FOR REFERENCE

Do Not Take From This Room

ESHBACH'S HANDBOOK OF ENGINEERING FUNDAMENTALS

ESHBACH'S HANDBOOK OF ENGINEERING FUNDAMENTALS, FIFTH EDITION

Edited by
Myer Kutz

WILEY

John Wiley & Sons, Inc.

Published by John Wiley & Sons, Inc., Hoboken, New Jersey.
Published simultaneously in Canada.

For general information about our other products and services, please contact our Customer Care Department within the United States at (800) 762-2974, outside the United States at (317) 572-3993 or fax (317) 572-4002.

Wiley also publishes its books in a variety of electronic formats. Some content that appears in print may not be available in electronic books. For more information about Wiley products, visit our web site at www.wiley.com.

Library of Congress Cataloging-in-Publication Data:

Eshbach, Ovid W. (Ovid Wallace), 1893–1958.
  Eshbach's handbook of engineering fundamentals / edited by Myer Kutz.—5th ed.
    p. cm.
  Includes bibliograhical references.
  ISBN 978-0-470-08578-3 (cloth: alk. paper)
1. Engineering—Handbooks, manuals, etc. I. Kutz, Myer. II. Title. III. Title: Handbook of engineering fundamentals.
  TA151.E8 2009
  620—dc22

                                                        2008041561

Printed in the United States of America.

10 9 8 7 6 5 4 3 2 1

To Ovid W. Eshbach (1893–1958), educator and editor

CONTENTS

PREFACE

In the years 1934–1936, when Ovid Wallace Eshbach (1893–1958) was preparing the first edition of the handbook that still bears his name, he was employed as special assistant in the Personnel Relations Department of AT&T. An electrical engineering graduate with honors from Lehigh University in 1915, he was well known in engineering education circles, particularly at schools which offered a cooperative option to their undergraduates. He coordinated the Bell System–MIT Cooperative Plan, an option in the Electrical Engineering Department at MIT, which permitted selected students to alternate study terms at MIT with terms of work, either with the Bell System or with the General Electric Company. In a memoir (available on the Northwestern University web site), to which I am indebted for this information, Eshbach's son wrote that his father, in addition to interviewing, hiring, and placing students within the Bell System, monitored their progress, counseled them, and followed their careers. He was also an adjunct MIT professor and taught electrical engineering courses for students co-oping at Bell. Eshbach served on committees of the Society for the Promotion of Engineering Education and the American Institute of Electrical Engineers. He was a member of the Regional Accrediting Committee of the Engineers' Council for Professional Development as well as the Special Advisory Committee to the President's Committee on Civil Service Improvement. In 1932 he had directed a survey of adult technical education for the Chamber of Commerce of the State of New York.

Several years after he published his handbook, Eshbach was approached to become dean of the Northwestern engineering school. Northwestern had established a new engineering school in the early 1900s, initially as a department within the College of Liberal Arts. In the mid-1920s the College of Engineering became the autonomous School of Engineering, with faculty members devoted exclusively to engineering. There was a crisis in engineering education at Northwestern in 1937 when, after years of declining enrollments, the school was denied accreditation during a national survey of engineering schools carried out by the Engineers' Council for Professional Development. A major criticism was that the curriculum was too heavily weighted with nonprofessional courses. But in 1939, Walter P. Murphy, a wealthy inventor of railroad equipment, donated $6.7 million for the construction of Northwestern's Technological Institute building. When the construction of Tech, as the engineering school was then known, was completed in 1942, Northwestern received an additional bequest of $28 million from Murphy's estate to provide for an engineering school "second to none." Although Murphy insisted that the school not be named for him—he would not appear in public or on programs of ceremonies, such as at the cornerstone laying or the dedication of the new building—the cooperative engineering education program bears his name to commemorate his interest in "practical education." Over the next 45 years cooperative engineering education remained a constant requirement at Tech, now known as the Robert R. McCormick School of Engineering and Applied Science.

Eshbach remained Tech's dean for the rest of his life as far as I can tell. His son reports in his memoir that Eshbach always had himself assigned to teach an undergraduate quiz section, usually in physics. And his name lives on at Northwestern. There is the Ovid W. Eshbach Society, in which alumni and other donors provide funds to strengthen undergraduate engineering education through support for such needs as laboratory equipment, undergraduate research, design competitions, and instructional software. There is also the Ovid W. Eshbach Award, established in 1948 by Tech's first graduating class, which is awarded for overall excellence in scholarship and leadership. Each spring, nominations are accepted from the graduating class on who they feel most closely typifies the ideal engineering student.

The team that Ovid W. Eshbach put together for the first edition of his handbook, which was called *Handbook of Engineering Fundamentals*, included 40 representatives from academia, industry, and government, most of them based in the northeast and some in the midwest. The handbook was the first volume in the **Wiley Engineering Handbook Series**, which also included the eleventh edition of the two-volume *Kent's Mechanical Engineers' Handbook* (one volume covered power, the second design and shop practice); the third edition of the two-volume *Handbook for Electrical Engineers* (one volume covered electric power, the second communications and electronics); and the third edition of the one-volume *Mining Engineers' Handbook*. Tables of contents for all handbooks in the series were included on pages following the index of the Eshbach volume.

The *Handbook of Engineering Fundamentals*, published in 1936 jointly by Wiley in New York and Chapman & Hall in London, contains 13 sections (chapters) and 1081 pages. Eshbach wrote in the Editor's Preface: "This handbook has been prepared for the purpose of embodying in a single volume those fundamental laws and theories of science which are basic to engineering practice. It is essentially a summary of the principles of mathematics, physics, and chemistry, the properties and uses of engineering materials, the mechanics of solids and fluids, and the commonly used mathematics and physical tables, to which has been added a discussion of contract relations. Thus, with the exception of the technics of surveying and drawing, there is included the fundamental technology common to all engineering curricula."

The **second edition** of *Handbook of Engineering Fundamentals* was published in 1952. It was still part of the **Wiley Engineering Handbook Series**, to which had been added *Handbook of Mineral Dressing*. Again, it was jointly published by Wiley and Chapman & Hall. The copy that I have is from the fourth printing, May 1954. On the front cover, COLLEGE EDITION is stamped underneath the name ESHBACH.

Eshbach made numerous changes for the **second edition**. He went west to find contributors—one from Texas and four from California were among the 38 contributors to this edition. With a new section on aerodynamics, he increased the number of sections to 14. He expanded the contracts section and renamed it Engineering Law. In addition, he enlarged the engineering tables to include standard structural sizes for aluminum and data on tangents and offsets for civil engineers; revised the mathematics section to eliminate "simple and commonly known items previously introduced for completeness" and put greater stress on "statistics, determinants, and vector analysis"; thoroughly revised the sections on solid and fluid mechanics; completely revised the section on electricity and magnetism; and in the sections on metallic and nonmetallic materials, "much material, more detailed, and of interest to special groups, has been eliminated to keep the volume within practical size."

By 1975, when the **third edition** was published, Eshbach had been dead for 17 years. Dr. Mott Souders, a chemical engineer from Piedmont, California, had taken over the editorship, although Eshbach's name was the only one stamped on the spine and front cover of the book. Souders, too, had died, in 1974, before the book was published, this time solely by Wiley, which now had offices in London, Sydney, and Toronto as well as New York. The handbook was still part of the **Wiley Engineering Handbook Series**. The center of gravity of contributor locations had shifted further west. In addition to seven contributors from the West Coast and one from Texas, the roster of 40 contributors included 18 on the staff of the U.S. Air

Force Academy, who contributed a section of over 180 pages on aeronautics and astronautics.

The **third edition** has 16 sections and 1562 pages. In his Preface, written in February 1974, Souders noted that the handbook contained new sections on astronautics, heat transfer, electronics, automatic control, and engineering economy. The sections on aeronautics and chemistry had been completely rewritten. New material had been added to the sections on mathematical and physical tables; mathematics, including an article on elements of Fortran; physical units and standards; as well as radiation, light, and acoustics. In the single section on properties of materials, all text was eliminated to provide space for more charts and tables. Souders also eliminated the section on engineering law. But the **third edition** did feature, on two pages following the Preface, canons of ethics of engineers approved by the Engineer's Council for Professional Development on September 30, 1963.

By the latter 1980s, the handbook's editorship had passed to Byron Tapley, a professor in the Department of Aerospace Engineering and Engineering Mechanics at the University of Texas at Austin. The **fourth edition's** size and scope increased dramatically. Whereas the trim size of the previous three editions had been $5\frac{1}{2}$ by $8\frac{3}{8}$ inches, the new edition was 7 by 10. The number of sections remained the same, at 16, but the number of pages increased dramatically to close to 2100. The number of contributors nearly doubled, to 77 and included, for the first time, one from overseas, in Athens, Greece. The rest were located throughout the United States—the East Coast orientation of the first edition was a thing of the distant past. As a result of the increased scope and complexity of the undertaking, a recently retired Wiley employee, Thurman Poston, was brought on board to assist Tapley in preparing the new edition.

The **fourth edition**, published in 1990, also had a new name. It was now called *Eshbach's Handbook of Engineering Fundamentals*. Also, major topic areas were placed into "chapters" and the term "sections" was now being used for subtopics. The most important changes to the handbook were undertaken in "recognition," Tapley wrote in his Preface, in November 1989, "given to the dramatic change that computers and computer technology have made in the way we generate, receive, and display information." Tapley continued: "The handbook has been modified to account for this impact in three substantial ways: (1) the chapter on mathematical and trigonometric tables has been reduced substantially in recognition of the fact that both small handheld computers and desktop personal computers allow a rapid generation of much of the information contained in this chapter, (2) a specific chapter dealing with computers and computer science has been added, and (3) specific applications where computers are useful have been included in many of the chapters." Tapley added sections on differential equations and the finite-element method; expanded the control theory chapter; split the aeronautics and astronautics chapter into two distinct chapters (due, I have been told, to usage of the handbook by students at the U.S. Air Force Academy for some years); and extensively revised the chapters on electromagnetics and circuits, electronics, radiation, light, acoustics, and engineering economics. In addition, international standard units were adopted throughout the handbook.

My approach to the **fifth edition**, which is being published nearly two decades after the appearance of the previous edition, has been to revise or update the chapters where there has been substantial change over the intervening years, but the scope of those chapters does not require substantial expansion or alteration; add new chapters in areas where the scope was insufficient and engineers need more basic information; and eliminate chapters superseded by the ubiquity of the digital environment. So the overall goal has been to add more knowledge essential to engineers while reducing the size of the handbook. As a result, there are fewer pages but more chapters.

The chapters that have been substantially updated and revised, but where the scope has remained unaltered for the most part, include those on mechanics of incompressible fluids, electromagnetics and circuits, acoustic, and engineering economy. All except the electromagnetics and circuits chapter have new contributors.

There are numerous chapters that either cover topics new to the handbook or replace chapters, or sections of chapters, where more basic information is essential for practicing engineers and students at any level. These chapters include Selection of Metals for Design; Plastics: Thermoplastics, Thermosets, and Elastomers; Ceramics; Nondestructive Testing; Aerodynamics of Wings; Mathematical Models of Dynamic Physical Systems; Basic Control Systems Design; Thermodynamics Fundamentals; Heat Transfer Fundamentals; and Electronics (with sections on bipolar transistors, data acquisition and conversion, data analysis, diodes, electronic components, input devices, instruments, integrated circuits, microprocessors, oscilloscopes, and power devices).

I have eliminated the chapter on computers and computer science, inasmuch as contributors now routinely absorb the digital world into their work whenever appropriate, as well as the over 250 pages of materials properties data, which have been replaced by a chapter, Sources of Materials Data, which is a current description of where and how to find reliable materials properties data on the Internet, the standard practice in this digital age. In addition, I have left alone those chapters which contain basic and theoretical information that does not change.

Eshbach has gone through a great many iterations in its long life, yet the handbook remains true to its creator's original vision. My thanks to him as well as to the legion of contributors whose efforts have graced the pages of the five editions of this great reference work.

Myer Kutz

Delmar, New York

CONTRIBUTORS

Kate D. Abel School of Systems and Enterprises, Stevens Institute of Technology, Hoboken, New Jersey

Adrian Bejan Department of Mechanical Engineering and Materials Science, Duke University, Durham, North Carolina

Jonathan Blotter Department of Mechanical Engineering, Brigham Young University, Provo, Utah

Arbee L. P. Chen National Tsing Hua University, Hsinchu, Taiwan, Republic of China

Robert P. Colwell Intel Corporation, Hillsboro, Oregon

Robert L. Crane Air Force Research Laboratory, Materials Directorate, Wright Patterson Air Force Base, Dayton, Ohio

John D. Cressler Georgia Institute of Technology, Atlanta, Georgia

Clarence W. de Silva University of British Columbia, Vancouver, British Columbia, Canada

D. H. Daley Department of Aeronautics, United States Air Force Academy, Colorado Springs, Colorado

Matthew J. Donachie Rensselaer at Hartford, Hartford, Connecticut

Neil F. Enke Department of Engineering Mechanics, University of Wisconsin, Madison, Wisconsin

Halit Eren Curtin University of Technology, Bentley, Western Australia, Australia

Wallace Fowler Department of Aerospace Engineering and Engineering Mechanics, The University of Texas at Austin, Austin, Texas

Kent L. Gee Department of Mechanical Engineering, Brigham Young University, Provo, Utah

M. Parker Givens Institute of Optics, University of Rochester, Rochester, New York

Georges Grinstein University of Massachusetts Lowell, Lowell, Massachusetts

Ramesh Harjani University of Minnesota, Minneapolis, Minnesota

Alex Q. Huang Virginia Polytechnic Institute and State University, Blacksburg, Virginia

Wade W. Huebsch Department of Mechanical and Aerospace Engineering, College of Engineering and Mineral Resources, West Virginia University, Morgantown, West Virginia

R. Nathan Katz Department of Mechanical Engineering, Worcester Polytechnic Institute, Worcester, Massachusetts

J. G. Kaufman Kaufman Associates, Inc., Columbus, Ohio

Jeremy S. Knopp Air Force Research Laboratory, Materials Directorate, Wright Patterson Air Force Base, Dayton, Ohio

D. A. Kohl The University of Texas at Austin, Austin, Texas

J. G. Kaufman Kaufman Associates, Inc., Columbus, Ohio

Konstantinos Misiakos NCSR "Demokritos," Athens, Greece

Kavita Nair University of Minnesota, Minneapolis, Minnesota

Egemen Ol Ogretim Department of Civil and Environmental Engineering, College of Engineering and Mineral Resources, West Virginia University, Morgantown, West Virginia

William J. Palm III Department of Mechanical Engineering, University of Rhode Island, Kingston, Rhode Island

Edward N. Peters General Electric Company, Selkirk, New York

G. P. Peterson Rensselaer Polytechnic Institute, Troy, New York

Warren F. Phillips Department of Mechanical and Aerospace Engineering, Utah State University, Logan, Utah

Dennis Polla University of Minnesota, Minneapolis, Minnesota

N. Ranganathan University of South Florida, Tampa, Florida

J. N. Reddy Department of Mechanical Engineering, Texas A&M University, College Station, Texas

Albert J. Rosa Professor Emeritus, University of Denver, Denver, Colorado

Andrew Rusek Oakland University, Rochester, Michigan

Bela I. Sandor Department of Engineering Mechanics, University of Wisconsin, Madison, Wisconsin

Scott Sommerfeldt Department of Mechanical Engineering, Brigham Young University, Provo, Utah

Marjan Trutschl University of Massachusetts Lowell, Lowell, Massachusetts

Raju D. Venkataramana University of South Florida, Tampa, Florida

Jack H. Westbrook Ballston Spa, New York

K. Preston White, Jr. Department of Systems and Information Engineering, University of Virginia, Charlottesville, Virginia

J. B. Wissler Department of Aeronautics, United States Air Force Academy, Colorado Springs, Colorado

Yi-Hung Wu National Tsing Hua University, Hsinchu, Taiwan, Republic of China

Bo Zhang Virginia Polytechnic Institute and State University, Blacksburg, Virginia

Chris Zillmer University of Minnesota, Minneapolis, Minnesota

CHAPTER 1

MATHEMATICAL AND PHYSICAL UNITS, STANDARDS, AND TABLES*

Jack H. Westbrook
Ballston Spa, New York

*This chapter is a revision and extension of Sections 1 and 3 of the third edition, which were written by Mott Souders and Ernst Weber, respectively. Section 4.4 is derived principally from ASTM's *Standard for Metric Practice*, ASTM E380-82, Philadelphia, 1982 (with permission). Section 6.1 is derived from *MIS Newsletter*, General Electric Co., 1980 (with permission).

1 SYMBOLS AND ABBREVIATIONS

Table 1   Greek Alphabet

A	α	Alpha	H	η		Eta	N	ν		Nu	T	τ	Tau
B	β	Beta	Θ	ϑ	θ	Theta	Ξ	ξ		Xi	Υ	υ	Upsilon
Γ	γ	Gamma	I	ι		Iota	O	o		Omicron	Φ	ϕ	Phi
Δ	δ	Delta	K	κ		Kappa	Π	π		Pi	X	χ	Chi
E	ε	Epsilon	Λ	λ		Lambda	P	ρ		Rho	Ψ	ψ	Psi
Z	ζ	Zeta	M	μ		Mu	Σ	σ	ς	Sigma	Ω	ω	Omega

Table 2   Symbols for Mathematical Operations[a]

Addition and Subtraction

$a + b$, a plus b
$a - b$, a minus b
$a \pm b$, a plus or minus b
$a \mp b$, a minus or plus b

Multiplication and Division

$a \times b$, or $a \cdot b$, or ab, a times b
$a \div b$, or $\dfrac{a}{b}$, or a/b, a divided by b

Symbols of Aggregation

() parentheses
[] brackets
{} braces

– vinculum

Equalities and Inequalities

$a = b$, a equals b
$a \approx b$, a approximately equals b

$a \neq b$, a is not equal to b
$a > b$, a is greater than b
$a < b$, a is less than b
$a \gg b$, a much larger than b
$a \ll b$, a much smaller than b
$a \geqq b$, a equals or is greater than b
$a \leqq b$, a is less than or equals b
$a \equiv b$, a is identical to b
$a \rightarrow b$, or $a = b$, a approaches b as a limit

Proportion

$a/b = c/d$, or $a : b :: c : d$, a is to b as c is to d
$a \propto b$, $a \sim b$, a varies directly as b
%, percent

Powers and Roots

a^2, a squared
a^n, a raised to the nth power
$\sqrt{a}$, square root of a
$\sqrt[3]{a}$, cube root of a
$\sqrt[n]{a}$, or $a^{1/n}$, nth root of a

(Continues)

Table 2   *(Continued)*

$a^{-n}, 1/a^n$

$3.14 \times 10^4 = 31{,}400$

$3.14 \times 10^{-4} = 0.000314$

Miscellaneous

$\bar{a}$, mean value of a

$a!, = 1 \cdot 2 \cdot 3 \ldots a$, factorial a

$|a|$ = absolute value of a

$P(n, r) = n(n-1)(n-2) \cdots (n-r+1)$

$C(n, r) = \dfrac{P(n, r)}{r!} = \dbinom{n}{r}$ = binomial coefficients

i (or j) $= \sqrt{-1}$, imaginary unit

$\pi = 3.1416$, ratio of the circumference to the diameter of a circle

∞, infinity

Plane Geometry

$<$, angle

$\triangle$, triangle

$\parallel$, parallel

$\perp$, perpendicular

$\odot$, circle

$\square$, parallelogram

$\therefore$, therefore

$° \prime \prime\prime$, degree, minute, second

$\prime \prime\prime$, feet, inches

Logarithms and Exponentials

$\log a = \log_{10} a$, common logarithm of a or log of a to the base 10

$\ln a = \log_e a$, natural logarithm of a or log of a to the base e ($e = 2.718$)

$\log^{-1} a$, number whose log is a

lb x or $\log_2 x$ = binary logarithm of x exponential of x, $\exp x, e^x$

Trigonometry

$\left.\begin{array}{l} \sin, \cos, \tan \\[4pt] \mathrm{cosec}\ \text{or}\ \csc,\ \sec, \cot\ \text{or}\ \mathrm{ctn} \\[4pt] \mathrm{vers}, \mathrm{covers} \end{array}\right\}$ trigonometric functions

$\left.\begin{array}{l} \sin^{-1}, \cos^{-1}, \text{etc.} \\[4pt] \arcsin, \arccos \end{array}\right\}$ inverse of the functions

Analytic Geometry

$x, y, z; \xi, \eta, \zeta$, rectangular coordinates

ρ, s, intrinsic coordinates

ρ, radius of curvature

s, length of arc

r, θ, polar coordinates

ψ, angle from radius vector to tangent

r, θ, ϕ, spherical coordinates

θ, colatitude

ϕ, longitude

r, θ, z, cylindrical coordinates

e, eccentricity in conics

p, semi latus rectum in conics

$l = \cos\alpha, m = \cos\beta, n = \cos\gamma$, direction cosines

Calculus

$y = f(x), y$ is a function of x

$y' = f'(x) = \dfrac{dy}{dx} = D_x y$, derivative of $y = f(x)$ with respect to x

$y'' = f''(x) = \dfrac{d(y')}{dx} = D_x^2 y = \dfrac{d^2 y}{dx^2}$, second derivative of $y = f(x)$ with respect to x

$u = f(x, y), u$ is a function of x and y

$u_x = f_x(x, y) = D_x(u) = \dfrac{\partial u}{\partial x}$, partial derivative of $u = f(x, y)$ with respect to x

$u_{xy} = f_{xy}(x, y) = D_y(D_x u) = \dfrac{\partial^2 u}{\partial y \partial x}$, second partial derivative of $u = f(x, y)$ with respect to x and y

Δy, increment of y

dy, differential of y

δy, variation of y

$\sum\limits_{i=a}^{b}$, summation over i from a to b

$\lim_{x \to a}(y) = b, y \to b$ as $x \to a$

$\int$, integral of

$\int_a^b$, definite integral of

Vector Analysis

i, j, k, unit vectors along the axes (right-handed system)

$a \cdot b = (ab) = Sab$, scalar product of a and b

$a \times b = [ab] = Vab$, vector product of a and b

Vectors are indicated in print by boldfaced type.

$|A|, A$, absolute value

$\partial/\partial r, \nabla$, differential vector operator

grad $\varphi, \nabla\varphi$, gradient

div $A, \nabla \cdot A$, divergence

curl A, rot $A, \nabla \times A$, curl

$\Delta\varphi, \nabla^2\varphi$, Laplacian

$\square\varphi, \square^2 = \dfrac{1}{c^2}\dfrac{\partial^2}{\partial t^2} - \nabla^2$, D'Alembertian

Logic and Boolean Algebra

$a \in A, a$ is contained in set A

$A \cap B, A \cdot B$, logical multiplication. Intersection of set A and set B, A AND B

$A \cup B, A + B$, logical addition. Union of set A and set B. A OR B

$A \oplus B$, exclusive OR

$A \supset B$, logical inclusion. Inclusion of set B in set A

$A \ominus B$, complement of set B in set A

$\tilde{A}, \overline{A}$, logical complementation. NOT set A. Negation

$\emptyset, 0$, logical impossibility. Empty (null) set. Zero state

$I, 1$, logical certainty. Universal set. One state

---

[a] References: Mathematic signs and symbols for use in the physical sciences and technology, ANSI Y10.20 − 1975.

Table 3 Abbreviations[a] for Scientific and Engineering Terms[b]

Name of Term	Abbreviation	Name of Term	Abbreviation
absolute	abs	cubic centimeter	cu cm, cm^3 (liquid, meaning milliliter, ml)
acre	spell out		
acre-foot	acre-ft		
air horsepower	air hp		
alternating-current (as adjective)	a-c	cubic foot	cu ft
ampere	amp or A	cubic feet per minute	cfm
ampere-hour	amp-hr	cubic feet per second	cfs
amplitude, an elliptic function	am.	cubic inch	cu in.
Angstrom unit	A	cubic meter	cu m or m^3
antilogarithm	antilog	cubic micron	cu μ or μ^3 or cu mu
atmosphere	atm		
atomic weight	at. wt.	cubic millimeter	cu mm or mm^3
average	avg	cubic yard	cu yd
avoirdupois	avdp	current density	spell out
azimuth	az or α	cycles per second	spell out or cps or Hz
		cylinder	cyl
barometer	bar.		
barrel	bbl	day	spell out
Baumé	Bé	decibel	db
board feet (feet board measure)	fbm	degree[d]	deg or °
boiler pressure	spell out	degree centigrade	C
boiling point	bp	degree Fahrenheit	F
brake horsepower	bhp	degree Kelvin	K
brake horsepower-hour	bhp-hr	degree Rankine	R
Brinell hardness number	Bhn	delta amplitude, an elliptic function	dn
British thermal unit[c]	Btu or B		
bushel	bu	diameter	diam
		direct-current (as adjective)	d-c
calorie	cal	dollar	$
candle	c	dozen	doz
candle-hour	c-hr	dram	dr
candlepower	cp		
cent	c or ¢	efficiency	eff
center to center	c to ¢	electric	elec
centigram	cg	electromotive force	emf
centiliter	cl	elevation	el
centimeter	cm	equation	eq
centimeter-gram-second (system)	cgs	external	ext
chemical	chem		
chemically pure	cp	farad	spell out or F
circular	cir	feet board measure (board feet)	fbm
circular mils	cir mils	feet per minute	fpm
coefficient	coef	feet per second	fps
cologarithm	colog	fluid	fl
concentrate	conc	foot	ft
conductivity	cond	foot-candle	ft-c
constant	const	foot-Lambert	ft-L
continental horsepower	cont hp	foot-pound	ft-lb
cord	cd	foot-pound-second (system)	fps
cosecant	csc	foot-second (see cubic feet per second)	
cosine	cos		
cosine of the amplitude, an elliptic function	cn	franc	fr
		free aboard ship	spell out
cost, insurance, and freight	cif	free alongside ship	spell out
cotangent	cot	free on board	fob
coulomb	spell out or C	freezing point	fp
counter electromotive force	cemf	frequency	spell out
cubic	cu	fusion point	fnp

(Continues)

Table 3 *(Continued)*

Name of Term	Abbreviation	Name of Term	Abbreviation
gallons per minute	gpm	low-pressure (as adjective)	
gallons per second	gps		l-p
grain	spell out	lumen	lm
gram	g	lumen-hour	lm-hr
gram-calorie	g-cal	lumens per watt	lpw
greatest common divisor	gcd		
		mass	spell out
haversine	hav	mathematics (ical)	math
hectare	ha	maximum	max
henry	H	mean effective pressure	mep
high-pressure (adjective)	h-p	mean horizontal candlepower	mhcp
hogshead	hhd	megacycle	spell out
horsepower	hp	megohm	spell out
horsepower-hour	hp-hr	melting point	mp
hour	hr	meter	m
hour (in astronomical tables)	h	meter-kilogram	m-kg
hundred	C	mho	spell out
hundredweight (112 lb)	cwt	microampere	μ a or mu a
hyperbolic cosine	cosh	microfarad	μF
hyperbolic sine	sinh	microinch	μ in.
hyperbolic tangent	tanh	micromicron	$\mu\mu$ or mu mu
		micron	μ or mu
inch	in.	microvolt	μ v
inch-pound	in. lb	microwatt	μ w or mu w
inches per second	ips	mile	spell out
indicated horsepower	ihp	miles per hour	mph
indicated horsepower-hour	ihp-hr	miles per hour per second	mphps
inside diameter	ID	milliampere	ma
intermediate-pressure (adjective)	i-p	milligram	mg
		millihenry	mH
internal	int	millilambert	mL
		milliliter	ml
joule	J	millimeter	mm
		millimicron	m μ or m mu
kilocalorie	kcal	million	spell out
kilocycles per second	kcps	million gallons per day	mgd
kilogram	kg	millivolt	mV
kilogram-calorie	kg-cal	minimum	min
kilogram-meter	kg-m	minute	min
kilograms per cubic meter	kg per cu m or kg/m^3	minute (angular measure)	
		minute (time) (in astronomical tables)	m
kilograms per second	kgps	mole	spell out
kiloliter	kl	molecular weight	mol. wt
kilometer	km	month	spell out
kilometers per second	kmps		
kilovolt	kv	National Electrical Code	NEC
kilovolt-ampere	kva		
kilowatt	kw	ohm	spell out or Ω
kilowatthour	kwhr	ohm-centimeter	ohm-cm
		ounce	oz
lambert	L	ounce-foot	oz-ft
latitude	lat or ϕ	ounce-inch	oz-in.
least common multiple	lcm	outside diameter	OD
linear foot	lin ft		
liquid	liq	parts per million	ppm
lira	spell out	peck	pk
liter	L	penny (pence)	d
logarithm (common)	log	pennyweight	dwt
logarithm (natural)	$\log_e$ or ln		
longitude	long. or λ		

Table 3 *(Continued)*

Name of Term	Abbreviation	Name of Term	Abbreviation
per	(see Fundamental Rules)	sine of the amplitude, an elliptic function	sn
peso	spell out	specific gravity	sp gr
pint	pt	specific heat	sp ht
potential	spell out	spherical candle power	scp
potential difference	spell out	square	sq
pound	lb	square centimeter	sq cm or cm^2
pound-foot	lb-ft	square foot	sq ft
pound-inch	lb-in.	square inch	sq in.
pound sterling	£	square kilometer	sq km or km^2
pounds per brake		square meter	sq m or m^2
horsepower-hour	lb per bhp-hr	square micron	sq μ or sq mu or μ^2
pounds per cubic foot	lb per cu ft	square millimeter	sq mm or mm^2
pounds per square foot	psf	square root of mean square	rms
pounds per square inch	psi	standard	std
pounds per square inch absolute	psia	steradian	sr
power factor	spell out or pf	tangent	tan
		temperature	temp
quart	qt	tensile strength	ts
		thousand	M
radian	spell out	thousand foot-pounds	kip-ft
reactive kilovolt-ampere	kvar	thousand pound	kip
reactive volt-ampere	var	ton	spell out
revolutions per minute	rpm	ton-mile	spell out
revolutions per second	rps		
rod	spell out	versed sine	vers
root mean square	rms	volt	V
		volt-ampere	Va
secant	sec	volt-coulomb	spell out
second	sec		
second (angular measure)	"	watt	W
second-foot (see cubic feet per second)		watthour	Whr
		watts per candle	Wpc
second (time) (in astronomical tables)	s	week	spell out
		weight	wt
shaft horsepower	shp		
shilling	s	yard	yd
sine	sin	year	yr

^{a}These forms are recommended for readers whose familiarity with the terms used makes possible a maximum of abbreviations. For other classes of readers, editors may wish to use less contracted combinations made up from this list. For example, the list gives the abbreviation of the term "feet per second" "fps." To some readers ft/sec will be more easily understood.

^{b}This list of abbreviations is adapted from the recommendations of the American National Standards Institute [see ANSI Y1.1-1972 (R1984)].

^{c}Abbreviation recommended by the American Society of Mechanical Engineers (ASME) Power Test Codes Committee. B = 1 Btu, kB = 1000 Btu, mB = 1,000,000 Btu. The American Society of Heating, Refrigerating and Air-Conditioning Engineers (ASHRAE) recommends the use of Mb = 1000 Btu and Mbh = 1000 Btu/hr.

^{d}There are circumstances under which one or the other of these forms is preferred. In general the sign° is used where space conditions make it necessary, as in tabular matter, and when abbreviations are cumbersome, as in some angular measurements, i.e., 59°23′ 42″. In the interest of simplicity and clarity the Committee has recommended that the abbreviation for the temperature scale, °F, °C, K, etc., always be included in expressions for numerical temperatures, but, wherever feasible, the abbreviation for "degree" be omitted; as 69 F.

Table 4 Symbols for Physical Quantities[a,b]

Name of Quantity	Symbol	Name of Quantity	Symbol
Absorption factor	α	Circular frequency $(2\pi f)$	ω
Acceleration		Circulation, strength of single	
Angular	α	vortex	Γ
Linear, general	a	Coefficient	
Acceleration due to gravity		Absolute	C
General	g	General	C
International adopted standard	g_0	Of contraction	C_c
Local	g_L	Of discharge	C
Gravitational conversion factor	g_c	Of discharge	C_q
Activity	a	Of energy per unit weight in	
Activity coefficient, molal basis	γ	$C_e(V^2/29)$	C_e
Adiabatic factor	X	Of flow (Chézy)	C
Admittance	Y	Of friction (Weisbach–Darcy)	f
Advanced ratio of propeller	J	Of friction	μ, f
Altitude	h, z	Of momentum per unit weight	
Amplitude	A	in $C_m(V/g)$	C_m
Angle	α	Of roughness (Bazin)	m
Angle	$\beta\ \phi$	Of roughness (Kutter and	
Blade angle	β	Manning)	n
Effective helix	ψ	Of heat transfer overall	V
Dihedral	Γ	Of velocity	C_v
Helical angle of advance	ϕ	Compressibility factor	z
Of attack	α	Concentrated load	F, P, Q
Of downwash	ε	Concentration	C, c
Of radiation	θ	Concentration, volumetric	c
Of sideslip	β	Concentration factor, stress	K
Of sidewash	σ	Conductance	
Solid	ω	Electrical	G
Angular		Thermal	$1/R$
Acceleration	α	Per unit area	$1/RA$
Displacements	δ	Conductivity	
Frequency	ω	Electrical	γ, σ
Momentum	H	Equivalent	Λ
Velocity	ω	Thermal	k
Area	A	Contraction, coefficient of	C_c
Area[c]	S	Correlation coefficient	R
Aspect ratio	A, AR	Coupling coefficient	k
Atomic weight	A	Critical state or indicating critical	
Attack, angle of	α	value (subscript)	c
Attenuation	a	Current[d]	I
Axes			
Of aircraft (left handed)		Damping	
Earth-bound coordinate		Coefficient	c
system	x, y, z	Constant or coefficient	δ
Lateral	Y	Factor	λ
Longitudinal	X	Deflection	δ
Normal	Z	Of beam, maximum	δ
Bazin's coefficient of roughness	m	Density	ρ
Blade width (propellers)	b	Relative to standard air density	σ
Boundary layer thickness	δ	Depth	h
Breadth	b	Depth	y
Capacitance, capacity	C	Of flow, channels	y
Capacitivity	ε	Diameter	D
Of evacuated space	ε_v	Dielectric constant	ε
Relative	ε_r	Difference between values	Δ
Charge, electric or quantity of	Q	Difference of potential[d,e]	E, e
electricity		Diffusion coefficient	D_v
Charge density		Diffusivity	α
Line density of charge	λ	Diffusivity, thermal	α
Surface density of charge	σ	Of vapor	D_v
Volume density of charge	ρ	Discharge	
Chézy's coefficient	C	Coefficient of	C_q
Chord length	c	Coefficient of	C
		Rate of; or flow	Q

Table 4 *(Continued)*

Name of Quantity	Symbol	Name of Quantity	Symbol
Per unit width	q	Force	F
Displacement, electric	D	Electromotive[e]	E, e
Distance		Magnetomotive	$M, \mathscr{F}$
From center of gravity to	f	Moment of	M
center of pressure of		Normal	N
horizontal tail surface		Shearing force in beam	V
Linear	s	section	
Drag, absolute coefficient of	D	Total load	F
Dynamic (or impact) pressure	q	Forces or loads, concentrated	P, Q, F
		Fraction	
Eccentricity of application of	e	By volume	x_v
load		By weight	x_w
Efficiency	η	Free energy	
Elastance	S	Gibbs	G
Mutual	S_m, S_{rc}	Helmholtz	A
Self	S, S_{cc}	Frequency	f
Elasticity		Circular ($2\pi f$)	θ
Bulk modulus, of liquids	K	Of radiant energy	v
Kinematic K/ρ	e	Reduced (flutter)	k
Modulus of	E	Rotational	n
Elastivity	σ	Frequency, angular	ω
Electric potential[d,e]	E, e	Friction	
Electricity, quantity of	Q, q	Coefficient of sliding	f, μ
Electromotive force[d]	E, e	Factor used in expressing	f
Electronic charge, absolute value	e	pipeloss	
Electrostatic flux	ψ	In energy balance	F
Elevation		Fugacity	f
Above datum	Z		
Above stream bed	Z_0	Gas constant	R
Elongation, total	δ	Gibbs function, total potential	G, g
Emissivity, total	ε	function	
Energy	W	Gyration, radius of	k
Work total	E		
Energy	E	Head	
Internal; intrinsic[f]	U, u	Atmospheric	h_a
Kinetic	E_k, T	Lost[c]	h
Per unit time (power)	P	Potential	h_{pz}
Potential	E_p, V	Pressure	h_p
Enthalpy[f]	H or h	Velocity	h_v
Enthalpy	H	Heat	
Of dry saturated vapor	h_g	Content; enthalpy[f]	H, h
Of saturated liquid	h_f	Content of dry saturated	h_g
Per unit weight	h	vapor; enthalpy of dry	
Entropy	S, s	saturated vapor	
Error signal	ε	Content of saturated liquid;	h_f
Expansion, exponent of	n	enthalpy of saturated liquid	
polytropic		Equivalent of work	$1/J$
Cubical, thermal coefficient	β	Flow rate	q
Linear, thermal coefficient	α	Across a boundary surface	h
		Latent, of evaporation	λ, h_{fg}
Factor of safety	N	Mechanical equivalent of	J
Film thickness, effective	B	Of vaporization at constant	$H_{fg}, \lambda,$ or h_{fg}
Flow rate	w	pressure[f]	
In pounds per unit of time	w	Specific, at constant pressure	c_p
Volumetric	q	Specific, at constant volume	c_v
Fluidity	$1/\mu$	Ratio of specific heats	$\gamma, \kappa,$ or k
Flux		Transfer, overall coefficient of	U
Density, magnetic	B	Transfer, surface coefficient of	h
Displacement	ψ	Height	h
Magnetic	Φ	Crest, weirs	z
Force	F	Helix, effective angle	Φ

(Continues)

Table 4 *(Continued)*

Name of Quantity	Symbol	Name of Quantity	Symbol
Helmholtz free energy; internal potential function[f]	A, a	Per unit distance	w, q
		Total	W, P
Humidity	H	Mach	
Density of water vapor; weight of water vapor per unit of volume of space	ρ_H	Angle	μ
		Number	M
Density of water vapor at saturation	ρ_s	Magnetic	
		Flux	Φ
Enthalpy of the mixture minus the enthalpy of the liquid at the temperature of adiabatic saturation; carrier sigma function	h_Σ	Intensity	**H**
		Magnetomotive force	$M, \mathscr{F}$
		Mass	m
		Flow rate	w
		Velocity	G
Humid volume, volume of mixture per unit of weight of dry air	v_H	Mean free path	λ
		Mechanical equivalent of heat	J
		Microscale (turbulence)	λ
Partial pressure of water vapor	p_H	Modulus	
Percentage humidity by weight	w_H/w_s	Bulk, of elasticity of liquids	K
		Of elasticity	E
Relative humidity; ratio of an actual partial pressure of water vapor in air to the saturation partial pressure	H_R	Of elasticity in shear	G
		Section	Z
		Shear	G
		Molecular weight	M
Saturation pressure of water vapor	p_s	Moment	
		Electric	**p**
Saturation weight of water vapor per unit of weight of air	H_s, w_s	Magnetic	**m**
		Of any area about a given axis, statical	Q
Weight of water vapor per unit of weight of dry air	H, w_H	Of force, including bending moment	M
Hydraulic radius	R_H	Of inertia, polar	J
Mean in a reach	R_m	Rectangular	I
Of cross-sectional area	R	Mutual inductance	L_m
Hydraulic slope	S_w		
		Neutral axis, distance to extreme fiber	c
Impedance	Z	Nozzle divergence factor	λ
Impulse	I	Number in general	N
Inductance	L	Of conductors or turns	N
Magnetic	B	Of moles, pound-moles, kilogram-moles, etc.	n
Mutual	L_m		
Self	L, L_{cc}	Of phases	m
Inertia, moment of	I	Of poles	p
Inertia, moment of		Of revolutions per unit of time	n
Polar	J		
Rectangular	I	Perimeter, wetted, of a sectional area	P
Product moment of	I_{xy}		
Intensity		Period	T
Electric	**E, K**	Permeability	
Magnetic	**H**	Magnetic	μ
Isentropic factor	X	Of evacuated space	μ_c
		Relative	μ_r
Joule–Thomson coefficient	μ	Permeance	$\mathscr{P}, \Lambda$
Kutter coefficient of roughness	n	Permittivity	ε
		Phase	
Length	L	Angle	ϕ
Length	l	Constant	β
Lift	L	Displacement	ϕ
Linear expansion, coefficient	α	Pitch, geometric	p
Linear velocity	v	Planck constant	h
Load		Poisson ratio	μ, ν
Concentrated	F, P, Q	Polarization, magnetic	B_i
Eccentricity of application of	e	Pole strength	m
Factor	n		

Table 4 *(Continued)*

Name of Quantity	Symbol	Name of Quantity	Symbol
Potential		Richness; equivalence ratio	R
Electric[d,e]	V	(combustion)	
Function	ϕ	Rotation	
Function, internal; Helmholtz	A, a	Rate of	n
free energy		Speed of	n
Function, total; Gibbs function	G, g	Safety factor	N
Magnetic	$\mathbf{M}, \mathscr{F}$	Saturation pressure of water	p_s
Magnetic vector	$\mathbf{A}$	vapor	
Retarded vector	$\mathbf{A}_r$	Section modulus	Z
Power		Self-inductance	L, L_{cc}
Active	P	Set of control surfaces, angle of [c]	δ
Apparent	S	Shape factor	S
Factor	F_p	Shearing force in beam section	V
Reactive	Q	Slip	s
Pressure		Slope	
Dynamic	q	Of channel bed	S_0
Intensity; force per unit area	p	Of cuts and embankments	s
Relative	δ	Of energy grade line	S
Saturation of water vapor	p_s	Of hydraulic grade line	S_w
Propagation constant	γ	Of lift curve	a
Poynting vector	Π	Solidity, propellers	σ
		Span	b
Q factor of a reactor	Q	Effectiveness	e
Quality of vapor	x	Specific	
Quantity		Gravity	G
Of electricity	Q	Heat	c
Of heat per unit mass or unit	q	Heat at constant pressure	c_p
weight		Molar	C_p
Of heat per unit time	q	Heat at constant volume	c_v
Of matter	W	Molar	C_v
Total, of a fluid, water, gas,	Q	Heats, ratio	γ or k or k
heat (by volume)		Volume	v
		Weight	γ
Radiant density	u	Speed	
Radiant energy	U	Linear	V, v, u
Radiant flux	Φ	Of rotation	n
Density	W	Spring constant	k
Radiant intensity	J	Stefan–Boltzmann constant	σ
Radiation, intensity of	N	Strain	
Radii	r, R	Normal	ε
Radius	r	Shear	γ
Of gyration	k	Stream function	ψ
Range	R	Stress	
Reactance	X	Concentration factor	K
Capacitive	X_c	Normal	σ, s
Inductive	X_L	Shear	τ, s_s
Mutual	X_m, X_{rc}	Supercompressibility factor	z
Self	X, X_{cc}	Surface coefficient of heat	h
Reactive factor	F_q	transfer	
Recovery factor	η_r	Surface per unit volume	a
Reduced frequency (flutter)	k	Surface tension	σ
Reflection factor	ρ	Kinematic σ/ρ	ω
Reluctance	$\mathscr{R}$	Susceptance	B
Reluctivity	ν	Susceptibility	
Resistance		Dielectric	η
Electrical	R	Magnetic	κ
Temperature coefficient	α	Sweepback angle	Λ
Thermal	R		
Per unit area	RA	Taper ratio	λ
Resistivity		Temperature	
Electrical	ρ	Absolute[g] (°F abs or K)	T or Θ
Thermal	$1/k$	Ordinary[g] (°F or °C)	t or θ
Revolutions per unit time	n		
Reynolds number	R		

(Continues)

Table 4 (*Continued*)

Name of Quantity	Symbol	Name of Quantity	Symbol
Ratio	θ	Of wave celerity	c
Thermal		Relative	v
Conductance	$1/R$	Temporal means of	$\bar{u}, \bar{v}, \bar{w}$
Per unit area; "unit	$1/RA$	components	
conductance"		Vibration constant	p
Conductivity	k	Viscosity	
Diffusivity	α	Absolute; coefficient of	μ
Resistance	R	Kinematic	ν
Resistance of unit area	RA	Relative (to absolute viscosity	μ/μ_w
Resistivity	$1/k$	of water)	
Transfer factor	j	Relative kinematic	use ν/ν_w
Transmission	q	Voltage[d]	E, e
Thickness	$d, t,$ or h	Volume	V
Thrust		Molar	V, V_m
Stream	F	Specific	v
Propeller	T	Total	V, V_L
Time	t	Volume rate; discharge by	q, Q
Time[h]	t or τ	volume, fluid rate of flow by	
Time constant	τ	volume	
Torque	Q	Wavelength	λ
Torque	T or M	Constant	β
Transmission		Weight	
Factor	τ	Molecular	M
Thermal	q	Per unit time per unit area of	G
Turbulence exchange,	ε	cross section; "mass velocity"	
coefficient		Per unit volume	γ
Turbulence scale	L	Rate; per unit of power; for	w
		unit of time	
Vaporization, heat of, at	H_{fg}, h_{fg}, λ	Rate of flow per unit of breadth	Γ
constant pressure		Specific, with g_c	γ
Velocity	V	Total	W
Velocity	V or v	Weirs	
Acoustic	V_a	Crest height	z
Angular	ω	Crest length	b
Average	V	Degree of submergence	N
Belanger critical	V_c	Wetted perimeter	L_p
Components in x, y, z	u, v, w	Width (same as breadth)	b
directions, respectively		Of stream bed	b
Linear	v	Width, channel surface	b_w
Local	u	Wing setting, angle of (angle	i_w
Mass, mass flow, per unit	G	between the wing chord and	
cross-sectional area, per		the thrust line)	
unit time		Work	W
Mean (Q/A)	V	External	W_e
Of light	c	Heat equivalent of	$1/J$ or A
Of sound	a, c	Per unit weight	w, w_k
Of uniform flow	V_0		

[a]The most frequently used American Standard and Tentative Standard Symbols are included in this table. Sources used are publications of the American National Standards Institute shown in the bibliography below.

[b]Where possible, capital letters denote total quantities and small letters denote specific quantities, or quantities per unit.

[c]Use with appropriate subscript.

[d]Where distinctions between maximum, instantaneous, effective (root-mean-square), and average values are necessary, E_m, I_m, P_m are recommended for maximum values; e, i, p for instantaneous values; E, I for effective (rms) values; and P for average value.

[e]Where a distinction between electromotive force and difference of electric potential is desirable, the symbols $E, e,$ and $V, v,$ respectively, may be used.

[f]In each instance uppercase italics may be used optionally for values in general or per mole. Molal values may have subscript M. Lowercase italics are to be used for specific values (per pound, gram, liter, etc.). Molecular values may be represented by lowercase italics or by lowercase italics with subscript m.

[g]θ is preferable only when t is used for time in the same discussion. Θ is preferable only when θ is used for ordinary temperature.

[h] τ should be used only when t is used for ordinary temperature in the same discussion.

BIBLIOGRAPHY FOR LETTER SYMBOLS

Acoustics, Letter Symbols and Abbreviations for Quantities Used in, ANSI/ASME Y10.11-1984.

Aeronautical Sciences, Letter Symbols for, ANSI Y10.7-1954.

Chemical Engineering, Letter Symbols for, ANSI Y10.12-1955(R1973).

Glossary of Terms Concerning Letter Symbols, ANSI Y10.1-1972.

Heat and Thermodynamics, Letter Symbols for, ANSI Y10.4-1982.

Hydraulics, Letter Symbols for, ANSI Y10.2-1958.

Illuminating Engineering, Letter Symbols for, ANSI Y10.18-1967(R1977).

Letter Symbols for SI Units and Certain Other Units of Measurement, ANSI/IEEE 260-1978.

Mathematic signs and Symbols for Use in Physical Sciences and Technology (includes supplement ANSI Y10.20a-1975), ANSI Y10.20-1975.

Mechanics and Time-Related Phenomena, ANSI/ASME Y10.3M-1984.

Meteorology, Letter Symbols for, ANSI Y10.10-1953(R1973).

Quantities Used in Electrical Science and Electrical Engineering, Letter Symbols for, ANSI/IEEE 280-1985.

Selecting Greek Letters Used as Letter Symbols for Engineering Mathematics, Guide for, ANSI Y10.17-1961 (R1973).

Table 5 Graphic Symbols (after Dreyfus, 1972)

Symbols are a graphical referent to information and have been used for millennia as devices for convenient shorthand notation, to restrict interpretation only to *cognoscenti*, or for compression of data. Examples from several engineering fields are shown here. ISO recommendations are indicated by ¶ and ISO draft recommendations by ¶¶.

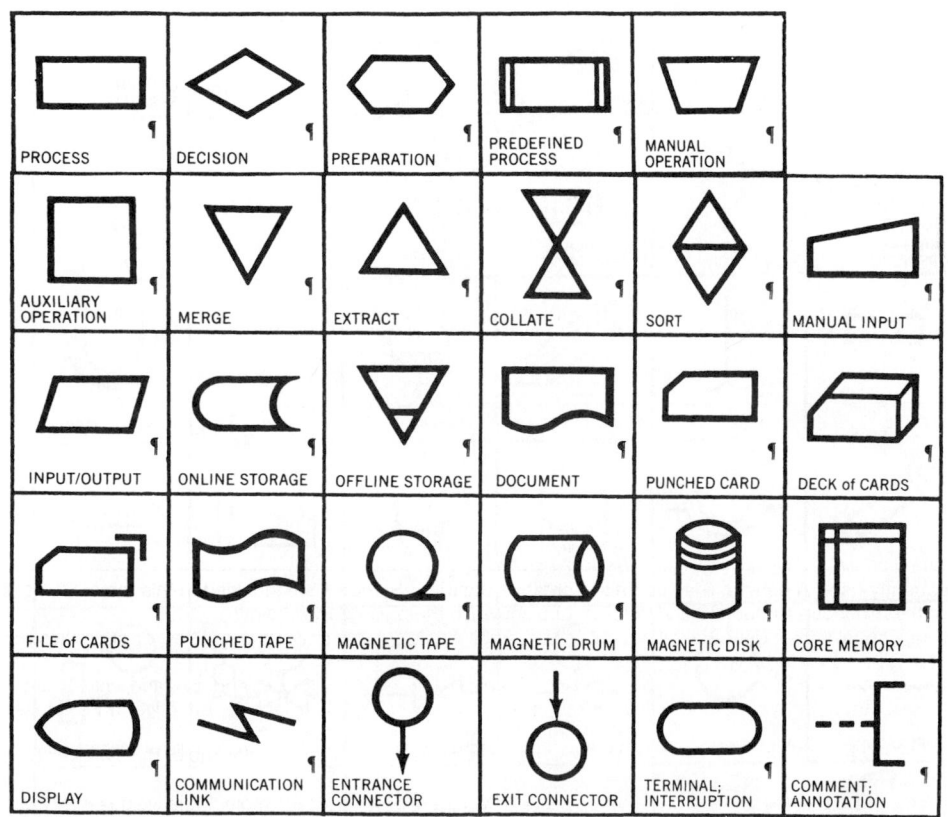

(Continues)

Table 5   (*Continued*)

JACKETED REACTOR, Stirred	NUCLEAR REACTOR	PACKED COLUMN	PLATE COLUMN	SECTIONED COLUMN	DISK and DONUT COLUMN
FIXED BED REACTOR	FLUIDIZED BED REACTOR	AUTOCLAVE	CENTRIFUGAL PUMP	RECIPROCATING PUMP	
REBOILER		HEAT EXCHANGER	WATER COOLER	COOLING TOWER	SPRAY DRYER
BLOWER; FAN	BELT CONVEYOR; SHAKER	BUCKET CONVEYOR	SCREW FEEDER	CENTRIFUGE	CYCLONE SEPARATOR
SINGLE–EFFECT EVAPORATOR	BAROMETRIC CONDENSER	ELECTRICAL PRECIPITATOR	PLATE and FRAME FILTER	ROTARY VACUUM FILTER	THICKENER
JET MIXER; EJECTOR	MIXER	SCREENER	BALL MILL	ROLLER CRUSHER	JACKETED VESSEL
ROTARY DRUM DRYER; KILN	ROTARY FILM DRYER; FLAKER	PRESSURE STORAGE TANK	BULK STORAGE TANK	GAS HOLDER STORAGE TANK	GAS FLOW (100 CFM)
TEMPERATURE (200 °F)	PRESSURE (100 psig)	ALL CONTROL VALVES	PRESSURE CONTROLLER (PC)	TEMPERATURE CONTROLLER (TC)	FLOW CONTROLLER (FC)
LEVEL CONTROLLER (LC)	SUMMATION POINT $R \xrightarrow{+} R - B$; B		OPERATIONAL BLOCK $M \rightarrow \boxed{G} \rightarrow MG$		

Table 5 *(Continued)*

C →●→ C ↓ C TAKE–OFF POINT				

Electrical Engineering

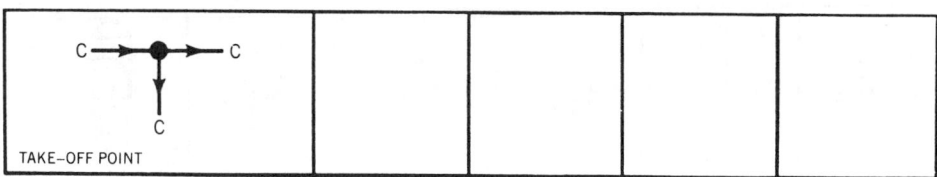

DIRECT CURRENT (DC)	ALTERNATING CURRENT (AC)	AUDIO FREQUENCY AC	SUPERAUDIO FREQUENCY AC	CROSSED CONDUCTORS	JOINED CONDUCTORS
SINGLE–PHASE	2–PHASE 3–WIRE	2–PHASE 4–WIRE	3–PHASE 3–WIRE (Delta)	3–PHASE 3–WIRE (Star)	3–PHASE 4–WIRE (Star)
2 and 3–PHASE TEE CONNECTED	3–PHASE, 3–WIRE VEE CONNECTED	6–PHASE; FORK with NEUTRAL	START of WINDING	VARIABLE CONTROL	VARIABLE CONTROL by STEPS
PRESET CONTROL	ADJUSTABLE TAPPING	PRESET TAPPING	NON–LINEAR VARIABILITY	SATURABLE PROPERTIES	EARTH (Ground)
CHASSIS of EQUIPMENT	INSULATED COUPLING	UNINSULATED COUPLING	SCREENED CONDUCTOR	RESISTOR	
NON–INDUCTIVE RESISTOR (Heater)	ADJUSTABLE CONTACT RESISTOR	INDUCTOR	TRANSFORMER	VACUUM TUBE (Triode)	DIODE
CONTROLLED RECTIFIER	TRANSISTOR (n–––n type)	TRANSISTOR (p–n–p type)	TRANSISTOR, Field–effect (n–channel)	FIXED CAPACITOR	ELECTROLYTIC CAPACITOR

(Continues)

Table 5   *(Continued)*

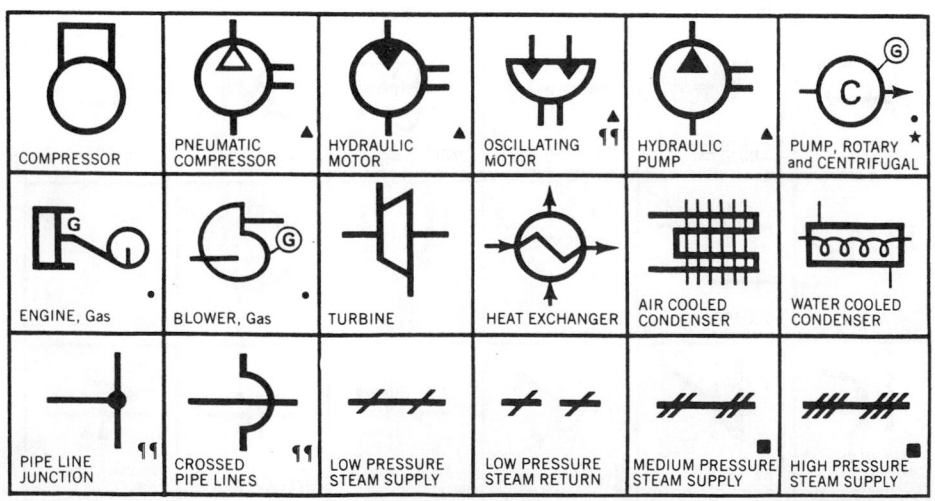

ALTERNATING CURRENT SOURCE	BATTERY (Direct Current Source)	PIEZOELECTRIC CRYSTAL UNIT	AMPLIFIER	LOUDSPEAKER ¶¶	MICROPHONE
CATHODE RAY TUBE (TV)	LAMP BULB ¶¶	INDICATOR ¶¶	LIGHTNING ARRESTER ¶¶	SWITCH (On/Off OR On/Off) ¶¶	
FUSE ¶¶	CIRCUIT BREAKER	RECORDING HEAD	PLAYBACK HEAD	ERASE HEAD	EQUIPMENT OUTLINE ¶¶
AERIAL (Antenna) General ¶¶ OR Dipole ¶¶	OR Loop ¶¶	OR Loop ¶¶	CONNECTOR Male Female ¶¶	Male OR Female	
PERMANENT MAGNET ¶¶	PHOTO-SENSITIVITY ¶¶	BELL ¶¶	BUZZER OR ¶¶ ¶¶		

¶¶ Draft ISO Recommendation

COMPRESSOR	PNEUMATIC COMPRESSOR ▲	HYDRAULIC MOTOR ▲	OSCILLATING MOTOR ¶¶	HYDRAULIC PUMP ▲	PUMP, ROTARY and CENTRIFUGAL ★
ENGINE, Gas	BLOWER, Gas	TURBINE	HEAT EXCHANGER	AIR COOLED CONDENSER	WATER COOLED CONDENSER
PIPE LINE JUNCTION ¶¶	CROSSED PIPE LINES ¶¶	LOW PRESSURE STEAM SUPPLY	LOW PRESSURE STEAM RETURN	MEDIUM PRESSURE STEAM SUPPLY	HIGH PRESSURE STEAM SUPPLY

Table 5 *(Continued)*

Mechanical Engineering (continued)

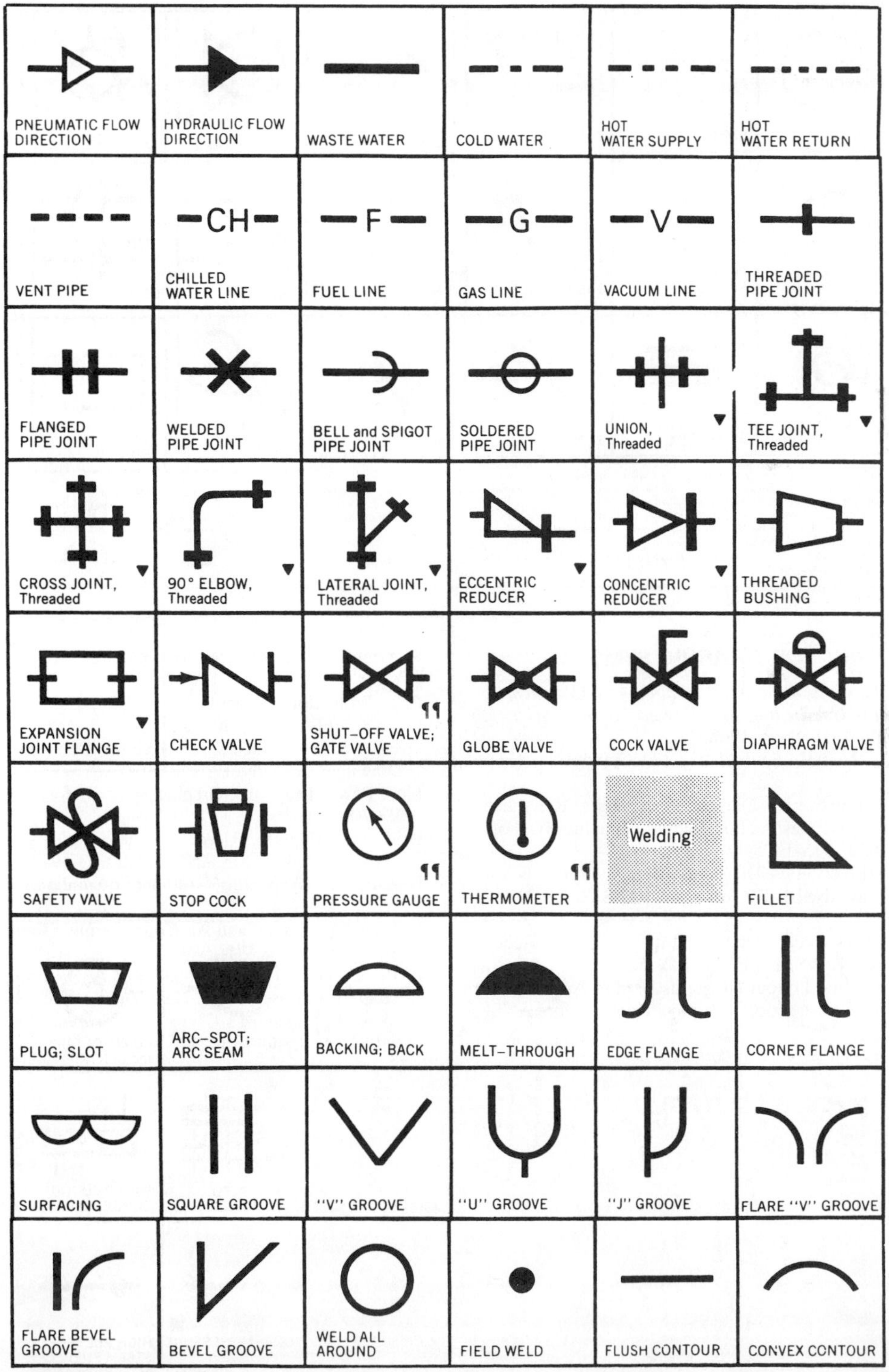

PNEUMATIC FLOW DIRECTION	HYDRAULIC FLOW DIRECTION	WASTE WATER	COLD WATER	HOT WATER SUPPLY	HOT WATER RETURN
VENT PIPE	CHILLED WATER LINE	FUEL LINE	GAS LINE	VACUUM LINE	THREADED PIPE JOINT
FLANGED PIPE JOINT	WELDED PIPE JOINT	BELL and SPIGOT PIPE JOINT	SOLDERED PIPE JOINT	UNION, Threaded	TEE JOINT, Threaded
CROSS JOINT, Threaded	90° ELBOW, Threaded	LATERAL JOINT, Threaded	ECCENTRIC REDUCER	CONCENTRIC REDUCER	THREADED BUSHING
EXPANSION JOINT FLANGE	CHECK VALVE	SHUT–OFF VALVE; GATE VALVE	GLOBE VALVE	COCK VALVE	DIAPHRAGM VALVE
SAFETY VALVE	STOP COCK	PRESSURE GAUGE	THERMOMETER	Welding	FILLET
PLUG; SLOT	ARC–SPOT; ARC SEAM	BACKING; BACK	MELT–THROUGH	EDGE FLANGE	CORNER FLANGE
SURFACING	SQUARE GROOVE	"V" GROOVE	"U" GROOVE	"J" GROOVE	FLARE "V" GROOVE
FLARE BEVEL GROOVE	BEVEL GROOVE	WELD ALL AROUND	FIELD WELD	FLUSH CONTOUR	CONVEX CONTOUR

(Continues)

Table 5   *(Continued)*

Mechanical Engineering (continued)

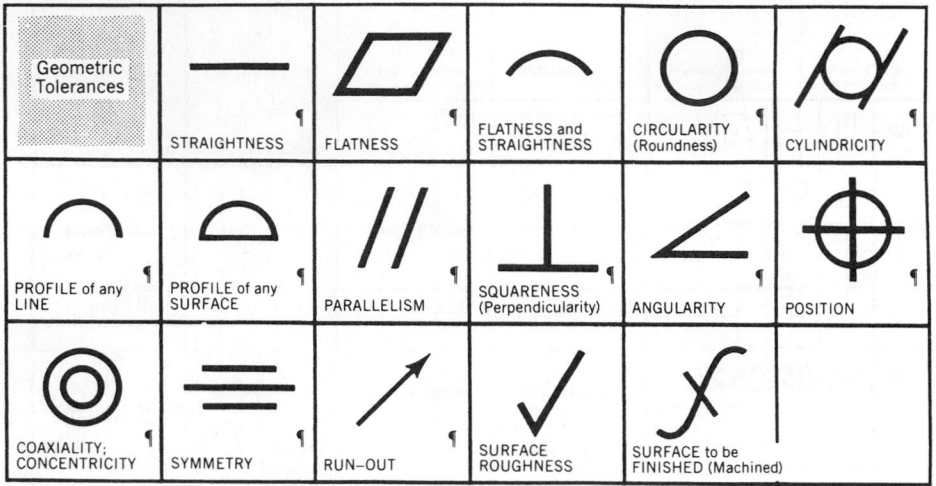

▲ Pneumatic machinery is indicated by △ , hydraulic machinery by ▲
• G indicates Gas. Different initial may be substituted to indicate other type of machine; e.g., D (diesel), M (motor), T (turbine), E (steam). Exception: (see Engine, Gas) Steam Engine is indicated by symbol without initial.
★ C indicates Circulating Water. Different initial indicates other type of machine or service; e.g., D (concentrate), F (boiler feed), O (oil), S (service), V (air).
■ "Return" indicated by broken line, as illustrated in Low Pressure Steam Return.
¶¶ Draft ISO Recommendation
▼ Flanged, Welded, Bell and Spigot, or Soldered Union indicated by substituting appropriate markings (see Joints). Example: ⫞⫟ Welded Union.
¶ ISO Recommendation

BIBLIOGRAPHY FOR GRAPHIC SYMBOLS

Arnell, A. *Standard Graphical Symbols—A Comprehensive Guide for Use in Industry, Engineering and Science*, McGraw-Hill, New York, 1963.

Dreyfus, H. *Symbol Sourcebook*, McGraw-Hill, New York, 1972.

Electrical and Electronics Diagrams (Including Reference Designation Class Designation Letters), Graphic Symbols for, ANSI/IEEE 315-1975.

Electrical and Electronics Diagrams, Graphic Symbols for (supplement to ANSI/IEEE 315-1975), ANSI/IEEE 315A-1986.

Electrical and Electronics Parts and Equipments, Reference Designations for, ANSI/IEEE 200-1975.

Electrical Wiring and Layout Diagrams Used in Architecture and Building Construction, Graphic Symbols for, ANSI Y32.9-1972.

Fire Fighting Operations, Symbols for, ANSI/NFPA 178-1980.

Fire Protection Symbols for risk Analysis Diagrams, ANSI/NFPA 174-1980.

Fire-Protection Symbols for Architectural and Engineering Drawings, ANSI/NFPA 172-1980.

Fluid Power Diagrams, Graphic Symbols for, ANSI Y32.10-1967(R1979).

Grid and Mapping Used in Cable Television Systems, Graphic Symbols for, ANSI/IEEE 623-1976.

Heat-Power Apparatus, Graphic Symbols for, ANSI Y32.2.6M-1984.

Heating, Ventilating, and Air Conditioning, Graphic Symbols for, ANSI Y32.2.4M-1984.

Polon, D. D. (Ed.), *Encyclopedia of Engineering Signs and Symbols*, Odyssey Press, New York, 1965.

Shepard, W., *Shepard's Glossary of Graphic Signs and Symbols*, Dent, London, 1971.

Table 6   Personal Computer Numeric Codes for Characters and Symbols

IBM PC Character Set (00–7F) Quick Reference

DECIMAL VALUE →	0	16	32	48	64	80	96	112
HEXADECIMAL VALUE →	0	1	2	3	4	5	6	7
0 / 0	BLANK (NULL)	►	BLANK (SPACE)	0	@	P	`	p
1 / 1	☺	◄	!	1	A	Q	a	q
2 / 2	☻	↕	"	2	B	R	b	r
3 / 3	♥	‼	#	3	C	S	c	s
4 / 4	♦	¶	$	4	D	T	d	t
5 / 5	♣	§	%	5	E	U	e	u
6 / 6	♠	▬	&	6	F	V	f	v
7 / 7	•	↨	'	7	G	W	g	w
8 / 8	◘	↑	(	8	H	X	h	x
9 / 9	○	↓	)	9	I	Y	i	y
10 / A	◙	→	*	:	J	Z	j	z
11 / B	♂	←	+	;	K	[	k	{
12 / C	♀	∟	,	<	L	\	l	|
13 / D	♪	↔	-	=	M	]	m	}
14 / E	♫	▲	.	>	N	^	n	~
15 / F	☼	▼	/	?	O	_	o	Δ

IBM PC Character Set (80–FF) Quick Reference

DECIMAL VALUE →	128	144	160	176	192	208	224	240
HEXADECIMAL VALUE →	8	9	A	B	C	D	E	F
0 / 0	Ç	É	á	░	└	╨	α	≡
1 / 1	ü	æ	í	▒	┴	╤	β	±
2 / 2	é	Æ	ó	▓	┬	╥	Γ	≥
3 / 3	â	ô	ú	│	├	╙	π	≤
4 / 4	ä	ö	ñ	┤	─	╘	Σ	⌠
5 / 5	à	ò	Ñ	╡	┼	╒	σ	⌡
6 / 6	å	û	ª	╢	╞	╓	µ	÷
7 / 7	ç	ù	º	╖	╟	╫	τ	≈
8 / 8	ê	ÿ	¿	╕	╚	╪	Φ	°
9 / 9	ë	Ö	⌐	╣	╔	┘	Θ	∙
10 / A	è	Ü	¬	║	╩	┌	Ω	·
11 / B	ï	¢	½	╗	╦	█	δ	√
12 / C	î	£	¼	╝	╠	▄	∞	ⁿ
13 / D	ì	¥	¡	╜	═	▌	φ	²
14 / E	Ä	₧	«	╛	╬	▐	ε	■
15 / F	Å	ƒ	»	┐	╧	▀	∩	BLANK FF

Source: Reprint courtesy of International Business Machines Corporation, copyright 1985 © by International Business Machines Corporation.

19

Table 7   Conversions for Number Systems of Different Bases

Radix 16 Hexadecimal	Radix 10 Decimal	Radix 8 Octal	Radix 2 Binary BIT 8765	Radix 2 Binary BIT 4321	Radix 16 Hexadecimal	Radix 10 Decimal	Radix 8 Octal	Radix 2 Binary BIT 8765	Radix 2 Binary BIT 4321
00	0	00	0000	0000	3A	58	72	0011	1010
01	1	01	0000	0001	3B	59	73	0011	1011
02	2	02	0000	0010	3C	60	74	0011	1100
03	3	03	0000	0011	3D	61	75	0011	1101
04	4	04	0000	0100	3E	62	76	0011	1110
05	5	05	0000	0101	3F	63	77	0011	1111
06	6	06	0000	0110	40	64	100	0100	0000
07	7	07	0000	0111	41	65	101	0100	0001
08	8	10	0000	1000	42	66	102	0100	0010
09	9	11	0000	1001	43	67	103	0100	0011
0A	10	12	0000	1010	44	68	104	0100	0100
0B	11	13	0000	1011	45	69	105	0100	0101
0C	12	14	0000	1100	46	70	106	0100	0110
0D	13	15	0000	1101	47	71	107	0100	0111
0E	14	16	0000	1110	48	72	110	0100	1000
0F	15	17	0000	1111	49	73	111	0100	1001
10	16	20	0001	0000	4A	74	112	0100	1010
11	17	21	0001	0001	4B	75	113	0100	1011
12	18	22	0001	0010	4C	76	114	0100	1100
13	19	23	0001	0011	4D	77	115	0100	1101
14	20	24	0001	0100	4E	78	116	0100	1110
15	21	25	0001	0101	4F	79	117	0100	1111
16	22	26	0001	0110	50	80	120	0101	0000
17	23	27	0001	0111	51	81	121	0101	0001
18	24	30	0001	1000	52	82	122	0101	0010
19	25	31	0001	1001	53	83	123	0101	0011
1A	26	32	0001	1010	54	84	124	0101	0100
1B	27	33	0001	1011	55	85	125	0101	0101
1C	28	34	0001	1100	56	86	126	0101	0110
1D	29	35	0001	1101	57	87	127	0101	0111
1E	30	36	0001	1110	58	88	130	0101	1000
1F	31	37	0001	1111	59	89	131	0101	1001
20	32	40	0010	0000	5A	90	132	0101	1010
21	33	41	0010	0001	5B	91	133	0101	1011
22	34	42	0010	0010	5C	92	134	0101	1100
23	35	43	0010	0011	5D	93	135	0101	1101
24	36	44	0010	0100	5E	94	136	0101	1110
25	37	45	0010	0101	5F	95	137	0101	1111
26	38	46	0010	0110	60	96	140	0110	0000
27	39	47	0010	0111	61	97	141	0110	0001
28	40	50	0010	1000	62	98	142	0110	0010
29	41	51	0010	1001	63	99	143	0110	0011
2A	42	52	0010	1010	64	100	144	0110	0100
2B	43	53	0010	1011	65	101	145	0110	0101
2C	44	54	0010	1100	66	102	146	0110	0110
2D	45	55	0010	1101	67	103	147	0110	0111
2E	46	56	0010	1110	68	104	150	0110	1000
2F	47	57	0010	1111	69	105	151	0110	1001
30	48	60	0011	0000	6A	106	152	0110	1010
31	49	61	0011	0001	6B	107	153	0110	1011
32	50	62	0011	0010	6C	108	154	0110	1100
33	51	63	0011	0011	6D	109	155	0110	1101
34	52	64	0011	0100	6E	110	156	0110	1110
35	53	65	0011	0101	6F	111	157	0110	1111
36	54	66	0011	0110	70	112	160	0111	0000
37	55	67	0011	0111	71	113	161	0111	0001
38	56	70	0011	1000	72	114	162	0111	0010
39	57	71	0011	1001	73	115	163	0111	0011

Table 7 *(Continued)*

Radix 16 Hexadecimal	Radix 10 Decimal	Radix 8 Octal	Radix 2 Binary BIT 8765	4321	Radix 16 Hexadecimal	Radix 10 Decimal	Radix 8 Octal	Radix 2 Binary BIT 8765	4321
74	116	164	0111	0100	D9	217	331	1101	1001
75	117	165	0111	0101	DA	218	332	1101	1010
76	118	166	0111	0110	DB	219	333	1101	1011
77	119	167	0111	0111	DC	220	334	1101	1100
78	120	170	0111	1000	DD	221	335	1101	1101
79	121	171	0111	1001	DE	222	336	1101	1110
7A	122	172	0111	1010	DF	223	337	1101	1111
7B	123	173	0111	1011	E0	224	340	1110	0000
7C	124	174	0111	1100	E1	225	341	1110	0001
7D	125	175	0111	1101	E2	226	342	1110	0010
7E	126	176	0111	1110	E3	227	343	1110	0011
7F	127	177	0111	1111	E4	228	344	1110	0100
80	128	200	1000	0000	E5	229	345	1110	0101
81	129	201	1000	0001	E6	230	346	1110	0110
82	130	202	1000	0010	E7	231	347	1110	0111
83	131	203	1000	0011	E8	232	350	1110	1000
84	132	204	1000	0100	E9	233	351	1110	1001
85	133	205	1000	0101	EA	234	352	1110	1010
86	134	206	1000	0110	EB	235	353	1110	1011
87	135	207	1000	0111	EC	236	354	1110	1100
88	136	210	1000	1000	ED	237	355	1110	1101
89	137	211	1000	1001	EE	238	356	1110	1110
8A	138	212	1000	1010	EF	239	357	1110	1111
8B	139	213	1000	1011	F0	240	360	1111	0000
8C	140	214	1000	1100	F1	241	361	1111	0001
8D	141	215	1000	1101	F2	242	362	1111	0010
8E	142	216	1000	1110	F3	243	363	1111	0011
8F	143	217	1000	1111	F4	244	364	1111	0100
90	144	220	1001	0000	F5	245	365	1111	0101
91	145	221	1001	0001	F6	246	366	1111	0110
92	146	222	1001	0010	F7	247	367	1111	0111
93	147	223	1001	0011	F8	248	370	1111	1000
94	148	224	1001	0100	F9	249	371	1111	1001
95	149	225	1001	0101	FA	250	372	1111	1010
96	150	226	1001	0110	FB	251	373	1111	1011
97	151	227	1001	0111	FC	252	374	1111	1100
98	152	230	1001	1000	FD	253	375	1111	1101
99	153	231	1001	1001	FE	254	376	1111	1110
9A	154	232	1001	1010	FF	255	377	1111	1111
9B	155	233	1001	1011	AB	171	253	1010	1011
9C	156	234	1001	1100	AC	172	254	1010	1100
9D	157	235	1001	1101	AD	173	255	1010	1101
9E	158	236	1001	1110	AE	174	256	1010	1110
9F	159	237	1001	1111	AF	175	257	1010	1111
A0	160	240	1010	0000	B0	176	260	1011	0000
A1	161	241	1010	0001	B1	177	261	1011	0001
A2	162	242	1010	0010	B2	178	262	1011	0010
A3	163	243	1010	0011	B3	179	263	1011	0011
A4	164	244	1010	0100	B4	180	264	1011	0100
A5	165	245	1010	0101	B5	181	265	1011	0101
A6	166	246	1010	0110	B6	182	286	1011	0110
A7	167	247	1010	0111	B7	183	267	1011	0111
A8	168	250	1010	1000	B8	184	270	1011	1000
A9	169	251	1010	1001	B9	185	271	1011	1001
AA	170	252	1010	1010	BA	186	272	1011	1010
D6	214	326	1101	0110	BB	187	273	1011	1011
D7	215	327	1101	0111	BC	188	274	1011	1100
D8	216	330	1101	1000	BD	189	275	1011	1101

(Continues)

Table 7 *(Continued)*

Radix 16 Hexadecimal	Radix 10 Decimal	Radix 8 Octal	Radix 2 Binary BIT		Radix 16 Hexadecimal	Radix 10 Decimal	Radix 8 Octal	Radix 2 Binary BIT	
			8765	4321				8765	4321
BE	190	276	1011	1110	CA	202	312	1100	1010
BF	191	277	1011	1111	CB	203	313	1100	1011
C0	192	300	1100	0000	CC	204	314	1100	1100
C1	193	301	1100	0001	CD	205	315	1100	1101
C2	194	302	1100	0010	CE	206	316	1100	1110
C3	195	303	1100	0011	CF	207	317	1100	1111
C4	196	304	1100	0100	D0	208	320	1101	0000
C5	197	305	1100	0101	D1	209	321	1101	0001
C6	198	306	1100	0110	D2	210	322	1101	0010
C7	199	307	1100	0111	D3	211	323	1101	0011
C8	200	310	1100	1000	D4	212	324	1101	0100
C9	201	311	1100	1001	D5	213	325	1101	0101

Table 8   Computer Graphics Codes and Standards

Modern computer-aided design (CAD), computer-aided manufacturing (CAM), and computer-aided engineering (CAE) are heavily dependent on computer graphics. A standard computer graphics metafile (CGM) is necessary in order to:

1. Allow picture information to be stored in an organized way on a graphical software system.
2. Facilitate transfer of picture information between different graphical software systems.
3. Enable picture information to be transferred between graphical devices.
4. Enable picture information to be transferred between different computer graphics installations.

More particularly, the CGM should provide these capabilities in a device-independent manner. To accomplish this, the standard defines the form (syntax) and functional behavior (semantics) of a set of elements that may occur in the CGM. There are eight classes of elements:

1. Delimiter Elements — delimit significant structures within the metafile.
2. Metafile Descriptor Elements — describe the functional content, default conditions, identification, and characteristics of the CGM.
3. Picture Descriptor Elements — set the interpretation modes of attribute elements for each picture.
4. Control Elements — allow picture boundaries and coordinate representation to be modified.
5. Graphical Primitive Elements — describe the visual components of a picture in the CGM.
6. Attribute Elements — describe the appearance of graphical primitive elements.
7. Escape Elements — describe device- or system-dependent elements used to construct a picture; however, the elements are not otherwise standardized.
8. External Elements — communicate information not directly related to the generation of a graphical image.

A computer graphics metafile is a collection of elements from this standardized set. The BEGIN METAFILE and END METAFILE elements each occur exactly once in a complete metafile; as many or as few of the elements in the other classes may occur as are needed. A metafile needs to be interpreted in order to display its pictorial content on a graphics device. The descriptor elements give the interpreter sufficient data to interpret metafile elements and to make informed decisions concerning the resources needed for display.

    A CGM contains delimiter elements; in addition it may include control elements for metafile interpretation, picture descriptor elements for declaring parameter modes of attribute elements, graphical primitive elements for defining graphical entities, attribute elements for defining the appearance of the graphical primitive elements, escape elements for accessing nonstandardized features of particular devices, and external elements for communication of information external to the definition of the pictures in the CGM.

    Full description and depiction of all the elements thus far defined in this standardized set is beyond the scope of this handbook. The interested reader is referred to ANSI standard X3.122-1986 and Smith, B. M., et al. "Initial Graphics Exchange Specification (IGES), Version 2.0," NBS (R82-2631) (AF) Feb. (1983) 26 pp.

2 MATHEMATICAL TABLES

Table 9 Certain Constants Containing e and π [a]

Powers of e			Multiples of π			Fractions of π		
e^n	Value	Logarithm	n_π	Value	Logarithm	π/n	Value	Logarithm
e	2.718282	0.434294	π	3.141593	0.497150	$\pi/2$	1.570780	0.196120
e^{-1}	0.367879	$\bar{1}$.565706	2π	6.283185	0.798180	$\pi/3$	1.047198	0.020029
e^2	7.389057	0.868589	3π	9.424778	0.974271	$\pi/4$	0.785398	$\bar{1}$.895090
e^{-2}	0.135335	$\bar{1}$.131411	4π	12.566371	1.099210	$\pi/180$	0.017453 [b]	2.241877
$e^{1/2}$	1.648721	0.217147	5π	15.707963	1.196120			

Reciprocals of π			Powers of π			Roots of π		
n/π	Value	Logarithm	$\pi^{\pm n}$	Value	Logarithm	$\pi^{\pm 1/n}$	Value	Logarithm
$1/\pi$	0.318310	$\bar{1}$.502850	π^2	9.869604	0.994300	$\sqrt{\pi}$	1.772454	0.248575
$2/\pi$	0.636620	$\bar{1}$.803880	$1/\pi^2$	0.101321	$\bar{1}$.005700	$1/\sqrt{\pi}$	0.564190	$\bar{1}$.751425
$3/\pi$	0.954930	$\bar{1}$.979971	π^3	31.006277	1.491450	$\sqrt[3]{\pi}$	1.464592	0.165717
$180/\pi$	57.295780 [c]	1.758123	$1/\pi^3$	0.032252	$\bar{2}$.508550	$1/\sqrt[3]{\pi}$	0.682784	$\bar{1}$.834283

[a] $e = 2.7182818285;\quad \pi = 3.1415926536;\quad M = \log_{10} e = 0.4342944819;\quad M^{-1} = \log_e 10 = 2.3025850930.$
[b] Number of radians per degree.
[c] Number of degrees per radian.

Table 10 Factorials

n	$n! = 1 \cdot 2 \cdot 3 \cdots n$	$1/n!$	n	$n! = 1 \cdot 2 \cdot 3 \cdots n$	$1/n!$
1	1	1.	11	$399{,}168 \times 10^2$	0.250521×10^{-7}
2	2	0.5	12	$479{,}002 \times 10^3$	0.208768×10^{-8}
3	6	0.166667	13	$622{,}702 \times 10^4$	0.160590×10^{-9}
4	24	0.416667×10^{-1}	14	$871{,}783 \times 10^5$	0.114707×10^{-10}
5	120	0.833333×10^{-2}	15	$130{,}767 \times 10^7$	0.764716×10^{-12}
6	720	0.138889×10^{-2}	16	$209{,}228 \times 10^8$	0.477948×10^{-13}
7	5,040	0.198413×10^{-3}	17	$355{,}687 \times 10^9$	0.281146×10^{-14}
8	40,320	0.248016×10^{-4}	18	$640{,}237 \times 10^{10}$	0.156192×10^{-15}
9	362,880	0.275573×10^{-5}	19	$121{,}645 \times 10^{12}$	0.822064×10^{-17}
10	3,628,800	0.275573×10^{-6}	20	$243{,}290 \times 10^{13}$	0.411032×10^{-19}

Table 11 Common and Natural Logarithms of Numbers

The common logarithm of a number is the index of the power to which the base 10 must be raised in order to equal the number.

The common logarithm of every positive number not an integral power of 10 consists of an *integral* and a *decimal part*. The integral part or whole number is called the *characteristic* and may be either *positive* or *negative*. The decimal or fractional part is a positive number called the *mantissa* and is the same for all numbers which have the same sequential digits.

The characteristic of the logarithm of any positive number greater than 1 is positive and is 1 less than the number of digits before the decimal point.

The characteristic of the logarithm of any positive number less than 1 is negative and is 1 more than the number of ciphers immediately after the decimal point.

A negative number or number less than zero has no real logarithm.

Examples: $\log_{10} 25{,}400. = 4.404834,\quad \log_{10} 0.0254 = \bar{2}.404834$, or $8.404834 - 10$

(Continues)

Table 11 *(Continued)*

The two systems of logarithms in general use are the common or Briggsian logarithms, introduced in 1615 by Henry Briggs, a contemporary of John Napier, the inventor of logarithms, and the natural or less appropriately termed Napierian or hyperbolic logarithms, which developed somewhat accidentally from Napier's original work. The latter have a base denoted by e, an irrational number, which is

$$\lim_{u=\infty} \left(1 + \frac{1}{u}\right)^{u} = 1 + 1 + \frac{1}{2!} + \frac{1}{3!} + \frac{1}{4!} + \cdots = 2.7182818$$

To obtain the natural logarithm, the common logarithm is multiplied by $\log_e 10$, which is 2.302585, or $\log_e N = 2.302585 \log_{10} N$.

The natural logarithm of a number is the index of the power to which the base e (= 2.7182818) must be raised in order to equal the number.

Example: $\log_e 4.12 = \ln 4.12 = 1.4159$.

Natural logarithms of numbers from 1.00 to 9.99 may be obtained directly; the natural logarithms of numbers outside of that range by the addition or subtraction of the natural logarithms of powers of 10.

Examples: $\log_e 679. \quad = \log_e 6.79 + \log_e 10^2 = 1.9155 + 4.6052 = 6.5207.$
$\log_e 0.0679 = \log_e 6.79 - \log_e 10^2 = 1.9155 - 4.6052 = -2.6897$

Natural Logarithms of Powers of 10

$$\log_e 10 \ = 2.302\ 585 \qquad \log_e 10^4 = 9.210\ 340 \qquad \log_e 10^7 = 16.118\ 096$$
$$\log_e 10^2 = 4.605\ 170 \qquad \log_e 10^5 = 11.512\ 925 \qquad \log_e 10^8 = 18.420\ 681$$
$$\log_e 10^3 = 6.907\ 755 \qquad \log_e 10^6 = 13.815\ 511 \qquad \log_e 10^9 = 20.723\ 266$$

To obtain the common logarithm, the natural logarithm is multiplied by $\log_{10} e$, which is 0.434294, or $\log_{10} N = 0.434294 \log_e N$.

A negative number or number less than zero has no real logarithm.

Tabulations of common and natural logarithms are no longer provided in this handbook because of ready access to them on modern pocket and desk calculators.

Values and Logarithms of Exponentials and Hyperbolic Functions

Many calculators directly give values of $e^x, e^{-x}, \sinh x, \cosh x$, and $\tanh x$ for any value of x. These quantities are therefore not tabulated here.

For values of x greater than 6, e^x may be computed from the relationship $e^x = \log^{-1}(x \log_{10} e) = \log^{-1} 0.43429x$; e^{-x} approaches zero; $\sinh x$ and $\cosh x$ are approximately equal and become $0.5e^x$; and $\tanh x$ and $\coth x$ have values approximately equal to unity.

Where more accurate values of the exponentials and functions are required they may be computed from the following relationships:

$$e = 2.7182818285 \qquad\qquad \tfrac{1}{e} = 0.3678794412$$

$$M = \log_{10} e = 0.4342944819 \qquad\qquad \frac{1}{M} = \log_e 10 = 2.3025850930$$

$$e^x = \log^{-1} Mx \qquad\qquad e^{-x} = \log^{-1}(-Mx)$$

$$\sin hx = \frac{e^x - e^{-x}}{2} \qquad\qquad \cos hx = \frac{e^x + e^{-x}}{2} \qquad\qquad \tan hx = \frac{e^x - e^{-x}}{e^x + e^{-x}}$$

$$\operatorname{csch} x = \frac{1}{\sin hx} \qquad\qquad \sec hx = \frac{1}{\cos hx} \qquad\qquad \cot hx = \frac{1}{\tan hx}$$

Table 11 (*Continued*)

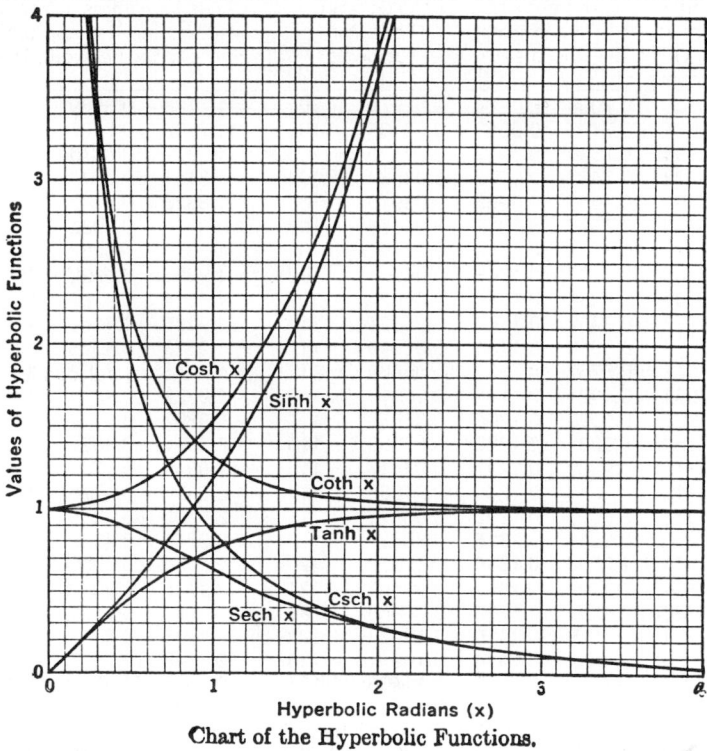

Chart of the Hyperbolic Functions.

Values of the hyperbolic functions are shown in the figure.

Table 12   Circular Arcs, Chords, and Segments

Central Angle in Degrees	Arc $\dfrac{}{R}$	Height $\dfrac{}{R}$	Chord $\dfrac{}{R}$	Height $\dfrac{}{\text{Chord}}$	Area $\dfrac{}{R^2}$	Central Angle in Degrees	Arc $\dfrac{}{R}$	Height $\dfrac{}{R}$	Chord $\dfrac{}{R}$	Height $\dfrac{}{\text{Chord}}$	Area $\dfrac{}{R^2}$
1	0.0175	0.0000	0.0175	0.0022	0.00000	21	0.3665	0.0167	0.3645	0.0459	0.00408
2	0.0349	0.0002	0.0349	0.0044	0.00000	22	0.3840	0.0184	0.3816	0.0481	0.00468
3	0.0524	0.0003	0.0524	0.0066	0.00001	23	0.4014	0.0201	0.3987	0.0503	0.00535
4	0.0698	0.0006	0.0698	0.0087	0.00003	24	0.4189	0.0219	0.4158	0.0526	0.00607
5	0.0873	0.0010	0.0872	0.0109	0.00006	25	0.4363	0.0237	0.4329	0.0548	0.00686
6	0.1047	0.0014	0.1047	0.0131	0.00010	26	0.4538	0.0256	0.4499	0.0570	0.00771
7	0.1222	0.0019	0.1221	0.0153	0.00015	27	0.4712	0.0276	0.4669	0.0592	0.00862
8	0.1396	0.0024	0.1395	0.0175	0.00023	28	0.4887	0.0297	0.4838	0.0614	0.00961
9	0.1571	0.0031	0.1569	0.0196	0.00032	29	0.5061	0.0319	0.5008	0.0636	0.01067
10	**0.1745**	**0.0038**	**0.1743**	**0.0218**	**0.00044**	**30**	**0.5236**	**0.0341**	**0.5176**	**0.0658**	**0.1180**
11	0.1920	0.0046	0.1917	0.0240	0.00059	31	0.5411	0.0364	0.5345	0.0680	0.01301
12	0.2094	0.0055	0.2091	0.0262	0.00076	32	0.5585	0.0387	0.5513	0.0703	0.01429
13	0.2296	0.0064	0.2264	0.0284	0.00097	33	0.5760	0.0412	0.5680	0.0725	0.01566
14	0.2443	0.0075	0.2437	0.0306	0.00121	34	0.5934	0.0437	0.5847	0.0747	0.01711
15	0.2618	0.0086	0.2611	0.0328	0.00149	35	0.6109	0.0463	0.6014	0.0770	0.01864
16	0.2793	0.0097	0.2783	0.0350	0.00181	36	0.6283	0.0489	0.6180	0.0792	0.02027
17	0.2967	0.0110	0.2956	0.0372	0.00217	37	0.6458	0.0517	0.6346	0.0814	0.02198
18	0.3142	0.0123	0.3129	0.0394	0.00257	38	0.6632	0.0545	0.6511	0.0837	00.2378
19	0.3316	0.0137	0.3301	0.0415	0.00302	39	0.6807	0.0574	0.6676	0.0859	0.02568
20	**0.3491**	**0.0152**	**0.3473**	**0.0437**	**0.00352**	**40**	**0.6981**	**0.0603**	**0.6840**	**0.0882**	**0.02767**

(Continues)

Table 12 (*Continued*)

Central Angle in Degrees	Arc R	Height R	Chord R	Height Chord	Area R²	Central Angle in Degrees	Arc R	Height R	Chord R	Height Chord	Area R²
41	0.7156	0.0633	0.7004	0.0904	0.02976	101	1.7628	0.3639	1.543	0.2358	0.39058
42	0.7330	0.0664	0.7167	0.0927	0.03195	102	1.7802	0.3707	1.554	0.2385	0.40104
43	0.7505	0.0696	0.7330	0.0949	0.03425	103	1.7977	0.3775	1.565	0.2412	0.41166
44	0.7679	0.0728	0.7492	0.0972	0.03664	104	1.8151	0.3843	1.576	0.2439	0.42242
45	0.7854	0.0761	0.7654	0.0995	0.03915	105	1.8326	0.3912	1.587	0.2466	0.43333
46	0.8029	0.0795	0.7815	0.1017	0.04176	106	1.8500	0.3982	1.597	0.2493	0.44439
47	0.8203	0.0829	0.7975	0.1040	0.04448	107	1.8675	0.4052	1.608	0.2520	0.45560
48	0.8378	0.0865	0.8135	0.1063	0.04731	108	1.8850	0.4122	1.618	0.2548	0.46695
49	0.8552	0.0900	0.8294	0.1086	0.05025	109	1.9024	0.4193	1.628	0.2575	0.47844
50	**0.8727**	**0.0937**	**0.8452**	**0.1108**	**0.05331**	**110**	**1.9199**	**0.4264**	**1.638**	**0.2603**	**0.49008**
51	0.8901	0.0974	0.8610	0.1131	0.05649	111	1.9373	0.4336	1.648	0.2631	0.50187
52	0.9076	0.1012	0.8767	0.1154	0.05978	112	1.9548	0.4408	1.658	0.2659	0.51379
53	0.9250	0.1051	0.8924	0.1177	0.06319	113	1.9722	0.4481	1.668	0.2687	0.52586
54	0.9425	0.1090	0.9080	0.1200	0.06673	114	1.9897	0.4554	1.677	0.2715	0.53807
55	0.9599	0.1130	0.9235	0.1223	0.07039	115	2.0071	0.4627	1.687	0.2743	0.55041
56	0.9774	0.1171	0.9389	0.1247	0.07417	116	2.0246	0.4701	1.696	0.2772	0.56289
57	0.9948	0.1212	0.9543	0.1270	0.07808	117	2.0420	0.4775	1.705	0.2800	0.57551
58	1.0123	0.1254	0.9696	0.1293	0.08212	118	2.0595	0.4850	1.714	0.2829	0.58827
59	1.0297	0.1296	0.9848	0.1316	0.08629	119	2.0769	0.4925	1.723	0.2858	0.60116
60	**1.0472**	**0.1340**	**1.0000**	**0.1340**	**0.09059**	**120**	**2.0944**	**0.5000**	**1.732**	**0.2887**	**0.61418**
61	1.0647	0.1384	1.015	0.1363	0.09502	121	2.1118	0.5076	1.741	0.2916	0.62734
62	1.0821	0.1428	1.030	0.1387	0.09958	122	2.1293	0.5152	1.749	0.2945	0.64063
63	1.0996	0.1474	1.045	0.1410	0.10428	123	2.1468	0.5228	1.758	0.2975	0.65404
64	1.1170	0.1520	1.060	0.1434	0.10911	124	2.1642	0.5305	1.766	0.3004	0.66759
65	1.1345	0.1566	1.075	0.1457	0.11408	125	2.1817	0.5383	1.774	0.3034	0.68125
66	1.1519	0.1613	1.089	0.1481	0.11919	126	2.1991	0.5460	1.782	0.3064	0.69505
67	1.1694	0.1661	1.104	0.1505	0.12443	127	2.2166	0.5538	1.790	0.3094	0.70897
68	1.1868	0.1710	1.118	0.1529	0.12982	128	2.2340	0.5616	1.798	0.3124	0.72301
69	1.2043	0.1759	1.133	0.1553	0.13535	129	2.2515	0.5695	1.805	0.3155	0.73716
70	**1.2217**	**0.1808**	**1.147**	**0.1576**	**0.14102**	**130**	**2.2689**	**0.5774**	**1.813**	**0.3185**	**0.75143**
71	1.2392	0.1859	1.161	0.1601	0.14683	131	2.2864	0.5853	1.820	0.3216	0.76584
72	1.2566	0.1910	1.176	0.1625	0.15279	132	2.3038	0.5933	1.827	0.3247	0.78034
73	1.2741	0.1961	1.190	0.1649	0.15889	133	2.3213	0.6013	1.834	0.3278	0.79497
74	1.2915	0.2014	1.204	0.1673	0.16514	134	2.3387	0.6093	1.841	0.3309	0.80970
75	1.3090	0.2066	1.218	0.1697	0.17154	135	2.3562	0.6173	1.848	0.3341	0.82454
76	1.3265	0.2120	1.231	0.1722	0.17808	136	2.3736	0.6254	1.854	0.3373	0.83949
77	1.3439	0.2174	1.245	0.1746	0.18477	137	2.3911	0.6335	1.861	0.3404	0.85455
78	1.3614	0.2229	1.259	0.1771	0.19160	138	2.4086	0.6416	1.867	0.3436	0.86971
79	1.3788	0.2284	1.272	0.1795	0.19859	139	2.4260	0.6498	1.873	0.3469	0.88497
80	**1.3963**	**0.2340**	**1.286**	**0.1820**	**0.20573**	**140**	**2.4435**	**0.6580**	**1.879**	**0.3501**	**0.90034**
81	1.4137	0.2396	1.299	0.1845	0.21301	141	2.4609	0.6662	1.885	0.3534	0.91580
82	1.4312	0.2453	1.312	0.1869	0.22045	142	2.4784	0.6744	1.891	0.3566	0.93135
83	1.4486	0.2510	1.325	0.1894	0.22804	143	2.4958	0.6827	1.897	0.3599	0.94700
84	1.4661	0.2569	1.338	0.1919	0.23578	144	2.5133	0.6910	1.902	0.3633	0.96274
85	1.4835	0.2627	1.351	0.1944	0.24367	145	2.5307	0.6993	1.907	0.3666	0.97858
86	1.5010	0.2686	1.364	0.1970	0.25171	146	2.5482	0.7076	1.913	0.3700	0.99449
87	1.5184	0.2746	1.377	0.1995	0.25990	147	2.5656	0.7160	1.918	0.3734	1.0105
88	1.5359	0.2807	1.389	0.2020	0.26825	148	2.5831	0.7244	1.923	0.3768	1.0266
89	1.5533	0.2867	1.402	0.2046	0.27675	149	2.6005	0.7328	1.927	0.3802	1.0428
90	**1.5708**	**0.2929**	**1.414**	**0.2071**	**0.28540**	**150**	**2.6180**	**0.7412**	**1.932**	**0.3837**	**1.0590**
91	1.5882	0.2991	1.427	0.2097	0.29420	151	2.6354	0.7496	1.936	0.3871	1.0753
92	1.6057	0.3053	1.439	0.2122	0.30316	152	2.6529	0.7581	1.941	0.3906	1.0917
93	1.6232	0.3116	1.451	0.2148	0.31226	153	2.6704	0.7666	1.945	0.3942	1.1082
94	1.6406	0.3180	1.463	0.2174	0.32152	154	2.6878	0.7750	1.949	0.3977	1.1247
95	1.6581	0.3244	1.475	0.2200	0.33093	155	2.7053	0.7836	1.953	0.4013	1.1413
96	1.6755	0.3309	1.486	0.2226	0.34050	156	2.7227	0.7921	1.956	0.4049	1.1580
97	1.6930	0.3374	1.498	0.2252	0.35021	157	2.7402	0.8006	1.960	0.4085	1.1747
98	1.7104	0.3439	1.509	0.2279	0.36008	158	2.7576	0.8092	1.963	0.4122	1.1915
99	1.7279	0.3506	1.521	0.2305	0.37009	159	2.7751	0.8178	1.967	0.4158	1.2084
100	**1.7453**	**0.3572**	**1.532**	**0.2332**	**0.38026**	**160**	**2.7925**	**0.8264**	**1.970**	**0.4195**	**1.2253**

Table 12 (*Continued*)

Central Angle in Degrees	Arc R	Height R	Chord R	Height Chord	Area R²	Central Angle in Degrees	Arc R	Height R	Chord R	Height Chord	Area R²
161	2.8100	0.8350	1.973	0.4233	1.2422	171	2.9845	0.9215	1.994	0.4622	1.4140
162	2.8274	0.8436	1.975	0.4270	1.2592	172	3.0020	0.9302	1.995	0.4663	1.4314
163	2.8449	0.8522	1.978	0.4308	1.2763	173	3.0194	0.9390	1.996	0.4704	1.4488
164	2.8623	0.8608	1.981	0.4346	1.2934	174	3.0369	0.9477	1.997	0.4745	1.4662
165	2.8798	0.8695	1.983	0.4385	1.3105	175	3.0543	0.9564	1.998	0.4786	1.4836
166	2.8972	0.8781	1.985	0.4424	1.3277	176	3.0718	0.9651	1.999	0.4828	1.5010
167	2.9147	0.8868	1.987	0.4463	1.3449	177	3.0892	0.9738	1.999	0.4871	1.5184
168	2.9322	0.8955	1.989	0.4502	1.3621	178	3.1067	0.9825	2.000	0.4914	1.5359
169	2.9496	0.9042	1.991	0.4542	1.3794	179	3.1241	0.9913	2.000	0.4957	1.5533
170	**2.9671**	**0.9128**	**1.992**	**0.4582**	**1.3967**	**180**	**3.1416**	**1.0000**	**2.000**	**0.5000**	**1.5708**

Table 13 Values of Degrees, Minutes, and Seconds in Radians[a]

Degrees	Radians Arc Length R = 1	Degrees	Radians Arc Length R = 1	Degrees	Radians Arc Length R = 1		Radians Arc Length R = 1 Minutes	Radians Arc Length R = 1 Seconds
0		**40**	**0.69813170**	**80**	**1.39626340**	**0**		
1	0.01745329	41	0.71558499	81	1.41371669	1	0.00029089	0.00000485
2	0.03490659	42	0.73303829	82	1.43116999	2	0.00058178	0.00000970
3	0.05235988	43	0.75049158	83	1.44862328	3	0.00087266	0.00001454
4	0.06981317	44	0.76794487	84	1.46607657	4	0.00116355	0.00001939
5	0.08726646	45	0.78539816	85	1.48352986	5	0.00145444	0.00002424
6	0.10471976	46	0.80285146	86	1.50098316	6	0.00174533	0.00002909
7	0.12217305	47	0.82030475	87	1.51843645	7	0.00203622	0.00003394
8	0.13962634	48	0.83775804	88	1.53588974	8	0.00232711	0.00003879
9	0.15707963	49	0.85521133	89	1.55334303	9	0.00261799	0.00004363
10	**0.17453293**	**50**	**0.87266463**	**90**	**1.57079633**	**10**	**0.00290888**	**0.00004848**
11	0.19198622	51	0.89011792	91	1.58824962	11	0.00319977	0.00005333
12	0.20943951	52	0.90757121	92	1.60570291	12	0.00349066	0.00005818
13	0.22689280	53	0.92502450	93	1.62315620	13	0.00378155	0.00006303
14	0.24434610	54	0.94247780	94	1.64060950	14	0.00407243	0.000068787
15	0.26179939	55	0.95993109	95	1.65806279	15	0.00436332	0.00007272
16	0.27925268	56	0.97738438	96	1.67551608	16	0.00465421	0.00007757
17	0.29670597	57	0.99483767	97	1.69296937	17	0.00494510	0.00008242
18	0.31415927	58	1.01229097	98	1.71042267	18	0.00523599	0.00008727
19	0.33161256	59	1.02974426	99	1.72787596	19	0.00552688	0.00009211
20	**0.34906585**	**60**	**1.04719755**	**100**	**1.74532925**	**20**	**0.00581776**	**0.00009696**
21	0.36651914	61	1.06465084	101	1.76278254	21	0.00610865	0.00010181
22	0.38397244	62	1.08210414	102	1.78023584	22	0.00639954	0.00010666
23	0.40142573	63	1.09955743	103	1.79768913	23	0.00669043	0.00011151
24	0.41887902	64	1.11701072	104	1.81514242	24	0.00698132	0.00011636
25	0.43633231	65	1.13446401	105	1.83259571	25	0.00727221	0.00012120
26	0.45378561	66	1.15191731	106	1.85004901	26	0.00756309	0.00012605
27	0.47123890	67	1.16937060	107	1.86750230	27	0.00785398	0.00013090
28	0.48869219	68	1.18682389	108	1.88495559	28	0.00814487	0.00013575
29	0.50614548	69	1.20427718	109	1.90240888	29	0.00843576	0.00014060
30	**0.52359878**	**70**	**1.22173048**	**110**	**1.91986218**	**30**	**0.00872665**	**0.00014544**
31	0.54105207	71	1.23918377	111	1.93731547	31	0.00901753	0.00015029
32	0.55850536	72	1.25663706	112	1.95476876	32	0.00930842	0.00015514
33	0.57595865	73	1.27409035	113	1.97222205	33	0.00959931	0.00015999
34	0.59341195	74	1.29154365	114	1.98967535	34	0.00989020	0.00016484
35	0.61086524	75	1.30899694	115	2.00712864	35	0.01018109	0.00016968
36	0.62831853	76	1.32645023	116	2.02458193	36	0.01047198	0.00017453
37	0.64577182	77	1.34390352	117	2.04203522	37	0.01076286	0.00017938
38	0.66322512	78	1.36135682	118	2.05948852	38	0.01105375	0.00018423
39	0.68067841	79	1.37881011	119	2.07694181	39	0.01134464	0.00018908

(*Continues*)

Table 13   *(Continued)*

Degrees	Radians Arc Length R = 1	Degrees	Radians Arc Length R = 1	Degrees	Radians Arc Length R = 1		Radians Arc Length R = 1	
							Minutes	Seconds
120	**2.09439510**	**140**	**2.44346095**	**160**	**2.79252680**	**40**	**0.01163553**	**0.00019393**
121	2.11184840	141	2.46091424	161	2.80998009	41	0.01192642	0.00019877
122	2.12930169	142	2.47836754	162	2.82743338	42	0.01221730	0.00020362
123	2.14675498	143	2.49582083	163	2.84488668	43	0.01250819	0.00020847
124	2.16420828	144	2.51327413	164	2.86233997	44	0.01279908	0.00021332
125	2.18166157	145	2.53072742	165	2.87979327	45	0.01308997	0.00021817
126	2.19911486	146	2.54818071	166	2.89724655	46	0.01338086	0.00022301
127	2.21656815	147	2.56563401	167	2.91469985	47	0.01367175	0.00022786
128	2.23402145	148	2.58308729	168	2.93215314	48	0.01396263	0.00023271
129	2.25147474	149	2.60054058	169	2.94960643	49	0.01425352	0.00023756
130	**2.26892803**	**150**	**2.61799388**	**170**	**2.96705972**	**50**	**0.01454441**	**0.00024241**
131	2.28638133	151	2.63544717	171	2.98451302	51	0.01483530	0.00024725
132	2.30383462	152	2.65290046	172	3.00196631	52	0.01512619	0.00025210
133	2.32128791	153	2.67035375	173	3.01941961	53	0.01541707	0.00025695
134	2.33874121	154	2.68780705	174	3.03687289	54	0.01570796	0.00026180
135	2.35619450	155	2.70526034	175	3.05432619	55	0.01599885	0.00026665
136	2.37364780	156	2.72271363	176	3.07177948	56	0.01628974	0.00027150
137	2.39110107	157	2.74016693	177	3.08923277	57	0.01658063	0.00027634
138	2.40855436	158	2.75762022	178	3.10668607	58	0.01687152	0.00028119
139	2.42600766	159	2.77507351	179	3.12413962	59	0.01716240	0.00028604
				180	**3.14159265**			

[a]Lengths of circular arcs, radius unity, for example:

$$\Theta = 30°20'10''$$

$$30° = 0.52359878$$

$$20' = 0.00581776$$

$$10'' = \underline{0.00004848}$$

Arc length = 0.52946502 radians

Table 14   **Values of Radians in Degrees**

Radian	0.00	0.01	0.02	0.03	0.04	0.05	0.06	0.07	0.08	0.09
0.0	0.0000	0.5730	1.1459	1.7189	2.2918	2.8648	3.4377	4.0107	4.5837	5.1566
0.1	5.7296	6.3025	6.8755	7.4485	8.0214	8.5944	9.1673	9.7403	10.3132	10.8862
0.2	11.4591	12.0321	12.6051	13.1780	13.7510	14.3239	14.8969	15.4699	16.0428	16.6158
0.3	17.1887	17.7617	18.3346	18.9076	19.4806	20.0535	20.6265	21.1994	21.7724	22.3454
0.4	22.9183	23.4913	24.0642	24.6372	25.2101	25.7831	26.3561	26.9290	27.5020	28.0749
0.5	28.6479	29.2208	29.7938	30.3668	30.9397	31.1527	32.0856	32.6586	33.2316	33.8045
0.6	34.3775	34.9504	35.5234	36.0963	36.6693	37.2423	37.8152	38.3882	38.9611	39.5341
0.7	40.1070	40.6800	41.2530	41.8259	42.3989	42.9718	43.5448	44.1178	44.6907	45.2637
0.8	45.8366	46.4096	46.9825	47.5555	48.1285	48.7014	49.2744	49.8473	50.4203	50.9932
0.9	51.5662	52.1392	52.7121	53.2851	53.8580	54.4310	55.0039	55.5769	56.1499	56.7228

1 rad = 57.29578°   2 rads = 114.59156°   3 rads = 171.88734°

Table 15   **Decimals of a Degree in Minutes and Seconds**

Decimal	0.00		0.01		0.02		0.03		0.04		0.05		0.06		0.07		0.08		0.09	
Decimal	Min	Sec	Min	Sec	Min	Sec	Min	Sec	Min	Sec	Min	Sec	Min	Sec	Min	Sec	Min	Sec	Min	Sec
0.0	0	0	0	36	1	12	1	48	2	24	3	0	3	36	4	12	4	48	5	24
0.1	6	0	6	36	7	12	7	48	8	24	9	0	9	36	10	12	10	48	11	24
0.2	12	0	12	36	13	12	13	48	14	24	15	0	15	36	16	12	16	48	17	24
0.3	18	0	18	36	19	12	19	48	20	24	21	0	21	36	22	12	22	48	23	24
0.4	24	0	24	36	25	12	25	48	26	24	27	0	27	36	28	12	28	48	29	24

(Continues)

Table 15 (*Continued*)

Decimal \ Decimal	0.00 Min	0.00 Sec	0.01 Min	0.01 Sec	0.02 Min	0.02 Sec	0.03 Min	0.03 Sec	0.04 Min	0.04 Sec	0.05 Min	0.05 Sec	0.06 Min	0.06 Sec	0.07 Min	0.07 Sec	0.08 Min	0.08 Sec	0.09 Min	0.09 Sec
0.5	30	0	30	36	31	12	31	48	32	24	33	0	33	36	34	12	34	48	35	24
0.6	36	0	36	36	37	12	37	48	38	24	39	0	39	36	40	12	40	48	41	24
0.7	42	0	42	36	43	12	43	48	44	24	45	0	45	36	46	12	46	48	47	24
0.8	48	0	48	36	49	12	49	48	50	24	51	0	51	36	52	12	52	48	53	24
0.9	54	0	54	36	55	12	55	48	56	24	57	0	57	36	58	12	58	48	59	24

Table 16   Minutes in Decimals of a Degree

Minutes	0	1	2	3	4	5	6	7	8	9
0	0.00000	0.01667	0.03333	0.05000	0.06667	0.08333	0.10000	0.11667	0.13333	0.15000
10	0.16667	0.18333	0.20000	0.21667	0.23333	0.25000	0.26667	0.28333	0.30000	0.31667
20	0.33333	0.35000	0.36667	0.38333	0.40000	0.41667	0.43333	0.45000	0.46667	0.48333
30	0.50000	0.51667	0.53333	0.55000	0.56667	0.58333	0.60000	0.61667	0.63333	0.65000
40	0.66667	0.68333	0.70000	0.71667	0.73333	0.75000	0.76667	0.78333	0.80000	0.81667
50	0.83333	0.85000	0.86667	0.88333	0.90000	0.91667	0.93333	0.95000	0.96667	0.98333

Table 17   Seconds in Decimals of a Degree

Seconds	0	1	2	3	4
0	0	0.0002778	0.0005555	0.0008333	0.0011111
10	0.0027778	0.0030555	0.0033333	0.0036111	0.0038888
20	0.0055555	0.0058333	0.0061111	0.0063888	0.0066667
30	0.0083333	0.0086111	0.0088888	0.0091667	0.0094444
40	0.0111111	0.0113888	0.0116667	0.0119444	0.0122222
50	0.0138888	0.0141667	0.0144444	0.0147222	0.0150000

Seconds	5	6	7	8	9
0	0.0013888	0.0016667	0.0019444	0.0022222	0.0024999
10	0.0041667	0.0044444	0.0047222	0.0050000	0.0052778
20	0.0069444	0.0072222	0.0075000	0.0077778	0.0080555
30	0.0097222	0.0100000	0.0102778	0.0105555	0.0108333
40	0.0125000	0.0127778	0.0130555	0.0133333	0.0136111
50	0.0152778	0.0155555	0.0158333	0.0161111	0.0163888

Table 18   Table of Integrals

Elementary Indefinite Integrals

1. $\int a\,dx = ax$

2. $\int (u + v + w + \cdots)\,dx = \int u\,dx + \int v\,dx + \int w\,dx + \cdots$

3. $\int u\,dv = uv - \int v\,du$, integration by parts

4. $\int f(x)\,dx = \int f[\phi(y)]\phi'(y)\,dy, \ x = \phi(y)$, change of variable

5. $\int x^n\,dx = \dfrac{x^{n+1}}{n+1} \quad (n \neq -1)$

6. $\int \dfrac{dx}{x} = \log_e x + c = \log_e c_1 x, \ [\log_e x = \log_e(-x) + (2k+1)\pi i]$

7. $\int e^{ax}\,dx = \dfrac{1}{a}e^{ax}$

(*Continues*)

Table 18   *(Continued)*

Elementary Indefinite Integrals (Continued)

8. $\int a^x\, dx = \dfrac{a^x}{\log_e a}$

9. $\int a^x \log_e a\, dx = a^x.$

10. $\int \sin ax\, dx = -\dfrac{1}{a} \cos ax$

11. $\int \cos ax\, dx = \dfrac{1}{a} \sin ax$

12. $\int \tan ax\, dx = -\dfrac{1}{a} \log_e \cos ax = \dfrac{1}{a} \log_e \sec ax$

13. $\int \cot ax\, dx = \dfrac{1}{a} \log_e \sin ax = -\dfrac{1}{a} \log_e \csc ax$

14. $\int \sec ax\, dx = \dfrac{1}{a} \log_e(\sec ax + \tan ax) = \dfrac{1}{a} \log_e \tan \left(\dfrac{ax}{2} + \dfrac{\pi}{4}\right)$

15. $\int \csc ax\, dx = \dfrac{1}{a} \log_e(\csc ax - \cot ax) = \dfrac{1}{a} \log_e \tan \dfrac{ax}{2}$

16. $\int \dfrac{dx}{\sqrt{a^2 - x^2}} = \sin^{-1} \dfrac{x}{a} = -\cos^{-1} \dfrac{x}{a} \quad (x^2 < a^2)$

17. $\int \dfrac{dx}{a^2 + x^2} = \dfrac{1}{a} \tan^{-1} \dfrac{x}{a} = -\dfrac{1}{a} \cot^{-1} \dfrac{x}{a}$

18. $\int \sinh ax\, dx = \dfrac{1}{a} \cosh ax$

19. $\int \cosh ax\, dx = \dfrac{1}{a} \sinh ax$

20. $\int \tanh ax\, dx = \dfrac{1}{a} \log_e(\cosh ax)$

21. $\int \coth ax\, dx = \dfrac{1}{a} \log_e(\sinh ax)$

22. $\int \operatorname{sech} ax\, dx = \dfrac{1}{a} \sin^{-1}(\tanh ax) = \dfrac{1}{a} \tan^{-1}(\sinh ax)$

23. $\int \operatorname{csch} ax\, dx = \dfrac{1}{a} \log_e \left(\tanh \dfrac{ax}{2}\right)$

24. $\int \sin^2 ax\, dx = \dfrac{1}{2}x - \dfrac{1}{2a} \sin ax \cos ax = \dfrac{1}{2}x - \dfrac{1}{4a} \sin 2ax$

25. $\int \cos^2 ax\, dx = \dfrac{1}{2}x + \dfrac{1}{2a} \sin ax \cos ax = \dfrac{1}{2}x + \dfrac{1}{4a} \sin 2ax$

26. $\int \tan^2 ax\, dx = \dfrac{1}{a} \tan ax - x$

27. $\int \cot^2 ax\, dx = -\dfrac{1}{a} \cot ax - x$

28. $\int \sec^2 ax\, dx = \dfrac{1}{a} \tan ax$

29. $\int \csc^2 ax\, dx = -\dfrac{1}{a} \cot ax$

30. $\int \sin^{-1} ax\, dx = x \sin^{-1} ax + \dfrac{1}{a}\sqrt{1 - a^2 x^2}$

31. $\int \cos^{-1} ax\, dx = x \cos^{-1} ax - \dfrac{1}{a}\sqrt{1 - a^2 x^2}$

32. $\int \tan^{-1} ax\, dx = x \tan^{-1} ax - \dfrac{1}{2a} \log_e(1 + a^2 x^2)$

Table 18 *(Continued)*

Elementary Indefinite Integrals (Continued)

33. $\int \cot^{-1} ax \, dx = x \cot^{-1} ax + \dfrac{1}{2a} \log_e(1 + a^2x^2)$

34. $\int \sec^{-1} ax \, dx = x \sec^{-1} ax - \dfrac{1}{a} \log_e(ax + \sqrt{a^2x^2 - 1})$

35. $\int \csc^{-1} ax \, dx = x \csc^{-1} ax + \dfrac{1}{a} \log_e(ax + \sqrt{a^2x^2 - 1})$

Integrals Involving ax + b

36. $\int (ax + b)^n \, dx = \dfrac{1}{a(n + 1)}(ax + b)^{n+1} \quad (n \neq -1)$

37. $\int \dfrac{dx}{ax + b} = \dfrac{1}{a} \log_e(ax + b)$

38. $\int x(ax + b)^n \, dx = \dfrac{1}{a^2(n + 2)}(ax + b)^{n+2} - \dfrac{b}{a^2(n + 1)}(ax + b)^{n+1} \quad (n \neq -1, -2)$

39. $\int \dfrac{x \, dx}{ax + b} = \dfrac{x}{a} - \dfrac{b}{a^2} \log_e(ax + b)$

40. $\int \dfrac{x \, dx}{(ax + b)^2} = \dfrac{b}{a^2(ax + b)} + \dfrac{1}{a^2} \log_e(ax + b)$

41. $\int \dfrac{x^2 \, dx}{ax + b} = \dfrac{1}{a^3} \left[\dfrac{1}{2}(ax + b)^2 - 2b(ax + b) + b^2 \log_e(ax + b) \right]$

42. $\int \dfrac{x^2 \, dx}{(ax + b)^2} = \dfrac{1}{a^3} \left[(ax + b) - 2b \log_e(ax + b) - \dfrac{b^2}{ax + b} \right]$

43. $\int \dfrac{x^2 \, dx}{(ax + b)^3} = \dfrac{1}{a^3} \left[\log_e(ax + b) + \dfrac{2b}{ax + b} - \dfrac{b^2}{2(ax + b)^2} \right]$

44. $\int \dfrac{dx}{x(ax + b)} = \dfrac{1}{b} \log_e \dfrac{x}{ax + b}$

45. $\int \dfrac{dx}{x^2(ax + b)} = -\dfrac{1}{bx} + \dfrac{a}{b^2} \log_e \dfrac{ax + b}{x}$

46. $\int \dfrac{dx}{x(ax + b)^2} = \dfrac{1}{b(ax + b)} - \dfrac{1}{b^2} \log_e \dfrac{ax + b}{x}$

47. $\int \dfrac{dx}{x^2(ax + b)^2} = -\dfrac{b + 2ax}{b^2x(ax + b)} + \dfrac{2a}{b^3} \log_e \dfrac{ax + b}{x}$

48. $\int \dfrac{dx}{x\sqrt{ax + b}} = \dfrac{1}{\sqrt{b}} \log_e \dfrac{\sqrt{ax + b} - \sqrt{b}}{\sqrt{ax + b} + \sqrt{b}} \quad (b \text{ positive})$

49. $\int \dfrac{dx}{x\sqrt{ax + b}} = \dfrac{2}{\sqrt{-b}} \tan^{-1} \sqrt{\dfrac{ax + b}{-b}} \quad (b \text{ negative})$

50. $\int \dfrac{\sqrt{ax + b}}{x} dx = 2\sqrt{ax + b} + \sqrt{b} \log_e \dfrac{\sqrt{ax + b} - \sqrt{b}}{\sqrt{ax + b} + \sqrt{b}} \quad (b \text{ positive})$

51. $\int \dfrac{\sqrt{ax + b}}{x} dx = 2\sqrt{ax + b} - 2\sqrt{-b} \tan^{-1} \sqrt{\dfrac{ax + b}{-b}} \quad (b \text{ negative})$

52. $\int \dfrac{dx}{x^2\sqrt{ax + b}} = -\dfrac{\sqrt{ax + b}}{bx} - \dfrac{a}{2b\sqrt{b}} \log_e \dfrac{\sqrt{ax + b} - \sqrt{b}}{\sqrt{ax + b} + \sqrt{b}} \quad (b \text{ positive})$

53. $\int \dfrac{dx}{x^2\sqrt{ax + b}} = -\dfrac{\sqrt{ax + b}}{bx} - \dfrac{a}{b\sqrt{-b}} \tan^{-1} \sqrt{\dfrac{ax + b}{-b}} \quad (b \text{ negative})$

54. $\int \dfrac{ax + b}{fx + g} dx = \dfrac{ax}{f} + \dfrac{bf - cg}{f^2} \log_e(fx + g)$

(Continues)

Table 18   *(Continued)*

Integrals Involving (ax + b) (Continued)

55. $\displaystyle\int \frac{dx}{(ax+b)(fx+g)} = \frac{1}{bf-ag} \log_e\left(\frac{fx+g}{ax+b}\right)$ $(ag \neq bf)$

56. $\displaystyle\int \frac{x\,dx}{(ax+b)(fx+g)} = \frac{1}{bf-ag}\left[\frac{b}{a}\log_e(ax+b) - \frac{g}{f}\log_e(fx+g)\right]$ $(ag \neq bf)$

57. $\displaystyle\int \frac{dx}{(ax+b)^2(fx+g)} = \frac{1}{bf-ag}\left(\frac{1}{ax+b} + \frac{f}{bf-ag}\log_e\frac{fx+g}{ax+b}\right)$ $(ag \neq bf)$

Integrals Involving $ax^n + b$

58. $\displaystyle\int (ax^2+b)^n x\,dx = \frac{1}{2a}\frac{(ax^2+b)^{n+1}}{n+1}$ $(n \neq -1)$

59. $\displaystyle\int \frac{dx}{ax^2+b} = \frac{1}{\sqrt{ab}}\tan^{-1}\left(x\sqrt{\frac{a}{b}}\right)$ *(a and b positive)*

60. $\displaystyle\int \frac{dx}{ax^2+b} = \frac{1}{2\sqrt{-ab}}\log_e\frac{x\sqrt{a}-\sqrt{-b}}{x\sqrt{a}+\sqrt{-b}}$ *(a positive, b negative)*

$\displaystyle\qquad\qquad = \frac{1}{2\sqrt{-ab}}\log_e\frac{\sqrt{b}+x\sqrt{-a}}{\sqrt{b}-x\sqrt{-a}}$ *(a negative, b positive)*

61. $\displaystyle\int \frac{dx}{x(ax^2+b)} = \frac{1}{2b}\log_e\frac{x^2}{ax^2+b}$

62. $\displaystyle\int \frac{dx}{(ax^2+b)^n} = \frac{1}{2(n-1)b}\frac{x}{(ax^2+b)^{n-1}}$

$\displaystyle\qquad\qquad + \frac{2n-3}{2(n-1)b}\int \frac{dx}{(ax^2+b)^{n-1}}$ *(n integer > 1)*

63. $\displaystyle\int \frac{x^2\,dx}{ax^2+b} = \frac{x}{a} - \frac{b}{a}\int \frac{dx}{ax^2+b}$

64. $\displaystyle\int \frac{x^2\,dx}{(ax^2+b)^n} = \frac{1}{2(n-1)a}\frac{x}{(ax^2+b)^{n-1}}$

$\displaystyle\qquad\qquad + \frac{1}{2(n-1)a}\int \frac{dx}{(ax^2+b)^{n-1}}$ *(n integer > 1)*

65. $\displaystyle\int \frac{dx}{x^2(ax^2+b)^n} = \frac{1}{b}\int \frac{dx}{x^2(ax^2+b)^{n-1}} - \frac{a}{b}\int \frac{dx}{(ax^2+b)^n}$ *(n = positive integer)*

66. $\displaystyle\int \sqrt{ax^2+b}\,dx = \frac{x}{2}\sqrt{ax^2+b} + \frac{b}{2\sqrt{a}}\log_e\frac{x\sqrt{a}+\sqrt{ax^2+b}}{\sqrt{b}}$ *(a positive)*

67. $\displaystyle\int \sqrt{ax^2+b}\,dx = \frac{x}{2}\sqrt{ax^2+b} + \frac{b}{2\sqrt{-a}}\sin^{-1}\left(x\sqrt{-\frac{a}{b}}\right)$ *(a negative)*

68. $\displaystyle\int \frac{dx}{\sqrt{ax^2+b}} = \frac{1}{\sqrt{a}}\log_e(x\sqrt{a}+\sqrt{ax^2+b})_n$ *(a positive)*

69. $\displaystyle\int \frac{dx}{\sqrt{ax^2+b}} = \frac{1}{\sqrt{-a}}\sin^{-1}\left(x\sqrt{-\frac{a}{b}}\right)$ *(a negative)*

70. $\displaystyle\int \frac{x\,dx}{\sqrt{ax^2+b}} = \frac{1}{a}\sqrt{ax^2+b}$

71. $\displaystyle\int \frac{\sqrt{ax^2+b}}{x}\,dx = \sqrt{ax^2+b} + \sqrt{b}\log_e\frac{\sqrt{ax^2+b}-\sqrt{b}}{x}$ *(b positive)*

72. $\displaystyle\int \frac{\sqrt{ax^2+b}}{x}\,dx = \sqrt{ax^2+b} - \sqrt{-b}\tan^{-1}\frac{\sqrt{ax^2+b}}{\sqrt{-b}}$ *(b negative)*

73. $\displaystyle\int x\sqrt{ax^2+b}\,dx = \frac{1}{3a}(ax^2+b)^{3/2}$

Table 18 *(Continued)*

Integrals Involving $(ax^n + b)$ (Continued)

74. $\displaystyle\int x^2 \sqrt{ax^2 + b}\, dx = \frac{x}{4a}(ax^2 + b)^{3/2} - \frac{bx}{8a}\sqrt{ax^2 + b}$

$\displaystyle\qquad\qquad - \frac{b^2}{8a\sqrt{a}} \log_e(x\sqrt{a} + \sqrt{ax^2 + b}) \quad (a \text{ positive})$

75. $\displaystyle\int x^2 \sqrt{ax^2 + b}\, dx = \frac{x}{4a}(ax^2 + b)^{3/2} - \frac{bx}{8a}\sqrt{ax^2 + b}$

$\displaystyle\qquad\qquad - \frac{b^2}{8a\sqrt{-a}} \sin^{-1}\left(x\sqrt{\frac{-a}{b}}\right) \quad (a \text{ negative})$

76. $\displaystyle\int \frac{dx}{x\sqrt{ax^2 + b}} = \frac{1}{\sqrt{b}} \log_e \frac{\sqrt{ax^2 + b} - \sqrt{b}}{x} \quad (b \text{ positive})$

77. $\displaystyle\int \frac{dx}{x\sqrt{ax^2 + b}} = \frac{1}{\sqrt{-b}} \sec^{-1}\left(x\sqrt{-\frac{a}{b}}\right) \quad (b \text{ negative})$

78. $\displaystyle\int \frac{x^2 dx}{\sqrt{ax^2 + b}} = \frac{x}{2a}\sqrt{ax^2 + b} - \frac{b}{2a\sqrt{a}} \log_e(x\sqrt{a} + \sqrt{ax^2 + b}) \quad (a \text{ positive})$

79. $\displaystyle\int \frac{x^2 dx}{\sqrt{ax^2 + b}} = \frac{x}{2a}\sqrt{ax^2 + b} - \frac{b}{2a\sqrt{-a}} \sin^{-1}\left(x\sqrt{-\frac{a}{b}}\right) \quad (a \text{ negative})$

80. $\displaystyle\int \frac{\sqrt{ax^2 + b}}{x^2}\, dx = -\frac{\sqrt{ax^2 + b}}{x} + \sqrt{a}\log_e(x\sqrt{a} + \sqrt{ax^2 + b}) \quad (a \text{ positive})$

81. $\displaystyle\int \frac{\sqrt{ax^2 + b}}{x^2}\, dx = -\frac{\sqrt{ax^2 + b}}{x} - \sqrt{-a}\sin^{-1}\left(x\sqrt{-\frac{a}{b}}\right) \quad (a \text{ negative})$

82. $\displaystyle\int \frac{dx}{x(ax^n + b)} = \frac{1}{bn} \log_e \frac{x^n}{ax^n + b}$

83. $\displaystyle\int \frac{dx}{x\sqrt{ax^n + b}} = \frac{1}{n\sqrt{b}} \log_e \frac{\sqrt{ax^n + b} - \sqrt{b}}{\sqrt{ax^n + b} + \sqrt{b}} \quad (b \text{ positive})$

84. $\displaystyle\int \frac{dx}{x\sqrt{ax^n + b}} = \frac{2}{n\sqrt{-b}} \sec^{-1}\sqrt{-\frac{ax^n}{b}} \quad (b \text{ negative})$

Integrals Involving $ax^2 + bx + d$

85. $\displaystyle\int \frac{dx}{ax^2 + bx + d} = \frac{1}{\sqrt{b^2 - 4ad}} \log_e \frac{2ax + b - \sqrt{b^2 - 4ad}}{2ax + b + \sqrt{b^2 - 4ad}} \quad (b^2 > 4ad)$

86. $\displaystyle\int \frac{dx}{ax^2 + bx + d} = \frac{2}{\sqrt{4ad - b^2}} \tan^{-1} \frac{2ax + b}{\sqrt{4ad - b^2}} \quad (b^2 < 4ad)$

87. $\displaystyle\int \frac{dx}{ax^2 + bx + d} = -\frac{2}{2ax + b} \quad (b^2 = 4ad)$

88. $\displaystyle\int \frac{dx}{\sqrt{ax^2 + bx + d}} = \frac{1}{\sqrt{a}} \log_e\left(2ax + b + 2\sqrt{a(ax^2 + bx + d)}\right) \quad (a \text{ positive})$

89. $\displaystyle\int \frac{dx}{\sqrt{ax^2 + bx + d}} = \frac{1}{\sqrt{-a}} \sin^{-1} \frac{-2ax - b}{\sqrt{b^2 - 4ad}} \quad (a \text{ negative})$

90. $\displaystyle\int \frac{x\, dx}{ax^2 + bx + d} = \frac{1}{2a} \log_e(ax^2 + bx + d) - \frac{b}{2a} \int \frac{dx}{ax^2 + bx + d}$

91. $\displaystyle\int \frac{x\, dx}{\sqrt{ax^2 + bx + d}} = \frac{\sqrt{ax^2 + bx + d}}{a} - \frac{b}{2a} \int \frac{dx}{\sqrt{ax^2 + bx + d}}$

92. $\displaystyle\int \frac{dx}{x\sqrt{ax^2 + bx + d}} = -\frac{1}{\sqrt{d}} \log_e\left(\frac{\sqrt{ax^2 + bx + d} + \sqrt{d}}{x} + \frac{b}{2\sqrt{d}}\right) \quad (d \text{ positive})$

(Continues)

Table 18   *(Continued)*

93. $\displaystyle\int \frac{dx}{x\sqrt{ax^2+bx+d}} = \frac{1}{\sqrt{-d}}\sin^{-1}\frac{bx+2d}{x\sqrt{b^2-4ad}}$ (*d* negative)

94. $\displaystyle\int \frac{dx}{x\sqrt{ax^2+bx}} = -\frac{2}{bx}\sqrt{ax^2+bx}$

95. $\displaystyle\int \sqrt{ax^2+bx+d}\,dx = \frac{2ax+b}{4a}\sqrt{ax^2+bx+d} + \frac{4ad-b^2}{8a}\int \frac{dx}{\sqrt{ax^2+bx+d}}$

96. $\displaystyle\int x\sqrt{ax^2+bx+d}\,dx = \frac{(ax^2+bx+d)^{3/2}}{3a} - \frac{b}{2a}\int \sqrt{ax^2+bx+d}\,dx$

Integrals Involving $\sin^n ax$

97. $\displaystyle\int \sin^3 ax\,dx = -\frac{1}{a}\cos ax + \frac{1}{3a}\cos^3 ax$

98. $\displaystyle\int \sin^4 ax\,dx = \frac{3}{8}x - \frac{1}{4a}\sin 2ax + \frac{1}{32a}\sin 4ax$

99. $\displaystyle\int \sin^n ax\,dx = -\frac{\sin^{n-1}ax\cos ax}{na} + \frac{n-1}{n}\int \sin^{n-2}ax\,dx$ (*n* = positive integer)

100. $\displaystyle\int x\sin ax\,dx = \frac{\sin ax}{a^2} - \frac{x\cos ax}{a}$

101. $\displaystyle\int x^2\sin ax\,dx = \frac{2x}{a^2}\sin ax - \left(\frac{x^2}{a} - \frac{2}{a^3}\right)\cos ax$

102. $\displaystyle\int x^3\sin ax\,dx = \left(\frac{3x^2}{a^2} - \frac{6}{a^4}\right)\sin ax - \left(\frac{x^3}{a} - \frac{6x}{a^3}\right)\cos ax$

103. $\displaystyle\int x^n\sin ax\,dx = -\frac{x^n}{a}\cos ax + \frac{n}{a}\int x^{n-1}\cos ax\,dx$ (*n* > 0)

104. $\displaystyle\int \frac{\sin ax}{x^n}dx = -\frac{1}{n-1}\frac{\sin ax}{x^{n-1}} + \frac{a}{n-1}\int \frac{\cos ax}{x^{n-1}}dx$

105. $\displaystyle\int \frac{dx}{\sin^n ax} = -\frac{1}{a(n-1)}\frac{\cos ax}{\sin^{n-1}ax} + \frac{n-2}{n-1}\int \frac{dx}{\sin^{n-2}ax}$ (*n* integer > 1)

106. $\displaystyle\int \frac{x\,dx}{\sin^2 ax} = -\frac{x}{a}\cot ax + \frac{1}{a^2}\log_e \sin ax$

107. $\displaystyle\int \frac{dx}{1+\sin ax} = -\frac{1}{a}\tan\left(\frac{\pi}{4} - \frac{ax}{2}\right)$

108. $\displaystyle\int \frac{dx}{1-\sin ax} = \frac{1}{a}\cot\left(\frac{\pi}{4} - \frac{ax}{2}\right)$

109. $\displaystyle\int \frac{x\,dx}{1+\sin ax} = -\frac{x}{a}\tan\left(\frac{\pi}{4} - \frac{ax}{2}\right) + \frac{2}{a^2}\log_e \cos\left(\frac{\pi}{4} - \frac{ax}{2}\right)$

110. $\displaystyle\int \frac{x\,dx}{1-\sin ax} = \frac{x}{a}\cot\left(\frac{\pi}{4} - \frac{ax}{2}\right) + \frac{2}{a^2}\log_e \sin\left(\frac{\pi}{4} - \frac{ax}{2}\right)$

111. $\displaystyle\int \frac{dx}{b+d\sin ax} = \frac{-2}{a\sqrt{b^2-d^2}}\tan^{-1}\left[\sqrt{\frac{b-d}{b+d}}\tan\left(\frac{\pi}{4} - \frac{ax}{2}\right)\right]$ ($b^2 > d^2$)

112. $\displaystyle\int \frac{dx}{b+d\sin ax} = \frac{-1}{a\sqrt{d^2-b^2}}\log_e \frac{d+b\sin ax + \sqrt{d^2-b^2}\cos ax}{b+d\sin ax}$ ($d^2 > b^2$)

113. $\displaystyle\int \sin ax\sin bx\,dx = \frac{\sin(a-b)x}{2(a-b)} - \frac{\sin(a+b)x}{2(a+b)}$ ($a^2 \neq b^2$)

Integrals Involving $\cos^n ax$

114. $\displaystyle\int \cos^3 ax\,dx = \frac{1}{a}\sin ax - \frac{1}{3a}\sin^3 ax$

115. $\displaystyle\int \cos^4 ax\,dx = \frac{3}{8}x + \frac{1}{4a}\sin 2ax + \frac{1}{32a}\sin 4ax$

Table 18 *(Continued)*

Integrals Involving $\cos^n ax$ (Continued)

116. $\displaystyle\int \cos^n ax\, dx = \frac{\cos^{n-1} ax \sin ax}{na} + \frac{n-1}{n}\int \cos^{n-2} ax\, dx$ (n = positive integer)

117. $\displaystyle\int x \cos ax\, dx = \frac{\cos ax}{a^2} + \frac{x \sin ax}{a}$

118. $\displaystyle\int x^2 \cos ax\, dx = \frac{2x}{a^2} \cos ax + \left(\frac{x^2}{a} - \frac{2}{a^3}\right) \sin ax$

119. $\displaystyle\int x^3 \cos ax\, dx = \left(\frac{3x^2}{a^2} - \frac{6}{a^4}\right) \cos ax + \left(\frac{x^3}{a} - \frac{6x}{a^3}\right) \sin ax$

120. $\displaystyle\int x^n \cos ax\, dx = \frac{x^n \sin ax}{a} - \frac{n}{a}\int x^{n-1} \sin ax\, dx$ ($n > 0$)

121. $\displaystyle\int \frac{\cos ax}{x^n}\, dx = -\frac{1}{n-1}\frac{\cos ax}{x^{n-1}} - \frac{a}{n-1}\int \frac{\sin ax}{x^{n-1}}\, dx$

122. $\displaystyle\int \frac{dx}{\cos^n ax} = \frac{1}{a(n-1)}\frac{\sin ax}{\cos^{n-1} ax} + \frac{n-2}{n-1}\int \frac{dx}{\cos^{n-2} ax}$ (n integer > 1)

123. $\displaystyle\int \frac{x\, dx}{\cos^2 ax} = \frac{x}{a} \tan ax + \frac{1}{a^2} \log_e \cos ax$

124. $\displaystyle\int \frac{dx}{1 + \cos ax} = \frac{1}{a} \tan \frac{ax}{2}$

125. $\displaystyle\int \frac{dx}{1 - \cos ax} = -\frac{1}{a} \cot \frac{ax}{2}$

126. $\displaystyle\int \frac{x\, dx}{1 + \cos ax} = \frac{x}{a} \tan \frac{ax}{2} + \frac{2}{a^2} \log_e \cos \frac{ax}{2}$

127. $\displaystyle\int \frac{x\, dx}{1 - \cos ax} = -\frac{x}{a} \cot \frac{ax}{2} + \frac{2}{a^2} \log_e \sin \frac{ax}{2}$

128. $\displaystyle\int \frac{dx}{b + d \cos ax} = \frac{2}{a\sqrt{b^2 - d^2}} \tan^{-1}\left(\sqrt{\frac{b-d}{b+d}} \tan \frac{ax}{2}\right)$ $(b^2 > d^2)$

129. $\displaystyle\int \frac{dx}{b + d \cos ax} = \frac{1}{a\sqrt{d^2 - b^2}} \log_e \frac{d + b \cos ax + \sqrt{d^2 - b^2} \sin ax}{b + d \cos ax}$ $(d^2 > b^2)$

130. $\displaystyle\int \cos ax \cos bx\, dx = \frac{\sin(a-b)x}{2(a-b)} + \frac{\sin(a+b)x}{2(a+b)}$ $(a^2 \neq b^2)$

Integrals Involving $\sin^n ax, \cos^n ax$

131. $\displaystyle\int \sin ax \cos bx\, dx = -\frac{1}{2}\left[\frac{\cos(a-b)x}{a-b} + \frac{\cos(a+b)x}{a+b}\right]$ $(a^2 \neq b^2)$

132. $\displaystyle\int \sin^2 ax \cos^2 ax\, dx = \frac{x}{8} - \frac{\sin 4ax}{32a}$

133. $\displaystyle\int \sin^n ax \cos ax\, dx = \frac{1}{a(n+1)} \sin^{n+1} ax$ ($n \neq -1$)

134. $\displaystyle\int \sin ax \cos^n ax\, dx = -\frac{1}{a(n+1)} \cos^{n+1} ax$ ($n \neq -1$)

135. $\displaystyle\int \sin^n ax \cos^m ax\, dx = -\frac{\sin^{n-1} ax \cos^{m+1} ax}{a(n+m)} + \frac{n-1}{n+m}\int \sin^{n-2} ax \cos^m ax\, dx$ (m, n pos)

136. $\displaystyle\int \frac{\sin^n ax}{\cos^m ax}\, dx = \frac{\sin^{n+1} ax}{a(m-1)\cos^{m-1} ax} - \frac{n-m+2}{m-1}\int \frac{\sin^n ax}{\cos^{m-2} ax}\, dx$ (m, n pos, $m \neq 1$)

137. $\displaystyle\int \frac{\cos^m ax}{\sin^n ax}\, dx = \frac{-\cos^{m+1} ax}{a(n-1)\sin^{n-1} ax} + \frac{n-m-2}{n-1}\int \frac{\cos^m ax}{\sin^{n-2} ax}\, dx$ (m, n pos, $n \neq 1$)

(Continues)

Table 18   (Continued)

Integrals Involving $\sin^n ax, \cos^n ax$ *(Continued)*

138. $\displaystyle\int \frac{dx}{\sin ax \cos ax} = \frac{1}{a} \log_e \tan ax$

139. $\displaystyle\int \frac{dx}{b \sin ax + d \cos ax} = \frac{1}{a\sqrt{b^2 + d^2}} \log_e \tan \frac{1}{2}\left(ax + \tan^{-1} \frac{d}{b}\right)$

140. $\displaystyle\int \frac{\sin ax}{b + d \cos ax} dx = -\frac{1}{ad} \log_e (b + d \cos ax)$

141. $\displaystyle\int \frac{\cos ax}{b + d \sin ax} dx = \frac{1}{ad} \log_e (b + d \sin ax)$

Integrals Involving $\tan^n ax, \cot^n ax, \sec^n ax, \csc^n ax$

142. $\displaystyle\int \tan^n ax \, dx = \frac{1}{a(n-1)} \tan^{n-1} ax - \int \tan^{n-2} ax \, dx \quad (n \text{ integer} > 1)$

143. $\displaystyle\int \cot^n ax \, dx = -\frac{1}{a(n-1)} \cot^{n-1} ax - \int \cot^{n-2} ax \, dx \quad (n \text{ integer} > 1)$

144. $\displaystyle\int \sec^n ax \, dx = \frac{1}{a(n-1)} \frac{\sin ax}{\cos^{n-1} ax} + \frac{n-2}{n-1} \int \sec^{n-2} ax \, dx \quad (n \text{ integer} > 1)$

145. $\displaystyle\int \csc^n ax \, dx = -\frac{1}{a(n-1)} \frac{\cos ax}{\sin^{n-1} ax} + \frac{n-2}{n-1} \int \csc^{n-2} ax \, dx \quad (n \text{ integer} > 1)$

146. $\displaystyle\int \frac{dx}{b + d \tan ax} = \frac{1}{b^2 + d^2}\left[bx + \frac{d}{a} \log_e (b \cos ax + d \sin ax)\right]$

147. $\displaystyle\int \frac{dx}{\sqrt{b + d \tan^2 ax}} = \frac{1}{a\sqrt{b - d}} \sin^{-1}\left[\sqrt{\frac{b-d}{b}} \sin ax\right] \quad (b \text{ pos}, b^2 > d^2)$

148. $\displaystyle\int \tan ax \sec ax \, dx = \frac{1}{a} \sec ax$

149. $\displaystyle\int \tan^n ax \sec^2 ax \, dx = \frac{1}{a(n+1)} \tan^{n+1} ax \quad (n \neq -1)$

150. $\displaystyle\int \frac{\sec^2 ax \, dx}{\tan ax} = \frac{1}{a} \log_e \tan ax$

151. $\displaystyle\int \cot ax \csc ax \, dx = -\frac{1}{a} \csc ax$

152. $\displaystyle\int \cot^n ax \csc^2 ax \, dx = -\frac{1}{a(n+1)} \cot^{n+1} ax \quad (n \neq -1)$

153. $\displaystyle\int \frac{\csc^2 ax}{\cot ax} dx = -\frac{1}{a} \log_e \cot ax$

Integrals Involving $b^{ax}, e^{ax}, \sin bx, \cos bx$

154. $\displaystyle\int x b^{ax} dx = \frac{x b^{ax}}{a \log_e b} - \frac{b^{ax}}{a^2 (\log_e b)^2}$

155. $\displaystyle\int x e^{ax} dx = \frac{e^{ax}}{a^2}(ax - 1)$

156. $\displaystyle\int x^n b^{ax} dx = \frac{x^n b^{ax}}{a \log_e b} - \frac{n}{a \log_e b} \int x^{n-1} b^{ax} dx \quad (n \text{ positive})$

157. $\displaystyle\int x^n e^{ax} dx = \frac{1}{a} x^n e^{ax} - \frac{n}{a} \int x^{n-1} e^{ax} dx \quad (n \text{ positive})$

158. $\displaystyle\int \frac{dx}{b + d e^{ax}} = \frac{1}{ab}[ax - \log_e (b + d e^{ax})]$

Table 18 *(Continued)*

Integrals Involving $b^{ax}, e^{ax}, \sin bx, \cos bx$ (Continued)

159. $\displaystyle\int \frac{e^{ax}\,dx}{b+de^{ax}} = \frac{1}{ad}\log_e(b+de^{ax})$

160. $\displaystyle\int \frac{dx}{be^{ax}+de^{-ax}} = \frac{1}{a\sqrt{bd}}\tan^{-1}\left(e^{ax}\sqrt{\frac{b}{d}}\right)$ (b and d positive)

161. $\displaystyle\int \frac{e^{ax}}{x}\,dx = \log_e x + ax + \frac{(ax)^2}{2\cdot 2!} + \frac{(ax)^3}{3\cdot 3!} + \cdots$

162. $\displaystyle\int \frac{e^{ax}}{x^n}\,dx = \frac{1}{n-1}\left(-\frac{e^{ax}}{x^{n-1}} + a\int \frac{e^{ax}}{x^{n-1}}\,dx\right)$ (n integer > 1)

163. $\displaystyle\int e^{ax}\sin bx\,dx = \frac{e^{ax}}{a^2+b^2}(a\sin bx - b\cos bx)$

164. $\displaystyle\int e^{ax}\cos bx\,dx = \frac{e^{ax}}{a^2+b^2}(a\cos bx + b\sin bx)$

165. $\displaystyle\int xe^{ax}\sin bx\,dx = \frac{xe^{ax}}{a^2+b^2}(a\sin bx - b\cos bx)$
$$- \frac{e^{ax}}{(a^2+b^2)^2}[(a^2-b^2)\sin bx - 2ab\cos bx]$$

166. $\displaystyle\int xe^{ax}\cos bx\,dx = \frac{xe^{ax}}{a^2+b^2}(a\cos bx + b\sin bx)$
$$- \frac{e^{ax}}{(a^2+b^2)^2}[(a^2-b^2)\cos bx + 2ab\sin bx]$$

Integrals Involving $\log_e ax$

167. $\displaystyle\int \log_e ax\,dx = x\log_e ax - x$

168. $\displaystyle\int (\log_e ax)^n dx = x(\log_e ax)^n - n(\log_e ax)^{n-1}dx$ (n positive)

169. $\displaystyle\int x^n \log_e ax\,dx = x^{n+1}\left(\frac{\log_e ax}{n+1}\right) - \frac{1}{(n+1)^2}$ $(n \neq -1)$

170. $\displaystyle\int \frac{(\log_e ax)^n}{x}\,dx = \frac{(\log_e ax)^{n+1}}{n+1}$ $(n \neq -1)$

171. $\displaystyle\int \frac{dx}{x\log_e ax} = \log_e(\log_e x)$

172. $\displaystyle\int \frac{dx}{\log_e ax} = \frac{1}{a}\left[\log_e(\log_e ax) + \log_e ax + \frac{(\log_e ax)^2}{2\cdot 2!} + \cdots\right]$

173. $\displaystyle\int x^m(\log_e ax)^n dx = \frac{x^{m+1}(\log_e ax)^n}{m+1} - \frac{n}{m+1}\int x^m(\log_e ax)^{n-1}dx$ (m, n ≠ 1)

174. $\displaystyle\int \frac{x^m dx}{(\log_e ax)^n} = -\frac{x^{m+1}}{(n-1)(\log_e ax)^{n-1}} + \frac{m+1}{n-1}\int \frac{x^m dx}{(\log_e ax)^{n-1}}$

Some Definite Integrals

1. $\displaystyle\int_0^a \sqrt{a^2-x^2}\,dx = \frac{\pi a^2}{4}$

2. $\displaystyle\int_0^a \sqrt{2ax-x^2}\,dx = \frac{\pi a^2}{4}$

(Continues)

Table 18   *(Continued)*

Some Definite Integrals (Continued)

3. $\int_0^\infty \dfrac{dx}{a+bx^2} = \dfrac{\pi}{2\sqrt{ab}}$ (*a* and *b* positive)

4. $\int_0^{\sqrt{a/b}} \dfrac{dx}{a+bx^2}\,dx = \int_{\sqrt{a/b}}^\infty \dfrac{dx}{a+bx^2} = \dfrac{\pi}{4\sqrt{ab}}$ (*a* and *b* positive)

5. $\int_0^{\sqrt{a/b}} \dfrac{dx}{\sqrt{a-bx^2}} = \dfrac{\pi}{2\sqrt{b}}$ (*a* and *b* positive)

6. $\int_0^\infty \dfrac{\sin bx}{x}\,dx = \begin{cases} \dfrac{\pi}{2} & (b>0) \\[2mm] 0 & (b=0) \\[2mm] -\dfrac{\pi}{2} & (b<0) \end{cases}$

7. $\int_0^\infty \dfrac{\tan x}{x}\,dx = \dfrac{\pi}{2}$

8. $\int_0^{\pi/2} \sin^{2n+1} x\,dx = \int_0^{\pi/2} \cos^{2n+1} x\,dx = \dfrac{2\cdot4\cdot6\cdots2n}{3\cdot5\cdot7\cdots(2n+1)}$ $(n>0)$

9. $\int_0^{\pi/2} \sin^{2n} x\,dx = \int_0^{\pi/2} \cos^{2n} x\,dx = \dfrac{1\cdot3\cdot5\cdots(2n-1)}{2\cdot4\cdot6\cdots2n}\cdot\dfrac{\pi}{2}$ $(n>0)$

10. $\int_0^\pi \sin ax \sin bx\,dx = \int_0^\pi \cos ax \cos bx\,dx = 0$ $(a\neq b)$

11. $\int_0^\pi \sin^2 ax\,dx = \int_0^\pi \cos^2 ax\,dx = \dfrac{\pi}{2}$

12. $\int_0^{\pi/2} \log_e \cos x\,dx = \int_0^{\pi/2} \log_e \sin x\,dx = -\dfrac{\pi}{2}\log_e 2$

13. $\int_0^\infty e^{-ax^2}\,dx = \dfrac{1}{2}\sqrt{\dfrac{\pi}{a}}$

14. $\int_0^\infty x^n e^{-ax}\,dx = \dfrac{n!}{a^{n+1}}$ $(a>0, n=1,2,3,\ldots)$

15. $\int_0^1 \dfrac{\log_e x}{1-x}\,dx = -\dfrac{\pi^2}{6}$

16. $\int_0^1 \dfrac{\log_e x}{1+x}\,dx = -\dfrac{\pi^2}{12}$

17. $\int_0^1 \dfrac{\log_e x}{1-x^2}\,dx = \dfrac{\pi^2}{8}$

Table 19   **Haversines**[a]

$\theta°$	Value	Log	$\theta°$	Value	Log	$\theta°$	Value	Log	$\theta°$	Value	Log
0	0.00000	—	**10**	0.00760	0.88059	**20**	0.03015	0.47934	**30**	0.06699	0.82599
1	0.00008	0.88168	11	0.00919	0.96315	21	0.03321	0.52127	31	0.07142	0.85380
2	0.00030	0.48371	12	0.01093	0.03847	22	0.03641	0.56120	32	0.07598	0.88068
3	0.00069	0.83584	13	0.01281	0.10772	23	0.03975	0.59931	33	0.08066	0.90668
4	0.00122	0.08564	14	0.01485	0.17179	24	0.04323	0.63576	34	0.08548	0.93187
5	0.00190	0.27936	15	0.01704	0.23140	25	0.04685	0.67067	35	0.09042	0.95628
6	0.00274	0.43760	16	0.01937	0.28711	26	0.05060	0.70418	36	0.09549	0.97996
7	0.00373	0.57135	17	0.02185	0.33940	27	0.05450	0.73637	37	0.10068	0.00295
8	0.00487	0.68717	18	0.02447	0.38867	28	0.05853	0.76735	38	0.10599	0.02528
9	0.00616	0.78929	19	0.02724	0.43522	29	0.06269	0.79720	39	0.11143	0.04699

Table 19 (*Continued*)

$\theta°$	Value	Log	$\theta°$	Value	Log	$\theta°$	Value	Log	$\theta°$	Value	Log
40	0.11698	0.06810	**80**	0.41318	0.61613	**120**	0.75000	0.87506	**160**	0.96985	0.98670
41	0.12265	0.08865	81	0.42178	0.62509	121	0.75752	0.87939	161	0.97276	0.98801
42	0.12843	0.10866	82	0.43041	0.63389	122	0.76496	0.88364	162	0.97553	0.98924
43	0.13432	0.12815	83	0.43907	0.64253	123	0.77232	0.88780	163	0.97815	0.99041
44	0.14033	0.14715	84	0.44774	0.65102	124	0.77960	0.89187	164	0.98063	0.99151
45	0.14645	0.16568	**85**	0.45642	0.65937	**125**	0.78679	0.89586	**165**	0.98296	0.99254
46	0.15267	0.18376	86	0.46512	0.66757	126	0.79389	0.89976	166	0.98515	0.99350
47	0.15900	0.20140	87	0.47383	0.67562	127	0.80091	0.90358	167	0.98719	0.99440
48	0.16543	0.21863	88	0.48255	0.68354	128	0.80783	0.90732	168	0.98907	0.99523
49	0.17197	0.23545	89	0.49127	0.69132	129	0.81466	0.91098	169	0.99081	0.99599
50	0.17861	0.25190	**90**	0.50000	0.69897	**130**	0.82139	0.91455	**170**	0.99240	0.99669
51	0.18534	0.26797	91	0.50873	0.70648	131	0.82803	0.91805	171	0.99384	0.99732
52	0.19217	0.28368	92	0.51745	0.71387	132	0.83457	0.92146	172	0.99513	0.99788
53	0.19909	0.29905	93	0.52617	0.72112	133	0.84100	0.92480	173	0.99627	0.99838
54	0.20611	0.31409	94	0.53488	0.72825	134	0.84733	0.92805	174	0.99726	0.99881
55	0.21321	0.32281	**95**	0.54358	0.73526	**135**	0.85355	0.93123	**175**	0.99810	0.99917
56	0.22040	0.34322	96	0.55226	0.74215	136	0.85967	0.93433	176	0.99878	0.99947
57	0.22768	0.35733	97	0.56093	0.74891	137	0.86568	0.93736	177	0.99931	0.99970
58	0.23504	0.37114	98	0.56959	0.75556	138	0.87157	0.94030	178	0.99970	0.99987
59	0.24248	0.38468	99	0.57822	0.76209	139	0.87735	0.94318	179	0.99992	0.99997
60	0.25000	0.39794	**100**	0.58682	0.76851	**140**	0.88302	0.94597	**180**	1.00000	0.00000
61	0.25760	0.41094	101	0.59540	0.77481	141	0.88857	0.94869			
62	0.26526	0.42368	102	0.60396	0.78101	142	0.89401	0.95134			
63	0.27300	0.43617	103	0.61248	0.78709	143	0.89932	0.95391			
64	0.28081	0.44842	104	0.62096	0.79306	144	0.90451	0.95641			
65	0.28869	0.46043	**105**	0.62941	0.79893	**145**	0.90958	0.95884			
66	0.29663	0.47222	106	0.63782	0.80470	146	0.91452	0.96119			
67	0.30463	0.48378	107	0.64619	0.81036	147	0.91934	0.96347			
68	0.31270	0.49512	108	0.65451	0.81592	148	0.92402	0.96568			
69	0.32082	0.50625	109	0.66278	0.82137	149	0.92858	0.96782			
70	0.32899	0.51718	**110**	0.67101	0.82673	**150**	0.93301	0.96989			
71	0.33722	0.52791	111	0.67918	0.83199	151	0.93731	0.97188			
72	0.34549	0.53844	112	0.68730	0.83715	152	0.94147	0.97381			
73	0.35381	0.54878	113	0.69537	0.84221	153	0.94550	0.97566			
74	0.36218	0.55893	114	0.70337	0.84718	154	0.94940	0.97745			
75	0.37059	0.56889	**115**	0.71131	0.85206	**155**	0.95315	0.97916			
76	0.37904	0.57868	116	0.71919	0.85684	156	0.95677	0.98081			
77	0.38752	0.58830	117	0.72700	0.86153	157	0.96025	0.98239			
78	0.39604	0.59774	118	0.73474	0.86613	158	0.96359	0.98389			
79	0.40460	0.60702	119	0.74240	0.87064	159	0.96679	0.98533			

[a] $\operatorname{hav} \theta = \frac{1}{2} \operatorname{vers} \theta = \frac{1}{2}(1 - \cos \theta) = \sin^2 \frac{1}{2}\theta$
$\operatorname{hav}(-\theta) = \operatorname{hav} \theta$
$\operatorname{hav}(180° - \theta) = \operatorname{hav}(180° + \theta) = 1 - \operatorname{hav} \theta$
Characteristics of the logarithms are omitted.

Table 20 Complete Elliptic Integrals[a]

$\sin^{-1} k$	K	$\log K$	E	$\log E$	$\sin^{-1} k$	K	$\log K$	E	$\log E$
0°	1.5708	0.196120	1.5708	0.196120	**5°**	1.5738	0.196947	1.5678	0.195293
1	1.5709	0.196153	1.5707	0.196087	6	1.5751	0.197312	1.5665	0.194930
2	1.5713	0.196252	1.5703	0.195988	7	1.5767	0.197743	1.5649	0.194500
3	1.5719	0.196418	1.5697	0.195822	8	1.5785	0.198241	1.5632	0.194004
4	1.5727	0.196649	1.5689	0.195591	9	1.5805	0.198806	1.5611	0.193442

(*Continues*)

Table 20 (*Continued*)

$\sin^{-1}k$	K	$\log K$	E	$\log E$	$\sin^{-1}k$	K	$\log K$	E	$\log E$
10	1.5828	0.199438	1.5589	0.192815	**50°**	1.9356	0.286811	1.3055	0.115790
11	1.5854	0.200137	1.5564	0.192121	51	1.9539	0.290895	1.2963	0.112698
12	1.5882	0.200904	1.5537	0.191362	52	1.9729	0.295101	1.2870	0.109563
13	1.5913	0.201740	1.5507	0.190537	53	1.9927	0.299435	1.2776	0.106386
14	1.5946	0.202643	1.5476	0.189646	54	2.0133	0.303901	1.2681	0.103169
15	1.5981	0.203615	1.5442	0.188690	**55**	2.0347	0.308504	1.2587	0.099915
16	1.6020	0.204657	1.5405	0.187668	56	2.0571	0.313247	1.2492	0.096626
17	1.6061	0.205768	1.5367	0.186581	57	2.0804	0.318138	1.2397	0.093303
18	1.6105	0.206948	1.5326	0.185428	58	2.1047	0.323182	1.2301	0.089950
19	1.6151	0.208200	1.5283	0.184210	59	2.1300	0.328384	1.2206	0.086569
20	1.6200	0.209522	1.5238	0.182928	**60**	2.1565	0.333753	1.2111	0.083164
21	1.6252	0.210916	1.5191	0.181580	61	2.1842	0.339295	1.2015	0.079738
22	1.6307	0.212382	1.5141	0.180168	62	2.2132	0.345020	1.1920	0.076293
23	1.6365	0.213921	1.5090	0.178691	63	2.2435	0.350936	1.1826	0.072834
24	1.6426	0.215533	1.5037	0.177150	64	2.2754	0.357053	1.1732	0.069364
25	1.6490	0.217219	1.4981	0.175545	**65**	2.3088	0.363384	1.1638	0.065889
26	1.6557	0.218981	1.4924	0.173876	66	2.3439	0.369940	1.1545	0.062412
27	1.6627	0.220818	1.4864	0.172144	67	2.3809	0.376736	1.1453	0.058937
28	1.6701	0.222732	1.4803	0.170348	68	2.4198	0.383787	1.1362	0.055472
29	1.6777	0.224723	1.4740	0.168489	69	2.4610	0.391112	1.1272	0.052020
30	1.6858	0.226793	1.4675	0.166567	**70**	2.5046	0.398730	1.1184	0.048589
31	1.6941	0.228943	1.4608	0.164583	71	2.5507	0.406665	1.1096	0.045183
32	1.7028	0.231173	1.4539	0.162537	72	2.5998	0.414943	1.1011	0.041812
33	1.7119	0.233485	1.4469	0.160429	73	2.6521	0.423596	1.0927	0.038481
34	1.7214	0.235880	1.4397	0.158261	74	2.7081	0.432660	1.0844	0.035200
35	1.7312	0.238359	1.4323	0.156031	**75**	2.7681	0.442176	1.0764	0.031976
36	1.7415	0.240923	1.4248	0.153742	76	2.8327	0.452196	1.0686	0.028819
37	1.7552	0.243575	1.4171	0.151393	77	2.9026	0.462782	1.0611	0.025740
38	1.7633	0.246315	1.4092	0.148985	78	2.9786	0.474008	1.0538	0.022749
39	1.7748	0.249146	1.4013	0.146519	79	3.0617	0.485967	1.0468	0.019858
40	1.7868	0.252068	1.3931	0.143995	**80**	3.1534	0.498777	1.0401	0.017081
41	1.7992	0.255085	1.3849	0.141414	81	3.2553	0.512591	1.0338	0.014432
42	1.8122	0.258197	1.3765	0.138778	82	3.3699	0.527613	1.0278	0.011927
43	1.8256	0.261406	1.3680	0.136086	83	3.5004	9.544120	1.0223	0.009584
44	1.8396	0.264716	1.3594	0.133340	84	3.6519	0.562514	1.0172	0.007422
45	1.8541	0.268127	1.3506	0.130541	**85**	3.8317	0.583396	1.0127	0.005465
46	1.8691	0.271644	1.3418	0.127690	86	4.0528	0.607751	1.0086	0.003740
47	1.8848	0.275267	1.3329	0.124788	87	4.3387	0.637355	1.0053	0.002278
48	1.9011	0.279001	1.3238	0.121836	88	4.7427	0.676027	1.0026	0.001121
49	1.9180	0.282848	1.3147	0.118836	89	5.4349	0.735192	1.0008	0.000326
					90	∞	∞	1.0000	0.000000

$\sin^{-1}$	k	K	$\log K$	$\sin^{-1}k$	k	K	$\log K$	$\sin^{-1}k$	k	K	$\log K$
89	**20**	5.840	0.76641	**89**	**40**	6.533	0.81511	**89**	**50**	7.226	0.85890
89	22	5.891	0.77019	89	41	6.584	0.81849	89	51	7.332	0.86522
89	24	5.946	0.77422	89	42	6.639	0.82210	89	52	7.449	0.87210
89	26	6.003	0.77837	89	43	6.696	0.82582	89	53	7.583	0.87984
89	28	6.063	0.78269	89	44	6.756	0.82969	89	54	7.737	0.88857
89	**30**	6.128	0.78732	**89**	**45**	6.821	0.83385	**89**	**55**	7.919	0.89867
89	32	6.197	0.79218	89	46	6.890	0.83822	89	56	8.143	0.91078
89	34	6.271	0.79734	89	47	6.964	0.84286	89	57	8.430	0.92583
89	36	6.351	0.80284	89	48	7.044	0.84782	89	58	8.836	0.94626
89	38	6.438	0.80875	89	49	7.131	0.85315	89	59	9.529	0.97905
								90	**0**	∞	∞

Table 21 Gamma Functions[a]

n	$\Gamma(n)$	n	$\Gamma(n)$	n	$\Gamma(n)$	n	$\Gamma(n)$
1.00	1.00000	1.25	0.90640	1.50	0.88623	1.75	0.91906
1.01	0.99433	1.26	0.90440	1.51	0.88659	1.76	0.92137
1.02	0.98884	1.27	0.90250	1.52	0.88704	1.77	0.92376
1.03	0.98355	1.28	0.90072	1.53	0.88757	1.78	0.92623
1.04	0.97844	1.29	0.89904	1.54	0.88818	1.79	0.92877
1.05	0.97350	1.30	0.89747	1.55	0.88887	1.80	0.93138
1.06	0.96874	1.31	0.89600	1.56	0.88964	1.81	0.93408
1.07	0.96415	1.32	0.89464	1.57	0.89049	1.82	0.93685
1.08	0.95973	1.33	0.89338	1.58	0.89142	1.83	0.93969
1.09	0.95546	1.34	0.89222	1.59	0.89243	1.84	0.94261
1.10	0.95135	1.35	0.89115	1.60	0.89352	1.85	0.94561
1.11	0.94739	1.36	0.89018	1.61	0.89468	1.86	0.94869
1.12	0.94359	1.37	0.88931	1.62	0.89592	1.87	0.95184
1.13	0.93993	1.38	0.88854	1.63	0.89724	1.88	0.95507
1.14	0.93642	1.39	0.88785	1.64	0.89864	1.89	0.95838
1.15	0.93304	1.40	0.88726	1.65	0.90012	1.90	0.96177
1.16	0.92980	1.41	0.88676	1.66	0.90167	1.91	0.96523
1.17	0.92670	1.42	0.88636	1.67	0.90330	1.92	0.96878
1.18	0.92373	1.43	0.88604	1.68	0.90500	1.93	0.97240
1.19	0.92088	1.44	0.88580	1.69	0.90678	1.94	0.97610
1.20	0.91817	1.45	0.88565	1.70	0.90864	1.95	0.97988
1.21	0.91558	1.46	0.88560	1.71	0.91057	1.96	0.98374
1.22	0.91311	1.47	0.88563	1.72	0.91258	1.97	0.98768
1.23	0.91075	1.48	0.88575	1.73	0.91466	1.98	0.99171
1.24	0.90852	1.49	0.88595	1.74	0.91683	1.99	0.99581
						2.00	1.00000

[a]Values of $\Gamma(n) = \int_0^\infty e^{-x} x^{n-1}\, dx$; $\Gamma(n+1) = n\Gamma(n)$.
For large positive integers, Stirling's formula gives an approximation in which the relative error decreases as n increases:

$$\Gamma(n+1) = (2\pi n)^{1/2} \left(\frac{n}{e}\right)^n$$

Source: From *CRC Standard Mathematical Tables*, Chemical Rubber Publishing Co., 12th ed., 1959. Used by permission.

Table 22 Bessel Functions

	$J_0(x)$ and $J_1(x)$[a]							
x	$J_0(x)$	$J_1(x)$	x	$J_0(x)$	$J_1(x)$	x	$J_0(x)$	$J_1(x)$
0.0	1.0000	0.0000	1.5	0.5118	0.5579	3.0	−0.2601	0.3391
0.1	0.9975	0.0499	1.6	0.4554	0.5699	3.1	−0.2921	0.3009
0.2	0.9900	0.0995	1.7	0.3980	0.5778	3.2	−0.3202	0.2613
0.3	0.9776	0.1483	1.8	0.3400	0.5815	3.3	−0.3443	0.2207
0.4	0.9604	0.1960	1.9	0.2818	0.5812	3.4	−0.3643	0.1792
0.5	0.9385	0.2423	2.0	0.2239	0.5767	3.5	−0.3801	0.1374
0.6	0.9120	0.2867	2.1	0.1666	0.5683	3.6	−0.3918	0.0955
0.7	0.8812	0.3290	2.2	0.1104	0.5560	3.7	−0.3992	0.0538
0.8	0.8463	0.3668	2.3	0.0555	0.5399	3.8	−0.4026	0.0128
0.9	0.8075	0.4059	2.4	0.0025	0.5202	3.9	−0.4018	−0.0272
1.0	0.7652	0.4401	2.5	−0.0484	0.4971	4.0	−0.3971	−0.0660
1.1	0.7196	0.4709	2.6	−0.0968	0.4708	4.1	−0.3887	−0.1033
1.2	0.6711	0.4983	2.7	−0.1424	0.4416	4.2	−0.3766	−0.1386
1.3	0.6201	0.5220	2.8	−0.1850	0.4097	4.3	−0.3610	−0.1719
1.4	0.5669	0.5419	2.9	−0.2243	0.3754	4.4	−0.3423	−0.2028

(Continues)

Table 22   *(Continued)*

$J_0(x)$ and $J_1(x)$[a]

x	$J_0(x)$	$J_1(x)$	x	$J_0(x)$	$J_1(x)$	x	$J_0(x)$	$J_1(x)$
4.5	−0.3205	−0.2311	6.0	0.1506	−0.2767	7.5	0.2663	0.1352
4.6	−0.2961	−0.2566	6.1	0.1773	−0.2559	7.6	0.2516	0.1592
4.7	−0.2693	−0.2791	6.2	0.2017	−0.2329	7.7	0.2346	0.1813
4.8	−0.2404	−0.2985	6.3	0.2238	−0.2081	7.8	0.2154	0.2014
4.9	−0.2097	−0.3147	6.4	0.2433	−0.1816	7.9	0.1944	0.2192
5.0	−0.1776	−0.3276	6.5	0.2601	−0.1538	8.0	0.1717	0.2346
5.1	−0.1443	−0.3371	6.6	0.2740	−0.1250	8.1	0.1475	0.2476
5.2	−0.1103	−0.3432	6.7	0.2851	−0.0953	8.2	0.1222	0.2580
5.3	−0.0758	−0.3460	6.8	0.2931	−0.0652	8.3	0.0960	0.2657
5.4	−0.0412	−0.3453	6.9	0.2981	−0.0349	8.4	0.0692	0.2708
5.5	−0.0068	−0.3414	7.0	0.3001	−0.0047	8.5	0.0419	0.2731
5.6	0.0270	−0.3343	7.1	0.2991	0.0252	8.6	0.0146	0.2728
5.7	0.0599	−0.3241	7.2	0.2951	0.0543	8.7	−0.0125	0.2697
5.8	0.0917	−0.3110	7.3	0.2882	0.0826	8.8	−0.0392	0.2641
5.9	0.1220	−0.2951	7.4	0.2786	0.1096	8.9	−0.0653	0.2559

$Y_0(x)$ and $Y_1(x)$

x	$Y_0(x)$	$Y_1(x)$	x	$Y_0(x)$	$Y_1(x)$	x	$Y_0(x)$	$Y_1(x)$
0.0	$(-\infty)$	$(-\infty)$	2.5	0.498	0.146	5.0	−0.309	0.148
0.5	−0.445	−1.471	3.0	0.377	0.325	5.5	−0.340	−0.024
1.0	0.088	−0.781	3.5	0.189	0.410	6.0	−0.288	−0.175
1.5	0.382	−0.412	4.0	−0.017	0.398	6.5	−0.173	−0.274
2.0	0.510	−0.107	4.5	−0.195	0.301	7.0	−0.026	−0.303

[a] $J_1(x) = 0$ for $x = 0,\ 3.832,\ 7.016,\ 10.173,\ 13.324, \ldots$
$J_0(x) = 0$ for $x = 2.405,\ 5.520,\ 8.654,\ 11.792, \ldots$

3   STATISTICAL TABLES*

Table 23   **Binomial Coefficients**

n	$\binom{n}{0}$	$\binom{n}{1}$	$\binom{n}{2}$	$\binom{n}{3}$	$\binom{n}{4}$	$\binom{n}{5}$	$\binom{n}{6}$	$\binom{n}{7}$	$\binom{n}{8}$	$\binom{n}{9}$	$\binom{n}{10}$
0	1										
1	1	1									
2	1	2	1								
3	1	3	3	1							
4	1	4	6	4	1						
5	1	5	10	10	5	1					
6	1	6	15	20	15	6	1				
7	1	7	21	35	35	21	7	1			
8	1	8	28	56	70	56	28	8	1		
9	1	9	36	84	126	126	84	36	9	1	
10	1	10	45	120	210	252	210	120	45	10	1
11	1	11	55	165	330	462	462	330	165	55	11
12	1	12	66	220	495	792	924	792	495	220	66
13	1	13	78	286	715	1287	1716	1716	1287	715	286
14	1	14	91	364	1001	2002	3003	3432	3003	2002	1001
15	1	15	105	455	1365	3003	5005	6435	6435	5005	3003
16	1	16	120	560	1820	4368	8008	11440	12870	11440	8008
17	1	17	136	680	2380	6188	12376	19448	24310	24310	19448
18	1	18	153	816	3060	8568	18564	31824	43758	48620	43758
19	1	19	171	969	3876	11628	27132	50388	75582	92378	92378
20	1	20	190	1140	4845	15504	38760	77520	125970	167960	184756

$$_nC_m = \binom{n}{m} = \frac{n!}{[(n-m)!m!]} = \binom{n}{n-m}, \binom{n}{0} = 1,$$

$$(p+q)^n = p^n + \binom{n}{1}p^{n-1}q + \cdots + \binom{n}{s}p^s q^t + \cdots + q^n, \; s+t = n.$$

Tables 23–25 from Burington, *Handbook of Math Tables and Formulas*, published by McGraw-Hill.

Probability Let p be the probability of an event e in one trial and q the probability of failure of e. The probability that, in n trials, the event e will occur exactly $n - t$ times is $\binom{n}{t} p^{n-t} q^t$. The probability that an event e will happen at least r times in n trials is $\sum_{t=0}^{t=n-r} \binom{n}{t} p^{n-t} q^t$; at most r times in n trials is $\sum_{t=n-r}^{t=n} \binom{n}{t} p^{n-t} q^t$.

In a *point binomial*, $(p + q)^n$, distribution, the mean number of favorable events is np; the mean number of unfavorable events is nq; the *standard deviation* is $\sigma = \sqrt{pqn}$; and $a_3 = (p - q)/\sigma$. The *mean deviation from the mean* MD is $\sigma\sqrt{2/\pi} = 0.7979\sigma$; the *semiquartile deviation from the mean* is $0.6745\sigma = 0.845$ MD.

The probability that a deviation of an individual measure from the average lies between $y = -a$ and $y = a$ is

$$\frac{1}{\sqrt{\pi}} \int_{y=-a}^{y=a} h e^{-h^2 y^2}\, dy = \frac{1}{\sigma\sqrt{2\pi}} \int_{y=-a}^{y=a} e^{-y^2/2\sigma^2}\, dy$$

$$= \frac{1}{\sqrt{2\pi}} \int_{x=-b}^{x=b} e^{-x^2/2}\, dx$$

where $x = hy\sqrt{2}$, $b = ha\sqrt{2}$, and $\sigma = 1/h\sqrt{2}$; h is called the *modulus of precision* and σ the *standard (quadratic mean) deviation*.

Table 24 Probability Functions

$$\tfrac{1}{2}(1 + \alpha) = \int_{-\infty}^{x} \Phi(x)\, dx = \text{area under } \Phi(x) \text{ from } -\infty \text{ to } x$$

$$\alpha = \int_{-x}^{x} \Phi(x)\, dx, \qquad \Phi(x) = \frac{1}{\sqrt{2\pi}} e^{-x^2/2} = \text{normal function}$$

$\Phi^{(2)}(x) = (x^2 - 1) \qquad \Phi(x) = \text{second derivative of } \Phi(x)$
$\Phi^{(3)}(x) = (3x - x^3) \qquad \Phi(x) = \text{third derivative of } \Phi(x)$
$\Phi^{(4)}(x) = (x^4 - 6x^2 + 3) \quad \Phi(x) = \text{fourth derivative of } \Phi(x)$

x	$\tfrac{1}{2}(1+\alpha)$	$\Phi(x)$	$\Phi^{(2)}(x)$	$\Phi^{(3)}(x)$	$\Phi^{(4)}(x)$	x	$\tfrac{1}{2}(1+\alpha)$	$\Phi(x)$	$\Phi^{(2)}(x)$	$\Phi^{(3)}(x)$	$\Phi^{(4)}(x)$
0.00	0.5000	0.3989	−0.3989	0.0000	1.1968	0.25	0.5987	0.3867	−0.3625	0.2840	1.0165
0.01	0.5040	0.3989	−0.3989	0.0120	1.1965	0.26	0.6026	0.3857	−0.3596	0.2941	1.0024
0.02	0.5080	0.3989	−0.3987	0.0239	1.1956	0.27	0.6064	0.3847	−0.3566	0.3040	0.9878
0.03	0.5120	0.3988	−0.3984	0.0359	1.1941	0.28	0.6103	0.3836	−0.3535	0.3138	0.9727
0.04	0.5160	0.3986	−0.3980	0.0478	1.1920	0.29	0.6141	0.3825	−0.3504	0.3235	0.9572
0.05	0.5199	0.3984	−0.3975	0.0597	1.1894	0.30	0.6179	0.3814	−0.3471	0.3330	0.9413
0.06	0.5239	0.3982	−0.3968	0.0716	1.1861	0.31	0.6217	0.3802	−0.3437	0.3423	0.9250
0.07	0.5279	0.3980	−0.3960	0.0834	1.1822	0.32	0.6255	0.3790	−0.3402	0.3515	0.9082
0.08	0.5319	0.3977	−0.3951	0.0952	1.1778	0.33	0.6293	0.3778	−0.3367	0.3605	0.8910
0.09	0.5359	0.3973	−0.3941	0.1070	1.1727	0.34	0.6331	0.3765	−0.3330	0.3693	0.8735
0.10	0.5398	0.3970	−0.3930	0.1187	1.1671	0.35	0.6368	0.3752	−0.3293	0.3779	0.8556
0.11	0.5438	0.3965	−0.3917	0.1303	1.1609	0.36	0.6406	0.3739	−0.3255	0.3864	0.8373
0.12	0.5478	0.3961	−0.3904	0.1419	1.1541	0.37	0.6443	0.3726	−0.3216	0.3947	0.8186
0.13	0.5517	0.3956	−0.3889	0.1534	1.1468	0.38	0.6480	0.3712	−0.3176	0.4028	0.7996
0.14	0.5557	0.3951	−0.3873	0.1648	1.1389	0.39	0.6517	0.3697	−0.3135	0.4107	0.7803
0.15	0.5596	0.3945	−0.3856	0.1762	1.1304	0.40	0.6554	0.3683	−0.3094	0.4184	0.7607
0.16	0.5636	0.3939	−0.3838	0.1874	1.1214	0.41	0.6591	0.3668	−0.3059	0.4259	0.7408
0.17	0.5675	0.3932	−0.3819	0.1986	1.1118	0.42	0.6628	0.3653	−0.3008	0.4332	0.7206
0.18	0.5714	0.3925	−0.3798	0.2097	1.1017	0.43	0.6664	0.3637	−0.2965	0.4403	0.7001
0.19	0.5753	0.3918	−0.3777	0.2206	1.0911	0.44	0.6700	0.3621	−0.2920	0.4472	0.6793
0.20	0.5793	0.3910	−0.3754	0.2315	1.0799	0.45	0.6736	0.3605	−0.2875	0.4539	0.6583
0.21	0.5832	0.3902	−0.3730	0.2422	1.0682	0.46	0.6772	0.3589	−0.2830	0.4603	0.6371
0.22	0.5871	0.3894	−0.3706	0.2529	1.0560	0.47	0.6808	0.3572	−0.2783	0.4666	0.6156
0.23	0.5910	0.3885	−0.3680	0.2634	1.0434	0.48	0.6844	0.3555	−0.2736	0.4727	0.5940
0.24	0.5948	0.3876	−0.3653	0.2737	1.0302	0.49	0.6879	0.3538	−0.2689	0.4785	0.5721

(Continues)

Table 24 *(Continued)*

x	$\frac{1}{2}(1+\alpha)$	$\Phi(x)$	$\Phi^{(2)}(x)$	$\Phi^{(3)}(x)$	$\Phi^{(4)}(x)$	x	$\frac{1}{2}(1+\alpha)$	$\Phi(x)$	$\Phi^{(2)}(x)$	$\Phi^{(3)}(x)$	$\Phi^{(4)}(x)$
0.50	0.6915	0.3521	−0.2641	0.4841	0.5501	0.95	0.8289	0.2541	−0.0248	0.5062	−0.4066
0.51	0.6950	0.3503	−0.2592	0.4895	0.5279	0.96	0.8315	0.2516	−0.0197	−0.521	−0.4228
0.52	0.6985	0.3485	−0.2543	0.4947	0.5056	0.97	0.8340	0.2492	−0.0147	0.4978	−0.4387
0.53	0.7019	0.3467	−0.2493	0.4996	0.4831	0.98	0.8365	0.2468	−0.0098	0.4933	−0.4541
0.54	0.7054	0.3448	−0.2443	0.5043	0.4605	0.99	0.8389	0.2444	−0.0049	0.4887	−0.4692
0.55	0.7088	0.3429	−0.2392	0.5088	0.4378	1.00	0.8413	0.2420	0.0000	0.4839	−0.4839
0.56	0.7123	0.3410	−0.2341	0.5131	0.4150	1.01	0.8438	0.2396	0.0048	0.4790	−0.4983
0.57	0.7157	0.3391	−0.2289	0.5171	0.3921	1.02	0.8461	0.2371	0.0096	0.4740	−0.5122
0.58	0.7190	0.3372	−0.2238	0.5209	0.3691	1.03	0.8485	0.2347	0.0143	0.4688	−0.5257
0.59	0.7224	0.3352	−0.2185	0.5245	0.3461	1.04	0.8508	0.2323	0.0190	0.4635	−0.5389
0.60	0.7257	0.3332	−0.2133	0.5278	0.3231	1.05	0.8531	0.2299	0.0236	0.4580	−0.5516
0.61	0.7291	0.3312	−0.2080	0.5309	0.3000	1.06	0.8554	0.2275	0.0281	0.4524	−0.5639
0.62	0.7324	0.3292	−0.2027	0.5338	0.2770	1.07	0.8577	0.2251	0.0326	0.4467	−0.5758
0.63	0.7357	0.3271	−0.1973	0.5365	0.2539	1.08	0.8599	0.2227	0.0371	0.4409	−0.5873
0.64	0.7389	0.3251	−0.1919	0.5389	0.2309	1.09	0.8621	0.2203	0.0414	0.4350	−0.5984
0.65	0.7422	0.3230	−0.1865	0.5411	0.2078	1.10	0.8643	0.2179	0.0458	0.4290	−0.6091
0.66	0.7454	0.3209	−0.1811	0.5431	0.1849	1.11	0.8665	0.2155	0.0500	0.4228	−0.6193
0.67	0.7486	0.3187	−0.1757	0.5448	0.1620	1.12	0.8686	0.2131	0.0542	0.4166	−0.6292
0.68	0.7517	0.3166	−0.1702	0.5463	0.1391	1.13	0.8708	0.2107	0.0583	0.4102	−0.6386
0.69	0.7549	0.3144	−0.1647	0.5476	0.1164	1.14	0.8729	0.2083	0.0624	0.4038	−0.6476
0.70	0.7580	0.3123	−0.1593	0.5486	0.0937	1.15	0.8749	0.2059	0.0664	0.3973	−0.6561
0.71	0.7611	0.3101	−0.1538	0.5495	0.0712	1.16	0.8770	0.2036	0.0704	0.3907	−0.6643
0.72	0.7642	0.3079	−0.1483	0.5501	0.0487	1.17	0.8790	0.2012	0.0742	0.3840	−0.6720
0.73	0.7673	0.3056	−0.1428	0.5504	0.0265	1.18	0.8810	0.1989	0.0780	0.3772	−0.6792
0.74	0.7704	0.3034	−0.1373	0.5506	0.0043	1.19	0.8830	0.1965	0.0818	0.3704	−0.6861
0.75	0.7734	0.3011	−0.1318	0.5505	−0.0176	1.20	0.8849	0.1942	0.0854	0.3635	−0.6926
0.76	0.7764	0.2989	−0.1262	0.5502	−0.0394	1.21	0.8869	0.1919	0.0890	0.3566	−0.6986
0.77	0.7794	0.2966	−0.1207	0.5497	−0.0611	1.22	0.8888	0.1919	0.0890	0.3566	−0.6986
0.78	0.7823	0.2943	−0.1153	0.5490	−0.0825	1.23	0.8907	0.1872	0.0960	0.3425	−0.7094
0.79	0.7852	0.2920	−0.1098	0.5481	−0.1037	1.24	0.8925	0.1849	0.0994	0.3354	−0.7141
0.80	0.7881	0.2897	−0.1043	0.5469	−0.1247	1.25	0.8944	0.1826	0.1027	0.3282	−0.7185
0.81	0.7910	0.2874	−0.0988	0.5456	−0.1455	1.26	0.8962	0.1804	0.1060	0.3210	−0.7224
0.82	0.7939	0.2850	−0.0934	0.5440	−0.1660	1.27	0.8980	0.1781	0.1092	0.3138	−0.7259
0.83	0.7967	0.2827	−0.0880	0.5423	−0.1862	1.28	0.8997	0.1758	0.1123	0.3065	−0.7291
0.84	0.7995	0.2803	−0.0825	0.5403	−0.2063	1.29	0.9015	0.1736	0.1153	0.2992	−0.7318
0.85	0.8023	0.2780	−0.0771	0.5381	−0.2260	1.30	0.9032	0.1714	0.1182	0.2918	−0.7341
0.86	0.8051	0.2756	−0.0718	0.5358	−0.2455	1.31	0.9049	0.1691	0.1211	0.2845	−0.7361
0.87	0.8078	0.2732	−0.0664	0.5332	−0.2646	1.32	0.9066	0.1669	0.1239	0.2771	−0.7376
0.88	0.8106	0.2709	−0.0611	0.5305	−0.2835	1.33	0.9082	0.1647	0.1267	0.2697	−0.7388
0.89	0.8133	0.2685	−0.0558	0.5276	−0.3021	1.34	0.9099	0.1626	0.1293	0.2624	−0.7395
0.90	0.8159	0.2661	−0.0506	0.5245	−0.3203	1.35	0.9115	0.1604	0.1319	0.2550	−0.7399
0.91	0.8186	0.2637	−0.0453	0.5212	−0.3383	1.36	0.9131	0.1582	0.1344	0.2476	−0.7400
0.92	0.8212	0.2613	−0.0401	0.5177	−0.3559	1.37	0.9147	0.1561	0.1369	0.2402	−0.7396
0.93	0.8238	0.2589	−0.0350	0.5140	−0.3731	1.38	0.9162	0.1539	0.1392	0.2328	−0.7389
0.94	0.8264	0.2565	−0.0299	0.5102	−0.3901	1.39	0.9177	0.1518	0.1415	0.2254	−0.7378

Table 24 *(Continued)*

x	$\frac{1}{2}(1+\alpha)$	$\Phi(x)$	$\Phi^{(2)}(x)$	$\Phi^{(3)}(x)$	$\Phi^{(4)}(x)$	x	$\frac{1}{2}(1+\alpha)$	$\Phi(x)$	$\Phi^{(2)}(x)$	$\Phi^{(3)}(x)$	$\Phi^{(4)}(x)$
1.40	0.9192	0.1497	0.1437	0.2180	−0.7364	1.85	0.9678	0.0721	0.1746	−0.0563	−0.4195
1.41	0.9207	0.1476	0.1459	0.2107	−0.7347	1.86	0.9686	0.0707	0.1740	−0.0605	−0.4095
1.42	0.9222	0.1456	0.1480	0.2033	−0.7326	1.87	0.9693	0.0694	0.1734	−0.0645	−0.3995
1.43	0.9236	0.1435	0.1500	0.1960	−0.7301	1.88	0.9699	0.0681	0.1727	−0.0685	−0.3894
1.44	0.9251	0.1415	0.1519	0.1887	−0.7274	1.89	0.9706	0.0669	0.1720	−0.0723	−0.3793
1.45	0.9265	0.1394	0.1537	0.1815	−0.7243	1.90	0.9713	0.0656	0.1713	−0.0761	−0.3693
1.46	0.9279	0.1374	0.1555	0.1742	−0.7209	1.91	0.9719	0.0644	0.1705	−0.0797	−0.3592
1.47	0.9292	0.1354	0.1572	0.1670	−0.7172	1.92	0.9726	0.0632	0.1697	−0.0832	−0.3492
1.48	0.9306	0.1344	0.1588	0.1599	−0.7132	1.93	0.9732	0.0620	0.1688	−0.0867	−0.3392
1.49	0.9319	0.1315	0.1604	0.1528	−0.7089	1.94	0.9738	0.0608	0.1679	−0.0900	−0.3292
1.50	0.9332	0.1295	0.1619	0.1457	−0.7043	1.95	0.9744	0.0596	0.1670	−0.0933	−0.3192
1.51	0.9345	0.1276	0.1633	0.1387	−0.6994	1.96	0.9750	0.0584	0.1661	−0.0964	−0.3093
1.52	0.9357	0.1257	0.1647	0.1317	−0.6942	1.97	0.9756	0.0573	0.1651	−0.0994	−0.2994
1.53	0.9370	0.1238	0.1660	0.1248	−0.6888	1.98	0.9761	0.0562	0.1641	−0.1024	−0.2895
1.54	0.9382	0.1219	0.1672	0.1180	−0.6831	1.99	0.9767	0.0551	0.1630	−0.1052	−0.2797
1.55	0.9394	0.1200	0.1683	0.1111	−0.6772	2.00	0.9772	0.0540	0.1620	−0.1080	−0.2700
1.56	0.9406	0.1182	0.1694	0.1044	−0.6710	2.01	0.9778	0.0529	0.1609	−0.1106	−0.2603
1.57	0.9418	0.1163	0.1704	0.0977	−0.6646	2.02	0.9783	0.0519	0.1598	−0.1132	−0.2506
1.58	0.9429	0.1145	0.1714	0.0911	−0.6580	2.03	0.9788	0.0508	0.1586	−0.1157	−0.2411
1.59	0.9441	0.1127	0.1722	0.0846	−0.6511	2.04	0.9793	0.0498	0.1575	−0.1180	−0.2316
1.60	0.9452	0.1109	0.1730	0.0781	−0.6441	2.05	0.9798	0.0468	0.1563	−0.1203	−0.2222
1.61	0.9463	0.1092	0.1738	0.0717	−0.6368	2.06	0.9803	0.0478	0.1550	−0.1225	−0.2129
1.62	0.9474	0.1074	0.1745	0.0654	−0.6293	2.07	0.9808	0.0468	0.1538	−0.1245	−0.2036
1.63	0.9484	0.1057	0.1751	0.0591	−0.6216	2.08	0.9812	0.0459	0.1526	−0.1265	−0.1945
1.64	0.9495	0.1040	0.1757	0.0529	−0.6138	2.09	0.9817	0.0449	0.1513	−0.1284	−0.1854
1.65	0.9505	0.1023	0.1762	0.0468	−0.6057	2.10	0.9821	0.0440	0.1500	−0.1302	−0.1765
1.66	0.9515	0.1006	0.1766	0.0408	−0.5975	2.11	0.9821	0.0440	0.1500	−0.1302	−0.1765
1.67	0.9525	0.0989	0.1770	0.0349	−0.5891	2.12	0.9830	0.0422	0.1474	−0.1336	−0.1588
1.68	0.9535	0.0973	0.1773	0.0290	−0.5806	2.13	0.9834	0.0413	0.1460	−0.1351	−0.1502
1.69	0.9545	0.0957	0.1776	0.0233	−0.5720	2.14	0.9838	0.0404	0.1446	−0.1366	−0.1416
1.70	0.9554	0.0940	0.1778	0.0176	−0.5632	2.15	0.9842	0.0395	0.1433	−0.1380	−0.1332
1.71	0.9564	0.0925	0.1779	0.0120	−0.5542	2.16	0.9846	0.0387	0.1419	−0.1393	−0.1249
1.72	0.9573	0.0909	0.1780	0.0065	−0.5452	2.17	0.9850	0.0379	0.1405	−0.1405	−0.1167
1.73	0.9582	0.0893	0.1780	0.0011	−0.5360	2.18	0.9854	0.0371	0.1391	−0.1416	−0.1086
1.74	0.9591	0.0878	0.1780	−0.0042	−0.5267	2.19	0.9857	0.0363	0.1377	−0.1426	−0.1006
1.75	0.9599	0.0863	0.1780	−0.0094	−0.5173	2.20	0.9861	0.0355	0.1362	−0.1436	−0.0927
1.76	0.9608	0.0848	0.1778	−0.0146	−0.5079	2.21	0.9864	0.0347	0.1348	−0.1445	−0.0850
1.77	0.9616	0.0833	0.1777	−0.0196	−0.4983	2.22	0.9868	0.0339	0.1333	−0.1453	−0.0774
1.78	0.9625	0.0818	0.1774	−0.0245	−0.4887	2.23	0.9871	0.0332	0.1319	−0.1460	−0.0700
1.79	0.9633	0.0804	0.1772	−0.0294	−0.4789	2.24	0.9875	0.0325	0.1304	−0.1467	−0.0626
1.80	0.9641	0.0790	0.1769	−0.0341	−0.4692	2.25	0.9878	0.0317	0.1289	−0.1473	−0.0554
1.81	0.9649	0.0775	0.1765	−0.0388	−0.4593	2.26	0.9881	0.0310	0.1275	−0.1478	−0.0484
1.82	0.9656	0.0761	0.1761	−0.0433	−0.4494	2.27	0.9884	0.0303	0.1260	−0.1483	−0.0414
1.83	0.9664	0.0748	0.1756	−0.0477	−0.4395	2.28	0.9887	0.0297	0.1245	−0.1486	−0.0346
1.84	0.9671	0.0734	0.1751	−0.0521	−0.4295	2.29	0.9890	0.0290	0.1230	−0.1490	−0.0279

(Continues)

Table 24 (*Continued*)

x	$\frac{1}{2}(1+\alpha)$	$\Phi(x)$	$\Phi^{(2)}(x)$	$\Phi^{(3)}(x)$	$\Phi^{(4)}(x)$	x	$\frac{1}{2}(1+\alpha)$	$\Phi(x)$	$\Phi^{(2)}(x)$	$\Phi^{(3)}(x)$	$\Phi^{(4)}(x)$
2.30	0.9893	0.0283	0.1215	−0.1492	−0.0214	2.80	0.9974	0.0079	0.0541	−0.1073	0.1379
2.31	0.9896	0.0277	0.1200	−0.1494	−0.0150	2.81	0.9975	0.0077	0.0531	−0.1059	0.1383
2.32	0.9898	0.0270	0.1185	−0.1495	−0.0088	2.82	0.9976	0.0075	0.0520	−0.1045	0.1386
2.33	0.9901	0.0264	0.1170	−0.1496	−0.0027	2.83	0.9977	0.0073	0.0510	−0.1031	0.1389
2.34	0.9904	0.0258	0.1155	−0.1496	0.0033	2.84	0.9977	0.0071	0.0500	−0.1017	0.1390
2.35	0.9906	0.0252	0.1141	−0.1495	0.0092	2.85	0.9978	0.0069	0.0490	−0.1003	0.1391
2.36	0.9909	0.0246	0.1126	−0.1494	0.0149	2.86	0.9979	0.0067	0.0480	−0.0990	0.1391
2.37	0.9911	0.0241	0.1111	−0.1492	0.0204	2.87	0.9979	0.0065	0.0470	−0.0976	0.1391
2.38	0.9913	0.0235	0.1096	−0.1490	0.0258	2.88	0.9980	0.0063	0.0460	−0.0962	0.1389
2.39	0.9916	0.0229	0.1081	−0.1487	0.0311	2.89	0.9981	0.0061	0.0451	−0.0948	0.1388
2.40	0.9918	0.0224	0.1066	−0.1483	0.0362	2.90	0.9981	0.0060	0.0441	−0.0934	0.1385
2.41	0.9920	0.0219	0.1051	−0.1480	0.0412	2.91	0.9982	0.0058	0.0432	−0.0920	0.1382
2.42	0.9922	0.0213	0.1036	−0.1475	0.0461	2.92	0.9982	0.0056	0.0423	−0.0906	0.1378
2.43	0.9925	0.0208	0.1022	−0.1470	0.0508	2.93	0.9983	0.0055	0.0414	−0.0893	0.1374
2.44	0.9927	0.0203	0.1007	−0.1465	0.0554	2.94	0.9984	0.0053	0.0405	−0.0879	0.1369
2.45	0.9929	0.0198	0.0992	−0.1459	0.0598	2.95	0.9984	0.0051	0.0396	−0.0865	0.1364
2.46	0.9931	0.0194	0.0978	−0.1453	0.0641	2.96	0.9985	0.0050	0.0388	−0.0852	0.1358
2.47	0.9932	0.0189	0.0963	−0.1446	0.0683	2.97	0.9985	0.0048	0.0379	−0.0838	0.1352
2.48	0.9934	0.0184	0.0949	−0.1439	0.0723	2.98	0.9986	0.0047	0.0371	−0.0825	0.1345
2.49	0.9936	0.0180	0.0935	−0.1432	0.0762	2.99	0.9986	0.0046	0.0363	−0.0811	0.1337
2.50	0.9938	0.0175	0.0920	−0.1424	0.0800	3.00	0.9987	0.0044	0.0355	−0.0798	0.1330
2.51	0.9940	0.0171	0.0906	−0.1416	0.0836	3.01	0.9987	0.0043	0.0347	−0.0785	0.1321
2.52	0.9941	0.0167	0.0892	−0.1408	0.0871	3.02	0.9987	0.0042	0.0339	−0.0771	0.1313
2.53	0.9943	0.0163	0.0878	−0.1399	0.0905	3.03	0.9988	0.0040	0.0331	−0.0758	0.1304
2.54	0.9945	0.0158	0.0864	−0.1389	0.0937	3.04	0.9988	0.0039	0.0324	−0.0745	0.1294
2.55	0.9946	0.0154	0.0850	−0.1380	0.0968	3.05	0.9989	0.0038	0.0316	−0.0732	0.1285
2.56	0.9948	0.0151	0.0836	−0.1370	0.0998	3.06	0.9989	0.0037	0.0309	−0.0720	0.1275
2.57	0.9949	0.0147	0.0823	−0.1360	0.1027	3.07	0.9989	0.0036	0.0302	−0.0707	0.1264
2.58	0.9951	0.0143	0.0809	−0.1350	0.1054	3.08	0.9990	0.0035	0.0295	−0.0694	0.1254
2.59	0.9952	0.0319	0.0796	−0.1339	0.1080	3.09	0.9990	0.0034	0.0288	−0.0682	0.1243
2.60	0.9953	0.0136	0.0782	−0.1328	0.1105	3.10	0.9990	0.0033	0.0281	−0.0669	0.1231
2.60	0.9953	0.0136	0.0782	−0.1328	0.1105	3.11	0.9991	0.0032	0.0275	−0.0657	0.1220
2.62	0.9956	0.0129	0.0756	−0.1305	0.1152	3.12	0.9991	0.0031	0.0268	−0.0645	0.1208
2.63	0.9957	0.0126	0.0743	−0.1294	0.1173	3.13	0.9991	0.0030	0.0262	−0.0633	0.1196
2.64	0.9959	0.0122	0.0730	−0.1282	0.1194	3.14	0.9992	0.0029	0.0256	−0.0621	0.1184
2.65	0.9960	0.0119	0.0717	−0.1270	0.1213	3.15	0.9992	0.0028	0.0249	−0.0609	0.1171
2.66	0.9961	0.0116	0.0705	−0.1258	0.1231	3.16	0.9992	0.0027	0.0243	−0.0598	0.1159
2.67	0.9962	0.0113	0.0692	−0.1245	0.1248	3.17	0.9992	0.0026	0.0237	−0.0586	0.1146
2.68	0.9963	0.0110	0.0680	−0.1233	0.1264	3.18	0.9993	0.0025	0.0232	−0.0575	0.1133
2.69	0.9964	0.0107	0.0668	−0.1220	0.1279	3.19	0.9993	0.0025	0.0226	−0.0564	0.1120
2.70	0.9965	0.0104	0.0656	−0.1207	0.1293	3.20	0.9993	0.0024	0.0220	−0.0552	0.1107
2.71	0.9966	0.0101	0.0644	−0.1194	0.1306	3.21	0.9993	0.0023	0.0215	−0.0541	0.1093
2.72	0.9967	0.0099	0.0632	−0.1181	0.1317	3.22	0.9994	0.0022	0.0210	−0.0531	0.1080
2.73	0.9968	0.0096	0.0620	−0.1168	0.1328	3.23	0.9994	0.0022	0.0204	−0.0520	0.1066
2.74	0.9969	0.0093	0.0608	−0.1154	0.1338	3.24	0.9994	0.0021	0.0199	−0.0509	0.1053
2.75	0.9970	0.0091	0.0597	−0.1141	0.1347	3.25	0.9994	0.0020	0.0194	−0.0499	0.1039
2.76	0.9971	0.0088	0.0585	−0.1127	0.1356	3.26	0.9994	0.0020	0.0189	−0.0488	0.1025
2.77	0.9972	0.0086	0.0574	−0.1114	0.1363	3.27	0.9995	0.0019	0.0184	−0.0478	0.1011
2.78	0.9973	0.0084	0.0563	−0.1100	0.1369	3.28	0.9995	0.0018	0.0180	−0.0468	0.0997
2.79	0.9974	0.0081	0.0552	−0.1087	0.1375	3.29	0.9995	0.0018	0.0175	−0.0458	0.0983

Table 24 (*Continued*)

x	$\frac{1}{2}(1+\alpha)$	$\Phi(x)$	$\Phi^{(2)}(x)$	$\Phi^{(3)}(x)$	$\Phi^{(4)}(x)$	x	$\frac{1}{2}(1+\alpha)$	$\Phi(x)$	$\Phi^{(2)}(x)$	$\Phi^{(3)}(x)$	$\Phi^{(4)}(x)$
3.30	0.9995	0.0017	0.0170	−0.0449	0.0969	3.75	0.9999	0.0004	0.0046	−0.0146	0.0410
3.31	0.9995	0.0017	0.0106	−0.0439	0.0955	3.76	0.9999	0.0003	0.0045	−0.0142	0.0401
3.32	0.9996	0.0016	0.0102	−0.0429	0.0941	3.77	0.9999	0.0003	0.0043	−0.0138	0.0392
3.33	0.9996	0.0016	0.0157	−0.0420	0.0927	3.78	0.9999	0.0003	0.0042	−0.0134	0.0382
3.34	0.9996	0.0015	0.0153	−0.0411	0.0913	3.79	0.9999	0.0003	0.0041	−0.0131	0.0373
3.35	0.9996	0.0015	0.0149	−0.0402	0.0899	3.80	0.9999	0.0003	0.0039	−0.0127	0.0365
3.36	0.9996	0.0014	0.0145	−0.0393	0.0885	3.81	0.9999	0.0003	0.0038	−0.0123	0.0356
3.37	0.9996	0.0014	0.0141	−0.0384	0.0871	3.82	0.9999	0.0003	0.0037	−0.0120	0.0347
3.38	0.9996	0.0013	0.0138	−0.0376	0.0857	3.83	0.9999	0.0003	0.0036	−0.0116	0.0339
3.39	0.9997	0.0013	0.0134	−0.0367	0.0843	3.84	0.9999	0.0003	0.0034	−0.0113	0.0331
3.40	0.9997	0.0012	0.0130	−0.0359	0.0829	3.85	0.9999	0.0002	0.0033	−0.0110	0.0323
3.41	0.9997	0.0012	0.0127	−0.0350	0.0815	3.86	0.9999	0.0002	0.0032	−0.0107	0.0315
3.42	0.9997	0.0012	0.0123	−0.0342	0.0801	3.87	1.0000	0.0002	0.0031	−0.0104	0.0307
3.43	0.9997	0.0011	0.0120	−0.0334	0.0788	3.88	1.0000	0.0002	0.0030	−0.0100	0.0299
3.44	0.9997	0.0011	0.0116	−0.0327	0.0774	3.89	1.0000	0.0002	0.0029	−0.0098	0.0292
3.45	0.9997	0.0010	0.0113	−0.0319	0.0761	3.90	1.0000	0.0002	0.0028	−0.0095	0.0284
3.46	0.9997	0.0010	0.0110	−0.0311	0.0747	3.91	1.0000	0.0002	0.0027	−0.0092	0.0277
3.47	0.9997	0.0010	0.0107	−0.0304	0.0734	3.92	1.0000	0.0002	0.0026	−0.0089	0.0270
3.48	0.9998	0.0009	0.0104	−0.0297	0.0721	3.93	1.0000	0.0002	0.0026	−0.0086	0.0263
3.49	0.9998	0.0009	0.0101	−0.0290	0.0707	3.94	1.0000	0.0002	0.0025	−0.0084	0.0256
3.50	0.9998	0.0009	0.0098	−0.0283	0.0694	3.95	1.0000	0.0002	0.0024	−0.0081	0.0250
3.51	0.9998	0.0008	0.0095	−0.0276	0.0681	3.96	1.0000	0.0002	0.0023	−0.0079	0.0243
3.52	0.9998	0.0008	0.0093	−0.0269	0.0669	3.97	1.0000	0.0002	0.0022	−0.0076	0.0237
3.53	0.9998	0.0008	0.0090	−0.0262	0.0656	3.98	1.0000	0.0001	0.0022	−0.0074	0.0230
3.54	0.9998	0.0008	0.0087	−0.0256	0.0643	3.99	1.0000	0.0001	0.0021	−0.0072	0.0224
3.55	0.9998	0.0007	0.0085	−0.0249	0.0631	4.00	1.0000	0.0001	0.0020	−0.0070	0.0218
3.56	0.9998	0.0007	0.0082	−0.0243	0.0618	4.05	1.0000	0.0001	0.0017	−0.0059	0.0190
3.57	0.9998	0.0007	0.0080	−0.0237	0.0606	4.10	1.0000	0.0001	0.0014	−0.0051	0.0165
3.58	0.9998	0.0007	0.0078	−0.0231	0.0594	4.15	1.0000	0.0001	0.0012	−0.0043	0.0143
3.59	0.9998	0.0006	0.0075	−0.0225	0.0582	4.20	1.0000	0.0001	0.0010	−0.0036	0.0123
3.60	0.9998	0.0006	0.0073	−0.0219	0.0570	4.25	1.0000	0.0001	0.0008	−0.0031	0.0105
3.61	0.9999	0.0006	0.0071	−0.0214	0.0559	4.30	1.0000	0.0000	0.0007	−0.0026	0.0090
3.62	0.9999	0.0006	0.0069	−0.0208	0.0547	4.35	1.0000	0.0000	0.0006	−0.0022	0.0077
3.63	0.9999	0.0006	0.0067	−0.0203	0.0536	4.40	1.0000	0.0000	0.0005	−0.0018	0.0065
3.64	0.9999	0.0005	0.0065	−0.0198	0.0524	4.45	1.0000	0.0000	0.0004	−0.0015	0.0055
3.65	0.9999	0.0005	0.0063	−0.0192	0.0513	4.50	1.0000	0.0000	0.0003	−0.0012	0.0047
3.66	0.9999	0.0005	0.0061	−0.0187	0.0502	4.55	1.0000	0.0000	0.0003	−0.0010	0.0039
3.67	0.9999	0.0005	0.0059	−0.0182	0.0492	4.60	1.0000	0.0000	0.0002	−0.0009	0.0033
3.68	0.9999	0.0005	0.0057	−0.0177	0.0481	4.65	1.0000	0.0000	0.0002	−0.0007	0.0027
3.69	0.9999	0.0004	0.0056	−0.0173	0.0470	4.70	1.0000	0.0000	0.0001	−0.0006	0.0023
3.70	0.9999	0.0004	0.0054	−0.0168	0.0460	4.75	1.0000	0.0000	0.0001	−0.0005	0.0019
3.71	0.9999	0.0004	0.0052	−0.0164	0.0450	4.80	1.0000	0.0000	0.0001	−0.0004	0.0016
3.72	0.9999	0.0004	0.0051	−0.0159	0.0440	4.85	1.0000	0.0000	0.0001	−0.0003	0.0013
3.73	0.9999	0.0004	0.0049	−0.0155	0.0430	4.90	1.0000	0.0000	0.0001	−0.0003	0.0011
3.74	0.9999	0.0004	0.0048	−0.0150	0.0420	4.95	1.0000	0.0000	0.0000	−0.0002	0.0009

The sum of those terms of

$$(p+q)^n \equiv \sum_{t=0}^{n} \binom{n}{t} p^{n-t} q^t \qquad p+q=1$$

in which t ranges from a to b inclusive, a and b being integers ($a \leqq t \leqq b$), is (if n is large enough) approximately

$$\int_{x_1}^{x_2} \phi(x)\ dx + \left[\frac{q-p}{6\sigma} \phi^{(2)}(x) \right.$$

$$\left. + \frac{1}{24}\left(\frac{1}{\sigma^2} - \frac{6}{n}\right)\phi^{(3)}(x) \right]_{x_1}^{x_2}$$

where $x_1 = (a - \frac{1}{2} - qn)/\sigma$, $x_2 = (b + \frac{1}{2} - qn)/\sigma$.

The sum of the first $t+1$ terms of

$$(p+q)^n \equiv \sum_{t=0}^{n} \binom{n}{t} p^{n-t} q^t \qquad p+q=1$$

is approximately

$$\int_x^\infty \phi(x)\ dx + \frac{q-p}{6\sigma}\phi^{(2)}(x) - \frac{1}{24}\left(\frac{1}{\sigma^2} - \frac{6}{n}\right)\phi^{(3)}(x)$$

where $x = (s - \frac{1}{2} - np)/\sigma$, $s = n - t$. The sum of the last $s+1$ terms is approximately

$$\int_x^\infty \phi(x)\,dx - \frac{q-p}{6\sigma}\phi^{(2)}(x) - \frac{1}{24}\left(\frac{1}{\sigma^2} - \frac{6}{n}\right)\phi^{(3)}(x)$$

where $x = (t - \frac{1}{2} - nq)/\sigma$.

Table 25 Factors for Computing Probable Errors

n	$\dfrac{1}{\sqrt{n}}$	$\dfrac{1}{\sqrt{n(n-1)}}$	$\dfrac{0.6745}{\sqrt{n-1}}$	$\dfrac{0.6745}{\sqrt{n(n-1)}}$	$\dfrac{0.8453}{n\sqrt{n-1}}$	$\dfrac{0.8453}{\sqrt{n(n-1)}}$
2	0.707 107	0.707 107	0.6745	0.4769	0.4227	0.5978
3	0.577 350	0.408 248	0.4769	0.2754	0.1993	0.3451
4	0.500 000	0.288 675	0.3894	0.1947	0.1220	0.2440
5	0.447 214	0.223 607	0.3372	0.1508	0.0845	0.1890
6	0.408 248	0.182 574	0.3016	0.1231	0.0630	0.1543
7	0.377 964	0.154 303	0.2754	0.1041	0.0493	0.1304
8	0.353 553	0.133 631	0.2549	0.0901	0.0399	0.1130
9	0.333 333	0.117 851	0.2385	0.0795	0.0332	0.0996
10	0.316 228	0.105 409	0.2248	0.0711	0.0282	0.0891
11	0.301 511	0.095 346	0.2133	0.0643	0.0243	0.0806
12	0.288 675	0.087 039	0.2034	0.0587	0.0212	0.0736
13	0.277 350	0.080 064	0.1947	0.0540	0.0188	0.0677
14	0.267 261	0.074 125	0.1871	0.0500	0.0167	0.0627
15	0.258 199	0.069 007	0.1803	0.0465	0.0151	0.0583
16	0.250 000	0.064 550	0.1742	0.0435	0.0136	0.0546
17	0.242 536	0.060 634	0.1686	0.0409	0.0124	0.0513
18	0.235 702	0.057 166	0.1636	0.0386	0.0114	0.0483
19	0.229 416	0.054 074	0.1590	0.0365	0.0105	0.0457
20	0.223 607	0.051 299	0.1547	0.0346	0.0097	0.0434
21	0.218 218	0.048 795	0.1508	0.0329	0.0090	0.0412
22	0.213 201	0.046 524	0.1472	0.0314	0.0084	0.0393
23	0.208 514	0.044 455	0.1438	0.0300	0.0078	0.0376
24	0.204 124	0.042 563	0.1406	0.0287	0.0073	0.0360
25	0.200 000	0.040 825	0.1377	0.0275	0.0069	0.0345
26	0.196 116	0.039 223	0.1349	0.0265	0.0065	0.0332
27	0.192 450	0.037 743	0.1323	0.0255	0.0061	0.0319
28	0.188 982	0.036 370	0.1298	0.0245	0.0058	0.0307
29	0.185 695	0.035 093	0.1275	0.0237	0.0055	0.0297
30	0.182 574	0.033 903	0.1252	0.0229	0.0052	0.0287
31	0.179 605	0.032 791	0.1231	0.0221	0.0050	0.0277
32	0.176 777	0.031 750	0.1211	0.0214	0.0047	0.0268
33	0.174 078	0.030 773	0.1192	0.0208	0.0045	0.0260
34	0.171 499	0.029 854	0.1174	0.0201	0.0043	0.0252

Table 25 *(Continued)*

n	$\dfrac{1}{\sqrt{n}}$	$\dfrac{1}{\sqrt{n(n-1)}}$	$\dfrac{0.6745}{\sqrt{n-1}}$	$\dfrac{0.6745}{\sqrt{n(n-1)}}$	$\dfrac{0.8453}{n\sqrt{n-1}}$	$\dfrac{0.8453}{\sqrt{n(n-1)}}$
35	0.169 031	0.028 989	0.1157	0.0196	0.0041	0.0245
36	0.166 667	0.028 172	0.1140	0.0190	0.0040	0.0238
37	0.164 399	0.027 400	0.1124	0.0185	0.0038	0.0232
38	0.162 221	0.026 669	0.1109	0.0180	0.0037	0.0225
39	0.160 128	0.025 976	0.1094	0.0175	0.0035	0.0220
40	0.158 114	0.025 318	0.1080	0.0171	0.0034	0.0214
41	0.156 174	0.024 693	0.1066	0.0167	0.0033	0.0209
42	0.154 303	0.024 098	0.1053	0.0163	0.0031	0.0204
43	0.152 499	0.023 531	0.1041	0.0159	0.0030	0.0199
44	0.150 756	0.022 990	0.1029	0.0155	0.0029	0.0194
45	0.149 071	0.022 473	0.1017	0.0152	0.0028	0.0190
46	0.147 442	0.021 979	0.1005	0.0148	0.0027	0.0186
47	0.145 865	0.021 507	0.0994	0.0145	0.0027	0.0182
48	0.144 338	0.021 054	0.0984	0.0142	0.0026	0.0178
49	0.142 857	0.020 620	0.0974	0.0139	0.0025	0.0174
50	0.141 421	0.020 203	0.0964	0.0136	0.0024	0.0171
51	0.140 028	0.019 803	0.0954	0.0134	0.0023	0.0167
52	0.138 675	0.019 418	0.0945	0.0131	0.0023	0.0164
53	0.137 361	0.019 048	0.0935	0.0129	0.0022	0.0161
54	0.136 083	0.018 692	0.0927	0.0126	0.0022	0.0158
55	0.134 840	0.018 349	0.0918	0.0124	0.0021	0.0155
56	0.133 631	0.018 019	0.0910	0.0122	0.0020	0.0152
57	0.132 453	0.017 700	0.0901	0.0119	0.0020	0.0150
58	0.131 306	0.017 392	0.0893	0.0117	0.0019	0.0147
59	0.130 189	0.017 095	0.0886	0.0115	0.0019	0.0145
60	0.129 099	0.016 807	0.0878	0.0113	0.0018	0.0142
61	0.128 037	0.016 529	0.0871	0.0112	0.0018	0.0140
62	0.127 000	0.016 261	0.0864	0.0110	0.0018	0.0138
63	0.125 988	0.016 001	0.0857	0.0108	0.0017	0.0135
64	0.125 000	0.015 749	0.0850	0.0106	0.0017	0.0133
65	0.124 035	0.015 504	0.0843	0.0105	0.0016	0.0131
66	0.123 091	0.015 268	0.0837	0.0103	0.0016	0.0129
67	0.122 169	0.015 038	0.0830	0.0101	0.0016	0.0127
68	0.121 268	0.014 815	0.0824	0.0100	0.0015	0.0125
69	0.120 386	0.014 599	0.0818	0.0099	0.0015	0.0123
70	0.119 523	0.014 389	0.0812	0.0097	0.0015	0.0122
71	0.118 678	0.014 185	0.0806	0.0096	0.0014	0.0120
72	0.117 851	0.013 986	0.0801	0.0094	0.0014	0.0118
73	0.117 041	0.013 793	0.0795	0.0093	0.0014	0.0117
74	0.116 248	0.013 606	0.0789	0.0092	0.0013	0.0115
75	0.115 470	0.013 423	0.0784	0.0091	0.0013	0.0113
76	0.114 708	0.013 245	0.0779	0.0089	0.0013	0.0112
77	0.113 961	0.013 072	0.0773	0.0088	0.0013	0.0111
78	0.113 228	0.012 904	0.0769	0.0087	0.0012	0.0109
79	0.112 509	0.012 739	0.0764	0.0086	0.0012	0.0108
80	0.111 803	0.012 579	0.0759	0.0085	0.0012	0.0106
81	0.111 111	0.012 423	0.0754	0.0084	0.0012	0.0105
82	0.110 432	0.012 270	0.0749	0.0083	0.0012	0.0104
83	0.109 764	0.012 121	0.0745	0.0082	0.0011	0.0103
84	0.109 109	0.011 976	0.0740	0.0081	0.0011	0.0101

(Continues)

Table 25   *(Continued)*

n	$\dfrac{1}{\sqrt{n}}$	$\dfrac{1}{\sqrt{n(n-1)}}$	$\dfrac{0.6745}{\sqrt{n-1}}$	$\dfrac{0.6745}{\sqrt{n(n-1)}}$	$\dfrac{0.8453}{n\sqrt{n-1}}$	$\dfrac{0.8453}{\sqrt{n(n-1)}}$
85	0.108 465	0.011 835	0.0736	0.0080	0.0011	0.0100
86	0.107 833	0.011 696	0.0732	0.0079	0.0011	0.0099
87	0.107 211	0.011 561	0.0727	0.0078	0.0011	0.0098
88	0.106 600	0.011 429	0.0723	0.0077	0.0010	0.0097
89	0.106 000	0.011 300	0.0719	0.0076	0.0010	0.0096
90	0.105 409	0.011 173	0.0715	0.0075	0.0010	0.0094
91	0.104 828	0.011 050	0.0711	0.0075	0.0010	0.0093
92	0.104 257	0.010 929	0.0707	0.0074	0.0010	0.0092
93	0.103 695	0.010 811	0.0703	0.0073	0.0010	0.0091
94	0.103 142	0.010 695	0.0699	0.0072	0.0009	0.0090
95	0.102 598	0.010 582	0.0696	0.0071	0.0009	0.0089
96	0.102 062	0.010 471	0.0692	0.0071	0.0009	0.0089
97	0.101 535	0.010 363	0.0688	0.0070	0.0009	0.0088
98	0.101 015	0.010 257	0.0685	0.0069	0.0009	0.0087
99	0.100 504	0.010 152	0.0681	0.0069	0.0009	0.0086
100	0.100 000	0.010 050	0.0678	0.0068	0.0008	0.0085

Table 26   Statistics and Probability Formulas

$p(x) = dP(x)/dx$	Differential probability (density) function of random variable x; univariate frequency function
$P(x) = \int_{-\infty}^{x} p(x')\,dx'$	Cumulative probability function of random variable x; univariate distribution function
$P(A < x < B)$	Cumulative probability that x is between A and B
$P(E \cap F)$	Probability of simultaneous (joint) occurrence of E and F
$P(E \cup F)$	Probability of occurrence of E or F or both
$P(E|F) = P(E \cap F)/P(F)$	Conditional probability; probability of occurrence of E provided F has occurred
$E[f(x)] = \int_{-\infty}^{\infty} f(x)p(x)\,dx$	Expected value of function of a random variable x
$E(x) = \bar{x} = \int_{-\infty}^{\infty} xp(x)\,dx$	Expected (mean) value of random variable x
$\alpha_r = E(x^r)$	rth moment of random variable x; rth moment about the origin
$\mu_r = E(x - \bar{x})^r$	rth moment of random variable x from mean value; rth central moment
$\text{Var } x = E[(x - \bar{x})^2] = \overline{(x - \bar{x})^2}$ $= \overline{x^2} - \bar{x}^2$	Variance value of random variable x
$\sigma = (\text{Var } x)^{1/2}$	Standard deviation of random variable x
$M_x(s) = E(e^{sx})$	Moment generating function associated with random variable x
$\psi_x(q) = E(e^{jqx})$	Characteristic function associated with random variable x
$\psi_g(q) = E[e^{jqg(x)}]$	Characteristic function of $g(x)$ with random variable x
$p(x, y) = d^2 P(x, y)/dx\,dy$	Differential probability (density) function of random variables x and y; bivariate frequency function
$P(x, y) = \int_{-\infty}^{x} \int_{-\infty}^{y} p(x', y')\,dx'\,dy'$	Cumulative probability function of random variables x and y; bivariate distribution function
$P(A < x < B, C < y < D)$	Cumulative probability that x is between A and B and that also y is between C and D; cumulative joint probability
$\text{Cov }(x, y) = E[(x - \bar{x})(y - \bar{y})]$ $= \overline{(x - \bar{x})(y - \bar{y})}$	Covariance value of random variables x and y
$\rho(x, y) = \text{Cov }(x, y)/\sigma_x \sigma_y$	Correlation coefficient of random variables x and y

Source: Giacoletto, *Electronic Designers' Handbook*, Copyright © 1977 by McGraw-Hill, pp. 1–8.

The *probable error* of a single observation in a series of n measures, $t_1, t_2, \ldots, t_n$, the arithmetic mean of which is m, is

$$e = \frac{0.6745}{\sqrt{n-1}}\sqrt{(m-t_1)^2 + (m-t_2)^2 + \cdots + (m-t_n)^2}$$

the probable error of the mean is

$$E = \frac{0.6745}{\sqrt{n(n-1)}}$$
$$\times\sqrt{(m-t_1)^2 + (m-t_2)^2 + \cdots + (m-t_n)^2}$$

Approximate values of e and E are

$$e = 0.8453\frac{\sum_{i=1}^{n} d_i}{\sqrt{n(n-1)}} \qquad E = 0.8453\frac{\sum_{i=1}^{n} d_i}{n\sqrt{n-1}}$$

where $\sum_{i=1}^{n} d_i$ is the sum of the deviations $d_i = |t_i - m|$.

4 UNITS AND STANDARDS

4.1 Physical Quantities and Their Relations

Mathematics is concerned with relations between numerical quantities, either constant or varying in a specified manner over a specified range of values. The numerical values are unique, absolute, and the same all over the world, being the expression of a fundamental perception of the mind. Any *mathematical equation* defines the values of one numerical quantity, known as the dependent, in terms of constants and one or more other numerical quantities, known as the independent variables, as, for example,

$$z = r^2 + 3x + 4 \qquad y = c \cdot \int_0^x \frac{x^2}{\cos x}\, dx \quad (1)$$

where z and y are dependent variables, r and x independent variables, and c a constant.

Physics, comprising the knowledge of inanimate nature and its laws, is concerned fundamentally with the measuring of the various quantities founded or created by definition, as, for example, *length*, *mass*, and *electric charge*. In order to specify a *physical quantity* it is not sufficient to state merely a number. The value of a physical quantity can be determined only by comparison of the sample with a known amount of the same quantity by the process of *measuring*. The reference amount is called a unit, and the result of any measurement must be a statement of "how many times the sample was found to contain the reference amount." Thus a physical quantity Q naturally appears to be the product of a numerical value N and a unit U,

$$Q = N \cdot U \quad (2)$$

as, for example: The length of a particular rod is 3.5 ft, or the rod is $3\frac{1}{2}$ times the length of 1 ft. Obviously, the reproduction of a unit must be possible at any time in order to facilitate correct measurements. This is being done by means of the "standards," which are simply a set of fundamental unit quantities kept under normalized conditions in order to preserve their values as accurately as facilities permit.

Any physical relation must be the result of a more or less obvious measurement, so that equations in physics are not merely numerical relations but express dependencies between physical quantities. Mathematics does not know "standards"; physics cannot be without "standards." The fact that physics often uses the methods of mathematics must not lead to the identification of the two sciences; it is merely an overlapping in the border regions.

Relations between Units. A unit is a particular amount of the physical quantity to be measured defined in terms of a standard. The choice of a unit depends on convenience, facility of reproduction, and easy subdivision so as to obtain smaller units if desired. The value of a physical quantity Q must be independent of the units used, so that for two different units of the same type

$$Q = N_1 \cdot U_1 = N_2 \cdot U_2 \quad (3)$$

The size of the unit and the numerical value of the quantity are inversely related: the larger the unit the smaller the number of units.

A unit relation is an equation between two different units of the same type,

$$U_1 = N_{12} \cdot U_2 \quad (4)$$

and serves to convert from one unit U_1 to a different one U_2. The conversion is achieved by replacing U_1, taken as a factor, by its equivalent according to Eq. (4) so that

$$Q = N_1 \cdot U_1 = N_1 \cdot (N_{12} \cdot U_2) = (N_1 \cdot N_{12}) \cdot U_2 \quad (5)$$

As an example, express the length 3.5 ft in centimeters. The unit relation is 1 ft = 30.5 cm, and therefore $l = 3.5$ ft $= 3.5 \times (30.5 \text{ cm}) = 106.75$ cm. No error is possible if this rule is followed properly.

Physical Equations. Relations between physical quantities are usually given in the form of equations. It is always possible, by the proper use of unit relations (see previous paragraph), to express each side in the same units. Since units are to be considered as factors, they may be canceled and a numerical identity must result. This fact always can be used to check the proper numerical relations and the consistency of the units used.

There are two fundamental types of physical equations:

1. The **mathematical definition** of a physical quantity determines a new quantity uniquely in terms of known quantities. An example is Newton's definition of mass by $f = m \cdot a$, where f is the force and a the acceleration of a moving body. If f and a are measured, m can be computed as a physical quantity with numerical value $N(f)/N(a)$ and unit $U(f)/U(a) = U(m)$. A definition should be in agreement with all the other known relations in a particular field of science; it can only be of restricted value if it contradicts other relations (see later the "absolute" electric systems).

2. The **statement of proportionality** defines one physical quantity as linearly depending on a combination of other, known quantities. It is always the result of an experimental investigation. An example is Newton's law of the gravitational force $f = k(m_1 m_2/r^2)$, where m_1 and m_2 are the two masses, r their center distance, and k the proportionality factor. In the case of a proportionality it is permissible to choose arbitrary units for all measurable physical quantities involved and to use the equation as a definition of the proportionality constant that, in general, will be a physical constant with numerical value and unit. In the example the value of k would be

$$\frac{N(f) \cdot N(r^2)}{N(m_1) \cdot N(m_2)} \times \frac{U(f) \cdot U(r^2)}{U(m_1) \cdot U(m_2)} = N(k) \cdot U(k)$$

Most of the fundamental laws of physics are statements of proportionalities, leading to universal physical constants, as, for instance, the gravitational constant k, the Planck constant h, the gas constant R, the absolute permeability of free space μ_v, and the absolute dielectric constant of free space ε_u. It may be observed that each branch of physics is represented by at least one fundamental proportionality constant.

Derived physical quantities are, in general, the result of mathematical definitions. The units of derived quantities are expressed from the combinations of the units used in the definition. All proportionality constants are ordinarily considered as derived physical quantities.

Fundamental Physical Quantities.
The physical quantities, arbitrarily chosen to define new quantities or derived quantities, are called fundamental physical quantities. Their number may vary according to needs and convenience. There is no possibility to designate any physical quantity as absolutely fundamental, or a priori fundamental. Quantities that appear to be fundamental in one special field may be derived quantities in some other field.

4.2 Dimensions and Dimension Systems

Definition of Dimension.
To choose a unit for a physical quantity one has an infinity of possibilities. The numerous units of length that were in use about 100 years ago present a good practical illustration. Yet all these units have in common the quality of being a distinct length and not, for example, a volume. It is convenient to state this fact by representing with the notation $[L]$ any unit of length whatsoever. The measurement of a physical quantity Q, therefore, leads to the statement

$$Q = N \cdot [Q] \tag{6}$$

where N is a numeric denoting the number of general units $[Q]$ that constitute the total quantity Q. According to Fourier, who first introduced this concept into the literature, $[Q]$ is called the "dimension" of the quantity Q. Be it clearly understood that dimension is simply the expression of a general unit and therefore a characteristic peculiarity of physical quantities not occurring in mathematics. Each new physical quantity gives rise to a new "dimension", as, for instance, time $[T]$, force $[F]$, mass $[M]$, and so on. There are as many dimensions, or general units, as there are kinds of physical quantities.

Derived Dimensions.
Many physical quantities have been introduced by mathematical definition. Velocity, for example, is defined as $v = ds/dt$, where s is the length of the path measured from a definite origin and t is the time. A possible expression for the dimension of velocity would be $[V]$. It is customary and convenient, however, to make use of the mathematical definition that is but the rule for the measurement of velocity and to express the dimension in terms of the more familiar dimensions of length and time as a derived dimension $[V] = [L]/[T] = [L][T]^{-1}$. [Read: velocity is of $+1$ dimension in length and -1 dimension in time.] The use of mathematical definitions, leading to derived dimensions of a composite nature, reduces the number of symbols. Thus the measurement of volume, if scientifically conducted, gives $[\text{Vol}] = [L]^3$, or in words, "volume is of $+3$ dimensions in length $[L]$."

Proportionality constants of physics have, in general, *derived dimensions*, as they are defined by the corresponding physical equations.

Fundamental Dimensions.
The more familiar dimensions used to express derived dimensions are referred to as fundamental dimensions. It is advantageous to use as few of these fundamental dimensions as possible, not because the physical relations become simpler or clearer, but merely as a matter of economy in symbols. In fact, any dimension can be chosen to be a fundamental dimension in a particular field and a derived dimension in some other field of physics. No fundamental dimension can be made a starting point of natural philosophy.

Dimensional Equations. Since a physical equation constitutes in fact two equations, one for the units and one for the numerics, one can disregard the numerical factors entirely and write the general units or dimensions only, arriving thus at a dimensional equation. For instance, the law of gravitation would read $[F] = [k][M]^2[L]^{-2}$ using $[F]$, $[k]$, $[M]$, and $[L]$ as dimensions for force, gravitation constant, mass, and length, respectively. From this dimensional equation a derived dimension can be obtained for any quantity involved. Conversely, dimensional equations are used to check the correctness of physical relations if all dimensions can be made to cancel. Finally, the validity of dimensional equations leads to the method of dimensional analysis.

A *set of fundamental dimensions* is any group of fundamental dimensions convenient and useful to express all the physical quantities of a particular field in terms of derived dimensions. The number of fundamental dimensions to make a set may vary according to the field of application. Whether or not a set of fundamental dimensions can be used beyond the field for which it was originally intended will depend upon its suitability as a dimension system. (See next paragraph.) In no case should it be used where it can lead to confusion.

A set of fundamental dimensions is *incomplete* when the number of fundamental dimensions composing it is less than the number required for a dimension system. Incomplete sets of fundamental dimensions should not be used outside the very restricted field for which they are defined; they necessarily would lead to confusing relations.

A *dimension system* is composed of the smallest number of fundamental dimensions that will form a consistent and complete set for a field of science. Since each relation between physical quantities can be split up into one relation of numerics and another one of dimensions (as general units), it is possible to combine all known relations of dimensions. In setting up these relations, all proportionality factors must be taken as physical quantities. If there are m independent relations known, $m + p$ dimensions may be involved, of which m dimensions can be expressed by any p "fundamental" dimensions chosen arbitrarily.

This set of p "fundamental" dimensions is then called a dimension system. From the theory of numbers, therefore, it is known that one generally has a choice of $\binom{m + p}{p}$ possible dimension systems. Thus, if $p = 3$, $m = 3$, then one has $\binom{6}{3} = 20$ different possibilities. A necessary condition, however, is that *each* independent relation involve at least $p + 1$ dimensions. If this is not the case, then the number of possible dimension systems is less, so that $\binom{m + p}{p}$ indicates the *upper* limit.

Any dimension system chosen in the described manner is consistent, as well as correct, and never leads to ambiguity with respect to the expression of physical quantities. Complete dimension systems in mechanics must have three, in thermodynamics four, and in electromagnetism four fundamental dimensions. It seems, according to present knowledge, that five fundamental dimensions suffice for the entire range of physics, namely, the three fundamental dimensions of mechanics, an additional one for thermodynamics, and another additional one for electromagnetism.

All the known dimension systems use length $[L]$ and time $[T]$ as primary fundamental dimensions, adding various fundamental dimensions from the available physical quantities of the fields of physics. The choice of $[L]$ and $[T]$ reduces at once the maximum number of possible dimension systems to $\binom{m + p - 2}{p - 2}$.

Why Dimension Systems? Since the proper choice of units is the ultimate goal of any critical analysis of physical quantities, the question may be asked: Why is it necessary to discuss dimension of systems? The answer is that each physical quantity may be measured by an infinite variety of units but has only one dimension within a given dimension system. The process of deciding upon the fundamental dimensions before fixing the units within the scope of the fundamental dimensions is, therefore, essentially a matter of economy and logic.

4.3 Dimension and Unit Systems

In the past different dimension systems were introduced for various fields of technology (mechanics, heat, electromagnetism) and based on different choices of fundamental dimensions, for example, for mechanics, length and time plus mass or force or energy or gravitational constant gave potentially four different dimension system classes. In turn, for a given dimensional system, a unit system could be developed, choosing for each fundamental dimension a unit desirably related to a fundamental standard or standards. In seeking to define units with appropriate size values, relationships, and so on, many different unit systems, for example, centimeter–gram–second (cgs), meter–kilogram–second (mks), "absolute," "technical," and so on, have been introduced over the years. (See O. W. Eshbach and M. Souders, *Handbook of Engineering Fundamentals*, 3rd ed., Wiley, New York, 1975, for a detailed exposition of the subject.)

In recent years a major step toward simplification and standardization has been taken by the increasing adoption of the International System of Units (SI).

4.4 The International System of Units

The SI system, composed of six fundamental units, has been adopted by the Conference Générale (BIPM Sevres, Paris, 1954 and 1960) to cover the whole range of physics and one in which all international reports are to be expressed.

Quantity	Unit	Symbol

Fundamental Units

Quantity	Unit	Symbol
Length	Meter	m
Mass	Kilogram	kg
Time	Second	s
Intensity of electric current	Ampere	A
Thermodynamic temperature	Degree kelvin	K
Luminous intensity	Candela	cd
Amount of substance	Mole	mol

Derived Units with Special Names

Quantity	Unit	Symbol
Area	Square meter	m^2
Volume	Cubic meter	m^3
Frequency	Hertz	Hz
Density (mass density)	Kilogram per cubic meter	kg/m^3
Velocity	Meter per second	m/s
Angular velocity	Radian per second	rad/s
Acceleration	Meter per square second	m/s^2
Angular acceleration	Radian per square second	rad/s^2
Force	Newton	$N, kg.m/s^2$
Pressure, stress	Newton per square meter	N/m^2
Kinematic viscosity	Square meter per second	m^2/s
Dynamic viscosity	Newton-second per square meter	$N.s/m^2$
Work, energy, heat (quantity of heat)	Joule	$J, N \cdot m$
Power, radiant flux	Watt	W, J/s
Plane angle	Radian	rad
Solid angle	Steradian	sr
Electric charge	Coulomb	$C, A \cdot s$
Electric potential, potential difference, electromotive force	Volt	V, W/A
Electric field strength	Volt per meter	V/m
Resistance (to direct current)	Ohm	$\Omega, V/A$
Electric conductance	Siemens	S, A/V
Capacitance	Farad	$F, A \cdot s/V$
Magnetic flux	Weber	$Wb, V \cdot s$
Inductance	Henry	$H, V \cdot s/A$
Magnetic flux density (magnetic induction)	Tesla	$T, Wb/m^2$
Magnetic field strength	Ampere per meter	A/m
Magnetomotive force	Ampere	A
Luminous flux	Lumen	$lm, cd \cdot sr$
Luminance	Candela per square meter	cd/m^2
Illumination	Lux	$lx, lm/m^2$
Activity (of a radionuclide)	Becquerel	Bq, l/S
Absorbed dose	Gray	Gy, J/kg
Dose equivalent	Sievert	Sv, J/kg

Other Common Derived Units

Quantity	Unit	Symbol
Absorbed dose rate	Gray per second	Gy/s
Acceleration	Meter per second squared	m/s^2
Angular acceleration	Radian per second squared	rad/s^2
Angular velocity	Radian per second	rad/s
Area	Square meter	m^2
Concentration (of amount of substance)	Mole per cubic meter	mol/m^3
Current density	Ampere per square meter	A/m^2

Quantity	Unit	Symbol
	Other Common Derived Units (Continued)	
Density, mass	Kilogram per cubic meter	kg/m^3
Electric charge density	Coulomb per cubic meter	C/m^3
Electric field strength	Volt per meter	V/m
Electric flux density	Coulomb per square meter	C/m^2
Energy density	Joule per cubic meter	J/m^3
Entropy	Joule per kelvin	J/K
Exposure (X and gamma rays)	Coulomb per kilogram	C/kg
Heat capacity	Joule per kelvin	J/K
Heat flux density	Watt per square meter	W/m^2
Irradiance	Watt per square meter	W/m^2
Luminance	Candela per square meter	cd/m^2
Magnetic field strength	Ampere per meter	A/m
Molar energy	Joule per mole	J/mol
Molar entropy	Joule per mole kelvin	$J/(mol \cdot K)$
Molar heat capacity	Joule per mole kelvin	$J/(mol \cdot K)$
Moment of force	Newton meter	$N \cdot m$
Permeability (magnetic)	Henry per meter	H/m
Permittivity	Farad per meter	F/m
Power density	Watt per square meter	W/m^2
Radiance	Watt per square meter steradian	$W/(m^2 \cdot sr)$
Radiant intensity	Watt per steradian	W/sr
Specific heat capacity	Joule per kilogram kelvin	$J/(kg \cdot K)$
Specific energy	Joule per kilogram	J/kg
Specific entropy	Joule per kilogram kelvin	$J/(kg \cdot K)$
Specific volume	Cubic meter per kilogram	m^3/kg
Surface tension	Newton per meter	N/m
Thermal conductivity	Watt per meter kelvin	$W/(m \cdot K)$
Velocity	Meter per second	m/s
Viscosity, dynamic	Pascal second	$Pa \cdot s$
Viscosity, kinematic	Square meter per second	m^2/s
Volume	Cubic meter	m^3
Wave number	1 per meter	$1/m$

Definitions of Derived Units of the International System Having Special Names

Quantity	Unit and Definition
1. Absorbed dose	The *gray* is the absorbed dose when the energy per unit mass imparted to matter by ionizing radiation is one joule per kilogram. *Note*: The gray is also used for the ionizing radiation quantities: specific energy imparted, kerma, and absorbed dose index, which have the SI unit joule per kilogram.
2. Activity	The *becquerel* is the activity of a radionuclide decaying at the rate of one spontaneous nuclear transition per second.
3. Celsius temperature	The *degree Celsius* is equal to the kelvin and is used in place of the kelvin for expressing Celsius temperature (symbol t) defined by the equation $t = T - T_0$, where T is the thermodynamic temperature and $T_0 = 273.15$ K by definition.
4. Dose equivalent	The *sievert* is the dose equivalent when the absorbed dose of ionizing radiation multiplied by the dimensionless factors Q (quality factor) and N (product of any other multiplying factors) stipulated by the International Commission on Radiological Protection is one joule per kilogram.

(Continues)

Quantity	Unit and Definition
5. Electric capacitance	The *farad* is the capacitance of a capacitor between the plates of which there appears a difference of potential of one volt when it is charged by a quantity of electricity equal to one coulomb.
6. Electric conductance	The *siemens* is the electric conductance of a conductor in which a current of one ampere is produced by an electric potential difference of one volt.
7. Electric inductance	The *henry* is the inductance of a closed circuit in which an electromotive force of one volt is produced when the electric current in the circuit varies uniformly at a rate of one ampere per second.
8. Electric potential difference, electromotive force	The *volt* (unit of electric potential difference and electromotive force) is the difference of electric potential between two points of a conductor carrying a constant current of one ampere when the power dissipated between these points is equal to one watt.
9. Electric resistance	The *ohm* is the electric resistance between two points of a conductor when a constant difference of potential of one volt, applied between these two points, produces in this conductor a current of one ampere, this conductor not being the source of any electromotive force.
10. Energy	The *joule* is the work done when the point of application of a force of one newton is displaced a distance of one meter in the direction of the force.
11. Force	The *newton* is that force that, when applied to a body having a mass of one kilogram, gives it an acceleration of one meter per second squared.
12. Frequency	The *hertz* is the frequency of a periodic phenomenon of which the period is one second.
13. Illuminance	The *lux* is the illuminance produced by a luminous flux of one lumen uniformly distributed over a surface of one square meter.
14. Luminous flux	The *lumen* is the luminous flux emitted in a solid angle of one steradian by a point source having a uniform intensity of one candela.
15. Magnetic flux	The *weber* is the magnetic flux that, linking a circuit of one turn, produces in it an electromotive force of one volt as it is reduced to zero at a uniform rate in one second.
16. Magnetic flux density	The *tesla* is the magnetic flux density given by a magnetic flux of one weber per square meter.
17. Power	The *watt* is the power that gives rise to the production of energy at the rate of one joule per second.
18. Pressure or stress	The *pascal* is the pressure or stress of one newton per square meter.
19. Quantity of electricity	The *coulomb* is the quantity of electricity transported in one second by a current of one ampere.

Prefixes.   The SI system has adopted the following standard set of prefixes:

Multiplication Factor	Prefix	Symbol
$1\ 000\ 000\ 000\ 000\ 000\ 000 = 10^{18}$	Exa	E
$1\ 000\ 000\ 000\ 000\ 000 = 10^{15}$	Peta	P
$1\ 000\ 000\ 000\ 000 = 10^{12}$	Tera	T
$1\ 000\ 000\ 000 = 10^{9}$	Giga	G
$1\ 000\ 000 = 10^{6}$	Mega	M
$1\ 000 = 10^{3}$	Kilo	k
$100 = 10^{2}$	Hecto[a]	h
$10 = 10^{1}$	Deka[a]	da
$0.1 = 10^{-1}$	Deci[a]	d
$0.01 = 10^{-2}$	Centi[a]	c
$0.001 = 10^{-3}$	Milli	m
$0.000\ 001 = 10^{-6}$	Micro	μ
$0.000\ 000\ 001 = 10^{-9}$	Nano	n
$0.000\ 000\ 000\ 001 = 10^{-12}$	Pico	p
$0.000\ 000\ 000\ 000\ 001 = 10^{-15}$	Femto	f
$0.000\ 000\ 000\ 000\ 000\ 001 = 10^{-18}$	Atto	a

Application of SI Prefixes

General. In general the SI prefixes should be used to indicate orders of magnitude, thus eliminating non-significant digits and leading zeros in decimal fractions and providing a convenient alternative to the powers-of-10 notation preferred in computation. For example,

$$12{,}300 \text{ mm becomes } 12.3 \text{ m}$$

$$12.3 \times 10^3 \text{ m becomes } 12.3 \text{ km}$$

$$0.00123 \ \mu\text{A becomes } 1.23 \text{ nA}$$

Selection. When expressing a quantity by a numerical value and a unit, a prefix should preferably be chosen so that the numerical value lies between 0.1 and 1000. To minimize variety, it is recommended that prefixes representing 1000 raised to an integral power be used. However, three factors may justify deviation:

1. In expressing area and volume, the prefixes hecto-, deka-, deci-, and centi- may be required, for example, square hectometer, cubic centimeter.
2. In tables of values of the same quantity or in a discussion of such values within a given context, it is generally preferable to use the same unit multiple throughout.
3. For certain quantities in particular applications, one particular multiple is customarily used. For example, the millimeter is used for linear dimensions in mechanical engineering drawings even when the values lie far outside the range 0.1–1000 mm; the centimeter is often used for body measurements and clothing sizes.

Prefixes in Compound Units.* It is recommended that only one prefix be used in forming a multiple of a compound unit. Normally the prefix should be attached to a unit in the numerator. One exception to this is when the kilogram occurs in the denominator. For example,

$$\text{V/m, } not \text{ mV/mm, and MJ/kg, } not \text{ kJ/g}$$

Compound Prefixes. Compound prefixes formed by the juxtaposition of two or more SI prefixes are not to be used. For example, use

$$1 \text{ nm, } not \text{ 1 m} \mu\text{m}$$

$$1 \text{ pF, } not \text{ 1} \mu\mu\text{ F}$$

If values are required outside the range covered by the prefixes, they should be expressed by using powers of 10 applied to the base unit.

---

*A compound unit is a derived unit that is expressed in terms of two or more units rather than by a single special name.

Powers of Units. An exponent attached to a symbol containing a prefix indicates that the multiple or submultiple of the unit (the unit with its prefix) is raised to the power expressed by the exponent. For example,

$$1 \text{ cm}^3 = (10^{-2} \text{ m})^3 = 10^{-6} \text{ m}^3$$

$$1 \text{ ns}^{-1} = (10^{-9} \text{ s})^{-1} = 10^9 \text{ s}^{-1}$$

$$1 \text{ mm}^2/\text{s} = (10^{-3} \text{ m})^2/\text{s} = 10^{-6} \text{ m}^2/\text{s}$$

Calculations. Errors in calculations can be minimized if the base and the coherent derived SI units are used and the resulting numerical values are expressed in powers-of-10 notation instead of using prefixes.

Other Units

Units from Different Systems. To assist in preserving the advantage of SI as a coherent system, it is advisable to minimize the use with it of units from other systems. Such use should be limited to units listed in this section.

A following section presents conversion factors to and from SI units.

4.5 Length, Mass, and Time

English Units and Standards

Units of Length. The *foot* (ft) is the *fundamental* unit of length in the foot–pound–second (fps) system. It equals, by definition, one-third of a *yard* (yd), which is the English legalized *standard* unit of length. The *U.S. yard* was defined by Act of Congress, July 28, 1866, as 3600/3937 the length of the *meter*. (See discussion of metric system for definitions of metric length.)

In Great Britain, the *Imperial yard* is measured by a bronze bar preserved in the Standards Office, Westminster. Its length, in terms of the *international prototype meter*, is 3600/3937.0113 m. For engineering purposes, the U.S. and British *yards* may be considered identical.

As subunits, the *inch* (in.) is defined as one-twelfth of one standard foot, and the *mil* as the one-thousandth part of one inch. The *nautical mile* (mi) is defined as one minute of arc on the earth's surface at the equator, whereas the U.S. mile (U.S. mi statute) is exactly 5280 ft and practically identical with the British mile.

Unit of Capacity (Dry). The *bushel* (bu) is the *standard* unit of *dry* capacity. The *Winchester bushel* (U.S. standard) has a volume of 2150.42 in.^{3}

In Great Britain, the *Imperial bushel* (bu) is defined as the volume of 80 lb of pure water at 62°F weighed against brass weights in air at the same temperature as the water and with the barometer at 30 in. Its volume is approximately 2219.36 in.^{3}

Unit of Capacity (Liquid). The *gallon* (gal) is the *standard* unit of *liquid* capacity. The *U.S. gallon* has a volume of 231 in.^{3}

In Great Britain, the *Imperial gallon* is defined as the volume of 10 lb of pure water at 62°F weighed against brass weights in air at the same temperature as the water and with the barometer at 30 in. Its volume is approximately 277.420 in.³ The Imperial gallon (liquid measure) equals exactly one-eighth of the Imperial bushel (dry measure). Subunits are the quart (qt), which is one-fourth of the standard gallon, and the pint (pt), which is $\frac{1}{2}$ qt.

Units of Mass.
The *pound* (*avoirdupois*) (lb avdp) is the *fundamental* unit of mass in the fps system.* It is also the English legalized *standard* unit of mass. The *U.S. pound* (*avoirdupois*) was defined by Act of Congress, 1866, as 1/2.2046 kg, but since 1895 there has been used, for greater accuracy, a value that agrees with that given by law as far as the latter is given, namely, 453.5924277 g. This value is now used by the Bureau of Standards as an exact definition and is the basis of the customary U.S. weights (Circular 47, Bureau of Standards).

In Great Britain, the *Imperial pound (avoirdupois)* is the mass of a *platinum cylinder* preserved in the Standards Office, Westminster. Its legal equivalent is 453.59243 g. For engineering purposes, the U.S. and British pounds (avoirdupois) may be considered as identical.

Subunits of mass are the grain (gr), defined as $\frac{1}{7000}$ of the standard pound (avoirdupois) and the ounce (avoirdupois) (oz-avdp), which is $\frac{1}{16}$ of the standard pound (avoirdupois). The grain was used as a fundamental unit in the so-called foot–gram–second (fgs) system of units prior to 1873.

Weight versus Mass.
Unfortunately, the word "weight" is used in two different senses, namely, (1) by the layman (as well as loosely by the scientist) to designate a given *mass* or quantity of matter and (2) by the scientist to designate the *pull* in standard gravitational force units that is exerted by the earth upon a piece of matter. The result of the commercial act of "weighing" a specific quantity is independent of the local gravitational pull of the earth, since both spring scales and balances are calibrated locally by comparison with standard masses.

Auxiliary Fundamental Units.
Auxiliary Fundamental Units and their principal derived units are defined and discussed under the sections of this handbook pertaining to the topics to which they apply. In general, however, conversion factors are included in the tables of Section 5.

For an interesting and rather complete history see *British Weights and Measures*, London, 1910, by Sir C. M. Watson.

Metric (or SI) Units and Standards
The development of the SI system and the operations of the international bodies (BIPM, CIPM, and CGPM) having cognizance over weights and measures are described in appendices to ASTM's *Standard for Metric Practice* (ASTM E380-82, American Society for for Testing and Materials, Philadelphia, 1982).

Units of Length.
The *centimeter* (cm) is the *fundamental* unit of length in the cgs system. It equals, by definition, $\frac{1}{100}$ of a *meter* (m). The meter has been standardized by international agreement as 1,650,763.73 times the wavelength in vacuum of the unperturbed transition $(2p_{10}-5d_5)$ of krypton 86. The *basic* meter for international comparisons is the *international prototype meter*, which is the distance, at zero degrees Centigrade, between two lines on a platinum–iridium bar located at the International Bureau of Weights and Measures at Sèvres, France. This meter is the nearest to a duplicate ever constructed of the *original* meter, which was constructed and deposited in the Archives of the French Republic in 1799. The meter is very nearly equal to one ten-millionth of the distance, measured at sea level, from the equator to either pole.

An interesting history of the development of the *international prototype meter* (as well as the *international prototype kilogram*—see discussion on unit of mass) is given by W. Parry, National Bureau of Standards, in *Merriman's Civil Engineers' Handbook*, as follows:

The use of the meter as the basis of geodetic surveys had become so general throughout Europe that a conference was called in Paris, France, in 1870, for the purpose of establishing a central bureau where the standards of the different countries could be compared. As a result of this conference an International Bureau of Weights and Measures was established near Paris in 1875, by the concurrent action of the principal nations of the world. One of the first tasks undertaken by the Bureau was the construction of exact copies of the meter and kilogram deposited in the Archives. Thirty-one standard meters of iridio-platinum and forty kilograms of the same alloy were constructed and carefully compared with the standards of the Archives and with one another. This great work was completed in 1889, and the meter and kilogram which agreed most nearly with the original standards were called international prototypes, and were deposited at the International Bureau, where they are maintained today subject to the authority of the International Committee on Weights and Measures. The remaining meters and kilograms were distributed by lot to the different nations which contributed to the support of the Bureau. The United States secured two copies of the meter and two copies of the kilogram, which are in the custody of

---

*The slug of mass, which is extensively used by engineers and physicists, is (in the English system) the mass to which an acceleration of one foot per second per second would be given by the application of a one-pound force. Under any gravity conditions, 1 slug of mass = 32.1739 lb of mass.

the Bureau of Standards at Washington. One of the meters, known as No. 27, and one kilogram, No. 20, were selected as the United States standards, while the other meter and kilogram are used as secondary standards. It was the declared intention of the International Committee that the various national prototypes should be returned to the International Bureau at regular intervals for the purpose of recomparing them with the international standards and with one another. In this way all measurements based upon metric standards throughout the world are ultimately referred to the international meter and kilogram.

Unit of Capacity. The *liter* (L) is the *standard* unit of capacity. It is defined as the volume of one kilogram of pure water at the temperature of maximum density (4°C) under a pressure of 76 cm of mercury. For all practical purposes, the liter may be regarded as the equivalent of the cubic decimeter, although the former is actually slightly greater, in the amount of less than three parts in one hundred thousand.

Unit of Mass. The *gram* (g) is the *fundamental* unit of mass in the cgs system.* It equals, by definition, $\frac{1}{1000}$ of a *kilogram* (kg), which is the *standard* unit of mass. The *basic* kilogram for international comparisons is the *international prototype kilogram*, which is a cylinder of platinum–iridium located at the International Bureau of Weights and Measures at Sèvres, France. This mass is the nearest to a duplicate ever constructed of the *original* kilogram, which was constructed and deposited in the Archives of the French Republic in 1799. The latter was made as nearly as possible equal to the mass of a cube of pure water at 4°C, the sides of the cube being one-tenth the length of the original meter.

An interesting history of the development of the *international prototype kilogram* was given under the discussion on units of length.

Weight versus Mass. See discussion under this same subheading of the English units and standards.

Auxiliary Fundamental Units. Auxiliary fundamental units and their principal derived units are defined and discussed under the sections of this handbook pertaining to the topics to which they apply. In general, however, conversion factors are included in Tables 27–64.

Standard of Time

Unit of Time. The *second* has been standardized by international agreement as 1/31,556,925.9747 of the tropical year at 12 hr, ephemeris time, January 0 for the year 1900.0. (This definition has been retained for the time being as an astronomical time standard—the following atomic standard of time interval is 100 times more precise.) The second has been standardized by international agreement as the time taken for 9,192,631,770.0 vibrations of the unperturbed hyperfine transition 4,0–3,0 for the $^2S_{1/2}$ fundamental state of the cesium 133 atom. The ^{133}Cs standard has been adopted provisionally (see resolution 5 of the 12th General Conference of Weights and Measures, BIPM, Sèvres, Paris, Oct. 1964). A more accurate hydrogen maser standard may be available in the near future that is 100 times more accurate than the ^{133}Cs standard.

Measures of Time. A *solar day* is measured by the rotation of the earth about its axis with respect to the sun. In *astronomical computations* and in *nautical time* the day commences at noon, and in the former it is counted throughout the 24 hr. In *civil computations* the day commences at midnight and is divided into two parts of 12 hr each.

A *solar year* is the time in which the earth makes one revolution around the sun. Its average time, called the *mean solar year*, is 365 days, 5 hr, 48 min, and 45.9747 sec, or nearly $365\frac{1}{4}$ days.

4.6 Force, Energy, and Power

Dynamical and Gravitational Units. According to the use of two different dimension and unit systems, the dynamical (or physical, or "absolute") system and the gravitational (or technical) system, two different sets of units of force, energy, power, and derived quantities are defined in both the English and the metric systems. *One dynamical unit of force* produces an acceleration of unity on unit standard mass. The *gravitational unit of force* is defined as that force required to give a unit standard mass an acceleration equal to that produced by the gravitational pull of the earth. As the acceleration due to gravity, *g*, varies with location and altitude,[†] the gravitational unit of force is not constant, and, therefore, its relation to the dynamical unit of force will vary. By international agreement, the value $g_0 = 980.665$ cm/sec sec = 32.1739 ft/sec sec (British) has been chosen as the standard acceleration of gravity to make invariant the gravitational unit of force.

English Units

Units of Force. The dynamical or physical unit of force is the *poundal*, defined as the force required to

---

*The slug of mass, which is extensively used by engineers and physicists, is (in the metric system) the mass to which an acceleration of one meter per second per second would be given by the application of a one-kilogram force. Under any gravity conditions, 1 slug of mass = 9.80665 kg of mass.

[†]The variation of *g* with latitude ϕ and altitude *H* is given approximately by (ϕ in degrees, *H* in meters) $g = 978.039\ (1 + 0.005295 \sin^2 \phi) - 0.000307H$. See *International Critical Tables*, Vol. 1, p. 395.

give a mass of one pound an acceleration of one foot per second per second.

The *pound-force* (or weight of the pound mass) is the gravitational or technical unit of force. It is, by definition, the force required to give a mass of one pound an acceleration of 32.1739 ft/sec sec. If a force is measured by "weighing," the result in pounds weight must be multiplied by g/g_0, the ratio of local to standard acceleration of gravity, in order to obtain the absolute value in pound-force units. For engineering purposes this correction can usually be neglected.

Unit of Pressure.

This is defined as the unit of force acting upon a unit area. The most commonly employed unit is the *pound* (force) *per square inch*.

Standard atmospheric pressure is defined to be the force exerted by a column of mercury 760 mm (29.92 in.) high at $0°$. This corresponds to 0.101325 MPa or 14.695 psi. Reference or fixed points for pressure calibration exist and are analogous to the phase changes used for temperature standards. These pressure references are based on phase changes or resistance jumps in selected materials.

Units of Work or Energy.

The foot-poundal is the physical unit of work or energy and is defined as the work done by a force of one poundal in moving a body through the distance of one foot in the direction of the force.

The *foot-pound* (*force*) is the technical unit of work or energy and is defined as the work required to raise a mass (or weight) of one pound through a vertical distance of one foot at standard acceleration of gravity g_0. If measurements are made in places where the local value of the acceleration of gravity g is different from g_0, a correction factor g/g_0 must be applied if the exact value of work or energy is desired.

The *British thermal unit* (Btu) is the quantity of heat required to raise the temperature of a one-pound mass of water either at $39°F$ (at its maximum density) or at $60°F$ and standard pressure through $1°F$. The mean British thermal unit is defined as the $\frac{1}{180}$ part of the heat required to raise the temperature of a one-pound mass of water from 32 to $212°F$ at standard pressure. It is obvious that the reference temperature must be indicated with the unit used.

Units of Power.

Power is the time rate at which work is done. Its physical unit is the *foot-poundal per second*, its technical units are the *foot-pound* (*force*) *per second*, or the *British thermal unit per second*. The *horsepower* (hp or Hp) is defined as 33,000 ft-lb (force) per minute or 550 ft-lb (force) per second.

Units of Torque.

Torque is the effectiveness of a force to produce rotation. It is defined as the product of the force and the perpendicular distance from its line of action to the instantaneous center of rotation. Its physical unit is the poundal-foot, and its technical unit the pound (force)-foot. (Note the reversal of force and

length units in the designation of the units of torque as compared with the units of energy or work.)

Metric Units

Units of Force.

The dynamical, or physical, unit of force is the *dyne*, defined as the force required to give a mass of one gram an acceleration of one centimeter per second per second.

The *newton* is the SI unit of force. It is the force required to give a mass of one kilogram an acceleration of one meter per second per second.

The *kilogram force* (or weight of the kilogram mass) is the gravitational or technical unit of force. It is, by definition, the force required to give a mass of one kilogram an acceleration of 980.665 cm/sec sec. If a force is measured by "weighing," the result in kilograms weight must be multiplied by g/g_0, the ratio of local to standard acceleration of gravity, in order to obtain the absolute value in kilogram-force units. For engineering purposes this correction can usually be neglected.

In the electrotechnical system of units the systematic unit of force is defined as the *joule per meter*, based on the fundamental definition of the joule. (See discussion on metric units of energy.)

Unit of Pressure.

This is defined as the unit of force acting upon a unit of area.

The *newton per square meter* is the SI unit of pressure and is called the *pascal*.

The *kilogram force per square meter* is the technical unit of pressure. With respect to correction for local gravity, see discussion on force versus weight.

Pressure is measured also by the height in centimeters of the column of water at $4°C$, or of the column of mercury at $0°C$, which it supports. (See conversion Table 41.)

The *normal atmosphere* (*at*), or the standard atmospheric pressure, is defined as the pressure exerted by a column of 76 cm of mercury at sea level and $0°C$ at standard acceleration of gravity g_0. It is equal to 1.01321 bars or 1.0332 kg/cm² force and is used extensively in the engineering literature. Some confusion exists since the unit of 1 kg/cm² is occasionally called 1 practical atmosphere.

Units of Work or Energy.

The *joule* is the physical or so-called absolute unit of work or energy. It is defined as the work done by a force of one newton acting through the distance of one meter. A larger unit is the *theoretical* or "absolute" *joule* defined as 10^7 ergs; it is a systematic unit in the practical electrical unit systems that is based on the theoretical unit systems. (See discussion on electrical units.)

The *international joule* is defined as the energy expended during one second by an electric current of one international ampere flowing through a resistance of one international ohm. (See discussion on electrical units.) The latest value of the international joule is equal to 1.000165 theoretical joules.[*]

The *kilowatt-hour* is the practical unit of energy in electrical metering. It is defined as a theoretical or an international unit (see definition of joule already given) and is equal to 3.6 megajoules.

The *meter-kilogram force* (commonly referred to as the kilogram-meter) is the technical unit of work or energy. It is defined as the work required to raise the mass (or weight) of one kilogram through a vertical distance of one meter at standard acceleration of gravity g_0. If measurements are made in places where the local value of the acceleration of gravity g is different from g_0, a correction factor g/g_0 has to be applied, if the exact value of work or energy is desired. (See discussion on force versus weight.)

The *gram-calorie* or small calorie is the physical unit of heat energy. It is defined as the quantity of heat required to raise the temperature of one gram mass of water either from 14.5 to 15.5°C or from 19.5 to 20.5°C at standard pressure. The two values are designated as 15 and 20°C cal, respectively. The mean gram-calorie is defined as $\frac{1}{100}$ part of the quantity of heat required to raise the temperature of one gram mass of water from 0 to 100°C at standard pressure. The same definitions apply to kilogram-calorie, or large calorie, if the kilogram mass is used as reference standard mass.

The *Ostwald calorie* is the quantity of heat required to raise the temperature of one gram mass from 0 to 100°C. This unit is frequently used by electrochemists and is equal to 100 mean gram-calories.

The *international kilocalorie* or international steam-table calorie (IT cal) is defined as the $\frac{1}{860}$ part of the international kilowatt-hour. This new unit avoids any reference to the thermal properties of water and was recommended for international adoption at the first International Steam Table Conference (1929).[†] Its value is very nearly equal to the mean kilocalorie, 1 IT cal = 1.00037 kilogram-calories (mean).

Units of Power. Power is the time rate at which work is done. Its physical unit is the watt, defined as the power which gives rise to the production of energy at the rate of one joule per second.

The *international watt* is defined as the power expended by an electric current of one international ampere flowing through a resistance of one international ohm. (See discussion on electrical units.) The latest value of the international watt is equal to 1.000165 theoretical watts.[‡]

The *electrical horsepower* is defined as 746 absolute watts and is commonly used in the United States and in England in rating electrical machinery.

The *meter-kilogram force per second* (commonly referred to as the kilogram-meter per second) is the technical unit of power. The *metric horsepower* is defined as 75 kg-m/sec and is the most common mechanical unit of power.

Units of Torque. Torque is the effectiveness of a force to produce rotation. It is defined as the product of the force and the perpendicular distance from its line of action to the instantaneous center of rotation. Its physical unit is the dyne-centimeter, and its technical unit the kilogram force meter. (Note the reversal of force and length units in the designation of the units of torque as compared with the units of energy and work.)

4.7 Thermal Units and Standards

Temperature

Definition of Temperature. The *temperature* of a body may be defined as its thermal state considered from the standpoint of its ability to communicate heat to other bodies. When two bodies are placed in thermal communication, the one that loses heat to the other is said to be at the higher temperature.

Standard Temperatures. Certain thermal states or "temperatures" may be reproduced and recognized by the fact that definite physical phenomena occur at these temperatures. Such thermal states are called "fixed points," and they may, quite apart from any temperature scale, be specified by the physical phenomena characteristic of those temperatures. The two fundamental fixed points are the ice point and the steam point.

The *ice point* is defined as the temperature of melting ice, which is realized experimentally as the temperature at which pure finely divided ice is in equilibrium with pure, air-saturated water under standard atmospheric pressure. The effect of increased pressure is to lower the freezing point to the extent of 0.007°C per atmosphere.

The *steam point* is defined as the temperature of condensing water vapor at standard atmospheric pressure, and it is realized experimentally by the use of a hypsometer so constructed as to avoid superheat of the vapor around the thermometer or contamination with air or other impurities. If the desired conditions have been attained, the observed temperature should be independent of the rate of heat supply to the boiler, except as this may affect the pressure within the hypsometer, and of the length of time the hypsometer has been in operation.

Definition of Temperature Scale. The purpose of establishing a temperature scale is to assign a number to every thermal state or temperature and to provide a means for determining the temperature of any particular body.

---

[*]*Mechanical Engineering*, Feb., 1930, pp. 122, 139.

[†]*Mechanical Engineering*, Nov., 1935, p. 710.

[‡]*Announcement of Changes in Electrical and Photometric Units*, Circular of National Bureau of Standards C459, Washington, DC, 1947.

A *temperature scale* may be defined by (1) selecting definite numbers for certain fixed points, (2) selecting some physical property of a definite substance that varies with temperature, and (3) selecting a mathematical law expressing temperatures on the scale in question in terms of the selected property of the thermometric substance. For example, on the Centigrade mercury-in-glass scale, the ice and steam points are numbered 0 and 100, respectively, the relative or "apparent" expansion of a volume of mercury enclosed in glass of a definite kind is the property used, and the mathematical relation used to express temperature on this scale is that equal increments of apparent volume of the mercury in this glass correspond to equal increments of temperature. If some other substance is substituted for mercury, or if glass of a different kind is used, another scale is obtained that agrees with it at 0 and 100 but not at other temperatures.

Although, in general, a temperature scale depends on the thermometric substance as well as on the expression for the temperature in terms of some property of this substance, Lord Kelvin has shown that, if the property selected is the availability of energy, the scale so defined is wholly independent of the substance and depends only on the mathematical relation chosen. Any scale so defined is known as a thermodynamic scale.

Kelvin Temperature Scale. The temperature scale finally chosen by Lord Kelvin is the one on which the temperature interval from the ice point to the steam point is $100°$ and the ratio of the values of any two temperatures is equal to the ratio of the heat taken in to the heat rejected by a reversible thermodynamic engine working with a source and refrigerator at the higher and lower temperatures, respectively. On this scale, which is also known as the absolute thermodynamic scale, the lowest attainable temperature is 0 and the ice point is found experimentally to be $273.16°$. The steam point therefore is $373.16°$ or $100°$ higher.

The *degree Kelvin* ($°K$) or degree of absolute temperature is the absolute unit of temperature and is, for practical purposes, identical with the degree Centigrade ($°C$) of the international temperature scale.

Thermodynamic Centigrade Scale. This is derived by subtracting from the Kelvin scale a constant number of the proper magnitude to make the ice point $0°$. On this scale, therefore, the ice and steam points are 0 and $100°$, respectively, and the so-called absolute zero is $-273.16°$.

International Centigrade Scale. This is a practical representation of the thermodynamic Centigrade scale to such a degree of accuracy as is possible with present-day apparatus and methods. It was adopted at the General Conference on Weights and Measures at Sèvres, France, in 1927 and is subject to revision and amendment as improved and more accurate methods of measurement are evolved.

The unit of temperature on the international scale is the *degree Centigrade* ($°$, or $°C$ int) and is very nearly equal to $\frac{1}{100}$ the difference between the temperature of melting ice and the temperature of condensing water vapor under standard atmospheric pressure. (See discussion on metric units for pressure.)

The standard of the international temperature scale between -190 and $+660°C$ is deduced from the electrical resistance of a standard platinum resistance thermometer by means of a formula connecting the resistance R_t at any temperature $t°C$ within the above range with the resistance R_0 at $0°C$. The purity of the platinum of which the thermometer is made should be such that the ratio R_t/R_0 for certain fixed temperatures is within specified limits. See also U.S. Bureau of Standards, *Journal of Research*, Vol. 1, p. 636, 1928.

The degree Centigrade is most widely used in scientific publications and increasingly also in the engineering literature. In many countries in Europe it is the common everyday temperature unit. The subdivision into a hundred degrees of the temperature interval between the ice point and the steam point was first used by Celsius, a German, in 1742; therefore, in the European literature "$°C$" is read "degree Celsius."

Fahrenheit Temperature Scale. This scale subdivides the temperature interval between the ice point and the steam point into 180 parts, one part of which is chosen as the unit of temperature and named *degree Fahrenheit* ($°F$). The ice point is assigned the value $32°F$, so that the steam point has a temperature of $212°F$.

The Fahrenheit unit of temperature is in common everyday use in the English-speaking countries. It was first introduced in England about 1665 by the physicist Fahrenheit; the choice of $32°F$ for the ice point has its explanation in the fact that Fahrenheit chose as zero the lowest temperature attainable by means of a salt–ice mixture.

Rankine Absolute Temperature Scale ($°R$). This is the thermodynamic Fahrenheit scale where absolute zero is $0°R$ ($-459.69°F$). The ice point is assigned the value $491.69°R$ and the steam point $671.69°R$.

Relations between Temperature Scales. The following table shows the interrelations between the various temperature scales in the form of equations.

Temperature Interrelationships

$x°F =$		$9/5(t°K - 273.16) + 32$	$9/5(t°C) + 32$
$x°K =$	$5/9(t°F - 32) + 273.16$		$(t°) + 273.16$
$x°C =$	$5/9(t°F - 32)$	$(t°K) - 273.16$	
$x°R =$	$(t°F) + 459.699$	$9/5(t°K)$	$9/5(t°C) + 491.69$

Here, X indicates the unknown number of chosen temperature units and t the known number of given temperature units:

Quantity of Heat and Some Derived Quantities

Units of Quantity of Heat. Quantity of heat is defined as the energy transferred from one body to another by a thermal process, that is, by radiation or conduction. The units for the quantity of heat are the *British thermal unit* and the *calorie* as specific thermal units and the *erg* and *joule* as general physical units (see discussion on units of energy, metric and English system of units).

Thermal Capacity or Specific Heat of a Substance. This is the quantity of heat required to produce a unit change in temperature in a unit of mass of the substance. The common English unit is the British thermal unit per degree Fahrenheit per pound mass (Btu per °F per lb); the usual metric unit is the gram-calorie per degree Centigrade per gram mass (cal per °C per g); and the general physical unit used in the scientific literature is the erg per degree Centigrade per gram mass (erg per °C per g). In the technical literature thermal capacity of a substance is often expressed in watt-seconds (or joules) per degree Centigrade per kilogram mass (W-sec per °C per kg) on account of the easy comparison with other technical units.

Calorimetric or Water Equivalent. This is the quantity of heat required to produce a unit change in temperature of a body or system. It is numerically equivalent to the mass of water (in units as involved in the definition of the unit of quantity of heat used) that could be raised a unit temperature by the same total quantity of heat. The thermal capacity is expressed in British thermal units per degree Fahrenheit (Btu per °F), calories per degree Centigrade (cal per °C), or watt-seconds per degree Centigrade (W-sec per °C).

Thermal Conductivity. This is the time rate of heat transfer through a unit area across a unit thickness per unit difference in temperature between the end surfaces. It is measured in British thermal units per second per degree Fahrenheit per inch thickness per square inch cross section (Btu per sec per °F per in. per in.²), in calories per second per degree Centigrade per centimeter thickness per square centimeter cross section (cal per sec per °C per cm per cm²), or in watts per degree Centigrade per meter thickness per square meter cross section (W per °C per m per m²).

Thermal Transmittance. The surface coefficient of transfer is the time rate of heat emitted by a unit area for a unit difference in temperature between the surface in question and the surroundings. It is measured in British thermal units per second per degree Fahrenheit per square inch (Btu per sec per °F per in.²), in calories per second per degree Centigrade per square centimeter (cal per sec per °C per cm²), or in watts per degree Centigrade per square meter (Wa per °C per m²).

Joule Equivalent The **Joule equivalent** is defined as the number of foot-pounds of energy per Btu. The numerical values for the various energy units used in the English and metric systems are shown in the table below.

4.8 Chemical Units and Standards

Atomic Weight. The present definition of atomic weights (1961) is based on ^{12}C, which is the most abundant isotope of carbon and whose atomic weight is defined as exactly 12.

Standard Cell Potential. A very large class of chemical reactions are characterized by the transfer of protons or electrons. Substances losing electrons in a reaction are said to be oxidized, those gaining electrons are said to be reduced. Many such reactions can be carried out in a galvanic cell that forms a natural basis for the concept of the half-cell, that is, the overall cell is conceptually the sum of two half-cells, one corresponding to each electrode. The half-cell potential measures the tendency of one reaction, for example, oxidation, to proceed at its electrode; the other half-cell of the pair measures the corresponding tendency for reduction to proceed at the other electrode. Measurable cell potentials are the sum of the two half-cell potentials. Standard cell potentials refer to the tendency of reactants in their standard state to form products in their standard states. The standard conditions are 1 M concentration for solutions, 101.325 kPa (1 atm) for gases, and for solids, their most stable form at 25 °C.

	Joules "Absolute"	Foot-pounds (force)	Foot-poundals	Meter-kilogram (force)	Kilowatt-hour "international"
1 British thermal unit (Btu) (mean) =	1055.18	778.26	25.040	107.599	2.93019×10^{-4}
1 gram-calorie (cal) (mean) =	4.1873	3.0884	99.366	0.42699	1.16279×10^{-6}
1 international kilocalorie (IT cal) =	4187.3	3088.4	99.366	426.99	1.16279×10^{-3}
1 Ostwald calorie =	418.73	308.84	9936.6	42.699	1.16279×10^{-4}

Since half-cell potentials cannot be measured directly, numerical values are obtained by assigning the hydrogen gas–hydrogen ion half reaction the half-cell potential of 0 V. Thus, by a series of comparisons referred directly or indirectly to the standard hydrogen electrode, values for the strength of a number of oxidants or reductants can be obtained, and standard reduction potentials can be calculated from established values.

Standard cell potentials are meaningful only when they are calibrated against an electromotive force (emf) scale. To achieve an absolute value of emf, electrical quantities must be referred to the basic metric system of mechanical units. If the current unit ampere and the resistance unit ohm can be defined, then the volt may be defined by Ohm's law as the voltage drop across a resistor of one standard ohm (Ω) when passing one standard ampere (A) of current. In the ohm measurement, a resistance is compared to the reactance of an inductor or capacitor at a known frequency. This reactance is calculated from the measured dimensions and can be expressed in terms of the meter and second. The ampere determination measures the force between two interacting coils while they carry the test current. The force between the coils is opposed by the force of gravity acting on a known mass; hence, the ampere can be defined in terms of the meter, kilogram, and second. Such a means of establishing a reference voltage is inconvenient for frequent use and reference is made to a previously calibrated standard cell.

Ideally, a standard cell is constructed simply and is characterized by a high constancy of emf, a low temperature coefficient of emf, and an emf close to 1 v. The Weston cell, which uses a standard cadmium sulfate electrolyte and electrodes of cadmium amalgam and a paste of mercury and mercurous sulfate, essentially meets these conditions. The voltage of the cell is 1.0183 V at 20 °C. The alternating current (ac) Josephson effect, which relates the frequency of a superconducting oscillator to the potential difference between two superconducting components, is used by the National Bureau of Standards to maintain the unit of emf, but the definition of the volt remains the Ω/A derivation described.

Concentration. The basic unit of concentration in chemistry is the mole, which is the amount of substance that contains as many entities, for example, atoms, molecules, ions, electrons, protons, and so on, as there are atoms in 12 g of ^{12}C, that is, Avogadro's number $N_A = 6.022045 \times 10^{23}$. Solution concentrations are expressed on either a weight or volume basis. *Molality* is the concentration of a solution in terms of the number of moles of solute per kilogram of solvent. *Molarity* is the concentration of a solution in terms of the number of moles of solute per liter of solution.

A particular concentration measure of acidity of aqueous solutions is pH, which, usually, is regarded as the common logarithm of the reciprocal of the hydrogen ion concentration (qv). More precisely, the potential difference of the hydrogen electrode in normal acid and in normal alkali solution (-0.828 V at 25 °C) is divided into 14 equal parts or pH units; each pH unit is 0.0591 V. Operationally, pH is defined by pH = pH (soln) + E/K, where E is the emf of the cell:

$$H_2|\text{solution of unknown pH}\|\text{saturated KCl}\|$$

$$\text{solution of known pH}|H_2$$

and $K = 2.303\ RT/F$, where R is the gas constant, 8.314 J/(mol/K) [1.987 cal/(mol · K)], T is the absolute temperature, and F is the value of the Faraday, 9.64845×10^4C/mol. pH usually is equated to the negative logarithm of the hydrogen ion activity, although there are differences between these two quantities outside the pH range 4.0–9.2:

$$-\log q_{H^+} \, m_{H^+} = \begin{cases} \text{pH} + 0.014\,(\text{pH} - 9.2) & \text{for pH} > 9.2 \\ \text{pH} + 0.009\,(4.0 - \text{pH}) & \text{for pH} < 4.0 \end{cases}$$

4.9 Theoretical, or Absolute, Electrical Units

With the general adoption of SI as the form of metric system that is preferred for all applications, further use of cgs units of electricity and magnetism is deprecated. Nonetheless, for historical reasons as well as for comprehensiveness, a brief review is included in this section and section 4.10.

The definitions of the *theoretical*, or *"absolute,"* **units** are based on a particular choice of the numerical value of either k_e, the constant in Coulomb's, electrostatic force law, or k_m, the constant in Ampere's electrodynamic force law. The designation absolute units is generally used because of historical tradition; an interesting account of the history can be found in Glazebrook's *Handbook for Applied Physics*, Vol. II, "Electricity," pp. 211 ff., 1922. Because of the theoretical background of the unit definitions, they have also been designated as "theoretical" units, which is in good contradistinction to practical units based on physical standards.

Theoretical Electrostatic Units The *theoretical electrostatic units* are based on the cgs system of mechanical units and the choice of the numerical value unity for k_{ev} in Coulomb's law. They are frequently referred to as the *cgs electrostatic units*, but no specific unit names are available. In order to avoid the cumbersome writing, for example, one "theoretical electrostatic unit of charge," it had been proposed to use the theoretical "practical" unit names and prefix them with either stat or E.S. as, for example, statcoulomb, or E.S. coulomb. The first alternative will be used here.

The *absolute dielectric constant (permittivity)* of free space is the reciprocal of the Coulomb constant k_{ev} and is chosen as the fourth fundamental quantity in the theoretical electrostatic system of units. Its numerical value is defined as unity, and it is identical with one statfarad per centimeter if use is made of prefixing the corresponding unit of the "practical" series.

The theoretical electrostatic unit of *charge*, or the *statcoulomb*, is defined as the quantity of electricity that, when concentrated at a point and placed at one centimeter distance from an equal quantity of electricity similarly concentrated, will experience a mechanical force of one dyne in free space. An alternative definition, based on the concept of field lines, gives the theoretical electrostatic unit of charge as a positive charge from which in free space exactly 4π displacement lines emerge.

The theoretical electrostatic unit of *displacement flux (dielectric flux)* is the "line of displacement flux," or $\frac{1}{4}\pi$ of the theoretical electrostatic unit of charge. This definition provides the basis for graphical field mapping insofar as it gives a definite rule for the selection of displacement lines to represent the distribution of the field quantitatively.

The theoretical electrostatic unit of *displacement*, or *dielectric flux density*, is chosen as one displacement line per square centimeter area perpendicular to the direction of the displacement lines. It can be given also as $\frac{1}{4}\pi$ statcoulomb per square centimeter (according to Gauss's law). In isotropic media the displacement has the same direction as the potential gradient, and the surfaces perpendicular to the field lines become the equipotential surfaces; the theoretical electrostatic unit of displacement can then be defined as one displacement line per square centimeter of equipotential surface.

The theoretical electrostatic unit of *electrostatic potential*, or the *statvolt*, is defined as existing at a point in an electrostatic field, if the work done to bring the theoretical electrostatic unit of charge, or the statcoulomb, from infinity to this point equals one erg. This customary definition implies, however, that the potential vanishes at infinite distances and has, therefore, only restricted validity. As it is fundamentally impossible to give absolute values of potential, the use of potential difference and its unit (see below) should be preferred.

The theoretical electrostatic unit of *electrical potential difference* or *voltage*, is the *statvolt* and is defined as existing between two points in space if the work done to bring the theoretical electrostatic unit of charge, or the statcoulomb, from one of these points to the other equals one erg. Potential difference is counted positive in the direction in which a negative quantity of electricity would be moved by the electrostatic field.

The theoretical electrostatic unit of *capacitance*, or the *statfarad*, is defined as the capacitance that maintains an electrical potential difference of one statvolt between two conductors charged with equal and opposite electrical charges of one statcoulomb. In the older literature, the cgs electrostatic unit of capacitance is identified with the "centimeter"; this was replaced by statfarad to avoid confusion.

The theoretical electrostatic unit of *electric potential gradient*, or *field strength* (field intensity), is defined to exist at a point in an electric field if the mechanical force exerted upon the theoretical electrostatic unit of charge concentrated at this point is equal to one dyne. It is expressed as one statvolt per centimeter.

The theoretical electrostatic unit of *current*, or the *statampere*, is defined as the time rate of transfer of the theoretical electrostatic unit of charge and is identical with the statcoulomb per second.

The theoretical electrostatic unit of *electrical resistance*, or the *statohm*, is defined as the resistance of a conductor in which a current of one statampere is produced if a potential difference of one statvolt is applied at its ends.

The theoretical electrostatic unit of *electromotive force (emf)* is defined as equivalent to the theoretical electrostatic unit of potential difference if it produces a current of one statampere in a conductor of one statohm resistance. It is identical with the statvolt but, according to its concept, requires an independent definition.

The theoretical electrostatic unit of *magnetic intensity* is defined as the magnetic intensity at the center of a circle of 4π centimeters diameter in which a current of one statampere is flowing. This unit is equal to 4π statamperes per centimeter but has no name as the factor 4π excludes the possibility of using the prefixed "practical" unit name.

The theoretical electrostatic unit of *magnetic flux*, or the *statweben*, is defined as the magnetic flux whose time rate of change through a linear conductor loop (linear conductor is used to designate a conductor of infinitely small cross section) produces in this loop an emf of one statvolt.

The theoretical electrostatic unit of *magnetic flux density*, or *induction*, is defined as the electrostatic unit of magnetic flux per square centimeter area, or the statweber per square centimeter.

The *absolute magnetic permeability of free space* is defined as the ratio of magnetic induction to the magnetic intensity. Its unit is the stathenry per centimeter as a derived unit.

The theoretical electrostatic unit of *inductance*, or the *stathenry*, is defined as connected with a conductor loop carrying a steady current of one statampere that produces a magnetic flux of one statweber. A more general definition, applicable to varying fields with nonlinear relation between magnetic flux and current, gives the stathenry as connected with a conductor loop in which a time rate of change in the current of one statcoulomb produces a time rate of change in the magnetic flux of one statweber per second.

Theoretical Electromagnetic Units   The *theoretical electromagnetic units* are based on the cgs system of mechanical units and Coulomb's law of mechanical force action between two isolated magnetic quantities m_1 and m_2 (approximately true for very long bar magnets) that must be written as

$$F_m = \frac{k_m}{2} \frac{m_1 m_2}{r^2} \qquad (7)$$

where k_m is the proportionality constant of Ampere's law for force action between parallel currents that is more basic, and amenable to much more accurate measurement, than (7). The factor $\frac{1}{2}$ appears here because of the three-dimensional character of the field distribution around point magnets as compared with the two-dimensional field of two parallel currents.

The theoretical electromagnetic units are obtained by defining the numerical value of $k_{mv}/2$ (for vacuum) as unity; they are frequently referred to as the cgs electromagnetic units. Only a few specific unit names are available. In order to avoid cumbersome writing, for example, one "theoretical electromagnetic unit of charge," it had been proposed to use the theoretical "practical" unit names and prefix them with either ab- or E.M. as, for example, abcoulomb, or E.M. coulomb. The first alternative will be used here.

The *absolute magnetic permeability* of free space is the value $k_{mv}/2$ in (7) and is chosen as the fourth fundamental quantity in the theoretical electromagnetic system of units. Its numerical value is assumed as unity, and it is identical with one abhenry per centimeter if use is made of prefixing the corresponding unit of the "practical" series.

The theoretical electromagnetic unit of *magnetic quantity* is defined as the magnetic quantity that, when concentrated at a point and placed at one centimeter distance from an equal magnetic quantity similarly concentrated, will experience a mechanical force of one dyne in free space. An alternative definition, based on the concept of magnetic intensity lines, gives the theoretical electromagnetic unit of magnetic quantity as a positive magnetic quantity from which, in free space, exactly 4π magnetic intensity lines emerge.

The theoretical electromagnetic unit of *magnetic moment* is defined as the magnetic moment possessed by a magnet formed by two theoretical electromagnetic units of magnetic quantity of opposite sign, concentrated at two points one centimeter apart. As a vector, its positive direction is defined from the negative to the positive magnetic quantity along the center line.

The theoretical electromagnetic unit of *magnetic induction (magnetic flux density)*, or the *gauss*, is defined to exist at a point in a magnetic field, if the mechanical torque exerted upon a magnet with theoretical electromagnetic unit of magnetic moment and directed perpendicular to the magnetic field is equal to one dyne-centimeter. The lines to which the vector of

magnetic induction is tangent at every point are called induction lines or magnetic flux lines; on the basis of this flux concept, magnetic induction is identical with magnetic flux density.

The theoretical electromagnetic unit of *magnetic flux*, or the *maxwell*, is the "field line" or line of magnetic induction. In free space, the theoretical electromagnetic unit of magnetic quantity issues 4π induction lines; the unit of magnetic flux, or the maxwell, is then $1/4\pi$ of the theoretical electromagnetic unit of magnetic quantity times the absolute permeability of free space.

The theoretical electromagnetic unit of *magnetic intensity (magnetizing force)*, or the *oersted*, is defined to exist at a point in a magnetic field in free space where one measures a magnetic induction of one gauss.

The theoretical electromagnetic unit of *current*, or the *abampere*, is defined as the current that flows in a circle of one centimeter diameter and produces at the center of this circle a magnetic intensity of one oersted.

The theoretical electromagnetic unit of *inductance*, or the *abhenry*, is defined as connected with a conductor loop in which a time rate of change of one maxwell per second in the magnetic flux produces a time rate of change in the current of one abampere per second. In the older literature, the cgs electromagnetic unit of inductance is identified with the "centimeter"; this should be replaced by a henry to avoid confusion.

The theoretical electromagnetic unit of *magnetomotive force (mmf)* is defined as the magnetic driving force produced by a conductor loop carrying a steady current of $\frac{1}{4}\pi$ abamperes; it has the name one gilbert. The concept of magnetomotive force as the driving force in a "magnetic circuit" permits an alternative definition of the gilbert as the magnetomotive force that produces a uniform magnetic intensity of one oersted over a length of one centimeter in the magnetic circuit. Obviously, one gilbert equals one oersted-centimeter.

The theoretical electromagnetic unit of *magnetostatic potential* is defined as the potential existing at a point in a magnetic field if the work done to bring the theoretical electromagnetic unit of magnetic quantity from infinity to this point equals one erg. This customary definition implies, however, that the potential vanishes at infinite distances, and the definition has therefore only restricted validity. The unit, thus defined, is identical with one gilbert. The difference in magnetostatic potential between any two points is usually called magnetomotive force (mmf).

The theoretical electromagnetic unit of *reluctance* is defined as the reluctance of a magnetic circuit in which a magnetomotive force of one gilbert produces a magnetic flux of one maxwell.

The theoretical electromagnetic unit of *electric charge*, or the *abcoulomb*, is defined as the quantity of electricity that passes through any section of an electric circuit in one second if the current is one abampere.

The theoretical electromagnetic unit of *displacement flux (dielectric flux)* is the "line of displacement flux," or $\frac{1}{4}\pi$ of the theoretical electromagnetic unit of electric charge. This definition provides the basis for graphical field mapping insofar as it gives a definite rule for the selection of displacement lines to represent the character of the field.

The theoretical electromagnetic unit of *displacement*, or *dielectric flux density*, is chosen as one displacement line per square centimeter area perpendicular to the direction of the displacement lines. It can also be given as $\frac{1}{4}\pi$ abcoulombs per square centimeter (according to Gauss's law). In isotropic media the theoretical electromagnetic unit of displacement can be defined as one displacement line per square centimeter of equipotential surface. (See discussion on theoretical electrostatic unit of displacement.)

The theoretical electromagnetic unit of *electrical potential difference*, or *voltage*, is the *abvolt* and is defined as the potential difference existing between two points in space if the work done in bringing the theoretical electromagnetic unit of charge, or the abcoulomb, from one of these points to the other equals one erg. Potential difference is counted positive in the direction in which a negative quantity of electricity would be moved by the electrostatic field.

The theoretical electromagnetic unit of *capacitance*, or the *abfarad*, is defined as the capacitance that maintains an electrical potential difference of one abvolt between two conductors charged with equal and opposite electrical quantities of one abcoulomb.

The theoretical electromagnetic unit of *potential gradient*, or *field strength* (field intensity), is defined to exist at a point in an electric field if the mechanical force exerted upon the theoretical electromagnetic unit of charge concentrated at this point is equal to one dyne. It is expressed as one abvolt per centimeter.

The theoretical electromagnetic unit of *resistance*, or the *abohm*, is defined as the resistance of a conductor in which a current of one abampere is produced if a potential difference of one abvolt is applied at its ends.

The theoretical electromagnetic unit of *electromotive force (emf)* is defined as the electromotive force acting in an electric circuit in which a current of one abampere is flowing and electrical energy is converted into other kinds of energy at the rate of one erg per second. This unit is identical with the abvolt.

The *absolute dielectric constant of free space* is defined as the ratio of displacement to the electric field intensity. Its unit is the abfarad per centimeter, a derived unit.

Theoretical Electrodynamic Units The theoretical electrodynamic units are based on the cgs system of mechanical units and are therefore frequently referred to as the *cgs electrodynamic units*. In contradistinction to the theoretical electromagnetic units, these units are derived from a significant experimental law, Ampere's

experiment on the mechanical force between two parallel currents. The units as proposed by Ampere and used by W. Weber differ from the electromagnetic units by factors of 2 and multiples thereof. They can be made to coincide with the theoretical electromagnetic units by proper definition of the fundamental unit of current. Some of the important definitions will be given for this latter case only.

For the *absolute magnetic permeability* of free space, see discussion on theoretical electromagnetic units.

The theoretical electrodynamic unit of *current*, or the *abampere*, is defined as the current flowing in a circuit consisting of two infinitely long parallel wires one centimeter apart when the electrodynamic force of repulsion between the two wires is *two* dynes per centimeter length in free space. If the more natural choice of *one* dyne per centimeter length is made, the original proposal of Ampere is obtained and the unit of current becomes $1/\sqrt{2}$ abampere.

The theoretical electrodynamic unit of *magnetic induction* is defined as the magnetic induction inducing an electromotive force of one abvolt in a conductor of one centimeter length and moving with a velocity of one centimeter per second if the conductor, its velocity, and the magnetic induction are mutually perpendicular. The unit thus defined is called one gauss.

The theoretical electrodynamic unit of *magnetic flux*, or the *maxwell*, is defined as the magnetic flux represented by a uniform magnetic induction of one gauss over an area of one square centimeter perpendicular to the direction of the magnetic induction.

The theoretical electrodynamic unit of *magnetic intensity*, or the *oersted*, is defined as the magnetic intensity at the center of a circle of 4π centimeters diameter in which a current of one abampere is flowing.

All the other unit definitions, which do not pertain to magnetic quantities, are identical with the definitions for the theoretical electromagnetic units.

4.10 Internationally Adopted Electrical Units and Standards

In October 1946, at Paris, the International Committee on Weights and Measures decided to abandon the so-called international practical units based on physical standards (see below) and to adopt effective January 1, 1948, the so-called absolute practical units for international use.

Adopted Absolute Practical Units By a series of international actions, the "absolute" practical electrical units are defined as exact powers of 10 of corresponding theoretical electrodynamic and electromagnetic units because they are based on the choice of the proportionality constant in Ampère's law for free space as $k_{mv} = 2 \times 10^{-7}$ H/m

The *absolute practical unit of current*, or the absolute is defined as the current flowing in a circuit consisting of two very long parallel thin wires spaced 1 m apart in free space if the electrodynamic force action between the wires is 2×10^{-7} N $= 0.02$ dyne per meter length. It is 10^{-1} of the theoretical or absolute electrodynamic or electromagnetic unit of current and was adopted internationally in 1881.

The absolute practical unit of *electric charge*, or the *absolute coulomb*, is defined as the quantity of electricity that passes through a cross-sectional surface in one second if the current is one absolute ampere. It is 10^{-1} of the theoretical or absolute electromagnetic unit of electric charge and was adopted internationally in 1881.

The absolute practical unit of *electric potential difference*, or the *absolute volt*, is defined as the potential difference existing between two points in space if the work done in bringing an electric charge of one absolute coulomb from one of these points to another is equal to one absolute joule $= 10^7$ ergs. It is 10^8 of the theoretical or absolute electromagnetic unit of potential difference and was adopted internationally in 1881.

The absolute practical unit of *resistance*, or the *absolute ohm*, is defined as the resistance of a conductor in which a current of one absolute ampere is produced if a potential difference of one absolute volt is applied at its ends. It is 10^9 of the theoretical or absolute electromagnetic unit of resistance and was adopted internationally in 1881.

The absolute practical unit of *magnetic flux*, or the *absolute weber*, is defined to be linked with a closed loop of thin wire of total resistance one absolute ohm if upon removing the wire loop from the magnetic field a total charge of one absolute coulomb is passed through any cross section of the wire. It is 10^8 of the theoretical or absolute electromagnetic unit of magnetic flux, the maxwell, and was adopted internationally in 1933.

The absolute practical unit of *inductance*, or the *absolute henry*, is defined as connected with a closed loop of thin wire in which a time rate of change of one absolute weber per second in the magnetic flux produces a time rate of change in the current of one absolute ampere. It is 10^9 of the theoretical or absolute electromagnetic unit of inductance and was adopted internationally in 1893.

The absolute practical unit of *capacitance*, or the *absolute farad*, is defined as the capacitance that maintains an electric potential difference of one absolute volt between two conductors charged with equal and opposite electrical quantities of one coulomb. It is 10^{-9} of the theoretical or absolute electromagnetic unit of capacitance and was adopted internationally in 1881.

Abandoned International Practical Units The International System of electrical and magnetic units is a system for electrical and magnetic quantities that takes as the four fundamental quantities resistance, current, length, and time. The units of resistance and current are defined by physical standards that were originally aimed to be exact replicas of the "absolute" practical units, namely the absolute ampere and the absolute ohm. On account of long-range variations in the physical standards, it proved impossible to rely upon them for international use and they recently have been replaced by the absolute practical units.

The international practical standards are defined as follows:

The *international ohm* is the resistance at $0\,^\circ$C of a column of mercury of uniform cross section having a length of 106.300 cm and a mass of 14.4521 g.

The *international ampere* is defined as the current that will deposit silver at the rate of 0.00111800 g/sec.

From these fundamental units, all other electrical and magnetic units can be defined in a manner similar to the absolute practical units. Because of the inconvenience of the silver voltameter as a standard, the various national laboratories actually used a volt, defining its value in terms of the other two standards.

At its conference in October 1946 in Paris, the International Committee on Weights and Measures accepted as the best relations between the international and the absolute practical units the following:

1 mean international ohm $= 1.00049$ absolute ohms

1 mean international volt $= 1.00034$ absolute volts

These mean values are the averages of values measured in six different national laboratories. On the basis of these mean values, the specific unit relation for converting international units appearing on certificates of the National Bureau of Standards, Washington, DC, into absolute practical units are as follows:

1 international ampere $= 0.999835$ absolute ampere

1 international coulomb $= 0.999835$ absolute coulomb

1 international henry $= 1.000495$ absolute henries

1 international farad $= 0.999505$ absolute farad

1 international watt $= 1.000165$ absolute watts

1 international joule $= 1.000165$ absolute joules

BIBLIOGRAPHY FOR UNITS
AND MEASUREMENTS

Cohen, E. R., and Taylor, B. N., "The 1986 Adjustment of the Fundamental Physical Constants," *Report of the CODATA Task Group on Fundamental Constants, November 1986*, CODATA Bulletin No. 63, International Council of Scientific Unions, Committee on Data for Science and Technology, Pergamon, 1986.

Hvistendahl, H. S., *Engineering Units and Physical Quantities*, Macmillan, London, 1964.

Jerrard, H. G., and McNeill, D. B., *A Dictionary of Scientific Units*, 2nd ed., Chapman & Hall, London, 1964.

Letter Symbols for Units of Measurement, ANSI/IEEE Std. 260-1978, Institute of Electrical and Electronic Engineers, New York, 1978.

Quantities, Units, Symbols, Conversion Factors, and Conversion Tables, ISO Reference 31, 15 sections, International Organization for Standardization Geneva, 1973–1979.

Standard for Metric Practice, ASTM E 380-82, American Society for Testing and Materials, Philadelphia, 1982.

Young, L., *System of Units in Electricity and Magnetism*, Oliver and Boyd, Edinburgh, 1969.

Young, L., *Research Concerning Metrology and Fundamental Constants*, National Academy Press, Washington, DC, 1983.

5   TABLES OF CONVERSION FACTORS*

J. G. Brainerd
(revised and extended by J. H. Westbrook)

Table 27   Temperature Conversion

$$^\circ F = (^\circ C \times \tfrac{9}{5}) + 32 = (^\circ C + 40) \times \tfrac{9}{5} - 40$$

$$^\circ C = (^\circ F - 32) \times \tfrac{5}{9} = (^\circ F + 40) \times \tfrac{5}{9} - 40$$

$$^\circ R = {}^\circ F + 459.69$$

$$^\circ K = C + 273.16$$

---

*Boldface units in Tables 28–63 are SI.

Table 28 Length [L]

Multiply Number of → / by → ; to Obtain →	Centimeters	Feet	Inches	Kilometers	Nautical Miles	**Meters**	Mils	Miles	Millimeters	Yards
Centimeters	1	30.48	2.540	10^5	1.853×10^5	100	2.540×10^{-3}	1.609×10^5	0.1	91.44
Feet	3.281×10^{-2}	1	8.333×10^{-2}	3281	6080.27	3.281	8.333×10^{-5}	5280	3.281×10^{-3}	3
Inches	0.3937	12	1	3.937×10^4	7.296×10^4	39.37	0.001	6.336×10^4	3.937×10^{-2}	36
Kilometers	10^{-5}	3.048×10^{-4}	2.540×10^{-5}	1	1.853	0.001	2.540×10^{-8}	1.609	10^{-6}	9.144×10^{-4}
Nautical Miles	—	1.645×10^{-4}	—	0.5396	1	5.396×10^{-4}	—	0.8684	—	4.934×10^{-4}
Meters	0.01	0.3048	2.540×10^{-2}	1000	1853	1	—	1609	0.001	0.9144
Mils	393.7	1.2×10^4	1000	3.937×10^7	—	3.937×10^4	1	—	39.37	3.6×10^4
Miles	6.214×10^{-6}	1.894×10^{-4}	1.578×10^{-5}	0.6214	1.1516	6.214×10^{-4}	—	1	6.214×10^{-7}	5.682×10^{-4}
Millimeters	10	304.8	25.40	10^6	—	1000	2.540×10^{-2}	—	1	914.4
Yards	1.094×10^{-2}	0.3333	2.778×10^{-2}	1094	2027	1.094	2.778×10^{-5}	1760	1.094×10^{-3}	1

Length

Land Measure

7.92 inches = 1 link

25 links = 1 rod = 16.5 feet = 5.5 yards (1 rod = 1 pole = 1 perch)

4 rods = 1 chain (Gunther's) = 66 feet = 22 yards = 100 links

10 chains = 1 furlong = 660 feet = 220 yards = 1000 links = 40 rods

8 furlongs = 1 mile = 5280 feet = 1760 yards = 8000 links = 320 rods = 80 chains

Ropes and Cables

2 yards = 1 fathom   120 fathoms = 1 cable length

Nautical Measure

6080.27 feet = 1 nautical mile = 1.15156 statute miles

3 nautical miles = 1 league (U.S.) 3 statute miles = 1 league (Gr. Britain)

(*Note:* A nautical mile is the length of a minute of longitude of the earth at the equator at sea level. The British Admiralty uses the round figure of 6080 feet. The word "knot" is used to denote "nautical miles per hour.")

Miscellaneous

3 inches = 1 palm   9 inches = 1 span

4 inches = 1 hand $2\frac{1}{2}$ feet = 1 military pace

Table 29 Area [L^2]

Multiply Number of → by →, to Obtain ↓

to Obtain ↓	Acres	Circular Mils	Square Centimeters	Square Feet	Square Inches	Square Kilometers	Square Meters	Square Miles	Square Millimeters	Square Yards
Acres	1	—	—	2.296×10^{-5}	—	247.1	2.471×10^{-4}	640	—	2.066×10^{-4}
Circular Mils	—	1	1.973×10^5	1.833×10^8	1.273×10^6	—	1.973×10^9	—	1973	8361
Square Centimeters	—	5.067×10^{-6}	1	929.0	6.452	10^{10}	10^4	2.590×10^{10}	0.01	9
Square Feet	4.356×10^4	—	1.076×10^{-3}	1	6.944×10^{-3}	1.076×10^7	10.76	2.788×10^7	1.076×10^{-5}	1296
Square Inches	6,272,640	7.854×10^{-7}	0.1550	144	1	1.550×10^9	1550	4.015×10^9	1.550×10^{-3}	8.361×10^{-7}
Square Kilometers	4.047×10^{-3}	—	10^{-10}	9.290×10^{-8}	6.452×10^{-10}	1	10^{-6}	2.590	10^{-12}	0.8361
Square Meters	4047	—	0.0001	9.290×10^{-2}	6.452×10^{-4}	10^6	1	2.590×10^6	10^{-6}	3.228×10^{-7}
Square Miles	1.562×10^{-3}	—	3.861×10^{-11}	3.587×10^{-8}	—	0.3861	3.861×10^{-7}	1	3.861×10^{-13}	8.361×10^5
Square Millimeters	—	5.067×10^{-4}	100	9.290×10^4	645.2	10^{12}	10^6	—	1	1
Square Yards	4840	—	1.196×10^{-4}	0.1111	7.716×10^{-4}	1.196×10^6	1.196	3.098×10^6	1.196×10^{-6}	1

Area

Land Measure

$30\frac{1}{4}$ square yards = 1 square rod = $272\frac{1}{4}$ square feet

16 square rods = 1 square chain = 484 square yards = 4356 square feet

$2\frac{1}{2}$ square chains = 1 rood = 40 square rods = 1210 square yards

4 roods = 1 acre = 10 square chains = 160 square rods

640 acres = 1 square mile = 2560 roods = 102,400 square rods

1 section of land = 1 square mile; 1 quarter section = 160 acres

Architect's Measure

100 square feet = 1 square

Circular Inch and Circular Mil   A circular inch is the area of a circle 1 inch in diameter = 0.7854 square inch

1 square inch = 1.2732 circular inches

A circular mil is the area of a circle 1 mil (or 0.001 inch) in diameter = 0.7854 square mil

1 square mil = 1.2732 circular mils

1 circular inch = 10^6 circular mils = 0.7854×10^6 square mils

1 square inch = 1.2732×10^6 circular mils = 10^6 square mils

73

Table 30 Volume [L^3]

to Obtain ↓ / Multiply Number of → by →	Bushels (Dry)	Cubic Centimeters	Cubic Feet	Cubic Inches	Cubic Meters	Cubic Yards	Gallons (Liquid)	Liters	Pints (Liquid)	Quarts (Liquid)
Bushels (Dry)	1	—	0.8036	4.651×10^{-4}	28.38	—	—	2.838×10^{-2}		
Cubic Centimeters	3.524×10^4	1	2.832×10^4	16.39	10^6	7.646×10^5	3785	1000	473.2	946.4
Cubic Feet	1.2445	3.531×10^{-5}	1	5.787×10^{-4}	35.31	27	0.1337	3.531×10^{-2}	1.671×10^{-2}	3.342×10^{-2}
Cubic Inches	2150.4	6.102×10^{-2}	1728	1	6.102×10^4	46,656	231	61.02	28.87	57.75
Cubic Meters	3.524×10^{-2}	10^{-6}	2.832×10^{-2}	1.639×10^{-5}	1	0.7646	3.785×10^{-3}	0.001	4.732×10^{-4}	9.464×10^{-4}
Cubic Yards	—	1.308×10^{-6}	3.704×10^{-2}	2.143×10^{-5}	1.308	1	4.951×10^{-3}	1.308×10^{-3}	6.189×10^{-4}	1.238×10^{-3}
Gallons (Liquid)	—	2.642×10^{-4}	7.481	4.329×10^{-3}	264.2	202.0	1	0.2642	0.125	0.25
Liters	35.24	0.001	28.32	1.639×10^{-2}	1000	764.6	3.785	1	0.4732	0.9464
Pints (Liquid)	—	2.113×10^{-3}	59.84	3.463×10^{-2}	2113	1616	8	2.113	1	2
Quarts (Liquid)	—	1.057×10^{-3}	29.92	1.732×10^{-2}	1057	807.9	4	1.057	0.5	1

Volume

Cubic Measure

1 cord of wood = pile cut 4 feet long piled 4 feet

high and 8 feet on the ground

= 128 cubic feet

1 perch of stone = quantity$1\frac{1}{2}$ feet thick,

1 foot high, and$16\frac{1}{2}$ feet long

= $24\frac{3}{4}$ cubic feet

(*Note:* A perch of stone is, however, often computed differently in different localities; thus, in most if not all of the states west of the Mississippi, stonemasons figure rubble by the perch of $16\frac{1}{2}$ cubic feet. In Philadelphia, 22 cubic feet is called a perch. In Chicago, stone is measured by the cord of 100 cubic feet. Check should be made against local practice.)

Board Measure. In board measure, boards are assumed to be one inch in thickness. Therefore, feet board measure of a stick of square timber = length in feet × breadth in feet × thickness in inches.

Shipping Measure. For register tonnage or measurement of the entire internal capacity of a vessel, it is arbitrarily assumed, to facilitate computation, that

100 cubic feet = 1 register ton

For the measurement of cargo:

40 cubic feet = 1 U.S. shipping ton

= 32.143 U.S. bushels

42 cubic feet = 1 British shipping ton

= 32.703 Imperial bushels

Dry Measure. One U.S. Winchester bushel contains 1.2445 cubic feet or 2150.42 cubic inches. It holds 77.601 pounds distilled water at 62°F.

(*Note:* This is a *struck* bushel. A *heaped* bushel in general equals $1\frac{1}{4}$ struck bushels, although for apples and pears it contains 1.2731 struck bushels = 2737.72 cubic inches.)

One U. S. gallon (dry measure) = $\frac{1}{8}$ bushel and contains 268.8 cubic inches.

(*Note:* This is not a legal U.S. *dry measure* and therefore is given for comparison only.)

One British Imperial bushel contains 1.2843 cubic feet or 2219.36 cubic inches. It holds 80 pounds distilled water at 62°F.

1 British Imperial gallon = $\frac{1}{8}$ Imperial bushel

and contains

277.42 cubic inches.

1 Winchester bushel = 0.9694 Imperial bushel

1 Imperial bushel = 1.032 Winchester bushels

Same relations as before maintain for gallons (dry measure).

[*Note:* 1 U.S. gallon (dry) = 1.164 U. S. gallons (liquid)).]

U.S. UNITS*

2 pints = 1 quart = 67.2 cubic inches
4 quarts = 1 gallon = 8 pints = 268.8 cubic inches
2 gallons = 1 peck = 16 pints = 8 quarts =
 537.6 cubic inches
4 pecks = 1 bushel = 64 pints = 32 quarts = 8 gallons =
 2150.42 cubic inches
1 cubic foot contains 6.428 gallons (dry measure)

Liquid Measure. One U.S. gallon (liquid measure) contains 231 cubic inches. It holds 8.336 pounds distilled water at 62°F.

One British Imperial gallon contains 277.42 cubic inches. It holds 10 pounds distilled water at 62°F.

1 U.S. gallon (liquid) = 0.8327 Imperial gallon

1 Imperial gallon = 1.201 U.S. gallons (liquid)

[*Note:* 1 U.S. gallon (liquid) = 0.8594 U.S. gallon (dry).]

U.S. UNITS

4 gills = 1 pint = 16 fluid ounces
2 pints = 1 quart = 8 gills = 32 fluid ounces
4 quarts = 1 gallon = 32 gills = 8 pints = 128 fluid ounces
1 cubic foot contains 7.4805 gallons (liquid measure)

Apothecaries' Fluid Measure

60 minims = 1 fluid drachm

8 drachms = 1 fluid ounce

In the United States a fluid ounce is the 128th part of a U.S. gallon, or 1.805 cubic inches or 29.58 cubic centimeters. It contains 455.8 grains of water at 62°F. In Great Britain the fluid ounce is 1.732 cubic inches and contains 1 ounce avoirdupois (or 437.5 grains) of water at 62°F.

---

*The *gallon* is not a U.S. legal *dry measure*.

Table 31 Plane Angle (No Dimensions)

Multiply Number of → by → to Obtain →

	Degrees	Minutes	Quadrants	Radians[a]	Revolutions[a] (Circumferences)	Seconds
Degrees	1	1.667×10^{-2}	90	57.30	360	2.778×10^{-4}
Minutes	60	1	5400	3438	2.16×10^{4}	1.667×10^{-2}
Quadrants	1.111×10^{-2}	1.852×10^{-4}	1	0.6366	4	3.087×10^{-6}
Radians[a]	1.745×10^{-2}	2.909×10^{-4}	1.571	1	6.283	4.848×10^{-6}
Revolutions[a] (Circumferences)	2.778×10^{-3}	4.630×10^{-5}	0.25	0.1591	1	7.716×10^{-7}
Seconds	3600	60	3.24×10^{5}	2.063×10^{5}	1.296×10^{6}	1

[a] 2π rad = 1 circumference = 360° by definition.

Table 32 Solid Angle (No Dimensions)

Multiply Number of → by → to Obtain →

	Hemispheres	Spheres[a]	Spherical Right Angles	Steradians[b]
Hemispheres	1	2	0.25	0.1592
Spheres[a]	0.5	1	0.125	7.958×10^{-2}
Spherical Right Angles	4	8	1	0.6366
Steradians[b]	6.283	12.57	1.571	1

[a] A sphere is the total solid angle about a point.
[b] 4π steradians = 1 sphere by definition.

Table 33 Time [T]

	Days	Hours	Minutes	Months (Average)[a]	Seconds	Weeks
Days	1	4.167×10^{-2}	6.944×10^{-4}	30.42	1.157×10^{-5}	7
Hours	24	1	1.667×10^{-2}	730.0	2.778×10^{-4}	168
Minutes	1440	60	1	4.380×10^{4}	1.667×10^{-2}	1.008×10^{4}
Months (Average)[a]	3.288×10^{-2}	1.370×10^{-3}	2.283×10^{-5}	1	3.806×10^{-7}	0.2302
Seconds	8.64×10^{4}	3600	60	2.628×10^{6}	1	6.048×10^{5}
Weeks	0.1429	5.952×10^{-3}	9.921×10^{-5}	4.344	1.654×10^{-6}	1

Multiply Number of → by →
to Obtain →

[a]One common year = 365 days; one leap year = 366 days; one average month = $\frac{1}{12}$ of a common year.

Table 34 Linear Velocity [LT^{-1}]

To Obtain ↓ \ Multiply Number of → by →	Centimeters per Second	Feet per Minute	Feet per Second	Kilometers per Hour	Kilometers per Minute	Knots[a]	Meters per Minute	**Meters per Second**	Miles per Hour	Miles per Minute
Centimeters per Second	1	0.5080	30.48	27.78	1667	51.48	1.667	100	44.70	2682
Feet per Minute	1.969	1	60	54.68	3281	101.3	3.281	196.8	88	5280
Feet per Second	3.281×10^{-2}	1.667×10^{-2}	1	0.9113	54.68	1.689	5.468×10^{-2}	3.281	1.467	88
Kilometers per Hour	0.036	1.829×10^{-2}	1.097	1	60	1.853	0.06	3.6	1.609	96.54
Kilometers per Minute	0.0006	3.048×10^{-4}	1.829×10^{-2}	1.667×10^{-2}	1	3.088×10^{-2}	0.001	0.06	2.682×10^{-2}	1.609
Knots[a]	1.943×10^{-2}	9.868×10^{-3}	0.5921	0.5396	32.38	1	3.238×10^{-2}	1.943	0.8684	52.10
Meters per Minute	0.6	0.3048	18.29	16.67	1000	30.88	1	60	26.82	1609
Meters per Second	0.01	5.080×10^{-3}	0.3048	0.2778	16.67	0.5148	1.667×10^{-2}	1	0.4470	26.82
Miles per Hour	2.237×10^{-2}	1.136×10^{-2}	0.6818	0.6214	37.28	1.152	3.728×10^{-2}	2.237	1	60
Miles per Minute	3.728×10^{-4}	1.892×10^{-4}	1.136×10^{-2}	1.036×10^{-2}	0.6214	1.919×10^{-2}	6.214×10^{-4}	3.728×10^{-2}	1.667×10^{-2}	1

[a]Nautical miles per hour.

Linear Velocity

The Miner's Inch. The miner's inch is used in measuring flow of water. An act of the California legislature, May 23, 1901, makes the standard miner's inch 1.5 ft^3/min, measured through any aperture or orifice.

The term miner's inch is more or less indefinite, for the reason that California water companies do not all use the same head above the center of the aperture,

and the inch varies from 1.36 to 1.73 ft^3/min, but the most common measurement is through an aperture 2 in. high and whatever length is required and through a plank $1\frac{1}{4}$ in. thick. The lower edge of the aperture should be 2 in. above the bottom of the measuring box and the plank 5 in. high above the aperture, thus making a 6-in. head above the center of the stream. Each square inch of this opening represents a miner's inch, which is equal to a flow of 1.5 ft^3/min.

Table 35   Angular Velocity [T^{-1}]

to Obtain ↓ / Multiply Number of → by →	Degrees per Second	Radians per Second	Revolutions per Minute	Revolutions per Second
Degrees per Second	1	57.30	6	360
Radians per Second	1.745×10^{-2}	1	0.1047	6.283
Revolutions per Minute	0.1667	9.549	1	60
Revolutions per Second	2.778×10^{-3}	0.1592	1.667×10^{-2}	1

Table 36   Linear Acceleration[a] [LT^{-2}]

to Obtain ↓ / Multiply Number of → by →	Centimeters per Second per Second	Feet per Second per Second	Kilometers per Hour per Second	Meters per Second per Second	Miles per Hour per Second
Centimeters per Second per Second	1	30.48	27.78	100	44.70
Feet per Second per Second	3.281×10^{-2}	1	0.9113	3.281	1.467
Kilometers per Hour per Second	0.036	1.097	1	3.6	1.609
Meters per Second per Second	0.01	0.3048	0.2778	1	0.4470
Miles per Hour per Second	2.237×10^{-2}	0.6818	0.6214	2.237	1

[a]The (standard) acceleration due to gravity $(g_0) = 980.7$ cm/sec sec, $= 32.17$ ft/sec sec $= 35.30$ km/hr sec $= 9.807$ m/sec sec $= 21.94$ mph/sec.

Table 37   Angular Acceleration [T^{-2}]

to Obtain ↓ / Multiply Number of → by →	Radians per Second per Second	Revolutions per Minute per Minute	Revolutions per Minute per Second	Revolutions per Second per Second
Radians per Second per Second	1	1.745×10^{-3}	0.1047	6.283
Revolutions per Minute per Minute	573.0	1	60	3600
Revolutions per Minute per Second	9.549	1.667×10^{-2}	1	60
Revolutions per Second per Second	0.1592	2.778×10^{-4}	1.667×10^{-2}	1

Table 38 Mass [M] and Weight[a]

to Obtain → / Multiply Number of →	Grains	Grams	**Kilograms**	Milligrams	Ounces[b]	Pounds[b]	Tons (Long)	Tons (Metric)	Tons (Short)
Grains	1	15.43	1.543×10^4	1.543×10^{-2}	437.5	7000			
Grams	6.481×10^{-2}	1	1000	0.001	28.35	453.6	1.016×10^6	$\times 10^6$	9.072×10^5
Kilograms	6.481×10^{-5}	0.001	1	10^{-6}	2.835×10^{-2}	0.4536	1016	1000	907.2
Milligrams	64.81	1000	10^6	1	2.835×10^4	4.536×10^5	1.016×10^9	10^9	9.072×10^8
Ounces[b]	2.286×10^{-3}	3.527×10^{-2}	35.27	3.527×10^{-5}	1	16	3.584×10^4	3.527×10^4	3.2×10^4
Pounds[b]	1.429×10^{-4}	2.205×10^{-3}	2.205	2.205×10^{-6}	6.250×10^{-2}	1	2240	2205	2000
Tons (Long)	—	9.842×10^{-7}	9.842×10^{-4}	9.842×10^{-10}	2.790×10^{-5}	4.464×10^{-4}	1	0.9842	0.8929
Tons (Metric)	—	10^{-6}	0.001	10^{-9}	2.835×10^{-5}	4.536×10^{-4}	1.016	1	0.9072
Tons (Short)	—	1.102×10^{-6}	1.102×10^{-3}	1.102×10^{-9}	3.125×10^{-5}	0.0005	1.120	1.102	1

[a]These same conversion factors apply to the *gravitational* units of force having the corresponding names. The dimensions of these units when used as gravitational units of force are MLT^{-2}; see Table 40.
[b]Avoirdupois pounds and ounces.

Table 39 Density or Mass per Unit Volume [ML^{-3}]

to Obtain ↓ / Multiply Number of → / by	Grams per Cubic Centimeter	Kilograms per Cubic Meter	Pounds per Cubic Foot	Pounds per Cubic Inch
Grams per Cubic Centimeter	1	0.001	1.602×10^{-2}	27.68
Kilograms per Cubic Meter	1000	1	16.02	2.768×10^{4}
Pounds per Cubic Foot	62.43	6.243×10^{-2}	1	1728
Pounds per Cubic Inch	3.613×10^{-2}	3.613×10^{-5}	5.787×10^{-4}	1
Pounds per Mil Foot[a]	3.405×10^{-7}	3.405×10^{-10}	5.456×10^{-9}	9.425×10^{-6}

[a]Unit of volume is a volume one foot long and one circular mil in cross-sectional area.

Avoirdupois Weight. Used Commercially.

27.343 grains = 1 drachm

16 drachms = 1 ounce (oz) = 437.5 grains

16 ounces = 1 pound (lb) = 7000 grains

28 pounds = 1 quarter (qr)

4 quarters = 1 hundredweight (cwt)

= 112 pounds

20 hundredweight = 1 gross or long ton*

200 pounds = 1 net or short ton

14 pounds = 1 stone100 pounds

= 1 quintal

Troy Weight. Used in weighing gold or silver.

24 grains = 1 pennyweight (dwt)

20 pennyweights = 1 ounce (oz) = 480 grains

12 ounces = 1 pound (lb) = 5760 grains

The grain is the same in avoirdupois, troy, and apothecaries' weights. A carat, for weighing diamonds, = 3.086 grains = 0.200 gram (International Standard, 1913.)

1 pound troy = 0.8229 pound avoirdupois

1 pound avoirdupois = 1.2153 pounds troy

Apothecaries' Weight. Used in compounding medicines.

20 grains = 1 scruple()

3 scruples = 1 drachm() = 60 grains

8 drachms = 1 ounce() = 480 grains

12 ounces = 1 pound(lb) = 5760 grains

The grain is the same in avoirdupois, troy, and apothecaries' weights.

1 pound apothecaries = 0.82286 pound avoirdupois

1 pound avoirdupois = 1.2153 pounds apothecaries

---

*The long ton is used by the U.S. custom houses in collecting duties upon foreign goods. It is also used in freighting coal and selling it wholesale.

Table 40 Force[a] $[MLT^{-2}]$ or $[F]$

to Obtain ↓ / Multiply Number of → by →	Dynes	Grams	Joules per Centimeter	Newtons, or Joules per Meter	Kilograms	Pounds	Poundals
Dynes	1	980.7	10^7	10^5	9.807×10^5	4.448×10^5	1.383×10^4
Grams	1.020×10^{-3}	1	1.020×10^4	102.0	1000	453.6	14.10
Joules per Centimeter	10^{-7}	9.807×10^{-5}	1	0.01	9.807×10^{-2}	4.448×10^{-2}	1.383×10^{-3}
Newtons, or Joules per Meter	10^{-5}	9.807×10^{-3}	100	1	9.807	4.448	0.1383
Kilograms	1.020×10^{-6}	0.001	10.20	0.1020	1	0.4536	1.410×10^{-2}
Pounds	2.248×10^{-6}	2.205×10^{-3}	22.48	0.2248	2.205	1	3.108×10^{-2}
Poundals	7.233×10^{-5}	7.093×10^{-2}	723.3	7.233	70.93	32.17	1

[a] Conversion factors between absolute and gravitational units apply only under standard acceleration due to gravity conditions. (See Section 4.)

Table 41 Pressure or Force per Unit Area $[ML^{-1}T^{-2}]$ or $[FL^{-2}]$

Multiply Number of → by → ; to Obtain →

to Obtain →	Atmospheres[a]	Baryes or Dynes per Square Centimeter	Centimeters of Mercury at 0°C[b]	Inches of Mercury at 0°C[b]	Inches of Water at 4°C	Kilograms per Square Meter[c]	Pounds per Square Foot	Pounds per Square Inch	Tons (Short) per Square Foot	Pascal
Atmospheres[a]	1	9.869×10^{-7}	1.316×10^{-2}	3.342×10^{-2}	2.458×10^{-3}	9.678×10^{-5}	4.725×10^{-4}	6.804×10^{-2}	0.9450	9.869×10^{-6}
Baryes or Dynes per Square Centimeter	1.013×10^{6}	1	1.333×10^{4}	3.386×10^{4}	2.491×10^{-3}	98.07	478.8	6.895×10^{4}	9.576×10^{5}	10
Centimeters of Mercury at 0°C[b]	76.00	7.501×10^{-5}	1	2.540	0.1868	7.356×10^{-3}	3.591×10^{-2}	5.171	71.83	7.501×10^{-4}
Inches of Mercury at 0°C[b]	29.92	2.953×10^{-5}	0.3937	1	7.355×10^{-2}	2.896×10^{-3}	1.414×10^{-2}	2.036	28.28	2.953×10^{-4}
Inches of Water at 4°C	406.8	4.015×10^{-4}	5.354	13.60	1	3.937×10^{-2}	0.1922	27.68	384.5	4.015×10^{-8}
Kilograms per Square Meter[c]	1.033×10^{4}	1.020×10^{-2}	136.0	345.3	25.40	1	4.882	703.1	9765	0.1020
Pounds per Square Foot	2117	2.089×10^{-3}	27.85	70.73	5.204	0.2048	1	144	2000	2.089×10^{-2}
Pounds per Square Inch	14.70	1.450×10^{-5}	0.1934	0.4912	3.613×10^{-2}	1.422×10^{-3}	6.944×10^{-3}	1	13.89	1.450×10^{-4}
Tons (Short) per Square Foot	1.058	1.044×10^{-6}	1.392×10^{-2}	3.536×10^{-2}	2.601×10^{-3}	1.024×10^{-4}	0.0005	0.072	1	1.044×10^{-5}
Pascal	1.013×10^{5}	10^{-1}	1.333×10^{3}	3.386×10^{3}	2.491×10^{-4}	9.807	47.88	6.895×10^{3}	9.576×10^{4}	1

[a]Definition: One atmosphere (standard) = 76 cm of mercury at 0°C.

[b]To convert height h of a column of mercury at t degrees Centigrade to the equivalent height h_0 at 0°C use $h_0 = h\{1 - (m - l)t/(1 + mt)\}$, where $m = 0.0001818$ and $l = 18.4 \times 10^{-6}$ if the scale is engraved on brass; $l = 8.5 \times 10^{-6}$ if on glass. This assumes the scale is correct at 0°C; for other cases (any liquid) see *International Critical Tables*, Vol. 1, p. 68.

[c]1 g/cm² = 10 kg/m².

Table 42   Torque or Moment of Force $[ML^2T^{-2}]$ or $[FL]$[a]

to Obtain ↓ Multiply Number of → by ↘	Dyne-Centimeters	Gram-Centimeters	Kilogram-Meters	Pound-Feet	**Newton-Meter**
Dyne-Centimeters	1	980.7	9.807×10^7	1.356×10^7	10^7
Gram-Centimeters	1.020×10^{-3}	1	10^5	1.383×10^4	1.020×10^4
Kilogram-Meters	1.020×10^{-8}	10^{-5}	1	0.1383	0.1020
Pound-Feet	7.376×10^{-8}	7.233×10^{-5}	7.233	1	0.7376
Newton-Meter	10^{-7}	9.807×10^{-4}	9.807	1.356	1

[a]Same dimensions as energy; more properly torque should be expressed as newton-meters per radian to avoid this confusion.

Table 43   Moment of Inertia $[ML^2]$

to Obtain ↓ Multiply Number of → by ↘	Gram-Centimeters Squared	**Kilogram-Meters Squared**	Pound-Inches Squared	Pound-Feet Squared	Slug-Feet Squared
Gram-Centimeters Squared	1	10^7	2.9266×10^3	4.21434×10^5	1.3559×10^7
Kilogram-Meters Squared	10^{-7}	1	2.9266×10^{-4}	4.21434×10^{-2}	1.3559
Pound-Inches Squared	3.4169×10^{-4}	3.4169×10^3	1	144	4.63304×10^3
Pound-Feet Squared	2.37285×10^{-6}	23.7285	6.944×10^{-3}	1	32.1739
Slug-Feet Squared	7.37507×10^{-8}	0.737507	2.15841×10^{-4}	3.10811×10^{-2}	1

Table 44 Energy, Work and Heat[a] $[ML^2T^{-2}]$ or $[FL]$

to Obtain → / Multiply Number of → by →	British Thermal Units[b]	Centimeter-Grams	Ergs or Centimeter-Dynes	Foot-Pounds	Horsepower-Hours	Joules,[c] or Watt-Seconds	Kilogram-Calories[b]	Kilowatt-Hours	Meter-Kilograms	Watt-Hours
British Thermal Units[b]	1	9.297×10^{-8}	9.480×10^{-11}	1.285×10^{-3}	2545	9.480×10^{-4}	3.969	3413	9.297×10^{-3}	3.413
Centimeter-Grams	1.076×10^7	1	1.020×10^{-3}	1.383×10^4	2.737×10^{10}	1.020×10^4	4.269×10^7	3.671×10^{10}	10^5	3.671×10^7
Ergs or Centimeter-Dynes	1.055×10^{10}	980.7	1	1.356×10^7	2.684×10^{12}	10^7	4.186×10^{10}	3.6×10^{13}	9.807×10^7	3.6×10^{10}
Foot-Pounds	778.0	7.233×10^{-5}	7.367×10^{-8}	1	1.98×10^6	0.7376	3087	2.655×10^6	7.233	2655
Horsepower-Hours	3.929×10^{-4}	3.654×10^{-11}	3.722×10^{-14}	5.050×10^{-7}	1	3.722×10^{-7}	1.559×10^{-3}	1.341	3.653×10^{-6}	1.341×10^{-3}
Joules,[c] or Watt-Seconds	1054.8	9.807×10^{-5}	10^{-7}	1.356	2.684×10^6	1	4186	3.6×10^6	9.807	3600
Kilogram-Calories[b]	0.2520	2.343×10^{-8}	2.389×10^{-11}	3.239×10^{-4}	641.3	2.389×10^{-4}	1	860.0	2.343×10^{-3}	0.8600
Kilowatt-Hours	2.930×10^{-4}	2.724×10^{-11}	2.778×10^{-14}	3.766×10^{-7}	0.7457	2.778×10^{-7}	1.163×10^{-3}	1	2.724×10^{-6}	0.001
Meter-Kilograms	107.6	10^{-5}	1.020×10^{-8}	0.1383	2.737×10^5	0.1020	426.9	3.671×10^5	1	367.1
Watt-Hours	0.2930	2.724×10^{-8}	2.778×10^{-11}	3.766×10^{-4}	745.7	2.778×10^{-4}	1.163	1000	2.724×10^{-3}	1

[a] See note at the bottom of Table 45.
[b] Mean calorie and Btu used throughout. One gram-calorie = 0.001 kilogram-calorie; one Ostwald calorie = 0.1 kilogram-calorie.
  The IT cal, 1000 international steam table calories, has been defined as the 1/860th part of the international kilowatthour (see *Mechanical Engineering*, Nov. 1935, p. 710). Its value is very nearly equal to the mean kilogram-calorie, 1 IT cal-1.00037 kilogram-calories (mean). 1 Btu = 251.996 IT cal.
[c] Absolute joule, defined as 10^7 ergs. The international joule, based on the international ohm and ampere, equals 1.0003 absolute joules.

Table 45 Power or Rate of Doing Work^a $[ML^2T^{-3}]$ or $[FLT^{-1}]$

Multiply Number of → by → to Obtain →	British Thermal Units per Minute	Ergs per Second	Foot-Pounds per Minute	Foot-Pounds per Second	Horsepower^a	Kilogram-Calories per Minute	Kilowatts	Metric Horsepower	Watts
British Thermal Units per Minute	1	5.689×10^{-9}	1.285×10^{-3}	7.712×10^{-2}	42.41	3.969	56.89	41.83	5.689×10^{-2}
Ergs per Second	1.758×10^{8}	1	2.259×10^{5}	1.356×10^{7}	7.457×10^{9}	6.977×10^{8}	10^{10}	7.355×10^{9}	10^{7}
Foot-Pounds per Minute	778.0	4.426×10^{-6}	1	60	3.3×10^{4}	3087	4.426×10^{4}	3.255×10^{4}	44.26
Foot-Pounds per Second	12.97	7.376×10^{-8}	1.667×10^{-2}	1	550	51.44	737.6	542.5	0.7376
Horsepower^a	2.357×10^{-2}	1.341×10^{-10}	3.030×10^{-5}	1.818×10^{-3}	1	9.355×10^{-2}	1.341	0.9863	1.341×10^{-3}
Kilogram-Calories per Minute	0.2520	1.433×10^{-9}	3.239×10^{-4}	1.943×10^{-2}	10.69	1	14.33	10.54	1.433×10^{-2}
Kilowatts	1.758×10^{-2}	10^{-10}	2.260×10^{-5}	1.356×10^{-3}	0.7457	6.977×10^{-2}	1	0.7355	10^{-3}
Metric Horsepower	2.390×10^{-2}	1.360×10^{-10}	3.072×10^{-5}	1.843×10^{-3}	1.014	9.485×10^{-2}	1.360	1	1.360×10^{-3}
Watts	17.58	10^{-7}	2.260×10^{-2}	1.356	745.7	69.77	1000	735.5	1

Note:

1 Cheval-vapeur = 75 kilogram-meters per second

1 Poncelet = 100 kilogram-meters per second

^{a}The "horsepower" used in these tables is equal to 550 foot-pounds per second by definition. Other definitions are one horsepower equals 746 watts (U.S. and Great Britain) and one horsepower equals 736 watts (continental Europe). Neither of these latter definitions is equivalent to the first; the "horsepowers" defined in these latter definitions are widely used in the rating of electrical machinery.

Table 46   Quantity of Electricity and Dielectric Flux [Q]

to Obtain ↓ / Multiply Number of → by	Abcoulombs	Ampere-Hours	**Coulombs**	Faradays	Stat coulombs
Abcoulombs	1	360	0.1	9649	3.335×10^{-11}
Ampere-Hours	2.778×10^{-3}	1	2.778×10^{-4}	26.80	9.259×10^{-14}
Coulombs	10	3600	1	9.649×10^4	3.335×10^{-10}
Faradays	1.036×10^{-4}	3.731×10^{-2}	1.036×10^{-5}	1	3.457×10^{-15}
Statcoulombs	2.998×10^{10}	1.080×10^{13}	2.998×10^9	2.893×10^{14}	1

Table 47   Charge per Unit Area and Electric Flux Density [QL⁻²]

to Obtain ↓ / Multiply Number of → by	Abcoulombs per Square Centimeter	Coulombs per Square Centimeter	Coulombs per Square Inch	Statcoulombs per Square Centimeter	**Coulombs per Square Meter**
Abcoulombs per Square Centimeter	1	0.1	1.550×10^{-2}	3.335×10^{-11}	10^{-5}
Coulombs per Square Centimeter	10	1	0.1550	3.335×10^{-10}	10^{-4}
Coulombs per Square Inch	64.52	6.452	1	2.151×10^{-9}	6.452×10^{-4}
Statcoulombs per Square Centimeter	2.998×10^{10}	2.998×10^9	4.647×10^8	1	2.998×10^5
Coulombs per Square Meter	10^5	10^4	1550	3.335×10^{-6}	1

Table 48   Electric Current [QT⁻¹]

to Obtain ↓ / Multiply Number of → by	Abamperes	**Amperes**	Statamperes
Abamperes	1	0.1	3.335×10^{-11}
Amperes	10	1	3.335×10^{-10}
Statamperes	2.998×10^{10}	2.998×10^9	1

Table 49   Current Density $[QT^{-1}L^{-2}]$

Multiply Number of → to Obtain ↓ by →	Abamperes per Square Centimeter	Amperes per Square Centimeter	Amperes per Square Inch	Statamperes per Square Centimeter	**Amperes per Square Meter**
Abamperes per Square Centimeter	1	0.1	1.550×10^{-2}	3.335×10^{-11}	10^{-5}
Amperes per Square Centimeter	10	1	0.1550	3.335×10^{-10}	10^{-4}
Amperes per Square Inch	64.52	6.452	1	2.151×10^{-9}	6.452×10^{-4}
Statamperes per Square Centimeter	2.998×10^{10}	2.998×10^{9}	4.647×10^{8}	1	2.998×10^{5}
Amperes per Square Meter	10^{5}	10^{4}	1550	3.335×10^{-6}	1

Table 50   Electric Potential and Electromotive Force $[MQ^{-1}L^{2}T^{-2}]$ or $[FQ^{-1}L]$

Multiply Number of → to Obtain ↓ by →	Abvolts	Microvolts	Millivolts	Statvolts	**Volts**
Abvolts	1	100	10^{5}	2.998×10^{10}	10^{8}
Microvolts	0.01	1	1000	2.998×10^{8}	10^{6}
Millivolts	10^{-5}	0.001	1	2.998×10^{5}	1000
Statvolts	3.335×10^{-11}	3.335×10^{-9}	3.335×10^{-6}	1	3.335×10^{-3}
Volts	10^{-8}	10^{-6}	0.001	299.8	1

Table 51 Electric Field Intensity and Potential Gradient $[MQ^{-1}LT^{-2}]$ **or** $[FQ^{-1}]$

Multiply Number of → / to Obtain ↓	Abvolts per Centimeter	Microvolts per Meter	Millivolts per Meter	Statvolts per Centimeter	Volts per Centimeter	Kilovolts per Centimeter	Volts per Inch	Volts per Mil	Volts per Meter
Abvolts per Centimeter	1	1	1000	2.998×10^{10}	10^8	10^{11}	3.937×10^7	3.937×10^{10}	10^6
Microvolts per Meter	1	1	1000	2.998×10^{10}	10^8	10^{11}	3.937×10^7	3.937×10^{10}	10^6
Millivolts per Meter	0.001	0.001	1	2.998×10^7	10^5	10^8	3.937×10^4	3.937×10^7	1000
Statvolts per Centimeter	3.335×10^{-11}	3.335×10^{-11}	3.335×10^{-8}	1	3.335×10^{-3}	3.335	1.313×10^{-3}	1.313	3.335×10^{-5}
Volts per Centimeter	10^{-8}	10^{-8}	10^{-5}	299.8	1	1000	0.3937	393.7	10^{-2}
Kilovolts per Centimeter	10^{-11}	10^{-11}	10^{-8}	0.2998	0.001	1	3.937×10^{-4}	0.3937	10^{-5}
Volts per Inch	2.540×10^{-8}	2.540×10^{-8}	2.540×10^{-5}	761.6	2.540	2540	1	1000	2.540×10^{-2}
Volts per Mil	2.540×10^{-11}	2.540×10^{-11}	2.540×10^{-8}	0.7616	2.540×10^{-3}	2.540	0.001	1	2.540×10^{-5}
Volts per Meter	10^{-6}	10^{-6}	10^{-3}	2.998×10^4	100	10^5	39.37	3.937×10^4	1

Table 52   Electric Resistance $[MQ^{-2}L^2T^{-1}]$ or $[FQ^{-2}LT]$

to Obtain ↓ / Multiply Number of → by →	Abohms	Megohms	Microhms	**Ohms**	Statohms
Abohms	1	10^{15}	1000	10^9	8.988×10^{20}
Megohms	10^{-15}	1	10^{-12}	10^{-6}	8.988×10^5
Microhms	0.001	10^{12}	1	10^6	8.988×10^{17}
Ohms	10^{-9}	10^6	10^{-6}	1	8.988×10^{11}
Statohms	1.112×10^{-21}	1.112×10^{-6}	1.112×10^{-18}	1.112×10^{-12}	1

Note: Electric Conductance $[F^{-1}Q^2L^{-1}T^{-1}]$. 1 Siemens = 1 mho = 1 ohm^{-1} = 10^{-6} megmho = 10^6 micromho.

Table 53   Electric Resistivity[a] $[MQ^{-2}L^3T^{-1}]$ or $[FQ^{-2}L^2T]$

to Obtain ↓ / Multiply Number of → by →	Abohm-Centimeters	Microhm-Centimeters	Microhm-Inches	Ohms (Mil, Foot)	Ohms (Meter, Gram)[b]	**Ohm-Meters**
Abohm-Centimeters	1	1000	2540	166.2	$10^5/\delta$	10^{11}
Microhm-Centimeters	0.001	1	2.540	0.1662	$100/\delta$	10^8
Microhm-Inches	3.937×10^{-4}	0.3937	1	6.545×10^{-2}	$39.37/\delta$	3.937×10^7
Ohms (Mil, Foot)	6.015×10^{-3}	6.015	15.28	1	$601.5/\delta$	6.015×10^8
Ohms (Meter, Gram)[b]	$10^{-5}\delta$	0.01δ	$2.540 \times 10^{-2}\delta$	$1.662 \times 10^{-3}\delta$	1	$10^{-6}\delta$
Ohm-Meters	10^{-11}	10^{-8}	2.540×10^{-8}	1.662×10^{-9}	$10^{-6}/\delta$	1

[a] In this table δ is density in grams per cubic-centimeters. The following names, corresponding respectively to those at the tops of columns, are sometimes used: abohms per centimeter cube; microhms per centimeter cube; microhms per inch cube; ohms per milfoot; ohms per meter-gram. The first four columns are headed by units of *volume* resistivity, the last by a unit of *mass* resistivity. The dimensions of the latter are $Q^{-2}L^6T^{-1}$, not those given in the heading of the table.
[b] One ohm (meter, gram) = 5710 ohms (mile, pound).

Table 54   Electric Conductivity[a] $[M^{-1}Q^2L^{-3}T]$ or $[F^{-1}Q^2L^{-2}T^{-1}]$

to Obtain ↓ / Multiply Number of → by →	Abmhos per Centimeter	Mhos (Mil, Foot)	Mhos (Meter, Gram)	Micromhos per Centimeter	Micromhos per Inch	**Siemens per Meter**
Abmhos per Centimeter	1	6.015×10^{-3}	$10^{-5}\delta$	0.001	3.937×10^{-4}	10^{-11}
Mhos (Mil, Foot)	166.2	1	$1.662 \times 10^{-3}\delta$	0.1662	6.524×10^{-2}	1.662×10^{-9}
Mhos (Meter, Gram)	$10^5/\delta$	$601.5/\delta$	1	$100/\delta$	$39.37/\delta$	$10^{-6}/\delta$
Micromhos per Centimeter	1000	6.015	0.01δ	1	0.3937	10^{-8}
Micromhos per Inch	2540	15.28	$2.540 \times 10^{-2}\delta$	2.540	1	2.54×10^{-8}
Siemens per Meter	10^{11}	6.015×10^8	$10^6\delta$	10^8	3.937×10^7	1

[a] See footnote of Table 53. Names sometimes used are abmho per centimeter cube, mho per mil-foot, etc. Dimensions of mass conductivity are $Q^2L^{-6}T$.

Table 55   Capacitance $[M^{-1}Q^2L^{-2}T^2]$ or $[F^{-1}Q^2L^{-1}]$

Multiply Number of → to Obtain ↓ by →	Abfarads	Farads	Microfarads	Statfarads
Abfarads	1	10^{-9}	10^{-15}	1.112×10^{-21}
Farads	10^9	1	10^{-6}	1.112×10^{-12}
Microfarads	10^{15}	10^6	1	1.112×10^{-6}
Statfarads	8.988×10^{20}	8.988×10^{11}	8.988×10^5	1

Table 56   Inductance $[MQ^{-2}L^2]$ or $[FQ^{-2}LT^2]$

Multiply Number of → to Obtain ↓ by →	Abhenries[a]	Henries	Microhenries	Millihenries	Stathenries
Abhenries[a]	1	10^9	1000	10^6	8.988×10^{20}
Henries	10^{-9}	1	10^{-6}	0.001	8.988×10^{11}
Microhenries	0.001	10^6	1	1000	8.988×10^{17}
Millihenries	10^{-6}	1000	0.001	1	8.988×10^{14}
Stathenries	1.112×10^{-21}	1.112×10^{-12}	1.112×10^{-18}	1.112×10^{-15}	1
					1

[a] An abhenry is sometimes called a "centimeter."

Table 57   Magnetic Flux $[MQ^{-1}L^2T^{-1}]$ or $[FQ^{-1}LT]$

Multiply Number of → to Obtain ↓ by →	Kilolines	Maxwells (or Lines)	Webers
Kilolines	1	0.001	10^5
Maxwells (or Lines)	1000	1	10^8
Webers	10^{-5}	10^{-8}	1

Table 58   Magnetic Flux Density $[MQ^{-1}T^{-1}]$ or $[FQ^{-1}L^{-1}T]$

to Obtain ↓ Multiply Number of → by	Gausses (or Lines per Square Centimeter)	Lines per Square Inch	Webers per Square Centimeter	Webers per Square Inch	**Tesla** (Webers per Square Meter)
Gausses (or Lines per Square Centimeter)	1	0.1550	10^8	1.550×10^7	10^4
Lines per Square Inch	6.452	1	6.452×10^8	10^8	6.452×10^4
Webers per Square Centimeter	10^{-8}	1.550×10^{-9}	1	0.1550	10^{-4}
Webers per Square Inch	6.452×10^{-8}	10^{-8}	6.452	1	6.452×10^{-4}
Tesla (Webers per Square Meter)	10^{-4}	1.550×10^{-5}	10^4	1550	1

Table 59   Magnetic Potential and Magnetomotive Force $[QT^{-1}]$

to Obtain ↓ Multiply Number of → by	Abampere-Turns	**Ampere-Turns**	Gilberts
Abampere-Turns	1	0.1	7.958×10^{-2}
Ampere-Turns	10	1	0.7958
Gilberts	12.57	1.257	1

Table 60   Magnetic Field Intensity, Potential Gradient, and Magnetizing Force $[QL^{-1}T^{-1}]$

to Obtain ↓ Multiply Number of → by	Abampere-Turns per Centimeter	Ampere-Turns per Centimeter	Ampere-Turns per Inch	Oersteds (Gilberts per Centimeter)	**Ampere-Turns per Meter**
Abampere-Turns per Centimeter	1	0.1	3.937×10^{-2}	7.958×10^{-2}	10^{-3}
Ampere-Turns per Centimeter	10	1	0.3937	0.7958	10^{-2}
Ampere-Turns per Inch	25.40	2.540	1	2.021	2.54×10^{-2}
Oersteds (Gilberts per Centimeter)	12.57	1.257	0.4950	1	1.257×10^{-2}
Ampere-Turns per Meter	10^3	10^2	39.37	79.58	1

Table 61   Specific Heat $[L^2T^{-2}t^{-1}]$ (t = temperature)

To change specific heat in gram-calories per gram per degree Centigrade to the units given in any line of the following table, multiply by the factor in the last column.

Unit of Heat or Energy	Unit of Mass	Temperature Scale[a]	Factor
Gram-calories	Gram	Centigrade	1
Kilogram-calories	Kilogram	Centigrade	1
British thermal units	Pound	Centigrade	1.800
British thermal units	Pound	Fahrenheit	1.000
Joules	Gram	Centigrade	4.186
Joules	Pound	Fahrenheit	1055
Joules	**Kilogram**	**Kelvin**	4.187×10^3
Kilowatt-hours	Kilogram	Centigrade	1.163×10^{-3}
Kilowatt-hours	Pound	Fahrenheit	2.930×10^{-4}

[a]Temperature conversion formulas:

$$t_c = \text{temperature in Centigrade degrees}$$

$$t_f = \text{temperature in Fahrenheit degrees}$$

$$t_K = \text{temperature in Kelvin degrees}$$

$$1\,\text{F} = \tfrac{5}{9}{}^{\circ}\text{C}$$

$$1\,\text{K} = 1\,^{\circ}\text{C}$$

$$t_c = \tfrac{5}{9}(t_f - 32)$$

$$t_f = \tfrac{9}{5}t_c + 32$$

$$t_K = t_c + 273$$

Table 62 Thermal Conductivity[a] $[LMT^{-3}t^{-1}]$

to Obtain (Multiply Number of → by →)	$\dfrac{\text{Btu}\cdot\text{ft}}{\text{h}\cdot\text{ft}^2\cdot{}^\circ\text{F}}$	$\dfrac{\text{Btu}\cdot\text{in.}}{\text{h}\cdot\text{ft}^2\cdot{}^\circ\text{F}}$	$\dfrac{\text{Btu}\cdot\text{in.}}{\text{sec}\cdot\text{ft}^2\cdot{}^\circ\text{F}}$	$\dfrac{\text{J}}{\text{m}\cdot\text{s}\cdot{}^\circ\text{C}}$	$\dfrac{\text{kcal}}{\text{m}\cdot\text{h}\cdot{}^\circ\text{C}}$	$\dfrac{\text{erg}}{\text{cm}\cdot\text{s}\cdot{}^\circ\text{C}}$	$\dfrac{\text{kcal}}{\text{m}\cdot\text{s}\cdot{}^\circ\text{C}}$	$\dfrac{\text{cal}}{\text{cm}\cdot\text{s}\cdot{}^\circ\text{C}}$	$\dfrac{\text{W}}{\text{ft}\cdot{}^\circ\text{C}}$	$\dfrac{\text{W}}{\text{m}\cdot\text{K}}$
$\text{Btu}\cdot\text{ft/h}\cdot\text{ft}^2\cdot{}^\circ\text{F}$	1	8.333×10^{-2}	3.0×10^{2}	5.778×10^{-1}	6.720×10^{-1}	5.778×10^{-6}	2.419×10^{3}	2.419×10^{2}	1.895	5.778×10^{-1}
$\text{Btu}\cdot\text{in./h}\cdot\text{ft}^2\cdot{}^\circ\text{F}$	12	1	3.6×10^{3}	6.933	8.064	6.933×10^{-5}	2.903×10^{4}	2.903×10^{3}	2.275×10^{1}	6.933
$\text{Btu}\cdot\text{in./s}\cdot\text{ft}^2\cdot{}^\circ\text{C}$	3.333×10^{-3}	2.778×10^{-4}	1	1.926×10^{-3}	2.240×10^{-3}	1.926×10^{-8}	8.064	8.064×10^{-1}	6.319×10^{-3}	1.926×10^{-3}
$\text{J/m}\cdot\text{s}\cdot{}^\circ\text{C}$	1.731	1.442×10^{-1}	5.192×10^{2}	1	1.163	1.000×10^{-5}	4.187×10^{3}	4.187×10^{2}	3.281	1.0
$\text{kcal/m}\cdot\text{h}\cdot{}^\circ\text{C}$	1.483	1.240×10^{-1}	4.465×10^{2}	8.599×10^{-1}	1	8.599×10^{-6}	3.6×10^{3}	3.6×10^{2}	2.821	8.599×10^{-1}
$\text{erg/cm}\cdot\text{s}\cdot{}^\circ\text{C}$	1.731×10^{5}	1.442×10^{4}	5.192×10^{7}	1.0×10^{5}	1.163×10^{5}	1	4.187×10^{8}	4.187×10^{7}	3.281×10^{5}	1.0×10^{5}
$\text{kcal/m}\cdot\text{s}\cdot{}^\circ\text{C}$	4.134×10^{-4}	3.445×10^{-5}	1.240×10^{-1}	2.388×10^{-4}	2.778×10^{-4}	2.388×10^{-9}	1	1.0×10^{-1}	7.835×10^{-4}	2.388×10^{-4}
$\text{cal/cm}\cdot\text{s}\cdot{}^\circ\text{C}$	4.134×10^{-3}	3.445×10^{-4}	1.240	2.388×10^{-3}	2.778×10^{-3}	2.388×10^{-8}	10	1	7.835×10^{-3}	2.388×10^{-3}
$\text{W/ft}\cdot{}^\circ\text{C}$	5.276×10^{-1}	4.395×10^{-2}	1.582×10^{2}	3.048×10^{-1}	3.545×10^{-1}	3.048×10^{-6}	1.276×10^{3}	1.276×10^{2}	1	3.048×10^{-1}
$\text{W/m}\cdot\text{K}$	1.731	1.442×10^{-1}	5.192×10^{2}	1.0	1.163	1.00×10^{-5}	4.187×10^{3}	4.187×10^{2}	3.281	1

[a] International Table Btu = 1.055056×10^{3} joules and International Table cal = 4.1868 J are used throughout.

Table 63 Photometric Units

	Common Unit	Multiply by	to Get **SI Unit**
Luminous intensity	International candle	9.81×10^{-1}	**cd**
Luminance	cd/in.^{2}	1.550×10^3	**cd/m^2**
	cd/cm^2	1×10^4	**cd/m^2**
	Foot · lambert	3.4263	**cd/m^2**
Luminous flux	cd · sr	1.0000	**lm**
	Candle power (spher.)	12.566	**lm**
Quantity of light flux			**lm·**
Luminous exitance[a]			**lm/m^2**
Illuminance[b]	lm	3.103×10^3	**cd/m^2**
	Foot candles	1.0764×10	**lm/m^2**
	lmft^2	1.0764×10	**lm/m^2**
	lx	1.000	**lm/m^2**
	Phots	1×10^4	**lm/m^2**
Luminous efficacy			**lm/W**

[a] Luminous emittance.
[b] Luminous flux density.

Table 64 Specific Gravity Conversions

Specific Gravity 60°/60°	°Be	°API	lb/gal 60°F, wt in air	lb/ft^3 at 60°F, wt in air	Specific Gravity 60°/60°	°Be	°API	lb/gal 60°F, wt in air	lb/ft^3 at 60°F, wt in air
0.600	103.33	104.33	4.9929	37.350	0.745	57.92	58.43	6.2020	46.394
0.605	101.40	102.38	5.0346	37.662	0.750	56.67	57.17	6.2437	46.706
0.610	99.51	100.47	5.0763	37.973	0.755	55.43	55.92	6.2854	47.018
0.615	97.64	98.58	5.1180	38.285	0.760	54.21	54.68	6.3271	47.330
0.620	95.81	96.73	5.1597	38.597	0.765	53.01	53.47	6.3688	47.642
0.625	94.00	94.90	5.2014	39.910	0.770	51.82	52.27	6.4104	47.953
0.630	92.22	93.10	5.2431	39.222	0.775	50.65	51.08	6.4521	48.265
0.635	90.47	91.33	5.2848	39.534	0.780	49.49	49.91	6.4938	48.577
0.640	88.75	89.59	5.3265	39.845	0.785	48.34	48.75	6.5355	48.889
0.645	87.05	87.88	5.3682	40.157	0.790	47.22	47.61	6.5772	49.201
0.650	85.38	86.19	5.4098	40.468	0.795	46.10	46.49	6.6189	49.513
0.655	83.74	84.53	5.4515	40.780					
0.660	82.12	82.89	5.4932	41.092	0.800	45.00	45.38	6.6606	49.825
0.665	80.53	81.28	5.5349	41.404	0.805	43.91	44.28	6.7023	50.137
0.670	78.96	79.69	5.5766	41.716	0.810	42.84	43.19	6.7440	50.448
0.675	77.41	78.13	5.6183	42.028	0.815	41.78	42.12	6.7857	50.760
0.680	75.88	76.59	5.6600	42.340	0.820	40.73	41.06	6.8274	51.072
0.685	74.38	75.07	5.7017	42.652	0.825	39.70	40.02	6.8691	51.384
0.690	72.90	73.57	5.7434	42.963	0.830	38.67	38.98	6.9108	51.696
0.695	71.44	72.10	5.7851	43.275	0.835	37.66	37.96	6.9525	52.008
					0.840	36.67	36.95	6.9941	52.320
0.700	70.00	70.64	5.8268	43.587	0.845	35.68	35.96	7.0358	52.632
0.705	68.58	69.21	5.8685	43.899	0.850	34.71	34.97	7.0775	52.943
0.710	67.18	67.80	5.9101	44.211	0.855	33.74	34.00	7.1192	53.225
0.715	65.80	66.40	5.9518	44.523	0.860	32.79	33.03	7.1609	53.567
0.720	64.44	65.03	5.9935	44.834	0.865	31.85	32.08	7.2026	53.879
0.725	63.10	63.67	6.0352	45.146	0.870	30.92	31.14	7.2443	54.191
0.730	61.78	62.34	6.0769	45.458	0.875	30.00	30.21	7.2860	54.503
0.735	60.48	61.02	6.1186	45.770	0.880	29.09	29.30	7.3277	54.815
0.740	59.19	59.72	6.1603	46.082	0.885	28.19	28.38	7.3694	55.127

(*Continues*)

Table 64 (*Continued*)

Specific Gravity 60°/60°	°Be	°TW	lb/gal 60°F, wt in air	lb/ft³ at 60°F, wt in air	Specific Gravity 60°/60°	°Be	°TW	lb/gal 60°F, wt in air	lb/ft³ at 60°F, wt in air
0.890	27.30	27.49	7.4111	55.438	1.140	17.81	28	9.4957	71.032
0.895	26.42	26.60	7.4528	55.750	1.145	18.36	29	9.5374	71.344
					1.150	18.91	30	9.5790	71.656
0.900	25.76	25.72	7.4944	56.062	1.155	19.46	31	9.6207	71.968
0.905	24.70	24.85	7.5361	56.374	1.160	20.00	32	9.6624	72.280
0.910	23.85	23.99	7.5777	56.685	1.165	20.54	33	9.7041	72.592
0.915	23.01	23.14	7.6194	56.997	1.170	21.07	34	9.7458	72.904
0.920	22.17	22.30	7.6612	57.410	1.175	21.60	35	9.7875	73.216
0.925	21.35	21.47	7.7029	57.622	1.180	22.12	36	9.8292	73.528
0.930	20.54	20.65	7.7446	57.934	1.185	22.64	37	9.8709	73.840
0.935	19.73	19.84	7.7863	58.246	1.190	23.15	38	9.9126	74.151
0.940	18.94	19.03	7.8280	58.557	1.195	23.66	39	9.9543	74.463
0.945	18.15	18.24	7.8697	58.869					
0.950	17.37	17.45	7.9114	59.181	1.200	24.17	40	9.9960	74.775
0.955	16.60	16.67	7.9531	59.493	1.205	24.67	41	10.0377	75.087
0.960	15.83	15.90	7.9947	59.805	1.210	25.17	42	10.0793	75.399
0.965	15.08	15.13	8.0364	60.117	1.215	25.66	43	10.1210	75.711
0.970	14.33	14.38	8.0780	60.428	1.220	26.15	44	10.1627	76.022
0.975	13.59	13.63	8.1197	60.740	1.225	26.63	45	10.2044	76.334
0.980	12.86	12.89	8.1615	61.052	1.230	27.11	46	10.2461	76.646
0.985	12.13	12.15	8.2032	61.364	1.235	27.59	47	10.2878	76.958
0.990	11.41	11.43	8.2449	61.676	1.240	28.06	48	10.3295	77.270
0.995	10.70	10.71	8.2866	61.988	1.245	28.53	49	10.3712	77.582
					1.250	29.00	50	10.4129	77.894
					1.255	29.46	51	10.4546	78.206
					1.260	29.92	52	10.4963	78.518
Specific Gravity 60°/60°	°Be	°TW	lb/gal 60°F, wt in air	lb/ft³ at 60°F, wt in air	1.265	30.38	53	10.5380	78.830
					1.270	30.83	54	10.5797	79.141
					1.275	31.27	55	10.6214	79.453
1.000	10.00	10.00	8.3283	62.300	1.280	31.72	56	10.6630	79.765
1.005	0.72	1	8.3700	62.612	1.285	32.16	57	10.7047	80.077
1.010	1.44	2	8.4117	62.924	1.290	32.60	58	10.7464	80.389
1.015	2.14	3	8.4534	63.236	1.295	33.03	59	10.7881	80.701
1.020	2.84	4	8.4950	63.547					
1.025	3.54	5	8.5367	63.859	1.300	33.46	60	10.8298	81.013
1.030	4.22	6	8.5784	64.171	1.305	33.89	61	10.8715	81.325
1.035	4.90	7	8.6201	64.483	1.310	34.31	62	10.9132	81.636
1.040	5.58	8	8.6618	64.795	1.315	34.73	63	10.9549	81.948
1.045	6.24	9	8.7035	65.107	1.320	35.15	64	10.9966	82.260
1.050	6.91	10	8.7452	65.419	1.325	35.57	65	11.0383	82.572
1.055	7.56	11	8.7869	65.731	1.330	35.98	66	11.0800	82.884
1.060	8.21	12	8.8286	66.042	1.335	36.39	67	11.1217	83.196
1.065	8.85	13	8.8703	66.354	1.340	36.79	68	11.1634	83.508
1.070	9.49	14	8.9120	66.666	1.345	37.19	69	11.2051	83.820
1.075	10.12	15	8.9537	66.978	1.350	37.59	70	11.2467	84.131
1.080	10.74	16	8.9954	67.290	1.355	37.99	71	11.2884	84.443
1.085	11.36	17	9.0371	67.602	1.360	38.38	72	11.3301	84.755
1.090	11.97	18	9.0787	67.914	1.365	38.77	73	11.3718	85.067
1.095	12.58	19	9.1204	68.226	1.370	39.16	74	11.4135	85.379
					1.375	39.55	75	11.4552	85.691
1.100	13.18	20	9.1621	68.537	1.380	39.93	76	11.4969	86.003
1.105	13.78	21	9.2038	68.849	1.385	40.31	77	11.5386	86.315
1.110	14.37	22	9.2455	69.161	1.390	40.68	78	11.5803	86.626
1.115	14.96	23	9.2872	69.473	1.395	41.06	79	11.6220	86.938
1.120	15.54	24	9.3289	69.785					
1.125	16.11	25	9.3706	70.097	1.400	41.43	80	11.6637	87.250
1.130	16.68	26	9.4123	70.409	1.405	41.80	81	11.7054	87.562
1.135	17.25	27	9.4540	70.721	1.410	42.16	82	11.7471	87.874

Table 64 (*Continued*)

Specific Gravity 60°/60°	°Be	°TW	lb/gal 60°F, wt in air	lb/ft³ at 60°F, wt in air	Specific Gravity 60°/60°	°Be	°TW	lb/gal 60°F, wt in air	lb/ft³ at 60°F, wt in air
1.415	42.53	83	11.7888	88.186	1.67	58.17	134	13.915	104.09
1.420	42.89	84	11.8304	88.498	1.68	58.69	136	13.998	104.72
1.425	43.25	85	11.8721	88.810	1.69	59.20	138	14.082	105.34
1.430	43.60	86	11.9138	89.121	1.70	59.71	140	14.165	105.96
1.435	43.95	87	11.9555	89.433	1.71	60.20	142	14.249	106.59
1.440	44.31	88	11.9972	89.745	1.72	60.70	144	14.332	107.21
1.445	44.65	89	12.0389	90.057	1.73	61.18	146	14.415	107.83
1.450	45.00	90	12.0806	90.369	1.74	61.67	148	14.499	108.46
1.455	45.34	91	12.1223	90.681	1.75	62.14	150	14.582	109.08
1.460	45.68	92	12.1640	90.993	1.76	62.61	152	14.665	109.71
1.465	46.02	93	12.2057	91.305	1.77	63.08	154	14.749	110.32
1.470	46.36	94	12.2473	91.616	1.78	63.54	156	14.832	110.95
1.475	46.69	95	12.2890	91.928	1.79	63.99	158	14.916	111.58
1.480	47.03	96	12.3307	92.240					
1.485	47.36	97	12.3724	92.552	1.80	64.44	160	14.999	112.20
1.490	47.68	98	12.4141	92.864	1.81	64.89	162	15.082	112.82
					1.82	65.33	164	15.166	113.45
1.495	48.01	99	12.4558	93.176	1.83	65.77	166	15.249	114.07
1.500	48.33	100	12.4975	93.488	1.84	66.20	168	15.333	114.70
1.51	48.97	102	12.581	94.11	1.85	66.62	170	15.416	115.31
1.52	49.61	104	12.644	94.79	1.86	67.04	172	15.499	115.94
1.53	50.23	106	12.748	95.36	1.87	67.46	174	15.583	116.56
1.54	50.84	108	12.831	95.98	1.88	67.87	176	15.666	117.19
1.55	51.45	110	12.914	96.61	1.89	68.28	178	15.750	117.81
1.56	52.05	112	12.998	97.23					
1.57	52.64	114	13.081	97.85	1.90	68.68	180	15.832	118.43
1.58	53.23	116	13.165	98.48	1.91	69.08	182	15.916	119.06
1.59	53.81	118	13.248	99.10	1.92	69.48	184	16.000	119.68
					1.93	69.87	186	16.083	120.31
1.60	54.38	120	13.331	99.73	1.94	70.26	188	16.166	120.93
1.61	54.94	122	13.415	100.35	1.95	70.64	190	16.250	121.56
1.62	55.49	124	13.498	100.97	1.96	71.02	192	16.333	122.18
1.63	56.04	126	13.582	101.60	1.97	71.40	194	16.417	122.80
1.64	56.59	128	13.665	102.22	1.98	71.77	196	16.500	123.43
1.65	57.12	130	13.748	102.84	1.99	72.14	198	16.583	124.05
1.66	57.65	132	13.832	103.47	2.00	72.50	200	16.667	124.68

[a] Baumé scale.
[b] Twaddell scale.

6 STANDARD SIZES

6.1 Preferred Numbers

Selection of standard sizes or ratings of many diverse products can be performed advantageously through the use of a geometrically based progression introduced by C. Renard. He originally adopted as a basis a rule that would yield a 10th multiple of the value a after every 5th step of the series:

$$a \times q^5 = 10a \quad or \quad q = \sqrt[5]{10}$$

where the numerical series $a, a[\sqrt[5]{10}], a[\sqrt[5]{10}]^2, a[\sqrt[5]{10}]^3, a[\sqrt[5]{10}]^4, 10a$, the values of which, to five significant figures, are a, $1.5849a$, $2.5119a$, $3.9811a$, $6.309a$, $10a$.

Renard's idea was to substitute, for these values, more rounded but more practical values. He adopted as a a power of 10, positive, nil, or negative, obtaining the series 10, 16, 25, 40, 63, 100, which may be continued in both directions.

From this series, designated by the symbol R5, the R10, R20, R40 series were formed, each adopted ratio being the square root of the preceding one: $\sqrt[10]{10}$, $\sqrt[20]{10}$, $\sqrt[40]{10}$. Thus each series provided Renard with twice as many steps in a decade as the preceding one.

Preferred numbers are immediately applicable to commercial sizes and ratings of products. It is advantageous to minimize the number of initial sizes and also to have adequate provision for logical expansion if and when additional sizes are required. By making the initial sizes correspond to a coarse series such as R5, unnecessary expense can be avoided if subsequent demand for the product is disappointing. If, on the other hand, the product is accepted, intermediate sizes may be selected in a rational manner by using the next finer series R10, and so on. Such a procedure assures a justifiable relationship between successive sizes and is a decided contrast to haphazard selection.

The application of preferred numbers to raw material sizes and to the dimensions of parts also has enormously important potentialities. Under present conditions, commercial sizes of material are the result of a great many dissimilar gauge systems. The current trend in internationally acceptable metric sizing is to use preferred numbers. Even here, though, in the midst of the greatest opportunity for worldwide standardization through the acceptance of Renard series, we have fallen prey to our individualistic nature. The preferred number 1.6 is used by most nations as a standard 1.6 mm material thickness. German manufacturers, however, like 1.5 mm of the International Organization for Standardization (ISO) 497 for a more rounded preferred number. Similarly in metric screw sizes, 6.3 mm is consistent with the preferred number series; yet, 6.0 mm (more rounded) has been adopted as a standard fastener diameter.

The International Electrochemical Commission (IEC) used preferred numbers to establish standard current ratings in amperes as follows: 1, 1.25, 1.6, 2.5, 3.15, 4.5, 6.3. Notice that R10 series is used except for 4.5, which is a third step R20 series.

The American Wire Gauge size for copper wire is based on a geometric series. However, instead of using 1.1220, the rounded value of $\sqrt[20]{10}$, in $a \times q^{20} = 10a$, the q chosen is 1.123.

A special series of preferred numbers is used for designating the characteristic values of capacitors, resistors, inductors, and other electronic products. Instead of using the Renard series R5, R10, R20, R40, R80 as derived from the geometric series of numbers $10^{N/5}, 10^{N/10}, 10^{N/20}, 10^{N/40}, 10^{N/80}$, the geometric series used is $10^{N/6}, 10^{N/12}, 10^{N/24}, 10^{N/48}, 10^{N/96}, 10^{N/192}$, which are designated respectively E6, E12, E24, E48, E96, E192.

It should be evident that any series of preferred numbers can be generated to serve any specific case. Examples taken from the American National Standards Institute (ANSI) and ISO standards are reproduced in Tables 65–68.

Table 65   Basic Series of Preferred Numbers: R5, R10, R20, and R40 Series

R5	R10	R20	R40	Theoretical Values		Differences between Basic Series and Calculated Values (%)
				Mantissas of Logarithms	Calculated Values	
1.00	1.00	1.00	1.00	000	1.0000	0
			1.06	025	1.0593	+0.07
		1.12	1.12	050	1.1220	−0.18
			1.18	075	1.1885	−0.71
	1.25	1.25	1.25	100	1.2589	−0.71
			1.32	125	1.3335	−1.01
		1.40	1.40	150	1.4125	−0.88
			1.50	175	1.4962	+0.25
1.60	1.60	1.60	1.60	200	1.5849	+0.95
			1.70	225	1.6788	+1.26
		1.80	1.80	250	1.7783	+1.22
			1.90	275	1.8836	+0.87
	2.00	2.00	2.00	300	1.9953	+0.24
			2.12	325	2.1135	+0.31
		2.24	2.24	350	2.2387	+0.06
			2.36	375	2.3714	−0.48
2.50	2.50	2.50	2.50	400	2.5119	−0.47
			2.65	425	2.6607	−0.40
		2.80	2.80	450	2.8184	−0.65
			3.00	475	2.9854	+0.49
	3.15	3.15	3.15	500	3.1623	−0.39
			3.35	525	3.3497	+0.01
		3.55	3.55	550	3.5481	+0.05
			3.75	575	3.7584	−0.22

Table 65 (*Continued*)

R5	R10	R20	R40	Theoretical Values		Differences between Basic Series and Calculated Values (%)
				Mantissas of Logarithms	Calculated Values	
4.00	4.00	4.00	4.00	600	3.9811	+0.47
			4.25	625	4.2170	+0.78
		4.50	4.50	650	4.4668	+0.74
			4.75	675	4.7315	+0.39
	5.00	5.00	5.00	700	5.0119	−0.24
			5.30	725	5.3088	−0.17
		5.60	5.60	750	5.6234	−0.42
			6.00	775	5.9566	+0.73
6.30	6.30	6.30	6.30	800	6.3096	−0.15
			6.70	825	6.6834	+0.25
		7.10	7.10	850	7.0795	+0.29
			7.50	875	7.4989	+0.01
	8.00	8.00	8.00	900	7.9433	+0.71
			8.50	925	8.4140	+1.02
		9.00	9.00	950	8.9125	+0.98
			9.50	975	9.4406	+0.63
10.00	10.00	10.00	10.00	000	10.0000	0

Table 66 Basic Series of Preferred Numbers: R80 Series

1.00	1.80	3.15	5.60
1.03	1.85	3.25	5.80
1.06	1.90	3.35	6.00
1.09	1.95	3.45	6.15
1.12	2.00	3.55	6.30
1.15	2.06	3.65	6.50
1.18	2.12	3.75	6.70
1.22	2.18	3.87	6.90
1.25	2.24	4.00	7.10
1.28	2.30	4.12	7.30
1.32	2.36	4.25	7.50
1.36	2.43	4.37	7.75
1.40	2.50	4.50	8.00
1.45	2.58	4.62	8.25
1.50	2.65	4.75	8.50
1.55	2.72	4.87	8.75
1.60	2.80	5.00	9.00
1.65	2.90	5.15	9.25
1.70	3.00	5.20	9.50
1.75	3.07	5.45	9.75

Table 67 Expansion of R5 Series

Preferred Number	Divided by 10	Multiplied by 10	Multiplied by 100	Multiplied by 1000
1.0	0.10	10	100	1000
1.6	0.16	16	160	1600
2.5	0.25	25	250	2500
4.0	0.40	40	400	4000
6.3	0.63	63	630	6300

Table 68 Rounding of Preferred Numbers[a]

Preferred Number	First Rounding	Second Rounding
1.12	1.1	1.1
1.25	1.25	1.2
1.60	1.6	1.5[a]
2.24	2.2	2.2
3.15	3.2	3.0
3.55	3.6	3.5
5.60	5.6	5.5
6.30	6.3	6.0
7.10	7.1	7.0

[a] Rounded only when using the R5 or R10 series.

Applicable Documents Adoption of Renard's preferred number system by international standardization bodies resulted in a host of national standards being generated for particular applications. The current organization in the United States that is charged with generating American national standards is the ANSI. Accordingly, the following national and international standards are in use in the United States.

ANSI Z17.1-1973	American National Standard for Preferred Numbers
ANSI C83.2-1971	American National Standard Preferred Values for Components for Electronic Equipment
EIA Standard RS-385	Preferred Values for Components for Electronic Equipment (issued by the Electronics Industries Association; Same as ANSI C83.2-1971)
ISO 3-1973	Preferred numbers—series of preferred numbers
ISO 17-1973	Guide to the use of preferred numbers and of series of preferred numbers

| ISO 497-1973 | Guide to the choice of series of preferred numbers and of series containing more rounded values of preferred numbers |

Table 67 shows the expansibility of preferred numbers in the positive direction. The same expansibility can be made in the negative direction. Table 68 shows a deviation by roundings for cases where adhering to a basic preferred number would be absurd as in 31.5 teeth in a gear when clearly 32 makes sense.

6.2 Gages

Table 69 U.S. Standard Gage[a] for Sheet and Plate Iron and Steel and Its Extension[b]

| Gage Number | Weight per Square Foot | | Weight per Square Meter | Approximate Thickness | | | |
| | | | | Wrought Iron, 480 lb/ft^3 | | Steel and open-hearth Iron, 489.6 lb/ft^3 | |
	oz.	lb	kg	in.	mm	in.	mm
0000000	320	20.00	97.65	0.500	12.70	0.490	12.45
000000	300	18.75	91.55	0.469	11.91	0.460	11.67
00000	280	17.50	85.44	0.438	11.11	0.429	10.90
0000	260	16.25	79.34	0.406	10.32	0.398	10.12
000	240	15.00	73.24	0.375	9.52	0.368	9.34
00	220	13.75	67.13	0.344	8.73	0.337	8.56
0	200	12.50	61.03	0.312	7.94	0.306	7.78
1	180	11.25	54.93	0.2812	7.14	0.2757	7.00
2	170	10.62	51.88	0.2656	6.75	0.2604	6.62
3	160	10.00	48.82	0.2500	6.35	0.2451	6.23
4	150	9.375	45.77	0.2344	5.95	0.2298	5.84
5	140	8.750	42.72	0.2188	5.56	0.2145	5.45
6	130	8.125	39.67	0.2031	5.16	0.1991	5.06
7	120	7.500	36.62	0.1875	4.76	0.1838	4.67
8	110	6.875	33.57	0.1719	4.37	0.1685	4.28
9	100	6.250	30.52	0.1562	3.97	0.1532	3.89
10	90	5.625	27.46	0.1406	3.57	0.1379	3.50
11	80	5.000	24.41	0.1250	3.18	0.1225	3.11
12	70	4.375	21.36	0.1094	2.778	0.1072	2.724
13	60	3.750	18.31	0.0938	2.381	0.0919	2.335
14	50	3.125	15.26	0.0781	1.984	0.0766	1.946
15	45	2.812	13.73	0.0703	1.786	0.0689	1.751
16	40	2.500	12.21	0.0625	1.588	0.0613	1.557
17	36	2.250	10.99	0.0562	1.429	0.0551	1.400
18	32	2.000	9.765	0.0500	1.270	0.0490	1.245
19	28	1.750	8.544	0.0438	1.111	0.0429	1.090
20	24	1.500	7.324	0.0375	0.952	0.0368	0.934
21	22	1.375	6.713	0.0344	0.873	0.0337	0.856
22	20	1.250	6.103	0.0312	0.794	0.0306	0.778
23	18	1.125	5.493	0.0281	0.714	0.0276	0.700
24	16	1.000	4.882	0.0250	0.635	0.0245	0.623
25	14	0.8750	4.272	0.0219	0.556	0.0214	0.545
26	12	0.7500	3.662	0.0188	0.476	0.0184	0.467

Table 69 (*Continued*)

Gage Number	Weight per Square Foot		Weight per Square Meter	Approximate Thickness			
				Wrought Iron, 480 lb/ft^3		Steel and open-hearth Iron, 489.6 lb/ft^3	
	oz.	lb	kg	in.	mm	in.	mm
27	11	0.6875	3.357	0.0172	0.437	0.0169	0.428
28	10	0.6250	3.052	0.0156	0.397	0.0153	0.389
29	9	0.5625	2.746	0.0141	0.357	0.0138	0.350
30	8	0.5000	2.441	0.0125	0.318	0.0123	0.311
31	7	0.4375	2.136	0.0109	0.278	0.0107	0.272
32	$6\frac{1}{2}$	0.4062	1.983	0.0102	0.258	0.0100	0.253
33	6	0.3750	1.831	0.0094	0.238	0.0092	0.233
34	$5\frac{1}{2}$	0.3438	1.678	0.0086	0.218	0.0084	0.214
35	5	0.3125	1.526	0.0078	0.198	0.0077	0.195
36	$4\frac{1}{2}$	0.2812	1.373	0.0070	0.179	0.0069	0.175
37	$4\frac{1}{4}$	0.2656	1.297	0.0066	0.169	0.0065	0.165
38	4	0.2500	1.221	0.0062	0.159	0.0061	0.156
39	$3\frac{3}{4}$	0.2344	1.144	0.0059	0.149	0.0057	0.146
40	$3\frac{1}{2}$	0.2188	1.068	0.0055	0.139	0.0054	0.136
41	$3\frac{3}{8}$	0.2109	1.030	0.0053	0.134	0.0052	0.131
42	$3\frac{1}{4}$	0.2031	0.9917	0.0051	0.129	0.0050	0.126
43	$3\frac{1}{8}$	0.1953	0.9536	0.0049	0.124	0.0048	0.122
44	3	0.1875	0.9155	0.0047	0.119	0.0046	0.117

[a] For the Galvanized Sheet Gage, add 2.5 oz to the weight per square foot as given in the table. Gage numbers below 8 and above 34 are not used in the Galvanized Sheet Gage. [b] Gage numbers greater than 38 were not in the standard as set up by law but are in general use.

Table 70 American Wire Gage: Weights of Copper, Aluminum, and Brass Sheets and Plates

Gage Number	Thickness		Approximate Weight,[a] lb/ft^2		
	in.	mm	Copper	Aluminum	Commercial (High) Brass
0000	0.4600	11.68	21.27	6.49	20.27
000	0.4096	10.40	18.94	5.78	18.05
00	0.3648	9.266	16.87	5.14	16.07
0	0.3249	8.252	15.03	4.58	14.32
1	0.2893	7.348	13.38	4.08	12.75
2	0.2576	6.544	11.91	3.632	11.35
3	0.2294	5.827	10.61	3.234	10.11
4	0.2043	5.189	9.45	2.880	9.00
5	0.1819	4.621	8.41	2.565	8.01
6	0.1620	4.115	7.49	2.284	7.14
7	0.1443	3.665	6.67	2.034	6.36
8	0.1285	3.264	5.94	1.812	5.66
9	0.1144	2.906	5.29	1.613	5.04
10	0.1019	2.588	4.713	1.437	4.490
11	0.0907	2.305	4.195	1.279	3.996
12	0.0808	2.053	3.737	1.139	3.560
13	0.0720	1.828	3.330	1.015	3.172
14	0.0641	1.628	2.965	0.904	2.824
15	0.0571	1.450	2.641	0.805	2.516

(*Continues*)

Table 70 *(Continued)*

Gage Number	Thickness		Approximate Weight,[a] lb/ft^2		
	in.	mm	Copper	Aluminum	Commercial (High) Brass
16	0.0508	1.291	2.349	0.716	2.238
17	0.0453	1.150	2.095	0.639	1.996
18	0.0403	1.024	1.864	0.568	1.776
19	0.0359	0.9116	1.660	0.506	1.582
20	0.0320	0.8118	1.480	0.451	1.410
21	0.0285	0.7230	1.318	0.402	1.256
22	0.0253	0.6438	1.170	0.3567	1.115
23	0.0226	0.5733	1.045	0.3186	0.996
24	0.0201	0.5106	0.930	0.2834	0.886
25	0.0179	0.4547	0.828	0.2524	0.789
26	0.0159	0.4049	0.735	0.2242	0.701
27	0.0142	0.3606	0.657	0.2002	0.626
28	0.0126	0.3211	0.583	0.1776	0.555
29	0.0113	0.2859	0.523	0.1593	0.498
30	0.0100	0.2546	0.4625	0.1410	0.4406
31	0.00893	0.2268	0.4130	0.1259	0.3935
32	0.00795	0.2019	0.3677	0.1121	0.3503
33	0.00708	0.1798	0.3274	0.0998	0.3119
34	0.00630	0.1601	0.2914	0.0888	0.2776
35	0.00561	0.1426	0.2595	0.0791	0.2472
36	0.00500	0.1270	0.2312	0.0705	0.2203
37	0.00445	0.1131	0.2058	0.0627	0.1961
38	0.00397	0.1007	0.1836	0.0560	0.1749
39	0.00353	0.0897	0.1633	0.0498	0.1555
40	0.00314	0.0799	0.1452	0.0443	0.1383

[a] Assumed specific gravities or densities in grams per cubic centimeter; copper, 8.89; aluminum, 2.71; brass, 8.47.

Wire Gages The sizes of wires having a diameter less than $\frac{1}{2}$ in. are usually stated in terms of certain arbitrary scales called "gages." The size or gage number of a solid wire refers to the cross section of the wire perpendicular to its length; the size or gage number of a stranded wire refers to the total cross section of the constituent wires, irrespective of the pitch of the spiraling. Larger wires are usually described in terms of their area expressed in circular mils. A circular mil is the area of a circle 1 mil in diameter, and the area of any circle in circular mils is equal to the square of its diameter in mils.

Table 71 **Comparison of Wire Gage Diameters in Mils**[a]

Gage No.	American Wire Gage (Brown & Sharpe)	Steel Wire Gage	Birmingham Wire Gage (Stubs')	Old English Wire Gage (London)	Stubs' Steel Wire Gage	(British) Standard Wire Gage	Metric Gage[b]
7–0	—	490.0	—	—	—	500	—
6–0	—	461.5	—	—	—	464	—
5–0	—	430.5	—	—	—	432	—
4–0	460	393.8	454	454	—	400	—
3–0	410	362.5	425	425	—	372	—
2–0	365	331.0	380	380	—	348	—
0	325	306.5	340	340	—	324	—
1	289	283.0	300	300	227	300	3.94

Table 71 (*Continued*)

Gage No.	American Wire Gage (B. & S.)	Steel Wire Gage	Birmingham Wire Gage (Stubs')	Old English Wire Gage (London)	Stubs' Steel Wire Gage	(British) Standard Wire Gage	Metric Gage[b]
2	258	262.5	284	284	219	276	7.87
3	229	243.7	259	259	212	252	11.8
4	204	225.3	238	238	207	232	15.7
5	182	207.0	220	220	204	212	19.7
6	162	192.0	203	203	201	192	23.6
7	144	177.0	180	180	199	176	27.6
8	128	162.0	165	165	197	160	31.5
9	114	148.3	148	148	194	144	35.4
10	102	135.0	134	134	191	128	39.4
11	91	120.5	120	120	188	116	—
12	81	105.5	109	109	185	104	47.2
13	72	91.5	95	95	182	92	—
14	64	80.0	83	83	180	80	55.1
15	57	72.0	72	72	178	72	—
16	51	62.5	65	65	175	64	63.0
17	45	54.0	58	58	172	56	—
18	40	47.5	49	49	168	48	70.9
19	36	41.0	42	42	164	40	—
20	32	34.8	35	35	161	36	78.7
21	28.5	31.7	32	31.5	157	32	—
22	25.3	28.6	28	29.5	155	28	—
23	22.6	25.8	25	27.0	153	24	—
24	20.1	23.0	22	25.0	151	22	—
25	17.9	20.4	20	23.0	148	20	98.4
26	15.9	18.1	18	20.5	146	18	—
27	14.2	17.3	16	18.75	143	16.4	—
28	12.6	16.2	14	16.50	139	14.8	—
29	11.3	15.0	13	15.50	134	13.6	—
30	10.0	14.0	12	13.75	127	12.4	118
31	8.9	13.2	10	12.25	120	11.6	—
32	8.0	12.8	9	11.25	115	10.8	—
33	7.1	11.8	8	10.25	112	10.0	—
34	6.3	10.4	7	9.50	110	9.2	—
35	5.6	9.5	5	9.00	108	8.4	138
36	5.0	9.0	4	7.50	106	7.6	—
37	4.5	8.5	—	6.50	103	6.8	—
38	4.0	8.0	—	5.75	101	6.0	—
39	3.5	7.5	—	5.00	99	5.2	—
40	3.1	7.0	—	4.50	97	4.8	157
41	—	6.6	—	—	95	4.4	—
42	—	6.2	—	—	92	4.0	—
43	—	6.0	—	—	88	3.6	—
44	—	5.8	—	—	85	3.2	—
45	—	5.5	—	—	81	2.8	177
46	—	5.2	—	—	79	2.4	—
47	—	5.0	—	—	77	2.0	—
48	—	4.8	—	—	75	1.6	—
49	—	4.6	—	—	72	1.2	—
50	—	4.4	—	—	69	1.0	197

[a] Bureau of Standards, Circulars No. 31 and No. 67.
[b] For diameters corresponding to metric gage numbers, 1.2, 1.4, 1.6, 1.8, 2.5, 3.5, and 4.5, divide those of 12, 14, etc., by 10.

6.3 Paper Sizes

Table 72 Standard Engineering Drawing Sizes[a]

Flat Sizes[b]

Size Designation	Width[c] (Vertical)	Length (Horizontal)	Margin Horizontal	Margin Vertical
A (horizontal)	8.5	11.0	0.38	0.25
A (vertical)	11.0	8.5	0.25	0.38
B	11.0	17.0	0.38	0.62
C	17.0	22.0	0.75	0.50
D	22.0	34.0	0.50	1.00
E	34.0	44.0	1.00	0.50
F	28.0	40.0	0.50	0.50

Roll Sizes

Size Designation	Width[b] (Vertical)	Length[c] (Horizontal) Min	Length[c] (Horizontal) Max	Margin[c] Horizontal	Margin[c] Vertical
G	11.0	22.5	90.0	0.38	0.50
H	28.0	44.0	143.0	0.50	0.50
J	34.0	55.0	176.0	0.50	0.50
K	40.0	55.0	143.0	0.50	0.50

[a] See ANSI Y14.1-1980.
[b] All dimensions are in inches.
[c] Not including added protective margins.

International Paper Sizes Countries that are committed to the International System of Units (SI) have a standard series of paper sizes for printing, writing, and drafting. These paper sizes are called the "international paper sizes."

The advantages of the international paper sizes are as follows:

1. The ratio of width to length remains constant for every size, namely:

$$\frac{\text{Width}}{\text{Length}} = \frac{1}{\sqrt{2}} \quad \text{or} \quad \frac{1}{1.414} \text{ approximately}$$

Since this is the same ratio as the D aperture in the unitized 35-mm microfilm frame, the advantages are apparent.

2. If a sheet is cut in half, that is, if the $\sqrt{2}$ length is cut in half, the two halves retain the constant width-to-length ratio of $1/\sqrt{2}$. No other ratio could do this.

3. All international sizes are created from the A-0 size by single cuts without waste. In storing or stacking they fit together like parts of a jigsaw puzzle—without waste.

Table 73 Eleven International Paper Sizes

International Paper Size	Millimeters	Inches, Approximate
A-0	841 × 1189	$33\frac{1}{8} \times 46\frac{3}{4}$
A-1	594 × 841	$23\frac{3}{8} \times 33\frac{1}{8}$
A-2	420 × 594	$16\frac{1}{2} \times 23\frac{3}{8}$
A-3	297 × 420	$11\frac{3}{4} \times 16\frac{1}{2}$
A-4	210 × 297	$8\frac{1}{4} \times 11\frac{3}{4}$
A-5	148 × 210	$5\frac{7}{8} \times 8\frac{1}{4}$
A-6	105 × 148	$4\frac{1}{8} \times 5\frac{7}{8}$
A-7	74 × 105	$2\frac{7}{8} \times 4\frac{1}{8}$
A-8	52 × 74	$2 \times 2\frac{7}{8}$
A-9	37 × 52	$1\frac{1}{2} \times 2$
A-10	26 × 37	$1 \times 1\frac{1}{2}$

6.4 Sieve Sizes

Table 74 Tyler Standard Screen Scale Sieves

This screen scale has as its base an opening of 0.0029 in., which is the opening in 200-mesh 0.0021-in. wire, the standard sieve, as adopted by the Bureau of Standards of the U.S. government, the openings increasing in the ratio of the square root of 2 or 1.414.

Where a closer sizing is required, column 5 shows the Tyler Standard Screen Scale with intermediate sieves. In this series the sieve openings increase in the ratio of the fourth root of 2, or 1.189.

Tyler Standard Screen Scale $\sqrt{2}$ or 1.414 Openings (in.) (1)	Every Other Sieve from 0.0029 to 0.742 in., Ratio of 2 to 1 (2)	Every Other Sieve from 0.0041 to 1.050 in., Ratio of 2 to 1 (3)	Every Fourth Sieve from 0.0029 to 0.742 in., Ratio of 4 to 1 (4)	For Closer Sizing Sieves from 0.0029 to 1.050 in., Ratio $\sqrt[4]{2}$ or 1.189 (5)	openings (mm) (6)	Openings in Fractions of inch (approx.) (7)	Mesh (8)	Diameter of Wire (9)
1.050	—	1.050	—	1.050	26.67	1	—	0.148
—	—	—	—	0.883	22.43	$\frac{7}{8}$	—	0.135
0.742	0.742	—	0.742	0.742	18.85	$\frac{3}{4}$	—	0.135
—	—	—	—	0.624	15.85	$\frac{5}{8}$	—	0.120
0.525	—	0.525	—	0.525	13.33	$\frac{1}{2}$	—	0.105
—	—	—	—	0.441	11.20	$\frac{7}{16}$	—	0.105
0.371	0.371	—	—	0.371	9.423	$\frac{3}{8}$	—	0.092
—	—	—	—	0.312	7.925	$\frac{5}{16}$	$2\frac{1}{2}$	0.088
0.263	—	0.263	—	0.263	6.680	$\frac{1}{4}$	3	0.070
—	—	—	—	0.221	5.613	$\frac{7}{32}$	$3\frac{1}{2}$	0.065
0.185	0.185	—	0.185	0.185	4.699	$\frac{3}{16}$	4	0.065
—	—	—	—	0.156	3.962	$\frac{5}{32}$	5	0.044
0.131	—	0.131	—	0.131	3.327	$\frac{1}{8}$	6	0.036
—	—	—	—	0.110	2.794	$\frac{7}{64}$	7	0.0328
0.093	0.093	—	—	0.093	2.362	$\frac{3}{32}$	8	0.032
—	—	—	—	0.078	1.981	$\frac{5}{84}$	9	0.033
0.065	—	0.065	—	0.065	1.651	$\frac{1}{16}$	10	0.035
—	—	—	—	0.055	1.397	—	12	0.028
0.046	0.046	—	0.046	0.046	1.168	$\frac{3}{64}$	14	0.025
—	—	—	—	0.0390	0.991	—	16	0.0235
0.0328	—	0.0328	—	0.0328	0.833	$\frac{1}{32}$	20	0.0172
—	—	—	—	0.0276	0.701	—	24	0.0141
0.0232	0.0232	—	—	0.0232	0.589	—	28	0.0125
—	—	—	—	0.0195	0.495	—	32	0.0118
0.0164	—	0.0164	—	0.0164	0.417	$\frac{1}{64}$	35	0.0122
—	—	—	—	0.0138	0.351	—	42	0.0100
0.0116	0.0116	—	0.0116	0.0116	0.295	—	48	0.0092
—	—	—	—	0.0097	0.246	—	60	0.0070
0.0082	—	0.0082	—	0.0082	0.208	—	65	0.0072
—	—	—	—	0.0069	0.175	—	80	0.0056
0.0058	0.0058	—	—	0.0058	0.147	—	100	0.0042
—	—	—	—	0.0049	0.124	—	115	0.0038
0.0041	—	0.0041	—	0.0041	0.104	—	150	0.0026
—	—	—	—	0.0035	0.088	—	170	0.0024
0.0029	0.0029	—	0.0029	0.0029	0.074	—	200	0.0021

Table 75   Nominal Dimensions, Permissible Variations, and Limits for Woven Wire Cloth of Standard Sieves, U.S. Series, ASTM Standard[a]

Size or Sieve Designation		Sieve Opening		Permissible Variations in Average Opening (±%)	Permissible Variations in Maximum Opening (±%)	Wire Diameter	
μm	No.	mm	in. (approx. equivalents)			mm	in. (approx. equivalents)
5660	$3\frac{1}{2}$	5.66	0.233	3	10	1.28–1.90	0.050–0.075
4760	4	4.76	0.187	3	10	1.14–1.68	0.045–0.066
4000	5	4.00	0.157	3	10	1.00–1.47	0.039–0.058
3360	6	3.36	0.132	3	10	0.87–1.32	0.034–0.052
2830	7	2.83	0.111	3	10	0.80–1.20	0.031–0.047
2380	8	2.38	0.0937	3	10	0.74–1.10	0.0291–0.0433
2000	10	2.00	0.0787	3	10	0.68–1.00	0.0268–0.0394
1680	12	1.68	0.0661	3	10	0.62–0.90	0.0244–0.0354
1410	14	1.41	0.0555	3	10	0.56–0.80	0.0220–0.0315
1190	16	1.19	0.0469	3	10	0.50–0.70	0.0197–0.0276
1000	18	1.00	0.0394	5	15	0.43–0.62	0.0169–0.0244
840	20	0.84	0.0331	5	15	0.38–0.55	0.0150–0.0217
710	25	0.71	0.0280	5	15	0.33–0.48	0.0130–0.0189
590	30	0.59	0.0232	5	15	0.29–0.42	0.0114–0.0165
500	35	0.50	0.0197	5	15	0.26–0.37	0.0102–0.0146
420	40	0.42	0.0165	5	25	0.23–0.33	0.0091–0.0130
350	45	0.35	0.0138	5	25	0.20–0.29	0.0079–0.0114
297	50	0.297	0.0117	5	25	0.170–0.253	0.0067–0.0100
250	60	0.250	0.0098	5	25	0.149–0.220	0.0059–0.0087
210	70	0.210	0.0083	5	25	0.130–0.187	0.0051–0.0074
177	80	0.177	0.0070	6	40	0.114–0.154	0.0045–0.0061
149	100	0.149	0.0059	6	40	0.096–0.125	0.0038–0.0049
125	120	0.125	0.0049	6	40	0.079–0.103	0.0031–0.0041
105	140	0.105	0.0041	6	40	0.063–0.087	0.0025–0.0034
88	170	0.088	0.0035	6	40	0.054–0.073	0.0021–0.0029
74	200	0.074	0.0029	7	60	0.045–0.061	0.0018–0.0024
62	230	0.062	0.0024	7	90	0.039–0.052	0.0015–0.0020
53	270	0.053	0.0021	7	90	0.035–0.046	0.0014–0.0018
44	325	0.044	0.0017	7	90	0.031–0.040	0.0012–0.0016
37	400	0.037	0.0015	7	90	0.023–0.035	0.0009–0.0014

[a] For sieves from the 1000-μm (No. 18) to the 37-μm (No. 400) size, inclusive, not more than 5% of the openings shall exceed the nominal opening by more than one-half of the permissible variation in the maximum opening.

6.5   Standard Structural Sizes — Steel

Steel Sections.   Tables 76–83 give the dimensions, weights, and properties of *rolled steel* structural sections, including wide-flange sections, American standard beams, channels, angles, tees, and zees. The values for the various structural forms, taken from the eighth edition, 1980, of *Steel Construction*, by the kind permission of the publisher, the American Institute of Steel Construction, give the section specifications required in designing steel structures. The theory of design is covered in Section 4—Mechanics of Deformable Bodies.

Most of the sections can be supplied promptly steel mills. Owing to variations in the rolling practice of the different mills, their products are not identical, although their divergence from the values given in the tables is practically negligible. For standardization, only the lesser values are given, and therefore they are on the side of safety.

Further information on sections listed in the tables, together with information on other products and on the requirements for placing orders, may be gathered from mill catalogs.

Table 76 Properties of Wide-Flange Sections

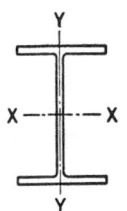

Nominal Size (in.)	Weight per Foot (lb)	Area (in.^{2})	Depth (in.)	Flange Width (in.)	Flange Thickness (in.)	Web Thickness (in.)	Axis X–X I (in.^{4})	Axis X–X S (in.^{3})	Axis X–X r (in.)	Axis Y–Y I (in.^{4})	Axis Y–Y S (in.^{3})	Axis Y–Y r (in.)
$36 \times 16\frac{1}{2}$	300	88.17	36.72	16.655	1.680	0.945	20290.2	1105.1	15.17	1225.2	147.1	3.73
	280	82.32	36.50	16.595	1.570	0.885	18819.3	1031.2	15.12	1127.5	135.9	3.70
	260	76.56	36.24	16.555	1.440	0.845	17233.8	951.1	15.00	1020.6	123.3	3.65
	245	72.03	36.06	16.512	1.350	0.802	16092.2	892.5	14.95	944.7	114.4	3.62
	230	67.73	35.88	16.475	1.260	0.765	14988.4	835.5	14.88	870.9	105.7	3.59
36×12	194	57.11	36.48	12.117	1.260	0.770	12103.4	663.6	14.56	355.4	58.7	2.49
	182	53.54	36.32	12.072	1.180	0.725	11281.5	621.2	14.52	327.7	54.3	2.47
	170	49.98	36.16	12.027	1.100	0.680	10470.0	579.1	14.47	300.6	50.0	2.45
	160	47.09	36.00	12.000	1.020	0.653	9738.8	541.0	14.38	275.4	45.9	2.42
	150	44.16	35.84	11.972	0.940	0.625	9012.1	502.9	14.29	250.4	41.8	2.38
$33 \times 15\frac{3}{4}$	240	70.52	33.50	15.865	1.400	0.830	13585.1	811.1	13.88	874.3	110.2	3.52
	220	64.73	33.25	15.810	1.275	0.775	12312.1	740.6	13.79	782.4	99.0	3.48
	200	58.79	33.00	15.750	1.150	0.715	11048.2	669.6	13.71	691.7	87.8	3.43
$33 \times 11\frac{1}{2}$	152	44.71	33.50	11.565	1.055	0.635	8147.6	486.4	13.50	256.1	44.3	2.39
	141	41.51	33.31	11.535	0.960	0.605	7442.2	446.8	13.39	229.7	39.8	2.35
	130	38.26	33.10	11.510	0.855	0.580	6699.0	404.8	13.23	201.4	35.0	2.29
30×15	210	61.78	30.38	15.105	1.315	0.775	9872.4	649.9	12.64	707.9	93.7	3.38
	190	55.90	30.12	15.040	1.185	0.710	8825.9	586.1	12.57	624.6	83.1	3.34
	172	50.65	29.88	14.985	1.065	0.655	7891.5	528.2	12.48	550.1	73.4	3.30
$30 \times 10\frac{1}{2}$	132	38.83	30.30	10.551	1.000	0.615	5753.1	379.7	12.17	185.0	35.1	2.18
	124	36.45	30.16	10.521	0.930	0.585	5347.1	354.6	12.11	169.7	32.3	2.16
	116	34.13	30.00	10.500	0.850	0.564	4919.1	327.9	12.00	153.2	29.2	2.12
	108	31.77	29.82	10.484	0.760	0.548	4461.0	299.2	11.85	135.1	25.8	2.06
27×14	177	52.10	27.31	14.090	1.190	0.725	6728.6	492.8	11.36	518.9	73.7	3.16
	160	47.04	27.08	14.023	1.075	0.658	6018.6	444.5	11.31	458.0	65.3	3.12
	145	42.68	26.88	13.965	0.975	0.600	5414.3	402.9	11.26	406.9	58.3	3.09
27×10	114	33.53	27.28	10.070	0.932	0.570	4080.5	299.2	11.03	149.6	29.7	2.11
	102	30.01	27.07	10.018	0.827	0.518	3604.1	266.3	10.96	129.5	25.9	2.08
	94	27.65	26.91	9.990	0.747	0.490	3266.7	242.8	10.87	115.1	23.0	2.04
24×14	160	47.04	24.72	14.091	1.135	0.656	5110.3	413.5	10.42	492.6	69.9	3.23
	145	42.62	24.49	14.043	1.020	0.608	4561.0	372.5	10.34	434.3	61.8	3.19
	130	38.21	24.25	14.000	0.900	0.565	4009.5	330.7	10.24	375.2	53.6	3.13
24×12	120	35.29	24.31	12.088	0.930	0.556	3635.3	299.1	10.15	254.0	42.0	2.68
	110	32.36	24.16	12.042	0.855	0.510	3315.0	274.4	10.12	229.1	38.0	2.66
	100	29.43	24.00	12.000	0.775	0.468	2987.3	248.9	10.08	203.5	33.9	2.63
24×9	94	27.63	24.29	9.061	0.872	0.516	2683.0	220.9	9.85	102.2	22.6	1.92
	84	24.71	24.09	9.015	0.772	0.470	2364.3	196.3	9.78	88.3	19.6	1.89
	76	22.37	23.91	8.985	0.682	0.440	2096.4	175.4	9.68	76.5	17.0	1.85
21×13	142	41.76	21.46	13.132	1.095	0.659	3403.1	317.2	9.03	385.9	58.8	3.04
	127	37.34	21.24	13.061	0.985	0.588	3017.2	284.1	8.99	338.6	51.8	3.01
	112	32.93	21.00	13.000	0.865	0.527	2620.6	249.6	8.92	289.7	44.6	2.96
21×9	96	28.21	21.14	9.038	0.935	0.575	2088.9	197.6	8.60	109.3	24.2	1.97
	82	24.10	20.86	8.962	0.795	0.499	1752.4	168.0	8.53	89.6	20.0	1.93

(Continues)

Table 76 (*Continued*)

Nominal Size (in.)	Weight per Foot (lb)	Area (in.^{2})	Depth (in.)	Flange Width (in.)	Flange Thickness (in.)	Web Thickness (in.)	Axis X–X I (in.^{4})	Axis X–X S (in.^{3})	Axis X–X r (in.)	Axis Y–Y I (in.^{4})	Axis Y–Y S (in.^{3})	Axis Y–Y r (in.)
$21 \times 8\frac{1}{4}$	73	21.46	21.24	8.295	0.740	0.455	100.3	150.7	8.64	66.2	16.0	1.76
	68	20.02	21.13	8.270	0.685	0.430	1478.3	139.9	8.59	60.4	14.6	1.74
	62	18.23	20.99	8.240	0.615	0.400	1326.8	126.4	8.53	53.1	12.9	1.71
$18 \times 11\frac{3}{4}$	114	33.51	18.48	11.833	0.991	0.595	2033.8	220.1	7.79	255.6	43.2	2.76
	105	30.86	18.32	11.792	0.911	0.554	1852.5	202.2	7.75	231.0	39.2	2.73
	96	28.22	18.16	11.750	0.831	0.512	1674.7	184.4	7.70	206.8	35.2	2.71
$18 \times 8\frac{3}{4}$	85	24.97	18.32	8.838	0.911	0.526	1429.9	156.1	7.57	99.4	22.5	2.00
	77	22.63	18.16	8.787	0.831	0.475	1286.8	141.7	7.54	88.6	20.2	1.98
	70	20.56	18.00	8.750	0.751	0.438	1153.9	128.2	7.49	78.5	17.9	1.95
	64	18.80	17.87	8.715	0.686	0.403	1045.8	117.0	7.46	70.3	16.1	1.93
$18 \times 7\frac{1}{2}$	60	17.64	18.25	7.558	0.695	0.416	984.0	107.8	7.47	47.1	12.5	1.63
	55	16.19	18.12	7.532	0.630	0.390	889.9	98.2	7.41	42.0	11.1	1.61
	50	14.71	18.00	7.500	0.570	0.358	800.6	89.0	7.38	37.2	9.9	1.59
$16 \times 11\frac{1}{2}$	96	28.22	16.32	11.533	0.875	0.535	1355.1	166.1	6.93	207.2	35.9	2.71
	88	25.87	16.16	11.502	0.795	0.504	1222.6	151.3	6.87	185.2	32.2	2.67
$16 \times 8\frac{1}{2}$	78	22.92	16.32	8.586	0.875	0.529	1042.6	127.8	6.74	87.5	20.4	1.95
	71	20.86	16.16	8.543	0.795	0.486	936.9	115.9	6.70	77.9	18.2	1.93
	64	18.80	16.00	8.500	0.715	0.443	833.8	104.2	6.66	68.4	16.1	1.91
	58	17.04	15.86	8.464	0.645	0.407	746.4	94.1	6.62	60.5	14.3	1.88
16×7	50	14.70	16.25	7.073	0.628	0.380	655.4	80.7	6.68	34.8	9.8	1.54
	45	13.24	16.12	7.039	0.563	0.346	583.3	72.4	6.64	30.5	8.7	1.52
	40	11.77	16.00	7.000	0.503	0.307	515.5	64.4	6.62	26.5	7.6	1.50
	36	10.59	15.85	6.992	0.428	0.299	446.3	56.3	6.49	22.1	6.3	1.45
14×16	426	125.25	18.69	16.695	3.033	1.875	6610.3	707.4	7.26	2359.5	282.7	4.34
	398	116.98	18.31	16.590	2.843	1.770	6013.7	656.9	7.17	2169.7	261.6	4.31
	370	108.78	17.94	16.475	2.658	1.655	5454.2	608.1	7.08	1986.0	241.1	4.27
	342	100.59	17.56	16.365	2.468	1.545	4911.5	559.4	6.99	1806.9	220.8	4.24
	314	92.30	17.19	16.235	2.283	1.415	4399.4	511.9	6.90	1631.4	201.0	4.20
	287	84.37	16.81	16.130	2.093	1.310	3912.1	465.5	6.81	1466.5	181.8	4.17
	264	77.63	16.50	16.025	1.938	1.205	3526.0	427.4	6.74	1331.2	166.1	4.14
	246	72.33	16.25	15.945	1.813	1.125	3228.9	397.4	6.68	1226.6	153.9	4.12
	237	69.69	16.12	15.910	1.748	1.090	3080.9	382.2	6.65	1174.8	147.7	4.11
	228	67.06	16.00	15.865	1.688	1.045	2942.4	367.8	6.62	1124.8	141.8	4.10
	219	64.36	15.87	15.825	1.623	1.005	2798.2	352.6	6.59	1073.2	135.6	4.08
	211	62.07	15.75	15.800	1.563	0.980	2671.4	339.2	6.56	1028.6	130.2	4.07
	202	59.39	15.63	15.750	1.503	0.930	2538.8	324.9	6.54	979.7	124.4	4.06
	193	56.73	15.50	15.710	1.438	0.890	2402.4	310.0	6.51	930.1	118.4	4.05
	184	54.07	15.38	15.660	1.378	0.840	2274.8	295.8	6.49	882.7	112.7	4.04
	176	51.73	15.25	15.640	1.313	0.820	2149.6	281.9	6.45	837.9	107.1	4.02
	167	49.09	15.12	15.600	1.248	0.780	2020.8	267.3	6.42	790.2	101.3	4.01
	158	46.47	15.00	15.550	1.188	0.730	1900.6	253.4	6.40	745.0	95.8	4.00
	150	44.08	14.88	15.515	1.128	0.695	1786.9	240.2	6.37	702.5	90.6	3.99
	142	41.85	14.75	15.500	1.063	0.680	1672.2	226.7	6.32	660.1	85.2	3.97
	320[a]	94.12	16.81	16.710	2.093	1.890	4141.7	492.8	6.63	1635.1	195.7	4.17
$14 \times 14\frac{1}{2}$	136	39.98	14.75	14.740	1.063	0.660	1593.0	216.0	6.31	567.7	77.0	3.77
	127	37.33	14.62	14.690	0.998	0.610	1476.7	202.0	6.29	527.6	71.8	3.76
	119	34.99	14.50	14.650	0.938	0.570	1373.1	189.4	6.26	491.8	67.1	3.75
	111	32.65	14.37	14.620	0.873	0.540	1266.5	176.3	6.23	454.9	62.2	3.73
	103	30.26	14.25	14.575	0.813	0.495	1165.8	163.6	6.21	419.7	57.6	3.72
	95	27.94	14.12	14.545	0.748	0.465	1063.5	150.6	6.17	383.7	52.8	3.71
	87	25.56	14.00	14.500	0.688	0.420	966.9	138.1	6.15	349.7	48.2	3.70
14×12	84	24.71	14.18	12.023	0.778	0.451	928.4	130.9	6.13	225.5	37.5	3.02
	78	22.94	14.06	12.000	0.718	0.428	851.2	121.1	6.09	206.9	34.5	3.00
14×10	74	21.76	14.19	10.072	0.783	0.450	796.8	112.3	6.05	133.5	26.5	2.48
	68	20.00	14.06	10.040	0.718	0.418	724.1	103.0	6.02	121.2	24.1	2.46
	61	17.94	13.91	10.000	0.643	0.378	641.5	92.2	5.98	107.3	21.5	2.45

Table 76 *(Continued)*

Nominal Size (in.)	Weight per Foot (lb)	Area (in.^{2})	Depth (in.)	Flange Width (in.)	Flange Thickness (in.)	Web Thickness (in.)	Axis X–X I (in.^{4})	Axis X–X S (in.^{3})	Axis X–X r (in.)	Axis Y–Y I (in.^{4})	Axis Y–Y S (in.^{3})	Axis Y–Y r (in.)
14 × 8	53	15.59	13.94	8.062	0.658	0.370	542.1	77.8	5.90	57.5	14.3	1.92
	48	14.11	13.81	8.031	0.593	0.339	484.9	70.2	5.86	51.3	12.8	1.91
	43	12.65	13.68	8.000	0.528	0.308	429.0	62.7	5.82	45.1	11.3	1.89
14 × 6¾	38	11.17	14.12	6.776	0.513	0.313	385.3	54.6	5.87	24.6	7.3	1.49
	34	10.00	14.00	6.750	0.453	0.287	339.2	48.5	5.83	21.3	6.3	1.46
	30	8.81	13.86	6.733	0.383	0.270	289.6	41.8	5.73	17.5	5.2	1.41
12 × 12	190	55.86	14.38	12.670	1.736	1.060	1892.5	263.2	5.82	589.7	93.1	3.25
	161	47.38	13.88	12.515	1.486	0.905	1541.8	222.2	5.70	486.2	77.7	3.20
	133	39.11	13.38	12.365	1.236	0.755	1221.2	182.5	5.59	389.9	63.1	3.16
	120	35.31	13.12	12.320	1.106	0.710	1071.7	163.4	5.51	345.1	56.0	3.13
	106	31.19	12.88	12.230	0.986	0.620	930.7	144.5	5.46	300.9	49.2	3.11
	99	29.09	12.75	12.190	0.921	0.580	858.5	134.7	5.43	278.2	45.7	3.09
	92	27.06	12.62	12.155	0.856	0.545	788.9	125.0	5.40	256.4	42.2	3.08
	85	24.98	12.50	12.105	0.796	0.495	723.3	115.7	5.38	235.5	38.9	3.07
	79	23.22	12.38	12.080	0.736	0.470	663.0	107.1	5.34	216.4	35.8	3.05
	72	21.16	12.25	12.040	0.671	0.430	597.4	97.5	5.31	195.3	32.4	3.04
	65	19.11	12.12	12.000	0.606	0.390	533.4	88.0	5.28	174.6	29.1	3.02
12 × 10	58	17.06	12.19	10.014	0.641	0.359	476.1	78.1	5.28	107.4	21.4	2.51
	53	15.59	12.06	10.000	0.576	0.345	426.2	70.7	5.23	96.1	19.2	2.48
12 × 8	50	14.71	12.19	8.077	0.641	0.371	394.5	64.7	5.18	56.4	14.0	1.96
	45	13.24	12.06	8.042	0.576	0.336	350.8	58.2	5.15	50.0	12.4	1.94
	40	11.77	11.94	8.000	0.516	0.294	310.1	51.9	5.13	44.1	11.0	1.94
12 × 6½	36	10.59	12.24	6.565	0.540	0.305	280.8	45.9	5.15	23.7	7.2	1.50
	31	9.12	12.09	6.525	0.465	0.265	238.4	39.4	5.11	19.8	6.1	1.47
	27	7.97	11.95	6.500	0.400	0.240	204.1	34.1	5.06	16.6	5.1	1.44
10 × 10	112	32.92	11.38	10.415	1.248	0.755	718.7	126.3	4.67	235.4	45.2	2.67
	100	29.43	11.12	10.345	1.118	0.685	625.0	112.4	4.61	206.6	39.9	2.65
	89	26.19	10.88	10.275	0.998	0.615	542.4	99.7	4.55	180.6	35.2	2.63
	77	22.67	10.62	10.195	0.868	0.535	457.2	86.1	4.49	153.4	30.1	2.60
	72	21.18	10.50	10.170	0.808	0.510	420.7	80.1	4.46	141.8	27.9	2.59
	66	19.41	10.38	10.117	0.748	0.457	382.5	73.7	4.44	129.2	25.5	2.58
	60	17.66	10.25	10.075	0.683	0.415	343.7	67.1	4.41	116.5	23.1	2.57
	54	15.88	10.12	10.028	0.618	0.368	305.7	60.4	4.39	103.9	20.7	2.56
	49	14.40	10.00	10.000	0.558	0.340	272.9	54.6	4.35	93.0	18.6	2.54
10 × 8	45	13.24	10.12	8.022	0.618	0.350	248.6	49.1	4.33	53.2	13.3	2.00
	39	11.48	9.94	7.990	0.528	0.318	209.7	42.2	4.27	44.9	11.2	1.98
	33	9.71	9.75	7.964	0.433	0.292	170.9	35.0	4.20	36.5	9.2	1.94
10 × 5¾	29	8.53	10.22	5.799	0.500	0.289	157.3	30.8	4.29	15.2	5.2	1.34
	25	7.35	10.08	5.762	0.430	0.252	133.2	26.4	4.26	12.7	4.4	1.31
	21	6.19	9.90	5.750	0.340	0.240	106.3	21.5	4.14	9.7	3.4	1.25
8 × 8	67	19.70	9.00	8.287	0.933	0.575	271.8	60.4	3.71	88.6	21.4	2.12
	58	17.06	8.75	8.222	0.808	0.510	227.3	52.0	3.65	74.9	18.2	2.10
	48	14.11	8.50	8.117	0.683	0.405	183.7	43.2	3.61	60.9	15.0	2.08
	40	11.76	8.25	8.077	0.558	0.365	146.3	35.5	3.53	49.0	12.1	2.04
	35	10.30	8.12	8.027	0.493	0.315	126.5	31.1	3.50	42.5	10.6	2.03
	31	9.12	8.00	8.000	0.433	0.288	109.7	27.4	3.47	37.0	9.2	2.01
8 × 6½	28	8.23	8.06	6.540	0.463	0.285	97.8	24.3	3.45	21.6	6.6	1.62
	24	7.06	7.93	6.500	0.398	0.245	82.5	20.8	3.42	18.2	5.6	1.61
8 × 5¼	20	5.88	8.14	5.268	0.378	0.248	69.2	17.0	3.43	8.5	3.2	1.20
	17	5.00	8.00	5.250	0.308	0.230	56.4	14.1	3.36	6.7	2.6	1.16

[a] Column core section.

Table 77   Properties of American Standard Beams

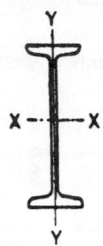

Nominal Size (in.)	Weight per Foot (lb)	Area (in.^{2})	Depth (in.)	Flange Width (in.)	Flange Thickness (in.)	Web Thickness (in.)	Axis X–X I (in.^{4})	Axis X–X S (in.^{3})	Axis X–X r (in.)	Axis Y–Y I (in.^{4})	Axis Y–Y S (in.^{3})	Axis Y–Y r (in.)
$24 \times 7\frac{7}{8}$	120.0	35.13	24.00	8.048	1.102	0.798	3010.8	250.9	9.26	84.9	21.1	1.56
	105.9	30.98	24.00	7.875	1.102	0.625	2811.5	234.3	9.53	78.9	20.0	1.60
24×7	100.0	29.25	24.00	7.247	0.871	0.747	2371.8	197.6	9.05	48.4	13.4	1.29
	90.0	26.30	24.00	7.124	0.871	0.624	2230.1	185.8	9.21	45.5	12.8	1.32
	79.9	23.33	24.00	7.000	0.871	0.500	2087.2	173.9	9.46	42.9	12.2	1.36
20×7	95.0	27.74	20.00	7.200	0.916	0.800	1599.7	160.0	7.59	50.5	14.0	1.35
	85.0	24.80	20.00	7.053	0.916	0.653	1501.7	150.2	7.78	47.0	13.3	1.38
$20 \times 6\frac{1}{4}$	75.0	21.90	20.00	6.391	0.789	0.641	1263.5	126.3	7.60	30.1	9.4	1.17
	65.4	19.08	20.00	6.250	0.789	0.500	1169.5	116.9	7.83	27.9	8.9	1.21
18×6	70.0	20.46	18.00	6.251	0.691	0.711	917.5	101.9	6.70	24.5	7.8	1.09
	54.7	15.94	18.00	6.000	0.691	0.460	795.5	88.4	7.07	21.2	7.1	1.15
$15 \times 5\frac{1}{2}$	50.0	14.59	15.00	5.640	0.622	0.550	481.1	64.2	5.74	16.0	5.7	1.05
	42.9	12.49	15.00	5.500	0.622	0.410	441.8	58.9	5.95	14.6	5.3	1.08
$12 \times 5\frac{1}{4}$	50.0	14.57	12.00	5.477	0.659	0.687	301.6	50.3	4.55	16.0	5.8	1.05
	40.8	11.84	12.00	5.250	0.659	0.460	268.9	44.8	4.77	13.8	5.3	1.08
12×5	35.0	10.20	12.00	5.078	0.544	0.428	227.0	37.8	4.72	10.0	3.9	0.99
	31.8	9.26	12.00	5.000	0.544	0.350	215.8	36.0	4.83	9.5	3.8	1.01
$10 \times 4\frac{5}{8}$	35.0	10.22	10.00	4.944	0.491	0.594	145.8	29.2	3.78	8.5	3.4	0.91
	25.4	7.38	10.00	4.660	0.491	0.310	122.1	24.4	4.07	6.9	3.0	0.97
8×4	23.0	6.71	8.00	4.171	0.425	0.441	64.2	16.0	3.09	4.4	2.1	0.81
	18.4	5.34	8.00	4.000	0.425	0.270	56.9	14.2	3.26	3.8	1.9	0.84
$7 \times 3\frac{5}{8}$	20.0	5.83	7.00	3.860	0.392	0.450	41.9	12.0	2.68	3.1	1.6	0.74
	15.3	4.43	7.00	3.660	0.392	0.250	36.2	10.4	2.86	2.7	1.5	0.78
$6 \times 3\frac{3}{8}$	17.25	5.02	6.00	3.565	0.359	0.465	26.0	8.7	2.28	2.3	1.3	0.68
	12.5	3.61	6.00	3.330	0.359	0.230	21.8	7.3	2.46	1.8	1.1	0.72
5×3	14.75	4.29	5.00	3.284	0.326	0.494	15.0	6.0	1.87	1.7	1.0	0.63
	10.0	2.87	5.00	3.000	0.326	0.210	12.1	4.8	2.05	1.2	0.82	0.65
$4 \times 2\frac{5}{8}$	9.5	2.76	4.00	2.796	0.293	0.326	6.7	3.3	1.56	0.91	0.65	0.58
	7.7	2.21	4.00	2.660	0.293	0.190	6.0	3.0	1.64	0.77	0.58	0.59
$3 \times 2\frac{3}{8}$	7.5	2.17	3.00	2.509	0.260	0.349	2.9	1.9	1.15	0.59	0.47	0.52
	5.7	1.64	3.00	2.330	0.260	0.170	2.5	1.7	1.23	0.46	0.40	0.53

Table 78 Properties of American Standard Channels

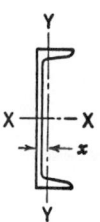

Nominal Size (in.)	Weight per Foot (lb)	Area (in.²)	Depth (in.)	Flange Width (in.)	Flange Average Thickness (in.)	Web Thickness (in.)	Axis X–X I (in.⁴)	Axis X–X S (in.³)	Axis X–X r (in.)	Axis Y–Y I (in.⁴)	Axis Y–Y S (in.³)	Axis Y–Y r (in.)	x (in.)
18 × 4[a]	58.0	16.98	18.00	4.200	0.625	0.700	670.7	74.5	6.29	18.5	5.6	1.04	0.88
	51.9	15.18	18.00	4.100	0.625	0.600	622.1	69.1	6.40	17.1	5.3	1.06	0.87
	45.8	13.38	18.00	4.000	0.625	0.500	573.5	63.7	6.55	15.8	5.1	1.09	0.89
	42.7	12.48	18.00	3.950	0.625	0.450	549.2	61.0	6.64	15.0	4.9	1.10	0.90
15 × 3⅜	50.0	14.64	15.00	3.716	0.650	0.716	401.4	53.6	5.24	11.2	3.8	0.87	0.80
	40.0	11.70	15.00	3.520	0.650	0.520	346.3	46.2	5.44	9.3	3.4	0.89	0.78
	33.9	9.90	15.00	3.400	0.650	0.400	312.6	41.7	5.62	8.2	3.2	0.91	0.79
12 × 3	30.0	8.79	12.00	3.170	0.501	0.510	161.2	26.9	4.28	5.2	2.1	0.77	0.68
	25.0	7.32	12.00	3.047	0.501	0.387	143.5	23.9	4.43	4.5	1.9	0.79	0.68
	20.7	6.03	12.00	2.940	0.501	0.280	128.1	21.4	4.61	3.9	1.7	0.81	0.70
10 × 2⅝	30.0	8.80	10.00	3.033	0.436	0.673	103.0	20.6	3.42	4.0	1.7	0.67	0.65
	25.0	7.33	10.00	2.886	0.436	0.526	90.7	18.1	3.52	3.4	1.5	0.68	0.62
	20.0	5.86	10.00	2.739	0.436	0.379	78.5	15.7	3.66	2.8	1.3	0.70	0.61
	15.3	4.47	10.00	2.600	0.436	0.240	66.9	13.4	3.87	2.3	1.2	0.72	0.64
9 × 2½	20.0	5.86	9.00	2.648	0.413	0.448	60.6	13.5	3.22	2.4	1.2	0.65	0.59
	15.0	4.39	9.00	2.485	0.413	0.285	50.7	11.3	3.40	1.9	1.0	0.67	0.59
	13.4	3.89	9.00	2.430	0.413	0.230	47.3	10.5	3.49	1.8	0.97	0.67	0.61
8 × 2¼	18.75	5.49	8.00	2.527	0.390	0.487	43.7	10.9	2.82	2.0	1.0	0.60	0.57
	13.75	4.02	8.00	2.343	0.390	0.303	35.8	9.0	2.99	1.5	0.86	0.62	0.56
	11.5	3.36	8.00	2.260	0.390	0.220	32.3	8.1	3.10	1.3	0.79	0.63	0.58
7 × 2⅛	14.75	4.32	7.00	2.299	0.366	0.419	27.1	7.7	2.51	1.4	0.79	0.57	0.53
	12.25	3.58	7.00	2.194	0.366	0.314	24.1	6.9	2.59	1.2	0.71	0.58	0.53
	9.8	2.85	7.00	2.090	0.366	0.210	21.1	6.0	2.72	0.98	0.63	0.59	0.55
6 × 2	13.0	3.81	6.00	2.157	0.343	0.437	17.3	5.8	2.13	1.1	0.65	0.53	0.52
	10.5	3.07	6.00	2.034	0.343	0.314	15.1	5.0	2.22	0.87	0.57	0.53	0.50
	8.2	2.39	6.00	1.920	0.343	0.200	13.0	4.3	2.34	0.70	0.50	0.54	0.52
5 × 1¾	9.0	2.63	5.00	1.885	0.320	0.325	8.8	3.5	1.83	0.64	0.45	0.49	0.48
	6.7	1.95	5.00	1.750	0.320	0.190	7.4	3.0	1.95	0.48	0.38	0.50	0.49
4 × 1⅝	7.25	2.12	4.00	1.720	0.296	0.320	4.5	2.3	1.47	0.44	0.35	0.46	0.46
	5.4	1.56	4.00	1.580	0.296	0.180	3.8	1.9	1.56	0.32	0.29	0.45	0.46
3 × 1½	6.0	1.75	3.00	1.596	0.273	0.356	2.1	1.4	1.08	0.31	0.27	0.42	0.46
	5.0	1.46	3.00	1.498	0.273	0.258	1.8	1.2	1.12	0.25	0.24	0.41	0.44
	4.1	1.19	3.00	1.410	0.273	0.170	1.6	1.1	1.17	0.20	0.21	0.41	0.44

[a] Car and Shipbuilding Channel; not an American standard.

Table 79   Properties of Angles with Equal Legs

 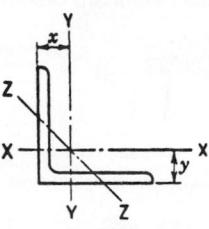

Size (in.)	Thickness (in.)	Weight per Foot (lb)	Area (in.^{2})	Axis X–X and Axis Y–Y				Axis Z–Z
				I (in.^{4})	S (in.^{3})	r (in.)	x or y (in.)	r (in.)
8×8	$1\frac{1}{8}$	56.9	16.73	98.0	17.5	2.42	2.41	1.56
	1	51.0	15.00	89.0	15.8	2.44	2.37	1.56
	$\frac{7}{8}$	45.0	13.23	79.6	14.0	2.45	2.32	1.57
	$\frac{3}{4}$	38.9	11.44	69.7	12.2	2.47	2.28	1.57
	$\frac{5}{8}$	32.7	9.61	59.4	10.3	2.49	2.23	1.58
	$\frac{9}{16}$	29.6	8.68	54.1	9.3	2.50	2.21	1.58
	$\frac{1}{2}$	26.4	7.75	48.6	8.4	2.50	2.19	1.59
6×6	1	37.4	11.00	35.5	8.6	1.80	1.86	1.17
	$\frac{7}{8}$	33.1	9.73	31.9	7.6	1.81	1.82	1.17
	$\frac{3}{4}$	28.7	8.44	28.2	6.7	1.83	1.78	1.17
	$\frac{5}{8}$	24.2	7.11	24.2	5.7	1.84	1.73	1.18
	$\frac{9}{16}$	21.9	6.43	22.1	5.1	1.85	1.71	1.18
	$\frac{1}{2}$	19.6	5.75	19.9	4.6	1.86	1.68	1.18
	$\frac{7}{16}$	17.2	5.06	17.7	4.1	1.87	1.66	1.19
	$\frac{3}{8}$	14.9	4.36	15.4	3.5	1.88	1.64	1.19
	$\frac{5}{16}$	12.5	3.66	13.0	3.0	1.89	1.61	1.19
5×5	$\frac{7}{8}$	27.2	7.98	17.8	5.2	1.49	1.57	0.97
	$\frac{3}{4}$	23.6	6.94	15.7	4.5	1.51	1.52	0.97
	$\frac{5}{8}$	20.0	5.86	13.6	3.9	1.52	1.48	0.98
	$\frac{1}{2}$	16.2	4.75	11.3	3.2	1.54	1.43	0.98
	$\frac{7}{16}$	14.3	4.18	10.0	2.8	1.55	1.41	0.98
	$\frac{3}{8}$	12.3	3.61	8.7	2.4	1.56	1.39	0.99
	$\frac{5}{16}$	10.3	3.03	7.4	2.0	1.57	1.37	0.99
4×4	$\frac{3}{4}$	18.5	5.44	7.7	2.8	1.19	1.27	0.78
	$\frac{5}{8}$	15.7	4.61	6.7	2.4	1.20	1.23	0.78
	$\frac{1}{2}$	12.8	3.75	5.6	2.0	1.22	1.18	0.78
	$\frac{7}{16}$	11.3	3.31	5.0	1.8	1.23	1.16	0.78
	$\frac{3}{8}$	9.8	2.86	4.4	1.5	1.23	1.14	0.79
	$\frac{5}{16}$	8.2	2.40	3.7	1.3	1.24	1.12	0.79
	$\frac{1}{4}$	6.6	1.94	3.0	1.1	1.25	1.09	0.80
$3\frac{1}{2} \times 3\frac{1}{2}$	$\frac{1}{2}$	11.1	3.25	3.6	1.5	1.06	1.06	0.68
	$\frac{7}{16}$	9.8	2.87	3.3	1.3	1.07	1.04	0.68
	$\frac{3}{8}$	8.5	2.48	2.9	1.2	1.07	1.01	0.69
	$\frac{5}{16}$	7.2	2.09	2.5	0.98	1.08	0.99	0.69
	$\frac{1}{4}$	5.8	1.69	2.0	0.79	1.09	0.97	0.69
3×3	$\frac{1}{2}$	9.4	2.75	2.2	1.1	0.90	0.93	0.58
	$\frac{7}{16}$	8.3	2.43	2.0	0.95	0.91	0.91	0.58
	$\frac{3}{8}$	7.2	2.11	1.8	0.83	0.91	0.89	0.58
	$\frac{5}{16}$	6.1	1.78	1.5	0.71	0.92	0.87	0.59

Table 79 *(Continued)*

Size (in.)	Thickness (in.)	Weight per Foot (lb)	Area (in.²)	Axis X–X and Axis Y–Y				Axis Z–Z
				I (in.⁴)	S (in.³)	r (in.)	x or y (in.)	r (in.)
$2\frac{1}{2} \times 2\frac{1}{2}$	$\frac{1}{4}$	4.9	1.44	1.2	0.58	0.93	0.84	0.59
	$\frac{3}{16}$	3.71	1.09	0.96	0.44	0.94	0.82	0.59
	$\frac{1}{2}$	7.7	2.25	1.2	0.72	0.74	0.81	0.49
	$\frac{3}{8}$	5.9	1.73	0.98	0.57	0.75	0.76	0.49
	$\frac{5}{16}$	5.0	1.47	0.85	0.48	0.76	0.74	0.49
	$\frac{1}{4}$	4.1	1.19	0.70	0.39	0.77	0.72	0.49
	$\frac{3}{16}$	3.07	0.90	0.55	0.30	0.78	0.69	0.49
2×2	$\frac{3}{8}$	4.7	1.36	0.48	0.35	0.59	0.64	0.39
	$\frac{5}{16}$	3.92	1.15	0.42	0.30	0.60	0.61	0.39
	$\frac{1}{4}$	3.19	0.94	0.35	0.25	0.61	0.59	0.39
	$\frac{3}{16}$	2.44	0.71	0.27	0.19	0.62	0.57	0.39
	$\frac{1}{8}$	1.65	0.48	0.19	0.13	0.63	0.55	0.40
$1\frac{3}{4} \times 1\frac{3}{4}$	$\frac{1}{4}$	2.77	0.81	0.23	0.19	0.53	0.53	0.34
	$\frac{3}{16}$	2.12	0.62	0.18	0.14	0.54	0.51	0.34
	$\frac{1}{8}$	1.44	0.42	0.13	0.10	0.55	0.48	0.35
$1\frac{1}{2} \times 1\frac{1}{2}$	$\frac{1}{4}$	2.34	0.69	0.14	0.13	0.45	0.47	0.29
	$\frac{3}{16}$	1.80	0.53	0.11	0.10	0.46	0.44	0.29
	$\frac{1}{8}$	1.23	0.36	0.08	0.07	0.47	0.42	0.30
$1\frac{1}{4} \times 1\frac{1}{4}$	$\frac{1}{4}$	1.92	0.56	0.08	0.09	0.37	0.40	0.24
	$\frac{3}{16}$	1.48	0.43	0.06	0.07	0.38	0.38	0.24
	$\frac{1}{8}$	1.01	0.30	0.04	0.05	0.38	0.36	0.25
1×1	$\frac{1}{4}$	1.49	0.44	0.04	0.06	0.29	0.34	0.20
	$\frac{3}{16}$	1.16	0.34	0.03	0.04	0.30	0.32	0.19
	$\frac{1}{8}$	0.80	0.23	0.02	0.03	0.30	0.30	0.20

Table 80 Properties of Angles with Unequal Legs

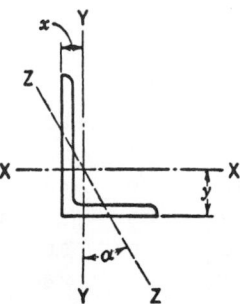

Size (in.)	Thickness (in.)	Weight per Foot (lb)	Area (in.²)	Axis X–X				Axis Y–Y				Axis Z–Z	
				I (in.⁴)	S (in.³)	r (in.)	y (in.)	I (in.⁴)	S (in.³)	r (in.)	x (in.)	r (in.)	$\tan\alpha$
9×4	1	40.8	12.00	97.0	17.6	2.84	3.50	12.0	4.0	1.00	1.00	0.83	0.203
	$\frac{7}{8}$	36.1	10.61	86.8	15.7	2.86	3.45	10.8	3.6	1.01	0.95	0.84	0.208
	$\frac{3}{4}$	31.3	9.19	76.1	13.6	2.88	3.41	9.6	3.1	1.02	0.91	0.84	0.212
	$\frac{5}{8}$	26.3	7.73	64.9	11.5	2.90	3.36	8.3	2.6	1.04	0.86	0.85	0.216
	$\frac{9}{16}$	23.8	7.00	59.1	10.4	2.91	3.33	7.6	2.4	1.04	0.83	0.85	0.218
	$\frac{1}{2}$	21.3	6.25	53.2	9.3	2.92	3.31	6.9	2.2	1.05	0.81	0.85	0.220

(Continues)

Table 80   (*Continued*)

Size (in.)	Thickness (in.)	Weight per Foot (lb)	Area (in.^{2})	Axis X–X				Axis Y–Y				Axis Z–Z	
				I (in.^{4})	S (in.^{3})	r (in.)	y (in.)	I (in.^{4})	S (in.^{3})	r (in.)	x (in.)	r (in.)	$\tan\alpha$
8 × 6	1	44.2	13.00	80.8	15.1	2.49	2.65	38.8	8.9	1.73	1.65	1.28	0.543
	$\frac{7}{8}$	39.1	11.48	72.3	13.4	2.51	2.61	34.9	7.9	1.74	1.61	1.28	0.547
	$\frac{3}{4}$	33.8	9.94	63.4	11.7	2.53	2.56	30.7	6.9	1.76	1.56	1.29	0.551
	$\frac{5}{8}$	28.5	8.36	54.1	9.9	2.54	2.52	26.3	5.9	1.77	1.52	1.29	0.554
	$\frac{9}{16}$	25.7	7.56	49.3	9.0	2.55	2.50	24.0	5.3	1.78	1.50	1.30	0.556
	$\frac{1}{2}$	23.0	6.75	44.3	8.0	2.56	2.47	21.7	4.8	1.79	1.47	1.30	0.558
	$\frac{7}{16}$	20.2	5.93	39.2	7.1	2.57	2.45	19.3	4.2	1.80	1.45	1.31	0.560
8 × 4	1	37.4	11.00	69.6	14.1	2.52	3.05	11.6	3.9	1.03	1.05	0.85	0.247
	$\frac{7}{8}$	33.1	9.73	62.5	12.5	2.53	3.00	10.5	3.5	1.04	1.00	0.85	0.253
	$\frac{3}{4}$	28.7	8.44	54.9	10.9	2.55	2.95	9.4	3.1	1.05	0.95	0.85	0.258
	$\frac{5}{8}$	24.2	7.11	46.9	9.2	2.57	2.91	8.1	2.6	1.07	0.91	0.86	0.262
	$\frac{9}{16}$	21.9	6.43	42.8	8.4	2.58	2.88	7.4	2.4	1.07	0.88	0.86	0.265
	$\frac{1}{2}$	19.6	5.75	38.5	7.5	2.59	2.86	6.7	2.2	1.08	0.86	0.86	0.267
	$\frac{7}{16}$	17.2	5.06	34.1	6.6	2.60	2.83	6.0	1.9	1.09	0.83	0.87	0.269
7 × 4	$\frac{7}{8}$	30.2	8.86	42.9	9.7	2.20	2.55	10.2	3.5	1.07	1.05	0.86	0.318
	$\frac{3}{4}$	26.2	7.69	37.8	8.4	2.22	2.51	9.1	3.0	1.09	1.01	0.86	0.324
	$\frac{5}{8}$	22.1	6.48	32.4	7.1	2.24	2.46	7.8	2.6	1.10	0.96	0.86	0.329
	$\frac{9}{16}$	20.0	5.87	29.6	6.5	2.24	2.44	7.2	2.4	1.11	0.94	0.87	0.332
	$\frac{1}{2}$	17.9	5.25	26.7	5.8	2.25	2.42	6.5	2.1	1.11	0.92	0.87	0.335
	$\frac{7}{16}$	15.8	4.62	23.7	5.1	2.26	2.39	5.8	1.9	1.12	0.89	0.88	0.337
	$\frac{3}{8}$	13.6	3.98	20.6	4.4	2.27	2.37	5.1	1.6	1.13	0.87	0.88	0.339
6 × 4	$\frac{7}{8}$	27.2	7.98	27.7	7.2	1.86	2.12	9.8	3.4	1.11	1.12	0.86	0.421
	$\frac{3}{4}$	23.6	6.94	24.5	6.3	1.88	2.08	8.7	3.0	1.12	1.08	0.86	0.428
	$\frac{5}{8}$	20.0	5.86	21.1	5.3	1.90	2.03	7.5	2.5	1.13	1.03	0.86	0.435
	$\frac{9}{16}$	18.1	5.31	19.3	4.8	1.90	2.01	6.9	2.3	1.14	1.01	0.87	0.438
	$\frac{1}{2}$	16.2	4.75	17.4	4.3	1.91	1.99	6.3	2.1	1.15	0.99	0.87	0.440
	$\frac{7}{16}$	14.3	4.18	15.5	3.8	1.92	1.96	5.6	1.9	1.16	0.96	0.87	0.443
	$\frac{3}{8}$	12.3	3.61	13.5	3.3	1.93	1.94	4.9	1.6	1.17	0.94	0.88	0.446
	$\frac{5}{16}$	10.3	3.03	11.4	2.8	1.94	1.92	4.2	1.4	1.17	0.92	0.88	0.449
6 × 3$\frac{1}{2}$	$\frac{1}{2}$	15.3	4.50	16.6	4.2	1.92	2.08	4.3	1.6	0.97	0.83	0.76	0.344
	$\frac{3}{8}$	11.7	3.42	12.9	3.2	1.94	2.04	3.3	1.2	0.99	0.79	0.77	0.350
	$\frac{5}{16}$	9.8	2.87	10.9	2.7	1.95	2.01	2.9	1.0	1.00	0.76	0.77	0.352
	$\frac{1}{4}$	7.9	2.31	8.9	2.2	1.96	1.99	2.3	0.85	1.01	0.74	0.78	0.355
5 × 3$\frac{1}{2}$	$\frac{3}{4}$	19.8	5.81	13.9	4.3	1.55	1.75	5.6	2.2	0.98	1.00	0.75	0.464
	$\frac{5}{8}$	16.8	4.92	12.0	3.7	1.56	1.70	4.8	1.9	0.99	0.95	0.75	0.472
	$\frac{1}{2}$	13.6	4.00	10.0	3.0	1.58	1.66	4.1	1.6	1.01	0.91	0.75	0.479
	$\frac{7}{16}$	12.0	3.53	8.9	2.6	1.59	1.63	3.6	1.4	1.01	0.88	0.76	0.482
	$\frac{3}{8}$	10.4	3.05	7.8	2.3	1.60	1.61	3.2	1.2	1.02	0.86	0.76	0.486
	$\frac{5}{16}$	8.7	2.56	6.6	1.9	1.61	1.59	2.7	1.0	1.03	0.84	0.76	0.489
	$\frac{1}{4}$	7.0	2.06	5.4	1.6	1.61	1.56	2.2	0.83	1.04	0.81	0.76	0.492
5 × 3	$\frac{1}{2}$	12.8	3.75	9.5	2.9	1.59	1.75	2.6	1.1	0.83	0.75	0.65	0.357
	$\frac{7}{16}$	11.3	3.31	8.4	2.6	1.60	1.73	2.3	1.0	0.84	0.73	0.65	0.361
	$\frac{3}{8}$	9.8	2.86	7.4	2.2	1.61	1.70	2.0	0.89	0.84	0.70	0.65	0.364
	$\frac{5}{16}$	8.2	2.40	6.3	1.9	1.61	1.68	1.8	0.75	0.85	0.68	0.66	0.368
	$\frac{1}{4}$	6.6	1.94	5.1	1.5	1.62	1.66	1.4	0.61	0.86	0.66	0.66	0.371
4 × 3$\frac{1}{2}$	$\frac{5}{8}$	14.7	4.30	6.4	2.4	1.22	1.29	4.5	1.8	1.03	1.04	0.72	0.745
	$\frac{1}{2}$	11.9	3.50	5.3	1.9	1.23	1.25	3.8	1.5	1.04	1.00	0.72	0.750

Table 80 *(Continued)*

Size (in.)	Thickness (in.)	Weight per Foot (lb)	Area (in.²)	Axis X–X I (in.⁴)	S (in.³)	r (in.)	y (in.)	Axis Y–Y I (in.⁴)	S (in.³)	r (in.)	x (in.)	Axis Z–Z r (in.)	tan α
	7/16	10.6	3.09	4.8	1.7	1.24	1.23	3.4	1.4	1.05	0.98	0.72	0.753
	3/8	9.1	2.67	4.2	1.5	1.25	1.21	3.0	1.2	1.06	0.96	0.73	0.755
	5/16	7.7	2.25	3.6	1.3	1.26	1.18	2.6	1.0	1.07	0.93	0.73	0.757
	1/4	6.2	1.81	2.9	1.0	1.27	1.16	2.1	0.81	1.07	0.91	0.73	0.759
4 × 3	5/8	13.6	3.98	6.0	2.3	1.23	1.37	2.9	1.4	0.85	0.87	0.64	0.534
	1/2	11.1	3.25	5.1	1.9	1.25	1.33	2.4	1.1	0.86	0.83	0.64	0.543
	7/16	9.8	2.87	4.5	1.7	1.25	1.30	2.2	1.0	0.87	0.80	0.64	0.547
	3/8	8.5	2.48	4.0	1.5	1.26	1.28	1.9	0.87	0.88	0.78	0.64	0.551
	5/16	7.2	2.09	3.4	1.2	1.27	1.26	1.7	0.73	0.89	0.76	0.65	0.554
	1/4	5.8	1.69	2.8	1.0	1.28	1.24	1.4	0.60	0.90	0.74	0.65	0.558
3½ × 3	1/2	10.2	3.00	3.5	1.5	1.07	1.13	2.3	1.1	0.88	0.88	0.62	0.714
	7/16	9.1	2.65	3.1	1.3	1.08	1.10	2.1	0.98	0.89	0.85	0.62	0.718
	3/8	7.9	2.30	2.7	1.1	1.09	1.08	1.9	0.85	0.90	0.83	0.62	0.721
	5/16	6.6	1.93	2.3	0.95	1.10	1.06	1.6	0.72	0.90	0.81	0.63	0.724
	1/4	5.4	1.56	1.9	0.78	1.11	1.04	1.3	0.59	0.91	0.79	0.63	0.727
3½ × 2½	1/2	9.4	2.75	3.2	1.4	1.09	1.20	1.4	0.76	0.70	0.70	0.53	0.486
	7/16	8.3	2.43	2.9	1.3	1.09	1.18	1.2	0.68	0.71	0.68	0.54	0.491
	3/8	7.2	2.11	2.6	1.1	1.10	1.16	1.1	0.59	0.72	0.66	0.54	0.496
	5/16	6.1	1.78	2.2	0.93	1.11	1.14	0.94	0.50	0.73	0.64	0.54	0.501
	1/4	4.9	1.44	1.8	0.75	1.12	1.11	0.78	0.41	0.74	0.61	0.54	0.506
3 × 2½	1/2	8.5	2.50	2.1	1.0	0.91	1.00	1.3	0.74	0.72	0.75	0.52	0.667
	7/16	7.6	2.21	1.9	0.93	0.92	0.98	1.2	0.66	0.73	0.73	0.52	0.672
	3/8	6.6	1.92	1.7	0.81	0.93	0.96	1.0	0.58	0.74	0.71	0.52	0.676
	5/16	5.6	1.62	1.4	0.69	0.94	0.93	0.90	0.49	0.74	0.68	0.53	0.680
	1/4	4.5	1.31	1.2	0.56	0.95	0.91	0.74	0.40	0.75	0.66	0.53	0.684
3 × 2	1/2	7.7	2.25	1.9	1.0	0.92	1.08	0.67	0.47	0.55	0.58	0.43	0.414
	7/16	6.8	2.00	1.7	0.89	0.93	1.06	0.61	0.42	0.55	0.56	0.43	0.421
	3/8	5.9	1.73	1.5	0.78	0.94	1.04	0.54	0.37	0.56	0.54	0.43	0.428
	5/16	5.0	1.47	1.3	0.66	0.95	1.02	0.47	0.32	0.57	0.52	0.43	0.435
	1/4	4.1	1.19	1.1	0.54	0.95	0.99	0.39	0.26	0.57	0.49	0.43	0.440
	3/16	3.07	0.90	0.84	0.41	0.97	0.97	0.31	0.20	0.58	0.47	0.44	0.446
2½ × 2	3/8	5.3	1.55	0.91	0.55	0.77	0.83	0.51	0.36	0.58	0.58	0.42	0.614
	5/16	4.5	1.31	0.79	0.47	0.78	0.81	0.45	0.31	0.58	0.56	0.42	0.620
	1/4	3.62	1.06	0.65	0.38	0.78	0.79	0.37	0.25	0.59	0.54	0.42	0.626
	3/16	2.75	0.81	0.51	0.29	0.79	0.76	0.29	0.20	0.60	0.51	0.43	0.631
2½ × 1½	3/8	4.7	1.36	0.82	0.52	0.78	0.92	0.22	0.20	0.40	0.42	0.32	0.340
	5/16	3.92	1.15	0.71	0.44	0.79	0.90	0.19	0.17	0.41	0.40	0.32	0.349
	1/4	3.19	0.94	0.59	0.36	0.79	0.88	0.16	0.14	0.41	0.38	0.32	0.357
	3/16	2.44	0.72	0.46	0.28	0.80	0.85	0.13	0.11	0.42	0.35	0.33	0.364
2 × 1½	1/4	2.77	0.81	0.32	0.24	0.62	0.66	0.15	0.14	0.43	0.41	0.32	0.543
	3/16	2.12	0.62	0.25	0.18	0.63	0.64	0.12	0.11	0.44	0.39	0.32	0.551
	1/8	1.44	0.42	0.17	0.13	0.64	0.62	0.09	0.08	0.45	0.37	0.33	0.558
1¾ × 1¼	1/4	2.34	0.69	0.20	0.18	0.54	0.60	0.09	0.10	0.35	0.35	0.27	0.486
	3/16	1.80	0.53	0.16	0.14	0.55	0.58	0.07	0.08	0.36	0.33	0.27	0.496
	1/8	1.23	0.36	0.11	0.09	0.56	0.56	0.05	0.05	0.37	0.31	0.27	0.506

Table 81 Properties and Dimensions of Tees

 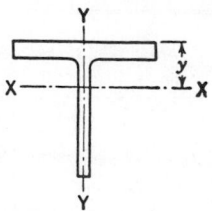

Tees are seldom used as structural framing members. When so used they are generally employed on short spans in flexure. This table lists a few selected sizes, the range of whose section moduli will cover all ordinary conditions. For sizes not listed, the catalogs of the respective rolling mills should be consulted.

Section Number	Weight per Foot (lb)	Area (in.²)	Depth of Tee (in.)	Flange Width (in.)	Flange Average Thickness (in.)	Stem Thickness (in.)	Axis X–X I (in.⁴)	Axis X–X S (in.³)	Axis X–X r (in.)	Axis X–X y (in.)	Axis Y–Y I (in.⁴)	Axis Y–Y S (in.³)	Axis Y–Y r (in.)
ST 18 WF[a]	150	44.09	18.36	16.655	1.680	0.945	1222.7	85.9	5.27	4.13	612.6	73.6	3.73
	140	41.16	18.25	16.595	1.570	0.885	1133.3	79.9	5.25	4.07	563.7	67.9	3.70
	130	38.28	18.12	16.555	1.440	0.845	1059.2	75.4	5.26	4.07	510.3	61.6	3.65
	122.5	36.01	18.03	16.512	1.350	0.802	994.3	71.1	5.25	4.04	472.3	57.2	3.62
	115	33.86	17.94	16.475	1.260	0.765	935.8	67.2	5.26	4.02	435.5	52.9	3.59
ST 18 WF	97	28.56	18.24	12.117	1.260	0.770	904.0	67.3	5.63	4.81	177.7	29.3	2.49
	91	26.77	18.16	12.072	1.180	0.725	844.0	63.0	5.61	4.77	163.9	27.1	2.47
	85	24.99	18.08	12.027	1.100	0.680	784.7	58.8	5.60	4.74	150.3	25.0	2.45
	80	23.54	18.00	12.000	1.020	0.653	741.0	56.0	5.61	4.76	137.7	22.9	2.42
	75	22.08	17.92	11.972	0.940	0.625	696.7	53.0	5.62	4.79	125.2	20.9	2.38
ST 16 WF	120	35.26	16.75	15.865	1.400	0.830	822.5	63.2	4.83	3.73	437.2	55.1	3.52
	110	32.36	16.63	15.810	1.275	0.775	754.1	58.4	4.83	3.71	391.2	49.5	3.48
	100	29.40	16.50	15.750	1.150	0.715	683.6	53.3	4.82	3.67	345.8	43.9	3.43
ST 16 WF	76	22.35	16.75	11.565	1.055	0.635	591.9	47.4	5.15	4.26	128.1	22.1	2.39
	70.5	20.76	16.66	11.535	0.960	0.603	551.8	44.7	5.16	4.30	114.9	19.9	2.35
	65	19.13	16.55	11.510	0.855	0.580	513.0	42.1	5.18	4.37	100.7	17.5	2.29
ST 15 WF	105	30.89	15.19	15.105	1.315	0.775	578.0	48.7	4.33	3.31	354.0	46.9	3.38
	95	27.95	15.06	15.040	1.185	0.710	520.4	44.1	4.31	3.26	312.3	41.5	3.34
	86	25.32	14.94	14.985	1.065	0.655	471.0	40.2	4.31	3.23	275.1	36.7	3.30
ST 15 WF	66	19.41	15.15	10.551	1.000	0.615	420.7	37.4	4.66	3.90	92.5	17.5	2.18
	62	18.22	15.08	10.521	0.930	0.585	394.8	35.3	4.65	3.90	84.8	16.1	2.16
	58.0	17.07	15.00	10.500	0.850	0.564	371.8	33.6	4.67	3.94	76.6	14.6	2.12
	54.0	15.88	14.91	10.484	0.760	0.548	349.5	32.1	4.69	4.03	67.6	12.9	2.06
ST 13 WF	88.5	26.05	13.66	14.090	1.190	0.725	391.8	36.7	3.88	2.97	259.4	36.8	3.16
	80	23.72	13.54	14.023	1.075	0.658	351.4	33.1	3.87	2.91	229.0	32.7	3.12
	72.5	21.34	13.44	13.965	0.975	0.600	316.3	29.9	3.85	2.85	203.5	29.1	3.09
ST 13 WF	57	16.77	13.64	10.070	0.932	0.570	288.9	28.3	4.15	3.42	74.8	14.9	2.11
	51	15.01	13.53	10.018	0.827	0.518	257.7	25.4	4.14	3.39	64.8	12.9	2.08
	47	13.83	13.45	9.990	0.747	0.490	238.5	23.7	4.15	3.41	57.5	11.5	2.04
ST 12 WF	80	23.54	12.36	14.091	1.135	0.656	271.6	27.6	3.40	2.51	246.3	35.0	3.23
	72.5	21.31	12.24	14.043	1.020	0.608	246.2	25.2	3.40	2.48	217.1	30.9	3.19
	65	19.11	12.13	14.000	0.900	0.565	222.6	23.1	3.41	2.47	187.6	26.8	3.13
ST 12 WF	60	17.64	12.16	12.088	0.930	0.556	213.6	22.4	3.48	2.62	127.0	21.0	2.68
	55	16.18	12.08	12.042	0.855	0.510	195.2	20.5	3.47	2.57	114.5	19.0	2.66
	50	14.71	12.00	12.000	0.775	0.468	176.7	18.7	3.46	2.54	101.8	17.0	2.63
ST 12 WF	47	13.81	12.15	9.061	0.872	0.516	185.9	20.3	3.67	2.99	51.1	11.3	1.92
	42	12.35	12.04	9.015	0.772	0.470	165.9	18.3	3.66	2.97	44.2	9.8	1.89
	38	11.18	11.95	8.985	0.682	0.440	151.1	16.9	3.68	3.00	38.3	8.5	1.85
ST 10 WF	71	20.88	10.73	13.132	1.095	0.659	177.3	20.8	2.91	2.18	193.0	29.4	3.04
	63.5	18.67	10.62	13.061	0.985	0.588	155.8	18.3	2.89	2.11	169.3	25.9	3.01
	56	16.47	10.50	13.000	0.865	0.527	136.4	16.2	2.88	2.06	144.8	22.3	2.96
ST 10 WF[a]	48	14.11	10.57	9.038	0.935	0.575	137.1	17.1	3.11	2.55	54.7	12.1	1.97
	41	12.05	10.43	8.962	0.795	0.499	115.4	14.5	3.09	2.48	44.8	10.0	1.93

Table 81 (*Continued*)

Section Number	Weight per Foot (lb)	Area (in.²)	Depth of Tee (in.)	Flange Width (in.)	Flange Average Thickness (in.)	Stem Thickness (in.)	Axis X–X I (in.⁴)	Axis X–X S (in.³)	Axis X–X r (in.)	Axis X–X y (in.)	Axis Y–Y I (in.⁴)	Axis Y–Y S (in.³)	Axis Y–Y r (in.)
ST 10 WF	36.5	10.73	10.62	8.295	0.740	0.455	110.2	13.7	3.21	2.60	33.1	7.98	1.76
	34	10.01	10.57	8.270	0.685	0.430	102.8	12.9	3.20	2.59	30.2	7.30	1.74
	31	9.12	10.49	8.240	0.615	0.400	93.7	11.9	3.21	2.59	26.6	6.45	1.71
ST 9 WF	57	16.77	9.24	11.833	0.991	0.595	102.6	13.9	2.47	1.85	127.8	21.6	2.76
	52.5	15.43	9.16	11.792	0.911	0.554	93.9	12.8	2.47	1.82	115.5	19.6	2.73
	48	14.11	9.08	11.750	0.831	0.512	85.3	11.7	2.46	1.78	103.4	17.6	2.71
ST 9 WF	42.5	12.49	9.16	8.838	0.911	0.526	84.4	11.9	2.60	2.05	49.7	11.3	2.00
	38.5	11.32	9.08	8.787	0.831	0.475	75.3	10.6	2.58	1.99	44.3	10.1	1.98
	35	10.28	9.00	8.750	0.751	0.438	68.1	9.67	2.57	1.96	39.2	8.97	1.95
	32	9.40	8.94	8.715	0.686	0.403	61.8	8.82	2.56	1.93	35.2	8.07	1.93
ST 9 WF	30	8.82	9.12	7.558	0.695	0.416	64.8	9.32	2.71	2.17	23.5	6.23	1.63
	27.5	8.09	9.06	7.532	0.630	0.390	59.6	8.63	2.71	2.16	21.0	5.57	1.61
	25	7.35	9.00	7.500	0.570	0.358	53.9	7.85	2.71	2.14	18.6	4.96	1.59
ST 8 WF	48	14.11	8.16	11.533	0.875	0.535	64.7	9.82	2.14	1.57	103.6	18.0	2.71
	44	12.94	8.08	11.502	0.795	0.504	59.5	9.11	2.14	1.55	92.6	16.1	2.67
ST 8 WF	39	11.46	8.16	8.586	0.875	0.529	60.0	9.45	2.28	1.81	43.8	10.2	1.95
	35.5	10.43	8.08	8.543	0.795	0.486	54.0	8.57	2.28	1.77	38.9	9.11	1.93
	32	9.40	8.00	8.500	0.715	0.443	48.3	7.71	2.27	1.73	34.2	8.05	1.91
	29	8.52	7.93	8.464	0.645	0.407	43.6	7.00	2.26	1.70	30.2	7.14	1.88
ST 8 WF	25	7.35	8.13	7.073	0.628	0.380	42.2	6.77	2.40	1.89	17.4	4.92	1.54
	22.5	6.62	8.06	7.039	0.563	0.346	37.8	6.10	2.39	1.87	15.2	4.33	1.52
	20	5.88	8.00	7.000	0.503	0.307	33.2	5.37	2.37	1.82	13.3	3.79	1.50
	18	5.30	7.93	6.992	0.428	0.299	30.7	5.10	2.41	1.90	11.1	3.17	1.45
ST 7 WF	105.5	31.04	7.88	15.800	1.563	0.980	102.2	16.2	1.81	1.57	514.3	65.1	4.07
	101	29.70	7.82	15.750	1.503	0.930	95.7	15.2	1.80	1.53	489.8	62.2	4.06
	96.5	28.36	7.75	15.710	1.438	0.890	90.1	14.4	1.78	1.49	465.1	59.2	4.05
	92	27.04	7.69	15.660	1.378	0.840	83.9	13.4	1.76	1.45	441.4	56.4	4.04
	88	25.87	7.63	15.640	1.313	0.820	80.2	12.9	1.76	1.42	418.9	53.6	4.02
	83.5	24.55	7.56	15.600	1.248	0.780	75.0	12.1	1.75	1.39	395.1	50.7	4.01
	79	23.24	7.50	15.550	1.188	0.730	69.3	11.3	1.73	1.34	372.5	47.9	4.00
	75	22.04	7.44	15.515	1.128	0.695	64.9	10.6	1.72	1.31	351.3	45.3	3.99
	71	20.92	7.38	15.500	1.063	0.680	62.1	10.2	1.72	1.29	330.1	42.6	3.97
ST 7 WF	68	19.99	7.38	14.740	1.063	0.660	60.0	9.89	1.73	1.31	283.9	38.5	3.77
	63.5	18.67	7.31	14.690	0.998	0.610	54.7	9.04	1.71	1.26	263.8	35.9	3.76
	59.5	17.49	7.25	14.650	0.938	0.570	50.4	8.36	1.70	1.22	245.9	33.6	3.75
	55.5	16.33	7.19	14.620	0.873	0.540	46.7	7.80	1.69	1.19	227.4	31.1	3.73
	51.5	15.13	7.13	14.575	0.813	0.495	42.4	7.10	1.67	1.15	209.9	28.8	3.72
	47.5	13.97	7.06	14.545	0.748	0.465	39.1	6.58	1.67	1.12	191.9	26.4	3.71
	43.5	12.78	7.00	14.5	0.688	0.420	34.9	5.88	1.65	1.08	174.8	24.1	3.70
ST 7 WF	42	12.36	7.09	12.023	0.778	0.451	37.4	6.36	1.74	1.21	112.7	18.8	3.02
	39	11.47	7.03	12.000	0.718	0.428	34.8	5.96	1.74	1.19	103.5	17.2	3.00
ST 7 WF	37	10.88	7.10	10.072	0.783	0.450	36.1	6.26	1.82	1.32	66.7	13.3	2.48
	34	10.00	7.03	10.040	0.718	0.418	33.0	5.74	1.81	1.29	60.6	12.1	2.46
	30.5	8.97	6.96	10.000	0.643	0.378	29.2	5.13	1.80	1.25	53.6	10.7	2.45
ST 7 WF	26.5	7.79	6.97	8.062	0.658	0.370	27.7	4.95	1.88	1.38	28.8	7.14	1.92
	24	7.06	6.91	8.031	0.593	0.339	24.9	4.49	1.88	1.35	25.6	6.38	1.91
	21.5	6.32	6.84	8.000	0.528	0.308	22.2	4.02	1.87	1.33	22.6	5.64	1.89
ST 7 WF[a]	19	5.59	7.06	6.776	0.513	0.313	23.5	4.27	2.05	1.56	12.3	3.64	1.49
	17	5.00	7.00	6.750	0.453	0.287	21.1	3.86	2.05	1.55	10.6	3.15	1.46
	15	4.41	6.93	6.733	0.383	0.270	19.0	3.55	2.08	1.59	8.77	2.61	1.41
ST 6 WF	80.5	23.69	6.94	12.515	1.486	0.905	62.6	11.5	1.63	1.47	243.1	38.9	3.20
	66.5	19.56	6.69	12.365	1.236	0.755	48.4	9.03	1.57	1.33	195.0	31.5	3.16
	60	17.65	6.56	12.320	1.106	0.710	43.4	8.22	1.57	1.28	172.5	28.0	3.13
	53	15.59	6.44	12.230	0.986	0.620	36.7	7.01	1.53	1.20	150.4	24.6	3.11
	49.5	14.54	6.38	12.190	0.921	0.580	33.7	6.46	1.52	1.16	139.1	22.8	3.09

Table 81 (*Continued*)

Section Number	Weight per Foot (lb)	Area (in.²)	Depth of Tee (in.)	Flange Width (in.)	Flange Average Thickness (in.)	Stem Thickness (in.)	Axis X–X I (in.⁴)	Axis X–X S (in.³)	Axis X–X r (in.)	Axis X–X y (in.)	Axis Y–Y I (in.⁴)	Axis Y–Y S (in.³)	Axis Y–Y r (in.)
	46	13.53	6.31	12.155	0.856	0.545	31.0	5.98	1.51	1.13	128.2	21.1	3.08
	42.5	12.49	6.25	12.105	0.796	0.495	27.8	5.38	1.49	1.08	117.7	19.5	3.07
	39.5	11.61	6.19	12.080	0.736	0.470	25.8	5.02	1.48	1.06	108.2	17.9	3.05
	36	10.58	6.13	12.040	0.671	0.430	23.1	4.53	1.48	1.02	97.6	16.2	3.04
	32.5	9.55	6.06	12.000	0.606	0.390	20.6	4.06	1.47	0.98	87.3	14.6	3.02
ST 6 WF	29	8.53	6.10	10.014	0.641	0.359	19.0	3.75	1.49	1.03	53.7	10.7	2.51
	26.5	7.80	6.03	10.000	0.576	0.345	17.7	3.54	1.51	1.02	48.0	9.60	2.48
ST 6 WF	25	7.36	6.10	8.077	0.641	0.371	18.7	3.80	1.60	1.17	28.2	6.98	1.96
	22.5	6.62	6.03	8.042	0.576	0.336	16.6	3.40	1.59	1.13	25.0	6.20	1.94
	20	5.89	5.97	8.000	0.516	0.294	14.4	2.94	1.56	1.08	22.0	5.50	1.94
ST 6 WF	18	5.29	6.12	6.565	0.540	0.305	15.3	3.14	1.70	1.26	11.9	3.62	1.50
	15.5	4.56	6.04	6.525	0.465	0.265	13.0	2.69	1.69	1.22	9.9	3.04	1.47
	13.5	3.98	5.98	6.500	0.400	0.240	11.4	2.39	1.69	1.21	8.3	2.55	1.44
ST 6 WF	7	2.07	5.96	3.970	0.224	0.200	7.66	1.83	1.92	1.76	1.13	0.57	0.74
ST 6 I[b]	25	7.29	6.00	5.477	0.660	0.687	25.2	6.05	1.85	1.84	7.85	2.87	1.03
	20.4	5.92	6.00	5.250	0.660	0.460	18.8	4.26	1.77	1.57	6.77	2.58	1.06
ST 6 I	17.5	5.10	6.00	5.078	0.544	0.428	17.2	3.95	1.83	1.65	4.93	1.94	0.98
	15.9	4.63	6.00	5.000	0.544	0.350	14.9	3.31	1.78	1.51	4.68	1.87	1.00
ST 5 I	17.5	5.11	5.00	4.944	0.491	0.594	12.5	3.63	1.56	1.56	4.18	1.69	0.90
	12.7	3.69	5.00	4.660	0.491	0.310	7.81	2.05	1.45	1.20	3.39	1.46	0.95
ST 4 I	11.5	3.36	4.00	4.171	0.425	0.441	5.03	1.77	1.22	1.15	2.15	1.03	0.80
	9.2	2.67	4.00	4.000	0.425	0.270	3.50	1.14	1.14	0.94	1.86	0.93	0.83
ST 3.5 I	10	2.92	3.50	3.860	0.392	0.450	3.36	1.36	1.07	1.04	1.58	0.82	0.73
	7.65	2.22	3.50	3.660	0.392	0.250	2.18	0.81	0.99	0.81	1.32	0.72	0.77
ST 3 I	8.625	2.51	3.00	3.565	0.359	0.465	2.13	1.02	0.92	0.91	1.15	0.65	0.67
	6.25	1.81	3.00	3.330	0.359	0.230	1.27	0.55	0.83	0.69	0.93	0.56	0.71
ST 5 WF	56	16.46	5.69	10.415	1.248	0.755	28.8	6.42	1.32	1.21	117.7	22.6	2.67
	50	14.72	5.56	10.345	1.118	0.685	24.8	5.62	1.30	1.14	103.3	20.0	2.65
	44.5	13.09	5.44	10.275	0.998	0.615	21.3	4.88	1.28	1.07	90.3	17.6	2.63
	38.5	11.33	5.31	10.195	0.868	0.535	17.7	4.10	1.25	1.00	76.7	15.1	2.60
	36	10.59	5.25	10.170	0.808	0.510	16.4	3.83	1.24	0.97	70.9	13.9	2.59
	33	9.70	5.19	10.117	0.748	0.457	14.5	3.39	1.22	0.92	64.6	12.8	2.58
	30	8.83	5.13	10.075	0.683	0.415	12.8	3.02	1.21	0.88	58.2	11.6	2.57
	27	7.94	5.06	10.028	0.618	0.368	11.2	2.64	1.18	0.84	51.95	10.4	2.56
	24.5	7.20	5.00	10.000	0.558	0.340	10.1	2.40	1.18	0.81	46.5	9.30	2.54
ST 5 WF	22.5	6.62	5.06	8.022	0.618	0.350	10.3	2.48	1.25	0.91	26.6	6.63	2.00
	19.5	5.74	4.97	7.990	0.528	0.318	8.96	2.19	1.25	0.88	22.5	5.62	1.98
	16.5	4.85	4.88	7.964	0.433	0.292	7.80	1.95	1.27	0.88	18.2	4.58	1.94
ST 5 WF[a]	14.5	4.27	5.11	5.799	0.500	0.289	8.38	2.07	1.40	1.05	7.61	2.62	1.34
	12.5	3.67	5.04	5.762	0.430	0.252	7.12	1.77	1.39	1.02	6.34	2.20	1.31
	10.5	3.10	4.95	5.750	0.340	0.240	6.31	1.62	1.43	1.06	4.87	1.69	1.25
ST 4 WF	33.5	9.85	4.50	8.287	0.933	0.575	10.94	3.07	1.05	0.94	44.3	10.7	2.12
	29	8.53	4.38	8.222	0.808	0.510	9.11	2.60	1.03	0.87	37.5	9.10	2.10
	24	7.06	4.25	8.117	0.683	0.405	6.92	2.00	0.99	0.78	30.45	7.50	2.08
	20	5.88	4.13	8.077	0.558	0.365	5.80	1.71	0.99	0.74	24.5	6.05	2.04
	17.5	5.15	4.06	8.027	0.493	0.315	4.88	1.45	0.97	0.69	21.25	5.30	2.03
	15.5	4.56	4.00	8.000	0.433	0.288	4.31	1.30	0.97	0.67	18.5	4.60	2.01
ST 4 WF	14	4.11	4.03	6.540	0.463	0.285	4.22	1.28	1.01	0.73	10.8	3.30	1.62
	12	3.53	3.97	6.500	0.398	0.245	3.53	1.08	1.00	0.70	9.10	2.80	1.61
ST 4 WF	10	2.94	4.07	5.268	0.378	0.248	3.66	1.13	1.12	0.83	4.25	1.61	1.20
	8.5	2.50	4.00	5.250	0.308	0.230	3.21	1.01	1.13	0.84	3.36	1.28	1.16

Table 81 *(Continued)*

Nominal Size (in.)	Weight per Foot (lb)	Area (in.^{2})	Depth (in.)	Width Flange (in.)	Minimum Flange (in.)	Thickness Stem (in.)	Axis X–X I (in.^{4})	S (in.^{3})	r (in.)	y (in.)	Axis Y–Y I (in.^{4})	S (in.^{3})	r (in.)
$5 \times 3\frac{1}{8}$	13.6	4.00	$3\frac{3}{8}$	5	$\frac{1}{2}$	$\frac{13}{32}$	2.7	1.1	0.82	0.76	5.2	2.1	1.14
5×3	11.5	3.37	3	5	$\frac{3}{8}$	$\frac{13}{32}$	2.4	1.1	0.84	0.76	3.9	1.6	1.10
$4 \times 4\frac{1}{2}$	11.2	3.29	$4\frac{1}{2}$	4	$\frac{3}{8}$	$\frac{3}{8}$	6.3	2.0	1.39	1.31	2.1	1.1	0.80
4×4	13.5	3.97	4	4	$\frac{1}{2}$	$\frac{1}{2}$	5.7	2.0	1.20	1.18	2.8	1.4	0.84
4×3	9.2	2.68	3	4	$\frac{3}{8}$	$\frac{3}{8}$	2.0	0.90	0.86	0.78	2.1	1.1	0.89
$4 \times 2\frac{1}{2}$	8.5	2.48	$2\frac{1}{2}$	4	$\frac{3}{8}$	$\frac{3}{8}$	1.2	0.62	0.69	0.62	2.1	1.0	0.92
3×3	7.8	2.29	3	3	$\frac{3}{8}$	$\frac{3}{8}$	1.84	0.86	0.89	0.88	0.89	0.60	0.63
3×3	6.7	1.97	3	3	$\frac{5}{16}$	$\frac{5}{16}$	1.61	0.74	0.90	0.85	0.75	0.50	0.62
$3 \times 2\frac{1}{2}$	6.1	1.77	$2\frac{1}{2}$	3	$\frac{5}{16}$	$\frac{5}{16}$	0.94	0.51	0.73	0.68	0.75	0.50	0.65
$2\frac{1}{2} \times 2\frac{1}{2}$	6.4	1.87	$2\frac{1}{2}$	$2\frac{1}{2}$	$\frac{3}{8}$	$\frac{3}{8}$	1.0	0.59	0.74	0.76	0.52	0.42	0.53
$2\frac{1}{2} \times 2\frac{1}{2}$	4.6	1.33	$2\frac{1}{2}$	$2\frac{1}{2}$	$\frac{1}{4}$	$\frac{1}{4}$	0.74	0.42	0.75	0.71	0.34	0.27	0.51
$2\frac{1}{4} \times 2\frac{1}{4}$	4.1	1.19	$2\frac{1}{4}$	$2\frac{1}{4}$	$\frac{1}{4}$	$\frac{1}{4}$	0.52	0.32	0.66	0.65	0.25	0.22	0.46
2×2	4.3	1.26	2	2	$\frac{5}{16}$	$\frac{5}{16}$	0.44	0.31	0.59	0.61	0.23	0.23	0.43
2×2	3.56	1.05	2	2	$\frac{1}{4}$	$\frac{1}{4}$	0.37	0.26	0.59	0.59	0.18	0.18	0.42

[a] WF indicates structural tee cut from wide-flange section.
[b] I indicates structural tee cut from standard beam section.

Table 82 Properties and Dimensions of Zees

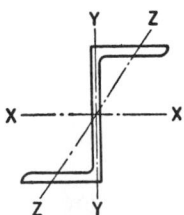

Zees are seldom used as structural framing members. When so used they are generally employed on short spans in flexure. This table lists a few selected sizes, the range of whose section moduli will cover all ordinary conditions. For sizes not listed, the catalogs of the respective rolling mills should be consulted.

Nominal Size (in.)	Weight per Foot (lb)	Area (in.^{2})	Depth (in.)	Width of Flange (in.)	Thickness (in.)	Axis X–X I (in.^{4})	S (in.^{3})	r (in.)	Axis Y–Y I (in.^{4})	S (in.^{3})	r (in.)	Axis Z–Z r (in.)
$6 \times 3\frac{1}{2}$	21.1	6.19	$6\frac{1}{8}$	$3\frac{5}{8}$	$\frac{1}{2}$	34.4	11.2	2.36	12.9	3.8	1.44	0.84
	15.7	4.59	6	$3\frac{1}{2}$	$\frac{3}{8}$	25.3	8.4	2.35	9.1	2.8	1.41	0.83
$5 \times 3\frac{1}{4}$	17.9	5.25	5	$3\frac{1}{4}$	$\frac{1}{2}$	19.2	7.7	1.91	9.1	3.0	1.31	0.74
	16.4	4.81	$5\frac{1}{8}$	$3\frac{3}{8}$	$\frac{7}{16}$	19.1	7.4	1.99	9.2	2.9	1.38	0.77
	14.0	4.10	$5\frac{1}{16}$	$3\frac{5}{16}$	$\frac{3}{8}$	16.2	6.4	1.99	7.7	2.5	1.37	0.76
	11.6	3.40	5	$3\frac{1}{4}$	$\frac{5}{16}$	13.4	5.3	1.98	6.2	2.0	1.35	0.75
$4 \times 3\frac{1}{16}$	15.9	4.66	$4\frac{1}{16}$	$3\frac{1}{8}$	$\frac{1}{2}$	11.2	5.5	1.55	8.0	2.8	1.31	0.67
	12.5	3.66	$4\frac{1}{8}$	$3\frac{3}{16}$	$\frac{3}{8}$	9.6	4.7	1.62	6.8	2.3	1.36	0.69
	10.3	3.03	$4\frac{1}{16}$	$3\frac{1}{8}$	$\frac{5}{16}$	7.9	3.9	1.62	5.5	1.8	1.34	0.68
	8.2	2.41	4	$3\frac{1}{16}$	$\frac{1}{4}$	6.3	3.1	1.62	4.2	1.4	1.33	0.67
$3 \times 2\frac{11}{16}$	12.6	3.69	3	$2\frac{11}{16}$	$\frac{1}{2}$	4.6	3.1	1.12	4.9	2.0	1.15	0.53
	9.8	2.86	3	$2\frac{11}{16}$	$\frac{3}{8}$	3.9	2.6	1.16	3.9	1.6	1.17	0.54
	6.7	1.97	3	$2\frac{11}{16}$	$\frac{1}{4}$	2.9	1.9	1.21	2.8	1.1	1.19	0.55

Table 83   Properties and Dimensions of H Bearing Piles

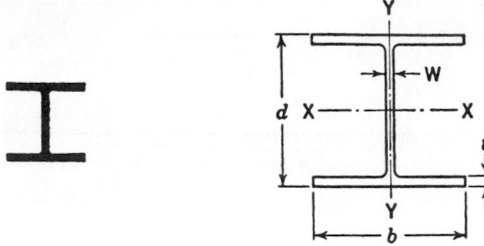

Section Number and Nominal Size	Weight per Foot (lb)	Area A (in.^{2})	Depth d (in.)	Flange Width b (in.)	Flange Thickness t (in.)	Web Thickness W (in.)	Axis X–X I (in.^{4})	Axis X–X S (in.^{3})	Axis X–X r (in.)	Axis Y–Y I' (in.^{4})	Axis Y–Y S' (in.^{3})	Axis Y–Y r' (in.)
BP 14,	117	34.44	14.234	14.885	0.805	0.805	1228.5	172.6	5.97	443.1	59.5	3.59
$14 \times 14\frac{1}{2}$	102	30.01	14.032	14.784	0.704	0.704	1055.1	150.4	5.93	379.6	51.3	3.56
	89	26.19	13.856	14.696	0.616	0.616	909.1	131.2	5.89	326.2	44.4	3.53
	73	21.46	13.636	14.586	0.506	0.506	733.1	107.5	5.85	261.9	35.9	3.49
BP 12,	74	21.76	12.122	12.217	0.607	0.607	566.5	93.5	5.10	184.7	30.2	2.91
12×12	53	15.58	11.780	12.046	0.436	0.436	394.8	67.0	5.03	127.3	21.2	2.86
BP 10,	57	16.76	10.012	10.224	0.564	0.564	294.7	58.9	4.19	100.6	19.7	2.45
10×10	42	12.35	9.720	10.078	0.418	0.418	210.8	43.4	4.13	71.4	14.2	2.40
BP 8,												
8×8	36	10.60	8.026	8.158	0.446	0.446	119.8	29.9	3.36	40.4	9.9	1.95

Table 84   Square and Round Bars[a]

Size (in.)	Square Weight/ft (lb)	Square Area (in.^{2})	Round Weight/ft (lb)	Round Area (in.^{2})	Size (in.)	Square Weight/ft (lb)	Square Area (in.^{2})	Round Weight/ft (lb)	Round Area (in.^{2})
0					$\frac{5}{16}$	5.857	1.7227	4.600	1.3530
$\frac{1}{16}$	0.013	0.0039	0.010	0.0031	$\frac{3}{8}$	6.428	1.8906	5.049	1.4849
$\frac{1}{8}$	0.053	0.0156	0.042	0.0123	$\frac{7}{16}$	7.026	2.0664	5.518	1.6230
$\frac{3}{16}$	0.120	0.0352	0.094	0.0276	$\frac{1}{2}$	7.650	2.2500	6.008	1.7671
$\frac{1}{4}$	0.213	0.0625	0.167	0.0491	$\frac{9}{16}$	8.301	2.4414	6.519	1.9175
$\frac{5}{16}$	0.332	0.0977	0.261	0.0767	$\frac{5}{8}$	8.978	2.6406	7.051	2.0739
$\frac{3}{8}$	0.478	0.1406	0.376	0.1105	$\frac{11}{16}$	9.682	2.8477	7.604	2.2365
$\frac{7}{16}$	0.651	0.1914	0.511	0.1503	$\frac{3}{4}$	10.413	3.0625	8.178	2.4053
$\frac{1}{2}$	0.850	0.2500	0.668	0.1963	$\frac{13}{16}$	11.170	3.2852	8.773	2.5802
$\frac{9}{16}$	1.076	0.3164	0.845	0.2485	$\frac{7}{8}$	11.953	3.5156	9.388	2.7612
$\frac{5}{8}$	1.328	0.3906	1.043	0.3068	$\frac{15}{16}$	12.763	3.7539	10.024	2.9483
$\frac{11}{16}$	1.607	0.4727	1.262	0.3712	2	13.600	4.0000	10.681	3.1416
$\frac{3}{4}$	1.913	0.5625	1.502	0.4418	$\frac{1}{16}$	14.463	4.2539	11.359	3.3410
$\frac{13}{16}$	2.245	0.6602	1.763	0.5185	$\frac{1}{8}$	15.353	4.5156	12.058	3.5466
$\frac{7}{8}$	2.603	0.7656	2.044	0.6013	$\frac{3}{16}$	16.270	4.7852	12.778	3.7583
$\frac{15}{16}$	2.988	0.8789	2.347	0.6903	$\frac{1}{4}$	17.213	5.0625	13.519	3.9761
1	3.400	1.0000	2.670	0.7854	$\frac{5}{16}$	18.182	5.3477	14.280	4.2000
$\frac{1}{16}$	3.838	1.1289	3.015	0.8866	$\frac{3}{8}$	19.178	5.6406	15.062	4.4301
$\frac{1}{8}$	4.303	1.2656	3.380	0.9940	$\frac{7}{16}$	20.201	5.9414	15.866	4.6664
$\frac{3}{16}$	4.795	1.4102	3.766	1.1075	$\frac{1}{2}$	21.250	6.2500	16.690	4.9087
$\frac{1}{4}$	5.313	1.5625	4.172	1.2272	$\frac{9}{16}$	22.326	6.5664	17.534	5.1572

Table 84 (*Continued*)

Size (in.)	Square Weight/ft (lb)	Area (in.^{2})	Round Weight/ft (lb)	Area (in.^{2})	Size (in.)	Square Weight/ft (lb)	Area (in.^{2})	Round Weight/ft (lb)	Area (in.^{2})
$\frac{5}{8}$	23.428	6.8906	18.400	5.4119	$\frac{3}{8}$	98.23	28.891	77.15	22.691
$\frac{11}{16}$	24.557	7.2227	19.287	5.6727	$\frac{7}{16}$	100.53	29.566	78.95	23.221
$\frac{3}{4}$	25.713	7.5625	20.195	5.9396	$\frac{1}{2}$	102.85	30.250	80.78	23.758
$\frac{13}{16}$	26.895	7.9102	21.123	6.2126	$\frac{9}{16}$	105.20	30.941	82.62	24.301
$\frac{7}{8}$	28.103	8.2656	22.072	6.4918	$\frac{5}{8}$	107.58	31.641	84.49	24.850
$\frac{15}{16}$	29.338	8.6289	23.042	6.7771	$\frac{11}{16}$	109.98	32.348	86.38	25.406
3	30.60	9.000	24.03	7.069	$\frac{3}{4}$	112.41	33.063	88.29	25.967
$\frac{1}{16}$	31.89	9.379	25.05	7.366	$\frac{13}{16}$	114.87	33.785	90.22	26.535
$\frac{1}{8}$	33.20	9.766	26.08	7.670	$\frac{7}{8}$	117.35	34.516	92.17	27.109
$\frac{3}{16}$	34.54	10.160	27.13	7.980	$\frac{15}{16}$	119.86	35.254	94.14	27.688
$\frac{1}{4}$	35.91	10.563	28.21	8.296	6	122.40	36.000	96.13	28.274
$\frac{5}{16}$	37.31	10.973	29.30	8.618	$\frac{1}{16}$	124.96	36.754	98.15	28.866
$\frac{3}{8}$	38.73	11.391	30.42	8.946	$\frac{1}{8}$	127.55	37.516	100.18	29.465
$\frac{7}{16}$	40.18	11.816	31.55	9.281	$\frac{3}{16}$	130.17	38.285	102.23	30.069
$\frac{1}{2}$	41.65	12.250	32.71	9.621	$\frac{1}{4}$	132.81	39.063	104.31	30.680
$\frac{9}{16}$	43.15	12.691	33.89	9.968	$\frac{5}{16}$	135.48	39.848	106.41	31.296
$\frac{5}{8}$	44.68	13.141	35.09	10.321	$\frac{3}{8}$	138.18	40.641	108.53	31.919
$\frac{11}{16}$	46.23	13.598	36.31	10.680	$\frac{7}{16}$	140.90	41.441	110.66	32.548
$\frac{3}{4}$	47.81	14.063	37.55	11.045	$\frac{1}{2}$	143.65	42.250	112.82	33.183
$\frac{13}{16}$	49.42	14.535	38.81	11.416	$\frac{9}{16}$	146.43	43.066	115.00	33.824
$\frac{7}{8}$	51.05	15.016	40.10	11.793	$\frac{5}{8}$	149.23	43.891	117.20	34.472
$\frac{15}{16}$	52.71	15.504	41.40	12.177	$\frac{11}{16}$	152.06	44.723	119.43	35.125
4	54.40	16.000	42.73	12.566	$\frac{3}{4}$	154.91	45.563	121.67	35.785
$\frac{1}{16}$	56.11	16.504	44.07	12.962	$\frac{13}{16}$	157.79	46.410	123.93	36.450
$\frac{1}{8}$	57.85	17.016	45.44	13.364	$\frac{7}{8}$	160.70	47.266	126.22	37.122
$\frac{3}{16}$	59.62	17.535	46.83	13.772	$\frac{15}{16}$	163.64	48.129	128.52	37.800
$\frac{1}{4}$	61.41	18.063	48.23	14.186	7	166.60	49.000	130.85	38.485
$\frac{5}{16}$	63.23	18.598	49.66	14.607	$\frac{1}{16}$	169.59	49.879	133.19	39.175
$\frac{3}{8}$	65.08	19.141	51.11	15.033	$\frac{1}{8}$	172.60	50.766	135.56	39.871
$\frac{7}{16}$	66.95	19.691	52.58	15.466	$\frac{3}{16}$	175.64	51.660	137.95	40.574
$\frac{1}{2}$	68.85	20.250	54.07	15.904	$\frac{1}{4}$	178.71	52.563	140.36	41.282
$\frac{9}{16}$	70.78	20.816	55.59	16.349	$\frac{5}{16}$	181.81	53.473	142.79	41.997
$\frac{5}{8}$	72.73	21.391	57.12	16.800	$\frac{3}{8}$	184.93	54.391	145.24	42.718
$\frac{11}{16}$	74.71	21.973	58.67	17.257	$\frac{7}{16}$	188.07	55.316	147.71	43.445
$\frac{3}{4}$	76.71	22.563	60.25	17.721	$\frac{1}{2}$	191.25	56.250	150.21	44.179
$\frac{13}{16}$	78.74	23.160	61.85	18.190	$\frac{9}{16}$	194.45	57.191	152.72	44.918
$\frac{7}{8}$	80.80	23.766	63.46	18.665	$\frac{5}{8}$	197.68	58.141	155.26	45.664
$\frac{15}{16}$	82.89	24.379	65.10	19.147	$\frac{11}{16}$	200.93	59.098	157.81	46.415
5	85.00	25.000	66.76	19.635	$\frac{3}{4}$	204.21	60.063	160.39	47.173
$\frac{1}{16}$	87.14	25.629	68.44	20.129	$\frac{13}{16}$	207.52	61.035	162.99	47.937
$\frac{1}{8}$	89.30	26.266	70.14	20.629	$\frac{7}{8}$	210.85	62.016	165.60	48.707
$\frac{3}{16}$	91.49	26.910	71.86	21.135	$\frac{15}{16}$	214.21	63.004	168.24	49.483
$\frac{1}{4}$	93.71	27.563	73.60	21.648	8	217.60	64.000	170.90	50.265
$\frac{5}{16}$	95.96	28.223	75.36	22.166					

[a] One cubic inch of rolled steel is assumed to weigh 0.2833 lb.

Table 85 Dimensions of Ferrous Pipe

Nominal Pipe Size (in.)	Outside Diameter (in.)	Schedule No.	Wall Thickness (in.)	Inside Diameter (in.)	Cross-Sectional Area Metal (in.²)	Cross-Sectional Area Flow (ft²)	Circumference, ft, or surface, ft²/ft of Length Outside	Circumference, ft, or surface, ft²/ft of Length Inside	Capacity at 1 ft/sec Velocity U.S. gal/min	Capacity at 1 ft/sec Velocity lb/hr water	Weight of Plain-End Pipe (lb/ft)
$\frac{1}{8}$	0.405	10S	0.049	0.307	0.055	0.00051	0.106	0.0804	0.231	115.5	0.19
		40ST, 40S	0.068	0.269	0.072	0.00040	0.106	0.0705	0.179	89.5	0.24
		80XS, 80S	0.095	0.215	0.093	0.00025	0.106	0.0563	0.113	56.5	0.31
$\frac{1}{4}$	0.540	10S	0.065	0.410	0.097	0.00092	0.141	0.107	0.412	206.5	0.33
		40ST, 40S	0.088	0.364	0.125	0.00072	0.141	0.095	0.323	161.5	0.42
		80XS, 80S	0.119	0.302	0.157	0.00050	0.141	0.079	0.224	112.0	0.54
$\frac{3}{8}$	0.675	10S	0.065	0.545	0.125	0.00162	0.177	0.143	0.727	363.5	0.42
		40ST, 40S	0.091	0.493	0.167	0.00133	0.177	0.129	0.596	298.0	0.57
		80XS, 80S	0.126	0.423	0.217	0.00098	0.177	0.111	0.440	220.0	0.74
$\frac{1}{2}$	0.840	5S	0.065	0.710	0.158	0.00275	0.220	0.186	1.234	617.0	0.54
		10S	0.083	0.674	0.197	0.00248	0.220	0.176	1.112	556.0	0.67
		40ST, 40S	0.109	0.622	0.250	0.00211	0.220	0.163	0.945	472.0	0.85
		80XS, 80S	0.147	0.546	0.320	0.00163	0.220	0.143	0.730	365.0	1.09
		160	0.188	0.464	0.385	0.00117	0.220	0.122	0.527	263.5	1.31
		XX	0.294	0.252	0.504	0.00035	0.220	0.066	0.155	77.5	1.71
$\frac{3}{4}$	1.050	5S	0.065	0.920	0.201	0.00461	0.275	0.241	2.072	1036.0	0.69
		10S	0.083	0.884	0.252	0.00426	0.275	0.231	1.903	951.5	0.86
		40ST, 40S	0.113	0.824	0.333	0.00371	0.275	0.216	1.665	832.5	1.13
		80XS, 80S	0.154	0.742	0.433	0.00300	0.275	0.194	1.345	672.5	1.47
		160	0.219	0.612	0.572	0.00204	0.275	0.160	0.917	458.5	1.94
		XX	0.308	0.434	0.718	0.00103	0.275	0.114	0.461	230.5	2.44
1	1.315	5S	0.065	1.185	0.255	0.00768	0.344	0.310	3.449	1725	0.87
		10S	0.109	1.097	0.413	0.00656	0.344	0.287	2.946	1473	1.40
		40ST, 40S	0.133	1.049	0.494	0.00600	0.344	0.275	2.690	1345	1.68
		80XS, 80S	0.179	0.957	0.639	0.00499	0.344	0.250	2.240	1120	2.17
		160	0.250	0.815	0.836	0.00362	0.344	0.213	1.625	812.5	2.84
		XX	0.358	0.599	1.076	0.00196	0.344	0.157	0.878	439.0	3.66
$1\frac{1}{4}$	1.660	5S	0.065	1.530	0.326	0.01277	0.435	0.401	5.73	2865	1.11
		10S	0.109	1.442	0.531	0.01134	0.435	0.378	5.09	2545	1.81
		40ST, 40S	0.140	1.380	0.668	0.01040	0.435	0.361	4.57	2285	2.27
		80XS, 80S	0.191	1.278	0.881	0.00891	0.435	0.335	3.99	1995	3.00
		160	0.250	1.160	1.107	0.00734	0.435	0.304	3.29	1645	3.76
		XX	0.382	0.896	1.534	0.00438	0.435	0.235	1.97	985	5.21

Nominal size	OD	Schedule									
$1\frac{1}{2}$	1.900	5S	0.065	1.770	0.375	0.01709	0.497	7.67	0.463	3835	1.28
		10S	0.109	1.682	0.614	0.01543	0.497	6.94	0.440	3465	2.09
		40ST, 40S	0.145	1.610	0.800	0.01414	0.497	6.34	0.421	3170	2.72
		80ST, 80S	0.200	1.500	1.069	0.01225	0.497	5.49	0.393	2745	3.63
		160	0.281	1.338	1.429	0.00976	0.497	4.38	0.350	2190	4.86
		XX	0.400	1.100	1.885	0.00660	0.497	2.96	0.288	1480	6.41
2	2.375	5S	0.065	2.245	0.472	0.02749	0.622	12.34	0.588	6170	1.61
		10S	0.109	2.157	0.776	0.02538	0.622	11.39	0.565	5695	2.64
		40ST, 40S	0.154	2.067	1.075	0.02330	0.622	10.45	0.541	5225	3.65
		80ST, 80S	0.218	1.939	1.477	0.02050	0.622	9.20	0.508	4600	5.02
		160	0.344	1.687	2.195	0.01552	0.622	6.97	0.436	3485	7.46
		XX	0.436	1.503	2.656	0.01232	0.622	5.53	0.393	2765	9.03
$2\frac{1}{2}$	2.875	5S	0.083	2.709	0.728	0.04003	0.753	17.97	0.709	8985	2.48
		10S	0.120	2.635	1.039	0.03787	0.753	17.00	0.690	8500	3.53
		40ST, 40S	0.203	2.469	1.704	0.03322	0.753	14.92	0.647	7460	5.79
		80XS, 80S	0.276	2.323	2.254	0.02942	0.753	13.20	0.608	6600	7.66
		160	0.375	2.125	2.945	0.02463	0.753	11.07	0.556	5535	10.01
		XX	0.552	1.771	4.028	0.01711	0.753	7.68	0.464	3840	13.70
3	3.500	5S	0.083	3.334	0.891	0.06063	0.916	27.21	0.873	13,605	3.03
		10S	0.120	3.260	1.274	0.05796	0.916	26.02	0.853	13,010	4.33
		40ST, 40S	0.216	3.068	2.228	0.05130	0.916	23.00	0.803	11,500	7.58
		80XS, 80S	0.300	2.900	3.016	0.04587	0.916	20.55	0.759	10,275	10.25
		160	0.438	2.624	4.213	0.03755	0.916	16.86	0.687	8430	14.31
		XX	0.600	2.300	5.466	0.02885	0.916	12.95	0.602	6475	18.58
$3\frac{1}{2}$	4.0	5S	0.083	3.834	1.021	0.08017	1.047	35.98	1.004	17.990	3.48
		10S	0.120	3.760	1.463	0.07711	1.047	34.61	0.984	17,305	4.97
		40ST, 40S	0.226	3.548	2.680	0.06870	1.047	30.80	0.929	15,400	9.11
		80XS, 80S	0.318	3.364	3.678	0.06170	1.047	27.70	0.881	13,850	12.51
4	4.5	5S	0.083	4.334	1.152	0.10245	1.178	46.0	1.135	23,000	3.92
		10S	0.120	4.260	1.651	0.09898	1.178	44.4	1.115	22,200	5.61
		40ST, 40S	0.237	4.026	3.17	0.08840	1.178	39.6	1.054	19,800	10.79
		80XS, 80S	0.337	3.826	4.41	0.07986	1.178	35.8	1.002	17,900	14.98
		120	0.438	3.624	5.58	0.07170	1.178	32.2	0.949	16,100	19.01
		160	0.531	3.438	6.62	0.06647	1.178	28.9	0.900	14,450	22.52
		XX	0.674	3.152	8.10	0.05419	1.178	24.3	0.825	12,150	27.54

(Continues)

Table 85 *(Continued)*

Nominal Pipe Size (in.)	Outside Diameter (in.)	Schedule No.	Wall Thickness (in.)	Inside Diameter (in.)	Cross-Sectional Area Metal (in.²)	Cross-Sectional Area Flow (ft²)	Circumference, ft, or surface, ft²/ft of Length Outside	Circumference, ft, or surface, ft²/ft of Length Inside	Capacity at 1 ft/sec Velocity U.S. gal/min	Capacity at 1 ft/sec Velocity lb/hr water	Weight of Plain-End Pipe (lb/ft)
5	5.563	5S	0.109	5.345	1.87	0.1558	1.456	1.399	69.9	34,950	6.36
		10S	0.134	5.295	2.29	0.1529	1.456	1.386	68.6	34,300	7.77
		40ST, 40S	0.258	5.047	4.30	0.1390	1.456	1.321	62.3	31,150	14.62
		80XS, 80S	0.375	4.813	6.11	0.1263	1.456	1.260	57.7	28,850	20.78
		120	0.500	4.563	7.95	0.1136	1.456	1.195	51.0	25,500	27.04
		160	0.625	4.313	9.70	0.1015	1.456	1.129	45.5	22,750	32.96
		XX	0.750	4.063	11.34	0.0900	1.456	1.064	40.4	20,200	38.55
6	6.625	5S	0.109	6.407	2.23	0.2239	1.734	1.677	100.5	50,250	7.60
		10S	0.134	6.357	2.73	0.2204	1.734	1.664	98.9	49,450	9.29
		40ST, 40S	0.280	6.065	5.58	0.2006	1.734	1.588	90.0	45,000	18.97
		80XS, 80S	0.432	5.761	8.40	0.1810	1.734	1.508	81.1	40,550	28.57
		120	0.562	5.501	10.70	0.1650	1.734	1.440	73.9	36,950	36.42
		160	0.719	5.187	13.34	0.1467	1.734	1.358	65.9	32,950	45.34
		XX	0.864	4.897	15.64	0.1308	1.734	1.282	58.7	29,350	53.16
8	8.625	5S	0.109	8.407	2.915	0.3855	2.258	2.201	173.0	86,500	9.93
		10S	0.148	8.329	3.941	0.3784	2.258	2.180	169.8	84,900	13.40
		20	0.250	8.125	6.578	0.3601	2.258	2.127	161.5	80,750	22.36
		30	0.277	8.071	7.260	0.3553	2.258	2.113	159.4	79,700	24.70
		40ST, 40S	0.322	7.981	8.396	0.3474	2.258	2.089	155.7	77,850	28.55
		60	0.406	7.813	10.48	0.3329	2.258	2.045	149.4	74,700	35.66
		80XS, 80S	0.500	7.625	12.76	0.3171	2.258	1.996	142.3	71,150	43.39
		100	0.594	7.437	14.99	0.3017	2.258	1.947	135.4	67,700	50.93
		120	0.719	7.187	17.86	0.2817	2.258	1.882	126.4	63,200	60.69
		140	0.812	7.001	19.93	0.2673	2.258	1.833	120.0	60,000	67.79
		XX	0.875	6.875	21.30	0.2578	2.258	1.800	115.7	57,850	72.42
		160	0.906	6.813	21.97	0.2532	2.258	1.784	113.5	56,750	74.71
10	10.75	5S	0.134	10.842	4.47	0.5993	2.814	2.744	269.0	134,500	15.23
		10S	0.165	10.420	5.49	0.5922	2.814	2.728	265.8	132,900	18.70
		20	0.250	10.250	8.25	0.5731	2.814	2.685	257.0	128,500	28.04
		30	0.307	10.136	10.07	0.5603	2.814	2.655	252.0	126,000	34.24
		40ST, 40S	0.365	10.020	11.91	0.5475	2.814	2.620	246.0	123,000	40.48
		80S, 60XS	0.500	9.750	16.10	0.5185	2.814	2.550	233.0	116,500	54.74
		80	0.594	9.562	18.95	0.4987	2.814	2.503	223.4	111,700	64.40
		100	0.719	9.312	22.66	0.4729	2.814	2.438	212.3	106,150	77.00
		120	0.844	9.062	26.27	0.4479	2.814	2.372	201.0	100,500	89.27
		140, XX	1.000	8.750	30.63	0.4176	2.814	2.291	188.0	94,000	104.13
		160	1.125	8.500	34.02	0.3941	2.814	2.225	177.0	88,500	115.65

Nominal	OD	Schedule	Wall								
12	12.75	5S	0.156	12.438	6.17	0.8438	3.338	3.26	378.7	189,350	22.22
		10S	0.180	12.390	7.11	0.8373	3.338	3.24	375.8	187,900	24.20
		20	0.250	12.250	9.82	0.8185	3.338	3.21	367.0	183,500	33.38
		30	0.330	12.090	12.88	0.7972	3.338	3.17	358.0	179,000	43.77
		ST, 40S	0.375	12.000	14.58	0.7854	3.338	3.14	352.5	176,250	49.56
		40	0.406	11.938	15.74	0.7773	3.338	3.13	349.0	174,500	53.56
		XS, 80S	0.500	11.750	19.24	0.7530	3.338	3.08	338.0	169,000	65.42
		60	0.562	11.626	21.52	0.7372	3.338	3.04	331.0	165,500	73.22
		80	0.688	11.374	26.07	0.7056	3.338	2.98	316.7	158,350	88.57
		100	0.844	11.062	31.57	0.6674	3.338	2.90	299.6	149,800	107.29
		120, XX	1.000	10.750	36.91	0.6303	3.338	2.81	283.0	141,500	125.49
		140	1.125	10.500	41.09	0.6013	3.338	2.75	270.0	135,000	139.68
		160	1.312	10.126	47.14	0.5592	3.338	2.65	251.0	125,500	160.33
14	14	5S	0.156	13.688	6.78	1.0219	3.665	3.58	459	229,500	22.76
		10S	0.188	13.624	8.16	1.0125	3.665	3.57	454	227,000	27.70
		10	0.250	13.500	10.80	0.9940	3.665	3.53	446	223,000	36.71
		20	0.312	13.376	13.42	0.9750	3.665	3.50	438	219,000	45.68
		30, ST	0.375	13.250	16.05	0.9575	3.665	3.47	430	215,000	54.57
		40	0.438	13.124	18.66	0.9397	3.665	3.44	422	211,000	63.37
		XS	0.500	13.000	21.21	0.9218	3.665	3.40	414	207,000	72.09
		60	0.594	12.812	25.02	0.8957	3.665	3.35	402	201,000	85.01
		80	0.750	12.500	31.22	0.8522	3.665	3.27	382	191,000	106.13
		100	0.938	12.124	38.49	0.8017	3.665	3.17	360	180,000	130.79
		120	1.094	11.812	44.36	0.7610	3.665	3.09	342	171,000	150.76
		140	1.250	11.500	50.07	0.7213	3.665	3.01	324	162,000	170.22
		160	1.406	11.188	55.63	0.6827	3.665	2.93	306	153,000	189.12
16	16	5S	0.165	15.670	8.18	1.3393	4.189	4.10	601	300,500	27.87
		10S	0.188	15.624	9.34	1.3314	4.189	4.09	598	299,000	31.62
		10	0.250	15.500	12.37	1.3104	4.189	4.06	587	293,500	42.05
		20	0.312	15.376	15.38	1.2985	4.189	4.03	578	289,000	52.36
		30, ST	0.375	15.250	18.41	1.2680	4.189	3.99	568	284,000	62.58
		40, XS	0.500	15.000	24.35	1.2272	4.189	3.93	550	275,000	82.77
		60	0.656	14.688	31.62	1.1766	4.189	3.85	528	264,000	107.54
		80	0.844	14.312	40.19	1.1171	4.189	3.75	501	250,500	136.58
		100	1.031	13.938	48.48	1.0596	4.189	3.65	474	237,000	164.86
		120	1.219	13.562	56.61	1.0032	4.189	3.55	450	225,000	192.40
		140	1.438	13.124	65.79	0.9394	4.189	3.44	422	211,000	223.57
		160	1.594	12.812	72.14	0.8953	4.189	3.35	402	201,000	245.22

(Continues)

Table 85 *(Continued)*

Nominal Pipe Size (in.)	Outside Diameter (in.)	Schedule No.	Wall Thickness (in.)	Inside Diameter (in.)	Cross-Sectional Area Metal (in.²)	Flow (ft²)	Circumference, ft, or surface, ft²/ft of Length Outside	Inside	Capacity at 1 ft/sec Velocity U.S. gal/min	lb/hr water	Weight of Plain-End Pipe (lb/ft)
18	18	5S	0.165	17.670	9.25	1.7029	4.712	4.63	764	382,000	31.32
		10S	0.188	17.624	10.52	1.6941	4.712	4.61	760	379,400	35.48
		10	0.250	17.500	13.94	1.6703	4.712	4.58	750	375,000	47.39
		20	0.312	17.376	17.34	1.6468	4.712	4.55	739	369,500	59.03
		ST	0.375	17.250	20.76	1.6230	4.712	4.52	728	364,000	70.59
		30	0.438	17.124	24.16	1.5993	4.712	4.48	718	359,000	82.06
		XS	0.500	17.000	27.49	1.5763	4.712	4.45	707	353,500	93.45
		40	0.562	16.876	30.79	1.5533	4.712	4.42	697	348,500	104.76
		60	0.750	16.500	40.64	1.4849	4.712	4.32	666	333,000	138.17
		80	0.938	16.124	50.28	1.4180	4.712	4.22	636	318,000	170.75
		100	1.156	15.688	61.17	1.3423	4.712	4.11	602	301,000	208.00
		120	1.375	15.250	71.82	1.2684	4.712	3.99	569	284,500	244.14
		140	1.562	14.876	80.66	1.2070	4.712	3.89	540	270,000	274.30
		160	1.781	14.438	90.75	1.1370	4.712	3.78	510	255,000	308.55
20	20	5S	0.188	19.624	11.70	2.1004	5.236	5.14	943	471,500	39.76
		10S	0.218	19.564	13.55	2.0878	5.236	5.12	937	467,500	45.98
		10	0.250	19.500	15.51	2.0740	5.236	5.11	930	465,000	52.73
		20, ST	0.375	19.250	23.12	2.0211	5.236	5.04	902	451,000	78.60
		30, XS	0.500	19.000	30.63	1.9689	5.236	4.97	883	441,500	104.13
		40	0.594	18.812	36.21	1.9302	5.236	4.92	866	433,000	123.06
		60	0.812	18.376	48.95	1.8417	5.236	4.81	826	413,000	166.50
		80	1.031	17.938	61.44	1.7550	5.236	4.70	787	393,500	208.92
		100	1.281	17.438	75.33	1.6585	5.236	4.57	744	372,000	256.15
		120	1.500	17.000	87.18	1.5763	5.236	4.45	707	353,500	296.37
		140	1.750	16.500	100.3	1.4849	5.236	4.32	665	332,500	341.10
		160	1.969	16.062	111.5	1.4071	5.236	4.21	632	316,000	379.14

	Schedule No.									
24	5S	0.218	23.564	16.29	3.0285	6.283	6.17	1359	679,500	55.08
	10, 10S	0.250	23.500	18.65	3.012	6.283	6.15	1350	675,000	63.41
	20, ST	0.375	23.250	27.83	2.948	6.283	6.09	1325	662,500	94.62
	XS	0.500	23.000	36.90	2.885	6.283	6.02	1295	642,500	125.49
	30	0.562	22.876	41.39	2.854	6.283	5.99	1281	640,500	140.80
	40	0.688	22.624	50.39	2.792	6.283	5.92	1253	626,500	171.17
	60	0.969	22.062	70.11	2.655	6.283	5.78	1192	596,000	238.29
	80	1.219	21.562	87.24	2.536	6.283	5.64	1138	569,000	296.53
	100	1.531	20.938	108.1	2.391	6.283	5.48	1073	536,500	367.45
	120	1.812	20.376	126.3	2.264	6.283	5.33	1016	508,000	429.50
	140	2.062	19.876	142.1	2.155	6.283	5.20	965	482,500	483.24
	160	2.344	19.312	159.5	2.034	6.283	5.06	913	456,500	542.09
30	5S	0.250	29.500	23.37	4.746	7.854	7.72	2130	1,065,000	79.43
	10, 10S	0.312	29.376	29.10	4.707	7.854	7.69	2110	1,055,000	99.08
	ST	0.375	29.250	34.90	4.666	7.854	7.66	2094	1,048,000	118.65
	20, XS	0.500	29.000	46.34	4.587	7.854	7.59	2055	1,027,000	157.53
	30	0.625	28.750	57.68	4.508	7.854	7.53	2020	1,010,000	196.08

Schedule Nos. 5S, 10S, and 40S American National Standards Institute (ANSI)/American Society of Mechanical Engineers (ASME) B.36.19-1985, "Stainless Steel Pipe." ST = standard wall, XS = extra strong wall, XX = double extra strong wall are all taken from ANSI/ASME, B.36.10M-1985, "Welded and Seamless Wrought-steel Pipe." Wrought-iron pipe has slightly thicker walls, approximately 3%, but the same weight per foot, because of lower density. Decimal thicknesses for respective pipe sizes represent their nominal or average wall dimensions. Mill tolerances as high as $12\frac{1}{2}\%$ are permitted.

Plain-end pipe is produced by a square cut. Pipe is also shipped from the mills threaded, with a threaded coupling on one end, or with the ends beveled for welding, or grooved or sized for patented couplings. Weights per foot for threaded and coupled pipe are slightly greater because of the weight of the coupling, but it is not available larger than 12 in., or lighter than Schedule 30 sizes 8 through 12 in., or Schedule 40 6 in. and smaller.

Source: From Chemical Engineer's Handbook, 4th ed., New York, McGraw-Hill, 1963. Used by permission.

Table 86 Properties and Dimensions of Steel Pipe[a]

	Dimensions						Couplings			Properties		
				Weight per Foot (lb)								
Nominal Diameter (in.)	Outside Diameter (in.)	Inside Diameter (in.)	Thickness (in.)	Plain Ends	Thread and Coupling	Threads per Inch	Outside Diameter (in.)	Length (in.)	Weight (lb)	I (in.^{4})	A (in.^{2})	k (in.)
					Schedule 40ST							
$\frac{1}{8}$	0.405	0.269	0.068	0.24	0.25	27	0.562	$\frac{7}{8}$	0.03	0.001	0.072	0.12
$\frac{1}{4}$	0.540	0.364	0.088	0.42	0.43	18	0.685	1	0.04	0.003	0.125	0.16
$\frac{3}{8}$	0.675	0.493	0.091	0.57	0.57	18	0.848	$1\frac{1}{8}$	0.07	0.007	0.167	0.21
$\frac{1}{2}$	0.840	0.622	0.109	0.85	0.85	14	1.024	$1\frac{3}{8}$	0.12	0.017	0.250	0.26
$\frac{3}{4}$	1.050	0.824	0.113	1.13	1.13	14	1.281	$1\frac{5}{8}$	0.21	0.037	0.333	0.33
1	1.315	1.049	0.133	1.68	1.68	$11\frac{1}{2}$	1.576	$1\frac{7}{8}$	0.35	0.087	0.494	0.42
$1\frac{1}{4}$	1.660	1.380	0.140	2.27	2.28	$11\frac{1}{2}$	1.950	$2\frac{1}{8}$	0.55	0.195	0.669	0.54
$1\frac{1}{2}$	1.900	1.610	0.145	2.72	2.73	$11\frac{1}{2}$	2.218	$2\frac{3}{8}$	0.76	0.310	0.799	0.62
2	2.375	2.067	0.154	3.65	3.68	$11\frac{1}{2}$	2.760	$2\frac{5}{8}$	1.23	0.666	1.075	0.79
$2\frac{1}{2}$	2.875	2.469	0.203	5.79	5.82	8	3.276	$2\frac{7}{8}$	1.76	1.530	1.704	0.95
3	3.500	3.068	0.216	7.58	7.62	8	3.948	$3\frac{1}{8}$	2.55	3.017	2.228	1.16
$3\frac{1}{2}$	4.000	3.548	0.226	9.11	9.20	8	4.591	$3\frac{5}{8}$	4.33	4.788	2.680	1.34
4	4.500	4.026	0.237	10.79	10.89	8	5.091	$3\frac{5}{8}$	5.41	7.233	3.174	1.51
5	5.563	5.047	0.258	14.62	14.81	8	6.296	$4\frac{1}{8}$	9.16	15.16	4.300	1.88
6	6.625	6.065	0.280	18.97	19.19	8	7.358	$4\frac{1}{8}$	10.82	28.14	5.581	2.25
8	8.625	8.071	0.277	24.70	25.00	8	9.420	$4\frac{5}{8}$	15.84	63.35	7.265	2.95
8	8.625	7.981	0.322	28.55	28.81	8	9.420	$4\frac{5}{8}$	15.84	72.49	8.399	2.94
10	10.750	10.192	0.279	31.20	32.00	8	11.721	$6\frac{1}{8}$	33.92	125.4	9.178	3.70
10	10.750	10.136	0.307	34.24	35.00	8	11.721	$6\frac{1}{8}$	33.92	137.4	10.07	3.69
10	10.750	10.020	0.365	40.48	41.13	8	11.721	$6\frac{1}{8}$	33.92	160.7	11.91	3.67
12	12.750	12.090	0.330	43.77	45.00	8	13.958	$6\frac{1}{8}$	48.27	248.5	12.88	4.39
12	12.750	12.000	0.375	49.56	50.71	8	13.958	$6\frac{1}{8}$	48.27	279.3	14.38	4.38
					Schedule 80XS							
$\frac{1}{8}$	0.405	0.215	0.095	0.31	0.32	27	0.582	$1\frac{1}{8}$	0.05	0.001	0.093	0.12
$\frac{1}{4}$	0.540	0.302	0.119	0.54	0.54	18	0.724	$1\frac{3}{8}$	0.07	0.004	0.157	0.16
$\frac{3}{8}$	0.675	0.423	0.126	0.74	0.75	18	0.898	$1\frac{5}{8}$	0.13	0.009	0.217	0.20
$\frac{1}{2}$	0.840	0.546	0.147	1.09	1.10	14	1.085	$1\frac{7}{8}$	0.22	0.020	0.320	0.25
$\frac{3}{4}$	1.050	0.742	0.154	1.47	1.49	14	1.316	$2\frac{1}{8}$	0.33	0.045	0.433	0.32
1	1.315	0.957	0.179	2.17	2.20	$11\frac{1}{2}$	1.575	$2\frac{3}{8}$	0.47	0.106	0.639	0.41
$1\frac{1}{4}$	1.660	1.278	0.191	3.00	3.05	$11\frac{1}{2}$	2.054	$2\frac{7}{8}$	1.04	0.242	0.881	0.52
$1\frac{1}{2}$	1.900	1.500	0.200	3.63	3.69	$11\frac{1}{2}$	2.294	$2\frac{7}{8}$	1.17	0.391	1.068	0.61
2	2.375	1.939	0.218	5.02	5.13	$11\frac{1}{2}$	2.870	$3\frac{5}{8}$	2.17	0.868	1.477	0.77
$2\frac{1}{2}$	2.875	2.323	0.276	7.66	7.83	8	3.389	$4\frac{1}{8}$	3.43	1.924	2.254	0.92
3	3.500	2.900	0.300	10.25	10.46	8	4.014	$4\frac{1}{8}$	4.13	3.894	3.016	1.14
$3\frac{1}{2}$	4.000	3.364	0.318	12.51	12.82	8	4.628	$4\frac{5}{8}$	6.29	6.280	3.678	1.31
4	4.500	3.826	0.337	14.98	15.39	8	5.233	$4\frac{5}{8}$	8.16	9.610	4.407	1.48
5	5.563	4.813	0.375	20.78	21.42	8	6.420	$5\frac{1}{8}$	12.87	20.67	6.112	1.84
6	6.625	5.761	0.432	28.57	29.33	8	7.482	$5\frac{1}{8}$	15.18	40.49	8.405	2.20
8	8.625	7.625	0.500	43.39	44.72	8	9.596	$6\frac{1}{8}$	26.63	105.7	12.76	2.88
10	10.750	9.750	0.500	54.74	56.94	8	11.958	$6\frac{5}{8}$	44.16	211.9	16.10	3.63
12	12.750	11.750	0.500	65.42	68.02	8	13.958	$6\frac{5}{8}$	51.99	361.5	19.24	4.34

Table 86 *(Continued)*

	Dimensions						Couplings			Properties		
					Weight per Foot (lb)							
Nominal Diameter (in.)	Outside Diameter (in.)	Inside Diameter (in.)	Thickness (in.)	Plain Ends	Thread and Coupling	Threads per Inch	Outside Diameter (in.)	Length (in.)	Weight (lb)	I (in.^{4})	A (in.^{2})	k (in.)
				Schedule XX								
$\frac{1}{2}$	0.840	0.252	0.294	1.71	1.73	14	1.085	$1\frac{7}{8}$	0.22	0.024	0.504	0.22
$\frac{3}{4}$	1.050	0.434	0.308	2.44	2.46	14	1.316	$2\frac{1}{8}$	0.33	0.058	0.718	0.28
1	1.315	0.599	0.358	3.66	3.68	$11\frac{1}{2}$	1.575	$2\frac{3}{8}$	0.47	0.140	1.076	0.36
$1\frac{1}{4}$	1.660	0.896	0.382	5.21	5.27	$11\frac{1}{2}$	2.054	$2\frac{7}{8}$	1.04	0.341	1.534	0.47
$1\frac{1}{2}$	1.900	1.100	0.400	6.41	6.47	$11\frac{1}{2}$	2.294	$2\frac{7}{8}$	1.17	0.568	1.885	0.55
2	2.375	1.503	0.436	9.03	9.14	$11\frac{1}{2}$	2.870	$3\frac{5}{8}$	2.17	1.311	2.656	0.70
$2\frac{1}{2}$	2.875	1.771	0.552	13.70	13.87	8	3.389	$4\frac{1}{8}$	3.43	2.871	4.028	0.84
3	3.500	2.300	0.600	18.58	18.79	8	4.014	$4\frac{1}{8}$	4.13	5.992	5.466	1.05
$3\frac{1}{2}$	4.000	2.728	0.636	22.85	23.16	8	4.628	$4\frac{5}{8}$	6.29	9.848	6.721	1.21
4	4.500	3.152	0.674	27.54	27.95	8	5.233	$4\frac{5}{8}$	8.16	15.28	8.101	1.37
5	5.563	4.063	0.750	38.55	39.20	8	6.420	$5\frac{1}{8}$	12.87	33.64	11.34	1.72
6	6.625	4.897	0.864	53.16	53.92	8	7.482	$5\frac{1}{8}$	15.18	66.33	15.64	2.06
8	8.625	6.875	0.875	72.42	73.76	8	9.596	$6\frac{1}{8}$	26.63	162.0	21.30	2.76
				Large Outside Diameter Pipe								

Pipe 14 in. and larger is sold by actual outside step diameter and thickness.

Sizes 14, 15, and 16 in. are available regularly in thicknesses varying by $\frac{1}{16}$ in. from $\frac{1}{4}$ to 1 in., inclusive.

All pipe is furnished random length unless otherwise ordered, viz: 12–22 ft with privilege of furnishing 5 % in 6–12-ft lengths. Pipe railing is most economically detailed with slip joints and random lengths between couplings.

[a] *Steel Construction*, 1980, A.I.S.C.

6.6 Standard Structural Shapes — Aluminum*

Table 87 **Aluminum Association Standard Channels — Dimensions, Areas, Weights, and Section Properties**[a]

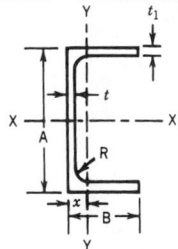

Size				Flange Thickness	Web Thickness	Fillet Radius	Section Properties[d]						
							Axis X–X			Axis Y–Y			
Depth	Width												
A (in.)	B (in.)	Area[b] (in.^{2})	Weight[c] (lb/ft)	t_1 (in.)	t (in.)	R (in.)	I (in.^{4})	S (in.^{3})	r (in.)	I (in.^{4})	S (in.^{3})	r (in.)	x (in.)
2.00	1.00	0.491	0.557	0.13	0.13	0.10	0.288	0.288	0.766	0.045	0.064	0.303	0.298
2.00	1.25	0.911	1.071	0.26	0.17	0.15	0.546	0.546	0.774	0.139	0.178	0.391	0.471
3.00	1.50	0.965	1.135	0.20	0.13	0.25	1.41	0.94	1.21	0.22	0.22	0.47	0.49
3.00	1.75	1.358	1.597	0.26	0.17	0.25	1.97	1.31	1.20	0.42	0.37	0.55	0.62
4.00	2.00	1.478	1.738	0.23	0.15	0.25	3.91	1.95	1.63	0.60	0.45	0.64	0.65
4.00	2.25	1.982	2.331	0.29	0.19	0.25	5.21	2.60	1.62	1.02	0.69	0.72	0.78
5.00	2.25	1.881	2.212	0.26	0.15	0.30	7.88	3.15	2.05	0.98	0.64	0.72	0.73
5.00	2.75	2.627	3.089	0.32	0.19	0.30	11.14	4.45	2.06	2.05	1.14	0.88	0.95

(Continues)

Tables 87–101 are from *Aluminum Standards and Data*. Copyright © 1984 The Aluminum Association.

Table 87 *(Continued)*

| Size | | | | Flange Thickness | Web Thickness | Fillet Radius | Section Properties[d] | | | | | | |
| Depth | Width | | | | | | Axis X–X | | | Axis Y–Y | | | |
A (in.)	B (in.)	Area[b] (in.²)	Weight[c] (lb/ft)	t₁ (in.)	t (in.)	R (in.)	I (in.⁴)	S (in.³)	r (in.)	I (in.⁴)	S (in.³)	r (in.)	x (in.)
6.00	2.50	2.410	2.834	0.29	0.17	0.30	14.35	4.78	2.44	1.53	0.90	0.80	0.79
6.00	3.25	3.427	4.030	0.35	0.21	0.30	21.04	7.01	2.48	3.76	1.76	1.05	1.12
7.00	2.75	2.725	3.205	0.29	0.17	0.30	22.09	6.31	2.85	2.10	1.10	0.88	0.84
7.00	3.50	4.009	4.715	0.38	0.21	0.30	33.79	9.65	2.90	5.13	2.23	1.13	1.20
8.00	3.00	3.526	4.147	0.35	0.19	0.30	37.40	9.35	3.26	3.25	1.57	0.96	0.93
8.00	3.75	4.923	5.789	0.41	0.25	0.35	52.69	13.17	3.27	7.13	2.82	1.20	1.22
9.00	3.25	4.237	4.983	0.35	0.23	0.35	54.41	12.09	3.58	4.40	1.89	1.02	0.93
9.00	4.00	5.927	6.970	0.44	0.29	0.35	78.31	17.40	3.63	9.61	3.49	1.27	1.25
10.00	3.50	5.218	6.136	0.41	0.25	0.35	83.22	16.64	3.99	6.33	2.56	1.10	1.02
10.00	4.25	7.109	8.360	0.50	0.31	0.40	116.15	23.23	4.04	13.02	4.47	1.35	1.34
12.00	4.00	7.036	8.274	0.47	0.29	0.40	159.76	26.63	4.77	11.03	3.86	1.25	1.14
12.00	5.00	10.053	11.822	0.62	0.35	0.45	239.69	39.95	4.88	25.74	7.60	1.60	1.61

[a] Users are encouraged to ascertain current availability of particular structural shapes through inquiries to their suppliers.
[b] Areas listed are based on nominal dimensions.
[c] Weights per foot are based on nominal dimensions and a density of 0.098 lb/in.³, which is the density of alloy 6061.
[d] I = moment of inertia; S = section modulus; r = radius of gyration.

Table 88　Aluminum Association Standard I Beams — Dimensions, Areas, Weights, and Section Properties[a]

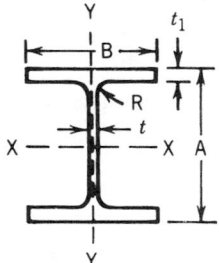

| Size | | | | Flange Thickness | Web Thickness | Fillet Radius | Section Properties[d] | | | | | |
| Depth | Width | | | | | | Axis X–X | | | Axis Y–Y | | |
A (in.)	B (in.)	Area[b] (in.²)	Weight[c] (lb/ft)	t₁ (in.)	t (in.)	R (in.)	I (in.⁴)	S (in.³)	r (in.)	I (in.⁴)	S (in.³)	r (in.)
3.00	2.50	1.392	1.637	0.20	0.13	0.25	2.24	1.49	1.27	0.52	0.42	0.61
3.00	2.50	1.726	2.030	0.26	0.15	0.25	2.71	1.81	1.25	0.68	0.54	0.63
4.00	3.00	1.965	2.311	0.23	0.15	0.25	5.62	2.81	1.69	1.04	0.69	0.73
4.00	3.00	2.375	2.793	0.29	0.17	0.25	6.71	3.36	1.68	1.31	0.87	0.74
5.00	3.50	3.146	3.700	0.32	0.19	0.30	13.94	5.58	2.11	2.29	1.31	0.85
6.00	4.00	3.427	4.030	0.29	0.19	0.30	21.99	7.33	2.53	3.10	1.55	0.95
6.00	4.00	3.990	4.692	0.35	0.21	0.30	25.50	8.50	2.53	3.74	1.87	0.97
7.00	4.50	4.932	5.800	0.38	0.23	0.30	42.89	12.25	2.95	5.78	2.57	1.08
8.00	5.00	5.256	6.181	0.35	0.23	0.30	59.69	14.92	3.37	7.30	2.92	1.18
8.00	5.00	5.972	7.023	0.41	0.25	0.30	67.78	16.94	3.37	8.55	3.42	1.20
9.00	5.50	7.110	8.361	0.44	0.27	0.30	102.02	22.67	3.79	12.22	4.44	1.31
10.00	6.00	7.352	8.646	0.41	0.25	0.40	132.09	26.42	4.24	14.78	4.93	1.42
10.00	6.00	8.747	10.286	0.50	0.29	0.40	155.79	31.16	4.22	18.03	6.01	1.44
12.00	7.00	9.925	11.672	0.47	0.29	0.40	255.57	42.60	5.07	26.90	7.69	1.65
12.00	7.00	12.153	14.292	0.62	0.31	0.40	317.33	52.89	5.11	35.48	10.14	1.71

[a] Users are encouraged to ascertain current availability of particular structural shapes through inquiries to their suppliers.
[b] Areas listed are based on nominal dimensions.
[c] Weights per foot are based on nominal dimensions and a density of 0.098 lb/in.³, which is the density of alloy 6061.
[d] I = moment of inertia; S = section modulus; r = radius of gyration.

Table 89 Standard Structural Shapes — Equal Angles [a]

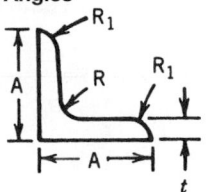

A	t	R	R_1	Area[b] (in.^{2})	Weight per Foot[c] (lb)
$\frac{3}{4}$	$\frac{1}{8}$	$\frac{1}{8}$	$\frac{3}{32}$	0.171	0.201
$\frac{3}{4}$	$\frac{3}{16}$	$\frac{1}{8}$	$\frac{3}{32}$	0.246	0.289
1	$\frac{3}{32}$	$\frac{1}{8}$	$\frac{3}{32}$	0.179	0.211
1	$\frac{1}{8}$	$\frac{1}{8}$	$\frac{3}{32}$	0.234	0.275
1	$\frac{3}{16}$	$\frac{1}{8}$	$\frac{3}{32}$	0.340	0.400
1	$\frac{1}{4}$	$\frac{1}{8}$	$\frac{3}{32}$	0.437	0.514
$1\frac{1}{4}$	$\frac{1}{8}$	$\frac{3}{16}$	$\frac{1}{8}$	0.292	0.343
$1\frac{1}{4}$	$\frac{3}{16}$	$\frac{3}{16}$	$\frac{1}{8}$	0.434	0.510
$1\frac{1}{4}$	$\frac{1}{4}$	$\frac{3}{16}$	$\frac{1}{8}$	0.558	0.656
$1\frac{1}{2}$	$\frac{1}{8}$	$\frac{3}{16}$	$\frac{1}{8}$	0.360	0.423
$1\frac{1}{2}$	$\frac{3}{16}$	$\frac{3}{16}$	$\frac{1}{8}$	0.529	0.619
$1\frac{1}{2}$	$\frac{1}{4}$	$\frac{3}{16}$	$\frac{1}{8}$	0.688	0.809
$1\frac{3}{4}$	$\frac{1}{8}$	$\frac{3}{16}$	$\frac{1}{8}$	0.423	0.497
$1\frac{3}{4}$	$\frac{3}{16}$	$\frac{3}{16}$	$\frac{1}{8}$	0.622	0.731
$1\frac{3}{4}$	$\frac{1}{4}$	$\frac{3}{16}$	$\frac{1}{8}$	0.813	0.956
$1\frac{3}{4}$	$\frac{5}{16}$	$\frac{3}{16}$	$\frac{1}{8}$	0.996	1.171
2	$\frac{1}{8}$	$\frac{1}{4}$	$\frac{1}{8}$	0.491	0.577
2	$\frac{3}{16}$	$\frac{1}{4}$	$\frac{1}{8}$	0.723	0.850
2	$\frac{1}{4}$	$\frac{1}{4}$	$\frac{1}{8}$	0.944	1.110
2	$\frac{5}{16}$	$\frac{1}{4}$	$\frac{1}{8}$	1.160	1.364
2	$\frac{3}{8}$	$\frac{1}{4}$	$\frac{1}{8}$	1.366	1.606
$2\frac{1}{2}$	$\frac{1}{8}$	$\frac{1}{4}$	$\frac{1}{8}$	0.616	0.724
$2\frac{1}{2}$	$\frac{3}{16}$	$\frac{1}{4}$	$\frac{1}{8}$	0.910	1.070
$2\frac{1}{2}$	$\frac{1}{4}$	$\frac{1}{4}$	$\frac{1}{8}$	1.194	1.404
$2\frac{1}{2}$	$\frac{5}{16}$	$\frac{1}{4}$	$\frac{1}{8}$	1.470	1.729
$2\frac{1}{2}$	$\frac{3}{8}$	$\frac{1}{4}$	$\frac{1}{8}$	1.714	2.047
3	$\frac{3}{16}$	$\frac{5}{16}$	$\frac{1}{4}$	1.084	1.275
3	$\frac{1}{4}$	$\frac{5}{16}$	$\frac{1}{4}$	1.432	1.684
3	$\frac{5}{16}$	$\frac{5}{16}$	$\frac{1}{4}$	1.770	2.082
3	$\frac{3}{8}$	$\frac{5}{16}$	$\frac{1}{4}$	2.104	2.474
3	$\frac{7}{16}$	$\frac{5}{16}$	$\frac{1}{4}$	2.428	2.855
3	$\frac{1}{2}$	$\frac{5}{16}$	$\frac{1}{4}$	2.744	3.227
$3\frac{1}{2}$	$\frac{1}{4}$	$\frac{3}{8}$	$\frac{1}{4}$	1.691	1.989
$3\frac{1}{2}$	$\frac{5}{16}$	$\frac{3}{8}$	$\frac{1}{4}$	2.093	2.461
$3\frac{1}{2}$	$\frac{3}{8}$	$\frac{3}{8}$	$\frac{1}{4}$	2.488	2.926
$3\frac{1}{2}$	$\frac{1}{2}$	$\frac{3}{8}$	$\frac{1}{4}$	3.253	3.826

(Continues)

Table 89 *(Continued)*

A	t	R	R_1	Area[b] (in.²)	Weight per Foot[c] (lb)
4	$\frac{1}{4}$	$\frac{3}{8}$	$\frac{1}{4}$	1.941	2.283
4	$\frac{5}{16}$	$\frac{3}{8}$	$\frac{1}{4}$	2.406	2.829
4	$\frac{3}{8}$	$\frac{3}{8}$	$\frac{1}{4}$	2.862	3.366
4	$\frac{7}{16}$	$\frac{3}{8}$	$\frac{1}{4}$	3.310	3.893
4	$\frac{1}{2}$	$\frac{3}{8}$	$\frac{1}{4}$	3.753	4.414
4	$\frac{9}{16}$	$\frac{3}{8}$	$\frac{1}{4}$	4.187	4.924
4	$\frac{5}{8}$	$\frac{3}{8}$	$\frac{1}{4}$	4.613	5.425
4	$\frac{11}{16}$	$\frac{3}{8}$	$\frac{1}{4}$	5.032	5.918
4	$\frac{3}{4}$	$\frac{3}{8}$	$\frac{1}{4}$	5.441	6.399
5	$\frac{3}{8}$	$\frac{1}{2}$	$\frac{3}{8}$	3.603	4.237
5	$\frac{7}{16}$	$\frac{1}{2}$	$\frac{3}{8}$	4.177	4.912
5	$\frac{1}{2}$	$\frac{1}{2}$	$\frac{3}{8}$	4.743	5.578
5	$\frac{5}{8}$	$\frac{1}{2}$	$\frac{3}{8}$	5.853	6.883
6	$\frac{3}{8}$	$\frac{1}{2}$	$\frac{3}{8}$	4.353	5.119
6	$\frac{7}{16}$	$\frac{1}{2}$	$\frac{3}{8}$	5.052	5.941
6	$\frac{1}{2}$	$\frac{1}{2}$	$\frac{3}{8}$	5.743	6.754
6	$\frac{5}{8}$	$\frac{1}{2}$	$\frac{3}{8}$	7.102	8.352
8	$\frac{1}{2}$	$\frac{5}{8}$	$\frac{3}{8}$	7.773	9.141
8	$\frac{3}{4}$	$\frac{5}{8}$	$\frac{3}{8}$	11.461	13.478
8	1	$\frac{5}{8}$	$\frac{3}{8}$	15.023	17.667

[a] Users are encouraged to ascertain current availability of particular structural shapes through inquiries to their suppliers.
[b] Areas listed are based on nominal dimensions.
[c] Weights per foot are based on nominal dimensions and a density of 0.098 lb/in.³, which is the density of alloy 6061.

Table 90 Standard Structural Shapes — Unequal Angles[a]

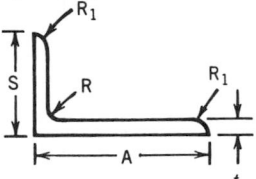

A	B	t	R	R_1	Area[b] (in.²)	Weight per Foot[c] (lb)
$1\frac{1}{4}$	$\frac{3}{4}$	$\frac{3}{32}$	$\frac{3}{32}$	$\frac{3}{64}$	0.180	0.212
$1\frac{1}{4}$	1	$\frac{1}{8}$	$\frac{1}{8}$	$\frac{1}{16}$	0.267	0.314
$1\frac{1}{2}$	$\frac{3}{4}$	$\frac{1}{8}$	$\frac{1}{8}$	$\frac{1}{16}$	0.267	0.314
$1\frac{1}{2}$	$\frac{3}{4}$	$\frac{3}{16}$	$\frac{1}{8}$	$\frac{3}{32}$	0.386	0.454
$1\frac{1}{2}$	1	$\frac{5}{32}$	$\frac{5}{32}$	$\frac{5}{64}$	0.368	0.433
$1\frac{1}{2}$	1	$\frac{1}{4}$	$\frac{3}{16}$	$\frac{1}{8}$	0.563	0.662
$1\frac{1}{2}$	$1\frac{1}{4}$	$\frac{1}{8}$	$\frac{3}{16}$	$\frac{1}{8}$	0.329	0.387
$1\frac{1}{2}$	$1\frac{1}{4}$	$\frac{3}{16}$	$\frac{3}{16}$	$\frac{1}{8}$	0.481	0.566
$1\frac{1}{2}$	$1\frac{1}{4}$	$\frac{1}{4}$	$\frac{3}{16}$	$\frac{1}{8}$	0.624	0.734
$1\frac{3}{4}$	$1\frac{1}{4}$	$\frac{1}{8}$	$\frac{3}{16}$	$\frac{1}{8}$	0.358	0.421

Table 90 *(Continued)*

A	B	t	R	R_1	Area[b] (in.²)	Weight per Foot[c] (lb)
$1\frac{3}{4}$	$1\frac{1}{4}$	$\frac{3}{16}$	$\frac{3}{16}$	$\frac{1}{8}$	0.528	0.621
$1\frac{3}{4}$	$1\frac{1}{4}$	$\frac{1}{4}$	$\frac{3}{16}$	$\frac{1}{8}$	0.688	0.809
2	$1\frac{1}{2}$	$\frac{1}{8}$	$\frac{3}{16}$	$\frac{1}{8}$	0.422	0.496
2	$1\frac{1}{2}$	$\frac{3}{16}$	$\frac{3}{16}$	$\frac{1}{8}$	0.622	0.731
2	$1\frac{1}{2}$	$\frac{1}{4}$	$\frac{3}{16}$	$\frac{1}{8}$	0.813	0.956
2	$1\frac{1}{2}$	$\frac{3}{8}$	$\frac{3}{16}$	$\frac{1}{8}$	1.172	1.378
$2\frac{1}{2}$	$1\frac{1}{2}$	$\frac{3}{16}$	$\frac{1}{4}$	$\frac{1}{8}$	0.723	0.850
$2\frac{1}{2}$	$1\frac{1}{2}$	$\frac{1}{4}$	$\frac{1}{4}$	$\frac{1}{8}$	0.944	1.110
$2\frac{1}{2}$	$1\frac{1}{2}$	$\frac{5}{16}$	$\frac{3}{16}$	$\frac{1}{8}$	1.152	1.355
$2\frac{1}{2}$	2	$\frac{1}{8}$	$\frac{1}{4}$	$\frac{1}{8}$	0.554	0.652
$2\frac{1}{2}$	2	$\frac{3}{16}$	$\frac{1}{4}$	$\frac{1}{8}$	0.817	0.961
$2\frac{1}{2}$	2	$\frac{1}{4}$	$\frac{1}{4}$	$\frac{1}{8}$	1.069	1.257
$2\frac{1}{2}$	2	$\frac{5}{16}$	$\frac{1}{4}$	$\frac{1}{8}$	1.314	1.545
$2\frac{1}{2}$	2	$\frac{3}{8}$	$\frac{1}{4}$	$\frac{1}{8}$	1.554	1.828
3	2	$\frac{3}{16}$	$\frac{5}{16}$	$\frac{3}{16}$	0.911	1.071
3	2	$\frac{1}{4}$	$\frac{5}{16}$	$\frac{3}{16}$	1.193	1.403
3	2	$\frac{5}{16}$	$\frac{5}{16}$	$\frac{3}{16}$	1.471	1.730
3	2	$\frac{3}{8}$	$\frac{5}{16}$	$\frac{3}{16}$	1.740	2.046
3	2	$\frac{7}{16}$	$\frac{5}{16}$	$\frac{3}{16}$	2.001	2.353
3	$2\frac{1}{2}$	$\frac{1}{4}$	$\frac{5}{16}$	$\frac{1}{4}$	1.307	1.537
3	$2\frac{1}{2}$	$\frac{5}{16}$	$\frac{5}{16}$	$\frac{1}{4}$	1.614	1.898
3	$2\frac{1}{2}$	$\frac{3}{8}$	$\frac{5}{16}$	$\frac{1}{4}$	1.916	2.253
$3\frac{1}{2}$	$2\frac{1}{2}$	$\frac{1}{4}$	$\frac{5}{16}$	$\frac{1}{4}$	1.432	1.684
$3\frac{1}{2}$	$2\frac{1}{2}$	$\frac{5}{16}$	$\frac{5}{16}$	$\frac{1}{4}$	1.770	2.082
$3\frac{1}{2}$	$2\frac{1}{2}$	$\frac{3}{8}$	$\frac{5}{16}$	$\frac{1}{4}$	2.104	2.474
$3\frac{1}{2}$	$2\frac{1}{2}$	$\frac{1}{2}$	$\frac{5}{16}$	$\frac{1}{4}$	2.744	3.227
$3\frac{1}{2}$	3	$\frac{1}{4}$	$\frac{3}{8}$	$\frac{1}{4}$	1.566	1.842
$3\frac{1}{2}$	3	$\frac{5}{16}$	$\frac{3}{8}$	$\frac{1}{4}$	1.937	2.278
$3\frac{1}{2}$	3	$\frac{3}{8}$	$\frac{3}{8}$	$\frac{1}{4}$	2.300	2.705
$3\frac{1}{2}$	3	$\frac{1}{2}$	$\frac{3}{8}$	$\frac{1}{4}$	3.003	3.532
4	3	$\frac{1}{4}$	$\frac{3}{8}$	$\frac{1}{4}$	1.691	1.988
4	3	$\frac{5}{16}$	$\frac{3}{8}$	$\frac{1}{4}$	2.091	2.459
4	3	$\frac{3}{8}$	$\frac{3}{8}$	$\frac{1}{4}$	2.488	2.926
4	3	$\frac{7}{16}$	$\frac{3}{8}$	$\frac{1}{4}$	2.874	3.380
4	3	$\frac{1}{2}$	$\frac{3}{8}$	$\frac{1}{4}$	3.253	3.826
4	3	$\frac{5}{8}$	$\frac{3}{8}$	$\frac{1}{4}$	3.988	4.690
4	$3\frac{1}{2}$	$\frac{3}{8}$	$\frac{3}{8}$	$\frac{5}{16}$	2.660	3.128
4	$3\frac{1}{2}$	$\frac{1}{2}$	$\frac{3}{8}$	$\frac{5}{16}$	3.488	4.102
5	3	$\frac{3}{8}$	$\frac{3}{8}$	$\frac{5}{16}$	2.848	3.349
5	3	$\frac{1}{2}$	$\frac{3}{8}$	$\frac{5}{16}$	3.738	4.396
5	$3\frac{1}{2}$	$\frac{5}{16}$	$\frac{7}{16}$	$\frac{5}{16}$	2.558	3.008
5	$3\frac{1}{2}$	$\frac{3}{8}$	$\frac{7}{16}$	$\frac{5}{16}$	3.046	3.582
5	$3\frac{1}{2}$	$\frac{7}{16}$	$\frac{7}{16}$	$\frac{5}{16}$	3.527	4.148

(Continues)

Table 90 *(Continued)*

A	B	t	R	R_1	Area[b] (in.²)	Weight per Foot[c] (lb)
5	$3\frac{1}{2}$	$\frac{1}{2}$	$\frac{7}{16}$	$\frac{5}{16}$	4.000	4.704
5	$3\frac{1}{2}$	$\frac{5}{8}$	$\frac{7}{16}$	$\frac{5}{16}$	4.921	5.787
6	$3\frac{1}{2}$	$\frac{5}{16}$	$\frac{1}{2}$	$\frac{5}{16}$	2.878	3.385
6	$3\frac{1}{2}$	$\frac{3}{8}$	$\frac{1}{2}$	$\frac{5}{16}$	3.433	4.037
6	$3\frac{1}{2}$	$\frac{1}{2}$	$\frac{1}{2}$	$\frac{5}{16}$	4.512	5.306
6	4	$\frac{3}{8}$	$\frac{1}{2}$	$\frac{3}{8}$	3.603	4.237
6	4	$\frac{7}{16}$	$\frac{1}{2}$	$\frac{3}{8}$	4.179	4.915
6	4	$\frac{1}{2}$	$\frac{1}{2}$	$\frac{3}{8}$	4.743	5.578
6	4	$\frac{9}{16}$	$\frac{1}{2}$	$\frac{3}{8}$	5.298	6.230
6	4	$\frac{5}{8}$	$\frac{1}{2}$	$\frac{3}{8}$	5.853	6.883
6	4	$\frac{3}{4}$	$\frac{1}{2}$	$\frac{3}{8}$	6.931	8.151
8	6	$\frac{5}{8}$	$\frac{1}{2}$	$\frac{5}{16}$	8.371	9.844
8	6	$\frac{11}{16}$	$\frac{1}{2}$	$\frac{3}{8}$	9.152	10.763
8	6	$\frac{3}{4}$	$\frac{1}{2}$	$\frac{3}{8}$	9.931	11.679

[a] Users are encouraged to ascertain current availability of particular structural shapes through inquiries to their suppliers.
[b] Areas listed are based on nominal dimensions.
[c] Weights per foot are based on nominal dimensions and a density of 0.098lb/in.³, which is the density of alloy 6061.

Table 91   Channels, American Standard[a]

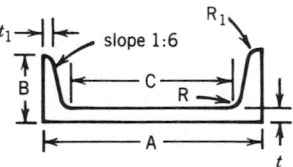

A	B	C	t	t_1	R	R_1	Area[b] (in.²)	Weight per Foot[c] (lb)
3	1.410	$1\frac{3}{4}$	0.170	0.170	0.270	0.100	1.205	1.417
3	1.498	$1\frac{3}{4}$	0.258	0.170	0.270	0.100	1.470	1.729
3	1.596	$1\frac{3}{4}$	0.356	0.170	0.270	0.100	1.764	2.074
4	1.580	$2\frac{3}{4}$	0.180	0.180	0.280	0.110	1.570	1.846
4	1.647	$2\frac{3}{4}$	0.247	0.180	0.280	0.110	1.838	2.161
4	1.720	$2\frac{3}{4}$	0.320	0.180	0.280	0.110	2.129	2.504
5	1.750	$3\frac{3}{4}$	0.190	0.190	0.290	0.110	1.969	2.316
5	1.885	$3\frac{3}{4}$	0.325	0.190	0.290	0.110	2.643	3.108
5	2.032	$3\frac{3}{4}$	0.472	0.190	0.290	0.110	3.380	3.975
6	1.920	$4\frac{1}{2}$	0.200	0.200	0.300	0.120	2.403	2.826
6	1.945	$4\frac{1}{2}$	0.225	0.200	0.300	0.120	2.553	3.002
6	2.034	$4\frac{1}{2}$	0.314	0.200	0.300	0.120	3.088	3.631
6	2.157	$4\frac{1}{2}$	0.437	0.200	0.300	0.120	3.825	4.498
7	2.110	$5\frac{1}{2}$	0.230	0.210	0.310	0.130	3.011	3.541
7	2.194	$5\frac{1}{2}$	0.314	0.210	0.310	0.130	3.599	4.232
7	2.299	$5\frac{1}{2}$	0.419	0.210	0.310	0.130	4.334	5.097
8	2.290	$6\frac{1}{4}$	0.250	0.220	0.320	0.130	3.616	4.252
8	2.343	$6\frac{1}{4}$	0.303	0.220	0.320	0.130	4.040	4.751
8	2.435	$6\frac{1}{4}$	0.395	0.220	0.320	0.130	4.776	5.617
8	2.527	$6\frac{1}{4}$	0.487	0.220	0.320	0.130	5.514	6.484

Table 91   (Continued)

A	B	C	t	t_1	R	R_1	Area[b] (in.^{2})	Weight per Foot[c] (lb)
9	2.430	$7\frac{1}{4}$	0.230	0.230	0.330	0.140	3.915	4.604
9	2.648	$7\frac{1}{4}$	0.448	0.230	0.330	0.140	5.877	6.911
10	2.600	$8\frac{1}{4}$	0.240	0.240	0.340	0.140	4.488	5.278
10	2.886	$8\frac{1}{4}$	0.526	0.240	0.340	0.140	7.348	8.641
12	2.960	10	0.300	0.280	0.380	0.170	6.302	7.411
12	3.047	10	0.387	0.280	0.380	0.170	7.346	8.639
12	3.170	10	0.510	0.280	0.380	0.170	8.822	10.374
15	3.400	$12\frac{3}{8}$	0.400	0.400	0.500	0.240	9.956	11.708
15	3.716	$12\frac{3}{8}$	0.716	0.400	0.500	0.240	14.696	17.282

[a] Users are encouraged to ascertain current availability of particular structural shapes through inquiries to their suppliers.
[b] Areas listed are based on nominal dimensions.
[c] Weights per foot are based on nominal dimensions and a density of 0.098 lb/in.^{3}, which is the density of alloy 6061.

Table 92   Channels, Shipbuilding, and Carbuilding[a]

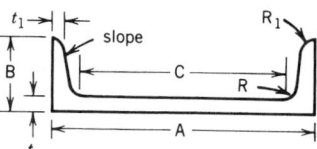

A	B	C	t	t_1	R	R_1	Slope	Area[b] (in.^{2})	Weight per Foot[c] (lb)
3	2	$1\frac{3}{4}$	0.250	0.250	0.250	0	12:12.1	1.900	2.234
3	2	$1\frac{7}{8}$	0.375	0.375	0.188	0.375	0	2.298	2.702
4	$2\frac{1}{2}$	$2\frac{3}{8}$	0.318	0.313	0.375	0.125	1:34.9	2.825	3.322
5	$2\frac{7}{8}$	3	0.438	0.438	0.250	0.094	1:9.8	4.950	5.821
6	3	$4\frac{1}{2}$	0.500	0.375	0.375	0.250	0	4.909	5.773
6	$3\frac{1}{2}$	4	0.375	0.412	0.480	0.420	1:49.6	5.044	5.932
8	3	$5\frac{3}{4}$	0.380	0.380	0.550	0.220	1:14.43	5.600	6.586
8	$3\frac{1}{2}$	$5\frac{3}{4}$	0.425	0.471	0.525	0.375	1:28.5	6.682	7.858
10	$3\frac{1}{2}$	$7\frac{1}{2}$	0.375	0.375	0.625	0.188	1:9	7.298	8.581
10	$3\frac{9}{16}$	$7\frac{1}{2}$	0.438	0.375	0.625	0.188	1:9	7.928	9.323
10	$3\frac{5}{8}$	$7\frac{1}{2}$	0.500	0.375	0.625	0.188	1:9	8.548	10.052

Table 93   H Beams[a]

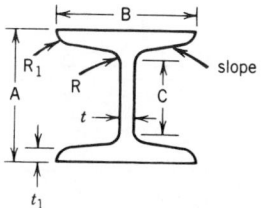

A	B	C	t	t_1	R	R_1	Slope	Area[b] (in.^{2})	Weight per Foot[c] (lb)
4	4	$2\frac{3}{8}$	0.313	0.290	0.313	0.145	1:11.3	4.046	4.758
5	5	$3\frac{3}{8}$	0.313	0.330	0.313	0.165	1:13.6	5.522	6.494
6	5.938	$4\frac{3}{8}$	0.250	0.360	0.313	0.180	1:15.6	6.678	7.853
8	7.938	$6\frac{1}{4}$	0.313	0.358	0.313	0.179	1:18.9	9.554	11.263
8	8.125	$6\frac{1}{4}$	0.500	0.358	0.313	0.179	1:18.9	11.050	12.995

[a] Users are encouraged to ascertain current availability of particular structural shapes through inquiries to their suppliers.
[b] Areas listed are based on nominal dimensions.
[c] Weights per foot are based on nominal dimensions and a density of 0.098 lb/in.^{3}, which is the density of alloy 6061.

Table 94 I Beams[a]

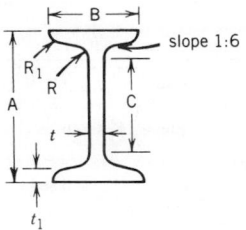

A	B	C	t	t_1	R	R_1	Area[b] (in.²)	Weight per Foot[c] (lb)
3	2.330	$1\frac{3}{4}$	0.170	0.170	0.270	0.100	1.669	1.963
3	2.509	$1\frac{3}{4}$	0.349	0.170	0.270	0.100	2.203	2.591
4	2.660	$2\frac{3}{4}$	0.190	0.190	0.290	0.110	2.249	2.644
4	2.796	$2\frac{3}{4}$	0.326	0.190	0.290	0.110	2.792	3.283
5	3	$3\frac{1}{2}$	0.210	0.210	0.310	0.130	2.917	3.430
5	3.284	$3\frac{1}{2}$	0.494	0.210	0.310	0.130	4.337	5.100
6	3.330	$4\frac{1}{2}$	0.230	0.230	0.330	0.140	3.658	4.302
6	3.443	$4\frac{1}{2}$	0.343	0.230	0.330	0.140	4.336	5.099
7	3.755	$5\frac{1}{4}$	0.345	0.250	0.350	0.150	5.147	6.053
8	4	$6\frac{1}{4}$	0.270	0.270	0.370	0.160	5.398	6.348
8	4.262	$6\frac{1}{4}$	0.532	0.270	0.370	0.160	7.494	8.813
10	4.660	8	0.310	0.310	0.410	0.190	7.452	8.764
12	5	$9\frac{3}{4}$	0.350	0.350	0.450	0.210	9.349	10.994

[a] Users are encouraged to ascertain current availability of particular structural shapes through inquiries to their suppliers.
[b] Areas listed are based on nominal dimensions.
[c] Weights per foot are based on nominal dimensions and a density of 0.098 lb/in.³, which is the density of alloy 6061.

Table 95 Wide-Flange Beams[a]

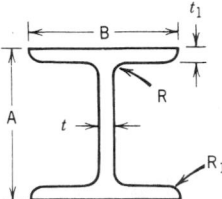

A	B	t	t_1	R	R_1	Area[b] (in.²)	Weight per Foot[c] (lb)
6.000	4.000	0.230	0.279	0.250	—	3.538	4.161
6.000	6.000	0.240	0.269	0.250	—	4.593	5.401
8.000	5.250	0.230	0.308	0.320	—	5.020	5.904
8.000	6.500	0.245	0.398	0.400	—	7.076	8.321
8.000	8.000	0.288	0.433	0.400	—	9.120	10.725
9.750	7.964	0.292	0.433	0.500	—	9.706	11.414
9.900	5.750	0.240	0.340	0.312	0.031	6.205	7.297
11.940	8.000	0.294	0.516	0.600	—	11.772	13.844
12.060	10.000	0.345	0.576	0.600	—	15.593	18.337

[a] Users are encouraged to ascertain current availability of particular structural shapes through inquiries to their suppliers.
[b] Areas listed are based on nominal dimensions.
[c] Weights per foot are based on nominal dimensions and a density of 0.098 lb/in.³, which is the density of alloy 6061.

Table 96 Tees[a]

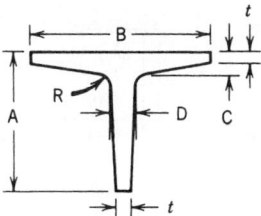

A	B	C	D	t	R	Area[b] (in.²)	Weight per Foot[c] (lb)
2	2	0.312	0.312	0.250	0.250	1.071	1.259
$2\frac{1}{4}$	$2\frac{1}{4}$	0.312	0.312	0.250	0.250	1.208	1.421
$2\frac{1}{2}$	$2\frac{1}{2}$	0.375	0.375	0.312	0.250	1.626	1.912
3	3	0.438	0.438	0.375	0.312	2.310	2.717
4	4	0.438	0.438	0.375	0.500	3.183	3.743

[a] Users are encouraged to ascertain current availability of particular structural shapes through inquiries to their suppliers.
[b] Areas listed are based on nominal dimensions.
[c] Weights per foot are based on nominal dimensions and a density of 0.098 lb/in.³, which is the density of alloy 6061.

Table 97 Zees[a]

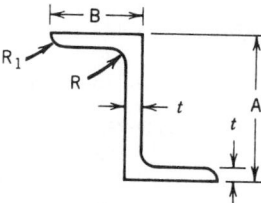

A	B	t	R	R_1	Area[b] (in.²)	Weight per Foot[c] (lb)
3	$2\frac{11}{16}$	0.250	0.312	0.250	1.984	2.333
3	$2\frac{11}{16}$	0.375	0.312	0.250	2.875	3.381
4	$3\frac{1}{16}$	0.250	0.312	0.250	2.422	2.848
$4\frac{1}{16}$	$3\frac{1}{8}$	0.312	0.312	0.250	3.040	3.575
$4\frac{1}{4}$	$3\frac{3}{16}$	0.375	0.312	0.250	3.672	4.318
5	$3\frac{1}{4}$	0.500	0.312	0.250	5.265	6.192
$5\frac{1}{16}$	$3\frac{5}{16}$	0.375	0.312	0.250	4.093	4.813

[a] Users are encouraged to ascertain current availability of particular structural shapes through inquiries to their suppliers.
[b] Areas listed are based on nominal dimensions.
[c] Weights per foot are based on nominal dimensions and a density of 0.098 lb/in.³, which is the density of alloy 6061.

Table 98   Aluminum Pipe — Diameters, Wall Thicknesses, and Weights

Nominal Pipe Size[a] (in.)	Schedule Number[a]	Outside Diameter (in.)			Inside Diameter (in.)	Wall Thickness (in.)			Weight per Foot (lb)	
		Nom[a]	Min[b,c]	Max[b,c]	Nom	Nom[a]	Min[b]	Max[b]	Nom[d]	Min[b,d]
$\frac{1}{8}$	40	0.405	0.374	0.420	0.269	0.068	0.060	—	0.085	0.091
	80	0.405	0.374	0.420	0.215	0.095	0.083	—	0.109	0.118
$\frac{1}{4}$	40	0.540	0.509	0.555	0.364	0.088	0.077	—	0.147	0.159
	80	0.540	0.509	0.555	0.302	0.119	0.104	—	0.185	0.200
$\frac{3}{8}$	40	0.675	0.644	0.690	0.493	0.091	0.080	—	0.196	0.212
	80	0.675	0.644	0.690	0.493	0.091	0.080	—	0.196	0.212
	5	0.840	0.809	0.855	0.710	0.065	0.053	0.077	0.186	—
	10	0.840	0.809	0.855	0.674	0.083	0.071	0.095	0.232	—
$\frac{1}{2}$	40	0.840	0.809	0.855	0.622	0.109	0.095	—	0.294	0.318
	80	0.840	0.809	0.855	0.546	0.147	0.129	—	0.376	0.406
	160	0.840	0.809	0.855	0.464	0.188	0.164	—	0.453	0.489
	5	1.050	1.019	1.065	0.920	0.065	0.053	0.077	0.237	—
	10	1.050	1.019	1.065	0.884	0.083	0.071	0.095	0.297	—
$\frac{3}{4}$	40	1.050	1.019	1.065	0.824	0.113	0.099	—	0.391	0.422
	80	1.050	1.019	1.065	0.742	0.154	0.135	—	0.510	0.551
	160	1.050	1.019	1.065	0.612	0.219	0.192	—	0.672	0.726
	5	1.315	1.284	1.330	1.185	0.065	0.053	0.077	0.300	—
	10	1.315	1.284	1.330	1.097	0.109	0.095	0.123	0.486	—
1	40	1.315	1.284	1.330	1.049	0.133	0.116	—	0.581	0.627
	80	1.315	1.284	1.330	0.957	0.179	0.157	—	0.751	0.811
	160	1.315	1.284	1.330	0.815	0.250	0.219	—	0.984	1.062
	5	1.660	1.629	1.675	1.530	0.065	0.053	0.077	0.383	—
	10	1.660	1.629	1.675	1.442	0.109	0.095	0.123	0.625	—
$1\frac{1}{4}$	40	1.660	1.629	1.675	1.380	0.140	0.122	—	0.786	0.849
	80	1.660	1.629	1.675	1.278	0.191	0.167	—	1.037	1.120
	160	1.660	1.629	1.675	1.160	0.250	0.219	—	1.302	1.407
	5	1.900	1.869	1.915	1.770	0.065	0.053	0.077	0.441	—
	10	1.900	1.869	1.915	1.682	0.109	0.095	0.123	0.721	—
$1\frac{1}{2}$	40	1.900	1.869	1.915	1.610	0.145	0.127	—	0.940	1.015
	80	1.900	1.869	1.915	1.500	0.200	0.175	—	1.256	1.357
	160	1.900	1.869	1.915	1.338	0.281	0.246	—	1.681	1.815
	5	2.375	2.344	2.406	2.245	0.065	0.053	0.077	0.555	—
	10	2.375	2.344	2.406	2.157	0.109	0.095	0.123	0.913	—
2	40	2.375	2.351	2.399	2.067	0.154	0.135	—	1.264	1.365
	80	2.375	2.351	2.399	1.939	0.218	0.191	—	1.737	1.876
	160	2.375	2.351	2.399	1.687	0.344	0.301	—	2.581	2.788
	5	2.875	2.844	2.906	2.709	0.083	0.071	0.095	0.856	—
	10	2.875	2.844	2.906	2.635	0.120	0.105	0.135	1.221	—
$2\frac{1}{2}$	40	2.875	2.846	2.904	2.469	0.203	0.178	—	2.004	2.164
	80	2.875	2.846	2.904	2.323	0.276	0.242	—	2.650	2.862
	160	2.875	2.846	2.904	2.125	0.375	0.328	—	3.464	3.741
	5	3.500	3.469	3.531	3.334	0.083	0.071	0.095	1.048	—
	10	3.500	3.469	3.531	3.260	0.120	0.105	0.135	1.498	—
3	40	3.500	3.465	3.535	3.068	0.216	0.189	—	2.621	2.830
	80	3.500	3.465	3.535	2.900	0.300	0.262	—	3.547	3.830
	160	3.500	3.465	3.535	2.624	0.438	0.383	—	4.955	5.351
$3\frac{1}{2}$	5	4.000	3.969	4.031	3.834	0.083	0.071	0.095	1.201	—
	10	4.000	3.969	4.031	3.760	0.120	0.105	0.135	1.720	—
	40	4.000	3.960	4.040	3.548	0.226	0.198	—	3.151	3.403
	80	4.000	3.960	4.040	3.364	0.318	0.278	—	4.326	4.672

Table 98 (*Continued*)

Nominal Pipe Size[a] (in.)	Schedule Number[a]	Outside Diameter (in.)			Inside Diameter (in.)	Wall Thickness (in.)			Weight per Foot (lb)	
		Nom[a]	Min[b,c]	Max[b,c]	Nom	Nom[a]	Min[b]	Max[b]	Nom[d]	Min[b,d]
4	5	4.500	4.469	4.531	4.334	0.083	0.071	0.095	1.354	—
	10	4.500	4.469	4.531	4.160	0.120	0.105	0.135	1.942	—
	40	4.500	4.455	4.545	4.026	0.237	0.207	—	3.733	4.031
	80	4.500	4.455	4.545	3.826	0.337	0.295	—	5.183	5.598
	120	4.500	4.455	4.545	3.624	0.438	0.383	—	6.573	7.099
	160	4.500	4.455	4.545	3.438	0.531	0.465	—	7.786	8.409
5	5.563	5.532	5.625	5.345	0.109	0.095	0.123	2.196	—	—
	10	5.563	5.532	5.625	5.295	0.134	0.117	0.151	2.688	—
	40	5.563	5.507	5.619	5.047	0.258	0.226	—	7.057	5.461
	80	5.563	5.507	5.619	4.813	0.375	0.328	—	7.188	7.763
	120	5.563	5.507	5.619	4.563	0.500	0.438	—	9.353	10.10
	160	5.563	5.507	5.619	4.313	0.625	0.547	—	11.40	12.31
6	5	6.625	6.594	6.687	6.407	0.109	0.095	0.123	2.624	—
	10	6.625	6.594	6.687	6.357	0.134	0.117	0.151	3.213	—
	40	6.625	6.559	6.691	6.065	0.280	0.245	—	6.564	7.089
	80	6.625	6.559	6.691	5.761	0.432	0.378	—	9.884	10.67
	120	6.625	6.559	6.691	5.501	0.562	0.492	—	12.59	13.60
	160	6.625	6.559	6.691	5.187	0.719	0.629	—	15.69	16.94
	5	8.625	8.594	8.718	8.407	0.109	0.095	0.123	3.429	—
	10	8.625	8.594	8.718	8.329	0.148	0.130	0.166	4.635	—
	20	8.625	8.539	8.711	8.125	0.250	0.219	—	7.735	8.354
	30	8.625	8.539	8.711	8.071	0.277	0.242	—	8.543	9.227
	40	8.625	8.539	8.711	7.981	0.322	0.282	—	9.878	10.67
8	60	8.625	8.539	8.711	7.813	0.406	0.355	—	12.33	13.31
	80	8.625	8.539	8.711	7.625	0.500	0.438	—	15.01	16.21
	100	8.625	8.539	8.711	7.437	0.594	0.520	—	17.62	19.03
	120	8.625	8.539	8.711	7.187	0.719	0.629	—	21.00	22.68
	140	8.625	8.539	8.711	7.001	0.812	0.710	—	23.44	25.31
	160	8.625	8.539	8.711	6.813	0.906	0.793	—	25.84	27.90
10	5	10.750	10.719	10.843	10.482	0.134	0.117	0.151	5.256	—
	10	10.750	10.719	10.843	10.420	0.165	0.144	0.186	6.453	—
	20	10.750	10.642	10.858	10.250	0.250	0.219	—	9.698	10.47
	30	10.750	10.642	10.858	10.136	0.307	0.269	—	11.84	12.69
	40	10.750	10.642	10.858	10.020	0.365	0.319	—	14.00	15.12
	60	10.750	10.642	10.858	9.750	0.500	0.438	—	18.93	24.07
	80	10.750	10.642	10.858	9.562	0.594	0.520	—	22.29	28.78
	100	10.750	10.642	10.858	9.312	0.719	0.629	—	26.65	28.78
	5	12.750	12.719	12.843	12.438	0.156	0.136	0.176	7.258	—
	10	12.750	12.719	12.843	12.390	0.180	0.158	0.202	8.359	—
	20	12.750	12.622	12.878	12.250	0.250	0.219	—	11.55	12.47
12	30	12.750	12.622	12.878	12.090	0.330	0.289	—	15.14	16.35
	40	12.750	12.622	12.878	11.938	0.406	0.355	—	18.52	20.00
	60	12.750	12.622	12.878	11.626	0.562	0.492	—	25.31	27.33
	80	12.750	12.622	12.878	11.374	0.688	0.602	—	30.66	33.11

[a] In accordance with ANSI Standards B36.10 and B36.19.
[b] Based on standard tolerances for pipe.
[c] For schedules 5 and 10 these values apply to mean outside diameters.
[d] Based on nominal dimensions, plain ends, and a density of 0.098 lb/in.^{3}, the density of 6061 alloy. For alloy 6063 multiply by 0.99, and for alloy 3003 multiply by 1.01.

Table 99   Aluminum Electrical Conduit — Designed Dimensions and Weights

Nominal or Trade Size of Conduit (in.)	Nominal Inside Diameter (in.)	Outside Diameter (in.)	Nominal Wall Thickness (in.)	Length without Coupling (ft and in.)	Minimum Weight of 10 Unit Lengths with Couplings Attached (lb)
$\frac{1}{4}$	0.364	0.540	0.088	$9-11\frac{1}{2}$	13.3
$\frac{3}{8}$	0.493	0.675	0.091	$9-11\frac{1}{2}$	17.8
$\frac{1}{2}$	0.622	0.840	0.109	$9-11\frac{1}{4}$	27.4
$\frac{3}{4}$	0.824	1.050	0.113	$9-11\frac{1}{4}$	36.4
1	1.049	1.315	0.133	$9-11$	53.0
$1\frac{1}{4}$	1.380	1.660	0.140	$9-11$	69.6
$1\frac{1}{2}$	1.610	1.900	0.145	$9-11$	86.2
2	2.067	2.375	0.154	$9-11$	115.7
$2\frac{1}{2}$	2.469	2.875	0.203	$9-10\frac{1}{2}$	182.5
3	3.068	3.500	0.216	$9-10\frac{1}{2}$	238.9
$3\frac{1}{2}$	3.548	4.000	0.226	$9-10\frac{1}{4}$	287.7
4	4.026	4.500	0.237	$9-10\frac{1}{4}$	340.0
5	5.047	5.563	0.258	$9-10$	465.4
6	6.065	6.625	0.280	$9-10$	612.5

Table 100   Equivalent Resistivity Values

Volume Conductivity, Percent International Amended Copper Standard at 68°F	Equivalent Resistivity at 68°F Volume	
	Ohm — Circular Mil/ft	Microhm — in.
52.5	19.754	1.2929
53.5	19.385	1.2687
53.8	19.277	1.2617
53.9	19.241	1.2593
54.0	19.206	1.2570
54.3	19.099	1.2501
55.0	18.856	1.2341
56.0	18.520	1.2121
56.5	18.356	1.2014
57.0	18.195	1.1908
59.0	17.578	1.1505
59.5	17.430	1.1408
61.0	17.002	1.1128
61.2	16.946	1.1091
61.3	16.918	1.1073
61.4	16.891	1.1055
61.5	16.863	1.1037
61.8	16.782	1.0983
62.0	16.727	1.0948
62.1	16.700	1.0931
62.2	16.674	1.0913
62.3	16.647	1.0896
62.4	16.620	1.0878

Table 101 Property Limits — Wire (Up to 0.374 in. Diameter)

Alloy and Temper	Ultimate Strength (ksi)		Electrical Conductivity[a] percent IACS at 68°F min
	Min	Max	
1350			
1350-O	8.5	14.0	61.8
1350-H12 and H22	12.0	17.0	61.0
1350-H14 and H24	15.0	20.0	61.0
1350-H16 and H26	17.0	22.0	61.0
8017			
8017-H212[b]	15.0	21.0	61.0
8030			
8030-H221	15.0	22.0	61.0
8176			
8176-H24	15.0	20.0	61.0
8177			
8177-H221	15.0	22.0	61.0

Alloy and Temper	Specified Diameter (in.)	Ultimate Strength (ksi min)		Elongation Percent min in 10 in.		Electrical Conductivity[a] min percent IACS at 68°F
		Individual[a]	Average[d]	Individual[a]	Average[d]	
1350						
1350-H19	0.0105–0.0500	23.0	25.0	—	—	61.0
	0.0501–0.0600	27.0	29.0	1.2	1.4	
	0.0601–0.0700	27.0	28.5	1.3	1.5	
	0.0701–0.0800	26.5	28.0	1.4	1.6	
	0.0801–0.0900	26.0	27.5	1.5	1.6	
	0.0901–0.1000	25.5	27.0	1.5	1.6	
	0.1001–0.1100	24.5	26.0	1.5	1.6	
	0.1101–0.1200	24.0	25.5	1.6	1.7	
	0.1201–0.1400	23.5	25.0	1.7	1.8	
	0.1401–0.1500	23.5	24.5	1.8	1.9	
	0.1501–0.1800	23.0	24.0	1.9	2.0	
	0.1801–0.2100	23.0	24.0	2.0	2.1	
	0.2101–0.2600	22.5	23.5	2.2	2.3	
5005						
5005-H19	0.0601–0.0700	38.0	40.0	1.3	—	53.5
	0.0701–0.0800	37.5	39.5	1.4	—	
	0.0801–0.0900	37.0	39.0	1.5	—	
	0.0901–0.1000	36.5	38.5	1.5	—	
	0.1001–0.1100	36.0	38.0	1.5	—	
5005-H19	0.1101–0.1200	35.5	37.5	1.6	—	
	0.1201–0.1400	35.0	37.0	1.7	—	
	0.1401–0.1500	35.0	36.5	1.8	—	
	0.1501–0.1600	34.5	36.0	1.9	—	
	0.1601–0.2100	32.5	34.0	2.0	—	
	0.2101–0.2600	31.5	33.0	2.2	—	
6201						
6201-T81	0.0612–0.1327	46.0	48.0	3.0	—	52.5
	0.1328–0.1878	44.0	46.0	3.0	—	
8176						
8176-H24	0.0500–0.2040	15.0	17.0	10.0	—	61.0

[a] To convert conductivity to maximum resistivity use Table 100.
[b] Applicable up to 0.250 in.
[c] Any test in a lot.
[d] Average of all tests in a lot.

7 STANDARD SCREWS*

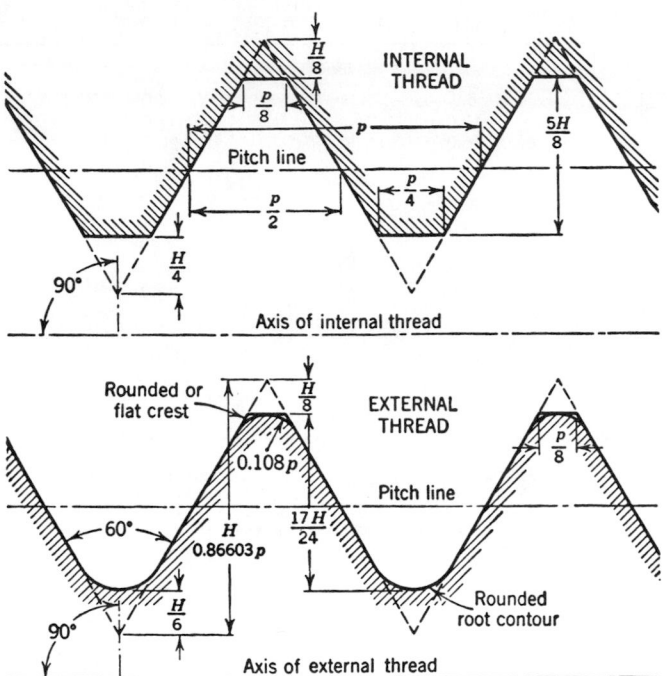

Standard Screw Threads The Unified and American Screw Threads included in Table 102 are taken from the publication of the American Standards Association, ASA B1.1—1949. The *coarse-thread series* is the former United States Standard Series. It is recommended for general use in engineering work where conditions do not require the use of a fine thread. The *fine-thread series* is the former "Regular Screw Thread Series" established by the Society of Automotive Engineers (SAE). The *fine-thread series* is recommended for general use in automotive and aircraft work and where special conditions require a fine thread. The *extra-fine-thread series* is the same as the former SAE fine series and the present SAE extra-fine series. It is used particularly in aircraft and aeronautical equipment where (a) thin-walled material is to be threaded; (b) thread depth of nuts clearing ferrules, coupling flanges, and so on, must be held to a minimum; and (c) a maximum practicable number of threads is required within a given thread length.

The method of designating a screw thread is by the use of the initial letters of the thread series, preceded by the nominal size (diameter in inches or the screw number) and number of threads per inch, all in Arabic numerals, and followed by the classification designation, with or without the pitch diameter tolerances or limits of size. An example of an external thread designation and its meaning is as follows:

Example 1

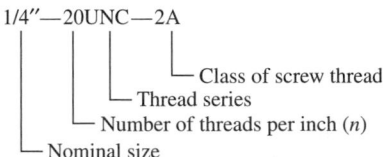

A left-hand thread must be identified by the letters LH following the class designation. If no such designation is used, the thread is assumed to be right hand.

Classes of thread are distinguished from each other by the amounts of tolerance and allowance specified in ASA B1.1—1949.

Table 102 Standard Screw Threads

Sizes	Basic Major Diameter D (in.)	Threads per Inch n	Basic Pitch Diameter[a] E (in.)	Minor Diameter External Threads K_s (in.)	Minor Diameter Internal Threads K_n (in.)	Section at Minor Diameter at $D - 2h_b$) (in.^{2})	Stress Area[b] (in.^{2})
			Coarse-thread Series — UNC and NC (Basic Dimensions)				
1 (0.073)	0.0730	64	0.0629	0.0538	0.0561	0.0022	0.0026
2 (0.086)	0.0860	56	0.0744	0.0641	0.0667	0.0031	0.0036
3 (0.099)	0.0990	48	0.0855	0.0734	0.0764	0.0041	0.0048
4 (0.112)	0.1120	40	0.0958	0.0813	0.0849	0.0050	0.0060
5 (0.125)	0.1250	40	0.1088	0.0943	0.0979	0.0067	0.0079
6 (0.138)	0.1380	32	0.1177	0.0997	0.1042	0.0075	0.0090
8 (0.164)	0.1640	32	0.1437	0.1257	0.1302	0.0120	0.0139
10 (0.190)	0.1900	24	0.1629	0.1389	0.1449	0.0145	0.0174
12 (0.216)	0.2160	24	0.1889	0.1649	0.1709	0.0206	0.0240
$\frac{1}{4}$	0.2500	20	0.2175	0.1887	0.1959	0.0269	0.0317
$\frac{5}{16}$	0.3125	18	0.2764	0.2443	0.2524	0.0454	0.0522
$\frac{3}{8}$	0.3750	16	0.3344	0.2983	0.3073	0.0678	0.0773
$\frac{7}{16}$	0.4375	14	0.3911	0.3499	0.3602	0.0933	0.1060
$\frac{1}{2}$	0.5000	13	0.4500	0.4056	0.4167	0.1257	0.1416
$\frac{1}{2}$	0.5000	12	0.4459	0.3978	0.4098	0.1205	0.1374
$\frac{9}{16}$	0.5625	12	0.5084	0.4603	0.4723	0.1620	0.1816
$\frac{5}{8}$	0.6250	11	0.5660	0.5135	0.5266	0.2018	0.2256
$\frac{3}{4}$	0.7500	10	0.6850	0.6273	0.6417	0.3020	0.3340
$\frac{7}{8}$	0.8750	9	0.8028	0.7387	0.7547	0.4193	0.4612
1	1.0000	8	0.9188	0.8466	0.8647	0.5510	0.6051
$1\frac{1}{8}$	1.1250	7	1.0322	0.9497	0.9704	0.6931	0.7627
$1\frac{1}{4}$	1.2500	7	1.1572	1.0747	1.0954	0.8898	0.9684
$1\frac{3}{8}$	1.3750	6	1.2667	1.1705	1.1946	1.0541	1.1538
$1\frac{1}{2}$	1.5000	6	1.3917	1.2955	1.3196	1.2938	1.4041
$1\frac{3}{4}$	1.7500	5	1.6201	1.5046	1.5335	1.7441	1.8983
2	2.0000	$4\frac{1}{2}$	1.8557	1.7274	1.7594	2.3001	2.4971
$2\frac{1}{4}$	2.2500	$4\frac{1}{2}$	2.1057	1.9774	2.0094	3.0212	3.2464
$2\frac{1}{2}$	2.5000	4	2.3376	2.1933	2.2294	3.7161	3.9976
$2\frac{3}{4}$	2.7500	4	2.5876	2.4433	2.4794	4.6194	4.9326
3	3.0000	4	2.8376	2.6933	2.7294	5.6209	5.9659
$3\frac{1}{4}$	3.2500	4	3.0876	2.9433	2.9794	6.7205	7.0992
$3\frac{1}{2}$	3.5000	4	3.3376	3.1933	3.2294	7.9183	8.3268
$3\frac{3}{4}$	3.7500	4	3.5876	3.4433	3.4794	9.2143	9.6546
4	4.0000	4	3.8376	3.6933	3.7294	10.6084	11.0805

(Continues)

Table 102 *(Continued)*

Sizes	Basic Major Diameter D (in.)	Threads per Inch n	Basic Pitch Diameter[a] E (in.)	Minor Diameter External Threads K_s (in.)	Minor Diameter Internal Threads K_n (in.)	Section at Minor Diameter at $D - 2h_b$) (in.^{2})	Stress Area[b] (in.^{2})
			Fine-Thread Series — UNF and NF				
			(Basic Dimensions)				
0 (0.060)	0.0600	80	0.0519	0.0447	0.0465	0.0015	0.0018
1 (0.073)	0.0730	72	0.0640	0.0560	0.0580	0.0024	0.0027
2 (0.086)	0.0860	64	0.0759	0.0668	0.0691	0.0034	0.0039
3 (0.099)	0.0990	56	0.0874	0.0771	0.0797	0.0045	0.0052
4 (0.112)	0.1120	48	0.0985	0.0864	0.0894	0.0057	0.0065
5 (0.125)	0.1250	44	0.1102	0.0971	0.1004	0.0072	0.0082
6 (0.138)	0.1380	40	0.1218	0.1073	0.1109	0.0087	0.0101
8 (0.164)	0.1640	36	0.1460	0.1299	0.1339	0.0128	0.0146
10 (0.190)	0.1900	32	0.1697	0.1517	0.1562	0.0175	0.0199
12 (0.216)	0.2160	28	0.1928	0.1722	0.1773	0.0226	0.0257
$\frac{1}{4}$	**0.2500**	**28**	**0.2268**	**0.2062**	**0.2113**	**0.0326**	**0.0362**
$\frac{5}{16}$	**0.3125**	**24**	**0.2854**	**0.2614**	**0.2674**	**0.0524**	**0.0579**
$\frac{3}{8}$	**0.3750**	**24**	**0.3479**	**0.3239**	**0.3299**	**0.0809**	**0.0876**
$\frac{7}{16}$	**0.4375**	**20**	**0.4050**	**0.3762**	**0.3834**	**0.1090**	**0.1185**
$\frac{1}{2}$	**0.5000**	**20**	**0.4675**	**0.4387**	**0.4459**	**0.1486**	**0.1597**
$\frac{9}{16}$	**0.5625**	**18**	**0.5264**	**0.4943**	**0.5024**	**0.1888**	**0.2026**
$\frac{5}{8}$	**0.6250**	**18**	**0.5889**	**0.5568**	**0.5649**	**0.2400**	**0.2555**
$\frac{3}{4}$	**0.7500**	**16**	**0.7094**	**0.6733**	**0.6823**	**0.3513**	**0.3724**
$\frac{7}{8}$	**0.8750**	**14**	**0.8286**	**0.7874**	**0.7977**	**0.4805**	**0.5088**
1	**1.0000**	**12**	**0.9459**	**0.8978**	**0.9098**	**0.6245**	**0.6624**
$1\frac{1}{8}$	**1.1250**	**12**	**1.0709**	**1.0228**	**1.0348**	**0.8118**	**0.8549**
$1\frac{1}{4}$	**1.2500**	**12**	**1.1959**	**1.1478**	**1.1598**	**1.0237**	**1.0721**
$1\frac{3}{8}$	**1.3750**	**12**	**1.3209**	**1.2728**	**1.2848**	**1.2602**	**1.3137**
$1\frac{1}{2}$	**1.5000**	**12**	**1.4459**	**1.3978**	**1.4098**	**1.5212**	**1.5799**
			Extra-Fine-Thread Series — NEF				
			(Basic Dimensions)				
12 (0.216)	0.2160	32	0.1957	0.1777	0.1822	0.0242	0.0269
$\frac{1}{4}$	0.2500	32	0.2297	0.2117	0.2162	0.0344	0.0377
$\frac{5}{16}$	0.3125	32	0.2922	0.2742	0.2787	0.0581	0.0622
$\frac{3}{8}$	0.3750	32	0.3547	0.3367	0.3412	0.0878	0.0929
$\frac{7}{16}$	0.4375	28	0.4143	0.3937	0.3988	0.1201	0.1270
$\frac{1}{2}$	0.5000	28	0.4768	0.4562	0.4613	0.1616	0.1695
$\frac{9}{16}$	0.5625	24	0.5354	0.5114	0.5174	0.2030	0.2134
$\frac{5}{8}$	0.6250	24	0.5979	0.5739	0.5799	0.2560	0.2676
$\frac{11}{16}$	0.6875	24	0.6604	0.6364	0.6424	0.3151	0.3280
$\frac{3}{4}$	0.7500	20	0.7175	0.6887	0.6959	0.3685	0.3855
$\frac{13}{16}$	0.8125	20	0.7800	0.7512	0.7584	0.4388	0.4573
$\frac{7}{8}$	0.8750	20	0.8425	0.8137	0.8209	0.5153	0.5352
$\frac{15}{16}$	0.9375	20	0.9050	0.8762	0.8834	0.5979	0.6194

Table 102 *(Continued)*

Sizes	Basic Major Diameter D (in.)	Threads per Inch n	Basic Pitch Diameter[a] E (in.)	Minor Diameter External Threads K_s (in.)	Minor Diameter Internal Threads K_n (in.)	Section at Minor Diameter at $D - 2h_b$) (in.²)	Stress Area[b] (in.²)
			Fine-Thread Series — UNF and NF (Basic Dimensions)				
1	1.0000	20	0.9675	0.9387	0.9459	0.6866	0.7095
$1\frac{1}{16}$	1.0625	18	1.0264	0.9943	1.0024	0.7702	0.7973
$1\frac{1}{8}$	1.1250	18	1.0889	1.0568	1.0649	0.8705	0.8993
$1\frac{3}{16}$	1.1875	18	1.1514	1.1193	1.1274	0.9770	1.0074
$1\frac{1}{4}$	1.2500	18	1.2139	1.1818	1.1899	1.0895	1.1216
$1\frac{5}{16}$	1.3125	18	1.2764	1.2443	1.2524	1.2082	1.2420
$1\frac{3}{8}$	1.3750	18	1.3389	1.3068	1.3149	1.3330	1.3684
$1\frac{7}{16}$	1.4375	18	1.4014	1.3693	1.3774	1.4640	1.5010
$1\frac{1}{2}$	1.5000	18	1.4639	1.4318	1.4399	1.6011	1.6397
$1\frac{9}{16}$	1.5625	18	1.5264	1.4943	1.5024	1.7444	1.7846
$1\frac{5}{8}$	1.6250	18	1.5889	1.5568	1.5649	1.8937	1.9357
$1\frac{11}{16}$	1.6875	18	1.6514	1.6193	1.6274	2.0493	2.0929
$1\frac{3}{4}$	1.7500	16	1.7094	1.6733	1.6823	2.1873	2.2382
2	2.0000	16	1.9594	1.9233	1.9323	2.8917	2.9501

Note: Bold type indicates unified threads — UNC and UNF.
[a] British: effective diameter.
[b] The stress area is the assumed area of an externally threaded part which is used for the purpose of computing the tensile strength.

Table 103 ASA[a] Standard Bolts and Nuts

Nominal Size	Across Flats (in.)	Across Square Corners (in.)	Across Hex Corners (in.)	Thickness Unfinished (in.)	Thickness Semifinished (in.)
			Regular Bolt Heads		
$\frac{1}{4}$	$\frac{3}{8}$	0.498	0.413	$\frac{11}{64}$	$\frac{5}{32}$
$\frac{5}{16}$	$\frac{1}{2}$	0.665	0.552	$\frac{13}{64}$	$\frac{3}{16}$
$\frac{3}{8}$	$\frac{9}{16}$	0.747	0.620	$\frac{1}{4}$	$\frac{15}{64}$
$\frac{7}{16}$	$\frac{5}{8}$	0.828	0.687	$\frac{19}{64}$	$\frac{9}{32}$
$\frac{1}{2}$	$\frac{3}{4}$	0.995	0.826	$\frac{21}{64}$	$\frac{19}{64}$
$\frac{9}{16}$	$\frac{7}{8}$	1.163	0.966	$\frac{3}{8}$	$\frac{11}{32}$
$\frac{5}{8}$	$\frac{15}{16}$	1.244	1.033	$\frac{27}{64}$	$\frac{25}{64}$
$\frac{3}{4}$	$1\frac{1}{8}$	1.494	1.240	$\frac{1}{2}$	$\frac{15}{32}$
$\frac{7}{8}$	$1\frac{5}{16}$	1.742	1.447	$\frac{19}{32}$	$\frac{9}{16}$
1	$1\frac{1}{2}$	1.991	1.653	$\frac{21}{32}$	$\frac{19}{32}$
$1\frac{1}{8}$	$1\frac{11}{16}$	2.239	1.859	$\frac{3}{4}$	$\frac{11}{16}$

(Continues)

Table 103   (*Continued*)

Nominal Size	Across Flats (in.)	Across Square Corners (in.)	Across Hex Corners (in.)	Thickness Unfinished (in.)	Thickness Semifinished (in.)
$1\frac{1}{4}$	$1\frac{7}{8}$	2.489	2.066	$\frac{27}{32}$	$\frac{25}{32}$
$1\frac{3}{8}$	$2\frac{1}{16}$	2.738	2.273	$\frac{29}{32}$	$\frac{27}{32}$
$1\frac{1}{2}$	$2\frac{1}{4}$	2.986	2.480	1	$\frac{15}{16}$
$1\frac{5}{8}$	$2\frac{7}{16}$	3.235	2.686	$1\frac{3}{32}$	$1\frac{1}{32}$
$1\frac{3}{4}$	$2\frac{5}{8}$	3.485	2.893	$1\frac{5}{32}$	$1\frac{3}{32}$
$1\frac{7}{8}$	$2\frac{13}{16}$	3.733	3.100	$1\frac{1}{4}$	$1\frac{3}{16}$
2	3	3.982	3.306	$1\frac{11}{32}$	$1\frac{7}{32}$
$2\frac{1}{4}$	$3\frac{3}{8}$	4.479	3.719	$1\frac{1}{2}$	$1\frac{3}{8}$
$2\frac{1}{2}$	$3\frac{3}{4}$	4.977	4.133	$1\frac{21}{32}$	$1\frac{17}{32}$
$2\frac{3}{4}$	$4\frac{1}{8}$	5.476	4.546	$1\frac{53}{64}$	$1\frac{11}{16}$
3	$4\frac{1}{2}$	5.973	4.959	2	$1\frac{7}{8}$

Heavy Bolt Heads

Nominal Size	Across Flats (in.)	Across Square Corners (in.)	Across Hex Corners (in.)	Thickness Unfinished (in.)	Thickness Semifinished (in.)
$\frac{1}{2}$	$\frac{7}{8}$	1.167	0.969	$\frac{7}{16}$	$\frac{13}{32}$
$\frac{9}{16}$	$\frac{15}{16}$	1.249	1.037	$\frac{15}{32}$	$\frac{7}{16}$
$\frac{5}{8}$	$1\frac{1}{16}$	1.416	1.175	$\frac{17}{32}$	$\frac{1}{2}$
$\frac{3}{4}$	$1\frac{1}{4}$	1.665	1.383	$\frac{5}{8}$	$\frac{19}{32}$
$\frac{7}{8}$	$1\frac{7}{16}$	1.914	1.589	$\frac{23}{32}$	$\frac{11}{16}$
1	$1\frac{5}{8}$	2.162	1.796	$\frac{13}{16}$	$\frac{3}{4}$
$1\frac{1}{8}$	$1\frac{13}{16}$	2.411	2.002	$\frac{29}{32}$	$\frac{27}{32}$
$1\frac{1}{4}$	2	2.661	2.209	1	$\frac{15}{16}$
$1\frac{3}{8}$	$2\frac{3}{16}$	2.909	2.416	$1\frac{3}{32}$	$1\frac{1}{32}$
$1\frac{1}{2}$	$2\frac{3}{8}$	3.158	2.622	$1\frac{3}{16}$	$1\frac{1}{8}$
$1\frac{5}{8}$	$2\frac{9}{16}$	3.406	2.828	$1\frac{9}{32}$	$1\frac{7}{32}$
$1\frac{3}{4}$	$2\frac{3}{4}$	3.655	3.036	$1\frac{3}{8}$	$1\frac{5}{16}$
$1\frac{7}{8}$	$2\frac{15}{16}$	3.905	3.242	$1\frac{15}{32}$	$1\frac{13}{32}$
2	$3\frac{1}{8}$	4.153	3.449	$1\frac{9}{16}$	$1\frac{7}{16}$
$2\frac{1}{4}$	$3\frac{1}{2}$	4.652	3.862	$1\frac{3}{4}$	$1\frac{5}{8}$
$2\frac{1}{2}$	$3\frac{7}{8}$	5.149	4.275	$1\frac{15}{16}$	$1\frac{13}{16}$
$2\frac{3}{4}$	$4\frac{1}{4}$	5.646	4.688	$2\frac{1}{8}$	2
3	$4\frac{5}{8}$	6.144	5.102	$2\frac{5}{16}$	$2\frac{3}{16}$

Nominal Size	Width Across Flats (in.)	Width Across Corners Square (in.)	Hex (in.)	Thickness Unfinished, Regular Nuts (in.)	Jam Nuts (in.)	Thickness Semifinished, Regular Nuts (in.)	Jam Nuts (in.)

Regular Nuts and Regular Jam Nuts

Nominal Size	Width Across Flats (in.)	Square (in.)	Hex (in.)	Nuts (in.)	Jam Nuts (in.)	Nuts (in.)	Jam Nuts (in.)
$\frac{1}{4}$	$\frac{7}{16}$	0.584	0.484	$\frac{7}{32}$	$\frac{5}{32}$	$\frac{13}{64}$	$\frac{9}{64}$
$\frac{5}{16}$	$\frac{9}{16}$	0.751	0.624	$\frac{17}{64}$	$\frac{3}{16}$	$\frac{1}{4}$	$\frac{11}{64}$
$\frac{3}{8}$	$\frac{5}{8}$	0.832	0.691	$\frac{21}{64}$	$\frac{7}{32}$	$\frac{5}{16}$	$\frac{13}{64}$
$\frac{7}{16}$	$\frac{3}{4}$	1.000	0.830	$\frac{3}{8}$	$\frac{1}{4}$	$\frac{23}{64}$	$\frac{15}{64}$
$\frac{1}{2}$	$\frac{13}{16}$	1.082	0.898	$\frac{7}{16}$	$\frac{5}{16}$	$\frac{27}{64}$	$\frac{19}{64}$
$\frac{9}{16}$	$\frac{7}{8}$	1.163	0.966	$\frac{1}{2}$	$\frac{11}{32}$	$\frac{31}{64}$	$\frac{21}{64}$
$\frac{5}{8}$	1	1.330	1.104	$\frac{35}{64}$	$\frac{3}{8}$	$\frac{17}{32}$	$\frac{23}{64}$
$\frac{3}{4}$	$1\frac{1}{8}$	1.494	1.240	$\frac{21}{32}$	$\frac{7}{16}$	$\frac{41}{64}$	$\frac{27}{64}$

Table 103 (*Continued*)

Nominal Size	Width Across Flats (in.)	Width Across Corners		Thickness Unfinished, Regular		Thickness Semifinished, Regular	
		Square (in.)	Hex (in.)	Nuts (in.)	Jam Nuts (in.)	Nuts (in.)	Jam Nuts (in.)
$\frac{7}{8}$	$1\frac{5}{16}$	1.742	1.447	$\frac{49}{64}$	$\frac{1}{2}$	$\frac{3}{4}$	$\frac{31}{64}$
1	$1\frac{1}{2}$	1.991	1.653	$\frac{7}{8}$	$\frac{9}{16}$	$\frac{55}{64}$	$\frac{35}{64}$
$1\frac{1}{8}$	$1\frac{11}{16}$	2.239	1.859	1	$\frac{5}{8}$	$\frac{31}{32}$	$\frac{39}{64}$
$1\frac{1}{4}$	$1\frac{7}{8}$	2.489	2.066	$1\frac{3}{32}$	$\frac{3}{4}$	$1\frac{1}{16}$	$\frac{23}{32}$
$1\frac{3}{8}$	$2\frac{1}{16}$	2.738	2.273	$1\frac{13}{64}$	$\frac{13}{16}$	$1\frac{11}{64}$	$\frac{25}{32}$
$1\frac{1}{2}$	$2\frac{1}{4}$	2.986	2.480	$1\frac{5}{16}$	$\frac{7}{8}$	$1\frac{9}{32}$	$\frac{27}{32}$
$1\frac{5}{8}$	$2\frac{7}{16}$	3.235	2.686	$1\frac{27}{64}$	$\frac{15}{16}$	$1\frac{25}{64}$	$\frac{29}{32}$
$1\frac{3}{4}$	$2\frac{5}{8}$	3.485	2.893	$1\frac{17}{32}$	1	$1\frac{1}{2}$	$\frac{31}{32}$
$1\frac{7}{8}$	$2\frac{13}{16}$	3.733	3.100	$1\frac{41}{64}$	$1\frac{1}{16}$	$1\frac{39}{64}$	$1\frac{1}{32}$
2	3	3.982	3.306	$1\frac{3}{4}$	$1\frac{1}{8}$	$1\frac{23}{32}$	$1\frac{3}{32}$
$2\frac{1}{4}$	$3\frac{3}{8}$	4.479	3.719	$1\frac{31}{32}$	$1\frac{1}{4}$	$1\frac{59}{64}$	$1\frac{13}{64}$
$2\frac{1}{2}$	$3\frac{3}{4}$	4.977	4.133	$2\frac{3}{16}$	$1\frac{1}{2}$	$2\frac{9}{64}$	$1\frac{29}{64}$
$2\frac{3}{4}$	$4\frac{1}{8}$	5.476	4.546	$2\frac{13}{32}$	$1\frac{5}{8}$	$2\frac{23}{64}$	$1\frac{37}{64}$
3	$4\frac{1}{2}$	5.973	4.959	$2\frac{5}{8}$	$1\frac{3}{4}$	$2\frac{37}{64}$	$1\frac{45}{64}$
Heavy Nuts and Heavy Jam Nuts							
$\frac{1}{4}$	$\frac{1}{2}$	0.670	0.556	$\frac{1}{4}$	$\frac{3}{16}$	$\frac{15}{64}$	$\frac{11}{64}$
$\frac{5}{16}$	$\frac{19}{32}$	0.794	0.659	$\frac{5}{16}$	$\frac{7}{32}$	$\frac{19}{64}$	$\frac{13}{64}$
$\frac{3}{8}$	$\frac{11}{16}$	0.919	0.763	$\frac{3}{8}$	$\frac{1}{4}$	$\frac{23}{64}$	$\frac{15}{64}$
$\frac{7}{16}$	$\frac{25}{32}$	1.042	0.865	$\frac{7}{16}$	$\frac{9}{32}$	$\frac{27}{64}$	$\frac{17}{64}$
$\frac{1}{2}$	$\frac{7}{8}$	1.167	0.969	$\frac{1}{2}$	$\frac{5}{16}$	$\frac{31}{64}$	$\frac{19}{64}$
$\frac{9}{16}$	$\frac{15}{16}$	1.249	1.037	$\frac{9}{16}$	$\frac{11}{32}$	$\frac{35}{64}$	$\frac{21}{64}$
$\frac{5}{8}$	$1\frac{1}{16}$	1.416	1.175	$\frac{5}{8}$	$\frac{3}{8}$	$\frac{39}{64}$	$\frac{23}{64}$
$\frac{3}{4}$	$1\frac{1}{4}$	1.665	1.382	$\frac{3}{4}$	$\frac{7}{16}$	$\frac{47}{64}$	$\frac{27}{64}$
$\frac{7}{8}$	$1\frac{7}{16}$	1.914	1.589	$\frac{7}{8}$	$\frac{1}{2}$	$\frac{55}{64}$	$\frac{31}{64}$
1	$1\frac{5}{8}$	2.162	1.796	1	$\frac{9}{16}$	$\frac{63}{64}$	$\frac{35}{64}$
$1\frac{1}{8}$	$1\frac{13}{16}$	2.411	2.002	$1\frac{1}{8}$	$\frac{5}{8}$	$1\frac{7}{64}$	$\frac{39}{64}$
$1\frac{1}{4}$	2	2.661	2.209	$1\frac{1}{4}$	$\frac{3}{4}$	$1\frac{7}{32}$	$\frac{23}{32}$
$1\frac{3}{8}$	$2\frac{3}{16}$	2.909	2.416	$1\frac{3}{8}$	$\frac{13}{16}$	$1\frac{11}{32}$	$\frac{25}{32}$
$1\frac{1}{2}$	$2\frac{3}{8}$	3.158	2.622	$1\frac{1}{2}$	$\frac{7}{8}$	$1\frac{15}{32}$	$\frac{27}{32}$
$1\frac{5}{8}$	$2\frac{9}{16}$	3.406	2.828	$1\frac{5}{8}$	$\frac{15}{16}$	$1\frac{19}{32}$	$\frac{29}{32}$
$1\frac{3}{4}$	$2\frac{3}{4}$	3.656	3.035	$1\frac{3}{4}$	1	$1\frac{23}{32}$	$\frac{31}{32}$
$1\frac{7}{8}$	$2\frac{15}{16}$	3.905	3.242	$1\frac{7}{8}$	$1\frac{1}{16}$	$1\frac{27}{32}$	$1\frac{1}{32}$
2	$3\frac{1}{8}$	4.153	3.449	2	$1\frac{1}{8}$	$1\frac{31}{32}$	$1\frac{3}{32}$
$2\frac{1}{4}$	$3\frac{1}{2}$	4.652	3.862	$2\frac{1}{4}$	$1\frac{1}{4}$	$2\frac{13}{64}$	$1\frac{13}{64}$
$2\frac{1}{2}$	$3\frac{7}{8}$	5.149	4.275	$2\frac{1}{2}$	$1\frac{1}{2}$	$2\frac{29}{64}$	$1\frac{29}{64}$
$2\frac{3}{4}$	$4\frac{1}{4}$	5.646	4.688	$2\frac{3}{4}$	$1\frac{5}{8}$	$2\frac{45}{64}$	$1\frac{37}{64}$
3	$4\frac{5}{8}$	6.144	5.102	3	$1\frac{3}{4}$	$2\frac{61}{64}$	$1\frac{45}{64}$
$3\frac{1}{4}$	5	6.643	5.515	$3\frac{1}{4}$	$1\frac{7}{8}$	$3\frac{3}{16}$	$1\frac{13}{16}$
$3\frac{1}{2}$	$5\frac{3}{8}$	7.140	5.928	$3\frac{1}{2}$	2	$3\frac{7}{16}$	$1\frac{15}{16}$
$3\frac{3}{4}$	$5\frac{3}{4}$	7.637	6.341	$3\frac{3}{4}$	$2\frac{1}{8}$	$3\frac{11}{16}$	$2\frac{1}{16}$
4	$6\frac{1}{8}$	8.135	6.755	4	$2\frac{1}{4}$	$3\frac{15}{16}$	$2\frac{3}{16}$

Table 103 *(Continued)*

| Nominal Size | Regular Slotted Nuts Semifinished | | | Heavy Slotted Nuts Semifinished | | | Slot | |
| | Width | | | Width | | | | |
	Across Flats (in.)	Across Corners (in.)	Thickness (in.)	Across Flats (in.)	Across Corners (in.)	Thickness (in.)	Width (in.)	Depth (in.)
$\frac{1}{4}$	$\frac{7}{16}$	0.485	$\frac{13}{64}$	$\frac{1}{2}$	0.556	$\frac{15}{64}$	$\frac{5}{64}$	$\frac{3}{32}$
$\frac{5}{16}$	$\frac{9}{16}$	0.624	$\frac{1}{4}$	$\frac{19}{32}$	0.659	$\frac{19}{64}$	$\frac{3}{32}$	$\frac{3}{32}$
$\frac{3}{8}$	$\frac{5}{8}$	0.691	$\frac{5}{16}$	$\frac{11}{16}$	0.763	$\frac{23}{64}$	$\frac{1}{8}$	$\frac{1}{8}$
$\frac{7}{16}$	$\frac{3}{4}$	0.830	$\frac{23}{64}$	$\frac{25}{32}$	0.865	$\frac{27}{64}$	$\frac{1}{8}$	$\frac{5}{32}$
$\frac{1}{2}$	$\frac{13}{16}$	0.898	$\frac{27}{64}$	$\frac{7}{8}$	0.969	$\frac{31}{64}$	$\frac{5}{32}$	$\frac{5}{32}$
$\frac{9}{16}$	$\frac{7}{8}$	0.966	$\frac{31}{64}$	$\frac{15}{16}$	1.037	$\frac{35}{64}$	$\frac{5}{32}$	$\frac{3}{16}$
$\frac{5}{8}$	1	1.104	$\frac{17}{32}$	$1\frac{1}{16}$	1.175	$\frac{39}{64}$	$\frac{3}{16}$	$\frac{7}{32}$
$\frac{3}{4}$	$1\frac{1}{8}$	1.240	$\frac{41}{64}$	$1\frac{1}{4}$	1.382	$\frac{47}{64}$	$\frac{3}{16}$	$\frac{1}{4}$
$\frac{7}{8}$	$1\frac{5}{16}$	1.447	$\frac{3}{4}$	$1\frac{7}{16}$	1.589	$\frac{55}{64}$	$\frac{3}{16}$	$\frac{1}{4}$
1	$1\frac{1}{2}$	1.653	$\frac{55}{64}$	$1\frac{5}{8}$	1.796	$\frac{63}{64}$	$\frac{1}{4}$	$\frac{9}{32}$
$1\frac{1}{8}$	$1\frac{11}{16}$	1.859	$\frac{31}{32}$	$1\frac{13}{16}$	2.002	$1\frac{7}{64}$	$\frac{1}{4}$	$\frac{11}{32}$
$1\frac{1}{4}$	$1\frac{7}{8}$	2.066	$1\frac{1}{16}$	2	2.209	$1\frac{7}{32}$	$\frac{5}{16}$	$\frac{3}{8}$
$1\frac{3}{8}$	$2\frac{1}{16}$	2.273	$1\frac{11}{64}$	$2\frac{3}{16}$	2.416	$1\frac{11}{32}$	$\frac{5}{16}$	$\frac{3}{8}$
$1\frac{1}{2}$	$2\frac{1}{4}$	2.480	$1\frac{9}{32}$	$2\frac{3}{8}$	2.622	$1\frac{15}{32}$	$\frac{3}{8}$	$\frac{7}{16}$
$1\frac{5}{8}$	$2\frac{7}{16}$	0.686	$1\frac{25}{64}$	$2\frac{9}{16}$	2.828	$1\frac{19}{32}$	$\frac{3}{8}$	$\frac{7}{16}$
$1\frac{3}{4}$	$2\frac{5}{8}$	2.893	$1\frac{1}{2}$	$2\frac{3}{4}$	3.035	$1\frac{23}{32}$	$\frac{7}{16}$	$\frac{1}{2}$
$1\frac{7}{8}$	$2\frac{13}{16}$	3.100	$1\frac{39}{64}$	$2\frac{15}{16}$	3.242	$1\frac{27}{32}$	$\frac{7}{16}$	$\frac{9}{16}$
2	3	3.306	$1\frac{23}{32}$	$3\frac{1}{8}$	3.449	$1\frac{31}{32}$	$\frac{7}{16}$	$\frac{9}{16}$
$2\frac{1}{4}$	$3\frac{3}{8}$	3.719	$1\frac{59}{64}$	$3\frac{1}{2}$	3.862	$2\frac{13}{64}$	$\frac{7}{16}$	$\frac{9}{16}$
$2\frac{1}{2}$	$3\frac{3}{4}$	4.133	$2\frac{9}{64}$	$3\frac{7}{8}$	4.275	$2\frac{29}{64}$	$\frac{9}{16}$	$\frac{11}{16}$
$2\frac{3}{4}$	$4\frac{1}{8}$	4.546	$2\frac{23}{64}$	$4\frac{1}{4}$	4.688	$2\frac{45}{64}$	$\frac{9}{16}$	$\frac{11}{16}$
3	$4\frac{1}{2}$	4.959	$2\frac{37}{64}$	$4\frac{5}{8}$	5.102	$2\frac{61}{64}$	$\frac{5}{8}$	$\frac{3}{4}$

[a] ANSI standards B18.2.1 — 1981, B18.2.2 — 1972 (R1983), B18.6.3 — 1972 (R1983).

Selection of Screws By definition, a *screw* is a fastener that is intended to be torqued by the head. Screws are the most widely used method of assembly despite recent technical advances of adhesives, welding, and other joining techniques. Use of screws is essential in those applications that require ease of disassembly for normal maintenance and service. There is no real economy if savings made in factory installation create service problems later. There are many types of screws, and each variety will be treated separately. Material selection is generally common to all types of screws.

Material. Not all materials are suitable for the processes used in the manufacture of fasteners. Large-volume users or those with critical requirements can be very selective in their choice of materials. Low-volume users or those with noncritical applications would be wise to permit a variety of materials in a general category in order to improve availability and lower cost. For example, it is usually desirable to specify low-carbon steel or 18-8-type stainless steel* rather than ask for a specific grade.

Low-carbon steel is widely used in the manufacture of fasteners where lowest cost is desirable and tensile strength requirements are ∼50,000 psi. If corrosion is a problem, these fasteners can be plated either electrically or mechanically. Zinc or cadmium plating is used in most applications. Other finishes include nickel, chromium, copper, tin, and silver electroplating; electroless nickel and other immersion coatings; hot dip galvanizing; and phosphate coatings.

---

*Manufacturer may use UNS—S30200, S30300, S30400, S30500 (AISI type 302, 303, 304, or 305) depending upon quantity, diameter, and manufacturing process.

Medium-carbon steel, quenched, and tempered is widely used in applications requiring tensile strengths from 90,000 to 120,000 psi. Alloy steels are used in applications requiring tensile strengths from 115,000 to 180,000 psi, depending on the grade selected. Where better corrosion resistance is required, 300 series stainless steel can be specified. The 400 series stainless steel is used if it is necessary to have a corrosion-resistant material that can be hardened and tempered by heat treatment.

For superior corrosion resistance, materials such as brass, bronze, aluminum, or nickel are sometimes used in the manufacture of fasteners. If strength is no problem, plastics such as nylons are used in severe corrosion applications.

Drivability. When selecting a screw, thought must be given to the means of driving for assembly and disassembly as well as the head shape. Most screw heads provide a slot, a recess, or a hexagon shape as a means of driving. The slotted screw is the least preferred driving style and serves only when appearance must be combined with ease of disassembly with a common screwdriver. Only a limited amount of torque can be applied with a screwdriver. A slot can become inoperative after repeated disassembly destroys the edge of the wall that the blade of the screwdriver bears against. The hexagon head is preferred for the following reasons:

Least likely to accidentally spin out (thereby marring the surface of the product)

Lowest initial cost

Adaptable to high-speed power drive

Minimum worker fatigue

Ease of assembly in difficult places

Permits higher driving torque, especially in large sizes where strength is important

Contains no recess to become clogged with dirt and interfere with driving

Contains no recess to weaken the head

Unless frequent field disassembly is required, use of the unslotted hex head is preferred.

Appearance is the major disadvantage of the hex head, and this one factor is judged sufficient to eliminate it from consideration for the front or top of products.

The recessed head fastener is widely used and becomes the first choice for appearance applications. It usually costs more than a slot or a hexagon shape. There are many kinds of recesses. The Phillips and Phillips POZIDRIV are most widely used. To a lesser extent the Frearson, clutch-type, hexagonal, and fluted socket heads are used. For special applications, proprietary types of tamper-resistant heads can be selected (Fig. 1).

The recessed head has some of the same advantages as the hex head (see preceding list). It also has improved appearance. The Phillips POZIDRIV is slowly replacing the Phillips recess. The POZIDRIV recess can be readily identified by four radial lines centered between each recess slot. These slots are a slight modification of the conventional Phillips recess. This change improves the fit between the driver and the recess, thus minimizing the possibility of marring a surface from accidental spinout of the driver as well as increasing the life of the driver. The POZIDRIV design is recommended in high-production applications requiring high driving torques. The POZIDRIV

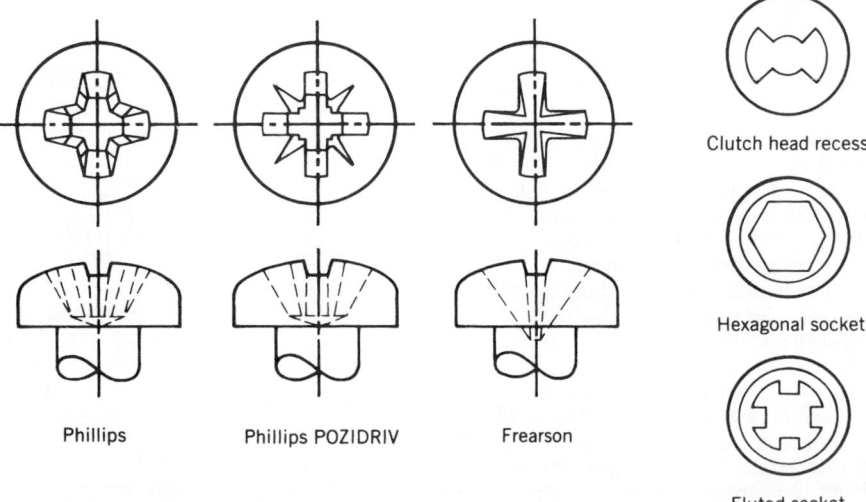

Clutch head recess

Hexagonal socket

Fluted socket

Phillips   Phillips POZIDRIV   Frearson

Fig. 1 Recessed head fasteners.

recess usually sells at a high-production applications requiring high driving torques. The POZIDRIV recess usually sells at a slightly higher price than the conventional Phillips recess, but some suppliers will furnish either at the same price. The savings resulting from longer tool life will usually justify the higher initial cost.

A conventional Phillips driver could be used to install or disassemble a POZIDRIV screw. However, a POZIDRIV driver should be used with a POZIDRIV screw in order to take advantage of the many features inherent in the new design. To avoid confusion, it should be clearly understood that a POZIDRIV driver cannot be used to install or remove a conventional Phillips head screw.

A Frearson recess is a somewhat different design than a Phillips recess and has the big advantage that one driving tool can be used for all sizes whereas a Phillips may require four driving tools in the range from no. 2 (0.086-in.) to 3/8 (0.375-in.) screw size. This must be balanced against the following disadvantages:

Limited availability.

Greater penetration of the recess means thinner walls between the bottom of the recess and the outer edge of the screw, which tends to weaken the head.

The sharp point of the driver can easily scratch or otherwise mar the surface of the product if it accidently touches.

Although one driver can be used for all sizes, for optimum results, different size drivers are recommended for installing various screw sizes, thus minimizing the one real advantage of the Frearson recess.

The hexagon and fluted socket head cap screws are only available in expensive high-strength alloy steel. Its unique small outside diameter or cylindrical head is useful on flanges, counterbored holes, or other locations where clearances are restricted. Such special applications may justify the cost of a socket head cap screw. Appreciable savings can be made in other applications by substitution of a hexagon head screw.

Despite any claims to the contrary, the dimensional accuracy of hexagon socket head cap screws is no better than that of other cold-headed products, and there is no merit in close-thread tolerances, which are advocated by some manufacturers of these products. The high prices, therefore, should be justified solely on the basis of possible space savings in using the cylindrical head.

The fluted socket is not as readily available and should only be considered in the very small sizes where a hexagon key tends to round out the socket. The fluted socket offers spline design so that the key will neither slip nor be subject to excessive wear.

Many types of special recesses are tamper resistant. In most of these designs, the recess is an unusual shape requiring a special tool for assembly and disassembly. A readily available driving tool such as a screwdriver or hexagon key would not fit the recess. The purpose of a tamper-resistant fastener is to prevent unauthorized removal of parts and equipment. Their protection is needed on any product located in public places to discourage vandalism and thievery. They may also be necessary on some consumer products as a safety measure to protect the amateur repairman from injury or to prevent him from causing serious damage to equipment. With product liability mania what it is today, the term "tamperproof" has all but disappeared. Now the fasteners are called "tamper resistant." They are the same as they were under their previous name, but the new term better reflects their true capabilities. Any skilled thief with ample time and proper tools can saw, drill, blast, or otherwise disassemble any tamper-resistant fastener. Therefore, these fasteners are intended only to discourage the casual thief or amateur tinkerer and make it more difficult for a skilled professional. Whatever the choice of fastener design, it is essential that hardened material be specified. No fastener is ever truly tamperproof, but hardened steel helps. Fasteners made of soft material can be disassembled easily by sawing a slot, hammering with a chisel, or drilling a hole and using an extraction bit.

Head Shapes   The following information is equally applicable to all types of recesses as well as a slotted head. For simplification only slotted screws are shown.

The pan head is the most widely used and is intended to replace the round, binding, and truss heads in order to keep varieties to a minimum. It is preferred because it presents the best combination of appearance with adequate head height to minimize weakness due to depth of penetration of the recess (Fig. 2).

The round head was widely used in the past (Fig. 3). It has since been delisted as an American National Standard. Give preference to pan heads on all new designs. Figure 4 shows the superiority of the pan head: The high edge of the pan head at its periphery,

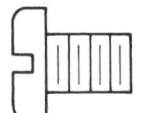

Fig. 2   Pan head.

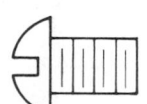

Fig. 3   Round head.

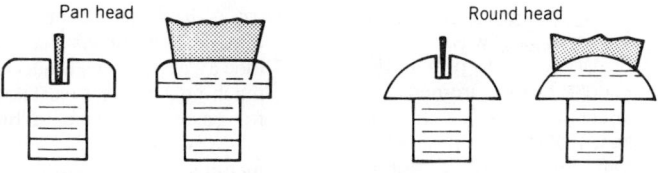

Fig. 4   Drive-slot engagement.

where driving action is most effective, provides superior driver-slot engagement and reduces the tendency to chew away the metal at the edge of the slot.

The flat head is used where a flush surface is required. The countersunk section aids in centering the screw (Fig. 5).

The oval head is similar to a flat head except that instead of a flush surface it presents a low silhouette that improves the appearance (Fig. 6).

The truss head is similar to the round head except that the head is shallower and has a larger diameter. It is used where extra bearing surface is required for extra holding power or where the clearance hole is oversized or the material is soft. It also presents a low silhouette that improves the appearance (Fig. 7).

The binding head is similar to the pan head and is commonly used for electrical connections where an undercut is usually specified to bind and prevent the fraying of stranded wire (Fig. 8).

The fillister head has the smallest diameter for a given shank size. It also has a deep slot that allows a higher torque to be applied during assembly. It is not as readily available or as widely used as some of the other head styles (Fig. 9).

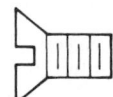

Fig. 5   Flat head.

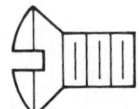

Fig. 6   Oval head.

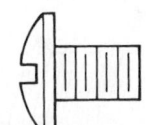

Fig. 7   Truss head.

Fig. 8   Binding head.

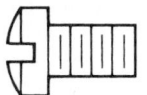

Fig. 9   Fillister head.

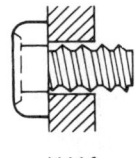

(1) Indented hex head
(2) Slotted indented hex head

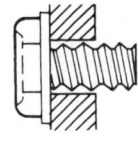

(3) Indented hex washer head
(4) Slotted indented hex washer head

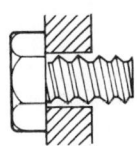

(5) Trimmed hex head
(6) Slotted trimmed hex head
(7) Trimmed hex washer head
(8) Slotted trimmed hex washer head

Fig. 10   Hex head.

The advantages of a hex head are listed in the discussion on drivability. This type head is available in eight variations (Fig. 10).

The indented design is lowest cost as the hex is completely cold upset in a counterbore die and possesses an identifying depression in the top surface of the head.

The trimmed design requires an extra operation to produce clean sharp corners with no indentation. Appearance is improved and there is no pocket on top to collect moisture.

The washer design has a larger bearing surface to spread the load over a wider area. The washer is an integral part of the head and also serves to protect the finish of the assembly from wrench disfigurement.

The slot is used to facilitate field service. It adds to the cost, weakens the head, and limits the amount of tightening torque that can be applied. A slot is unnecessary in high-production factory installation.

Any given location should standardize on one or possibly two of the eight variations.

Types of Screws

Machine Screws. Machine screws are meant to be assembled in tapped holes, either into a product or into a nut. The screw threads of a machine screw are readily available in American National Standard Unified Inch Coarse and Fine Thread series. They are generally considered for applications where the material to be joined is too hard, too weak, too brittle, or too thick to take a tapping screw. It is also used in applications where the assembly requires a fastener made of a material that cannot be hardened enough to make its own thread, such as brass or nylon machine screws. Applications requiring freedom from dust or particles of any kind cannot use thread-cutting screws and, therefore, must be joined by machine screws or a tapping screw which forms or rolls a thread.

There are many combinations of head styles, shapes, and materials.

Self-Tapping Screws. There are many different types of self-tapping screws commercially available. The following three types are capable of creating an internal thread by being twisted into a smooth hole:

1. Thread-forming screws
2. Thread-cutting screws
3. Thread-rolling screws

The following two types create their own opening before generating the thread:

4. Self-drilling and tapping screws
5. Self-extruding and tapping screws

1. **Thread-Forming Screws.** Thread-forming screws create an internal thread by forming or squeezing material. They rely on the pressure of the screw thread to force a mating thread into the workpiece. They are applicable in materials where large internal stresses are permissible or desirable to increase resistance to loosening. They are generally used to fasten sheet metal parts. They cannot be used to join brittle materials, such as plastics, because the stresses created in the workpiece can cause cracking. The following types of thread-forming screws are commonly used:

Types A and AB. Type AB screws have a spaced thread. This means that each thread is spaced further away from its adjacent thread than the popular machine screw series. They also have a gimlet point for ease in entering a predrilled hole. This type of screw is primarily intended to be used in sheet metal with a thickness from 0.015 in. (0.38 mm) to 0.05 in. (1.3 mm), resin-impregnated plywood, natural woods, and asbestos compositions.

Type AB screws were introduced several years ago to replace the type A screws. The type A screw is the same as the type AB except for a slightly wider spacing of the threads. Both are still available and can be used interchangeably. The big advantage of the type AB screw is that its threads are spaced exactly as the type B screws to be discussed later. In the interest of standardization it is recommended that type AB screws be used in place of either the type A or the type B series (Fig. 11).

Type B. Type B screws have the same spacing as type AB screws. Instead of a gimlet point, they have a blunt point with incomplete threads at the point. This point makes the type B more suitable for thicker metals and blind holes. The type B screws can be used in any of the applications listed under type AB. In addition the type B screw can be used in sheet metal up to a thickness of 0.200 in. (5 mm) and in nonferrous castings (Fig. 12).

Type C. Type C screws look like type B screws except that threads are spaced to be exactly the same as a machine screw thread and may be used to replace a machine screw in the field. They are recommended for general use in metal 0.030–0.100 in. (0.76–2.54 mm) thick. It should be recognized that in specific applications, involving long thread engagement or hard materials, this type of screw requires extreme driving torques.

2. **Thread-Cutting Screws.** Thread-cutting screws create an internal thread by actual removal of material from the internal hole. The design of the cavity to provide space for the chips and the design of the cutting edge differ with each type. They are used in place of the thread-forming type for applications

Fig. 11   Type AB.

Fig. 12   Type B.

in materials where disruptive internal stresses are undesirable or where excessive driving torques are encountered. The following types of thread-cutting screws are commonly used:

Type BT (Formerly Known as Type 25). Type BT screws have a spaced thread and a blunt point similar to the type B screw. In addition they have one cutting edge and a wide chip cavity. These screws are primarily intended for use in very friable plastics such as urea compositions, asbestos, and other similar compositions. In these materials, a larger space between threads is required to produce a satisfactory joint because it reduces the buildup of internal stresses that fracture brittle plastic when a closer spaced thread is used. The wide cutting slot creates a large cutting edge and permits rapid deflection of the chips to produce clean mating threads. For best results all holes should be counterbored to prevent fracturing the plastic. Use of this type screw eliminates the need to use tapped metallic inserts in plastic materials (Fig. 13).

Type ABT. Type ABT screws are the same as type BT screws except that they have a gimlet point similar to a type AB screw. This design is not recognized as an American National Standard and should only be selected for large-volume applications (over 50,000 pieces of one size and type). It is primarily intended for use in plastic for the same reasons as listed for type BT screws (Fig. 14).

Type D (Formerly Known as Type 1). Type D screws have threads of machine screw diameter–pitch combinations approximating unified form with a blunt point and tapered entering threads. In addition a slot is cut off center with one side on the center line. This radial side of the slot creates the sharp serrated cutting edge such as formed on a tap. The slot leaves a thinner section on one side of the screw that collapses and helps concentrate the pressure on the cutting edge. This screw is suitable for use in all thicknesses of metals (Fig. 15).

Type F. Type F screws are identical to type D except that instead of one slot there are several slots cut at a slight angle to the axis of the thread. This screw is

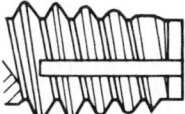

Fig. 15   Type D.

Fig. 16   Type F.

suitable for use in all thicknesses of metals and can be used interchangeably with a type D screw in many applications. However, the type F screw is superior to the type D screw for tapping into cast iron and permits the use of a smaller pilot hole (Fig. 16).

Type D or Type F. Because in many applications these two types can be used interchangeably with the concomitant advantages of simpler inventory and increased availability, a combined specification is often issued permitting the supplier to furnish either type.

Type T (Formerly Known as Type 23). Type T screws are similar to type D and type F except that they have an acute rake angle cutting edge. The cut in the end of the screw is designed to eliminate a pocket that confines the chips. The shape of the slot is such that the chips are forced ahead of the screw as it is driven. This screw is suitable for plastics and other soft materials when a standard machine screw series thread is desired. It is used in place of type D and type F when more chip room is required because of deep penetration (Fig. 17).

Type BF. Type BF screws are intended for use in plastics. The wide thread pitch reduces the buildup of internal stresses that fracture brittle plastics when a smaller thread pitch is used. The screw has a blunt point and tapered entering threads with several cutting edges and chip cavity (Fig. 18).

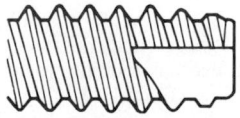

Fig. 13   Type BT.

Fig. 14   Type ABT.

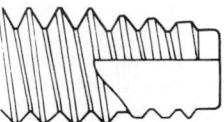

Fig. 17   Type T.

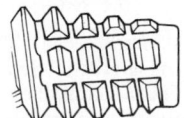

Fig. 18   Type BF.

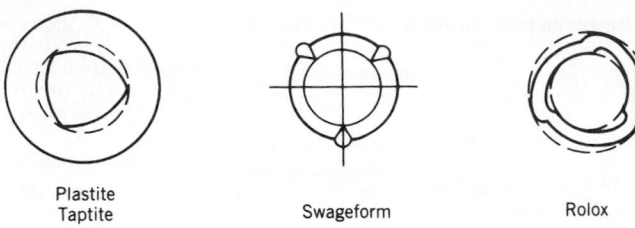

Plastite
Taptite

Swageform

Rolox

Fig. 19   Thread-rolling screws.

3. **Thread-Rolling Screws.** Thread-rolling screws (see Fig. 19) form an internal thread by flowing metal and thus do not cut through or disrupt the grain flow lines of materials as do thread-cutting screws. The screw compacts and work hardens the material, thereby forming strong, smoothly burnished internal threads. The screws have the threads of machine screw diameter–pitch combinations. This type screw is ideal for applications where chips can cause electrical shorting of equipment or jamming of delicate mechanism. Freedom from formation of chips eliminates the costly problem of cleaning the product of chips and burrs as would otherwise be required.

The ratio of driving torque to stripping torque is approximately 1 : 8 for a thread-rolling screw as contrasted to 1 : 3 for a conventional tapping screw. This higher ratio permits the driver torque release to be set well over the required driving torque and yet safely below the stripping torque. This increased ratio minimizes poor fastening due to stripped threads or inadequate seating of the screws.

Plastite is intended for use in filled or unfilled thermoplastics and some of the thermosetting plastics. The other three types are intended for use in metals. At present, there are no data to prove the superiority of one type over another.

4. **Self-Drilling and Tapping Screws.** The self-drilling and tapping screw (Fig. 20) drills its own hole and forms a mating thread, thus making a complete fastening in a single operation. Assembly labor is reduced by eliminating the need to predrill holes at assembly and by solving the problem of hole alignment. These screws must complete their metal-drilling function and fully penetrate the material before the screw thread can engage and begin its advancement. In order to meet this requirement, the unthreaded point length must be equal to or greater than the material thickness to be drilled. Therefore, there is a strict limitation on minimum and maximum material thickness that varies with screw size. There are many different styles and types of self-drilling and tapping screws to meet specific needs.

5. **Self-Extruding Screws.** Self-extruding screws provide their own extrusion as they are driven into an inexpensively produced punched hole. The resulting extrusion height is several times the base material thickness. This type screw is suitable for material in

Fig. 20   Self-drilling and tapping screws.

Fig. 21   Self-extruding screw.

thicknesses up to 0.048 in. (1.2 mm). By increasing the thread engagement, these screws increase the differential between driving and stripping torque and provide greater pull-out strength. Since they do not produce chips, they are excellent for grounding sheet metal for electrical connections (Fig. 21).

There is almost no limit to the variety of head styles, thread forms, and screw materials that are available commercially. The listing only shows representative examples. Users should attempt to keep varieties to a minimum by carefully selecting those variations that best meet the needs of their type of product.

Set Screws. Set screws are available in various combinations of head and point style as well as material and are used as locking, locating, and adjustment devices. The common head styles are slotted headless, square head, hexagonal socket, and fluted socket. The slotted headless has the lowest cost and can be used in a counterbored hole to provide a flush surface. The square head is applicable for location or adjustment of static parts where the projecting head is not objectionable. Its use should be avoided on all rotating parts. The hexagonal socket head can be used in a counterbored hole to provide a flush surface. It permits greater torque to be applied than with a slotted headless design. Fluted sockets are useful in very small

Table 104   Holding Power of Flat or Cup Point Set Screws

d (in.)	$\frac{1}{4}$	$\frac{5}{16}$	$\frac{3}{8}$	$\frac{7}{16}$	$\frac{1}{2}$	$\frac{9}{16}$	$\frac{5}{8}$	$\frac{3}{4}$	$\frac{7}{8}$	1	$1\frac{1}{8}$	$1\frac{1}{4}$
P (lb)	100	168	256	366	500	658	840	1280	1830	2500	3388	4198

diameters, that is, no. 6 (0.138 in.) and under, where hexagon keys tend to round out the socket in hexagonal socket set screws. Set screws should not be used to transmit large amounts of torque, particularly under shock torsion loads. Increased torsion loads may be carried by two set screws located 120° apart.

The following points are available with the head styles discussed: The cup point (Table 104) is the standard stock point for all head shapes and is recommended for all general locking purposes. Flats are recommended on round shafts when close fits are used and it is desirable to avoid interference in disassembling parts because of burrs produced by action of the cup point or when the flats are desired to increase torque transmission. When flats are not used, it is recommended that the minimum shaft diameter be not less than four times the cup diameter since otherwise the whole cup may not be in contact with the shaft. The self-locking cup point has limited availability. It has counterclockwise knurls to prevent the screw from working loose even in poorly tapped holes (Fig. 22).

When oval points are used, the surface it contacts should be grooved or spotted to the same general contour as the point to assure good seating. It is used where frequent adjustment is necessary without excessive deformation of the part against which it bears (Fig. 23).

When flat points are used, it is customary to grind a flat on the shaft for better point contact. This point is preferred where wall thickness is thin and on top of plugs made of any soft material (Fig. 24).

When the cone point is used, it is recommended that the angle of countersink be as nearly as possible the angle of screw point for the best efficiency. Cone point

Fig. 22   Cup point.

Fig. 23   Oval point.

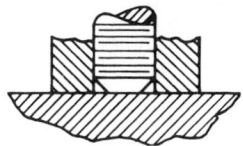

Fig. 24   Flat point.

Fig. 25   Cone point.

Fig. 26   Half-dog point.

set screws have some application as pivot points. It is used where permanent location of parts is required. Because of penetration, it has the highest axial and torsional holding power of any point (Fig. 25).

The half-dog point should be considered in lieu of full-dog points when the usable length of thread is less than the nominal diameter. It is also more readily obtained than the full-dog point. It can be used in place of dowel pins and where end of thread must be protected (Fig. 26).

Lag Screws.   Lag screws (Table 105) are usually used in wood but also can be used in plastics and with expansion shields in masonry. A 60° gimlet point is the most readily available type. A 60° cone point, not covered in these drawings, is also available. Some suppliers refer to this item as a lag bolt (Fig. 27).

A lag screw is normally used in wood when it is inconvenient or objectionable to use a through bolt and nut. To facilitate the insertion of the screw especially in denser types of wood, it is advisable to use a lubricant on the threads. It is important to have a pilot hole of proper size and following are some recommended hole sizes for commonly used types of wood. Hole sizes for other types of wood should be in proportion

Table 105 Lag Screws

Diameter of screw (in.)	$\frac{1}{4}$	$\frac{5}{16}$	$\frac{3}{8}$	$\frac{7}{16}$	$\frac{1}{2}$	$\frac{5}{8}$	$\frac{3}{4}$	$\frac{7}{8}$	1
No. of threads per inch	10	9	7	7	6	5	$4\frac{1}{2}$	4	$3\frac{1}{2}$
Across flats of hexagon and square heads (in.)	$\frac{3}{8}$	$\frac{15}{32}$	$\frac{9}{16}$	$\frac{21}{32}$	$\frac{3}{4}$	$\frac{15}{16}$	$1\frac{1}{8}$	$1\frac{5}{16}$	$1\frac{1}{2}$
Thickness of hexagon and square heads (in.)	$\frac{3}{16}$	$\frac{1}{4}$	$\frac{5}{16}$	$\frac{3}{8}$	$\frac{7}{16}$	$\frac{17}{32}$	$\frac{5}{8}$	$\frac{3}{4}$	$\frac{7}{8}$

Length of Threads for Screws of All Diameters							
Length of screw (in.)	$1\frac{1}{2}$	2	$2\frac{1}{2}$	3	$3\frac{1}{2}$	4	$4\frac{1}{2}$
Length of thread (in.)	To head	$1\frac{1}{2}$	2	$2\frac{1}{4}$	$2\frac{1}{2}$	3	$3\frac{1}{2}$
Length of screw (in.)	5	$5\frac{1}{2}$	6	7	8	9	10–12
Length of thread (in.)	4	4	$4\frac{1}{2}$	5	6	6	7

Table 106 Recommended Diameters of Pilot Hole for Types of Wood[a]

Screw Diameter (in.)	White Oak	Southern Yellow Pine, Douglas Fir	Redwood, Northern White Pine
0.250	0.160	0.150	0.100
0.312	0.210	0.195	0.132
0.375	0.260	0.250	0.180
0.438	0.320	0.290	0.228
0.500	0.375	0.340	0.280
0.625	0.485	0.437	0.375
0.750	0.600	0.540	0.480

[a] Pilot holes should be slightly larger than listed when lag screws of excessive lengths are to be used.

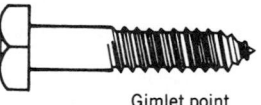

Gimlet point

Fig. 27 Lag screws.

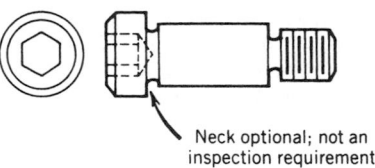

Neck optional; not an inspection requirement

Fig. 28 Shoulder screw.

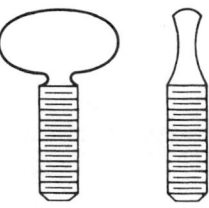

Fig. 29 Thumb screws.

to the relative specific gravity of that wood to the ones listed in Table 106.

Shoulder Screws.   These screws are also referred to as "stripper bolts." They are used mainly as locators or retainers for spring strippers in punch and die operations and have found some application as fulcrums or pivots in machine designs that involve links, levers, or other oscillating parts. Consideration should be given to the alternative use of a sleeve bearing and a bolt on the basis of both cost and good design (Fig. 28).

Thumb Screws.   Thumb screws have a flattened head designed for manual turning without a driver or a wrench. They are useful in applications requiring frequent disassembly or screw adjustment (Fig. 29).

Weld Screws.   Weld screws come in many different head configurations, all designed to provide one or more projections for welding the screw to a part.

Overhead projections are welded directly to the part. Underhead projections go through a pilot hole. The designs in Figs. 30 and 31 are widely used.

In projection welding of carbon steel screws, care should be observed in applications, since optimum weldability is obtained when the sum, for either parent metal or screw, of one-fourth the manganese content

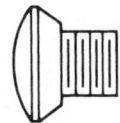

Fig. 30 Single-projection weld screw.

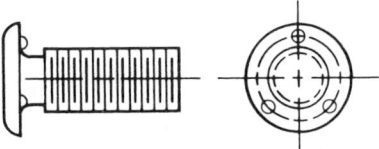

Fig. 31 Underhead weld screws.

plus the carbon content does not exceed 0.38. For good weldability with the annular ring type, the height of the weld projection should not exceed half the parent metal thickness as a rule of thumb.

Copper flash plating is provided for applications where cleanliness of the screw head is necessary in obtaining good welds.

Wood Screws. Wood screws are (Table 107) readily available in lengths from $\frac{1}{4}$ to 5 in. for steel and from $\frac{1}{4}$ to $3\frac{1}{2}$ in. for brass. Consideration should be given to the use of type AB thread-forming screws, which are lower in cost and more efficient than wood screws for use in wood. Wood screws are made with flat, round, or oval heads.

The *resistance of wood screws to withdrawal* from side grain of seasoned wood is given by the formula $P = 2850G^2D$, where P is the allowable load on the screw (lb/in. penetration of the threaded portion), G is specific gravity of oven-dry wood, and D is the diameter of the screw (in.). Wood screws should not be designed to be loaded in withdrawal from the end grain.

The *allowable safe lateral resistance* of wood screws embedded seven diameters in the side grain

of seasoned wood is given by the formula $P = KD^2$, where P is the lateral resistance per screw (lb), D is the diameter (in.), and K is 4000 for oak (red and white), 3960 for Douglas fir (coast region) and southern pine, and 3240 for cypress (southern) and Douglas fir (inland region).

The following rules should be observed: (a) The size of the lead hole in soft (hard) woods should be about 70% (90%) of the core or root diameter of the screw; (b) lubricants such as soap may be used without great loss in holding power; (c) long, slender screws are preferable generally, but in hardwood too slender screws may reach the limit of their tensile strength; and (d) in the screws themselves, holding power is favored by thin sharp threads, rough unpolished surface, full diameter under the head, and shallow slots.

SEMS. The machine and tapping screws can be purchased with washers or lock washers as an integral part of the purchased screws. When thus joined together, the part is known as a SEMS unit. The washer is assembled on a headed screw blank before the threads are rolled. The inside diameter of the washer is of a size that will permit free rotation and yet prevent disassembly from the screw after the threads are rolled. If these screws and washers were purchased separately, there would be an initial cost savings over the preassembled units. However, these preassembled units reduce installation time because only one hand is needed to position them, leaving the other hand free to hold the driving tool. The time required to assemble a loose washer is eliminated. In addition, these assemblies act to minimize installation errors and inspection time because the washer is in place, correctly oriented. Also the use of a single unit, rather than two separate parts, simplifies bookkeeping, handling, inventory, and other related operations.

7.1 Nominal and Minimum Dressed Sizes of American Standard Lumber

Table 108 applies to boards, dimensional lumber, and timbers. The thicknesses apply to all widths and all widths to all thicknesses.

Table 107 American Standard Wood Screws[a]

Number	0	1	2	3	4	5	6	7	8
Threads per inch	32	28	26	24	22	20	18	16	15
Diameter (in.)	0.060	0.073	0.086	0.099	0.112	0.125	0.138	0.151	0.164
Number	9	10	11	12	14	16	18	20	24
Threads per inch	14	13	12	11	10	9	8	8	7
Diameter (in.)	0.177	0.190	0.203	0.216	0.242	0.268	0.294	0.320	0.372

[a]Included angle of flathead = 82°; see Fig. 18.

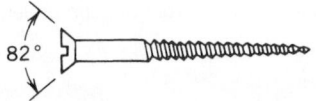

82°

Table 108   Nominal and Minimum Dressed Sizes of American Standard Lumber

	Thicknesses			Face Widths		
		Minimum Dressed			Minimum Dressed	
Item	Nominal	Dry[a] (in.)	Green (in.)	Nominal	Dry[a] (in.)	Green (in.)
Boards[b]				2	$1\frac{1}{2}$	$1\frac{9}{16}$
				3	$2\frac{1}{2}$	$2\frac{9}{16}$
				4	$3\frac{1}{2}$	$3\frac{9}{16}$
				5	$4\frac{1}{2}$	$4\frac{5}{8}$
				6	$5\frac{1}{2}$	$5\frac{5}{8}$
	1	$\frac{3}{4}$	$\frac{25}{32}$	7	$6\frac{1}{2}$	$6\frac{5}{8}$
	$1\frac{1}{4}$	1	$1\frac{1}{32}$	8	$7\frac{1}{4}$	$7\frac{1}{2}$
	$1\frac{1}{2}$	$1\frac{1}{4}$	$1\frac{9}{32}$	9	$8\frac{1}{4}$	$8\frac{1}{2}$
				10	$9\frac{1}{4}$	$9\frac{1}{2}$
				11	$10\frac{1}{4}$	$10\frac{1}{2}$
				12	$11\frac{1}{4}$	$11\frac{1}{2}$
				14	$13\frac{1}{4}$	$13\frac{1}{2}$
				16	$15\frac{1}{4}$	$15\frac{1}{2}$
Dimension				2	$1\frac{1}{2}$	$1\frac{9}{16}$
				3	$2\frac{1}{2}$	$2\frac{9}{16}$
				4	$3\frac{1}{2}$	$3\frac{9}{16}$
	2	$1\frac{1}{2}$	$1\frac{9}{16}$	5	$4\frac{1}{2}$	$4\frac{5}{8}$
	$2\frac{1}{2}$	2	$2\frac{1}{16}$	6	$5\frac{1}{2}$	$5\frac{5}{8}$
	3	$2\frac{1}{2}$	$2\frac{9}{16}$	8	$7\frac{1}{4}$	$7\frac{1}{2}$
	$3\frac{1}{2}$	3	$3\frac{1}{16}$	10	$9\frac{1}{4}$	$9\frac{1}{2}$
				12	$11\frac{1}{4}$	$11\frac{1}{2}$
				14	$13\frac{1}{4}$	$13\frac{1}{2}$
				16	$15\frac{1}{4}$	$15\frac{1}{2}$
Dimension				2	$1\frac{1}{2}$	$1\frac{9}{16}$
				3	$2\frac{1}{2}$	$2\frac{9}{16}$
				4	$3\frac{1}{2}$	$3\frac{9}{16}$
				5	$4\frac{1}{2}$	$4\frac{5}{8}$
	4	$3\frac{1}{2}$	$3\frac{9}{16}$	6	$5\frac{1}{2}$	$5\frac{5}{8}$
	$4\frac{1}{2}$	4	$4\frac{1}{16}$	8	$7\frac{1}{4}$	$7\frac{1}{2}$
				10	$9\frac{1}{4}$	$9\frac{1}{2}$
				12	$11\frac{1}{4}$	$11\frac{1}{2}$
				14		$13\frac{1}{2}$
				16		$15\frac{1}{2}$
Timbers	5 and thicker		$\frac{1}{2}$ off	5 and wider		$\frac{1}{2}$ off

[a]Maximum moisture content of 19 % or less.

[b]Boards less than the minimum thickness for 1 in. nominal but $\frac{5}{8}$ in. or greater thickness dry ($\frac{11}{16}$ in. green) may be regarded as American Standard Lumber, but such boards shall be marked to show the size and condition of seasoning at the time of dressing. They shall also be distinguished from 1-in. boards on invoices and certificates.

Source: From *American Softwood Lumber Standard*, NBS 20–70, National Bureau of Standards, Washington, DC, 1970, amended 1986 (available from Superintendent of Documents).

CHAPTER 2
MATHEMATICS*

J. N. Reddy
Department of Mechanical Engineering
Texas A&M University
College Station, Texas

*This chapter is a revision and extension of Section 2 of the third edition, which was written by John L. Barnes.

The names of Greek letters are found in Table 1, standard mathematical symbols in Table 2, and abbreviations for engineering terms in Table 3 in Section 4.5.

1 ARITHMETIC

1.1 Roman Numerals

Roman Notation. This uses seven letters and a bar; a letter with a bar placed over it represents a thousand times as much as it does without the bar. The letters and rules for combining them to represent numbers are as follows:

I	V	X	L	C	D	M	$\overline{\text{L}}$
1	5	10	50	100	500	1000	50,000

Rule 1 If no letter precedes a letter of greater value, add the numbers represented by the letters.

Example 1 XXX represents 30; VI represents 6.

Rule 2 If a letter precedes a letter of greater value, subtract the smaller from the greater; add the remainder or remainders thus obtained to the numbers represented by the other letters.

Example 2 IV represents 4; XL represents 40; CXLV represents 145.

Other illustrations:

IX	XIII	XIV	LV	XLII	XCVI	MDCI	$\overline{\text{IV}}$ CCXL
9	13	14	55	42	96	1601	4240

1.2 Roots of Numbers

Roots can be found by use of Table 7, or logarithms, in Section 2.9.

To find an nth root by arithmetic, use a method indicated by the binomial theorem expansion of $(a + b)^n$:

$$(a + b)^n = a^n + na^{n-1}b + \frac{n(n-1)}{2}a^{n-2}b^2$$
$$+ \frac{n(n-1)(n-2)}{3 \cdot 2}a^{n-3}b^3 + \cdots + b^n$$
$$= a^n + bD$$

where $D = na^{n-1} + \frac{1}{2}n(n-1)a^{n-2}b + \cdots + b^{n-1}$.

1. Point off the given number into periods of n figures each, starting at the decimal point and going both ways.

2. Find the largest nth power in the left-hand period and use its root as the first digit of the result. Subtract this nth power from the left-hand period and bring down the next period.

3. Use the quantity D, in which a is 10 times the first digit since the first digit occupies a higher place than the second, as the divisor to obtain the second digit b. As a trial divisor to estimate b, use the first term in D, since it is the largest. Multiply D by b, subtract, and bring down the next period.

4. To get the next digit use 10 times the first two digits as a and proceed as before.

Example 3
See the tabulation for Example 3 below.

1.3 Approximate Computation

Standard Notation. $N = a \cdot 10^b$, N is a given number; $1 \le a < 10$, the figures in a being the *significant figures* in N; b is an integer, positive or negative or zero.

Example 4   If $N = 2,953,000$, in which the first five figures are significant, then $N = 2.9530 \times 10^6$.

A number is *rounded* to contain fewer significant figures by dropping figures from the right-hand side. If the figures dropped amount to more than $\frac{1}{2}$ in the last figure kept, this last figure is increased by 1. If the figures dropped amount to $\frac{1}{2}$, the last figure may or may not be increased.

Since the last significant figure used in making a measurement, an estimate, and so on, is not exact but is usually the nearer of two consecutive figures, an approximate number may represent any value in a range from $\frac{1}{2}$ less in its last significant figure to $\frac{1}{2}$ more. The *absolute error* in an approximate number may be as much as $\frac{1}{2}$ in the last significant figure.

Example 5   If $N = 2.9530 \times 10^6$ is an approximate number, then $2.95295 \times 10^6 \le N \le 2.95305 \times 10^6$.

The absolute error is between -0.00005×10^6 and 0.00005×10^6.

The size of the absolute error depends on the location of the decimal point.

The *relative error* is the ratio of the absolute error to the number. Its size depends on the number of significant figures.

Example 6   The relative error in Example 5 is at most $0.00005 \times 10^6 / 2.9530 \times 10^6$, or about 1 in 60,000; the percentage error is at most $100 \times (0.00005/2.9530)$, or less than 0.002%.

In the result of a computation with approximate numbers, some figures on the right are doubtful and should be rounded off. It is always possible, by using the bounds of the ranges that approximate numbers represent, to compute exactly the bounds of the range in which the result lies and then round off the uncertain figures.

Example 7   Divide the approximate number 536 by the approximate number 217.4:

	At least	At most
$\dfrac{536}{217.4} = 2.47-$	$\dfrac{535.5}{217.45} = 246+$	$\dfrac{536.5}{217.35} = 2.47-$

Tabulation for Example 3

1. Square root of 302.980652:

$$3'02.'98'06'52' \quad \big|\; 17.406 +$$

$$\begin{array}{rl}
 & 1 \\
D = 2a + b = 27 & \overline{\,202\,} \\
 & 189 \\
344 & \overline{\,1398\,} \\
 & 1376 \\
34,806 & \overline{\,220,652\,} \\
 & 208,836
\end{array}$$

2. Cube root of 1,58,252.632929: $158'252'.632'929$ $\big|\; 54.09$

$$\begin{array}{rl}
5^3 = & 125 \\
\text{Trial divisor} = 3a^2 = 3 \times 50^2 = \quad 7,500 & \overline{\,33,252\,} \\
3ab = 3 \times 50 \times 4 = \quad 600 & \\
b^2 = 4^2 = \quad 16 & \\
D = 3a^2 + 3ab + b^2 = \quad \overline{8,116} & 32,464 \\
3 \times 5400^2 = 8,7480,000 & \overline{\,788,632,929\,} \\
3 \times 5400 \times 9 = \quad 145,800 & \\
9^2 = \quad 81 & \\
\overline{87,625,881} & 788,632,929
\end{array}$$

In the quotient the third figure may be in error. It is useless to carry the division further.

The following rules usually give the largest number of significant figures that it is reasonable to keep.

Addition and Subtraction.
Keep as the last significant figure in the result the figure in the last full column. The absolute accuracy of the result is determined by the least absolutely accurate number.

Example 8

$$2.953xx$$
$$0.8942x$$
$$\underline{0.06483}$$
$$3.912xx$$

Multiplication, Division, Powers, and Roots.
Keep no more significant figures in the result than the fewest in any number involved. The relative accuracy of the result is determined by that of the least relatively accurate number. Shortcuts as shown in the examples may be used.

Example 9

1.

$$2953 \times 413$$

$$2953$$
$$413$$
$$\overline{118\,|\,12}$$
$$3\,|\,0$$
$$\overline{9}$$
$$\overline{122\quad xxxx = 1.22 \times 10^6}$$

2.

$$(1.22 \times 10^6)/2953$$
$$413$$
$$2953\,\overline{|\,1{,}220{,}000}$$
$$11{,}812$$
$$\overline{295\,|\,388}$$
$$295$$
$$\overline{30\,|\,93}$$
$$90$$

In intermediate results keep one additional figure.

If there is much difference in the relative accuracy, that is, the number of significant figures, of the numbers involved in a computation, round all of them to one more significant figure than the least accurate number has. This procedure may introduce a small error in the last figure kept in the result. A three-digit number beginning with 8 or 9 has about the same relative accuracy as a four-digit number beginning with 1.

Use of Tables.
In using a table to find the value of a function corresponding to an approximate value of an argument, it is usually advisable to retain no more significant figures in the function than there are in the argument, although the accuracy of the function varies considerably, depending inversely on the slope of the curve representing the function. However, there is no need for many-place tables if the values of the argument are known only to a few significant figures.

Example 10
$\frac{1}{52} = 0.019$; $\cos 61.3° = 0.877$; $\log 3.74 = 0.573$.

To investigate the behavior of the error for any given function, the differential approximation is useful. If $y = f(x)$, then $dy = f'(x)\,dx$ approximates the absolute error, and $dy/y = f'(x)\,dx/f(x)$ the relative error.

For particular approximate values of the arguments, the bounds of the ranges of the functions can be found directly from a table with arguments given to one additional place.

1.4 Interpolation

Gregory–Newton Interpolation Formula.
Let $f(x)$ be a tabulated function of the argument x, Δx the constant difference between values of x for which the function is tabulated, and p a proper fraction. To find $f(x + p\Delta x)$ use the formula

$$f(x + p\,\Delta x) = f(x) + p\,\Delta f +_p C_2\,\Delta_2 f$$
$$+ _pC_3\,\Delta_3 f + \cdots$$

in which

$$_pC_r = \frac{p(p-1)\cdots(p-r+1)}{r!}$$

and $\Delta_r f = r$ th functional difference.

Binomial coefficients for interpolation:

p	$_pC_2$	$_pC_3$	$_pC_4$	$_pC_5$
0.1	−0.0450	0.0285	−0.0207	0.0161
0.2	−0.0800	0.0480	−0.0336	0.0255
0.3	−0.1050	0.0595	−0.0402	0.0297
0.4	−0.1200	0.0640	−0.0416	0.0300
0.5	−0.1250	0.0625	−0.0391	0.0273
0.6	−0.1200	0.0560	−0.0336	0.0228
0.7	−0.1050	0.0455	−0.0262	0.0173
0.8	−0.0800	0.0320	−0.0176	0.0113
0.9	−0.0450	0.0165	−0.0087	0.0054

In ordinary linear interpolation the first two terms of the formula are used.

Example 11 Find $\sqrt{15.4}$.

x	$f(x) = \sqrt{x}$	Δf	$\Delta_2 f$	$\Delta_3 f$
15	3.8730			
		0.1270		
16	4.0000		−0.0039	
		0.1231		0.0003
17	4.1231		−0.0036	
		0.1195		
18	4.2426			

$$\Delta x = 1 \qquad p = 0.4$$

$$f(15 + 0.4 \times 1) = 3.8730 + 0.4 \times 0.1270 + 0.1200$$

$$\times 0.0039 + 0.0640 \times 0.0003$$

$$= 3.9243$$

2 ALGEBRA
2.1 Numbers
Classification

1. *Real* (*positive* and *negative*).
   (a) *Rational*, expressible as the quotient of two integers.
      i. *Integers*, as $-1, 2, 53$.
      ii. *Fractions*, as $\frac{3}{4}, -\frac{5}{2}$.
   (b) *Irrational*, not expressible as the quotient of two integers, as $\sqrt{2}, \pi$.
2. *Imaginary*, a product of a real number and the *imaginary unit* $i(= \sqrt{-1})$. Electrical engineers use j to avoid confusion with i for current. Example: $\sqrt{-2} = \sqrt{2}i$.
3. *Complex*, a sum of a real number and an imaginary number, as $a + bi$ (a and b real), $-3 + 0.5i$. A real number may be regarded as a complex number in which $b = 0$ and an imaginary number as one in which $a = 0$.

The Absolute Value of:

1. A *real number* is the number itself if the number is positive and the number with its sign changed if it is negative, as, for example, $|3| = |-3| = 3$.
2. A *complex number* $a + bi$ is $\sqrt{a^2 + b^2}$, as, for example, $|-3 + 0.5i| = \sqrt{9 + \frac{1}{4}} = 3.04$.

2.2 Identities
Powers

1. $(-a)^n = a^n$ if n is even
2. $(-a)^n = -a^n$ if n is odd
3. $a^m \cdot a^n = a^{m+n}$
4. $a^m / a^n = a^{m-n}$

5. $(ab)^n = a^n b^n$
6. $(a/b)^n = a^n / b^n = (b/a)^{-n} = b^{-n}/a^{-n}$
   $= a^n b^{-n}$
7. $a^{-n} = (1/a)^n = \frac{1}{a^n}$
8. $(a^m)^n = a^{mn}$
9. $a^0 = 1; 0^n = 0; 0^0$ is meaningless

Roots

1. $\sqrt[n]{a} = a^{1/n}$
2. $(\sqrt[n]{a})^n = \sqrt[n]{a^n} = a$
3. $\sqrt[n]{ab} = \sqrt[n]{a}\sqrt[n]{b}$
4. $\sqrt[n]{a/b} = \sqrt[n]{a}/\sqrt[n]{b}$
5. $\sqrt[m]{a}\sqrt[n]{a} = a^{(1/m)+(1/n)} = \sqrt[mn]{a^{m+n}}$
6. $\sqrt[m]{a^n} = (\sqrt[m]{a})^n = a^{n/m}$
7. $\sqrt[m]{\sqrt[n]{a}} = \sqrt[mn]{a} = \sqrt[n]{\sqrt[m]{a}} = (a^{1/m})^{1/n} = a^{1/mn}$
8. $\sqrt{a} + \sqrt{b} = \sqrt{a + b + 2\sqrt{ab}}$

Products

1. $(a \pm b)^2 = a^2 \pm 2ab + b^2$
2. $(a + b)(a - b) = a^2 - b^2$.
3. $(a + b + c)^2 = a^2 + b^2 + c^2 + 2ab + 2ac + 2bc$
4. $(a \pm b)^3 = a^3 \pm 3a^2 b + 3ab^2 \pm b^3$
5. $a^3 \pm b^3 = (a \pm b)(a^2 \mp ab + b^2)$

Quotients

1. $(a^n - b^n)/(a - b) = a^{n-1} + a^{n-2}b + a^{n-3}b^2 + \cdots + ab^{n-2} + b^{n-1}$ if $a \neq b$
2. $(a^n + b^n)/(a + b) = a^{n-1} - a^{n-2}b + a^{n-3}b^2 - \cdots - ab^{n-2} + b^{n-1}$ if n is odd
3. $(a^n - b^n)/(a + b) = a^{n-1} - a^{n-2}b + a^{n-3}b^2 - \cdots + ab^{n-2} - b^{n-1}$ if n is even

Fractions

Signs:

$$\frac{a}{b} = \frac{-a}{-b} = \frac{-a}{b} = -\frac{a}{-b}.$$

Addition and subtraction:

$$\frac{a}{c} \pm \frac{b}{d} = \frac{ad \pm bc}{cd}, \frac{a}{c} \pm \frac{b}{c} = \frac{a \pm b}{c}, \frac{a}{c} \pm \frac{a}{d}$$

$$= \frac{a(d \pm c)}{cd} \quad \frac{a}{def} + \frac{b}{e^3 g} - \frac{c}{df^2}$$

$$= \frac{ae^2 fg + bdf^2 - ce^3 g}{de^3 f^2 g}$$

Multiplication:

$$\frac{a}{b} \times \frac{c}{d} = \frac{ac}{bd}, \; \frac{a}{b} = \frac{ac}{bc}$$

Division:

$$\frac{a/b}{c/d} = \frac{a}{b} \times \frac{d}{c} = \frac{ad}{bc}, \; \frac{a}{b} = \frac{a/c}{b/c}$$

Series

1. $1 + 2 + 3 + 4 + \cdots + (n - 1) + n = \frac{1}{2}n(n + 1)$

2. $p + (p + 1) + (p + 2) + \cdots + (q - 1) + q =$ $\frac{1}{2}(q + p)(q - p + 1)$

3. $2 + 4 + 6 + 8 + \cdots + (2n - 2) + 2n =$ $n(n + 1)$

4. $1 + 3 + 5 + 7 + \cdots + (2n - 3) + (2n - 1)$ $= n^2$

5. $1^2 + 2^2 + 3^2 + 4^2 + \cdots + (n - 1)^2 + n^2 =$ $\frac{1}{6}n(n + 1)(2n + 1)$

6. $1^3 + 2^3 + 3^3 + 4^3 + \cdots + (n - 1)^3 + n^3 =$ $\frac{1}{4}n^2(n + 1)^2$

7. $1^4 + 2^4 + 3^4 + 4^4 + \cdots + (n - 1)^4 + n^4 =$ $\frac{1}{30}n(n + 1)(2n + 1)(3n^2 + 3n - 1)$

2.3 Binomial Theorem

$$(a \pm b)^n = a^n \pm na^{n-1}b + \frac{n(n - 1)}{1 \cdot 2}a^{n-2}b^2$$

$$\pm \frac{n(n - 1)(n - 2)}{1 \cdot 2 \cdot 3}a^{n-3}b^3 + \cdots$$

$$+ (\pm 1)^r \frac{n(n - 1) \cdots (n - r + 1)}{r!}a^{n-r}b^r + \cdots$$

in which the last term shown is the $(r + 1)$th; $r!$, called *r factorial*, equals $1 \cdot 2 \cdot 3 \cdots (r - 1) \cdot r$; and $0! = 1$.

If n is a positive integer, the series is finite; it has $n + 1$ terms, the last being b^n; and it holds for all values of a and b. If n is fractional or negative, the series is infinite; it converges only for $|b| < |a|$ (see Section 9.4).

The coefficients $n, n(n - 1)/2!, n(n - 1)(n - 2)/3!, \ldots$ are called *binomial coefficients*. For brevity the coefficient $n(n - 1) \cdots (n - r + 1)/r!$ of the $(r + 1)$th

terms is written $\binom{n}{r}$ or $_nC_r$. If n is a positive integer,

the coefficients of the rth term from the beginning and the rth from the end are equal.

For any value of n and $-1 < x < 1$,

$$(1 \pm x)^n = 1 \pm nx + \frac{n(n - 1)}{1 \cdot 2}x^2 \pm \frac{n(n - 1)(n - 2)}{1 \cdot 2 \cdot 3}x^3$$

$$+ \frac{n(n - 1)(n - 2)(n - 3)}{1 \cdot 2 \cdot 3 \cdot 4}x^4 \pm \cdots$$

$$\frac{1}{1 \pm x} = (1 \pm x)^{-1} = 1 \mp x + x^2 \mp x^3 + x^4 \mp x^5 + \cdots$$

$$\sqrt{1 \pm x} = (1 \pm x)^{1/2} = 1 \pm \frac{1}{2}x - \frac{1}{2 \cdot 4}x^2 \pm \frac{1 \cdot 3}{2 \cdot 4 \cdot 6}x^3$$

$$- \frac{1 \cdot 3 \cdot 5}{2 \cdot 4 \cdot 6 \cdot 8}x^4 \pm \frac{1 \cdot 3 \cdot 5 \cdot 7}{2 \cdot 4 \cdot 6 \cdot 8 \cdot 10}x^5 - \cdots$$

$$\frac{1}{\sqrt{1 \pm x}} = (1 \pm x)^{-1/2} = 1 \mp \frac{1}{2}x + \frac{1 \cdot 3}{2 \cdot 4}x^2 \mp \frac{1 \cdot 3 \cdot 5}{2 \cdot 4 \cdot 6}x^3$$

$$+ \cdots$$

2.4 Approximate Formulas

(a) If $|x|$ and $|y|$ are small compared with 1:

1. $(1 \pm x)^2 = 1 \pm 2x$

2. $(1 \pm x)^{1/2} = 1 \pm \frac{1}{2}x$

3. $1/(1 \pm x) = 1 \mp x$

4. $(1 + x)(1 + y) = 1 + x + y$

5. $(1 + x)(1 - y) = 1 + x - y$

6. $e^x = 1 + x + \frac{1}{2}x^2$
   (where $e = 2.71828$)

7. $\log_e(1 \pm x)$
   $= \pm x - x^2/2 \pm x^3/3$

8. $\log_e\left(\frac{1 + x}{1 - x}\right)$
   $= 2\left(x + \frac{1}{3}x^3 + \frac{1}{5}x^5\right)$

(Last term often may be omitted.)

(b) If $|x|$ is small compared with a and $a > 0$:

9. $a^x = 1 + x \log_e a + \frac{1}{2}x^2(\log_e a)^2$. (Last term often may be omitted.)

(c) If a and b are nearly equal and both > 0:

10. $\sqrt{ab} = \frac{1}{2}(a + b)$

(d) If b is small compared with a and both > 0:

11. $\sqrt{a^2 \pm b} = a \pm b/2a$

12. $\sqrt{a^2 \pm b} = a \pm b/3a^2$

13. $\sqrt{a^2 + b^2} = 0.960a + 0.398b$. This is within 4% of the true value if $a > b$.

A closer approximation is $\sqrt{a^2 + b^2} = 0.9938a + 0.0703b + 0.3567(b^2/a)$.

14. $\sqrt{a^2 + b^2 + c^2} = 0.939a + 0.389b + 0.297c$. This is within 6% of the true value if $a > b > c$. For instance, for the numbers 43, 42, and 41, the error is $< 5.2\%$.

(e) If $|x|$ is less than $\pi/18$:

15. $\sin x = x - \frac{1}{6}x^3$

16. $\cos x = 1 - \frac{1}{2}x^2$

17. $\tan x = x + \frac{1}{3}x^3$

(Last term often may be omitted.)

(Note: If $x = 8° = 8\pi/180 = 0.13963$, $\sin x = x - \frac{1}{6}x^3 = 0.13918$, which is one unit in error in the fifth decimal place. If the absolute value of the angle is less than $5°$, the values of x and $\sin x$ do not differ more than one unit in the fourth decimal place.)

(f) If $|y|'$ is less than $\pi/36$ and small compared with $|x|$:

18. $\sin(x \pm y) = \sin x \pm y \cos x$

19. $\cos(x \pm y) = \cos x \mp y \sin x$

20. $\tan(x \pm y) = \tan x \pm y/\cos^2 x$

(g) If $|n| > 1$:

21. $e^{1/n} = 1 + 1/(n - 0.5)$

22. $e^{-1/n} = 1 - 1/(n + 0.5)$

(h) As $n \to \infty$:

23. $\dfrac{1 + 2 + 3 + 4 + 5 \cdots + n}{n^2} \to \dfrac{1}{2}$

24. $\dfrac{1 + 2^2 + 3^2 + 4^2 + \cdots + n^2}{n^3} \to \dfrac{1}{3}$

25. $\dfrac{1 + 2^3 + 3^3 + 4^3 + \cdots + n^3}{n^4} \to \dfrac{1}{4}$

2.5 Inequalities

Laws of Inequalities for Positive Quantities

(a) If $a > b$, then

$$a + c > b + c \qquad b < a$$
$$a - c > b - c \qquad c - a < c - b$$
$$ac > bc \qquad -ca < -cb$$
$$\frac{a}{c} > \frac{b}{c} \qquad \frac{c}{a} < \frac{c}{b}$$

Corollary: If $a - c > b$, then $a > b + c$.

(b) If $a > b$ and $c > d$, then $a + c > b + d$; $ac > bd$; but $a - c$ may be greater than, equal to, or less than $b - d$; a/c may be greater than, equal to, or less than b/d.

2.6 Ratio and Proportion

Laws of Ratio and Proportion

(a) If $a/b = c/d$, then

$$\frac{a}{c} = \frac{b}{d} \qquad\qquad ad = bc$$

$$\frac{ma + nb}{pa + qb} = \frac{mc + nd}{pc + qd} \qquad \left(\frac{a}{b}\right)^n = \left(\frac{c}{d}\right)^n$$

If also $e/f = g/h$, then, $ae/bf = cg/dh$.

(b) If $a/b = c/d = e/f = \cdots$, then

$$\frac{a}{b} = \frac{c}{d} = \frac{e}{f} = \cdots = \frac{pa + qc + re + \cdots}{pb + qd + rf + \cdots}$$

Variation

If $y = kx$, y varies directly as x; that is, y is directly proportional to x.

If $y = k/x$, y varies inversely as x; that is, y is inversely proportional to x.

If $y = kxz$, y varies jointly as x and z.

If $y = k(x/z)$, y varies directly as x and inversely as z.

The constant k is called the proportionality factor.

2.7 Progressions

Arithmetic Progression. This is a sequence in which the *difference d* of any two consecutive terms is a constant. If n = number of terms, a = first term, l = last term, s = sum of n terms, then $l = a + (n - 1)d$, and $s = (n/2)(a + l)$. The *arithmetic mean A* of two quantities m, n is the quantity that placed between them makes with them an arithmetic progression; $A = (m + n)/2$.

Example 12 Given the series $3 + 5 + 7 + \cdots$ to 10 terms. Here $n = 10$, $a = 3$, $d = 2$; hence $l = 3 + (10 - 1) \times 2 = 21$ and $s = (10/2)(3 + 21) = 120$.

Geometric Progression. This is a sequence in which the *ratio r* of any two consecutive terms is a constant. If n = number of terms, a = first term, l = last term, s = sum of n terms, then $l = ar^{n-1}$, $s = (rl - a)/(r - 1) = a(1 - r^n)/(1 - r)$. The *geometric mean G* of two quantities m, n is the quantity that placed between them makes with them a geometric progression; $G = \sqrt{mn}$.

Example 13 Given the series $3 + 6 + 12 + \cdots$ to six terms. Here $n = 6$, $a = 3$, $r = 2$; hence $l = 3 \times 2^{6-1} = 96$ and $s = (2 \times 96 - 3)/(2 - 1) = 3(1 - 2^6)/(1 - 2) = 189$.

If $|r| < 1$ then, as $n \to \infty$, $s \to a/(1 - r)$.

Example 14 Given the infinite series $\frac{1}{2} + \frac{1}{4} + \frac{1}{8} + \cdots$. Here $a = \frac{1}{2}$ and $r = \frac{1}{2}$; hence $s \to (\frac{1}{2})/(1 - \frac{1}{2}) = 1$ as $n \to \infty$.

Harmonic Progression. This is a sequence in which the reciprocals of the terms form an arithmetic progression. The *harmonic mean H* of two quantities m, n is the quantity that placed between them makes with them a harmonic progression; $H = 2mn/(m + n)$.

The relation among the arithmetic, geometric, and harmonic means of two quantities is $G^2 = AH$.

2.8 Partial Fractions

A *proper* algebraic fraction is one in which the numerator is of lower degree than the denominator. An improper fraction can be changed to the sum of a polynomial and a proper fraction by dividing the numerator by the denominator.

A proper fraction can be resolved into *partial fractions*, the denominators of which are factors, prime to each other, of the denominator of the given fraction.

CASE 1: The denominator can be factored into real linear factors $P, Q, R, \ldots$ all different. Let

$$\frac{\text{Num}}{PQR\cdots} = \frac{A}{P} + \frac{B}{Q} + \frac{C}{R} + \cdots$$

Example 15

$$\frac{6x^2 - x + 1}{x^3 - x} = \frac{A}{x} + \frac{B}{x-1} + \frac{C}{x+1}$$

Clearing fractions,

$$6x^2 - x + 1 = A(x-1)(x+1) + Bx(x+1) \\ + Cx(x-1) \quad (1)$$

(a) *Substitution method.* Letting $x = 0$, $A = -1$; $x = 1$, $B = 3$; and $x = -1$, $C = 4$ yields

$$\frac{6x^2 - x + 1}{x^3 - x} = -\frac{1}{x} + \frac{3}{x-1} + \frac{4}{x+1}$$

(b) *Method of undetermined coefficients.* Rewriting Eq. (1),

$$6x^2 - x + 1 = (A + B + C)x^2 + (B - C)x - A$$

Equating coefficients of like powers of x, $A + B + C = 6$, $B - C = -1$, $-A = 1$. Solving this system of equations, $A = -1$, $B = 3$, $C = 4$.

CASE 2: The denominator can be factored into real linear factors, $P, Q, \ldots$, one or more repeated. Let

$$\frac{\text{Num}}{p^2 Q^3} = \frac{A}{P} + \frac{B}{p^2} + \frac{C}{Q} + \frac{D}{Q^2} + \frac{E}{Q^3} + \cdots$$

Example 16

$$\frac{x+1}{x(x-1)^3} = \frac{A}{x} + \frac{B}{x-1} + \frac{C}{(x-1)^2} + \frac{D}{(x-1)^3}$$

Clearing fractions,

$$x + 1 = A(x-1)^3 + Bx(x-1)^2 + Cx(x-1) + Dx$$

A and D can be found by substituting $x = 0$ and $x = 1$. After inserting these numerical values for A and D, B and C can be found by the method of undetermined coefficients.

CASE 3: The denominator can be factored into quadratic factors, $P, Q, \ldots$, all different, which cannot be factored into real linear factors. Let

$$\frac{\text{Num}}{PQ\cdots} = \frac{Ax + B}{P} + \frac{Cx + D}{Q} + \cdots$$

Example 17

$$\frac{3x^2 - 2}{(x^2 + x + 1)(x + 1)} = \frac{Ax + B}{x^2 + x + 1} + \frac{C}{x+1}$$

Clearing fractions,

$$3x^2 - 2 = (Ax + B)(x + 1) + C(x^2 + x + 1)$$
$$= (A + C)x^2 + (A + B + C)x + (B + C)$$

Use the method of undetermined coefficients to find A, B, C.

CASE 4: The denominator can be factored into quadratic factors, $P, Q, \ldots$, one or more repeated, which cannot be factored into real linear factors. Let

$$\frac{\text{Num}}{P^2 Q^3 \cdots} = \frac{Ax + B}{P} + \frac{Cx + D}{P^2} + \frac{Ex + F}{Q}$$
$$+ \frac{Gx + H}{Q^2} + \frac{Ix + J}{Q^3} + \cdots$$

Example 18

$$\frac{5x^2 - 4x + 16}{(x-3)(x^2 - x + 1)^2} = \frac{A}{x-3} + \frac{Bx + C}{x^2 - x + 1}$$
$$+ \frac{Dx + E}{(x^2 - x + 1)^2}$$

Clearing fractions,

$$5x^2 - 4x + 16 = A(x^2 - x + 1)^2 + (Bx + C)(x - 3) \\ \times (x^2 - x + 1) + (Dx + E)(x - 3)$$

Find A by substituting $x = 3$. Then use the method of undetermined coefficients to find B, C, D, E.

2.9 Logarithms

If $N = b^x$, then x is the *logarithm* of the number N to the *base b*. For computation, *common*, or Briggs, logarithms to the base 10 (abbreviated $\log_{10}$ or log) are used. For theoretical work involving calculus, *natural*, or Naperian, logarithms to the irrational base

$e = 2.71828\cdots$ (abbreviated ln, $\log_e$, or log) are used. The relation between logarithms of the two systems is

$$\log_e n = \frac{\log_{10} n}{\log_{10} e} = \frac{\log_{10} n}{0.4343} = 2.303 \log_{10} n$$

The integral part of a common logarithm, called the *characteristic*, may be positive, negative, or zero. The decimal part, called the *mantissa* and given in tables, is always positive.

To find the common logarithm of a number, first find the mantissa from Table 10 in Section 9.4, disregarding the decimal point of the number. Then from the location of the decimal point find the characteristic as follows. If the number is greater than 1, the characteristic is positive or zero. It is 1 less than the number of figures preceding the decimal point. For a number expressed in standard notation the characteristic is the exponent of 10.

Example 19 $\log 6.54 = 0.8156$, $\log 6540 = \log(6.54 \times 10^3) = 3.8156$.

If the number is less than 1, the characteristic is negative and is numerically 1 greater than the number of zeros immediately following the decimal point. To avoid having a negative integral part and a positive decimal part, the characteristic is written as a difference.

Example 20 $\log 0.654 = \log(6.54 \times 10^{-1}) = \overline{1}.8156 = 9.8156 - 10$, $\log \quad 0.000654 = \log(6.54 \times 10^{-4}) = \overline{4}.8156 = 6.8156 - 10$.

To find a number whose logarithm is given, each of the preceding steps is reversed.

The cologarithm of a number is the logarithm of its reciprocal. Hence, $\operatorname{colog} N = \log 1/N = \log 1 - \log N = -\log N$.

Use of Logarithms in Computation

To multiply a and b	$\log ab = \log a + \log b$
To divide a by b	$\log a/b = \log a - \log b$
To raise a to the nth power	$\log a^n = n \log a$
To find the nth root of a	$\log a^{1/n} = (1/n) \log a$

Example 21

1. $68.31 \times 0.2754 = 18.81$:

$$\log 68.31 = \quad 1.8345$$

$$\log 0.2754 = \quad \underline{9.4400 - 10}$$
$$11.2745 - 10 = 1.2745 = \log 18.81$$

2. $0.6841^{1.53} = 0.5582$:

$$\log 0.6831 = 9.8345 - 10$$

$$1.53 \times (9.8345 - 10) = 15.0468 - 15.3$$

To subtract 15.3 from 15.0468, add 10 to 15.0468 and subtract 10 from it:

$$25.0468 - 10$$
$$\underline{15.3}$$
$$9.7468 - 10 = \log 0.5582$$

3. $\sqrt[5]{0.6831} = 0.9266$:

$$\log 0.6831 = 9.8345 - 10$$

$$\tfrac{1}{5}(49.8345 - 50) = 9.9669 - 10 = \log 0.9266$$

To solve a simple exponential equation of the form $a^x = b$, equate the logarithms of the two sides of the equation:

$$x \log a = \log b$$

from which

$$x = \frac{\log b}{\log a} \quad \text{and} \quad \log x = \log(\log b) - \log(\log a)$$

Example 22 $0.6831^x = 27.54$.

$$x = \frac{\log 27.54}{\log 0.6831} = \frac{1.4400}{9.8345 - 10} = \frac{1.4400}{-0.1655} = -8.701$$

2.10 Equations

The equation $f(x) = a_0 x^n + a_1 x^{n-1} + a_2 x^{n-2} + \cdots + a_n = 0$, a_i real, is a *polynomial equation* of *degree n* in one variable.

For $n = 1$, the equation $f(x) = ax + b = 0$ is *linear*. It has one root, $x_1 = -b/a$.

Quadratic Equation For $n = 2$, the equation $f(x) = ax^2 + bx + c = 0$ is *quadratic*. It has two roots, both real or both complex, given by the formulas

$$x_1, \; x_2 = \frac{-b \pm \sqrt{b^2 - 4ac}}{2a} = \frac{2c}{-b \mp \sqrt{b^2 - 4ac}}$$

To avoid loss of precision if $\sqrt{b^2 - 4ac}$ and $|b|$ are nearly equal, use the form that does not involve the difference.

If the quantity $b^2 - 4ac$, called the *discriminant*, is greater than zero, the roots are real and unequal; if it equals zero, the roots are real and equal; if it is less than zero, the roots are complex.

Cubic Equation For $n = 3$, the equation $f(x) = a_0 x^3 + a_1 x^2 + a_2 x + a_3 = 0$ is *cubic*. It has three roots, all real or one real and two complex.

Algebraic Solution. Write the equation in the form $ax^3 + 3bx^2 + 3cx + d = 0$. Let

$$q = ac - b^2 \quad \text{and} \quad r = \tfrac{1}{2}(3abc - a^2 d) - b^3$$

Also let

$$s_1 = (r + \sqrt{q^3 + r^2})^{1/3} \quad \text{and}$$

$$s_2 = (r - \sqrt{q^3 + r^2})^{1/3}$$

Then the roots are

$$x_1 = \frac{(s_1 + s_2) - b}{a}$$

$$x_2 = \frac{-\frac{1}{2}(s_1 + s_2) + \frac{1}{2}\sqrt{-3}(s_1 - s_2) - b}{a}$$

$$x_3 = \frac{-\frac{1}{2}(s_1 + s_2) - \frac{1}{2}\sqrt{-3}(s_1 - s_2) - b}{a}$$

If $q^3 + r^2 > 0$, there are one real and two complex roots. If $q^3 + r^2 = 0$, there are three real roots of which at least two are equal. If $q^3 + r^2 < 0$, there are three real roots, but the numerical solution leads to finding the cube roots of complex quantities. In such a case the trigonometric solution is employed.

Example 23　Given the equation $x^3 + 12x^2 + 45x + 54 = 0$. Here $a = 1, b = 4, c = 15, d = 54$. Let $q = 15 - 16 = -1$; $r = \frac{1}{2}(180 - 54) - 64 = 1$; $q^3 + r^2 = -1 + 1 = 0$, $s_1 = s_2 = (-1)^{1/2} = -1$; $s_1 + s_2 = -2$; $s_1 - s_2 = 0$. Hence the roots are $x_1 = (-2-4) = -6$; $x_2 = x_3 = [-\frac{1}{2}(-2) - 4] = -3$.

Trigonometric Solution.　Write the equation in the form $ax^3 + 3bx^2 + 3cx + d = 0$. Let $q = ac - b^2$ and $r = \frac{1}{2}(3abc - a^2d) - b^3$ (as in algebraic solution). Then the roots are

$$x_1 = \frac{y_1 - b}{a} \qquad x_2 = \frac{y_2 - b}{a} \qquad x_3 = \frac{y_3 - b}{a}$$

where $y_1, y_2,$ and y_3 have the following values (upper of alternative signs being used when r is positive and the lower when r is negative):

CASE 1: If q is negative and $q^3 + r^2 \le 0$:

$$y_1 = \pm 2\sqrt{-q}\cos\left(\frac{1}{3}\cos^{-1}\frac{\pm r}{\sqrt{-q^3}}\right)$$

$$y_2 = \pm 2\sqrt{-q}\cos\left(\frac{1}{3}\cos^{-1}\frac{\pm r}{\sqrt{-q^3}} + \frac{2\pi}{3}\right)$$

$$y_3 = \pm 2\sqrt{-q}\cos\left(\frac{1}{3}\cos^{-1}\frac{\pm r}{\sqrt{-q^3}} + \frac{4\pi}{3}\right)$$

CASE 2: If q is negative and $q^3 + r^2 \ge 0$:

$$y_1 = \pm 2\sqrt{-q}\cosh\left(\frac{1}{3}\cosh^{-1}\frac{\pm r}{\sqrt{-q^3}}\right)$$

$$y_2 = \mp\sqrt{-q}\cosh\left(\frac{1}{3}\cosh^{-1}\frac{\pm r}{\sqrt{-q^3}}\right)$$

$$+ i\sqrt{-3q}\sinh\left(\frac{1}{3}\cosh^{-1}\frac{\pm r}{\sqrt{-q^3}}\right)$$

$$y_3 = \mp\sqrt{-q}\cosh\left(\frac{1}{3}\cosh^{-1}\frac{\pm r}{\sqrt{-q^3}}\right)$$

$$- i\sqrt{-3q}\sinh\left(\frac{1}{3}\cosh^{-1}\frac{\pm r}{\sqrt{-q^3}}\right)$$

CASE 3: If q is positive:

$$y_1 = \pm 2\sqrt{q}\sinh\left(\frac{1}{3}\sinh^{-1}\frac{\pm r}{\sqrt{q^3}}\right)$$

$$y_2 = \mp\sqrt{q}\sinh\left(\frac{1}{3}\sinh^{-1}\frac{\pm r}{\sqrt{q^3}}\right)$$

$$+ i\sqrt{3q}\cosh\left(\frac{1}{3}\sinh^{-1}\frac{\pm r}{\sqrt{q^3}}\right)$$

$$y_3 = \mp\sqrt{q}\sinh\left(\frac{1}{3}\sinh^{-1}\frac{\pm r}{\sqrt{q^3}}\right)$$

$$- i\sqrt{3q}\cosh\left(\frac{1}{3}\sinh^{-1}\frac{\pm r}{\sqrt{q^3}}\right)$$

Example 24　Given the equation $x^3 + 6x^2 - 9x - 54 = 0$.

Here $a = 1, b = 2, c = -3, d = -54$; $q = -3 - 4 = -7$; $r = \frac{1}{2}(-18 + 54) - 8 = 10$; $q^3 + r^2 = -343 + 100 = -243$. Note that q is negative; $q^3 + r^2 < 0$; r is positive. Therefore use Case 1 with upper signs:

$$y_1 = 2\sqrt{7}\cos\left(\frac{1}{3}\cos^{-1}\frac{10}{\sqrt{343}}\right) = 2\sqrt{7}\cos 19.1° = 5$$

Hence, one root is $x_1 = 5 - 2 = 3$. The other roots can be similarly determined.

Quartic Equation　For $n = 4$, the equation $f(x) = a_0x^4 + a_1x^3 + a_2x^2 + a_3x + a_4 = 0$ is *quartic*. It has four roots, all real, all complex, or two real and two complex.

To solve, first divide the equation by a_0 to put it in the form $x^4 + ax^3 + bx^2 + cx + d = 0$. Find any real root y_1 of the cubic equation:

$$8y^3 - 4by^2 + 2(ac - 4d)y - [c^2 + d(a^2 - 4b)] = 0$$

Then the four roots of the quartic equation are given by the roots of the two quadratic equations:

$$x^2 + \left(\tfrac{1}{2}a + \sqrt{\tfrac{1}{4}a^2 + 2y_1 - b}\right)x + (y_1 + \sqrt{y_1^2 - d}) = 0$$

$$x^2 + \left(\tfrac{1}{2}a - \sqrt{\tfrac{1}{4}a^2 + 2y_1 - b}\right)x + (y_1 - \sqrt{y_1^2 - d}) = 0$$

nth-Degree Equation

Properties of $f(x) = a_0x^n + a_1x^{n-1} + \cdots + a_n = 0$.
Assume a_n's are real.

1. *Remainder Theorem.* If $f(x)$ is divided by $x - r$ until a remainder independent of x is obtained, this remainder is equal to $f(r)$, the value of $f(x)$ for $x = r$.
2. *Factor Theorem.* If and only if $x - r$ is a factor of $f(x)$, then $f(r) = 0$.
3. The equation $f(x) = 0$ has n roots, not necessarily distinct. Complex roots occur in conjugate pairs, $a + bi$ and $a - bi$. If n is odd, there is at least one real root.
4. The sum of the roots is $-a_1/a_0$, the sum of the products of the roots taken two at a time is a_2/a_0, the sum of the products of the roots taken three at a time is $-a_3/a_0$, and so on. The product of all the roots is $(-1)^n a_n/a_0$.
5. If the a_i are integers and p/q is a rational root of $f(x) = 0$ reduced to its lowest terms, then p is a divisor of a_n and q of a_0. If a_0 is 1, the rational roots are integers.

6. If x is replaced by (a) y/m, (b) $-y$, (c) $y + h$, the roots of the resulting equation $\phi(y) = 0$ are (a) m times, (b) the negatives of, (c) less by h than the corresponding roots of $f(x) = 0$.
7. *Descartes' Rule of Signs.* A *variation* of sign occurs in $f(x) = 0$ if two consecutive terms have unlike signs. The number of positive roots is either equal to the number of variations of sign or is less by a positive even integer. For negative roots apply the rule to $f(-x) = 0$.
8. If, for two real numbers a and b, $f(a)$ and $f(b)$ have opposite signs, there is an odd number of roots between a and b.
9. If k is the exponent of the first term with a negative coefficient and G the greatest of the absolute values of the negative coefficients, then an upper bound of the real roots is $1 + \sqrt[n-k]{G/a_0}$.
10. *Sturm's Theorem.* Let the equation $f(x) = 0$ have no multiple roots. With $f_0 = f(x)$ and $f_1 = f'(x)$, form the sequence $f_0, f_1, f_2, \ldots, f_n$ as follows:

$$f_0 = q_1f_1 - f_2 \qquad f_1 = q_2f_2 - f_3$$
$$f_2 = q_3f_3 - f_4, \ldots, f_{n-2} = q_{n-1}f_{n-1} - f_n$$

At any step, a function f_i may be multiplied by a positive number to avoid fractions. Let a and b be real numbers, $a < b$ such that $f(a) \neq 0$, $f(b) \neq 0$, and let $V(a)$ be the number of variations of sign in the nonzero members of the sequence $f_0(a), f_1(a), \ldots, f_n$. Then the number of real roots between a and b is $V(a) - V(b)$.

If $f(x) = 0$ has multiple roots, the sequence terminates with the function $f_m, m < n$, when $f_{m-1} = q_m f_m$. For this sequence, $V(a) - V(b)$ is the number of distinct real roots between a and b.

Example 25 See the tabulation for Example 25 below.

Tabulation for Example 25

1. Locate the real roots of $x^3 - 7x - 7 = 0$.

$x =$	-2	-1	0	1	2	3	4		$3x - \dfrac{9}{2}$		x	
$f_0 = x^3 - 7x - 7$	$-$	$-$	$-$	$-$	$-$	$-$	$+$	$2x + 3$	$3x^2$	-7	$x^3 - 7x - 7$	
$f_1 = 3x^2 - 7$	$+$	$-$	$-$	$-$	$+$	$+$	$+$		$6x^2$	-17	$3x^3 - 21x - 21$	
$f_2 = 2x + 3$	$-$	$+$	$+$	$+$	$+$	$+$	$+$		$6x^2 + 9x$		$3x^2 - 7x$	
$f_3 = 1$	$+$	$+$	$+$	$+$	$+$	$+$	$+$		$-9x - 14$		$-14x - 21$	
$V(x) =$	3	1	1	1	1	1	1		$-9x - \dfrac{27}{2}$		$2x + 3 = f_2$	
									$-\dfrac{1}{2}$			
									$1 = f_3$			

$$V(-2) - V(-1) = 2 \qquad\qquad V(3) - V(4) = 1$$
$$-2 < r_1r_2 < -1 \qquad\qquad 3 < r_3 < 4$$

Tabulation for Example 25 (continued)

2. Locate the real roots of $4x^3 - 3x - 1 = 0$.

$x =$	-1	0	1	2		$2x - 1x$		
$f_0 = 4x^3 - 3x - 1$	$-$	$-$	0	$+$	$2x + 1$	$4x^2 - 1$	$4x^3 - 3x - 1$	
$f_1 = 3(4x^2 - 1)$	$+$	$-$		$+$		$4x^2 + 2x$	$4x^3 - x$	
$f_2 = 2x + 3$	$-$	$+$		$+$		$-2x - 1$	$-2x - 1$	
$V(x) =$	2	1	0			$-2x - 1$	$2x - 1 = f_2$	

$$V(-1) - V(0) = 2 \qquad V(3) - V(4) = 1$$
$$-1 < r_1 < 0 \qquad\qquad 0 < r_2 < 2$$

Then r_1 can be found to be a double root.

Synthetic Division. To divide a polynomial $f(x)$ by $x - a$, proceed as in Example 25. Divide $f(x) = 4x^3 - 7x + 1$ by $x + 2$. Arrange the coefficients in order of descending powers of x, supplying zeros for missing powers. Place $a(= -2)$ to the left. Bring down the first coefficient, multiply it by a, and add the product to the next coefficient. Multiply the sum by a, add the product to the next coefficient, and continue thus:

$$
\begin{array}{r}
-2\,|\,4 + 0 - 7 + 1 \\
\underline{-8 - 16 - 18} \\
4 - 8 + 9\,\underline{|-17}
\end{array}
$$

The last number is the remainder. It is the value of the polynomial $f(x) = 4x^3 - 7x + 1$ for $x = -2$, or $f(-2) = -17$. The other numbers in the last line are the coefficients of the quotient $4x^2 - 8x + 9$, a polynomial of one degree less than the dividend.

Rational Roots. Possible integral and fractional roots can be found by property 5 and tested by synthetic division. If a rational root r is found, then the remaining roots are roots of $q(x) \equiv f(x)/(x - r) = 0$.

Irrational Roots.

Horner's Method This consists of diminishing a root repeatedly toward zero and adding together the amounts by which it is diminished. This sum approximates the original root. The method is explained by an example.

A root of $x^3 + 4x - 7 = 0$ is located between the successive integers 1 and 2, graphically or by synthetic division, using property 8. First, the roots are diminished by 1 (property 6c) to give an equation $f(y + 1) \equiv \phi(y) = 0$, which has a root between 0 and 1. The method of obtaining the coefficients of $\phi(y)$ by use of successive synthetic divisions is illustrated. The remainders are the required coefficients. The root between 0 and 1 of $\phi(y) = 0$ is then located between successive tenths. Since its value is small, the last two terms set equal to 0 suffice to estimate that it is between 0.2 and 0.3. Next, diminish the roots by 0.2 to obtain

an equation with a root between 0 and 0.1. To check that the root was between 0.2 and 0.3, note that the first remainder, which is the value of $\phi(0.2)$, remains negative when $\phi(y)$ is divided by $y - 0.2$, and that the remainder would be found to be positive if $\phi(y)$ were divided by $y - 0.3$. Repeat the process, using the last two terms to estimate that the root of the new equation is between 0.05 and 0.06, and then diminish by 0.05. At the next stage it is frequently possible to estimate two more figures by using the last two terms.

$$
\begin{array}{llll}
1 + 0 & +4 & - & 7\,|\,1 \\
\underline{+1} & \underline{+1} & \underline{+} & 5 \\
1 + 1 & +5 & - & 2 \\
\underline{+1} & \underline{+2} & & \\
1 + 2 & +7 & & \\
\underline{+1} & & & \\
1 + 3 & + & 7 & - & 2\,|\,0.2 \\
\underline{+0.2} & \underline{+} & 0.64 & \underline{+} & 1.528 \\
1 + 3.2 & + & 7.64 & - & 0.472 \\
\underline{+0.2} & \underline{+} & 0.68 & \underline{-} & \\
1 + 3.4 & + & 8.32 & & \\
\underline{+0.2} & & & & \\
1 + 3.6 & + & 8.32 & -0.472 & |\,0.05 \\
\underline{+0.05} & \underline{+} & 0.1825 & \underline{+0.425125} & \\
1 + 3.65 & + & 8.5025 & -0.046875 & \\
\underline{+0.05} & \underline{+} & 0.185 & & \\
1 + 3.70 & \underline{+8.6875} & & & \\
\underline{+0.05} & & & & \\
1 + 3.75 & & & &
\end{array}
$$

$$8.6875x - 0.046875 = 0$$
$$x = 0.0054$$

The root is 1.2554.

To find a *negative* irrational root $-r$ by Horner's method, replace x in $f(x) = 0$ by $-y$, find the positive root r of $\phi(y) = f(-y) = 0$, and change its sign.

Newton's Method This can be used to find a root of either an algebraic or a transcendental equation. The

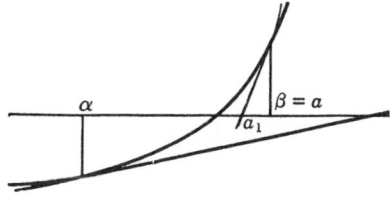

Fig. 1

root is first located graphically between α and β, $f(\alpha)$ and $f(\beta)$ having unlike signs (Fig. 1). Assume that there is no maximum, minimum, or inflection point in the interval (α, β), that is, that neither $f'(x)$ nor $f''(x)$ equals zero for any point in (α, β). Take as a first approximation a the endpoint α or β for which $f(x)$ and $f''(x)$ have the same sign, that is, if the curve is concave up, take the endpoint at which $f(x)$ is positive, and, if concave down, the endpoint at which $f(x)$ is negative. The point $a_1 = a - f(a)/f'(a)$, at which the tangent to the curve at $[a, f(a)]$ intersects the x axis, is between a and the root. Then, by using a_1 instead of a, a still better approximation a_2 is obtained, and so forth. If the endpoint for which $f(x)$ and $f''(x)$ have opposite signs were used, it could happen that the approximation obtained would be better than a_1, but it might be much worse since the tangent would not cross the x axis between the endpoint used and the root (Fig. 1).

Example 26 Find the real root of $x^3 + 4x - 7 = 0$.

$$f(x) = x^3 + 4x - 7$$

$$f'(x) = 3x^2 + 4$$

$$f''(x) = 6x$$

Graphically (Fig. 2), $\alpha = 1.2$, $\beta = 1.3$. Since $f(1.2) = -0.472$ and $f(1.3) = 0.397$, and $f''(x)$ is positive in the interval, then $a = 1.3$.

$$a_1 = a - \frac{f(a)}{f'(a)} = 1.3 - \frac{0.397}{9.07} = 1.3 - 0.044 = 1.256$$

$$a_2 = 1.256 - \frac{0.005385}{8.7326} = 1.256 - 0.00062 = 1.25538$$

If Newton's method of using the tangent is not applicable, either because of the presence of a maximum, minimum, or inflection point or because of difficulty in finding $f'(x)$, the interpolation method using the chord joining $[\alpha, f(\alpha)]$ and $[\beta, f(\beta)]$ can be used. The chord crosses the x axis at $a = \alpha - f(\alpha)(\beta - \alpha)/[f(\beta) - f(\alpha)]$, a better approximation than either α or β. Note that this formula differs from Newton's only in having the difference quotient, which is the slope of the chord, in place of the derivative, which is the slope of the tangent. To get a still better approximation, repeat the procedure, using as one endpoint a

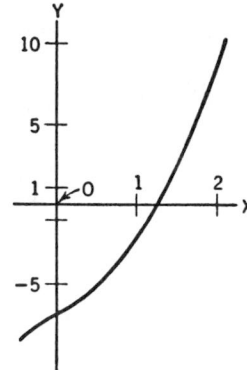

Fig. 2

and as the other either α or β, chosen so that $f(x)$ has opposite signs at the endpoints of the new interval.

Graphical Method of Solution This can be used to solve any kind of equation if it gives sufficient accuracy. To solve the equation $f(x) = 0$, graph the function $y = f(x)$. The x coordinates of the points at which the graph intersects the x axis are roots of $f(x) = 0$. Another method is to set $f(x)$ equal to any convenient difference $f_1(x) - f_2(x)$ and graph the functions $y = f_1(x)$ and $y = f_2(x)$ on the same axes. The x coordinates of the points of intersection of the two graphs are real roots of $f(x) = 0$. Also, see section 2.12.

Graeffe's Method for Real and Complex Roots Let $x_1, x_2, \ldots, x_n$ be the roots of the equation $a_0x^n + a_1x^{n-1} + \cdots + a_n = 0$, arranged in descending order of absolute values. Form a sequence of equations such that the roots of each are the negatives of the squares of the roots of the preceding equation. Using the negatives of the squares gives more uniform formulas.

Let A_i be a coefficient of the equation being formed, and a_i a coefficient of the preceding equation:

$$A_0 = a_0 = 1$$

$$A_1 = a_1^2 - 2a_0a_2 = a_1^2 - 2a_2$$

$$A_2 = a_2^2 - 2a_1a_3 + 2a_4$$

$$A_3 = a_3^2 - 2a_2a_4 + 2a_1a_5 - 2a_6$$

$$\vdots$$

$$A_{n-1} = a_{n-1}^2 - 2a_{n-2}a_n$$

$$A_n = a_n^2$$

Each coefficient is the sum of the square of the preceding and twice the product of all pairs of equidistant coefficients in the preceding equation, taken with alternately minus and plus signs. Missing coefficients are zero. The process is ended when further steps do not affect the nonfluctuating coefficients to the accuracy desired in the roots.

As the successive equations are formed, various cases arise depending on the behavior of the coefficients. Among them are:

CASE 1: Each coefficient approaches the square of the preceding. The roots are real and unequal in absolute value. Let A_i be a coefficient of the equation whose roots are $-x_1^p, -x_2^p, \ldots, -x_n^p$. Then, approximately, $x_1 = \pm \sqrt[p]{A_1}, x_2 = \pm \sqrt[p]{A_2/A_1}, \ldots, x_n = \pm \sqrt[p]{A_n/A_{n-1}}$. The signs of the roots are determined by substitution in the original equation. It is usually sufficient to find successive integers between which a root is located.

Example 27 $f(x) = x^3 - 2x^2 - 5x + 4 = 0$. See the tabulation for Example 27 below.

Using synthetic substitution,

$$
\begin{array}{r}
3 \,\underline{|\, 1 - 2 - 5 + 4} \\
+ 3 + 3 - 6 \\
\hline
1 + 1 - 2 - 2
\end{array}
\qquad
\begin{array}{r}
4 \,\underline{|\, 1 - 2 - 5 + 4} \\
+ 4 + 8 + 12 \\
\hline
1 + 2 - 3 - 16
\end{array}
$$

we have $f(3) = -2$, $f(4) = 16$. Therefore there is a root between 3 and 4, and $x_1 = 3.177$.

$$
\log x_2 = \tfrac{1}{16}(\log 2.136 \times 10^{12} - \log 1.080 \times 10^8)
$$

$$
= \tfrac{1}{16}(12.3296 - 8.0334)
$$

$$
= \tfrac{1}{16} \times 4.2962 = 0.2685
$$

$$
x_2 = \pm 1.856
$$

Using synthetic substitution, $f(-2) = -2$, $f(-1) = 6$. Therefore $x_2 = -1.856$.

$$
\log x_3 = \tfrac{1}{16}(\log 4.295 \times 10^9 - \log 2.136 \times 10^{12})
$$

$$
= \tfrac{1}{16}(9.6330 - 12.3296)
$$

$$
= \tfrac{1}{16}(157.3034 - 160) = 9.8315 - 10
$$

$$
x_3 = \pm 0.678
$$

Since $x_1 + x_2 + x_3 = 2$, $x_3 = 0.678$.

CASE 2: A coefficient fluctuates in sign. There is a pair of complex roots. If the sign of A_i fluctuates, then $x_i = u + iv$ and $x_{i+1} = u - iv$ are complex. Let $r^2 = u^2 + v^2$. Then $r^2 = \sqrt[p]{A_{i+1}/A_{i-1}}$, $2u = -a_1 - (\text{sum of real roots})$, $v = \sqrt{r^2 - u^2}$.

Example 28 $f(x) = x^4 - 2x^3 - 4x^2 + 5x - 7 = 0$. See the tabulation for Example 28 on next page.

If, for a fourth-degree equation, alternate coefficients, that is, the second and fourth, fluctuate in sign, all four roots are complex. Let the roots be $u_1 \pm iv_1, u_2 \pm iv_2$. Then $r_1^2 = \sqrt[p]{A_2}, r_2^2 = \sqrt[p]{A_4/A_2}, 2(u_1 + u_2) = -a_1, 2(r_2^2 u_1 + r_1^2 u_2) = -a_3$.

Tabulation for Example 27

		x^3	x^2	x	x^0
1st		1	-2	-5	4
		1	4	25	16
			10	16	
2nd		1	14	41	16
		1	196	1.681×10^3	256
			-82	-0.448×10^3	
4th		1	1.14×10^2	1.233×10^3	256
		1	1.300×10^4	1.520×10^6	6.554×10^4
			-0.247×10^4	-0.058×10^6	
8th		1	1.053×10^4	1.462×10^6	6.554×10^4
		1	1.109×10^8	2.137×10^{12}	4.295×10^9
			-0.029×10^8	-0.001×10^{12}	
16th		1	1.080×10^8	2.136×10^{12}	4.295×10^9

$$
\log x_1 = \tfrac{1}{16}\log 1.080 \times 10^8 = \tfrac{1}{16} \times 8.0334 = 0.5021
$$

$$
x_1 = \pm 3.177
$$

Tabulation for Example 28

	x^4	x^3	x^2	x	x^0
1st	1	-2	-4	5	-7
	1	4	16	25	49
		8	20	-56	
			-14		
2nd	1	12	22	-31	49
	1	144	484	961	2401
		-44	744	-2156	
			98		
4th	1	100	1326	-1195	2401
	1	1.0000×10^4	1.758×10^6	1.428×10^6	5.765×10^6
		-0.2652×10^4	0.239×10^6	-6.367×10^6	
			0.005×10^6		
8th	1	7.348×10^3	2.002×10^6	-4.939×10^6	5.765×10^6
	1	5.399×10^7	4.008×10^{12}	2.439×10^{13}	3.324×10^{13}
		-0.400×10^7	0.073×10^{12}	-2.308×10^{13}	
16th	1	4.999×10^7	4.081×10^{12}	1.31×10^{12}	3.324×10^{13}
	1	2.499×10^{15}	1.665×10^{25}	0.017×10^{26}	1.105×10^{27}
		-0.001×10^{15}		-2.713×10^{26}	
32nd	1	2.498×10^{15}	1.665×10^{25}	-2.696×10^{26}	1.105×10^{27}

Since the sign of A_3 fluctuates, x_3 and x_4 are complex.

$$x_1 = \pm \sqrt[32]{2.498 \times 10^{15}} = \pm 3.028 \qquad f(3) = - \quad f(4) = +$$

$$\frac{\log(2.498 \times 10^{15})}{32} = \frac{15.3976}{32} = 0.4812 \qquad x_1 = 3.028$$

$$x_2 = \pm \sqrt[32]{\frac{1.665 \times 10^{25}}{2.498 \times 10^{15}}} = \pm 2.028 \qquad f(-3) = + \quad f(-2) = -$$

$$\begin{array}{c} 25.2214 \\ 15.3976 \\ \hline \dfrac{9.8238}{32} = 0.3070 \end{array} \qquad x_2 = -2.028$$

$$r^2 = \sqrt[32]{\frac{1.105 \times 10^{27}}{1.665 \times 10^{25}}} = 1.140 \qquad u = \frac{2 - (3.028 - 2.028)}{2} = 0.500$$

$$\begin{array}{c} 27.0434 \\ 25.2214 \\ \hline \dfrac{1.8220}{32} = 0.05694 \end{array} \qquad v = \sqrt{1.140 - 0.250} = \sqrt{0.890} = 0.943$$

$$x_3, x_4 = 0.5 \pm 0.943 i$$

Tabulation for Example 29

		x^4	x^3	x^2	x	x^0
1st		1	-3	-1	4	14
		1	9	1	16	196
			2	24	28	
				28		
2nd		1	11	53	44	196
		1	121	2809	1936	38416
			-106	-968	-20776	
				392		
4th		1	15	2.233×10^3	-1.884×10^4	3.842×10^4
		1	0.225×10^3	4.986×10^6	3.549×10^8	1.476×10^9
			-4.466×10^3	0.565×10^6	-1.716×10^8	
				0.077×10^6		
8th		1	-4.241×10^3	5.628×10^6	1.833×10^8	1.476×10^9
		1	1.799×10^7	3.1674×10^{13}	3.360×10^{16}	2.178×10^{18}
			-1.126×10^7	0.1555×10^{13}	-1.661×10^{16}	
				0.0003×10^{13}		
16th		1	0.675×10^7	3.323×10^{13}	1.699×10^{16}	2.178×10^{18}

Since A_1 and A_3 fluctuate in sign, there are four complex roots.

$$r_1^2 = \sqrt[16]{3.323 \times 10^{13}} = 7.000$$

$$\frac{\log(3.323 \times 10^{13})}{16} = \frac{13.5215}{16} = 0.8451$$

$$r_2^2 = \sqrt[16]{\frac{2.178 \times 10^{18}}{3.323 \times 10^{13}}} = 2.000$$

$$\begin{array}{c} 18.3380 \\ 13.5215 \\ \hline 4.8165 \\ \hline 16 \end{array} = 0.3010$$

$$2(u_1 + u_2) = 3$$
$$2(2u_1 + 7u_2) = -4$$
$$u_2 = -1 \quad u_1 = 2.5$$

$$v_2 = \sqrt{r_2^2 - u_2^2} = \sqrt{2 - 1} = 1$$

$$v_1 = \sqrt{r_1^2 - u_1^2} = \sqrt{7 - 6.25} = \sqrt{0.75} = 0.866$$

$$x_1, x_2 = 2.5 \pm 0.866i$$

$$x_3, x_4 = -1 \pm i$$

Example 29 $f(x) = x^4 - 3x^3 - x^2 + 4x + 14 = 0$.
See the tabulation for Example 29 above.

CASE 3: A coefficient approaches one-half the square of the preceding. There is a double real root or there are two real roots of equal absolute value. If A_i approaches one-half the square of the preceding coefficient, then $|x_i| = |x_{i+1}| = \sqrt[2p]{A_{i+1}/A_{i-1}}$.

Example 30 $f(x) = x^3 + 2.20x^2 - 2.95x + 0.80 = 0$.
See the tabulation for Example 30 below.

Tabulation for Example 30

		x^3	x^2	x	x^0
1st		1	2.20	-2.95	0.80
		1	4.84	8.703	0.64
			5.90	-3.52	
2nd		1	10.74	5.183	0.64
		1	1.1535×10^2	2.686×10	0.4096
			-0.1037×10^2	-1.375×10	
4th		1	1.050×10^2	1.311×10	0.4096
		1	1.1025×10^4	1.719×10^2	0.1678
			-0.0026×10^4	-0.860×10^2	
8th		1	1.100×10^4	0.859×10^2	0.1678

Tabulation for Example 30

Since A_2 approaches one-half the square of the preceding coefficient, $|x_2| = |x_3|$.

$$x_1 = \pm\sqrt[8]{1.100 \times 10^4} = \pm 3.20$$

$$\frac{\log(1.100 \times 10^4)}{8} = \frac{4.0414}{8} = 0.5052$$

$$|x_2| = |x_3| = \sqrt[16]{\frac{0.1678}{1.100 \times 10^4}} = 0.50$$

$$\begin{array}{c} 9.2248 - 10 \\ 4.0414 \\ \hline 155.1834 - 160 \\ \hline 16 \end{array} = 9.6990 - 10$$

$f(-4) = -$ $\quad f(-3) = +$

$f(0.5) = 0$ $\quad f(-0.5) \neq 0$

$x_1 = -3.20$

$x_2 = x_3 = 0.50$

For a more extensive treatment of Graeffe's method, see mathworld.wolfram.com/GraeffesMethod. html (August 2008) and math.fullerton.edu/mathews/ n2003/GraeffesMethodMod/html (August 2008).

2.11 Matrices and Determinants

Definitions

1. A *matrix* is a system of mn quantities, called *elements*, arranged in a rectangular array of m rows and n columns:

$$A = \begin{pmatrix} a_{11} & a_{12} & \cdots & a_{1n} \\ a_{21} & a_{22} & \cdots & a_{2n} \\ \vdots & \vdots & \ddots & \vdots \\ a_{m1} & a_{m2} & \cdots & a_{mn} \end{pmatrix} = \begin{Vmatrix} a_{11} & a_{12} & \cdots & a_{1n} \\ a_{21} & a_{22} & \cdots & a_{2n} \\ \vdots & \vdots & \ddots & \vdots \\ a_{m1} & a_{m2} & \cdots & a_{mn} \end{Vmatrix}$$

$$= (a_{ij}) = \|a_{ij}\|$$

$$i = 1, \ldots, m \qquad j = 1, \ldots, n$$

2. If $m = n$, then A is a *square matrix of order n*.

3. Two matrices are *equal* if and *only if* they have the same number of rows and of columns and corresponding elements are equal.

4. Two matrices are *transposes* (sometimes called *conjugates*) of each other if either is obtained from the other by interchanging rows and columns.

5. The *complex conjugate* of a matrix (a_{ij}) with complex elements is the matrix $(\bar{a}_{ij})$. See Section 13.1.

6. A matrix is *symmetric* if it is equal to its transpose, that is, if $a_{ij} = a_{ji}, i, \ j = 1, \ldots, n$.

7. A matrix is *skew symmetric*, or *antisymmetric*, if $a_{ij} = -a_{ji}, i, \ j = 1, \ldots, n$. The diagonal elements $a_{ii} = 0$.

8. A matrix all of whose elements are zero is a *zero matrix*.

9. If the nondiagonal elements $a_{ij}, i \neq j$, of a square matrix A are all zero, then A is a *diagonal matrix*. If, furthermore, the diagonal elements are all equal, the matrix is a *scalar matrix*; if they are all 1, it is an *identity* or *unit matrix*, denoted by I.

10. The *determinant* $|A|$ of a square matrix $(a_{ij}), i, \ j = 1, \ldots, n$, is the sum of the $n!$ products $a_{1r_1} a_{2r_2} \cdots a_{nr_n}$, in which $r_1, r_2, \ldots, r_n$ is a permutation of $1, 2, \ldots, n$, and the sign of each product is plus or minus according as the permutation is obtained from $1, 2, \ldots, n$ by an even or an odd number of interchanges of two numbers.

Symbols used are

$$|A| = \begin{vmatrix} a_{11} & a_{12} & \cdots & a_{1n} \\ a_{21} & a_{22} & \cdots & a_{2n} \\ \vdots & \vdots & \ddots & \vdots \\ a_{n1} & a_{n2} & \cdots & a_{nn} \end{vmatrix} = |a_{ij}| \qquad i, j = 1, \ldots, n$$

11. A square matrix (a_{ij}) is *singular* if its determinant $|a_{ij}|$ is zero.

12. The determinants of the square submatrices of any matrix A, obtained by striking out certain rows or columns or both, are called the *determinants* or *minors* of A. A matrix is of *rank r* if it has at least one r-rowed determinant that is not zero while all its determinants of order higher than r are zero. The *nullity d* of a square matrix of order n is $d = n - r$. The zero matrix is of rank 0.

13. The minor D_{ij} of the element a_{ij} of a square matrix is the determinant of the submatrix obtained by striking out the row and column in which a_{ij} lies. The *cofactor* A_{ij} of the element a_{ij} is $(-1)^{i+j} D_{ij}$. A *principal minor* is the minor obtained by striking out the same rows as columns.

14. The *inverse* of the square matrix A is

$$A^{-1} = \begin{pmatrix} \dfrac{A_{11}}{|A|} & \cdots & \dfrac{A_{n1}}{|A|} \\ \vdots & \ddots & \vdots \\ \dfrac{A_{1n}}{|A|} & \cdots & \dfrac{A_{nn}}{|A|} \end{pmatrix}$$

$$A A^{-1} = A^{-A} = I$$

15. The *adjoint* of A is

$$\text{adj } A = \begin{pmatrix} A_{11} & \cdots & A_{n1} \\ \vdots & \ddots & \vdots \\ A_{1n} & \cdots & A_{nn} \end{pmatrix}$$

16. *Elementary transformations* of a matrix are:

    a. The interchange of two rows or of two columns

    b. The addition to the elements of a row (or column) of any constant multiple of the corresponding elements of another row (or column)

    c. The multiplication of each element of a row (or column) by any nonzero constant

17. Two $m \times n$ matrices A and B are *equivalent* if it is possible to pass from one to the other by a finite number of elementary transformations.

    a. The matrices A and B are equivalent if and only if there exist two nonsingular square matrices E and F, having m and n rows, respectively, such that $EAF = B$.

    b. The matrices A and B are equivalent if and only if they have the same rank.

Matrix Operations

Addition and Subtraction. The sum or difference of two matrices (a_{ij}) and (b_{ij}) is the matrix $(a_{ij} \pm b_{ij})$, $i = 1, \ldots, m$, $j = 1, \ldots, n$.

Scalar Multiplication. The product of the scalar k and the matrix (a_{ij}) is the matrix (ka_{ij}).

Matrix Multiplication. The product (p_{ik}), $i = 1, \ldots, m$, $k = 1, \ldots, q$, of two matrices (a_{ij}), $i = 1, \ldots, m$, $j = 1, \ldots, n$, and (b_{jk}), $j = 1, \ldots, n$, $k = 1, \ldots, q$, is the matrix whose elements are

$$p_{ik} = \sum_{j=1}^{n} a_{ij} b_{jk} = a_{i1} b_{1k} + a_{i2} b_{2k} + \cdots + a_{in} b_{nk}$$

The element in the ith row and kth column of the product is the sum of the n products of the n elements of the ith row of (a_{ij}) by the corresponding n elements of the kth column of (b_{jk}).

Example 31

$$\begin{pmatrix} a_{11} & a_{12} \\ a_{21} & a_{22} \end{pmatrix} \begin{pmatrix} b_{11} & b_{12} & b_{13} \\ b_{21} & b_{22} & b_{23} \end{pmatrix}$$

$$= \begin{pmatrix} a_{11}b_{11} + a_{12}b_{21} & a_{11}b_{12} + a_{12}b_{22} & a_{11}b_{13} + a_{12}b_{23} \\ a_{21}b_{11} + a_{22}b_{21} & a_{21}b_{12} + a_{22}b_{22} & a_{21}b_{13} + a_{22}b_{23} \end{pmatrix}$$

All the laws of ordinary algebra hold for the addition and subtraction of matrices and for scalar multiplication.

Multiplication of matrices is not in general commutative, but it is associative and distributive.

If the product of two or more matrices is zero, it does not follow that one of the factors is zero. The factors are *divisors of zero*.

Example 32

$$\begin{pmatrix} a & 0 \\ b & 0 \end{pmatrix} \begin{pmatrix} 0 & 0 \\ c & d \end{pmatrix} = \begin{pmatrix} 0 & 0 \\ 0 & 0 \end{pmatrix}$$

Linear Dependence

1. The quantities $l_1, l_2, \ldots, l_n$ are *linearly dependent* if there exist constants $c_1, c_2, \ldots, c_n$, not all zero, such that

$$c_1 l_1 + c_2 l_2 + \cdots + c_n l_n = 0$$

If no such constants exist, the quantities are *linearly independent*.

2. The linear functions

$$l_i = a_{i1} x_1 + a_{i2} x_2 + \cdots + a_{in} x_n \qquad i = 1, 2, \ldots, m$$

are linearly dependent if and only if the matrix of the coefficients is of rank $r < m$. Exactly r of the l_i form a linearly independent set.

3. For $m > n$, any set of m linear functions are linearly dependent.

Consistency of Equations

1. The system of homogeneous linear equations

$$a_{i1} x_1 + a_{i2} x_2 + \cdots + a_{in} x_n = 0 \qquad i = 1, 2, \ldots, m$$

has solutions not all zero if the rank r of the matrix (a_{ij}) is less than n.

If $m < n$, there always exist solutions not all zero. If $m = n$, there exist solutions not all zero if $|a_{ij}| = 0$.

If r of the equations are so selected that their matrix is of rank r, they determine uniquely r of the variables as homogeneous linear functions of the remaining $n - r$ variables. A solution of the system is obtained by assigning arbitrary values to the $n - r$ variables and finding the corresponding values of the r variables.

2. The system of linear equations

$$a_{i1} x_1 + a_{i2} x_2 + \cdots + a_{in} x_n = k_i \qquad i = 1, 2, \ldots, m$$

is consistent if and only if the *augmented* matrix derived from (a_{ij}) by annexing the column $k_1, \ldots, k_m$ has the same rank r as (a_{ij}).

As in the case of a system of homogeneous linear equations, r of the variables can be expressed in terms of the remaining $n - r$ variables.

Linear Transformations

1. If a linear transformation

$$x_i' = a_{i1}x_1 + a_{i2}x_2 + \cdots + a_{in}x_n \qquad i = 1, 2, \ldots, n$$

with matrix (a_{ij}) transforms the variables x_i into the variables x_i' and a linear transformation

$$x_i' = b_{i1}'x_1 + b_{i2}'x_2 + \cdots + b_{in}'x_n \qquad i = 1, 2, \ldots, n$$

with matrix (b_{ij}) transforms the variables x_i' into the variables x_i'', then the linear transformation with matrix $(b_{ij})(a_{ij})$ transforms the variables x_i into the variables x_i'' directly.

2. A real *orthogonal* transformation is a linear transformation of the variables x_i into the variables x_i' such that

$$\sum_{i=1}^{n} x_i^2 = \sum_{i=1}^{n} x_i'^2$$

A transformation is orthogonal if and only if the transpose of its matrix is the inverse of its matrix.

3. A *unitary* transformation is a linear transformation of the variables x_i into the variables x_i' such that

$$\sum_{i=1}^{n} x_i \bar{x}_i = \sum_{i=1}^{n} x_i' \bar{x}_i'$$

A transformation is uni'tary if and only if the transpose of the conjugate of its matrix is the inverse of its matrix.

Quadratic Forms A *quadratic form* in n variables is

$$\sum_{i,j=1}^{n} a_{ij}x_ix_j = a_{11}x_1^2 + a_{12}x_1x_2 + \cdots + a_{1n}x_1x_n$$

$$+ a_{21}x_2x_1 + a_{22}x_2^2 + \cdots + a_{2n}x_2x_n$$
$$+ a_{n1}x_nx_1 + a_{n2}x_nx_2 + \cdots + a_{nn}x_n^2$$

in which $a_{ji} = a_{ij}$. The symmetric matrix (a_{ij}) of the coefficients is the matrix of the quadratic form and the rank of (a_{ij}) is the rank of the quadratic form.

A real quadratic form of rank r can be reduced by a real nonsingular linear transformation to the *normal form*

$$x_1^2 + \cdots + x_p^2 - x_{p+1}^2 - \cdots - x_r^2$$

in which the *index p* is uniquely determined.

If $p = r$, a quadratic form is *positive*, and if $p = 0$, it is *negative*. If, furthermore, $r = n$, both are *definite*. A quadratic form is positive definite if and only if the determinant and all the principal minors of its matrix are positive.

A method of reducing a quadratic form to its normal form is illustrated.

Example 33 See the tabulation for Example 33 below.

The transformation

$$x' = 3x + 2y - z \qquad y' = -\tfrac{16}{3}y + \tfrac{8}{3}z \qquad z' = z$$

reduces q to $\tfrac{1}{3}x'^2 - \tfrac{3}{16}y'^2$.

The transformation

$$x'' = \sqrt{3}x' \qquad y'' = \frac{4}{\sqrt{3}}y' \qquad z'' = z'$$

further reduces q to the normal form $x''^2 - y''^2$ of rank 2 and index 1.

Expressing x, y, z in terms of x'', y'', z'', the real nonsingular linear transformation that reduces q to the normal form is

---

Tabulation for Example 33

$$q = 3x^2 - 4y^2 - z^2 + 4xy - 2xz + 4yz$$

$$q = \left\{ \begin{array}{l} 3x^2 + 2xy - xz \\ +2xy - 4y^2 + 2yz \\ -xz + 2yz - z^2 \end{array} \right\} \begin{array}{l} = \tfrac{1}{3}(3x + 2y - z)^2 + q_1, \text{ in which the quantity} \\ \text{in parentheses is obtained by factoring } x \text{ out} \\ \text{of the first row} \\ = \tfrac{1}{3}(9x^2 + 4y^2 + z^2 + 12xy - 6xz - 4yz) + q_1 \end{array}$$

$$q_1 = -\tfrac{4}{3}y_1^2 - \tfrac{1}{3}z^2 + \tfrac{4}{3}yz - 4y^2 + 4yz - z^2$$

$$= \left\{ \begin{array}{l} -\tfrac{16}{3}y^2 + \tfrac{8}{3}yz \\ +\tfrac{8}{3}yz - \tfrac{4}{3}z^2 \end{array} \right\} = -\tfrac{3}{16}(-\tfrac{16}{3}y + \tfrac{8}{3}z)^2 + q_2$$

$$q_2 = 0$$

$$x = \frac{\sqrt{3}}{3}x'' + \frac{1}{2\sqrt{3}}y''$$

$$y = -\frac{\sqrt{3}}{4}y'' + \tfrac{1}{2}z''$$

$$z = z''$$

Hermitian Forms A *Hermitian form* in n variables is

$$\sum_{i,j=1}^{n} a_{ij}x_i\overline{x}_j \qquad a_{ji} = \overline{a}_{ij}$$

The matrix (a_{ij}) is a *Hermitian matrix*. Its transpose is equal to its conjugate. The rank of (a_{ij}) is the rank of the Hermitian form.

A Hermitian form of rank r can be reduced by a nonsingular linear transformation to the normal form

$$x_1\overline{x}_1 + \cdots + x_p\overline{x}_p - x_{p+1}\overline{x}_{p+1} - \cdots - x_r\overline{x}_r$$

in which the index p is uniquely determined.

If $p = r$, the Hermitian form is positive, and, if $p = 0$, it is negative. If, furthermore, $r = n$, both are definite

Determinants Second- and third-order determinants are formed from their square symbols by taking diagonal products, down from left to right being positive and up negative:

$$\begin{vmatrix} a_{11} & a_{12} \\ a_{21} & a_{22} \end{vmatrix} = a_{11}a_{22} - a_{21}a_{12}$$

$$\begin{vmatrix} a_{11} & a_{12} & a_{13} \\ a_{21} & a_{22} & a_{23} \\ a_{31} & a_{32} & a_{33} \end{vmatrix} = a_{11}a_{22}a_{33} + a_{12}a_{23}a_{31} + a_{13}a_{32}a_{21}$$
$$- a_{31}a_{22}a_{13} - a_{32}a_{23}a_{11} - a_{33}a_{12}a_{21}$$

Third- and higher order determinants are formed by selecting any row or column and taking the sum of the products of each element and its cofactor. This process is continued until second- or third-order cofactors are reached:

$$\begin{vmatrix} a_{11} & a_{12} & a_{13} \\ a_{21} & a_{22} & a_{23} \\ a_{31} & a_{32} & a_{33} \end{vmatrix} = a_{11}\begin{vmatrix} a_{22} & a_{23} \\ a_{32} & a_{33} \end{vmatrix} - a_{21}\begin{vmatrix} a_{12} & a_{13} \\ a_{32} & a_{33} \end{vmatrix}$$
$$+ a_{31}\begin{vmatrix} a_{12} & a_{13} \\ a_{22} & a_{23} \end{vmatrix}$$

The determinant of a matrix A is:

1. Zero if two rows or two columns of A have proportional elements

2. Unchanged if:
   a. The rows and columns of A are interchanged
   b. To each element of a row or column of A is added a constant multiple of the corresponding element of another row or column

3. Changed in sing if two rows or two columns of A are interchanged

4. Multiplied by c if each element of any row or column of A is multiplied by c

5. The sum of the determinants of two matrices B and C if A, B, and C have all the same elements except that in one row or column each element of A is the sum of the corresponding elements of B and C

Example 34 See the tabulation for Example 34 below.

Tabulation for Example 34

$$\begin{vmatrix} 2 & 9 & 9 & 4 \\ 2 & -3 & 12 & 8 \\ 4 & 8 & 3 & -5 \\ 1 & 2 & 6 & 4 \end{vmatrix} = \begin{vmatrix} 2 & 5 & 9 & 4 \\ 2 & -7 & 12 & 8 \\ 4 & 0 & 3 & -5 \\ 1 & 0 & 6 & 4 \end{vmatrix} = 3\begin{vmatrix} 2 & 5 & 3 & 4 \\ 2 & -7 & 4 & 8 \\ 4 & 0 & 1 & -5 \\ 1 & 0 & 2 & 4 \end{vmatrix}$$

Multiply 1st column   Factor 3 out of
by -2 and add to 2nd   3rd column

$$= 3 \times (-5)\begin{vmatrix} 2 & 4 & 8 \\ 4 & 1 & -5 \\ 1 & 2 & 4 \end{vmatrix} + 3 \times (-7)\begin{vmatrix} 2 & 3 & 4 \\ 4 & 1 & -5 \\ 1 & 2 & 4 \end{vmatrix} = \qquad 0 \qquad -21\begin{vmatrix} 1 & 1 & 0 \\ 4 & 1 & -5 \\ 1 & 2 & 4 \end{vmatrix}$$

Expand according to 2nd column   1st and 3rd   Subtract 3rd
     rows are proportional   row from 1st

$$= -21\begin{vmatrix} 1 & -5 \\ 2 & 4 \end{vmatrix} - (-21)\begin{vmatrix} 4 & -5 \\ 1 & 4 \end{vmatrix} = -21[(4+10) - (16+5)] = +147$$

Expand according to 1st row

2.12 Systems of Equations

Linear Systems (also see Section 11.6)

Homogeneous. $a_{i1}x_1 + \cdots + a_{in}x_n = 0, i = 1, \ldots,$ m. Let $r = $ rank of(a_{ij}).

For $m = n$:

$r = n, |a_{ij}| \neq 0$; one solution, $x_1 = \cdots = x_n = 0$.
$r < n, |a_{ij}| = 0$; infinite number of solutions.

Nonhomogeneous. $a_{i1}x_1 + \cdots + a_{in}x_n = k_i, i = 1, \ldots, m$.

Let $a = (a_{ij})$, an $m \times n$ matrix.

$$k = \text{augmented matrix} = \begin{pmatrix} a_{11} & \cdots & a_{1n}k_1 \\ \vdots & \ddots & \vdots \\ a_{m1} & \cdots & a_{mn}k_m \end{pmatrix},$$

an $m \times (n+1)$ matrix.
$r_a = $ rank of a.
$r_k = $ rank of k.

For $m = n$:

$r_a = r_k$; consistent.

    (a) $r_a = r_k = n, |a_{ij}| \neq 0$; independent. One solution.

    (b) $r_a = r_k < n, |a_{ij}| = 0$; dependent. Infinite number of solutions.

$r_a < r_k$; inconsistent. No solution.

Methods of Solution *Elimination* is a practical method of solution for a system of two or three linear equations in as many variables.

Example 35

1. By addition and subtraction, solve

$$2x + y + 3z = 9 \qquad (2)$$
$$x - 2y + z = -2 \qquad (3)$$
$$3x + 2y + 2z = 7 \qquad (4)$$

(3) + (4)   gives $4x + 3z = 5$ (5)

$2 \times (2) + (3)$ gives $5x + 7z = 16$ (6)

$5 \times (5) - 4 \times (6)$ gives $-13z = -39$ or

$$z = 3$$

Putting $z = 3$ in (5) or (6) gives $x = -1$. Then from (2), (3), or (4), $y = 2$.

2. By substitution, solve

$$\begin{cases} x + 2y - z = 5 & (7) \\ x - y = 2 & (8) \\ 2x + z = 1 & (9) \end{cases}$$

From (8), $y = x - 2$, and from (9), $z = -2x + 1$. Substituting for y and z in (7), $x - 2x - 4 + 2x - 1 = 5$, from which $x = 2$. Then $y = 2 - 2 = 0, z = -4 + 1 = -3$.

Determinants can be used to solve a system of n nonhomogeneous linear equations in n variables for which $|a_{ij}| \neq 0$. To solve for x_j, form a fraction the denominator of which is the determinant $|a_{ij}|$ and the numerator the determinant obtained from $|a_{ij}|$ by replacing its jth column by the constants k_i.

Example 36   Solve

$$2x + y + 3z = 9$$
$$x - 2y + z = -2$$
$$3x + 2y + 2z = 7$$

$$x = \frac{\begin{vmatrix} 9 & 1 & 3 \\ -2 & -2 & 1 \\ 7 & 2 & 2 \end{vmatrix}}{\begin{vmatrix} 2 & 1 & 3 \\ 1 & -2 & 1 \\ 3 & 2 & 2 \end{vmatrix}} = \frac{\begin{vmatrix} 9 & 1 & 3 \\ -2 & -2 & 1 \\ 5 & 0 & 3 \end{vmatrix}}{\begin{vmatrix} 2 & 1 & 3 \\ 1 & -2 & 1 \\ 4 & 0 & 3 \end{vmatrix}} = \frac{\begin{vmatrix} 9 & 1 & 3 \\ 16 & 0 & 7 \\ 5 & 0 & 3 \end{vmatrix}}{\begin{vmatrix} 2 & 1 & 3 \\ 5 & 0 & 7 \\ 4 & 0 & 3 \end{vmatrix}}$$

$$= \frac{-(48 - 35)}{-(15 - 28)} = -1$$

Miscellaneous Systems   To be solvable a system of equations must have as many independent equations as variables. A system of two polynomial equations of degrees m and n has mn solutions, real or complex. For systems in general no statement can be made regarding the number of solutions.

Graphical Method of Solution.   This is a general method for systems of two equations in two variables. It consists of graphing both equations on the same axes and reading the pairs of coordinates of the points of intersection of the graphs as solutions of the system. This method gives real solutions only.

Example 37   Solve

$$y = \sin x \qquad x^2 + y^2 = 2$$

Solution from the graph (Fig. 3) gives $x = 1.1, y = 0.9$. From symmetry, $x = -1.1, y = -0.9$, is also a solution.

Method of Elimination of Variables.   This is a general method that can be applied to systems composed of any kinds of equations, algebraic or transcendental. However, except in fairly simple cases, practical difficulties are frequently encountered.

Example 38   Solve

$$y = \sin x \qquad x^2 + y^2 = 2$$

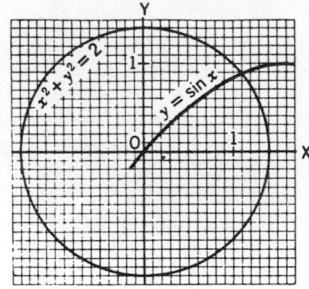

Fig. 3

Squaring both sides of the first equation and subtracting it from the second to eliminate y, $x^2 = 2 - \sin^2 x$. This equation can be solved by Newton's method. Extraneous solutions introduced by squaring can be eliminated by reference to the graph.

There are numerous devices for eliminating variables in special systems. For example, to solve the system of *two general quadratics*

$$a_1 x^2 + b_1 xy + c_1 y^2 + d_1 x + e_1 y + f_1 = 0 \quad (10)$$

$$a_2 x^2 + b_2 xy + c_2 y^2 + d_2 x + e_2 y + f_2 = 0 \quad (11)$$

eliminate x^2 by multiplying (10) by a_2 and (11) by a_1 and subtracting, solve the resulting equation for x, substitute this expression in either of the given equations, and clear fractions. The resulting fourth-degree equation in y can be solved by Horner's method. In a similar manner y could have been eliminated instead of x.

2.13 Permutations and Combinations

Fundamental Principle. If in a sequence of s events the first event can occur in n_1 ways, the second in $n_2, \ldots$, the s th in n_s, then the number of different ways in which the sequence can occur is $n_1 n_2 \ldots n_s$.

A *permutation* of n objects taken r at a time is an arrangement of any r objects selected from the n objects. The number of permutations of n objects taken r at a time is

$$_n P_r = n(n-1)(n-2)\cdots(n-4+1) = \frac{n!}{(n-r)!}$$

In particular, $_n P_1 = n$, $_n P_n = n!$. Cyclic permutations are

$$_n P_r^c = \frac{n!}{r(n-r)!} \qquad _n P_n^c = (n-1)!$$

If the n objects are divided into s sets each containing n_i objects that are alike, the distinguishable permutations are

$$n = n_1 + n_2 + \cdots + n_s \qquad _n P_n = \frac{n!}{n_1! n_2! \cdots n_s!}$$

A *combination* of n objects taken r at a time is an unarranged selection of any r of the n objects. The number of combinations of n objects taken r at a time is

$$_n C_r = \frac{_n P_r}{r!} = \frac{n!}{r!(n-r)!} = {_n C_{n-r}}$$

In particular, $_n C_1 = n$, $_n C_n = 1$. Combinations taken any number at a time, $_n C_1 + _n C_2 + \cdots + _n C_n = 2^n - 1$.

2.14 Probability

If, in a set M of m events that are mutually exclusive and equally likely, one event will occur, and if in the set M there is a subset N of n events ($n \leq m$), then the *a priori probability* p that the event that will occur is one of the subset N is n/m. The probability q that the event that will occur does not belong to N is $1 - n/m$.

Example 39 If the probability of drawing one of the 4 aces from a deck of 52 cards is to be found, then $m = 52$, $n = 4$, and $p = \frac{4}{52} = \frac{1}{13}$. The probability of drawing a card that is not an ace is $q = 1 - \frac{1}{13} = \frac{12}{13}$.

If, out of a large number r of observations in which a given event might or might not occur, the event has occurred s times, then a useful approximate value of the *experimental*, or *a posteriori, probability* of the occurrence of the event under the same conditions is s/r.

Example 40 From the American Experience Mortality Table, out of 100,000 persons living at age 10 years 749 died within a year. Here $r = 100,000$, $s = 749$, and the probability that a person of age 10 will die within a year is 749/100,000.

If p is the probability of receiving an amount A, then the *expectation* is pA.

Addition Rule (either or). The probability that any one of several mutually exclusive events will occur is the sum of their separate probabilities.

Example 41 The probability of drawing an ace from a deck of cards is $\frac{1}{13}$, and the probability of drawing a king is the same. Then the probability of drawing either an ace or a king is $\frac{1}{13} + \frac{1}{13} = \frac{2}{13}$.

Multiplication Rule (both and). (a) The probability that two (or more) independent events will both (or all) occur is the product of their separate probabilities.

(b) If p_1 is the probability that an event will occur, and if, after it has occurred, p_2 is the probability that another event will occur, then the probability that both will occur in the given order is $p_1 p_2$. This rule can be extended to more than two events.

Example 42 (a) The probability of drawing an ace from a deck of cards is $\frac{1}{13}$, and the probability of drawing a king from another deck is $\frac{1}{13}$. Then the probability that an ace will be drawn from the first deck and a king from the second is $\frac{1}{13} \cdot \frac{1}{13} = \frac{1}{169}$. (b) After an ace has been drawn from a deck of cards, the probability of drawing a king is $\frac{4}{51}$. If two cards are drawn in succession without the first being replaced, the probability that the first is an ace and the second a king is $\frac{1}{13} \cdot \frac{4}{51} = \frac{4}{663}$.

Repeated Trials. If p is the probability that an event will occur in a single trial, then the probability that it will occur exactly s times in r trials is the *binomial*, or *Bernoulli*, distribution function

$$_rC_s\, p^s (1 - p)^{r-s}$$

The probability that it will occur at least s times is

$$p^r + {_rC_{r-1}}\, p^{r-1}(1 - p) + {_rC_{r-2}}\, p^{r-2}(1 - p)^2$$
$$+ \cdots + {_rC_s}\, p^s (1 - p)^{r-s}$$

Example 43 If five cards are drawn, one from each of five decks, the probability that exactly three will be aces is $_5C_3(\frac{1}{13})^3(\frac{12}{13})^2$. The probability that at least three will be aces is $(\frac{1}{13})^5 + {_5C_4}(\frac{1}{13})^4(\frac{12}{13}) + {_5C_3}(\frac{1}{13})^3(\frac{12}{13})^2$.

3 SET ALGEBRA

3.1 Sets

A *set* is a collection of objects called *elements* that are distinguished by a particular characteristic. Examples are a set of engineers, a set of integers, a set of points. Element e belongs to set $\mathscr{S}$ is written $e \in \mathscr{S}$. If not, $e \notin \mathscr{S}$. A set can be denoted by including the listed elements, or merely by a typical element, in curly brackets: $\{2, 4, 6\}$; $\{e_1, e_2\}$, $\{e\}$. A set with no elements is called the *null set* and is denoted by $\emptyset$. A set with one element e_1 is denoted by $\{e_1\}$; and to

avoid a paradox in logic, these two ideas must be kept distinct.

Two sets $\mathscr{S}_1, \mathscr{S}_2$ may be compared as follows. If every element of set $\mathscr{S}_1$ is also an element of $\mathscr{S}_2$, then $\mathscr{S}_1$ is contained in $\mathscr{S}_2$. This is written $\mathscr{S}_1 \subset \mathscr{S}_2$ and is read "$\mathscr{S}_1$ *is contained in* $\mathscr{S}_2$" or "$\mathscr{S}_1$ is a *subset* of $\mathscr{S}_2$." If, in addition, $\mathscr{S}_2 \subset \mathscr{S}_1$, then their relation is written $\mathscr{S}_1 = \mathscr{S}_2$. On the other hand, if $\mathscr{S}_4$ has at least one element not contained in $\mathscr{S}_3$ but $\mathscr{S}_3 \subset \mathscr{S}_4$, $\mathscr{S}_3$ is a *proper subset of* $\mathscr{S}_4$. If $\mathscr{S}_5$ can contain all the elements of $\mathscr{S}_6$, this can be stressed by writing $\mathscr{S}_5 \subseteq \mathscr{S}_6$. Evidently $\emptyset \subset \mathscr{S}$ for every set $\mathscr{S}$.

If S, called the *space*, is the largest set concerned in a particular discussion, all the other sets are subsets of S. Thus set $\mathscr{A} \subset S$. The *complement* of $\mathscr{A}$, $\mathscr{A}^c$, with respect to space S is the set of elements in S that are not elements of $\mathscr{A}$.

Binary Operations for Sets. The *union*, $\mathscr{S}_a \cup \mathscr{S}_b$, of sets $\mathscr{S}_a$ and $\mathscr{S}_b$ is the set of elements in $\mathscr{S}_a$ or $\mathscr{S}_b$ or in both. Note that union differs from the idea of sum since in the union the common elements are counted only once. The *intersection*, $\mathscr{S}_a \cap \mathscr{S}_b$, of sets $\mathscr{S}_a$ and $\mathscr{S}_b$ is the set of elements in both $\mathscr{S}_1$ and $\mathscr{S}_2$. See the tabulation below.

Let $\mathscr{S}_a, \mathscr{S}_b, \mathscr{S}_c$ have their elements in space S.

Boolean algebra has as one representation the following:

UNICITY. Unique union $\mathscr{S}_a \cup \mathscr{S}_b \subset S$. Unique intersection $\mathscr{S}_a \cap \mathscr{S}_b \subset S$.

COMMUTATIVITY. $\mathscr{S}_a \cup \mathscr{S}_b = \mathscr{S}_b \cup \mathscr{S}_a$, $\mathscr{S}_a \cap \mathscr{S}_b = \mathscr{S}_b \cap \mathscr{S}_a$.

ASSOCIATIVITY. $\mathscr{S}_a \cup (\mathscr{S}_b \cup \mathscr{S}_c) = (\mathscr{S}_a \cup \mathscr{S}_b) \cup \mathscr{S}_c$, $\mathscr{S}_a \cap (\mathscr{S}_b \cap \mathscr{S}_c) = (\mathscr{S}_a \cap \mathscr{S}_b) \cap \mathscr{S}_c$.

DISTRIBUTIVITY. $\mathscr{S}_a \cup (\mathscr{S}_b \cap \mathscr{S}_c) = (\mathscr{S}_a \cup \mathscr{S}_b) \cap (\mathscr{S}_a \cup \mathscr{S}_c)$, $\mathscr{S}_a \cap (\mathscr{S}_b \cup \mathscr{S}_c) = (\mathscr{S}_a \cap \mathscr{S}_b) \cup (\mathscr{S}_a \cap \mathscr{S}_c)$.

IDEMPOTENCY. $\mathscr{S}_a \cup \mathscr{S}_a = \mathscr{S}_a$, $\mathscr{S}_a \cap \mathscr{S}_a = \mathscr{S}_a$.

SPACE. $\mathscr{S}_a \cup S = S$, $\mathscr{S}_a \cap S = \mathscr{S}_a$.

NULL SET. $\mathscr{S}_a \cup \emptyset = \mathscr{S}_a$, $\mathscr{S}_a \cap \emptyset = \emptyset$.

SUBSET. $\emptyset \subset \mathscr{S}_a \subset S$, $\mathscr{S}_a \subset (\mathscr{S}_a \cup \mathscr{S}_b)$, $(\mathscr{S}_a \cap \mathscr{S}_b) \subset \mathscr{S}_a$, $\mathscr{S}_a \subset \mathscr{S}_b \Rightarrow \mathscr{S}_a \cup \mathscr{S}_b = \mathscr{S}_b$, and $\mathscr{S}_a \cap \mathscr{S}_b = \mathscr{S}_a$.

Tabulation for Binary Operations for Sets

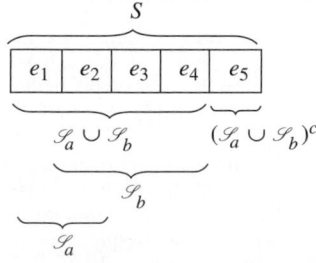

Union

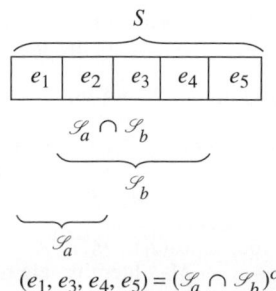

$(e_1, e_3, e_4, e_5) = (\mathscr{S}_a \cap \mathscr{S}_b)^c$

Intersection

COMPLEMENT. To $\mathscr{S}_a \subset S$ there corresponds unique $\mathscr{S}_{a^c} \subset S$; $\mathscr{S}_a \cup \mathscr{S}_{a^c} = S$, $\mathscr{S}_a \cap \mathscr{S}_{a^c} = \emptyset$.

DE MORGAN'S RELATIONS. $(\mathscr{S}_a \cup \mathscr{S}_b)^c = \mathscr{S}_{a^c} \cap \mathscr{S}_{b^c}$, $(\mathscr{S}_a \cap \mathscr{S}_b)^c = \mathscr{S}_{a^c} \cup \mathscr{S}_{b^c}$.

INVARIANT under the *duality transformation*, $\cup \leftrightarrow \cap$, $\subset \leftrightarrow \supset$, $S \leftrightarrow \emptyset$, are all the preceding relations.

3.2 Groups

A *group* is a system composed of a *set of elements* $\{a\}$ and a *rule of combination* of any two of them to form a product, such that:

1. The *product* of any ordered pair of elements and the square of each element are elements of the set.
2. The *associative* law holds.
3. The set contains an *identity* element I such that $Ia = aI = a$ for any element a of the set.
4. For any element a of the set there is in the set an *inverse* a^{-1} such that $aa^{-1} = a^{-1}a = I$.
5. If, in addition, the *commutative* law holds, the group is *commutative*, or *Abelian*.

The *order* of a group is the number n of elements in the group.

3.3 Rings, Integral Domains, and Fields

Rings. Space S consists of a set of elements $e_1, e_2, e_3, \ldots$. These elements are compared for *equality* and *order* and combined by the operations of *addition* and *multiplication*. These terms are partially defined by the following sets of assumptions.

Equality is a term from logic and means that if two expressions have this relation, then one may be substituted for the other.

Assumptions of equality E_1. Unicity: either $e_1 = e_2$ or $e_1 \neq e_2$. E_2. Reflexivity: $E_1 = e_1$, E_3. Symmetry: $e_1 = e_2 \Rightarrow e_2 = e_1$. E_4. Transitivity: $e_1 = e_2, e_2 = e_3 \Rightarrow e_1 = e_3$.

Assumptions of addition A_1. Closure: $e_1 + e_2 \subset S$. A_2; $e_1 = e_2 \Rightarrow e_1 + e_3 = e_2 + e_3$ and $e_3 + e_1 = e_3 + e_2$. (Invariance under addition.) A_3. Associativity: $e_1 + (e_2 + e_3) = (e_1 + e_2) + e_3$. A_4. Identity element: There exists an element $z \subset S$ such that $e_1 + z = e_1, z + e_1 = e_1$. A_5. Commutativity: $e_1 + e_2 = e_2 + e_1$.

Theorem 1: z is unique.

Negative. To each $e \subset S$, there corresponds an $e' \subset S$ such that $e + e' = z$; e' is called the negative of e and written $-e$.

Theorem 2: e' or $-e$ is unique. Theorem 3: $-(-e) = e$. Theorem 4: $-z = z$. Theorem 5: Equation $x + e_1 = e_2$ has the solution $x = e_2 - e_1$. Theorem 6: $e_1 + e_3 = e_1 \Rightarrow e_3 = z$.

Assumptions of multiplication M_1. Closure: $e_1 \cdot e_2 \subset S$. M_2. $e_1 = e_2 \Rightarrow e_1 \cdot e_3 = e_2 \cdot e_3$ and $e_3 \cdot e_1 = e_3 \cdot e_2$. (Invariance under multiplication.) M_3. Associativity: $e_1(e_2 \cdot e_3) = (e_1 \cdot e_2)e_3$. M_4. Identity element:

There exists an element $u \subset S$ such that $e_1 \cdot u = e_1, u \cdot e_1 = e_1$. M_5. Commutativity: $e_1 \cdot e_2 = e_2 \cdot e_1$.

Theorem 7: u is unique.

Reciprocal. To each element $e \subset S$ except z there corresponds an $e'' \subset S$ such that $e \cdot e'' = u$; e'' is called the reciprocal of e and written e^{-1}.

Theorem 8: e'' or e^{-1} is unique.

M_7. Distributivity: $e_1(e_2 + e_3) = e_1 \cdot e_2 + e_1 \cdot e_3$. Theorem 9: $e \cdot z = z$. Theorem 10: $e_1(-e_2) = -(e_1 \cdot e_2) = (-e_1)e_2$. Theorem 11: $(-e_1)(-e_2) = e_1 \cdot e_2$. Theorem 12: If S contains an element besides z, then it is $u \neq z$. Theorem 13: $e_1 \cdot e_2 = z \Rightarrow$ either $e_1 = z$ or $e_2 = z$.

A *ring* is a space S having at least two elements for which assumptions E_1 to E_4, A_1 to A_6, M_1 to M_5, and M_7 hold. An example is a residue system modulo 4.

Integral Domain. An *integral domain* is a ring for which, as an assumption, Theorem 13 holds. An example is the set of all integers.

Field. A *field* is an integral domain for which M_6 holds. An example of a field is the set of algebraic numbers.

Assumptions of (linear) order O_1. (Contains E_1.) If $e_1, e_2 \subset S$, then either $e_1 < e_2, e_1 = e_2$, or $e_2 < e_1$. O_2. $e_1 < e_2 \Rightarrow e_1 + e_3 < e_2 + e_3$. (Invariance under addition.) O_3. Transitivity: $e_1 < e_2, e_2 < e_3 \Rightarrow e_1 < e_3$.

Negative. If $e_1 < z$, then e_1 is called negative.

Positive. If $z < e_2$, then e_2 is called positive.

O_4. $z < e_2 z < e_3 \Rightarrow z < e_2 \cdot e_3$.

An *ordered integral domain* is an integral domain for which O_1 to O_4 hold. An example is the set of all integers.

An *ordered field* is an ordered integral domain for which M_6 holds. An example is the set of all rational numbers.

If an additional order assumption, O_5, known as the Dedekind assumption—see a book on real analysis—is included, then the space S for which assumptions E_1 to E_4, A_1 to A_6, M_1 to M_7, and O_1 to O_5 hold is called the real number space. An example is the set of real numbers. Here z is denoted 0, and u is denoted 1. Another example is the set of points on the real line.

4 STATISTICS AND PROBABILITY

4.1 Frequency Distributions of One Variable

Definitions A *frequency distribution* of statistical data consisting of N values of a variable x is a tabulation by intervals, called *classes,* showing the number f_i, called the *frequency* or *weight,* in each class; $N = \sum f_i$.

The midvalue x_i of a class is the *class mark.*

For equal classes, the *class interval* is $c = x_{i+1} - x_i$.

The *cumulative frequency*, cum f, at any class is the sum of the frequencies of all classes up to and including the given class.

Graphs

Frequency Polygon. Plot the points (x_i, f_i) and draw a broken line through them.

Histogram. Draw a set of rectangles using as bases intervals representing the classes marked off on a straight line and using altitudes proportional to the frequencies.

Frequency Curve. Draw a continuous curve approximating a frequency polygon or such that the region under the curve approximates a histogram. As the class interval c is taken smaller and the total frequency N larger, the approximation becomes better.

Ogive. This is a graph of cumulative frequencies.

Averages

Arithmetic Mean

$$\text{AM} = \bar{x} = \frac{1}{N} \sum_{i=1}^{k} f_i x_i$$

in which $N = \sum_{i=1}^{k} f_i$.

Geometric Mean

$$\text{GM} = (x_1^{f_1} \cdot x_2^{f_2} \cdots x_k^{f_k})^{1/N}$$

$$\log \text{ GM} = \frac{1}{N} \sum_{i=1}^{k} f_i \log x_i$$

Harmonic Mean

$$\text{HM} = \frac{N}{\sum_{i=1}^{k} (f_i / x_i)}$$

Root-Mean-Square

$$\text{rms} = \sqrt{\frac{\sum_{i=1}^{k} f_i x_i^2}{N}}$$

Median.   (a) For continuously varying data, the value of x for which cum $f = N/2$; (b) for discrete data, the value of x such that there is an equal number of values larger and smaller; for N odd, $N = 2k - 1$, the median is x_k; for N even, $N = 2k$, the median may be taken as $\frac{1}{2}(x_k + x_{k+1})$.

Mode.   The value of x that occurs most frequently.

Moments
1. About x_0. In x units

$$\nu_r = \frac{1}{N} \sum_{i=1}^{k} f_i (x_i - x_0)^r \qquad r = 0, 1, \ldots$$

If $x_0 = 0$, $\nu_1 = \bar{x}$, which is the *arithmetic mean*. In u units

$$\nu_r = \frac{1}{N} \sum_{i=1}^{k} f_i u_i^r \qquad r = 0, 1, \ldots \qquad u = \frac{x - x_0}{c}$$

$c = $ class interval

2. About the *mean*. In x units

$$\mu_r = \frac{1}{N} \sum_{i=1}^{k} f_i (x_i - \bar{x})^r \qquad r = 0, 1, \ldots$$

$$\bar{x} = \nu_1 \text{ in } x \text{ units}$$

In u units

$$\mu_r = \frac{1}{N} \sum_{i=1}^{k} f_i (u_i - \bar{u})^r \qquad r = 0, 1, \ldots$$

$$\bar{u} = \nu_1 \text{ in } u \text{ units}$$

In either x or u units, the μ's as functions of the ν's are

$$\mu_0 = 1$$

$$\mu_1 = 0$$

$$\mu_2 = \nu_2 - \nu_1^2$$

$$\mu_3 = \nu_3 - 3\nu_1\nu_2 + 2\nu_1^3$$

$$\mu_4 = \nu_4 - 4\nu_1\nu_3 + 6\nu_1^2\nu_2 - 3\nu_1^4$$

$$\mu_r (\text{in } x \text{ units}) = c^r \mu_r (\text{in } u \text{ units}) \qquad r = 0, 1, \ldots.$$

In x units, μ_2 is the *variance*; $\sqrt{\mu_2}$ is the *standard deviation* σ. Both are used as measures of dispersion. To compute σ,

$$\sigma = c \sqrt{\frac{\sum_{i=1}^{k} f_i u_i^2}{N} - \bar{u}^2}$$

Probable error $= 0.6745\sigma$.

3. In *standard (deviation) units*,

$$\alpha_1 = 0 \qquad \alpha_2 = 1$$

$$\alpha_3 = \frac{\mu_3}{\sigma^3} \quad \text{(a measure of skewness)}$$

$$\alpha_4 = \frac{\mu_4}{\sigma^4} \quad \text{(a measure of kurtosis)}$$

The *moment-generating function,* or arbitrary-range inverse real Laplace transform, is

$$M(\theta) = \int_a^b e^{\theta x} f(x) \; dx$$

The rth moment is

$$\mu_r = \left. \frac{d^r M}{d\theta^r} \right|_{\theta=0} \qquad r = 0, 1, 2, \ldots$$

M Tiles The rth *quartile* Q_r is the value of x for which cum $f/N = r/4$. The rth *percentile* P_r is the value of x for which cum $f/N = r/100$. For $r = 10s$, $P_r = D_s$, the sth *decile*.

Other Measures of Shape

Dispersion

1. *Range* of x, the difference between the largest and the smallest values of x.
2. *Mean deviation*, $(1/N) \sum_{i=1}^{k} f_i |x_i - \overline{x}|$.
3. *Semi-interquartile range*, or *quartile deviation*,

$$Q = \tfrac{1}{2} |Q_3 - Q_1|$$

Skewness. *Quartile coefficient of skewness*, $(Q_3 - 2Q_2 + Q_1)/Q$.

Statistical Hypotheses A hypothesis concerning one or more statistical distribution parameters is a *statistical hypothesis*. A *test* of such a hypothesis is a procedure leading to a decision to accept or reject the hypothesis. The *significance level* is the probability value below which a hypothesis is rejected.

A *type 1 error* is made if the hypothesis is correct but the test rejects the hypothesis. A *type 2 error* is made if the hypothesis is false but the test accepts the hypothesis.

If the variable x has a distribution function $f(x; \theta)$, with parameter θ, then the *likelihood* function, that is, distribution function of a random sample of size n, is $P(\theta) = f(x_1; \theta) f(x_2; \theta) \cdots f(x_n; \theta)$. The use of $P_{\max}(\theta)$ in the estimation of population parameters is the *method of maximum likelihood*. It often consists of solving $dP/d\theta = 0$ for θ.

Random Sampling A set $x_1, x_2, \ldots, x_n$ of values of x with distribution function $f(x)$ is a *sample of size n* drawn from the population described by $f(x)$. If repeated samples of size n drawn from the population have the x_r's independently distributed in the probability sense and each x_r has the same distribution as the population, then the sampling is *random*.

Normal and Nonnormal Distributions The normal distribution function in analytic and tabular form is found in Section 4.5. A linear combination of independent normal variables is normally distributed.

The *Poisson distribution*, $P(x) = e^{-m} m^x / x!$, is the limit approached by the binomial distribution (Section) if the probability p that an event will occur in a single trial approaches zero and the number of trials r

becomes infinite in such a way that $rp = m$ remains constant.

If m is the mean of a nonnormal distribution of x, σ the standard deviation, and if the moment-generating function exists, then the variable $(\overline{x} - m)n^{1/2}/\sigma$, in which $\overline{x}$ is the mean of a sample of size n, has a distribution that approaches the normal distribution as $n \to \infty$.

Nonparametric methods are those that do not involve the estimation of parameters of a distribution function. Tchebycheff's inequality (Section) provides nonparametric tests for the validity of hypotheses. It leads to the *law of large numbers*. Let p be the probability of an event occurring in one trial and p_n the ratio of the number of occurrences in n trials to the number n. The probability that $|p_n - p| > \varepsilon$ is $\leq pq/n\varepsilon$; this can be made arbitrarily small, however small ε is, by taking n large enough. The ratio p_n converges *stochastically* to the probability p.

Two numbers L_1, L_2 between which a large fraction of a population is expected to lie are *tolerance limits*. If z is the fraction of the population of a variable with a continuous distribution that lies between the extreme values of a random sample of size n from this population, then the distribution of z is $f(z) = n(n-1)z^{n-2}(1-z)$.

Statistical Control of Production Processes A chart on which percentage defective in a sample is graphed as a function of output time can be used for control of an industrial process. Horizontal lines are drawn through the mean m and the controls $m \pm 3\sigma/n^{1/2}$. The behavior of the graph with respect to these control lines is used as an error signal in a feedback system that controls the process. If the graph goes out of the band bounded by the control lines, the process is stopped until the trouble is located and removed.

4.2 Correlation

To discover whether there is a simple relation between two variables, corresponding pairs of values are used as coordinates to plot the points of a *scatter diagram*. The simplest relation exists if the scatter diagram can be approximated more or less closely by a straight line.

Least-Square Straight Line. This line, which minimizes the sum of the squares of the y deviations of the points, is

$$\hat{y} - \overline{y} = M(x - \overline{x})$$

in which

$$M = \frac{\Sigma(x - \overline{x})y}{\Sigma(x - \overline{x})^2}$$

(x, y) is a plotted point, and $(x, \hat{y})$ is a point on this *line of regression* of y on x. The *correlation coefficient*

$$r = \pm \left[1 - \frac{\Sigma(y - \hat{y})^2}{\Sigma(y - \overline{y})^2} \right]^{1/2}$$

is a measure of the usefulness of the regression line. If $r = 0$, the line is useless; if $r = \pm 1$, the line gives a perfect estimate. The percentage of the variance of y that has been accounted for by y's relation to x is equal to r^2.

Polynomial of Degree $n-1$. This can be passed through n points (x_i, y_i). The method of doing this by *divided differences* is as follows:

Example 44 Find the polynomial through (1, 5), (3, 11), (4, 31), (6, 3).

Using the first three values of x, assume the polynomial to be of the form $y = a_1 + a_2(x - 1) + a_3(x - 1)(x - 3) + a_4(x - 1)(x - 3)(x - 4)$. The a_i are the last four numbers in the top diagonal of the following:

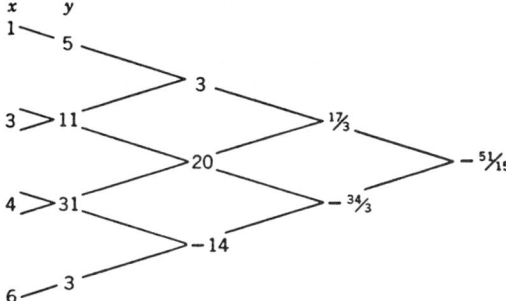

To form the graphic, put the given (x_i, y_i) in the first two columns. To find a number in any other column, divide the difference of the two numbers just above and below it immediately to the left by the difference of the x's in the two diagonals through it. The polynomial is $y = 5 + 3(x - 1) + \frac{17}{3}(x - 1)(x - 3) - \frac{51}{15}(x - 1)(x - 3)(x - 4)$.

Power Formula. $y = ax^n$ fits well if the points (x_i, y_i) lie approximately on a straight line when plotted on logarithmic (log scales on both horizontal and vertical axes) graph paper. To find a and n use two of the points (x_1, y_1) and (x_2, y_2), preferably far apart:

$$n = \frac{\log y_2 - \log y_1}{\log x_2 - \log x_1}$$

$$\log a = \log y_1 - n \log x_1$$

Exponential Formula. $y = ae^{nx}$ fits well if the points (x_i, y_i) lie approximately on a straight line when plotted on semilogarithmic (log scale on vertical axis) graph paper. To find a and n use two of the points (x_1, y_1) and (x_2, y_2), preferably far apart:

$$n = \frac{\ln y_1 - \ln y_2}{x_1 - x_2} = 2.3026 \frac{\log y_1 - \log y_2}{x_1 - x_2}$$

$$\ln a = \ln y_1 - nx_1 \quad \text{or} \quad \log a = \log y_1 - 0.4343nx_1$$

4.3 Statistical Estimation by Small Samples

A statistic is an *unbiased estimate* of a population parameter if its expected value is equal to the population parameter.

In the problem of estimating a population parameter, such as the mean or variance, the interval within which c percent of the sample parameter values lies is the *c percent confidence interval* for the parameter.

The χ^2 *distribution function* for v degrees of freedom is

$$f(\chi^2) = \frac{1}{2^{v/2}\Gamma(v/2)}(\chi^2)^{(v-2)/2}e^{-\chi^2/2}$$

and its moment-generating function is

$$M(\phi) = (1 - 2\theta)^{-v/2}$$

The sum of the squares of n random sample values of x has a χ^2 distribution with n degrees of freedom if x has a normal distribution with zero mean and unit variance.

The *binomial index of dispersion* is

$$\chi^2 = \sum_{r=1}^{k} \frac{(x_r - \overline{x})^2}{\overline{x}(1 - \overline{x}/n)}$$

For p small and n large, this reduces to the *Poisson index of dispersion* $\sum_{r=1}^{k}(x_r - \overline{x})^2/\overline{x}$.

These indices are used to test the hypothesis that k sample frequencies x_r came from the same binomial or Poisson population, respectively.

Student's t distribution for the variable $t = uv^{1/2}/v$ is

$$f(t) = c\left(1 + \frac{t^2}{v}\right)^{-(v+1)/2}$$

v degrees of freedom, c constant, if u has a normal distribution with zero mean and unit variance and v^2 has a χ^2 distribution with v degrees of freedom.

The *F distribution* for the variable $F = (u/v_1)/(v/v_2)$ is

$$f(F) = \frac{cF^{(v_1-2)/2}}{(v_2 + v_1 F)^{(v_1+v_2)/2}}$$

v_1 and v_2 degrees of freedom, c constant, if u and v have independent χ^2 distributions with v_1 and v_2 degrees of freedom, respectively.

Analysis of Variance *Experimental error* is the variation in the basic variable remaining after the effects of controlled variables have been removed (Section 4.5). The *analysis of variance* means the resolution of the basic sum of squares into the component that measures the part of the variation being tested and the component that measures the experimental error.

4.4   Statistical Design of Experiments

To get *valid* conclusions from an experiment, there is need for proper control of the other variables besides those being investigated and also for sufficiently large and random samples.

Sampling Inspection.   To make an inspection *efficient*, the cost and usually the amount of sampling should be minimized.

It is a common practice in industry for a consumer to accept or reject a lot on the basis of a sample drawn from the lot. There is a maximum fraction of defectives that the consumer will tolerate. This is the *lot tolerance fraction defective* p_t. A random sample of n pieces is selected from a lot of N pieces. The maximum allowable number of defective pieces in an acceptable sample is c. *Single sampling* means: (a) Inspect a sample of n pieces. (b) Accept the lot if the number of defective pieces is c or less; otherwise inspect the remainder of the lot. (c) Replace all defective pieces found by nondefective ones.

The *consumer's risk*, that is, the probability that a consumer will accept a lot of quality lower than p_t, is

$$P_c = \sum_{x=0}^{c} \frac{\binom{Np_t}{x}\binom{N-Np_t}{n-x}}{\binom{N}{n}} \qquad (12)$$

If a producer has standardized quality at a fractional value $\bar{p}$, the *process average fraction defective*, then the *producer's risk*, that is, the probability that a lot will be erroneously rejected, is

$$P_p = 1 - \sum_{x=0}^{c} \frac{\binom{N\bar{p}}{x}\binom{N-N\bar{p}}{n-x}}{\binom{N}{n}} \qquad (13)$$

These two risks correspond to errors of type 2 and type 1, respectively.

The average number of pieces inspected per lot for single sampling is $I = n + (N - n)P_p$. The amount of inspection and ordinarily the cost are minimized by finding the pair of values of n and c that satisfy (a) above for an assigned value of P_c and minimize I.

Sequential Analysis.   An improvement on the fixed-size sampling methods already described results in greater efficiency if the inspection can be conducted on an accumulation-of-information basis. Such sequential methods operate on successive terms of a sequence of observations as they are received. They involve two steps: (a) to accept or reject the hypothesis under test and (b) to continue taking additional observations if the hypothesis is rejected.

For a more extensive treatment of the elementary theory of statistics, see *Applied Statistics and Probability for Engineers* by D. C. Montgomery and G. C. Runger (Wiley, Hoboken, New Jersey, 2007).

4.5   Precision of Measurements

Observations and Errors   The *error of an observation* is $e_i = m_i - m, i = 1, 2, \ldots, n$, where the m_i are the observed values, the e_i the errors, and m the mean value, that is, the arithmetic mean of a very large number (theoretically infinite) of observations.

In a large number of measurements *random errors* are as often negative as positive and have little effect on the arithmetic mean. All other errors are classed as *systematic*. If due to the same cause, they affect the mean in the same sense and give it a definite bias.

Best Estimate and Measured Value.   If all systematic errors have been eliminated, it is possible to consider the sample of individual repeated measurements of a quantity with a view to securing the "best" estimate of the mean value m and assessing the degree of reproducibility that has been obtained. The final result will then be expressed in the form $E \pm L$, where E is the best estimate of m and L the characteristic limit of variation associated with a certain risk. Not merely E but the entire result $E \pm L$ is the value measured.

Arithmetic Mean.   If a large number of measurements have been made to determine directly the mean m of a certain quantity, all measurements having been made with equal skill and care, the best estimate of m from a sample of n is the arithmetic mean $\bar{m}$ of the measurements in the sample,

$$\bar{m} = \frac{1}{n}\sum_{i=1}^{n} m_i$$

Standard deviation is the *root-mean-square* of the deviations e_i of a set of observations from the mean,

$$\sigma = \left(\frac{1}{n}\sum_{i=1}^{n} e_i^2\right)^{1/2}$$

Since neither the mean m nor the errors of observation e_i are ordinarily known, the deviations from the arithmetic mean, or the *residuals*, $x_i = m_i - \bar{m}, i = 1, 2, \ldots, n$, will be referred to as errors.

Likewise, for σ the unbiased value

$$\sigma = (n-1)^{-1/2}\left[\sum_{i=1}^{n}(m_i - \bar{m})^2\right]^{1/2}$$

$$= (n-1)^{1/2}\left(\sum_{i=1}^{n} e_i^2\right)^{1/2}$$

will be used, in which n is replaced by $n-1$ since one degree of freedom is lost by using $\overline{m}$ instead of m, $\overline{m}$ being related to the m_i.

Normal Distribution

Relative Frequency of Errors.
The *Gauss–Laplace*, or *normal*, distribution of frequency of errors is (Fig. 4)

$$y = \frac{1}{\sigma\sqrt{2\pi}}e^{-x^2/2\sigma^2}$$

or

$$y = \frac{1}{\sqrt{\pi}}he^{-h^2x^2}$$

where $2h^2\sigma^2 = 1$, or $h = 1/(\sqrt{2}\sigma)$, and y represents the proportionate number of errors of value x. The area under the curve is unity. The dotted curve is also an error distribution curve with a greater value of the precision index h, which measures the concentration of observations about their mean.

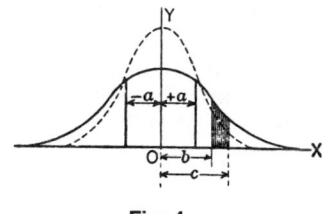

Fig. 4

Probability.
The fraction of the total number of errors whose values lie between $x = -a$ and $x = a$ is

$$P = \frac{h}{\sqrt{\pi}}\int_{-a}^{+a}e^{-h^2x^2}\,dx = \frac{2}{\sqrt{\pi}}\int_{0}^{ha}e^{-h^2x^2}\,d\,(hx)$$

(14)

that is, P is the probability of an error x having a value between $-a$ and a (see Table 1). Similarly, the shaded area represents the probability of errors between b and c.

Table 1 Values of $P = \dfrac{2}{\sqrt{\pi}}\displaystyle\int_{0}^{ha}e^{-h^2x^2}d(hx)$

ha^a	0	1	2	3	4	5	6	7	8	9
0.0		0.01128	0.02256	0.03384	0.04511	0.05637	0.06762	0.07886	0.09008	0.10128
0.1	0.11246	0.12362	0.13476	0.14587	0.15695	0.16800	0.17901	0.18999	0.20094	0.21184
0.2	0.22270	0.23352	0.24430	0.25502	0.26570	0.27633	0.28690	0.29742	0.30788	0.31828
0.3	0.32863	0.33891	0.34913	0.35928	0.36936	0.37938	0.38933	0.39921	0.40901	0.41874
0.4	0.42839	0.43797	0.44747	0.45689	0.46623	0.47548	0.48466	0.49375	0.50275	0.51167
0.5	0.52050	0.52924	0.53790	0.54646	0.55494	0.56332	0.57162	0.57982	0.58792	0.59594
0.6	0.60386	0.61168	0.61941	0.62705	0.63459	0.64203	0.64938	0.65663	0.66378	0.67084
0.7	0.67780	0.68467	0.69143	0.69810	0.70468	0.71116	0.71754	0.72382	0.73001	0.73610
0.8	0.74210	0.74800	0.75381	0.75952	0.76514	0.77067	0.77610	0.78144	0.78669	0.79184
0.9	0.79691	0.80188	0.80677	0.81156	0.81627	0.82089	0.82542	0.82987	0.83423	0.83851
1.0	**0.84270**	0.84681	0.85084	0.85478	0.85865	0.86244	0.86614	0.86977	0.87333	0.87680
1.1	0.88021	0.88353	0.88679	0.88997	0.89308	0.89612	0.89910	0.90200	0.90484	0.90761
1.2	0.91031	0.91296	0.91553	0.91805	0.92051	0.92290	0.92524	0.92751	0.92973	0.93190
1.3	0.93401	0.93606	0.93807	0.94002	0.94191	0.94376	0.94556	0.94731	0.94902	0.95067
1.4	0.95229	0.95385	0.95538	0.95686	0.95830	0.95970	0.96105	0.96237	0.96365	0.96490
1.5	0.96611	0.96728	0.96841	0.96952	0.97059	0.97162	0.97263	0.97360	0.97455	0.97546
1.6	0.97635	0.97721	0.97804	0.97884	0.97962	0.98038	0.98110	0.98181	0.98249	0.98315
1.7	0.98379	0.98441	0.98500	0.98558	0.98613	0.98667	0.98719	0.98769	0.98817	0.98864
1.8	0.98909	0.98952	0.98994	0.99035	0.99074	0.99111	0.99147	0.99182	0.99216	0.99248
1.9	0.99279	0.99309	0.99338	0.99366	0.99392	0.99418	0.99443	0.99466	0.99489	0.99511
2.0	**0.99532**	0.99552	0.99572	0.99591	0.99609	0.99626	0.99642	0.99658	0.99673	0.99688
2.1	0.99702	0.99715	0.99728	0.99741	0.99753	0.99764	0.99775	0.99785	0.99795	0.99805
2.2	0.99814	0.99822	0.99831	0.99839	0.99846	0.99854	0.99861	0.99867	0.99874	0.99880
2.3	0.99886	0.99891	0.99897	0.99902	0.99906	0.99911	0.99915	0.99920	0.99924	0.99928
2.4	0.99931	0.99935	0.99938	0.99941	0.99944	0.99947	0.99950	0.99952	0.99955	0.99957
2.5	0.99959	0.99961	0.99963	0.99965	0.99967	0.99969	0.99971	0.99972	0.99974	0.99975
2.6	0.99976	0.99978	0.99979	0.99980	0.99981	0.99982	0.99983	0.99984	0.99985	0.99986
2.7	0.99987	0.99987	0.99988	0.99989	0.99989	0.99990	0.99991	0.99991	0.99992	0.99992
2.8	0.99992	0.99993	0.99993	0.99994	0.99994	0.99994	0.99995	0.99995	0.99995	0.99996
2.9	0.99996	0.99996	0.99996	0.99997	0.99997	0.99997	0.99997	0.99997	0.99997	0.99998
3.0	**0.99998**	1.0000	1.0000	1.0000						

^{a} $ha = 0.47694(a/r) = (1/\sqrt{2})(a/\sigma)$.

Probable Error.   Results of measurements are sometimes expressed in the form $E \pm r$, where r is the *probable error* of a single observation and is defined as the number that the actual error may with equal probability be greater or less than. From (14)

$$\frac{2}{\sqrt{\pi}} \int_0^{hr} e^{-h^2 x^2} \, d(hx) = 0.50$$

and

$$hr = 0.47694$$

or

$$r = 0.4769 \times \sqrt{2}\sigma = 0.6745\sigma$$

Similarly, 5% of the errors x are greater than 2σ and less than 1% are greater than 3σ.

For rapid comparisons the following approximate formula due to Peters is useful:

$$r \approx 0.8453[n(n-1)]^{-1/2} \sum_{i=1}^{n} |x_i|$$

The *standard deviation of the arithmetic mean*, $\sigma_{\overline{m}}$, as calculated from data, is related to the standard deviation σ by the formula

$$\sigma_{\overline{m}} = n^{-1/2}\sigma = [n(n-1)]^{-1/2} \left(\sum_{i=1}^{n} x_i^2 \right)^{1/2}$$

From this formula and Tables 1 and 2 the limits corresponding to given risks can be determined as indicated previously. It is evident that the stability of the mean increases with n, that is, the effect of the erratic behavior of single cases decreases with increase of n.

The *probable error of the arithmetic mean* as calculated from data, $r_{\overline{m}}$, is then given by

$$r_{\overline{m}} = 0.6745[n(n-1)]^{-1/2} \left(\sum_{i=1}^{n} x_i^2 \right)^{1/2}$$

and Peters's formula for the approximate value is

$$r_{\overline{m}} \approx 0.8453[n^2(n-1)]^{-1/2} \sum_{i=1}^{n} |x_i|$$

Example 45.   The following are 10 measurements, m_i, of the length of a baseline. The values of the residuals, x_i, and their squares are given: $m = 455.35$, $455.35, 455.20, 455.05, 455.75, 455.40, 455.10, 455.30, 455.50, 455.30$.

Arithmetic mean $\overline{m} = 455.330$.
x_i: $0.02, 0.02, -0.13, -0.28, 0.42, 0.07, -0.23, -0.03, -0.17, -0.03$.

x_i^2: $0.0004, 0.0004, 0.0169, 0.0784, 0.1764, 0.0049, 0.0529, 0.0009, 0.0289, 0.0009$.

Hence

$$\sum_{i=1}^{10} x_i^2 = 0.3610 \qquad \text{and} \qquad \sum_{i=1}^{10} |x_i| = 1.40$$

So by the standard formulas, $r = 0.6745(9)^{-1/2} (0.3610)^{1/2} = 0.13$, $r_{\overline{m}} = (10)^{-1/2} r = 0.042$. By the approximate formulas, $r \approx 0.8453(90)^{-1/2}(1.40) = 0.12$, $r_{\overline{m}} \approx 0.039$.

For the best estimate of the baseline, the result is 455.330 with probable error ± 0.042 (using result given by the standard formula), usually written 455.330 ± 0.042. In any considerable number of observations it should be the case, as it is here, that half of the residuals are less than the probable error.

Rounded Numbers.   It can be shown that the standard deviation σ of a *rounded number* (Section 1.3) due to rounding is $\sigma = 0.2887 \, w$, where w is a unit in the last place retained. Consequently, the probable error of a rounded number due to rounding is

$$r = 0.6745 \times 0.2887 \, w = 0.1947 \, w$$

Weighted   Observations.   Sometimes, notwithstanding the care with which observations are taken, there are reasons for believing that certain observations are better than others. In such cases the observations are given different weights, that is, are counted a different numbers of times, the weights or numbers expressing their relative practical worth. If there are n weighted observations m_i with weights p_i, these being made directly on the same quantity, then the best estimate of the mean value m of the quantity is the ***weighted arithmetic mean*** $\overline{m}$ of the sample,

$$\overline{m} \equiv \frac{\sum_{i=1}^{n} p_i m_i}{\sum_{i=1}^{n} p_i}$$

For the set of weighted observations we have

$$r = 0.6745(n-1)^{-1/2} \left(\sum_{i=1}^{n} p_i x_i^2 \right)^{1/2}$$

as the probable error of an observation of unit weight and

$$r_{\overline{m}} = 0.6745 \left[(n-1) \sum_{i=1}^{n} p_i \right]^{-1/2} \left(\sum_{i=1}^{n} p_i x_i^2 \right)^{1/2}$$

Table 2 Values of Functions of n and $n - 1$ Factors for Computing Actual and Approximate Values of r and $r_{\overline{m}}$

n	$\dfrac{0.6745}{\sqrt{n-1}}$	$\dfrac{0.6745}{\sqrt{n(n-1)}}$	$\dfrac{0.8453}{\sqrt{n(n-1)}}$	$\dfrac{0.8453}{n\sqrt{n-1}}$	n	$\dfrac{0.6745}{\sqrt{n-1}}$	$\dfrac{0.6745}{\sqrt{n(n-1)}}$	$\dfrac{0.8453}{\sqrt{n(n-1)}}$	$\dfrac{0.8453}{n\sqrt{n-1}}$
1					51	0.0954	0.0134	0.0167	0.0023
2	0.6745	0.4769	0.5978	0.4227	52	0.0944	0.0131	0.0164	0.0023
3	0.4769	0.2754	0.3451	0.1993	53	0.0935	0.0128	0.0161	0.0022
4	0.3894	0.1947	0.2440	0.1220	54	0.0926	0.0126	0.0158	0.0022
5	0.3372	0.1508	0.1890	0.0845	55	0.0918	0.0124	0.0155	0.0021
6	0.3016	0.1231	0.1543	0.0630	56	0.0909	0.0122	0.0152	0.0020
7	0.2754	0.1041	0.1304	0.0493	57	0.0901	0.0119	0.0150	0.0020
8	0.2549	0.0901	0.1130	0.0399	58	0.0893	0.0117	0.0147	0.0019
9	0.2385	0.0795	0.0996	0.0332	59	0.0886	0.0115	0.0145	0.0019
10	**0.2248**	**0.0711**	**0.0891**	**0.0282**	**60**	**0.0878**	**0.0113**	**0.0142**	**0.0018**
11	0.2133	0.0643	0.0806	0.0243	61	0.0871	0.0111	0.0140	0.0018
12	0.2034	0.0587	0.0736	0.0212	62	0.0864	0.0110	0.0137	0.0017
13	0.1947	0.0540	0.0677	0.0188	63	0.0857	0.0108	0.0135	0.0017
14	0.1871	0.0500	0.0627	0.0167	64	0.0850	0.0106	0.0133	0.0017
15	0.1803	0.0465	0.0583	0.0151	65	0.0843	0.0105	0.0131	0.0016
16	0.1742	0.0435	0.0546	0.0136	66	0.0837	0.0103	0.0129	0.0016
17	0.1686	0.0409	0.0513	0.0124	67	0.0830	0.0101	0.0127	0.0016
18	0.1636	0.0386	0.0483	0.0114	68	0.0824	0.0100	0.0125	0.0015
19	0.1590	0.0365	0.0457	0.0105	69	0.0818	0.0098	0.0123	0.0015
20	**0.1547**	**0.0346**	**0.0434**	**0.0097**	**70**	**0.0812**	**0.0097**	**0.0122**	**0.0015**
21	0.1508	0.0329	0.0412	0.0090	71	0.0806	0.0096	0.0120	0.0014
22	0.1472	0.0314	0.0393	0.0084	72	0.0800	0.0094	0.0118	0.0014
23	0.1438	0.0300	0.0376	0.0078	73	0.0795	0.0093	0.0117	0.0014
24	0.1406	0.0287	0.0360	0.0073	74	0.0789	0.0092	0.0115	0.0013
25	0.1377	0.0275	0.0345	0.0069	75	0.0784	0.0091	0.0113	0.0013
26	0.1349	0.0265	0.0332	0.0065	76	0.0779	0.0089	0.0112	0.0013
27	0.1323	0.0255	0.0319	0.0061	77	0.0774	0.0088	0.0111	0.0013
28	0.1298	0.0245	0.0307	0.0058	78	0.0769	0.0087	0.0109	0.0012
29	0.1275	0.0237	0.0297	0.0055	79	0.0764	0.0086	0.0108	0.0012
30	**0.1252**	**0.0229**	**0.0287**	**0.0052**	**80**	**0.0759**	**0.0085**	**0.0106**	**0.0012**
31	0.1231	0.0221	0.0277	0.0050	81	0.0754	0.0084	0.0105	0.0012
32	0.1211	0.0214	0.0268	0.0047	82	0.0749	0.0083	0.0104	0.0011
33	0.1192	0.0208	0.0260	0.0045	83	0.0745	0.0082	0.0102	0.0011
34	0.1174	0.0201	0.0252	0.0043	84	0.0740	0.0081	0.0101	0.0011
35	0.1157	0.0196	0.0245	0.0041	85	0.0736	0.0080	0.0100	0.0011
36	0.1140	0.0190	0.0238	0.0040	86	0.0732	0.0079	0.0099	0.0011
37	0.1124	0.0185	0.0232	0.0038	87	0.0727	0.0078	0.0098	0.0010
38	0.1109	0.0180	0.0225	0.0037	88	0.0723	0.0077	0.0097	0.0010
39	0.1094	0.0175	0.0220	0.0035	89	0.0719	0.0076	0.0096	0.0010
40	**0.1080**	**0.0171**	**0.0214**	**0.0034**	**90**	**0.0715**	**0.0075**	**0.0094**	**0.0010**
41	0.1066	0.0167	0.0209	0.0033	91	0.0711	0.0075	0.0093	0.0010
42	0.1053	0.0163	0.0204	0.0031	92	0.0707	0.0074	0.0092	0.0010
43	0.1041	0.0159	0.0199	0.0030	93	0.0703	0.0073	0.0091	0.0009
44	0.1029	0.0155	0.0194	0.0029	94	0.0699	0.0072	0.0090	0.0009
45	0.1017	0.0152	0.0190	0.0028	95	0.0696	0.0071	0.0089	0.0009
46	0.1005	0.0148	0.0186	0.0027	96	0.0692	0.0071	0.0089	0.0009
47	0.0994	0.0145	0.0182	0.0027	97	0.0688	0.0070	0.0088	0.0009
48	0.0984	0.0142	0.0178	0.0026	98	0.0685	0.0069	0.0087	0.0009
49	0.0974	0.0139	0.0174	0.0025	99	0.0681	0.0068	0.0086	0.0009
50	**0.0964**	**0.0136**	**0.0171**	**0.0024**	**100**	**0.0678**	**0.0068**	**0.0085**	**0.0009**

Tabulation for Example 46

p_i	5	4	1	4	3	4
m_i	178.26	176.30	181.06	177.95	176.20	180.85
$p_i m_i$	891.30	705.20	181.06	711.80	528.60	723.40
x_i	0.10	1.86	2.90	0.21	1.96	2.69
x_i^2	0.010	3.460	8.410	0.441	3.842	7.230
$p_i x_i^2$	0.05	13.84	8.41	0.18	11.53	28.94

as the probable error of the arithmetic mean of weighted items, in which

$$x_i \equiv \frac{m_i - \sum_{i=1}^{n} p_i m_i}{\sum_{i=1}^{n} p_i}$$

Example 46. Let six observations on the same quantity be made with weights p_i, the sum of these weights being 21 (see the tabulation above). The sum of the weighted observations, $\sum_{i=1}^{6} p_i m_i$, is 3741.36. The best estimate of the value of m for the observed quantity is $\overline{m} = 3741.36/21 = 178.16$. Subtracting this from each m_i gives the residuals x_i. The sum of the weighted squares of the residuals, $\sum_{i=1}^{6} p_i x_i^2$, is 62.95. Then the preceding formulas give the probable error of an observation of weight unity as $r = 2.39$ and the probable error of the weighted mean as $r_{\overline{m}} = 0.52$. The final result then is 178.16 ± 0.52.

Probable Error in a Result Calculated from Means of Several Observed Quantities.
Let Z be a sum of n means of observed independent quantities, each taken with a plus or a minus sign. Then, if $r_j, j = 1, 2, \ldots, n$, are the probable errors in these means, the probable error in Z is $\left(\sum_{j=1}^{n} r_j^2 \right)^{1/2}$.

Let $Z = Az$, where z is the mean of an observed quantity with probable error r and A an exact number. Then the probable error in Z is Ar.

Let Z be any differentiable function of the means of independently observed quantities z_j with probable errors r_j. Then the probable error in Z is $\left[\sum_{j=1}^{m} (\partial Z/\partial z_j)^2 r_j^2 \right]^{1/2}$. For example, if $Z = z_1 z_2$, the probable error in Z is $(z_1^2 r_2^2 + z_2^2 r_1^2)^{1/2}$.

Conditions of Applicability.
The theory underlying the foregoing development depends on the following assumptions: (a) The sample consists of a large number of observations. (b) The observations have been made with equal care and skill so that (i) there are approximately an equal number of readings above and below the mean (except in the case of weighted items), (ii) the individual deviations from the mean are small in most cases, and (iii) the number of deviations diminishes rapidly as their size increases.

The extent to which the observed data satisfy these assumptions is a measure of the extent to which we are justified in using the Gauss error distribution curve, which is consistent with the statement that $\overline{m}$ is the best estimate of the mean value m and which leads to the factor 0.6745 used in computing probable error. Even if we were not justified in assuming the Gaussian distribution of errors, the arithmetic mean still remains the best estimate we have for m. Therefore, there is little difficulty in this regard, especially since "errors" appear to follow the Gaussian distribution as closely as any other we know. Our difficulties enter in connection with the factor 0.6745 and the accuracy of the σ, as estimated from the data.

If the number of observations n in a sample is small, the estimate of the standard deviation of the possible infinity of observations with mean m is itself subject to considerable error. For example, for $n = 3$ the standard error of the standard deviation is as large as the standard deviation itself, and hence the probable error calculated from $r = 0.6745\sigma$ would not be very reliable. Table 3 will illustrate this. The second and third columns give the probability that the probable error of a single observation should be out 20 and 50%, respectively.

From Table 3 it is clear that with 10 observations the odds are only 3:2 that the calculated probable error is within 20% of the correct value and about 30:1 that it is within 50% of the correct value. Of course, the probable error of the mean will be correspondingly out.

The use of Table 2.3 is quite legitimate for $100 < n$, and for $30 < n < 100$ the table may be used provided σ is multiplied by $(n - 3)^{-1/2}$. For $n < 30$, a rough estimate can be obtained from the fact that the percentage of cases lying outside the range, $m \pm k\sigma$, is $< 100k^{-2}$ for $1 < k$. A striking property of this inequality due to Tchebycheff is that it is *nonparametric*,

Table 3 Combination of Observations

n	20%	50%	n	20%	50%
5	0.64	0.24	30	0.12	0.00014
10	0.40	0.034	40	0.076	8×10^{-6}
15	0.29	0.008	50	0.047	6×10^{-7}
20	0.21	0.0002	100	0.0050	

Source: D. Brunt, *The Combination of Observations*, Cambridge University Press, 1917.

which means independent of the nature of the distribution assumed.

5   GEOMETRY

5.1   Geometric Concepts

Plane Angles   A *degree* (°) is $\frac{1}{360}$ of a revolution (or *perigon*) and is divided into 60 units called *minutes* (′) that in turn are divided into 60 units called *seconds* (″).

A *radian* is a *central angle* that intercepts a *circular arc* equal to its *radius*. One radian, therefore, equals $360/2\pi$ degrees, or 57.295779513°, and 1° = 0.017453293 radian.

An *angle* of 90° is a *right angle*, and the lines that form it are *perpendicular*. An angle less than a right angle is *acute*. An angle greater than a right angle but less than 180° is *obtuse*. If the sum of two angles equals 90°, they are *complementary* to each other, and if their sum is 180°, *supplementary* to each other.

Polygons   A *polygon*, or *plane rectilinear figure*, is a closed broken line.

A *triangle* is a polygon of three sides. It is *isosceles* if two sides (and their opposite angles) are equal; it is *equilateral* if all three sides (and all three angles) are equal.

A *quadrilateral* is a polygon of four sides. This classification includes the *trapezium*, having no two sides parallel; the *trapezoid*, having two opposite sides parallel (*isosceles trapezoid* if the nonparallel sides are equal); and the *parallelogram*, having both pairs of opposite sides parallel and equal. The parallelogram includes the *rhomboid*, having no right angles and, in general, adjacent sides not equal; the *rhombus*, having no right angles but all sides equal; the *rectangle*, having only right angles and, in general, adjacent sides not equal; and the *square*, having only right angles and all sides equal.

Similar polygons have their respective angles equal and their corresponding sides proportional.

A *regular polygon* has all sides equal and all angles equal. An equilateral triangle and a square are regular polygons.

Other polygons classified according to number of sides are (5) *pentagon*, (6) *hexagon*, (7) *heptagon*, (8) *octagon*, (9) *enneagon*, or *nonagon*, (10) *decagon*, and (12) *dodecagon*. Two regular polygons of the same number of sides are *similar*.

Properties of Triangles

General Triangle.   The sum of the angles equals 180°. $\angle XAB$ (Fig. 5) is an *exterior angle* of $\triangle ABC$ and equals the sum of the opposite *interior angles* (i.e., $\angle XAB = \angle B + \angle C$). A *median* of a triangle is a line joining a vertex to the midpoint of the opposite side. The three medians meet at the *center of gravity*, G, and G trisects each median (e.g., $AG = \frac{2}{3}AD$). *Bisectors of angles* of a triangle (Fig. 6) meet in a point M equidistant from all sides. M is the center of the *inscribed circle* (tangent to all sides), or the *incenter* of the triangle. An angle bisector divides the opposite side into segments proportional to the adjacent sides of the angle (e.g. $AK/KC = AB/BC$). An *altitude* of a triangle is a perpendicular from a vertex to the opposite side. The three altitudes meet in a point called the *orthocenter*. The *perpendicular bisectors* of the sides of a triangle (Fig. 7) meet in a point O equidistant from all vertices. O is the center of the *circumscribed circle* (passing through all vertices), or the *circumcenter* of the triangle. The longest side of a triangle is opposite the largest angle, and vice versa. The line joining the midpoints of two sides of a triangle is parallel to the third side and half its length. If two triangles are mutually equiangular, they are similar, and their corresponding sides are proportional.

Orthogonal Projection.   In Figs. 8 and 9, AE is the orthogonal projection of AB on AC, BE being perpendicular to AC. The square of the side opposite an acute angle equals the sum of the squares of the other two sides diminished by twice the product of one of those sides by the orthogonal projection of the other side upon it. In Fig. 8, $a^2 = b^2 + c^2 - 2b \cdot AE$. The

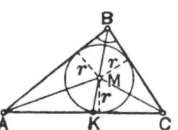

Fig. 6

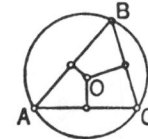

Fig. 7

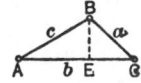

Fig. 8

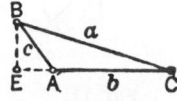

Fig. 9

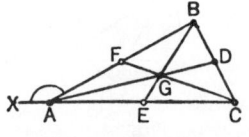

Fig. 5

square of the side opposite an obtuse angle equals the sum of the squares of the other two sides increased by twice the product of one of those sides by the orthogonal projection of the other side upon it. In Fig. 9, $a^2 = b^2 + c^2 + 2b \cdot AE$.

Right Triangle.   In Fig. 10, let h be the *altitude* drawn from the vertex of right angle C to the *hypotenuse* c. Then $\angle A + \angle B = 90°$; $c^2 = a^2 + b^2$; $h^2 = mn$; $b^2 = cm$; $a^2 = cn$; median from $C = c/2$.

Isosceles Triangle.   Two sides are equal and their opposite angles are equal. If a straight line from the vertex at which the equal sides meet bisects the base, it also bisects the angle at the vertex and is perpendicular to the base.

Circles   A *circle* is a closed plane curve, all the points of which are equidistant from a *center* point. A *chord* is a straight line joining two points on a curve, that is, joining the extremities of an *arc*. A *segment* of a circle is the part of its plane included between a concave arc and its chord. An angle *intercepts* an arc cut off by its sides; the arc *subtends* the angle. A *central angle* of a circle is one whose vertex is at the center and whose sides are two radii. A *sector* of a circle is the part of its plane that is included between an arc and two radii drawn to its extremities. A *secant* of a circle is a straight line intersecting it in two points. Parallel secants (or tangents) intercept equal arcs. A tangent line meets a circle in only one point and is perpendicular to the radius to that point. If a radius is perpendicular to a chord, it bisects both the chord and the arc intercepted by the chord. If two circles are tangent to each other, the line of centers passes through the point of contact; if the circles intersect, the line of centers bisects the common chord at right angles. In Fig. 11, the product of linear segments AC and AE equals the product of linear segments AB and AF. In Fig. 12, the product of the whole secant AB and its external segment AE equals the product of the whole

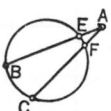

Fig. 12

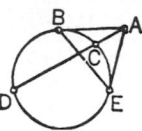

Fig. 13

secant AC and its external segment AF. In Fig. 13, the product of the whole secant AD and its external segment AC equals the square of tangent AB (or AE). Also $\angle ABE = \angle AEB$.

Angle Measurement.   Considering the arc of a circle to be expressed in terms of the central angle that it subtends, the arc may be said to contain a certain number of degrees and hence be used to express the measurement of other angles related to the circle. On this basis, an entire circle equals $360°$. The *inscribed angle* formed by two chords intersecting on a circle equals half the arc intercepted by it. Thus, in Fig. 14, $\angle BAC = \frac{1}{2}$ arc BC. An angle inscribed in a semicircle is a right angle. The angle formed by a tangent to a circle and a chord having one extremity at the point of contact equals half the arc intercepted by the chord. In Fig. 14, $\angle BAT = \frac{1}{2}$ arc BCA. The angle formed by two chords intersecting within a circle equals half the sum of the intercepted arcs. In Fig. 11, $\angle BAC$ (or $\angle EAF$) $= \frac{1}{2}$(arc BC + arc EF). The angle formed by two secants, or two tangents, or a secant and a tangent, intersecting outside a circle, equals half the difference of the intercepted arcs. In Fig 12, $\angle BAC = \frac{1}{2}$(arc BC − arc EF). In Fig. 13, $\angle BAE = \frac{1}{2}$(arc BDE − arc BCE), and $\angle BAD = \frac{1}{2}$(arc BD − arc BC).

Coaxal Systems

Types

1. A set of nonintersecting circles having collinear centers and orthogonal to a given circle with

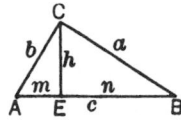

Fig. 10

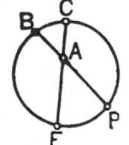

Fig. 11

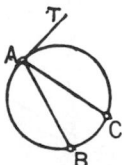

Fig. 14

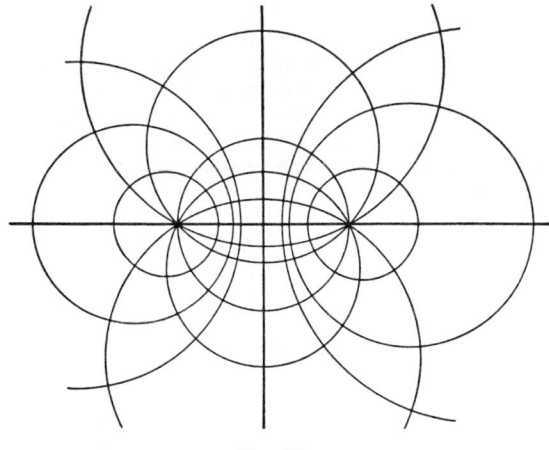

Fig. 15

center also collinear. The endpoints of the diameter of the given circle on the line of centers are the limiting points of the system (Fig. 15, centers on horizontal line).

2. A set of circles through two given points (Fig. 15, centers on vertical line).
3. A set of circles with a common point of tangency.
4. A set of concentric circles.
5. A set of concurrent lines.
6. A set of parallel lines.

Conjugate Systems.   Two coaxal systems whose members are mutually orthogonal are *conjugate*. A conjugate pair may consist of (a) a system of type 1 and one of type 2, with the limiting points of one the common points of the other (Fig. 15); (b) two systems of type 3; (c) a system of type 4 and one of type 5; (d) two systems of type 6.

Inversion   If the point O is the center of a circle c of radius r, if P and P' are collinear with O, and if $OP \cdot OP' = r^2$, then P and P' are *inverse* to each other with respect to the circle c (Fig. 16). The point O is the *center of inversion*.

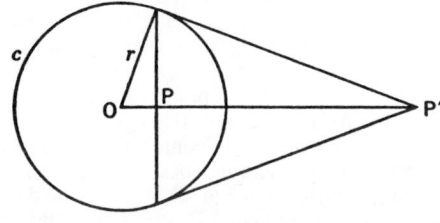

Fig. 16

The inverse of a circle not passing through the center of inversion is a circle, the inverse of a circle through the center is a straight line not through the center, and the inverse of a straight line through the center is itself.

Two intersecting curves invert into curves intersecting at the same angle.

Nonplanar Angles   A *dihedral angle* is the opening between two intersecting planes. In Fig. 17, P–BD–Q is a dihedral angle of which the two planes are the *faces* and their line of intersection DB is the *edge*. A *plane angle* that measures a dihedral angle is an angle formed by two lines, one in each face, drawn perpendicular to the edge at the same point (as $\angle ABC$). A *right dihedral angle* is one whose plane angle is a right angle. Through a given line oblique or parallel to a given plane, one and only one plane can be passed perpendicular to the given plane. The line of intersection CD (Fig. 18) is the *orthogonal projection* of line AB upon plane P. The *angle between a line and a plane* is the angle that the line (produced if necessary) makes with its orthogonal projection on the plane. This angle is the least angle that the line makes with any line in the plane.

A *polyhedral angle* is the opening of three or more planes that meet in a common point. In Fig. 19, O–$ABCDE$ is a polyhedral angle of which the intersections of the planes, as OA, OB, and so on, are the edges; the portions of the planes lying between the edges are the *faces*; and the common point O is the vertex. Angles formed by adjacent edges, as angles AOB, BOC, and so on, are *face angles*. A polyhedral angle

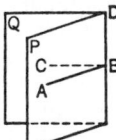

Fig. 17

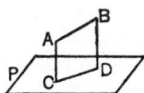

Fig. 18

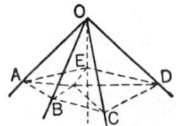

Fig. 19

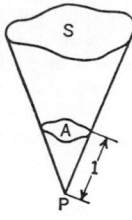

Fig. 20

is called a *trihedral angle* if it has three faces; a *tetrahedral angle* if it has four faces; and so on.

A *solid angle* measures the opening between surfaces, either planar or nonplanar, which meet in a common point. The polyhedral angle is a special case. In Fig. 20 the solid angle at any point P, subtended by any surface S, is equal numerically to the portion. A of the surface of a sphere of unit radius that is cut out by a conical surface with vertex at P and having the boundary of S for base. The *unit solid angle* is the *steradian* and equals the central solid angle that intercepts a spherical area (of any shape) equal to the radius squared. The total solid angle about a point equals 4π steradians.

A *spherical angle* is the opening between two arcs of great circles drawn on a sphere from the same point (vertex) and is measured by the plane angle formed by tangents to its sides at its vertex. If the planes of the great circles are perpendicular, the angle is a *right spherical angle*.

Polyhedrons   A *polyhedron* is a convex closed surface consisting of parts of four or more planes, called its faces; its faces intersect in straight lines, called its edges; its edges at points, called its vertices.

A *prism* is a polyhedron of which two faces (the *bases*) are congruent polygons in parallel planes and the other (*lateral*) faces are parallelograms whose planes intersect in the *lateral edges*. Prisms are *triangular, rectangular, quadrangular*, and so on, according as their bases are triangles, rectangles, quadrilaterals, and so on. A *right prism* has its lateral edges perpendicular to its bases. A prism whose bases are parallelograms is a *parallepiped*; if in addition the edges are perpendicular to the bases, it is a *right parallelepiped*. A *rectangular parallelepiped* is a right parallelepiped whose bases are rectangles. A *cube* is a parallelepiped whose six faces are squares. A *truncated prism* is that part of a prism included between a base and a section made by a plane oblique to the base. A *right section* of a prism is a section made by a plane that cuts all the lateral edges perpendicularly.

A *prismatoid* is a polyhedron of which two faces (the bases) are polygons in parallel planes and the other (lateral) faces are triangles or trapezoids with one side common with one base and the opposite vertex or side common with the other base.

A *pyramid* is a polyhedron of which one face (the base) is a polygon and the other (lateral) faces are triangles meeting in a common point called the vertex of the pyramid and intersecting one another in its lateral edges. Pyramids are triangular, quadrangular, and so on, according as their bases are triangles, quadrilaterals, and so on. A *regular pyramid* (or *right pyramid*) has for its base a regular polygon whose center coincides with the foot of the perpendicular dropped from the vertex to the base. A *frustum of a pyramid* is the portion of a pyramid included between its base and a section parallel to the base. If the section is not parallel to the base, a *truncated pyramid* results.

A *regular polyhedron* has all faces formed of congruent regular polygons and all polyhedral angles equal. The only regular polyhedrons possible are the five types discussed in the mensuration table (Table 4).

A *tetrahedron* is a polyhedron of four faces. It may be described also as a triangular pyramid, and any one of its four triangular faces may be considered as the base. The four perpendiculars erected at circumcenters of the four faces meet in a point equidistant from all vertices, which is the center of the circumscribed sphere. The four *medians*, joining each vertex with the center of gravity of the opposite face, meet in a point, which is the *center of gravity* of the tetrahedron. This point is three-fourths of the distance from each vertex along a median. The four altitudes meet in a point, called the *orthocenter* of the tetrahedron. The six planes bisecting the six dihedral angles meet in a point equidistant from all faces, this being the center of the inscribed sphere.

Solids Having Curved Surfaces   A *cylinder* is a solid bounded by two parallel plane surfaces (the bases) and a cylindrical lateral surface. A *cylindrical surface* is a surface generated by the movement of a straight line (the *generatrix*) which constantly is parallel to a fixed straight line and touches a fixed curve (the *directrix*) not in the plane of the fixed straight line. The generatrix in any position is an *element* of the cylindrical surface. A *circular cylinder* is one having circular bases. A *right cylinder* is one whose elements are perpendicular to its bases. A *truncated cylinder* is the part of a cylinder included between a base and a section made by a plane oblique to the base. A *right section* of a cylinder is a section made by a plane which cuts all the elements perpendicularly.

A *cone* is a solid bounded by a conic lateral surface and a plane (the base) that cuts all the elements of the conic surface. A *conic surface* is a surface generated by the movement of a straight line (the generatrix) that constantly touches a fixed plane curve (the directrix) and passes through a fixed point (the vertex) not in the plane of the fixed curve. The generatrix in any position is an *element* of the conic surface. A *circular cone* is one having a circular base. A *right cone* is a circular cone whose center of the base coincides with the foot of the perpendicular dropped from the vertex to the base. A *frustum of a cone* is the portion of a

Table 4 Mensuration Formulas

Approximate Decimal Equivalents (for reference)

$\pi = 3.1416$

$\pi/2 = 1.5708$
$\pi/4 = 0.7854$
$\pi/180 = 0.01745$
$\pi/360 = 0.00873$

$\dfrac{1}{\pi} = 0.318$
$1/2\pi = 0.159$
$1/4\pi = 0.080$
$180/\pi = 57.296$
$360/\pi = 114.592$

$\sqrt{2} = 1.414$
$\sqrt{3} = 1.732$
$1/\sqrt{2} = 0.707$
$1/\sqrt{3} = 0.577$

1a. Plane Rectilinear Figures

Notation. Lines, $a, b, c, \ldots$; angles, $\alpha, \beta, \gamma, \ldots$; altitude (perpendicular height), h; side, l; diagonals, $d, d_1, \ldots$; perimeter, p; radius of inscribed circle, r; radius of circumscribed circle, R; area, A.

1. Right triangle

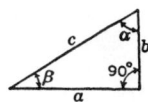

(One angle 90°)

$$p = a + b + c; c^2 = a^2 + b^2;$$
$$A = \tfrac{1}{2}ab = \tfrac{1}{2}a^2 \tan \beta = \tfrac{1}{4}c^2 \sin 2\beta = c\tfrac{1}{4}c^2 \sin 2\alpha.$$

For additional formulas, see *general triangle* below and also trigonometry.

2. General triangle (and equilateral triangle)

For general triangle:

$$p = a + b + c. \text{ Let } s = \tfrac{1}{2}(a + b + c).$$

$$r = \frac{\sqrt{s(s-a)(s-b)(s-c)}}{s}; \; R = \frac{a}{2\sin\alpha} = \frac{abc}{4rs};$$

$$A = \frac{ah}{2} = \frac{ab}{2}\sin\gamma = \frac{b^2\sin\gamma\sin\alpha}{2\sin\beta} = rs = \frac{abc}{4R}.$$

Length of median to side $c = \tfrac{1}{2}\sqrt{2(a^2 + b^2) - c^2}$.

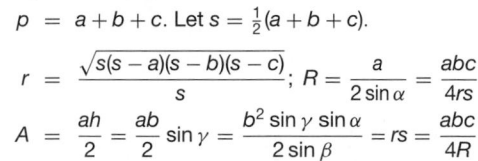

Length of bisector of angle $\gamma = \dfrac{\sqrt{ab[(a+b)^2 - c^2]}}{a+b}$.

For equilateral triangle ($a = b = c = l$ and $\alpha = \beta = \gamma = 60°$):

(Equal sides and equal angles)

$$p = 3l, \; r = \frac{l}{2\sqrt{3}}; \; R = \frac{l}{\sqrt{3}} = 2r;$$

$$h = \frac{l\sqrt{3}}{2}; \; l = \frac{2h}{\sqrt{3}}; \; A = \frac{l^2\sqrt{3}}{4}.$$

For additional formulas, see trigonometry.

3. Rectangle (and square)

For rectangle:

$$p = 2(a + b); \; d = \sqrt{a^2 + b^2}; \; A = ab.$$

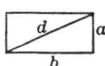

For square ($a = b = l$):

$$p = 4l; \; d = l\sqrt{2}; \; l = \frac{d}{\sqrt{2}}; \; A = l^2 = \frac{d^2}{2}.$$

4. General parallelogram (and rhombus)

For general parallelogram (rhomboid):
(Opposite sides parallel)

$$p = 2(a + b); \; d_1 = \sqrt{a^2 + b^2 - 2ab \, \cos\gamma};$$

$$d_2 = \sqrt{a^2 + b^2 + 2ab \, \cos\gamma}; \; d_1^2 + d_2^2 = 2(a^2 + b^2);$$

$$A = ah = ab \, \sin\gamma.$$

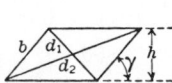

(Continues)

Table 4   *(Continued)*

1a. Plane Rectilinear Figures (Continued)

For rhombus ($a = b = l$):

(Opposite sides parallel and all sides equal)

$p = 4l$; $d_1 = 2l \sin \frac{1}{2}\gamma$; $d_2 = 2l \cos \frac{1}{2}\gamma$; $d_1^2 + d_2^2 = 4l^2$;

$d_1 d_2 = 2l^2 \sin \gamma$; $A = lh = l^2 \sin \gamma = \frac{1}{2}(d_1 d_2)$.

5. General trapezoid (and isosceles trapezoid)

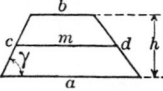

Let midline bisecting nonparallel sides $= m$. Then $m = \frac{1}{2}(a + b)$.

For general trapezoid:

(Only one pair of opposite sides parallel)

$p = a + b + c + d$; $A = \frac{1}{2}(a + b)h = mh$.

For isosceles trapezoid ($d = c$):

(Nonparallel sides equal)

$$A = \frac{1}{2}(a + b)h = mh = \frac{1}{2}(a + b)c \sin \gamma$$

$$= (a - c \cos \gamma)c \sin \gamma = (b + c \cos \gamma)c \sin \gamma.$$

6. General quadrilateral (trapezium)

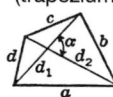

(No sides parallel)

$$p = a + b + c + d.$$

$A = \frac{1}{2}d_1 d_2 \sin \alpha =$ sum of areas of the two triangles formed by either diagonal and the four sides.

7. Quadrilateral inscribed in circle

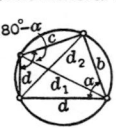

(Sum of opposite angles $= 180°$)

$ac + bd = d_1 d_2$.

Let $s = \frac{1}{2}(a + b + c + d) = \frac{1}{2}p$ and $\alpha =$ angle between sides a and b.

$A = \sqrt{(s - a)(s - b)(s - c)(s - d)} = \frac{1}{2}(ab + cd) \sin \alpha$.

8. Regular polygon (and general polygon)

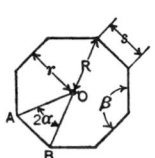

For regular polygon:

(Equal sides and equal angles)

Let $n =$ number of sides.

Central angle $= 2\alpha = \dfrac{2\pi}{n}$ radians;

Vertex angle $= \beta = \dfrac{n - 2}{n}\pi$ radians.

$p = ns$; $s = 2r \tan \alpha = 2R \sin \alpha$;

$r = \frac{1}{2}s \cot \alpha$; $R = \frac{1}{2}s \csc \alpha$;

$A = \frac{1}{2}nsr = nr^2 \tan \alpha = \frac{1}{2}nR^2 \sin 2\alpha = \frac{1}{4}ns^2 \cot \alpha =$ sum of areas of the n equal triangles such as OAB.

For general polygon:

$A =$ sum of areas of constituent triangles into which it can be divided.

Table 4   (*Continued*)

1b. Plane Curvilinear Figures

Notation. Lines, $a, b, \ldots$; radius, r; diameter, d; perimeter, p; circumference, c; central angle n radians, θ; arc, s; chord of arc s, l; chord of half arc $s/2$, l'; rise, h; area, A.

9. Circle (and circular arc)

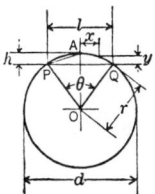

For circle:

$$d = 2r; \quad c = 2\pi r = \pi d; \quad A = \pi r^2 = \frac{\pi d^2}{4} = \frac{c^2}{4\pi}.$$

For circular arc:

Let arc $PAQ = s$; and chord $PA = l'$. Then, $s = r\theta = \frac{1}{2}d\theta$; $s = \frac{1}{3}(8l' - l)$. (The latter equation is Huygen's approximate formula. For θ small; error is very small; for $\theta = 120°$, error is about 0.25%; for $\theta = 180°$, error is less than 1.25%.)

$$l = 2r \sin \tfrac{1}{2}\theta; \quad l = 2\sqrt{2hr - h^2} \text{ (approximate formula)}$$

$$r = \frac{s}{\theta} = \frac{l}{2\sin(\theta/2)}; \quad r = \frac{4h^2 + l^2}{8h} \text{ (approximate formula)}$$

$$h = r \mp \sqrt{r^2 - \tfrac{1}{4}l^2}(- \text{ if } \theta \leq 180° + \text{ if } \theta \geq 180°) = r(1 - \cos \tfrac{1}{2}\theta)$$

$$= r \operatorname{versin} \tfrac{1}{2}\theta = 2r \sin^2 \tfrac{1}{4}\theta = \tfrac{1}{2}l \tan \tfrac{1}{4}\theta = r + y - \sqrt{r^2 - x^2}.$$

Side ordinate $y = h - r + \sqrt{r^2 - x^2}$.

10. Circular sector (and semicircle)

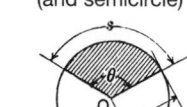

For circular sector:

$$A = \tfrac{1}{2}\theta r^2 = \tfrac{1}{2}sr.$$

For semicircle:

$$A = \tfrac{1}{2}\pi r^2.$$

11. Circular segment

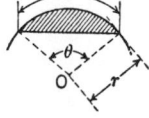

$$A = \tfrac{1}{2}r^2(\theta - \sin \theta)$$

$$= \tfrac{1}{2}[sr \mp l(r - h)](- \text{ if } h \leq r; + \text{ if } h \geq r).$$

$$A = \frac{2lh}{3} \text{ or } \frac{h}{15}(8l' + 6l). \text{ (Approximate formulas. For } h \text{ small compared with } r, \text{ error is very small; for } h = \tfrac{1}{4}r, \text{ first formula errs about 3.5% and second less than 1.0%.)}$$

12. Annulus

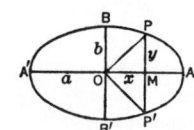

(Region between two concentric circles)

$$A = \pi(r_1^2 - r_2^2) = \pi(r_1 + r_2)(r_1 - r_2);$$

$$A \text{ of sector } ABCD = \tfrac{1}{2}\theta(r_1^2 - r_2^2) = \tfrac{1}{2}\theta(r_1 + r_2)(r_1 - r_2)$$

$$= \tfrac{1}{2}t(s_1 + s_2).$$

13. Ellipse

$$p = \pi(a + b)\left(1 + \frac{R^2}{4} + \frac{R^4}{64} + \frac{R^4}{256} + \cdots\right) \text{ where } R = \frac{a - b}{a + b}.$$

$$p = \pi(a + b)\frac{64 - 3R^4}{64 - 16R^2} \text{ (approximate formula).}$$

$$A = \pi ab; \quad A \text{ of quadrant } AOB = \tfrac{1}{4}\pi ab;$$

$$A \text{ of sector } AOP = \frac{ab}{2}\cos^{-1}\frac{x}{a}; \quad A \text{ of sector } POB = \frac{ab}{2}\sin^{-1}\frac{x}{a};$$

(*Continues*)

Table 4   (*Continued*)

1b. Plane Curvilinear Figures (Continued)

A of section $BPP'B' = xy + ab \sin^{-1} \dfrac{x}{a}$;

A of segment $PAP'P = -xy + ab \cos^{-1} \dfrac{x}{a}$.

For additional formulas, see analytic geometry.

14. Parabola

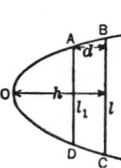

Arc $BOC = s = \frac{1}{2}\sqrt{l^2 + 16h^2} + \dfrac{l^2}{8h} \log_e \dfrac{4h + \sqrt{l^2 + 16h^2}}{l}$.

Let $R = \dfrac{h}{l}$. Then:

$s = l\left(1 + \dfrac{8R^2}{3} - \dfrac{32R^4}{5} + \cdots\right)$ (approximate formula).

$d = \dfrac{h}{l^2}(l^2 - l_1^2); \ l_1 = l\sqrt{\dfrac{h-d}{h}}; \ h = \dfrac{dl^2}{l^2 - l_1^2}$;

A of segment $BOC = \dfrac{2hl}{3}$;

A of section $ABCD = \dfrac{2}{3}d\left(\dfrac{l^3 - l_1^3}{l^2 - l_1^2}\right)$.

For additional formulas, see analytic geometry.

15. Hyperbola

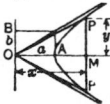

A of figure $OPAP'O = ab \log_e\left(\dfrac{x}{a} + \dfrac{y}{b}\right) = ab \cosh^{-1} \dfrac{x}{a}$;

A of segment $PAP' = xy - ab \log_e\left(\dfrac{x}{a} + \dfrac{y}{b}\right) = xy - ab \cosh^{-1} \dfrac{x}{a}$.

For additional formulas, see analytic geometry.

16. Cycloid

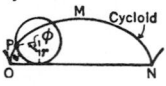

Arc $OP = s = 4r(1 - \cos \frac{1}{2}\phi)$; arc $OMN = 8r$;

A under curve $OMN = 3\pi r^2$.

For additional formulas, see analytic geometry.

17. Epicycloid

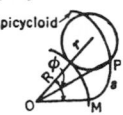

Arc $MP = s = \dfrac{4r}{R}(R + r)\left(1 - \cos \dfrac{R\phi}{2r}\right)$;

Area $MOP = A = \dfrac{r}{2R}(R + r)(R + 2r)\left(\dfrac{R\phi}{r} - \sin \dfrac{R\phi}{r}\right)$.

For additional formulas, see analytic geometry.

18. Hypocycloid

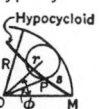

Arc $MP = s = \dfrac{4r}{R}(R - r)\left(1 - \cos \dfrac{R\phi}{2r}\right)$;

Area $MOP = A = \dfrac{r}{2R}(R - r)(R - 2r)\left(\dfrac{R\phi}{r} - \sin \dfrac{R\phi}{r}\right)$.

For additional formulas, see analytic geometry.

19. Catenary

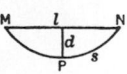

If d is small compared with l:

Arc $MPN = s = l\left[1 + \dfrac{2}{3}\left(\dfrac{2d}{l}\right)^2\right]$ (approximately).

For additional formulas, see analytic geometry:

20. Helix (a skew curve)

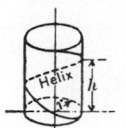

Let length of helix $= s$; radius of coil ($=$ radius of cylinder in figure) $= r$; distance advanced in one revolution $=$ pitch $= h$; and number of revolutions $= n$. Then:

$s = n\sqrt{(2\pi r)^2 + h^2}$.

Table 4 *(Continued)*

1b. Plane Curvilinear Figures (Continued)

21. Spiral of Archimedes

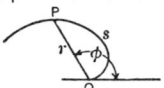

Let $a = \dfrac{r}{\phi}$. Then:

Arc $OP = s = \frac{1}{2}a[\phi\sqrt{1 + \phi^2} + \log_e(\phi + \sqrt{1 + \phi^2})]$.

For additional formulas, see analytic geometry.

22. Irregular figure

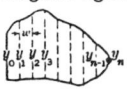

Divide the figure into an *even* number n of strips by means of $n + 1$ ordinates y_i spaced equal distances w. The area can then be determined approximately by any of the following approximate formulas, which are presented in the order of usual increasing approach to accuracy. In any of the first three cases, the greater the number of strips used, the more nearly accurate will be the result:

Trapezoidal rule

$A = w[\frac{1}{2}(y_0 + y_n) + y_1 + y_2 + \cdots + y_{n-1}]$;

Durand's rule

$A = w[0.4(y_0 + y_n) + 1.1(y_1 + y_{n-1}) + y_2 + y_3 + \cdots + y_{n-2}]$;

Simpson's rule (*n must* be even)

$A = \frac{1}{3}w[(y_0 + y_n) + 4(y_1 + y_3 + \cdots + y_{n-1})$
$\qquad + 2(y_2 + y_4 + \cdots + y_{n-2})]$;

Weddle's rule (for 6 strips only)

$A = \dfrac{3w}{10}[5(y_1 + y_5) + 6y_3 + y_0 + y_2 + y_4 + y_6]$.

Areas of irregular regions can often be determined more quickly by such methods as plotting on squared paper and counting the squares; graphical coordinate representation (see analytic geometry); or use of a planimeter.

1c. Solids Having Plane Surfaces

Notation. Lines, a, b, c,...; altitude (perpendicular height), h; slant height, s; perimeter of base, p_h or p_B; perimeter of a right section, p_r; area of base, A_b or A_B; area of a right section, A_r; total area of lateral surfaces, A_l; total area of all surfaces, A_t; volume, V.

23. Wedge (and right triangular prism)

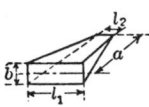

For wedge:

(Narrow-side rectangular); $V = \frac{1}{6}ab(2l_1 + l_2)$.

For right triangular prism (or wedge having parallel triangular bases perpendicular to sides): $l_2 = l_1 = l$:

$V = \frac{1}{2}abl$.

24. Rectangular prism (or rectangular parallelepiped) (and cube)

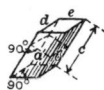

For rectangular prism or rectangular parallelepiped:

$A_l = 2c(a + b)$; $A_t = 2(de + ac + bc)$;

$V = A_r c = abc$.
For cube (letting $b = c = a$):

$A_t = 6a^2$; $V = a^3$; diagonal $= a\sqrt{3}$.

25. General prism

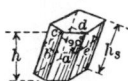

$A_l = hp_b = sp_r = s(a + b + \cdots + n)$;

$V = hA_b = sA_r$.

26. General truncated prism (and truncated triangular prism)

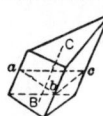

For general truncated prism:

$V = A_r \cdot$ (length of line BC joining centers of gravity of bases).

For truncated triangular prism:

$V = \frac{1}{3}A_r(a + b + c)$.

27. Prismatoid

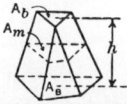

Let area of midsection $= A_m$.

$V = \frac{1}{6}h(A_B + A_b + 4A_m)$.

Table 4 *(Continued)*

1c. Solids Having Plane Surfaces (Continued)

28. Right regular pyramid (and prustum of right regular pyramid)

For right regular pyramid:

$$A_l = \tfrac{1}{2} s p_B; \quad V = \tfrac{1}{3} h A_B.$$

For prustum of right regular pyramid:

$$A_l = \tfrac{1}{2} s(p_B + p_b); \quad V = \tfrac{1}{3} h(A_B + A_b + \sqrt{A_B A_b}).$$

29. General pyramid (and prustum of pyramid)

For general pyramid:

$$V = \tfrac{1}{3} h A_B.$$

For prustum of general pyramid:

$$V = \tfrac{1}{3} h(A_B + A_b + \sqrt{A_B A_b}).$$

30. Regular polyhedrons

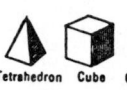

Tetrahedron   Cube   Octahedron

Dodecahedron   Icosahedron

Let edge $= a$, and radius of inscribed sphere $= r$. Then:

$$r = \frac{3V}{A_t} \text{ and:}$$

Number of Faces	Form of Faces	Total Area A_t	Volume V
4	Equilateral triangle	$1.7321a^2$	$0.1179a^3$
6	Square	$6.0000a^2$	$1.0000a^3$
8	Equilateral triangle	$3.4641a^2$	$0.4714a^3$
12	Regular pentagon	$20.6457a^2$	$7.6631a^3$
20	Equilateral triangle	$8.6603a^2$	$2.1817a^3$

(Factors shown only to four decimal places.)

1d. Solids Having Curved Surfaces

Notation. Lines, $a, b, c, \ldots$; altitude (perpendicular height), $h, h_1, \ldots$; slant height, s; radius, r; perimeter of base, p_b; perimeter of a right section, p_r; angle in radians, ϕ; arc, s; chord of segment, l; rise, h; area of base, A_b or A_B; area of a right section, A_r; total area of convex surface, A_l; total area of all surfaces, A_t; volume, V.

31. Right circular cylinder (and truncated right circular cylinder)

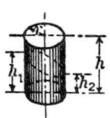

For right circular cylinder:

$$A_l = 2\pi r h; A_t = 2\pi r(r + h);$$

$$V = \pi r^2 h.$$

For truncated right circular cylinder:

$$A_l = \pi r(h_1 + h_2); \quad A_t = \pi r[h_1 + h_2 + r + \sqrt{r^2 + \tfrac{1}{2}(h_1 - h_2)^2}];$$

$$V = \tfrac{1}{2}\pi r^2(h_1 + h_2).$$

32. Ungula (wedge) of right circular cylinder

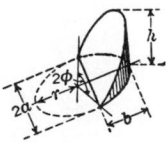

$$A_l = \frac{2rh}{b}[a + (b - r)\phi];$$

$$V = \frac{h}{3b}[a(3r^2 - a^2) + 3r^2(b - r)\phi]$$

$$= \frac{hr^3}{b}\left[\sin\phi - \frac{\sin^3\phi}{3} - \phi\cos\phi\right].$$

For semicircular base (letting $a = b = r$):

$$A_l = 2rh; \quad V = \frac{2r^2 h}{3}.$$

33. General cylinder

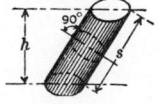

$$A_l = p_b h = p_r s;$$

$$V = A_b h = A_r s.$$

Table 4 *(Continued)*

1d. Solids Having Curved Surfaces (Continued)

34. Right circular cone (and frustum of right circular cone)

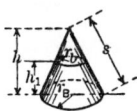

For right circular cone:

$$A_l = \pi r_B s = \pi r_B \sqrt{r_B^2 + h^2}; \; A_t = \pi r_B (r_B + s);$$

$$V = \tfrac{1}{3}\pi r_B^2 h.$$

For frustum of right circular cone:

$$s = \sqrt{h_1^2 + (r_B - r_b)^2}; \; A_l = \pi s (r_B + r_b);$$

$$V = \tfrac{1}{3}\pi h_1 (r_B^2 + r_b^2 + r_B r_b).$$

35. General cone (and frustum of general cone)

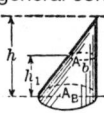

For general cone:

$$V = \tfrac{1}{3} A_B h.$$

For frustum of general cone:

$$V = \tfrac{1}{3} h_1 (A_B + A_b + \sqrt{A_B A_b}).$$

36. Sphere

Let diameter $= d$.

$$A_t = 4\pi r^2 = \pi d^2;$$

$$V = \frac{4\pi r^3}{3} = \frac{\pi d^3}{6}.$$

37. Spherical sector (and hemisphere)

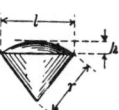

For spherical sector:

$$A_t = \frac{\pi r}{2}(4h + l); \; V = \frac{2\pi r^2 h}{3}.$$

For hemisphere (letting $h = \tfrac{1}{2}l = r$):

$$A_t = 3\pi r^2; V = \frac{2\pi r^3}{3}.$$

38. Spherical zone (and spherical segment)

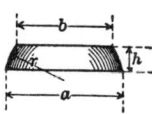

For spherical zone bounded by two planes:

$$A_l = 2\pi rh; A_t = \tfrac{1}{4}\pi(8rh + a^2 + b^2).$$

For spherical zone bounded by one plane ($b = 0$):

$$A_l = 2\pi rh = \tfrac{1}{4}\pi(4h^2 + a^2);$$

$$A_t = \tfrac{1}{4}\pi(8rh + a^2) = \tfrac{1}{2}\pi(2h2 + a^2).$$

For spherical segment with two bases:

$$V = \tfrac{1}{24}\pi h(3a^2 + 3b^2 + 4h^2).$$

For spherical segment with one base ($b = 0$):

$$V = \tfrac{1}{24}\pi h(3a^2 + 4h^2) = \pi h^2(r - \tfrac{1}{3}h).$$

39. Spherical polygon (and spherical triangle)

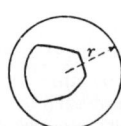

For spherical polygon:
Let sum of angles in radians $= \theta$ and number of sides $= n$.

$$A = [\theta - (n - 2)\pi]r^2$$

[The quantity $\theta - (n - 2)\pi$ is called "spherical excess."]

For spherical triangle ($n = 3$):

$$A = (\theta - \pi)r^2$$

For additional formulas, see trigonometry.

(Continues)

Table 4   *(Continued)*

1d. Solids Having Curved Surfaces (Continued)

40. Torus

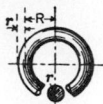

$$A_l = 4\pi^2 Rr;$$

$$V = 2\pi^2 Rr^2.$$

41. Ellipsoid (and spheroids)

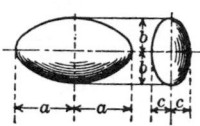

For ellipsoid:

$$V = \tfrac{4}{3}\pi abc.$$

For prolate spheroid:

Let $c = b$ and $\dfrac{\sqrt{a^2 - b^2}}{a} = e.$

$$A_t = 2\pi b^2 + 2\pi ab\,\frac{\sin^{-1} e}{e}; \quad V = \frac{4}{3}\pi ab^2.$$

For oblate spheroid:

Let $c = a$ and $\dfrac{\sqrt{a^2 - b^2}}{a} = e.$

$$A_t = 2\pi a^2 + \frac{\pi b^2}{e}\ln\left(\frac{1+e}{1-e}\right); \quad V = \frac{4}{3}\pi a^2 b.$$

42. Paraboloid of revolution

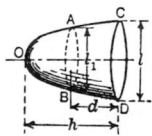

$$A_l \text{ of segment } DOC = \frac{2\pi l}{3h^2}\left[\left(\frac{l^2}{16} + h^2\right)^{3/2} - \left(\frac{l}{4}\right)^3\right].$$

For paraboloidal segment with two bases:

$$V \text{ of } ABCD = \frac{\pi d}{8}(l^2 + l_1^2).$$

For paraboloidal segment with one base ($l_1 = 0$ and $d = h$):

$$V \text{ of } DOC = \frac{\pi h l^2}{8}.$$

43. Hyperboloid of revolution

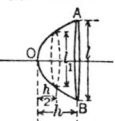

$$V \text{ of segment } AOB = \frac{\pi h}{24}(l^2 + 4l_1^2).$$

44. Surface and solid of revolution

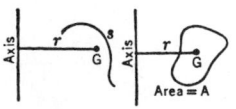

Let perpendicular distance from axis to center of gravity (G) of curve (or surface) $= r$. Curve (or surface) must not cross axis. Then,

Area of surface generated by curve revolving about axis:

$$A_l = 2\pi rs.$$

Volume of solid generated by surface revolving about axis:

$$V = 2\pi rA.$$

45. Irregular solid

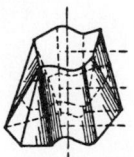

One of the following methods can often be employed to determine the volume of an irregular solid with a reasonable approach to accuracy:

(a) Divide the solid into prisms, cylinders, etc., and sum their individual volumes.

(b) Divide one surface into triangles after replacing curved lines by straight ones and curved surfaces by plane ones. Then multiply the area of each triangle by the mean depth of the section beneath it (which generally approximates the average of the depths at its corners). Sum the volumes thus obtained.

(c) If two surfaces are parallel, replace any curved lateral surfaces by plane surfaces best suited to the contour and then employ the prismatoidal formula.

cone included between its base and a section parallel to the base.

A *sphere* is a solid bounded by a surface all points of which are equidistant from a point within called the *center*. Every plane section of a sphere is a circle. This circle is a *great circle* if its plane passes through the center of the sphere; otherwise, it is a *small circle*. *Poles* of such a circle are the extremities of the diameter of the sphere that is perpendicular to the plane of the circle. Through two points on a spherical surface, not extremities of a diameter, one great circle can be passed. The shortest line that can be drawn on the surface of a sphere between two such points is an arc of a great circle less than a semicircumference joining those points. If two spherical surfaces intersect, their line of intersection is a circle whose plane is perpendicular to the line of centers and whose center lies on this line.

A *spherical sector* is the portion of a sphere generated by the revolution of a circular sector about a diameter of the circle of which the sector is a part. A *hemisphere* is half of a sphere.

A *spherical segment* is the portion of a sphere contained between two parallel plane sections (the bases), one of which may be tangent to the sphere (in which case there is only one base). The term "segment" also is applied in an analogous manner to various solids of revolution, the planes in such cases being perpendicular to an axis. A *zone* is the portion of a spherical surface included between two parallel planes.

A *spherical polygon* is a figure on a spherical surface bounded by three or more arcs of great circles. The sum of the angles of a spherical triangle (polygon of three sides) is greater than two right angles and less than six right angles.

Other solids appearing in the mensuration table (Table 4), if not sufficiently defined by their figures, may be found discussed in the section on analytic geometry.

5.2   Mensuration

Perimeters of similar figures are proportional to their respective linear dimensions, areas to the squares of their linear dimensions, and volumes of similar solids to the cubes of their linear dimensions (see Table 4).

5.3   Constructions

Lines

1.   *To draw a line parallel to a given line.*

CASE 1: At a given distance from the given line (Fig. 21). With the given distance as radius and with any centers *m* and *n* on the given line *AB*, describe arcs *xy* and *zw*, respectively. Draw *CD* touching these arcs. *CD* is the required parallel line.

CASE 2: Through a given point (Fig. 22). Let *C* be the given point and *D* be any point on the given line *AB*. Draw *CD*. With equal radii draw arcs *bf* and *ce*

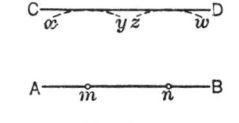

Fig. 21

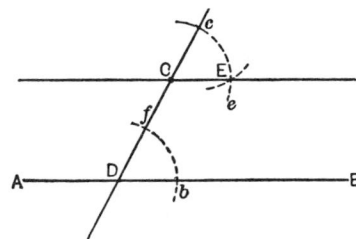

Fig. 22

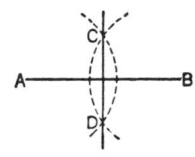

Fig. 23

with *D* and *C*, respectively, as centers. With radius equal to chord *bf* and with *c* as center draw an arc cutting arc *ce* at *E*. *CE* is the required parallel line.

2.   *To bisect a given line* (Fig. 23). Let *AB* be the given line. With any radius greater than 0.5 *AB* describe two arcs with *A* and *B* as centers. The line *CD*, through points of intersection of the arcs, is the perpendicular bisector of the given line.

3.   *To divide a given line into a given number of equal parts* (Fig. 24). Let *AB* be the given line and let the number of equal parts be five. Draw line *AC* at any convenient angle with *AB*, and step off with dividers five equal lengths from *A* to *b*. Connect *b* with *B*, and draw parallels to *Bb* through the other points in *AC*. The intersections of these parallels with *AB* determine the required equal parts on the given line.

4.   *To divide a given line into segments proportional to a number of given unequal parts.* Follow the same procedure as under item 3 except make the

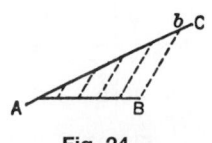

Fig. 24

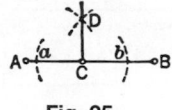

Fig. 25

lengths on *AC* equal to (or proportional to) the lengths of the given unequal parts.

5. *To erect a perpendicular to a given line at a given point in the line.*

CASE 1: Point *C* is at or near the middle of the line *AB* (Fig. 25). With *C* as center, describe arcs of equal radii intersecting *AB* at *a* and *b*. With *a* and *b* as centers, and any radius greater than *Ca*, describe arcs intersecting at *D*. *CD* is the required perpendicular.

CASE 2: Point *C* is at or near the extremity of the line *AB* (Fig. 26). With any point *O* as center and radius *OC*, describe an arc intersecting *AB* at *a*. Extend *aO* to intersect the arc at *D*. *CD* is the required perpendicular.

6. *To erect a perpendicular to a given line through a given point outside the line.*

CASE 1: Point *C* is opposite, or nearly opposite, the middle of the line *AB* (Fig. 27). With *C* as center, describe an arc intersecting *AB* at *a* and *b*. With *a* and *b* as centers, describe arcs of equal radii intersecting at *D*. *CD* is the required perpendicular.

CASE 2: Point *C* is opposite, or nearly opposite, the extremity of the line *AB* (Fig. 28). Through *C*, draw any line intersecting *AB* at *a*. Divide line *Ca* into two equal parts, *ab* and *bC* (method given previously). With *b* as center and radius *bC*, describe an arc intersecting *AB* at *D*. *CD* is the required perpendicular.

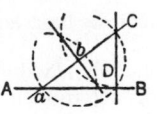

Fig. 28

Angles

7. *To bisect a given angle.*

CASE 1: Vertex *B* is accessible (Fig. 29). Let *ABC* be the given angle. With *B* as center and a large radius, describe an arc intersecting *AB* and *BC* at *a* and *c*, respectively. With *a* and *c* as centers, describe arcs of equal radii intersecting at *D*. *DB* is the required bisector.

CASE 2: The vertex is inaccessible (Fig. 30). Let the given angle be that between lines *AB* and *BC*. Draw lines *ab* and *bc* parallel to the given lines, and at equal distances from them, intersecting at *b*. Construct *Db* bisecting angle *abc* (method given previously). *Db* is the required bisector.

8. *To construct an angle equal to a given angle if one new side and the new vertex are given* (Fig. 31). Let *ABC* be the given angle, *DE* the new side, and *E* the new vertex. With center *B* and a convenient radius, describe arc *ac*. With the same radius and center *E*, draw arc *df*. With radius equal to chord *ac* and with center *d* draw an arc cutting the arc *df* at *F*. Draw *EF*. Then *DEF* is the required angle.

9. *To construct angles of 60° and 30°* (Fig. 32). About any point *A* on a line *AB*, describe with a

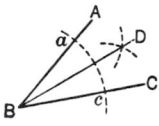

Fig. 29

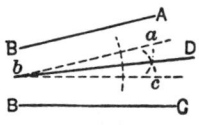

Fig. 30

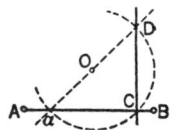

Fig. 26

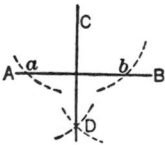

Fig. 27

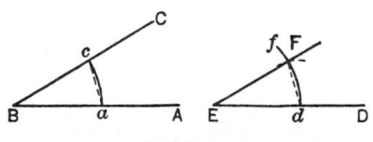

Fig. 31

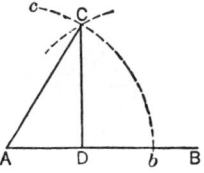

Fig. 32

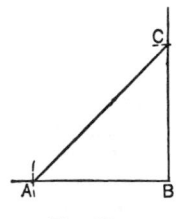

Fig. 33

convenient radius the arc *bc*. From *b*, using an equal radius, describe an arc cutting the former one at *C*. Draw *AC*, and drop a perpendicular *CD* from *C* to line *AB*. Then *CAD* is a 60° angle and *ACD* is a 30° angle.

10. *To construct an angle of 45°* (Fig. 33). Set off any distance *AB*; draw *BC* perpendicular and equal to *AB* and join *CA*. Angles *CAB* and *ACB* are each 45°.

11. *To draw a line making a given angle with a given line* (Fig. 34). Let *AB* be the given line. With *A* as the center and with as large a radius as convenient, describe arc *bc*. Determine from Table 12 in Chapter 1, the length of chord to radius 1, corresponding to the given angle. Multiply this chord by the length of *Ab*, and with the product as a new radius and *b* as a center, describe an arc cutting *bc* at *C*. Draw *AC*. This line makes the required angle with *AB*.

Circles

12. *To describe through two given points an arc of a circle having a given radius* (Fig. 35). Let *A* and *B*

be the given points. With the given radius and these points as centers, describe arcs cutting each other at *C*. From *C*, with the same radius, describe arc *AB*, which is the required arc.

13. *To bisect a given arc of a circle.* Draw the perpendicular bisector of the chord of the arc. The point in which this bisector meets the arc is the required midpoint.

14. *To locate the center of a given circle or circular arc* (Fig. 36). Select three points *A, B, C* on the circle (or arc) located well apart. Draw chords *AB* and *BC* and erect their perpendicular bisectors. The point *O*, where the bisectors intersect, is the required center.

15. *To draw a circle through three given points not in the same straight line.*

CASE 1: Radius small and center accessible (Fig. 36). Let *A, B, C* be the given points. Draw lines *AB* and *BC* and erect their perpendicular bisectors. From point *O*, where the bisectors intersect, describe a circle of radius *OA* that is the required circle.

CASE 2: Radius very long or center inaccessible (Fig. 37). Let *A, O, A'* be the given points (*O* not necessarily midpoint of *AOA'*). Draw arcs *Aa* and *A'a* with centers at *A'* and *A*, respectively; extend *AO* to determine *a* and *A'O* to determine *a'*; point off from *a* on *aA'* equal parts *ab, bc*, and so on; lay off *a'b', b'c'*, and so on, equal to *ab*; join *A* with any point as *b* and *A'* with the corresponding point *b'*; the intersection *P* of these joining lines is a point on the required circle.

16. *To lay out a circular arc without locating the center of the circle, given the chord and the rise* (Fig. 37). Let *AA'* be the chord and *QO* the rise. (In this case, *O* is the midpoint of *AOA'*.) The arc can be

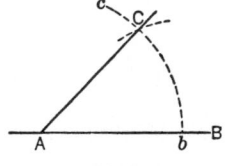

Fig. 34

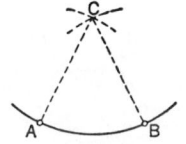

Fig. 35

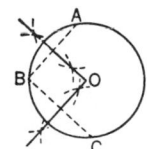

Fig. 36

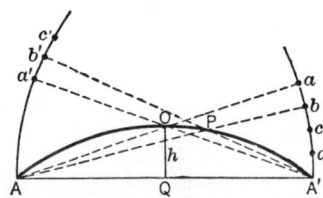

Fig. 37

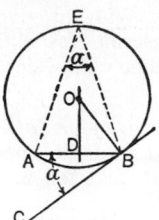

Fig. 38

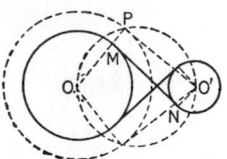

Fig. 41

constructed through the points *A, O, A'*, as under item 15, Case 2.

17. *To construct, upon a given chord, a circle in which a given angle can be inscribed* (Fig. 38). Let *AB* be the given chord and α the given angle. Construct angle *ABC* equal to angle α. Bisect line *AB* by the perpendicular at *D*. Draw a perpendicular to *BC* from point *B*. With *O*, the point of intersection of the perpendiculars, as center and *OB* as radius, describe a circle. The angle *AEB*, with vertex *E* located anywhere on the arc *AEB*, equals α, and therefore the circle just drawn is the one required.

18. *To draw a tangent to a given circle through a given point.*

CASE 1: Point *A* is on the circle (Fig. 39). Draw radius *OA*. Through *A*, perpendicular to *OA*, draw *BAC*, the required tangent.

CASE 2: Point *A* is outside the circle (Fig. 40). Two tangents can be drawn. Join *O* and *A*. Bisect *OA* at *D*, and with *D* as center and *DO* as radius, describe an arc intersecting the given circle at *B* and *C*. *BA* and *CA* are the required tangents.

19. *To draw a common tangent to two given circles.* Let the circles have centers *O* and *O'* and corresponding radii *r* and *r'* (*r* > *r'*).

CASE 1: Common internal tangents (when circles do not intersect) (Fig. 41). Construct a circle having the same center *O* as the larger circle and a radius equal to the sum of the radii of the given circles (*r* + *r'*). Construct a tangent *O'P* from center *O'* of the smaller circle to this circle. Construct *O'N* perpendicular to this tangent. Draw *OP*. The line *MN* joining the extremities of the radii *OM* and *O'N* is a common tangent. The figure shows two such common internal tangents.

CASE 2: Common external tangents (Fig. 42). Construct a circle having the same center *O* as the larger circle and radius equal to the difference of the radii (*r* − *r'*). Construct a tangent to this circle from the center of the smaller circle. The line joining the extremities *M, N* of the radii of the given circles perpendicular to this tangent is a required common tangent. There are two such tangents.

20. *To draw a circle with a given radius that will be tangent to two given circles.* (Fig. 43). Let *r* be the given radius and *A* and *B* the given circles. About center of circle *A* with radius equal to *r* plus radius of *A*, and about center of *B* with radius equal to *r* plus radius of *B*, draw two arcs cutting each other in *C*, which is the center of the required circle.

21. *To describe a circular arc touching two given circles, one of them at a given point.* (Fig. 44). Let *AB*, *FG* be the given circles and *F* the given point. Draw the radius *EF*, and produce it both ways. Set off *FH*

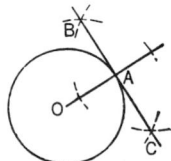

Fig. 39

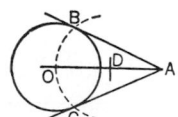

Fig. 40

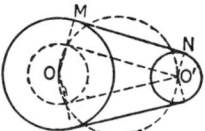

Fig. 42

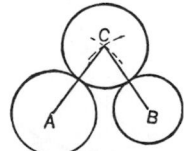

Fig. 43

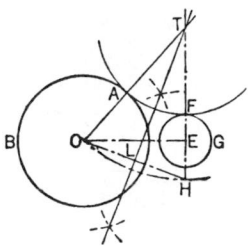

Fig. 44

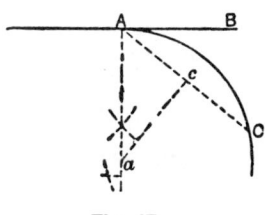

Fig. 47

equal to the radius *AC* of the other circle; join *CH*, and bisect it by the perpendicular *LT*, cutting *EF* at *T*. About center *T*, with radius *TF*, describe arc *FA* as required.

22. *To draw a circular arc that will be tangent to two given lines inclined to one another, one tangential point being given* (Fig. 45). Let *AB* and *CD* be the given lines and *E* the given point. Draw the line *GH*, bisecting the angle formed by *AB* and *CD*. From *E* draw *EF* at right angles to *AB*; then *F*, its intersection with *GH*, is the center of the required circular arc.

23. *To connect two given parallel lines by a reversed curve composed of two circular arcs of equal radius, the curve being tangent to the lines at given points* (Fig. 46). Let *AD* and *BE* be the given lines and *A* and *B* the given points. Join *A* and *B*, and bisect the connecting line at *C*. Bisect *CA* and *CB* by perpendiculars. At *A* and *B* erect perpendiculars to the given lines, and the intersections *a* and *b* are the centers of the arcs composing the required curve.

24. *To describe a circular arc that will be tangent to a given line at a given point and pass through another given point outside the line* (Fig. 47). Let *AB* be the given line, *A* the given point on the line, and *C* the given point outside it. Draw from

A a line perpendicular to the given line. Connect *A* and *C* by a straight line, and bisect this line by the perpendicular *ca*. The point *a* where these two perpendiculars intersect is the center of the required circular arc.

25. *To draw a circular arc joining two given relatively inclined lines, tangent to the lines, and passing through a given point on the line bisecting their included angle* (Fig. 48). Let *AB* and *DE* be the given lines and *F* the given point on the line *FC* that bisects their included angle. Through *F* draw *DA* at right angles to *FC*; bisect the angles *A* and *D* by lines intersecting at *C*, and about *C* as a center, with radius *CF*, draw the arc *HFG* required.

26. *To draw a series of circles between two given relatively inclined lines touching the lines and touching each other* (Fig. 49). Let *AB* and *CD* be the given lines. Bisect their included angle by the line *NO*. From a point *P* in this line draw the perpendicular *PB* to the line *AB*, and on *P* describe the circle *BD*, touching the given lines and cutting the center line at *E*. From *E* draw *EF* perpendicular to the center line, cutting *AB* at *F*; and about *F* as a center describe an arc *EG*, cutting *AB* at *G*. Draw *GH* parallel to *BP*, giving *H*, the center of the next circle, to be described with the radius *HE*; and so on for the next circle *IN*.

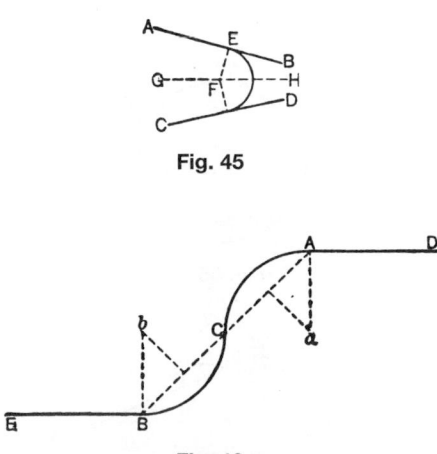

Fig. 45

Fig. 46

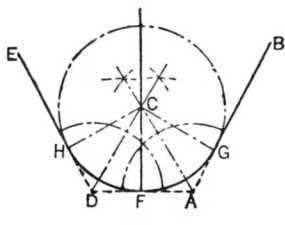

Fig. 48

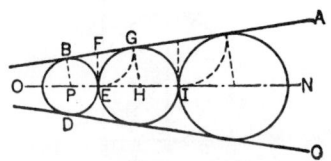

Fig. 49

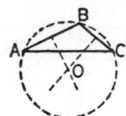

Fig. 50

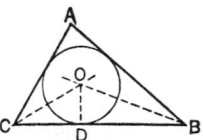

Fig. 51

27. *To circumscribe a circle about a given triangle* (Fig. 50). Construct perpendicular bisectors of two sides. Their point of intersection *O* is the center (circumcenter) of the required circle.

28. *To inscribe a circle in a given triangle* (Fig. 51). Draw bisectors of two angles intersecting in *O* (incenter). From *O* draw *OD* perpendicular to *BC*. Then the circle with center *O* and radius *OD* is the required circle.

29. *To circumscribe a circle about a given square* (Fig. 52). Let *ACBD* be the given square. Draw diagonals *AB* and *CD* of the square intersecting at *E*. On center *E*, with radius *AE*, describe the required circle. The same procedure can be used for circumscribing a circle about a given rectangle.

30. *To inscribe a circle in a given square* (Fig. 53). Let *ACBD* be the given square. Draw diagonals *AB* and *CD* of the square intersecting at *E*. Drop a perpendicular *EF* from *E* to one side. On center *E*, with radius *EF*, describe the required circle.

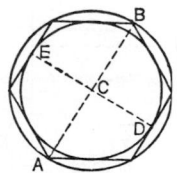

Fig. 54

31. *To circumscribe a circle about a given regular polygon.*

CASE 1: The polygon has an even number of sides (Fig. 54). Draw a diagonal *AB* joining two opposite vertices. Bisect the diagonal by a perpendicular line *DE*, which is another diagonal or a line bisecting two opposite sides, depending on whether the number of sides is or is not divisible by 4. With the midpoint *C* as the center and radius *CA*, describe the required circle.

CASE 2: The polygon has an odd number of sides (Fig. 55). Bisect two of the sides at *D* and *E* by the perpendicular lines *DB* and *EA* which pass through the respective opposite vertices and intersect at a point *C*. With *C* as the center and radius *CA*, describe the required circle.

32. *To inscribe a circle in a given regular polygon* (Figs. 54 and 55). Locate the center *C* as in item 31. With *C* as center and radius *CD*, describe the required circle.

Polygons

33. *To construct a triangle on a given base, the lengths of the sides being given* (Fig. 56). Let *AB* be the given base and *a*, *b* the given lengths of sides. With

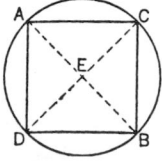

Fig. 52

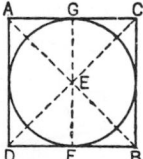

Fig. 53

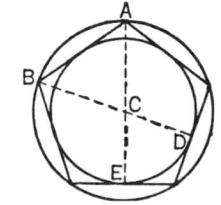

Fig. 55

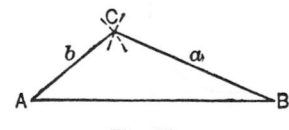

Fig. 56

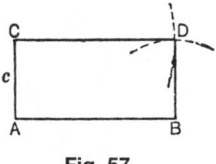

Fig. 57

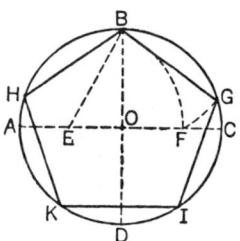

Fig. 60

A and *B* as centers and *b* and *a* as respective radii, describe arcs intersecting at *C*. Draw *AC* and *BC* to complete the required triangle.

34. *To construct a rectangle of given base and given height* (Fig. 57). Let *AB* be the base and *c* the height. Erect the perpendicular *AC* equal to *c*. With *C* and *B* as centers and *AB* and *c* as respective radii, describe arcs intersecting at *D*. Draw *BD* and *CD* to complete the required rectangle.

35. *To construct a square with a given diagonal* (Fig. 58). Let *AC* be the given diagonal. Draw a circle on *AC* as diameter and erect the diameter *BD* perpendicular to *AC*. Then *ABCD* is the required square.

36. *To inscribe a square in a given circle* (Fig. 58). Draw perpendicular diameters *AC* and *BD*. Their extremities are the vertices of an inscribed square.

37. *To circumscribe a square about a given circle* (Fig. 59). Draw perpendicular diameters *AC* and *BD*. With *A, B, C, D* as centers and the radius of the circle as radius, describe the four semicircular arcs shown. Their outer intersections are the vertices of the required square.

38. *To inscribe a regular pentagon in a given circle* (Fig. 60). Draw perpendicular diameters *AC* and *BD* intersecting at *O*. Bisect *AO* at *E* and, with *E* as center

and *EB* as radius, draw an arc cutting *AC* at *F*. With *B* as center and *BF* as radius, draw an arc cutting the circle at *G* and *H*; also with the same radius, step around the circle to *I* and *K*. Join the points thus found to form the pentagon.

39. *To inscribe a regular hexagon in a given circle* (Fig. 61). Step around the circle with compasses set to the radius and join consecutive divisions thus marked off.

40. *To circumscribe a regular hexagon about a given circle* (Fig. 62). Draw a diameter *ADB* and, with center *A* and radius *AD*, describe an arc cutting the circle at *C*. Draw *AC* and bisect it with the radius *DE*. Through *E*, draw *FG* parallel to *AC*, cutting diameter *AB* extended at *F*. With center *D* and radius *DF*, describe the circumscribing circle *FH*; within this circle inscribe a regular hexagon as under item 39. This hexagon circumscribes the given circle, as required.

41. *To construct a regular hexagon having a side of given length* (Fig. 61). Draw a circle with radius equal to the given length of side and inscribe a regular hexagon (see item 39).

42. *To construct a regular octagon having a side of given length* (Fig. 63). Let *AB* be the given side. Produce *AB* in both directions, and draw perpendiculars

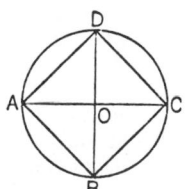

Fig. 58

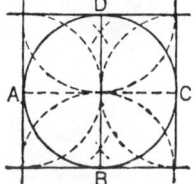

Fig. 59

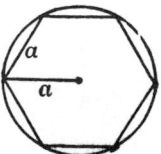

Fig. 61

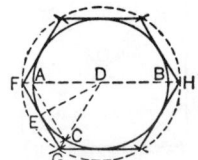

Fig. 62

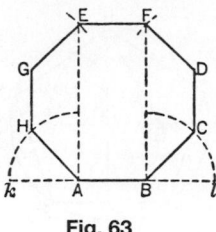

Fig. 63

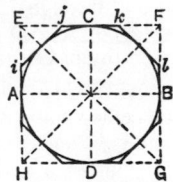

Fig. 66

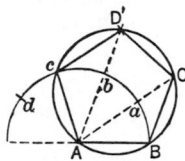

Fig. 67

AE and *BF*. Bisect the external angles at *A* and *B* by the lines *AH* and *BC* making them equal to *AB*. Draw *CD* and *HG* parallel to *AE* and equal to *AB*; from the centers *G, D*, with the radius *AB*, draw arcs cutting the perpendiculars at *E, F*, and draw *EF* to complete the octagon.

43. *To inscribe a regular octagon in a given circle* (Fig. 64). Draw perpendicular diameters *AC* and *BD*. Bisect arcs *AB, BC*, ... and join *Ae, eB*, ... to form the octagon.

44. *To inscribe a regular octagon in a given square* (Fig. 65). Draw diagonals of the given square intersecting at *O*. With *A, B, C, D* as centers and *AO* as radius, describe arcs cutting the sides of the square at *gn, fk, hm*, and *ol*. Join the points thus found to form the octagon.

45. *To circumscribe a regular octagon about a given circle* (Fig. 66). Describe a square about the given circle. Draw perpendiculars *ij, kl*, and so on, to the diagonals of the squares, touching the circle. Then *ij, jk, kl*, and so on, form the octagon.

46. *To describe a regular polygon of any given number of sides when one side is given* (Fig. 67). Let

AB be the given side and let the number of sides be five. Produce the line *AB*, and with *A* as center and *AB* as radius, describe a semicircle. Divide this into as many equal parts as there are to be sides of the polygon—in this case, five. Draw lines from *A* through the division points *a, b*, and *c* (omitting the last). With *B* and *c* as centers and *AB* as radius, cut *Aa* at *C* and *Ab* at *D*. Draw *cD, DC*, and *CB* to complete the polygon.

47. *To inscribe a regular polygon of a given number of sides in a given circle*. Determine the central angle subtended by any side by dividing 360° by the number of sides. Lay off this angle successively round the center of the circle by means of a protractor. The radii thus drawn intersect the circle at vertices of the required polygon.

Ellipse An *ellipse* is a curve for which the sum of the distances of any point on it from two fixed points (the *foci*) is constant.

48. *To describe an ellipse for which the axes are given* (Fig. 68). Let *AB* be the *major* and *RS* the *minor* axis (*AB* > *RS*). With *O* as center and *OB* and *OR* as radii, describe circles. From *O* draw any radial line intersecting the circles at *M* and *N*. Through *M* draw

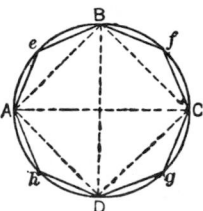

Fig. 64

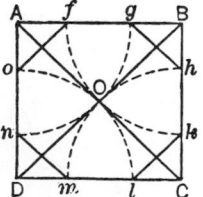

Fig. 65

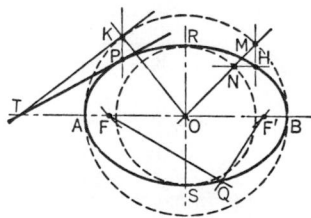

Fig. 68

a line parallel to *OR* and through *N* a line parallel to *OB*. These lines intersect at *H*, a point on the ellipse. Repeat the construction to obtain other points.

49. *To locate the foci of an ellipse, given the axes* (Fig. 68). With *R* as center and radius equal to *AO*, describe arcs intersecting *AB* at *F* and *F'*, the required foci.

50. *To describe an ellipse mechanically, given an axis and the foci* (Fig. 68). A cord of length equal to the major axis is pinned or fixed at its ends to the foci *F* and *F'*. With a pencil inside the loop, keeping the cord taut so as to guide the pencil point, trace the outline of the ellipse (*Q* represents the pencil point and length *FQF'* the cord). If the minor axis *RS* is given rather than the major axis *AB*, the length *AB* (for the cord) is readily determined as *FR* + *RF'*.

51. *To draw a tangent to a given ellipse through a given point.*

CASE 1: Point *P* is on the curve (Fig. 68). With *O* as center and *OB* as radius, describe a circle. Through *P* draw a line parallel to *OR* intersecting the circle at *K*. Through *K* draw a tangent to the circle intersecting the major axis at *T*. *PT* is the required tangent.

CASE 2: Point *P* is not on the curve (Fig. 69). With *P* as center and radius *PF'*, describe an arc. With *F* as center and radius *AB*, describe an arc intersecting the first arc at *M* and *N*. Draw *FM* and *FN* intersecting the ellipse at *E* and *G*. *PE* and *PG* are the required tangents.

52. *To describe an ellipse approximately by means of circular arcs of three radii* (Fig. 70). On the major

axis *AB* draw the rectangle *BG* of altitude equal to half the minor axis, *OC*; to the diagonal *AC* draw the perpendicular *GHD*; set off *OK* equal to *OC*, and describe a semicircle on *AK*; produce *OC* to *L*; set off *OM* equal to *CL*, and from *D* describe an arc with radius *DM*; from *A*, with radius *OL*, draw an arc cutting *AB* at *N*; from *H*, with radius *HN*, draw an arc cutting arc *ab* at *a*. Thus the five centers *H*, *a*, *D*, *b*, *H'* are found, from which the arcs *AR*, *RP*, *PQ*, *QS*, *SB* are described. The part of the ellipse below axis *AB* can be constructed in like manner.

Parabola A *parabola* is a curve for which the distance of any point on it from a fixed line (the *directrix*) is equal to its distance from a fixed point (the *focus*). For a general discussion of its properties, see the section on analytic geometry.

53. *To describe a parabole for which the vertex, the axis, and a point of the curve are given* (Fig. 71). Let *A* be the given vertex, *AB* the given axis, and *M* the given point. Construct the rectangle *ABMC*. Divide *MC* and *CA* into the same number of equal parts (say four), numbering the divisions consecutively in the manner shown. Connect *A*1, *A*2, and *A*3. Through 1', 2', 3', draw parallels to the axis *AB*. The intersections I, II, and III of these lines are points on the required curve. A similar construction below the axis will give the other symmetric branch of the curve.

54. *To locate the focus and directrix of a parabola, given the vertex, the axis, and a point of the curve* (Fig. 71). Let *A* be the given vertex, *AB* the given axis, and *M* the given point. Drop the perpendicular *MB* from *M* to *AB*. Bisect it at *E* and draw *AE*. Draw *ED* perpendicular to *AE* at *E* and intersecting the axis at *D*. With *A* as center and *BD* as radius, describe arcs cutting the axis at *F* and *J*. Then *F* is the focus and the line *GH*, perpendicular to the axis through *J*, is the directrix.

55. *To describe a parabola mechanically, given the focus and directrix* (Fig. 72). Let *F* be the given focus and *EN* the given directrix. Place a straight edge to the directrix *EN*, and apply to it a square, *LEG*. Fasten to the end *G* one end of a cord equal in length to the edge *EG*, and attach the other end to the focus *F*; slide the square along the straight edge, holding the cord taut

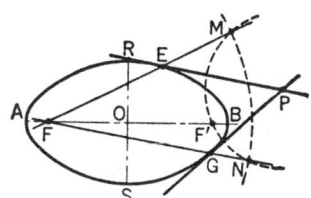

Fig. 69

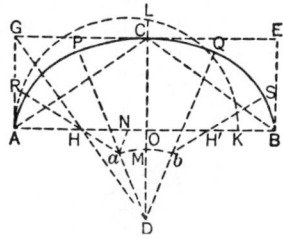

Fig. 70

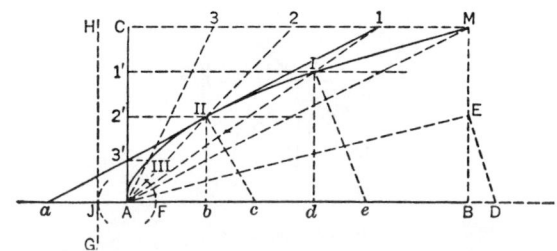

Fig. 71

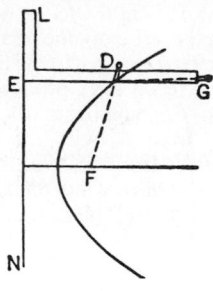

Fig. 72

against the edge of the square by a pencil D, by which the parabolic curve is described.

56. *To draw a tangent to a given parabola through a given point.*

CASE 1: The point is on the curve (Fig. 71). Let II be the given point. Drop a perpendicular from II to the axis, cutting it at b. Make Aa equal to Ab. Then a line through a and II is the required tangent. The line II c perpendicular to the tangent at II is the *normal* at that point; bc is the *subnormal*. All subnormals of a given parabola are equal to the distance from the directrix to the focus and hence equal to each other. Thus the subnormal at I is de equal to bc, where d is the foot of the perpendicular dropped from I. The tangent at I can be drawn as a perpendicular to Ie through I.

CASE 2: The point is off the curve (on the convex side) (Fig. 73). Let P be the given point and F the focus of the parabola. With P as center and PF as radius, draw arcs intersecting the directrix at B and D. Through B and D draw lines parallel to the axis intersecting the parabola at E and H. PE and PH are the required tangents.

Hyperbola   A *hyperbola* is a curve for which the difference of the distances of any point on it from two fixed points (the foci) is constant. It has two distinct branches.

57. *To describe a hyperbola for which the foci and the difference of the focal radii are given* (Fig. 74). Let

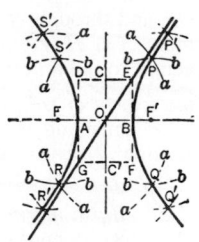

Fig. 74

F and F' be the given foci and AOB the given difference of the focal radii. Lay out AOB (the transverse axis) so that $AF = F'B$ and $AO = OB$. A and B are points on the required curve. With centers F and F' and any radius greater than FB or $F'A$, describe arcs aa. With the same centers and radius equal to the difference between the first radius and the transverse axis AOB, describe arcs bb, intersecting arcs aa at P, Q, R, and S, points on the required curve. Repeat the construction for additional points.

Make $BC = BC' = OF = OF'$, and construct the rectangle $DEFG$; CC' is the conjugate axis. The diagonals DF and EG, produced, are called *asymptotes*. The hyperbola is tangent to its asymptotes at infinity.

58. *To locate the foci of a hyperbola, given the axes* (Fig. 74). With O as center and radius equal to BC, describe arcs intersecting AB extended at F and F', the required foci.

59. *To describe a hyperbola mechanically, having given the foci and the difference of the focal radii* (Fig. 75). Let F and F' be the given foci and AB the given difference of focal radii. Using a ruler longer than the distance $F'F$, fasten one of its extremities at the focus F'. At the other extremity H attach a cord of such a length that the length of the ruler exceeds the length of the cord by the given distance AB. Attach the other extremity of the cord at the focus F. Press a pencil P against the ruler, and keep the cord constantly taut while the ruler is turned around F' as a center. The point of the pencil will describe one branch of the curve, and the other can be obtained in like manner.

60. *To draw a tangent to a given hyperbola through a given point.*

CASE 1: Point P is on the curve (Fig. 76). Draw lines connecting P with the foci. Bisect the angle $F'PF$. The bisecting line TP is the required tangent.

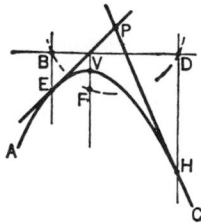

Fig. 73

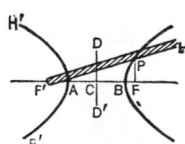

Fig. 75

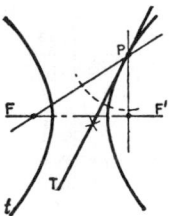

Fig. 76

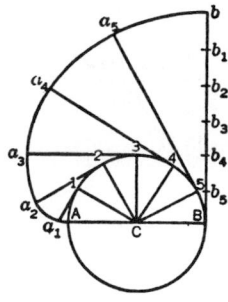

Fig. 79

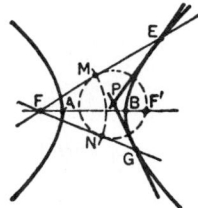

Fig. 77

CASE 2: Point P is off the curve on the convex side (Fig. 77). With P as center and radius PF', describe an arc. With F as center and radius AB, describe an arc intersecting the first arc at M and N. Produce lines FM and FN to intersect the curve at E and G. PE and PG are the required tangents.

Cycloid   A *cycloid* is a curve generated by a point on a circle rolling on a straight line.

61. *To describe a cycloid for which the generating circle is given* (Fig. 78). Let A be the generating point. Divide the circumference of the generating circle into an even number of equal arcs, as $A1, 1\text{–}2, \ldots$, and set off the rectified arcs on the base. Through the points $1, 2, 3, \ldots$ on the circle, draw horizontal lines, and on them set off distances $1a = A1$, $2b = A2$, $3c = A3, \ldots$. The points $A, a, b, c, \ldots$ are points of the cycloid.

An *epicycloid* is a curve generated by a point on one circle rolling on the *outside* of another circle. A *hypocycloid* is a curve generated by the point if the generating circle rolls on the *inside* of the second circle.

Involute of a Circle   An *involute of a circle* is a curve generated by the free end of a taut string as it is unwound from a circle.

62. *To describe an involute of a given circle* (Fig. 79). Let AB be the given circle. Through B draw Bb perpendicular to AB. Make Bb equal in length to half the circumference of the circle. Divide Bb and the semicircumference into the same number of equal parts, say six. From each point of division $1, 2, 3, \ldots$ of the circumference, draw lines to the center C of the circle. Then draw $1a_1$ perpendicular to $C1$, $2a_2$ perpendicular to $C2$, and so on. Make $1a_1$ equal to bb_1; $2a_2$ equal to bb_2; $3a_3$ equal to bb_3; and so on. Join the points A, a_1, a_2, a_3, etc., by a curve; this curve is the required involute.

6   TRIGONOMETRY

6.1   Circular Functions of Plane Angles

Definitions and Values

Trigonometric Functions.   The angle α in Fig. 80 is measured in degrees or radians, as defined in Section 5.1. The ratio of any two of the quantities x, y, or r determines the extent of the opening between the lines OP and OX. Since these ratios are functions of the angle, they may be used to measure or construct it.

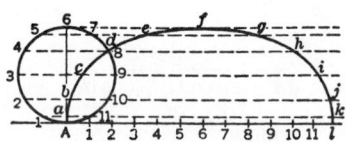

Fig. 78

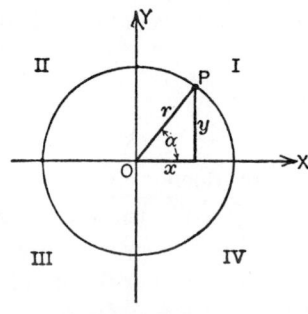

Fig. 80

The definitions and terms used to designate the functions are as follows:

$$\text{Sine } \alpha = \frac{y}{r} = \sin \alpha$$

$$\text{Cosine } \alpha = \frac{x}{r} = \cos \alpha$$

$$\text{Tangent } \alpha = \frac{y}{x} = \tan \alpha$$

$$\text{Cotangent } \alpha = \frac{x}{y} = \cot \alpha$$

$$\text{Secant } \alpha = \frac{r}{x} = \sec \alpha$$

$$\text{Cosecant } \alpha = \frac{r}{y} = \csc \alpha$$

$$\text{Versine } \alpha = \frac{r - x}{r} = \text{vers } \alpha = 1 - \cos \alpha$$

$$\text{Coversine } \alpha = \frac{r - y}{r} = \text{covers } \alpha = 1 - \sin \alpha$$

$$\text{Haversine } \alpha = \frac{r - x}{2r} = \text{hav } \alpha = \tfrac{1}{2}\text{vers } \alpha$$

Values of Trigonometric Functions. An angle α (Fig. 80), if measured in a *counterclockwise* direction, is said to be *positive*; if measured *clockwise, negative*. Following the convention that x is positive if measured along OX to the right of the OY axis and negative if measured to the left, and similarly, y is positive if measured along OY above the OX axis and negative if measured below, the signs of the trigonometric functions are different for angles in the quadrants I, II, III, and IV (Table 5).

Values of trigonometric functions are *periodic*, the period of the sin, cos, sec, csc being 2π radians, and that of the tan and cot, π radians (Tables 6–8). For example, in Fig. 81 (n an integer)

$$\sin(\alpha + 2\pi n) = \sin \alpha$$

$$\tan(\alpha + \pi n) = \tan \alpha$$

Inverse, or Antifunctions. The symbol $\sin^{-1} x$ means the angle whose sine is x and is read inverse sine of x, antisine of x, or arc sine x. Similarly for $\cos^{-1} x, \tan^{-1} x, \cot^{-1} x, \sec^{-1} x, \csc^{-1} x, \text{vers}^{-1}x$, the last meaning an angle α such that $1 - \cos \alpha = x$.

Table 5   Signs of Trigonometric Functions

Quadrant	sin	cos	tan	cot	sec	csc
I	+	+	+	+	+	+
II	+	−	−	−	−	+
III	−	−	+	+	−	−
IV	−	+	−	−	+	−

Table 6   Functions of Angles in Any Quadrant in Terms of Angles in First Quadrant

	$-\alpha$	$90° \pm \alpha$	$180° \pm \alpha$	$270° \pm \alpha$	$360° \pm \alpha$
sin	$-\sin\alpha$	$+\cos\alpha$	$\mp\sin\alpha$	$-\cos\alpha$	$\pm\sin\alpha$
cos	$+\cos\alpha$	$\mp\sin\alpha$	$-\cos\alpha$	$\pm\sin\alpha$	$+\cos\alpha$
tan	$-\tan\alpha$	$\mp\cot\alpha$	$\pm\tan\alpha$	$\mp\cot\alpha$	$\pm\tan\alpha$
cot	$-\cot\alpha$	$\mp\tan\alpha$	$\pm\cot\alpha$	$\mp\tan\alpha$	$\pm\cot\alpha$
sec	$+\sec\alpha$	$\mp\csc\alpha$	$-\sec\alpha$	$\pm\csc\alpha$	$+\sec\alpha$
csc	$-\csc\alpha$	$+\sec\alpha$	$\mp\csc\alpha$	$-\sec\alpha$	$\pm\csc\alpha$

While the direct functions (e.g., sine) are single valued, the indirect are many valued; thus $\sin 30° = 0.5$, but $\sin^{-1} 0.5 = 30°, 150°, \ldots$.

Functional Relations Identities

Functions of the Sum and Difference of Two Angles

$$\sin(\alpha \pm \beta) = \sin\alpha \cos\beta \pm \cos\alpha \sin\beta$$

$$\cos(\alpha \pm \beta) = \cos\alpha \cos\beta \mp \sin\alpha \sin\beta$$

$$\tan(\alpha \pm \beta) = 1 \mp \tan\alpha \tan\beta / \tan\alpha \pm \tan\beta$$

$$\cot(\alpha \pm \beta) = \cot\beta \pm \cot\alpha / \cot\beta \cot\alpha \mp 1$$

If x is small, say $3°$ or $4°$, then the following are close approximations, in which the quantity x is to be expressed in radians ($1° = 0.01745$ rad):

$$\sin\alpha \approx \alpha \qquad \cos\alpha \approx 1 \qquad \tan\alpha \approx \alpha$$

$$\sin(\alpha \pm x) \approx \sin\alpha \pm x \cos\alpha$$

$$\cos(\alpha \pm x) \approx \cos\alpha \mp x \sin\alpha$$

Functions of Half-Angles

$$\sin\tfrac{1}{2}\alpha = \sqrt{\tfrac{1}{2}(1 - \cos\alpha)}$$

$$= \tfrac{1}{2}\sqrt{1 + \sin\alpha} - \tfrac{1}{2}\sqrt{1 - \sin\alpha}$$

$$\cos\tfrac{1}{2}\alpha = \sqrt{\tfrac{1}{2}(1 + \cos\alpha)}$$

$$= \tfrac{1}{2}\sqrt{1 + \sin\alpha} + \tfrac{1}{2}\sqrt{1 - \sin\alpha}$$

$$\tan\tfrac{1}{2}\alpha = \sqrt{1 + \cos\alpha / 1 - \cos\alpha} = \frac{1 - \cos\alpha}{\sin\alpha}$$

$$= \frac{\sin\alpha}{1 + \cos\alpha}$$

$$\cot\tfrac{1}{2}\alpha = \sqrt{1 + \cos\alpha / 1 - \cos\alpha} = \frac{1 + \cos\alpha}{\sin\alpha}$$

$$= \frac{\sin\alpha}{1 - \cos\alpha}$$

Table 7   Functions of Certain Angles

	0°	30°	45°	60°	90°	180°	270°	360°
sin	0	$\frac{1}{2}$	$\frac{1}{2}\sqrt{2}$	$\frac{1}{2}\sqrt{3}$	1	0	-1	0
cos	1	$\frac{1}{2}\sqrt{3}$	$\frac{1}{2}\sqrt{2}$	$\frac{1}{2}$	0	-1	0	1
tan	0	$1/3\sqrt{3}$	1	$\sqrt{3}$	∞	0	∞	0
cot	∞	$\sqrt{3}$	1	$1/3\sqrt{3}$	0	∞	0	∞
sec	1	$2/3\sqrt{3}$	$\sqrt{2}$	2	∞	-1	∞	1
csc	∞	2	$\sqrt{2}$	$2/3\sqrt{3}$	1	∞	-1	∞

Table 8   Functions of an Angle in Terms of Each of the Others[a]

	$\sin\alpha = a$	$\cos\alpha = a$	$\tan\alpha = a$	$\cot\alpha = a$	$\sec\alpha = a$	$\csc\alpha = a$
sin	a	$\sqrt{1-a^2}$	$\dfrac{a}{\sqrt{1+a^2}}$	$\dfrac{1}{\sqrt{1+a^2}}$	$\dfrac{\sqrt{a^2-1}}{a}$	$\dfrac{1}{a}$
cos	$\sqrt{1-a^2}$	a	$\dfrac{1}{\sqrt{1+a^2}}$	$\dfrac{a}{\sqrt{1+a^2}}$	$\dfrac{1}{a}$	$\dfrac{\sqrt{a^2-1}}{a}$
tan	$\dfrac{a}{\sqrt{1-a^2}}$	$\dfrac{\sqrt{1-a^2}}{a}$	a	$\dfrac{1}{a}$	$\sqrt{a^2-1}$	$\dfrac{1}{\sqrt{a^2-1}}$
cot	$\dfrac{\sqrt{1-a^2}}{a}$	$\dfrac{a}{\sqrt{1-a^2}}$	$\dfrac{1}{a}$	a	$\dfrac{1}{\sqrt{a^2-1}}$	$\sqrt{a^2-1}$
sec	$\dfrac{1}{\sqrt{1-a^2}}$	$\dfrac{1}{a}$	$\sqrt{1+a^2}$	$\dfrac{\sqrt{1+a^2}}{a}$	a	$\dfrac{a}{\sqrt{a^2-1}}$
csc	$\dfrac{1}{a}$	$\dfrac{1}{\sqrt{1-a^2}}$	$\dfrac{\sqrt{1+a^2}}{a}$	$\sqrt{1+a^2}$	$\dfrac{a}{\sqrt{a^2-1}}$	a

[a]The sign of the radical is to be determined by the quadrant.

Functions of Multiples of Angles

$$\sin 2\alpha = 2\sin\alpha\cos\alpha$$

$$\tan 2\alpha = \frac{2\tan\alpha}{1-\tan^2\alpha}$$

$$\cos 2\alpha = \cos^2\alpha - \sin^2\alpha = 2\cos^2\alpha - 1 = 1 - 2\sin^2\alpha$$

$$\cot 2\alpha = \frac{\cot^2\alpha - 1}{2\cot\alpha}$$

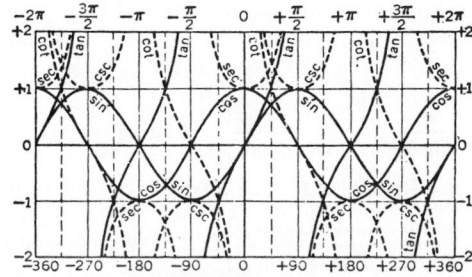

Fig. 81

$$\sin 3\alpha = 3\sin\alpha - 4\sin^3\alpha$$

$$\cos 3\alpha = 4\cos^3\alpha - 3\cos\alpha$$

$$\sin 4\alpha = 8\cos^3\alpha\sin\alpha - 4\cos\alpha\sin\alpha$$

$$\cos 4\alpha = 8\cos^4\alpha - 8\cos^2\alpha + 1$$

$$\sin n\alpha = 2\sin(n-1)\alpha\cos\alpha - \sin(n-2)\alpha$$

$$= n\sin\alpha\cos^{n-1}\alpha - {}_nC_3\sin^3\alpha\cos^{n-3}\alpha$$

$$+ {}_nC_5\sin^5\alpha\cos^{n-5}\alpha - \cdots$$

$$\cos n\alpha = 2\cos(n-1)\alpha\cos\alpha - \cos(n-2)\alpha$$

$$= \cos^n\alpha - {}_nC_2\sin^2\alpha\cos^{n-2}\alpha$$

$$+ {}_nC_4\sin^4\alpha\cos^{n-4}\alpha - \cdots$$

(For ${}_nC_r$, see p. 164.)

Products and Powers of Functions

$$\sin\alpha\sin\beta = \tfrac{1}{2}\cos(\alpha-\beta) - \tfrac{1}{2}\cos(\alpha+\beta)$$

$$\cos\alpha\cos\beta = \tfrac{1}{2}\cos(\alpha-\beta) + \tfrac{1}{2}\cos(\alpha+\beta)$$

$$\sin\alpha\cos\beta = \tfrac{1}{2}\sin(\alpha-\beta) + \tfrac{1}{2}\sin(\alpha+\beta)$$

$$\tan\alpha\cot\alpha = \sin\alpha\csc\alpha = \cos\alpha\sec\alpha = 1$$

$$\sin^2\alpha = \tfrac{1}{2}(1-\cos 2\alpha) \quad \cos^2\alpha = \tfrac{1}{2}(1+\cos 2\alpha)$$

$$\sin^3\alpha = \tfrac{1}{4}(3\sin\alpha - \sin\beta\alpha) \quad \cos^3\alpha$$

$$= \tfrac{1}{4}(3\cos\alpha + \cos 3\alpha)$$

$$\sin^4\alpha = \tfrac{1}{8}(3\sin 4\cos 2\alpha + \cos 4\alpha) \quad \cos^4\alpha$$

$$= \tfrac{1}{8}(3 + 4\cos 2\alpha + \cos 4\alpha)$$

$$\sin^5\alpha = \tfrac{1}{16}(10\sin\alpha - 5\sin 3\alpha + \sin 5\alpha)$$

$$\sin^6\alpha = \tfrac{1}{32}(10 - 15\cos 2\alpha + 6\cos 4\alpha - \cos 6\alpha)$$

$$\cos^5\alpha = \tfrac{1}{16}(10\cos\alpha + 5\cos 3\alpha + \cos 5\alpha)$$

$$\cos^6\alpha = \tfrac{1}{32}(10 + 15\cos 2\alpha + 6\cos 4\alpha + \cos 6\alpha)$$

Sums and Differences of Functions

$$\sin\alpha + \sin\beta = 2\sin\tfrac{1}{2}(\alpha+\beta)\cos\tfrac{1}{2}(\alpha-\beta)$$

$$\sin\alpha - \sin\beta = 2\cos\tfrac{1}{2}(\alpha+\beta)\sin\tfrac{1}{2}(\alpha-\beta)$$

$$\cos\alpha + \cos\beta = 2\cos\tfrac{1}{2}(\alpha+\beta)\cos\tfrac{1}{2}(\alpha-\beta)$$

$$\cos\alpha - \cos\beta = -2\sin\tfrac{1}{2}(\alpha+\beta)\sin\tfrac{1}{2}(\alpha-\beta)$$

$$\tan\alpha + \tan\beta = \frac{\sin(\alpha+\beta)}{\cos\alpha\cos\beta} \quad \cot\alpha + \cot\beta$$

$$= \frac{\sin(\alpha+\beta)}{\sin\alpha\sin\beta}$$

$$\tan\alpha - \tan\beta = \frac{\sin(\alpha-\beta)}{\cos\alpha\cos\beta} \quad \cot\alpha - \cot\beta$$

$$= -\frac{\sin(\alpha-\beta)}{\sin\alpha\sin\beta}$$

$$\sin^2\alpha - \sin^2\beta = \sin(\alpha+\beta)\sin(\alpha-\beta)$$

$$\cos^2\alpha - \cos^2\beta = -\sin(\alpha+\beta)\sin(\alpha-\beta)$$

$$\cos^2\alpha - \sin^2\beta = \cos(\alpha+\beta)\cos(\alpha-\beta)$$

Antitrigonometric or Inverse Functional Relations.

In the following formulas the periodic constant is omitted:

$$\sin^{-1}x = -\sin^{-1}(-x) = \frac{\pi}{2} - \cos^{-1}x$$

$$= \cos^{-1}\sqrt{1-x^2} = \tan^{-1}\frac{x}{\sqrt{1-x^2}}$$

$$= \cot^{-1}\frac{\sqrt{1-x^2}}{x} = \csc^{-1}\frac{1}{x}$$

$$= \sec^{-1}\frac{1}{\sqrt{1-x^2}}$$

$$\cos^{-1}x = \pi - \cos^{-1}(-x) = \frac{\pi}{2} - \sin^{-1}x$$

$$= \tfrac{1}{2}\cos^{-1}(2x^2-1) = \sin^{-1}\sqrt{1-x^2}$$

$$= \tan^{-1}\frac{\sqrt{1-x^2}}{x} = \cot^{-1}\frac{x}{\sqrt{1-x^2}}$$

$$= \sec^{-1}\frac{1}{x} = \csc^{-1}\frac{1}{\sqrt{1-x^2}}$$

$$\tan^{-1}x = -\tan^{-1}(-x) = \frac{\pi}{2} - \cot^{-1}x$$

$$= \sin^{-1}\frac{x}{\sqrt{1+x^2}} = \cos^{-1}\frac{1}{\sqrt{1+x^2}}$$

$$= \cot^{-1}\frac{1}{x} = \sec^{-1}\sqrt{1+x^2}\,\csc^{-1}\frac{\sqrt{1+x^2}}{x}$$

$$\cot^{-1}x = \tan^{-1}\frac{1}{x} \quad\quad \sec^{-1}x = \cos^{-1}\frac{1}{x}$$

$$\csc^{-1}x = \sin^{-1}\frac{1}{x}$$

$$\sin^{-1}x \pm \sin^{-1}y = \sin^{-1}(x\sqrt{1-y^2} \pm y\sqrt{1-x^2})$$

$$\cos^{-1}x \pm \cos^{-1}y = \cos^{-1}[xy \mp \sqrt{(1-x^2)(1-y^2)}]$$

$$\sin^{-1}x \pm \cos^{-1}y = \sin^{-1}[xy \pm \sqrt{(1-x^2)(1-y^2)}]$$

$$= \cos^{-1}(y\sqrt{1-x^2} \mp x\sqrt{1-y^2})$$

$$\tan^{-1}x \pm \tan^{-1}y = \tan^{-1}\frac{x\pm y}{1\mp xy}$$

$$\tan^{-1}x \pm \cot^{-1}y = \tan^{-1}\frac{xy\pm 1}{y\mp x} = \cot^{-1}\frac{y\mp x}{xy\pm 1}$$

6.2 Solution of Triangles

Relations between Angles and Sides of Plane Triangles.

Let a, b, c = sides of triangle; α, β, γ = angles opposite a, b, c, respectively; A = area of triangle; $s = \tfrac{1}{2}(a+b+c)$; r = radius of inscribed circle (Fig. 82).

$$\frac{a}{\sin\alpha} = \frac{b}{\sin\beta} = \frac{c}{\sin\gamma} \quad \text{(law of sines)}$$

$$a^2 = b^2 + c^2 - 2bc\cos\alpha \quad \text{(law of cosines)}$$

$$\frac{a-b}{a+b} = \frac{\tan\tfrac{1}{2}(\alpha-\beta)}{\tan\tfrac{1}{2}(\alpha+\beta)} \quad \text{(law of tangents)}$$

$$\alpha + \beta + \gamma = 180°$$

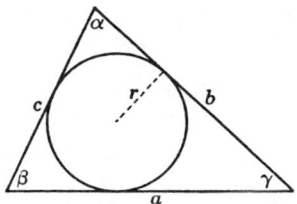

Fig. 82

$$a = b\cos\gamma + c\cos\beta \quad b = c\cos\alpha + a\cos\gamma$$
$$c = a\cos\beta + b\cos\alpha$$
$$A = \sqrt{s(s-a)(s-b)(s-c)}$$
$$\sin\alpha = \frac{2}{bc}A \quad \sin\beta = \frac{2}{ca}A \quad \sin\gamma = \frac{2}{ab}A$$

$$\sin\frac{\alpha}{2} = \sqrt{\frac{(s-b)(s-c)}{bc}}$$

$$\sin\frac{\beta}{2} = \sqrt{\frac{(s-c)(s-a)}{ca}}$$

$$\sin\frac{\gamma}{2} = \sqrt{\frac{(s-a)(s-b)}{ab}}$$

$$\cos\frac{\alpha}{2} = \sqrt{\frac{s(s-a)}{bc}}$$

$$\cos\frac{\beta}{2} = \sqrt{\frac{s(s-b)}{ca}} \qquad \cos\frac{\gamma}{2} = \sqrt{\frac{s(s-c)}{ab}}$$

$$\tan\frac{\alpha}{2} = \sqrt{\frac{(s-b)(s-c)}{s(s-a)}}$$

$$\tan\frac{\beta}{2} = \sqrt{\frac{(s-c)(s-a)}{s(s-b)}}$$

$$\tan\frac{\gamma}{2} = \sqrt{\frac{(s-a)(s-b)}{s(s-c)}}$$

Solution of Plane Oblique Triangles. Given a, b, c (if logarithms are to be used, use 1):

1. $$r = \sqrt{\frac{(s-a)(s-b)(s-c)}{s}},$$
    $$A = \sqrt{s(s-a)(s-b)(s-c)} = rs,$$
    $$\tan\frac{\alpha}{2} = \frac{r}{s-a} \qquad \tan\frac{\beta}{2} = \frac{r}{s-b},$$
    $$\tan\frac{\gamma}{2} = \frac{r}{s-c}.$$

2. $$\cos\alpha = \frac{b^2+c^2-a^2}{2bc},$$
    $$\cos\beta = \frac{a^2+c^2-b^2}{2ac},$$
    $$\cos\gamma = \frac{a^2+b^2-c^2}{2ab} \quad \text{or}$$
    $$\gamma = 180° - (\alpha+\beta).$$

Given a, b, α:

$$\sin\beta = \frac{b\sin\alpha}{a}$$

(if $a > b$, $\beta < \pi/2$ and has only one value; if $b > a$, β has two values, β_1, $\beta_2 = 180° - \beta_1$); $\gamma = 180° - (\alpha+\beta)$; $c = a\sin\gamma/\sin\alpha$; $A = \frac{1}{2}ab\,\sin\gamma$.
Given a, α, β:

$$b = \frac{a\sin\beta}{\sin\alpha} \quad \gamma = 180° - (\alpha+\beta)$$

$$c = \frac{a\sin\gamma}{\sin\alpha} \qquad A = \frac{1}{2}ab\,\sin\gamma$$

Given a, b, γ (if logarithms are to be used, use 1):

1. $$\tan\frac{1}{2}(\alpha-\beta) = \frac{a-b}{a+b}\cot\frac{1}{2}\gamma,$$
    $$\frac{1}{2}(\alpha+\beta) = 90° - \frac{1}{2}\gamma, \quad c = \frac{a\sin\gamma}{\sin\alpha},$$
    $$A = \frac{1}{2}ab\,\sin\,\gamma.$$

2. $$c = \sqrt{a^2+b^2-2ab\,\cos\gamma}, \quad \sin\alpha = \frac{a\sin\gamma}{c},$$
    $$\beta = 180° - (\alpha+\gamma).$$

3. $$\tan\alpha = \frac{a\sin\gamma}{b-a\cos\gamma}, \quad \beta = 180° - (\alpha+\gamma),$$
    $$c = \frac{a\sin\gamma}{\sin\alpha}.$$

Mollweide's Check Formulas

$$\frac{a-b}{c} = \frac{\sin\frac{1}{2}(\alpha-\beta)}{\cos\frac{1}{2}\gamma} \qquad \frac{a+b}{c} = \frac{\cos\frac{1}{2}(\alpha-\beta)}{\sin\frac{1}{2}\gamma}$$

Solution of Plane Right Triangles. Let $\gamma = 90°$ and c be the hypotenuse. Given any two sides or one side and an acute angle α:

$$a = \sqrt{c^2-b^2} = \sqrt{(c+b)(c-b)} = b\tan\alpha = c\sin\alpha$$
$$b = \sqrt{c^2-a^2} = \sqrt{(c+a)(c-a)} = \frac{a}{\tan\alpha} = c\cos\alpha$$
$$c = \sqrt{a^2+b^2} = \frac{a}{\sin\alpha} = \frac{b}{\cos\alpha}$$
$$\alpha = \sin^{-1}\frac{a}{c} = \cos^{-1}\frac{b}{c} = \tan^{-1}\frac{a}{b} \qquad \beta = 90° - \alpha$$
$$A = \frac{ab}{2} = \frac{a^2}{2\tan\alpha} = \frac{b^2\tan\alpha}{2} = \frac{c^2\sin 2\alpha}{4}$$

6.3   Spherical Trigonometry

Spherical Trigonometry.   Let O be the center of the sphere and a, b, c the sides of a triangle on the surface with opposite angles α, β, γ, respectively, the sides being measured by the angle subtended at the center of the sphere. Let $s = \frac{1}{2}(a + b + c)$, $\sigma = \frac{1}{2}(\alpha + \beta + \gamma)$, $E = \alpha + \beta + \gamma - 180°$, the spherical excess. The following formulas are valid usually only for triangles of which the sides and angles are all between $0°$ and $180°$. To each such triangle there is a polar triangle whose sides are $180° - \alpha$, $180° - \beta$, $180° - \gamma$ and whose angles are $180° - a$, $180° - b$, $180° - c$.

General Formulas

$$\frac{\sin a}{\sin \alpha} = \frac{\sin b}{\sin \beta} = \frac{\sin c}{\sin \gamma} \quad \text{(law of sines)}$$

$$\cos a = \cos b \cos c + \sin b \sin c \cos \alpha$$

$$\text{(law of cosines)}$$

$$\cos \alpha = -\cos \beta \cos \gamma + \sin \beta \sin \gamma \cos a$$

$$\text{(law of cosines)}$$

$$\cos a \sin b = \sin a \cos b \cos \gamma + \sin c \cos \alpha$$

$$\cot a \sin b = \sin \gamma \cot \alpha + \cos \gamma \cos b$$

$$\cos \alpha \sin \beta = \sin \gamma \cos a - \sin \alpha \cos \beta \cos c$$

$$\cot \alpha \sin \beta = \sin c \cot a - \cos c \cos \beta$$

$$\sin \frac{a}{2} = \sqrt{\frac{-\cos \sigma \cos(\sigma - \alpha)}{\sin \beta \sin \gamma}}$$

$$\sin \frac{\alpha}{2} = \sqrt{\frac{\sin(s - b)\sin(s - c)}{\sin b \sin c}}$$

$$\cos \frac{a}{2} = \sqrt{\frac{\cos(\sigma - \beta)\cos(\sigma - \gamma)}{\sin \beta \sin \gamma}}$$

$$\cos \frac{\alpha}{2} = \sqrt{\frac{\sin s \sin(s - a)}{\sin b \sin c}}$$

$$\tan \frac{a}{2} = \sqrt{\frac{-\cos \sigma \cos(\sigma - \alpha)}{\cos(\sigma - \beta)\cos(\sigma - \gamma)}}$$

$$\tan \frac{\alpha}{2} = \sqrt{\frac{\sin(s - b)\sin(s - c)}{\sin s \sin(s - a)}}$$

$$\tan \frac{E}{4}$$
$$= \sqrt{\tan \frac{s}{2}\tan \frac{(s - a)}{2}\tan \frac{(s - b)}{2}\tan \frac{(s - c)}{2}}$$

$$\cot \frac{E}{2} = \frac{\cot(a/2)\cot(b/2) + \cos \gamma}{\sin \gamma}$$

$$\tan \left(\frac{a + b}{2}\right) = \frac{\cos[(\alpha - \beta)/2]}{\cos[(\alpha + \beta)/2]}\tan \frac{c}{2}$$

$$\tan \left(\frac{a - b}{2}\right) = \frac{\sin[(\alpha - \beta)/2]}{\sin[(\alpha + \beta)/2]}\tan \frac{c}{2}$$

$$\tan \left(\frac{\alpha + \beta}{2}\right) = \frac{\cos[(a - b)/2]}{\cos[(a + b)/2]}\cot \frac{\gamma}{2}$$

$$\tan \left(\frac{\alpha - \beta}{2}\right) = \frac{\sin[(a - b)/2]}{\sin[(a + b)/2]}\cot \frac{\gamma}{2}$$

$$\cos \left(\frac{\alpha + \beta}{2}\right)\cos \frac{c}{2} = \cos \left(\frac{a + b}{2}\right)\sin \frac{\gamma}{2}$$

$$\sin \left(\frac{\alpha + \beta}{2}\right)\cos \frac{c}{2} = \cos \left(\frac{a - b}{2}\right)\cos \frac{\gamma}{2}$$

$$\cos \left(\frac{\alpha - \beta}{2}\right)\sin \frac{c}{2} = \sin \left(\frac{a + b}{2}\right)\sin \frac{\gamma}{2}$$

$$\sin \left(\frac{\alpha - \beta}{2}\right)\sin \frac{c}{2} = \sin \left(\frac{a - b}{2}\right)\cos \frac{\gamma}{2}$$

The Right Spherical Triangle.   Let $\gamma = 90°$ and c be the hypotenuse.

$$\cos c = \cos a \cos b = \cot \alpha \cot \beta \qquad \cos a = \frac{\cos \alpha}{\sin \beta}$$

$$\cos b = \frac{\cos \beta}{\sin \alpha}$$

$$\sin \alpha = \frac{\sin a}{\sin c} \qquad\qquad \cos \alpha = \frac{\tan b}{\tan c}$$

$$\tan \alpha = \frac{\tan a}{\sin b}$$

6.4   Hyperbolic Trigonometry

Hyperbolic Angles.   These are defined in a manner similar to circular angles but with reference to an *equilateral hyperbola.* The comparative relations are shown in Figs. 83 and 84. A *circular angle* is a central angle measured in radians by the ratio s/r or the ratio $2A/r^2$, where A is the area of the sector included by the angle α and the arc s (Fig. 83). For the *hyperbola* the radius ρ is not constant and only the value of the *differential hyperbolic angle* $d\theta$ is defined by the ratio ds/ρ. Thus

$$\theta = \int \frac{ds}{\rho} = \frac{2A}{a^2}$$

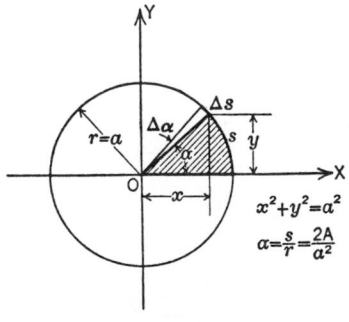

Fig. 83

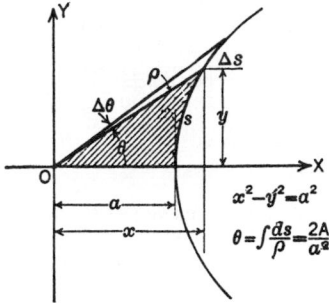

Fig. 84

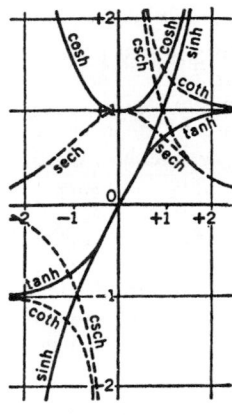

Fig. 85

where A represents the shaded area in Fig. 84. If both s and ρ are measured in the same units, the angle is expressed in *hyperbolic radians*.

Hyperbolic Functions. These are defined by ratios similar to those defining functions of circular angles and also named similarly. Their names and abbreviations are

$$\text{Hyperbolic sine } \theta = \frac{y}{a} = \sinh\ \theta$$

$$\text{Hyperbolic cosine } \theta = \frac{x}{a} = \cosh\ \theta$$

$$\text{Hyperbolic tangent } \theta = \frac{y}{x} = \tanh\ \theta$$

$$\text{Hyperbolic cotangent } \theta = \frac{x}{y} = \coth\ \theta$$

$$\text{Hyperbolic secant } \theta = \frac{a}{x} = \text{sech}\ \theta$$

$$\text{Hyperbolic cosecant } \theta = \frac{a}{y} = \text{csch}\ \theta$$

Values and Exponential Equivalents. The values of hyperbolic functions may be computed from their exponential equivalents. The graphs are shown in Fig. 85. Values for increments of 0.01 rad are given in Table 18.

$$\sinh\theta = \frac{e^{\theta} - e^{-\theta}}{2} \qquad \cosh\theta = \frac{e^{\theta} + e^{-\theta}}{2}$$

$$\tanh\theta = \frac{e^{\theta} - e^{-\theta}}{e^{\theta} + e^{-\theta}}$$

If θ is extremely small, $\sinh\theta \approx \theta$, $\cosh\theta \approx 1$, and $\tanh\theta \approx \theta$. For large values of θ, $\sinh\theta \approx \cosh\theta$ and $\tanh\theta \approx \coth\theta \approx 1$.

Fundamental Identities

$$\text{csch } \theta = \frac{1}{\sinh\theta} \qquad \text{sech } \theta = \frac{1}{\cosh\theta}$$

$$\coth\theta = \frac{1}{\tanh\theta}$$

$$\cosh^2\theta - \sinh^2\theta = 1 \qquad\qquad \text{sech}^2\,\theta = 1 - \tanh^2\theta$$

$$\text{csch}^2\,\theta = \coth^2\theta - 1$$

$$\cosh\theta + \sinh\theta = e^{\theta} \qquad \cosh\theta - \sinh\theta = e^{-\theta}$$

$$\sinh(-\theta) = -\sinh\theta \qquad \cosh(-\theta) = \cosh\theta$$

$$\tanh(-\theta) = -\tanh\theta \qquad \coth(-\theta) = -\coth\theta$$

$$\sinh(\theta_1 \pm \theta_2) = \sinh\theta_1 \cosh\theta_2 \pm \cosh\theta_1 \sinh\theta_2$$

$$\cosh(\theta_1 \pm \theta_2) = \cosh\theta_1 \cosh\theta_2 \pm \sinh\theta_1 \sinh\theta_2$$

$$\tanh(\theta_1 \pm \theta_2) = \frac{\tanh\theta_1 \pm \tanh\theta_2}{1 \pm \tanh\theta_1 \tanh\theta_2}$$

$$\coth(\theta_1 \pm \theta_2) = \frac{1 \pm \coth\theta_1 \coth\theta_2}{\coth\theta_1 \pm \coth\theta_2}$$

$$\sinh 2\theta = 2 \sinh\theta \cosh\theta = \frac{2\tanh\theta}{1 - \tanh^2\theta}$$

$$\cosh 2\theta = \sinh^2\theta + \cosh^2\theta = 1 + 2\sinh^2\theta$$

$$= 2\cosh^2\theta - 1 = \frac{1 + \tanh^2\theta}{1 - \tanh^2\theta}$$

$$\tanh 2\theta = \frac{2\tanh\theta}{1 + \tanh^2\theta}$$

$$\coth 2\theta = \frac{1 + \coth^2\theta}{2\coth\theta}$$

$$\sinh \tfrac{1}{2}(0) = \sqrt{\tfrac{1}{2}(\cosh\theta - 1)}$$

$$\cosh \tfrac{1}{2}(0) = \sqrt{\tfrac{1}{2}(\cosh\theta + 1)}$$

$$\tanh \frac{\theta}{2} = \sqrt{\frac{\cosh\theta - 1}{\cosh\theta + 1}} = \frac{\sinh\theta}{\cosh\theta + 1}$$

$$= \frac{\cosh\theta - 1}{\sinh\theta}$$

$$\sinh\theta_1 \pm \sinh\theta_2 = 2\sinh\tfrac{1}{2}(\theta_1 \pm \theta_2)\cosh\tfrac{1}{2}(\theta_1 \mp \theta_2)$$

$$\cosh\theta_1 + \cosh\theta_2 = 2\cosh\tfrac{1}{2}(\theta_1 + \theta_2)\cosh\tfrac{1}{2}(\theta_1 - \theta_2)$$

$$\cosh\theta_1 - \cosh\theta_2 = 2\sinh\tfrac{1}{2}(\theta_1 + \theta_2)\sinh\tfrac{1}{2}(\theta_1 - \theta_2)$$

$$\tanh\theta_1 \pm \tanh\theta_2 = \frac{\sinh(\theta_1 \pm \theta)}{\cosh\theta_1 \cosh\theta_2}$$

$$(\cosh\theta \pm \sinh\theta)^n = \cosh n\theta \pm \sinh n\theta$$

Antihyperbolic or Inverse Functions.

The inverse hyperbolic sine of u is written $\sinh^{-1} u$. Values of the inverse functions may be computed from their logarithmic equivalents:

$$\sinh^{-1}u = \log_e(u + \sqrt{u^2 + 1})$$
$$\cosh^{-1}u = \log_e(u + \sqrt{u^2 + 1})$$
$$\tanh^{-1}u = \tfrac{1}{2}\log_e\frac{1+u}{1-u}$$
$$\coth^{-1}u = \tfrac{1}{2}\log_e\frac{u+1}{u-1}$$

6.5 Functions of Imaginary and Complex Angles

Relations of Hyperbolic to Circular Functions.

By comparison of the exponential equivalents of hyperbolic and circular functions, the following identities are established ($i = \sqrt{-1}$):

$$\begin{array}{ll}
\sin\alpha = -i\sinh i\alpha & \sinh\beta = -i\sin i\beta \\
\cos\alpha = \cosh i\alpha & \cosh\beta = \cos i\beta \\
\tan\alpha = -i\tanh i\alpha & \tanh\beta = -i\tan i\beta \\
\cot\alpha = i\coth i\alpha & \coth\beta = i\cot i\beta \\
\sec\alpha = \operatorname{sech} i\alpha & \operatorname{sech}\beta = \sec i\beta \\
\csc\alpha = i\operatorname{csch} i\alpha & \operatorname{csch}\beta = i\csc i\beta
\end{array}$$

Relations between Inverse Functions

$$\begin{array}{ll}
\sin^{-1} A = -i\sinh^{-1} iA & \sinh^{-1} B = -i\sinh^{-1} iB \\
\cos^{-1} A = -i\cosh^{-1} A & \cosh^{-1} B = -i\cos^{-1} B \\
\tan^{-1} A = -i\tanh^{-1} iA & \tanh^{-1} B = -i\tan^{-1} iB \\
\cot^{-1} A = i\coth^{-1} iA & \coth^{-1} B = i\cot^{-1} iB \\
\sec^{-1} A = -i\operatorname{sech}^{-1} A & \operatorname{sech}^{-1} B = i\sec^{-1} B \\
\csc^{-1} A = i\operatorname{csch}^{-1} iA & \operatorname{csch}^{-1} B = i\csc^{-1} iB
\end{array}$$

Functions of a Complex Angle.

In complex, notation $c = a + ib = |c|(\cos\theta + i\sin\theta) = |c|e^{i\theta}$, where $|c| = \sqrt{a^2 + b^2}$, $i = \sqrt{-1}$, and $\theta = \tan^{-1} b/a$. Frequently $|c|e^{i\theta}$ is written $c\angle\theta$.

$\operatorname{Log}_e |c|e^{i\theta} = \log|c| + i(\theta + 2k\pi)$ and is infinitely many valued. By its principal part will be understood $\log_e |c| + i\theta$. Some convenient identities are

$$\log_e 1 = 0 \qquad \log_e(-1) = i\pi$$

$$\log_e i = i\frac{\pi}{2} \qquad \log_e(-i) = i\frac{3\pi}{2}$$

$$(\cos\theta \pm i\sin\theta)^n = \cos n\theta \pm i\sin n\theta$$

$$\sqrt[n]{\cos\theta \pm i\sin\theta} = \cos\frac{\theta + 2\pi k}{n} \pm i\sin\frac{\theta + 2\pi k}{n}$$

The use of complex angles occurs frequently in electric circuit problems where it is often necessary to express the functions of them as a complex number:

$$\sin(\alpha \pm i\beta) = \sin\alpha\cosh\beta \pm i\cos\alpha\sinh\beta$$
$$= \sqrt{\cosh^2\beta - \cos^2\alpha}\,e^{\pm i\theta}$$

where $\theta = \tan^{-1}\cot\alpha\tanh\beta$;

$$\cos(\alpha \pm i\beta) = \cos\alpha\cosh\beta \mp i\sin\alpha\sinh\beta$$
$$= \sqrt{\cosh^2\beta - \sin^2\alpha}\,e^{\pm i\theta}$$

where $\theta = \tan^{-1}\tan\alpha\tanh\beta$;

$$\sinh(\alpha \pm i\beta) = \sinh\alpha\cos\beta \pm i\cosh\alpha\sin\beta$$
$$= \sqrt{\sinh^2\alpha + \sin^2\beta}\,e^{\pm i\theta}$$
$$= \sqrt{\cosh^2\alpha + \cos^2\beta}\,e^{\pm i\theta}$$

where $\theta = \tan^{-1}\coth\alpha\tan\beta$;

$$\cosh(\alpha \pm i\beta) = \cosh\alpha\cos\beta \pm i\sinh\alpha\sin\beta$$
$$= \sqrt{\sinh^2\alpha + \cos^2\beta}\,e^{\pm i\theta}$$
$$= \sqrt{\cosh^2\alpha + \sin^2\beta}\,e^{\pm i\theta}$$

where $\theta = \tan^{-1}\tanh\alpha\tan\beta$; and

$$\tan(\alpha \pm i\beta) = \frac{\sin 2\alpha \pm i\sinh 2\beta}{\cos 2\alpha + \cosh 2\beta}$$

$$\tanh(\alpha \pm i\beta) = \frac{\sinh 2\alpha \pm i\sin 2\beta}{\cosh 2\alpha + \cos 2\beta}$$

The hyperbolic sine and cosine have the period $2\pi i$; the hyperbolic tangent has the period πi:

$$\sinh(\alpha + 2k\pi i) = \sinh\alpha \qquad \cosh(\alpha + 2k\pi i) = \cosh\alpha$$

$$\tanh(\alpha + k\pi i) = \tanh\alpha \qquad \coth(\alpha + k\pi i) = \coth\alpha$$

Inverse Functions of Complex Numbers

$\sin^{-1}(A \pm iB)$

$$= \sin^{-1}\left[\frac{1}{2}(\sqrt{B^2 + (1 + A)^2} - \sqrt{B^2 + (1 - A)^2})\right]$$

$$\pm i \cosh^{-1}\left[\frac{1}{2}(\sqrt{B^2 + (1 + A)^2} + \sqrt{B^2 + (1 - A)^2})\right]$$

$\cos^{-1}(A \pm iB)$

$$= \cos^{-1}\left[\frac{1}{2}(\sqrt{B^2 + (1 + A)^2} - \sqrt{B^2 + (1 - A)^2})\right]$$

$$\mp i \cosh^{-1}\left[\frac{1}{2}(\sqrt{B^2 + (1 + A)^2} + \sqrt{B^2 + (1 - A)^2})\right]$$

$\tan^{-1}(A \pm iB)$

$$= \left[\frac{1}{2}\left(\pi - \tan^{-1}\frac{A}{\pm B - 1} + \tan^{-1}\frac{A}{\pm B + 1}\right)\right]$$

$$\pm i\frac{1}{4}\log_e \frac{A^2 + (1 \pm B)^2}{A^2 + (1 \mp B)^2}$$

$\sinh^{-1}(A \pm iB)$

$$= \cosh^{-1}\left[\frac{1}{2}(\sqrt{A^2 + (1 + B)^2} - \sqrt{A^2 + (1 - B)^2})\right]$$

$$\pm i \sin^{-1}\left[\frac{1}{2}(\sqrt{A^2 + (1 + B)^2} - \sqrt{A^2 + (1 - B)^2})\right]$$

$\cosh^{-1}(A \pm iB)$

$$= \cosh^{-1}\left[\frac{1}{2}(\sqrt{B^2 + (1 + A)^2} + \sqrt{B^2 + (1 - A)^2})\right]$$

$$\pm i \cos^{-1}\left[\frac{1}{2}(\sqrt{B^2 + (1 + A)^2} - \sqrt{B^2 + (1 - A)^2})\right]$$

$\tanh^{-1}(A \pm iB)$

$$= \frac{1}{2}\tanh^{-1}\frac{2A}{1 + A^2 + B^2} + i\frac{1}{2}\tan^{-1}\frac{\pm 2B}{1 - 2A - B^2}$$

7 PLANE ANALYTIC GEOMETRY

7.1 Point and Line

Coordinates. The position of a point P_1 in a plane is determined if its distance and direction from each of two lines or axes OX and OY which are perpendicular to each other are known. The distances x and y (Fig. 86) perpendicular to the axes are called the *Cartesian* or *rectangular coordinates* of the point. The directions to the right of OY and above OX are called *positive* and opposite directions *negative*. The point O of intersection of OY and OX is called the *origin*.

The position of a point P is also given by its radial distance r from the origin and the angle θ between the radius r and the horizontal axis OX (Fig. 87). These coordinates r, θ are called *polar coordinates*.

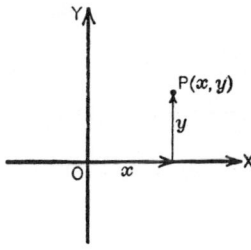

Fig. 86

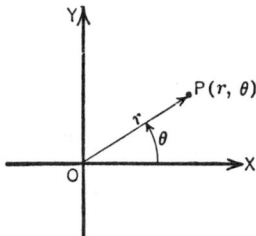

Fig. 87

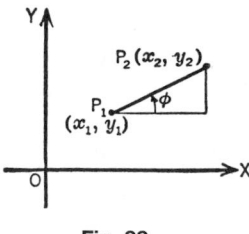

Fig. 88

The distance s between two points $P_1(x_1, y_1)$ and $P_2(x_2, y_2)$ (Fig. 88) on a straight line is

$$s = \sqrt{(x_2 - x_1)^2 + (y_2 - y_1)^2} \qquad (15)$$

In polar coordinates the distance s between $P_1(r_1, \theta_1)$ and $P_2(r_2, \theta_2)$ is

$$s = \sqrt{r_1^2 + r_2^2 - 2r_1r_2\cos(\theta_2 - \theta_1)} \qquad (16)$$

The slope m of the line $P_1 P_2$ is defined as the tangent of the angle ϕ which the line makes with OX:

$$m = \tan\phi = \frac{y_2 - y_1}{x_2 - x_1} \qquad (17)$$

To divide the segment $P_1 P_2$ in the ratio c_1/c_2, internally or externally,

$$x = \frac{c_2 x_1 \pm c_1 x_2}{c_2 \pm c_1} \qquad y = \frac{c_2 y_1 \pm c_1 y_2}{c_2 \pm c_1}$$

The midpoint of $P_1 P_2$ is

$$x = \tfrac{1}{2}(x_1 + x_2) \qquad y = \tfrac{1}{2}(y_1 + y_2)$$

Equation of a Straight Line. In Cartesian coordinates the equation of a straight line is of the first degree and is expressed as

$$Ax + By + C = 0 \tag{18}$$

where A, B, and C are constants.

Other forms of the equation are

$$y = mx + b \tag{19}$$

where m is the slope and b is the y intercept;

$$y - y_1 = m(x - x_1) \tag{20}$$

where m is the slope and (x_1, y_1) is a point on the line;

$$\frac{x - x_1}{y - y_1} = \frac{x_1 - x_2}{y_1 - y_2} \tag{21}$$

where (x_1, y_1) and (x_2, y_2) are two points on the line;

$$\frac{x}{a} + \frac{y}{b} = 1 \tag{22}$$

where a and b are the x and y intercepts, respectively;

$$x \cos\alpha + y \sin\alpha - p = 0 \tag{23}$$

where α is the angle between OX and the perpendicular from the origin to the line and p is the length of the perpendicular (Fig. 89). This is called the *perpendicular form* and is obtained by dividing the general form

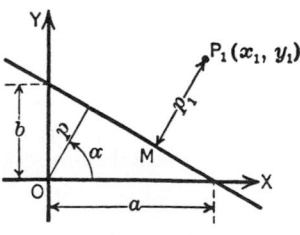

Fig. 89

$Ax + By + C = 0$ by $\pm\sqrt{A^2 + B^2}$. The sign before the radical is taken opposite to that of C if $C \neq 0$ and the same as that of B if $C = 0$.

Equations of lines parallel to the x and y axes, respectively, are

$$y = k \qquad x = k \tag{24}$$

The *perpendicular distance of a point* $P_1(x_1, y_1)$ (Fig. 89) from the line $Ax + By + C = 0$ is

$$p_1 = \frac{Ax_1 + By_1 + C}{\pm\sqrt{A^2 + B^2}} \tag{25}$$

where the sign before the radical is opposite to that of C if $C \neq 0$ and the same as B if $C = 0$.

Parallel Lines. The two lines $y = m_1 x + b_1$, $y = m_2 x + b_2$ are parallel if $m_1 = m_2$. For the form $A_1 x + B_1 y + C_1 = 0$, $A_2 x + B_2 y + C_2 = 0$, the lines are parallel if

$$\frac{A_1}{A_2} = \frac{B_1}{B_2} \tag{26}$$

The equation of a line through the point (x_1, y_1) and parallel to the line $Ax + By + C = 0$ is

$$A(x - x_1) + B(y - y_1) = 0 \tag{27}$$

Perpendicular Lines. The two lines $y = m_1 x + b_1$ and $y = m_2 x + b_2$ are perpendicular if

$$m_1 = -\frac{1}{m_2} \tag{28}$$

For the form $A_1 x + B_1 y + C_1 = 0$, $A_2 x + B_2 y + C_2 = 0$, the lines are perpendicular if

$$A_1 A_2 + B_1 B_2 = 0 \tag{29}$$

The equation of a line through the point (x_1, y_1) perpendicular to the line $Ax + By + C = 0$ is

$$B(x - x_1) - A(y - y_1) = 0 \tag{30}$$

Intersecting Lines. Let $A_1 x + B_1 y + C_1 = 0$ and $A_2 x + B_2 y + C_2 = 0$ be the equations of two intersecting lines and λ an arbitrary real number. Then

$$(A_1 x + B_1 y + C_1) + \lambda(A_2 x + B_2 y + C_2) = 0 \tag{31}$$

represents the system of lines through the point of intersection.

The three lines $A_1x + B_1y + C_1 = 0$, $A_2x + B_2y + C_2 = 0$, $A_3x + B_3y + C_3 = 0$ meet in a point if

$$\begin{vmatrix} A_1 & B_1 & C_1 \\ A_2 & B_2 & C_2 \\ A_3 & B_3 & C_3 \end{vmatrix} = 0 \qquad (32)$$

The *angle θ between two lines* with equations $A_1x + B_1y + C_1 = 0$ and $A_2x + B_2y + C_2 = 0$ can be found from

$$\sin\theta = \frac{A_1 B_2 - A_2 B_1}{\sqrt{(A_1^2 + B_1^2)(A_2^2 + B_2^2)}}$$

$$\cos\theta = \frac{A_1 A_2 + B_1 B_2}{\sqrt{(A_1^2 + B_1^2)(A_2^2 + B_2^2)}} \qquad (33)$$

$$\tan\theta = \frac{A_1 B_2 - A_2 B_1}{A_1 A_2 - B_1 B_2}$$

The signs of $\tan\theta$ and $\cos\theta$ determine whether the acute or obtuse angle is meant. If the equations are in the form $y = m_1 x + b_1$, $y = m_2 x + b_2$, then

$$\sin\theta = \frac{m_2 - m_1}{\sqrt{(1 + m_1^2)(1 + m_2^2)}}$$

$$\cos\theta = \frac{1 + m_1 m_2}{\sqrt{(1 + m_1^2)(1 + m_2^2)}} \qquad (34)$$

$$\tan\theta = \frac{m_2 - m_1}{1 + m_1 m_2}$$

7.2   Transformation of Coordinates

Change of Origin O to O′.   Let (x, y) denote the coordinates of a point P with respect to the old axes and (x', y') the coordinates with respect to the new axes (Fig. 90). Then, if the coordinates of the new origin O' with respect to the old axes are $x = h$, $y = k$,

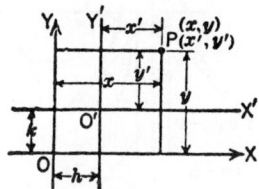

Fig. 90

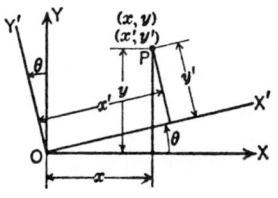

Fig. 91

the relations between the old and the new coordinates under transformation are

$$x = x' + h \qquad y = y' + k \qquad (35)$$

Rotation of Axes about the Origin.   Let θ (Fig. 91) be the angle through which the axes are rotated. Then

$$x = x' \cos\theta - y \sin\theta \qquad y = x' \sin\theta + y \cos\theta \qquad (36)$$

If the axes are both translated and rotated,

$$x = x' \cos\theta - y' = \sin\theta + h$$

$$y = x' \sin\theta + y' \cos\theta + k \qquad (37)$$

Coordinate Transformation.   The relations between the rectangular coordinates x, y and the polar coordinates r, θ are

$$x = r\cos\theta \qquad y = r\sin\theta \qquad r = \sqrt{x^2 + y^2}$$

$$\theta = \tan^{-1}\frac{y}{x} \qquad (38)$$

7.3   Conic Sections

Conic Section.   This is a curve traced by a point P moving in a plane so that the distance PF of the point from a fixed point (*focus*) is in constant ratio to the distance PM of the point from a fixed line (*directrix*) in the plane of the curve. The ratio $e = PF/PM$ is called the *eccentricity*. If $e < 1$, the curve is an *ellipse*; $e = 1$, a *parabola*; $e > 1$, a *hyperbola*; and $e = 0$, a *circle*, which is a special case of an ellipse.

Circle.   The equation is

$$(x - x_0)^2 + (y - y_0)^2 = r^2 \qquad (39)$$

where (x_0, y_0) is the center and r the radius. If the center is at the origin,

$$x^2 + y^2 = r^2 \qquad (40)$$

Another form is

$$x^2 + y^2 + 2gx + 2fy + c = 0 \qquad (41)$$

with center $(-g, -f)$ and radius $\sqrt{g^2 + f^2 - c}$.

The equation of the tangent to (41) at a point $P_1(x_1, y_1)$ is

$$xx_1 + yy_1 + g(x + x_1) + f(y + y_1) + c = 0 \quad (42)$$

Ellipse (Fig. 92). The equation is

$$\frac{(x - x_0)^2}{a^2} + \frac{(y - y_0)^2}{b^2} = 1 \qquad (43)$$

where (x_0, y_0) is the center, a the semimajor axis, and b the semiminor axis. In Fig. 92, $(x_0, y_0) = (0, 0)$.

Coordinates of foci are $F_1 = (-ae, 0)$, $F_2 = (ae, 0)$; $e^2 = (F_1 P)^2 / (MP)^2 = 1 - b^2/a^2 < 1$; and the directrices are the lines $x = -a/e, x = a/e$.

The chord LL' through F is called the *latus rectum* and has the length $2b^2/a = 2a(1 - e^2)$. If P_1 is any point on the ellipse, $F_1 P_1 = a + ex_1$, $F_2 P_1 = a + ex_1$, and $F_1 P_1 + F_2 P_1 = 2a$ (a constant).

The area of the ellipse with semiaxes a and b is

$$A = \pi ab \qquad (44)$$

The equation of the tangent to the ellipse (Fig. 92) at the point (x_1, y_1) is

$$\frac{xx_1}{a^2} + \frac{yy_1}{b^2} = 1 \qquad (45)$$

the equation of the tangent with slope m is

$$y = mx \pm \sqrt{a^2 m^2 + b^2} \qquad (46)$$

The equation of the normal to the ellipse at the point (x_1, y_1) is

$$a^2 y_1 (x - x_1) - b^2 x_1 (y - y_1) = 0 \qquad (47)$$

Conjugate Diameters. A line through the center of an ellipse is a *diameter*; if the slopes m and m' of the two diameters $y = mx$ and $y = m'x$ are such that $mm' = -b^2/a^2$ each diameter bisects all chords parallel to the other and the diameters are called *conjugate.*

Other Forms of the Equation of the Ellipse

$$\frac{x^2}{a^2} + \frac{y^2}{a^2(1 - e^2)} = 1 \qquad (48)$$

$$ax^2 + by^2 + 2gx + 2fy + c = 0 \qquad (49)$$

If a, b, and $g^2/a + f^2/b - c$ have the same sign, (49) is an ellipse whose axes are parallel to the coordinate axes.

The parametric form is

$$x = a \cos \phi \qquad y = b \sin \phi \qquad (50)$$

Hyperbola (Fig. 93). The equation is

$$\frac{(x - x_0)^2}{a^2} - \frac{(y - y_0)^2}{b^2} = 1 \qquad (51)$$

where (x_0, y_0) is the center, $AA' = 2a$ is the transverse axis, and $BB' = 2b$ is the conjugate axis. In Fig. 93, $(x_0, y_0) = (0, 0)$;

$$e^2 = \frac{(F_1 P)^2}{(PM)^2} = 1 + \frac{b^2}{a^2} > 1$$

the coordinates of the foci are $F_1 = (-ae, 0)$, $F_2 = (ae, 0)$; and the directrices are the lines $x = -a/e, x = a/e$.

The chord LL' through F is called the latus rectum and has the length $2b^2/a = 2a(e^2 - 1)$. If P_1 is any point on the curve, $F_1 P_1 = ex_1 - a$, $F_2 P_1 = ex_1 + a$, and $|F_2 P_1 - F_1 P_1| = 2a$ (a constant).

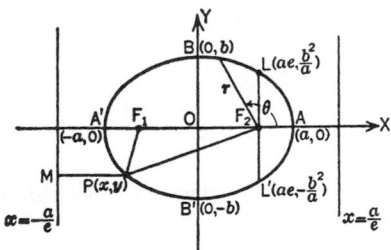

Fig. 92

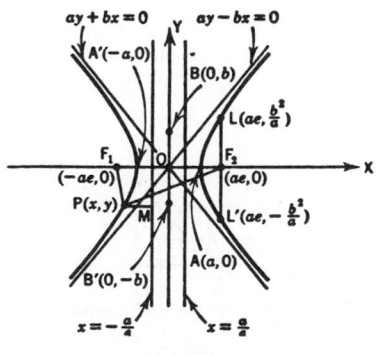

Fig. 93

The equation of the tangent to the hyperbola (Fig. 93) at the point (x_1, y_1) is

$$\frac{xx_1}{a^2} - \frac{yy_1}{b^2} = 1 \qquad (52)$$

The equation of the tangent whose slope is m is

$$y = mx \pm \sqrt{a^2m^2 - b^2} \qquad (53)$$

The equation of the normal to the hyperbola at the point (x_1, y_1) is

$$a^2y_1(x - x_1) + b^2x_1(y - y_1) = 0 \qquad (54)$$

Conjugate Hyperbolas and Diameters. The two hyperbolas

$$\frac{x^2}{a^2} - \frac{y^2}{b^2} = 1 \qquad \frac{y^2}{b^2} - \frac{x^2}{a^2} = 1$$

are conjugate. The transverse axis of each is the conjugate axis of the other.

If the slopes of the two lines $y = mx$ and $y = m_1 x$ through the center O are connected by the relation $mm_1 = b^2/a^2$, each of these lines bisects all chords of the hyperbola that are parallel to the other line. Two such lines are called *conjugate diameters*. The equation of the hyperbola referred to its conjugate diameters as oblique axes is

$$\frac{x'^2}{a_1^2} - \frac{y'^2}{b_1^2} = 1 \qquad (55)$$

where $2a_1$ and $2b_1$ are the conjugate axes.

Asymptotes. The lines $y = (b/a)x$ and $y = -(b/a)x$ are the *asymptotes* of the hyperbola $x^2/a^2 - y^2/b^2 = 1$. The asymptotes are two tangents whose points of contact with the curve are at an infinite distance from the center. The equation of the hyperbola when referred to its asymptotes as oblique axes is

$$4x'y' = a^2 + b^2 \qquad (56)$$

If $a = b$, the asymptotes are the perpendicular lines $y = x, y = -x$; the corresponding hyperbola

$$x^2 - y^2 = a^2 \qquad (57)$$

is called the *rectangular* or *equilateral hyperbola*.

Other Forms of the Equation of the Hyperbola

$$\frac{x^2}{a^2} - \frac{y^2}{a^2(e^2 - 1)} = 1 \qquad (58)$$

$$ax^2 + by^2 + 2gx + 2fy + c = 0 \qquad (59)$$

If a and b have unlike signs, (59) is a hyperbola with axes parallel to the coordinate axes.

The parametric form is

$$x = a \sec \phi \quad y = a \tan \phi \qquad (60)$$

Parabola. The equation of the parabola is

$$(y - y_0)^2 = 4a(x - x_0) \qquad (61)$$

If $(x_0, y_0) = (0, 0)$, the vertex is at the origin (Fig. 94); the focus F is on OX, called the *axis of the parabola*, and has the coordinates $(a, 0)$; and the directrix is $x = -a$. The chord LL' through F is the latus rectum and has the length $4a$. The eccentricity $e = FP/PM = 1$.

The tangent to the parabola $y^2 = 4ax$ at the point (x_1, y_1) is

$$yy_1 = 2a(x + x_1) \qquad (62)$$

The equation of the tangent whose slope is m is

$$y = mx + \frac{a}{m} \qquad (63)$$

The normal to the parabola at the point (x_1, y_1) is

$$2a(y - y_1) + y_1(x - x_1) = 0 \qquad (64)$$

A diameter of the curve is a straight line parallel to the axis. It bisects all chords parallel to the tangent at the point where the diameter meets the parabola.

If $P_1 T$ is tangent to the curve at (x_1, y_1), then $TQ = 2x_1$ is the *subtangent*, and $QN = 2a$ (a constant) is the *subnormal*, where $P_1 N$ is perpendicular to $P_1 T$.

The equation of the form $y^2 + 2gx + 2fy + c = 0$, where $g \neq 0$, is a parabola whose axis is parallel to OX; and the equation $x^2 + 2gx + 2fy + c = 0$, where $f \neq 0$, is a parabola whose axis is parallel to OY.

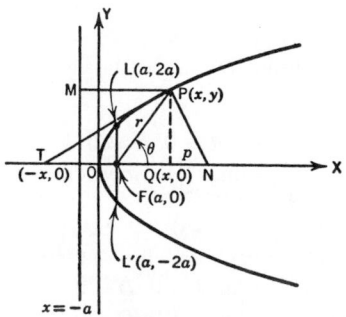

Fig. 94

The parabola referred to the tangents at the extremities of its latus rectum as axes of coordinates is

$$x^{1/2} \pm y^{1/2} = b^{1/2} \qquad (65)$$

where b is the distance from the origin to each point of tangency.

Polar Equations of the Conics. If e is the eccentricity, the directrix is vertical, the focus is at a distance p to the right or left of it, respectively, and the polar origin is taken at the focus, the polar equation is

$$r = \begin{cases} \dfrac{ep}{1 \mp e \cos\theta} & \text{for ellipse, hyperbola, or parabola} \\[2mm] & \qquad\qquad\qquad (66) \\[2mm] \dfrac{a(1 - e^2)}{1 \mp e \cos\theta} & \text{for ellipse or circle} \qquad (67) \\[2mm] \dfrac{a(e^2 - 1)}{1 \mp e \cos\theta} & \text{for hyperbola} \qquad\quad (68) \end{cases}$$

If the directrix is horizontal and the focus is at a distance p above or below it, respectively, the polar equation is

$$r = \begin{cases} \dfrac{ep}{1 \mp e \sin\theta} & \text{for ellipse, hyperbola, or parabola} \\[2mm] & \qquad\qquad\qquad (69) \\[2mm] \dfrac{a(1 - e^2)}{1 \mp e \sin\theta} & \text{for ellipse or circle} \qquad (70) \\[2mm] \dfrac{a(e^2 - 1)}{1 \mp e \sin\theta} & \text{for hyperbola} \qquad\quad (71) \end{cases}$$

General Equation of a Conic Section. This equation has the form

$$ax^2 + 2hxy + by^2 + 2gx + 2fy + c = 0 \qquad (72)$$

Let

$$D = \begin{vmatrix} a & h & g \\ h & b & f \\ g & f & c \end{vmatrix} \qquad d = \begin{vmatrix} a & h \\ h & b \end{vmatrix} \qquad \delta = a + b$$

$$(73)$$

Then the following is a classification of conic sections:

1. A parabola for $d = 0, D \neq 0$
2. Two parallel lines (possibly coincident or imaginary) for $d = 0, D = 0$
3. An ellipse for $d > 0, \delta D < 0$
4. No locus (imaginary ellipse) for $d > 0, \delta D > 0$
5. Point ellipse for $d > 0, D = 0$
6. A hyperbola for $d < 0, D \neq 0$
7. Two intersecting lines for $d < 0, D = 0$

Let $A + B = a + b$, $AB = ab - h^2 = d$, and $A - B$ have the same sign as h. Let $c' = D/d$; then the equation of the conic referred to its axes is

$$\frac{x^2}{-c'/A} + \frac{y^2}{-c'/B} = 1 \qquad (74)$$

To find the center (x_0, y_0) of the conic solve the equations

$$ax_0 + hy_0 + g = 0 \qquad hx_0 + by_0 + f = 0 \qquad (75)$$

To remove the term in xy from (64), rotate the axes about the origin through an angle θ such that $\tan 2\theta = 2h/(a - b)$.

7.4 Higher Plane Curves

Plane Curves. The point (x, y) describes a plane curve if x and y are continuous functions of a variable t (parameter), as $x = x(t)$, $y = y(t)$. The elimination of t from the two equations gives $F(x, y) = 0$ or in explicit form $y = f(x)$. The angle τ, which a tangent to the curve makes with OX, can be found from

$$\sin\tau = \frac{dy}{ds} \qquad \cos\tau = \frac{dx}{ds} \qquad \tan\tau = \frac{dy}{dx} = y' \quad (76)$$

where ds is the element of arc length:

$$ds = \sqrt{dx^2 + dy^2} = \sqrt{1 + y'^2}\, dx \qquad (77)$$

In polar coordinates,

$$ds = \sqrt{dr^2 + r^2\, d\theta^2} = \sqrt{\left(\frac{dr}{d\theta}\right)^2 \theta p + r^2} \qquad (78)$$

From (Fig. 95), it may be seen that

$$\sin\psi = \frac{r\, d\theta}{ds} \qquad \cos\psi = \frac{dr}{ds} \qquad \tan\Psi = \frac{r\, d\theta}{dr} \qquad (79)$$

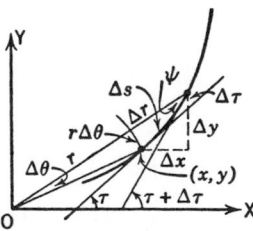

Fig. 95

The equation of the tangent to the curve $F(x, y) = 0$ at the point (x_1, y_1) is

$$\left(\frac{\partial F}{\partial x}\right)_{x=x_1, y=y_1}(x - x_1)$$

$$+ \left(\frac{\partial F}{\partial y}\right)_{x=x_1, y=y_1}(y - y_1) = 0 \qquad (80)$$

The equation of the normal to the curve $F(x, y) = 0$ at the point (x_1, y_1) is

$$\left(\frac{\partial F}{\partial y}\right)_{x=x_1, y=y_1}(x - x_1)$$

$$- \left(\frac{\partial F}{\partial x}\right)_{x=x_1, y=y_1}(y - y_1) = 0 \qquad (81)$$

The equation of the tangent to the curve $y = f(x)$ at the point (x_1, y_1) is

$$y - y_1 = \left(\frac{dy}{dx}\right)_{x=x_1}(x - x_1) \qquad (82)$$

The equation of the normal to the curve $y = f(x)$ at the point (x_1, y_1) is

$$y - y_1 = -\frac{1}{(dy/dx)_{x=x_1}}(x - x_1) \qquad (83)$$

The radius of curvature of the curve at the point (x, y) is

$$\rho = \frac{ds}{d\tau} = \frac{[1 + (dy/dx)^2]^{3/2}}{d^2y/dx^2} = \frac{[1 + y'^2]^{3/2}}{y''} \qquad (84)$$

The reciprocal $1/\rho$ is called the *curvature of the curve* at (x, y).

The coordinates (x_0, y_0) of the center of curvature for the point (x, y) on the curve [the center of the circle of curvature tangent to the curve at (x, y) and of radius ρ] are

$$x_0 = x - \rho\frac{dy}{ds} = x - y'\frac{[1 + y'^2]}{y''}$$

$$y_0 = y + \rho\frac{dx}{ds} = y + \frac{[1 + y'^2]}{y''}$$

A curve has a *singular point* if, simultaneously,

$$F(x, y) = 0 \qquad \frac{\partial F}{\partial x} = 0 \qquad \frac{\partial F}{\partial y} = 0 \qquad (85)$$

Let

$$D = \left(\frac{\partial^2 F}{\partial x \partial y}\right)^2 - \frac{\partial^2 F}{\partial x^2}\frac{\partial^2 F}{\partial y^2} \qquad (86)$$

Then for $D > 0$, the curve has a *double point* with two real different tangents. For $D = 0$, the curve has a *cusp* with two coincident tangents. For $D < 0$, the curve has an *isolated point* with no real tangents. See Figs. 96–100 for special curves.

For l large compared with d,

$$s \approx l\left[1 + \frac{2}{3}\left(\frac{2d}{l}\right)^2\right]$$

Semicubic, or Neil's, Parabola

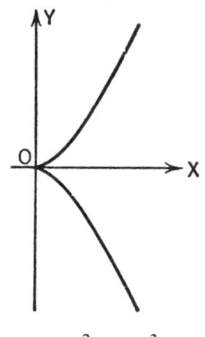

$$y^2 = ax^3$$

Fig. 96

Logarithmic Curve

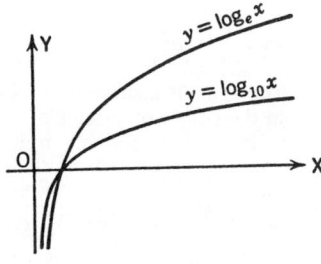

$$y = \log_b x$$

Fig. 97

Exponential Curve

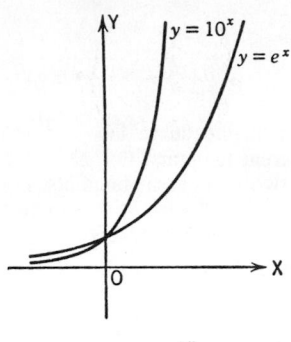

$$y = b^x$$

Fig. 98

Catenary

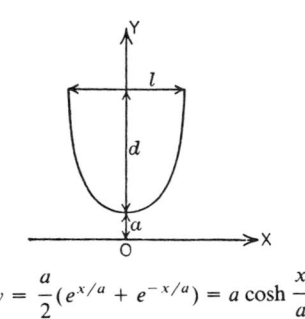

$$y = \frac{a}{2}(e^{x/a} + e^{-x/a}) = a \cosh \frac{x}{a}$$

Fig. 99

Damped Wave

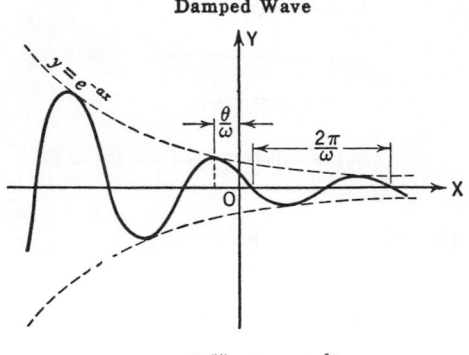

$$y = e^{-ax}\cos(\omega x + \theta)$$

Fig. 100

Cycloid
$a = b$

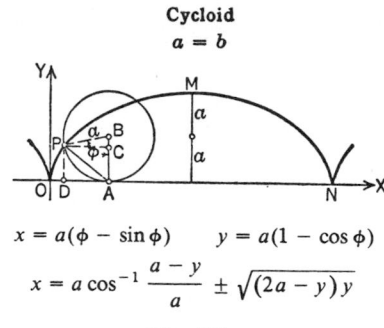

$$x = a(\phi - \sin \phi) \qquad y = a(1 - \cos \phi)$$

$$x = a \cos^{-1} \frac{a - y}{a} \pm \sqrt{(2a - y)y}$$

Fig. 101

Prolate Cycloid
$a < b$

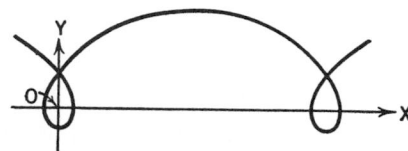

Fig. 102

Trochoid. This is a curve traced by a point at a distance b from the center of a circle of radius a as the circle rolls on a straight line:

$$x = a\phi - b \sin \phi \qquad y = a - b \cos \phi$$

See Figs. 101–103 for cycloids. For one arch, arc length $= 8a$, area $= 3\pi a^2$.

Hypotrochoid. This is a curve traced by a point at a distance b from the center of a circle of radius a as the circle rolls on the inside of a fixed circle of radius R:

$$x = (R - a) \cos \phi + b \cos \frac{R - a}{a} \phi$$

$$y = (R - a) \sin \phi - b \sin \frac{R - a}{a} \phi$$

Hypocycloid. $b = a$ (Fig. 104).

Epitrochoid. This is a curve traced by a point at a distance b from the center of a circle of radius a as the circle rolls on the outside of a fixed circle of radius R. See Figs. 105 and 106.

Other forms of the right-hand side of the equation, $b + 2a \sin \theta, b - 2a \cos \theta, b - 2a \sin \theta$, give curves rotated through 1, 2, 3 right angles, respectively. See Figs. 107–110.

In Fig. 111, as $\theta \to \infty, r \to 0$. The curve winds an indefinite number of times around the origin. As

Curtate Cycloid
$a > b$

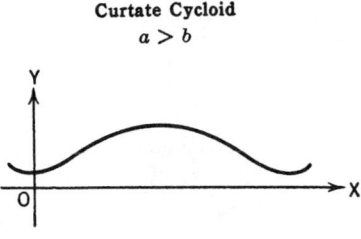

Fig. 103

Hypocycloid of Four Cusps, or Astroid
$b = a = 1/4\,R$

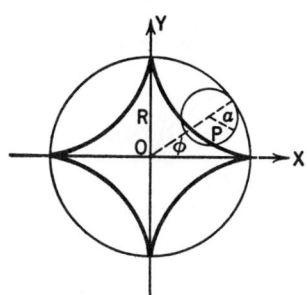

$$x = R \cos^3\phi, \quad y = R \sin^3\phi$$
$$x^{2/3} + y^{2/3} = R^{2/3}$$

Fig. 104

Epicycloid
$b = a$

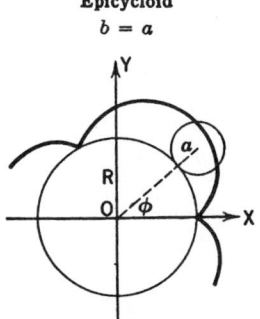

Fig. 105

Limacon of Pascal

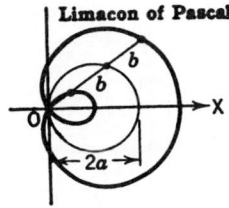

$b < 2a$
$r = b + 2a\cos\theta$

Fig. 106

Limacon of Pascal

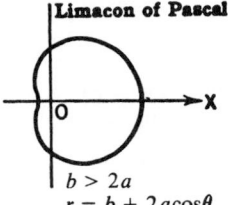

$b > 2a$
$r = b + 2a\cos\theta$

Fig. 107

Cardioid

Limaçon in which $b = 2a$
Epicycloid in which $R = a$

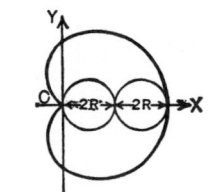

$$r = 2a(1 + \cos\theta)$$
$$(x^2 + y^2 - 2ax)^2 = 4a^2(x^2 + y^2)$$

Fig. 108

$\theta \to 0, r \to \infty$. The curve has an asymptote parallel to the polar axis at a distance a.

In Fig. 112, the tangent to the curve at any point makes a constant angle $\alpha(= \cot^{-1} m)$ with the radius vector. As $\theta \to -\infty, r \to 0$. The curve winds an indefinite number of times around the origin.

Figure 113 illustrates the locus of a point P, the product of whose distances from two fixed points F_1 and F_2 is equal to the square of half the distance between them, $r_1 \cdot r_2 = c^2$.

The roses $r = a \sin n\theta$ and $r = a \cos n\theta$ have, for n even, $2n$ leaves; for n odd, n leaves.

In Fig. 118, the locus of point P is such that $OP = AB$.

In Fig. 119, if the line AB rotates about A, intersecting the y axis at B, and if $PB = BP' = OB$, the locus of P and P' is the strophoid.

Figure 123 illustrates the locus of one end P of tangent line of length a as the other end Q is moved along the x axis.

In Fig. 126, $y = \cos \pi/2t^2$, $\Omega(t) = \pi t$.

Involute of a Circle

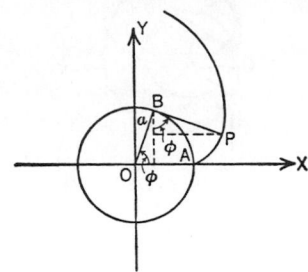

$$x = a(\cos\phi + \phi\sin\phi)$$

$$y = a(\sin\phi - \phi\cos\phi)$$

$$\theta = \sqrt{r^2/a^2 - 1} - \tan^{-1}\sqrt{r^2/a^2 - 1}$$

Spiral traced by the end of a taut string
unwinding from a circle.

Fig. 109

8 SOLID ANALYTIC GEOMETRY

8.1 Coordinate Systems

Right-Hand Rectangular   (Fig. 127). The position
of a point $P(x, y, z)$ is fixed by its distances x, y, z
from the mutually perpendicular planes yz, xz, and xy,
respectively.

Spherical, or Polar   (Fig. 128). The position of
a point $P(r, \theta, \phi)$ is fixed by its distance from a
given point O, the origin, and its direction from O,
determined by the angles θ and ϕ.

Cylindrical   (Fig. 128). The position of a point
$P(\rho, \phi, z)$ is fixed by its distance z from a given plane
and the polar coordinates (ρ, ϕ) of the projection Q
of P on the given plane.

Hyperbolic, or Reciprocal, Spiral

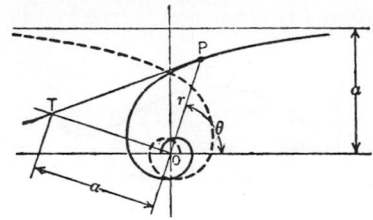

$$r\theta = a$$

Polar subtangent $OT = -a$

Fig. 111

Logarithmic, or Equiangular, Spiral

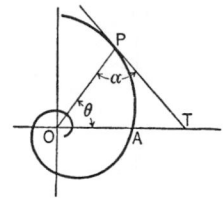

$$r = ae^{m\theta} \qquad m > 0$$

$$\ln\frac{r}{a} = m\theta$$

Fig. 112

Relations among coordinates of the three systems
are

$$x = r\sin\theta\cos\phi = \rho\cos\phi \tag{88}$$

$$y = r\sin\theta\sin\phi = \rho\sin\phi \tag{89}$$

$$z = r\cos\theta \tag{90}$$

$$\rho = \sqrt{x^2 + y^2} = r\sin\theta \tag{91}$$

Spiral of Archimedes

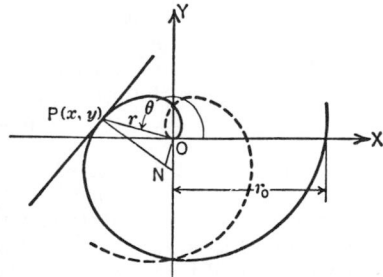

$$r = a\theta$$

Polar subnormal $ON = a$

Length of arc $OP = s = \frac{1}{2}a(\theta\sqrt{1 + \theta^2}$
$$+ \sinh^{-1}\theta)$$

For many turns, $s \approx \frac{1}{2}a\theta^2$

Fig. 110

Lemniscate of Bernoulli

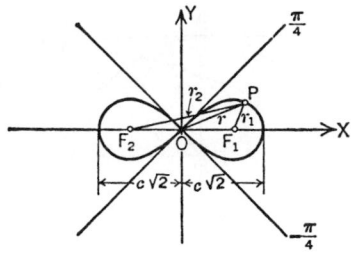

$$(x^2 + y^2)^2 + 2c^2(y^2 - x^2) = 0$$
$$r^2 = 2c^2\cos 2\theta$$

Fig. 113

Three-leaved Roses

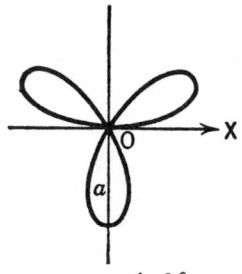

$$r = a \sin 3\theta$$

Fig. 114

Three-leaved Roses

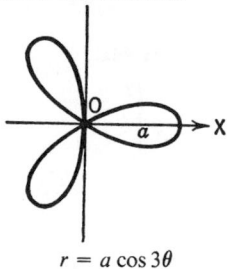

$$r = a \cos 3\theta$$

Fig. 115

Four-leaved Roses

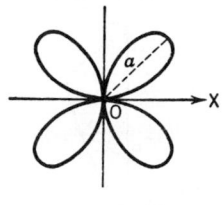

$$r = a \sin 2\theta$$

Fig. 116

Four-leaved Roses

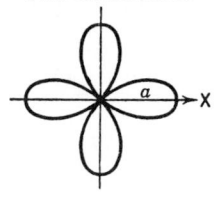

$$r = a \cos 2\theta$$

Fig. 117

Cissoid of Diocles

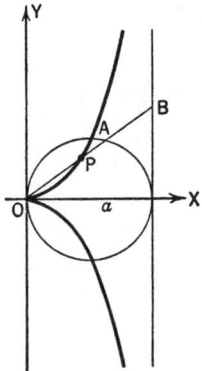

$$y^2 = \frac{x^3}{a - x}$$
$$r = a(\sec \theta - \cos \theta)$$

Fig. 118

$$\phi = \tan^{-1} \frac{y}{x} \qquad (92)$$

$$r = \sqrt{x^2 + y^2 + z^2} = \sqrt{\rho^2 + z^2} \qquad (93)$$

$$\theta = \tan^{-1} \frac{\sqrt{x^2 + y^2}}{z} = \tan^{-1} \frac{\rho}{z} \qquad (94)$$

8.2 Point, Line, and Plane

Euclidean Distance between Two Points. This distance between $P_1(x_1, y_1, z_1)$ and $P_2(x_2, y_2, z_2)$ is

$$s = \sqrt{(x_2 - x_1)^2 + (y_2 - y_1)^2 + (z_2 - z_1)^2} \qquad (95)$$

Strophoid

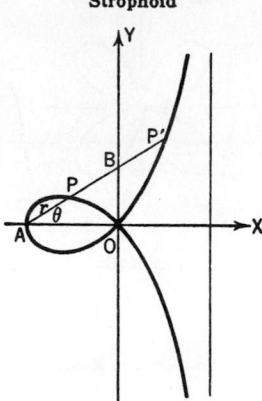

$$y^2 = \frac{x^2(a+x)}{a-x}$$

$$r = a(\sec\theta - \tan\theta)$$

Fig. 119

Conchoid of Nicomedes

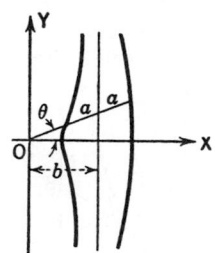

$$(x^2+y^2)(x-b)^2 = a^2 x^2$$
$$r = b\sec\theta - a$$

Fig. 120

Witch of Agnesi

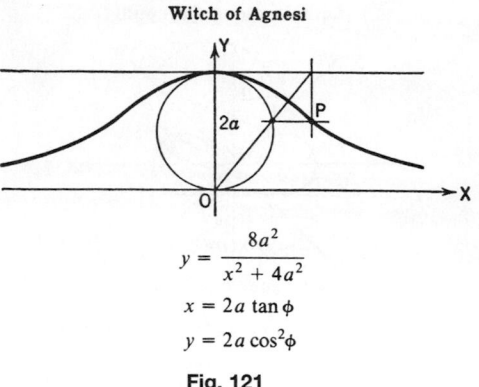

$$y = \frac{8a^2}{x^2 + 4a^2}$$

$$x = 2a\tan\phi$$

$$y = 2a\cos^2\phi$$

Fig. 121

Folium of Descartes

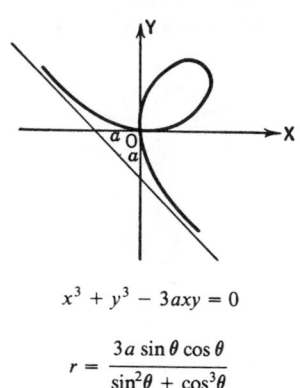

$$x^3 + y^3 - 3axy = 0$$

$$r = \frac{3a\sin\theta\cos\theta}{\sin^2\theta + \cos^3\theta}$$

Fig. 122

Tractrix

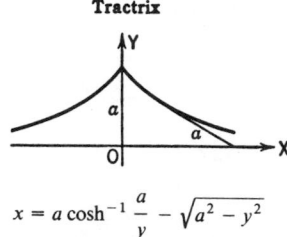

$$x = a\cosh^{-1}\frac{a}{y} - \sqrt{a^2 - y^2}$$

Fig. 123

To divide the segment $P_1 P_2$ in the ratio c_1/c_2, internally or externally,

$$x = \frac{c_2 x_1 \pm c_1 x_2}{c_2 \pm c_1} \qquad y = \frac{c_2 y_1 \pm c_1 y_2}{c_2 \pm c_1}$$

$$z = \frac{c_2 z_1 \pm c_1 z_z}{c_2 \pm c_1} \tag{96}$$

The *midpoint* of $P_1 P_2$ is

$$x = \tfrac{1}{2}(x_1 + x_2) \qquad y = \tfrac{1}{2}(y_1 + y_2)$$

$$z = \tfrac{1}{2}(z_1 + z_2) \tag{97}$$

Angles.   The angles α, β, γ that the line $P_1 P_2$ makes with the coordinate directions x, y, z, respectively, are the *direction angles* of $P_1 P_2$. The consines

$$\cos\alpha = \frac{x_2 - x_1}{s} \qquad \cos\beta = \frac{y_2 - y_1}{s}$$

$$\cos\gamma = \frac{z_2 - z_1}{s} \tag{98}$$

Circles in Polar Coordinates

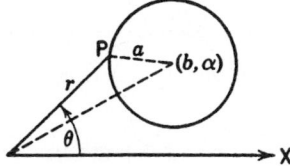

$$r^2 + b^2 - 2rb\cos(\theta - \alpha) = a^2$$

Center at (b, α), radius a

Fig. 124

Frequency-modulated Wave

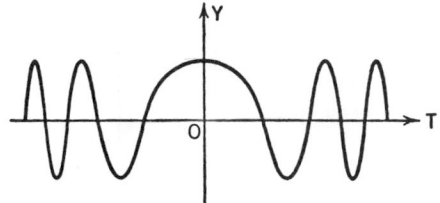

$$y = k\cos[\phi(t)] \qquad \text{instantaneous frequency} = \Omega(t) = \frac{d\phi}{dt}$$

Fig. 126

Circles in Polar Coordinates

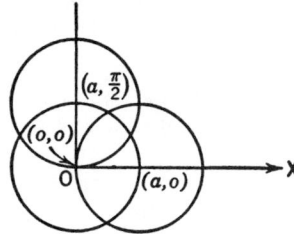

Center $(0, 0)$ $r = a$

Center $(a, 0)$ $r = 2a\cos\theta$

Center $\left(a, \dfrac{\pi}{2}\right)$ $r = 2a\sin\theta$

Fig. 125

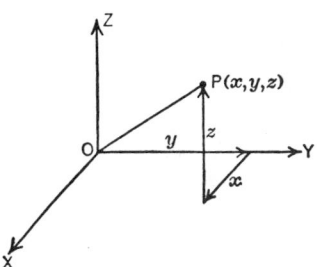

Fig. 127

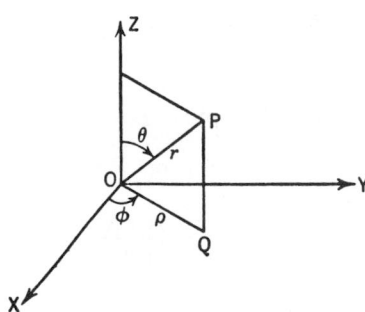

Fig. 128

are the *direction cosines* of P_1P_2, and

$$\cos^2 x + \cos^2 \beta + \cos^2 \gamma = 1 \qquad (99)$$

If $\ell : m : n = \cos\alpha : \cos\beta : \cos\gamma$, then

$$\cos\alpha = \frac{l}{\sqrt{l^2 + m^2 + n^2}} \qquad \cos\beta = \frac{m}{\sqrt{l^2 + m^2 + n^2}}$$

$$\cos\gamma = \frac{n}{\sqrt{l^2 + m^2 + n^2}} \qquad (100)$$

The angle θ between two lines in terms of their direction angles $\alpha_1, \beta_1, \gamma_1$ and $\alpha_2, \beta_2, \gamma_2$ is obtained from

$$\cos\theta = \cos\alpha_1\cos\alpha_2 + \cos\beta_1\cos\beta_2 + \cos\gamma_1\cos\gamma_2 \qquad (101)$$

If $\cos\theta = 0$, the lines are perpendicular to each other.

Planes. A plane is represented by

$$Ax + By + Cz + D = 0 \qquad (102)$$

If one of the variables is missing, the plane is parallel to the axis of the missing variable. For example, $Ax + By + D = 0$ represents a plane parallel to the z axis. If two of the variables are missing, the plane is parallel to the plane of the missing variables. For example, $z = k$ represents a plane parallel to the xy plane and k units from it.

A plane through *three points* $P_1(x_1, y_1, z_1)$, $P_2(x_2, y_2, z_2)$, and $P_3(x_3, y_3, z_3)$ has the equation

$$\begin{vmatrix} x & y & z & 1 \\ x_1 & y_1 & z_1 & 1 \\ x_2 & y_2 & z_2 & 1 \\ x_3 & y_3 & z_3 & 1 \end{vmatrix} = 0 \qquad (103)$$

The equation of a plane whose x, y, z intercepts are, respectively, a, b, c (Fig. 129) is

$$\frac{x}{a} + \frac{y}{b} + \frac{z}{c} = 1 \qquad (104)$$

The perpendicular form of the equation of a plane, where $OP = p$ is the perpendicular distance of the plane from the origin O and has the direction angles α, β, γ, is

$$x \cos \alpha + y \cos \beta + z \cos \gamma - p = 0 \qquad (105)$$

To bring the general form $Ax + By + Cz + D = 0$ into the perpendicular form, divide it by $\pm\sqrt{A^2 + B^2 + C^2}$, where the sign before the radical is opposite to that of D.

The coefficients A, B, C are proportional to the direction cosines λ, μ, ν of a line perpendicular to the plane. Therefore,

$$A(x - x_1) + B(y - y_1) + C(z - z_1) = 0 \qquad (106)$$

is a plane through $P_1(x_1, y_1, z_1)$ and perpendicular to a line with direction cosines λ, μ, ν proportional to A, B, C.

Perpendicular Distance between Point and Plane. The distance between point P_1 from a plane $Ax + By + Cz + D = 0$ is given by

$$PP_1 = \frac{Ax_1 + By_1 + Cz_1 + D}{\pm\sqrt{A^2 + B^2 + C^2}} \qquad (107)$$

where the sign before the radical is opposite to that of D.

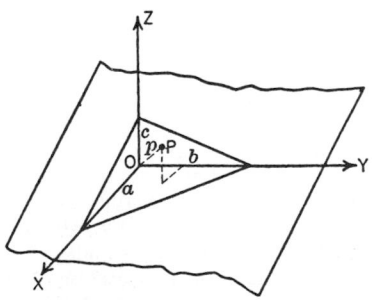

Fig. 129

Parallel Planes. Two planes $A_1x + B_1y + C_1z + D_1 = 0$ and $A_2x + B_2y + C_2z + D_2 = 0$ are parallel if $A_1 : B_1 : C_1 = A_2 : B_2 : C_2$;

$$A(x - x_1) + B(y - y_1) + C(z - z_1) = 0 \qquad (108)$$

is a plane through the point $P_1(x_1, y_1, z_1)$ and parallel to the plane $Ax + By + Cz + D = 0$.

Angle θ between Two Planes. The angle between $Ax + By + Cz + D = 0$ and $A_1x + B_1y + C_1z + D_1 = 0$ is the angle between two intersecting lines, each perpendicular to one of the planes:

$$\cos \theta = \frac{AA_1 + BB_1 + CC_1}{\pm\sqrt{(A^2 + B^2 + C^2)(A_1^2 + B_1^2 + C_1^2)}} \qquad (109)$$

The two planes are perpendicular if $AA_1 + BB_1 + CC_1 = 0$.

Points, Planes, and Lines. *Four points*, $P_k(x_k, y_k, z_k)(k = 1, 2, 3, 4)$, lie in the same plane if

$$\begin{vmatrix} 1 & x_1 & y_1 & z_1 \\ 1 & x_2 & y_2 & z_2 \\ 1 & x_3 & y_3 & z_3 \\ 1 & x_4 & y_4 & z_4 \end{vmatrix} = 0 \qquad (110)$$

Four planes, $A_kx + B_ky + C_kz + D_k = 0$ $(k = 1, 2, 3, 4)$, pass through the same point if

$$\begin{vmatrix} A_1 & B_1 & C_1 & D_1 \\ A_2 & B_2 & C_2 & D_2 \\ A_3 & B_3 & C_3 & D_3 \\ A_4 & B_4 & C_4 & D_4 \end{vmatrix} = 0 \qquad (111)$$

A *straight line* is represented as the intersection of two planes by two first-degree equations

$$A_1x + B_1y + C_1z + D_1 = 0$$
$$A_2x + B_2y + C_2z + D_2 = 0 \qquad (112)$$

The three planes through the line perpendicular to the coordinate planes are its *projecting planes*. The equation of the *xy* projecting plane is found by eliminating z between the two given equations, and so on. The line can be represented by any two of its projecting planes, for example,

$$y = m_1x + b_1 \qquad z = m_2x + b_2 \qquad (113)$$

If the line goes through a point $P_1(x_1, y_1, z_1)$ and has the direction angles α, β, γ, then

$$\frac{x - x_1}{\cos \alpha} = \frac{y - y_1}{\cos \beta} = \frac{z - z_1}{\cos \gamma} \qquad (114)$$

and

$$m_1 = \frac{\cos \beta}{\cos \alpha} \qquad m_2 = \frac{\cos \gamma}{\cos \alpha}$$

The equations of a line through two points (x_1, y_1, z_1) and (x_2, y_2, z_2) are

$$\frac{x - x_1}{x_2 - x_1} = \frac{y - y_1}{y_2 - y_1} = \frac{z - z_1}{z_2 - z_1} \qquad (115)$$

A line through a point P_1 perpendicular to a plane $Ax + By + Cz + D = 0$ has the equations

$$\frac{x - x_1}{A} = \frac{y - y_1}{B} = \frac{z - z_1}{C} \qquad (116)$$

Line of Intersection of Two Planes. The direction cosines λ, μ, ν of the line of intersection of two planes $Ax + By + Cz + D = 0$ and $A_1x + B_1y + C_1z + D_1 = 0$ are found from the ratios

$$\lambda : \mu : \nu = \begin{vmatrix} B & C \\ B_1 & C_1 \end{vmatrix} : \begin{vmatrix} C & A \\ C_1 & A_1 \end{vmatrix} : \begin{vmatrix} A & B \\ A_1 & B_1 \end{vmatrix} \qquad (117)$$

8.3 Transformation of Coordinates

Changing the Origin. Let the coordinates of a point P with respect to the original axes be x, y, z and with respect to the new axes x', y', z'. For a parallel displacement of the axes with x_0, y_0, z_0 the coordinates of the new origin

$$x = x_0 + x' \qquad y = y_0 + y' \qquad z = z_0 + z' \quad (118)$$

Rotation of the Axes about the Origin. Let the cosines of the angles of the new axes x', y', z' with the x axis be λ_1, μ_1, ν_1, with the y axis be λ_2, μ_2, ν_2, with the z axis be λ_3, μ_3, ν_3. Then

$$
\begin{aligned}
x &= \lambda_1 x' + \mu_1 y' + \nu_1 z' & x' &= \lambda_1 x + \lambda_2 y + \lambda_3 z \\
y &= \lambda_2 x' + \mu_2 y' + \nu_2 z' & y' &= \mu_1 x + \mu_2 y + \mu_3 z \\
z &= \lambda_3 x' + \mu_3 y' + \nu_3 z' & z' &= \nu_1 x + \nu_2 y + \nu_3 z
\end{aligned}
$$
$$(119)$$

The following relations exist:

$$
(1) \quad
\begin{aligned}
\lambda_1^2 + \mu_1^2 + \nu_1^2 &= 1 \\
\lambda_2^2 + \mu_2^2 + \nu_2^2 &= 1 \\
\lambda_3^2 + \mu_3^2 + \nu_3^2 &= 1
\end{aligned}
$$

$$
(2) \quad
\begin{aligned}
\lambda_1^2 + \lambda_2^2 + \lambda_3^2 &= 1 \\
\mu_1^2 + \mu_2^2 + \mu_3^2 &= 1 \\
\nu_1^2 + \nu_2^2 + \nu_3^2 &= 1
\end{aligned}
$$

$$
(3) \quad
\begin{aligned}
\lambda_1\lambda_2 + \mu_1\mu_2 + \nu_1\nu_2 &= 0 \\
\lambda_2\lambda_3 + \mu_2\mu_3 + \nu_2\nu_3 &= 0 \\
\lambda_3\lambda_1 + \mu_3\mu_1 + \nu_3\nu_1 &= 0
\end{aligned}
$$

$$
(4) \quad
\begin{aligned}
\lambda_1\mu_1 + \lambda_2\mu_2 + \lambda_3\mu_3 &= 0 \\
\mu_1\nu_1 + \mu_2\nu_2 + \mu_3\nu_3 &= 0 \\
\nu_1\lambda_1 + \nu_2\lambda_2 + \nu_3\lambda_3 &= 0
\end{aligned}
$$

$$
(5) \quad
\begin{aligned}
\lambda_1 &= \mu_2\nu_3 - \nu_2\mu_3 \\
\mu_1 &= \nu_2\lambda_3 - \lambda_2\nu_3 \\
\nu_1 &= \lambda_2\mu_3 - \mu_2\lambda_3
\end{aligned}
$$

$$
(6) \quad
\begin{aligned}
\lambda_2 &= \nu_1\mu_3 - \mu_1\nu_3 \\
\mu_2 &= \lambda_1\nu_3 - \nu_1\lambda_3 \\
\nu_2 &= \mu_1\lambda_3 - \lambda_1\mu_3
\end{aligned}
$$

$$
(7) \quad
\begin{aligned}
\lambda_3 &= \mu_1\nu_2 - \nu_1\mu_2 \\
\mu_3 &= \nu_1\lambda_2 - \lambda_1\nu_2 \\
\nu_3 &= \lambda_1\mu_2 - \mu_1\lambda_2
\end{aligned}
$$

$$
(8) \quad
\begin{vmatrix}
\lambda_1 & \mu_1 & \nu_1 \\
\lambda_2 & \mu_2 & \nu_2 \\
\lambda_3 & \mu_3 & \nu_3
\end{vmatrix} = 1
$$

For a combination of displacement and rotation, apply the corresponding equations simultaneously.

8.4 Quadric Surfaces

The *general form* of the *equation of a surface of the second degree* is

$$
\begin{aligned}
F(x, y, z) \equiv{}& a_{11}x^2 + 2a_{12}xy + 2a_{13}xz + a_{22}y^2 \\
&+ 2a_{23}yz + a_{33}z^2 + 2a_{14}x \\
&+ 2a_{24}y + 2a_{34}z + a_{44} = 0 \qquad (120)
\end{aligned}
$$

where the a_{ik} are constants and $a_{ik} = a_{ki}$, that is, $a_{12} = a_{21}$, and so on. Let

$$
D = \begin{vmatrix}
a_{11} & a_{12} & a_{13} & a_{14} \\
a_{21} & a_{22} & a_{23} & a_{24} \\
a_{31} & a_{32} & a_{33} & a_{34} \\
a_{41} & a_{42} & a_{43} & a_{44}
\end{vmatrix}
\qquad
d = \begin{vmatrix}
a_{11} & a_{12} & a_{13} \\
a_{21} & a_{22} & a_{23} \\
a_{31} & a_{32} & a_{33}
\end{vmatrix}
$$

Let $I \equiv a_{11} + a_{22} + a_{33}$ and $J \equiv a_{22}a_{33} + a_{33}a_{11} + a_{11}a_{22} - a_{23}^2 - a_{13}^2 - a_{12}^2$. Here, D, d, I, and J are invariant under coordinate transformation. The following is a classification of the quadratic surfaces, so far as they are real and do not degenerate into curves in one plane:

Ellipsoid for $D < 0, Id > 0, J > 0$

Hyperboloid of two sheets for $D < 0, Id$ and J not both >0

Hyperboloid of one sheet for $D > 0, Id$ and J not both >0

Cone for $D = 0, d \neq 0, Id$ and J not both >0

Elliptic paraboloid for $D < 0, d = 0, J > 0$

Hyperbolic paraboloid for $D > 0, d = 0, J < 0$

Cylinder for $D = 0, d = 0$

Ellipsoid and Hyperboloids. Consider the center of the quadric as the origin and the principal axes of the quadric as the orthogonal coordinate axes. Then

$$\frac{x^2}{a^2} + \frac{y^2}{b^2} + \frac{z^2}{c^2} = 1 \quad [\textit{ellipsoid (Fig. 130)}] \quad (121)$$

$$\frac{x^2}{a^2} + \frac{y^2}{b^2} - \frac{z^2}{c^2} = 1 \quad [\textit{hyperboloid of one sheet (Fig. 131)}] \quad (122)$$

$$\frac{x^2}{a^2} + \frac{y^2}{b^2} + \frac{z^2}{c^2} = -1 \quad [\textit{hyperboloid of two sheets (Fig. 132)}] \quad (123)$$

where a, b, c are the semiaxes.

The length of the semiaxis is found from

$$a^2 = -\frac{D}{\lambda_1 d} \qquad b^2 = -\frac{D}{\lambda_2 d} \qquad c^2 = -\frac{D}{\lambda_3 d} \quad (124)$$

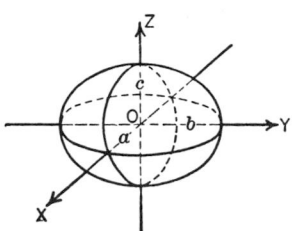

Fig. 130

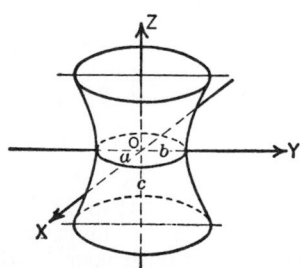

Fig. 131

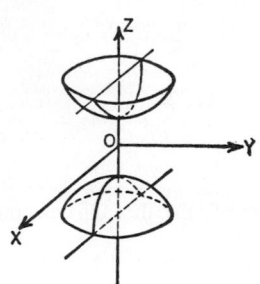

Fig. 132

where $\lambda_1, \lambda_2, \lambda_3$ are the real roots of the cubic equation

$$\begin{vmatrix} a_{11} - \lambda & a_{12} & a_{13} \\ a_{12} & a_{22} - \lambda & a_{23} \\ a_{13} & a_{23} & a_{33} - \lambda \end{vmatrix} = 0 \quad (125)$$

Cone. The equation

$$ax^2 + by^2 + cz^2 + 2hxy + 2gxz + 2fyz = 0 \quad (126)$$

represents a cone with vertex at the origin. If the cross section of the cone is an ellipse with axes $2a$ and $2b$ whose plane is parallel to the xy plane and at a distance c from the origin, then the equation of the cone with vertex at the origin is

$$\frac{x^2}{a^2} + \frac{y^2}{b^2} - \frac{z^2}{c^2} = 0 \quad (127)$$

If $a = b$, the cross section is circular and the cone is a cone of revolution.

Sphere. An equation of the form

$$x^2 + y^2 + z^2 + ax + by + cz + d = 0 \quad (128)$$

represents a sphere with radius

$$r = \tfrac{1}{2}\sqrt{a^2 + b^2 + c^2 - 4d} \quad (129)$$

and center

$$x_0 = -\tfrac{1}{2}a \qquad y_0 = -\tfrac{1}{2}b \qquad z_0 = -\tfrac{1}{2}c \quad (130)$$

If (x_0, y_0, z_0) are the coordinates of the center and r is the radius, then the equation of the sphere is

$$(x - x_0)^2 + (y - y_0)^2 + (z - z_0)^2 = r^2 \quad (131)$$

If $x_0 = 0, y_0 = 0, z_0 = 0$, then the equation is

$$x^2 + y^2 + z^2 = r^2 \quad (132)$$

Paraboloids. The equation

$$\frac{x^2}{a^2} + \frac{y^2}{b^2} = 2cz \qquad (133)$$

represents an *elliptic paraboloid* (Fig. 133).
If $a = b$, the equation is of the form

$$x^2 + y^2 = 2cz \quad (paraboloid\ of\ revolution) \qquad (134)$$

The equation

$$\frac{x^2}{a^2} - \frac{y^2}{b^2} = 2cz \quad [hyperbolic\ paraboloid\ (Fig.134)] \qquad (135)$$

Cylinder. The equation of a cylinder perpendicular to the *yz*, *xz*, or *xy* plane is the same as the equation of a section of the cylinder in the corresponding plane. Thus

$$\frac{x^2}{a^2} + \frac{y^2}{b^2} = 1 \qquad (136)$$

$$\frac{x^2}{a^2} - \frac{y^2}{b^2} = 1 \qquad (137)$$

$$y^2 = 4ax \qquad (138)$$

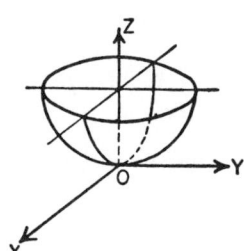

Fig. 133

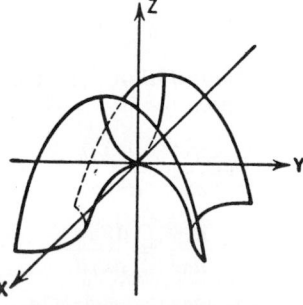

Fig. 134

are elliptic, hyperbolic, and parabolic cylinders, respectively, with elements or generators parallel to *OZ*.

Tangent Plane. The equation of the tangent plane to any quadric

$$F(x, y, z) \equiv a_{11}x^2 + 2a_{12}xy + 2a_{13}xz + a_{22}y^2$$
$$+ 2a_{23}yz + a_{33}z^2 + 2a_{14}x$$
$$+ 2a_{24}y + 2a_{34}z + a_{44} = 0 \qquad (139)$$

at the point (x_1, y_1, z_1) is

$$\left(\frac{\partial F}{\partial x}\right)_{x=x_1, y=y_1, z=z_1} (x - x_1)$$

$$+ \left(\frac{\partial F}{\partial y}\right)_{x=x_1, y=y_1, z=z_1} (y - y_1)$$

$$+ \left(\frac{\partial F}{\partial z}\right)_{x=x_1, y=y_1, z=z_1} (z - z_1) = 0 \qquad (140)$$

Example 47. Find the tangent plane to the hyperboloid of one sheet at point (x_1, y_1, z_1). Given $x^2/a^2 + y^2/b^2 - z^2/c^2 = 1$. Then

$$\left(\frac{\partial F}{\partial x}\right)_{x=x_1, y=y_1, z=z_1} (x - x_1)$$

$$+ \left(\frac{\partial F}{\partial y}\right)_{x=x_1, y=y_1, z=z_1} (y - y_1)$$

$$+ \left(\frac{\partial F}{\partial z}\right)_{x=x_1, y=y_1, z=z_1} (z - z_1)$$

$$= \frac{2x_1(x - x_1)}{a^2} + \frac{2y_1(y - y_1)}{b^2} - \frac{2z_1(z - z_1)}{c^2} = 0$$

$$\frac{xx_1}{a^2} + \frac{yy_1}{b^2} - \frac{zz_1}{c^2} - \frac{x_1^2}{a^2} - \frac{y_1^2}{b^2} + \frac{z_1^2}{c^2}$$

$$= \frac{xx_1}{a^2} + \frac{yy_1}{b^2} - \frac{zz_1}{c^2} - 1 = 0 \quad (tangent\ plane)$$

The Normal. The line through a point P_1 on a surface and perpendicular to the tangent plane at P_1 is called the normal to the surface at P_1.
The equations of the normal to the surface $F(x, y, z) = 0$ at the point (x_1, y_1, z_1) are

$$\frac{x - x_1}{\left(\dfrac{\partial F}{\partial x}\right)_{x=x_1, y=y_1, z=z_1}} = \frac{y - y_1}{\left(\dfrac{\partial F}{\partial y}\right)_{x=x_1, y=y_1, z=z_1}}$$

$$= \frac{z - z_1}{\left(\dfrac{\partial F}{\partial z}\right)_{x=x_1, y=y_1, z=z_1}} \qquad (141)$$

9 DIFFERENTIAL CALCULUS

9.1 Functions and Derivatives

Function.   If two variables x and y are so related that to each value of x in a given domain there corresponds a value of y, then y is a *function* of x in that domain. The variable x is the *independent* variable and y the *dependent* variable. The symbols $F(x)$, $f(x)$, $\phi(x)$, and so on, are used to represent functions of x; the symbol $f(a)$ represents the value of $f(x)$ for $x = a$.

Limit, Derivative, Differential.   The function $f(x)$ approaches the limit 1 as x approaches a if the difference $|f(x) - 1|$ can be made arbitrarily small for all values of x except a within a sufficiently small interval with a as midpoint. In symbols, $\lim_{x \to a} f(x) = 1$.

The symbols $\lim_{x \to a} f(x) = \infty$ or $\lim_{x \to a} f(x) = -\infty$ mean that, for all values of x except a within a sufficiently small interval with a as midpoint, the values of $f(x)$ can be made arbitrarily large positively or negatively, respectively.

The symbols $\lim_{x \to \infty} f(x) = 1$ and $\lim_{x \to -\infty} f(x) = 1$ mean that the difference $|f(x) - 1|$ can be made arbitrarily small for all values of x sufficiently large positively or negatively, respectively.

A change in x is called an *increment* of x and is denoted by Δx. The corresponding change in y is denoted by Δy. If

$$\lim_{\Delta x \to 0} \frac{f(x + \Delta x) - f(x)}{\Delta x}$$

exists, it is called the *derivative* of y with respect to x and is denoted by dy/dx, $f'(x)$, or $D_x y$.

The geometric interpretation of $f'(x)$ is

$$f'(x) = \frac{dy}{dx} = \tan \theta \qquad (142)$$

or $f'(x)$ is equal to the slope of the tangent to the curve $y = f(x)$ at the point $P(x, y)$ (Fig. 135):

$$\frac{RS}{PR} = \lim_{PR \to 0} \frac{RQ}{PR} = \lim_{\Delta x \to 0} \frac{\Delta y}{\Delta x}$$

$$= \lim_{\Delta x \to 0} \frac{f(x + \Delta x) - f(x)}{\Delta x} \qquad (143)$$

$$= \frac{dy}{dx} = f'(x) = \tan \theta$$

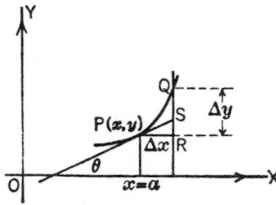

Fig. 135

The differentials of x and y, respectively, are

$$dx = \Delta x \qquad dy = f'(x)\ dx$$

Continuity.   A function is *continuous* at $x = b$ if it has a definite value at b and approaches that value as a limit whenever x approaches b as a limit. The notion of continuity at a point suggests that the graph of the function can be drawn without lifting pencil from paper at the point. The analytic conditions that $f(x)$ be continuous at b are that $f(b)$ have a definite value and that for an arbitrarily small positive number ε there exist a $\delta(\varepsilon)$ such that

$$|f(x) - f(b)| < \varepsilon \quad \text{for all values of } x$$

$$\text{for which } |x - b| < \delta(\varepsilon)$$

$$(144)$$

A function that is continuous at each point of an interval is said to be continuous in that interval. An example of a continuous function is $f(x) = x^2$. The function $\phi(x) = 1/(x - a)$ is continuous for all values of x except $x = a$, at which point it becomes infinite. Every differentiable function is continuous, although the reverse is not always true.

If, in the preceding definition of continuity, the number δ can be chosen the same for all points in the interval, the function is said to be *uniformly continuous* in that interval.

Derivatives of Higher Order.   The derivative of the *first derivative* of y with respect to x is called the *second derivative* of y with respect to x and is denoted by

$$\frac{d}{dx}\left(\frac{dy}{dx}\right) = \frac{d^2 y}{dx^2} = f''(x) = D_x^2 y \qquad (145)$$

By successive differentiations the nth derivative

$$\frac{d^n y}{dx^n} = f^{(n)}(x) = D_x^n y \qquad (146)$$

is obtained.

The nth differential of y is denoted by

$$d^n y = f^{(n)}(x)\ dx^n \qquad (147)$$

Parametric Differentiation.   To find the derivatives of y with respect to x if $y = y(t)$ and $x = x(t)$:

$$y' = \frac{dy}{dx} = \frac{dy/dt}{dx/dt} \qquad (148)$$

$$y'' = \frac{d^2 y}{dx^2} = \frac{dy'/dt}{dx/dt} \qquad (149)$$

$$y^{(n)} = \frac{d^n y}{dx^n} = \frac{dy^{(n-1)}/dt}{dx/dt} \qquad (150)$$

Example 48. Find the derivatives of y with respect to x for the ellipse $x = a \cos t$, $y = b \sin t$:

$$y' = \frac{dy}{dx} = \frac{b \cos t}{-a \sin t} = -\frac{b}{a} \cot t$$

$$y'' = \frac{dy'}{dx} = \frac{(b/a) \csc^2 t}{-a \sin t} = -\frac{b}{a^2} \csc^3 t$$

$$y''' = \frac{dy''}{dx} = \frac{(3b/a^2) \csc^3 t \cot t}{-a \sin t} = -\frac{3b}{a^3} \csc^4 t \cot t$$

Logarithmic Differentiation for Products and Quotients. If

$$y = \frac{u' v^m}{w^n} \qquad (151)$$

take the logarithms of both sides before differentiating:

$$\ln y = l \ln u + m \ln v - n \ln w \qquad (152)$$

$$\frac{1}{y} \frac{dy}{dx} = \frac{l}{u} \frac{du}{dx} + \frac{m}{v} \frac{dv}{dx} - \frac{n}{w} \frac{dw}{dx} \qquad (153)$$

$$\frac{dy}{dx} = y \left(\frac{l}{u} \frac{du}{dx} + \frac{m}{v} \frac{dv}{dx} - \frac{n}{w} \frac{dw}{dx} \right) \qquad (154)$$

Example 49. Find dy/dx if

$$y = \frac{\sqrt{x^2 - 25}}{(x-1)^3 (x+5)^2}$$

$$\ln y = \tfrac{1}{2} \ln(x^2 - 25) - 3 \ln(x-1) - 2 \ln(x+5)$$

$$\frac{1}{y} \frac{dy}{dx} = \frac{2x}{2(x^2 - 25)} - \frac{3}{x-1} - \frac{2}{x+5}$$

$$\frac{dy}{dx} = \frac{y(-4x^2 + 11x + 65)}{(x^2 - 25)(x-1)}$$

Mean Value Theorem. If $f(x)$ is single valued, continuous in the interval $a \le x \le b$, and has a derivative for all values of x between a and b, then there is a value $x = \xi, a < \xi < b$, such that

$$f(b) - f(a) = (b-a) f'(\xi) \qquad (155)$$

Another form is

$$f(x+h) = f(x) + h f'(x + \theta h) \qquad 0 < \theta < 1 \qquad (156)$$

Indeterminate Forms If a function $f(x)$ for $x = a$ (where a can also be ∞) has no determined value but

appears in one of the meaningless forms

$$\frac{0}{0} \quad \frac{\infty}{\infty} \quad 0 \cdot \infty \quad \infty - \infty \quad 0^\circ \quad \infty^\circ \quad 0^\infty \quad 1^\infty$$

then it may happen that $\lim f(x)$ has a definite value. For the determination of this limiting value, if it exists, the following rules can be used:

0/0. If $f(x) = \phi(x)/\psi(x), \phi(a) = 0$, and $\psi(a) = 0$, then

$$\lim_{x \to a} f(x) = \lim_{x \to a} \frac{\phi'(x)}{\psi'(x)} \quad (1' \text{ Hospital's rule})$$

$$(157)$$

If, however, $\phi'(a) = 0$ and $\psi'(a) = 0$, the rule is applied again, with the result

$$\lim_{x \to a} \frac{\phi(x)}{\psi(x)} = \lim_{\xi \to a} \frac{\phi'(\xi)}{\psi'(\xi)} = \frac{\phi''(a)}{\psi''(a)} \qquad (158)$$

unless $\phi''(a) = 0$ and $\psi''(a) = 0$. In this case, the rule is applied again, and so forth.

Example 50. Find the value of $\sin x / x$ for $x = 0$:

$$\lim_{x \to 0} \frac{\sin x}{x} = \lim_{x \to 0} \frac{\cos x}{1} = 1$$

∞/∞. If $f(x) = \phi(x)/\psi(x), \phi(a) = \infty$, and $\psi(a) = \infty$, then

$$\lim_{x \to a} \frac{\phi(x)}{\psi(x)} = \lim_{x \to a} \frac{\phi'(x)}{\psi'(x)} \qquad (159)$$

as before.

0 · ∞. If $f(x) = \phi(x) \cdot \psi(x), \phi(a) = 0$, and $\psi(a) = \infty$, then place $1/\psi(x) = \omega(x)$ and obtain the previous case 0/0.

∞ − ∞. If $f(x) = \phi(x) - \psi(x), \phi(a) = \infty$, and $\psi(a) = \infty$, then place $\phi(x) = 1/u(x), \psi(x) = 1/v(x)$ and obtain

$$f(x) = \frac{v(x) - u(x)}{u(x) v(x)} \qquad (160)$$

which takes the form 0/0.

$0^\circ, \infty^\circ, 0^\infty, 1^\infty$. An expression of the type $[\psi(x)]^{\phi(x)}$ may, for $x = a$, give rise to the forms $0^\circ, \infty^\circ, 0^\infty, 1^\infty$.
Such an expression may be reduced to a type 0/0 or ∞/∞ by the use of logarithms. Thus,

$$u = [\psi(x)]^{\phi(x)} \qquad \log_e u = \phi(x) \log_e \psi(x)$$

$$(161)$$

If $\lim_{x \to a} \phi(x) \log_e \psi(x)$ can be found by the previous methods, the limit approached by u can be found.

Example 51. $u = (1 - x)^{1/x}$ for $x = 0$:

$$\log_e u = \frac{\log_e(1 - x)}{x}$$

$$\lim_{x \to 0} \frac{\log_e(1 - x)}{x} = \lim_{x \to 0} \frac{-1/(1 - x)}{1} = -1$$

Therefore $\lim_{x \to 0} \log_e u = -1$ and $\lim_{x \to 0} u = e^{-1}$

9.2 Differentiation Formulas

These are listed in Table 9.

9.3 Partial Derivatives

Functions of Two Variables. If three variables $f(x, y)$, x, y are so related that to each pair of values of x and y in a given domain there corresponds a value of $f(x, y)$, then $f(x, y)$ is a function of x and y in that domain. If x is considered as the only variable while y is taken as constant, then the derivative of $f(x, y)$

Table 9　Differentiation Formulas

Let $u, v, w, \ldots$ be functions of x; a and n be constants; and e be the base of the natural or Naperian logarithms. Then $e = 2.7183$.

$$\frac{d}{dx}a = 0$$

$$\frac{d}{dx}(u + v + w + \cdots) = \frac{du}{dx} + \frac{dv}{dx} + \frac{dw}{dx} + \cdots$$

$$\frac{d}{dx}au = a\frac{du}{dx}$$

$$\frac{d}{dx}uv = u\frac{dv}{dx} + v\frac{du}{dx}$$

$$\frac{d}{dx}(uvw\cdots)$$

$$= \left(\frac{1}{u} \cdot \frac{du}{dx} + \frac{1}{v}\frac{dv}{dx} + \frac{1}{w}\frac{dw}{dx} + \cdots\right)(uvw\cdots)$$

$$\frac{d}{dx}f(u) = \frac{df(u)}{du} \cdot \frac{du}{dx}$$

$$\frac{d^2f(u)}{dx^2} = \frac{df(u)}{du} \cdot \frac{d^2u}{dx^2} + \frac{d^2f(u)}{du^2}\left(\frac{du}{dx}\right)^2$$

$$\frac{d}{dx}\sin u = \cos u\frac{du}{dx}$$

$$\frac{d}{dx}\cos u = -\sin u\frac{du}{dx}$$

$$\frac{d}{dx}\tan u = \sec^2 u\frac{du}{dx}$$

$$\frac{d}{dx}\cot u = -\csc^2 u\frac{du}{dx}$$

$$\frac{d}{dx}\sec u = \sec u\tan u\frac{du}{dx}$$

$$\frac{d}{dx}\csc u = -\csc u\cot u\frac{du}{dx}$$

$$\frac{d}{dx}\sin^{-1} u = \frac{1}{\sqrt{1 - u^2}}\frac{du}{dx}\left(-\frac{\pi}{2} \leq \sin^{-1} u \leq \frac{\pi}{2}\right)$$

$$\frac{d}{dx}\cos^{-1} u = -\frac{1}{\sqrt{1 - u^2}}\frac{du}{dx}(0 \leq \cos^{-1} u \leq \pi)$$

$$\frac{d}{dx}\tan^{-1} u = \frac{1}{1 + u^2}\frac{du}{dx}$$

$$\frac{d}{dx}\cot^{-1} u = -\frac{1}{1 + u^2}\frac{du}{dx}$$

$$\frac{d}{dx}\sec^{-1} u = \frac{1}{u\sqrt{u^2 - 1}}\frac{du}{dx}{}^a$$

$$\frac{d}{dx}\left(\frac{u}{v}\right) = \frac{v(du/dx) - u(dv/dx)}{v^2}$$

$$\frac{d}{dx}u^n = nu^{n-1}\frac{du}{dx}$$

$$\frac{d}{dx}\log_e u = \frac{1}{u}\frac{du}{dx}$$

$$\frac{d}{dx}\log_{10} u = \frac{1}{u}\frac{du}{dx}\log_{10} e = (0.4343)\frac{1}{u}\frac{du}{dx}$$

$$\frac{d}{dx}e^u = e^u\frac{du}{dx}$$

$$\frac{d}{dx}u^v = vu^{v-1}\frac{du}{dx} + u^v\frac{dv}{dx}\log_e u$$

$$\frac{d}{dx}\csc^{-1} u = -\frac{1}{u\sqrt{u^2 - 1}}\frac{du}{dx}{}^a$$

$$\frac{d}{dx}\sinh u = \cosh u\frac{du}{dx}$$

$$\frac{d}{dx}\cosh u = \sinh u\frac{du}{dx}$$

$$\frac{d}{dx}\tanh u = \text{sech}^2 u\frac{du}{dx}$$

$$\frac{d}{dx}\coth u = -\text{csch}^2 u\frac{du}{dx}$$

$$\frac{d}{dx}\text{sech } u = -\text{sech } u\tanh u\frac{du}{dx}$$

$$\frac{d}{dx}\text{csch } u = -\text{csch } u\coth u\frac{du}{dx}$$

$$\frac{d}{dx}\sinh^{-1} u = \frac{1}{\sqrt{u^2 + 1}}\frac{du}{dx}$$

$$\frac{d}{dx}\cosh^{-1} u = \frac{1}{\sqrt{u^2 - 1}}\frac{du}{dx}$$

$$\frac{d}{dx}\tanh^{-1} u = \frac{1}{1 - u^2}\frac{du}{dx}$$

$$\frac{d}{dx}\coth^{-1} u = \frac{1}{1 - u^2}\frac{du}{dx}$$

$$\frac{d}{dx}\text{sech}^{-1} u = -\frac{1}{u\sqrt{1 - u^2}}\frac{du}{dx}$$

$$\frac{d}{dx}\text{csch}^{-1} u = -\frac{1}{u\sqrt{u^2 + 1}}\frac{du}{dx}$$

^{a}For angles in the first and third quadrants. Use the opposite sign in the second and fourth quadrants.

with respect to x is called the *partial derivative* of f with respect to x and is denoted by

$$\frac{\partial f}{\partial x} = f_x = \lim_{\Delta x \to 0} \frac{f(x + \Delta x, y) - f(x, y)}{\Delta x} \quad (162)$$

Likewise, the partial derivative of f with respect to y is obtained by considering x to be constant while y varies:

$$\frac{\partial f}{\partial y} = f_y = \lim_{\Delta y \to 0} \frac{f(x, y + \Delta y) - f(x, y)}{\Delta y} \quad (163)$$

If $\partial f/\partial x$ and $\partial f/\partial y$ are again differentiable, partial derivatives of second order may be found:

$$\frac{\partial}{\partial x}\left(\frac{\partial f}{\partial x}\right) = \frac{\partial^2 f}{\partial x^2} = f_{xx}$$

$$\frac{\partial}{\partial y}\left(\frac{\partial f}{\partial y}\right) = \frac{\partial^2 f}{\partial y^2} = f_{yy}$$

$$\frac{\partial}{\partial x}\left(\frac{\partial f}{\partial y}\right) = \frac{\partial^2 f}{\partial x\, \partial y} = f_{yx} \quad (164)$$

$$\frac{\partial}{\partial y}\left(\frac{\partial f}{\partial x}\right) = \frac{\partial^2 f}{\partial y\, \partial x} = f_{xy}$$

If the derivatives in question are continuous, the order of differentiation is immaterial, that is,

$$\frac{\partial^2 f}{\partial y\, \partial x} = \frac{\partial^2 f}{\partial x\, \partial y} \quad (165)$$

Similarly, the third and higher partial derivatives of $f(x, y)$ may be found. The third partial derivatives, if continuous, are four in number:

$$\frac{\partial}{\partial x}\left(\frac{\partial^2 f}{\partial x^2}\right) = \frac{\partial^3 f}{\partial x^3} \quad \frac{\partial}{\partial x}\left(\frac{\partial^2 f}{\partial y^2}\right) = \frac{\partial}{\partial y}\left(\frac{\partial^2 f}{\partial x\, \partial y}\right)$$

$$= \frac{\partial^2}{\partial y^2}\left(\frac{\partial f}{\partial x}\right) = \frac{\partial^3 f}{\partial x\, \partial y^2}$$

$$\frac{\partial}{\partial y}\left(\frac{\partial^2 f}{\partial y^2}\right) = \frac{\partial^3 f}{\partial y^3} \quad \frac{\partial}{\partial y}\left(\frac{\partial^2 f}{\partial x^2}\right) = \frac{\partial}{\partial x}\left(\frac{\partial^2 f}{\partial x\, \partial y}\right)$$

$$= \frac{\partial^2}{\partial x^2}\left(\frac{\partial f}{\partial y}\right) = \frac{\partial^3 f}{\partial x^2\, \partial y} \quad (166)$$

Functions of N Variables. The preceding formulas may be generalized to the case where f is a function of more than two variables, that is, there corresponds a value of $f(x, y, z, \ldots)$ to every set of values of $x, y, z, \ldots$.

If the increments $\Delta x, \Delta y, \Delta z, \ldots$ are assigned to $x, y, z, \ldots$ in $f(x, y, z, \ldots)$, the *total increment* of f is

$$\Delta f = f(x + \Delta x, y + \Delta y, z + \Delta z, \ldots) \\ - f(x, y, z, \ldots) \quad (167)$$

The *total differential* of f is

$$df = \frac{\partial f}{\partial x} dx + \frac{\partial f}{\partial y} dy + \frac{\partial f}{\partial z} dz + \cdots \quad (168)$$

The *second total differential* of f is

$$d^2 f = \frac{\partial^2 f}{\partial x^2}(dx)^2 + \frac{\partial^2 f}{\partial y^2}(dy)^2 + \frac{\partial^2 f}{\partial z^2}(dz)^2 + \cdots \\ + 2\frac{\partial^2 f}{\partial x\, \partial y} dx\, dy + \cdots \quad (169)$$

In general,

$$d^n f = \left(\frac{\partial}{\partial x} dx + \frac{\partial}{\partial y} dy + \frac{\partial}{\partial z} dz + \cdots\right)^n \\ \times f(x, y, z, \ldots) \quad (170)$$

Exact Differential. In order for the expression $P(x, y)\, dx + Q(x, y)\, dy$ to be the *exact* or *complete differential* of a function of two variables, it is necessary and sufficient that

$$\frac{\partial Q}{\partial x} = \frac{\partial P}{\partial y} \quad \text{(integrability condition)} \quad (171)$$

For three variables, $P\, dx + Q\, dy + R\, dz$, the corresponding conditions are

$$\frac{\partial Q}{\partial z} = \frac{\partial R}{\partial y} \quad \frac{\partial R}{\partial x} = \frac{\partial P}{\partial z} \quad \frac{\partial P}{\partial y} = \frac{\partial Q}{\partial x} \quad (172)$$

Differentiation of Composite Functions. If $u = f(x, y, z, \ldots, w)$, and $x, y, z, \ldots, w$ are functions of a single variable t, then

$$\frac{du}{dt} = \frac{\partial u}{\partial x}\frac{dx}{dt} + \frac{\partial u}{\partial y}\frac{dy}{dt} + \cdots + \frac{\partial u}{\partial w}\frac{dw}{dt} \quad (173)$$

which is the total derivative of u with respect to t.

Example 52. Given $u = x^2 + y^2 + 3xy$, $x = t^2$, $y = 1/t$. Then

$$\frac{dx}{dt} = 2t \qquad \frac{dy}{dt} = -\frac{1}{t^2}$$

and

$$\frac{du}{dt} = \left(2t^2 + \frac{3}{t}\right)2t - \left(\frac{2}{t} + 3t^2\right)\frac{1}{t^2}$$

The equation reduces to

$$\frac{du}{dt} = 4t^3 + 3 - \frac{2}{t^3}$$

which expresses the rate of change of u with respect to t as a function of t.

Implicit Functions. The equation $F(x, y) = 0$ defines y as an *implicit* function of x and x as an implicit function of y. If the equation is solved for y in terms of x, $y = f(x)$, then y is called an *explicit* function of x.

Example 53. Implicit function: $F(x, y) = x^2 + y^2 - r^2 = 0$. Explicit function: $y = \pm\sqrt{r^2 - x^2}$. To find dy/dx, either differentiate $y = f(x)$ or use

$$\frac{dy}{dx} = -\frac{\partial F/\partial x}{\partial F/\partial y} \qquad \left(\frac{\partial F}{\partial y} \neq 0\right) \tag{174}$$

$$\frac{d^2y}{dx^2} = -\frac{\dfrac{\partial^2 F}{\partial x^2}\left(\dfrac{\partial F}{\partial y}\right)^2 - 2\dfrac{\partial^2 F}{\partial x \partial y}\dfrac{\partial F}{\partial x}\dfrac{\partial F}{\partial y} + \dfrac{\partial^2 F}{\partial y^2}\left(\dfrac{\partial F}{\partial x}\right)^2}{(\partial F/\partial y)^3}$$

$$\left(\frac{\partial F}{\partial y} \neq 0\right) \tag{175}$$

9.4 Infinite Series

Let $a_1, a_2, \ldots, a_n, \ldots$ be a sequence of numbers formed according to some rule. The indicated sum

$$\sum_{n=1}^{\infty} a_n = a_1 + a_2 + \cdots + a_n + \cdots \tag{176}$$

is called an *infinite series*. Let $s_n = a_1 + a_2 + \cdots + a_n$. If the partial sums s_n approach a limit S as $n \to \infty$, then the series is *convergent* and S is the *sum* or *value* of the series. A series that is not convergent is *divergent*.

If the series of absolute values $|a_1| + |a_2| + \cdots + |a_n| + \cdots$ is convergent, then series (176) is *absolutely convergent*. A series that converges, but not absolutely, is *conditionally convergent*. The sum of an absolutely convergent series is not changed by rearrangement of its terms.

Tests for Convergence

Comparison Test. A comparison test for series of positive terms. If there is a convergent series of positive terms $c_1 + c_2 + \cdots + c_n + \cdots$ such that $a_n \leq c_n$ for every n from some term on, then series (176) converges. If there is a divergent series of positive terms $d_1 + d_2 + \cdots + d_n + \cdots$ such that $a_n \geq d_n$ for every n from some term on, then series (176) diverges. Two useful comparison series are:

1. The *geometric* series $a + ar + ar^2 + \cdots + ar^{n-1} + \cdots$, which converges for $|r| < 1$ and diverges for $|r| \geq 1$
2. The *p* series $1 + 1/2^p + 1/3^p + \cdots + 1/n^p + \cdots$, which converges for $p > 1$ and diverges for $p \leq 1$

Ratio Test. Let

$$L = \lim_{n \to \infty}\left|\frac{a_{n+1}}{a_n}\right| \tag{177}$$

If $L < 1$, series (176) converges absolutely; if L does not exist or if $L > 1$, series (176) diverges; if $L = 1$, the test fails.

Example 54

$$(1) \qquad 10 + \frac{10^2}{2!} + \frac{10^3}{3!} + \cdots + \frac{10^n}{n!} + \cdots$$

Since

$$L = \lim_{n \to \infty}\frac{10^{n+1}/(n+1)!}{10^n/n!} = \lim_{n \to \infty}\frac{10}{n+1} = 0$$

the series converges.

$$(2) \qquad \frac{1+1}{1+3} + \frac{(1+1)(2+1)}{(1+3)(2+3)} + \cdots$$

$$+ \frac{(1+1)(2+1)\cdots(n+1)}{(1+3)(2+3)\cdots(n+3)} + \cdots$$

Since

$$L = \lim_{n \to \infty}\frac{(n+1)+1}{(n+1)+3} = 1$$

the test fails. Raabe's test can be used. See Eq. (179).

Root Test. Let

$$L = \lim_{n \to \infty}|a_n|^{1/n} \tag{178}$$

If $L < 1$, series (176) converges; if $L > 1$, series (176) diverges; if $L = 1$, the test fails.

Example 55

$$1 + \frac{1}{(\log 2)^2} + \frac{1}{(\log 3)^3} + \cdots + \frac{1}{(\log n)^n} + \cdots$$

Since

$$L = \lim_{n \to \infty} \frac{1}{\log n} = 0$$

the series converges.

Integral Test. Let $f(n) = a_n$. If $f(x)$ is a positive nonincreasing function of x for $x > k$, then series (176) converges or diverges with the improper integral $\int_k^\infty f(x)\,dx$.

Example 56

$$1 + \frac{1}{2(\log 2)^3} + \frac{1}{3(\log 3)^3} + \cdots + \frac{1}{n(\log n)^3} + \cdots$$

Then

$$f(x) = \frac{1}{x(\log x)^3} \quad \text{for } x \geq 2$$

and

$$\int_2^\infty \frac{dx}{x(\log x)^3} = \lim_{n \to \infty} \frac{1}{2}\left(\frac{1}{(\log 2)^2} - \frac{1}{(\log n)^2}\right)$$

$$= \frac{1}{2(\log 2)^2}$$

Since the integral is convergent, the series is also.

Raabe's Test. Let

$$L = \lim_{n \to \infty} n\left(\frac{a_n}{a_{n+1}} - 1\right) \tag{179}$$

If $L > 1$, series (176) converges; if $L < 1$, series (176) diverges; if $L = 1$, the test fails.

Example 57

$$\frac{1+1}{1+3} + \frac{(1+1)(2+1)}{(1+3)(2+3)} + \cdots$$

$$+ \frac{(1+1)(2+1)\cdots(n+1)}{(1+3)(2+3)\cdots(n+3)} + \cdots$$

Since

$$L = \lim_{n \to \infty} n\left(\frac{(n+1)+3}{(n+1)+1} - 1\right) = \lim_{n \to \infty} \frac{2n}{n+2} = 2 > 1$$

the series converges.

Convergence of an Alternating Series. A series

$$a_1 - a_2 + a_3 - + \cdots + (-1)^{n+1}a_n + \cdots \tag{180}$$

in which the terms are alternately positive and negative is an *alternating series*. If, from some term on, $|a_{n+1}| \leq |a_n|$ and $a_n \to 0$ as $n \to \infty$, the series converges. The sum of the first n terms differs numerically from the sum of the series by less than $|a_{n+1}|$.

Series of Functions A *power series* is a series of the form

$$\sum_{n=0}^\infty a_n x^n = a_0 + a_1 x + a_2 x^2 + \cdots + a_n x^n + \cdots \tag{181}$$

If $\lim_{n \to \infty} |a_{n-1}/a_n| = r$, the power series converges absolutely for all values of x in the interval $-r < x < r$. For $|x| = r$, it is necessary to use one of the convergence tests for series of numerical terms.

Example 58

$$1 - \frac{x}{1 \cdot 2} + \frac{x^2}{2 \cdot 2^2} - \frac{x^3}{3 \cdot 2^3} + \cdots + (-1)^n \frac{x^n}{n \cdot 2^n} + \cdots$$

Since

$$\lim_{n \to \infty} \frac{n \cdot 2^n}{(n-1)2^{n-1}} = 2$$

the interval of convergence is $-2 < x < 2$. For $x = 2$, the series is a convergent alternating series. For $x = -2$, it is a divergent p series.

Taylor's Series. If $f(x)$ has continuous derivatives in the neighborhood of a point $x = a$, then

$$f(x) = f(a) + \frac{f'(a)}{1!}(x - a) + \frac{f''(a)}{2!}(x - a)^2 + \cdots$$

$$+ \frac{f^{(n-1)}(a)}{(n-1)!}(x - a)^{n-1} + \cdots \tag{182}$$

with the remainder after n terms

$$R_n = \frac{f^{(n)}(\xi)}{n!}(x - a)^n$$

$$\xi = a + \theta(x - a) \qquad 0 < \theta < 1 \tag{183}$$

Another form of Taylor's series is

$$f(x + h) = f(x) + \frac{h}{1!}f'(x) + \frac{h^2}{2!}f''(x) + \cdots$$

$$+ \frac{h^{n-1}}{(n-1)!}f^{(n-1)}(x) + \cdots \tag{184}$$

with the remainder after n terms

$$R_n = \frac{h^n}{n!} f^{(n)}(\xi) \qquad \xi = x + \theta h \qquad 0 < \theta < 1 \tag{185}$$

Maclaurin's Series.
If $a = 0$ in Eq. (182),

$$f(x) = f(0) + \frac{f'(0)}{1!} x + \frac{f''(0)}{2!} x^2 + \cdots$$

$$+ \frac{f^{(n-1)}(0)}{(n-1)!} x^{n-1} + \cdots \tag{186}$$

with the remainder after n terms

$$R_n = \frac{f^{(n)}(\xi)}{n!} x^n \qquad \xi = \theta x \qquad 0 < \theta < 1 \tag{187}$$

A Taylor or Maclaurin series represents a function in an interval if and only if $R_n \to 0$ as $n \to \infty$.

Example 59. Expand e^{ax} in powers of x:

$$
\begin{aligned}
f(x) &= e^{ax} & f'(x) &= ae^{ax} \\
f''(x) &= a^2 e^{ax} & f'''(x) &= a^3 e^{ax}, \ldots \\
f(0) &= 1 & f'(0) &= a \\
f''(0) &= a^2 & f'''(0) &= a^3, \ldots
\end{aligned}
$$

$$f(x) = e^{ax} = 1 + \frac{a}{1!} x + \frac{a^2}{2!} x^2 + \frac{a^3}{3!} x^3 + \cdots$$

Since

$$\lim_{n \to \infty} \frac{a^{n-1}/(n-1)!}{a^n/n!} = \lim_{n \to \infty} \frac{n}{a} = \infty$$

the series converges for all values of x.

Taylor's Series for Two Variables

$$f(x + h, y + k)$$

$$= f(x, y) + \frac{1}{1!} \left(h \frac{\partial}{\partial x} + k \frac{\partial}{\partial y} \right) f(x, y)$$

$$+ \frac{1}{2!} \left(h \frac{\partial}{\partial x} + k \frac{\partial}{\partial y} \right)^2 f(x, y) + \cdots$$

$$+ \frac{1}{(n-1)!} \left(h \frac{\partial}{\partial x} + k \frac{\partial}{\partial y} \right)^{n-1} f(x, y) + \cdots \tag{188}$$

with the remainder

$$R_n = \frac{1}{n!} \left(h \frac{\partial}{\partial x} + k \frac{\partial}{\partial y} \right)^n$$

$$\times f(x + \theta h, y + \theta k) \qquad 0 < \theta < 1 \tag{189}$$

Fourier Series.
If $f(x)$ is of bounded variation over an interval of length $2l$, that is, if it can be expressed as the difference of two nondecreasing or nonincreasing bounded functions, then

$$f(x) = \frac{a_0}{2} + \sum_{n=1}^{\infty} \left(a_n \cos \frac{n\pi x}{l} + b_n \sin \frac{n\pi x}{l} \right)$$

$$= \frac{a_0}{2} + a_1 \cos \frac{\pi x}{l} + a_2 \cos \frac{2\pi x}{l}$$

$$+ \cdots + b_1 \sin \frac{\pi x}{l} + b_2 \sin \frac{2\pi x}{l} + \cdots \tag{190}$$

in which

$$a_n = \frac{1}{l} \int_k^{k+2l} f(x) \cos \frac{n\pi x}{l} \, dx$$

$$b_n = \frac{1}{l} \int_k^{k+2l} f(x) \sin \frac{n\pi x}{l} \, dx \qquad n = 0, 1, 2, \ldots \tag{191}$$

In exponential form

$$f(x) = \sum_{n=-\infty}^{\infty} c_n e^{in\pi x/l}$$

$$c_n = \frac{1}{2l} \int_k^{k+2l} f(x) e^{-in\pi x/l} \, dx$$

$$n = \ldots, -2, -1, 0, 1, 2, \ldots \tag{192}$$

At a point of discontinuity, a Fourier series gives the value at the midpoint of the jump.

Example 60. Expand e^x in the interval 0 to 2π:

$$a_0 = \frac{1}{\pi} \int_0^{2\pi} e^x \, dx = \frac{1}{\pi} (e^{2\pi} - 1)$$

$$a_n = \frac{1}{\pi} \int_0^{2\pi} e^x \cos nx \, dx = \frac{e^{2\pi} - 1}{\pi (n^2 + 1)}$$

$$b_n = -\frac{n(e^{2\pi} - 1)}{\pi (n^2 + 1)}$$

Hence

$$e^x = \frac{1}{\pi} (e^{2\pi} - 1) \left[\frac{1}{2} + \frac{1}{1^2 + 1} \cos x + \frac{1}{2^2 + 1} \cos 2x \right.$$

$$\left. + \frac{1}{3^2 + 1} \cos 3x + \cdots \right]$$

$$- \frac{1}{\pi} (e^{2\pi} - 1) \left[\frac{1}{1^2 + 1} \sin x + \frac{2}{2^2 + 1} \sin 2x \right.$$

$$\left. + \frac{3}{3^2 + 1} \sin 3x + \cdots \right]$$

The expansion is valid only in the interval from 0 to 2π; outside that interval the series repeats itself owing to the periodic property of $\sin nx$ and $\cos nx$.

Fourier Series for Even or Odd Functions.
If $f(-x) = f(x)$, it is an *even* function. Then

$$a_n = \frac{2}{l} \int_0^l f(x) \cos \frac{n\pi x}{l} \, dx \qquad n = 0, 1, 2, \ldots$$

$$b_n = 0 \tag{193}$$

If $f(-x) = -f(x)$, it is an *odd* function. Then

$$a_n = 0$$

$$b_n = \frac{2}{l} \int_0^l f(x) \sin \frac{n\pi x}{l} \, dx \qquad n = 0, 1, 2, \ldots \tag{194}$$

Example 61. Expand $f(x) = x$ in a cosine series in the interval $(0, \pi)$. Here

$$\frac{1}{2} a_0 = \frac{1}{\pi} \int_0^\pi x \, dx = \frac{\pi}{2}$$

$$a_n = \frac{2}{\pi} \int_0^\pi x \cos nx \, dx$$

$$= \frac{2}{\pi} \left(\left[\frac{x \sin nx}{n} \right]_0^\pi - \int_0^\pi \frac{\sin nx}{n} \, dx \right)$$

$$= \frac{2}{\pi} \left[\frac{1}{n^2} \cos nx \right]_0^\pi = \frac{2}{\pi n^2} (\cos n\pi - 1)$$

Therefore

$$x = \frac{\pi}{2} - \frac{4}{\pi} \left(\cos x + \frac{\cos 3x}{3^2} + \frac{\cos 5x}{5^2} + \cdots \right)$$

$$(0 < x < \pi)$$

If $x = 0$, the sum of the series is 0; if $x = \pi$, the sum of the series is π.

Uniform Convergence.
Let $R_n(x)$ be the remainder after n terms of the series of functions

$$\sum_{n=1}^\infty u_n(x) = u_1(x) + u_2(x) + \cdots + u_n(x) + \cdots \tag{195}$$

The series is uniformly convergent in the interval $a \le x \le b$ if, for any $\varepsilon > 0$, there exists an N dependent on ε but not on x such that $|R_n(x)| < \varepsilon$ for $n > N$.

If a power series converges in the interval $-r < x < r$, then it converges uniformly in any interval within this interval.

The sum of a uniformly convergent series of continuous functions is also a continuous function.

Weierstrass M Test. If $\sum_{n=1}^\infty u_n(x)$ is a series of functions defined in an interval, $\sum_{n=1}^\infty M_n$ is a series of positive constants, and $|u_n(x)| \le M_n$ for all values of x in the interval, then $\sum_{n=1}^\infty u_n(x)$ is absolutely and uniformly convergent in the interval.

Operations

Term-by-Term Differentiation If $f(x) = \sum_{n=1}^\infty u_n(x)$ is a convergent series of differentiable functions in an interval and $\sum_{n=1}^\infty u'_n(x)$ is a series of continuous functions that converges uniformly in the interval, then

$$\sum_{n=1}^\infty u'_n(x) = f'(x)$$

Term-by-Term Integration If

$$f(x) = \sum_{n=1}^\infty u_n(x)$$

converges uniformly in an interval and a and x are any two values in the interval, then $\sum_{n=1}^\infty \int_a^x u_n(t) \, dt$ converges to $\int_a^x f(t) \, dt$. It converges uniformly with respect to x for each fixed value of a.

Two power series can be *added, subtracted*, or *multiplied* term by term, and the result is a power series that converges when both of the first do and represents the *sum, difference*, or *product*, respectively, of the two series. The product of two power series $\sum_{n=0}^\infty a_n x^n$ and $\sum_{n=0}^\infty b_n x^n$ is

$$a_0 b_0 + (a_0 b_1 + a_1 b_0)x + (a_0 b_2 + a_1 b_1 + a_2 b_0)x^2$$

$$+ \cdots + (a_0 b_n + a_1 b_{n-1} + \cdots + a_n b_0)x^n + \cdots \tag{196a}$$

The *quotient* of two convergent power series $\sum_{n=0}^\infty a_n x^n$ and $\sum_{n=0}^\infty b_n x^x$, $b_0 \ne 0$, is

$$\sum_{n=0}^\infty q_n x^n = \frac{a_0}{b_0} + \frac{a_1 b_0 - a_0 b_1}{b_0^2} x$$

$$+ \frac{a_2 b_0^2 - a_1 b_0 b_1 + a_0 b_1^2 - a_0 b_0 b_2}{b_0^3} x^2 + \cdots \tag{196b}$$

The interval of convergence of the quotient series must be determined. To obtain $q_3, q_4, \ldots$, solve the equations

$$a_0 = q_0 b_0$$

$$a_1 = q_0 b_1 + q_1 b_0$$

$$a_2 = q_0 b_2 + q_1 b_1 + q_2 b_0$$

$$a_n = q_0 b_n + q_1 b_{n-1} + \cdots + q_n b_0 \tag{197}$$

See Table 10 for series expansions of various functions.

Table 10 Functions Expanded in Series (log = log$_e$)

$$(a+x)^n = a^n + na^{n-1}x + \frac{n(n-1)}{2!}a^{n-2}x^2 + \frac{n(n-1)(n-2)}{3!}a^{n-3}x^3 + \cdots \qquad (x^2 < a^2)$$

$$e^x = 1 + x + \frac{x^2}{2!} + \frac{x^3}{3!} + \frac{x^4}{4!} + \cdots \qquad (-\infty < x < \infty)$$

$$a^x = 1 + x\log a + \frac{(x\log a)^2}{2!} + \frac{(x\log a)^3}{3!} + \cdots \qquad (-\infty < x < \infty)$$

$$e^{-x^2} = 1 - x^2 + \frac{x^4}{2!} - \frac{x^6}{3!} + \frac{x^8}{4!} - \cdots \qquad (-\infty < x < \infty)$$

$$e^{\sin x} = 1 + x + \frac{x^2}{2!} - \frac{3x^4}{4!} - \frac{8x^5}{5!} - \frac{3x^6}{6!} + \frac{56x^7}{7!} + \cdots \qquad (-\infty < x < \infty)$$

$$e^{\cos x} = e\left(1 - \frac{x^2}{2!} + \frac{4x^4}{4!} - \frac{31x^6}{6!} + \cdots\right) \qquad (-\infty < x < \infty)$$

$$e^{\tan x} = 1 + x + \frac{x^2}{2!} + \frac{3x^3}{3!} + \frac{9x^4}{4!} + \frac{37x^5}{5!} + \cdots \qquad \left(-\frac{\pi}{2} < x < \frac{\pi}{2}\right)$$

$$\log x = \frac{x-1}{x} + \frac{1}{2}\left(\frac{x-1}{x}\right)^2 + \frac{1}{3}\left(\frac{x-1}{x}\right)^3 + \cdots \qquad \left(x > \frac{1}{2}\right)$$

$$\log x = 2\left[\frac{x-1}{x+1} + \frac{1}{3}\left(\frac{x-1}{x+1}\right)^3 + \frac{1}{5}\left(\frac{x-1}{x+1}\right)^5 + \cdots\right] \qquad (x > 0)$$

$$\log(1+x) = x - \frac{x^2}{2} + \frac{x^3}{3} - \frac{x^4}{4} + \cdots \qquad (-1 < x < 1)$$

$$\log\left(\frac{1+x}{1-x}\right) = 2\left[x + \frac{x^3}{3} + \frac{x^5}{5} + \frac{x^7}{7} + \cdots\right] \qquad (-1 < x < 1)$$

$$\log\left(\frac{x+1}{x-1}\right) = 2\left[\frac{1}{x} + \frac{1}{3x^3} + \frac{1}{5x^5} + \cdots\right] \qquad (x^2 > 1)$$

$$\log\sin x = \log x - \frac{x^2}{6} - \frac{x^4}{180} - \frac{x^6}{2835} - \cdots \qquad (-\pi < x < \pi)$$

$$\log\cos x = -\frac{x^2}{2} - \frac{x^4}{12} - \frac{x^6}{45} - \frac{17x^8}{2520} - \cdots \qquad \left(-\frac{\pi}{2} < x < \frac{\pi}{2}\right)$$

$$\log\tan x = \log x + \frac{x^2}{3} + \frac{7x^4}{90} + \frac{62x^6}{2835} + \cdots \qquad \left(-\frac{\pi}{2} < x < \frac{\pi}{2}\right)$$

$$\sin x = x - \frac{x^3}{3!} + \frac{x^5}{5!} - \frac{x^7}{7!} + \cdots \qquad (-\infty < x < \infty)$$

$$\cos x = 1 - \frac{x^2}{2!} + \frac{x^4}{4!} - \frac{x^6}{6!} + \cdots \qquad (-\infty < x < \infty)$$

$$\tan x = x + \frac{x^3}{3} + \frac{2x^5}{15} + \frac{17x^7}{315} + \frac{62x^9}{2835} + \cdots \qquad \left(-\frac{\pi}{2} < x < \frac{\pi}{2}\right)$$

$$\cot x = \frac{1}{x} - \frac{x}{3} - \frac{x^3}{45} - \frac{2x^5}{945} - \frac{x^7}{4725} - \cdots \qquad (-\pi < x < \pi)$$

$$\sec x = 1 + \frac{x^2}{2!} + \frac{5x^4}{4!} + \frac{61x^6}{6!} + \cdots \qquad \left(-\frac{\pi}{2} < x < \frac{\pi}{2}\right)$$

$$\csc x = \frac{1}{x} + \frac{x}{3!} + \frac{7x^3}{3\cdot 5!} + \frac{31x^5}{3\cdot 7!} + \cdots \qquad (-\pi < x < \pi)$$

$$\sin^{-1}x = x + \frac{x^3}{2\cdot 3} + \frac{3x^5}{2\cdot 4\cdot 5} + \frac{3\cdot 5x^7}{2\cdot 4\cdot 6\cdot 7} + \cdots \qquad (-1 \le x \le 1)$$

$$\cos^{-1}x = \frac{\pi}{2} - \sin^{-1}x$$

$$\tan^{-1}x = \frac{\pi}{2} - \frac{1}{x} + \frac{1}{3x^3} - \frac{1}{5x^5} + \cdots \qquad (x^2 \ge 1)$$

$$= x - \frac{x^3}{3} + \frac{x^5}{5} - \frac{x^7}{7} + \cdots \qquad (-1 \le x \le 1)$$

(Continues)

Table 10   (Continued)

$$\cot^{-1}x = \frac{\pi}{2} - \tan^{-1}x$$

$$\sec^{-1}x = \frac{\pi}{2} - \frac{1}{x} - \frac{1}{2 \cdot 3x^3} - \frac{3}{2 \cdot 4 \cdot 5x^5} - \frac{3 \cdot 5}{2 \cdot 4 \cdot 6 \cdot 7x^7} - \cdots \qquad (x^2 > 1)$$

$$\csc^{-1}x = \frac{\pi}{2} - \sec^{-1}x$$

$$\sinh x = x + \frac{x^3}{3!} + \frac{x^5}{5!} + \frac{x^7}{7!} + \cdots \qquad (-\infty < x < \infty)$$

$$\cosh x = 1 + \frac{x^2}{2!} + \frac{x^4}{4!} + \frac{x^6}{6!} + \frac{x^8}{8!} + \cdots \qquad (-\infty < x < \infty)$$

$$\tanh x = x - \frac{x^3}{3} + \frac{2x^5}{15} - \frac{17x^7}{315} + \cdots \qquad \left(-\frac{\pi}{2} < x < \frac{\pi}{2}\right)$$

$$\coth x = \frac{1}{x} + \frac{x}{3} - \frac{x^3}{45} + \frac{2x^5}{945} - \frac{x^7}{4725} + \cdots \qquad (-\pi < x < \pi)$$

$$\operatorname{sech} x = 1 - \frac{x^2}{2!} + \frac{5x^4}{4!} - \frac{61x^6}{6!} + \frac{1385x^8}{8!} - \cdots \qquad \left(-\frac{\pi}{2} < x < \frac{\pi}{2}\right)$$

$$\operatorname{csch} x = \frac{1}{x} - \frac{x}{6} + \frac{7x^3}{360} - \frac{31x^5}{15{,}120} + \cdots \qquad (-\pi < x < \pi)$$

$$\sinh^{-1}x = x - \frac{x^3}{2 \cdot 3} + \frac{3x^5}{2 \cdot 4 \cdot 5} - \frac{3 \cdot 5x^7}{2 \cdot 4 \cdot 6 \cdot 7} + \cdots \qquad (-1 < x < 1)$$

$$\sinh^{-1}x = \log 2x + \frac{1}{2 \cdot 2x^2} - \frac{3}{2 \cdot 4 \cdot 4x^4} + \frac{3 \cdot 5}{2 \cdot 4 \cdot 6 \cdot 6 \cdot x^6} + \cdots \qquad (x^2 > 1)$$

$$\cosh^{-1}x = \pm\left(\log 2x - \frac{1}{2 \cdot 2x^2} - \frac{1 \cdot 3}{2 \cdot 4 \cdot 4x^4} - \frac{1 \cdot 3 \cdot 5}{2 \cdot 4 \cdot 6 \cdot 6x^6} - \cdots\right) \qquad (x > 1)$$

$$\tanh^{-1}x = x + \frac{x^3}{3} + \frac{x^5}{5} + \frac{x^7}{7} + \cdots \qquad (-1 < x < 1)$$

$$\coth^{-1}x = \frac{1}{x} + \frac{1}{3x^3} + \frac{1}{5x^5} + \frac{1}{7x^7} + \cdots \qquad (x^2 > 1)$$

$$\operatorname{sech}^{-1}x = \pm\left(\log \frac{2}{x} - \frac{1}{2 \cdot 2}x^2 - \frac{1 \cdot 3}{2 \cdot 4 \cdot 4}x - \frac{1 \cdot 3 \cdot 5}{2 \cdot 4 \cdot 6 \cdot 6}x^6 - \cdots\right) \qquad (0 < x < 1)$$

$$\operatorname{csch}^{-1}x = \frac{1}{x} - \frac{1}{2 \cdot 3x^3} + \frac{3}{2 \cdot 4 \cdot 5x^5} - \frac{3 \cdot 5}{2 \cdot 4 \cdot 6 \cdot 7x^7} + \cdots \qquad (x^2 > 1)$$

9.5   Maxima and Minima

Function of One Variable.   A function $f(x)$ has a relative *maximum* (*minimum*) at a point $x = a$ if at every point in some neighborhood of $x = a$ the values of $f(x)$ are all less (greater) than $f(a)$. Either a maximum or a minimum is an *extreme*. If the derivative exists at a relative extreme, it must be zero, that is, the tangent must be parallel to the x axis. To locate possible extreme points solve the equation $f'(x) = 0$. A solution $x = a$ gives a maximum (minimum) value of $f(x)$ if and only if the derivative is positive (negative) for $x < a$ and negative (positive) for $x > a$. If the derivative does not change sign, $x = a$ gives a *point of inflection*.

A solution $x = a$ can be tested also by using the higher derivatives of $f(x)$. Let $f^{(n)}(x)$ be the first derivative that does not equal zero for $x = a$, $f^{(n)}(a) \neq 0$. If n is even, there is a maximum (Fig. 136) if $f^{(n)}(a) < 0$ and a minimum (Fig. 137)

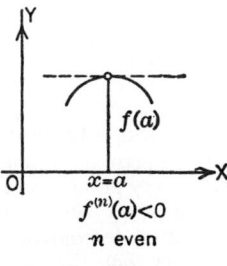

Fig. 136

if $f^{(n)}(a) > 0$. If n is odd, there is a point of inflection (Figs. 138 and 139). In many problems physical considerations make testing unnecessary.

Example 62.   A piece of wire of length 30 in. is bent into a rectangle. Find the maximum area. Let $x =$ the base, then $\frac{1}{2}(30 - 2x) =$ the altitude. The area

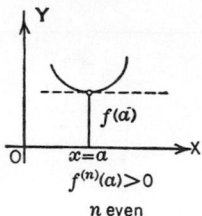

$f(a)$

$x = a$

$f^{(n)}(a) > 0$

n even

Fig. 137

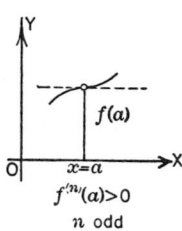

$f(a)$

$x = a$

$f^{(n)}(a) > 0$

n odd

Fig. 138

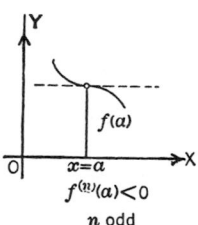

$f(a)$

$x = a$

$f^{(n)}(a) < 0$

n odd

Fig. 139

$A = x(15 - x) = 15x - x^2$. For a maximum or minimum, $dA/dx = 15 - 2x = 0$, $x = 7.5$. Then $A = 7.5 (15 - 7.5) = 56.25$ in.^{2} To find whether the area is maximum or minimum, $d^2A/dx^2 = -2$, which is less than 0, and therefore the area 56.25 in.^{2} is a maximum.

Function of Two or More Variables. A function $f(x, y)$ has a relative maximum (minimum) at a point $(x, y) = (a, b)$ if at every point in some neighborhood of (a, b) the values of $f(x, y)$ are all less (greater) than $f(a, b)$. If the first partial derivatives exist at a relative extreme, it is necessary that

$$\frac{\partial f}{\partial x} = \frac{\partial f}{\partial y} = 0 \tag{198}$$

If, furthermore, at the point (a, b)

$$\frac{\partial^2 f}{\partial x^2} \frac{\partial^2 f}{\partial y^2} - \left(\frac{\partial^2 f}{\partial x \, \partial y} \right)^2 > 0 \tag{199}$$

then $f(a, b)$ is an extreme value, which is a maximum if

$$\frac{\partial^2 f}{\partial x^2} < 0 \quad \left(\text{and consequently } \frac{\partial^2 f}{\partial y^2} < 0 \right) \tag{200}$$

and a minimum if

$$\frac{\partial^2 f}{\partial x^2} > 0 \tag{201}$$

If

$$\frac{\partial^2 f}{\partial x^2} \frac{\partial^2 f}{\partial y^2} - \left(\frac{\partial^2 f}{\partial x \, \partial y} \right)^2 < 0 \tag{202}$$

then $f(x, y)$ does not have an extreme value but has a saddle point. If

$$\frac{\partial^2 f}{\partial x^2} \frac{\partial^2 f}{\partial y^2} - \left(\frac{\partial^2 f}{\partial x \, \partial y} \right)^2 = 0 \tag{203}$$

the test gives no information.

For a function of several variables $f(x, y, z, \ldots)$, necessary conditions for an extreme value are

$$\frac{\partial f}{\partial x} = \frac{\partial f}{\partial y} = \frac{\partial f}{\partial z} = \cdots = 0 \tag{204}$$

10 INTEGRAL CALCULUS

10.1 Integration

The operation of *integration* is the inverse of differentiation. If $F'(x) = f(x)$, then $F(x)$ is an *indefinite integral* of $f(x)$. Any other indefinite integral of $f(x)$ differs from $F(x)$ at most by an additive constant. In symbols, $F(x) = \int f(x) \, dx + c$; or, more precisely, $F(x) = \int_a^x f(\xi) \, d\xi + c$, where a and c are constants (see Section 10.2). The function $f(x)$ is the *integrand*.

The integrals of many functions are given in Table 18 of Chapter 1. Table 11 is a convenient short table of fundamental integrals. The constant of integration is usually omitted in tables.

In general, with the exception of the square root of polynomials of the second degree, integrals containing fractional powers of polynomials above the first degree cannot be integrated in terms of the elementary integral forms.

Elliptic Integrals. An elliptic integral has the form

$$\int R[x, \sqrt{f(x)}] \, dx \tag{205}$$

where R represents a rational function and $f(x) = a + bx + cx^2 + dx^3 + ex^4$, an algebraic function of the third or fourth degree.

Table 11 Fundamental Integrals

1. $\int u^n\,du = \dfrac{u^{n+1}}{n+1}$

2. $\int \dfrac{du}{u} = \ln u$

3. $\int a^u\,du = \dfrac{a^u}{\ln u}$

4. $\int e^u\,du = e^u$

5. $\int \cos u\,du = \sin u$

6. $\int \sin u\,du = -\cos u$

7. $\int \sec^2 u\,du = \tan u$

8. $\int \csc^2 u\,du = -\cot u$

9. $\int \sec u \tan u\,du = \sec u$

10. $\int \csc u \cot u\,du = -\csc u$

11. $\int \tan u\,du = \ln \sec u$

$\qquad = -\ln \cos u$

12. $\int \cot u\,du = \ln \sin u$

$\qquad = -\ln \csc u$

13. $\int \sec u\,du = \ln(\sec u + \tan u)$

$\qquad = \ln \tan\left(\dfrac{\pi}{4} + \dfrac{u}{2}\right)$

14. $\int \csc u\,du = \ln(\csc u - \cot u)$

$\qquad = \ln \tan \dfrac{u}{2}$

15. $\int \dfrac{du}{u^2 + a^2} = \dfrac{1}{a}\tan^{-1}\dfrac{u}{a}$

$\qquad = -\dfrac{1}{a}\cot^{-1}\dfrac{u}{a}$

16. $\int \dfrac{du}{u^2 - a^2} = \dfrac{1}{2a}\ln\dfrac{u-a}{u+a} \qquad (u^2 > a^2)$

$\qquad = \dfrac{1}{2a}\ln\dfrac{a-u}{a+u} \qquad (u^2 < a^2)$

$\qquad = -\dfrac{1}{a}\tanh^{-1}\dfrac{u}{a} \qquad (u^2 < a^2)$

$\qquad = -\dfrac{1}{a}\coth^{-1}\dfrac{u}{a} \qquad (u^2 > a^2)$

17. $\int \dfrac{du}{\sqrt{a^2 - u^2}} = \sin^{-1}\dfrac{u}{a}$

$\qquad = -\cos^{-1}\dfrac{u}{a}$

18. $\int \dfrac{du}{\sqrt{u^2 \pm a^2}} = \ln(u + \sqrt{u^2 \pm a^2})$

$\int \dfrac{du}{\sqrt{u^2 + a^2}} = \sinh^{-1}\dfrac{u}{a}$

$\int \dfrac{du}{\sqrt{u^2 - a^2}} = \cosh^{-1}\dfrac{u}{a}$

19. $\int \dfrac{du}{u\sqrt{u^2 - a^2}} = \dfrac{1}{a}\sec^{-1}\dfrac{u}{a}$

$\qquad = -\dfrac{1}{a}\csc^{-1}\dfrac{u}{a}$

20. $\int \dfrac{du}{\sqrt{2au - u^2}} = \text{vers}^{-1}\dfrac{u}{a}$

$\qquad = \cos^{-1}\left(1 - \dfrac{u}{a}\right)$

21. $\int \sinh u\,du = \cosh u$

22. $\int \cosh u\,du = \sinh u$

23. $\int \tanh u\,du = \ln \cosh u$

24. $\int \coth u\,du = \ln \sinh u$

25. $\int \text{sech}\,u\,du = 2\tan^{-1} e^u$

26. $\int \text{csch}\,u\,du = \ln \tanh \dfrac{u}{2}$

The constant of integration is omitted in the above integrals.

Elliptic integral of the first kind:

$$F(\phi, k) = \int_0^\phi \frac{d\theta}{\sqrt{1 - k^2 \sin^2 \theta}}$$

$$= \int_0^x \frac{d\xi}{\sqrt{(1 - \xi^2)(1 - k^2\xi^2)}}$$

$$x = \sin\phi \qquad k^2 < 1 \qquad (206)$$

Elliptic integral of the second kind:

$$E(\phi, k) = \int_0^\phi \sqrt{1 - k^2 \sin^2 \theta}\,d\theta$$

$$= \int_0^x \frac{\sqrt{1 - k^2\xi^2}}{\sqrt{1 - \xi^2}}\,d\xi$$

$$x = \sin\phi \qquad k^2 < 1 \qquad (207)$$

Elliptic integral of the third kind:

$$\Pi(\phi, n, k) = \int_0^\phi \frac{d\theta}{(1 + n\sin^2\theta)\sqrt{1 - k^2 \sin^2\theta}}$$

$$= \int_0^x \frac{d\xi}{(1 + n\xi^2)\sqrt{(1 - \xi^2)(1 - k^2\xi^2)}}$$

$$x = \sin\phi \qquad k^2 < 1 \qquad (208)$$

The "complete" integrals are

$$K = F\left(\frac{\pi}{2}, k\right) = \frac{\pi}{2}\left[1 + \left(\frac{1}{2}\right)^2 k^2 + \left(\frac{3}{2 \cdot 4}\right)^2 k^4\right.$$

$$\left. + \left(\frac{3 \cdot 5}{2 \cdot 4 \cdot 6}\right)^3 k^6 + \cdots\right] \tag{209}$$

$$E = E\left(\frac{\pi}{2}, k\right) = \frac{\pi}{2}\left[1 - \left(\frac{1}{2^2}\right) k^2 - \left(\frac{3}{2^2 \cdot 4^2}\right) k^4\right.$$

$$\left. - \left(\frac{3^2 \cdot 5}{2^2 \cdot 4^2 \cdot 6^2}\right) k^6 - \cdots\right] \tag{210}$$

$$K = F\left(\frac{\pi}{2}, \sqrt{1 - k^2}\right) \qquad E' = E\left(\frac{\pi}{2}, \sqrt{1 - k^2}\right) \tag{211}$$

They are connected by the relation

$$KE' + EK' - KK' = \tfrac{1}{2}\pi \tag{212}$$

The inverse function of $u = F(\phi, k)$ is $\phi = \text{am } u$ (am = amplitude),

$$x \equiv \sin\phi \equiv \text{sn } u = u - (1 + k^2)\frac{u^3}{3!}$$

$$+ (1 + 14k^2 + k^4)\frac{u^5}{5!} - \cdots \tag{213}$$

$$\cos\phi \equiv \text{cn } u = 1 - \frac{u^2}{2!} + (1 + 4k^2)\frac{u^4}{4!}$$

$$- (1 + 44k^2 + 16k^4)\frac{u^6}{6!} + \cdots \tag{214}$$

$$\sqrt{1 - k^2 x^2} \equiv \Delta\phi \cong \text{dn } u = 1 - k^2\frac{u^2}{2!} + k^2(4 + k^2)\frac{u^4}{4!}$$

$$- k^2(16 + 44k^2 + k^4)\frac{u^6}{6!} + \cdots \tag{215}$$

Methods of Integration

Integration by Parts. The formula

$$\int u \, dv = uv - \int v \, du \quad (u \text{ and } v \text{ functions of } x)$$
$$\tag{216}$$

is useful in integrating a product if factors of the product are a function of x and the derivative of another function of x.

Example 63. To find $\int x \sin x \, dx$, let $u = x, dv = \sin x \, dx$. Then $du = dx, v = -\cos x$, and $\int x \sin x \, dx = -x \cos x + \int \cos x \, dx = -x \cos x + \sin x + c$.

Integration of Rational Fractions. If the quotient of two polynomials

$$R(x) = \frac{P_n(x)}{P_d(x)} \tag{217}$$

is not a proper fraction, that is, if the degree of the numerator is not less than that of the denominator, $R(x)$ can be changed, by dividing as indicated, to the sum of a polynomial, which is immediately integrable, and a proper fraction. If the proper fraction cannot be integrated by reference to Table 18, use the methods of Section 2.8 to resolve it, if possible, into partial fractions. These can be integrated from the table.

Irrational Functions. These can sometimes be put into integrable forms by rationalizing them by a change of variable.

Form	Substitution
$f[(ax + b)^{p/q}] \, dx$	Let $ax + b = y^q$
$f[(ax + b)^{p/q}(ax + b)^{r/s}] \, dx$	Let $ax + b = y^n$, where n is the LCM of q, s
$f[x, \sqrt{x^3 + ax + b}] \, dx$	Let $\sqrt{x^2 + ax + b}$ $= y - x$
$f[x, \sqrt{-x^2 + ax + b}] \, dx$	Let $\sqrt{-x^2 + ax + b}$ $= \sqrt{(\alpha - x)(\beta + x)}$ $= (\alpha - x)y$ or $= (\beta + x)y$
$f[\sin x, \cos x] \, dx$	Let $\tan\dfrac{x}{2} = y$
$f[x, \sqrt{a^2 - x^2}] \, dx$	Let $x = a \sin y$
$f[x, \sqrt{x^2 - a^2}] \, dx$	Let $x = a \sec y$ or $x = a \cosh y$
$f[x, \sqrt{x^2 + a^2}] \, dx$	Let $x = a \tan y$ or $x = a \sinh y$

10.2 Definite Integrals

The *definite integral* of $f(x)$ from a to b is

$$\int_a^b f(x) \, dx = \lim_{\substack{n \to \infty \\ \max \Delta x_\nu \to 0}} \sum_{\nu=1}^n f(\xi_\nu) \, \Delta x_\nu \tag{218}$$

in which the interval $a \le x \le b$ is divided into n arbitrary parts $\Delta x_\nu, \nu = 1, 2, \ldots, n$, and ξ_ν is an arbitrary point in Δx_ν (Fig. 140). A sufficient condition that this integral exists is that $f(x)$ be continuous. However, it is necessary and sufficient only that $f(x)$ be bounded and that its points of discontinuity form a set of Lebesgue measure 0. A set of points is of Lebesgue measure 0 if the points can be enclosed in a set of intervals $I_\nu, \nu = 1, 2, 3, \ldots$, finite or infinite in number, such that, for any $\varepsilon > 0$, the sum of the lengths of the I_ν is $<\varepsilon$.

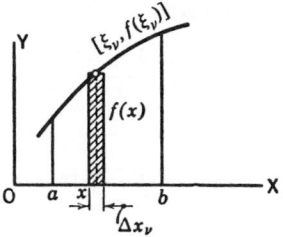

Fig. 140

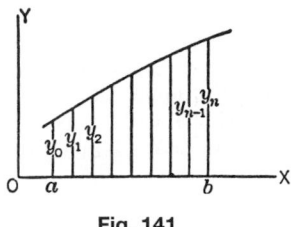

Fig. 141

A definite integral of a given function over a given interval is a number. The geometric interpretation of the definite integral of $f(x)$, $f(x) \geq 0$, is the area bounded by the curve $y = f(x)$, the x axis, and the ordinates at $x = a$ and $x = b$.

To evaluate a definite integral, if $F(x) = \int f(x)\,dx$, then $\int_a^b f(x)\,dx = F(b) - F(a)$.

Example 64

$$\int_3^5 x^2\,dx = \frac{x^3}{3}\Big|_3^5 = \frac{125}{3} - 9 = \frac{98}{3}$$

Some Fundamental Theorems

$$\frac{d}{dx}\int_a^x f(\xi)\,d\xi = f(x)$$

$$\int_a^b c f(x)\,dx = c\int_a^b f(x)\,dx$$

$$\int_a^b [f_1(x) + f_2(x) + \cdots + f_n(x)]\,dx$$

$$= \int_a^b f_1(x)\,dx + \int_a^b f_2(x)\,dx$$

$$+ \cdots + \int_a^b f_n(x)\,dx$$

$$\int_a^b f(x)\,dx = -\int_b^a f(x)\,dx$$

$$\int_a^b f(x)\,dx = \int_a^c f(x)\,dx + \int_c^b f(x)\,dx$$
$$a \leq c \leq b$$

$$\int_a^b f(x)\,dx = (b-a)f(\xi)$$

for some ξ such that $a \leq \xi \leq b$ (mean value theorem).

Simpson's Rule for Approximate Integration. To evaluate $\int_a^b f(x)\,dx$ approximately, divide the interval from a to b into an even number n of equal parts with endpoints $x_0 = a, x_1, \ldots, x_n = b$ and let $y_i = f(x_i)$ (Fig. 141). Then

$$\int_a^b f(x)\,dx \approx \frac{b-a}{3n}(y_0 + 4y_1 + 2y_2 + 4y_3 + 2y_4$$

$$+ \cdots + 4y_{n-1} + y_n) \qquad (219)$$

Improper Integrals. If one limit is *infinite*,

$$\int_a^\infty f(x)\,dx = \lim_{b\to\infty}\int_a^b f(x)\,dx \qquad (220)$$

The integral exists, or converges, if there is a number $k > 1$ and a number M independent of x such that $x^k|f(x)| < M$ for arbitrarily large values of x. If $x|f(x)| > m$, an arbitrary positive number, for sufficiently large values of x, the interval diverges.

Example 65. The integral $\int_0^\infty x\,dx/(x + x^2)^{3/2}$ exists, since, for $k = 2$ and $M = 1$,

$$x^2\left|\frac{x}{(x+x^2)^{3/2}}\right| = \left(\frac{x^2}{x+x^2}\right)^{3/2} < 1$$

no matter how large the value of x.

If the integrand is infinite at the upper limit

$$\int_a^b f(x)\,dx = \lim_{\varepsilon\to 0}\int_a^{b-\varepsilon} f(x)\,dx \quad 0 < \varepsilon < (b-a)$$
$$(221)$$

The integral exists if there is a number $k < 1$ and a number M independent of x such that $(b - x)^k|f(x)| < M$ for $a \leq x < b$. If there is a number $k \geq 1$ and a number m such that $(b - x)^k|f(x)| > m$ for $a \leq x < b$, the integral diverges.

Example 66. The integral $\int_0^1 dx/(1 - x)$ diverges, since, for $k = 1$ and $m = \frac{1}{2}$, $(1 - x)/(1 - x) = 1 > \frac{1}{2}$.

If the integrand is infinite at the lower limit, the tests are analogous. If the integrand is infinite at an intermediate point, use the point to divide the interval into two subintervals and apply the preceding tests.

Multiple Integrals. Let $f(x, y)$ be defined in the region R of the xy plane. Divide R into subregions $\Delta R_1, \Delta R_2, \ldots, \Delta R_n$ of areas $\Delta A_1, \Delta A_2, \ldots, \Delta A_n$. Let (ξ_i, η_i) be any point in ΔR_i. If the sum $\sum_{i=1}^{n} f(\xi_i, \eta_i)\, \Delta A_i$ has a limit as $n \to \infty$ and the maximum diameter of the subregions ΔR_i approaches zero, then

$$\int_R f(x, y)\, dA = \lim_{n \to \infty} \sum_{i=1}^{n} f(\xi_i, \eta_i)\, \Delta A_i \qquad (222)$$

The double integral is evaluated by two successive single integrations, first with respect to y holding x constant between variable limits of integration and then with respect to x between constant limits (Fig. 142). If $f(x, y)$ is continuous, the order of integration can be reversed,

$$\int_R f(x, y)\, dA = \int_a^b \int_{y_1(x)}^{y_2(x)} f(x, y)\, dy\, dx$$

$$= \int_c^d \int_{x_1(y)}^{x_2(y)} f(x, y)\, dx\, dy \qquad (223a)$$

In polar coordinates,

$$\int_R F(r, \theta)\, dA = \int_\alpha^\beta \int_{r_1(\theta)}^{r_2(\theta)} F(r, \theta) r\, dr\, d\theta$$

$$= \int_k^l \int_{\theta_1(r)}^{\theta_2(r)} F(r, \theta) r\, d\theta\, dr \qquad (223b)$$

The analogous triple integrals are evaluated by three single integrations. In rectangular coordinates,

$$\int_R f(x, y, z)\, dV = \int \int \int f(x, y, z)\, dx\, dy\, dz \qquad (224)$$

In spherical coordinates,

$$\int_R F(r, \theta, \phi)\, dV = \int \int \int F(r, \theta, \phi) r^2 \sin\theta\, dr\, d\theta\, d\phi \qquad (225)$$

In cylindrical coordinates,

$$\int_R G(\rho, \phi, z)\, dV = \int \int \int G(\rho, \phi, z)\rho\, d\rho\, d\phi\, dz \qquad (226)$$

Integrals Containing a Parameter. If $f(x, y)$ is a continuous function of x and y in the closed rectangle $x_0 \le x \le x_1$, $y_0 \le y \le y_1$, and if $f(x, y)$ is integrated with respect to x, with y regarded as fixed and called a *parameter*, then

$$\int_{x_0}^{x_1} f(x, y)\, dx = \phi(y) \qquad (227)$$

is a continuous function of y. Geometrically, the function $f(x, y)$ may be plotted as a surface $z = f(x, y)$. Then the value of $\phi(y_i)$ is the area of the section under the surface made by the plane $y = y_i$ (Fig. 143). If the limits of integration are continuous functions of y instead of constants, then $\phi(y)$ is continuous.

Differentiation under the Integral Sign. If $\partial f/\partial y$ is a continuous function of x and y in a closed rectangle, then

$$\frac{d\phi}{dy} = \int_{x_0}^{x_1} \frac{\partial f(x, y)}{\partial y}\, dx \qquad (228)$$

If $\partial f/\partial y$ is continuous and the limits of integration are differentiable functions of y, then

$$\frac{d\phi}{dy} = \frac{d}{dy} \int_{x_0=g_0(y)}^{x_1=g_1(y)} f(x, y)$$

$$dx = \int_{g_0(y)}^{g_1(y)} \frac{\partial f(x, y)}{\partial y}\, dx - f(g_0, y)\frac{dg_0}{dy}$$

$$+ f(g_1, y)\frac{dg_1}{dy} \qquad (229)$$

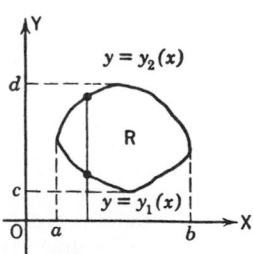

Fig. 142

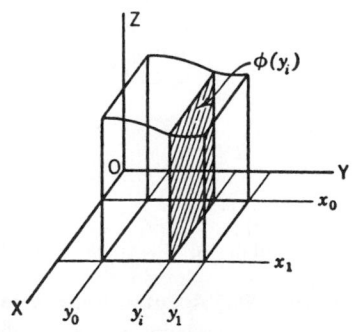

Fig. 143

If $f(x)$ is integrable in the interval $a \leq x \leq b$ and continuous at a point within the interval, then at that point the function $F(x) = \int_a^x f(\xi)\, d\xi$ has a derivative $F'(x) = f(x)$.

Uniform Convergence and Change of Order of Integration.

The improper integral

$$\phi(y) = \int_{x_0}^{\infty} f(x, y)\, dx \qquad (230)$$

converges uniformly in y in the interval $y_0 \leq y \leq y_1$ if for any $\varepsilon > 0$ there exists an L dependent on ε but not on y such that

$$\left| \int_l^{\infty} f(x_1 y)\, dx \right| < \varepsilon \quad \text{for } l \geq L \qquad (231)$$

If $\int_{x_0}^{\infty} f(x, y)\, dx$ is uniformly convergent for $y_0 \leq y \leq y_1$, then

$$\int_{y_0}^{y_1} \int_{x_0}^{\infty} f(x, y)\, dx\, dy = \int_{x_0}^{\infty} \int_{y_0}^{y_1} f(x, y)\, dy\, dx \qquad (232)$$

Stieltjes Integral.

If $f(x)$ and $\phi(x)$ are defined in the interval (a, b), the Stieltjes integral of $f(x)$ with respect to $\phi(x)$ is

$$\int_a^b f(x)\, d\phi(x)$$

$$= \lim_{\substack{n \to \infty \\ \max \Delta x_\nu \to 0}} \sum_{\nu=1}^{n} f(\xi_\nu)[\phi(x_\nu) - \phi(x_{\nu-1})] \qquad (233)$$

in which the interval (a, b) is divided into n arbitrary parts $\Delta x_\nu = x_\nu - x_{\nu-1}$ by the points $x_0 = a, x_1, \ldots, x_n = b$, and ξ_ν is an arbitrary point in Δx_ν. This limit exists if $f(x)$ is continuous and $\phi(x)$ is of bounded variation, that is, can be expressed as the difference of two nonincreasing or two nondecreasing bounded functions. However, it is not necessary that $f(x)$ be continuous, but only that the variation of $\phi(x)$ over the set of points of discontinuity of $f(x)$ be zero.

Lebesgue Integral.

Let S be a set of points in the interval (a, b), and $C(S)$ the complement of S, that is, the set of all the points of (a, b) that do not belong to S. Enclose the points of S in a set of intervals I_ν, $\nu = 1, 2, 3, \ldots$, finite or infinite in number, and let the sum of the lengths of the I_ν be L. The greatest lower bound of all possible values of L is the exterior measure $\overline{m}(S)$ of S. The interior measure of S is $m(S) = (b - a) - \overline{m}[C(S)]$. If $\overline{m}(S) = m(S)$, the set S is measurable and its *measure* is $m(S) = \overline{m}(S)$.

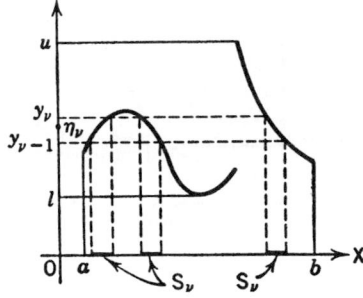

Fig. 144

A function $f(x)$ defined in the interval (a, b) is *measurable* if the set of points x for which $y_0 \leq f(x) < y_1$ is measurable for any values of y_0 and y_1.

Let u and l be the upper and lower bounds of a measurable function $f(x)$ defined in the interval (a, b) (Fig. 144). Divide the interval (u, l) into n arbitrary subintervals Δy_ν by the points $y_0 = 1, y_1, \ldots, y_n = u$. Let S_ν be the set of points for which $y_{\nu-1} \leq f(x) < y_\nu$ and η_ν any point in the interval Δy_ν. Then the Lebesgue integral of $f(x)$ in the interval (a, b) is

$$\int_a^b f(x)\, dx = \lim_{\substack{n \to \infty \\ \max \Delta y_\nu \to 0}} \sum_{\nu=1}^{n} \eta_\nu \cdot m(S_\nu) \qquad (234)$$

If the Riemann integral in the interval (a, b), defined on p. 250, exists, the Lebesgue integral does also, and the two are equal, but not conversely.

10.3 Line, Surface, and Volume Integrals

Line Integrals.

Let $P(x, y)$ and $Q(x, y)$ be functions continuous at all points of a continuous curve C joining the points A and B in the xy plane. Divide the curve C into n arbitrary parts Δs_ν by the points (x_ν, y_ν), let (ξ_ν, η_ν) be an arbitrary point on Δs_ν, and let Δx_ν and Δy_ν be the projections of ΔS_ν on the x and the y axes (Fig. 145). The line integral is

$$\int_A^B [P(x, y)\, dx + Q(x, y)\, dy] = \lim_{\substack{n \to \infty \\ \max \Delta x_\nu, \Delta y_\nu \to 0}}$$

$$\times \sum_{\nu=1}^{n} [P(\xi_\nu, \eta_\nu)\, \Delta x_\nu + Q(\xi_\nu, \eta_\nu)\, \Delta y_\nu] \qquad (235)$$

If the equation of the curve C is $y = f(x)$, $x = \phi(y)$, or the parametric equations $x = x(t)$, $y = y(t)$, the line integral can be evaluated as a definite integral in the one variable x, y, or t, respectively.

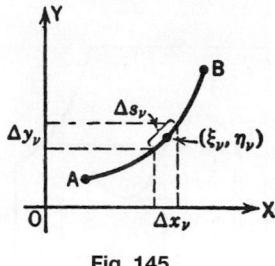

Fig. 145

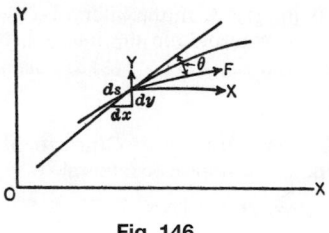

Fig. 146

Example 67. Find the value of $\int_{0,0}^{1,3}[y^2\,dx + (xy - x^2)\,dy]$ along the paths (a) $y = 3x$, (b) $y^2 = 9x$.

(a) Substitute $y = 3x$, $dy = 3\,dx$ and obtain

$$\int_0^1 [9x^2 + (3x^2 - x^2)3]\,dx$$

$$= \int_0^1 15x^2\,dx = 5$$

(b) Substitute $y^2 = 9x$, $2y\,dy = 9\,dx$, and obtain

$$\int_0^3 \left[\frac{2}{9}y^3 + \left(\frac{y^3}{9} - \frac{y^4}{81} \right) \right] dy$$

$$= \left[\frac{1}{12}y^4 - \frac{y^5}{405} \right]_0^3 = 6\frac{3}{20}$$

A line integral in the xyz space

$$\int_A^B [P(x, y, z)\,dx + Q(x, y, z)\,dy$$

$$+ R(x, y, z)\,dz] \tag{236}$$

is defined similarly.

Applications

Work. The work done by a constant force F acting on a particle that moves a distance s along a straight line inclined at an angle θ to the force is $W = Fs\cos\theta$. If the path is a curve C and the force variable, the differential of work is $dW = F\cos\theta\,ds$, where ds is the differential of the path. Then

$$W = \int dW = \int_C F\cos\theta\,ds = \int_C (X\,dx + Y\,dy)$$

$$\tag{237}$$

where X and Y are the x and y components of F (Fig. 146).

Area. The area of a region bounded by a closed curve C such that a line parallel to the x or y axis meets C in no more than two points is

$$A = \frac{1}{2}\int_C (x\,dy - y\,dx) \tag{238}$$

The formula can be applied to any region that can be divided by a finite number of lines into regions satisfying the preceding condition.

Surface Integrals. Let $P(x, y, z)$ be a function continuous at all points of a region S (bounded by a simple closed curve) of a surface $z = f(x, y)$ which has a continuously turning tangent plane except possibly at isolated points or lines. Let A be the projection of S on the xy plane. Divide S into arbitrary subregions ΔS_ν and let $(\xi_\nu, \eta_\nu, \zeta_\nu)$ be an arbitrary point in ΔS_ν (Fig. 147). The surface integral is

$$\lim_{\substack{n \to \infty \\ \text{max diam}\Delta S_\nu \to 0}} \sum_{\nu=1}^n P(\xi_\nu, \eta_\nu, \zeta_\nu)\,\Delta S_\nu$$

$$= \int_S P(x, y, z)\,dS$$

$$= \iint_A P(x, y, z)\sqrt{1 + \left(\frac{\partial z}{\partial x} \right)^2 + \left(\frac{\partial z}{\partial y} \right)^2}\,dx\,dy$$

$$\tag{239}$$

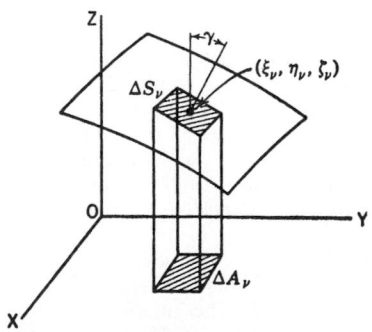

Fig. 147

If α, β, γ are the direction angles of the normal to S, the form of the surface integral analogous to the line integral (235) is

$$\iint_A (P\,dy\,dz + Q\,dz\,dx + R\,dx\,dy)$$

$$= \int_S (P\cos\alpha + Q\cos\beta + R\cos\gamma)\,dS \quad (240)$$

Green's Theorem. Let $P(x, y)$ and $Q(x, y)$ be continuous functions with continuous partial derivatives $\partial P/\partial y$ and $\partial Q/\partial x$ in a simply connected region R bounded by a simple closed curve C. Then

$$\iint_R \left(\frac{\partial Q}{\partial x} - \frac{\partial P}{\partial y}\right) dx\,dy = \int_C (P\,dx + Q\,dy) \quad (241)$$

A region is *simply connected* if any closed curve in the region can be shrunk to a point without passing outside the region.

Stokes's Theorem. Let $P(x, y, z)$, $Q(x, y, z)$, $R(x, y, z)$ be continuous functions with continuous first partial derivatives, S a region (bounded by a simple closed curve C) of a surface $z = f(x, y)$, continuous with continuous first partial derivatives. Then

$$\iint_S \left[\left(\frac{\partial R}{\partial y} - \frac{\partial Q}{\partial z}\right) dy\,dz + \left(\frac{\partial P}{\partial z} - \frac{\partial R}{\partial x}\right) dz\,dx\right.$$

$$\left. + \left(\frac{\partial Q}{\partial x} - \frac{\partial P}{\partial y}\right) dx\,dy\right]$$

$$= \int_C (P\,dx + Q\,dy + R\,dz) \quad (242)$$

The signs are such that an observer standing on the surface with head in the direction of the normal will see the integration around C taken in the positive direction.

Divergence, or Gauss's, Theorem. Let $P(x, y, z)$, $Q(x, y, z)$, $R(x, y, z)$ be continuous functions with continuous first partial derivatives. Let V be a region in the xyz space bounded by a closed surface S with a continuously turning tangent plane except possibly at isolated points or lines. Then

$$\iiint_V \left(\frac{\partial P}{\partial x} + \frac{\partial Q}{\partial y} + \frac{\partial R}{\partial z}\right) dx\,dy\,dz$$

$$= \iint_S (P\,dy\,dz + Q\,dz\,dx + R\,dx\,dy) \quad (243)$$

Example 68. Evaluate $\int (x\,dy\,dz + y\,dz\,dx + z\,dx\,dy)$ over the cylinder

$$x^2 + y^2 = a^2 \qquad z = \pm b$$

Since

$$P = x \qquad Q = y \qquad R = z$$

$$\frac{\partial P}{\partial x} = 1 \qquad \frac{\partial Q}{\partial y} = 1 \qquad \frac{\partial R}{\partial z} = 1$$

and

$$\int_{-a}^{a} \int_{-\sqrt{a^2-x^2}}^{\sqrt{a^2-x^2}} \int_{-b}^{+b} 3\,dz\,dy\,dx = 6\pi a^2 b$$

Independence of Path and Exact Differential. Under the conditions of Green's theorem, the following statements are equivalent:

1. $\int_C (P\,dx + Q\,dy) = 0$ for any closed curve C in the region R.
2. The value of $\int_{(a,b)}^{(\xi,\eta)} (P\,dx + Q\,dy)$ is independent of the curve connecting (a, b) and (ξ, η), any points in R.
3. $\partial P/\partial y = \partial Q/\partial x$ at all points of R.
4. There exists a function $F(x, y)$ such that $dF = P\,dx + Q\,dy$.

Under the conditions of Stokes's theorem, the corresponding statements for three dimensions are:

1. $\int_C (P\,dx + Q\,dy + R\,dz) = 0$ for any closed curve C in the region S.
2. The value of $\int_{(a,b,c)}^{(\xi,\eta,\zeta)} (P\,dx + Q\,dy + R\,dz)$ is independent of the curve connecting (a, b, c) and (ξ, η, ζ), any points in S.
3. $\partial P/\partial y = \partial Q/\partial x$, $\partial Q/\partial z = \partial R/\partial y$, $\partial R/\partial x = \partial P/\partial z$ at all points of S.
4. There exists a function $F(x, y, z)$ such that $dF = P\,dx + Q\,dy + R\,dz$.

10.4 Applications of Integration

Length of Arc of a Curve. The length s of the arc of a plane curve $y = f(x)$ from the point (a, b) to the point (c, d) is

$$s = \int_a^c \sqrt{1 + \left(\frac{dy}{dx}\right)^2}\,dx = \int_b^d \sqrt{1 + \left(\frac{dx}{dy}\right)^2}\,dy$$
$$(244)$$

If the equation of the curve is in polar coordinates, $r = f(\theta)$, then the length of the arc from the point (r_1, θ_1) to the point (r_2, θ_2) is

$$s = \int_{\theta_1}^{\theta_2} \sqrt{r^2 + \left(\frac{dr}{d\theta}\right)^2}\,d\theta = \int_{r_1}^{r_2} \sqrt{1 + r^2 \left(\frac{d\theta}{dr}\right)^2}\,dr$$
$$(245)$$

If the curve is in three dimensions, represented by the equations $y = f_1(x), z = f_2(x)$, the length of arc from $x_1 = a$ to $x_2 = b$ is

$$s = \int_a^b \sqrt{1 + \left(\frac{dy}{dx}\right)^2 + \left(\frac{dz}{dx}\right)^2}\, dx \qquad (246)$$

Plane Area. The area bounded by the curve $y = f(x)$, the x axis, and the ordinates at $x = a, x = b$ is

$$A = \int_a^b f(x)\, dx \qquad (247)$$

where y has the same sign for all values of x between a and b.

In polar coordinates, the area bounded by the curve $r = f(\theta)$ and the two radii $\theta = \alpha, \theta = \beta$ (Fig. 148) is

$$A = \frac{1}{2} \int_\alpha^\beta r^2\, d\theta \qquad (248)$$

In rectangular coordinates, if the area is bounded by the two curves $y_2 = f(x), y_1 = \phi(x)$ and the lines $x_2 = b, x_1 = a$ (Fig. 149), then

$$A = \int_a^b dx \int_{\phi(x)}^{f(x)} dy \qquad (249)$$

If the area is bounded by the two curves $x_2 = \psi(y), x_1 = \xi(y)$ and the lines $y_2 = d, y_1 = c$, then

$$A = \int_c^d dy \int_{\xi(y)}^{\psi(y)} dx \qquad (250)$$

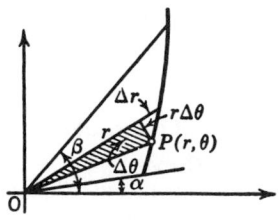

Fig. 148

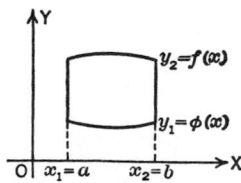

Fig. 149

If expressed in polar coordinates, the area by double integration is

$$A = \int_{\theta_1}^{\theta_2} d\theta \int_{r_1=f_1(\theta)}^{r_2=f_2(\theta)} r\, dr \qquad \text{or}$$

$$\int_{r_1}^{r_2} r\, dr \int_{\theta_1=\phi_1(r)}^{\theta_2=\phi_2(r)} d\theta \qquad (251)$$

Area of a Surface Revolution. The area of the surface of a solid of revolution generated by revolving the curve $y = f(x)$ between $x = a$ and $x = b$ is given as

$$2\pi \int_a^b y \sqrt{1 + \left(\frac{dy}{dx}\right)^2}\, dx \quad \text{about the } x \text{ axis} \quad (252)$$

$$2\pi \int_c^d x \sqrt{1 + \left(\frac{dx}{dy}\right)^2}\, dy \quad \text{about the } y \text{ axis} \quad (253)$$

where $c = f(a)$ and $d = f(b)$.

Volume. By triple integration,

$$V = \begin{cases} \iiint dx\, dy\, dz & \text{(rectangular coordinates)} \\ & \qquad\qquad\qquad (254) \\ \iiint r^2 \sin\theta\, d\theta\, d\phi\, dr & \text{(spherical coordinates)} \\ & \qquad\qquad\qquad (255) \\ \iiint \rho\, d\rho\, d\phi\, dz & \text{(cylindrical coordinates)} \\ & \qquad\qquad\qquad (256) \end{cases}$$

(the limits of integration to be supplied).

Volume of a Solid of Revolution. The volume of a solid of revolution generated by revolving the region bounded by the x axis and the curve $y = f(x)$ between $x = a$ and $x = b$ is

$$\pi \int_a^b y^2\, dx \quad \text{about the } x \text{ axis} \qquad (257a)$$

$$\pi \int_c^d x^2\, dy \quad \text{about the } y \text{ axis} \qquad (257b)$$

where $c = f(a)$ and $d = f(b)$.

Surfaces. If the equation of a surface is written in the parametric form $x = f_1(u, v), y = f_2(u, v), z = f_3(u, v)$, the length of arc of a curve $u = u(t), v = v(t)$ on the surface is

$$s = \int \sqrt{E\left(\frac{du}{dt}\right)^2 + 2F \frac{du}{dt}\frac{dv}{dt} + G\left(\frac{dv}{dt}\right)^2}\, dt \qquad (258)$$

The area S of a region on the surface is

$$S = \int \int \sqrt{EG - F^2} \, du \, dv \qquad (259)$$

where

$$E = \left(\frac{\partial x}{\partial u}\right)^2 + \left(\frac{\partial y}{\partial u}\right)^2 + \left(\frac{\partial z}{\partial u}\right)^2$$

$$F = \frac{\partial x}{\partial u}\frac{\partial x}{\partial v} + \frac{\partial y}{\partial u}\frac{\partial y}{\partial v} + \frac{\partial z}{\partial u}\frac{\partial z}{\partial v}$$

$$G = \left(\frac{\partial x}{\partial v}\right)^2 + \left(\frac{\partial y}{\partial v}\right)^2 + \left(\frac{\partial z}{\partial v}\right)^2$$

If the equation of the surface is written as $x = u$, $y = v$, $z = f(u, v) = f(x, y)$, the arc length is given as

$$s = \int \sqrt{(1+p^2)\left(\frac{dx}{dt}\right)^2 + 2pq\frac{dx}{dt}\frac{dy}{dt} + (1+q^2)\left(\frac{dy}{dt}\right)^2} \, dt \qquad (260)$$

and the area as

$$S = \int \int \sqrt{1 + p^2 + q^2} \, dx \, dy \quad \text{where}$$

$$p = \frac{\partial z}{\partial x} \qquad q = \frac{\partial z}{\partial y} \qquad (261)$$

(the limits of integration to be supplied).

Moment. The moments of a mass m about the yz, xz, and xy planes are respectively

$$M_{yz} = \int x \, dm \quad M_{xz} = \int y \, dm \quad M_{xy} = \int z \, dm \qquad (262)$$

(the limits of integration to be supplied).

Center of Gravity. The coordinates of the center of gravity of a mass m are

$$x = \frac{\int x \, dm}{\int dm} \qquad y = \frac{\int y \, dm}{\int dm} \qquad z = \frac{\int z \, dm}{\int dm} \qquad (263)$$

(the limits of integration to be supplied).

Moment of Inertia. The moments of inertia I are

$$I_x = \int y^2 \, ds \quad I_y = \int x^2 \, ds \quad I_0 = \int (x^2 + y^2) \, ds \qquad (264)$$

for a plane curve about the x and y axis and about the origin, respectively;

$$I_x = \int y^2 \, dA \quad I_y = \int x^2 \, dA \quad I_0 = \int (x^2 + y^2) \, dA \qquad (265)$$

for a plane area about the x and y axis and about the origin, respectively; and

$$I_{yz} = \int x^2 \, dm$$

$$I_{xz} = \int y^2 \, dm \qquad (266)$$

$$I_{xy} = \int z^2 \, dm$$

$$I_x = I_{xz} + I_{xy}, \text{etc.}$$

for a solid of mass m about the yz, xz, and xy plane and about the x axis, respectively (the limits of integration to be supplied).

Fluid Pressure. The total force F against a plane surface perpendicular to the surface of the liquid and between the depths a and b is

$$F = \int_{y=a}^{y=b} \rho y \, dA = \int_a^b \rho yx \, dy \qquad (267)$$

where ρ is the weight of the liquid per unit volume and y is the depth beneath the surface of the liquid of a horizontal element of area dA. Usually, $dA = x \, dy$, where x is the width of the vertical surface expressed as a function of y.

Center of Pressure. The depth $\overline{y}$ of the center of pressure against a surface perpendicular to the surface of the liquid and between the depths a and b is

$$\overline{y} = \frac{\displaystyle\int_{y=a}^{y=b} \rho y^2 \, dA}{\displaystyle\int_{y=a}^{y=b} \rho y \, dA} \qquad (268)$$

Work. The work W done in moving a particle from $s = a$ to $s = b$ against a force whose component expressed as a function of s in the direction of motion is $F(s)$ is

$$W = \int_{s=a}^{s=b} F(s) \, ds \qquad (269)$$

11 DIFFERENTIAL EQUATIONS

11.1 Definitions

A *differential equation* is an equation containing an unknown function of a set of variables and its derivatives. If the equation has derivatives with respect to one variable only, it is an *ordinary differential equation*, otherwise it is a *partial differential equation*.

Example 69

$$\frac{d^2 y}{dx^2} + k^2 y = 0 \tag{270}$$

$$\frac{d^2 y}{dx^2} = \sqrt{1 + y^2 + \frac{dy}{dx}} \tag{271}$$

$$y \frac{\partial^2 z}{\partial x^2} + zx \frac{\partial^2 z}{\partial x\, \partial y} - \frac{\partial z}{\partial y} = xyz \tag{272}$$

$$y - x \frac{dy}{dx} + 3 \frac{dx}{dy} = 0 \tag{273}$$

Equations (270), (271), and (273) are ordinary differential equations and (272) is a partial differential equation.

The *order* of a differential equation is the order of the highest derivative involved. Thus in Eqs. (270)–(272), the order is 2; in (273), the order is 1.

The *degree* of a differential equation is the exponent of the highest order appearing in the equation after it is rationalized and cleared of fractions with respect to the derivatives. The degree of (270), (272), and (273) is 1; that of (271) is 2.

A *solution* or *integral* of a differential equation is a relation among the variables that satisfies the equation identically.

A *general solution* of an ordinary differential equation of the *n*th order is one that contains *n* independent constants. Thus, $y = \sin x + c$ is a general solution of the equation $dy/dx = \cos x$.

A *particular solution* is one that is derivable from a general solution by assigning fixed values to the arbitrary constants. Thus, $y_1 = \sin x$, $y_2 = \sin x + 4$ are two particular solutions of the preceding equation.

11.2 First-Order Equations

Separation of Variables. A differential equation of the *first order*,

$$f\left(x, y, \frac{dy}{dx}\right) = 0 \tag{274}$$

can be brought into the form

$$P(x, y)\, dx + Q(x, y)\, dy = 0 \tag{275}$$

For the special case where P is a function of x only and Q a function of y only,

$$P(x)\, dx + Q(y)\, dy = 0 \tag{276}$$

the variables are separated. The solution is

$$\int P(x)\, dx + \int Q(y)\, dy = c \tag{277}$$

Example 70. Solve

$$\frac{dy}{dx} = -\frac{x}{y}$$

This can be written as $x\, dx + y\, dy = 0$ and has the solution

$$\int x\, dx + \int y\, dy = \tfrac{1}{2} x^2 + \tfrac{1}{2} y^2 = c$$

If $c = r^2/2$, then $x^2 + y^2 = r^2$, a set of concentric circles. There are an infinite number of solutions depending on the value of r. Through each point in the plane there passes one circle and only one.

Homogeneous Equations. A function $f(x, y)$ is homogeneous of the *n*th degree in x and y if $f(kx, ky) = k^n f(x, y)$. An equation

$$P(x, y)\, dx + Q(x, y)\, dy = 0 \tag{278}$$

is homogeneous if the functions $P(x, y)$ and $Q(x, y)$ are homogeneous in x and y. By substituting $y = vx$, the variables can be separated.

Example 71. Solve $(x^2 + y^2)\, dx - 2xy\, dy = 0$.

This is of the form $P(x, y)\, dx + Q(x, y)\, dy = 0$, where P and Q are homogeneous functions of the second degree. Making the substitution $y = vx$, the equation becomes $(1 + v^2)\, dx - 2v(x\, dv + v\, dx) = 0$.

Separating variables,

$$\frac{dx}{x} - \frac{2v}{1 - v^2}\, dv = 0$$

Integrating, $\log_e x(1 - v^2) = \log_e c$; replacing $v = y/x$, $\log(1 - y^2/x^2)x = \log_e c$; and taking exponentials, $x^2 - y^2 = cx$.

Linear Differential Equation. The differential equation

$$\frac{dy}{dx} + P(x)y = Q(x) \tag{279}$$

in which y and dy/dx appear only in the first degree and P and Q are functions of x is a *linear equation of the first order*. This has the general solution

$$y = e^{-\int p(x)\,dx}\left[\int Q(x)e^{\int P(x)\,dx}\,dx + c\right] \qquad (280)$$

Example 72. An equation in the theory of electric networks is

$$L\frac{di}{dt} + Ri = E$$

where i is the current, L the inductance (a constant), R the resistance (a constant), and E the electromotive force, a function of time or constant. If $E = E(t)$,

$$i = e^{-(R/L)t}\left[\int \frac{E}{L}e^{(R/L)t}\,dt + c\right]$$

If E is constant and if $i = 0$ at $t = 0$, then

$$i = \frac{E}{R}(1 - e^{-(R/L)t})$$

Bernoulli Equation. This is

$$\frac{dy}{dx} + P(x)Y = Q(x)y^n \qquad (281)$$

in which $n \neq 1$. By making the substitution $z = y^{1-n}$, a linear equation is obtained and the general solution is

$$y = e^{-\int P(x)\,dx}\left[(1-n)\int e^{(1-n)\int P(x)\,dx}Q(x)\,dx + c\right]^{1/1-n} \qquad (282)$$

Example 73. Solve the equation

$$\frac{dy}{dx} - xy = xy^2$$

Substitute $z = y^{-1}$ and obtain $dz/dx + xz = -x$. The general integral is

$$z = ce^{-x^2/2} - 1 \qquad \text{or} \qquad y = \frac{1}{ce^{-x^2/2}-1}$$

Exact Differential Equation. The equation

$$P(x, y)\,dx + Q(x, y)\,dy = 0 \qquad (283)$$

is an *exact differential equation* if its left side is an exact differential

$$du = P\,dx + Q\,dy \qquad (284)$$

that is, if $\partial P/\partial y = \partial Q/\partial x$. Then,

$$\int P\,dx + \int \left[Q - \frac{\partial \int P\,dx}{\partial y}\right]dy = c \qquad (285)$$

is a solution.

Example 74. Solve $(x^2 - 4xy - y^2)\,dx + (y^2 - 2xy - 2x^2)\,dy = 0$.

This is an exact equation because $\partial P/\partial y = -4x - 2y = \partial Q/\partial x$,

$$\int (x^2 - 4xy - y^2)\,dx = \tfrac{1}{3}x^3 - 2x^2y - xy^2$$

$$\int [(y^2 - 2xy - 2x^2) - (-2x^2 - 2xy)]\,dy = \tfrac{1}{3}y^3$$

The general solution is

$$\tfrac{1}{3}x^3 - 2x^2y - xy^2 + \tfrac{1}{3}y^3 = c$$

Integrating Factor. If the left member of the differential equation $P(x, y)\,dx + Q(x, y)\,dy = 0$ is not an exact differential, look for a factor $v(x, y)$ such that $du = v(P\,dx + Q\,dy)$ is an exact differential. Such an *integrating factor* satisfies the equation

$$Q\frac{\partial v}{\partial x} - P\frac{\partial v}{\partial y} + \left(\frac{\partial Q}{\partial x} - \frac{\partial P}{\partial y}\right)v = 0 \qquad (286)$$

Example 75. The equation $(xy^2 - y^3)\,dx + (1 - xy^2)\,dy = 0$ when multiplied by $v = 1/y^2$ becomes $(x - y)\,dx + (1/y^2 - x)\,dy = 0$, of which the left side $du = (x - y)\,dx + (1/y^2 - x)\,dy$ is an exact differential since $\partial P/\partial y = \partial Q/\partial x$. The integration gives $u = x^2/2 - xy - 1/y$. The general solution is $u = c$ or $x^2y - 2xy^2 - 2cy - 2 = 0$.

Riccati's Equation. This is

$$\frac{dy}{dx} + P(x)y^2 + Q(x)y + R(x) = 0 \qquad (287)$$

If a particular integral y_1 is known, place $y = y_1 + 1/z$ and obtain a linear equation in z.

11.3 Second-Order Equations

The differential equation

$$F\left(x, y, \frac{dy}{dx}, \frac{d^2y}{dx^2}\right) = 0 \qquad (288)$$

is of the *second order*. If some of these variables are missing, there is a straightforward method of solution.

CASE 1: With y and dy/dx missing,

$$\frac{d^2y}{dx^2} = f(x) \qquad (289)$$

This has the solution

$$y = \int dx \int f(x)\, dx + cx + c_1 \qquad (290)$$

CASE 2: With x and dy/dx missing,

$$\frac{d^2y}{dx^2} = f(y) \qquad (291)$$

Multiply both sides by 2 dy/dx and obtain

$$x = \int \frac{dy}{\sqrt{c + 2\int f(y)\, dy}} + c_1 \qquad (292)$$

as a solution.

CASE 3: With x and y missing,

$$\frac{d^2y}{dx^2} = f\left(\frac{dy}{dx}\right) \qquad (293)$$

Place

$$\frac{dy}{dx} = p \qquad \frac{d^2y}{dx^2} = \frac{dp}{dx} \qquad (294)$$

Then

$$x = \int \frac{dp}{f(p)} + c$$

Solve for p, replace p by dy/dx, and solve the resulting first-order equation.

Example 76. The differential equation of the catenary is

$$a\frac{d^2y}{dx^2} = \sqrt{1 + \left(\frac{dy}{dx}\right)^2}$$

Let $p = dy/dx$. Then

$$a\frac{dp}{dx} = \sqrt{1 + p^2}$$

By separating variables,

$$\frac{dp}{\sqrt{1 + p^2}} = \frac{dx}{a}$$

which has the solution

$$\sinh^{-1} p = \frac{x + c}{a} \qquad \text{or} \qquad p = \frac{dy}{dx} = \sinh\frac{x + c}{a}$$

Integrating this latter,

$$y = a\,\cosh\frac{x + c}{a} + c_1$$

CASE 4: With y missing,

$$\frac{d^2y}{dx^2} = f\left(\frac{dy}{dx}, x\right) \qquad (295)$$

Place $dy/dx = p$ and obtain the first-order equation $dp/dx = f(p, x)$. If this can be solved for p, then

$$y = \int p(x)\, dx + c \qquad (296)$$

CASE 5: With x missing,

$$\frac{d^2y}{dx^2} = f\left(\frac{dy}{dx}, y\right) \qquad (297)$$

Place $dy/dx = p$ and obtain the first-order equation $p\, dp/dy = f(p, y)$. If this can be solved for p, then

$$x = \int \frac{dy}{p(y)} + c \qquad (298)$$

11.4 Bessel Functions

Wherever the mathematics of problems having circular or cylindrical symmetry appears, it is usually appropriate to consider the solutions of Bessel's differential equation (299). Such applications include radiation from a cylindrical antenna, eddy current losses in a cylindrical wire, and sinusoidal angle modulations including phase and frequency modulation,

$$x^2\frac{d^2y}{dx^2} + x\frac{dy}{dx} + (x^2 - n^2)y = 0 \qquad (299)$$

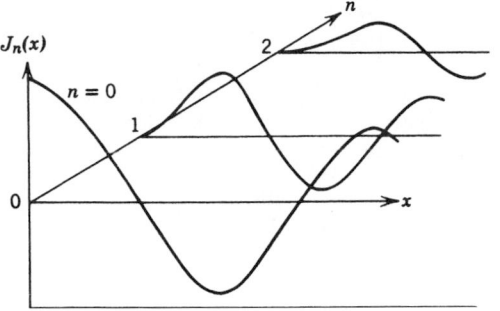

Fig. 150 Bessel functions of first kind.

where n is real, possibly integral or fractional, or complex, and the solution $y(x)$ is said to be of the *first kind* and denoted $J_n(x)$ for $0 \le n$ an integer. Tables of $J_0(x)$ and $J_1(x)$ are available in Table 22 in Chapter 1. Graphs of these are shown in Fig. 150.

Bessel functions $J_n(x)$ are *almost periodic functions* that for increasing x have a zero-crossing half "period" approaching π from below. A sequence of these functions can be used to construct an orthogonal series much in the same way that periodic functions, sine, and cosine waves make up a Fourier series.

For an extensive set of tables of Bessel functions of many types, *Essentials of Mathematical Methods in Science and Engineering*, by S. S. Bayin (Wiley, Hoboken, New Jersey, 2008).

11.5 Linear Equations

General Theorem. The differential equation

$$\frac{d^n y}{dx^n} + P_1(x)\frac{d^{n-1} y}{dx^{n-1}} + \cdots + P_{n-1}(x)\frac{dy}{dx} + P_n(x)y$$
$$= F(x) \tag{300}$$

is called the general nth-order linear differential equation. If $F(x) = 0$, the equation is *homogeneous*; otherwise it is *nonhomogeneous*. If $\phi(x)$ is a solution of the nonhomogeneous equation and $y_1, y_2, \ldots, y_n$ are linearly independent solutions of the homogeneous equation, then the *general solution* of (300) is

$$y = c_1 y_1 + c_2 y_2 + \cdots + c_n y_n + \phi(x) \tag{301}$$

The part $\phi(x)$ is called the *particular integral*, and the part $c_1 y_1 + \cdots + c_n y_n$ is the *complementary function*.

Homogeneous Differential Equation with Constant Coefficients

$$\frac{d^n y}{dx^n} + a_1\frac{d^{n-1} y}{dx^{n-1}} + \cdots + a_{n-1}\frac{dy}{dx} + a_n y = 0 \tag{302}$$

A solution of this equation is

$$y_k = ce^{r_k x} \tag{303}$$

if r_k is a root of the algebraic equation,

$$r^n + a_1 r^{n-1} + \cdots + a_{n-1} r + a_n = 0 \tag{304}$$

If all the n roots $r_1, r_2, \ldots, r_n$ of (304) are different, then

$$y = c_1 e^{r_1 x} + c_2 e^{r_2 x} + \cdots + c_n e^{r_n x} \tag{305}$$

is a general solution of (302). If k of the roots are equal, $r_1 = r_2 = \cdots = r_k$ while $r_{k+1}, \ldots, r_n$ are different, then

$$y = (c_1 + c_2 x + \cdots + c_k x^{k-1})e^{r_1 x} + c_{k+1}e^{r_{k+1} x}$$
$$+ \cdots + c_n e^{r_n x} \tag{306}$$

is a general solution. If $r_1 = p + iq, r_2 = p - iq$ are conjugate complex roots of (304), then

$$c_1 e^{r_1 x} + c_2 e^{r_2 x} = e^{px}(C_1 \cos qx + C_2 \sin qx) \tag{307}$$

Example 77

$$\frac{d^2 y}{dx^2} + 13\frac{dy}{dx} + 40y = 0$$

has the solution $y = c_1 e^{-5x} + c_2 e^{-8x}$.

Example 78

$$\frac{d^2 y}{dx^2} + 6\frac{dy}{dx} + 34y = 0$$

has the solution $y = (c_1 \cos 5x + c_2 \sin 5x)e^{-3x}$.

Nonhomogeneous Differential Equation with Constant Coefficients

$$\frac{d^n y}{dx^n} + a_1\frac{d^{n-1} y}{dx^{n-1}} + \cdots + a_{n-1}\frac{dy}{dx} + a_n y = F(x) \tag{308}$$

The complementary function is found as previously. To find the particular integral, replace

$$\frac{dy}{dx} \text{ by } D, \quad \frac{d^2 y}{dx^2} \text{ by } D^2, \ldots, \quad \frac{d^n y}{dx^n} \text{ by } D^n \tag{309}$$

$$P(D)y = (D^n + a_1 D^{n-1} + \cdots + a_{n-1} D + a_n)y$$
$$= F(x) \tag{310}$$

Particular integrals y_p, in which B_i, A, B are undetermined coefficients, to be determined by substituting y_p in (310) and equating coefficients of like terms, are:

(a) If $F(x) = x^n + b_1 x^{n-1} + \cdots + b_{n-1}x + b_n$, then $y_p = x^n + B_1 x^{n-1} + \cdots + B_{n-1}x + B_n$. If D^m is a factor of $P(D)$, then $y_p = (x^n + B_1 x^{n-1} + \cdots + B_{n-1}x + B_n)x^m$.

(b) If $F(x) = b \sin ax$ or $b \cos ax$, then $y_p = A \sin ax + B \cos ax$. If $(D^2 + a^2)^m$ is a factor of $P(D)$, then $y_p = (A \sin ax + B \cos ax)x^m$.

(c) If $F(x) = ce^{axam}$ then $y_p = Ae^{axam}$. If $(D - a)^m$ is a factor of $P(D)$, then $y_p = x^m Ae^{ax}$.

(d) If $F(x) = g(x)e^{axam}$, place $y_p = e^{ax}w$ in (308), divide out e^{ax}, and solve the equation for w_p as a function of x.

(e) If $F(x)$ is the sum of a number of these functions, then y_p is the sum of the particular integrals corresponding to each of the functions.

(f) If $F(x)$ is not of the type (e), try the method of Laplace transformation (Section 13).

Example 79. $d^2y/dx^2 + 4y = x^2 + \cos x$ can be written as $(D^2 + 4)y = (D + 2i)(D - 2i) = x^2 + \cos x$. By (307), the complementary function is $y = c_1 \cos 2x + c_2 \sin 2x$. For a particular integral take $y_p = ax^2 + bx + c + f \sin x + g \cos x$ [by (a), (b), (e)].
Then

$$\frac{d^2 y_p}{dx^2} = 2a - f \sin x - g \cos x$$

and substituting in the original equation

$$\frac{d^2 y_p}{dx^2} + 4y_p = 2a - f \sin x - g \cos x + 4ax^2$$

$$+ 4bx + 4c + 4f \sin x + 4g \cos x$$

$$= x^2 + \cos x$$

Equating coefficients, $a = \frac{1}{4}, b = 0, c = -\frac{1}{8}, f = 0$, $g = \frac{1}{3}$ and the general solution is $y = c_1 \cos 2x + c_2 \sin 2x + x^2/4 - \frac{1}{8} + \frac{1}{3} \cos x$.

Euler's Homogeneous Equation

$$x^n \frac{d^n y}{dx^n} + ax^{n-1}\frac{d^{n-1}y}{dx^{n-1}} + \cdots + a_{n-1}x\frac{dy}{dx} + a_n y = 0 \tag{311}$$

Place $x = e^t$, and since

$$x\frac{dy}{dx} = \frac{dy}{dt} \qquad x^2 \frac{d^2 y}{dx^2} = \left[\frac{d}{dt}\left(\frac{d}{dt} - 1\right)\right]y$$

$$x^3 \frac{d^3 y}{dx^3} = \left[\frac{d}{dt}\left(\frac{d}{dt} - 1\right)\left(\frac{d}{dt} - 2\right)\right]y, \ldots \tag{312}$$

(311) is transformed into a linear homogeneous differential equation with constant coefficients.

Depression of Order. If a particular integral of a linear homogeneous differential equation is known, the order of the equation can be lowered. If y_1 is a particular integral of

$$\frac{d^n y}{dx^n} + P_1(x)\frac{d^{n-1}y}{dx^{n-1}} + \cdots + P_{n-1}(x)\frac{dy}{dx} + P_n(x) = 0 \tag{313}$$

substitute $y = y_1 z$. The coefficient of z will be zero, and then by placing $dz/dx = u$, the equation is reduced to the $(n-1)$st order.

Example 80. Given

$$\frac{d^2 y}{dx^2} + p(x)\frac{dy}{dx} + q(x)y = 0$$

and y_1, a particular integral of this equation.
Let $y = y_1 z$. Then

$$\frac{dy}{dx} = y_1 \frac{dz}{dx} + z\frac{dy_1}{dx}$$

$$\frac{d^2 y}{dx^2} = y_1 \frac{d^2 z}{dx^2} + 2\frac{dy_1}{dx}\frac{dz}{dx} + z\frac{d^2 y_1}{dx^2}$$

Substituting in the original equation

$$y_1 \frac{d^2 z}{dx^2} + 2\frac{dy_1}{dx}\frac{dz}{dx} + z\frac{d^2 y_1}{dx^2}$$

$$+ p\left[y_1 \frac{dz}{dx} + z\frac{dy_1}{dx}\right] + qy_1 z = 0$$

and since the coefficient of z is zero, this reduces to

$$y_1 \frac{d^2 z}{dx^2} + \left(2\frac{dy_1}{dx} + py_1\right)\frac{dz}{dx} = 0$$

Writing

$$\frac{dz}{dx} = u \qquad \frac{du}{u} + \left(2\frac{dy_1}{dx} + py_1\right)\frac{dx}{y_1} = 0$$

By integrating,

$$\log_e u + \int p\,dx + \log_e y_1^2 = \log_e c \qquad \text{or}$$

$$u = \frac{c}{y_1^2}\exp\left(-\int P\,dx\right)$$

Another integration gives z. Then

$$y = y_1 \int \frac{c}{y_1^2} \exp\left(-\int P \, dx\right) dx + c_1$$

Systems of Linear Differential Equations with Constant Coefficients.

For a system of n linear equations with constant coefficients in n dependent variables and one independent variable t, the symbolic algebraic method of solution may be used. If $n = 2$,

$$(D^n + a_1 D^{n-1} + \cdots + a_n)x$$
$$+ (D^m + b_1 D^{m-1} + \cdots + b_m)y = R(t)$$
$$(D^p + c_1 D^{p-1} + \cdots + c_p)x \tag{314}$$
$$+ (D^q + d_1 D^{q-1} + \cdots d_q)y = S(t)$$

where $D = d/dt$. The equations may be written as

$$P_1(D)x + Q_1(D)y = R \qquad P_2(D)x + Q_2(D)y = S \tag{315}$$

Treating these as algebraic equations, eliminate either x or y and solve the equation thus obtained.

Example 81. Solve the system,

$$(a) \quad \frac{dx}{dt} + \frac{dy}{dt} + 2x + y = 0$$

$$(b) \quad \frac{dy}{dt} + 5x + 3y = 0$$

By using the symbol D these equations can be written

$$(D + 2)x + (D + 1)y = 0 \qquad 5x + (D + 3)y = 0$$

Eliminating x, $(D^2 + 1)y = 0$. From (307) (a) this has the solution $y = c_1 \cos t + c_2 \sin t$. Substituting this in (b),

$$x = -\frac{3c_1 + c_2}{5} \cos t + \frac{c_1 - 3c_2}{5} \sin t$$

11.6 Linear Algebraic Equations

Consider the set of linear algebraic equations

$$\sum_{i=1}^{n} a_{ki} \alpha_i = f_k \qquad (k = 1, 2, \ldots, m) \tag{316}$$

Equation (316) contains m linear algebraic equations in n unknowns, α_i.

Any system of linear equations in which all f_k are zero is called *homogeneous*. Consider the following equations associated with the matrix operator A:

$$A\alpha = 0 \quad \text{(homogeneous equation)} \tag{317}$$

$$A^*\beta = 0 \quad \text{(adjoint homogeneous equation)} \tag{318}$$

where A^* is the adjoint of A. For the linear algebraic equations, $A^* = A^T$, the transpose of A.

The homogeneous adjoint equations can also be written in the form

$$(\mathbf{A}_i, \beta) = 0 \qquad (i = 1, 2, \ldots, n) \tag{319}$$

where $(\cdot, \cdot)$ denotes the inner product in Euclidean space and A_i are the column vectors of the matrix $[A]$. From Eqs. (316) and (319), we deduce the following result, known as the *solvability condition:* The nonhomogeneous equation $A\alpha = \mathbf{f}$ possesses a solution α if and only if the vector $\mathbf{f}$ is orthogonal to all vectors β that are the solutions of the homogeneous adjoint equation, $A^*\beta = \mathbf{0}$. In analytical form this statement can be expressed as

$$(\mathbf{f}, \beta) = 0 \tag{320}$$

We now consider two cases of linear equations and discuss the existence and uniqueness of solutions of linear equations.

1. If (317) has only the trivial (i.e., zero) solution, it follows that $\det A \neq 0$ (otherwise, the trivial solution cannot be determined) and hence $\det A^* \neq 0$. Therefore, the adjoint homogeneous equation (318) also has only the trivial solution. Moreover, the solvability conditions are automatically satisfied for any $\mathbf{f}$ [since the only solution of (318) is $\beta = \mathbf{0}$], and the nonhomogeneous equation 316 has one and only one solution, $\alpha = A^{-1}\mathbf{f}$, where A^{-1} is the inverse of the matrix A.

2. If (317) has nontrivial solutions, then $\det A = 0$. This in turn implies that the rows (or columns) of A are linearly dependent. If these linear dependencies are also reflected in the column vector $\mathbf{f}$ (e.g., if the third row of A is the sum of the first and second rows, we must have $f_3 = f_1 + f_2$ in order to have any solutions), then there is a hope of having a solution to the system. If there are $r (\geq n)$ number of independent solutions to (316), A is said to have a *r-dimensional null space* (i.e., nullity of A is r). It can be shown that A^* also has a r-dimensional null space, which is in general different from that of A. A necessary and sufficient condition for (316) to have solutions is provided by the solvability condition

$$(\mathbf{f}, \beta) \equiv \sum_{i=1}^{n} f_i \beta_i = 0$$

where β is the solution of Eq. (318).

Example 82. This example has three cases:

1. Consider the following pair of equations in two unknowns α_1 and α_2:

$$3\alpha_1 - 2\alpha_2 = 4 \qquad 2\alpha_1 + \alpha_2 = 5$$

or

$$\begin{bmatrix} 3 & -2 \\ 2 & 1 \end{bmatrix} \begin{Bmatrix} \alpha_1 \\ \alpha_2 \end{Bmatrix} = \begin{Bmatrix} 4 \\ 5 \end{Bmatrix} \qquad (A\alpha = f)$$

We note that $\det A = 3 + 4 = 7 \neq 0$. The solution is then given by

$$\begin{Bmatrix} \alpha_1 \\ \alpha_2 \end{Bmatrix} = \begin{bmatrix} \frac{1}{7} & \frac{2}{7} \\ -\frac{2}{7} & \frac{3}{7} \end{bmatrix} \begin{Bmatrix} 4 \\ 5 \end{Bmatrix} = \begin{Bmatrix} 2 \\ 1 \end{Bmatrix}$$

The solution of the adjoint equations is trivial, $\beta = 0$, and therefore, the solvability condition is identically satisfied.

2. Next consider the pair of equations

$$6\alpha_1 + 4\alpha_2 = 4 \qquad 3\alpha_1 + 2\alpha_2 = 2$$

or

$$\begin{bmatrix} 6 & 4 \\ 3 & 2 \end{bmatrix} \begin{Bmatrix} \alpha_1 \\ \alpha_2 \end{Bmatrix} = \begin{Bmatrix} 4 \\ 2 \end{Bmatrix} \qquad A\alpha = f$$

We have $\det A = 0$, because row 1 (R_1) is equal to 2 times row 2 (R_2). However, we also have $2f_2 = f_1$. Consequently, we have one linearly independent solution, say $\alpha^{(1)}$, and the other depends on $\alpha^{(1)}$:

$$\alpha^{(1)} = (2, -2)$$

Note that there are many dependent solutions to the pair. For example, $(2, -2), (4, -5), (-2, 4)$, and so on, are solutions of $A\alpha = f$. The solution to the adjoint homogeneous equation

$$\begin{bmatrix} 6 & 3 \\ 4 & 2 \end{bmatrix} \begin{Bmatrix} \beta_1 \\ \beta_2 \end{Bmatrix} = \begin{Bmatrix} 0 \\ 0 \end{Bmatrix}$$

is given by $\beta_2 = -2\beta_1$. Note that $(f, \beta) \equiv f_1\beta_1 + f_2\beta_2 = 4(-\frac{1}{2}\beta_2) + 2\beta_2 = 0$; hence the solvability condition is satisfied.

3. Finally, consider the pair of equations

$$6\alpha_1 + 4\alpha_2 = 3 \qquad 3\alpha_1 + 2\alpha_2 = 2$$

or

$$\begin{bmatrix} 6 & 4 \\ 3 & 2 \end{bmatrix} \begin{Bmatrix} \alpha_1 \\ \alpha_2 \end{Bmatrix} = \begin{Bmatrix} 3 \\ 2 \end{Bmatrix}$$

We note that $\det A = 0$, because $2R_2 = R_1$. However, $2f_2 \neq f_1$. Hence the pair of equations is inconsistent, and therefore no solutions exist.

Geometrically, we can interpret these three pairs of equations as pairs of straight lines in R^2 with $\alpha_i = x_i, i = 1, 2$ (see Fig. 151). In part 1, the lines represented by the two equations intersect at the point $(x_1, x_2) = (2, 1)$. In part 2, the lines coincide, or intersect, at an infinite number of points, and hence many solutions exist. In part 3, the lines do not intersect at all showing that no solutions exist. From this geometric interpretation, one can see that the lines are nearly parallel (i.e., the angle θ is nearly zero), the determinant of A is nearly zero [because $\tan \theta = (a_{11}a_{22} - a_{12}a_{21})/(a_{11}a_{21} + a_{12}a_{22})$], and therefore it is difficult to obtain an accurate numerical solution. In such cases the system of equations is said to be *ill conditioned*. While these observations can be generalized to a system of n equations, the geometric interpretation becomes complicated.

Example 83. This example has two cases:

1. Consider the following set of three equations in three unknowns:

$$\alpha_1 + \alpha_2 + \alpha_3 = 2$$
$$\alpha_1 - \alpha_2 - 3\alpha_3 = 3 \qquad \text{or} \qquad A\alpha = f$$
$$3\alpha_1 + \alpha_2 - \alpha_3 = 1$$

The adjoint homogeneous equations become

$$\beta_1 + \beta_2 + 3\beta_3 = 0$$
$$\beta_1 - \beta_2 + \beta_3 = 0 \qquad \text{or} \qquad A^*\beta = 0$$
$$\beta_1 - 3\beta_2 - \beta_3 = 0$$

Solving for β, we obtain $\beta_1 = 2\beta_2 = -2\beta_3$. Hence, the null space of A^* is defined by

$$\mathcal{N}(A^*) = \{(2a, a, -a), a \text{ is a real number}\}$$

Clearly $\mathcal{N}(A^*)$ is one dimensional. The null space of A is given by

$$\mathcal{N}(A) = \{(a, -2a, a), a \text{ is a real number}\}$$

Note that $\mathcal{N}(A^*) \neq \mathcal{N}(A)$, but their dimension is the same. Clearly $(2, 1, -1)$ is a solution of $A^*\beta = 0$ while $(1, -2, 1)$ is a solution of $A\alpha = 0$. The solvability condition gives

$$(f, \beta) = 2 \times 2 + 3 \times 1 + 1 \times (-1) \neq 0$$

and therefore $A\alpha = f$ has *no* solution.

2. Reconsider the preceding linear equations with $f = \{-1, 3, 1\}^T$. Then the solvability condition

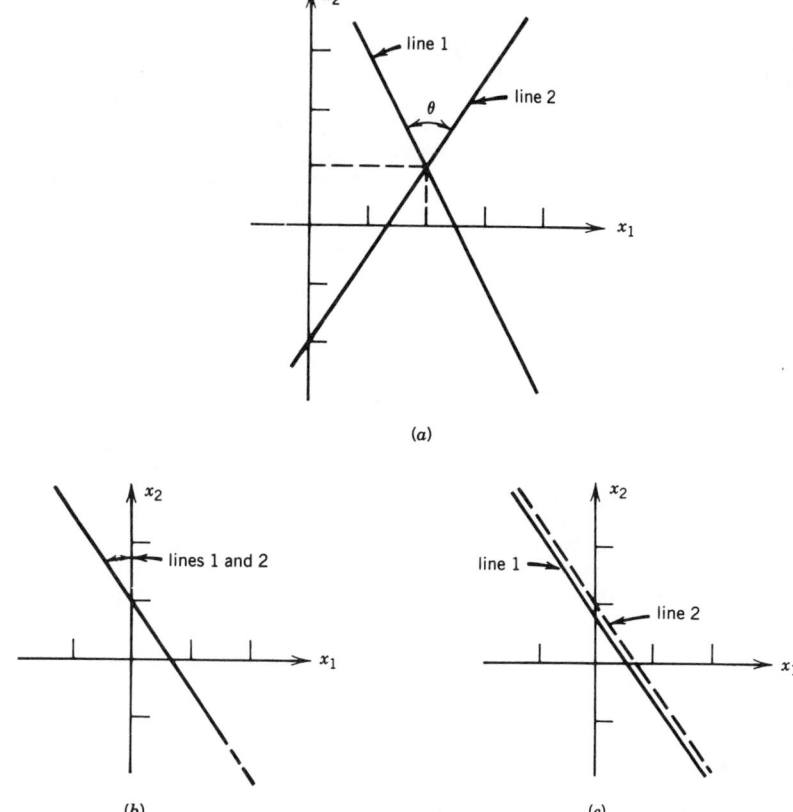

(a)

(b)   (c)

Fig. 151 Geometric interpretation of the solution of two simultaneous algebraic equations in a plane: (a) unique solution; (b) many solutions; (c) no solution.

is clearly satisfied. Hence there is one linearly independent solution to $A\boldsymbol{\alpha} = \mathbf{f}$ (note that $-2R_1 + R_3 = R_2$ and $-2f_1 + f_3 = f_2$).

$$\boldsymbol{\alpha} \equiv (\alpha_1, \alpha_2, \alpha_3) = (1, -2, 0)$$

Only one of the three α's is arbitrary (not determined), and the remaining two α's are given in terms of the arbitrary α. For example, if $\boldsymbol{\alpha_1}$ is arbitrary, we have

$$\alpha_1 + \alpha_2 + \alpha_3 = 1$$
$$\alpha_1 + \alpha_2 - 3\alpha_3 = 2 \quad \text{or} \quad A\boldsymbol{\alpha} = \mathbf{f}$$
$$3\alpha_1 + \alpha_2 - \alpha_3 = 3$$

We have $\det A \neq 0$. It can be easily verified that $(A) = (A^*) = \{(0, 0, 0)\}$. The unique solution to $A\boldsymbol{\alpha} = \mathbf{f}$ is given by

$$\alpha_1 = \tfrac{3}{4} \qquad \alpha_2 = \tfrac{1}{2} \qquad \alpha_3 = -\tfrac{1}{4}.$$

11.7 Partial Differential Equations
First Order

Definition. If $x_1, x_2, \ldots, x_n$ are n independent variables, $z = z(x_1, x_2, \ldots, x_n)$ the dependent variable, and

$$\frac{\partial z}{\partial x_1} = p_1, \ldots, \frac{\partial z}{\partial x_n} = p_n \qquad (321)$$

then

$$F(x_1, x_2, \ldots, x_n, z, p_1, p_2, \ldots, p_n) = 0 \qquad (322)$$

is a partial differential equation of the first order. An equation

$$f(x_1, x_2, \ldots, x_n, z, c_1, \ldots, c_n) = 0 \qquad (323)$$

with n independent constants is a *complete integral* of (322) if the elimination of the constants by partial differentiation gives the differential equation (322).

Example 84

$$F = z^2 \left[\left(\frac{\partial z}{\partial x} \right)^2 + \left(\frac{\partial z}{\partial y} \right)^2 + 1 \right] - c^2 = 0$$

Then

$$f = (x - h)^2 + (y - k)^2 + z^2 - c^2 = 0$$

is a solution, since by differentiating it with respect to x and y,

$$(x - h) + z \frac{\partial z}{\partial x} = 0 \qquad (y - k) + z \frac{\partial z}{\partial y} = 0$$

and substituting the values of $x - h$, $y - k$ from the last two equations in f, expression F is obtained.

If the eliminant obtained by eliminating $c_1, \dots, c_n$ from the equations $f = 0$, $\partial f / \partial c_1 = 0, \dots, \partial f / \partial c_n = 0$ satisfies the differential equation, it is a *singular solution*. This differs from a particular integral in that it is usually not obtainable from the complete integral by giving particular values to the constants.

Suppose that the equation

$$F \left(x, y, z, \frac{\partial z}{\partial x}, \frac{\partial z}{\partial y} \right) = 0$$

has the complete integral $f(x, y, z, a, b) = 0$. Let one of the constants $b = \phi(a)$; then $f[x, y, z, a, \phi(a)] = 0$. The *general integral* is the set of solutions found by eliminating a between $f[x, y, z, a, \phi(a)] = 0$ and $d\phi / da = 0$ for all choices of ϕ.

Linear Differential Equations

$$P(x, y, z)p + Q(x, y, z)q = R(x, y, z) \qquad \text{where}$$

$$p = \frac{\partial z}{\partial x} \qquad q = \frac{\partial z}{\partial y} \qquad (324)$$

is a linear partial differential equation. From the system of ordinary equations

$$\frac{dx}{P} = \frac{dy}{Q} = \frac{dz}{R}$$

the two independent solutions $u(x, y, z) = c_1$, $v(x, y, z) = c_2$ are obtained. Then $\Phi(u, v) = 0$, where Φ is an arbitrary function, is the *general solution* of $Pp + Qq = R$.

Example 85. Given $xp + yq = z$. The system $dx/x = dy/y = dz/z$ has the solution $u = y/x = c_1$, $v = z/x = c_2$. Then the general solution is

$$\Phi(u, v) = \Phi \left(\frac{y}{x}, \frac{z}{x} \right) = 0$$

Example 86. Given $(ny - mz)p + (lz - nx)q = mx - ly$. From

$$\frac{dx}{ny - mz} = \frac{dy}{lz - nx} = \frac{dz}{mx - ly}$$

by using the multipliers l, m, n and adding the fraction $(l\, dx + m\, dy + n\, dz)/0$ is obtained. Therefore $l\, dx + m\, dy + n\, dz = 0$. This has the solution $lx + my + nz = c_1$. Similarly, $x\, dx + y\, dy + z\, dz = 0$, or $x^2 + y^2 + z^2 = c_2$. Then the general solution is $\Phi(x^2 + y^2 + z^2, lx + my + nz) = 0$.

General Method of Solution. Given $F(x, y, z, p, q) = 0$, the partial differential equation to be solved. Since z is a function of x and y, it follows that $dz = p\, dx + q\, dy$. If another relation can be found among x, y, z, p, q, such as $f(x, y, z, p, q) = 0$, then p and q can be eliminated. The solution of the ordinary differential equation thus formed involving x, y, z will satisfy the given equation, $F(x, y, z, p, q) = 0$. The unknown function f must satisfy the following linear partial differential equation:

$$\frac{\partial F}{\partial p} \frac{\partial f}{\partial x} + \frac{\partial F}{\partial q} \frac{\partial f}{\partial y} + \left(p \frac{\partial F}{\partial p} + q \frac{\partial F}{\partial q} \right) \frac{\partial f}{\partial z}$$

$$- \left(\frac{\partial F}{\partial x} + p \frac{\partial F}{\partial z} \right) \frac{\partial f}{\partial p} - \left(\frac{\partial F}{\partial y} + q \frac{\partial F}{\partial z} \right) \frac{\partial f}{\partial q} = 0$$

$$(325)$$

which is satisfied by any of the solutions of the system

$$\frac{\partial x}{\partial F / \partial p} = \frac{\partial y}{\partial F / \partial q} = \frac{dz}{p\, \partial F / \partial p + q\, \partial F / \partial q}$$

$$= \frac{-dp}{\partial F / \partial x + p\, \partial F / \partial z} = \frac{-dq}{\partial F / \partial y + q\, \partial F / \partial z}$$

$$(326)$$

Example 87. Solve $p(q^2 + 1) + (b - z)q = 0$. Here Eqs. (326) reduce to

$$\frac{dp}{pq} = \frac{dp}{q^2} = \frac{dz}{3pq^2 + p + (b - z)q} = \frac{dx}{q^2 + 1}$$

$$= \frac{dy}{-z + b + 2pq}$$

The third fraction, by virtue of the given equation, reduces to $dz/2pq^2$. From the first two fractions, by integration, $q = cp$. This and the original equation determine the values of p and q, namely,

$$p = \frac{\sqrt{c_1(z - b) - 1}}{c_1} \qquad q = \sqrt{c_1(z - b) - 1}$$

Substitution of these values in $dz = p\,dx + q\,dy$ gives

$$dz = \left(\frac{dx}{c_1} + dy\right)\sqrt{c_1(z-b)-1}$$

In this equation the variables are separable; this on integration gives the complete integral $2\sqrt{c_1(z-b)-1} = x + c_1 y + c_2$. There is no singular solution. In this work, had another pair of ratios been chosen, say $dq/q^2 = dx/(q^2+1)$, another complete integral would have been obtained, namely,

$$(z-b)\left\{\tfrac{1}{2}(x+k_1) - \sqrt{\left[\tfrac{1}{2}(x+k_1)\right]^2 + 1}\right\}$$
$$+ y + k_2 = 0$$

Second Order

Definitions. A linear partial differential equation of the second order with two independent variables is of the form

$$L = Ar + 2Bs + Ct + Dp + Eq + Fz = f(x, y) \tag{327}$$

where

$$r = \frac{\partial^2 z}{\partial x^2} \qquad s = \partial^2 z/(\partial x\,\partial y) \qquad t = \partial^2 z/\partial y^2$$
$$p = \partial z/\partial x \qquad q = \partial z/\partial y$$

The coefficients $A, \ldots, F$ are real continuous functions of the real variables x and y. Let $\xi = \xi(x, y), \eta = \eta(x, y)$ be two solutions of the following homogeneous partial differential equation of the first order:

$$Ap^2 + 2Bpq + Cq^2 = 0 \tag{328}$$

If $B^2 - AC = 0$, the homogeneous form of (327), $L = 0$, is called the *parabolic* type and has the normal form

$$\frac{\partial^2 z}{\partial \xi^2} + a\frac{\partial z}{\partial \xi} + b\frac{\partial z}{\partial \eta} + cz = 0 \tag{329}$$

where a, b, c are functions of ξ and η. An example is the equation of heat flow, $\partial u/\partial t = a^2 \partial^2 u/\partial t^2$, where $u = u(x, t)$ is the temperature, t is the time, a^2 is constant. If $B^2 - AC > 0$ in (328), the homogeneous form of (327) is the *hyperbolic* type that has as its two normal forms

$$\frac{\partial^2 z}{\partial \xi\,\partial \eta} + a\frac{\partial z}{\partial \xi} + b\frac{\partial z}{\partial \eta} + cz = 0 \tag{330}$$

$$\frac{\partial^2 z}{\partial \xi^2} - \frac{\partial^2 z}{\partial \eta^2} + a\frac{\partial z}{\partial \xi} + b\frac{\partial z}{\partial \eta} + cz = 0 \tag{331}$$

An example is the equation of a vibrating string,

$$\frac{\partial^2 z}{\partial t^2} = a^2\frac{\partial^2 z}{\partial x^2}$$

where z is the transverse displacement of a point on the string, with abscissa x at time t and a^2 is constant. If $B^2 - AC < 0$, the equation is of the *elliptic* type that has the normal form

$$\frac{\partial^2 z}{\partial \xi^2} + \frac{\partial^2 z}{\partial z^2} + a\frac{\partial z}{\partial \xi} + b\frac{\partial z}{\partial \eta} + cz = 0 \tag{332}$$

An example is Laplace's equation

$$\frac{\partial^2 z}{\partial \xi^2} + \frac{\partial^2 z}{\partial \eta^2} = 0$$

usually written $\nabla^2 z = 0$. The two solutions of (328) are real in the hyperbolic case and conjugate complex in the elliptic case. That is, in the latter case, $\xi = \tfrac{1}{2}(\alpha + i\beta), \eta = \tfrac{1}{2}(\alpha - i\beta)$, where α and β are real, and

$$\frac{\partial^2 z}{\partial \xi\,\partial \eta} = \frac{1}{4}\left(\frac{\partial^2 z}{\partial \alpha^2} + \frac{\partial^2 z}{\partial \beta^2}\right)$$

As in ordinary linear equations, the whole solution consists of the complementary function and the particular integral. Also, if $z = z_1, z = z_2, \ldots, z = z_n$ are solutions of the homogeneous equation (327), $L = 0$, then $z = c_1 z_1 + c_2 z_2 + \cdots + c_n z_n$ is again a solution.

Equations Linear in the Second Derivatives. The general type of second-order equation linear in the second derivatives may be written in the form

$$Ar + Bs + Ct = V \tag{333}$$

where A, B, C, V are functions of x, y, z, p, q. From the equations

$$A\,dy^2 - B\,dx\,dy + C\,dx^2 = 0 \tag{334}$$
$$A\,dp\,dy + C\,dq\,dx - V\,dx\,dy = 0 \tag{335}$$
$$p\,dx + q\,dy = dz \tag{336}$$

it may be possible to derive either one or two relations between x, y, z, p, q, called intermediary integrals, and from these to deduce the solution of (333). To obtain an intermediary integral, resolve (334), supposing the left member is not a perfect square, into the two equations $dy - n_1\,dx = 0, dy - n_2\,dx = 0$. From the first of these and from (335) combined, if necessary, with (336), obtain the two integrals $u_1(x, y, z, p, q) = a, v_1(x, y, z, p, q) = b$; then $u_1 = f_1(v_1)$, where f_1 is an arbitrary function, is now an intermediary integral.

In the same way, from $dy - n_2\,dx = 0$, obtain another pair of integrals $u_2 = a_1$, $v_2 = b_1$; then $u_2 = f_2(v_2)$ is an intermediary integral. For the final integral, if $n_1 = n_2$, the intermediary integral may be integrated. If $n_1 \neq n_2$, solve the two intermediary integrals for p and q, substitute in $p\,dx + q\,dy = dz$, and integrate for the solution.

Example 88. Solve

$$r^2 - a^2 t = 0 \qquad (337)$$

the equation for a vibrating string.

The auxiliary equations are

$$dy - a\,dx = 0 \qquad dy + a\,dx = 0$$

$$dp\,dy - a^2\,dx\,dq = 0 \qquad (338)$$

Hence $y + ax = c_1$, $y - ax = c_2$. Combining $y + ax = c_1$ with (338), $dp + a\,dq = 0$ is obtained, whereupon $p + aq = c_3 = f_1(y + ax)$. Combining $y - ax = c_1$ with (338), $dp - a\,dq = 0$ is obtained, whereupon $p - aq = c_4 = f_2(y - ax)$.

Solving for p and q,

$$p = \tfrac{1}{2}[f_1(y + ax) + f_2(y - ax)]$$

$$q = \frac{1}{2a}[f_1(y + ax) - f_2(y - ax)]$$

Substituting these in $p\,dx + q\,dy = dz$,

$$dz = \frac{1}{2a}[f_1(y + a)(dy + a\,dx)$$
$$- f_2(y - ax)(dy - a\,dx)]$$

which is an exact differential. Integration gives $z = \phi(y + ax) + \psi(y - ax)$.

Homogeneous Equation with Constant Coefficients

$$\frac{\partial^2 z}{\partial x^2} + A_1 \frac{\partial^2 z}{\partial x\,\partial y} + A_2 \frac{\partial^2 z}{\partial y^2} = 0 \qquad (339)$$

This equation is equivalent to

$$\left(\frac{\partial}{\partial x} - m_1 \frac{\partial}{\partial y}\right)\left(\frac{\partial}{\partial x} - m_2 \frac{\partial}{\partial y}\right) z = 0 \qquad (340)$$

where m_1 and m_2 are roots of the auxiliary equation $X^2 + A_1 X + A_2 = 0$. The general solution of (340) is

$$z = f_1(y + m_1 x) + f_2(y + m_2 x) \qquad (341)$$

Example 89. Solve

$$8\frac{\partial^2 z}{\partial x^2} + 2\frac{\partial^2 z}{\partial x\,\partial y} - 15\frac{\partial^2 z}{\partial y^2} = 0$$

The auxiliary equation is $8X^2 + 2X - 15 = (2X + 3)(4X - 5) = 0$. Hence $m_1 = -\tfrac{3}{2}$, $m_2 = \tfrac{5}{4}$. The general solution is $z = f_1(2y - 3x) + f_2(4y + 5x)$.

If the auxiliary equation has multiple factors, the general solution is $z = f_1(y + m_1 x) + x f_2(y + m_1 x)$.

Example 90. Solve

$$\frac{\partial^2 z}{\partial x^2} + 6\frac{\partial^2 z}{\partial x\,\partial y} + 9\frac{\partial^2 z}{\partial y^2} = 0$$

The auxiliary equation is $X^2 + 6X + 9 = (X + 3)(X + 3) = 0$. The general solution is $z = f_1(y - 3x) + x f_2(y - 3x)$.

If the coefficients in Eq. (339) are real, the complex roots of the auxiliary equation occur in conjugate pairs. Then the general solution will have the form

$$z = f(y + \alpha x + i\beta x) + g(y + \alpha x - i\beta x)$$

Example 91. Solve

$$\frac{\partial^2 z}{\partial x^2} - 2\frac{\partial^2 z}{\partial x\,\partial y} + 2\frac{\partial^2 z}{\partial y^2} = 0$$

The auxiliary equation is $X^2 - 2X + 2 = 0$ and $m = 1 \pm i$. The general solution is $z = f(y + x + ix) + g(y + x - ix)$, which can be written as $z = f_1(y + x + ix) + f_1(y + x - ix) + i[g_1(y + x + ix) - g_1(y + x - ix)]$, where f_1 and g_1 are any twice-differentiable real functions. If, in particular, $f_1 = \cos u$ and $g_1 = c^u$, it can be shown that $z = 2\cos(x + y)\cosh x - 2e^{x+y}\sin x$.

Method of Separation of Variables. As an example of this method, the solution will be given to Laplace's equation

$$\nabla^2 u = \frac{\partial^2 u}{\partial x^2} + \frac{\partial^2 u}{\partial y^2} = 0 \qquad (342)$$

Assume that

$$u = X(x) \cdot Y(y) \qquad (343)$$

where X is a function of x only and Y a function of y only. By substitution and dividing by $X \cdot Y$, (342) becomes

$$\frac{1}{X}\frac{d^2 X}{dx^2} = -\frac{1}{Y}\frac{d^2 Y}{dy^2} \qquad (344)$$

Since the left side does not contain y, the right side does not contain x, and the two sides are equal, they must equal a constant, say $-k^2$:

$$\frac{1}{X}\frac{d^2X}{dx^2} = -k^2 \qquad \frac{1}{Y}\frac{d^2Y}{dy^2} = k^2 \qquad (345)$$

The solutions of these homogeneous linear differential equations with constant coefficients are

$$X = c_1 \cos kx + c_2 \sin kx \qquad Y = c_3 e^{ky} + c_4 e^{-ky} \qquad (346)$$

Hence, from (343),

$$u = (c_1 \cos kx + c_2 \sin kx)(c_3 e^{ky} + c_4 e^{-ky})$$

$$= e^{ky}(k_1 \cos kx + k_2 \sin kx)$$

$$+ e^{-ky}(k_3 \cos kx + k_4 \sin kx) \qquad (347)$$

Since (342) is linear, the sum of any number of solutions is again a solution. An infinite number of solutions may be taken provided the series converges and may be differentiated term by term. Then

$$u = \sum_{n=0}^{\infty}[e^{ky}(A_n \cos kx + B_n \sin kx)$$

$$+ e^{-ky}(D_n \cos kx + E_n \sin kx)] \qquad (348)$$

is a solution of (342). The coefficients of (348) are determined by using the series as a Fourier series to fit the boundary conditions.

Functions that satisfy Laplace's equation are *harmonic*. In polar coordinates (342) becomes

$$\nabla^2 u = \frac{\partial^2 u}{\partial r^2} + \frac{1}{r^2}\frac{\partial^2 u}{\partial \theta^2} + \frac{1}{r}\frac{\partial u}{\partial r} = 0 \qquad (349)$$

In three dimensions, Laplace's equation in rectangular coordinates is

$$\nabla^2 u = \frac{\partial^2 u}{\partial x^2} + \frac{\partial^2 u}{\partial y^2} + \frac{\partial^2 u}{\partial z^2} = 0 \qquad (350)$$

In cylindrical coordinates,

$$\nabla^2 u = \frac{\partial^2 u}{\partial \rho^2} + \frac{1}{\rho}\frac{\partial u}{\partial \rho} + \frac{1}{\rho^2}\frac{\partial^2 u}{\partial \phi^2} + \frac{\partial^2 u}{\partial z^2} = 0 \qquad (351)$$

In spherical coordinates,

$$\nabla^2 u = \frac{1}{r^2}\frac{\partial}{\partial r}\left(r^2\frac{\partial u}{\partial r}\right) + \frac{1}{r^2 \sin^2 \theta}\frac{\partial^2 u}{\partial \phi^2}$$

$$+ \frac{1}{r^2 \sin \theta}\frac{\partial}{\partial \theta}\left(\sin \theta \frac{\partial u}{\partial \theta}\right) \qquad (352)$$

12 FINITE-ELEMENT METHOD

12.1 Introduction

The finite-element method is a powerful numerical technique that uses variational methods and interpolation theory for solving differential and integral equations of initial and boundary-value problems. The method is so general that it can be applied to a wide variety of engineering problems, including heat transfer, fluid mechanics, solid mechanics, chemical processing, electrical systems, and a host of other fields. The method is also so systematic and modular that it can be implemented on a digital computer and can be utilized to solve a wide range of practical engineering problems by merely changing the data input to the program. The method is naturally suited for the description of complicated geometries and the modeling and simulation of most physical phenomena.

Basic Features The finite-element method is characterized by two distinct features: First, the *domain* of the problem is viewed as a collection of simple subdomains, called *finite elements*. By the word domain we refer to a physical structure, system, or region over which the governing equations are to be solved. The collection of the elements is called the *finite-element mesh*. Second, over each element, the solution of the equations being solved is approximated by interpolation polynomials. The first feature, dividing a whole into parts, called *discretization of the domain*, allows the analyst to represent any complex system as one of numerous smaller connected elements, each element being of a simpler shape that permits approximation of the solution by a linear combination of algebraic polynomials. The second feature, *elementwise polynomial approximation*, enables the analyst to represent the solution on an element by polynomials so that the numerical evaluation of integrals becomes easy. The polynomials are typically interpolants of the solution at a preselected number of points, called *nodes*, in the element. The number and location of the nodes in an element depends on the geometry of the element and the degree of the polynomial, which in turn depends on the equation being solved. Since the solution is represented by polynomials on each element, a continuous approximation of the solution of the whole can only be obtained by imposing the continuity of the finite-element solution, and possibly its derivatives, at element interfaces (i.e., at the nodes common to two elements). The procedure of putting the elements together is called the *connectivity* or *assembly*.

Finite-Element Approximation Beyond the two features already described, the finite-element method is a variational method, like the Ritz, Galerkin, and weighted-residual methods, in which the approximate solution is sought in the form

$$u \approx U_N = \sum_{j=1}^{N} c_j \phi_j$$

where ϕ_j are preselected functions and c_j are parameters that are determined using a *variational statement* of the equation governing u. However, the finite-element method typically entails the solution of a very large number of equations for the nodal values of the function being sought. The number of equations is equal to the number of unknown nodal values. In most practical problems the number of unknown nodal values are so large that it is practical only if the calculations are carried on an electronic computer.

12.2 One-Dimensional Problems

The finite-element analysis consists of dividing a domain into simple parts (i.e., elements) that are easier to work with. Over each element the method involves representing the solution in terms of its nodal values and the development of a relationship between the nodal values and their counterparts by means of a variational method. Assembly of these relations and solution of the equations after imposing known boundary and initial conditions completes the analysis.

Evaluation of an Integral Consider the evaluation of the integral

$$I = \int_a^b f(x)\,dx \qquad (353)$$

where $f(x)$ is a complicated function whose integration by conventional methods (e.g., exact integration) is not possible. A step-by-step procedure of the numerical evaluation of the integral I by the finite-element method is given later.

Discretization of Domain. The area can be approximated by representing the interval (domain) $\Omega = (a, b)$ as a finite set of subintervals (see Fig. 152). A typical subinterval (element), $\Omega^e = (x_e, x_{e+1})$, is of length $h_e \equiv x_{e+1} - x_e$, with $x_1 = a$, and $x_{N+1} = b$, where N is the number of elements.

Approximation of Solution. Over each element, the function $f(x)$ is approximated using polynomials of a desired degree. The accuracy increases with increasing N and degree of the approximating polynomial. Over each element Ω^e, the function $f(x)$ can be approximated by a linear polynomial (see Fig. 153)

$$f(x) \approx F_e(x) = c_1^e + c_2^e x \qquad (354)$$

where c_1^e and c_2^e are constants that can be determined in terms of the values of the function f at the endpoints, x_e and x_{e+1}, called the nodes. Let F_1^e and F_2^e denote the values of $F_e(x)$ at nodes 1 and 2 of element Ω^e:

$$F_1^e = F_e(x_e) \qquad F_2^e = F_e(x_{e+1}) \qquad (355)$$

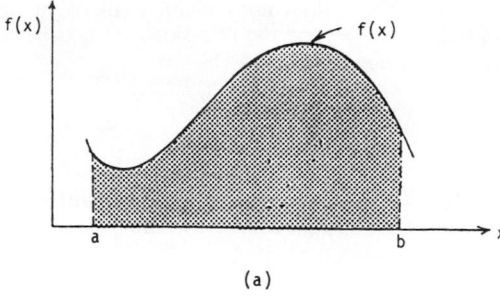

(a)

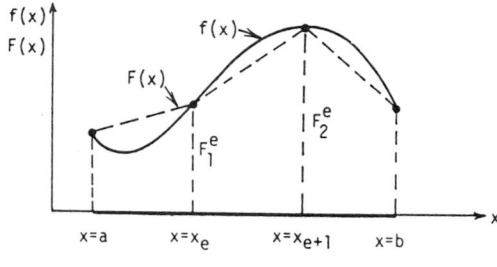

(b)

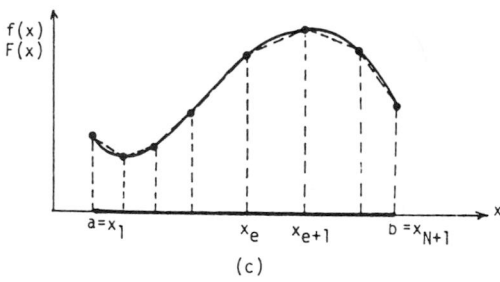

(c)

Fig. 152 Piecewise approximation of integral of function by polynomials.

Now $F_e(x)$ can be expressed in terms of its values at the nodes as

$$F_e(x) = \frac{x_{e+1} - x}{h_e} F_1^e + \frac{x - x_e}{h_e} F_2^e = \sum_{j=1}^{2} F_j^e \psi_j^e \qquad (356)$$

where ψ_j^e are called the *element interpolation functions* (see Fig. 153),

$$\psi_1^e = \frac{x_{e+1} - x}{h_e} \qquad \psi_2^e = \frac{x - x_e}{h_e} \qquad (357)$$

Let the approximation of the area I over a typical element Ω^e be denoted by I_e,

$$I_e = \int_{x_e}^{x_{e+1}} F_e(x)\,dx \qquad (358)$$

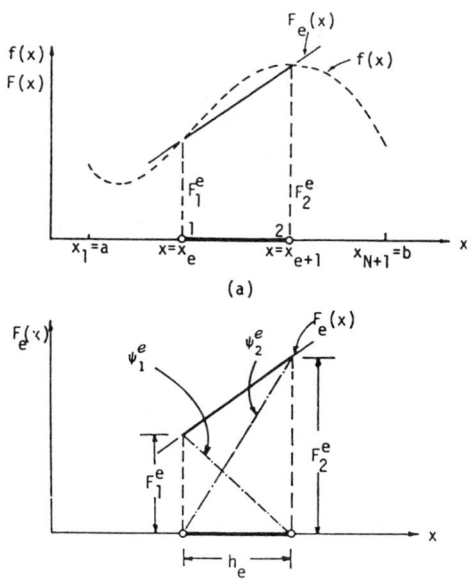

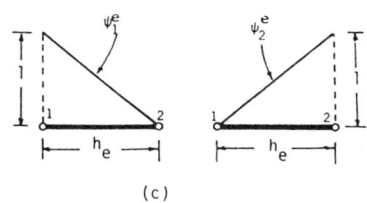

Fig. 153 Finite-element approximation of function $f(x)$ over typical element.

Substituting Eq. (356) into (358) and integrating, one obtains

$$
I_e = \sum_{j=1}^{2} F_j^e \int_{x_e}^{x_{e+1}} \psi_j^e \, dx
$$

$$
= \frac{1}{h_e} \left\{ F_1^e \left[h_e x_{e+1} - \frac{h_e}{2}(x_{e+1} + x_e) \right] \right.
$$

$$
\left. + F_2^e \left[\frac{h_e}{2}(x_{e+1} + x_e) - h_e x_e \right] \right\}
$$

$$
= \frac{h_e}{2}(F_1^e + F_2^e) \tag{359}
$$

Thus, the area under the function $F_e(x)$ over the element Ω^e is given by the area of the trapezoid of sides F_1^e and F_2^e and width h_e (see Fig. 153b).

Assembly of Equations. An approximation of the total area I is given by the sum of the areas $I_e, e = 1, 2, \ldots, N$:

$$
I = \sum_{e=1}^{N} \int_{x_e}^{x_{e+1}} f(x) \, dx
$$

$$
\approx \sum_{e=1}^{N} \int_{x_e}^{x_{e+1}} F_e(x) \, dx
$$

$$
= \sum_{e=1}^{N} I_e = \sum_{e=1}^{N} \frac{h_e}{2}(F_1^e + F_2^e) \tag{360}
$$

Incidentally, Eq. (360) is known as the *trapezoidal rule.*

The accuracy of the approximation can be improved by increasing the number of elements N (see Fig. 152c) or by using higher order approximation of $f(x)$ over each element. Note that the accuracy can also be improved by using unequal intervals, with smaller elements in areas where function $f(x)$ varies rapidly.

The quadratic interpolation of $f(x)$ over Ω^e is given by

$$
f(x) \approx F_e(x) = F_1^e \psi_1^e + F_2^e \psi_2^e + F_3^e \psi_3^e
$$

$$
= \sum_{j=1}^{3} F_j^e \psi_j^e \tag{361}
$$

where ψ_j^e are the quadratic interpolation functions

$$
\psi_1^e = \frac{(x - \xi_2)(x - \xi_3)}{(\xi_1 - \xi_2)(\xi_1 - \xi_3)}
$$

$$
\psi_2^e = \frac{(x - \xi_1)(x - \xi_3)}{(\xi_2 - \xi_1)(\xi_2 - \xi_3)} \tag{362}
$$

$$
\psi_3^e = \frac{(x - \xi_1)(x - \xi_2)}{(\xi_3 - \xi_1)(\xi_3 - \xi_2)}
$$

and $\xi_1, \xi_2,$ and ξ_3 are the coordinates of the three nodes in Ω^e. If nodes are equally spaced within each element (see Fig. 154), (ξ_1, ξ_2, ξ_3) take the values

$$
\xi_1 = x_{2e-1} \qquad \xi_2 = x_{2e} \qquad \xi_3 = x_{2e+1}
$$

$$
(e = 1, 2, \ldots, N)
$$

Then Eqs. (362) become

$$
\psi_1^e = \left(\frac{2\overline{x}}{h_e} - 1 \right)\left(\frac{\overline{x}}{h_e} - 1 \right) \qquad \psi_2^e = -\frac{4\overline{x}}{h_e}\left(\frac{\overline{x}}{h_e} - 1 \right)
$$

$$
\psi_3^e = \frac{\overline{x}}{h_e}\left(\frac{2\overline{x}}{h_e} - 1 \right) \tag{363}
$$

where $\overline{x} = x - x_{2e-1}$ and h_e is the length of the element Ω^e.

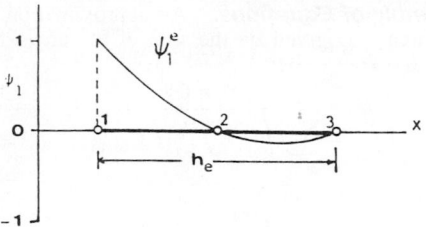

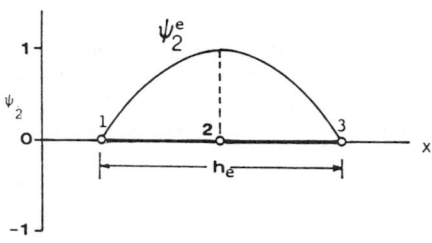

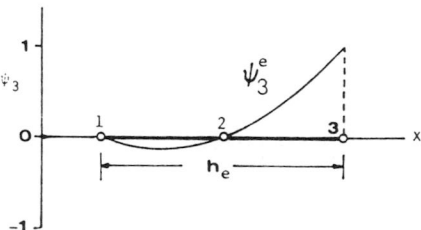

Fig. 154 One-dimensional quadratic interpolation functions.

In general, the interpolation functions ψ_j^e satisfy the properties

$$\psi_j^e(\xi_i) = \begin{cases} 0 & \text{if } i \neq j \\ 1 & \text{if } i = j \end{cases} \qquad (364)$$

Substituting Eq. (361) into (358) and integrating, one obtains

$$I_e = \sum_{j=1}^{3} F_j^e \int_{x_{2e-1}}^{x_{2e+1}} \psi_j^e(x)\, dx$$

$$= \sum_{j=1}^{3} F_j^e \int_{0}^{h_e} \psi_j^e(\bar{x})\, d\bar{x}$$

$$= \tfrac{1}{6} h_e (F_1^e + 4F_2^e + F_3^e)$$

The total area is given by

$$I \approx \sum_{e=1}^{N} I_e = \sum_{e=1}^{N} \frac{h_e}{6}(F_1^e + 4F_2^e + F_3^e)$$

This equation is known as the *one-third Simpson's rule.*

Example 92. Consider the integral of the function

$$f(x) = \sin(2\cos x)\sin^2 x$$

over the domain $\Omega = (0, \pi/2)$. Table 12 contains the finite-element solutions obtained using linear and quadratic interpolation. It is clear that the accuracy improves as the number of elements or the degree of polynomial is increased.

Solution of a Differential Equation

Model Equation. Consider the differential equation

$$-\frac{d}{dx}\left[a(x)\frac{du}{dx}\right] - f(x) = 0 \qquad 0 < x < L \quad (365)$$

which arises in connection with heat transfer in a heat exchanger fin, where $a(x) = kA$, k is the thermal conductivity, A is the cross-sectional area of the fin, $f(x)$ is the heat source, and $u = u(x)$ is temperature to be determined. Equation (365) also arises in many fields of engineering. In addition to Eq. (365), the function u is required to satisfy certain boundary conditions (i.e., conditions at points $x = 0$ and $x = L$). Equation (365), in general, has the following types of boundary conditions:

Specify either u or $(a\, du/dx)$ at a boundary point

Discretization. The domain $\Omega = (0, L)$ is represented as a collection of line elements, each element having at least two end nodes so that it can be connected to adjacent elements. A two-node element with one unknown per node requires, uniquely, a linear polynomial approximation of the variable over the element (see Fig. 155).

Approximation. Over a typical element $\Omega^e = (x_e, x_{e+1})$, the function $u(x)$ is approximated by $U_e(x)$, which is assumed to be of the form

$$U_e(x) = \sum_{j=1}^{n} U_j^e \psi_j^e(x) \qquad (366)$$

where U_j^e denotes the value of $U_e(x)$ at the jth node and ψ_j^e are the linear [see Eq. (357)], quadratic [see Eq. (363)], or higher order interpolation functions. The values U_j^e are to be determined such that Eq. (365), with appropriate boundary conditions, is satisfied in integral sense.

Table 12 Finite-Element Solutions Using Linear and Quadratic Interpolation

Number of Elements	Linear Interpolation		Quadratic Interpolation		Exact
	$\bar{I}$	Error[a] (%)	$\bar{I}$	Error (%)	
2 (1)[b]	0.38790	23.6	0.51719	−1.8	0.50797
4 (2)	0.48149	5.2	0.51268	−0.9	0.50797
6 (3)	0.49640	2.3	0.50865	−0.1	0.50797
8 (4)	0.50150	1.3	0.50817	−0.04	0.50797
10 (5)	0.50384	0.8	0.50805	−0.02	0.50797

[a]$(1 − \bar{I}/I(100)$.
[b]Numbers in parentheses indicate number of equivalent quadratic elements.

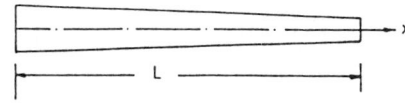

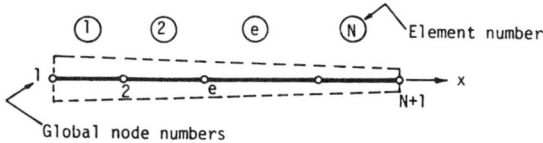

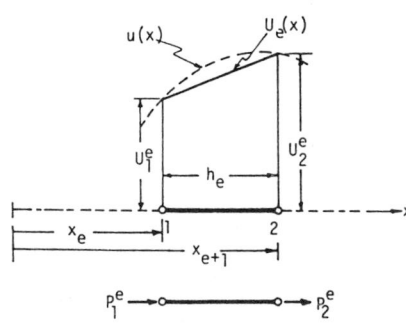

Fig. 155 One-dimensional domain, finite-element discretization, and finite-element approximation over an element.

Need for a Variational Statement.

The difference between the numerical evaluation of an integral and the numerical solution of a differential equation is that in the case of a differential equation one is required to determine a function that satisfies a given differential equation and boundary conditions. It is possible to recast the differential equation as an integral statement, called a *variational statement*. The variational statement of Eq. (333), with the aid of a variational method of approximation, gives the same number of algebraic equations as the number of unknowns (*n*) in the approximation (334).

Variational Formulation.

The variational statement of Eq. (333) over an element $\Omega^e = (x_e, x_{e+1})$ (see Fig. 155) is constructed as follows. Multiply Eq. (333) with an arbitrary but continuous function W and integrate over the domain of the element to obtain

$$0 = \int_{x_A}^{x_B} W \left[-\frac{d}{dx}\left(a\frac{dU}{dx}\right) - f \right] dx \qquad (367)$$

The (Ritz) finite-element model uses a *weak form* that can be obtained from Eq. (367) by trading differentiation between the weight function W and the variable of approximation U equally:

$$0 = \int_{x_A}^{x_B} \left(a\frac{dW}{dx}\frac{dU}{dx} - Wf \right) dx - \left[W\left(a\frac{dU}{dx}\right) \right]_{x_A}^{x_B} \qquad (368)$$

which is obtained by integrating the first term in Eq. (367) by parts. The term weak form is appropriate because the solution U of Eq. (368) requires weaker continuity conditions on ψ_i than U of Eq. (367). Also, the weak formulation allows the incorporation of the boundary conditions of the "flux" type, dU/dx (the coefficient of the weight function W in the boundary term, called the *natural boundary condition*), into the variational statement (368). Boundary conditions on U in the same form as the weight function in the boundary terms are called the *essential boundary conditions*.

Identifying the coefficients of the weight function in the boundary terms (i.e., fluxes) as the *dual variables*,

$$\left(a\frac{dU}{dx}\right)\Bigg|_{x=x_e} = -P_1^e \qquad \left(a\frac{dU}{dx}\right)\Bigg|_{x=x_{e+1}} = P_2^e$$

Eq. (368) can be written as

$$0 = \int_{x_e}^{x_{e+1}} \left(a\frac{dW}{dx}\frac{dU}{dx} - Wf \right) dx$$
$$- W(x_e)P_1^e - W(x_{e+1})P_2^e \qquad (369)$$

Equation (369) represents the variational statement of Eq. (365) for the (Ritz) finite-element model.

As a general rule, the essential boundary conditions of the variational form of a problem indicate what interelement continuity conditions are to be imposed on the function U and its derivatives. This in turn dictates the type and degree of approximation and hence the element type. For example, Eq. (369) indicates that U must be continuous in the interval (x_e, x_{e+1}). A *complete* continuous polynomial in x is a linear polynomial

$$U_e(x) = c_1^e + c_2^e x$$

The constants c_1^e and c_2^e are expressed in terms of the values of U_e at nodes 1 and 2,

$$U_e = \sum_{j=i}^{2} U_j^e \psi_j^e(x)$$

For the $(n-1)$st-degree polynomial approximation, U_e is of the form

$$U_e(x) = \sum_{j=1}^{n} U_j^e \psi_j^e(x)$$

(Ritz) Finite-Element Model. In the Ritz model U_j^e Eq. (369) is satisfied for each $W = \psi_i^e (i = 1, 2, \ldots, n)$. For each choice of W, an algebraic equation can be obtained:

$$0 = \int_{x_e}^{x_{e+1}} \left[a \frac{d\psi_i^e}{dx} \left(\sum_{j=1}^{n} U_j^e \frac{d\psi_j^e}{dx} \right) - \psi_1^e f \right] dx$$
$$- \psi_1^e(x_e) P_1^e - \psi_1^e(x_{e+1}) P_2^e$$

$$0 = \int_{x_e}^{x_{e+1}} \left[a \frac{d\psi_2^e}{dx} \left(\sum_{j=1}^{n} U_j^e \frac{d\psi_j^e}{dx} \right) - \psi_2^e f \right] dx$$
$$- \psi_2^e(x_e) P_1^e - \psi_2^e(x_{e+1}) P_2^e$$

$$\vdots$$

$$0 = \int_{x_e}^{x_{e+1}} \left[a \frac{d\psi_n^e}{dx} \left(\sum_{j=1}^{n} U_j^e \frac{d\psi_j^e}{dx} \right) - \psi_n^e f \right] dx$$
$$- \psi_n^e(x_e) P_1^e - \psi_n^e(x_{e+1}) P_2^e$$

The ith equation can be written in compact form as

$$0 = \sum_{j=1}^{n} K_{ij}^e U_j^e - F_i^e \tag{370a}$$

where

$$K_{ij}^e = \int_{x_e}^{x_{e+1}} a \frac{d\psi_i^e}{dx} \frac{d\psi_j^e}{dx} dx$$

$$F_i^e = \int_{x_e}^{x_{e+1}} f \psi_i^e \, dx + \psi_i^e(x_e) P_1^e + \psi_i^e(x_{e+1}) P_2^e \tag{370b}$$

To be more specific, let ψ_i^e be the linear interpolation functions of Eq. (357). Because of the interpolation property (364) of ψ_j^e, the F_i^e of Eq. (371) can be written as

$$F_i^e = \int_{x_e}^{x_{e+1}} f \psi_j^e \, dx + P_i^e \equiv f_i^e + P_i^e$$

For elementwise constant values of a and f, the element coefficient matrix $[K^e]$ and source vector $\{f^e\}$ become

$$[K^e] = \frac{a_e}{h_e} \begin{bmatrix} 1 & -1 \\ -1 & 1 \end{bmatrix} \qquad \{f^e\} = \frac{h_e f_e}{2} \begin{Bmatrix} 1 \\ 1 \end{Bmatrix}$$

Assembly of Elements. The element equations (370) must be put together to obtain the equations of the whole domain. Geometrically, the elements are connected together by noting that the second node of element Ω^e is the same as the first node of element Ω^{e+1}. Since the solution and hence its approximation are single valued throughout the domain, the geometric continuity also implies the continuity of the approximate solution (see Fig. 156):

$$U_2^e = U_1^{e+1} \qquad e = 1, 2, \ldots, N$$

In addition to the continuity of U_e, the balance of the dual variables P_i at interelement nodes is also enforced:

$$P_2^e + P_1^{e+1} = 0 \qquad e = 1, 2, \ldots, N$$

Note that this *does not* guarantee the continuity of $a \, dU_e/dx$ at interelement nodes.

The finite-element approximation on the entire domain $\Omega = \sum_{e=1}^{N} \Omega^e$ is given by

$$U = \sum_{e=1}^{N} U_e = \sum_{e=1}^{N} \sum_{j=1}^{2} U_j^e \psi_j^e(x)$$

In view of the continuity conditions and the elementwise definition of the interpolation functions ψ_j^e, the finite-element approximation can be written as

$$U = \sum_{J=1}^{N+1} U_J \Phi_J(x) \tag{371}$$

where U_J denotes the value of $U(x)$ at the Jth (global) node of the mesh and Φ_J are the *global interpolation*

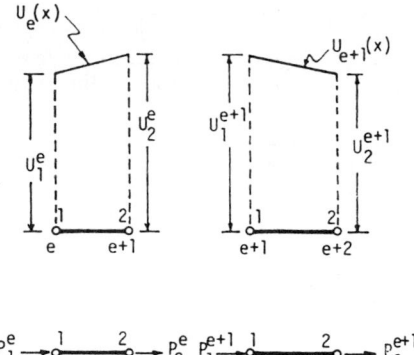

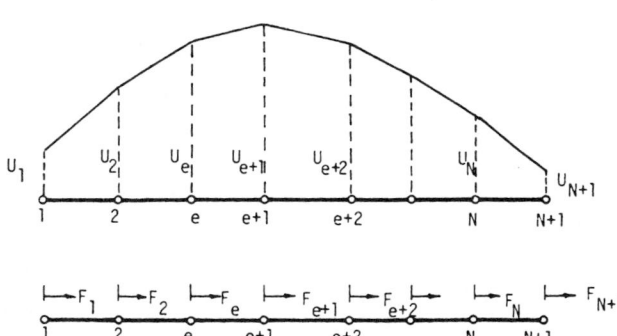

Fig. 156 Assembly of finite elements using continuity of finite-element approximation between elements.

functions, related to the local (or element) interpolation functions by

$$\Phi_1 = \psi_1^1 \qquad 0 = x_1 \le x \le x_2$$

$$\Phi_J = \begin{cases} \psi_2^{J-1} & x_{J-1} \le \ x \le x_J \\ \psi_1^J & x_J \le x \ \ \le x_{J+1} \end{cases}$$

$$(J = 2, 3, \ldots, N)$$

$$\Phi_{N+1} = \psi_2^N \qquad x_N \le x \le x_{N+1} = L$$

Note that Φ_J are continuous and defined only on the two elements connected at the global node J.

Analogous to the variational form (369) for an element Ω^e, a variational form for the entire domain can be derived as

$$0 = \int_0^L \left(a \frac{dW}{dx} \frac{dU}{dx} - Wf \right) dx - \left[W \left(a \frac{dU}{dx} \right) \right]_{x=0}^{x=L} \tag{372}$$

Substitution of Eq. (371) for U and $W = \Phi_I (I = 1, 2, \ldots, N + 1)$ into Eq. (372) gives

$$0 = \int_0^L \left[a \frac{d\Phi_I}{dx} \left(\sum_{j=1}^{N+1} U_J \frac{d\Phi_J}{dx} \right) \right.$$

$$\left. - \Phi_I f \right] dx - \left[\Phi_f \sum_{e=1}^N a \frac{dU_e}{dx} \right]_{x=0}^{x=L}$$

Since each Φ_I is defined on two neighboring elements, this equation becomes

$$0 = \int_{x_{I-1}}^{X_I} \left[a \frac{d\psi_2^{I-1}}{dx} \left(U_{I-1} \frac{d\psi_1^{I-1}}{dx} + U_I \frac{d\psi_2^{I-1}}{dx} \right) \right.$$

$$\left. - \psi_2^{I-1} f \right] dx - \psi_2^{I-1}(L) P_2^{I-1}$$

$$+ \int_{x_I}^{x_{I+1}} \left[a \frac{d\psi_1^I}{dx} \left(U_I \frac{d\psi_1^I}{dx} + U_{I+1} \frac{d\psi_2^I}{dx} \right) \right.$$

$$\left. - \psi_1^I f \right] dx - \psi_1^I(a) P_1^I$$

$$= K_{21}^{I-1} U_{I-1} + (K_{22}^{I-1} + K_{11}^I) U_I + K_{12}^I U_{I+1}$$

$$- (f_2^{I-1} + f_1^I) - \psi_2^{I-1}(L) P_2^{I-1} - \psi_1^I(0) P_1^I$$
$$(373)$$

Thus, the equations of the connected elements (i.e., the finite-element equations of the entire domain) are given by setting $I = 1, 2, \ldots, N + 1$ in Eq. (373) (set $K_{IJ}^0 = F_I^0 = P_I^0 = 0$):

$$K_{11}^1 U_1 + K_{12}^1 U_2 = f_1^1 + P_1^1$$

$$K_{21}^1 U_1 + (K_{22}^1 + K_{11}^2) U_2 + K_{12}^2 U_3 = f_2^1 + f_1^2 + \underbrace{(P_2^1 + P_1^2)}_{0}$$

$$K_{21}^2 U_2 + (K_{22}^2 + K_{11}^3) U_3 + K_{12}^3 U_4 = f_2^2 + f_1^3 + \underbrace{(P_2^2 + P_1^3)}_{0}$$

$$\vdots$$

$$K_{21}^N U_N + K_{22}^N U_{N+1} = f_2^N + P_2^N$$

or, in matrix form,

$$\begin{bmatrix} K_{11}^1 & K_{12}^1 & 0 & 0 & 0 \\ K_{21}^1 & K_{22}^1 + K_{11}^2 & K_{12}^2 & 0 & 0 \\ 0 & K_{21}^2 & K_{22}^2 + K_{11}^3 & K_{12}^3 & 0 \\ \vdots & \vdots & \vdots & \vdots & \vdots \\ 0 & 0 & 0 & K_{21}^N & K_{22}^N \end{bmatrix}$$

$$\times \begin{Bmatrix} U_1 \\ U_2 \\ U_3 \\ \vdots \\ U_{N+1} \end{Bmatrix} = \begin{Bmatrix} f_1^1 + P_1^1 \\ f_2^1 + f_1^2 \\ f_2^2 + f_1^3 \\ \vdots \\ f_2^N + P_2^N \end{Bmatrix} \qquad (374)$$

One does not repeat the connectivity procedure described in Eqs. (371)–(374) for every problem but uses the pattern implied in the final equations (374) for all problems described by Eq. (365).

Example 93. Heat conduction in a long radially symmetric coaxial cylindrical cable can be described by

$$-\frac{d}{dr} \left[a(r) \frac{du}{dr} \right] = 0 \qquad (375)$$

where u denotes the temperature and $a = 2\pi r k$, k being the thermal conductivity of the medium.

Equation (375) is in the same form as Eq. (365). Therefore, Eqs. (370) and (374) describe the element and global finite-element models of Eq. (375). For the choice of linear interpolation functions, we have

$$K_{ij}^e = \int_{r_e}^{r_{e+1}} 2\pi k_e r \frac{d\psi_i^e}{dr} \frac{d\psi_i^e}{dr} dr$$

where

$$\psi_i^e = \frac{r_{e+1} - r}{r_{e+1} - r_e} \qquad \psi_2^e = \frac{r - r_e}{r_{e+1} - r_e}$$

$$h_e = r_{e+1} - r_e$$

For example, K_{11}^e is given by

$$K_{11}^e = 2\pi k_e \int_{r_e}^{r_{e+1}} r \left(-\frac{1}{h_e} \right)^2 dr$$

$$= \frac{\pi k_e}{h_e} (r_{e+1} + r_e)$$

We have

$$\frac{\pi k_e}{h_e} (r_{e+1} + r_e) \begin{bmatrix} 1 & -1 \\ -1 & 1 \end{bmatrix} \begin{Bmatrix} u_1^e \\ u_2^e \end{Bmatrix} = \begin{Bmatrix} P_1^e \\ P_2^e \end{Bmatrix}$$

where P_i^e denote the internal heats,

$$P_1^e = -2\pi k_e \left. \left(r \frac{dU}{dr} \right) \right|_{r=r_e}$$

$$P_2^e = 2\pi k_e \left. \left(r \frac{dU}{dr} \right) \right|_{r=r_{e+1}}$$

The assembled equations for an N-element case are shown in the tabulation at the top of page 277.

We now impose the boundary conditions of the problem. Suppose that the domain is the cross section of a coaxial cylinder with two materials (i.e., with different thermal conductivities), as shown in Fig. 157. Let the internal and external radii be $r_1 = 20$ mm and $r_{N+1} = 50$ mm and let the thickness of the first material be 11.6 mm and that of the second material be 18.4 mm and the associated material constants (k) be 5 and 1. We assume the boundary conditions to be $u(20) = 100°C$ and $u(50) = 0.0$. These conditions translate to

$$U_1 = 100.0 \qquad U_{N+1} = 0.0$$

$$P_2^1 + P_1^2 = 0, \ldots \qquad P_2^{N-1} + P_1^N = 0$$

For a nonuniform mesh of four elements ($h_1 = 5.1, h_2 = 6.5, h_3 = 8.2, h_4 = 10.2$; equivalently, $r_1 = 20, r_2 = 25.1, r_3 = 31.6, r_4 = 39.8,$ and $r_5 = 50.0$),

Tabulation for Example 93

$$
\begin{bmatrix}
\dfrac{K_1}{h_1} & -\dfrac{K_1}{h_1} & & 0 & & & \\[2mm]
-\dfrac{K_1}{h_1} & \dfrac{K_1}{h_1}+\dfrac{K_2}{h_2} & -\dfrac{K_2}{h_2} & & & & \\[2mm]
0 & -\dfrac{K_2}{h_2} & \dfrac{K_2}{h_2}+\dfrac{K_3}{h_3} & \ddots & & & \\[2mm]
. & . & \ddots & \cdots & -\dfrac{K_N}{h_N} & 0 & \\[2mm]
. & . & & -\dfrac{K_N}{h_N} & \dfrac{K_N}{h_N}+\dfrac{K_{N+1}}{h_{N+1}} & -\dfrac{K_{N+1}}{h_{N+1}} & \\[2mm]
. & . & & 0 & -\dfrac{K_{N+1}}{h_{N+1}} & \dfrac{K_{N+1}}{h_{N+1}}
\end{bmatrix}
\begin{Bmatrix}
U_1 \\ U_2 \\ U_3 \\ \vdots \\ U_N \\ U_{N+1}
\end{Bmatrix}
=
\begin{Bmatrix}
P_1^1 \\ P_2^1 + P_1^2 \\ P_2^2 + P_1^3 \\ \vdots \\ P_2^{N-1}+P_1^N \\ P_2^N
\end{Bmatrix}
$$

where $K_i = k_i(r_{i+1} + r_i)\pi$.

---

the assembled equations become

$$
2\pi
\begin{bmatrix}
22.108 & -22.108 & 0 & 0 & 0 \\
-22.108 & 43.916 & -21.808 & 0 & 0 \\
0 & -21.808 & 26.162 & -4.354 & 0 \\
0 & 0 & -4.354 & 8.756 & -4.402 \\
0 & 0 & 0 & -4.402 & 4.402
\end{bmatrix}
$$

$$
\times
\begin{Bmatrix}
U_1 \\ U_2 \\ U_3 \\ U_4 \\ U_5
\end{Bmatrix}
=
\begin{Bmatrix}
P_1^2 \\ P_2^1 + P_1^2 \\ P_2^2 + P_1^3 \\ P_2^3 + P_1^4 \\ P_2^4
\end{Bmatrix}
$$

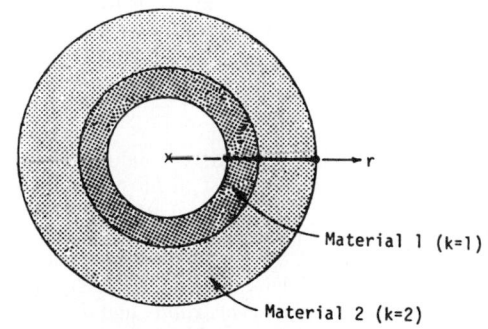

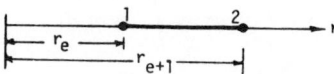

Fig. 157 Finite-element representation of radially symmetric problem with two different materials.

The boundary and continuity conditions are

$$
U_1 = 100.0 \qquad U_5 = 0.0 \qquad P_2^1 + P_1^2 = 0
$$
$$
P_2^2 + P_1^3 = 0 \qquad P_2^3 + P_1^3 = 0
$$

The solution for U_2, U_3, and U_4 is obtained by solving the second, third, and fourth equations of the assembled system:

$$
\begin{bmatrix}
43.916 & -21.808 & 0 \\
-21.808 & 26.162 & -4.354 \\
0 & -4.354 & 8.756
\end{bmatrix}
\begin{Bmatrix}
U_2 \\ U_3 \\ U_4
\end{Bmatrix}
$$
$$
=
\begin{Bmatrix}
22.108 U_1 \\ 0 \\ 0
\end{Bmatrix}
$$

or

$$
U_2 = 91.745^\circ C \quad U_2 = 83.377^\circ C \quad U_4 = 41.458^\circ C
$$

Table 13 contains a comparison of the finite-element solutions obtained by three different nonuniform meshes with the analytical solution. The numerical convergence and accuracy are apparent from the results.

12.3 Two-Dimensional Problems

As a model equation, consider the following second-order equation in two dimensions:

$$
-\frac{\partial}{\partial x}\left(a_{11}\frac{\partial u}{\partial x}\right) - \frac{\partial}{\partial y}\left(a_{22}\frac{\partial u}{\partial y}\right) + a_0 u = f \quad \text{in } \Omega
$$

$$(376)$$

The coefficients a_{11}, a_{22}, and a_0 and the source term f are known functions of position (x, y) in the domain Ω.

Table 13   Finite-Element Solutions Obtained by Various Nonuniform Meshes

r	Two Elements	Four Elements	Eight Elements	Analytical Solution
20.0	100.000	100.000	100.00	100.000
22.6	—	—	95.559	95.559
25.1	—	91.745	91.746	91.746
28.4	—	—	87.258	87.257
31.6	83.375	83.377	83.377	83.377
35.7	—	—	61.213	61.210
39.8	—	41.458	41.457	41.457
44.9	—	—	19.551	19.549
50.0	0.000	0.000	0.000	0.000

Equation (376) arises in the study of a number of engineering problems, including heat transfer, irrotational flow of a fluid, transverse deflection of a membrane, and torsion of a cylindrical member. Also, the Stokes flow and plane elasticity problems are described by a pair of equations of the same form as the model equation. Thus, the finite-element procedure to be described for Eq. (376) is applicable to *any* problem that can be formulated as one of solving equations of the form of (376).

While the basic ideas are the same as described before, the mathematical complexity for two-dimensional problems increases because of the partial differential equations on two-dimensional domains with possibly curved boundaries. It is necessary to approximate not only the solution of a partial differential equation but also the domain by a suitable finite-element mesh. This latter property is what made the finite-element method a more attractive practical analysis tool over other competing methods.

Discretization of a Domain   Two-dimensional domains can be represented by more than one type of geometric shape. For example, a plane curved domain can be represented by triangular elements or rectangular elements. Without reference to a specific geometric shape, we simply denote a typical element by Ω^e and proceed to discuss the approximation of Eq. (376).

The choice of the finite-element mesh depends both on the element characteristics (convergence, computational simplicity, etc.) and the ability to represent the domain accurately. The concept of so-called *isoparametric formulations* allows the representation of the element geometry by the same interpolation as that used in the approximation of the dependent variables. Thus, by identifying nodes on the boundary of the domain, one can approximate the domain by suitable collection of elements to a desired accuracy.

Element Equations

Variational Formulation.   Consider a typical finite element Ω^e from the finite-element mesh of the domain $\overline{\Omega}$ (see Fig. 158). Let $\psi_i^e \, (i = 1, 2, \ldots, n)$ denote the interpolation functions used to approximate u on Ω^e. Multiply Eq. (376) with a weight function W, integrate

over the element domain Ω^e, and use the Green–Gauss theorem to trade differentiation to W to obtain the weak variational form

$$
0 = \int_{\Omega^e} \left[\frac{\partial W}{\partial x} \left(a_{11}^e \frac{\partial U}{\partial x} \right) + \frac{\partial W}{\partial y} \left(a_{22}^e \frac{\partial U}{\partial y} \right) \right.
$$
$$
\left. + a_0^e W U - W f_e \right] dx \, dy - \int_{\Gamma^e} W \left[n_x \left(a_{11}^e \frac{\partial U}{\partial x} \right) \right.
$$
$$
\left. + n_y \left(a_{22}^e \frac{\partial U}{\partial y} \right) \right] ds \tag{377}
$$

where n_x and n_y are the components (i.e., direction cosines) of the unit normal $\hat{n}$,

$$
\hat{n} = n_x \hat{i} + n_y \hat{j} = \cos \alpha \hat{i} + \sin \alpha \hat{j}
$$

on the boundary Γ^e and ds is the elemental arc length along the boundary of the element. From an inspection of the boundary term in Eq. (377), it follows that the specification of the coefficient of W,

$$
q_n^e \equiv n_x \left(a_{11}^e \frac{\partial U}{\partial x} \right) + n_y \left(a_{22}^e \frac{\partial U}{\partial y} \right) \tag{378}
$$

constitutes the natural boundary condition. The variable q_n is of physical interest in most problems. For example, in the case of the heat transfer through an anisotropic medium (where a_{ij} denotes the conductivities of the medium), q_n denotes the heat flux across the boundary of the element (see Fig. 158). The variable U is called the *primary variable* and q_n (heat flux) is termed the *secondary variable*.

The variational form in Eq. (377) now becomes

$$
0 = \int_{\Omega^e} \left[\frac{\partial W}{\partial x} \left(a_{11}^e \frac{\partial U}{\partial x} \right) + \frac{\partial W}{\partial y} \left(a_{22}^e \frac{\partial U}{\partial y} \right) \right.
$$
$$
\left. + a_{00}^e W U - W f_e \right] dx \, dy - \int_{\Gamma^e} W q_n^e \, ds \tag{379}
$$

This variational equation forms the basis of the Ritz finite-element model. The boundary term indicates that W should be continuous at interelement boundaries.

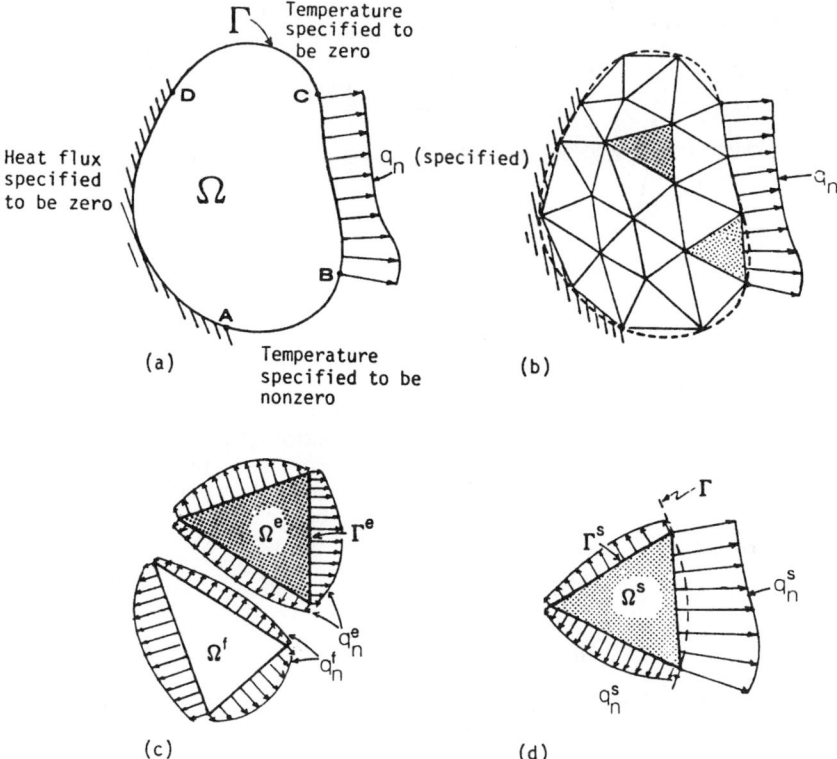

Fig. 158 Finite-element representation of two-dimensional domain with various types of boundary conditions.

Finite-Element Formulation. The variational form in (379) indicates that the approximation chosen for u should be at least bilinear in x and y so that $\partial u/\partial x$ and $\partial u/\partial y$ are nonzero and the interelement continuity of u can be imposed. Suppose that the temperature is approximated by the expression

$$u \approx U_e = \sum_{j=1}^{n} U_j^e \psi_j^e \qquad (380)$$

where U_j^e are the values of U_e at the point (x_j, y_j) in Ω^e and ψ_j^e are the interpolation functions with the property

$$\psi_i^e(x_j, y_j) = \delta_{ij}$$

The specific form of ψ_i^e will be derived later for linear triangular and rectangular elements.

Substituting Eq. (380) for U_e and ψ_i^e for W into the variational form (379), the ith algebraic equation of the model is obtained,

$$\sum_{j=1}^{n} K_{ij}^e U_j^e = F_i^e \qquad (i = 1, 2, \ldots, n) \qquad (381)$$

where

$$K_{ij}^e = \int_{\Omega^e} \left[\frac{\partial \psi_i^e}{\partial x} \left(a_{11}^e \frac{\partial \psi_j^e}{\partial x} \right) + \frac{\partial \psi_i^e}{\partial y} \left(a_{22}^e \frac{\partial \psi_j^e}{\partial y} \right) \right.$$
$$\left. + a_0^e \psi_i^e \psi_j^e \right] dx\, dy$$

$$F_i^e = \int_{\Omega^e} f_e \psi_i^e \, dx\, dy + \int_{\Gamma^e} q_n^e \psi_i^e \, ds \equiv f_i^e + P_i^e$$
$$(382)$$

Note that $K_{ij}^e = K_{ji}^e$ (i.e., $[K^e]$ is symmetric). Equation (381) is called the finite-element model of Eq. (376).

Assembly of Elements The assembly of finite-element equations is based on the same principle as that employed in one-dimensional problems. We illustrate the procedure by considering a finite-element mesh consisting of two triangular elements (see Fig. 159). Let K_{ij}^e and $K_{ij}^f (i, j = 1, 2, 3)$ denote the coefficient matrices and $\{F^e\}$ and $\{F^f\}$ denote the column vectors of three-node triangular elements Ω^e and Ω^f. From the finite-element mesh shown in Fig. 159, the following correspondence between the global and

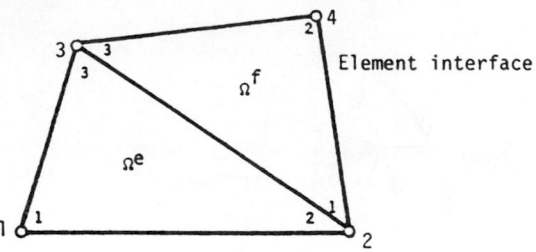

Fig. 159 Assembly (or connectivity) of linear triangular elements.

element nodal values of the temperature is noted:

$$U_1 = U_1^e \qquad U_2 = U_2^e = U_1^f$$

$$U_3 = U_3^e = U_e^f \qquad U_4 = U_2^f$$

The continuity of U at the interelement nodes guarantees its continuity along the *entire* interelement boundary. To see this, consider two linear triangular elements (see Fig. 159). The finite-element solution for U is linear along the boundaries of the elements. The interelement boundary is along the line connecting global nodes 2 and 3. Since U_e is linear along side 2–3 of element Ω^e, it is uniquely determined by the two values U_2^e and U_3^e. Similarly, U_f is uniquely determined along side 1–3 of element Ω^f by the two values U_1^f and U_3^f. Since $U_2^e = U_1^f$ and $U_3^e = U_3^f$, it follows that $U_e = U_f$ along the interface. Similar arguments can be presented for higher order elements.

The coefficient K_{ij}^e is a representation of a physical property of node i with respect to node j of element Ω^e. The assembled coefficient matrix also represents the same property among the global nodes. But the global property comes from the element nodes shared by the global nodes. For example, the coefficient K_{23} of the global coefficient matrix is the sum of the contributions from nodes 2 and 3 of Ω^e and nodes 1 and 3 of Ω^f (see Fig. 159):

$$K_{23} = K_{23}^e + K_{13}^f \qquad K_{32} = K_{32}^e + K_{31}^f$$

Similarly,

$$K_{22} = K_{22}^e + K_{11}^f \qquad K_{33} = K_{33}^e + K_{33}^f, \dots$$

If the global nodes I and J do not correspond to nodes in the same element, then $K_{IJ} = 0$. For example, K_{14} is zero because global nodes 1 and 4 do not belong to the same element. The column vectors can be assembled using the same logic:

$$F_2 = F_2^e + F_1^f \qquad F_3 = F_3^e + F_3^f, \dots$$

The complete assembled equations for the two-element mesh is given by

$$\begin{bmatrix} K_{11}^e & K_{12}^e & K_{13}^e & 0 \\ K_{21} & K_{22}^e + K_{11}^f & K_{23}^e + K_{13}^f & K_{12}^f \\ K_{31}^e & K_{32}^e + K_{31}^f & K_{33}^e + K_{33}^f & K_{32}^f \\ 0 & K_{21}^f & K_{23}^f & K_{32}^f \end{bmatrix} \begin{Bmatrix} U_1 \\ U_2 \\ U_3 \\ U_4 \end{Bmatrix}$$

$$= \begin{Bmatrix} F_1^e \\ F_2^e + F_1^f \\ F_3^e + F_3^f \\ F_2^f \end{Bmatrix}$$

Imposition of Boundary Conditions   The boundary conditions on the primary variables (temperatures) and secondary variables (heats) are imposed on the assembled equations in the same way as in the one-dimensional problems. To understand the physical significance of the P's [see Eq. (382)], take a closer look at the definition,

$$P_i^e \equiv \int_{\Gamma^e} q_n^e \psi_i^e(s)\, ds \qquad (383)$$

where $\psi_i^e(s)$ is the value of $\psi_i^e(x, y)$ on the boundary Γ^e. The heat flux q_n^e [see Eq. (378)] is an unknown when Ω^e is an interior element of the mesh (see Fig. 158a). However, when the element equations are assembled, the contribution of the heat flux q_n^e to the nodes (namely, P_i^e) of Ω^e get canceled by similar contributions from the adjoining elements (see Fig. 158b). If the element Ω^r has any of its sides on the boundary Γ of the domain Ω (see Fig. 158c), then on that side the heat flux q_n^r is either specified or unspecified. If q_n^r is specified, then the heat P_i^r at the nodes on that side can be computed using Eq. (383). If q_n^r is not specified, then the primary variable U_r is known on that portion of the boundary.

The remaining steps of the analysis do not differ from those of one-dimensional problems.

Interpolation Functions

Linear Triangular Element.   The simplest finite element in two dimensions is the triangular element. Since a triangle is defined uniquely by three points that form its vertices, the vertex points are chosen as the nodes (see Fig. 160a). These nodes will be connected to the nodes of adjoining elements in a finite-element mesh.

A polynomial in x and y that is uniquely defined by three constants is of the form $p(x, y) = c_0 + c_1 x + c_2 y$. Hence, assume approximation of u_e in the form

$$U_e = c_0^e + c_1^e x + c_2^e y \qquad (384)$$

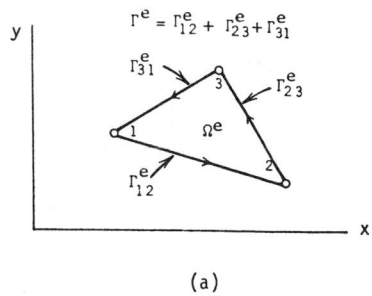

(a)

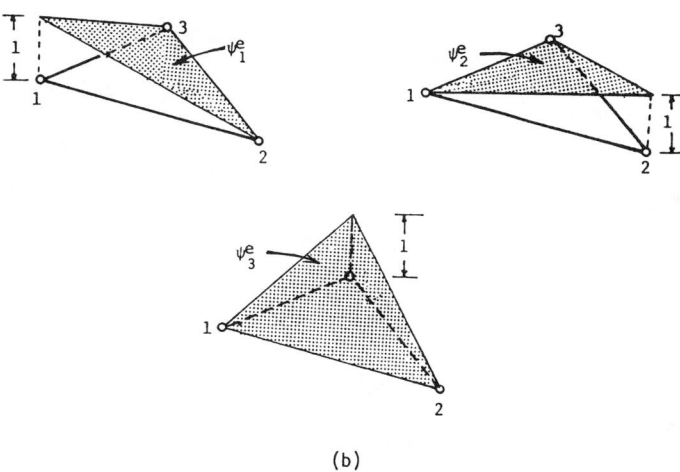

(b)

Fig. 160 Typical linear triangular element and associated finite-element interpolations function.

Proceeding as in the case of one-dimensional elements, write

$$U_i^e \equiv U_e(x_i, y_i) = c_0^e + c_1^e x_i + c_2^e y_i \qquad i = 1, 2, 3$$

where (x_i, y_i) denote the global coordinates of the element node i in Ω^e. In explicit form this equation becomes

$$\begin{Bmatrix} U_1^e \\ U_2^e \\ U_3^e \end{Bmatrix} = \begin{bmatrix} 1 & x_1 & y_1 \\ 1 & x_2 & y_2 \\ 1 & x_3 & y_3 \end{bmatrix} \begin{Bmatrix} c_0^e \\ c_1^e \\ c_2^e \end{Bmatrix}$$

Note that the element nodes are numbered counterclockwise. Upon solving for c's and substituting back into Eq. (384), one obtains

$$U_e = \sum_{i=1}^{3} U_i^e \psi_i^e(x, y)$$

$$\psi_i^e = \frac{1}{2A_e}(\alpha_i^e + \beta_i^e x + \gamma_i^e y)$$

where A_e represents the area of the triangle, and

$$\alpha_i^e = x_j y_k - x_k y_j \qquad \beta_i^e = y_j - y_k \qquad \gamma_i^e = x_k - x_j$$
$$i \neq j \neq k \qquad i, j, k = 1, 2, 3$$

and the indices on α_i^e, β_i^e, and γ_i^e permute in a natural order. For example, α_1^e is given by setting $i = 1, j = 2$, and $k = 3$:

$$\alpha_1^e = x_2 y_3 - x_3 y_2$$

The sign of the determinant changes if the node numbering is changed to clockwise. The interpolation functions ψ_i^e satisfy the interpolation properties listed in Eq. (364). The shape of these functions is shown in Fig. 160b.

Note that the derivative of ψ_i^e with respect to x or y is a constant. Hence, the derivatives of the solution evaluated in the postcomputation would be elementwise constant. Also, the coefficient matrix

$$K_{ij}^e = \int_{\Omega^e} \left(a_{11}^e \frac{\partial \psi_i^e}{\partial x} \frac{\partial \psi_j^e}{\partial x} + a_{22}^e \frac{\partial \psi_i^e}{\partial y} \frac{\partial \psi_j^e}{\partial y} \right) dx \, dy$$

$$(385)$$

can be easily evaluated for the linear interpolation functions for a triangle. We have

$$\frac{\partial \psi_i^e}{\partial x} = \frac{\beta_i^e}{2A_e} \qquad \frac{\partial \psi_i^e}{\partial y} = \frac{\gamma_i^e}{2A_e}$$

and, for elementwise constant values of a_{11}^e and a_{22}^e, the coefficients of K_{ij}^e become

$$K_{ij}^e = \frac{1}{4A_e^2}(a_{11}^e \beta_i^e \beta_j^e + a_{22}^e \gamma_i^e \gamma_j^e)\left(\int_{\Omega^e} dx\,dy\right)$$

$$= \frac{1}{4A_e}(a_{11}^e \beta_i^e \beta_j^e + a_{22}^e \gamma_i^e \gamma_j^e)$$

Linear Rectangular Element. A rectangular element is uniquely defined by the four corner points (see Fig. 161). Therefore, the four-term polynomial can be used to derive the interpolation functions. Express u_e in the form

$$u_e = c_0^e + c_1^e x + c_2^e y + c_3^e xy \qquad (386)$$

and obtain

$$\begin{Bmatrix} U_1^e \\ U_2^e \\ U_3^e \\ U_4 \end{Bmatrix} = \begin{bmatrix} 1 & x_1 & y_1 & x_1 y_1 \\ 1 & x_2 & y_2 & x_2 y_2 \\ 1 & x_3 & y_3 & x_3 y_3 \\ 1 & x_4 & y_4 & x_4 y_4 \end{bmatrix} \begin{Bmatrix} c_0^e \\ c_1^e \\ c_2^e \\ c_3^e \end{Bmatrix}$$

By inverting the equations for c's and substituting into Eq. (386), one obtains

$$\psi_1^e = \left(1 - \frac{\xi}{a}\right)\left(1 - \frac{\eta}{b}\right) \qquad \psi_2^e = \frac{\xi}{a}\left(1 - \frac{\eta}{b}\right)$$

$$\psi_3^e = \frac{\xi}{a}\frac{\eta}{b} \qquad \psi_4^e = \left(1 - \frac{\xi}{a}\right)\frac{\eta}{b}$$

where (ξ, η) are the element coordinates,

$$\xi = x - x_1 \qquad \eta = y - y_1$$

The functions are geometrically represented in Fig. 161. In calculating element matrices, one finds that the use

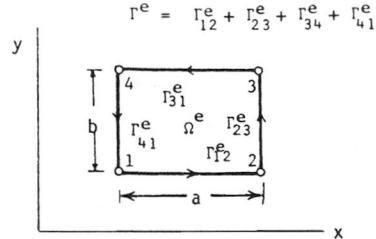

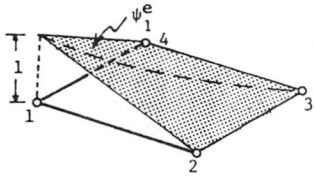

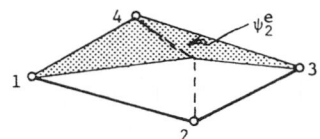

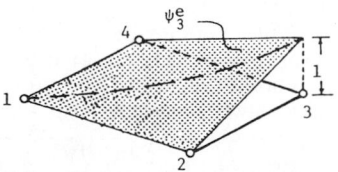

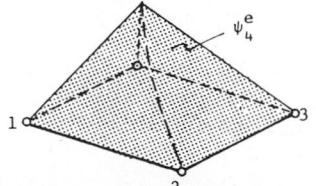

Fig. 161 Typical linear rectangular element and associated finite-element interpolation functions.

of the local coordinate system (ξ, η) is more convenient than using the global coordinates (x, y).

For the linear rectangular element, the derivatives of the shape functions are not constant within the element:

$$\frac{\partial \psi_i^e}{\partial x} = \text{linear in } y \qquad \frac{\partial \psi_i^e}{\partial y} = \text{linear in } x$$

The integration of polynomial expressions over a rectangular element is made simple by the fact that $\Omega^e = (0, a) \times (0, b)$:

$$\int_{\Omega^e} f(x, y)\, dx\, dy = \int_b^a \int_0^b f(x, y)\, dx\, dy$$

The coefficients in Eq. (385) can be easily evaluated over a linear rectangular element for elementwise constant values of a_{11}^e and a_{22}^e:

$$[K^e] = \frac{b}{6a}\begin{bmatrix} 2 & -2 & -1 & 1 \\ -2 & 2 & 1 & -1 \\ -1 & 1 & 2 & -2 \\ 1 & -1 & -2 & 2 \end{bmatrix}$$

$$+ \frac{a}{6b}\begin{bmatrix} 2 & 1 & -1 & -2 \\ 1 & 2 & -2 & -1 \\ -1 & -2 & 2 & 1 \\ -2 & -1 & 1 & 2 \end{bmatrix}$$

Example 94. Consider a computational example of Eq. (376) for the case where $a_{11} = a_{22} = 1$, $f = 0$, and Ω is a unit square. Let the boundary conditions be as follows (see Fig. 162a):

$$u(0, y) = u(1, y) = 0 \qquad u(x, 0) = 0$$

$$\frac{\partial u}{\partial y}(x, 1) = x$$

The finite-element model is given by Eq. (381), with

$$K_{ij}^e = \int_{\Omega^e} \left(\frac{\partial \psi_i^e}{\partial x} \frac{\partial \psi_j^e}{\partial x} + \frac{\partial \psi_i^e}{\partial y} \frac{\partial \psi_j^e}{\partial y} \right) dx\, dy \qquad f_i^e = 0$$

Triangular Elements. The 2×2 mesh of triangular elements is shown in Fig. 162b. The element

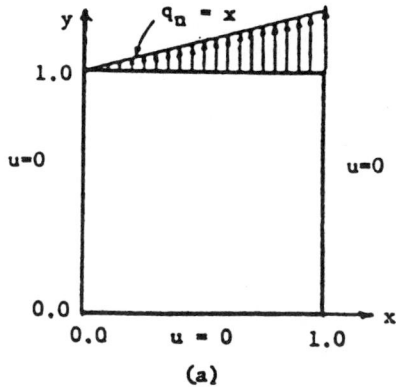

(a)

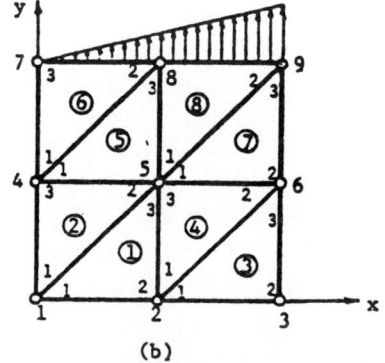

(b)

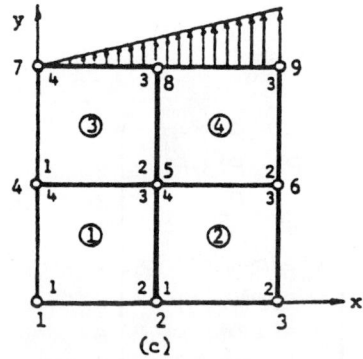

(c)

Fig. 162 Domain, boundary conditions, and finite-element meshes.

coefficient matrices are given by

$$[K^1] = [K^3] = [K^5] = [K^7] = \frac{1}{2}\begin{bmatrix} 1 & -1 & 0 \\ -1 & 2 & -1 \\ 0 & -1 & 1 \end{bmatrix}$$

$$[K^2] = [K^4] = [K^6] = [K^8] = \frac{1}{2}\begin{bmatrix} 1 & 0 & -1 \\ 0 & 1 & -1 \\ -1 & -1 & 2 \end{bmatrix}$$

The assembled equations are given by (refer to Fig. 162b) the tabulation below where $\times$ denotes a zero due to disconnectivity (e.g., $K_{13} = 0$ because global nodes 1 and 3 do not belong to the same element).

The boundary conditions on the primary variables (i.e., U's) are

$$U_1 = U_2 = U_3 = U_4 = U_6 = U_7 = U_9 = 0$$

The known secondary variables are (correspond to nodes 5 and 8)

$$P_3^1 + P_2^2 + P_3^4 + P_2^5 + P_1^7 + P_1^8 + = 0$$

(because no flux is specified at node 5)

$$P_3^5 + P_2^6 + P_3^8 = \int_0^{0.5} q\psi_2^6(x, 1)\, dx$$

$$+ \int_{0.5}^{1.0} q\psi_3^8(x, 1)\, dx \tag{387}$$

Note that the individual fluxes P_i^e in Eqs. (387) are not zero, but their sum is equal to the values indicated. For example, consider P_2^6:

$$P_2^6 = \int_{S^6} q\psi_2^6\, ds$$

$$= \int_{0.5}^{1.0} q\psi_2^6(y - 0.5, y)\left(\frac{1}{\sqrt{2}}\, dy\right)$$

$$+ \int_{0.5}^0 q\psi_2^6(x, 1)(-dx) + \int_{1.0}^{0.5} q\psi_2^6(0, y)(-dy)$$

where $\psi_2^6(x, y) = 2x$. The first integral is nonzero but gets canceled by a similar but negative contribution from P_3^5, the second integral is nonzero and can be evaluated since $q = x$ is known, and the third integral is zero because $\psi_2^6(0, y) = 0$.

Evaluating the integral in Eq. (387) [with $\psi_2^6(x, y) = 2x$ and $\psi_3^8 = 2(y - x)$], we obtain

$$P_3^5 + P_2^6 + P_3^8 = \int_0^{0.5} x2x\, dx + \int_{0.5}^{1.0} 2x(1 - x)\, dx$$

$$= \tfrac{2}{3}(0.5)^3 + (1)^2 - (0.5)^2$$

$$- \tfrac{2}{3}[(1)^3 - (0.5)^3]$$

$$= \tfrac{1}{12} + \tfrac{1}{6} = \tfrac{1}{4}$$

Tabulation for Triangular Elements

	1	2	3	4	5	6	7	8	9
	$1+1$	-1	$\times$	-1	0	$\times$	$\times$	$\times$	$\times$
		$2+1+1$	-1	$\times$	$-1-1$	$0+0$	$\times$	$\times$	$\times$
			2	$\times$	$\times$	-1	$\times$	$\times$	$\times$
				$2+1+1$	$-1-1$	$\times$	-1	$0+0$	$\times$
$\frac{1}{2}$					$2+2+1$ $+1+1+1$	$-1-1$	$\times$	$-1-1$	$0+0$
						$2+1+1$	$\times$	$\times$	-1
	Symmetric						2	-1	$\times$
								$2+1+1$	-1
									$1+1$

$$\times \begin{Bmatrix} U_1 \\ U_2 \\ U_3 \\ U_4 \\ U_5 \\ U_6 \\ U_7 \\ U_8 \\ U_9 \end{Bmatrix} = \begin{Bmatrix} P_1^1 + P_1^2 \\ P_2^1 + P_1^3 + P_1^4 \\ P_2^3 \\ P_3^2 + P_1^5 + P_1^6 \\ (P_3^1 + P_2^2 + P_3^4 + P_2^5 \\ + P_1^7 + P_1^8) \\ P_3^3 + P_2^4 + P_2^7 \\ P_3^6 \\ P_3^5 + P_2^6 + P_3^8 \\ P_3^7 + P_2^8 \end{Bmatrix} \begin{matrix} 1 \\ 2 \\ 3 \\ 4 \\ 5 \\ \\ 6 \\ 7 \\ 8 \\ 9 \end{matrix}$$

To solve for the unknowns U_5 and U_8, equations (5) and (8) of the assembled equations are used. This choice is dictated by the fact that the remaining equations contain additional unknowns in P's. The solution is given by

$$U_5 = \tfrac{1}{28} \qquad U_8 = \tfrac{2}{14}$$

The internal heat P_i^e can be determined from either the element equations (381) or by definition (383). In general, the values computed by the two methods are not the same because P_i^e determined from the element equations is the internal heat in equilibrium with the heat from the neighboring elements, whereas P_i^e computed from the gradient of the approximate temperature field is not.

Rectangular Elements. For the 2×2 mesh of rectangular elements shown in Fig. 162c, the element matrices are given by

$$[K^-] = [K^2] = \frac{1}{6} \begin{bmatrix} 4 & -1 & -2 & -1 \\ -1 & 4 & -1 & -2 \\ -2 & -1 & 4 & -1 \\ -1 & -2 & -1 & 4 \end{bmatrix}$$

The assembled equations are shown in the tabulation below.

The boundary conditions are given by

$$P_3^1 + P_4^2 + P_2^3 + P_1^4 = 0 \qquad P_3^3 + P_4^4 = 0.25$$

The condensed equations become

$$\frac{1}{6} \begin{bmatrix} 16 & -2 \\ -2 & 8 \end{bmatrix} \begin{Bmatrix} U_5 \\ U_8 \end{Bmatrix} = \begin{Bmatrix} 0 \\ \tfrac{1}{4} \end{Bmatrix}$$

and the solution is given by

$$U_5 = \tfrac{3}{124} \qquad U_8 = \tfrac{6}{31}$$

The exact solution of the problem is given by

$$u(x, y) = \frac{2}{\pi^2} \sum_{n=1}^{\infty} \frac{(-1)^{n+1}}{n^2 \cosh n\pi} \sin n\pi x \sinh n\pi y$$

A comparison of the finite-element solutions obtained with 2×2 and 4×4 meshes of linear rectangular and triangular elements with the series solution is presented in Table 14. The finite-element solution improves as the mesh is refined.

Tabulation for Rectangular Elements

Table 14   Comparison of Finite-Element Solutions

		Triangles		Rectangles		Series
x	y	2×2	4×4	2×2	4×4	Solution
0.25	0.25	—	0.0101	—	0.0095	0.0103
0.50	0.25	—	0.0151	—	0.0136	0.0152
0.75	0.25	—	0.0114	—	0.0097	0.0112
0.25	0.50	—	0.0253	—	0.0254	0.0264
0.50	0.50	0.0357	0.0387	0.0242	0.0370	0.0400
0.75	0.50	—	0.0305	—	0.0270	0.0308
0.25	0.75	—	0.0525	—	0.0552	0.0555
0.50	0.75	—	0.0840	—	0.0882	0.0894
0.75	0.75	—	0.0719	—	0.0675	0.0765
0.25	1.00	—	0.1007	—	0.1059	0.1057
0.50	1.00	0.1429	0.1729	0.1936	0.1851	0.1846
0.75	1.00	—	0.1729	—	0.2027	0.1990

Table 15   Basic Steps in Finite-Element Analysis of Typical Problem

1. *Discretization of a Domain*. Represent the given domain as a collection of a finite number of simple subdomains, called *finite elements*. The number, shape, and type of element depend on the domain and differential equation being solved. The principal parts of this step include:

   (a) Number the nodes (see step 2) and elements of the collection, called the *finite-element mesh*.

   (b) Generate the coordinates of the nodes in the mesh and the relationship between the element nodes to global nodes (called *the connectivity matrix*, which indicates the relative position of each element in the mesh).

2. *Approximation of the Solution*

   (a) *Derivation of the Approximating Functions*. For each element in the mesh, derive the approximation functions needed in the variational method. These functions are generally algebraic polynomials generated by interpolating the unknown function in terms of its values at preselected points, the *nodes*, of the element.

   (b) *Variational Approximation of the Equation*. Using the functions derived in step 2a and any appropriate variational method, derive the algebraic equations among the nodal values of the primary and secondary variables.

3. *Connectivity (or Assembly) of Elements*. Combine the algebraic equations of all elements in the mesh by imposing the continuity of the primary nodal variables (i.e., the values of the primary variables at a node shared by two or more elements are the same). This can be viewed as putting the elements (which were isolated in steps 2a and 2b from the mesh to derive the algebraic equations) back into their original places. This gives the algebraic equations governing the whole problem.

4. *Imposition of Boundary Conditions*. Impose the boundary conditions, both on primary and secondary variables of the assembled equations.

5. *Solution of Equations*. Solve the equations for the unknown nodal values of the primary variables.

6. *Computation of Additional Quantities*. Using the nodal values of the primary variables, compute the secondary variables (via constitutive equations).

In summary, the finite-element method is a numerical technique of solving field problems of engineering. It is endowed with two unique features: The domain in which the equations are defined is represented by a collection of simple parts (finite elements), and over each element the problem is approximated using any one of the variational methods with polynomials for the approximation functions. The first feature allows approximate representation of geometrically complicated domains by simple geometric shapes, while the second feature enables the approximation of the field variables, evaluation of the coefficient matrices, and solution of the finite-element equations on a computer. A list of basic steps of the finite-element analysis is presented in Table 15.

13   LAPLACE TRANSFORMATION

13.1   Transformation Principles

The Laplace and Fourier transformation methods and the Heaviside operational calculus are in essence different aspects of the same method. This method simplifies the solving of linear constant-coefficient integrodifferential equations and convolution-type integral equations. For brevity the conditions under which the steps of the method may be validly applied will be omitted. Hence the correctness of a final result should be checked in each case by showing that the formal solution satisfies the given equation and conditions.

1. *Direct Laplace Transformation*. Let t be a real variable, s a complex variable (Section 14.2),
$f(t)$ a real function of t that equals zero for $t < 0$, $F(s)$ a function of s, and e the base of the natural logarithms. If the Lebesgue integral

$$\int_0^\infty e^{-st} f(t)\, dt = F(s) \qquad (388)$$

then $F(s)$ is the *direct Laplace transform* of $f(t)$; in simpler notation

$$\mathscr{L}[f(t)] = F(s) \qquad (389)$$

2. *Inverse Laplace Transformation*. Under certain conditions the direct transformation can be inverted, giving as one explicit representation

$$\frac{1}{2\pi i} \int_{c-i\infty}^{c+i\infty} e^{ts} F(s)\, ds (=) f(t) \qquad (390)$$

in which c is a real constant chosen so that the path of integration lies to the right of all the singularities of $F(s)$, and $(=)$ means equals except possibly for a set of values of t of measure zero. If this relation holds, then $f(t)$ is the *inverse Laplace transform* of $F(s)$. In simpler notation the transformation is written

$$\mathscr{L}^{-1}[F(s)] (=) f(t) \qquad (391)$$

3. *Transformation of nth Derivative.* If $\mathscr{L}[f(t)] = F(s)$, then

$$\mathscr{L}\left[\frac{d^n f(t)}{dt^n}\right] = s^n F(s) - \sum_{k=0}^{n-1} f^{(k)}(0+) \cdot s^{n-1-k}$$

(392)

where $f^{(2)}(0+)$ means $d^2 f(t)/dt^2$ evaluated for $t \to 0$ and $f^{(0)}(0+)$ means $f(0+)$ and $n = 1, 2, 3, \ldots$.

4. *Transformation of nth Integral.* If $\mathscr{L}[f(t)] = F(s)$, then

$$\mathscr{L}\left[\overbrace{\int\!\!\int \cdots \int}^{n} f(t)\, dt\right]$$

$$= s^{-n} F(e) + \sum_{k=-1}^{-n} f^{(k)}(0+) \cdot s^{-n-1-k}$$

(393)

where $n = 1, 2, 3, \ldots$. For example, $f^{(-2)}(0+)$ means $\int\int f(t)\, dt\, dt$ evaluated for $t \to 0$.

5. *Inverse Transformation of Product.* If

$$\mathscr{L}^{-1}[F_1(s)] = f_1(t) \qquad \mathscr{L}^{-1}[F_2(s)] = f_2(t)$$

(394)

then

$$\mathscr{L}^{-1}[F_1(s) \cdot F_2(s)] = \int_0^t f_1(t - \lambda) \cdot f_2(\lambda)\, d\lambda$$

(395)

6. *Linear Transformations $\mathscr{L}$ and $\mathscr{L}^{-1}$.* Let k_1, k_2 be real constants. Then

$$\mathscr{L}[k_1 f_1(t) + k_2 f_2(t)] = k_1 \mathscr{L}[f_1(t)] + k_2 \mathscr{L}[f_2(t)]$$

(396)

and

$$\mathscr{L}^{-1}[k_1 F_2(s) + k_2 F_2(s)]$$

$$= k_1 \mathscr{L}^{-1}[F_1(s)] + k_2 \mathscr{L}^{-1}[F_2(s)] \quad (397)$$

13.2   Procedure

To illustrate the application of the rules of procedure the following simple initial-value problem will be solved. Given the equation

$$k_1 \frac{dy(t)}{dt} + k_2 y(t) + k_3 \int y(t)\, dt = u(t)$$

and initial values $y(0)$, $y^{(-1)}(0)$ where $u(t) = 0$ for $t < 0$ and $u(t) = 1$ for $0 < t$ and k_1, k_2, k_3 are real constants. Assume that $y(t)$ has a Laplace transform $Y(s)$, that is, $\mathscr{L}[y(t)] = Y(s)$.

Step A. Find the Laplace transform of the equation to be solved and express it in terms of the transform of the unknown function.

Thus,

$$\mathscr{L}\left[k_1 \frac{dy(t)}{dt} + k_2 y(t) + k_3 \int y(t)\, dt\right] = \mathscr{L}[u(t)]$$

By (396) this becomes

$$k_1 \mathscr{L}\left[\frac{dy(t)}{dt}\right] + k_2 \mathscr{L}[y(t)] + k_3 \mathscr{L}\left[\int y(t)\, dt\right]$$

$$= \mathscr{L}[u(t)]$$

By (392) and (393) and the given initial conditions of the problem the equation becomes

$$k_1[sY(s) - y(0)] + k_2 Y(s) + k_3[s^{-1} Y(s)$$

$$+ y^{(-1)}(0) \cdot s^{-1}] = \mathscr{L}[u(t)]$$

Step B. Solve the resulting equation for the transform of the unknown function. Thus,

$$Y(s) = \frac{\mathscr{L}[u(t)] + k_1 y(0) - y^{(-1)}(0) \cdot s^{-1}}{k_1 s + k_2 + k_3 s^{-1}}$$

Step C. Evaluate the direct transform of the given function (right member) in the original equation. Since

$$\mathscr{L}[u(t)] = \frac{1}{s}$$

$$Y(s) = \frac{k_1 y(0) \cdot s - y^{(-1)}(0) + 1}{k_1 s^2 + k_2 s + k_3}$$

Step D. Obtain the solution of the problem by evaluating the inverse Laplace transform of the function obtained by the preceding steps.

One way to carry out step D is to find the inverse transform from the table of Laplace transforms in Section 12.3. To use the table, the denominator of the fraction should be factored:

$$y(t) = \mathscr{L}^{-1}[Y(s)] = \mathscr{L}^{-1}\left[\frac{k_1 y(0) \cdot s - y^{(-1)}(0) + 1}{k_1 s^2 + k_2 s + k_3}\right]$$

$$= \mathscr{L}^{-1}\left[\frac{k_1 y(0) \cdot s - y^{(-1)}(0) + 1}{k_1 (s + K_1)(s + K_2)}\right]$$

in which

$$K_1 \equiv \frac{k_2}{2k_1} - \frac{1}{2k_1}(k_2^2 - 4k_1 k_3)^{1/2}$$

$$K_2 \equiv \frac{k_2}{2k_1} + \frac{1}{2k_1}(k_2^2 - 4k_1 k_3)^{1/2}$$

To find the result it is necessary to distinguish between two cases.

CASE 1: If $K_1 \neq K_2$,

$$y(t) = \frac{[k_1 y(0)K_1 + y^{(-1)}(0) - 1]e^{-K_1 t} - [k_1 y(0)K_2 + y^{(-1)}(0) - 1]e^{-K_2 t}}{k_1(K_1 - K_2)}$$

for $0 < t$ and $y(t) = 0$ for $t < 0$.

CASE 2: If $K_1 = K_2 = K$, then $K = k_2/2k_1$, and

$$y(t) = \mathscr{L}^{-1}\left[\frac{k_1 y(0) \cdot s - y^{(-1)}(0) + 1}{k_1(s + K)^2} \right]$$

From Table 16,

$$y(t) = \frac{k_1 y(0)e^{-Kt} - [y^{(-1)}(0) - 1 + k_1 y(0)K]te^{-Kt}}{k_1}$$

for $0 < t$ and $y(t) = 0$ for $t < 0$.

The solutions can be shown to satisfy the original equation and initial conditions.

The use of step C can be avoided by using steps E, F, and G in place of steps C and D in the following way.

Step E. Factor the transform of the unknown function obtained by step B and evaluate the inverse Laplace transform of each factor.

Note. The inverse transform of a rational fraction can be found only if it is a proper fraction.

Thus

$$Y(s) = \frac{k_1 y(0) \cdot s - y^{(-1)}(0)}{k_1 s^2 + k_2 s + k_3} + \frac{s\mathscr{L}[u(t)]}{k_1 s^2 + k_2 s + k_3}$$

Let

$$y_1(t) \equiv \mathscr{L}^{-1}\left[\frac{k_1 y(0) \cdot s - y^{(-1)}(0)}{k_1(s + K_1)(s + K_2)} \right]$$

$$= [k_1(y)(0)K_1 + y^{(-1)}(0)]e^{-K_1 t} - \frac{[k_1 y(0)K_2 + y^{(-1)}(0)]e^{-K_2 t}}{k_1(K_1 - K_2)}$$

for $0 < t$ and $y_1(t) = 0$ for $t < 0$. Also

$$\mathscr{L}^{-1}\left[\frac{s}{k_1(s + K_1)(s + K_2)} \right] = \frac{K_1 e^{-K_1 t} - K_2 e^{-K_2 t}}{k_1(K_1 - K_2)}$$

for $0 < t$ and $y_1(t) = 0$ for $t < 0$. Finally, $\mathscr{L}^{-1}\{\mathscr{L}[u(t)]\} = u(t)$.

Step F. Use condition 5 to find the inverse transform of the product.

Thus, by condition 6 and step F,

$$y(t) = y_1(t) + [k_1(K_1 - K_2)]^{-1}$$
$$\times \int_0^t [K_1 e^{-K_1(t-\tau)} - K_2 e^{-K_2(t-\tau)}]u(\tau)\,d\tau$$

Step G. Evaluate the (convolution) integral arising from step F. Thus,

$$y(t) = y_1(t) + [k_1(K_1 - K_2)]^{-1}(e^{-k_2 t} - e^{-K_1 t})$$

for $0 < t$ and $y(t) = 0$ for $t < 0$.

For the particular problem treated it is much simpler to use steps C and D than steps E, F, and G. However, for a more complicated right member of the original equation it could happen that step G would be easier to carry out than step C, in which case the second method (A, B, E, F, G) should be used rather than the first (A, B, C, D).

One physical representation of the initial-value problem that we have used for illustration is the problem of finding the current response of a series electric circuit containing constant lumped inductance, resistance, and capacitance to an applied electromotive force $u(t)$, with an initial current in the inductance and an initial charge on the condenser.

The complete method (of which only a part has been given) is not restricted in its field of application to linear equations with constant coefficients, but the solution of this type of equation is most simplified.

13.3 Transform Pairs

The Laplace transforms in Table 16 are applicable in the solution of ordinary integrodifferential and difference equations.

14 COMPLEX ANALYSIS

14.1 Complex Numbers

A *complex number* A is a combination of two real numbers a_1, a_2 in the ordered pair $(a_1, a_2) = A = a_1 + ia_2$, where $i = (-1)^{1/2}$. Real and imaginary numbers are special cases of complex numbers obtained by placing $(a_1, 0) = a_1$, $(0, a_2) = ia_2$ (see Fig. 163).

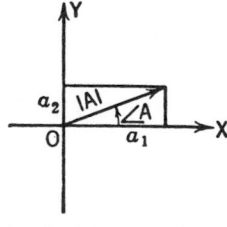

Fig. 163

Table 16   Laplace Transforms[a]

Unilateral Laplace Operation Transform Pairs

Name	$f(t), 0 \leq t$	$F(s)$
Linearity	$af(t),$ a is constant or variable independent of t, $f_1(t) \pm f_2(t)$	$aF(s),$ a is constant or variable independent of s, $F_1(s) \pm F_2(s)$
Real differentiation	$\dfrac{df(t)}{dt} \triangleq f'(t)$	$sF(s) - f(0+)$
Multiplication by s	$f'(t)$ if $f(0+) = 0$	$sF(s)$
Real integration	$\displaystyle\int f(t)\,dt \triangleq f^{(-1)}(t)$	$\dfrac{F(s)}{s} + \dfrac{f^{(-1)}(0+)}{s}$
Division by s	$\displaystyle\int_0^t f(t)\,dt \triangleq f^{(-1)}(t) - f^{(-1)}(0+)$	$\dfrac{F(s)}{s}$

Scale change	$f\left(\dfrac{t}{a}\right),$
	
	a is positive constant or positive variable
	    independent of t
		$aF(as),$
		
		a is positive constant or positive variable
		    independent of s

Complex multiplication	$\displaystyle\int_0^1 f_1(t-\tau)f_2(\tau)\,d\tau \triangleq f_1(t) * f_2(t)$	$F_1(s)F_2(s)$
Real translation	$f(t-a)$ if $f(t-a) = 0, 0 < t < a,$ $f(t+a)$ if $f(t+a) = 0, -a < t < 0,$ a is a nonnegative real number.	$e^{-as}F(s),$ $e^{as}F(s)$

Complex translation	$e^{-at}f(t),$
	$e^{at}f(t),$
	a is complex number with nonnegative real
	    part
		$F(s+a),$
		$F(s-a),$
		a is complex number with nonnegative real
		    part.

Second independent variable	$\displaystyle\lim_{a \to a_0} f(t, a),$ a is second variable independent of t	$\displaystyle\lim_{a \to a_0} F(s, a),$ a is second variable independent of s

Differentiation with respect to second independent variable	$\dfrac{\partial}{\partial a} f(t, a),$
	
	    a is second variable independent of t
		$\dfrac{\partial}{\partial a} F(s, a),$
		
		    a is second variable independent of t

Final value	$\displaystyle\lim_{t \to \infty} f(t) = \lim_{s \to 0} s(s)$
	if $sF(s)$ is analytic on axis of imaginaries and
	    in right half-plane
		$\displaystyle\lim_{t \to \infty} f(t) = \lim_{s \to 0} s(s)$
		if $sF(s)$ is analytic on axis of imaginaries and
		    in right half-plane

Initial value	$\displaystyle\lim_{t \to 0} f(t) = \lim_{s \to \infty} sF(s)$	$\displaystyle\lim_{t \to 0} f(t) = \lim_{s \to \infty} sF(s)$
Complex differentiation	$tf(t)$	$-\dfrac{d}{ds}F(s)$
Complex integration	$\dfrac{1}{t}f(t)$	$\displaystyle\int_s^\infty F(s)\,ds$

Integration with respect to second independent variable	$\displaystyle\int_{a_0}^a f(t, a)\,da,$
	
	    a is second variable independent of t
		$\displaystyle\int_{a_0}^a F(s, a)\,da,$
		
		    a is second variable independent of s

(Continues)

Table 16   (*Continued*)

Real multiplication	$f_1(t)f_2(t)$	$\dfrac{1}{2\pi j}\displaystyle\int_{c_1-j\infty}^{c_2+j\infty} F_1(s-w)F_2(w)dw \triangleq$

$$F_1(s) \otimes F_2(s),\ \max(\sigma_{a1},\sigma_{a2},\sigma_{a1}+\sigma_{a2}) < \sigma,\sigma_{a2} < c_2 < \sigma - \sigma_{a2}$$

$$\sum_{k-1}^{q} \frac{A_1(s_k)}{B_1'(s_k)}F_2(s-s_k)$$

if $F_1(s) \triangleq \dfrac{A_1(s)}{B_1(s)}$ is rational algebraic
fraction having only first-order poles

$$\sum_{k-1}^{n}\sum_{j-1}^{m_k} \frac{(-1)^{m_k-j}K_{kj}}{(m_k-j)!}\left[\frac{d^{m_k-j}}{ds^{m_k-j}}F_2(s)\right]_{s=s-s_k},$$

$$K_{kj} \triangleq \frac{1}{(j-1)!}\left[\frac{d^{j-1}}{ds^{j-1}}(s-s_k)^{m_k}F_1(s)\right]_{s=s_k}$$

if $F_1(s)$ is rational algebraic fraction having
multiple-order poles

Unilateral Laplace Function Transform Pairs

$F(s)$	$f(t), 0 \le t$
$\dfrac{A(s)}{B(s)}$; rational proper fraction; first-order poles only	$\displaystyle\sum_{k-1}^{q}\frac{A(s_k)}{B'(s_k)}e^{s_k i}$
$\dfrac{A(s)}{B(s)}$; rational proper fraction; higher order poles, general case	$\displaystyle\sum_{k-1}^{n}\sum_{j-1}^{mk}\frac{K_{kj}}{(m_k-j)^t}t^{m_k-j}e^{s_k i}\ m_1+m_2+\cdots+m_n=q$
	$K_{kj} \triangleq \dfrac{1}{(j-1)!}\left[\dfrac{d^{j-1}}{ds^{j-1}}\dfrac{(s-s_k)^{m_k}A(s)}{B(s)}\right]_{s=s_k}$
	$B(s) \triangleq (s-s_1)^{m_1}(s-s_2)^{m_2}\cdots(s-s_k)^{m_k}\cdots(s-s_n)^{m_n}$
1	$u_1(t) \triangleq \displaystyle\lim_{a\to 0}\frac{u(t)-u(t-a)}{a}$, unit impulse at $t=0$
s	$u_2(t) \triangleq \displaystyle\lim_{a\to 0}\frac{u(t)-2u(t-a)+u(t-2a)}{a^2}$, unit doublet impulse at $t=0$
$\dfrac{1}{s}$	1, or $u(t)$, unit step at $t=0$
$\dfrac{1}{s+\alpha}$	$e^{-\alpha t}$
$\dfrac{1}{(s+\alpha)(s+\gamma)}$	$\dfrac{e^{-\alpha t}-e^{-\gamma t}}{\gamma-\alpha}$

Table 16 (Continued)

Unilateral Laplace Function Transform Pairs

$F(s)$	$f(t), 0 \leq t$
$\dfrac{s + a_0}{(s + \alpha)(s + \gamma)}$	$\dfrac{(a_0 - \alpha)e^{-\alpha t} - (a_0 - \gamma)e^{-\gamma t}}{\gamma - \alpha}$
$\dfrac{1}{s(s + \alpha)(s + \gamma)}$	$\dfrac{1}{\alpha\gamma} + \dfrac{\gamma e^{-\alpha t} - \alpha e^{-\gamma t}}{\alpha\gamma(\alpha - \gamma)}$
$\dfrac{s + a_0}{s(s + \alpha)(s + \gamma)}$	$\dfrac{a_0}{\alpha\gamma} + \dfrac{a_0 - \alpha}{\alpha(\alpha - \gamma)}e^{-\alpha t} + \dfrac{a_0 - \gamma}{\gamma(\gamma - \alpha)}e^{-\gamma t}$
$\dfrac{s^2 + a_1 s + a_0}{s(s + \alpha)(s + \gamma)}$	$\dfrac{a_0}{\alpha\gamma} + \dfrac{\alpha^2 - a_1\alpha + a_0}{\alpha(\alpha - \gamma)}e^{-\alpha t} - \dfrac{\gamma^2 - a_1\gamma + a_0}{\gamma(\alpha - \gamma)}e^{-\gamma t}$
$\dfrac{1}{(s + \alpha)(s + \gamma)(s + \delta)}$	$\dfrac{e^{-\alpha t}}{(\gamma - \alpha)(\delta - \alpha)} + \dfrac{e^{-\gamma t}}{(\alpha - \gamma)(\delta - \gamma)} + \dfrac{e^{-\delta t}}{(\alpha - \delta)(\gamma - \delta)}$
$\dfrac{s + a_0}{(s + \alpha)(s + \gamma)(s + \delta)}$	$\dfrac{a_0 - \alpha}{(\gamma - \alpha)(\delta - \alpha)}e^{-\alpha t} + \dfrac{a_0 - \gamma}{(\alpha - \gamma)(\delta - \gamma)}e^{-\gamma t} + \dfrac{a_0 - \delta}{(\alpha - \delta)(\gamma - \delta)}e^{-\delta t}$
$\dfrac{s^2 + a_1 s + a_0}{(s + \alpha)(s + \gamma)(s + \delta)}$	$\dfrac{\alpha^2 - a_1\alpha + a_0}{(\gamma - \alpha)(\delta - \alpha)}e^{-\alpha t} + \dfrac{\gamma^2 - a_1\gamma + a_0}{(\alpha - \gamma)(\delta - \gamma)}e^{-\gamma t} + \dfrac{\delta^2 - a_1\delta + a_0}{(\alpha - \delta)(\gamma - \delta)}e^{-\delta t}$
$\dfrac{1}{s^2 + \beta^2}$	$\dfrac{1}{\beta}\sin\beta t$
$\dfrac{1}{s^2 - \beta^2}$	$\dfrac{1}{\beta}\sinh\beta t$
$\dfrac{s}{s^2 + \beta^2}$	$\cos\beta t$
$\dfrac{s}{s^2 - \beta^2}$	$\cosh\beta t$
$\dfrac{s + a_0}{s^2 + \beta^2}$	$\dfrac{1}{\beta}(a_0 + \beta^2)^{1/2}\sin(\beta t + \psi)$ $\psi \overset{\Delta}{=} \tan^{-1}\dfrac{\beta}{a_0}$
$\dfrac{1}{s(s^2 + \beta^2)}$	$\dfrac{1}{\beta^2}(1 - \cos\beta t)$
$\dfrac{s + a_0}{s(s^2 + \beta^2)}$	$\dfrac{a_0}{\beta^2} - \dfrac{(a_0^2 + \beta^2)^{1/2}}{\beta^2}\cos(\beta t + \psi)$ $\psi \overset{\Delta}{=} \tan^{-1}\dfrac{\beta}{a_0}$
$\dfrac{s^2 + a_1 s + a_0}{s(s^2 + \beta^2)}$	$\dfrac{a_0}{\beta^2} - \dfrac{[(a_0 - \beta^2)^2 + a_1^2\beta^2]^{1/2}}{\beta^2}\cos(\beta t + \psi)$ $\psi \overset{\Delta}{=} \tan^{-1}\dfrac{a_1\beta}{a_0 - \beta^2}$
$\dfrac{s + a_0}{(s + \alpha)(s^2 + \beta^2)}$	$\dfrac{a_0 - \alpha}{\alpha^2 + \beta^2}e^{-\alpha t} + \dfrac{1}{\beta}\left[\dfrac{a_0^2 + \beta^2}{\alpha^2 + \beta^2}\right]^{1/2}\sin(\beta t + \psi)$ $\psi \overset{\Delta}{=} \tan^{-1}\dfrac{\beta}{a_0} - \tan^{-1}\dfrac{\beta}{\alpha}$

(Continues)

Table 16 *(Continued)*

Unilateral Laplace Function Transform Pairs (Continued)

$F(s)$	$f(t), 0 \leq t$
$\dfrac{s^2 + a_1 s + a_0}{(s + \alpha)(s^2 + \beta^2)}$	$\dfrac{\alpha^2 - a_1\alpha + a_0}{\alpha^2 + \beta^2} e^{-\alpha t} + \dfrac{1}{\beta}\left[\dfrac{(a_0 - \beta^2)^2 + a_1^2\beta^2}{\alpha^2 + \beta^2}\right]^{1/2} \sin(\beta t + \psi)$
	$\psi \stackrel{\Delta}{=} \tan^{-1}\dfrac{a_1\beta}{a_0 - \beta^2} - \tan^{-1}\dfrac{\beta}{\alpha}$
$\dfrac{s + a_0}{s(s + \alpha)(s^2 + \beta^2)}$	$\dfrac{a_0}{\alpha\beta^2} + \dfrac{\alpha - a_0}{\alpha(\alpha^2 + \beta^2)} e^{-\alpha t} - \dfrac{1}{\beta^2}\left[\dfrac{a_0^2 + \beta^2}{\alpha^2 + \beta^2}\right]^{1/2} \cos(\beta t + \psi)$
	$\psi \stackrel{\Delta}{=} \tan^{-1}\dfrac{\beta}{a_0} - \tan^{-1}\dfrac{\beta}{\alpha}$
$\dfrac{s^2 + a_1 s + a_0}{s(s + \alpha)(s^2 + \beta^2)}$	$\dfrac{a_0}{\alpha\beta^2} - \dfrac{\alpha^2 - a_1\alpha + a_0}{\alpha(\alpha^2 + \beta^2)} e^{-\alpha t} - \dfrac{1}{\beta^2}\left[\dfrac{(a_0 - \beta^2)^2 + a_1^2\beta^2}{\alpha^2 + \beta^2}\right]^{1/2} \cos(\beta t + \psi)$
	$\psi \stackrel{\Delta}{=} \tan^{-1}\dfrac{a_1\beta}{a_0 - \beta^2} - \tan^{-1}\dfrac{\beta}{\alpha}$
$\dfrac{s^2 + a_1 s + a_0}{(s + \alpha)(s + \gamma)(s^2 + \beta^2)}$	$\dfrac{\alpha^2 - a_1\alpha + a_0}{(\gamma - \alpha)(\alpha^2 + \beta^2)} e^{-\alpha t} + \dfrac{\gamma^2 + a_1\gamma + a_0}{(\alpha - \gamma)(\gamma^2 + \beta^2)} e^{-\gamma t}$
	$\qquad + \dfrac{1}{\beta}\left[\dfrac{(a_0 - \beta^2)^2 + a_1^2\beta^2}{(\alpha^2 + \beta^2)(\gamma^2 + \beta^2)}\right]^{1/2} \sin(\beta t + \psi)$
	$\psi \stackrel{\Delta}{=} \tan^{-1}\dfrac{a_1\beta}{a_0 - \beta^2} - \tan^{-1}\dfrac{\beta}{\alpha} - \tan^{-1}\dfrac{\beta}{\gamma}$
$\dfrac{s^3 + a_2 s^2 + a_1 s + a_0}{(s + \alpha)(s + \gamma)(s^2 + \beta^2)}$	$\dfrac{-\alpha^3 + a_2\alpha^2 - a_1\alpha - a_0}{(\gamma - \alpha)(\alpha^2 + \beta^2)} e^{-\alpha t} + \dfrac{-\gamma^3 + a_2\gamma^2 - a_1\gamma + a_0}{(\alpha - \gamma)(\gamma^2 + \beta^2)} e^{-\gamma t}$
	$\qquad + \dfrac{1}{\beta}\left[\dfrac{(a_0 - a_2\beta^2)^2 + \beta^2(a_1 - \beta^2)^2}{(\alpha^2 + \beta^2)(\gamma^2 + \beta^2)}\right]^{1/2} \sin(\beta t + \psi)$
	$\psi \stackrel{\Delta}{=} \tan^{-1}\dfrac{\beta(a_1 - \beta^2)}{a_0 - a_2\beta^2} - \tan^{-1}\dfrac{\beta}{\alpha} - \tan^{-1}\dfrac{\beta}{\gamma}$
$\dfrac{s}{(s^2 + \beta^2)(s^2 + \gamma^2)}$	$\dfrac{\cos\beta t - \cos\lambda t}{\lambda^2 - \beta^2}$
$\dfrac{s}{s^2 + (\beta + \lambda)^2} \times$ $\dfrac{1}{s^2 + (\beta - \lambda)^2}$	$\dfrac{1}{2\lambda\beta} \sin\lambda t \sin\beta t$

Table 16   *(Continued)*

Unilateral Laplace Function Transform Pairs (Continued)

$F(s)$	$f(t), 0 \le t$

$$\frac{s^2 + a_1 s + a_0}{(s^2 + \beta^2)(s^2 + \lambda^2)}$$

$$\frac{[(a_0 - \beta^2)^2 + a_1^2 \beta^2]^{1/2}}{\beta(\lambda^2 - \beta^2)} \sin(\beta t + \psi_1)$$

$$+ \frac{[(a_0 - \lambda^2)^2 + a_1^2 \lambda^2]^{1/2}}{\lambda(\beta^2 - \lambda^2)} \sin(\lambda t + \psi_2)$$

$$\psi_1 \triangleq \tan^{-1} \frac{a_1 \beta}{a_0 - \beta_2}$$

$$\psi_2 \triangleq \tan^{-1} \frac{a_1 \lambda}{a_0 - \lambda^2}$$

$$\frac{s^3 + a_2 s^2 + a_1 s + a_0}{(s^2 + \beta^2)(s^2 + \lambda^2)}$$

$$\frac{[(a_0 - a_2 \beta^2)^2 + \beta^2(a_1 - \beta^2)^2]^{1/2}}{\beta(\lambda^2 - \beta^2)} \sin(\beta t + \psi_1)$$

$$+ \frac{[(a_0 - a_2 \lambda^2)^2 + \lambda^2(a_1 - \lambda^2)^2]^{1/2}}{\lambda(\beta^2 - \lambda^2)} \sin(\lambda t + \psi_2)$$

$$\psi_1 \triangleq \tan^{-1} \frac{\beta(a_1 - \beta^2)}{a_0 - a_2 \beta^2}$$

$$\psi_2 \triangleq \tan^{-1} \frac{\lambda(a_1 - \lambda^2)}{a_0 - a_2 \lambda^2}$$

$$\frac{1}{(s + \alpha)^2 + \beta^2}$$

$$\frac{1}{\beta} e^{-\alpha t} \sin \beta t$$

$$\frac{s + a_0}{(s + \alpha)^2 + \beta^2}$$

$$\frac{1}{\beta}[(a_0 - \alpha)^2 + \beta^2]^{1/2} e^{-\alpha t} \sin(\beta t + \psi)$$

$$\psi \triangleq \tan^{-1} \frac{\beta}{a_0 - \alpha}$$

$$\frac{s + \alpha}{(s + \alpha)^2 + \beta^2}$$

$$e^{-\alpha t} \cos \beta t$$

$$\frac{1}{s[(s + \alpha)^2 + \beta^2]}$$

$$\frac{1}{\beta_0^2} + \frac{1}{\beta_0 \beta} e^{-\alpha t} \sin(\beta t - \psi)$$

$$\psi \triangleq \tan^{-1} \frac{\beta}{-\alpha}$$

$$\beta_0^2 \triangleq \alpha^2 + \beta^2$$

$$\frac{s^2 + a_1 s + a_0}{s[(s + \alpha)^2 + \beta^2]}$$

$$\frac{a_0}{\beta_0^2} + \frac{1}{\beta \beta_0}[(\alpha^2 - \beta^2 - a_1 \alpha + a_0)^2 + \beta^2(a_1 - 2\alpha)^2]^{1/2} e^{-\alpha t} \sin(\beta t + \psi)$$

$$\psi \triangleq \tan^{-1} \frac{\beta(a_1 - 2\alpha)}{\alpha^2 - \beta^2 - a_1 \alpha + a_0} - \tan^{-1} \frac{\beta}{-\alpha}$$

$$\beta_0^2 \triangleq \beta^2 + \alpha^2$$

(Continues)

Table 16 *(Continued)*

Unilateral Laplace Function Transform Pairs (Continued)

$F(s)$	$f(t), 0 \leq t$
$\dfrac{1}{(s+\gamma)[(s+\alpha)^2+\beta^2]}$	$\dfrac{e^{-\gamma t}}{(\gamma-\alpha)^2+\beta^2} + \dfrac{1}{\beta[(\gamma-\alpha)^2+\beta^2]^{1/2}}e^{-\alpha t}\sin(\beta t - \psi)$ $\psi \triangleq \tan^{-1}\dfrac{\beta}{\gamma-\alpha}$
$\dfrac{s+a_0}{(s+\gamma)[(s+\alpha)^2+\beta^2]}$	$\dfrac{a_0-\gamma}{(\alpha-\gamma)^2+\beta^2}e^{-\gamma t} + \dfrac{1}{\beta}\left[\dfrac{(a_0-\alpha)^2+\beta^2}{(\gamma-\alpha)^2+\beta^2}\right]^{1/2}e^{-\alpha t}\sin(\beta t + \psi)$ $\psi \triangleq \tan^{-1}\dfrac{\beta}{a_0-\alpha} - \tan^{-1}\dfrac{\beta}{\gamma-\alpha}$
$\dfrac{s^2+a_1 s+a_0}{(s+\gamma)[(s+\alpha)^2+\beta^2]}$	$\dfrac{\gamma^2-a_1\gamma+a_0}{(\alpha-\gamma)^2+\beta^2}e^{-\gamma t}$ $+\dfrac{1}{\beta}\left[\dfrac{(\alpha^2-\beta^2-a_1\alpha+a_0)^2+\beta^2(a_1-2\alpha)^2}{(\gamma-\alpha)^2+\beta^2}\right]^{1/2}e^{-\alpha t}\sin(\beta t+\psi)$ $\psi \triangleq \tan^{-1}\dfrac{\beta(a_1-2\alpha)}{\alpha^2-\beta^2-a_1\alpha+a_0} - \tan^{-1}\dfrac{\beta}{\gamma-\alpha}$
$\dfrac{1}{s(s+\gamma)[(s+\alpha)^2+\beta^2]}$	$\dfrac{1}{\gamma\beta_0^2} - \dfrac{1}{\gamma[(\alpha-\gamma)^2+\beta^2]}e^{-\gamma t}$ $+\dfrac{1}{\beta\beta_0[(\gamma-\alpha)^2+\beta^2]^{1/2}}e^{-\alpha t}\sin(\beta t - \psi)$ $\gamma \triangleq \tan^{-}\dfrac{\beta}{-\alpha} + \tan^{-1}\dfrac{\beta}{\gamma-\alpha}$ $\beta_0^2 \triangleq \alpha^2+\beta^2$
$\dfrac{s+a_0}{s(s+\gamma)[(s+\alpha)^2+\beta^2]}$	$\dfrac{a_0}{\gamma\beta_0^2} + \dfrac{\gamma-a_0}{\gamma[(\alpha-\gamma)^2+\beta^2]}e^{-\gamma t}$ $+\dfrac{1}{\beta\beta_0}\left[\dfrac{(a_0-\alpha)^2+\beta^2}{(\gamma-\alpha)^2+\beta^2}\right]^{1/2}e^{-\alpha t}\sin(\beta t+\psi)$ $\psi \triangleq \tan^{-1}\dfrac{\beta}{a_0-\alpha} - \tan^{-1}\dfrac{\beta}{\gamma-\alpha} - \tan^{-1}\dfrac{\beta}{-\alpha}$ $\beta_0^2 \triangleq \alpha^2+\beta^2$
$\dfrac{s^2+a_1 s+a_0}{(s+\gamma)(s+\delta)[(s+\alpha)^2+\beta^2]}$	$\dfrac{\gamma^2-a_1\gamma+a_0}{(\delta-\gamma)[(\alpha-\gamma)^2+\beta^2]}e^{-\gamma t} + \dfrac{\delta^2-a_1\delta+a_0}{(\gamma-\delta)[(\alpha-\delta)^2+\beta^2]}e^{-\delta t}$ $+\dfrac{1}{\beta}\left\{\dfrac{(\alpha^2-\beta^2-a_1\alpha+a_0)^2+\beta^2(a_1-2\alpha)^2}{[(\delta-\alpha)^2+\beta^2][(\gamma-\alpha)^2+\beta^2]}\right\}^{1/2}e^{-\alpha t}\sin(\beta t+\psi)$ $\psi \triangleq \tan^{-1}\dfrac{\beta(a_1-2\alpha)}{\alpha^2-\beta^2-a_1\alpha+a_0} - \tan^{-1}\dfrac{\beta}{\gamma-\alpha} - \tan^{-1}\dfrac{\beta}{\delta-\alpha}$

Table 16 (Continued)

Unilateral Laplace Function Transform Pairs (Continued)

$F(s)$	$f(t), 0 \leq t$
$\dfrac{1}{(s^2 + \lambda^2)[(s + \alpha)^2 + \beta^2]}$	$\dfrac{1}{[(\beta_0^2 - \lambda^2)^2 + 4\alpha^2\lambda^2]^{1/2}} \left[\dfrac{1}{\lambda} \sin(\lambda t - \psi_1) + \dfrac{1}{\beta} e^{-\alpha t} \sin(\beta t - \psi_2) \right]$

$$\psi_1 \overset{\Delta}{=} \tan^{-1} \frac{2\alpha\lambda}{\beta_0^2 - \lambda^2}$$

$$\psi_2 \overset{\Delta}{=} \frac{-2\alpha\beta}{\alpha^2 - \beta^2 + \lambda^2}; \quad \beta_0^2 \overset{\Delta}{=} \alpha^2 + \beta^2$$

$\dfrac{s + a_0}{(s^2 + \lambda^2)[(s + \alpha)^2 + \beta^2]}$	$\dfrac{1}{\lambda} \left[\dfrac{a_0^2 + \lambda^2}{(\beta_0^2 - \lambda^2)^2 + 4\alpha^2\lambda^2} \right]^{1/2} \sin(\lambda t + \psi_1)$

$$+ \frac{1}{\beta} \left[\frac{(a_0 - \alpha)^2 + \beta^2}{(\beta_0^2 - \lambda^2)^2 + 4\alpha^2\lambda^2} \right]^{1/2} e^{-\alpha t} \sin(\beta t + \psi_2)$$

$$\psi_1 \overset{\Delta}{=} \tan^{-1} \frac{\lambda}{a_0} - \tan^{-1} \frac{2\alpha\lambda}{\beta_0^2 - \lambda^2}$$

$$\psi_2 \overset{\Delta}{=} \tan^{-1} \frac{\beta}{a_0 - \alpha} - \tan^{-1} \frac{-2\alpha\beta}{\alpha^2 - \beta^2 + \lambda^2}$$

$$\beta^2 \overset{\Delta}{=} \alpha^2 + \beta_0^2$$

$\dfrac{s^2 + a_1 s + a_0}{(s^2 + \lambda^2)[(s + \alpha)^2 + \beta^2]}$	$\dfrac{1}{\lambda} \left[\dfrac{(a_0 - \lambda^2)^2 + a_1^2\lambda^2}{(\beta_0^2 - \lambda^2)^2 + 4\alpha^2\lambda^2} \right]^{1/2} \sin(\lambda t + \psi_1)$

$$+ \frac{1}{\beta} \left[\frac{(\alpha^2 - \beta^2 - a_1\alpha + a_0)^2 + \beta^2(a_1 - 2\alpha)^2}{(\beta_0^2 - \lambda^2)^2 + 4\alpha^2\lambda^2} \right]^{1/2} e^{-\alpha t} \sin(\beta t + \psi_2)$$

$$\psi_1 \overset{\Delta}{=} \tan^{-1} \frac{a_1\lambda}{a_0 - \lambda^2} - \tan^{-1} \frac{2\alpha\lambda}{\beta_0^2 - \lambda^2}$$

$$\psi_2 \overset{\Delta}{=} \tan^{-1} \frac{\beta(a_1 - 2\alpha)}{\alpha^2 - \beta^2 - a_1\alpha + a_0} - \tan^{-1} \frac{-2\alpha\beta}{\alpha^2 - \beta^2 + \lambda^2}$$

$$\beta_0^2 \overset{\Delta}{=} \alpha^2 + \beta^2$$

$\dfrac{s + a_0}{(s + \gamma)(s^2 + \lambda^2)}$ $\times \dfrac{1}{(s + \alpha)^2 + \beta^2}$	$\dfrac{a_0 - \gamma}{(\lambda^2 + \gamma^2)[(\alpha - \gamma)^2 + \beta^2]} e^{-\gamma t}$

$$+ \frac{1}{\lambda} \left\{ \frac{a_0^2 + \lambda^2}{(\gamma^2 + \lambda^2)[(\beta_0^2 - \lambda^2)^2 + 4\alpha^2\lambda^2]} \right\}^{1/2} \sin(\lambda t + \psi_1)$$

$$+ \frac{1}{\beta} \left\{ \frac{(a_0 - \alpha)^2 + \beta^2}{[(\gamma - \alpha)^2 + \beta^2][(\beta_0^2 - \lambda^2)^2 + 4\alpha^2\lambda^2]} \right\}^{1/2} e^{-\alpha t} \sin(\beta t + \psi_2)$$

$$\psi_1 \overset{\Delta}{=} \tan^{-1} \frac{\lambda}{a_0} - \tan^{-1} \frac{\lambda}{\gamma} - \tan^{-1} \frac{2\alpha\lambda}{\beta_0^2 - \lambda^2}$$

$$\gamma^2 \overset{\Delta}{=} \tan^{-1} \frac{\beta}{a_0 - \alpha} - \tan^{-1} \frac{\beta}{\gamma - \alpha} - \tan^{-1} \frac{-2\alpha\beta}{\alpha^2 - \beta^2 + \lambda^2}$$

$$\beta_0^2 \overset{\Delta}{=} \alpha^2 + \beta^2$$

(Continues)

Table 16 *(Continued)*

Unilateral Laplace Function Transform Pairs (Continued)

$F(s)$	$f(t), 0 \leq t$
$\dfrac{1}{s^2}$	t
$\dfrac{1}{s^n}$	$\dfrac{1}{(n-1)!}t^{n-1}$, n is a positive integer
$\dfrac{1}{(s+\alpha)s^2}$	$\dfrac{e^{-\alpha t}+\alpha t-1}{\alpha^2}$
$\dfrac{s+a_0}{(s+\alpha)s^2}$	$\dfrac{a_0-\alpha}{\alpha^2}e^{-\alpha t}+\dfrac{a_0}{\alpha}t+\dfrac{\alpha-a_0}{\alpha^2}$
$\dfrac{s^2+a_1 s+a_0}{(s+\alpha)s^2}$	$\dfrac{\alpha^2-a_1\alpha+a_0}{\alpha^2}e^{-\alpha t}+\dfrac{a_0}{\alpha}t+\dfrac{a_1\alpha-a_0}{\alpha^2}$
$\dfrac{1}{(s+\alpha)^2}$	$te^{-\alpha t}$
$\dfrac{s+a_0}{(s+\alpha)^2}$	$[(a_0-\alpha)t+1]e^{-\alpha t}$
$\dfrac{1}{(s+\alpha)^n}$	$\dfrac{1}{(n-1)!}t^{n-1}e^{-\alpha t}$, n is a positive integer
$\dfrac{s^n}{(s+\alpha)^{n+1}}$	$e^{-\alpha t}\displaystyle\sum_{k-0}^{n}\dfrac{n!(-\alpha)^k}{(n-k)!(k!)^2}t^k$, n is a nonnegative integer
$\dfrac{1}{s(s+\alpha)^2}$	$\dfrac{1-(1+\alpha t)e^{-\alpha t}}{\alpha^2}$
$\dfrac{s+a_0}{s(s+\alpha)^2}$	$\dfrac{a_0}{\alpha^2}+\left(\dfrac{\alpha-a_0}{\alpha}t-\dfrac{a_0}{\alpha^2}\right)e^{-\alpha t}$
$\dfrac{s^2+a_1 s+a_0}{s(s+\alpha)^2}$	$\dfrac{a_0}{\alpha^2}+\left(\dfrac{a_1\alpha-a_0-\alpha^2}{\alpha}t+\dfrac{\alpha^2-a_0}{\alpha^2}\right)e^{-\alpha t}$
$\dfrac{1}{(s+\gamma)(s+\alpha)^2}$	$\dfrac{1}{(\gamma-\alpha)^2}e^{-\gamma t}+\dfrac{(\gamma-\alpha)t-1}{(\gamma-\alpha)^2}e^{-\alpha t}$
$\dfrac{s+a_0}{(s+\gamma)(s+\alpha)^2}$	$\dfrac{a_0-\gamma}{(\alpha-\gamma)^2}e^{-\gamma t}+\left[\dfrac{a_0-\alpha}{\gamma-\alpha}t+\dfrac{\gamma-a_0}{(\gamma-\alpha)^2}\right]e^{-\alpha t}$
$\dfrac{s^2+a_1 s+a_0}{(s+\gamma)(s+\alpha)^2}$	$\dfrac{\gamma^2+a_1\gamma+a_0}{(\alpha-\gamma)^2}e^{-\gamma t}$
	$+\left[\dfrac{\alpha^2-a_1\alpha+a_0}{\gamma-\alpha}t+\dfrac{\alpha^2-2\alpha\gamma+a_1\gamma-a_0}{(\gamma-\alpha)^2}\right]e^{-\alpha t}$
$\dfrac{s+a_0}{(s+\gamma)(s+\alpha)^3}$	$\dfrac{a_0-\gamma}{(\alpha-\gamma)^3}e^{-\gamma t}+\left[\dfrac{a_0-\alpha}{2(\gamma-\alpha)}t^2+\dfrac{\gamma-a_0}{(\gamma-\alpha)^2}t+\dfrac{a_0-\gamma}{(\gamma-\alpha)^3}\right]e^{-\alpha t}$
$\dfrac{s+a_0}{s(s+\gamma)(s+\alpha)^2}$	$\dfrac{a_0}{\gamma\alpha^2}+\dfrac{\gamma-a_0}{\gamma(\alpha-\gamma)^2}e^{-\gamma t}+\left[\dfrac{a_0-\alpha}{\alpha(\alpha-\gamma)}t+\dfrac{2a_0\alpha-\alpha^2-a_0\gamma}{\alpha^2(\alpha-\gamma)^2}\right]e^{-\alpha t}$
$\dfrac{s^2+a_1 s+a_0}{s(s+\gamma)(s+\alpha)^2}$	$\dfrac{a_0}{\gamma\alpha^2}-\dfrac{\gamma^2-a_1\gamma+a_0}{\gamma(\alpha-\gamma)^2}e^{-\gamma t}$
	$+\left[\dfrac{\alpha^2-a_1\alpha+a_0}{\alpha(\alpha-\gamma)}t+\dfrac{(\gamma-a_1)\alpha^2+(2\alpha-\gamma)a_0}{\alpha^2(\alpha-\gamma)^2}\right]e^{-\alpha t}$

Table 16 (Continued)

Unilateral Laplace Function Transform Pairs (Continued)

$F(s)$	$f(t), 0 \le t$
$\dfrac{s + a_0}{(s + \gamma)(s + \delta)(s + \alpha)^2}$	$\dfrac{a_0 - \gamma}{(\delta - \gamma)(\alpha - \gamma)^2}e^{-\gamma t} + \dfrac{a_0 - \delta}{(\gamma - \delta)(\alpha - \delta)^2}e^{-\delta t}$ $+ \left[\dfrac{a_0 - \alpha}{(\gamma - \alpha)(\delta - \alpha)}t + \dfrac{2a_0\alpha - \alpha^2 - a_0(\gamma + \delta) + \gamma\delta}{(\gamma - \alpha)^2(\delta - \alpha)^2}\right]e^{-\alpha t}$
$\dfrac{s + a_0}{(s + \alpha)(s + \gamma)s^2}$	$\dfrac{a_0 - \alpha}{\alpha^2(\gamma - \alpha)}e^{-\alpha t} + \dfrac{a_0 - \gamma}{\gamma^2(\alpha - \gamma)}e^{\gamma - t} + \dfrac{a_0}{\alpha\gamma}t + \dfrac{\alpha\gamma - a_0(\alpha + \gamma)}{\alpha^2\gamma^2}$
$\dfrac{s^2 + a_1 s + a_0}{(s + \alpha)(s + \gamma)s^2}$	$\dfrac{\alpha^2 - a_1\alpha + a_0}{\alpha^2(\gamma - \alpha)}e^{-\alpha t} + \dfrac{\gamma^2 - a_1\gamma + a_0}{\gamma^2(\alpha - \gamma)}e^{-\gamma t} + \dfrac{a_0}{\alpha\gamma}t$ $+ \dfrac{a_1\alpha\gamma - a_0(\alpha + \gamma)}{\alpha^2\gamma^2}$
$\dfrac{s^2 + a_1 s + a_0}{(s + \alpha)^2 s^2}$	$\left[\dfrac{\alpha^2 - a_1\alpha + a_0}{\alpha^2}t + \dfrac{2a_0 - a_1\alpha}{\alpha^3}\right]e^{-\alpha t} + \dfrac{a_0}{\alpha^2}t + \dfrac{a_1\alpha - 2a_0}{\alpha^3}$
$\dfrac{s + a_0}{(s + \alpha)^2(s + \gamma)^2}$	$\left[\dfrac{a_0 - \alpha}{(\gamma - \alpha)^2}t + \dfrac{\alpha + \gamma - 2a_0}{(\gamma - \alpha)^3}\right]e^{-\alpha t}$ $+ \left[\dfrac{a_0 - \gamma}{(\alpha - \gamma)^2}t + \dfrac{\alpha + \gamma - 2a_0}{(\alpha - \gamma)^3}\right]e^{-\gamma t}$
$\dfrac{s^2 + a_1 s + a_0}{(s + \alpha)^2(s + \gamma)^2}$	$\left[\dfrac{\alpha^2 - a_1\alpha + a_0}{(\gamma - \alpha)^2}t + \dfrac{a_1(\alpha + \gamma) - 2(\alpha\gamma + a_0)}{(\gamma - \alpha)^3}\right]e^{-\alpha t}$ $+ \left[\dfrac{\gamma^2 - a_1\gamma + a_0}{(\gamma - \alpha)^2}t - \dfrac{a_1(\alpha + \gamma) - 2(\alpha\gamma + a_0)}{(\gamma - \alpha)^3}\right]e^{-\gamma t}$
$\dfrac{s^2 + a_1 s + a_0}{(s + \alpha)^3 s^2}$	$\left(\dfrac{\alpha^2 - a_1\alpha + a_0}{2\alpha^2}t^2 + \dfrac{-a_1\alpha + 2a_0}{\alpha^3}t + \dfrac{-a_1\alpha + 3a_0}{\alpha^4}\right)e^{-\alpha t}$ $+ \dfrac{a_0}{\alpha^3}t + \dfrac{a_1\alpha - 3a_0}{\alpha^4}$
$\dfrac{1}{(s^2 + \beta^2)s^2}$	$\dfrac{1}{\beta^2}t - \dfrac{1}{\beta^3}\sin\beta t$
$\dfrac{1}{(s^2 - \beta^2)s^2}$	$\dfrac{1}{\beta^3}\sinh\beta t - \dfrac{1}{\beta^2}t$
$\dfrac{s + a_0}{(s^2 + \beta^2)s^2}$	$\dfrac{a_0}{\beta^2}t + \dfrac{1}{\beta^2} - \dfrac{1}{\beta^3}(a_0^2 + \beta^2)^{1/2}\sin(\beta t + \psi)$ $\psi \triangleq \tan^{-1}\dfrac{\beta}{a_0}$

(Continues)

Table 16 (*Continued*)

Unilateral Laplace Function Transform Pairs (Continued)

$F(s)$	$f(t),\ 0 \le t$
$\dfrac{s^2 + a_1 s + a_0}{(s^2 + \beta^2)s^2}$	$\dfrac{a_0}{\beta^2}t + \dfrac{a_1}{\beta^2} - \dfrac{1}{\beta^3}[(a_0 - \beta^2)^2 + a_1^2\beta^2]^{1/2}\sin(\beta t + \psi)$
	$\psi \overset{\Delta}{=} \tan^{-1}\dfrac{a_1\beta}{a_0 - \beta^2}$
$\dfrac{1}{(s^2 + \beta^2)s^3}$	$\dfrac{1}{\beta^4}(\cos\beta t - 1) + \dfrac{1}{2\beta^2}t^2$
$\dfrac{1}{(s^2 - \beta^2)s^3}$	$\dfrac{1}{\beta^4}(\cosh\beta t - 1) - \dfrac{1}{2\beta^2}t^2$
$\dfrac{1}{(s^2 + \beta^2)(s + \alpha)^2}$	$\dfrac{1}{\beta(\alpha^2 + \beta^2)}\sin(\beta t - \psi) + \left[\dfrac{1}{a^2 + \beta^2}t + \dfrac{2\alpha}{(\alpha^2 + \beta^2)^2}\right]e^{-\alpha t}$
	$\psi \overset{\Delta}{=} 2\tan^{-1}\dfrac{\beta}{\alpha}$
$\dfrac{s + a_0}{(s^2 + \beta^2)(s + \alpha)^2}$	$\dfrac{(a_0^2 + \beta^2)^{1/2}}{\beta(\alpha^2 + \beta^2)}\sin(\beta t + \psi) + \left[\dfrac{a_0 - \alpha}{\alpha^2 + \beta^2}t + \dfrac{2a_0\alpha + \beta^2 - \alpha^2}{(\alpha^2 + \beta^2)^2}\right]e^{-\alpha t}$
	$\psi \overset{\Delta}{=} \tan^{-1}\dfrac{\beta}{a_0} - 2\tan^{-1}\dfrac{\beta}{\alpha}$
$\dfrac{s^2 + a_1 s + a_0}{(s^2 + \beta^2)(s + \alpha)^2}$	$\dfrac{[(a_0 - \beta^2)^2 + a_1^2\beta^2]^{1/2}}{\beta(\alpha^2 + \beta^2)}\sin(\beta t + \psi)$
	$\quad + \left[\dfrac{\alpha^2 - a_1\alpha + a_0}{\alpha^2\beta^2}t + \dfrac{a_1(\beta^2 - \alpha^2) + 2\alpha(a_0 - \beta^2)}{(\alpha^2 + \beta^2)^2}\right]e^{-\alpha t}$
	$\psi \overset{\Delta}{=} \tan^{-1}\dfrac{a_1\beta}{a_0 - \beta^2} - 2\tan^{-1}\dfrac{\beta}{\alpha}$
$\dfrac{s + a_0}{s(s^2 + \beta^2)(s + \alpha)^2}$	$\dfrac{a_0}{\beta^2\alpha^2} - \dfrac{(a_0^2 + \beta^2)^{1/2}}{\beta^2(\alpha^2 + \beta^2)}\cos(\beta t + \psi)$
	$\quad + \left[\dfrac{\alpha - a_0}{\alpha(\alpha^2 + \beta^2)}t + \dfrac{2\alpha^3 - 3a_0\alpha^2 - a_0\beta^2}{\alpha^2(\alpha^2 + \beta^2)^2}\right]e^{-\alpha t}$
	$\psi \overset{\Delta}{=} \tan^{-1}\dfrac{\beta}{a_0} - 2\tan^{-1}\dfrac{\beta}{\alpha}$
$\dfrac{s^2 + a_1 s + a_0}{(s + \gamma)(s^2 + \beta^2)(s + \alpha)^2}$	$\dfrac{\gamma^2 - a_1\gamma + a_0}{(\gamma^2 + \beta^2)(\alpha - \gamma)^2}e^{-\gamma t} + \dfrac{[(a_0 - \beta^2)^2 + a^2\beta^2]^{1/2}}{\beta(\gamma^2 + \beta^2)^{1/2}(\alpha^2 + \beta^2)}\sin(\beta t + \psi)$
	$\quad + \dfrac{\alpha^2 - a_1\alpha + a_0}{(\gamma - \alpha)(\alpha^2 + \beta^2)}te^{-\alpha t}$
	$\quad + \dfrac{(\gamma - \alpha)(\alpha^2 + \beta^2)(a_1 - 2\alpha) - (\alpha^2 - a_1\alpha + a_0)(3\alpha^2 + \beta^2 - 2\alpha\gamma)}{(\gamma - \alpha)^2(\alpha^2 + \beta^2)^2}$
	$\quad \times e^{-\alpha t}$
	$\psi \overset{\Delta}{=} \tan^{-1}\dfrac{a_1\beta}{a_0 - \beta^2} - \tan^{-1}\dfrac{\beta}{\gamma} - 2\tan^{-1}\dfrac{\beta}{\alpha}$

Table 16   (Continued)

Unilateral Laplace Function Transform Pairs (Continued)

$F(s)$	$f(t), 0 \le t$
$\dfrac{1}{(s^2+\beta^2)^2}$	$\dfrac{1}{2\beta^3}(\sin\beta t - \beta t\cos\beta t)$
$\dfrac{s}{(s^2+\beta^2)^2}$	$\dfrac{1}{2\beta}t\sin\beta t$
$\dfrac{s^2}{(s^2+\beta^2)^2}$	$\dfrac{1}{2\beta}(\sin\beta t + \beta t\cos\beta t)$
$\dfrac{s^2-\beta^2}{(s^2+\beta^2)^2}$	$t\cos\beta t$
$\dfrac{1}{s(s^2+\beta^2)^2}$	$\dfrac{1}{\beta^4}(1-\cos\beta t) - \dfrac{1}{2\beta^3}t\sin\beta t$
$\dfrac{s^2+a_1 s+a_0}{s(s^2+\beta^2)^2}$	$\dfrac{a_0}{\beta^4} - \dfrac{[(a_0-\beta^2)^2+a_1^2\beta^2]^{1/2}}{2\beta^3}t$
	$\quad \sin(\beta t+\psi_1) - \dfrac{(4a_0^2+a_1^2\beta^2)^{1/2}}{2\beta^4}\cos(\beta t+\psi_2)$
	$\psi_1 \overset{\Delta}{=} \tan^{-1}\dfrac{a_1\beta}{a_0-\beta^2}$
	$\psi_2 \overset{\Delta}{=} \tan^{-1}\dfrac{a_1\beta}{2a_0}$
$\dfrac{1}{[(s+\alpha)^2+\beta^2]s^2}$	$\dfrac{1}{\beta_0^2}\left[t - \dfrac{2\alpha}{\beta_0^2} + \dfrac{1}{\beta}e^{-\alpha t}\sin(\beta t-\psi)\right]$
	$\psi \overset{\Delta}{=} 2\tan^{-1}\dfrac{\beta}{-\alpha}$
	$\beta_0^2 \overset{\Delta}{=} \alpha^2+\beta^2$
$\dfrac{1}{(s+\gamma)^2[(s+\alpha)^2+\beta^2]}$	$\dfrac{1}{(\alpha-\gamma)^2+\beta^2}\left[te^{-\gamma t} + \dfrac{2(\gamma-\alpha)}{(\alpha-\gamma)^2+\beta^2}e^{-\gamma t} + \dfrac{1}{\beta}e^{-\alpha t}\sin(\beta t-\psi)\right]$
	$\psi \overset{\Delta}{=} 2\tan^{-1}\dfrac{\beta}{\gamma-\alpha}$
$\dfrac{s^2+a_1 s+a_0}{(s+\gamma)^2[(s+\alpha)^2+\beta^2]}$	$\dfrac{\gamma^2-a_1\gamma+a_0}{(\alpha-\gamma)^2+\beta^2}te^{-\gamma t}$
	$\quad + \dfrac{[(\alpha-\gamma)^2+\beta^2](a_1-2\gamma)-2(\alpha-\gamma)(\gamma^2-a_1\gamma+a_0)}{[(\alpha-\gamma)^2+\beta^2]^2}e^{-\gamma t}$
	$\quad + \dfrac{[(\alpha^2-\beta^2-a_1\alpha+a_0)^2+\beta^2(a_1-2\alpha)^2]^{1/2}}{\beta[(\gamma-\alpha)^2+\beta^2]}e^{-\alpha t}\sin(\beta t+\psi)$
	$\psi \overset{\Delta}{=} \tan^{-1}\dfrac{\beta(a_1-2\alpha)}{\alpha^2-\beta^2-a_1\alpha+a_0} - 2\tan^{-1}\dfrac{\beta}{\gamma-\alpha}$

(Continues)

Table 16   (*Continued*)

Unilateral Laplace Function Transform Pairs (Continued)

$F(s)$	$f(t), 0 \le t$
$\dfrac{1}{[(s+\alpha)^2 + \beta^2]^2}$	$\dfrac{1}{2\beta^3} e^{-\alpha t}(\sin \beta t - \beta t \cos \beta t)$
$\dfrac{s+\alpha}{[(s+\alpha)^2 + \beta^2]^2}$	$\dfrac{1}{2\beta} te^{-\alpha t}\sin \beta t$
$\dfrac{s^2 + a_0}{[(s+\alpha)^2 + \beta^2]^2}$	$\dfrac{\beta_0^2 + a_0}{2\beta^3} e^{-\alpha t}\sin \beta t - \dfrac{[(\alpha^2 - \beta^2 + a_0)^2 + 4\alpha^2\beta^2]^{1/2}}{2\beta^2} te^{-\alpha t}\cos(\beta t + \psi)$
	$\psi \triangleq \tan^{-1}\dfrac{-2\alpha\beta}{\alpha^2 - \beta^2 + a_0}$
	$\beta_0^2 \triangleq \alpha^2 + \beta^2$
$\dfrac{(s+\alpha)^2 - \beta^2}{[(s+\alpha)^2 + \beta^2]^2}$	$te^{-\alpha t}\cos \beta t$
$\tan^{-1}\dfrac{\beta}{s}$	$\dfrac{\sin \beta t}{t}$
$\ln\dfrac{s+\beta}{s+\alpha}$	$\dfrac{e^{-\alpha t} - e^{-\beta t}}{t}$
$e^{s^2/4a}\operatorname{cerf}\dfrac{s}{2\sqrt{a}}$	$2\sqrt{\dfrac{a}{\pi}}e^{-\alpha t^2}$
	$\operatorname{cerf} y \triangleq 1 - \operatorname{erf} y \triangleq 1 - \dfrac{2}{\sqrt{\pi}}\displaystyle\int_0^y e^{-x^2}\,dx$
$\dfrac{1}{\sqrt{s^2 + \alpha^2}}$	$J_0(\alpha t)$
$\dfrac{1}{\sqrt{s^2 + \alpha^2}(\sqrt{s^2 + \alpha^2} + s)}$	$\dfrac{1}{\alpha}J_1(\alpha t)$
$\dfrac{1}{\sqrt{s^2 + \alpha^2}(\sqrt{s^2 + \alpha^2} + s)^n}$	$\dfrac{1}{\alpha^n}J_n(\alpha t)$, n is a nonnegative integer
$\dfrac{1}{s\sqrt{s^2 + \alpha^2}(\sqrt{s^2 + \alpha^2} + s)^n}$	$\dfrac{1}{\alpha^n}\displaystyle\int_0^t J_n(\alpha t)\,dt$, n is a nonnegative integer
$\dfrac{1}{\sqrt{s^2 + \alpha^2} + s}$	$\dfrac{1}{\alpha}\dfrac{J_1(\alpha t)}{t}$
$\dfrac{1}{(\sqrt{s^2 + \alpha^2} + s)^n}$	$\dfrac{n}{\alpha^n}\dfrac{J_n(\alpha t)}{t}$, n is a positive integer
$\dfrac{1}{s(\sqrt{s^2 + \alpha^2} + s)^n}$	$\dfrac{n}{\alpha^n}\displaystyle\int_0^t \dfrac{J_n(\alpha t)}{t}\,dt$, n is a positive integer
$\dfrac{1}{s}e^{-as}$	$u(t-a)$
$\dfrac{1}{s^2}e^{-as}$	$(t-a)u(t-a)$
$\left(\dfrac{a}{s} + \dfrac{1}{s^2}\right)e^{-as}$	$tu(t-a)$
$\left(\dfrac{2}{s^3} + \dfrac{2a}{s^2} + \dfrac{a^2}{s}\right)e^{-as}$	$t^2 u(t-a)$

Table 16   *(Continued)*

Unilateral Laplace Function Transform Pairs (Continued)

$F(s)$	$f(t), 0 \leq t$
$\dfrac{1}{s}(e^{-as} - e^{-bs})$ $a < b$	$u(t - a) - u(t - b)$
$\left(\dfrac{1 - e^{-s}}{s}\right)^2$	$\begin{cases} t, & 0 < t < 1 \\ 2 - t, & 1 < t < 2 \\ 0, & 2 < t \end{cases}$
$\left(\dfrac{1 - e^{-s}}{s}\right)^3$	$\begin{cases} 0.5t^2, & 0 < t < 1 \\ 0.75 - (t - 1.5)^2, & 1 < t < 2 \\ 0.5(t - 3)^2, & 2 < t < 3 \\ 0, & 3 < t \end{cases}$
$\dfrac{1}{s^2}(1 - e^{-s})$	$\begin{cases} t, & 0 < t < 1 \\ 1, & 1 < t \end{cases}$
$\dfrac{1}{s^3}(1 - e^{-s})^2$	$\begin{cases} 0.5t^2, & 0 < t < 0 \\ 1 - 0.5(t - 2)^2, & 1 < t < 2 \\ 1, & 2 < t \end{cases}$
$\dfrac{1}{s(1 + e^{-s})}$	$\displaystyle\sum_{k=0}^{\infty}(-1)^k u(t - k)$
$\dfrac{1}{s \sinh s}$	$\displaystyle 2\sum_{k=0}^{\infty} u(t - 2k - 1)$
$\dfrac{1}{s \cosh s}$	$\displaystyle 2\sum_{k=0}^{\infty}(-1)^k u(t - 2k - 1)$
$\dfrac{1}{s} \tanh s$	$\displaystyle u(t) + 2\sum_{k=1}^{\infty}(-1)^k u(t - 2k)$
	or $\displaystyle\sum_{k=0}^{\infty}(-1)^k u(t - 2k)u(2k + 2 - t)$
$\dfrac{e^s - s - 1}{s^2(e^s - 1)}$	$\displaystyle t - \sum_{k=1}^{\infty} u(t - k)$
	or $\displaystyle\sum_{k=0}^{\infty}(t - k)u(t - k)u(k + 1 - t)$

[a] $\alpha, \beta, \gamma, \delta$, and λ are real numbers.

Source: From M. F. Gardner and J. L. Barnes, *Transients in Linear Systems*, Wiley, New York, 1942. Used with permission.

1. If $a_1 + ia_2 = 0$, then $a_1 = 0, a_2 = 0$.
2. If $a_1 + ia_2 = b_1 + ib_2$, then $a_1 = b_1, a_2 = b_2$.
3. $a_1 + ia_2$ and $a_1 - ia_2$ are *conjugate* complex numbers. The complex conjugate of A is $\overline{A}$ or A^*.
4. $A + B = (a_1 + ia_2) + (b_1 + ib_2) = (a_1 + b_1) + i(a_2 + b_2)$.
5. $a_1 + ia_2 = |A|(\cos \angle A + i \sin \angle A) = |A|e^{i\angle A}$ and $a_1 - ia_2 = |A|(\cos \angle A - i \sin \angle A) = |A|e^{-i\angle A}$, where $|A| = \sqrt{a_1^2 + a_2^2}$, $\sin \angle A = a_2/|A|$, $\cos \angle A = a_1/|A|$, $|A|$ is the *absolute value* (*modulus*), and $\angle A$ is the *angle* of A.
6. $AB = (a_1 + ia_2)(b_1 + ib_2) = (a_1b_1 - a_2b_2) + i(a_2b_1 + a_1b_2) = |A||B|e^{i(\angle A + \angle B)}$.
7. $A\overline{A} = (a_1 + ia_2)(a_1 - ia_2) = a_1^2 + a_2^2 = |A|^2$.
8. $\dfrac{A}{B} = \dfrac{a_1 + ia_2}{b_1 + ib_2} = \dfrac{(a_1 + ia_2)(b_1 - ib_2)}{b_1 + ib_2)(b_1 - ib_2)}$

$= \dfrac{a_1b_1 + a_2b_2}{b_1^2 + b_2^2} + i\dfrac{a_2b_1 - a_1b_2}{b_1^2 + b_2^2}$

$= \dfrac{|A|}{|B|}e^{i(\angle A - \angle B)}$.

9. $A^n = (a_1 + ia_2)^n = [|A|(\cos \angle A + i \sin \angle A)]^n$

$= |A|^n e^{in\angle A}$

$= |A|^n(\cos n\angle A + i \sin n\angle A)$.

$\overline{A}^n = (a_1 - ia_2)^n = [|A|(\cos \angle A - i \sin \angle A)]^n$

$= |A|^n e^{-in\angle A}$

$= |A|^n(\cos n\angle A - i \sin n\angle A)$.

10. $\sqrt[n]{A} = \sqrt[n]{a_1} + ia_2$

$= \sqrt[n]{|A|}\left(\cos \dfrac{\angle A + 2k\pi}{n}\right.$

$\left. +i \sin \dfrac{\angle A + 2k\pi}{n}\right)$

$= R|A|ne^{i\angle A + 2k\pi/n}$,

where k is an integer. For $k = 0, 1, 2, \ldots, n - 1$, all of the n roots are obtained.

14.2   Complex Variables

Analytic Functions of a Complex Variable.   A function $w = f(z), z = x + iy$, which has a derivative

$$\frac{df}{dz} = f'(z) = \lim_{h \to 0} \frac{f(z + h) - f(z)}{h}$$

at a point z independent of the manner of approach of $z + h$ to z, is *analytic* at z and may be expanded in a convergent power series there. A function that is analytic at every point of a region is *analytic in the region*. If $f(z) = w = u(x, y) + iv(x, y)$ and $f(z)$ is

analytic at z, then the Cauchy–Riemann differential equations

$$\frac{\partial u}{\partial x} = \frac{\partial v}{\partial y}, \frac{\partial u}{\partial y} = -\frac{\partial v}{\partial x}$$

hold at z. If in a neighborhood of a point z these four partial derivatives exist and are continuous and if the Cauchy–Riemann equations hold, then $f(z)$ is analytic at z. The functions u and v satisfy Laplace's equation

$$\frac{\partial^2 \phi}{\partial x^2} + \frac{\partial^2 \phi}{\partial y^2} = 0$$

Example 95.   Examples of analytic functions are $z, 1/z, e^z$, and $\sin z$. An example of a nonanalytic function is $w = x - iy$.

Conformal Mapping.   The function $w = f(z)$, analytic in a region R_z of the z plane, *conformally* maps each point in R_z on a point of the w plane in the region R_w if $f'(z) \neq 0$ at all points of R_z. This mapping is also *isogonal*, that is, the angle between two curves starting at z_0 is equal to the angle between their mapped curves starting at w_0.

Example 96

1. $w = z + b, b$ complex, is a *translation* of magnitude $|b|$ in the direction $\angle b$.
2. $w = az, a$ complex, is a *rotation* through $\angle a$ and a *magnification* by $|a|$.
3. $w = az + b, a, b$ complex, the *integral linear transformation*, is a combination of 1 and 2.
4. $w = 1/z$, the *inversion transformation*, carries the origin of the z plane into the point at infinity in the *enlarged w* plane.
5. $w = (az + b)/(cz + d), ad - bc \neq 0$, the *general linear* or *bilinear transformation*, can be resolved into two linear integral and one inversion transformations.

Integrals of Analytic Functions.   If $dF(z)/dz = f(z)$ in a simply connected region R_z, then $F(z) = \int f(z) dz$ is analytic throughout R_z. If $f_1(z), f_2(z)$ are analytic in R_z and the path of integration is in R_z, then:

1. $\int_{z_0}^{z_0} f_1(z) dz = 0$
2. $\int_{z_0}^{z_1} [k_1 f_1(z) + k_2 f_2(z)] dz = k_1 \int_{z_0}^{z_1} f_1(z) dz + k_2 \int_{z_0}^{z_1} f_2(z) dz$
3. $\int_{z_1}^{z_0} f_1(z) dz = -\int_{z_0}^{z_1} f_1(z) dz$
4. $\int_{z_0}^{z_1} f_1(z) dz + \int_{z_1}^{z_2} f_1(z) dz = \int_{z_0}^{z_2} f_1(z) dz$

Cauchy's Integral Theorem.   If $f(z)$ is analytic and single valued on and within a simple closed contour C, then $\int_C f(z) dz = 0$.

A *contour* is a continuous curve made up of a finite number of elementary arcs.

If $f(z)$ is continuous on a simple closed contour C and analytic in the region bounded by C, then *Cauchy's integral formula*

$$f(z) = \frac{1}{2\pi i} \int_C \frac{f(\zeta)}{\zeta - z}\, d\zeta$$

holds; also

$$f^{(n)}(z) = \frac{n!}{2\pi i} \int_C \frac{f(\zeta)}{(\zeta - z)^{n+1}}\, d\zeta$$

Laurent Series. A function $f(z)$ has a *zero of order* n at z_1 if it can be put in the form $f(z) = (z - z_1)^n f_1(z)$, n a positive integer, $f_1(z_1) \neq 0$. A function $f(z)$ has a *pole of order n* at z_1 if it can be put in the form $f(z) = f_2(z)/(z - z_1)^n$, n a positive integer, $f_2(z_1) \neq 0$.

If $f(z)$ is analytic in a ring between and on two concentric circles C_1 and C_2 with radii R_1 and R_2, $R_1 < R_2$, and center z_1, then the *Laurent series*

$$f(z) = \sum_{n=-\infty}^{\infty} c_n (z - z_1)^n \qquad R_1 < |z - z_1| < R_2$$

is convergent everywhere in the ring, and

$$c_n = \frac{1}{2\pi i} \int_C \frac{f(z)}{(z - z_1)^{n+1}}\, dz$$

C is circle $|z - z_1| = r$, $R_1 < r < R_2$.

If a single-valued analytic function $f(z)$ is expanded in a Laurent series in the neighborhood of an isolated singularity z_1, then the *residue* of $f(z)$ at z_1 is

$$c_{-1} = \frac{1}{2\pi i} \int_C f(z)\, dz$$

C is any circle with center at z_1 that excludes all other singularities of $f(z)$.

A function is *meromorphic* in a region if it is analytic in the region except for a finite number of poles. If $f(z)$ is analytic on and inside a contour C, except for a finite number of poles, and has no zeros on C, then

$$\frac{1}{2\pi i} \int_C \frac{f'(z)}{f(z)}\, dz = N - P$$

N the total order of the zeros and P the total order of the poles within the contour.

15   VECTOR ANALYSIS

15.1   Vector Algebra

A *scalar* is a quantity that has magnitude, such as mass, density, and temperature. A *vector* is a quantity that has magnitude and direction, such as force, velocity, and acceleration. A vector may be represented geometrically by an oriented line segment.

Two vectors $\mathbf{A}$ and $\mathbf{B}$ are equal if they have the same magnitude and direction. A vector may be displaced parallel to itself provided it retains the same magnitude and direction. A vector having the same magnitude but direction opposite to that of $\mathbf{A}$ is the negative of $\mathbf{A}$ and is written $-\mathbf{A}$. If $\mathbf{A}$ is a vector of magnitude, or length, a, then $|\mathbf{A}| = a$. A vector parallel to $\mathbf{A}$ but with magnitude equal to the reciprocal of the magnitude of $\mathbf{A}$ is written $\mathbf{A}^{-1} = 1/\mathbf{A}$. A unit vector $\mathbf{A}/|\mathbf{A}|(\mathbf{A} \neq 0)$ has the direction of $\mathbf{A}$ and magnitude 1.

The *sum* of two vectors $\mathbf{A}$ and $\mathbf{B}$ is $\mathbf{A} + \mathbf{B}$ (Fig. 164). Similarly, the sum of three or more vectors can be found by adding them end to end.

The sum of $\mathbf{A}$ and $-\mathbf{B}$ is $\mathbf{A} - \mathbf{B}$ (Fig. 164), the *difference* of two vectors.

Let $\mathbf{A}$, $\mathbf{B}$, $\mathbf{C}$ be vectors and p, q scalars.

$p\mathbf{A} = \mathbf{A}p$, a vector p times as long as $\mathbf{A}$ with the same direction as $\mathbf{A}$ if p is positive and opposite if p is negative,

$$(p + q)\mathbf{A} = p\mathbf{A} + q\mathbf{A} \qquad p(\mathbf{A} + \mathbf{B}) = p\mathbf{A} + p\mathbf{B}$$

$$\mathbf{A} + \mathbf{B} = \mathbf{B} + \mathbf{A}$$

$$\mathbf{A} + (\mathbf{B} + \mathbf{C}) = (\mathbf{A} + \mathbf{B}) + \mathbf{C}$$

$|\mathbf{A} + \mathbf{B}| \leq |\mathbf{A}| + |\mathbf{B}|$, where the equality sign holds only for $\mathbf{A}$ parallel to $\mathbf{B}$.

Rectangular Coordinates. Figure 165 shows a right-hand coordinate system. Let $\mathbf{i}$, $\mathbf{j}$, $\mathbf{k}$ be unit vectors with the directions OX, OY, OZ, respectively. The vector $\mathbf{R}$ with initial point O and endpoint $P(x, y, z)$ can be expressed as the sum of its components,

$$\mathbf{R} = \mathbf{i}x + \mathbf{j}y + \mathbf{k}z$$

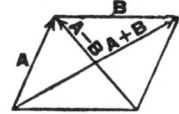

Fig. 164

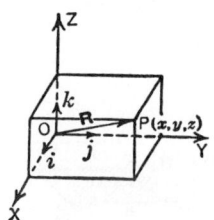

Fig. 165

If $\mathbf{A} = \mathbf{i}a_1 + \mathbf{j}a_2 + \mathbf{k}a_3$ and $\mathbf{B} = \mathbf{i}b_1 + \mathbf{j}b_2 + \mathbf{k}b_3$, then

$$\mathbf{A} + \mathbf{B} = \mathbf{i}(a_1 + b_1) + \mathbf{j}(a_2 + b_2) + \mathbf{k}(a_3 + b_3)$$

The *scalar, inner,* or *dot product* of two vectors $\mathbf{A}$ and $\mathbf{B}$ is $\mathbf{A} \cdot \mathbf{B} = |\mathbf{A}||\mathbf{B}| \cos \theta$ (Fig. 166),

$$\mathbf{A} \cdot \mathbf{B} = \mathbf{B} \cdot \mathbf{A}$$

$$\mathbf{A} \cdot (\mathbf{B} + \mathbf{C}) = \mathbf{A} \cdot \mathbf{B} + \mathbf{A} \cdot \mathbf{C}$$

$$\mathbf{A} \cdot \mathbf{A} = \mathbf{A}^2 = |\mathbf{A}|^2$$

$$\mathbf{i} \cdot \mathbf{i} = \mathbf{j} \cdot \mathbf{j} = \mathbf{k} \cdot \mathbf{k} = 1$$

$$\mathbf{i} \cdot \mathbf{j} = \mathbf{j} \cdot \mathbf{k} = \mathbf{k} \cdot \mathbf{i} = 0$$

If $\mathbf{A} \cdot \mathbf{B} = 0$, then either $\mathbf{A} = 0, \mathbf{B} = 0$, or $\mathbf{A}$ is perpendicular to $\mathbf{B}$.

If $\mathbf{A} = \mathbf{i}a_1 + \mathbf{j}a_2 + \mathbf{k}a_3$ and $\mathbf{B} = \mathbf{i}b_1 + \mathbf{j}b_2 + \mathbf{k}b_3$, then $\mathbf{A} \cdot \mathbf{B} = a_1b_1 + a_2b_2 + a_3b_3$.

The *vector, restricted outer,* or *cross product* of two vectors $\mathbf{A}$ and $\mathbf{B}$ is $\mathbf{A} \times \mathbf{B} = \mathbf{C}$, where $\mathbf{C}$ is perpendicular to the plane of $\mathbf{A}$ and $\mathbf{B}$ with the magnitude $|\mathbf{C}| = |\mathbf{A}||\mathbf{B}| \sin \theta$ (the area of the parallelogram made by $\mathbf{A}$ and $\mathbf{B}$, Fig. 167) and so directed that a right-hand rotation of less than $180°$ carries $\mathbf{A}$ into $\mathbf{B}$,

$$\mathbf{A} \times \mathbf{B} = -\mathbf{B} \times \mathbf{A}$$

$$\mathbf{A} \times (\mathbf{B} + \mathbf{C}) = \mathbf{A} \times \mathbf{B} + \mathbf{A} \times \mathbf{C}$$

$$(\mathbf{B} + \mathbf{C}) \times \mathbf{A} = \mathbf{B} \times \mathbf{A} + \mathbf{C} \times \mathbf{A}$$

$$\mathbf{i} \times \mathbf{i} = \mathbf{j} \times \mathbf{j} = \mathbf{k} \times \mathbf{k} = 0$$

$$\mathbf{i} \times \mathbf{j} = \mathbf{k} = -\mathbf{j} \times \mathbf{i}$$

$$\mathbf{j} \times \mathbf{k} = \mathbf{i} = -\mathbf{k} \times \mathbf{j}$$

$$\mathbf{k} \times \mathbf{i} = \mathbf{j} = -\mathbf{i} \times \mathbf{k}$$

If $\mathbf{A} \times \mathbf{B} = 0$, then either $\mathbf{A} = 0, \mathbf{B} = 0$, or $\mathbf{A}$ is parallel to $\mathbf{B}$. If $\mathbf{A} = \mathbf{i}a_1 + \mathbf{j}a_2 + \mathbf{k}a_3$ and $\mathbf{B} = \mathbf{i}b_1 + \mathbf{j}b_2 + $

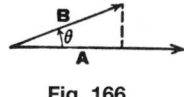

Fig. 166

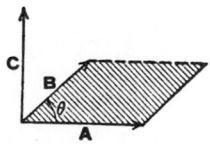

Fig. 167

kb_3, then

$$\mathbf{A} \times \mathbf{B} = \begin{vmatrix} \mathbf{i} & \mathbf{j} & \mathbf{k} \\ a_1 & a_2 & a_3 \\ b_1 & b_2 & b_3 \end{vmatrix}$$

If $\mathbf{A} = \mathbf{i}a_1 + \mathbf{j}a_2 + \mathbf{k}a_3, \mathbf{B} = \mathbf{i}b_1 + \mathbf{j}b_2 + \mathbf{k}b_3, \mathbf{C} = \mathbf{i}c_1 + \mathbf{j}c_2 + \mathbf{k}c_3$, then the *scalar triple product*

$$\mathbf{A} \cot(\mathbf{B} \times \mathbf{C}) = (\mathbf{A} \times \mathbf{B}) \cdot \mathbf{C} = \mathbf{B} \cdot (\mathbf{C} \times \mathscr{A})$$

$$= (\mathbf{ABC}) = \begin{vmatrix} a_1 & a_2 & a_3 \\ b_1 & b_2 & b_3 \\ c_1 & c_2 & c_3 \end{vmatrix}$$

and is equal to the volume of a parallelepiped whose three determining edges are $\mathbf{A}, \mathbf{B}, \mathbf{C}$,

$$(\mathbf{A} \times \mathbf{B}) \times \mathbf{C} = (\mathbf{A} \cdot \mathbf{C})\mathbf{B} - (\mathbf{B} \cdot \mathbf{C})\mathbf{A}$$

$$= -\mathbf{C} \times (\mathbf{A} \times \mathbf{B})$$

$$(\mathbf{A} \times \mathbf{B}) \cdot (\mathbf{C} \times \mathscr{D}) = (\mathbf{A} \cdot \mathbf{C})(\mathbf{B} \cdot \mathbf{D})$$

$$- (\mathbf{A} \cdot \mathbf{D})(\mathbf{B} \cdot \mathbf{C})$$

15.2 Differentiation and Integration of Vectors

Differentiation. A vector function of one or more scalar variables is called a *variable vector* or *field vector*. The derivative is

$$\frac{d\mathbf{F}}{dt} = \mathbf{F}'(t) = \lim_{\Delta t \to 0} \frac{\mathbf{F}(t + \Delta t) - \mathbf{F}(t)}{\Delta t}$$

$$= \lim_{\Delta t \to 0} \frac{\Delta \mathbf{F}}{\Delta t} \text{ (Fig. 168)}$$

If the length of $\mathbf{F}$ remains unaltered, then $\mathbf{F} \cdot d\mathbf{F} = 0$. If the direction of $\mathbf{F}$ remains unaltered, then $\mathbf{F} \times d\mathbf{F} = 0$,

$$d(\mathbf{A} + \mathbf{B}) = d\mathbf{A} + d\mathbf{B}$$

$$d(\mathbf{A} \cdot \mathbf{B}) = \mathbf{A} \cdot d\mathscr{B} + \mathscr{B} \cdot d\mathscr{A}$$

$$d(\mathbf{A} \times \mathbf{B}) = d\mathbf{A} \times \mathbf{B} + \mathbf{A} \times d\mathbf{B}$$

$$= \mathbf{A} \times d\mathbf{B} - \mathbf{B} \times d\mathbf{A}$$

$$d(\mathbf{A} \cdot \mathbf{B} \cdot \mathbf{C}) = \mathbf{A} \cdot \mathbf{B} \cdot d\mathscr{C} + \mathscr{B} \cdot \mathscr{C} \cdot d\mathscr{A}$$

$$+ \mathbf{C} \cdot \mathbf{A} \cdot d\mathscr{B}$$

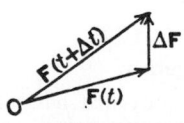

Fig. 168

Derivative Operators

$$\nabla = \text{del} = \mathbf{i}\frac{\partial}{\partial x} + \mathbf{j}\frac{\partial}{\partial y} + \mathbf{k}\frac{\partial}{\partial z}$$

$$\nabla^2 = \text{Laplacian} = \frac{\partial^2}{\partial x^2} + \frac{\partial^2}{\partial y^2} + \frac{\partial^2}{\partial z^2}$$

If V is a scalar function, then

$$\nabla V = \text{grad}\, V = \mathbf{i}\frac{\partial V}{\partial x} + \mathbf{j}\frac{\partial V}{\partial y} + \mathbf{k}\frac{\partial V}{\partial z}$$

If $\mathbf{A}$ is a vector function with components $\mathbf{A}_x, \mathbf{A}_y, \mathbf{A}_z$, then

$$\nabla \cdot \mathbf{A} = \text{div } \mathbf{A} = \frac{\partial \mathbf{A}_x}{\partial x} + \frac{\partial \mathbf{A}_y}{\partial y} + \frac{\partial \mathbf{A}_z}{\partial z}$$

$$\nabla \times \mathbf{A} = \text{curl } \mathbf{A} = \text{rot } \mathbf{A} = \begin{vmatrix} \mathbf{i} & \mathbf{j} & \mathbf{k} \\ \frac{\partial}{\partial x} & \frac{\partial}{\partial y} & \frac{\partial}{\partial z} \\ \mathbf{A}_x & \mathbf{A}_y & \mathbf{A}_z \end{vmatrix}$$

Formulas for Differentiation.

Let U and V be scalar functions and $\mathbf{A}$ and $\mathbf{B}$ be vector functions of x, y, z. Then (see *Advanced Engineering Analysis* by J. N. Reddy and M. L. Rasmussen, Wiley, New York, 1982)

$$\nabla(U + V) = \nabla U + \nabla V \qquad \nabla \cdot (\mathbf{A} + \mathbf{B})$$

$$= \nabla \cdot \mathbf{A} + \nabla \cdot \mathbf{B}$$

$$\nabla \times (\mathbf{A} + \mathbf{B}) = \nabla \times \mathbf{A} + \nabla \times \mathbf{B}$$

$$\nabla(UV) = V\nabla U + U\nabla V$$

$$\nabla \cdot (U\mathbf{A}) = U\nabla \cdot \mathbf{A} + \mathbf{A} \cdot \nabla U$$

$$\nabla \times (U\mathbf{A}) = \nabla U \times \mathbf{A} + U\nabla \times \mathbf{A}$$

$$\nabla \cdot (\mathbf{A} \times \mathbf{B}) = \mathbf{B} \cdot \nabla \times \mathbf{A} - \mathbf{A} \cdot \nabla \times \mathbf{B}$$

$$\nabla(\mathbf{A} \cdot \mathbf{B}) = \mathbf{A} \cdot \nabla \mathbf{B} + \mathbf{B} \cdot \nabla \mathbf{A} + \mathbf{A} \times (\nabla \times \mathbf{B})$$

$$+ \mathbf{B} \times (\nabla \times \mathbf{A})$$

$$\nabla \times (\mathbf{A} \times \mathbf{B}) = \mathbf{B} \cdot \nabla \mathbf{A} - \mathbf{A} \cdot \nabla \mathbf{B} + \mathbf{A}(\nabla \cdot \mathbf{B})$$

$$- \mathbf{B}(\nabla \cdot \mathbf{A})$$

$$\nabla \times (\nabla \times \mathbf{A}) = \nabla(\nabla \cdot \mathbf{A}) - \nabla^2 \mathbf{A}$$

$$\nabla \cdot (\nabla \times \mathbf{A}) = 0$$

$$\nabla \times (\nabla U) = 0$$

If $\mathbf{R} = \mathbf{i}x + \mathbf{j}y + \mathbf{k}z$ (Fig. 165), then

$$\nabla \cdot \mathbf{R} = 3 \qquad \nabla \times \mathbf{R} = \mathbf{0} \qquad \mathbf{A} \cdot \nabla |\mathbf{R}| = |\mathbf{A}|$$

$$\nabla \cdot \frac{1}{|\mathbf{R}|} = -\frac{\mathbf{R}}{|\mathbf{R}|^3} \qquad \nabla^2 \frac{1}{|\mathbf{R}|} = 0$$

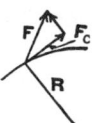

Fig. 169

Integration.

The line integral of a vector $\mathbf{F}$ along a curve AB denotes the integral of the tangential component of the vector along the curve; thus

$$\int_A^B \mathbf{F} \cdot d\mathbf{R} = \int_A^B |\mathcal{F}_c|\, ds \quad \text{(Fig. 169)}$$

where $d\mathbf{R} = \mathbf{i}\,dx + \mathbf{j}\,dy + \mathbf{k}\,dz$.

If $\mathbf{F} = \nabla U$ is the gradient of a single-valued continuous function $U(x, y, z)$, the line integral of $\mathbf{F}$ depends only on the endpoints. Conversely, if $\mathbf{F}(x, y, z)$ is continuous and $\int_C \mathbf{F} \cdot d\mathbf{R} = 0$ for any closed path C in a three-dimensional region, there is a function $U(x, y, z)$ such that $\mathbf{F} = \nabla U$.

15.3   Theorems and Formulas

Let $\mathbf{n}$ be the vector of unit length perpendicular to a surface at a point P and extending on the positive side (the outward normal), dS the element of surface, and dv the element of volume.

Divergence (Gauss) Theorem.

If a field vector $\mathbf{F}$ and its first derivatives are continuous at all points in a region of volume v bounded by a closed elementary surface S, then

$$\iint_S \mathbf{F} \cdot \mathbf{n}\, dS = \iiint_v \nabla \cdot \mathcal{F}\, dv$$

Stokes's Theorem.

If a field vector $\mathbf{F}$ and its first derivatives are continuous at all points in a region of area S bounded by a closed curve C, then

$$\iint_S \nabla \times \mathbf{F} \cdot \mathbf{n}\, dS = \int_C \mathcal{F} \cdot d\mathbf{R}$$

Green's Theorem.

Under the conditions of the divergence theorem,

$$\iint_S \mathbf{n} \cdot U\nabla V\, dS = \iiint_v U\nabla^2 V\, dv$$

$$+ \iiint_v (\nabla U \cdot \nabla V)\, dv$$

$$\iint_S \mathbf{n} \cdot (U\nabla V - V\nabla U)\, dS = \iiint_v (U\nabla^2 V$$

$$- V\nabla^2 U)\, dv$$

Cylindrical Coordinates

$$x = r \cos \theta \qquad y = r \sin \theta \qquad z = z$$

The element of volume $dv = r\, dr\, d\theta\, dz$. The unit vectors $\mathbf{u}_r, \mathbf{u}_\theta, \mathbf{u}_z$ are perpendicular to each other,

$$\text{grad } V = \nabla V = \frac{\partial V}{\partial r}\mathbf{u}_r + \frac{1}{r}\frac{\partial V}{\partial \theta}\mathbf{u}_\theta + \frac{\partial V}{\partial z}\mathbf{u}_z$$

$$\text{div } \mathbf{F} = \nabla \cdot \mathbf{F} = \frac{1}{r}\frac{\partial}{\partial r}(r\mathbf{F}_r) + \frac{1}{r}\frac{\partial}{\partial \theta}(\mathbf{F}_\theta) + \frac{\partial}{\partial z}(\mathbf{F}_s)$$

$$\text{curl } \mathbf{F} = \nabla \times \mathbf{F} = \begin{vmatrix} \dfrac{\mathbf{u}_r}{r} & \mathbf{u}_\theta & \dfrac{\mathbf{u}_z}{r} \\ \dfrac{\partial}{\partial r} & \dfrac{\partial}{\partial \theta} & \dfrac{\partial}{\partial z} \\ \mathbf{F}_r & r\mathbf{F}_\theta & \mathbf{F}_s \end{vmatrix}$$

$$\nabla^2 V = \frac{1}{r}\frac{\partial V}{\partial r} + \frac{\partial^2 V}{\partial r^2} + \frac{1}{r^2}\frac{\partial^2 V}{\partial \theta^2} + \frac{\partial^2 V}{\partial z^2}$$

Spherical Coordinates

$$x = r \cos\phi \sin\theta \qquad y = r\sin\phi\sin\theta \qquad z = r\cos\theta$$

The unit vectors $\mathbf{u}_r, \mathbf{u}_\phi, \mathbf{u}_\theta$ are perpendicular to each other,

$$\text{grad } V. = \nabla V = \frac{\partial V}{\partial r}\mathbf{u}_r + \frac{1}{r\sin\theta}\frac{\partial V}{\partial \phi}\mathbf{u}_\phi$$

$$+ \frac{1}{r}\frac{\partial V}{\partial \theta}\mathbf{u}_\theta$$

$$\text{div } \mathbf{F} = \nabla \cdot \mathbf{F} = \frac{1}{r^2}\frac{\partial}{\partial r}(r^2\mathbf{F}_r)$$

$$+ \frac{1}{r\sin\theta}\frac{\partial \mathbf{F}_\phi}{\partial \phi} + \frac{1}{r\sin\theta}\frac{\partial}{\partial \phi}(\sin\theta\mathbf{F}_\theta)$$

$$\text{curl } \mathbf{F} = \nabla \times \mathbf{F} = \begin{vmatrix} \dfrac{\mathbf{u}_r}{r^2\sin\theta} & \dfrac{\mathbf{u}_\theta}{r\sin\theta} & \dfrac{\mathbf{u}_\phi}{r} \\ \dfrac{\partial}{\partial r} & \dfrac{\partial}{\partial \theta} & \dfrac{\partial}{\partial \phi} \\ \mathbf{F}_r & r\mathbf{F}_\theta & r\sin\theta\mathbf{F}_\phi \end{vmatrix}$$

$$\nabla^2 V = \frac{1}{r^2}\frac{\partial}{\partial r}\left(r^2\frac{\partial V}{\partial r}\right) + \frac{1}{r^2\sin^2\theta}\frac{\partial^2 V}{\partial \phi^2}$$

$$+ \frac{1}{r^2\sin\theta}\frac{\partial}{\partial \theta}\left(\sin\theta\frac{\partial V}{\partial \theta}\right)$$

Solid Angle. The lines joining the point P to points of a surface S generate a solid angle. If a is the area intercepted by these lines on a sphere of center P with radius r, then

$$\omega = \frac{a}{r^2}$$

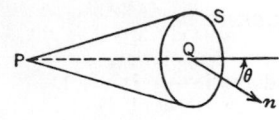

Fig. 170

is the measure of the solid angle (Fig. 170). If S is a surface that does not pass through P, $\cos\theta$ is nowhere zero, and $\mathbf{n}$ is everywhere continuous on the surface, then

$$\omega = \int_S \frac{\mathbf{R} \cdot \mathbf{n}\, da}{r^3}$$

where $\mathbf{R} = PQ$. If S forms the complete boundary of a three-dimensional region, the total solid angle subtended by S at P is zero if P lies outside the region and 4π if P lies inside the region.

BIBLIOGRAPHY

Beaumont, R., and Pierce, R., *The Algebraic Foundations of Mathematics*, Addison-Wesley, Reading, MA, 1963.

Beckenbach, E., and Bellman, R., *Introduction to Inequalities*, Random House, New York, 1962.

Becker, E. B., Carey, G. F., and Oden, J. T., *Finite Elements: An Introduction*, Prentice-Hall, Englewood Cliffs, NJ, 1981.

Boole, G., *An Investigation of the Laws of Thoughts*, Dover, New York, 1951.

Brebbia, C. A., and Connor, J. J., *Fundamentals of Finite Element Techniques for Structural Engineers*, Butterworth, London, 1975.

Carnahan, B., Luther, H. A., and Wilkes, J. O., *Applied Numerical Methods*, Wiley, New York, 1969.

Carrier, G. F., and Pearson, C. E., *Ordinary Differential Equations*, Blaisdell, Waltham, MA, 1968.

Chapra, S. C., and Canale, R. P., *Numerical Methods for Engineers with Personal Computer Applications*, McGraw-Hill, New York, 1985.

Churchill, R. V., *Fourier Series and Boundary Values Problems*, McGraw-Hill, New York, 1941.

Churchill, R. V., *Complex Variables with Applications*, McGraw-Hill, New York, 1960.

Collatz, L., *The Numerical Treatment of Numerical Differential Equations*, Springer-Verlag, Berlin, 1960.

Courant, R., and Hilbert, D., *Methods of Mathematical Physics*, Vols. I and II, Wiley, New York, 1968.

Davis, A. S., *The Finite Element Method, a First Approach*, Clarendon, Oxford, 1980.

Feller, W., *An Introduction to Probability Theory and Its Applications*, Wiley, New York, 1968.

Forsyth, A. R., *Theory of Differential Equations*, Cambridge University Press, New York, 1906; Dover, New York, 1959.

Gantmacher, F. R., *The Theory of Matrices*, Chelsea, New York, 1959.

Garabedian, P., *Partial Differential Equations*, Wiley, New York, 1964.

Hashisaki, J., and Peterson, J., *Theory of Arithmetic*, Wiley, New York, 1971.

Hildebrand, F. B., *Methods of Applied Mathematics*, 2nd ed., Prentice-Hall, Englewood Cliffs, NJ, 1965.

Hildebrand, F. B., *Introduction to Numerical Analysis*, McGraw-Hill, New York, 1974.

Hinton, E., and Owen, D. R. J., *An Introduction to Finite Element Computations*, Pineridge Press, Swansea, 1979.

Householder, A. S., *The Theory of Matrices in Numerical Analysis*, Blaisdell, New York, 1964.

Isaacson, E., and Keller, H. B., *Analysis of Numerical Methods*, Wiley, New York, 1966.

Leeds, H. D., and Weinberg, G. M., *Computer Programming Fundamentals*, McGraw-Hill, New York, 1961.

McCormick, E. M., *Digital Computer Primer*, McGraw-Hill, New York, 1949.

McCracken, D. D., *A Guide to FORTRAN IV Programming*, Wiley, New York, 1965.

McCrea, W., *Analytic Geometry of Three Dimensions*, Interscience, New York, 1948.

Morse, P. M., and Feshbach, H., *Methods of Theoretical Physics*, McGraw-Hill, New York, 1953.

Noble, B., *Applied Linear Algebra*, Prentice-Hall, Englewood-Cliffs, NJ, 1969.

Olmsted, J., *The Real Number System*, Appleton-Century-Crofts, New York, 1962.

Pearson, C. E., *Handbook of Applied Mathematics*, Van Nostrand Reinhold, New York, 1974.

Pipes, L. A., and Harvill, L. R., *Applied Mathematics for Engineers and Physicists*, 3rd ed., McGraw-Hill, New York, 1970.

Ralston, A., and Wilf, H. S., *Mathematical Methods for Digital Computers*, Wiley, New York, 1967.

Reddy, J. N., *An Introduction to the Finite Element Method*, McGraw-Hill, New York, 1984a.

Reddy, J. N., *Energy and Variational Methods in Applied Mechanics* (with an Introduction to the Finite Element Method), Wiley, New York, 1984b.

Reddy, J. N., *Applied Functional Analysis and Variational Methods in Engineering*, McGraw-Hill, New York, 1986.

Reddy, J. N., and Rasmussen, M. L., *Advanced Engineering Analysis*, Wiley, New York, 1982.

Rice, J. R., *Numerical Methods, Software and Analysis*, McGraw-Hill, New York, 1983.

Salmon, G., *A Treatise on Conic Sections*, Chelsea, New York, 1954.

Titscmarsh, E. C., *Theory of Fourier Integrals*, 2nd ed., Oxford University Press, New York, 1948.

Whittaker, E. T., and Watson, G. N., *Modern Analysis*, Cambridge University Press, New York, 1946.

Wilks, S. S., *Mathematical Statistics*, Wiley, New York, 1962.

Zienkiewicz, O. C., *The Finite Element Method in Engineering Science*, McGraw-Hill, London, 1969.

CHAPTER 3
MECHANICS OF RIGID BODIES

Wallace Fowler
Department of Aerospace Engineering and Engineering Mechanics
The University of Texas at Austin
Austin, Texas

1 DEFINITIONS

Mechanics is that branch of science that treats forces and motion.

Statics is that branch of mechanics that deals with the equilibrium of forces on bodies *at rest* (or moving at a uniform velocity in a straight line).

Kinematics is that branch of mechanics that deals with the motion of bodies without consideration of the character of the bodies or of the influence of forces upon their motion. It considers only concepts of *geometry* and *time*.

Kinetics (or *dynamics*) is that branch of mechanics that deals with the effect of unbalanced external forces in *modifying the motion* of bodies.

Mass and weight, in the *gravitational system of units* employed by English engineers, are related by the formula $W = Mg$, where W = weight, M = mass, and g = acceleration due to gravity. For a thorough discussion of these terms, see Chapter 1.

Force is that which changes or tends to change the state of rest or motion of a body.

Inertia is that property of a body by virtue of which it tends to continue in the state of rest or motion in which it may be placed until acted on by some force.

Reaction is that *equal and opposite force* exerted by a body in opposing another force acting upon it.

Newton's Laws of Motion

First Law. If a body is at rest, it will remain at rest, or if in motion, it will move uniformly in a straight line until acted on by some force.

Second Law. The time rate of change of the linear momentum of a particle (defined as the product of its mass and its acceleration, a vector) is proportional to the force (a vector) acting on the particle.

Third Law. If a force acts to change the state of a body with respect to rest or motion, the body will offer a resistance equal and directly opposed to the force. Or, to every action there is opposed an equal and opposite reaction.

Special terms such as *hydrostatics*, *aerodynamics*, and so on, are used to denote the theory of statics as applied to *liquid bodies*, the theory of dynamics as applied to *gaseous bodies*, and so on. *Mechanics of materials* considers, in addition to *external forces*, the *internal forces*, or *stresses*, between molecules of a body. Subjects of these types are covered in other sections of this handbook. The present chapter on mechanics is confined, in general, to the discussion of motion of, and external forces applied to, *rigid bodies*.

2 STATICS

2.1 Graphical Representation and Classification of Forces

Graphical Representation of Force A *force* is completely specified by its *magnitude, direction*, and *point of application*. The word *sense* as applied to a force refers to one of the two directions along the line of action of the force. The effect of any force applied to a rigid body at rest is the same, no matter where in its own line of action the force is applied. This is known as the principle of the *transmissibility of force*. A force may be represented graphically in magnitude and direction by a straight line drawn parallel to its line of action, the length being proportional to the magnitude of the force; its sense is indicated by an arrowhead placed on the line. The English engineers' unit of force is the pound, or the earth's pull on a mass of 1 lb. A drawing that indicates the lines of action of the various forces acting on a machine or structure is called a *space diagram*; one in which vectors are drawn to represent the magnitudes and directions of the forces is a *vector diagram*. A force is indicated on a space diagram by two lowercase letters placed on opposite sides of the line of action of the force; the vector, representing its magnitude and direction, by the same capital letters placed at the ends. Thus, in Fig. 1, *AB* represents the magnitude and direction of

the force *W* and *ab* its action line. The vector being read as *AB* indicates a downward sense; read as *BA*, an upward sense.

Classification of Systems of Forces A *system of forces* consists of any number of forces taken collectively.

Classification of systems of forces is made according to the arrangement of their *action lines*. If the action lines lie in the same plane, the system is *coplanar*, otherwise *noncoplanar*. If they pass through the same point, the system is *concurrent*, otherwise *nonconcurrent*. If two or more forces have the same action line, they are *collinear*. A system of two equal forces parallel, opposite in sense, and having different action lines is a *couple*. Two or more forces equivalent to a single force are components of the single force. *Resolution* is the operation of replacing a single force by a system of components. The single force is the *resultant* of its components. In general, the resultant of a system of forces is the simplest equivalent system. This may be a *single force*, a *single couple*, or a *noncoplanar force and couple* (or *two skewed forces*). When the resultant is a single force, the equilibrant is a force equal in magnitude, having the same line of action but opposite sense. *Composition* is the operation of replacing a system of forces by its resultant.

2.2 Addition and Resolution of Concurrent Forces

Addition of Two Concurrent Forces*

Parallelogram Law. If magnitudes, lines of action, and senses of two concurrent forces acting on a rigid body are represented by *OA* and *OB* (Fig. 2), the magnitude, line of action, and sense of their resultant are represented by the diagonal *OC* of the parallelogram *OABC*. The points of application of the forces may be anywhere on the body in the lines *OA, OB*, and *OC* or their extensions. The arrowheads on the lines *OA, OB*, and *OC* all point toward or all away from the point of concurrence *O*.

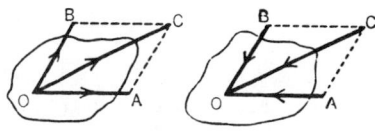

Fig. 2 Concurrent forces.

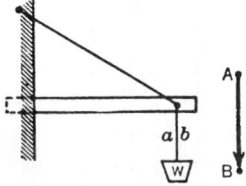

Fig. 1 Space diagram.

---

*It is evident that a pair of concurrent forces and their resultant are necessarily coplanar.

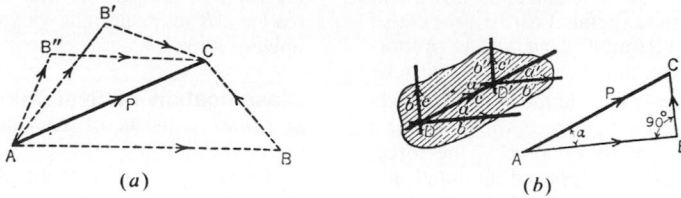

Fig. 3 Force components.

Triangle Law.   The triangle law for graphical vector addition of two concurrent forces follows directly from the parallelogram law. In the parallelogram law, vectors are drawn tail to tail and a parallelogram is created. If either half of the parallelogram in Fig. 2 (either above or below line OC, with each "half" containing line OC) is examined, it can be seen that each triangle thus considered contains both of the forces being composed (added) laid out in a head-to-tail sequence.

Resolution into Two Concurrent Forces   A force may be resolved into an infinite number of pairs of components by constructing triangles with the given force as one of the sides. In each such triangle, the other two sides of the triangle are the components of the given force in the directions defined by the other two sides of the triangle, as shown in Fig. 3a. Of special interest are those cases in which the given force is the hypotenuse of a right triangle, in which case the sides of the triangle are rectangular components of the given force, as shown in Fig. 3b. Algebraically, the rectangular components of the force P in Fig. 3b in the x and y directions are given by $P_x = P \cos \alpha$ and $P_y = P \sin \alpha$.

Addition of More Than Two Coplanar Concurrent Forces

Graphic Method.   In Fig. 4, consider body G acted on by the four forces shown. Construct a *force polygon* as follows: Plot AB parallel to ab, and scale it to represent 60 lb; from B plot BC parallel to bc, and

scale it to represent 80 lb; in like manner plot CD and DE, so that the arrows lead *confluently* from A to E. The resultant of the system is AE in magnitude and sense and equals 114 lb. Its action line is ae. The resultant will be the same regardless of the order in which the forces are plotted. Note particularly that the resultant is not confluent with the component forces.

Algebraic Method.   Choose rectangular axes OX and OY. Referring to Fig. 4 resolve each force into its x and y components, considering components acting upward or to the right as positive and those acting downward or to the left as negative. Arrange the results in tabular form, placing the forces in the first column, the x components in the second, and the y components in the third. $\sum F_x$ = algebraic sum of x components and $\sum F_y$ = algebraic sum of y components.

F (lb)	F_x (lb)	F_y (lb)
$ab = 60$	$-60 \times 0.707 = -42.4$	$+60 \times 0.707 = +42.4$
$bc = 80$	$+80 \times 2/\sqrt{5} = +71.4$	$+80 \times 1/\sqrt{5} = +35.7$
$cd = 120$	$+120 \times 0.866 = +104$	$-120 \times 0.5 = -60$
$de = 40$	$-40 \times 0.5 = -20$	$-40 \times 0.866 = -34.6$
	$\sum F_r = +113$	$\sum F_y = -16.5$

Then $R = \sqrt{\overline{\sum F_x^2} + \overline{\sum F_y^2}} = \sqrt{13{,}041} = 114$ lb. Sense is downward and to the right (Fig. 5):

$$\tan \theta = \frac{\sum F_y}{\sum F_x} = -0.146 \qquad \theta = -8°20'$$

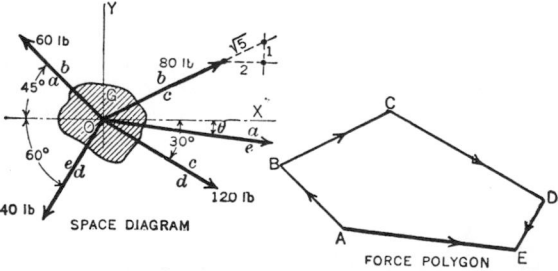

Fig. 4 Force polygon.

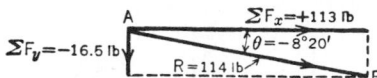

Fig. 5 Rectangular components.

Addition of Noncoplanar Forces

Graphical Methods. Graphical methods for the addition of noncoplanar forces are unnecessarily cumbersome in view of the ease with which the process can be handled algebraically using rectangular components.

Algebraic Method. Consider any number of concurrent forces (e.g., the three forces **F, G,** and **H** acting at point O as shown in Fig 6). Each such force can be resolved into its components with respect to a set of rectangular axes by using the direction angles of the line of action of the force (and their cosines, called direction cosines) as shown in Fig. 6. The rectangular components of the force **F** are given by

$$F_x = F \cos \alpha_F \qquad F_y = F \cos \beta_F \qquad F_2 = F \cos \gamma_F$$

and the force **F** can be written as

$$\mathbf{F} = F_x \mathbf{i} + F_y \mathbf{j} + F_z \mathbf{k}$$

where **i, j,** and **k** are unit vectors in the $x,\ y,$ and z directions, respectively. Similar forms can be written for the forces **G** and **H**. The resultant or sum of the three forces **F, G,** and **H** is obtained by adding corresponding components of the forces. Thus, the resultant, **R,** is given by

$$\mathbf{R} = R_x \mathbf{i} + R_y \mathbf{j} + R_z \mathbf{k}$$

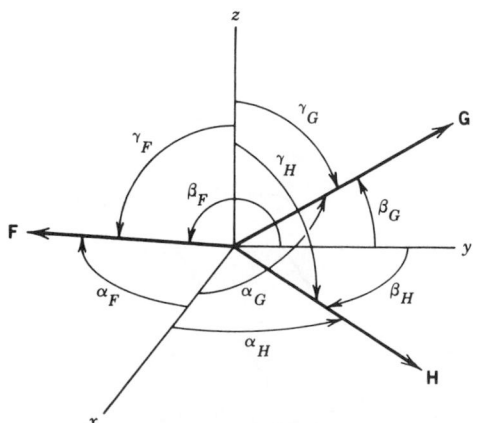

Fig. 6 Direction angles.

where

$$R_x = F_x + G_x + H_x$$
$$R_y = F_y + G_y + H_y$$
$$R_z = F_z + G_z + H_z$$

Nature of Resultant of Concurrent Forces The *resultant* of any system of concurrent forces that are not in equilibrium is a single force.

2.3 Moments and Couples

Moment (or Torque) of a Force about a Point
Moment, or *torque*, of a force about a point is the product of the force magnitude and the distance from the point to its action line. This perpendicular distance is called the *arm* of the force, and the point is the *origin or center of moments*. The product is the measure of the rotational tendency of the force. The name of the unit of moment is a combination of the names of force and distance units, as foot-pound, inch-ton, and so on. (Some writers use pound-foot as a unit of moment of a force to distinguish from foot-pound as the unit of work or energy, similar distinction being made for the other units.)

The computation of moments is most easily carried out by introducing a coordinate system and then taking the vector cross product,

$$\mathbf{M} = \mathbf{R} \times \mathbf{F}$$

where **R** is the vector from the reference point to the point of action of the force and **F** is the vector representation of the force in the same coordinate system. The moment vector **M** is perpendicular to the plane of **R** and **F**. This vector operation, in effect, takes into account the angle between **R** and **F** and can be used in two- and three-dimensional situations.

Moment (or Torque) of a Force about a Line
The moment of a force **F** about a line is most easily calculated using vectors. Let O be any point on the line and let A be any other point on the line. Let P be the point of application of the force **F**. The moment of the force **F** about the line OA measures the torque of the force about the axis OA. In order to calculate this torque, it is most convenient to break **F** into two components, with one of the components parallel to OA. The component of **F** parallel to OA does not exert a torque around the axis OA. Thus, the torque of the force **F** around the axis OA can be written as

$$\mathbf{M} \ (\text{about } OA) = \mathbf{R} \times (\mathbf{F} - \mathbf{F} \cdot \mathbf{U}_{OA})$$

where **R** is the vector from O to P, **F** is the force vector, and $\mathbf{U}_{OA}$ is a unit vector directed along the axis OA. The quantity $\mathbf{F} \cdot \mathbf{U}_{OA}$ is the component of **F** parallel to the axis OA.

Principle of Moments

For a Point. The moment of the resultant of any set of forces about a point is the vector sum of the moments of the individual forces about the point.

For an Axis. The moment of the resultant of any set of forces about an axis is the vector sum of the moments of the individual forces about the axis.

Couples

Nature of Couples. Two equal and parallel forces of opposite sense are called a couple. The tendency of a couple is to produce rotation only. Since a couple has no single resultant, no single force can balance it. To prevent the rotation of a body acted upon by a couple, the application of two other forces is required, forming a second couple.

The arm of a couple is the perpendicular distance between the lines of action of the forces. The moment of a couple is constant and independent of the origin of moments; it is equal to one of the forces times the arm of the couple. Its sense is positive or negative according as rotational tendency is counterclockwise or clockwise. Couples of equal moments, in the same or parallel planes, are *equivalent* and may be replaced one by the other. Further, the *center of rotation* for a couple may be anywhere in its plane. Hence, a couple may be turned about in its own plane or moved to a parallel plane or replaced by another couple (having an arm of any given length but the same moment) without altering its effect on a rigid body.

Resultant of Couples. The resultant of any number of coplanar couples is a couple. Its moment and sense are determined by the algebraic sum of the moments of the individual couples.

In vector form, the moment of a couple can be calculated by choosing a point O on the line of action of one of the forces and determining the moment of the other force of the couple about that point. Thus, the moment of a couple is calculated from

$$\mathbf{M} = \mathbf{R} \times \mathbf{F}$$

The resultant of any number of couples is the vector sum of the moments of the individual forces (each calculated as shown), and this resultant is itself a couple.

Composition of Single Force and Couple. A single force and couple in the *same plane* (or *parallel planes*) may be composed into *another* single force equal and parallel to the original force, at a distance from it equal to the moment of the couple divided by the magnitude of the force and so situated that the moment of the resultant about the point of application of the original force is of the same sign as the moment of the couple. The couple may be brought into the position shown in Fig. 7. The resultant of P, $-Q$, and

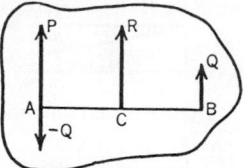

Fig. 7 Composition of a force and a couple.

Q is $R(= P)$ acting in a line through point C so that $(P - Q) \times AC = Q \times BC$. From this it follows that

$$AC = \frac{Q(AC + BC)}{P} = \frac{\text{moment of couple}}{P}$$

Resolution into Single Force through Chosen Point and Couple. A single force may be resolved into another single force acting through a *chosen point and a couple* (the new force being equal and parallel to the original force). In Fig. 8, P_1 is the given force and O the chosen point. Through O apply a pair of forces, opposite in sense, equal and parallel to P_1. As P_2 and P_3 balance, no change is produced in the motion of the body due to the addition. P_1 and P_3 constitute a couple of moment $P \times a$, which is the same as moment of P_1 about O; P_2 is a force just like P_1 but acting through the chosen point O.

2.4   Addition of Nonconcurrent Forces and Moments

The addition of nonconcurrent forces involves the determination of the force and moment resultants of the forces. The procedure requires that a reference point O for the moments be defined and that all position vectors and force vectors be written in terms of vector components. The procedure works for all force and moment sets:

Force Resultant. The force resultant of a set of non-concurrent forces is the vector sum of the individual forces:

$$\mathbf{R} = \sum_{i=1}^{n} \mathbf{F}_i = \sum_{i=1}^{n} (\mathbf{F}_{ix})\mathbf{i} + \sum_{i=1}^{n} (\mathbf{F}_{iy})\mathbf{j} + \sum_{i=1}^{n} (\mathbf{F}_{iz})\mathbf{k}$$

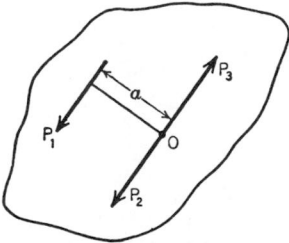

Fig. 8 Resolution of a force into a force plus couple.

Moment Resultant. The moment resultant of a set of nonconcurrent moments about a point O is the vector sum of the moments of the individual forces about O:

$$\mathbf{M} = \sum_{i=1}^{n} \mathbf{M}_i = \sum_{i=1}^{n} (\mathbf{r}_i \times \mathbf{F}_i)$$

2.5 Principles of Equilibrium

Forces in Equilibrium. A system of forces is in equilibrium if their combined action produces no change in motion of the body to which they are applied., There is no change in motion if the body remains at rest or moves in a straight line at constant speed. When a force system is in equilibrium, its resultant must be zero. This statement may be called the *general condition of equilibrium.* It implies both zero force and zero couple.

In all cases, the conditions of equilibrium can be written in vector form as

$$\sum \mathbf{F} = 0 \qquad \sum \mathbf{M} = 0$$

and in scalar form in rectangular Cartesian coordinates as

$$\sum F_x = 0 \qquad \sum M_x = 0$$

$$\sum F_y = 0 \qquad \sum M_y = 0$$

$$\sum F_z = 0 \qquad \sum M_z = 0$$

Note: If a problem is planar (e.g., in the xy plane), then the force equation in the z direction and the moment equations in the x and y directions are identically zero and may be ignored.

Body in Equilibrium. A rigid body is in equilibrium if it remains at rest or moves in a straight line at constant speed, that is, if its state of motion does not change. This condition obtains if all the external forces acting upon it (including those due to pull of gravity, friction, etc.) form a system in equilibrium.

Conditions of Equilibrium In a problem in statics a body is known to be in equilibrium; hence the system composed of all the external forces acting upon it must be in equilibrium. In such a case, tests are not needed to ascertain if equilibrium exists, but they are used to set up relations involving unknown forces, distances, or angles, and the unknown elements are then computed provided their number does not exceed the number of independent equations that may be set up by means of the equilibrium conditions. When the number of unknown elements exceeds the number of independent equations, the problem is said to be statically indeterminate.

Special Conditions. If three forces are in equilibrium, they must be coplanar and concurrent or parallel; if concurrent, each force is proportional to the sine of the angle between the other two; if parallel, each force is proportional to the distance between the other two. If a force system is in equilibrium, the resultant of any part must balance the resultant of the other part. It follows that if four coplanar nonconcurrent nonparallel forces are in equilibrium, the resultant of any two is concurrent with the other two.

Stability of Equilibrium. When a body (or collection of bodies) is in equilibrium and the state is such that when displaced slightly in any way the body returns of itself to its original position, the equilibrium is *stable*; when displaced slightly the body moves further from its original position, the equilibrium is *unstable*; and when displaced slightly it remains in that displaced position, the equilibrium is *neutral* or *indifferent.* The body or collection is also said to be stable, unstable, or neutral (or indifferent) under these respective conditions. When the body is stable or unstable, the system of forces is changed by the slight displacement and is no longer in equilibrium—hence the further displacement. Only when the stability is neutral is the equilibrium of the force system undisturbed by a slight displacement of the body.

2.6 Equilibrium Problems
General Principles

1. It frequently happens that the external force system acting on a body as a whole cannot be solved directly owing to the presence of more unknown elements (forces, distances, and angles) than there are conditions of equilibrium. In such cases, endeavor to separate the original body into simpler parts that will permit solutions, making use of the unknowns thus determined in solving the force systems acting on other sections until the complete solution, if obtainable, has been found.

2. To facilitate computations, it is desirable if resolution equations are used, to resolve perpendicular to one of the unknown forces; if a moment origin is used, to select it on the action line of an unknown force; if moment axes are used, to select them so as to intersect some of the unknown forces.

3. Assume senses for unknown forces. A plus answer then indicates the sense to have been correctly assumed; a minus answer, incorrect assumption.

4. When force polygons are used, letter action lines of wholly known forces first and those of the remainder last. Draw the force polygon to the end of the last known vector. Vectors required to close it (remembering that the senses must read confluently from the starting point back to the same point) determine unknown magnitudes and senses and/or lines of action.

Typical Problems

I. System of Coplanar Concurrent Forces in Equilibrium.

*In this system all forces are known except two whose action lines only are known. The magnitudes and senses of these two forces are to be determined.**

Example 1. Two smooth cylinders rest upon a 30° plane and against a vertical wall as shown in Fig. 9. Determine all forces acting on each cylinder (a) The forces involved are 100 lb, 200 lb, P, Q, R, and S, the last four being normal to the surfaces of contact (smooth surfaces). (b) Consider the two cylinders as a single free body. The external force system is 100 lb, 200 lb, P, R, and S (Q_1 and Q_2 are internal). The system is nonconcurrent and so does not come under a typical problem I. Consider the large cylinder as a free body. The external force system is 100 lb, Q, R, and S. While this system is concurrent, it cannot be solved because there are more than two unknown quantities. Next consider the small cylinder as a free body. The force system is 200 lb, P, and Q, and this is a typical problem I.

Algebraic Solution. Choose X and Y directions parallel and perpendicular to the plane. $\sum F_x = 0 = Q_1(\sqrt{60}/8) - 200\sin 30°$. Hence $Q_1 = 800/\sqrt{60} = 103.3$ lb. $\sum F_y = 0 = P - 200\cos 30° - 103.3 \times 2/8$. Hence $P = 199$ lb. Consider the large cylinder as a free body. $Q_1 = Q_2 = 103.3$ lb. Use the same X and Y directions. $\sum F_x = 0 = S\cos 30° - 103.3 \times (\sqrt{60}/8) - 100\sin 30°$. Hence, $S = 173.2$ lb. $\sum F_y = 0 = R - 100\cos 30° - 173.2\sin 30° + 103.3 \times 2/8$. Hence $R = 147.4$ lb.

II. System of Coplanar Parallel Forces in Equilibrium.

In this system all forces are known except two whose action lines only are known. The magnitudes and senses of these two forces are to be determined.

Example 2. A beam is loaded as shown in Fig. 10 and supported at the points P and Q. Determine the

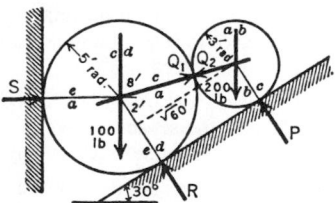

Fig. 9 Space diagram of cylinders.

---

*This is a common problem in the determination of the stresses of a roof or bridge truss.

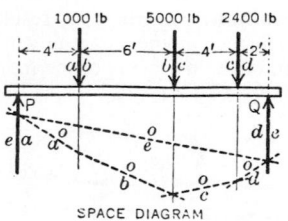

SPACE DIAGRAM

Fig. 10 Space diagram of beam.

reactions of the supports. Consider the beam as a free body. The external force system consists of the forces 1000 lb, 5000 lb, 2400 lb, P, and Q. This is a coplanar parallel system and is a typical problem II.

Algebraic Solution. Assume senses for the reactions,

$$\sum M_p = 0 = -4 \times 1000 - 10 \times 5000 - 14 \times 2400 + 16Q$$

Hence $Q = 5475$ lb; the correct sense was assumed,

$$\sum M_q = 0 = 2 \times 2400 + 6 \times 5000 + 12 \times 1000 - 16P$$

Hence $P = 2925$ lb; the correct sense was assumed. As a check, apply a third equilibrium condition, $\sum F = 0$,

$$\sum F = 0 = -2925 + 1000 + 5000 + 2400 - 5475$$

III. System of Coplanar Nonparallel Nonconcurrent Forces in Equilibrium.

In this system all forces are known except two, of which the action line of one and a point in the action line of the other are known. The magnitude and sense of the one and the magnitude, sense, and angular direction of the other are to be determined.†

Example 3. A roof truss is loaded as in Fig. 11. The left end of the truss rests on a smooth horizontal support. The right end is secured to a wall by means of a pin. Determine the reactions. The external forces acting on the truss are the given loads, the left reaction P (vertical, on account of the smooth support), and the right reaction Q (inclined, through point M). The unknown quantities are the reactions P and Q. This is a typical problem III.

---

†This is a common problem in the determination of the reactions on a roof truss sustaining wind pressures, the truss being fixed at one end and resting on rollers at the other.

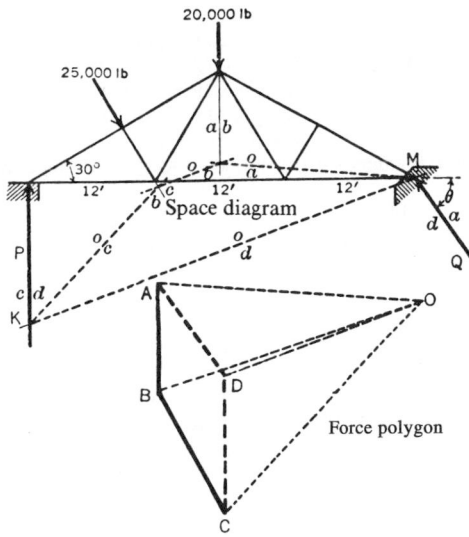

Fig. 11 Space diagram of roof truss.

Algebraic Solution. Assume P upward,

$$\sum M_M = 0 = 20,000 \times 18 + 25,000$$
$$\times 24 \cos 30° - 36P$$

hence, $P = 24,430$ lb; the correct sense was assumed.

Assume Q upward to the left at angle θ with horizontal,

$$\sum F_x = 0 = 25,000 \sin 30° - Q \cos \theta$$

$$\sum F_y = 0 = -25,000 \cos 30° - 20,000 + 24,430$$
$$+ Q \sin \theta$$

Solving simultaneously, $Q = 21,300$ lb and $\theta = 54°$. Sense and direction were correctly assumed; hence Q acts upward to the left at $54°$ to the horizontal. As a check, apply condition $\sum M_P = 0$,

$$\sum M_P = 0 = -25,000 \times 12 \cos 30° - 20,000 \times 18$$
$$+ 21,300 \times 36 \sin 54°$$

Graphic Solution. ab and bc are the action lines of the given loads, cd of the reaction P, and da of the reaction Q. Draw the vectors AB and BC and a line through C parallel to cd. Choose a pole and draw the rays. Construct the funicular polygon, drawing oa through M, and draw closing string od from K to M. Draw OD through O parallel to od to intersect CD at D. Draw DA. Vectors CD and DA represent the two

unknown forces $P = 24,430$ lb and $Q = 21,300$ lb. The action line of Q is da, making the angle with the horizontal $54°$.

Special Case. A case coming under the preceding classification that requires a variation in treatment when employing the graphic method is one in which the action lines of all three forces are known but their magnitudes and senses are unknown. The procedure is in general similar to methods employed before except that, in the graphic solution, two of the unknown forces that are concurrent must be replaced by their unknown resultant acting through their point of concurrency along an unknown action line. After the magnitude and sense of this resultant have been determined (by the method employed in Example 3), it is resolved into its two components along the action lines of the two unknown forces that it had replaced. These components represent the magnitudes and senses of this pair of forces.

IV. System of Noncoplanar Nonparallel Noncon-current Forces in Equilibrium. *In this system one force is completely known and action lines (or a point in the action line) of the others are known.* All unknown force magnitudes, action lines, and senses are to be determined.

Example 4. The crane (Fig. 12) is supported by a socket at the foot of the post at D, is kept from overturning by the backstays AB and AC, and carries a load of 600 lb (E, A, F, G, D are in the vertical XY plane). Determine the axial components of the reaction on the post at D and the tensions in the backstays. The external forces acting on the post are the load, the reaction at D, and the tensions in the backstays at A. This is a typical problem IV. Moment equations are the most convenient to apply for this solution:

$$\sum M_{BC} = 0 = 600 \times 40 - 20D_y \qquad D_y = 1200 \text{ lb}$$
$$\sum M_{Z_A} = 0 = 600 \times 20 - 16D_x \qquad D_x = 750 \text{ lb}$$

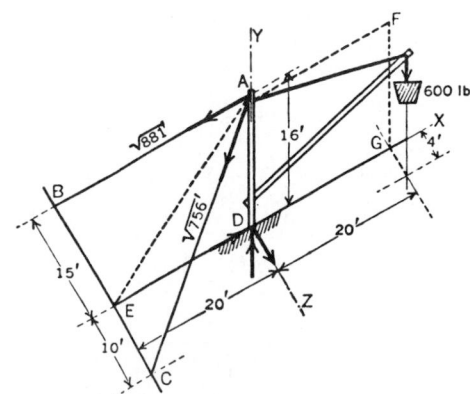

Fig. 12 Space diagram of truss.

$$\sum M_{X_A} = 0 = 600 \times 4 - 16D_z \qquad D_z = 150 \text{ lb}$$

$$\sum M_{X_C} = 0 = AB \times \frac{16}{\sqrt{881}} \times 25 \qquad AB = 622 \text{ lb}$$
$$-1200 \times 10 + 600 \times 6$$

$$\sum M_{X_B} = 0 = AC \times \frac{16}{\sqrt{756}} \times 25 \qquad AC = 452 \text{ lb}$$
$$+ 600 \times 19 - 1200 \times 15$$

The senses of all forces are as shown in Fig. 12.

Truss Analysis A *truss* is a framework* for carrying loads, each *member* of which is subjected only to tension or compression loads. The members are usually pin jointed with loads applied only at the joints.

The *stress in a member* at any section is the force that either of its two parts exerts internally on the other part as a result of the external forces acting on the member. Longitudinal stresses, like external longitudinal forces, may be either tensile or compressive.

The *analysis of a truss* under a given loading condition refers to the determination of the stresses in its members due to the loads.

Analysis by Method of Sections. First, determine the reactions on the truss due to the loads; second, imagine the truss separated into two distinct parts (i.e., pass a section through the truss) so that the member under consideration is one of the members cut and so that the system of forces, including stresses, acting on either part of the truss is solvable for the desired stress; third, solve the system.

To pass the section, suppose the stress in *HI* (Fig. 13a) is required, the truss being supported at its ends and bearing five loads *L* and one *P*, and suppose the reactions determined. Trying section $1-1$, the force system on the left part of the truss (Fig. 13b) is a nonconcurrent one of seven forces and includes four unknown stresses S_1, S_2, S_3, and S_4; it is not solvable

---

Redundant frames (i.e., ones having more members than necessary to preserve their shapes under the loading conditions) are not considered in this section, since the stresses in them cannot be determined by elementary static methods.

for the desired stress S_1. Trying section $2-2$, the force system on the lower part (Fig. 13c) is a concurrent one and includes four unknown stresses S_1, S_2, S_5, and S_6; it is not solvable. Trying section $3-3$, the force system on the left part (Fig. 13d) is nonconcurrent with three unknown stresses S_1, S_7, and S_8; it is solvable. In some instances different sections may be used, each leading to a solution.

S_1 having been determined, the force system of Fig. 13b becomes solvable, and then, with S_2 also determined, the force system of Fig. 13c may be solved.

Algebraic Solution. Following the general method of procedure outlined above, determine the various stresses by employing algebraic conditions of equilibrium in manners similar to those illustrated in Section 2.6.

In making the imaginary separations of the truss, care should be taken to cut not more than three members in which the stresses are unknown. It is advantageous to make the separation so that not more than two such members are cut. If this is done, a single force polygon will determine the two unknowns, whereas if three are cut, a force polygon and an equilibrium polygon, or the equivalent, are necessary for determining the three unknowns.

Analysis by Method of Joint Resolution.[†] Consider the pin at each joint as a body acted upon by forces in equilibrium.

Algebraic Solution. Determine the various stresses by employing algebraic conditions of equilibrium in manners similar to those illustrated in Section 2.6.

Graphic Solution. For each joint, draw the force polygon. In doing so, it will be advantageous to represent the forces in the order in which they occur about the joint. A force polygon so drawn will be called a polygon for the joint; for brevity, if the order taken

---

[†]This method is not usually so convenient for determining algebraically the stress in a single specified member.

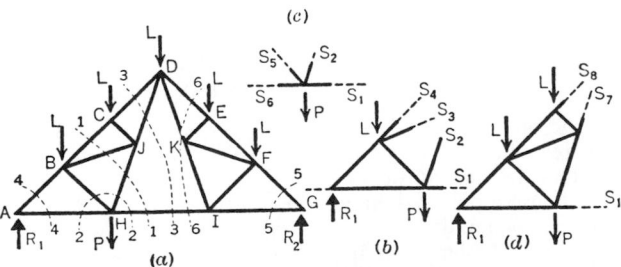

Fig. 13 Truss sections.

is clockwise, the polygon will be called a clockwise polygon, and if counterclockwise, it will be called a counterclockwise polygon. If the polygons for all the joints of a truss are drawn separately, the stress in each member will have been represented twice. It is possible to combine the polygons so that it will not be necessary to represent the stress in any member more than once, thus reducing the number of lines to be drawn. Such a combination of force polygons is called a *stress diagram*. Each triangular space in the truss diagram is marked by a small letter; also the space between consecutive action lines of the loads and reactions. Then the two letters on opposite sides of any line serve to designate that line, and the same large letters are used to designate the magnitude of the corresponding force.

To construct a stress diagram for a truss under given loads: (a) Determine the reactions. (b) Letter the truss diagram as directed. (c) Construct a force polygon for all the external forces applied to the truss (loads and reactions), representing them in the order in which their application points occur about the truss, clockwise or counterclockwise. (d) On the sides of that polygon construct the polygons for all the joints. They must be clockwise or counterclockwise according as the polygon for the loads and reactions was drawn clockwise or counterclockwise. (The first polygon drawn must be for a joint at which only two members are fastened; the joints at the supports are usually such. Next, that joint is considered, and its polygon is drawn, at which not more than two stresses are unknown.)

Example 5. Figure 14 represents a roof truss sustaining loads of 600, 1000, 1200, and 1800 lb; the right reaction is 2100 lb, and the left 2500 lb. *ABCDEFA* is a polygon for the loads and reactions, these being represented in the order in which their points of application occur about the truss. The polygon for joint 1 is *FABGF*; the force *BG* acts toward the joint, hence *bg* is under compression, and *GF* acts away from the joint, hence *gf* is in tension. The polygon for joint 2 is *CDEHC*; the force *EH* acts away

from the joint, hence *eh* is in tension; and *HC* acts toward the joint, hence *hc* is in compression. The polygon for joint 3 is *HEFGH*; the force *GH* acts away from the joint and hence *gh* is in tension. If the work has been done correctly, *GH* is parallel to *gh*. (In Fig. 14*a* all the polygons are clockwise, and in Fig. 14*b* counterclockwise.)

2.7 Center of Gravity

Definitions The *centroid* of a system of parallel forces having fixed application points is the point through which their resultant will always pass regardless of how the forces may be turned, provided they remain parallel.

The *center of gravity* of a body[*] or system of bodies is the *centroid of the forces of gravitation*[†] acting upon all the particles thereof. Referring the application points of such a force system to a set of coordinate axes, the coordinates of the centroid, or center of gravity (CG), are

$$\bar{x} = \frac{\sum F_i \cdot x_i}{\sum F_i} = \frac{\int x\, dF}{F}$$

$$\bar{y} = \frac{\sum F_i \cdot y_i}{\sum F_i} = \frac{\int y\, dF}{F}$$

$$\bar{z} = \frac{\sum F_i \cdot z_i}{\sum F_i} = \frac{\int z\, dF}{F}$$

in which F_i represents the force on (or weight of) one particle and x_i, y_i, z_i are the coordinates of its application point. If a group of bodies is involved, the coordinates of the center of gravity of the group are

$$\bar{x} = \frac{\sum W_i \cdot \bar{x}_i}{\sum W_i} \qquad \bar{y} = \frac{\sum W_i \cdot \bar{y}_i}{\sum W_i} \qquad \bar{z} = \frac{\sum W_i \cdot \bar{z}_i}{\sum W_i}$$

in which W_i represents the weight of one body and $\bar{x}_i, \bar{y}_i, \bar{z}_i$ are the coordinates of its center of gravity.[‡] A body (or system of bodies), if supported at its center of gravity, will remain at rest in any position.

The center of gravity of part of a body may be located by the rule that its moment, with respect to any plane, equals the moment of the whole minus, algebraically, the moment of the remainder.

The center of gravity of a line, surface, or volume is that point that would be the center of gravity if

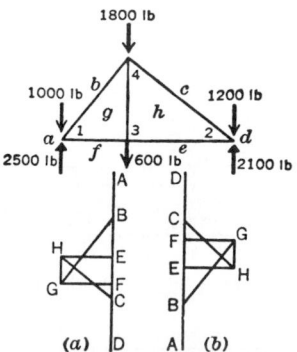

Fig. 14 Roof truss/polygon solution.

---

[*]Sometimes called *center of mass* or *center of inertia*.

[†]For practical purposes, the forces of gravitation may be considered as parallel.

[‡]As F (in the gravitational system of units) equals W, these symbols may be used interchangeably in the two sets of formulas. Also, any one of the expressions may be read as "the sum of the moments divided by the sum of the forces (or weights)."

the line were replaced by a homogeneous rod of infinitesimal diameter, the surface by a homogeneous plate of infinitesimal thickness, or the volume by a homogeneous body.

Symmetry. Two points are symmetrical with respect to a third point if the line joining the two is bisected by the third. Two points are symmetrical with respect to a line or a plane if the line joining them is perpendicular to the given line or plane and is bisected by it. A body, line, surface, or volume is symmetrical with respect to a point, a line, or a plane if all the points of the body, line, surface, or volume can be paired so that each pair is symmetrical with respect to the point, line, or plane. If a homogeneous body or a line, surface, or volume is symmetrical with respect to a point, line, or plane, its center of gravity is at the point, in the line, or in the plane.

The *static moment* of a body (having weight), a line (having length), a surface (having area), or a solid* (having volume) with respect to any plane is the product of the weight, length, area, or volume and the distance of the center of gravity of the body, line, surface, or solid from the plane. The static moment of a plane line or plane surface with respect to a straight line in the plane is the product of the length or area and the distance of the center of gravity of the line or surface from the reference line. A static moment is regarded as positive or negative according as the corresponding center of gravity is on the positive or negative side of the reference plane or line.

Determination of Center of Gravity Location

When practicable, determination of center of gravity location by algebraic or integration methods, based on dividing the sum of the moments by the sum of the forces, is generally the simplest process. For some bodies of nonhomogeneous nature or of very irregular shape, one of the following methods of procedure may be necessary or at least preferable:

Graphic Method. For application to plane figures.[†] Referring to Fig. 15, take a point O and a line bb on opposite sides of the figure at any convenient distance m apart; project any width of the figure parallel to bb as aa on bb, connect the projections bb with O, and note the intersections cc; determine other points cc and draw a smooth curve through them as shown; measure the area A' within the curve cc; then A'm is the static

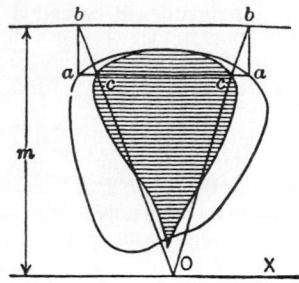
Fig. 15 Center-of-gravity determination.

moment of the given figure with respect to OX; if A is the area of the given figure and y the distance of its center of gravity from OX, $y = A'm/A$. In a similar way the distance of the center of gravity from a line perpendicular to OX can be determined and its exact position thus definitely located.

Suspension Method. For application to plane figures.[‡] Suspend the body (or a model representing it) from a point near its edge and mark on it the direction of a plumb line hung from that point. Repeat this operation using a second suspension point. The center of gravity is at (or behind) the intersection of the two markings.

Weighing Method. Generally applied where location of the center of gravity in one plane only is required. Determine weight W of the body and then support it on a knife edge (Fig. 16) and on a point support resting upon a platform scale. Weigh reaction R of the point support and measure the horizontal distance a between the point and the knife edge. Then the horizontal distance from the knife edge to the center of gravity is $\overline{x} = Ra/W$.

Balancing Method. For general application. Balance the body (or a model representing it) on a straightedge, marking on the body the vertical plane containing the edge, Repeat for two more balancing positions of the body. The center of gravity is at the point common to the three planes thus determined.

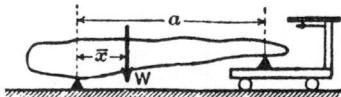
Fig. 16 Weighing/knife edge method.

---

*Herein the word "solid" denotes "that which has volume." Care should be taken to distinguish this from a "body," which has "mass" (as well as "volume"). Some writers use the word solid to denote at various times either volume or mass, which is sometimes confusing. In this section, the word volume is frequently used even in preference to solid to avoid the possibility of confusion with mass.
†Including areas or flat homogeneous bodies of uniform thickness.

‡Including areas or flat bodies of uniform homogeneous thickness.

2.8 Moment of Inertia

Plane Surfaces — Definitions The *moment of inertia* of a plane surface (or figure) with respect to (or about) a line (or axis) is the sum of the products obtained by multiplying the area of each element of the surface by the square of its distance from the line*. Letting I_x denote the moment of inertia about an X axis,

$$I_x = \int y^2 \, dA$$

in which A is the total area and y is the perpendicular distance of any element of area dA from the axis. The moment of inertia of a surface is obviously the sum of the moments of inertia of its parts. The moment of inertia of a plane surface is *rectangular* if the axis used is *in* the plane of the area; it is *polar* if the axis is *perpendicular* to the plane of the area.

The *radius of gyration* of a plane surface with respect to a line is the length whose square multiplied by the area of the surface equals the moment of inertia of the surface with respect to the line. Letting k denote the radius of gyration,

$$I = k^2 A \qquad \text{or} \qquad k = \sqrt{I/A}$$

in which I is the moment of inertia and A the area.

The *product of inertia* of a plane surface with respect to a pair of coordinate axes in the plane is the algebraic sum of the products obtained by multiplying the area of each element of the surface by its coordinates[†] Letting U_{xy} denote the product of inertia with respect to X and Y axes,

$$U_{xy} = \int xy \, dA$$

in which A is the total area and x and y are the coordinates of any element of area dA.

The *principal axes of inertia* of a plane surface at a particular point in the plane are the two axes about which the moments of inertia are greater and less than for any other axis through the point in the plane.[‡] The corresponding moments of inertia are called the *principal moments of inertia* of the surface at the point.

The principal axes are always at right angles to each other. The product of inertia with respect to them is zero.

The *customary engineer's unit* for both moment and product of inertia of a surface is biquadratic inches (in.[4]).

Determination of Moment of Inertia of Plane Surfaces When practicable, determination of moment of inertia with respect to an axis by algebraic or integration methods is generally the simplest process. For some surfaces of very irregular shape, the following graphic method of procedure may be necessary or at least preferable.

Graphic Method. Let *aaaa* (Fig. 17) be the outline and XX' the axis with respect to which the moment of inertia is desired; at any convenient distance m from XX' draw two parallels (but if XX' does not cut the figure, only one parallel, the one on the opposite side of the figure from XX'); draw any line as *aa* parallel to XX' and project the points *aa* on the nearer parallel; join the projections *bb* to any point O in XX', and note the intersections *cc* on *aa*; project the same parallel; join the projections *dd* with O, and note the intersections *ee* on *aa*. In a similar manner determine points like *ee* for other widths like *aa*, and connect all points *e* as shown. Then measure the area of the loops OPO and OQO; denoting this combined area by A'', $I = A'' m^2$. (There will be only one loop if only one parallel *bb* is used.)

Transformation Formulas — Plane Surfaces

Parallel-Axes Theorems. Let $I =$ moment of inertia (either rectangular or polar) of a plane figure with respect to any line or axis, $\bar{I} =$ that with respect to a parallel axis passing through the center of gravity of the figure, $d =$ distance between the axes, k, $k =$ radii of gyration with respect to the same area, respectively, and $A =$ area of the figure; then

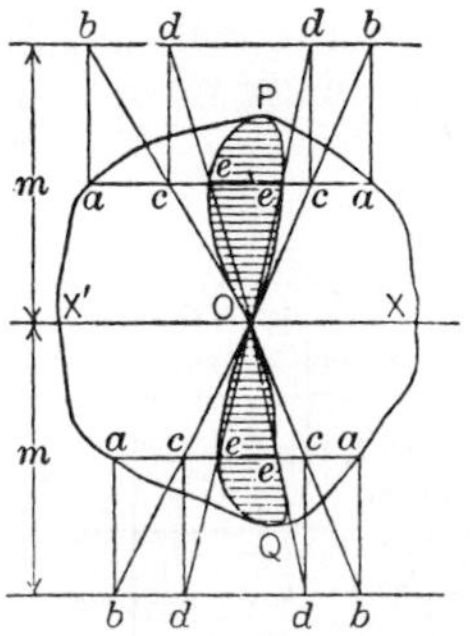

---

*Moment of inertia is always positive and never zero. Product of inertia may be positive, zero, or negative, depending on the distribution of the area with respect to the axes. If a surface has an axis of symmetry, its product of inertia with respect to that axis and one perpendicular thereto is zero.

[†]In certain special cases, as for axes through the point in the center of a circular area, the moment of inertia is the same for any axis and therefore there is no principal axis through that point.

[‡]If U_{xy} and $I_y - I_x$ are both zero, there is no principal axis through the point.

Fig. 17 Moments of inertia/graphic method.

$$I = \overline{I} + Ad^2 \qquad \text{and} \qquad k^2 = \overline{k}^2 + d^2$$

These show that with respect to all parallel axes the moment of inertia and the radius of gyration are least for the one passing through the center of gravity of the figure.

Similarly, let U = product of inertia of a plane figure with respect to a pair of coordinate axes in the plane and $\overline{U}$ = that with respect to a parallel pair whose origin is at the center of gravity; $\overline{x}$, $\overline{y}$ the coordinates of the center of gravity referred to the first pair, and A the area of the figure; then $U = \overline{U} + A\overline{x}\overline{y}$.

Relation of Rectangular and Polar Moments of Inertia.
Let I_x, I_y, J_z = moments of inertia of a plane figure with respect to x, y, and z axes, respectively, the axes being *at right angles* to each other and the x and y axes in the plane; and let k_x, k_y, k_z = corresponding radii of gyration; then $J_z = I_x + I_y$, $k_z^2 = k_x^2 + k_y^2$.

Rotated-Axes Theorem.
Let XOY and UOV (Fig. 18) be two sets of rectangular coordinate axes with a common origin and in a given plane figure; I_x, I_y, I_u, I_v = moments of inertia of the figure with respect to x, y, u, and v axes, respectively; U_{xy}, U_{uv} = its products of inertia with respect to the sets of axes, respectively; α = angle through which x axis must be rotated to bring it into u axis, regarded as positive or negative according as the turning is counterclockwise or clockwise. Then $I_u + I_v = I_x + I_y$ and

$$I_u = I_x \cos^2 \alpha + I_y \sin^2 \alpha - U_{xy} \sin 2\alpha$$

$$U_{uv} = \tfrac{1}{2}(I_x - I_y)\sin 2\alpha + U_{xy}\cos 2\alpha$$

If OU and OV are principal axes (see definition already given), $U_{uv} = 0$ and therefore $\tan 2\alpha = 2U_{xy}/(I_y - I_x)$. Hence, the principal axes of a figure at a point can be readily found if the moments of inertia and the product of inertia of the figure with respect to two rectangular axes through the point and in the plane are known. The principal moments of inertia are then I_u from the preceding formula and I_v from the same formula after replacing α by $\alpha + \pi/2$. As a check, $I_u + I_v = I_x + I_y$.

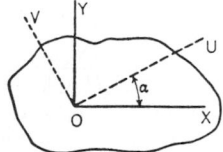

Fig. 18 Rotated-axes theorem.

Graphic Transformations — Plane Surfaces
The *inertia circle* is a device for determining *graphically*, the moment of inertia of a plane figure with respect to any line of the plane through a given point and the principal axes and principal moments of inertia for the same point. To construct the circle, it is necessary to know the moments of inertia and the product of inertia with respect to two rectangular axes through the point in the plane figure. Suppose I_x, I_y, and U_{xy} given for the shaded area in Fig. 19. To convenient scale, plot OX' and OY' to represent I_x and I_y and $Y'A$ to represent U_{xy} (downward if negative and upward if positive). Center C is midway between X' and Y'. With CA as radius, describe the inertia circle. To find I_u, draw chord AB parallel to axis OU; draw perpendicular BU'. OU' (to scale) = I_u, and BU' (to scale) = U_{uv}. OM, parallel to $A2$, is axis of least I; and a parallel to $A1$, through O, is axis of greatest I. $O2$ (to scale) is the value of least $I = I_2$; and $O1$, value of greatest $I = I_1$. Least radius of gyration for an axis through point $O = \sqrt{I_2/\text{area}}$.

Bodies — Definitions
The moment of inertia of a body with respect to (or about) a line (or axis) is the sum of the products obtained by multiplying the mass of each elementary part by the square of its distance from the line.* Letting I_x denote the moment of inertia about an X axis,

$$I_x = \int y^2\, dm$$

in which m is the total mass and y is the perpendicular distance of any element of mass dm from the

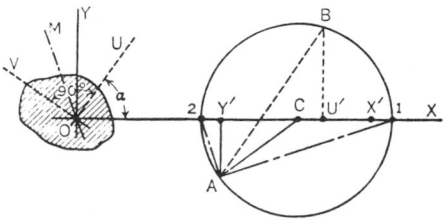

Fig. 19 Inertia circle.

---

*Moment of inertia is always positive and never zero. Product of inertia may be positive, zero, or negative, depending on the distribution of the mass with respect to the coordinate planes. If a body has a plane of symmetry, its product of inertia with respect to that plane and one perpendicular thereto is zero.

axis.* The moment of inertia of a body is obviously the sum of the moments of inertia of its parts.

The *center of gyration* of a body with respect to a line is a point at such a distance from the line that, if the entire mass of the body were concentrated there, its moment of inertia would be the same as that of the body.

The **radius of gyration of a body** with respect to a line is the distance from the center of gyration to the line. Letting k denote the radius of gyration,

$$I = k^2 m \qquad \text{or} \qquad k = \sqrt{\frac{I}{m}}$$

in which I is the moment of inertia and m the mass.

The **product of inertia of a body** with respect to a pair of coordinate planes is the algebraic sum of the products obtained by multiplying the mass of each element of the body by its coordinates with reference to those planes. Thus with respect to *YOZ* and *ZOX* (Fig. 20), *ZOX* and *XOY*, and *XOY* and *YOZ* planes, the products of inertia are, respectively,

$$U_{xy} = \int xy \, dm \quad U_{yz} = \int yz \, dm \quad U_{zx} = \int zx \, dm$$

Principal Axes of Inertia of a Body at a Particular Point.

The values of moments of inertia of a body for all axes through a given point are in general unequal; for one axis the moment of inertia is greater and for another it is less than for any other axis through the point. These two axes are at right angles, and they together with one at right angles to their plane and passing through the point are *principal axes of inertia* of the body at the point; the corresponding moments of inertia are the *principal moments of inertia* of the body at the point. If the point is the center of gravity of the body, the axes and moments are called *central principal axes* and *central principal moments of inertia*. For a set of principal axes, the

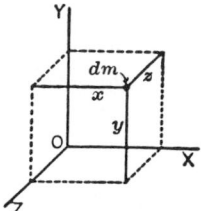

Fig. 20 Products of inertia.

three products of inertia, with respect to the *principal planes* determined by them, are zero.

The **customary engineer's unit** for both moment and product of inertia is slug-ft.^{2}

Transformation Formulas — Bodies

Parallel-Axes Theorems.

Let I = moment of inertia of a body with respect to any line or axis, $\bar{I}$ = that with respect to a parallel axis passing through the center of gravity of the body, d = distance between the axes, $k, \bar{k}$ = radii of gyration with respect to the same axes, respectively, and m = mass of the body; then

$$I = \bar{I} + md^2 \qquad \text{and} \qquad k^2 = \bar{k}^2 + d^2$$

Rotated-Axes Theorems.

Let I_x, I_y, and I_z denote the moments of inertia of a body with respect to rectangular axes x, y, and z, respectively; U_{xy}, U_{yz}, and U_{zx} its products of inertia with respect to yz and zx planes, zx and xy planes, and xy and yz planes, respectively; I the moment of inertia of the body with respect to a line through the origin of coordinates having direction angles α, β, and γ; then

$$I = I_x \cos^2 \alpha + I_y \cos^2 \beta + I_z \cos^2 \gamma - 2U_{yz} \cos \beta$$
$$\times \cos \gamma - 2U_{zx} \cos \gamma \cos \alpha - 2U_{xy} \cos \alpha \cos \beta$$

If $U_{xy} = U_{yz} = 0$, the y axis is a principal axis at the origin. If $U_{yz} = U_{zx} = 0$, the z axis is a principal axis at the origin. If $U_{zx} = U_{xy} = 0$, the x axis is a principal axis at the origin.

If a homogeneous body has a plane of symmetry, any perpendicular to the plane is a principal axis of the body at the point where the line pierces the plane. If it has two planes of symmetry at right angles to each other, their intersection is a principal axis at any point of the intersection, the other two being in the planes of symmetry. If it has three planes of symmetry, their lines of intersection are the central principal axes of the body.

Properties of Various Lines, Surfaces, Volumes, and Bodies

Symbols.

I_i = rectangular moment of inertia; k_i = corresponding radius of gyration; J_o = polar moment of inertia about axis through O perpendicular to plane; k_o = corresponding radius of gyration; m = mass = W/g, where W = weight and g = acceleration due to gravity. Moments of inertia of bodies are given in terms of mass. For their values in terms of weight, replace m by W in the formulas.

---

*Strictly speaking, this is the moment of inertia of the mass of a body. If m be replaced by w = weight, the result is the moment of inertia of the weight of a body.

Decimal Equivalents (for reference in using Table 1):

$$\pi = 3.1416 \qquad \sqrt{2} = 1.414 \qquad \frac{1}{\sqrt{2}} = 0.707$$

$$\frac{\pi}{2} = 1.5708 \qquad \sqrt{3} = 1.732 \qquad \frac{1}{\sqrt{3}} = 0.577$$

$$\frac{\pi}{4} = 0.7854 \qquad \sqrt{5} = 2.236 \qquad \frac{1}{\sqrt{5}} = 0.447$$

$$\frac{\pi}{8} = 0.3927 \qquad \sqrt{6} = 2.449 \qquad \frac{1}{\sqrt{6}} = 0.408$$

$$\frac{\pi}{32} = 0.0982 \qquad \sqrt{8} = 2.828 \qquad \frac{1}{\sqrt{8}} = 0.354$$

$$\frac{\pi}{64} = 0.0491 \qquad \sqrt{10} = 3.162 \qquad \frac{1}{\sqrt{10}} = 0.316$$

$$\frac{\pi}{128} = 0.0245 \qquad \sqrt{12} = 3.464 \qquad \frac{1}{\sqrt{12}} = 0.289$$

$$\frac{1}{\pi} = 0.318 \qquad \sqrt{18} = 4.242 \qquad \frac{1}{\sqrt{18}} = 0.238$$

3 KINEMATICS

3.1 Motions of a Particle

Definitions *Motion* of a particle is the continual change of its position with respect to some base point or coordinate set. Particle motion is characterized by a time-dependent position vector $\mathbf{r}(t)$ that describes the location of the particle with respect to the base point or coordinates.

Rectilinear motion is motion along a straight line.

Curvilinear motion is motion along a curved path that may be planar or in three dimensions.

Displacement of a particle is its change of position and is a vector quantity. The displacement vector associated with the times t_1 and t_2, $\Delta \mathbf{r}$, is given by

$$\Delta \mathbf{r} = \mathbf{r}(t_2) - \mathbf{r}(t_1)$$

Velocity of a particle is the time rate of change of its displacement and is a vector quantity. If the time interval $t_2 - t_1$ is denoted as Δt, then

$$\mathbf{v} = \lim_{\Delta t \to 0} \frac{\Delta \mathbf{r}}{\Delta t} = \frac{d\mathbf{r}}{dt}$$

Acceleration of a particle is the time rate of change of its velocity vector and is itself a vector quantity:

$$\mathbf{a} = \lim_{\Delta t \to 0} \frac{\Delta \mathbf{v}}{\Delta t} = \frac{d\mathbf{v}}{dt}$$

3.2 Rectilinear Motion

Velocity. Let $s =$ distance measured along the path of a particle, $s_1 =$ distance from origin at time t_1, $s_2 =$ distance at a later time t_2, $\Delta s = s_2 - s_1 =$ displacement* in time interval $\Delta t = t_2 - t_1$. Then *average velocity* $= \Delta s / \Delta t$. If the position changes at at uniform rate (which implies no change in sense), actual velocity at any time is $\Delta s / \Delta t$. For every case, *instantaneous velocity* is

$$v = \frac{ds}{dt} = \lim_{\Delta t \to 0} \frac{\Delta s}{\Delta t}$$

The *unit of velocity* is any distance unit divided by any time unit. Units commonly used are feet per second, meters per second, miles per hour, and so on.

Acceleration. Let $v_1 =$ velocity of particle at time t_1, $v_2 =$ velocity at a later time t_2, $\Delta_v = v_2 - v_1 =$ change in velocity in time interval $\Delta t = t_2 - t_1$. Then *average acceleration* $= \Delta_v / \Delta_t$. If the velocity changes at a uniform rate, the actual acceleration at any time is Δ_v / Δ_t. For every case, *instantaneous acceleration* is

$$a = \frac{dv}{dt} = \frac{d^2 s}{dt^2} = \lim_{\Delta t \to 0} \frac{\Delta_v}{\Delta_t}$$

The *unit of acceleration* is the distance unit divided by the square of any time unit. Units commonly used are feet per second per second (feet per second squared), meters per second squared, and so on.)

Formulas for Determination of a, v, s, t. If s is given algebraically in terms of t, then v and a may be determined in terms of t by differentiation as indicated previously. If a is given algebraically in terms of t, then v and s may be determined in terms of t by integration. Other relations not involving t may be determined by similar methods. The common formulas are

$$v = \frac{ds}{dt} \qquad a = \frac{dv}{dt} = \frac{d^2 s}{dt^2} \qquad \frac{a}{v} = \frac{dv}{ds}$$

$$s_2 - s_1 = \int_{t_1}^{t_2} v \, dt \qquad v_2 - v_1 = \int_{t_1}^{t_2} a \, dt$$

$$t_2 - t_1 = \int_{s_1}^{s_2} \frac{ds}{v} = \int_{v_1}^{v_2} \frac{dv}{a} \qquad v_2^2 - v_1^2 = 2\int_{s_1}^{s_2} a \, ds$$

---

*The difference in distances along the path equals the displacement *only* when the path is a straight line. (See definition of displacement.)

Table 1   Centroids of Lines and Plane Areas

Lines	
Figure	**Centroid Location**

1. Any plane curve

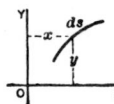

CG is at point having coordinates $\bar{x}$, $\bar{y}$, where

$$\bar{x} = \frac{\int x\,ds}{\text{length}} \quad \text{where } ds = \sqrt{1 + \left(\frac{dy}{dx}\right)^2}\,dx$$

$$\bar{y} = \frac{\int y\,ds}{\text{length}} \quad \text{where } ds = \sqrt{1 + \left(\frac{dx}{dy}\right)^2}\,dy$$

2. Circular arc

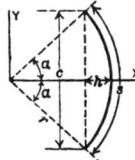

CG is on axis of symmetry at $\bar{x} = r\sin\alpha/\alpha = rc/s$. If α is small, distance from CG to chord = approx. $2h/3$. (Error is small even for $\alpha = 45°$.)

For semicircle $\bar{x} = 2r/\pi$
For quadrant: $\bar{x} = 2r\sqrt{2}/\pi$ and distance from radius drawn to either end of arc = $2r/\pi$

3. Any Plane surface

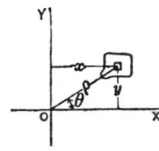

CG is at point having coordinates $\bar{x}$, $\bar{y}$, where

$$\bar{x} = \frac{\iint x\,dx\,dy}{\text{area}} = \frac{\iint \rho^2\cos\theta\,d\rho\,d\theta}{\text{area}}$$

$$\bar{y} = \frac{\iint y\,dx\,dy}{\text{area}} = \frac{\iint \rho^2\sin\theta\,d\rho\,d\theta}{\text{area}}$$

$$I_x = \iint y^2\,dx\,dy; \quad I_y = \iint x^2\,dx\,dy; \quad J_0 = \iint \rho^3\,d\rho\,d\theta = I_x + I_y$$

$$k_x = \sqrt{\frac{I_x}{\text{area}}}; \quad k_y = \sqrt{\frac{I_y}{\text{area}}}; \quad k_0 = \sqrt{\frac{J_0}{\text{area}}} = \sqrt{\frac{I_x + I_y}{\text{area}}}$$

4. Triangle

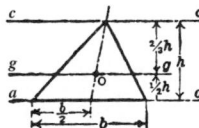

CG is at O = intersection of medians. Perpendicular distance from $a - a = \frac{1}{3}h$

$$I_g = \frac{bh^3}{36}; \quad I_a = \frac{bh^3}{12}; \quad I_c = \frac{bh^3}{4}$$

$$k_g = \frac{h}{3\sqrt{2}}; \quad k_a = \frac{h}{\sqrt{6}}; \quad k_c = \frac{h}{\sqrt{2}}$$

5. Solid rectangle (or square)

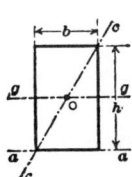

CG is at O = intersection of diagonals.
For rectangle:

$$I_g = \frac{bh^3}{12}; \quad I_a = \frac{bh^3}{3}; \quad I_c = \frac{b^3h^3}{6(b^2 + h^2)}; \quad J_0 = \frac{bh(b^2 + h^2)}{12}$$

$$k_g = \frac{h}{2\sqrt{3}}; \quad k_a = \frac{h}{\sqrt{3}}; \quad k_c = \frac{bh}{\sqrt{6(b^2 + h^2)}}; \quad k_0 = \sqrt{\frac{b^2 + h^2}{12}}$$

For square (letting $b = h = s$):

$$I_g = \frac{s^4}{12}; \quad I_a = \frac{s^4}{3}; \quad I_c = \frac{s^4}{12}; \quad J_0 = \frac{s^4}{6}$$

$$k_g = \frac{s}{2\sqrt{3}}; \quad k_a = \frac{s}{\sqrt{3}}; \quad k_c = \frac{s}{2\sqrt{3}}; \quad k_0 = \frac{s}{\sqrt{6}}$$

Table 1 *(Continued)*

Lines	
Figure	Centroid Location

6. Hollow rectangle (or square)

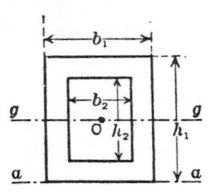

CG is at. O = intersection of diagonals. For hollow rectangle:

$$I_g = \frac{(b_1h_1^3 - b_2h_2^3)}{12}; \qquad I_a = \frac{b_1h_1^3}{3} - \frac{b_2h_2(3h_1^2 + h_2^2)}{12}$$

$$k_g = \sqrt{\frac{b_1h_1^3 - b_2h_2^3}{12(b_1h_1 - b_2h_2)}}; \qquad J_0 = \frac{b_1h_1(b_1^2 + h_1^2) - b_2h_2(b_2^2 + h_2^2)}{12}$$

For hollow square (letting $b_1 = h_1 = s_1$ and $b_2 = h_2 = s_2$):

$$I_g = \frac{1}{12}(s_1^4 - s_2^4); \qquad I_a = \frac{1}{3}s_1^4 - \frac{1}{12}s_2^2(3s_1^2 + s_2^2)$$

$$k_g = \sqrt{\frac{1}{12}(s_1^2 + s_2^2)}; \qquad J_0 = \frac{1}{6}(s_1^4 - s_2^4)$$

(Note: For a diagonal $c - c, I_c = I_g$ and $k_c = k_g$)

7. Trapezoid

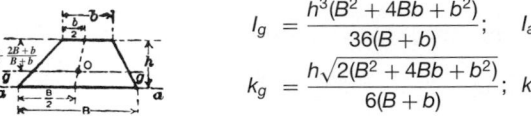

CG is at O, located as shown.

$$I_g = \frac{h^3(B^2 + 4Bb + b^2)}{36(B + b)}; \qquad I_a = \frac{h^3(B + 3b)}{12}$$

$$k_g = \frac{h\sqrt{2(B^2 + 4Bb + b^2)}}{6(B + b)}; \qquad k_a = \frac{h}{\sqrt{6}}\sqrt{\frac{B + 3b}{B + b}}$$

8. Quadrilateral

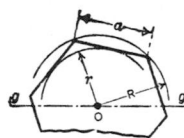

CG is at O, located as follows: Divide the sides into thirds and construct the paralleogram with sides passing through the third points as shown. The intersection of the diagonals of this parallelogram is the desired centroid.

9. Regular polygon

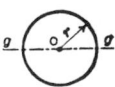

CG is at O = geometric center. Let $g - g$ be any axis through O and in plane of polygon. Then

$$I_g = \frac{1}{24}\text{area} \cdot (6R^2 - a^2) = \frac{1}{48}\text{area} \cdot (12r^2 + a^2)$$

$$J_0 = \frac{1}{12}\text{area} \cdot (6R^2 - a^2) = \frac{1}{24}\text{area} \cdot (12r^2 + a^2)$$

$$k_g = \sqrt{\frac{1}{24}(6R^2 - a^2)} = \sqrt{\frac{1}{48}(12r^2 + a^2)}$$

$$k_0 = \sqrt{\frac{1}{12}(6R^2 - a^2)} = \sqrt{\frac{1}{24}(12r^2 + a^2)}$$

10. Circle

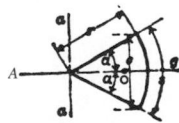

CG is at O = geometric center.

$$I_g = \frac{1}{4}\pi r^4 = \frac{1}{64}\pi d^4; \qquad J_0 = \frac{1}{2}\pi r^4 = \frac{1}{32}\pi d^4$$

$$k_g = \frac{1}{2}r = \frac{1}{4}d; \qquad k_0 = \frac{r}{\sqrt{2}} = \frac{d}{\sqrt{8}}$$

11. Circular sector

CG is on axis of symmetry at O. Distance from $a - a = 2r\sin\alpha/(3\alpha) = 2rc/(3s)$.

$$A = \text{area} = r^2\alpha$$

$$I_g = \frac{Ar^2}{4}\left(1 - \frac{\sin\alpha\cos\alpha}{\alpha}\right); \qquad I_\alpha = \frac{Ar^2}{4}\left(1 + \frac{\sin\alpha\cos\alpha}{\alpha}\right)$$

$$k_g = \frac{r}{2}\sqrt{1 - \frac{\sin\alpha\cos\alpha}{\alpha}}; \qquad k_\alpha = \frac{r}{2}\sqrt{1 + \frac{\sin\alpha\cos\alpha}{\alpha}}$$

(Continues)

Table 1 (*Continued*)

Plane Surfaces	
Figure	Centroid Location; Moments of Inertia; Radii of Gyration

12. Semicircle

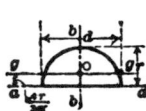

CG is on axis of symmetry at O. Distance from $a - a = 4r/(3\pi) = 0.424r$.

$$I_g = \frac{d^4(9\pi^2 - 64)}{1152\pi} = \frac{r^4(9\pi^2 - 64)}{72\pi} = 0.1098r^4$$

$$I_a = I_b = \frac{\pi d^4}{128} = \frac{\pi r^4}{4}; J_o = r^4\left(\frac{\pi}{4} - \frac{8}{9\pi}\right) = 0.5025r^4$$

$$k_g = \frac{d\sqrt{9\pi^2 - 64}}{12\pi} = \frac{r\sqrt{9\pi^2 - 64}}{6\pi} = 0.264r; \ k_a = k_b = \frac{d}{4} = \frac{r}{2}$$

$$k_o = r\sqrt{\frac{1}{2} - \frac{16}{9\pi^2}} = 0.566r$$

13. Circular segment

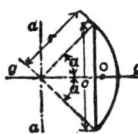

CG is on axis of symmetry at O. Distance from

$$a - a = \frac{2r^3\sin^3 ga}{3A} = \frac{c^3}{12A} \text{ where } A = \text{area} = \frac{r^2(2\alpha - \sin 2\alpha)}{2}.$$

$$I_g = \frac{Ar^2}{4}\left(1 - \frac{2\sin^3\alpha\cos\alpha}{3(\alpha - \sin\alpha\cos\alpha)}\right); \quad I_a = \frac{Ar^2}{4}\left(1 + \frac{2\sin^3\alpha\cos\alpha}{\alpha - \sin\alpha\cos\alpha}\right)$$

$$k_g = \frac{r}{2}\sqrt{1 - \frac{2\sin^3\alpha\cos\alpha}{3(\alpha - \sin\alpha\cos\alpha)}}; \quad k_a = \frac{r}{2}\sqrt{1 + \frac{2\sin^3\alpha\cos\alpha}{\alpha - \sin\alpha\cos\alpha}}$$

14. Annulus

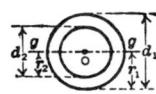

CG is at O = geometric center.

$$I_g = \frac{1}{64}\pi(d_1^4 - d_2^4) = \frac{1}{4}\pi(r_1^4 - r_2^4); \quad J_o = \frac{1}{32}\pi(d_1^4 - d_2^4) = \frac{1}{2}\pi(r_1^4 - r_2^4)$$

$$k_g = \sqrt{\frac{1}{4}(d_1^2 + d_2^2)} = \sqrt{\frac{1}{2}(r_1^2 + r_2^2)}; \quad k_o = \sqrt{\frac{1}{8}(d_1^2 + d_2^2)} = \sqrt{\frac{1}{2}(r_1^2 + r_2^2)}$$

15. Ellipse

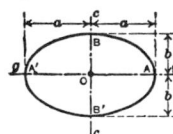

CG is at O = geometric center. *For semiellipse ABB′, CG is on OA at distance to right of $c - c = 4a/(3\pi)$. For quarter-ellipse ABO, CG is at distance to right of $c - c = 4a/(3\pi)$ and at distance above $g - g = 4b/(3\pi)$.*

$$I_g = \frac{1}{4}\pi ab^3 = \frac{1}{4}Ab^2; \quad I_c = \frac{1}{4}\pi a^3 b = \frac{1}{4}Aa^2; \quad J_o = \frac{1}{4}A(a^2 + b^2)$$

$$k_g = \frac{1}{2}b; \quad k_c = \frac{1}{2}a; \quad k_o = \sqrt{\frac{1}{2}(a^2 + b^2)}$$

16. Parabolic segment

CG is on axis of symmetry at O. Distance from $c - c = 3a/5$.

$$I_g = \frac{4ab^3}{15}; \quad I_c = \frac{4a^3 b}{7}$$

$$k_g = \frac{b}{\sqrt{5}} = 0.447b; \quad k_c = a\sqrt{\frac{3}{7}} = 0.654a$$

Homogeneous Bodies (Including Nonplanar Surfaces)[a]

17. Any surface or body of revolution

Let axis of revolution be X axis. Then generating curve is $y = f(x)$. CG is at point having coordinates $\bar{x}, \bar{y}, \bar{z}$.
For surface:

$$\bar{x} = \frac{\int 2\pi xy\, ds}{\int 2\pi y\, ds} = \frac{\int xy\sqrt{1 + (dy/dx)^2}\, dx}{\int y\sqrt{1 + (dy/dx)^2}\, dx}; \quad \bar{y} = 0; \quad \bar{z} = 0$$

For body (letting δ = density = m/volume):

$$\bar{x} = \frac{\int \pi xy^2\, dx}{\int \pi y^2\, dx}; \quad \bar{y} = 0; \quad \bar{z} = 0$$

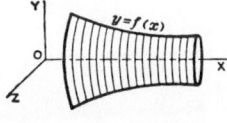

$$I_x = \frac{1}{2}\pi\delta\int y^4\, dx; \quad I_y = I_z = \pi\delta\int\left(\frac{1}{4}y^4 + x^2 y^2\right) dx$$

$$k_x = \sqrt{\frac{I_x}{m}}; \quad k_y = k_z = \sqrt{\frac{I_y}{m}} = \sqrt{\frac{I_z}{m}}$$

Table 1   *(Continued)*

Homogeneous Bodies *(Continued)*
Figure	Centroid Location; Moments of Inertia; Radii of Gyration

For thin shell having mass: CG. coordinates are same as for surface.

$$I_x = 2\pi\delta \int y^3 \, ds = 2\pi\delta \int y^3 \sqrt{1 + \left(\frac{dy}{dx}\right)^2} \, dx; \quad k_x = \sqrt{\frac{I_x}{m}}$$

18. Thin straight rod

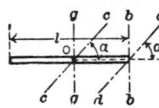

CG is at O = geometric center. For body:

$$I_g = \frac{1}{12} m l^2; \quad I_b = \frac{1}{3} m l^2; \quad I_c = \frac{1}{12} m l^2 \sin^2 a; \quad I_d = \frac{1}{3} m l^2 \sin^2 \alpha$$

$$k_g = \frac{l}{\sqrt{12}}; \quad k_b = \frac{l}{\sqrt{3}}; \quad k_c = \frac{l \sin \alpha}{\sqrt{12}}; \quad k_d = \frac{l \sin \alpha}{\sqrt{3}}$$

19. Thin rod bent into circular arc

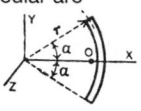

CG is on axis of symmetry at $\bar{x} = \frac{(r \sin\alpha)}{\alpha}$. For body:

$$I_x = \frac{mr^2}{2}\left(1 - \frac{\sin\alpha\cos\alpha}{\alpha}\right); I_y = \frac{mr^2}{2}\left(1 + \frac{\sin\alpha\cos\alpha}{\alpha}\right); I_z = mr^2$$

$$k_x = r\sqrt{\frac{1}{2} - \frac{\sin\alpha\cos\alpha}{2\alpha}}; k_y = r\sqrt{\frac{1}{2} + \frac{\sin\alpha\cos\alpha}{2\alpha}} \quad k_z = r$$

20. Rectangular paral lelepiped (or cube)

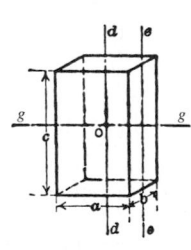

CG is at O = geometric center.

For parallelepiped:

$$I_g = \frac{1}{12}m(b^2 + c^2); \quad I_d = \frac{1}{12}m(a^2 + b^2); \quad I_e = \frac{1}{12}m(4a^2 + b^2)$$

$$k_g = \sqrt{\frac{1}{12}(b^2 + c^2)}; \quad k_d = \sqrt{\frac{1}{12}(a^2 + b^2)}; \quad k_e = \sqrt{\frac{1}{12}(4a^2 + b^2)}$$

For cube (letting $a = b = c = s$):

$$I_g = I_d = \frac{ms^2}{6}; \quad I_e = \frac{5ms^2}{12}$$

$$k_g = k_d = \frac{s}{\sqrt{6}}; \quad k_e = s\sqrt{\frac{5}{12}}$$

21. Right rectangular pyramid

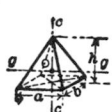

CG is on axis of symmetry at O. Distance from base = $h/4$. Drawing g–g axis through O parallel to side a:

$$I_g = \frac{m}{20}\left(b^2 + \frac{3h^2}{4}\right); \quad I_c = \frac{m}{20}(a^2 + b^2)$$

$$k_g = \sqrt{\frac{4b^2 + 3h^2}{80}}; \quad k_c = \sqrt{\frac{a^2 + b^2}{20}}$$

22. Pyramid (or frustum of pyramid)

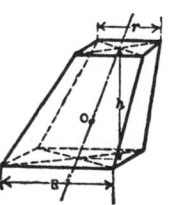

For surface of any pyramid: CG of surface (base excluded) is on line joining apex with centroid of perimeter of base at a distance two-thirds its length from the apex.

For body of any pyramid: CG of body is on line joining apex with centroid of base at a distance three-fourths its length from the apex.

For surface of frustum of pyramid having rectangular bases: Letting R and r be the lengths of sides of the larger and smaller bases respectively, and h the altitude: CG of surface (bases excluded) is at distance from larger base = $h(R + 2r)/[3(R + r)]$.

For body of frustum of any pyramid: Letting A and a be the areas of the larger and smaller bases, respectively, and h the altitude: CG of body is at distance from larger base = $h(A + 2\sqrt{Aa} + 3a)/[4(A + \sqrt{Aa} + a)]$.

(Continues)

Table 1 (*Continued*)

Homogeneous Bodies (*Continued*)
Figure   Centroid Location; Moments of Inertia; Radii of Gyration

23. Right elliptical cylinder (or circular cylinder)

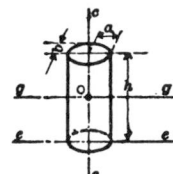

CG is at O = geometric center.

For right elliptical cylinder:

$$I_g = \frac{1}{12}m(3b^2 + h^2); \qquad I_c = \frac{1}{4}m(a^2 + b^2); \qquad I_e = \frac{1}{12}m(3r^2 + 4h^2)$$

$$k_g = \sqrt{\frac{1}{12}(3b^2 + h^2)}; \qquad k_c = \sqrt{\frac{1}{2}(a^2 + b^2)}; \qquad k_e = \sqrt{\frac{1}{12}(3r^2 + 4h^2)}$$

For right circular cylinder (letting $a = b = r$):

$$I_g = \frac{1}{12}m(3r^2 + h^2); \qquad I_c = \frac{1}{2}mr^2$$

$$k_g = \sqrt{\frac{3r^2 + h^2}{12}}; \qquad k_c = \frac{r}{\sqrt{2}}$$

24. Hollow right circular cylinder

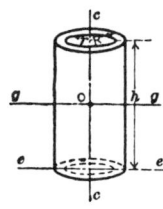

CG is at O = geometric center.

$$I_g = \frac{1}{12}m(3R^2 + 3r^2 + h^2); \quad I_c = \frac{1}{2}m(R^2 + r^2); \quad I_e = \frac{1}{12}m(3R^2 + 3r^2 + 4h^2)$$

$$k_g = \sqrt{\frac{1}{12}(3R^2 + 3r^2 + h^2)}; \quad k_c = \sqrt{\frac{1}{2}(R^2 + r^2)}; \quad k_e = \sqrt{\frac{1}{12}(3R^2 + 3r^2 + 4h^2)}$$

For thin shell (radius R):

$$I_g = \frac{1}{12}m(6R^2 + h^2); \quad I_c = mR^2; \quad I_e = \frac{1}{6}m(3R^2 + 2h^2)$$

$$k_g = \sqrt{\frac{1}{12}(6R^2 + h^2)}; \quad k_c = R; \quad k_e = \sqrt{\frac{1}{6}(3R^2 + 2h^2)}$$

25. Right circular cone

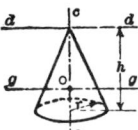

CG is on axis of symmetry at O. Distance from base = $h/4$. Drawing g–g axis through O and d–d axis through apex, both parallel to base:

$$I_g = \frac{3m}{20}\left(r^2 + \frac{h^2}{4}\right); \quad I_c = \frac{3mr^2}{10}; \quad I_d = \frac{3m}{20}(r^2 + 4h^2)$$

$$k_g = \sqrt{\frac{3}{80}(4r^2 + h^2)}; \quad k_c = \frac{3r}{\sqrt{30}}; \quad k_d = \sqrt{\frac{3}{20}(r^2 + 4h^2)}$$

26. Frustum of right circular cone

CG is on axis of symmetry at O.

$$\text{Distance from base} = \frac{h(R^2 + 2Rr + 3r^2)}{4(R^2 + Rr + r^2)}$$

$$I_c = \frac{3m(R^5 - r^5)}{10(R^3 - r^3)}; \quad k_c = \sqrt{\frac{3(R^5 - r^5)}{10(R^3 - r^3)}}$$

27. Cone (or frustum of cone)

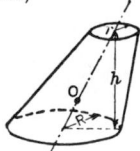

For surface of any cone: CG of surface (base excluded) is on line joining apex with centroid of perimeter of base at a distance two-thirds its length from the apex.

For body of any cone: CG of body is on line joining apex with centroid of base at a distance three-fourths its length from the apex.

For surface of frustum of a circular cone: Letting R and r be the radii of the larger and smaller bases, respectively, and h the altitude: CG of surface (bases excluded) is at distance from larger base = $h(R + 2r)/[3(R + r)]$.

For body of frustum of a circular cone: Letting R and r be the radii of the larger and smaller bases, respectively, and h the altitude: CG of body is at distance from larger base = $h(R^2 + 2Rr + 3r^2)/[4(R^2 + Rr + r^2)]$.

Table 1 *(Continued)*

Homogeneous Bodies *(Continued)*	
Figure	Centroid Location; Moments of Inertia; Radii of Gyration

28. Thin circular lamina

CG is at O = geometric center.

$$I_g = \frac{1}{4}mr^2; \quad I_c = \frac{1}{2}mr^2 \text{(where } c\text{-}c \text{ axis is perpendicular to the plane)}$$

$$k_g = \frac{1}{2}r; \quad k_c = \frac{r}{\sqrt{2}}$$

29. Sphere

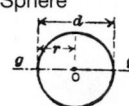

CG is at O = geometric center.

$$I_g = \frac{2mr^2}{5}; k_g = \frac{2r}{\sqrt{10}}$$

30. Hollow sphere

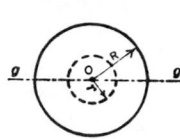

CG is at O = geometric center.

$$I_g = \frac{2m}{5}\left(\frac{R^5 - r^5}{R^3 - r^3}\right); \quad k_g = \sqrt{\frac{2}{5}\left(\frac{R^5 - r^5}{R^3 - r^3}\right)}$$

For thin shell (radius R):

$$I_g = \frac{2mR^2}{3}; \quad k_g = \frac{2R}{\sqrt{6}}$$

31. Spherical sector

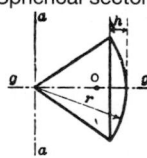

CG is on axis of symmetry at O. Distance from center of sphere = $3(2r - h)/8$.

$$I_g = \frac{m}{5}(3rh - h^2); k_g = \sqrt{\frac{3rh - h^2}{5}}$$

32. Hemisphere

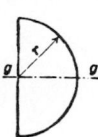

For surface: CG is on axis of symmetry at distance from center of sphere = $r/2$.
For body: CG is on axis of symmetry at distance from center of sphere = $3r/8$.

$$I_g = \frac{2mr^2}{5}; \quad k_g = \frac{2r}{\sqrt{10}}$$

33. Spherical segment

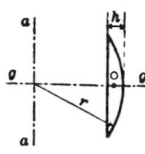

CG is on axis of symmetry at distance from center of
  sphere = $3(2r - h)^2/[4(3r - h)]$.

$$I_g = m\left(r^2 - \frac{3rh}{4} + \frac{3h^2}{20}\right)\frac{2h}{(3r - h)}$$

$$k_g = \sqrt{\left(r^2 - \frac{3rh}{4} + \frac{3h^2}{20}\right)\frac{2h}{3r - h}}$$

34. Torus

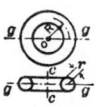

CG is at O = geometric center.

$$I_g = \frac{1}{8}m(4R^2 + 5r^2); \quad I_c = \frac{1}{4}m(4R^2 + 3r^2)$$

$$k_g = \sqrt{\frac{1}{8}(4R^2 + 5r^2)}; \quad k_c = \sqrt{\frac{1}{2}(4R^2 + 3r^2)}$$

35. Ellipsoid

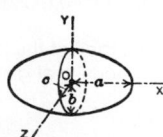

CG is at O = geometric center. CG of *one octant* is at point having coordinates:

$$\bar{x} = \frac{3a}{8}; \quad \bar{y} = \frac{3b}{8}; \quad \bar{z} = \frac{3c}{8}$$

For complete ellipsoid:

$$I_x = \frac{1}{5}m(b^2 + c^2); \quad I_y = \frac{1}{5}m(a^2 + c^2); \quad I_z = \frac{1}{5}m(a^2 + b^2)$$

$$k_x = \sqrt{\frac{1}{5}(b^2 + c^2)}; \quad k_y = \sqrt{\frac{1}{5}(a^2 + c^2)}; \quad k_x = \sqrt{\frac{1}{5}(a^2 + b^2)}$$

(Continues)

Table 1 (*Continued*)

Homogeneous Bodies (*Continued*)	
Figure	Centroid Location; Moments of Inertia; Radii of Gyration

36. Paraboloid

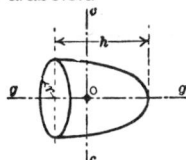

CG is on axis of symmetry at O. Distance from base = $h/3$.

$$I_g = \frac{1}{3}mr^2; \quad I_c = \frac{1}{18}m(3r^2 + h^2)$$

$$k_g = \frac{r}{\sqrt{3}}; \quad k_c = \sqrt{\frac{3r^2 + h^2}{18}}$$

---

[a] "Body" is to be understood unless "Surface" is indicated.

For *uniform acceleration*, a = constant; $v = at + v_0$; $s = \frac{1}{2}at^2 + v_0t + s_0$; $v^2 = 2a(s - s_0) + v_0^2$; v_0 being initial velocity and s_0 initial distance.

If algebraic relations between a, v, s, and t are not given but a number of pairs of corresponding values of two of the variables are known, curves may be plotted for the approximate determination of other corresponding pairs of values and of other unknowns within the range of the data. Such curves are discussed below under "Motion Graphs."

Examples of Rectilinear Motion

Falling Body.* If a body *falls from rest* in a vacuum, $v_0 = 0$, $s_0 = 0$, and $a = g = 32.2$ ft/sec² (approx). Hence $v = gt = \sqrt{2gs}$; $s = \frac{1}{2}gt^2$. If a body is *projected upward* at an initial velocity v_0, $a = -g$ and the formulas become $v = -gt + v_0 = \sqrt{-2gs + v_0^2}$, $s = -\frac{1}{2}gt^2 + v_0t$. Total ascent (to highest position) is $v_0^2/2g$ and time required v_0/g.

Crank and Connecting Rod Mechanism. The problem is to find expressions for the velocity and acceleration of any point in the crosshead, as A in Fig. 21. Such a point describes rectilinear motion. Let $c = r/l$, n = revolutions per second (assumed constant), ω = radians of angle described by crank per second, and s = distance of A from its extreme left position, all distances expressed in feet. Then.

$$s = (l + r) - l(1 - c^2 \sin^2 \theta)^{1/2} - r\cos\theta$$

$$v = r\omega\left(\sin\theta + \frac{c\sin 2\theta}{2(1 - c^2\sin^2\theta)^{1/2}}\right)$$

$$a = r\omega^2\left(\cos\theta + \frac{c\cos 2\theta + c^3\sin^4\theta}{(1 - c^2\sin^2\theta)^{3/2}}\right)$$

---
*Rotation is disregarded and the body is considered as a particle.

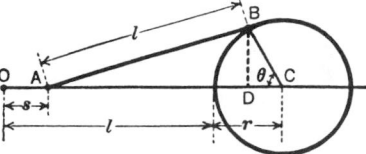

Fig. 21 Motion graph.

These formulas are exact; close approximations are

$$s = r(1 - \cos\theta) + \frac{1}{4}cr(1 - \cos 2\theta)$$

$$v = r\omega(\sin\theta + \frac{1}{2}c\sin 2\theta)$$

$$a = r\omega^2(\cos\theta + c\cos 2\theta)$$

Motion Graphs Space–time, velocity–time, acceleration–time, velocity–space, and acceleration–space curves for a particle are graphs showing the relations between magnitudes of s and t, v and t, a and t, v and s, and a and s, respectively. Figures 22–26 illustrate such graphs but do not correspond to the same motion.

Space–Time Diagram. In Fig. 22, the *slope* of the curve at any point represents the magnitude of the velocity. If AB and BC are measured by the s and t scales of the drawing, respectively, the slope equals the velocity magnitude; thus if $AB = 0.2$ in. and $BC = 0.4$ in., $v = 0.4/4 = 0.1$ ft/sec.

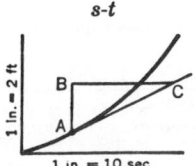

Fig. 22 Motion graph.

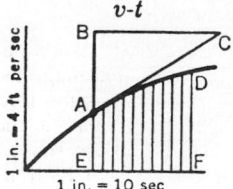

Fig. 23   Velocity versus time.

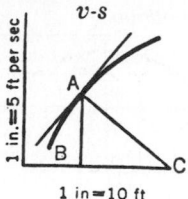

Fig. 25   Velocity versus distance.

Velocity–Time Diagram.   In Fig. 23, the slope of the curve at any point represents the magnitude of the acceleration.* If AB and BC are measured by the v and t scales, respectively, the slope equals the acceleration magnitude; thus if $AB = 0.3$ in. and $BC = 0.5$ in., $a = 1.2/5 = 0.24$ ft/sec^2. The *area* included between any two ordinates (as AE and DF), the curve, and the t axis represents the displacement† of the moving point in the time EF. If the area is below the time axis, it is considered minus. If the area is computed by multiplying its average ordinate measured by the velocity scale (this being the average velocity) by EF measured by the time scale, the product equals the displacement; thus if the average ordinate is 0.35 in. and EF is 0.4 in., the displacement is $1.4 \times 4 = 5.6$ ft.

Acceleration–Time Diagram.   In Fig. 24, the slope represents the rate at which the acceleration is changing. The area (plus above and minus below time axis) included between any two ordinates (as AE and DF), the curve, and the t axis represents the velocity change in the time EF.‡ Thus if the average ordinate is 0.3 in. and EF is 0.2 in., the velocity change is $6 \times 2 = 12$ ft/sec.

Velocity–Space Diagram.   In Fig. 25, the subnormal represents the acceleration.§ If the length of the subnormal is multiplied by the square of the velocity scale number and the product is divided by the space scale number, the result will equal the acceleration; thus suppose that the subnormal $BC = \frac{1}{3}$in., then $a = (\frac{1}{3} \times 25)/10 = 0.83$ ft/sec^2.

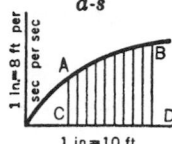

Fig. 26   Acceleration versus distance.

Acceleration–Space Diagram.   In Fig. 26, the area (plus above and minus below the space axis) included between two ordinates (as AC and BD), the curve, and the s axis represents the change in the velocity square. If the area is computed by multiplying the mean ordinate measured by the acceleration scale by CD measured by the space scale, the product times 2 equals the change in the velocity square; thus if the average ordinate is 0.3 in. and $CD = 0.4$ in., the change is $2.4 \times 4 \times 2 = 19.2$ ft^2/ sec^2.

Simple Harmonic Motion

Simple Harmonic Motion and Its Motion Graphs.
These have wide application in physics and engineering. If a point P moves in a circular path of radius r at uniform speed, its projection on any diameter has *simple harmonic motion*. The radius r is called the *amplitude*. The *period* is the time required for the projection to go from one end of the diameter to the other and back. The *frequency* is the number of periods per unit time, which makes it the reciprocal of the period. Angle XOP (Fig. 27) (considered as less than 2π radians) is the *phase angle. The* displacement at any time is the distance of the point having simple harmonic motion from the center of its path or range.

When $t = 0$, let P be at P_0. ε is called the *lead angle* (*lag*, if negative). For simple harmonic motion (SHM) of V in the vertical diameter, $y = r\sin(\theta + \varepsilon) = r\sin(\omega t + \varepsilon)$, in which $\omega = d\theta/dt$ in radians per unit time (i.e., 2π times the frequency):

$$v_y = r\omega \cos(\omega t + \varepsilon) = \omega x$$

$$a_y = -r\omega^2 \sin(\omega t + \varepsilon) = -\omega^2 y$$

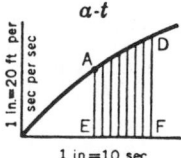

Fig. 24   Acceleration versus time.

---

*For curvilinear motion, this is tangential acceleration only.
†For curvilinear motion, this is distance along the path (not displacement).
‡For rectilinear motion only.
§For curvilinear motion, this is tangential acceleration only.

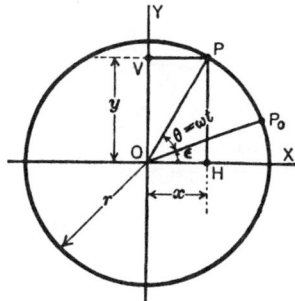

Fig. 27 Simple harmonic motion.

For SHM of H in horizontal diameter,

$$x = r \cos(\theta + \varepsilon) = r \cos(\omega t + \varepsilon)$$

$$v_x = -r\omega \sin(\omega t + \varepsilon) = -\omega y$$

$$a_x = -r\omega^2 \cos(\omega t + \varepsilon) = -\omega^2 x$$

If the time is reckoned from the instant when V is in its midposition and moving upward, $\varepsilon = 0$. The three curves (Fig. 28) $OA, O'B$, and OC are the space–time, velocity–time, and acceleration–time curves, respectively, for one complete period of a simple harmonic motion; $\varepsilon = 0$; Ot represents the period; the values of y, v, and a marked are for position Q, shown. In Fig. 28a the curve is the velocity–space curve and the inclined line the acceleration–space curve. They show how v and a vary with the displacement of the moving point; thus for the position Q (Fig. 28b), v and a have values as marked.

From the preceding equations and curves, it will be noted that simple harmonic motion may be defined also as any rectilinear motion in which the acceleration is always directed toward a fixed point in the path and is proportional to the distance between that point and the moving point.*

---

*A common example of simple harmonic motion is the motion of a weight suspended from an elastic spring.

3.3 Curvilinear Motion

Velocity. If s is distance measured along the curved path of a particle, then the magnitude of velocity (speed) at any instant is ds/dt; the linear direction of the velocity is tangent to the path at the instantaneous position of the particle; and the sense of the velocity corresponds to the direction of motion of the particle at the instant.

The velocity vector changes in magnitude and direction. In Fig. 29, let A, B, C represent positions of particle P in its curved path; s distance along the path; and v_1, v_2, v_3 velocity vectors at A, B, C. Plot velocity vectors $O'A'$, $O'B'$, $O'C'$, and so on, from any origin O' to represent the velocities at A, B, C, and so on. The curve $A'B'C'$, drawn through the ends of the vectors, is called a *hodograph* for the motion. For every position of P in its path, there is a corresponding position P' in the hodograph; and P' describes distances s' on the hodograph while P describes distance s on the path. Vector $O'P'$ represents the velocity of P. In time Δt, P moves from A to C, its velocity changes from $O'A'$ to $O'C'$, and the velocity change is $A'C'$.

Acceleration. Referring to the hodograph (Fig. 29), average acceleration for interval Δt, during which particle P moves from A to C, is vector $A'C'/\Delta t$, and it has the direction of the chord $A'C'$. The instantaneous acceleration of P at $A = a =$ limit of the average acceleration as Δt approaches zero,

$$a = \lim_{\Delta t \to 0} \left(\frac{\text{vector } A'C'}{\Delta t} \right)$$

$$= \lim_{\Delta t \to 0} = \frac{\text{arc } A'B'C'}{\Delta t} = \frac{ds'}{dt} = \text{speed of } P'$$

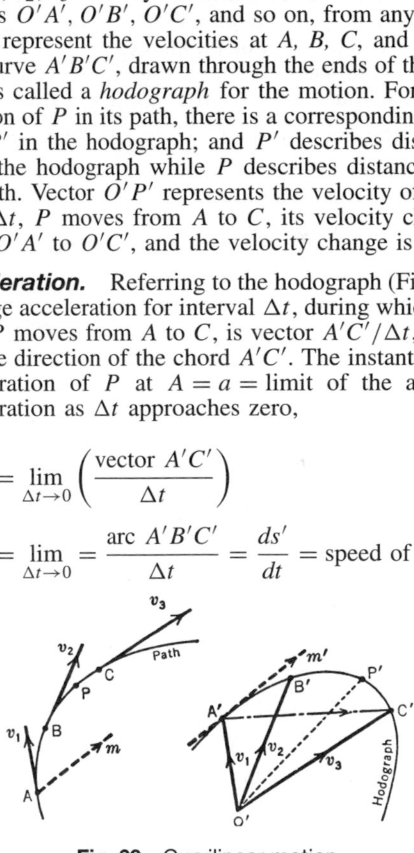

Fig. 29 Curvilinear motion.

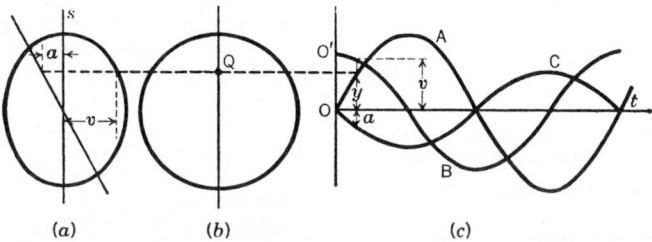

Fig. 28 (a) Space–time curve; (b) velocity–time curve; (c) acceleration–time curve.

on the hodograph. The direction of a is along the tangent $A'm'$, and as P' is moving clockwise, the sense is as indicated by arrow at m'. Hence acceleration at A is Am, parallel to $A'M'$ and $= ds'/dt$. Its tangential component is ds/dt' and its normal component is v^2/ρ, ρ being the radius of curvature at A. *The unit of acceleration is any velocity unit divided by any time unit.*

Components of Velocity and Acceleration

Components of Velocity and Acceleration of a Particle for Any Curved Path (*not necessarily planar*).
The position of the particle P being defined by its coordinates, x, y, z, the axial components of velocity are $v_x = dx/dt$, $v_y = dy/dt$, and $v_z = dz/dt$. Resultant velocity $v = \sqrt{v_x^2 + v_y^2 + v_z^2}$, and its direction cosines are $\cos\theta_x = v_x/v$, $\cos\theta_y = v_y/v$, and $\cos\theta_z = v_z/v$. The axial components of acceleration are

$$a_x = \frac{dv_x}{dt} = \frac{d^2x}{dt^2} \qquad a_y = \frac{dv_y}{dt} = \frac{d^2y}{dt^2}$$

$$a_z = \frac{dv_z}{dt} = \frac{d^2z}{dt^2}$$

The resultant acceleration $a = \sqrt{a_x^2 + a_y^2 + a_z^2}$ and its direction cosines are

$$\cos\phi_x = \frac{a_x}{a} \qquad \cos\phi_y = \frac{a_y}{a} \qquad \cos\phi_z = \frac{a_z}{a}$$

The tangential and normal components of acceleration are $a_t = dv/dt = d^2s/dt^2$ and $a_n = v^2/\rho$, ρ being the radius of curvature. The resultant acceleration is

$$a = \sqrt{a_t^2 + a_n^2} = \sqrt{a_x^2 + a_y^2 + a_z^2}$$

If the path is a plane curve, $v_z = 0$ and $a_z = 0$.

The preceding discussion shows that velocities and accelerations (like forces) may be composed or resolved according to the parallelogram and parallelepiped laws.

Motion of a Projectile

Projectile*Describing Plane Curvilinear Motion.
In the following formulas air resistance is neglected; $v_0 = $ velocity of projection; $\theta = $ angle of projection

---

*Rotation is disregarded and the body is considered as a particle.

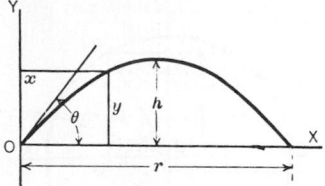

Fig. 30   Projectile motion.

(Fig. 30); x, $y = $ coordinates of the projectile at any time t after projection; $v = $ velocity; v_x and v_y are x and y components, respectively, of v; $r = $ range on the horizontal plane through O; $\theta_1 = $ value of θ for maximum r; $h = $ greatest height attained; and $T = $ time of flight. The path of the projectile, or the trajectory, is a parabola as represented, and a set of parametric equations for it are

$$x = v_0 \cos\theta \cdot t \qquad y = v_0 \sin\theta \cdot t - \tfrac{1}{2}gt^2$$

from which

$$y = x \tan\theta - \frac{gx^2}{2v_0^2 \cos^2\theta}$$

Also,

$$v_x = v_0 \cos\theta \qquad\qquad r = \frac{\sin 2\theta \cdot v_0^2}{g}$$

$$v_y = v_0 \sin\theta - gt \qquad \theta_1 = 45°$$

$$v = \sqrt{v_0^2 - 2gy} \qquad\quad T = \frac{2v_0 \sin\theta}{g}$$

$$h = \frac{\sin^2\theta \cdot v_0^2}{2g}$$

If the direction of projection is horizontal, $\theta = 0$; the equation of the path is $y = -gx^2/2v_0^2$; and $x = v_0t$ and $y = -\tfrac{1}{2}gt^2$.

The fact that the horizontal component of velocity is constant indicates that the hodograph of the motion of a projectile is a straight vertical line.

Motion Graphs   Motion graphs similar to those previously discussed for rectilinear motion may be constructed for curvilinear motion of a particle. Great care must be exercised, however, in interpreting the significance of slopes, areas, and subnormals when acceleration or distance is involved. In this connection, reference should be made to the footnotes referred to in the previous discussion.

In general, accelerations obtained are tangential components only, while "displacements" must be replaced by "distances along the curve." Thus, in the velocity–time graph (Fig. 23), the slope of the curve represents the magnitude of the *tangential component* of the acceleration,* while the area under the curve represents the *distance along* the curve.

3.4 Motions of a Body

Translation of a *rigid body* is a motion such that each straight line in it remains fixed in direction. The paths of all particles of the body are exactly alike, straight or curved (not necessarily plane curves); the displacements of all particles during a given time are the same; the velocities of all particles at any instant are the same; and their accelerations at any instant are the same. For these reasons, it is customary to use the expressions "velocity of the body" and "acceleration of the body." The motion is described by the same formulas as those previously derived for rectilinear and curvilinear motions of a particle.

Rotation of a rigid body is a motion such that one line of the body, or of its extension, remains fixed. The fixed line is the *axis*. The plane through the mass center perpendicular to the axis is the *plane of rotation*.

Plane motion of a rigid body is a motion such that each particle of the body moves in a plane at a constant distance from a fixed plane through the mass center (called the *plane of motion*), while each line of the body parallel to the plane of motion turns through the same angle in the same time interval.

Three-dimensional motion of a rigid body is a term covering all types of motion in three-dimensional space, including pure translation along a skewed curve as a special case. Even in the most general case, any three-dimensional motion of a rigid body may be regarded as consisting of two components: one a translation equal to that of the mass center and the other a rotation about some axis through the mass center.

Angular displacement of a rigid body is the change of angular position of any line in the plane of motion. The angular displacement of a rigid body *cannot* be expressed as a vector quantity.

Angular velocity of a rigid body is its time rate of angular displacement (i.e., rate of change of angular position). The angular velocity of a rigid body can be expressed as a vector quantity.

Angular acceleration of a rigid body is its time rate of change of angular velocity. The angular acceleration of a rigid body can be expressed as a vector quantity.

3.5 Rotation

Angular Velocity. The paths of all particles are circles with centers on the axis. Since all lines of the body parallel to the plane of rotation sweep out equal angles in equal times, it is customary to describe rotation by the behavior of one radial line. In Fig. 31,

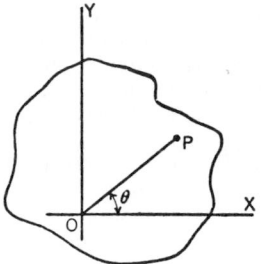

Fig. 31 Angular displacement.

let θ be the angle from the x axis to the radial line OP. $\Delta\theta = \theta_2 - \theta_1$ is the angular displacement of the body in the time $\Delta t = t_2 - t_1$ and is expressed in any angular unit,

$$\text{Average angular velocity} = \frac{\theta_2 - \theta_1}{t_2 - t_1} = \frac{\Delta\theta}{\Delta t}$$

If the angle changes at a uniform rate, actual velocity at any time equals $\Delta\theta/\Delta t$. For every case, the instantaneous angular velocity is

$$\omega = \lim_{\Delta t \to 0}\left(\frac{\Delta\theta}{\Delta t}\right) = \frac{d\theta}{dt}$$

The sign depends on the numerator of the fraction, or the way in which θ is changing. The *unit of angular velocity* is any angular displacement unit divided by any time unit, such as radians per second, revolutions per minute, and so on.

Angular velocity is generally expressed as a vector quantity. The direction of the vector is along the instantaneous axis of rotation, and the magnitude of the vector is equal to the instantaneous angular velocity. The sense along the axis is such that the right-hand rule applies (if the right hand grasps the axis of rotation with the fingers curled in the direction of rotation, the right thumb points in the direction of the angular velocity vector).

Angular Acceleration

$$\text{Average angular acceleration} = \frac{\omega_2 - \omega_1}{t_2 - t_1} = \frac{\Delta\omega}{\Delta t}$$

If the angular velocity changes at a uniform rate, actual angular acceleration at any time equals $\Delta\omega/\Delta t$. For every case, the instantaneous angular acceleration is

$$\alpha = \lim_{\Delta t \to 0}\left(\frac{d\omega}{\Delta t}\right) = \frac{d\omega}{dt} = \frac{d^2\theta}{dt^2}$$

---

*However, if a velocity–time graph were made for motion along the hodograph of the original motion, the slope of the curve would represent the magnitude of the *total* acceleration of the original particle along the original path.

The sign of α depends on the numerator of the fraction or on the way in which ω is changing. The *unit of angular acceleration* is any angular velocity unit divided by any time unit, as radians per second per second (i.e., radians per second squared).

Formulas for Determination of $\alpha, \omega, \theta, t$***.*** The formulas are exactly analogous to those previously derived for rectilinear motion, a, v, and s being replaced by α, ω, and θ, respectively. The formulas are

$$\omega = \frac{d\theta}{dt} \qquad \alpha = \frac{d\omega}{dt} = \frac{d^2\theta}{dt^2} \qquad \frac{\alpha}{\omega} = \frac{d\omega}{d\theta}$$

$$\theta_2 - \theta_1 = \int_{t_1}^{t_2} \omega\, dt \qquad \omega_2 - \omega_1 = \int_{t_1}^{t_2} \alpha\, dt$$

$$t_2 - t_1 = \int_{\theta_1}^{\theta_2} \frac{d\theta}{\omega} = \int_{\omega_1}^{\omega_2} \frac{d\omega}{\alpha}$$

$$\omega_2^2 - \omega_1^2 = 2 \int_{\theta_1}^{\theta_2} \alpha\, d\theta$$

Angular acceleration is generally expressed as a vector quantity. The magnitude of the vector is equal to the instantaneous angular acceleration. The direction of the angular acceleration vector is perpendicular to the plane formed by successive values of the angular velocity vector. The sense along the axis is such that the right-hand rule applies (if the right hand grasps the axis of rotation with the fingers curled in the direction of increasing angular velocity, the right thumb points in the direction of the angular acceleration vector).

In the case of planar motion, there is no plane formed by successive values of the angular velocity vector. In this case, the angular velocity vector and the angular acceleration vector both lie perpendicular to the plane of motion.

Relations between Rectilinear and Angular Velocities and Accelerations. Let ω and α, respectively. be instantaneous angular velocity and acceleration of a rotating body and v and a the corresponding instantaneous rectilinear velocity and acceleration of a point P of the body located at distance r from the axis of rotation. Then

$$v = r\omega \quad a_t = r\alpha \quad a_n = r\omega^2 \quad a = r\sqrt{\alpha^2 + \omega^4}$$

The sense of v must agree with the sense of ω, and the sense of a_t with the sense of α. The sense of a_n is always toward the axis.

Motion Graphs   Motion graphs analogous to those previously discussed for rectilinear motion may be constructed to show the relations between angular displacement, velocity and acceleration, and time. θ, ω, and α correspond to s, v, and a, respectively.

3.6   Plane Motion

Any displacement resulting from plane motion may be accomplished by a translation of the body that will bring any one line of it, which is perpendicular to the plane of motion, into final position, followed by a rotation of the body about that line into final position. The necessary amount of translation depends on the line of the body selected as axis of the rotation; the amount of the rotation does not. The *state of motion* of a body at any instant may be regarded as consisting of two components, a translational motion and a rotational motion. Thus a plane motion may be traced by giving the history of the movement of one point of the body (called a base point) in its own curved path and a description of the rotation of the body about the selected base point.* The point selected as base should be one for which the motion is readily specified. For a wheel rolling along a straight path, the center would be selected as a base point.

Velocity of any point P of the body at any instant with respect to a fixed point O is the vector sum of the velocity of base point A with respect to O and of the velocity of P with respect to A due to rotation about A. Thus (Fig. 32) O is the fixed point, A the moving base point, and P any other point of the body at distance r from A; v_1 is the velocity of A with respect to O, and $v_2 = r\omega$ is the velocity of P with respect to A. The resultant velocity of P with respect to O is v, or $v_{ptoO} = v_{PtoA} + v_{AtoO}$. In vector from this relation can be written as the vector cross product:

$$\mathbf{V}_P = \mathbf{V}_A + \boldsymbol{\omega} \times \mathbf{r}_P$$

Acceleration of any point P with respect to a fixed point O at any instant has two components; one is that of the base point A with respect to O and the other that of P with respect to base A. Acceleration of P with respect to A is rotational and is conveniently replaced by its tangential and normal components, $a_t = r\alpha$ and $a_n = r\omega^2$. Then resultant acceleration of

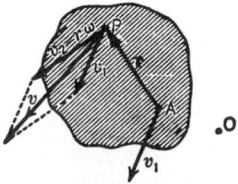

Fig. 32   Velocity of a point on a rigid body.

---

*To simplify matters, "points" are referred to throughout this and the following discussion but "lines" through the points perpendicular to the plane of motion should be understood. Thus, in Fig. 32 and 33, the parallel lines through P, A, and O perpendicular to the plane of motion move relative to each other.

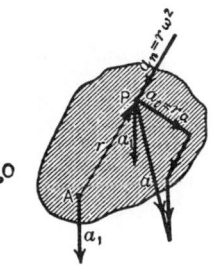

Fig. 33 Acceleration of a point on a rigid body.

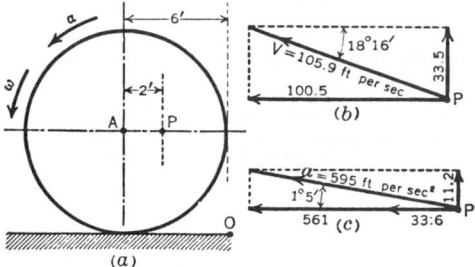

Fig. 34 Instantaneous axis of rotation.

P, with respect to O is the vector sum of $r\alpha$, $r\omega^2$, and acceleration of A with respect to O. Thus (Fig. 33) a_1 is acceleration of A to O and acceleration P to A is resultant of a_t and a_n. Acceleration of P to O is a, the vector sum of a_1, a_t, and a_n. In vector form this relation can be written as three vector cross products:

$$\mathbf{a}_P = \mathbf{a}_A + \boldsymbol{\alpha} \times \mathbf{r} + \boldsymbol{\omega} \times (\boldsymbol{\omega} \times \mathbf{r})$$

Instantaneous Axis. For a body having plane motion, there is always one point in it (or in its extension) at each instant for which the velocity with respect to A (Fig. 32) is equal and opposite to velocity of A with respect to O; that is, its velocity is zero at the instant. This point Q is called the *instantaneous (or instant) center* of rotation, and a line through Q perpendicular to the plane of motion is called the *instantaneous axis*. Since Q is at rest for the instant, the resultant velocities of all points at the instant are purely rotational about the instant axis. The instant center is the intersection of two lines drawn from any two points, C and D, in the plane of the motion perpendicular to their velocities. If the velocity of the point C is known, ω for the body is determined by dividing v_C by the distance of C from Q, or by r_C. The velocity of any other point E is $\boldsymbol{\omega} \times \mathbf{r}_E$, perpendicular to the radius $\mathbf{r}_E$.

The position of Q in the body (or in its extension) is continually changing; its locus is a line (usually curved) fixed in the body and moving with it, called the *body centrode*. The locus of the positions of Q in the fixed plane of motion is a line (usually curved) called the *space centrode*. The plane motion may be considered as produced by the rolling, without slipping, of the body centrode upon the space centrode.

Example 36. Rolling Wheel Describing Plane Motion. A wheel of 6 ft radius rolls along a straight horizontal path, and at a certain instant the point P, 2 ft from the center of the wheel, is in the position shown in Fig. 34a. At this instant $\omega = 16.75$ rad/sec and $\alpha = 5.6$ rad/sec^2. Determine the velocity and acceleration of point P with respect to fixed point O at the specified instant.

Solution. Select center A as base point. From relations between v and a of any point of a rotating body and ω and α of the body, $v_{A \text{ to } O} = r\omega = 6 \times 16.75 = 100.5$ ft/sec, horizontally toward left, and $v_{P \text{ to } A} = r\omega = 2 \times 16.75 = 33.5$ ft/sec, vertically upward. Therefore, $v_{P \text{ to } O} = 105.9$ ft/sec, upward to left, at $18°16'$ to horizontal (Fig. 34b).

Then $a_{A \text{ to} O} = r\alpha = 6 \times 5.6 = 33.6$ ft/sec^2, horizontally toward left; $a_{t_P \text{ toA}} = r\alpha = 2 \times 5.6 = 11.2$ ft/sec^2, vertically upward; and $a_{n_P \text{ toA}} = rw^2 = 2 \times (16.75)^2 = 561$ ft/sec^2, horizontally toward left. Therefore, $\alpha_{P \text{ to } O} = 595$ ft/sec^2, upward to left, at $1°5'$ to horizontal (Fig. 34c).

4 KINETICS

4.1 Basic Quantities

System of Units* The *unit of force* is the pound force in the English system and is the newton in the SI system: The pound force is defined as the force required to give one slug of mass an acceleration of one foot per second per second. The newton is defined as the force required to give one kilogram of mass an acceleration of one meter per second per second.

The *unit of acceleration* in the English system is the foot per second per second. In the SI system, the unit is the meter per second per second.

The *unit of mass* in the English system is the slug, which is the mass of 32.1739 standard pounds. The standard pound is a platinum standard kept at the National Bureau of Standards in Washington, D.C. Its mass is 0.45359243 kg. The SI unit of mass is the standard kilogram, a platinum standard kept at the International Bureau of Weights and Measures near Paris, France.

The *relation* between force, mass, and acceleration of a constant mass particle, using either of the preceding systems of units, is expressed by the relation

$$\mathbf{F} = m\mathbf{a} \quad \left(\text{or } \mathbf{F} = \frac{W}{g}\mathbf{a} \right)$$

---

*The units herein defined are those of the English gravitational system. For a complete discussion of English and metric gravitational and absolute units, see Section 2.2.

where **F** is the force in pounds or newtons, m is the mass in slugs or kilograms, and a is acceleration in feet per second per second or meters per second per second.*

4.2 Derived Quantities and Relations

Work, Power, and Energy

Work. The *work of a force*, if constant, is the product of the force and the effective displacement of its application point. *Effective displacement* of the application point is the component of the displacement parallel to the force. The body exerting the force is also said to do work. In Fig. 35, the work of force F as the application point describes path $AB = F \times AC$. Since $F \cdot (AB \cos \alpha) = (F \cos \alpha) \cdot AB$, the work is also equal to displacement of the application point times the component of force parallel to the displacement. *Work of a variable force* in moving a body through distance $\Delta S = (s_2 - s_1)$ is $W = \int_{s_1}^{s_2} F \cos \alpha \ ds = \int_{s_1}^{s_2} F_t \ ds$, in which F is the variable force, ds is the elementary length of path, α is the angle between the force and element ds, and F_t is the tangential component of force. The *sign of work* is positive if force and effective displacement have the same sense; it is negative if they differ in sense. Work done by a body against a force is equal and opposite to work done by the force on the body.

In vector form, work is calculated by integrating the dot product $\mathbf{F} \cdot d\mathbf{r}$ along the path of the application point. In equation form,

$$ W = \int_{s_1}^{s_2} \mathbf{F} \cdot d\mathbf{r} $$

where **F** is the vector representation of the force and $d\mathbf{r}$ is the differential displacement along the path of the application point. Work is a scalar quantity.

The *unit of work* is any force unit times any distance unit (such as the foot-pound or newton-meter) or any power unit times any time unit (such as the watt-hour).

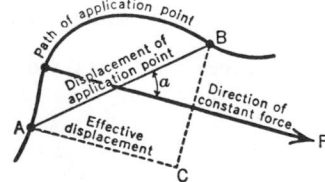

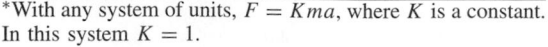

Fig. 35 Effective displacement of a force.

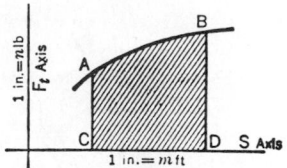

Fig. 36 Work of a force.

In a *work diagram* (Fig. 36), plot values of F_t as ordinates, corresponding values of s as abscissas; draw curve AB through ends of ordinates. Area $ABDC$ times mn equals work, in foot-pounds, done by F_t over distance $s_2 - s_1$.

The *work of gravity* on a body in any motion equals the product of weight and the change in height of the mass center. The *work of a central force F* (one always directed toward a fixed point), in any displacement of its application point, is $\int_{r_1}^{r_2} F \ dr$, in which r_2 and r_1 are the distances of the application point from the center at the beginning and end of the displacement.

The *work of a torque T* on a rotating body for an angular displacement $\theta = (\theta_2 - \theta_1)$ radians is $W = \int_{\theta_1}^{\theta_2} T \ d\theta$. If T is constant, $W = T(\theta_2 - \theta_1)$.

The *mechanical efficiency* of a machine is the ratio of useful output to total input of work. Let W_u = useful work performed, W_f = useless work required to overcome friction or air or any other type of resistance, W_a = work applied to the machine. Then $W_a = W_u + W_f$, and mechanical efficiency = W_u / W_a.

Power. The *power of a force* is its time rate of doing work. The body exerting the force is also said to have power. Let P = power and W = work. Then instantaneous power $P = dW/dt = F_t(ds/dt) = F_1 v$, where v is the instantaneous velocity of an application point of force F.

The *unit of power* is any work unit divided by any time unit (as foot-pound per second). One horsepower = 550 ft-lb/sec = 33,000 ft-lb/min = 0.7457 W.

The *power of a torque* at any instant is $P = dW/dt = T(d\theta/dt) = T\omega$, where ω is the instantaneous angular velocity of the body.

Energy. The *energy*[†] of a body (or system of bodies) is the amount of work it can do, by virtue of its motion or position, against forces applied to it while changing to a standard state.

The *potential energy* (PE) of a body is that possessed by virtue of its configuration. Thus, a body of weight W located at a height above the earth's surface such that its mass center can descend h feet has a potential energy PE = Wh.

The *kinetic energy* (KE) of a body is that possessed by virtue of its velocity, and the standard state is zero velocity, KE of a *body in translation* is $\frac{1}{2}mv^2$. KE of a *rotating body* is $\frac{1}{2}I\omega^2 = \frac{1}{2}mk^2\,\omega^2$, I, k, and ω being the moment of inertia, radius of gyration, and angular velocity, respectively, about the axis of rotation. KE of a body having plane motion is $\frac{1}{2}I\omega^2 = \frac{1}{2}mk^2\,\omega^2 = \frac{1}{2}m\bar{v}^2 + \frac{1}{2}\bar{I}\omega^2$, in which I and k are referred to instantaneous axis, $\bar{v}$ = velocity of mass center, and I = moment of inertia about axis through mass center perpendicular to plane of motion. The *unit of energy* is the same as the unit of work. For KE in foot-pounds, use m in slugs, v in feet per second, ω in radians per second, and k in feet.

Principle of Conservation of Energy.

If a body or system of bodies is isolated so that it neither receives nor gives out energy, its total store of energy, all forms included, remains constant; there may be a transfer of energy from one part of the system to another, but the total gain or loss in one part is exactly equivalent to the loss or gain in the remainder. This is the principle of conservation of energy.

Principle of Work and Kinetic Energy.

The total work of the applied forces acting on any body or on any system of connected bodies equals the change in the kinetic energy of the body or bodies. (This assumes no work is converted into nonmechanical types of energy.) Work done $=\Delta$KE. ΔKE in translation is $\frac{1}{2}m(v_2^2 - v_1^2)$, v_1 and v_2 being initial and final velocities. ΔKE in rotation is $\frac{1}{2}I(\omega_2^2 - \omega_1^2) = \frac{1}{2}mk^2(\omega_2^2 - \omega_1^2)$, ω_1 and ω_2 being initial and final angular velocities. In plane motion, change in KE is

$$\Delta\text{KE} = \tfrac{1}{2}I\left(\omega_2^2 - \omega_1^2\right) = \tfrac{1}{2}mk^2\left(\omega_2^2 - \omega_1^2\right)$$
$$= \tfrac{1}{2}m\left(\bar{v}_2^2 - \bar{v}_1^2\right) + \tfrac{1}{2}\bar{I}\left(\omega_2^2 - \omega_1^2\right)$$
$$= \tfrac{1}{2}m\left(\bar{v}_2^2 - \bar{v}_1^2\right) + \tfrac{1}{2}m\bar{k}^2\left(\omega_2^2 - \omega_1^2\right)$$

in which I and k are referred to instantaneous axis and $\bar{I}$ and $\bar{k}$ to a parallel axis through the mass center and $\bar{v}$ is velocity of mass center.

Example 7.

Water falling from a height of 120 ft at the rate of 1000 ft^3/min drives a turbine directly connected to an electric generator at 120 rpm. If the total resisting torque due to friction is 250 lb-ft and the water leaves the turbine blades with a velocity of 15 ft/sec, find the power developed by the generator.

Solution.

This is a problem in the conversion of potential energy to work that in turn is converted to useful kinetic energy, wasted kinetic energy, and wasted thermal energy, the total energy of the system

of course remaining constant. Assume that 1 ft^3 of water weighs 62.5 lb and $g = 32$ ft/sec^2. In 1 min,

$$\Delta\text{PE} = Wh$$
$$= 1000 \times 62.5 \times 120$$
$$= 7{,}500{,}000 \text{ ft-lb}$$

$$\text{Wasted } \Delta\text{KE} = \tfrac{1}{2}mv^2$$
$$= \frac{1000 \times 62.5 \times \overline{15}^2}{2 \times 32}$$
$$= 219{,}700 \text{ ft-lb}$$

$$\text{Wasted friction (thermal) } \Delta\text{TE} = T\theta$$
$$= 250 \times 2\pi \times 120$$
$$= 188{,}500 \text{ ft-lb}$$

Therefore

$$\text{Useful } \Delta\text{KE} = \Delta\text{PE} - \text{wasted } \Delta\text{KE} - \text{wasted } \Delta\text{TE}$$
$$= 7{,}500{,}000 - 219{,}700 - 188{,}500$$
$$= 7{,}091{,}800 \text{ ft-lb}$$
$$P = \frac{7{,}091{,}800}{33{,}000} = 215 \text{ hp or } 215 \times 0.7457$$
$$= 160 \text{ kW}$$

Impulse, Momentum, and Impact of a Force

Impulse.

Linear impulse of a force is the integral of the force over the time of application of the force. The integral can be separated into components along convenient coordinate directions. Linear impulse is a vector quantity. Thus the linear impulse of a force $\mathbf{F}$, can be calculated as

$$\int_{t_1}^{t_2} \mathbf{F}\,dt = \int_{t_1}^{t_2} F_x\,dt\,\mathbf{i} + \int_{t_1}^{t_2} F_y\,dt\,\mathbf{j} + \int_{t_1}^{t_2} F_z\,dt\,\mathbf{k}$$

The direction cosines of the resultant vector are determined in the usual manner. The *unit of impulse* is any unit force times any unit time, as pound (force) seconds.

Angular impulse of a force about a point is the time integral of the moment of that force about the point. If $\mathbf{r}$ is the vector from the point O to any point on the line of application of the force, $\mathbf{F}$, then the angular impulse of $\mathbf{F}$ about O is given by

$$\int_{t_1}^{t_2} (\mathbf{r} \times \mathbf{F})\,dt$$

The angular impulse of a force about a line is the time integral of the moment of the force about the line. If the line connects points O and A, and if $\mathbf{r}$ is the vector from O to any point on the line of action of $\mathbf{F}$, then the moment of the force $\mathbf{F}$ about OA is given by $\mathbf{M}_{OA} = \mathbf{r} \times (\mathbf{F} - \mathbf{F} \cdot \mathbf{U}_{OA})$, and the angular impulse of $\mathbf{F}$ about OA is

$$\int_{t_1}^{t_2} \mathbf{r} \times (\mathbf{F} - \mathbf{F} \cdot \mathbf{U}_{OA})\, dt$$

where $\mathbf{U}_{OA}$ is a unit vector along the OA line selected to be positive when directed from O to A. The *unit of angular impulse* is unit torque times unit time, as pound (force) feet seconds.

Momentum.   *Linear momentum* of a particle is the product of its mass and velocity. It is a vector quantity and has the sense and direction of the velocity. The *unit of momentum* is the same as the unit of impulse. Linear momentum of a body is the resultant, or vector sum, of the momentums of its particles. In any motion the linear momentum of a body is mv, m being the mass of the body and $\bar{v}$ the velocity of its mass center.

Angular momentum of a particle about a point is the moment of its linear momentum about that point. If the linear momentum of the particle is $\mathbf{P} = m\mathbf{v}$, O is the reference point, and $\mathbf{r}$ is the vector from O to the particle, then the angular momentum of the particle about O is

$$\mathbf{h}_O = \mathbf{r} \times \mathbf{P} = \mathbf{r} \times m\mathbf{U}$$

Angular momentum is a vector quantity. The *unit of angular momentum* is mass times velocity times distance.

Angular momentum of a particle about an axis is the moment of the particle's linear momentum component which is not parallel to the axis about any point on the axis. If OA is the axis, then the angular momentum of the particle about the axis is

$$\mathbf{h}_{OA} = \mathbf{r} \times (\mathbf{P} - \mathbf{P} \cdot \mathbf{U}_{OA})$$

where $\mathbf{U}_{OA}$ is a unit vector along the line OA.

In Fig. 37, let $mv =$ momentum of particle P. Resolve the momentum into components parallel and perpendicular to the axis. DE is the perpendicular distance from the axis to line AP. The angular momentum of $P = mv \cos\alpha \times DE$. The angular momentum of a body about an axis is the algebraic sum of the angular momentums of its particles. The angular momentum of a rotating body about the axis of rotation is $I\omega = mk^2\,\omega$, I and k being the moment of inertia and radius of gyration, respectively, about the axis of rotation and ω the angular velocity. The unit of angular momentum is the same as the unit of angular impulse.

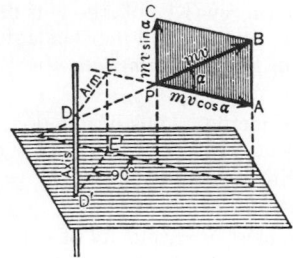

Fig. 37   Angular momentum of a particle.

Principle of Conservation of Linear and Angular Momentum.   When no external forces are acting upon a body or system of bodies, the component linear momentum along any line and the angular momentum about any line remain constant; this is the principle of conservation of linear and angular momentum.

Principle of Impulse and Momentum.   For linear momentum, the impulse of the resultant force acting for an infinitesimal time upon a body is equal to the change in linear momentum of its mass center during that time parallel to the direction of the force. Referred to coordinate axes, the change in the component of linear momentum parallel to any axis x for any length of time $t_2 - t_1$ equals the algebraic sum of the components of the impulses of the applied forces parallel to the axis in the same time, or, more briefly, $\Delta(m\bar{v}_x) = \sum \int_{t_1}^{t_2} F_x\, dt$. Similarly, the change in the angular momentum about any axis y in the time $t_2 - t_1$ equals the algebraic sum of the angular impulses of the applied forces about the axis in the same time, or, more briefly, $\Delta(I_y\omega) = \sum \int_{t_1}^{t_2} T_y\, dt$.

Example 8.   A jet of water strikes a concave vessel with a velocity of 80 ft/sec and leaves it with a velocity that has the same magnitude but makes an angle of $120°$ with the original direction. If the diameter of the jet is 1 in., find the force necessary to hold the vessel in position.

Solution.   The sustaining force $\mathbf{F}$ must bisect the acute angle between the lines representing the original and final velocities. Let the line of action of $\mathbf{F}$ (Fig. 38) be taken as the X axis. There is no change in the Y component of momentum. The impulse of the force in the X direction in t seconds is $F \times t$ pound-seconds. The weight of water deflected in t seconds is $W = 80\pi \times 62.5t/576$ lb. The component of the original momentum in the X direction is $-80\,W \cos 30°/g$ pound-seconds. The component of the final momentum in the X direction is $80\,W \cos 30°/g$ pound-seconds. The change in momentum in the X direction is $160\,W \cos 30°/g = 5\,W \cos 30°$ pound-seconds. The fundamental relation gives $F \times t = 5 \times 80\pi \times 62.5 \times \cos 30° \times t/576$, whence $F = 118$ lb. Observe that the sustaining force F does no mechanical work and that the water suffers no loss of kinetic energy.

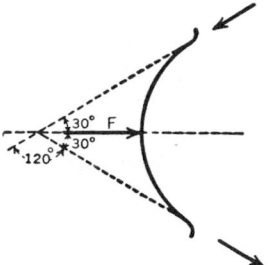

Fig. 38   Deflected water jet.

Impact.   *Impact* occurs when two bodies collide. It is *direct* when the motion is perpendicular to the striking surfaces; otherwise it is *oblique*. It is *central* if the forces that the bodies exert on each other are directed along the line joining the mass centers; otherwise it is *eccentric*. In any collision, the forces that the two bodies exert on each other are equal and opposite at each instant; hence the total impulses of these forces during the collision are equal and opposite, and according to the principle of impulse and momentum the changes in the momentum of the bodies produced by the collision must be equal and opposite; or, otherwise stated, the total momentum of the two bodies is unchanged by the collision. Or, for *direct central impact*,

$$m_1 v_1 + m_2 v_2 = m_1 V_1 + m_2 V_2$$

wherein m_1, m_2 are the masses of the bodies, v_1, v_2 their velocities before, and V_1, V_2 their velocities after the collision; but in numerical substitution, velocities in one direction are given the same sign and those in the other direction the opposite sign.

Experiments on direct central impact of spherical bodies have shown that the relative velocities of spheres after impact are always less than before the impact and that these relative velocities are opposite in direction. The ratio of the relative velocities after impact to that before impact is called *coefficient of restitution*; it seems to depend only on the material of the impinging spheres. For glass the coefficient is $\frac{15}{16}$, for steel and cork $\frac{5}{9}$, ivory $\frac{8}{9}$, wood about $\frac{1}{2}$, clay and putty 0. If e = coefficient, then

$$V_1 - V_2 = -e(v_1 - v_2)$$

This equation and the preceding one solved simultaneously show that

$$V_1 = v_1 - \frac{(1+e)m_2}{m_1 + m_2}(v_1 - v_2)$$

$$V_2 = v_2 - \frac{(1+e)m_1}{m_1 + m_2}(v_2 - v_1)$$

During impact there is, in general, loss of kinetic energy*; the loss is $\frac{1}{2}(v_1 - v_2)^2(1 - e^2)m_1 m_2/(m_1 + m_2)$. Bodies for which $e = 0$ are said to be *inelastic*, and those for which e is nearly 1 are said to be nearly perfectly *elastic*. When a sphere is dropped on a horizontal surface of a large body from a height h, if H = height of rebound, then $H = e^2 h$. This equation furnishes a means of computing e.

Example 9. Ballistic Pendulum.   This is a device for determining the velocity of a bullet. The bullet is imbedded in soft material, such as clay, for which $e = 0$. Referring to Fig. 39, let m_1 = mass of bullet, m_2 = mass of pendulum, k = radius of gyration about axis of suspension O, v_1 = velocity of bullet (to be determined), $v_2 = r\omega$ = velocity of bullet after impact, and ω = angular velocity of pendulum after impact and assume the pendulum is stationary before impact. Then the angular momentum of the system before impact is $m_1 v_1 r$, and the angular momentum of the system just after impact is $m_1 r^2 \omega + m_2 k^2 \omega$. Since the total momentum of the system remains constant, $m_1 v_1 r = m_1 r^2 \omega + m_2 k^2 \omega$. If the mass center of the pendulum rises to height h, $k^2 \omega^2 = 2gh$. Combining the last two equations to eliminate ω and solving for v_1, $v_1 = (m_1 r^2 + m_2 k^2)\sqrt{2gh}/m_1 rk$. The quantities on the right-hand side of this equation are easily determined experimentally.

4.3   Kinematic and Kinetic Formulas

Symbols. s = distance along path of motion; x, y, z = coordinates of any point; $\bar{x}, \bar{y}, \bar{z}$ = coordinates of mass center; t = time; a = resultant linear acceleration of any point; $\bar{a}$ = resultant linear acceleration of mass center; $a_{x,y,z}$ = components of resultant acceleration along x, y, z axes; a_t = resultant tangential acceleration of any point; a_n = resultant normal acceleration of any point; $v, \bar{v}, v_{x,y,z}$ = linear velocities having corresponding significances; θ = angular displacement; α = angular acceleration;

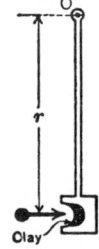

Fig. 39   Ballistic pendulum.

*As energy can be neither created nor destroyed, the total energy remains constant. The "lost" kinetic energy is simply converted into other forms, as, e.g., into work done in distorting the bodies' thermal (heat) energy.

ω = angular velocity; n = revolutions per unit time; r = radius (of curvature); g = acceleration of gravity = 32.2 ft/sec^2 (approx.); m = weight/g = mass; F = resultant force; $F_{x,y,z}$ = components of resultant force along x, y, z axes; F_t = resultant tangential force; F_n = resultant normal force; W = work; Eff = efficiency; P = power; KE = kinetic energy; Imp = linear impulse;   Mom = linear momentum, T = resultant torque about axis of rotation; $T_{x,y,z}$ = torques about x, y, z axes; Ang Imp = angular impulse;   Ang Mom = angular momentum; I = moment of inertia (for mass) about axis of rotation; $I_{x,y,z}$ = moments of inertia about x, y, z axes; $\overline{I}$ = moment of inertia about axis through mass center; $k, k_{x,y,z}, \overline{k}$ = corresponding radii of gyration; d = distance between axes; U = product of inertia; U_{xy} = product of inertia with respect to YOZ and ZOX planes; (U_{yz}, U_{zx} have corresponding significances); Δ indicates "change in."

Gravitation and Inertia Functions.
Mass center has coordinates:

$$\overline{x}= \frac{\int x\,dm}{m} \qquad \overline{y}= \frac{\int y\,dm}{m} \qquad \overline{z}= \frac{\int z\,dm}{m}$$

$$\overline{I}= \int r^2\,dm \qquad I= \overline{I} + md^2 \qquad k^2= \overline{k}^2 + d^2$$

$$\overline{k}= \sqrt{\frac{\overline{I}}{m}} \qquad k= \sqrt{\frac{I}{m}}$$

$$U_{xy}= \int xy\,dm \qquad U_{yz}= \int yz\,dm \qquad U_{zx}= \int zx\,dm$$

Translation* — (Rectilinear Motion)

$$v= \frac{ds}{dt} \quad a= \frac{dv}{dt} = \frac{d^2s}{dt^2} \quad \frac{a}{v} = \frac{dv}{ds}$$

$$\Delta s = \int_{t_1}^{t_2} v\,dt \qquad\qquad \Delta v = \int_{t_1}^{t_2} a\,dt$$

$$\Delta t = \int_{s_1}^{s_2} \frac{ds}{v} = \int_{v_1}^{v_2} \frac{dv}{a} \qquad \Delta v^2 = 2\int_{s_1}^{s_2} a\,ds$$

$$F = m\overline{a} \qquad\qquad \Delta W = \int_{s_1}^{s_2} F\,ds$$

$$\text{KE} = \tfrac{1}{2}m\overline{v}^2 \qquad\qquad \Delta W = \Delta\text{KE}$$

$$P = \frac{dW}{dt} = Fv \qquad \Delta\text{Imp} = \int_{t_1}^{t_2} F\,dt$$

$$\text{Mom} = m\overline{v} \qquad\qquad \Delta\text{Imp} = \Delta\text{Mom}$$

---

*For a rigid body in translation, accelerations and velocities of all particles are equal. However, $\overline{a}$ and $\overline{v}$ are indicated in certain of the kinetic translation formulas to make them applicable also to nonrigid bodies.

Translation — (curvilinear Motion)

$$v = \frac{ds}{dt} \qquad v_x = \frac{dx}{dt} \qquad v_y = \frac{dy}{dt} \qquad v_z = \frac{dz}{dt}$$

$$v = \sqrt{v_x^2 + v_y^2 + v_z^2}$$

Directional cosines of v are $\cos\theta_x = v_x/v$, $\cos\theta_y = v_y/v$, and $\cos\theta_z = v_z/v$:

$$a_x = \frac{dv_x}{dt} = \frac{d^2x}{dt^2} \qquad a_y = \frac{dv_y}{dt} = \frac{d^2y}{dt^2}$$

$$a_z = \frac{dv_z}{dt} = \frac{d^2z}{dt^2} \qquad a = \sqrt{a_x^2 + a_y^2 + a_z^2}$$

Directional cosines of a are $\cos\phi_x = a_x/a$, $\cos\phi_y = a_y/a$, and $\cos\phi_z = a_z/a$:

$$a_t = \frac{dv}{dt} = \frac{d^2s}{dt^2} \qquad a_n = \frac{v^2}{r} \qquad a = \sqrt{a_t^2 + a_n^2}$$

$$\frac{a_t}{v} = \frac{dv}{ds}$$

$$\Delta s = \int_{t_1}^{t_2} v\,dt \qquad \Delta v = \int_{t_1}^{t_2} a_t\,dt$$

$$\Delta t = \int_{s_1}^{s_2} \frac{ds}{v} = \int_{v_1}^{v_2} \frac{dv}{a_t} \qquad \Delta v^2 = 2\int_{s_1}^{s_2} a_t\,ds$$

$$\mathbf{F} = \mathbf{m\overline{a}} \qquad F_x = m\overline{a}_x \qquad F_y = m\overline{a}_y$$

$$F_z = m\overline{a}_z \qquad F_t = m\overline{a}_t \qquad F_n = m\overline{a}_n$$

$$F = \sqrt{F_x^2 + F_y^2 + F_z^2} \quad\text{or}\quad F = \sqrt{F_t^2 + F_n^2}$$

$$\Delta W = \int_{s_1}^{s_2} F_t\,ds \qquad \text{KE} = \tfrac{1}{2}m\overline{v}^2 \quad \Delta W = \Delta\text{KE}$$

$$P = \frac{dW}{dt} = F_t v$$

$$\Delta\text{Imp}_x = \int_{t_1}^{t_2} F_x\,dt \qquad \Delta\text{Imp}_y = \int_{t_1}^{t_2} F_y\,dt$$

$$\Delta\text{Imp}_z = \int_{t_1}^{t_2} F_z\,dt \quad \text{Imp} = \sqrt{\text{Imp}_x^2 + \text{Imp}_y^2 + \text{Imp}_z^2}$$

$$\text{Mom}_x = m\overline{v}_x \qquad \text{Mom}_y = m\overline{v}_y \qquad \text{Mom}_z = m\overline{v}_z$$

$$\text{Mom} = \sqrt{\text{Mom}_x^2 + \text{Mom}_y^2 + \text{Mom}_z^2}$$

$$\Delta\text{Imp}_x = \Delta\text{Mom}_x \qquad \Delta\text{Imp}_y = \Delta\text{Mom}_y$$

$$\Delta\text{Imp}_z = \Delta\text{Mom}_z \qquad \Delta\text{Imp} = \Delta\text{Mom}$$

Directional cosines of $\Delta\mathbf{Imp} = \Delta\mathbf{Mom}$ are $\cos\psi_x = \Delta\overline{v_x}/\Delta\overline{v}$, $\cos\psi = \Delta\overline{v_y}/\Delta\overline{v}$, and $\cos\psi_z = \Delta/\overline{v}_z/\Delta\overline{v}$.

For kinetic formulas applying to a translated body for rotation about an axis not fixed in the body or its extension, use formulas applying to "rotation of a particle" about its axis, considering the entire mass of the body as concentrated at the mass center.

Rotation*

$$\omega = \frac{d\theta}{dt} \qquad \alpha = \frac{d\omega}{dt} = \frac{d^2\theta}{dt^2} \qquad \frac{\alpha}{\omega} = \frac{d\omega}{d\theta}$$

$$\Delta\theta = \int_{t_1}^{t_2} \omega \, dt \qquad \Delta\omega = \int_{t_1}^{t_2} \alpha \, dt$$

$$\Delta t = \int_{\theta_1}^{\theta_2} \frac{d\theta}{\omega} = \int_{\omega_1}^{\omega_2} \frac{d\omega}{\alpha} \qquad \Delta\omega^2 = 2\int_{\theta_1}^{\theta_2} \alpha \, d\theta$$

For a "particle" ($\bar{I}$ *infinitesimal* compared with I for finite body):

$$s = r\theta \quad v = r\omega \quad a_t = r\alpha \quad a_n = r\omega^2$$

$$a = r\sqrt{\alpha^2 + \omega^4}$$

$$T = F_t r = mr^2\alpha \quad F_t = mr\alpha \quad F_n = mr\omega^2$$

$$F = mr\sqrt{\alpha^2 + \omega^4}$$

$$\Delta W = \int_{\theta_1}^{\theta_2} T \, d\theta \quad KE = \tfrac{1}{2}mr^2 \, \omega^2 \quad \Delta W = \Delta KE$$

$$P = \frac{dW}{dt} = T\omega$$

$$\Delta\mathbf{AngImp} = \int_{t_1}^{t_2} \mathbf{T} \, dt \quad \text{Ang Mom} = mr^2 \, \omega$$

$$\Delta\mathbf{AngImp} = \Delta\mathbf{AngMom}$$

For a body:

$$T = I\alpha = mk^2\alpha \quad \Delta W = \int_{\theta_1}^{\theta_2} T \, d\theta \quad \Delta W = \Delta KE$$

$$P = \frac{dW}{dt} = T\omega \qquad KE = \tfrac{1}{2}I\omega^2 = \tfrac{1}{2}mk^2 \, \omega^2$$

$$\Delta\mathbf{AngImp} = \int_{t_1}^{t_2} \mathbf{T} \, dt \qquad \mathbf{AngMom} = I\omega = mk^2\omega$$

$$\Delta\mathbf{AngImp} = \Delta\mathbf{AngMom}$$

*Formulas are for rigid bodies.

Constrained Rotation†   Assume the plane of rotation is fixed above and parallel to the horizontal XZ plane; the vertical Y axis of rotation is not passing through the mass center (except as a special case). All previous rotation formulas apply if T is replaced by T_y. Additional formulas are:

For a particle ($\bar{I}$ *infinitesimal* compared with I for a finite body):

$$F_x = m\bar{z}\alpha - m\bar{x}\omega^2 \qquad F_y = 0$$

$$F_z = -m\bar{x}\alpha - m\bar{z}\omega^2 \qquad F = \sqrt{F_x^2 + F_y^2 + F_z^2}$$

$$T_x = F_z\bar{y} = -m\overline{xy}\alpha - m\overline{yz}\omega^2$$

$$T_y = F_x\bar{z} = m\bar{z}^2\alpha - m\overline{xz}\omega^2 = mr^2\alpha$$

$$\quad = -F_z\bar{x} = m\bar{x}^2\alpha + m\overline{zx}\omega^2 = mr^2\alpha$$

$$T_z = -F_x\bar{y} = -m\overline{yz}\alpha + m\overline{xy}\omega^2$$

(θ, ω, α, positive for counterclockwise rotation facing origin from plus point on axis)

For a body:

$$T_x = -U_{xy}\alpha - U_{yz}\omega^2 \qquad T_y = I_y\alpha$$

$$T_z = -U_{yz}\alpha + U_{xy}\omega^2 \quad \text{(sign convention as before)}$$

The **center of percussion and center of oscillation** of a pendulum are located at a distance from the center of suspension, given as $k^2/\bar{z}$, where $\bar{z}$ is distance from the center of suspension to the mass center.

Plane and Three-Dimensional Motions **For translation of mass center**, consider the entire mass as concentrated at the mass center. Refer motion to a set of fixed axes located outside the body. To determine acceleration of the mass center, apply formulas for translation.

For rotation about mass center, consider the mass center as fixed and the resultant of forces as a couple. Refer motion to a set of central principal axes. To determine the components of angular acceleration about these axes, use the formulas

$$T_x = I_x\alpha_x + (I_z - I_y)\omega_y\omega_z$$

$$T_y = I_y\alpha_y + (I_x - I_z)\omega_z\omega_x$$

$$T_z = I_z\alpha_z + (I_y - I_x)\omega_x\omega_y$$

For complete resultant motion, combine the motion of translation of mass center with the motion of rotation about mass center.

---

†In obtaining total forces and torques, the effect of the weight of a body (this effect depending on position of plane of rotation) must not be neglected.

Work and kinetic energy changes also are equal to the respective sums of the corresponding changes under the preceding component motions.

4.4 Translation

Kinetic formulas for motion of translation follow directly from the kinematic formulas applying to such motion and the previous discussion on kinetic quantities and their relations. The formulas are summarized in Section 4.3 under the heading "Translation." For the solution of a specific problem, careful choice of formulas will often facilitate the computations. As there is no rotation, the resultant force acts through the mass center and there is no couple.

Example 10. Motion of Parallel Rod of a Locomotive.

The problem is to find the forces acting upon the parallel rod when it is in any position with respect to the wheels. Assume the velocity of the locomotive constant at 60 mph on a level track, driver diameter 5.5 ft, crank length 1 ft, and weight of rod 275 lb. The forces acting on the rod are its weight and the pressures of the crank pins at its ends; the latter are represented (Fig. 40) by their horizontal and vertical components. Since the resultant of all these forces acts through the mass center, $V_1 = V_2$; also $2V_1 - 275 = (W/g)a_y = 8.55a_y$ and $H_1 - H_2 = (W/g)a_x = 8.55a_x$. To determine a_x and a_y, The velocity of the center of either crank pin relative to the locomotive is $(88 \times 1)/2.75 = 32$ ft/sec(60 mph = 88 ft/sec), and the relative motion of the pin being circular at constant velocity, the relative acceleration is toward the center of the crank pin circle at all times and equals $32^2/1 = 1024$ ft/sec^2. This is also the absolute acceleration of the crank pin, since the locomotive is assumed to have no acceleration. But the rod has the same acceleration as the crank pin; hence $a_x = 1024 \sin\theta$ and $a_y = 1024 \cos\theta$. Thus $V = \frac{1}{2}(8755\cos\theta + 275)$ and $H_1 - H_2 = 8755\sin\theta$. In the lowest position of the rod, $\theta = 0$, $a_x = 0$, $a_y = 1024$, $H_1 = H_2$, and $V = \frac{1}{2}(8755 + 275) = 4515$. In a midposition, when $\theta = 90°$, $a_x = 1024$, $a_y = 0$, $H_1 - H_2 = 8755$, and $V = \frac{1}{2}(275) = 137.5$. In the highest position, $\theta = 180°$, $a_x = 0$, $a_y = -1024$, $H_1 = H_2$, and $V = \frac{1}{2}(275 - 8755) = -4240$, the negative sign meaning that V acts downward on the rod.

4.5 Rotation

Kinetic formulas for motion of rotation follow directly from the kinematic formulas applying to such motion

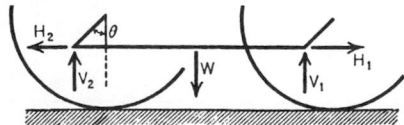

Fig. 40 Translational motion/locomotive parallel rod.

and the previous discussion on kinetic quantities and their relations. The formulas are summarized in Section 4.3 under the heading "Rotation." For the solution of a specific problem, careful choice of formulas will often facilitate the computations. As there is no translation, the resultant force is zero but there is a couple.

Example 11.

A punch is required to exert a force of 100,000 lb through a distance of $\frac{1}{4}$ in., and the work is to be supplied by a flywheel of radius of gyration 1.5 ft making 120 rpm. Find the weight of the wheel if the speed is not to be reduced below 100 rpm.

Solution. $\omega_1 = 120$ rpm $= 4\pi$ rad/sec, $\omega_2 = 100$ rpm $= (10/3)\pi$ rad/sec. Work done by punch $= 100,000/48$ ft-lb $=$ reduction in KE of flywheel.

Change in KE $= \frac{1}{2}mk^2 \ \Delta\omega^2 = W \times 2.25(\omega_1^2 - \omega_2^2)/64 = W \times 2.25(\omega_1 - \omega_2)(\omega_1 + \omega_2)/64$. Hence $(W \times 2.25)/64(2\pi/3)(22\pi/3) = 100,000/48$; hence $W = 1230$ lb $=$ minimum weight of flywheel.

Constrained Rotation *Constrained rotation* refers to rotation of a body about a fixed axis that does not pass through its mass center. Such an axis, since it constrains the motion,* must be held by forces (exerted by bearings) to keep it from shifting position. These bearing reactions depend on the weight of the body, the manner in which the mass of body is distributed about the axis, the applied forces, the angular velocity ω, and the angular acceleration α. Generally, the resultant of the applied forces for such a body is not a single force, but a single force at a selected origin and a couple. Selecting the origin on the axis of rotation, the axial components of the single force and axial components of the couple are given by the following six equation.†

$$\Sigma F_x = m\bar{z}\alpha - m\bar{x}\omega^2$$

$$\Sigma T_x = -\alpha \int xy \, dm - \omega^2 \int yz \, dm$$

$$= -U_{xy}\alpha - U_{yz}\omega^2$$

$$\Sigma F_y = 0 \qquad \Sigma T_y = I_y\alpha$$

$$\Sigma F_z = -m\bar{x}\alpha - m\bar{z}\omega^2$$

$$\Sigma T_z = -\alpha \int yz \, dm + \omega^2 \int xy \, dm$$

$$= -U_{yz}\alpha + U_{xy}\omega^2$$

In these equations, the axis of rotation is fundamentally the y axis; $\bar{x}, \bar{y}, \bar{z}$ are the instantaneous coordinates of the mass center; $\Sigma F_x, \Sigma F_y, \Sigma F_z$ are the sums of

---

*In certain cases, the "physical path" itself constrains the rotation, as the action of the track on a train rounding a curve.
†In obtaining total forces and torques, the effect of the weight of the body must not be neglected.

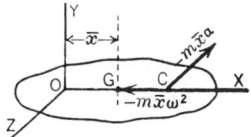

Fig. 41 Constrained rotation.

components of all applied forces in the axial directions; ΣT_x, ΣT_y, ΣT_z are the sums of moments of all applied forces about the axes; and the convention of signs for moments of forces and senses of θ, ω, and α are that counterclockwise rotation, facing the origin from any plus point on an axis, is positive.

The equations are simultaneous at each instant. They are used more often to determine the forces exerted by the bearings on the axle than to determine the resultant.

Special Cases (Fig. 41). Choose the x axis through an instantaneous location of the mass center and let XZ be a plane of symmetry of a homogeneous body. The resultant is a single force in the plane of symmetry having the Z component $-m\bar{x}\alpha$ and the X component $-m\bar{x}\omega^2$ acting at point C, $\overline{OC} = k_y^2/\bar{x}$, k_y being the radius of gyration about the y axis. If $\bar{x} = 0$, the resultant becomes a couple in the XZ plane of moment $\Sigma T_y = I_y\alpha$. If $\alpha = 0$ and $\bar{x} \neq 0$, the resultant is $-m\bar{x}\omega^2$ in the sense CO. If $\alpha = 0$ and $\bar{x} = 0$, the resultant vanishes.

Centrifugal Force. Let any particle of mass m move in a circular path of radius r about a fixed y axis. The resultant of all forces acting on the particle has a normal component $mr\omega^2$ and a tangential component $mr\alpha$. The component $mr\alpha$ increases or decreases the speed of the particle; the component $mr\omega^2$ continually changes the direction of the linear velocity. The resultant of such forces for all the particles of the body is equivalent to the resultant specified by the preceding general equations. If ω is constant and $\alpha = 0$, the resultant force acting on the particle to make it rotate in its circular path is $mr\omega^2$ toward the axis and is called *centripetal force. Centrifugal force* for the particle is equal and opposite to centripetal force and is exerted by the particle upon its neighboring particles or upon the axis of rotation. The *centrifugal resultant* for a body is the resultant of the centrifugal forces of all its particles. Generally, this resultant is not a single force; it may be computed from the general equations by making $\alpha = 0$ and reversing senses of resultant force and couple.

Center of Percussion. A prismatic bar. (Fig. 42) is suspended on a horizontal y axis at O and G is the mass center. If a force P, parallel to the x axis, is applied to the body, the axle reaction OD will generally be inclined to the z axis at some angle $\pm\beta$, the angle depending on the distance h of P from

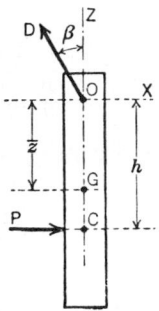

Fig. 42 Center of percussion.

the axis of rotation. If $h = k_y^2/\bar{z}$, in which k_y is the radius of gyration about the y axis, P will cause no x component of axle reaction; that is, β will be zero, and the point C, where the action line of P intersects OG, is the *center of percussion*. In impact testing machines, heavy pendulums are used to deliver blows, and proper design requires the striking point to coincide with the center of percussion in order to avoid shock to the axle and detrimental vibration of the pendulum itself.

Examples of Constrained Rotation

Simple Pendulum. This consists of a small heavy bob on a light string (Fig. 43).* The forces acting on it are the weight W and tension T. The resultant force along the tangent is $-W\sin\theta$; the resultant force along the normal is $T - W\cos\theta$. The force equations are $Wa_t/g = -W\sin\theta$, $Wa_n/g = T - W\cos\theta$. Since $a_n = l\omega^2$, tension $T = W(\cos\theta + l\bar{\omega}^2/g)$.

To determine the motion, $a_t/g = l\alpha/g = -\sin\theta$.

The solution of this equation leads to elliptic functions. An approximate solution for small oscillations can be obtained by putting $\sin\theta = \theta$. (The difference between θ and $\sin\theta$ is less than 1% if θ is less than 14°.) The differential equation becomes $\omega\, d\omega/d\theta = -g\theta/l$. If the pendulum is at the end of its swing

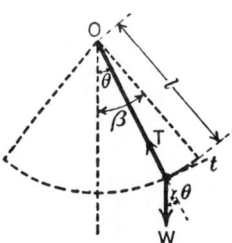

Fig. 43 Simple pendulum.

---

*The radius of gyration of the bob about its axis through its mass center parallel to the axis of rotation is considered negligible compared with the radius of its path.

when $t = 0$, then $\theta = \beta$, $\omega = 0$. Integrating, $\omega^2 = g(\beta^2 - \theta^2)/l$ and $\omega = d\theta/dt = \pm\sqrt{(g/l)(\beta^2 - \theta^2)}$. Integrating, $\theta = \beta \cos \sqrt{(g/l)}t$. Period of oscillation $= 2\pi \sqrt{l/g}$.

Conical Pendulum.* This consists of a small heavy bob suspended from a fixed point by a light string go that it can be made to rotate about the vertical axis through the fixed point (Fig. 44). If the bob rotates with constant angular velocity ω, the quantities ϕ, r, h are constants. Since there is no vertical acceleration, $T \cos \phi = W$. The force acting inward on the bob is $T \sin \phi$. Hence the force equation gives $T \sin \phi = Wa_n/g = Wv^2/gr$ and $\tan \phi = v^2/gr = r\omega^2/g$. Also $h = g/\omega^2$, $T = Wl\omega^2/g$, period of one revolution $= 2\pi/\omega = 2\pi \sqrt{h/g}$.

Compound (or Physical) Pendulum. This is any rigid body suspended from a horizontal axis about which it may rotate under the action of its own weight. The forces acting on the body are its weight, acting downward at G (Fig. 45), and the reaction of the axis at O. Let $\bar{r} = $ distance OG; $k = $ radius of gyration about O. The torque equation gives $Wk^2\alpha/g = -W\bar{r} \sin \theta$, whence $\alpha = -\bar{r}g \sin \theta/k^2$. This is the equation of a simple pendulum (discussed previously) of length $l = k^2/\bar{r}$ called the length of the equivalent simple pendulum. The motion of a compound pendulum is the same as the motion of the equivalent simple pendulum. The point on the compound pendulum located at the distance $k^2/\bar{r}$ from the axis of rotation is called the *center of oscillation*. It coincides with the center of percussion (discussed previously).

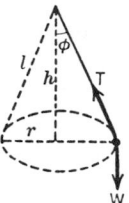

Fig. 44 Conical pendulum.

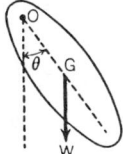

Fig. 45 Compound pendulum.

*The principle of the conical pendulum is employed in the Watt governor for steam engines.

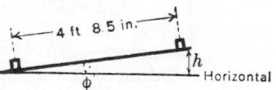

Fig. 46 Banking of track on a curve.

Superelevation of Outer Rail of a Railroad Track. This is determined as follows (Fig. 46): Let $r = $ radius of curvature in feet and $v = $ speed in feet per second of car. Then horizontal centrifugal force is Wv^2/gr and vertical force is W acting through mass center.[†] For the resultant to be perpendicular to the track and thus impose no side load on the rails, $\tan \Phi = v^2/gr$. For small angles, the sine instead of the tangent may be used and, if $h = $ superelevation of the outer rail in inches, $h = 56.5v^2/gr$.

Skidding and Tipping. Suppose a car (Fig. 47) is taking a curve of radius r feet at a speed of v feet per second, G is mass center, N_1 is the vertical and F_1 the horizontal pressure on the outer wheel, and f is the coefficient of friction. The problem becomes one of statics by introducing $Wv^2/gr = F_1 + F_2 = F = fW$. If $f < v^2/gr$, the car will skid. Suppose $f ¿ v^2/gr$; then $N_1 = W \left(\frac{1}{2} + v^2h/dgr\right)$, $N_2 = W \left(\frac{1}{2} - v^2h/dgr\right)$. The critical speed is $v_1 = \sqrt{dgr}/2h$ when the total weight is borne on the outer wheel. If this critical speed is exceeded, the car will tip over.

Note on Use of Rotation Formulas. In practice, nearly all problems of the nature illustrated by the car problems are solved by the use of formulas applying to rotation of a *particle* about an axis. It should be realized, however, that the assumption thus made that the mass is concentrated at the mass center is not strictly correct except in the event that the body has a true motion of translation (as exemplified by the motion of the parallel rod of a locomotive). Seldom is this the case in practice as usually every line of the body lying in the plane of motion makes one complete revolution for each revolution of the body about the center of rotation of its path. Therefore the formulas

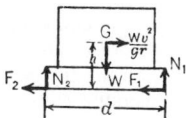

Fig. 47 Skidding and tipping on a curve.

[†]Here and in the next problem, the radius of gyration of the car about its axis through its mass center parallel to the axis of rotation is considered negligible compared with the radius of its path.

applying to rotation of a body are the only ones giving absolutely correct results.

For example, in the problem above on the motion of a simple pendulum, considering the path as that of the mass center, the actual torque $T = Wk^2\alpha/g$, where k is radius of gyration of the bob about the horizontal axis of rotation through O. Let $\bar{k}$ = radius of gyration of bob about horizontal axis through mass center parallel to axis through O. Then $k^2 = l^2 + \bar{k}^2$ and actual torque $T = W(l^2 + k^2)\alpha/g$. But the approximate assumption made in the problem that $F_1 = Wl\alpha/g$ gives $T = Wl^2\alpha/g$, which is too small by the amount $W\bar{k}^2\alpha/g$. However, when $\bar{k}$ is very small compared with l, the results obtained are sufficiently near accurate.

4.6   Plane Motion

Kinetic formulas for plane motion are a combination of those for motions of translation and rotation. The procedure for solution of a problem was summarized under the heading "Plane and Three-Dimensional Motions." The general formula given there for determination of angular accelerations reduces to $T_x = I_x\alpha = mk_x^2\alpha$ the x axis being perpendicular to the plane of motion and passing through the mass center. The theory forming the basis for the assumptions regarding mass concentration and arrangement of forces is explained under the general case of "Three-Dimensional Motion."

Example 12. Wheel on Inclined Plane.   In Fig. 48, $\tan\beta = \frac{3}{4}$, the wheel weighs 100 lb, diameter = 4 ft, radius of gyration = 1.6 ft. (a) Find the acceleration of the center if the wheel rolls without slipping. (b) Find the least coefficient of friction to prevent slipping. (c) If the coefficient of friction is 0.1, find the acceleration of the center and the number of turns made while the center moves 20 ft.

Solution.   (a) The forces acting to move the wheel are $W \sin\beta = 60$ lb and friction F. The equation of motion of the center is $100a/32 = 60 - F$. The force acting to turn the wheel is F. The torque equation is $100 \times 1.6 \times 1.6\alpha/32 = 2F$. Since the wheel does not slip, $a = 2\alpha$. Elimination of F and α gives $a = 11.7$ ft/sec^2.

(b) Friction = minimum coefficient of friction × normal pressure, or $F = fW \cos\beta = $ minimum $f \times$

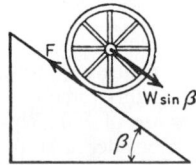

Fig. 48   Wheel on an inclined plane.

80. From the equation above, $F = 23.4$ lb; thus minimum $f = 0.29$.

(c) The relation between a and α is not known when the wheel slips. $F = 80 \times 0.1 = 8$ lb. The equation of motion of the center is $100a/32 = 60 - 8 = 52$, whence $a = 16.6$ ft/sec^2. Distance moved by center, $x = 8.3t^2$. Time to move 20 ft is given by $t^2 = 20/8.3$. Torque equation is $100 \times 1.6 \times 1.6\alpha/32 = 2 \times 8$, hence $\alpha = 2$ rad/sec^2. The angle turned through, $\theta = t^2 = 20/8.3 = 2.41$ rad = 0.38 rev.

4.7   Three-Dimensional Motion

Kinetic formulas for three-dimensional motion are a combination of those for motions of translation and rotation. The procedure for the solution of a problem was summarized under the heading "Plane and Three-Dimensional Motions."

Any motion of a body may be regarded as consisting of two components: one a translation equal to that of the mass center and the other a rotation about some axis through the mass center. These motions may be said to be produced independently by the forces acting on the body: thus (a) the acceleration of the mass center is the same as if the whole mass were concentrated there and acted upon by forces equal in magnitude to and in the same in direction as the actual external forces and (b) the angular acceleration is the same as if the mass center were fixed and the actual external forces applied. The reasonableness of this will be seen from the following: Imagine each force acting on the body replaced by a force acting at the mass center G and a couple; the resultant of all the forces acting at G is a single force R, and the resultant of all the couples is a single couple C; R cannot turn the body but gives it a motion of translation only, and C cannot move G but merely turns the body about some line through G. In general, C does not cause turning about a line perpendicular to the plane of C, only so if the plane of C is perpendicular to one of the principal central axes of the body. To determine the acceleration of the mass center, take fixed x, y, and z axes outside the body and resolve all external forces F_1, F_2, and so on, into x, y, and z components; then

$$\Sigma F_x = m\bar{a}_x \qquad \Sigma F_y = m\bar{a}_y \qquad \Sigma F_z = m\bar{a}_z$$

m denoting the mass of the body. To determine the angular acceleration of the body, take moments of all the forces F_1, F_2, and so on, about the three central principal axes; calling the sums of the moments about these axes ΣT_1, ΣT_2, and ΣT_3, the components of the angular acceleration α_1, α_2, and α_3, and the components of the angular velocity ω_1, ω_2, and ω_3,

$$\Sigma T_1 = I_1\alpha_1 + (I_3 - I_2)\,\omega_2\,\omega_3$$

$$\Sigma T_2 = I_2\alpha_2 + (I_1 - I_3)\,\omega_3\,\omega_1$$

$$\Sigma T_3 = I_3\alpha_3 + (I_2 - I_1)\,\omega_1\,\omega_2$$

wherein I_1, I_2, and I_3 denote the three central principal moments of inertia of the body. In any motion of a body, the kinetic energy may be computed in two parts: (a) the kinetic energy of the whole body moving with a velocity equal to that of the mass center and (b) the sum of the kinetic energies of the constituent particles of the body due to their velocities relative to an axis through the mass center.

4.8   Moving Axes

A *frame of reference* is a set of coordinate axes or coordinate curves with respect to which linear and angular positions, velocities, and accelerations are measured. For brevity, the word *frame* will denote a frame of reference, and a measurement with respect to a frame will be said to be made *in* that frame. A frame may move relative to another frame. In so doing, it has the same freedom of motion as a rigid body.

Forces can be specified and measured in a manner independent of the observer's frame of reference, as, for example, by spring extensions. It follows that Newton's third law, which is concerned with forces only, holds in all frames: in the mechanical interaction of two bodies, their mutual forces are equal but opposite and in the same line of action.

An *inertial frame* is a frame in which the first law of Newton holds, that is, that a particle free from forces is unaccelerated. This defines an inertial frame, while experiment shows that inertial frames exist. The second law of Newton, that net force equals mass times acceleration, is experimentally true in all inertial frames and only in inertial frames. There is no particular inertial frame that may be called absolute; but rather there are infinitely many possible inertial frames, and all of them are equivalent. Suppose that a frame R_1 is inertial. Other inertial frames may have any position of their origin in R_1 and their axes may have any orientation in R_1. Another frame R_2 is inertial if and only if its origin has no acceleration in R_1 and in addition its orientation in R_1 is fixed. In other words, R_2 is inertial if and only if its motion in R_1 is at most a pure translation without acceleration. These conditions are necessary and sufficient in order that all accelerations should appear the same in R_2 as in R_1; they are therefore the conditions that, if R_1 is inertial, R_2 is also. Acceleration in an inertial frame will be called *true acceleration*.

Noninertial frames are of practical interest, for all terrestrial experiments are performed in such frames. Any frame fixed in the earth is noninertial because the earth is both spinning and accelerating relative to an inertial frame. Nevertheless, the observation of phenomena caused by the noninertial character of an earth-fixed frame is a matter of some delicacy, so that to a certain approximation such a frame may be considered inertial. An earth-fixed frame will be referred to simply as *the earth*.

The effects of the noninertial character of a frame are apparent accelerations or forces not accounted for by actual forces. Let a_x, a_y, and a_z be the components

of the apparent acceleration of a point mass m observed in a noninertial frame S. Let the true acceleration components in the same directions be a'_x, a'_y, and a'_z. Define components of *acceleration difference*, the difference between apparent and true accelerations:

$$g_x \equiv a_x - a'_x \qquad g_y \equiv a_y - a'_y \qquad g_z \equiv a_z - a'_z$$

The acceleration difference involves the motion of S relative to an inertial frame R, but it also involves the position x, y, z and the velocity v_x, v_y, v_z of m in S. Now, if the mass is free from actual forces, it is observed in S to have the acceleration g_x, g_y, g_z; or, if the actual forces on it are specified, it is observed to have this acceleration in addition to that predicted from Newton's second law by the observer in S. If, on the other hand, the observer in S specifies the motion of the particle relative to his frame, he finds it necessary to exert upon m, in addition to the force that he computes from Newton's second law, the force

$$f_x = -mg_x \qquad f_y = -mg_y \qquad f_z = -mg_z$$

The particle m thus exerts a reaction force having no apparent physical cause:

$$r_x = mg_x \qquad r_y = mg_y \qquad r_z = mg_z$$

The following paragraphs classify the possible acceleration differences.

Translational acceleration of frame S with components a_{0x}, a_{0y}, a_{0z} relative to an inertial frame R produces an acceleration difference

$$g_x = -a_{0x} \qquad g_y = -a_{0y} \qquad g_z = -a_{0z}$$

Thus, in a train that has acceleration a_{0x} along a straight track in the forward direction, a body free to move would have an apparent backward acceleration $g_x = -a_{0x}$. A mass m constrained to remain at rest in the train would exert a backward reaction $-ma_{0x}$.

Centripetal acceleration causes an acceleration difference when the system S rotates relative to R. If the rotation of S is about its z axis with an angular speed ω, then a point mass m at x, y, z has an acceleration difference $\omega^2\sqrt{x^2 + y^2}$ directed perpendicularly away from the z axis. The components of acceleration difference are

$$g_x = \omega^2 x \qquad g_y = \omega^2 y \qquad g_z = 0$$

If m is fixed in S, its reaction force is $r_x = m\omega^2 x$, $r_y = m\omega^2 y$. This is called centrifugal force. The rider on a merry-go-round exerts this force on his mount.

Angular acceleration of S in R produces an acceleration difference. Consider first the case when the axis of rotation remains in a fixed direction, and let this coincide with the z axis. Let $\alpha = d\omega/dt$ be the rate of change of the magnitude of the angular velocity ω. The resulting acceleration difference for a mass point m at x, y, z is given by $\alpha\sqrt{x^2 + y^2}$ in a direction tangent to the circle of radius $\sqrt{x^2 + y^2}$ about the z axis. The components of the acceleration difference are

$$g_x = \alpha y \qquad g_y = -\alpha x \qquad g_z = 0$$

If a speck of dust is to cling to a phonograph record as the turntable starts or stops, the frictional force on it must be $-m\alpha y m \alpha x$.

Precessional acceleration occurs when the frame S is rotating in R in such a way that its angular velocity vector changes direction. This change of direction may be described at a given instant as the rotation of the angular velocity vector about some axis perpendicular to itself with an angular velocity of precession Ω. Let the axes of S and R coincide at a particular instant, let the angular velocity ω of S be directed along positive z, and let the precession be along positive y. The right-handed-screw convention may be used to refer both angular velocities to their axis directions. The components of the acceleration difference for a mass point at x, y, z are then

$$g_z = 0 \qquad g_y = \Omega \omega z \qquad g_z = -\Omega \omega y$$

Its magnitude is $\Omega \omega \sqrt{y^2 + z^2}$, and its direction is tangent to a circle about the x axis. The force required to be exerted on mass m is then

$$f_y = -m\Omega \omega z \qquad f_z = m\Omega \omega y$$

If a collection of particles fixed in S forms a flywheel, there must be exerted on the flywheel a torque about the positive x axis in order to supply these forces and produce the precession. This is the well-known gyroscopic torque.

Coriolis acceleration is the only difference acceleration that depends on the velocity v_x, v_y, v_z of a point m in a rotating frame S. It occurs when m has a component of velocity in S that is perpendicular to the axis of rotation. Coriolis acceleration is always perpendicular to the velocity of m in S and is independent of the location of m. Let the rotation of S be about the z axis in the positive sense. The components of the Coriolis acceleration are then

$$g_x = 2\omega v_y \qquad g_y = -2\omega v_x \qquad g_z = 0$$

Coriolis acceleration appears in a body falling freely on the earth, causing it to deviate eastward from the plumb line. If the velocity of fall is v and the latitude of the place is λ, the acceleration difference is $2\omega v \cos \lambda$. As another illustration, let a train run north on a horizontal track with velocity v at the same (north) latitude λ. The train then exerts a Coriolis reaction force eastward, on the track, of magnitude $r = 2\omega v \sin \lambda$.

The general case may involve all the above types of acceleration simultaneously. These accelerations simply add vectorially.

4.9 Gyroscopic Motion

A *gyroscope* is essentially a symmetrical rotor that spins rapidly about its axis. The moment of inertia about the axis is made as large as possible within the limitations of weight and size of the instrument. Gyroscopic phenomena are only those that relate changes of direction of the spin axis to applied torque.

Precession is a term for rotation of the axis direction.

The *angular momentum* of the gyroscope about the spin axis is its basic characteristic quantity. If the moment of inertia about the axis is I_1 and the component of the angular velocity parallel to the axis is ω, the component of angular momentum parallel to the axis is

$$\rho = I_1 \omega \tag{1}$$

When precession occurs, the total angular momentum vector is not parallel to the axis; also the value of ω may possibly be affected. In practice, however, the spin is very large compared with precession velocities. It is then a good approximation to consider that the total angular momentum vector is always parallel to the spin axis and has a constant magnitude ρ. This vector will be called ρ.

It is necessary to adopt a convention for the sense of ρ. The right-handed-screw convention will be adopted for both angular momentum and torque. Thus ρ has the direction in which a right-handed screw spinning with the rotor would advance. Likewise torque is in the direction of advance of a right-handed screw to which it is applied.

The theory of the gyroscope rests upon the theorem of mechanics that the rate of change of the angular momentum vector equals the applied torque vector. Thus, if $\mathbf{L}$ is the vector torque applied to a gyroscope,

$$\mathbf{L} = \frac{d\rho}{dt} \tag{2}$$

Since ρ is assumed to have a constant magnitude in a gyroscope problem, $d\rho/dt$ must always be perpendicular to ρ. Thus the gyroscopic torques that can be applied to a gyroscope, and hence its reaction torques, are always perpendicular to the axis. If the vector ρ is drawn from a fixed origin, its tip moves always in the direction of the torque vector. Thus, suppose a gyroscope to have its ρ vector pointed toward an observer and to be mounted so as to be free to precess about its center of mass. See Fig. 49. If the observer exerts a downward force on the end of the axis projecting toward him, the torque direction is horizontally to the right. Hence, this end of the axis precesses horizontally to the right. If the gyroscope is constrained so that horizontal precession is prevented, no downward

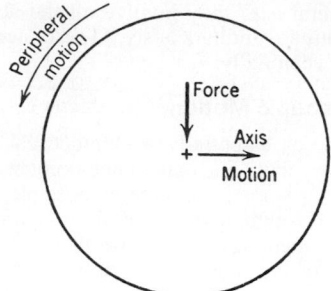

Fig. 49   Force/motion relations for gyroscope.

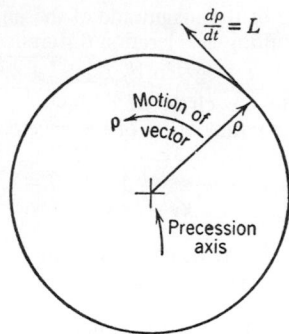

Fig. 51   Necessary condition for steady precession.

force (except that required for ordinary angular acceleration) can be exerted on the end of the axis: the axis yields freely. The constraints are then providing torque to produce this downward precession of the end of the axis.

In *steady precession* about a fixed axis, the tip of the ρ vector drawn from a fixed origin describes a circle about this axis, and the ρ vector sweeps out a half cone about the axis. See Fig. 50. In the simple case where the cone degenerates into a plane, that is, when the ρ vector is perpendicular to the precession axis, the magnitude of the torque is given by

$$L = \rho\Omega = I_1\,\omega\,\Omega \qquad (3)$$

where Ω is the angular velocity of precession. See Fig. 51. To maintain the precession, the torque axis must rotate with the ρ vector. In the general case of a cone of half angle θ, the torque is given, to the approximation that has been made, by

$$L = (\rho\sin\theta)\Omega \qquad (4)$$

It should be noticed that the axis of precession is always perpendicular to the torque vector.

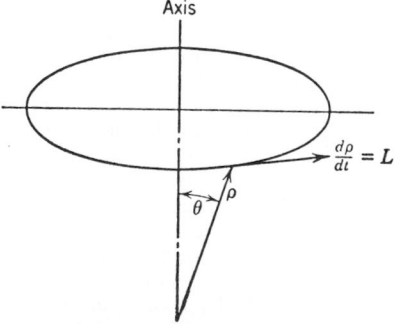

Fig. 50   Steady precession.

Example 13.   A gyroscope has a rotor that weighs 8 lb, has a radius of gyration of 3 in., and spins at 3000 rpm. A torque of 2 lb-ft is applied, and precession occurs about an axis perpendicular to the spin. Find the velocity of precession.

$$I_1 = mk^2 = \tfrac{8}{32}\text{slugs} \times \left(\tfrac{1}{4}\ \text{ft}\right)^2 = \tfrac{1}{64}\text{slug-ft}^2$$

$$\omega = 3000\ \text{rpm} = 100\pi\ \text{rad/sec}$$

Therefore

$$\rho = \frac{100\pi}{64}\text{slug-ft}^2/\ \text{sec} = \frac{100\pi}{64}\text{lb-ft/sec}$$

$$L = 2\ \text{lb-ft}$$

Therefore

$$\Omega = \frac{2\ \text{lb-ft}}{100\pi/64\,\text{lb-ft/sec}} = 0.408\ \text{rad/sec}$$

The "*resistance*" of a gyroscope to change of its axis direction involves two different types of motion: rotation about the precession axis and rotation about the torque axis. Suppose a torque **L** is to be applied to a gyroscope that is free to precess. Let **L** be exerted always about the axis of the angle θ between the spin and precession axes. See Fig. 52. A transient rotation about the torque axis occurs when the torque is first applied. After the transient motion dies out, there is a final steady deflection produced by the torque. The gyroscope behaves as if there were a strong spring opposing rotation about the torque axis. The transient motion may be analyzed by means of the general equations for rotation of a body about its mass center. The result is simple if the increment $\alpha \equiv \theta - \theta_0$ is sufficiently small, θ_0 being the value of θ for no applied torque and no precession. The equation is

$$L = I_2\frac{d^2\alpha}{dt^2} + \left(\frac{I_1^2\,\omega^2}{I_2}\right)\alpha \qquad (5)$$

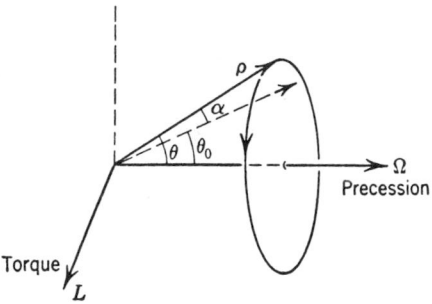

Fig. 52 Gyroscopic resistance.

Here L is torque related by the right-handed-screw rule to positive sense of α and I_2 is the moment of inertia of the rotor about a transverse axis. Solutions of (7) lead to simple harmonic motion when L is constant. When the friction that has been omitted from this equation has damped out the oscillations, the resulting deflection is

$$\alpha_{\text{final}} = \frac{I_2 L}{I_1^2\, \omega^2} \qquad (6)$$

Thus the gyroscope acts like a spring whose torsional stiffness is $I_1^2\, \omega^2 / I_2 = \rho^2 / I_2$. The stiffness may be very large in practice.

Example 14. For the gyroscope of Example 13, find the torsional stiffness, assuming $I_2 = 0.80 I_1$:

$$\rho = \frac{100\pi}{64}\,\text{lb-ft/sec} = 4.90\ \text{lb-ft/sec}$$

$$\rho^2 = 24\ \text{lb}^2 - \text{ft}^2/\ \text{sec}^2$$

$$I_2 = \frac{0.80}{64}\,\text{slug-ft}^2 = \frac{1}{80}\,\text{slug-ft}^2 = \frac{1}{80}\,\text{lb-ft/sec}^2$$

Therefore

$$\frac{\rho^2}{I_2} = \frac{24}{1/80}\,\text{lb-ft/rad} = 1920\ \text{lb-ft/rad}$$

Here the angle unit of the radian has had to be supplied.

The total deflection of the gyroscope about the precession axis depends on the angular impulse received about the torque axis. If the angle α of Fig. 52 is small, then in the steady state the precession velocity is

$$\Omega = \frac{\rho}{I_2 \sin \theta_0}\alpha \qquad (7)$$

If the torque acts for a time Δt and if $d\alpha/dt$ is the same at the beginning and end of Δt, the resulting deflection about the precession axis is

$$\Delta\psi = \frac{1}{\rho \sin \theta_0} \int_0^{\Delta t} L\, dt$$

$\Delta\psi$ may be made small for a given angular impulse by making ρ large. Thus the gyroscope yields but little in any direction to an applied impulse, provided it is free to precess. In time, however, a steady torque produces an indefinitely large deflection about the precession axis.

Applications of the gyroscope are numerous and important. A few follow:

1. *Maintenance of a fixed orientation in space for a brief time.* The gyroscope's axis has this property when it is mounted in very friction-free gimbals and when it is balanced to remove gravitational torque. Such a free gyroscope is used to control a torpedo or a rocket.

2. *Indication of vertical in an airplane.* This cannot be done by a free gyroscope for any length of time because of rotation of the earth and motion over the earth's surface. Also the irreducible minimum of friction in the mounting would produce precession after long operation. Small correcting torques are applied to the gyroscope to cause it to precess very slowly toward the *apparent* vertical. In this way the time average of the apparent vertical is indicated; and if accelerations are random and relatively short-lived, this average will approximate the true vertical.

3. *Ship stabilization.* A huge gyroscope is mounted so that it can precess about a horizontal axis transverse to the ship. The precession is limited to a small range of angle so that the spin axis remains near the ship's vertical. The torque resulting from this precession is then about a fore-and-aft axis, so that it can combat rolling of the ship. The precession is not free but is motor driven and is controlled by a small gyroscope so as to be most effective.

4. *Bicycle operation* depends on gyroscopic action. Turning is a precession about a vertical axis and requires the torque produced by unbalancing the weight toward the inside of the turn. On the other hand, torque applied to the handlebars about a roughly vertical axis produces precession about a horizontal axis, which is tipping of the bicycle. This assists the rider to control her balance.

5. *Indication of the meridian.* The gyrocompass makes use of the rotation of the earth together with the gravitational field in such a way that the gyroscope axis seeks north.

6. *Indication of rate of turn of an airplane.* The gyroscope is mounted with its axis horizontal and is forced to turn with the airplane. This precession produces a reaction torque that works against a spring and actuates an indicator.

4.10 Generalized Coordinates

The *configuration* of a mechanical system is that characteristic that is determined by the position and orientation of all its parts. If the system is composed of particles so small relative to the whole that their rotations are irrelevant to the mechanical problem, the configuration of the system may be specified by the Cartesian coordinates of all its particles in some inertial frame of reference. In problems of ordinary mechanics, the atoms of matter of which the system is composed may be treated as particles. The concept of configuration does not involve velocities or accelerations. A mechanical system is solved when its configuration is known as a function of time. From the time dependence of the configuration, all the kinematic properties, such as velocity and acceleration, may be deduced. Generalized coordinates and Lagrange's equations provide a systematic procedure for solving mechanical systems.

A *displacement* of a mechanical system is a change of its configuration.

Constraints are conditions imposed upon a system to limit its possible displacements. For example, a particle may be constrained to remain on a given surface or on a given line. The atoms composing a rigid body may for many purposes be considered to be constrained to remain at fixed distances from one another. A flywheel may be constrained to rotate in fixed bearings. Sometimes constraints change with time, as when a bead is constrained to slide on a moving wire. Such constraints will be called time dependent. Their motion will be supposed to be given; otherwise, the constraining body must be considered a part of the mechanical system to be solved.

Generalized coordinates are quantities describing the configuration of a system. They describe the configuration completely if the constraints are not time dependent; otherwise, the time is needed explicitly, together with the generalized coordinates, to specify the configuration. It is required for what follows that the set of generalized coordinates of a system be so chosen as to contain the least number of quantities necessary to describe the configuration. Thus, for example, the position of three noncollinear points of a rigid body is not a satisfactory set of generalized coordinates for the body, for it requires nine coordinates to locate these points, whereas only six coordinates are required for a rigid body. A true set of generalized coordinates is the location of one point of the body together with a set of three angles, the Euler angles, describing the orientation of the body. A set of generalized coordinates for a mechanical system will be denoted by $q_1, q_2, \ldots, q_j, \ldots, q_n$.

The *number of degrees of freedom* of a system is equal to the number n of generalized coordinates required to specify its configuration *provided* all n of the increments δq_j are independent. The criterion of independence of the increments δq_j is that any one of them can be specified arbitrarily while all the others are zero. When this is the case, the system is called *holonomic*. This discussion will be limited to holonomic systems. Nonholonomic systems are characterized by the existence of nonintegrable relations between infinitesimal increments δq_j so that they are not independent.

Generalized forces are quantities Q_j such that the work done by the actual forces of the system during an infinitesimal displacement δq_j of only one generalized coordinate is

$$\delta W_j = Q_j \, \delta q_j \tag{8}$$

The Q_j are ordinarily functions of the generalized coordinates, but they may also involve the time derivatives of the coordinates.

It is important to notice that any actual forces that do not work during possible displacements of the system may be omitted completely from consideration. This kind of force is often present owing to constraints. For example, a weightless, rigid rod connecting two masses may exert forces along its length, but these forces do no work since the length of the rod is constant. Forces that constrain a body to remain in contact with a surface are workless.

A generalized force does not have the dimensions of force unless its corresponding generalized coordinate has the dimensions of length. For example, the generalized force is a torque when its corresponding coordinate is an angle.

The generalized forces of a system can ordinarily be obtained directly by computing the work done by the applied forces for infinitesimal changes of the generalized coordinates. There is a formula for Q_j in terms of the Cartesian components of force, X_i, Y_i, and Z_i, acting on the particles of the system whose coordinates are x_i, y_i, and z_i:

$$Q_i = \sum_{i=1}^{N} \left(X_i \frac{\partial x_i}{\partial q_j} + Y_i \frac{\partial y_i}{\partial q_j} + Z_i \frac{\partial z_i}{\partial q_j} \right) \tag{9}$$

This is summed over all N particles of the system.

In the case of moving constraints, the motion of the constraint is not considered in computing the generalized forces. The displacements δq_j of Eq. 8 are assumed to take place in zero time, that is, before the constraint has time to change.

The kinetic energy of a mechanical system may be expressed in terms of the generalized coordinates and their time derivatives. The time appears explicitly only if the constraints are time dependent. The usual notation for the time derivatives of the generalized coordinates is

$$\dot{q}_j \equiv \frac{dq_j}{dt} \tag{10}$$

The kinetic energy function is then written in general as

$$T(q_1, q_2 \cdots q_n; \dot{q}_1, \dot{q}_2 \cdots \dot{q}_n; t) \tag{11}$$

If the constraints are not time dependent, this function becomes a homogeneous quadratic form in the $\dot{q}_j$:

$$T = \sum_{j=1}^{n} \sum_{k=1}^{n} A_{jk}(q_1, q_2 \cdots q_n) \dot{q}_j \dot{q}_k \qquad (12)$$

Equation (12) is quite easy to prove, starting with the kinetic energy equation in terms of the ultimate particles of the system,

$$T = \tfrac{1}{2} \sum_{l=1}^{N} m_l \left(\dot{x}_l^2 + \dot{y}_l^2 + \dot{z}_l^2 \right) \qquad (13)$$

and using the functional dependence of the Cartesian coordinates upon the generalized coordinates in the case of fixed constraints:

$$x_l = x_l(q_1 \cdots q_n) \qquad (14)$$

In an actual problem, the kinetic energy function (11) or (12) is found not by working from Eq. (13) but rather by inspection of the system and addition of the kinetic energies of its large-scale members.

Lagrange Equations *Lagrange equations* of motion are second-order differential equations in the generalized coordinates, time being the independent variable. They have the same form for any holonomic system, and there are n such equations of this form for a given system:

$$\frac{d}{dt}\frac{\partial T}{\partial \dot{q}_j} - \frac{\partial T}{\partial q_j} = Q_j \qquad j = 1, 2, \ldots, n \qquad (15)$$

The solution of any holonomic problem is thus systematized. The process involves assigning generalized coordinates; finding the Q_j functions and the T function; use of Eq (15) to obtain n differential equations; simultaneous solution and integration of the differential equations; and consideration of boundary conditions.

In the use of Eq. (15), the meaning of the partial derivatives must be clearly understood. In the process of finding $\partial T/\partial q_j$, all other q's besides q_j are considered constants; all the $\dot{q}$'s are considered constant, including $\dot{q}_j$; and, if time appears explicitly, it too is considered constant. Similarly, to find $\partial T/\partial \dot{q}_j$, $\dot{q}_j$ is considered the only variable. It is understood, of course, that T has been expressed in the form of Eq. (15) to start with.

Two examples of the use of Lagrange's equations will illustrate most of the ideas involved. Consider first the Atwood machine shown in Fig. 53. This consists of a frictionless pulley of groove radius R and moment of inertia I over which is passed a light, inextensible cord tied to two masses m_1 and m_2. The problem is simplest when the masses are constrained by frictionless guides to move only vertically. There is then only one degree of freedom. Let the generalized coordinate q be the

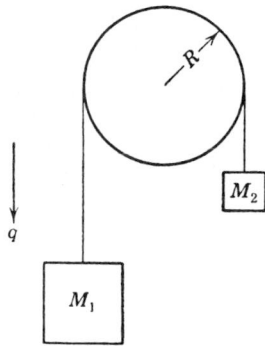

Fig. 53 Generalized coordinates/pulley system.

downward vertical coordinate to m_1. The work done by gravity for a displacement δq is then

$$\delta W = (m_1 g - m_2 g)\delta q$$

whence the generalized force is

$$Q = (m_1 - m_2)g$$

The kinetic energy is the sum of the kinetic energies of the pulley and the weights:

$$T = \frac{1}{2}m_1 \dot{q}^2 + \frac{1}{2}m_2 \dot{q}^2 + \frac{1}{2}I\left(\frac{\dot{q}}{R}\right)^2$$

The Lagrange equation is thus

$$\frac{d}{dt}\left[\left(m_1 + m_2 + \frac{I}{R^2}\right)\dot{q}\right] = (m_1 - m_2)g$$

Therefore

$$\frac{d^2 q}{dt^2} = \frac{(m_1 - m_2)q}{m_1 + m_2 + I/R^2}$$

As a second example, consider the problem of a point mass m constrained to move in a plane under a central force $F = kr$ toward the origin, where r is the distance from the origin to m and k is constant. Let us choose polar coordinates r, θ as the generalized coordinates, as in Fig. 54,

$$q_1 = r \qquad q_2 = \theta$$

For displacement δq_1, the force F does work

$$W = -F \ \delta r = -kr \ \delta r = -kq_1 \ \delta q_1$$

Thus the generalized force Q_1 is $-kq_1$. No work is done during a rotation $\delta\theta$, so $Q_2 = 0$. The kinetic energy $\frac{1}{2}mv^2$ is readily expressed in polar coordinates:

$$T = \tfrac{1}{2}m\left(\dot{r}^2 + r^2\dot{\theta}^2\right) = \tfrac{1}{2}m\left(\dot{q}_1^2 + q_1^2\dot{q}_2^2\right)$$

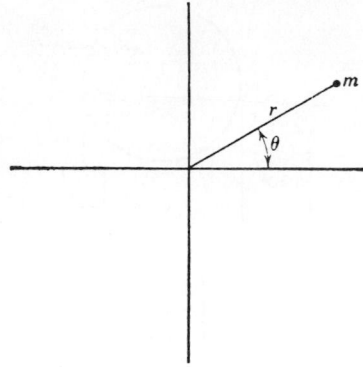

Fig. 54 Generalized coordinates/central force motion.

The two Lagrange equations of the motion are therefore

$$\frac{d}{dt}(m\dot{q}_1) - mq_1\dot{q}_2^2 = -kq_1$$

and

$$\frac{d}{dt}(mq_1^2\dot{q}_2) = 0$$

Potential energy may be used in the case of conservative systems. In such systems, there is a potential energy function V such that, during any displacement of the system, the work done by the force is

$$\delta W = -\delta V \qquad (16)$$

Since potential energy is energy of position or configuration, it must be expressible as a function $V(q_1, q_2, \ldots, q_n)$ of the generalized coordinates. Then we may write for an infinitesimal displacement

$$\delta V = -\delta W = \left(\frac{\partial V}{\partial q_1}\delta_{q_1} + \frac{\partial V}{\partial q_2}\delta_{q_2} + \cdots + \frac{\partial V}{\partial q_n}\delta_{q_n}\right)$$

It follows immediately from definition (8) that the generalized forces are

$$Q_1 = -\frac{\partial V}{\partial q_1} \cdots$$

or in general

$$Q_j = -\frac{\partial V}{\partial q_j} \qquad (17)$$

If the Lagrangian function

$$L(q_1 \cdots q_n; \dot{q}_1 \cdots \dot{q}_n; t)$$

$$\equiv T(q_1 \cdots q_n; \dot{q}_1 \cdots \dot{q}_n; t) - V(q_1 \cdots q_n) \qquad (18)$$

is formed, it may be seen immediately that for a conservative system the Lagrange equation 15 may be written in the compact form

$$\frac{d}{dt}\frac{\partial L}{\partial \dot{q}_j} - \frac{\partial L}{\partial \dot{q}_j} = 0 \qquad j = 1, 2, \ldots, n \qquad (19)$$

5 FRICTION

5.1 Static and Kinetic Friction

A *smooth surface* is one that offers no resistance to the sliding of a body upon it. A *rough surface* does offer resistance to such motion. The *total reaction R* (Fig. 55) of the surface of one body upon another body is its resultant force. *Friction F* is that component of the total reaction R that is tangent to the surface. *Normal reaction N* is that component that is normal to the surface.

Static frictional force F is that friction that opposes motion when there is no slipping. Its value varies as the need for it to prevent motion is developed. *Limiting frictional force F'* is the value of static frictional force when slipping impends. *Coefficient of static friction* (f) is the ratio F'/N. *Angle of static friction* (ϕ) is defined by $\tan\phi = F'/N = f$. *Angle of repose* is that angle that the surface of one body makes with the horizontal when slipping of another body upon it impends. It applies to the particular rubbing surfaces in contact. It equals the angle of static friction.

Kinetic frictional force F_k opposes motion when one body is slipping on the surface of the other. Its value is usually less than that of the limiting friction. *Coefficient of kinetic friction* (f_k) is the ratio F_k/N. *Angle of kinetic friction* (ϕ_k) is defined by $\tan\phi_k = F_k/N = f_k$.

Laws of Friction for Dry Surfaces

1. Friction between two given bodies is directly proportional to the pressure; the coefficient of friction is constant for all pressures.
2. The coefficient and amount of friction for given pressures are independent of the area of contact.
3. The coefficient of friction is independent of the relative velocity, although static friction is greater than kinetic friction.

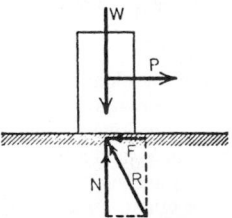

Fig. 55 Force diagram/friction.

Table 2 Coefficients of Static and Kinetic Friction

Materials	Condition	Sliding Friction		Static Friction	
		ϕ	f	ϕ	f
Cast iron on cast iron or bronze	Wet	$17\frac{1}{4}°$	0.31	—	—
	Greased	$4\frac{1}{2}°-5\frac{3}{4}°$	0.08–0.10	$9°$	0.16
Cast iron on oak (fibers parallel)	Dry	$16\frac{3}{4}°-26\frac{1}{2}°$	0.30–0.50	—	—
	Wet	$12\frac{1}{2}°$	0.22	$33°$	0.65
	Greased	$10\frac{3}{4}°$	0.19	—	—
Earth on earth	—	—	—	$14°-45°$	0.25–1.0
Earth on earth (clay)	Damp	—	—	$45°$	1.0
	Wet	—	—	$17\frac{1}{4}°$	0.31
Hemp-rope on rough wood	Dry	$26\frac{1}{2}°$	0.50	$26\frac{1}{2}°-38\frac{3}{4}°$	0.50–0.80
Hemp-rope on polished wood	Dry	—	—	$18\frac{1}{4}°$	0.33
Leather on oak	Dry	$16\frac{3}{4}°-26\frac{1}{2}°$	0.30–0.50	$26\frac{1}{2}°-31°$	0.50–0.60
Leather on cast iron	Dry	$29\frac{1}{4}°$	0.56	$16\frac{3}{4}°-26\frac{1}{2}°$	0.30–0.50
Oak on oak (fibers parallel)	Dry	$25\frac{3}{4}°$	0.48	$31\frac{3}{4}°$	0.62
Oak on oak (fibers crossed)	Dry	$18\frac{3}{4}°$	0.34	$28\frac{1}{4}°$	0.54
	Wet	$14°$	0.25	$35\frac{1}{4}°$	0.71
Oak on oak (fibers perpendicular)	Dry	$10\frac{3}{4}°$	0.19	$23\frac{1}{4}°$	0.43
Steel on ice	Dry	—	0.01	$1\frac{1}{2}°$	0.027
Steel on steel	Dry	Velocity 10 ft/sec — 0.09 / Velocity 100 ft/sec — 0.03		$8\frac{1}{2}°$	0.15
Stone masonry on concrete	Dry	—	—	$37\frac{1}{4}°$	0.76
Stone masonry on undisturbed ground	Dry	—	—	$33°$	0.65
	Wet	—	—	$16\frac{3}{4}°$	0.30
Wrought iron on wrought iron	Dry	$23\frac{3}{4}°$	0.44	—	—
	Greased	$4\frac{1}{2}°-5\frac{3}{4}°$	0.08–10.10	$6\frac{1}{2}°$	0.11
Wrought iron on cast iron or bronze	Dry	$10\frac{1}{4}°$	0.18	$10\frac{3}{4}°$	0.19
	Greased	—	—	$4°-4\frac{1}{2}°$	0.07–0.08

Source: Hudson, Ralph G., *The Engineer's Manual*, John Wiley & Sons, New York, 1939, page 102.

The preceding laws are only approximately true. The coefficient of friction is slightly greater for small pressures upon large areas than for great pressures upon small areas. The coefficient of friction decreases as the speed increases.*

Coefficients of Static and Kinetic Friction These are affected by the preceding laws of friction and also by the characters of the surfaces, the kinds of material, and the nature of any lubricant used. Rough averages for a number of materials and conditions are given in Table 2.

5.2 Axle Friction and Lubricated Surfaces

Axle Friction, Nonlubricated Axle friction is the friction that opposes the turning of an axle in its bearing. For a dry bearing, the axle (Fig. 56) will obviously move to the right. It will climb until a point is reached, at some angle ϕ, where the friction

---

*Recent experiments have proved also that time of contact of the surfaces affects the coefficient of static friction.

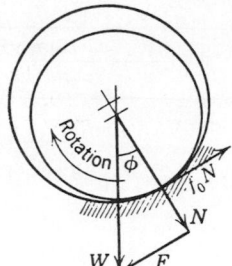

Fig. 56   Force diagram/axle friction.

force f_0N on the shaft or journal is balanced by the tangential component F of the vertical load W, that is, $f_0N = F$. Normal force N is equal to $W \cos \phi$, and f_0 is the coefficient of friction for the surfaces in question. Friction force F is equal to $W \sin \phi$, so that the usual relationship $f_0 = \tan \phi$ for friction prevails. It is customary, however, to define the coefficient of friction f for a journal bearing as the ratio of load W to the tangential friction force F, or $f = F/W = \sin \phi$. The relationship between f_0 and f is then represented by $f = f_1 \cos \phi$.

Friction of Lubricated Surfaces   The friction of lubricated surfaces is characterized by two types of sliding motion:

1. Hydrodynamic or thick-film lubrication
2. Boundary of thin-film lubrication

Hydrodynamic lubrication occurs with a plentiful supply of oil when the thickness of the film is large as compared with the height of the irregularities of the surfaces. Metallic contact or wear does not occur, and the frictional resistance is independent of the kinds of materials composing the bodies.

Although somewhat involved mathematically, the laws for this type of friction are well understood. Figure 57 shows a weightless plate of area A being pushed along at velocity U by force F as it rests on an oil film of thickness h. Newton observed that force F was directly proportional to area A and velocity U and inversely proportional to the film thickness h. This can be written in the form of an equation, $F = \mu UA/h$, where the constant of proportionality μ is called the coefficient of viscosity of the lubricant. It has dimensions of lb-sec/in^2 or dyne sec/cm^2.

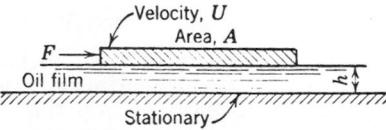

Fig. 57   Lubricated surface friction.

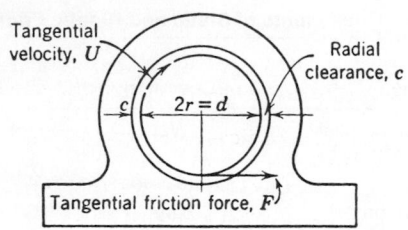

Fig. 58   Journal friction.

Example 15.   Find the force necessary to maintain a velocity of 20 ft/sec for a 3 in. × 3 in. plate if the film has a thickness of 0.003 in. A heavy oil of viscosity 0.000005 lb-sec/in.^{2} is used.

Solution.   Substitution in Newton's equation gives $F = 0.000005 \times 240 \times 9/0.003 = 3.6$ lb.

Journal Friction, Lubricated   Newton's equation can be adapted to the journal bearing of Fig. 58 of radius r and axial length l by substituting $A = 2\pi rl$ and $U = 2\pi rn_s$, where n_s is in revolutions per second. Film thickness h is replaced by radial clearance c. The result,

$$F = 4\pi^2 \mu r^2 h_s \frac{l}{c}$$

is known as *Petroff's equation*. It is valid only for very lightly loaded bearings where the shaft is centrally located in the clearance space. The friction is practically unchanged when the bearing is loaded, and an equation for the coefficient of friction, $f = F/W$, is obtained by dividing both sides of Petroff's equation by the load W. The result, $f = 2\pi^2 \mu h_s r/pc$, is obtained by the substitution $p = W/2rl$, where p is the load per square inch of projected area of the bearing.

Example 16.   Find the coefficient of friction for a 2-in.-diameter by 3-in.-long journal bearing carrying a load of 600 lb at a speed of 1200 rpm. Radial clearance is 0.002 in. Viscosity of the lubricant is 0.000003 lb-sec/in.^{2}

Solution.   Substitution gives

$$f = \frac{2\pi^2 0.000003 \times 20 \times 1}{100 \times 0.002} = 0.0059$$

This example shows that the coefficient of friction under conditions of flooded lubrication is very much smaller than the value of the coefficient when the surfaces are dry. It should be noted that the coefficient of friction for lubricated surfaces varies directly with the velocity and inversely with the pressure. This is in contrast with the coefficient of friction for dry surfaces, which is independent of both velocity and pressure.

The plate of Fig. 57 could carry a downward load provided that the forward edge would be tipped at a small upward angle. The oil would be drawn into the wedge-shaped opening, and a hydrostatic pressure would be developed in the film sufficient to support the load. When a lubricated journal bearing carries a load, the shaft center will shift sidewise in the direction opposite to that for a dry bearing. The oil film will converge in such a manner as to produce the pressure necessary to carry the load. The mathematical treatment of these phenomena may be found in almost any treatise on lubrication.

Boundary Friction Boundary friction occurs between surfaces that are slightly oily or greasy but where the oil supply is insufficient to maintain a hydrodynamic film. Although the bodies approach each other closely, the load is carried by the absorbed layers of lubricant attached to the surfaces. The chemical and physical properties of the lubricant and of the materials composing the bodies are now of importance. At the present time, available knowledge for this type of friction is highly specific and cannot be formulated into general laws. It would appear, however, that the friction is relatively independent of the area in contact and the velocity of sliding. The value for the coefficient of friction is usually intermediate between that for dry friction and that for hydrodynamic friction.

In boundary lubrication one type of oil may have less friction than another of the same viscosity. This

Table 3 Pivot Friction[a]

Type of Pivot	Torque T (lb-in.)	Power P Lost by Friction (ft-lb/sec)
Shafts and Journals (180° bearing)	$T = fWr$	$P = \dfrac{2\pi n}{12} fWr$
Flat pivot	$T = \tfrac{2}{3} fWr$	$P = \dfrac{4\pi n}{3 \times 12} fWr$
Collar bearing	$T = \dfrac{2}{3} fW \dfrac{R^3 - r^3}{R^2 - r^2}$	$P = \dfrac{4\pi n}{3 \times 12} fW \dfrac{R^3 - r^3}{R^2 - r^2}$
Conical pivot	$T = \dfrac{2}{3} fW \dfrac{r}{\sin \alpha}$	$P = \dfrac{4\pi n fWr}{3 \times 12 \sin \alpha}$
Truncated-cone pivot	$T = \dfrac{2}{3} fW \dfrac{R^3 - r^3}{(R^2 - r^2)\sin \alpha}$	$P = \dfrac{4\pi n fW(R^3 - r^3)}{3 \times 12(R^2 - r^2)\sin \alpha}$

Source: Hudson, Ralph G., *The Engineer's Manual*, John Wiley & Sons, New York, 1939, page 105.
[a]f = coefficient of friction, W = load in pounds, T = torque of friction about the axis of the shaft.
r = radius in inches, n = revolutions per second.

property has been given the name of *oiliness*. Certain animal and vegetable oils and compounds are superior to mineral oils with respect to oiliness. They are therefore used as additives to mineral oils to decrease the friction.

If the load on a bearing is increased until the absorbed boundary layers of lubricant can no longer be maintained, metal-to-metal contact will occur at the high points of the sliding surfaces and wear will take place. The exact physical and metallurgical process of wear is not clearly understood. The harder metal may gouge off particles of the softer metal. Experimental evidence exists that the surfaces weld together at the points of metallic contact because of the extremely high localized pressures. The welds are broken as soon as made and may account for the major portion of the resistance to the motion. The amount of welding decreases with decrease of solid solubility between the metals of the two bodies. There is thus justification for the long-established rule that the two members of a bearing should be made of dissimilar metals to decrease friction and the possibility of galling and failure. The *running-in* of a bearing is a form of wear process in which an improvement is effected in the surface conditions.

In addition to a scanty oil supply, a bearing operating with boundary friction will be adversely affected by a decrease in the viscosity of the lubricant, a decrease in the speed, or an increase in the pressure. An unfavorable combination of these qualities may cause the coefficient of friction to increase until the bearing overheats and failure becomes imminent.

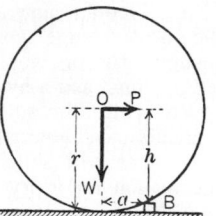

Fig. 59   Rolling friction.

5.3   Rolling Friction

Rolling friction is that friction developed when one body rolls over the surface of another and depends on the hardness of the surfaces in contact and the radius of the rolling surface. The theory is based on the idea that surfaces are slightly deformed at the place of contact and that the effect of rolling friction is the same as if the surfaces were not deformed and the rolling body passed constantly over a small obstruction. Let P (Fig. 59) be the horizontal force required to overcome the small obstruction B. Then $hP = aW$, and, since h is nearly equal to r, $P = aW/r$ (approximately). The *coefficient of rolling friction* is a and, as an analogy with definitions of static and kinetic friction coefficients, might be defined as the ratio T_r/W, where $T_r = Pr$ is the torque resisting rolling motion and W is the normal force (in this case, the weight of the body). The coefficient is a linear distance and is usually given in inches. The values of the coefficient of rolling

Table 4   **Maximum Ratio T_1/T_2 (Slipping Impending)**

α Radians	Values of f (Coefficient of Friction)								
2π	0.10	0.15	0.20	0.25	0.30	0.35	0.40	0.45	0.50
0.1	1.06	1.1	1.13	1.17	1.21	1.25	1.29	1.33	1.37
0.2	1.13	1.21	1.29	1.37	1.46	1.55	1.65	1.76	1.87
0.3	1.21	1.32	1.45	1.60	1.76	1.93	2.13	2.34	2.57
0.4	1.29	1.46	1.65	1.87	2.12	2.41	2.73	3.10	3.51
0.425	1.31	1.49	1.70	1.95	2.23	2.55	2.91	3.33	3.80
0.45	1.33	1.53	1.76	2.03	2.34	2.69	3.10	3.57	4.11
0.475	1.35	1.56	1.82	2.11	2.45	2.84	3.30	3.83	4.45
0.5	1.37	1.60	1.87	2.19	2.57	3.00	3.51	4.11	4.81
0.525	1.39	1.64	1.93	2.28	2.69	3.17	3.74	4.41	5.20
0.55	1.41	1.68	2.00	2.37	2.82	3.35	3.98	4.74	5.63
0.6	1.46	1.76	2.13	2.57	3.10	3.74	4.52	5.45	6.59
0.7	1.52	1.93	2.41	3.00	3.74	4.66	5.81	7.24	9.02
0.8	1.65	2.13	2.73	3.51	4.52	5.81	7.47	9.60	12.35
0.9	1.76	2.34	3.10	4.11	5.45	7.24	9.60	12.74	16.90
1.0	1.87	2.57	3.51	4.81	6.59	9.02	12.35	16.90	23.14
1.5	2.57	4.11	6.59	10.55	16.90	27.08	43.38	69.49	111.32
2.0	3.51	6.59	12.35	23.14	43.38	81.31	152.40	285.68	535.49
2.5	4.81	10.55	23.14	50.75	111.32	244.15	535.49	1,174.5	2,575.9
3.0	6.59	16.90	43.38	111.32	285.68	733.14	1,881.5	4,828.5	12,391.
3.5	9.02	27.08	81.31	244.15	733.14	2,199.9	6,610.7	19,851.	59,608.
4.0	12.35	43.38	152.40	535.49	1,881.5	6,610.7	23,227.	81,610.	286,744.

Table 5 Coefficients of Friction for Belts and Pulley Materials

Belt Material	Pulley Material					
	Iron–Steel	Wood	Paper	Wet Iron	Greasy Iron	Oily Iron
Oak-tanned leather	0.25	0.30	0.35	0.20	0.15	0.12
Mineral-tanned leather	0.40	0.45	0.50	0.35	0.25	0.20
Canvas stitched	0.20	0.23	0.25	0.15	0.12	0.10
Balata	0.32	0.35	0.40	0.20	—	—
Cotton woven	0.22	0.25	0.28	0.15	0.12	0.10
Camel hair	0.35	0.40	0.45	0.25	0.20	0.15
Rubber friction	0.30	0.32	0.35	0.18	—	—
Rubber covered	0.32	0.35	0.38	0.15	—	—
Rubber on fabric	0.35	0.38	0.40	0.20	—	—

Source: *Machinery*, Vol. 37, 1931, p. 560-A.

friction given by various investigators are not in close agreement and should be used with caution.

The following are some reported values of coefficients of rolling friction:

Lignum vitae roller on oak track	0.019 in.
Elm roller on oak track	0.032 in.
Cast-iron wheel (20 in. diameter) on cast-iron rail	0.018–0.019 in.
Railroad wheels (39.4 in. diameter)	0.020–0.022 in.
Iron or steel wheels on wood track	0.06–0.010 in.

5.4 Pivot Friction

Pivot friction is that friction that opposes the turning of the end of a vertical, or inclined, shaft in its bearing. Some examples of pivot friction, with friction torque and power formulas applying, are shown in Table 3.

5.5 Belt Friction

Belt, or *coil*, *friction* is that friction that opposes the slipping of a belt, rope, brake band, or similar article coiled about a pulley, sheave, post, capstan, or similar device. When power is being transmitted, say by a belt driving a pulley, the tension T_1 on the driving side of the belt is greater than the tension T_2 on the driven side. Neglecting the effect of centrifugal force, which is small at low speeds, the tensions are related by the formula $T_1/T_2 = e^{f\alpha}$, where $e = 2.718+$ (i.e., base of natural logarithms), f = coefficient of friction between belt and pulley, and α = angle of contact between belt and pulley (Fig. 60). Values of T_1/T_2 for various values of f and α are shown in Table 4.

Power transmitted is given by $P = (T_1 - T_2)v$, where P is power in foot-pounds per second and v

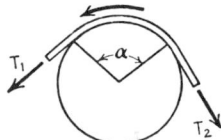

Fig. 60 Belt friction.

is velocity in feet per second. Coefficients of friction for belts and pulley material are shown in Table 5.

ACKNOWLEDGMENTS

This chapter was originally a revision of material in previous handbooks published by John Wiley & Sons, most of which was written by C. H. Burnside and E. R. Maurer for *Merriam's Civil Engineers' Handbook* and *Peele's Mining Engineers' Handbook*. Later revisions by Janvier M. Rice appeared in the *Handbook of Engineering Fundamentals*, third edition, a part of the Wiley Handbook Series. The present chapter was revised by Fowler, with the primary revisions being focused on the introduction of vector mechanics.

BIBLIOGRAPHY

Beer, F. P., and Russell Johnston, E. Jr., *Vector Mechanics for Engineers, Statics and Dynamics*, 4th ed., McGraw-Hill, New York, 1984.

Ginsberg, J. H., and Genin, J., *Statics and Dynamics*, Combined Version, Wiley, New York, 1984.

Hibbeler, R. C., *Engineering Mechanics, Statics and Dynamics*, 3rd ed., Macmillan, New York, 1983.

Meriam, J. L., *Dynamics*, 2nd ed., SI Version, Wiley, New York, 1975.

Shames, I. H., *Engineering Mechanics*, Vol. I, *Statics*, 2nd ed., Prentice-Hall, Englewood Cliffs, NJ, 1966a.

Shames, I. H., *Engineering Mechanics*, Vol. II, *Dynamics*, 2nd ed., Prentice-Hall, Englewood Cliffs, NJ, 1966b.

CHAPTER 4

SELECTION OF METALS FOR STRUCTURAL DESIGN

Matthew J. Donachie
Rensselaer at Hartford
Hartford, Connecticut

1 INTRODUCTION

1.1 Metals and Alloys

Metals are a unique class of elements which provide special properties not available in naturally occurring materials such as wood, cement, and ceramics or in human-produced nonmetallic polymeric materials. (*Note:* Some metals occur in elemental form in nature but most do not.) Since metals are frequently combined (alloyed) with other metallic and some nonmetallic elements to produce alloys (metal combinations with certain desired properties), we will refer to all metallic materials in this chapter as "alloys" with the understanding that often pure or nearly pure metal elements may be used in design.

1.2 Purpose

The purpose of this chapter is to create a sufficient understanding of alloys so that selection of them for specific designs will be appropriate. The primary intent will be to cover alloy selection for structural purposes, that is, where the alloy must support its weight and, probably, the additional loads of a design. The chapter will provide sufficient information for the selector to work with designers to create successful components as well as to evaluate the capability of alloy providers and manufacturers to meet the required end uses of the design. To this end, mechanical, physical, and environmental property behavior that can influence alloy selection will be described.

There is nocook book for alloy selection. Proprietary alloys and/or proprietary/restricted processing can create conditions and properties not listed in a handbook or catalog of available alloys. Critical applications may require a selector to work with one or more manufacturers to develop an understanding of what is available and to determine what one can expect from a chosen alloy. At times it may be necessary to develop new or adapt different manufacturing processes for the utilization of alloys for specific applications.

2 COMMON ALLOY SYSTEMS

While there are many metallic elements which are used to produce alloys, the more industrially important alloys in structural design are iron and its alloys (called ferrous alloys) or other (nonferrous) alloys based on copper, aluminum, magnesium, nickel, cobalt, titanium, and a handful of other metallic elements. Some special elements (e.g., zirconium, beryllium) also may find structural use in applications such as nuclear reactors or other devices. Precious metals (e.g., gold, platinum, silver) are used for applications from jewelry to jet engines! Zinc-based alloys are important in die casting of small articles. [*Note:* The words *article* or *component* may be used interchangeably to indicate a designed item which, in service, will likely be part of a collection of items (e.g., a gas turbine). A gas turbine high-pressure turbine blade would be an article or component.]

3 WHAT ARE ALLOYS AND WHAT AFFECTS THEIR USE?

Alloys, for purposes of this chapter, are solid entities of a given chemical composition, crystal structure, and metallurgical structure at the temperature of use. The range of temperature use can be from near absolute zero to many thousands of degrees. Alloys are made from metallic elements (and some nonmetallic ones added in minor amounts) which are recovered from various ores. In the process of recovery from ores and transfer to desired products, alloys most often are subject to melting (liquification), casting, and/or mechanical deformation processes. This chapter is not concerned with extraction from ores or the subsequent purification of alloy elements prior to the creation of specific alloys for structural use.

Alloys normally exist as a crystalline arrangement of individual atoms (see Fig. 1). Aggregates of multiples of the basic crystalline individual units (e.g., face-centered cubic, body-centered cubic) are termed *grains*. These grains have the same arrangement/orientation of unit cells of the given alloy across their dimension. The mechanical properties of alloys are normally affected by chemistry and intrinsic unit-cell capability, grain size, grain shape and grain orientation, temperature, and certain other factors. Typically, properties of interest are tensile properties (Fig. 2), fatigue properties (Fig. 3), and high-temperature time-dependent properties such as creep rupture (Fig. 4). Modulus of elasticity is another mechanical property used in design and additional properties such as toughness or crack propagation characteristics can be incorporated into structural design, though such properties are not always available.

The methods of preparing alloys can significantly affect their resultant properties. The preparation of an alloy normally will result from melting of appropriate elements/alloys followed by pouring (casting) the molten alloy into a mold to produce an electrode for further primary melting or an ingot for remelting or for deformation processing. Generic mold shapes may be used to produce ingots if the cast alloy is to be remelted and cast into a specific mold shape (e.g., a blade for use in a gas turbine engine). Specific mold shapes may be used if articles are to be created directly by casting. The electrode, ingot, or component produced will show a "cast" structure which will be influenced by the casting process and subsequent treatments applied to create the desired alloy properties.

Often, an alloy may be processed by one or more deformation processes, that is, a billet cut from an ingot may be formed to shape or mostly to shape by hot rolling, forging, and so on, to produce a "wrought" structure. There are a variety of possible cast structures as well as a variety of wrought structures. By examining a component under a microscope, microstructure (appearance at 100x magnification and up) can be viewed. Alloys also can be examined at low magnifications for macrostructure (probably 1x to 10x magnification). Subsequent to production of a

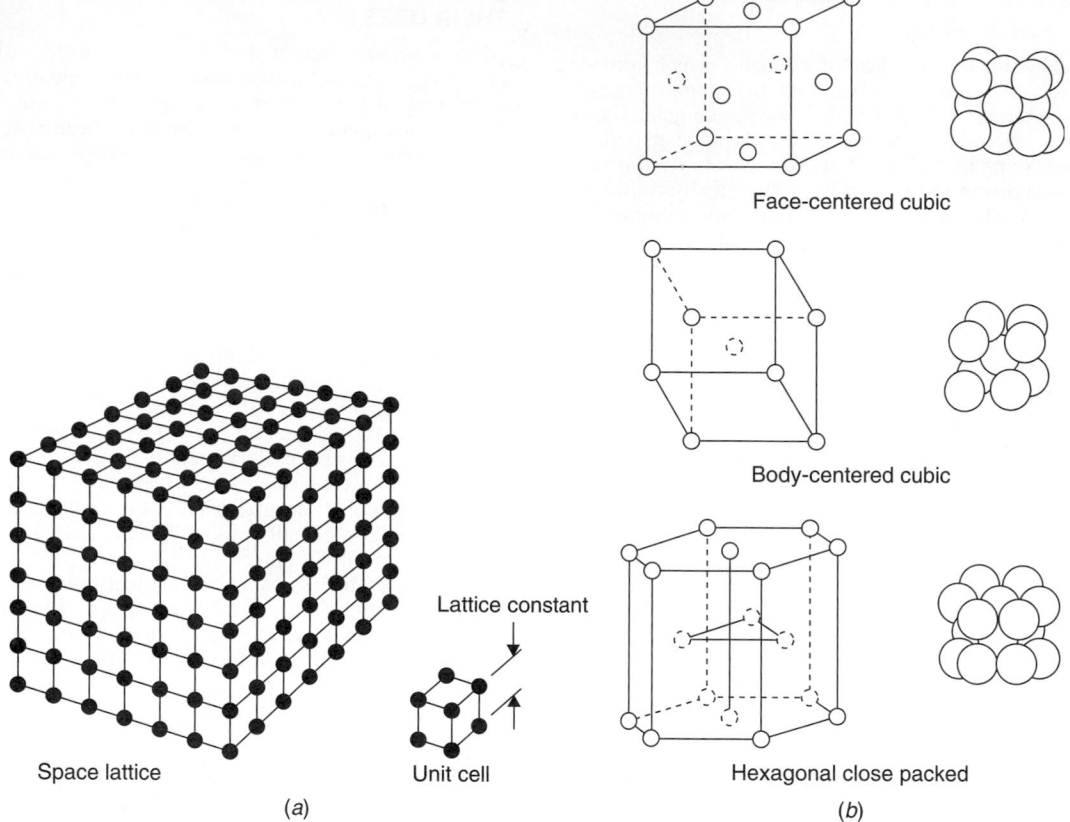

Face-centered cubic

Body-centered cubic

Lattice constant

Space lattice   Unit cell   Hexagonal close packed

(a)   (b)

Fig. 1 (a) Simple cubic lattice. Atoms are represented by solid spheres and, in this lattice, atoms are located at the corners of cubes which repeat indefinitely in three-dimensional space. (b) Common lattice structures for most metals/alloys.

component, the application of special heating cycles (heat treatment) often is used to convey certain property values or characteristics to alloys. In-process heat treatments may be used during the shaping of the component.

4   WHAT ARE THE PROPERTIES OF ALLOYS AND HOW ARE ALLOYS STRENGTHENED?

Metals are crystalline as noted above and, in the solid state, the atoms of a metal or alloy may arrange themselves in various crystallographic groupings, often occurring as cubic structures. Some crystal structures tend to be associated with better property characteristics than others. Crystalline structure normally is a characteristic of the major (base) alloy element in an alloy. In addition to basic crystal structure, crystalline aggregates of atoms have orientation relationships in space. As noted, crystalline aggregates are called grains and, in an alloy, there are usually many grains with random orientation directions within each

manufactured article. Metal alloy articles with multiple random grain directions are known as polycrystalline (frequently referred to as having equiaxed grain structure), often having roughly equal dimensions in all directions. However, columnar-shaped grains (with one relatively longer axis) are common in polycrystalline cast products.

Properties of alloys may be mechanical, physical, or chemical. Mechanical behavior encompasses the strength properties determined in appropriate tests and tend to be the more important properties for selection of alloys for structural applications. Some mechanical properties of alloys are tensile or compressive strength (e.g., onset of plasticity, yielding strength, fracture strength under normal stress loading), fatigue strength (i.e., cycles to first crack or fracture or stress for a specific number of failure cycles), modulus of elasticity, and so on, which are measured primarily at lower to intermediate temperatures of operation. In addition, high-temperature mechanical properties such as creep or creep rupture strength, fatigue strength, and thermally induced fatigue capability are of concern but

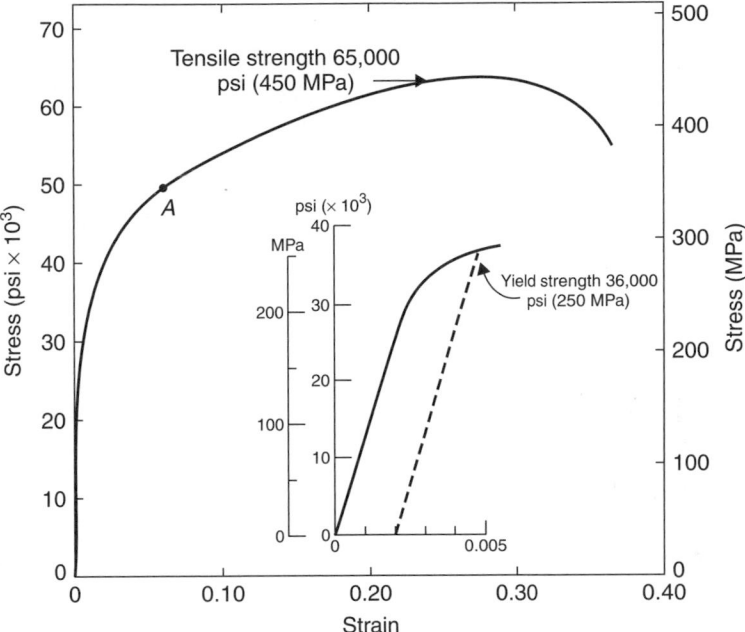

Fig. 2 Ultimate tensile strength, UTS or TS, and defined yield strength at 0.2% offset (inset), TYS or YS. Young's modulus is the slope of the elastic region before plastic deformation begins.

primarily at temperatures above about 0.5 of the absolute melting point of an alloy. At high temperatures, mechanical properties may be influenced by sequences of cyclic loading and time as well as frequency of load application in addition to temperature of testing.

Physical properties include heat capacity, electrical conductivity, thermal conductivity, and magnetic properties. Chemical properties may be of most concern in environmental attack when alloys for structural applications are considered. Intrinsic chemical properties of alloys are modified (as are mechanical properties) by chemical changes during alloy invention and by heat treatment of alloys during processing. In addition, surface chemical properties frequently are modified by surface changes induced, for example, by coatings.

While the physical properties to a major extent depend upon the chemistry of the alloy, the mechanical properties are dependent not only on the chemistry of an alloy but more importantly on the microstructure. Microstructure is assumed to mean grain shape, orientation, grain size, and so on, and the presence, absence, type, and location of new crystal phases in the alloy. The mechanical properties result from the interaction of imperfections such as dislocations with the microstructural components.

The peripheral surface of a grain is called a grain boundary. Aggregates of atoms without grain boundaries are rarely created in nature. However, alloys without internal grain boundaries (i.e., single crystals) or alloys with aligned boundaries (i.e., columnar grained

structures) can be produced in some systems by appropriate manufacturing techniques. The introduction of different atom types and new crystal phases and/or the manipulation of grain boundaries plus the use of stored energy of deformation enable inhibition of the movement of the imperfections (e.g., dislocations) that enable deformation to occur.

It is quite important for the engineer selecting alloys to have a realistic understanding of the strengthening process in alloys as the mechanical properties of alloys can be modified considerably by processing to manipulate the strengthening level achieved. The energy of mechanical deformation (working) can be stored in an alloy (work hardening), producing higher strengths than the unworked (e.g., soft, annealed) alloy. Alloys such as steels (ferrous alloys) can have increased strengths produced by the introduction of fine structure from transformations of crystal lattices. This fine structure usually results from the formation of nonequilibrium martensite phases (in preference to the expected equilibrium phases). The fine structure produced by the martensitic reaction acts to restrict deformation by dislocation movements. This type of hardening often produces very high strengths (as in iron alloys) but much reduced ductility. Subsequent heat treatment (tempering) normally is used to enhance ductility at the expense of strength in transformation hardened alloys. Industrial use of martensite to strengthen alloys is most prevalent with iron-base alloys (containing carbon) or in titanium alloys.

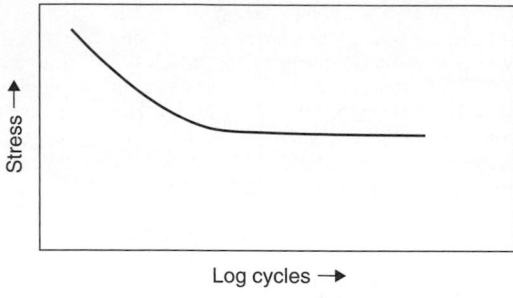

Typical fatigue curve at low temperature

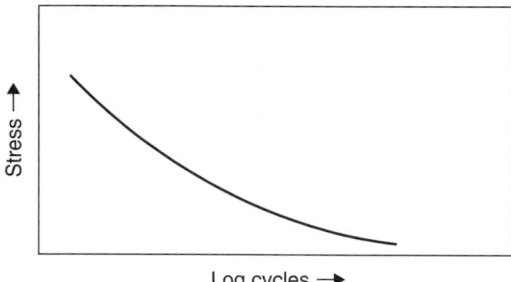

Typical fatigue curve at high temperature

(a)

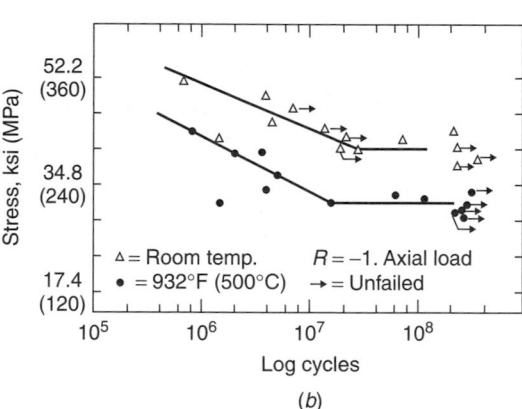

(b)

Fig. 3 (a) Typical fatigue response at high temperature versus that at low temperature. (b) Actual fatigue curves at room and one elevated temperature for specific nickel-base superalloy.

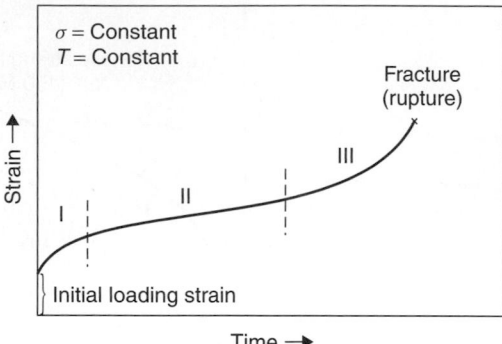

Fig. 4 Time-dependent deformation under constant load at constant high temperature followed by rupture. All loads below short-time yield strength.

Dislocation movement at lower temperatures is reduced not only by fine structure but also by smaller grain sizes. Ultimate tensile strength is not particularly affected by grain size changes, although ductility effects are noticed. As operating temperatures increase, finer grain size material usually becomes less strong than coarse-grained material. Increased temperatures and time at high temperatures may cause recrystallization (new grain formation and subsequent loss of stored energy) and/or grain growth. Consequently, work hardening to increase strength has upper limits of application since the high-strength benefits of (cold) work may be lost with time at higher operating temperatures.

Some alloys derive their strength from solid-solution hardeners (elements dissolved in the basis metal) as well as from secondary precipitates that form in the matrix (principal) phase and produce precipitation (age) hardening from the dispersed particles. Precipitates may be phases or they may be zones of enrichment in one of the alloy elements present. In lower temperature use alloys, such as aluminum, age hardening may be from zones or phases. In higher temperature use alloys, precipitated phases are the norm. Principal strengthening precipitate phases vary with alloy but in nickel-base and iron–nickel-base superalloys they are ordered particles of γ', η, and/or γ''. Phases such as carbides may provide limited direct strengthening (e.g., through dispersion hardening) or, more commonly, indirectly (e.g., by stabilizing grain boundaries against movement at high temperature). In some instances, ceramic (e.g., silica, yttria) or other phases may be dispersed mechanically instead of metallurgically in a matrix to produce dispersion hardening. Thus the hardening of alloys can consist of:

- Solid-solution hardening
- Work hardening
- Dispersion/precipitate hardening (sometimes with ordered precipitates)
- Transformation (martensite) hardening
- Grain size/morphology (shape/orientation) hardening

The results of hardening vary by alloy and by the temperature of testing. Improvements in lower temperature short-time strength may not necessarily be transferred to high-temperature ranges of operation.

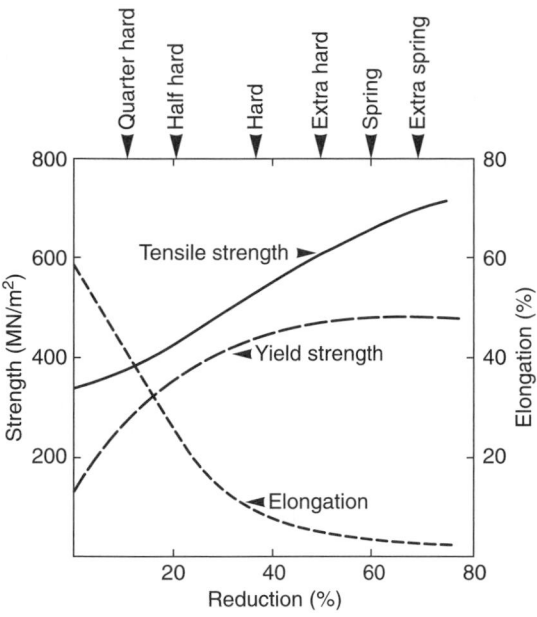

Fig. 5 Terms used in copper industry to indicate degree of cold working are shown at top of plot.

Control of grain structure can significantly influence mechanical properties. The extent to which the secondary phases contribute directly to strengthening depends on the alloy and its processing. It should be noted that improper distributions of phases such as carbides and precipitate phases can be detrimental to properties. Work hardening is an important characteristic of alloys since work hardening may be used to strengthen an alloy (Fig. 5) but also may contribute to the need for intermediate heating (in-process annealing) during forging/forming of an article to prevent cracking of the article or damage to fabrication equipment.

5 MANUFACTURE OF ALLOY ARTICLES

Appropriate compositions of most alloy types can be worked, that is, they can be forged, rolled to sheet, or otherwise deformed mechanically into a variety of shapes. More highly alloyed compositions may be processed as castings. Castings may be less costly than equivalent forgings. Large and small castings can be made. Fabricated alloy structures can be built up by combining (joining) separate article pieces, particularly by welding or brazing, but the more highly alloyed the composition, the more difficult it may be to join an alloy.

Many alloys may be available and used in cast or wrought form. In the latter situation, alloys may be available in extruded, forged, or rolled form. Bar stock, strip, sheet, plate, wire, and other wrought forms/sizes may be obtained. Deformation of alloys may be performed hot or cold. Often, higher strength or "exotic"

alloys are only able to be produced and used in the cast condition. Powder metallurgy processing is an accepted method to produce articles of difficult-to-process articles of certain higher strength or demanding alloys in "wrought" conditions. However, simple components of ordinary alloys, particularly of ferrous metals, are often produced by powder metallurgy for reduced cost compared to conventional cast or wrought processing.

As noted above cold deformation (work hardening), for example, by cold rolling, may be used to increase short-time strength properties for applications at low to intermediate temperatures. Energy stored by deformation processing is an accepted method for improving alloy mechanical properties below a use temperature of about 0.5 of the absolute melting point. As noted above, heat treatment (heating of components in sometimes multiple and complex steps) is frequently used to obtain desired mechanical property values.

6 ALLOY INFORMATION

There is no substitute for consultation with alloy producers or manufacturers about the forms (cast, wrought, powder metallurgy) which can be provided and the exact chemistries/properties available in alloys. It should be understood that not all alloys are readily available as off-the-shelf items. While many thousands of alloy compositions have been evaluated, some for over a century, only a relative handful are routinely produced. Moreover, some alloys are not available for use in all forms and sizes. Sometimes the highest strength alloys will be useful only as powder metal products or as castings. In many instances, specific alloy compositions may only be available at limited times in a production cycle.

Design data for alloys are not intended to be conveyed here, but Tables 1–9 offer limited information on some typical properties for a few alloy classes. Design properties should be obtained from internal testing if possible. Data from producers or other validated sources may be substituted if sufficient test data are not available in-house. Typical properties are merely a guide for comparison. Exact alloy chemistry, article section size, heat treatment, and other processing steps must be known to generate adequate property values for design.

The alloy selector needs to be aware that processing treatments such as forging conditions, heat treatment, and coatings for corrosion protection dramatically affect properties of alloys. All data should be reconciled with the actual manufacturing specifications and processing conditions expected. Alloy selectors should work with competent metallurgical engineers to establish the validity of data intended for design as well as to specify the processing conditions that will be used for component production.

Application of design data must take into consideration the probability of components containing locally inhomogeneous regions under some circumstances. Short-time properties such as tensile strength

Table 1 Elastic Moduli and Tensile Strengths of Selected Representative Metals and Alloys

Material	Elastic Modulus (GPa)		Tensile strengths (MPa)
	Absolute	Specific	
Pure metals			
Aluminum	70	26	70
Beryllium	295	160	290
Copper	130	14	250
Iron	210	27	420
Magnesium	45	26	180
Molybdenum	270	25	700
Nickel	220	25	340
Titanium	120	26	300
Alloys			
Aluminum	70	26	80–400
Copper	130	14	250–750
Iron	210	27	400–2000
Titanium	120	26	750–1500

Table 2 Mechanical Characteristics and Typical Applications for Carbon and Alloy Steels[a]

Hot-Rolled Material

AISI/SAE or ASTM Number	Tensile Strength [psi × 10^3 (MPa)]	Yield Strength [psi × 10^3 (MPa)]	Ductility (% Elongation in 2 in.)	Typical Applications
Plain low-carbon steels				
1010	47 (325)	26 (180)	28	Automobile panels, nails, and wire
1020	55 (380)	30 (205)	25	Pipe; structural and sheet steel
A36	58 (400)	32 (220)	23	Structural (bridges and buildings)
A516 Grade 70	70 (485)	38 (260)	21	Low-temperature pressure vessels
High-strength, low-alloy steels				
A440	63 (435)	42 (290)	21	Structures that are bolted or riveted
A633 Grade E	75 (520)	55 (380)	23	Structures used at low ambient temperatures
A656 Grade 1	95 (655)	80 (552)	15	Truck frames and railway cars

Oil-Quenched and Tempered

AISI Number	UNS Number	Mechanical Property Ranges			Typical Applications
		Tensile Strength [psi × 10^3 (MPa)]	Yield Strength [psi × 10^3 (MPa)]	Ductility (% EL in 2 in.)	
Plain carbon steels					
1040	G10400	88–113 (605–780)	62–85 (430–585)	33–19	Crankshafts, bolts
1080[b]	G10800	116–190 (800–1310)	70–142 (480–980)	24–13	Chisels, hammers
1095[b]	G10950	110–186 (760–1280)	74–120 (510–830)	26–10	Knives, hacksaw blades
Alloy steels					
4063	G40630	114–345 (786–2380)	103–257 (710–1770)	24–4	Springs, hand tools
4340	G43400	142–284 (980–1960)	130–228 (895–1570)	21–11	Bushings, aircraft tubing
6150	G61500	118–315 (815–2170)	108–270 (745–1860)	22–7	Shafts, pistons, gears

[a] AISI, American Iron and Steel Institute; SAE, Society of Automotive Engineers; ASTM, American Society for Testing and Materials; UNS, Unified Numbering System.

[b] Classified as high-carbon steels.

Source: Data for hot-rolled material adapted from *Metals Handbook: Properties and Selection, Irons and Steels*, Vol. 1, 9th ed., B. Bardes, Ed., American Society for Metals, Materials Park, OH, 1978, pp. 190, 192, 405, 406.

Table 3 Designations, Compositions, Mechanical Properties; and Typical Applications for Austenitic, Ferritic, Martensitic, and Precipitation-Hardenable Stainless Steels

AISI Number	UNS Number	Composition (wt %)[a]				Condition[b]	Mechanical Properties			Typical Applications
		C	Cr	Ni	Other		Tensile Strength [psi×10³ (MPa)]	Yield Strength [psi×10³ (MPa)]	Ductility (% EL in 2 in.)	
Ferritic										
409	S40900	0.08	11		1.0Mn, 0.75Ti	Annealed	65 (448)	35 (240)	25	Automotive exhaust
446	S44600	0.20	25		1.5Mn	Annealed	80 (552)	50 (345)	20	Valves (high temperature), glass molds
Austenitic										
304	S30400	0.08	19	9	2.0Mn	Annealed	85 (586)	35 (240)	55	Food processing
316L	S31603	0.03	17	12	2.0Mn, 2.5Mo	Annealed	80 (552)	35 (240)	50	Welding construction
Martensitic										
410	S41000	0.15	12.5		1.0Mn	Annealed / Q and T	70 (483) / 140 (965)	40 (275) / 100 (690)	30	Rifle barrels, cutlery
440A	S44002	0.70	17		1.0Mn, 0.75Mo	Annealed / Q and T	105 (724) / 260 (1790)	60 (414) / 240 (1655)	23 / 20 / 5	Cutlery, surgical tools
Precipitation Hardenable										
17-7PH	S17700	0.09	17	7	1.0Mn, 1.0Al	Solution treated / Precipitation hardened	130 (897) / 215 (1480)	40 (275) / 195 (1345)	35 / 9	Knives, springs

[a]The balance of the composition is iron.

[b]"Q and T" denotes quenched and tempered.

Source: Adapted from Metal Progress 1982 Materials and Processing Databook. Copyright © 1982 American Society for Metals.

Table 4 Designations, Minimum Mechanical Properties, Approximate Compositions, and Typical Applications for Various Gray, Nodular, and Malleable Cast Irons

Grade	UNS Number	Composition (wt %)[a] C	Si	Other	Matrix Structure	Mechanical Properties Tensile Strength [psi×10³ (MPa)]	Yield Strength [psi×10³ (MPa)]	Ductility (% EL in 2 in.)	Typical Applications
Gray Iron									
SAE G2500	F10005	3.3	2.2	0.7Mn	Pearlite + ferrite	25 (173)	—	—	Engine blocks, brake drums
SAE G4000	F10008	3.2	2.0	0.8Mn	Pearlite + ferrite	40 (276)	—	—	Engine cylinders and pistons
Ductile (Nodular) Iron									
ASTM A536									
60-40-18	F32800	3.5–3.8	2.0–2.8	0.05Mg, <0.20Ni, <0.10Mo	Ferrite	60 (414)	40 (276)	18	Valve and pump bodies
100-70-03	F34800				Pearlite	100 (690)	70 (483)	3	High-strength gears
120-90-02	F36200				Tempered martensite	120 (828)	90 (621)	2	Gears, rollers
Malleable Iron									
32510	F22200	2.3–2.7	1.0–1.75	<0.55Mn	Ferrite	50 (345)	32 (224)	10	General engineering service at room and elevated temperatures
45006	—	2.4–2.7	1.25–1.55	<0.55Mn	Ferrite + pearlite	65 (448)	45 (310)	6	

[a]The balance of the composition is iron.

Source: Adapted from *Metals Handbook: Properties and Selection: Irons and Steels*, Vol. I, 9th ed., B. Bardes, Ed., American Society for Metals Materials Park, 6H, 1978.

Table 5 Compositions, Mechanical Properties, and Typical Applications for Eight Copper Alloys

Alloy Name	UNS Number	Composition (wt %)				Condition	Mechanical Properties			Typical Applications
		Cu	Zn	Sn	Other		Tensile Strength [psi×10³ (MPa)]	Yield Strength [psi×10³ (MPa)]	Ductility (% EL in 2 in.)	
Wrought Alloys										
Electrolytic tough pitch	C11000	99.9	—	—	0.04O	Annealed	32 (220)	10 (69)	55	Roofing, rivets, radiators
Beryllium–copper	C17200	97.9	—	—	1.9Be, 0.2Co	Annealed	68 (470)	25 (172)	48	Springs, diaphragms
						Precipitation hardened	165 (1140)	145 (1000)	7	
Cartridge brass	C26000	70	30	—	—	Annealed	44 (303)	11 (76)	66	Ammunition components
Phosphor bronze, 5% A	C51000	95	—	5	Trace P	Annealed	47 (324)	19 (131)	64	Bellows, welding rods
Copper–nickel, 30%	C71500	70	—	—	30Ni	Annealed	54 (372)	20 (138)	45	Saltwater piping
Cast Alloys										
Leaded yellow brass	C85400	67	29	1	3Pb	As cast	34 (234)	12 (83)	35	Battery clamps, fittings
Tin bronze	C90500	88	2	10	—	As cast	45 (310)	22 (152)	25	Bearings, bushings
Aluminum bronze	C95400	85	—	—	4Fe, 11Al	As cast	85 (586)	35 (241)	18	Gears, valve seats

Source: *Metal Progress 1980 Databook*. Copyright © 1980 American Society for Metals.

Table 6 Compositions, Mechanical Properties, and Typical Applications for Eight Common Aluminum Alloys

Aluminum Association Number	UNS Number	Composition (wt %)[a]				Condition	Mechanical Properties			Typical Applications
		Cu	Mg	Mn	Other		Tensile Strength [psi×10^3 (MPa)]	Yield Strength [psi×10^3 (MPa)]	Ductility (% EL in 2 in.)	
Wrought, Non-Heat-Treatable Alloys										
1100	A91100	0.12	—	—	—	Annealed	13 (90)	5 (34)	35	Sheet metal work
3003	A93003	0.12	—	1.2	—	Annealed	16 (110)	6 (42)	30	Cooking utensils
5052	A95052	—	2.5	—	0.25Cr	Annealed	28 (195)	13 (90)	25	Bus, truck uses
Wrought, Heat-Treatable Alloys										
2014	A92014	4.4	0.5	0.8	0.8Si	Heat treated	70 (485)	60 (415)	13	General structures
6061	A96061	0.3	1.0	—	0.6Si, 0.2Cr	Heat treated	45 (310)	40 (275)	12	Trucks, towers, furniture
7075	A97075	1.6	2.5	—	5.6Zn, 0.23Cr	Heat treated	83 (570)	73 (505)	11	Aircraft structural parts
Cast, Heat-Treatable Alloys										
295.0	A02950	4.5	—	—	1.1Si	Heat treated	36 (250)	24 (165)	5	Crankcases, aircraft wheels
356.0	A03560	—	0.3	—	7.0Si	Heat treated	33 (230)	24 (165)	4	Water-cooled cylinder blocks

[a]The balance of the composition is aluminum.

Source: Adapted from *Metals Handbook: Properties and Selection: Nonferrous Alloys and Pure Metals*, Vol. 2, 9th ed., H. Baker, Managing Editor, American Society for Metals, Materials Park, OH, 1979.

Table 7 Compositions, Mechanical Properties, and Typical Applications for Six Common Magnesium Alloys

| ASTM Number | UNS Number | Composition (wt %)[a] | | | | Condition | Mechanical Properties | | | Typical Applications |
		Al	Mn	Zn	Other		Tensile Strength [psi×10³ (MPa)]	Yield Strength [psi×10³ (MPa)]	Ductility (% EL in 2 in.)	
Wrought Alloys										
AZ80A	M11800	8.5	0.12	0.5	—	As extruded	49 (340)	36 (250)	11	Highly stressed extrusions
HM31A	M13312	—	1.20	—	3.0Th	Artifically aged	37 (255)	26 (179)	4	Missile and aircraft use to 425°C
ZK60A	M16600	—	—	5.5	0.45Zr	Artifically aged	51 (350)	41 (285)	11	Forgings of maximum strength for aircraft
Cast Alloys										
AZ92A	M11920	9.0	0.10	2.0	—	As cast	25 (170)	14 (97)	2	Pressure-tight castings
EZ33A	M12330	—	—	2.6	3.2 Rare earths, 0.7Zr	Artifically aged	23 (160)	16 (110)	3	Pressure-tight castings for use between 175 and 250°C
AZ91A	M11910	9.0	0.13	0.7	—	As cast	33 (230)	24 (165)	3	Parts for cars, lawnmowers, luggage

[a]The balance of the composition is magnesium.

Source: Adapted from *Metals Handbook: Properties and Selection: Nonferrous Alloys and Pure Metals*, Vol. 2, 9th ed., H. Baker, Managing Editor, American Society for Metals, Materials Park, OH, 1979.

Table 8 Compositions, Mechanical Properties, and Typical Applications for Four Common Titanium Alloys

Alloy Type	UNS Number	Composition (wt %)	Condition	Mechanical Properties			Typical Applications
				Tensile Strength [psi×10³ (MPa)]	Yield Strength [psi×10³ (MPa)]	Ductility (% EL in 2 in.)	
Commercially pure	R50550	99.1Ti	Annealed	75 (517)	65 (448)	25	Chemical, marine, aircraft parts
α	R54521	5Al, 2.5Sn, balance Ti	Annealed	125 (862)	117 (807)	16	Aircraft engine compressor blades
α − β	R56401	6Al, 4V, balance Ti	Annealed	144 (993)	134 (924)	14	Rocket motor cases
β	R58010	13V, 11Cr, 3Al, balance Ti	Precipitation hardened	177 (1220)	170 (1172)	8	High-strength fasteners

Source: Adapted from *Metal Progress 1978 Databook.* Copyright © 1978 American Society for Metals.

Table 9 Characteristics of Selected Superalloys

Designation	Alloy Type	Form	Condition	Ultimate TS, 1000°F(540°C)	0.2% YS, 1000°F(540°C)	10³ hr Rupture Life, 1400°F(760°C)	10³ hr Rupture Life, 1600°F(870°C)	Applications
Haynes 188	Cobalt base	Sheet	Prob. annealed	107 (740)	44 (305)	24 (165)	10 (70)	GT burners
Hastelloy X	Nickel base	Sheet	Prob. annealed	94 (650)	42 (290)	15 (105)	6 (40)	GT burners
Inconel X-750	Nickel base	Bar	Age hardened	152 (1050)	105 (725)	—	7 (50)	GT/ST general
A 286	Iron base	Bar	Age hardened	131 (905)	88 (605)	15 (105)	—	GT/ST disks
Waspaloy	Nickel base	Bar	Age hardened	170 (1170)	105 (725)	42 (290)	16 (110)	GT disks, hardware
Udimet 700	Nickel base	Bar	Age hardened	185 (1275)	130 (895)	62 (425)	29 (200)	GT blades
Astroloy	Nickel base	Bar	Age hardened	180 (1240)	140 (965)	62 (425)	25 (170)	GT disks
Inconel 718	Nickel base	Bar	Age hardened	185 (1275)	154 (1065)	28 (195)	—	GT disks
Rene 95	Nickel base	Bar	Age hardened	224 (1550)	182 (1255)	—	—	GT disks
Inconel 718	Nickel base	Cast	Age hardened	—		—	—	GT cases
IN 713 C	Nickel base	Cast	Age hardened	125 (860)	102 (705)	44 (305)[a]	31 (215)	GT blades
IN 792	Nickel base	Cast	Age hardened	—		55 (380)[a]	38 (260)	GT blades
Rene 80	Nickel base	Cast	Age hardened	—		—	35 (240)	GT blades
PWA 1480	Nickel base	Cast (SC)	Age hardened	164 (1130)	131 (905)	—	—	GT blades
CMSX-2	Nickel base	Cast (SC)	Age hardened	188 (1295)	181 (1245)	—	50 (345)	GT blades
X-40	Cobalt base	Cast	As cast	80 (550)	40 (275)	—	15 (105)	GT vanes
WI 52	Cobalt base	Cast	As cast	108 (745)	64 (440)	—	22 (150)	GT vanes
Mar-M 509	Cobalt base	Cast	Prob. as cast	83 (570)	58 (400)	—	20 (140)	GT vanes

Note: TS = tensile strength, YS = yield strength, SC = single crystal, GT = gas turbine, ST = steam turbine, ksi (mPa) = strength units.
[a] 1500°F (815°C).

may be little affected by local inhomogeneities. Time- or cyclic-dependent properties at high temperatures and cyclic properties at low to intermediate temperatures can be detrimentally affected by inhomogeneities. For wrought alloys, the probability of occurrence of such regions (which are highly detrimental to fatigue life) is principally dependent upon the melting method selected to produce the cast ingot for subsequent processing. For alloys, whether used as cast or in the wrought form, the degree of inhomogeneity and the likelihood of defects such as porosity are related to the alloy composition, the casting technique used, and the complexity of the final component. Defects may be detected by metallurgical surface examination, radiography, or sonic inspection, but all methods have limitations on the defect size that can be detected.

For sources of property data other than data from the producers (e.g., melters, forgers) of an alloy or from an alloy selector's own institution, one may refer to organizations such as ASM International which publish compilations of data that may form a basis for

the development of design allowables for many alloys. A list of some trade and professional organizations where alloy information may be obtained is provided at the end of this chapter.

Standards organizations such as the American Society for Testing and Materials (ASTM) publish information about alloys, but that information does not ordinarily contain design data. It is important to note that the same nominal alloy chemistry may have some composition modifications made from one manufacturer or customer to another. Sometimes this extends from one country to another. Tweaking of the casting or "wroughting" processes or the heat treatment often associated with what seem to be minor composition changes can cause significant variations in properties. All facets of chemistry and processing need to be considered when selecting an alloy for an application.

There are instances when different heat treatments may be applied to the same nominal alloy composition with resultant differences in properties. An illustration of this is shown in Fig. 6 where a wrought superalloy

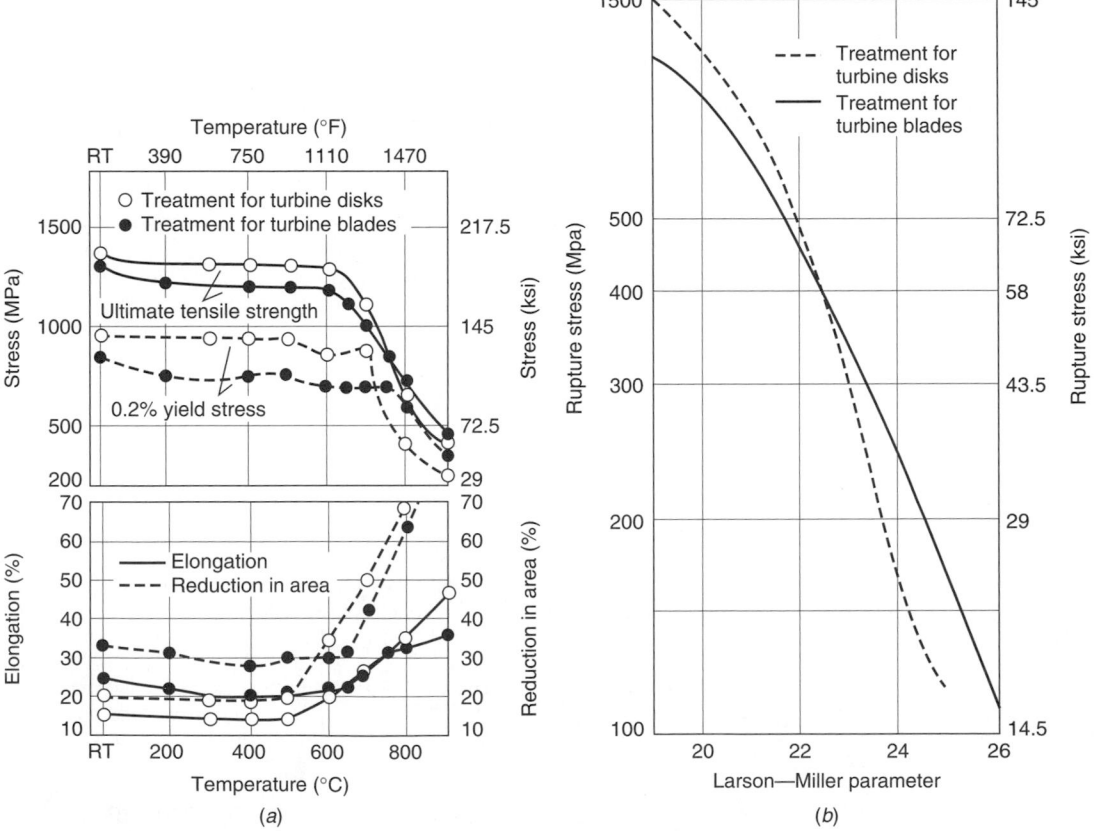

Fig. 6 Heat treatment for two different applications showing how (a) tensile properties and (b) rupture properties differ; $P_{LM} = T(C + \log t)$, where $C = 20$ is Larson–Miller constant, T is absolute temperature (K), and t is time in hours. RT = room temperature.

(Waspaloy) was produced for two different gas turbine engine parts, a disk and a turbine blade. Note the changes in properties. It also is not unusual for the heat treatments applied to cast parts of the same alloy composition to differ from the heat treatments for wrought products of the same alloy.

7 METALS AT LOWER TEMPERATURES

7.1 General

Alloy strengths at lower temperatures (usually considerably below $0.5T_m$) are not a function of time under normal conditions. Environmental attack (general overall corrosion, selective corrosion, etc.) can be significant and will be a function of time, temperature, and environment. Presence of notches and/or cyclic conditions may cause premature failure of a component. Consequently, in addition to evaluation of tensile (e.g., ultimate, yield strengths) properties, possible environmental considerations should be checked and the fatigue (low cycle/high cycle, smooth/notched) strengths, crack propagation characteristics, and toughness behavior should be reviewed.

7.2 Mechanical Behavior

Temperature or time of load application may influence the property behavior, including failure mode, of an article in mechanical testing. Failure may be defined as initiation of a crack or of fracture of an article into two or more pieces. In the case of short-time tensile properties (yield strength, ultimate strength), strengths usually are reduced as temperatures increase from absolute zero to higher temperatures near about 0.5 of the absolute melting temperature. Figure 7 illustrates the trends in tensile strength and ductility of alloys as temperatures increase.

Ductility of an alloy, as measured by elongation at fracture or reduction of area in cross section after fracture, tends to increase with temperature. Sometimes ductility or toughness of an alloy is measured by energy absorbed during impact tests. Strength properties tend to appear in design calculations while ductility values do not. Ductility is considered important in actual alloy use since limited ductility may cause an alloy article to fail with no immediate prior indication of likelihood of failure. The "forgiveness" of higher ductility alloys makes designers more comfortable with application of some alloys. Attempts to build "ductility" considerations into alloy selection and design resulted in various types of fracture toughness property concepts being developed and applied to alloy selection. Fracture toughness (critical fracture toughness) is one concept for which design data may be available for alloys. Higher fracture toughness is most desirable, as is higher ductility.

Some alloys may exhibit a "ductile-to-brittle" transition. This concept occurs when an alloy undergoes a rapid transition to much lower failure ductility over a very small temperature range on cooling of the alloy.

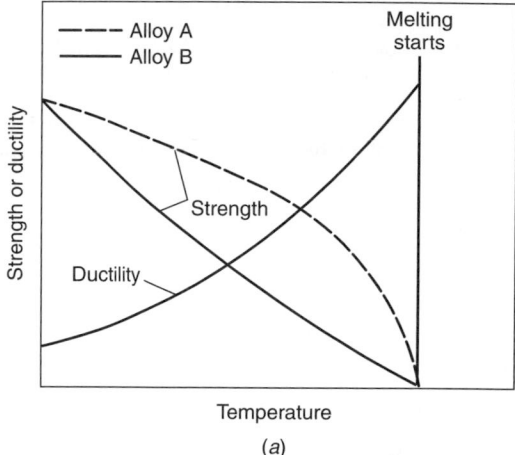

(a)

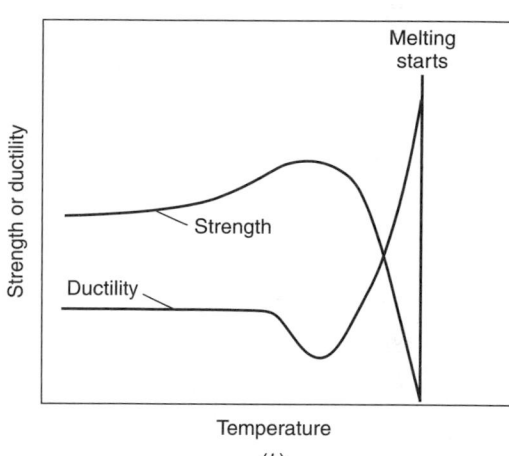

(b)

Fig. 7 (a) Most metals and alloys show a decrease in strength and increase in ductility as temperatures increase. (b) Superalloys precipitation hardened by γ' particles often show peak in strength and minimum in ductility as temperatures increase.

Steels, for example, may show a ductile-to-brittle transition, particularly at temperatures below room temperature. Figure 8 illustrates how toughness, as measured by ductility, for steels drops as temperature decreases (and carbon content increases).

Hardness properties (usually measured by means of standard indentation tests) may be used to judge an alloy's strength capability at lower temperatures and confirm the results of heat treatment but play no role in design.

8 METALS AT HIGH TEMPERATURES

8.1 General

While material strengths at low temperatures are usually not a function of time, at high temperatures

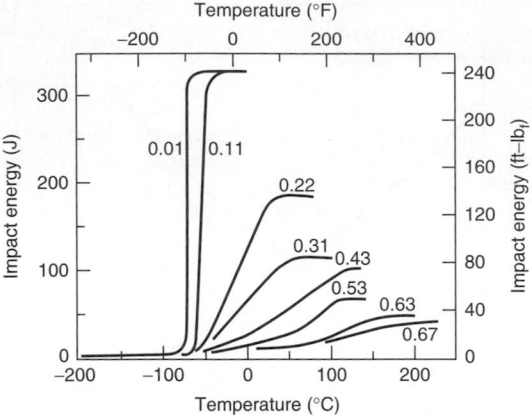

Fig. 8 Charpy V-notch energy versus temperature behavior for steel.

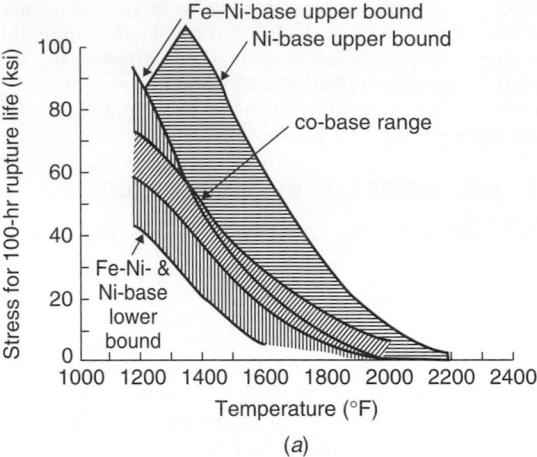

(a)

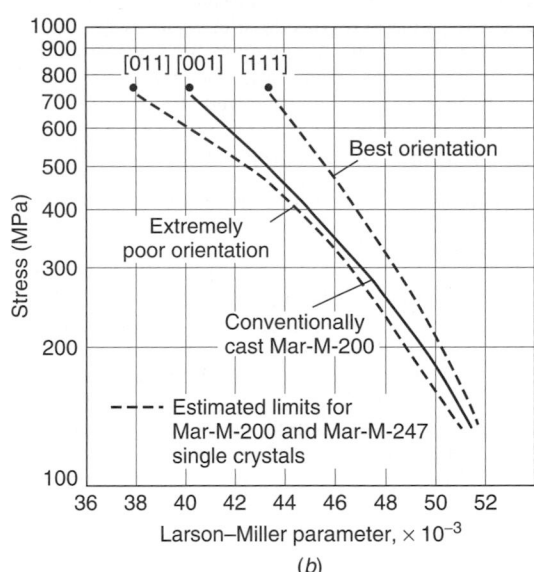

(b)

Fig. 9 (a) Stress for 100-hr rupture life for three classes of superalloys. (b) Log failure stress versus Larson–Miller parameter for single-crystal nickel-base superalloys in three possible test orientations (P_{LM} = actual $P_{LM} \times 1.8$). $P_{LM} = T(C + \log t)$, where $C = 20$ is Larson–Miller constant, T is absolute temperature (K), and t is time in hours.

the time of load application becomes very significant for mechanical properties. Concurrently, environmental attack by oxygen and/or other elements at high temperatures accelerates the conversion of some of the metal atoms to oxides or other compounds. Environmental attack proceeds much more rapidly at high temperatures than at room or lower temperatures. Environmental considerations can vary significantly with different types of environment and different temperatures, in some instances, showing a greater rate of attack at one temperature than the attack of the same alloy at a higher temperature. As in the case of lower temperatures, presence of notches and/or cyclic conditions may cause premature failure of a component. Thus, in addition to evaluation of short-time tensile (e.g., ultimate, yield strengths) properties, creep rupture properties, fatigue, and thermomechanical fatigue (smooth/notched) strengths, crack propagation characteristics, toughness behavior, and possible environmental considerations should be checked.

8.2   Mechanical Behavior

In the case of short-time tensile properties (yield strength, ultimate strength), the mechanical behavior of alloys at higher temperatures is similar to that at room temperature, but with alloys becoming weaker as the temperature increases. However, when steady loads below the normal yield or ultimate strength (determined in short-time tests) are applied for prolonged times at higher temperatures, the situation is different. Figure 4 illustrates the way in which most alloys respond to steady extended-time loads at high temperatures.

Owing to the higher temperature, a time-dependent extension (creep) is noticed under load. If the alloy is exposed for a long time, the alloy eventually fractures (ruptures). The degradation process is called creep or, in the event of failure, creep rupture (sometimes stress rupture) and alloys for elevated temperature use

are selected on their ability to resist creep and creep rupture failure. Data for superalloys frequently are provided as the stress which can be sustained for a fixed time (e.g., 100-hr rupture) versus the temperature. Figure 9 shows such a plot with ranges of expected performance for various superalloy families.

One of the contributory aspects to elevated-temperature failure is that alloys tend to come apart at the grain boundaries when tested for long times above about 0.5 of their absolute melting temperature. Thus, fine-grained alloys which are usually favored for lower temperature applications may not be the best materials for creep-rupture-limited applications at high temperatures. Elimination or reorientation/alignment of grain boundaries is sometimes a key factor in maximizing the higher temperature life of an alloy.

A factor in mechanical behavior not often recognized is that the static modulus (e.g., Young's modulus E) is affected by increases in temperature. This should be obvious from the above discussion about creep. The dynamic modulus of alloys shows a decrease with increasing temperatures. However, dependent on rate of load application, as test temperatures increase, static moduli determined by measurements from a (short-time) tensile test tend to gradually fall below moduli determined by dynamic means. Since moduli are used in design and may affect predictions of life and durability, every effort should be made to determine the dynamic, not just static, moduli of alloys for high-temperature applications. Table 10 shows dynamic moduli of cast superalloys for illustration.

Moduli of cast alloys also may be affected dramatically by the orientation of grains or by use of single crystals. Thus, a columnar grain nickel-base superalloy will show a dynamic modulus parallel to the growth direction of the columnar grains that is around 18×10^6 psi (124.1 GPa) at 70°F (21°C). This is much lower than the polycrystalline (random-orientation) value of around 30 x 10^6 psi (206.8 GPa). A more accurate method of depicting moduli would be to use the appropriate single-crystal elastic constants (three needed) of an alloy, but these are rarely available.

Cyclically applied loads that cause failure (fatigue) at lower temperatures also cause failures in shorter times (lesser cycles) at high temperatures. For example, Fig. 3 shows schematically how the cyclic resistance is degraded at high temperatures when the locus of failure is plotted as stress versus applied cycles ($S-N$) of load. From the $S-N$ curves shown, it should be clear that there is not necessarily an endurance limit for metals and alloys at high temperatures.

Cyclic loads can be induced not only by mechanical loads in a structure but also by thermal changes. The combination of thermally induced and mechanically induced loads leads to failure in thermomechanical fatigue (TMF). TMF failures occur in a relatively low number of cycles. Thus TMF is a low-cycle fatigue (LCF) process (less than about 10^5 cycles). LCF can be induced by high repeated mechanical loads while lower stress repeated mechanical loads lead to fatigue failure in a high number of cycles (HCF greater than 10^6 cycles). Dependent on application, LCF failures in structures can be either mechanically induced or TMF type. In airfoils for the hot section of gas turbines, TMF is a major concern. In highly mechanically loaded parts such as gas turbine disks, mechanically induced LCF is the major concern. HCF normally is not a problem with alloys unless a design error occurs and a component is subjected to a high-frequency vibration that forces rapid accumulation of fatigue cycles. Although $S-N$ plots are common, the designer should be aware that data for LCF and TMF behavior frequently are gathered as plastic strain (ε_p or $\Delta\varepsilon_p$) versus applied cycles. The final component application will determine the preferred method of depicting fatigue behavior.

While life under cyclic load ($S-N$ behavior) is a common criterion for design, resistance to crack propagation is an increasingly desired property. Thus, the crack growth rate (da/dn) versus a fracture toughness parameter is required. The toughness parameter in this instance is the stress intensity factor K range ΔK over an incremental distance which a crack has grown. A

Table 10 Dynamic Modulus of Elasticity for Selected Cast Superalloys

Alloy	Dynamic Modulus of Elasticity					
	At 21°C (70°F)		At 538°C (1000°F)		At 1093°C (2000°F)	
	GPa	10^6 psi	GPa	10^6 psi	GPa	10^6 psi
Nickel base						
IN-713 C	206	29.9	179	26.2		
IN-713 LC	197	28.6	172	25.0		
B-1900	214	31.0	183	27.0		
IN-100	215	31.2	187	27.1		
IN-162	197	28.5	172	24.9		
IN-738	201	29.2	175	25.4		
MAR-M-200	218	31.6	184	26.7		
MAR-M-246	205	29.8	178	25.8	145	21.1
MAR-M-247	—	—	—	—	—	—
MAR-M-421	203	29.4	—	—	141	20.4
Rene 80	208	30.2	—	—		
Cobalt base						
Haynes 1002	210	30.4	173	25.1		
MAR-M-509	225	32.7				

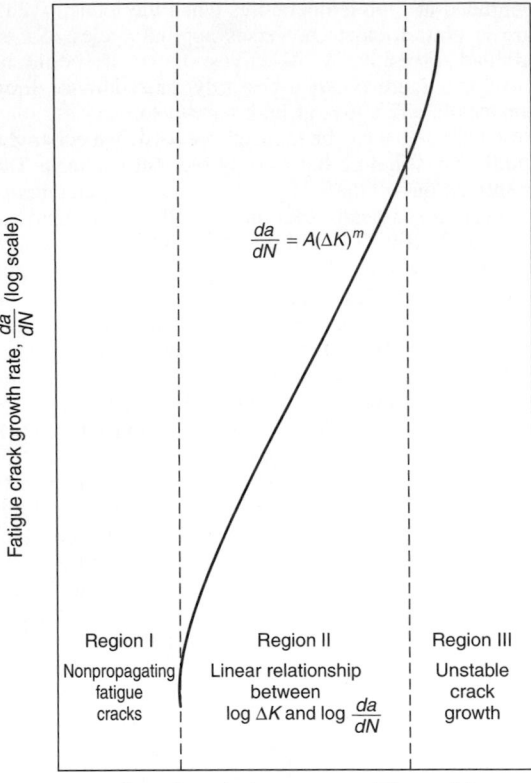

$$\frac{da}{dN} = A(\Delta K)^m$$

Fatigue crack growth rate, $\frac{da}{dN}$ (log scale)

Region I	Region II	Region III
Nonpropagating fatigue cracks	Linear relationship between $\log \Delta K$ and $\log \frac{da}{dN}$	Unstable crack growth

Stress intensity factor range, ΔK (log scale)

Fig. 10 Three regions of crack growth response.

plot of the resultant type (da/dn vs. ΔK) is shown schematically in Fig. 10.

9 MELTING AND CASTING PRACTICES

9.1 Melting

In order to produce most alloys, the appropriate chemistries are made liquid; mixing occurs (if more than one element or master alloy is present) and then the liquid is poured into containers (molds) to produce the desired shape. Most alloys have melting points sufficiently low that they can be melted and cast to shapes economically and with acceptable purity and quality. Melting consists of providing a furnace with an appropriate (usually ceramic-lined) crucible to contain the molten-metal charge and a power source (usually gas or electricity) to produce the appropriate temperatures for melting. For common industrial alloys (white metals, magnesium, aluminum, copper, steels, and superalloys), the melting range needed probably varies from around 800°F to about 2500°F. Melting temperatures for some alloys such as titanium alloys range higher than this.

As indicated earlier, there are occasions when a primary alloy to eventually be used in a final article

shape is cast into an electrode or an ingot one or more times before an article is produced. Electrodes are the precursors to final forged or deformation processed articles. They are intended to be remelted by electrical energy to refine chemistry, microstructure, and so on, prior to being cast as ingots to be delivered to the manufacturer who may cut the ingots to billets from which to make the (raw shape at least) final article by deformation processing. Cast ingots may be remelted without additional processing for use in the direct casting of articles. Alloys produced using the variety of available furnaces in ambient air environments satisfy most common needs. Vacuum melting and similar processes produce higher quality alloys and usually are the production melting methods of choice for applications such as those in aerospace.

For metals such as titanium alloys and superalloys, melting is done under vacuum to retain and/or enhance an alloy's properties. Ordinarily, most alloys to be cast into articles or ingots (which are then processed into billets for wrought processing) are melted open to the atmosphere (air) except for the use of slags on the molten-alloy surface to minimize pickup of gases or loss of elements from the molten alloy or to introduce elements to the alloy from the slag. Furnaces may be resistance wire heated, gas fire heated, arc heated, or induction heated. For more sophisticated melting, especially where vacuum is required, alloys may be heated by induction, arc, electron beam, or plasma. The latter three processes can produce higher local heat input and so may be used to melt alloys such as titanium. The more sophisticated the melting techniques, the higher the cost but, generally, the better the quality.

Vacuum induction melting (VIM) is used for superalloys and, in some instances, stainless steels. The VIM furnace consists of a ceramic-lined crucible built up around water-cooled induction coils. The crucible is mounted in a vacuum chamber. The vacuum chamber may have several vacuum ports (of various sizes) built into it so that, without breaking vacuum,

- charge material may be introduced into the crucible,
- molds may be introduced into the chamber, and
- systems for removing slag from the pour stream (tundish) may be introduced.

Sampling of the molten metal is required for chemistry control in primary alloy production and when the VIM unit is producing ingot for subsequent processing. VIM is more costly than electric arc furnace or other routine processes for melting alloys. A major advantage of VIM is excellent control of chemistry and reduction of impurities such as gases (oxygen, nitrogen) and detrimental tramp elements. Sometimes, in lieu of VIM, an alloy (e.g., of steel) may be vacuum degassed (VD) after melting but before pouring into molds. Improved properties result from VIM or VD processing. The essence of selecting a melting process to create the desired composition and electrode/ingot/cast article structure hinges on the correct choice and adaptation of melting principles and attention to detail in processing.

Consumable vacuum arc remelting (CVAR, most often referred to as VAR) is used to primary melt and to remelt titanium and to remelt superalloys. High heat inputs can be achieved and the alloys are cast into water-cooled molds. No slag is used but rather the ultrahigh vacuum acts to protect the alloy. An electrode is used to strike an arc between the consumable electrode of the alloy and a striker plate in the water-cooled mold. The molten metal produced from the electrode by the arc drips into a pool and solidifies in characteristic ways related to alloy composition, melt rate, mold size, and so on. Generally the mold is larger than the electrode. Molds for VAR processing are circular. Sometimes multiple VAR is applied to an alloy.

Another melting process is electroslag remelting (ESR), which uses an arc within the slag to produce heat that melts the face of an electrode. In ESR, current is applied to an electrode situated inside a water-cooled crucible containing a molten-slag charge. The intended circuit of current is from the electrode, through the molten slag, through the solidifying ingot, through the water-cooled stool, and back to the electrode. The molten droplets from the electrode face fall through the slag to a molten pool and then solidify. Changes in chemistry can be effected in ESR to a greater degree than in VAR processing. There are subtle aspects of the processing for both VAR and ESR which may be discussed with a knowledgeable alloy producer. For superalloy primary melting, where larger ingot sizes of highly alloyed materials are needed, VAR is used in preference to ESR.

VAR and ESR are secondary melting processes (following primary VIM melting to produce electrodes). Several producers of critical rotating superalloy components in the gas turbine industry have adopted the use of a hybrid secondary melt process. Primary melting by VIM to ensure production of a low-oxygen, precise chemistry, initial electrode is followed by ESR. The ESR electrode will be clean and sound. The clean, sound electrode is then remelted by VAR. The improved cleanliness and soundness of the electrode facilitate VAR control. This product is referred to as "triple melt" and has a much reduced frequency of defect occurrence compared to double-melt (VIM + VAR) superalloy product. The cost of the triple-melt process is clearly higher than double melt (VIM + VAR or VIM + ESR).

As noted, alloy-melting practices may be classified as either primary (the initial melt of elemental materials and/or scrap which sets the composition) or secondary (remelt of a primary melt for the purpose of making larger electrodes, controlling the solidification structure, etc.). The melt type or combination of melt types selected depends upon both the alloy composition and intended application (mill form and size desired, properties desired, sensitivity of the final component to localized inhomogeneity in the alloy, economics, etc.).

Although not widely used owing to economic considerations, the use of electron beams or plasma playing on the surface of an electrode and on the molten pool provides an alternative source for clean melting of some specialty alloys.

9.2 Casting to Prepare for Subsequent Processing

When alloys are cast, the resulting object may be (a) an electrode for remelting, (b) an ingot to be melted (in whole or in part) later and then cast into articles, (c) a cast article of commerce, or (d) an ingot for processing into billet. This section discusses the casting of ingot for later use, not directly into articles of commerce.

From the billet, wrought shaped articles may be formed. Also from billet stock, mill products may be produced. Mill products would be bar, plate, sheet, wire, and so on, which might later be deformation processed into a shape or used directly. For example, bar stock might be machined into fasteners while wire might be sold and used with no further processing. Alloys usually not only are available in standard mill forms (plate, sheet, and bar, from which components may be machined) but also may be produced as specific component shapes by the use of forging (or casting). It is critically necessary that the property requirements of a component be fully understood in order to maximize properties and minimize costs. For instance, a component selected to resist corrosive attack may not need to be melted by a process that assures maximum fatigue resistance.

The most critical feature of melt process selection is the size of the component to be manufactured from an ingot. Larger components require larger ingot sizes. The higher the melting point of an alloy and the more highly alloyed the chemistry of the alloy, the greater may be the restrictions of maximum ingot size. For forged components consideration must be made not only of the final component size but also of the capability of the ingot to be sufficiently deformed so as to develop the properties characteristic of a forged structure. Internal ingot cleanliness is also a consideration in selecting the melting route.

Many alloys will have no special cyclic requirements and routine melting and casting to ingots is the rule. However, for more sensitive applications (e.g., aircraft structures), more stringent requirements will exist. Generally, for fatigue-sensitive components, a melt practice is selected which will guarantee the best structure. The needs of sensitive applications may require the generation of specific structure in an ingot cast and machined to billet. For example, a specific grain size may be required by the wrought processor in order to obtain the ultimate customer's specified properties after a forging process and subsequent heat treatment. Specific grain sizes and/or ingot surface finish may be required in order to maximize the effectiveness of ultrasonic or etch inspections. The processes for producing ingot for sensitive wrought-product applications will require that the alloy selector work closely with the melter and forger on the melt-to-ingot-to-billet-to-article sequence. Those selecting materials for less critical articles may be able to rely on standard "mill" practices to generate adequate product.

9.3   Casting Practices for Producing Articles

Generally, cast articles are made by remelting ingots of the desired alloy. Articles may range from grams (ounces) in weight for jewelry (e.g., rings) to tens or hundreds of pounds (machinery, vehicle applications, power turbine components, aircraft gas turbine components). The casting invariably is done in a foundry. The alloy generally is melted in smaller versions of the furnaces used to do primary/secondary melting to produce ingots. Melting in the furnace is done in crucibles which are almost always ceramic-lined containers. In order to cast an article after it has been designed, a copy of the design must be replicated as a negative cavity in a mold. Obviously the mold must be resistant to attack by the molten alloy to be poured into it.

Multiple molds generally are prepared, dependent on the weight/size of the intended cast article. The molds must be connected in such a way that transfer of liquid alloy to them is easy. Design of molds is critical to successful casting. The cavity in the mold must be slightly larger than the intended article to provide for the shrinkage that occurs in the solidification of alloys. There will be a basin or tundish into which the molten metal is poured and from which the metal is dispersed to the individual molds. In order to transfer the metal and properly feed the cavities, various channels (sprues, risers, runners, etc.) are needed. The final mold assembly must have sufficient rigidity to be stored and to be moved into place for casting at the appropriate time.

Different media are used for molds. Sand is a common and inexpensive molding material and plaster and other similar compositions may be used. Surface finish and dimensions are affected by the molding material as is the cost of the casting process. In some instances, molten metal can be melted and injected into water-cooled metal dies producing fine definition with die-cast articles. The alloys used for such processing are lower melting alloys but work has been done to cast high-temperature alloys this way. Cost will be a factor and die casting is more commonly a process for large runs of small- to moderate-size articles such as medals cast in moderate-melting-range alloys.

Complex shapes can be produced by casting. Often the article shape is such that cores must be added to the negative cavity to produce desired passages and holes in the finished casting. Most routine casting will not require extra thin article walls or extra high tolerances. However, for aircraft or power gas turbine parts, in particular, very high tolerances are required

and wall thicknesses (nickel-base superalloys) may drop to 0.040 in. (0.101 cm) or so.

The principal casting practice for high-precision casting is investment casting (also known as the lost–wax process). A reverse-cast model of the desired component is made and wax is solidified in the resultant die. Then a series of these wax models are joined to a central wax-pouring stem. The assembly is coated (invested) with appropriate ceramic, processed to remove wax, and fired to strengthen the invested ceramic mold. A small percentage of a (VIM) alloy ingot is remelted and cast into the mold. Upon solidification, a series of components in the desired form are created, attached to the central pouring stem. These objects, frequently gas turbine hot-section airfoil components, are removed and then machined and processed to desired dimensions. Superalloy investment-cast cases now are available up to about 40 in. (101.6 cm) in diameter. Titanium alloys similarly can be cast in large sizes using investment-casting techniques. Turbine airfoils can now be cast not only in smaller sizes for aircraft gas turbines but also with airfoil lengths of several feet for land-based power turbines.

Most investment-cast articles are of conventional, polycrystalline structure with more or less randomly oriented grains. Mechanical properties are nominally random but may show some directionality. Increased high-temperature property strength levels have been achieved in nickel-base superalloys by columnar grain directional solidification (CGDS), which removes grain boundaries that are perpendicular to the applied principal load in turbine airfoils. The ultimate solution for optimum high-temperature strength is single-crystal directional solidification (SCDS) to produce a superalloy with no grain boundaries. Maximum creep rupture strength in nickel-base superalloys now is achieved with SCDS alloys. SCDS has been applied not only to aircraft gas turbine engines but also to large-frame land-based gas turbines. Most alloy systems used commercially have not shown a need for increases in the high-temperature strength which might be produced by single-crystal casting. Such processing would considerably increase cost and property improvements with SCDS depend on alloy chemistry. Certain nickel-base superalloys are the only alloys where acceptable cost–benefit trades are possible.

9.4   Casting Considerations in Alloy Selection

Alloy selection for cast articles generally will be on the basis of alloy type needed (e.g., copper, aluminum) and then on specific alloys within the selected type so as to provide the desired properties. Not all alloys are equally good for casting. In many alloys systems, compositions of casting alloys differ from those of alloys used for wrought applications. Depending on the application, there may be more or less tolerance for lesser quality castings. Need for pressure tightness, for example, might dictate a casting process different from that for another application. Casting defects occur and vary with alloy composition, article shape, and dimensions as well as casting practice. Oxide (dross) carried over from the alloy-melting production process may cause surface and internal inclusion problems in some alloy systems. Porosity is another major concern, especially with large castings. Porosity may be controlled by mold design and pouring modifications. In some alloy systems where articles for critical applications are required, non-surface-connected porosity may be closed by hot isostatic pressing (HIP) of the cast articles. Surface porosity in large castings may be repaired by welding. Other casting concerns may include surface finish, intergranular attack (IGA) caused during chemical removal of molding materials, selectively located coarse grains in PC materials, and so on. Sometimes alloys may be modified in chemistry to optimize product yield but with a possible compromise in properties. Product yield is an important determinator of final component cost.

10 FORGING, FORMING, POWDER METALLURGY, AND JOINING OF ALLOYS

10.1 Forging and Forming

Forging is one of the most common methods of producing modest production lots of wrought articles for alloy applications. Mill products such as bar stock and wire generally are produced in greater quantities than forgings but, for the most part, are used to make noncomplex shapes. The most demanding applications use wrought forged ingot metallurgy components. One or more intermediate shape stages usually are involved when conventional forging is practiced. Special processing by isothermal (usually superplastic) forging may enable the forging to go directly from billet to final stage in one step using a closed die.

A billet is the precursor to a forging in large-sized articles. Billets are transformed (forged) to shapes approximating the final desired shapes with the approximation dimensions being dependent on the alloy system and known issues in the forging process. Multiple reductions may be needed and blocker dies (open dies) may be needed before closed die forging to final dimension/shape. Forging may be accomplished with one or a group of dies and in open or closed dies. While closed dies better define the finish form of a forging, the cost of using such dies exceeds that of open dies. Some dies may be just flat platens that are used to transfer the forging load. The more complex the dies, the more difficult it becomes to extract heat. Significant costs can be incurred for die design and procurement.

Forging is usually a process carried out at elevated temperatures with the temperature being determined by alloy composition, amount of reduction needed, and so on. Forging always requires significant deformation of the alloy as it is transformed from billet to article. Sometimes the forging deformation is retained to increase strength; sometimes forging is used to modify the microstructure of the article when forging is combined with appropriate heat treatments. In the case of high-performance superalloys, the objectives of a forging cycle may be:

- Uniform grain refinement
- Control of second-phase morphology
- Controlled grain flow
- Structurally sound component

Hot forging is a multistep process with several to many cycles of heating to forge temperature, then forging, followed again by heating, forging, and so on.

Forming, while containing elements of deformation, generally restricts itself mostly to sheet that, for example, may be bent or formed to a mandrel or other forming die. Forming of automotive hoods from sheet metal would be an example. Deformation induced in forming may not be extreme and forming is not used to control microstructure and properties of alloys. Basically, forming produces relatively two-dimensional shapes while forging produces three-dimensional shapes with distinct properties. Roll forming is often used to produce small shapes from bar stock or rod.

There are two major types of forging: open die and closed die. Open-die forgings are less expensive but in general require more stock removal by machining to get to the finish dimensions. Closed-die forgings are more costly but reduce the machining needed to achieve finish dimensions in the article. In forging, alloy billets are squeezed between two platens or dies using an appropriately sized press. Several to many forging steps may be needed to bring the billet to the approximate article shape and dimensions. A great deal of skill is necessary not only in die design but by the operators of the forge presses in order to achieve optimum properties and cost balances. As alloy strengths increase, it becomes increasingly difficult to move an alloy around during forging. When higher forging pressures are required, defects become more probable.

Superplastic forming became available in the 1960s and was first used to produce aluminum alloy door panels, for example, for high-end automotive vehicles. Superplastic forging was introduced for superalloys when the strengths of alloys were bumping up against metallurgical limits and forging was becoming too difficult. A process called superplastic forming/diffusion bonding was developed in the air frame business to produce large articles with better properties than existing technology could produce. This is a high-technology process without applicability to processing of routine materials.

10.2 Powder Metallurgy Processing

Powder metallurgy (PM) has been in use for many decades, dating back over 50 years. The principal intent of PM was/is to produce alloy articles at less cost. However, since the latter part of the twentieth century, PM also has meant a way to create billets of the highest strength alloys (without casting) and subsequent forging of billets to final parts. Different powder production manufacturing processes exist for alloys. A common practice is gas atomization where a prealloyed ingot is remelted and then blasted by a jet of air/argon/helium to produce powder in a chamber. The powder is recovered and classified for size, for example. Powder production generally is a batch process.

Molds (dies) for making PM parts are made to the shape of the planned article and then filled with powder, tapped, pressed, and sintered (high-temperature heated) or otherwise metallurgically bonded (by heat/pressure) to increase density and properties. The process often can be automated. Dimensions of dies will need to reflect the expansion coefficient of the die material and the alloy powder. There are different process variables in PM, including multiple methods of powder production. Compaction consists of processes such as cold pressing or vacuum hot pressing or HIP, among others. Some high-strength superalloys are made only by PM methods. Because powder articles made by other than HIP are not 100% dense, there will be porosity in the form of very small voids in the powder article compact. For the most part, a small number of defects such as holes do not significantly affect the properties of alloys, at least in static tests. For fatigue-limited applications, higher quality production processes would need to be used.

A relatively new aspect of the use of PM is in the superalloy field where alloys, which can be cast in relatively small parts such as turbine blades, are not workable by cogging, forging, and so on, to produce a billet for disk forging. By using powder of the desired alloy, a compact of the desired article can be made by HIP or extrusion and then the compact can subsequently be worked (usually forged) to final shape and machined to produce a higher strength wrought article than can be produced from conventional wrought-alloy ("ingot metallurgy") procedures.

10.3 Forging/Working Considerations in Alloy Selection

The stronger the alloy to be utilized in design, the more difficult it will be to manipulate by mechanical working forces. In addition to working difficulties, problems with ordinary forgings can arise from many sources. Poor grain size control, grain size banded areas, poor second-phase morphology/distribution, internal cracking, and surface cracking are among the sources for rejection of forged parts. Limits exist on the capability of an alloy to be worked without cracking, encountering other defects, or stalling a forge or extrusion press. Some shaping equipment such as rotary forging devices may be better able to change the shape of the stronger alloys than other equipment. Powder metallurgy offers an alternative processing route for high-strength alloys such as superalloys used for rotating disks, shafts, hubs, and so on, in aircraft and power generation turbines. Powder production reflects an art

that is not always directly transferable from one producer or one alloy to another. Powder components are best made by producing an ingot from powder and then forging the ingot to component shape. Extensive work with an alloy melt shop and an associated powder producer may be necessary to create a satisfactory metal powder of a high-strength/high-performance alloy selected for design.

10.4 Joining

Welding, soldering, and brazing are used to manufacture articles in many instances. Soldering is a lower temperature, less than 800°F (427°C), process of letting molten metal be drawn into a very tight crevice between the components to be joined and then allowing it to be solidified. Brazing is accomplished in the same general way, but the braze alloys melt above about 800°F (427°C). Braze alloys produce stronger bonds than solder alloys. Solder finds extensive use in the electronics industry. Brazing is more often found in structures which require relatively high strength. Such structures might be aircraft gas turbine stator assembly parts, for example.

Welding produces a metallurgical bond between components of an article. There are many welding processes. The simplest definition is solid-state bonding versus fusion welding. Relative to fusion welding, resistance welding techniques are low-cost methods to join parts such as tabs on sheet metal. Arc welding is perhaps the most common fusion welding technique with higher quality associated with gas–tungsten arc or gas–metal arc welding. There are other techniques used for massive structures and also specialized techniques that produce maximum-quality welds. Those techniques use electron beams, laser beams, or plasmas to heat the weld metal. Relative to solid-state welding, a number of techniques exist, including friction or inertia bonding, diffusion bonding, and ultrasonic bonding. The success of these techniques is dependent on clean surfaces and optimum bonding areas.

Many complex articles in devices such as gas or steam turbines were/are combinations of components joined to produce a larger article. In some applications, when the joined article is not too large or complex, efforts have been made to replace such articles with cast components. However, surface-connected defects often are found in large superalloy castings. In addition to large superalloy castings, many articles produced in industry can have production defects. If defects can be repaired, articles may be acceptable for service. Consequently, repair welding is employed with great frequency, dependent on the dollar value of the article and other economic and availability aspects.

10.5 Considerations in Joining Process Selection

An alloy selector will need to consider the weldability or joinability of an alloy before suggesting initial production use of joining techniques and/or their use in repair. Many nickel-base superalloys are not weldable. High-performance alloys such as titanium alloys and sheet metal of cobalt-base, solid-solution-strengthened nickel-base and lower hardener content nickel-base alloys can be joined as can iron–nickel-base alloys. Aluminum alloys can be welded with high-heat input welding techniques. Some techniques such as electron beam welding (EBW) may be best for high-performance materials owing to the vacuum environments of the processes and the very narrow fusion zone that is created in the process. High-vacuum EBW and similar joining processes may not find wide use in production of large quantities of consumer goods owing to the cost of the process. On the other hand, the use of nonvacuum electron bean welding (NVEBW) has been shown to be cost effective for steels used in auto frames.

11 SURFACE PROTECTION OF MATERIALS

11.1 Intrinsic Corrosion Resistance

Most alloys are not intrinsically resistant to surface attack by corrosion. Precious metal alloys containing gold or platinum may be chemically stable. Most alloys consist of metallic elements which want to revert to compounds (e.g., oxides) in chemical states such as those in which the elements are found naturally. Corrosion is the interaction of elements in an alloy with the environment to produce chemical compounds. Surface attack occurs most frequently since the surface is the primary interface with the surroundings of an article in use. Most typically, articles "see" oxygen from the air and a water-based environment owing to humidity in the air. The processes of uniform corrosion in aqueous environments and with selected other environments are fairly well understood. Corrosion also can take place in nonuniform ways through the action of or development of pits, crevices, or cracks. Under some circumstances, individual alloy elements may be selectively removed from an alloy.

Using an understanding of the nature of the corrosion process, metallurgists have developed alloys which have improved resistance to corrosion in a variety of environments. In fact, many alloys containing certain elements such as chromium or aluminum demonstrate natural resistance to corrosion in oxidizing environments. However, alloys which are corrosion resistant in oxidizing environments often are not resistant in reducing environments, especially when attacked in the presence of halide elements such as chlorine. Some alloys/elements are more noble with respect to other alloys/elements. One representation of the corrosion sensitivity of alloys is the "galvanic series of alloys/elements." Alloys/metals that do not corrode or are more corrosion resistant in aqueous environments are called noble or cathodic while more corrosion prone alloys/elements are called active or anodic. Table 11 shows a galvanic series for some commercial alloys and metals in seawater. Note that some more common alloys other than those of the precious elements are near the top. Specifically, titanium

Table 11 Galvanic Series of Selected Commercial Metals and Alloys in Seawater[a]

Noble, or cathodic	Platinum
	Gold
	Graphite
	Titanium
	Silver
	Stainless steel, austenitic, P (18% Cr, 8% Ni, low C)
	Stainless steel, ferritic, P (10–30% Cr, high C)
	Nickel–chromium–iron alloy, P (80% Ni, 13% Cr, 7% Fe)
	Nickel, P
	Silver solder
	Nickel–copper alloy (70% Ni, 30% Cu)
	Copper–nickel alloys (60–90% Cu)
	Copper–tin bronzes
	Copper–zinc brasses
	Nickel–chromium–iron alloy, A (80% Ni, 13% Cr, 7% Fe)
	Nickel, A
	Tin
	Lead
	Lead–tin solders
	Stainless steel, austenitic, A (18% Cr, 8% Ni, low C)
	Stainless steel, ferritic, A (10–30% Cr, high C)
	Cast iron
	Steel
	Aluminum alloys, precipitation hardened
	Cadmium
	Aluminum, commercially pure
	Zinc
Active, or anodic	Magnesium and magnesium alloys

[a]P indicates passive condition; A indicates active condition.

and its alloys are very resistant to aqueous corrosion. Similarly, titanium alloys display excellent corrosion resistance in body fluids. Consequently, titanium alloys are often choices for applications such as prostheses for bone implants or in dental work. Commercially pure titanium finds extensive use in the chemical processing industry.

Although cast iron and steel are near the bottom of the series and so are active (steels rust in most instances of environmental exposure), chromium additions to iron can produce a passive layer that inhibits oxidation. Hence alloys of iron with sufficient chromium additions became "stainless." Stainless steels are near the top (cathodic end) of the galvanic series and perform very well in a variety of commercial applications. Some construction steels are alloyed to produce a rust-resistant finish and are used for uncoated/unpainted steel work on bridges, guard rails, and so on.

Different alloys may display varying behavior dependent on the environment. In some instances, alloys that display good aqueous corrosion resistance (general corrosion attack resistance) show tendencies to selective attack such as that which occurs in stress corrosion. Connections between anodic and cathodic alloys in service invariably lead to accelerated attack of the anodic element/alloy.

Some alloys have moderate resistance to corrosion. That is, they form discolored surface layers but do not have surface recession at a great rate. A case in point would be copper or copper alloy flashing often seen on roofs at chimney lines. The alloys degrade owing to formation of a surface patina, but there is no significant surface recession and the copper retains its cross section and hence strength integrity for a long time.

Some alloys show adequate corrosion resistance so long as the surface of the article is not degraded. However, surface erosion or inadvertent scratching or similar degradation can cause corrosion to accelerate. Pitting and similar corrosion owe their existence to such tendencies.

Some alloy elements confer better corrosion resistance in certain temperature/environment regimes than in others. Nickel-base superalloys oxidize at elevated temperatures. Nickel was alloyed with chromium in the early twentieth century to produce oxidation-resistant materials used for electrical resistance wires. In the mid-twentieth century, the nickel–chromium alloys were adapted for use as superalloys for gas turbine engines. In the early years of use, these chromium-containing alloys showed excellent oxidation resistance. Later, as temperatures went higher, alloys with higher amounts of aluminum (and reduced chromium) than in the early superalloys showed improved high-temperature oxidation resistance.

Generally, as temperatures increase, corrosion attack increases. However, for some forms of environmental attack, the kinetics of the attack process may be such that a specific type of attack begins at a certain temperature and increases with increased temperature but then drops again to lower rates at still higher temperatures. Nickel-base superalloys exposed to sulfur and/or halide-containing environments may show lower temperature hot-corrosion rates that are greater than the hot-corrosion attack seen at higher temperatures.

In addition to general corrosion, grain boundary oxidation/corrosion or the selective attack of some secondary phases in an alloy can create notches. Coatings will help protect against such attack. Other types of chemical attack are more subtle. For example, brass (copper–zinc) alloys can be susceptible to stress corrosion cracking. In this instance, the attack may be on/in the grains and cause cracking owing to a residual or applied stress on an article made of the alloy. In the early days of manufacture of cartridge brass (70 Cu and 30 Zn), cases would crack or be embrittled by stress corrosion cracking and would burst on firing. Stress corrosion cracking can be found in titanium alloys and some stainless steels among other materials. Reduced residual stresses help to alleviate the problem for brass.

11.2 Coatings for Protection

From early use of steel, the concept of rust was prevalent. At first no alloys were made that were rust resistant, but it was discovered that, if one could protect the iron or low-carbon steels from the environment, then there was no attack. In the automotive industry, this concept was applied to bumpers by electroplating coatings onto the bumpers. First a nickel plate was applied, then a high-reflectivity chromium was plated over the nickel. So long as the nickel did not develop a pore or scratch or suffer a similar breakdown, the steel bumper did not rust. Similarly, prior to the widespread use of polymer trash cans, steel cans were coated with a sacrificial layer of zinc, an element anodic to iron. The trash cans did not rust because the zinc corroded preferentially to the steel. These two illustrations show the use of impervious coatings or anodic coatings to protect materials.

Corrosion at lower temperatures is not the same as corrosion at high temperatures where the interaction of an alloy with oxygen can cause an oxide to form with consequent reduction in the cross-sectional area of an alloy. If the oxide itself is essentially protective, e.g., chromium oxide at temperatures below about $1500-1600°F$ ($186-871°C$) or aluminum oxide above about $1800°F$ ($982°C$), then the alloy is protected from oxygen attack for some time. It was reasoned that the production of a deliberately introduced chromium-oxide-forming or aluminum-oxide-forming coating on alloys for high-temperature use would increase their resistance to oxidation.

Such coatings were introduced and applied by a number of methods. The early coatings were diffusion coatings on cobalt- or nickel-base superalloys and were created by pack aluminizing or slurry application. The chemistry of the coating was determined by the chemistry of the alloy. Some high-temperature protective aluminide coatings were applied by painting or by spray processes. Later (second-generation) coatings were produced by overlaying a specific chemistry of a protective nature on the surface of the component using physical vapor deposition (evaporation of elements of an alloy by using an electron beam and redeposition of the elements on the target alloy's surface). Overlay coatings are generally more expensive than diffusion coatings. However, vapor-transported diffusion-type coatings can coat internal (non-line-of-sight) surfaces while overlay coatings can only coat external line-of-sight surfaces that can be seen by the coating apparatus. Some commercial diffusion coating processes are available, but most overlay coating processes are proprietary, having been developed by users such as aircraft gas turbine manufacturers.

Of course, for lower temperature use, spray or electrodeposited coatings may provide adequate protection for many alloys and overlay coatings might be overkill.

Despite the coatings on high-temperature superalloys, oxidation (or other corrosion) continues but at a markedly lower rate. Since the coating is being used up over time, eventually the coating will not be sufficiently protective and the surface of the alloy must be recoated. This concept of eventual coating degradation will apply at both high temperatures in oxidation and lower temperature corrosion (e.g., aqueous corrosion). Depending on the alloy's structural application, the cost to replace the article that is coated, and the cost of recoating, a decision will need to be made about continued use of the article in its application. Generally, replating or recoating of articles is more apt to occur if the cost of the article's replacement is very high. So, gas turbine airfoils which may cost in the $100–$1000 range to make as castings are more apt to be recoated and reused than a car bumper that has rusted.

11.3 Coating Selection

Coating selection is based on knowledge of oxidation/corrosion behavior in laboratory, pilot-plant, and field tests. Attributes that are required for successful coating selection include:

- High resistance to general oxidation/corrosion
- Ductility sufficient to provide adequate resistance to thermal mechanical fatigue if the article sees high temperatures since coatings are not particularly ductile and coating cracking/spalling could occur
- Compatibility with the base alloy to be coated (important anywhere but especially for high-temperature operation)
- Low rate of interdiffusion with the base alloy (if used at high temperatures)
- Ease of application and low cost relative to improvement in component life
- Ability to be stripped and reapplied without significant reduction of base-metal dimensions or degradation of base-metal properties

12 POSTSERVICE REFURBISHMENT AND REPAIR

One important aspect of modern alloy use is the concern for maximizing service life of articles. Manufacturers do not wish articles or devices to fail in less than the warranted life. Selection of alloys for design should include concern for potential early removal and repair or refurbishment. Surface degradation and mechanical property loss are major economic factors in applications of articles. These factors have become of greater interest as the base cost of materials and subsequent components has risen dramatically over the past 40 years. (As this chapter is being written, for example, the price of nickel, a common base material or alloy element, has risen 10-fold in 3 years!)

Although initial cost has usually prevailed in alloy selection for consumer devices and articles, as the cost of machinery (nonelectronic) has increased owing to various factors, more attention in alloy selection has been given not only to initial ability to produce a viable article but also with regard to future life and refurbishment/repair concerns. In practice, when public safety consideration limits or product integrity design limits are reached, costly components may be withdrawn from use. Some components may appear to be unaffected by service time. For economic reasons there may be incentives to return these components to service.

Other components after service may have visible changes in appearance. For example, a gas turbine high-pressure turbine blade may be missing a coating in the hottest regions of the airfoil or a crack may develop in a vane airfoil. Seals may be worn. It is highly desirable that damage can be repaired so that the costly parts can be returned to service.

If possible, alloys would be selected on the basis of restoration of capability after initial capability has (apparently) been reduced by service exposure of the component. However, most applications do not permit alloy selection on that basis. Many applications have limited lifetimes to reduce initial cost (and, possibly, to improve the likelihood that new products will be purchased). The best alloy from a property and initial economic viewpoint is usually the choice. However, it is common practice with certain applications to refurbish or repair many components which have visible external changes.

Stripping and recoating of turbine airfoils is one example of refurbishment and repair practice. Oxidation- and corrosion-resistant coatings and thermal barrier coatings may be reapplied (after appropriate surface-cleaning treatments) to restore resistance of the surface to heat and gaseous environments. When a high-performance article made of a superalloy is to be refurbished by recoating, all traces of the original coating should be removed before recoating is attempted. In the case of missing, eroded, cracked, and/or routed material, welding traditionally has been used to fill the gaps for some alloys and components. Care must be taken to assure that no additional alloy degradation occurs owing to the refurbishment and repair practice.

The restoration of mechanical properties degraded by creep and/or fatigue is not clear-cut. In the laboratory, reheat treatment has been shown to restore the mechanical properties of some superalloys after service exposure. The degree of restoration is a function of the mechanical history of the component. Results of reheat treatment of service exposed parts are variable. Most postservice procedures for high-cost flight safety articles do not provide for mechanical property restoration.

It is important to recognize that refurbishment or repair may not result in cost-effective performance.

13 ALLOY SELECTION: A LOOK AT POSSIBILITIES

13.1 General

Selection of alloys for design may require a comprehensive review of the design and potential materials or, as noted below, may rely on prior use to dictate alloy selection. On the assumption that a design will start from scratch with no preconceived notions of alloy usage, several questions need to be answered:

- What is the expected temperature of use?
- What is the expected environment (gas, liquid, moving, static, etc.)?
- What strength levels are required?
- Is there a cyclic component to any loads?
- Are elastic properties likely to be a criterion?
- Will loading be uniform or will some directions within the article see different loads?
- Are there special characteristics at issue, for example, heat capacity, electrical, magnetic?
- For how long must the article to be designed and manufactured last?
- Will successful operation depend on special cooling or other conditions?
- What weight or dimensional levels must not be exceeded?
- Is there an aesthetic component to the application of the desired article?

Assuming that the above questions can be answered in a definitive way, referral to general alloy characteristics will narrow the choices available. For example, operating temperatures above about 1000°F (538°C) will eliminate aluminum, magnesium, and zinc alloys since alloys of those metals will be molten at such operating temperatures. Elastic modulus concerns might further limit alloys. For example, if it would appear that materials for a given article would need Young's modulus of near 30×10^6 psi (206.8 GPa), then titanium alloys would be eliminated since their moduli are less than 20×10^6 psi (137.9 GPa). Would the application be in oxidizing gases above 1000°F (538°C)? Then iron would be ruled out owing to its lower oxidation resistance compared to other alloys at those temperatures. Stainless steels might be possible but creep rupture strength might be too low, thus moving the likely candidates to superalloys, those of cobalt, iron–nickel, or nickel base.

Now, attention would need to be directed to the manufacturing ability of alloys to be considered. If the alloy is to be mechanically deformed to reach its near final shape, ductility/workability will need to be considered. Is the article complex, with internal cooling passages and thin walls? Then investment casting may be the best production method! Consideration must also be given to the fact that some processing methods produce or permit improved property levels over other processing methods. For example, investment-cast superalloys are superior in creep rupture strength to wrought superalloys and SCDS investment-cast superalloys are superior to polycrystal investment-cast superalloys.

At this point, alloy families will have been reduced considerably from the total matrix of alloys to one or two alloy types, for example, cobalt-base or nickel-base superalloys that can be investment cast. Design now will hinge on strength (static and/or dynamic-cyclic). A review of strength requirements might show creep rupture capability requirements on the order of 24 hr life at 1850°F/36 ksi (1010°C/248.2 MPa). Cobalt alloys cannot reach those levels so nickel-base superalloys become the choice. Now the selection may hinge on specific alloy chemistry, past experience with an alloy, and investment-casting experience with the alloys from which selection needs to be made. Economics of the basic alloy chemistry, economics of the investment-casting route (polycrystalline, columnar grain, single crystal) to be used, ease of applying surface coatings (if required), minimum alloy properties which are needed, and perhaps other factors need to be considered.

The final selection will need to be verified by actual manufacturing and device testing in simulated or (preferably) actual service. Refurbishment procedures may need to be considered if the article is particularly costly and can have its life extended by suitable and timely refurbishment.

It may be seen that the alloy selection in this instance is not exactly a case of working from "ground zero" to scientifically evaluate the likely candidates. Rather, the selection has depended on real engineering data and judgment and would have involved potential alloy suppliers and manufacturing specialists as soon as possible. A great deal of "alloy selection" that appears in the literature has been done in retrospect, that is after some particular article has been manufactured (e.g., a skateboard), scientific approaches to the evaluation of the actual article show the requirements which were met to produce it. There is nothing in principle wrong with scientific alloy selection. However, the general lack of data (not enough, maybe virtually no, property data), no or limited manufacturing experience, insufficient awareness of the "shop" climate (what furnaces might be available for heat treating the articles and how would the treatment impact the shop schedule), and so on, can cause unacceptable delays or even inability to manufacture an acceptable article.

Assuming that minimum alloy property data will be available, the key to alloy selection is to find knowledgeable engineers in-house and engage them and other knowledgeable persons from the manufacturing pipeline in conversation early on in the selection process.

13.2 Impact of Materials' Data Validity on Selection

There ought to be one basic rule for alloy selection and that is: "Do not believe all you see or hear" (from your fellow employees or from persons purveying new alloys, coatings, or processes). The most common type of statement from alloy/process developers with whom one talks (the "sellers") at the start of new alloy (or process) evaluation is: "This is our average product/property." Translated, this statement should probably be treated by the alloy selectors (the "buyers" in this instance) as meaning: "It's the best we have gotten!" One needs to be skeptical and inquisitive about statements from the seller!

One should be especially wary of selective use of data. There was a situation where an alloy (never used in production) was claimed to have property levels up to 150°F (94°C) better than any other similar alloy then existent. Unfortunately, when a laboratory program was run at a different laboratory (lab 2) than the laboratory (lab 1) of alloy invention, results were troublingly low. On inquiry to the "inventor of the alloy," the question was raised as to how he got such good results. The answer? The inventor only cut out good material to test.

At that point, it was reported to the manager of lab 2 that, with careful work in the laboratory, one might get a "100°F (62.5°C)" improvement over the best existing alloy but never 150°F (94°C). And, it was said that, in a production shop, the best improvement might be no better than 50°F (31°C). The latter projection was proven true when maximum alloy improvement turned out in production to be only 47°F (29.4°C).

Engineers have been known to discard data that are not as good as what they expect (and occasionally get). There need to be valid reasons for excluding data. When much data (properties, results of processing, etc.) get discarded, the discarded data often come back to haunt the user of the alloy/production process. Dig into the database for anything you use in your alloy selection. Assure yourself that there are no hidden items which may appear again in production.

In an instance of mechanical property tests of a new wrought nickel-base superalloy to be used for a gas turbine disk, low creep rupture test results were routinely discarded. The alloy (as a disk) went into production and most of the first production run began to fail the creep rupture acceptance tests in the shop, with test values almost identical to those which had been routinely discarded during alloy development. A modification of heat treatment was necessary to bring the alloy creep rupture life back to the range promised in development and much time and money were spent in the recovery operation.

14 LEVEL OF PROPERTY DATA

Different organizations have different ways to treat property data. Generally minimum and typical properties are desired by design engineers. Minima can vary in definition (-2σ, -3σ deviations are used). One of the cardinal rules of alloy property development is that enough data will never be generated to truly determine a statistically valid property level over all possible property space. Data cost money! Estimates are made and may often be used in design. If these estimates are conservative and can be justified by alloy data on the alloy of interest and similar alloys, satisfactory designs with the selected alloy usually result.

Not all properties require minimum values. Some property values used for design are typical values. While minima may be used in design, estimates of typical behavior of a component made in order to track its operation may require typical property values.

15 THOUGHTS ON ALLOY SYSTEMS

15.1 General

Melting temperatures for homogenous alloys range near but generally below the melting temperature for the major (base) element. Natural segregation of alloy elements during solidification can lead to even lower melting temperatures. The incipient melting temperatures (lowest melting point of an alloy after solidification) can be many tens of degrees less than an alloy's nominal melting temperature. While some homogenization may occur with heat treatment or with wrought processing and heat treatment, it is possible for alloys, particularly cast alloys, to still show incipient melting. This factor needs to be kept in mind during processing and use of a given alloy. Some alloy elements cause significant depression of the melting temperatures of an alloy system while others show relatively little effect. For example, boron, carbon, zirconium, and hafnium depress the melting temperatures of nickel-base superalloys. Since these elements were added primarily to enhance grain boundary strength/ductility in cast versions of many alloys, the introduction of single-crystal nickel-base superalloys enabled most or all of those elements to be removed from the alloy chemistry. The resultant melting temperature increase enabled enhanced heat treatments and optimized strength properties in these alloys.

Another aspect of alloys is the effect of alloy elements on the density of the base metal of an alloy. When much lighter elements than the base metal are introduced, significant density reductions occur. Such additions may or may not be beneficial. For example, early nickel-base superalloys tended to have densities of about 0.3 lb/in.^{3} (8.3 g/cm^3). Some inventors added vanadium, a relatively light metal, to an alloy and reduced the density of the alloy to about 0.28 lb/in.^{3} (7.75 g/cm^3). When first introduced, the vanadium-containing alloy seemed destined to become an excellent turbine blade airfoil performer. However, the introduction of vanadium made the alloy sensitive to hot-corrosion degradation and it was dropped from consideration. Major changes in alloy density for an application may often be achieved but often only by selecting a different base alloy.

Alloy costs today are much more variable than they have been historically. Alloy selection needs to consider the possibility of alloy and processing cost increases during the lifetime of a component design. Alloy costs can be significantly impacted by local conditions in areas of the world where the base-metal ore is found. Dramatic increases in prices have occurred when wars, embargoes, or significant new manufacturing restrictions are imposed. Sometimes such cost increases have led to changes in alloy selection for a component or to chemistry modifications to reduce the amount of costly or strategic element used in the alloy. In the mid-1970s, the price of cobalt shot up by a factor of 10 (the price eventually dropped). Cobalt was a principal ingredient in many superalloys used for gas turbines. Corporate managements were appalled by this state of affairs and decreed that cobalt be removed as soon as possible from the manufacturing process. As a result, a particular iron–nickel-base superalloy, IN 718 (alloy with no cobalt), became the world standard for applications which previously employed cobalt-containing alloys. It behooves the alloy selector to consider strategic and economic conditions when recommending an alloy for design.

As mentioned earlier, there are limited numbers of alloy systems available for structural applications. Iron is the most common and least expensive. Aluminum, copper, magnesium, and titanium follow iron in volume of structural alloy application. Nickel and superalloys are widely used as well. Alloys from all of the systems mentioned can be welded, although some alloys or systems may require special attention. High-strength, high-temperature nickel-base superalloys generally are not weldable owing to a tendency to cracking, particularly in the heat-affected zone.

15.2 Iron

Iron alloys consist of cast iron, wrought iron, low-carbon steel, alloy steel, stainless steel, precipitation-hardening steels, and so on. Iron has been around for several millennia. Steel, the name commonly applied to many iron alloys, is the most widely used metallic material. The range of property levels achievable in iron alloys is quite broad and the general availability of iron makes it a relatively inexpensive alloy base for structural design. Steels are hardenable by solid-solution strengthening, grain size hardening, work hardening, precipitation hardening, and transformation hardening (martensite or bainite formation). Very high short-time strengths, up to over 200 ksi (1379 MPa), can be produced in steels. Stainless steels provide significant corrosion protection in many environments. Many steels are forgeable and formable with much steel sheet going into vehicle production. Automobiles can contain nearly 60% of their weight in steel. Steels are ubiquitous, being found in tall buildings, aircraft, cars, trucks, small and large internal combustion engines, hand tools and large machine tools, garden implements,

kitchen utensils, pots, pans, and so on. Iron-base super-alloys are used in gas and steam turbines. Much iron wire was used to fence the western United States in the late 1800s and iron boilers, rails, and rail cars were standard on trains. Iron has a density of 0.285 lb/in.^{3} (7.89 g/cm^3) but alloys can range as high as 0.3 lb/in.^{3} (8.3 g/cm^3) or above. Steels have elastic moduli in the range of about 30×10^6 psi (207 GPa). Iron melts at 1220°F (660°C) but steels melt at lower temperatures. Steels can be plated, painted, or coated to enhance corrosion protection. The alloys can be joined with relative ease depending on alloy composition, and economics favors iron alloys as materials of choice in many instances.

Other than for certain stainless grades, steel normally is considered an alloy of iron and carbon plus other elements. Now some modern steels incorporate very important changes, for example, interstitial-free steels where carbon is considered an impurity and vacuum-degassed steels with extra low oxygen content. The application of carbon and alloy steels is related largely to the ability to get the desired mechanical properties ("hardness") where we want it. The selection of steels will need to consider the hardenability (maximum thickness at which one can achieve the desired hardness/strength) of the selected alloy with respect to the requirements for strength in the article being designed. Combinations of alloy content and processing can almost always enable the strength (hardness) to be achieved. There are more steels from which to choose than any other alloy class. Strength in steels is frequently referred to in terms of hardness (e.g., on a Rockwell C scale or some other scale). There are sophisticated tools available to determine the likelihood of success in getting an alloy with the correct hardenability.

The properties of steels are modified not only by chemistry changes but also by heating for various times at various temperatures. This heat treatment of steels is effective because iron has a phase transition as it heats up and the lower (room temperature) body-centered-cubic ferrite phase changes to the face-centered-cubic austenite phase. Of course, carbides continue to exist in steels as appropriate to the temperatures and times at which the steel has been held. On cooling, the steel can change back to the preferred lower temperature structure in many ways (not all ways are possible with all steels). Essentially the austenite, which absorbs a lot of carbon at its high temperature of formation, rejects the carbon and tries to transform back to ferrite. Sometimes the austenite forms a nonequilibrium body-centered-tetragonal phase, martensite. Martensite then exists along with carbides. Sometimes austenite transforms to bainite and carbides and sometimes austenite transforms to pearlite (a special form of ferrite and carbide aggregate). The result of these various possibilities is the ability of steels to possess a range of possible properties. Selection of a particular steel is an exercise in estimating the strength needs for a particular property and finding one steel from the multitude of available steels that will get the properties where they are wanted, along with any other desired properties such as corrosion resistance, and at the lowest cost.

Plain carbon steels are used for routine and lower temperature applications. For elevated-temperature applications (e.g., tubing/piping in a steam turbine), higher alloy contents are needed and carbon content is only a part of the selection process. Chromium is often added in small amounts to low-alloy steels used at moderate temperatures. With sufficient chromium, scaling of steels can be prevented. When corrosion is meant to be almost entirely prevented, chromium is added above about 12 wt %, sometimes going up above 20 wt % for certain stainless steels.

Stainless steels are intended to promote integrity of the surface while retaining reasonable strength at high temperatures. The chromium oxide film which forms on stainless steels conveys a passivity and provides corrosion resistance to the alloy. Stainless steels not only see high-temperature service but also find much use at lower temperatures in aqueous or other nongaseous environments. To an extent, stainless steels for elevated-temperature service form a continuum with the iron–nickel- and nickel-base superalloys as well as the heat-resistant and nickel-base corrosion-resistant alloys. The so-called precipitation-hardened steels are chromium-containing steels with special additives to bring out precipitates for hardening, much like aluminum alloys or nickel-base superalloys. It is important to note that the stainless nature of stainless steels is promoted by chemically oxidizing environments. Reducing environments, particularly environments containing halogens, can have devastating effects on stainless steels.

15.3 Copper

Copper is a system which also has been around for millennia. The bronze age was the age of tin alloyed with copper to produce strong (by the standards of the day) alloys. The bronze age was followed by the iron age. There are many benefits to copper compared to iron-base alloys and other systems. The copper alloy systems do not compete solely on a strength basis since copper alloys normally are weaker than iron alloys. However, there are areas where copper alloys show sufficient strength and unique other properties (corrosion resistance, nonsparking characteristics when struck, etc.) to compete with steels. A major property of pure or relatively pure copper alloys is their high electrical and thermal conductivity. Specially strengthened copper alloys find use as electrodes and similar current-carrying articles where high temperatures and surface wear resistance are significant. Copper has a moderately high melting point, 1981°F (1083°C); this is less than the melting points of iron alloys but much greater than the melting points of aluminum and magnesium alloys.

Copper is strengthened by solid-solution hardening, grain size hardening, work hardening, and precipitation hardening. Copper is moderately abundant and copper alloys such as the brasses offer reasonable strength with good formability and corrosion resistance. The modulus of copper is 17×10^6 psi (115 GPa); this is less than the modulus of steels and nickel alloys, higher than the modulus of aluminum, and comparable to the moduli of titanium alloys. Most wrought alloys are available in work-hardened conditions and typically find use in moderate to small parts such as springs, fasteners, hardware, gears, and cams. Copper is very formable and forgeable though forming is more apt to be used than forging to manufacture articles of copper alloys. Casting can be used to create copper articles as well. A major use is in conductivity devices (e.g., wire). Copper alloys find wide use in plumbing. Copper, brasses, bronzes, and cupronickels are used for pipes, valves, and fittings in transporting water. The density of copper alloys is around 0.32 lb/in.^{3} (8.86 g/cm^3) for many compositions.

15.4 Aluminum

The aluminum industry is relatively new and got its start in the late 1800s. Aluminum alloys became available in the early part of the twentieth century. Aluminum melts at 1220°F (660°C) and owes its popularity to corrosion-resistant, moderately high strength alloys with a substantially lower density, 0.098 lb/in.^{3} (2.77 g/cm^3), than most other structural alloys. Strong aluminum alloys can be produced for operation up to about several hundred degrees Fahrenheit. Aluminum is strengthened by solid solution and work hardening but the highest strength aluminum alloys are precipitation (age) hardened. Cast aluminum is often used. Aluminum alloys have become synonymous with lightweight articles of acceptable strength when age-hardened aluminum is used. The density is considerably less than the density of steel. Figure 11 shows the effect of density on the specific strength of several alloy systems. Note that, despite the much greater strength of iron and nickel alloys, at room temperature the specific strength of aluminum alloys makes them competitive with iron alloys.

Aluminum is thermodynamically a reactive metal, as is magnesium. However, it has excellent resistance to corrosion in water (including salt water), atmospheric environments, oils, and many chemicals. Aluminum owes its excellent corrosion resistance to the aluminum oxide films that form naturally but can be enhanced by artificial means. When the excellent corrosion resistance of aluminum alloys in many environments plus the excellent formability/forgeability of aluminum alloys is considered, their high specific strength can make them alloys of choice. In weight-sensitive applications at low to moderate temperatures, aluminum alloys perform very well. The modulus of aluminum alloys is about 10×10^6 psi (69 GPa) and so the alloys are not as stiff as iron, copper, or nickel

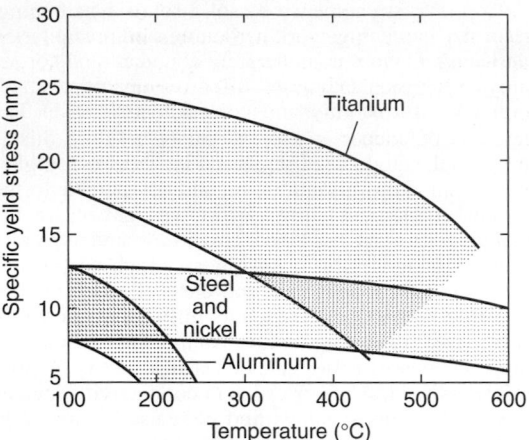

Fig. 11 In certain circumstances lower densities can make somewhat less strong alloys the materials of choice.

alloys. They find use in aircraft structures, however, and aluminum beams have been used in construction. Aluminum is nontoxic and is found as pots/pans, utensils, and beverage cans and in other uses.

Although aluminum alloys generally are corrosion resistant in atmospheric environments, some alloys are corroded in special localized environments. Exfoliation, intergranular corrosion, stress corrosion cracking, and similar problems can exist. Some alloy types within the aluminum alloy family are less corrosion resistant than other aluminum alloys. However, alloy choice, the availability of anodizing and similar finishing processes, plus the process of cladding aluminum alloys with pure aluminum can provide good corrosion resistance for aluminum alloy systems. Alclad aluminum alloys generally are sheet and tube. Because of the nature of artificially formed aluminum oxide coatings on aluminum alloys, the anodized alloys can be colored. In addition to household goods, aluminum siding provides colorful protection for many homes while resisting denting. Aluminum finds use as ladders and in many outdoor activities where its intrinsic corrosion-resistant oxide surface skin protects it better than iron alloys. Aluminum alloys are extensively used in aircraft structures.

15.5 Magnesium

Magnesium alloys became available in the early twentieth century. Magnesium alloy applications are fueled by its density of 0.063 lb/in.^{3} (1.74 g/cm^3), which is the lowest of any commercially used metal. The melting point of magnesium of 1202°F (650°C) is competitive with that of aluminum. The modulus of magnesium is about 6.5 x 10^6 psi (45 GPa), which is considerably less than aluminum or steel. However, owing to its low density, magnesium can have better specific rigidity than the latter alloys. For rectangular steel, aluminum, and magnesium sections of equal

rigidity, calculations show that the magnesium article will weigh only about 71% as much as the aluminum article and 41% as much as the steel article. Magnesium alloys have good strength and very good strength when their density is factored into the selection process. Magnesium alloys generally are forged/formed at elevated temperatures and forgings can be produced in the same general variety of shapes and sizes as forgings of other metals. However, most magnesium articles are produced as castings. Commercial applications have included luggage, ladders, materials handling equipment, as well as aerospace applications.

Magnesium is the most active element in the galvanic series of elements/alloys and so normally should not be coupled with any more cathodic alloy in an application. (Magnesium is used in hot-water tanks to provide cathodic protection.) Under extreme conditions of galvanic corrosion, rapid corrosion results. However, magnesium alloys generally are used with limited corrosion problems. Magnesium rapidly forms an oxide layer when exposed to corrosion conditions such as atmospheric moisture. The oxide is not too protective but neither is alloy section loss excessive. Use of high-purity magnesium reduces problems and anodizing, chromate coatings, and polymeric coatings have been developed to increase magnesium alloy protection from the environment.

15.6 Titanium

Titanium became commercially available at the start of the last half of the twentieth century. Titanium owes its industrial use to two significant factors: It has exceptional room temperature resistance to a variety of corrosive media and a relatively low density of 0.163 lb/in.^{3} (4.51 g/cm^3) and can be strengthened to achieve outstanding properties when compared with competitive materials on a strength-to-density basis. Limited numbers of titanium alloys have been invented and marketed, perhaps owing to the greater cost of the product compared to the metal systems covered thus far. Titanium is produced by a batch process and must be further processed by expensive vacuum-melting techniques if its use is desired for aerospace applications.

There are three main classes of titanium alloys, α, $\alpha-\beta$, and β alloys. Property levels and processing capability vary with class of alloy in addition to specific composition. Titanium has a melting point of 3035°F (1668°C). The high melting point of titanium induced inventors to assume that high-strength alloys comparable to those of iron-, nickel-, and cobalt-base alloys would be capable of being produced. Unfortunately, titanium alloys are not able to operate at very high temperatures. Thus, while titanium and its alloys have melting points higher than those of steels, their maximum upper useful temperatures for structural applications generally range from as low as 800°F (427°C) to the region of about 1000–1100°F (538–595°C), depending on composition. Normally the alloys are used well below about

1000°F (538°C). However, titanium alloys are capable of being strengthened to high levels and, combined with their reduced densities relative to iron alloys and excellent corrosion resistance, they were able to replace iron alloys extensively in aircraft gas turbines.

The modulus of titanium is 15.0 x 10^6 psi (103 GPa). However, titanium undergoes an allotropic transformation on heating and changes from a low-temperature hexagonal-close-packed crystal structure (α) to one that is body-centered cubic (β). On cooling back to lower temperatures, a martensitic reaction can occur or the β can transform to α with varying degrees of fine structure. If a titanium alloy has a β structure at room temperature, the modulus of the alloy will differ from the modulus of an α alloy. The modulus will vary with chemistry and processing. Texturing in rolling processes can further change the effective modulus depending on grain alignment and the amount of β or α present. In addition to the variation of modulus with chemistry and processing, titanium alloy density will vary with the amount and type of alloy elements in a given composition. Many β-forming elements are significantly heavier than titanium while the α-forming element, aluminum, is lighter.

One alloy, Ti–6Al–4V, has been the workhorse of the titanium industry and, for many years, claimed over 50% of the titanium market. This alloy can be used up to the order of 500°F (260°C) to 600°F (316°C). Ti–6Al–4V is an α–β alloy with good all-around properties. Generally α–β alloys are the most used titanium alloys. However, titanium α alloys have better high-temperature properties and β alloys have superior tensile properties and improved fabricability. Titanium alloys have found use in the chemical processing industry, as golf clubs and eyeglass frames, in dental applications, and as structural implants (e.g., hip replacements) in the human body, in addition to their widespread aerospace use. The alloys have greater high-temperature capability than aluminum alloys and are better in specific strengths than iron alloys, facts which ensure their continued use. A large amount of titanium production is used in aerospace applications. Large fan disks in high-bypass-ratio gas turbines and wing spars for aircraft are made of these alloys.

15.7 Nickel

Nickel has been available as an alloy element since the late-nineteenth century. It has exceptional value as an alloy element in steels and other elements as well as being the basis for the high-performance superalloys and heat-resistant alloys used in the aerospace and nuclear industries. Superalloys are covered below. Nickel alloys other than superalloys tend to be more or less relatively pure nickels used in food processing equipment, chemical processing, or some aerospace components. There is another class of nickel alloys which have applications other than those just mentioned—specifically, nickel–copper alloys (tradenames begin with Monel), which are characterized by high strength, good weldability, and excellent corrosion resistance at room temperatures in aqueous conditions. Some of these alloys are age hardened. Monels have applications in pumps, valves, storage tanks, heat exchangers, waveguides, screw machine products, fasteners for nuclear applications, and other uses.

15.8 Superalloys

Superalloys came into early use about 1940, with use motivated in large part by military concerns. Stainless steels and cobalt alloys were modified to provide improved creep rupture capability at higher temperatures such as those found in piston engine superchargers. Concurrently, a few nickel-base superalloys such as Inconel X were created and made available. The terminology "superalloys" did not take hold until the late 1940s when the more widespread use of gas turbine engines prompted a search for improved alloys. Eventually, the addition of more precipitation-creating elements such as aluminum into the basic Nichrome (80 nickel–20 chromium) chemistries plus continued work on modified stainless steels with titanium and other additions led to a class of alloys generally called superalloys. Superalloys are iron-, iron–nickel-, cobalt-, or nickel-based alloys which have exceptional high-temperature strength properties from about 1000°F (538°C) to about 2000°F (1093°C),

Cast nickel-base superalloys have the best strength in creep rupture at high temperatures above about 1300°F (704°C) while wrought iron–nickel and nickel-base superalloys have the best combination of properties below about 1300°F (704°C). Iron-base superalloys have acceptable properties below about 1000°F (538°C) and are the least expensive superalloys. Cobalt-base superalloys are used for long-time, high-temperature, and somewhat lower stress applications. For highest tensile properties, the nickel- or iron–nickel-base alloys stand out. For best forgeability, iron–nickel-base superalloys such as IN718 are outstanding.

Superalloy articles up to 48 in. (122 cm.) in length are cast as blades for gas turbines by the investment casting process, sometimes using SCDS technology. CGDS blades are made as well. Both types are stronger than polycrystalline castings but decidedly more expensive. Large relatively thin walled case castings up to nearly 48 in. (122 cm) in diameter are made of IN718. Gas turbine disks up to nearly 36 in. ($\sim$ 1 m) in diameter are forged or powder metallurgy processed and forged from IN718 and nickel-base alloys. Oxidation/corrosion-resistant coatings are used to protect surfaces of the hottest parts (turbine airfoils) from environmental attack. Superalloys are normally melted with various vacuum-melting techniques using induction or other high-technology furnaces. They may have as many as a dozen or more alloy elements controlled in the alloy and are subject (as is titanium) to extensive surveillance during the melting, casting, and forging operations.

Parts as small as thumb size or as large as gas turbine diffuser cases can be cast in superalloys. Investment-casting techniques produce exceptionally high quality alloy articles with maximum property capability. Most alloys are available to potential customers through normal channels. However, certain manufacturers of gas turbine engine hardware have invented proprietary alloys and these most likely would not be available for public purchase in the foreseeable future.

16 SELECTED ALLOY INFORMATION SOURCES

16.1 General

In the selection of alloys for structural applications, selectors may have access to prior related designs from company records or data from public records that may be applicable to the planned design. Consequently, at least a class of alloys and, possibly, a specific alloy or two may be readily apparent as alloy selection starts. However, more information on alloy types, properties, economics, and availability will most certainly be required eventually. It is beyond the scope of this chapter to provide information on all alloy classes, let alone specific compositions which might be used for alloy selection within the alloy classes. However, many organizations, technical and sales, exist to promote specific elements and alloys of such elements as well as to promote processes for the manufacture of specific alloys into useful articles.

The number of organizations available to provide alloys, alloy property data, produce articles, and so on, is very large. The following list of websites is provided as a starting point to help alloy selectors determine possible alloy types, properties, economics, and availability. This list almost certainly will change with time and should only be used as a guide to locate potential information providers. Hopefully the resources at these and similar sites will enable a selector and the designer to come up with an appropriate alloy for a planned design.

It is vital to remember in alloy selection that many alloys are not off-the-shelf consumer items and that properties may vary with material specifications for the alloy. Alloys often are made to user specifications which may vary from user to user and time to time. While a selector may be able to use common readily available alloys, often he or she may not be able to do so. When working with any specialty alloys, diligence in working with the producers will pay dividends in obtaining optimum properties and reducing difficulties for users of alloys. This concept is more costly than buying off the shelf but invariably leads to best quality and more reliable articles in design.

16.2 Selected Websites for Alloy Selection

The following is only a partial list. More information as to websites, locations, and so on, may be available. Not all companies or institutions active in any given technology are represented. No recommendation is made or implied by this list.

General Information

ASM International: www.asminternational.org

The Minerals, Metals & Materials Society of the American Institute of Mining, Metallurgical and Petroleum Engineers: www.tms.org

Institute of Materials, Minerals and Mining: www.iom3.org

Specifications/Handbooks/Property Data

American Society for Testing and Materials: www.astm.org

ASM International: www.asminternational.org

Military property handbooks: http://projects.battelle.org/mmpds/

Fundamental Approaches to Alloy Selection

Granta Design: www.grantadesign.com

Specific Elements/Alloys

Nickel (Nickel Development Institute): www.nickel-institute.org

Cobalt (Cobalt Development Institute): www.thecdi.com

Chromium (International Chromium Development Institute): www.chromium-assoc.com

Superalloys/special steels (Specialty Steel Industry of North America): www.ssina.com

Copper (Copper Development Association): www.copper.org

Copper (International Copper Association): www.copperinfo.com

Titanium (Titanium Development Association): www.titanium.org

Titanium (Titanium Information Group): www.titaniuminfogroup.co.uk

Aluminum (Aluminum Association): www.aluminum.org

Magnesium (Magnesium Association): www.intlmag.org

Iron/Steel (American Iron & Steel Institute): www.steel.org

Gold (*Gold Bulletin*): www.goldbulletin.org

Platinum (*Platinum Metals Review*): www.platinum-metalsreview.com

Silver (Silver Institute): www.silverinstitute.org

Zinc (American Zinc Association): www.zinc.org

Tungsten (International Tungsten Institute): www.itia.org.uk

Rare earth metals (Metall): www.metall.com.cn

Molybdenum (International Molybdenum Organization: www.imoa.org.uk

Molybdenum (Climax Molybdenum Corp.): www.climaxmolybdenum.com

Tantalum/niobium (Tantalum-Niobium International Study Center): www.tanb.org

Manufacturing Industry Organizations

Casting (American Foundry Society): www.afsinc.org

Forging (Forging Industry Association): www.forging.org

Forming (Institute for Metal Forming): www.lehigh.edu/~inimf

Joining (American Welding Society): www.aws.org

Heat Treating (Heat Treating Society): www.asminternational.org

BIBLIOGRAPHY

Data Related

Budinsky, K. G., and Budinsky M. K., *Engineering Materials, Properties and Selection*, 6th ed., Prentice-Hall, Englewood Cliffs, NJ, 1999.

Kutz, M. (Ed), *Handbook of Materials Selection*, Wiley, New York, 2002.

Various authors/editors, *ASM Metals Handbooks,* 10th ed., 20 vols. covering properties, chemistry, surface effects, metals manufacturing, etc., ASM International, Materials Park, OH, continuously updated, latest updated volumes 2006.

Selection Procedure Related

Ashby, M. F., *Materials Selection in Mechanical Design,* 3rd ed., Butterworth Heinemann, Oxford, 2005.

Ashby, M. F. and Jones, D. R. H., *Engineering Materials,* Vol. 1, 3rd ed., Butterworth Heinemann, Oxford, 2007.

Ashby, M. F., and Johnson, K., *Materials and Design, the Art and Science of Material Selection in Product Design*, Butterworth Heinemann, Oxford, 2002.

Ashby, M. F., Shercliff, H., and Cebon, D., *Materials: Engineering, Science, Processing and Design*, Butterworth Heinemann, Oxford, 2007.

Charles, J. A., Crane, F. A. A., and Furness, J. A. G., *Selection and Use of Engineering Materials,* 3rd ed., Butterworth Heinemann, Oxford, 1997.

Dieter, G. E., *Engineering Design, a Materials and Processing Approach,* 2nd ed., McGraw-Hill, New York, 1991.

Farag, M. M., *Selection of Materials and Manufacturing Processes for Engineering Design*, Prentice-Hall, Englewood Cliffs, NJ, 1989.

Lewis, G., *Selection of Engineering Materials*, Prentice-Hall, Englewood Cliffs, NJ, 1990.

CHAPTER 5

PLASTICS: INFORMATION AND PROPERTIES OF POLYMERIC MATERIALS

Edward N. Peters
General Electric Company
Selkirk, New York

Reprinted from *Mechanical Engineers' Handbook*, Vol. 1, Wiley, New York, 2006, with permission of the publisher.

1 INTRODUCTION

Plastics or polymers are ubiquitous. Through human ingenuity and necessity natural polymers have been modified to improve their utility and synthetic polymers developed. Synthetic polymers in the form of plastics, fibers, elastomers, adhesives, and coatings have come on the scene as the result of a continual search for man-made substances that can perform better or can be produced at a lower cost than natural materials such as wood, glass, and metal, which require mining, refining, processing, milling, and machining. The use of plastics can also increase productivity by producing finished parts and consolidating parts. For example, an item made of several metal parts that require separate fabrication and assembly can often be consolidated to one or two plastic parts. Such increases in productivity have led to fantastic growth in macromolecules. Indeed, the use of plastics has increased almost 20-fold in the last 30 years. Today there is a tremendous range of polymers and formulated plastics available which offer a wide range of performance. This chapter presents concise information on their synthesis, structure, properties, and applications.

1.1 Classification of Plastics

The availability of plastic materials for use in various applications is huge.[1-5] There are various ways to categorize or classify plastics that facilitate understanding similarities and difference between materials. The two major classifications are thermosetting material and thermoplastics materials.[5-7] As the name implies, thermosetting plastics, or thermosets, are set, cured, or hardened into a permanent shape.[5,6] The curing, which usually occurs rapidly under heat or ultraviolet (UV) light, leads to an irreversible crosslinking of the polymer. Thermoplastics differ from thermosetting materials in that they do not set or cure under heat.[5] When heated, thermoplastics merely soften to a mobile, flowable state where they can be shaped into useful objects. Upon cooling, thermoplastics harden and hold their shape. Thermoplastics can be repeatedly softened by heat and shaped.

Another major classification of thermoplastics is as amorphous or semicrystalline plastics.[5,6] Most polymers are either completely amorphous or have an amorphous component as part of a semicrystalline polymer. Amorphous polymers are hard, rigid glasses below a fairly sharply defined temperature, which is known as the glass transition temperature, or T_g. Above the T_g the amorphous polymer becomes soft and flexible and can be shaped. Mechanical properties show profound changes near the glass transition temperature.

Semicrystalline polymers have crystalline melting points T_m that are above their glass transition temperature. The degree of crystallinity and the morphology of the crystalline phase have an important effect on mechanical properties. Crystalline plastics will become

less rigid above their glass transition temperature but will not flow until the temperature is above its T_m.

Many mechanical and physical properties of plastics are related to their structure. In general, at ambient temperatures, amorphous polymers have greater transparency and greater dimensional stability over a wide temperature range. In semicrystalline plastics the ordering and packing of the polymer chains make them denser, stiffer, stronger, and harder. Semicrystalline plastics also have better inherent lubricity and chemical resistance and during molding will tend to shrink more and have a greater tendency to warp.

Another important class of polymeric resins is elastomers. Elastomeric materials are rubberlike polymers with glass transition temperatures below room temperature. Below that glass transition temperature an elastomer will become rigid and lose its rubbery characteristics.

1.2 Chemical/Solvent Resistance

The solvent or chemical resistance of plastics is a measure of the resin's ability to withstand chemical interaction with minimal change in appearance, weight, dimensions, and mechanical properties over a period of time.[8] When polymers are exposed to chemical environments, they can dissolve in the solvent, absorb the solvent, become plasticized by the solvent, react with the solvent or chemical, or be stress cracked.[6] Hence mechanical properties can be reduced and dimensions of molded parts changed.

Qualitative comments are made in this chapter about the chemical resistance of various plastics. Generally, these comments are based on laboratory screening that the manufacturers have reported. The stress level of the test part has a very significant influence on the performance of the material. When the stress level increases, resistance to chemicals and solvents decreases. Sources of stress can occur from injection molding, applied loads, forming, and assembly operations.

When an application requires exposure to or immersion in chemicals or other harsh environments, it is of prime importance to use prototypes of suitable stressed material for testing under actual operating conditions.[8]

1.3 Plastics Additives

The wide variety of synthetic polymers offers a huge range of performance attributes.[9-13] Additives play a key role in optimizing the performance of polymers and transforms them into commercially viable products. Additives are materials that are added to a polymer to produce a desired change in material properties or characteristics. A wide variety of additives are currently used in thermoplastics to enhance material properties, expand processability, modify aesthetics, and/or increase environmental resistance. Enhancement of properties include thermal stability, flame retardancy, impact resistance, and UV light stability. Reinforcing fibers (i.e., high-strength, inert fibrous material such as glass and carbon fibers) can be incorporated in the polymer matrix to improve modulus and strength while lowering the coefficient of thermal expansion and unusually lowering impact and increasing density. Particulates of flaked fillers usually increase stiffness as well as lower costs.[13] Plasticizers lower modulus and enhance flexibility. Thus additives are found in most commercial plastics and are an essential part of the formulation in which performance is tailored for specific applications.[9-13]

This chapter focuses on properties of the base resins with very few additives other than stabilizers. Properties for plastics that contain glass fiber (GF) are listed for those plastics that routinely use such reinforcing fibers.

1.4 Properties

Plastic materials can encounter mechanical stress, impact, flexure, elevated temperatures, different environments, and the like. Hence various properties are measured and reported to give an indication of a material's ability to perform under various conditions. Properties are determined on standard test parts and are useful for comparative purposes. A description of properties used in the chapter would be instructive and are listed below.

Density: The mass per unit volume of a material at $73°F$ ($23°C$).

Tensile Modulus: An indicator of the stiffness of a part. It is basically the applied tensile stress, based on the force and cross-sectional area, divided by the observed strain at that stress level.

Tensile Strength: The amount of force required to elongate the plastic by a defined amount. Higher values mean the material is stronger.

Elongation at Break: The increase in length of a specimen under tension before it breaks under controlled test conditions. Usually expressed as a percentage of the original length. Also called "strain."

Flexural Modulus: The ratio of applied stress to the deflection caused in a bending test. It is a measure of the stiffness of a material during the first part of the bending process.

Flexural Strength: The maximum stress that can be applied to a beam in pure bending before permanent deformation occurs.

Izod Impact Strength: The amount of impact energy required to cause a test specimen to break completely. The specimen may be notched or unnotched.

Heat Deflection Temperature (HDT, also called deflection temperature under load, or DTUL): Gives an indication of a material's ability to perform at higher temperatures while supporting a load. It shows the temperature at which a test specimen will deflect a given distance under a given load in flexure under specified test conditions, usually at 1.82 and/or 0.42-MPa loads.

Vicat Softening Temperature: A measure of the temperature at which a plastic starts to soften at specified test conditions according to International Organization for Standardization (ISO) 306. It gives an indication of a material's ability to withstand limited short-term contact with a heated object.

Relative Thermal Index (RTI, formerly called continuous-use temperature): The continuous operating temperature of plastics materials used in electrical applications, as specified by Underwriter's Laboratories (UL). It is the maximum temperature at which the material retains at least 50% of the original value of all properties tested after the specified amount of time. The RTI tests are important if the final product is to receive UL recognition.

UL94 VO: Classification of flammability of plastic materials that follows the UL 94 standard, vertical burning test. The VO rating is where the total burn time for the five specimens after two 10-sec applications of flame is ≤ 50 sec. Sample thicknesses on VO-rated test specimens are reported. A rating on thin samples would suggest possible utility in applications requiring thin walls.

Oxygen Index: A flammability test that determines the minimum volumetric concentration of oxygen that is necessary to maintain combustion of a specimen after it has been ignited.

Hardness: The resistance of a material to indentation under standardized conditions. A hard indenter or standard shape is pressed into the surface of the material under a specified load. The resulting area of indentation or the depth of indentation is measured and assigned a numerical value. For plastics, the most widely used methods are Rockwell and Shore methods and Ball hardness.

Coefficient of Thermal Expansion (CTE): A measure of how much a material will lengthen (or shorten upon cooling) based on its original length and the temperature difference to which it is exposed. It becomes important when part dimensions are critical or when two different materials with different CTEs are attached to each other.

Shrinkage: The percentage of reduction in overall part dimensions during injection molding. The shrinkage occurs during the cooling phase of the process.

Water Absorption/Moisture Absorption: The percentage weight gain of a material after immersion in water for a specified time at a specified temperature. Low values are preferred and are important in applications requiring good dimensional stability. As water is absorbed, dimensions of the part tend to increase and physical properties can deteriorate. In addition, low water absorption is very important for most electrical properties.

Relative Permittivity (formerly called dielectric constant): A measure of the amount of electrical energy stored in a material. It is equal to the capacitance of a material divided by the capacitance of a dimensionally equivalent vacuum. Important for high-frequency or power applications in order to minimize power losses. Low values indicate a good insulator. Moisture, frequency, and temperature increases may have adverse effects.

Dissipation Factor (also called "loss tangent"): A measure of the dielectric characteristics of a plastic resin. Moisture, frequency, and temperature increases may have adverse effects.

2 POLYOLEFINIC THERMOPLASTICS

2.1 Polyethylene

Polyethylene (PE) plastics are lightweight, semicrystalline thermoplastics that are prepared by the catalytic polymerization of ethylene.[14–16] Depending on the temperature, pressure, catalyst, and use of a comonomer, three basic types of polyethylene can be produced: high-density polyethylene (HDPE), low-density polyethylene (LDPE), and linear low-density polyethylene (LLDPE). LDPE and LLDPE contain branching. This branching results in decreased crystallinity and lower density. Most properties of PEs are a function of their density and molecular weight. As density decreases, the strength, modulus, and hardness decrease and flexibility, impact, and clarity increase. Hence HDPE exhibits greater stiffness, rigidity, improved heat resistance, and increased resistance to permeability than LDPE and LLDPE.

LDPE is prepared under more vigorous conditions, which results in short-chain branching. The amount of branching and the density can be controlled by the polymerization conditions.

LLDPE is prepared by using an α-olefin comonomer during polymerization. Hence branching is introduced in a controlled manner and where the chain length of the branching is uniform. In general, the comonomers are 1-butene, 1-hexene, 1-octene, and 4-methyl-1-pentene (4M1P).

Polyethylene polymers are versatile and inexpensive resins that have become the largest volume use of any plastics. They exhibit toughness, near-zero moisture absorption, excellent chemical resistance, excellent electrical insulating properties, low coefficient of friction, and ease of processing. They are not high-performance resins. Their heat deflection temperatures are reasonable but not high. Specialty grades of PE include very low density (VLDPE), medium-density (MDPE), and ultrahigh-molecular-weight PE (UHMWPE)

Some typical properties of HDPE and LDPE are listed in Table 1. Properties of the various copolymers of LLDPE appear in Table 2.

Uses HDPE's major use is in blow-molded containers (milk, water, antifreeze, household chemical bottles), shipping drums, carboys, automotive gasoline tanks; injection-molded material-handling pallets, crates, totes, trash and garbage containers, and household and automotive parts; and extruded pipe products (corrugated, irrigation, sewer/industrial, and gas distribution pipes).

LDPE/LLDPEs find major applications in film form for food packaging as a vapor barrier film, including stretch and shrink wrap; for plastic bags such as

Table 1 Typical Properties of HPDE, LDPE, LLDPE, and UHMWPE

Property	HDPE	LDPE	LLDPE	UHMWPE
Density (g/cm^3)	0.94–0.97	0.915–0.935	0.91–0.92	0.93
Tensile modulus (GPa)	0.76–1.0	0.14–0.31	0.13–0.19	110
Tensile strength (MPa)	22–32	7–17	14–21	30
Elongation at break (%)	200–1000	100–700	200–1100	300
Flexural modulus (GPa)	0.7–1.6	0.24–0.33	0.25–0.37	—
HDT, 0.45 MPa ($^\circ$C)	65–90	43	—	79
Vicat softening temperature ($^\circ$C)	120–130	90–102	80–94	136
Brittle temperature ($^\circ$C)	< -75	< -75	< -75	—
Hardness (Shore)	D60–69	D45–60	D45–50	—
CTE ($10^{-5}/^\circ$C)	15	29	18	0.1.6
Shrinkage (in./in.)	0.007–0.009	0.015–0.035	—	—
Water absorption, 24 hr (%)	<0.01	<0.01	<0.01	<0.01
Relative permittivity at 1 MHz	2.3	2.2	2.3	2.3
Dissipation factor at 1 MHz	0.0003	0.0003	0.0004	0.0005

Table 2 Typical Properties of LLDPE Copolymers

Property	1-Butene	1-Hexene	1-Octene	4M1P
Density (g/cm^2)	0.919	0.919	0.920	—
Tensile modulus (GPa)	0.185	0.206	0.200	0.277
Tensile strength (MPa)	41	39	58	42
Elongation at break (%)	430	430	440	510

grocery, laundry, and dry cleaning bags; for extruded wire and cable insulation; and for bottles, closures, and toys.

2.2 Polypropylene

Polypropylene (PP) is a versatile semicrystalline thermoplastic offering a useful balance of properties that are typically higher than for HDPE. It is prepared by the catalyzed polymerization of propylene.[17,18] Crystallinity is key to the properties of PP. The degree of crystallinity is a function of the degree of geometric orientation (stereoregularity) of the methyl groups on the polymer chain (backbone). There are three possible geometric forms of PP. Isotactic PP has predominantly all the methyl groups aligned on the same side of the chain (above or below the chain). In syndiotactic PP the methyl groups alternate above and below the polymer chain. Finally, atactic PP has the methyl groups randomly positioned along the chain. Both isotactic and syndiotactic PP will crystallize when cooled from the molten states; however, commercial PP resins are generally isotactic.

Isotactic PP is highly crystalline thermoplastic that exhibits low density; rigidity; good chemical resistance to hydrocarbons, alcohols, and oxidizing agents; negligible water absorption; excellent electrical properties; and excellent impact/stiffness balance. PP has the highest flexural modulus of the commercially available polyolefins. In general, PP has poor impact resistance. However, PP–elastomer blends have improved impact strength. Unfilled PP has poor flame resistance and is degraded by sunlight. Flame-retardant and UV-stabilized grades are available. Typical properties appear in Table 3.

Uses   End uses for PP are in blow-molding bottles and automotive parts, injection-molded closures, appliances (washer agitators, dishwasher components), house wares, automotive parts, luggage, syringes, storage battery cases, and toys. PP can be extruded into fibers and filaments for use in carpets, rugs, and cordage. In addition, PP can be extruded into film for packaging applications.

2.3 Polymethylpentane

Polymethylpentane (PMP) is prepared by the catalytic polymerization of 4-methyl-1-pentene.[14] PMP is a semicrystalline thermoplastic that exhibits transparency (light transmission up to 90%), very low density, chemical resistance, negligible water absorption, and good electrical properties. PMP resins are available as homopolymers and copolymers with higher α-olefins. The homopolymer is more rigid and the copolymers have increased ductility. PMP is degraded by sunlight and high-energy radiation. Moreover, strong oxidizing acids attack it. PMP grades are available with increased radiation resistance as well as with reinforcing fillers. Properties appear in Table 4.

Uses   Applications for PMP resins cover a wide spectrum of end uses. These include medical (hypodermic syringes, disposable cuvettes, blood-handling equipment, respiration equipment, laboratory ware), packaging (microwave and hot-air-oven cookware, service trays, coated paper plates), and miscellaneous (transparent ink cartridges for printers, light covers, sight glasses, lenses, liquid level and flow indicators, fluid reservoirs).

3   SIDE-CHAIN-SUBSTITUTED VINYL THERMOPLASTICS

3.1   Polystyrene

The catalytic or thermal polymerization of styrene yields general-purpose or crystal polystyrene (PS).[19,20] The designation of "crystal" refers to its high clarity. PS is an amorphous polymer (atactic geometric configuration) and exhibits high stiffness, good dimensional stability, moderately high heat deflection temperature, and excellent electrical insulating properties. However, it is brittle under impact and exhibits very poor resistance to surfactants and a wide range of solvents.

The three main categories of commercial PS resins are crystal PS, impact PS, and expanded PS foam.

Table 3   Typical Properties of PP

Density (g/cm^3)	0.90–0.91
Tensile modulus (GPa)	1.4–1.8
Tensile strength (MPa)	22–35
Elongation at break (%)	10–60
Flexural modulus (GPa)	1.0–1.39
Notched Izod (kJ/m)	0.03–0.13
HDT at 1.81 MPa ($^\circ$C)	60–65
HDT at 0.45 MPa ($^\circ$C)	75–107
Vicat softening temperature ($^\circ$C)	130–148
CTE ($10^{-5}/^\circ$C)	9.0
Hardness (Shore)	D76
Hardness (Rockwell)	R60–R90
Shrinkage (in./in.)	0.01–0.02
Water absorption, 24 hr (%)	<0.01
Relative permittivity at 1 MHz	2.25
Dissipation factor at 1 MHz	0.0003

Table 4   Typical Properties of PMP

Density (g/cm^3)	0.84
Tensile modulus (GPa)	1.75
Tensile strength (MPa)	20.5
Elongation at break (%)	18
Flexural modulus (GPa)	1.55
Flexural strength (MPa)	30.0
Notched Izod (kJ/m)	0.150
HDT at 1.81 MPa ($^\circ$C)	49
HDT at 0.45 MPa ($^\circ$C)	85
Vicat softening temperature ($^\circ$C)	173
CTE ($10^{-5}/^\circ$C)	3.8
Hardness (Rockwell)	L85
Relative permittivity at 100 Hz	2.12

Copolymerization of styrene with a rubber, polybutadiene, results in the rubber being grafted onto the PS. This impact-modified PS has increased impact strength, which is accompanied by a decrease in rigidity and heat deflection temperature. Depending on the levels of rubber, impact polystyrene (IPS) or high-impact polystyrene (HIPS) can be prepared. These materials are translucent to opaque and generally exhibit poor weathering characteristics.

PS finds wide use in styrofoam. Typically, styrofoam is produced from expandable styrene (EPS) beads, which contain a blowing agent. When heated, the blowing agent vaporizes and expands the PS and forms a low-density foam. The density of the foam is controlled by the amount of blowing agent.

Typical properties for PS, IPS, and HIPS appear in Table 5.

Uses Ease of processing, rigidity, clarity, and low cost combine to support applications in toys, displays, consumer goods (television housings), medical (labware, tissue culture flasks), and housewares such as food packaging, audio/video consumer electronics, office equipment, and medical devices.

EPS can readily be prepared and is characterized by excellent low thermal conductivity, high strength-to-weight ratio, low water absorption, and excellent energy absorption. These attributes have made EPS of special interest as insulation boards for construction (structural foam sections for insulating walls), protective packaging materials (foamed containers for meat and produce), insulated drinking cups and plates, and flotation devices.

3.2 Syndiotactic Polystyrene

Syndiotactic polystyrene (SPS) is a semicrystalline polymer and is produced via metallocene-catalyzed polymerization of styrene.[21] By comparison to general-purpose, amorphous polystyrene (PS, HIPS), SPS has stereoregularity in its structure, which facilitates crystallization. In SPS the phenyl groups alternate above and below the polymer chain. Hence SPS has a high crystalline melt point and good chemical resistance.

The slow rate of crystallization and high T_g of SPS typically require an oil-heated tool when injection molding and longer cycle times to maximize physical propereties. Typical properties appear in Table 6.

Uses SPS is targeted at automotive under-the-hood and specialty electronic applications. It will compete with polyamides, thermoplastic polyesters, and polyphenylene sulfide (PPS).

3.3 Styrene/Acrylonitrile Copolymer

Copolymerization of styrene with a moderate amount of acrylonitrile (AN) provides a clear, amorphous polymer (SAN).[20] The addition of the polar AN group gives increased heat deflection temperature and chemical resistance compared to PS. Other benefits of SAN are stiffness, strength, and clarity. Like PS the impact resistance is still poor. SAN is chemically resistant to hydrocarbons, bases, most detergents, and battery acid. However, SAN has poor resistance to chlorinated and aromatic solvents, esters, ketones, and aldehydes. The composition and molecular weight can be varied to change properties. An increase in AN will increase SAN's physical properties but will make melt processing more difficult and will decrease the transparency. In general, the AN level in SAN does not exceed 30% for molding applications. Typical properties appear in Table 7.

Uses SAN is utilized in typical PS-type applications where a slight increase in heat deflection temperature and/or chemical resistance is needed. Such applications include appliances (refrigerator compartments, knobs, blender and mixer bowls), electronics (cassette cases, tape windows, meter lenses), packaging (bottle jars, cosmetic containers, closures), medical (syringe

Table 5 Typical Properties of PS, IPS, and HIPS

Property	PS	IPS	HIPS
Density (g/cm^3)	1.04	1.04	1.04
Tensile modulus (GPa)	3.14	2.37	1.56
Tensile strength (MPa)	51.1	26.9	15.0
Elongation at break (%)	21	40	65
Flexural modulus (GPa)	3.54	2.61	1.68
Flexural strength (MPa)	102	56.5	30
Notched Izod (kJ/m)	0.021	0.112	0.221
HDT at 1.81 MPa (°C)	77	79	73
HDT at 0.42 MPa (°C)	89	88	79
Vicat softening temperature (°C)	105	104	96
Oxygen index (%)	17.8	—	—
Hardness (Rockwell)	R130	R110	R75
CTE (10^{-5}/°C)	9.0	9.0	9.0
Shrinkage (in./in.)	0.005	0.005	0.005
Relative permittivity at 1 kHz	2.53	—	—

Table 6 Typical Properties of SPS

Property	SPS	SPS + 30% GF
Density (g/cm^3)	1.05	1.25
Tensile modulus (GPa)	3.44	10.0
Tensile strength (MPa)	41	121
Elongation at break (%)	1	1.5
Flexural modulus (GPa)	3.9	9.7
Flexural strength (MPa)	71	166
Notched Izod (kJ/m)	0.011	0.096
HDT at 1.81 MPa (°C)	99	249
HDT at 0.42 MPa (°C)	–	263
CTE (10^{-5}/°C)	9.2	2.5
Water absorption, 24 hr (%)	0.04	0.05
Relative permittivity, 1 MHz	2.6	2.9
Dissipation factor at 1 MHz	<0.001	<0.001

Table 7 Typical Properties of SAN

Property	SAN	SAN + 30% GF
Density (g/cm^3)	1.07	1.22
Tensile modulus (GPa)	3.45	11.0
Tensile strength (MPa)	76	139
Elongation at break (%)	2.5	1.6
Flexural modulus (GPa)	3.65	9.60
Flexural strength (MPa)	128	—
Notched Izod (kJ/m)	0.016	0.060
HDT at 1.81 MPa (°C)	87	100
HDT at 0.42 MPa (°C)	103	108
Vicat softening temperature (°C)	111	—
RTI (°C)	50	74
Oxygen index (%)	19.0	—
Hardness (Rockwell)	R125	R123
CTE (10^{-5}/°C)	6.6	1.9
Shrinkage (in./in.)	0.005	—
Water absorption, 24 hr (%)	—	0.15
Relative permittivity at 1 MHz	2.9	3.6
Dissipation factor at 1 MHz	0.009	0.008

Table 8 Typical Properties of ABS Resins

Property	General Purpose	High Heat	High Impact
Density (g/cm^3)	1.05	1.05	1.04
Tensile modulus (GPa)	2.28	2.28	2.00
Tensile strength (MPa)	43	47	39
Elongation at break (%)	—	—	26
Flexural modulus (GPa)	2.48	2.41	2.07
Flexural strength (MPa)	77	83	65
Notched Izod (kJ/m)	0.203	0.214	0.454
HDT at 1.81 MPa (°C)	81	99	81
HDT at 0.42 MPa (°C)	92	110	97
Vicat softening temperature (°C)	—	—	99
RTI (°C)	60	—	60
Hardness (Rockwell)	—	R110	—
CTE (10^{-15}/°C)	8.82	7.92	9.45
Shrinkage (in./in.)	0.006	0.006	0.006

components, dialyzer housings), and other (glazing, battery cases, pen barrels).

3.4 Acrylonitrile/Butadiene/Styrene Polymers

ABS is a generic name for a family of amorphous ter polymer prepared from the combination of acrylonitrile, butadiene, and styrene monomers.[22] Acrylonitrile (A) provides chemical resistance, hardness, rigidity, and fatigue resistance and increases the heat deflection temperature. Butadiene (B) provides toughness and low-temperature ductility but lowers the heat resistance and rigidity. Styrene (S) provides rigidity, hardness, gloss, aesthetics, and processing ease. This three-monomer system offers a lot of flexibility in tailoring ABS resins. Optimization of these monomers can enhance a desired performance profile. Because of the possible variations in composition, properties can vary over a wide range.

Typical resin grades in this product family consist of a blend of an elastomeric component and an amorphous thermoplastic component. Typically the elastomeric component is the polybutadiene-based copolymer (ABS with high B). The amorphous thermoplastic component is SAN.

ABS plastics offer a good balance of properties centering on toughness, hardness, high gloss, dimensional stability, and rigidity. Compared to PS, ABS offers good impact strength, improved chemical resistance, and similar heat deflection temperature. ABS is also opaque. ABS has good chemical resistance to acids and bases but poor resistance to aromatic compounds, chlorinated solvents, esters, ketones, and aldehydes. ABS has poor resistance to UV light. Typical properties are shown in Table 8.

Uses ABS materials are suitable for tough consumer products (refrigerator door liners, luggage, telephones, business machine housings, power tools, small appliances, toys, sporting goods, personal care devices), automotive (consoles, door panels, various interior trim, exterior grills and lift gates), medical (clamps, stopcocks, blood dialyzers, check valves), and building and construction (pipes, fittings, faucets, conduit, shower heads, bathtubs).

3.5 Acrylonitrile/Styrene/Acrylate Polymers

Acrylonitrile/styrene/acrylate (ASA) ter polymers are amorphous thermoplastics and are similar to ABS resins. However, the butadiene rubber has been replaced by an acrylate-based elastomer, which has excellent resistance to sunlight. Hence ASA offers exceptional durability in weather-related environments without painting. In outdoor applications, ASA resins retain color stability under long-term exposure to UV, moisture, heat, cold, and impact. In addition, ASA polymers offer high gloss and mechanical properties similar to those of ABS resins. ASA has good chemical resistance to oils, greases, and salt solutions. However, resistance to aromatic and chlorinated hydrocarbons, ketones, and esters is poor.

ASA resins exhibit good compatibility with other polymers. This facilitates its use in polymer blends and alloys and as a cost-effective, cap-stock (overlayer) to protect PS, polyvinyl chloride (PVC), or ABS in outdoor applications.

Various grades are available for injection molding, profile and sheet extrusion, thermoforming, and blow molding. Typical properties appear in Table 9.

Uses ASA resins have applications in automotive/transportation (body moldings, bumper parts, side-view mirror housings, truck trailer doors, roof luggage containers), building/construction (window lineals, door profiles, downspouts, gutters, house

Table 9   Typical Properties of ASA Resins

Property	ASA	ASA/PVC	ASA/Polycarbonate
Density (g/cm^3)	1.06	1.21	1.15
Tensile modulus (GPa)	1.79	—	—
Tensile strength (MPa)	41	39	62
Elongation at break (%)	40	30	25
Flexural modulus (GPa)	1.79	1.93	2.52
Flexural strength (MPa)	59	48	88
Notched Izod (kJ/m)	0.320	0.961	0.320
Notched Izod, $-30°$C (kJ/m)	0.059	0.107	0.080
HDT at 1.81 MPa ($°$C)	77	74	104
HDT at 0.42 MPa ($°$C)	88	82	116
Vicat softening temperature ($°$C)	99	—	—
RTI ($°$C)	50	50	—
Hardness (Rockwell)	R86	R102	R110
CTE (10^{-5}/$°$C)	9.0	8.46	7.2
Shrinkage (in./in.)	0.006	0.004	0.006
Water absorption, 24 hr (%)	—	0.11	0.25
Relative permittivity at 1 MHz	3.2	—	—
Dissipation factor at 1 MHz	0.026	—	—

siding, windows, mail boxes, shutters, fencing, wall fixtures), sporting goods (snowmobile and all-terrain vehicle housings, small water craft, camper tops, windsurfer boards), and consumer items (garden hose fittings and reels, lawnmower components, outdoor furniture, telephone handsets, covers for outdoor lighting, spa and swimming pool steps and pumps, housings for garden tractors).

3.6   Poly(methyl methacrylate)

The catalytic or thermal polymerization of methyl methacrylate yields poly(methyl methacrylate) (PMMA). It is often referred to as acrylic. PMMA is a strong, rigid, clear, amorphous polymer. The optical clarity, rigidity, colorability, and ability to resist sunlight and other environmental stresses make PMMA ideal for glass replacement.

In addition, PMMA has low water absorption and good electrical properties.

Acrylics have fair chemical resistance to many chemicals. However, resistance to aromatic and chlorinated hydrocarbons, ketones, and esters is poor. PMMA properties appear in Table 10.

Uses   PMMA is used in construction (glazing, lighting diffusers, domed skylights, enclosures for swimming pools and buildings), automotive (exterior lighting lenses in cars and trucks, nameplate, medallions, lenses on instrument panels), household (laboratory, vanity and counter tops, tubs), medical (filters, blood pumps), and others (appliances, aviation canopies and window, outdoor signs, display cabinets).

3.7   Styrene/Maleic Anhydride Copolymer

Styrene/maleic anhydride (SMA) resins are copolymers of styrene and maleic anhydride (MA) and offer

Table 10   Typical Properties of PMMA, SMMA, and SMA Resins

Property	PMMA	SMMA	SMA
Density (g/cm^3)	1.19	1.09	1.08
Tensile modulus (GPa)	3.10	3.50	—
Tensile strength (MPa)	70	57.2	48.3
Elongation at break (%)	6	2	—
Flexural modulus (GPa)	3.10	3.50	3.61
Flexural strength (MPa)	103	103	115.8
Notched Izod (kJ/m)	0.016	0.020	0.011
HDT at 1.81 MPa ($°$C)	93	98	96
HDT at 0.42 MPa ($°$C)	94	—	—
Vicat softening temperature ($°$C)	103	—	118
RTI ($°$C)	90	—	—
Hardness (Rockwell)	M91	M64	L108
CTE (10^{-5}1/$°$C)	7.6	7.92	6.3
Shrinkage (in./in.)	0.004	0.006	0.005
Water absorption, 24 hr (%)	0.3	0.15	0.10
Relative permittivity at 1 kHz	3.3	—	—

increased heat deflection temperatures, strength, solvent reistance, and density compared to PS.[20] SMA resins are usually produced via catalyzed bulk polymerization. There is a strong tendency to form the 1 : 1 styrene–MA copolymer. Random SMA resins containing 5–12% MA are produced via starve feeding, i.e., keeping the MA concentration low during the polymerization. SMA resins are brittle and have poor UV resistance. Impact-modified grades and ter polymers (grafting rubber into polymer during polymerization) are available. Typical properties for SMA prepared with 9% MA appear in Table 10.

Uses SMA copolymers have been used in automotive (instrument panels, headliners) food service items, plumbing, and electrical applications.

3.8 Styrene/Methyl Methacrylate Copolymer

Styrene/methyl methacrylate (SMMA) copolymers are prepared by the catalyzed polymerization of styrene and methyl methacrylate (MMA). The advantages of SMMA over PS include improved outdoor weathering and light stability, better clarity, increased chemical resistance, and improved toughness. Properties for SMMA prepared with 30% MMA appear in Table 10.

Uses Applications for SMMA plastics are in small appliances and kitchen and bathroom accessories.

3.9 Polyvinyl Chloride

The catalytic polymerization of vinyl chloride yields polyvinyl chloride (PVC).[23] It is commonly referred to as PVC or vinyl and is second only to PE in volume use. Normally, PVC has a low degree of crystallinity and good transparency. The high chlorine content of the polymer produces advantages in flame resistance, fair heat deflection temperature, and good electrical properties. The chlorine also makes PVC difficult to process. The chlorine atoms have a tendency to split out under the influence of heat during processing and heat and light during end use in finished products, producing discoloration and embrittlement. Therefore, special stabilizer systems are often used with PVC to retard degradation.

PVC has good chemical resistance to alcohols, mild acids and bases, and salts. However, PVC is attacked by halogenated solvents, ketones, esters, aldehydes, ethers, and phenols.

There are two major classifications of PVC: rigid and flexible (plasticized). In addition, there are also foamed PVC and PVC copolymers. Typical properties of PVC resins appear in Table 11.

Rigid PVC PVC alone is a fairly rigid polymer, but it is difficult to process and has low impact strength. Both of these properties are improved by the addition of elastomers or impact-modified graft copolymers, such as ABS and ASA resins. These improve the melt flow during processing and improve the impact strength without seriously lowering the rigidity or the heat deflection temperature.

Uses With this improved balance of properties, rigid PVCs are used in such applications as construction (door and window frames, water supply, pipe, fittings, conduit, building panels and siding, rainwater gutters and downspouts, interior molding and flooring), packaging (bottles, food containers, films for food wrap), consumer goods (credit cards, furniture parts), and other (agricultural irrigation and chemical processing piping).

Plasticized PVC The rigid PVC is softened by the addition of compatible, nonvolatile, liquid plasticizers. The plasticizers are usually used in > 20 parts per hundred resins. It lowers the crystallinity in PVC and acts as internal lubricant to give clear, flexible plastics. Plasticized PVC is also available in liquid formulations known as plastisols or organosols.

Uses Plasticized PVC is used for construction (wire and cable insulation, interior wall covering), consumer goods (outdoor apparel, rainwear, upholstery, garden hose, toys, shoes, tablecloths, sporting goods, shower curtains), medical (clear tubing, blood and solution bags, connectors), and automotive (seat covers). Plastisols are used in coating fabric, paper, and metal and rotationally cast into balls, dolls, and the like.

Foamed PVC Rigid PVC can be foamed to a low-density cellular material that is used for decorative moldings and trim. Foamed PVC is also available via foamed plastisols. Foamed PVC adds greatly to the softness and energy absorption characteristics already inherent in plasticized PVC. In addition, it gives rich, warm, leatherlike material.

Table 11 Typical Properties of PVC Materials

Property	General Purpose	Rigid	Rigid Foam	Plasticized	Copolymer
Density (g/cm^3)	1.40	1.34–1.39	0.75	1.29–1.34	1.37
Tensile modulus (GPa)	3.45	2.41–2.45	—	—	3.15
Tensile strength (MPa)	56.6	37.2–42.4	> 13.8	14–26	52–55
Elongation at break (%)	85	—	> 40	250–400	—
Notched Izod (kJ/m)	0.53	0.74–1.12	> 0.06	—	0.02
HDT at 1.81 MPa (°C)	77	73–77	65	—	65
Brittle temperature (°C)	—	—	—	−60 to −30	
Hardness	D85 (Shore)	R107–R122 (Rockwell)	D55 (Shore)	A71–A96 (Shore)	
CTE (10^{-5}/°C)	7.00	5.94	5.58		
Shrinkage (in./in.)	0.003				
Relative permittivity at 1 kHz	3.39				
Dissipation factor at 1 kHz	0.081				

Uses Upholstery, clothing, shoe fabrics, handbags, luggage, and auto door panels and energy absorption for quiet and comfort in flooring, carpet backing, auto headliners, and so forth.

PVC Copolymers Copolymerization of vinyl chloride with 10–15% vinyl acetate gives a vinyl polymer with improved flexibility and less crystallinity than PVC, making such copolymers easier to process without detracting seriously from the rigidity and heat deflection temperature. These copolymers find primary applications in flooring and solution coatings.

3.10 Poly(vinylidene chloride)

Poly(vinylidene chloride) (PVDC) is prepared by the catalytic polymerization of 1,1-dichloroethylene. This crystalline polymer exhibits high strength, abrasion resistance, high melting point, better than ordinary heat resistance ($100°C$ maximum service temperature), and outstanding impermeability to oil, grease, water vapor, oxygen, and carbon dioxide. It is used in films, coatings, and monofilaments.

When the polymer is extruded into film, quenched, and oriented, the crystallinity is fine enough to produce high clarity and flexibility. These properties contribute to widespread use in packaging film, especially for food products that require impermeable barrier protection.

PVDC and/or copolymers with vinyl chloride, alkyl acrylate, or acrylonitrile are used in coating paper, paperboard, or other films to provide more economical, impermeable materials. Properties appear in Table 12.

Table 12 Typical Properties of PVDC

Density (g/cm^3)	1.65–1.72
Tensile strength (MPa)	25
Elongation at break (%)	120
Notched Izod (kJ/m)	0.04
Hardness (Rockwell)	M50–M65

Uses PVDC is used in food packaging were barrier properties are needed. Applications for injection-molding grades are fittings and parts in the chemical industry. PVDC pipe is used in the disposal of waste acids. PVDC is extruded into monofilament and tape that is used in outdoor furniture upholstery.

4 POLYURETHANE AND CELLULOSIC RESINS

4.1 Polyurethanes

Polyurethanes (PUs) are prepared from polyols and isocyanates.[24,25] The isocyanate groups react with the hydroxyl groups on the polyol to form a urethane bond. The polyol can be a low-molecular-weight polyether or polyester. The isocyanate can be aliphatic or aromatic and in the preparation of linear PU is typically difunctional. However, isocyanates with greater functionality are used in preparing rigid foam PUs. The family of PU resins is very complex because of the enormous variation in the compositional features of the polyols and isocyanates. This variety results in a large number of polymer structures and performance profiles. Indeed, PUs can be rigid solids or soft and elastomeric or a have a foam (cellular) structure.

The majority of PU resins are thermoset (PUR). However, there are also important thermoplastic polyurethane resins (TPU). Polyurethanes offer high impact strength, even at low temperatures, good abrasion resistance, excellent heat resistance, excellent resistance to nonpolar solvents, fuels, and oils, and resistance to ozone, oxidation, and humidity.

TPUs are generally processed by injection-molding techniques. PURs are processed by reaction injection molding (RIM) and various foaming techniques. A major use of PUR is in the production of flexible, semirigid, and rigid foams. In addition, PUs can be used as fibers, sealants, adhesives, and coatings.

Typical properties of PUs appear in Table 13.

Uses Typical applications for PUs are in consumer goods (furniture padding, bedding, skateboard and roller blade wheels, shoe soles, athletic shoes,

Table 13 Typical Properties of PUs

Property	TPU Polyester	TPU Polyether	PUR–RIM Foam	PUR–RIM Solid	PUR–RIM Elastomer
Density (g/cm^3)	1.21	1.18	0.56	1.13	1.04
Tensile strength (MPa)	41	38	13.8	39	24
Elongation at break (%)	500	250	10	10	250
Flexural modulus (GPa)	0.14	0.54	0.86	0.14	0.36
Flexural strength (MPa)	—	—	27.6	60	—
Notched Izod (kJ/m)	—	—	—	0.05	0.6
HDT at 0.45 MPa ($°C$)	59	45	70	—	—
Vicat softening temperature ($°C$)	167	140	—	—	—
Brittle temperature ($°C$)	<−68	<−70	—	—	—
Hardness (Shore)	D55	D70	D65	D76	D58
CTE ($10^{-5}/°C$)	13	11.5	7.9	10	11
Shrinkage (in./in.)	0.008	0.008	0.8	1.1	1.3

ski booths, backing on carpets and tiles), automotive (padding, seals, fascias, bumpers, structural door panels), and miscellaneous (tubing, membranes, bearings, nuts, seals, gaskets).

4.2 Cellulosic Polymers

Cellulose-based plastics are manufactured by the chemical modification of cellulose.[26–29] Cellulose does not melt and hence is not a thermoplastic material. However, esterification of cellulose gives organic esters of cellulose, which are thermoplastic. These include cellulose acetate (CA), cellulose acetate butyrate (CAB), and cellulose proprionate (CP). Cellulosics are noted for their wide range of properties, which include clarity (up to 80% light transmission), abrasion resistance, stress crack resistance, high gloss, and good electrical properties. CA offers good rigidity and hardness. CAB and CP offer good weatherability, low-temperature impact strength, and dimensional stability.

In general, cellulosic esters are resistant to aliphatic hydrocarbons, ethylene glycol, bleach, and various oils. However, alkaline materials attack them. Cellulosic esters have high water absorption and low continuous-use temperatures. Typical properties appear in Table 14.

Uses Typical applications for cellulosic plastics are automotive (steering wheels, trim), films (photographic, audio tape, pressure-sensitive tape), home furnishings (table edging, Venetian blind wands), packaging (tubular containers, thermoformed containers for nuts, bolts, etc.), and miscellaneous (tool and brush handles, toys, filaments for toothbrushes, eye glass frames, lighting fixtures).

5 ENGINEERING THERMOPLASTICS: CONDENSATION POLYMERS

Engineering thermoplastics comprise a special-performance segment of synthetic plastics that offer enhanced properties.[2,3] When properly formulated, they may be shaped into mechanically functional, semiprecision parts or structural components. Mechanically functional implies that the parts may be subjected to mechanical stress, impact, flexure, vibration, sliding friction, temperature extremes, hostile environments, and the like, and continue to function.

As substitutes for metal in the construction of mechanical apparatus, engineering plastics offer advantages such as transparency, light weight, self-lubrication, and economy in fabrication and decorating. Replacement of metals by plastic is favored as the physical properties and operating temperature ranges of plastics improve and the cost of metals and their fabrication increases.

5.1 Thermoplastic Polyesters

Thermoplastic polyesters are prepared from the condensation polymerization of a diol and typically a dicarboxylic acid. Usually the dicarboxylic acid is aromatic, that is, terephthalic acid. As a family of polymers thermoplastic polyesters are typically semicrystalline and hence have good chemical resistance. An important attribute of a semicrystalline polymer is a fast rate of crystallization, which facilitates short injection-molding cycles.

Poly(butylene terephthalate) Poly(butylene terephthalate) (PBT) is prepared from butanediol with dimethyl terephthalate.[29–31] PBT is a semicrystalline polymer that has a fast rate of crystallization and rapid molding cycles. PBT has a unique and favorable balance of properties between polyamides and polyacetals. PBT combines high mechanical, thermal, and electrical properties with low moisture absorption, extremely good self-lubricity, fatigue resistance, very good chemical resistance, very good dimensional stability, and good maintenance of properties at elevated temperatures. Dimensional stability and electrical properties are unaffected by high-humidity conditions.

PBT has good chemical resistance to water, ketones, alcohols, glycols, ethers, and aliphatic and chlorinated hydrocarbons at ambient temperatures. In addition, PBT has good resistance to gasoline, transmission fluid, brake fluid, greases, and motor oil. At ambient temperatures PBT has good resistance to dilute acids and bases, detergents, and most aqueous salt solutions. It is not recommended for use in strong bases or aqueous media at temperatures above 50°C.

PBT grades range from unmodified to glass-fiber-reinforced to combinations of glass fiber and mineral fillers that enhance strength, modulus, and heat deflection temperature. Both filled and unfilled grades of PBT offer a range of UL and other agency compliance ratings.

A high-density PBT combines the inherent characteristics of PBT with the advantages of high levels of mineral reinforcement. This combination provides

Table 14 Typical Properties of Cellulosic Materials

Property	CA	CAB	CP
Density (g/cm^3)	1.28	1.19	1.19
Tensile modulus (GPa)	2.17	1.73	1.73
Tensile strength (MPa)	40	34.5	35
Elongation at break (%)	25	50	60
Flexural modulus (GPa)	2.4	1.8	—
Flexural strength (MPa)	66	60	—
Notched Izod (kJ/m)	0.16	0.187	0.41
HDP at 1.81 MPa (°C)	61	65	72
HDP at 0.42 MPa (°C)	72	72	80
Vicat softening temperature (°C)	—	—	100
Hardness (Rockwell)	R82	R75	R70
CTE (10^{-5}/°C)	13.5	13.5	14
Relative permittivity at 1 kHz	3.6	3.6	3.6

Table 15 Typical Properties of PBT

Property	PBT	PBT + 30% GF	PBT + 40% GF
Density (g/cm^3)	1.31	1.53	1.63
Tensile strength (MPa)	52	119	128
Elongation at break (%)	300	3	—
Flexural modulus (GPa)	2.34	7.58	8.27
Flexural strength (MPa)	83	190	200
Notched Izod (kJ/m)	0.053	0.085	0.096
Unnotched Izod (kJ/m)	1.602	0.801	0.961
HDT at 1.81 MPa ($^\circ$C)	54	207	204
HDT at 0.42 MPa ($^\circ$C)	154	216	216
RTI ($^\circ$C)	—	140	—
Hardness (Rockwell)	R117	R118	R118
CTE (10^{-5}/$^\circ$C)	8.1	2.5	2.7
Shrinkage (in./in.)	0.006	0.004	0.004
Water absorption, 24 hr (%)	0.08	0.06	0.05
Relative permittivity at 1 MHz	3.1	3.7	4.0
Dissipation factor at 1 MHz	0.419	0.02	0.02

a balance of mechanical, thermal, and electrical properties, broad chemical and stain resistance, low water absorption, and dimensional stability. In addition, the smooth, satin finish and heavy weight offer the appearance and quality feel of ceramics while providing design flexibility, injection-molding processing advantages, and recycling opportunities which are common in engineering thermoplastics. Properties appear in Table 15.

Uses Applications of PBT include automotive components (brake system parts, fuel injection modules, grill-opening panels), electrical/electronic components (connectors, smart network interface devices, power plugs, electrical components, switches, relays, fuse cases, light sockets, television tuners, fiber-optic tubes), medical (check valves, catheter housings, syringes), consumer goods (hair dryer and power tool housings, iron and toaster housings, food processor blades, cooker-fryer handles), and miscellaneous (gears, rollers, bearing, housings for pumps, impellers, pulleys, industrial zippers). High-density PBT is being used in kitchen and bath sinks, countertops, wall tiles, shower heads, speaker housings, medical equipment, and consumer goods.

PBT/PC Alloy PBT/PC resins are thermoplastic alloys of PBT and polycarbonate (PC). The amorphous PC provides impact resistance and toughness while the PBT provides enhanced chemical resistance and thermal stability. Impact modification completes the balanced performance profile by providing both low- and high-temperature durability. Hence high levels of impact strength are achieved at low temperatures, below −40°C.

PBT/PC resins offer a balance of performance characteristics unique among engineering thermoplastics with their optimal combination of toughness, chemical resistance, dimensional stability, lubricity, and

high heat distortion temperature. In addition, this family of resins offers very good aesthetics, lubricity, UV resistance, and color retention. Originally developed for the automotive industry, PBT/PC resins are designed to provide resistance to various automotive fluids—gasoline, greases, oils, and the like. In general, the higher the amount of PBT in the resin blend, the higher the resin's chemical resistance. Hence resistance to gasoline may vary from grade to grade. PBT/PC resins generally are not very hydrolytically stable. Properties appear in Table 16.

Uses Applications for PBT/PC resins include automotive bumpers/fascia, tractor hoods and panels, components on outdoor recreational vehicles, lawn mower decks, power tool housings, material-handling pallets, and large structural parts.

Table 16 Typical Properties of PBT/PC Alloy

Property	PBT/PC	PBT/PC +30% GF
Density (g/cm^3)	1.21	1.44
Tensile strength (MPa)	59	92
Elongation at break (%)	120	4
Flexural modulus (GPa)	2.04	5.38
Flexural strength (MPa)	85	138
Notched Izod (kJ/m)	0.710	0.171
Notched Izod, −30°C (kJ/m)	3.204	0.112
Unnotched Izod (kJ/m)	0.299	0.641
HDT at 1.81 MPa ($^\circ$C)	99	149
HDT at 0.42 MPa ($^\circ$C)	106	204
Hardness (Rockwell)	R112	R109
CTE (10^{-5}/$^\circ$C)	8.4	2.3
Shrinkage (in./in.)	0.009	0.005
Water absorption, 24 hr (%)	0.12	0.09
Relative permittivity at 1 MHz	3.04	3.9
Dissipation factor at 1 MHz	0.019	0.02

Poly(ethylene terephthalate) Poly (ethylene terephthalate) (PET) is prepared from the condensation polymerization of dimethyl terephthalate or terephthalic acid with ethylene glycol.[29-31] PET is a semicrystalline polymer that exhibits high modulus, high strength, high melting point, good electrical properties, and moisture and solvent resistance. The crystallization rate of PET is relatively slow. This slow crystallization is a benefit in blow-molding bottles where clarity is important. Indeed, a small amount of a comonomer is typically added during polymerization of PET. The function of the comonomer is to disrupt the crystallinity and lower the rate of crystallization. However, in injection-molding applications the slow rate of crystallization will increase the molding cycle time.

For most injection-molding applications PET generally contains glass fiber or mineral filler to enhance properties. Typical properties appear in Table 17.

Uses Primary applications of PET include blow-molded beverage bottles; fibers for wash-and-wear, wrinkle-resistant fabrics; and films that are used in food packaging, electrical applications (e.g., capacitors), magnetic recording tape, and graphic arts. Injection-molding application of PET include automotive (cowl vent grills, wiper blade supports) and electrical (computer fan blades, fuse holders and connectors).

Poly(trimethylene terephthalate) Poly(trimethylene terephthalate) (PTT) is prepared from 1,3-propanediol and terephthalic acid. PTT is a semicrystalline polymer that exhibits properties between PET and PBT. In particular, it offers high modulus, high strength, good electrical properties, and moisture and solvent resistance. Its crystallization rate is slower than PBT but faster than PET.[32] Properties appear in Table 18.

Table 17   Typical Properties of PET

Property	PET	PET + 30% GF
Density (g/cm^3)	1.41	1.60
Tensile modulus (GPa)	1.71	11.5
Tensile strength (MPa)	50	175
Elongation at break (%)	180	2
Flexural modulus (GPa)	2.0	—
Flexural strength (MPa)	—	225
Notched Izod (kJ/m)	0.090	—
HDT at 1.81 MPa (°C)	63	225
HDT at 0.42 MPa (°C)	71	—
Vicat softening temperature (°C)	—	260
Oxygen index (%)	—	21
Hardness (Rockwell)	R105	—
CTE (10^{-5}/°C)	9.1	2.0
Water absorption, 24 hr (%)	—	0.15
Relative permittivity at 1 MHz	3.3	4.2
Dissipation factor at 1 MHz	—	0.018

Table 18   Typical Properties of PTT

Property	PTT	PTT + 30% GF
Density (g/cm^3)	1.35	1.55
Flexural modulus (GPa)	2.76	10.3
Tensile strength (Mpa)	67	159
Notched Izod (kJ/m)	0.05	0.11
HDT at 1.81 MPa (°C)	59	216
Shrinkage (in./in.)	0.020	0.002
Dielectric constant at 1 MHz	3.0	—
Relative permittivity at 1 MHz	0.02	—

Uses Initial applications for PTT were in carpet fiber, where it offers a softer feel in combination with stain resistance and resiliency. Other fiber markets include textiles and monofilaments. Injection-molding applications would be various automotive parts.

5.2   Polyamides (Nylon)

Polyamides, commonly called nylons, are produced by the condensation polymerization of dicarboxylic acids and diamines or the catalytic polymerization of a lactam monomer (a cyclic amide).[33,34] In general, polyamides are semicrystalline thermoplastics. Polyamides are a class of resins characterized by broad chemical resistance, high strength, and toughness. In addition, polyamides absorb high levels of water. Moisture from the atmosphere diffuses into the polymer and hydrogen bonds to the amide groups. This absorbed water causes dimensional changes where the molded part will increase in size and weight. The higher the amount of amide groups in the polymer, the greater the moisture uptake. In addition, the water acts as a plasticizer and lowers the rigidity and strength. Polar solvents such as alcohols are also absorbed into the nylon.

There are numerous dicarboxylic acids and diamines that can be used to make polyamides. The shorthand method for describing the various types of polyamides uses a number to designate the number of carbon atoms in each starting monomer(s). In the case of terephthalic and isophthalic acids T and I are used, respectively.

Polyamide 6/6 and 6 The two major types of polyamides are polyamide 6/6, or nylon 6/6 (PA6/6), and polyamide 6, or nylon 6 (PA6).[33,34] PA6/6 is made from a six-carbon dicarboxylic acid and a six-carbon diamine—that is, adipic acid and hexamethylene diamine. PA6 is prepared from caprolactam. Both PA6/6 and PA6 are semicrystalline resins.

Key features of nylons include toughness, fatigue resistance, and chemical resistance. Nylons do exhibit a tendency to creep under applied load. Nylons are resistant to many chemicals, including ketones, esters, fully halogenated hydrocarbons, fuels, and brake fluids. Nylons have a relatively low heat deflection temperature. However, glass fibers or mineral fillers are used to enhance the properties of polyamides. In

Table 19   Typical Properties of PA6 and PA6/6

Property	PA6	PA6 + 33% GF	PA6/6	PA6/6 + 40% GF
Density (g/cm^3)	1.13	1.46	1.14	1.44
Tensile modulus (GPa)	—	—	3.30	—
Tensile strength (MPa)	79	200	86	—
Elongation at break (%)	70	3	45	—
Flexural modulus (GPa)	2.83	9.38	2.90	9.3
Flexural strength (MPa)	108	276	—	219
Notched Izod (kJ/m)	0.053	0.117	0.059	0.14
HDT at 1.81 MPa (°C)	64	210	90	250
HDT at 0.45 MPa (°C)	165	220	235	260
RTI (°C)	105	120	—	—
Hardness (Rockwell)	R119	R121	—	M119
CTE (10^{-5}/°C)	8.28	2.16	8.10	3.42
Shrinkage (in./in.)	0.013	0.003	0.0150	0.0025
Water absorption, 24 hr (%)	—	—	1.2	—
Relative permittivity at 1 MHz	—	3.8	3.6	—
Dissipation factor at 1 MHz	—	0.022	0.02	—

addition to increasing the heat deflection temperature, the fibers and fillers lessen the effect of moisture and improve the dimensional stability. Properties of PA6/6 and PA6 appear in Table 19.

Uses   The largest application of nylons is in fibers. Molded applications include automotive components (electrical connectors, wire jackets, fan blades, valve covers, emission control valves, light-duty gears), electronic (connectors, cable ties, plugs, terminals, coil forms), related machine parts (gears, cams, pulleys, rollers, boat propellers), and appliance parts.

Polyamide/PPE Alloys   Polyamide/polyphenylene ether (PA/PPE) alloys are compatible blends of amorphous PPE and a semicrystalline PA that have a microstructure in which the PA is the continuous phase and the PPE is the discrete phase.[3,35-37] The PPE acts as an organic reinforcing material. This technology combines the inherent advantages of PPE (dimensional stability, very low water absorption, and high heat resistance) with the chemical resistance and ease of processing of the PA. This combination results in a chemically resistant material with the stiffness, impact resistance, dimensional stability, and heat performance required for automotive body panels that can undergo online painting.

PA/PPE alloys offer broad environmental resistance to commonly used automotive fuels, greases, and oils. In addition, this family of alloys is resistant to detergents, alcohols, aliphatic and aromatic hydrocarbons, and alkaline chemicals.

Since PPE does not absorb any significant amount of moisture, the effect of moisture on properties is reduced. Indeed, the moisture uptake in PA/PPE alloys is lower. Hence PA/PPE alloys minimize the effect of moisture on rigidity, strength, and dimensional stability vis-à-vis PA.[35,36] In addition, heat deflection temperatures have been enhanced by the PPE. Properties are shown in Table 20.

Uses   PA/PPE alloys are used in automotive body panels (fenders and quarter panels), automotive wheel covers, exterior truck parts, under-the-hood automotive parts (air intake resonators, electrical junction boxes and connectors), and fluid-handling applications (e.g., pumps).

Table 20   Typical Properties of PPE/PA6/6 Alloys

Property	Unfilled		10% GF		30% GF	
	PA	PPE/PA	PA	PPE/PA	PA	PPE/PA
Density (g/cm^3)	1.14	1.10	1.204	1.163	1.37	1.33
Flexural modulus (GPa)						
Dry as molded	2.8	2.2	4.5	3.8	8.3	8.1
100% Relative humidity	0.48	0.63	2.3	2.6	4.1	5.8
At 150°C	0.21	0.70	0.9	1.6	3.2	4.3
Flexural strength (MPa)						
Dry as molded	96	92	151	146	275	251
100% Relative humidity	26	60	93	109	200	210
At 150°C	14	28	55	60	122	128

Table 21 Typical Properties of Nylon 4/6

Property	PA4/6	PA4/6 + 30% GF
Density (g/cm^3)	1.18	1.41
Melting point (°C)	295	—
Glass transition temperature (°C)	75	—
Tensile modulus (GPa)	3.3	10.0
Tensile strength (MPa)	100	210
Elongation at break (%)	15	4
Flexural modulus (GPa)	3.1	—
Flexural strength (MPa)	149.6	—
Notched Izod (kJ/m)	0.096	0.069
HDT at 1.81 MPa (°C)	190	290
HDT at 0.42 MPa (°C)	280	290
Vicat softening temperature (°C)	290	290
Hardness (Shore)	D85	—
CTE (10^{-5}/°C)	9.0	5.0
Water absorption, 24 hr (%)	3.7	2.6
Relative permittivity at 1 kHz	3.83	—

Polyamide 4/6 (PA4/6)   Nylon 4/6 is prepared from the condensation polymerization of 1,4-diaminobutane and adipic acid.[33] PA4/6 has a higher crystalline melting point, greater crystallinity, and a faster rate of crystallization than other PAs. In comparison to PA6/6 and PA6, PA4/6 offers high tensile strength and heat deflection temperature; however, it has higher moisture uptake, which can affect properties and dimensional stability. Properties appear in Table 21.

Uses   Application for PA4/6 include under-the-hood automotive parts, gears, bearings, and electrical parts.

Semiaromatic Polyamide (PA6/6T, PA6I/6T)
Semiaromatic PAs have been developed in order to increase the performance over that of PA6/6 and

PA6.[33] In general, these resins are modified copolymers based on poly(hexamethylene terephthalate), or PA6/T. Pure nylon 6/T exhibits a very high T_m of 370°C and a T_g of 180°C. This high T_m is above its decomposition temperature. PA6/T copolymers have been prepared using additional monomers such as isophthalic acid, adipic acid, caprolactam, or 1,5-hexyl diamine, which lower the crystalline melting point to useful ranges for melt processing. These ter polymers exhibit T_m values of 290–320°C and T_g values of 100–125°C and offer enhanced performance (i.e., stiffer, stronger, greater thermal and dimensional stability) over PA6/6 and 6. Melt processing requires higher temperatures and often an oil-heated mold (> 100°C). These semicrystalline PAs have good chemical resistance, good dielectric properties, and lower moisture absorption and are more dimensionally stable in the presence of moisture than PA6 and PA6/6. Properties appear in Table 22.

Uses   High heat application, automotive (radiator ventilation and fuel supply systems), electrical/electronic (housings, plugs, sockets, connectors), recreational (tennis rackets, gold clubs), mechanical (industrial and chemical processing equipment, bearings, gears), appliance and plumbing parts, and aerospace components.

Aromatic Polyamides   Polyamides prepared from aromatic diamines and aromatic diacids give very high heat aromatic nylons or aramides. Examples are poly(p-phenyleneterephthalamide) (PPTA) and poly (m-phenyleneisophthalamide) (MPIA). These wholly aromatic PAs have high strength, high modulus, high chemical resistance, high toughness, excellent dimensional stability, and inherent flame resistance. MPIA has a T_g of 280°C and is difficult to melt process. Typically it is spun into fibers. PPTA has very high T_g and T_m of 425 and 554°C, respectively. In addition,

Table 22 Typical Properties of Semiaromatic Polyamides

Property	PA6/6T	PA6/6T + 35% GF	PA6T/6I	PA6T/6I + 40% GF
Density (g/cm^3)	1.16	1.43	1.21	1.46
Tensile modulus (GPa)	3.20	12.0	2.44	11.1
Tensile strength (MPa)	100	210	108	187
Elongation at break (%)	11.5	3	5	2
Flexural modulus (GPa)	—	—	3.43	10.8
Flexural strength (MPa)	—	—	157	284
Notched Izod (kJ/m)	0.070	—	0.049	0.079
Unnotched Izod (kJ/m)	—	—	0.395	0.592
HDT at 1.81 MPa (°C)	100	270	140	295
HDT at 0.45 MPa (°C)	120	—	—	—
RTI (°C)	125	—	120	120
CTE (10^{-5}/°C)	7.0	1.5	—	—
Shrinkage (in./in.)	0.0065	0.0035	0.006	0.002
Water absorption, 24 hr (%)	1.8		0.3	0.2
Dielectric constant at 1 MHz	4.0	4.2	—	—
Dissipation factor at 1 MHz	0.030	0.020	—	—

Table 23 Typical Properties of Aromatic PAs

Property	PPTA	MPIA
Density (g/cm^3)	—	1.38
Melting point (°C)	554	—
Glass transition temperature (°C)	425	280
Tensile modulus (GPa)	80–125	—
Tensile strength (MPa)	1500–2500	61
Elongation at break (%)	2	25
Flexural modulus (GPa)	—	3.1
Oxygen index (%)	29	28
CTE (10^{-5}/°C)	−0.32	0.62

PPTA exhibits liquid crystalline behavior and is spun into highly oriented, very high modulus, crystalline fibers. Properties appear in Table 23.

Uses PPTA fibers have uses in bullet-resistant apparel, composites, brake and transmission parts, gaskets, ropes and cables, sporting goods, tires, belts, and hoses.

MPIA fibers have uses in heat-resistant and flame-retardant apparel, electrical insulation, and composite structures.

5.3 Polyacetals

Polyacetal homopolymer, or polyoxymethylene (POM), is prepared via the polymerization of formaldehyde followed by capping each end of the polymer chain with an ester group for thermal stability. Polyacetal copolymer (POM-Co) is prepared by copolymerizing trioxane with relatively small amounts of a comonomer such as ethylene oxide.[29] The comonomer functions to stabilize the polymer to reversion reactions. POM and POM-Co are commonly referred to as acetals and are also semicrystalline resins.

Polyacetals exhibit rigidity, high strength, excellent creep resistance, fatigue resistance, toughness, self-lubricity/wear resistance, and solvent resistance. Acetals are resistant to gasoline, oils, greases, ethers,

alcohols, and aliphatic hydrocarbons. They are not recommended for use with strong acids.

Properties are enhanced by the addition of glass fiber or mineral fillers. Properties of POM and POM-Co appear in Table 24.

Uses Applications of polyacetals include moving parts in appliances and machines (gears, bearings, bushings, rollers, springs, valves, conveying equipment), automobiles (door handles, fasteners, knobs, fuel pumps, housings), plumbing and irrigation (valves, pumps, faucet underbodies, shower heads, impellers, ball cocks), industrial or mechanical products (rollers, bearings, gears, conveyer chains, housings), consumer products (A/V cassette components, toiletry articles, zippers, pen barrels, disposable lighters, toy parts), and electronic parts (key tops, buttons, switches).

5.4 Polycarbonate

Most commercial PCs are derived from the reaction of bisphenol A and phosgene.[29,38–40] PCs are amorphous resins that have a unique combination of outstanding clarity and high impact strength. In addition, PCs offer high dimensional stability, resistance to creep, and excellent electrical insulating characteristics. Indeed, PCs are among the stronger, tougher, and more rigid thermoplastics available. PC is a versatile material and a popular blend material used to enhance the performance of ABS, ASA, and polyesters (PBT).

PCs offer limited resistance to chemicals. PC properties are shown in Table 25.

Uses Applications of PC include glazing (safety glazing, safety shields, nonbreakable windows), automotive parts (lamp housings and lenses, exterior parts, instrument panels), packaging (large water bottles, reusable bottles), food service (mugs, food processor bowls, beverage pitchers), ophthalmic (optical

Table 24 Typical Properties of Polyacetals

Property	POM	POM-Co	POM + 25% GF	POM-Co + 30% GF
Density (g/cm^3)	1.42	1.41	1.58	1.60
Tensile modulus (GPa)	3.12	2.83	9.50	9.20
Tensile strength (MPa)	68.9	60.6	140	135
Elongation at break (%)	50	60	3	2.5
Flexural modulus (GPa)	2.83	2.58	8.00	—
Flexural strength (MPa)	97	90	—	—
Notched Izod (kJ/m)	0.074	0.054	0.096	—
HDT at 1.81 MPa (°C)	136	110	172	160
HDT at 0.45 MPa (°C)	172	158	176	—
Vicat softening point (°C)	—	151	178	158
Hardness (Rockwell)	M94	M80	—	—
CTE (10^{-5}/°C)	11.1	11.0	5.6	6.0
Shrinkage (in./in.)	0.02	—	0.008	—
Water absorption, 24 hr (%)	0.2	0.2	0.17	0.17
Relative permittivity at 1 MHz	3.7	4.0	—	4.3
Dissipation factor at 1 MHz	—	0.005	—	0.006

Table 25   Typical Properties of PCs

Property	PC	PC + 30% GF
Density (g/cm^3)	1.20	1.43
Tensile modulus (GPa)	2.38	8.63
Tensile strength (MPa)	69	131
Elongation at break (%)	130	3
Flexural modulus (GPa)	2.35	7.59
Flexural strength (MPa)	98	158
Notched Izod (kJ/m)	0.905	0.105
Unnotched Izod (kJ/m)	3.20	1.06
HDT at 1.81 MPa ($^\circ$C)	132	146
HDT at 0.45 MPa ($^\circ$C)	138	152
Vicat softening point ($^\circ$C)	154	165
RTI ($^\circ$C)	121	120
Hardness (Rockwell)	R118	R120
CTE (10^{-5}/$^\circ$C)	6.74	1.67
Shrinkage (in./in.)	0.006	0.002
Water absorption, 24 hr (%)	0.15	0.14
Relative permittivity at 1 MHz	2.96	3.31

lenses, corrective eyewear, sun wear lenses), medical/laboratory ware (filter housings, tubing connectors, eyewear, health care components), consumer (various appliance parts and housings, power tool housings, cellular phone housings, food processor bowls, lighting), media storage [compact discs (CDs), digital video discs (DVDs)], and miscellaneous (electrical relay covers, aircraft interiors, building and construction). Extruded PC film is used in membrane switches.

5.5   Polycarbonate/ABS Alloy

Polycarbonate/ABS alloys are amorphous blends of PC and ABS resins. They offer a unique balance of properties that combines the most desirable properties of both resins.[41] The addition of ABS improves the melt processing of PC/ABS blends, which facilitates filling large, thin-walled parts. Moreover, the ABS enhances the toughness of PC, especially at low temperatures, while maintaining the high strength and rigidity of the PC. In addition, PC/ABS offers excellent UV stability, high dimensional stability at ambient and elevated temperatures, and the ability for chlorine/bromine-free flame retardance.

The properties are a function of the ABS-to-PC ratio. Properties appear in Table 26.

Uses   Automotive (interior and exterior automotive applications as instrument panels, pillar, bezel, grills, interior and exterior trim), business machines (enclosures and internal parts of products such as lap- and desk-top computers, copiers, printers, plotters, and monitors), telecommunications [mobile telephone housings, accessories, and smart cards (GSM SIM cards)], electrical (electronic enclosures, electricity meter covers and cases, domestic switches, plugs and sockets and extruded conduits), and appliances (internal and external parts of appliances such as washing machines, dryers, and microwave ovens).

5.6   Polyestercarbonates

Polyestercarbonate (PEC) resins have iso- and terephthalate units incorporated into standard bisphenol A polycarbonate.[38] This modification of the polymer enhanced the performance between that of PC and polyarylates. Thus, PECs have properties similar to PC but with higher heat deflection temperature and better hydrolytic performance. Higher levels of iso- and terephthalate units result in higher heat deflection temperatures and continuous-use temperatures. Properties of PECs with different heat deflection temperatures appear in Table 27.

Uses   PEC is marketed into typical polycarbonate applications that require slightly higher heat deflection temperature and continuous-use temperatures, such as consumer goods, electrical/electronic, and automotive parts.

5.7   Polyarylates

The homopolymers of bisphenol A and isophthalic acid or terephthalic acids are semicrystalline.[42] The semicrystalline polyarylates (PARs) have a high crystalline melting point and very slow crystallization rates. Hence oil-heated molds and log cycle times would have made commercialization unattractive. Amorphous PARs are prepared from a mixture of isophthalic and terephthalic acids and bisphenol A and can be melt processed without difficulty. They are

Table 26   Typical Properties of PC/ABS Blends

Property	At Selected PC/ABS Ratio (wt/wt)			
	0/100	50/50	80/20	100/0
Density (g/cm^3)	1.06	1.13	1.17	1.20
Tensile modulus (GPa)	1.8	1.9	2.5	2.4
Tensile strength (MPa)	40	57	60	65
Elongation at break (%)	20	70	150	110
Notched Izod, 25°C (kJ/m)	0.30	0.69	0.75	0.86
Notched Izod, −20°C (kJ/m)	0.11	0.32	0.64	0.15
HDT at 1.81 MPa ($^\circ$C)	80	100	113	132

Table 27 Typical Properties of PECs

Property	PEC-1	PEC-2
Density (g/cm^3)	1.20	1.20
Tensile strength (MPa)	71.8	78.0
Elongation at break (%)	120	75
Flexural modulus (GPa)	2.03	2.33
Flexural strength (MPa)	95	96.6
Notched Izod (kJ/m)	0.534	0.534
Unnotched Izod (kJ/m)	3.20	3.22
HDT at 1.81 MPa (°C)	151	162
HDT at 0.42 MPa (°C)	160	174
RTI (°C)	125	130
Hardness (Rockwell)	R122	R127
CTE (10^{-5}/°C)	5.1	4.5
Linear mold shrinkage (in./in.)	0.007	0.009
Water absorption, 24 hr (%)	0.16	0.19
Relative permittivity at 1 MHz	3.19	3.45

clear, slightly yellow in color, dimensionally stable, and resistant to creep, have excellent electrical properties, are rigid, and have good impact strength.

PARs have poor chemical resistance to ketones, esters, and aromatic and chlorinated hydrocarbons. Typical properties appear in Table 28.

Uses PARs are marketed into applications requiring a higher heat deflection temperature than PC. These include electrical/electronic and automotive applications.

5.8 Modified Polyphenylene Ether

Poly(2,6-dimethylphenylene ether), or PPE, is produced by the oxidative coupling of 2,6-dimethyl phenol.[43–46] PPE is an amorphous thermoplastic. The PPE polymer has a very high heat deflection temperature, good inherent flame resistance, outstanding dimensional stability, and outstanding electrical properties. In addition, PPE has one of the lowest moisture absorption rates found in any engineering plastic.

PPE by itself is difficult to process; hence, PPE is commonly blended with styrenics (e.g., HIPS, ABS) to form a family of modified PPE-based resins.[44–46] What is truly unique about the PPE/PS blends is that the PPE and PS form a miscible, single-phased blend. Most polymers have limited solubility in other polymers. Modified PPE resins cover a wide range of compositions and properties.

Modified PPE resins are characterized by ease of processing, high impact strength, outstanding dimensional stability at elevated temperatures, long-term stability under load (creep resistance), and excellent electrical properties over a wide range of frequencies and temperatures. Another unique feature of modified PPE resins is their ability to make chlorine/bromine-free flame-retardant grades.

Modified PPE resins are especially noted for their outstanding hydrolytic stability—they do not contain any hydrolyzable bonds. Their low water absorption rates—both at ambient and elevated temperatures—promote the retention of properties and dimensional stability in the presence of water and high humidity and even in steam environments. In addition, modified PPE resins are also virtually unaffected by a wide variety of aqueous solutions—slats, detergents, acids, and bases. PPE is a versatile material and is used in alloys with PA to enhance the performance and decrease the moisture absorbance of the PA.

The chemical compatibility with oils and greases is limited. It is not recommended for contact with ketones, esters, toluene, and halogenated solvents. Typical properties appear in Table 29.

Table 28 Typical Properties of PARs

Property	PAR	PAR + 40% GF
Density (g/cm^3)	1.21	1.40
Tensile modulus (GPa)	2.0	8.28
Tensile strength (MPa)	66	149
Elongation at break (%)	50	0.8
Flexural modulus (GPa)	2.14	7.59
Flexural strength (MPa)	86	220
Notched Izod (kJ/m)	0.22	0.085
HDT at 1.81 MPa (°C)	174	—
Oxygen index (%)	34	—
UL94 VO (mm)	1.60	—
RTI (°C)	120	—
Hardness (Rockwell)	R100	R125
CTE (10^{-5}/°C)	5.6	3.0
Shrinkage (in./in.)	0.009	0.002
Water absorption (%)	0.27	0.18
Relative permittivity at 1 MHz	2.62	3.8

Uses Applications include automotive (instrument panels, trim, spoilers, under-the-hood components, grills), telecommunication equipment (TV cabinets, cable splice boxes, wire board frames, electrical connectors, structural and interior components in electrical/electronic equipment), plumbing/water handling (pumps, plumbing fixtures), consumer goods (microwavable food packaging, appliance parts), medical, and building and construction.

6 HIGH-PERFORMANCE MATERIALS

High-performance resins arbitrarily comprise the high end of engineering plastics and offer premium performance. Typically high-performance resins will be used in tough metal replacement applications or replacement of ceramic materials. Such materials offer greater resistance to heat and chemicals, high RTIs, high strength and stiffness, and inherent flame resistance. These resins have higher cost and the processing can be more challenging.

Table 29 Typical Properties of Modified PPE Resins

Property	190 Grade	225 Grade	300 Grade
Density (g/cm^3)	1.12	1.11	1.12
Tensile modulus (GPa)	2.70	—	—
Tensile strength (MPa)	61	67	76
Elongation at break (%)	18	17	20
Flexural modulus (GPa)	2.35	2.49	2.50
Flexural strength (MPa)	93	99	110
Notched Izod (kJ/m)	0.029	0.019	0.023
Unnotched Izod (kJ/m)	0.721	—	—
HDT at 1.81 MPa (°C)	88	99	145
HDT at 0.42 MPa (°C)	96	109	156
Vicat softening temperature (°C)	113	129	—
RTI (°C)	95	95	105
UL94 VO (mm)	1.5	1.5	1.5
Oxygen index (%)	39	—	—
Hardness (Rockwell)	R120	—	R119
CTE (10^{-5}/°C)	8.8	—	5.4
Shrinkage (in. / in.)	0.006	0.006	0.006
Water absorption, 24 hr (%)	0.08	—	0.06
Relative permittivity at 1 MHz	2.60	2.55	2.63
Dissipation factor at 1 MHz	0.0055	0.007	0.009

6.1 Polyphenylene Sulfide

The condensation polymerization of 1,4-dichlorobenzene and sodium sulfide yields a semicrystalline polymer, polyphenylene sulfide (PPS).[47] It is characterized by high heat resistance, rigidity, excellent chemical resistance, dimensional stability, low friction coefficient, good abrasion resistance, and electrical properties. PPS has good mechanical properties, which remain stable during exposure to elevated temperatures. Water absorption for PPS is very low and hydrolytic stability is very high. PPS resins are inherently flame resistant.

PPS has excellent chemical resistance to a wide range of chemicals. Indeed, even at elevated temperatures PPS can withstand exposure to a wide range of chemicals, such as mineral and organic acids and alkali. However, it is attacked by chlorinated hydrocarbons.

Depending on the polymerization process, PPS can be a linear or branched polymer. The branched polymer is somewhat more difficult to process due to the very high melting temperature and relatively poor flow characteristics.

PPS resins normally contain glass fibers or mineral fillers. Properties appear in Table 30.

Uses Applications for PPS resins include industrial (parts requiring heat and chemical resistance, submersible, vane, and gear-type pump components), electrical/electronic (high-voltage electrical components), automotive (electrical connectors, under-the-hood components), appliance parts (hair dryers, small cooking appliances, range components), medical (hydraulic components, bearing cams, valves), and aircraft/aerospace.

6.2 Polyarylsulfones

Polyarylsulfones are a class of amorphous, high-end-use-temperature thermoplastics that characteristically exhibit excellent thermo-oxidative stability, good solvent resistance, creep resistance, transparency, and high heat deflection temperatures.[48–50] Polyarylsulfones have excellent hydrolytic stability even in saturated steam; they can be repeatedly steam sterilized. Typically the polyarylsulfone families of resins have low resistance to weathering and are degraded by UV light. There are three major categories of aromatic polyarylsulfone resins: polysulfone (PSU), polyphenylsulfone (PPSU), and polyethersulfone (PES).

PSU is prepared by nucleophilic aromatic displacement of the chloride on bis(p-chlorophenyl)sulfone by the anhydrous disodium salt of bisphenol A. This amorphous polymer has a T_g of 190°C.

PPSU is higher performing than PSU. This wholly aromatic resin is prepared from biphenol and bis(p-chlorophenyl)sulfone.[50] The lack of aliphatic groups in this all-aromatic polymer and the presence of the biphenyl moiety impart enhanced chemical/solvent resistance, outstanding toughness, greater resistance to combustion, greater thermo-oxidative stability, and a T_g of 220°C. PPSU has excellent resistance to mineral acids, caustic, salt solution, and various automotive fluids. Exposure to esters, ketones, and polar aromatic solvents should be avoided.

PES consists of a diphenyl sulfone unit linked through an ether (oxygen) unit. Again the lack of aliphatic groups results in higher thermo-oxidative stability. PES offers high heat (T_g of 230°C), chemical/solvent resistance, and improved toughness over PSU. PES is chemically resistant to most inorganic

Table 30 Typical Properties of PPS

Property	Branched	Branched + 40% GF	Linear	Linear + 40% GF
Density (g/cm^3)	1.35	1.60	1.35	1.65
Tensile modulus (GPa)	—	14.5	4.0	15.7
Tensile strength (MPa)	65	150	66	150
Elongation at break (%)	2	1.2	12	1.7
Flexural modulus (GPa)	3.85	15.0	3.90	15.0
Flexural strength (MPa)	104	153	130	230
Notched Izod (kJ/m)	0.080	0.578	0.139	0.241
Unnotched Izod (kJ/m)	0.107	0.482	0.167	0.589
HDT at 1.81 MPa (°C)	115	>260	115	265
RTI (°C)	220	—	220	—
UL94 VO (mm)	—	0.8	—	0.6
Oxygen index (%)	44	46.5	44	47
Hardness (Rockwell)	R120	R123	M95	M100
CTE (10^{-5}/°C)	4.9	4.0	5.3	4.1
Shrinkage (in./in.)	—	0.004	0	0.004
Water absorption, 24 hr (%)	—	0.03	—	0.03
Relative permittivity at 1 MHz	—	3.9	—	4.7
Dissipation factor at 1 MHz	—	0.0014	—	0.020

Table 31 Typical Properties of Polyarylsulfones

Property	PSU	PES	PPSU
Density (g/cm^3)	1.24	1.37	1.29
Tensile modulus (GPa)	2.48	2.41	2.35
Tensile strength (MPa)	69	82	70
Elongation at break (%)	75	50	60
Flexural modulus (GPa)	2.55	2.55	2.42
Flexural strength (MPa)	102	110	91
Notched Izod (kJ/m)	0.080	0.075	0.694
Unnotched Izod, −40°C (kJ/m)	0.064	—	0.425
HDT at 1.81 MPa (°C)	174	203	204
HDT at 0.42 MPa (°C)	180	210	—
Vicat softening temperature (°C)	188	226	—
RTI (°C)	160	180	180
UL94 VO (mm)	—	0.46	0.58
Oxygen index (%)	—	38	—
Hardness (Rockwell)	M69	M88	—
CTE (10^{-5}/°C)	5.1	5.5	5.5
Shrinkage (in./in.)	0.005	0.006	0.006
Water absorption, 24 hr (%)	0.3	0.43	0.37
Relative permittivity at 1 MHz	3.19	3.45	3.5
Dissipation factor at 1 MHz	—	0.0076	—

gasoline. However, esters, ketones, methylene chloride, and polar aromatic solvents attack PES.

Typical properties of PSU, PPSU, and PES appear in Table 31.

Uses Typical applications of polyarylsulfones include medical/laboratory (surgical equipment, laboratory equipment, life support parts, autoclavable tray systems, suction bottles, tissue culture bottles, surgical hollow shapes), food handling (microwave cookware, coffee makers, hot-water and food-handling equipment, range components), electrical/electronic (components, multipin connectors, coil formers, printed circuit boards), chemical processing equipment (pump housings, bearing cages), and miscellaneous (radomes, alkaline battery cases).

6.3 Liquid Crystalline Polyesters

Liquid crystalline polymers (LCPs) have a rigid rod-like aromatic structure. The rodlike molecules arrange themselves in parallel domains in both the melt and solid states. In the molten state the molecules readily slide over one another, giving the resin very high flow under shear. Most commercially important polyester LCPs are based on *p*-hydroxybenzoic acid (HBA). An example would be the copolyester of HBA and hydroxynaphthanoic acid (HNA) with a molar ratio of 73/27.

LCPs are highly crystalline, thermotropic (melt-orienting) thermoplastics. Because of the possibility of melt orientation during molding, LCPs can have anisotropic properties; that is, properties can differ in the direction of flow and perpendicular to the flow direction. LCPs offer high strength, rigidity, dimensional stability, inherent flame resistance, and high heat resistance. LCPs have high flow and can deliver exceptionally precise and stable dimensions in very thin walled applications.

LCPs are resistant to weathering, burning, γ-radiation, steam autoclaving, and most chemical sterilization methods. They have outstanding strength at elevated temperatures.

LCPs are resistant to virtually all chemicals, including acids, organic hydrocarbons, and boiling water. It is attacked by concentrated, boiling caustic but is unaffected by milder solutions. Typical properties of the HBA–HNA LCPs appear in Table 32.

Uses Typical applications of LCPs include electrical/electronic (stator insulation, rotors, boards for motors, burn-in sockets, interface connectors, bobbins, switches, chip carriers and sensors), medical (surgical instruments, needleless syringes, dental tools, sterilizable trays and equipment, drug delivery systems and diagnostics), industrial (chemical process and oil field

equipment), and packaging (food packaging requiring barrier properties).

6.4 Polyimides

Polyimides are a class of polymers prepared from the condensation reaction of a carboxylic acid anhydride with a diamine.[51] Polyimides are among the most heat-resistant polymers. Poly(pyromellitimide-1,4-diphenyl ether) is prepared from pyromellitic anhydride (PMDA) and 4, 4′-oxydianiline (ODA). PMDA–ODA has a T_g of 360°C or higher. This very high T_g does not lend itself to standard melt-processing techniques. PMDA–ODA resins are available as films and direct-formed parts from formulated resin. PMDA–ODA has excellent thermal stability, useful mechanical properties over a very wide temperature range, creep resistance, high toughness, and excellent wear resistance. PMDA–ODA had excellent electrical properties to a broad range of chemicals. However, it will be attacked by 10% and stronger solutions of sodium hydroxide. Typical properties appear in Table 33.

Uses PMDA–ODA films are used as wire and cable wrap, motor-slot liners, flexible printed circuit boards, transformers, and capacitors. Molded parts are used in application requiring resistance to thermally harsh environments such as automotive transmission parts, thermal and electrical insulators, valve seats, rotary seal rings, thrust washers and discs, bushings, and the like.

Table 32   Typical Properties of LCP Resins (HBA–HNA)

Property	LCP	LCP + 15% GF	LCP + 30% GF
Density (g/cm^3)	1.40	1.50	1.62
Tensile modulus (GPa)	10.6	12.0	15.0
Tensile strength (MPa)	182	190	200
Elongation at break (%)	3.4	3.1	2.1
Flexural modulus (GPa)	9.10	12.0	15.0
Flexural strength (MPa)	158	240	280
Notched Izod (kJ/m)	0.75	0.35	0.17
Unnotched Izod (kJ/m)	2.0	0.48	0.23
HDT at 0.45 MPa (°C)	—	250	250
HDT at 1.81 MPa (°C)	187	230	235
Vicat softening point (°C)	145	162	160
Oxygen index (%)	—	—	45
Hardness (Rockwell)	—	M80	M85
CTE (10^{-5}/°C)			
Parallel to flow	0.4	1.0	0.6
Perpendicular to flow	3.62	1.8	2.3
Shrinkage (in./in.)			
Parallel to flow	0.0	0.001	0.002
Perpendicular to flow	0.007	0.004	0.004
Water absorption, 24 hr (%)	0.03	—	—
Relative permittivity at 1 MHz	3.0	3.0	3.7
Dissipation factor at 1 MHz	0.02	0.018	0.018

Table 33   Typical Properties of Polyimide

Property	PMDA–ODA	15% Graphite	40% Graphite
Density (g/cm^3)	1.43	1.51	1.65
Tensile strength (MPa)	86.2	65.5	51.7
Elongation at break (%)	7.5	4.5	3.0
Flexural modulus (GPa)	3.10	3.79	4.83
Flexural strength (MPa)	110	110	90
Notched Izod (kJ/m)	0.043	0.043	—
Unnotched Izod (kJ/m)	0.747	0.320	—
HDT at 2 MPa ($^\circ$C)	360	360	—
Oxygen index (%)	53	49	—
Hardness (Rockwell)	—	M80	M85
CTE ($10^{-5}/^\circ$C)	5.4	4.9	3.8
Water absorption, 24 hr (%)	0.24	0.19	0.14
Relative permittivity at 1 MHz	3.55	13.4	—
Dissipation factor at 1 MHz	0.0034	0.0106	—

6.5   Polyetherimides

Polyetherimides (PEIs) are prepared from the polymerization of aromatic diamines and etherdianhydrides.[52,53] PEI is an amorphous thermoplastic that contains repeating aromatic imide and ether units. The rigid aromatic imide units provide PEI with its high-performance properties at elevated temperatures, while the ether linkages provide it with the chain flexibility necessary to have good melt flow and processability by injection molding and other melt-processing techniques.

PEI resins offer high strength, rigidity, high heat resistance, inherent flame resistance, low smoke generation, excellent processability with very tight molding tolerances, heat resistance up to 200°C, RTI of 170°C, excellent dimensional stability (low creep sensitivity and low coefficient of thermal expansion), superior torque strength and torque retention and stable dielectric constant and dissipation factor over a wide range of temperatures and frequencies.

PEIs have excellent hydrolytic stability, UV stability, and radiation resistance. They are well suited for repeated steam, hot air, ethylene oxide gas, and cold chemical sterilizations. In addition, PEIs have proven property retention through over 1000 cycles in industrial washing machines with detergents and compliancy with Food and Drug Administration (FDA), European Union (EU), and national food contact regulations.

In addition PEI resins are resistant to a wide range of chemicals, including alcohols, hydrocarbons, aqueous detergents and bleaches, strong acids, mild bases, and most automotive fuels, fluids, and oils. Chlorinated hydrocarbons and aromatic solvents can attack PEI resins. Typical properties appear in Table 34.

Uses   Automotive (transmission components, throttle bodies, ignition components, sensors, thermostat housings), automotive lighting (headlight reflectors, fog light reflectors, bezels, light bulb sockets), telecommunications (molded interconnect devices, electrical control units, computer components, mobile phone internal antennae, radio frequency (RF) duplexers or microfilters, fiber-optic connectors), electrical (lighting, connectors to reflectors), heating, ventilation, and air conditioning (HVAC)/fluid handling (water pump impellers, expansion valves, hot-water reservoirs, heat exchange systems), tableware/catering (food trays, soup mugs, steam insert pans or gastronome containers, cloches, microwavable bowls, ovenware, cooking utensils, reusable airline casseroles), medical (reusable medical devices such as sterilization trays, stopcocks, dentist devices, and pipettes), and aircraft (air and fuel valves, food tray containers, steering wheels, interior cladding parts, semistructural components).

6.6   Polyamide Imides

Polyamide imides (PAIs) are prepared from trimellitic anhydride and various aromatic diamines.[54] PAIs are tough, high-modulus thermoplastics capable of high continuous-use temperatures. In addition, they have inherent flame resistance and very high strength. PAIs have very broad chemical resistance but are attacked by aqueous caustic and amines. PAIs have high melt temperature and require very high processing temperatures. PAIs contain an amide group and hence will absorb moisture, which can affect dimensional stability. Properties appear in Table 35.

Uses   Automotive (housings, connectors, switches, relays, thrust washers, valve seats, bushings, wear rings, rollers, thermal insulators), composites (printed wiring boards, radomes), and aerospace (replacement of metal parts).

6.7   Aromatic Polyketones

Aromatic polyketones are a family of semicrystalline high-performance thermoplastics that consist of a combination of ketone, ether, and aromatic units.[55] The absence of aliphatic groups results in outstanding high-temperature performance, exceptional thermal stability, excellent environmental resistance, high

Table 34 Typical Properties of PEIs

Property	PEI	10% GF + PEI	20% GF + PEI	30% GF + PEI
Density (g/cm³)	1.27	1.34	1.42	1.51
Tensile modulus (GPa)	3.59	4.69	6.89	9.31
Tensile strength (MPa)	110	116	131	169
Elongation at break (%)	60	6	4	3
Flexural modulus (GPa)	3.52	5.17	6.89	8.96
Flexural strength (MPa)	165	200	228	228
Notched Izod (kJ/m)	0.053	0.053	0.064	0.085
Unnotched Izod (kJ/m)	1.335	0.481	0.481	0.427
HDT at 1.81 MPa (°C)	201	209	210	210
HDT at 0.42 MPa (°C)	210	210	210	212
Vicat softening temperature (°C)	219	223	220	228
RTI (°C)	170	170	170	180
UL94 VO (mm)	—	0.4	0.4	0.3
Oxygen index (%)	47	47	50	50
Hardness (Rockwell)	M109	M114	M114	M114
CTE (10^{-5}/°C)	5.5	3.24	2.52	1.98
Shrinkage (in./in.)	0.006	0.005	0.004	0.003
Water absorption, 24 hr (%)	0.25	0.21	0.19	0.16
Dissipation factor at 1 kHz	0.0012	0.0014	0.0015	0.0015
Relative permittivity at 1 kHz	3.15	3.5	3.5	3.7

Table 35 Typical Properties of PAIs

Property	PAI	PAI + 30% GF
Density (g/cm³)	1.38	1.42
Tensile modulus (GPa)	5.2	20.0
Tensile strength (MPa)	117	207
Elongation at break (%)	15	4
Flexural modulus (GPa)	3.59	17.9
Flexural strength (MPa)	189	316.7
Notched Izod (kJ/m)	0.136	0.07
Unnotched Izod (kJ/m)	1.088	—
HDT at 1.81 MPa (°C)	254	275
RTI (°C)	—	220
Hardness (Rockwell)	E78	E94
Oxygen index (%)	41	49
CTE (10^{-5}/°C)	3.6	4.9
Relative permittivity at 1 MHz	4.0	3.8
Dissipation factor at 1 MHz	0.009	—
Water absorption, 24 hr (%)	1.0	0.8

mechanical properties, resistance to chemical environments at elevated temperatures, inherent flame resistance, excellent friction and wear resistance, and impact resistance. In addition, aromatic polyketones have high chemical purity suitable for applications in silicon chip manufacturing.

There are two major aromatic polyketone resins: polyetherketone (PEK) and polyetheretherketone (PEEK). Typical properties appear in Table 36.

Uses Applications are in the chemical process industry (compressor plates, valve seats, pump impellers, thrust washers, bearing cages), aerospace (aircraft fairings, radomes, fuel valves, ducting),

and electrical/electronic (wire coating, semiconductor wafer carriers).

7 FLUORINATED THERMOPLASTICS

Fluoropolymers or fluoroplastics are a family of fluorine-containing thermoplastics that exhibit some unusual properties.[56,57] These properties include inertness to most chemicals, resistance to high temperatures, extremely low coefficient of friction, weather resistance, and excellent dielectric properties. Mechanical properties are normally low but can be enhanced with glass or carbon fiber or molybdenum disulfide fillers. Properties are shown in Table 37.

7.1 Poly(tetrafluoroethylene)

Poly(tetrafluoroethylene) (PTFE) is a highly crystalline polymer produced by the polymerization of tetrafluoroethylene. PTFE offers very high heat resistance (up to 250°C), exceptional chemical resistance, and outstanding flame resistance.[56] The broad chemical resistance includes strong acids and strong bases. In addition, PTFE has the lowest coefficient of friction of any polymer.

PTFE has a high melting point and extremely high melt viscosity and hence cannot be melt processed by normal techniques. Therefore PTFE has to be processed by unconventional techniques (PTFE powder is compacted to the desired shape and sintered).

Uses Typically PTFE is used in applications requiring long-term performance in extreme-service environments. PTFE applications include electrical (high-temperature, high-performance wire and cable insulation, sockets, pins, connectors), mechanical

Table 36 Typical Properties of Aromatic Polyketones

Property	PEEK	PEEK + 30% GF	PEEK + 30% Carbon Fiber	PEK	PEK + 30% GF
Density (g/cm^3)	1.32	1.49	1.44	1.30	1.53
Tensile modulus (GPa)	3.6	9.7	13.0	4.0	10.5
Tensile strength (MPa)	192	157	208	104	160
Elongation at break (%)	50	2.2	1.3	5	4
Flexural modulus (GPa)	3.7	10.3	13.0	3.7	9.0
Flexural strength (MPa)	170	233	318	—	—
Notched Izod (kJ/m)	0.08	0.1	0.09	0.08	0.1
Unnotched Izod (kJ/m)	—	0.73	0.75	—	—
HDT at 1.81 MPa (JC)	160	315	315	165	340
HDT at 0.42 MPa (°C)	—	—	315	—	—
RTI (°C)	—	—	—	260	260
UL94 VO (mm)	—	—	—	1.6	1.6
Oxygen index (%)	—	—	—	40	46
Hardness (Rockwell)	M126	R124	R124	R126	R126
CTE (10^{-5}/°C)	16	11.5	7.9	5.7	1.7
Shrinkage (in./in.)	0.004	0.005	0.001	0.006	0.004
Water absorption, 24 hr (%)	0.5	0.11	0.06	0.11	0.08
Relative permittivity at 1 MHz	3.2	3.7	—	3.4	3.9

Table 37 Typical Properties of Fluoropolymers

Property	PTFE	PCTFE	FEP	PVDF	ECTFE
Density (g/cm^3)	2.29	2.187	2.150	1.77	1.680
Tensile modulus (GPa)	4.10	14.0	—	1.194	—
Tensile strength (MPa)	27.6	40	20.7	44	48.3
Elongation at break (%)	~275	150	~300	43	200
Flexural modulus (GPa)	5.2	1.25	—	—	—
Flexural strength (MPa)	—	74	—	—	—
Notched Izod (kJ/m)	0.19	0.27	0.15	—	—
HDT at 0.45 MPa (JC)	132	126	—	104	116
HDT at 1.81 MPa (°C)	60	75	—	71	77
RTI (°C)	260	199	204	—	150–170
Hardness	D42 (Shore)	D77 (Shore)	D55 (Shore)	D75 (Shore)	R93 (Rockwell)
Dielectric constant at 10^2 Hz	2.1	3.0	2.1	9	2.5
Dissipation factor at 10^3 Hz	0.0003	—	—	0.03	—
CTE (10^{-5}/°C)	12.0	4.8	9.3	11	—

(bushings, rider rings, seals, bearing pads, valve seats, chemical resistance processing equipment and pipe, nonlubricated bearings, pump parts, gaskets, and packings), nonstick coatings (home cookware, tools, food-processing equipment), and miscellaneous (conveyor parts, packaging, flame-retardant laminates).

7.2 Poly(chlorotrifluoroethylene)

Poly(chlorotrifluoroethylene) (PCTFE) is less crystalline and exhibits higher rigidity and strength than PTFE.[57] Poly(chlorotrifluoroethylene) has excellent chemical resistance and heat resistance up to 200°C. Unlike PTFE, PCTFE can be molded and extruded by conventional processing techniques.

Uses PCTFE applications include electrical/electronic (insulation, cable jacketing, coilforms), industrial (pipe and pump parts, gaskets, seals, diaphragms, coatings for corrosive process equipment and in cryogenic systems), and pharmaceutical packaging.

7.3 Fluorinated Ethylene–Propylene

Copolymerization of tetrafluoroethylene with some hexafluoropropylene produces fluorinated ethylene–propylene polymer (FEP), which has less crystallinity, lower melting point, and improved impact strength than PTFE. FEP can be molded by normal thermoplastic techniques.[57]

Uses FEP applications include wire insulation and jacketing, high-frequency connectors, coils, gaskets, and tube sockets.

7.4 Polyvinylidene Fluoride

Polyvinylidene fluoride (PVDF) is made by emulsion and suspension polymerization and has better ability to be processed but less thermal and chemical resistance than FEP, CTFE, and PTFE.[57] PVDF has excellent mechanical properties and resistance to severe environmental stresses and good chemical resistance.

Uses Polyvinylidene fluoride applications include electronic/electrical (wire and cable insulation, films), fluid handling (solid and lined pipes, fittings, valves, seals and gaskets, diaphragms, pumps, microporous membranes), and coatings (metal finishes, exterior wall panels, roof shingles).

7.5 Poly(ethylene chlorotrifluoroethylene)

The copolymer of ethylene and chlorotrifluoroethylene (ECTFE) has high strength and chemical and impact resistance. ECTFE can be processed by conventional techniques.

Uses Poly(ethylene chlorotrifluoroethylene) applications include wire and cable coatings, chemical-resistant coatings and linings, molded laboratory ware, and medical packing.

7.6 Poly(vinyl fluoride)

Poly(vinyl fluoride) (PVF) is the least chemical resistant fluoropolymer.[57] PVF offers weathering and UV resistance, antisoiling, durability, and chemical resistance.

Uses Poly(vinyl fluoride) uses include protective coatings (aircraft interior, architectural fabrics, presurfaced exterior building panels, wall coverings, glazing, lighting).

8 THERMOSETS

Thermosetting resins are used in molded and laminated plastics.[58] They are first polymerized into a low-molecular-weight linear or slightly branched polymer or oligomers, which are still soluble, fusible, and highly reactive during final processing. Thermoset resins are generally highly filled with mineral fillers and glass fibers. Thermosets are generally catalyzed and/or heated to finish the polymerization reaction, crosslinking them to almost infinite molecular weight. This step is often referred to as curing. Such cured polymers cannot be reprocessed or reshaped.

The high filler loading and the high crosslink density of thermoset resins result in high densities and low ductility but high rigidity and good chemical resistance.

8.1 Phenolic Resins

Phenolic resins combine the high reactivity of phenol and formaldehyde to form prepolymers and oligomers called resoles and novolacs. These materials can be used as adhesives and coatings or combined with fibrous fillers to give phenolic resins, which when heated undergo a rapid, complete crosslinking to give a highly cured structure. The curing typically requires heat and pressure. Fillers include cellulosic materials and glass and mineral fibers/fillers. The high-crosslinked aromatic structure has high hardness, dimensional stability, rigidity, strength, heat resistance, chemical resistance, and good electrical properties.

Phenolic resins have good chemical resistance to hydrocarbons, phenols, and ethers. However, they can undergo chemical reaction with acids and bases.

Uses Phenolic applications include appliance parts (handles, knobs, bases, end panels), automotive uses (parts in electric motors, brake linings, fuse blocks, coil towers, solenoid covers and housings, ignition parts), electrical/electronic (high-performance connectors and coil bobbins, circuit breakers, terminal switches and blocks, light sockets, receptacles), and miscellaneous (adhesives in laminated materials such as plywood, coatings).

8.2 Epoxy Resins

Epoxy resins are materials whose molecular structure contains an epoxide or oxirane ring. The most common epoxy resins are prepared from the reaction of bisphenol A and epichlorohydrin to yield low-molecular-weight resins that are liquid either at room temperature or on warming. Other epoxy resins are available—these include novalac and cycloaliphatic epoxy resins. Each polymer chain usually contains two or more epoxide groups. The high reactivity of the epoxide groups with aromatic and aliphatic amines, anhydrides, carboxylic acids, and other curing agents provides facile conversion into highly crosslinked materials. The large number of variations possible in epoxy structure and curing agent coupled with their ability to be formulated with a wide variety of fillers and additives results in broad application potential. Cured epoxy resins exhibit hardness, strength, heat resistance, good electrical properties, and broad chemical resistance.

Most epoxies are resistant to numerous chemicals, including hydrocarbons, esters, and bases. However, epoxies can exhibit poor resistance to phenols, ketones, ethers, and concentrated acids.

Uses Epoxy resins are used in composites (glass-reinforced, high-strength composites in aerospace, pipes, tasks, pressure vessels), coatings (marine coatings, protective coatings on appliances, chemical scrubbers, pipes), electrical/electronic (encapsulation or casting of various electrical and electronic components, printed wiring boards, switches, coils, insulators, bushings), and miscellaneous (adhesives, solder mask, industrial equipment, sealants).

8.3 Unsaturated Polyesters

Thermoset polyesters are prepared by the condensation polymerization of various diols and maleic anhydride to give a very viscous unsaturated polyester oligomer that is dissolved in styrene monomer. The addition of styrene lowers the viscosity to a level suitable for impregnation and lamination of glass fibers. The low-molecular-weight polyester has numerous maleate/fumarate ester units that provide facile reactivity with styrene monomer. Unsaturated polyesters can be prepared from a variety of different monomers.

Properly formulated glass-reinforced unsaturated polyesters are commonly referred to as sheet-molding compound (SMC) or reinforced plastics. The combination of glass fibers in the cured resins offers outstanding strength, high rigidity, impact resistance, high strength-to-weight ratio, and chemical resistance. SMC typically is formulated with 50% calcium carbonate filler, 25% long glass fiber, and 25% unsaturated polyester. The highly filled nature of SMC results in a high density and a brittle, easily pitted surface.

Bulk-molding compounds (BMCs) are formulated similar to SMCs except $\frac{1}{4}$-in. chopped glass is typically used. The shorter glass length gives easier processing but lower strength and impact.

Thermoset polyesters have good chemical resistance, which includes exposure to alcohols, ethers, and organic acids. They exhibit poor resistance to hydrocarbons, ketones, esters, phenols, and oxidizing acids.

Uses The prime use of unsaturated polyesters is in combination with glass fibers in high-strength composites and in SMC and BMC materials. The applications include transportation markets (large body parts for automobiles, trucks, trailers, buses, and aircraft, automotive ignition components), marine markets (small- to medium-sized boat hulls and associated marine equipment), building and construction (building panels, housing and bathroom components—bathtub and shower stalls), and electrical/electronic (components, appliance housing, switch boxes, breaker components, encapsulation).

8.4 Vinyl Esters

Vinyl esters are part of the unsaturated polyester family. They are prepared by the reaction of an epoxy resin with methacrylic acid. Thus the epoxide group is converted into a methacrylate ester. Vinyl esters offer an enhancement in properties over unsaturated polyesters with greater toughness and better resistance to corrosion to a wide range of chemicals. This chemical resistance includes halogenated solvents, acids, and bases.

Uses Applications for vinyl esters are similar to those for unsaturated polyesters but where added

toughness and chemical resistance are required, that is, electrical equipment, flooring, fans, adsorption towers, process vessels, and piping.

8.5 Alkyd Resins

Alkyd resins are unsaturated polyesters based on branched prepolymers from polyhydridic alcohols (e.g., glycerol, pentaerythritol, ethylene glycol), polybasic acids (e.g., phthalic anhydride, maleic anhydride, fumaric acid), and fatty acids and oils (e.g., oleic, stearic, palmitic, linoleic, linolenic, lauric, and licanic acids). They are well suited for coating with their rapid drying, good adhesion, flexibility, mar resistance, and durability. Formulated alkyd molding resins have excellent heat resistance, are dimensionally stable at high temperatures and have excellent dielectric strength, high resistance to electrical leakage, and excellent arc resistance.

Alkyl resins can be hydrolyzed under alkaline conditions.

Uses Alkyd resin applications include coatings (drying oils in paints, enamels, and varnish) and molding compounds when formulated with reinforcing fillers for electrical applications (circuit breaker insulation, encapsulation of capacitors and resistors, coil forms).

8.6 Diallyl Phthalate

Diallyl phthalate (DAP) is the most widely used compound in the allylic ester family. The neat resin is a medium-viscosity liquid. These low-molecular-weight prepolymers can be reinforced and compression molded into highly crosslinked, completely cured products.

The most outstanding properties of DAP are excellent dimensional stability and high insulation resistance. In addition, DAP has high dielectric strength and excellent arc resistance. DAP has excellent resistance to aliphatic hydrocarbons, oils, and alcohols but is not recommended for use with phenols and oxidizing acids.

Uses. DAP applications include electrical/electronic (parts, connectors, bases and housings, switches, transformer cases, insulators, potentiometers) and miscellaneous (tubing, ducting, radomes, junction bases, aircraft parts). DAP is also used as a coating and an impregnating resin.

8.7 Amino Resins

The two main members of the amino resin family of thermosets are the melamine–formaldehyde and urea–formaldehyde resins. They are prepared from the condensation reaction of melamine and urea with formaldehyde. In general, these materials exhibit extreme hardness, inherent flame resistance, heat resistance, scratch resistance, arc resistance, chemical resistance, and light fastness.

Table 38 Properties of General-Purpose Elastomers

Rubber	ASTM[a] Nomenclature	Outstanding Characteristic	Property Deficiency	Temperature Use Range (°C)
Butadiene rubber	BR	Very flexible; resistance to wear	Sensitive to oxidation; poor resistance to fuels and oil	−100–90
Natural rubber	NR	Similar to BR but less resilient	Similar to BR	−50–80
Isoprene rubber	IR	Similar to BR but less resilient	Similar to BR	−50–80
Isobutylene–isoprene rubber (butyl rubber)	IIR	High flexibility; low-permeability air		−45–150
Chloroprene	CR	Flame resistant; fair fuel and oil resistance; increased resistance to oxygen, ozone, heat, and light	Poor low-temperature flexibility	−40–115
Nitrile–butadiene	NBR	Good resistance to fuels, oils, and solvents; improved abrasion resistance	Lower resilience; higher hysteresis; poor electrical properties; poorer low-temperature flexibility	−45–80
Styrene–butadiene rubber	SBR	Relatively low cost	Less resilience; higher hysteresis; limited low-temperature flexibility	−45–80
Ethylene–propylene copolymer	EPDM	Resistance to ozone and weathering	Poor hydrocarbon and oil resistance	−50 to <175
Polysulfide	T	Chemical resistance; resistance to ozone and weathering	Creep; low resilience	−45–120

[a]American Society for Testing and Materials.

Table 39 Properties of Specialty Elastomers

Elastomer	ASTM Nomenclature	Temperature Use Range (°C)	Outstanding Characteristic	Typical Applications
Silicones (polydimethylsiloxane)	MQ	−100–300	Wide temperature range; resistance to aging, ozone, and sunlight; very high gas permeability	Seals, molded and extruded goods; adhesive, sealants; biomedical; personal care products
Fluoroelastomers	CFM	−40–200	Resistance to heat, oils, and chemicals	Seals such as O-rings, corrosion-resistant coatings
Acrylic	AR	−40–200	Oil, oxygen, ozone, and sunlight resistance	Seals, hose
Epichlorohydrin	ECO	−18–150	Resistance to oil and fuels; some flame resistance; low gas permeability	Hose, tubing, coated fabrics, vibration isolators
Chlorosulfonated	CSM	−40–150	Resistance to oil, ozone weathering, and oxidizing chemicals	Automotive hose, wire and cable, linings for reservoirs
Chlorinated polyethylene	CM	−40–150	Resistance to oils, ozone, chemicals	Impact modifier, automotive applications
Ethylene–acrylic	—	−40–175	Resistance to ozone and weathering	Seals, insulation, vibration damping
Propylene oxide	—	−6–150	Low-temperature properties	Motor mounts

Uses Melamine–formaldehyde resins find use in colorful, rugged dinnerware, decorative laminates (countertops, tabletops, furniture surfacing), electrical applications (switchboard panels, circuit breaker parts, arc barriers, armature and slot wedges), and adhesives and coatings.

Urea–formaldehyde resins are used in particleboard binders, decorative housings, closures, electrical parts, coatings, and paper and textile treatment.

9 GENERAL-PURPOSE ELASTOMERS

Elastomers are polymers that can be stretched substantially beyond their original length and can retract rapidly and forcibly to essentially their original dimensions (on release of the force).[59,60]

The optimum properties and/or economics of many rubbers are obtained through formulating with reinforcing agents, fillers, extending oils, vulcanizing agents, antioxidants, pigments, and the like. End-use markets for formulated rubbers include automotive tire products (including tubes, retread applications, valve stems, and inner liners), adhesives, cements, caulks, sealants, latex foam products, hose (automotive, industrial, and consumer applications), belting (V-conveyor and trimming), footwear (heels, soles, slab stock, boots, and canvas), and molded, extruded, and calendered products (athletic goods, flooring, gaskets, household products, O-rings, blown sponge, thread, and rubber sundries). A summary of general-purpose elastomers and properties is provided in Table 38.

10 SPECIALTY ELASTOMERS

Specialty rubbers offer higher performance over general-purpose rubbers and find use in more demanding applications.[61] They are more costly and hence are produced in smaller volume. Properties and uses are summarized in Table 39.

ACKNOWLEDGMENT

The author wishes to thank John Wiley & Sons for permission to adapt the content on pages 115–129 from *Mechanical Engineer's Handbook*, 2nd ed., 1988, edited by M. Kutz.

REFERENCES

1. Peters, E. N., "Plastics: Thermoplastics, Thermosets, and Elastomers," in *Handbook of Materials Selection*, M. Kutz (Ed.), Wiley-Interscience, New York, 2002, Chapter 11, pp. 335–355.
2. Fox, D. W., and Peters, E. N., "Engineering Thermoplastics: Chemistry and Technology," in *Applied Polymer Science*, 2nd ed., R. W. Tess and G. W. Poehlein (Eds.), American Chemical Society, Washington, DC, 1985, Chapter 21, pp. 495–514.
3. Peters, E. N., and Arisman, R. K., "Engineering Thermoplastics," in *Applied Polymer Science—21st Century*, C. D. Craver and C. E. Carraher (Eds.), Elsevier, New York, 2000, pp. 177–196.
4. Peters, E. N., "Inorganic High Polymers," in *Kirk-Othmer Encyclopedia of Chemical Technology*, 3rd ed., Vol. 13, M. Grayson (Ed.), Wiley, New York, 1981, pp. 398–413.
5. Peters, E. N., "Introduction to Polymer Characterization," in *Comprehensive Desk Reference of Polymer Characterization and Analysis*, R. F. Brady, Jr. (Ed.), Oxford University Press, New York, 2003, Chapter 1, pp. 3–29.
6. Goodman, S. H. (Ed.), *Handbook of Thermoset Plastics*, 2nd ed., Plastics Design Library, Brookfield, CT, 1999.
7. Olabisi, O. (ed.), *Handbook of Thermoplastics*, Marcel Dekker, New York, 1998.
8. Peters, E. N., "Behavior in Solvents," in *Comprehensive Desk Reference of Polymer Characterization and Analysis*, R. F. Brady, Jr. (Ed.), Oxford University Press, New York, 2003, Chapter 20, pp. 535–554.
9. Zweifel, H., *Plastics Additives Handbook*, 5th ed. Hanser/Gardner, Cincinnati, OH, 2001.
10. Stepek, J., and Daoust, H., *Additives for Plastics*, Springer-Verlag, New York, 1983.
11. Pritchard, G., *Plastics Additives: An A-Z Reference*, Kluwer Academic, London, 1998.
12. Lutz, J. T. (ed.), *Thermoplastic Polymer Additives: Theory and Practice*, Marcel Dekker, New York, 1989.
13. Wypych, G., *Handbook of Fillers: The Definitive User's Guide and Databook on Properties, Effects, and Users*, 2nd ed., Plastics Design Library, Brookfield, CT, 2000.
14. Benedikt, G. M., and Goodall, B. L. (ed.), *Metallocene-Catalyzed Polymers—Materials, Properties, Processing and Markets*, Plastics Design Library, Brookfield, CT, 1998.
15. Shah, V. (ed.), *Handbook of Polyolefins*, 2nd ed., Wiley, New York, 1998.
16. Peacock, A. J., *Handbook of Polyethylene: Structures, Properties, and Applications*, Marcel Dekker, New York, 2000.
17. Karian, H. G. (ed.), *Handbook of Polypropylene and Polypropylene Composites*, Marcel Dekker, New York, 1999.
18. Maier, C., and Calafut, T., *Polypropylene—The Definitive User's Guide*, Plastics Design Library, Brookfield, CT, 1998.
19. Smith, D. A. (ed.), *Addition Polymers: Formation and Characterization*, Plenum, New York, 1968.
20. Scheirs, J., and Priddy, D. (eds.), *Modern Styrenic Polymers*, Wiley, New York, 2003.
21. Bank, D., Brentin, R., and Hus, M., "SPS Crystalline Polymer: A New Material for Automotive Interconnect Systems," paper presented at the SAE Conference, SAE, Troy, MI, Vol. 71, 1997, pp. 305–309.
22. Vernaleken, H., "Polycarbonates," in *Interfacial Synthesis*, Vol. 2, F. Millich and C. Carraher (Eds.), Marcel Dekker, New York, 1997, Chapter 13, pp. 65–124.
23. Wickson, E. J. (ed.), *Handbook of PVC Formulating*, Wiley, New York, 1993.
24. Randall, D., and Lee, S., *The Polyurethanes Book*, Wiley, New York, 2003.
25. Uhlig, K., *Discovering Polyurethanes*, Hanser Gardner, New York, 1999.

26. Pigman, W., and Horton, D. (eds.), *The Carbohydrates*, 2nd ed., Academic, New York, 1970.

27. Aspinall, G. O. (ed.), *The Polysaccharides*, Vol. 1, Academic, New York, 1982.

28. Gilbert, R. (ed.), *Cellulosic Polymers*, Hanser Gardner, Cincinnati, OH, 1993.

29. Bottenbruch, L. (Ed.), *Engineering Thermoplastics: Polycarbonates–Polyacetals–Polyesters–Cellulose Esters*, Hanser Gardner, New York, 1996.

30. Fakirov, S. (Ed.), *Handbook of Thermoplastic Polyesters*, Wiley, New York, 2002.

31. Scheirs, J., and Long, T. E. (eds.), *Modern Polyesters*, Wiley, New York, 2003.

32. Chisholm, B. J., and Zimmer, J. G., "Isothermal Crystallization Kinetics of Commercially Important Polyalkylene Terephthalates," *J. Appl. Polym. Sci.*, **76,** 1296–1307 (2000).

33. Kohan, M. I. (Ed.), *Nylon Plastics Handbook SPE Monograph*, Hanser Gardner, Cincinnati, OH, 1995.

34. Ahorani, S. M., *n-Nylons: Their Synthesis, Structure and Properties*, Wiley, New York, 1997.

35. Gallucci, R. R., "Polyphenylene Ether-Polyamide Blends," in *Conference Proceedings for the Society of Plastics Engineers, Inc., 44th Annual Technical Conference*, Society of Plastics Engineers, Washington, DC, 1986, pp. 48–50.

36. Peters, E. N., "High Performance Polyamides. I. Long Glass PPE/PA66 Alloys," in *Conference Proceedings for the Society of Plastics Engineers, Inc., 55th Annual Technical Conference*, Society of Plastics Engineers, Washington, DC, 1997, pp. 2322–2326.

37. Majumdar, B., and Paul, D. R., "Reactive Compatibilization," in *Polymer Blends: Formulation and Performance*, Vol. 2, D. R. Paul and C. P. Bucknall (Eds.), Wiley, New York, 1999, pp. 539–580.

38. LeGrand, D. G., and Bendler, J. T. (eds.), *Polycarbonates: Science and Technology*, Marcel Dekker, New York, 1999.

39. Christopher, W. F., and Fox, D. W., *Polycarbonates*, Reinhold, New York, 1962.

40. Schnell, H., *Chemistry and Physics of Polycarbonates*, Wiley-Interscience, New York, 1964.

41. Peters, E. N., "Plastics and Elastomers," in *Mechanical Engineer's Handbook*, 2nd ed., M. Kutz (Ed.), Wiley-Interscience, New York, 1998, Chapter 8, pp. 115 129.

42. Robeson, L. M., "Polyarylate," in *Handbook of Plastic Materials and Technology*, I. I. Rubin (Ed.), Wiley-Interscience, New York, 1990, Chapter 21, pp. 237–246.

43. Hay, A. S., "Polymerization by Oxidative Coupling. II. Oxidation of 2,6-Disubstituted Phenols," *J. Polym. Sci.*, **58,** 581–591 (1962).

44. White, D. M., "Polyethers, Aromatic," in *Kirk-Othmer Encyclopedia of Chemical Technology*, 5th ed., Wiley, New York, 2006.

45. Cizek, E. P., U.S. Patent 3,338,435, 1968.

46. Peters, E. N., "Polyphenylene Ether Blends and Alloys," in *Engineering Plastics Handbook*, J. M. Margolis (Ed.), McGraw-Hill, New York, 2005.

47. Gardner, J. W., and Boeke, P. J., "Poly(phenylene sulfide) (PPS)," in *Handbook of Plastic Materials and Technology*, I. I. Rubin (Ed.), Wiley-Interscience, New York, 1990, Chapter 37, pp. 417–432.

48. Johnson, R. N., Farnham, A. G., Clendinning, R. A., Hale, W. F., and Merriam, C. N., "Poly(aryl ethers) by Nucleophilic Aromatic Substitution. I. Synthesis and Properties," *J. Polym. Sci., Part A-1*, **5,** 2375–2398 (1967).

49. Harris, J. E., "Polysulfone (PSO)," in *Handbook of Plastic Materials and Technology*, I. I. Rubin (Ed.), Wiley-Interscience, New York, 1990, Chapter 40, pp. 487–500.

50. Robeson, L. M., "Poly(phenyl) Sulfone," in *Handbook of Plastic Materials and Technology* I. I. Rubin (Ed.), Wiley-Interscience, New York, 1990, Chapter 34, pp. 385–394.

51. Sroog, C. E., "Polyimides," *J. Polym. Sci., Macromol. Rev.*, **11,** 161–208 (1976).

52. Takekoshi, T., "Polyimides," *Adv. Polym. Sci.*, **94,** 1–25 (1990).

53. Serfaty, I. W., "Polyetherimide (PEI)," in *Handbook of Plastic Materials and Technology*, I. I. Rubin (Ed.), Wiley-Interscience, New York, 1990, Chapter 24, pp. 263–276.

54. Thorne, J. L., "Polyamid-Imid (PAI)," in *Handbook of Plastic Materials and Technology*, I. I. Rubin (Ed.), Wiley-Interscience, New York, 1990, Chapter 20, pp. 225–236.

55. Haas, T. W., "Polyetheretherketone (PEEK)," in *Handbook of Plastic Materials and Technology*, I. I. Rubin (Ed.), Wiley-Interscience, New York, 1990, Chapter 25, pp. 277–294.

56. Ebnesajjad, S., *Fluoroplastics*, Vol. I: *Non-Melt Processible Fluoroplastics*, Plastics Design Library, Brookfield, CT, 2000.

57. Ebnesajjad, S., *Fluoroplastics*, Vol. 2: *Melt-Processible Fluoroplastics*, Plastics Design Library, Brookfield, CT, 2002.

58. Goodman, S. H. (Ed.), *Handbook of Thermoset Plastics*, 2nd ed., Plastics Design Library, Brookfield, CT, 1999.

59. Ciullo, P. A., and Hewitt, N., *The Rubber Formulary*, Plastics Design Library, Brookfield, CT, 1999.

60. Morton, M. (Ed.), *Rubber Technology*, Van Norstrand Reinhold, New York, 1973.

61. Zeigler, J. M., and F. W. Gordan Fearon (Eds.), in *Silicon-Based Polymer Science: A Comprehensive Resource*, American Chemical Society, Washington, DC, 1990.

CHAPTER 6

OVERVIEW OF CERAMIC MATERIALS, DESIGN, AND APPLICATION

R. Nathan Katz
Department of Mechanical Engineering
Worcester Polytechnic Institute
Worcester, Massachusetts

1 INTRODUCTION

Engineering ceramics possess unique combinations of physical, chemical, electrical, optical, and mechanical properties. Utilizing the gains in basic materials science understanding and advances in processing technology accrued over the past half century, it is now frequently possible to custom tailor the chemistry, phase content, and microstructure to optimize applications-specific combinations of properties in ceramics (which include glasses, single crystals, and coatings technologies, in addition to bulk polycrystalline materials). This capability in turn has led to many important, new applications of these materials over the past few decades. Indeed, in many of these applications the new ceramics and glasses are the key enabling technology.

Ceramics include materials that have the highest melting points, highest elastic moduli, highest hardness, highest particulate erosion resistance, highest thermal conductivity, highest optical transparency, lowest thermal expansion, and lowest chemical reactivity known. Counterbalancing these beneficial factors are brittle behavior and vulnerability to thermal shock and impact. Over the past three decades major progress has been made in learning how to design to mitigate the brittleness and other undesirable behaviors associated with ceramics and glasses. Consequently, many exciting new applications for these materials have emerged over the past several decades.

Among the major commercial applications for these materials are:

- Passive electronics (capacitors and substrates)
- Optronics/photonics (optical fibers)
- Piezoceramics (transducers)
- Mechanical (bearings, cutting tools)
- Biomaterials (hard-tissue replacement)
- Refractories (furnace linings, space vehicle thermal protection)
- Electrochemical (sensors, fuel cells)
- Transparencies (visible, radar)

This chapter will provide a brief overview of how ceramics are processed and the ramifications of processing on properties. Next a short discussion of the

Reprinted from *Handbook of Materials Selection*, Wiley, New York, 2002, with permission of the publisher.

special issues that one encounters in mechanical design with brittle materials is provided. Short reviews of several of the above engineering applications of ceramics and glasses, which discuss some of the specific combinations of properties that have led design engineers to the selected material(s), follow. A section on how to obtain information on materials sources is provided. Tables listing typical properties of candidate materials for each set of applications are included throughout. Finally, some areas of future potential will be discussed.

2 PROCESSING OF ADVANCED CERAMICS

The production of utilitarian ceramic artifacts via the particulate processing route outlined in Fig. 1 actually commenced about 10,000 years ago.[1] Similarly, glass melting technology goes back about 3500 years, and as early as 2000 years ago optical glass was being produced.[1] While many of the basic unit processes for making glasses and ceramics are still recognizable across the millennia, the level of sophistication in equipment, process control, and raw material control have advanced by "light years." In addition, the past 50 years has created a fundamental understanding of the materials science principles that underlie the processing–microstructure–property relationships. Additionally, new materials have been synthesized that possess extraordinary levels of performance for specific applications. These advances have led to the use of advanced ceramics and glasses in roles that were unimaginable 50 or 60 years ago. For example, early Egyptian glass ca. 2000 BC had an optical loss of $\sim 10^7$ dB/km, compared to an optical loss of $\sim 10^{-1}$ in mid-1980s glass optical fibers,[2] a level of performance that has facilitated the fiber-optic revolution in telecommunications. Similarly, the invention of barium titanate and lead zirconate titanate ceramics, which

have much higher piezoelectric moduli and coupling coefficients than do naturally occurring materials, has enabled the existence of modern sonar and medical ultrasound imaging.[3]

The processing of modern ceramics via the particulate route, shown in Fig. 1, is the way that $\sim 99\%$ of all polycrystalline ceramics are manufactured. Other techniques for producing polycrystalline ceramics, such as chemical vapor deposition[4] or reaction forming,[5] are of growing importance but still represent a very small fraction of the ceramic industry. There are three basic sets of unit processes in the particulate route (and each of these three sets of processes may incorporate dozens of subprocesses). The first set of processes involves powder synthesis and treatment. The second set of processes involves the consolidation of the treated powders into a shaped preform, known as a "green" body. The green body typically contains about 50 vol % porosity and is extremely weak. The last set of unit processes utilizes heat, or heat and pressure combined, to bond the individual powder particles, remove the free space and porosity in the compact via diffusion, and create a fully dense, well-bonded ceramic with the desired microstructure. (See Ref. 4, chapters 9–11.) If only heat is used, this process is called sintering. If pressure is also applied, the process is then referred to as hot pressing (unidirectional pressure) or hot isostatic pressing [(HIP), which applies uniform omnidirectional pressure].

Each of the above steps can introduce processing flaws that can diminish the intrinsic properties of the material. For example, chemical impurities introduced during the powder synthesis and treatment steps may adversely affect the optical, magnetic, dielectric, or thermal properties of the material. Alternatively, the impurities may segregate in the grain boundary of

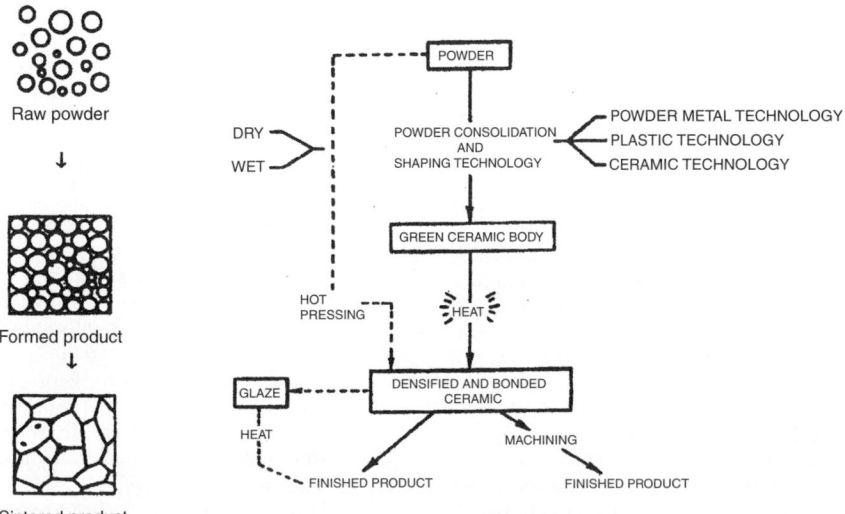

Fig. 1 Processing of polycrystalline ceramics via particular route.

the sintered ceramic and negatively affect its melting point, high-temperature strength, dielectric properties, or optical properties. In green-body formation, platey or high-aspect-ratio powders may align with a preferred orientation, leading to anisotropic properties. Similarly, hot pressing may impose anisotropic properties on a material. Since ceramics are not ductile materials, they can (usually) not be thermomechanically modified after primary fabrication. Thus, the specific path by which a ceramic component is fabricated can profoundly affect its properties. The properties encountered in a complex-shaped ceramic part are often quite different than those encountered in a simply shaped billet of material. This is an important point of which a design engineer specifying a ceramic component needs to be constantly mindful.

3   BRITTLENESS AND BRITTLE MATERIALS DESIGN

Even when ceramics are selected for other than mechanical applications, in most cases some levels of strength and structural integrity are required. It is therefore necessary to briefly discuss the issue of brittleness and how one designs with brittle materials before proceeding to discuss applications and the various ceramic and glass materials families and their properties.

The main issues in designing with a brittle material are that a very large scatter in strength (under tensile stress), a lack of capacity for mitigating stress concentrations via plastic flow, and relatively low-energy absorption prior to failure dominate the mechanical behavior. Each of these issues is a result of the presence of one or more flaw distribution within or at the surface of the ceramic material and/or the general lack of plastic flow available in ceramics. As a consequence, ceramic and glass components that are subjected to tensile stresses are not designed using a single valued strength (*deterministic design*) as commonly done with metals. Rather, ceramic components are designed to a specified probability of failure (*probabilistic design*) that is set at acceptably low values. The statistics of failure of brittle materials whose strength is determined by a population of varying sized flaws are similar to modeling the statistics of a chain failing via its weakest link. The statistics utilized are known as Weibull statistics. A Weibull probability of failure distribution is characterized by two parameters, the characteristic stress and the Weibull modulus.[6] Computer programs for incorporating Weibull statistical distributions into finite-element design codes have been developed that facilitate the design of ceramic components optimized for low probabilities of failure.[7] The effectiveness of such probabilistic design methodology has been demonstrated by the reliable performance of ceramics in many highly stressed structural applications, such as bearings, cutting tools, turbocharger rotors, missile guidance domes, and hip prosthesis.

Flaws (strength-limiting features) can be intrinsic or extrinsic to the material and processing route by which a test specimen or a component is made. Intrinsic strength-limiting flaws are generally a consequence of the processing route and may include features such as pores, aggregations of pores, large grains, agglomerates, and shrinkage cracks. While best processing practices will eliminate or reduce the size and frequency of many of these flaws, it is inevitable that some will still persist. Extrinsic flaws can arise from unintended foreign material entering the process stream, that is, small pieces of debris from the grinding media or damage (cracks) introduced in machining a part to final dimensions. Exposure to a service environment may bring new flaw populations into existence, that is, oxidation pits on the surface of nonoxide ceramics exposed to high temperatures, or may cause existing flaws to grow larger as in the case of static fatigue of glass. In general, one can have several flaw populations present in a component at any time, and the characteristics of each population may change with time. As a consequence of these constantly changing flaw populations, at the present time the state of the art in life prediction of ceramic components for use in extreme environments significantly lags the state of the art in component design. As in most fields of engineering, there are some rules of thumb that one can apply to ceramic design.[8] While these are not substitutes for a carefully executed probabilistic finite-element design analysis, they are very useful in spotting pitfalls and problems when a full-fledged design cannot be executed due to financial or time constraints.

Rules of Thumb for Design with Brittle Materials

1. Point loads should be avoided to minimize stress where loads are transferred. It is best to use areal loading (spherical surfaces are particularly good); line loading is next best.
2. Structural compliance should be maintained by using compliant layers or springs or radiusing of mating parts (to avoid lockup).
3. Stress concentrators—sharp corners, rapid changes in section size, undercuts and holes—should be avoided or minimized. Generous radii and chamfers should be used.
4. The impact of thermal stresses should be minimized by using the smallest section size consistent with other design constraints. The higher the symmetry, the better (a cylinder will resist thermal shock better than a prism), and breaking up complex components into subcomponents with higher symmetry may help.
5. Components should be kept as small as possible—the strength and probability of failure at a given stress level are dependent on size; thus minimizing component size increases reliability.
6. The severity of impact should be minimized. Where impact (i.e., particulate erosion) cannot be avoided, low-angle impacts ($20°-30°$) should be designed for. Note this is very different than

the case of metals, where minimum erosion is at $90°$.

7. Avoid surface and subsurface damage. Grinding should be done so that any residual grinding marks are parallel, not perpendicular, to the direction of principal tensile stress during use. Machining-induced flaws are often identified to be the strength-limiting defect.

4 APPLICATIONS

The combinations of properties available in many advanced ceramics and glasses provide the designers of mechanical, electronic, optical, and magnetic systems a variety of options for significantly increasing systems performance. Indeed, in some cases the increase in systems performance is so great that the use of ceramic materials is considered an enabling technology. In the applications examples provided below the key properties and combinations of properties required will be discussed, as well as the resultant systems benefits.

4.1 Ceramics in Wear Applications

In the largest number of applications where modern ceramics are used in highly stressed mechanical applications, they perform a wear resistance function. This is true of silicon nitride used as balls in rolling element bearings, silicon carbide journal bearings or water pump seals, alumina washers in faucets and beverage dispensing equipment, silicon nitride and alumina-based metal-cutting tools, zirconia fuel injector components, or boron carbide sand blast nozzles, to cite some typical applications and materials.

Wear is a systems property rather than a simple materials property. As a systems property, wear depends upon what material is rubbing, sliding, or rolling over what material, upon whether the system is lubricated or not, upon what the lubricant is, and so forth. To the extent that the wear performance of a material can be predicted, the wear resistance is usually found to be a complex function of several parameters. Wear of ceramic materials is often modeled using an abrasive wear model where the material removed per length of contact with the abrasive is calculated. A wide variety of such models exist, most of which are of the form

$$V \propto P^{0.8} K_{Ic}^{-0.75} H^{-0.5} N \qquad (1)$$

where V is the volume of material worn away, P is the applied load, K_{Ic} is the fracture toughness, H is the indentation hardness, and N is the number of abrasive particles contacting the wear surface per unit length. Even if there are no external abrasives particles present, the wear debris of the ceramics themselves act as abrasive particles. Therefore, the functional relationships that predict that wear resistance should increase as fracture toughness and hardness increase are, in fact, frequently observed in practice.

Even though the point contacts that occur in abrasive wear produce primarily hertzian compressive stresses, in regions away from the hertzian stress field tensile stresses will be present and strength is, thus, a secondary design property. In cases where inertial loading or weight is a design consideration, density may also be a design consideration. Accordingly, Table 1 lists typical values of the fracture toughness, hardness, Young's modulus, four-point modulus of rupture (MOR) in tension, and the density for a variety of advanced ceramic wear materials. Several successful applications of ceramics to challenging wear applications are described below.

Bearings Rolling element bearings, for use at very high speeds or in extreme environments, are limited in performance by the density, compressive strength, corrosion resistance, and wear resistance of traditional high-performance bearing steels. The key screening test to assess a material's potential as a bearing element is rolling contact fatigue (RCF). RCF tests on a variety of alumina, SiC, Si_3N_4, and zirconia materials, at loads representative of high-performance bearings demonstrated that only fully dense silicon nitride (Si_3N_4) could outperform bearing steels.[9] This behavior has been linked to the high fracture toughness of silicon nitride, which results from a unique "self-reinforced" microstructure combined with a high hardness. Additionally, the low density of silicon nitride creates a reduced centrifugal stress on the outer races at high speeds. Fully dense Si_3N_4 bearing materials have demonstrated RCF lives 10 times that of high-performance bearing steel. This improved RCF behavior translates into DN (DN = bearing bore diameter in millimeters × shaft rpm) ratings for hybrid ceramic bearings (Si_3N_4 balls running in steel races, the most common ceramic bearing configuration) about 50% higher than the DN rating of

Table 1 Key Properties for Wear-Resistant Ceramics

Material	K_{Ic}(MPa·m$^{1/2}$)	H(kg/mm^2)	E (GPa)	MOR (MPa)	ρ(g/cm^3)
Al_2O_3 99%	3.9–4.5	1900	360–395	350–560	3.9
B_4C	–	3000	445	300–480	2.5
Diamond	6–10	8000	800–925	800–1400	3.5
SiC	2.6–4.6	2800	380–445	390–550	3.2
Si_3N_4	4.2–7	1600	260–320	450–1200	3.3
TiB_2	5–6.5	2600	550	240–400	4.6
ZrO_2 (Y-TZP)	7–12	1000	200–210	800–1400	5.9

steel bearings. Other benefits of silicon nitride hybrid bearings include an order-of-magnitude less wear of the inner race, excellent performance under marginal lubrication, survival under lubrication starvation conditions, lower heat generation than comparable steel bearings, and reduced noise and vibration.

Another important plus for Si_3N_4 is its failure mechanism. When Si_3N_4 rolling elements fail, they do not fail catastrophically; instead they spall—just like bearing steel elements (though by a different microstructural mechanism). Thus, the design community only had to adapt their existing practices, instead of developing entirely new practices to accommodate new failure modes. The main commercial applications of silicon nitride bearing elements are listed in Table 2.

Cutting Tool Inserts While ceramic cutting tools have been in use for over 60 years, it is only within the past two decades that they have found major application, principally in turning and milling cast iron and nickel-based superalloys and finishing of hardened steels. In these areas ceramics based on aluminum oxide and silicon nitride significantly outperform cemented carbides and coated carbides. High-speed cutting tool tips can encounter temperatures of $1000°C$ or higher. Thus, a key property for an efficient cutting tool is hot hardness. Both the alumina and Si_3N_4 families of materials retain a higher hardness at temperatures between 600 and $1000°C$ than either tool steels or cobalt-bonded WC cermets. The ceramics are also more chemically inert.

The combination of hot hardness and chemical inertness means that the ceramics can run hotter and longer with less wear than the competing materials. Historic concerns with ceramic cutting tools have focused on low toughness, susceptibility to thermal shock, and unpredictable failure times. Improvements in processing together with microstructural modifications to increase fracture toughness have greatly increased the reliability of the ceramics in recent years

Alumina-based inserts are reinforced (toughened) with zirconia, TiC, or TiN particles or SiC whiskers. The thermal shock resistance of alumina–SiC^w is sufficiently high, so that cooling fluids can be used when cutting Ni-based alloys. Silicon-nitride-based inserts include fully dense Si_3N_4 and SiAlON's, which are solid solutions of alumina in Si_3N_4. Fully dense Si_3N_4

can have a fracture toughness of $6-7$ $MPa \cdot m^{1/2}$, almost as high as cemented carbides (~ 9 $MPa \cdot m^{1/2}$), a high strength (greater than symbol 1000 MPa), and a low thermal expansion that yields excellent thermal shock behavior. Silicon nitride is the most efficient insert for the turning of gray cast iron and is also used for milling and other interrupted cut operations on gray iron. Because of its thermal shock resistance, coolant may be used with silicon nitride for turning applications. SiAlON's are typically more chemically stable than the Si_3N_4's but not quite as tough or thermal shock resistant. They are mainly used in rough turning of Ni-based superalloys.

Ceramic inserts are generally more costly than carbides ($1.5-2$ times more), but their metal removal rates are $\sim 3-4$ times greater. However, that is not the entire story. Ceramic inserts also demonstrate reduced wear rates. The combination of lower wear and faster metal removal means many more parts can be produced before tools have to be indexed or replaced. In some cases this enhanced productivity is truly astonishing. In the interrupted single-point turning of the outer diameter counterweights on a gray cast-iron crankshaft, a SiAlON tool was substituted for a coated carbide tool. This change resulted in the metal removal rate increasing 150% and the tool life increasing by a factor of 10. Each tool now produced 10 times as many parts and in much less time. A gas turbine manufacturer performing a machining operation on a Ni-based alloy using a SiAlON tool for roughing and a tungsten carbide tool for finishing required a total of 5 hr. Changing to SiC-whisker-reinforced alumina inserts for both operations reduced the total machining time to only 20 min. This yielded a direct savings of $250,000 per year, freed up 3000 hr of machine time per year, and avoided the need to purchase a second machine tool.

Ceramic Wear Components in Automotive and Light-Truck Engines Several engineering ceramics have combinations of properties that make them attractive materials for a variety of specialized wear applications in automotive engines.

The use of structural ceramics as wear components in commercial engines began in Japan in the early 1980s. Table 3 lists many of the components that have been manufactured, the engine company that first

Table 2   Commercial Applications of Si_3N_4 Hybrid Bearings

Machine tool spindles	The first and largest application, its main benefits are higher speed and stiffness, hence greater throughput and tighter tolerances
Turbomolecular pump shaft	Presently the industry standard, the main benefits are improved pump reliability and marginal lubrication capability, which provide increased flexibility in pump-mounting orientation
Dental drill shaft	The main benefit is sterilization by autoclaving
Aircraft wing flap actuators	Wear and corrosion resistance are the main benefits
In-line skates/mountain bikes	Wear and corrosion resistance are the main benefits
Space Shuttle main-engine oxygen fuel pump	Here, the bearing is lubricated by liquid oxygen. Steel bearings are rated for one flight; Si_3N_4 hybrid bearings are rated for five.

Table 3   Ceramic Wear Components in Automotive and Light-Truck Engines

Component	Engine Manufacturer	Engine Type	Ceramic	Year of Introduction
Rocker arm insert	Mitsubishi	Spark Ignited	Si_3N_4	1984
Tappet	Nissan	Diesel	Si_3N_4	1993
Fuel injector link	Cummins	Diesel	Si_3N_4	1989
Injector shim	Yanmar	Diesel	Si_3N_4	1991
Cam roller	Detroit Diesel	Diesel	Si_3N_4	1992
Fuel injector timing plunger	Cummins	Diesel	ZrO_2	1995
Fuel pump roller	Cummins	Diesel	Si_3N_4	1996

introduced the component, the material, and the year of introduction. In some of these applications several companies have introduced a version of the component into one or more of their engines.

Many of these applications are driven by the need to control the emissions of heavy-duty diesels. Meeting current emissions requirements creates conditions within the engine fuel delivery system that increase wear of lubricated steel against steel. One of these conditions is increased injection pressure, another is an increase in the soot content of engine lubricating oils. Strategic utilization of ceramic components within the fuel delivery systems of many heavy-duty truck engines has enabled the engines to maintain required performance for warranties of 500,000 miles and more. The fuel injector link introduced by Cummins in 1989 is still in production. Well over a million of these components have been manufactured. And many of these have accumulated more than a million miles of service with so little wear that they can be reused in engine rebuilds. In a newer model electronic fuel injector, Cummins introduced a zirconia timing plunger. The part has proved so successful that a second zirconia component was added to the timing plunger assembly several years later. Increasingly stringent emissions requirements for heavy diesels have increased the market for ceramic components in fuel injectors and valve train components. Many of these heavy-duty engine parts are manufactured at rates of 20,000 up to 200,000 per month.

Perhaps the largest remaining problem for this set of applications is cost. Ceramic parts are still more expensive than generally acceptable for the automotive industry. Reluctance of designers to try ceramic solutions still exists, but it is greatly diminishing thanks to the growing list of reliable and successful applications of structural ceramic engine components.

4.2   Thermostructural Applications

Due to the nature of their chemical bond, many ceramics maintain their strength and hardness to higher temperatures than metals. For example, at temperatures above 1200°C, silicon carbide and silicon nitride ceramics are considerably stronger than any superalloy. As a consequence, structural ceramics have been considered and utilized in a number of demanding applications where both mechanically imposed tensile stresses and thermally imposed tensile stresses are present. One dramatic example is the ceramic (silicon nitride) turbocharger that has been in commercial production for automobiles in Japan since 1985. Over one million of these have been manufactured and driven with no recorded failure. This is a very demanding application, as the service temperature can reach 900°C, stresses at ~700°C can reach 325 MPa, and the rotor must also endure oxidative and corrosive exhaust gases that may contain erosion-inducing rust and soot particles. Silicon nitride gas turbine nozzle vanes have been flying for several years in aircraft auxiliary power units. Other applications include heat exchangers and hot-gas valving. Recently, ceramic matrix composites have been introduced as disks for disk breaks in production sports cars by two European manufacturers. A major future market for structural ceramics may be high-performance automotive valves. Such valves are currently undergoing extensive, multiyear fleet tests in Germany.

This class of applications requires a focus on the strength, Weibull modulus, m (the higher the m, the narrower the distribution of observed strength values), thermal shock resistance, and often the stress rupture (strength decrease over time at temperature) and/or creep (deformation with time at temperature) behavior of the materials. Indeed, as shown in Fig. 2, the stress rupture performance of current structural ceramics represents a significant jump in materials performance over superalloys.

The thermal shock resistance of a ceramic is a systems property rather than a fundamental materials property. Thermal shock resistance is given by the maximum temperature change a component can sustain, ΔT:

$$\Delta T = \frac{\sigma(1-\mu)}{\alpha E} \frac{k}{r_m h} S \qquad (2)$$

where σ is strength, μ is Poisson's ratio, α is the coefficient of thermal expansion (CTE), E is Young's modulus, k is thermal conductivity, r_m is the half-thickness for heat flow, h is the heat transfer coefficient, and S is a shape factor totally dependent on component geometry.[10] Thus it can be seen that thermal shock resistance, ΔT, is made up of terms wholly dependent on materials properties and dependent on heat transfer conditions and geometry. It is the role of the

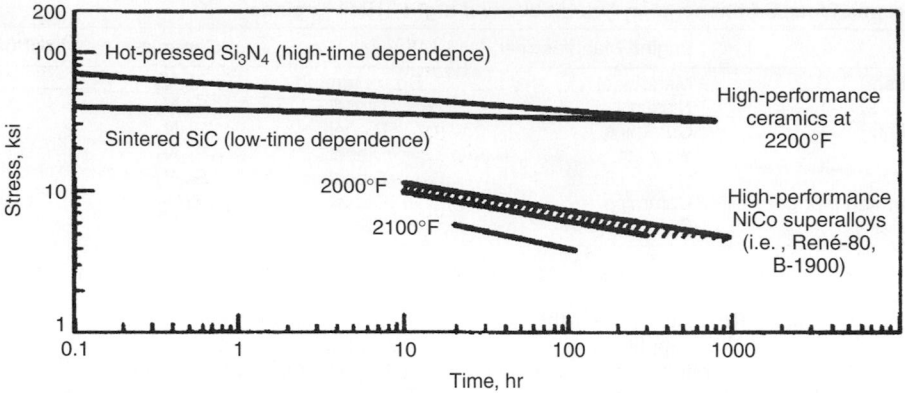

Fig. 2   Stress rupture performance of nonoxide structural ceramics compared to superalloys (oxidizing atmosphere).

Table 4   Calculated Thermal Shock Resistance of Various Ceramics

Material	σ(MPa)	μ	CTE (cm/cm · K)	E (GPa)	R (K)
Al_2O_3 (99 %)	345	0.22	7.4×10^{-6}	375	97
AlN	350	0.24	4.4×10^{-6}	350	173
SiC (sintered)	490	0.16	4.2×10^{-6}	390	251
PSZ	1000	0.3	10.5×10^{-6}	205	325
Si_3N_4 (sintered)	830	0.3	2.7×10^{-6}	290	742
LAS (glass CERAMIC)	96	0.27	0.5×10^{-6}	68	2061
Al titanate	41	0.24	1.0×10^{-6}	11	2819

ceramic engineer to maximize the former and of the design engineer to maximize the latter two terms. It has become usual practice to report the materials-related thermal shock resistance as the instantaneous thermal shock parameter R, which is equal to

$$R = \frac{\sigma(1 - \mu)}{\alpha E} \qquad (3)$$

The value of R for selected ceramics is presented in Table 4. Another frequently used parameter is R', the thermal shock resistance where some heat flow occurs: R' is simply R multiplied by the thermal conductivity, k. For cases where heat transfer environments are complex, Ref. 11 lists 22 figures of merit for selecting ceramics to resist thermal stress.

4.3   Corrosion Resistance

Many advanced structural ceramics such as alumina, silicon nitride, or SiC have strong atomic bonding that yields materials that are highly resistant to corrosion by acidic or basic solutions at room temperature (the notable exception being glass or glass-bonded ceramics attacked by HF). This corrosion resistance has led to many applications. Carbonated soft drinks are acidic, and alumina valves are used to meter and dispense these beverages at refreshment stands. The chemical

industry utilizes a wide variety of ceramic components in pumps and valves for handling corrosive materials. For example, the outstanding corrosion resistance of fully dense SiC immersed in a variety of hostile environments is given in Table 5. There are many cases where corrosion and particulate wear are superimposed, as in the handling of pulp in papermaking or transporting slurries in mineral processing operations, and ceramics find frequent application in such uses.

Table 5   Weight Loss of Fully Dense SiC in Acids and Bases[a]

Reagent (wt %)	Test Temperature (°C)	Weight Loss (mg/cm² · yr)
98% H_2SO_4	100	1.5
50% NaOH	100	2.5
53% HF	100	<0.2
85% H_3PO_4	100	<0.2
45% KOH	100	<0.2
25% HCl	100	<0.2
10% HF + 57% HNO_3	25	<0.2

[a]Specimens submerged 125–300 hr, continuously stirred.
Source: Data Courtesy of ESK-Wacker, Adrian, MI.

Table 6 Key Properties for Electronic Substrates and Packages

Material	CTE (10^{-6}/K)	Thermal Conductivity (W/mK)	Resistivity	Dielectric Constant
Al_2O_3 (96%)	6.8	26	$>10^{14}$	9.5
Al_2O_3 (99%)	6.7	35	$>10^{14}$	10
AlN	4.5	140–240	$>10^{14}$	9
BeO	6.4	250	$>10^{14}$	6.5
Diamond	2	2000	$>10^{14}$	5.5
Silicon	2.8	150		

Note: CTE and thermal conductivity are at room temperature, and the dielectric constant is at 1 MHz.

4.4 Passive Electronics

The role of passive electronics is to provide insulation (prevent the flow of electrons) either on a continuous basis (as in the case of substrates or packages for microelectronics) or on an intermittent basis, as is the case for ceramic capacitors (which store electric charge and hence need a high polarizability). These applications constitute two of the largest current markets for advanced ceramics. For electronic substrates and packages key issues include the minimization of thermal mismatch stresses between the Si (or GaAS) chip and the package material (so the CTE will be important) and dissipation of the heat generated as electrons flow through the millions of transistors and resistors that comprise modern microelectronic chips; hence the thermal conductivity is a key property. All other things being equal, the delay time for electrons to flow in the circuit is proportional to the square root of the dielectric constant of the substrate (or package) material. Additionally, the chip or package must maintain its insulating function, so resistivities of $>10^{14}$ are required. Most high-performance packages for computer chips are alumina. With the advent of microwave integrated circuits (e.g., cell phones) aluminum nitride substrates are beginning to be utilized for high thermal conductivity. The environmental drawbacks to machining BeO have tended to favor the use of AlN to replace or avoid the use of BeO. Synthetic diamond is an emerging substrate material for special applications. Isotopically "pure" synthetic, single-crystal diamond has values of thermal conductivity approaching 10,000 W/mK. Typical values of the above properties for each of these materials are given in Table 6, along with selected properties of silicon for comparison. For design purposes exact values for specific formulations of the materials should be obtained from the manufacturers.

Over a billion ceramic capacitors or multilayer ceramic capacitors (MLCCs) are made every day.[12] Since electrons do not flow through capacitors, they are considered passive electronic components. However, the insulators from which ceramic capacitors are made polarize, thereby separating electric charge. This separated charge can be released and flow as electrons, but the electrons do not flow through the dielectric

Table 7 Dielectric Constants for Various Ceramic Capacitor Materials

Material	Dielectric Constant at RT
Tantalum oxide (Ta_2O_5)	~25
Barium titanate	~5,000
Barium–zirconium titanate	~20,000
Lead–zirconium titanate (PZT)	~2,000
PZT with W or Mg additives	~9,000
Lead magnesiun niobate (PMN)	~20,000
Lead zinc niobate (PZN)	~20,000

material of which the capacitor is composed. Thus, the materials parameter, which determines the amount of charge that can be stored, the dielectric constant k, is the key parameter for design and application. Table 7 lists the approximate dielectric constant at room temperature (RT) for several families of ceramics used in capacitor technology. The dielectric constant varies with both temperature and frequency. Thus, for actual design precise curves of materials performance over a relevant range of temperatures and frequencies are often utilized. Many ceramics utilized as capacitors are ferroelectrics, and the dielectric constant of these materials is usually a maximum at or near the Curie temperature.

4.5 Piezoceramics

Piezoceramics are a multi-billion-dollar market.[13] Piezoceramics are an enabling material for sonar systems, medical ultrasonic imaging, micromotors and micropositioning devices, the timing crystals in our electronic watches, and numerous other applications. A piezoelectric material will produce a charge (or a current) if subjected to pressure (the *direct* piezoelectric effect) or, if a voltage is applied, the material will produce a strain (the *converse* piezoelectric effect). Upon the application of a stress, a polarization charge, P, per unit area is created that equals $d\sigma$, where σ is the applied stress and d is the piezoelectric modulus. This modulus, which determines piezoelectric behavior, is a third-rank tensor[14] that is thus highly dependent on directions along which the crystal is stressed. For example, a quartz crystal stressed in the [100] direction will produce a voltage, but one stressed in the

[001] direction will not. In a polycrystalline ceramic the random orientation of the grains in an as-fired piezoceramic will tend to minimize or zero out any net piezoelectric effects. Thus, polycrystalline piezoceramics have to undergo a postsintering process to align the electrically charged dipoles within the polycrystalline component. This process is known as poling and it requires the application of a very high electric field. If the piezoceramic is taken above a temperature, known as the Curie temperature, a phase transformation occurs and piezoelectricity will disappear. The piezoelectric modulus and the Curie temperature are thus two key materials selection parameters for piezoceramics.

The ability of piezoceramics to almost instantaneously convert electrical current to mechanical displacement, and vice versa, makes them highly useful as transducers. The efficiency of conversion between mechanical and electrical energy (or the converse) is measured by a parameter known as the coupling coefficient. This is a third key parameter that guides the selection of piezoelectric materials.

Although piezoelectricity was discovered by Pierre and Jacques Curie in 1880, piezoceramics were not widely utilized until the development of polycrystalline barium titanate in the 1940s and PZTs in the 1950s. Both of these materials have high values of d and thus develop a high voltage for a given applied stress. PZT has become widely used because, in addition to a high d value, it also has a very high coupling coefficient. Sonar, in which ultrasonic pulses are emitted and reflected "echoes" are received, is used to locate ships and fish and map the ocean floor by navies, fishermen, and scientists all over the globe. Medical ultrasound utilizes phased arrays of piezoceramic transducers to image organs and fetuses noninvasively and without exposure to radiation. A relatively new application that has found significant use in the microelectronics industry is the use of piezoceramics to drive micropositioning devices and micromotors. Some of these devices can control positioning to a nanometer or less. Piezoceramic transducers are combined with sophisticated signal detection and generation electronics to create "active" noise and vibration damping devices. In such devices the electronics detect and quantify a noise spectrum and then drive the transducers to provide a spectrum $180°$ out of phase with the noise, thereby effectively canceling it.

Many of the current high-performance applications of piezoceramics are based on proprietary modifications of PZT, which contain additions of various dopants or are solid solutions with perovskite compounds of Pb with Mg, Mn, Nb, Sn, Mo, or Ni. Table 8 lists the range of several key piezoceramic selection parameters for proprietary PZT compositions from one manufacturer.

4.6 Transparencies

Transparent ceramics (which include glasses and single-crystal and polycrystalline ceramics) have been

Table 8 Key Properties for PZT-Based Piezoceramics

Material	Piezoelectric Modulus, d_{33} (m/V)	Curie Temperature (°C)	Coupling Coefficient, k_{33}
A	226×10^{-12}	320	0.67
B	635×10^{-12}	145	0.68
C	417×10^{-12}	330	0.73

used as optical transparencies or lenses for millennia. Glass windows were in commercial production in first-century Rome, but it was not until the 1800s, with the need for precision optics for microscopes, telescopes, and ophthalmic lenses, that glasses and other optical materials became the object of serious scientific study. As noted in the introduction, progress in glass science and technology, coupled with lasers, has led to the current broadband digital data transmission revolution via optical fibers. Various ceramic crystals are used as laser hosts and specialty optical lenses and windows. A significant fraction of supermarket scanner windows combine the scratch resistance of sapphire (single-crystal alumina) with its ability to transmit the red laser light that we see at the checkout counter. While such windows are significantly more costly than glass, their replacement rate is so low that they have increased profitability for several supermarket chains. For the same reason the crystal in many high-end watches are scratch-resistant man-made sapphire. Polycrystalline translucent (as opposed to fully transparent) alumina is used as containers (envelopes) for the sodium vapor lamps that light our highways and industrial sites.

Not all windows have to pass visible light. Radar or mid- to far-infrared transparencies look opaque to the human eye but are perfectly functional windows at their design wavelengths. The most demanding applications for such transparencies is for the guidance domes of missiles. Materials that can be used for missile radomes include slip-cast fused silica, various grades of pyroceram (glass ceramics), and silicon-nitride-based materials. Infrared (IR) windows and missile domes include MgF_2 and ZnSe. Requirements exist for having missile guidance domes that can transmit in the visible, IR, and radar frequencies (multimode domes). Ceramic materials that can provide such functionality include sapphire and aluminum oxynitride spinel (AlON). In addition to optical properties, missile domes must be able to take high aerothermal loading (have sufficient strength) and be thermal shock resistant (a high-speed missile encountering a rain cloud can have an instant ΔT of minus several hundred degrees kelvin).

Key properties for visible and IR optical materials include the index of refraction, n (which will be a function of wavelength), and absorption or loss. For radar transparencies key parameters are dielectric constant (which can be thought of as analogous to the index of refraction) and dielectric loss.

5 INFORMATION SOURCES

5.1 Manufacturers and Suppliers

There are hundreds of manufacturers of advanced ceramics and glasses. Locating ones that already have the material that is needed and can produce it in the configuration required can be a daunting task. There are two resources published annually that make this task much easier. The American Ceramic Society publishes a directory of suppliers of materials, supplies, and services that can help locate such information quickly. It is called *Ceramic Source*. This directory can also be accessed on the web at www.ceramics.org. A similar *Buyers Guide* is published by *Ceramic Industry Magazine*, and this can also be viewed online at www.ceramicindustry.com. Once a likely source for your need has been identified, a visit to the supplier's website can often provide a great deal of background information and specific data, which can make further contacts with the supplier much more meaningful and informative.

5.2 Data

Manufacturer's literature, both hard copy and posted on the web, is an invaluable source of data. The handbooks, textbooks, and encyclopedias listed below are also excellent sources of data. However, before committing to a finalized design or to production, it is advisable to develop your own test data in conformance with your organization's design practice. Such data should be acquired from actual components made by the material, processing route, and manufacturer that have been selected for the production item.

ASM, *Ceramics and Glasses*, Vol. 4: *Engineered Materials Handbook*, ASM International, Materials Park, OH, 1991.

Brook, R. J. (Ed.), *Concise Encyclopedia of Advanced Ceramic Materials*, Pergamon, Oxford, 1991.

Campbell, C. X., and El-Rahaiby, S. K. (Eds.), *Databook on Mechanical and Thermophysical Properties of Whisker-Reinforced Ceramic Matrix Composites*, Ceramics Information Analysis Center, Purdue University, W. Lafayette, IN, and The American Ceramic Society, Westerville, OH, 1995.

Shackelsford, J. F., Alexander, W., and Park, J. (Eds.), *Materials Science and Engineering Handbook*, 2nd ed., CRC Press, Boca Raton, FL, 1994.

5.3 Standards and Test Methods

To reliably design, procure materials, and assure reliability, it is necessary to have common, agreed-upon, and authoritative test standards, methods, and practices. Institutions such as the American Society for Testing and Materials (ASTM), the Japanese Institute for Standards (JIS), the German Standards Organization (DIN), and the International Standards Organization (ISO) all provide standards for their various constituencies. The following are a sampling of standards available from the ASTM and JIS for advanced ceramics and ceramic matrix composites. One can reach these organizations at the following addresses.

American Society for Testing and Materials, 100 Barr Harbor Drive, Conshohocken, PA 19428-2959:

- C-177-85(1993), Test Method for Steady State Heat Flux and Thermal Transmission by Means of the Gradient-Hot-Plate Apparatus
- C-1161-90, Test Method for Flexural Strength of Advanced Ceramics at Ambient Temperature
- C-1211-92, Test Method for Flexural Strength of Advanced Ceramics at Elevated Temperature
- C-1259-94, Test Method for Dynamic Young's Modulus, Shear Modulus and Poisson's Ratio for Advanced Ceramics by Impulse Excitation of Vibration
- C-1286-94, Classification for Advanced Ceramics
- C-1292-95A, Test Method for Shear Strength of Continuous Fiber-Reinforced Ceranic Composites (CFCCs) at Ambient Temperatures
- C-1337-96, Test Method for Creep and Creep-Rupture of CFFFs under Tensile Loading at Elevated Temperature
- C-1421-99, Standard Test Method for Determination of Fracture Toughness of Advanced Ceramics at Ambient Temperature
- C-1425-99, Test Method for Interlaminar Shear Strength of 1-D and 2D CFCCs at Elevated Temperature
- E-228-85(1989), Test Method for Linear Thermal Expansion of Solid Materials with a Vitreous Silica Dilatometer
- E-1269-94, Test Method for Determining Specific Heat Capacity by Differential Scanning Calorimetry
- E-1461-92, Test Method for Thermal Diffusion of Solids by the Flash Method

The Japanese Standards Association, 1-24, Akasaka 4, Minato-ku, Tokyo 107 Japan:

- Testing Methods for Elastic Modulus of High Performance Ceramics at Elevated Temperatures; JIS R 1605-(1989)
- Testing Methods for Tensile Strength of High Performance Ceramics at Room and Elevated Temperatures; JIS R 1606-(1990)
- Testing Methods for Fracture Toughness of High Performance Ceramics; JIS R 1607-(1990)
- Testing Methods for Compressive Strength of High Performance Ceramics; JIS R 1608-(1990)
- Testing Methods for Oxidation Resistance of Non-Oxide of High Performance Ceramics; JIS R 1609-(1990)
- Testing Methods for Vickers Hardness of High Performance Ceramics; JIS R 1610-(1991)
- Testing Methods of Thermal Diffusivity, Specific Heat Capacity, and Thermal Conductivity for High Performance Ceramics by Laser Flash Method; JIS R 1611-(1991)

In addition to testing standards, it is possible to obtain standard materials with certified properties to calibrate several of these new standards against your own tests. Such standard materials can be obtained from the National Institute of Science and Technology (NIST). For example, a standard material to calibrate ASTM C-1421-99 is available. Materials standards are not available for all of the above tests.

5.4 Design Handbooks

It has been widely recognized that procedural handbooks that provide methodology on how to design with advanced ceramics and that can provide high-quality evaluated design data are sorely needed for ceramic materials. The ceramics matrix composites (CMCs) community has taken the initiative to begin the process of creating such a handbook for its constituency. The activity is sponsored by various U.S. governmental agencies, including the Department of Defense, the Department of Energy, the Federal Aviation Administration, and the National Aeronautics and Space Administration, and is entitled MIL-Handbook-17. This activity brings together materials suppliers, materials testers, designers, and end users who are engaged in developing a handbook that will provide design tools and guidance; provide guidelines on data generation, documentation, and use; and provide an authoritative source of design quality data. This is a work in progress and its completion is many years off, if ever. Nevertheless, much guidance in design and testing of advanced CMCs has already resulted from this activity. Progress can be followed by periodically accessing the handbook websites at http://mil-17.udel.edu or http://www.materials-sciences.com/MIL17/. Unfortunately, no similar activity exists for monolithic ceramics.

6 FUTURE TRENDS

It has been estimated that in the United States advanced ceramics of the type discussed above are an over $8-billion-a-year industry with a growth rate of ~8% per year.[15]

The largest segment of this growth will come from the electronics area. Not only will there be significant growth in the "traditional" roles of ceramics as insulators, packages, substrates, and capacitors, but structural ceramics will play a major role in the equipment used in semiconductor manufacturing. This trend will be especially driven by the resistance of ceramics such as SiC, AIN, silicon nitride, and alumina to the erosive and corrosive environments within high-energy plasma chambers used in single-wafer processing operations.

The intertwined global issues of energy sufficiency and environmental protection will see commercial use of advanced ceramics in energy systems as diverse as solid oxide fuel cells and pebble-bed modular reactors (nuclear). As more and more industries move toward "green" (pollution-free) manufacturing, there will be growth in wear- and corrosion-resistant ceramics for industrial machinery. There will also be substantial growth potential for ceramic filters and membranes. One major environmentally driven opportunity will be particulate traps for diesel trucks and industrial power sources. This technology is just beginning to be commercialized, and it is certain to see rapid growth as emissions requirements for diesel engines grow more stringent. Not all progress in these areas will create increased markets for ceramics; some will reduce them. For example, the rapid growth of energy-efficient light-emitting diode technology for illumination will create a significant growth opportunity for producers of single-crystal SiC substrates and GaN materials. However, this will come at a cost to the ceramics industry of a significant decrease in glass envelopes for incandescent bulbs and fluorescent tubes. Another area of growth will be filters and membranes for filtration of hot or corrosive, or both, gases and liquids.

The explosive growth of fiber-optic- and microwave-based digital communications technology has produced significant opportunities and markets for advanced ceramics and glasses and will continue to do so into the foreseeable future.

Medical applications are sure to grow, in the areas both of diagnostics and prosthetics.

At the entrance to the Pohang Steel complex in Pohang, Republic of Korea, is a wonderful sign. It proclaims, "Resources are Limited—Creativity is Unlimited." This thought certainly applies to the global future of advanced ceramics. Creatively utilized advanced ceramics will effectively expand our resources, protect our environment, and create new technological opportunities. The potential opportunities go far beyond the few discussed in this chapter.

REFERENCES

1. Vandiver, P. B., "Reconstructing and Interpreting the Technologies of Ancient Ceramics," in *Materials Issues in Art and Archaeology*, Materials Res. Soc. Symposium Proceed., Vol. 123, Materials Research Socity, Pittsburgh, PA, 1988, pp. 89–102.

2. *Materials Science and Engineering for the 1990's*, National Academy Press, Washington, DC, 1989, p. 24.

3. Katz, R. N., "Piezoceramics," *Ceramic Industry*, p. 20 (Aug. 20, 2000).

4. Richerson, D. W., *Modern Ceramic Engineering*, 2nd ed., Marcel Dekker, New York, 1992, pp. 582–588.

5. Haggerty, J. S., and Chiang, Y. M., "Reaction-Based Processing Methods for Materials and Composites," *Ceramic Eng. Sci. Proc.*, **11** (7–8), 757–781 (1990).

6. McLean, A. F., and Hartsock, D., "Design with Structural Ceramics," in *Treatise on Materials Science and Technology*, Vol. 29, J. B. Wachtman (Ed.), Academic, Boston, 1989, pp. 27–95.

7. Nemeth, N. N., and Gyekenyesi, J. P., "Probabilistic Design of Ceramic Components with the NASA/CARES Computer Program," in *Ceramics and*

Glasses, Vol. 4, *Engineered Materials Handbook*, ASM International, Metals Park, OH, 1991, pp. 700–708.

8. Katz, R. N., "Application of High Performance Ceramics in Heat Engine Design," *Mater. Sci. Eng.*, **71**, 227–249 (1985).

9. Katz, R. N., "Ceramic Materials for Roller Element Bearing Application," in *Friction and Wear of Ceramics*, S. Jahanmir (Ed.), Marcel Dekker, New York, 1994, pp. 313–328.

10. Kingery, W. D., Bowen, H. K., and Uhlmann, D. R., *Introduction to Ceramics*, 2nd ed., Wiley, New York, 1976.

11. Hasselman, D. P. H., "Figures-of-Merit for the Thermal Stress Resistance of High Temperature Brittle Materials: A Review," *Ceramurgia Int.*, **4**(4), 147–150 (1998).

12. Richerson, D. W., *The Magic of Ceramics*, American Ceramic Society, Westerville, OH, 2000, p. 141.

13. NSF Workshop Report, Fundamental Research Needs in Ceramics, Washington, DC, April 1999, p. 9.

14. Nye, J. F., *Physical Properties of Crystals*, Oxford University Press, London, 1964.

15. Abraham, T., "US Advanced Ceramics Growth Continues," *Ceramic Industry*, 23–25 (Aug. 2000).

CHAPTER 7

MECHANICS OF DEFORMABLE BODIES

Neal F. Enke and Bela I. Sandor
Department of Engineering Mechanics
University of Wisconsin
Madison, Wisconsin

1 INTRODUCTION TO STRESS AND STRAIN

1.1 Definitions of Stress and Strain

Stress at a Point Stress is the description of the intensity of force. For a general state of loading, such as represented in Fig. 1, the stress vector **S** acting on the section D at point Q is given by

$$\mathbf{S} = \lim_{\Delta A \to 0} \frac{\Delta \mathbf{P}}{\Delta A}$$

where $\Delta \mathbf{P}$ is the force acting on the area ΔA. The stress vector **S** can be further divided into a component

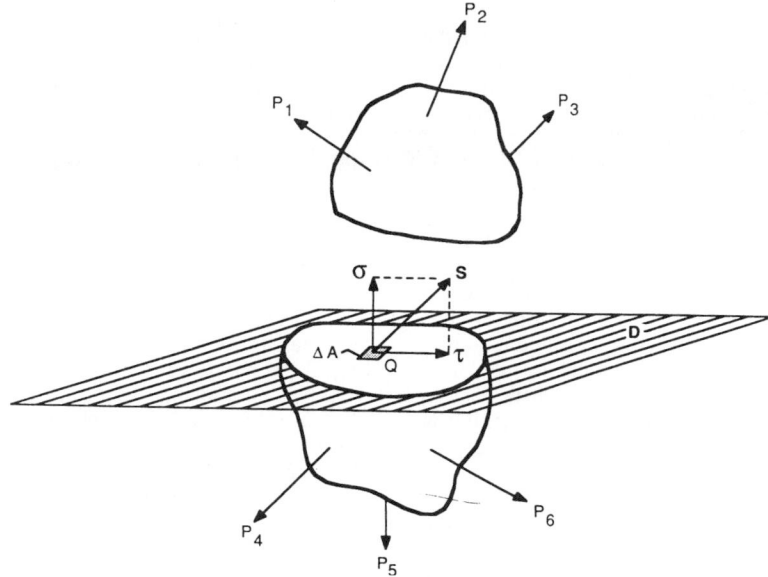

Fig. 1 Normal and shear components of stress.

normal to the section and a component tangential to the section. The normal component is referred to as the normal stress and is usually denoted by σ. The tangential component is referred to as the shear (or shearing) stress and is commonly denoted by τ. It is often useful to further divide τ into two orthogonal components, thus giving an orthogonal triad of stress components. Tensile stresses are normal stresses that tend to increase the length of a member. Compressive stresses are normal stresses that tend to decrease the length of a member. In the U.S. customary system of units, stress is expressed in pounds per square inch (psi). Stress is also commonly expressed in kilopounds per square inch (ksi).

In general, if different sections are considered through the same point in a body under load, different stress vectors will be obtained. In order to completely specify the state of stress at a point, it is sufficient to consider the stresses acting on three mutually orthogonal sections at that point. The common procedure is to envision an infinitesimally small cube at the point of interest (see Fig. 2). In the figure, σ_{ij} is the stress acting on the i face in the j direction, where i and j can be any of x, y, or z. The state of stress at a point is often represented in matrix form as

$$[\sigma_{ij}] = \begin{bmatrix} \sigma_{xx} & \sigma_{xy} & \sigma_{xz} \\ \sigma_{yx} & \sigma_{yy} & \sigma_{yz} \\ \sigma_{zx} & \sigma_{zy} & \sigma_{zz} \end{bmatrix} \tag{1}$$

The three diagonal terms ($\sigma_{xx}, \sigma_{yy}, \sigma_{zz}$) are the normal stress components in the x, y, and z directions, respectively. The off-diagonal terms represent the shear stress components. The i face is considered positive if its outward normal acts in the same direction as the positive i axis; otherwise, it is considered negative. The stress

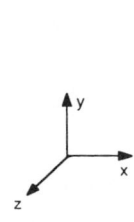

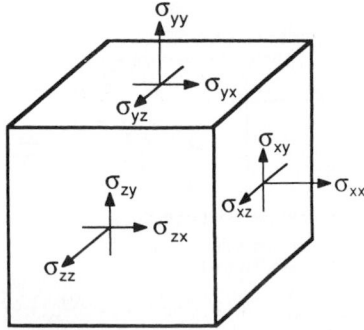

Fig. 2 General state of stress.

σ_{ij} is considered positive if it acts on the positive i face in the positive j direction or if it acts on the negative i face in the negative j direction. Thus, the three faces shown in Fig. 2 are positive as are the stress components acting on them. A positive normal stress is a tensile stress while a negative normal stress is a compressive stress.

If the point of interest is in a state of static equilibrium, it can be shown (by summing moments about a small element) that the cross-shearing stresses must be equal. That is, $\sigma_{xy} = \sigma_{yx}, \sigma_{xz} = \sigma_{zx}$, and $\sigma_{yz} = \sigma_{zy}$. The total number of stress components needed to completely define the state of stress at a point is thus reduced to six in this case. This so-called complementary property of shear is not valid for dynamic situations. It is also invalid in situations where surface or body couples are significant, but this condition is rarely encountered in practice.

Strain at a Point Deformation is the movement of points in a body relative to each other. Strain is the description of the intensity of deformation. The normal strain at a point Q in the x direction, ε_{xx}, is given by

$$\varepsilon_{xx} = \lim_{L \to 0} \frac{\Delta L}{L}$$

Here, L is the original length of a line segment in the x direction centered at Q, and ΔL is the change in length due to the deformation (Fig. 3). Normal strains are considered positive if they cause an increase in the length of the line segment. The shear (or shearing) strain associated with the orthogonal x and y directions, γ_{xy}, is defined as the change in angle between infinitesimal line segments originally in the x and y directions before deformation (Fig. 4). Shear strains are considered positive if they bring about a reduction in angle. By this definition of shear strain, it is apparent that $\gamma_{xy} = \gamma_{yx}$, provided that no discontinuities (such as void formation) occur at the point of interest.

Tensorial shear strains ε_{ij} are defined to be equal to one-half the ordinary shear strains γ_{ij}. That is, $\varepsilon_{ij} = \gamma_{ij}/2$ for $i \neq j$. Tensorial shear strains are often used in the theory of elasticity and continuum mechanics because they result in algebraically simpler equations than if ordinary shear strains are employed. Utilizing the tensorial shear strains, the state of strain at a point can be represented in matrix form as

$$[\varepsilon_{ij}] = \begin{bmatrix} \varepsilon_{xx} & \varepsilon_{xy} & \varepsilon_{xz} \\ \varepsilon_{yx} & \varepsilon_{yy} & \varepsilon_{yz} \\ \varepsilon_{zx} & \varepsilon_{zy} & \varepsilon_{zz} \end{bmatrix} \quad (2)$$

The three diagonal terms are the normal strain components while the off-diagonal terms are the tensorial shear strain components. Although strain is a dimensionless quantity, it is common to express it in terms of inches per inches or similar units.

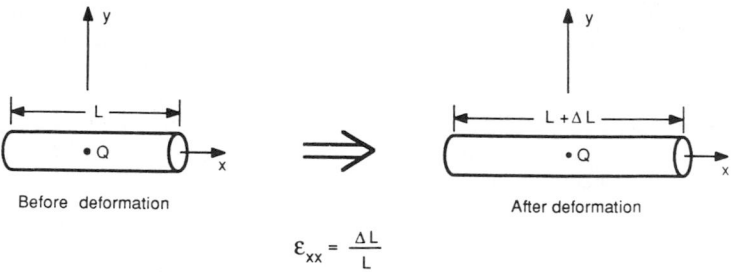

$$\varepsilon_{xx} = \frac{\Delta L}{L}$$

Fig. 3 Definition of normal strain.

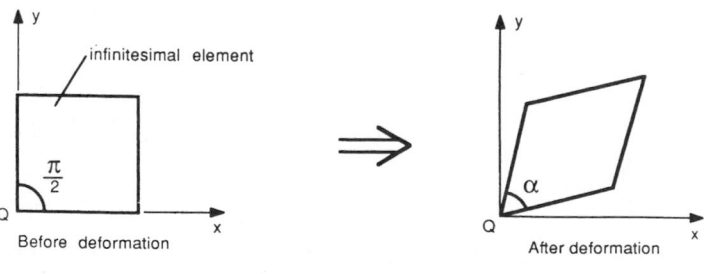

$$\gamma_{xy} = \frac{\pi}{2} - \alpha$$

Fig. 4 Definition of shear strain.

Strain–Displacement Relations Mathematically, strains are defined in terms of derivatives of displacements. Assume a general point Q in a body undergoes a displacement to Q' when the body deforms under application of load. Let u, v, and w represent the displacement components of Q in the x, y, and z directions, respectively. Using the previously given definitions of normal and shear strains, the strain components at the point Q can be shown to be[1]

$$\varepsilon_{xx} = \sqrt{1 + 2\frac{\partial u}{\partial x} + \left(\frac{\partial u}{\partial x}\right)^2 + \left(\frac{\partial v}{\partial x}\right)^2 + \left(\frac{\partial w}{\partial x}\right)^2} - 1$$

$$\varepsilon_{yy} = \sqrt{1 + 2\frac{\partial v}{\partial y} + \left(\frac{\partial u}{\partial y}\right)^2 + \left(\frac{\partial v}{\partial y}\right)^2 + \left(\frac{\partial w}{\partial y}\right)^2} - 1$$

$$\varepsilon_{zz} = \sqrt{1 + 2\frac{\partial w}{\partial z} + \left(\frac{\partial u}{\partial z}\right)^2 + \left(\frac{\partial v}{\partial z}\right)^2 + \left(\frac{\partial w}{\partial z}\right)^2} - 1 \quad (3)$$

$$\gamma_{xy} = \gamma_{yx} = \sin^{-1}\left[\frac{\frac{\partial u}{\partial y} + \frac{\partial v}{\partial x} + \frac{\partial u}{\partial x}\frac{\partial u}{\partial y} + \frac{\partial v}{\partial x}\frac{\partial v}{\partial y} + \frac{\partial w}{\partial x}\frac{\partial w}{\partial y}}{(1 + \varepsilon_{xx})(1 + \varepsilon_{yy})}\right]$$

$$\gamma_{xz} = \gamma_{zx} = \sin^{-1}\left[\frac{\frac{\partial w}{\partial x} + \frac{\partial u}{\partial z} + \frac{\partial w}{\partial z}\frac{\partial w}{\partial x} + \frac{\partial u}{\partial z}\frac{\partial u}{\partial x} + \frac{\partial v}{\partial x}\frac{\partial v}{\partial x}}{(1 + \varepsilon_{xx})(1 + \varepsilon_{zz})}\right]$$

$$\gamma_{yz} = \gamma_{zy} = \sin^{-1}\left[\frac{\frac{\partial v}{\partial z} + \frac{\partial w}{\partial y} + \frac{\partial v}{\partial y}\frac{\partial v}{\partial z} + \frac{\partial w}{\partial y}\frac{\partial w}{\partial z} + \frac{\partial u}{\partial y}\frac{\partial u}{\partial z}}{(1 + \varepsilon_{yy})(1 + \varepsilon_{zz})}\right]$$

Equations (3) give the strain components for arbitrary displacements. They are known as the strain–displacement relations for large displacements. Slight variations of these equations, due to simplifying assumptions or slightly different definitions of strain, are often found in the literature (see, e.g., Refs. 2, Chapter 33, and 3).

If the displacements and strains are sufficiently small, the higher-order terms in Eqs. (3) can be neglected. The equations then reduce to

$$\varepsilon_{xx} = \frac{\partial u}{\partial x} \qquad \gamma_{xy} = \gamma_{yx} = \frac{\partial v}{\partial x} + \frac{\partial u}{\partial y}$$

$$\varepsilon_{yy} = \frac{\partial v}{\partial y} \qquad \gamma_{xz} = \gamma_{zx} = \frac{\partial w}{\partial x} + \frac{\partial u}{\partial z} \quad (4)$$

$$\varepsilon_{zz} = \frac{\partial w}{\partial z} \qquad \gamma_{yz} = \gamma_{zy} = \frac{\partial w}{\partial y} + \frac{\partial v}{\partial z}$$

These equations are known as the strain–displacement relations for small displacements.

Equations of Compatibility Equations (4) allow for the determination of six strain components using only three displacement components. Physically, this implies that the strains must be compatible; that is, the infinitesimal elements of the deformed member must fit together without any voids or overlap being present. Mathematically, this implies that the strains are not arbitrary functions of the coordinates. There must exist three equations relating the strains to one another. These equations of compatibility are as follows:

$$\frac{\partial^2 \gamma_{xy}}{\partial x\,\partial y} = \frac{\partial^2 \varepsilon_{xx}}{\partial y^2} + \frac{\partial^2 \varepsilon_{yy}}{\partial x^2}$$

$$\frac{\partial^2 \gamma_{yz}}{\partial y\,\partial z} = \frac{\partial^2 \varepsilon_{yy}}{\partial z^2} + \frac{\partial^2 \varepsilon_{zz}}{\partial y^2} \quad (5a)$$

$$\frac{\partial^2 \gamma_{zx}}{\partial z\,\partial x} = \frac{\partial^2 \varepsilon_{zz}}{\partial x^2} + \frac{\partial^2 \varepsilon_{xx}}{\partial z^2}$$

Three additional compatibility equations are often used,

$$2\frac{\partial^2 \varepsilon_{xx}}{\partial y\,\partial z} = \frac{\partial}{\partial x}\left(-\frac{\partial \gamma_{yz}}{\partial x} + \frac{\partial \gamma_{zx}}{\partial y} + \frac{\partial \gamma_{xy}}{\partial z}\right)$$

$$2\frac{\partial^2 \varepsilon_{yy}}{\partial z\,\partial x} = \frac{\partial}{\partial y}\left(\frac{\partial \gamma_{yz}}{\partial x} - \frac{\partial \gamma_{zx}}{\partial y} + \frac{\partial \gamma_{xy}}{\partial z}\right) \quad (5b)$$

$$2\frac{\partial^2 \varepsilon_{zz}}{\partial x\,\partial y} = \frac{\partial}{\partial z}\left(\frac{\partial \gamma_{yz}}{\partial x} + \frac{\partial \gamma_{zx}}{\partial y} - \frac{\partial \gamma_{xy}}{\partial z}\right)$$

It can be shown that Eqs. (5b) are not independent of Eqs. (5a).[2] The equations of compatibility are used in the theory of elasticity. They are also useful in verifying the accuracy of experimentally determined strains such as in the Moiré method of strain analysis.

Stress Equations of Equilibrium By considering the static equilibrium of an infinitesimal element, it can be shown that the following equations must hold:

$$\frac{\partial \sigma_{xx}}{\partial x} + \frac{\partial \sigma_{yx}}{\partial y} + \frac{\partial \sigma_{zx}}{\partial z} + F_x = 0$$

$$\frac{\partial \sigma_{xy}}{\partial x} + \frac{\partial \sigma_{yy}}{\partial y} + \frac{\partial \sigma_{zy}}{\partial z} + F_y = 0 \quad (6)$$

$$\frac{\partial \sigma_{xz}}{\partial x} + \frac{\partial \sigma_{yz}}{\partial y} + \frac{\partial \sigma_{zz}}{\partial z} + F_z = 0$$

where F_x, F_y, and F_z are the body forces per unit volume in the x, y, and z directions, respectively. Equations (6) are known as the stress equations of equilibrium.

1.2 Linear Elastic Stress–Strain Relationships

Stresses and strains in a solid body are generally not independent but rather can be related to each other. The exact form of these relationships depends on whether the body is behaving in an elastic, plastic, viscoelastic, or some other fashion. The relationships also depend on whether the material is isotropic or anisotropic. An isotropic material displays the same properties in all directions. An anisotropic material displays different properties in different directions. Only isotropic response will be considered here. Anisotropic response is discussed in Section 7 on composite materials.

A member is said to experience elastic response if it returns to its original shape upon removal of loads. If, in addition, the relationship between stress and strain can be written in a linear form, the member is said to be experiencing linear elastic response.

Axial Loading For the case of uniaxial loading of a prismatic member, the stress–strain relationship is

$$\sigma_a = \frac{P}{A} = E\varepsilon_a \tag{7}$$

where P is the applied axial force, A is the cross-sectional area of the member, ε_a is the axial strain, and E is a material constant known as the modulus of elasticity or Young's modulus. Equation (7) is commonly referred to as Hooke's law. The transverse strain, ε_t, is given by

$$\varepsilon_t = -\mu\varepsilon_a = \frac{-\nu\sigma_a}{E} \tag{8}$$

where ν is a material constant known as Poisson's ratio. Its value is always between 0 and $\frac{1}{2}$ for linear elastic, isotropic response.

If the member has a length L, the axial elongation, δ, is given by $\delta = \varepsilon_a L$. In view of Eq. (7) this becomes $\delta = PL/AE$. If P, A, or E varies along the length of the member, the deflection must be found by integrating $\varepsilon_a\, dx$ over the length of the member:

$$\delta = \int_0^L \frac{P\, dx}{AE}$$

Multiaxial State of Stress For the general case of a three-dimensional state of stress, the stress–strain relationships are

$$\varepsilon_{xx} = \frac{1}{E}[\sigma_{xx} - \nu(\sigma_{yy} + \sigma_{zz})] \qquad \gamma_{xy} = \frac{\sigma_{xy}}{G}$$

$$\varepsilon_{yy} = \frac{1}{E}[\sigma_{yy} - \nu(\sigma_{xx} + \sigma_{zz})] \qquad \gamma_{xy} = \frac{\sigma_{xz}}{G} \tag{9}$$

$$\varepsilon_{zz} = \frac{1}{E}[\sigma_{zz} - \nu(\sigma_{xx} + \sigma_{yy})] \qquad \gamma_{yz} = \frac{\sigma_{yz}}{G}$$

where G is a material constant known as the shear modulus or modulus of rigidity. Equations (9) are often referred to as the generalized Hooke's law. These equations can be inverted to give the stresses in terms of the strains:

$$\sigma_{xx} = \frac{E}{(1+\nu)(1-2\nu)}[(1-\nu)\varepsilon_{xx} + \nu(\varepsilon_{yy} + \varepsilon_{zz})]$$

$$\sigma_{xy} = G\gamma_{xy}$$

$$\sigma_{yy} = \frac{E}{(1+\nu)(1-2\nu)}[(1-\nu)\varepsilon_{yy} + \nu(\varepsilon_{xx} + \varepsilon_{zz})]$$

$$\sigma_{xz} = G\gamma_{xz}$$

$$\sigma_{zz} = \frac{E}{(1+\nu)(1-2\nu)}[(1-\nu)\varepsilon_{zz} + \nu(\varepsilon_{xx} + \varepsilon_{yy})]$$

$$\sigma_{yz} = G\gamma_{yz} \tag{10}$$

The material constants G, E, and ν are not independent of one another. It is shown in the theory of elasticity that they are related by the equation $G = E/2(1+\nu)$.

The dilatation, e, represents the change in volume per unit volume. For small strains it is given by $e = \varepsilon_{xx} + \varepsilon_{yy} + \varepsilon_{zz}$. For the case of pure hydrostatic stress, $\sigma_{xx} = \sigma_{yy} = \sigma_{zz} = -p$ and $\sigma_{xy} = \sigma_{xz} = \sigma_{yz} = 0$, where p is the uniform pressure acting on the member. In this case the dilatation is linearly related to the pressure by $e = -p/k$, where k is a material constant known as the bulk modulus. The bulk modulus is related to E and ν by the equation $k = E/3(1-2\nu)$.

One other elastic constant that arises in the theory of elasticity is Lamé's constant, λ. Unlike the other constants (E, ν, G, and k), λ has no physical significance. That is, there is no mechanical test that can be used to directly measure λ. Since only two material constants are required to completely describe linear elastic isotropic response, the constants E, ν, G, k, and λ can be interrelated (see Ref. 1, p. 44, for details).

Plane Stress Plane stress, or the two-dimensional state of stress, exists at a point if an orientation can be found such that the stress in one of the three coordinate directions is zero. If this direction of zero stress is arbitrarily taken to be the z direction (i.e., $\sigma_{zz} = \sigma_{xz} = \sigma_{yz} = 0$), then Eqs. 9 reduce to

$$\varepsilon_{xx} = \frac{1}{E}(\sigma_{xx} - \nu\sigma_{yy}) \qquad \gamma_{xy} = \frac{\sigma_{xy}}{G}$$

$$\varepsilon_{yy} = \frac{1}{E}(\sigma_{yy} - \nu\sigma_{xx}) \qquad \gamma_{xz} = 0 \tag{11}$$

$$\varepsilon_{zz} = \frac{-\nu}{E}(\sigma_{xx} + \sigma_{yy}) \qquad \gamma_{yz} = 0$$

In addition, ε_{zz} can be written as

$$\varepsilon_{zz} = \frac{-\nu}{1-\nu}(\varepsilon_{xx} + \varepsilon_{yy})$$

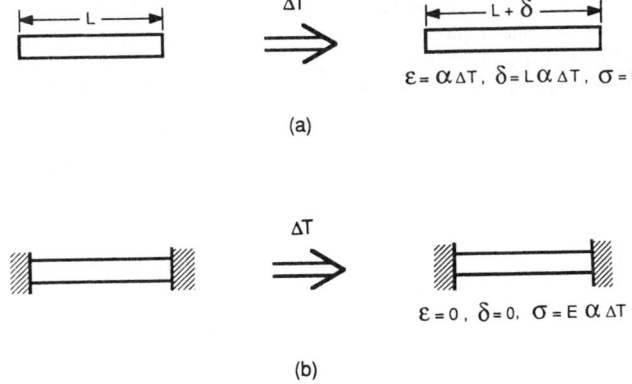

$$\varepsilon = \alpha \, \Delta T, \quad \delta = L \alpha \, \Delta T, \quad \sigma = 0$$

(a)

$$\varepsilon = 0, \quad \delta = 0, \quad \sigma = E \alpha \, \Delta T$$

(b)

Fig. 5 Thermally induced stresses and strains: (a) member free to expand; (b) member fixed at both ends.

Equations (10) reduce to

$$\sigma_{xx} = \frac{E}{1-v^2}(\varepsilon_{xx} + v\varepsilon_{yy}) \qquad \sigma_{xy} = G\gamma_{xy}$$

$$\sigma_{yy} = \frac{E}{1-v^2}(\varepsilon_{yy} + v\varepsilon_{xx}) \qquad \sigma_{xz} = 0 \tag{12}$$

$$\sigma_{zz} = 0 \qquad\qquad\qquad \sigma_{yz} = 0$$

Plane Strain Plane strain occurs at a point if an orientation can be found such that the strain in one of the three coordinate directions is zero. If this direction is arbitrarily taken to be the z direction, then $\varepsilon_{zz} = \gamma_{xz} = \gamma_{yz} = 0$. Equations (9) reduce to

$$\varepsilon_{xx} = \frac{1+v}{E}[(1-v)\sigma_{xx} - v\sigma_{yy}] \qquad \gamma_{xy} = \frac{\sigma_{xy}}{G}$$

$$\varepsilon_{xx} = \frac{1+v}{E}[(1-v)\sigma_{yy} - v\sigma_{xx}] \qquad \gamma_{xz} = 0 \tag{13}$$

$$\varepsilon_{zz} = 0 \qquad\qquad\qquad\qquad \gamma_{yz} = 0$$

Equations (10) reduce to

$$\sigma_{xx} = \frac{E}{(1+v)(1-2v)}[(1-v)\varepsilon_{xx} + v\varepsilon_{yy}]$$

$$\sigma_{xy} = G\gamma_{xy}$$

$$\sigma_{yy} = \frac{E}{(1+v)(1-2v)}[(1-v)\varepsilon_{yy} + v\varepsilon_{xx}]$$

$$\sigma_{xz} = 0 \tag{14}$$

$$\sigma_{zz} = \frac{vE}{(1+v)(1-2v)}[\varepsilon_{xx} + \varepsilon_{yy}]$$

$$\gamma_{yz} = 0$$

In addition, σ_{zz} can be written as $\sigma_{zz} = v(\sigma_{xx} + \sigma_{yy})$.

Thermal Stresses and Strains Thermal stresses and strains can occur when a member is subjected to a temperature change. For the case of a uniform, unrestrained slender rod of length L, the total increase in length, δ, due to a temperature change ΔT is given by (Fig. 5a)

$$\delta = L \int_{T}^{T+\Delta T} \alpha(T)\, dT$$

where α is the material's coefficient of thermal expansion. For many materials, α can be considered to be constant over a wide temperature range. In this case δ is given by $\delta = \alpha L \, \Delta T$, and the thermal strain is given by $\varepsilon = \delta/L = \alpha \, \Delta T$. If the slender rod is fixed at both ends so that no deformation can occur in the axial direction as the temperature is changed, an internal axial stress will develop (Fig. 5b). If the magnitude of this stress remains within the linear elastic range, its value is given by

$$|\sigma| = E \int_{T}^{T+\Delta T} \alpha(T)\, dT$$

If the temperature change is small enough so that α can be considered a constant, then $\sigma = E\alpha \, \Delta T$. Thus, the thermal stress is equal to E times the thermal strain that would develop if the member were free to expand. The determination of thermal stresses and strains for members of more complex geometries can be a formidable task. In such cases, assuming the stresses remain within the elastic range, it is necessary to apply the theory of thermal elasticity (see, e.g., Ref. 4).

1.3 Transformations of Stress and Strain

General Equations of Transformation Given the complete state of stress or strain in a given orientation, it is often desirable to obtain the state of stress or strain in some new orientation. Using the notation that

$\cos(i', j)$ represents the direction cosine between the original j coordinate axis and the new i' coordinate axis, the matrix equation for stress transformation from the $Oxyz$ coordinate system to the $Ox'y'z'$ coordinate system is given by

$$[\sigma'] = [\alpha]^T[\sigma][\alpha] \qquad (15)$$

In this equation, $[\sigma]$ is the stress matrix for the original xyz coordinate system [see Eq. (11)], $[\sigma']$ is the stress matrix for the new $x'y'z'$ coordinate system, and $[\alpha]$ is the matrix of direction cosines given by

$$[\alpha] = \begin{bmatrix} \cos(x', x) & \cos(y', x) & \cos(z', x) \\ \cos(x', y) & \cos(y', y) & \cos(z', y) \\ \cos(x', z) & \cos(y', z) & \cos(z', z) \end{bmatrix} \qquad (16)$$

Finally, $[\alpha]^T$ is the matrix transpose of $[\alpha]$.

The equation for strain transformation is similarly given by

$$[\varepsilon'] = [\alpha]^T[\varepsilon][\alpha] \qquad (17)$$

Here, $[\varepsilon]$ is the strain matrix for the original xyz coordinate system [see Eq. (2)], $[\varepsilon']$ is the strain matrix

for the new $x'y'z'$ coordinate system, and $[\alpha]$ is again given by Eq. (16).

Transformation about a Fixed Axis If the new orientation is obtained by rotating through an angle θ about the z axis (i.e., $z' = z$), Eqs. (15) and (16) can be greatly simplified. Using the notation of Fig. 6b, where θ is measured positive counterclockwise, it follows that

$$\cos(x', x) = \cos(y', y) = \cos\theta$$

$$\cos(y', x) = \cos(90° + \theta)$$

$$\cos(x', y) = \cos(90° - \theta)$$

$$\cos(z', x) = \cos(x', z) = \cos(z', y) = \cos(y', z) = 0$$

$$\cos(z', z) = 1$$

Equation (15) reduces to

$$\sigma'_{xx} = \tfrac{1}{2}(\sigma_{xx} + \sigma_{yy}) + \tfrac{1}{2}(\sigma_{xx} - \sigma_{yy})\cos 2\theta$$
$$+ \sigma_{xy}\sin 2\theta$$
$$\sigma'_{yy} = \tfrac{1}{2}(\sigma_{xx} + \sigma_{yy}) - \tfrac{1}{2}(\sigma_{xx} - \sigma_{yy})\cos 2\theta$$
$$- \sigma_{xy}\sin 2\theta \qquad (18)$$

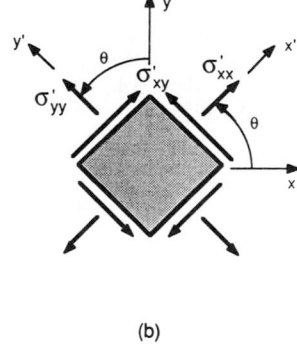

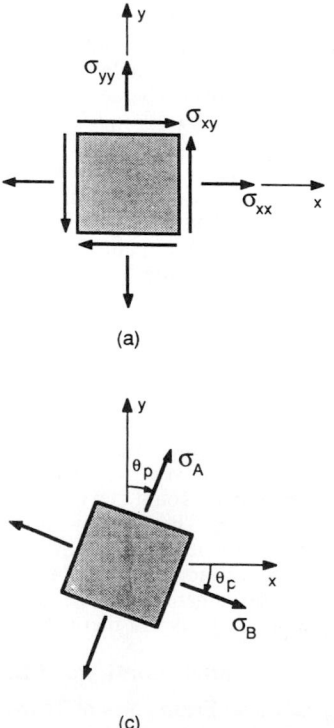

Fig. 6 Two-dimensional coordinate rotation: (a) original configuration; (b) arbitrary rotation; (c) rotation to principal coordinates.

$$\sigma'_{xy} = -\tfrac{1}{2}(\sigma_{xx} - \sigma_{yy})\sin 2\theta - \sigma_{xy}\cos 2\theta$$

$$\sigma'_{zz} = \sigma_{zz} \qquad \sigma'_{xz} = \sigma_{xz} \qquad \sigma'_{yz} = \sigma_{yz}$$

Similarly, Eq. (17) reduces to

$$\varepsilon'_{xx} = \tfrac{1}{2}(\varepsilon_{xx} + \varepsilon_{yy}) + \tfrac{1}{2}(\varepsilon_{xx} - \varepsilon_{yy})\cos 2\theta$$
$$\qquad + \varepsilon_{xy}\sin 2\theta$$

$$\varepsilon'_{yy} = \tfrac{1}{2}(\varepsilon_{xx} + \varepsilon_{yy}) - \tfrac{1}{2}(\varepsilon_{xx} - \varepsilon_{yy})\cos 2\theta \qquad (19)$$
$$\qquad - \varepsilon_{xy}\sin 2\theta$$

$$\varepsilon'_{xy} = -\tfrac{1}{2}(\varepsilon_{xx} - \varepsilon_{yy})\sin 2\theta + \varepsilon_{xy}\cos 2\theta$$

$$\varepsilon'_{zz} = \varepsilon_{zz} \qquad \varepsilon'_{xz} = \varepsilon_{xz} \qquad \varepsilon'_{yz} = \varepsilon_{yz}$$

Mohr's Circle for Stress and Strain Through further mathematical manipulation, Eqs. (18) can be written as

$$(\sigma'_{xx} - \sigma_{avg})^2 + (\sigma'_{xy})^2 = R^2$$

where

$$\sigma_{avg} = \frac{\sigma_{xx} + \sigma_{yy}}{2} \quad \text{and}$$

$$R = \left[\left(\frac{\sigma_{xx} - \sigma_{yy}}{2}\right)^2 + \sigma_{xy}^2\right]^{1/2}$$

This is the equation of a circle in the $(\sigma_{xx}, \sigma_{xy})$ plane known as Mohr's circle for stress. The horizontal and vertical axes are chosen to represent the applied normal and shear stresses, respectively (Fig. 7a). Normal stresses are plotted positive to the right. Shear stresses are plotted positive downward. With this convention, the points $(\sigma_{xx}, \sigma_{xy})$ and $(\sigma_{yy}, -\sigma_{xy})$ should be plotted on the Mohr's circle in order that the positive direction of θ is the same in Figs. 6b and 7a. Mohr's circle for stress thus consists of the points $(\sigma_{xx}, \sigma_{xy})$ and $(\sigma_{yy}, -\sigma_{xy})$ as 2θ ranges from 0° to 360°. Note that exactly the same Mohr's circle would be obtained if shear stresses were plotted positive upward and the points $(\sigma_{xx}, -\sigma_{xy})$ and $(\sigma_{yy}, \sigma_{xy})$ were plotted. A variety of sign conventions for the shear stress can be found in the literature, but all

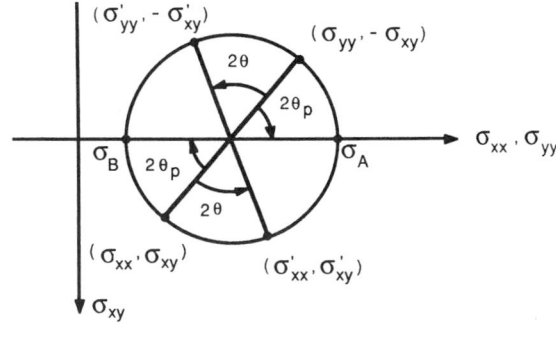

(a)

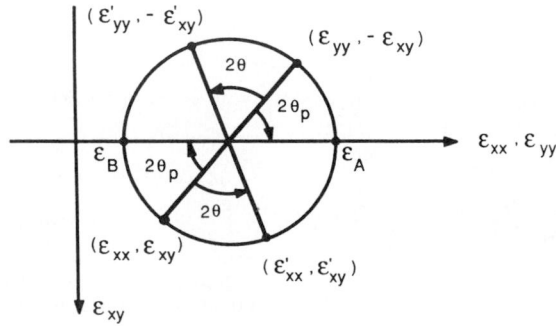

(b)

Fig. 7 Mohr's circles for stress and strain: (a) Mohr's circle for stress; (b) Mohr's circle for strain.

of them will lead to the same Mohr's circle when used consistently in each system.

It is important to keep in mind that a rotation of the element in Fig. 6a through an angle θ to bring about the orientation of Fig. 6b corresponds to a rotation of 2θ on the Mohr's circle in Fig. 7a. It is also important to realize that the application of Mohr's circle relies only on the fact that the old and new coordinate systems have a common axis of rotation (e.g., the z axis). It is not required that the element also be in a state of plane stress as is often stated in elementary strength of materials texts.

Mohr's circle for strain can be constructed in a manner analogous to Mohr's circle for stress. The appropriate equation is

$$(\varepsilon'_{xx} - \varepsilon_{avg})^2 + (\varepsilon'_{xy})^2 = R^2$$

where

$$\varepsilon_{avg} = \frac{\varepsilon_{xx} + \varepsilon_{yy}}{2} \quad \text{and}$$

$$R = \left[\left(\frac{\varepsilon_{xx} - \varepsilon_{yy}}{2} \right)^2 + \varepsilon_{xy}^2 \right]^{1/2}$$

Normal strains are plotted horizontally with positive values to the right. Tensorial shear strains are plotted vertically with positive values downward. Mohr's circle for strain thus consists of the points $(\varepsilon_{xx}, \varepsilon_{xy})$ and $(\varepsilon_{yy}, -\varepsilon_{xy})$ as 2θ ranges from $0°$ to $360°$ (Fig. 7b).

Principal Stresses and Strains   For any state of stress at a point there exists an orientation such that the shear stresses vanish. In this orientation the stress matrix is given by

$$[\sigma] = \begin{bmatrix} \sigma_1 & 0 & 0 \\ 0 & \sigma_2 & 0 \\ 0 & 0 & \sigma_3 \end{bmatrix}$$

where σ_1, σ_2, and σ_3 are known as the principal stresses. It is common to assume that the orientation has been taken so that σ_1 is the largest algebraic stress and σ_3 is the smallest algebraic stress. The principal stresses represent the solutions to the cubic equation

$$\sigma_n^3 - I_1\sigma_n^2 + I_2\sigma_n - I_3 = 0 \tag{20}$$

where $I_1 = \sigma_{xx} + \sigma_{yy} + \sigma_{zz}$
$I_2 = \sigma_{xx}\sigma_{yy} + \sigma_{xx}\sigma_{zz} + \sigma_{yy}\sigma_{zz}$
$\quad - \sigma_{xy}^2 - \sigma_{xz}^2 - \sigma_{yz}^2$
$I_3 = \sigma_{xx}\sigma_{yy}\sigma_{zz} - \sigma_{xx}\sigma_{yz}^2 - \sigma_{yy}\sigma_{xz}^2 - \sigma_{zz}\sigma_{xy}^2$
$\quad + 2\sigma_{xy}\sigma_{xz}\sigma_{yz}$

I_1, I_2, and I_3 are known as the first, second, and third invariants of stress, respectively. These quantities are independent of the orientation being considered.

Once the principal stresses have been found, their orientations can be determined by substituting each principal stress, σ_n, individually into the following set of equations and solving for the direction cosines associated with that principal stress:

$$(\sigma_{xx} - \sigma_n)\cos(n, x) + \sigma_{yx}\cos(n, y) + \sigma_{zx}\cos(n, z) = 0$$

$$\sigma_{xy}\cos(n, x) + (\sigma_{yy} - \sigma_n)\cos(n, y) + \sigma_{zy}\cos(n, z) = 0$$

$$\sigma_{xz}\cos(n, x) + \sigma_{yz}\cos(n, y) + (\sigma_{zz} - \sigma_n)\cos(n, z) = 0$$

$$\cos^2(n, x) + \cos^2(n, y) + \cos^2(n, z) = 1$$

The last equation is based on the fact that a direction cosine vector has unit magnitude. It is required since the first three equations are not linearly independent and hence are insufficient to explicitly solve for the direction cosines.

Just as for stresses, an orientation can be found such that the shear strains vanish. The principal strains $(\varepsilon_1, \varepsilon_2,$ and $\varepsilon_3)$ represent the solutions to the cubic equation

$$\varepsilon_n^3 - J_1\varepsilon_n^2 + J_2\varepsilon_n - J_3 = 0 \tag{21}$$

where $J_1 = \varepsilon_{xx} + \varepsilon_{yy} + \varepsilon_{zz}$
$J_2 = \varepsilon_{xx} + \varepsilon_{yy} + \varepsilon_{xx}\varepsilon_{zz} + \varepsilon_{yy}\varepsilon_{zz}$
$\quad - \varepsilon_{xy}^2 - \varepsilon_{xz}^2 - \varepsilon_{yz}^2$
$J_3 = \varepsilon_{xx}\varepsilon_{yy}\varepsilon_{zz} - \varepsilon_{xx}\varepsilon_{yz}^2 - \varepsilon_{yy}\varepsilon_{xz}^2 - \varepsilon_{zz}\varepsilon_{xy}^2$
$\quad + 2\varepsilon_{xy}\varepsilon_{xz}\varepsilon_{yz}$

J_1, J_2, and J_3 are known as the first, second, and third invariants of strain, respectively. These quantities remain the same regardless of the orientation under consideration.

Once the principal strains have been found, their orientations can be determined by substituting each principal strain, ε_n, individually into the following set of equations and solving for the direction cosines associated with that principal strain:

$$(\varepsilon_{xx} - \varepsilon_n)\cos(n, x) + \varepsilon_{yx}\cos(n, y) + \varepsilon_{zx}\cos(n, z) = 0$$

$$\varepsilon_{xy}\cos(n, x) + (\varepsilon_{yy} - \varepsilon_n)\cos(n, y) + \varepsilon_{zy}\cos(n, z) = 0$$

$$\varepsilon_{xz}\cos(n, x) + \varepsilon_{yz}\cos(n, y) + (\varepsilon_{zz} - \varepsilon_n)\cos(n, z) = 0$$

$$\cos^2(n, x) + \cos^2(n, y) + \cos^2(n, z) = 1$$

Again, the last equation is required since the first three equations are not linearly independent.

For linear elastic isotropic response, $\sigma_{ij} = G\gamma_{ij} = 2G\varepsilon_{ij}$ for $i \neq j$. It thus follows in this case that $\varepsilon_{ij} = 0$ when $\sigma_{ij} = 0$. That is, the principal directions of stress and strain coincide. If the principal stresses are known, the principal strains can be found using Eqs. (9) rather than solving Eq. (21). Similarly, if the principal strains are known, the principal stresses can be found using Eqs. (10) rather than solving Eq. (20).

For anisotropic response, such as often occurs in composite materials, the principal strains may have a different orientation than the principal stresses. It is worth noting, however, that Eqs. (20) and (21) are still valid for anisotropic response, as are Mohr's circles for stress and strain.

For the case of plane stress ($\sigma_{zz} = \sigma_{xz} = \sigma_{yz} = 0$) in a material exhibiting linear elastic isotropic response, it is clear that the z direction is a principal direction. The other two principal directions lie in the xy plane. Since the horizontal axis of Mohr's circle represents a state of zero shear stress, the principal stresses σ_A and σ_B and their directions relative to the x axis (θ_p and $\theta_p + 90°$, respectively) can be found easily from Mohr's circle (Fig. 7a). The corresponding rotated element is shown in Fig. 6c. The principal strains ε_A and ε_B in this case can also be found easily from the corresponding Mohr's circle for strain (Fig. 7b). The principal strain in the z direction can be found from $\varepsilon_{zz} = (-\nu/E)(\sigma_{xx} + \sigma_{yy})$.

Similarly, for the case of plane strain ($\varepsilon_{zz} = \varepsilon_{xz} = \varepsilon_{yz} = 0$), the z direction is again a principal direction. The determination of the principal stresses and strains lying within the xy plane can once more be accomplished using Mohr's circles. The principal stress in the z direction can be found from

$$\sigma_{zz} = \frac{\nu E}{(1+\nu)(1-2\nu)}(\varepsilon_{xx} + \varepsilon_{yy})$$

Maximum Shear Stress The maximum shear stress at a point is given by $\tau_{max} = (\sigma_1 - \sigma_3)/2$, where σ_1 and σ_3 are the maximum and minimum principal stresses, respectively. The maximum shear stress acts in the 1–3 plane at an angle of $\pm 45°$ to the 1 axis.

1.4 Tension Test

The tension test is one of the most fundamental and important of mechanical tests. A wide and diversified set of material properties can be determined from it. Specimens used for tension tests commonly consist of round or rectangular cross sections with a gage region, a gripping region, and an intermediate fillet region. It is important that the gage region has a reduced cross-sectional area; otherwise, the specimen will often fail in or near the region of gripping. The *Annual Book of ASTM Standards*[5] is a multiple-volume reference containing adopted standards for a wide variety of tests. Included within this book are standards for tension tests on a number of different types of materials. These standards describe the required specimen dimensions, gripping arrangement, and testing procedure. In particular, see Volume 3.01, standard E8 for tension testing of metallic materials.

In general, to perform a tension test, the following procedure is used. First, the specimen is appropriately gripped in the testing system. This is usually a screw-driven tension testing machine or the more sophisticated closed-loop, servo-hydraulic mechanical testing system (see Section 9.1 for further discussion of these testing systems). Next, a displacement measuring device must be attached to the specimen. When specimen size and rigidity are sufficient, the attachment of an extensometer within the gage region of the specimen is desirable. Finally, the specimen is pulled apart at a constant displacement rate until failure occurs. While the test is proceeding, the load on the specimen and the displacement of the specimen are continuously recorded. The resulting load–displacement plot must be converted to an engineering stress–strain plot or a true stress–strain plot so that mechanical properties can be determined.

Engineering Stress and Strain Engineering stress S is defined as the instantaneous value of the load P on the specimen divided by the specimen's original cross-sectional area, A_0. That is, $S = P/A_0$. Engineering strain e is defined by $e = (L - L_0)/L_0$, where L is the instantaneous gage length and L_0 the initial gage length. Engineering stress–strain diagrams for some common materials are shown in Fig. 8. For some materials, the engineering stress does not monotonically increase to failure. This is due to the fact that the specimen necks down within the gage area. When necking down occurs, one small region of the specimen begins to deform at a much more rapid rate than the rest of the specimen. This region of the specimen thus has a much smaller cross-sectional area and hence a larger stress. The strain rate within this region is much higher than in the rest of the specimen; as a result, the load applied to the specimen must be lowered in order to maintain a constant displacement rate in the necked-down region. The state of stress within this region is no longer uniaxial, which further complicates matters. When engineering stress and strain are used, this necking-down phenomenon disguises the true nature of the material's mechanical response. In order to obtain a more realistic material response up to the point of failure, the true stress–strain diagram should be determined.

True Stress and Strain True stress σ is defined as $\sigma = P/A_i$, where A_i is the instantaneous minimal cross-sectional area and P is again the instantaneous value of the load acting on the specimen. Before necking down occurs, A_i will be approximately the same along the entire gage region. Once necking down begins, however, A_i must be measured at the point where the necking down is taking place.

True strain ε is defined by $\varepsilon = \log(L_i/L_0)$, where log represents the natural logarithm. For this equation to hold, the strain must be uniform within the region where L_i is being measured. This assumption is obviously invalid when necking down occurs within the gage region. To overcome this problem, it is often assumed that plastic deformation occurs with

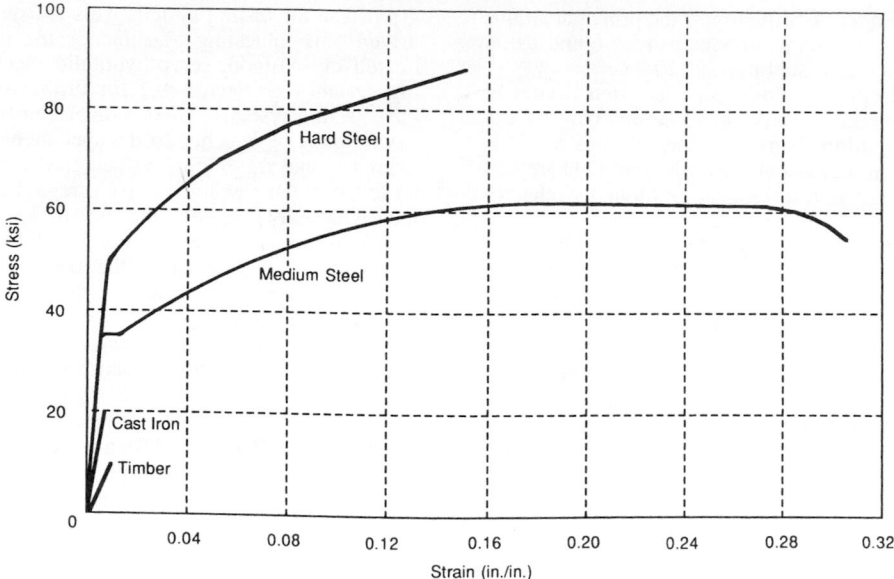

Fig. 8 Monotonic engineering stress–strain diagrams for some common materials.

no change in volume: $A_0 L_0 = A_i L_i$. This assumption is fairly reasonable for many ductile metals. By using this assumption, the true strain can be written as $\varepsilon = \log(A_0/A_i)$.

If the instantaneous minimal cross-sectional area can be measured during a test along with P and L_i, and if the constant-volume deformation assumption is valid while plastic deformation is occurring, then a true stress–strain diagram can be constructed. The construction should consist of using the equation $\varepsilon = \log(L_i/L_0)$ for true strain in the elastic and initial plastic regions and the equation $\varepsilon = \log(A_0/A_i)$ once significant plastic deformation has begun. In practice, however, the equation $\varepsilon = \log(A_0/A_i)$ is used for the entire strain range since the amount of error introduced is usually negligible. The equation $\sigma = P/A_i$ for true stress is valid throughout the entire test. Accurate construction of a true stress–strain diagram becomes exceedingly difficult if the plastic deformation cannot be assumed to occur at a constant volume. Finally, it should be mentioned that neither the true stress–strain diagram nor the engineering stress–strain diagram accounts for the fact that the state of stress within the necked-down region is multiaxial. There is no simple way to account for the effect of this multiaxial state of stress on the material's response, so it is customarily ignored. A schematic representation of an engineering stress–strain diagram and the corresponding true stress–strain diagram can be found in Fig. 9.

Relationships between Engineering and True Stress and Strain There are two useful equations for constructing a true stress–strain diagram if the engineering stress–strain diagram is known. The first equation is given by $\varepsilon = \log(1 + e)$. It is based on the assumption of homogeneous strains and is thus valid only up to the point in the tension test when necking down begins. The second equation, $\sigma = S(1 + e)$, assumes a constant-volume process as well as homogeneous strains and is thus valid between the onset of plastic deformation (assuming constant-volume plastic deformation) and the beginning of the necking-down process. In fact, this equation will be slightly inaccurate because of the fact that the elastic deformation that took place prior to the beginning of plastic flow was not a constant-volume process. However, because of the small size of the elastic strains, the error introduced can usually be ignored.

For small strains (such as those that occur within the elastic range of most metals), there is no significant difference between engineering stress–strain and true stress–strain, but for large strains the difference can be considerable. It becomes apparent that the true stress–strain diagram is considerably more difficult to obtain than the engineering stress–strain diagram. As a result, material properties based on the engineering stress–strain diagram are usually reported. However, the importance of the true stress–strain diagram in giving a more representative picture of a material's response should not be overlooked.

Mechanical Properties Determined from the Tension Test A variety of mechanical properties can be determined from a tension test. Of course, some properties such as the yield strength and the strain-hardening exponent only apply to ductile materials.

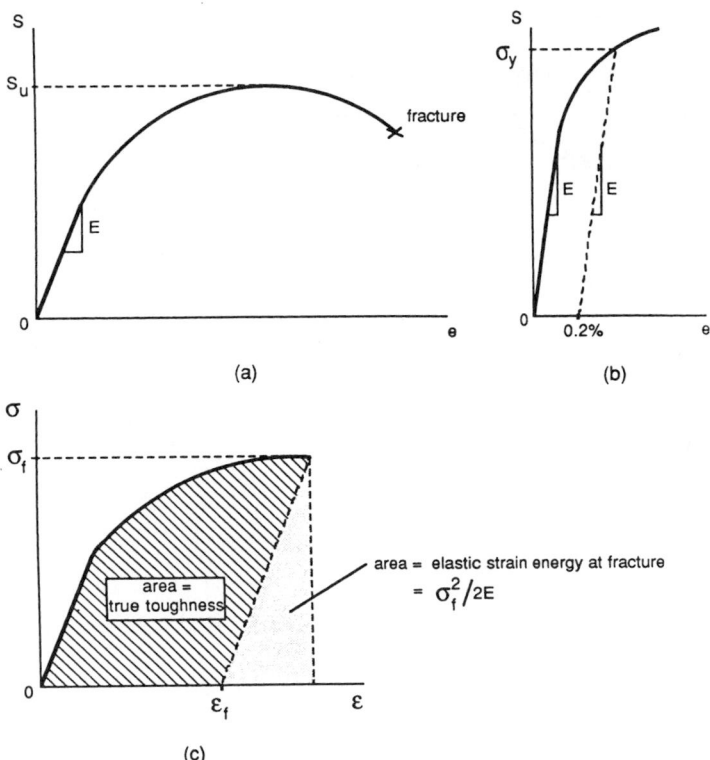

Fig. 9 Material properties determined from stress–strain diagrams: (a) engineering stress–strain diagram; (b) determination of σ_y by offset method; (c) true stress–strain diagram.

Table 1 lists mechanical properties for a wide variety of materials. Many of the mechanical properties that will be discussed are displayed graphically in Fig. 9.

The *modulus of elasticity E* is given by the slope of the initial straight-line portion of the stress–strain diagram.

The *proportional limit* is the largest stress for which the stress–strain relationship is linear. It is the largest stress for which Hooke's law can be applied for the material. This property depends on the judgment of the observer and is thus of limited value.

The *elastic limit* is the largest stress to which a material can be subjected and still return to its original form upon removal of the load. It can be determined by application and release of a series of increasing loads to the specimen until a permanent deformation is observed upon release of the load. With the advent of equipment that can measure extremely small strains, it has been found that most materials will experience a small amount of permanent deformation at stresses far below what was once thought to be the elastic limit. The determination of this property is thus dependent on the sensitivity of the testing equipment. Its use in design is strongly discouraged.

The *yield strength* σ_y is roughly defined as the stress at which the material begins to experience significant plasticity. In order to make this material property of use to the designer, the yield strength is now usually taken to be the stress that will cause a specified amount of permanent set (also called offset). The amount of permanent set specified is arbitrary, and hence the value used should be reported along with the yield strength. The most commonly used offset is 0.2%. The yield strength is determined by drawing a line of slope E on the engineering stress–strain diagram through the point on the strain axis representing the desired amount of offset. The stress corresponding to the intersection of this line with the stress–strain curve represents the yield strength (Fig. 9b).

The *ultimate strength* S_u is the maximum tensile load a specimen can resist, P_{max}, divided by the specimen's original cross-sectional area. Thus, $S_u = P_{max}/A_0$.

The *percent elongation* equals $100 (L_f - L_0)/L_0$, where L_f is the final gage length at fracture. The percent elongation varies depending on the initial gage length and is of somewhat limited value.

The *percent reduction of area* is given by %RA = $100(A_0 - A_f)/A_0$, where A_f is the cross-sectional

Table 1 Mechanical Properties of Selected Materials

Material	Specific Weight (lb/in.^{3})	E (10^6 psi)	G (10^6 psi)	σ_y (ksi)	σ_u (ksi)	Percentage of Elongation (in 2 in.)	α ($10^{-6}/°$F)
Aluminum							
2014 T6 (155 BHN)	0.10	10.5	4	67	74	13	12.8
2024 T4	0.1	10.5	4	44	69	20	12.9
6061 T6	0.098	10.0	3.7	37	42	17	13.1
7075 T6 (95 BHN)	0.1	10.3	4	68	84	11	13.1
Cast iron							
Gray # 20	0.251	14	—	—	20	—	0.60
Gray # 30	0.260	15.2	—	—	30	—	0.60
Gray # 40	0.260	18.3	—	—	40	—	0.60
Gray # 60	0.270	19	—	—	60	—	0.60
Malleable	0.266	26	8.8	32–45	50–65	—	0.75
Nodular	0.257	23.5	—	45–65	60–100	—	0.66
Concrete (compression)							
Medium strength	0.084	3.6	—	—	4.0	—	5.5
High strength	—	4.5	—	—	6.0	—	5.5
Copper							
High purity, annealed	0.323	17	6.4	10	33	50	9.8
High purity, cold worked	0.323	17	6.4	45	50	10	9.8
Glass (compression)							
98% silica	0.079	9.6	4.1	—	7	—	44
Magnesium: AZ80A-T5	0.065	6.5	2.4	38	55	7	16.0
Nickel 200	0.321	30	—	25	65	45	7.4
Nylon							
6 (cast)	0.04	0.30	—	—	8	50	45
Molded	0.04	0.55	—	13	13	20	44
Phosphor bronze							
Cold rolled (510)	0.320	15.9	5.9	75	81	10	9.9
Spring temper (524)	0.317	16	—	—	122	4	10.2
Polyethylene							
Medium density	0.033	—	—	—	2	200	120
Polystyrene (molded)	0.039	—	—	5	5	2–30	40
Rubber							
Natural	0.034	—	—	—	3	800	400
Neoprene	0.045	—	—	—	3	850	350
Steel							
SAE 1005 (90 BHN)	0.283	30	12	38	50	40	6.5
SAE 1005 (125 BHN)	0.283	30	12	60	65	30	6.5
SAE 1020 (90 BHN)	0.283	30	12	35	65	30	6.5
SAE 1045 (225 BHN)	0.283	29	12	92	105	—	6.5
SAE 1045 (450 BHN)	0.283	30	12	220	230	—	6.5
SAE 1045 (650 BHN)	0.283	30	12	270	325	—	6.5
SAE 4142 (670 BHN)	0.283	29	12	240	355	—	6.5
SAE 5160 (430 BHN)	0.283	28	12	222	242	—	6.5
SAE 9262 (260 BHN)	0.283	30	12	200	227	—	6.5
SAE 9262 (410 BHN)	0.281	29	12	300	375	—	6.5
Stainless steel							
AISI 304 (160 BHN)	0.29	27	—	37	108	—	9.6
18 NI Maraging (460 BHN)	0.29	27	—	260	270	—	6.3
17-7 PH (TH-1050)	0.281	29	—	182	193	10	6.0
Titanium							
6% AL, 4% V	0.161	16.5	—	120	130	10	5.3
5% AL, 2.5% Sn	0.161	17	6.2	110	115	—	5.7
8-1.1	0.170	17	—	150	160	15	4.7
Wood (loaded parallel to grain)							
Birch	0.026	2.1	—	8.3	2.0	—	1.1
Douglas fir	0.019	1.8	—	—	—	—	1.7–2.5
Eastern spruce	0.016	1.3	—	—	—	—	1.7–2.5
Southern pine	0.022	1.6	—	—	—	—	1.7–2.5
White oak	0.028	1.6	—	7.0	1.9	—	2.7

Note: σ_y is the monotonic yield strength. For some materials, the cyclic yield strength (discussed in Section 8.4) differs radically from σ_y. AISI, American Iron and Steel Institute; SAE, Society of Automotive Engineers; BHN, Brinell hardness number.

area of the specimen at the point of fracture. The %RA is not as strongly dependent on specimen geometry as the percent elongation. It is a good measure of a material's ductility.

The *fracture strength* σ_f represents the true stress just prior to fracture. It should not be confused with the ultimate strength.

The *fracture ductility* ε_f is the true plastic strain at fracture. If the plastic deformation occurs without volume change, ε_f is related to %RA by

$$\varepsilon_f = \log\left(\frac{A_0}{A_f}\right) = \log\frac{100}{100 - \%RA}$$

Toughness has several common definitions. In its simplest form, the toughness is the area under the engineering stress–strain diagram.

True toughness represents the irrecoverable work done on a material during plastic deformation. It is given by the area under a plot of true stress versus true plastic strain, ε_p. The true plastic strain is related to σ and ε by $\varepsilon_p = \varepsilon - \sigma/E$. The true toughness is equivalently given by the total area under the true stress–strain diagram minus the elastic strain energy at fracture, $\sigma_f^2/2E$ (Fig. 9c).

For ductile materials it is often found that a simple power law relationship exists between the true stress and the true plastic strain. This relationship takes the form

$$\sigma = K\varepsilon_p^n \tag{22}$$

where K is known as the *strength coefficient*, and n is the *strain-hardening exponent*. These two quantities are easily determined by plotting $\log \sigma$ versus $\log \varepsilon_p$. The slope of such a plot gives n, while K is given by the value of σ corresponding to $\varepsilon_p = 1$.

2 BEAMS AND BENDING

2.1 Shear and Bending Moment in Beams

Classification of Beams A *beam* is a structural member whose length is large compared to its transverse dimensions and is subjected to forces acting transverse to its longitudinal axis. The following categories of beams are simple abstractions that approximate actual beams used in practice.

A *simple beam*, or simply supported beam (Fig. 10a), has a roller support at one end and a pin support at the other. The ends of a simple beam cannot support a bending moment but can support upward and downward vertical loads. Stated differently, the ends are free to rotate but cannot translate in the vertical direction. The end with the roller support is free to translate in the axial direction.

A *cantilever beam* (Fig. 10b) has one end rigidly fixed and the other end free. The fixed end can neither translate nor rotate while the free end can do both.

A *continuously supported beam* (Fig. 10c) is a beam resting on more than two supports.

A *fixed beam* (Fig. 10d) is rigidly fixed at both ends.

A *restrained beam* (Fig. 10e) is rigidly fixed at one end and simply supported at the other.

An *overhanging beam* (Fig. 10f) projects beyond one or both ends of its supports.

Loads on a beam commonly include concentrated loads (Fig. 10a), distributed loads (Fig. 10b), concentrated moments (Fig. 10c), and combinations of these.

Relations among Load, Shear, and Bending Moment At any point along a beam, the shear force and bending moment acting on the beam cross section must be such as to balance the external loads on either side of the cross section. The commonly used sign convention for internal shear force and bending moment is shown in Fig. 11. It is important to determine the points along a beam where the shear force and bending moment are maximum since it is at these points that the shear stresses and bending stresses, respectively, reach their maximum values. These points are commonly referred to as critical points.

Consider an infinitesimal section of a beam that is under a general state of loading (Fig. 12). Equilibrium in the vertical direction requires that

$$\frac{dV}{dx} = -q(x) \tag{23a}$$

or

$$V_B - V_A = \Delta V_{A-B} = -\int_{x_A}^{x_B} q(x)\,dx \tag{23b}$$

where $q(x)$ is the applied distributed load, A and B are arbitrary points along the beam, V is the internal shear force, and ΔV_{A-B} is the change in shear force between points A and B. There is equilibrium of moments if

$$\frac{dM}{dx} = V(x) \tag{24a}$$

or

$$M_B - M_A = \Delta M_{A-B} = \int_{x_A}^{x_B} V(x)\,dx \tag{24b}$$

where ΔM_{A-B} is the change in bending moment between points A and B. Equation (23b) is valid only if no concentrated loads act between points A and B since discontinuities in the shear occur at points of application of concentrated loads. Similarly, Eq. (23b) is valid only if no concentrated moments act between points A and B.

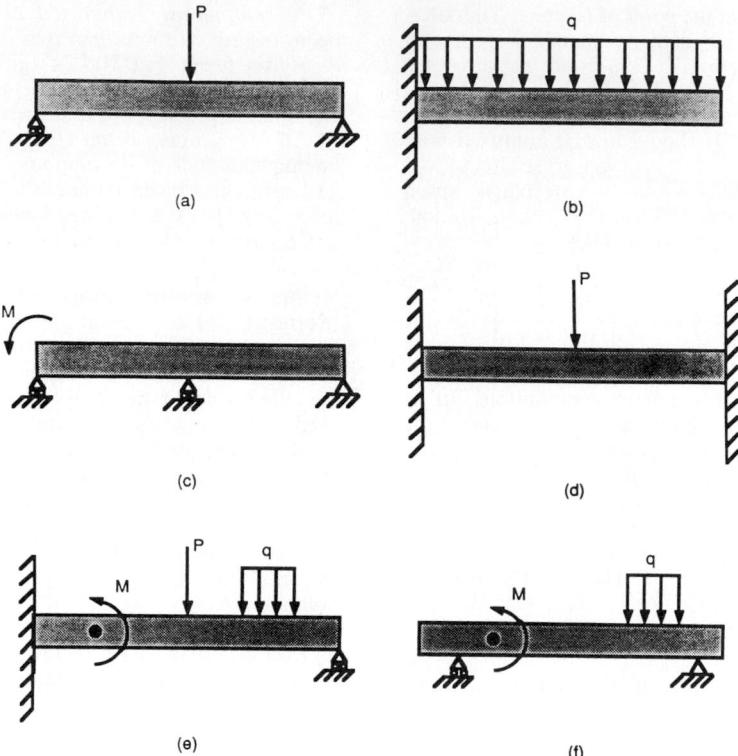

Fig. 10 Classification of beams and bending: (*a*) simple beam, concentrated load; (*b*) cantilever beam, distributed load; (*c*) continuous beam, concentrated moment; (*d*) fixed beam, concentrated load; (*e*) restrained beam, combined loading; (*f*) overhanging beam, combined loading.

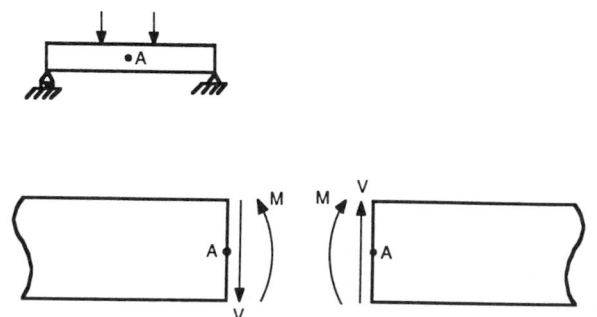

Fig. 11 Positive-sign convention for internal shear and bending moment.

Shear and Bending Moment Diagrams
Equations (23b) and (24b) provide a simple means for determining the shear and bending moment as a function of distance along the beam. In particular, if no distributed or concentrated loads act between two points on a beam, Eq. (23b) shows that the shear force is constant in this region. Equation (24b) shows that

the moment varies linearly in this region, provided that no concentrated moments exist.

A shear diagram is a graphical representation of the vertical shear at every point along a beam. A bending moment diagram is a graphical representation of the bending moment at every point along a beam. It is clear from Eq. (24b) that the bending moment at any

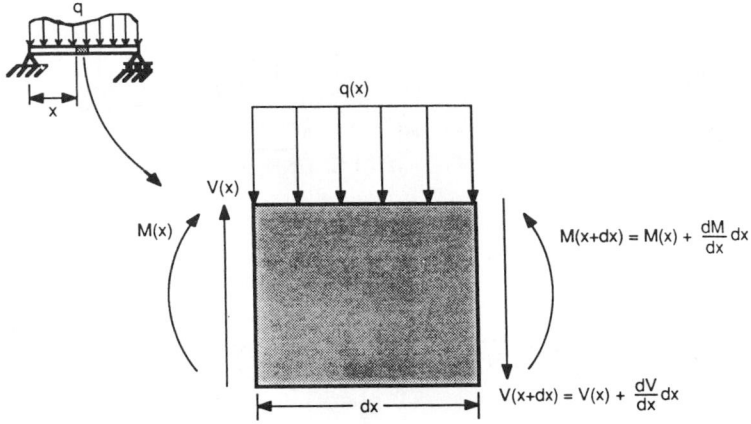

Fig. 12 Infinitesimal element of a beam.

point along a beam is equal to the area under the shear diagram up to that point, provided that no concentrated moments are present. If concentrated moments do exist, they will cause abrupt changes in the bending moment diagram, but this situation can be accounted for easily.

The general procedure for constructing shear and bending moment diagrams is described with the aid of Fig. 13:

1. Using the equations of static equilibrium ($\Sigma F_y = 0$ and $\Sigma M_z = 0$), determine the reactions at the supports (Fig. 13b). If the beam is statically indeterminate, the reactions can often be found by considering the deformations of the beam (see Section 2.4).

2. Determine the equations for shear and bending moment within each region of the beam in which the loading is continuous (Fig. 13c). Every discontinuity in loading, such as the application of a concentrated force, requires that another set of equations for shear and bending moment be determined.

3. Using the equations derived in step 2, draw the shear and bending moment diagrams (Fig. 13d).

Shear and bending moment diagrams are useful because they provide a rapid means for determining the critical points in a beam. These are the points at which failure is most likely to occur. With practice, it becomes possible to draw the diagrams directly without the need for first deriving the equations for shear and bending moment (at least for relatively simple loadings).

2.2 Theory of Flexure

For a straight beam possessing a plane of symmetry and subjected to moments acting in that plane of symmetry, it is assumed that cross sections perpendicular to the longitudinal axis remain plane. This assumption is very accurate except for short beams subjected to shear forces. It follows from this assumption that the beam deforms into a circular arc subtending an angle θ (Fig. 14a). The only stress acting in the beam will be a normal stress component in the longitudinal direction, σ_{xx}. On the concave side, the strain ε_{xx} and stress σ_{xx} are negative (compressive), while on the convex side they are positive (tensile); consequently, there exists a surface in the beam where ε_{xx} and σ_{xx} are zero. The neutral surface is the surface within which the normal stress and strain are zero. The neutral axis is the intersection of the neutral surface with any cross section perpendicular to the longitudinal axis (Fig. 14b).

Since the normal strain within the neutral surface is zero, it follows that $L = \rho\theta$. Here, L is the original beam length and ρ is the radius of the arc AB of the neutral surface (Fig. 14a). At a radial distance y from the neutral surface (where y is shown as positive on the concave side of the beam), the length of the beam is given by $L' = (\rho - y)\theta$. The strain along this arc is given by $\varepsilon_{xx} = (L' - L)/L$, or equivalently by $\varepsilon_{xx} = -y/\rho$. Thus, the longitudinal strain varies linearly with distance from the neutral axis.

Assuming linearly elastic isotropic response, the longitudinal stress distribution is given by $\sigma_{xx} = E\varepsilon_{xx}$, or

$$\sigma_{xx} = \frac{-Ey}{\rho} \qquad (25)$$

From statics, the integral of σ_{xx} over the cross-sectional area of the beam must equal zero since there is no applied longitudinal force. Placing the origin of the coordinate axes at the center of the neutral surface and integrating over the cross-sectional area gives

$$-\frac{E}{\rho} \int y \, dA = 0$$

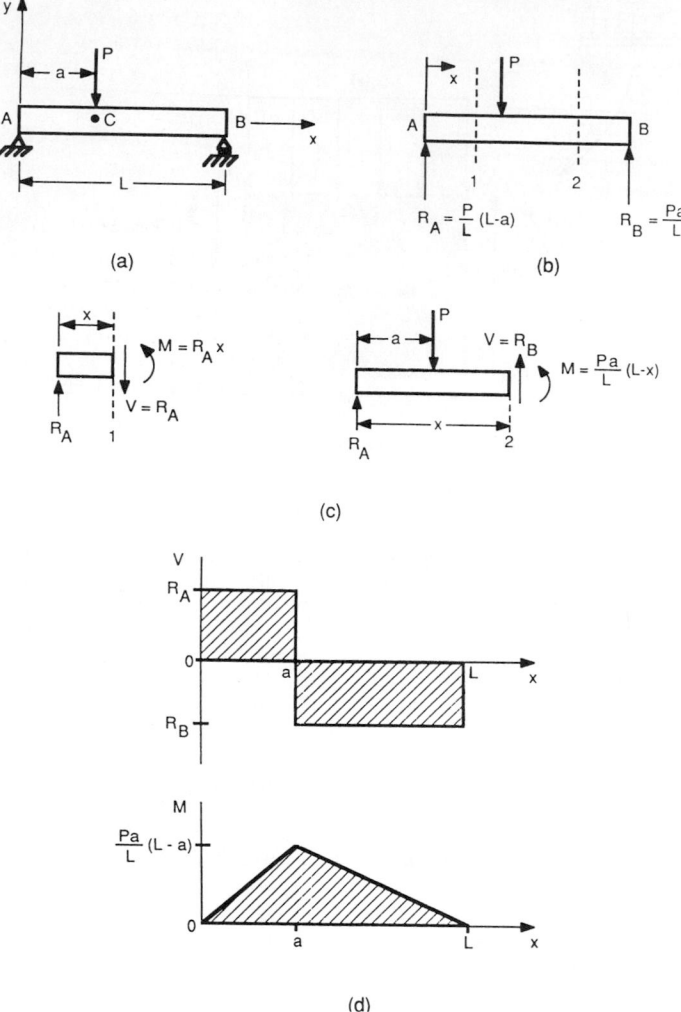

Fig. 13 Construction of shear and bending moment diagrams: (*a*) beam configuration; (*b*) static equilibrium of forces; (*c*) shear and bending moment in sections *AC* and *CB*; (*d*) shear and bending moment diagrams.

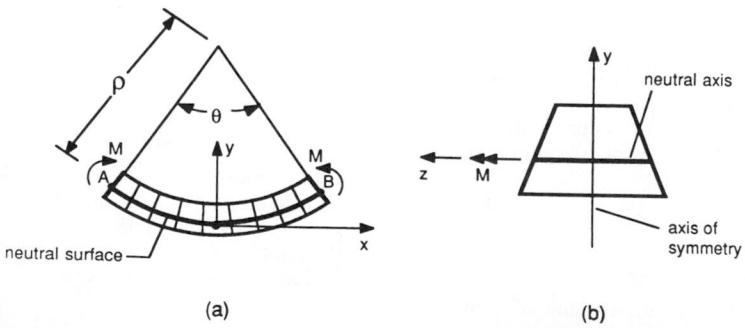

Fig. 14 Symmetric beam in pure bending: (*a*) side view; (*b*) enlarged cross-sectional view.

For finite ρ and nonzero E it follows that

$$\int y \, dA = 0$$

Stated in words, the first moment of the cross-sectional area with respect to the neutral axis must equal zero. The neutral axis therefore passes through the centroid of the cross section.

The applied moment M must be equal to the integral of $-y\sigma_{xx}$ over the cross-sectional area:

$$M = -\int y\sigma_{xx} \, dA$$

Substituting Eq. (25) into this expression gives

$$M = \frac{E}{\rho} \int y^2 \, dA$$

This can be rewritten as

$$\frac{1}{\rho} = \frac{M}{EI_{zz}} \tag{26}$$

where I_{zz} is the moment of inertia of the cross section with respect to the neutral axis (the z axis in Fig. 14b). Thus, $I_{zz} = \int y^2 \, dA$. Substituting for ρ from Eq. (25) into (26) leads to

$$\sigma_{xx} = \frac{-My}{I_{zz}} \tag{27}$$

This is known as the elastic flexure formula. The longitudinal stress is seen to have a maximum value at the farthest distance from the neutral axis. Denoting this distance by c, it follows that $|\sigma_{max}| = Mc/I_{zz}$.

The section modulus, S, is equal to I_{zz}/c. The maximum longitudinal stress in the beam can be written as $|\sigma_{max}| = M/S$. The section modulus represents the ability of a beam section to resist an applied bending moment. When designing beams, S should be made as large as is practical while minimizing the amount and cost of the material used. Table 2 gives values for the moment of inertia and section modulus of some common beam cross sections.

In summary, the following conditions must be satisfied if the elastic flexure formula is to be valid:

1. The beam must be straight or nearly straight.
2. The beam must possess a plane of symmetry and be subjected to moments acting in that plane of symmetry.
3. The material must be homogeneous and exhibit linear elastic, isotropic response.
4. The beam should be long enough so that shear forces do not cause warping of the cross section. A length-to-height ratio of 10 or better is normally sufficient to make shearing effects negligible.
5. The beam must have sufficient lateral stiffness so that it does not buckle under the applied loads (see Section 5.4 for details).

2.3 Shear Stresses in Beams

Symmetric beams that are subjected to a state of pure bending acting in the plane of symmetry experience only a flexural normal stress as given by Eq. 27. Thus, $\sigma_{xx} = -My/I_{zz}$, and $\sigma_{yz} = \sigma_{xz} = \sigma_{xy} = \sigma_{yy} = \sigma_{zz} = 0$ (using the coordinate axes given in Fig. 14). Beams that are subjected to transverse loads experience shear stresses σ_{xy} in addition to the flexural stress σ_{xx}. The average vertical shear stress throughout a cross section is given by $(\sigma_{xy})_{avg} = V/A$, where V is the resisting vertical shear force on the cross section, and A is the cross-sectional area.

In general, the shear stress is not uniformly distributed over the cross section. To determine the shear stress distribution, it must first be noted that the presence of a shear stress σ_{xy} implies the presence of a shear stress σ_{yx} of equal magnitude. This follows from the complementary property of shearing stresses (see Section 1.1).

Consider the beam shown in Fig. 15. For simplicity of demonstration, a rectangular cross section has been drawn, but Eq. (29), to be derived, is valid for an arbitrary cross-sectional shape provided there is at least one axis of symmetry, and the beam is loaded in that plane of symmetry. A portion CD of the beam is shown enlarged in Fig. 15c. Static equilibrium in the horizontal direction leads to

$$(\sigma_{yx})_{avg} L t = \int_{y=y_0}^{y=h} (\sigma_D - \sigma_C) \, dA \tag{28}$$

where L is the length of the beam element CD, t is the thickness of CD, and $(\sigma_{yx})_{avg}$ is the average horizontal shear stress acting on the bottom face of CD. From Eq. (24a), $M_D - M_C = VL$, since V is constant between C and D. Note that V is the shear force acting on the cross-sectional area of the whole beam at C, whereas V', shown in Fig. 15c, is only that portion of V that acts on the cross-sectional area of element CD. Also, recalling Eq. 27, $\sigma = -My/I$, and substituting $M_D - M_C = VL$ gives

$$\sigma_D - \sigma_C = -\frac{(M_D - M_C)y}{I} = -\frac{VLy}{I}$$

Finally, substituting this result into Eq. (28) and considering the limit as L goes to zero leads to

$$(\sigma_{yx})_{avg} = -\frac{VQ}{It} \tag{29}$$

Table 2   Properties for Common Beam Cross Sections

Beam Section	Moment of Inertia, I_{xx}	Section Modulus, $\dfrac{I_{xx}}{c}$
	$\dfrac{bd^3}{12}$	$\dfrac{bd^2}{6}$
	$\dfrac{b_1 d_1^3 - b_2 d_2^3}{12}$	$\dfrac{b_1 d_1^3 - b_2 d_2^3}{6d_1}$
	$\dfrac{bt_f^3}{6} + \dfrac{bt_f}{2}(d - t_f)^2$ $+ \dfrac{t_w}{12}(d - 2t_f)^3$	$\dfrac{2I_{xx}}{d}$
	$\dfrac{\pi d^4}{64}$	$\dfrac{\pi d^3}{32}$
	$\dfrac{\pi(d_1^4 - d_2^4)}{64}$	$\dfrac{\pi(d_1^4 - d_2^4)}{32d_1}$
	$\dfrac{bd^3}{36}$	$\dfrac{bd^2}{24}$

where

$$Q = \int_{y=y_0}^{y=h} y\, dA$$

The quantity Q represents the first moment of the shaded area in Fig. 15b with respect to the neutral axis. The minus sign in Eq. (29) implies that σ_{yx} is negative when V is positive (see Sections 1.1 and 2.1 for discussions of the sign conventions for σ_{yx} and V, respectively).

From the complementary property of shear (Section 1.1), it follows that $\sigma_{xy} = \sigma_{yx}$, provided that static equilibrium exists. In particular, along the top and bottom of the beam, $Q = 0$. Thus, $\sigma_{xy} = \sigma_{yx} = 0$ on these surfaces; consequently, the shear stress goes from zero at the top and bottom surfaces of the beam to some maximum value in the interior.

Shear Stresses in Rectangular Beams   For the case of a beam having a rectangular cross section, Eq. (29) can be greatly simplified. Denoting the height of the beam by $2h$, it can be shown that the shear stress is given by

$$\sigma_{xy} = \sigma_{yx} = \frac{3}{2}\frac{V}{A}\left(1 - \frac{y^2}{h^2}\right)$$

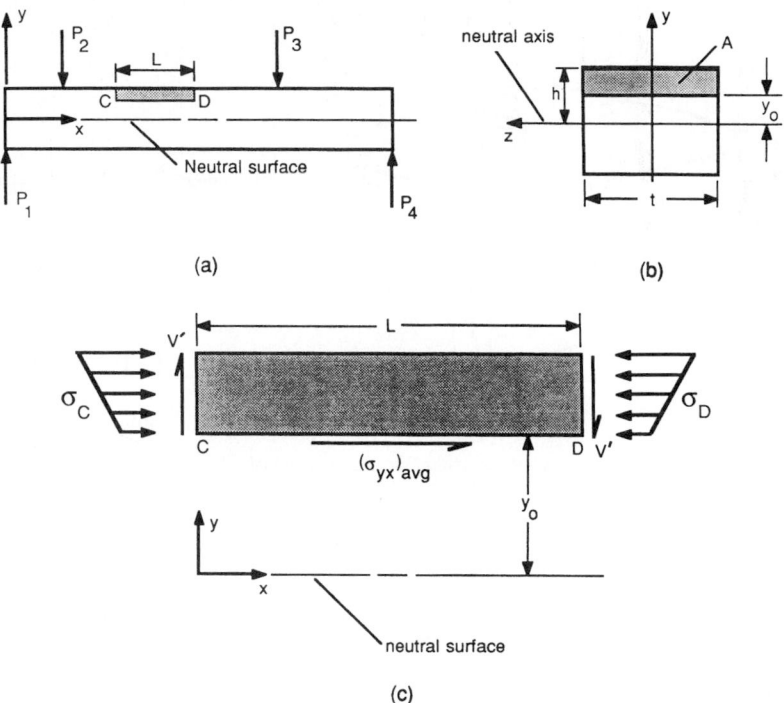

(a)

(b)

(c)

Fig. 15 Shear stresses in a beam: (a) side view; (b) cross-sectional view; (c) enlarged view of segment *CD*.

Clearly, the maximum shear stress is given by $(\sigma_{xy})_{\max} = 1.5(V/A)$. This maximum shear stress occurs at $y = 0$ (at the neutral axis, where the normal stress σ_{xx} is zero). The shear stress distribution for a beam of rectangular cross section is shown in Fig. 16.

Shear Stresses in Thin-Walled Beams It can be shown that Eq. (29) can be used to describe the shear stress on an arbitrary longitudinal cut. For example, consider a section of a **T** beam as shown in Fig. 17a. The shear stress σ_{xy} must equal zero on the top and bottom of the flange since these are free surfaces.

Provided the flange thickness is small, σ_{xy} will not become significant through the thickness of the flange. Note that on the bottom of the flange Q is nonzero. Equation (29) thus predicts that a shear stress $\sigma_{xy} = \sigma_{yx}$ would exist on the free surface. In reality, this is physically impossible. This demonstrates one of the limitations of Eq. (29). This equation also gives incorrect results along the free surface of beams of circular cross section, except along the neutral axis (see Ref. 6 for further details). For a proper analysis that leads to an answer of zero shear stress on the bottom of the flange, the theory of elasticity must be employed. Equation (29) should thus be used with care.

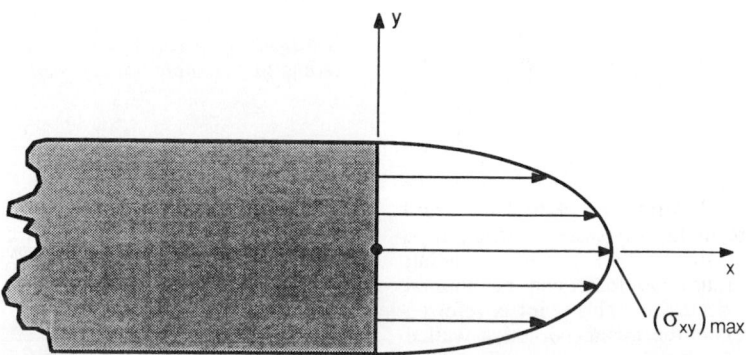

Fig. 16 Shear stress distribution in a rectangular beam subjected to transverse loads.

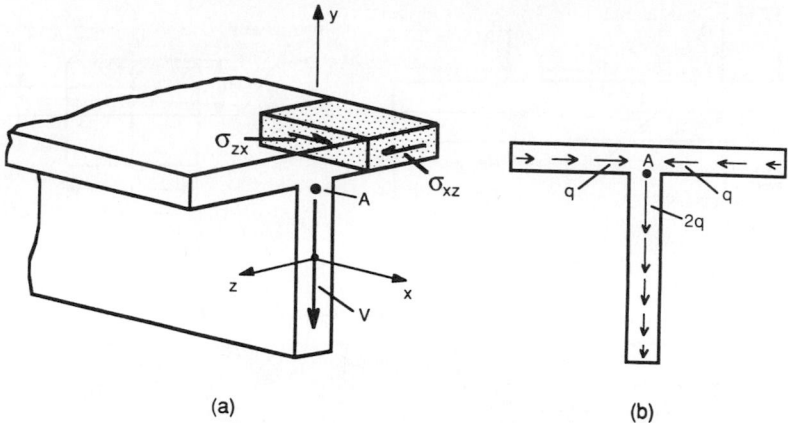

Fig. 17 Shear flow in a thin-walled beam: (*a*) isometric view of a beam; (*b*) sketch of shear flow.

The shear stresses $\sigma_{xz} = \sigma_{zx}$ in the flange can become large. Their value is again given by Eq. (29). As the web is reached (point A in Fig. 17*a*), the horizontal shear stress, σ_{xz}, becomes a vertical shear stress, σ_{xy}. The shear flow, q, is defined as

$$q = \frac{VQ}{I_{zz}} = \tau_{\text{avg}}t$$

where V is again the internal resisting shear force, Q is the first moment of area with respect to the neutral axis, I_{zz} is the moment of inertia of the beam cross section, and τ_{avg} is the average horizontal shear stress acting in the web. Shear flow in a thin-walled beam is analogous to the flow of water through a channel. If the beam thickness decreases, τ_{avg} will increase so as to keep q constant. This is analogous to conservation of mass in fluid flow through a channel of variable diameter. Similarly, as the point A is approached in Fig. 17*b*, the shear flows from each flange, q, combine to give a shear flow of $2q$ in the web. This is analogous to water flow through a T section.

The total shear flow acting over the area of the cross section must be such as to equal the resisting shearing force, V. For example, if no external horizontal forces are present, horizontal forces due to shear flow must cancel each other out. This is obvious for the T section of Fig. 17*b* since the horizontal shear flow in one flange exactly cancels the horizontal shear flow in the other flange.

Shear Center for Beams   In general, a beam will undergo both bending and twisting when it is subjected to transverse loads. There does exist a point, however, at which transverse loads can be applied without any twisting occurring. This point is referred to as the shear center of the beam. For thick-walled or solid cross sections, the shear center will usually be located close to the centroid of the cross section. Knowledge of the precise location of the shear center in beams with thick or solid cross sections is generally unimportant since such beams will exhibit considerable resistance to torsional loads. If such a beam is loaded through a point slightly away from the shear center, no significant amount of twisting occurs.

For thin-walled open-section beams, the situation is quite different. Such beams exhibit very little torsional resistance; therefore, it is important that the application of transverse loads occurs at the shear center. The shear center is found by balancing out the moments due to the shear flows q in the beam and the resisting shear force V. Details of the procedure can be found in Refs. 7 and 8. Table 3 gives the location of the shear center for some common thin-walled cross sections. More extensive tables can be found in Refs. 8 and 9.

2.4   Deflection of Beams

A variety of techniques exist for determining the elastic deflection of beams. The techniques to be discussed here are only valid for small deflections. In addition, the assumptions used in deriving the elastic flexure formula must also be satisfied (see Section 2.2).

Method of Integration   For small deflections the radius of curvature, ρ, can be written as

$$\frac{1}{\rho} = \frac{d^2y(x)}{dx^2}$$

where $y(x)$ represents the deflection of the beam at a longitudinal distance, x, from the origin of the coordinates. Substituting this result into Eq. (26) leads to

$$\frac{d^2y(x)}{dx^2} = \frac{M(x)}{EI_{zz}} \tag{30}$$

Table 3   Shear Centers for Common Sections

Beam Cross Section	Location of Q
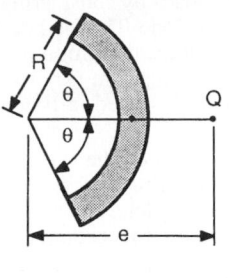	$e = \dfrac{4R(\sin\theta - \theta\cos\theta)}{2\theta - \sin 2\theta}$
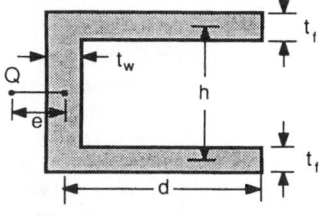	$e = \dfrac{3t_f d^2}{6t_f d + t_w h}$
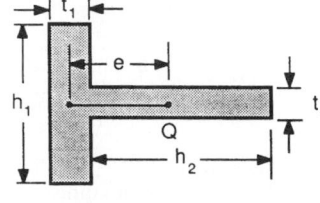	$e = \dfrac{(t_1 + h^2)(t_2^3 h_2)}{2(t_2^3 h_2 + h_1^3 t_1)}$
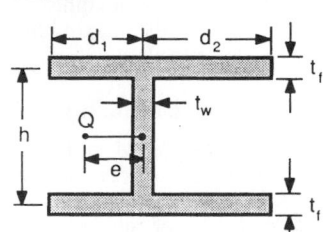	$e = \dfrac{3t_f(d_2^2 - d_1^2)}{6t_f(d_1 + d_2) + t_w h}$ provided $d_1 < d_2$

Equation (30) is a second-order linear differential equation for y. Its solution requires a knowledge of the bending moment at every point along the beam as well as two boundary conditions. The equation is only valid provided that no sudden changes in M, E, or I_{zz} occur. When there is a sudden change, such as the application of a concentrated force, it is necessary to divide the beam into several portions and solve Eq. (30) within each portion. The following example illustrates the method of integration.

Example 1. For the beam shown in Fig. 13*a*, derive the deflection equation by the method of integration. Consider the case where $a = L/2$.

Solution. Since the beam contains a concentrated load at $x = L/2$, it is necessary to write two separate differential equations for the two halves of the beam. Using the moment equations derived in Fig. 13*c* and substituting them into Eq. (30), with $a = L/2$, leads to

$$\frac{d^2y_1}{dx^2} = \frac{P}{2EI}x \qquad 0 \le x \le \frac{L}{2}$$

and

$$\frac{d^2y_2}{dx^2} = \frac{-P}{2EI}x + \frac{PL}{2EI} \qquad \frac{L}{2} \le x \le L$$

Integrating these equations gives

$$\frac{dy_1}{dx} = \frac{Px^2}{4EI} + C_1$$

$$\frac{dy_2}{dx} = \frac{-Px^2}{4EI} + \frac{PL}{2EI}x + C_3$$

$$y_1 = \frac{Px^3}{12EI} + C_1x + C_2$$

$$y_2 = \frac{-Px^3}{12EI} + \frac{PLx^2}{4EI} + C_3x + C_4$$

Finally, the boundary conditions must be applied. Clearly, four boundary conditions are needed to solve for the four constants of integration, C_1 through C_4. These boundary conditions are as follows:

At $x = 0$: $y_1 = 0 \Rightarrow C_2 = 0$

At $x = L$: $y_2 = 0 \Rightarrow C_4 = \dfrac{-PL^3}{6EI} - C_3L$

At $x = \dfrac{L}{2}$: $\dfrac{dy_1}{dx} = \dfrac{dy_2}{dx} \Rightarrow \dfrac{PL^2}{16EI} + C_1$

$$= \frac{3PL^2}{16EI} + C_3$$

At $x = \dfrac{L}{2}$: $y_1 = y_2 \Rightarrow \dfrac{PL^3}{96EI} + C_1\dfrac{L}{2}$

$$= \frac{5PL^3}{96EI} + C_3\frac{L}{2} + C_4$$

The last two conditions are due to the fact that the slope and deflection must be continuous at $x = L/2$. Solving these equations for the constants gives

$$C_1 = \frac{-PL^2}{16EI} \qquad C_2 = 0$$

$$C_3 = \frac{-3PL^2}{16EI} \qquad C_4 = \frac{PL^3}{48EI}$$

The deflection equations for the two halves of the beam are thus given by

$$y_1 = \frac{P}{48EI}(4x^3 - 3L^2x) \qquad 0 \le x \le \frac{L}{2}$$

$$y_2 = \frac{P}{48EI}(-4x^3 + 12Lx^2 - 9L^2x + L^3)$$

$$\frac{L}{2} \le x \le L$$

In general, if n discontinuities in the loading occur between the end points of the beam, it will be necessary to write and solve $n + 1$ differential equations in order to describe the displacement along the entire length of the beam. It is possible to write a single differential equation for even the most general state of loading by using singularity functions.[10] Table 4 contains deflection equations for a variety of loading conditions.

Method of Superposition Since Eq. (30), the governing differential equation for beam deflection, is linear, it is possible to solve a beam deflection problem involving many loads by considering the effect of each load as if it acted alone. The total deflection is obtained by adding together the deflections caused by each individual load. The procedure (known as the method of superposition) is illustrated in Example 2.

Example 2. Consider the case of a simple beam of length L carrying a uniformly distributed load q, and a concentrated load P at its center. Find the center deflection of the beam using the method of superposition.

Solution. From Table 4, the center deflection for a simple beam carrying a concentrated load at its center is equal to $-PL^3/48EI$. Similarly, for a simple beam having a uniformly distributed load, the center deflection is equal to $-5qL^4/384EI$. Thus, applying the method of superposition, it follows that the center deflection for a simple beam having both a uniformly distributed load along its length and a concentrated load at its center is given by

$$Y_{x=L/2} = \frac{-PL^3}{48EI} - \frac{5qL^4}{384EI} = \frac{-L^3}{384EI}(8P + 5qL)$$

Since the center deflections were the maximum deflections for the individual loadings, it follows that the center deflection is the maximum deflection for the combined state of loading as well.

2.5 Unsymmetric Bending

Moments of Inertia of a Plane Area In order to determine the flexural stress in a beam undergoing unsymmetric bending, it is necessary to compute the moments of inertia of the cross-sectional area of the beam. Assuming the beam cross section lies in the xy plane, the required terms are as follows:

$$I_{xx} = \int y^2 \, dA \quad I_{yy} = \int x^2 \, dA \quad I_{xy} = \int xy \, dA$$

$$\tag{31}$$

The terms I_{xx} and I_{yy} are known as the moments of inertia of the cross section, while the term I_{xy} is called the product of inertia of the cross section. Whereas I_{xx} and I_{yy} are always positive, I_{xy} can be positive, negative, or zero. For the cross sections shown in Table 2, the products of inertia are zero. This is because each of these cross sections has one or more axes of symmetry, and at least one of the two coordinate axes lies along an axis of symmetry. In fact, every cross section has some orientation for which $I_{xy} = 0$. A Mohr's circle for inertias can even be drawn. Details can be found in most statics and mechanics texts.

Table 4 Deflections and Slopes of Uniform Beams

Loading	Equation of Elastic Curve	Maximum Deflection	End Slopes
	$y = \dfrac{P}{48EI}(4x^3 - 3L^2x)$ for $0 \le x \le \dfrac{L}{2}$	$y_{max} = -\dfrac{PL^3}{48EI}$ at $x = \dfrac{L}{2}$	$\theta_{x=0} = -\dfrac{PL^2}{16EI}$ $\theta_{x=L} = \dfrac{PL^2}{16EI}$
	$y = \dfrac{Pb}{6EIL}[x^3 - (L^2 - b^2)x]$ for $0 \le x \le a$	$y_{max} = -\dfrac{Pa(L^2 - a^2)^{3/2}}{9\sqrt{3}EIL}$ at $x = \sqrt{\dfrac{(L^2 - a^2)}{3}}$ provided $a < b$	$\theta_{x=0} = -\dfrac{Pb(L^2 - b^2)}{6EIL}$ $\theta_{x=L} = \dfrac{Pa(L^2 - a^2)}{6EIL}$
	$y = \dfrac{P}{6EI}(x^3 - 3ax(L - a))$ for $0 \le x \le a$	$y_{max} = -\dfrac{Pa}{24EI}(3L^2 - 4a^2)$ at $x = \dfrac{L}{2}$	$\theta_{x=0} = -\dfrac{Pa}{2EI}(L - a)$ $\theta_{x=L} = \dfrac{Pa}{2EI}(L - a)$
	$y = -\dfrac{q}{24EI}(x^4 - 2Lx^3 + L^3x)$	$y_{max} = -\dfrac{5qL^4}{384EI}$ at $x = \dfrac{L}{2}$	$\theta_{x=0} = -\dfrac{qL^3}{24EI}$ $\theta_{x=L} = \dfrac{qL^3}{24EI}$
	$y = -\dfrac{M}{6EIL}(x^3 - L^2x)$	$y_{max} = \dfrac{ML^2}{9\sqrt{3}EI}$ at $x = \dfrac{L}{\sqrt{3}}$	$\theta_{x=0} = \dfrac{ML}{6EI}$ $\theta_{x=L} = -\dfrac{ML}{3EI}$
	$y = \dfrac{P}{6EI}(x^3 - 3Lx^2)$	$y_{max} = -\dfrac{PL^3}{3EI}$ at $x = L$	$\theta_{x=0} = 0$ $\theta_{x=L} = -\dfrac{PL^2}{2EI}$
	$y = -\dfrac{q}{24EI}(x^4 - 4Lx^3 + 6L^2x^2)$	$y_{max} = -\dfrac{qL^4}{8EI}$ at $x = L$	$\theta_{x=0} = 0$ $\theta_{x=L} = -\dfrac{qL^3}{6EI}$
	$y = -\dfrac{M}{2EI}x^2$	$y_{max} = -\dfrac{ML^2}{2EI}$ at $x = L$	$\theta_{x=0} = 0$ $\theta_{x=L} = -\dfrac{ML}{EI}$

Table 4 *(Continued)*

Loading	Equation of Elastic Curve	Maximum Deflection	End Slopes
	$y = \dfrac{Px^2}{48EI}(4x - 3L)$ for $0 \le x \le \frac{L}{2}$	$y_{max} = -\dfrac{PL^3}{192EI}$ at $x = \dfrac{L}{2}$	$\theta_{x=0} = 0$ $\theta_{x=L} = 0$
	$y = -\dfrac{qx^2}{24EI}(L - x)^2$	$y_{max} = -\dfrac{qL^4}{384EI}$ at $x = \dfrac{L}{2}$	$\theta_{x=0} = 0$ $\theta_{x=L} = 0$

Parallel Axis Theorem   The determination of the moments of inertia for a complex cross section using the definitions in Eq. (31) is often quite difficult. Instead, it is usually more expeditious to consider the cross section to be made up of several smaller sections of simple shapes and to determine the moments of inertia of these smaller sections separately. After this, the moments of inertia of the smaller sections must somehow be assembled to give the moment of inertia of the whole cross section. This assemblage is accomplished using the parallel axis theorem. This theorem states that the moments of inertia of a cross section with respect to an arbitrary xy coordinate system are related to the moments of inertia of that cross section with respect to a parallel, centroidal $x'y'$ coordinate system by the following:

$$I_{xx} = I'_{xx} + A\bar{y}^2 \qquad I_{yy} = I'_{yy} + A\bar{x}^2 \qquad I_{xy} = I'_{xy} + A\bar{x}\bar{y}$$

where A is the area of the cross section, and $\bar{x}$ and $\bar{y}$ are the distances from the origin of the xy coordinate system to the origin of the $x'y'$ coordinate system.

Example 3.   For the angle section shown in Fig. 18, determine the moments of inertia of the cross section with respect to its centroid.

Solution.   First determine the coordinates of the centroid C with respect to the given x^*y^* reference frame. The cross section is considered to be made up of two small rectangles as shown in Fig. 18. The centroid of the cross section is given by the coordinates x_C^* and y_C^*:

$$x_C^* = \frac{\sum(\bar{x}A)}{\sum A} = \frac{1(2)(4) + 3(2)(6)}{(2)(4) + (2)(6)} = 2.2 \text{ in.}$$

$$y_C^* = \frac{\sum(\bar{y}A)}{\sum A} = \frac{2(2)(4) + 5(2)(6)}{(2)(4) + (2)(6)} = 3.8 \text{ in.}$$

These equations are based on the definition of the centroid of a cross section. Next, place the xy coordinate system at the centroid C. Using the values for the moments of inertia of a rectangle with respect to its centroid (from Table 2) and applying the parallel axis theorem leads to

$$I_{xx} = \frac{6(2)^3}{12} + 6(2)(1.2)^2 + \frac{2(4)^3}{12} + 2(4)(-1.8)^2$$

$$= 57.87 \text{ in.}^4$$

$$I_{yy} = \frac{2(6)^3}{12} + 6(2)(0.8)^2 + \frac{4(2)^3}{12} + 2(4)(-1.2)^2$$

$$= 57.87 \text{ in.}^4$$

$$I_{xy} = 0 + 6(2)(1.2)(0.8) + 0 + 2(4)(-1.8)(-1.2)$$

$$= 28.80 \text{ in.}^4$$

Equation for Flexural Stress   When a beam cross section does not have a plane of symmetry, or if it has a plane of symmetry but the applied moments do not act in this plane, Eq. (27) is no longer valid. Provided that the origin of the xy coordinate axes is at the centroid of the cross section, the flexural stress at a point $P(x, y)$ is given by

$$\sigma_{zz} = \frac{(M_x I_{yy} + M_y I_{xy})y - (M_x I_{xy} + M_y I_{xx})x}{I_{xx}I_{yy} - I_{xy}^2}$$

$$(32)$$

where M_x and M_y are the components of the moment acting along the x and y axes, respectively.

The neutral axis can be found by setting $\sigma_{xx} = 0$ in Eq. (32). Since $x = y = 0$ satisfies $\sigma_{xx} = 0$, it is clear that the neutral axis passes through the centroid of the cross section. Let α be the angle of the neutral

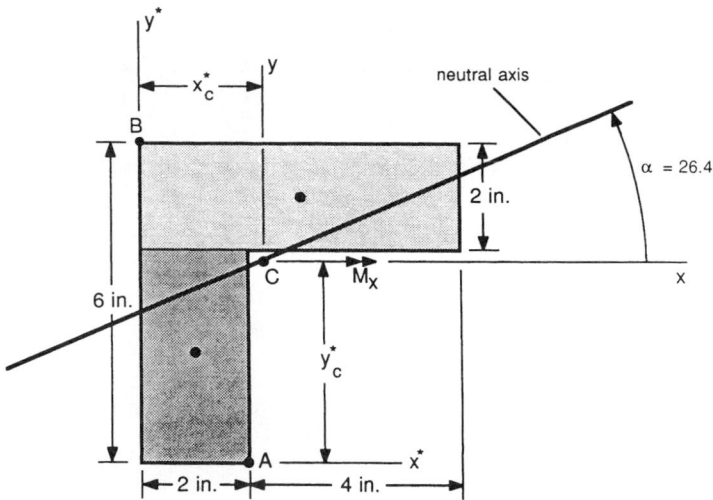

Fig. 18 Beam cross section for use with Examples 3 and 4.

axis with respect to the x axis. Then α satisfies the following:

$$\tan \alpha = \frac{y}{x} = \frac{M_x I_{xy} + M_y I_{xx}}{M_x I_{yy} + M_y I_{xy}} \qquad (33)$$

By eliminating M_y between Eqs. (32) and (33), a simpler expression for the flexural stress is obtained:

$$\sigma_{zz} = \frac{y - x \tan \alpha}{I_{xx} - I_{xy} \tan \alpha} M_x \qquad (34)$$

This equation is invalid if $\alpha = \pm 90°$. In such cases, Eq. (32) should be used.

Example 4. Suppose a beam has the cross-sectional dimensions given in Fig. 18. Assume the beam is acted upon by a 90,000-lb-in. moment that acts along the positive x axis. Determine the maximum flexural stress in the beam.

Solution. The moments of inertia for this cross section were previously found in Example 3. The values are $I_{xx} = 57.9$ in.[4], $I_{yy} = 57.9$ in.[4], and $I_{xy} = 28.8$ in.[4]. The orientation of the neutral axis is found using Eq. (33) with $M_y = 0$:

$$\tan \alpha = \frac{28.8}{57.9} \Rightarrow \alpha = 26.4°$$

The neutral axis has been shown in Fig. 18. From this figure it is logical to expect that the maximum stress will occur at A or B since these points are furthest

from the neutral axis. Using Eq. (34), the stresses at these points are given by

$$(\sigma_{zz})_A = \frac{(-3.8) - (-0.2)\tan 26.4°}{57.9 - 28.8 \tan 26.4°}(90,000)$$

$$= -7640 \text{ psi}$$

$$(\sigma_{zz})_B = \frac{(2.2) - (-2.2)\tan 26.4°}{57.9 - 28.8 \tan 26.4°}(90,000)$$

$$= 6800 \text{ psi}$$

Thus, the maximum tensile stress in the beam is 6800 psi and the maximum compressive stress is 7640 psi.

2.6 Curved Beams

Circumferential Stress The formula presented in Section 2.2 for flexural stress, $\sigma = -My/I$, is applicable for long, initially straight beams. It can still be used with reasonable accuracy for beams with some initial curvature provided the beam depth is small compared to the radius of curvature. When the beam's depth is of the same order of magnitude as its curvature, however, a new equation must be used. The derivation of this equation is too involved to be presented here. The details can be found in almost any modern text on strength of materials; in particular, see Refs. 7 and 8.

The equation for the circumferential stress distribution of a curved member with a plane of symmetry and subjected to moments acting in this plane of symmetry is given by (see Fig. 19):

$$\sigma = \frac{-My}{A(R - r_n)r} \qquad (35)$$

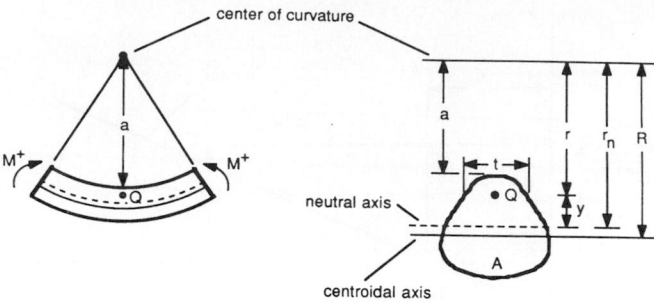

Fig. 19   Notation for theory of curved beams.

where r = distance from center of curvature to point under consideration

r_n = distance from center of curvature to neutral axis of cross section

y = distance from neutral axis to point Q under consideration

R = distance from center of curvature to centroidal axis of cross section

A = area of cross section

M = applied moment

The moment is considered positive if it tends to create compressive stresses on the concave side of the beam cross section. The coordinate y is taken to be positive on the concave side of the neutral axis.

It can be shown that r_n is given by

$$r_n = \frac{A}{\int dA/r}$$

Also, R is given by

$$R = \frac{1}{A} \int r\, dA$$

Thus, the centroidal axis and the neutral axis do not coincide. In general their difference is small; consequently, it is important to determine these distances accurately since their difference occurs in the denominator of Eq. (35). Table 5 gives required cross-sectional properties for application to curved beams. More complete tables can be found in Refs. 7–9.

Radial Stress   For an initially straight beam, only circumferential stresses develop upon application of a bending moment. For initially curved beams, however, significant radial stresses can develop. An approximate solution for the radial stress is given by[7]

$$\sigma_r = \frac{-M}{A(r - r_n)} tr \left(r_n \int_a^r \frac{dA}{r} - \int_a^r dA \right) \quad (36)$$

where a is the distance from the center of curvature to the edge of the cross section, and t is the thickness

of the cross section at the point under consideration. All other terms were previously defined in conjunction with Eq. (35). See Fig. 19 for details.

It should be noted that Eq. (36) neglects the effect of normal and shear forces on the radial stress. In general, these forces do not have a major effect on the radial stress and can be ignored for most applications.[7]

3   TORSION AND SHAFTS

3.1   Circular Shafts

Shear Stress Distribution   Circular shafts have the unique property that, when subjected to torsion, every cross section of the shaft remains plane and undistorted. This fact allows for the easy determination of the shear strain distribution. When subjected to torsion, a small square element, A, deforms into a rhombus, A', as shown in Fig. 20a. The angle γ shown in Fig. 20a is identical to the shear strain that the square element has gone through, Fig. 20b.

For small values of γ it is reasonable to assume that $\gamma = r\phi/L$, where L is the shaft length, r is the radius of the shaft at the point under consideration, and ϕ is the angle of twist of the shaft (see Fig. 20). Denoting the outer radius of the shaft by c, it follows that

$$\gamma_{max} = \frac{c\phi}{L} \quad (37)$$

and hence,

$$\gamma(r) = \frac{r}{c} \gamma_{max} \quad (38)$$

The only assumption made in deriving Eq. (38) was to assume that the shear strains are small. For linear elastic response, $\tau = G\gamma$. Substituting this relationship into Eq. (38) leads to

$$\tau(r) = \frac{r}{c} \tau_{max} \quad (39)$$

It follows that, for torsion of an isotropic linear solid elastic circular shaft, the shear stress distribution is linear with respect to radius. This is illustrated in

Table 5   Cross-Sectional Properties for Curved Beams

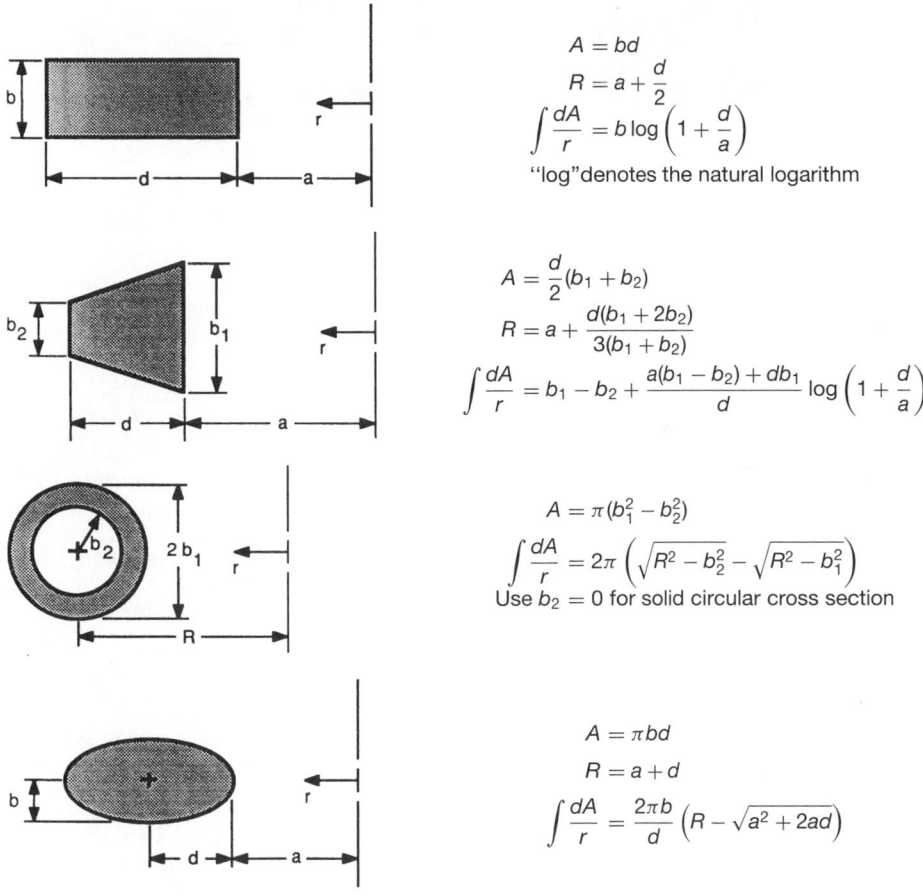

$$A = bd$$

$$R = a + \frac{d}{2}$$

$$\int \frac{dA}{r} = b \log\left(1 + \frac{d}{a}\right)$$

"log" denotes the natural logarithm

$$A = \frac{d}{2}(b_1 + b_2)$$

$$R = a + \frac{d(b_1 + 2b_2)}{3(b_1 + b_2)}$$

$$\int \frac{dA}{r} = b_1 - b_2 + \frac{a(b_1 - b_2) + db_1}{d} \log\left(1 + \frac{d}{a}\right)$$

$$A = \pi(b_1^2 - b_2^2)$$

$$\int \frac{dA}{r} = 2\pi\left(\sqrt{R^2 - b_2^2} - \sqrt{R^2 - b_1^2}\right)$$

Use $b_2 = 0$ for solid circular cross section

$$A = \pi bd$$

$$R = a + d$$

$$\int \frac{dA}{r} = \frac{2\pi b}{d}\left(R - \sqrt{a^2 + 2ad}\right)$$

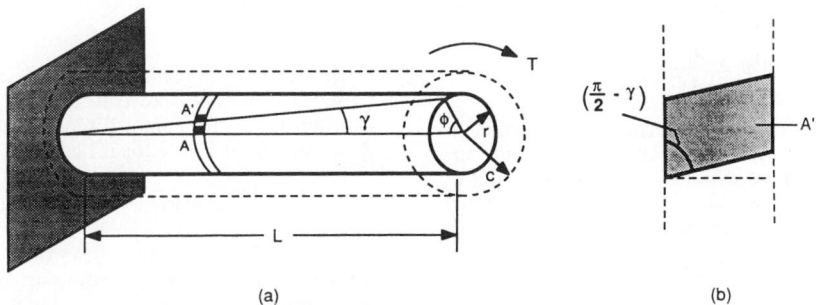

(a)   (b)

Fig. 20   Torsion of a circular shaft.

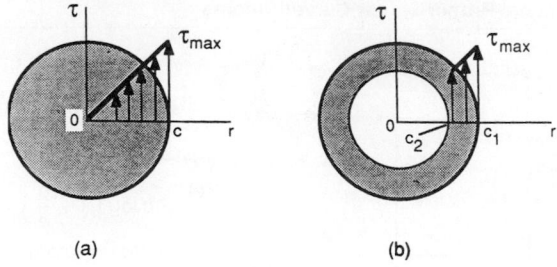

Fig. 21   Shear stress distribution in a circular shaft: (a) solid shaft; (b) hollow shaft.

Fig. 21a. The shear stress distribution of a hollow circular shaft (Fig. 21b) is also given by Eq. (39).

From statics, the net resisting torque generated by the internal shear stresses must equal the applied external torque, T:

$$T = \int r\tau \, dA \qquad (40)$$

Substituting for τ from Eq. (39) into (40) leads to

$$\tau_{max} = \frac{Tc}{J} \qquad (41)$$

where $J = \int r^2 \, dA$. The term J is known as the polar moment of inertia of the cross section with respect to its centroid. For a solid circular cross section, $J = \pi c^4/2$. For a hollow circular tube, $J = \pi(c_1^4 - c_2^4)/2$, where c_1 and c_2 are the outer radius and inner radius of the tube, respectively. Since the shear stress distribution is linear with respect to radius, it follows that the shear stress at a distance r from the centroid of a circular shaft is given by $\tau(r) = Tr/J$.

Angle of Twist   Using Hooke's law, Eq. (41) can be rewritten as

$$\gamma_{max} = \frac{\tau_{max}}{G} = \frac{Tc}{JG}$$

Equating this expression with Eq. (37) leads to

$$\phi = \frac{TL}{JG} \qquad (42)$$

Equation (42) applies to a uniform shaft subjected to a single torque at its end. If the shaft dimensions vary continuously with axial distance, ϕ must be found by integration:

$$\phi = \int_0^L \frac{T \, dx}{JG}$$

where x is the axial distance coordinate.

If several concentrated torques are applied to the shaft, or if the shaft has sudden changes in cross

section, the relative angle of twist must be found within each uniform shaft region. The total angle of twist is found by summing these relative angles of twist:

$$\phi_{total} = \sum_k \phi_k = \sum_k \frac{T_k L_k}{J_k G_k}$$

Design of Transmission Shafts   Circular shafts are often used for power transmission. The relation between the power transmitted, P, the angular speed of the shaft, ω, and the applied torque, T, is given by $P = T\omega$. If ω is expressed in revolutions per minute and P is expressed in horsepower,

$$T = \frac{63,030P}{\omega} \qquad (43)$$

where T has units of pound-inches. For a given power transmission and a given speed of rotation, it is important that the shaft dimensions are large enough so that the torque, as given by Eq. (43), does not cause the maximum shear stress in the shaft to exceed allowable limits. Since $\tau_{max} = Tc/J$, it is clear that the quantity c/J should be kept as small as possible.

3.2   Thin-Walled Closed Hollow Tubes

A tube is considered thin if the shear stress due to an applied torque does not vary significantly across the wall thickness. Put another way, the wall thickness must be small compared to the overall dimensions of the cross section.

Single Cells   Consider the single-celled, thin-walled, closed hollow tube shown in Fig. 22a. By considering equilibrium of horizontal forces for the small element A, as shown in Fig. 22b, it becomes apparent that

$$q = \tau_1 t_1 = \tau_2 t_2 = \text{const}$$

where q is known as the shear flow and t is the tube thickness. The following equation is obtained when equilibrium of the torsional forces is considered:

$$T = \oint qr \, ds \qquad (44)$$

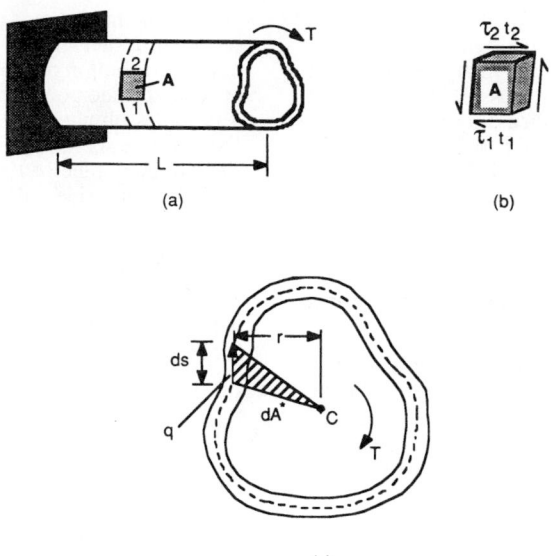

Fig. 22 Torsion of a single-celled, thin-walled, closed hollow tube.

where T is the applied torque, and r is the perpendicular distance from the centroid of the cross section to the shear flow q acting on the small section ds (Fig. 22c). The area of the shaded triangle in Fig. 22c is given by

$$dA^* = \tfrac{1}{2} r \, ds \qquad (45)$$

Substituting for $r \, ds$ from Eq. (45) into (44) leads to

$$T = 2q A^*$$

where A^* is the area bounded by the centerline of the wall. Recal that $q = \tau t$ gives

$$\tau = \frac{T}{2t A^*} \qquad (46)$$

Clearly, the maximum shear stress occurs at the thinnest section of the wall.

Using energy methods, it can be shown that the angle of twist is given by (see Ref. 7 for details):

$$\phi = \frac{TL}{4G(A^*)^2} \oint \frac{ds}{t} \qquad (47)$$

where L is the length of the tube, and G is the shear modulus of the material.

Multicells For a multicell section, the shear flow is no longer constant throughout the cross section.

Instead, the shear flow is analogous to the flow of water through pipes or the flow of current in an electric circuit. Consider the rectangular multicell section shown in Fig. 23. If the two sections are labeled 1 and 2 as shown, the shear flows around the outer walls of these sections are q_1 and q_2, respectively. The shear flow in the intermediate wall is $q_1 - q_2$.

The torque can be expressed in terms of the shear flows using a procedure similar to that used for single-celled tubes. In fact, it can be shown that for an n-celled tube:

$$T = \sum_{i=1}^{n} A_i^* q_i \qquad (48)$$

where A_i^* is the area contained within the centerline bounding the ith cell.

For an n-celled section there are $n + 1$ unknowns $(q_1, q_2, \ldots, q_n, \phi)$; consequently, $n + 1$ equations are

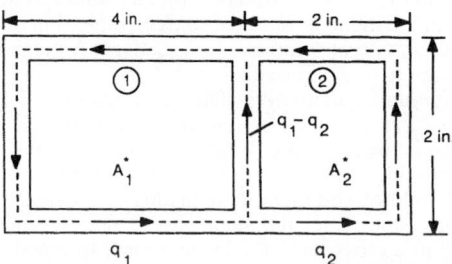

Fig. 23 Two-celled rectangular cross section for use with Example 5.

required in order to solve for these unknowns. Equation (48) provides one equation. The other n equations come about by noting that, according to Saint-Venant's torsion theory, cross sections do not warp within their own plane when a torque is applied. In other words, the angle of twist ϕ is the same for each cell. It follows that

$$\phi = \frac{L}{2G(A_i^*)} \left(\oint \frac{q\, ds}{t} \right)_i, \qquad i = 1, 2, \ldots, n \quad (49)$$

Equation (49) provides the other n equations needed to solve for all the unknowns.

Example 5. For the cross section shown in Fig. 23, assume the wall thickness is uniform at 0.25 in. If the tube length is 12 in. and a torque of 50,000 lb-in. is applied, find the maximum shear stress in the tube. Assume $G = 11.5 \times 10^6$ psi.

Solution. Writing Eq. (49) for the two cells,

$$2G\frac{\phi}{L} = \frac{1}{4(2)} \left[\frac{4+2+4}{0.25}q_1 + \frac{2}{0.25}(q_1 - q_2) \right]$$

$$= \frac{1}{2(2)} \left[\frac{2+2+2}{0.25}q_2 + \frac{2}{0.25}(q_2 - q_1) \right]$$

Using Eq. (48),

$$T = 50,000 = 4(2)q_1 + 2(2)q_2$$

Solving these equations for q_1, q_2, and ϕ gives $\phi = 0.0115$ rad, $q_1 = 4310$ lb/in., and $q_2 = 3830$ lb/in. Thus, $\tau_{max} = \tau_1 = 4310/0.25 = 17,250$ psi.

3.3   Torsion of Noncircular Cross Sections

Theories for analyzing the torsional response of circular shafts and thin-walled, closed hollow tubes were developed in Sections 3.1 and 3.2. For solid or thick-walled noncircular cross sections, it is usually necessary to use the theory of elasticity in order to determine torsion formulas. Because of the complexities involved, the most practical means for determining the torsion formula for a noncircular cross section is to find the appropriate formula in a handbook or other source. Table 6 gives formulas for several common cross sections. A much more thorough listing is provided in Ref. 9.

Although quantitative determination of the stress distribution for the torsion of a noncircular cross section is usually quite difficult, an analogy does exist that provides for a qualitative understanding. This is the Prandtl elastic-membrane analogy. In essence, this analogy states that the shear stress at a given point (x, y) in a cross section is proportional to the slope at the point (x, y) of a membrane (such as a soap film) made of the same cross-sectional dimensions, where the membrane is pressurized from one side. Thus, by

visualizing the shape of a membrane of the same cross-sectional dimensions, it is possible to obtain an idea of the distribution of shear stresses in the cross section of a shaft undergoing torsional loading. In particular, it becomes obvious that the shear stress is zero at the corners of a shaft made of a rectangular cross section since a membrane of the same geometry and pressurized from one side will have zero slope at its corners. See Refs. 7, 8, and 11 for further details.

4   PLATES, SHELLS, AND CONTACT STRESSES

4.1   Classical Theory of Elastic, Isotropic Plates

Flat plates are most often used to carry transverse pressure loads. Just as with beams, these pressure loads are carried by out-of-plane shear stresses, which in turn induce a bending moment distribution in the plate. Unlike beams, however, the bending moment distribution in a plate is two dimensional, and a torsional moment distribution is also usually present.

Assumptions Used   In deriving the equations of equilibrium, it is conventional to take the origin of coordinates at the middle surface of the plate, as shown in Fig. 24a. The commonly used sign convention for moments and shear forces is shown in Fig. 24b. A number of assumptions are used in deriving the classical theory of elastic isotropic plates. These assumptions are listed below:

1. The plate is initially flat.
2. The material is linear elastic, isotropic, and homogeneous.
3. The plate thickness, h, is small compared to its in-plane dimensions, a and b (see Fig. 24a). Generally, a and b should be at least 10 times greater than h.
4. Deflections are small and slopes of the deflected middle surface are small compared to unity.
5. The middle surface of the plate is a neutral surface; that is, it remains unstrained during bending.
6. Distortions due to transverse shear are negligible. Thus, initially straight lines normal to the middle surface remain straight and normal after deformation.
7. Stresses normal to the middle surface are negligible compared to other stresses.

Stresses and Moments in Terms of Curvature   The displacements in the x and y directions, u and v, respectively, depend on the z coordinate and the slope components:

$$u = -z\frac{\partial w}{\partial x} \qquad v = -z\frac{\partial w}{\partial y}$$

where w is the displacement component in the z direction. Using the strain–displacement relations,

Table 6 Torsion Formulas for Solid Sections

Cross Section	Maximum Shear Stress	Angle of Twist
Ellipse	$\tau_{max} = \dfrac{2T}{\pi ab^2}$ $(a > b)$	$\theta = \dfrac{T(a^2 + b^2)L}{\pi a^3 b^3 G}$
Equilateral Triangle	$\tau_{max} = \dfrac{20T}{a^3}$	$\theta = \dfrac{46.2TL}{a^4 G}$
Rectangle	$\tau_{max} = \dfrac{T}{K_1 ab^2}$	$\theta = \dfrac{TL}{K_2 ab^3 G}$

$\dfrac{a}{b}$	K_1	K_2
1.0	0.208	0.141
1.5	0.231	0.196
2.0	0.246	0.229
3.0	0.267	0.263
4.0	0.282	0.281
6.0	0.299	0.299
10.0	0.312	0.312
∞	0.333	0.333

Cross Section	Maximum Shear Stress	Angle of Twist
Regular Hexagon	$\tau_{max} = \dfrac{1.09T}{a^3}$	$\theta = \dfrac{0.967TL}{a^4 G}$

Note: T = applied torque, L = length of shaft, G = shear modulus.

Eqs. (4), the strains can be expressed in terms of the curvatures as follows:

$$\varepsilon_{xx} = -z\frac{\partial^2 w}{\partial x^2} \quad \varepsilon_{yy} = -z\frac{\partial^2 w}{\partial y^2} \quad \gamma_{xy} = -2z\frac{\partial^2 w}{\partial x \, \partial y}$$

$$(50)$$

For thin plates, plane stress conditions prevail ($\sigma_{zz} = \sigma_{xz} = \sigma_{yz} = 0$). Substituting from Eqs. (50) into (11) leads to

$$\sigma_{xx} = \frac{-Ez}{1 - \mu^2}\left(\frac{\partial^2 w}{\partial x^2} + v\frac{\partial^2 w}{\partial y^2}\right)$$

$$(51)$$

$$\sigma_{yy} = \frac{-Ez}{1 - v^2}\left(\frac{\partial^2 w}{\partial y^2} + v\frac{\partial^2 w}{\partial x^2}\right)$$

$$\sigma_{xy} = \frac{-Ez}{1 + v}\frac{\partial^2 w}{\partial x \, \partial y}$$

The moments per unit length are given by

$$M_x = \int_{-h/2}^{h/2} \sigma_{xx} z \, dz$$

$$M_y = \int_{-h/2}^{h/2} \sigma_{yy} z \, dz \qquad (52)$$

$$M_{yx} = M_{xy} = \int_{-h/2}^{h/2} \sigma_{xy} z \, dz$$

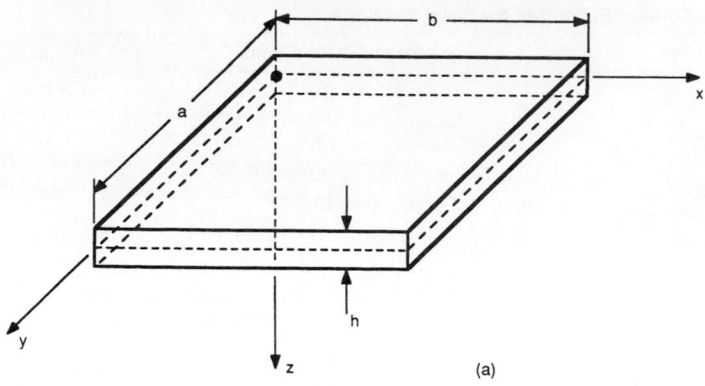

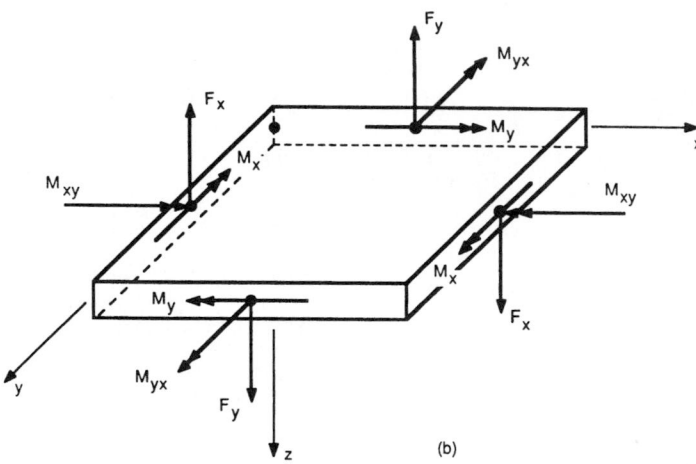

Fig. 24 Positive-sign convention for plate deformation: (*a*) plate dimensions; (*b*) positive-sign convention for forces and moments.

Substitution of Eqs. (51) into (52) gives

$$M_x = -D\left(\frac{\partial^2 w}{\partial x^2} + v\frac{\partial^2 w}{\partial y^2}\right)$$

$$M_y = -D\left(\frac{\partial^2 w}{\partial y^2} + v\frac{\partial^2 w}{\partial x^2}\right) \qquad (53)$$

$$M_{xy} = -D(1-v)\frac{\partial^2 w}{\partial x\,\partial y}$$

where

$$D = \frac{Eh^3}{12(1-v^2)}$$

The parameter D is known as the bending stiffness, or flexural rigidity, of the plate.

Equations of Equilibrium If an infinitesimal section of the plate is considered, equilibrium of moments about the x and y axes leads to

$$\frac{\partial M_x}{\partial x} + \frac{M_{xy}}{\partial y} - F_x = 0 \qquad \frac{\partial M_y}{\partial y} + \frac{M_{xy}}{\partial x} - F_y = 0$$
$$(54)$$

where F_x and F_y are the shear forces per unit length on the x and y faces, respectively (see Fig. 24*b*). These equations, when combined with Eqs. (53), lead to

$$F_x = -D\frac{\partial}{\partial x}\nabla^2 w \qquad F_y = -D\frac{\partial}{\partial y}\nabla^2 w \qquad (55)$$

where

$$\nabla^2 = \frac{\partial^2}{\partial x^2} + \frac{\partial^2}{\partial y^2}$$

Consideration of equilibrium in the z direction gives

$$\frac{\partial F_x}{\partial x} + \frac{\partial F_y}{\partial x} + p = 0 \qquad (56)$$

where p is the load per unit area in the z direction acting on the infinitesimal section. Finally, substituting Eqs. (55) into (56) gives the governing differential equation for equilibrium of a plate:

$$\nabla^2 \nabla^2 \, w = \nabla^4 \, w = \frac{p}{D} \qquad (57)$$

where

$$\nabla^4 = \frac{\partial^4}{\partial x^4} + 2\frac{\partial^4}{\partial x^2 \partial y^2} + \frac{\partial^4}{\partial y^4}$$

Boundary Conditions When attempting to solve Eq. (57) for a given pressure distribution $p(x, y)$, the boundary conditions of the plate edges must be known. The mathematical forms of the three commonly encountered boundary conditions follow for a plate edge parallel to the y axis and at a distance $x = a$.

1. Simply-supported-edge conditions

$$w = 0 \quad \text{at } x = a$$

$$\frac{\partial^2 w}{\partial x^2} + \nu \frac{\partial^2 w}{\partial y^2} = 0 \quad \text{at } x = a$$

2. Fixed-edge conditions

$$w = 0 \quad \text{at } x = a$$

$$\frac{\partial w}{\partial y} = 0 \quad \text{at } x = a$$

3. Free-edge conditions

$$\frac{\partial^2 w}{\partial x^2} + \nu \frac{\partial^2 w}{\partial y^2} = 0 \quad \text{at } x = a$$

$$\frac{\partial^3 w}{\partial x^3} + (2 - \nu)\frac{\partial^3 w}{\partial x \partial y^2} = 0 \quad \text{at } x = a$$

Strain Energy When the elastic energy of a component is written in terms of displacements, it is referred to as the strain energy. For thin plates undergoing small deflections, only the strain energy due to bending is significant. This quantity is given by

$$U = \frac{1}{2}\int_A D\left\{ \left(\frac{\partial^2 w}{\partial x^2}\right)^2 + \left(\frac{\partial^2 w}{\partial y^2}\right)^2 + 2\nu\frac{\partial^2 w}{\partial x^2}\frac{\partial^2 w}{\partial y^2} \right.$$

$$\left. + 2(1 - \nu)\left(\frac{\partial^2 w}{\partial x \partial y}\right)^2 \right\} dx \, dy$$

This expression can be useful for obtaining approximate solutions to plate problems.

Solutions to Plate Problems For relatively simple loadings and edge conditions, solutions for the lateral deflection, w, can be found using classical techniques, such as Fourier series analysis.[12,13] Approximate solutions can be obtained by means of Galerkin's method and energy methods such as the Rayleigh–Ritz method (Ref. 14, Chapter 5). In more recent years, the trend has been to apply numerical methods, most notably the finite-difference and finite-element methods.[15,16].

The most important quantities in plate design are usually the maximum stress and maximum deflection. Figure 25b and 26a contain this information in graphical form for a variety of loading and edge conditions. More extensive data bases can be found in Refs. 9 (Chapter 10) and 17.

4.2 Orthotropic Plates

The discussion of Section 4.1 assumed an isotropic plate. That is, the material properties are independent of direction. With the current widespread usage of composite materials, the problem of determining stresses and deflections in anisotropic plates has become quite prevalent. A good treatment of the general theory of anisotropic plates can be found in Ref. 18. Typical examples of anisotropic plates include asymmetrically laminated plates made from multiple plies of fiber-reinforced composite materials, and plates reinforced with stringers and ribs.

While the general theory of anisotropic plates is too complex to present here, one simpler plate configuration, which is easier to analyze, often occurs in practice. This is the case of an orthotropic plate. A material exhibits orthotropic response if there are two orthogonal planes of material property symmetry. These two directions of symmetry normally lie within the horizontal plane of the plate (the xy plane in Fig. 24). If these directions of symmetry coincide with the x and y directions, the governing differential equation for equilibrium of a plate becomes[15]

$$D_{xx}\frac{\partial^4 w}{\partial x^4} + 2H\frac{\partial^4 w}{\partial x^2 \, y^2} + D_{yy}\frac{\partial^4 w}{\partial y^4} = p \qquad (58)$$

where

$$D_{xx} = \tfrac{1}{12}h^3 Q_{xx} \quad D_{yy} = \tfrac{1}{12}h^3 Q_{yy} \qquad D_{xy} = \tfrac{1}{12}h^3 Q_{xy}$$

$$G_{xy} = \tfrac{1}{12}h^3 G \qquad H = D_{xy} + 2G_{xy}$$

where h is the plate thickness, and G is the shear modulus of the material. The quantities Q_{xx}, Q_{yy}, Q_{xy}, and G characterize the linearly elastic, orthotropic stress–strain response of the material as given by

$$\left\{ \begin{array}{c} \sigma_{xx} \\ \sigma_{yy} \\ \sigma_{xy} \end{array} \right\} = \left[\begin{array}{ccc} Q_{xx} & Q_{xy} & 0 \\ Q_{xy} & Q_{yy} & 0 \\ 0 & 0 & G \end{array} \right] \left\{ \begin{array}{c} \varepsilon_{xx} \\ \varepsilon_{yy} \\ \gamma_{xy} \end{array} \right\}$$

(See Section 7.2 for further discussion of orthotropic response.) The bending moments in terms of the curvatures have the form

$$M_x = -\left(D_{xx}\frac{\partial^2 w}{\partial x^2} + D_{xy}\frac{\partial^2 w}{\partial y^2}\right)$$

$$M_y = -\left(D_{yy}\frac{\partial^2 w}{\partial y^2} + D_{xy}\frac{\partial^2 w}{\partial x^2}\right)$$

$$M_{xy} = -2G_{xy}\frac{\partial^2 w}{\partial x\,\partial y}$$

It should be mentioned that the preceding equations are valid only for small deflection, linear elastic, orthotropic response. In addition to symmetrically stacked composite laminated plates, corrugated plates, plates reinforced by closely spaced ribs and stiffeners, and reinforced concrete slabs can also be modeled as

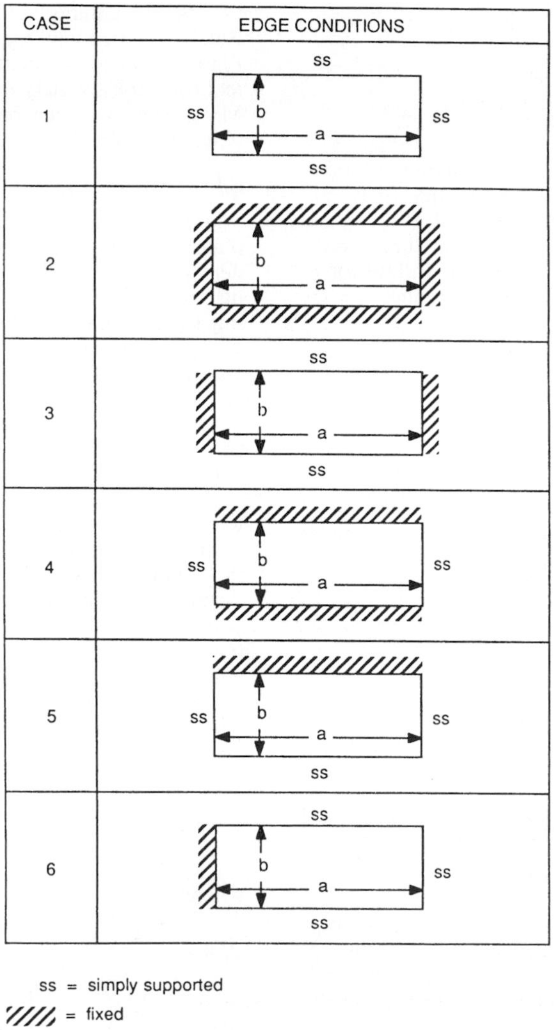
CASE
EDGE CONDITIONS
1
2
3
4
5
6
ss
a
b
ss = simply supported
= fixed

Fig. 25a Maximum stress and deflection of linear elastic, isotropic, rectangular plates for $\nu = 0.3$. Plate thickness $= h$. Elastic modulus $= E$. Each plate is uniformly loaded over its entire surface with a pressure P:

$$\sigma_{\max} = \frac{\alpha P b^2}{h^2} \qquad w_{\max} = \frac{\beta P b^4}{E h^3}$$

Values of α and β for the cases shown are plotted in charts A and B, respectively, in Fig. 25b*b*.

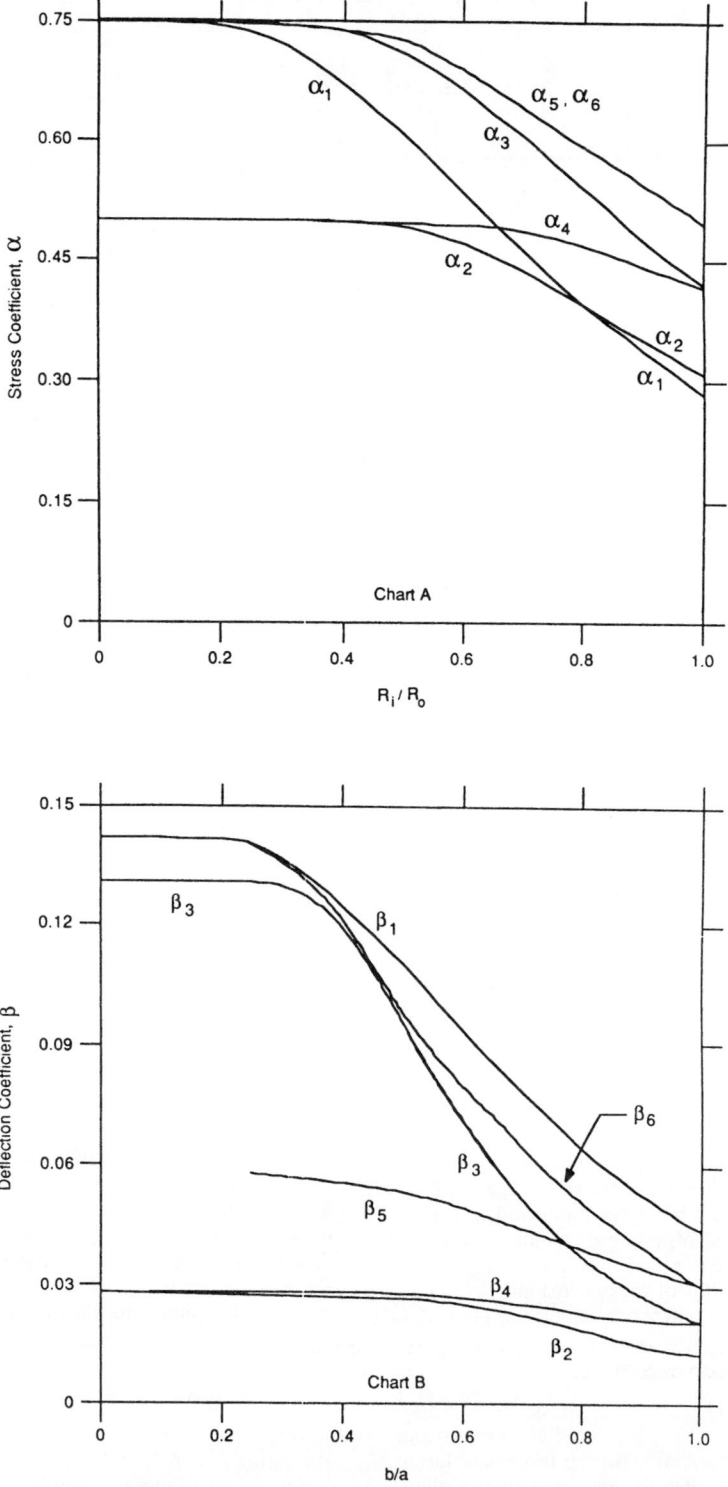

Fig. 25b Values of α and β for cases in Fig. 25ba.

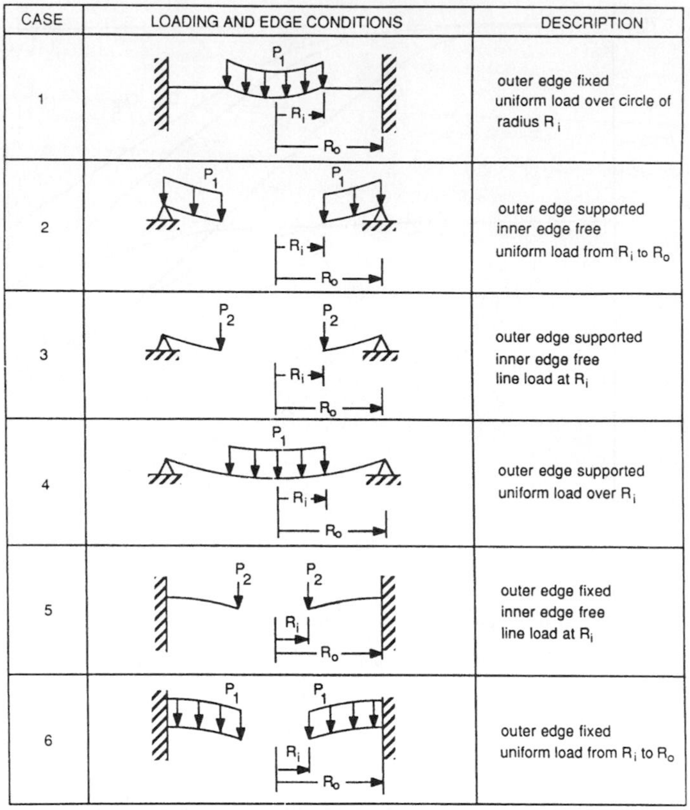

CASE	LOADING AND EDGE CONDITIONS	DESCRIPTION
1		outer edge fixed uniform load over circle of radius R_i
2		outer edge supported inner edge free uniform load from R_i to R_o
3		outer edge supported inner edge free line load at R_i
4		outer edge supported uniform load over R_i
5		outer edge fixed inner edge free line load at R_i
6		outer edge fixed uniform load from R_i to R_o

R_i = inside radius
R_o = outside radius

Fig. 26a Maximum stress and deflection of linear elastic, isotropic, circular plates for $\nu = 0.3$. Plate thickness $= h$. Elastic modulus $= E$:

$$\sigma_{max} = \frac{\alpha P}{h^2} \qquad w_{max} = \frac{\beta P R_0^2}{E h^3}$$

where P is the applied loading, either a pressure load (P_1) or a line load (P_2). Values of α and β for the case shown are plotted in charts A and B, respectively, in Fig. 26a*b*

orthotropic plates.[12] The problem of analyzing orthotropic plates reduces to determining D_{xx}, D_{yy}, D_{xy}, and G_{xy}. Once this has been accomplished, Eq. (58) can be solved by classical, energy, or numerical techniques just as is done for isotropic plates. Figure 27 gives approximate values of these constants for corrugated plates and rib and stiffener reinforced plates.

4.3 Buckling of Isotropic Plates

Critical Buckling Loads The classical isotropic plate theory discussed in Section 4.1 ignored the effects of in-plane forces. When such forces are large, they can significantly alter the response of the plate. If the in-plane forces are tensile, the plate will exhibit

an effective increase in stiffness with respect to lateral loads. If the in-plane forces are compressive, buckling of the plate may occur. Let N_x and N_y represent the in-plane normal forces per unit length in the x and y directions, respectively. Similarly, N_{xy} is the in-plane shear force per unit length. The governing differential equation for plate equilibrium takes the form

$$D\nabla^4 w = N_x \frac{\partial^2 w}{\partial x^2} + N_y \frac{\partial^2 w}{\partial y^2} + 2N_{xy} \frac{\partial^2 w}{\partial x\, \partial y} + p$$

(59)

By setting $p = N_{xy} = N_y = 0$, the critical buckling load for uniaxial compression can be obtained. Similarly, the critical buckling load for any combination of shear and

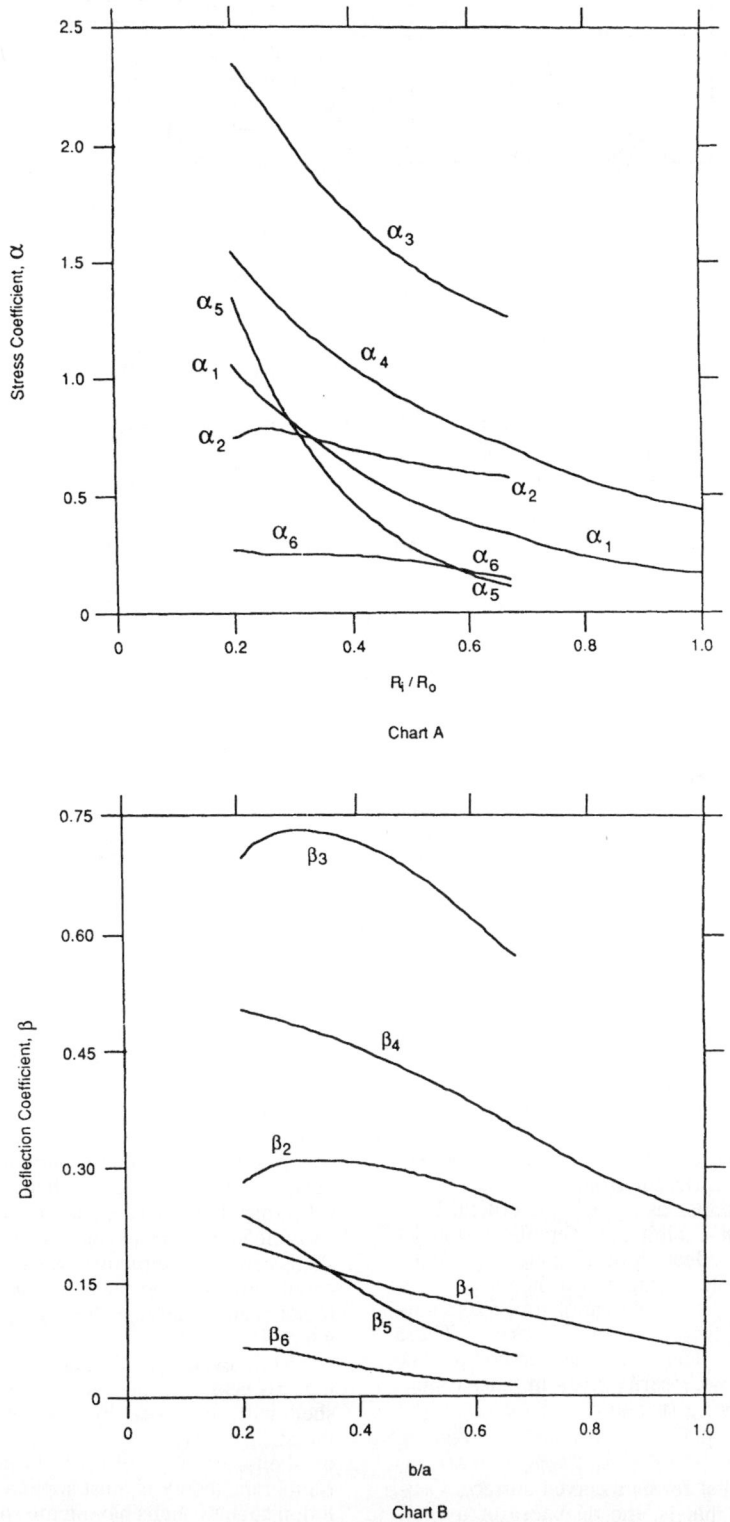

Fig. 26b Values of α and β for cases shown in Fig. 26a.

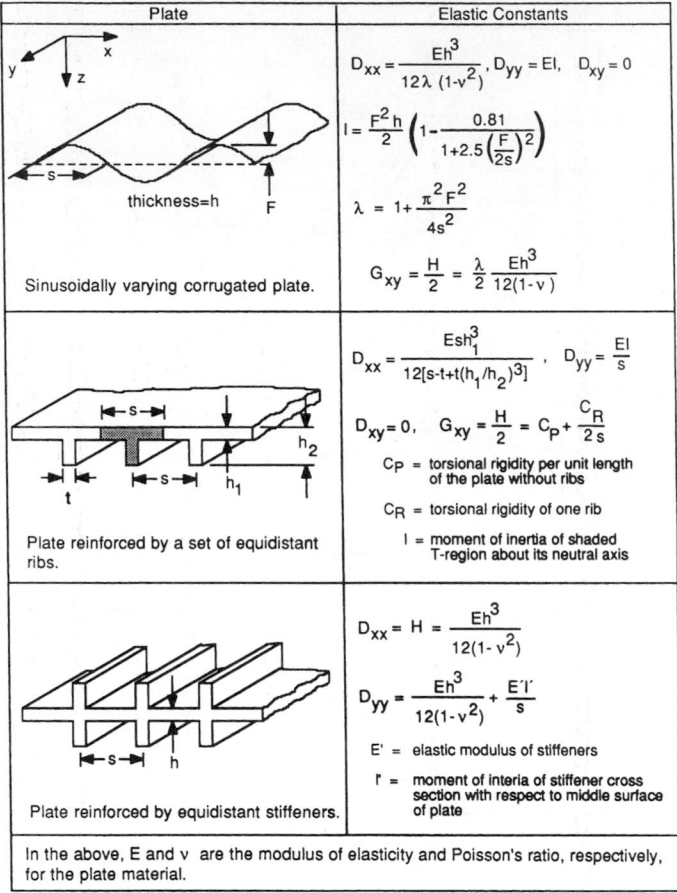

Plate	Elastic Constants
Sinusoidally varying corrugated plate.	$D_{xx} = \dfrac{Eh^3}{12\lambda(1-v^2)}$, $D_{yy} = EI$, $D_{xy} = 0$ $I = \dfrac{F^2 h}{2}\left(1 - \dfrac{0.81}{1+2.5\left(\frac{F}{2s}\right)^2}\right)$ $\lambda = 1 + \dfrac{\pi^2 F^2}{4s^2}$ $G_{xy} = \dfrac{H}{2} = \dfrac{\lambda}{2}\dfrac{Eh^3}{12(1-v)}$
Plate reinforced by a set of equidistant ribs.	$D_{xx} = \dfrac{Esh_1^3}{12[s-t+t(h_1/h_2)^3]}$, $D_{yy} = \dfrac{EI}{s}$ $D_{xy} = 0$, $G_{xy} = \dfrac{H}{2} = C_P + \dfrac{C_R}{2s}$ C_P = torsional rigidity per unit length of the plate without ribs C_R = torsional rigidity of one rib I = moment of inertia of shaded T-region about its neutral axis
Plate reinforced by equidistant stiffeners.	$D_{xx} = H = \dfrac{Eh^3}{12(1-v^2)}$ $D_{yy} = \dfrac{Eh^3}{12(1-v^2)} + \dfrac{E'I'}{s}$ E' = elastic modulus of stiffeners I' = moment of interia of stiffener cross section with respect to middle surface of plate
In the above, E and v are the modulus of elasticity and Poisson's ratio, respectively, for the plate material.	

Fig. 27 Stiffness coefficients for various orthotropic plates.

axial forces can be obtained. See Refs. 12–15 for details on Fourier series, energy, and numerical techniques for solving Eqs. (59). Figure 28 gives critical buckling stresses for plates with various boundary conditions. More extensive results can be found in Ref. 9.

Postbuckling Strength The postbuckling behavior of plates is quite different from that of columns. Whereas columns collapse as soon as the critical load is reached, plates can exhibit considerable postbuckling strength. As the deflection of the plate after buckling increases, the plate stiffness also increases. This behavior is due to the development of transverse tensile stresses subsequent to the start of buckling. Thus, while plate buckling is clearly an undesired component response, it does not necessarily result in catastrophic failure of the engineering structure.

4.4 Shells

A shell is an object that forms a curved surface. Usually, a shell is thin; that is, the thickness of a shell is small compared to its radius of curvature. It is important to note that this definition of a thin shell is only for classical mechanics analysis of average stresses and deformations. The absolute thickness of a so-called thin shell may be critically important in other areas such as embrittlement of a ductile material caused by thickness and notches or flaws (see Section 8.3). Examples of shells include pressure vessels, soap films, and stadium domes. Whereas plates carry load by the development of bending stresses, shells carry load directly. The strength of a shell comes about primarily because of the way it carries load rather than because of the strength of the material of which it is made.

While even very thin shells will have some bending stiffness, the majority of the loads acting on a shell will be carried by stresses acting tangential to the shell's surface. In the classical membrane theory of shells, bending stiffness is neglected completely. Membrane theory is most applicable for shells of revolution such as shells having the shapes of hemispheres,

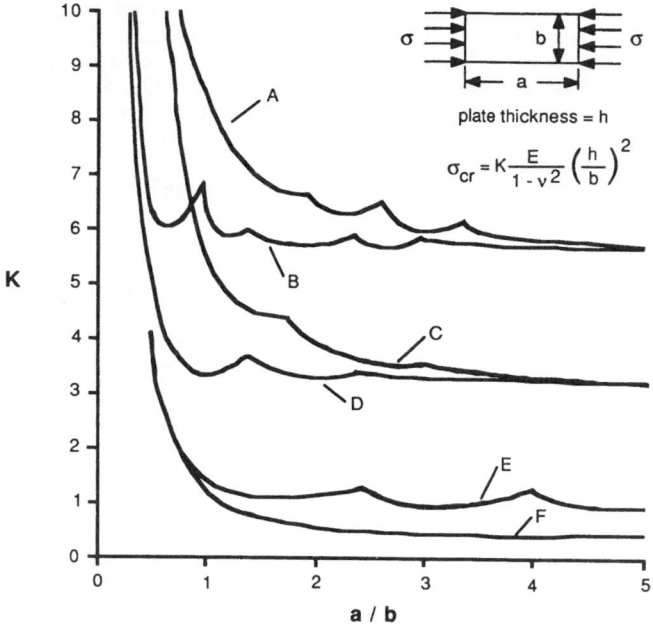

Fig. 28 Critical buckling stress for rectangular plates with various boundary conditions.

A - All edges clamped
B - Edges a clamped, edges b simply supported
C - Edges a simply supported, edges b clamped
D - All edges simply supported
E - One a edge clamped, one a edge free, b edges simply supported
F - One a edge free, one a edge and b edges simply supported

paraboloids, and cylinders. For shells of these shapes, classical solutions can be obtained.[12,19] Because of the variety of shell shapes and the loadings imposed on them, it is difficult to compile a useful yet brief table of formulas. One of the more complete listings of results can be found in Ref. 16. Because of the common use of cylindrical shells as pressure vessels, this particular shell type is discussed in detail in Section 4.5. The previously mentioned potential problem with absolute thickness is especially important for pressure vessels but cannot be discussed here in detail (see Section 8.3 for background on fracture mechanics).

4.5 Cylinders

Thin-Walled Cylinders A cylinder can be considered thin provided its wall thickness is less than one-tenth its radius. Under such conditions, only membrane stresses need to be considered when the cylinder is pressurized internally. In addition, these membrane stresses are essentially constant throughout the wall thickness. If the cylinder is end capped and pressurized with an internal pressure p, the longitudinal stress

is given by

$$\sigma_l = \frac{pr}{2t} \qquad (60)$$

where r is the inner radius of the cylinder and t is the wall thickness. Similarly, the transverse stress (hoop stress) is given by

$$\sigma_h = \frac{pr}{t} \qquad (61)$$

For a thin-walled spherical pressure vessel, the tensile stress in any direction is given by Eq. (60).

Equations (60) and (61) are derived easily by considering the equilibrium forces acting on a cross section of the pressure vessel.

Thick-Walled Cylinders Thick-walled cylinders are used in many industrial fields. Examples include pressure vessels, gun tubes, and pipes. When the wall thickness is a significant fraction of the cylinder's radius, the tangential stresses can no longer be considered constant through the wall thickness. For an end-capped cylinder, the stress distribution near the ends can become quite complex due to the effects of welds, joints, localized yielding, and so on. Thus, the stresses near the ends are commonly determined using experimental techniques. For sections far removed from the ends, however, analytical solutions are available. Assuming linear elastic isotropic response, the stresses are given by (Ref. 7, Chapter 3):

$$\sigma_l = \frac{p_1 r_1^2 - p_2 r_2^2}{r_2^2 - r_1^2} + \frac{F}{\pi(r_2^2 - r_1^2)} \tag{62a}$$

$$\sigma_h = \frac{p_1 r_1^2 - p_2 r_2^2}{r_2^2 - r_1^2} + \frac{r_1^2 r_2^2}{r^2(r_2^2 - r_1^2)}(p_1 - p_2) \tag{62b}$$

$$\sigma_r = \frac{p_1 r_1^2 - p_2 r_2^2}{r_2^2 - r_1^2} - \frac{r_1^2 r_2^2}{r^2(r_2^2 - r_1^2)}(p_1 - p_2) \tag{62c}$$

where σ_l = longitudinal stress
σ_h = hoop stress
σ_r = radial stress
p_1 = internal pressure
p_2 = external pressure
r_1 = inner radius of cylinder
r_2 = outer radius of cylinder
r = radial distance to point under consideration
F = longitudinal load applied to ends of cylinder

If temperature gradients exist through the thickness of the cylinder, the equations for the stresses are more complex (Ref. 8, Chapter 11).

Failure Criteria For thin-walled pressure vessels made of a ductile material, the maximum shear stress criterion or maximum distortion energy criterion is applicable (see Section 8.2). If a brittle material is used, or if the vessel is thick walled, it is important to consider the possibility of failure due to unstable crack propagation. Fracture-mechanics-based design should be used in such cases (see Section 8.3). If the vessel will be under considerable cyclic loading, then fatigue failure may occur (see Section 8.4). An extensive set of rules for the design of pressure vessels and piping has been prepared by the American Society of Mechanical Engineers (ASME).[20]

Compound Cylinders under Internal Pressure
It is apparent from Eqs. (62) that all the stresses cannot be made arbitrarily small by making the outer radius of the cylinder arbitrarily large. In particular, the hoop stress along the inner radius ($r = r_1$) is always greater than the magnitude of the internal pressure, $\sigma_h > p_1$, and the radial stress is always less than $-p_1$, $\sigma_h < -p_1$. Consequently, the maximum shear stress is always larger than p_1. This situation limits the allowable internal pressure. In order to make more

efficient use of material, it is common practice to pre-stress the cylinder. This can be done by shrink-fitting a larger cylinder over the original or prestressing the cylinder to high enough pressures so that plastic flow occurs.[7] Only the former technique will be discussed here.

In order to shrink-fit cylinders, the inner diameter of the larger cylinder must be slightly smaller than the outer diameter of the smaller cylinder. While at the same temperature, the smaller cylinder will not fit inside the larger. If the larger cylinder is heated, however, it will expand in size to the point where the smaller cylinder will fit inside. When the assembly is then brought back to room temperature, a compressive residual hoop stress will exist along the inner radius of the smaller cylinder. This compressive residual stress helps to counteract the tensile hoop stress that develops when internal pressure is applied to the assembly. Details of the procedures to be used for determining the final stress distribution and optimum design of compound cylinders can be found in Refs. 7 and 8.

4.6 Contact Stresses

Contact stresses are the stresses that develop when two solid bodies are pressed against each other. Knowledge of the magnitude of these stresses is important in the design of ball bearings, expansion rollers, and so on. The pioneering work in this field was performed by H. Hertz who is better known for his work with radio waves. The contact stress problem is a nonlinear one, and the analysis is difficult. The primary assumptions used in Hertz's theory are as follows:

1. The contacting bodies are linear elastic, homogeneous, and isotropic.
2. The zone of contact is small compared to the dimensions of the bodies.
3. No friction is present.

Hertz's formulas give the maximum compressive stress that occurs in the contact region, but not the maximum shear stress or maximum tensile stress. Table 7 gives formulas for the maximum stress and contact radius for some common cases. These formulas are based on the original work of Hertz. References 7–9 give more details on the problem of contact stresses.

5 NONLINEAR RESPONSE OF MATERIALS

5.1 Introduction to Plasticity

Plastic flow is said to have occurred in a component if the component does not return to its original shape upon removal of the applied loads. Theories of plasticity fall into two major categories: mathematical theories and physical theories. Mathematical theories provide useful, quantitative information on the plastic flow of materials from a macroscopic point of view. Physical theories help one explain why a material plastically flows; this is done by looking at

Table 7   Stresses and Contact Area for Two Surfaces in Contact

Surface Types	Maximum Compressive Stress, σ (psi)	Radius r or Width b of Contact Area (in.)
sphere on a sphere	$\sigma = 0.616\left[PE^2\left(\dfrac{D_1+D_2}{D_1D_2}\right)^2\right]^{1/3}$	$r = 0.881\left[\dfrac{P}{E}\left(\dfrac{D_1D_2}{D_1+D_2}\right)\right]^{1/3}$
sphere in a spherical socket	$\sigma = 0.616\left[PE^2\left(\dfrac{D_1-D_2}{D_1D_2}\right)^2\right]^{1/3}$	$r = 0.881\left[\dfrac{P}{E}\left(\dfrac{D_1D_2}{D_1-D_2}\right)\right]^{1/3}$
sphere on a plate	$\sigma = 0.616\left[\dfrac{PE^2}{D^2}\right]^{1/3}$	$r = 0.881\left[\dfrac{PD}{E}\right]^{1/3}$
cylinder on a cylinder	$\sigma = 0.591\sqrt{pE\left(\dfrac{D_1+D_2}{D_1D_2}\right)}$	$b = 2.15\sqrt{\dfrac{P}{E}\left(\dfrac{D_1D_2}{D_1+D_2}\right)}$
cylinder on a plate	$\sigma = 0.591\sqrt{\dfrac{pE}{D}}$	$b = 2.15\sqrt{\dfrac{pD}{E}}$

Note: Poisson's ratio $= 0.3$, $E =$ modulus of elasticity (psi), $P =$ load (lb), $p =$ load per unit length (lb/in.), r, b, D_1, D_2, and D in inches.

the material's microstructure. From the viewpoint of the design engineer, the phenomenological approach offered by mathematical theories is the more useful of the two approaches.

Idealized Response   The plastic flow of metals is extremely nonlinear, as is evident from the monotonic stress–strain curves shown later in Fig. 58. Accurate mathematical modeling of such material response is exceedingly difficult. In order to provide useful (but only approximate) predictions, several idealized stress–strain curves are often employed. These are illustrated in Fig. 29. The model of rigid, perfectly plastic behavior neglects elastic strains as well as strain-hardening effects. While this model may seem to be overly simplistic, it has proven useful in limit analysis as well as confined metal-forming processes

such as extrusion.[3] The model of elastic, perfectly plastic behavior is useful for representing the initial yielding response of mild steels. The model of rigid, linear strain-hardening plastic behavior and the model of elastic, linear strain-hardening plastic behavior are useful for representing large plastic strains in strain-hardening materials.

Bauschinger Effect   Low-carbon steels (mild steels) are among the most commonly used engineering materials. They are in a special class because of their sharp-yielding behavior and related phenomena described in the following.

If monotonic tension and compression tests are performed on homogeneous specimens of mild steel and the resulting true stress–strain curves are constructed, the two curves look almost identical. In particular,

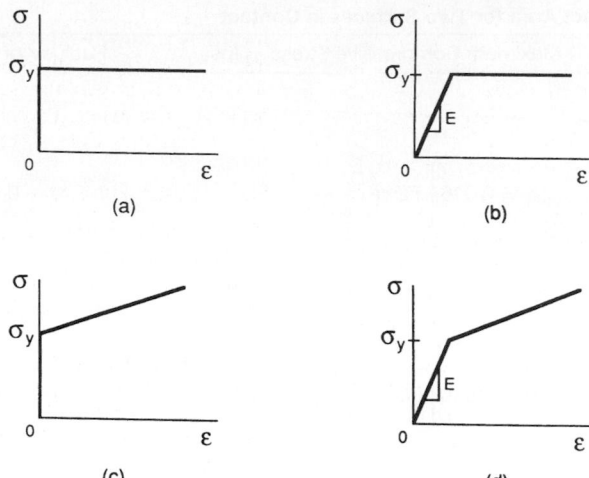

Fig. 29 Idealized stress–strain curves: (*a*) rigid, perfectly plastic; (*b*) elastic, perfectly plastic; (*c*) rigid, linear strain-hardening plastic; (*d*) elastic, linear strain-hardening plastic.

the yield strengths in tension and compression are essentially the same. If a specimen is first loaded in tension into the plastic range and then loaded in compression, the yield strength in compression (σ_{yc}) is generally much lower than the yield strength in tension (σ_{yt}) (Fig. 30). This phenomenon is known as the Bauschinger effect. It plays an important role in cyclic plasticity of these metals.

It is also worthwhile to note that plastic flow is an anisotropic phenomenon. If an initially isotropic material is loaded into the plastic regime and the load is then released, the material will no longer exhibit the same strength properties in all directions. For example, the yield strength of cold-rolled sheet will be substantially different in the thickness direction from that in the rolling direction.

Effects of Hydrostatic Pressure and Incompressibility It has been found that, except for extremely large pressures, hydrostatic pressure has a negligible effect on plastic deformations.[21] Many theories of plasticity assume that the hydrostatic component of stress, $\sigma_{xx} + \sigma_{yy} + \sigma_{zz}$, does not influence the yield behavior of the material. Instead of using the stress tensor [Eq. (1)], it is often found convenient to work with the deviatoric stress tensor defined by

$$[\sigma_d] = \begin{bmatrix} \sigma_{xx} + p & \sigma_{xy} & \sigma_{xz} \\ \sigma_{yx} & \sigma_{yy} + p & \sigma_{yz} \\ \sigma_{zx} & \sigma_{zy} & \sigma_{zz} + p \end{bmatrix}$$

where $p = -\frac{1}{3}(\sigma_{xx} + \sigma_{yy} + \sigma_{zz})$.

It is also commonly assumed in theories of plasticity that plastic deformation is a constant-volume process. Provided the strains are small enough so that

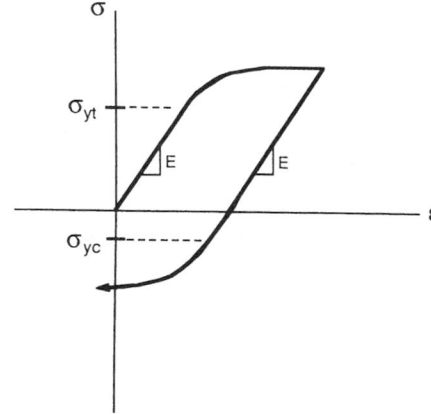

Fig. 30 Bauschinger effect.

product terms can be neglected, the assumption of incompressibility takes the form $\varepsilon_{xx} + \varepsilon_{yy} + \varepsilon_{zz} = 0$.

Solution procedures for general problems using mathematical theories of plasticity are beyond the scope of this work. Details of such solution procedures can be found in Refs. 2 (Chapter 49) and 22.

Multiaxial Loadings A variety of theories exist for the yielding of ductile metals under multiaxial loading conditions. The most popular of these are the distortion energy theory and the maximum shear stress theory. Details of these theories are given in Section 8.2. Reference 22 provides a comparison of these two theories with experimental data obtained by various researchers.

Limit Analysis Limit analysis refers to the determination of the magnitude of the load that will cause plastic collapse of a structure. The plastic collapse load is that load at which the increment of deflection per unit increment of load becomes large. This phenomenon is referred to as uncontained plastic flow. In such situations, enough of the engineering component is undergoing plastic flow so that the remaining elastic regions play only a minor role in sustaining the applied loads. Since failure due to plastic collapse is the governing design criterion for many structures, limit analysis has received increasing attention over the years. It represents a more efficient design criterion than, for example, design based on the yield strength. In many structures, of course, other failure modes such as fatigue, brittle fracture, or buckling govern the design. Limit analysis is most useful in the design of beams and rigid-jointed frames made of mild steels.

In order to obtain quantitative results, the following two assumptions are made:

1. The material exhibits elastic, perfectly plastic response.
2. Changes in geometry of the structure for loads less than the limit load are insignificant.

The advantages of limit load analysis, besides making more efficient use of material, is that the limit load can often be obtained through very simple calculations.

One of the biggest disadvantages is that the deflections for loads below the plastic collapse load are not determined. For some structures, these deflections may be so large as to govern the design, so that plastic collapse becomes irrelevant. Further details on limit analysis can be found in Refs. 2 and 22.

5.2 Plastic Response of Beams and Shafts

When analyzing the plastic response of beams and shafts, it is commonly assumed that the material to be used is elastic–plastic; that is, it behaves in an elastic, perfectly plastic manner (Fig. 29b).

Plastic Bending of a Rectangular Beam To illustrate the analysis procedures for determining plastic response of beams, the case of a beam of rectangular cross section undergoing pure bending is considered. The beam is assumed to have a width b and height $2h$. The moment of inertia is given by $I = b(2h)^3/12 = 2bh^3/3$. The maximum stress σ_m in the beam for elastic response occurs at the top and bottom edges ($y = \pm h$) (Fig. 31a). Its magnitude is given by Eq. (27):

$$|\sigma_m| = \frac{Mh}{I} = \frac{3M}{2bh^2}$$

The bending moment corresponding to the initiation of yield on the top and bottom edges of the beam,

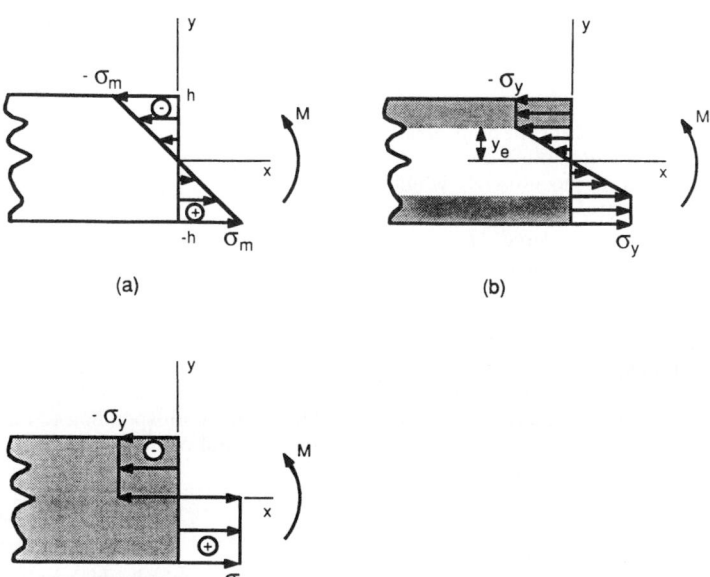

Fig. 31 Plastic deformations in a beam undergoing bending: (a) elastic response; (b) elastic–plastic response; (c) fully plastic response.

M_y, is given by $M_y = (\frac{2}{3})bh^2\sigma_y$, where σ_y is the yield strength of the material.

As the bending moment is further increased, plastic zones develop on the top and bottom of the beam (Fig. 31b). The stress is constant in these zones with a value of σ_y. Within the remaining elastic core, the stresses have the form

$$\sigma_x = -\frac{y}{y_e}\sigma_y$$

where y_e is half the height of the elastic core. The corresponding moment is given by

$$M = -2\int_0^{y_e} y\left(-\frac{y}{y_e}\sigma_y\right)(b\,dy)$$
$$-2\int_{y_e}^h y(-\sigma_y)(b\,dy)$$

Integrating this expression and simplifying it leads to

$$M = \frac{3}{2}M_y\left(1 - \frac{1}{3}\frac{y_e^2}{h^2}\right) \tag{63}$$

As y_e approaches zero, the beam becomes fully plastic, and plastic collapse is imminent (Fig. 31c). The bending moment for fully plastic collapse, M_p, is found by substituting $y_e = 0$ into Eq. (63): $M_p = 1.5M_y = bh^2\sigma_y$. By using limit analysis, this result can be found directly:

$$M_p = -2\int_0^h (-y\sigma_y)(b\,dy) = bh^2\sigma_y$$

Similar expressions can be derived for the limit loads of other kinds of beam loadings. It is often convenient to introduce the plastic section modulus Z, which satisfies the following equation: $M_p = \sigma_y Z$. For the rectangular cross section just considered, $Z = bh^2$. Plastic section moduli for common steel shapes can be found in the *Manual of Steel Construction*.[23]

Plastic Deformations in a Circular Shaft Undergoing Torsion The circular shaft is assumed to be solid with a radius c. The polar moment of inertia of the cross section is $J = 0.5\pi c^4$. The maximum shear stress for elastic response, τ_m, occurs on the outer surface of the shaft (Fig. 32a). Its value is given by Eq. (41):

$$\tau_m = \frac{Tc}{J} = \frac{2T}{\pi c^3}$$

Denoting the torque that will initiate yielding by T_y, it follows that

$$T_y = \frac{\pi}{2}c^3\tau_y$$

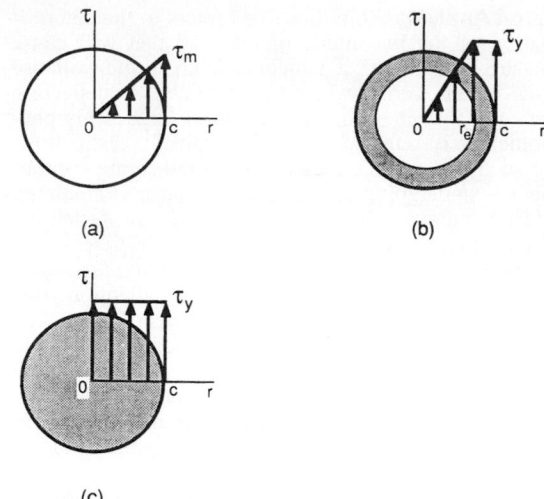

Fig. 32 Plastic deformations in a circular shaft undergoing torsion: (a) elastic response; (b) elastic–plastic response; (c) fully plastic response.

where τ_y is the yield strength in shear of the material.

As the torque is further increased, a plastic zone develops in the outer regions of the shaft (Fig. 32b). This plastic zone surrounds an elastic core. Within the plastic zone the shear stress is given by τ_y. In the elastic core, the shear stress has the form $\tau = (r/r_e)\tau_y$, where r is the radial distance to the point under consideration, and r_e is the radial distance to the edge of the elastic core. The corresponding torque can be found with the aid of Eq. (40):

$$T = \int_0^{r_e} r\left(\frac{r}{r_e}\tau_y\right)(2\pi r\,dr) + \int_{r_e}^c r(\tau_y)(2\pi r\,dr)$$

Integrating and simplifying this equation leads to

$$T = \frac{4}{3}T_y\left(1 - \frac{1}{4}\frac{r_e^3}{c^3}\right) \tag{64}$$

The torque corresponding to fully plastic deformation, T_p, is found by setting r_e equal to zero in Eq. (64):

$$T_p = \tfrac{4}{3}T_y = \tfrac{2}{3}\pi c^3\tau_y$$

Using limit analysis, this result can be obtained directly from Eq. (40) by setting $\tau = \tau_y$:

$$T_p = \int_0^c r(\tau_y)(2\pi r\,dr) = \tfrac{2}{3}\pi c^3\tau_y$$

5.3 Viscoelasticity

Viscoelasticity is the study of the time-dependent response of solid materials. Time, rate, and temperature effects must be considered when analyzing the viscoelastic response of materials. In particular, it is useful to know the homologous temperature T_h. This is defined as the ratio of the operating temperature T to the melting temperature of the material $T_m : T_h = T/T_m$. Both T and T_m must be expressed as absolute temperatures. For U.S. customary units, absolute temperature is measured in Rankines ($^\circ$R $=^\circ$ F $+ 459.67$). If T_h is below 0.3, viscoelastic effects can usually be neglected. If T_h is greater than 0.5, viscoelastic phenomena will almost certainly play a role in the mechanical response of the material.

Creep and the Stress Rupture Test Creep refers to the time-dependent accumulation of strain in a component subjected to a constant stress. In a typical creep test, a slender specimen is subjected to a constant load rather than a constant stress. A schematic diagram of a creep curve that results from a constant-load test is shown in Fig. 33. The creep process can usually be divided into three stages. During primary creep, the creep rate is a decreasing function of time. This response is brought about due to strain-hardening effects. During secondary creep, also called steady-state creep, the creep rate is constant. During this stage, strain-hardening effects are balanced by the effects of increasing true stress amplitude due to decreasing cross-sectional area. During tertiary creep, the decreasing cross-sectional area becomes the controlling feature, and the creep rate rises rapidly until failure occurs. This response is somewhat analogous to the decrease in load amplitude during a tension test due to the necking down of the specimen (see Section 1.4).

It is possible, although difficult, to perform a creep test in which the true stress is maintained constant rather than the load. For such tests, tertiary creep is not observed. This indicates that tertiary creep is indeed due to a mechanical instability rather than representing actual material response. It is often found that the strain at which tertiary creep begins in a constant-load creep test is roughly the same as the strain at which necking

down begins in a tensile test. Even secondary creep is probably a misconception. For carefully controlled constant-true-stress creep tests, the creep rate is usually found to continuously decrease with time.[24]

In spite of the aforementioned problems, constant-load creep testing, commonly referred to as stress rupture testing, is still the most common test method used to determine the viscoelastic properties of a material. In a typical testing series, the time to rupture is determined for a range of loads and temperatures. The data are then plotted as the engineering stress versus time to rupture on log-log coordinates.

Stress Relaxation Tests In a stress relaxation test, the strain is held constant, and the decrease in stress with time is recorded. The advantage of a stress relaxation test over a stress rupture test is that no mechanical instability arises. Thus, the stress relaxation test gives more representative material response than the stress rupture test. The primary disadvantage is that the stress must be continuously adjusted so as to keep the strain constant. This requires the use of sophisticated mechanical testing equipment. A schematic of the typical stress versus time response observed in a stress relaxation test is shown in Fig. 34. Unlike the three-stage response of a stress rupture test, a stress

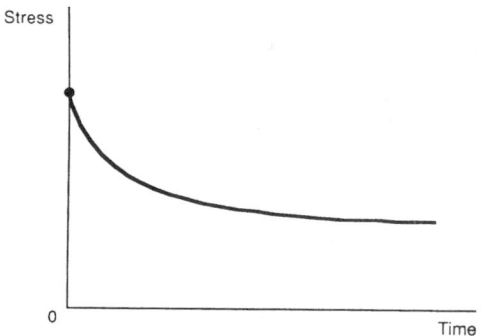

Fig. 34 Stress versus time for stress relaxation test.

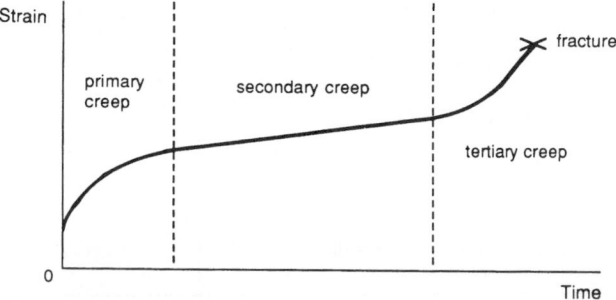

Fig. 33 Schematic diagram of typical creep curve.

relaxation test shows only one general response: a continuing decrease in the stress relaxation rate with time.

Linear Viscoelastic Models Viscoelastic response is sometimes modeled by using combinations of springs and dashpots. Two of the most commonly used of these models are the Maxwell model and the Kelvin–Voigt model (Fig. 35). In this figure, E is the stiffness of the spring, and η is the viscosity coefficient of the dashpot.

The Maxwell model consists of a spring and dashpot in series. This model predicts a linear creep rate (which is representative of secondary creep). The model predicts exponential decay of stress under constant strain conditions. This is the kind of response typically observed in stress relaxation tests of metals.

The Kelvin–Voigt model consists of a spring and dashpot in parallel. The creep response of this model is indicative of primary creep. This model does not exhibit stress relaxation, however.

It is clear that neither the Maxwell model nor the Kelvin–Voigt model is capable of completely describing the creep and stress relaxation response of common materials. Even so, each model is applicable for certain stages of the creep process. These models are not necessarily useful for their ability to accurately predict material response, but rather for the simple visualization of viscoelastic response that they provide. More sophisticated spring-dashpot models are sometimes used. A good discussion of these models can be found in Ref. 25.

In the continuum mechanics approach to viscoelasticity, time-dependent response is modeled by using constitutive equations containing convolution integrals. The stress response $\sigma(t)$ for an arbitrary strain input $\varepsilon(t)$ has the form

$$\sigma(t) = \varepsilon(0)G(t) + \int_0^t G(t - \tau) - \frac{d\varepsilon(\tau)}{d\tau} \, d\tau$$

where t represents time, $\varepsilon(0)$ is the strain corresponding to $t = 0$, and $G(t)$ is the relaxation modulus function (which is often a complicated function of time).[26] Similarly, the strain response due to an arbitrary stress

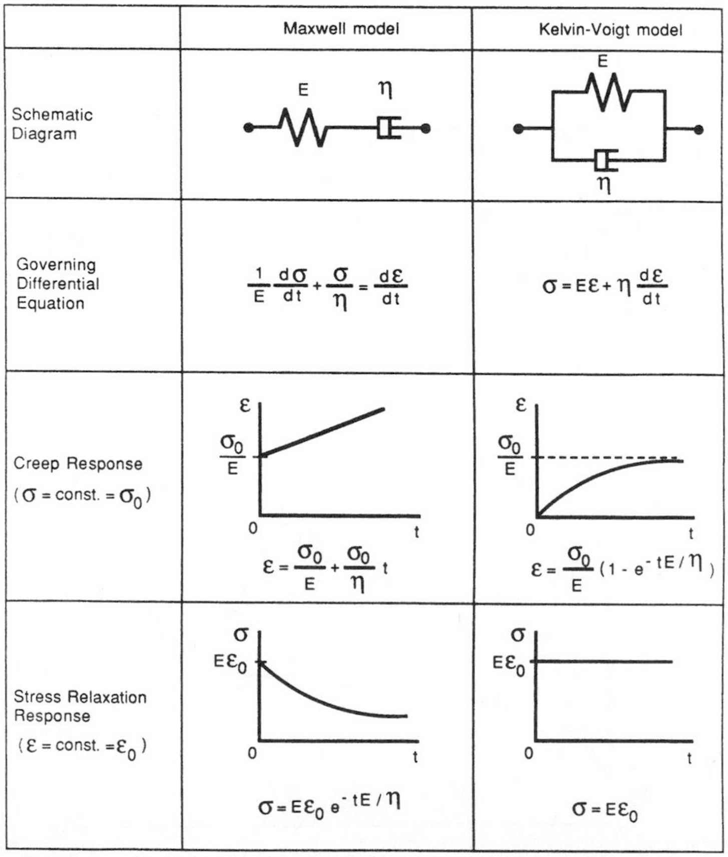

	Maxwell model	Kelvin-Voigt model
Schematic Diagram		
Governing Differential Equation	$\dfrac{1}{E}\dfrac{d\sigma}{dt} + \dfrac{\sigma}{\eta} = \dfrac{d\varepsilon}{dt}$	$\sigma = E\varepsilon + \eta\dfrac{d\varepsilon}{dt}$
Creep Response (σ = const. = σ_0)	$\varepsilon = \dfrac{\sigma_0}{E} + \dfrac{\sigma_0}{\eta}t$	$\varepsilon = \dfrac{\sigma_0}{E}(1 - e^{-tE/\eta})$
Stress Relaxation Response (ε = const. = ε_0)	$\sigma = E\varepsilon_0\, e^{-tE/\eta}$	$\sigma = E\varepsilon_0$

Fig. 35 Models for viscoelastic response.

input is given by

$$\varepsilon(t) = \sigma(0)J(t) + \int_0^t J(t-\tau)\frac{d\sigma(\tau)}{d\tau}\,d\tau$$

where $\sigma(0)$ is the stress corresponding to $t = 0$, and $J(t)$ is the creep compliance function. It can be shown that $G(t)$ and $J(t)$ satisfy

$$\int_0^t G(t-\tau)\frac{dJ(\tau)}{d\tau}\,d\tau = h(t)$$

where $h(t)$ is the unit step function. Thus, determination of $G(t)$ using a stress relaxation test automatically prescribes $J(t)$. Similarly, determining $J(t)$ using a constant-true-stress creep test automatically prescribes $G(t)$. Therefore, the continuum mechanics approach to viscoelasticity rests on the assumption that creep and stress relaxation are simply different manifestations of the same material response. While the continuum mechanics approach to viscoelasticity can lead to powerful methods of analysis, these methods are generally exceedingly difficult to implement in practice.

5.4 Elastic Stability and Column Buckling

When a slender column is subjected to small compressive loads, the column axially shortens according to $\delta = PL/AE$, where δ is the axial shortening, P is the applied load, A is the cross-sectional area, E is the elastic modulus of the material, and L is the column length. If continually larger loads are applied, a load is reached at which the column suddenly bows out sideways. This load is referred to as the critical or buckling load of the column. These sideways deformations are normally too large to be acceptable; consequently, the column is considered to have failed. For slender columns, the axial stress corresponding to

the critical load is generally below the yield strength of the material. Since the stresses in the column just prior to buckling are within the elastic range, the failure is referred to as elastic buckling. Of course, once the column has buckled, large sideways deformations may cause some plastic flow to occur. The term *elastic stability* is commonly used to designate the study of elastic buckling problems. For short columns, failure may be governed by yielding or rupture of the column while it is still axially straight. Failure of short columns may also be caused by inelastic buckling; that is, large sideways deformations that occur when the nominal axial stress is greater than the yield strength.

Besides columns, other structural components that are prone to buckling include plates, shells, and frames. The discussion here will be limited to columns. Buckling of plates is described in Section 4.3. Buckling of frames and shells, as well as torsional buckling phenomena, are discussed in Ref. 27.

Theory of Column Buckling The first studies of the elastic buckling of columns were performed by Euler in the mid-eighteenth century. The governing differential equation for the elastic stability of a pinned–pinned column is given by (see Fig. 36):

$$y'' = \frac{d^2 y}{dx^2} = \frac{M}{EI} \tag{65}$$

where E is the elastic modulus, M is the moment acting on the cross section, and I is the moment of inertia of the cross section. For this equation to govern the buckling of the column, I must be the minimum moment of inertia of the cross section. Also, in using this equation, one assumes linear elastic, isotropic response. From statics, $M = -Py$. Equation (65) thus becomes

$$EIy'' + Py = 0 \tag{66}$$

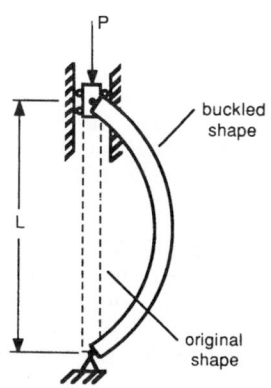

(a) Original and buckled shapes

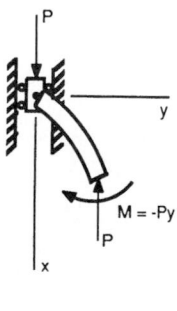

(b) Static equilibrium of buckled element

Fig. 36 Pinned–pinned Euler column: (a) original and buckled shapes; (b) static equilibrium of buckled element.

This is a second-order, homogeneous, linear differential equation with constant coefficients. Its general solution takes the form

$$y = A \sin \lambda x + B \cos \lambda x \qquad (67)$$

where $\lambda^2 = P/EI$ and A and B are constants.

Four boundary conditions can be applied to solve for A and B in Eq. (67). Two of these conditions are

$$y'' = \begin{cases} 0 & \text{at } x = 0 \\ 0 & \text{at } x = L \end{cases}$$

where the primes denote differentiation with respect to x. These conditions are known as natural boundary conditions. They result from the fact that the moments at the pin connections must be zero. Two other boundary conditions are

$$y = \begin{cases} 0 & \text{at } x = 0 \\ 0 & \text{at } x = L \end{cases}$$

These are known as forced boundary conditions. They result from the geometric constraints that exist. Using either the forced or the natural boundary conditions results in the same two possible solutions:

$$A = 0 \qquad \text{and} \qquad B = 0$$

or

$$\sin \lambda L = 0 \qquad \text{and} \qquad B = 0$$

The first of these solutions ($A = 0$ and $B = 0$) is known as the trivial solution. It represents the condition that the beam is still straight. The second solution leads to

$$\lambda L = n\pi \qquad n = 0, 1, 2, 3, \ldots$$

Using $\lambda^2 = P/EI$,

$$P = \frac{n^2 \pi^2 EI}{L^2} \qquad n = 0, 1, 2, 3, \ldots$$

Clearly, $n = 0$ corresponds to no load being applied to the column. The minimum load for which buckling will occur corresponds to $n = 1$. This critical buckling load P_{cr} is given by

$$P_{cr} = \frac{\pi^2 EI}{L^2} \qquad (68)$$

Equation (68) is often referred to as Euler's formula. If I is expressed in terms of the radius of gyration

and area of the cross section, $I = r^2 A$, Eq. (68) can be expressed in terms of the critical buckling stress:

$$\sigma_{cr} = \frac{\pi^2 E}{(L/r)^2}$$

The quantity L/r is referred to as the slenderness ratio of the column. The deflected shape is found from Eq. (67): $y = A \sin(\pi x/L)$. The deflected shape is hence sinusoidal in nature. Since the constant A cannot be prescribed, the magnitude of the deflected shape is indeterminate using this approach.

Equation (66) is valid only for a pinned–pinned column. For other end conditions, a different second-order differential equation applies. It has been shown by Timoshenko and Gere[28] that a single fourth-order differential equation applicable to all end conditions can be employed. This equation has the form

$$y^{iv} + \lambda^2 y'' = 0 \qquad (69)$$

This differential equation has the general solution

$$y = A \sin \lambda x + B \cos \lambda x + Cx + D \qquad (70)$$

where A, B, C, and D are constants. Application of the four available boundary conditions allows for determination of P_{cr} and three of the constants. One of the constants (either A or B) is indeterminate. Thus, the magnitude of the deflected shape cannot be determined using this classical technique either.

Effective Column Length   For end conditions other than pinned-pinned, it is common to express the Euler buckling load in the form

$$P_{cr} = \frac{\pi^2 EI}{L_e^2}$$

where L_e is the effective column length. Figure 37 provides effective column lengths and mathematical boundary conditions for a variety of end conditions. The mathematical boundary conditions are used in conjunction with Eq. (70) to obtain L_e. See Ref. 27 for further details.

Imperfect Columns   In reality, columns are never perfectly straight. Instead, they have some initial eccentricity, often denoted by e, as shown in exaggerated form in Fig. 38. If e is known or can be estimated, the maximum deflection as a function of load can be obtained as

$$y_{max} = e\left[\sec\left(\frac{L}{2}\sqrt{\frac{P}{EI}}\right) - 1\right]$$

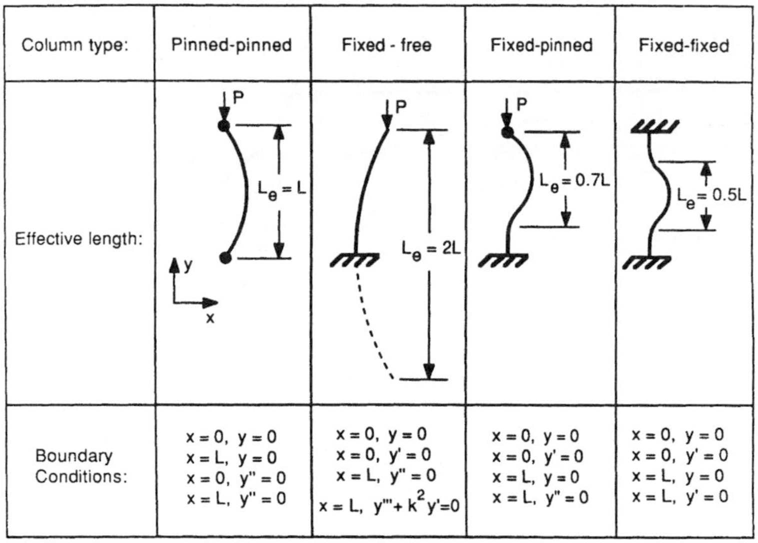

Column type:	Pinned-pinned	Fixed - free	Fixed-pinned	Fixed-fixed
Effective length:	$L_e = L$	$L_e = 2L$	$L_e = 0.7L$	$L_e = 0.5L$
Boundary Conditions:	$x = 0, \ y = 0$ $x = L, \ y = 0$ $x = 0, \ y'' = 0$ $x = L, \ y'' = 0$	$x = 0, \ y = 0$ $x = 0, \ y' = 0$ $x = L, \ y'' = 0$ $x = L, \ y''' + k^2 y' = 0$	$x = 0, \ y = 0$ $x = 0, \ y' = 0$ $x = L, \ y = 0$ $x = L, \ y'' = 0$	$x = 0, \ y = 0$ $x = 0, \ y' = 0$ $x = L, \ y = 0$ $x = L, \ y' = 0$

Fig. 37 Effective lengths and boundary conditions for various columns.

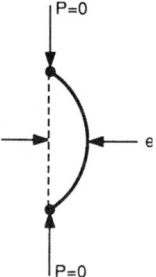

Fig. 38 Initially eccentric column.

The maximum compressive stress is given by

$$\sigma_{\max} = \frac{P}{A} \left[1 + \frac{ec}{r^2} \ \sec \left(\frac{L}{2} \sqrt{\frac{P}{EI}} \right) \right] \qquad (71)$$

where c is the distance from the neutral axis to the out-ermost compressed fiber of the cross section. All other symbols have been previously defined. Equation (71) is referred to as the secant formula. It gives the maximum stress that exists in a column for a given applied load (assuming elastic response throughout the column). For design purposes, the maximum stress is set equal to the yield strength (or to the fracture stress for brittle materials), and the corresponding critical load P_{cr} is found by trial and error. The appropriate factor of safety is applied to this load. The factor of safety must not be applied to the stress since the relation between stress and load in Eq. (71) is nonlinear. Figure 39 gives

a comparison of the secant and Euler formulas for a mild steel.

The secant formula can be used for columns that are eccentrically loaded if e is taken as the eccentricity due to the load plus the initial eccentricity of the column.

Inelastic Buckling For very short columns, the critical stress corresponds to the compressive strength of the material. For very long columns, the Euler critical stress is applicable. For intermediate length columns, failure often occurs at stresses below that predicted by the Euler critical stress or the compressive strength. The effects of residual stresses, especially in rolled carbon steel columns, play an important role in the buckling behavior of intermediate length columns. The tangent modulus theory is sometimes useful in dealing with intermediate length columns. In this theory the elastic modulus in Euler's formula, E, is replaced by the tangent modulus E_T. The tangent modulus corresponding to a given stress is the slope of the stress–strain curve at that stress level. For elastic response, E and E_T are identical. In equation form, the tangent modulus theory is given by

$$P_{cr} = \frac{\pi^2 E_T I}{L_e^2} \qquad (72)$$

where E_T is the slope of the stress–strain curve at the critical stress. Equation (72) is sometimes referred to as the Engesser equation. It is valid only for a rectangular cross section bent about its weak axis. Because use of the tangent modulus theory requires a detailed knowledge of a material's stress–strain curve, it is difficult to implement in practice. Simplified column formulas are more commonly used.

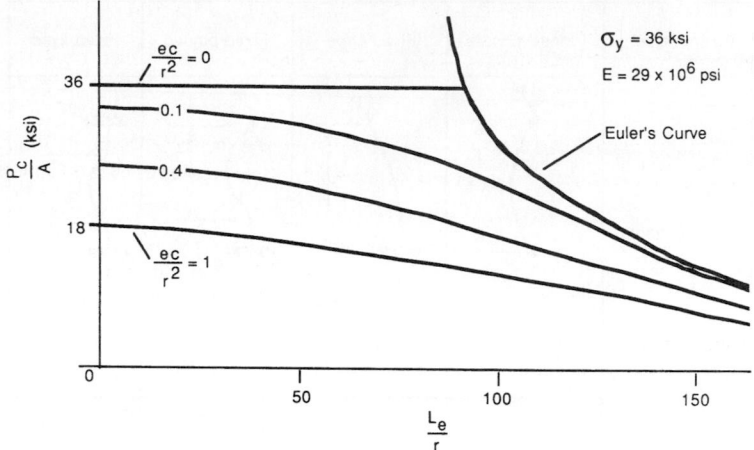

Fig. 39 Comparison of Euler and secant formulas.

Simplified Column Formulas Because of the complexities of the tangent modulus theory, a number of simpler formulas have been proposed:

$$\frac{P_c}{A} = \begin{cases} \sigma_0 - C\left(\dfrac{L_e}{r}\right)^2 & \text{(parabolic)} \quad (73) \\[2ex] \dfrac{\sigma_0}{1 + C\left(\dfrac{L_e}{r}\right)^2} & \text{(Gordon–Rankine)} \quad (74) \\[2ex] \sigma_0 - C\left(\dfrac{L_e}{r}\right) & \text{(straight line)} \quad (75) \end{cases}$$

where P_c/A is the average buckling stress, σ_0 is the compressive strength of the material, C is a material constant, L_e is the effective column length, and r is the radius of gyration of the cross section. The constants σ_0 and C may not be the same in each formula, even for the same material, since they are adjusted to provide a best fit to experimental data. Equations (73)–(75) are displayed graphically in Fig. 40.

Steel Columns Because of the variation in cooling rates of different regions of a hot rolled wide-flange member, large compressive residual stresses can develop. Cold rolling of steel members can also result in significant residual stresses. These residual stresses must be allowed for (by appropriate factors of safety) when designing against column buckling. The current American Institute of Steel Construction (AISC) specification for buckling of steel columns is as follows.[29]

For slenderness ratios less than C_c, where $C_c = (2\pi^2 E/\sigma_y)^{0.5}$, the column buckling formula is given

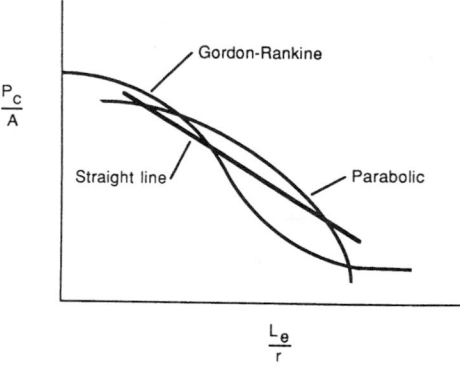

Fig. 40 Simplified column formulas.

by

$$\sigma_a = \frac{\sigma_y\left\{1 - [(\tfrac{1}{2}C_c^2)(L_e/r)^2]\right\}}{FS} \quad (76)$$

where σ_a = allowable compressive stress
L_e/r = slenderness ratio
$FS = \tfrac{5}{3} + 3(L_e/r)/8C_c - (L_e/r)^3/8C_c^3$

For slenderness ratios greater than C_c, the Euler equation is used with a built-in factor of safety of 1.92:

$$\sigma_a = \frac{\pi^2 E}{1.92(L_e/r)^2} \quad (77)$$

Aluminum Columns The Aluminum Association provides three formulas for use with each aluminum alloy.[30] For short columns, the allowable compressive stress is constant with respect to slenderness ratio. For intermediate columns, a straight-line relationship similar to Eq. (75) is used. For long columns, a Euler formula is used.

Timber Columns The American Institute of Timber Construction specifies the use of Euler's formula with a factor of safety of 2.727:

$$\sigma_a = \frac{\pi^2 E}{2.727(L_e/r)^2} \tag{78}$$

Of course, the allowable stress must not exceed the allowable compressive strength parallel to the grain.[31]

6 ENERGY METHODS

6.1 Strain Energy and Strain Energy Density

Axial Loading Consider the case of a uniform rod on which an axial load P is acting. Assume that the axis of the rod is in the x direction. The work done by the load P in extending the rod from its original length L to a length $L + \Delta L$ is given by

$$U = \int_0^{\Delta L} P \, dx$$

where U is known as the strain energy of the rod. It is normally expressed in foot-pounds or inch-pounds.

The strain energy density u is defined as the strain energy per unit volume. For the case of an axially loaded uniform rod,

$$u = \frac{U}{V} = \frac{1}{AL} \int_0^{\Delta L} P \, dx = \int_0^\varepsilon \sigma_{xx} \, d\varepsilon_{xx} \tag{79}$$

where V is the volume of the rod, and A is its cross-sectional area. It is apparent from Eq. (79) that the strain energy density for the case of axial loading of a rod represents the area under the stress–strain diagram. Strain energy density has units of inch-pounds per cubic inches.

For linear elastic, isotropic response, $\varepsilon_{xx} = \sigma_{xx}/E$, where E is the elastic modulus. Equation (79) becomes

$$u = \frac{\sigma_{xx}^2}{2E} \tag{80}$$

The total strain energy of the uniform rod is

$$U = uV = \frac{\sigma_{xx}^2}{2E}(AL) = \frac{P^2 L}{2AE}$$

If the cross section of the rod varies along the length of the rod, the total strain energy must be found by integration of Eq. (80):

$$U = \int \frac{\sigma_{xx}^2}{2E} \, dV \tag{81}$$

It should be kept in mind that Eqs. (80) and (81) apply to a linear elastic, isotropic rod undergoing axial loading.

The modulus of resilience u_y represents the area under the stress–strain diagram up to the yield strength (Fig. 41). Assuming that the yield strength and proportional limit coincide, u_y can be written as

$$u_y = \frac{\sigma_y^2}{2E} \tag{82}$$

where σ_y is the yield strength of the material. The modulus of resilience is important because it represents the energy per unit volume that a material can absorb without yielding.

General State of Loading Assuming linear elastic, isotropic response, the strain energy density can be expressed in several equivalent forms,

$$u = \begin{cases} \dfrac{1}{2}\{\sigma_{xx}\varepsilon_{xx} + \sigma_{yy}\varepsilon_{yy} + \sigma_{zz}\varepsilon_{zz} + \sigma_{xy}\gamma_{xy} \\ \qquad + \sigma_{yz}\gamma_{yz} + \sigma_{zx}\gamma_{zx}\} \hfill (83a) \\[4pt] \dfrac{1}{2E}\{\sigma_{xx}^2 + \sigma_{yy}^2 + \sigma_{zz}^2 \\ \qquad - 2\nu(\sigma_{xx}\sigma_{yy} + \sigma_{yy}\sigma_{zz} + \sigma_{zz}\sigma_{xx}) \\ \qquad + 2(1+\nu)(\sigma_{xy}^2 + \sigma_{yz}^2 + \sigma_{zx}^2)\} \hfill (83b) \\[4pt] G\left\{\varepsilon_{xx}^2 + \varepsilon_{yy}^2 + \varepsilon_{zz}^2 + \dfrac{\nu e^2}{1-2\nu} \right. \\ \qquad \left. + \dfrac{1}{2}(\gamma_{xy}^2 + \gamma_{yz}^2 + \gamma_{zx}^2)\right\} \hfill (83c) \end{cases}$$

where $e = \varepsilon_{xx} + \varepsilon_{yy} + \varepsilon_{zz}$. The total strain energy of a component is found by integrating any of Eqs. (83) over the volume of the component.

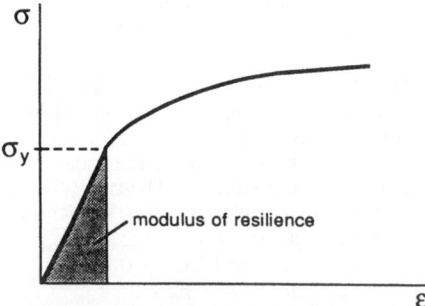

Fig. 41 Definition of modulus of resilience.

Strain Energy in Bending For the case of a beam in pure bending, the flexural stress is given by $\sigma_x = -My/I$, where M is the applied moment, I is the moment of inertia of the cross section, and y is the distance from the neutral axis to the point under consideration (see Section 2.2 for details). All other stresses are zero. Equation (83b) becomes

$$u = \frac{M^2 y^2}{2EI^2}$$

and thus

$$U = \int_0^L \frac{M^2}{2EI^2} \left(\int y^2 \, dA \right) dx$$

where L is the length of the beam, and dA represents an infinitesimal element of the cross-sectional area. Noting that $I = \int y^2 \, dA$,

$$U = \int_0^L \frac{M^2}{2EI} \, dx \qquad (84)$$

Strain Energy Due to Transverse Shear For a beam subjected to transverse loads, the total strain energy consists of strain energy due to bending [as given by Eq. (84)] and strain energy due to shear as given by

$$U = \int_0^L \frac{kV(x)^2}{2\,GA} \, dx \qquad (85)$$

where A is the cross-sectional area, $V(x)$ is the shear force at a distance x, and K is a constant that depends on the cross section of the beam. Values of K for various cross sections are as follows[7]:

Cross Section	K
Rectangular	1.20
Circular	≈ 1.11
Thin-walled cylinder	2.00
I section	1.20
Closed, thin-walled box section	1.00

For the I section, A should be the area of the flanges. For the closed, thin-walled box section, A should be the area of the webs.

Strain Energy in Torsion For the case of a circular rod of length L undergoing torsion, the strain energy can be expressed in the form

$$U = \int_0^L \frac{T^2}{2JG} \, dx \qquad (86)$$

where T is the applied torque, G is the shear modulus, and J is the polar moment of inertia of the cross section. If the cross section is uniform along the length of the rod, Eq. (86) reduces to

$$U = \frac{T^2 L}{2JG}$$

Example 6. Determine the strain energy of a cantilever beam of length L consisting of a uniform rectangular cross section (width $= b$, height $= h$), taking into account both shear and bending effects (see Fig. 42).

Solution. The moment of inertia is given by $I = bh^3/12$. The moment at a longitudinal distance x from the applied load ($0 \leq x \leq L$) takes the form $M = Px$. The shear force is constant along the length of the beam with a value of P. By Eq. (84), the strain energy due to bending is

$$U_b = \int_0^L \frac{P^2 x^2}{2E(\frac{1}{12}bh^3)} \, dx = \frac{2P^2 L^3}{Ebh^3}$$

From Eq. (85), with $K = 1.20$, the strain energy due to shear is

$$U_v = \int_0^L \frac{1.20 P^2}{2Gbh} \, dx = \frac{3P^2 L}{5Gbh}$$

The total strain energy of the beam is thus

$$U = U_b + U_v = U_b \left(1 + \frac{3Eh^2}{10GL^2} \right)$$

For isotropic materials, $E/G < 3$. Thus, provided $h/L < 0.1$, neglecting the strain energy due to shear results in an error of less than 1%. This is why shear effects in long, slender beams are normally ignored.

6.2 Castigliano's Theorems

Complementary Energy In Section 6.1, the strain energy density u for the case of axial loading was defined to be the area under the stress–strain curve.

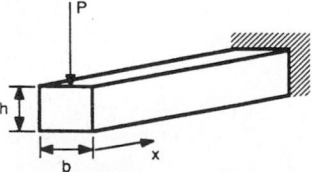

Fig. 42 Cantilever beam for use with Examples 6–9.

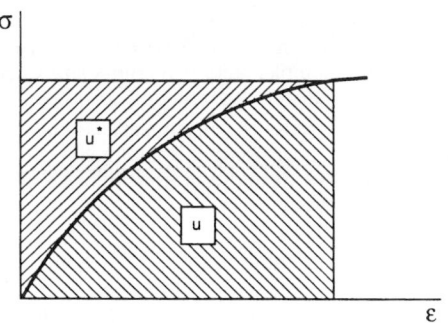

Fig. 43 Definitions of strain energy density and complementary energy density.

The complementary energy density u^* is defined as the area above the stress–strain diagram (Fig. 43). For multiaxial states of stress, the visualization of complementary energy is more difficult. The total complementary energy of a structure, U^*, is given by $U^* = \int u^* \, dV$. See Ref. 14 for details.

Castigliano's First Theorem Consider linear elastic structures that dissipate no energy and undergo small displacements. Also consider conservative forces, that is, forces that keep their orientations as the structure deforms. Castigliano's first theorem states that the load P_i that corresponds to a displacement D_i is given by the partial derivative of the strain energy with respect to D_i. In equation form,

$$P_i = \frac{\partial U}{\partial D_i} \qquad (87)$$

Castigliano's Second Theorem Once again consider linear elastic structures subjected to conservative forces and undergoing small displacements. Castigliano's second theorem states that the displacement D_i due to an applied load P_i equals the partial derivative of the complementary energy with respect to P_i. In equation form,

$$D_i = \frac{\partial U^*}{\partial P_i} \qquad (88)$$

For linear elastic response, $U = U^*$, so that U and U^* can be interchanged in Eqs. (87) and (88).

It can be shown that a similar relationship holds for moments and angular displacements. In equation form,

$$\theta_i = \frac{\partial U^*}{\partial M_i} \qquad (89)$$

where θ_i is the angular displacement at point i due to the applied moment M_i.

Example 7. Using Castigliano's second theorem, determine the end displacement of a cantilever beam loaded at its end. Neglect shear effects.

Solution. From Example 6, the strain energy is given by

$$U = U_b = \frac{P^2 L^3}{6EI}$$

For linear elastic response, $U = U^*$, thus

$$\Delta = \frac{\partial U}{\partial P} = \frac{P L^3}{3EI}$$

This answer agrees with the results obtained using beam deflection theory (Section 2.4).

Example 8. For the beam of Example 7, determine the end slope.

Solution. Since no end moment is applied, a fictitious moment, M, is created. Its value will be set equal to zero after differentiating the strain energy expression. The strain energy due to P and M is given by

$$U = \int_0^L \frac{(M + Px)^2}{2EI} \, dx$$

Differentiating this expression with respect to M gives

$$\theta = \frac{\partial U}{\partial M} = \int_0^L \frac{2(M + Px)}{2EI} \, dx$$

Setting $M = 0$ and integrating leads to

$$\theta = \int_0^L \frac{Px}{EI} \, dx = \frac{P L^2}{2EI}$$

Again, this answer agrees with beam deflection theory.

Castigliano's theorems can also be used to solve statically indeterminate problems. Details can be found in most books on advanced mechanics of materials.

6.3 Impact Stresses

An interesting application of energy methods is in the determination of stresses due to impact loadings, such as a weight dropped onto a beam. In order to solve impact problems using simple energy methods, the following two assumptions are made:

1. No energy is dissipated during impact.
2. The striking body does not bounce off the structure.

In essence, these two assumptions require that all of the kinetic energy of the striking object is transferred to the structure. To solve an impact problem, it is assumed that the kinetic energy of the striking object equals the strain energy induced in the structure.

Example 9.   Determine the maximum stress induced in a uniform cantilever beam of rectangular cross section (height $= h$, width $= b$) if a weight W is dropped from a height h_0 above the free end of the beam.

Solution.   Assuming the end deflection of the beam is negligible compared to h_0, the energy transferred to the beam by the weight is simply Wh_0. From Example 6, the strain energy of the beam (neglecting shear effects) is

$$U = \frac{P_w^2 L^3}{6EI}$$

where P_w is the static load that would bring about the same strain energy in the beam as does the weight. Thus,

$$Wh_0 = \frac{P_w^2 L^3}{6EI} \qquad \text{or} \qquad P_w = \sqrt{\frac{6EIWh_0}{L^3}}$$

The maximum stress in the beam is given by

$$\sigma_{\max} = \frac{Mc}{I} = \frac{P_w L(h/2)}{I} = \frac{h}{2}\sqrt{\frac{6EWh_0}{LI}}$$

In general, for a structure to efficiently resist an impact load, it should have the following properties[10]:

1. Possess a large volume.
2. Have a low modulus of elasticity and high yield strength.
3. Have as uniform a shape as possible so that the stresses are evenly distributed.

7 COMPOSITE MATERIALS

7.1 Introduction to Composite Materials

Definition of a Composite Material   There is no universally accepted definition of what constitutes a composite material. In everyday usage, a composite is something made up of several parts. In engineering, whether or not a given material should be considered a composite depends on the scale of viewing:

At the atomic level, any material consisting of more than one kind of atom could be considered a composite. With this definition, only pure elements would not be composites.

At the microscopic level, any material composed of more than one phase could be considered a composite. For example, steel is a composite since it is a multiphase alloy of carbon and iron.

At the macroscopic level, only materials consisting of two or more macroscopic constituents are called composites. An example is glass fibers embedded in an epoxy matrix.

The following definition of composite materials is provided by Schwartz[32]:

"A composite material is a material system composed of a mixture or combination of two or more macroconstituents differing in form and/or material composition and that are essentially insoluble in each other."

Classification of Composite Materials   Composites can be classified in a variety of ways such as by the matrix constituent (i.e., metal matrix, ceramic matrix, and resin matrix composites). More commonly, classification is based on the structural constituents. This leads to three common types of composite materials:

1. Fiber composites are composed of fibers embedded in a matrix.
2. Particulate composites consist of particles embedded in a matrix.
3. Laminated composites are composed of layers of various materials.

Advantages of Composites   Some of the advantages of composites over conventional materials include the following:

1. Higher specific strength (i.e., strength per unit weight)
2. Higher specific stiffness (i.e., stiffness per unit weight)
3. The ability to custom design the properties of a composite for a specific application

Many other advantages exist for certain composites as discussed in the technical literature.

7.2 Orthotropic Elasticity

While composites offer higher specific strength and stiffness than many conventional materials, they suffer from the fact that their mechanical behavior is more difficult to understand. In the most general case, a material can display anisotropic stress–strain response that requires 21 material constants to describe the elastic behavior. In comparison, only 2 elastic constants are needed to describe isotropic response. For unidirectional, fiber-reinforced materials such as shown in Fig. 44, there are 3 mutually orthogonal planes of material symmetry. This reduces the required number of elastic constants from 21 to 9. A material having

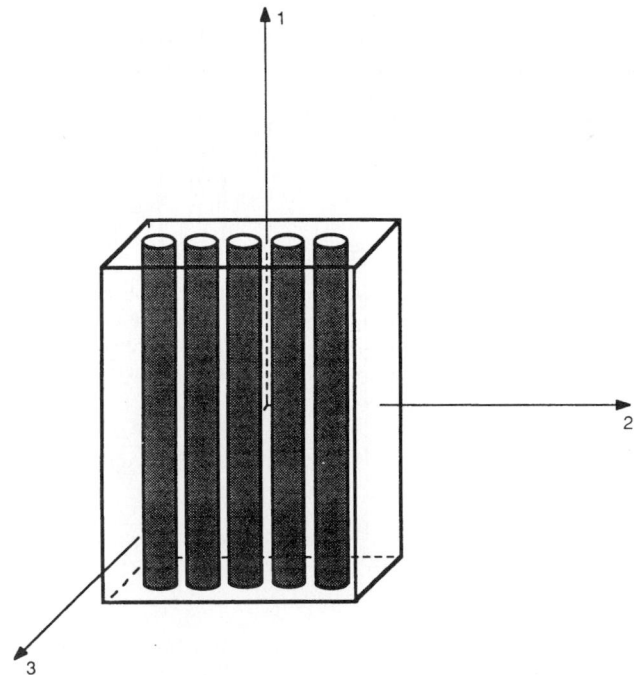

Fig. 44 Principal material directions for unidirectionally reinforced lamina.

such symmetry is said to be orthotropic. Assuming the 1, 2, and 3 directions are the principal directions of material symmetry, the stress–strain relationships are

$$
\begin{Bmatrix} \sigma_1 \\ \sigma_2 \\ \sigma_3 \\ \sigma_{23} \\ \sigma_{31} \\ \sigma_{12} \end{Bmatrix}
=
\begin{bmatrix}
C_{11} & C_{12} & C_{13} & 0 & 0 & 0 \\
C_{12} & C_{22} & C_{23} & 0 & 0 & 0 \\
C_{13} & C_{23} & C_{33} & 0 & 0 & 0 \\
0 & 0 & 0 & C_{44} & 0 & 0 \\
0 & 0 & 0 & 0 & C_{55} & 0 \\
0 & 0 & 0 & 0 & 0 & C_{66}
\end{bmatrix}
\begin{Bmatrix} \varepsilon_1 \\ \varepsilon_2 \\ \varepsilon_3 \\ \gamma_{23} \\ \gamma_{31} \\ \gamma_{12} \end{Bmatrix}
\tag{90}
$$

where $[C_{ij}]$ is known as the stiffness matrix. The shear strains (γ_{12}, γ_{23}, and γ_{31}) represent engineering shear strains. Equation 90 can be inverted to give strains in terms of stresses:

$$
\begin{Bmatrix} \varepsilon_1 \\ \varepsilon_2 \\ \varepsilon_3 \\ \gamma_{23} \\ \gamma_{31} \\ \gamma_{12} \end{Bmatrix}
=
\begin{bmatrix}
S_{11} & S_{12} & S_{13} & 0 & 0 & 0 \\
S_{12} & S_{22} & S_{23} & 0 & 0 & 0 \\
S_{13} & S_{23} & S_{33} & 0 & 0 & 0 \\
0 & 0 & 0 & S_{44} & 0 & 0 \\
0 & 0 & 0 & 0 & S_{55} & 0 \\
0 & 0 & 0 & 0 & 0 & S_{66}
\end{bmatrix}
\begin{Bmatrix} \sigma_1 \\ \sigma_2 \\ \sigma_3 \\ \sigma_{23} \\ \sigma_{31} \\ \sigma_{12} \end{Bmatrix}
\tag{91}
$$

where $[S_{ij}]$ is known as the compliance matrix. Note that both the stiffness and compliance matrices are symmetric. Also, they are matrix inverses of each other. The components of the stiffness matrix can be related to the components of the compliance matrix by

the following expressions:

$$
C_{11} = \frac{S_{22}S_{33} - S_{23}^2}{S} \qquad C_{12} = \frac{S_{13}S_{23} - S_{12}S_{33}}{S}
$$
$$
C_{22} = \frac{S_{33}S_{11} - S_{13}^2}{S} \qquad C_{13} = \frac{S_{12}S_{23} - S_{13}S_{22}}{S}
$$
$$
C_{33} = \frac{S_{11}S_{22} - S_{12}^2}{S} \qquad C_{23} = \frac{S_{12}S_{13} - S_{23}S_{11}}{S}
$$
$$
C_{44} = \frac{1}{S_{44}} \qquad C_{55} = \frac{1}{S_{55}} \qquad C_{66} = \frac{1}{S_{66}}
\tag{92}
$$

where

$$
S = S_{11}S_{22}S_{33} - S_{11}S_{23}^2 - S_{22}S_{13}^2 - S_{33}S_{12}^2
$$
$$
+ 2S_{12}S_{23}S_{13}
$$

The converse relationship (i.e., S_{ij}'s in terms of C_{ij}'s) can be obtained by interchanging the symbols C and S everywhere in Eqs. (92).

Engineering Constants   In order to predict material behavior, it is necessary to determine the material constants in either Eq. (90) or (92). It is convenient to introduce the following engineering constants[33]:

E_1, E_2, E_3 Elastic moduli in the 1, 2, and 3 directions, respectively. Thus, $E_i = \sigma_i / \varepsilon_i$ for $\sigma_i = \sigma$ and all other stresses being zero.

G_{12}, G_{23}, G_{31} Shear moduli in the 1–2, 2–3, and 3–1 planes, respectively.

v_{ij} Poisson's ratio for transverse strain in the j direction when the specimen is stressed in the i direction. Hence, $v_{ij} = -\varepsilon_j/\varepsilon_i$ for $\sigma_i = \sigma$ and all other stresses being zero.

The compliance matrix can be rewritten as

$$[S_{ij}] = \begin{bmatrix} \dfrac{1}{E_1} & -\dfrac{v_{21}}{E_2} & -\dfrac{v_{31}}{E_3} & 0 & 0 & 0 \\ -\dfrac{v_{12}}{E_1} & \dfrac{1}{E_2} & -\dfrac{v_{32}}{E_3} & 0 & 0 & 0 \\ -\dfrac{v_{13}}{E_1} & -\dfrac{v_{23}}{E_2} & \dfrac{1}{E_3} & 0 & 0 & 0 \\ 0 & 0 & 0 & \dfrac{1}{G_{23}} & 0 & 0 \\ 0 & 0 & 0 & 0 & \dfrac{1}{G_{31}} & 0 \\ 0 & 0 & 0 & 0 & 0 & \dfrac{1}{G_{12}} \end{bmatrix}$$

(93)

Because of the symmetry of the compliance matrix $(S_{ij} = S_{ji})$, it is apparent that

$$\frac{v_{ij}}{E_i} = \frac{v_{ji}}{E_j}$$

In terms of the engineering constants, the stiffness matrix components take the form

$$C_{11} = \frac{1 - v_{23}v_{32}}{E_2 E_3 \Delta}$$

$$C_{12} = \frac{v_{21} + v_{31}v_{23}}{E_2 E_3 \Delta} = \frac{v_{12} + v_{32}v_{13}}{E_1 E_3 \Delta}$$

$$C_{13} = \frac{v_{31} + v_{21}v_{32}}{E_2 E_3 \Delta} = \frac{v_{13} + v_{12}v_{23}}{E_1 E_2 \Delta}$$

$$C_{22} = \frac{1 - v_{31}v_{13}}{E_1 E_3 \Delta}$$

$$C_{23} = \frac{v_{32} + v_{12}v_{31}}{E_1 E_3 \Delta} = \frac{v_{23} + v_{21}v_{13}}{E_1 E_2 \Delta}$$

$$C_{33} = \frac{1 - v_{12}v_{21}}{E_1 E_2 \Delta}$$

$$C_{44} = G_{23} \qquad G_{55} = G_{31} \qquad C_{66} = G_{12}$$

where

$$\Delta = \frac{1 - v_{12}v_{21} - v_{23}v_{32} - v_{31}v_{13} - 2v_{21}v_{32}v_{13}}{E_1 E_2 E_3}$$

7.3 Plane Stress

For orthotropic response under plane stress conditions, the in-plane stress–strain relationships are given by

$$\begin{Bmatrix} \varepsilon_1 \\ \varepsilon_2 \\ \gamma_{12} \end{Bmatrix} = \begin{bmatrix} S_{11} & S_{12} & 0 \\ S_{12} & S_{22} & 0 \\ 0 & 0 & S_{66} \end{bmatrix} \begin{Bmatrix} \sigma_1 \\ \sigma_2 \\ \sigma_{12} \end{Bmatrix}$$

and

$$\begin{Bmatrix} \sigma_1 \\ \sigma_2 \\ \sigma_{12} \end{Bmatrix} = \begin{bmatrix} Q_{11} & Q_{12} & 0 \\ Q_{12} & Q_{22} & 0 \\ 0 & 0 & Q_{66} \end{bmatrix} \begin{Bmatrix} \varepsilon_1 \\ \varepsilon_2 \\ \gamma_{12} \end{Bmatrix}$$

where the compliance coefficients (S_{ij}) are the same as in Eq. (93). The components Q_{ij}, sometimes referred to as the reduced stiffnesses, can be expressed in terms of the compliance coefficients as

$$Q_{11} = \frac{S_{22}}{S_{11}S_{22} - S_{12}^2} \qquad Q_{22} = \frac{S_{11}}{S_{11}S_{22} - S_{12}^2}$$

$$Q_{12} = -\frac{S_{12}}{S_{11}S_{22} - S_{12}^2} \qquad Q_{66} = \frac{1}{S_{66}}$$

7.4 Stress–Strain Relationships for Arbitrary Orientations

In Section 7.2, the stress–strain relationships with respect to the principal material directions were presented. Unfortunately, the principal material directions often do not coincide with the coordinates that naturally fit the geometry of the structure. For plane stress conditions, Mohr's circle can be used to transform stresses or strains; however, the principal stresses and strains do not necessarily occur at the same orientation. For an xy coordinate system located at an angle θ with respect to the principal material coordinates (Fig. 45),

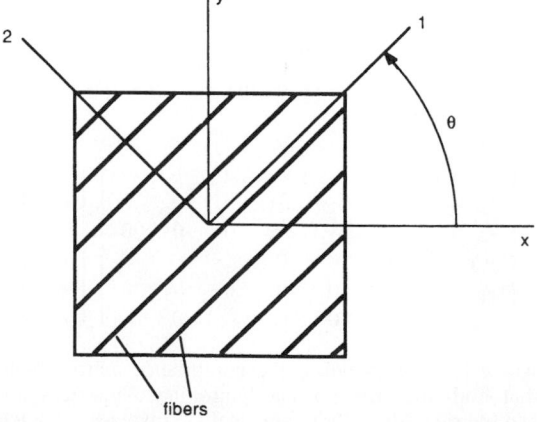

Fig. 45 Arbitrary in-plane orientation of xy coordinate axes.

the stress–strain relationships take the form

$$\left\{ \begin{array}{c} \sigma_x \\ \sigma_y \\ \sigma_{xy} \end{array} \right\} = \begin{bmatrix} \overline{Q}_{11} & \overline{Q}_{12} & \overline{Q}_{16} \\ \overline{Q}_{12} & \overline{Q}_{22} & \overline{Q}_{26} \\ \overline{Q}_{16} & \overline{Q}_{26} & \overline{Q}_{66} \end{bmatrix} \left\{ \begin{array}{c} \varepsilon_x \\ \varepsilon_y \\ \gamma_{xy} \end{array} \right\} \qquad (94)$$

where

$$\overline{Q}_{11} = Q_{11}\cos^4\theta + 2(Q_{12} + 2Q_{66})\sin^2\theta\cos^2\theta$$
$$+ Q_{22}\sin^4\theta$$

$$\overline{Q}_{12} = (Q_{11} + Q_{22} - 4Q_{66})\sin^2\theta\cos^2\theta$$
$$+ Q_{12}(\sin^4\theta + \cos^4\theta)$$

$$\overline{Q}_{22} = Q_{11}\sin^4\theta + 2(Q_{12} + 2Q_{66})\sin^2\theta\cos^2\theta$$
$$+ Q_{22}\cos^4\theta$$

$$\overline{Q}_{16} = (Q_{11} - Q_{12} - 2Q_{66})\sin\theta\cos^3\theta$$
$$+ (Q_{12} - Q_{22} + 2Q_{66})\sin^3\theta\cos\theta$$

$$\overline{Q}_{26} = (Q_{11} - Q_{12} - 2Q_{66})\sin^3\theta\cos\theta$$
$$+ (Q_{12} - Q_{22} + 2Q_{66})\sin\theta\cos^3\theta$$

$$\overline{Q}_{66} = (Q_{11} + Q_{22} - 2Q_{12} - 2Q_{66})\sin^2\theta\cos^2\theta$$
$$+ Q_{66}(\sin^4\theta + \cos^4\theta)$$

or alternatively

$$\left\{ \begin{array}{c} \varepsilon_x \\ \varepsilon_y \\ \gamma_{xy} \end{array} \right\} = \begin{bmatrix} \overline{S}_{11} & \overline{S}_{12} & \overline{S}_{16} \\ \overline{S}_{12} & \overline{S}_{22} & \overline{S}_{26} \\ \overline{S}_{16} & \overline{S}_{26} & \overline{S}_{66} \end{bmatrix} \left\{ \begin{array}{c} \sigma_x \\ \sigma_y \\ \sigma_{xy} \end{array} \right\} \qquad (95)$$

where

$$\overline{S}_{11} = S_{11}\cos^4\theta + (2S_{12} + S_{66})\sin^2\theta\cos^2\theta$$
$$+ S_{22}\sin^4\theta$$

$$\overline{S}_{12} = S_{12}(\sin^4\theta + \cos^4\theta)$$
$$+ (S_{11} + S_{12} - S_{66})\sin^2\theta\cos^2\theta$$

$$\overline{S}_{22} = S_{11}\sin^4\theta + (2S_{12} + S_{66})\sin^2\theta\cos^2\theta$$
$$+ S_{22}\cos^4\theta$$

$$\overline{S}_{16} = (2S_{11} - 2S_{12} - S_{66})\sin\theta\cos^3\theta$$
$$- (2S_{22} - 2S_{12} - S_{66})\sin^3\theta\cos\theta$$

$$\overline{S}_{26} = (2S_{11} - 2S_{12} - S_{66})\sin^3\theta\cos\theta$$
$$- (2S_{22} - 2S_{12} - S_{66})\sin\theta\cos^3\theta$$

$$\overline{S}_{66} = 2(2S_{11} + 2S_{22} - 4S_{12} - S_{66})\sin^2\theta\cos^2\theta$$
$$+ S_{66}(\sin^4\theta + \cos^4\theta)$$

If $\gamma_{xy} = 0$ in Eq. (94), the xy orientation corresponds to the principal strain directions. However, since $\sigma_{xy} = Q_{16}\varepsilon_{xx} + Q_{26}\varepsilon_{yy}$, this is seldom the direction of the principal stresses. Similarly, when $\sigma_{xy} = 0$ in Eq. (95), it is seldom true that $\gamma_{xy} = 0$. This fact is very important since some theories of lamina strength are based on principal strains, while others are based on principal stresses.

7.5 Biaxial Strength Theories

Several theories are available for predicting the strength of lamina under in-plane biaxial loading. Some of the simpler of these theories are now discussed.

Maximum-Strength Theory Maximum-strength theory states that the lamina is safe provided the stresses in the principal material directions are less than the ultimate strengths in these directions. Hence, the lamina is safe provided

$$\sigma_1 < X_t \qquad \sigma_2 < Y_t \qquad \sigma_{12} < S$$

for tension, and

$$|\sigma_1| < |X_c| \qquad |\sigma_2| < |Y_c|$$

for compression. In these inequalities, X_t is the tensile strength in the 1 direction, X_c is the compressive strength in the 1 direction, Y_t is the tensile strength in the 2 direction, Y_c is the compressive strength in the 2 direction, and S is the in-plane shear strength.

Maximum-Strain Theory Maximum-strain theory is similar to the maximum-strength theory except that the stresses are replaced by strains. Thus, the material is safe provided

$$\varepsilon_1 < \varepsilon_1^{ut} \qquad \varepsilon_2 < \varepsilon_2^{ut} \qquad |\gamma_{12}| < S_\varepsilon$$

for tension, and

$$|\varepsilon_1| < |\varepsilon_1^{uc}| \qquad |\varepsilon_2| < |\varepsilon_2^{uc}|$$

for compression. In these inequalities, ε_1^{ut} is the maximum tensile strain at failure for loading in the 1 direction, ε_1^{uc} is the maximum compressive strain at failure for loading in the 1 direction, ε_2^{ut} is the maximum tensile strain at failure for loading in the 2 direction, ε_2^{uc} is the maximum compressive strain at failure for loading in the 2 direction, and S_ε is the shear strain at failure for an in-plane shear test of the lamina.

Tsai–Hill Theory It should be noted that the maximum stress and maximum strain theories assume no interaction between modes of failure. As such, these theories can lead to inaccurate predictions of strength. The Tsai–Hill theory states that failure is imminent when the following inequality is satisfied:

$$\left(\frac{\sigma_1}{X}\right) - \left(\frac{\sigma_1\sigma_2}{X^2}\right) + \left(\frac{\sigma_2}{Y}\right)^2 + \left(\frac{\sigma_{12}}{S}\right)^2 \geq 1$$

where X is the ultimate strength in the 1 direction, Y is the ultimate strength in the 2 direction, and S is the in-plane shear strength of the lamina. The Tsai–Hill theory is easy to use and gives reasonably good results for many kinds of composite lamina. Note that it does not differentiate between tensile and compressive strengths, however.

More sophisticated lamina failure theories are available. A thorough review of these is given in Ref. 34.

Laminates A lamina is a single layer of composite material with fibers running in one direction. A laminate consists of several individual laminae stacked one on top of another and bonded together to form a single component. Theories for predicting the stress–strain relationships and failure strengths of composite laminates are available, but they are too complex to be presented here (see Ref. 34 for a thorough review of these theories). Further details on the mechanics of composite materials can be found in Refs. 33–38.

8 THEORIES OF STRENGTH AND FAILURE

8.1 Stress Concentrations

Large increases in stresses occur near abrupt changes in geometry such as holes, notches, fillets, and cracks. With regards to mechanics of materials, these geometrical discontinuities are referred to as stress raisers or stress concentrations. For some simple geometries, the effect of stress raisers on the stress distribution can be determined using the theory of elasticity. For complicated geometries, however, numerical or experimental techniques represent more practical means of determining the stress distribution.

The theoretical stress concentration factor, K_t, is defined by

$$K_t = \frac{\sigma_{\max}}{\sigma_{\text{avg}}} \tag{96}$$

where $\sigma_{\max}$ is the maximum stress that occurs in the vicinity of the stress concentration, and σ_{avg} is the average stress that acts at the minimal cross-sectional area. For example, consider the case of a hole in a semi-infinite plate under uniaxial loading (Fig. 46). Far away from the hole, the longitudinal stress is given by $\sigma = P/wt$, where P is the applied load, w is the plate width, and t is the plate thickness. Along the line AB in Fig. 46, the average longitudinal stress is higher

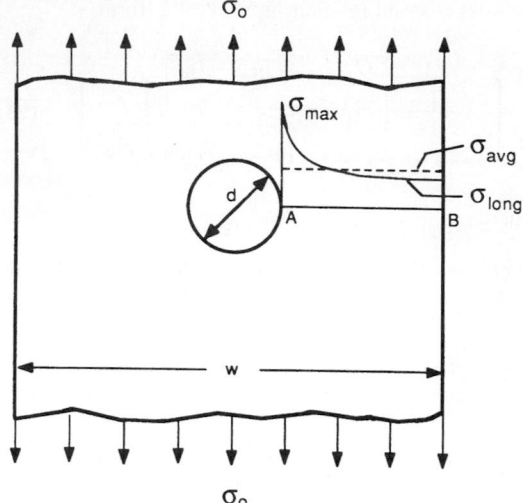

Fig. 46 Hole in a semi-infinite plate.

due to the reduced cross-sectional area caused by the presence of the hole. Thus,

$$\sigma_{\text{avg}} = \frac{P}{(w-d)t}$$

where d is the hole diameter. The actual stress distribution along the line AB is not constant at σ_{avg}, but rather it varies as shown in the figure. The maximum longitudinal stress occurs at the edge of the hole.

Stress concentration factors provide a quick method for estimating the severity of a given stress raiser. Extensive tables of these factors for a wide variety of situations are available in the literature.[9,39] Stress concentration factors should be used with caution in design, however, since several important issues are neglected:

1. Stress concentration factors are normally given for uniaxial states of loading. In many instances, there is a combined state of loading (e.g., biaxial loading of a plate with a hole).

2. Even when the far-field loading is uniaxial, the geometrical discontinuity induces a state of biaxial stresses in its immediate vicinity.

3. If the maximum stress exceeds the yield strength, a plastic zone develops. The maximum stress predicted by Eq. (96) is too high in such circumstances.

It is apparent from the last of these statements that the use of the stress concentration factor can lead to overly conservative design in the case of a ductile material. It is possible, however, for normally ductile materials to exhibit brittle behavior. This is especially true if the component thickness is large enough to induce a state of plane strain. In such cases, the use of the stress concentration factor in design can lead to nonconservative predictions. This apparent dilemma can be explained by the application of modern theories of fracture mechanics. Further details are provided in Sections 8.2 and 8.3.

8.2 Classical Theories of Static Failure

Ductile versus Brittle Material Behavior For materials exhibiting extremely ductile behavior, such as mild steel at room temperature, design against failure is often based on the yield strength. Roughly speaking, a material can be considered to be ductile if it exhibits an elongation of at least 5% in a standard tension test.[40] It is important to recognize, however, that materials are in fact neither ductile nor brittle. Instead, materials are ductile in some circumstances and brittle in others. For example, while most mild steels display considerable ductility at room temperature, the ductility tends to decrease as the temperature is lowered. In fact, a rapid decrease in ductility often occurs over a narrow temperature range (see Section 8.3 for further details). In general, several other factors besides temperature can induce brittle behavior in what are normally considered to be ductile materials. These include the presence of notches, corrosive environments, and very rapid rates of loading.[41]

Maximum-Shear-Stress Theory The maximum-shear-stress theory states that a region of material in a component yields when the maximum shear stress at that point equals the maximum shear stress at yield of a tensile specimen of the same material. From Section 1.3, $\tau_{max} = (\sigma_1 - \sigma_3)/2$, where σ_1 and σ_3 are the algebraically largest and smallest principal stresses, respectively. For the case of a uniaxial tension specimen at yield, $\sigma_3 = 0$ and $\sigma_1 = \sigma_y$. The maximum-shear-stress theory takes the form

$$\tau_{max} = \tfrac{1}{2}(\sigma_1 - \sigma_3) = \tfrac{1}{2}\sigma_y \quad \text{at yield} \quad (97)$$

This theory is relatively simple to use and normally results in conservative predictions. For biaxial states of loading, σ_1 and σ_3 are determined easily using Mohr's circle.

For a state of pure shear, $\tau_{max} = \tau_y$ at yield. The maximum shear stress theory thus predicts a value of 0.5 for the ratio of the yield strength in torsion to the yield strength in tension,

$$\frac{\tau_y}{\sigma_y} = 0.5$$

Distortion Energy Theory The distortion energy theory, also known as the von Mises–Hencky theory, is based on the assumption that yielding occurs at a

point in a component when the total distortional energy per unit volume at that point equals the distortional energy per unit volume in a tensile specimen at yield. The distortion energy per unit volume, u_d, is given by

$$u_d = \frac{1}{12G}[(\sigma_{xx} - \sigma_{yy})^2 + (\sigma_{yy} - \sigma_{zz})^2 \\ + (\sigma_{zz} - \sigma_{xx})^2 + 6(\sigma_{xy}^2 + \sigma_{zx}^2 + \sigma_{yz}^2)]$$

or, in terms of the principal stresses,

$$u_d = \frac{1}{6G}(\sigma_1^2 + \sigma_2^2 + \sigma_3^2 - \sigma_1\sigma_2 - \sigma_1\sigma_3 - \sigma_2\sigma_3)$$

For a tensile specimen at yield, $\sigma_1 = \sigma_y$ and $\sigma_2 = \sigma_3 = 0$. Accordingly, $u_d = \sigma_y^2/6G$ in this case. The distortion energy theory can therefore be rewritten as

$$\sigma_1^2 + \sigma_2^2 + \sigma_3^2 - \sigma_1\sigma_2 - \sigma_1\sigma_3 - \sigma_2\sigma_3 = \sigma_y^2 \quad \text{at yield} \quad (98)$$

For a state of pure shear, $\sigma_1 = \tau$, $\sigma_2 = 0$, and $\sigma_3 = -\tau$. Using the distortion energy theory, the relationship between yield in shear and tension is

$$\tau_y^2 + 0 + \tau_y^2 - 0 - 0 - 0 = \sigma_y^2$$

or

$$\frac{\tau_y}{\sigma_y} = 0.577$$

Comparison of Theories for Yielding For most ductile materials, the distortion energy theory gives more accurate results than the maximum-shear-stress theory.[22] For a biaxial state of loading, the out-of-plane principal stress is zero. Denoting the in-plane principal stresses by σ_a and σ_b, it is possible to graphically portray the two theories. This has been done in Fig. 47. It is seen that the maximum-shear-stress theory is represented by a hexagon, while the distortion energy theory is represented by an ellipse. In general, both theories give reasonably good results for plane stress conditions. While the distortion energy theory is usually more accurate, its use can result in slightly nonconservative predictions. On the other hand, use of the maximum shear stress theory almost always leads to slightly conservative predictions.

Maximum-Normal-Stress Theory The maximum-normal-stress theory is generally used for calculating the failure stress for brittle materials. It states that a component fails when its maximum principal stress, σ_1, exceeds the ultimate strength as determined by a tensile test. This theory should be used with caution. For plane stress situations, this theory gives reasonable results if both in-plane principal stresses are tensile. It can be grossly inaccurate, however, when the in-plane principal stresses are of opposite sign.[42]

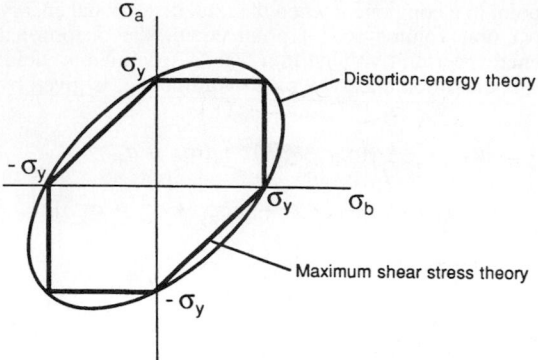

Fig. 47 Comparison of theories of yielding for plane stress conditions.

Coulomb–Mohr Theory The Coulomb–Mohr theory is used almost exclusively for predicting failure of brittle materials. It is based on the ultimate strengths of a material in tension and compression, σ_{ut} and σ_{uc}, respectively. In general, brittle materials exhibit a larger ultimate strength in compression than in tension. To apply the Coulomb–Mohr theory, Mohr's circles are drawn corresponding to the states of stress at failure for tension and compression specimens. The tangent lines to these circles are then drawn. The theory states that failure occurs for any state of stress that produces a Mohr's circle as large as the envelope of the σ_{ut} and σ_{uc} circles and their tangent lines. That is, if the Mohr's circle for the state of stress under consideration extends out to or beyond the shaded region in Fig. 48, failure is imminent.

Figure 48b shows a graphical representation of the Coulomb–Mohr theory for a state of plane stress. It should be noted that this theory is identical to the maximum-normal-stress theory in the first quadrant. In general, the Coulomb–Mohr theory gives conservative

results. It is the best classical theory available for estimating failure of brittle materials. In modern design, it is better to use the theory of fracture mechanics to estimate failure loads rather than the Coulomb–Mohr theory because the presence of even a very small flaw in a brittle material can lead to failure at loads much lower than those predicted by the Coulomb–Mohr theory.

8.3 Fracture Mechanics

Fracture mechanics is that branch of mechanics that deals with the ability of a material to resist crack propagation under a given set of loading and environmental conditions. Modern fracture mechanics can be divided into two major categories: linear elastic fracture mechanics (LEFM) and elastic–plastic fracture mechanics (EPFM). Before discussing these topics, some background information regarding the nature of brittle fracture must be presented.

Brittle Fracture Brittle fracture involves the rapid propagation of a crack leading to failure of the component. Little energy is absorbed by the component, and only a small amount of plastic deformation occurs. Crack propagation is conventionally considered under three basic modes of loading as shown in Fig. 49. Mode I loading involves an opening or tensile mode. It is the most common loading encountered in practice. Mode II loading represents an in-plane shearing mode. Mode III represents an out-of-plane shearing mode. Because of the prevalence of mode I loading in practice and the relative ease in experimentation and analysis, a majority of investigations have focused on this cracking mode. As mentioned in Section 8.2, several factors can cause a normally ductile material to behave in a brittle fashion. These include reduced temperatures, high strain rates, corrosive environments, and the presence of notches.

On a macroscopic level, for uniaxial states of loading, a brittle fracture results in a relatively flat fracture surface with only small amounts of slanted surfaces.

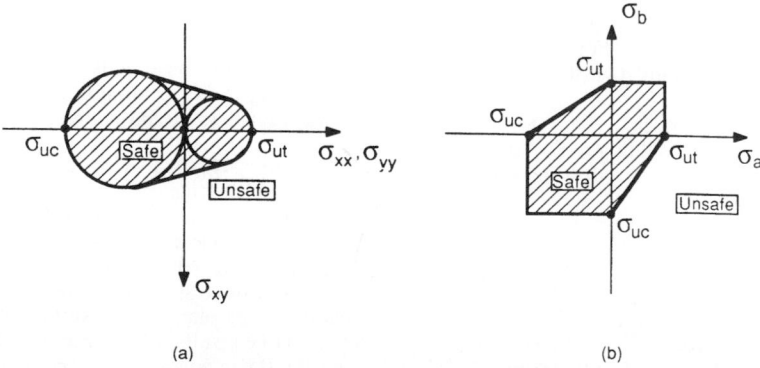

(a)   (b)

Fig. 48 Coulomb–Mohr theory for brittle failure.

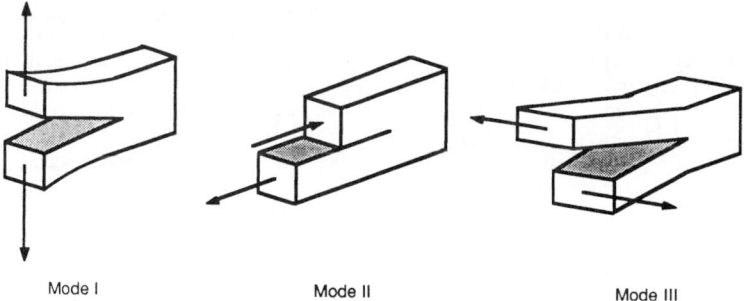

Mode I   Mode II   Mode III

Fig. 49 Basic modes of loading for crack propagation.

The slanted surfaces, commonly referred to as shear lips, are indicative of plastic deformation. In a uniaxial test specimen, these shear lips are at approximately a 45° angle to the applied load axis since the maximum shear stress occurs along planes at this angle. For a ductile fracture, the fracture surface is rougher, and considerable shear lip formations are present.

On a microscopic level, the fracture process can appear quite complicated. For isotropic homogeneous metals, three common microstructural failure modes are microvoid coalescence, cleavage fracture, and intergranular fracture. Cleavage and intergranular fracture are often associated with a brittle failure, while microvoid coalescence is most often associated with ductile failure. This categorization is not always accurate, however. Further information on microscopic fracture analysis can be found in Refs. 43–45.

Linear Elastic Fracture Mechanics The theory of linear elastic fracture mechanics is based on the assumption that the zone of plastic deformation at the tip of a crack is small. In this case, the stress distributions near the tip of a crack can be determined using the theory of elasticity. Assuming the crack tip is infinitely sharp, the stresses near the crack tip are given by (Fig. 50):

$$\sigma_{xx} = \frac{K}{\sqrt{2\pi r}} \cos\frac{\theta}{2} \left(1 - \sin\frac{\theta}{2} \sin\frac{3\theta}{2}\right)$$

$$\sigma_{yy} = \frac{K}{\sqrt{2\pi r}} \cos\frac{\theta}{2} \left(1 + \sin\frac{\theta}{2} \sin\frac{3\theta}{2}\right)$$

$$\sigma_{xy} = \frac{K}{\sqrt{2\pi r}} \left(\sin\frac{\theta}{2} \cos\frac{\theta}{2} \cos\frac{3\theta}{2}\right)$$

where the parameter K is known as the stress intensity factor. These equations were first derived by Westergaard.[46] It is apparent that these equations predict infinite stresses at the crack tip ($r = 0$). In reality, yielding occurs at the crack tip and limits the maximum stresses attained. The stress intensity factor K is a function of the crack geometry and the

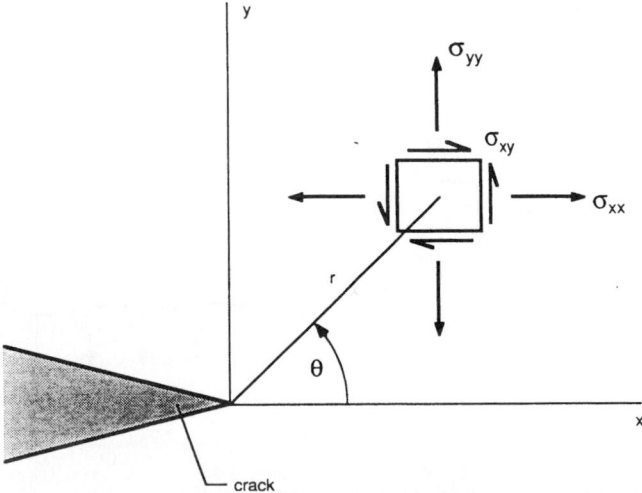

Fig. 50 Stresses near a crack tip.

applied loads. Extensive tabulations of K for a variety of geometries and loading conditions can be found in the literature.[47,48] For a center crack of length $2a$ in an infinite sheet, the stress intensity factor is given by $K = \sigma\sqrt{\pi a}$ (Fig. 51a). For an edge crack of length a in a semi-infinite sheet, $K \approx 1.1\sigma\sqrt{\pi a}$ (Fig. 51b). As the crack length increases, the stress intensity factor rises sharply. In essence, K is a measure of the severity of the crack tip (for a given component, loading, and crack length). If the applied load is increased high enough, rapid failure due to unstable crack propagation occurs. The value of the stress intensity factor at which failure occurs is known as the fracture toughness of the particular component, K_c. For a component of specified in-plane dimensions, the fracture toughness varies with thickness as shown in Fig. 52. For very thin specimens, there is little resistance to cracking, and the fracture toughness is low. For moderate thicknesses where considerable plastic flow can occur, plane stress conditions prevail, and maximum toughness is achieved. As the thickness is further increased, transverse stresses begin to develop near the crack tip. These stresses limit the ability of the material near the crack tip to yield, which in turn reduces the fracture toughness of the component. If the component is made thick enough, a state of plane strain exists in the midthickness near the crack tip. The fracture toughness reaches a minimum value at this point. Further increases in thickness do not change its value. This limiting value of toughness is known as the plane strain fracture toughness and is denoted by K_{Ic}. The subscript I indicates that the toughness value is for a

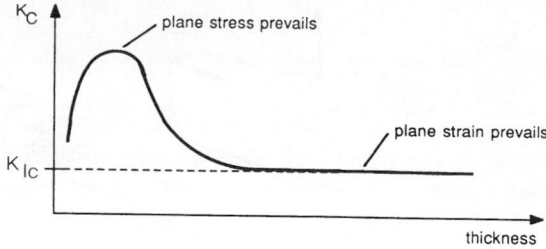

Fig. 52   Variation of fracture toughness with thickness.

mode I loading situation. In general, the fracture toughness, K_c, depends on both component geometry and material, whereas the plane strain fracture toughness is considered a material property when the thickness is sufficiently large. Since K_{Ic} represents a lower bound on the fracture toughness, its use in design should lead to conservative predictions. Table 8 lists values of plane strain fracture toughness for a variety of materials along with the associated yield strengths. It must be noted that the values given represent averages. Considerable variation in reported plane strain fracture toughness values is not uncommon. More extensive tables for K_{Ic} can be found elsewhere.[49,50] Standardized testing procedures for determining K_{Ic} values can be found in Ref. 51.

Transition Temperature Phenomenon   Face-centered-cubic metals, such as aluminum, copper, and

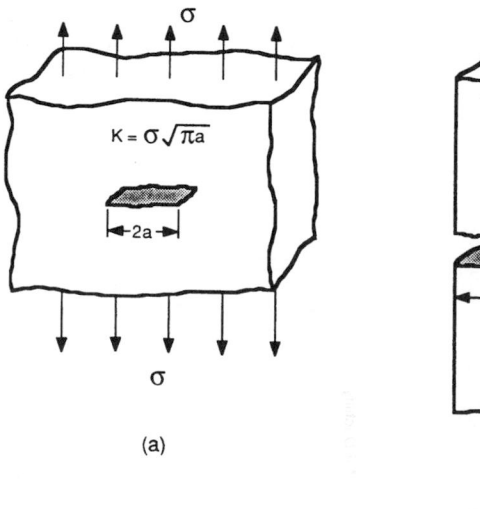

(a)

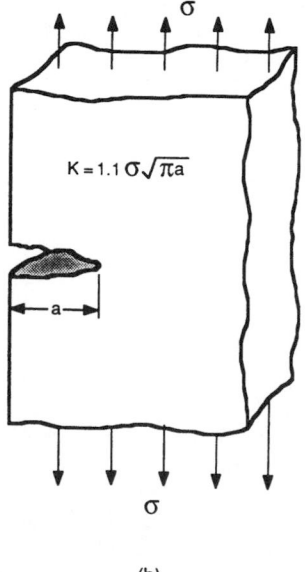

(b)

Fig. 51   Examples of stress intensity factors: (a) through-thickness crack in an infinite plate; (b) through-thickness edge crack in a semi-infinite plate.

Table 8   Plane Strain Fracture Toughness of Selected Alloys

Material	K_{Ic} (ksi $\sqrt{in.}$)	Yield Strength (ksi)
Aluminum		
2014-T651	22	66
2024-T851	24	66
7075-T651	27	72
Titanium		
Ti–6A1–4V	84	124
Ti–6A1–6V–2Sn; mill annealed (1300°F)	52	145
Ti–6A1–6V–2Sn; solution treated and aged (1050°F)	30	179
Steels		
4340; tempered at 400°F	50	235
4340; tempered at 800°F	76	204
300M; tempered at 600°F	49	260
18 Ni maraging (200); aged 900°F, 6 hr	100	210
Stainless steels		
PH 13-8Mo; H1000	98	215
PH 13-8Mo; H950	54	210

nickel, show little variation in their mechanical properties with moderate changes in temperature. In contrast, the strength and toughness of body-centered-cubic metals, such as ferritic alloys, are often quite sensitive to temperature and strain rate effects. In order to study temperature effects, some means of measuring toughness is required. While performing K_{Ic} tests over a range of temperatures would probably provide the most reliable data, the cost of performing these tests often precludes their use. The most common technique employed is the Charpy test. This test consists of dropping an impact hammer pendulum onto a small notched specimen. The energy absorbed by the specimen during fracture is given by $W(h_i - h_f)$ where W is the weight of the hammer, h_i is the initial height of the hammer, and h_f is the maximum height to which the hammer swings after impacting the specimen. The primary disadvantages of the Charpy test are that the small specimen size may allow for considerable plasticity at the crack tip and that it is a rather severe test (a single blow causing fracture). The occurrence of plasticity may result in overly optimistic toughness values. Other means of measuring toughness include the drop weight tear test and the dynamic tear test.[52]

Whatever technique is used for measuring toughness (and the choice is important), one common trend that is observed in low-strength ferritic steels and some other alloys is that a sudden drop in material toughness occurs over a fairly narrow temperature range (Fig. 53a). This is known as the transition temperature phenomenon. The temperature corresponding to the lower toughness end of this transition is known as the nil ductility temperature (NDT). In general, the measured value for the NDT depends on the testing technique used, specimen thickness, and material. For thicker specimens, plastic flow is constrained, and the NDT rises (Fig. 53b).

Elastic–Plastic Fracture Mechanics While LEFM has proven to be a very powerful design tool, its use is not always appropriate. For high-toughness materials or thin components experiencing plane stress conditions near a crack tip rather than

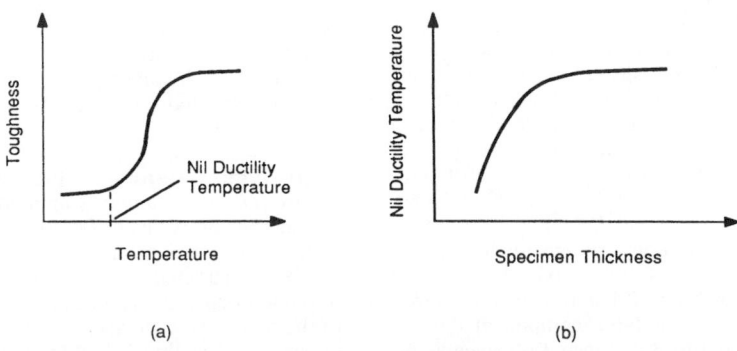

Fig. 53 Transition temperature phenomenon.

plane strain conditions, LEFM leads to overly conservative design restrictions. Also, large specimen sizes are often required to bring about plane strain conditions for K_{Ic} determination. These large specimens can be expensive to fabricate and difficult to test. For all these reasons and others, considerable research in the field of elastic–plastic fracture mechanics has been done during the past 20 years. The most notable work has been the J integral originally defined by Rice to be a path-independent line integral for two-dimensional crack problems.[53] It was later concluded that the value of the J integral, J, is a measure of the stress and strain singularity near a crack tip. In this regard, J for the plastic case is analogous to the stress intensity factor K for the linear elastic case.

It has been shown that, in the linear elastic range, the critical value of J required to cause unstable crack propagation, J_{Ic}, is related to the plane strain fracture toughness by

$$J_{Ic} = \frac{1 - \mu^2}{E} K_{Ic} \qquad (99)$$

where E and ν are Young's modulus and Poisson's ratio, respectively. The advantage of determining J_{Ic} instead of K_{Ic} is that much smaller specimen sizes can be used. A standard procedure for determining J_{Ic} has been adopted by the American Society for Testing and Materials.[54]

In addition to providing an alternative technique for determining plane strain fracture toughness, the J integral may be used to determine whether a given cracked component is stable with respect to crack initiation and propagation when the material near the crack tip deforms plastically (i.e., when the size of the plastic zone ahead of the crack tip is substantial).[55]

Fracture Mechanics Design   To apply fracture mechanics in design, it is necessary to know the loading conditions, specimen geometry, and crack configuration. If no cracks are known to be present, the maximum crack size that could have been overlooked by nondestructive inspection should be used. Once these conditions are known, the stress intensity factor K can be determined. This value of K is then compared to the critical stress intensity factor, K_c. If K is larger than K_c, the design is unsafe. In a more general sense, given any two of K_c, σ, and a (fracture toughness, applied stress, and crack length, respectively), the third can be determined since the three are related by the following condition: Unstable crack propagation occurs when the value of K reaches K_c, where $K = K_0 \sigma \sqrt{\pi a}$. Here, K_0 is a constant that depends on the specimen geometry and loading conditions. For example, $K_0 = 1$ for uniaxial loading of an infinite plate with a through-thickness center crack of length $2a$ (Fig. 51a). For large component thicknesses, K_c will be equal to K_{Ic}. Since K_c values as a function of thickness are not commonly available for most materials, it is common to use K_{Ic} values even when component thicknesses are too small to bring about plane strain conditions. This can lead to overly conservative design.

8.4 Fatigue

Fatigue investigations concern the effects of cyclic stresses and strains on the mechanical integrity of a structural component. If enough cycles are applied to a component, it is possible for failure to occur even though the maximum stress never exceeded the ultimate strength of the material. It is often useful to classify fatigue problems into two regimes: high-cycle fatigue (cycles to failure $> 10^4$) and low-cycle fatigue (cycles to failure $< 10^4$). This classification needs to be refined in some cases. It is often necessary to further classify the fatigue loading of a given component. If the operating temperature of the material is a sizable fraction of its melting temperature, creep of the material can become substantial. In such cases, theories for creep fatigue must be employed. Another important distinction is whether the cyclic loads imposed on the component result in uniaxial or multiaxial states of stress.

Looked at simplistically, the fatigue process can be divided into three categories: initiation of a dominant macrocrack, propagation of this crack, and final failure of the component. In rough terms, a macrocrack exists if the crack length is large compared to the microstructure grain size. For high-cycle fatigue, the majority of the life is spent in initiating a macrocrack. In low-cycle fatigue, propagation of a macrocrack can consume over half the life.

Two distinct approaches to fatigue analysis have evolved. In the classical approach, empirical models are developed that relate the imposed loading conditions to the cycles to failure. In the fracture mechanics approach, crack propagation rates are related to the cyclic change in stress intensity factors. The resulting relationship is then integrated to give the cycles to failure. In the application of either approach, two steps need to be performed. First of all, the expected cyclic loading history of the component should be estimated. The obtaining of these data is known as load spectrum analysis. Second, the response of the material to the applied cyclic loading must be known. These data are obtained by laboratory testing under controlled conditions. Finally, the laboratory results must be correlated with the expected service loads so that a life prediction can be made.

High-Cycle Fatigue   In high-cycle fatigue testing, a fully reversed loading is applied to a specimen, and the number of cycles to failure is recorded. When this is done for a variety of load amplitudes, an $S-N$ plot can be constructed (Fig. 54). In this diagram, the log of the applied stress amplitude is plotted versus the log of the reversals to failure. For every cycle applied, two reversals in the direction of applied loading must occur (one reversal when the maximum stress is obtained and

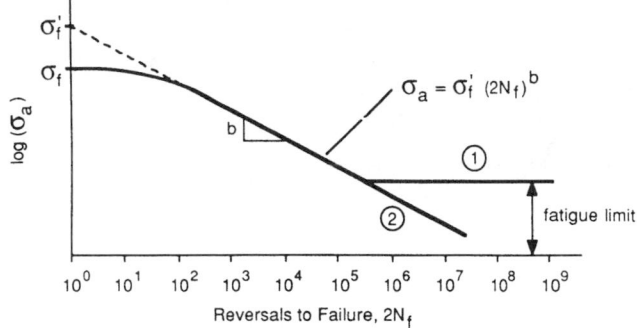

$\sigma_a = \sigma_f' \, (2N_f)^b$

① material exhibiting a fatigue limit

② material not exhibiting a fatigue limit

Fig. 54 Schematic representation of an *S–N* diagram.

the other when the minimum stress is obtained). Thus, if the number of cycles to fracture is designated by N_f, the number of reversals to failure is $2N_f$. For many materials, this $S-N$ plot shows a linear relationship over an extensive portion. Within this region, the applied true stress amplitude, σ_a, can be related to the number of reversals to failure by

$$\sigma_a = \sigma_f'(2N_f)^b \qquad (100)$$

where σ_f' and b are material constants known as the fatigue strength coefficient and fatigue strength exponent, respectively. The constant b is also commonly referred to as Basquin's exponent. For some materials, this linear relationship is valid from $2N_f = 1$ (which is equivalent to the tension test in the lower limit). In such cases, the fatigue strength coefficient, σ_f', is equal to the true fracture strength, σ_f.

Some materials exhibit a stress amplitude below which no failures occur. This stress amplitude is known as the fatigue limit or endurance limit of the material. This phenomenon is observed in mild steels, for example. Most nonferritic metals behave differently. In these cases the fatigue strength (instead of a fatigue limit) is defined as the stress amplitude required to bring about failure in a specified number of cycles (normally, $N_f = 10^8$).[56]

When relating the results from laboratory testing, as given by an $S-N$ diagram, to structural components, it is necessary to consider many additional factors such as stress concentrations, surface finish quality, residual stresses, and so on. Accurate accounting of the influence of these factors on the fatigue life of the component requires considerable experience. Further details can be found in Ref. 57.

Effect of Mean Stress on Fatigue Life In the previous discussion, it was assumed that the fatigue

loading was fully reversed as shown in Fig. 55*a*. It is common in actual components to find a mean stress σ_m superimposed on the reversed loading. Tensile mean stresses (Fig. 55*b*) tend to reduce the fatigue life while compressive mean stresses (Fig. 55*c*) tend to increase the life. Several empirical relationships have been devised to account for mean stress effects. The most commonly used of these is the modified Goodman diagram (Fig. 55*d*). In this diagram, the alternating stress amplitude is plotted versus the mean stress. The ordinate is chosen to be the stress amplitude for a given cyclic life. The abscissa is taken to be the ultimate strength of the material. The line connecting these two points represents combinations of stress amplitudes and superimposed mean stresses that give the same fatigue life as for the selected fully reversed stress amplitude. Thus, given the desired cyclic life and σ_a, the allowable mean stress can be determined. Similarly, given N_f and σ_m, σ_a can be determined. This empirical technique is easy to use and leads to reasonably accurate results for many materials.

Low-Cycle Fatigue When performing laboratory tests in the low-cycle fatigue regime ($N_f < 10^4$ cycles), it is common to control the strain amplitude rather than the stress amplitude. Considerable plastic flow can occur in low-cycle fatigue testing. If the stress–strain response for a complete cycle is plotted, the resulting diagram is known as a hysteresis loop (Fig. 56). A considerable amount of information is contained in this diagram. The area within the hysteresis loop represents the work per unit volume done on the material in a single cycle. Most of this work is dissipated as heat. The vertical distance between the tips of the hysteresis loop is the stress range $\Delta\sigma$. The horizontal distance between the tips of the hysteresis loop represents the applied strain range $\Delta\varepsilon$. The maximum width of the loop is known as the plastic strain range $\Delta\varepsilon_p$. The elastic strain range $\Delta\varepsilon_e$ is given by

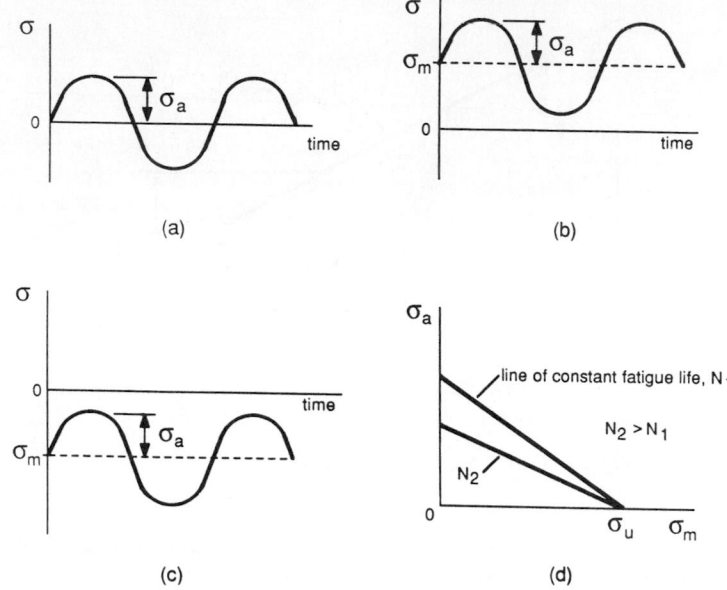

Fig. 55 Mean stress effects in fatigue: (a) fully reversed loading: (b) tensile mean stress; (c) compressive mean stress; (d) modified Goodman diagram.

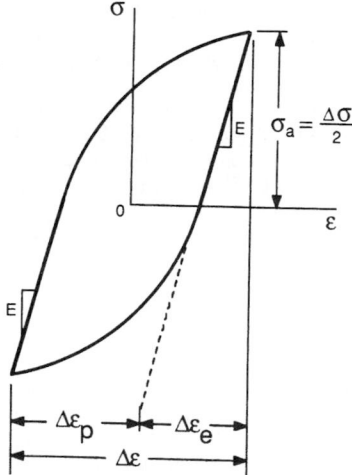

Fig. 56 Hysteresis loop.

$\Delta\varepsilon_e = \Delta\varepsilon - \Delta\varepsilon_p$. Young's modulus is given by the slope of the initial unloading portion of each half of the hysteresis loop. For many materials, the plastic strain range is not constant for each cycle. Instead, some materials show a cyclic hardening effect (i.e., $\Delta\varepsilon_p$ decreases with increasing cycles), while others show a cyclic softening effect (i.e., $\Delta\varepsilon_p$ increases with increasing cycles). Usually, a material stabilizes by the time the half-life is reached.

If the log of the total strain amplitude, $\log(\Delta\varepsilon/2)$, is plotted versus the log of the reversals to failure, a strain–life curve is obtained (Fig. 57). It is often instructive to include the plastic and elastic strain amplitudes, as measured at the half-life, as well. These elastic and plastic strain versus life curves are often straight lines. The plastic strain versus life line is described by the well-known Coffin–Manson equation.[58]

$$\frac{1}{2}\Delta\varepsilon_p = \varepsilon'_f (2N_f)^c \qquad (101)$$

where ε'_f is the fatigue ductility coefficient and c is the fatigue ductility exponent. For most materials, ε'_f is between $0.5\varepsilon_f$ and $1.5\varepsilon_f$, where ε_f is the true strain at fracture as measured in a tension test (see Section 1.4). The exponent c is usually found to be between -0.5 and -0.7. Thus, if low-cycle fatigue data are lacking, the assumptions $\varepsilon'_f = \varepsilon_f$ and $c = -0.6$ should give reasonable results.

The elastic strain versus life line is described by dividing Eq. (100) by Young's modulus E:

$$\frac{\Delta\varepsilon_e}{2} = \frac{\sigma_a}{E} = \frac{\sigma'_f}{E}(2N_f)^b \qquad (102)$$

By combining Eqs. (101) and (102), a relationship between total strain amplitude and reversals to failure

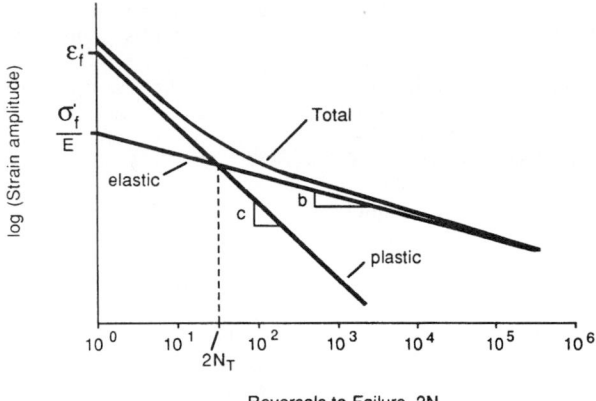

Fig. 57 Strain–life curve.

is obtained:

$$\frac{\Delta\varepsilon}{2} = \frac{\Delta\varepsilon_e}{2} + \frac{\Delta\varepsilon_p}{2} = \frac{\sigma'_f}{E}(2N_f)^b + \varepsilon'_f(2N_f)^c$$

The fatigue properties of some selected materials are given in Table 9.

Another important parameter obtained from the strain–life curve is the transition fatigue life, $2N_T$. This is the number of reversals corresponding to the intersection of the elastic and plastic strain versus life lines ($\Delta\varepsilon_e/2 = \Delta\varepsilon_p/2$). For lives less than $2N_T$, plastic strains dominate the fatigue process and a material's ductility is of great importance. For lives greater than $2N_T$, elastic strains are the dominant strains, and a material's strength is of greater importance than its ductility. It is not surprising that high-strength, low-ductility materials show a short transition life (sometimes N_T is at a few cycles) while lower-strength, ductile materials tend to have a high transition life.

By equating $\Delta\varepsilon_p/2$ to $\Delta\varepsilon_e/2$ in Eqs. (100) and (101), respectively, the following equation for the transition life is obtained:

$$2N_t = \left(\frac{\varepsilon'_f E}{\sigma'_f}\right)^{1/(b-c)}$$

The transition fatigue life is more satisfactory to distinguish the low- and high-cycle regimes of a particular material than an arbitrary number of cycles such as 10^4.

Cyclic Stress–Strain Diagram For components that undergo significant dynamic loading, design based on the monotonic stress–strain curve may lead to poor results. This is because many materials exhibit significant changes in strength properties due to cyclic loading. These changes are not necessarily due to the development of a macroscopic crack, but rather represent

Table 9 Fatigue Properties of Selected Alloys

Material	E (ksi)	σ'_f (ksi)	ε'_f	b	c
Steels					
SAE 1005-100	30×10^3	77	0.17	−0.066	−0.45
SAE 1015	30×10^3	120	0.96	−0.11	−0.64
SAE 1045-390 BHN	30×10^3	230	0.45	−0.074	−0.68
A1SI 4130-365 BHN	29×10^3	246	0.89	−0.081	−0.69
SAE 4142-380 BHN	30×10^3	265	0.45	−0.08	−0.75
SAE 4340-350 BHN	28×10^3	240	0.73	−0.076	−0.62
A1SI 304-327 BHN	25×10^3	330	0.89	−0.12	−0.69
Gray cast iron					
Pearlitic-210 BHN	16×10^3	59	0.008	−0.084	−0.385
Aluminum alloys					
1100	10×10^3	28	1.80	−0.106	−0.69
2024-T4	10.2×10^3	147	0.21	−0.11	−0.52
7075-T6	10.3×10^3	191	0.19	−0.126	−0.52

global changes in the material's microstructure. Some materials exhibit cyclic hardening while others exhibit cyclic softening and still others are relatively stable with respect to cyclic loading. Thus, when designing components that undergo significant dynamic loading, the cyclic stress–strain diagram should be used.

Several techniques exist for construction of a cyclic stress–strain curve. In one commonly used procedure, the tips of stabilized hysteresis loops from a number of strain-controlled fatigue tests are connected together to form a smooth curve. In another single-specimen technique, the specimen is cycled at higher and higher strain amplitudes. At each amplitude, enough cycles are performed to obtain a relatively stable hysteresis loop, but not so many cycles as to significantly damage the specimen. Again, the tips of the hysteresis loops so obtained are connected together to form a smooth

curve. Some examples of cyclic stress–strain diagrams and the corresponding monotonic curves are provided in Fig. 58.

If true stress and true strain are used to construct the cyclic stress–strain diagram, a relationship similar to Eq. (22) is often found to be valid:

$$\sigma_a = K'(\tfrac{1}{2}\Delta\varepsilon_p)^{n'}$$

where σ_a is the steady-state true stress amplitude, $\Delta\varepsilon_p/2$ is the corresponding plastic strain amplitude, K' is the cyclic strength coefficient, and n' is the cyclic strain-hardening exponent. The value of n' is usually between 0.10 and 0.20. In general, metals with high monotonic strain-hardening exponents ($n > 0.15$) cyclically harden while those with low values of

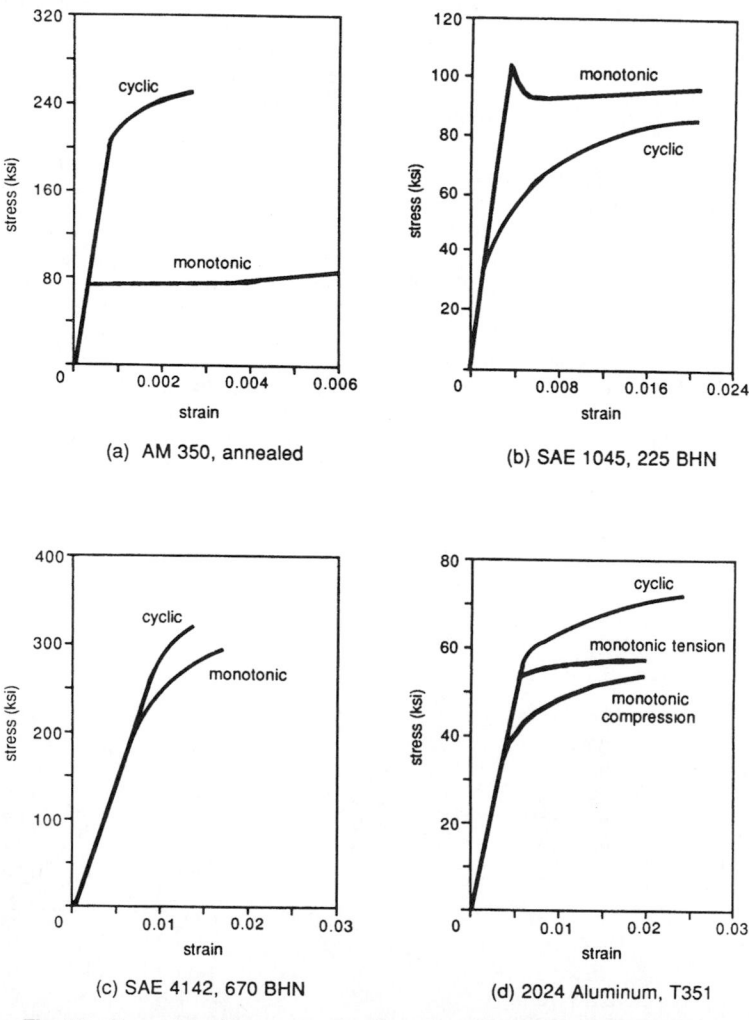

Fig. 58 Examples of some monotonic and cyclic stress–strain diagrams.

$n(n < 0.15)$ cyclically soften. It can be shown, by considering energy quantities, that[59]

$$n' = \frac{b}{c}$$

This equation provides a correlation between fatigue properties and cyclic stress–strain properties. If typical values for c and b of -0.6 and -0.9, respectively, are substituted into this equation, a value for n' of 0.15 results. This is in good agreement with the fact that n' is usually found to be between 0.1 and 0.2 for most metals.

Theories of Cumulative Damage Constant amplitude cyclic loading does not occur in most real situations. Instead, the loading is most often random (e.g., the loadings on a car axle) or consists of distinct blocks of loading at given amplitudes (e.g., machinery operated at several discrete speeds). In order to estimate fatigue life under these conditions, a theory of cumulative damage is required.

For simple block-loading situations, the most commonly used theory is the Palmgren–Miner rule (also known as Miner's rule, or as the linear cumulative damage theory). This theory states that the fraction of life exhausted by the ith block loading is given by n_i/N_{fi}, where n_i is the number of cycles applied in the ith block and N_{fi} is the number of cycles to failure of a virgin specimen cycled at the same load amplitude as used in the ith block. Clearly, an S–N or strain–life diagram is required to use this theory since N_f data are needed. Failure is assumed to occur when the total fraction of life exhausted becomes equal to one. That is,

$$\sum_i \frac{n_i}{N_{fi}} = \frac{n_1}{N_{f1}} + \frac{n_2}{N_{f2}} + \frac{n_3}{N_{f3}} + \cdots = 1 \quad \text{at failure}$$

In using the Palmgren–Miner rule, one assumes that fatigue damage accumulates in a linear fashion. In other words, the damage accumulated per cycle must be the same early and late in the cycling sequence for a given block loading. In reality, this is often not the case. Even so, this empirical rule or modified versions of it usually give satisfactory results.

For components undergoing random loading, it is necessary to obtain or estimate a typical portion of the loading history of the component. As previously mentioned, this is referred to as load spectrum analysis. Once this is done, more sophisticated models for fatigue damage accumulation, such as rain flow counting, can be applied. An excellent discussion of fatigue life estimation techniques is provided in Ref. 60.

Creep and Fatigue There are many situations in modern engineering in which a material must be subjected to cyclic loads at temperatures that are relatively near the melting point of the material. Some examples include turbine blades in aircraft jet engines,

nuclear pressure vessels, and solder joints used in electronics interconnection technology. Under these circumstances, the standard theories of fatigue presented previously are no longer adequate. Instead, theories for creep fatigue need to be applied. The two most commonly used theories are the Robinson–Taira theory[61] and the theory of strain range partitioning.[62]

The Robinson–Taira theory has the form

$$\sum_i \frac{n_i}{N_{fi}} + \sum_i \frac{t_i}{t_{ri}} = 1 \quad \text{at failure}$$

where n_i = number of cycles applied at ith loading
N_{fi} = number of cycles to failure at ith loading
t_i = time spent at ith stress amplitude σ_i
t_{ri} = time to creep failure at stress level of σ_i

Note the similarity between the Robinson–Taira equation and the Palmgren–Miner rule. In fact, the Robinson–Taira equation reduces to the Palmgren–Miner rule when creep damage is insignificant. The Robinson–Taira equation is displayed graphically in Fig. 59. For some materials (such as 304 stainless steel), this equation can lead to very nonconservative predictions. A modification of the Robinson–Taira equation, which forms the basis for an ASME code, helps alleviate this problem.

The theory of strain range partitioning (SRP) was developed at the National Aeronautics and Space Administration (NASA) Lewis Research Center in the early 1970s. It involves the partitioning of a hysteresis loop obtained from a region of a structural member into four basic strain components: plastic–plastic $(\Delta\varepsilon_{pp})$, creep–creep $(\Delta\varepsilon_{cc})$, plastic–creep $(\Delta\varepsilon_{pc})$, and creep–plastic $(\Delta\varepsilon_{cp})$. The idealized hysteresis loops for these four strain components are shown in Fig. 60a. Only one of $\Delta\varepsilon_{pc}$ and $\Delta\varepsilon_{cp}$ can be present in any real hysteresis loop. Strain–life plots for all four types

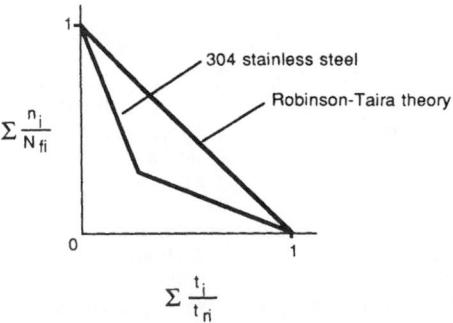

Fig. 59 Robinson–Taira theory for creep fatigue.

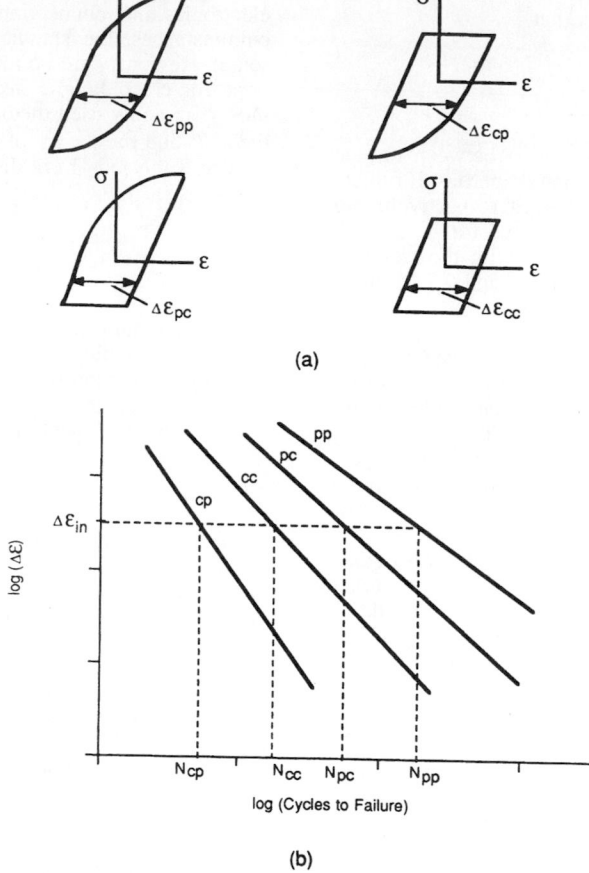

Fig. 60 Strain range partitioning components: (a) idealized hysteresis loops for four strain range components; (b) strain–life plots for four strain range components.

of strain components must be obtained from laboratory testing. Once these strain–life plots are obtained, the predicted life of the structural member, N_{pred}, is obtained using a linear interactive damage rule,

$$\frac{1}{N_{pred}} = \frac{F_{pp}}{N_{pp}} + \frac{F_{cc}}{N_{cc}} + \frac{F_{pc}}{N_{pc}} + \frac{F_{cp}}{N_{cp}}$$

where $F_{pp} = \Delta\varepsilon_{cc}/\Delta\varepsilon_{in}$, $F_{cc} = \Delta\varepsilon_{cc}/\Delta\varepsilon_{in}$, and so on. The quantity $\Delta\varepsilon_{in}$ is the total inelastic strain range of the structural member's hysteresis loop. That is, $\Delta\varepsilon_{in} = \Delta\varepsilon_{pp} + \Delta\varepsilon_{cc} + \Delta\varepsilon_{pc} + \Delta\varepsilon_{cp}$ (where one of $\Delta\varepsilon_{pc}$ and $\Delta\varepsilon_{cp}$ must be zero). N_{pp} is the number of cycles to failure if $\Delta\varepsilon_{in}$ were made up solely of plastic–plastic strains. Similar definitions hold for N_{cc}, N_{pc}, and N_{cp}. Figure 60b should provide for further understanding of these quantities.

For isothermal fatigue, the modified Robinson–Taira theory and the theory of strain range partitioning both give reasonable results. While the theory of strain range partitioning is more complicated to use and can involve considerable laboratory testing, it usually leads to life predictions that are within a factor of 2 or 3 of actual results.

For thermal fatigue situations (i.e., situations in which large temperature fluctuations occur within each cycle), no theories have yet gained broad acceptance. An excellent overview of highly regarded theories is given in Ref. 63.

Multiaxial Fatigue When attempting to predict fatigue lives under multiaxial loading conditions, many new factors must be considered. First of all, it must be known whether the loading is proportional (i.e., $\sigma_1/\sigma_2 =$ constant throughout time) as shown in Fig. 61a, or nonproportional. Nonproportional loading can be further divided into in-phase (Fig. 61b) and out-of-phase (Fig. 61c) situations. It must also be known if the cycling is periodic, as shown in Fig. 61, or of a block-loading or random-loading nature. If a point on the free surface of a component is being analyzed,

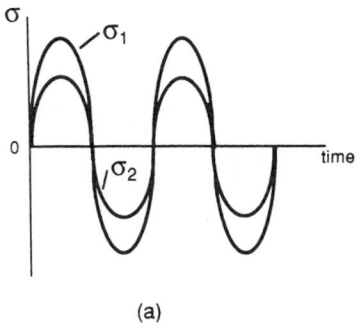

(a)

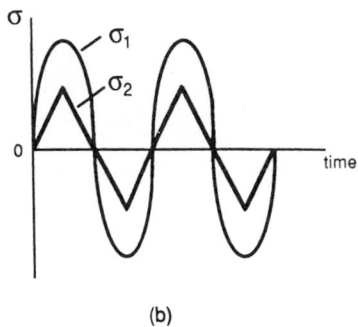

(b)

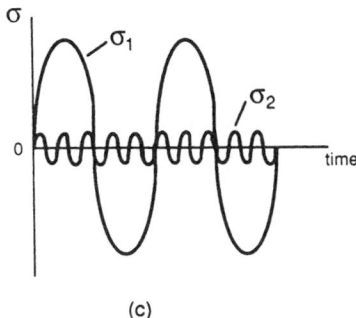

(c)

Fig. 61 Distinct examples of multiaxial cyclic loading: (a) proportional loading; (b) in-phase nonproportional loading; (c) out-of-phase nonproportional loading.

it should be noted that the orientation of the principal stresses (or strains) with respect to the free surface can have a major influence on the fatigue life. Only theories for periodic, in-phase, proportional loading are considered here. Excellent reviews of the available theories for multiaxial fatigue can be found in Refs. 64–66.

Some of the earliest work on multiaxial fatigue centered around modifying static yield theories for fatigue situations. The results were the maximum normal stress theory,

$$S_f = \sigma_{a1}$$

the maximum-shear-stress theory,

$$S_f = \tfrac{1}{2}(\sigma_{a1} - \sigma_{a3})$$

or

$$e_f = \tfrac{1}{2}(\varepsilon_{a1} - \varepsilon_{a3})$$

and the energy-of-distortion theory,

$$S_f = \tfrac{1}{2}\left[(\sigma_{a1} - \sigma_{a2})^2 + (\sigma_{a1} - \sigma_{a3})^2 + (\sigma_{a2} - \sigma_{a3})^2\right]^{1/3}$$

$$e_f = \tfrac{2}{3}\left[(\varepsilon_{a1} - \varepsilon_{a2})^2 + (\varepsilon_{a1} - \varepsilon_{a3})^2 + (\varepsilon_{a2} - \varepsilon_{a3})^2\right]^{1/2}$$

In these equations, σ_{ai} is the amplitude of the ith principal stress. Similarly, ε_{ai} is the amplitude of the ith principal strain, and S_f and e_f are stress and strain parameters, respectively, used to predict fatigue life. While each of these theories has been found to be applicable under certain loading conditions, each one can lead to serious errors in life prediction under other conditions.

A more successful approach that has been used by many investigators is to account for the effects of hydrostatic pressure on fatigue life. One of the more useful versions of these theories has the form

$$\left(\frac{\sigma_{a1} - \sigma_{a3}}{2}\right) + K_{N_f}\left(\frac{\sigma_{a1} + \sigma_{a3}}{2}\right) = C_{N_f}$$

where K_{N_f} and C_{N_f} are empirically determined parameters applicable to a certain fatigue life, N_f. The first term in parentheses represents the maximum-shear-stress amplitude, while the second term represents the normal stress acting on the plane of maximum shear. For low-cycle fatigue analysis, strain amplitudes should be used instead of stress amplitudes. Although this modern approach to multiaxial fatigue tends to give reasonably accurate results, considerable amounts of laboratory data must be acquired for its application.

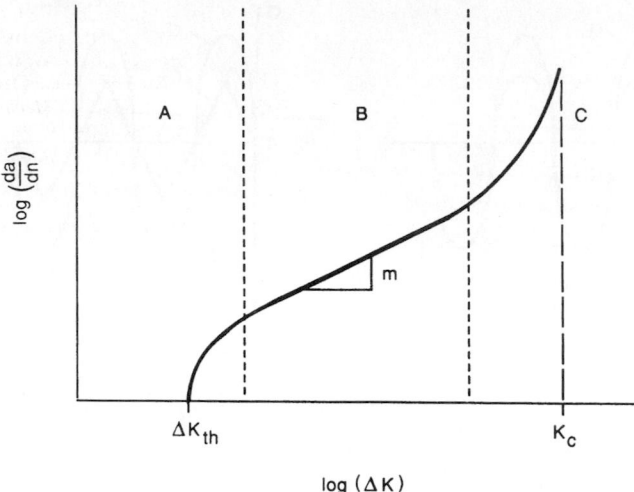

Fig. 62 Variation of fatigue crack growth rate with ΔK.

Damage-Tolerant Design The damage-tolerant design approach to fatigue analysis is based on LEFM. In the usual approach, an initial crack size is assumed or known via nondestructive testing. Using LEFM techniques, the number of cycles required to propagate the crack to a size at which component failure occurs is determined. The primary disadvantage of damage-tolerant design is that, in many high-cycle fatigue applications, the initiation of a fatigue crack may entail a majority of the component life. Thus, assuming a certain initial crack size when none is actually present may lead to an overly conservative design. Neglecting this disadvantage, damage-tolerant design can be useful in many applications:

1. Given the desired service life, the smallest initial flaw size that would lead to failure at the end of the service life can be determined. Nondestructive testing that can detect this flaw size must be employed to assure the minimum desired service life.

2. If a flaw is discovered during routine in-service inspection (such as occurs with inspection of aircraft), the remaining safe life of the component before replacement is necessary can be assessed.

The primary relationship used in damage-tolerant design is the Paris equation[67]

$$\frac{da}{dN} = A(\Delta K)^m \qquad (103)$$

where da/dN is the crack growth rate, ΔK is the change in stress intensity factor in a single cycle, and A and m are material constants. The constant m is usually

between 2 and 4 for most metals. Equation (103) is not valid for all values of ΔK; instead, it is applicable only within a certain range as shown in Fig. 62. Region A in the figure is often referred to as the threshold regime. Equation (103) is not valid in this region. In fact, for values of ΔK below the threshold intensity factor range, ΔK_{th}, no fatigue cracks should propagate. There is still some debate over the proper use of ΔK_{th} in design. In region B of Fig. 62, material response is relatively stable, and Eq. (103) is valid. In region C, crack propagation becomes unstable, and failure is imminent.

In order to apply the Paris equation for life prediction, the initial crack size and stress range must be such that $\Delta K > \Delta K_{th}$. Also, it must be possible to compute ΔK as a function of crack length. If these two things can be done, the Paris equation can be integrated to give the number of cycles until ΔK is so large that unstable crack propagation occurs. A wealth of information on damage-tolerant design can be found in Ref. 68.

REFERENCES

1. Dally, J. W., and Riley, W. F., *Experimental Stress Analysis*, 2nd ed., McGraw-Hill, New York, 1978.
2. Flügge, W. (Ed.), *Handbook of Engineering Mechanics*, McGraw-Hill, New York, 1962.
3. Malvern, L. E., *Introduction to the Mechanics of a Continuous Medium*, Prentice-Hall, Englewood Cliffs, NJ, 1969.
4. Timoshenko, S. P., and Goodier, J. N., *Theory of Elasticity*, 3rd ed., McGraw-Hill, New York, 1970.
5. American Society for Testing and Materials (ASTM), *Annual Book of ASTM Standards*, ASTM, Philadelphia, 1987.
6. Timoshenko, S., *Strength of Materials*, 2nd ed., Van Nostrand, New York, 1940.

7. Cook, R. D., and Young, W. C., *Advanced Mechanics of Materials*, Macmillan, New York, 1985.
8. Boresi, A. P., Sidebottom, O. M., Seely, F. B., and Smith, J. O., *Advanced Mechanics of Materials*, 3rd ed., Wiley, New York, 1978.
9. Roark, R. J., and Young, W. C., *Formulas for Stress and Strain*, 5th ed., McGraw-Hill, New York, 1982.
10. Beer, F. P., and Johnston, E. R., Jr., *Mechanics of Materials*, McGraw-Hill, New York, 1981.
11. Den Hartog, J. P., *Advanced Strength of Materials*, McGraw-Hill, New York, 1952.
12. Timoshenko, S. P., and Woinowsky-Krieger, S., *Theory of Plates and Shells*, 2nd ed., McGraw-Hill, New York, 1959.
13. Szilard, R., *Theory and Analysis of Plates*, Prentice-Hall, Englewood Cliffs, NJ, 1974.
14. Langhaar, H. L., *Energy Methods in Applied Mechanics*, Wiley, New York, 1962.
15. McFarland, D., Smith, B. L., and Bernhart, W. D., *Analysis of Plates*, Spartan Books, New York, 1972.
16. Cook, R. D., *Concepts and Applications of Finite Element Analysis*, 2nd ed., Wiley, New York, 1981, Chapter 9.
17. Griffel, W., *Plate Formulas*, Frederick Ungar, New York, 1968.
18. Lekhnetskii, S. G., *Anisotropic Plates*, Gordon & Breach, New York, 1968.
19. Vinson, J. R., *Structural Mechanics: The Behavior of Plates and Shells*, Wiley, New York, 1974.
20. *Boiler and Pressure Vessel Code*, Section III, Nuclear Power Plant Components, Division 1, The American Society of Mechanical Engineers, New York, 1980.
21. Bridgman, P. W., *Studies in Large Plastic Flow and Fracture with Special Emphasis on the Effects of Hydrostatic Pressure*, McGraw-Hill, New York, 1952.
22. Mendelson, A., *Plasticity: Theory and Application*, Macmillan, New York, 1968.
23. *Manual of Steel Construction* (AISC Handbook), 7th ed., American Institute of Steel Construction, New York, 1970.
24. Lubahn, J. D., and Felgar, R. P., *Plasticity and Creep of Metals*, Wiley, New York, 1961.
25. Flügge, W., *Viscoelasticity*, Blaisdell, Waltham, MA, 1967.
26. Christensen, R. M., *Theory of Viscoelasticity*, 2nd ed., Academic, New York, 1982.
27. Chajes, A., *Principles of Structural Stability Theory*, Prentice-Hall, Englewood Cliffs, NJ, 1974.
28. Timoshenko, S. P., and Gere, J. M., *Theory of Elastic Stability*, McGraw-Hill, New York, 1952.
29. *AISC Specification for the Design, Fabrication, and Erection of Structural Steel for Buildings*, American Institute of Steel Construction, Chicago, November 1978.
30. *Specifications for Aluminum Structures*, Aluminum Association, Washington, DC, 1976.
31. *Timber Construction Manual*, American Institute of Timber Construction, Wiley, New York, 1974.
32. Schwartz, M. M., *Composite Materials Handbook*, McGraw-Hill, New York, 1984.
33. Jones, R. M., *Mechanics of Composite Materials*, McGraw-Hill, New York, 1975.
34. Rowlands, R. E., "Strength (Failure) Theories and Their Experimental Correlation," in *Handbook of Composites*, Vol. 3, G. C. Sih and A. M. Skudra (Eds.), North-Holland, Amsterdam, 1985, pp. 71–125.
35. Christensen, R. M., *Mechanics of Composite Materials*, Wiley, New York, 1979.
36. Tewary, V. K., *Mechanics of Fibre Composites*, Wiley, New York, 1978.
37. Agarwal, B. D., and Broutman, L. J., *Analysis and Performance of Fiber Composites*, Wiley, New York, 1980.
38. Tsai, S. W., and Hahn, H. T., *Introduction to Composite Materials*, Technomic Publishing, Westport, CT, 1980.
39. Peterson, R. E., *Stress Concentration Design Factors*, Wiley, New York, 1953.
40. Blake, A. (Ed.), *Handbook of Mechanics, Materials, and Structures*, Wiley, New York, 1985, p. 268.
41. Kanninen, M. F., and Popelar, C. H., *Advanced Fracture Mechanics*, Oxford University Press, New York, 1985.
42. Blake, A. (Ed.), *Handbook of Mechanics, Materials, and Structures*, Wiley, New York, 1985, p. 270.
43. Hertzberg, R. W., *Deformation and Fracture Mechanics of Materials*, Wiley, New York, 1976.
44. Hellan, K., *Introduction to Fracture Mechanics*, McGraw-Hill, New York, 1984.
45. *Fatigue and Microstructure*, American Society for Metals, Metals Park, OH, 1979.
46. Westergaard, H. M., *Trans. ASME J. Appl. Mech.*, **61**, 49 (1939).
47. Sih, G. C., *Handbook of Stress Intensity Factors*, Lehigh University, Bethlehem, PA, 1973.
48. Rooke, D. P., and Cartwright, D. J., *Compendium of Stress Intensity Factors*, Her Majesty's Stationary Office, London, 1976.
49. Mathews, W. T., *Plain Strain Fracture Toughness (K_{Ic}) Data Handbook for Metals*, AD-773-673, Army Materials and Mechanics Research Center, 1974.
50. Campbell, J. E., "Plane-Strain Fracture Toughness Data for Selected Metals and Alloy", DMIC Report S-28, 1969.
51. American Society for Testing and Materials (ASTM), *Annual Book of ASTM Standards*, E339-83, Section 3, Vol. 03.01, ASTM, Philadelphia, 1983, pp. 518–553.
52. American Society for Testing and Materials (ASTM), *Annual Book of ASTM Standards*, E436-71T, Part 31, ASTM, Philadelphia, 1971, p. 1005.
53. Rice, J. R., "A Path Independent Integral and the Approximate Analysis of Strain Concentration by Notches and Cracks," *Trans. ASME J. Appl. Mech.*, **35**, 379–386 (1968).
54. American Society for Testing and Materials (ASTM), *Annual Book of ASTM Standards*, E813-81, Section 3, Vol. 03.01, ASTM, Philadelphia, 1983, pp. 762–780.
55. Kanninen, M. F., and Popelar, C. H., *Advanced Fracture Mechanics*, Oxford University Press, New York, 1985.
56. Sandor, B. I., *Fundamentals of Cyclic Stress and Strain*, The University of Wisconsin Press, Madison, WI, 1972.
57. Shigley, J. E., *Mechanical Engineering Design*, 2nd ed., McGraw-Hill, New York, 1972, pp. 243–259.

58. Coffin, L. F., Jr., "Low Cycle Fatigue: A Review," *Appl. Mater. Res.*, **1**, 129–141 (1962).

59. Morrow, J., "Cyclic Plastic Strain Energy and Fatigue of Metals," ASTM STP 378, American Society for Testing and Materials, Philadelphia, 1965, pp. 45–87.

60. Socie, D. F., "Fatigue Life Estimation Techniques," Datamyte Corporation, Minnetonka, MN, 1981.

61. Robinson, E. L., "Effect of Temperature Variation on the Long-Time Strength of Steels," *Trans. ASME*, **74**, 777–781 (1952).

62. Zamrik, S. Y. (Ed.), *Design for Elevated Temperature Environment*, American Society of Mechanical Engineers, New York, 1971, pp. 12–24.

63. Halford, G. R., "Low-Cycle Thermal Fatigue," NASA TM-87225, 1986.

64. Zamrik, S. Y., and Dietrich, D. (Eds.), *ASME Pressure Vessels and Piping: Design Technology—1982. A Decade of Progress*, American Society of Mechanical Engineers, New York, 1982, pp. 507–518.

65. Brown, M. W., and Miller, K. J., "Two Decades of Progress in the Assessment of Multiaxial Low-Cycle Fatigue Life," ASTM STP 770, American Society for Testing and Materials, Philadelphia, 1982, pp. 482–499.

66. Garud, Y. S., "Multiaxial Fatigue: A Survey of the State of the Art," *J. Test. Eval.*, **9**, 165–178 (1981).

67. Paris, P. C., and Erdogan, F., "A Critical Analysis of Crack Propagation Laws," *Trans. ASME J. Basic Eng.*, **85**, 528 (1963).

68. *USAF Damage Tolerant Design Handbook: Guidelines for the Analysis and Design of Damage Tolerant Aircraft Structures*, Wright-Patterson Air Force Base, OH, 1984.

CHAPTER 8

NONDESTRUCTIVE INSPECTION*

Robert L. Crane and Jeremy S. Knopp
Air Force Research Laboratory
Materials Directorate
Wright Patterson Air Force Base
Dayton, Ohio

1 INTRODUCTION

This chapter deals with the nondestructive inspection of materials, components, and structures. The term *nondestructive inspection* (NDI) or *nondestructive evaluation* (NDE) is defined as that class of physical and chemical tests that permit the detection and/or measurement of significant properties or the detection of defects in a material without impairing its usefulness. The inspection process is often complicated by the fact that many materials are anisotropic, and most NDI techniques were developed for isotropic materials such as metals. The added complication due to the anisotropy usually means that an inspection is more complicated than it would be with isotropic materials.

Inspection of complex materials and structures is frequently carried out by comparing the expected inspection data with a standard and noting any significant deviations. This means a well-defined standard must be available for calibration of the inspection

*Reprinted from *Mechanical Engineers' Handbook*, Vol. 1, Wiley, New York, 2006, with permission of the publisher.

instrumentation. Furthermore, standards also must contain implanted flaws that mimic those that naturally occur in the material or structure to be inspected. Without a well-defined standard to calibrate the inspection process, the analysis of NDI results can be significantly in error. For example, to estimate the amount of porosity in a cast component from ultrasonic measurements, standard calibration specimens with calibrated levels of porosity must be available to calibrate the instrumentation. Without such standards, estimation of porosity from ultrasonic data is a highly speculative process.

This chapter covers some important and some less well-known NDI tests. Since information on the less frequently used tests is not generally in standard texts, additional sources of information are listed in the References.

Inspection instrumentation must possess four qualities in order to receive widespread acceptance in the NDI community:

1. *Accuracy.* The instrument must accurately measure a property of the material or structure that can be used to infer either its properties or the presence of flaws.

2. *Reliability.* The instrument must be highly reliable, that is, it must consistently detect and quantify flaws or a property with a high degree of reliability. If an instrument is not reliable, then it may not detect flaws that can lead to failure of the component, or it may indicate the presence of a flaw where none exists. The detection of a phantom flaw can mean that an adequate component is rejected, which is a costly error.

3. *Simplicity.* The most frequently used instruments are those used by factory or repair technicians. The inspection community rarely uses highly skilled operators due to the cost constraints.

4. *Low Cost.* An instrument need not be low cost in an absolute sense. Instead, it must be inexpensive relative either to the value of the component under test or to the cost of a failure or aborted mission. For example, in the aircraft industry as much as 12% of the value of the component may be spent on inspection of a flight-critical aircraft component.

1.1 Information on Inspection Methods

To the engineer confronted by a new inspection requirement, there may arise the question of where to find pertinent information regarding an inspection procedure and its interpretation. Fortunately, many potential sources of information about instrumentation and techniques are available for NDI, and a brief examination of this literature is presented here. Many of these references were generated because of the demands of materials used in flight-critical aerospace structures. In this chapter, we will refer only to scientific and engineering books and journals that one would reasonably expect to find in a well-provisioned library.

With the rise of the Internet, there are now many electronic sources of information available on the World Wide Web. These include library catalogs, societal home pages, online journals devoted to inspection, home pages of instrument manufacturers with online demonstrations of their capabilities and inspection services, inspection software, and online forums devoted to solving inspection problems. The References provide many such sources. However, with new electronic sources appearing daily, it is only a brief snapshot of those available at the beginning of the twenty-first century. For those new to the technology, American Society for Testing and Materials (ASTM) standards are particularly valuable because they give very detailed directions on many NDE techniques. More importantly, they are widely accepted standards for inspections. The References also provides sources for those situations where standard inspection methods are not sufficient to detect the material condition of interest.

General NDE Reference Books General overviews to NDE techniques are provided in Refs. 1–22. The reader will note that some of these citations are not recent, but they are included because of their value to the engineer who does not possess formal training in the latest inspection technologies. Additionally, some older works were included because of their clarity of presentation, completeness, or usefulness to the inspection of complex structures.

NDE Journals The periodical literature is often a source of the latest research results for new or modified inspection methodologies.[5,23–28] Some excellent journals are no longer available but are still a valuable source of information or may contain data available nowhere else. Whenever possible, World Wide Web addresses are provided to give the reader ready access to this material.

1.2 Electronic References

There are many useful electronic references for those working in NDE technology. Only a few of the many useful sites on the World Wide Web are included here. Many sites contain links to other sites that contain information on a special topic of interest to the reader. Because the Web is constantly being updated, the list in the References represents a very brief snapshot of the information available to the NDE community. Some useful sites associated with government agencies were not included due to space limitations. The Web addresses provided are associated with NDE societies,[5,6,27,30–36] institutes,[7,8,10,38–40] government agencies,[25,38,41] and general-interest sites.[26,33,35,42–44] There are also many references for the reader interested in using or modifying existing NDE techniques.[31,45–47]

1.3 Future NDE Capabilities

At this point, the reader might be tempted to ask if there are new technologies on the horizon that will enable more cost-effective, anticipatory inspection or monitoring of materials and structures. The answer to this is an emphatic yes. There are new developments in solid-state detectors that should significantly affect both inspection capability and cost. For example, optical and X-ray detectors now give the inspector the ability to rapidly scan large areas of structures for defects. Many new developments in these areas are the outgrowth of advances in noninvasive medical imaging. By coupling this technology with computer algorithms that search an image, the inspection of large areas can be automated, providing more accurate inspections with much less operator fatigue. Hopefully, this technological advance will remove much of the drudgery of detecting the rather small number of flaws in an otherwise large population of satisfactory components.

The area of data fusion is just beginning to be explored in the NDE field. This means that data collected with one technique can be combined with another technique to detect a range of flaws not detected when either is used independently. Data from several techniques can then be coupled at the basic physics level to provide a more complete description of the microstructural details of a material than is now possible.

Finally, the development of new semiconductor-based devices microelectromechanical systems (MEMS) and radio-frequency identification (RFID) allows the implantation of monitoring devices into a material at the time of manufacture to enable real-time structural health monitoring. These devices will permit the inspector to detect and quantify material or structural degradation remotely. This should also enable management of the components and structures for optimum usage over their lifetimes. Remote inspection and tracking of material degradation should reduce the burden of inspection while giving the inspector the ability to examine areas of structure that are now called "hidden." For more information about this rapidly evolving area the reader is referred to the literature.[29,33,48–50]

A brief review of the commonly used NDI methods are listed in Table 1 along with types of flaws that each method detects and the advantages and disadvantages of each technique. For detailed information regarding the capabilities of any particular method, the reader is referred to the literature. A good place to start any search for the latest NDE technology is the home page of the American Society for Nondestructive Testing.[6]

2 LIQUID PENETRANTS

Liquid penetrants are used to detect surface-connected discontinuities, such as cracks, porosity, and laps, in solid, nonporous materials.[51] The method uses a brightly colored visible or fluorescent penetrating liquid that is applied to the surface of a cleaned part. During a specified "dwell time," the liquid enters the discontinuity and is then removed from the surface of the part in a separate step. The penetrant is drawn from the flaw to the surface by a developer to provide an

indication of surface-connected defects. This process is depicted schematically in Figs. 1–4. A penetrant indication of a flaw in a turbine blade is shown in Fig. 5.

2.1 Penetrant Process

Both technical societies and military specifications require a classification system for penetrants. Society documents (typically ASTM E165)[51] categorize penetrants into visible and fluorescent, depending on the type of dye used. In each category, there are three types, depending on how the excess penetrant is removed from the part. These are water washable, postemulsifiable, and solvent removable.

Fig. 1 Representation of part surface before cleaning for penetrant inspection.

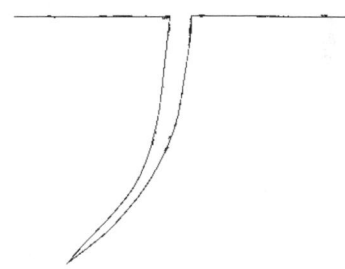

Fig. 2 Part surface after cleaning and before penetrant application.

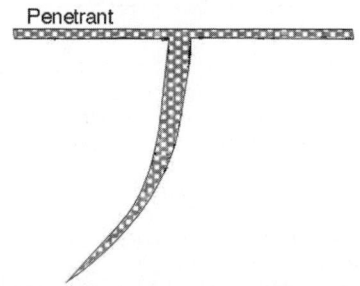

Fig. 3 Part after penetrant application.

Table 1   Capabilities of the Common NDI Methods

Method	Typical Flaws Detected	Typical Application	Advantages	Disadvantages
Radiography	Voids, porosity, inclusions, and cracks	Castings, forging, weldments, and structural assemblies	Detects internal flaws; useful on a wide variety of geometric shapes; portable; provides a permanent record	High cost; insensitive to thin laminar flaws, such as tight fatigue cracks and delaminations; potential health hazard
Liquid penetrants technique	Cracks, gouges, porosity, laps, and seams open to a surface	Castings, forging, weldments, and components subject to fatigue or stress–corrosion cracking	Inexpensive; easy to apply; portable; easily interpreted	Flaw must be open to an accessible surface, level of detectability operator dependent
Eddy current inspection	Cracks and variations in alloy composition or heat treatment, wall thickness, dimensions	Tubing, local regions of sheet metal, alloy sorting, and coating thickness measurement	Moderate cost; readily automated; portable	Detects flaws that change conductivity of metals; shallow penetration; geometry sensitive
Magnetic particles method	Cracks, laps, voids, porosity, and inclusions	Castings, forging, and extrusions	Simple; inexpensive; detects shallow subsurface flaws as well as surface flaws	Useful on ferromagnetic materials only; surface preparation required; irrelevant indications often occur; operator dependent
Thermal testing	Voids or disbonds in both metallic and nonmetallic materials, location of hot or cold spots in thermally active assemblies	Laminated structures, honeycomb, and electronic circuit boards	Produces a thermal image that is easily interpreted	Difficult to control surface emissivity and poor discrimination between flaw types
Ultrasonic methods	Cracks, voids, porosity, inclusions and delaminations, and lack of bonding between dissimilar materials	Composites, forgings, castings, and weldments and pipes	Excellent depth penetration; good sensitivity and resolution; can provide permanent record	Requires acoustic coupling to component; slow; interpretation of data is often difficult

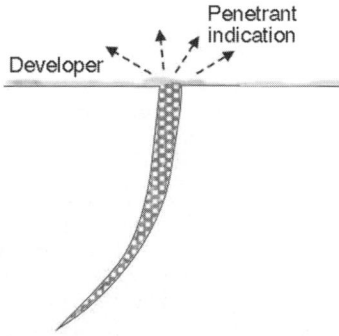

Fig. 4   Representation of part after excess penetrant has been removed and developer has been applied.

The first step in penetrant testing (PT) or inspection is to clean the part. This critical step is one of the most neglected phases of the PT procedure. Since PT only detects flaws that are open to the surface, the flaw and part surface must be free of dirt, grease, oil, water, chemicals, and other foreign materials that might block the penetrant's entrance into a defect. Typical cleaning procedures use vapor degreasers, ultrasonic cleaners, alkaline cleaners, or solvents.

After the surface is clean, a liquid penetrant is applied to the part by dipping, spraying, or brushing. In this step, the penetrant on the surface is wicked into the flaw. In the case of tight or narrow surface openings, such as fatigue cracks, the penetrant must be allowed to remain on the part for a minimum of 30 min to completely fill the flaw. High-sensitivity fluorescent dye penetrants are used for this type of inspection.

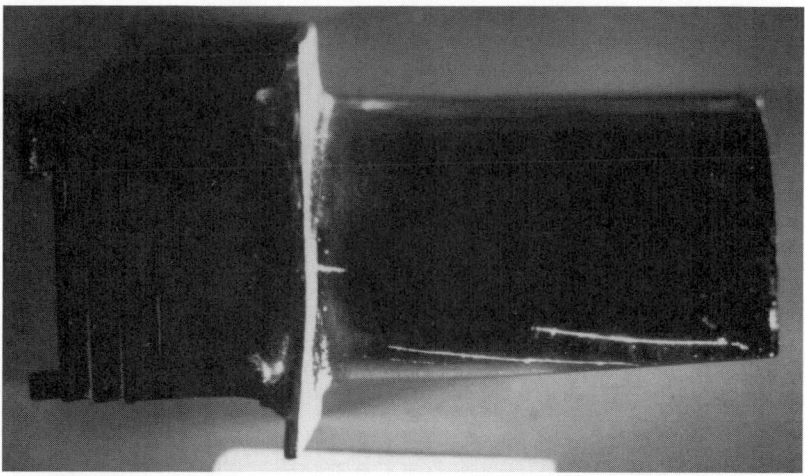

Fig. 5 Penetrant indication of crack running along edge of jet engine turbine blade. Ultraviolet illumination causes the extracted penetrant to fluoresce.

After the dwell time, excess penetrant is removed by one of the processes mentioned previously. For water-based penetrants an emulsifier is sprayed onto the part and again a dwell time is observed. Water is then used to remove the penetrant from the surface of the part. In some cases, the emulsifier is included in the penetrant, so one only needs to wash the part after the penetrant has had time to penetrate the flaw. These penetrants are therefore called "water washable." Of course, the emulsifier reduces the brightness of any flaw indication because it dilutes the penetrant. Ideally, only the surface penetrant is removed with the penetrant in the flaw left undisturbed.

The final step in a basic penetrant inspection is the application of a fine powder developer. This may be applied either wet or dry. The developer aids in wicking the penetrant from the flaw and provides a suitable background for its detection. The part is then viewed under a suitable illumination—either an ultraviolet or a visible source. A typical fluorescent penetrant indication for a crack in a jet engine turbine blade is shown in Fig. 5.

2.2 Reference Standards

Several reference standards are used to check the effectiveness of liquid penetrant systems. One of the oldest and most often used methods involves applying penetrant to hard chromium-plated brass panels. The panel is bent to place the chromium in tension, producing a series of cracks in the plating. These panels are available in sets containing fine, medium, and coarse cracks. The panels are used to classify penetrant materials by sensitivity and to detect degrading changes in the penetrant process.

2.3 Limitations of Penetrant Inspections

The major limitation of liquid penetrant inspection is that it can only detect flaws that are open to the surface. Other inspection methods must be used to detect subsurface defects. A factor that may inhibit the effectiveness of liquid penetrant inspection is surface roughness. Rough surfaces are likely to produce false indications by trapping penetrant, therefore, PT is not suited to the inspection of porous materials. Other penetrant-like methods are available for porous components—see the discussion of filtered particle inspection in Ref. 51.

3 RADIOGRAPHY

In radiography used in NDE, the projected X-ray attenuations of a multitude of paths through a specimen are recorded as a two-dimensional image on a recording media, usually film. One might ask if the newer solid-state X-ray imaging technologies used in medicine also apply to NDE. The answer is yes, as will be discussed in the latter part of this section. The most used recording medium is still film because it is the simplest to apply and provides a resolution of subtle details not currently available with solid-state detectors. However, this situation may not be the case much longer as rapid progress is being made in the development of solid-state detectors with significantly enhanced resolution capabilities. Therefore, since this chapter is written at the beginning of the twenty-first century, when film usage for inspection is still commonplace, this portion of the chapter approaches radiography from the standpoint of film-based recording. Since most quantitative relationships for film also apply to solid-state detectors, the material presented should be applicable for the near future.

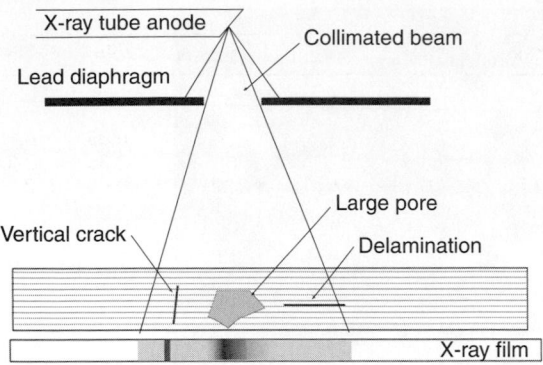

Fig. 6 Schematic radiograph with typical flaws.

The radiography testing (RT) process is shown schematically in Fig. 6. RT records any feature that changes the attenuation of the X-ray beam as it traverses the component. This local change in attenuation produces a change in the intensity of the X-ray beam, which translates into a change in the density, or darkness, on a film. This change in brightness may appear as a distinct shadow or in some cases a delicate shadow on the radiograph. The inspector is greatly aided in detecting a flaw or discrepancy in a part by his or her knowledge of part shape and its influence on the radiographic image. Flaws, which do not change the attenuation of the X-ray beam on passage through the part, are not recorded. For example, a delamination in a laminated specimen is not visible because there is no local change in attenuation of the X-ray beam as it transverses the part. Conversely, flaws that are oriented parallel to the X-ray path do not attenuate the beam as much, allowing more radiation to expose the film and appearing darker than the surrounding image. An example of a crack in the correct orientation to be visible on a radiograph of a piece of tubing is shown in Fig. 7.

3.1 Generation and Absorption of X Radiation

X radiation can be produced from a number of processes. The most common method of generating X rays is with an electron tube in which a beam of energetic electrons impacts a metal target. As the electrons are rapidly decelerated by this collision, a wide band of X radiation is produced, analogous to white light. This band of radiation is referred to as *Bremsstrahlung* or breaking radiation. These high-energy electrons produce short-wavelength energetic X rays. The relationship between the shortest wavelength radiation and the highest voltage applied to the tube is given by

$$\lambda = \frac{12,336}{\text{voltage}}$$

where λ is the wavelength in angstroms and is the shortest wavelength of the X radiation produced. The more energetic the radiation, the more penetrating powers it possesses, and very high energy radiation is used on dense materials such as metals. While it is possible to analytically predict what X-ray energy would provide the best image for a specific material and geometry, a simpler method of determining the optimum X-ray energy is shown in Fig. 8. Note that high-energy X-ray beams are used for dense materials, for example, steels, or for thick low-density materials, for example, large plastic parts. An alternative method to using this figure is to use the radiographic equivalence factors given in Table 2.[52] Aluminum is the standard material for X-ray tube voltages below 100 keV, while steel is the standard above this voltage. When radiographing another material, its thickness is multiplied by the factor in this table to obtain the equivalent thickness of the standard material. The radiographic parameters are set up for this thickness of aluminum or steel. When used in this manner, good radiographs can be obtained for most parts. For example, assume that one must radiograph a 0.75-in.-thick piece of brass with a 400-keV X-ray source. The inspector should multiply the 0.75 in. of brass by the factor of 1.3 to obtain 0.98. This means that an acceptable radiograph of the brass plates would be obtained with the same exposure parameters as would be used for 0.98 in. (approximately 1 in.) of steel.

Radiation for RT can also be obtained from the decay of radioactive sources. In this case, the process is usually referred to as gamma radiography. These radiation sources have several characteristics that differ from X-ray tubes. First, gamma radiation is very nearly

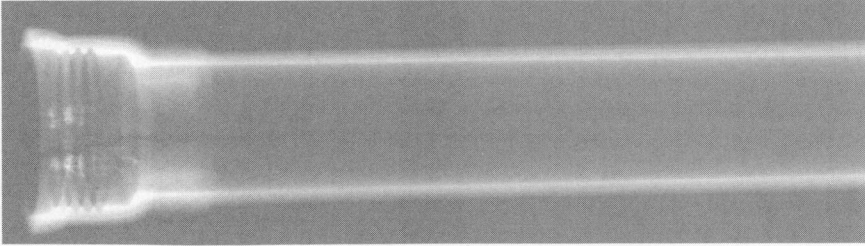

Fig. 7 Radiograph of crack in end of aluminum tubing.

Table 2   Approximate Radiographic Equivalence Factors

Energy Level	100 kV	150 kV	220 kV	250 kV	400 kV	1 MeV	2 MeV	4–25 MeV	^{192}Ir	^{60}Co
Metal										
Magnesium	0.05	0.05	0.08							
Aluminum	0.08	0.12	0.18						0.35	0.35
Aluminum alloy	0.10	0.14	0.18						0.35	0.35
Titanium	—	0.54	0.54		0.71	0.9	0.9	0.9	0.9	0.9
Iron/all steels	1.0	1.0	1.0	1.0	1.0	1.0	1.0	1.0	1.0	1.0
Copper	1.5	1.6	1.4	1.4	1.4	1.1	1.1	1.2	1.1	1.1
Zinc	—	1.4	1.3	—	1.3	—	—	1.2	1.1	1.0
Brass	—	1.4	1.3	—	1.3	1.2	1.1	1.0	1.1	1.0
Inconel X	—	1.4	1.3	—	1.3	1.3	1.3	1.3	1.3	1.3
Monel	1.7	—	1.2	—						
Zirconium	2.4	2.3	2.0	1.7	1.5	1.0	1.0	1.0	1.2	1.0
Lead	14.0	14.0	12.0	—	—	5.0	2.5	2.7	4.0	2.3
Halfnium		14.0	12.0	9.0	3.0	—				
Uranium		20.0	16.0	12.0	4.0	—		3.9	12.6	3.4

Source: From Ref. 52.

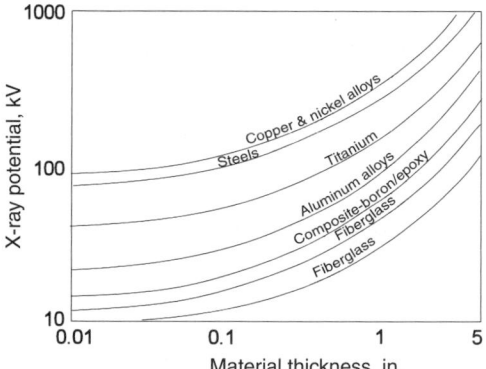

Fig. 8 Plot of X-ray tube voltage versus thickness of several industrial materials.

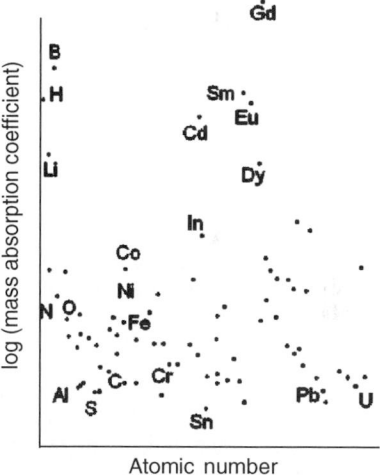

Fig. 9 Plot of mass absorption coefficient for neutron radiography versus atomic number.

monochromatic; that is, the spectrum of radiation contains only one or two dominant energies. Second, the energies of most sources are on the order of millions-of-volts range, making this source ideal for inspecting highly attenuating materials or very large structures. Third, the small size of these sources permits them to be used in tight locations where an X-ray tube could not fit. Fourth, since the gamma-ray source is continually decaying, adjustments to the exposure time must be made in order to achieve consistent results over time. Finally, the operator must always remember that the source is continually on and is therefore a persistent safety hazard! Aside from these differences, gamma radiography differs little from standard practice, so no further distinction between the two will be given.

3.2   Neutron Radiography

Neutron radiography[53] may be useful to inspect some materials and structures. Because the attenuation of

neutrons is not related to the elemental composition of the part, some elements can be more easily detected than others. While X rays are most heavily absorbed by high-atomic-number elements, this is not true of neutrons, as shown in Fig. 9. In Fig. 10 two aluminum panels are bonded with an epoxy adhesive. The reader can discern that hydrogen adsorbs neutrons more than aluminum does, and thus the missing adhesive is easily detectable.

Neutron radiography, however, does have several constraints. First, neutrons do not expose radiographic film and therefore a fluorescing medium is often used to produce light, which exposes the film. The image produced in this manner is not as sharp and well defined as that from X rays. Second, at present there is

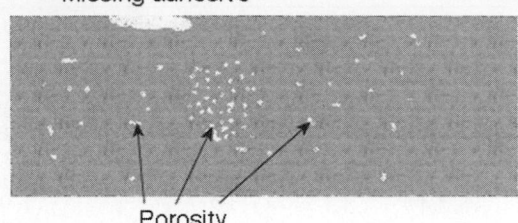

Fig. 10 Representation of neutron radiograph showing flaws in adhesive bond.

no portable high-flux source of neutrons. This means that a nuclear reactor is most often used to supply the neutron radiation. Although neutron radiography has these severe restrictions, at times there is no alternative, and the utility of this method outweighs its expense and complexity.

3.3 Attenuation of X Radiation

An appreciation of how radiographs are interpreted requires a fundamental understanding of X-ray absorption. The relationship governing this phenomenon is de Beer's law:

$$I = I_0 e^{-\mu x}$$

where I, I_0 = transmitted and incident X-ray beam intensities, respectively
μ = attenuation coefficient of material, cm^{-1}
x = thickness of specimen, cm

Since the attenuation coefficient is a function of both the composition of the specimen and the wavelength of the X rays, it would be necessary to calculate or measure it for each wavelength used in RT. However, it is possible to calculate the attenuation coefficient of a material for a specific X-ray energy using the mass absorption coefficient μ_m as defined below. The mass absorption coefficients for most elements are readily available for a variety of X-ray energies,[54]

$$\mu_m = \frac{\mu}{\rho}$$

where μ = attenuation coefficient of an element, cm^{-1}
ρ = density, g/cm^3

The mass absorption coefficient for the material is obtained, at a specific X-ray energy, by multiplying the μ_m of each element by its weight fraction in a material and summing these quantities. Multiplying this sum by the density of the material yields its attenuation coefficient for the material. This procedure is often

not used in practice because the results are valid only for a narrow band of wavelengths. Radiographic equivalency factors are used instead. This process points out that each element in a material contributes to the attenuation coefficient by an amount proportional to its amount in the material.

3.4 Film-Based Radiography

The classical method of recording an X-ray image is with film. Because of the continued importance of this medium of recording and the fact that much of the technology associated with it is applicable to newer solid-state recording methods, this section explores film radiography in some detail.

The relationship between the darkness produced on an X-ray film and the quantity of radiation impinging on it is shown by log-log plots of darkness, or film density, and relative exposure (Figs. 11 and 12). Varying the time of exposure, intensity of the beam, or specimen thickness changes the density, or darkness, of the image. The slope of the curve along its linear portion is referred to as the film gamma, γ. Film has characteristics that are analogous to electronic devices: The greater the gamma or amplification capability of the film, the smaller its dynamic range—the range of exposures over which density is linearly related to thickness. If it is necessary to use a high-gamma film to detect very subtle flaws in a part with a wide range of thicknesses, then it is necessary to use several different film types in the same cassette or package. In this way, each film will be optimized for flaw detection in a narrow thickness range of the part.

Using this information, one may calculate the minimum detectable flaw size for a specific RT inspection. A simple method is available to check the radiographic

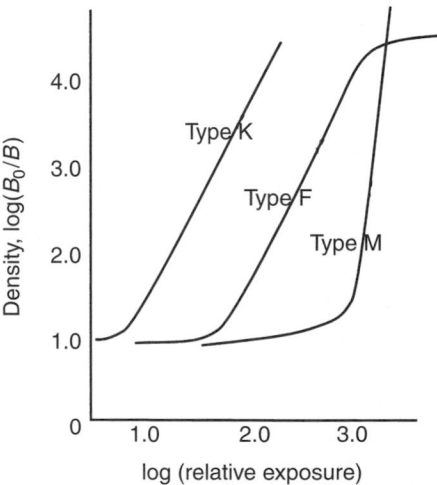

Fig. 11 Density or darkness of X-ray film versus relative exposure for three common films.

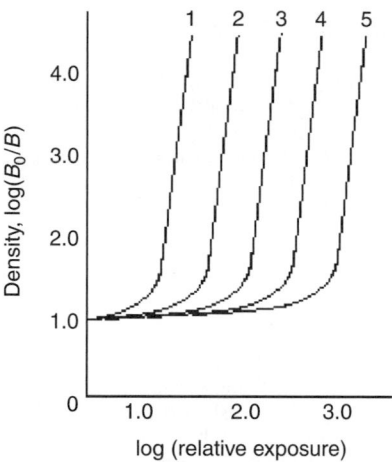

Fig. 12 Density versus relative exposure for films that could be used in multiple film exposure to obtain optimum flaw delectability in complex part.

procedure to determine if this detectability has been achieved on the film. This method does not ensure that the radiograph was taken with the specimen in the proper orientation; it merely provides a method of checking for proper execution of a radiographic procedure; see Section 3.5.

Using a knowledge of the minimum density difference that is detectable by the average radiographic inspector, the following equation relates the radiographic sensitivity, S, to radiographic parameters:

$$S = \frac{2.3}{\gamma \mu x}$$

where S is the radiographic sensitivity in percent, γ is the film gamma, μ is the attenuation coefficient of the specimen material, and x is the maximum thickness of the part associated with a particular radiographic film. The radiographer uses a penetrameter to determine if this sensitivity was achieved. Table 3 give the

Table 3 Radiographic Sensitivity with Thinnest Penetrameter and Smallest Hole Visible on Radiograph

Sensitivity, S (%)	Quality Level (%T − Hole Diameter)
0.7	1 − 1T
1.0	1 − 2T
1.4	2 − 1T
2.0	2 − 2T
2.8	2 − 4T
4.0	4 − 2T

sensitivity S in percent and the expected RT performance in penetrameter values (see Section 3.5).

3.5 Penetrameter

An example of a penetrameter is shown schematically in Fig. 13, while its image on a radiograph is shown in Fig. 14. While there are many types of penetrameters, this one was chosen because it is easily related to radiographic sensitivity. The penetrameter is simply a thin strip of metal or polymeric material[55] in which three holes of varying sizes are drilled or punched. It is composed of the same material as the specimen and has a thickness 1, 2, or 4% of maximum part thickness. The holes in the penetrameter have diameters that are $1T$, $2T$, and $4T$. The sensitivity achieved for each radiographic is determined by noting the smallest hole just visible in the thinnest penetrameter on a film and using Table 3 to determine the sensitivity achieved. By calculating the radiographic sensitivity and then noting the level achieved in practice, the radiographic process can be quantitatively evaluated. While this procedure does not offer any guarantee of flaw detection, it is useful in evaluating the effectiveness of the RT process.

Almost all variables of the radiographic process may be easily and rapidly changed with the aid of tables, graphs, and nomograms, which are usually provided by film manufacturers free of charge. For more information, the reader is referred to the commercial literature.

3.6 Real-Time Radiography

While film radiography represents the bulk of radiographic NDE performed at this time (the beginning

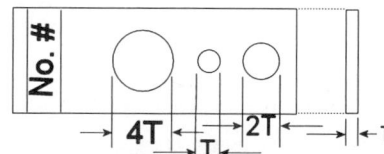

Fig. 13 Schematic of typical film penetrameter.

Fig. 14 Radiograph of penetrameter shown in Figure 13. The $1T$ hole is just visible, indicating the resolution obtained in the radiograph.

of the twenty-first century), new methods of both recording the data and analyzing it are coming into widespread usage. For example, filmless radiography (FR) and real-time radiography (RTR) use solid-state detectors and digital signal processing (DSP) software instead of film to record and enhance the radiographic image. These methods have many advantages, along with some disadvantages. For example, FR permits viewing a radiographic image while the specimen is being moved. This often permits the detection of flaws that would normally be missed in conventional film radiography because of the limited number of views or exposures taken—remember that the X-ray beam must pass along a crack or void for it to be detectable. Additionally, the motion of some flaws enhances their detectability because they present the inspector with a different image as a function of time. Additionally, image enhancement techniques can now be economically and rapidly applied to these images because of the availability of inexpensive, fast computing hardware. The disadvantage of RTR is its lower resolution compared to film. Typical resolution capabilities of RTR or FR systems are in the range of 4 to perhaps 20 line pairs/mm, while film resolution capabilities are in the range of 10–100 line pairs/mm. This means some very fine flaws may not be detectable with FR and the inspector must resort to film. However, in cases where resolution is not the limiting factor, the benefits of software image enhancement can be significant. While the images on film may also be enhanced using the image processing schemes, they cannot be performed in real or near real time, as can be done with an electronic system.

3.7 Computed Tomography

Another advance in industrial radiography has been the incorporation of computed tomography (CT) into the repertoire of the radiographer. Unfortunately, CT has not been exploited to it fullest extent principally due to the high cost of instrumentation. The capability of CT to link NDE measurements with engineering design and analysis gives this inspection a unique ability to provide quantitative estimates of performance not associated with NDE.

The principal advantage of this method is that it produces an image of a thin slice of the specimen under examination. This slice is parallel to the path of the X-ray beam that passes through the specimen, in contrast to the shadowgraph image produced by traditional radiography shown in Fig. 7. Whereas the shadowgraph image can be difficult to interpret, the computed CT image does not contain information from planes outside the thin slice.

A comparison between CT and traditional film radiography is best made with images from these two modalities. Figure 7 shows a typical radiograph where one can easily see the image of the top and bottom surfaces of the tube under inspection. The reader can contrast this with the image in Fig. 15, a CT image of a flashlight. The individual components of the flashlight are easily visible and any misplacement of its components or defects in its assembly can be easily detected. An image with a finer scale that reveals the microstructural details of a pencil is shown in Fig. 16. Clearly visible are not only the key features and even the growth rings of the wood. In fact, the details of the growth during each season are visible as

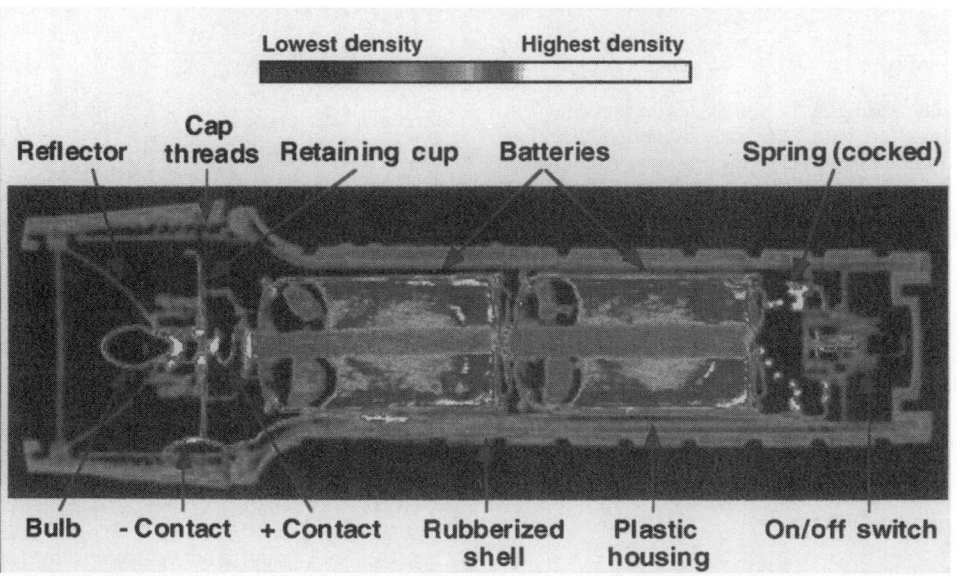

Fig. 15 Computed tomography image of flashlight showing details of internal structure.

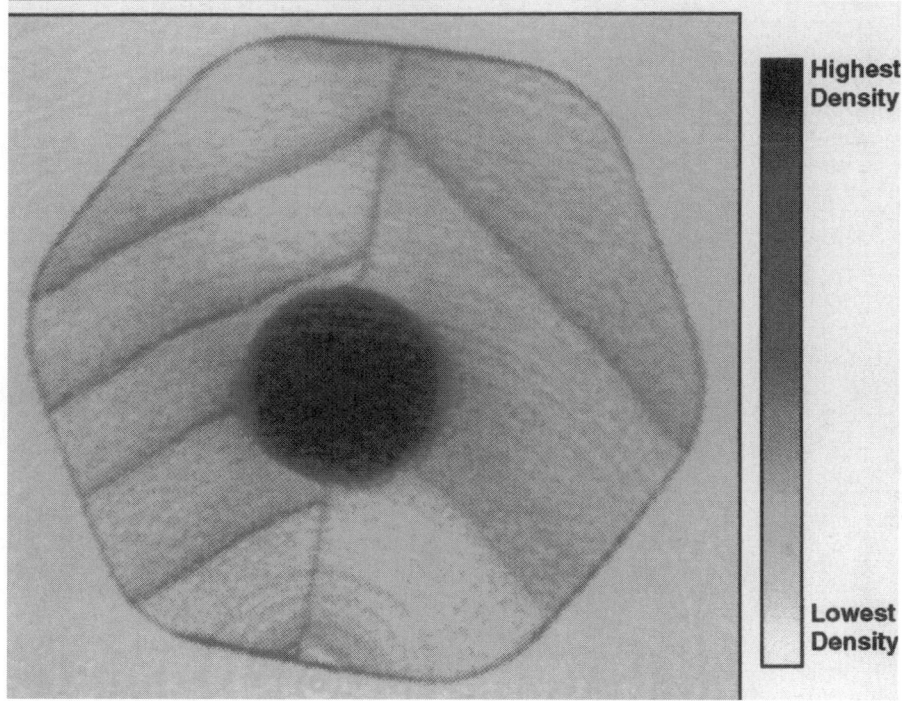

Fig. 16 Computed tomograph of pencil. The reader will note the yearly growth rings and even the growth variations within a single growing season.

rings within rings. The information in the CT image contrasted with conventional radiographs is striking. First, the detectability of a defect is independent of its position in the image. This is not the case with the classical radiograph, where the defect detectability decreases significantly with depth in the specimen because the defect represents a smaller change in the attenuation of the X-ray beam as the depth increases. Second, the defect detectability is very nearly independent of its orientation. This again is clearly not the case with classical radiography. New applications for CT are constantly being discovered. For example, with a digital CT image it is possible to search for various flaw conditions using computer analysis and relieve the inspector of much of the tedium of examining structures for the odd flaw. In addition, it is possible to link the digital CT image with finite-element analysis software to examine precisely how the flaws present will affect such parameters as stress distribution, heat flow, and the like. With little effort, one could analyze the full three-dimensional performance of many engineering structures.

4 ULTRASONIC METHODS

Ultrasonic inspection methods utilize high-frequency sound waves to inspect the interior of solid parts. Sound waves are mechanical or elastic disturbances or waves that propagate in fluid and solid media. Ultrasonic testing (UT) or inspection is similar to the angler who uses sonar to detect fish.[56] The government and various technical societies have developed standard practice specifications for UT. These include ASTM specifications 214-68, 428-71, and 494-75 and military specification MIL-1-8950H. Acoustic and ultrasonic testing can take many forms, from simple coin tapping to the transmission and reception of very high frequency or ultrasonic waves into a part to analyze its internal structure.

UT instruments operating in the frequency range between 20 and 500 kHz are referred to as *sonic* instruments, while those that operate above 500 kHz are called *ultrasonic*. To generate and receive ultrasonic waves, a piezoelectric transducer is employed to convert electrical signals to sound waves and back again. The usual form of a transducer is a piezoelectric crystal mounted in a waterproof housing that is electrically connected to a pulsar (transmitter) and a receiver. In the transmit mode a high-voltage, short-duration electrical spike is applied to the crystal, causing it to rapidly change shape and emit an acoustic pulse. In the receive mode, sound waves or returning echoes compress the piezoelectric crystal, producing an electrical signal that is amplified and processed by the receiver. This process is shown schematically in Fig. 17.

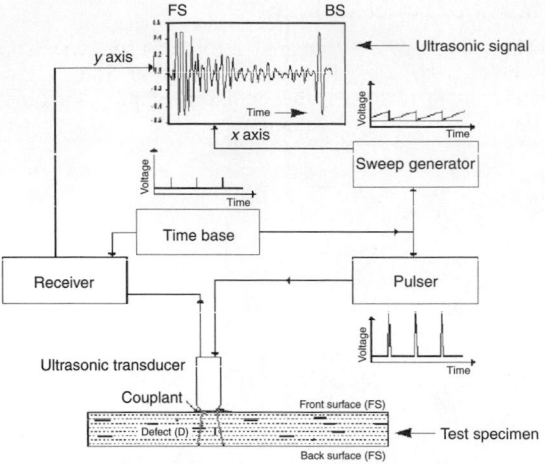

Fig. 17 Schematic of ultrasonic data collection and display in A-scan mode.

4.1 Sound Waves

Ultrasonic waves have physical characteristics such as wavelength (λ), frequency (f), velocity (v), pressure (p), and amplitude (a). The following relationship between wavelength, frequency, and sound velocity is valid for all sound waves:

$$f\lambda = v$$

For example, the wavelength of longitudinal ultrasonic waves of frequency 2 MHz propagating in steel is 3 mm and the wavelength of shear waves is about half this value, 1.6 mm. The relation between the sound pressure and the particle amplitude is

$$p = 2\pi f \rho v a$$

where p is density, f is the frequency of the sound wave, v is its velocity, and a the amplitude.

Ultrasonic waves are reflected from boundaries between different materials or media. Each medium has characteristic acoustic impedance and reflections occur in a manner similar to those observed with electrical signals. The acoustic impedance Z of any media capable of supporting sound waves is defined by

$$Z = \rho v$$

where ρ = density of medium, g/cm^3
$\quad\quad v$ = velocity of sound along direction of propagation

Materials with high acoustic impedance are often referred to as sonically hard, in contrast to sonically soft materials with low impedances. For example, steel

($Z = 7.7$ g/cm^3 $\times$ 5.9 km/s $= 45.4 \times 10^6$ kg/m$^2 \cdot$s) is sonically harder than aluminum ($Z = 2.7$ g/cm^3 $\times$ 6.3 km/s $= 17 \times 10^6$ kg/m$^2 \cdot$s). The Appendix at the end of this chapter lists the acoustic properties of many common materials.

4.2 Reflection and Transmission of Sound

Almost all acoustic energy incident on air–solid interfaces is reflected because of the large impedance mismatch between air and most solids. For this reason, a medium with impedance close to that of the part is used to couple the sonic energy from the transducer into the part. A liquid couplant has obvious advantages for parts with a complex geometry, and water is the couplant of choice for most inspection situations. The receiver, in addition to amplifying the returning echoes, also time gates the returning echoes between the front surface and rear surfaces of the component. Thus, any unusually occurring echo is displayed separately or used to set off an alarm, as shown in Fig. 17. This method of displaying the voltage amplitude of the returning pulse versus time or depth (if acoustic velocity is known) at a single point in the specimen is known as an A scan. In this figure, the first signal corresponds to a reflection from the front surface (FS) of the part and the last signal corresponds to the reflection from its back surface (BS). The signal or echo between the FS and BS is from the defect in the middle of the part.

The portion of sound energy that is either reflected from or transmitted through each interface is a function of the impedances of the medium on each side of that interface. The reflection coefficient R (ratio of the sound pressures or intensities of the reflected and incident waves) and the power reflection coefficient R_{pwr} (ratio of the power in the reflected and incident sound waves) for normally incident waves onto an interface are given as

$$R = \frac{p_r}{p_i} = \frac{Z_1 - Z_2}{Z_1 + Z_2} \quad R_{pwr} = \frac{I_r}{I_i} = \left(\frac{Z_1 - Z_2}{Z_1 + Z_2}\right)^2$$

Likewise, the transmission coefficients T and T_{pwr} are defined as

$$T = \frac{p_t}{p_i} = \frac{2Z_2}{Z_1 + Z_2} \quad T_{pwr} = \frac{I_t}{I_i} = \frac{4(Z_2/Z_1)}{[1 + (Z_2/Z_1)]^2}$$

where I_i, I_r and I_t are the incident, reflected, and transmitted acoustic field intensities, respectively; Z_1 is the acoustic impedance of the medium from which the sound wave is incident; and Z_2 is impedance of the medium into which the wave is transmitted. From these equations one can calculate the reflection and transmission coefficients for a planar flaw containing air, $Z_1 = 450$ kg/cm$^2 \cdot$s, located in a steel part, $Z_2 = 45.4 \times 10^6$ kg/m$^2 \cdot$s. In this case, the reflection coefficient for the flaw is virtually −1.0. The minus sign indicates a phase change of 180° for the reflected pulse (note that the defect signal in Fig. 17 is inverted or phase shifted by 180° from the FS signal). Effectively no acoustic energy is transmitted across an air gap, necessitating the use of water as a coupling media in ultrasonic testing. Using the acoustic properties of common materials given in the Appendix the reader can make a number of simple, yet informative, calculations.

Thus far, our discussion has involved only longitudinal waves. This is the only wave that travels through fluids such as air and water. The particle motion in this wave, if one could see it, is similar to the motion of a spring, or a Slinky toy, where the displacement and wave motion are collinear (the oscillations occur along the direction of propagation). The wave is called compressional or dilatational since both compressional and dilatational forces are active in it. Audible sound waves are compressional waves. This wave propagates in liquids and gases as well as in solids. However, a solid medium can also support additional types of waves such as shear and Rayleigh or surface waves. Shear or transverse waves have a particle motion that is analogous to what one sees in an oscillating rope. That is, the displacement of the rope is perpendicular to the direction of wave propagation. The velocity of this wave is about half that of compressional waves and is only found in solid media, as indicated in the Appendix. Shear waves are often generated when a longitudinal wave is incident on a fluid–solid interface at angles of incidence other than 90°. Rayleigh or surface waves have elliptical wave motion, as shown in Fig. 18, and penetrate the surface for about one wavelength; therefore, they can be used to detect surface and very near surface flaws. The velocity of Rayleigh waves is about 90% of the shear wave velocity. Their generation requires a special device, or wedge as shown in Fig. 18, which enables an incident ultrasonic wave on the sample at a specific angle that is characteristic of the material (Rayleigh angle). The reader can find more details in the scientific literature.[57–59]

4.3 Refraction of Sound

The direction of propagation of acoustic waves is governed by the acoustic equivalent of Snell's law. Referring to Fig. 19, the direction of propagation is determined with the equation

$$\frac{\sin \theta_i}{c_{\mathrm{I}}} = \frac{\sin \theta_r}{c_{\mathrm{I}}} = \frac{\sin \gamma_{\mathrm{r}}}{b_{\mathrm{I}}} = \frac{\sin \theta_t}{c_{\mathrm{II}}} = \frac{\sin \gamma_t}{b_{\mathrm{II}}}$$

where c_1 is the velocity of the incident longitudinal wave, c_{I} and b_{I} are the velocities of the longitudinal and shear reflected waves, and c_{II} and b_{II} are the velocities of the longitudinal and shear transmitted waves in solid II. In the water–steel interface, there is no reflected shear wave because these waves do not propagate in fluids such as water. In this case, the above relationship is simplified. Since the water has a lower longitudinal wave speed than either the longitudinal or shear wave speeds of the steel, the transmitted acoustic waves are refracted away from the normal. If the incident wave approaches the interface at increasing angles, there will be an angle above which there will be no transmitted acoustic wave in the higher wave speed material. This angle is referred to as a critical angle. At this angle, the refracted wave travels along the interface and does not enter the solid. A computer-generated curve is shown in Fig. 20 in which the normalized acoustic energy that is reflected and refracted at a water–steel interface is plotted as a function of the angle of the incident longitudinal wave. Note that a longitudinal or first critical angle for steel occurs at 14.5°. Likewise, the shear or second critical angle occurs at about 30°. If the angle of incidence is increased above the first critical angle but less than the second critical angle, only the shear wave is generated in the metal and travels at an angle of refraction described by Snell's law. Angles of incidence above the second critical angle produce a complete reflection of the incident acoustic wave; that is, no acoustic energy enters the solid. At a specific angle of incidence (Rayleigh angle) surface

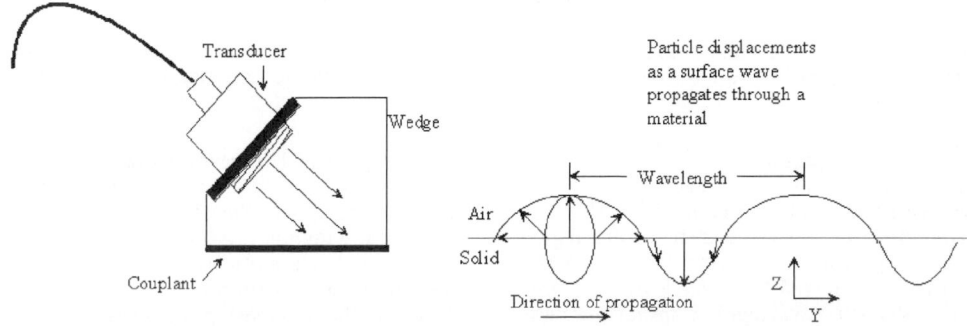

Fig. 18 Generation and propagation of surface waves in a material.

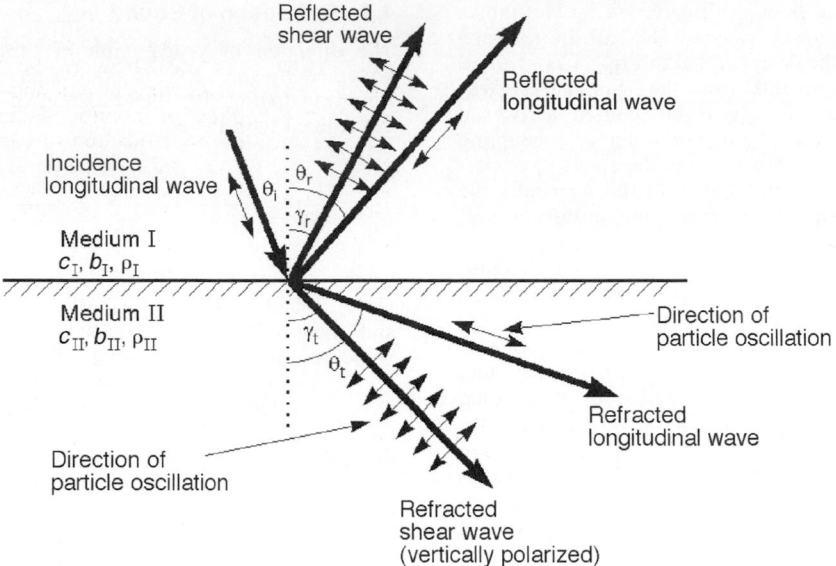

Fig. 19 Representation of Snell's law and mode conversion of longitudinal wave incident on solid–solid interface.

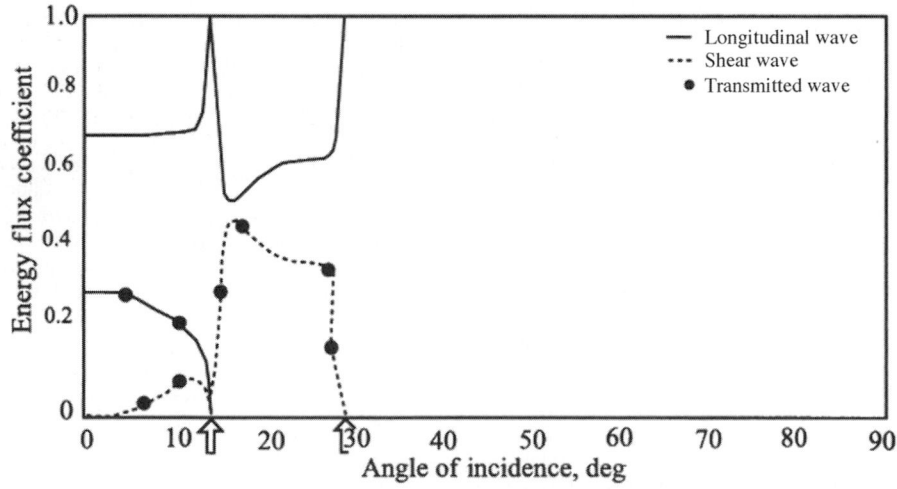

Fig. 20 Amplitude (energy flux) and phase of reflected coefficient and transmitted amplitude versus angle of incidence for longitudinal wave incident on water–steel interface. The arrows indicate the critical angles for the interface.

acoustic waves are generating on the material. The Rayleigh angle can be easily calculated from Snell's law by assuming that the refracted angle is 90°. The Rayleigh angle for steel occurs at 29.5°. In the region between the two critical angles, only the shear wave is generated and is referred to as *shear wave testing*. There are two distinct advantages to inspecting parts with this type of shear wave. First, with only one type of wave present, the ambiguity that would exist

concerning which type of wave is reflected from a defect does not occur. Second, the lower wave speed of the shear wave means that it is easier to resolve distances within the part. For these reasons, shear wave inspection is often chosen for inspection of thin metallic structures such as those in aircraft.

Using the relationships for the reflection and transmission coefficients, a great deal of information can be deduced about any ultrasonic inspection situation when the acoustic wave is incident at $90°$ to the surface. For other angles of incidence, computer software is often used to analyze the acoustic interactions. Analytic predictions of ultrasonic performance in complex materials such as fiber-reinforced composites require the use of more complex algorithms because more complicated modes of wave propagation can occur. Examples of these include Lamb waves (plate waves), Stoneley waves (interface waves), Love waves (guided in layers of a solid material coated onto another one), and others.

4.4   Inspection Process

Once the type of ultrasonic inspection has been chosen and the optimum experimental parameters determined, one must choose the mode of presentation of the data. If the principal dimension of the flaw is less than the diameter of the transducer, then the A-scan method may be chosen, as shown in Fig. 17. The acquisition of a series of A-scans obtained by scanning the transducer in one direction across the specimen and displaying the data as distance versus depth is referred to as a B scan. This is the mode most often used by medical ultrasound instrumentation. In the A-scan mode, the size of the flaw may be inferred by comparing the amplitude of the defect signal to a set of standard calibration blocks. Each block has a flat bottom hole (FBH) drilled from one end. Calibration blocks have FBH diameters that vary in $\frac{1}{64}$-in. increments, for example, a number 5 block has a $\frac{5}{64}$-in. FBH. By comparing the amplitude of the signal from a calibration block with one from a defect, the inspector may specify a defect size as equivalent to a certain size FBH. The equivalent size is meaningful only for smooth flaws that are nearly perpendicular to the path of the ultrasonic beam and is used in many industrial situations where a reference size is required by a UT procedure.

If the flaw size is larger than the transducer diameter, then the C-scan mode is usually selected. In this mode, shown in Fig. 21, the transducer is rastered back and forth across the part. In normal operation, a line is traced on a computer monitor or piece of paper. When a flaw signal is detected between the front and back surfaces, the line drawing ceases and a blank place appears on the paper or monitor. Using this mode of presentation, a planar projection of each flaw is presented to the viewer and its positional relationship to other flaws and to the component boundaries is easily ascertained. Unfortunately, the C-scan mode does not show depth information, unless an electronic gate is set to capture only information from within a specified time window or time gate in the part. With current computer capability, it is a rather simple matter to store all of the returning A-scan data and display only the data in a C-scan mode for a specific depth.

Depending on the structural complexity and the attenuation of the signal, cracklike flaws as small as 0.015 in. in diameter may be reliably detected and quantified with this method. An example of a typical C-scan printout of an adhesively bonded test panel is shown in Fig. 22. This panel was fabricated with a void-simulating Teflon implant and the numerous additional white areas indicate the presence of a great deal of porosity in the part.

Through Transmission versus Pulse–Echo

Thus far, the discussion of ultrasonic inspection methods has been concerned with the setup that uses a single transducer to send a signal into the part and to receive any returning echoes. This method is variously referred to as pulse–echo or pitch–catch inspection and is shown schematically in Fig. 23. The other frequently used inspection setup for many structures is called through transmission. With this setup two transducers are used, one to send ultrasonic pulses and the other placed on the opposite side of the part to receive the transmitted signals, as shown schematically in Fig. 24. In Figs. 23 and 24 a large number of reflections occur for the many individual layers in a composite part that can obscure subtle reflections from inclusions whose reflectivity is similar to that of the layered materials. An inclusion with an acoustic impedance very close to that of the part, for example, paper or peal-plys in polymer-based composites, is

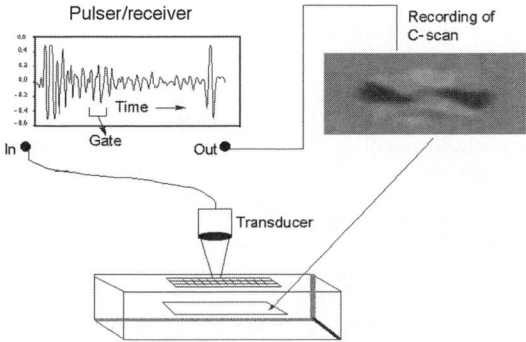

Fig. 21 Representation of ultrasonic data collection. The data are displayed using the C-scan mode. The image shows a defect located at a certain depth in the material.

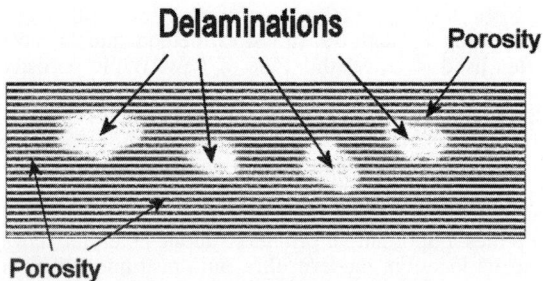

Fig. 22 Typical C-scan image of composite specimen showing delaminations and porosity.

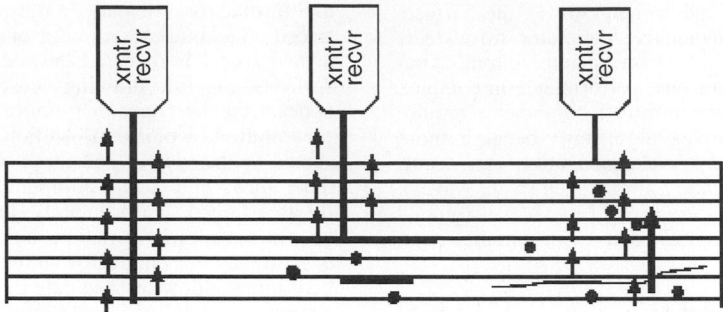

Fig. 23 Representation of pulse–echo mode of ultrasonic inspection.

Transmitting transducer

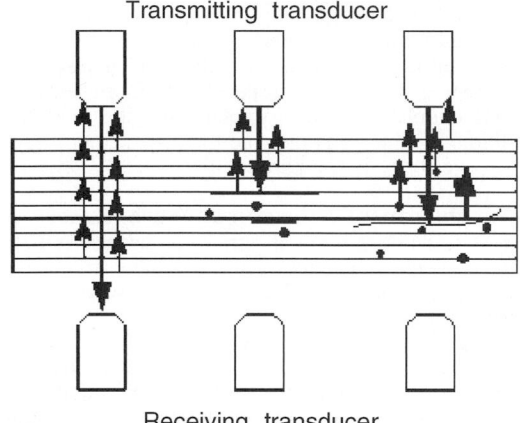

Receiving transducer

Fig. 24 Representation of through-transmission mode of inspection.

very difficult to detect with the through-transmission mode of inspection. In this case, the pulse–echo inspection mode is often used to detect these flaws. On the other hand, reflections from distributed flaws such as porosity, as shown on the right-hand side of Figs. 23 and 24, can be obscured by the general background noise present in an acoustic signal. Therefore, it is the loss in signal strength of the transmitted signal of the through-transmission method that is most often used to detect this type of flaw. While porosity is detectable in this manner, its location may not be determined. In this situation, the pulse–echo mode is required because the distance from the front or back surface to the flaw can be determined by the relative position of the reflections of the scattered porosity with respect to the surface reflection. Because each method supplies important information about potential flaws and its location, modern ultrasonic instrumentation is frequently equipped to perform both types of inspection nearly simultaneously.[60,61] In such a setup, two transducers are used to conduct a through-transmission

test and then each is used separately to conduct pulse–echo tests from opposite sides of the part. This method also helps ensure that a large flaw does not shadow a smaller one, as shown in Figs. 23 and 24.

Portable Ultrasonic Systems This ability to image defects on specific levels within a layered component, for example, a composite, is so important that C-scan instrumentation has been miniaturized for usage in the field. An example of one such system developed for aircraft inspection is shown in Fig. 25. The heart of this system is a computer that records the position of a hand-held transducer and the complete A-scan wave train at each point of the scan. Since the equipment tracks the motion of the transducer as it is scanned manually across a structure, the inspector can see which areas have been scanned. If areas are missed, he or she can return to "color them in," as shown in the image on the monitor. Additionally, computer manipulation of ultrasonic image data allows the inspector to select either one or a small number of layers for evaluation. In this way, an orderly assessment of the flaws in critical structures can be accomplished. This process of selecting flaws on a layer-by-layer basis for evaluation is shown schematically in Fig. 26 for the instrument depicted in Fig. 25.

4.5 Bond Testers

A great deal of the ultrasonic inspection literature is devoted to instruments that test adhesive bonds. There has been a recent resurgence of interest in the inspection of adhesive bonds due to concerns about the viability of bonded patches on our aging aircraft.[62] For an extensive treatment of most of the currently used instruments, the reader is referred to review articles.[63–67] However, while there may seem to be a large number of instruments, some with exaggerated claims of performance, most operate on the same physical principles.

Bond-testing instruments use a variety of means to excite sonic or low-frequency sound waves into the part. In these methods, a low-frequency acoustic transducer is attached to the structure through a couplant.

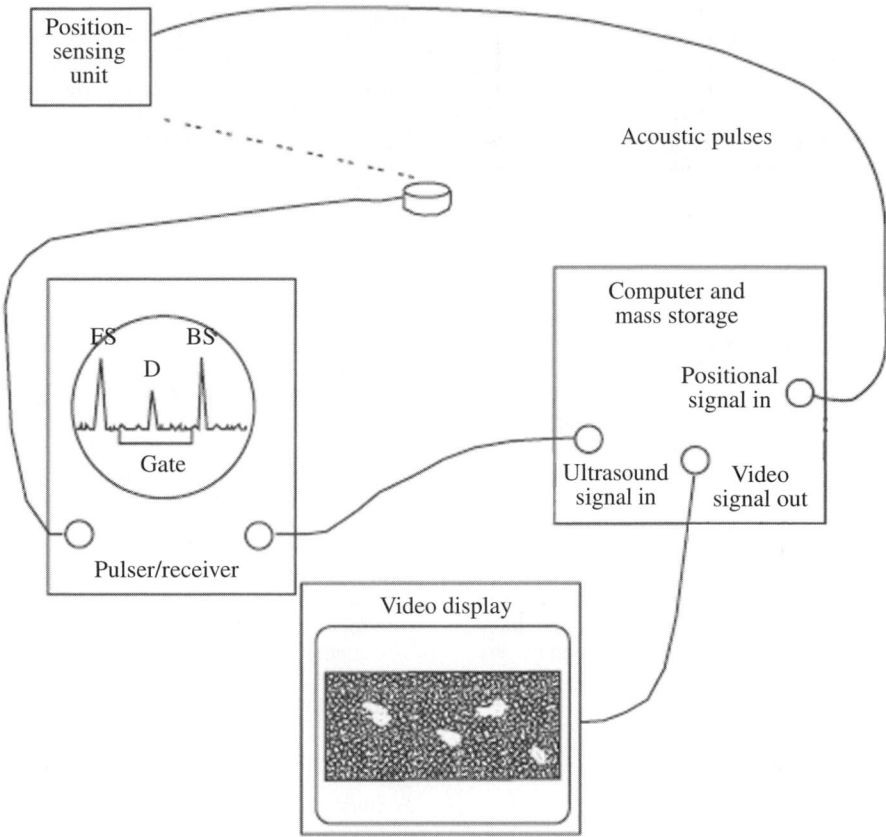

Fig. 25 Field level C-scan instrumentation that is capable of simultaneously tracking the motion of a handheld transducer and recording the ultrasonic information.

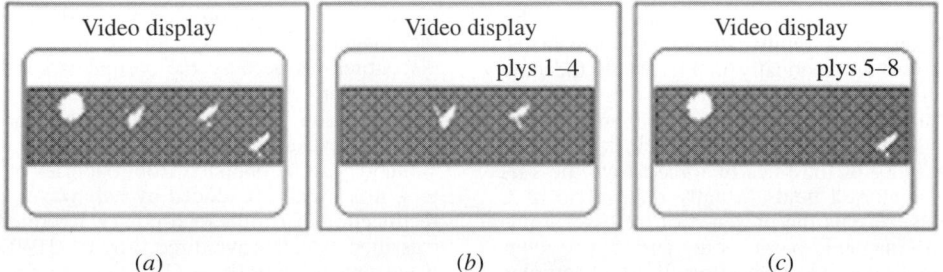

Fig. 26 Three different displays of delaminations in 16-ply composite obtained from field-level C-scan system. (a) Projection of all flaws through specimen. (b, c) Images from selected depths within specimen.

As the driving frequency of the transducer is varied, the amplitude and phase of the transducer oscillations change dramatically as it passes through a resonance. The phase and amplitude of these vibrations change very rapidly and reach a maximum as the driving frequency passes through the resonance frequency of the transducer. The effect of the structure is to dampen the resonant response of the transducer–block combination because of the transfer of acoustic energy into it. Defects such as delaminations and porosity in the adhesive bond layer increase the stiffness of the structure and lower the resonant frequency of the combination.

The amplitude of the resonance is increased since there is less material to adsorb the sound energy. These changes in the sharpness of the resonant response are easily detectable electronically.

An alternate method of detecting flaws in bonded components is with a low-frequency or sonic instrument that senses the change in the time of flight for sound waves in the layered structure due to the presence of planar delamination. In such instruments, the increased time of traveling from a transmitting to a receiving transducer is detected electronically. Several commercially available bond-testing instruments successfully exploit this principle. A clever adaptation of a commercial version of this instrument has recently been used to successfully test the joints of structures made from sheet molding compounds.[64]

Probably the most often used method of detecting delaminations in laminated structures is with a coin or tap hammer. This simple instrument is surprisingly effective in trained hands at detecting flaws since an exceedingly complex computer interprets the output signal, that is, the human brain. Consider, for a moment, that most parents can easily hear their child playing a musical instrument at a school concert. They can perform this task even though their child may have a minor part to play and all the other instruments are much louder than the one that their child is playing. With this powerful real-time signal-processing capability, inspectors can often detect flaws that cannot be detectable with current instrumentation and computers.

5 MAGNETIC PARTICLE METHOD

The magnetic particle method of nondestructive testing is used to locate surface and subsurface discontinuities in ferromagnetic materials.[68] An excellent reference for this NDE method is Ref. 21, especially Chapters 10–16. Magnetic particle inspection is based on the principle that magnetic lines of force, when present in a ferromagnetic material, are distorted by changes in material continuity, such as cracks or inclusions, as shown schematically in Fig. 27. If the flaw is open to the surface or close to it, the flux lines escape the surface at the site of the discontinuity. Near-surface flaws, such as nonmagnetic inclusions, cause the same bulging of the lines of force above the surface. These distorted fields, usually referred to as a leakage fields, reveals the presence of the discontinuity when fine magnetic particles are attracted to them during magnetic particle inspection. If these particles are fluorescent, their presence at a flaw will be visible under ultraviolet light, much like penetrant indications.

Magnetic particle inspection is used for steel components because it is fast and easily implemented and has rather simple flaw indications. The part is usually magnetized with an electric current and then a solution containing fluorescent particles is applied by flowing it over the part. The particles stick to the part, forming the indication of the flaw.

5.1 Magnetizing Field

The magnetizing field may be applied to a component by a number of methods. Its function is to generate a residual magnetic field at the surface of the part. The application of a magnetizing force (H) generates a magnetic flux (B) in the component, as shown schematically in Fig. 28. In this figure, the magnetic flux density B has units of newtons per ampere or webers per square meter, and the strength of the magnetic field or magnetic flux intensity H has units of oersteds or amperes per meter. Starting at the origin, a magnetizing force is applied and the magnetic field internal to the part increases in a nonlinear fashion along the path shown by the arrows. If the force is reversed, the magnetic field does not return to zero but follows the arrows around the curve as shown. The reader will note that once the magnetizing force is removed, the flux density does not return to zero but remains at an elevated value called the material's remanence. This is the point at which most magnetic particle inspections are performed. The reader will also note that an appreciable reverse magnetic force H must be applied before the internal field density is again zero. This point is referred to as the coercivity of the material. If the magnetizing force is applied and reversed, the material will respond by continually moving around this hysteresis loop.

Selection of the type of magnetizing current depends primarily on whether the defects are open to the surface or are wholly below it. Alternating-current (ac) magnetization is best for the detection of surface discontinuities because the current is concentrated in the near-surface region of the part. Direct-current (dc) magnetization is best suited for subsurface discontinuities because of its deeper penetration of the part. While dc can be obtained from batteries or dc generators, it is usually produced by half-wave or full-wave rectification of commercial power. Rectified current is classified as half-wave direct current (HWDC) or full-wave direct current (FWDC). Alternating-current fields are usually obtained from conventional power mains, but it is supplied to the part at reduced voltage for

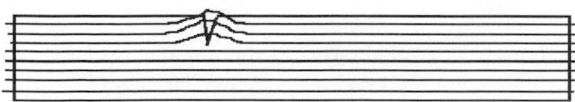

Fig. 27 Representation of magnetic lines of flux in ferromagnetic metal near a flaw. Small magnetic particles are attracted to the leakage field associated with the flaw.

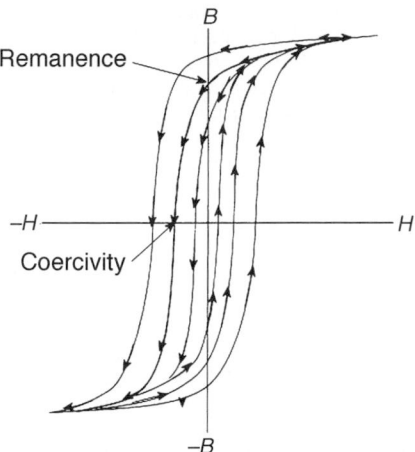

Fig. 28 Magnetic flux intensity *H* versus magnetic flux density *B* hysteresis curve for typical steel. Initial magnetization starts at the origin and progresses as shown by the arrows. Demagnetization follows the arrows of the smaller hysteresis loops.

reasons of safety and the high-current requirements of the magnetizing process.

Two general types of magnetic particles are available to highlight flaws. One type is low-carbon steel with high-permeability and low-retentivity particles, which are used dry and consist of different sizes and shapes to respond to leakage fields. The other type of is very fine particles of magnetic iron oxide that are suspended in a liquid (either a petroleum distillate or water). These particles are smaller and have a lower permeability than the dry particles. Their small mass permits them to be held by the weak leakage fields at very fine surface cracks. Magnetic particles are available in several colors to increase their contrast against different surfaces or backgrounds. Dry powders are typically gray, red, yellow, and black, while wet particles are usually red, black, or fluorescent.

5.2 Continuous versus Noncontinuous Fields

Because the field is always stronger while the magnetizing current is on, the continuous magnetizing method is generally preferred. Additionally, for specimens with low retentivity this continuous method is often preferred. In the continuous method, the current can be applied in short pulses, typically 0.5 sec. The magnetic particles are applied to the surface during this interval and are free to move to the site of the leakage fields. Liquid suspended fluorescent particles produces the most sensitive indications. For field inspections, the magnetizing current is often continuously applied during the test to give time for the powder to migrate to the defect site. In the residual method, the particles are applied after the magnetizing current is removed. This method is particularly well suited for production inspection of multiple parts.

The choice of direction of the magnetizing field within the part involves the nature of the flaw and its direction with respect to the surface and the major axis of the part. In circular magnetization, the field runs circumferentially around the part. It is induced into the part by passing current through it between two contacting electrodes. Since flaws perpendicular to the magnetizing lines are readily detectable, circular magnetization is used to detect flaws that are parallel or less than 45° to the surface of the long, circular specimens. Placing the specimen inside a coil to create a field running lengthwise through the part produces longitudinal magnetization. This induction method is used to detect transverse discontinuities to the axis of the part.

5.3 Inspection Process

The surface of the part to be examined should be clean, dry, and free of contaminants such as oil, grease, loose rust, loose sand, loose scale, lint, thick paint, welding flux, and weld splatter. Cleaning of the specimen may be accomplished with detergents, organic solvents, or mechanical means, such as scrubbing or grit blasting.

Portable and stationary equipment are available for this inspection process. Selection of the specific type of equipment depends on the nature and location of testing. Portable equipment is available in lightweight units (35–90 lb) that can be readily taken to the inspection site. Generally, these units operate at 115, 230, or 460 V ac and supply current outputs of 750–1500 A in half-wave ac.

5.4 Demagnetizing the Part

Once the inspection process is complete, the part must be demagnetized. This is done by one of several ways depending on the subsequent usage of the component. A simple method of demagnetizing to remove residual magnetism from small tools is to draw it through the loop-shaped coil tip of a soldering iron. This has the effect of retracing the hysteresis loop a large number of times, each time with a smaller magnetizing force. When completely withdrawn, the tool will then have a very small remnant magnetic field, which for all practical purposes is zero. This same process is accomplished with an industrial part by slowly reducing and reversing the magnetizing current until it is essentially zero, as shown schematically by the arrows in Fig. 28. Another method of demagnetizing a part is to heat it above its Curie temperature (about 550°C for iron), at which point all residual magnetism disappears. This last process is the best means of removing all residual magnetism, but it requires the expense and time of an elevated heat treatment.

6 THERMAL METHODS

Thermal nondestructive inspection methods involve the detection of infrared energy emitted from the surface of a test object.[69] This technique is used to detect the flow of thermal energy either into or out of a specimen and the effect of anomalies have on the surface temperature distribution. The material properties that influence this method are heat capacity, density, thermal conductivity, and emissivity. Defects that are usually detected include porosity, cracks, and delaminations that are parallel to the surface. The sensitivity of any thermal method is greatest for near-surface flaws that impede heat flow and degrades rapidly for deeply buried flaws in high-conductivity materials. Materials with lower thermal conductivity yield better resolution because they allow larger thermal gradients.

6.1 Infrared Cameras

All objects emit infrared (IR) radiation with a temperature above absolute zero. At room temperature, the thermal radiation is predominately IR with a wavelength of approximately 10 μm. IR cameras are available that can produce images from this radiation and are capable of viewing large areas by scanning. Since the IR images are usually captured and stored in digital form, image processing is easily performed and the enhanced images are stored on magnetic or optical media. For many applications, an uncalibrated thermal image of a specimen is sufficient to detect flaws. However, if absolute temperatures are required, the IR instrumentation must be calibrated to account for the surface emissivity of the test subject.

The ability of thermography to detect flaws is often affected by the type of flaw and its orientation with respect to the surface of the object. To have a maximum effect on the surface temperature, the flaw must interrupt heat flow to the surface. Since a flaw can occur at any angle to the surface, the important parameter is its projected area to the camera. Subsurface flaws such as cracks parallel to the surface of the object, porosity, and debonding of a surface layer are easily detected. Cracks that are perpendicular to the surface can be very difficult or impossible to detect using thermography.

Most thermal NDE methods do not have good spatial resolution due to spreading of thermal energy as it diffuses to the surface. The greatest advantage of thermography is that it can be a noncontact, remote-viewing technique requiring only line-of-sight access to one side of a test specimen. Large areas can be viewed rapidly, since scan rates for IR cameras run between 16 and 30 frames per second. Temperature differences of 0.02°C or less can be detected in a controlled environment.

6.2 Thermal Paints

A number of contact thermal methods are available for inspection purposes. These usually involve applying a coating to the sample and observing a color change as the specimen is thermally cycled. Several different types of coatings are available that cover a wide temperature ranges. Temperature-sensitive pigments in the form of paints have been made to cover a temperature range of 40–1600°C. Thermal phosphors emit visible light when exposed to ultraviolet (UV) radiation. (The amount of visible light is inversely proportional to temperature.) Thermochromic compounds and cholesteric liquid crystals change color over large temperature ranges. The advantages of these approaches are the simplicity of application and relatively low cost if only small areas are scanned.

6.3 Thermal Testing

Excellent results may be achieved for thermographic inspections performed in dynamic environments where the transient effects of heat flow in the test object can be monitored. This enhances detection of areas where different heat transfer rates are present. Applications involving steady-state conditions are more limited. Thermography has been successfully used in several different areas of testing. In medicine, it is used to detect subsurface tumors. In aircraft manufacturing and maintenance, it may be used to detect debonding in layered materials and structures. In the electronics industry, it is used to detect poor thermal performance of circuit board components. Recently thermography has been used to detect stress-induced thermal gradients around defects in dynamically loaded test samples. For more information on thermal NDE methods, the reader is referred to Refs. 69–71.

7 EDDY CURRENT METHODS

7.1 Eddy Current Inspection

Eddy current (EC) methods are used to inspect electrically conducting components for flaws. Flaws that cause a change in electrical conductivity or magnetic permeability such as surface-breaking cracks, subsurface cracks, voids, and errors in heat treatment are detectable using EC methods. Thickness measurements and the thickness of nonconducting coatings on metal substrates can also be determined with EC methods.[72] Quite often, several of these conditions can be monitored simultaneously if instrumentation capable of measuring the phase of the EC signal is used.

This inspection method is based on the principle that eddy currents are induced in a conducting material when a coil (probe) is excited with an alternating or transient electric current that is placed in close proximity to the surface of a conductor. The induced currents create an electromagnetic field that opposes the field of the inducing coil in accordance with Lenz's law. The eddy currents circulate in the part in closed, continuous paths, and their magnitude depends on many variables. These include the magnitude and frequency of the current in the inducing coil, the coil's shape and position relative to the surface of the part, electrical conductivity, magnetic permeability, shape of the part, and presence of discontinuities or inhomogeneities within the material. Therefore, EC inspection is useful for measuring the electrical properties of materials and detecting discontinuities or variations in the geometry of components.

Skin Effect Eddy current inspections are limited to the near-surface region of the conductor by the skin effect. Within the material, the EC density decreases with the depth. The density of the EC field falls off exponentially with depth and diminishes to a value of about 37% of the surface value at a depth referred to as the standard depth of penetration (SDP). The SDP in meters is calculated with the formula

$$\text{SDP} = \frac{1}{\sqrt{\pi f \sigma \mu}}$$

where f = test frequency, Hz
   σ = test material's electrical conductivity, mho/m
   μ = permeability, H/m

The latter quantity is the product of the relative permeability of the specimen, 1.0 for nonmagnetic materials, and the permeability of free space, $4\pi \times 10^{-7}$ H/m.

Impedance Plane While the SDP is used to give an indication of the depth from which useful information can be obtained, the choice of the independent variables in most test situations is usually made using the impedance plane diagram suggested by Förster.[73] It is theoretically possible to calculate the optimum inspection parameters from numerical codes based on Maxwell's equations, but this is a laborious task that is justified in special situations.

The eddy currents induced at the surface of a material are time varying and have amplitude and phase. The complex impedance of the coil used in the inspection of a specimen is a function of a number of variables. The effect of changes in these variables can be conveniently displayed with the impedance diagram, which shows the variations in amplitude and phase of the coil impedance as functions of the dependent variables specimen conductivity, thickness, and distance between the coil and specimen, or lift-off. For the case of an encircling coil on a solid cylinder, shown schematically in Fig. 29, the complex impedance plane is displayed in Fig. 30. The reader will note that the ordinate and abscissa are normalized by the inductive reactance of the empty coil. This eliminates the effect of the geometry of the coil and specimen. The numerical values shown on the large curve, which are called reference numbers, are used to combine the effects of the conductivity, size of the test specimen, and frequency of the measurement into a single parameter. This yields a diagram that is useful for most test conditions. The reference numbers shown on the outermost curve are obtained with the following relationship for nonmagnetic materials:

$$\text{Reference number} = r\sqrt{2\pi f \mu \sigma}$$

where r = radius of bar, m
   f = frequency of test, Hz
   m = magnetic permeability of free space, $4\pi \times 10^{-7}$ H/m
   σ = conductivity of specimen, mho/m

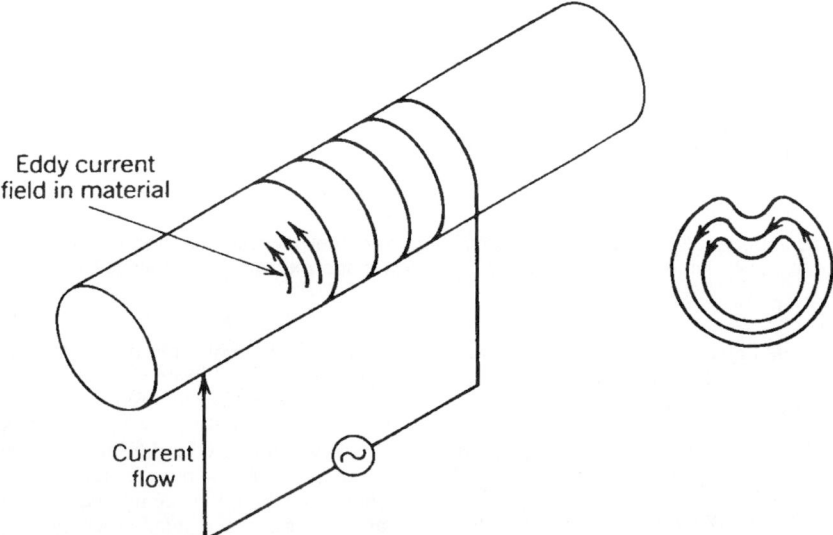

Fig. 29 Representations of eddy current inspection of solid cylinder. Also shown are the eddy current paths within the cross section of the cylinder near the crack.

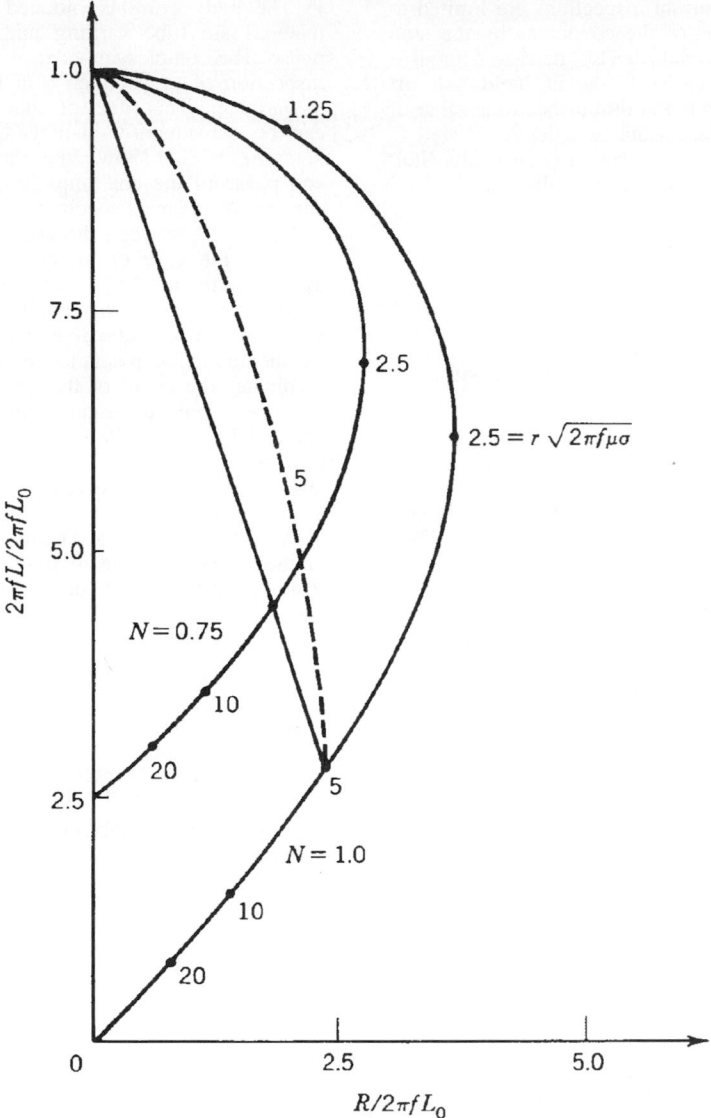

Fig. 30 Normalized impedance diagram for long encircling coil on solid, nonferromagnetic cylinder. For $N = 1$ the coil and cylinder have the same diameter, while for $N = 0.75$ the coil is approximately 1.155 times larger than the cylinder.

The outer curve in Figs. 30 and 31 is useful only for the case where the coil is the same size as the solid cylinder, which can never happen. For those cases where the coil is larger than the test specimen, which is usually the case, a coil-filling factor is calculated. This is quite easily accomplished with the formula

$$N = \frac{\text{diameter}_{\text{specimen}}}{\text{diameter}_{\text{coil}}}$$

Figure 30 shows the impedance plane with a curve for specimen/coil inspection geometry with a fill factor of 0.75. Note that the reference numbers on the curves representing the different fill factors can be determined by projecting a straight line from point 1.0 on the ordinate to the reference number of interest, as is shown for the reference number 5.0. Both the fill factor and the reference number change when the size of either the specimen or coil changes. Assume that a reference number of 5.0 is appropriate to a specific test with $N = 1.0$; if the coil diameter is changed so that

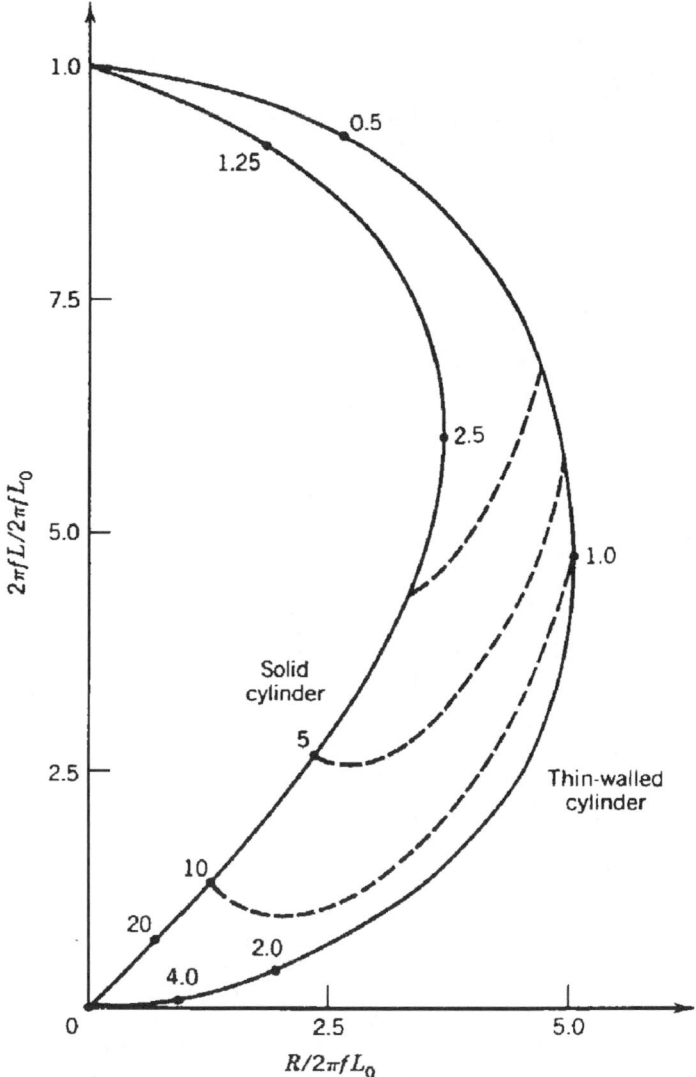

Fig. 31 Normalized impedance diagram for long encircling coil on both solid and thin-walled conductive but nonferromagnetic cylinders. The dashed lines represent the effects of varying wall thicknesses.

the fill factor becomes 0.75, then the new reference number will be equal to approximately 7. While the actual change in reference number for this case follows the path indicated by the dashed line in Fig. 30, we have estimated the change along a straight line. This yields a small error in optimizing the test setup but is sufficient for most purposes. For a more detailed treatment of the impedance plane, the reader is referred to Ref. 72. The inspection geometry discussed thus far has been for a solid cylinder. The other geometry of general interest is the thin-walled tube. In this case the

skin effect limits the thickness of the metal that may be effectively inspected.

For an infinitely thin-walled tube, the impedance plane is shown in Fig. 31, which includes the curve for a solid cylinder. The dashed lines that connect these two cases are for thin-walled cylinders of varying thicknesses. The semicircular curve for the thin cylinder is used in the same manner as described above for the solid cylinder.

Lift-off of Inspection Coil from Specimen In most inspection situations, the only independent

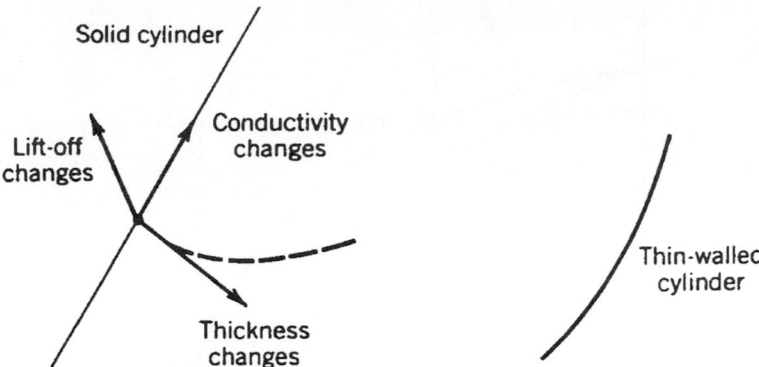

Fig. 32 Effects of various changes in inspection conditions on local signal changes in impedance plane of Figure 31. Phase differentiation is relatively easily accomplished with current instrumentation.

variables are frequency and lift-off. High-frequency excitations are frequently used for detecting defects such as surface-connected cracks or corrosion, while low frequencies are used to detect subsurface flaws. It is also possible to change the coil shape and measurement configuration to enhance detectability, but the discussion of these more complex parameters is beyond the scope of this chapter and the reader is referred to the literature. The relationships discussed so far may be applied by examining Fig. 32, where changes in thickness, lift-off, and conductivity are represented by vectors. These vectors all point in different directions representing the phases of the different possible signals. Instrumentation with phase discrimination circuitry can differentiate between these signals and therefore is often capable of detecting two changes in specimen condition at once. Changes in conductivity can arise from several different conditions. For example, aluminum alloys can have different conductivities depending on their heat treatment. Changes in apparent conductivity are also due to the presence of cracks or voids. A crack decreases the apparent conductivity of the specimen because the eddy currents must travel a longer distance to complete their circuit within the material. Lift-off and wall thinning are also shown in Fig. 32. Thus, two different flaw conditions can be rapidly detected. There are situations where changes in wall thickness and lift-off result in signals that are very nearly out of phase and therefore the net change is not detectable. If this situation is suspected, then inspection at two different frequencies is warranted. There are other inspection situations that cannot be covered in this brief description. These include the inspection of ferromagnetic alloys, plate, and sheet stock and the measurement of film thicknesses on metal substrates. For a treatment of

these and other special applications of EC inspection, the reader is referred to Ref. 72.

7.2 Probes and Sensors

In some situations, it may be advantageous to have a core with a high magnetic permeability inside the coil. Magnetic fields will pass through the medium with the highest permeability if possible. Therefore, materials with high permeability can be placed in different geometric configurations to enhance the sensitivity of a probe. One example of this is the cup-core probe where the coil has a ferrite core, shield, and cap.[74]

For low-frequency or transient tests, using the inductive coil as a sensor is not sufficient since it responds to the time change in magnetic field and not the direct magnetic field. It is necessary to induce a magnetic field sensor that responds well at the lower frequencies. Sensors such as the Hall effect sensor and giant magnetoresistive (GMR) sensors have been used to accomplish this.[75]

There are numerous methods of making eddy current NDE measurements. Two of the more common methods are shown schematically in Fig. 33. In the absolute coil arrangement, very accurate measurements can be made of the differences between the two samples. In the differential coil method, it is the differences between the two variables at two slightly different locations that may be detected. For this arrangement, slightly varying changes in dimensions and conductivity are not sensed, while singularities such as cracks or voids are highlighted, even in the presence of other slowly changing variables. Since the specific electronic circuitry used to accomplish this task can vary dramatically, depending on the specific inspection situation, the reader is referred to the current NDE and instrumentation Refs. 73, 76, and 77.

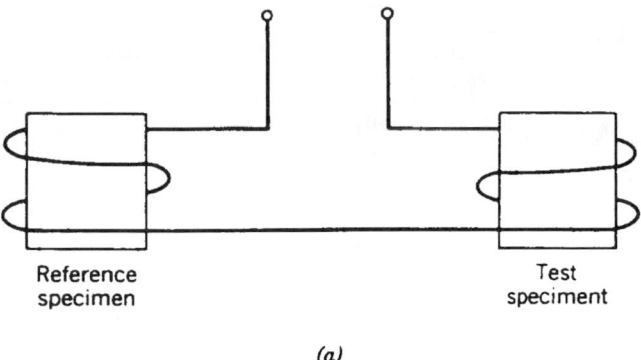

(a)

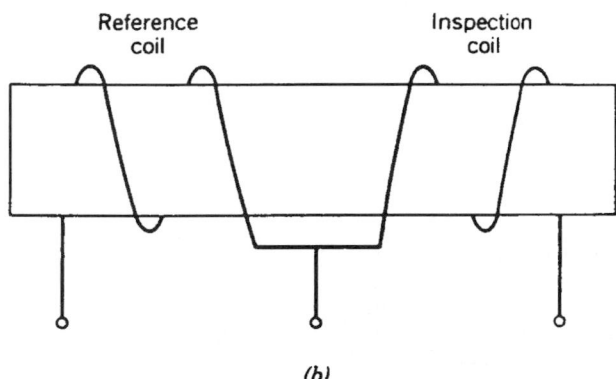

(b)

Fig. 33 Representation of (*a*) absolute versus (*b*) differential coil configurations used in eddy current testing.

APPENDIX: ULTRASONIC PROPERTIES OF COMMON MATERIALS

Material	Comments	Density (g/cm^3)	V_1 (km/s)	V_s (km/s)	Impedance (MRayl)	Attenuation (dB/cm/MHz)	Attenuation (dB/cm at 5 MHz)
Gases							
Alcohol vapor			0.23				
Air	25 atm		0.33				
	50 atm		0.34				
	100 atm		0.35				
			0.34				
			0.39				
			0.55				
Ammonia			0.42				
Argon		0.00178	0.32				
Carbon monoxide		0.34					
Carbon dioxide			0.26				
Carbon disulfide			0.19				

Material	Comments	Density (g/cm^3)	V_1 (km/s)	V_s (km/s)	Impedance (MRayl)	Attenuation (dB/cm/MHz)	Attenuation (dB/cm at 5 MHz)
Chlorine			0.21				
Ether vapor			0.18				
Ethylene			0.31				
Helium		0.00018	0.97				
Hydrogen		0.00009	1.28				
Methane		0.00074	0.43				
Neon		0.0009	0.43				
Nitric oxide			0.33				
Nitrogen		0.00125	0.33				
		0.00116	0.35				
Nitrous oxide			0.26				
Oxygen		0.00142	0.32				
		0.00132	0.33				
Water vapor			0.4				
			0.41				
			0.42				

Gases—Cryogenic

Material	Comments	Density (g/cm^3)	V_1 (km/s)	V_s (km/s)	Impedance (MRayl)	Attenuation (dB/cm/MHz)	Attenuation (dB/cm at 5 MHz)
Argon		1.404	0.84			1.18	
		1.424	0.86			1.23	
Helium		0.125	0.18			0.023	
		0.146	0.23			0.034	
Helium-4	Liquid at 2 K	0.15	0.18			0.027	
Hydrogen	Liquid at 20 K	0.07	1.19			0.08	
		0.355	1.13			0.401	
Nitrogen		0.815	0.87			0.708	
		0.843	0.93			0.783	
Oxygen		1.143	0.97			1.04	
		1.272	1.13			1.44	
		1.149	0.95			1.09	

Liquids

Material	Comments	Density (g/cm^3)	V_1 (km/s)	V_s (km/s)	Impedance (MRayl)	Attenuation (dB/cm/MHz)	Attenuation (dB/cm at 5 MHz)
Acetate butyl (n)		0.871	1.17			1.02	
Acetate ethyl		0.9	1.18			1.06	
Acetate methyl		0.928	1.15			1.07	
Acetate propyl		0.891	1.18			1.05	
Acetone		0.79	1.17			0.92	
		0.791	1.16	—	0.92	—	0.0469
Acetonitrile		0.783	1.29			1.01	
Acetonyl acetone		0.729	1.4			1.36	
Acetylene dichloride		1.26	1.02			1.29	
Adiprene	CW-520	0.79	1.68	—	1.33	—	0.0469
Alcohol, butyl		0.81	1.24			1	
Alcohol, ethyl		0.789	1.18			0.93	
Alcohol, furfuryl		1.135	1.45			1.65	
Alcohol, isopropyl		0.79	1.17	—	0.92	—	0.08
Alcohol, methyl		0.792	1.12			0.89	
Alcohol, propyl (i)		0.786	1.17			0.92	
Alcohol, propyl (n)		0.804	1.22			0.98	

Material	Comments	Density (g/cm^3)	V_1 (km/s)	V_s (km/s)	Impedance (MRayl)	Attenuation (dB/cm/MHz)	Attenuation (dB/cm at 5 MHz)
Alcohol, *t*-amyl		0.81	1.2			0.97	
Alkazene 13		0.86	1.32			1.14	
Analine		1.022	1.69			1.68	
A-Spirit	Ethanol >96%	0.79	1.18			0.93	
Benzene		0.87	1.3			1.13	
	C_6H_6	0.88	1.31			1.15	
Benzol		0.878	1.33			1.17	
Benzol ethyl		0.868	1.34			1.16	
Bromo-benzene	C_6H_5Br	1.52	1.17			1.78	
Bromoform		2.89	0.92			2.66	
Butanol	Butyl	0.71	1.27	—	0.9		
		0.81	1.27			1.03	
Butoxyethanol	(2*n*-)		1.31				
tert-Butyl chloride		0.84	0.98			0.82	
Butylene glycol (2.3)		1.019	1.48			1.51	
Butyrate ethyl		0.877	1.17			1.03	
Carbitol		0.988	1.46			1.44	
Carbon disulfide			1.16				
		1.26	1.15			1.45	
Carbon tetrachloride	CCl_4	1.595	0.93			1.48	
Cerechlor 42		1.26	1.43			1.8	
Chinolin		1.09	1.57			1.71	
Chlorobenzene	C_6H_5Cl	1.1	1.3			1.43	
Chloroform		1.49	0.99			1.47	
Chlorohexanol	>98%	0.95	1.42			1.35	
Cyclohexanol		0.962	1.45			1.39	
	Freon		1.2				
	DTE 21 oil		1.39				
	Glycerol		1.52				
Decahydro-naphtaline		0.948	1.42			1.39	
	$C_{10}H_{18}$	0.89	1.42			1.27	
	Paraffin		1.41				
Diacetyl		0.99	1.24			1.22	
Diamine propane	(1.3) >99%	0.89	1.66			1.47	
Dichloro isobutane (1.3)		1.14	1.22			1.39	
Diethylamine	$(C_2H_5)_2$ NH	0.7	1.13			0.8	
Diethylene glycol		1.116	1.58			1.76	
Diethyl ketone		0.813	1.31			1.07	
Dimethyl phthalate		1.2	1.46			1.75	
Dioxane		1.033	1.38			1.43	
Diphenyl	Diphenyl oxide		1.5				
Dodecanol		0.83	1.41			1.16	
DTE 21	Mobil		1.39				
DTE 24	Mobil		1.42				

Material	Comments	Density (g/cm³)	V_1 (km/s)	V_s (km/s)	Impedance (MRayl)	Attenuation (dB/cm/MHz)	Attenuation (dB/cm at 5 MHz)
DTE 26	Mobil		1.43				
Dubanol	Shell		1.43				
			1.44				
	Water		1.55				
Ethanol	C_2H_4OH	0.79	1.13	—	0.89	—	0.0421
Ethanol amide		1.018	1.72			1.75	
Ethyl acetate		0.9	1.19			1.07	
Ethylenclycolol	Sodium benzoate	—	1.67				
Ethylene diamine	> 99.5%	0.9	1.69			1.52	
Ethylene glycol	> 99.5%	1.11	1.69			1.88	
	1.2-Ethanediol	1.112	1.67			1.86	
		1.113	1.66			1.85	
	H_2O 1:4		1.6				
	H_2O 2:3		1.68				
	H_2O 3:2		1.72				
	H_2O 4:1		1.72				
Ethyl ether		0.713	0.99			0.7	
Fluorinert	H_2O-72	1.68	0.51			0.86	
	FC-104	1.76	0.58			1.01	
	FC-75	1.76	0.59			1.02	
	FC-77	1.78	0.6			1.05	
	FC-43	1.85	0.66			1.21	
	FC-40	1.86	0.64			1.19	
	FC-70	1.94	0.69			1.33	
Fluoro-benzene	C_6H_5F	1.024	1.18			1.21	
Formamide		1.134	1.62			1.84	
Freon			0.68				
Freon MF 21.1		1.485	0.8			1.19	
Freon TF 21.1		1.574	0.97			1.52	
Furfural		1.157	1.45			1.68	
Gasoline		0.803	1.25			1	
Glycerine	CH_2OHCHO HCH_2OH	1.23	1.9			2.34	
	Glycerol <98%	1.26	1.88			2.37	
		1.26	1.92			2.42	
Glycerol	Water	1.22	1.88			2.29	
	Butanol		1.45				
	Ethanol		1.52				
			1.56				
	Isopropanol		1.57				
Glycerol trioleate		0.91	1.44			1.31	
Glycol	Polyethylene	1.06	1.62			1.71	
		1.087	1.62			1.75	
	Ethylene	1.108	1.59			1.76	
		1.112	1.67			1.86	
Hexane	$(n\text{-})C_6H_{14}$	0.659	1.1			0.727	
n-Hexanol		0.819	1.3			1.06	
Honey	Sue Bee Orange	1.42	2.03			2.89	

Material	Comments	Density (g/cm^3)	V_1 (km/s)	V_s (km/s)	Impedance (MRayl)	Attenuation (dB/cm/MHz)	Attenuation (dB/cm at 5 MHz)
Iodobenzene	C$_6$H$_5$I	1.183	1.1			2.01	
Isopentane		0.62	0.99			0.62	
Isopropanol			1.14				
Isopropyl alcohol		0.786	1.17			0.92	
	Propylene glycol	0.79	1.14			0.9	0
		0.84	1.21			1.01	0.14
		0.88	1.24			1.09	0.4
		0.88	1.28			1.13	0.23
		0.92	1.3			1.2	0.44
		0.94	1.35			1.27	0.33
		0.96	1.36			1.31	0.53
		0.97	1.36			1.32	1.12
		0.99	1.43			1.41	0.47
		1	1.42			1.42	1.08
		1	1.43			1.43	0.67
		1.03	1.48			1.52	1.13
		1.03	1.51			1.56	0.72
		1.04	1.54			1.64	1.2
		1.07	1.58			1.7	0.93
		1.09	1.6			1.75	1.4
		1.13	1.65			1.86	1.68
		1.16	1.75			2.03	2.57
		1.2	1.82			2.19	2.12
		1.25	1.92			2.39	4.55
Jeffox WL-1400			1.53				
Kerosene		0.81	1.32			1.07	
Linalool		0.884	1.4			1.24	
Mercury	Hg	13.6	1.45			19.7	
Mercury, 20°C			1.42			19.7	
Mesityloxide		0.85	1.31			1.11	
Methanol	CH$_3$OH	0.796	1.09	—	0.87	0.87	0.0262
Methyl acetate		0.934	1.21			1.13	
Methylene iodide			0.98				
Methylethyl ketone		0.805	1.21			0.97	
Methyl napthalene		1.09	1.51			1.65	
Methyl salicylate		1.16	1.38			1.6	
Modinet P40		1.06	1.38			1.47	
Monochloro-benzene		1.107	1.27			1.41	
Morpholine		1	1.44			1.44	
M-xylol		0.864	1.32			1.14	
NaK	Mix	0.64	1.66				
		0.713	1.72				
		0.714	1.84				
		0.73	1.95				
		0.736	1.77				
		0.738	1.89				
		0.754	2				

Material	Comments	Density (g/cm^3)	V_1 (km/s)	V_s (km/s)	Impedance (MRayl)	Attenuation (dB/cm/MHz)	Attenuation (dB/cm at 5 MHz)
		0.759	1.82				
		0.761	1.99				
		0.778	2.05				
		0.781	1.88				
		0.784	1.99				
		0.801	2.1				
		0.804	1.93				
		0.807	2.04				
		0.825	2.15				
		0.826	1.98				
		0.83	2.09				
		0.848	2.2				
		0.849	2.04				
		0.853	2.14				
		0.871	2.25				
		0.876	2.19				
		0.893	2.31				
Nicotine	$C_{10}H_{14}N_2$	1.01	1.49			1.51	
Nitrobenzene		1.2	1.46			1.75	
Nitrogen	N_2	0.8	0.86			0.68	
Nitromethane		1.13	1.33			1.5	
Oil, baby		0.821	1.43			1.17	
Oil, castor	Jeffox WL-1400	—	1.52				
	Castor	0.95	1.54			1.45	
	Ricinus oil	0.969	1.48			1.43	
Oil, corn		0.922	1.46			1.34	
Oil, cutting	64 AS (red)		1.4				
Oil, diesel			1.25				
Oil, fluorosilicone	Dow FS-1265		0.76				
Oil, grape seed	Cerechlor	0.92	1.43				
	Castor oil	0.936	1.44			1.35	
Oil, gravity fuel AA		0.99	1.49			1.48	
Oil, linseed		0.922	1.77			1.63	
		0.94	1.46			1.34	
Oil, mineral (heavy)		0.843	1.46			1.37	
Oil, mineral (light)		0.825	1.44			1.19	
Oil, motor (2-cycle)			1.43				
Oil, motor (SAE 20)		0.87	1.74			1.51	
Oil, motor (SAE 30)		0.88	1.7			1.5	
Oil, olive			1.43				
		0.918	1.45			1.32	
		0.948	1.43			1.39	
Oil, paraffin			1.28				
			1.43				
		0.835	1.42			1.86	
Oil, peanut		0.914	1.44			1.31	
		0.936	1.46			1.37	
Oil, safflower		0.92	1.45			1.34	

Material	Comments	Density (g/cm^3)	V_1 (km/s)	V_s (km/s)	Impedance (MRayl)	Attenuation (dB/cm/MHz)	Attenuation (dB/cm at 5 MHz)
Oil, silicone	Dow 710 fluid		1.35				
	Silicone 200	0.818	0.96			0.74	
		0.94	0.97			0.91	
		0.972	0.99			0.96	
	30 cP	0.993	0.99			0.983	
		1.1	1.37			1.5	
Oil, soybean		0.93	1.43			1.32	
Oil, sperm		0.88	1.44			1.27	
Oil, sun	Nivea		1.41				
Oil, sunflower		0.92	1.45			1.34	
Oil, synthetic		0.98	1.27			1.33	
Oil, transformer		0.92	1.39			1.28	
Oil, transmission	Dexron (red)		1.42				
Oil, velocite	Mobil		1.3				
Oil, wheat germ		0.94	1.49			1.39	
Paraffin		1.5	1.5			2.3	
d-Penchone		0.94	1.32			1.24	
Pentane		0.621	1.01			0.63	
	(n-)C$_5$H$_{12}$	0.626	1.03			0.64	
Petroleum		0.825	1.29			1.06	
Polypropylene glycol	Polyglycol P-400		1.3				
	Polyglycol P-1200		1.3				
	Polyglycol E-200		1.57				
Polypropylene oxide	Ambiflo		1.37				
Potassium		0.662	1.49				
		0.685	1.55				
		0.707	1.6				
		0.729	1.65				
		0.751	1.71				
		0.773	1.76				
		0.796	1.81				
		0.818	1.86				
Propane diol	(1.3) >97%	1.05	1.62			1.7	
Pyridine		0.982	1.41			1.38	
Sodium		0.759	2.15				
		0.784	2.21				
		0.809	2.26				
		0.833	2.31				
		0.857	2.37				
		0.881	2.42				
		0.904	2.48				
		0.926	2.53				
Solvesso #3		0.877	1.37			0.201	
Sonotrack	Coupling gel	1.4	1.62			1.68	
Span 20			1.48				
Span 85			1.46				
Tallow			0.39				
Tetraethylene glycol		1.12	1.58			1.77	

Material	Comments	Density (g/cm^3)	V_1 (km/s)	V_s (km/s)	Impedance (MRayl)	Attenuation (dB/cm/MHz)	Attenuation (dB/cm at 5 MHz)
Tetrahydro-naphtaline	(1.2.3.4)	0.97	1.47			1.42	
Trichloro-ethylene		1.05	1.05			0.41	
Triethylene glycol		1.12	1.61			1.81	
		1.123	1.61			1.98	
Trithylamine	$(C_2H_5)_3N$	0.73	1.12			0.81	
Turpentine		0.87	1.25			1.11	
		0.893	1.28			1.14	
			1.27				
Ucon 75H450			1.54				
Univis 800		0.87	1.35			1.19	
Water–salt solution	10%		1.47				
	15%		1.53				
	20%		1.6				
Water, sea		1.025	1.53			1.57	
		1.026	1.5			1.54	
Water		1	1.51			1.51	
		1	1.55			1.55	
	Propylene glycol	1	1.5			1.5	
		1.01	1.61			1.63	0.021
		1.02	1.69			1.72	0.038
		1.03	1.51			1.56	0.669
		1.03	1.62			1.66	0.213
		1.03	1.69			1.73	0.088
		1.05	1.6			1.69	
		1.06	1.69			1.79	0.059
		1.07	1.58			1.7	1.025
		1.07	1.66			1.78	0.395
		1.07	1.71			1.83	0.174
		1.07	1.73			1.84	0.112
		1.11	1.71			1.89	0.086
		1.11	1.75			1.95	0.321
		1.11	1.76			1.94	0.117
		1.11	1.77			1.97	0.182
		1.12	1.66			1.86	1.744
		1.12	1.71			1.91	0.582
		1.16	1.75			2.03	2.57
		1.16	1.78			2.06	1.242
		1.16	1.8			2.09	0.175
		1.16	1.81			2.09	0.648
		1.16	1.82			2.11	0.241
		1.16	1.82			2.11	0.397
		1.2	1.82			2.19	2.12
		1.2	1.85			2.23	1.469
		1.2	1.85			2.22	2.033
		1.2	1.86			2.24	1.023
		1.2	1.87			2.24	0.731
		1.2	1.88			2.25	0.544
		1.25	1.92			2.39	4.55
Water	UCON 50HB400	0.79	1.16			0.91	0.11
		0.83	1.25			1.04	0.06

Material	Comments	Density (g/cm³)	V_1 (km/s)	V_s (km/s)	Impedance (MRayl)	Attenuation (dB/cm/MHz)	Attenuation (dB/cm at 5 MHz)
		0.83	1.27			1.06	0.06
		0.83	1.28			1.07	0.07
		0.83	1.29			1.07	0.07
		0.84	1.21			1.01	0.15
		0.84	1.24			1.03	0.08
		0.87	1.41			1.23	0.3
		0.88	1.31			1.16	0.15
		0.88	1.34			1.18	0.15
		0.88	1.37			1.2	0.2
		0.88	1.4			1.22	0.24
		0.89	1.26			1.12	0.26
		0.91	1.54			1.4	0.55
		0.92	1.52			1.4	0.74
		0.93	1.44			1.33	0.4
		0.93	1.48			1.38	0.57
		0.94	1.32			1.24	0.35
		0.94	1.4			1.31	0.35
		0.96	1.63			1.57	1.38
		0.96	1.64			1.57	0.03
		0.97	1.54			1.5	1.06
		0.97	1.59			1.55	1.54
		0.99	1.38			1.37	0.52
		0.99	1.49			1.46	0.7
		1	1.48			1.48	
		1	1.5			1.5	0
		1	1.5			1.5	0.04
		1.01	1.61			1.63	0.13
		1.01	1.61			1.63	0.13
		1.01	1.63			1.65	0
		1.01	1.69			1.71	0.4
		1.02	1.64			1.69	2.72
		1.02	1.66			1.7	2.11
		1.02	1.66			1.7	2.11
		1.02	1.66			1.7	2.11
		1.02	1.68			1.72	0.84
		1.02	1.69			1.72	1.5
		1.02	1.69			1.73	0.17
		1.02	1.69			1.73	0.29
		1.02	1.7			1.73	0.08
		1.02	1.71			1.74	0.44
		1.03	1.5			1.55	0.92
		1.03	1.5			1.55	0.92
		1.03	1.53			1.58	0.72
		1.03	1.53			1.58	0.72
		1.03	1.54			1.6	0.72
		1.03	1.54			1.6	0.72
		1.03	1.56			1.61	0.73
		1.03	1.56			1.61	0.73
		1.03	1.57			1.63	2.13
		1.03	1.57			1.61	0.92
		1.03	1.57			1.62	0.58
		1.03	1.57			1.63	2.13
		1.03	1.57			1.61	0.92
		1.03	1.57			1.62	0.58
		1.03	1.57			1.63	2.13
		1.03	1.58			1.63	1.25

Material	Comments	Density (g/cm^3)	V_1 (km/s)	V_s (km/s)	Impedance (MRayl)	Attenuation (dB/cm/MHz)	Attenuation (dB/cm at 5 MHz)
		1.03	1.58			1.63	1.25
		1.03	1.6			1.65	0.47
		1.03	1.6			1.65	0.47
		1.03	1.61			1.65	0.56
		1.03	1.61			1.65	0.56
		1.03	1.62			1.67	0.38
		1.03	1.62			1.67	0.38
		1.03	1.63			1.68	0.9
		1.03	1.63			1.68	0.9
		1.03	1.64			1.69	2.72
		1.03	1.65			1.7	1.53
		1.03	1.65			1.7	0.46
		1.03	1.65			1.7	0.32
		1.03	1.65			1.7	1.53
		1.03	1.65			1.7	0.46
		1.03	1.65			1.7	0.32
		1.04	1.44			1.5	0.94
		1.04	1.44			1.5	0.94
		1.04	1.46			1.52	1.05
		1.04	1.48			1.53	1.02
		1.04	1.51			1.57	0.85
Water, D$_2$O		1.104	1.4			1.55	
Xylene hexaflouride		1.37	0.88			1.21	
Solids (Metals and Alloys)							
Aluminum		2.7	6.32	3.1		17.1	
	Duraluminum	2.71	6.32	3.1		17.1	
Al 1100-0	2S0	2.71	6.35	3.1	2.9	17.2	
Al 2014	14S	2.8	6.32	3.1	—	17.7	
Al 2024 T4	24ST	2.77	6.37	3.2	2.95	17.6	
Al 2117 T4	17ST	2.8	6.5	3.1		18.2	
Antimony	Sb		3.4				
Bearing Babbit		10.1	2.3	—	—	23.2	
Beryllium		1.82	12.9	8.9	7.87	23.5	
Bismuth		9.8	2.18	1.1		21.4	
Brass	70% Cu-30% Zn	8.64	4.7	2.1		40.6	
		8.56	4.28	2		36.6	
	Half hard	8.1	3.83	2.1		31.0	
	Naval	8.42	4.43	2.1	1.95	37.3	
Bronze	Phospho	8.86	3.53	2.2	2.01	31.3	
Cadmium	Cd	8.6	2.8	1.5		42.0	
		8.64	2.78	1.5		24.0	
Cesium		1.88	0.97			1.82	
Columbium		8.57	4.92	2.1		42.2	
Constantan		8.88	5.24	2.6		46.5	
Copper		8.93	4.66	2.3	1.93	41.6	
Copper, rolled	Cu	8.9	5.01	2.3		44.6	
E-Solder		2.71	1.9	1		5.14	
Gallium		5.95	2.74	—	—	16.3	
Germanium		5.47	5.41	—	—	29.6	
Gold	Hard drawn	19.32	3.24	1.2	—	62.6	
Hafnium			3.84	—			

Material	Comments	Density (g/cm^3)	V_1 (km/s)	V_s (km/s)	Impedance (MRayl)	Attenuation (dB/cm/MHz)	Attenuation (dB/cm at 5 MHz)
Inconel		8.25	5.72	3	2.79	64.5	
Indium		7.3	2.22			16.2	
Iron		7.7	5.9	3.2	2.79	45.4	
	Cast	7.22	4.6	2.6	—	33.2	
Lead		11.4	2.16	0.7	0.63	24.6	
	5% Antimony	10.9	2.17	0.8	0.74	23.7	
Magnesium		1.74	6.31			11.0	
	AM-35	1.74	5.79	3.1	2.87	10.1	
	FS-1	1.69	5.47	3		9.2	
	J-1	1.7	5.67	3		9.6	
	M	1.75	5.76	3.1		10.1	
	O-1	1.82	5.8	3		10.6	
	ZK-60A-TS	1.83	5.71	3.1		10.4	
		1.72	5.8	3		10.0	
Manganese		7.39	4.66	2.4		34.4	
Molybdenum		10.2	6.29	3.4	3.11	64.2	
Monel		8.83	6.02	2.7	1.96	53.2	
Nickel		8.88	5.63	3	2.64	50.0	
Nickel–silver		11.2	3.58	2.2		40.0	
Platinum		21.4	3.96	1.7		84.7	
Plutonium			1.79			28.2	
	1% Gallium		1.82			28.6	
Potassium		0.83	1.82			1.51	
Rubidium		1.53	1.26			1.93	
Silver		10.5	3.6	1.6		37.8	
	Nickel	8.75	4.62	2.3	1.69	40.4	
	Germanium	8.7	4.76			41.4	
Steel	302 Cres	8.03	5.66	3.1	3.12	45.4	
	347 Cres	7.91	5.74	3.1	—	45.4	
	410 Cres	7.67	7.39	3	2.16	56.7	
	1020	7.71	5.89	3.2		45.4	
	1095	7.8	5.9	3.2		51.0	
	4150	7.84	5.86	2.8		45.9	
		7.82	5.89	3.2		46.1	
		7.81	5.87	3.2		45.8	
		7.8	5.82	2.8		45.4	
	4340	7.8	5.85	3.2		51.0	
	Mild	7.8	5.9	3.2		46.00	
	Stainless 347	7.89	5.79	3.1		45.70	
Tantalum		16.6	4.1	2.9		54.8	
Thallium		11.9	1.62			19.3	
Thorium		11.3	2.4	1.6		33.2	
Tin		7.3	3.3	1.7		24.1	
Titanium		4.5	6.07	3.1		27.3	
		4.48	6.1	3.1		27.3	
Tungsten		19.25	5.18	2.9	2.65	99.7	
Uranium		18.5	3.4	2		63.0	
Vanadium		6.03	6	2.8		36.2	
Zinc		7.1	4.17	2.4		29.6	
Zircalloy			4.72	2.4		44.2	
Zirconium		6.48	4.65	2.3		30.1	
Solids (Ceramics)							
Ammonium dihydrogen	502/118.9 : 1	1.35	2.73	—	3.69		

Material	Comments	Density (g/cm^3)	V_1 (km/s)	V_s (km/s)	Impedance (MRayl)	Attenuation (dB/cm/MHz)	Attenuation (dB/cm at 5 MHz)
phosphate (ADP)							
	502/118.5 : 1	1.35	2.67	—	3.60		
			3.28				
Arsenic trisulfide		3.2	2.58	1.4		8.25	
Barium titanate		5.55	5.64	2.9		33.5	
Boron carbide		2.4	11			26.4	
Brick		1.7	4.3			7.40	
		3.6	3.65	2.6		15.3	
Calcium fluoride	CaFl. X-cut		6.74				
Clay rock		2.5	3.48	3.4		14.2	
Concrete		2.6	3.1			8.00	
Flint		3.6	4.26	3		18.9	
Glass	Crown	2.24	5.1	2.8		11.4	
	205 Sheet	2.49	5.66			14.1	
	FK3	2.26	4.91	2.9		11.1	
	FK6	2.28	4.43	2.5		10.1	
	Flint	3.6	4.5			16.0	
	Macor	2.54	5.51			14.0	
	Plate	2.75	5.71			10.7	
	Pyrex	2.24	5.64	3.3		13.1	
	Quartz	2.2	5.57	3.4		14.5	
	Silica	2.2	5.9			13.0	
	Soda lime	2.24	6			13.4	
	T1K	2.38	4.38			10.5	
	Window		6.79	3.4			
Glass crown	Reg.	2.6	5.66	3.5		14.5	
Granite		4.1	6.5			26.8	
Graphite	Pyrolytic	1.46	4.6			6.60	
	Pressed	1.8	2.4			4.10	
Hydrogen	Solid at 4.2 K	0.089	2.19			0.19	
Ice		0.92	3.6			3.20	
		2.65	3.99	3.3		16.4	
Ivory		2.17	3.01			10.4	
Leadmeta niobate	$PbNbO_3$	6.2	3.3			20.5	
	K-81	6.2	3.3			20.5	
	K-83	4.3	5.33			22.9	
	K-85	5.5	3.35			18.4	
Lead zirconate titanate	$PbZrTiO_3$	7.75	3.28			29.3	
		7.5	4			30.0	
		7.45	4.2			31.3	
		7.43	4.44			33.0	
		7.95	4.72			37.5	
Lithium niobate	46 Rot. Y-cut	4.7	7.08			33.0	
	Z-cut	4.64	7.33			34.0	
	Y-cut		6.88				
Lithium sulfate	Y-cut	2.06	5.46			11.2	
Marble		2.8	3.8			10.5	
Porcelain		2.3	5.9			13.50	
Potassium bromide			3.38				

Material	Comments	Density (g/cm^3)	V_1 (km/s)	V_s (km/s)	Impedance (MRayl)	Attenuation (dB/cm/MHz)	Attenuation (dB/cm at 5 MHz)
Potassium chloride			4.14				
Potassium sodium niobate		4.46	6.94			31.0	
PZT-2		7.6	4.41	1.7		31.3	
PZT-4		7.5	4.6	1.9		34.5	
PZT-5A		7.75	4.35	1.7		33.7	
PZT-5H		7.5	4.56	1.8		34.2	
Quartz		6.82	5.66			15.2	
	X-cut	2.65	5.75			15.3	
Salt	NaCl	2.17	4.85			10.5	
Salt, rochelle		2.2	5.36	3.8		13.1	
	KNaC$_4$H4$_{(6}$		2.47				
Salt, rock	X dir		4.78				
Sapphire		2.6	9.8			11.7	
	Al$_2$O$_3$	3.98	11.2			44.5	
Silica, fused		2.2	5.96	3.8		13.1	
Silicon	Anisotropic	2.33	9			21.0	
Silicon carbide		13.8	6.66			91.8	
Silicon nitride		3.27	11	6.3		36.0	
Slate			4.5				
		3	4.5			13.5	
Sodium bismuth titanate		6.5	4.06			26.4	
Sodium bromide	NaBr		2.79				
Sulfur			1.35				
Titanium carbide		5.15	8.27	5.2		42.6	
Tourmaline	Z-cut	3.1	7.54			23.4	
Uranium oxide	UO$_2$		5.18			56.7	
Zinc oxide		5.68	6.4	3		36.4	
Solids (Polymer)							
ABS	Acrylonitrile	1.04	2.11	—	2.20		
Acrylic		1.2	2.7	—	3.24		
Acrylic resin		1.18	2.67	1.1		3.15	
Araldite	502/956	1.16	2.62	—	4.04		
Bakelite		1.4	2.59			3.63	
		1.9	1.9			4.80	
Butyl rubber		1.11	1.8			2.00	
Carbon, pyrolytic	Soft	2.21	3.31			7.31	
Carbon, vitreous		1.47	4.26	2.7		6.26	
Celcon	Acetal copolymer	1.41	2.51			3.54	
Cellulose acetate		1.3	2.45			3.19	
Cycolac	Acrylonitrile–butadiene–styrene		2.27			2.49	

Material	Comments	Density (g/cm³)	V_1 (km/s)	V_s (km/s)	Impedance (MRayl)	Attenuation (dB/cm/MHz)	Attenuation (dB/cm at 5 MHz)
Delrin		1.36	2.47			3.36	
	Acetal homopolymer	1.42	2.52			3.57	
DER317	10.5PHR DEH20	1.18	2.75			3.25	
		2.23	2.07			4.61	
	13.5PHR MPDA	1.6	2.4			3.84	
		2.03	2.19			4.44	
		3.4	1.86	0.9		6.40	
	9PHR DEH20	7.27	1.5	—		10.9	
		2.23	2.03	1		4.53	
		2.37	1.93	—		4.58	
DER332	10PHR DEH20	1.76	3.18	1.6		5.58	
		1.2	2.6			3.11	
	10.5PHR DEH20	1.29	2.65			3.41	
		1.26	2.61			3.29	
		1.37	2.75			3.78	
	11PHR DEH20	1.72	2.35			4.05	
		1.29	2.71			3.49	
	14PHR MPDA	1.25	2.59			3.24	
	15PHR MPDA	1.54	2.78	1.5		4.27	
		1.49	2.8	1.4		4.18	
		1.24	2.66	—		3.30	
		1.24	2.55	1.2		3.16	
		2.15	3.75			8.06	
		2.24	3.9			8.74	
		6.45	1.75			11.3	
	64PHR V140	1.13	2.36			2.65	
	75PHR V140	1.12	2.35			2.62	
	100PHR V140	1.1	2.32			2.55	
		1.13	2.27			2.55	
		1.16	2.36			2.74	
ECHOGEL 1265	100PHA of B	9.19	1.32			12.2	
		1.4	1.7			2.38	
		1.1	1.71			1.90	
EPON 828	MPDA	1.21	2.83	1.2		3.40	
EPOTEK 301		1.08	2.64			2.85	
EPOTEK 330		1.14	2.57			2.94	
EPOTEK H70S		1.68	2.91			4.88	
EPOTEK V6	10PHA of B	1.23	2.55			3.14	
		1.23	2.61			3.21	
		1.26	2.55			3.22	
		1.25	2.6			3.25	
Epoxy	Silver	3.098	1.89			5.85	
		3.383	1.87			6.31	
EPX-1 or EPX-2	100PHA of B	1.1	2.44			2.68	
Ethyl vinyl acetate		0.94	1.8			1.69	
		0.95	1.68			1.60	
		0.93	1.86			1.72	

Material	Comments	Density (g/cm³)	V_l (km/s)	V_s (km/s)	Impedance (MRayl)	Attenuation (dB/cm/MHz)	Attenuation (dB/cm at 5 MHz)
Glucose		1.56	3.2			5.00	
Hysol	C8-4143/3404	1.58	2.85			4.52	
	C9-4183/3561	3.17	2.16			7.04	
		2.14	2.49			5.33	
		1.8	2.62			4.70	
		1.48	2.92			4.30	
		2.66	2.3			6.10	
	C8-4412	1.68	2.02			3.39	
		1.5	2.32			3.49	
Hysol	R9-2039/3404						
Ivory			3.01				
Kel-F			1.79				
Kydex		1.35	2.22			2.99	
Lucite	Polymethyla-crylate	1.29	2.72			3.50	
		1.18	2.68	1.3		3.16	
		1.15	2.7	1.1		3.10	
Marlex 5003	High-density polyethylene	0.95	2.56			2.43	
Melopas		1.7	2.9			4.93	
Micarta	Linen base		3				
Mylar		1.18	2.54			3.00	
Neoprene		1.31	1.6			2.10	
Noryl	Polyphenylene oxide	1.08	2.27			2.45	
Nylon 6-6		1.12	2.6	1.1		2.90	
Penton	Chlorinated polyether	1.4	2.57			3.60	
	Syntactic foam (33 lb/ft³)	0.53	2.57			1.36	
Phenolic		1.34	1.42			1.90	
Plexiglas	UVA	1.27	2.76			3.51	
	UVAII	1.18	2.73	1.4		3.22	
Polyamide			2.6			2.90	
Polycarbonate	Lexan	1.18	2.3			2.71	
Polyester	Casting resin	1.07	2.29			2.86	
Polyethylene	Low density	0.92	2.06	—	1.90	22	26.5
		1.1	2.67			2.80	
	TCI		1.6				
	HD. LB-861	0.96	2.43			2.33	
Polyisobutylene			1.49				
	mol. wt. 200		1.85				
Polypropylene	Profax 6423	0.901	2.49			2.24	
		0.88	2.74			2.40	
Polysulfone		1.24	2.24			2.78	
Polystyrene		1.1	2.67			2.80	
	Styron 666	1.05	2.4			2.52	
Polyurethane	RP-6400	1.04	1.5			1.56	
	RP-6401	1.07	1.71	—	1.83	35	73
		1.07	1.63			1.74	
	RP-6402	1.08	1.77			1.91	
	RP-6403	1.1	1.87			2.05	
	RP-6405	1.3	2.09			2.36	
	RP-6410	1.04	1.71	—	1.78	36	73

Material	Comments	Density (g/cm^3)	V_1 (km/s)	V_s (km/s)	Impedance (MRayl)	Attenuation (dB/cm/MHz)	Attenuation (dB/cm at 5 MHz)
		1.04	1.33			1.38	
	RP-6413	1.04	1.71		1.78	21	35.2
		1.04	1.65			1.66	
	RP-6414	1.05	1.78			1.86	
		1.05	1.85		1.94	18	35.2
	RP-6422	1.04	1.6			1.66	
	EN-9	1.01	1.68			1.70	
	REN plastic	1.07	1.71			1.83	35
		1.04	1.49			1.55	36
		1.04	1.62			1.69	15
		1.04	1.71			1.78	21
		1.05	1.85			1.92	18
	RP6422	1.04	1.62		1.69	14	27.6
Polyvinyl chloride (PVC)		1.45	2.27			3.31	
Polyvinylbutyral	Butracite	1.11	2.35			2.60	
Polyvinylidene difluoride		1.79	2.3			4.20	
Profax	Polypropylene	0.9	2.79			2.51	
Refrasil		1.73	3.75			6.49	
Rubber	BFG#6063-19-71	0.97	1.53			1.56	
	BFG#35080						
	Hard	1.1	1.45			2.64	
	Rho-C	1	1.55			1.55	
	Soft	0.95	0.07			1.00	
Scotchcast	XR2535	1.49	2.48			3.70	
Scotchply XP241	Syntactic foam (42 lb/ft^3)	0.65	2.84			1.84	
	Syntactic foam (38 lb/ft^3)	0.61	2.81			1.71	
Scotchply	XP241	0.65	2.84			1.84	
	SP1002	1.94	3.25			6.24	
Scotch tape	2.5 mils thick	1.16	1.9			2.08	
Silicon rubber	Sylgard 170	1.38	0.97			1.34	
	Sylgard 182	1.05	1.03			1.07	
		1.12	1.03			1.15	
	Sylgard 184	1.03	1.03			1.04	
	RTV-11	1.18	1.05			1.24	
	RTV-21	1.31	1.01			1.32	
	RTV-30	1.45	0.97			1.41	
	RTV-41	1.31	1.01			1.32	
	RTV-60	1.47	0.96			1.41	
	RTV-77	1.33	1.02			1.36	
	RTV-90	1.5	0.96			1.44	
	RTV-112	1.05	0.94			0.99	
	RTV-511	1.18	1.11			1.31	
	RTV-116	1.1	1.02			1.12	
	RTV-118	1.04	1.03			1.07	
	RTV-577	1.35	1.08			1.46	
	RTV-560	1.42	1.03			1.46	
	RTV-602	1.02	1.16			1.18	
	RTV-615	1.02	1.08			1.10	
	RTV-616	1.22	1.06			1.29	

Material	Comments	Density (g/cm^3)	V_1 (km/s)	V_s (km/s)	Impedance (MRayl)	Attenuation (dB/cm/MHz)	Attenuation (dB/cm at 5 MHz)
	RTV-630	1.24	1.05			1.30	
	PRC 1933-2	1.48	0.95			1.40	
Silly Putty		1	1			1.00	
Stycast	1251-40	1.67	2.9	1.5		4.83	
		1.63	2.95			4.82	
		1.57	2.88			4.53	
		1.5	2.77			4.16	
	1264	1.19	2.22			2.64	
	2741	1.17	2.29			2.68	
	CPC-41	1.01	1.52			1.54	
	CPC-39	1.06	1.53			1.63	
Styrene 50D	Polystyrene	1.04	2.33			2.43	
Styron	Modified polystyrene	1.03	2.24			2.31	
Surlyn	1555 Ionomer	0.95	1.91			1.81	
Tapox	Epoxy	1.11	2.48			2.76	
Techform	EA700	1.2	2.63			3.14	
Teflon		2.14	1.39			2.97	
		2.2	1.35			2.97	
TPX	DX845	0.83	2.22		1.84	4.2	5.8
Tracon	2135 D	1.03	2.45			1.52	
	2143 D	1.05	2.37			2.50	
	2162 D	1.19	2.02			2.41	
	3011	1.2	2.12			2.54	
	401 ST	1.62	2.97			4.82	
Uvex			2.11				
WR 106-1	Fluoro elastomer		0.87				
Zytel-101	Nylon-101	1.14	2.71			3.08	
Solids (Natural)							
Ash	Along fiber		4.67				
Beech	Along fiber		3.34				
Beef			1.55			1.68	
Brain			1.49			1.55	
Cork			0.5				
Douglas Fir	Cross grain		1.4				
	With grain		4.8				
Elm			1.4			0.798	
Human			1.47			1.58	
Kidney			1.54			1.62	
Liver			1.54			1.65	
Maple	Along fiber		4.11				
Oak			4.47			3.60	
Pine	Along fiber		3.32				
Poplar	Along fiber		4.28				
Spleen			1.5			1.60	
Sycamore	Along fiber		4.46				
Water		0.88	4	2		3.50	
Wood	Cork	0.24	0.5			0.12	
	Elm		4.1				
	Oak	0.72	4			1.57	
	Pine	0.45	3.5			1.57	

REFERENCES

1. *Metals Handbook*, 3rd ed., Vol. 17: *Nondestructive Evaluation and Quality Control*, ASM International, Metals Park, OH, 1989.
2. Mitchell, M. R., and Buck, O. (Eds.), *Cyclic Deformation, Fracture, and Nondestructive Evaluation of Advanced Materials*, American Society for Testing and Materials, Philadephlia, PA, 1992.
3. Shapuk, H. J. (Ed.), *Annual Book of ASTM Standards: E-7, Nondestructive Testing*, American Society for Testing and Materials, West Conshohocken, PA, 1997.
4. Green, R. E. Jr. (Ed.), *Nondestructive Characterization of Materials*, Vol. 8, International Symposium on Nondestructive Characterization of Materials, Plenum, New York, 1998.
5. *British Journal of Nondestructive Testing*, no longer published.
6. American Society for Nondestructive Testing, http://www.asnt.org.
7. Center for Nondestructive Evaluation, http://www.cnde.iastate.edu.
8. Center for Quality Engineering & Failure Prevention, http://www.cqe.nwu.edu.
9. *Nondestructive Evaluation System Reliability Assessment*, http://www.ihserc.com.
10. Nondestructive Testing Information Analysis Center, http://www.ntiac.com.
11. Altergott, W., and Henneke, E. (eds.), *Characterization of Advanced Materials*, Plenum, New York, 1990.
12. Halmshaw, R., *Nondestructive Testing Handbook*, Chapman & Hall, London, 1991.
13. Kline, R. A., *Nondestructive Characterization of Materials*, Technomic Publishing, Lancaster, PA, 1992.
14. Mallick, P. K., "Nondestructive Tests," in *Composites Engineering Handbook*, P. K. Mallick (Ed.), Marcel Dekker, New York, 1997.
15. McGonnagle, W., *Nondestructive Testing*, Gordon Breach, New York, 1961.
16. Ruud, C. O., et al. (Eds.), *Nondestructive Characterization of Materials*, Vols. I–IV, Plenum, New York, 1986.
17. Sharpe, R. S., *Research Techniques in Nondestructive Testing*, Academic, New York, 1984.
18. Summerscales, J., "Manufacturing Defects in Fibre-Reinforced Plastic Composites," *Insight*, **36**(12), 936–942 (1994).
19. Thompson, D. O., and Chimenti, D. E. (eds.), *Review of Progress in Quantitative Nondestructive Evaluation*, Plenum, New York, 1982–2000.
20. Boogaard, J., and van Dijk, G. M. (eds.), *Nondestructive Testing: Proceedings of the 12th World Conference on Nondestructive Testing*, Elsevier Science, New York, 1989.
21. Bray, D. E., and Stanley, R. K., *Nondestructive Evaluation, a Tool for Design, Manufacturing, and Service*, McGraw-Hill, New York, 1989.
22. Geier, M. H., *Quality Handbook for Composite Materials*, Chapman & Hall, London, 1994.
23. Online Journal Publication Service, http://ojps.aip.org.
24. *Journal of Composite Materials*, 2004.
25. Electronic journals, http://lib-www.lanl.gov/cgi-bin/ejrnlsrch.cgi.
26. Elsevier Science, http://www.elsevier.com/homepage/elecserv.htt.
27. *Japanese Journal of Nondestructive Inspection*, http://sparc5.kid.ee.cit.nihon-u.ac.jp/homepage_Eng.html.
28. *Journal of Nondestructive Evaluation*, Kluwer Academic, Norwell, MA, 2004.
29. *Journal of Micromechanics and Microengineering*, 2004.
30. British Institute of Non-Destructive Testing, http://www.bindt.org/, 1999.
31. IFANT, International Foundation for the Advancement of Nondestructive Testing, http://www.ifant.org.
32. Japan JSNDI, http://sparc5.kid.ee.cit.nihonu.ac.jp/homepage_Eng.html.
33. SPIE, http://spie.org/.
34. Institute of Electrical and Electronic Engineers, http://www.ieee.org/.
35. *IEEE-ASME, Journal of Microelecromechanical Systems*, Vol. 2000, 2004.
36. American Society of Mechanical Engineers, http://www.asme.org/.
37. Center for Nondestructive Evaluation, http://www.cnde.com.
38. Airport and Aircraft Safety Research & Development, http://www.asp.tc.faa.gov.
39. Fraunhofer IZFP, http://www.fhg.de/english/profile/institute/izfp/index.html.
40. Stasuk Testing & Inspection, http://www.nde.net.
41. AFRL electronic journals, http://www.wrs.afrl.af.mil/infores/library/ejournals.htm.
42. Link, Springer Verlag, http://link.springer-ny.com/.
43. IBM Intellectual Property Network, http://www.patents.ibm.com.
44. Lavender International NDT, in *Lavender International*, 2004.
45. *Trends in NDE Science and Technology, Proceedings of the 14th World Conferences on Nondestructive Testing*, Ashgate Publishing, Brookfield, VT, 1997.
46. Rose, J. L., and Tseng, A. A. (Eds.), *New Directions in Nondestructive Evaluation of Advanced Materials*, American Society of Mechanical Engineers, New York, 1988.
47. *Journal of Intelligent Material Systems and Structures*, http://www.techpub.com.
48. Smart Structures, http://www.adaptive-ss.com/.
49. Smart Materials and Structures, http://www.adaptive-ss.com/, 2001.
50. Smart Structures—Harvard, http://iti.acns.nwu.edu/clear/infr/imat_smart.html.
51. Tracy, N. (Ed.), *Liquid Penetrant Testing*, 3rd ed., Vol. 2 of *Nondestructive Testing Handbook*, P. Moore (Ed.), American Society for Nondestructive Testing, Columbus, OH, 1999.
52. Quinn, R. A., *Industrial Radiography—Theory and Practice*, Eastman Kodak, Rochester, NY, 1980.
53. Burger, H., *Neutron Radiography; Methods, Capabilities and Applications*, Elsevier Science, New York, 1965.
54. Bossi, R. H., Iddings, F. A., and Wheeler, G. C. (Eds.), *Radiographic Testing*, 3rd ed., Vol. 4 of *Nondestructive Testing Handbook*, P. Moore (Ed.), American Society for Nondestructive Testing, Columbus, OH, 2002.

55. Fassbender, R. H., and Hagemaier, D. J., "Low-Kilovoltage Radiography of Composites," *Mater. Eval.*, **41**(7), 381–838 (1983).

56. Birks, A. S., and Green, J., *Ultrasonic Testing*, 2nd ed., Vol. 7 of *Nondestructive Testing Handbook*, P. Intire (Ed.), American Society for Nondestructive Testing, Columbus, OH, 1991.

57. Ash, E. A., and Paige, E. G. S., *Rayleigh Wave Theory and Application*, Springer Series on Wave Phenomena, Vols. 1 and 2, Springer-Verlag, Berlin, 1985.

58. Viktorov, I. A., *Rayleigh and Lamb Waves*, Plenum, New York, 1967.

59. Krautkramer, J., and Krautkramer, H., *Ultrasonic Testing of Materials*, 3rd ed., Springer-Verlag, New York, 1983.

60. Jones, R. B., and Stone, D. E. W., "Toward an Ultrasonic-Attenuation Technique to Measure Void Content in Carbon-Fibre Composites," *Nondestructive Testing*, **9**(3), 71–79 (1976).

61. Jones, T. S., "Inspection of Composites Using the Automated Ultrasonic Scanning System (AUSS)," *Mater. Eval.*, **43**(5), 746–753 (1985).

62. Hsu, D. K., and Patton, T. C., "Development of Ultrasonic Inspection for Adhesive Bonds in Aging Aircraft," *Mater. Eval.*, **51**(12), 1390–1397 (1993).

63. Swamy, R. N., and Ali, A. M. A. H., "Assessment of In Situ Concrete Strength by Various Non-Destructive Tests," *NDT Int.*, **17**(3), 139–146 (1984).

64. Papadakis, E. P., and Chapman II, G. B., "Modification of a Commercial Ultrasonic Bond Tester for Quantitative Measurements in Sheet-Molding Compound Lap Joints," *Mater. Eval.*, **51**(4), 496–500 (1993).

65. Hagemier, D. J., "Bonded Joints and Nondestructive Testing—1," *Nondestructive Testing*, **4**(12), 401–406 (1971).

66. Hagemier, D. J., "Bonded Joints and Nondestructive Testing—2," *Nondestructive Testing*, **5**(2), 38–47 (1972).

67. Hagemier, D. J., "Nondestructive Testing of Bonded Metal-to-Metal Joints—2," *Nondestructive Testing*, **5**(6), 144–153 (1972).

68. Schmidt, J. T., and Skeie, K. (Eds.), *Magnetic Particle Testing*, 2nd ed., Vol. 6 of *Nondestructive Testing Handbook*, P. McIntire (Ed.), American Society for Nondestructive Testing, Columbus, OH, 2001.

69. Maldague, X. P. V. (Ed.), *Infrared and Thermal Testing*, 3rd ed., Vol. 3 of *Nondestructive Testing Handbook*, P. Moore (Ed.), American Society for Nondestructive Testing, Columbus, OH, 2001.

70. Stanley, R. K. (Ed.), *Special Nondestructive Testing Methods*, 2nd ed., Vol. 9 of *Nondestructive Testing Handbook*, P. O. Moore and P. McIntire (Eds.), American Society for Nondestructive Testing, Columbus, OH, 1995.

71. Zorc, T. B. (Ed.), *Nondestructive Evaluation and Quality Control*, 9th ed., Vol. 17 of *Metals Handbook*, ASM International, Metals Park, OH, 1989.

72. Udpa, S. S. (Ed.), *Electromagnetic Testing*, 3rd ed., Vol. 5 of *Nondestructive Testing Handbook*, P. Moore (Ed.), American Society for Nondestructive Testing, Columbus, OH, 2004.

73. Förster, F., "Theoretische und experimentalle Grundlagen der zerstörungfreien Werkstoffprufung mit Wirbelstromverfahren, I. Das Tastpulverfahern," *Zeits. Metall.*, **43**, 163–171 (1952).

74. Vernon, S. N., "Parametric Eddy Current Defect Depth Model and Its Application to Graphite Epoxy," *NDT Int.*, **22**(3), 139–148 (1989).

75. Wincheski, R. A., et al., "Development of Giant Magnetoresistive Inspection System for Detection of Deep Fatigue Cracks under Airframe Fastners," in *Review of Progress in Quantitative Nondestructive Evaluation*, 2002.

76. Blitz, J., *Electrical and Magnetic Method of Non-Destructive Testing*, Chapman & Hall, London, 1997.

77. Libby, H. L., *Introduction to Electromagnetic Non-destructive Test Methods*, Wiley-Interscience, New York, 1971.

CHAPTER 9

MECHANICS OF INCOMPRESSIBLE FLUIDS

Egemen Ol Ogretim
Department of Civil and Environmental Engineering
College of Engineering and Mineral Resources
West Virginia University
Morgantown, West Virginia

Wade W. Huebsch
Department of Mechanical and Aerospace Engineering
College of Engineering and Mineral Resources
West Virginia University
Morgantown, West Virginia

1 INTRODUCTION

We experience and interact with fluids, such as air, water, oil, and gas, in many aspects of our daily life. These fluids can be categorized in different ways depending on the level of analysis required. An obvious first categorization is that they can be viewed as liquid or gas; but they can be classified depending on their behavior in a particular application as well. Some common fluid classifications are viscous or inviscid, compressible or incompressible, laminar or turbulent, Newtonian or non-Newtonian, and steady or unsteady. These classifications are sometimes idealizations (approximations) and sometimes exact descriptions. Overall, they offer information on fluid properties and characteristics, the state of the flow field, and how to approach an engineering solution to the relevant flow problem. This chapter will cover the basics on many of these classifications but will primarily restrict a bulk of the discussion to the category of Newtonian incompressible fluids.

All substances are compressible to some extent and fluids are no exception. However, in many cases, fluids may be treated as incompressible without introducing unacceptable inaccuracies in either computations or measurements. In the case of liquids, the vast majority of problems may be addressed as incompressible flow problems. Even in situations where pressure changes are significant enough to cause small changes in density (water hammer), incompressible flow techniques are applied to solve problems. For gases where flow velocities are low compared to the local speed of sound (low Mach number) or where density changes in the system are small, incompressible flow theory may be used to good approximation.

1.1 Definition of a Fluid

A *fluid*, which can be a liquid or gas, is defined as a collection of molecules of a substance (or several substances) that cannot support a shearing stress of any size (magnitude) without undergoing permanent and continual angular deformation. Even though the fluid is a collection of molecules, it is generally considered to be a continuous substance without voids, referred to as a continuum (typical liquid has on the order of 10^{21} molecules/mm^3). Newtonian fluids are those whose rate of angular deformation is linear with respect to the magnitude of the deforming shear stress. Most common liquids and gases are Newtonian fluids (e.g., air, water, oil). Non-Newtonian fluids are those whose rate of angular deformation bears a nonlinear relationship to the applied shear stress. Examples of non-Newtonian fluids are blood, paints, and suspensions.

1.2 System of Units and Dimension

The scientific world is in the midst of a conversion from various systems of units to SI (Système International d'Unités) units. However, the English FSS (foot-slug-second) system is still in widespread use throughout the United States. Hence, this work will use both the English system of units as well as the SI equivalent in the text, tables, and numerical examples. Table 1 provides the units used to identify the basic dimensional quantities in each system, and Table 2 gives a conversion table for fluid properties and other commonly used quantities.

2 FLUID PROPERTIES

Properties of fluids generally encountered in engineering practice are presented in tables and figures in this section. Included are the commonly used properties of density, specific gravity, bulk modulus (compressibility), viscosity, surface tension, and vapor pressure.

2.1 Density, Specific Weight, and Specific Gravity

Density is defined as the mass per unit volume ρ. Table 3 provides values for a selection of liquids at standard atmospheric pressure (14.7 lb/in.^{2} or 101325 Pa), and Table 4 provides properties of some common gases. Density varies with temperature and pressure, but for liquids, variation with pressure is generally negligible. Also included in these tables are the specific weight γ, which is the density multiplied by the gravitational acceleration, which produces a measure of weight per unit volume (pounds per cubic feet or newtons per cubic meter).

For a gas, the density depends heavily on the pressure and temperature. Most gases closely follow the ideal gas equation of state

$$\rho = \frac{p}{RT} \tag{1}$$

where p is the absolute pressure, T is the absolute temperature, and R is the specific gas constant. This equation should be used with discretion under conditions of very high temperature or very low pressure or when the gas approaches a liquid. Because all gases at the same temperature and pressure contain the same number of molecules per unit volume (Avogadro's law), the specific gas constant can be calculated to good accuracy by dividing the universal gas constant, 8314 J/mol-K, by the molecular weight of the gas. For example, oxygen has the molecular weight of 32 kg/mol. The specific gas constant of oxygen is calculated as $R = 8314/32 = 259.8$ J/kg-K (compare with Table 4).

Another widely used fluid property is the specific gravity, SG, which is defined as the ratio of the fluid density to the density of water at a specified temperature (typically 39.2°F, or 4°C, which gives a density of water as 1.94 slugs/ft^3 or 1000 kg/m^3). So the specific gravity can be found from

$$SG = \frac{\rho}{\rho_{water}} \tag{2}$$

Table 1 Basic Dimensions and Abbreviations in English and SI Units

Abbreviations

BTU = British thermal unit	m = meter (SI) = mile (FSS)
cfs = cubic feet per second	mb = millibar = 10^{-3} bar
fps. ft/sec = feet per second	mm = millimeter = 10^{-3} meter
ft = foot	mm^2 = square millimeter
gpm = gallons per minute	mph = miles per hour
hp = horsepower	mps, m/s = meters per second
hr, h = hour	N = newton
Hz = hertz	Pa = pascal = N/m^2
in. = inch	psi = pounds per square inch
J = joule = N-m	sec, s = second
kg = kilogram = 10^3 grams	W = watt = J/s
lb = pound force	

Units

Quantity	SI Unit Name (Symbol)	FSS Unit Name (Symbol)
Basic units		
Length	Meter (m)	Foot (ft)
Mass	Kilogram (kg)	Slug (slug)
Time	Second (s)	Second (sec)
Temperature	Kelvin (K)	Rankine (°R)
	[Celsius (°C)]	[Fahrenheit (°F)]
Derived units		
Energy	Joule (J)	Foot-pound (ft-lb)
Force	Newton (N)	Pound (lb)
Frequency	Hertz (Hz)	Hertz (Hz)
Power	Watt (W)	Horsepower (hp)
Pressure	Pascal (Pa)	—

Source: From Ref. 1.

This provides a direct measure of whether the fluid is less dense or more dense than water and by what degree; for example, the specific gravity of mercury is 13.6, so it is 13.6 times denser than water.

2.2 Compressibility

The *compressibility* of fluids is confined to their behavior in the mode of compression. The definition of the bulk modulus of elasticity stems from the relative change in volume and provides a measure of the compressibility of a fluid:

$$E_v = -\frac{\Delta p}{\Delta V/V} \qquad (3)$$

where Δp is the pressure increment which causes a relative decrease in volume, $\Delta V/V$. The E_v value has the dimensions of pressure, and it depends on

the pressure and temperature of a liquid. For a gas, the E_v value depends on the thermodynamic process governing the change in volume resulting from the pressure increment. For example, if the process is isentropic, $E_v = kp$, where k is the ratio of the specific heats given in Table 4. If the process is isothermal, $E_v = p$. Values of E_v for a variety of liquids at standard temperature and pressure are given in Table 3.

As expected, the values of E_v for liquids are large, which indicates the fluid can be considered incompressible. Figure 1 is provided to demonstrate the variation of E_v with pressure for pure water. Entrained gas in a liquid can drastically affect the E_v value; Fig. 2 illustrates the effect of entrained air on the compressibility of water for relatively small amounts of air.

One of the common applications of the property of fluid compressibility is in the computation of the acoustic wave speed (speed of sound) in a fluid. The

Table 2 Conversion Factors for English FSS and SI Units

Absolute viscosity: 1 slug/ft-sec = 1 lb-sec/ft^2 = 47.88 N-s/m^2 = 47.88 kg/m-s = 478.8 P

Acceleration due to gravity: 32.174 ft/sec^2 = 9.80665 m/s^2

Area: 1 ft^2 = 0.0929 m^2

    1 in.^{2} = 645.2 mm^2

Density: 1 slug/ft^3 = 515.4 kg/m^3

Energy: 1 ft-lb = 1.356 J = 1.356 Nm = 3.77 × 10^{-7} kWhr

    1 Btu = 778.2 ft-lb = 1055 J = 2.93 × 10^{-4} kWhr

Flow rate: 1 ft^3/sec = 0.02832 m^3/s = 28.32 liters/s

    1 mgd = 1.55 cfs = 0.0438 m^3/s = 43.8 liters/s

Force: 1 lb = 4.448 N

Frequency: 1 cycle/s = 1 Hz

Kinematic viscosity: 1 ft^2/sec = 0.0929 m^2/s = 929 stokes

Length: 1 in. = 25.4 mm

    1 ft = 0.3048 m

    1 mile = 1.609 km

Mass: 1 slug = 14.59 kg

Power: 1 ft-lb/sec = 1.356 W = 1.356 J/s

    1 hp = 550 ft-lb/sec = 745.7 W

Pressure: 1 psi = 6895 N/m^2 = 6895 Pa   1 atm = 14.70 psi

    1 in. Hg = 25.4 mm Hg = 3386 N/m^2 = 29.92 in. Hg = 760 mm Hg

    1 in. H$_2$O = 249.1 N/m^2 = 101.325 kN/m^2

    1 lb/ft^2 = 47.88 N/m^2 = 47.88 Pa = 0.4788 mb 1 bar = 14.504 psi

                                               = 10^5 N/m^2 = 100 kN/m^2

Specific heat, engineering gas constant: 1 ft-lb/slug °R = 0.1672 Nm/kg °K

Specific weight: 1 lb/ft^3 = 157.1 N/m^3

Temperature: 1°C = 1 K = 1.8°F = 1.8°R

Velocity: 1 fps = 0.3048 mps = 0.3048 m/s

    1 mph = 1.609 km/h = 0.447/m/s

    1 knot = 1.152 mph = 1.689 fps = 0.5155 m/s

Volume: 1 ft^3 = 0.02832 m^3

    1 U.S. gallon = 0.1337 ft^3 = 0.003785 m^3 = 3.785 liters

Source: From Ref. 1.

equation applicable to both liquids and gases is

$$a = \sqrt{\frac{E_v}{\rho}} \qquad (4)$$

where a is the acoustic wave speed. The value of E_v is obtained from a table (e.g., Table 3) or, in the case of gases, computed using the isentropic thermodynamic process. For reference, the speed of sound is much higher in water than in air.

2.3 Viscosity

Viscosity is a fluid's resistance to angular deformation under action of a shearing stress. It is a measure of how easy or difficult a fluid "flows." For example, pancake syrup has a higher viscosity than water. Viscosity is a property that changes according to the thermodynamic conditions; thus, it strongly depends on temperature but is relatively unaffected by pressure.

In liquids, viscosity depends on the strength of cohesive forces between fluid molecules; hence an increase in temperature results in a decrease in viscosity. For gases, viscosity depends on the momentum exchange between layers of gas moving at different velocities. An increase in temperature provides an increase in molecular activity and an increase in viscosity.

All real fluids have some level of viscosity. However, for certain types of flows or specific regions of the flow field, it is sometimes reasonable to assume that the effects of viscosity are negligible. In these cases, the flow is classified as inviscid and the viscosity is set

Table 3 Approximate Physical Properties of Some Common Liquids

British Gravitational Units

Liquid	Temperature ($^\circ$F)	Density, ρ (slugs/ft^3)	Specific Weight, γ (lb/ft^3)	Dynamic Viscosity, μ (lb-sec/ft^2)	Kinematic Viscosity, ν (ft^2/ sec)	Surface Tension,[a] σ (lb/ft)	Vapor Pressure, p_v [lb/in.^{2}(abs)]	Bulk Modulus,[b] E_v (lb/in.^{2})
Carbon tetrachloride	68	3.09	99.5	2.00×10^{-5}	6.47×10^{-6}	1.84×10^{-3}	1.9	1.91×10^5
Ethyl alcohol	68	1.53	49.3	2.49×10^{-5}	1.63×10^{-5}	1.56×10^{-3}	0.85×10^{-1}	1.54×10^5
Gasoline[c]	60	1.32	42.5	6.5×10^{-6}	4.9×10^{-6}	1.5×10^{-3}	8.0	1.9×10^5
Glycerin	68	2.44	78.6	3.13×10^{-2}	1.28×10^{-2}	4.34×10^{-3}	2.0×10^{-6}	6.56×10^5
Mercury	68	26.3	847	3.28×10^{-5}	1.25×10^{-6}	3.19×10^{-2}	2.3×10^{-5}	4.14×10^6
SAE 30 oil[c]	60	1.77	57.0	8.0×10^{-3}	4.5×10^{-3}	2.5×10^{-3}	—	2.2×10^5
Seawater	60	1.99	64.0	2.51×10^{-5}	1.26×10^{-5}	5.03×10^{-3}	0.226	3.39×10^5
Water	60	1.94	62.4	2.34×10^{-5}	1.21×10^{-5}	5.03×10^{-3}	0.226	3.12×10^5

SI Units

Liquid	Temperature ($^\circ$C)	Density, ρ (kg/m^3)	Specific Weight, γ (kN/m^3)	Dynamic Viscosity, μ (N-s/m^2)	Kinematic Viscosity, ν (m^2/s)	Surface Tension,[a] σ (N/m)	Vapor Pressure, p_v [N/m^2(abs)]	Bulk Modulus,[b] E_v (N/m^2)
Carbon tetrachloride	20	1,590	15.6	9.58×10^{-4}	6.03×10^{-7}	2.69×10^{-2}	1.3×10^4	1.31×10^9
Ethyl alcohol	20	789	7.74	1.19×10^{-3}	1.51×10^{-6}	2.28×10^{-2}	$5.9 \times 10^{+3}$	1.06×10^9
Gasoline[c]	15.6	680	6.67	3.1×10^{-4}	4.6×10^{-7}	2.2×10^{-2}	5.5×10^4	1.3×10^9
Glycerin	20	1,260	12.4	1.50	1.19×10^{-3}	6.33×10^{-2}	1.4×10^{-2}	4.52×10^9
Mercury	20	13,600	133	1.57×10^{-3}	1.15×10^{-7}	0.466	0.16	2.85×10^{10}
SAE 30 oil[c]	15.6	912	8.95	0.38	4.2×10^{-4}	3.6×10^{-2}	—	1.5×10^9
Seawater	15.6	1,030	10.1	1.20×10^{-3}	1.17×10^{-6}	7.34×10^{-2}	$1.77 \times 10^{+3}$	2.34×10^9
Water	15.6	999	9.80	1.12×10^{-3}	1.12×10^{-6}	7.34×10^{-2}	$1.77 \times 10^{+3}$	2.15×10^9

[a] In contact with air.
[b] Isentropic bulk modulus calculated from speed of sound.
[c] Typical values. Properties of petroleum products vary.
Source: From Ref. 2.

Table 4   Approximate Physical Properties of Some Common Gases at Standard Atmospheric Pressure

BG Units

Gas	Temperature (°F)	Density ρ (slugs/ft³)	Specific Weight γ (lb/ft³)	Dynamic Viscosity μ (lb-sec/ft²)	Kinematic Viscosity ν (ft²/sec)	Gas Constant,[a] R (ft-lb/slug-°R)	Specific Heat Ratio,[b] k
Air (standard)	59	2.38×10^{-3}	7.65×10^{-2}	3.74×10^{-7}	1.57×10^{-4}	1.716×10^{3}	1.40
Carbon dioxide	68	3.55×10^{-3}	0.114	3.07×10^{-7}	8.65×10^{-5}	1.130×10^{3}	1.30
Helium	68	3.23×10^{-4}	1.04×10^{-2}	4.09×10^{-7}	1.27×10^{-3}	1.242×10^{4}	1.66
Hydrogen	68	1.63×10^{-4}	5.25×10^{-3}	1.85×10^{-7}	1.13×10^{-3}	2.466×10^{4}	1.41
Methane (natural gas)	68	1.29×10^{-3}	4.15×10^{-2}	2.29×10^{-7}	1.78×10^{-4}	3.099×10^{3}	1.31
Nitrogen	68	2.26×10^{-3}	7.28×10^{-2}	3.68×10^{-7}	1.63×10^{-4}	1.775×10^{3}	1.40
Oxygen	68	2.58×10^{-3}	8.31×10^{-2}	4.25×10^{-7}	1.65×10^{-4}	1.554×10^{3}	1.40

SI Units

Gas	Temperature (°C)	Density ρ (kg/m³)	Specific Weight γ (N/m³)	Dynamic Viscosity μ (N-s/m²)	Kinematic Viscosity ν (m²/s)	Gas Constant,[a] R (J/kg-K)	Specific Heat Ratio,[b] k
Air (standard)	15	1.23	12	1.79×10^{-5}	1.46×10^{-5}	2.869×10^{2}	1.40
Carbon dioxide	20	1.83	18	1.47×10^{-5}	8.03×10^{-6}	1.889×10^{2}	1.30
Helium	20	0.166	1.63	1.94×10^{-5}	1.15×10^{-4}	2.077×10^{3}	1.66
Hydrogen	20	0.0838	0.822	8.84×10^{-6}	1.05×10^{-4}	4.124×10^{3}	1.41
Methane (natural gas)	20	0.667	6.54	1.10×10^{-5}	1.65×10^{-5}	5.183×10^{2}	1.31
Nitrogen	20	1.16	0.114	1.76×10^{-5}	1.52×10^{-5}	2.968×10^{2}	1.40
Oxygen	20	1.33	13	2.04×10^{-5}	1.53×10^{-5}	2.598×10^{2}	1.40

[a]Values of the gas constant are independent of temperature.
[b]Values of the specific heat ratio depend only slightly on temperature.
Source: From Ref. 2.

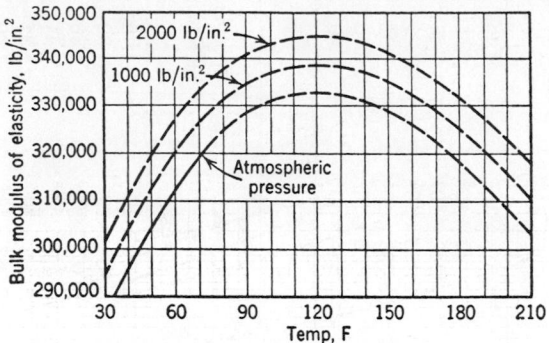

Fig. 1 Bulk modulus of elasticity for pure water as function of temperature and pressure. By permission from H. Rouse, *Fluid Mechanics for Hydraulic Engineers*, McGraw-Hill, New York, 1938. Copyright © 1938 McGraw-Hill.

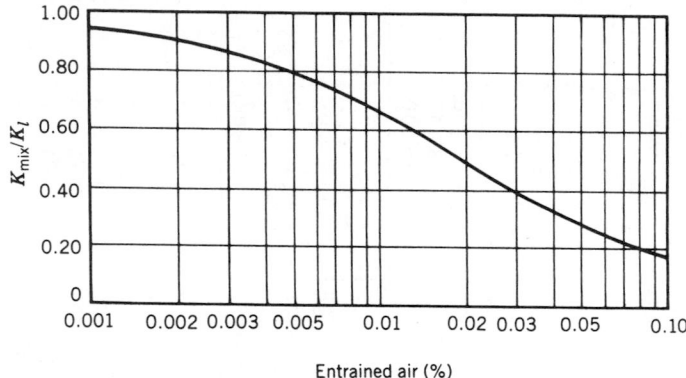

Fig. 2 Effect of entrained air on elasticity of water.

to zero. If the flow cannot be assumed inviscid, then you must account for the effects of viscosity.

A fundamental device for measuring fluid viscosity leads to the definition of viscosity. Figure 3 illustrates the flow situation occurring when a moving plate, under the action of a shearing force, slides over a fixed plate separated by a fluid. If the movement is sufficiently slow so that the flow between the plates

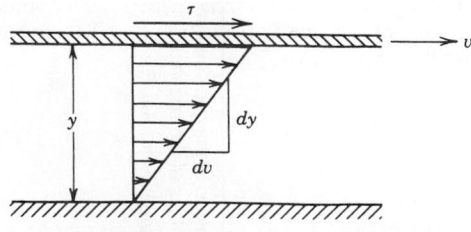

Fig. 3 Definition sketch for absolute viscosity.

is laminar, the shear stress is related to the velocity difference by the equation

$$\tau = \mu \frac{du}{dy} \tag{5}$$

where τ is the shear stress, du/dy is the velocity gradient (or rate of angular strain), and μ is the coefficient of viscosity.

In a more precise sense, μ is known as the absolute (or dynamic) viscosity and has the units pound-seconds per foot squared in the English FSS system. The values of μ for various fluids are listed in Tables 3 and 4. Because the viscosity shows up in many instances divided by the density, the ratio $\nu = \mu/\rho$ is defined as the kinematic viscosity with the English FSS units square feet per second. In the metric system, absolute viscosity is commonly given in poises (dyne-second per square centimeter) where 1 lb-sec/ft^2 = 478.8 P. The metric equivalent of kinematic viscosity is the stoke (square centimeters per second) where 1 ft^2/ sec = 929 stokes.

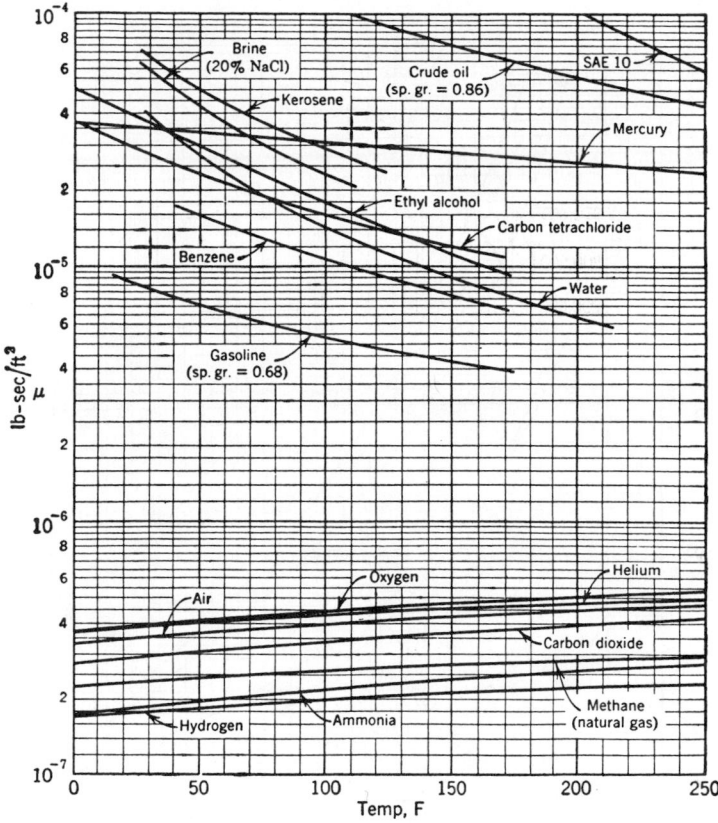

Fig. 4 Dynamic viscosity versus temperature for common gases and liquids. (Courtesy of Hunter Rouse, State University of Iowa, Iowa City.)

Figures 4 and 5 give absolute and kinematic viscosities for a wide range of liquids and gases. The English FSS values may be converted to the SI values using the conversions

$$\mu = 1 \text{ lb-sec/ft}^2 = 47.88 \text{ N-s/m}^2$$

$$\nu = 1 \text{ ft}^2/\text{sec} = 0.0929 \text{ m}^2/\text{s}$$

2.4 Surface Tension and Capillarity

Molecules of a liquid exert a mutual cohesive force that causes those molecules in the interior of a liquid to be in a balanced state of cohesive forces that is attracted equally in all directions. However, liquid molecules at a surface are attracted by the interior molecules but experience no balancing attraction from above the surface. This imbalance of forces causes the liquid surface to behave as though it were covered with an elastic membrane, hence the term *surface tension*. This phenomenon is also apparent at the interface between immiscible liquids. This is why surface tension values for a liquid must always specify

what fluid lies across the interface. As a result of its dependence on molecular cohesion, surface tension σ (in pounds per foot or newtons per meter) decreases with increasing temperature. Table 3 gives values for a variety of liquids.

Surface tension manifests itself in free liquid jets, bubbles, small waves, and capillary action in small conduits. At the interface between two fluids, surface tension can sustain a pressure discontinuity across the interface. The magnitude of the pressure difference is a function of the surface curvature with the higher pressure on the concave side of the interface.

Capillary rise in a circular tube is an important application of surface tension (Fig. 6). The equation for capillary rise h is

$$h = \frac{2\sigma \cos \theta}{\gamma r} \tag{6}$$

The value of h depends heavily on the contact angle Θ, which in turn depends strongly on the liquid and the tube material as well as the cleanliness of the tube

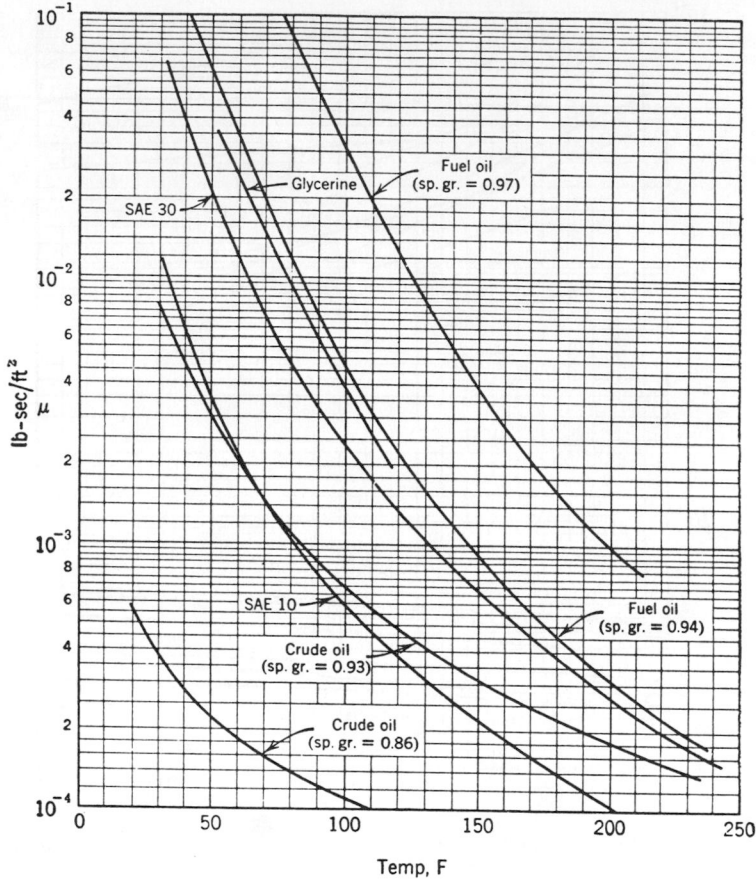

Fig. 5 Dynamic viscosity versus temperature for typical grades of oil. (Courtesy of Hunter Rouse, State University of Iowa, Iowa City.)

surface. This equation should not be used unless the capillary tube is small enough that the liquid surface is approximately spherical in shape (constant curvature).

2.5 Vapor Pressure

A liquid exhibiting a free surface is continually ejecting and absorbing molecules of the liquid across the free surface. Some of the molecules at the surface have enough energy due to local collisions that they can break free of the intermolecular cohesive forces. If more molecules leave than return, the liquid is said to be evaporating. In an equilibrium situation, the number of molecules expelled equals the number returning. The molecules of the liquid in the overlying gaseous fluid exert a force on the liquid surface, in conjunction with the other gas molecules bombarding the surface, to make up the total surface pressure. That portion of the surface pressure generated by the vapor molecules from the liquid is called the partial pressure, or *vapor pressure*, p_v, of the liquid. An alternative view is if you

started with a completely liquid-filled container. You were then able to move one end of the container outward, which increased the volume, but did not allow air to enter the container. The free space between the liquid and the movable end would become filled with a vapor from the liquid and impose a pressure on the liquid surface equal to the vapor pressure of the liquid. Table 3 provides values of the vapor pressure p_v for a variety of liquids.

The higher the liquid temperature, the more vigorous the molecular activity and the larger the fraction of the total pressure contributed by the liquid vapor. When the liquid temperature is elevated to a level where the vapor pressure is equal to the total pressure, boiling occurs (formation of vapor bubbles in liquid). Boiling is also experienced when the total pressure is reduced to the liquid vapor pressure, for example, in liquid cavitation in pumps, valves, and propellers. Therefore, boiling can occur by either raising the temperature to increase the vapor pressure to

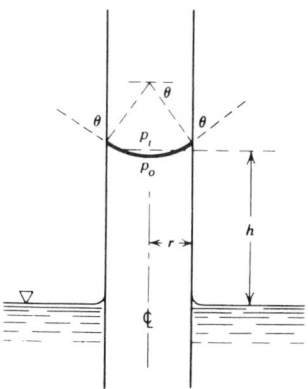

Fig. 6 Capillary rise in a small circular tube. By permission from J. K. Vennard and R. L. Street, *Elementary Fluid Mechanics*, 5th ed., Wiley, New York, 1975.

meet the local pressure or reducing the local pressure to meet the vapor pressure at the current temperature. In fluid mechanics, the latter is the most common phenomenon; flow conditions and geometry contribute to a significant reduction in local pressure. If the pressure is low enough to reach the vapor pressure of the liquid, vapor bubbles will form. In some instances, these vapor bubbles travel downstream and suddenly collapse due to higher pressure regions, which causes cavitation to occur. Cavitation can have enough intensity to cause structural damage to the system.

3 FLUID STATICS

3.1 Pressure Variation in a Fluid at Rest

By definition, a Newtonian fluid at rest has no internal shear stresses, only normal stresses. As a consequence, pressure does not vary horizontally in a homogeneous fluid, only vertically. The equation describing the vertical variation in pressure is

$$\frac{dp}{dz} = -\rho g = -\gamma \tag{7}$$

where z is measured vertically upward as shown in Fig. 7. If a fluid has variable density, this expression must be integrated to determine differences in pressure. However, for an incompressible fluid (i.e., constant density), the commonly used form of this equation is the hydrostatic pressure variation equation

$$\Delta p = \gamma \, \Delta h \tag{8}$$

where Δh is the difference in elevation between two points and Δp is the corresponding pressure difference. Note that the pressure is increased linearly with increasing depth; the fluid at a lower level needs to support all the fluid above it so the pressure is higher.

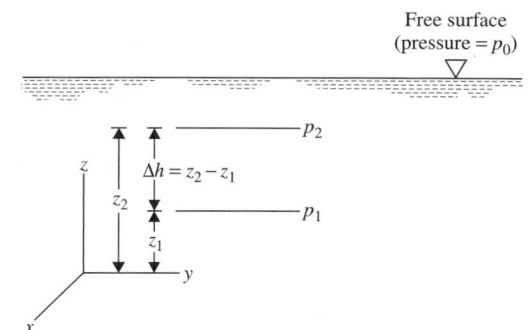

Fig. 7 Notation for pressure variation in a fluid at rest. From Ref. 2.

3.2 Basic Pressure-Measuring Devices

Pressure is generally measured from one of two datum locations—absolute zero or the surrounding environment (gage). Pressures measured above absolute zero, for example, barometric pressure, are known as absolute pressure. Those measured above the local atmospheric pressure are known as gage pressures. Absolute pressure always equals gage pressure plus atmospheric pressure:

$$p_{\text{absolute}} = p_{\text{gauge}} + p_{\text{atmospheric}} \tag{9}$$

It is possible to have negative gage pressure, since it is measured with respect to a datum. But, it is impossible to have negative absolute pressure, since the minimum you can achieve is a perfect vacuum, which is zero absolute pressure.

A fundamental pressure-measuring device, which is based on the hydrostatic pressure variation equation, is the manometer (Fig. 8). Because one end of this manometer is open to the atmosphere, it is known as an open-end manometer. By making use of the fact that pressure does not vary in the horizontal direction for a fluid at rest and making use of the hydrostatic pressure equation at various points, the gage pressure at A is calculated as

$$p_A = \gamma_2 y_2 - \gamma_1 y_1 \tag{10}$$

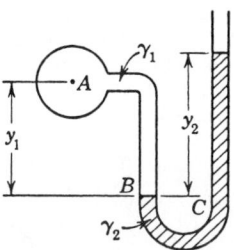

Fig. 8 Simple manometer.

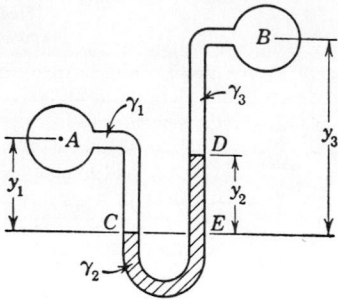

Fig. 9   Differential manometer.

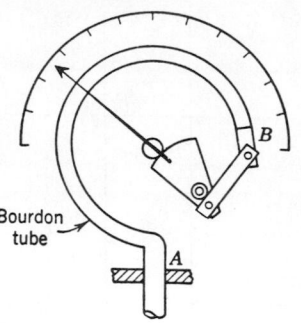

Fig. 10   Essential features of Bourdon gage.

Figure 9 illustrates a differential manometer. The difference in pressure between A and B is calculated as

$$p_B - p_A = \gamma_1 y_1 - \gamma_2 y_2 - \gamma_3(y_3 - y_2) \qquad (11)$$

Another commonly used mechanical device for measuring both gage pressure and absolute pressure is the Bourdon-type gage (Fig. 10).

3.3   Fluid Forces on Plane Surfaces

The magnitude of the total fluid force exerted on a plane (flat) surface can be calculated with the equation (refer to Fig. 11)

$$F_R = p_c A \qquad (12)$$

where p_c is the pressure at the centroid of the surface and A is the total area of the surface. Note that pressure always acts normal to the surface.

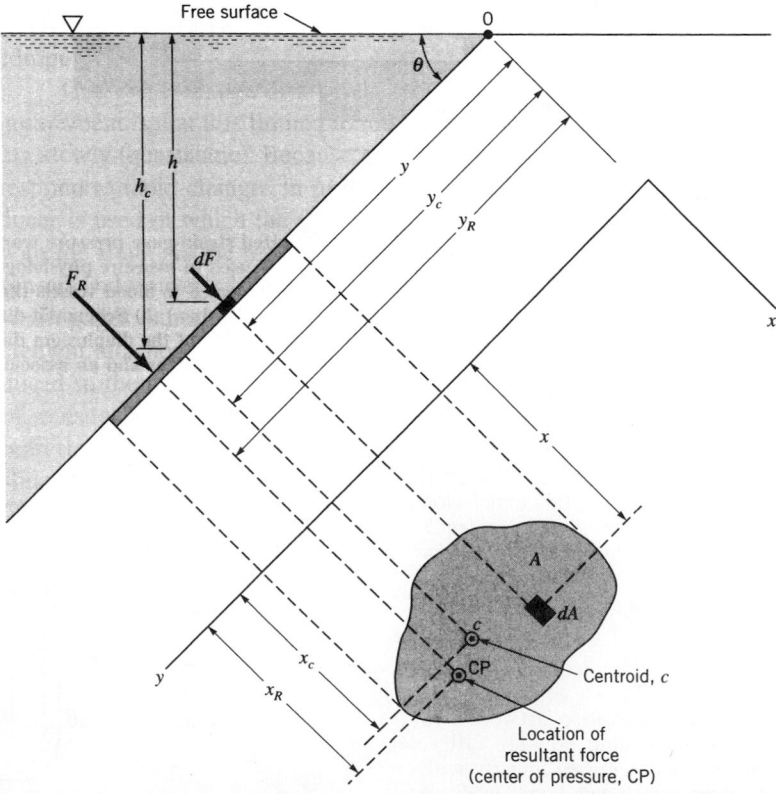

Fig. 11   Notation for pressure force calculation on submerged plane of arbitrary shape. From Ref. 2.

If the fluid is a liquid with a free surface open to the atmosphere,

$$p_c = \gamma h_c \qquad (13)$$

where h_c is the distance the centroid lies below the free surface. Locations of centroids of variously shaped areas are shown in Table 5.

The location of the resultant of the pressure forces acting on a plane area is known as the center of pressure. For a plane area lying below a liquid surface, the center-of-pressure location is computed with the equation

$$y_R - y_c = \frac{I_c}{y_c A} \qquad (14)$$

where y_R is measured in the plane of the area (slant distance) from the free surface to the center of pressure, y_c is the distance (in the same plane) from the free surface to the centroid of the area, and I_c is the second moment of the area about a horizontal axis through the centroid and in the plane of the area. Values of I_c for common geometric shapes are given in Table 5.

Two special cases of this application are the horizontal-plane and the vertical-plane cases. For a horizontal plane ($\theta = 0°$ of inclination), the centroid of the surface and the center of pressure coincide, since the pressure distribution over the entire surface area is uniform. For the vertical case ($\theta = 90°$ of inclination), the center of pressure will always be below the centroid of the surface since the pressure acting on the surface increases with depth. Note that for deeply submerged plane areas, the center of pressure approaches the centroid of the area regardless of the angle of inclination.

3.4 Fluid Forces on Curved Surfaces

Pressure forces exerted on curved surfaces vary in direction and must usually be divided into horizontal and vertical components to facilitate calculation. The resultants of the horizontal and vertical components of the pressure forces are determined with assistance from the equations developed for plane surfaces. Figure 12 illustrates a common situation. The fluid mass ABC is in equilibrium under the action of pressure forces; hence the vertical component of the resultant of the pressure forces is equal to the weight of the fluid mass plus the force $\mathbf{F}_{AC}$ exerted on the upper surface of the fluid mass by the overlying fluid. The horizontal component of the resultant of the pressure forces is equal to the force on a vertical projection of the curved surface. The line of action and point of application (center of pressure) for the resultant force is determined by taking moments of the contributing forces about an axis of convenience. The force $\mathbf{F}_{BC}$ is located using the formula for plane surfaces. The weight W_{ABC} is concentrated at the centroid of the area ABC (see Table 5).

3.5 Buoyancy and Stability

An object floating or submerged in a fluid at rest experiences pressure forces from the surrounding fluid. Consider a square block of material that is submerged in a liquid; due to hydrostatic pressure variation with depth, the pressure on the bottom surface will be greater than on the top surface (pressure on the two sides is equal and opposite). The resultant of these pressure forces is known as the buoyant force, which acts in the upward direction, and is equal in magnitude to the weight of the displaced fluid. The line of action of the buoyant force is vertical through the centroid of the volume of displaced fluid (center of buoyancy). The magnitude of the buoyant force can be found from the following equation:

$$F_B = \gamma \mathbb{V} \qquad (15)$$

where $\mathbb{V}$ is the volume of the displaced fluid. If the object is totally submerged in the fluid, then the volume of the displaced fluid equals the volume of the object; otherwise, the displaced volume equals only the volume of the submerged portion of the object.

The overturning stability (Fig. 13) of a floating or submerged body is affected by the relative position of the center of gravity of the body and the center of buoyancy. When the center of gravity lies below the center of buoyancy, the body is always *stable*. If the center of buoyancy is below the center of gravity, the body may or may not be stable in terms of overturning. If the geometric shape of the body is such that a rocking motion results in a shift of the buoyancy center that in turn causes a righting action, the body is stable. For example, a 2×4 board floating on its side would be stable. However, the same board floating on its edge would be unstable.

3.6 Accelerated Fluid Masses without Relative Motion

There are situations when a fluid body is accelerated with no relative motion among fluid particles (zero-shear stress). This is true only for constant accelerations. As in a fluid at rest, the pressure will still vary linearly with elevation, but the pressure gradient will generally not equal the specific weight. Further, there will be a variation in pressure in the horizontal plane in this situation.

For horizontal acceleration of a container of liquid (Fig. 14), the free surface will tilt just enough to provide the unbalanced hydrostatic force on the opposite sides of the container required to impart a constant acceleration to the fluid mass. As shown in the figure, this will produce lines of constant pressure that are parallel to the titled free surface. Under these conditions the slope of the free surface is given as

$$\text{Slope} = \frac{a_x}{g} \qquad (16)$$

where a_x is the constant horizontal acceleration.

Table 5 Properties of Areas and Volumes

	Sketch	Area or Volume	Location of Centroid	I or I_c
Rectangle		bh	$y_c = \dfrac{h}{2}$	$I_c = \dfrac{bh^2}{12}$
Triangle		$\dfrac{bh}{2}$	$y_c = \dfrac{h}{3}$	$I_c = \dfrac{bh^2}{36}$
Circle		$\dfrac{\pi d^2}{4}$	$y_c = \dfrac{d}{2}$	$I_c = \dfrac{\pi d^4}{64}$
Semicircle[a]		$\dfrac{\pi d^2}{8}$	$y_c = \dfrac{4r}{3\pi}$	$I = \dfrac{\pi d^4}{128}$
Ellipse		$\dfrac{\pi bh}{4}$	$y_c = \dfrac{h}{2}$	$I_c = \dfrac{\pi bh^2}{64}$
Semiellipse		$\dfrac{\pi bh}{4}$	$y_c = \dfrac{4h}{3\pi}$	$I = \dfrac{\pi bh^2}{16}$
Parabola		$\dfrac{2}{3}bh$	$y_c = \dfrac{3h}{5}$ $x_c = \dfrac{3b}{8}$	$I = \dfrac{2bh^2}{7}$
Cylinder		$\dfrac{\pi d^2 h}{4}$	$y_c = \dfrac{h}{2}$	
Cone		$\dfrac{1}{3}\left(\dfrac{\pi d^2 h}{4}\right)$	$y_c = \dfrac{h}{4}$	
Paraboloid of revolution		$\dfrac{1}{2}\left(\dfrac{\pi d^2 h}{4}\right)$	$y_c = \dfrac{h}{3}$	
Sphere		$\dfrac{\pi d^2}{6}$	$y_c = \dfrac{d}{2}$	
Hemisphere		$\dfrac{\pi d^2}{12}$	$y_c = \dfrac{3r}{8}$	

[a]For the quarter-circle, the respective values are $\pi d^2/16$, $4r/3\pi$, and $\pi d^4/256$.

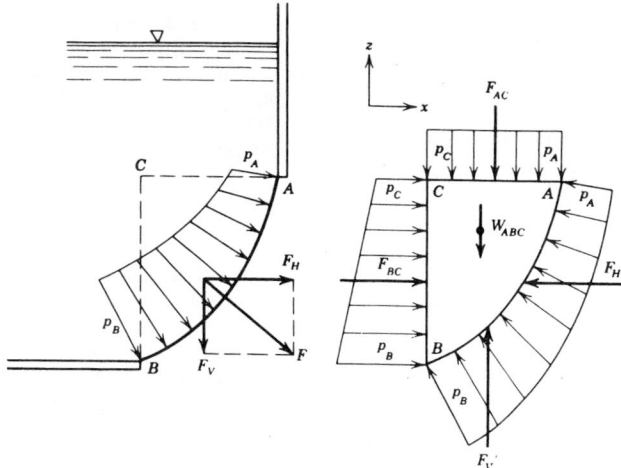

Fig. 12 Forces acting on curved surface. By permission from J. K. Vennard and R. L. Street, *Elementary Fluid Mechanics*, 5th ed., Wiley, New York, 1975.

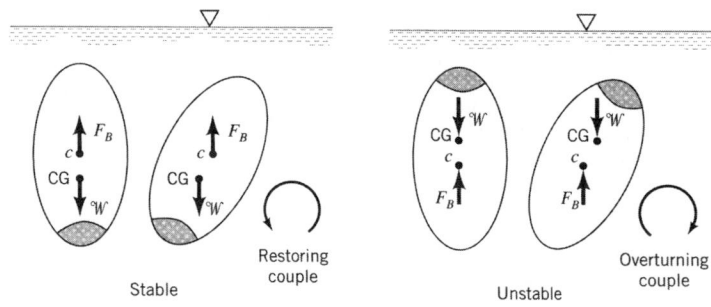

Fig. 13 Stable and unstable settings for totally submerged body. From Ref. 2.

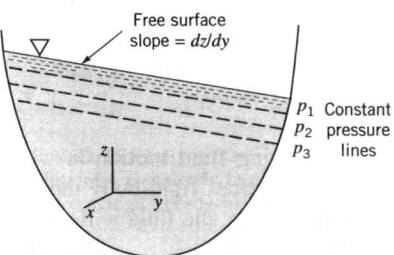

Fig. 14 Horizontal acceleration of liquid with free surface. From Ref. 2.

For vertical acceleration, the free surface remains horizontal, but the pressure variation with depth is given by the equation

$$\frac{dp}{dz} = -\gamma \left(1 + \frac{a_z}{g}\right) \qquad (17)$$

Note that if $a_z = -g$ (free fall), the pressure gradient throughout the fluid is zero.

Another situation occurs for a container of liquid undergoing a constant rotational angular velocity (Fig. 15). When the relative motion between fluid particles ceases, the free surface forms the shape of a parabola described by the equation

$$z = \frac{\omega^2}{2g} r^2 \qquad (18)$$

where z is the elevation of the free surface at radius r above that at the center of rotation and ω is the angular velocity of rotation of the container. Pressure in the fluid varies parabolically with radius,

$$p = p_o + \frac{\gamma \omega^2}{2g} r^2 \qquad (19)$$

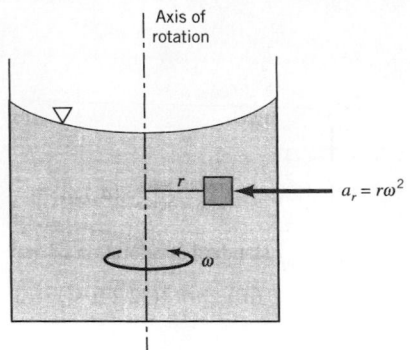

Fig. 15 Rotation of tank filled with liquid. From Ref. 2.

where p is the pressure at radius r in a horizontal plane and p_o is the pressure at the center of rotation in the same horizontal plane. Because there is no vertical acceleration, pressure variation in the vertical direction is the some as if the fluid were at rest.

4 IDEAL (INVISCID) FLUID DYNAMICS

An *ideal fluid* is a hypothetical substance exhibiting all the characteristics of a real fluid except that it is a continuum (no molecular spaces), it has no viscosity (frictionless) or inviscid, and it can accept pressures down to negative infinity. These characteristics greatly simplify the mathematical treatment of fluid flow and, in many cases, give acceptable engineering results for real fluid situations where viscous effects are not significant.

Although not restricted to ideal fluid flow, some definitions are common to all flow situations. Steady flow occurs when all conditions in the flow at any point do not change with time. In turbulent flows, this definition is extended to include temporal mean values of all conditions. Uniform flow is defined as the condition where the velocity is everywhere the same at a given time. For one-dimensional flow, this definition refers to the average velocity at any cross section of the flow. In nonuniform flow, the velocity varies with position in the flow domain at any given time.

Qualitative aspects of a fluid flow situation can be illustrated effectively through the use of streamlines. For steady flow, *streamlines* are tracks of fluid particles that are everywhere tangent to fluid velocity vectors. As a consequence, flow cannot occur across streamlines; so a collection of streamlines defining a closed surface (a stream tube) would act like a conduit within the flow. Further, regions of flow where streamlines are close together represent locally high velocities (and low pressures). Regions where the streamlines are straight and parallel identify situations where there is no acceleration normal to the flow; therefore, the pressure variation normal to the flow is hydrostatic.

4.1 One-Dimensional Flow

Most real flows are three dimensional in nature. However, in order to simplify the analysis and obtain solutions to engineering problems, some assumptions can be made. Some types of flows have a dominant flow direction, with the remaining two spatial directions only contributing secondary effects. Therefore, making the assumption that the flow is one dimensional can simplify the analysis and provide reasonable results under certain conditions. The introduction of the various conservation laws (mass, momentum, and energy) for fluid flow is also simplified by using a one-dimensional analysis, which is given below.

Conservation of Mass (Continuity) Conservation of mass is an analog of the conservation of energy. It states that in a fluid flow accumulation of mass within a control volume (a volume whose boundaries are specified) is equal to the difference of the net inflow and the net outflow of mass. For steady state, the accumulation, by definition, is zero. Consequently, conservation of mass simplifies to a balance of net inflow and net outflow of mass.

For one-dimensional flow, the control volume is generated by the cross-sectional area of the conduit at two specified locations, where the streamlines are essentially straight and parallel to each other and normal to the cross-sectional area, and the walls of the conduit between these two locations (see Fig. 16).

For steady flow, the mass flow rate $\dot{m}$ (slugs per second or kilograms per second) is a constant along the conduit and is written

$$\dot{m} = \rho\, AV = \rho Q = \text{const} \qquad (20)$$

where Q is the volume flow rate (cubic feet per second), A is the cross-sectional area of the conduit, and V is the average velocity, Q/A. In cases where the flow enters the control volume at section 1 and leaves at section 2 (see Fig. 16), conservation of mass states that the mass inflow rate equals mass outflow rate, or

$$\rho_1 A_1 V_1 = \rho_1 A_2 V_2 \qquad (21)$$

In the case of incompressible flow, the density will not vary and therefore will drop from the above expression, yielding constant volume flow rate:

$$Q_1 = Q_2 \qquad (22)$$

Euler's Equations Application of Newton's second law to a fluid particle gives Euler's equations of motion, assuming inviscid flow. Along a streamline (see Fig. 16)

$$\frac{dp}{\gamma} + \frac{v\,dv}{g} + dz = 0 \qquad (23)$$

where v is the local velocity. Normal to the streamline,

$$\frac{dp}{\gamma} - \frac{v^2}{gr}\,dr + dz = 0 \qquad (24)$$

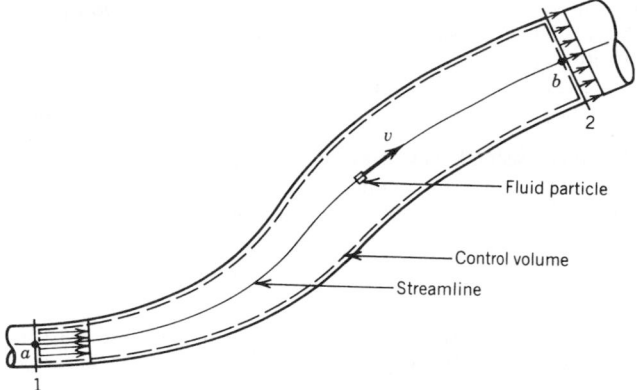

Fig. 16 One-dimensional inviscid flow.

where r is the local radius of curvature of the streamline along which the fluid particle is moving.

Bernoulli Equation Integration of the Euler equation along a streamline for constant density yields the Bernoulli equation

$$\frac{p}{\gamma} + \frac{v^2}{2g} + z = \text{const} \qquad (25)$$

If the integration is performed along a streamline between points a and b (Fig. 16), the Bernoulli equation takes the form

$$z_a + \frac{P_a}{\gamma} + \frac{v_a^2}{2g} = z_b + \frac{P_b}{\gamma} + \frac{v_b^2}{2g} \qquad (26)$$

Multiplying through by the specific weight, the Bernoulli equation can also be written as

$$p_a + \tfrac{1}{2}\rho v_a^2 + \gamma z_a = p_b + \tfrac{1}{2}\rho v_b^2 + \gamma z_b \qquad (27)$$

This states that the combination of these three terms remains constant everywhere along the streamline.

The Bernoulli equation can be "expanded" to apply to a full-sized one-dimensional flow (as opposed to a single streamline) if the sections at 1 and 2 (Fig. 16) are regions where the streamlines are essentially straight and parallel and the velocity profile uniform (irrotational flow),

$$p_1 + \tfrac{1}{2}\rho V_1^2 + \gamma z_1 = p_2 + \tfrac{1}{2}\rho V_2^2 + \gamma z_2 \qquad (28)$$

where V is the average velocity. This is true because at a section where the streamlines are straight and parallel the velocity is constant over the cross section and the pressure variation is hydrostatic ($z + p/\gamma = \text{const}$). Hence the sum of the three terms on each side of the equation is a constant regardless of the streamline

selected. Because 1 and 2 are sections and not points, the quantities z and p/γ (or γz and P) are usually evaluated at some arbitrary point in the cross section, for example, at the centerline for a cylindrical pipe.

A further conclusion extends from the uniform-flow section concept. If the sum of the three terms is the same for all streamlines at the special cross section, the sum is everywhere constant throughout the flow:

$$p_1 + \tfrac{1}{2}\rho V_1^2 + \gamma z_1$$

$$= \text{const throughout flow field for irrotational flow} \qquad (29)$$

Work–Energy Equation An equation similar to the Bernoulli equation can be derived from work–energy principles. However, the work–energy equation has broader application because it can include thermal energy and heat transfer. Utilizing the principle that the work done on a fluid system is equal to the change in energy of the system and applying this principle to the fluid in the control volume of Fig. 16,

$$z_1 + \frac{p_1}{\gamma} + \frac{V_1^2}{2g} = z_2 + \frac{p_2}{\gamma} + \frac{V_2^2}{2g} \qquad (30)$$

This equation is identical to the Bernoulli equation; however, we have not considered other energies beyond mechanical energies. Further, the terms in the work–energy equation have the following meanings:

z = potential energy per pound of fluid flowing
$V^2/(2g)$ = kinetic energy per pound of fluid flowing
p/γ = "flow work" done on fluid in control volume per pound of fluid flowing (also known as "pressure energy")

Based on the fact that we can obtain the Bernoulli equation from both Newton's second law and the conservation of energy, we can conclude that it is a very powerful equation, which is also witnessed by its numerous applications in nature.

Momentum Equation

Linear Momentum. Newton's second law was applied to a fluid particle moving along a streamline to produce the Euler equations. If the second law is applied to a finite fluid mass within a control volume (e.g., Fig. 16), another set of useful equations are derived. All internal forces cancel so that only external forces due to pressure, shear (for real fluids), and gravity need be considered. The resulting linear momentum equations are vector equations and are written as

$$\sum \mathbf{F}_{ext} = \sum (Q\rho\mathbf{V})_{out} - \sum (Q\rho\mathbf{V})_{in} \qquad (31)$$

where $\mathbf{F}_{ext}$ are the external forces acting on the fluid in the control volume and summations are used if there are multiple inlets and outlets. Again, the control volume is generally selected in such a way that the sections where the momentum enters and leaves the control volume are regions where the streamlines are straight and parallel.

In applying the linear momentum equations to a problem, it is customary to use the vector component equations with the orthogonal axes chosen in a convenient orientation. For example, the scalar equation in the x direction would be

$$\sum (F_x)_{ext} = \sum (Q\rho V_x)_{out} - \sum (Q\rho V_x)_{in} \qquad (32)$$

It should be noted that this equation can be used to determine the resultant force but not the distribution of that force. Also note that in the calculation of Q the velocity that is normal to the cross section is used [also see explanation for Eq. (20)], whereas the V_x term only refers to the component of that velocity in the x direction. Therefore, when dealing with the y and z component of this equation, the Q term is always calculated with the velocity that is normal to the cross section, and the V_y or V_z is calculated by considering the respective component of velocity in those directions. When considering the component magnitudes, it is also important to pay attention to the sign of the magnitudes as $(+)$ or $(-)$.

Moment of Momentum. In problems involving fluid flow through a conduit, the linear momentum equation provides the necessary force information. However, for some problems the moment of a force, or the torque, is important; for example, fluid flow through a rotary water sprinkler. In these cases, a moment-of-momentum equation is needed that relates torque to angular momentum.

For the derivation of this equation, a similar approach to that which resulted in the linear momentum equation is used; but this time, the moments of the vectors are used instead. The resultant moment-of-momentum equation is

$$\sum (\mathbf{r} \times \mathbf{F})_{ext} = \sum [Q\rho(\mathbf{r} \times \mathbf{V})]_{out}$$
$$- \sum [Q\rho(\mathbf{r} \times \mathbf{V})]_{in} \qquad (33)$$

where the terms in parentheses are vector cross products.

A more directly usable form of the equation is that applicable in two dimensions,

$$\sum (M_o)_{ext} = \sum (Q\rho r V_t)_{out} - \sum (Q\rho r V_t)_{in} \quad (34)$$

where M_o are moments of the external forces about an axis through o and V_t is the value of the velocity normal to a vector $\mathbf{r}$ extending from the moment center o to the location of the V_t. This equation is vastly used in the design of turbomachinery such as pumps and turbines.

4.2 Two- and Three-Dimensional Flow*

The equations describing two- and three-dimensional ideal fluid flow are derived using a small cubic fluid element or control volume of sides dx, dy, and dz.

Conservation of Mass (Continuity)   The continuity equation for an ideal fluid states that under steady-state conditions the net flow rate into any small volume must be zero. In equation form

$$\frac{\partial u}{\partial x} + \frac{\partial v}{\partial y} + \frac{\partial w}{\partial z} = 0 \qquad (35)$$

or in vector notation

$$\nabla \cdot \mathbf{V} = 0 \qquad (36)$$

that is, the divergence of the velocity vector $\mathbf{V}$ is everywhere zero (see Chapter 2 for vector analysis).

Euler's Equation of Motion   Consideration of body and pressure forces on a fluid element and neglecting the viscous contributions lead to the Euler equations:

$$X - \frac{1}{\rho}\frac{\partial p}{\partial x} = u\frac{\partial u}{\partial x} + v\frac{\partial u}{\partial y} + w\frac{\partial u}{\partial z} + \frac{\partial u}{\partial t} \qquad (37)$$

$$Y - \frac{1}{\rho}\frac{\partial p}{\partial y} = u\frac{\partial v}{\partial x} + v\frac{\partial v}{\partial y} + w\frac{\partial v}{\partial z} + \frac{\partial v}{\partial t} \qquad (38)$$

$$Z - \frac{1}{\rho}\frac{\partial p}{\partial z} = u\frac{\partial w}{\partial x} + v\frac{\partial w}{\partial y} + w\frac{\partial w}{\partial z} + \frac{\partial w}{\partial t} \qquad (39)$$

---

*Illustrations and material extracted all or in part from the previous edition by Victor L. Streeter.

where X, Y, Z are the components of the body (extraneous) forces per unit mass in the xyz directions, respectively; ρ is the mass density; p is the pressure at a point (independent of direction); and u, v, w are velocity components in the xyz directions at any point x, y, z. The terms on the right-hand side of the equations are acceleration components, the first three of which are known as the *convective* acceleration (account for fluid convecting in space) and the fourth as the *local* acceleration (accounts for local unsteadiness).

Boundary Conditions A kinematic boundary condition must be satisfied at every solid boundary. For real fluids (i.e., with viscosity), the widely known "no-slip" condition states that the fluid particles adjacent to a solid boundary move with the same velocity as the boundary. Therefore, in the case of wind over a stationary flat plate, for example, the air particles adjacent to the plate will not move, while their fellow particles that are further into the air will move with the wind. Therefore, both u and v components of the velocity at the solid boundary are zero according to the no-slip condition for real fluids. However, the Euler equations are derived for inviscid cases, and so, for Euler solutions to flow problems, there will be a slip velocity tangent to the solid boundary. Therefore, the u component of the velocity, in the case of wind over a flat plate, will be equal to the free-stream velocity, whereas the normal component of the velocity, v, of the air particles at the solid boundary will be zero (since there is no suction or blowing from the surface and it is not a porous surface).

Irrotational Flow-Velocity Potential Rotation of a fluid element may be represented by a vector that has a length proportional to the magnitude of the rotation (radians per second) and in a direction parallel to the instantaneous axis of rotation. The right-handed rule is adopted; that is, the positive direction of the vector is the direction a right-handed screw would progress when rotating in the same sense as the element. In vector notation the *curl* of the velocity vector is twice the rotation vector, which is called vorticity ($\mathbf{\Omega}$):

$$\nabla \times \mathbf{V} = 2\boldsymbol{\omega} = \mathbf{\Omega} \qquad (40)$$

Scalar components of the rotation vector $\omega_x, \omega_y, \omega_z$ in the directions of the xyz axes may be used in place of the vector itself. Defining a rotation component of an element about an axis as the average angular velocity of two infinitesimal line segments through the point, mutually perpendicular to themselves and the axis,

$$\omega_x = \frac{1}{2}\left(\frac{\partial w}{\partial y} - \frac{\partial v}{\partial z}\right) \qquad (41)$$

$$\omega_y = \frac{1}{2}\left(\frac{\partial u}{\partial z} - \frac{\partial w}{\partial x}\right) \qquad (42)$$

$$\omega_z = \frac{1}{2}\left(\frac{\partial v}{\partial x} - \frac{\partial u}{\partial y}\right) \qquad (43)$$

For an irrotational flow, each of these rotational components must be zero, or

$$\nabla \times \mathbf{V} = 0 \qquad (44)$$

Then,

$$\frac{\partial w}{\partial y} = \frac{\partial v}{\partial z} \qquad \frac{\partial u}{\partial z} = \frac{\partial w}{\partial x} \qquad \frac{\partial v}{\partial x} = \frac{\partial u}{\partial y} \qquad (45)$$

A visual concept of irrotational flow may be obtained by considering as a free body a small element of fluid in the form of a sphere. As the fluid is frictionless, no tangential stresses or forces may be applied to its surface. The pressure forces act normal to its surface and hence through its center. Extraneous, or body, forces act through its mass center, which is also its geometric center for constant density. Hence, it is evident that no torque may be applied about any diameter of the sphere. The angular acceleration of the sphere must always be zero. If the sphere is initially at rest, it cannot be set in rotation by any means whatsoever; if it is initially in rotation, there is no means of changing its rotation. As this applies to every point in the fluid, one may visualize the fluid elements as being pushed around by boundary movements but not being rotated if initially at rest. Rotation or lack of rotation of the fluid particles is a property of the fluid itself and not its position in space.

The velocity potential Φ is a scalar function of space such that its rate of change with respect to any direction is the velocity component in that direction. In vector notation

$$\mathbf{V} = \nabla\Phi \qquad (46)$$

or in terms of Cartesian coordinates

$$u = \frac{\partial\Phi}{\partial x} \qquad v = \frac{\partial\Phi}{\partial y} \qquad w = \frac{\partial\Phi}{\partial z} \qquad (47)$$

The assumption of irrotational flow is equivalent to the assumption of a velocity potential.

Laplace Equation The Laplace equation results when the continuity equation is written in terms of the velocity potential. By using Eqs. (36) and (46), we get

$$\nabla \cdot \mathbf{V} = \nabla \cdot (\nabla\Phi) = \nabla^2\Phi = 0 \quad (48)$$

$$\frac{\partial^2\Phi}{\partial x^2} + \frac{\partial^2\Phi}{\partial y^2} + \frac{\partial^2\Phi}{\partial z^2} = 0 \qquad (49)$$

For plane polar coordinates (r, θ)

$$\frac{\partial \Phi}{\partial r} + r\frac{\partial^2 \Phi}{\partial r^2} + \frac{1}{r}\frac{\partial^2 \Phi}{\partial \theta^2} = 0 \qquad (50)$$

For spherical polar coordinates (r, θ, γ), where r is the distance from the origin, θ is the polar angle and γ is the meridian angle,

$$\frac{\partial}{\partial r}\left(r^2 \frac{\partial \Phi}{\partial r}\right) + \frac{1}{\sin \theta}\frac{\partial}{\partial \theta}\left(\sin \theta \frac{\partial \Phi}{\partial \theta}\right)$$

$$+ \frac{1}{\sin^2 \theta}\frac{\partial^2 \Phi}{\partial y^2} = 0 \qquad (51)$$

Once the velocity field is found by use of the potential function, the pressure distribution may be found by the Bernoulli equation.

Two-Dimensional Flow   In two-dimensional flow all lines of motion are parallel to a fixed plane, say the xy plane, and the flow patterns are identical in all planes to this plane. Application of this assumption to the Laplace equation in Cartesian coordinates gives

$$\nabla^2 \Phi = \frac{\partial^2 \Phi}{\partial x^2} + \frac{\partial^2 \Phi}{\partial y^2} = 0 \qquad (52)$$

In addition to streamlines as a means of visualizing the flow, equipotential lines are also useful in two dimensions. Equipotential lines are defined as the lines connecting the points in the flow domain where the value of the potential function Φ remains constant. Both the streamlines and the equipotential lines are utilized in the following sections of this chapter.

Stream Function.   A streamline is a continuous line drawn through the fluid in such a way that at every point it is tangent to the velocity vector; therefore, there is no flow across a streamline. The stream function ψ is a scalar function of space whose value remains constant along a streamline. The volume rate of flow between two streamlines is given by the difference in values of the stream function along these streamlines, as shown in Fig. 17. So,

$$\psi_2 - \psi_1 = q \qquad (53)$$

This notion of the stream function suggests that for the no-slip condition in real fluids the value of the stream function is zero at the stationary solid boundary; as for Euler solutions where viscous effects are neglected, the stream function will have a nonzero value proportional to the slip velocity tangent to the solid boundary. However, the important point is not the actual value used for the stream function at the surface, but rather the difference in stream function values between two streamlines, indicating the flow rate at that location.

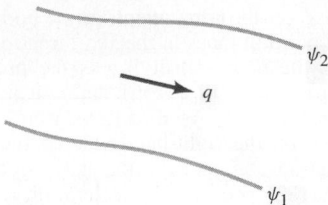

Fig. 17   Flow between two streamlines. From Ref. 2.

The velocity components u, v in the x and y directions, respectively, can be expressed in terms of stream function ψ as

$$u = \frac{\partial \psi}{\partial y} \qquad v = -\frac{\partial \psi}{\partial x} \qquad (54)$$

From the definition of the stream function, it is subject to addition of an arbitrary constant. Substituting these values of u and v into the continuity equation for two-dimensional flow,

$$\frac{\partial u}{\partial x} + \frac{\partial v}{\partial y} = \frac{\partial}{\partial x}\left(\frac{\partial \psi}{\partial y}\right) + \frac{\partial v}{\partial y}\left(-\frac{\partial \psi}{\partial x}\right) = 0 \quad (55)$$

which exactly satisfies the conservation of mass.

Application of Complex Variables to Irrotational Flow.   The idea of combining the stream function and the velocity potential function into a single variable can be realized through the use of complex variables. This method of expression provides simplicity in the formulations.

Let $z = x + iy$ be a complex variable where x and y are real and $i = \sqrt{-1}$. Any function of z

$$w = f(z) = f(x + iy) = \phi + i\psi \qquad (56)$$

that is defined throughout a region and that has a derivative throughout the region gives rise to two possible irrotational flow cases, since it may be shown that

$$\nabla^2 w = \nabla^2 \phi + i\,\nabla^2 \psi = 0 \qquad (57)$$

and hence $\nabla^2 \phi = 0$ and $\nabla^2 \psi = 0$. Here, w is termed the complex potential, ϕ is the real part of w, and ψ is the pure imaginary part of w. The complex velocity

$$\frac{dw}{dz} = -u + iv \qquad (58)$$

provides a simple method of finding velocity components from the complex potential. Stagnation points occur at those points in the flow where the velocity is zero, that is,

$$\frac{dw}{dz} = 0 \qquad (59)$$

Conformal Mapping. Each of the two complex variables w and z introduced in the preceding paragraphs may be represented by plotting on a graph. The w plane is a graph having ϕ as abscissa and ψ as ordinate, showing values of $\phi = \pm nc$, $\psi = \pm nc$, where n is every integer and c is a constant. Figure 18 is a flow net of the simplest form showing equipotential and streamlines as series of parallel straight lines. The z plane is a plot of the equipotential and streamlines of the w plane, with x as abscissa and y as ordinate. The particular flow net in the z plane depends entirely on the functional relation between w and z, that is, $w = f(z)$. Since

$$w = f(z) = \phi + i\psi \qquad z = x + iy \qquad (60)$$

where ϕ, ψ, x, y are real,

$$\phi = \phi(x, y) \qquad \psi = \psi(x, y) \qquad (61)$$

The values of $\phi = \text{const}$ and $\psi = \text{const}$ can be plotted on the z plane from the functional relations of ϕ and ψ with x, y.

Examples of Two-Dimensional Flow. Since the Laplace equation is linear in ϕ, the sum of two solutions is also a solution, and the product of a solution by a constant is a solution. Several of the important solutions are given.

Rectilinear Flow. Uniform straight-line flow is given by

$$w = Uz - iVz \quad \phi = Ux + Vy \quad \psi = -Vx + Uy \qquad (62)$$

where the character of flow is easily seen from the complex velocity

$$\frac{dw}{dz} = u + iv = U + iV \qquad u = U \qquad v = V \qquad (63)$$

and U, V are the $+x, y$ components of the velocity, respectively.

Source or Sink. A source in two-dimensional flow is a point from which fluid flows outward uniformly in all directions. A sink is a negative source; that is, the flow is into the point:

$$\phi = -\mu \ln r \qquad \psi = -\theta \qquad (64)$$

where $2\pi\mu$ is the outward flow from the line per unit length, known as the *strength*, with r and Θ shown in Fig. 19. In this figure, streamlines are the radial lines and equipotential lines are the concentric circles. When μ is negative, the equations become those for a sink.

Vortex. A vortex defines a rotation around a point. The relevant equations for a vortex are

$$\phi = -\mu\theta \qquad \psi = \mu \ln r \qquad (65)$$

Figure 20 is the flow net for a vortex with the radial lines now equipotential lines and the concentric circles streamlines. The vortex causes circulation about itself. Circulation about any closed curve is defined as the line integral of the velocity around the curve. The circulation about the vortex is $\kappa = 2\pi\mu$, considered

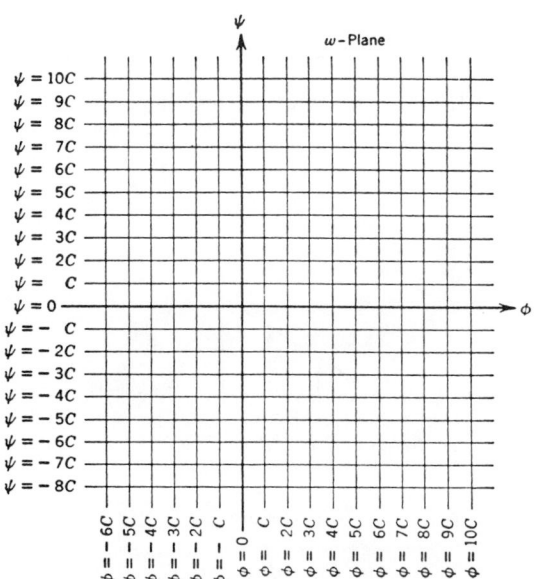

Fig. 18 Flow net in w plane.

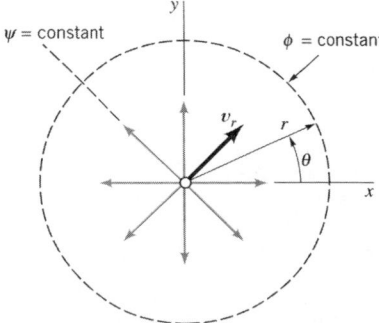

Fig. 19 Streamline pattern for source (or sink). From Ref. 2.

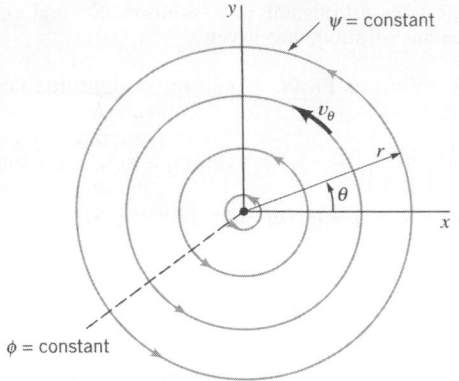

Fig. 20 Flow net for vortex. From Ref. 2.

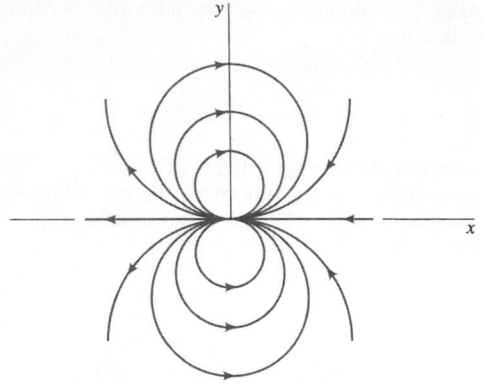

Fig. 22 Streamlines for doublet. From Ref. 2.

positive in the counterclockwise direction around the vortex.

Based on the radial variation of the strength of the rotation, two major types of vortices are assumed: forced vortex and free vortex (Fig. 21). In the forced vortex, which is also known as solid-body rotation, the angular velocity is the same at all points, and the tangential velocity increases linearly with distance from the center of the vortex. In the free vortex, however, both the angular velocity and the tangential velocity decay with distance from the vortex center.

Doublet The doublet is defined as the limiting case of a source and a sink of equal strength that approach each other such that the product of the strength by the distance between them remains a constant. The constant μ is called the strength of the doublet:

$$\phi = \frac{\mu x}{x^2 + y^2} \qquad \psi = \frac{\mu y}{x^2 + y^2} \qquad (66)$$

Figure 22 shows the flow net. The axis of the doublet is in the direction from sink to source and is parallel to the $+x$ axis as given here. The equipotential lines are circles having their centers on the x axis; the streamlines are circles having centers on the y axis.

Flow around a Circular Cylinder The superposition of a uniform flow in the $-x$ direction on a doublet with axis in the $+x$ direction results in flow around a circular cylinder. Adding the two flows, taking $\mu = Ua^2$, where a is the radius of the cylinder,

$$\phi = U\left(r + \frac{a^2}{r}\right)\cos\theta \qquad \psi = U\left(r - \frac{a^2}{r}\right)\sin\theta \qquad (67)$$

The streamline $\psi = 0$ is given by $\Theta = 0, \pi$ and by $r = a$; hence the circle $r = a$ may be taken as a solid boundary and the flow pattern of Fig. 23 is obtained.

Three-Dimensional Flow Three-dimensional flows in general are more difficult to handle than those in two

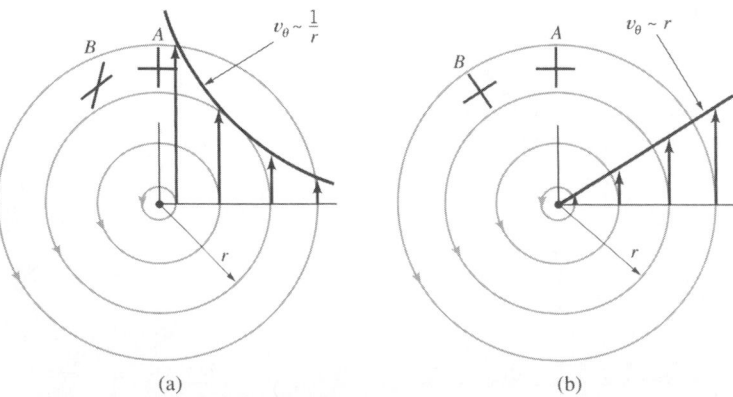

(a)   (b)

Fig. 21 Depiction for (a) free vortex and (b) forced vortex. From Ref. 2.

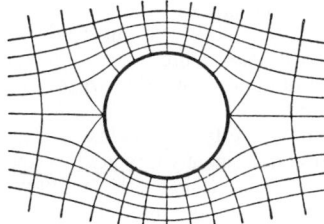

Fig. 23 Flow pattern for uniform flow around circular cylinder without circulation.

dimensions, primarily owing to a lack of methods comparable to the use of complex variables and conformal mapping. A velocity potential must be found that satisfies the Laplace equation and the boundary conditions. When the velocity potential is *single valued*, it may be shown that the solution is unique. For a body moving through an infinite fluid, otherwise at rest, a necessary condition is that the solution be such that the fluid at infinity remains at rest. Flow cases are usually found by investigating solutions of the Laplace equation to determine the particular boundary conditions that they satisfy.

Stokes's Stream Function.

The Stokes stream function is defined only for those three-dimensional flow cases that have axial symmetry, that is, where the flow is in a series of planes passing through a given line and where the flow pattern is identical in each of these planes. The intersection of these planes is the axis of symmetry.

In any one of these planes through the axis of symmetry select two points A, P such that A is fixed and P is variable. Draw a line connecting AP. The flow through the surface generated by rotating AP about the axis of symmetry is a function of the position of P. Let this function be $2\pi\psi$, x the axis the axis of symmetry, r the distance from the origin, and Θ the angle the radius vector makes with the x axis. The meridian angle is not needed because of axial symmetry. Then

$$v_r = -\frac{1}{r^2 \sin\theta}\frac{\partial\psi}{\partial\theta} \qquad v_\theta = \frac{1}{r\sin\theta}\frac{\partial\psi}{\partial r} \qquad (68)$$

and

$$\frac{1}{\sin\theta}\frac{\partial\psi}{\partial\theta} = r^2\frac{\partial\phi}{\partial r} \qquad \frac{\partial\psi}{\partial r} = -\sin\theta\frac{\partial\phi}{\partial\theta} \qquad (69)$$

The surfaces $\psi = $ const are stream surfaces. Since A is an arbitrary point, the stream function is always subject to the addition of an arbitrary constant. These equations are useful in dealing with flow about spheres, ellipsoids, and discs and through apertures. Stokes's stream function has the dimensions *volume per unit time*.

Examples of Three-Dimensional Flow.

Two examples of three-dimensional flow with axial symmetry follow.

Source in a Uniform Stream.

A point source is a point from which fluid issues at a uniform rate in all directions. Its strength m is the flow rate from the point, and the velocity potential and stream function for a source at the origin are

$$\phi = \frac{m}{4\pi r} \qquad \psi = \frac{Ur^2}{2}\sin^2\theta \qquad (70)$$

The flow net is shown in Fig. 24. Superposing a uniform flow on a point source results in *a half body*, with the flow equations

$$\phi = \frac{m}{4\pi r} + Ur\cos\theta \qquad \psi = \frac{m}{4\pi}\cos\theta + \frac{Ur^2}{2}\sin^2\theta \qquad (71)$$

The resulting flow net is shown in Fig. 25. The body extends to infinity in the downstream direction and has the asymptotic cylinder whose radius is

$$R_{\text{asymp}} = \sqrt{\frac{m}{\pi U}} \qquad (72)$$

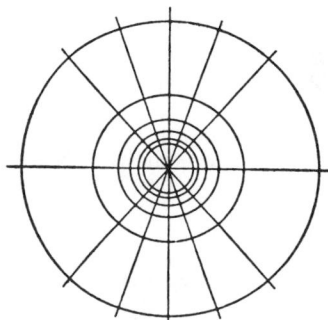

Fig. 24 Streamlines and equipotential lines for source.

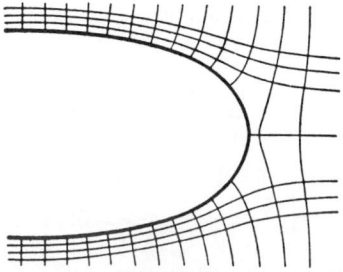

Fig. 25 Streamlines and equipotential lines for half body.

The equation of the half body is

$$r = \frac{1}{2}\sqrt{\frac{m}{\pi U}}\ \sec\frac{\theta}{2} \tag{73}$$

The pressure on the half body is given by

$$p = \frac{\rho}{2}U^2\left(\frac{3m^2}{16\pi^2 r^4 U^2} - \frac{m}{2\pi r^2 U}\right) \tag{74}$$

showing that the dynamic pressure drops to zero at great distances downstream along the body.

Flow around a Sphere. A doublet is defined as the limiting case as a source and sink approach each other such that the product of their strength by the distance between them remains a constant. The doublet has directional properties. Its axis is positive from the sink toward the source. For a doublet at the origin with axis in the $+x$ direction

$$\phi = \frac{\mu}{r^2}\cos\theta \qquad \psi = -\frac{\mu}{r}\sin^2\theta \tag{75}$$

where μ is the strength of the doublet.

Superposing a uniform flow, $u = -U$, on the doublet results in flow around a sphere,

$$\phi = \frac{Ua^3}{2r^2}\cos\theta + Ur\cos\theta$$

$$\psi = -\frac{Ua^3}{2r}\sin^2\theta + \frac{Ur^2}{2}\sin^2\theta \tag{76}$$

where $\mu = Ua^3/2$. This is shown to be the case, since the streamline $\psi = 0$ is satisfied by $\Theta = 0, \pi$ and $r = a$.

The flow net is shown in Fig. 26,

5 VISCOUS FLUID DYNAMICS

The additional consideration of viscosity in the flow of fluids greatly complicates analysis and understanding of flow situations. In the fundamental sense, viscosity only introduces the possibility of shear stress into the flow process. However, the effects are far-reaching and for many years have caused researchers and designers

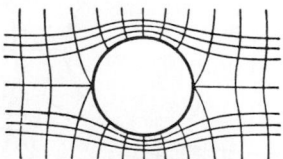

Fig. 26 Streamlines and equipotential lines for uniform flow about sphere at rest.

to rely heavily on experimental techniques. In recent years the wide availability of supercomputers has led to an increasing use of numerical analysis to predict the behavior of fluid flows (known as computational fluid dynamics, or CFD).

Viscous flows are typically classified as laminar or turbulent, although in any given flow there may be regions of both laminar and turbulent flow. In laminar flow, fluid particles slide smoothly over one another with no mixing except that which normally occurs as a result of molecular activity. Shear stress τ in laminar flow depends directly on the local velocity gradient and the fluid viscosity,

$$\tau = \mu\frac{du}{dy} \tag{77}$$

Shear stress is not affected by boundary roughness in laminar flow so long as the boundary roughness is small in comparison to the flow cross section.

In turbulent flow, a great deal of mixing occurs with random motion of fluid masses of various sizes occurring. Shear stress in turbulent flow depends strongly on the momentum exchange occurring as the result of turbulent mixing. An equivalent expression for shear stress in turbulent flow is

$$\tau = \varepsilon\frac{du}{dy} \tag{78}$$

where ε is the eddy viscosity, which is a function of both fluid viscosity and the local structure of the turbulent flow. This expression was unsatisfactory because it was impossible to quantify ε. An analysis of the turbulent momentum exchange process by Reynolds yielded a shear stress equation directly related to the flow turbulence:

$$\tau = -\rho\overline{u'v'} \tag{79}$$

where u', v' are the turbulent velocity fluctuations about the mean values and $\overline{u'v'}$ is the time average of the product of the fluctuations in the x and y directions. This expression is also not very useful in a practical sense. Research eventually led to the Prandtl–von Karman equation for shear stress:

$$\tau = \rho\kappa^2\frac{(dv/dy)^4}{(d^2v/dy^2)^2} \tag{80}$$

where κ is a dimensionless turbulence constant. This practical form of the shear stress equation can be integrated to generate velocity profiles.

5.1 Internal Flows

Flow in pipes and ducts under the action of viscosity requires some changes in basic concepts as well as in the equations for one-dimensional flow. Specifically, the velocity is no longer uniform in the cross section where the streamlines are straight and parallel. The velocity will be zero at the wall due to the "no-slip" condition, and increase to a maximum near the center of the conduit. This nonuniform velocity profile requires correction coefficients in the work–energy and momentum equations.

In addition to the need for compensating for the nonuniform velocity profile in one-dimensional flow, it is necessary to account for the energy converted to heat through the frictional processes caused by viscosity. If heat transfer is of no interest or concern, then energy conversion to heat is considered a loss in usable energy. The amount of useful energy converted to heat per pound of fluid flowing is designated by h_L, or the head loss. The one-dimensional work–energy equation now has the form

$$\frac{P_2}{\gamma} + \alpha_2 \frac{\overline{V}_2^2}{2g} + z_2 = \frac{P_1}{\gamma} + \alpha_1 \frac{\overline{V}_1^2}{2g} + z_1 - h_{L_{1-2}} \quad (81)$$

where α_1, α_2 are the correction factors for a nonuniform velocity profile ($\alpha \geq 1$ with $\alpha = 1$ for uniform profile) and $\overline{V}$ is an average velocity. Looking at Fig. 16, this equation states that the frictional effects reduce the amount of incoming available energy [i.e., $P_1/\gamma + \alpha_1 \overline{V}_1^2/(2g) + z_1$] by an amount h_L, resulting in a reduced amount of available energy in the fluid leaving the control volume. In actuality, the h_L term reveals itself in an increase in internal energy (temperature) of the fluid or a heat transfer from the fluid. It should be noted that the Bernoulli equation (same as above except for the h_L term) is a statement that the amount of incoming and outgoing available energy are equal since a primary assumption is inviscid flow, or no frictional losses.

Another phenomenon stemming directly from the effect of friction is flow separation. Separation occurs when the fluid is forced to decelerate too rapidly or change direction too quickly. The result is regions of reversed flow, formation of wakes and eddies, poor pressure recovery, and excessive friction losses (see Fig. 27a).

The nonuniform velocity profile occurring in viscous flow also causes vortices and cross currents to occur at certain locations. Vortices form in corners and at abrupt changes in cross section or flow direction. Cross currents or secondary flows occur at bends in the conduit. These secondary flows are the consequence of equal pressure gradients in the bend acting on fluid particles moving at different speeds in the flow cross section. The result is a spiral flow(s) superposed on the main flow (Fig. 27b). Both vortexes and secondary flows generate additional head losses and may cause some additional problems in hydraulic machinery

5.2 External Flows

External flows are generally associated with lift and drag on solid bodies in a fluid of large extent. Although for ideal fluids there is no drag, this is not the case for real fluids. Frictional effects on the surface of the body create skin friction drag. That is, the no-slip fluid condition at the body surface creates a velocity profile and the resulting shear stress at the surface to occur. The frictional effects are confined to the "boundary layer" near the body where the flow may be laminar, turbulent, or both. Theoretical analysis of skin friction drag has been relatively successful. Some specifics on lift and drag are presented in a subsequent section.

Separation in external flows is also caused by "adverse" pressure gradients that are forcing the flow to decelerate faster than it can and still remain "attached" to the boundary. Separation manifests itself as a turbulent wake behind the body (see Fig. 28), which creates considerable drag on the body largely resulting from the pressure difference between the front and rear. Smooth bodies with gradually changed form tend to generate less separation and lower drag than blunt or sharp-cornered objects.

5.3 Navier–Stokes Equations*

Application of Newton's second law to an incompressible small fluid particle of dimension dx, dy, dz

---

*Illustrations and material extracted all or in part from the previous edition by Victor L. Streeter.

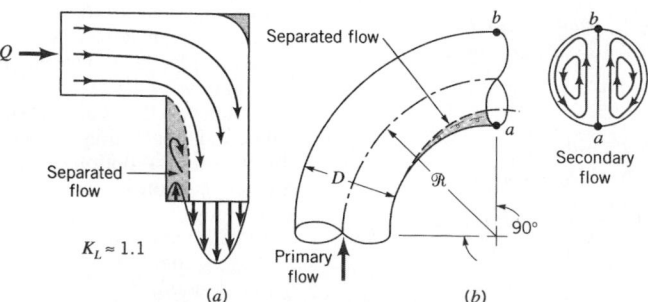

Fig. 27 Character of flow in 90°-turn elbow and subsequent viscous phenomena: (a) separated flow; (b) secondary flow. From Ref. 2.

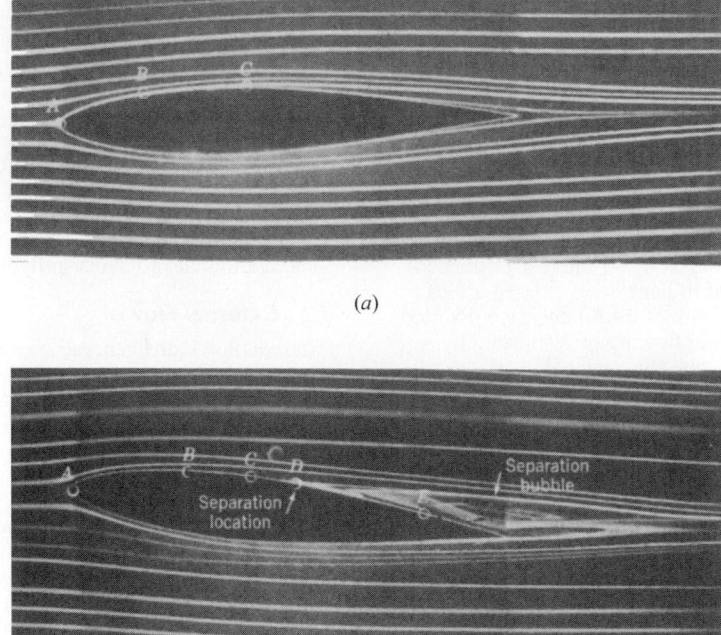

(a)

(b)

Fig. 28 Flow visualization past an airfoil at (a) 0° angle of attack with no flow separation and (b) 5° angle of attack with flow separation. From Ref. 2.

yields a set of equations comparable to the Euler equation previously derived, but now viscous effects are included. The equations are referred to as the Navier–Stokes equations, and they take the following form for laminar flow:

$$X - \frac{1}{\rho}\frac{\partial p}{\partial x} + v\left(\frac{\partial^2 u}{\partial x^2} + \frac{\partial^2 u}{\partial y^2} + \frac{\partial^2 u}{\partial z^2}\right)$$
$$= u\frac{\partial u}{\partial x} + v\frac{\partial u}{\partial y} + w\frac{\partial u}{\partial z} + \frac{\partial u}{\partial t} \tag{82}$$

$$Y - \frac{1}{\rho}\frac{\partial p}{\partial y} + v\left(\frac{\partial^2 v}{\partial x^2} + \frac{\partial^2 v}{\partial y^2} + \frac{\partial^2 v}{\partial z^2}\right)$$
$$= u\frac{\partial v}{\partial x} + v\frac{\partial v}{\partial y} + w\frac{\partial v}{\partial z} + \frac{\partial v}{\partial t} \tag{83}$$

$$Z - \frac{1}{\rho}\frac{\partial p}{\partial z} + v\left(\frac{\partial^2 w}{\partial x^2} + \frac{\partial^2 w}{\partial y^2} + \frac{\partial^2 w}{\partial z^2}\right)$$
$$= u\frac{\partial w}{\partial x} + v\frac{\partial w}{\partial y} + w\frac{\partial w}{\partial z} + \frac{\partial w}{\partial t} \tag{84}$$

where X, Y, Z are external force components per unit mass in the xyz directions, respectively; ρ is the mass

density, considered constant for incompressible flow; p is the average pressure at a point; v is the kinematic viscosity; and u, v, w are the velocity components in the xyz directions.

These simultaneous, nonlinear, differential equations cannot be integrated except for extremely simple flow cases where many of the terms are neglected. They contain the basic assumption that the stresses on a particle may be expressed as the most general linear function of the velocity gradients.

Boundary Conditions A real fluid in contact with a solid boundary must have a velocity exactly equal to the velocity of the boundary; this is much more restrictive than for an inviscid fluid, where no restrictions are placed on tangential velocity components at a boundary.

When two fluids are flowing side by side, a dynamical boundary condition arises at the interface. Applying the equation of motion to a thin layer of fluid enclosing a small portion of the interface shows that the terms containing mass are of higher order of smallness than the surface stress intensities and hence that the stresses must be continuous through the surface.

In general, the viscous boundary conditions at a solid surface give rise to rotational flow. Although the Navier–Stokes equations are satisfied by a velocity potential, since the viscous terms drop out, the viscous boundary conditions cannot be satisfied.

Continuity   The continuity equation must hold, as in the case of inviscid flow. It is

$$\frac{\partial u}{\partial x} + \frac{\partial v}{\partial y} + \frac{\partial w}{\partial z} = 0 \qquad (85)$$

Several examples of flow at low Reynolds numbers are given. It is assumed that in each case any turbulent fluctuations are completely damped out by viscous action, and the dominant flow is one dimensional with incompressible conditions.

Flow between Parallel Boundaries: Pressure-Driven Flow.   For steady flow between fixed parallel boundaries at low Reynolds numbers, the Navier–Stokes equations can be greatly reduced. Taking the coordinates as shown in Fig. 29 and assuming a pressure gradient exists in the x direction which drives the flow, the differential equations reduce to

$$\frac{\partial}{\partial x}(p + \gamma h) = \mu \frac{\partial^2 u}{\partial z^2} \qquad (86)$$

where h is measured vertically upward and γ is the specific weight of the fluid.

Integrating and introducing the boundary conditions $u = 0$ for $z = \pm b$, which are the no-slip conditions for this pressure-driven flow,

$$u = \frac{z^2 - b^2}{2\mu} \frac{\partial}{\partial x}(p + \gamma h) \qquad (87)$$

As the z^2 term implies, and as is shown in Fig. 29, such a velocity profile is a parabolic one. It has the maximum value at the center of the pipe—hence maximum velocity. Given a linear pressure gradient, which is an acceptable assumption for such a laminar flow, the derivative term becomes a constant, and the velocity profile becomes only a function of radial distance from the center; thus it remains the same along the length of the pipe (assuming the pipe does not change in direction, diameter, or makeup). Keep in

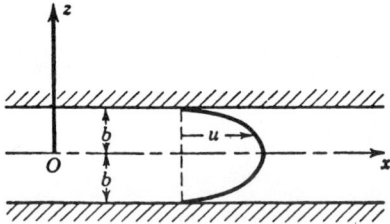

Fig. 29   Viscous flow between fixed parallel boundaries.

mind that the sustainability of laminar flow in a pipe is subject to various conditions. Therefore, once the flow characteristics change (e.g., transition to turbulence, development of secondary flows), this formula is no longer applicable.

Although the velocity potential function is essentially for irrotational flows (i.e., inviscid cases), for this special case, it is possible to express the parabolic velocity profile in terms of a velocity potential

$$u = \frac{z^2 - b^2}{2\mu} \frac{\partial}{\partial x}(p + \gamma h)$$

$$= \frac{\partial}{\partial x}\left[(p + \gamma h)\left(\frac{z^2 - b^2}{2\mu}\right)\right] = \frac{\partial \phi}{\partial x} \qquad (88)$$

where ϕ, the velocity potential, is the quantity in brackets.

Flow between Parallel Boundaries: Shear-Driven Flow.   A similar approach can be used for flow between parallel plates, one of which is moving with a constant speed U in the x direction. In this case, the motion of the fluid is due to the viscous force between the moving plate and the adjacent fluid layer. Thus, despite the absence of a pressure gradient, there will be a flow. Without the pressure gradient, the simplified version of the Navier–Stokes equations further simplifies to

$$\frac{d^2 u}{dz^2} = 0 \qquad (89)$$

The relevant boundary conditions for this case are $u = 0$ for $z = -b$ and $u = U$ for $z = +b$. Again integrating and using the boundary conditions, the velocity profile becomes

$$u = \frac{U}{2}\left(1 + \frac{z}{b}\right) \qquad (90)$$

This results in a linear velocity profile between the plates.

Flow between Parallel Boundaries: Combined Effects of Pressure Gradient and Shear.   One can also have the effect of both pressure gradient and shear for flow between parallel plates. For that case, the simplified version of the Navier–Stokes equations is the same as that for the pressure-driven flow, but the boundary conditions are those for the shear driven flow. The resultant velocity profile is

$$u = \frac{U}{2}\left(1 + \frac{z}{b}\right) + \frac{z^2 - b^2}{2\mu} \frac{\partial}{\partial x}(p + \gamma h) \qquad (91)$$

This is still a parabolic velocity profile, but now the maximum velocity has been displaced from the middle plane.

Theory of Lubrication The equations for two-dimensional viscous flow are applicable to the case of a slider bearing and can be applied to journal bearings. The simple case of a bearing of unit width is developed here, under the assumption that there is no flow out of the sides of the block, that is, normal to the plane of Fig. 30, where the clearance b is shown to a greatly exaggerated scale. The motion of a bearing block sliding over a plane surface, inclined slightly so that fluid is crowded between the two surfaces, develops large supporting forces normal to the surfaces. The angle of inclination is very small; therefore the differential equations of the preceding example apply. Since elevation changes also are very small and flow is in the x direction only, the equations reduce to

$$\frac{\partial p}{\partial x} = \mu \frac{\partial^2 u}{\partial z^2} \qquad \frac{\partial p}{\partial z} = 0 \qquad (92)$$

Considering the inclined block stationary and the plane surface in motion and taking the pressure at the two ends of the block as zero, the boundary conditions become $x = 0, x = L, \ p = 0; \ z = 0, \ u = U$; and $z = b, \ u = 0$. Integrating the equations and considering unit width normal to the figure, the discharge Q and pressure distribution are determined:

$$Q = \frac{U b_1 b_2}{b_1 + b_2} \qquad p = \frac{6\mu U x (b - b_2)}{b^2 (b_1 + b_2)} \qquad (93)$$

The last relation shows that b must be greater than b_2 for positive pressure buildup in the bearing. The point of maximum pressure and its value are

$$x\big|_{p\max} = \frac{b_1 L}{b_1 + b_2} \qquad p_{\max} = \frac{3}{2}\frac{\mu U L}{b_1 b_2}\frac{b_1 - b_2}{b_1 + b_2} \qquad (94)$$

The force P which the bearing will sustain is

$$P = \int_0^L p\, dx = \frac{6\mu U L^2}{b_2^2 (k-1)^2}\left[\ln k - \frac{2(k-1)}{k+1}\right] \qquad (95)$$

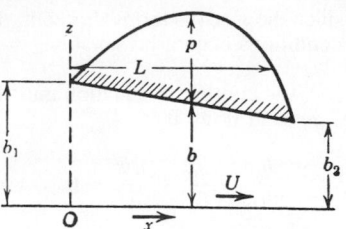

Fig. 30 Slider bearing.

The maximum bearing load is obtained for $k = 2.2$, yielding

$$P = 0.16\mu\frac{UL^2}{b_2^2} \qquad D = 0.75\mu\frac{UL}{b_2} \qquad (96)$$

The ratio

$$\frac{P}{D} = 0.21\frac{L}{b_2} \qquad (97)$$

can be made very large since b_2 is small. For $k = 2.2$, the line of action of the bearing load is at $x = 0.58L$. In general the line of action is given by

$$\bar{x} = \frac{L}{2}\left[\frac{2k}{k-1} - \frac{k^2 - 1 - 2k\ln k}{(k^2 - 1)\ln k - 2(k-1)^2}\right] \qquad (98)$$

Journal bearings are computed in an analogous manner. In general the clearances are so small compared with the radius of curvature of the bearing surface that the equations for plane motion can be applied.

Flow through Circular Tubes For steady flow through circular tubes with values of the Reynolds number, $UD\rho/\mu$, less than 2000, the motion is laminar. Equilibrium conditions on a cylindrical element concentric with the tube axis (Fig. 31) show that the shear stress varies linearly from zero at the pipe axis to a maximum $\Delta p r_0 / 2\Delta L$ at the wall, where $\Delta p/\Delta L$ is the drop in pressure per unit length of tube. In one-dimensional laminar flow the relation between shear

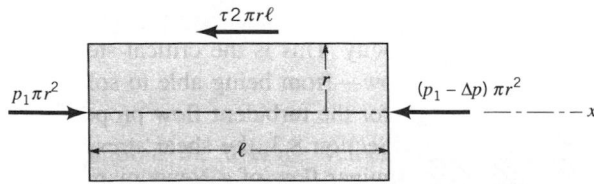

Fig. 31 Free-body diagram of cylinder of fluid in steady laminar flow through circular pipe. From Ref. 2.

stress and the velocity gradient is given by Newton's law of viscosity,

$$\tau = -\mu \frac{du}{dr} \qquad (99)$$

Using this relation for the value of shear stress, the velocity distribution is found to be

$$u = \frac{\Delta p}{\Delta L} \frac{r_0^2 - r^2}{4\mu} \qquad (100)$$

where r is the distance from the tube axis. The maximum velocity is at the axis:

$$u_{max} = \frac{\Delta p r_0^2}{\Delta L 4\mu} \qquad (101)$$

The average velocity is half the maximum velocity:

$$V = \left(\frac{\Delta p}{\Delta L}\right)\left(\frac{r_0^2}{8\mu}\right) \qquad (102)$$

The discharge, obtained by multiplying average velocity by cross-sectional area, is

$$Q = \left(\frac{\Delta p}{\Delta L}\right)\left(\frac{\pi D^4}{128\mu}\right) \qquad (103)$$

which is the *Hagen–Poiseuille* law. The preceding equations are independent of the surface condition in the tube, and they therefore hold for either rough or smooth tubes as long as the flow is laminar.

Viscous Flow around a Sphere: Stokes's Law
The flow of an infinite viscous fluid around a sphere at very low Reynolds numbers (Re $\ll$ 1, which states that viscous forces are dominant compared to inertial forces) has been solved by Stokes.[3] The Navier–Stokes equations with the acceleration terms omitted must be satisfied, as well as continuity and the boundary condition that the velocity vanish at the surface of the sphere. Stokes's solution is

$$u = U\left[\frac{3}{4}\frac{ax^3}{r^3}\left(\frac{a^2}{r^2} - 1\right) + 1 - \frac{1}{4}\frac{a}{r}\left(3 + \frac{a^2}{r^2}\right)\right] \qquad (104)$$

$$v = U\frac{3}{4}\frac{axy}{r^3}\left(\frac{a^2}{r^2} - 1\right) \qquad (105)$$

$$w = U\frac{3}{4}\frac{axz}{r^3}\left(\frac{a^2}{r^2} - 1\right) \qquad (106)$$

$$p = -\frac{3}{2}\frac{\mu U ax}{r^3} \qquad (107)$$

The radius of a sphere is a; the undisturbed velocity is in the x direction, $u = U$; and p is the average dynamic pressure at a point. The drag on a sphere is made up from the pressure difference over the surface of the sphere and from the shear stress. The viscous drag due to shear stress is twice as great as that due to the pressure difference. The total drag is

$$D = 6\pi a \mu U \qquad (108)$$

This is known as *Stokes's law*.

The settling velocity of small spheres may be obtained by writing the equation for drag, weight of particle, and buoyant force. Solving for the settling velocity,

$$U = \frac{2}{9}\frac{a^2}{\mu}(\gamma_s - \gamma) \qquad (109)$$

where γ_s is the specific weight of the solid particle. Stokes's law has been found by experiment to hold for Reynolds numbers below 1, that is,

$$\frac{2a\rho U}{\mu} < 1 \qquad (110)$$

6 SIMILITUDE AND DIMENSIONAL ANALYSIS

Similitude and dimensional analysis are inextricably tied to experimental testing and the analysis of experimental data. Principles of similitude permit the prediction of prototype behavior based on the performance of models. Techniques of dimensional analysis permit the efficient and logical presentation of test data in dimensionless form. Even in recent times when CFD techniques are supplementing laboratory work at an increasing pace (note: CFD will never totally replace experiments), many flow phenomena are so complex as to warrant experimental investigation.

6.1 Similitude

For complete similitude the model and the prototype must be geometrically, kinematically, and dynamically similar. In this context, the prototype is the physical system for which you are trying to predict the behavior, while the model (typically a scaled version of the prototype) is the device used for testing in the laboratory. Geometric similarity exists when the model is a photographic reduction (or enlargement) of the prototype. Kinematic similarity occurs when the streamline pattern in the model is a photographic reduction (or enlargement) of that of the prototype. Also ratios of velocity vectors at corresponding points are constant. Dynamic similarity requires that the ratios of similar forces at all corresponding points in the flow be constant. In general, if geometric similarity and dynamic similarity exist, then kinematic similarity is guaranteed.

Quantitative relationships for dynamic similarity are obtained by considering the contributing forces acting on a fluid particle. Potential contributing forces include viscosity (shear), pressure, gravity, elasticity, and surface tension. These forces all contribute to the acceleration of the fluid particle in the prototype flow:

$$\mathbf{F}_v + \mathbf{F}_p + \mathbf{F}_g + \mathbf{F}_e + \mathbf{F}_s = m\mathbf{a}$$

$$= -\mathbf{F}_I \quad \text{(inertial force)} \tag{111}$$

Dividing by the inertial force and considering the prototype (P),

$$\left(\frac{\mathbf{F}_v}{\mathbf{F}_I}\right)_P + \left(\frac{\mathbf{F}_p}{\mathbf{F}_I}\right)_P + \left(\frac{\mathbf{F}_g}{\mathbf{F}_I}\right)_P + \left(\frac{\mathbf{F}_e}{\mathbf{F}_I}\right)_P$$
$$+ \left(\frac{\mathbf{F}_s}{\mathbf{F}_I}\right)_P = -1 \tag{112a}$$

If the same process is followed in the model (m),

$$\left(\frac{\mathbf{F}_v}{\mathbf{F}_I}\right)_m + \left(\frac{\mathbf{F}_p}{\mathbf{F}_I}\right)_m + \left(\frac{\mathbf{F}_g}{\mathbf{F}_I}\right)_m + \left(\frac{\mathbf{F}_e}{\mathbf{F}_I}\right)_m$$
$$+ \left(\frac{\mathbf{F}_s}{\mathbf{F}_I}\right)_m = -1 \tag{112b}$$

This scaling procedure renders the force polygons at the corresponding points in the model and prototype as congruent provided the force ratios for each source of force are the same in the model and prototype. In fact, the scaled polygons are congruent if *all but one* of the force ratios are equal. This relationship can be summarized as follows:

Viscous Forces:

$$\left(\frac{\rho V l}{\mu}\right)_m = \left(\frac{\rho V l}{\mu}\right)_P \tag{113}$$

where $\rho V l/\mu$ = Reynolds number, Re = inertia force/ viscous force.

Pressure Forces:

$$\left(\frac{P}{\rho V^2}\right)_m = \left(\frac{P}{\rho V^2}\right)_P \tag{114}$$

where $P/\rho V^2$ = Euler number, E = pressure force/ inertia force.

Gravity Forces:

$$\left(\frac{V^2}{gl}\right)_m = \left(\frac{V^2}{gl}\right)_P \tag{115}$$

where $V/\sqrt{gl}$ = Froude number, F = inertia force/ gravitational force.

Elastic Forces:

$$\left(\frac{V}{a}\right)_m = \left(\frac{V}{a}\right)_P \tag{116}$$

where a is the speed of sound and V/a = Mach number, M = inertia force/compression force.

Surface Tension Forces:

$$\left(\frac{\rho V^2 l}{\sigma}\right)_m = \left(\frac{\rho V^2 l}{\sigma}\right)_P \tag{117}$$

where $\rho V^2 l/\sigma$ = Weber number, W = inertia force/ surface tension force.

In most fluid problems, some of these forces are negligible and can be ignored in model–prototype relationships. For example, if there is no free surface and flows are well below the sonic velocity, then only viscous and pressure forces are important. In this case, equality of Reynolds numbers would guarantee similarity. Typically, pressure forces are always in existence and the Euler number is the force ratio omitted in specifying similarity (as noted earlier).

Example 1. Model testing of a new submarine is to be conducted in a wind tunnel using standard air at sea level conditions. Due to the limitations on the test section size, the model needs to be a one-twentieth-scale version of the prototype. If the actual submarine will have a velocity of $V_p = 0.5$ ft/sec in sea water at $20°C$, what must the air velocity be in the wind tunnel for the model to achieve dynamic similarity?

For dynamic similarity, $\text{Re}_P = \text{Re}_m$, or $(\rho V l/\mu)_m = (\rho V l/\mu)_P$. Solving for model velocity V_m yields

$$V_m = \frac{l_P}{l_m} \frac{\rho_P}{\rho_m} \frac{\mu_m}{\mu_P} V_P$$

The necessary values are found as follows:

$$\rho_m = 0.00238 \text{ slug/ft}^3 \qquad \mu_m = 3.74 \times 10^{-7} \text{ lb-sec/ft}^2$$

$$\rho_P = 2.0 \text{ slugs/ft}^3 \qquad \mu_P = 2.23 \times 10^{-5} \text{ lb-sec/ft}^2$$

So, the required wind tunnel velocity is found from

$$V_m = \left(\frac{20}{1}\right) \left(\frac{2.0}{0.00238}\right) \left(\frac{3.74 \times 10^{-7}}{2.23 \times 10^{-5}}\right) (0.5)$$

$$= 141 \text{ ft/sec}$$

There are situations where complete similarity cannot be practically achieved. For example, in the modeling of ship hulls, if water is used as the modeling fluid, the model must be as large as the prototype. This is true because viscous and gravity forces predominate and both Reynolds and Froude numbers must be equal to guarantee similarity. In this case, to get around the problem, the model–prototype relation is determined by the Froude number equality and the viscous effects on hull drag are determined through other means. This situation is known as incomplete similarity or distorted model.

In other flow situations such as rivers and harbors, the lateral dimensions are so large compared to the vertical dimensions that geometric scaling would make the model so shallow as to be under strong surface tension effects. Also the boundary roughness would be reduced to supersmoothness. In these situations, two different model scales are used: one for horizontal dimensions, another for vertical dimensions. These models generally have to be adjusted for "scale effects" by employing artificial roughness to properly represent prototype behavior.

6.2 Dimensional Analysis

Dimensional analysis is based on the principle that meaningful physical relationships between quantities must be dimensionally homogeneous; that is, both sides of an equation must have the same dimensions. The four basic dimensions used in fluid mechanics are force (F), mass (M), length (L), and time (T). In fact, the dimension of force is related to the dimensions of mass, length, and time through Newton's second law:

$$F = M \frac{L}{T^2} \qquad (118)$$

The units of force are said to be *equivalent* to the units of mass × length/time squared.

The principle of dimensional homogeneity can be used to develop a formal procedure for establishing relations between physical quantities and, most importantly, dimensionless groups. Dimensionless groups of quantities are extremely valuable in the analysis and presentation of experimental data. Their use permits large amounts of data to be presented on simple graphs, graphs with parametric relationships, or in some more complex cases multicorrelational graphs.

For example, consider the case of the drag force on a sphere falling at constant velocity in a very viscous liquid. Drag *(D)* is a function of velocity (V), size (d), and viscosity (μ):

$$D = f(V, d, \mu)$$

In basic dimensions,

$$F = f'\left(\frac{L}{T}, L, \frac{FT}{L^2}\right)$$

The only combination of quantities on the right-hand side that will make the relationship dimensionally correct is

$$D = KVd\mu \qquad (119)$$

where K is a dimensionless constant. Thus, by dimensional reasoning alone, Stokes's law has been deduced. Sometimes the relationship does not fall out so easily, quantities may need to have positive or negative exponents to create a combination that is dimensionally correct.

However, formal dimensional analysis does not always offer a unique approach to generating the desired dimensionless groups. A more modern and intuitive method, called the Buckingham Π theorem, is more useful. This theorem informs the user as to how many dimensionless groups may be formed and leaves the configuration of each group up to the user. This freedom, coupled with knowledge of the important similarity principles, permits the user to make a good choice of the proper groups.

According to the Buckingham Π theorem, a problem described by k variables can be expressed in terms of $k - r$ nondimensional Π terms, where r is the minimum number of reference dimensions involved in the problem. For example, when establishing the quantities affecting the drag on a ship's hull, the list includes

$$D(\text{drag}) = f(l, V, \mu, \rho, g) \qquad (120)$$

or

$$f'(D, l, V, \mu, \rho, g) = 0 \qquad (121)$$

The Π theorem allows the formation of three dimensionless groups ($k - r = 6 - 3$). Knowing that viscous and gravitational effects are most significant in affecting drag, the dimensionless groups of Reynolds number and Froude number are selected. Then the drag force D is combined into a third group. The final relationship states that the nondimensional drag (i.e., drag coefficient) is a function of the Reynolds and Froude numbers:

$$\frac{D}{\rho l^2 V^2} = \phi\left(\frac{\rho V l}{\mu}, \frac{V^2}{gl}\right) \qquad (122)$$

or

$$C_D = \phi(\text{Re}, F^2) \qquad (123)$$

The use of this kind of nondimensional expression allows the experimenter to make use of scale-size models and can perform experiments more easily and affordably by reducing the number of parameters that need to be varied. Once the experiments are performed on the model, the experimenter can plot all data on one graph using the nondimensional values; for example, $D/\rho l^2 V^2$ as the ordinate, Re as the abscissa, and F as the parametric variable.

7 FLOW IN CLOSED CONDUITS

Flow through a conduit may be steady or unsteady, uniform or nonuniform, and laminar or turbulent. Steady flow refers to flow at constant rate, uniform flow to prismatic sections of conduit, and laminar flow to those cases where viscous forces predominate and the losses are a linear function of the velocity; whereas for turbulent flow the losses vary as the velocity to some power (1.7–2.0), depending in part upon the Reynolds number.

The classical methods of hydrodynamics applying to an ideal fluid are of little value in solving flow problems in conduits, although they are extremely useful in connection with flow around immersed bodies (see Section 9). The nature of turbulent flow, on the other hand, is not sufficiently well understood to permit computation of the energy losses for given boundary conditions and rates of flow, and hence recourse must generally be taken to experimentation.

The work–energy equation is of first importance in solving flow problems. Momentum relationships are of use in certain cases in which the forces acting on the fluid are known or are desired. In special situations where both the energy and momentum equations are applicable, the energy loss may be computed without recourse to experimentation. The continuity equation in steady flow usually states that the flow past every cross section is the same. When the fluid is compressible, this is a statement of mass or weight flow, but for liquids it is sufficient to deal with volume rates only. The three types of equations (energy, momentum, and continuity), together with the experimentally determined loss relationships, provide the general framework for solving closed-conduit problems.

7.1 Velocity Distribution*

The velocity distribution for established laminar flow through round tubes and between parallel plates has been discussed under laminar fluid motion, Section 5.

---

*Illustrations and material extracted all or in part from previous edition by Victor L. Streeter.

Development of Flow The velocity distribution for fully developed uniform flow in a closed conduit is determined by the relationship between the radial velocity gradient and the shear stress. In turbulent flow the velocity distribution cannot be derived exactly, although much has been accomplished in recent years in the analytical approach to rational velocity–distribution equations.

Downstream from any change m cross section or direction, there is a length over which the velocity distribution regains its characteristic form (entrance region), depending on the shape of cross section, the wall roughness, and the Reynolds number. For example (follow the description from Fig. 32), when the flow passes from a reservoir through a rounded entrance into a conduit, the velocity is practically constant over the section at the upstream end of the conduit. Such flow during the initial stages is therefore practically irrotational, since the boundary layer is very thin, The effect of boundary resistance, however, is to retard the fluid in the wall vicinity, resulting, through lateral transmission of shear, in a continuous growth of the boundary layer with distance from the inlet. Since the mean velocity must nevertheless remain constant, the central portion of the fluid is simultaneously accelerated until the forces of shear and pressure gradient reach equilibrium as the velocity distribution of uniform flow becomes fully developed some distance downstream. If the flow experiences a change in cross section or direction, it will undergo another developing flow region until fully developed flow is once again achieved (see Fig. 32).

For laminar flow, experiments by Nikuradse give the entranee length l for development of flow as

$$\frac{l_e}{D} = 0.06 \, \text{Re} \qquad (124)$$

where D is the pipe diameter and the Reynolds number is based on average velocity and diameter.

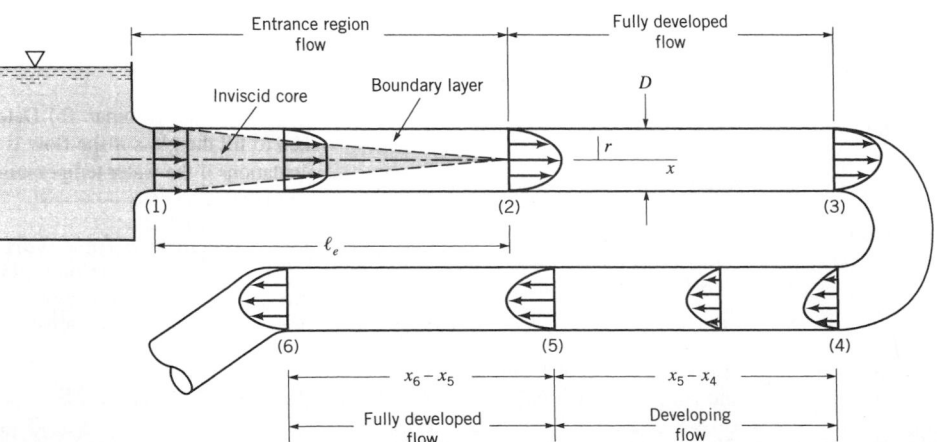

Fig. 32 Entrance region, developing flow, and fully developed flow in pipe system. From Ref. 2.

In turbulent flow, the transition to the established velocity distribution is effected in a much shorter reach, because of the pronounced mixing action that then prevails. Nikuradse's experiments indicate that a distance of 25–40 diameters is sufficient and that the length is not so dependent on the Reynolds number.

Rational Formulas for Fully Developed Turbulent Flow

In turbulent flow, the ratio of shear stress to velocity gradient depends not only on the physical properties of the fluid but also on the characteristics of the flow.

Stanton first stated that the turbulent velocity distribution in the central portions of a conduit has a form that is independent of the wall roughness and viscous effects provided that the wall shear remains the same. In equation form

$$\frac{v_{max} - v}{\sqrt{\tau_o/\rho}} = F\left(\frac{r}{r_o}\right) \tag{125}$$

where v_{max} is the velocity at the pipe axis, v is the velocity at the distance r from the axis, r_o is the pipe radius, τ_o is the wall shear, ρ is the mass density of fluid, and F is an unknown function. The proof is evident by an inspection of the Nikuradse data on smooth and sand-roughened pipes, given in Fig. 33; k is the diameter of sand grains cemented to the pipe walls.

Based on the preceding, von Karman[4] obtained the formula for smooth pipes,

$$\frac{v}{\sqrt{\tau_o/\rho}} = C_1 + \frac{1}{\kappa} \ln\left(\sqrt{\frac{\tau_o}{\rho}} \frac{y}{v}\right) \tag{126}$$

where κ is a universal constant having the value 0.40 and v is the kinematic viscosity. Figure 34, based on Nikuradse's tests, shows the value of C_1 to be 5.5 for best agreement with the data. In the immediate vicinity of the pipe wall, through a film called the *laminar sublayer*, the velocity is given closely by $v = y\tau_o/\mu$. This may be written as

$$\frac{v}{\sqrt{\tau_o/\rho}} = \sqrt{\frac{\tau_o}{\rho}} \frac{y}{v} \tag{127}$$

and is plotted in Fig. 34. The intersection of the two curves may be taken arbitrarily as the border between the two types of flow, although actually there is a transition phase from the laminar to the turbulent zone. From the figure, the laminar film has the thickness

$$\delta = \frac{11.6v}{\sqrt{\tau_o/\rho}} \tag{128}$$

For rough pipes von Karman obtained the formula

$$\frac{v}{\sqrt{\tau_o/\rho}} = C_2 + \frac{1}{\kappa} \ln\left(\frac{y}{\kappa}\right) \tag{129}$$

Figure 35 shows the Nikuradse sand-roughened pipe tests. From these data, $C_2 = 8.5$.

The two logarithmic equations do not give a zero slope of the velocity distribution curve at the center line. This is a defect in the formulas that, nevertheless, has little significance from a practical viewpoint. The equations actually portray the true velocity distribution in the central region of the flow very well, although they were derived for the region near the wall.

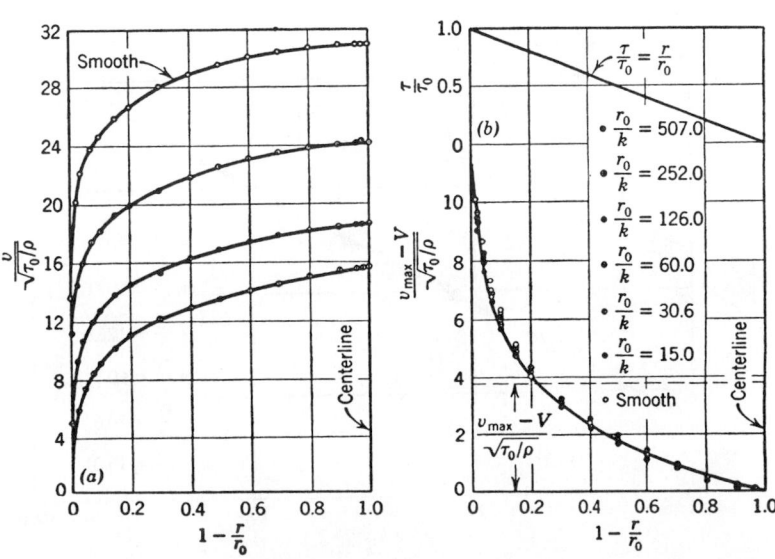

Fig. 33 Generalized plot of velocity distribution for smooth and rough pipes.

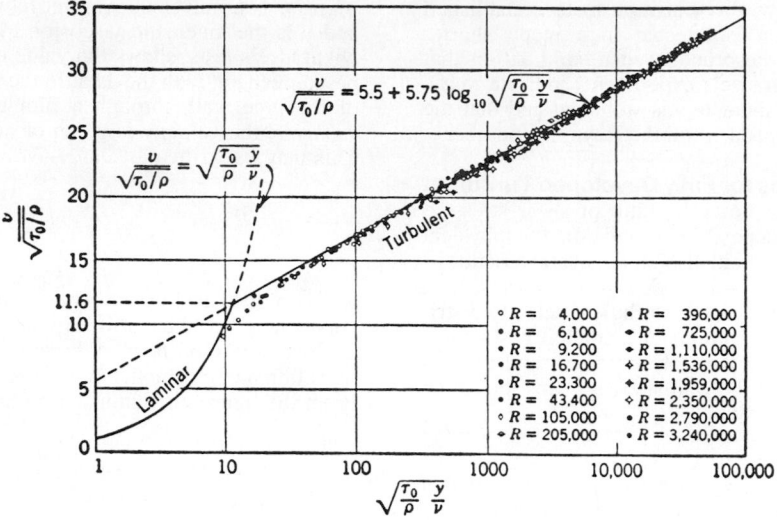

Fig. 34 Universal velocity distribution for smooth pipes.

Energy and Momentum Correction Factors In writing the work–energy equation between two cross sections of a conduit, it is usually satisfactory to express the mean kinetic energy per unit weight simply as $V^2/2g$, where V is the average velocity at a section. As discussed earlier, however, this is strictly true only when the velocity is constant over both cross sections. For laminar flow, the correction α by which $V^2/2g$ must be multiplied to give the true mean value is 2.0, and for sections where there is back flow the factor may be even larger. Values of α and β based on Prandtl–Karman turbulent velocity distribution equations are given in Fig. 36.

7.2 Pipe Friction*

A general equation for solving one-dimensional pipe flow problems is the work–energy equation

$$\frac{P_1}{\gamma} + \alpha_1 \frac{V_1^2}{2g} + z_1 = \frac{P_2}{\gamma} + \alpha_2 \frac{V_2^2}{2g} + z_2 + \sum h_{L_{1-2}}$$
(130)

---

*Material and illustrations used by permission from G. Z. Watters, *Analysis and Control of Unsteady Flow in Pipelines* Butterworth, Stoneham, MA, 1984.

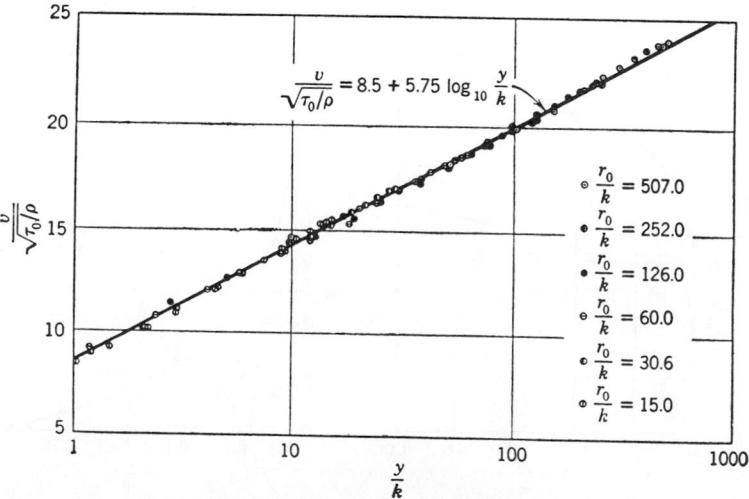

Fig. 35 Universal velocity distribution for rough pipes.

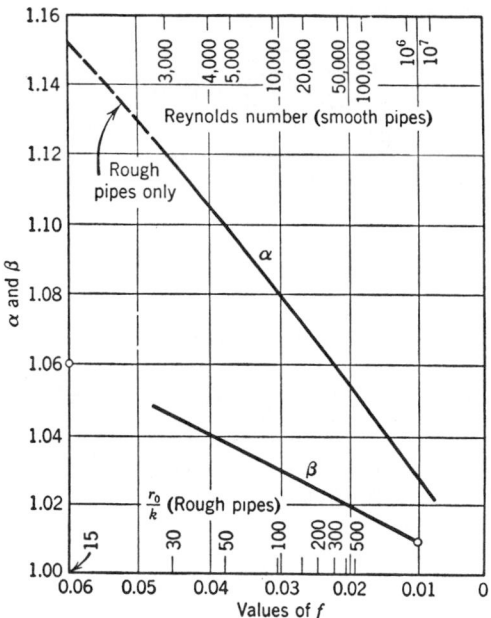

Fig. 36 Energy and momentum correction factors for smooth and rough pipes.

where $\sum h_{L_{1-2}}$ is the total of all the friction losses between cross sections 1 and 2. These losses can be divided into two general categories: pipe friction and minor losses. Pipe friction losses are those caused by the continuing viscous action along the conduit. Minor losses are the result of the additional friction losses over and above the pipe friction. These losses are caused by flow separation, eddies, and wakes that are generated by changes in flow direction or cross section and pipe components such as a valve. This section deals with pipe friction.

Pipe friction losses for circular cylindrical pipes depend on the flow velocity, pipe size, fluid viscosity, and wall roughness. In regard to wall roughness, friction loss specifically depends on *relative* roughness, that is, the *absolute* roughness compared with the pipe

diameter. The various formulas for pipe friction loss differ in how they incorporate wall roughness into the head loss equation. The formulas in most common use today are the following:

$$h_f = f \frac{L}{D} \frac{V^2}{2g} \qquad \text{(Darcy–Weisbach)} \qquad (131)$$

$$V = 0.55 C D^{0.63} S^{0.54} \quad \text{(Hazen–Williams)} \quad (132)$$

$$V = \frac{0.59}{n} D^{2/3} S^{1/2} \qquad \text{(Manning)} \qquad (133)$$

where f, C, n are friction coefficients; S is head loss per unit length of pipe, D is inside pipe diameter, and L is pipe length. The Darcy–Weisbach formula, commonly referred to as the Darcy formula, is the most general in application. It can be used for a variety of liquids and gases, for laminar and turbulent flow, and for rough or smooth pipes. Its main disadvantage is the fact that the friction factor f is often dependent on one of the design unknowns (pipe diameter or discharge) and an iterative solution results. However, engineers are increasingly using this formula because of its breadth of application.

The Hazen–Williams formula was developed for the computation of friction losses for water flowing in distribution system pipes. It works well for moderately smooth pipes (such as cast iron), but it is not accurate for rough pipes, small pipes, or laminar low.

The main task is to find the friction factor f. The value depends on two parameters: Reynolds number and relative roughness. The Reynolds number measures the effect of viscosity on f and is defined as

$$\text{Re} = \frac{VD}{\nu} \qquad (134)$$

where ν is the kinematic viscosity of the fluid. The relative roughness measures the roughness of the pipe wall relative to the pipe diameter and is expressed as k/D, where k is a measure of pipe wall roughness.

For convenience, Table 6 was compiled from several sources to provide assistance in selecting roughness values for various pipe materials. It should be

Table 6 Roughness Values for Commercial Pipes

Pipe Material	k (in.)	C (Hazen–Williams)	n (Manning)
Riveted steel	0.036–0.36	110	0.013–0.017
Concrete	0.012–0.12	120–140	0.011–0.014
Cast iron (new)	0.010	130	0.013
Cast iron (old)	—	100	0.015–0.035
Galvanized iron	0.0060	—	0.016
Asphalted iron	0.0048	—	0.013
Welded steel	0.0018	120	0.012
Asbestos cement	—	140	0.011
Copper, aluminum tube	Smooth	150	0.010
PVC, plastic	Smooth	150	0.009

noted that, typically, there are several inconsistencies between the roughness values given in Table 6 for the different formulas and those generated using the following equations.

The Reynolds number and the relative roughness are used in conjunction with the Moody diagram (Fig. 37) to find the f value. The use of the Moody diagram can best be shown by example.

Example 2. Compute the friction factor f for the flow of 1500 gpm of water at normal temperature in a 10-in. cast-iron pipe.

$$V = \frac{1500}{449 \frac{\pi}{4} \left(\frac{10}{12}\right)^2}$$

$$= 6.13 \text{ ft/sec}$$

From Table 6

$$\frac{k}{D} = \frac{0.01}{10} = 0.001$$

Using $v = 1.22 \times 10^{-5}$ ft^2/sec, from Eq. (133)

$$\text{Re} = \frac{6.13 \times (10/12)}{1.22 \times 10^{-5}}$$

$$= 418,000$$

From Fig. 37, $f = 0.020$.

Both the Hazen–Williams and Manning formulas can be manipulated into the same form as the Darcy formula. The resulting f expressions may be compared

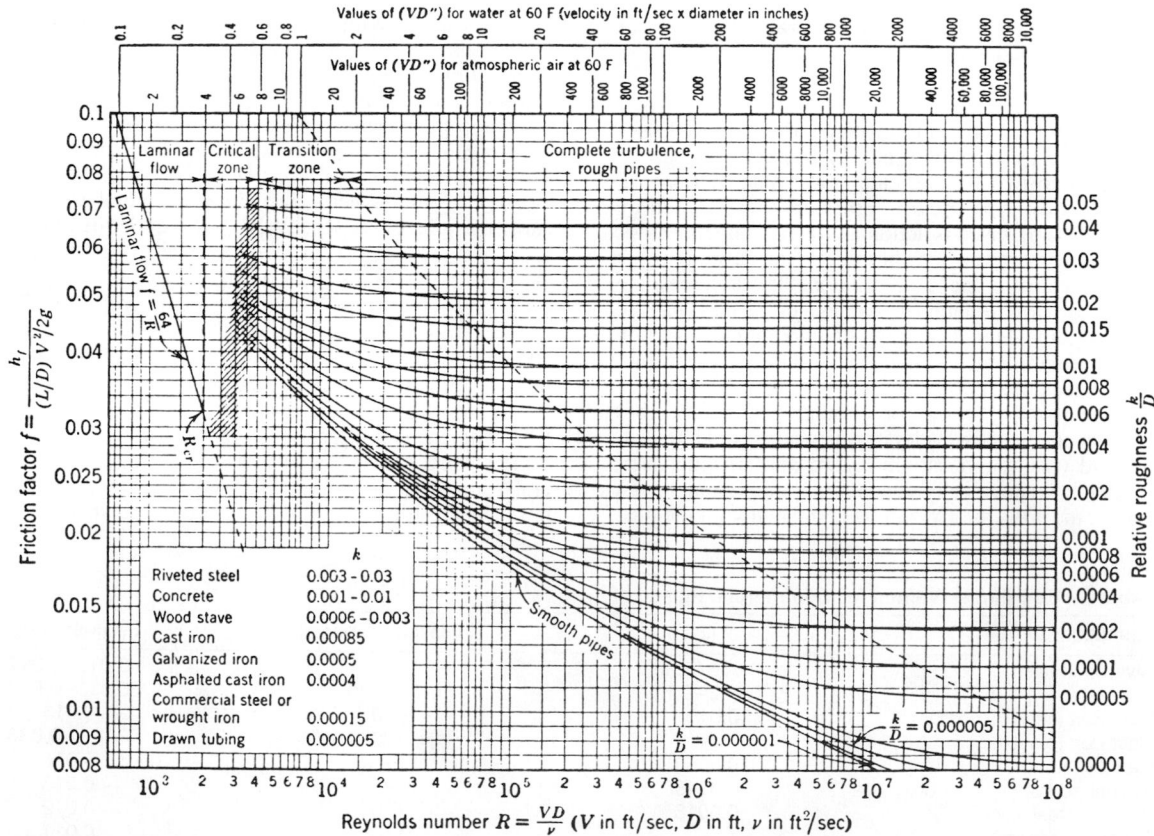

Fig. 37 Resistance diagram. Reproduced with permission from L. F. Moodes, "Friction Factor for Pipe Flow," *Transactions ASME*, November 1944, Princeton, NJ.

with the Darcy f value to deduce the situations where these formulas may be confidently applied:

$$h_f = \left(\frac{1090}{C^{1.85}\text{Re}^{0.15}} \right) \frac{L}{D} \frac{V^2}{2g} \quad \text{(Hazen–Williams)}$$

(135)

$$h_f = \left(\frac{185n^2}{D^{1/3}} \right) \frac{L}{D} \frac{V^2}{2g} \quad \text{(Manning)} \quad (136)$$

Both C and n are roughness values that can be obtained from tables such as Table 6. For additional information on C values, see Davis and Sorensen[5] for n values, see Chow.[6]

The areas of applicability of the three head loss formulas can be deduced. It is clear that the Manning formula is indeed a rough pipe formula because the Reynolds number does not appear. On the other hand, the Hazen–Williams formula is a relatively smooth pipe formula because the head loss varies with discharge in the same manner as smooth pipes. Their use should be strictly limited to these specific hydraulic conditions. However, in actual design use of the formulas, the results are not quite so dramatic. Consider the following example.

Example 3. Select a pipe to convey 20 ft³/sec of water between two reservoirs 5 miles apart and 200 ft different in elevation. Use welded steel pipe.

Hazen–Williams: From Table 6, $C = 120$. From Eq. (131),

$$Q = AV = \frac{1}{4}\pi D^2 \times 0.55CD^{0.63}S^{0.54}$$

$$20 = \frac{1}{4}\pi D^2 \times 0.55 \times 120$$

$$\times D^{0.63} \left(\frac{200}{5 \times 5280} \right)^{0.54}$$

$D = 1.90$ ft or 22.8 in. Design $D = 24$ in.

Equivalent f value $= 0.019$

Manning: From Table 6, $n = 0.012$. From Eq. (132),

$$Q = AV = \frac{1}{4}\pi D^2 \times \frac{0.59}{n} D^{2/3}S^{1/2}$$

$$20 = \frac{1}{4}\pi D^2 \times \frac{0.59}{0.012} D^{2/3} \left(\frac{200}{5 \times 5280} \right)^{1/2}$$

$D = 1.95$ ft or 23.4 in. Design $D = 24$ in.

Equivalent f value $= 0.021$

Darcy–Weisbach: This is an iterative solution because the Reynolds number and relative roughness cannot be found unless the diameter is known. Because

we know roughly what the diameter is from the previous solution, we will use that as an estimate. From an estimated $D \approx 24$ in., from Table 6, $k/D = 0.000075$. Then

$$V = \frac{20}{(\pi/4)2^2} = 6.4 \text{ ft/sec}$$

$$\text{Re} = \frac{6.4 \times 2}{1.2 \times 10^{-5}} = 1.07 \times 10^6$$

From Fig. 37,

$$f = 0.0135$$

$$D = 1.85 \text{ ft or 22 in.} \quad \text{Design } D = 22 \text{ in.}$$

Even though there is a rather dramatic variance in the f value (variation of almost 50%), the resulting effect on the design pipe diameter is quite a lot less. This is a consequence of the fact that diameter varies at about the $\frac{1}{5}$ power of the f value; so, dramatic differences in f value have greatly reduced impact.

For purposes of computer application, it is advantageous to express the information on the Moody diagram in algebraic form. This is most commonly done with the equations of Colebrook[7]:

$$\frac{1}{\sqrt{f}} = -2\log_{10}\left(\frac{2.51}{\text{Re}\sqrt{f}} \right) \quad \text{(smooth)} \quad (137)$$

$$\frac{1}{\sqrt{f}} = 1.74 - 2\log_{10}\left(2\frac{k}{D} + \frac{18.7}{\text{Re}\sqrt{f}} \right) \quad \text{(transition)}$$

(138)

$$\frac{1}{\sqrt{f}} = -2\log_{10}\left(0.266\frac{k}{D} \right) \quad \text{(rough)} \quad (139)$$

These equations can then be solved iteratively by computer to determine f for a given Re value and k/D value.

Effects of Aging* The values of k given for the various pipe materials of Fig. 37 and Table 6 are for new, clean pipes. In general, pipes become increasingly rough with age, owing to deposition or corrosion. Colebrook and White have determined an approximately linear increase in absolute roughness with time, which may be expressed as

$$k = k_0 + \alpha t \quad (140)$$

where k_0 is the absolute roughness of the new material, α is a constant, and k is the absolute roughness at time t.

---

*Material extracted from third edition by Victor L. Streeter.

Example 4. After 10 years of service, a 10-in cast-iron pipe line in water service has a drop of 3.13 psi per 1000 ft for a flow of 1000 gpm. What is the estimated pressure drop for 1200 gpm after 20 years of service?

$$V = \frac{1000}{7.48 \times 60 \frac{\pi}{4} \left(\frac{5}{6}\right)^2} = 4.1 \text{ ft/sec} \qquad \frac{V^2}{2g} = 0.262 \text{ ft}$$

$$h_f = \frac{3.13 \times 144}{62.4} = f \frac{1000}{5/6} 0.262 \qquad f = 0.023$$

Taking $v = 1.2 \times 10^{-5} \text{ft}^2/\text{sec}$,

$$\text{Re} = \frac{4.1 \times 5}{6 \times 1.2 \times 10^{-5}} = 285,000$$

From Fig. 37, for the above calculated values of f and Re, we get $k/D = 0.0017$ and $k = 0.00142$ ft. For new cast iron take $k_0 = 0.001$ ft; hence for 10 years,

$$0.00142 = 0.001 + 10\alpha \qquad \alpha = 0.000042 \text{ ft/yr}$$

and for 20 years,

$$k = 0.001 + 20 \times 0.000042 = 0.00184 \text{ ft}$$

$$\frac{k}{D} = 0.0022 \qquad V = \frac{1200}{1000} \times 4.1 = 4.92 \text{ ft/sec}$$

$$\text{Re} = \frac{4.92}{4.1} \times 285,000 = 342,000$$

From Fig. 37, $f = 0.025$; then

$$h_f = 0.025 \times \frac{1000}{5/6} \frac{4.92^2}{64.4} = 11.3 \text{ ft}$$

$$\Delta p = \frac{11.3 \times 62.4}{144} = 4.9 \text{ psi}$$

Conduits of Noncircular Cross Section The Darcy–Weisbach equation may also be applied to noncircular conduits if the diameter D is replaced by some equivalent linear measure of the cross section. The hydraulic radius R, widely used in open-channel equations, can be related to D for the circular cross section; this relationship is usually assumed to be a valid replacement of D in the pipe formula. The hydraulic radius is defined as the ratio of the cross-sectional area to the wetted perimeter. For a circular cross section

$$R = \frac{\pi D^2/4}{\pi D} = \frac{D}{4} \qquad (141)$$

Hence the diameter may be replaced by four times the hydraulic radius in the Reynolds number, the relative roughness, and the resistance equation; the resistance equation becomes

$$h_f = f \frac{L}{4R} \frac{V^2}{2g} \qquad (142)$$

Although satisfactory for conduits which are reasonably comparable to pipes in cross-sectional form, this equation cannot be expected to give accurate results for cross sections that are at great variance therefrom.

Example 5. Find the head loss per 1000 ft for a flow of 200 gpm of water through a clear cast-iron conduit of rectangular cross section 3 in. by 6 in.

$$R = \frac{3 \times 6}{12 + 6} = 1 \text{ in.} = \tfrac{1}{2} \text{ ft}$$

$$Q = \frac{200}{7.48 \times 60} = 0.446 \text{ ft}^3/\text{sec}$$

$$V = \frac{0.446}{18} \times 144 = 3.57 \text{ ft/sec}$$

Taking $v = 1.2 \times 10^{-5} \text{ ft}^2/\text{sec}$

$$\text{Re} = \frac{V 4R}{v} \approx 99,000$$

$$\frac{k}{D} = \frac{k}{4R} = \frac{0.00085}{4/12} = 0.00255$$

From Fig. 37, $f = 0.025$; hence

$$h_f = f \frac{L}{4R} \frac{V^2}{2g} = \frac{0.025 \times 1000}{4 \times \frac{1}{12}} \frac{3.57^2}{64.4} = 14.88 \text{ ft}$$

7.3 Minor Losses

Minor losses are the second type of friction losses and are caused by flow separation, eddies, and excessive turbulence beyond that occurring as a result of normal pipe friction. While the losses occur over a finite length of pipe, they are usually assumed to be concentrated at the location of the causative valve, fitting, and so on. This definition is depicted in Fig. 38.

The formula for minor losses of all kinds is generally of the form

$$h_m = K_L \frac{V^2}{2g} \qquad (143)$$

where h_m is the minor head loss, K_L is the loss coefficient whose value depends on the device causing the loss, and V is the velocity of flow in the smaller pipe. Virtually all hydraulics or fluid mechanics textbooks have tables of loss coefficients for various types of devices. In addition, Davis and Sorensen[5] list several tables of K_L values. The reference book published by

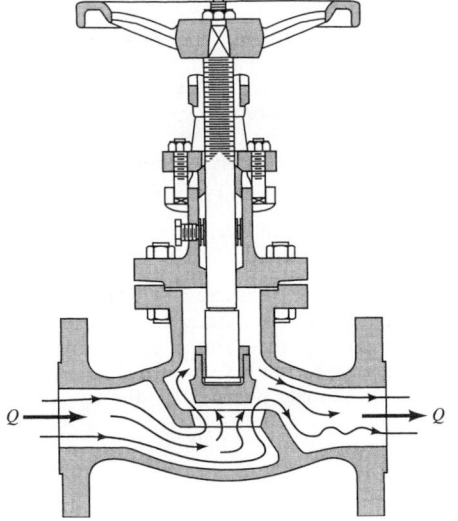

Fig. 38 Depiction of flow through valve illustrating causes of minor losses. From Ref. 2.

the Crane Company[8] is also a good source for loss coefficients, and a selection of K_L values for common situations has been presented in Tables 7 and 8. For an exhaustive collection of loss coefficients, see Idelchik.[9]

To demonstrate the use of the tables and a complete friction loss situation in pipeline analysis, the following example is presented.

Example 6. A series of pipes connecting two reservoirs composed of 150-ft-long 12-in. pipeline and 8-ft-long 6-in. pipeline conveys water at 50°F between two reservoirs. The pipes are asphalted cast iron and the entrance to the pipe is sharp edged. Compute the discharge in the pipeline.

Before starting the solution to the problem, it is useful to classify the various losses that are going to be acting on the flow:

$h_{m_{en}}$ = minor loss at sharp-edged entry to pipeline
$h_{f_{12}}$ = friction loss in 12-in. pipe
h_{m_c} = minor loss at connection between 12- and 6-in. pipes
h_{f_6} = friction loss in 6-in. pipe
$h_{m_{ex}}$ = minor loss at exit from pipeline

When the discharge is unknown, a flow velocity must be estimated to permit calculating Re and finding an f value.

Assume the velocity in the 12-in. pipe is 7 ft/sec. From Table 6

$$\nu = 1.41 \times 10^{-5} \text{ ft}^2/\text{sec}$$

$$Re_{12} = \frac{VD}{\nu} = \frac{7 \times 1.0}{1.41 \times 10^{-5}} = 496,000$$

$$Re_6 = \frac{28 \times 0.5}{1.41 \times 10^{-5}} = 993,000$$

Table 7 Loss Coefficients for Enlargements and Contractions Based on Velocity in Small Pipe[a]

Diameter ratio	0	0.1	0.2	0.3	0.4	0.5	0.6	0.7	0.8	0.9	1.0
Contraction K_L	0.5	0.48	0.45	0.42	0.38	0.34	0.29	0.22	0.12	0.04	0.0
Enlargement K_L	1.0	0.98	0.92	0.83	0.70	0.57	0.40	0.26	0.12	0.04	0.0

[a]Courtesy of Crane Company.[8]

Table 8 Selected Minor Loss Coefficients[a]

Minor Loss Device	K_L.	Minor Loss Device	K_L.
Pipe entrances		Valves	
Inward projecting	0.78	Globe	$340f$[b]
Sharp edged	0.50	Angle	$145f$
Slightly rounded	0.23	Ball or plug	$3f$
Well rounded	0.04	Butterfly	$40f$
Pipe exits (all types)	1.00	Gate (fully open)	$13f$
Bends		(75% open)	$35f$
90° miter bends	$58f$[b]	(50% open)	$160f$
45° miter bends	$15f$	(25% open)	$900f$
Fittings			
90° standard elbow	$30f$[b]		
45° standard elbow	$16f$		

[a]Courtesy of Crane Company.[8]
[b]f is friction factor for pipe.

From Table 6

$$k = 0.0048 \text{ in.}$$

$$\left(\frac{k}{D}\right)_{12} = \frac{0.0048}{12} = 0.0004$$

$$\left(\frac{k}{D}\right)_{6} = \frac{0.0048}{6} = 0.0008$$

From Fig. 36

$$f_{12} = 0.017 \qquad f_6 = 0.019$$

From the work–energy equation,

$$z_1 + \frac{p_1}{\gamma} + \frac{V_1^2}{2g} = z_2 + \frac{p_2}{\gamma} + \frac{V_2^2}{2g} + \sum h_{L_{1-2}}$$

$$4230 + 0 + 0 = 4200 + 0 + 0 + \sum h_{L_{1-2}}$$

$$\sum h_{L_{1-2}} = 30$$

Summarizing frictional head losses,

$$\sum h_{L_{1-2}} = h_{m_{en}} + h_{f_{12}} + h_{m_c} + h_{f_6} + h_{m_{ex}}$$

From Table 7

$$h_{m_{en}} = 0.5\frac{V_{12}^2}{2g} \qquad h_{m_r} = 0.34\frac{V_6^2}{2g}$$

$$h_{m_{ex}} = 1.0\frac{V_6^2}{2g}$$

$$\sum h_{L_{1-2}} = 0.5\frac{V_{12}^2}{2g} + 0.017 \times \frac{150}{1.0}\frac{V_{12}^2}{2g} + 0.34\frac{V_6^2}{2g}$$

$$+ 0.019 \times \frac{8}{6/12}\frac{V_6^2}{2g} + 1.0\frac{V_6^2}{2g}$$

From continuity considerations,

$$V_6 = 4V_{12}$$

$$\sum h_{L_{1-2}} = (0.5 + 2.55 + 5.44 + 4.86 + 16.0)\frac{V_{12}^2}{2g}$$

Solving for V_{12},

$$29.35\frac{V_{12}^2}{2g} = 30$$

$$V_{12} = 8.11 \text{ fps} \qquad V_6 = 32.45 \text{ fps} \qquad Q = 6.37 \text{ cfs}$$

Now a check must be made on the accuracy of the initial assumption used to find the f values:

$$\text{Re}_{12} = 575,000 \qquad f_{12} = 0.0169$$
$$\text{Re}_{6} = 1,150,000 \qquad f_{6} = 0.0190$$

Because these values are as close to the initial estimates as one could reasonably expect, we will consider the first estimates as the final values.

Another form of expressing minor losses is often used by the manufacturers of valves. This formula is

$$Q = C_v\sqrt{\Delta p} \qquad (144)$$

where C_v is the flow coefficient for the valve in question and Δp and Q are expressed in psi (pounds per square inch) and gpm (gallons per minute), respectively. Of course C_v and K_L measure the same thing. They are approximately related by the equation

$$K_L = \frac{890}{C_v^2}D^4 \qquad (145)$$

for the units already described with diameter expressed in inches.

Transitions are sections of conduit that connect one prismatic portion to another by a gradual change in cross section. Since, owing to the inherent stability of accelerated flow, losses are small in gradual contractions, transition design is usually determined by factors other than energy loss. For example, it is often important that the pressure decrease continuously to that of the reduced section, so that the sections will be both *separation proof* and *cavitation proof*.

In expanding transitions or diffusers, wherein kinetic energy is converted into potential energy, it is even more essential that separation be avoided. The slowly moving fluid near the wall, which is retarded by surface resistance, is also retarded by the adverse pressure gradient due to the flow expansion. If the adverse pressure gradient acts over a sufficient length, it is certain to result in boundary layer separation. Once separation occurs, with the backflow and eddies that accompany it, the losses become high. A series of experiments was conducted by Gibson[10] on conical diffusers, the results of which are shown in Fig. 39.

7.4 Steady-State Pipeline Analysis

The design situation in pipelines is generally that of selecting a pipe diameter that will convey a prescribed amount of discharge between two locations of known elevation with specified reservoir surface elevations or pressure requirements. The attractiveness of the Hazen–Williams and Manning formulas is immediately clear because, if minor losses are neglected, both formulas lead to direct solutions for the required diameter. Use of the Darcy formula requires several iterations beginning with an assumed diameter and ending with a final check on f, Re, and k/D to ensure all are consistent in the final solution. Techniques and skills can be developed to streamline this Darcy equation to determine head loss in a variety of systems with the purpose of refamiliarizing the reader with the techniques.

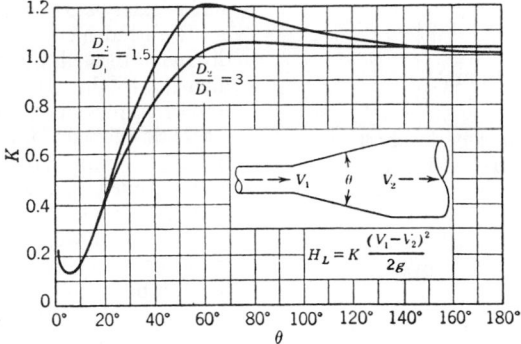

Fig. 39 Loss coefficients for conical diffusers.

Single Pipelines Single-pipeline problems are solved by applying the work–energy equation between two points of known energy levels. An example best illustrates the procedure.

Example 7 An 18-in. welded steel pipeline 1200 ft long connects two reservoirs differing in elevation by 20 ft (see Fig. 40). The pipe has a sharp-edged entrance and a wide-open globe valve at the downstream end neat the location where it enters the lower reservoir. Find the flow rate in the pipe. From the work–energy equation,

$$z_1 + \frac{p_1}{\gamma} + \frac{V_1^2}{2g} = z_2 + \frac{p_2}{\gamma} + \frac{V_2^2}{2g} + \sum h_{L_{1-2}}$$

Using the lower reservoir surface as a datum,

$$20 + 0 + 0 = 0 + 0 + 0 + \sum h_{L_{1-2}}$$

where $\sum h_{L_{1-2}} = $ (entrance + friction + valve + exit) losses. From Tables 7 and 8,

$$\sum h_{L_{1-2}} = 0.5\frac{V^2}{2g} + f\frac{L}{D}\frac{V^2}{2g} + 340f\frac{V^2}{2g} + \frac{V^2}{2g}$$

$$= \left(0.5 + f\frac{L}{D} + 340f + 1.0\right)\frac{V^2}{2g}$$

From Table 6,

$$k = 0.0018 \qquad \frac{k}{D} = 0.0001$$

From Table 3

$$V \approx 8 \text{ ft/see} \quad \text{Re} = \frac{8 \times 1.5}{1.2 \times 10^{-5}} = 1 \times 10^6$$

From Fig. 36, $f = 0.0135$. Substituting into the preceding work–energy equation,

$$20 = 0 + \left(0.5 + 0.0135 \times \frac{1200}{1.5} + 340\right.$$

$$\left. \times 0.0135 + 1.0\right)\frac{V^2}{2g}$$

$$= (0.5 + 10.8 + 4.6 + 1.0)\frac{V^2}{2g}$$

$$= 16.9\frac{V^2}{2g}$$

$$V = 8.73 \text{ ft/sec}$$

Checking on initial assumptions,

$$\text{Re} = \frac{8.73 \times 1.5}{1.2 \times 10^{-5}} = 1.1 \times 10^6$$

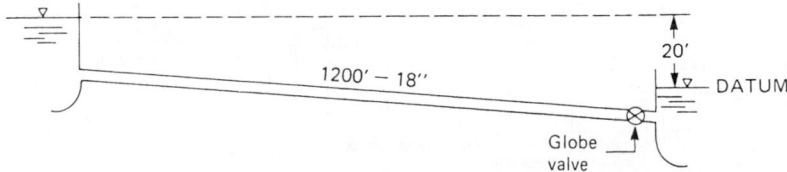

Fig. 40 Diagram for Example 7. By permission from G. Z. Watters, *Analysis and Control of Unsteady Flow in Pipelines*, Stoneham, MA, 1984.

$f = 0.0135$, so solution is acceptable

$$Q = AV = 0.7854 \times 1.5^2 \times 8.73$$

$$Q = 15.4 \, \text{ft}^3/\text{sec or 6927 gpm}$$

Series pipelines, that is, pipelines of different diameters connected end to end, are handled in a manner similar to the previous example. It is only required that the single pipe friction term be replaced by a summation of the pipe friction losses in each of the pipes, including the additional minor losses between pipes.

Single Pipelines with Pumps The pumped pipeline is a common design situation with which engineers are confronted. Pumps occur at the upstream end of pipelines (source pumps) and at intermediate locations in the pipeline (booster pumps). In each case, the discharge through the pump and the head increase across the pump are affected by the pipe system in which the pump is installed.

To analyze pumped pipelines, the work–energy equation must be modified to include the energy added by the pump:

$$z_1 + \frac{p_1}{\gamma} + \frac{V_1^2}{2g} + H_p = z_2 + \frac{p_2}{\gamma} + \frac{V_2^2}{2g} + \sum h_{L_{1-2}}$$
(146)

where H_p is the energy added to each pound of liquid passing through the pump. In the case of series or multistaged pumps, H_p is the sum of the head increases across each pump or stage.

The head increase across a pump is a function of discharge through the pump and is determined experimentally by the manufacturer. The information is presented graphically on a diagram known as the characteristic diagram. Information on the power requirements and pump efficiencies at varying discharges is also included. An example of a typical pump characteristic diagram is shown in Fig. 41.

Because power requirements for a pumping situation are of interest, it is important to establish the relation between energy added (H_p) and brake horsepower required. The power added to the liquid by the pump can be expressed as

$$\text{WHP} = \frac{Q\gamma H_p}{550}$$
(147)

where WHP is the horsepower added to the water. Of course, a greater amount of power must be added to the pump shaft because of friction and other losses in the pumping process. The power that must be supplied to a pump shaft (brake horsepower, BHP) in order to provide a given WHP is related to the hydraulic parameters by the equation

$$\text{BHP} = \frac{Q\gamma H_p}{550\eta}$$
(148)

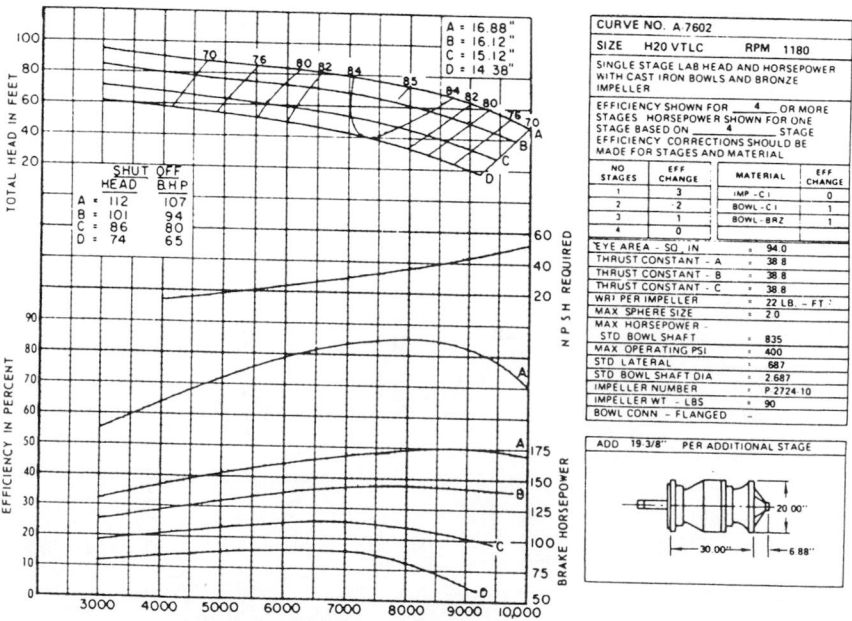

Fig. 41 Typical pump characteristics diagram for vertical turbine pump. (Courtesy of Allis-Chalmers.) By permission from G. Z. Watters, *Analysis and Control of Unsteady Flow in Pipelines*, Butterworth, Stoneham, MA, 1984.

where η is the overall pump efficiency. Both overall pump efficiency and brake horsepower are displayed on the pump characteristic diagram in Fig. 41.

Two examples follow that demonstrate the use of the characteristic diagrams in pipeline analysis.

Example 8. A single-stage pump with the characteristics shown in Fig. 41 (curve A) is used to pump water from a reservoir of elevation 1350 ft to another reservoir at elevation 1400 ft. The line is 6000 ft long and 24 in. in diameter with an f value of 0.021.

Neglecting minor losses, compute the discharge in the pipeline. From the work–energy equation

$$z_1 + \frac{p_1}{\gamma} + \frac{V_1^2}{2g} + H_p = z_2 + \frac{p_2}{\gamma} + \frac{V_2^2}{2g} + \sum h_{L_{1-2}}$$

$$1350 + 0 + 0 + H_p = 1400 + 0 + 0 + f\frac{L}{D}\frac{V^2}{2g}$$

$$H_p = 50 + 0.021 \times \frac{6000}{2}\frac{V^2}{2g}$$

$$= 50 + 63.0\frac{V^2}{2g}$$

This equation must be solved by trial in conjunction with the head-versus-discharge characteristic for the pump given in Fig. 41. In the solution process we neglect losses in the pump discharge column and head, which would normally be included in an analysis.

The solution is best approached using a trial solution table:

Assume Q (gpm)	V (ft/sec)	$\frac{V^2}{2g}$	$63\frac{V^2}{2g}$	RHS[a]	H_p (ft)[a]
8000	5.67	0.50	31.5	81.5	73
7000	4.96	0.38	24.1	74.1	79
7500	5.32	0.44	27.7	77.7	76
7400	5.25	0.43	26.9	76.9	77

[a]Right-hand side $(50 + 63. 0V^2/2g)$.

The solution is $Q = 7400$ gpm.

Example 9. Solve the problem of Example 8 if two three-stage parallel pumps with curve C characteristics were employed.

The pipeline analysis would remain the same:

$$H_p = 50 + 63.0\frac{V^2}{2g}$$

However, H_p is now the total head put out by the three stages in each pump. Also the pipeline discharge must

be halved to obtain the amount passing through each pump.

The resulting trial solution table is as follows:

Assume Pump Q (gpm)	Q_{pipe}	V (ft/sec)	$\frac{V^2}{2g}$	RHS	$H_p/$ stage	H_p (ft)
7000	14,000	9.93	1.53	146	54	159
7500	15,000	10.63	1.76	161	50	150
7250	14,500	10.28	1.64	153	51	153

The solution is $Q = 7250$ gpm.

7.5 Pipe Network Analysis

The steady-state analysis of flows in pipe networks can be a very complex problem. Devices such as pressure-reducing valves, minor losses, booster pumps, and supply pumps, as well as reservoirs, all serve to complicate the analysis. The subject is covered comprehensively by Jeppson[11] and the reader is referred to that text for detailed information. The presentation here will be introductory and will apply only to relatively simple systems. However, we shall discuss all three of the most popular analysis methods—the Hardy Cross method, the linear theory method, and the Newton—Raphson method. In addition, a good summary of the application of these three methods is given by Wood and Rayes.[12]

Hardy Cross Method Because of its simplicity of application, its easily understood theory, and its amenability to hand calculation, the Hardy Cross method has enjoyed (and still enjoys) considerable popularity among practicing engineers.

The first step is to estimate flow rates in all the pipes in a network so that continuity is satisfied at each junction (node). Of course it is unlikely that the energy line (EL) and hydraulic grade line (HGL) are continuous throughout the network because the original estimates of the flow rates are always erroneous to some degree. This method assumes that there can be found a unique flow rate adjustment that can be applied to each loop in the network that will cause the EL–HGL to be continuous around each loop. In hydraulic terms this continuity is expressed as

$$\sum_{i=1}^{N} h_{L_i} = 0 \qquad (149)$$

around each loop where i is the pipe number and N is the number of pipes in the loop. Assuming that the head loss can be written in the form

$$h_{L_i} = K_i Q_i^n \qquad (150)$$

and assuming a correction ΔQ is being added to each pipe flow in the loop to satisfy the requirement that the sum of the head loss equals zero, this equation becomes

$$\sum_{i=1}^{N} = K_i(Q_1 + \Delta Q)^n = 0 \qquad (151)$$

where Q_i is the most recent estimate for the discharge in each pipe in the loop. It remains only to solve for ΔQ.

Because n is generally a noninteger, the preceding is generally expanded by the binomial theorem to yield an equation for ΔQ. Retaining only the first two terms of the binomial expansion, the following equation for ΔQ is derived:

$$\Delta Q = \frac{\sum_{i=1}^{N} K_i Q_i^n}{-n \sum_{i=1}^{N} K_i Q_i^{n-1}} \qquad (152)$$

To produce the proper sign on ΔQ, the denominator is kept negative and the terms in the summation in the numerator are positive or negative, depending on whether one moves with or against the flow while proceeding clockwise around the loop.

Once the ΔQ is computed for each loop, it is added (or subtracted) from the flow rates in each member of the loop to get a better estimate of the true flow rate. Because the decomposition of $(Q_i + \Delta Q)^n$ with the binomial theorem was not exact and because pipes that are common to more than one loop have multiple ΔQ corrections, the calculated ΔQ's will not be correct. Therefore, the process is iterative and must be continued until the error is acceptably small (or no convergence to a solution occurs).

Although this numerical method is not so sophisticated as the other methods, the results are just as valid, provided convergence is obtained. Actually, a more careful investigation would reveal that Hardy Cross analysis is a decoupled Newton–Raphson analysis.

Linear Theory Method The linear theory method is a technique for solving a set of network equations, some of which are nonlinear, for the unknown flow rates in the pipes. The equations are generated by writing continuity equations for flow into and out of each junction and by specifying that the algebraic sum of the head losses around each loop is zero. Solving a set of nonlinear equations is an iterative process and there are many techniques for doing this. In the linear theory approach the nonlinear equations for the sum of the head losses around each loop are linearized. Then the complete set of linear equations (the continuity equations are already linear) is solved.

To understand how the procedure works, look at the equations involved. For each loop in the network the following equation is valid:

$$\sum_{i=1}^{N} h_{L_i} = \sum_{i=1}^{N} K_i Q_i^n = 0 \qquad (153)$$

where N is the number of pipes in the loop. To linearize this equation, Q_i^n is decomposed into two parts so that this equation becomes

$$\sum_{i=1}^{N} (K_i Q_i^{n-1}) Q_i = \sum_{i=1}^{N} K_i' Q_i = 0 \qquad (154)$$

Of course K_i' is now a function of Q_i so the process is still iterative.

As the set of linear-plus-linearized equations is successively solved, the estimate of K_i' is revised after each solution. After several iterations the values of Q_i and K_i' should converge to their final values. The mathematical form of the iteration equation is

$$\sum_{i=1}^{N} K_i [Q_i^{(j-1)}]^{n-1} Q_i^j = 0 \qquad (155)$$

where j is the iteration number. For example, if we are making the eighth iteration ($j = 8$), then we would calculate K_i''s from the results of the seventh iteration.

Experience with the linear theory has shown that the numerical solution tends to oscillate around the final values. To damp out this numerical oscillation, the iteration equation is altered to include the last two iterations for Q_i in computing K_i':

$$\sum_{i=1}^{N} K_i \left[\frac{Q_i^{(j-1)} + Q_i^{(j-2)}}{2} \right]^{n-1} Q_i^{(j)} = 0 \qquad (156)$$

When starting an analysis, only the direction of flow (not the quantity) has to be specified. This is a substantial savings in effort over the Hardy Cross approach. For the first iteration $K_i'^{(1)}$ is assumed to be equal to K_i. For the second iteration $K_i'^{(2)}$ will equal $K_i[Q_i^{(1)}]^{n-1}$. Thereafter, Eq. (156) will be used for each loop.

Newton–Raphson Method The Newton–Raphson technique has the same conceptual basis as the Hardy Cross method. Flow rates in each pipe are assumed that satisfy continuity, and these flow rates are corrected so that the sum of the head losses around each loop approaches zero. In the Hardy Cross method the flow rates in each pipe are corrected after each ΔQ computation. In the Newton–Raphson method the equations containing ΔQ are written for each loop; then this nonlinear set of equations is solved successively for the final value of ΔQ in each loop. When the solution is complete, only then are the initial flow rates in each pipe adjusted to their final value.

The method gets its name from the technique used to solve the nonlinear set of equations. The Newton–Raphson technique is a frequently used, powerful method of numerical analysis. In operation, it adjusts successive approximations to the solution by computing the way the solution is moving with respect to each variable and then, based on that computation, calculates new trial values for the unknowns.

The Newton–Raphson technique in two or more dimensions (two or more equations with two or more unknowns) is most conveniently expressed in matrix form,

$$\{F^{(j-1)}\} + [\mathbf{J}^{(j-1)}]\{x^{(j)} - x^{(j-1)}\} = \{0\} \qquad (157)$$

where $\mathbf{J}$ is a $K \times K$ matrix of $\partial F^{(j-1)}/\partial x_k$ known as the Jacobian. Converting this to another form,

$$\{x^{(j)}\} = \{x^{(j-1)}\} - [\mathbf{J}^{(j-1)}]^{-1}\{F^{(j-1)}\} \qquad (158)$$

Because all the F_k can be differentiated, the Jacobian can be evaluated at each new approximation for x_i and the inverse computed. However, this is a very large computational task for large systems of equations; hence, a slightly different approach is employed when working with hydraulic networks.

In hydraulic networks,

$$F_k = \sum_{i=1}^{N} K_i Q_i^n = 0 \qquad (159)$$

However, because the Q_i's are unknown, a value in each pipe must be estimated and a search for ΔQ's, which will correct the Q_0's to the proper value, must be made. The loop equations now are of the form

$$F_k = \sum_{i=1}^{N} K_{L_i} \left[Q_{0_i} + \sum_{l=1}^{K} \Delta Q_l \right]^n = 0 \qquad k = 1, \ K$$

$$(160)$$

because any pipe in a given loop may be a member of other loops and their ΔQ's must be included. Then, in general,

$$F_k(\Delta Q_1, \Delta Q_2, \ldots, \Delta Q_k) = 0 \qquad k = 1, \ K \quad (161)$$

If $\Delta Q^{(j)} - \Delta Q^{(j-1)}$ is now represented as $\delta Q^{(j)}$ in each loop, then

$$[\mathbf{J}^{(j-1)}]\{\delta Q^{(j)}\} = -\{F^{(j-1)}\} \qquad (162)$$

This equation is now solved for $\{\delta Q^{(j)}\}$ and

$$\{\Delta Q^{(j)}\} = \{\Delta Q^{(j-1)}\} + \{\delta Q^{(j)}\} \qquad (163)$$

after each iteration. When the δQ's become small enough, an acceptable solution has been obtained.

It should be noted here that this method only requires the solution to a set of equations equal in number to the number of loops. The linear theory must solve a set of equations equal in number to the number of unknown flow rates. Consequently, the

Newton–Raphson technique may require substantially less storage space for solution.

7.6 Unsteady Flow in Pipe Systems

Unsteady flow in pipe systems is important because it can result in serious problems. Some of these are:

1. Pipe rupture
2. Pipe collapse
3. Vibration
4. Excessive pipe displacements
5. Pipe fitting and support deformation or failure
6. Vapor cavity formation (cavitation, column separation)

Some of the primary causes of unsteady flow are:

1. Valve closure (or opening)
2. Flow demand changes
3. Pump shutdown (or power failure to the pump)
4. Pump startup
5. Air venting from lines
6. Failure of flow or pressure regulators
7. Pipe rupture

Unsteady flow analysis in pipe systems is generally divided into two categories:

1. Rigid water column theory (surge theory) where the fluid and pipe are inelastic, pressure changes propagate instantaneously, and the differential equation of motion is "ordinary"
2. Elastic theory (water hammer) where the elasticity of fluid and pipe affect pressure changes, pressure changes propagate with wave speed $\mathbf{a}$ (1000–4700 ft/sec), and the differential equations of motion are partial and nonlinear

A simple problem is used to demonstrate phenomena and introduce concepts. A steady flow situation is shown in Fig. 42 where velocity V is caused by head H in reservoir. Friction is neglected and the EL and HGL are coincident because water hammer pressures are large compared to velocity head.

The valve is closed suddenly causing a pressure head to propagate upstream at speed $\mathbf{a}$. The sequence of events shown in Fig. 42 are as follows:

1. Pressure head increase ΔH reaches reservoir at $L/\mathbf{a}$ seconds. Velocity $= 0$ and pressure head $= H + \Delta H$ throughout pipe. Pipe is stretched. Water is compressed.
2. High pressure in pipe ejects water into reservoir. At $2L/\mathbf{a}$ seconds, velocity $= -V$, pressure head $= H$ throughout pipe.

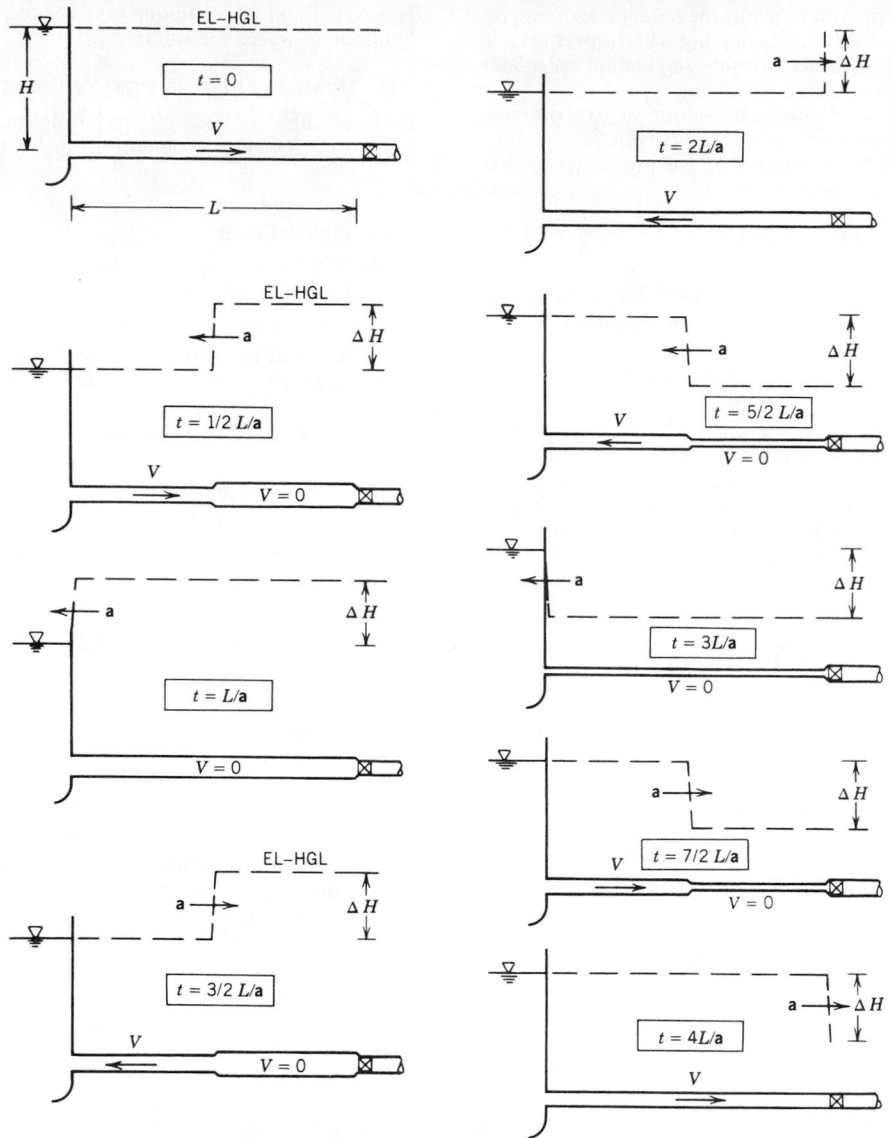

Fig. 42 Pressure wave propagation in simple pipe system. By permission from G. Z. Watters, *Analysis and Control of Unsteady Flow in Pipelines*, Butterworth, Stoneham, MA, 1984.

3. Upstream flow suddenly stopped at valve. Negative wave propagates upstream. At 3 L/a seconds, velocity = 0, pressure head = $H - \Delta H$ throughout pipe.

4. High pressure in reservoir forces water into pipe giving downstream flow. At 4 L/a, velocity = V, pressure = H, and wave period is complete.

5. As long as valve remains closed, these cycles repeat. Any friction in system would cause damping of ΔH until it eventually disappears.

The following are important ideas:

1. L/a is an important time parameter in water hammer situations.

2. Pressure head at the valve reaches its maximum if the valve is closed in any time less than 2 L/a seconds. Valve need not be suddenly closed to create maximum water hammer pressures.

Column Separation When pressure change is severe enough to drop pressure in the pipe to the vapor pressure of water, "column separation" occurs. Dissolved gases come out of solution, water vapor cavities occur, and liquid columns "separate." Eventually cavity closure causes water hammer pressure "shocks" of a magnitude difficult to calculate and potentially destructive.

Column separation could be caused by simple valve closure. In Fig. 42, if ΔH were large enough, column separation would occur on both sides of the valve, although at different times.

Rigid Water Column Theory   This analysis uses Newton's second law, $F = ma$. For unsteady flow the equation is

$$\frac{P_1}{\gamma} - \frac{P_2}{\gamma} - \frac{fL}{2gD}V^2 = \frac{L}{g}\frac{dV}{dt} \qquad (164)$$

where p = pressure, f = Darcy–Weisbach friction factor, L = pipe length, γ = specific weight of water, D = pipe diameter, V = flow velocity, g = acceleration of gravity, and dV/dt = liquid acceleration.

Example 10   Flow establishment in a pipe. The physical situation is shown in Fig. 43 with free discharge at the valve.

The equation applied to this problem is

$$H_0 - \frac{fL}{2gD}V^2 = \frac{L}{g}\frac{dV}{dt}$$

This expression can be integrated in closed form. The time to reach steady flow is infinite. To reach 99% of V,

$$t_{99} = 2.65\frac{LV_0}{gH_0}$$

Pipe systems may not be single, constant-diameter pipes. The equivalent pipe technique may be used to reduce complex pipe systems to single pipes. Criteria for equivalence is friction loss equality and similar dynamic behavior. For series pipes

$$\left[\frac{fL}{D^5}\right]_{eq} = \sum_{i=1}^{N}\left[\frac{F_iL_i}{D_i^5}\right] \qquad (165)$$

$$\left[\frac{L}{D^2}\right]_{eq} = \sum_{i=1}^{N}\left[\frac{L_i}{D_i^2}\right] \qquad (166)$$

For parallel pipes:

$$\left[\frac{D_{eq}^5}{F_{eq}L_{eq}}\right]^{1/2} = \sum_{i=1}^{N}\left[\frac{D_i^5}{f_iL_i}\right]^{1/2} \qquad (167)$$

$$\left[\frac{D_{eq}^2}{L_{eq}}\right] = \sum_{i=1}^{N}\left[\frac{D_i^2}{L_i}\right] \qquad (168)$$

Example 11   A three-unit pumped storage facility is shown in Fig. 44. Flow through the turbines is shut down so that penstock velocities decrease from 60 ft/sec to zero linearly in 30 sec. Compute p_{max} if the f value is the same for all pipes.

Using the parallel-pipe equations,

$$\frac{D_{eq}^2}{L_{eq}} = 3\left(\frac{8^2}{800}\right) = 0.240 \qquad \left[\frac{D_{eq}^5}{L_{eq}}\right]^{1/2} = 3\left[\frac{8^5}{800}\right]^{1/2}$$

$$= 19.20$$

The parallel pipes can be replaced by a single pipe $D = 11.54$ ft and $L = 555$ ft:

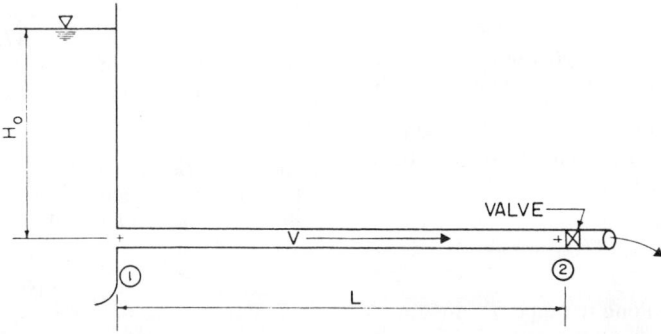

L = 2000'   L = 555'

D = 24'   D = 11.54'

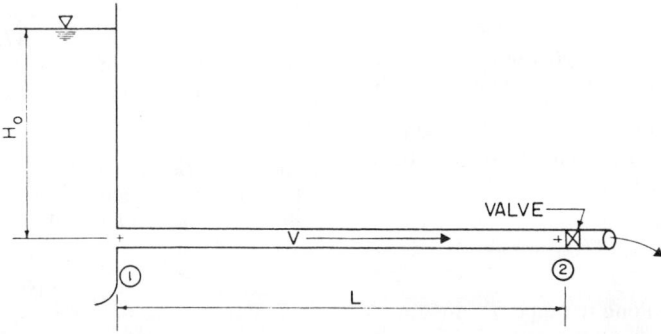

Fig. 43   Simple system for applying rigid water column theory. By permission from G. Z. Watters, *Analysis and Control of Unsteady Flow in Pipelines*, Butterworth, Stoneham, MA, 1984.

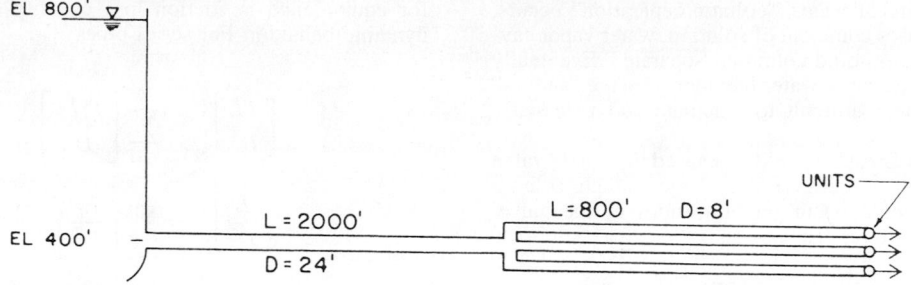

Fig. 44 Schematic sketch of a pumped storage facility. By permission from G. Z. Watters, *Analysis and Control of Unsteady Flow in Pipelines*, Butterworth, Stoneham, Mass., 1984.

Using the series-pipe equation,

$$\frac{L_{eq}}{D_{eq}^2} = \frac{2000}{24^2} + \frac{555}{11.54^2} = 7.64$$

$$\frac{L_{eq}}{D_{eq}^5} = \frac{2000}{24^5} + \frac{555}{11.54^5} = 0.00296$$

This series pipe is replaced by the single pipe $D = 13.71$ ft and $L = 1437$ ft. Steady flow velocity in the equivalent pipe is

$$V_{eq} = \frac{Q}{A_{eq}} = 61.3 \text{ ft/sec}$$

From the unsteady flow equation,

$$400 - \frac{p_2}{\gamma} - \frac{fLV^2}{2gD} = \left(\frac{1437}{32.2}\right)\frac{0 - 61.3}{30}$$

$$\frac{p_2}{\gamma} = 400 + 91.2 - \frac{fLV^2}{2gD}$$

A maximum pressure head of 491 ft occurs at the instant the valve is completely closed.

Elastic Theory Elastic theory includes the effect of water and pipe elasticity on pressures and velocities. The impulse–momentum equation and the conservation-of-mass equation are used.

The impulse–momentum equation is used to develop the equation for ΔH. The x component of the equation is

$$\left(\sum \mathbf{F}_{ext}\right)_x = Q\rho(V_{out} - V_{in}) \qquad (169)$$

where the x direction is along the pipe, F_x are forces in the x direction, and Q is discharge.

To make the unsteady case steady, the coordinate system is moved along the pipe at the wave speed so

the wave appears to be standing still. The resulting analysis gives

$$\Delta H = \frac{\mathbf{a}}{g}\Delta V \qquad (170)$$

It is clear that ΔH depends on the wave speed **a**. Conservation of mass is used to find an equation for wave speed. The result for thin-walled pipes is

$$\mathbf{a} = \frac{[K/\rho]^{1/2}}{\left[1 + \frac{K}{E}\frac{D}{e}(C)\right]^{1/2}} \qquad (171)$$

where K = bulk modulus of elasticity of liquid, E = modulus of elasticity of pipe, D = pipe diameter, e = pipe wall thickness, and C = restraint coefficient:

$C = \frac{5}{4} - \mu$ if pipe is free to stretch in longitudinal direction as a pressure vessel.

$C = 1 - \mu^2$ if no longitudinal stretching occurs.

$C = 1.0$ if functioning expansion joints occur, where μ is Poisson's ratio.

Thin-walled pipes have D/e greater than about 40. Suggested values of E and μ are shown in Table 9.

For water $[K/p]^{1/2} = 4720$ ft/sec. Most buried-pipe situations are close to $C = 1 - \mu^2$; however, for $\mu \approx 0.3$, the result is about the same regardless of restraint.

Example 12 A 10,000-ft pipe has $V = 10$ ft/sec and a wave speed of 3220 ft/sec. Compute head increase at a valve for sudden valve closure:

$$\Delta H = \frac{\mathbf{a}}{g}\Delta V = \frac{3220}{32.2}10 = 1000 \quad \text{or} \quad 433 \text{ psi}$$

Note that $L/a = 3.1$ sec. If the valve is closed in less than 6 sec, full water hammer pressure is developed.

Table 9 E and μ Values for Common Pipe Materials

Steel	$E = 30 \times 10^6$ psi	$\mu \approx 0.30$
Ductile cast iron	$E = 24 \times 10^6$ psi	$\mu \approx 0.28$
Copper	$E = 16 \times 10^6$ psi	$\mu \approx 0.36$
Brass	$E = 15 \times 10^6$ psi	$\mu \approx 0.34$
Aluminum	$E = 10.5 \times 10^6$ psi	$\mu \approx 0.33$
PVC	$E = 4 \times 10^5$ psi	$\mu \approx 0.45$
Fiberglass-reinforced plastic (FRP)	$E_2 = 4.0 \times 10^6$ psi	$\mu_2 = 0.27 - 0.30$
	$E_1 = 1.3 \times 10^6$ psi	$\mu_1 = 0.20 - 0.24$
Asbestos cement	$E \approx 3.4 \times 10^6$ psi	$\mu \approx 0.30$
Concrete	$E = 57{,}000\sqrt{f_c'}{}^a$	$\mu \approx 0.24$ (dynamically)

^{a} Where $f_c' = $ 28-day strength.

The previous C values for thin-walled pipes must be modified when D/e is less than 40. For homogeneous pipes,

Case (a): $\quad C = \dfrac{1}{1 + e/D}\left[\left(\dfrac{5}{4} - \mu\right)\right.$

$$\left. + 2\dfrac{e}{D}(1+\mu)\left(1 + \dfrac{e}{D}\right)\right] \qquad (172)$$

Case (b): $\quad C = \dfrac{1}{1 + e/D}\left[(1 - \mu^2)\right.$

$$\left. + 2\dfrac{e}{D}(1+\mu)\left(1 + \dfrac{e}{D}\right)\right] \qquad (173)$$

Case (c): $\quad C = \dfrac{1}{1 + e/D}$

$$\times \left[1 + 2\dfrac{e}{D}(1+\mu)\left(1 + \dfrac{e}{D}\right)\right] \qquad (174)$$

These C values are used in the wave speed equation to compute wave speed.

When air or other dissolved gases come out of solution and form small bubbles, wave speed is affected dramatically. This occurs because:

1. Low pressure at pipeline summit allows air release.
2. Pump sump is aerated by improper inflow design.

If the fraction of the air volume is known, the wave speed can be estimated from the following equation:

$$\mathbf{a} = \dfrac{\sqrt{K_l/\rho_{\text{ave}}}}{\sqrt{1 + \frac{K_l}{E}\frac{D}{e}C + (\text{void fraction})\frac{K_l}{K_a}}} \qquad (175)$$

Amounts of air as low as 0.5% can reduce wave velocity to 25% of its unaerated value.

The previous approach permits calculation of pressure head increase at a point where velocity changes suddenly. Generally, it is necessary to find head H and velocity V at any section of pipe system at any time t.

To accomplish this Newton's second law and conservation of mass are applied to a differential length of pipe through which the water hammer wave is passing. The result is two partial differential equations,

$$\dfrac{dV}{dt} + g\dfrac{\partial H}{\partial s} + \dfrac{1}{2}\dfrac{F}{D}V|V| = 0 \qquad (176)$$

$$\dfrac{\mathbf{a}^2}{g}\dfrac{\partial V}{\partial s} + V\left[\dfrac{\partial H}{\partial s} - \dfrac{\partial z}{\partial s}\right] + \dfrac{\partial H}{\partial t} = 0 \qquad (177)$$

where $s = $ location along pipe and $z = $ elevation above datum.

This type of equation can be solved using the "method of characteristics." First, the equations are simplified by replacing dV/dt with $\partial V/\partial t$ and deleting $\partial H/\partial s$ in Eqs. (176) and (177). This approximation had been shown to have negligible effects on accuracy. If independent variables s and t follow certain relationship in Eqs. (176) and (177), namely

$$\dfrac{ds}{dt} = \pm\mathbf{a} \qquad (178)$$

Then *partial* differential equations can be written as *total* differential equations:

$$C^+ : \quad \dfrac{dV}{dt} + \dfrac{g}{\mathbf{a}}\dfrac{dH}{dt} + \dfrac{f}{2D}V|V| = 0 \qquad (179)$$

if

$$\dfrac{ds}{dt} = \mathbf{a} \qquad (180)$$

and

$$C^- : \quad \dfrac{dV}{dt} = \dfrac{g}{\mathbf{a}}\dfrac{dH}{dt} + \dfrac{f}{2D}V|V| = 0 \qquad (181)$$

if

$$\dfrac{ds}{dt} = -\mathbf{a} \qquad (182)$$

Equations (180) and (182) called the characteristics of the C^+ and C^- equations, respectively.

The physical meaning of C^+, C^- and characteristic equations is that changes in pressure caused by disturbances (valve closing) propagate at wave speeds upstream and downstream in the pipe $(ds/dt = \pm\mathbf{a})$. If this rule is followed, the partial differential equations become total differential equations. The equations are solved numerically as described in Watters,[13] Wylie and Streeter,[14] and Chaudhry.[15]

8 FLOW IN OPEN CHANNELS*

Flow in open channels is similar to that in pipes in that flow can be laminar or turbulent, can have smooth or rough boundaries, and be uniform or nonuniform. Open-chanel flow has the one unique characeristic that the pressure is zero on the free surface. Laminar flow is quite rare and will not be addressed here. This work will consider primarily uniform and nonuniform flow in the turbulent rough boundary mode.

8.1 Uniform Flow

In steady uniform flow, the slope fo channel bottom, free surface (hydraulic grade line), and energy grade line are the same (tan θ, Fig. 45). For very wide channels, the shear stress varies linearly with distance from the free surface y, given by $\tau = y\gamma \sin\theta$. For other channels, the average shear stress τ_0 at the solid boundary is $\tau_0 = \gamma R \sin\theta$, where R is the hydraulic radius, which is defined as the ratio of area of cross section A to wetted perimeter P. The liquid velocity at the solid boundary is zero; it increase generally with distance from a boundary. The maximum velocity is usually below the free surface.

The *Manning formula* is the most commonly used open-channel formula,

$$V = \frac{1.49}{n} R^{2/3} S^{1/2} \tag{183}$$

where V is the average velocity, R is the hydraulic radius, $S = \sin\theta$ (Fig. 45), and n is an absolute rough-

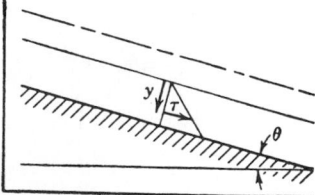

Fig. 45 Uniform flow.

---

*Material and illustrations extracted in whole or in part from previous edition by Victor L. Streeter.

ness factor, having the dimensions $L^{1/6}$, whose values for different surfaces are determined experimentally. Table 10 lists many of these values. Since the constant in the Manning formula is not dimensionless, it is necessary to use the foot-pound-second system of units:

Multiplying the formula by A,

$$Q = \frac{1.49}{n} A R^{2/3} S^{1/2} \tag{184}$$

When the cross section is known, the equation may be solved directly for any one of the other quantities that is unknown. For determination of depth of flow in a given section, with Q, n, S given, the solution is effected by trial.

Example 13 Find the depth of flow in a trapezoidal channel of roughness 0.012, bottom width 10 ft, and side slopes $1:1$ for 650 ft³/sec. The channel slope is 0.0009.

Writing

$$AR^{2/3} = \frac{A^{5/3}}{P^{2/3}}$$

Table 10 Average Manning n Values for Selected Boundaries

Closed Conduits Flowing Partially Full	
Welded steel	0.012
Coated cast iron	0.013
Uncoated cast iron	0.014
Corrugated metal storm drain	0.024
Cement mortar	0.013
Concrete culvert	0.011
Finished concrete	0.012
Unfinished concrete (smooth wood form)	0.014
Vitrified sewer pipe	0.014
Lined Open Channels	
Painted steel	0.013
Cement mortar	0.013
Planed, untreated wood	0.012
Unfinished concrete	0.017
Gunite concrete (good)	0.019
Glazed brick	0.013
Cemented rubble	0.025
Smooth asphalt	0.013
Excavated Channels	
Earth, straight, uniform and clean	0.018
Gravel, straight, uniform and clean	0.025
Earth with short grass and a few weeds	0.027
Dredged channel	0.028
Smooth rock cuts	0.035
Jagged rock cuts	0.040

Source: V. T. Chow, *Open Channel Hydraulics*, McGraw-Hill, New York, 1959.

from the Manning formula

$$\frac{Qn}{1.49S^{1/2}} = \frac{A^{5/3}}{P^{2/3}} = \frac{650 \times 0.012}{1.49 \times 0.03} = 174.7$$

$A = 10D + D^2$, $P = 10 + 2\sqrt{2}\,D$; hence

$$f(D) = \frac{(100 + D^2)^{5/3}}{(10 + 2\sqrt{2}D)^{2/3}} = 174.7$$

Trying $D = 5$, $f(D) = 160$; hence D must be larger. Trying $D = 5.5$, $f(D) = 191$. By straight-line interpolation, $D = 5.24$, $f(D) = 174$, which is a satisfactory check. Hence $D = 5.24$ ft is the answer sought.

The cross section having the least perimeter for given conditions is called the *most efficient* cross section. The semicircular section is the most efficient of all cross sections since it has the least perimeter for a given area. The most efficient *rectangular* channel has a bottom width twice the depth. The most efficient *trapezoidal* channel is half of a hexagon.

Specific Energy — Critical Depth   The mechanical energy per unit weight, with elevation datum taken as the bottom of the channel, is called *specific energy*. It is simply the sum of depth of flow and velocity head. In steady uniform flow, when all cross sections are identical, the specific energy is constant along the channel.

Referring to Fig. 46, the specific energy is

$$E = y + \frac{V^2}{2g} \tag{185}$$

assuming uniform distribution of velocity over the cross section. For a given discharge Q, the specific energy varies with the depth of flow. Substituting $V = Q/A$, where A is the cross-sectional area and a function of y,

$$E = y + \frac{Q^2}{2gA^2} \tag{186}$$

For a unit width of rectangular channel, with q the discharge per unit width,

$$E = y + \frac{q^2}{2gy^2} \tag{187}$$

A plot of specific energy against depth, Fig. 47, for a constant q, reveals that a certain minimum specific energy is required for the flow, found by setting $dE/dy = 0$. Calling this depth y_c the *critical depth*, we have

$$y_c = \left(\frac{q^2}{g}\right)^{1/3} \tag{188}$$

In terms of the velocity, $V_c = \sqrt{gy_c}$. Hence the critical depth is the depth at which the velocity of flow V_c is just equal to the velocity of an elementary wave $\sqrt{gy}$ in still liquid. Greater specific energy is required for both greater and lesser depths of flow. It is obvious from Fig. 47 that there are two depths at which the flow has the same specific energy.

For nonrectangular channels the critical depth occurs when

$$\frac{Q^2}{g} = \frac{A^3}{b} \tag{189}$$

where b is the top width of the cross section at the liquid surface.

8.2   Steady, Nonuniform Flow

Gradually varied channel flow is steady flow in which changes in depth, section, slope, and roughness with respect to length along the channel are small. By assuming that the energy loss at any section is the same as in uniform flow at the same discharge and

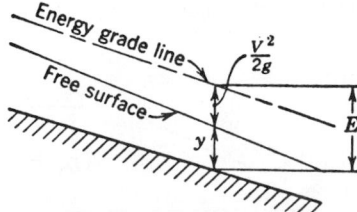

Fig. 46   Specific energy.

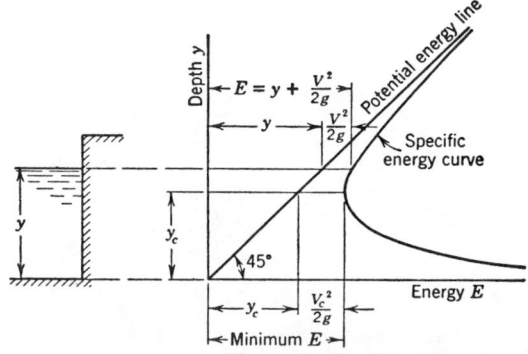

Fig. 47   Specific energy diagram. (Courtesy of R. A. Dodge, University of Michigan, Ann Arbor.)

the same depth, a differential equation for change in depth as a function of distance along the channel can be developed:

$$\frac{dy}{dl} = \frac{S_0 - n^2 Q^2 / [(1.49)^2 A^2 R^{4/3}]}{1 - Q^2 b / (g A^3)} \qquad (190)$$

where y is the depth, l the distance along the channel, S_0 the since of the angle the bottom makes with the horizontal, n the Manning roughness factor, Q the discharge, A the cross-sectional area, R the hydraulic radius, and b the width of cross section at the liquid surface. Solving for l,

$$l = \int \frac{1 - Q^2 b / (g A^3)}{S_0 - n^2 Q^2 / [(1.49)^2 A^2 R^{4/3}]} \, dy \qquad (191)$$

For constant S_0 and n, the integrand is a function of y only, and l may be determined as a function of y, usually by numerical integration. When the integrand is zero, $Q^2 b / g A^3 = 1$, which is the condition for critical depth. Hence for a change in depth there is no change in l; that is, neglecting the effects of the curvature of streamlines and the nonhydrostatic pressure distribution, the liquid surface is vertical as the flow goes through critical. When the denominator is zero, uniform flow occurs, and there is no change in depth along the channel.

The various possible free-surface profiles given by the preceding equation are shown in Fig. 48. In each case the flow is from left to right. Here, y_0 is the normal depth, that is, the depth given by the Manning

uniform flow equation; y_c is the critical depth. When the normal depth is greater than the critical depth, the slope of the channel is *mild*; when normal depth equals *critical* depth; the slope is critical; when normal depth is less than critical depth, the slope is *steep*. The two other cases are *horizontal* and *adverse*.

The determination of surface profiles from the equation is effected by starting the numerical integration at a *control* section. When the flow is above critical depth, the control is always downstream, and the depth is evaluated first for the control section, and then use is made of the gradually varied flow equation. Writing the equation in the form

$$l = \int F(y) \, dy$$

a plot of $F(y)$ as ordinate against y as abscissa is made, starting with the control depth and varying y in the direction indicated by the characteristic curves. This plot (Fig. 49) gives the value of l from the control section to the new depth y as the area under the curve between the values of y. In this manner the whole profile may be worked out.

When the depth of flow is less than critical, the control section is upstream (i.e., flow is out from under a gate), and the integration is handled in a similar fashion for determination of the profile downstream from the control.

A phenomenon known as the *hydraulic jump* occurs under certain conditions in channel flow. The flow prior to the jump must always be below the critical depth, and when the downstream depth is such that the momentum equation is satisfied for the liquid contained in the jump, the hydraulic jump will occur. The momentum equation applied to the liquid between sections 1 and 2 of Fig. 50 for a rectangular channel

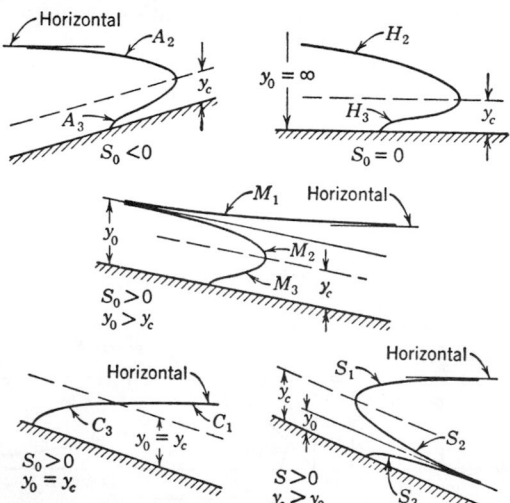

Fig. 48 Surface profiles on adverse, horizontal, mild, critical, and steep slopes. (Courtesy of Hunter Rouse, State University of Iowa, Iowa City.)

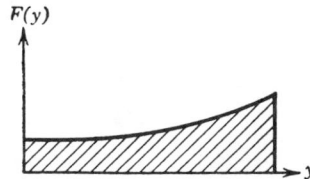

Fig. 49 Plot for determination of liquid surface profile.

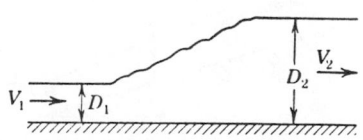

Fig. 50 Hydraulic jump.

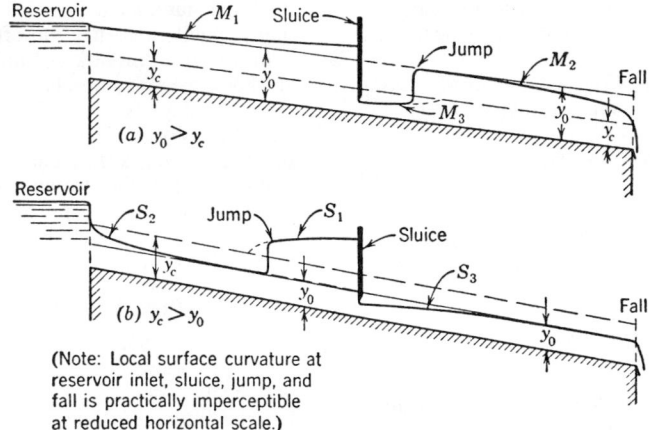

Fig. 51 Examples of surface profiles. (Courtesy of Hunter Rouse, State University of Iowa, Iowa City.)

yields the relation between depths

$$D_2 = -\frac{D_1}{2} + \sqrt{\frac{2D_1 V_1^2}{g} + \frac{D_1^2}{4}} \qquad (192)$$

Examples of occurrence of the surface profiles, including situations where the jump results, are given in Fig. 51.

8.3 Unsteady, Nonuniform Flow

Any change in discharge in an open channel results in an unsteady nonuniform flow. The changes in flow result in gravity waves moving through the system. In certain cases, waves of fixed form propagate along the channel. The most common is the surge wave depicted in Fig. 52. In a rectangular channel, the celerity of the surge can be computed by the equation

$$c = \sqrt{gy_1}\left[\frac{1}{2}\frac{y_2}{y_1}\left(\frac{y_2}{y_1}+1\right)\right]^{1/2} \qquad (193)$$

where c is the surge celerity relative to the undisturbed fluid velocity V_1.

There are other waves of fixed form, including the monoclinal rising flood wave, the solitary wave, and roll waves. All of these are the result of special flow or channel conditions and can be found in the literature.

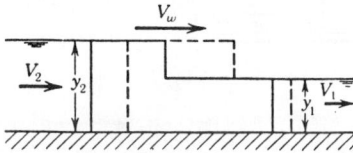

Fig. 52 Surge wave.

The most general approach to open-channel unsteady flow is through a procedure similar to that used in pipe flow. The principles of continuity and momentum are employed to develop a pair of nonlinear partial differential equations:

$$\frac{\partial A}{\partial t} + V\frac{\partial A}{\partial x} + A\frac{\partial V}{\partial x} = q \qquad (194)$$

$$g\frac{\partial y}{\partial x} + \frac{\partial V}{\partial t} + V\frac{\partial V}{\partial x} = g(S_0 - S_f) - \frac{V}{A}q \qquad (195)$$

where A = cross-sectional area, V = velocity of flow, y = depth of flow, S = slope of energy gradient, and q = lateral inflow per unit length.

In a manner similar to that in pipe flow these two equations can be transformed into ordinary differential equations so that

$$\frac{dV}{dt} + \frac{g}{c}\frac{dy}{dt} + g(S - S_0) + \frac{q}{A}(V + c) = 0 \qquad (196)$$

if

$$\frac{dx}{dt} = V + c \qquad (197)$$

and

$$\frac{dV}{dt} - \frac{g}{c}\frac{dy}{dt} + g(S - S_0) + \frac{q}{A}(V - c) = 0 \qquad (198)$$

if

$$\frac{dx}{dt} = V - c \qquad (199)$$

where $c = \sqrt{gA/b}$ and b = surface width of the channel.

The equations can now be solved numerically by finite-difference methods. This application of the method of characteristics requires that considerable care be exercised to guarantee that the Courant condition-relating time step, length step, and wave celerity,

$$\Delta t \leq \frac{\Delta x}{|V| + c} \tag{200}$$

be satisfied. The work of Wylie and Streeter[14] detail this general approach to the analysis of unsteady flow in open channels.

9   FLOW ABOUT IMMERSED OBJECTS

Flow about immersed objects has been discussed in Section 4 for the case of an ideal (frictionless) fluid. This section considers the effects of viscosity.

When a viscous fluid flows past an object, the fluid exerts a shear stress on the surface of the object as well as a normal pressure force. If the components of the surface shear and pressure in the direction of the flow are summed, the resulting force is known as the *drag*. The drag consists of a contribution from shear (skin friction drag) and pressure (form drag). In the case of well-formed bodies, the skin friction drag is the most significant. For blunt bodies, form drag dominates.

If the components of shear and pressure forces normal to the oncoming flow are summed, the *lift* on the body results. Typically, shear forces play a minor *direct* role on lift. Pressure is the dominant contributor. However, viscous forces can have a considerable indirect effect on lift and drag by causing boundary layer separation. It is common to represent lift L and drag D in terms of lift and drag coefficients:

$$D = C_D \tfrac{1}{2} \rho A V_o^2 \tag{201}$$

$$L = C_L \tfrac{1}{2} \rho A V_o^2 \tag{202}$$

where C_D, C_L = drag and lift coefficients, respectively, ρ = fluid density, A = frontal area of object, and V_o = free-stream velocity.

In 1904 Prandtl developed the concept of the boundary layer and thus forged the link between ideal fluid mechanics and viscous flow. For fluids of relatively small viscosity, the effects of fluid friction are confined to a thin layer of fluid adjacent to the boundary known as the boundary layer. The flow outside the boundary layer can be determined with the tools of ideal fluid flow analysis. It is important to note that the boundary layer is thin and there is little normal acceleration; hence, the pressure variation along the body is determined, for all practical purposes, by the ideal fluid flow. This revelation led to the first analytical approach to the calculation of drag.

9.1   Flat-Plate Boundary Layer

Flow across a flat plate parallel to the flow direction is subject to boundary layer growth. The forward portion of the plate develops a laminar boundary layer. The laminar boundary layer then "breaks down", forming a turbulent boundary layer that continues downstream indefinitely (see Fig. 53). The laminar boundary layer exists until a Reynolds number of 3900 occurs:

$$\text{Re} = \frac{V\delta}{\nu} \tag{203}$$

where δ = boundary layer thickness. Note that if the approaching flow is turbulent or the leading edge of the plate is rough, the laminar boundary layer may be considerably shorter.

Analysis of the flow on a flat plate provides the following values of shear stress and drag for the laminar flow portion:

$$\tau_o = c_f \tfrac{1}{2} \rho V_o^2 \tag{204}$$

where

$$c_f = \sqrt{\frac{8}{15 \text{Re}_x}} \quad \text{and} \quad \text{Re}_x = \frac{V_o x}{\nu} \tag{205}$$

where x = distance from forward edge of the plate. The total drag force D is given by

$$D = C_f \tfrac{1}{2} A \rho V_o^2 \tag{206}$$

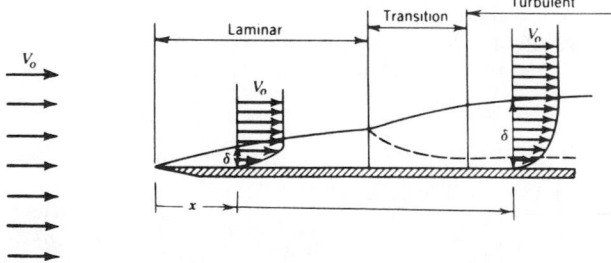

Fig. 53 Boundary layers on flat plate. By permission from J. K. Vennard and R. L. Street, *Elementary Fluid Mechanics*, 5th ed., Wiley, New York, 1975.

where

$$C_f = \sqrt{\frac{32}{15\mathrm{Re}_x}} \qquad (207)$$

These formulas are valid for Reynolds numbers based on Re_x up to 500,000 where

$$\mathrm{Re}_x = \frac{Vx}{\nu} \qquad (208)$$

If the flow over the flat plate is largely turbulent, then the drag coefficient C_f can be expressed as

$$\frac{1}{\sqrt{C_f}} = 4.13 \log C_f \mathrm{Re}_x \qquad (209)$$

In general the C_f values in Fig. 54 can be used in the drag force equation to compute total drag on a smooth flat plate.

9.2 Drag on Immersed Objects

The total drag on an immersed object may be dominated by skin friction or form drag but, in any case, is generally obtained by experiment. The data are presented as drag coefficients and plotted as C_D versus Reynolds number. The total drag force can then be calculated from Eq. (201). Data on drag coefficients for common objects are presented in Figs. 55 and 56.

9.3 Lift

Lift is also presented through graphs of C_L. However, Chapter 6 adequately covers lift, and the reader should refer to that section for more detailed information.

10 FLUID MEASUREMENTS*

In spite of the advances in numerical analysis, analytical techniques, and computer power in recent years, the complex phenomena of fluid flow must still be addressed empirically. The measurement of pressure, shear, discharge, velocity, and so forth remain necessary skills to the serious fluid mechanician. Many techniques and devices remain relatively unchanged over the years. Others, particularly in velocity measurement, reflect the recent advances in technology. This section presents an abbreviated review of the range of devices available.

10.1 Fluid Property Measurement

Specific Weight or Density Measurement of the specific weight of a liquid relies on fundamental concepts and devices that generally need no calibration. Figure 57 illustrates three methods that depend on buoyancy calculations and one (Fig. 57d) that utilizes hydrostatics. Probably the simplest method is not illustrated—simply weighing a known volume of liquid. The device or method selected depends on the availability of the required equipment.

Figure 57a utilizes the submerged weight of a known volume of liquid. Figure 57b, in a somewhat reversed approach, submerges a known weight and volume into the unknown liquid. Figure 57c uses calibrated hydrometers. Figure 57d illustrates the U-tube method, which requires no calibration.

Viscosity Measurement Viscosity-measuring devices (viscometers) generally employ one of three

---

*This section follows closely the treatment in Vennard and Street.[1]

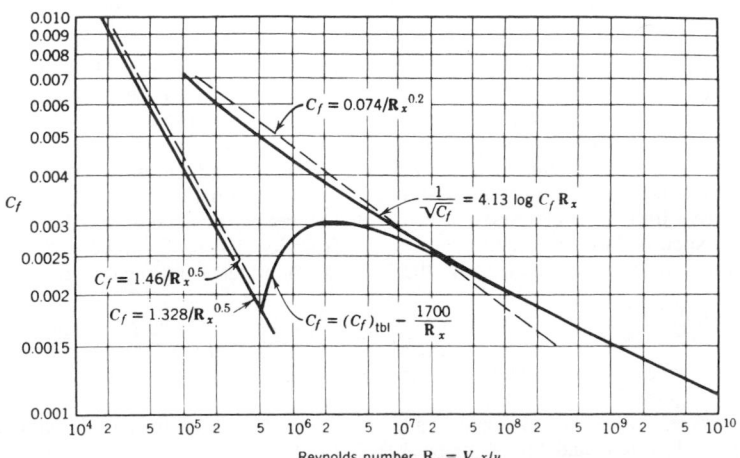

Fig. 54 Drag coefficients for smooth, flat plates. By permission from J. K. Vennard and R. L. Street, *Elementary Fluid Mechanics*, 5th ed., Wiley, New York, 1975.

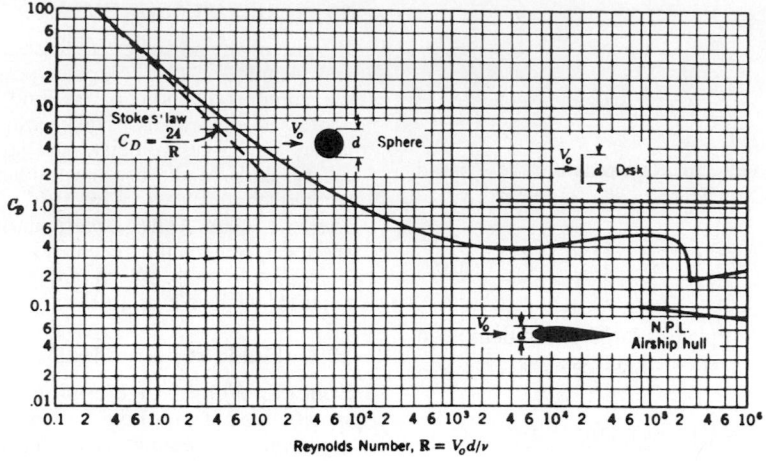

Fig. 55 Drag coefficients for sphere, disk, and streamlined body. By permission from J. K. Vennard and R. L. Street, *Elementary Fluid Mechanics*, 5th ed., Wiley, New York, 1975.

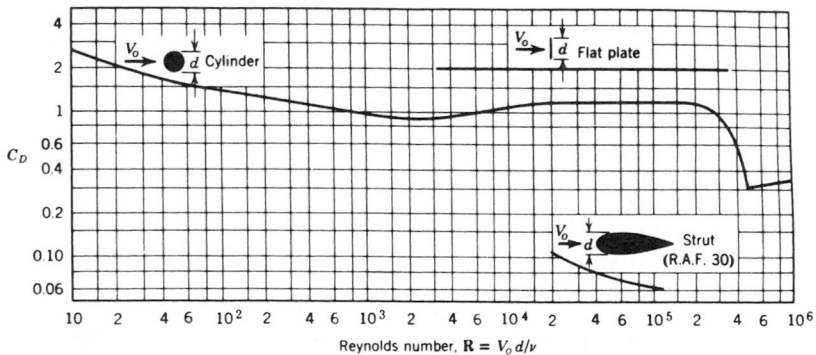

Fig. 56 Drag coefficients for circular cylinders, flat plates, and streamlined struts of infinite length. By permission from J. K. Vennard and R. L. Street, *Elementary Fluid Mechanics*, 5th ed., Wiley, New York, 1975.

approaches to measuring viscosity—the falling sphere, the flowing tube, or the rotating cylinder. All the devices require that laminar flow be maintained throughout the measurement period.

The falling-sphere device is illustrated in Fig. 58. Stokes's law for a sphere falling in a viscous liquid is employed. The time required for a sphere of known size and weight to fall a specified distance is measured. The absolute viscosity can be calculated from the equation

$$\mu = \frac{d^2(\gamma_s - \gamma_l)}{18V} \qquad (210)$$

where d = sphere diameter, V = sphere velocity, and γ_s, γ_l = specific weight of sphere and liquid, respectively.

The flowing-tube devices (Ostwald, Saybolt, Bingham, Redwood, and Engler viscometers) are typified by

the Ostwald and Saybolt devices in Fig. 59. All these devices depend on the laminar unsteady flow of a liquid. The time required for a given volume of liquid to flow through a small tube is measured, and an empirical formula based on laminar flow principles is used to calculate kinematic viscosity. For example, in the Saybolt device, the time required for the liquid level to drop from level B to C is recorded. The kinematic viscosity is calculated from the equation

$$\nu(\text{ft}^2/\text{sec}) = 0.000002365t - \frac{0.001935}{t} \qquad (211)$$

where t = time in seconds required for liquid level to drop.

The other devices will not be discussed here as the principles are similar and instructions and calibrated equations are supplied with the devices.

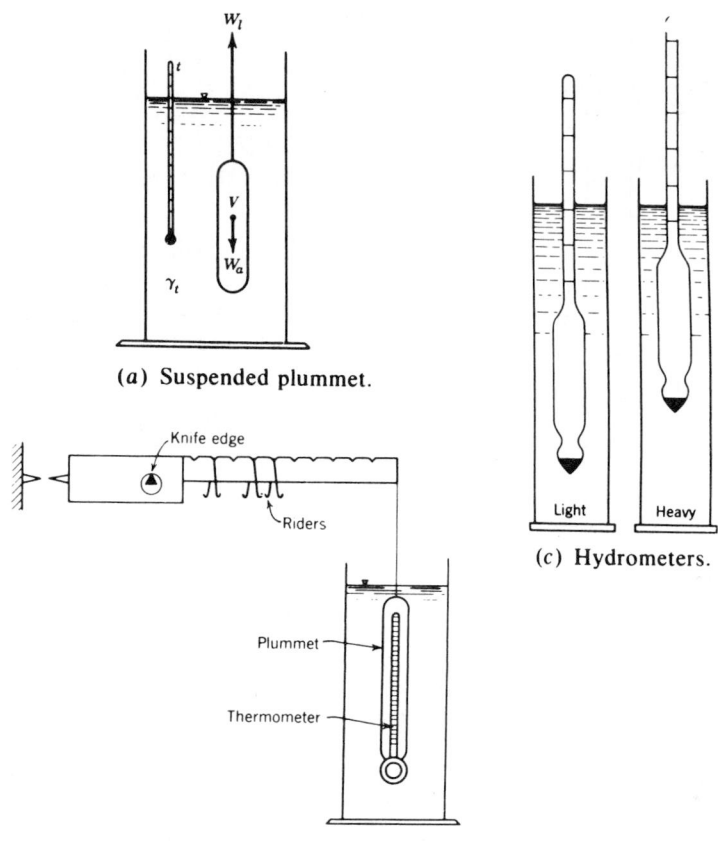

(a) Suspended plummet.

(c) Hydrometers.

(b) Westphal balance.

Fig. 57 Devices for density measurement. By permission from J. K. Vennard and R. L. Street, *Elementary Fluid Mechanics*, 5th ed., Wiley, New York, 1975.

The rotating-cylinder devices (Stormer, Mac-Michael, Brookfield) generally employ a fixed cylindrical container and a rotating inner cylinder (see Fig. 60). The space between the two cylinders and the speed of rotation is purposely kept small enough to maintain laminar flow. Measurement of the time required to complete a given number of revolutions under a constant torque leads to a calculation of absolute viscosity. The fundamental equation used is

$$T = \frac{2\pi R^2 h \mu V}{\Delta R} + \frac{\pi R^3 \mu V}{2\Delta h} \qquad (212)$$

10.2 Pressure Measurement

Pressure measurement in a fluid at rest is relatively easy to accomplish. The manometers and Bourdon gauges discussed in Section 3 are common devices used for this purpose. If the fluid is moving, pressure measurement is more difficult.

Of primary concern is the pressure-sensing connection between the fluid and the pressure-measuring system. In measuring *static* pressure (pressure unaffected by the velocity of flow), an opening is made in the conduit wall or the pressure probe, such as a pitot-static tube (see Fig. 61), so that the pressure in the flow at the surface may be conducted to the pressure-measuring device. The surface opening must be small (less than 1 mm) and well finished (square edged, no burr) so that the flow is not disturbed. In the case of a probe, the device must be small enough to not disturb the flow and alter the pressure situation and properly oriented to produce the true static pressure.

The pressure-measuring devices used in fluid flows may be manometers or gauges. However, for electronic recording as well as measurement of fluctuating pressures, a pressure transducer is commonly used. A typical pressure transducer is a small diaphragm to which is attached a strain gauge to measure deflection of the diaphragm. The gauge is part of a Wheatstone bridge circuit and is calibrated to an electrical output that can be processed to produce a plot, directed to a computer for further analysis, or simply stored on some electronic device.

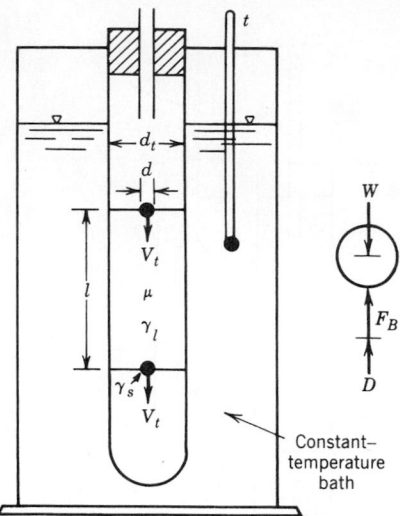

Fig. 58 Falling-sphere viscometer. By permission from J. K. Vennard and R. L. Street, *Elementary Fluid Mechanics*, 5th ed., Wiley, New York, 1975.

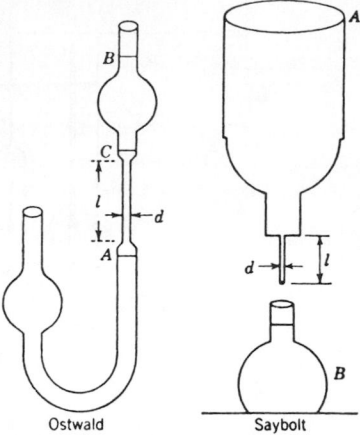

Fig. 59 Tube viscometers. By permission from J. K. Vennard and R. L. Street, *Elementary Fluid Mechanics*, 5th ed., Wiley, New York, 1975.

Piezoelectric transducers are also used wherein the sensing device is a piezoelectric crystal that produces an electric field when deformed. Care must be taken to install the sensors in a shock-free environment as they are very sensitive to any vibration of the facility hardware.

Pressure-sensitive paint (PSP) is a special paint material that adjusts its color based on the local pressure acting on the paint. Due to its nature, PSP provides nonintrusive, global surface pressure measurements. Therefore, PSP can be applied to the model surface being tested, and through special imaging equipment,

a quantitative picture of surface pressure values can be produced. Although PSP technology was initially developed in the 1970s, its use for flow visualization was first proposed in 1980. Since then, many different chemical formulations have been developed and have been used in all fields of aerodynamics, including low subsonic, high subsonic, transonic, and supersonic.

The working mechanism of PSP is simple but it is not reflected in its name. Actually, the paint is sensitive not to the local pressure but to the local partial pressure of oxygen. If the fluid that is used in the experiment is air and at constant composition (nonreacting), then the partial pressure of oxygen is directly proportional with the pressure of the air. Therefore, the higher the air pressure, the higher the "quenching effect." As a

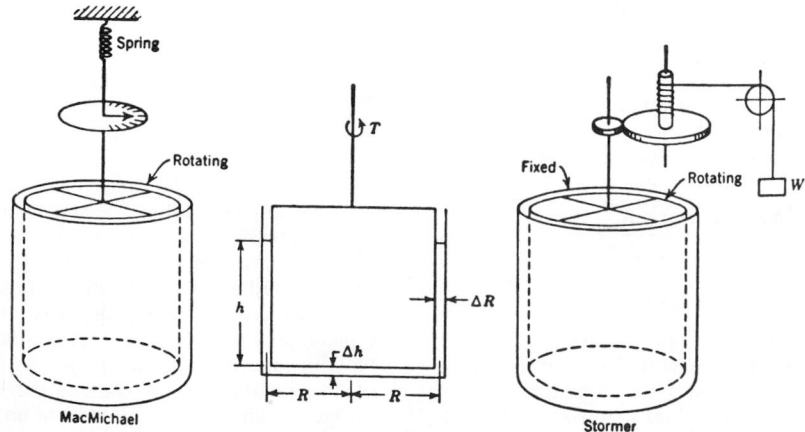

Fig. 60 Rotational viscometers (schematic). By permission from J. K. Vennard and R. L. Street, *Elementary Fluid Mechanics*, 5th ed., Wiley, New York, 1975.

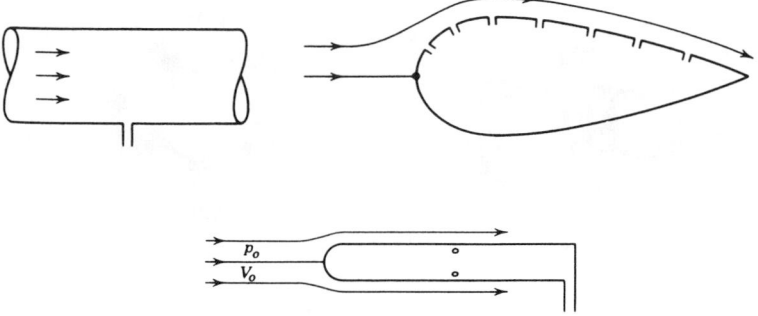

Fig. 61 Static tube. By permission from J. K. Vennard and R. L. Street, *Elementary Fluid Mechanics*, 5th ed., Wiley, New York, 1975.

result of this quenching effect, the energy of the fluorescent light coming from the PSP decreases, which translates to an increase in the wavelength of the fluorescent light. Modified wavelength leads to variable color based on the magnitude of the quenching, and hence the local pressure. Once calibrated for different wavelength–pressure correspondence, PSP can be used on the model object.

A PSP system consists of a light source to excite the paint for luminescence, the model painted with the PSP, a high-definition camera, and a processing unit for matching the light intensity values to the pressure values. As in all other measurement devices, PSPs also have their own error sources and they require a calibration for different operating conditions.

10.3 Velocity Measurement

One of the commonest and simplest devices for the measurement of velocity is a pitot-static tube (see Fig. 62). The device senses the difference in pressure between the tip of the pitot-static tube and the side of the tube. This pressure difference can be used to calculate the velocity from the equation

$$V_o = \sqrt{\frac{2(p_s - p_0)}{\rho}} = \sqrt{\frac{2g(p_s - p_o)}{\gamma}} \qquad (213)$$

where $p_s - p_o$ is the aforementioned difference in pressure.

In very low velocity flows, the anemometer or current meter is often used (see Fig. 63). These are

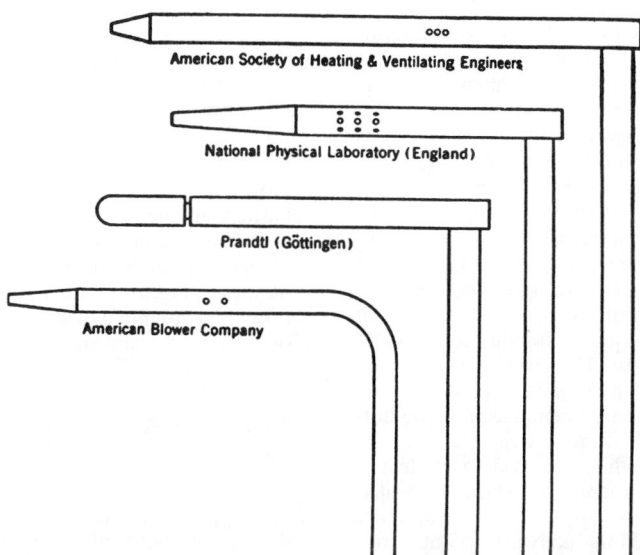

Fig. 62 Pitot-static tubes (to scale). By permission from J. K. Vennard and R. L. Street, *Elementary Fluid Mechanics*, 5th ed., Wiley, New York, 1975.

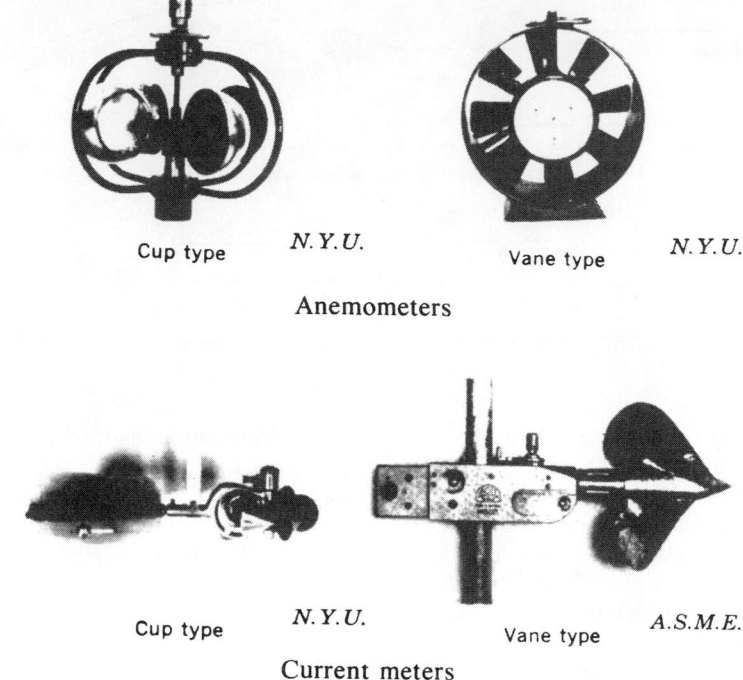

Cup type   *N.Y.U.*   Vane type   *N.Y.U.*

Anemometers

Cup type   *N.Y.U.*   Vane type   *A.S.M.E.*

Current meters

Fig. 63   By permission from J. K. Vennard and R. L. Street, *Elementary Fluid Mechanics*, 5th ed., Wiley, New York, 1975.

calibrated devices that relate flow velocity to the number of revolutions per minute of the rotating element in the meter. Anemometers are typically used to measure wind speeds, and current meters are employed to measure water velocities in rivers.

Measurement of rapidly fluctuating velocities in air and water require devices that can respond quickly to changes in velocity. The hot-wire anemometer is commonly used to measure velocity fluctuations in air. A thin wire connected between two supports passes an electric current that heats the wire (see Fig. 64a). Air moving perpendicular to the wire cools the wire in proportion to the flow velocity. The electronics of the system measures the increased voltage necessary to keep the wire temperature constant and relates that voltage to the flow velocity. In some devices the current flowing through the probe is kept constant and the voltage change required for this to occur is related to the flow velocity. However, the constant-temperature device is by far the most popular.

A variation on the hot-wire anemometer is the hot-film anemometer. The hot-film device is similar to the hot wire except the wire is coated to protect it from contaminated environments. In addition to coated wires, hot-film anemometry employs probes of other shapes that are sturdier and less likely to trap impurities in the flow (lint, small pieces of organic material, etc.). Figure 65 illustrates a few of these different types of probes. Hot-film devices are commonly used in liquid

flows. Both these devices are capable of measuring velocity fluctuation frequencies of better than 300 kHz.

Another technique for measuring velocities is laser-Doppler velocimetry (LDV). The fundamental basis for the technique is the Doppler shift of light that is scattered from extremely small particles in a moving fluid. One valuable attribute of LDV is that it is non intrusive, that is, the sensing device is not in the flow. It only needs visual access to the flow for the required light beams. Refer to Goldstein[16] for a thorough treatment of the subject. One of the most recent advances in measuring fluid velocities is particle image velocimetry (PIV), which also makes use of the laser and seed particles but works with a planar laser light sheet. This provides the advantage of measuring an entire two-dimensional plane of the flow field simultaneously, as compared to LDV, which only offers a point measurement. For more details on PIV, see Raffel et al.[17]

10.4   Flow Rate Measurement

Methods of measuring flow rate or discharge can generally be categorized into total quantity measurements, pressure drop or pressure difference measuring devices, tracer transport techniques, and devices that induce critical flow conditions or simply changes in water surface depth in open channels. Another indirect approach is the measurement of velocity at several

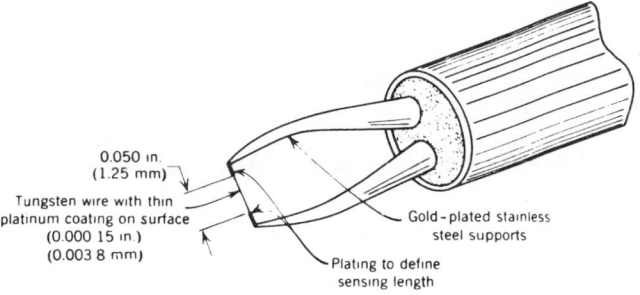

0.050 in.
(1.25 mm)

Tungsten wire with thin
platinum coating on surface
(0.000 15 in.)
(0.003 8 mm)

Gold-plated stainless
steel supports

Plating to define
sensing length

(a) Hot-wire sensor and support needles.

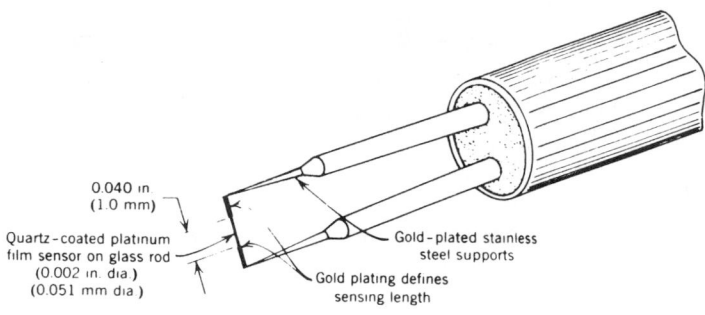

0.040 in
(1.0 mm)

Quartz-coated platinum
film sensor on glass rod
(0.002 in. dia.)
(0.051 mm dia.)

Gold-plated stainless
steel supports

Gold plating defines
sensing length

(b) Hot-film sensor and support needles.

Fig. 64 Anemometer sensors. Reproduced from TB5, Thermo-Systems, Inc., 2500 Cleveland Ave. North, St. Paul, Minnesota, 55113. By permission from J. K. Vennard and R. L. Street, *Elementary Fluid Mechanics*, 5th ed., Wiley, New York, 1975.

points and a numerical integration of velocity times area to calculate discharge.

Total Quantity Methods The success of these methods depends on the availability of a means to collect and measure (or weigh) the amount of liquid captured in the container in a given time period. Advantages are the lack of any need to calibrate a device. This approach is often used to calibrate other flow-rate-measuring devices.

Pressure Difference Methods All the devices that create pressure differences in a fluid flow to permit calculation of flow rate also create friction losses in the flow. One of the devices that has the lowest friction loss is the venturi meter (see Fig. 66). In the venturi meter the Bernoulli effect is used to generate a pressure difference between the entrance and the throat of the device. This pressure difference is related to the flow rate through the equation

$$Q = \frac{C_v A_2}{\sqrt{1 - (A_2/A_1)^2}} \sqrt{2g \left(\frac{p_1}{\gamma} + z_1 - \frac{p_2}{\gamma} - z_2 \right)}$$

(214)

where A is the cross-sectional area of the meter and C_v is a calibration coefficient shown in Fig. 66. Venturi meters are relatively expensive but produce very little friction loss and are quite accurate.

A device based on a similar technique to the venturi meter is the flow nozzle. This device resembles the upstream portion of a venturi meter (see Fig. 67). Pressure recovery experienced in the downstream section of the venturi meter is not realized here so the flow nozzle creates a larger friction loss than the venturi meter. The calculation for discharge is made with the same equation used for the venturi meter with the C_v value taken from Fig. 67.

One further step in the direction of simplicity (and lower cost) is the orifice meter (see Fig. 68). This device is simply a circular plate with a hole cut out of its center and installed in a pipe, generally in a flanged connection. The orifice meter creates considerably more friction loss than the venturi meter or flow nozzle but its low cost makes it attractive in many instances. Flow rates through an orifice meter may be calculated with the equation

$$Q = CA \sqrt{2g \left(\frac{p_1}{\gamma} + z_1 - \frac{p_2}{\gamma} - z_2 \right)}$$

(215)

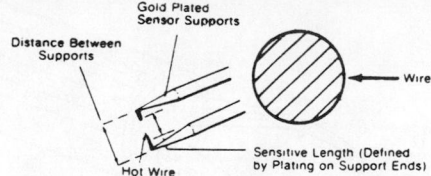

(a) CROSS SECTION OF HOT-WIRE SENSOR

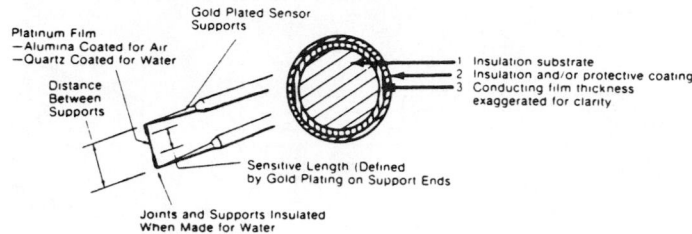

(b) CROSS SECTION OF HOT-FILM SENSOR

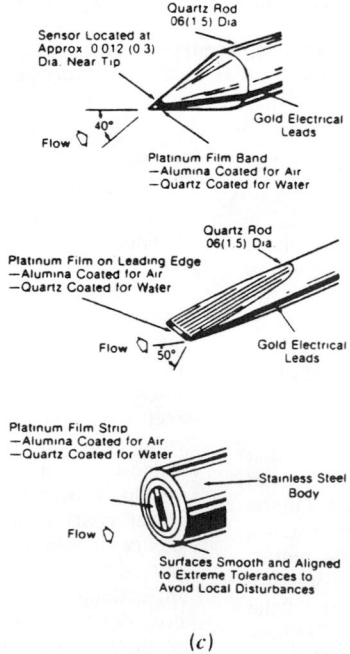

(c)

Fig. 65 Thermal sensor configuration. By permission from R. J. Goldstein, *Fluid Mechanics Measurements*, Hemisphere, New York, 1983.

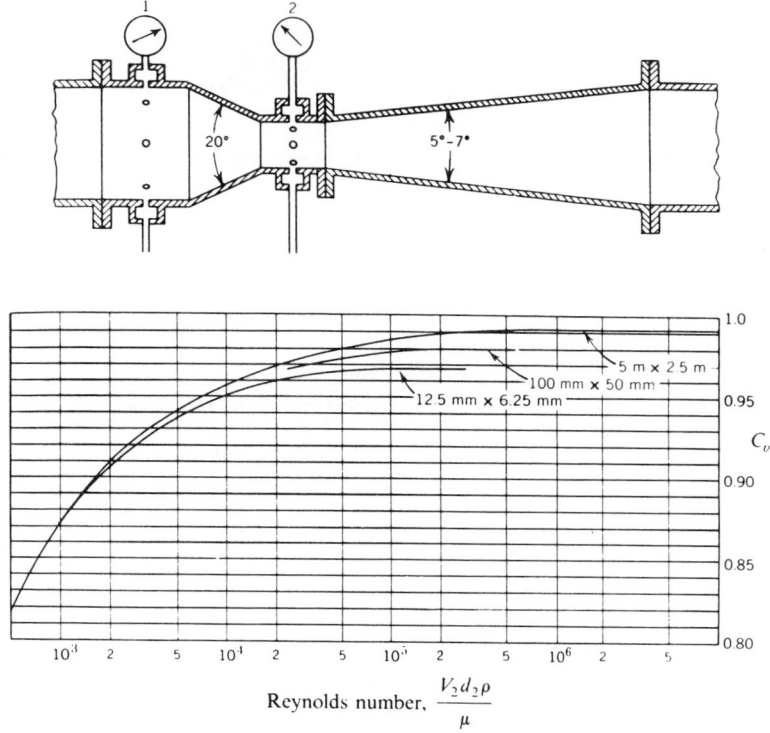

Fig. 66 Venturi meter and coefficients. By permission from J. K. Vennard and R. L. Street, *Elementary Fluid Mechanics*, 5th ed., Wiley, New York, 1975.

The value of C for a given flow rate is found from Fig. 69.

There are certain special cases of "orifice meters" shown in Fig. 70 where discharges through the orifice can be related to liquid levels on one (or two) side(s) of the opening. The equations used to compute the discharge are

$$Q = A_2 V_2 = C_c C_v A \sqrt{2g(h_1 - h_2)}$$

$$= CA\sqrt{2g(h_1 - h_2)} \tag{216}$$

$$= C_c C_v A \sqrt{2gh} = CA\sqrt{2gh} \tag{217}$$

All the C values for various types of orifices can be found in Fig. 71. More precise data for C values of sharp-edged orifices over a large range of heads are given in Fig. 72.

Probably the most inexpensive technique for calculating flow rate by the pressure difference method is the elbow meter (see Fig. 73). The only requirements are two pressure taps on the inner and outer portions of the elbow. The flow rate is calculated from the equation

$$Q = CA\sqrt{2g\left(\frac{P_o}{\gamma} + z_o - \frac{p_i}{\gamma} - z_i\right)} \tag{218}$$

Because the value of C depends strongly on the elbow geometry, it must generally be determined through a calibration procedure.

Tracer Transport Techniques This technique is based on the premise that the discharge in a conduit can be measured by the time of travel of the average concentration of a tracer between two points. The tracer could be as simple as salt where concentration is measured via conductivity instruments (see Fig. 74), although any other tracer may be employed. Fluorescent dyes are commonly used because they have little effect on water quality and are detectable in minute quantities. From Fig. 74, the average velocity is computed from dividing the length l between sensors by the time t required for the centroid of the concentration curve to travel over the distance. This approach is particularly useful in open channels of relatively constant cross section and conduits where, for one reason or another, other techniques cannot be used.

Open-Channel Flow-Measuring Devices In free-surface flows, the common techniques for flow measurement center on either (a) constricting the flow to create differences in water surface elevations that can be related to flow rate or (b) creating critical flow

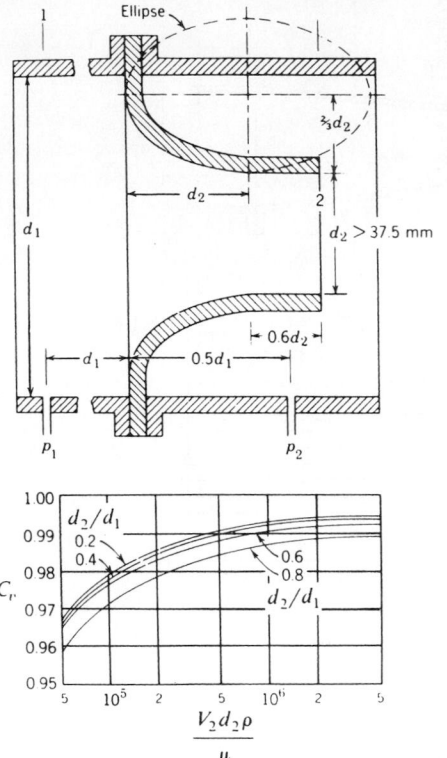

Fig. 67 ASME flow nozzle and coefficients. By permission from J. K. Vennard and R. L. Street, *Elementary Fluid Mechanics*, 5th ed., Wiley, New York, 1975.

conditions (see Section 8) that provide a strong analytical connection to discharge calculation.

The simplest devices are venturi flumes, which are the open-channel version of venturi meters. The drop in water surface in the flume throat is measured and related to the flow rate. Although not extremely accurate, this approach does not create large friction losses.

A simpler version similar to the flow nozzle in closed conduits is the cutthroat flume where the diverging recovery section is not present.

A somewhat more complicated version of the venturi flume, which causes critical or near-critical flow to occur, is the Parshall flume (see Fig. 75). This device has been commonly used in irrigation systems for well over 70 years. The discharge is calculated from the equation

$$Q = 4Bh_a^{1.522B^{0.26}} \tag{219}$$

Another category of open-channel flow-measuring devices is weirs. These devices generally generate more friction loss than the previously described devices, but they are relatively simple and accurate. A typical sharp-crested rectangular weir is shown in Fig. 76. The other types of weirs are broad-crested (Fig. 77) and triangular (Fig. 78). The general form of the discharge equation per foot of width for weirs with two-dimensional flow is

$$q = C_w \frac{2}{3} \sqrt{2g} H^{3/2}. \tag{220}$$

where H is defined in Figs. 76 and 77 as the head on the weir. The value of C_w for sharp-crested well-ventilated weirs (Fig. 76) is given by the equation

$$C_w = 0.605 + 0.08\frac{H}{P} + \frac{1}{305H} \tag{221}$$

For broad-crested weirs (Fig. 77), the flow rate per foot of width is given by

$$q = \sqrt{g\left(\frac{2E}{3}\right)^3} = \left(\frac{2}{3}\right)^{3/2} \sqrt{g}\, E^{3/2} \tag{222}$$

or Eq. (220) where

$$C_w = \frac{1}{\sqrt{3}}\left(\frac{E}{H}\right)^{3/2} \tag{223}$$

where E is defined in Fig. 77.

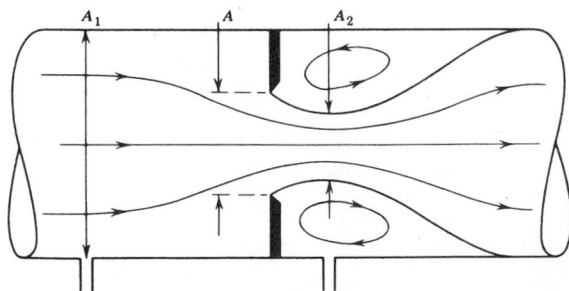

Fig. 68 Definition sketch for orifice meter. By permission from J. K. Vennard and R. L. Street, *Elementary Fluid Mechanics*, 5th ed., Wiley, New York, 1975.

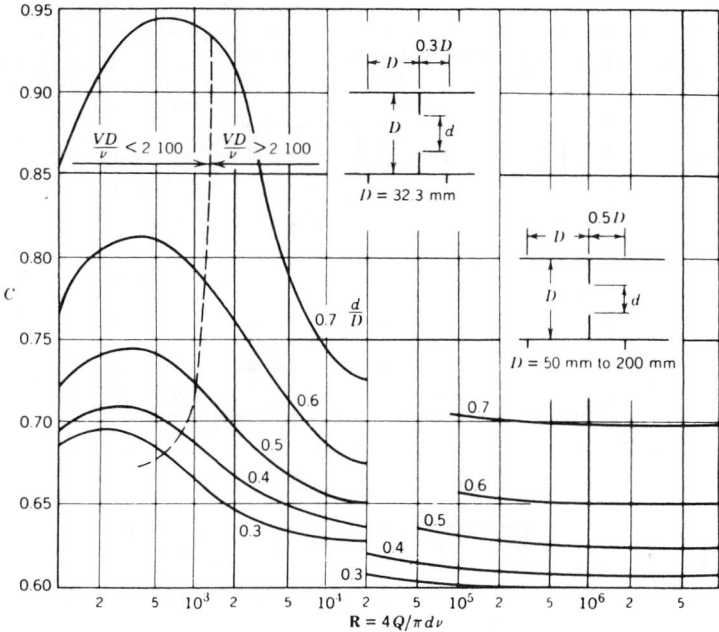

Fig. 69 Orifice meter coefficients. By permission from J. K. Vennard and R. L. Street, *Elementary Fluid Mechanics*, 5th ed., Wiley, New York, 1975.

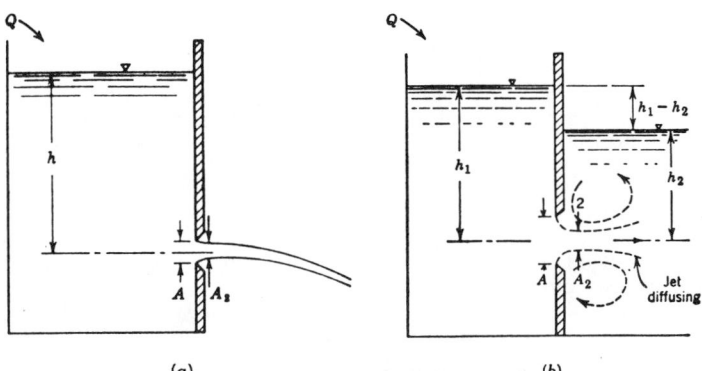

Fig. 70 (a) Orifice discharging freely; (b) submerged orifice. By permission from J. K. Vennard and R. L. Street, *Elementary Fluid Mechanics*, 5th ed., Wiley, New York, 1975.

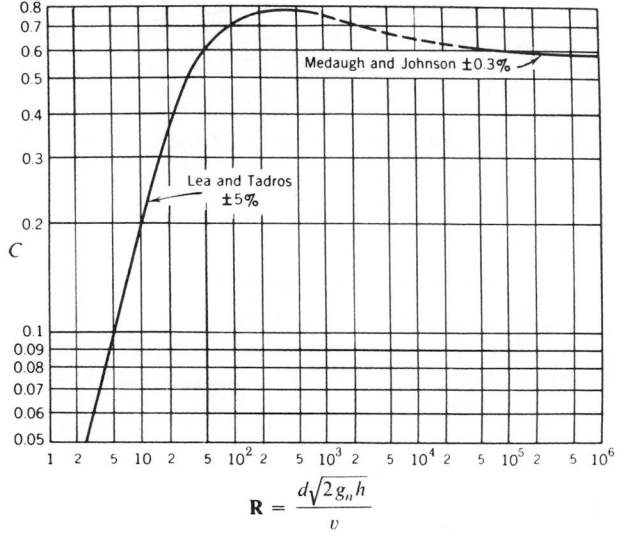

Orifices and their Nominal Coefficients				
	Sharp edged	Rounded	Short tube	Borda
C	0.61	0.98	0.80	0.51
C_c	0.62	1.00	1.00	0.52
C_v	0.98	0.98	0.80	0.98

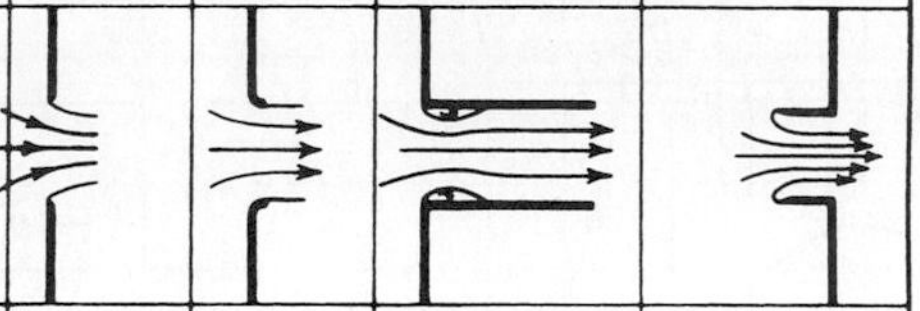

Fig. 71 By permission from J. K. Vennard and R. L. Street, *Elementary Fluid Mechanics*, 5th ed., Wiley, New York, 1975.

$$\mathbf{R} = \frac{d\sqrt{2g_n h}}{v}$$

Fig. 72 Coefficient for sharp-edged orifices under static head ($h/d > 5$). By permission from J. K. Vennard and R. L. Street, *Elementary Fluid Mechanics*, 5th ed., Wiley, New York, 1975.

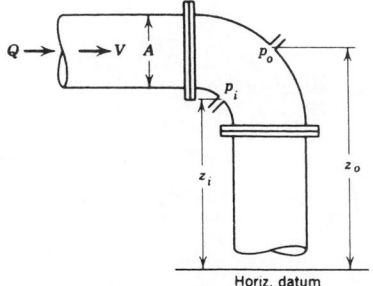

Fig. 73 Elbow meter. By permission from J. K. Vennard and R. L. Street, *Elementary Fluid Mechanics*, 5th ed., Wiley, New York, 1975.

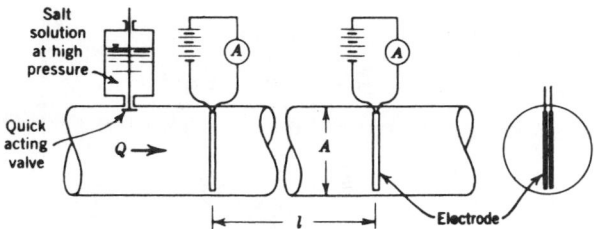

Fig. 74 Salt dilution discharge measuring. By permission from J. K. Vennard and R. L. Street, *Elementary Fluid Mechanics*, 5th ed., Wiley, New York, 1975.

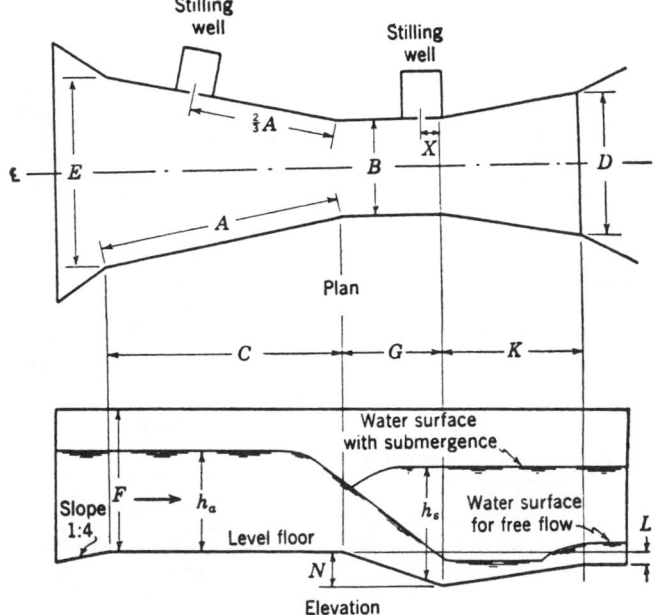

Fig. 75 Parshall measuring flume. From R. K. Linsley and J. B. Franzini, *Elements of Hydraulic Engineering*, McGraw-Hill, New York, 1955. Copyright 1955 McGraw-Hill. Reprinted by permission.

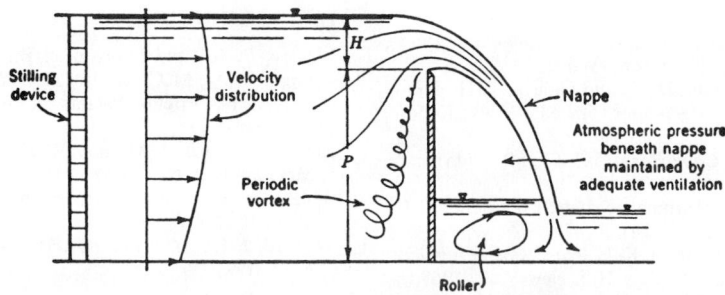

Fig. 76 Weir flow (actual). By permission from J. K. Vennard and R. L. Street, *Elementary Fluid Mechanics*, 5th ed., Wiley, New York, 1975.

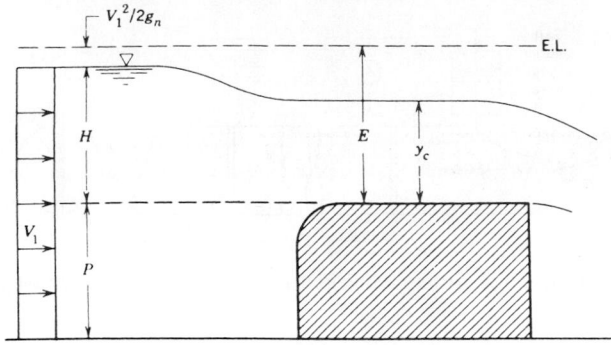

Fig. 77 Broad-crested weir. By permission from J. K. Vennard and R. L. Street, *Elementary Fluid Mechanics*, 5th ed., Wiley, New York, 1975.

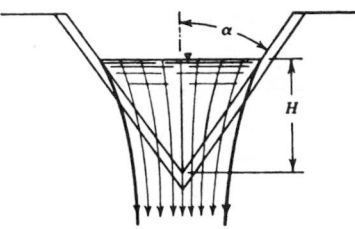

Fig. 78 Triangular weir. By permission from J. K. Vennard and R. L. Street, *Elementary Fluid Mechanics*, 5th ed., Wiley, New York, 1975.

For triangular notch weirs (Fig. 78) the equation for computing discharge is

$$Q = C_w \frac{8}{15} \tan \alpha \sqrt{2g} \; H^{5/2} \qquad (224)$$

where C_w depends on the notch angle α. For $\alpha = 90°$,

$$C_w = 0.56 + \frac{0.70}{R^{0.165} \; W^{0.170}} \qquad (225)$$

REFERENCES

1. Vennard, J. K., and Street, R. L., *Elementary Fluid Mechanics*, 5th ed., New York, Wiley, 1975.
2. Munson, B. R., Young, D. F., and Okiishi, T. H., *Fundamentals of Fluid Mechanics*, 5th ed., Wiley, Hoboken, NJ, 2006.
3. Stokes, G., *Trans. Cambridge Phil. Soc.*, **8** (1845) and **9** (1851).
4. Kármán, Th. von, "Turbulence and Skin Friction," *J. Aeronaut. Sci.*, **1**(1), 1 (1934).
5. Davis, C. V., and Sorenson, K. E., *Handbook of Applied Hydraulics*, 3rd ed., McGraw-Hill, New York, 1969.
6. Chow, V. T., *Open Channel Hydraulics*, McGraw-Hill, New York, 1959.
7. Colebrook, C. F., "Turbulent Flow in Pipes, with Particular Reference to the Transition Region between the Smooth and Rough Pipe Laws," *J. Inst. Civil Eng. London*, **11** (1938–1939).
8. Crane Company, "Flow of Fluids through Valves, Fittings and Pipe," Tech. Paper No. 410, Crane Co., New York, 1969.
9. Idelchik, J. E., *Handbook of Hydraulic Resistance*, 2nd ed., Hemisphere Publishing, New York, 1986.
10. Gibson, A. H., *Hydraulics and Its Applications*, Constable, London, 1912.
11. Jeppson, R. W., *Analysis of Flow in Pipe Networks*, Ann Arbor Science, Ann Arbor, MI, 1976.
12. Wood, D. J., and Rayes, A. G., "Reliability of Algorithms for Pipe Network Analysis," *J. Hydraulics Div., ASCE*, **107**(10) (1981).
13. Watters, G. Z., *Analysis and Control of Unsteady Flow in Pipelines*, Butterworth, Stoneham, MA, 1984.
14. Wylie, E. B., and Streeter, V. L., *Fluid Transients*, McGraw-Hill, New York, 1978.
15. Chaudhry, M. H., *Applied Hydraulic Transients*, Van Nostrand Reinhold, New York, 1979.
16. Goldstein, R. J., *Fluid Mechanics Measurements*, Hemisphere Publishing, New York, 1983.
17. Raffel, M., Wifferk, C., and Kompenhans, J., *Particle Image Velocimetry: A Practical Guide*, Springer, Berlin, 1998.

BIBLIOGRAPHY

Linsley, R. K., and Franzini, J. B., *Elements of Hydraulic Engineering*, McGraw-Hill, New York, 1955.

Moody, L. F., "Friction Factors for Pipe Flow," *Trans. ASME*, **66** (1944).

Rouse, H., *Fluid Mechanics for Hydraulic Engineers*, McGraw-Hill, New York, 1938.

Streeter, V. L., *Fluid Dynamics*, 3rd ed., McGraw-Hill, New York, 1962.

Vennard, J. K., *Elementary Fluid Mechanics*, 4th ed., Wiley, New York, 1961.

CHAPTER 10

AERODYNAMICS OF WINGS

Warren F. Phillips
Department of Mechanical and Aerospace Engineering
Utah State University
Logan, Utah

1 INTRODUCTION AND NOTATION

Aerodynamics is the science of predicting and controlling the forces and moments that act on an object moving through the atmosphere. The aerodynamic forces and moments acting on any such object originate from only two sources:

1. The pressure distribution over the body surface
2. The shear stress distribution over the body surface

A resultant aerodynamic force $\mathbf{F}_a$ and a resultant aerodynamic moment $\mathbf{M}_a$ are the net effects of the pressure and shear stress distributions integrated over the entire surface of the body. To express these two vectors in terms of components, we must define a coordinate system. While several different coordinate systems are used in aeronautics, the coordinate system commonly used in the study of aerodynamics is referred to here as *Cartesian aerodynamic coordinates*. When considering flow over a body such as an airfoil, wing, or airplane, the x axis of this particular coordinate system is aligned with the body axis or *chord line*, pointing in the general direction of relative airflow. The origin is located at the front of the body or *leading edge*. The y axis is chosen normal to the x axis in an upward direction. Choosing a conventional right-handed coordinate system requires the z axis to be pointing in the spanwise direction from right to left, as shown in Fig. 1. Here, the components of the resultant aerodynamic force and moment, described in this particular coordinate system, are denoted as

$$\mathbf{F}_a = A\mathbf{i}_x + N\mathbf{i}_y + B\mathbf{i}_z$$
$$\mathbf{M}_a = -l\mathbf{i}_x - n\mathbf{i}_y - m\mathbf{i}_z$$

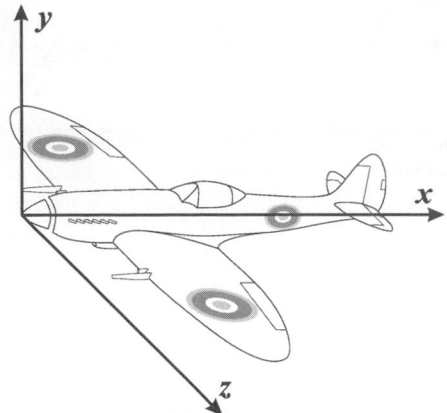

Fig. 1 Cartesian aerodynamic coordinate system used in the study of aerodynamics.

$\tilde{A} \equiv$ section axial force $\equiv$ axial force per unit span (parrallel to chord)

$\tilde{N} \equiv$ section normal force $\equiv$ normal force per unit span (perpendicular to chord)

$\tilde{m} \equiv$ section moment $\equiv$ pitching moment per unit span (positive nose up)

where the chord is a line extending from the leading edge to the trailing edge of the body. The chord length, c, is the length of this chord line.

The aerodynamic forces and moments are usually expressed in terms of dimensionless force and moment coefficients. For example,

$$C_D \equiv \text{drag coefficient} \equiv \frac{D}{\frac{1}{2}\rho_\infty V_\infty^2 S}$$

$$C_L \equiv \text{lift coefficient} \equiv \frac{L}{\frac{1}{2}\rho_\infty V_\infty^2 S}$$

$$C_A \equiv \text{axial force coefficient} \equiv \frac{A}{\frac{1}{2}\rho_\infty V_\infty^2 S}$$

$$C_N \equiv \text{normal force coefficient} \equiv \frac{N}{\frac{1}{2}\rho_\infty V_x^2 S}$$

$$C_m \equiv \text{pitching moment coefficient} \equiv \frac{m}{\frac{1}{2}\rho_\infty V_\infty^2 Sc}$$

where ρ_∞ is the free-stream density, S is the reference area, and c is the reference length. For a streamlined body such as a wing, S is the planform area and c is the mean chord length. For a bluff body, the frontal area is used as the reference. For 2D flow, the section aerodynamic coefficients per unit span are defined:

$$\tilde{C}_D \equiv \text{section drag coefficient} \equiv \frac{\tilde{D}}{\frac{1}{2}\rho_\infty V_\infty^2 c}$$

$$\tilde{C}_L \equiv \text{section lift coefficient} \equiv \frac{\tilde{L}}{\frac{1}{2}\rho_\infty V_\infty^2 c}$$

$$\tilde{C}_A \equiv \text{section axial force coefficient} \equiv \frac{\tilde{A}}{\frac{1}{2}\rho_\infty V_\infty^2 c}$$

$$\tilde{C}_N \equiv \text{section normal force coefficient} \equiv \frac{\tilde{N}}{\frac{1}{2}\rho_\infty V_\infty^2 c}$$

$$\tilde{C}_m \equiv \text{section moment coefficient} \equiv \frac{\tilde{m}}{\frac{1}{2}\rho_\infty V_\infty^2 c^2}$$

where $\mathbf{i}_x$, $\mathbf{i}_y$, and $\mathbf{i}_z$ are the unit vectors in the x, y, and z directions, respectively. The terminology that describes these components is

$A \equiv$ aftward axial force $\equiv$ x component of $\mathbf{F}_a$ (parallel to chord)

$N \equiv$ upward normal force $\equiv$ y component of $\mathbf{F}_a$ (normal to chord and span)

$B \equiv$ leftward side force $\equiv$ z component of $\mathbf{F}_a$ (parallel with span)

$\ell \equiv$ rolling moment (positive right wing down)

$n \equiv$ yawing moment (positive nose right)

$m \equiv$ pitching moment (positive nose up)

The traditional definitions for the moments in roll, pitch, and yaw do not follow the right-hand rule in this coordinate system. It is often convenient to split the resultant aerodynamic force into only two components,

$D \equiv$ drag $\equiv$ component of $\mathbf{F}_a$ parallel to $\mathbf{V}_\infty$ ($D = \mathbf{F}_a \cdot \mathbf{i}_\infty$)

$L \equiv$ lift $\equiv$ component of $\mathbf{F}_a$ perpendicular to $\mathbf{V}_\infty$ ($L = |\mathbf{F}_a - D\mathbf{i}_\infty|$)

where $\mathbf{V}_\infty$ is the free-stream velocity or *relative wind* far from the body and $\mathbf{i}_\infty$ is the unit vector in the direction of the free stream.

For two-dimensional flow (2D), it is often advantageous to define the *section force* and *section moment* to be the force and moment per unit span. For these definitions the notation used here will be

$\tilde{D} \equiv$ section drag $\equiv$ drag force per unit span (parallel to $\mathbf{V}_\infty$)

$\tilde{L} \equiv$ section lift $\equiv$ lift force per unit span (perpendicular to $\mathbf{V}_\infty$)

The resultant aerodynamic force acting on a 2D airfoil section is completely specified in terms of either lift and drag or axial and normal force. These two equivalent descriptions of the resultant aerodynamic force are related to each other through the angle of attack, as shown in Fig. 2:

$\alpha \equiv$ angle of attack $\equiv$ angle from $\mathbf{V}_\infty$ to chord line (positive nose up)

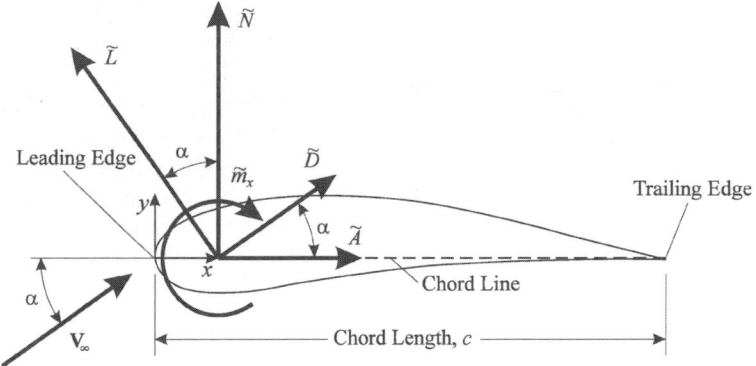

Fig. 2 Section forces and moment.

If the normal and axial coefficients are known, the lift and drag coefficients can be found from the relations

$$\widetilde{C}_L = \widetilde{C}_N \cos\alpha - \widetilde{C}_A \sin\alpha \qquad (1)$$

$$\widetilde{C}_D = \widetilde{C}_A \cos\alpha + \widetilde{C}_N \sin\alpha \qquad (2)$$

and when the lift and drag coefficients are known, the normal and axial coefficients are found from

$$\widetilde{C}_N = \widetilde{C}_L \cos\alpha + \widetilde{C}_D \sin\alpha \qquad (3)$$

$$\widetilde{C}_A = \widetilde{C}_D \cos\alpha - \widetilde{C}_L \sin\alpha \qquad (4)$$

Because the lift is typically much larger than the drag, from Eq. (4) we see why the axial force is sometimes negative even though the angle of attack is small.

The resultant aerodynamic force and moment acting on a body must have the same effect as the distributed loads. Thus, the resultant moment will depend on where the resultant force is placed on the body. For example, let x be the coordinate measured along the chord line of an airfoil, from the leading edge toward the trailing edge. If we place the resultant force and moment on the chord line, the value of the resultant moment will depend on the x location of the resultant force. The resultant moment about some arbitrary point on the chord line a distance x from the leading edge, $\widetilde{m}_x$, is related to the resultant moment about the leading edge, $\widetilde{m}_{le}$, according to

$$\widetilde{m}_{le} = \widetilde{m}_x - x\widetilde{N}$$

or in terms of dimensionless coefficients,

$$\widetilde{C}_{m_{le}} = \widetilde{C}_{m_x} - \frac{x}{c}\widetilde{C}_N \qquad (5)$$

Two particular locations along the chord line are of special interest:

$x_{cp} \equiv$ center of pressure $\equiv$ point about which resultant moment is zero

$x_{ac} \equiv$ aerodynamic center $\equiv$ point about which resultant moment is independent of α

Using the definition of center of pressure in Eq. (5), the section pitching moment coefficient relative to the leading edge can be written as

$$\widetilde{C}_{m_{le}} = \widetilde{C}_{m_x} - \frac{x}{c}\widetilde{C}_N = -\frac{x_{cp}}{c}\widetilde{C}_N$$

or, after solving for the *center of pressure*,

$$\frac{x_{cp}}{c} = \frac{x}{c} - \frac{\widetilde{C}_{m_x}}{\widetilde{C}_N} = -\frac{\widetilde{C}_{m_{le}}}{\widetilde{C}_N} \qquad (6)$$

Using the aerodynamic center in Eq. (5), we can write

$$\widetilde{C}_{m_x} - \frac{x}{c}\widetilde{C}_N = \widetilde{C}_{m_{ac}} - \frac{x_{ac}}{c}\widetilde{C}_N \qquad (7)$$

Equation (7) must hold for any angle of attack and any value of x. Thus, at the angle of attack that gives a normal force coefficient of zero, the moment coefficient about any point on the chord line is equal to the moment coefficient about the aerodynamic center, which by definition does not vary with angle of attack. Thus, the section pitching moment coefficient relative to the aerodynamic center can be expressed as

$$\widetilde{C}_{m_{ac}} = \left(\widetilde{C}_{m_x}\right)_{\widetilde{C}_N = 0} \qquad (8)$$

Using Eq. (8) in Eq. (7) and solving for the location of the *aerodynamic center* results in

$$\frac{x_{ac}}{c} = \frac{x}{c} + \frac{\left(\widetilde{C}_{m_x}\right)_{\widetilde{C}_N=0} - \widetilde{C}_{m_x}}{\widetilde{C}_N} = \frac{\left(\widetilde{C}_{m_{le}}\right)_{\widetilde{C}_N=0} - \widetilde{C}_{m_{le}}}{\widetilde{C}_N} \qquad (9)$$

Thus, to determine the aerodynamic center of an airfoil section, one can evaluate the normal force and pitching moment coefficients for any point on the chord line, as a function of angle of attack. The moment coefficient is then plotted as a function of the normal force coefficient. The moment axis intercept is the moment coefficient about the aerodynamic center. For any nonzero normal force coefficient, the location of the aerodynamic center can be determined according to Eq. (9) from knowledge of the normal force coefficient and the moment coefficient about some arbitrary point.

2 BOUNDARY LAYER CONCEPT

As pointed out in Section 1, there are only two types of aerodynamic forces acting on a body moving through the atmosphere: pressure forces and viscous shear forces. The Reynolds number provides a measure of the relative magnitude of the pressure forces in relation to the viscous shear forces. For the airspeeds typically encountered in flight, Reynolds numbers are quite high and viscous forces are usually small compared to pressure forces. This does not mean that viscous forces can be neglected. However, it does allow us to apply the simplifying concept of boundary layer theory.

For flow over a streamlined body at low angle of attack and high Reynolds number, the effects of viscosity are essentially confined to a thin layer adjacent to the surface of the body, as shown in Fig. 3. Outside the boundary layer, the shear forces can be neglected and since the boundary layer is thin, the change in pressure across the thickness of this layer is insignificant. With this flow model, the pressure forces can be determined from the inviscid flow outside the boundary layer, and the shear forces can be obtained from a solution to the boundary layer equations.

While boundary layer theory provides a tremendous simplification over the complete Navier–Stokes equations, solutions to the boundary layer equations are far from trivial, especially for the complex geometry that is often encountered in an aircraft. A thorough review of boundary layer theory is beyond the intended scope of this chapter and is not a prerequisite to an understanding of the fundamental principles of wing theory. However, there are some important results of boundary layer theory that the reader should know and understand:

1. For the high-Reynolds-number flows typically encountered in flight, the viscous shear forces are small compared to the pressure forces.
2. For flow at high Reynolds number, the pressure forces acting on a body can be closely approximated from an inviscid flow analysis, outside the boundary layer.
3. For 2D flow about streamlined bodies at high Reynolds numbers and low angles of attack, pressure forces do not contribute significantly to drag.
4. For bluff bodies and streamlined bodies at high angles of attack, boundary layer separation occurs, as shown in Fig. 4, and pressure forces dominate the drag.

Boundary layer separation in the flow over airfoils and wings is commonly referred to as *stall*. An understanding of stall and its effect on lift and drag is critical in the study of aeronautics. As the angle of attack for a wing is increased from zero, at first the boundary layer remains attached, as shown in Fig. 3, the lift coefficient increases as a nearly linear function of angle of attack, and the drag coefficient increases approximately with the angle of attack squared. As angle of attack continues to increase, the positive pressure gradient on the aft portion of the upper surface of the wing also increases. At some angle of attack this adverse pressure gradient may result in local boundary layer separation and the increase in lift with angle of attack will begin to diminish. At a slightly higher angle of attack, boundary layer separation becomes complete, as shown in Fig. 4, and the lift rapidly decreases as the angle of attack is increased further. Boundary layer separation also greatly accelerates the increase in drag with angle of attack. The maximum lift coefficient and the exact shape of the lift curve for angles of attack near stall depend substantially on the airfoil section geometry. As the angle of attack increases beyond stall, the lift and drag coefficients become less sensitive to section geometry and for angles of attack beyond about 25°, the lift and drag coefficients are nearly independent of the airfoil section.

The maximum lift coefficient that can be attained on a given wing before stall is quite important in aeronautics. This parameter not only determines the maximum

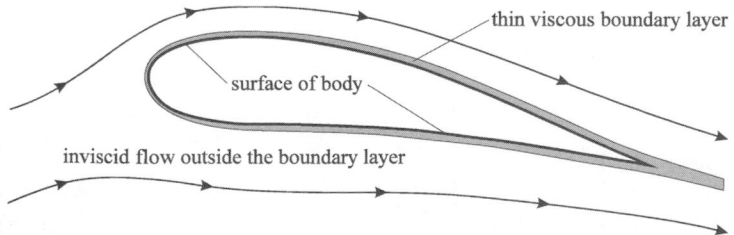

Fig. 3 Boundary layer flow over a streamlined body at a low angle of attack.

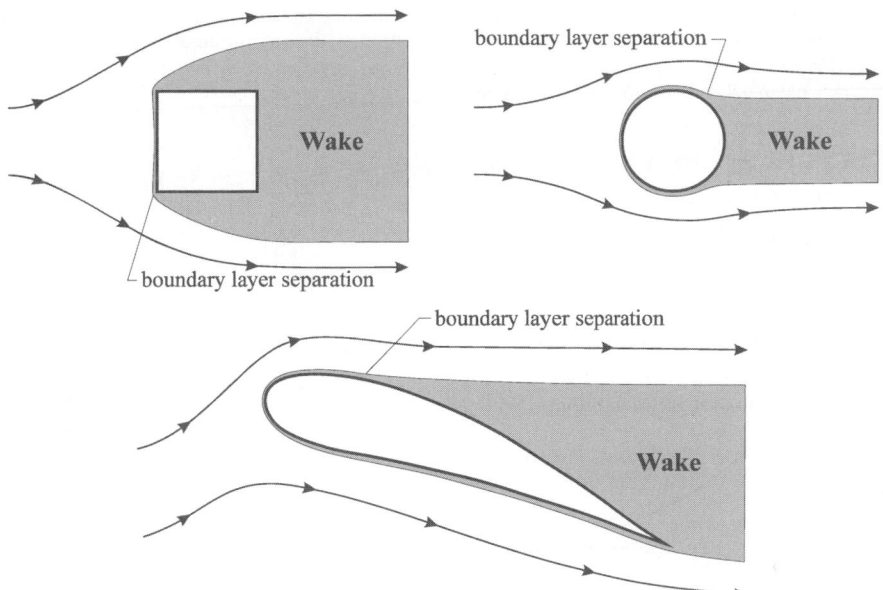

Fig. 4 Examples of boundary layer separation..

weight that can be carried by a given wing at a particular airspeed, it also affects takeoff distance, turning radius, and other measures of airplane performance.

3 INVISCID AERODYNAMICS

Since the pressure forces acting on an aircraft at high Reynolds number can be closely estimated from an inviscid analysis of the flow outside the boundary layer, inviscid aerodynamics is an important tool used in the estimation of the forces acting on an aircraft in flight. Written in vector notation, the general equations that govern inviscid fluid flow are the continuity equation

$$\frac{\partial \rho}{\partial t} + (\mathbf{V} \cdot \nabla)\rho + \rho \nabla \cdot \mathbf{V} = 0 \qquad (10)$$

and the momentum equation

$$\frac{\partial \mathbf{V}}{\partial t} + (\mathbf{V} \cdot \nabla)\mathbf{V} = -\frac{\Delta p}{\rho} - g\, \nabla H \qquad (11)$$

where $\mathbf{V}$ is the fluid velocity vector and t is time. Using the important mathematical identity

$$(\mathbf{V} \cdot \nabla)\mathbf{V} = \nabla\left(\frac{1}{2}V^2\right) - \mathbf{V} \times (\nabla \times \mathbf{V})$$

the momentum equation for inviscid flow can be written

$$\frac{\partial \mathbf{V}}{\delta t} + \nabla\left(\frac{1}{2}V^2\right) + \frac{\nabla p}{\rho} + g\, \nabla H = \mathbf{V} \times \boldsymbol{\Omega} \qquad (12)$$

where $\boldsymbol{\Omega}$ is the curl of the velocity vector, traditionally called the *vorticity*:

$$\boldsymbol{\Omega} \equiv \nabla \times \mathbf{V} \qquad (13)$$

It can be shown from the application of vector calculus to Eqs. (10) and (12) that vorticity cannot be generated in an inviscid flow. In the aerodynamics problem associated with flow over an aircraft in flight, the fluid velocity far upstream from the aircraft is uniform. From the definition of vorticity given in Eq. (13), this means that the vorticity far upstream from the aircraft is zero. Since vorticity cannot be generated in an inviscid flow, the vorticity must be zero everywhere in the inviscid flow outside the boundary layer. Vorticity is generated in the viscous boundary layer next to the skin of the aircraft and thus there is also vorticity in the boundary layer wake that trails behind an aircraft in flight. However, outside the boundary layer and trailing wake, the flow can be assumed to be inviscid and free of vorticity. Thus, for this flow, the right-hand side of Eq. (12) will be zero. This type of flow is called *inviscid irrotational flow*.

In summary, inviscid irrotational flow can be assumed to exist outside the boundary layer and trailing wake of an aircraft in flight. For this flow, the continuity and momentum equations are

$$\frac{\partial \rho}{\partial t} + (\mathbf{V} \cdot \nabla)\rho + \rho \nabla \cdot \mathbf{V} = 0 \qquad (14)$$

$$\frac{\partial \mathbf{V}}{\partial t} + \nabla\left(\frac{1}{2}V^2\right) + \frac{\nabla p}{\rho} + g\,\nabla H = 0 \qquad (15)$$

For the important special case of an aircraft in steady flight, the time derivatives are zero and Eq. (15) reduces to

$$\nabla\left(\frac{1}{2}V^2\right) + \frac{\nabla p}{\rho} + g\,\nabla H = 0 \qquad (16)$$

For the free-stream flow far from the aircraft, Eq. (16) gives

$$\nabla\left(\frac{1}{2}V_\infty^2\right) + \frac{\nabla p_\infty}{\rho\infty} + g\,\nabla H = 0$$

Since the free-stream velocity is uniform, the gradient of V_∞^2 is zero, which gives

$$g\,\nabla H = -\frac{\nabla p_\infty}{\rho_\infty} \qquad (17)$$

This is simply a form of the hydrostatic equation. Using Eq. (17) in Eq. (16) to eliminate the geometric elevation gradient, the momentum equation for the inviscid flow about an aircraft in steady flight is

$$\nabla\left(\frac{1}{2}V^2\right) + \frac{\nabla p}{\rho} - \frac{\nabla p_\infty}{\rho_\infty} = 0 \qquad (18)$$

where p_∞ and ρ_∞ are evaluated at the same elevation as p and ρ.

A special case of Eq. (18) that is of particular interest is the case of flow with negligible variation in air density. This approximation can be applied to flight at low Mach numbers. Assuming that ρ is constant, Eq. (18) reduces to

$$\nabla\left(\frac{1}{2}V^2 + \frac{p - p_\infty}{\rho}\right) = 0 \qquad (19)$$

Integrating this result from the free stream to some arbitrary point in the flow and solving for the local static pressure, the momentum equation for *incompressible inviscid flow* about an aircraft in steady flight yields

$$p = p_\infty + \frac{1}{2}\rho(V_\infty^2 - V^2) \qquad (20)$$

Once the velocity field has been determined in some manner, Eq. (20) can be used to determine the pressure at any point in the flow. The net contribution that these pressure forces make to the resultant aerodynamic force is then computed from

$$\mathbf{F}_p = -\iint_S p\mathbf{n}\,d\mathscr{S} = -\iint_S p_\infty \mathbf{n}\,d\mathscr{S}$$
$$+ \frac{1}{2}\rho V_\infty^2 \iint_S \left[\left(\frac{V}{V_\infty}\right)^2 - 1\right]\mathbf{n}\,d\mathscr{S} \qquad (21)$$

where $\mathbf{n}$ is the unit outward normal and $\mathscr{S}$ is the surface area. The first integral on the far right-hand side of Eq. (21) is the buoyant force, and the second integral is the vector sum of the pressure contribution to the lift and drag. For most conventional airplanes the buoyant force is small and can be neglected.

From Eq. (14), the continuity equation for incompressible flow is

$$\nabla \cdot \mathbf{V} = 0 \qquad (22)$$

Because the curl of the gradient of any scalar function is zero, a flow field is irrotational if the velocity field is written as the gradient of a scalar function

$$\mathbf{V} = \nabla\phi \qquad (23)$$

where ϕ is normally called the *velocity potential* and could be any scalar function of space and time that satisfies the continuity equation and the required boundary conditions. Using Eq. (23) in Eq. (22), this requires that

$$\nabla \cdot \mathbf{V} = \nabla \cdot \nabla\phi = \nabla^2\phi = 0 \qquad (24)$$

Thus the incompressible velocity potential must satisfy Laplace's equation and some appropriate boundary conditions. The far-field boundary condition is uniform flow. At the surface of the aircraft the normal component of velocity must go to zero. However, since the flow is inviscid, we do not require zero tangential velocity at the surface.

For flight at higher Mach numbers the result predicted by Eq. (18) is more complex. When the density of a fluid changes with pressure, the temperature changes as well. Thus, in a compressible fluid, the pressure gradients that accompany velocity gradients also produce temperature gradients. In general, temperature gradients result in heat transfer. However, the thermal conductivity of air is extremely low. Thus, it is commonly assumed that the flow outside the boundary layer is adiabatic as well as inviscid. Inviscid adiabatic flow is isentropic. From the fundamentals of thermodynamics recall that for isentropic flow of an ideal gas, density is related to pressure according to the relation

$$\rho = \rho_\infty \left(\frac{p}{p_\infty}\right)^{1/\gamma} \qquad (25)$$

where again γ is the ratio of specific heats ($\gamma = 1.4$ for air). Substituting Eq. (25) into Eq. (18) and neglecting the gradients of p_∞ and ρ_∞ yields

$$\nabla \left(\frac{1}{2} V^2 + \frac{\gamma}{\gamma - 1} \frac{p}{\rho} \right) = 0 \qquad (26)$$

Integrating Eq. (26) gives

$$\frac{1}{2} V^2 + \frac{\gamma}{\gamma - 1} \frac{p}{\rho} = \text{const} \qquad (27)$$

Equation (27) must apply at the stagnation state, where $V = 0$. Thus, the constant in Eq. (27) can be expressed in terms of the stagnation pressure and density, p_0 and ρ_0.

$$\frac{1}{2} V^2 + \frac{\gamma}{\gamma - 1} \frac{p}{\rho} = \frac{\gamma}{\gamma - 1} \frac{p_0}{\rho_0} \qquad (28)$$

For an ideal gas $p/\rho = RT$ and $a^2 = \gamma RT$, where R is the gas constant and a is the speed of sound. Thus, Eqs. (25) and (28) written in terms of Mach number, M, require that

$$T = T_0 \left(1 + \frac{\gamma - 1}{2} M^2 \right)^{-1} \qquad (29)$$

$$p = p_0 \left(1 + \frac{\gamma - 1}{2} M^2 \right)^{-\gamma/(\gamma-1)} \qquad (30)$$

$$\rho = \rho_0 \left(1 + \frac{\gamma - 1}{2} M^2 \right)^{-1/(\gamma-1)} \qquad (31)$$

Stagnation conditions are readily evaluated by applying Eqs. (29)–(31) to the free-stream flow:

$$T_0 = T_\infty \left(1 + \frac{\gamma - 1}{2} M_\infty^2 \right) \qquad (32)$$

$$p_0 = p_\infty \left(1 + \frac{\gamma - 1}{2} M_\infty^2 \right)^{\gamma/(\gamma-1)} \qquad (33)$$

$$\rho_0 = \rho_\infty \left(1 + \frac{\gamma - 1}{2} M_\infty^2 \right)^{1/(\gamma-1)} \qquad (34)$$

Equation (23) applies to compressible flow as well as to incompressible flow. For any irrotational flow, the velocity vector field can always be expressed as the *gradient of a scalar potential field*. All such flows are called *potential flows*. The only requirement for potential flow is that the flow be irrotational. There are no further restrictions. It can be shown mathematically that the curl of the gradient of any scalar function is zero. Thus, it follows that every potential flow is an *irrotational flow*. It can also be shown mathematically that if the curl of any vector field is zero, that vector field can be expressed as the gradient of some scalar function. Therefore, every irrotational flow is a potential flow. For this reason the terms potential flow and irrotational flow are used synonymously. For complete coverage of potential flow see Karamcheti.[1]

4 INCOMPRESSIBLE FLOW OVER AIRFOILS

An airfoil is any 2D cross section of a wing or other lifting surface that lies in a plane perpendicular to the spanwise coordinate. An airfoil section is completely defined by the geometric shape of its boundary. However, the aerodynamic properties of an airfoil section are most profoundly affected by the shape of its centerline; see Abbott and Von Doenhoff.[2] This centerline is midway between the upper and lower surfaces of the airfoil and is called the camber line. If the airfoil is not symmetric, the camber line is not a straight line but rather a planar curve.

Because the shape of the camber line is such an important factor in airfoil design, it is critical that the reader understand exactly how the camber line is defined. In addition, there are several other designations that will be used throughout this chapter when referring to the geometric attributes of airfoil sections. The reader should be sure that he or she understands the following nomenclature as it applies to airfoil geometry such as that shown in Fig. 5:

The *camber line* is the locus of points midway between the upper and lower surfaces of an airfoil section as measured perpendicular to the camber line itself.

The *leading edge* is the most forward point on the camber line.

The *trailing edge* is the most rearward point on the camber line.

The *chord line* is a straight line connecting the leading edge and the trailing edge.

The *chord length*, often referred to simply as the *chord*, is the distance between the leading edge and the trailing edge as measured along the chord line.

The *maximum camber*, often referred to simply as the *camber*, is the maximum distance between the chord line and the camber line as measured perpendicular to the chord line.

The *local thickness*, at any point along the chord line, is the distance between the upper and lower surfaces as measured perpendicular to the camber line.

The *maximum thickness*, often referred to simply as the *thickness*, is the maximum distance between the upper and lower surfaces as measured perpendicular to the camber line.

The **upper and lower surface coordinates** for an airfoil can be obtained explicitly from the camber line

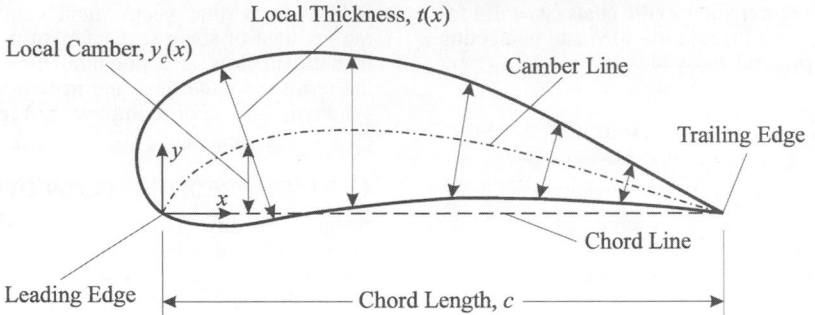

Fig. 5 Airfoil coordinates and nomenclature.

geometry $y_c(x)$ and the thickness distribution $t(x)$:

$$x_u(x) = x - \frac{t(x)}{2\sqrt{1 + (dy_c/dx)^2}} \frac{dy_c}{dx} \qquad (35)$$

$$y_u(x) = y_c(x) + \frac{t(x)}{2\sqrt{1 + (dy_c/dx)^2}} \qquad (36)$$

$$x_l(x) = x + \frac{t(x)}{2\sqrt{1 + (dy_c/dx)^2}} \frac{dy_c}{dx} \qquad (37)$$

$$y_l(x) = y_c(x) - \frac{t(x)}{2\sqrt{1 + (dy_c/dx)^2}} \qquad (38)$$

4.1 Thin Airfoil Theory

For airfoils with a maximum thickness of about 12% or less, the inviscid aerodynamic force and moment are only slightly affected by the thickness distribution. The resultant aerodynamic force and moment acting on such an airfoil depend almost exclusively on the angle of attack and the shape of the camber line. For this reason, the inviscid aerodynamics for these airfoils can be closely approximated by assuming that the airfoil thickness is zero everywhere along the camber line. Thus, airfoils with a thickness of about 12% or less can be approximated by combining a uniform flow with a vortex sheet placed along the camber line, as shown

schematically in Fig. 6. The strength of this vortex sheet is allowed to vary with the distance, s, measured along the camber line. The variation in this strength, $\gamma(s)$, is determined so that the camber line becomes a streamline for the flow.

In the development of thin airfoil theory it is shown that the vortex strength distribution necessary to make the camber line a streamline is related to the camber line geometry according to

$$\frac{1}{2\pi} \int_{x_o=0}^{c} \frac{\gamma(x_o)}{x - x_o} \, dx_o = V_\infty \left(\alpha - \frac{dy_c}{dx} \right) \qquad (39)$$

This is the fundamental equation of thin airfoil theory. Any vortex strength distribution, $\gamma(x_o)$, which satisfies Eq. (39) will make the camber line a streamline of the flow, at least within the accuracy of the approximations used in thin airfoil theory. For a given airfoil at a given angle of attack, the only unknown in Eq. (39) is the vortex strength distribution, $\gamma(x_o)$. This equation is subject to a boundary condition known as the *Kutta condition*, which requires that

$$\gamma(c) = 0 \qquad (40)$$

Development of the general solution to Eq. (39), subject to (40), is presented in most undergraduate

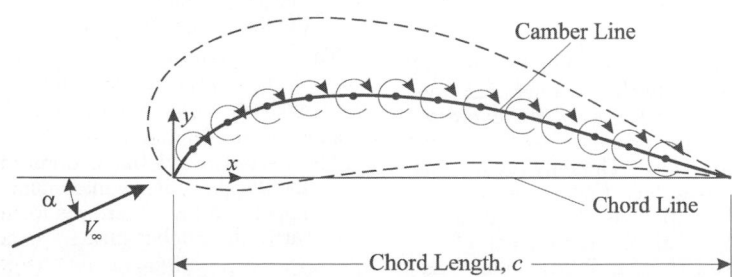

Fig. 6 Synthesis of a thin airfoil section from a uniform flow and a curved vortex sheet distributed along the camber line.

engineering textbooks on aerodynamics. This solution is found by using the change of variables, $x = c(1 - \cos\theta)/2$. Only the final result is presented here. The solution is in the form of an infinite series for the vortex strength distribution,

$$\gamma(\theta) = 2V_\infty \left(A_0 \frac{1 + \cos\theta}{\sin\theta} + \sum_{n=1}^{\infty} A_n \sin(n\theta) \right) \tag{41}$$

$$A_0 = \alpha - \frac{1}{\pi} \int_{\theta=0}^{\pi} \frac{dy_c}{dx} d\theta \tag{42}$$

$$A_n = \frac{2}{\pi} \int_{\theta=0}^{\pi} \frac{dy_c}{dx} \cos(n\theta) d\theta \tag{43}$$

$$x(\theta) \equiv \frac{c}{2}(1 - \cos\theta) \tag{44}$$

The aerodynamic force per unit span can be predicted from the *Kutta–Joukowski law*, which applies to all 2D potential flows. This requires that the net aerodynamic force is always normal to the free stream and equal to

$$\tilde{L} = \rho V_\infty \Gamma \tag{45}$$

A direct consequence of the Kutta–Joukowski law is that the pressure drag for any 2D flow without boundary layer separation is zero.

Applying the Kutta–Joukowski law to a differential segment of the vortex sheet that is used to synthesize a thin airfoil gives

$$d\tilde{L} = \rho V_\infty d\Gamma = \rho V_\infty \gamma(x_o) dx_o$$

$$= \frac{\rho V_\infty c}{2} \gamma(\theta_o) \sin\theta_o d\theta_o$$

Applying Eq. (41) for the vortex strength distribution, this result can be used to evaluate the lift and moment coefficients as well as the center of pressure.

$$\tilde{C}_l = 2\pi \left(A_0 + \frac{1}{2}A_1 \right)$$

$$= 2\pi \left(\alpha - \frac{1}{\pi} \int_{\theta=0}^{\pi} \frac{dy_c}{dx}(1 - \cos\theta) d\theta \right)$$

$$= 2\pi(\alpha - \alpha_{l,0}) \tag{46}$$

$$\tilde{C}_{m_{le}} = -\frac{\tilde{C}_L}{4} + \frac{\pi}{4}(A_2 - A_1)$$

$$= -\frac{\tilde{C}_L}{4} + \frac{1}{2} \int_{\theta=0}^{\pi} \frac{dy_c}{dx}[\cos(2\theta) - \cos\theta] d\theta \tag{47}$$

$$\tilde{C}_{m_{c/4}} = \frac{\pi}{4}(A_2 - A_1)$$

$$= \frac{1}{2} \int_{\theta=0}^{\pi} \frac{dy_c}{dx}[\cos(2\theta) - \cos\theta] d\theta \tag{48}$$

$$\frac{x_{cp}}{c} = \frac{1}{4} + \frac{\pi}{4\tilde{C}_L}(A_1 - A_2)$$

$$= \frac{1}{4} + \frac{1}{2\tilde{C}_L} \int_{\theta=0}^{\pi} \frac{dy_c}{dx}[\cos\theta - \cos(2\theta)] d\theta \tag{49}$$

Note from Eq. (46) that thin airfoil theory predicts a section lift coefficient that is a linear function of angle of attack, and that the change in lift coefficient with respect to angle of attack is 2π per radian. Also note that the lift coefficient at zero angle of attack is a function only of the shape of the camber line. Thus, for thin airfoils,

$$\text{Section lift slope} \equiv \frac{d\tilde{C}_L}{d\alpha} \equiv \tilde{C}_{L,\alpha} = 2\pi$$

$$\text{Zero lift angle of attack} \equiv \alpha_{L0}$$

$$= \frac{1}{\pi} \int_{\theta=0}^{\pi} \frac{dy_c}{dx}(1 - \cos\theta) d\theta$$

Also notice that the leading-edge moment coefficient and center of pressure both depend on lift coefficient and hence on angle of attack. The quarter-chord moment coefficient, on the other hand, is independent of angle of attack and depends only on the shape of the camber line. Since the aerodynamic center is defined to be the point on the airfoil where the moment is independent of angle of attack, for incompressible flow, the quarter chord is the *aerodynamic center* of a thin airfoil. Thus, the quarter chord is usually referred to as the *theoretical* or *ideal aerodynamic center* of the cambered airfoil. Viscous effects and airfoil thickness can cause the quarter-chord moment coefficient to vary slightly with angle of attack, but this variation is small and the aerodynamic center is always close to the quarter chord for subsonic flow. For this reason, airfoil section moment data are usually reported in terms of the quarter-chord moment coefficient. For an airfoil with no camber, Eq. (49) shows that the quarter chord is also the *center of pressure* for incompressible flow about a thin symmetric airfoil.

As is the case with all 2D potential flow, thin airfoil theory predicts a net aerodynamic force that is normal to the free stream. Thus, thin airfoil theory predicts a section drag coefficient that is exactly zero. This is not a function of the thin airfoil approximation. Numerical panel methods will also predict zero section drag, including the effects of thickness. Section drag in any

2D subsonic flow results entirely from viscous effects, which are neglected in the potential flow equations. The viscous forces also have some effect on lift, but this effect is relatively small.

4.2 Vortex Panel Method

Potential flow over an airfoil of arbitrary shape can be synthesized by combining uniform flow with a curved vortex sheet wrapped around the surface of the airfoil, as shown in Fig. 7. The vortex strength must vary along the surface such that the normal component of velocity induced by the entire sheet and the uniform flow is zero everywhere along the surface of the airfoil. In most cases, the strength distribution necessary to satisfy this condition is very difficult or impossible to determine analytically. However, for numerical computations, such a sheet can be approximated as a series of flat vortex panels wrapped around the surface of the airfoil. In the limit as the panel size becomes very small, the panel solution approaches that for the curved vortex sheet. For complete coverage of the vortex panel method see Katz and Plotkin.[3]

To define the vortex panels, a series of nodes is placed on the airfoil surface. For best results the nodes should be clustered more tightly near the leading and trailing edges. The most popular method for attaining this clustering is called *cosine clustering*. For this method we use the change of variables

$$\frac{x}{c} = \frac{1}{2}(1 - \cos\theta)$$

which is the same change of variables as that used in thin airfoil theory. Distributing the nodes uniformly in θ will provide the desired clustering in x. For best results near the leading edge, an even number of nodes should always be used. For this particular distribution, the nodal coordinates are computed from the algorithm

$$\delta\theta = \frac{2\pi}{n-1} \quad (50)$$

$$\begin{Bmatrix} x_N\left(\frac{1}{2}n+i\right) \\ y_N\left(\frac{1}{2}n+i\right) \\ x_N\left(\frac{1}{2}n+1-i\right) \\ y_N\left(\frac{1}{2}n+1-i\right) \end{Bmatrix} = \begin{Bmatrix} x_u \\ y_u \\ x_l \\ y_l \end{Bmatrix},$$

$$\frac{x}{c} = 0.5\{1 - \cos[(i-0.5)\delta\theta]\}$$

$$i = 1, \tfrac{1}{2}n \quad (51)$$

where n is the total number of nodes and the upper and lower surface coordinates are computed from Eqs. (35)–(38). A nodal distribution using cosine clustering with 12 nodes is shown in Fig. 8. Notice that both the first and last nodes are placed at the trailing edge. Between 50 and 200 nodes should be used for computation.

We can now synthesize an airfoil using $n-1$ vortex panels placed between these n nodes on the airfoil surface. The panels start at the trailing edge, are

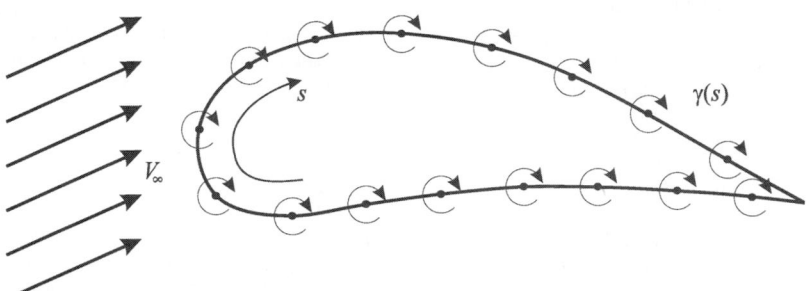

Fig. 7 Synthesis of an arbitrary airfoil section from a uniform flow and a vortex sheet.

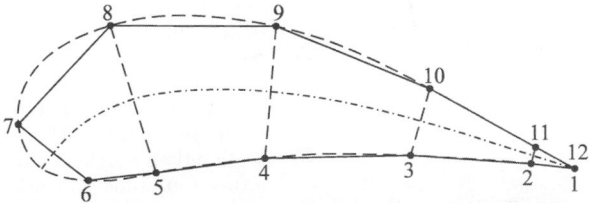

Fig. 8 Vortex panel distribution with cosine clustering and an even number of nodes.

spaced forward along the lower surface, are wrapped up around the leading edge, and then run back along the upper surface to the trailing edge. The last panel ends at the trailing edge where the first panel began. The strength of each vortex panel is assumed to be linear along the panel and is required to be continuous from one panel to the next, that is, the strength at the end of one panel must equal the strength at the beginning of the next panel. The strength is not required to be continuous across the trailing edge, that is, $\gamma_1 \neq \gamma_n$. Each panel is assigned a local panel coordinate system (ξ, η), as shown in Fig. 9. The velocity induced by each of these panels is expressed in panel coordinates, and the induced velocity components are then transformed to airfoil coordinates.

To solve for the n unknown nodal vortex strengths, control points are placed at the center of each of the $n - 1$ panels. The coordinates of these control points are given by

$$\begin{Bmatrix} x_C(i) \\ y_C(i) \end{Bmatrix} = \begin{Bmatrix} \dfrac{x_N(i) + x_N(i+1)}{2} \\ \dfrac{y_N(i) + y_N(i+1)}{2} \end{Bmatrix} \quad i = 1, n-1$$

(52)

The normal velocity at each control point, induced by all $n - 1$ panels and the uniform flow, must be zero. This gives $n - 1$ equations for the n unknown nodal vortex strengths.

For the remaining equation, we know that the flow must leave the airfoil from the trailing edge. This means that the velocity just above the trailing edge must equal the velocity just below the trailing edge. If the angle between the upper and lower surfaces at the trailing edge is nonzero, the trailing edge is a stagnation point. If the angle between the upper and lower surfaces at the trailing edge is zero, the velocity at the trailing edge can be finite but the velocity must be continuous across the trailing edge. Since there is discontinuity in tangential velocity across a vortex sheet, we know that at the trailing edge, the discontinuity across the upper surface is equal and opposite to the discontinuity across the lower surface. Thus, the net discontinuity across both surfaces will be zero. Because the discontinuity in tangential velocity across any vortex sheet is equal to the local strength of the sheet, at the trailing edge, the strength of the upper surface must be exactly equal and opposite to the strength of the lower surface. That is,

$$\gamma_1 + \gamma_n = 0$$

This is called the *Kutta condition* and it provides the remaining equation necessary to solve for the n unknown nodal strengths.

The 2×2 panel coefficient matrix in airfoil coordinates, $[\mathbf{P}]_{i(x,y)}$, for the velocity induced at the arbitrary point (x,y) by panel i, extending from the node at (x_i, y_i) to the node at (x_{i+1}, y_{i+1}), is computed from the algorithm

$$l_i = \sqrt{(x_{i+1} - x_i)^2 + (y_{i+1} - y_i)^2}$$

(53)

$$\begin{Bmatrix} \xi \\ \eta \end{Bmatrix} = \frac{1}{l_i} \begin{bmatrix} (x_{i+1} - x_i) & (y_{i+1} - y_i) \\ -(y_{i+1} - y_i) & (x_{i+1} - x_i) \end{bmatrix} \begin{Bmatrix} (x - x_i) \\ (y - y_i) \end{Bmatrix}$$

(54)

$$\Phi = \text{atan2}(\eta l_i, \eta^2 + \xi^2 - \xi l_i)$$

(55)

$$\Psi = \frac{1}{2} \ln \left[\frac{\xi^2 + \eta^2}{(\xi - l_i)^2 + \eta^2} \right]$$

(56)

$$[\mathbf{P}]_{i(x,y)} = \frac{1}{2\pi l_i^2} \begin{bmatrix} (x_{i+1} - x_i) & -(y_{i+1} - y_i) \\ (y_{i+1} - y_i) & (x_{i+1} - x_i) \end{bmatrix}$$

$$\times \begin{bmatrix} [(l_i - \xi)\Phi + \eta\Psi] & (\xi\Phi - \eta\Psi) \\ [\eta\Phi - (l_i - \xi)\Psi - l_i] & (-\eta\Phi - \xi\Psi - l_i) \end{bmatrix}$$

(57)

The $n \times n$ airfoil coefficient matrix, $[\mathbf{A}]$, is generated from the 2×2 panel coefficient matrix in airfoil coordinates, $[\mathbf{P}]_{ji}$, for the velocity induced at control point i by panel j, extending from node j to node $j + 1$, using the algorithm

$$A_{ij} = 0 \quad i = 1, n \quad j = 1, n \quad (58)$$

$$\left. \begin{aligned} A_{ij} &= A_{ij} + \frac{x_{i+1} - x_i}{l_i} P_{21_{ji}} - \frac{y_{i+1} - y_i}{l_i} P_{11_{ji}} \\ A_{ij+1} &= A_{ij+1} + \frac{x_{i+1} - x_i}{l_i} P_{22_{ji}} - \frac{y_{i+1} - y_i}{l_i} P_{12_{ji}} \end{aligned} \right\}$$

$$i = 1, n-1 \quad j = 1, n-1 \quad (59)$$

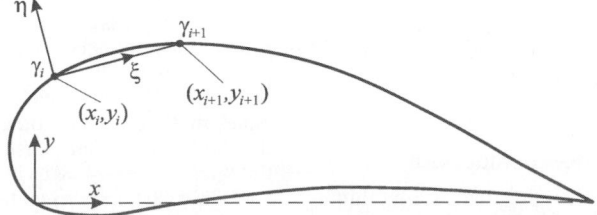

Fig. 9 A vortex panel on the surface of an arbitrary airfoil section.

$$A_{n1} = 1 \qquad (60)$$

$$A_{nn} = 1 \qquad (61)$$

The *n nodal vortex strengths* $\gamma_1, \ldots, \gamma_n$ are then obtained by numerically solving the $n \times n$ linear system

$$[A] \begin{Bmatrix} \gamma_1 \\ \gamma_2 \\ \vdots \\ \gamma_{n-1} \\ \gamma_n \end{Bmatrix}$$

$$= V_\infty \begin{Bmatrix} [(y_2 - y_1)\cos\alpha - (x_2 - x_1)\sin\alpha]/l_1 \\ [(y_3 - y_2)\cos\alpha - (x_3 - x_2)\sin\alpha]/l_2 \\ \vdots \\ [(y_n - y_{n-1})\cos\alpha - (x_n - x_{n-1})\sin\alpha]/l_{n-1} \\ 0 \end{Bmatrix}$$

$$(62)$$

Once the nodal strengths are known, the velocity and pressure at any point in space can be computed by adding the velocity induced by all $n - 1$ vortex panels to the free-stream velocity,

$$\begin{Bmatrix} V_x \\ V_y \end{Bmatrix} = V_\infty \begin{Bmatrix} \cos\alpha \\ \sin\alpha \end{Bmatrix} + \sum_{i=1}^{n-1} [P]_{i(x,y)} \begin{Bmatrix} \gamma_i \\ \gamma_{i+1} \end{Bmatrix} \qquad (63)$$

$$V^2 = V_x^2 + V_y^2 \qquad (64)$$

$$C_p \equiv \frac{p - p_\infty}{\frac{1}{2}\rho V_\infty^2} = 1 - \frac{V^2}{V_\infty^2} \qquad (65)$$

The lift and moment coefficients for the entire airfoil are the sum of those induced by all of the $n - 1$ vortex panels,

$$\tilde{C}_L = \sum_{i=1}^{n-1} \frac{l_i}{c} \frac{\gamma_i + \gamma_{i+1}}{V_\infty} \qquad (66)$$

$$\tilde{C}_{m_{le}} = -\frac{1}{3} \sum_{i=1}^{n-1} \frac{l_i}{c}$$

$$\left[\frac{2x_i\gamma_i + x_i\gamma_{i+1} + x_{i+1}\gamma_i + 2x_{i+1}\gamma_{i+1}}{cV_\infty} \cos(\alpha) \right.$$

$$\left. + \frac{2y_i\gamma_i + y_i\gamma_{i+1} + y_{i+1}\gamma_i + 2y_{i+1}\gamma_{i+1}}{cV_\infty} \sin(\alpha) \right]$$

$$(67)$$

4.3 Comparison with Experimental Data

Section lift and moment coefficients predicted by thin airfoil theory and panel codes are in good agreement with experimental data for low Mach numbers and small angles of attack. In Fig. 10, the inviscid lift coefficient for an NACA 2412 airfoil, as predicted by thin airfoil theory, is compared with the inviscid lift coefficient predicted by the vortex panel method and with experimental data for total lift coefficient as reported by Abbott and Von Doenhoff.[2] Thin airfoil theory predicts a section lift coefficient that is independent of the thickness distribution and dependent only on angle of attack and camber line shape. At small angles of attack, the thin airfoil approximation agrees closely with experimental data based on total lift, for airfoils as thick as about 12%. The agreement seen in Fig. 10 is quite typical. For airfoils much thicker than about 12%, viscous effects become increasingly important, even at fairly low angles of attack, and the inviscid lift coefficient predicted by thin airfoil theory begins to deviate more from experimental observations based on total lift. This can be seen in Fig. 11.

The reason why thin airfoil theory predicts total lift so well over such a wide range of thickness is that thickness tends to increase the lift slope slightly, while viscous effects tend to decrease the lift slope. Since both thickness effects and viscous effects are neglected in thin airfoil theory, the resulting errors tend to cancel, giving the theory a broader range of applicability than would otherwise be expected. Coincidentally, because of these opposing errors, thin airfoil theory actually predicts a lift slope that agrees more closely with experimental data for total lift than does that predicted by the more elaborate vortex panel method. The vortex panel method accurately predicts the pressure distribution around the airfoil, including the effects of thickness. However, since the vortex panel method provides a potential flow solution, it does not account for viscous effects in any way.

After seeing that thin airfoil theory predicts total section lift better than the vortex panel method, one may wonder why we should ever be interested in the vortex panel method. For the answer, we must remember that both the vortex panel solution and the thin airfoil solution come from potential flow theory. Thus, neither of these two solutions can be expected to predict viscous forces. However, potential flow solutions are often used as boundary conditions for viscous flow analysis. Potential flow is used to predict the pressure distribution around the airfoil. The viscous forces are then computed from boundary layer theory. However, since boundary layer flow is greatly affected by the pressure distribution around the airfoil, our potential flow solution must accurately predict the section pressure distribution. Because thin airfoil theory does not account for airfoil thickness, it cannot be used to predict the surface pressure distribution on an airfoil section with finite thickness. The vortex panel method, on the other hand, accurately predicts the inviscid pressure distribution for airfoils of any thickness.

While the experimental data shown in Figs. 10 and 11 are based on total section lift, section lift data have been obtained that are based on pressure

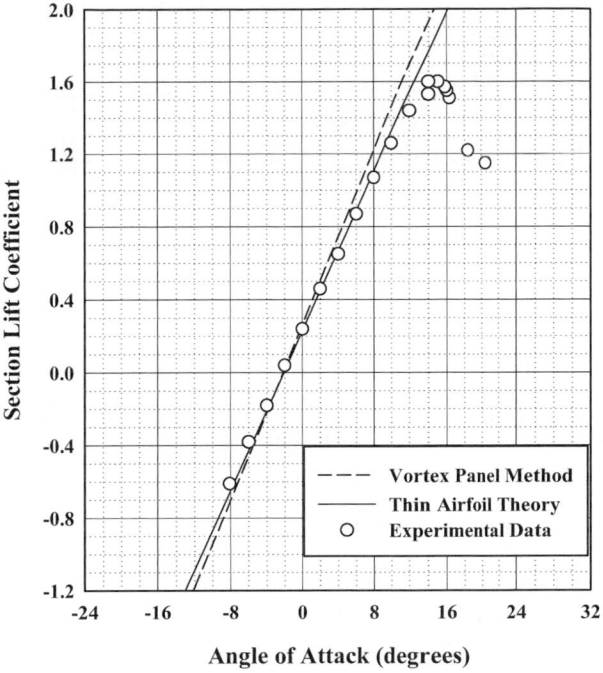

Fig. 10 Section lift coefficient comparison among thin airfoil theory, the vortex panel method, and experimental data based on total lift, for the NACA 2412 airfoil.

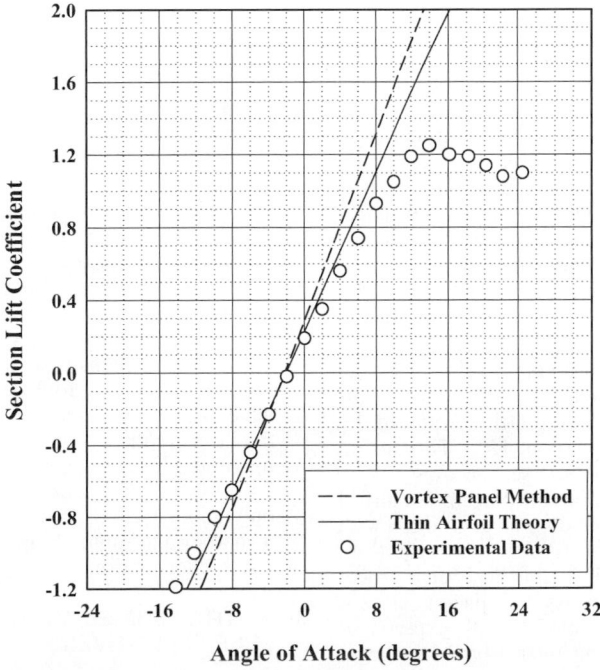

Fig. 11 Section lift coefficient comparison among thin airfoil theory, the vortex panel method, and experimental data based on total lift, for the NACA 2421 airfoil.

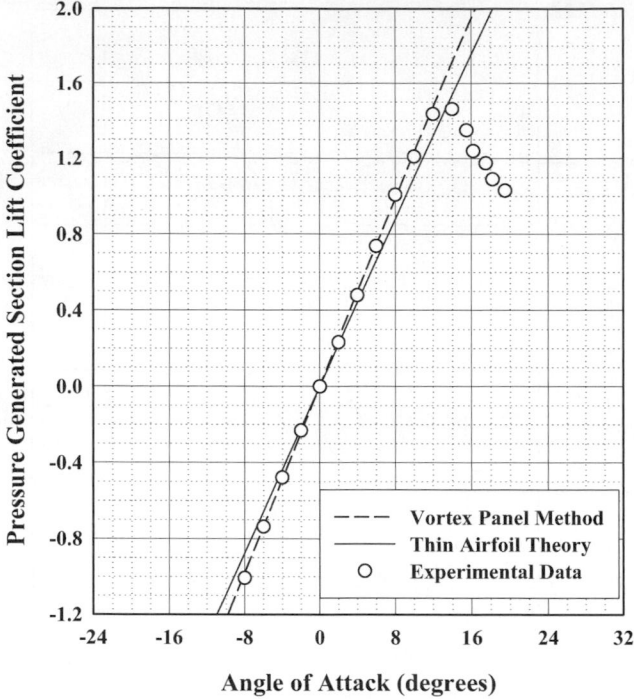

Fig. 12 Section lift coefficient comparison among thin airfoil theory, the vortex panel method, and experimental data based on pressure lift, for the NACA 0015 airfoil.

forces only. This is usually accomplished by spacing a large number of static pressure taps around the circumference of an airfoil. The net pressure force acting on the airfoil section is derived by numerically integrating the forces obtained from these pressure measurements. The lift coefficient predicted by thin airfoil theory and that predicted by the vortex panel method are compared with experimental data of this type in Fig. 12. Notice that the vortex panel method agrees very closely with these experimental data, while thin airfoil theory predicts a lift coefficient that is somewhat low.

In summary, thin airfoil theory can be used to obtain a first approximation for the total section lift coefficient, at small angles of attack, produced by airfoils of thickness less than about 12%. Thin airfoil theory gives no prediction for the section drag.

To improve on the results obtained from thin airfoil theory, we can combine a boundary layer solution with the velocity and pressure distribution obtained from the vortex panel method. Since this analytical procedure accounts for both thickness and viscous effects, it produces results that agree closely with experimental data for total lift and drag over a broad range of section thickness. Another alternative available with today's high-speed computers is the use of computational fluid dynamics (CFD). However, a CFD solution will increase the required computation time by several orders of magnitude.

In Figs. 10–12 it should be noticed that at angles of attack near the zero-lift angle of attack, the result predicted by thin airfoil theory agrees very well with that predicted by the vortex panel method and with that observed from all experimental data. Thus, we see that the thickness distribution has little effect on the lift produced by an airfoil at angles of attack near the zero-lift angle of attack. The thickness distribution does, however, have a significant effect on the maximum lift coefficient and on the stall characteristics of the airfoil section. This is seen by comparing Fig. 10 with Fig. 11. The two airfoils described in these two figures have exactly the same camber line. They differ only in thickness. Notice that the NACA 2412 section has a maximum lift coefficient of about 1.6, while the NACA 2421 airfoil produces a maximum lift coefficient of only about 1.2. Also notice that the thinner section has a sharper and more abrupt stall than that displayed by the thicker section.

5 TRAILING-EDGE FLAPS AND SECTION FLAP EFFECTIVENESS

An attached trailing-edge flap is formed by hinging some aft portion of the airfoil section so that it can be deflected, either downward or upward, by rotating the

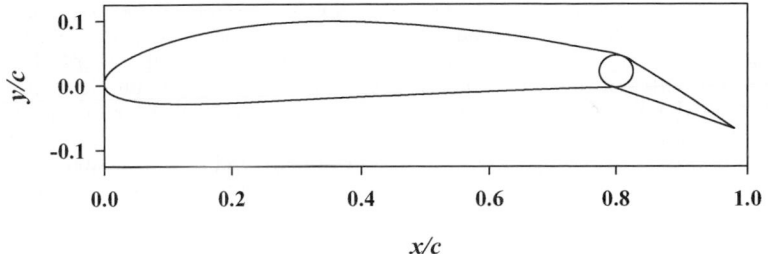

Fig. 13 NACA 4412 airfoil section with a 20% trailing-edge flap.

flap about the hinge point as shown in Fig. 13. The deflection of a trailing-edge flap effectively changes the camber of the airfoil section, and in so doing changes the aerodynamic characteristics of the section. A downward deflection of the flap increases the effective camber and is thus usually considered to be a positive deflection.

For small angles of attack and small flap deflections, thin airfoil theory can be applied to an airfoil section with a deflected trailing-edge flap. Let y_d be the y position of the section camber line with the trailing-edge flap deflected as shown in Fig. 14. All of the results obtained from thin airfoil theory must apply to this modified camber line geometry. Thus from Eq. (46), we can write

$$\tilde{C}_L = 2\pi(\alpha - \alpha_{L0})$$
$$= 2\pi\left(\alpha - \frac{1}{\pi}\int_{\theta=0}^{\pi} \frac{dy_d}{dx}(1 - \cos\theta)\,d\theta\right) \quad (68)$$

However, within the small-angle approximation, the slope of the deflected camber line geometry can be related to the slope of the undeflected geometry according to

$$\frac{dy_d}{dx} = \begin{cases} \dfrac{dy_c}{dx} & x \leq c - c_f \\ \dfrac{dy_c}{dx} - \delta & x \geq c - c_f \end{cases} \quad (69)$$

where y_c is the undeflected camber line ordinate, c_f is the flap chord length, and δ is the deflection of the flap in radians, with positive deflection being downward.

Using Eq. (69) in Eq. (68), we have

$$\tilde{C}_L = 2\pi(\alpha - \alpha_{L0})$$
$$= 2\pi\left(\alpha - \frac{1}{\pi}\int_{\theta=0}^{\pi} \frac{dy_c}{dx}(1 - \cos\theta)\,d\theta\right.$$
$$\left. + \frac{\delta}{\pi}\int_{\theta=\theta_f}^{\pi}(1 - \cos\theta)\,d\theta\right) \quad (70)$$

or

$$\alpha_{L0} = \frac{1}{\pi}\int_{\theta=0}^{\pi} \frac{dy_c}{dx}(1 - \cos\theta)\,d\theta$$
$$- \frac{\delta}{\pi}\int_{\theta=\theta_f}^{\pi}(1 - \cos\theta)\,d\theta \quad (71)$$

where α_{L0} is the zero-lift angle of attack for the airfoil section with the flap deflected and θ_f is given by

$$\theta_f = \cos^{-1}\left(2\frac{c_f}{c} - 1\right) \quad (72)$$

From Eqs. (70) and (71), we see that the thin airfoil lift coefficient is affected by flap deflection only through a change in the zero-lift angle of attack. The lift coefficient is still a linear function of the angle of

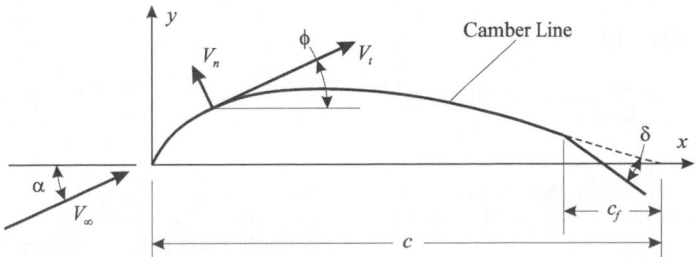

Fig. 14 Camber line geometry for an airfoil section with attached trailing-edge flap and positive flap deflection.

attack and the lift slope is not affected by the flap deflection. The first integral on the right-hand side of Eq. (71) is simply the zero-lift angle of attack with no flap deflection. The second integral in this equation is readily evaluated to yield what is commonly called the *ideal section flap effectiveness*,

$$\varepsilon_{fi} \equiv -\frac{\partial \alpha_{L0}}{\partial \delta} = \frac{1}{\pi} \int_{\theta=\theta_f}^{\pi} (1 - \cos\theta)\, d\theta$$

$$= 1 - \frac{\theta_f - \sin\theta_f}{\pi} \qquad (73)$$

Notice that the ideal section flap effectiveness depends only on the ratio of flap chord to total chord and is independent of both camber line geometry and flap deflection. Using this definition in Eq. (71), the zero-lift angle of attack for a thin airfoil with an ideal trailing-edge flap is found to vary linearly with flap deflection,

$$\alpha_{L0}(\delta) = \alpha_{L0}(0) - \varepsilon_{fi}\delta \qquad (74)$$

In a similar manner we can predict the quarter-chord moment coefficient for a thin airfoil with a deflected trailing-edge flap. From Eq. (48), we have

$$\widetilde{C}_{m_{c/4}} = \frac{1}{2} \int_{\theta=0}^{\pi} \frac{dy_d}{dx} [\cos(2\theta) - \cos\theta]\, d\theta \qquad (75)$$

Using Eq. (69) in Eq. (75) gives

$$\widetilde{C}_{m_{c/4}} = \frac{1}{2} \int_{\theta=0}^{\pi} \frac{dy_c}{dx} [\cos(2\theta) - \cos\theta]\, d\theta$$

$$- \frac{\delta}{2} \int_{\theta=\theta_f}^{\pi} [\cos(2\theta) - \cos\theta]\, d\theta \qquad (76)$$

From Eq. (76) we see that the section quarter-chord moment coefficient for a thin airfoil is also a linear function of flap deflection. The first integral on the right-hand side of Eq. (76) is the quarter-chord moment coefficient for the airfoil section with no flap deflection, and the second integral can be evaluated to yield the section quarter-chord moment slope with respect to flap deflection. Thus, the quarter-chord moment coefficient for a thin airfoil section with an ideal trailing-edge flap can be written as

$$\widetilde{C}_{m_{c/4}}(\delta) = \widetilde{C}_{m_{c/4}}(0) + \widetilde{C}_{m,\delta}\delta \qquad (77)$$

where the change in the section quarter-chord moment coefficient with respect to flap deflection is given by

$$\widetilde{C}_{m,\delta} \equiv \frac{\partial \widetilde{C}_{m_{c/4}}}{\partial \delta} = -\frac{1}{2} \int_{\theta=\theta_f}^{\pi} [\cos(2\theta) - \cos\theta]\, d\theta$$

$$= \frac{\sin(2\theta_f) - 2\sin\theta_f}{4} \qquad (78)$$

Notice that the change in moment coefficient with respect to flap deflection depends only on the ratio of flap chord to total chord. Thus, as was the case with the ideal section flap effectiveness, the ideal section quarter-chord moment slope with respect to flap deflection is independent of both camber line geometry and flap deflection.

In summary, at angles of attack below stall, the lift coefficient for an airfoil section with a deflected trailing-edge flap is found to be very nearly a linear function of both the airfoil angle of attack, α, and the flap deflection, δ. This linear relation can be written

$$\widetilde{C}_L(\alpha, \delta) = \widetilde{C}_{L,\alpha}[\alpha - \alpha_{L0}(0) + \varepsilon_f \delta] \qquad (79)$$

where $\widetilde{C}_{L,\alpha}$ is the section lift slope, $\alpha_{L0}(0)$ is the zero-lift angle of attack with no flap deflection, and ε_f is called the *section flap effectiveness*. Previously, we found that thin airfoil theory predicts a section lift slope of 2π. However, solutions obtained using the vortex panel method and experimental measurements have shown that the actual section lift slope can vary somewhat from this value. The zero-lift angle of attack with no flap deflection, as predicted by thin airfoil theory, was previously shown to be in excellent agreement with both the vortex panel method and experimental data. As we shall see, the section flap effectiveness predicted by thin airfoil theory agrees with results predicted using the vortex panel method, but deviates somewhat from experimental observation.

The actual section flap effectiveness is always less than the ideal section flap effectiveness given by Eq. (73). The hinge mechanism in a real trailing-edge flap always reduces the flap effectiveness. This reduction results from local boundary layer separation and from flow leakage through the hinge from the high-pressure side to the low-pressure side. In addition, at flap deflections greater than about $\pm 10°$, the error associated with the small-angle approximation used to obtain Eq. (73) begins to become significant. This results in an additional decrease in the section flap effectiveness for larger flap deflections. A comparison among the section flap effectiveness predicted by thin airfoil theory, typical results predicted by the vortex panel method, and results observed experimentally is shown in Fig. 15. The data shown in this figure are from Abbott and Von Doenhoff.[2]

The discrepancy between the theoretical results and the experimental data shown in Fig. 15 is only about 7% with a flap chord fraction of 0.4, but at a flap chord fraction of 0.1 this discrepancy is nearly 25%. The deviation between actual section flap effectiveness and the ideal section flap effectiveness continues to increase as the flap chord fraction becomes smaller. The poor agreement at low flap chord fraction is attributed to the thickness of the boundary layer, which is much larger near the trailing edge. The trailing-edge flaps used to generate the data shown in Fig. 15 all had flap hinges that were sealed to prevent leakage from

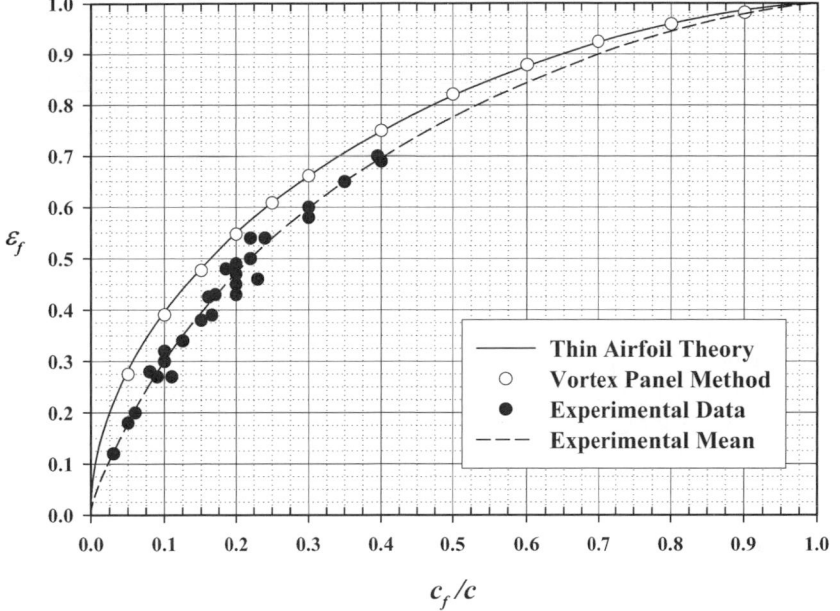

Fig. 15 Section flap effectiveness comparison among thin airfoil theory, the vortex panel method, and experimental data.

the high-pressure side to the low-pressure side. For unsealed trailing-edge flaps, an additional decrease in section flap effectiveness of about 20% is observed.

The data shown in Fig. 15 were all taken at flap deflections of less than $10°$. For larger deflections, actual flap effectiveness deviates even more from the ideal, as a result of errors associated with the small-angle approximation. The actual section flap effectiveness for a trailing-edge flap can be expressed in terms of the ideal section flap effectiveness, ε_{fi}, given by Eq. (73), according to

$$\varepsilon_f = \eta_h \eta_d \varepsilon_{fi} \qquad (80)$$

where η_h and η_d are, respectively, the section flap hinge efficiency and the section flap deflection efficiency.

For well-designed sealed flaps, the section hinge efficiency can be approximated from Fig. 16. This figure is based on the mean line for the experimental data shown in Fig. 15. If the gap between the main wing and the trailing-edge flap is not sealed, it is recommended that the values found from Fig. 16 be reduced by 20%. Remember that the result given in this figure represents a mean experimental efficiency. The actual hinge efficiency for a specific trailing-edge flap installed in a particular airfoil section may deviate significantly from the value found in this figure. For flap deflections of more than $\pm10°$, Perkins and Hage[4] recommend the flap deflection efficiency shown in Fig. 17. For flap deflections of less

than $\pm10°$, a flap deflection efficiency of 1.0 should be used.

At angles of attack below stall, the pitching moment coefficient for an airfoil section with a deflected trailing-edge flap is also found to be very nearly a linear function of both angle of attack, α, and flap deflection, δ. This linear relation can be written in the form

$$\widetilde{C}_{m_{c/4}}(\alpha, \delta) = \widetilde{C}_{m_{c/4}}(0,0) + \widetilde{C}_{m,\alpha}\alpha + \widetilde{C}_{m,\delta}\delta \qquad (81)$$

where $\widetilde{C}_{m_{c/4}}(0,0)$, $\widetilde{C}_{m,\alpha}$, and $\widetilde{C}_{m,\delta}$ are, respectively, the section pitching moment coefficient at zero angle of attack and zero flap deflection, the moment slope with angle of attack, and the moment slope with flap deflection.

Thin airfoil theory predicts that the quarter chord is the aerodynamic center of an airfoil section. Thus, thin airfoil theory predicts a zero quarter-chord moment slope with respect to angle of attack. In reality, solutions obtained from the vortex panel method and experimental observations have shown that the quarter chord is not exactly the aerodynamic center of all airfoil sections. Thus, in general, we should allow for a finite quarter-chord moment slope with angle of attack. However, for preliminary design, the quarter-chord moment slope with angle of attack is usually taken to be zero.

The section quarter-chord moment slope with respect to flap deflection is shown in Fig. 18. In this figure the ideal quarter-chord moment slope, as predicted by thin airfoil theory in Eq. (78), is compared with typical results predicted from the vortex panel

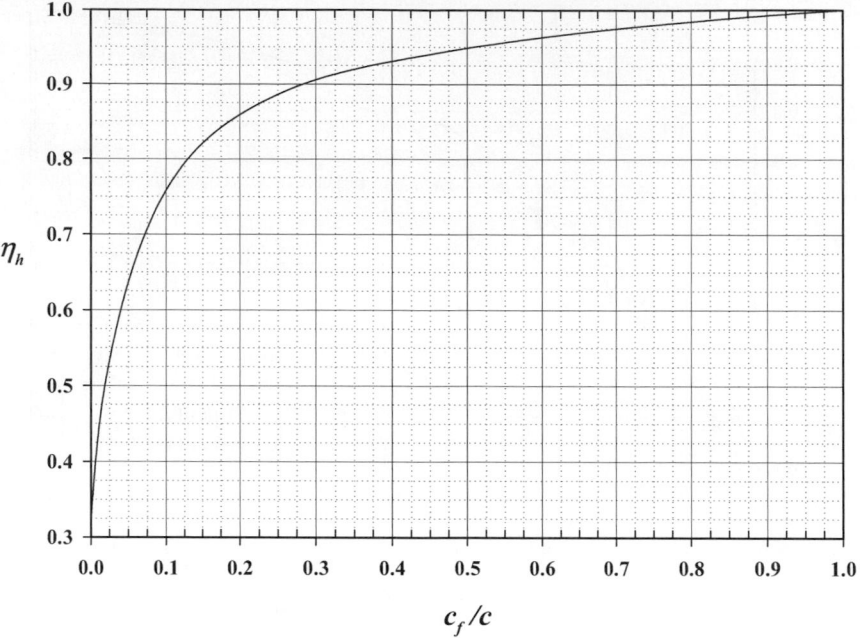

Fig. 16 Section flap hinge efficiency for well-designed and sealed trailing-edge flaps. For unsealed flaps this hinge efficiency should be decreased by about 20%.

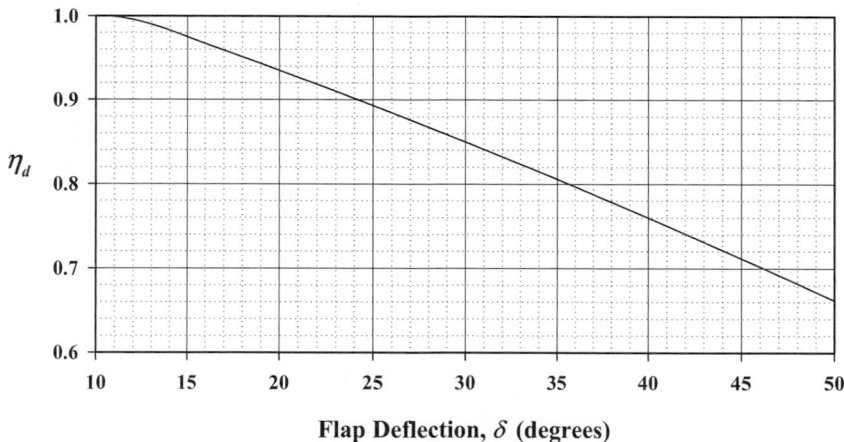

Fig. 17 Section flap deflection efficiency for trailing-edge flaps with flap deflections of more than ±10°. For flap deflections of less than ±10° a deflection efficiency of 1.0 should be used.

method and with limited experimental data. Airfoil section thickness tends to increase the magnitude of the negative quarter-chord moment slope with respect to flap deflection. The hinge effects, on the other hand, tend to lessen the magnitude of this moment slope. The result from thin airfoil theory is often used for preliminary design.

6 INCOMPRESSIBLE FLOW OVER FINITE WINGS

In Section 4 we reviewed the aerodynamic properties of airfoils, which are the same as the aerodynamic properties of wings with infinite span. A wing of constant cross section and infinite span would have no variation in aerodynamic forces in the spanwise

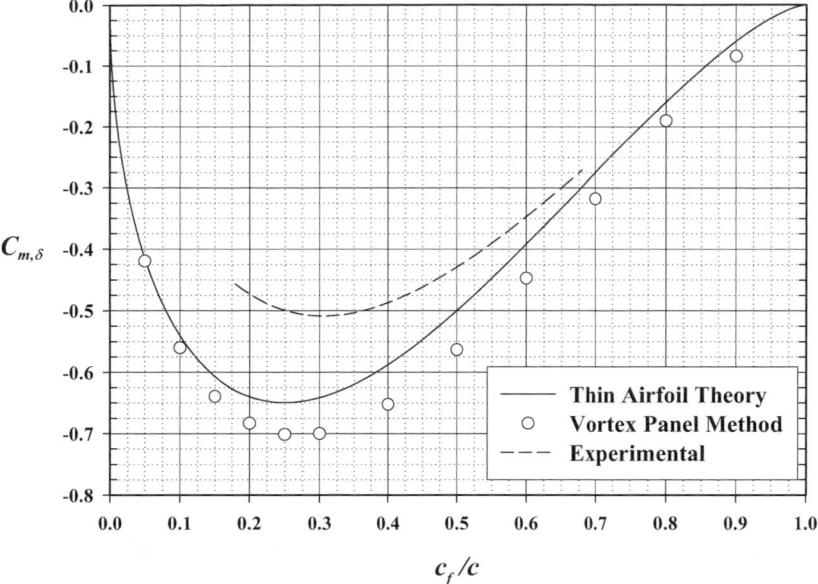

Fig. 18 Quarter-chord moment slope with respect to flap deflection.

direction. The aerodynamic forces per unit span acting on such a wing, at any point along the span, would be the same as those for an airfoil of the same cross section. An airfoil or wing of infinite span is synthesized using vortex sheets that are made up of straight vortex filaments that extend to $\pm\infty$ in the direction of span. The vortex strength can vary over the sheet as we move in a chordwise direction from one vortex filament to another. However, there is no variation in vortex strength as we move along a vortex filament in the direction of span.

Any real wing, of course, must have finite span. At the tips of a finite wing, the air on the lower surface

of the wing comes in direct contact with the air on the upper surface of the wing. Thus, at the tips of a finite wing, the pressure difference between the upper and lower surfaces must always go to zero. As a result, the lift on any finite wing must go to zero at the wingtips, as shown schematically in Fig. 19.

The pressure difference between the upper and lower surfaces of a finite wing is reduced near the wingtips because some of the air from the high-pressure region below the wing spills outward, around the wingtip, and back inward toward the low-pressure region above the wing. Thus, while the flow around an infinite wing is entirely in the plane of the airfoil

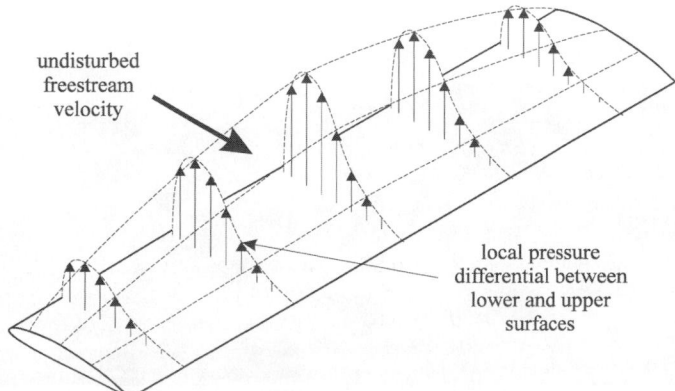

Fig. 19 Lift distribution on a finite wing.

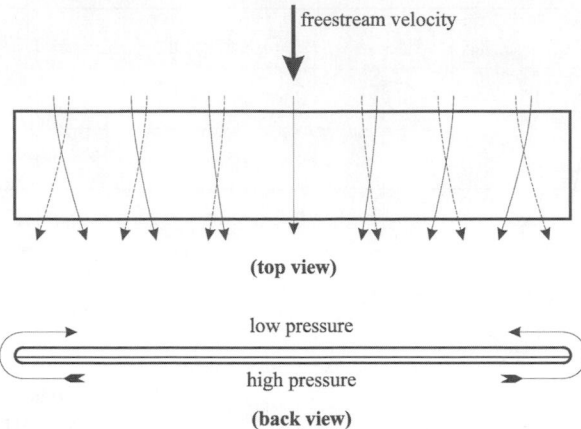

Fig. 20 Airflow around a finite wing.

section, the flow around a finite wing is three dimensional.

As air flows over a finite wing, the air below the wing moves outward toward the wingtip and the air above the wing moves inward toward the root, as shown in Fig. 20. Where the flows from the upper and lower surfaces combine at the trailing edge, the difference in spanwise velocity generates a trailing vortex sheet. Because this vortex sheet is free and not bound to the wing's surface, the flow field induced by the sheet tends to change the shape of the sheet as it moves downstream from the wing. As can be seen in Fig. 21, this vortex sheet rolls up around an axis trailing slightly inboard from each wingtip. At some distance behind the wing, the sheet becomes completely rolled up to form two large vortices, one trailing aft of each wingtip. For this reason, these vortices are referred to as *wingtip vortices*, even though they are generated over the full span of the wing. The downward velocity component that is induced between the wingtip vortices is called *downwash*. Potential

flow theory predicts that the wingtip vortices must trail behind the wing for an infinite distance with no reduction in strength. In reality, viscous effects will eventually dissipate the energy in these vortices, but this is a slow process. These vortices will still have significant energy several miles behind a large aircraft and are of sufficient strength to cause control loss or structural damage to other aircraft following too closely.

From the discussion above, we see that a wing of finite span cannot be synthesized with a vortex sheet made up of vortex filaments that are always perpendicular to the airfoil sections of the wing. We can, however, still synthesize a finite wing with a vortex sheet, but the sheet must be made up of horseshoe-shaped vortex filaments. These filaments run out along the wing in the direction of span, curving back and eventually leaving the wing from the trailing edge at some point inboard of the wingtip. This is seen schematically in the plan view of the elliptic wing shown in Fig. 22. Because the strength of a vortex filament cannot vary

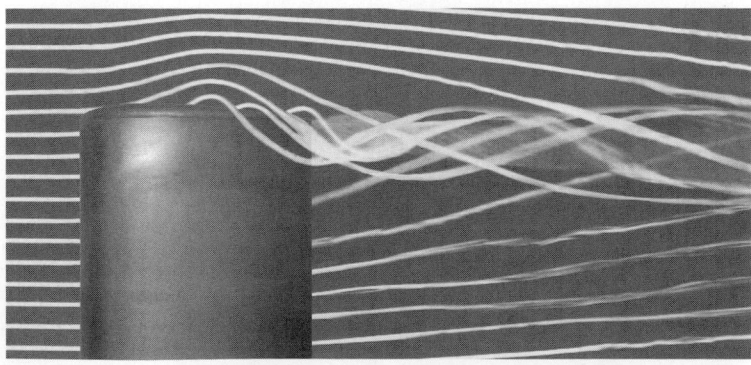

Fig. 21 Vorticity trailing aft of a finite wing.

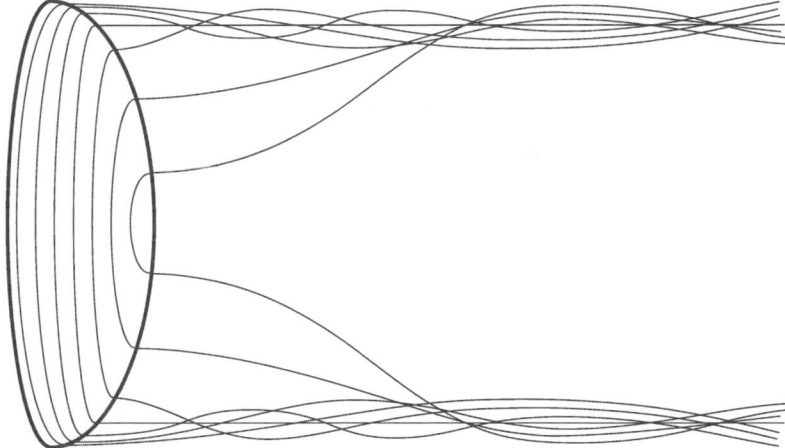

Fig. 22 Schematic of the vorticity distribution on a finite wing with elliptic planform shape.

along its length, the decrease in the circulation about the wing as we move out from the root toward the tip must result entirely from the vorticity that leaves the wing in the trailing vortex sheet.

The classical lifting-line theory developed by Ludwig Prandtl[5] was the first analytical method to satisfactorily predict the performance of a finite wing. While a general 3D vortex lifting law was not available at the time of Prandtl's development, the 2D vortex lifting law of Kutta[6] and Joukowski[7] was well known. Prandtl's lifting-line theory is based on the hypothesis that each spanwise section of a finite wing has a section lift that is equivalent to that acting on a similar section of an infinite wing having the same section circulation. However, to reconcile this theory with reality, the undisturbed free-stream velocity in the Kutta–Joukowski law was intuitively replaced with the vector sum of the free-stream velocity and the velocity induced by the trailing vortex sheet.

Today we know that Prandtl's hypothesis was correct. This can be shown as a direct consequence of the *vortex lifting law*, which is a 3D counterpart to the 2D Kutta–Joukowski law. This vortex lifting law states that for any potential flow containing vortex filaments, the force per unit length exerted on the surroundings at any point along a vortex filament is given by

$$\mathbf{dF} = \rho \Gamma \mathbf{V} \times \mathbf{dl} \qquad (82)$$

where $\mathbf{dF}$ is the differential aerodynamic force vector and ρ, Γ, $\mathbf{V}$, and $\mathbf{dl}$ are, respectively, the fluid density, vortex strength, local fluid velocity, and the directed differential vortex length vector (see Saffman[8]).

The force computed from Eq. (82) is called the *vortex force* or *Kutta lift*. The vortex lifting law is a very useful tool in the study of aerodynamics. It provides the basis for much of finite wing theory. There are two important consequences of Eq. (82).

First, for a *bound vortex filament*, the vortex force is always perpendicular to both the local velocity vector and the vortex filament. Second, since a free vortex filament can support no force, *free vortex filaments* must be aligned everywhere with the streamlines of the flow.

6.1 Prandtl's Classical Lifting-Line Theory

Prandtl's lifting-line theory gives good agreement with experimental data for straight wings of aspect ratio greater than about 4. Development of this theory is presented in any undergraduate engineering textbook on aerodynamics and will not be repeated here. Only summary results are presented in this section.

The model used by Prandtl to approximate the bound vorticity and trailing vortex sheet is shown in Fig. 23. All bound vortex filaments are assumed to follow the wing quarter-chord line, and all trailing vortex filaments are assumed to be straight and parallel with the free stream. Rollup of the trailing vortex sheet is ignored.

The foundation of lifting-line theory is the requirement that for each cross section of the wing, the lift predicted from the vortex lifting law must be equal to that predicted from airfoil section theory, that is;

$$\widetilde{C}_L(z) = \widetilde{C}_{L,\alpha}[\alpha_{\mathrm{eff}}(z) - \alpha_{L0}(z)]$$

where α_{eff} is the local section angle of attack, including the effects of velocity induced by the trailing vortex sheet. Because of the downwash induced on the wing by the trailing vortex sheet, the local relative wind is inclined at an angle, α_i, to the free stream, as shown in Fig. 24. This angle is called the *induced angle of attack*. Since the lift is always perpendicular to the local relative wind, the downwash tilts the lift vector back, creating a component of lift parallel to the free stream. This is called *induced drag*.

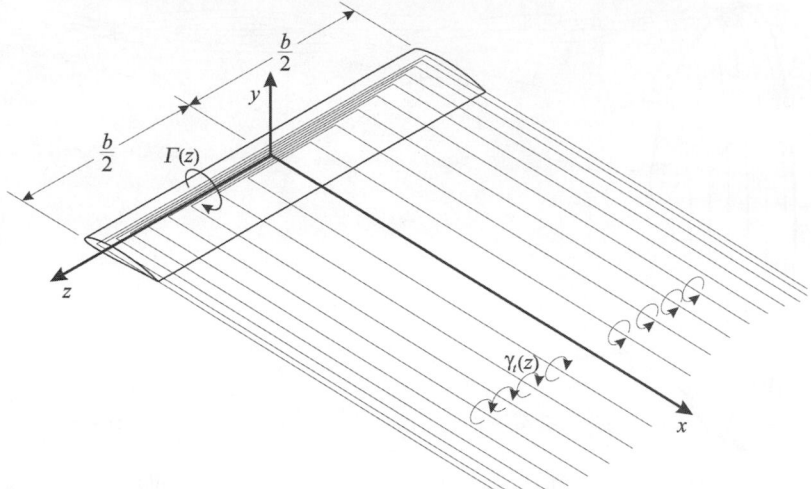

Fig. 23 Prandtl's model for the bound vorticity and the trailing vortex sheet generated by a wing of finite span.

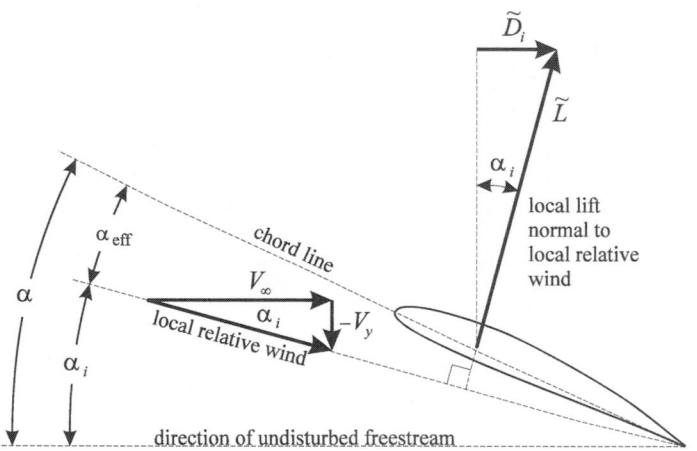

Fig. 24 Induced angle of attack.

When the downwash is accounted for, Prandtl's hypothesis requires that

$$\frac{2\Gamma(z)}{V_\infty c(z)} + \frac{\widetilde{C}_{L,\alpha}}{4\pi V_\infty} \int_{\zeta=-b/2}^{b/2} \frac{1}{z-\zeta} \left(\frac{d\Gamma}{dz}\right)_{z=\zeta} d\zeta$$
$$= \widetilde{C}_{L,\alpha}[\alpha(z) - \alpha_{L0}(z)] \qquad (83)$$

Equation (83) is the fundamental equation of Prandtl's lifting-line theory. It is a nonlinear integrodifferential equation that involves only one unknown, the local section circulation as a function of the spanwise position, $\Gamma(z)$. All other parameters are known for a given wing design at a given geometric angle of attack and

a given free-stream velocity. Remember that the chord length, c, the geometric angle of attack, α, and the zero-lift angle of attack, α_{L0}, are all allowed to vary in the spanwise direction. The section lift slope could also vary with z but is usually assumed to be constant.

There is certain terminology associated with spanwise variation in wing geometry with which the reader should be familiar. If a wing has spanwise variation in geometric angle of attack as shown in Fig. 25, the wing is said to have *geometric twist*. The tip is commonly at a lower angle of attack than the root, in which case the geometric twist is referred to as *washout*. A spanwise variation in zero-lift angle of attack, like that shown in Fig. 26, is called *aerodynamic twist*. Deflecting a trailing-edge flap that extends over only part of

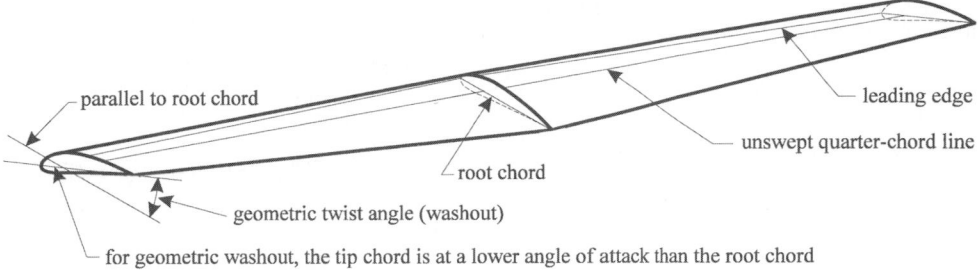

Fig. 25   Geometric twist in an unswept rectangular wing.

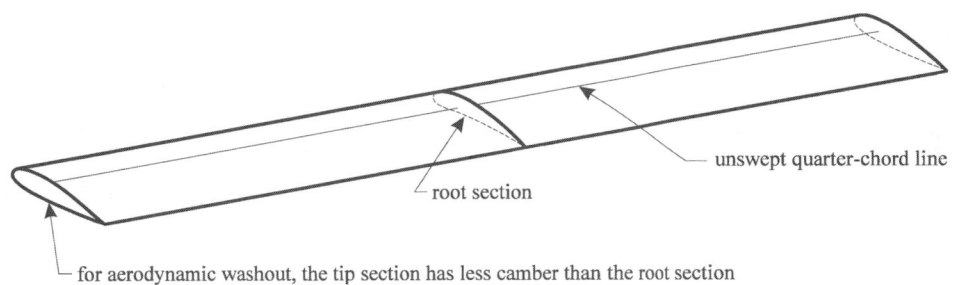

Fig. 26   Aerodynamic twist in an unswept rectangular wing.

the span is another form of aerodynamic twist (see Fig. 27).

For a single finite wing with no sweep or dihedral, an analytical solution to Prandtl's lifting-line equation can be obtained in terms of a Fourier sine series. For a given wing design, at a given angle of attack, the planform shape, $c(z)$, the geometric twist, $\alpha(z)$, and the aerodynamic twist, $\alpha_{L0}(z)$, are all known as functions of spanwise position. The circulation distribution is written as a Fourier series, the series is truncated to a finite number of terms (N), and the Fourier coefficients are determined by forcing the lifting-line equation to be satisfied at N specific sections along the span of the wing. From this solution the circulation distribution is given by

$$\Gamma(\theta) = 2bV_\infty \sum_{n=1}^{N} A_n \sin(n\theta) \qquad \theta = \cos^{-1}\left(\frac{-2z}{b}\right)$$
(84)

where the Fourier coefficients, A_n, are obtained from

$$\sum_{n=1}^{N} A_n \left[\frac{4b}{\widetilde{C}_{L,\alpha} c(\theta)} + \frac{n}{\sin(\theta)} \right] \sin(n\theta) = \alpha(\theta) - \alpha_{L0}(\theta)$$
(85)

Once the circulation distribution has been determined, the section lift distribution can be obtained

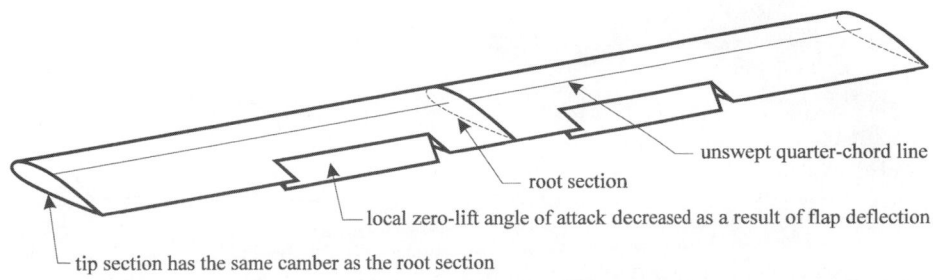

Fig. 27   Aerodynamic twist resulting from deflection of a trailing-edge flap spanning only a portion of the wing.

from the vortex lifting law. The resulting lift and induced drag coefficients for the finite wing are

$$C_L = \pi R_A A_1 \qquad R_A \equiv \frac{b^2}{S} \qquad (86)$$

$$C_{D_i} = \pi R_A \sum_{n=1}^{N} n A_n^2 = \frac{C_L^2}{\pi R_A e_s} \qquad e_s \equiv \frac{1}{1 + \sigma}$$

$$\sigma \equiv \sum_{n=2}^{N} n \left(\frac{A_n}{A_1} \right)^2 \qquad (87)$$

The wingspan squared divided by the planform area, R_A, is called the *aspect ratio* and the parameter e_s is called the *span efficiency factor*.

Lifting-line theory predicts that an elliptic wing with no geometric or aerodynamic twist produces minimum possible induced drag for a given lift coefficient and aspect ratio. This planform has a chord that varies with the spanwise coordinate according to

$$c(z) = \frac{4b}{\pi R_A} \sqrt{1 - (2z/b)^2} \qquad \text{or}$$

$$c(\theta) = \frac{4b}{\pi R_A} \sin(\theta)$$

As shown in Fig. 28, an unswept elliptic wing has a straight quarter-chord line. This gives the leading edge less curvature and the trailing edge more curvature than

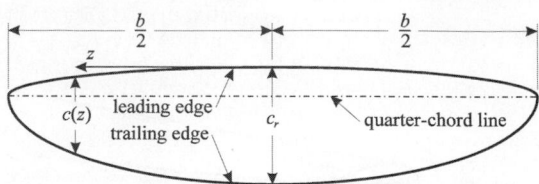

Fig. 28 Planform shape of an unswept elliptic wing with an aspect ratio of 6.

a conventional ellipse, which has a straight half-chord line. Several aircraft have been designed and built with elliptic wings. One of the best known is the British Spitfire, shown in Fig. 29.

The lift and induced drag coefficients predicted from Eqs. (86) and (87) for an elliptic wing with no geometric or aerodynamic twist are

$$C_L = C_{L,\alpha}(\alpha - \alpha_{L0}) \qquad C_{L,\alpha} = \frac{\widetilde{C}_{L,\alpha}}{1 + \widetilde{C}_{L,\alpha}/(\pi R_A)} \qquad (88)$$

$$C_{D_i} = \frac{C_L^2}{\pi R_A} \qquad (89)$$

where $C_{L,\alpha}$ is the lift slope for the finite wing. Equation (88) shows that the lift slope for an untwisted elliptic wing is less than the section lift slope for the airfoil from which the wing was generated. However, as the aspect ratio for the wing becomes large, the lift slope

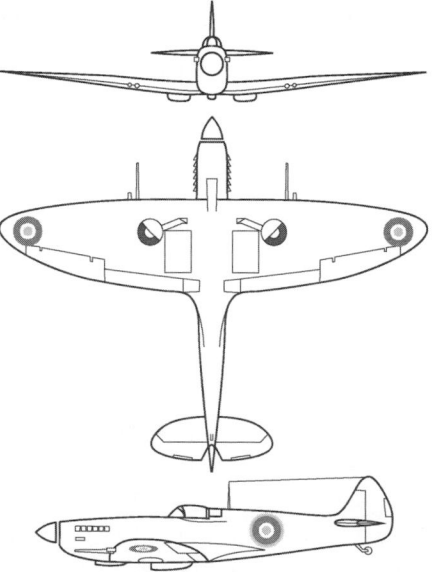

Fig. 29 Elliptic wing used on the famous British Spitfire. (Photograph by Barry Santana.)

for the finite wing approaches that of the airfoil section and the induced drag becomes small. At a given angle of attack, the untwisted elliptic wing produces more lift than any other untwisted wing of the same aspect ratio. Planform shape affects both the induced drag and the lift slope of a finite wing. However, the effect of planform shape is small compared to that of aspect ratio.

While untwisted elliptic wings produce minimum possible induced drag, they are more expensive to manufacture than simple rectangular wings. Untwisted rectangular wings are easy to manufacture, but they generate induced drag at a level that is less than optimum. The untwisted tapered wing, shown in Fig. 30, has commonly been used as a compromise. Tapered wings have a chord length that varies linearly from the root to the tip. They are nearly as easy to manufacture as rectangular wings, and they can be designed to produce induced drag close to the optimum value of an elliptic wing.

The chord length for a tapered wing varies with the spanwise coordinate according to the relation

$$c(z) = \frac{2b}{R_A(1 + R_T)}\left[1 - (1 - R_T)\left|\frac{2z}{b}\right|\right] \quad \text{or}$$

$$c(\theta) = \frac{2b}{R_A(1 + R_T)}[1 - (1 - R_T)|\cos\theta|]$$

where R_T is the taper ratio, related to the root chord, c_r, and the tip chord, c_t, by

$$R_T \equiv \frac{c_t}{c_r}$$

For a wing of any planform having no sweep, dihedral, geometric twist, or aerodynamic twist, the circulation distribution predicted from Prandtl's lifting-line theory can be written in terms of a Fourier series with coefficients, a_n, that are independent of angle of attack. Under these conditions, Eq. (84) can be rearranged to give

$$\Gamma(\theta) = 2bV_\infty(\alpha - \alpha_{L0})\sum_{n=1}^{N} a_n \sin(n\theta)$$

$$\theta = \cos^{-1}\left(\frac{-2z}{b}\right) \quad (90)$$

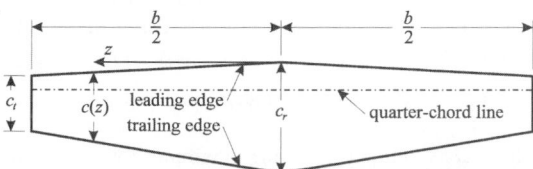

Fig. 30 Planform shape of an unswept tapered wing with a taper ratio of 0.5 and an aspect ratio of 6.

where the Fourier coefficients are obtained from

$$\sum_{n=1}^{N} a_n\left[\frac{4b}{\widetilde{C}_{L,\alpha}c(\theta)} + \frac{n}{\sin(\theta)}\right]\sin(n\theta) = 1 \quad (91)$$

The lift and induced drag coefficients predicted from Eqs. (86) and (87) for a wing with no geometric or aerodynamic twist can be written as

$$C_L = C_{L,\alpha}(\alpha - \alpha_{L0})$$

$$C_{L,\alpha} = \frac{\widetilde{C}_{L,\alpha}}{[1 + \widetilde{C}_{L,\alpha}/(\pi R_A)](1 + \kappa_L)} \quad (92)$$

$$C_{D_i} = \frac{C_L^2}{\pi R_A e_s} \qquad e_s = \frac{1}{1 + \kappa_D} \quad (93)$$

where the lift slope factor, κ_L, and the induced drag factor, κ_D, are given by

$$\kappa_L = \frac{1 - (1 + \pi R_A/\widetilde{C}_{L,\alpha})a_1}{(1 + \pi R_A/\widetilde{C}_{L,\alpha})a_1} \quad (94)$$

$$\kappa_D = \sum_{n=2}^{N} n\left(\frac{a_n}{a_1}\right)^2 \quad (95)$$

For untwisted tapered wings, numerical results obtained for these two factors are shown in Figs. 31 and 32. These results were generated using a section lift slope of 2π.

Glauert[9] first presented results similar to those shown in Fig. 32. Such results have sometimes led to the conclusion that a tapered wing with a taper ratio of about 0.4 always produces significantly less induced drag than a rectangular wing of the same aspect ratio developing the same lift. As a result, tapered wings are often used as a means of reducing induced drag. However, this reduction in drag usually comes at a price. Because a tapered wing has a lower Reynolds number at the wingtips than at the root, a tapered wing with no geometric or aerodynamic twist tends to stall first in the region of the wingtips. This wingtip stall commonly leads to poor handling qualities during stall recovery.

The results shown in Fig. 32 can be misleading if one loses sight of the fact that these results apply only for the special case of wings with no geometric or aerodynamic twist. This is only one of many possible washout distributions that could be used for a wing of any given planform. Furthermore, it is not the washout distribution that produces minimum induced drag with finite lift, except for the case of an elliptic wing. When the effects of washout are included, it can be shown that the conclusions sometimes reached from consideration of only those results shown in Fig. 32 are erroneous.

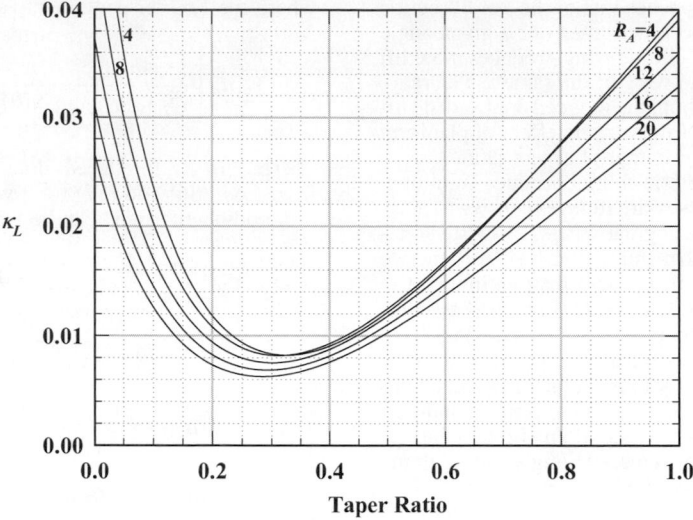

Fig. 31   Lift slope factor for untwisted tapered wings from Prandtl's lifting-line theory.

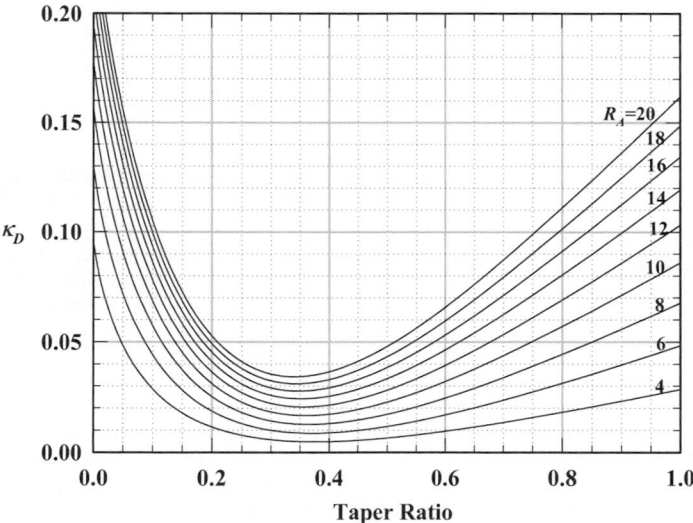

Fig. 32   Induced drag factor for untwisted tapered wings from Prandtl's lifting-line theory.

6.2   Effects of Wing Twist

For a wing with geometric and/or aerodynamic twist, the solution expressed in the form of Eqs. (84)–(87) is cumbersome for the evaluation of traditional wing properties because the Fourier coefficients depend on angle of attack and must be reevaluated for each operating point studied. Furthermore, the definition of κ_D that is given in Eq. (87) is not practical for use at an arbitrary angle of attack because the value becomes singular for a twisted wing when the lift coefficient

approaches zero. A more useful form of the solution was obtained by Phillips et al.[10–13] using the change of variables

$$\alpha(\theta) - \alpha_{L0}(\theta) \equiv (\alpha - \alpha_{L0})_{\text{root}} - \Omega\omega(\theta) \quad (96)$$

where Ω is defined to be the maximum total washout, geometric plus aerodynamic,

$$\Omega \equiv (\alpha - \alpha_{L0})_{\text{root}} - (\alpha - \alpha_{L0})_{\text{max}} \quad (97)$$

and $\omega(\theta)$ is the local washout distribution function, which is normalized with respect to maximum total washout

$$\omega(\theta) \equiv \frac{\alpha(\theta) - \alpha_{L0}(\theta) - (\alpha - \alpha_{L0})_{\text{root}}}{(\alpha - \alpha_{L0})_{\text{max}} - (\alpha - \alpha_{L0})_{\text{root}}} \quad (98)$$

The normalized washout distribution function, $\omega(\theta)$, is independent of angle of attack and always varies from 0.0 at the root to 1.0 at the point of maximum washout, which is commonly at the wingtips.

Using Eq. (96) in Eq. (85) gives

$$\sum_{n=1}^{N} A_n \left[\frac{4b}{\widetilde{C}_{L,\alpha} c(\theta)} + \frac{n}{\sin(\theta)} \right] \sin(n\theta)$$
$$= (\alpha - \alpha_{L0})_{\text{root}} - \Omega \omega(\theta) \quad (99)$$

The Fourier coefficients in Eq. (99) for a wing with geometric and/or aerodynamic twist can be conveniently written as

$$A_n \equiv a_n (\alpha - \alpha_{L0})_{\text{root}} - b_n \Omega \quad (100)$$

where

$$\sum_{n=1}^{N} a_n \left[\frac{4b}{\widetilde{C}_{L,\alpha} c(\theta)} + \frac{n}{\sin(\theta)} \right] \sin(n\theta) = 1 \quad (101)$$

$$\sum_{n=1}^{N} b_n \left[\frac{4b}{\widetilde{C}_{L,\alpha} c(\theta)} + \frac{n}{\sin(\theta)} \right] \sin(n\theta) = \omega(\theta) \quad (102)$$

Comparing Eq. (101) with Eq. (91), we see that the Fourier coefficients defined by Eq. (101) are those corresponding to the solution for a wing of the same planform shape but with no geometric or aerodynamic twist. The solution to Eq. (102) can be obtained in a similar manner and is also independent of angle of attack.

Using Eq. (100) in Eq. (86), the lift coefficient for a wing with washout can be expressed as

$$C_L = \pi R_A A_1 = \pi R_A [a_1 (\alpha - \alpha_{L0})_{\text{root}} - b_1 \Omega] \quad (103)$$

Using Eq. (100) in Eq. (87), the induced drag coefficient is given by

$$C_{D_i} = \pi R_A \sum_{n=1}^{N} n A_n^2 = \pi R_A [a_1 (\alpha - \alpha_{L0})_{\text{root}} - b_1 \Omega]^2$$

$$+ \pi R_A \sum_{n=2}^{N} n [a_n (\alpha - \alpha_{L0})_{\text{root}} - b_n \Omega]^2$$

or after using Eq. (103) to express the first term on the right-hand side in terms of the lift coefficient.

$$C_{D_i} = \frac{C_L^2}{\pi R_A} + \pi R_A \sum_{n=2}^{\infty} n [a_n^2 (\alpha - \alpha_{L0})_{\text{root}}^2$$
$$- 2a_n b_n (\alpha - \alpha_{L0})_{\text{root}} \Omega + b_n^2 \Omega^2] \quad (104)$$

Equations (103) and (104) can be algebraically rearranged to yield a convenient expression for the lift and induced drag developed by a finite wing with geometric and/or aerodynamic twist:

$$C_L = C_{L,\alpha}[(\alpha - \alpha_{L0})_{\text{root}} - \varepsilon_\Omega \Omega] \quad (105)$$

$$C_{D_i} = \frac{C_L^2 (1 + \kappa_D) - \kappa_{DL} C_L C_{L,\alpha} \Omega + \kappa_{D\Omega} (C_{L,\alpha} \Omega)^2}{\pi R_A} \quad (106)$$

$$C_{L,\alpha} = \pi R_A a_1 = \frac{\widetilde{C}_{L,\alpha}}{[1 + \widetilde{C}_{L,\alpha}/(\pi R_A)](1 + \kappa_L)} \quad (107)$$

$$\kappa_L \equiv \frac{1 - (1 + \pi R_A / \widetilde{C}_{L,\alpha}) a_1}{(1 + \pi R_A / \widetilde{C}_{L,\alpha}) a_1} \quad (108)$$

$$\varepsilon_\Omega \equiv \frac{b_1}{a_1} \quad (109)$$

$$\kappa_D \equiv \sum_{n=2}^{N} n \frac{a_n^2}{a_1^2} \quad (110)$$

$$\kappa_{DL} \equiv 2 \frac{b_1}{a_1} \sum_{n=2}^{N} n \frac{a_n}{a_1} \left(\frac{b_n}{b_1} - \frac{a_n}{a_1} \right) \quad (111)$$

$$\kappa_{D\Omega} \equiv \left(\frac{b_1}{a_1} \right)^2 \sum_{n=2}^{N} n \left(\frac{b_n}{b_1} - \frac{a_n}{a_1} \right)^2 \quad (112)$$

Comparing Eqs. (105)–(112) with Eqs. (92) and (95), we see that washout increases the zero-lift angle of attack for any wing but the lift slope for a wing of arbitrary planform shape is not affected by washout. Notice that the induced drag for a wing with washout is not zero at zero lift. In addition to the usual component of induced drag, which is proportional to the lift coefficient squared, a wing with washout produces a component of induced drag that is proportional to the washout squared, and this results in induced drag at zero lift. There is also a component of induced drag that varies with the product of the lift coefficient and the washout.

A commonly used washout distribution that is easy to implement is linear washout. For the special case of a linear variation in washout from the root to the tip, the normalized washout distribution function is simply

$$\omega(z) = |2z/b| \quad \text{or} \quad \omega(\theta) = |\cos(\theta)| \quad (113)$$

For tapered wings, the variation in chord length is also linear:

$$c(z) = \frac{2b}{R_A(1 + R_T)}[1 - (1 - R_T)|2z/b|] \quad \text{or}$$

$$c(\theta) = \frac{2b}{R_A(1 + R_T)}[1 - (1 - R_T)|\cos(\theta)|] \quad (114)$$

Using Eqs. (113) and (114) in Eqs. (101) and (102) yields the results for a tapered wing with linear washout,

$$\sum_{n=1}^{N} a_n \left[\frac{2R_A(1 + R_T)}{\widetilde{C}_{L,\alpha}[1 - (1 - R_T)|\cos(\theta)|]} + \frac{n}{\sin(\theta)} \right]$$
$$\times \sin(n\theta) = 1 \quad (115)$$

$$\sum_{n=1}^{N} b_n \left[\frac{2R_A(1 + R_T)}{\widetilde{C}_{L,\alpha}[1 - (1 - R_T)|\cos(\theta)|]} + \frac{n}{\sin(\theta)} \right]$$
$$\times \sin(n\theta) = |\cos(\theta)| \quad (116)$$

The solution obtained from Eq. (115) for the Fourier coefficients, a_n, is the familiar result that was used to produce Figs. 31 and 32. Induced drag for a wing with washout is readily predicted from Eq. (106) with the definitions given in Eqs. (110)–(112). For tapered wings with linear washout, the Fourier coefficients, b_n, can be obtained from Eq. (116) in exactly the same manner as the coefficients, a_n, are obtained from Eq. (115). Using the Fourier coefficients so obtained in Eqs. (109), (111), and (112) produces the results

shown in Figs. 33–35. Notice from examining either Eq. (112) or Fig. 35 that $\kappa_{D\Omega}$ is always positive. Thus, the third term in the numerator on the right-hand side of Eq. (106) always contributes to an increase in induced drag. However, from the results shown in Fig. 34, we see that the second term in the numerator on the right-hand side of Eq. (106) can either increase or decrease the induced drag, depending on the signs of κ_{DL} and Ω. This raises an important question regarding wing efficiency. What level and distribution of washout will result in minimum induced drag for a given wing planform and lift coefficient?

6.3 Minimizing Induced Drag with Washout

For a wing of any given planform shape having a fixed washout distribution, induced drag can be minimized with washout as a result of the trade-off between the second and third terms in the numerator on the right-hand side of Eq. (106). Thus, Eq. (106) can be used to determine the optimum value of total washout, which will result in minimum induced drag for any washout distribution function and any specified lift coefficient. Differentiating Eq. (106) with respect to total washout at constant lift coefficient gives

$$\frac{\partial C_{D_i}}{\partial \Omega} = \frac{-\kappa_{DL} C_L C_{L,\alpha} + 2\kappa_{D\Omega} C_{L,\alpha}^2 \Omega}{\pi R_A} \quad (117)$$

Setting the right-hand side of Eq. (117) to zero and solving for the total washout, it can be seen that minimum induced drag is attained for any given wing planform, $c(z)$, any given washout distribution, $\omega(z)$, and any given design lift coefficient, C_{Ld}, by

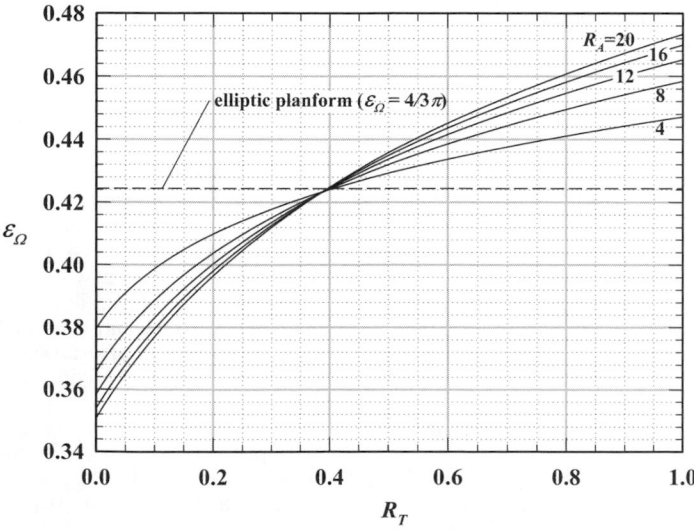

Fig. 33 Washout effectiveness for tapered wings with linear washout.

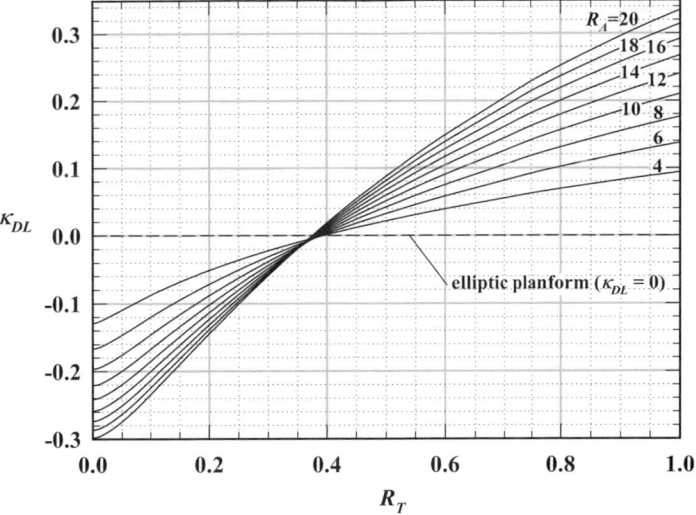

Fig. 34 Lift–washout contribution to induced drag for tapered wings with linear washout.

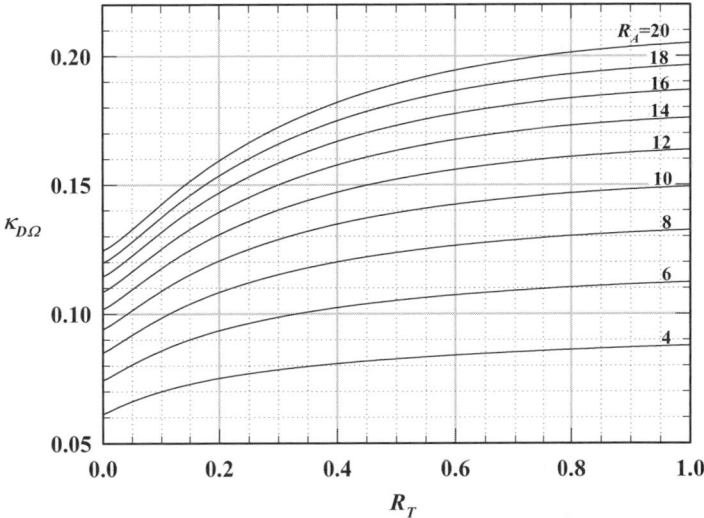

Fig. 35 Washout contribution to induced drag for tapered wings with linear washout.

using an optimum total washout, Ω_{opt}, given by the relation

$$\Omega_{opt} = \frac{\kappa_{DL}C_{Ld}}{2\kappa_{D\Omega}C_{L,\alpha}} \qquad (118)$$

For the elliptic planform, all of the Fourier coefficients, a_n, are zero for n greater than 1. Thus, Eq. (111) shows that κ_{DL} is zero for an elliptic wing. As a result, elliptic wings are always optimized with no washout. From

consideration of Eq. (118) together with the results shown in Figs. 34 and 35, we see that tapered wings with linear washout and taper ratios greater than 0.4 are optimized with positive washout, whereas those with taper ratios less than about 0.4 are optimized with negative washout.

Using the value of optimum washout from Eq. (118) in the expression for induced drag coefficient given by Eq. (106), we find that the induced drag coefficient for a wing of arbitrary planform with a fixed washout distribution and optimum total washout is given by

$$(C_{D_i})_\text{opt} = \frac{C_L^2}{\pi R_A}\left[1 + \kappa_D - \frac{\kappa_{DL}^2}{4\kappa_{D\Omega}}\left(2 - \frac{C_{Ld}}{C_L}\right)\frac{C_{Ld}}{C_L}\right]$$
(119)

From Eq. (119) it can be seen that a wing with optimum washout will always produce less induced drag than a wing with no washout having the same planform and aspect ratio, provided that the actual lift coefficient is greater than one-half the design lift coefficient. When the actual lift coefficient is equal to the design lift coefficient, the induced drag coefficient for a wing with optimum washout is

$$(C_{D_i})_\text{opt} = \frac{C_L^2}{\pi R_A}(1 + \kappa_{D_o}) \qquad \kappa_{D_o} \equiv \kappa_D - \frac{\kappa_{DL}^2}{4\kappa_{D\Omega}}$$
(120)

For tapered wings with linear washout, the variations in κ_{D_o} with aspect ratio and taper ratio are shown in Fig. 36. For comparison, the dashed lines in this figure show the same results for wings with no washout. Notice that when linear washout is used to further optimize tapered wings, taper ratios near 0.4 correspond closely to a maximum in induced drag, not to a minimum.

The choice of a linear washout distribution, which was used to generate the results shown in Fig. 36, is as arbitrary as the choice of no washout. While a linear variation in washout is commonly used and simple to implement, it is not the optimum washout distribution for wings with linear taper. Minimum possible induced drag for a finite lift coefficient always occurs when the local section lift varies with the spanwise coordinate in proportion to $\sin(\theta)$. This results in uniform downwash and requires that the product of the local chord length

and local aerodynamic angle of attack, $\alpha - \alpha_{L0}$, varies elliptically with the spanwise coordinate, that is,

$$\frac{c(z)[\alpha(z) - \alpha_{L0}(z)]}{\sqrt{1 - (2z/b)^2}} = \frac{c(\theta)[\alpha(\theta) - \alpha_{L0}(\theta)]}{\sin(\theta)} = \text{const}$$
(121)

There are many possibilities for wing geometry that will satisfy this condition. The elliptic planform with no geometric or aerodynamic twist is only one such geometry. Since the local aerodynamic angle of attack decreases along the span in direct proportion to the increase in washout, Eq. (121) can only be satisfied if the washout distribution satisfies the relation

$$\frac{c(z)[1 - \omega(z)]}{\sqrt{1 - (2z/b)^2}} = \frac{c(\theta)[1 - \omega(\theta)]}{\sin(\theta)} = \text{const}$$
(122)

Equation (122) is satisfied by the optimum washout distribution

$$\omega_\text{opt} = 1 - \frac{\sqrt{1 - (2z/b)^2}}{c(z)/c_\text{root}} = 1 - \frac{\sin(\theta)}{c(\theta)/c_\text{root}}$$

$$\Omega_\text{opt} = \frac{4bC_L}{\pi R_A \widetilde{C}_{L,\alpha} c_\text{root}}$$
(123)

For wings with linear taper, this gives

$$\omega_\text{opt} = 1 - \frac{\sqrt{1 - (2z/b)^2}}{1 - (1 - R_T)|2z/b|}$$

$$= 1 - \frac{\sin(\theta)}{1 - (1 - R_T)|\cos(\theta)|}$$

$$\Omega_\text{opt} = \frac{2(1 + R_T)C_L}{\pi \widetilde{C}_{L,\alpha}}$$
(124)

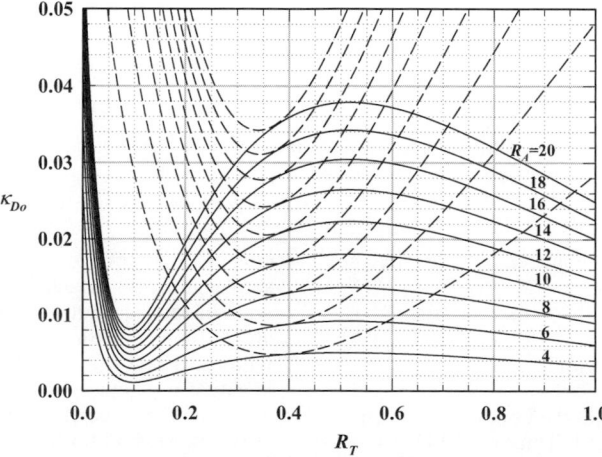

Fig. 36 Optimum induced drag factor for tapered wings with linear washout.

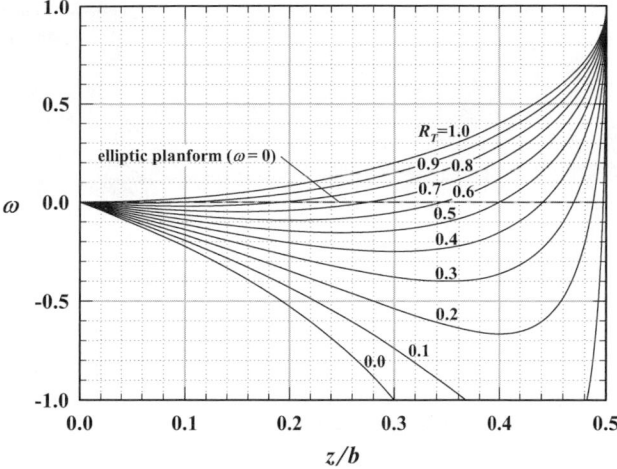

Fig. 37 Optimum washout distribution that results in production of minimum induced drag for wings with linear taper, as defined in Eq. (124).

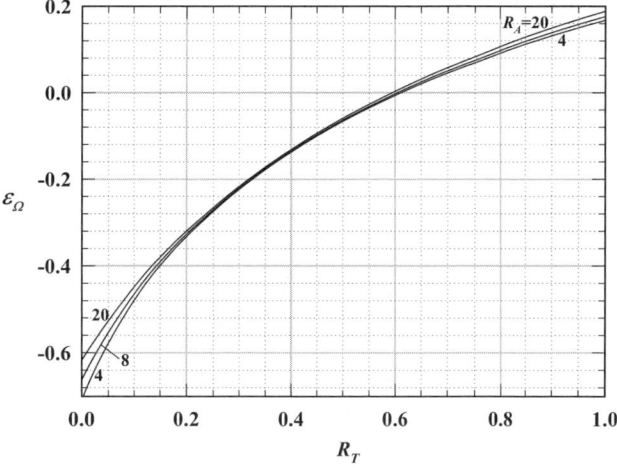

Fig. 38 Washout effectiveness for wings with linear taper and optimum washout.

This optimum washout distribution is shown in Fig. 37 for several values of taper ratio. Results obtained for tapered wings having this washout distribution are presented in Figs. 38–40. Notice that for tapered wings with the optimum washout distribution, κ_{DL} is positive for all taper ratios. Thus, we see from Eq. (118) that tapered wings with optimum washout always have positive washout at the wingtips.

When an unswept wing of arbitrary planform has the washout distribution specified by Eq. (123), the value of κ_{D_o} as defined in Eq. (120) is always identically zero. With this washout distribution and the total washout set according to Eq. (118), an unswept wing

of any planform shape can be designed to operate at a given lift coefficient with the same induced drag as that produced by an untwisted elliptic wing with the same aspect ratio and lift coefficient.

6.4 Solution with Control Surface Deflection and Rolling Rate

Trailing-edge flaps extending over only some portion of the wingspan are commonly used as control surfaces on an airplane wing. A spanwise symmetric control surface deflection can be used to provide pitch control, and spanwise asymmetric control surface deflection can be used to provide roll control. The

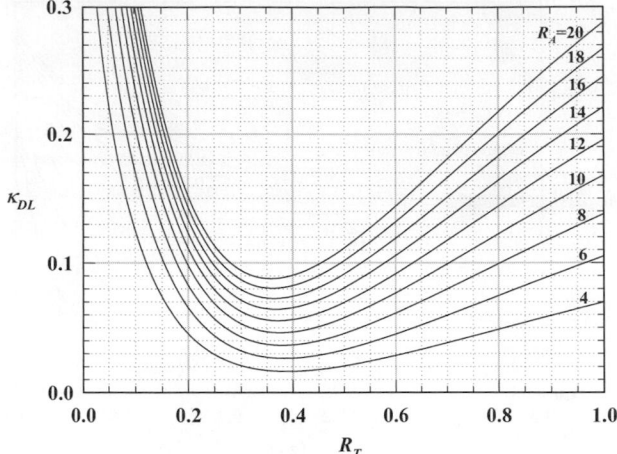

Fig. 39　Lift–washout contribution to the induced drag factor for wings with linear taper and the optimum washout distribution specified by Eq. (124).

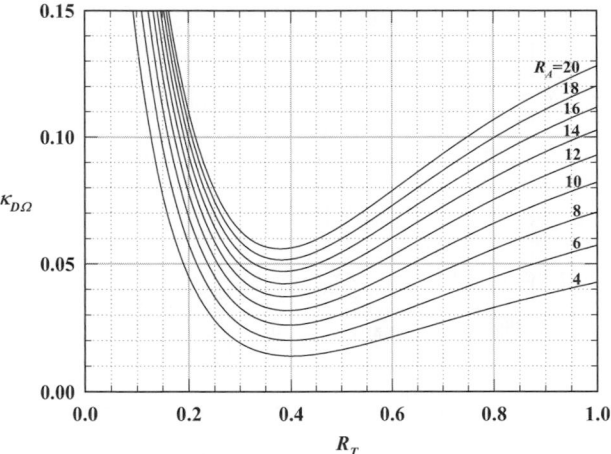

Fig. 40　Washout contribution to the induced drag factor for wings with linear taper and the optimum washout distribution specified by Eq. (124).

control surfaces commonly used to provide roll control are called *ailerons*. These are small trailing-edge flaps located in the outboard sections of the wing. The ailerons are deflected asymmetrically to change the rolling moment. One aileron is deflected downward and the other is deflected upward, as shown in Fig. 41. This increases lift on the semispan with the downward-deflected aileron and decreases lift on the semispan with the upward-deflected aileron. Aileron deflection is given the symbol δ_a, and the rolling moment is denoted as ℓ. The sign convention that is commonly used for aileron deflection is shown in Fig. 41. Aileron deflection is assumed positive when the right aileron is deflected downward and the left aileron is deflected

upward. The traditional sign convention used for the rolling moment is positive to the right (i.e., a moment that would roll the right wing down). With these two sign conventions, a positive aileron deflection produces a rolling moment to the left (i.e., negative ℓ). The two ailerons are not necessarily deflected the same magnitude. The aileron angle, δ_a, is usually defined to be the average of the two angular deflections.

The rolling moment produced by aileron deflection results in a rolling acceleration, which in turn produces a rolling rate. This rolling rate changes the spanwise variation in geometric angle of attack as shown in Fig. 42. The symbol traditionally used to represent rolling rate is p. As shown in Fig. 42, p is taken to be

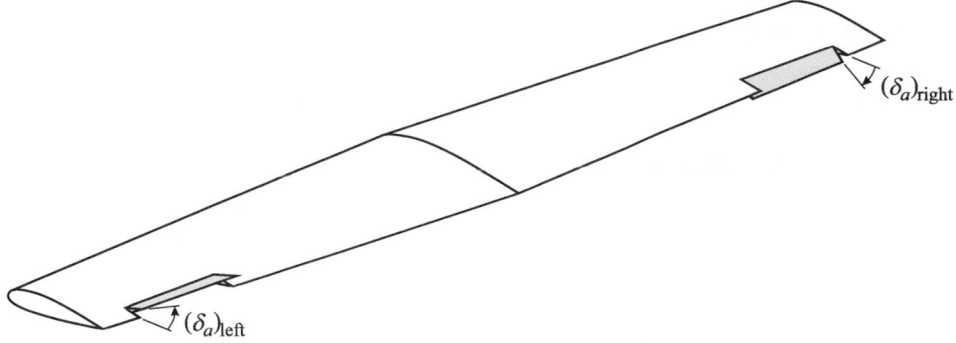

Fig. 41 Asymmetric aileron deflection typically used for roll control.

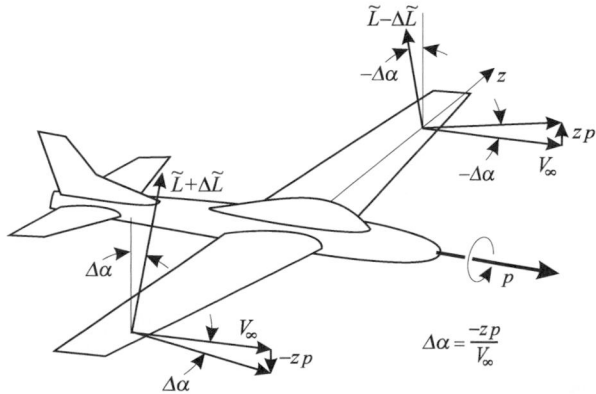

Fig. 42 Change in local section angle of attack resulting from a rolling rate.

positive when the right semispan is moving downward and the left semispan is moving upward. Thus, a positive rolling rate increases the local geometric angle of attack on the right semispan and decreases the local geometric angle of attack on the left semispan. This produces a negative rolling moment, which opposes the rolling rate.

When a small positive aileron deflection is first initiated, a negative rolling moment is produced by the asymmetric change in the wing's local aerodynamic angle of attack. This rolling moment imbalance results in negative rolling acceleration, which produces a negative rolling rate. As the magnitude of the rolling rate increases, an opposing rolling moment develops as a result of the asymmetric change in the local geometric angle of attack. This reduces the moment imbalance and slows the rolling acceleration. At some point, the positive rolling moment produced by the negative rolling rate will just balance the negative rolling moment produced by the positive aileron deflection, and a constant rolling rate will develop that is proportional to the aileron deflection.

Roll in a conventional airplane is what is referred to as *heavily damped motion*. This means that, when a typical fixed-wing aircraft responds to aileron input, the period of rolling acceleration is very short and the airplane quickly approaches a steady rolling rate, which is proportional to aileron deflection. This gives the pilot the perception that aileron input commands the airplane's rolling rate. Thus, analysis of an airplane's roll response must include consideration of the rolling moment produced by the rolling rate as well as that produced by aileron deflection. Lifting-line theory provides the capability to do just that.

In general, a wing's spanwise variation in local aerodynamic angle of attack can be expressed as the value at the wing root plus the changes due to washout, control surface deflection, and rolling rate,

$$\alpha(z) - \alpha_{L0}(z) \equiv (\alpha - \alpha_{L0})_{\text{root}} - \Omega\omega(z)$$

$$+ \delta_a \chi(z) - \frac{pz}{V_\infty} \qquad (125)$$

where Ω is defined to be the maximum total symmetric washout for the wing, geometric plus aerodynamic,

$$\Omega \equiv [(\alpha - \alpha_{L0})_{\text{root}} - (\Delta\alpha - \Delta\alpha_{L0})_{\text{max}}]_{\text{washout}} \quad (126)$$

$\omega(z)$ is the symmetric washout distribution function,

$$\omega(z) \equiv \left[\frac{\Delta\alpha(z) - \Delta\alpha_{L0}(z) - (\alpha - \alpha_{L0})_{\text{root}}}{(\Delta\alpha - \Delta\alpha_{L0})_{\text{max}} - (\alpha - \alpha_{L0})_{\text{root}}} \right]_{\text{washout}} \quad (127)$$

and $\chi(z)$ is the control surface distribution function,

$$\chi(z) \equiv \left[\frac{\Delta\alpha(z) - \Delta\alpha_{L0}(z) - (\alpha - \alpha_{L0})_{\text{root}}}{\delta_a} \right]_{\text{control}} \quad (128)$$

For example, ailerons extend from the spanwise coordinate z_{ar} to z_{at} give

$$\chi(z) \equiv \begin{cases} 0 & z < -z_{at} \\ \varepsilon_f(z) & -z_{at} < z < -z_{ar} \\ 0 & -z_{ar} < z < z_{ar} \\ -\varepsilon_f(z) & z_{ar} < z < z_{at} \\ 0 & z > z_{at} \end{cases} \quad (129)$$

where ε_f is the local section flap effectiveness.

Using the definition of θ from Eq. (84) together with Eq. (125) in the relation for the Fourier coefficients, A_n, specified by Eq. (85), gives

$$\sum_{n=1}^{N} A_n \left[\frac{4b}{\widetilde{C}_{L,\alpha} c(\theta)} + \frac{n}{\sin(\theta)} \right] \sin(n\theta) = (\alpha - \alpha_{L0})_{\text{root}}$$
$$- \Omega\omega(\theta) + \delta_a \chi(\theta) + \overline{p}\cos(\theta) \quad (130)$$

where $\overline{p}$ is a dimensionless rolling rate defined as $\overline{p} = pb/2V_\infty$. The Fourier coefficients in Eq. (130) can be conveniently written as

$$A_n = a_n(\alpha - \alpha_{L0})_{\text{root}} - b_n\Omega + c_n\delta_a + d_n\overline{p} \quad (131)$$

where

$$\sum_{n=1}^{N} a_n \left[\frac{4b}{\widetilde{C}_{L,\alpha} c(\theta)} + \frac{n}{\sin(\theta)} \right] \sin(n\theta) = 1 \quad (132)$$

$$\sum_{n=1}^{N} b_n \left[\frac{4b}{\widetilde{C}_{L,\alpha} c(\theta)} + \frac{n}{\sin(\theta)} \right] \sin(n\theta) = \omega(\theta) \quad (133)$$

$$\sum_{n=1}^{N} c_n \left[\frac{4b}{\widetilde{C}_{L,\alpha} c(\theta)} + \frac{n}{\sin(\theta)} \right] \sin(n\theta) = \chi(\theta) \quad (134)$$

$$\sum_{n=1}^{N} d_n \left[\frac{4b}{\widetilde{C}_{L,\alpha} c(\theta)} + \frac{n}{\sin(\theta)} \right] \sin(n\theta) = \cos(\theta) \quad (135)$$

Equations (132) and (133) are exactly Eqs. (101) and (102) and their solutions are obtained in the same manner. The solutions to Eqs. (134) and (135) can be obtained in a similar manner and are both independent of angle of attack and rolling rate.

Once the Fourier coefficients are determined from Eqs. (131)–(135), the spanwise circulation distribution is known from Eqs. (84) and the spanwise section lift distribution is given by

$$\widetilde{L}(z) = \rho V_\infty \Gamma(z)$$

Thus, in view of Eq. (84), the rolling moment coefficient can be evaluated from

$$C_\ell = \frac{1}{\frac{1}{2}\rho V_\infty^2 Sb} \int_{z=-b/2}^{b/2} \widetilde{L}(z) z \, dz$$
$$= \frac{2}{V_\infty Sb} \int_{z=-b/2}^{b/2} \Gamma(z) z \, dz$$
$$= -\frac{b^2}{S} \sum_{n=1}^{N} A_n \int_{\theta=0}^{\pi} \sin(n\theta) \cos(\theta) \sin(\theta) \, d\theta$$

or after applying the trigonometric identity, $\sin(2\theta) = 2\sin(\theta)\cos(\theta)$, along with the definition of aspect ratio

$$C_\ell = -\frac{R_A}{2} \sum_{n=1}^{N} A_n \int_{\theta=0}^{\pi} \sin(n\theta) \sin(2\theta) \, d\theta \quad (136)$$

The integral in Eq. (136) is evaluated from

$$\int_{\theta=0}^{\pi} \sin(m\theta) \sin(n\theta) \, d\theta = \begin{cases} 0 & n \neq m \\ \pi/2 & n = m \end{cases} \quad (137)$$

After applying Eqs. (131) and (137), Eq. (136) becomes

$$C_\ell = -\frac{\pi R_A}{4} A_2 = -\frac{\pi R_A}{4} [a_2(\alpha - \alpha_{L0})_{\text{root}}$$
$$- b_2\Omega + c_2\delta_a + d_2\overline{p}] \quad (138)$$

For a wing with a *spanwise symmetric planform* and *spanwise symmetric washout*, the solutions to Eqs. (132) and (133) give

$$a_n = b_n = 0 \qquad n \text{ even} \quad (139)$$

and Eq. (138) reduces to

$$C_\ell = C_{\ell,\delta_a}\delta_a + C_{\ell,\overline{p}}\overline{p} \tag{140}$$

where

$$C_{\ell,\delta_a} = -\frac{\pi R_A}{4}c_2 \tag{141}$$

$$C_{\ell,\overline{p}} = -\frac{\pi R_A}{4}d_2 \tag{142}$$

The Fourier coefficients obtained from Eq. (134) depend on control surface geometry as well as the planform shape of the wing. Thus, the change in rolling moment coefficient with respect to aileron deflection depends on the size and shape of the ailerons and the wing planform. On the other hand, the Fourier coefficients that are evaluated from Eq. (135) are functions of only wing planform. For an elliptic planform, the spanwise variation in chord length is given by

$$c(y) = \frac{4b}{\pi R_A}\sqrt{1 - (2y/b)^2} \quad \text{or}$$

$$c(\theta) = \frac{4b}{\pi R_A}\sin(\theta)$$

and Eq. (135) reduces to

$$\sum_{n=1}^{N} d_n\left(\frac{\pi R_A}{\widetilde{C}_{L,\alpha}} + n\right)\sin(n\theta) = \sin(\theta)\cos(\theta) \tag{143}$$

The solution to Eq. (143) is given by the Fourier integral

$$d_n\left(\frac{\pi R_A}{\widetilde{C}_{L,\alpha}} + n\right) = \frac{2}{\pi}\int_0^\pi \sin(\theta)\cos(\theta)\sin(n\theta)\,d\theta$$

$$= \frac{1}{\pi}\int_0^\pi \sin(2\theta)\sin(n\theta)\,d\theta \tag{144}$$

which is readily evaluated from Eq. (137) to give

$$d_n = \begin{cases} \dfrac{\widetilde{C}_{L,\alpha}}{2(\pi R_A + 2\widetilde{C}_{L,\alpha})} & n = 2 \\ 0 & n \ne 2 \end{cases} \tag{145}$$

Using Eq. (145) in Eq. (142) gives

$$C_{\ell,\overline{p}} = \frac{-\widetilde{C}_{L,\alpha}}{8(1 + 2\widetilde{C}_{L,\alpha}/\pi R_A)}$$

Thus, in view of Eq. (88), the *change in rolling moment coefficient* with respect to dimensionless rolling rate for an elliptic wing can be written as

$$C_{\ell,\overline{p}} = -\frac{\kappa_{\ell\overline{p}}C_{L,\alpha}}{8} \qquad \kappa_{\ell\overline{p}} \equiv \frac{1 + \widetilde{C}_{L,\alpha}/\pi R_A}{1 + 2\widetilde{C}_{L,\alpha}/\pi R_A} \tag{146}$$

Similarly, in the general case of a wing with arbitrary planform,

$$C_{\ell,\overline{p}} = -\frac{\kappa_{\ell\overline{p}}C_{L,\alpha}}{8} \qquad \kappa_{\ell\overline{p}} \equiv \frac{2d_2}{a_1} \tag{147}$$

Using an airfoil lift slope of 2π to determine a_1 and d_2 from Eqs. (132) and (135), the results shown in Fig. 43 are obtained for tapered wings.

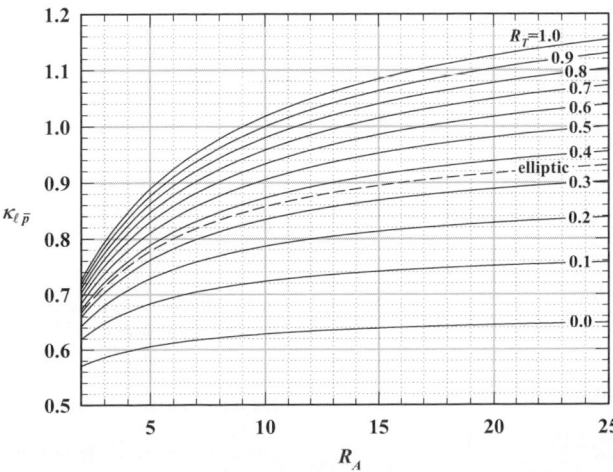

Fig. 43 Roll damping factor for wings with linear taper from Eq. (147).

The asymmetric spanwise variations in aerodynamic angle of attack, which result from aileron deflection and roll, can also produce a yawing moment. The aerodynamic yawing moment is given the symbol n, and the traditional sign convention for yaw is positive to the right (i.e., a moment that would yaw the airplane's nose to the right). The yawing moment develops as a direct result of an asymmetric spanwise variation in drag. Thus, the yawing moment coefficient can be written as

$$
C_n \equiv \frac{n}{\frac{1}{2}\rho V_\infty^2 Sb} = \frac{1}{\frac{1}{2}\rho V_\infty^2 Sb} \int_{z=-b/2}^{b/2} (-z)\widetilde{D}(z)\, dz
$$

(148)

As shown in Figs. 24 and 42, the asymmetric section drag results from tilting the section lift vector through the induced angle, α_i, and the roll angle, zp/V_∞. Thus, after expressing the section lift in terms of the section circulation, we have

$$
\widetilde{D}(z) = \widetilde{L}\sin\left(\alpha_i + \frac{zp}{V_\infty}\right) \cong \widetilde{L}\left(\alpha_i + \frac{zp}{V_\infty}\right)
$$

$$
= \rho V_\infty \Gamma(z)\left[\alpha_i(z) + \frac{zp}{V_\infty}\right]
$$

(149)

The induced angle of attack as predicted by lifting-line theory is

$$
a_i(z) = \frac{1}{4\pi V_\infty}\int_{\zeta=-b/2}^{b/2} \frac{1}{z-\zeta}\left(\frac{d\Gamma}{dz}\right)_{z=\zeta} d\zeta
$$

(150)

Using Eqs. (149) and (150) in Eq. (148) gives

$$
C_n = \int_{z=-b/2}^{b/2} \frac{-2z\Gamma(z)}{V_\infty^2 Sb}
$$

$$
\left[\frac{1}{4\pi}\int_{\zeta=-b/2}^{b/2}\frac{1}{z-\zeta}\left(\frac{d\Gamma}{dz}\right)_{z=\zeta} d\zeta + zp\right] dz
$$

(151)

From Eq. (84) we have

$$
\Gamma(\theta) = 2bV_\infty \sum_{n=1}^{N} A_n \sin(n\theta) \qquad \cos(\theta) = -\frac{2z}{b}
$$

$$
\frac{d\Gamma}{dz} = 4V_\infty \sum_{n=1}^{N} A_n \frac{n\cos(n\theta)}{\sin(\theta)}
$$

and Eq. (151) can be written as

$$
C_n = R_A \int_{\theta=0}^{\pi} \cos(\theta)\left[\sum_{n=1}^{N} A_n \sin(n\theta)\right]
$$

$$
\times \left[\sum_{n=1}^{N}\frac{nA_n}{\pi}\int_{\phi=0}^{\pi}\frac{\cos(n\phi)\, d\phi}{\cos(\phi)-\cos(\theta)}\right]\sin(\theta)\, d\theta
$$

$$
- R_A\overline{p}\sum_{n=1}^{N} A_n \int_{\theta=0}^{\pi}\cos^2(\theta)\sin(\theta)\sin(n\theta)\, d\theta
$$

Integrating in ϕ and using the trigonometric relation, $\sin(2\theta) = 2\cos(\theta)\sin(\theta)$, yields

$$
C_n = R_A \sum_{n=1}^{N}\sum_{m=1}^{N} nA_n A_m
$$

$$
\times \int_{\theta=0}^{\pi}\cos(\theta)\sin(n\theta)\sin(m\theta)\, d\theta
$$

$$
- R_A\overline{p}\sum_{n=1}^{N} A_n \frac{1}{2}\int_{\theta=0}^{\pi}\cos(\theta)\sin(2\theta)\sin(n\theta)\, d\theta
$$

(152)

Since m and n are positive integers, the integrals with respect to θ are evaluated from

$$
\int_{\theta=0}^{\pi}\cos(\theta)\sin(n\theta)\sin(m\theta)\, d\theta = \begin{cases} \frac{1}{4}\pi & m = n\pm 1 \\ 0 & m \neq n\pm 1 \end{cases}
$$

and after some rearranging, we obtain

$$
C_n = \frac{\pi R_A}{4}\sum_{n=2}^{N}(2n-1)A_{n-1}A_n - \frac{\pi R_A\overline{p}}{8}(A_1 + A_3)
$$

(153)

After applying Eq. (86), we find that the *yawing moment coefficient* for a wing of arbitrary geometry is given by

$$
C_n = \frac{C_L}{8}(6A_2 - \overline{p}) + \frac{\pi R_A}{8}(10A_2 - \overline{p})A_3
$$

$$
+ \frac{\pi R_A}{4}\sum_{n=4}^{N}(2n-1)A_{n-1}A_n
$$

(154)

where the Fourier coefficients, A_n, are related to Ω, δ_a, and $\overline{p}$ through Eq. (131).

If the wing planform and washout distribution are both spanwise symmetric, all even coefficients in both a_n and b_n are zero. The change in aerodynamic angle of attack that results from roll is always a spanwise odd function. Thus, all odd coefficients in d_n are zero for a spanwise symmetric wing. If the ailerons produce an equal and opposite change on each semispan, the control surface distribution function is a spanwise odd function as well and all odd coefficients in c_n are also zero. With this wing symmetry the Fourier coefficients must satisfy the relations $a_n = b_n = 0$ for n even and $c_n = d_n = 0$ for n odd. Thus, Eq. (131) reduces to

$$A_n = \begin{cases} a_n(\alpha - \alpha_{L0})_{\text{root}} - b_n\Omega & n \text{ odd} \\ c_n\delta_a + d_n\overline{p} & n \text{ even} \end{cases} \quad (155)$$

Using the Fourier coefficients from Eq. (155) in Eq. (154), the yawing moment coefficient with the wing symmetry described above is given by

$$C_n = \frac{C_L}{8}[6c_2\delta_a - (1 - 6d_2)\overline{p}] + \frac{\pi R_A}{8}[10c_2\delta_a$$
$$- (1 - 10d_2)\overline{p}][a_3(\alpha - \alpha_{L0})_{\text{root}} - b_3\Omega]$$
$$+ \frac{\pi R_A}{4}\left[\sum_{n=4}^{N}(2n - 1)[a_{n-1}(\alpha - \alpha_{L0})_{\text{root}}\right.$$
$$\left. - b_{n-1}\Omega](c_n\delta_a + d_n\overline{p})\right]_{n \text{ even}}$$
$$+ \frac{\pi R_A}{4}\left[\sum_{n=5}^{N}(2n - 1)(c_{n-1}\delta_a + d_{n-1}\overline{p})\right.$$
$$\left. \times [a_n(\alpha - \alpha_{L0})_{\text{root}} - b_n\Omega]\right]_{n \text{ odd}} \quad (156)$$

For the special case of a symmetric wing operating with optimum washout, which is specified by Eqs. (118) and (123), $a_n(\alpha - \alpha_{L0})_{\text{root}} - b_n\Omega$ is always zero for $n > 1$ and

$$C_n = C_{n,\delta_a}\delta_a + C_{n,\overline{p}}\overline{p} \quad C_{n,\delta_a} = \frac{3C_L}{4}c_2$$
$$C_{n,\overline{p}} = -\frac{C_L}{8}(1 - 6d_2) \quad (157)$$

For wings without optimum washout, the higher-order terms in Eq. (156) are not too large and the linear relation from Eq. (157) can be used as a rough approximation.

By comparing Eq. (140) with Eq. (157) we find that within the accuracy of Eq. (157), the yawing moment can be expressed in terms of the rolling moment and the rolling rate,

$$C_n = -\frac{3C_L}{\pi R_A}C_\ell - \frac{C_L}{8}\overline{p} \quad (158)$$

When an airplane responds to aileron input, the rolling rate quickly reaches the steady-state value, which results in no rolling moment. Thus, we see from Eq. (158) that the magnitude of the steady yawing moment produced by aileron deflection is proportional to the lift coefficient developed by the wing and the steady rolling rate that develops as a result of the

aileron deflection. From Eq. (158), we can also see that the sign of the yawing moment, which is induced on a wing by aileron deflection, is opposite to that of the rolling moment induced by the same aileron deflection. Positive aileron deflection induces a wing rolling moment to the left and a wing yawing moment to the right. For this reason, the yawing moment induced on the wing by aileron deflection is commonly called *adverse yaw*. We can also see from Eq. (158) that adverse yaw is more pronounced at low airspeeds, which require higher values of C_L.

6.5 Wing Aspect Ratio and Mean Chord Length

We have seen that the aspect ratio of a finite wing has a profound effect on the wing's performance. As defined in Eq. (86), the *aspect ratio* for a wing of arbitrary planform can be computed as the square of the wingspan divided by the planform area:

$$R_A \equiv \frac{b^2}{S} \quad (159)$$

In a more general sense, aspect ratio is defined to be the ratio of the longer to the shorter dimension for any 2D shape. The aspect ratio of a rectangle is simply the length of the long side divided by the length of the short side. Thus, the aspect ratio of a rectangular wing having a constant chord, c and wingspan, b, can be defined as

$$R_A \equiv \frac{b}{c} \quad (160)$$

Because the planform area of a rectangular wing is simply the wingspan multiplied by the chord length, $S = bc$, the definitions given in Eqs. (159) and (160) are equivalent. For the more general case of a wing having a chord length that varies with the spanwise coordinate, z, the aspect ratio can be written as the ratio of the wingspan to the average or *mean chord length*, $\overline{c}$,

$$R_A \equiv \frac{b}{\overline{c}} \quad (161)$$

where, by comparison of Eqs. (159) and (161), the mean chord length is defined as

$$\overline{c} \equiv \frac{S}{b} = \frac{1}{b}\int_{z=-b/2}^{b/2} c(z)\,dz \quad (162)$$

Because the wing aspect ratio defined by Eq. (159) has such a profound effect on wing performance, both the wingspan and the mean chord length defined by Eq. (162) are important characteristic dimensions for a finite wing.

For a wing with linear taper, the planform area is one half the sum of the root chord, c_r, and the tip chord, c_t, multiplied by the wingspan, b. Thus, the mean chord length for a wing with linear taper is

$$\bar{c} \equiv \frac{S}{b} = \frac{1}{b} \int_{z=-b/2}^{b/2} \left[c_r - (c_r - c_t) \left| \frac{2z}{b} \right| \right] dz$$

$$= \frac{c_r + c_t}{2} = \frac{1 + R_T}{2} c_r \qquad (163)$$

For an elliptic wing the mean chord length is

$$\bar{c} \equiv \frac{S}{b} = \frac{1}{b} \int_{z=-b/2}^{b/2} c_r \sqrt{1 - \left(\frac{2z}{b} \right)^2}\, dz = \frac{\pi}{4} c_r \quad (164)$$

7 FLOW OVER MULTIPLE LIFTING SURFACES

Prandtl's classical lifting-line equation, expressed in Eq. (83), applies only to a single lifting surface with no sweep or dihedral. In the development of this relation it was assumed that the only downwash induced on the wing was that induced by the trailing vortex sheet, which is shed from the wing itself. Airplanes are usually designed with more than one lifting surface. A typical configuration could be a wing combined with horizontal and vertical stabilizers. Each of these surfaces can generate lift and vorticity. The vorticity that is generated by one lifting surface will alter the flow about another. For example, a lifting wing will induce downwash on an aft horizontal stabilizer, and a lifting aft stabilizer will induce upwash on the main wing. Such lifting surface interactions are not accounted for in Eq. (83).

Even a flying wing cannot be analyzed using Eq. (83) if the wing has sweep or dihedral. In a wing with sweep and/or dihedral, the quarter-chord line on one semispan makes an angle with the quarter-chord line of the other semispan, as show in Fig. 44. A wing is said to have *positive sweep* if the wing is swept back from the root to the tip, with the wingtips being aft of the root. A wing has *positive dihedral* if the wing slopes up from the root to the tip, with the wingtips being above the root. *Negative dihedral* is usually called *anhedral*. Because a wing can also have taper and/or geometric twist, defining the sweep and dihedral angles can be somewhat ambiguous. For a conventional tapered wing, the sweep angle for the leading edge is greater than that for the trailing edge. For a wing with geometric washout, the dihedral angle for the trailing edge is greater than that for the leading edge. Furthermore, neither the sweep nor the dihedral must remain constant over the semispan of a lifting surface. For this reason, local sweep and dihedral angles are defined in terms of the wing quarter-chord line and the orientation of the local streamwise airfoil section, as shown in Fig. 44.

When constant dihedral is added to a wing, each side of the wing is rotated about the root as a solid body. The rotation of the left-hand side is opposite to that of the right-hand side, so that both wingtips are raised above the root and brought closer together. Since dihedral is a solid-body rotation, the local airfoil sections rotate with the quarter-chord line. For a wing with no dihedral, a vector drawn normal to a local airfoil section is also normal to the aircraft plane of symmetry. As dihedral is added to the wing, both the quarter-chord and the airfoil section normal are rotated upward, so that the normal to the local airfoil section always forms an angle equal to the dihedral angle with the normal to the aircraft plane of symmetry. For example, the vertical stabilizer of a conventional airplane usually has a dihedral angle of $90°$. For the rarely encountered case of an aircraft that does not have a plane of symmetry, the dihedral angle must be defined relative to some other defined reference plane.

Typically, for an airplane with fixed wings, constant sweep is not a solid-body rotation of each semispan. For a wing with no sweep, the normal to each local airfoil section is aligned with the wing quarter-chord line, and the quarter-chord line is straight. As constant positive sweep is added to a wing, the quarter-chord

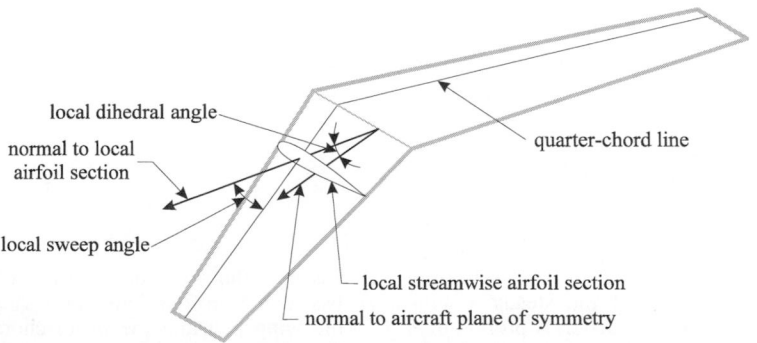

Fig. 44 Local sweep and dihedral angles.

line for each semispan is rotated back, moving the wingtips aft of the root. However, for constant planform area and aspect ratio, the local airfoil sections are not rotated with the quarter-chord line. Each airfoil section is translated straight back to the new quarter-chord location and the length of the quarter-chord line is increased, so that the distance from wingtip to wingtip does not change. A local sweep angle can be defined as the angle between the normal to the streamwise airfoil section and the quarter-chord line. Thus, the sweep angle for a wing with no dihedral is measured relative to a horizontal line, while the sweep angle for a vertical stabilizer with $90°$ dihedral is measured relative to a vertical line.

Supersonic airplanes are sometimes designed with wings that can be rotated to vary the sweep angle in flight. This type of sweep variation is solid-body rotation. However, varying sweep in this manner changes not only the sweep but the shape of the streamwise airfoil section, the wingspan, and usually even the planform area of the wing.

Most airplanes are designed with some sweep and/or dihedral in the wing and/or the horizontal and vertical stabilizers. While the reasons for using sweep and dihedral will be left for a later discussion, it is clear from looking at the myriad of airplanes parked at any airport that a means for analyzing the effects of sweep and dihedral is needed. In addition, the downwash created by one lifting surface, such as the wing, has a dramatic effect on the performance of other lifting surfaces, such as the horizontal stabilizer. Thus, a means for predicting the effects of such interactions is also needed.

To predict the aerodynamic forces and moments acting on a complete airplane in flight, 3D panel codes and CFD analysis are often used. However, these methods are very computationally intensive and may not be suitable for use in all phases of aircraft design and analysis. An alternative method that provides excellent accuracy at very low computational cost is the numerical lifting-line method, which is presented here.

7.1 Numerical Lifting-Line Method

In developing Prandtl's lifting-line equation, a single lifting surface with no sweep or dihedral was assumed. The lifting line was confined to the z axis and the trailing vortex sheet was assumed to be in the $x-z$ plane (see Fig. 23). This significantly simplified the expressions for downwash and induced angle of attack. In order to use lifting-line theory to study the effects of sweep, dihedral, and/or the interactions between lifting surfaces, the theory must be generalized to allow for an arbitrary position and orientation of both the lifting line and the trailing vortex sheet. Here we shall examine a numerical lifting-line method that can be used to obtain a potential flow solution for the forces and moments acting on a system of lifting surfaces, each with arbitrary position and orientation. For a detailed development of this numerical method and a comparison with panel codes, CFD, and experimental data, see Phillips and Snyder.[14]

A first-order numerical lifting-line method can be obtained by synthesizing a system of lifting surfaces using a composite of horseshoe-shaped vortices. The continuous distribution of bound vorticity over the surface of each lifting surface, as well as the continuous distribution of free vorticity in the trailing vortex sheets, is approximated by a finite number of discrete horseshoe vortices, as shown in Fig. 45.

The bound portion of each horseshoe vortex is placed coincident with the wing quarter-chord line and is, thus, aligned with the local sweep and dihedral. The trailing portion of each horseshoe vortex is aligned with the trailing vortex sheet. In Fig. 45, a small gap

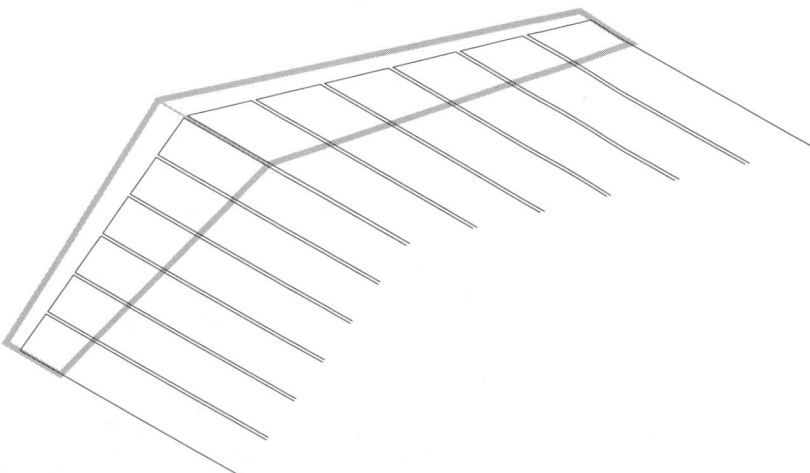

Fig. 45 Horseshoe-shaped vortices distributed along the lifting line of a finite wing with sweep and dihedral.

is shown between the left-hand trailing segment of one horseshoe vortex and the right-hand trailing segment of the next. This is for display purposes only. In reality, the left-hand corner of one horseshoe and the right-hand corner of the next are both placed on the same nodal point. Thus, except at the wingtips, each trailing vortex segment is coincident with another trailing segment from the adjacent vortex. If two adjacent vortices have exactly the same strength, then the two coincident trailing segments exactly cancel, since one has clockwise rotation and the other has counterclockwise rotation. The net vorticity that is shed from the wing at any internal node is simply the difference in strength of the two adjacent vortices that share the node.

Each horseshoe vortex is composed of three straight segments, a finite bound segment and two semi-infinite trailing segments. We can calculate the velocity induced at an arbitrary point in space (x, y, z) by a general horseshoe vortex. As shown in Fig. 46, a general horseshoe vortex is completely defined by two nodal points (x_1, y_1, z_1) and (x_2, y_2, z_2), a trailing unit vector $\mathbf{u}_\infty$, and a vortex strength Γ. The horseshoe vortex starts at the fluid boundary, an infinite distance downstream. The inbound trailing vortex segment is directed along the vector $-\mathbf{u}_\infty$ to node 1 at (x_1, y_1, z_1). The bound vortex segment is directed along the wing quarter-chord line from node 1 to node 2 at (x_2, y_2, z_2). The outbound trailing vortex segment is directed along the vector $\mathbf{u}_\infty$ from node 2 to the fluid boundary, an infinite distance back downstream. The velocity induced by the entire horseshoe vortex is simply the vector sum of the velocities induced by each of the three linear segments that make up the horseshoe.

The vorticity vector is assumed to point in the direction of $\mathbf{u}_\infty$ as determined by the right-hand rule. However, the inbound trailing vortex segment can be treated like an outbound segment with negative circulation. Thus, the *velocity vector induced at an arbitrary*

point in space by a complete horseshoe vortex is

$$\mathbf{V} = \frac{\Gamma}{4\pi}\left[\frac{\mathbf{u}_\infty \times \mathbf{r}_2}{r_2(r_2 - \mathbf{u}_\infty \cdot \mathbf{r}_2)} + \frac{(r_1 + r_2)(\mathbf{r}_1 \times \mathbf{r}_2)}{r_1 r_2(r_1 r_2 + \mathbf{r}_1 \cdot \mathbf{r}_2)} \right.$$
$$\left. - \frac{\mathbf{u}_\infty \times \mathbf{r}_1}{r_1(r_1 - \mathbf{u}_\infty \cdot \mathbf{r}_1)}\right] \qquad (165)$$

where $\mathbf{u}_\infty$ is the unit vector in the direction of the trailing vortex sheet, $\mathbf{r}_1$ is the spatial vector from (x_1, y_1, z_1) to (x, y, z), and $\mathbf{r}_2$ is the spatial vector from (x_2, y_2, z_2) to (x, y, z),

$$\mathbf{r}_1 = (x - x_1)\mathbf{i}_x + (y - y_1)\mathbf{i}_y + (z - z_1)\mathbf{i}_z \quad (166)$$
$$\mathbf{r}_2 = (x - x_2)\mathbf{i}_x + (y - y_2)\mathbf{i}_y + (z - z_2)\mathbf{i}_z \quad (167)$$

In obtaining the classical lifting-line solution for a single wing without sweep or dihedral, the trailing vortex sheet was assumed to be aligned with the wing chord. This was done to facilitate obtaining an analytical solution. In obtaining a numerical solution, there is little advantage in aligning the trailing vortex sheet with a vehicle axis such as the chord line. More correctly, the trailing vortex sheet should be aligned with the free stream. This is easily done in the numerical solution by setting $\mathbf{u}_\infty$ equal to the unit vector in the direction of the free stream. When using this method to predict the forces and moments on a single lifting surface, there is very little difference between the results obtained from slightly different orientations of the trailing vortex sheet.

When a system of lifting surfaces is synthesized using N of these horseshoe vortices, in a manner similar to that shown in Fig. 45, Eq. (165) can be used to determine the resultant velocity induced at any point in space if the vortex strength of each horseshoe vortex is known. However, these strengths are not known a priori. To compute the strengths of the N vortices, we must have a system of N equations relating these

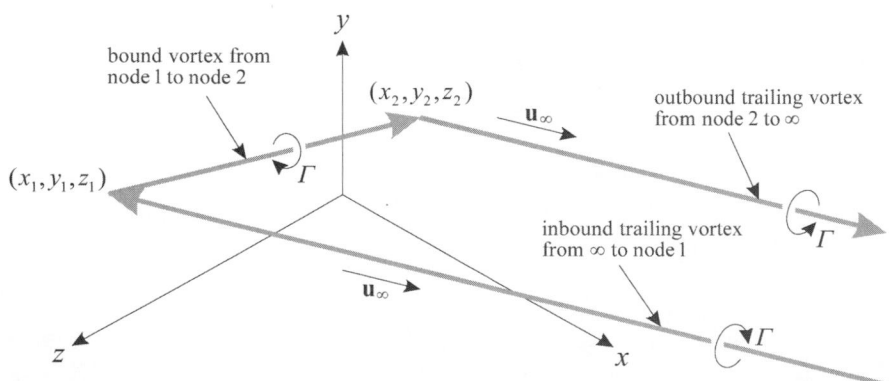

Fig. 46   General horseshoe vortex.

N strengths to some known properties of the wing. For this we turn to the 3D vortex lifting law given in Eq. (82).

If flow over a finite wing is synthesized from a uniform flow combined with horseshoe vortices placed along the wing quarter chord, from Eq. (165), the local velocity induced at a control point placed anywhere along the bound segment of horseshoe vortex i is

$$\mathbf{V}_i = \mathbf{V}_\infty + \sum_{j=1}^{N} \frac{\Gamma_j \mathbf{v}_{ji}}{\overline{c}_j} \qquad (168)$$

where $\mathbf{V}_\infty$ is the velocity of the uniform flow, Γ_j is the strength of horseshoe vortex j, $\mathbf{v}_{ji}$ is a dimensionless velocity that would be induced at control point i by horseshoe vortex j, having a unit strength

$$\mathbf{v}_{ji} = \begin{cases} \dfrac{\overline{c}_j}{4\pi}\left[\dfrac{\mathbf{u}_\infty \times \mathbf{r}_{j_2 i}}{r_{j_2 i}(r_{j_2 i} - \mathbf{u}_\infty \cdot \mathbf{r}_{j_2 i})} \right. \\ \quad + \dfrac{(r_{j_1 i} + r_{j_2 i})(\mathbf{r}_{j_1 i} \times \mathbf{r}_{j_2 i})}{r_{j_1 i} r_{j_2 i}(r_{j_1 i} r_{j_2 i} + \mathbf{r}_{j_1 i} \cdot \mathbf{r}_{j_2 i})} \qquad j \neq i \\ \quad \left. - \dfrac{\mathbf{u}_\infty \times \mathbf{r}_{j_1 i}}{r_{j_1 i}(r_{j_1 i} - \mathbf{u}_\infty \cdot \mathbf{r}_{j_1 i})} \right] \\ \dfrac{\overline{c}_j}{4\pi}\left[\dfrac{\mathbf{u}_\infty \times \mathbf{r}_{j_2 i}}{r_{j_2 i}(r_{j_2 i} - \mathbf{u}_\infty \cdot \mathbf{r}_{j_2 i})} \right. \\ \quad \left. - \dfrac{\mathbf{u}_\infty \times \mathbf{r}_{j_1 i}}{r_{j_1 i}(r_{j_1 i} - \mathbf{u}_\infty \cdot \mathbf{r}_{j_1 i})} \right] \qquad j = i \end{cases}$$
$$(169)$$

where $\mathbf{r}_{j_1 i}$ is the spatial vector from node 1 of horseshoe vortex j to the control point of horseshoe vortex i, $\mathbf{r}_{j_2 i}$ is the spatial vector from node 2 of horseshoe vortex j to the control point of horseshoe vortex i, and $\mathbf{u}_\infty$ is the unit vector in the direction of the free

stream. At this point, $\overline{c}_j$ could be any characteristic length associated with the wing section aligned with horseshoe vortex j. This characteristic length is simply used to nondimensionalize Eq. (169) and has no effect on the induced velocity. An appropriate choice for $\overline{c}_j$ will be addressed at a later point. The bound vortex segment is excluded from Eq. (169), when $j = i$, because a straight vortex segment induces no downwash along its own length. However, the second term in Eq. (169), for $j \neq i$, is indeterminate when used with $j = i$ because $r_{i_1 i} r_{i_2 i} + \mathbf{r}_{i_1 i} \cdot \mathbf{r}_{i_2 i} = 0$.

From Eqs. (82) and (168), the aerodynamic force acting on a spanwise differential section of the wing located at control point i is given by

$$d\mathbf{F}_i = \rho \Gamma_i \left(\mathbf{V}_\infty + \sum_{j=1}^{N} \frac{\Gamma_j}{\overline{c}_j} \mathbf{v}_{ji} \right) \times d\mathbf{l}_i \qquad (170)$$

Allowing for the possibility of flap deflection, the local section lift coefficient for the airfoil section located at control point i is a function of local angle of attack and local flap deflection,

$$\widetilde{C}_{Li} = \widetilde{C}_{Li}(\alpha_i, \delta_i) \qquad (171)$$

where $\widetilde{C}_{Li}$, α_i, and δ_i are, respectively, the local airfoil section lift coefficient, the local angle of attack, and the local flap deflection, all evaluated for the airfoil section aligned with control point i. Defining $\mathbf{u}_{ni}$ and $\mathbf{u}_{ai}$ to be the local unit normal and axial vectors for the airfoil section located at control point i, as shown in Fig. 47, the local angle of attack at control point i can be written as

$$\alpha_i = \tan^{-1}\left(\frac{\mathbf{V}_i \cdot \mathbf{u}_{ni}}{\mathbf{V}_i \cdot \mathbf{u}_{ai}} \right) \qquad (172)$$

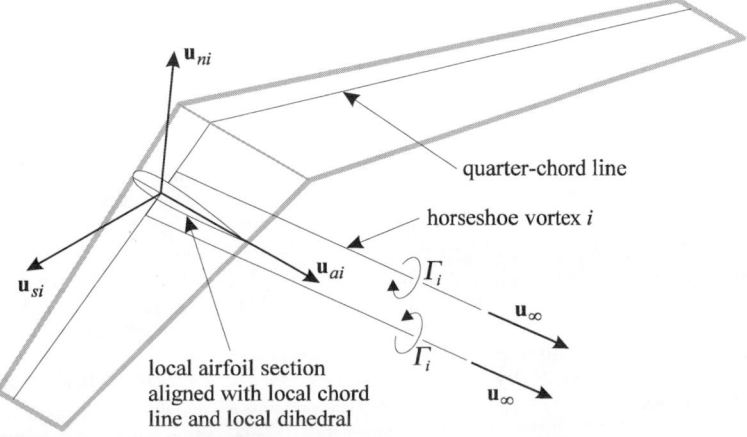

Fig. 47 Unit vectors describing the orientation of the local airfoil section.

If the relation implied by Eq. (171) is known at each section of the wing, the magnitude of the aerodynamic force acting on a spanwise differential section of the wing located at control point i can be written as

$$|\mathbf{dF}_i| = \tfrac{1}{2}\rho V_\infty^2 \widetilde{C}_{Li}(\alpha_i, \delta_i)\, dS_i \qquad (173)$$

where dS_i is a spanwise differential planform area element located at control point i. Setting the magnitude of the force obtained from Eq. (170) equal to that obtained from Eq. (173), applying Eqs. (168)–(172), and rearranging, we can write

$$2\left|\left(\mathbf{u}_\infty + \sum_{j=1}^{N} \mathbf{v}_{ji} G_j\right) \times \boldsymbol{\zeta}_i\right| G_i - \widetilde{C}_{Li}(\alpha_i, \delta_i) = 0 \qquad (174)$$

$$\mathbf{u}_\infty \equiv \frac{\mathbf{V}_\infty}{V_\infty} \qquad \boldsymbol{\zeta}_i \equiv \overline{c}_i \frac{\mathbf{dl}_i}{dS_i} \qquad G_i \equiv \frac{\Gamma_i}{\overline{c}_i V_\infty}$$

where the local section angle of attack is determined from

$$\alpha_i = \tan^{-1}\left[\frac{\left(\mathbf{u}_\infty + \sum_{j=1}^{N} \mathbf{v}_{ji} G_j\right) \cdot \mathbf{u}_{ni}}{\left(\mathbf{u}_\infty + \sum_{j=1}^{N} \mathbf{v}_{ji} G_j\right) \cdot \mathbf{u}_{ai}}\right] \qquad (175)$$

Equation (174) can be written for N different control points, one associated with each of the N horseshoe vortices used to synthesize the lifting surface or system of lifting surfaces. This provides a system of N nonlinear equations relating the N unknown dimensionless vortex strengths, G_i, to known properties of the wing. At angles of attack below stall, this system of nonlinear equations surrenders quickly to Newton's method.

To apply Newton's method, the system of equations is written in the vector form

$$\mathbf{f}(\mathbf{G}) = \mathbf{R} \qquad (176)$$

where

$$f_i(\mathbf{G}) = 2\left|\left(\mathbf{u}_\infty + \sum_{j=1}^{N} \mathbf{v}_{ji} G_j\right) \times \boldsymbol{\zeta}_i\right| G_i - \widetilde{C}_{Li}(\alpha_i, \delta_i) \qquad (177)$$

and $\mathbf{R}$ is a vector of residuals. We wish to find the vector of dimensionless vortex strengths, $\mathbf{G}$, that will make all components of the residual vector, $\mathbf{R}$, go to zero. Thus, we want the change in the residual vector to be $-\mathbf{R}$. We start with some initial estimate for the $\mathbf{G}$ vector and iteratively refine this estimate by applying the Newton corrector equation

$$[\mathbf{J}]\,\Delta\mathbf{G} = -\mathbf{R} \qquad (178)$$

where $[\mathbf{J}]$ is an $N \times N$ Jacobian matrix of partial derivatives, which is obtained by differentiating Eq. (177),

$$J_{ij} = \frac{\partial f_i}{\partial G_j}$$

$$= \begin{cases} \dfrac{2\mathbf{w}_i \cdot (\mathbf{v}_{ji} \times \boldsymbol{\zeta}_i)}{|\mathbf{w}_i|} G_i \\[2mm] \quad -\dfrac{\partial \widetilde{C}_{Li}}{\partial \alpha_i} \dfrac{v_{ai}(\mathbf{v}_{ji} \cdot \mathbf{u}_{ni}) - v_{ni}(\mathbf{v}_{ji} \cdot \mathbf{u}_{ai})}{v_{ai}^2 + v_{ni}^2} \quad j \neq i \\[4mm] 2|\mathbf{w}_i| + \dfrac{2\mathbf{w}_i \cdot (\mathbf{v}_{ji} \times \boldsymbol{\zeta}_i)}{|\mathbf{w}_i|} G_i \\[2mm] \quad -\dfrac{\partial \widetilde{C}_{Li}}{\partial \alpha_i} \dfrac{v_{ai}(\mathbf{v}_{ji} \cdot \mathbf{u}_{ni}) - v_{ni}(\mathbf{v}_{ji} \cdot \mathbf{u}_{ai})}{v_{ai}^2 + v_{ni}^2} \quad j = i \end{cases}$$

$$(179)$$

$$\mathbf{w}_i \equiv \mathbf{v}_i \times \boldsymbol{\zeta}_i \qquad v_{ni} \equiv \mathbf{v}_i \cdot \mathbf{u}_{ni} \qquad v_{ai} \equiv \mathbf{v}_i \cdot \mathbf{u}_{ai}$$

$$\mathbf{v}_i \equiv \mathbf{u}_\infty + \sum_{j=1}^{N} \mathbf{v}_{ji} G_j$$

Using Eq. (179) in Eq. (178), we compute the correction vector $\Delta\mathbf{G}$. This correction vector is used to obtain an improved estimate for the dimensionless vortex strength vector, $\mathbf{G}$, according to the relation

$$\mathbf{G} = \mathbf{G} + \Psi\,\Delta\mathbf{G} \qquad (180)$$

where Ψ is a relaxation factor. The process is repeated until the magnitude of the largest residual is less than some convergence criteria. For angles of attack below stall, this method converges very rapidly using almost any initial estimate for the $\mathbf{G}$ vector with a relaxation factor of unity. At angles of attack beyond stall, the method must be highly underrelaxed and is very sensitive to the initial estimate for the $\mathbf{G}$ vector.

For the fastest convergence of Newton's method, we require an accurate initial estimate for the dimensionless vortex strength vector. For this purpose, a linearized version of Eq. (174) is useful. Furthermore, for wings of high aspect ratio at small angles of attack, the nonlinear terms in Eq. (174) are quite small and the linearized system can be used directly to give an accurate prediction of aerodynamic performance. Applying the small-angle approximation to Eq. (174) and neglecting all other second-order terms, we obtain

$$2|\mathbf{u}_\infty \times \boldsymbol{\zeta}_i| G_i - \widetilde{C}_{Li,\alpha} \sum_{j=1}^{N} \mathbf{v}_{ji} \cdot \mathbf{u}_{ni} G_j$$

$$= \widetilde{C}_{Li,\alpha}(\mathbf{u}_\infty \cdot \mathbf{u}_{ni} - \alpha_{L0i} + \varepsilon_i \delta_i) \qquad (181)$$

The linearized system of equations given by Eq. (181) gives good results, at small angles of attack, for wings of reasonably high aspect ratio and little sweep. For larger angles of attack or highly swept wings, the nonlinear system given by Eq. (174) should be used. However, at angles of attack below stall, Eq. (181) still provides a reasonable initial estimate for the dimensionless vortex strength vector, to be used with Newton's method for obtaining a solution to the nonlinear system.

Once the vortex strengths are determined from either Eq. (174) or Eq. (181), the total aerodynamic force vector can be determined from Eq. (170). If the lifting surface or surfaces are synthesized from a large number of horseshoe vortices, each covering a small spanwise increment of one lifting surface, we can approximate the aerodynamic force as being constant over each spanwise increment. Then, from Eq. (170), the total aerodynamic force is given by

$$\mathbf{F}_a = \rho \sum_{i=1}^{N} \left(\Gamma_i \mathbf{V}_\infty + \sum_{j=1}^{N} \frac{\Gamma_i \Gamma_j}{\bar{c}_j} \mathbf{v}_{ji} \right) \times \delta \mathbf{l}_i \quad (182)$$

where $\delta \mathbf{l}$ is the spatial vector along the bound segment of horseshoe vortex i from node 1 to node 2. Nondimensionalizing this result, the *total nondimensional aerodynamic force vector* is

$$\frac{\mathbf{F}_a}{\frac{1}{2}\rho V_\infty^2 S_r} = 2 \sum_{i=1}^{N} \left(G_i \mathbf{u}_\infty + \sum_{j=1}^{N} G_i G_j \mathbf{v}_{ji} \right) \times \boldsymbol{\zeta}_i \frac{\delta S_i}{S_r} \quad (183)$$

where S_r is the global reference area and δS_i is the planform area of the spanwise increment of the lifting surface covered by horseshoe vortex i. If we assume a linear variation in chord length over each spanwise increment, we have

$$\delta S_i \equiv \int_{s=s_1}^{s_2} c \, ds = \frac{c_{i_1} + c_{i_2}}{2} (s_{i_2} - s_{i_1}) \quad (184)$$

where c is the local chord length and s is the spanwise coordinate.

The aerodynamic moment generated about the center of gravity is

$$\mathbf{M}_a = \rho \sum_{i=1}^{N} \left\{ \mathbf{r}_i \times \left[\left(\Gamma_i \mathbf{V}_\infty + \sum_{j=1}^{N} \frac{\Gamma_i \Gamma_j}{\bar{c}_j} \mathbf{v}_{ji} \right) \right. \right.$$
$$\left. \left. \times \delta l_i \right] + \delta \mathbf{M}_i \right\} \quad (185)$$

where $\mathbf{r}_i$ is the spatial vector from the center of gravity to control point i and $\delta \mathbf{M}_i$ is the quarterchord moment generated by the spanwise increment of the wing covered by horseshoe vortex i. If we

assume a constant section moment coefficient over each spanwise increment, then

$$\delta \mathbf{M}_i \cong -\frac{1}{2}\rho V_\infty^2 \tilde{C}_{mi} \int_{s=s_1}^{s_2} c^2 \, ds \, \mathbf{u}_{si} \quad (186)$$

where $\tilde{C}_{mi}$ is the local section moment coefficient and $\mathbf{u}_{si}$ is the local unit vector in the spanwise direction as shown in Fig. 47 and defined by

$$\mathbf{u}_{si} = \mathbf{u}_{at} \times \mathbf{u}_{ni} \quad (187)$$

Using Eq. (186) in Eq. (185) and nondimensionalizing gives

$$\frac{\mathbf{M}_a}{\frac{1}{2}\rho V_\infty^2 S_r l_r}$$
$$= \sum_{i=1}^{N} \left\{ 2\mathbf{r}_i \times \left[\left(G_i \mathbf{u}_\infty + \sum_{j=1}^{N} G_i G_j \mathbf{v}_{ji} \right) \times \boldsymbol{\zeta}_i \right] \right.$$
$$\left. - \frac{\tilde{C}_{mi}}{\delta S_i} \int_{s=s_1}^{s_2} c^2 \, ds \, \mathbf{u}_{si} \right\} \frac{\delta S_i}{S_r l_r} \quad (188)$$

where l_r is the global reference length.

To this point, the local characteristic length $\bar{c}_i$ has not been defined. It could be any characteristic length associated with the spanwise increment of the wing covered by horseshoe vortex i. From Eq. (88), the most natural choice for this local characteristic length is the integral of the chord length squared, with respect to the spanwise coordinate, divided by the incremental area. For a linear variation in chord length over each spanwise increment, this gives

$$\bar{c}_i = \frac{1}{\delta S_i} \int_{s=s_1}^{s_2} c^2 \, ds = \frac{2}{3} \frac{c_{i_1}^2 + c_{i_1} c_{i_2} + c_{i_2}^2}{c_{i_1} + c_{i_2}} \quad (189)$$

With this definition, the *dimensionless aerodynamic moment vector* is

$$\frac{\mathbf{M}_a}{\frac{1}{2}\rho V_\infty^2 S_r l_r} = \sum_{i=1}^{N} \left\{ 2\mathbf{r}_i \times \left[\left(G_i \mathbf{u}_\infty + \sum_{j=1}^{N} G_i G_j \mathbf{v}_{ji} \right) \right. \right.$$
$$\left. \left. \times \boldsymbol{\zeta}_i \right] - \tilde{C}_{mi} \bar{c}_i \mathbf{u}_{si} \right\} \frac{\delta S_i}{S_r l_r} \quad (190)$$

Once the dimensionless vortex strengths, G_i, are known, Eqs. (183) and (190) are used to evaluate the components of the aerodynamic force and moment.

Each lifting surface must, of course, be divided into spanwise elements, in a manner similar to that shown symbolically in Fig. 45. In this figure, the wing is divided into elements of equal spanwise increment. However, this is not the most efficient way in which to grid a wing. Since the spanwise derivative of shed vorticity is greater in the region near the wingtips, for best computational efficiency, the nodal points should be clustered more tightly in this region. Conventional cosine clustering has been found to be quite efficient for this purpose. For straight wings, clustering is only needed near the wingtips and the cosine distribution can be applied across the entire span of the wing. However, for wings with sweep or dihedral, there is a step change in the slope of the quarter-chord line at the root of the wing. This step change produces an increase in the spanwise variation of downwash in the region near the root. Thus, in general, it is recommended that cosine clustering be applied independently over each semispan of each lifting surface. This clusters the nodes more tightly at both the tip and root. This clustering is based on the change of variables,

$$\frac{s}{b} = \frac{1 - \cos(\theta)}{4} \qquad (191)$$

where s is the spanwise coordinate and b is twice the semispan. Over each semispan, θ varies from zero to π as s varies from zero to $b/2$. Distributing the nodes uniformly in θ will provide the desired clustering in s. If the total number of horseshoe elements desired on each semispan is n, the spanwise nodal coordinates are computed from

$$\frac{s_i}{b} = \frac{1}{4}\left[1 - \cos\left(\frac{i\pi}{n}\right)\right] \qquad 0 \le i \le n \qquad (192)$$

where the bound segment of horseshoe vortex i extends from node i to node $i - 1$ on any left semispan and from node $i - 1$ to node i on any right semispan. Using this nodal distribution with about 40 horseshoe elements per semispan gives the best compromise between speed and accuracy. Figure 48 shows a system of lifting surfaces overlaid with a grid of this type using 20 elements per semispan.

For maximum accuracy and computational efficiency, some attention must also be paid to the location of control points. At first thought, it would seem most reasonable to place control points on the bound segment of each vortex, midway between the two trailing legs. However, it has been found that this does not give the best accuracy. A significant improvement in accuracy, for a given number of elements, can be achieved by placing the control points midway in θ rather than midway in s. Thus, the spanwise control point coordinates should be computed from

$$\frac{s_i}{b} = \frac{1}{4}\left[1 - \cos\left(\frac{i\pi}{n} - \frac{\pi}{2n}\right)\right] \qquad 1 \le i \le n$$

$$(193)$$

This distribution places control points very near the spatial midpoint of each bound vortex segment over most of the wing. However, near the root and the tip, these control points are significantly offset from the spatial midpoint.

This numerical lifting-line method can be used to predict the aerodynamic forces and moments acting on a system of lifting surfaces with arbitrary position and orientation. Each lifting surface is synthesized by distributing horseshoe vortices along the quarter chord in the manner shown in Fig. 48. Because all of the horseshoe vortices used to synthesize the complete system of lifting surfaces are combined and forced to satisfy either Eq. (174) or Eq. (181) as a single system of coupled equations, all of the interactions between lifting surfaces are accounted for directly.

Unlike the closed-form solution to Prandtl's classical lifting-line theory, the numerical lifting-line method can be applied to wings with sweep and/or dihedral. To examine how well the numerical lifting-line method predicts the effects of sweep, Phillips and Snyder[14] compared results obtained from this method, a numerical panel method, and an inviscid CFD solution. These results were also compared with experimental data for two different wings. Some of the results from this comparison are shown in Fig. 49. The solid line and filled symbols correspond to a straight wing of aspect ratio 6.57, with experimental data obtained from McAlister and Takahashi.[15] The dashed line and open symbols are for a 45° swept wing of aspect ratio 5.0, having experimental data reported by Weber and Brebner.[16] Both wings have symmetric airfoil sections and constant chord throughout the span, with no geometric twist. The straight wing has a thickness of 15% and

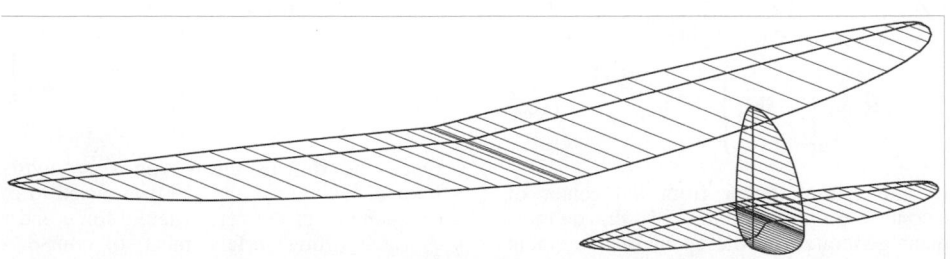

Fig. 48 Lifting-line grid with cosine clustering and 20 elements per semispan.

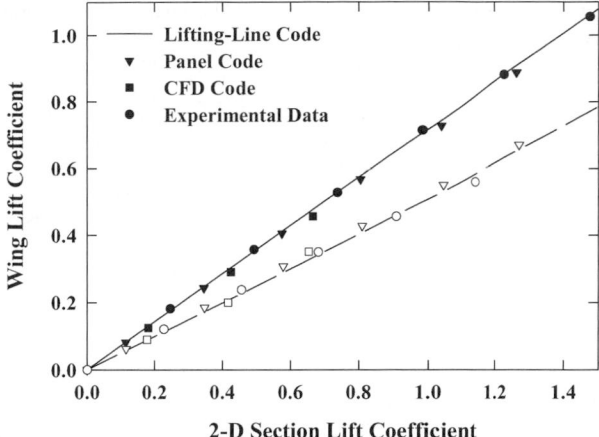

Fig. 49 Comparison between the lift coefficient predicted by the numerical lifting-line method, a numerical panel method, and an inviscid CFD solution with data obtained from wind tunnel tests for an unswept wing and a wing with 45° of sweep.

the swept wing has a thickness of 12%. From the results shown in Fig. 49, we see that the lift coefficient predicted by all four methods is in good agreement with experimental observations for both wings. However, the computational time required to obtain a solution using the numerical panel method was about 2.5×10^4 times that required for the lifting-line method, and the inviscid CFD solutions required approximately 2.7×10^6 times as long as the lift-line solutions. The accuracy of the numerical lifting-line method for predicting dihedral effects was also investigated by Phillips and Snyder,[14] and the results obtained were similar to those shown in Fig. 49.

The insight of Ludwig Prandtl (1875–1953) was nothing short of astonishing. This was never more dramatically demonstrated than in the development of his classical lifting-line theory, during the period 1911–1918. The utility of this simple and elegant theory is so great that it is still widely used today. Furthermore, with a few minor alterations and the use of a modern computer, the model proposed by Prandtl can be used to predict the inviscid forces and moments acting on lifting surfaces of aspect ratio greater than about 4 with an accuracy as good as that obtained from modern panel codes or CFD, but at a small fraction of the computational cost.

Like panel methods, lifting-line theory provides only a potential flow solution. Thus, the forces and moments computed from this method do not include viscous effects. In addition to this restriction, which also applies to panel methods, lifting-line theory imposes an additional restriction, which does not apply to panel methods. For lifting surfaces with low aspect ratio, Prandtl's hypothesis breaks down and the usual relationship between local section lift and local section angle of attack no longer applies. It has long been established that lifting-line theory gives good agreement with experimental data for lifting surfaces of aspect ratio greater than about 4. For lifting surfaces

of lower aspect ratio, panel methods or CFD solutions should be used.

The numerical lifting-line method contains no inherent requirement for a linear relationship between section lift and section angle of attack. Thus, the method can be applied, with caution, to account approximately for the effects of stall. The lifting-line method requires a known relationship for the section lift coefficient as a function of section angle of attack. Since such relationships are often obtained experimentally beyond stall, the numerical lifting-line method predicts stall by using a semiempirical correction to an otherwise potential flow solution. For this reason, the method should be used with extreme caution for angles of attack beyond stall.

The effects of viscous parasitic drag can be approximately accounted for in the present numerical lifting-line method through the vector addition of a parasitic drag component to the contribution from each segment in Eqs. (183) and (190). The parasitic segment contribution to the net aerodynamic force vector is

$$(\delta \mathbf{F}_i)_{\text{parasite}} \cong \tfrac{1}{2} \rho V_\infty^2 \widetilde{C}_{D_i}(\alpha_i) \, \delta S_i \mathbf{u}_i$$

where $\widetilde{C}_{D_i}(\alpha_i)$ represents a relation for the local section drag coefficient as a function of angle of attack and $\mathbf{u}_i$ is the unit vector in the direction of the local velocity vector,

$$\mathbf{u}_i \equiv \frac{\mathbf{V}_i}{V_i}$$

A simple polynomial fit to experimental airfoil section data could be used to describe the relation for section drag coefficient as a function of angle of attack. Similarly, the parasitic segment contribution to the net aerodynamic moment vector is

$$(\delta \mathbf{M}_i)_{\text{parasite}} \cong \frac{1}{2} \rho V_\infty^2 \widetilde{C}_{D_i}(\alpha_i) \, \delta S_i (\mathbf{r}_i \times \mathbf{u}_i)$$

Thus, the *dimensionless parasitic contributions to the force and moment vectors* are

$$\left[\frac{\mathbf{F}_a}{\frac{1}{2}\rho V_\infty^2 S_r}\right]_{\text{parasite}} = \sum_{i=1}^{N} \widetilde{C}_{D_i}(\alpha_i)\frac{\delta S_i}{S_r}\mathbf{u}_i \qquad (194)$$

$$\left[\frac{\mathbf{M}_a}{\frac{1}{2}\rho V_\infty^2 S_r l_r}\right]_{\text{parasite}} = \sum_{i=1}^{N} \widetilde{C}_{D_i}(\alpha_i)\frac{\delta S_i}{S_r l_r}(\mathbf{r}_i \times \mathbf{u}_i) \qquad (195)$$

In the study of aeronautics, we find it necessary to know how the aerodynamic force and moment components for a complete aircraft vary over a broad range of operating conditions. Ultimately, this information is commonly gathered from wind tunnel testing. However, such testing is very time consuming and expensive. Another alternative that is often used to obtain the required information is CFD computations. Although modern computers are very fast and the CFD algorithms available today are quite accurate, the volume of data needed to define the required aerodynamic parameters is large. Even with the fastest available computers, gathering the desired information from CFD computations requires considerable time. Thus, for preliminary design, it is important to have a more computationally efficient means for the estimation of aerodynamic force and moment components acting on a complete aircraft. When parasitic drag is included, the numerical lifting-line method described in this section provides such an analytical tool. This method reduces computation time by more than four orders of magnitude over that required for inviscid panel codes and by more than six orders of magnitude compared with CFD computations.

8 WING STALL AND MAXIMUM LIFT COEFFICIENT

Many aspects of aircraft design and performance analysis depend on the maximum lift coefficient that can be attained on a finite wing prior to stall. Because in general the local section lift coefficient is not constant along the span of a finite wing, it is of interest to know the value of the maximum section lift coefficient and the position along the span at which this maximum occurs. Such knowledge allows us to predict the onset of wing stall from known airfoil section properties, including the maximum airfoil section lift coefficient.

From Eq. (84) combined with the vortex lifting law, lifting-line theory predicts that the spanwise variation in local section lift for an unswept wing is given by

$$\widetilde{L}(\theta) = \rho V_\infty \Gamma(\theta) = 2b\rho V_\infty^2 \sum_{n=1}^{\infty} A_n \sin(n\theta)$$

$$\theta = \cos^{-1}\left(\frac{-2z}{b}\right) \qquad (196)$$

and the spanwise variation in local section lift coefficient is

$$\widetilde{C}_L(\theta) \equiv \frac{\widetilde{L}(\theta)}{\frac{1}{2}\rho V_\infty^2 c(\theta)} = \frac{4b}{c(\theta)}\sum_{n=1}^{\infty} A_n \sin(n\theta) \qquad (197)$$

The Fourier coefficients, A_n, in Eq. (197) can be written conveniently using the change of variables given by Eq. (100). Furthermore, solving Eq. (105) for the root aerodynamic angle of attack and applying Eqs. (107) and (109) gives

$$(\alpha - \alpha_{L0})_{\text{root}} = \frac{b_l}{a_l}\Omega + \frac{C_L}{\pi R_A a_1} \qquad (198)$$

Using Eq. (198) in Eq. (100) yields

$$A_n = \left(\frac{b_1 a_n}{a_1} - b_n\right)\Omega + \frac{a_n}{\pi R_A a_1}C_L \qquad (199)$$

Using this change of variables, Eq. (197) can be written as

$$\widetilde{C}_L(\theta) = \Omega \sum_{n=1}^{\infty} 4\left(\frac{b_1 a_n}{a_1} - b_n\right)\frac{\sin(n\theta)}{c(\theta)/b}$$

$$+ C_L \sum_{n=1}^{\infty} \frac{4a_n}{\pi R_A a_1}\frac{\sin(n\theta)}{c(\theta)/b} \qquad (200)$$

We see from Eq. (200) that the spanwise variation in local section lift coefficient can be divided conveniently into two components. The first term on the right-hand side of Eq. (200) is called the *basic section lift coefficient* and the second term is called the *additional section lift coefficient*. The basic section lift coefficient is independent of C_L and directly proportional to the total amount of wing twist, Ω. The additional section lift coefficient at any section of the wing is independent of wing twist and directly proportional to the net wing lift coefficient, C_L.

As can be seen from Eq. (200), the basic *section lift coefficient* is the spanwise variation in local section lift coefficient that occurs when the total net lift developed by the wing is zero. Examination of the first term on the right-hand side of Eq. (200) reveals that the basic section lift coefficient depends on all of the Fourier coefficients a_n and b_n. From Eq. (101) we see the Fourier coefficients a_n depend only on the wing planform. Equation (102) shows that the Fourier coefficients b_n depend on both the wing planform and the dimensionless twist distribution function, $\omega(\theta)$. Thus, the *spanwise variation* in the basic section lift coefficient depends on wing planform and wing twist but is independent of the wing angle of attack.

Examination of the second term on the right-hand side of Eq. (200) discloses that the additional section lift coefficient depends only on the wing planform and the Fourier coefficients a_n. From Eq. (101) we have seen that the coefficients a_n do not depend on wing twist. Thus, Eq. (200) exposes the important fact that the *additional* section lift coefficient is independent of wing twist. Because the basic section lift coefficient is zero for an untwisted wing, we see that the additional section lift coefficient is equivalent to the spanwise variation in local section lift coefficient that would be developed on an untwisted wing of the same planform operating at the same wing lift coefficient.

Figure 50 shows how the net section lift coefficient and its two components obtained from Eq. (200) vary along the span of a linearly tapered wing of aspect ratio 8.0 and taper ratio 0.5. This figure shows the spanwise variation in section lift coefficient for several values of total linear twist with the net wing lift coefficient held constant at 1.0. Similar results are shown in Fig. 51 for three different values of the net wing lift coefficient with the total linear twist held constant at $6°$.

Notice that the spanwise coordinates of the maximums in both the basic and additional section lift coefficients do not change with either the amount of wing twist or the net wing lift coefficient. Because the additional section lift coefficient is independent of wing twist, the spanwise position of the aerodynamic

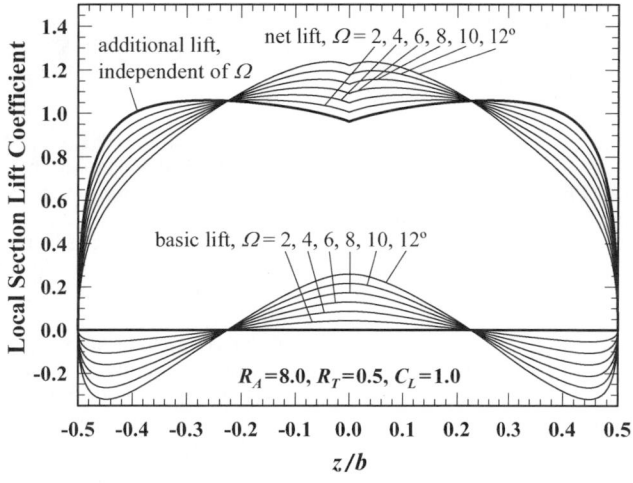

Fig. 50 Spanwise variation in local section lift coefficient as a function of the total amount of linear twist with the net wing lift coefficient held constant at $C_L = 1.0$.

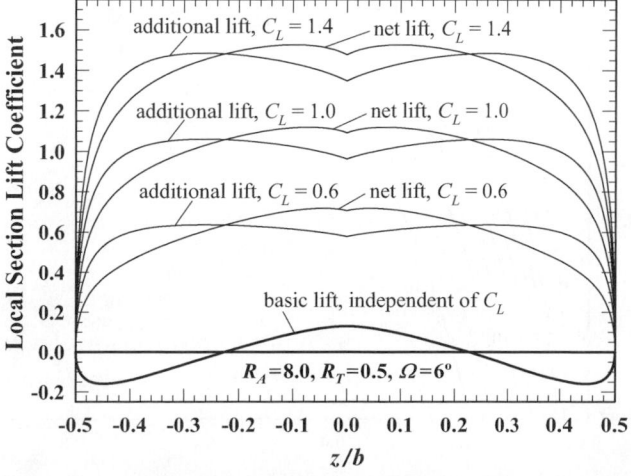

Fig. 51 Spanwise variation in local section lift coefficient as a function of the net wing lift coefficient with the total amount of linear twist held constant at $\Omega = 6°$.

center of each wing semispan is *not affected* by wing twist. However, the maximum in the net section lift coefficient moves inboard as the total amount of twist is increased, and for wings with positive twist (i.e., washout), this maximum moves outboard as the wing lift coefficient increases. Thus, the spanwise position of the center of pressure on each semispan of a twisted wing *varies* with both wing twist and angle of attack.

Combining terms from the basic and additional section lift coefficients on the right-hand side of Eq. (200) and rearranging, it can be shown that the maximum section lift coefficient occurs at a value of θ that satisfies the relation

$$\frac{d\widetilde{C}_L}{d\theta} = \frac{4C_L}{\pi R_A} \sum_{n=1}^{\infty} \left(\frac{a_n}{a_1} + \frac{b_1 a_n - a_1 b_n}{a_1^2} \frac{\pi R_A a_1 \Omega}{C_L} \right)$$

$$\times \left[b \frac{n \cos(n\theta)c(\theta) - \sin(n\theta)\, dc/d\theta}{c^2(\theta)} \right] = 0 \quad (201)$$

Following the development of Phillips and Alley[17] and applying Eq. (201), the spanwise location of the airfoil section that supports the largest section lift coefficient is found to be a root of the equation

$$\sum_{n=1}^{\infty} \left(a_n + \frac{b_1 a_n - a_1 b_n}{a_1} \frac{C_{L,\alpha}\Omega}{C_L} \right) \left[n \, \cos(n\theta) \frac{c(\theta)}{b} \right.$$

$$\left. - \sin(n\theta) \frac{d(c/b)}{d\theta} \right] = 0 \quad (202)$$

In the most general case, this root must be found numerically.

After finding the root of Eq. (202) to obtain the value of θ, which corresponds to the airfoil section that supports the maximum section lift coefficient, this value of θ can be used in Eq. (200) to determine the maximum section lift coefficient for the wing at the specified operating condition. Dividing Eq. (200) by the net wing lift coefficient and applying Eq. (107), the ratio of the local section lift coefficient to the total wing lift coefficient can be written as

$$\frac{\widetilde{C}_L(\theta)}{C_L} = \frac{4b}{\pi R_A c(\theta)} \left[\sum_{n=1}^{\infty} \frac{a_n}{a_1} \sin(n\theta) \right.$$

$$\left. + \frac{C_{L,\alpha}\Omega}{C_L} \sum_{n=2}^{\infty} \frac{b_1 a_n - a_1 b_n}{a_1^2} \sin(n\theta) \right] \quad (203)$$

Examination of Eqs. (202) and (203) reveals that, for $\Omega = 0$, these equations are independent of the net wing lift coefficient. This means that, for an untwisted wing of any planform, the ratio of the maximum section lift coefficient to the total wing lift coefficient and the position along the span at which this maximum occurs are *independent* of operating conditions and functions of the wing planform only. Figure 52 shows how the ratio of total wing lift coefficient to maximum section lift coefficient varies with aspect ratio and taper ratio for untwisted wings with linear taper.

The spanwise coordinate of the maximum section lift coefficient for such untwisted wings is presented in Fig. 53 as a function of aspect ratio and taper

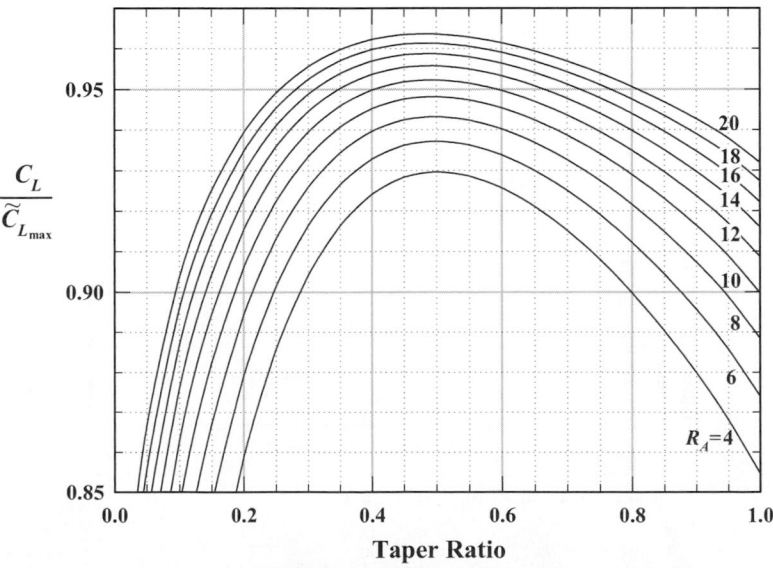

Fig. 52 Maximum lift coefficient for tapered wings with no sweep or twist.

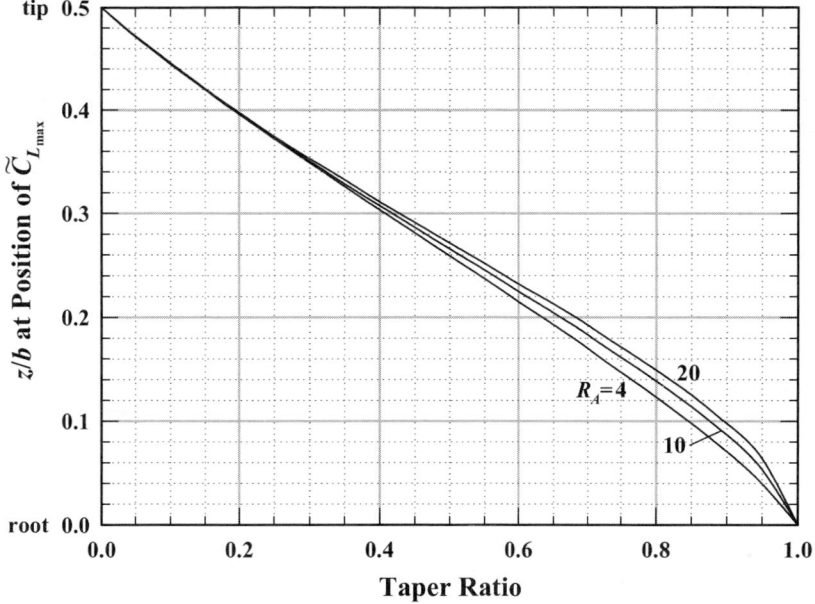

Fig. 53 Spanwise location of maximum section lift coefficient for tapered wings with no sweep or twist.

ratio. Notice that the spanwise coordinate of the airfoil section that supports the maximum section lift coefficient is quite insensitive to the aspect ratio and nearly a linear function of taper ratio.

For unswept wings of arbitrary planform and arbitrary twist, the maximum section lift coefficient is obtained by evaluating Eq. (203) at the value of θ corresponding to the root of Eq. (202). The result can be rearranged algebraically to solve for the ratio of the total wing lift coefficient to the maximum section lift coefficient, which yields

$$\left(\frac{C_L}{\widetilde{C}_{L_{\max}}}\right)_{\Lambda=0} = \left(\frac{C_L}{\widetilde{C}_{L_{\max}}}\right)_{\substack{\Omega=0 \\ \Lambda=0}} \left(1 - \kappa_{L\Omega}\frac{C_{L,\alpha}\Omega}{\widetilde{C}_{L_{\max}}}\right)$$

(204)

where

$$\left(\frac{C_L}{\widetilde{C}_{L_{\max}}}\right)_{\substack{\Omega=0 \\ \Lambda=0}} = \frac{\pi R_A c(\theta_{\max})/(4b)}{\sum_{n=1}^{\infty}(a_n/a_1)\sin(n\theta_{\max})}$$

(205)

$$\kappa_{L\Omega} \equiv \frac{4b}{\pi R_A c(\theta_{\max})}\sum_{n=2}^{\infty}\frac{b_1 a_n - a_1 b_n}{a_1^2}\sin(n\theta_{\max})$$

(206)

and $\theta_{\max}$ is the value of θ at the wing section that supports the maximum airfoil section lift coefficient, which is obtained from the root of Eq. (202).

For unswept rectangular wings with positive washout ($\Omega \geq 0$), the maximum section lift coefficient

occurs at the wing root ($\theta_{\max} = \pi/2$) and the constant chord length is given by $c(\theta) = b/R_A$. Using these results in Eqs. (205) and (206) and simplifying, for *unswept rectangular wings with positive washout* we obtain

$$\left(\frac{C_L}{\widetilde{C}_{L_{\max}}}\right)_{\substack{\Omega=0 \\ \Lambda=0}} = \frac{\pi/4}{\sum_{i=0}^{\infty}(-1)^i[(a_{2i+1})/a_1]}$$

$$\kappa_{L\Omega} = \frac{4}{\pi}\sum_{i=1}^{\infty}(-1)^i\frac{b_1 a_{2i+1} - a_1 b_{2i+1}}{a_1^2}$$

(207)

For an unswept elliptic wing with $\Omega \geq 0$, the maximum section lift coefficient occurs at the wing root ($\theta_{\max} = \pi/2$), the section chord length is $c(\theta) = [4b/(\pi R_A)]\sin(\theta)$, and the planform Fourier coefficients are $a_1 = \widetilde{C}_{L,\alpha}/(\pi R_A + \widetilde{C}_{L,\alpha})$ and $a_n = 0$ for $n > 1$. Using these results in Eqs. (205) and (206), Phillips and Alley[17] have shown that for *unswept elliptic wings with positive linear washout*

$$\left(\frac{C_L}{\widetilde{C}_{L_{\max}}}\right)_{\substack{\Omega=0 \\ \Lambda=0}} = 1$$

$$\kappa_{L\Omega} = \frac{1}{\pi}\sum_{i=1}^{\infty}\frac{4}{(2i+1)^2 - 4}\left(\frac{\pi R_A + \widetilde{C}_{L,\alpha}}{\pi R_A + (2i+1)\widetilde{C}_{L,\alpha}}\right)$$

(208)

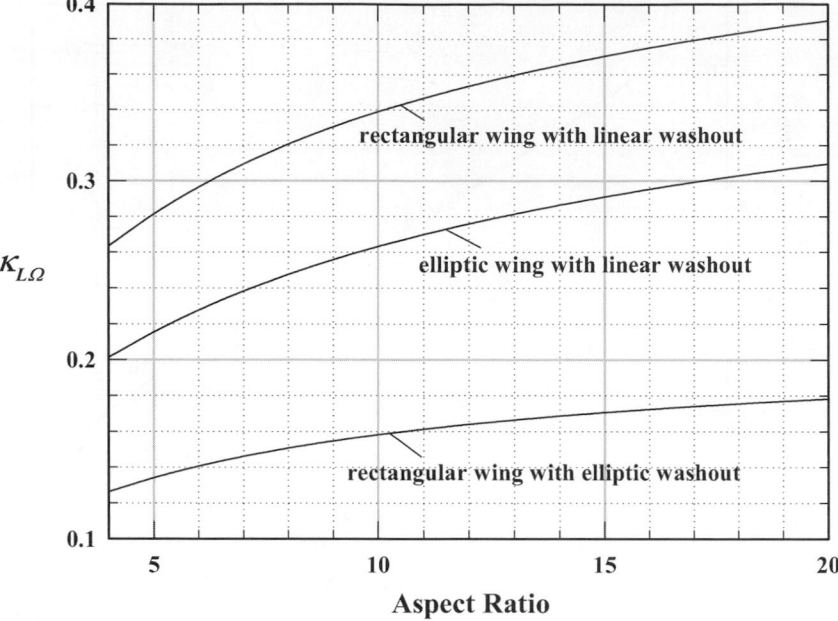

Fig. 54   Effect of wing twist on $\kappa_{L\Omega}$ for elliptic and rectangular wings.

Notice that for unswept rectangular and elliptic wings with positive washout, $\kappa_{L\Omega}$ is independent of both Ω and C_L and the ratio of wing lift coefficient to maximum section lift coefficient is a linear function of $C_{L,\alpha}\Omega/C_{L\max}$. Results predicted from Eqs. (207) and (208) are shown in Fig. 54 as a function of aspect ratio.

As was seen in Figs. 50 and 51, for wings of arbitrary planform the maximum section lift coefficient does not necessarily occur at the root. Thus, $\kappa_{L\Omega}$ may vary with $C_{L,\alpha}\Omega/C_{L\max}$ as well as the wing planform. The results plotted in Figs. 55 and 56 show how $\kappa_{L\Omega}$ varies with $C_{L,\alpha}\Omega/\widetilde{C}_{L\max}$ for wings with linear taper and linear washout.

For *unswept wings with twist optimized to produce minimum induced drag*, the optimum twist distribution and optimum total amount of twist are given in Eq. (123). For wings with linear taper, which are twisted in this manner, Phillips and Alley[17] obtained a closed-form solution to Eqs. (202) and (203). From this solution, we find that the z coordinate of the wing section that supports the maximum airfoil section lift coefficient is a linear function of the taper ratio and independent of operating conditions,

$$z_{\max} \equiv z(\theta_{\max}) = \pm\tfrac{1}{2}(1 - R_T)b \qquad (209)$$

The ratio of the total wing lift coefficient to the maximum section lift coefficient for such wings is found to be independent of operating conditions and given by

$$\left(\frac{C_L}{\widetilde{C}_{L\max}}\right)_{\Lambda=0} = \frac{\pi(2R_T - R_T^2)^{1/2}}{2(1 + R_T)} \qquad (210)$$

The reader is cautioned that results predicted from Eqs. (209) and (210) are only valid if the wing twist is maintained in proportion to the wing lift coefficient according to the relations provided by Eq. (123).

When Eq. (204) and Figs. 52–56 are used to estimate the total wing lift coefficient that corresponds to a given maximum airfoil section lift coefficient, the results apply only to wings without sweep. As a lifting wing of any planform is swept back, the lift near the root of each semispan is reduced as a result of the downwash induced by the bound vorticity generated on the opposite semispan. This tends to move the point of maximum section lift outboard. For wings with significant taper, this outboard shift causes the point of maximum section lift to occur at an airfoil section having a smaller section chord length, which increases the maximum airfoil section lift coefficient that is produced for a given wing lift coefficient. Because the series solution to Prandtl's lifting-line equation applies only to unswept wings, a numerical solution is required to predict the effects of wing sweep. Either the numerical lifting-line method presented in Section 7 or a numerical panel code could be used for this purpose.

The reader should also note that using the maximum airfoil section lift coefficient in Eq. (204) will

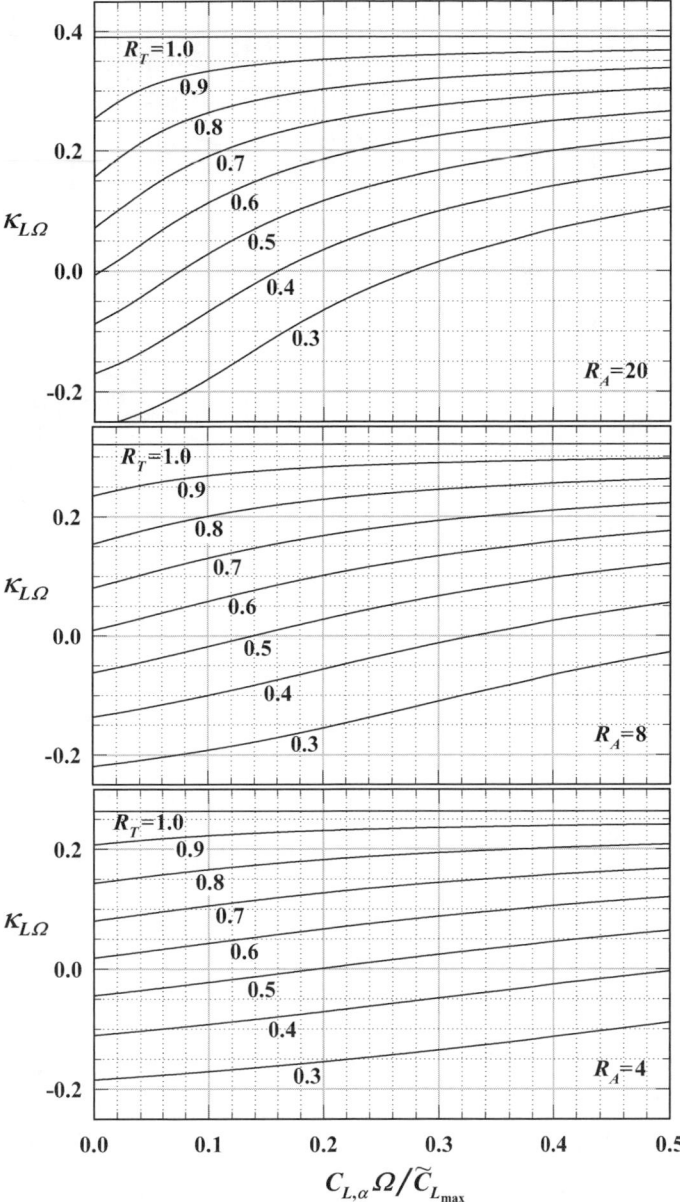

Fig. 55 Effect of wing twist and taper ratio on $\kappa_{L\Omega}$ for wings with linear taper and linear washout.

give an estimate for the wing lift coefficient at the onset of airfoil section stall. At higher angles of attack, separated flow will exist over some sections of the wing and drag will be substantially increased. However, the wing lift coefficient predicted from Eq. (204) is not exactly the maximum wing lift coefficient. Viscous interactions between adjacent sections of the wing can initiate flow separation at slightly lower angles of attack than

predicted by Eq. (204). Furthermore, as the angle of attack is increased somewhat beyond that which produces the onset of airfoil section stall, the section lift coefficient on the stalled section of the wing will decrease. However, the section lift coefficient on the unstalled sections of the wing will continue to increase with angle of attack until the maximum section lift coefficient is reached on these sections as well. Thus,

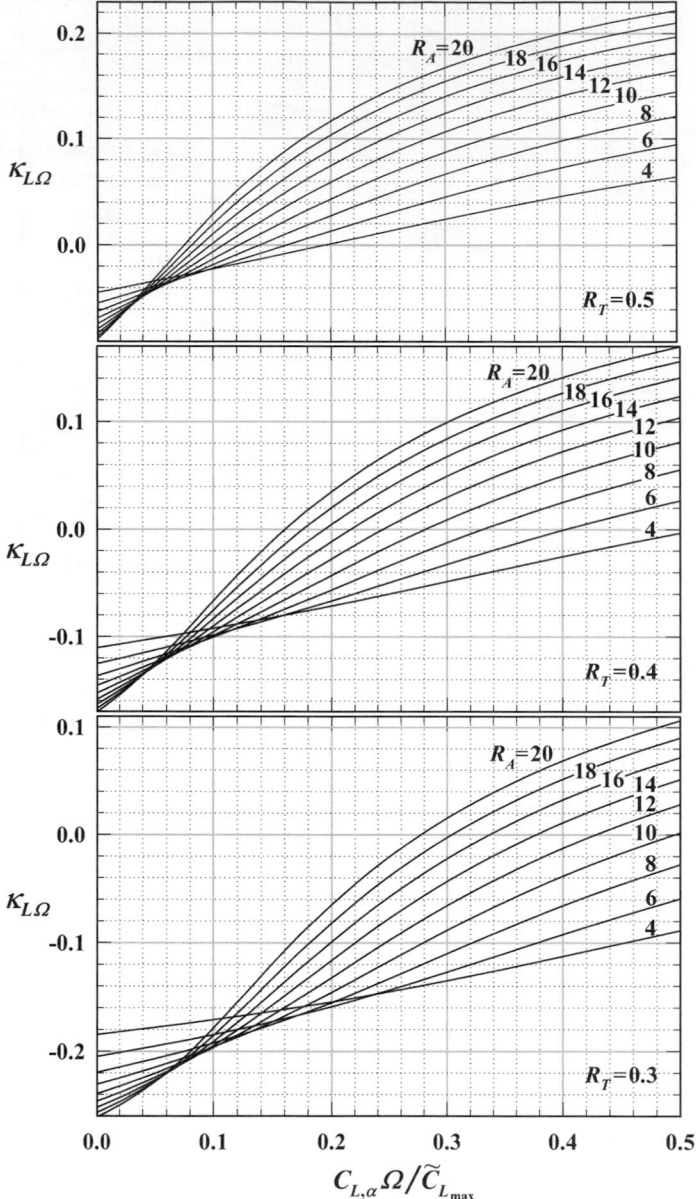

Fig. 56 Effect of wing twist and aspect ratio on $\kappa_{L\Omega}$ for wings with linear taper and linear washout.

the maximum wing lift coefficient could differ slightly from that which is predicted by Eq. (204). Because boundary layer separation is a viscous phenomenon, the maximum wing lift coefficient for a given wing geometry must be determined from experimental methods or computational fluid dynamics (CFD).

To account for the effects of wing sweep and stall, experimental data and results obtained from CFD

computations suggest that predictions from Eq. (204) can be modified by including sweep and stall correction factors,

$$C_{L_{\max}} = \left(\frac{C_L}{\widetilde{C}_{L_{\max}}} \right)_{\substack{\Omega=0 \\ A=0}} \kappa_{Ls} \kappa_{L\Lambda} (\widetilde{C}_{L_{\max}} - \kappa_{L\Omega} C_{L,\alpha} \Omega)$$

$$(211)$$

where the sweep factor $\kappa_{L\Lambda}$ depends on the wing sweep angle and planform, and the stall factor κ_{Ls} depends on wing aspect ratio and the wing twist parameter $C_{L,\alpha}\Omega/\widetilde{C}_{L_{\max}}$. For wings with linear taper, Phillips and Alley[17] presented numerical lifting-line results for the sweep factor, which correlate well with the approximation

$$\kappa_{L\Lambda} \cong 1 + \kappa_{\Lambda 1}\Lambda - \kappa_{\Lambda 2}\Lambda^{1.2} \qquad (212)$$

where Λ is the quarter-chord sweep angle in radians and the coefficients $\kappa_{\Lambda 1}$ and $\kappa_{\Lambda 2}$ depend on aspect ratio and taper ratio as shown in Figs. 57 and 58, respectively. For wings with linear taper and linear twist, CFD results for the stall factor obtained by Alley and co-workers[18] correlate well with the approximate relation

$$\kappa_{Ls} \cong 1 + (0.0042R_A - 0.068)\left(1 + \frac{2.3C_{L,\alpha}\Omega}{\widetilde{C}_{L_{\max}}}\right) \qquad (213)$$

Equation (213) was obtained from computations of the maximum wing lift coefficient for 25 different wing geometries. These wings had aspect ratios ranging from 4 to 20, taper ratios from 0.5 to 1.0, quarter-chord sweep angles from 0° to 30°, and linear geometric washout ranging from 0° to 8°.

When estimating a maximum wing lift coefficient from Eq. (211), it is essential that the maximum 2D

airfoil section lift coefficient be evaluated at the same Reynolds and Mach numbers as those for the 3D wing. For the case of a rectangular wing this is rather straightforward. However, for a tapered wing the section Reynolds number is not constant across the wingspan. This gives rise to an important question regarding what characteristic length should be used to define the Reynolds number associated with predicting the maximum lift coefficient for a wing of arbitrary planform. The simplest choice would be the mean chord length defined by Eq. (162). However, this does not provide a particularly suitable characteristic length for defining the Reynolds number associated with wing stall. A more appropriate characteristic length for this purpose is the chord length of the wing section supporting the maximum section lift coefficient.

The spanwise coordinate of the maximum airfoil section lift coefficient at the onset of stall depends on wing taper, twist, and sweep. For an untwisted rectangular wing with no sweep, the onset of airfoil section stall occurs at the wing root. As wing taper ratio is decreased from 1.0, the point of maximum section lift coefficient moves outboard from the root. Adding sweep to the wing also moves the point of maximum section lift coefficient outboard. On the other hand, adding washout to a wing with taper and/or sweep moves the point of maximum section lift coefficient inboard. Lifting-line theory can be used to predict the spanwise coordinate of the wing section that supports the maximum airfoil section lift coefficient. For tapered wings with no sweep or twist, the coordinate of

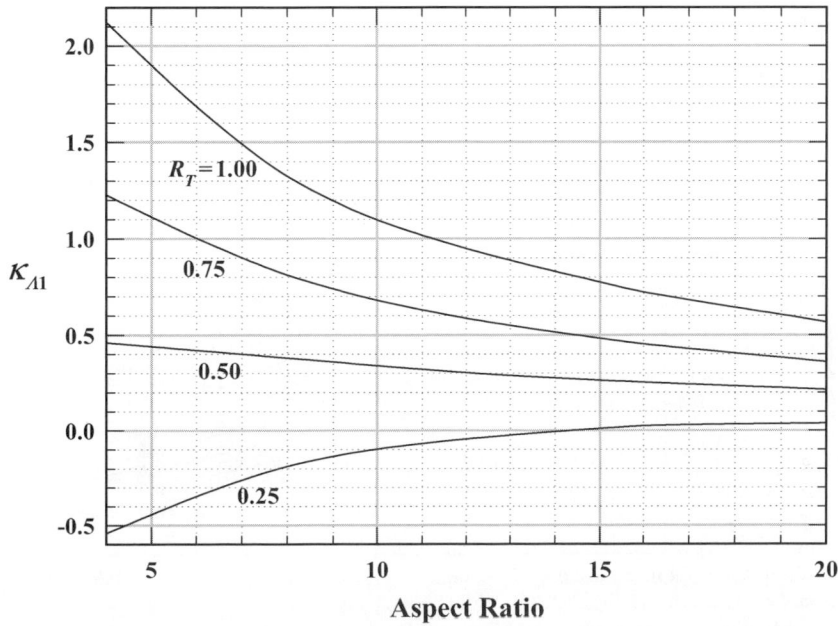

Fig. 57 Wing sweep coefficient $\kappa_{\Lambda 1}$ to be used in Eq. (212) for wings with linear taper.

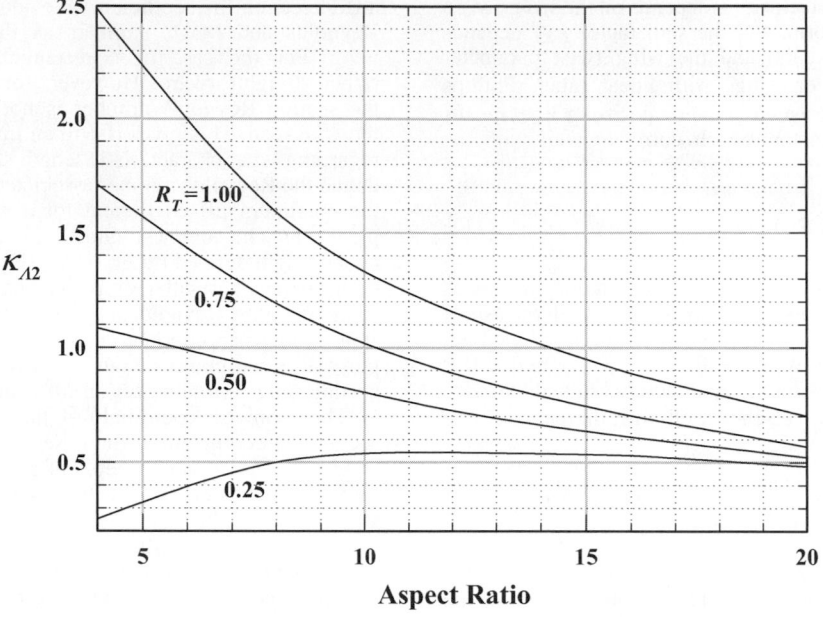

Fig. 58 Wing sweep coefficient $\kappa_{\Lambda 2}$ to be used in Eq. (212) for wings with linear taper.

the maximum section lift coefficient may be obtained from Fig. 53.

Results obtained from Fig. 53 can be modified to account for the effects of twist and sweep. The spanwise coordinate of the wing section that supports the maximum section lift coefficient can be estimated from

$$\frac{z_{max}}{b} = 0.5 - \kappa_{Z\Lambda}\left[0.5 - \kappa_{Z\Omega}\left(\frac{Z_{max}}{b}\right)_{\substack{\Omega=0 \\ \Lambda=0}}\right]$$

(214)

For wings with linear taper and linear twist, $\kappa_{Z\Omega}$ and $\kappa_{Z\Lambda}$ are obtained from Figs. 59 and 60, respectively

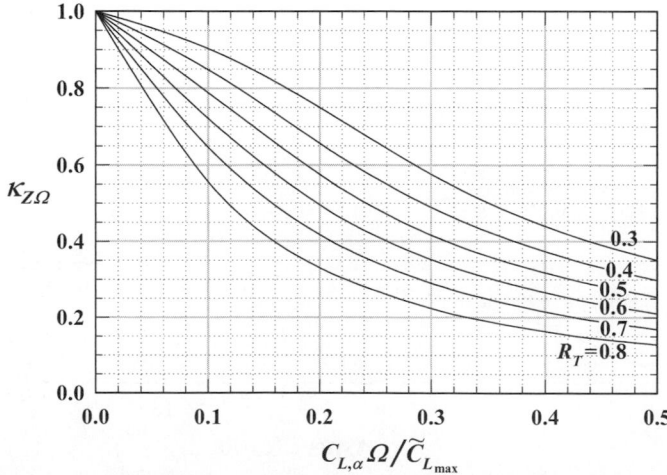

Fig. 59 Twist coefficient $\kappa_{Z\Omega}$ to be used in Eq. (214) for wings with linear twist.

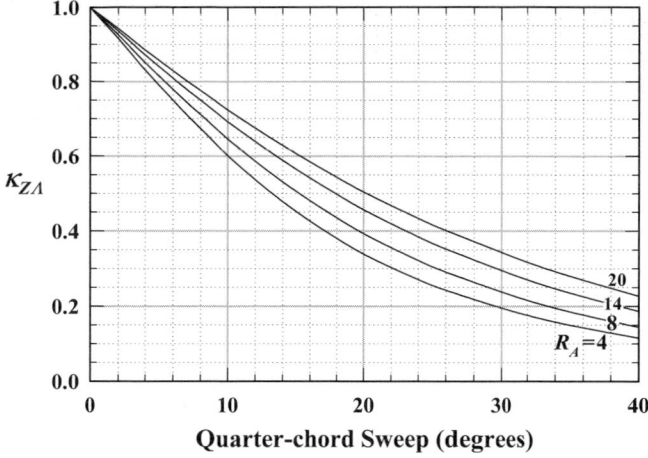

Fig. 60 Sweep coefficient $\kappa_{Z\Lambda}$ to be used in Eq. (214) for wings with linear taper.

9 INVISCID COMPRESSIBLE AERODYNAMICS

In the analytical methods that were reviewed in Sections 4–7, the variation in air density was assumed to be negligible. This is a reasonable approximation for flight Mach numbers less than about 0.3. For higher flight Mach numbers, compressibility effects become increasingly important. For flight speeds near or exceeding the speed of sound, the effects of compressibility vastly alter the airflow about an aircraft in flight. In this and the following sections, we review some of the important concepts associated with compressible aerodynamics.

As discussed in Section 3, inviscid flow over any body immersed in a uniform flow is irrotational. Any irrotational flow is a potential flow. There is no requirement that the flow be incompressible. Thus, the velocity field for inviscid flow about a body immersed in uniform flow can always be expressed according to Eq. (23):

$$\mathbf{V} = \nabla \phi \qquad (215)$$

where ϕ is the scalar velocity potential. From Eq. (14), the continuity equation for steady compressible flow is

$$(\mathbf{V} \cdot \nabla)\rho + \rho \nabla \cdot \mathbf{V} = 0 \qquad (216)$$

Inviscid adiabatic flow is isentropic. For an ideal gas this requires that

$$\frac{p}{p_0} = \left(\frac{\rho}{\rho_0}\right)^\gamma \qquad (217)$$

where p_0 and ρ_0 are the stagnation pressure and density. For steady isentropic flow of an ideal gas,

the momentum equation is expressed in Eq. (28):

$$\frac{1}{2}V^2 + \frac{\gamma}{\gamma - 1}\frac{p}{\rho} = \frac{\gamma}{\gamma - 1}\frac{p_0}{\rho_0} \qquad (218)$$

For an ideal gas

$$\frac{p}{\rho} = RT \qquad (219)$$

$$a^2 = \gamma RT \qquad (220)$$

Substituting Eqs. (219) and (220) into Eq. (217) gives

$$\frac{p}{p_0} = \frac{RT\rho}{RT_0\rho_0} = \frac{\gamma RT\rho}{\gamma RT_0\rho_0} = \frac{a^2\rho}{a_0^2\rho_0} = \left(\frac{\rho}{\rho_0}\right)^\gamma \qquad (221)$$

Similarly, applying Eqs. (219) and (220) to Eq. (218) yields

$$\frac{1}{2}V^2 + \frac{a^2}{\gamma - 1} = \frac{a_0^2}{\gamma - 1} \qquad (222)$$

Solving Eq. (221) for ρ and solving Eq. (222) for the speed of sound squared results in

$$\rho = \rho_0 \left(\frac{a^2}{a_0^2}\right)^{1/(\gamma - 1)} \qquad (223)$$

$$a^2 = a_0^2 - \frac{\gamma - 1}{2}V^2 \qquad (224)$$

Substituting Eq. (224) into Eq. (223), we have

$$\rho = \rho_0 \left(1 - \frac{\gamma - 1}{2}\frac{V^2}{a_0^2}\right)^{1/(\gamma - 1)} \qquad (225)$$

This specifies the local air density as a function of the local velocity and the stagnation conditions, which are known from the free-stream conditions according to Eqs. (32)–(34).

Applying Eq. (215) to express the velocity in Eq. (225) in terms of the velocity potential, the local air density is found to be a function only of ϕ and known stagnation properties of the flow:

$$\rho = \rho_0 \left(1 - \frac{\gamma - 1}{2a_0^2} \mathbf{V} \cdot \mathbf{V} \right)^{1/(\gamma - 1)}$$

$$= \rho_0 \left(1 - \frac{\gamma - 1}{2a_0^2} \nabla\phi \cdot \nabla\phi \right)^{1/(\gamma - 1)} \tag{226}$$

After using Eqs. (215) and (226) in (216), the *continuity equation for steady compressible potential flow* can be written as

$$\left(a_0^2 - \frac{\gamma - 1}{2} \nabla\phi \cdot \nabla\phi \right) \nabla^2\phi$$

$$- \frac{1}{2}(\nabla\phi \cdot \nabla)(\nabla\phi \cdot \nabla\phi) = 0 \tag{227}$$

This equation contains only one unknown, the velocity potential ϕ. Both γ and a_0 are known constants of the flow. Since Eq. (227) is a single scalar equation in only one scalar unknown, it provides a tremendous simplification over the more general Navier–Stokes equations. However, like the Navier–Stokes equations but unlike the Laplace equation that governs the velocity potential for incompressible flow, Eq. (227) is nonlinear. Once the velocity potential has been determined from Eq. (227), the velocity field can be evaluated from Eq. (215). With the velocity known, the speed of sound at any point in the flow can be determined from Eq. (224). Knowing the local velocity and speed of sound at every point in the flow allows us to compute the Mach number and then, using Eqs. (29)–(31), to evaluate the pressure, temperature, and air density.

It is sometimes useful to apply a change of variables in Eq. (227). Here we shall define a new velocity vector to be the difference between the local velocity and the free-stream velocity,

$$\mathbf{V}_p \equiv \mathbf{V} - \mathbf{V}_\infty \tag{228}$$

This is commonly called the *perturbation velocity*. From this definition, we also define the *perturbation velocity potential*,

$$\phi_p \equiv \phi - \phi_\infty \tag{229}$$

where ϕ_∞ is the velocity potential for the uniform flow. For the uniform flow potential we have

$$\nabla\phi_\infty = \mathbf{V}_\infty \tag{230}$$

$$\nabla\phi_\infty \cdot \nabla\phi_\infty = V_\infty^2 \tag{231}$$

$$\nabla^2\phi_\infty = 0 \tag{232}$$

Using Eqs. (229)–(232) in Eq. (227) results in

$$\left[a_0^2 - \frac{\gamma - 1}{2} V_\infty^2 \left(1 + 2\mathbf{u}_\infty \cdot \nabla\hat{\phi}_p \right. \right.$$

$$\left. \left. + \nabla\hat{\phi}_p \cdot \nabla\hat{\phi}_p \right) \right] \nabla^2\hat{\phi}_p$$

$$- \frac{1}{2} V_\infty^2 \left[(\nabla\hat{\phi}_p + \mathbf{u}_\infty) \cdot \nabla \right] (2\mathbf{u}_\infty \cdot \nabla\hat{\phi}_p$$

$$+ \nabla\hat{\phi}_p \cdot \nabla\hat{\phi}_p) = 0 \tag{233}$$

where $\mathbf{u}_\infty$ is the unit vector in the direction of the free stream

$$\mathbf{u}_\infty \equiv \frac{\mathbf{V}_\infty}{V_\infty} \tag{234}$$

and

$$\hat{\phi}_p \equiv \frac{\phi_p}{V_\infty} \tag{235}$$

Equation (233) is the general equation for the perturbation velocity potential. To this point in the development, no approximation has been made in going from Eq. (227) to Eq. (233). This result applies to any irrotational flow.

Equation (233) can be linearized under the assumption that the perturbation velocity is small compared to the free-stream velocity. This approximation results in

$$\left(a_0^2 - \frac{\gamma - 1}{2} V_\infty^2 \right) \nabla^2\hat{\phi}_p$$

$$- V_\infty^2 (\mathbf{u}_\infty \cdot \nabla)(\mathbf{u}_\infty \cdot \nabla\hat{\phi}_p) = 0 \tag{236}$$

Applying Eq. (224) to the free stream gives

$$a_\infty^2 = a_0^2 - \frac{\gamma - 1}{2} V_\infty^2 \tag{237}$$

Substituting Eq. (237) in Eq. (236) and dividing through by a_∞^2, the *linearized equation for the perturbation velocity potential* is

$$\nabla^2\phi_p - M_\infty^2 (\mathbf{u}_\infty \cdot \nabla)(\mathbf{u}_\infty \cdot \nabla\phi_p) = 0 \tag{238}$$

Notice that as the flight Mach number approaches zero this result reduces to Laplace's equation, which applies to incompressible flow. Equation (238) provides reasonable predictions for slender bodies at low angles of attack with subsonic Mach numbers ($M_\infty < 0.8$) and supersonic Mach numbers ($1.2 < M_\infty < 5$). It is not valid for transonic Mach numbers ($0.8 < M_\infty < 1.2$) or hypersonic Mach numbers ($M_\infty > 5$).

10 COMPRESSIBLE SUBSONIC FLOW

The aerodynamic theory of incompressible flow over airfoil sections was reviewed in Sections 4 and 5. The analytical methods that have been developed for incompressible flow over airfoils can be applied to subsonic compressible flow through a simple change of variables. However, the reader should recall that subsonic flow does not exist all the way to a flight Mach number of 1.0. For an airfoil section producing positive lift, the flow velocities just outside the boundary layer on the upper surface are greater than the free-stream velocity. Thus, at some flight speeds below Mach 1.0, supersonic flow will be encountered in some region above the upper surface of the airfoil. The flight Mach number at which sonic flow is first encountered at some point on the upper surface of the airfoil is called the *critical Mach number*. Here we review the theory of compressible flow over airfoils at flight speeds below the critical Mach number.

10.1 Prandtl–Glauert Compressibility Correction

For thin airfoils at small angles of attack and flight Mach numbers below critical, the linearized potential flow approximation given by Eq. (238) can be applied. For the special case of 2D flow and a coordinate system having the x axis aligned with the free-stream velocity vector, $\mathbf{u}_\infty$ is simply the unit vector in the x direction, and the continuity equation as expressed in Eq. (238) becomes

$$\left(\frac{\partial^2\phi_p}{\partial x^2} + \frac{\partial^2\phi_p}{\partial y^2}\right) - M_\infty^2\left(\frac{\partial}{\partial x}\right)\left(\frac{\partial\phi_p}{\partial x}\right) = 0$$

or after rearranging,

$$\frac{\partial^2\phi_p}{\partial x^2} + \frac{1}{1-M_\infty^2}\frac{\partial^2\phi_p}{\partial y^2} = 0 \qquad (239)$$

The nature of this partial differential equation depends on the sign of $1 - M_\infty^2$. Recall from your introductory course on differential equations that if the flight Mach number is greater than 1.0, Eq. (239) is hyperbolic. However, for subsonic Mach numbers this partial differential equation is elliptic.

For subsonic flight, Eq. (239) is simplified by using the change of independent variables

$$\hat{y} \equiv y\sqrt{1-M_\infty^2} \qquad (240)$$

With this change of variables, Eq. (239) becomes

$$\frac{\partial^2\phi_p}{\partial x^2} + \frac{\partial^2\phi_p}{\partial \hat{y}^2} = 0 \qquad (241)$$

which is exactly Laplace's equation that governs incompressible potential flow. Thus, with this coordinate

transformation, the incompressible flow solution can be used for compressible flow.

The local pressure coefficient obtained from any solution to Eq. (241) is given by

$$C_p = \frac{C_{pM0}}{\sqrt{1-M_\infty^2}} \qquad (242)$$

where C_{pM0} is the pressure coefficient for zero Mach number, which is the solution obtained for incompressible flow. Equation (242) relates the pressure coefficient for subsonic compressible flow to that for incompressible flow and is commonly referred to as the *Prandtl–Glauert compressibility correction*.

For inviscid flow, the section lift and moment coefficients can be found from simple integrals of the pressure distribution over the surface of the airfoil. As a result, the final result for section lift and moment coefficients look much like Eq. (242):

$$\widetilde{C}_L = \frac{\widetilde{C}_{LM0}}{\sqrt{1-M_\infty^2}} \qquad (243)$$

$$\widetilde{C}_m = \frac{\widetilde{C}_{mM0}}{\sqrt{1-M_\infty^2}} \qquad (244)$$

where $\widetilde{C}_{LM0}$ and $\widetilde{C}_{mM0}$ are the section lift and moment coefficients obtained for incompressible flow. The development of Eqs. (242)–(244) can be found in any undergraduate engineering textbook on aerodynamics and will not be repeated here.

10.2 Critical Mach Number

For potential flow over an airfoil producing positive lift, the flow velocities along some portion of the upper surface are greater than the free-stream velocity. Also recall that for incompressible potential flow, the ratio of local velocity to the free-stream velocity at any point in the flow does not vary with the magnitude of the free-stream velocity. In other words, if we double the free-stream velocity in an incompressible potential flow, the velocity at every point in the flow field will double as well. This is a direct result of the linear nature of the Laplace equation, which governs incompressible potential flow. Thus, within the approximation of Eq. (239), the position of the point of maximum velocity on the upper surface of the airfoil does not change with free-stream velocity.

From the definition of pressure coefficient, we can write

$$C_p \equiv \frac{p-p_\infty}{\frac{1}{2}\rho_\infty V_\infty^2} = \frac{2}{(\rho_\infty/p_\infty)V_\infty^2}\left(\frac{p}{p_\infty}-1\right)$$

$$= \frac{2}{(\gamma/\gamma RT_\infty)V_\infty^2}\left(\frac{p}{p_\infty}-1\right) = \frac{2}{\gamma M_\infty^2}\left(\frac{p}{p_\infty}-1\right) \qquad (245)$$

From Eq. (30),

$$\frac{p}{p_\infty} = \left(\frac{1 + [(\gamma - 1)/2]M_\infty^2}{1 + [(\gamma - 1)/2]M^2}\right)^{\gamma/(\gamma - 1)} \quad (246)$$

Substituting Eq. (246) into Eq. (245), the pressure coefficient can be expressed as a function of Mach number,

$$C_p = \frac{2}{\gamma M_\infty^2}\left[\left(\frac{1 + [(\gamma - 1)/2]M_\infty^2}{1 + [(\gamma - 1)/2]M^2}\right)^{\gamma/(\gamma - 1)} - 1\right] \quad (247)$$

This relation allows us to compute the local pressure coefficient at any point in the flow from the local Mach number and the free-stream Mach number.

The critical Mach number M_{cr} is the free-stream Mach number that results in a local Mach number of 1.0 at that point on the upper surface where the pressure coefficient is lowest. Using this fact together with Eqs. (242) and (247) gives the relation

$$\frac{(C_{pM0})_{min}}{\sqrt{1 - M_{cr}^2}}$$

$$= \frac{2}{\gamma M_{cr}^2}\left[\left(\frac{1 + [(\gamma - 1)/2]M_{cr}^2}{1 + [(\gamma - 1)/2]}\right)^{\gamma/(\gamma - 1)} - 1\right] \quad (248)$$

The minimum pressure coefficient on the airfoil surface for incompressible flow can be evaluated using a panel code. With this pressure coefficient known, Eq. (248) contains only one unknown, the critical Mach number. This is easily solved for the value of the critical Mach number by using the secant method.

10.3 Drag Divergence

While the lift and pitching moment produced on an airfoil section are generated primarily from pressure forces, at subsonic speeds the drag is attributed almost entirely to viscous forces generated in the boundary layer. Below the critical Mach number, airfoil section drag coefficient does not vary substantially with Mach number. However, beyond the critical Mach number the drag coefficient begins to increase very rapidly with increasing Mach number, reaching a maximum at Mach 1 as shown in Fig. 61. This figure is intended only to show the general shape of the drag curve that might be expected for some airfoil. The exact shape of this drag curve and the total drag rise for a particular airfoil is quite sensitive to the shape of the airfoil section. There is no simple theory capable of predicting the variation in section drag coefficient with Mach number in the transonic region. At Mach numbers near 1.0, the flow over an airfoil is extremely complex and very sensitive to section shape. In the transonic region we commonly rely on experimental data.

The large increase in drag that occurs at Mach numbers just below 1.0, called *drag divergence*, is due primarily to the shock waves that form in transonic flow and the premature boundary layer separation caused by these shocks. The dramatic increase in drag that is experienced by an airplane as it approaches Mach 1 is what gave rise to the concept and terminology of the *sound barrier*. Some early researchers believed that the drag would become infinite as the speed of sound was approached. Thus, it was thought by some that it would be impossible to exceed the speed of sound in an aircraft. As can be imagined from inspection of Fig. 61, the speed of sound does present a substantial barrier to an aircraft with limited thrust.

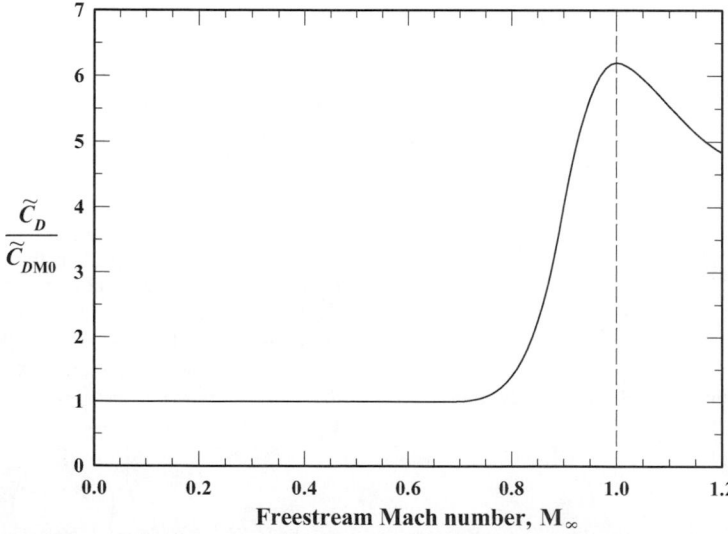

Fig. 61 Sketch of typical variation in section drag coefficient with Mach number.

Nevertheless, the drag at Mach 1 is finite, and like most things in this universe, the sound barrier will yield to sufficient power. It will yield more easily, however, if finesse is applied with the power.

In the last half-century a great deal of "engineering finesse" has been applied to help "lower the sound barrier" by increasing the critical Mach number and decreasing the peak drag experienced near Mach 1. Contributions that have led to the design of improved supersonic aircraft include the use of thinner airfoils and swept wings, the introduction of the *area rule* in the supersonic design philosophy, and development of the supercritical airfoil. All of these topics are covered in most aerodynamics textbooks and the reader is encouraged to review this material, if necessary.

11 SUPERSONIC FLOW

In an irrotational supersonic flow where the streamlines make only small angles with the free stream, the linearized potential flow approximation given by Eq. (238) can be applied. Again consider 2D flow with the x axis aligned with the free-stream velocity vector. These are exactly the conditions that were used in Section 10 to study subsonic compressible flow over an airfoil. Thus, Eq. (239) applies to supersonic flow at small angles as well as to subsonic flow. However, since our interest is now in Mach numbers greater than 1.0, Eq. (239) is more conveniently written as

$$\frac{\partial^2 \phi_p}{\partial x^2} - \frac{1}{M_\infty^2 - 1} \frac{\partial^2 \phi_p}{\partial y^2} = 0 \qquad (249)$$

Solutions to this hyperbolic partial differential equation are characteristically very different from solution to the elliptic equation encountered for subsonic flow. Thus, we should expect supersonic flow to be very different from subsonic flow.

In obtaining a solution to Eq. (249), a change of variables is again useful. We wish to find a variable change that will allow us to separate the independent variables, x and y, in Eq. (249). This is a common technique used in obtaining solutions to partial differential equations. We want to find a change of variables, say $\xi(x, y)$, that will make the perturbation potential a function of ξ only, that is, $\phi_p(\xi)$. With this in mind, Eq. (249) can be written as

$$\frac{d^2 \phi_p}{d\xi^2} \left(\frac{\partial \xi}{\partial x} \right)^2 + \frac{d\phi_p}{d\xi} \frac{\partial^2 \xi}{\partial x^2}$$

$$- \frac{1}{M_\infty^2 - 1} \left[\frac{d^2 \phi_p}{d\xi^2} \left(\frac{\partial \xi}{\partial y} \right)^2 + \frac{d\phi_p}{d\xi} \frac{\partial^2 \xi}{\partial y^2} \right] = 0$$

$$(250)$$

Equation (250) is satisfied for ϕ_p equal to any function of ξ if

$$\frac{\partial^2 \xi}{\partial x^2} = \frac{1}{M_\infty^2 - 1} \frac{\partial^2 \xi}{\partial y^2} \quad \text{and} \quad \frac{\partial \xi}{\partial x} = \pm \frac{1}{\sqrt{M_\infty^2 - 1}} \frac{\partial \xi}{\partial y}$$

Both of these relations are satisfied if we set both sides of the second relation equal to an arbitrary constant, say C_1. This gives

$$\frac{\partial \xi}{\partial x} = C_1 \quad \text{and} \quad \frac{\partial \xi}{\partial y} = \pm C_1 \sqrt{M_\infty^2 - 1}$$

These equations are easily integrated to yield

$$\xi(x, y) = C_1 x + f_1(y) \quad \text{and}$$

$$\xi(x, y) = \pm C_1 \sqrt{M_\infty^2 - 1}\, y + f_2(x)$$

which require that

$$\xi(x, y) = C_1 \left(x \pm \sqrt{M_\infty^2 - 1}\, y \right) + C_2$$

Because C_1 and C_2 are arbitrary, we can choose $C_1 = 1$ and $C_2 = 0$, which gives

$$\xi(x, y) = x \pm \sqrt{M_\infty^2 - 1}\, y \qquad (251)$$

Either of the two solutions for ξ that are given in Eq. (251) will provide the desired variable change. With either definition, any function $\phi_p(\xi)$ will satisfy Eq. (249). Thus, any problem associated with small-angle potential flow at supersonic Mach numbers reduces to that of finding the function $\phi_p(\xi)$ that will also satisfy the required boundary conditions.

From Eq. (53), a great deal can be deduced about supersonic potential flow at small angles without applying any boundary conditions. Since the perturbation potential is a function only of ξ, lines of constant perturbation potential are lines of constant ξ. Thus, the *constant perturbation potential lines* are described by the equation

$$x \pm \sqrt{M_\infty^2 - 1}\, y = \text{const} \qquad (252)$$

Since this is the equation of a straight line, the constant perturbation potential lines are all straight lines. Furthermore, solving Eq. (252) for y and differentiating with respect to x, we find that the *slope of any constant perturbation potential line* is

$$\left(\frac{\partial y}{\partial x} \right)_{\phi_p = \text{const}} = \frac{\mp 1}{\sqrt{M_\infty^2 - 1}} \qquad (253)$$

The angle that the constant perturbation potential lines makes with the free stream (i.e., the x axis) is called the *Mach angle* and is given by

$$\mu = \mp \tan^{-1} \left(\frac{1}{\sqrt{M_\infty^2 - 1}} \right) \qquad (254)$$

Notice that for supersonic potential flow at small angles, the Mach angle is independent of surface geometry. It depends only on the free-stream Mach number. Also notice that there are two solutions for the Mach angle and that the signs in Eqs. (253) and (254) are opposite to those in Eqs. (251) and (252). Thus, choosing the negative sign in Eq. (251) results in a positive Mach angle, and vice versa.

From the definition of perturbation potential, given in Eq. (229), the local fluid velocity vector is given by

$$\mathbf{V} = \nabla\phi = \nabla\phi_\infty + \nabla\phi_p = \mathbf{V}_\infty + \nabla\phi_p = V_\infty \mathbf{i}_x + \nabla\phi_p$$

Using the relation

$$\phi_p = \phi_p(\xi) = \phi_p\left(x \pm \sqrt{M_\infty^2 - 1}\, y\right)$$

we have

$$\mathbf{V} = \left(V_\infty + \frac{d\phi_p}{d\xi}\frac{\partial\xi}{\partial x}\right)\mathbf{i}_x + \frac{d\phi_p}{d\xi}\frac{\partial\xi}{\partial y}\mathbf{i}_y$$

$$= \left(V_\infty + \frac{d\phi_p}{d\xi}\right)\mathbf{i}_x \pm \sqrt{M_\infty^2 - 1}\,\frac{d\phi_p}{d\xi}\mathbf{i}_y \quad (255)$$

The slope of a streamline at any point in the flow can be evaluated from Eq. (255).

$$\frac{V_y}{V_x} = \pm\frac{\sqrt{M_\infty^2 - 1}\,(d\phi_p/d\xi)}{V_\infty + d\phi_p/d\xi}$$

Because we are considering the case of small perturbation velocity, only the first-order term in this result should be retained. Since the streamlines are tangent to the velocity vector, the *slope of the local streamline relative to the x axis* is

$$\left(\frac{\partial y}{\partial x}\right)_{\text{streamline}} = \pm\frac{\sqrt{M_\infty^2 - 1}}{V_\infty}\frac{d\phi_p}{d\xi} \quad (256)$$

From Eq. (255), the local velocity squared is

$$V^2 = \left(V_\infty + \frac{d\phi_p}{d\xi}\right)^2 + (M_\infty^2 - 1)\left(\frac{d\phi_p}{d\xi}\right)^2$$

$$= V_\infty^2 + 2V_\infty\frac{d\phi_p}{d\xi} + M_\infty^2\left(\frac{d\phi_p}{d\xi}\right)^2$$

Again, since we are considering the case of small perturbation velocity, the second-order term should be ignored and the *velocity at any point in the flow* is given by

$$V^2 = V_\infty^2 + 2V_\infty\frac{d\phi_p}{d\xi} \quad (257)$$

From the momentum equation for a compressible potential flow, which is given by Eq. (28), we can

express the local temperature in terms of the local velocity. Using the ideal gas law in Eq. (28) at an arbitrary point and the free stream gives

$$\frac{1}{2}V^2 + \frac{\gamma RT}{\gamma - 1} = \frac{1}{2}V_\infty^2 + \frac{\gamma RT_\infty}{\gamma - 1}$$

Solving this for temperature, using Eq. (257) to eliminate the velocity, and applying the relation $a^2 = \gamma RT$ gives the *temperature* at any point in the flow:

$$T = T_\infty\left[1 - (\gamma - 1)\frac{M_\infty}{a_\infty}\frac{d\phi_p}{d\xi}\right] \quad (258)$$

From the isentropic relations given in Eqs. (29) and (30), the pressure ratio is easily related to the temperature ratio and after applying Eq. (258), we obtain

$$\frac{p}{p_\infty} = \left(\frac{T}{T_\infty}\right)^{\gamma/(\gamma-1)}$$

$$= \left[1 - (\gamma - 1)\frac{M_\infty}{a_\infty}\frac{d\phi_p}{d\xi}\right]^{\gamma/(\gamma-1)}$$

Expanding this result in a Taylor series and retaining only the first-order term results in an expression for the *pressure* at any point in the flow:

$$p = p_\infty\left(1 - \gamma\frac{M_\infty}{a_\infty}\frac{d\phi_p}{d\xi}\right) \quad (259)$$

Using Eq. (259) in the definition of pressure coefficient as expressed in Eq. (245) results in

$$C_p = \frac{2}{\gamma M_\infty^2}\left(\frac{p}{p_\infty} - 1\right) = -\frac{2}{V_\infty}\frac{d\phi_p}{d\xi} \quad (260)$$

Solving Eq. (256) for the derivative of the perturbation potential and using Eq. (254) to eliminate the free-stream Mach number in favor of the Mach angle gives

$$\frac{d\phi_p}{d\xi} = \pm\frac{V_\infty}{\sqrt{M_\infty^2 - 1}}\left(\frac{\partial y}{\partial x}\right)_{\text{streamline}}$$

$$= -V_\infty\tan(\mu)\left(\frac{\partial y}{\partial x}\right)_{\text{streamline}} \quad (261)$$

Substituting Eq. (261) into Eq. (260), the *pressure coefficient at any point in a small-angle potential flow at supersonic Mach numbers* can be expressed as a function of only the Mach angle and the slope of the local streamline:

$$C_p = 2\,\tan(\mu)\left(\frac{\partial y}{\partial x}\right)_{\text{streamline}} \quad (262)$$

Remember that the Mach angle can be chosen as either positive or negative as needed to satisfy the required boundary conditions. However, the positive sign in Eq. (262) holds regardless of the sign chosen for the Mach angle. If the streamline slope has the same sign as the Mach angle, the pressure coefficient is positive. If these signs are opposite, the pressure coefficient is negative.

For those of us interested in computing aerodynamic forces, this is a very simple result in comparison to most solutions that have been obtained for fluid flow. In this respect, supersonic potential flow at small angles is less complex than subsonic flow at small angles. It rivals Bernoulli's equation for raw simplicity. The Mach angle depends only on the free-stream Mach number. Furthermore, at the surface of a solid body, the streamlines must be tangent to the surface. Thus, the slope of the surface streamlines depends only on the surface geometry and the angle of attack.

11.1 Supersonic Thin Airfoils

The supersonic small-angle potential flow equations can be used to predict the aerodynamic force and moment components acting on a thin airfoil at small angles of attack, provided that the airfoil has a sharp leading edge as shown in Fig. 62. This type of leading edge is commonly used on supersonic airfoils because a blunt leading edge produces a strong bow shock in supersonic flight.

Consider supersonic flow over the cambered airfoil shown in Fig. 62. Because the flow must be symmetric for a symmetric airfoil at zero angle of attack, the positive slope is chosen for the Mach lines above the airfoil and the negative slope is chosen for those Mach lines below the airfoil. The slope of the upper surface relative to the free stream can be expressed as the slope of the chord line relative to the free stream, plus the slope of the camber line relative to the chord line, plus the slope of the thickness line relative to the camber line. Thus, for small angles of attack, the *upper surface slope* is

$$\frac{dy_u}{dx} = -\alpha + \frac{dy_c}{dx} + \frac{dy_t}{dx} \tag{263}$$

where y_c is the local camber and y_t is one-half the local thickness. Similarly, the *lower surface slope* is

$$\frac{dy_l}{dx} = -\alpha + \frac{dy_c}{dx} - \frac{dy_t}{dx} \tag{264}$$

Since the streamlines must be tangent to the airfoil surface, the results above can be used in Eq. (162) to evaluate the pressure coefficient on the airfoil surface.

The contribution of pressure on the lower surface to the section lift is the surface pressure multiplied by a differential area per unit span projected on the x axis. Similarly, the upper surface contributes to the lift in just the opposite direction. Thus, the section lift coefficient is

$$\tilde{C}_L = \frac{\tilde{L}}{\frac{1}{2}\rho_\infty V_\infty^2 c} = \int_0^c \frac{p_l - p_u}{\frac{1}{2}\rho_\infty V_\infty^2} \frac{dx}{c}$$

$$= \int_0^c (C_{pl} - C_{pu}) \frac{dx}{c} \tag{265}$$

The slope of the Mach lines is positive on the upper surface and negative on the lower surface. Thus, from Eq. (254),

$$\tan(\mu_l) = \frac{-1}{\sqrt{M_\infty^2 - 1}} \quad \text{and} \quad \tan(\mu_u) = \frac{+1}{\sqrt{M_\infty^2 - 1}}$$

Using these results with Eqs. (262)–(264) applied to Eq. (265) gives

$$\tilde{C}_L = \int_0^c \left[2\frac{-1}{\sqrt{M_\infty^2 - 1}} \left(-\alpha + \frac{dy_c}{dx} - \frac{dy_t}{dx} \right) \right.$$
$$\left. - 2\frac{+1}{\sqrt{M_\infty^2 - 1}} \left(-\alpha + \frac{dy_c}{dx} + \frac{dy_t}{dx} \right) \right] \frac{dx}{c}$$

$$= \frac{4}{\sqrt{M_\infty^2 - 1}} \int_0^c \left(\alpha - \frac{dy_c}{dx} \right) \frac{dx}{c}$$

$$= \frac{4}{\sqrt{M_\infty^2 - 1}} \left[\alpha - \frac{y_c(c) - y_c(0)}{c} \right]$$

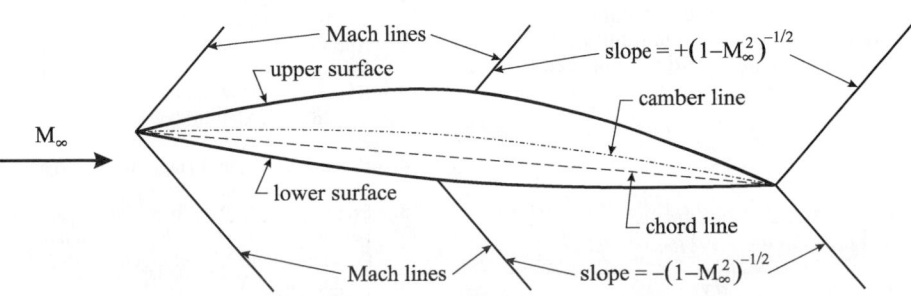

Fig. 62 Small-angle supersonic potential flow over an airfoil.

By definition, the camber of any airfoil is zero at both the leading and trailing edges. Thus, the *section lift coefficient* is

$$\widetilde{C}_L = \frac{4\alpha}{\sqrt{M_\infty^2 - 1}} = \widetilde{C}_{L,\alpha}\alpha \quad \text{where} \quad \widetilde{C}_{L,\alpha} = \frac{4}{\sqrt{M_\infty^2 - 1}}$$
(266)

From this result, we see that the small-angle potential flow equations predict that neither camber nor thickness makes a contribution to the lift at supersonic airspeeds.

Similarly, the pitching moment coefficient about the leading edge is evaluated from

$$\widetilde{C}_{m_{le}} = \frac{\widetilde{m}_{le}}{\frac{1}{2}\rho_\infty V_\infty^2 c^2} = \int_0^c (C_{pu} - C_{pl})\frac{x\,dx}{c^2}$$

which gives

$$\widetilde{C}_{m_{le}} = \frac{-4}{\sqrt{M_\infty^2 - 1}}\int_0^c \left(\alpha - \frac{dy_c}{dx}\right)\frac{x\,dx}{c^2}$$

$$= \frac{-2\alpha}{\sqrt{M_\infty^2 - 1}} + \frac{4}{\sqrt{M_\infty^2 - 1}}\left(\int_0^c \frac{dy_c}{dx}\frac{x\,dx}{c^2}\right)$$

or

$$\widetilde{C}_{m_{le}} = -\frac{1}{2}\widetilde{C}_L + \widetilde{C}_{L,\alpha}\int_0^c \frac{dy_c}{dx}\frac{x\,dx}{c^2}$$

From this result and the material that was reviewed in Section 1, it can be seen that the aerodynamic center is located at the half chord and the *pitching moment coefficient* about the aerodynamic center is

$$\widetilde{C}_{m_{ac}} = \widetilde{C}_{L,\alpha}\int_0^c \frac{dy_c}{dx}\frac{x\,dx}{c^2} \quad \text{where} \quad \frac{x_{ac}}{c} = \frac{1}{2} \quad (267)$$

The moment coefficient about the aerodynamic center only depends on the camber line shape. Since camber does not contribute to lift in small-angle supersonic potential flow, symmetric airfoils are often used for supersonic flight. Notice from Eq. (267) that a thin symmetric airfoil produces no moment about its half chord. Thus, the half chord is also the center of pressure for a thin symmetric airfoil in supersonic flight.

The section drag for supersonic potential flow over a thin airfoil is not zero as it is for subsonic flow. This section drag is called *wave drag* and is computed from

$$\widetilde{C}_D = \frac{\widetilde{D}}{\frac{1}{2}\rho_\infty V_\infty^2 c}$$

$$= \int_0^c \left(\frac{p_u - p_\infty}{\frac{1}{2}\rho_\infty V_\infty^2}\frac{dy_u}{dx} - \frac{p_l - p_\infty}{\frac{1}{2}\rho_\infty V_\infty^2}\frac{dy_l}{dx}\right)\frac{dx}{c}$$

$$= \int_0^c \left(C_{pu}\frac{dy_u}{dx} - C_{pl}\frac{dy_l}{dx}\right)\frac{dx}{c}$$

Applying Eqs. (254) and (262)–(264) gives

$$\widetilde{C}_D = \frac{4}{\sqrt{M_\infty^2 - 1}}$$

$$\times \left\{\alpha^2 + \int_0^c \left[\left(\frac{dy_c}{dx}\right)^2 + \left(\frac{dy_t}{dx}\right)^2\right]\frac{dx}{c}\right\}$$

This result can be written in terms of the lift slope that was evaluated previously and presented in Eq. (266). Thus, the *section wave drag coefficient* is

$$\widetilde{C}_D = \widetilde{C}_{L,\alpha}\alpha^2 + \widetilde{C}_{L,\alpha}\int_0^c \left[\left(\frac{dy_c}{dx}\right)^2 + \left(\frac{dy_t}{dx}\right)^2\right]\frac{dx}{c}$$
(268)

From the results presented in Eqs. (266)–(268), we see that small-angle potential flow theory predicts that there is no advantage to using camber in a supersonic airfoil. It only contributes to the pitching moment and the drag, making no predicted contribution to the lift. Furthermore, we see that airfoil thickness makes a predicted contribution to wave drag that is proportional to thickness squared. This is one reason why thin airfoil sections are commonly used on supersonic aircraft. Most modern airplanes that are designed to fly at Mach 1.5 and above use airfoils having a thickness on the order of 4%. This approximate linearized supersonic airfoil theory agrees reasonably well with experimental observations for airfoils as thick as about 10% and angles of attack as large as 20°. As should be expected, this inviscid theory underpredicts the drag and moment coefficients and overpredicts the lift coefficient. However, the discrepancy is only a few percent. Most of the discrepancy can be attributed to viscous effects in the boundary layer. The drag that is predicted by this linear theory includes only wave drag. If a contribution for viscous drag is included from boundary layer computations, the predictions are improved. However, the viscous drag is quite small compared to the wave drag and for many purposes can be ignored.

REFERENCES

1. Karamcheti, K., *Principles of Ideal-Fluid Aerodynamics*, Wiley, New York, 1966.
2. Abbott, I. H., and Von Doenhoff, A. E., *Theory of Wing Sections*, Dover Publications, New York, 1949.
3. Katz, J., and Plotkin, A., *Low-Speed Aerodynamics*, 2nd ed., Cambridge University Press, Cambridge, 2001.
4. Perkins, C. D., and Hage, R. E., *Airplane Performance Stability & Control*, Wiley, New York, 1949.
5. Prandtl, L., "Tragflügel Theorie," in *Nachricten von der Gesellschaf der Wisseschaften zu Göttingen*, Geschäeftliche Mitteilungen, Klasse, 1918.
6. Kutta, M. W., "Auftriebskräfte in Strömenden Flüssigkeiten," *Illustr. Aeronaut. Mitteil.*, **6**, 133 (1902).

7. Joukowski, N. E., "Sur les Tourbillons Adjionts," *Traraux Sect. Phys. Soci. Imperiale Amis Sci. Nat.*, **13**, 2 (1906).

8. Saffman, P. G., *Vortex Dynamics*, Cambridge University Press, Cambridge, 1992.

9. Glauert, H., *The Elements of Aerofoil and Airscrew Theory*, Cambridge University Press, London, 1926.

10. Phillips, W. F., *Mechanics of Flight*, Wiley, Hoboken, NJ, 2004.

11. Phillips, W. F., "Lifting-Line Analysis for Twisted Wings and Washout-Optimized Wings," *J. Aircraft*, **41**, 1 (2004).

12. Phillips, W. F., Alley, N. R., and Goodrich, W. D., "Lifting-Line Analysis of Roll Control and Variable Twist," *J. Aircraft*, **41**, 5 (2004).

13. Phillips, W. F., Fugal, S. R., and Spall, R. E., "Minimizing Induced Drag with Wing Twist, Computational-Fluid-Dynamics Validation," *J. Aircraft*, **43**, 2 (2006).

14. Phillips, W. F., and Snyder, D. O., "Modern Adaptation of Prandtl's Classic Lifting-Line Theory," *J. Aircraft*, **37**, 4 (2000).

15. McAlister, K. W., and Takahashi, R. K., "NACA 0015 Wing Pressure and Trailing Vortex Measurements," NASA TP 3151, 1991.

16. Weber, J., and Brebner, G. G., "Low-Speed Tests on 45-deg Swept-Back Wings, Part I: Pressure Measurements on Wings of Aspect Ratio 5," *Reports and Memoranda* 2882, British Aeronautical Research Council, London, 1958.

17. Phillips, W. F., and Alley, N. R., "Predicting Maximum Lift Coefficient for Twisted Wings Using Lifting-Line Theory," *J. Aircraft*, **44**, 3 (2007).

18. Alley, N. R., Phillips, W. F., and Spall, R. E., "Predicting Maximum Lift Coefficient for Twisted Wings Using Computational Fluid Dynamics," *J. Aircraft*, **44**, 3 (2007).

CHAPTER 11

STEADY ONE-DIMENSIONAL GAS DYNAMICS

D. H. Daley with contributions by J. B. Wissler
Department of Aeronautics
United States Air Force Academy
Colorado Springs, Colorado

1 GENERALIZED ONE-DIMENSIONAL GAS DYNAMICS

The steady one-dimensional flow of a chemically inert perfect gas with constant specific heats is conveniently described and governed by the following definitions and physical laws.[1]

1.1 Definitions

Perfect gas:
$$p = \rho R T \qquad (1)$$

Mach number:
$$M = \frac{u}{a} \qquad (2)$$

Stagnation temperature:
$$T_0 = T\left[1 + \tfrac{1}{2}(\gamma - 1)M^2\right] \qquad (3)$$

Stagnation pressure:
$$p_0 = p\left[1 + \tfrac{1}{2}(\gamma - 1)M^2\right]^{\gamma/\gamma - 1} \qquad (4)$$

where p = pressure, ρ = density, R = gas constant, T = temperature, M = Mach number, u = velocity, a = speed of sound, T_0 = stagnation temperature, γ = ratio of specific heat at constant pressure to specific heat at constant volume, and p_0 = stagnation pressure. It is conventional to denote those stream properties at the point in the flow where $M = 1$ by p^*, u^*, and so on.

1.2 Physical Laws

For one-dimensional flow through a control volume having the single inlet and exit flow Sections 1 and 2, respectively, we have

Continuity equation:
$$\rho_1 A_1 u_1 = \rho_2 A_2 u_2 \qquad (5)$$

Momentum equation:
$$F_{\text{frict}} = \left(pA + \rho A u^2\right)_1 - \left(pA + \rho A u^2\right)_2 \qquad (6)$$

Energy equation:
$$q = c_p(T_{02} - T_{01}) \qquad \text{(shaft work} = 0) \qquad (7)$$

Entropy equation:
$$s_2 \geq s_1 \qquad \text{(adiabatic flow)} \qquad (8)$$

where F_{frict} = frictional force of a solid control surface boundary on the flowing gas, A = the flow cross-sectional area normal to u, q = the heat flow per unit mass flow, c_p = specific heat at constant pressure, and s = entropy per unit mass flow.

The application of Eqs. (1)–(6) to flow in the presence of the simultaneous effects of area-change, heating, and friction (Fig. 1) results in the following set of equations:

Perfect gas:
$$\frac{dp}{p} - \frac{d\rho}{\rho} - \frac{dT}{T} = 0 \qquad (9)$$

Stagnation temperature:
$$\frac{dT}{T} + \frac{[(\gamma - 1)/2]M^2}{1 + [(\gamma - 1)/2]M^2} \times \frac{dM^2}{M^2} = \frac{dT_0}{T_0} \qquad (10)$$

Continuity:
$$\frac{d\rho}{\rho} + \frac{dA}{A} + \frac{du}{u} = 0 \qquad (11)$$

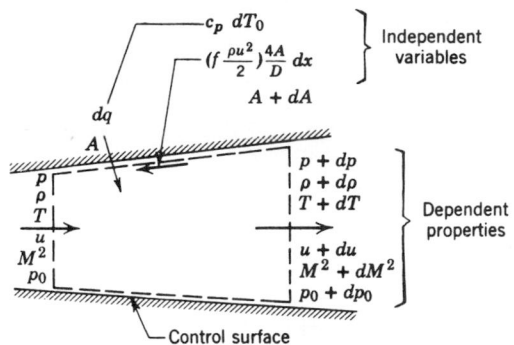

Fig. 1 Independent and dependent variables for generalized one-dimensional flow, where f = friction coefficient, D = hydraulic diameter.

Stagnation pressure:
$$\frac{dp}{p} + \frac{\gamma M^2/2}{1 + [(\gamma - 1)/2]M^2}\frac{dM^2}{M^2} = \frac{dp_0}{p_0} \tag{12}$$

Momentum:
$$\frac{dp}{p} + \gamma M^2 \frac{du}{u} + \frac{\gamma M^2}{2}\frac{4f\,dx}{D} = 0 \tag{13}$$

Mach number:
$$2\frac{du}{u} - \frac{dT}{T} = \frac{dM^2}{M^2} \tag{14}$$

In these equations heat effects are measured in terms of the stagnation temperature change according to Eq. (7). The entropy condition of Eq. (8) is also applicable if $dT_0 = 0$. If $dT_0 \neq 0$, then the entropy requirement is $ds \geq (dq/T)$.

The six dependent variables M^2, u, p, ρ, T, and p_0 in this set of six linear algebraic equations may be expressed in terms of the three independent variables, dA/A, dT_0/T_0, and $4f\,dx/D$. The solution of this set of equations is given in Table 1.

General conclusions can be made relative to the variation of the stream properties of the flow with each of the independent variables by the relations of Table 1.[1,2] An example of the relation given for (du/u) at the bottom of the table indicates that, in a constant area adiabatic flow, friction will increase the stream velocity in subsonic flow and will decrease the velocity in supersonic flow. Similar reasoning may be applied to determine the manner in which any dependent property varies with a single independent variable.

2   SIMPLE FLOWS

A simple flow is defined as one in which all but one of the independent variables in Table 1 are zero. Three types of simple flows are summarized in Fig. 2a by presenting for each simple flow (i) the independent effects present, (ii) a schematic of the flow situation, (iii) the locus on a temperature–entropy diagram of the possible states attained for each flow, and (iv) useful functions obtained by integration of the relations of Table 1 or the basic Eqs. (1)– (6). These useful functions are tabulated in Ref. 3.

In the temperature–entropy diagrams of Fig. 2a the path lines of states corresponding to simple area flow, simple heating flow, and simple friction flow, respectively, are shown. These path lines are called the isentrope, Rayleigh, and Fanno lines, respectively.

Table 1   Influence Coefficients for Steady One-Dimensional Flow

Dependent	Independent		
	$\dfrac{dA}{A}$	$\dfrac{dT_0}{T_0}$	$\dfrac{4f\,dx}{D}$
$\dfrac{dM^2}{M^2}$	$-\dfrac{2\left\{1 + [(\gamma - 1)/2]M^2\right\}}{1 - M^2}$	$\dfrac{(1 + \gamma M^2)\left\{1 + [(\gamma - 1)/2]M^2\right\}}{1 - M^2}$	$\dfrac{\gamma M^2\left\{1 + [(\gamma - 1)/2]M^2\right\}}{1 - M^2}$
$\dfrac{du}{u}$	$-\dfrac{1}{1 - M^2}$	$\dfrac{1 + [(\gamma - 1)/2]M^2}{1 - M^2}$	$\dfrac{\gamma M^2}{2(1 - M^2)}$
$\dfrac{dp}{p}$	$\dfrac{\gamma M^2}{1 - M^2}$	$\dfrac{-\gamma M^2\left\{1 + [(\gamma - 1)/2]M^2\right\}}{1 - M^2}$	$\dfrac{-\gamma M^2[1 + (\gamma - 1)M^2]}{2(1 - M^2)}$
$\dfrac{d\rho}{\rho}$	$\dfrac{M^2}{1 - M^2}$	$-\dfrac{1 + [(\gamma - 1)/2]M^2}{1 - M^2}$	$\dfrac{-\gamma M^2}{2(1 - M^2)}$
$\dfrac{dT}{T}$	$\dfrac{(\gamma - 1)M^2}{1 - M^2}$	$\dfrac{(1 - \gamma M^2)\left\{1 + [(\gamma - 1)/2]M^2\right\}}{1 - M^2}$	$\dfrac{-\gamma(\gamma - 1)M^4}{2(1 - M^2)}$
$\dfrac{dp_0}{p_0}$	0	$-\dfrac{\gamma M^2}{2}$	$-\dfrac{\gamma M^2}{2}$

Table is read:

$$\frac{du}{u} = \left[-\frac{1}{1 - M^2}\right]\frac{dA}{A} + \left[\frac{1 + [(\gamma - 1)/2]M^2}{1 - M^2}\right]\frac{dT_0}{T_0} + \left[\frac{\gamma M^2}{2(1 - M^2)}\right]\frac{4f\,dx}{D}$$

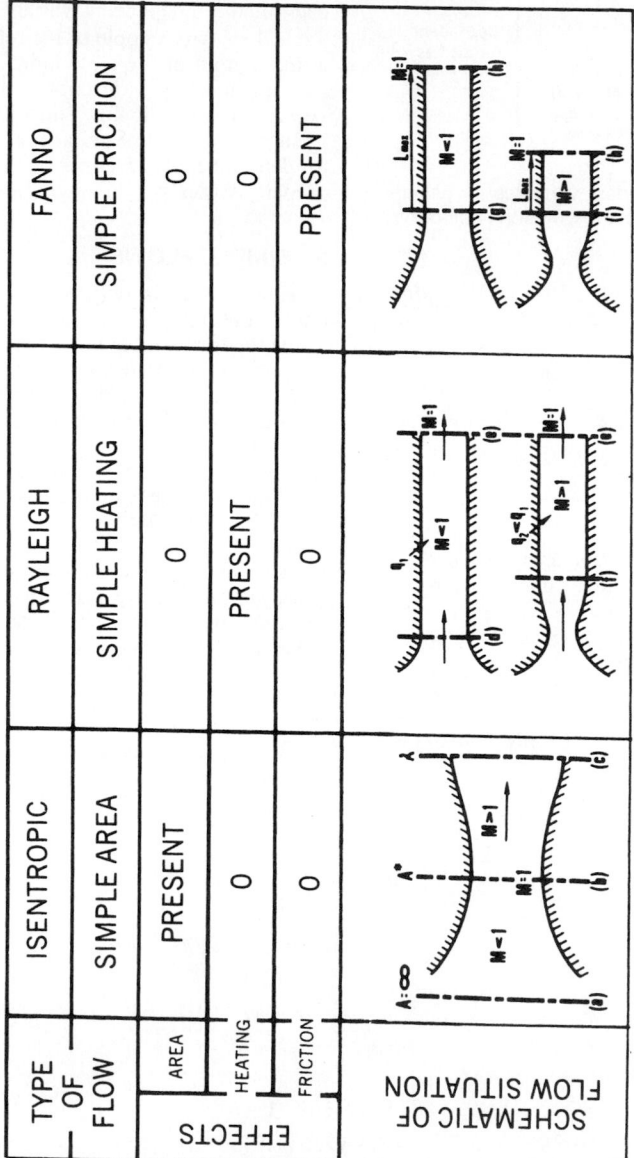

TYPE OF FLOW		ISENTROPIC	RAYLEIGH	FANNO
EFFECTS	AREA	SIMPLE AREA	SIMPLE HEATING	SIMPLE FRICTION
		PRESENT	0	0
	HEATING	0	PRESENT	0
	FRICTION	0	0	PRESENT
SCHEMATIC OF FLOW SITUATION				

Fig. 2a Simple flows.

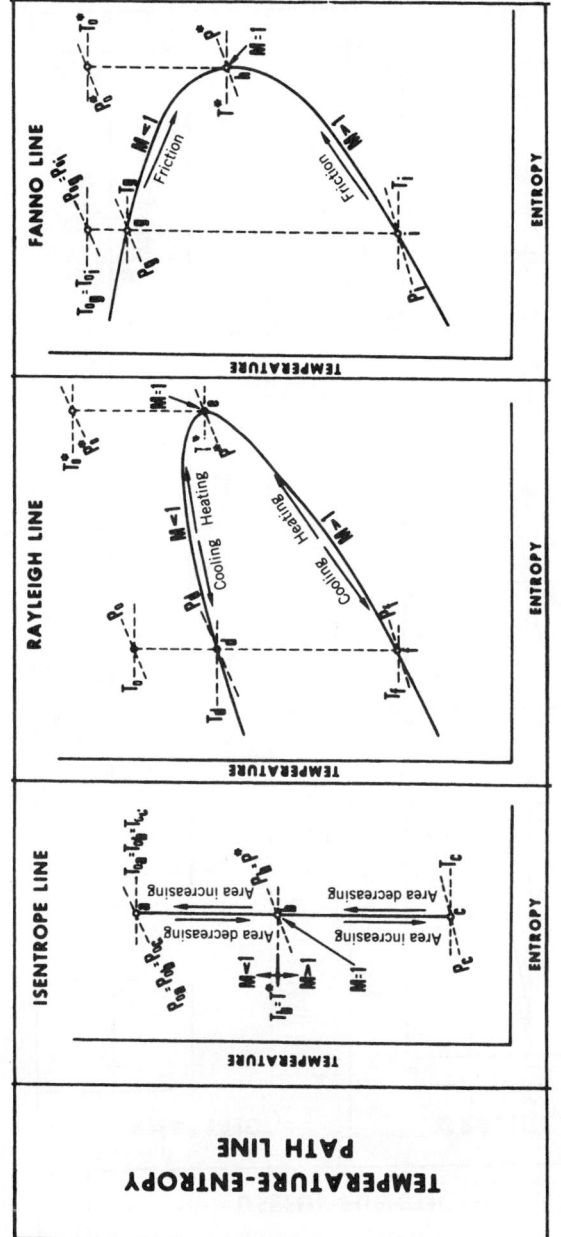

Fig. 2b (Continued)

685

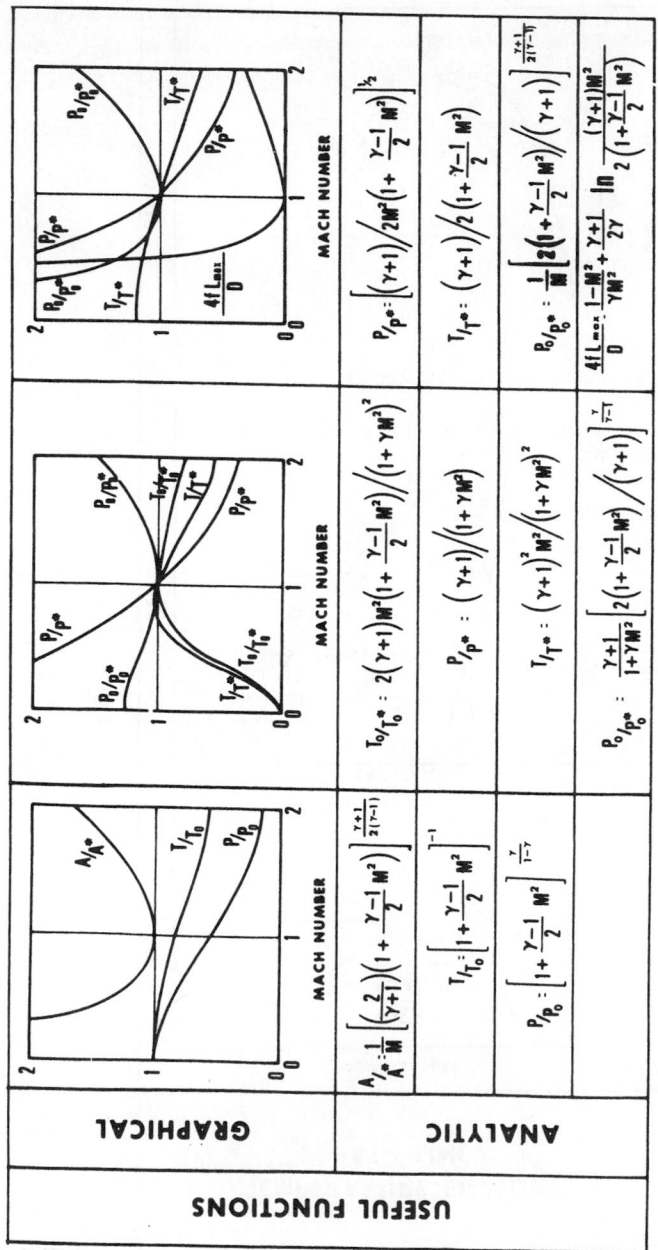

Fig. 2c (Continued)

686

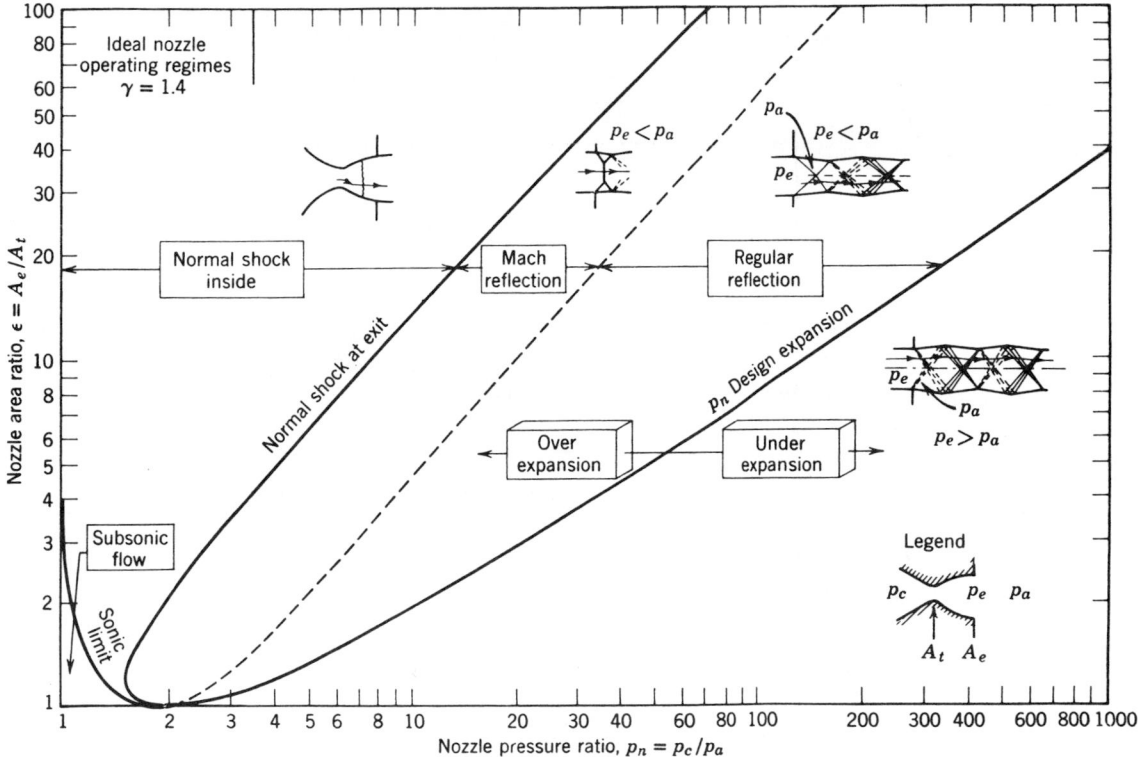

Fig. 3 Nozzle operating diagram (from Ref. 4).

To proceed downward along the *isentrope line* from point *a* of the diagram, the flow area is decreased until the sonic point at *b* is reached. The area must be increased after point *b* in order to continue down to point *c*. By proper adjustments in the flow area and boundary pressure, the flow may be made to proceed through point *b* in either direction along the isentrope line. Point *a* represents the isentropic stagnation condition for all points on the isentrope *ac*.

The *Rayleigh line* shows the series of possible states in a steady, frictionless, constant area flow. Motion along the Rayleigh line is caused by changes in the stagnation temperature produced by heating effects that, in turn, produce entropy changes in the manner indicated on the line. Heating in an initially subsonic flow (point *d*) causes the flow Mach number

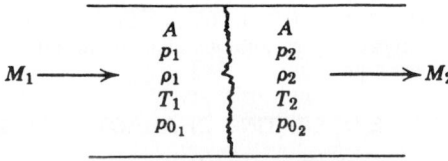

Fig. 4 Normal shock.

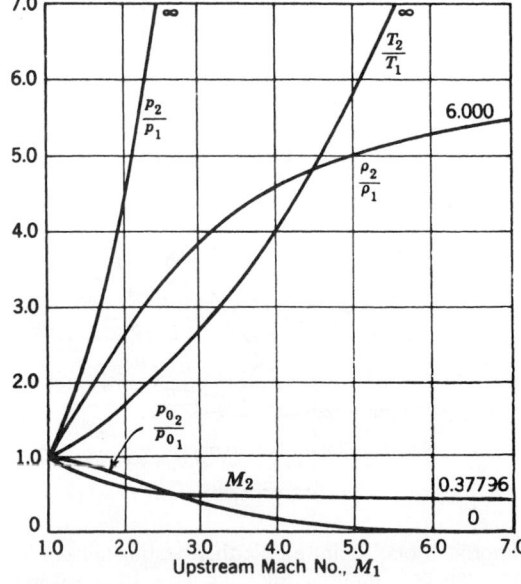

Fig. 5 Property variations across a normal shock.

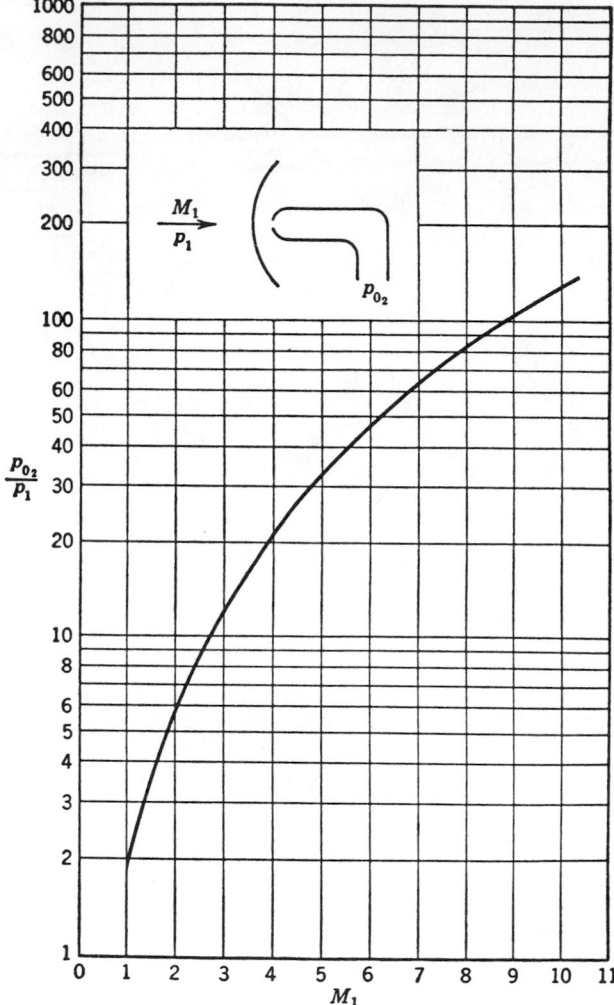

Fig. 6   Pitot-static pressure ratio for supersonic flow of air ($\gamma = 1.4$).

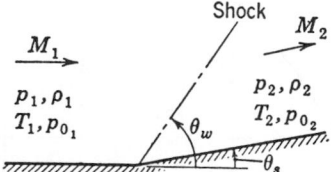

Fig. 7   Plane oblique shock wave.

The *Fanno line* represents the possible series of states in a steady, constant area, constant stagnation temperature flow. Frictional effects alone produce motion along the Fanno line. Consequently the flow progression along the line must always be one of increasing entropy toward the sonic point h. The flow is subsonic on the Fanno line above h and supersonic below. Since the entropy decreases along the Fanno line from point h, it is impossible in simple friction flow to proceed by continuous changes through sonic conditions at point h.

to approach one (point e). Neither heating nor cooling alone can continuously alter the flow from subsonic to supersonic speeds, or from supersonic to subsonic speeds.

3   NOZZLE OPERATING CHARACTERISTICS

The operating characteristics of a nozzle are governed by the ratio of its exit area to throat area (A_e/A_t) and by the ratio of its reservoir chamber pressure to exhaust region ambient ressure (p_ε/p_a). The operating regimes of a nozzle for isentropic (except for shock waves), steady one-dimensional flow of the perfect gas air arew depicted in Fig. 3 on a diagram with nozzle pressure ratio and nozzle area ratio as the coordinate axes.

The curve in Fig. 3 with the two branches labeled "design expansion" and "sonic limit" is obtained from the simple area isentropic functions of Fig. 2a by plotting A/A^* versus the reciprocal of p/p_0. The design expansion line corresponds to sets of values of nozzle pressure ratio and nozzle area ratio for which the flow is shock free with the nozzlen exit section Mach number supersonic and with the exit pressure p_e equal to the exhaust region ambient pressure p_a. For any given area ratio the sonic limit line locates the nozzle pressure ratio, which will produce sonic conditions at the throat and subsonic flow elsewhere. Alternatively, the sonic limit branch corresponds to the minimum nozzle pressure ratio that will give the maximum flow through a given area ratio nozzle with specified throat area and reservoir conditions. The flow is subsonic in a nozzle that has its operating point located below the sonic limit line. In the region to the right of the design expansion line $p_e > p_a$, the nozzle is underexpanded, and the flow expands to ambient conditions in the exhaust jet.

The area between the sonic limit and the design expansion curves is divided into two regions wherein shock waves occur inside the nozzle or in the jet exhaust. The common boundary of these two regions, labeled "normal shock at exit," is the locus of pressure ratios, which will produce a normal shock at the nozzle exit. Between the shock exit line and the sonic limit line, normal shock waves occur inside the nozzle. Between the shock exit and the design expansion lines oblique shocks, as shown, occur in the jet exhaust.

4 NORMAL SHOCK WAVES

A discontinuous change in stream properties can be sustained in a one-dimensional flow. When this phenomenon occurs normal to the flow direction, it is called a normal shock wave (Fig. 4). The process in this case is irreversible adiabatic and, therefore, with increasing entropy. The velocity of the flow entering

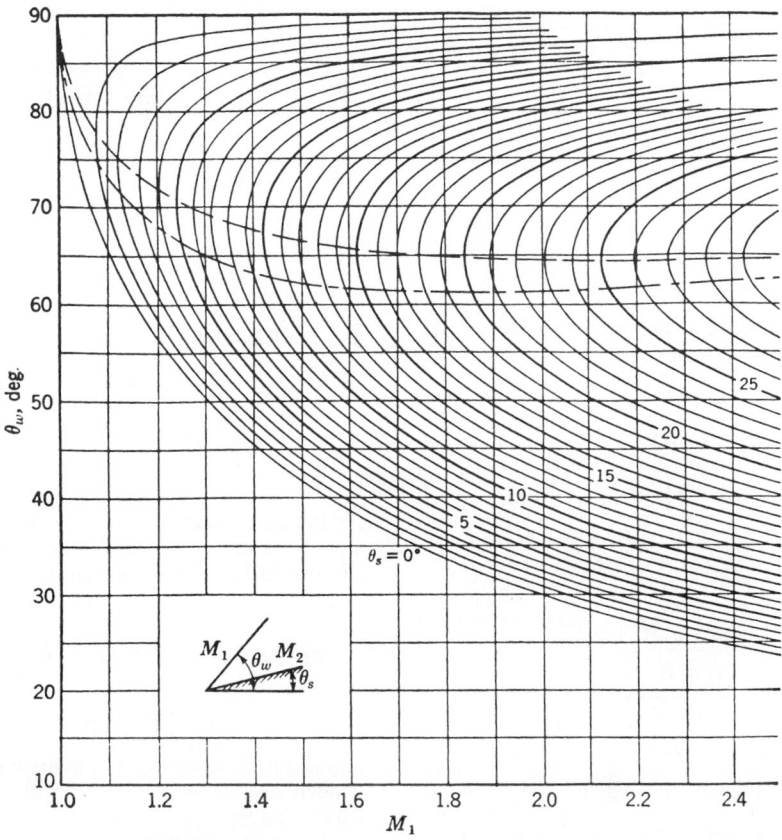

Fig. 8a   Wave angle for a plane oblique shock wave ($\gamma = 1.4$).

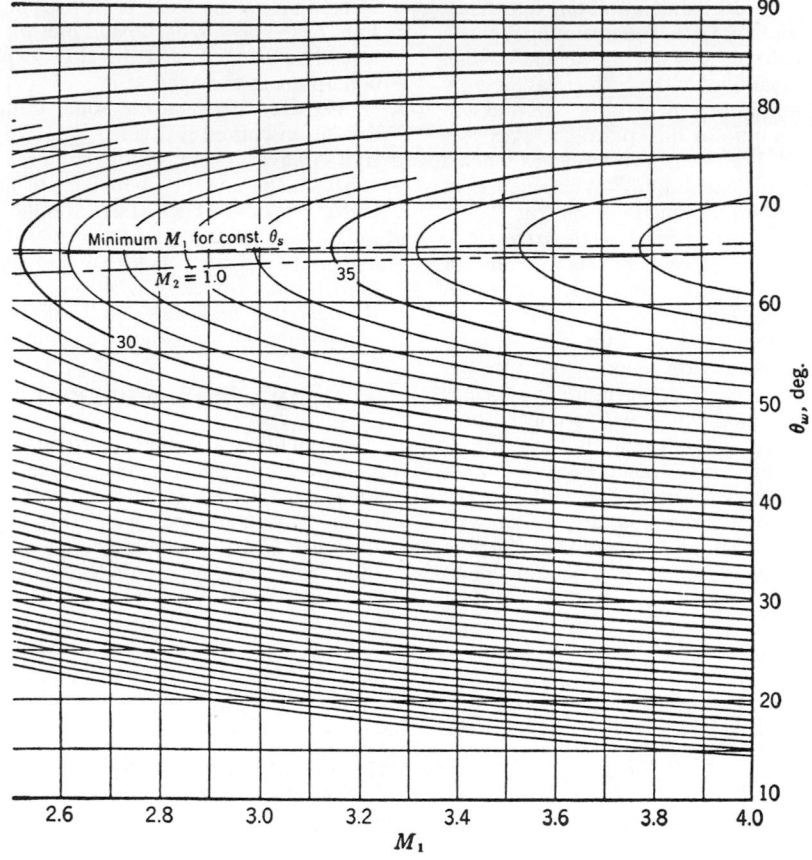

Fig. 8b   (*Continued*)

the shock is supersonic, and that leaving is subsonic. Various important relations are

$$\frac{p_2}{p_1} = \left(\frac{2\gamma}{\gamma+1}\right)M_1^2 - \left(\frac{\gamma-1}{\gamma+1}\right) \tag{15}$$

$$\frac{T_2}{T_1} = \frac{\left\{[2\gamma/(\gamma-1)]M_1^2 - 1\right\}\left\{1 + [(\gamma-1)/2]M_1^2\right\}}{\dfrac{(\gamma+1)^2}{2(\gamma-1)}M_1^2} \tag{16}$$

$$\frac{\rho_2}{\rho_1} = \frac{u_1}{u_2} = \left[\frac{\gamma-1}{\gamma+1} + \frac{2}{(\gamma+1)M_1^2}\right]^{-1} \tag{17}$$

$$\frac{p_{0_2}}{p_{0_1}} = \frac{A_1^*}{A_2^*} = \left[\frac{[(\gamma+1)/2]M_1^2}{1 + [(\gamma-1)/2]M_1^2}\right]^{\gamma/\gamma-1}$$

$$\times \left[\frac{2\gamma}{\gamma+1}M_1^2 - \frac{\gamma-1}{\gamma+1}\right]^{1/1-\gamma} \tag{18}$$

$$M_2 = \left[\frac{M_1^2 + 2/(\gamma-1)}{[2\gamma/(\gamma-1)]M_1^2 - 1}\right]^{1/2} \tag{19}$$

and the Rayleigh pitot formula,

$$\frac{p_{0_2}}{p_1} = \left[\frac{\gamma+1}{2}M_1^2\right]^{\gamma/\gamma-1}\left[\frac{2\gamma}{\gamma+1}M_1^2 - \frac{\gamma-1}{\gamma+1}\right]^{1/1-\gamma} \tag{20}$$

Numerical values for Eqs. (11)– (20) are given in Refs. 3 and 5–7. Equations (7)– (9) are plotted in Fig. 5 and Eq. (20) is shown in Fig. 6.

5   PLANE OBLIQUE SHOCK WAVES

A plane shock wave oblique to the oncoming flow will turn the flow sharply into the oncoming flow. Figure 7 shows an oblique shock wave turning the flow at a sudden change in wall direction. As with a normal shock, the flow process is nonisentropic.

The stream deflection angle θ_s, the wave angle θ_w, and the approach Mach number M_1 of an oblique

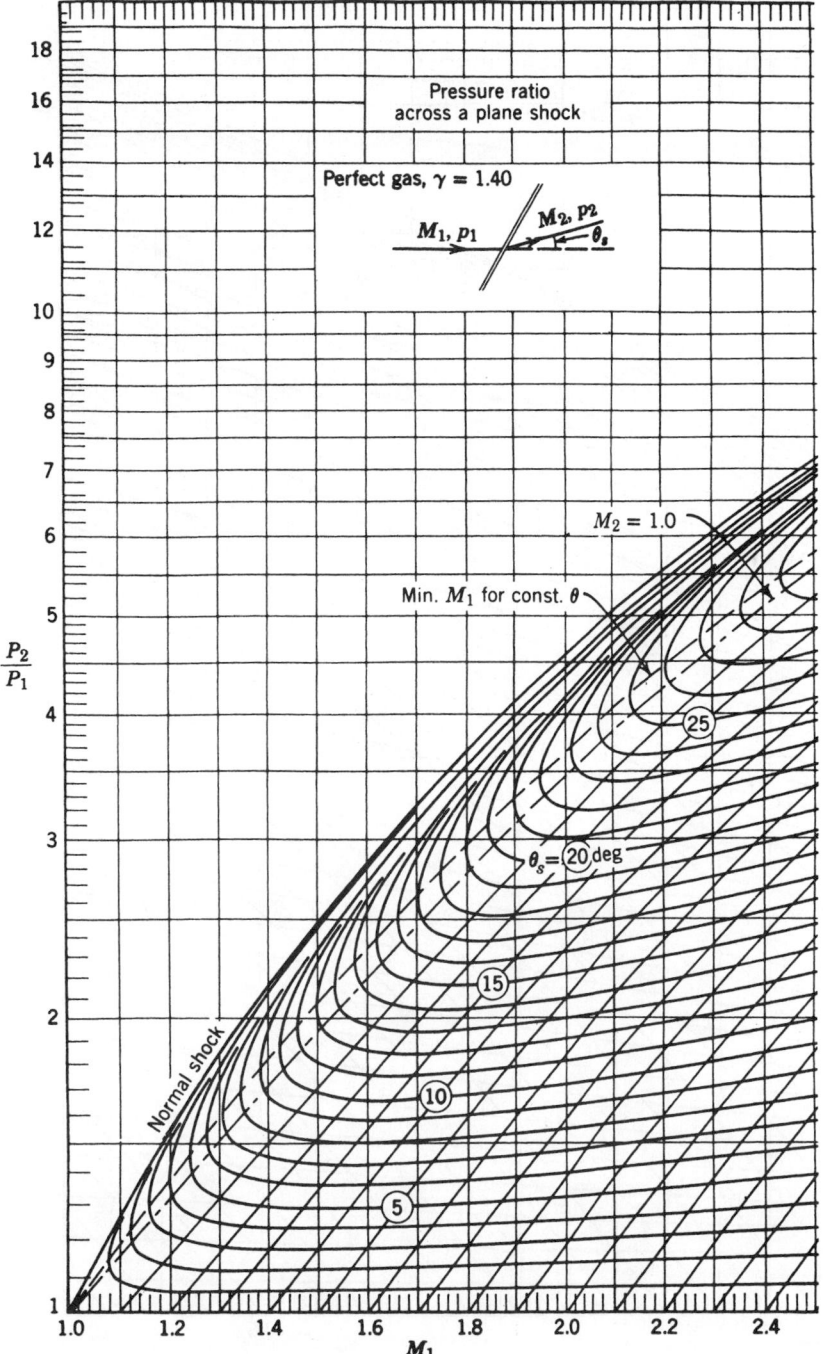

Fig. 9a Variation of pressure ratio and downstream Mach number with flow deflection angle and upstream Mach number M (from Ref. 6).

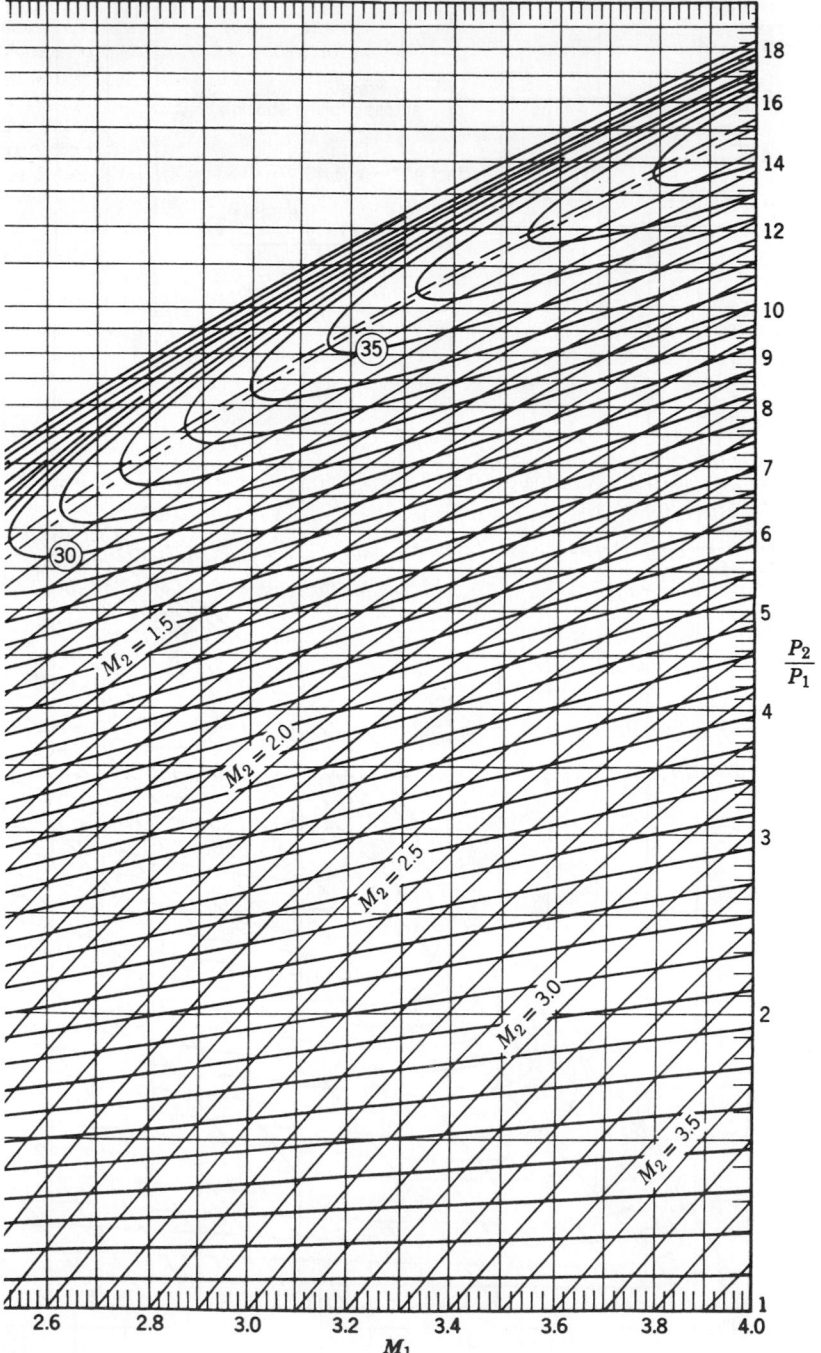

Fig. 9b   (Continued)

shock wave are related as follows:

$$\tan(\theta_w - \theta_s) = \frac{2/(\gamma + 1) + [(\gamma - 1)/(\gamma + 1)]M_1^2 \sin^2 \theta_w}{M_1^2 \sin \theta_w \cos \theta_w}$$

(21)

This relation is plotted in Fig. 8a. Observe from the figure that two waves are possible. However, when the shock is attached to the corner (Fig. 7), only the wave with the smaller wave angle, the so-called weak wave, can exist. In detached shocks, such as the curved shock in front of a blunt wedge, both solutions, strong and weak, occur. The line of minimum M_1 for constant θ_s is the locus of shock wave detachment.

Relations that relate the pressures, temperatures, densities, stagnation pressures, and Mach numbers before and after plane oblique shocks can be obtained in terms of M_1 and θ_w. In fact, when the products $M_1 \sin \theta_w$ and $M_2 \sin(\theta_w - \theta_s)$ are substituted for M_1 and M_2, respectively, in the normal shock equations [(15)–(19)], they become directly applicable to flow through plane oblique shocks. Thus by entering the graph of Fig. 5 with $M_1 \sin \theta_w$ (where θ_w may be obtained from Fig. 8a for known values of M_1 and θ_s), the values of the property ratios plotted correspond to those for a plane oblique shock wave. Similarly, the curve labeled M_2 in Fig. 5 gives $M_2 \sin(\theta_w - \theta_s)$ for a plane oblique shock wave. The variation of pressure ratio, downstream Mach number, and total pressure

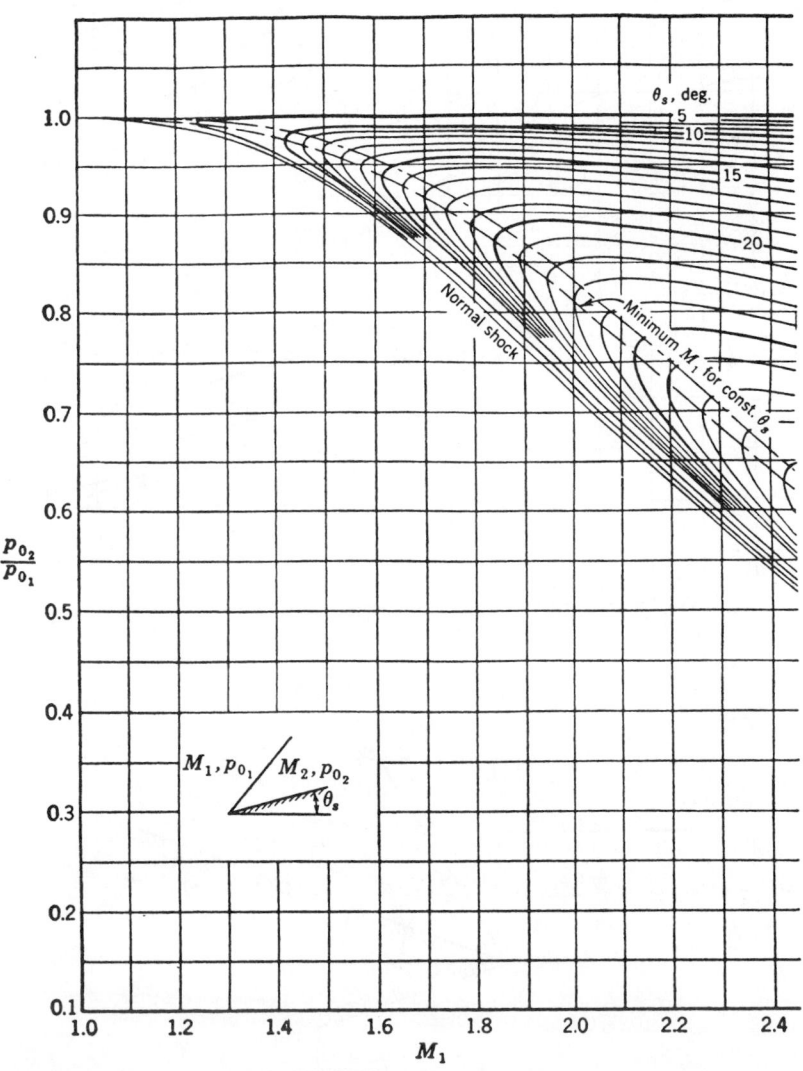

Fig. 10a Stagnation pressure ratio across a plane oblique shock wave ($\gamma = 1.4$).

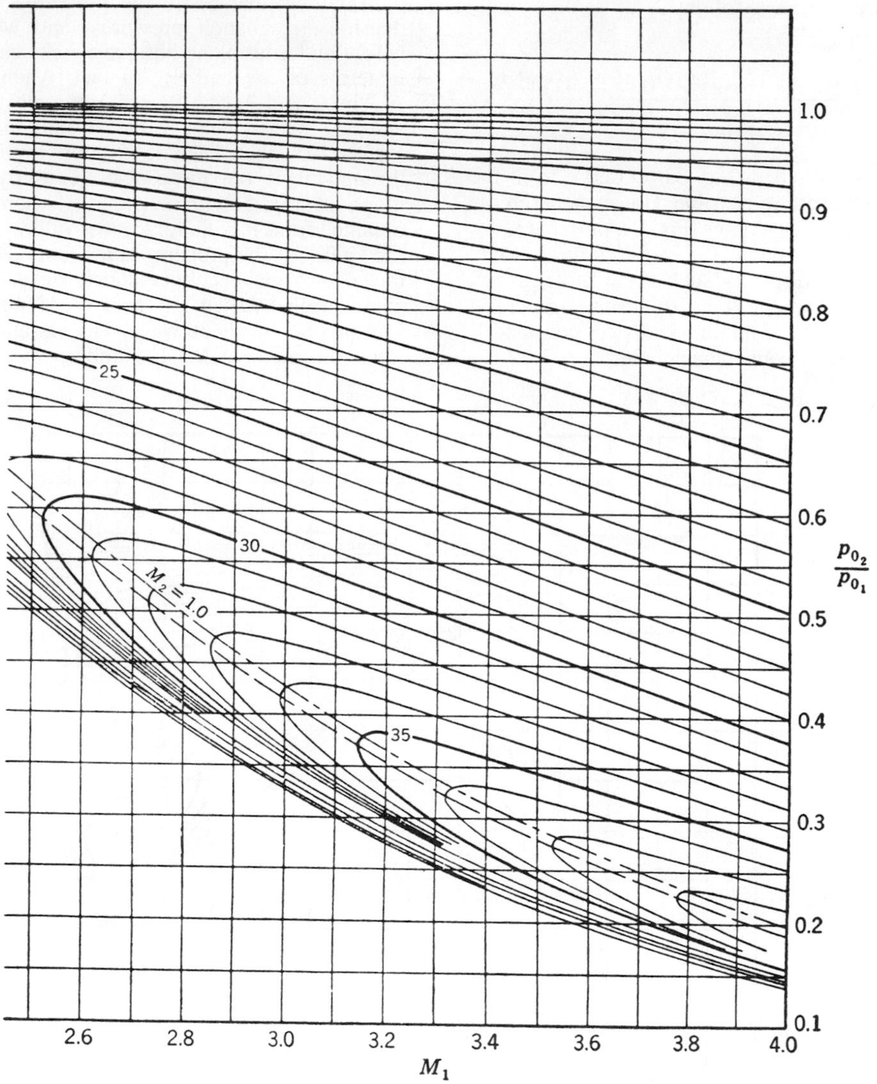

$$\frac{p_{0_2}}{p_{0_1}}$$

M_1

Fig. 10b   (*Continued*)

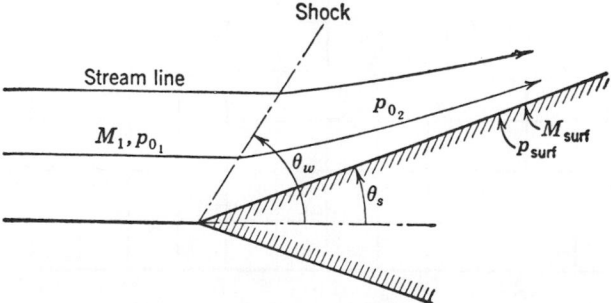

Fig. 11   Conical shock wave.

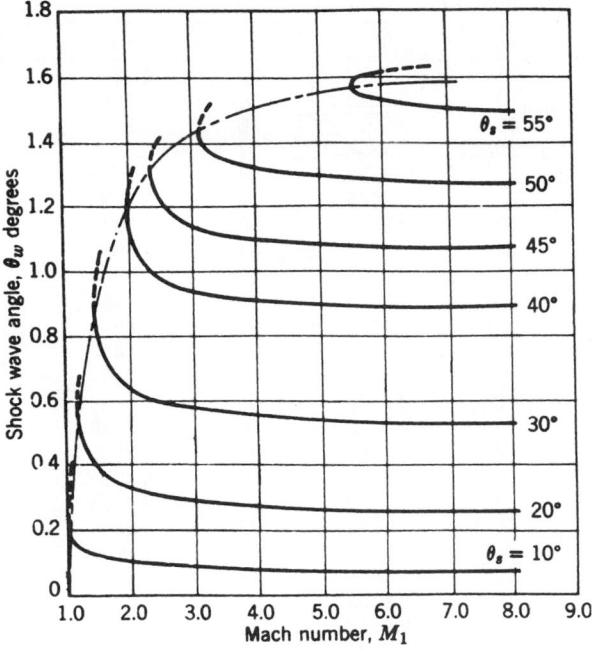

Fig. 12 Conical shock-wave angle versus Mach number for various cone angles.

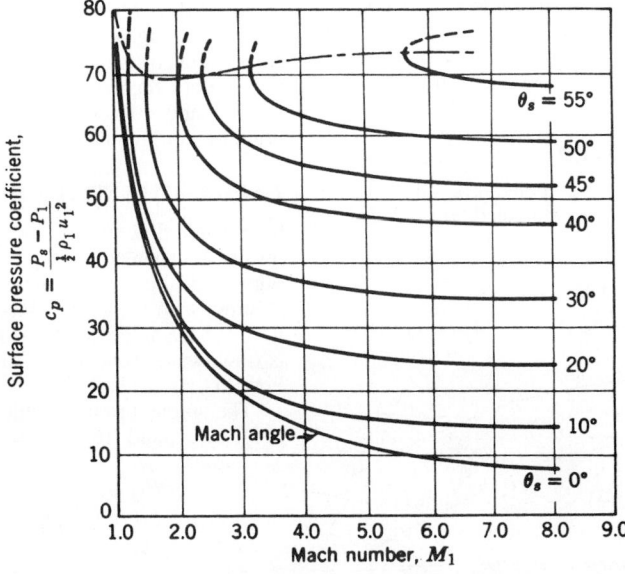

Fig. 13 Surface pressure coefficient versus Mach number for various cone angles.

Table 2 Mach Number Variation in Expanding Flow

θ_s	M_2	θ_s	M_2	θ_s	M_2	θ_s	M_2
0	1.0000	27	2.0222	54	3.2293	81	5.470
1	1.0808	28	2.0585	55	3.2865	82	5.595
2	1.1328	29	2.0957	56	3.3451	83	5.724
3	1.1770	30	2.1336	57	3.4055	84	5.867
4	1.2170	31	2.1723	58	3.4675	85	6.008
5	1.2554	32	2.2105	59	3.5295	86	6.155
6	1.2935	33	2.2492	60	3.5937	87	6.311
7	1.3300	34	2.2885	61	3.6610	88	6.472
8	1.3649	35	2.3288	62	3.7288	89	6.643
9	1.4005	36	2.3688	63	3.7980	90	6.820
10	1.4350	37	2.4108	64	3.8690	91	7.008
11	1.4688	38	2.4525	65	3.9417	92	7.202
12	1.5028	39	2.4942	66	4.0164	93	7.407
13	1.5365	40	2.5372	67	4.0940	94	7.623
14	1.5710	41	2.5810	68	4.1738	95	7.852
15	1.6045	42	2.6254	69	4.2543	96	8.093
16	1.6380	43	2.6716	70	4.3385	97	8.350
17	1.6723	44	2.7179	71	4.4257	98	8.622
18	1.7061	45	2.7643	72	4.5158	99	8.907
19	1.7401	46	2.8120	73	4.6086	100	9.210
20	1.7743	47	2.8610	74	4.7031	101	9.539
21	1.8090	48	2.9105	75	4.7979	102	9.887
22	1.8445	49	2.9616	76	4.9032	103	10.260
23	1.8795	50	3.0131	77	5.009	104	10.658
24	1.9150	51	3.0660	78	5.119	105	11.081
25	1.9502	52	3.1193	79	5.232		
26	1.9861	53	3.1737	80	5.349		

ratio with flow deflection and upstream Mach number is plotted in Figs. 9 and 10. References 3, 5, 7, and 8 present the oblique shock functions in tabular or graphical form. Reference 8 also contains numerous examples.

6 CONICAL SHOCK WAVES

Conditions on the surface of a cone, Fig. 11, with attached shock wave have been calculated on the basis of the Taylor–Maccoll theory.[9] All fluid properties are constant along radial lines through the nose of the cone. Data for the surface stream properties on a cone with an attached conical shock wave are tabulated in Refs. 5 and 10. The shock wave angles and surface pressure coefficients are plotted in Figs. 12 and 13.

7 PRANDTL–MEYER EXPANSION

Supresonic flow turning away from the oncoming flow results in expansion of a gas to a higher Mach number. Figure 14 shows such an expansion around a corner starting with an initial Mach number of unity. The flow is here assumed two dimensional and the process is downward from $M = 1$ along the isentrope line of Fig. 2a.

The angle through which a stream must turn in order to expand from $M_1 = 1$ to M_2 is

$$\theta_s = \sqrt{\frac{\gamma + 1}{\gamma - 1}} \tan^{-1} \sqrt{\frac{\gamma - 1}{\gamma + 1}(M_2^2 - 1)}$$

$$- \tan^{-1} \sqrt{M_2^2 - 1} \qquad (22)$$

Numerical values of Eq. (22) are given in Refs. 3 and 5 and Table 2 for air ($\gamma = 1.4$).

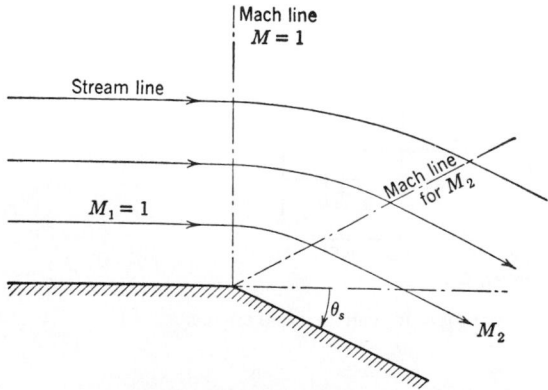

Fig. 14 Prandtl–Meyer expansion at a corner.

Values given in Table 2 can be applied to isentropic flow along any convex curved surface. The downstream Mach number M_2 after expanding M_1 through an angle $\Delta\theta$ can be found by adding $\Delta\theta$ to the θ_s corresponding to M_1 and then finding the M_2 corresponding to $\theta_s + \Delta\theta$ in Table 2.

REFERENCES

1. Ames Research Staff, "Equations, Tables, and Charts for Compressible Flow," National Advisory Committee for Aeronautics report NACA Report 1135, Ames Aeronautical Laboratory, Moffett Field, CA, 1953.
2. Chapman, S., and Cowling, T. G., *The Mathematical Theory of Non-Uniform Gases*, Cambridge University Press, London, 1952.
3. Keenan, J. H., and Kaye, J., *Gas Tables*, Wiley, New York, 1957.
4. Daley, D. H., "A Nozzle Operating Diagram," *Bull. Mech. Eng. Educ.*, **6**, 293–300 (1967).
5. *Equations, Tables and Charts for Compressible Flow*, NACA Report 1135, Government Printing Office, Washington, DC, 1953.
6. Dailey and Wood, *Computation Curves for Compressible Fluid Problems*, Wiley, New York, 1949.
7. Zucrow, M. J., and Hoffman, J. D., *Gas Dynamics*, Vol. I, Wiley, New York, 1976.
8. Dennard, J. S., and Spencer, P. B., "Ideal-Gas Tables for Oblique Shock Flow Parameters in Ai Mach Numbers from 1.05 to 12.0," NASA TN D-2221, March 1964.
9. Taylor, G. I., and Maccoll, J. W., "The Air Pressure on a Cone Moving at High Speeds," *Proc. Roy. Soc. (Lond.)*, Ser. A, **139**(838) (Feb. 1933).
10. Kopal, Z., *Tables and Supersonic Flow around Cones*, MIT Press, Cambridge, MA, 1947.

CHAPTER 12

MATHEMATICAL MODELS OF DYNAMIC PHYSICAL SYSTEMS

K. Preston White, Jr.
Department of Systems and Information Engineering
University of Virginia
Charlottesville, Virginia

1 RATIONALE

The design of modern control systems relies on the formulation and analysis of mathematical models of dynamic physical systems. This is simply because a model is more accessible to study than the physical system the model represents. Models typically are less costly and less time consuming to construct and test. Changes in the structure of a model are easier to implement, and changes in the behavior of a model are easier to isolate and understand. A model often can be used to achieve insight when the corresponding physical system cannot because experimentation with the actual system is too dangerous or too demanding. Indeed, a model can be used to answer "what if" questions about a system that has not yet been realized or actually cannot be realized with current technologies.

The type of model used by the control engineer depends upon the nature of the system the model represents, the objectives of the engineer in developing the model, and the tools the engineer has at his or her disposal for developing and analyzing the model. A mathematical model is a description of a system in terms of equations. Because the physical systems of primary interest to the control engineer are dynamic in nature, the mathematical models used to represent these systems most often incorporate difference or differential equations. Such equations, based on physical laws and observations, are statements of the fundamental relationships among the important variables that describe the system. Difference and differential equation models are expressions of the way in which the current values assumed by the variables combine to determine the future values of these variables.

Mathematical models are particularly useful because of the large body of mathematical and computational theory that exists for the study and solution of equations. Based on this theory, a wide range of techniques has been developed specifically for the study of control systems. In recent years, computer programs have been written that implement virtually all of these techniques. Computer software packages are now widely available for both simulation and computational assistance in the analysis and design of control systems.

It is important to understand that a variety of models can be realized for any given physical system. The choice of a particular model always represents a trade-off between the fidelity of the model and the effort required in model formulation and analysis. This trade-off is reflected in the nature and extent of simplifying assumptions used to derive the model. In general, the more faithful the model is as a description of the physical system modeled, the more difficult it is to obtain general solutions. In the final analysis, the best engineering model is not necessarily the most accurate or precise. It is, instead, the simplest model that yields the information needed to support a decision. A classification of various types of models commonly encountered by control engineers is given in Section 8.

A large and complicated model is justified if the underlying physical system is itself complex, if the individual relationships among the system variables are well understood, if it is important to understand the system with a great deal of accuracy and precision, and if time and budget exist to support an extensive study. In this case, the assumptions necessary to formulate the model can be minimized. Such complex models cannot be solved analytically, however. The model itself must be studied experimentally, using the techniques of computer simulation. This approach to model analysis is treated in Section 7.

Simpler models frequently can be justified, particularly during the initial stages of a control system study. In particular, systems that can be described by linear difference or differential equations permit the use of powerful analysis and design techniques. These include the transform methods of classical control theory and the state-variable methods of modern control

theory. Descriptions of these standard forms for linear systems analysis are presented in Sections 4, 5, and 6.

During the past several decades, a unified approach for developing lumped-parameter models of physical systems has emerged. This approach is based on the idea of idealized system elements, which store, dissipate, or transform energy. Ideal elements apply equally well to the many kinds of physical systems encountered by control engineers. Indeed, because control engineers most frequently deal with systems that are part mechanical, part electrical, part fluid, and/or part thermal, a unified approach to these various physical systems is especially useful and economic. The modeling of physical systems using ideal elements is discussed further in Sections 2, 3, and 4.

Frequently, more than one model is used in the course of a control system study. Simple models that can be solved analytically are used to gain insight into the behavior of the system and to suggest candidate designs for controllers. These designs are then verified and refined in more complex models, using computer simulation. If physical components are developed during the course of a study, it is often practical to incorporate these components directly into the simulation, replacing the corresponding model components. An iterative, evolutionary approach to control systems analysis and design is depicted in Fig. 1.

2 IDEAL ELEMENTS

Differential equations describing the dynamic behavior of a physical system are derived by applying the appropriate physical laws. These laws reflect the ways in which energy can be stored and transferred within the system. Because of the common physical basis provided by the concept of energy, a general approach to deriving differential equation models is possible. This approach applies equally well to mechanical, electrical, fluid, and thermal systems and is particularly useful for systems that are combinations of these physical types.

2.1 Physical Variables

An idealized *two-terminal* or *one-port* element is shown in Fig. 2. Two *primary physical variables* are associated with the element: a through variable $f(t)$ and an across variable $v(t)$. *Through variables* represent quantities that are transmitted through the element, such as the force transmitted through a spring, the current transmitted through a resistor, or the flow of fluid through a pipe. Through variables have the same value at both ends or terminals of the element. *Across variables* represent the difference in state between the terminals of the element, such as the velocity difference across the ends of a spring, the voltage drop across a resistor, or the pressure drop across the ends of a pipe. *Secondary physical variables* are the integrated through variable $h(t)$ and the integrated across variable $x(t)$. These represent the accumulation of quantities within an element as a result of the integration of the

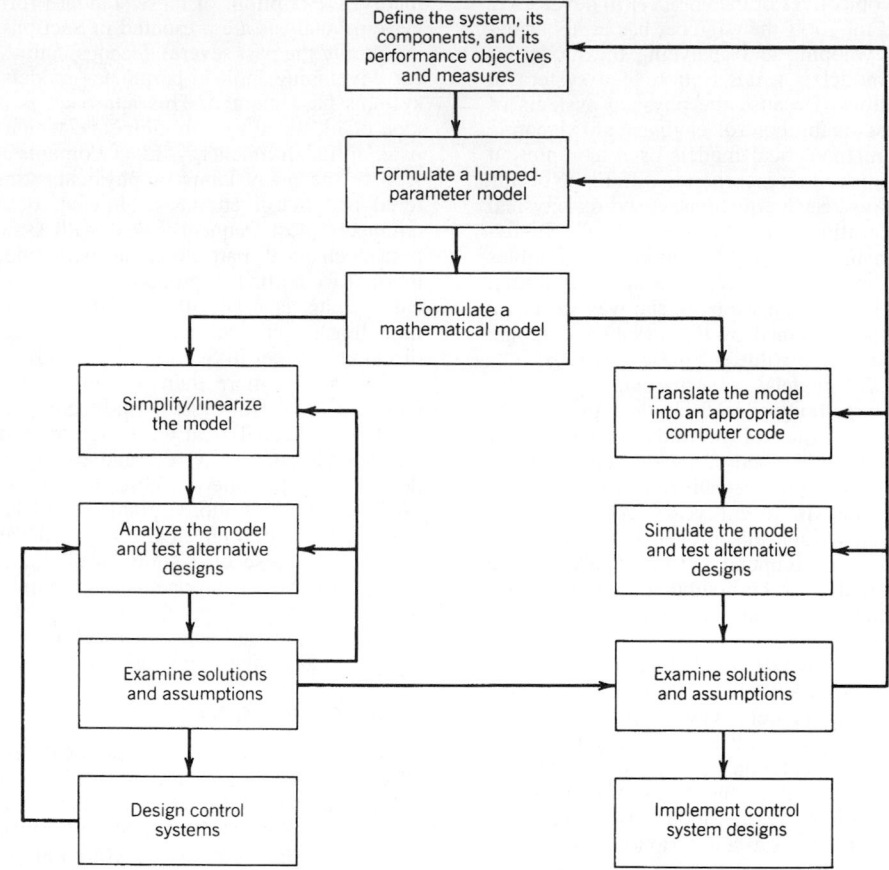

Fig. 1 Iterative approach to control system design, showing the use of mathematical analysis and computer simulation.

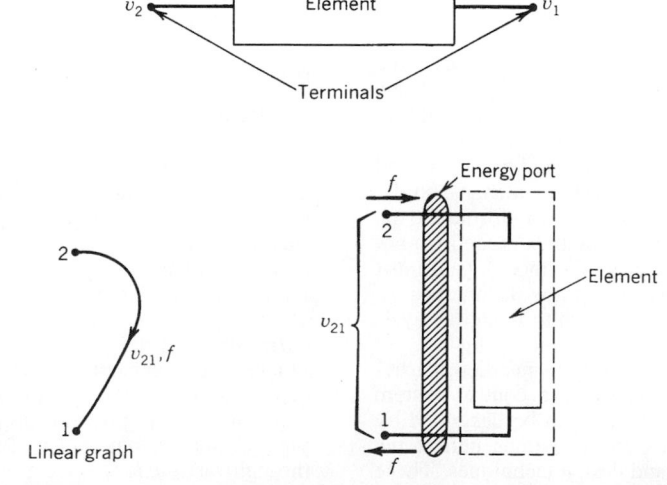

Fig. 2 Two-terminal or one-port element, showing through and across variables.[1]

Table 1 Primary and Secondary Physical Variables for Various Systems

System	Through Variable f	Integrated through Variable h	Across Variable v	Integrated across Variable x
Mechanical–translational	Force F	Translational momentum p	Velocity difference v_{21}	Displacement difference x_{21}
Mechanical–rotational	Torque T	Angular momentum h	Angular velocity difference Ω_{21}	Angular displacement difference Θ_{21}
Electrical	Current i	Charge q	Voltage difference v_{21}	Flux linkage λ_{21}
Fluid	Fluid flow Q	Volume V	Pressure difference P_{21}	Pressure–momentum Γ_{21}
Thermal	Heat flow q	Heat energy $\mathscr{H}$	Temperature difference θ_{21}	Not used in general

Source: From Ref. 1.

associated through and across variables. For example, the momentum of a mass is an integrated through variable, representing the effect of forces on the mass integrated or accumulated over time. Table 1 defines the primary and secondary physical variables for various physical systems.

2.2 Power and Energy

The flow of *power* $P(t)$ into an element through terminals 1 and 2 is the product of the through variable $f(t)$ and the difference between the across variables $v_2(t)$ and $v_1(t)$. Suppressing the notation for time dependence, this may be written as

$$P = f(v_2 - v_1) = f v_{21}$$

A negative value of power indicates that power flows out of the element. The *energy* $E(t_a, t_b)$ transferred to the element during the time interval from t_a to t_b is the integral of power, that is,

$$E = \int_{ta}^{tb} P \, dt = \int_{ta}^{tb} f v_{21} \, dt$$

A negative value of energy indicates a net transfer of energy out of the element during the corresponding time interval.

Thermal systems are an exception to these generalized energy relationships. For a thermal system, power is identically the through variable $q(t)$, heat flow. Energy is the integrated through variable $\mathscr{H}(t_a, t_b)$, the amount of heat transferred.

By the *first law of thermodynamics*, the net energy stored within a system at any given instant must equal the difference between all energy supplied to the system and all energy dissipated by the system. The generalized classification of elements given in the following sections is based on whether the element stores or dissipates energy within the system, supplies energy to the system, or transforms energy between parts of the system.

2.3 One-Port Element Laws

Physical devices are represented by idealized system elements, or by combinations of these elements. A physical device that exchanges energy with its environment through one pair of across and through variables is called a *one-port* or *two-terminal* element. The behavior of a one-port element expresses the relationship between the physical variables for that element. This behavior is defined mathematically by a *constitutive relationship*. Constitutive relationships are derived empirically, by experimentation, rather than from any more fundamental principles. The *element law*, derived from the corresponding constitutive relationship, describes the behavior of an element in terms of across and through variables and is the form most commonly used to derive mathematical models.

Table 2 summarizes the element laws and constitutive relationships for the one-port elements. Passive elements are classified into three types. *T-type* or *inductive storage* elements are defined by a single-valued constitutive relationship between the through variable $f(t)$ and the integrated across-variable difference $x_{21}(t)$. Differentiating the constitutive relationship yields the element law. For a linear (or ideal) T-type element, the element law states that the across-variable difference is proportional to the rate of change of the through variable. Pure translational and rotational compliance (springs), pure electrical inductance, and pure fluid inertance are examples of T-type storage elements. There is no corresponding thermal element.

A-type or *capacitive storage elements* are defined by a single-valued constitutive relationship between the across-variable difference $v_{21}(t)$ and the integrated through variable $h(t)$. These elements store energy by virtue of the across variable. Differentiating the constitutive relationship yields the element law. For a linear A-type element, the element law states that the through variable is proportional to the derivative of the across-variable difference. Pure translational and rotational inertia (masses) and pure electrical, fluid, and thermal capacitance are examples.

Table 2 Element Laws and Constitutive Relationships for Various One-Port Elements

Type of element	Physical element	Linear graph	Diagram	Constitutive relationship	Energy or power function	Ideal elemental equation	Ideal energy or power
T-type energy storage $\varepsilon \geq 0$ Pure: $x_{21}=f(f)$, $\varepsilon=\int_0^f f\,dx_{21}$ Idea: $x_{21}=Lf$, $\varepsilon=\frac{1}{2}Lf^2$	Translational spring			$x_{21}=f(F)$	$\varepsilon=\int_0^F f\,dx_{21}$	$v_{21}=\dfrac{1}{k}\dfrac{dF}{dt}$	$\varepsilon=\dfrac{1}{2}\dfrac{F^2}{k}$
	Rotational spring			$\Theta_{21}=f(T)$	$\varepsilon=\int_0^T T\,d\Theta_{21}$	$\Omega_{21}=\dfrac{1}{K}\dfrac{dT}{dt}$	$\varepsilon=\dfrac{1}{2}\dfrac{T^2}{K}$
	Inductance			$\lambda_{21}=f(I)$	$\varepsilon=\int_0^i i\,d\lambda_{21}$	$v_{21}=L\dfrac{di}{dt}$	$\varepsilon=\dfrac{1}{2}LI^2$
	Fluid inertance			$\Gamma_{21}=f(Q)$	$\varepsilon=\int_0^Q Q\,d\Gamma_{21}$	$P_{21}=I\dfrac{dQ}{dt}$	$\varepsilon=\dfrac{1}{2}IQ^2$
A-type energy storage $\varepsilon \geq 0$ Pure: $h=f(v_{21})$, $\varepsilon=\int_0^{v_{21}} v_{21}\,dh$ Idea: $h=Cv_{21}$, $\varepsilon=\frac{1}{2}Cv_{21}^2$	Translational mass			$p=f(v_2)$	$\varepsilon=\int_0^{v_2} v_2\,dp$	$F=m\dfrac{dv_2}{dt}$	$\varepsilon=\dfrac{1}{2}mv_2^2$
	Inertia			$h=f(\Omega_2)$	$\varepsilon=\int_0^{\Omega_2} \Omega_2\,dh$	$T=J\dfrac{d\Omega_2}{dt}$	$\varepsilon=\dfrac{1}{2}J\Omega_2^2$
	Electrical capacitance			$q=f(v_{21})$	$\varepsilon=\int_0^{v_{21}} v_{21}\,dq$	$i=C\dfrac{dv_{21}}{dt}$	$\varepsilon=\dfrac{1}{2}Cv_{21}^2$
	Fluid capacitance			$V=f(P_2)$	$\varepsilon=\int_0^{P_2} P_2\,dV$	$Q=C_f\dfrac{dP_2}{dt}$	$\varepsilon=\dfrac{1}{2}C_f P_2^2$
	Thermal capacitance			$\mathscr{H}=f(\theta_2)$	$\varepsilon=\int_0^{\theta_2} q\,dt=\mathscr{H}$	$q=C_t\dfrac{d\theta_2}{dt}$	$\varepsilon=C_t\theta_2$

(Continues)

Table 2 *(Continued)*

D-type energy dissipators $\mathcal{P} \geq 0$			$F = f(v_{21})$	$\mathcal{P} = Fv_{21}$	$F = bv_{21}$	$\mathcal{P} = bv_{21}^2$
Pure	$f = f(v_{21})$	$\mathcal{P} = v_{21}f(v_{21})$				
Idea	$f = \dfrac{1}{R}v_{21}$	$\mathcal{P} = \dfrac{1}{R}v_{21}^2 = Rf^2$				
Translational damper			$F = f(v_{21})$	$\mathcal{P} = Fv_{21}$	$F = bv_{21}$	$\mathcal{P} = bv_{21}^2$
Rotational damper			$T = f(\Omega_{21})$	$\mathcal{P} = T\Omega_{21}$	$T = B\Omega_{21}$	$\mathcal{P} = B\Omega_{21}^2$
Electrical resistance			$i = f(v_{21})$	$\mathcal{P} = iv_{21}$	$i = \dfrac{1}{R}v_{21}$	$\mathcal{P} = \dfrac{1}{R}v_{21}^2$
Fluid resistance			$Q = f(P_{21})$	$\mathcal{P} = QP_{21}$	$Q = \dfrac{1}{R_f}P_{21}$	$\mathcal{P} = \dfrac{1}{R_f}P_{21}^2$
Thermal resistance			$\mathbf{q} = f(\theta_{21})$	$\mathcal{P} = q$	$q = \dfrac{1}{R_f}\theta_{21}$	$\mathcal{P} = \dfrac{1}{R_f}\theta_{21}$
Energy sources						
A-type across-variable source $\mathcal{P} \gtrless 0$			$v_{21} = f(t)$	$\mathcal{P} = fv_{21}$		
T-type through-variable source $\varepsilon \gtrless 0$			$f = f(t)$	$\mathcal{P} = fv_{21}$		

Nomenclature:

λ = energy, $\mathcal{P}$ = power

f = generalized *throughvariable*, F = force, T = torque, i = current, Q = fluid flow rate, q = heat flow rate

h = generalized integrated through variable, p = translational momentum, h = angular momentum, q = charge, I' = fluid volume displaced, $\mathcal{H}$ = heat

v = generalized across variable, v = translational velocity, Ω = angular velocity, v = voltage, P = pressure, θ = temperature

x = generalized integrated across variable, x = translational displacement, Θ = angular displacement, λ = flux linkage, Γ = pressure-momentum

L = generalized ideal inductance, $1/k$ = reciprocal translational stiffness, $1/K$ = reciprocal rotational stiffness, L = inductance, I = fluid inertance

C = generalized ideal capacitance, m = mass, J = moment of inertia, C = capacitance, C_j = fluid capacitance, C_t = thermal capacitance

R = generalized ideal resistance, $1/b$ = reciprocal translational damping, $1/B$ = reciprocal rotational damping, R = electrical resistance,
R_j = fluid resistance, R_t = thermal resistance

Source: From Ref. 1.

It is important to note that when a nonelectrical capacitance is represented by an A-type element, one terminal of the element must have a constant (reference) across variable, usually assumed to be zero. In a mechanical system, for example, this requirement expresses the fact that the velocity of a mass must be measured relative to a noninertial (nonaccelerating) reference frame. The constant-velocity terminal of a pure mass may be thought of as being attached in this sense to the reference frame.

D-type or *resistive elements* are defined by a single-valued constitutive relationship between the across and the through variables. These elements dissipate energy, generally by converting energy into heat. For this reason, power always flows into a D-type element. The element law for a D-type energy dissipator is the same as the constitutive relationship. For a linear dissipator, the through variable is proportional to the across-variable difference. Pure translational and rotational friction (dampers or dashpots) and pure electrical, fluid, and thermal resistance are examples.

Energy storage and energy-dissipating elements are called *passive* elements because such elements do not supply outside energy to the system. The fourth set of one-port elements are *source elements*, which are examples of *active* or power-supplying elements. Ideal sources describe interactions between the system and its environment. A pure *A-type source* imposes an across-variable difference between its terminals, which is a prescribed function of time, regardless of the values assumed by the through variable. Similarly, a pure *T-type source* imposes a through-variable flow through the source element, which is a prescribed function of time, regardless of the corresponding across variable.

Pure system elements are used to represent physical devices. Such models are called *lumped-element models*. The derivation of lumped-element models typically requires some degree of approximation since (a) there rarely is a one-to-one correspondence between a physical device and a set of pure elements and (b) there

always is a desire to express an element law as simply as possible. For example, a coil spring has both mass and compliance. Depending on the context, the physical spring might be represented by a pure translational mass, or by a pure translational spring, or by some combination of pure springs and masses. In addition, the physical spring undoubtedly will have a nonlinear constitutive relationship over its full range of extension and compression. The compliance of the coil spring may well be represented by an ideal translational spring, however, if the physical spring is approximately linear over the range of extension and compression of concern.

2.4 Multiport Elements

A physical device that exchanges energy with its environment through two or more pairs of through and across variables is called a *multiport element*. The simplest of these, the idealized *four-terminal* or *two-port* element, is shown in Fig. 3. Two-port elements provide for transformations between the physical variables at different energy ports, while maintaining instantaneous continuity of power. In other words, net power flow into a two-port element is always identically zero:

$$P = f_a v_a + f_b v_b = 0$$

The particulars of the transformation between the variables define different categories of two-port elements.

A *pure transformer* is defined by a single-valued constitutive relationship between the integrated across variables or between the integrated through variables at each port:

$$x_b = f(x_a) \qquad \text{or} \qquad h_b = f(h_a)$$

For a linear (or ideal) transformer, the relationship is proportional, implying the following relationships

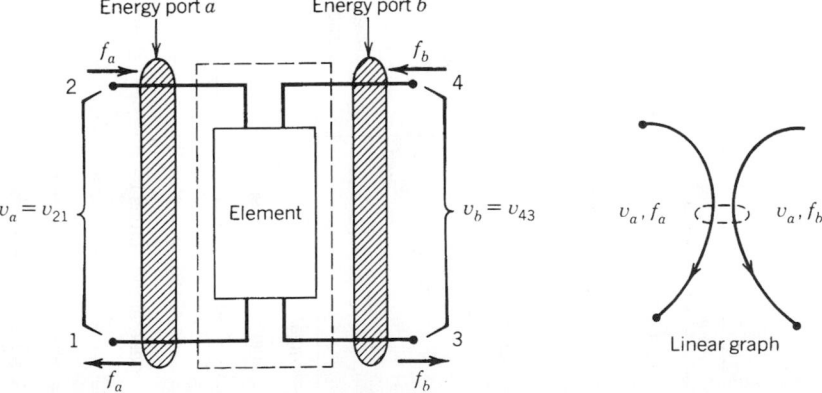

Fig. 3 Four-terminal or two-port element, showing through and across variables.

System	Symbol	Pure transformer	Ideal transformer	Transformation ratio
Mechanical translation (lever)	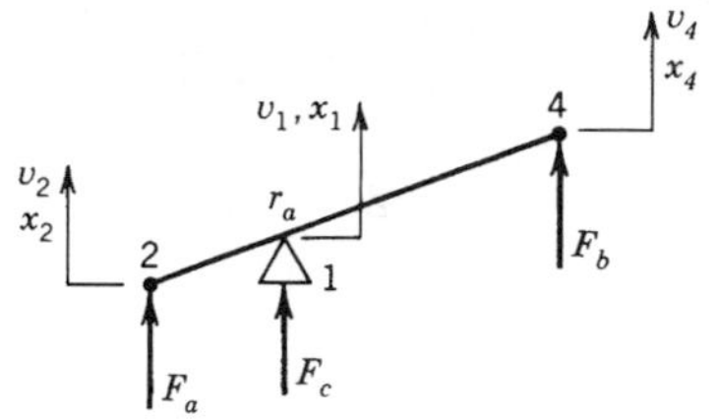	$x_{41} = f(x_{21})$	$v_{41} = nv_{21}$ $F_b = -\dfrac{1}{n} F_a$	$n = -\dfrac{r_b}{r_a}$ Lever ratio
Mechanical rotational (gears)	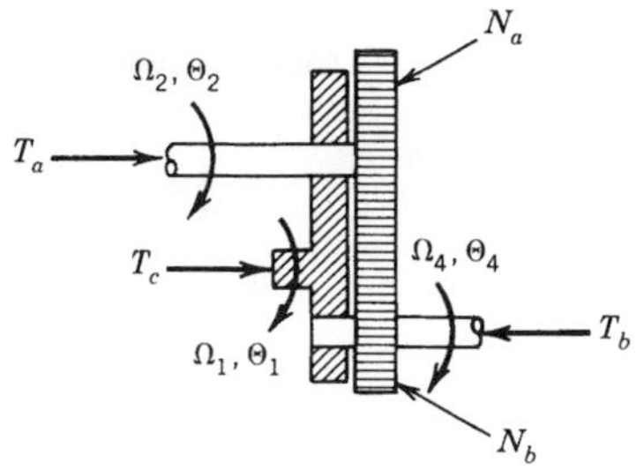	$\Theta_{41} = f(\Theta_2)$	$\Omega_{41} = n\Omega_{21}$ $T_b = -\dfrac{1}{n} T_a$	$n = -\dfrac{N_a}{N_b}$ Gear ratio
Electrical (magnetic)	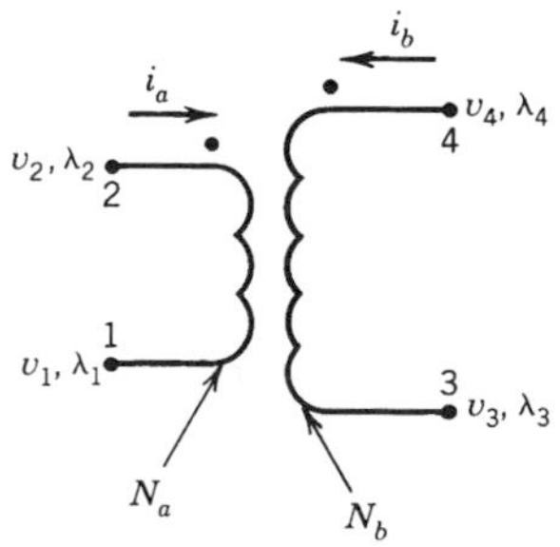	$\lambda_{43} = f(\lambda_{21})$	$v_{43} = nv_{21}$ $i_b = -\dfrac{1}{n} i_a$	$n = \dfrac{N_b}{N_a}$ Turns ratio
Fluid (differential piston)	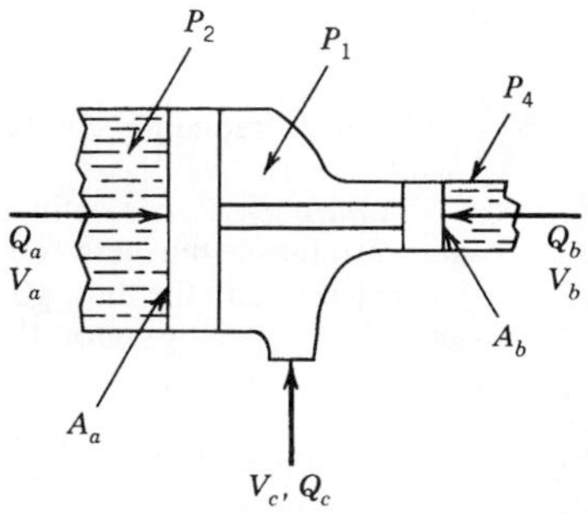	$V_b = f(V_a)$	$P_{41} = nP_{21}$ $Q_b = -\dfrac{1}{n} Q_a$	$n = \dfrac{A_a}{A_b}$ Area ratio

Fig. 4a Examples of transforms and transducers: (*a*) pure transformers (Ref. 1); (*b*) pure mechanical transformers and transforming transducers (Ref. 2).

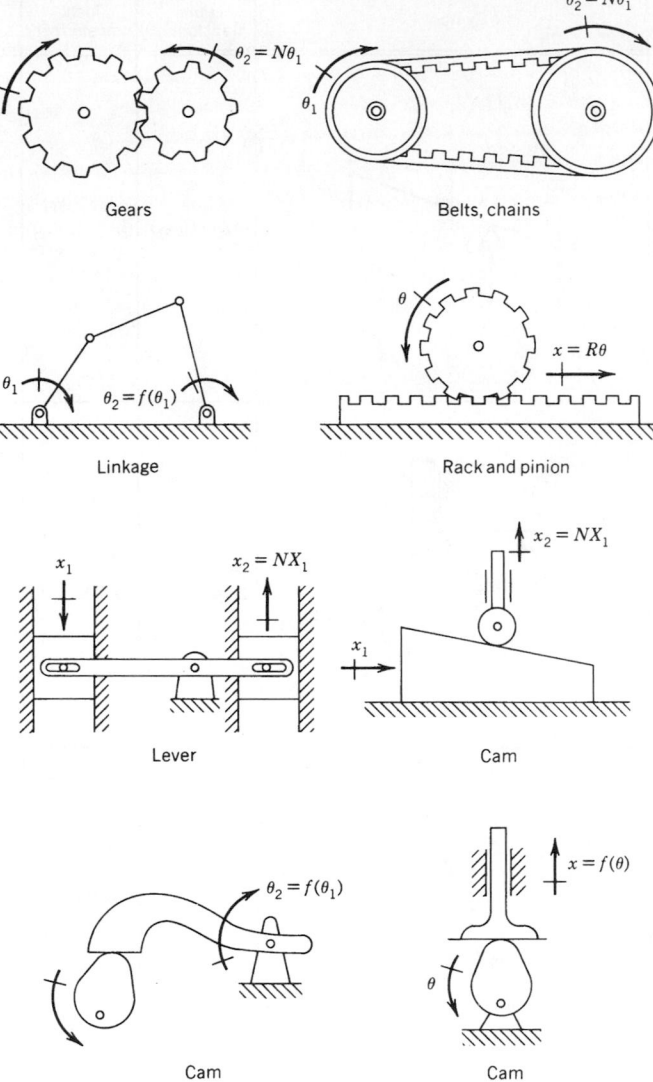

Fig. 4b (Continued)

between the primary variables:

$$v_b = nv_a \qquad f_b = -\frac{1}{n}f_a$$

where the constant of proportionality n is called the *transformation ratio*. Levers, mechanical linkages, pulleys, gear trains, electrical transformers, and differential-area fluid pistons are examples of physical devices that typically can be approximated by pure or ideal transformers. Figure 12.4 depicts some examples. *Pure transmitters*, which serve to transmit energy over

a distance, frequently can be thought of as transformers with $n = 1$.

A *pure gyrator* is defined by a single-valued constitutive relationship between the across variable at one energy port and the through variable at the other energy port. For a linear gyrator, the following relations apply:

$$v_b = rf_a \qquad f_b = \frac{-1}{r}v_a$$

where the constant of proportionality is called the *gyration ratio* or *gyrational resistance*. Physical devices that perform pure gyration are not as common as those performing pure transformation. A mechanical gyroscope is one example of a system that might be modeled as a gyrator.

In the preceding discussion of two-port elements, it has been assumed that the type of energy is the same at both energy ports. A *pure transducer*, on the other hand, changes energy from one physical medium to another. This change may be accomplished as either a transformation or a gyration. Examples of *transforming transducers* are gears with racks (mechanical rotation to mechanical translation) and electric motors and electric generators (electrical to mechanical rotation and vice versa). Examples of *gyrating transducers* are the piston-and-cylinder (fluid to mechanical) and piezoelectric crystals (mechanical to electrical).

More complex systems may have a large number of energy ports. A common *six-terminal* or *three-port element* called a *modulator* is depicted in Fig. 5. The flow of energy between ports *a* and *b* is controlled by the energy input at the modulating port *c*. Such devices inherently dissipate energy, since

$$P_a + P_c \geq P_b$$

although most often the modulating power P_c is much smaller than the power input P_a or the power output P_b. When port *a* is connected to a pure source element, the combination of source and modulator is called a *pure dependent source*. When the modulating power P_c is considered the input and the modulated power P_b is considered the output, the modulator is called an *amplifier*. Physical devices that often can be modeled as modulators include clutches, fluid valves and couplings, switches, relays, transistors, and variable resistors.

3 SYSTEM STRUCTURE AND INTERCONNECTION LAWS

3.1 A Simple Example

Physical systems are represented by connecting the terminals of pure elements in patterns that approximate the relationships among the properties of component devices. As an example, consider the mechanical–translational system depicted in Fig. 6a, which might represent an idealized automobile suspension system. The inertial properties associated with the masses of the chassis, passenger compartment, engine, and so on, all have been lumped together as the pure mass m_1. The inertial properties of the unsprung components (wheels, axles, etc.) have been lumped into the pure mass m_2. The compliance of the suspension is modeled as a pure spring with stiffness k_1 and the frictional effects (principally from the shock absorbers) as a pure damper with damping coefficient b. The road is represented as an input or source of vertical velocity, which is transmitted to the system through a spring of stiffness k_2, representing the compliance of the tires.

3.2 Structure and Graphs

The *pattern of interconnections* among elements is called the *structure* of the system. For a one-dimensional system, structure is conveniently represented by a *system graph*. The system graph for the idealized automobile suspension system of Fig. 6a is shown in Fig. 6b. Note that each distinct across variable (velocity) becomes a distinct *node* in the graph. Each distinct through variable (force) becomes a *branch* in the graph. Nodes coincide with the terminals of elements and branches coincide with the elements themselves. One node always represents *ground* (the constant velocity of the inertial reference frame v_g), and this is usually assumed to be zero for convenience. For nonelectrical systems, all the A-type elements (masses) have one terminal connection to the reference node. Because the masses are not physically connected to ground, however, the convention is to represent the corresponding branches in the graph by dashed lines.

System graphs are oriented by placing arrows on the branches. The orientation is arbitrary and serves to assign reference directions for both the

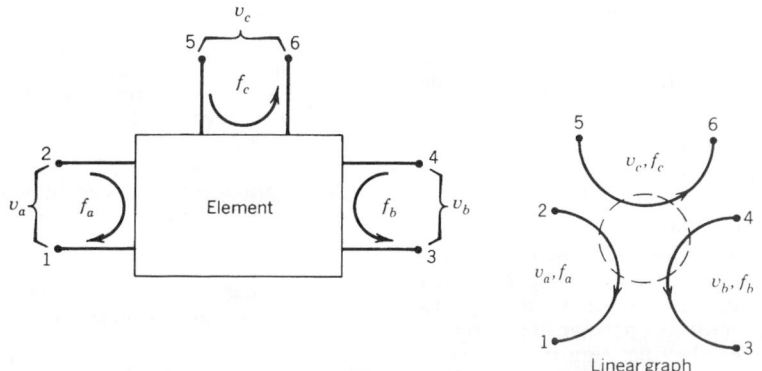

Fig. 5 Six-terminal or three-port element, showing through and across variables.

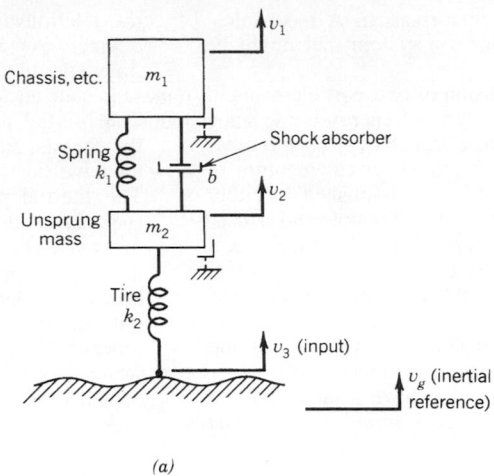

(a)

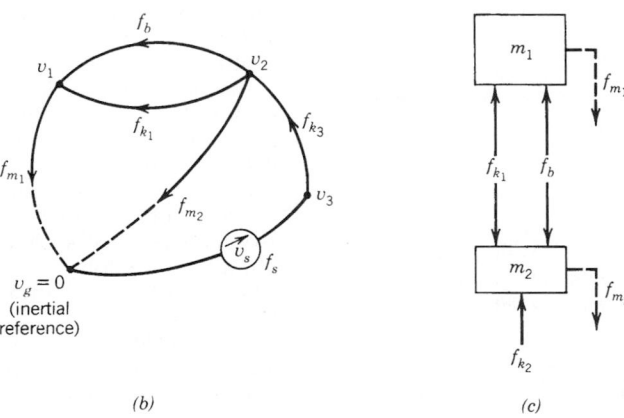

(b)   *(c)*

Fig. 6 Idealized model of automobile suspension system: (*a*) lumped-element model, (*b*) system graph, (*c*) free-body diagram.

through-variable and the across-variable difference. For example, the branch representing the damper in Fig. 6*b* is directed from node 2 (tail) to node 1 (head). This assigns $v_b = v_{21} = v_2 - v_1$ as the across-variable difference to be used in writing the damper elemental equation

$$f_b = bv_b = bv_{21}$$

The reference direction for the through variable is determined by the convention that power flow $P_b = f_b v_b$ into an element is positive. Referring to Fig. 6*a*, when v_{21} is positive, the damper is in compression. Therefore, f_b must be positive for compressive forces in order to obey the sign convention for power. By similar reasoning, tensile forces will be negative.

3.3 System Relations

The structure of a system gives rise to two sets of *interconnection laws* or *system relations*. Continuity relations apply to through variables and compatibility relations apply to across variables. The interpretation of system relations for various physical systems is given in Table 3.

Continuity is a general expression of dynamic equilibrium. In terms of the system graph, continuity states that the algebraic sum of all through variables entering a given node must be zero. Continuity applies at each node in the graph. For a graph with n nodes, continuity gives rise to n continuity equations, $n - 1$ of which are independent. For node i, the continuity equation is

$$\sum_j f_{ij} = 0$$

Table 3 System Relations for Various Systems

System	Continuity	Compatibility
Mechanical	Newton's first and third laws (conservation of momentum)	Geometrical constraints (distance is a scalar)
Electrical	Kirchhoff's current law (conservation of charge)	Kirchhoff's voltage law (potential is a scalar)
Fluid	Conservation of matter	Pressure is a scalar
Thermal	Conservation of energy	Temperature is a scalar

where the sum is taken over all branches (i, j) incident on i.

For the system graph depicted in Fig. 6b, the four continuity equations are

$$\text{Node 1:} \qquad f_{k_1} + f_b - f_{m_1} = 0$$

$$\text{Node 2:} \qquad f_{k_2} - f_{k_1} + f_b - f_{m_2} = 0$$

$$\text{Node 3:} \qquad f_s - f_{k_2} = 0$$

$$\text{Node g:} \qquad f_{m_1} + f_{m_2} - f_s = 0$$

Only three of these four equations are independent. Note, also, that the equations for nodes $1-3$ could have been obtained from the conventional *free-body diagrams* shown in Fig. 6c, where f_{m_1} and f_{m_2} are the *D'Alembert forces* associated with the pure masses. Continuity relations are also known as *vertex, node, flow*, and *equilibrium relations*.

Compatibility expresses the fact that the magnitudes of all across variables are scalar quantities. In terms of the system graph, compatibility states that the algebraic sum of the across-variable differences around any closed path in the graph must be zero. Compatibility applies to any closed path in the system. For convenience and to ensure the independence of the resulting equations, continuity is usually applied to the *meshes* or "windows" of the graph. A one-part graph with n nodes and b branches will have $b - n + 1$ meshes, each mesh yielding one independent compatibility equation. A planar graph with p separate parts (resulting from multiport elements) will have $b - n + p$ independent compatibility equations. For a closed path q, the compatibility equation is

$$\sum_q v_{ij} = 0$$

where the summation is taken over all branches (i, j) on the path.

For the system graph depicted in Fig. 6b, the three compatibility equations based on the meshes are

$$\text{Path } 1 \rightarrow 2 \rightarrow g \rightarrow 1: \qquad -v_b + v_{m_2} - v_{m_1} = 0$$

$$\text{Path } 1 \rightarrow 2 \rightarrow 1: \qquad -v_{k_1} + v_b = 0$$

$$\text{Path } 2 \rightarrow 3 \rightarrow g \rightarrow 2: \qquad -v_{k_2} - v_s - v_{m_2} = 0$$

These equations are all mutually independent and express apparent geometric identities. The first equation, for example, states that the velocity difference between the ends of the damper is identically the difference between the velocities of the masses it connects. Compatibility relations are also known as *path, loop*, and *connectedness* relations.

3.4 Analogs and Duals

Taken together, the element laws and system relations are a complete mathematical model of a system. When expressed in terms of generalized through and across variables, the model applies not only to the physical system for which it was derived, but also to any physical system with the same generalized system graph. Different physical systems with the same generalized model are called *analogs*. The mechanical rotational, electrical, and fluid analogs of the mechanical translational system of Fig. 6a are shown in Fig. 7. Note that because the original system contains an inductive storage element, there is no thermal analog.

Systems of the same physical type but in which the roles of the through variables and the across variables have been interchanged are called *duals*. The analog of a dual—or, equivalently, the dual of an analog—is sometimes called a *dualog*. The concepts of analogy and duality can be exploited in many different ways.

4 STANDARD FORMS FOR LINEAR MODELS

The element laws and system relations together constitute a complete mathematical description of a physical system. For a system graph with n nodes, b branches, and s sources, there will be $b - s$ element laws, $n - 1$ continuity equations, and $b - n + 1$ compatibility equations. This is a total of $2b - s$ differential and algebraic equations. For systems composed entirely of linear elements, it is always possible to reduce these $2b - s$ equations to either of two standard forms. The *input/output*, or *I/O, form* is the basis for *transform* or so-called *classical linear systems analysis*. The *state-variable form* is the basis for *state-variable* or so-called *modern linear systems analysis*.

4.1 I/O Form

The classical representation of a system is the "black-box," depicted in Fig. 8. The system has a set of p inputs (also called *excitations* or *forcing functions*), $u_j(t)$, $j = 1, 2, \ldots, p$. The system also has a set of

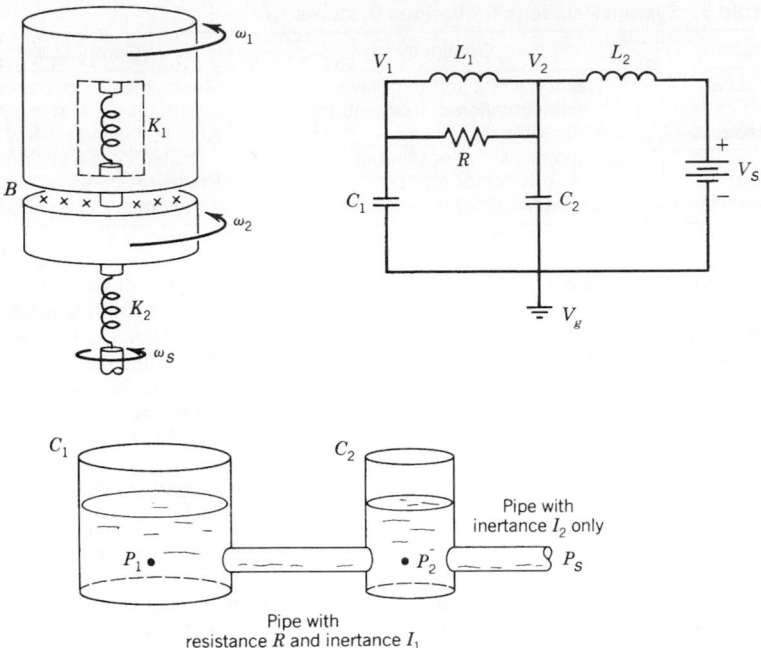

Fig. 7 Analogs of the idealized automobile suspension system depicted in Fig. 6.

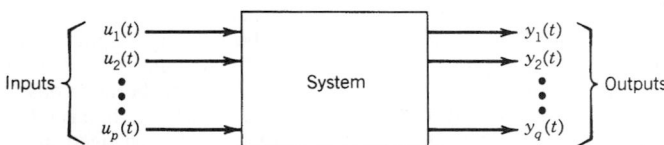

Fig. 8 Input/output (I/O), or blackbox, representation of a dynamic system.

q outputs (also called *response variables*), $y_k(t)$, $k = 1, 2, \ldots, q$. Inputs correspond to sources and are assumed to be known functions of time. Outputs correspond to physical variables that are to be measured or calculated.

Linear systems represented in I/O form can be modeled mathematically by *I/O differential equations*. Denoting as $y_{kj}(t)$ that part of the kth output $y_k(t)$ that is attributable to the jth input $u_j(t)$, there are $(p \times q)$ I/O equations of the form

$$\frac{d^n y_{kj}}{dt^n} + a_{n-1}\frac{d^{n-1} y_{kj}}{dt^{n-1}} + \cdots + a_1\frac{dy_{kj}}{dt} + a_0 y_{kj}(t)$$

$$= b_m\frac{d^m u_j}{dt^m} + b_{m-1}\frac{d^{m-1} u_j}{dt^{m-1}} + \cdots + b_1\frac{du_j}{dt} + b_0 u_j(t)$$

where $j = 1, 2, \ldots, p$ and $k = 1, 2, \ldots, q$. Each equation represents the dependence of one output and its derivatives on one input and its derivatives. By the

principle of superposition, the kth output in response to all of the inputs acting simultaneously is

$$y_k(t) = \sum_{j=1}^{p} y_{kj}(t)$$

A system represented by nth-order I/O equations is called an *nth-order system*. In general, the order of a system is determined by the number of *independent* energy storage elements within the system, that is, by the combined number of T-type and A-type elements for which the initial energy stored can be independently specified.

The coefficients $a_0, a_1, \ldots, a_{n-1}$ and $b_0, b_1, \ldots, b_m$ are parameter groups made up of algebraic combinations of the system physical parameters. For a system with constant parameters, therefore, these coefficients are also constant. Systems with constant parameters are called *time-invariant* systems and are the basis for classical analysis.

4.2 Deriving I/O Form: Example

I/O differential equations are obtained by combining element laws and continuity and compatibility equations in order to eliminate all variables except the input and the output. As an example, consider the mechanical system depicted in Fig. 9a, which might represent an idealized milling machine. A rotational motor is used to position the table of the machine tool through a rack and pinion. The motor is represented as a torque source T with inertia J and internal friction B. A flexible shaft, represented as a torsional spring K, is connected to a pinion gear of radius R. The pinion meshes with a rack, which is rigidly attached to the table of mass m. Damper b represents the friction opposing the motion of the table. The problem is to determine the I/O equation that expresses the relationship between the input torque T and the position of the table x.

The corresponding system graph is depicted in Fig. 9b. Applying continuity at nodes 1, 2, and 3 yields

$$\text{Node 1:} \quad T - T_J - T_B - T_K = 0$$

$$\text{Node 2:} \quad T_K - T_p = 0$$

$$\text{Node 3:} \quad -f_r - f_m - f_b = 0$$

Substituting the elemental equation for each of the one-port elements into the continuity equations and assuming zero ground velocities yield

$$\text{Node 1:} \quad T - J\dot{\omega}_1 - B\omega_1 - K\int(\omega_1 - \omega_2)\,dt = 0$$

$$\text{Node 2:} \quad K\int(\omega_1 - \omega_2)\,dt - T_p = 0$$

$$\text{Node 3:} \quad -f_r - m\dot{v} - bv = 0$$

Note that the definition of the across variables for each element in terms of the node variables, as above, guarantees that the compatibility equations are satisfied. With the addition of the constitutive relationships for the rack and pinion

$$\omega_2 = \frac{1}{R}v \quad \text{and} \quad T_p = -Rf_r$$

there are now five equations in the five unknowns $\omega_1, \omega_2, v, T_p$, and f_r. Combining these equations to eliminate all of the unknowns except v yields, after some manipulation,

$$a_3\frac{d^3v}{dt^3} + a_2\frac{d^2v}{dt^2} + a_1\frac{dv}{dt} + a_0v = b_1T$$

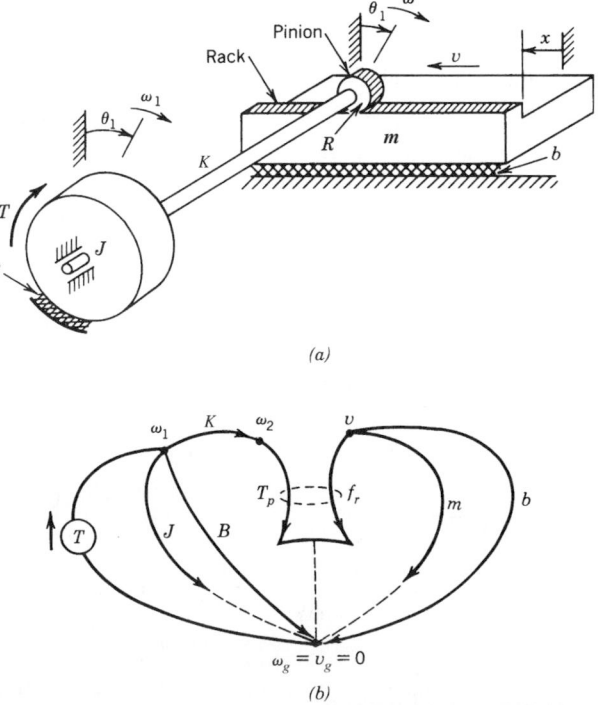

(a)

(b)

Fig. 9 Idealized model of milling machine: (a) lumped-element model (Ref. 3), (b) system graph.

where

$$a_3 = Jm \qquad a_1 = \frac{JK}{R^2} + Bb + mK \qquad b_1 = \frac{K}{R}$$

$$a_2 = Jb + mB \qquad a_0 = \frac{BK}{R^2} + Kb$$

Differentiating yields the desired I/O equation

$$a_3 \frac{d^3x}{dt^3} + a_2 \frac{d^2x}{dt^2} + a_1 \frac{dx}{dt} + a_0 x = b_1 \frac{dT}{dt}$$

where the coefficients are unchanged.

For many systems, combining element laws and system relations can best be achieved by ad hoc procedures. For more complicated systems, formal methods are available for the orderly combination and reduction of equations. These are the so-called *loop method* and *node method* and correspond to procedures of the same names originally developed in connection with electrical networks. The interested reader should consult Ref. 1.

4.3 State-Variable Form

For systems with multiple inputs and outputs, the I/O model form can become unwieldy. In addition, important aspects of system behavior can be suppressed in deriving I/O equations. The "modern" representation of dynamic systems, called the *state-variable form*, largely eliminates these problems. A state-variable model is the maximum reduction of the original element laws and system relations that can be achieved without the loss of any information concerning the behavior of a system. State-variable models also provide a convenient representation for systems with multiple inputs and outputs and for systems analysis using computer simulation.

State variables are a set of variables $x_1(t)$, $x_2(t), \dots, x_n(t)$ internal to the system from which any set of outputs can be derived, as depicted schematically in Fig. 10. A set of state variables is the minimum number of independent variables such that by knowing the values of these variables at any time t_0 and by knowing the values of the inputs for all time $t \geq t_0$, the

values of the state variables for all future time $t \geq t_0$ can be calculated. For a given system, the number n of state variables is unique and is equal to the order of the system. The definition of the state variables is not unique, however, and various combinations of one set of state variables can be used to generate alternative sets of state variables. For a physical system, the state variables summarize the *energy state* of the system at any given time.

A complete state-variable model consists of two sets of equations, the *state* or *plant equations* and the *output equations*. For the most general case, the state equations have the form

$$\dot{x}_1(t) = f_1[x_1(t), x_2(t), \dots, x_n(t), u_1(t), u_2(t), \dots, u_p(t)]$$

$$\dot{x}_2(t) = f_2[x_1(t), x_2(t), \dots, x_n(t), u_1(t), u_2(t), \dots, u_p(t)]$$

$$\vdots$$

$$\dot{x}_n(t) = f_n[x_1(t), x_2(t), \dots, x_n(t), u_1(t), u_2(t), \dots, u_p(t)]$$

and the output equations have the form

$$y_1(t) = g_1[x_1(t), x_2(t), \dots, x_n(t), u_1(t), u_2(t), \dots, u_p(t)]$$

$$y_2(t) = g_2[x_1(t), x_2(t), \dots, x_n(t), u_1(t), u_2(t), \dots, u_p(t)]$$

$$\vdots$$

$$y_q(t) = g_q[x_1(t), x_2(t), \dots, x_n(t), u_1(t), u_2(t), \dots, u_p(t)]$$

These equations are expressed more compactly as the two vector equations

$$\dot{x}(t) = f[x(t), u(t)] \qquad y(t) = g[x(t), u(t)]$$

where

$$\dot{x}(t) = n \times 1 \ \textit{state vector}$$

$$u(t) = p \times 1 \ \textit{input} \ \text{or} \ \textit{control vector}$$

$$y(t) = q \times 1 \ \textit{output} \ \text{or} \ \textit{response vector}$$

and f and g are vector-valued functions.

For linear systems, the state equations have the form

$$\dot{x}_1(t) = a_{11}(t)x_1(1) + \cdots + a_{1n}(t)x_n(t)$$
$$+ b_{11}(t)u_1(t) + \cdots + b_{1p}(t)u_p(t)$$

$$\dot{x}_2(t) = a_{21}(t)x_1(t) + \cdots + a_{2n}(t)x_n(t)$$
$$+ b_{21}(t)u_1(t) + \cdots + b_{2p}(t)u_p(t)$$

$$\vdots$$

$$\dot{x}_n(t) = a_{n1}(t)x_1(t) + \cdots + a_{nn}(t)x_n(t)$$
$$+ b_{n1}(t)u_1(t) + \cdots + b_{np}(t)u_p(t)$$

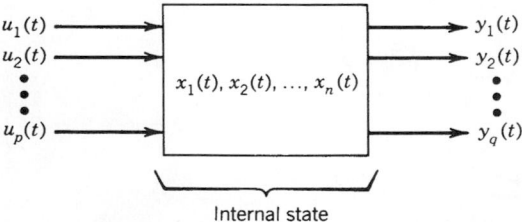

$$u_1(t) \longrightarrow \qquad \longrightarrow y_1(t)$$
$$u_2(t) \longrightarrow \qquad \longrightarrow y_2(t)$$
$$\vdots \qquad x_1(t), x_2(t), \dots, x_n(t) \qquad \vdots$$
$$u_p(t) \longrightarrow \qquad \longrightarrow y_q(t)$$

Internal state

Fig. 10 State-variable representation of a dynamic system.

and the output equations have the form

$$y_1(t) = c_{11}(t)x_1(t) + \cdots + c_{1n}(t)x_n(t)$$
$$+ d_{11}(t)u_1(t) + \cdots + d_{1p}(t)u_p(t)$$
$$y_2(t) = c_{21}(t)x_1(t) + \cdots + c_{2n}(t)x_n(t)$$
$$+ d_{21}(t)u_1(t) + \cdots + d_{2p}(t)u_p(t)$$

$$\vdots$$

$$y_n(t) = c_{q1}(t)x_1(t) + \cdots + c_{qn}(t)x_n(t)$$
$$+ d_{q1}(t)u_1(t) + \cdots + d_{qp}(t)u_p(t)$$

where the coefficients are groups of parameters. The linear model is expressed more compactly as the two linear vector equations

$$\dot{x}(t) = A(t)x(t) + B(t)u(t)$$
$$y(t) = C(t)x(t) + D(t)u(t)$$

where the vectors x, u, and y are the same as the general case and the matrices are defined as

$A = [a_{ij}]$ is the $n \times n$ system matrix

$B = [b_{jk}]$ is the $n \times p$ control, input, or

distribution matrix

$C = [c_{lj}]$ is the $q \times n$ output matrix

$D = [d_{lk}]$ is the $q \times p$ output distribution matrix

For a time-invariant linear system, all of these matrices are constant.

4.4 Deriving "Natural" State Variables

Procedure Because the state variables for a system are not unique, there are an unlimited number of alternative (but equivalent) state-variable models for the system. Since energy is stored only in generalized system storage elements, however, a natural choice for the state variables is the set of through and across variables corresponding to the independent T-type and A-type elements, respectively. This definition is sometimes called the set of *natural state variables* for the system.

For linear systems, the following procedure can be used to reduce the set of element laws and system relations to the natural state-variable model.

Step 1. For each independent T-type storage, write the element law with the derivative of the through variable isolated on the left-hand side, that is, $\dot{f} = L^{-1}v$.

Step 2. For each independent A-type storage, write the element law with the derivative of the across variable isolated on the left-hand side, that is, $\dot{v} = C^{-1}f$.

Step 3. Solve the compatibility equations, together with the element laws for the appropriate D-type and multiport elements, to obtain each of the across variables of the independent T-type elements in terms of the natural state variables and specified sources.

Step 4. Solve the continuity equations, together with the element laws for the appropriate D-type and multiport elements, to obtain the through variables of the A-type elements in terms of the natural state variables and specified sources.

Step 5. Substitute the results of step 3 into the results of step 1; substitute the results of step 4 into the results of step 2.

Step 6. Collect terms on the right-hand side and write in vector form.

Example The six-step process for deriving a natural state-variable representation, outlined in the preceding section, is demonstrated for the idealized automobile suspension depicted in Fig. 6:

Step 1

$$\dot{f}_{k_1} = k_1 v_{k_1} \qquad \dot{f}_{k_2} = k_2 v_{k_2}$$

Step 2

$$\dot{v}_{m_1} = m_1^{-1} f_{m_1} \qquad \dot{v}_{m_2} = m_2^{-1} f_{m_2}$$

Step 3

$$v_{k_1} = v_b = v_{m_2} - v_{m_1} \qquad v_{k_2} = -v_{m_2} - v_s$$

Step 4

$$f_{m_1} = f_{k_1} + f_b = f_{k_1} + b^{-1}(v_{m_2} - v_{m_1})$$
$$f_{m_2} = f_{k_2} - f_{k_1} - f_b = f_{k_2} - f_{k_1} - b^{-1}(v_{m_2} - v_{m_1})$$

Step 5

$$\dot{f}_{k_1} = k_1(v_{m_2} - v_{m_1})$$
$$\dot{v}_{m_1} = m_1^{-1}[f_{k_1} + b^{-1}(v_{m_2} - v_{m_1})]$$
$$\dot{f}_{k_2} = k_2(-v_{m_2} - v_s)$$
$$\dot{v}_{m_2} = m_2^{-1}[f_{k_2} - f_{k_1} - b^{-1}(v_{m_2} - v_{m_1})]$$

Step 6

$$\frac{d}{dt}\begin{bmatrix} f_{k_1} \\ f_{k_2} \\ v_{m_1} \\ v_{m_2} \end{bmatrix} = \begin{bmatrix} 0 & 0 & -k_1 & k_1 \\ 0 & 0 & 0 & -k_2 \\ 1/m_1 & 0 & -1/m_1 b & 1/m_1 b \\ -1/m_2 & 1/m_2 & 1/m_2 b & -1/m_2 b \end{bmatrix}$$

$$\times \begin{bmatrix} f_{k_1} \\ f_{k_2} \\ v_{m_1} \\ v_{m_2} \end{bmatrix} + \begin{bmatrix} 0 \\ -k_2 \\ 0 \\ 0 \end{bmatrix} v_s$$

4.5 Converting from I/O to Phase-Variable Form

Frequently, it is desired to determine a state-variable model for a dynamic system for which the I/O equation is already known. Although an unlimited number of such models is possible, the easiest to determine uses a special set of state variables called the *phase variables*. The phase variables are defined in terms of the output and its derivatives as follows:

$$x_1(t) = y(t)$$

$$x_2(t) = \dot{x}_1(t) = \frac{d}{dt}y(t)$$

$$x_3(t) = \dot{x}_2(t) = \frac{d^2}{dt^2}y(t)$$

$$\vdots$$

$$x_n(t) = \dot{x}_{n-1}(t) = \frac{d^{n-1}}{dt^{n-1}}y(t)$$

This definition of the phase variables, together with the I/O equation of Section 4.1, can be shown to result in a state equation of the form

$$\frac{d}{dt}\begin{bmatrix} x_1(t) \\ x_2(t) \\ \vdots \\ x_{n-1}(t) \\ x_n(t) \end{bmatrix} = \begin{bmatrix} 0 & 1 & 0 & \cdots & 0 \\ 0 & 0 & 1 & \cdots & 0 \\ \vdots & \vdots & \vdots & \ddots & \vdots \\ 0 & 0 & 0 & \cdots & 1 \\ -a_0 & -a_1 & -a_2 & \cdots & -a_{n-1} \end{bmatrix}$$

$$\times \begin{bmatrix} x_1(t) \\ x_2(t) \\ \vdots \\ x_{n-1}(t) \\ x_n(t) \end{bmatrix} + \begin{bmatrix} 0 \\ 0 \\ \vdots \\ 0 \\ 1 \end{bmatrix} u(t)$$

and an output equation of the form

$$y(t) = [b_0 \ b_1 \cdots b_m]\begin{bmatrix} x_1(t) \\ x_2(t) \\ \vdots \\ x_n(t) \end{bmatrix}$$

This special form of the system matrix, with 1's along the upper off-diagonal and 0's elsewhere except for the bottom row, is called a *companion matrix*.

5 APPROACHES TO LINEAR SYSTEMS ANALYSIS

There are two fundamental approaches to the analysis of linear, time-invariant systems. *Transform methods* use rational functions obtained from the Laplace transformation of the system I/O equations. Transform methods provide a particularly convenient algebra for combining the component submodels of a system and form the basis of so-called *classical control theory*. *State-variable methods* use the vector state and output equations directly. State-variable methods permit the adaptation of important ideas from linear algebra and form the basis for so-called *modern control theory*. Despite the deceiving names of "classical" and "modern," the two approaches are complementary. Both approaches are widely used in current practice and the control engineer must be conversant with both.

5.1 Transform Methods

A *transformation* converts a given mathematical problem into an equivalent problem, according to some well-defined rule called a *transform*. Prudent selection of a transform frequently results in an equivalent problem that is easier to solve than the original. If the solution to the original problem can be recovered by an inverse transformation, the three-step process of (a) transformation, (b) solution in the *transform domain*, and (c) inverse transformation may prove more attractive than direct solution of the problem in the original problem domain. This is true for fixed linear dynamic systems under the *Laplace transform*, which converts differential equations into equivalent algebraic equations.

Laplace Transforms: Definition The one-sided Laplace transform is defined as

$$F(s) = \mathscr{L}[f(t)] = \int_0^\infty f(t)e^{-st}\, dt$$

and the inverse transform as

$$f(t) = \mathscr{L}^{-1}[F(s)] = \frac{1}{2\pi j}\int_{\sigma-j\omega}^{\sigma+j\omega} F(s)e^{-st}\, ds$$

The Laplace transform converts the function $f(t)$ into the transformed function $F(s)$; the inverse transform recovers $f(t)$ from $F(s)$. The symbol $\mathscr{L}$ stands for the "Laplace transform of"; the symbol $\mathscr{L}^{-1}$ stands for "the inverse Laplace transform of."

The Laplace transform takes a problem given in the *time domain*, where all physical variables are functions of the *real variable* t, into the *complex-frequency domain*, where all physical variables are functions of the complex frequency $s = \sigma + j\omega$, where $j = \sqrt{-1}$ is the imaginary operator. Laplace transform pairs consist of the function $f(t)$ and its transform $F(s)$. Transform pairs can be calculated by substituting $f(t)$ into the defining equation and then evaluating the integral with s held constant. For a transform pair to exist, the corresponding integral must converge, that is,

$$\int_0^\infty |f(t)|e^{-\sigma^* t}\, dt < \infty$$

for some real $\sigma^* > 0$. Signals that are physically realizable always have a Laplace transform.

Tables of Transform Pairs and Transform Properties Transform pairs for functions commonly encountered in the analysis of dynamic systems rarely need to be calculated. Instead, pairs are determined by reference to a *table of transforms* such as that given in Table 4. In addition, the Laplace transform has a number of properties that are useful in determining the transforms and inverse transforms of functions in terms of the tabulated pairs. The most important of these are given in a *table of transform properties* such as that given in Table 5.

Table 4 Laplace Transform Pairs

$F(s)$	$f(t),\ t \geq 0$
1. 1	$\delta(t)$, the unit impulse at $t = 0$
2. $\dfrac{1}{s}$	1, the unit step
3. $\dfrac{n!}{s^{n+1}}$	t^n
4. $\dfrac{1}{s+a}$	e^{-at}
5. $\dfrac{1}{(s+a)^n}$	$\dfrac{1}{(n-1)!}t^{n-1}e^{-at}$
6. $\dfrac{a}{s(s+a)}$	$1 - e^{-at}$
7. $\dfrac{1}{(s+a)(s+b)}$	$\dfrac{1}{b-a}(e^{-at} - e^{-bt})$
8. $\dfrac{s+p}{(s+a)(s+b)}$	$\dfrac{1}{b-a}[(p-a)e^{-at} - (p-b)e^{-bt}]$
9. $\dfrac{1}{(s+a)(s+b)(s+c)}$	$\dfrac{e^{-at}}{(b-a)(c-a)} + \dfrac{e^{-bt}}{(c-b)(a-b)} + \dfrac{e^{-ct}}{(a-c)(b-c)}$
10. $\dfrac{s+p}{(s+a)(s+b)(s+c)}$	$\dfrac{(p-a)e^{-at}}{(b-a)(c-a)} + \dfrac{(p-b)e^{-bt}}{(c-b)(a-b)} + \dfrac{(p-c)e^{-ct}}{(a-c)(b-c)}$
11. $\dfrac{b}{s^2 + b^2}$	$\sin bt$
12. $\dfrac{s}{s^2 + b^2}$	$\cos bt$
13. $\dfrac{b}{(s+a)^2 + b^2}$	$e^{-at}\sin bt$
14. $\dfrac{s+a}{(s+a)^2 + b^2}$	$e^{-at}\cos bt$
15. $\dfrac{\omega_n^2}{s^2 + 2\zeta\omega_n s + \omega_n^2}$	$\dfrac{\omega_n}{\sqrt{1-\zeta^2}}e^{-\zeta\omega_n t}\sin\omega_n\sqrt{1-\zeta^2}\,t,\quad \zeta < 1$
16. $\dfrac{\omega_n^2}{s(s^2 + 2\zeta\omega_n s + \omega_n^2)}$	$1 + \dfrac{1}{\sqrt{1-\zeta^2}}e^{-\zeta\omega_n t}\sin(\omega_n\sqrt{1-\zeta^2}\,t + \phi)$
	$\phi = \tan^{-1}\dfrac{\sqrt{1-\zeta^2}}{\zeta} + \pi$
	(third quadrant)

Table 5   Laplace Transform Properties

$f(t)$	$F(s) = \int_0^\infty f(t)e^{-st}\,dt$
1. $af_1(t) + bf_2(t)$	$aF_1(s) + bF_2(s)$
2. $\dfrac{df}{dt}$	$sF(s) - f(0)$
3. $\dfrac{d^2f}{dt^2}$	$s^2F(s) - sf(0) - \dfrac{df}{dt}\Big|_{r=0}$
4. $\dfrac{d^nf}{dt^n}$	$s^nF(s) - \displaystyle\sum_{k=1}^{n} s^{n-k}g_{k-1}$
	$g_{k-1} = \dfrac{d^{k-1}f}{dt^{k-1}}\Big|_{r=0}$
5. $\displaystyle\int_0^t f(t)\,dt$	$\dfrac{F(s)}{s} + \dfrac{h(0)}{s}$
	$h(0) = \displaystyle\int f(t)\,dt\Big|_{r=0}$
6. $\begin{cases} 0 & t < D \\ f(t-D) & t \ge D \end{cases}$	$e^{-sD}F(s)$
7. $e^{-at}f(t)$	$F(s+a)$
8. $f\left(\dfrac{t}{a}\right)$	$aF(as)$
9. $f(t) = \displaystyle\int_0^t x(t-\tau)y(\tau)\,d\tau$	$F(s) = X(s)Y(s)$
$= \displaystyle\int_0^t y(t-\tau)x(\tau)\,d\tau$	
10. $f(\infty) = \displaystyle\lim_{s\to 0} sF(s)$	
11. $f(0+) = \displaystyle\lim_{s\to\infty} sF(s)$	

Poles and Zeros   The response of a dynamic system most often assumes the following form in the complex-frequency domain:

$$F(s) = \frac{N(s)}{D(s)} = \frac{b_m s^m + b_{m-1}s^{m-1} + \cdots + b_1 s + b_0}{s^n + a_{n-1}s^{n-1} + \cdots + a_1 s + a_0} \quad (1)$$

Functions of this form are called *rational functions* because these are the ratio of two polynomials $N(s)$ and $D(s)$. If $n \ge m$, then $F(s)$ is a *proper rational function;* if $n > m$, then $F(s)$ is a *strictly proper rational function.*

In factored form, the rational function $F(s)$ can be written as

$$F(s) = \frac{N(s)}{D(s)} = \frac{b_m(s - z_1)(s - z_2)\cdots(s - z_m)}{(s - p_1)(s - p_2)\cdots(s - p_n)} \quad (2)$$

The roots of the numerator polynomial $N(s)$ are denoted by z_j, $j = 1, 2, \ldots, m$. These numbers are

called the *zeros* of $F(s)$, since $F(z_j) = 0$. The roots of the denominator polynomial are denoted by p_i, 1, 2, $\ldots$, n. These numbers are called the *poles* of $F(s)$ since $\lim_{s\to p_i} F(s) = \pm\infty$.

Inversion by Partial-Fraction Expansion   The *partial-fraction expansion theorem* states that a strictly proper rational function $F(s)$ with *distinct (nonrepeated)* poles p_i, $i = 1, 2, \ldots, n$, can be written as the sum

$$F(s) = \frac{A_1}{s - p_1} + \frac{A_2}{s - p_2} + \cdots + \frac{A_n}{s - p_n} = \sum_{i=1}^{n} A_i\left(\frac{1}{s - p_i}\right) \quad (3)$$

where the A_i, $i = 1, 2, \ldots, n$, are constants called *residues.* The inverse transform of $F(s)$ has the simple form

$$f(t) = A_1 e^{p_1 t} + A_2 e^{p_2 t} + \cdots + A_n e^{p_n t} = \sum_{i=1}^{n} A_i e^{p_i t}$$

The *Heaviside expansion theorem* gives the following expression for calculating the residue at the pole p_i,

$$A_i = (s - p_i)F(s)|_{s=p_i} \quad \text{for } i = 1, 2, \ldots, n$$

These values can be checked by substituting into Eq. (3), combining the terms on the right-hand side of Eq. (3), and showing that the result yields the values for all the coefficients b_j, $j = 1, 2, \ldots, m$, originally specified in the form of Eq. (3).

Repeated Poles   When two or more poles of a strictly proper rational function are identical, the poles are said to be *repeated* or *nondistinct*. If a pole is repeated q times, that is, if $p_i = p_{i+1} = \cdots = p_{i+q-1}$, then the pole is said to be of *multiplicity q*. A strictly proper rational function with a pole of multiplicity q will contain q terms of the form

$$\frac{A_{i1}}{(s - p_i)^q} + \frac{A_{i2}}{(s - p_i)^{q-1}} + \cdots + \frac{A_{iq}}{s - p_i}$$

in addition to the terms associated with the distinct poles. The corresponding terms in the inverse transform are

$$\left(\frac{1}{(q - 1)!} A_{i1} t^{(q-1)} + \frac{1}{(q - 2)!} A_{i2} t^{(q-2)} + \cdots + A_{iq} \right) e^{p_i t}$$

The corresponding residues are

$$A_{i1} = (s - p_i)^q F(s)|_{s=p_i}$$

$$A_{i2} = \left(\frac{d}{ds} [(s - p_i)^q F(s)] \right)\Bigg|_{s=p_i}$$

$$\vdots$$

$$A_{iq} = \frac{1}{(q - 1)!} \left(\frac{d^{(q-1)}}{ds^{(q-1)}} [(s - p_i)^q F(s)] \right)\Bigg|_{s=p_i}$$

Complex Poles   A strictly proper rational function with complex-conjugate poles can be inverted using partial-fraction expansion. Using a method called *completing the square*, however, is almost always easier. Consider the function

$$F(s) = \frac{B_1 s + B_2}{(s + \sigma - j\omega)(s + \sigma + j\omega)}$$

$$= \frac{B_1 s + B_2}{s^2 + 2\sigma s + \sigma^2 + \omega^2}$$

$$= \frac{B_1 s + B_2}{(s + \sigma)^2 + \omega^2}$$

From the transform tables the Laplace inverse is

$$f(t) = e^{-\sigma t} [B_1 \cos \omega t + B_3 \sin \omega t]$$

$$= K e^{-\sigma t} \cos(\omega t + \phi)$$

where

$$B_3 = \left(\frac{1}{\omega} \right) (B_2 - a B_1) \qquad K = \sqrt{B_1^2 + B_3^2}$$

$$\phi = - \tan^{-1} \left(\frac{B_3}{B_1} \right)$$

Proper and Improper Rational Functions   If $F(s)$ is not a strictly proper rational function, then $N(s)$ must be divided by $D(s)$ using *synthetic division*. The result is

$$F(s) = \frac{N(s)}{D(s)} = P(s) + \frac{N^*(s)}{D(s)}$$

where $P(s)$ is a polynomial of degree $m - n$ and $N^*(s)$ is a polynomial of degree $n - 1$. Each term of $P(s)$ may be inverted directly using the transform tables. The strictly proper rational function $N^*(s)/D(s)$ may be inverted using partial-fraction expansion.

Initial-Value and Final-Value Theorems   The limits of $f(t)$ as time approaches zero or infinity frequently can be determined directly from the transform $F(s)$ without inverting. The *initial-value theorem* states that

$$f(0_+) = \lim_{s \to \infty} s F(s)$$

where the limit exists. If the limit does not exist (i.e., is infinite), the value of $f(0_+)$ is undefined. The *final-value theorem* states that

$$f(\infty) = \lim_{s \to 0} s F(s)$$

provided that (with the possible exception of a single pole at $s = 0$) $F(s)$ has no poles with nonnegative real parts.

Transfer Functions   The Laplace transform of the system I/O equation may be written in terms of the transform $Y(s)$ of the system response $y(t)$ as

$$Y(s) = \frac{G(s)N(s) + F(s)D(s)}{P(s)D(s)}$$

$$= \left(\frac{G(s)}{P(s)} \right) \left(\frac{N(s)}{D(s)} \right) + \frac{F(s)}{P(s)}$$

where (a) $P(s) = a_n s^n + a_{n-1} + \cdots + a_1 s + a_0$ is the *characteristic polynomial* of the system

(b) $G(s) = b_m s^m + b_{m-1} s^{m-1} + \cdots + b_1 s + b_0$ represents the *numerator dynamics* of the system

(c) $U(s) = N(s)/D(s)$ is the transform of the input to the system, $u(t)$, assumed to be a rational function

(d) $F(s) =$
$$a_n y(0)s^{n-1} + \left(a_n \frac{dy}{dt}(0) \right.$$
$$+ \left. a_{n-1}y(0) \right)s^{n-2} + \cdots$$
$$+ \left(a_n \frac{d^{n-1}y}{dt^{n-1}}(0) + a_{n-1}\frac{d^{n-2}y}{dt}(0) \right.$$
$$+ \cdots \left. + a_1 y(0) \right)$$

reflects the initial system state [i.e., the initial conditions on $y(t)$ and its first $n-1$ derivatives]

The transformed response can be thought of as the sum of two components,

$$Y(s) = Y_{zs}(s) + Y_{zi}(s)$$

where (e) $Y_{zs}(s) = [G(s)/P(s)][N(s)/D(s)] = H(s)U(s)$ is the transform of the *zero-state response*, that is, the response of the system to the input alone

(f) $Y_{zi}(s) = F(s)/P(s)$ is the transform of the *zero-input response*, that is, the response of the system to the initial state alone

The rational function

(g) $H(s) = Y_{zs}(s)/U(s) = G(s)/P(s)$ is the *transfer function* of the system, defined as the Laplace transform of the ratio of the system response to the system input, assuming zero initial conditions

The transfer function plays a crucial role in the analysis of fixed linear systems using transforms and can be written directly from knowledge of the system I/O equation as

$$H(s) = \frac{b_m s^m + \cdots + b_0}{a_n s^n + a_{n-1}s^{n-1} + \cdots + a_1 s + a_0}$$

Impulse Response Since $U(s) = 1$ for a unit impulse function, the transform of the zero-state response to a unit impulse input is given by the relation

(g) as

$$Y_{zs}(s) = H(s)$$

that is, the system transfer function. In the time domain, therefore, the unit *impulse response* is

$$h(t) = \begin{cases} 0 & \text{for } t \le 0 \\ \mathscr{L}^{-1}[H(s)] & \text{for } t > 0 \end{cases}$$

This simple relationship is profound for several reasons. First, this provides for a direct characterization of time domain response $h(t)$ in terms of the properties (poles and zeros) of the rational function $H(s)$ in the complex-frequency domain. Second, applying the convolution transform pair (Table 5) to relation (e) above yields

$$Y_{zs}(t) = \int_0^t h(\tau)u(t-\tau)\,d\tau$$

In words, the zero-state output corresponding to an arbitrary input $u(t)$ can be determined by convolution with the impulse response $h(t)$. In other words, the impulse response completely characterizes the system. The impulse response is also called the system *weighing function*.

Block Diagrams Block diagrams are an important conceptual tool for the analysis and design of dynamic systems because block diagrams provide a graphic means for depicting the relationships among system variables and components. A block diagram consists of unidirectional blocks representing specified system components or subsystems interconnected by arrows representing system variables. Causality follows in the direction of the arrows, as in Fig. 11, indicating that the output is caused by the input acting on the system defined in the block.

Combining transform variables, transfer functions, and block diagrams provides a powerful graphical means for determining the overall transfer function of a system when the transfer functions of its component subsystems are known. The basic blocks in such diagrams are given in Fig. 12. A block diagram

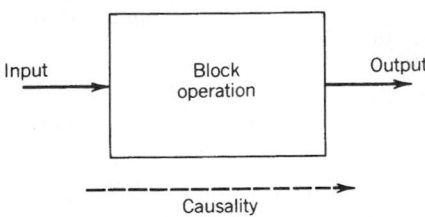

Fig. 11 Basic block diagram, showing assumed direction of causality or loading.

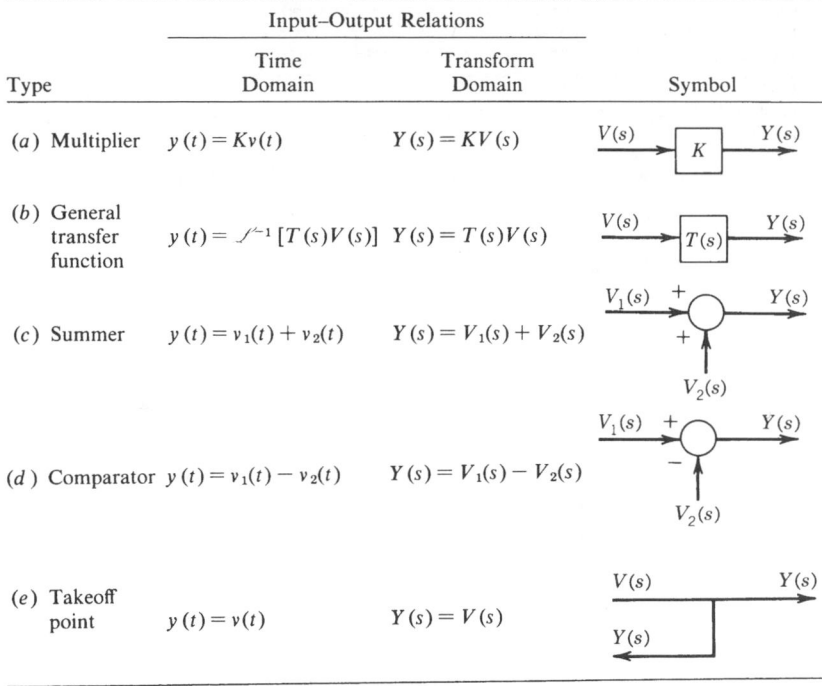

Type	Input–Output Relations		Symbol
	Time Domain	Transform Domain	
(a) Multiplier	$y(t) = Kv(t)$	$Y(s) = KV(s)$	
(b) General transfer function	$y(t) = \mathcal{L}^{-1}[T(s)V(s)]$	$Y(s) = T(s)V(s)$	
(c) Summer	$y(t) = v_1(t) + v_2(t)$	$Y(s) = V_1(s) + V_2(s)$	
(d) Comparator	$y(t) = v_1(t) - v_2(t)$	$Y(s) = V_1(s) - V_2(s)$	
(e) Takeoff point	$y(t) = v(t)$	$Y(s) = V(s)$	

Fig. 12 Basic block diagram elements (Ref. 4).

comprising many blocks and summers can be reduced to a single transfer function block by using the diagram transformations given in Fig. 13.

5.2 Transient Analysis Using Transform Methods

Basic to the study of dynamic systems are the concepts and terminology used to characterize system behavior or performance. These ideas are aids in *defining* behavior in order to consider for a given context those features of behavior that are desirable and undesirable; in *describing* behavior in order to communicate concisely and unambiguously various behavioral attributes of a given system; and in *specifying* behavior in order to formulate desired behavioral norms for system design. Characterization of dynamic behavior in terms of standard concepts also leads in many cases to analytical shortcuts since key features of the system response frequently can be determined without actually solving the system model.

Parts of the Complete Response A variety of names are used to identify terms in the response of a fixed linear system. The complete response of a system may be thought of alternatively as the sum of the following:

1. The *free response* (or complementary or homogeneous solution) and the *forced response* (or

particular solution). The free response represents the natural response of a system when inputs are removed and the system responds to some initial stored energy. The forced response of the system depends on the form of the input only.

2. The *transient response* and the *steady-state response*. The transient response is that part of the output that decays to zero as time progresses. The steady-state response is that part of the output that remains after all the transients disappear.

3. The *zero-state response* and the *zero-input response*. The zero-state response is the complete response (both free and forced responses) to the input when the initial state is zero. The zero-input response is the complete response of the system to the initial state when the input is zero.

Test Inputs or Singularity Functions For a stable system, the response to a specific input signal will provide several measures of system performance. Since the actual inputs to a system are not usually known a priori, characterization of the system behavior is generally given in terms of the response to one of a standard set of *test input signals*. This approach provides a common basis for the comparison of different systems. In addition, many inputs actually encountered can be approximated by some combination of standard inputs.

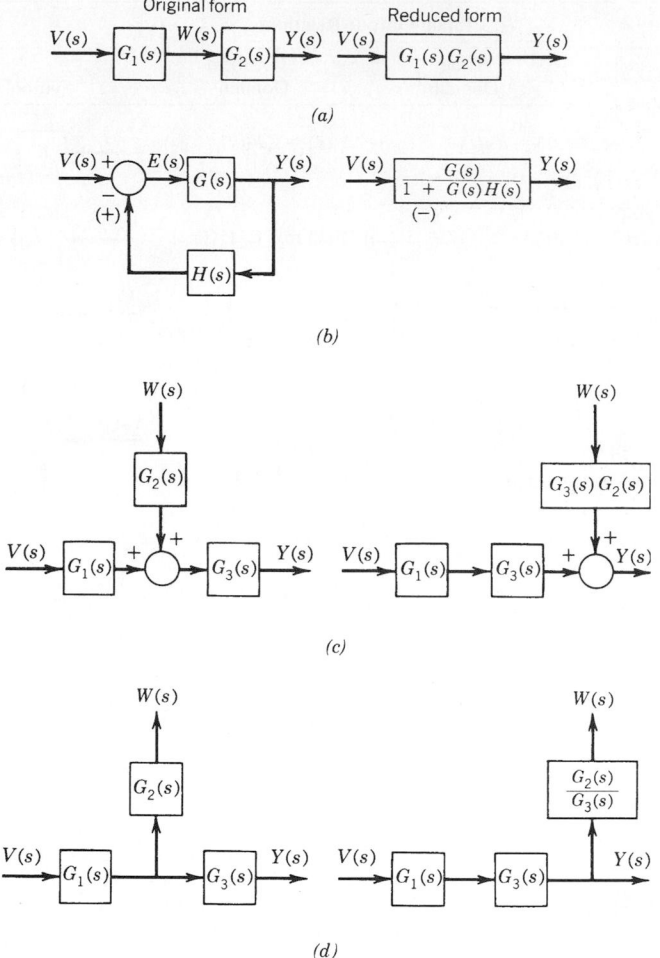

Fig. 13 Representative block diagram transformations: (a) series or cascaded elements, (b) feedback loop, (c) relocated summer, (d) relocated takeoff point (Ref. 4).

The most commonly used test inputs are members of the family of *singularity functions*, depicted in Fig. 14.

First-Order Transient Response   The standard form of the I/O equation for a first-order system is

$$\frac{dy}{dt} + \frac{1}{\tau}y(t) = \frac{1}{\tau}u(t)$$

where the parameter τ is called the system *time constant*. The response of this standard first-order system to three test inputs is depicted in Fig. 15, assuming zero initial conditions on the output $y(t)$. For all inputs, it is clear that the response approaches its steady state monotonically (i.e., without oscillations) and that the *speed of response* is completely characterized by the

time constant τ. The transfer function of the system is

$$H(s) = \frac{Y(s)}{U(s)} = \frac{1/\tau}{s + 1/\tau}$$

and therefore $\tau = -p^{-1}$, where p is the system pole. As the absolute value of p increases, τ decreases and the response becomes faster.

The response of the standard first-order system to a step input of magnitude u for arbitrary initial condition $y(0) = y_0$ is

$$y(t) = y_{ss} - [y_{ss} - y_0]e^{-t/r}$$

where $y_{ss} = u$ is the steady-state response. Table 6 and Fig. 16 record the values of $y(t)$ and $\dot{y}(t)$ for

Function	Graph	$u(t)$	$U(s)$
Impulse		$u(t) = \begin{cases} \lim\limits_{\epsilon \to 0} A/\epsilon, & 0_+ < t < \epsilon \\ 0, & \text{otherwise} \end{cases}$	$U(s) = A$
Step		$u(t) = \begin{cases} A, & 0_+ < t \\ 0, & \text{otherwise} \end{cases}$	$U(s) = \dfrac{A}{s}$
Ramp		$u(t) = \begin{cases} At, & 0_+ < t \\ 0, & \text{otherwise} \end{cases}$	$U(s) = \dfrac{A}{s^2}$
Parabolic		$u(t) = \begin{cases} At^2, & 0_+ < t \\ 0, & \text{otherwise} \end{cases}$	$U(s) = \dfrac{2A}{s^3}$
General		$u(t) = \begin{cases} At^n, & 0_+ < t \\ 0, & \text{otherwise} \end{cases}$	$U(s) = \dfrac{n!A}{s^{n+1}}$

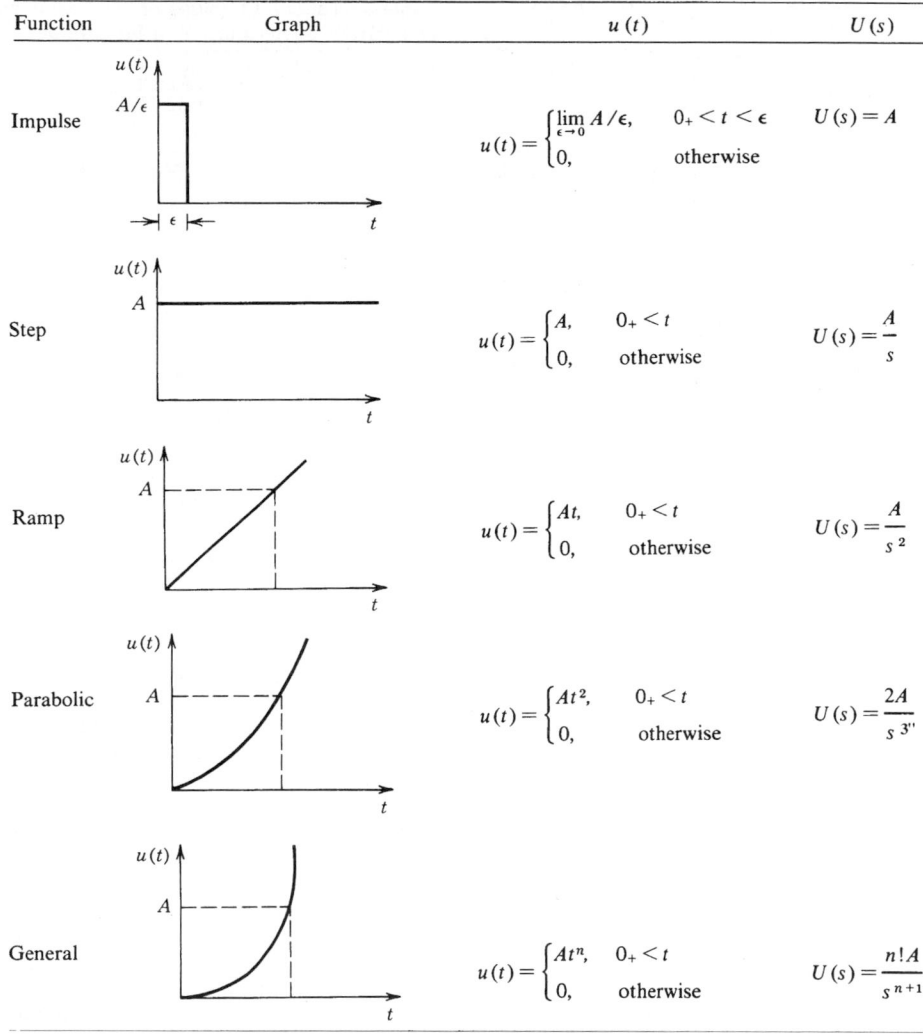

Fig. 14 Family of singularity functions commonly used as test inputs.

$t = k\tau, k = 0, 1, \ldots, 6$. Note that over any time interval of duration τ, the response increases approximately 63% of the difference between the steady-state value and the value at the beginning of the time interval, that is,

$$y(t + \tau) - y(t) \approx 0.63212[y_{ss} - y(t)]$$

Note also that the slope of the response at the beginning of any time interval of duration τ intersects the steady-state value y_{ss} at the end of the interval, that is,

$$\frac{dy}{dt}(t) = \frac{y_{ss} - y(t)}{\tau}$$

Finally, note that after an interval of four time constants, the response is within 98% of the steady-state value, that is,

$$y(4\tau) \approx 0.98168(y_{ss} - y_0)$$

For this reason, $T_s = 4\tau$ is called the (2%) *setting time*.

Second-Order Transient Response The standard form of the I/O equation for a second-order system is

$$\frac{d^2 y}{dt^2} + 2\zeta \omega_n \frac{dy}{dt} + \omega_n^2 y(t) = \omega_n^2 u(t)$$

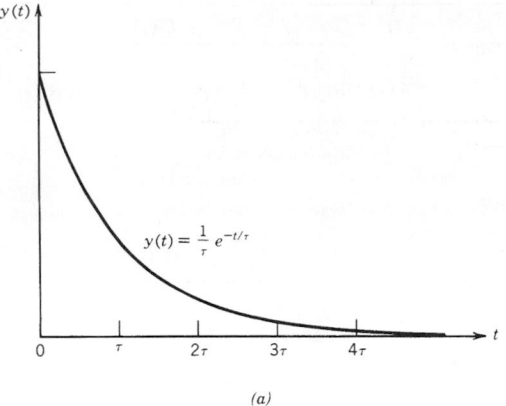

$$y(t) = \frac{1}{\tau} e^{-t/\tau}$$

(a)

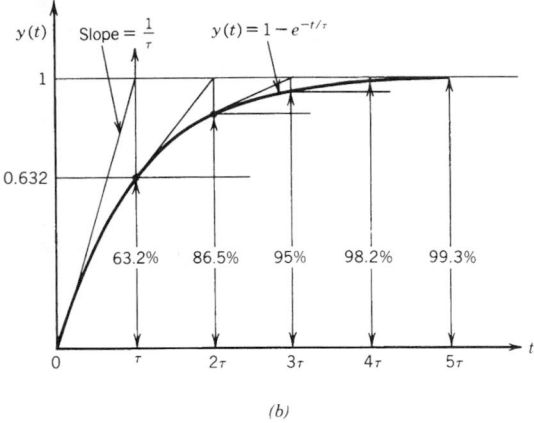

Slope = $\frac{1}{\tau}$ $y(t) = 1 - e^{-t/\tau}$

1

0.632

63.2%   86.5%   95%   98.2%   99.3%

(b)

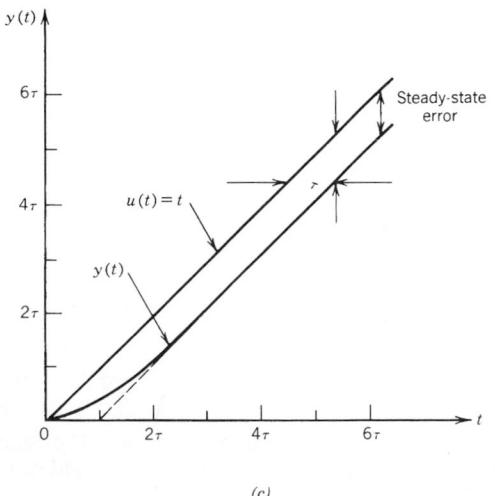

$u(t) = t$

Steady-state error

$y(t)$

(c)

Fig. 15 Response of a first-order system to (*a*) unit impulse, (*b*) unit step, and (*c*) unit ramp inputs.

Table 6   Tabulated Values of the Response of a First-Order System to a Unit Step Input

t	$y(t)$	$\dot{y}(t)$
0	0	τ^{-1}
τ	0.632	$0.368\tau^{-1}$
2τ	0.865	$0.135\tau^{-1}$
3τ	0.950	$0.050\tau^{-1}$
4τ	0.982	$0.018\tau^{-1}$
5τ	0.993	$0.007\tau^{-1}$
6τ	0.998	$0.002\tau^{-1}$

with transfer function

$$H(s) = \frac{Y(s)}{U(s)} = \frac{\omega_n^2}{s^2 + 2\zeta\omega_n s + \omega_n^2}$$

The system poles are obtained by applying the quadratic formula to the characteristic equation as

$$p_{1,2} = -\zeta\omega_n \pm j\omega_n\sqrt{1 - \zeta^2}$$

where the following parameters are defined: ζ is the *damping ratio*, ω_n is the *natural frequency*, and $\omega_d = \omega_n\sqrt{1 - \zeta^2}$ is the *damped natural frequency*.

The nature of the response of the standard second-order system to a step input depends on the value of the damping ratio, as depicted in Fig. 17. For a stable system, four classes of response are defined.

1. *Overdamped Response* ($\zeta > 1$). The system poles are real and distinct. The response of the second-order system can be decomposed into the response of two cascaded first-order systems, as shown in Fig. 18.
2. *Critically Damped Response* ($\zeta = 1$). The system poles are real and repeated. This is the limiting case of overdamped response, where the response is as fast as possible without overshoot.
3. *Underdamped Response* ($1 > \zeta > 0$). The system poles are complex conjugates. The response oscillates at the damped frequency ω_d. The magnitude of the oscillations and the speed with which the oscillations decay depend on the damping ratio ζ.
4. *Harmonic Oscillation* ($\zeta = 0$). The system poles are pure imaginary numbers. The response oscillates at the natural frequency ω_n and the oscillations are undamped (i.e., the oscillations are sustained and do not decay).

Complex *s* Plane   The location of the system poles (roots of the characteristic equation) in the *complex s plane* reveals the nature of the system response to test inputs. Figure 19 shows the relationship between the location of the poles in the complex plane and the parameters of the standard second-order system.

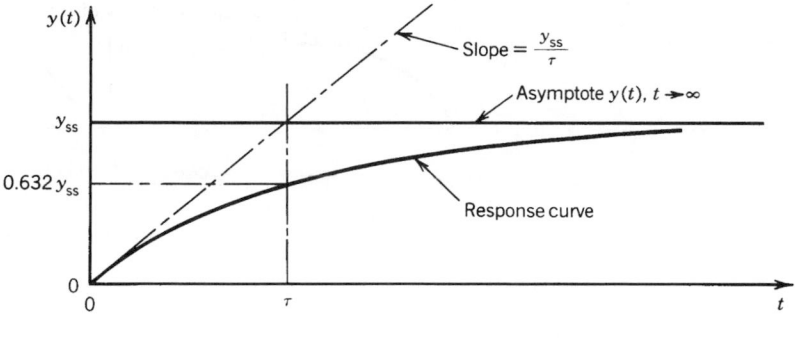

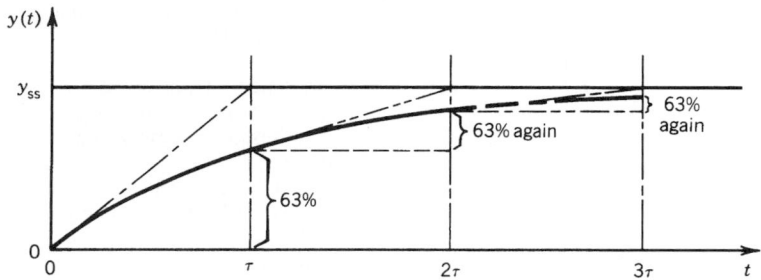

Fig. 16 Response of a first-order system to a unit step input, showing the relationship to the time constant.

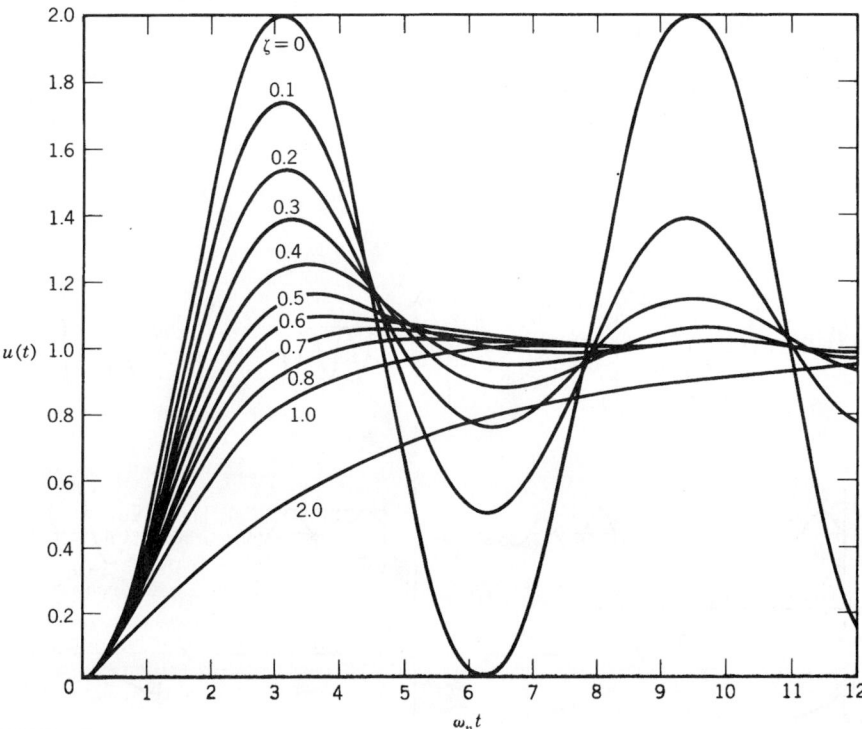

Fig. 17 Response of a second-order system to a unit step input for selected values of the damping ratio.

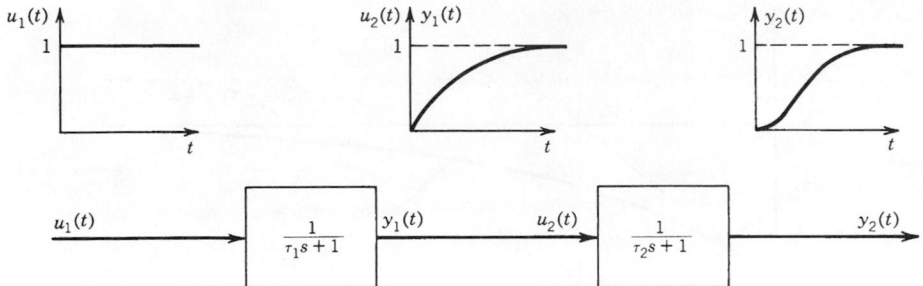

Fig. 18 Overdamped response of a second-order system decomposed into the responses of two first-order systems.

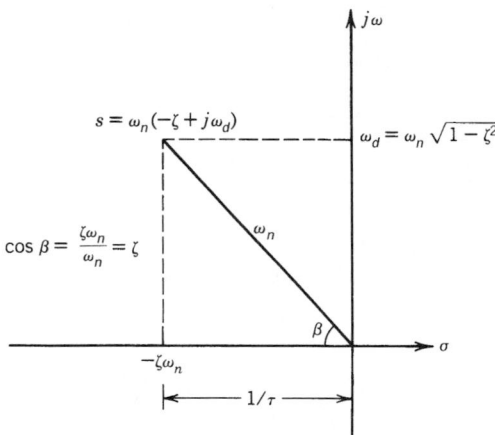

Fig. 19 Location of the upper complex pole in the s plane in terms of the parameters of the standard second-order system.

Figure 20 shows the unit impulse response of a second-order system corresponding to various pole locations in the complex plane.

Transient Response of Higher Order Systems
The response of third- and higher order systems to test inputs is simply the sum of terms representing component first- and second-order responses. This is because the system poles must either be real, resulting in first-order terms, or complex, resulting in second-order underdamped terms. Furthermore, because the transients associated with those system poles having the largest real part decay the most slowly, these transients tend to dominate the output. The response of higher order systems therefore tends to have the same form as the response to the *dominant poles*, with the response to the *subdominant poles* superimposed over it. Note that the larger the relative difference between the real parts of the dominant and subdominant poles, the more the output tends to resemble the dominant mode of response.

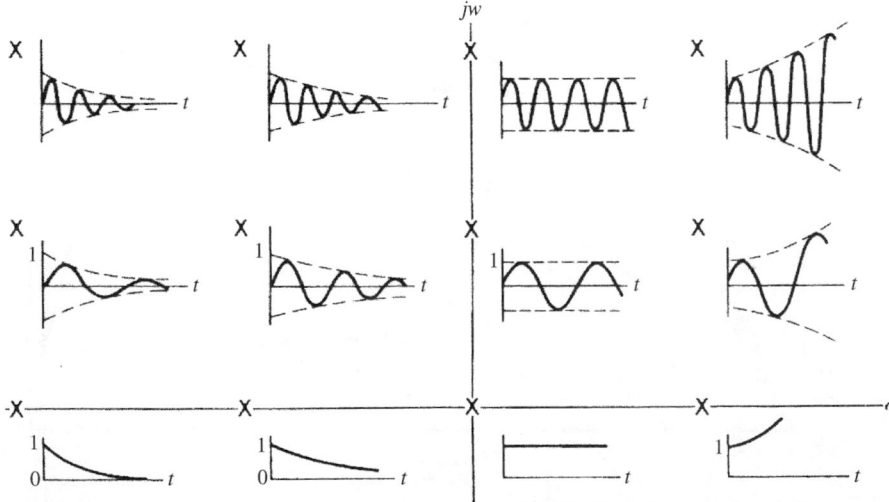

Fig. 20 Unit impulse response for selected upper complex pole locations in the s plane.

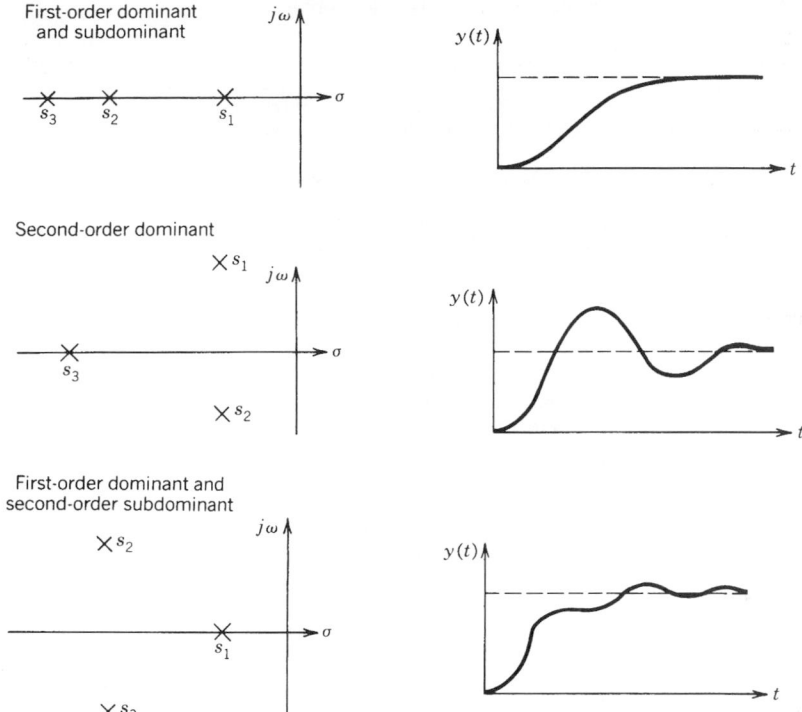

Fig. 21 Step response of a third-order system for alternative upper complex pole locations in the *s* plane.

For example, consider a fixed linear third-order system. The system has three poles. Either the poles may all be real or one may be real while the other pair is complex conjugate. This leads to the three forms of step response shown in Fig. 21, depending on the relative locations of the poles in the complex plane.

Transient Performance Measures The transient response of a system is commonly described in terms of the measures defined in Table 7 and shown in Fig. 22. While these measures apply to any output, for a second-order system these can be calculated exactly in terms of the damping ratio and natural frequency, as shown in column 3 of the table. A common practice in control system design is to determine an initial design with dominant second-order poles that satisfy the performance specifications. Such a design can easily be calculated and then modified as necessary to achieve the desired performance.

Effect of Zeros on Transient Response Zeros arise in a system transfer function through the inclusion of one or more derivatives of $u(t)$ among the inputs to the system. By sensing the rate(s) of change of $u(t)$, the system in effect *anticipates* the future values of $u(t)$. This tends to increase the speed of response of the system relative to the input $u(t)$.

The effect of a zero is greatest on the modes of response associated with neighboring poles. For example, consider the second-order system represented by the transfer function

$$H(s) = K \frac{s - z}{(s - p_1)(s - p_2)}$$

If $z = p_1$, then the system responds as a first-order system with $\tau = -p_2^{-1}$; whereas if $z = p_2$, then the system responds as a first-order system with $\tau = -p_1^{-1}$. Such *pole–zero cancellation* can only be achieved mathematically, but it can be approximated in physical systems. Note that by diminishing the residue associated with the response mode having the larger time constant, the system responds more quickly to changes in the input, confirming our earlier observation.

5.3 Response to Periodic Inputs Using Transform Methods

The response of a dynamic system to periodic inputs can be a critical concern to the control engineer. An input $u(t)$ is *periodic* if $u(t + T) = u(t)$ for all time t, where T is a constant called the period. Periodic inputs are important because these are ubiquitous: rotating unbalanced machinery, reciprocating pumps

Table 7   Transient Performance Measures Based on Step Response

Performance Measure	Definition	Formula for a Second-Order System
Delay time, t_d	Time required for the response to reach half the final value for the first time	
10–90% rise time, t_r	Time required for the response to rise from 10 to 90% of the final response (used for overdamped responses)	
0–100% rise time, t_r	Time required for the response to rise from 0 to 100% of the final response (used for underdamped responses)	$t_r = \dfrac{\pi - \beta}{\omega_d}$ where $\beta = \cos^{-1}\zeta$
Peak time, t_p	Time required for the response to reach the first peak of the overshoot	$t_p = \dfrac{\pi}{\omega_d}$
Maximum overshoot, M_p	The difference in the response between the first peak of the overshoot and the final response	$M_p = e^{-\zeta\pi/\sqrt{1-\zeta^2}}$
Percent overshoot (PO)	The ratio of maximum overshoot to the final response expressed as a percentage	$PO = 100e^{-\zeta\pi/\sqrt{1-\zeta^2}}$
Setting time, t_s	The time required for the response to reach and stay within a specified band centered on the final response (usually 2 or 5% of final response band)	$t_s = \dfrac{4}{\zeta\omega_n}$ (2% band) $t_s = \dfrac{3}{\zeta\omega_n}$ (5% band)

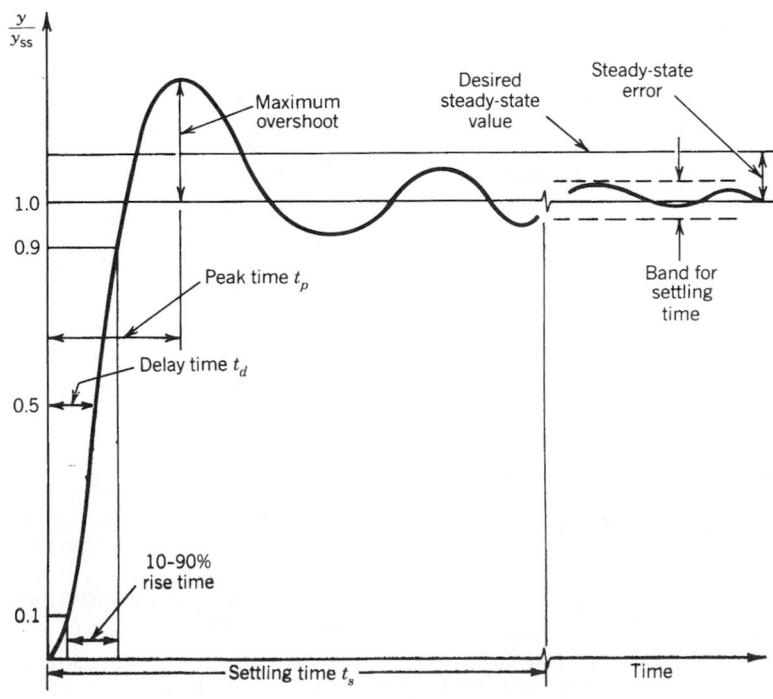

Fig. 22   Transient performance measures based on step response.

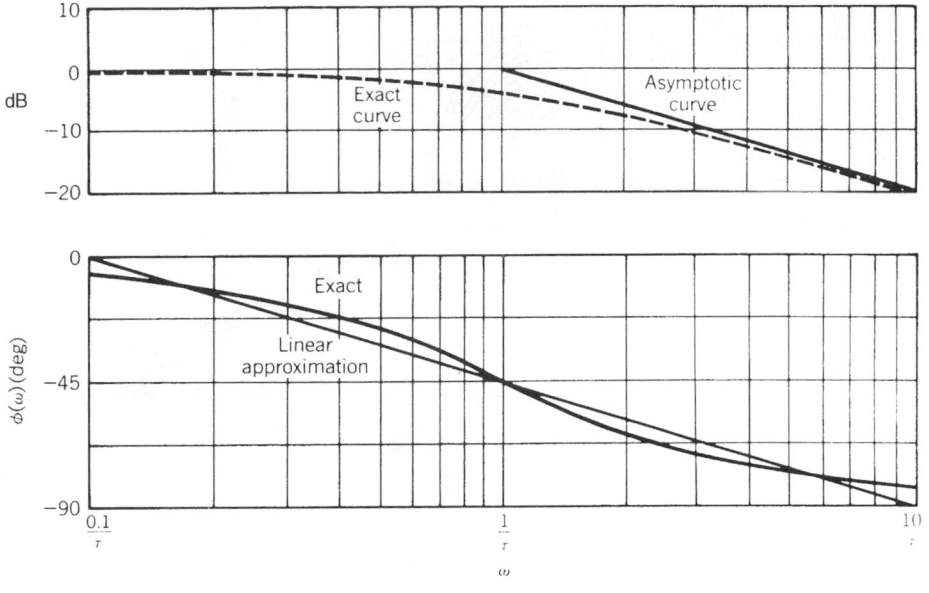

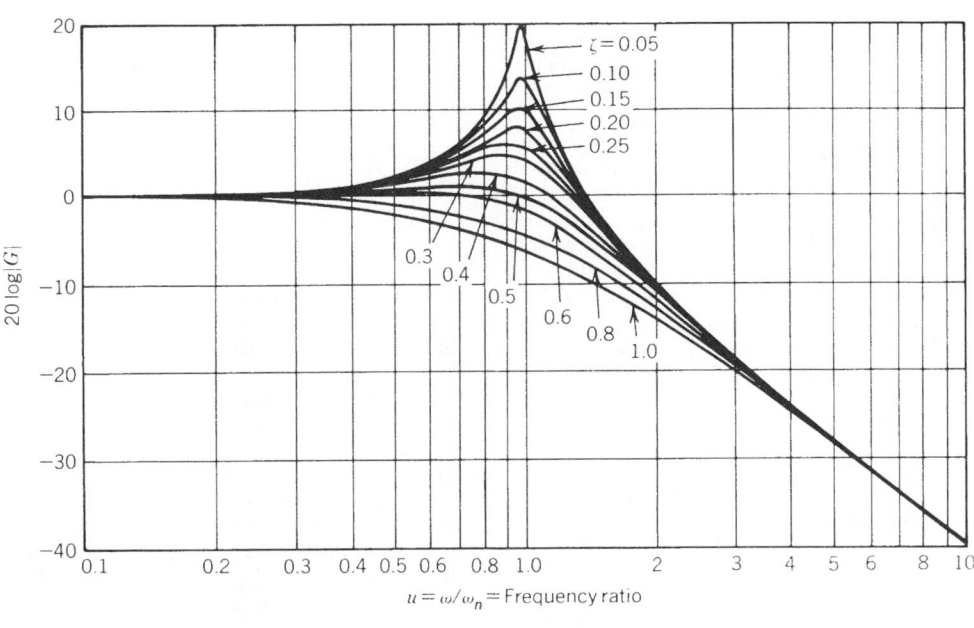

(a)

Fig. 23a Bode diagrams for normalized (*a*) first-order and (*b*) second-order systems.

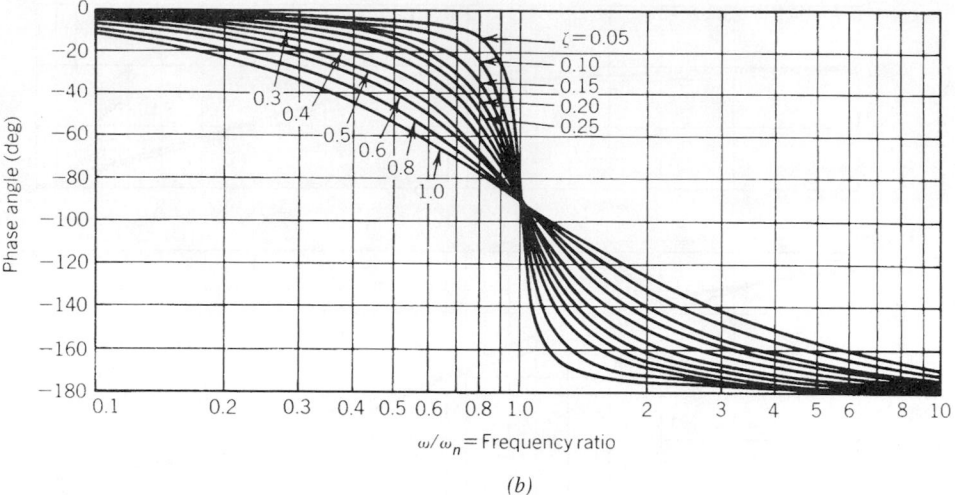

(b)

Fig. 23b (Continued)

and engines, alternating current (ac) electrical power, and a legion of noise and disturbance inputs can be approximated by periodic inputs. Sinusoids are the most important category of periodic inputs, because these are frequently occurring and easily analyzed and form the basis for analysis of general periodic inputs.

Frequency Response   The *frequency response* of a system is the steady-state response of the system to a sinusoidal input. For a linear system, the frequency response has the unique property that the response is a sinusoid of the same frequency as the input sinusoid, differing only in amplitude and phase. In addition, it is easy to show that the amplitude and phase of the response are functions of the input frequency, which are readily obtained from the system transfer function.

Consider a system defined by the transfer function $H(s)$. For an input

$$u(t) = A \sin \omega t$$

the corresponding steady-state output is

$$y_{ss}(t) = AM(\omega) \sin[\omega t + \phi(\omega)]$$

where $M(\omega) = H(j\omega)|$ is the *magnitude ratio*
$\phi(\omega) = \angle H(j\omega)$ is the *phase angle*
$H(j\omega) = H(s)|_{s=j\omega}$ is the *frequency transfer function*

The frequency transfer function is obtained by substituting $j\omega$ for s in the transfer function $H(s)$. If the complex quantity $H(j\omega)$ is written in terms

of its real and imaginary parts as $H(j\omega) = \text{Re}(\omega) + j \, \text{Im}(\omega)$, then

$$M(\omega) = [\text{Re}(\omega)^2 + \text{Im}(\omega)^2]^{1/2}$$

$$\phi(\omega) = \tan^{-1}\left[\frac{\text{Im}(\omega)}{\text{Re}(\omega)}\right]$$

and in polar form

$$H(j\omega) = M(\omega)e^{j\phi(\omega)}$$

Frequency Response Plots   The frequency response of a fixed linear system is typically represented graphically using one of three types of frequency response plots. A *polar plot* is simply a plot of the vector $H(j\omega)$ in the complex plane, where $\text{Re}(\omega)$ is the abscissa and $\text{Im}(\omega)$ is the ordinate. A *logarithmic plot* or *Bode diagram* consists of two displays: (a) the magnitude ratio in decibels $M_{\text{dB}}(\omega)$ [where $M_{\text{dB}}(\omega) = 20 \log M(\omega)$] versus $\log \omega$ and (b) the phase angle in degrees $\phi(\omega)$ versus $\log \omega$. Bode diagrams for normalized first- and second-order systems are given in Fig. 23a. Bode diagrams for higher order systems are obtained by adding these first- and second-order terms, appropriately scaled. A *Nichols diagram* can be obtained by cross plotting the Bode magnitude and phase diagrams, eliminating $\log \omega$. Polar plots and Bode and Nichols diagrams for common transfer functions are given in Table 8.

Frequency Response Performance Measures
Frequency response plots show that dynamic systems tend to behave like *filters*, "passing" or even amplifying certain ranges of input frequencies while blocking

Table 8 Transfer Function Plots for Representative Transfer Functions

$G(s)$	Polar Plot	Bode Diagram
1. $\dfrac{K}{S\tau_1 + 1}$		
2. $\dfrac{K}{(S\tau_1 + 1)(S\tau_2 + 1)}$		
3. $\dfrac{K}{(S\tau_1 + 1)(S\tau_2 + 1)(S\tau_3 + 1)}$		
4. $\dfrac{K}{S}$		
5. $\dfrac{K}{S(S\tau_1 + 1)}$		

Table 8   (*Continued*)

Nichols Diagram	Root Locus	Comments
		Stable; gain margin $= \infty$
		Elementary regulator; stable; gain margin $= \infty$
		Regulator with additional energy storage component; unstable, but can be made stable by reducing gain
		Ideal integrator; stable
		Elementary instrument servo; inherently stable; gain margin $= \infty$

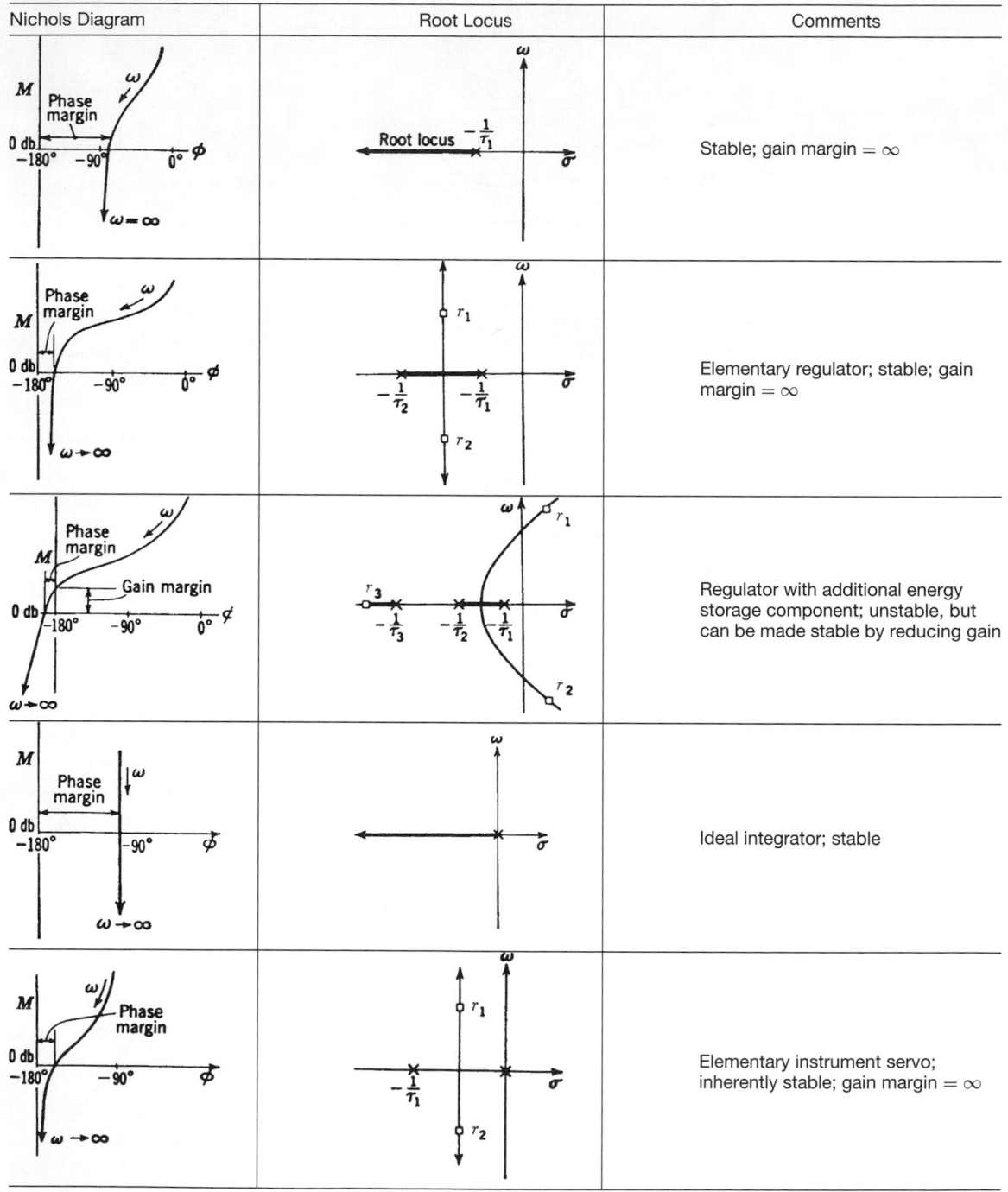

Table 8 *(Continued)*

$G(s)$	Polar Plot	Bode Diagram
6. $\dfrac{K}{S(S\tau_1 + 1)(S\tau_2 + 1)}$		
7. $\dfrac{K(S\tau_a + 1)}{S(S\tau_1 + 1)(S\tau_2 + 1)}$		
8. $\dfrac{K}{S^2}$		
9. $\dfrac{K}{S^2(S\tau_1 + 1)}$		
10. $\dfrac{K(S\tau_a + 1)}{S^2(S\tau_1 + 1)}$ $\tau_a > \tau_1$		

Table 8 (*Continued*)

Nichols Diagram	Root Locus	Comments
		Instrument servo with field control motor or power servo with elementary Ward-Leonard drive; stable as shown but may become unstable with increased gain
		Elementary instrument servo with phase lead (derivative) compensator; stable
		Inherently unstable; must be compensated
		Inherently unstable; must be compensated
		Stable for all gains

Table 8 *(Continued)*

$G(S)$	Polar Plot	Bode Diagram
11. $\dfrac{K}{S^3}$		
12. $\dfrac{K(S\tau_a + 1)}{S^3}$		
13. $\dfrac{K(S\tau_a + 1)(S\tau_b + 1)}{S^3}$		
14. $\dfrac{K(S\tau_a + 1)(S\tau_b + 1)}{(S\tau_1 + 1)(S\tau_2 + 1)(S\tau_3 + 1)(S\tau_4 + 1)}$		
15. $\dfrac{K(S\tau_a + 1)}{S^2(S\tau_1 + 1)(S\tau_2 + 1)}$		

Table 8 *(Continued)*

Nichols Diagram	Root Locus	Comments
		Inherently unstable
		Inherently unstable
		Conditionally stable; becomes unstable if gain is too low
		Conditionally stable; stable at low gain, becomes unstable as gain is raised, again becomes stable as gain is further increased, and becomes unstable for very high gains
		Conditionally stable; becomes unstable at high gain

Source: From Ref. 5.

or attenuating other frequency ranges. The range of frequencies for which the amplitude ratio is no less than 3 dB of its maximum value is called the *bandwidth* of the system. The bandwidth is defined by upper and lower *cutoff frequencies* ω_c, or by $\omega = 0$ and an upper cutoff frequency if $M(0)$ is the maximum amplitude ratio. Although the choice of "down 3 dB" used to define the cutoff frequencies is somewhat arbitrary, the bandwidth is usually taken to be a measure of the range of frequencies for which a significant portion of the input is felt in the system output. The bandwidth is also taken to be a measure of the system speed of response since attenuation of inputs in the higher frequency ranges generally results from the inability of the system to "follow" rapid changes in amplitude. Thus, a narrow bandwidth generally indicates a sluggish system response.

Response to General Periodic Inputs   The *Fourier series* provides a means for representing a general periodic input as the sum of a constant and terms containing sine and cosine. For this reason the *Fourier series*, together with the superposition principle for linear systems, extends the results of frequency response analysis to the general case of arbitrary periodic inputs. The Fourier series representation of a periodic function $f(t)$ with period $2T$ on the interval $t^* + 2T \geq t \geq t^*$ is

$$f(t) = \frac{a_0}{2} + \sum_{n=1}^{\infty} \left(a_n \, \cos \frac{n\pi t}{T} + b_n \, \sin \frac{n\pi t}{T} \right)$$

where

$$a_n = \frac{1}{T} \int_{t^*}^{t^*+2T} f(t) \, \cos \frac{n\pi t}{T} \, dt$$

$$b_n = \frac{1}{T} \int_{t^*}^{t^*+2T} f(t) \, \sin \frac{n\pi t}{T} \, dt$$

If $f(t)$ is defined outside the specified interval by a periodic extension of period $2T$ and if $f(t)$ and its first derivative are piecewise continuous, then the series converges to $f(t)$ if t is a point of continuity or to $1/2[f(t_+) + f(t_-)]$ if t is a point of discontinuity. Note that while the Fourier series in general is infinite, the notion of bandwidth can be used to reduce the number of terms required for a reasonable approximation.

6   STATE-VARIABLE METHODS

State-variable methods use the vector state and output equations introduced in Section 4 for analysis of dynamic systems directly in the time domain. These methods have several advantages over transform methods. First, state-variable methods are particularly advantageous for the study of multivariable (multiple-input/multiple-output) systems. Second, state-variable

methods are more naturally extended for the study of linear time-varying and nonlinear systems. Finally, state-variable methods are readily adapted to computer simulation studies.

6.1   Solution of State Equation

Consider the vector equation of state for a fixed linear system:

$$\dot{x}(t) = Ax(t) + Bu(t)$$

The solution to this system is

$$x(t) = \Phi(t)x(0) + \int_0^t \Phi(t - \tau)Bu(\tau) \, d\tau$$

where the matrix $\Phi(t)$ is called the *state transition matrix*. The state transition matrix represents the free response of the system and is defined by the matrix exponential series

$$\Phi(t) = e^{At} = I + At + \frac{1}{2!}A^2 t^2 + \cdots = \sum_{k=0}^{\infty} \frac{1}{k!}A^k t^k$$

where I is the identity matrix. The state transition matrix has the following useful properties:

$$\Phi(0) = I$$

$$\Phi^{-1}(t) = \Phi(-t)$$

$$\Phi^k(t) = \Phi(kt)$$

$$\Phi(t_1 + t_2) = \Phi(t_1)\Phi(t_2)$$

$$\Phi(t_2 - t_1)\Phi(t_1 - t_0) = \Phi(t_2 - t_0)$$

$$\dot{\Phi}(t) = A\Phi(t)$$

The Laplace transform of the state equation is

$$sX(s) - x(0) = AX(s) + BU(s)$$

The solution to the fixed linear system therefore can be written as

$$x(t) = \mathscr{L}^{-1}[X(s)]$$
$$= \mathscr{L}^{-1}[\Phi(s)]x(0) + \mathscr{L}^{-1}[\Phi(s)BU(s)]$$

where $\Phi(s)$ is called the *resolvent matrix* and

$$\Phi(t) = \mathscr{L}^{-1}[\Phi(s)] = \mathscr{L}^{-1}[sI - A]^{-1}$$

6.2   Eigenstructure

The internal structure of a system (and therefore its free response) is defined entirely by the system matrix A. The concept of matrix *eigenstructure*, as defined by the eigenvalues and eigenvectors of the system matrix, can provide a great deal of insight into the fundamental behavior of a system. In particular, the system eigenvectors can be shown to define a special set of first-order subsystems embedded within the system. These subsystems behave independently of one another, a fact that greatly simplifies analysis.

System Eigenvalues and Eigenvectors For a system with system matrix A, the system *eigenvectors* v_i and associated *eigenvalues* λ_i are defined by the equation

$$Av_i = \lambda_i v_i$$

Note that the eigenvectors represent a set of special directions in the state space. If the state vector is aligned in one of these directions, then the homogeneous state equation becomes $\dot{v}_i = A\dot{v}_i = \lambda v_i$, implying that each of the state variables changes at the *same* rate determined by the eigenvalue λ_i. This further implies that, in the absence of inputs to the system, a state vector that becomes aligned with an eigenvector will remain aligned with that eigenvector. The system eigenvalues are calculated by solving the nth-order polynomial equation

$$|\lambda I - A| = \lambda^n + a_{n-1}\lambda^{n-1} + \cdots + a_1\lambda + a_0 = 0$$

This equation is called the *characteristic equation*. Thus the system eigenvalues are the roots of the characteristic equation, that is, the system eigenvalues are identically the system poles defined in transform analysis.

Each system eigenvector is determined by substituting the corresponding eigenvalue into the defining equation and then solving the resulting set of simultaneous linear equations. Only $n-1$ of the n components of any eigenvector are independently defined, however. In other words, the magnitude of an eigenvector is arbitrary, and the eigenvector describes a direction in the state space.

Diagonalized Canonical Form There will be one linearly independent eigenvector for each distinct (nonrepeated) eigenvalue. If all of the eigenvalues of an nth-order system are distinct, then the n independent eigenvectors form a new basis for the state space. This basis represents new coordinate axes defining a set of state variables $z_i(t), i = 1, 2, \ldots, n$, called the *diagonalized canonical variables*. In terms of the diagonalized variables, the homogeneous state equation is

$$\dot{z}(t) = \Lambda z$$

where Λ is a diagonal system matrix of the eigenvectors, that is,

$$\Lambda = \begin{bmatrix} \lambda_1 & 0 & \cdots & 0 \\ 0 & \lambda_2 & \cdots & 0 \\ \vdots & \vdots & \ddots & \vdots \\ 0 & 0 & \vdots & \lambda_n \end{bmatrix}$$

The solution to the diagonalized homogeneous system is

$$z(t) = e^{\Lambda t} z(0)$$

where $e^{\Lambda t}$ is the diagonal state transition matrix

$$e^{\Lambda t} = \begin{bmatrix} e^{\lambda_1 t} & 0 & \cdots & 0 \\ 0 & e^{\lambda_2 t} & \cdots & 0 \\ \vdots & \vdots & \ddots & \vdots \\ 0 & 0 & \cdots & e^{\lambda_n t} \end{bmatrix}$$

Modal Matrix Consider the state equation of the nth-order system

$$\dot{x}(t) = Ax(t) + Bu(t)$$

which has real, distinct eigenvalues. Since the system has a full set of eigenvectors, the state vector $x(t)$ can be expressed in terms of the canonical state variables as

$$x(t) = v_1 z_1(t) + v_2 z_2(t) + \cdots + v_n z_n(t) = Mz(t)$$

where M is the $n \times n$ matrix whose columns are the eigenvectors of A, called the *modal matrix*. Using the modal matrix, the state transition matrix for the original system can be written as

$$\Phi(t) = e^{\Lambda t} = Me^{\Lambda t}M^{-1}$$

where $e^{\Lambda t}$ is the diagonal state transition matrix. This frequently proves to be an attractive method for determining the state transition matrix of a system with real, distinct eigenvalues.

Jordan Canonical Form For a system with one or more repeated eigenvalues, there is not in general a full set of eigenvectors. In this case, it is not possible to determine a diagonal representation for the system. Instead, the simplest representation that can be achieved is block diagonal. Let $L_k(\lambda)$ be the $k \times k$ matrix

$$L_k(\lambda) = \begin{bmatrix} \lambda & 1 & 0 & \cdots & 0 \\ 0 & \lambda & 1 & \cdots & 0 \\ \vdots & \vdots & \lambda & \ddots & 0 \\ \vdots & \vdots & \vdots & \ddots & 1 \\ 0 & 0 & 0 & 0 & \lambda \end{bmatrix}$$

Then for any $n \times n$ system matrix A there is certain to exist a nonsingular matrix T such that

$$T^{-1}AT = \begin{bmatrix} L_{k_1}(\lambda_1) & & & \\ & L_{k_2}(\lambda_2) & & \\ & & \ddots & \\ & & & L_{k_r}(\lambda_r) \end{bmatrix}$$

where $k_1 + k_2 + \cdots + k_r = n$ and $\lambda_i, i = 1, 2, \ldots, r$, are the (not necessarily distinct) eigenvalues of A. The matrix $T^{-1}AT$ is called the *Jordan canonical form*.

7 SIMULATION

7.1 Experimental Analysis of Model Behavior

Closed-form solutions for nonlinear or time-varying systems are rarely available. In addition, while explicit solutions for time-invariant linear systems can always be found, for high-order systems this is often impractical. In such cases it may be convenient to study the dynamic behavior of the system using *simulation*.

Simulation is the *experimental* analysis of model behavior. A *simulation run* is a controlled experiment in which a specific realization of the model is manipulated in order to determine the response associated with that realization. A *simulation study* comprises *multiple runs*, each run for a different combination of model parameter values and/or initial conditions. The generalized solution of the model must then be inferred from a finite number of simulated data points.

Simulation is almost always carried out with the assistance of computing equipment. *Digital simulation* involves the *numerical solution* of model equations using a digital computer. *Analog simulation* involves solving model equations by analogy with the behavior of a physical system using an analog computer. *Hybrid simulation* employs digital and analog simulation together using a hybrid (part digital and part analog) computer.

7.2 Digital Simulation

Digital continuous-system simulation involves the approximate solution of a state-variable model over successive time steps. Consider the general state-variable equation

$$\dot{x}(t) = f[x(t), u(t)]$$

to be simulated over the time interval $t_0 \le t \le t_K$. The solution to this problem is based on the repeated solution of the single-variable, single-step subproblem depicted in Fig. 24. The subproblem may be stated formally as follows:

Given:

1. $\Delta t(k) = t_k - t_{k-1}$, the length of the kth time step.
2. $x_i(t) = f_i[x(t), u(t)]$ for $t_{k-1} \le t \le t_k$, the ith equation of state defined for the state variable $x_i(t)$ over the kth time step.
3. $u(t)$ for $t_{k-1} \le t \le t_k$, the input vector defined for the kth time step.
4. $\tilde{x}(k - 1) \simeq x(t_{k-1})$, an initial approximation for the state vector at the beginning of the time step.

Find:

5. $\tilde{x}_i(k) \simeq x_i(t_k)$, a final approximation for the state variable $x_i(t)$ at the end of the kth time step.

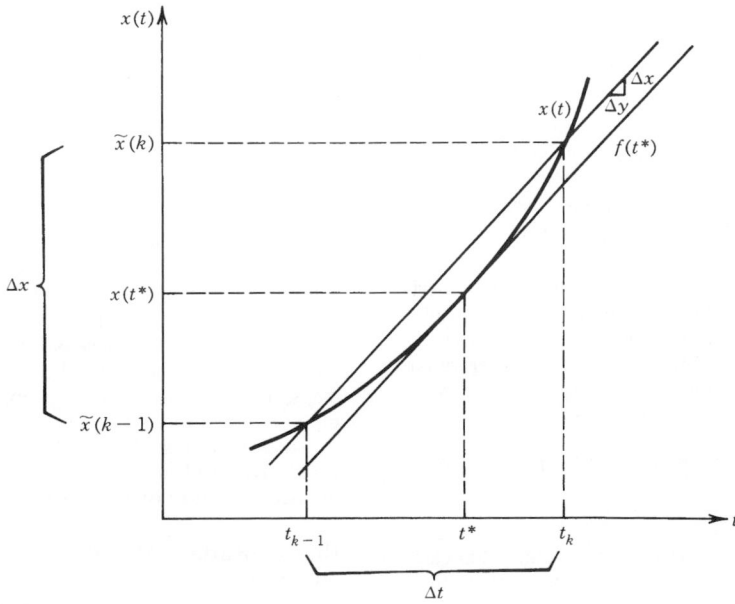

Fig. 24 Numerical approximation of a single variable over a single time step.

Solving this single-variable, single-step subproblem for each of the state variables $x_i(t), i = 1, 2, \ldots, n$, yields a final approximation for the state vector $\tilde{x}(k) \simeq x(t_k)$ at the end of the kth time step. Solving the complete single-step problem K times over K time steps, beginning with the initial condition $\tilde{x}(0) = x(t_0)$ and using the final value of $\tilde{x}(t_k)$ from the kth time step as the initial value of the state for the $(k+1)$st time step, yields a discrete succession of approximations $\tilde{x}(1) \simeq x(t_1), \tilde{x}(2) \simeq x(t_2), \ldots, \tilde{x}(K) \simeq x(t_k)$ spanning the solution time interval.

The basic procedure for completing the single-variable, single-step problem is the same regardless of the particular integration method chosen. It consists of two parts: (a) calculation of the average value of the ith derivative over the time step as

$$\dot{x}_i(t^*) = f_i[x(t^*), u(t^*)] = \frac{\Delta x_i(k)}{\Delta t(k)} \simeq \tilde{f}_i(k)$$

and (b) calculation of the final value of the simulated variable at the end of the time step as

$$\tilde{x}_i(k) = \tilde{x}_i(k-1) + \Delta x_i(k)$$

$$\simeq \tilde{x}_i(k-1) + \Delta t(k)\tilde{f}_i(k)$$

If the function $f_i[x(t), u(t)]$ is continuous, then t^* is guaranteed to be on the time step, that is, $t_{k-1} \leq t^* \leq t_k$. Since the value of t^* is otherwise unknown, however, the value of $x(t^*)$ can only be approximated as $\tilde{f}(k)$.

Different *numerical integration* methods are distinguished by the means used to calculate the approximation $f_i(k)$. A wide variety of such methods is available for digital simulation of dynamic systems. The choice of a particular method depends on the nature of the model being simulated, the accuracy required in the simulated data, and the computing effort available for the simulation study. Several popular classes of integration methods are outlined in the following sections.

Euler Method   The simplest procedure for numerical integration is the Euler method. The standard Euler method approximates the average value of the ith derivative over the kth time step using the derivative evaluated at the beginning of the time step, that is,

$$\tilde{f}_i(k) = f_i[\tilde{x}(k-1), u(t_{k-1})] \simeq f_i(t_{k-1})$$

$i = 1, 2, \ldots, n$ and $k = 1, 2, \ldots, K$. This is shown geometrically in Fig. 25 for the scalar single-step case. A modification of this method uses the newly calculated state variables in the derivative calculation as these new values become available. Assuming the state

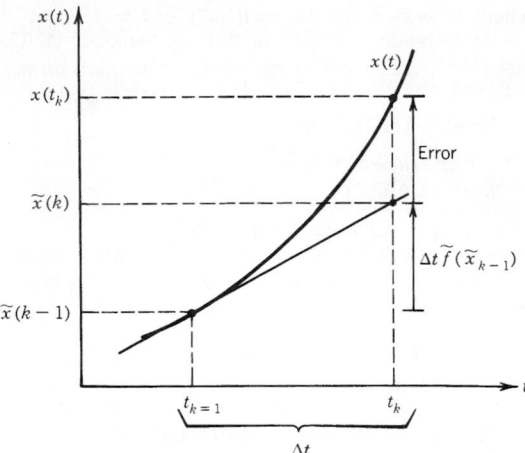

Fig. 25   Geometric interpretation of the Euler method for numerical integration.

variables are computed in numerical order according to the subscripts, this implies

$$\tilde{f}_i(k) = f_i[\tilde{x}_1(k), \ldots, \tilde{x}_{i-1}(k),$$

$$\tilde{x}_i(k-1), \ldots, \tilde{x}_n(k-1), u(t_{k-1})]$$

The modified Euler method is modestly more efficient than the standard procedure and, frequently, is more accurate. In addition, since the input vector $u(t)$ is usually known for the entire time step, using an average value of the input, such as

$$u(k) = \frac{1}{\Delta t(k)} \int_{t_{k-1}}^{t_k} u(\tau)\, d\tau$$

frequently leads to a superior approximation of $\tilde{f}_i(k)$.

The Euler method requires the least amount of computational effort per time step of any numerical integration scheme. Local truncation error is proportional to Δt^2, however, which means that the error within each time step is highly sensitive to step size. Because the accuracy of the method demands very small time steps, the number of time steps required to implement the method successfully can be large relative to other methods. This can imply a large computational overhead and can lead to inaccuracies through the accumulation of round-off error at each step.

Runge–Kutta Methods   *Runge–Kutta methods* precompute two or more values of $f_i[x(t), u(t)]$ in the time step $t_{k-1} \leq t \leq t_k$ and use some weighted average of these values to calculate $\tilde{f}_i(k)$. The *order* of a

Runge–Kutta method refers to the number of derivative terms (or *derivative calls*) used in the scalar single-step calculation. A Runge–Kutta routine of order N therefore uses the approximation

$$\tilde{f}_i(k) = \sum_{j=1}^{N} w_j f_{ij}(k)$$

where the N approximations to the derivative are

$$f_{i1}(k) = f_i[\tilde{x}(k-1), u(t_{k-1})]$$

(the Euler approximation) and

$$f_{ij} = f_i\left[\tilde{x}(k-1) + \Delta t \sum_{t=1}^{j-1} Ib_{jt} f_{il}, u\left(t_{k-1} + \Delta t \sum_{t=1}^{j-1} b_{jl}\right)\right]$$

where I is the identity matrix. The weighting coefficients w_j and b_{jl} are not unique, but are selected such that the error in the approximation is zero when $x_i(t)$ is some specified Nth-degree polynomial in t. Coefficients commonly used for Runge–Kutta integration are given in Table 9.

Among the most popular of the Runge–Kutta methods is fourth-order Runge–Kutta. Using the defining equations for $N = 4$ and the weighting coefficients from Table 9 yields the derivative approximation

$$\tilde{f}_i(k) = \tfrac{1}{6}[f_{i1}(k) + 2f_{i2}(k) + 2f_{i3}(k) + f_{i4}(k)]$$

based on the four derivative calls

$$f_{i1}(k) = f_i[\tilde{x}(k-1), u(t_{k-1})]$$

$$f_{i2}(k) = f_i\left[\tilde{x}(k-1) + \frac{\Delta t}{2} If_{i1}, u\left(t_{k-1} + \frac{\Delta t}{2}\right)\right]$$

$$f_{i3}(k) = f_i\left[\tilde{x}(k-1) + \frac{\Delta t}{2} If_{i2}, u\left(t_{k-1} + \frac{\Delta t}{2}\right)\right]$$

$$f_{i4}(k) = f_i[\tilde{x}(k-1) + \Delta t If_{i3}, u(t_k)]$$

where I is the identity matrix.

Because Runge–Kutta formulas are designed to be exact for a polynomial of order N, local truncation error is of the order Δt^{N+1}. This considerable improvement over the Euler method means that comparable accuracy can be achieved for larger step sizes. The penalty is that N derivative calls are required for each scalar evaluation within each time step.

Euler and Runge–Kutta methods are examples of *single-step methods* for numerical integration, so-called because the state $x(k)$ is calculated from knowledge of the state $x(k-1)$, without requiring knowledge of the state at any time prior to the beginning of the current time step. These methods are also referred to as *self-starting methods* since calculations may proceed from any known state.

Multistep Methods *Multistep methods* differ from the single-step methods previously described in that multistep methods use the stored values of two or more previously computed states and/or derivatives in order to compute the derivative approximation $\tilde{f}_i(k)$ for the current time step. The advantage of multistep methods over Runge–Kutta methods is that these

Table 9   Coefficients Commonly Used for Runge–Kutta Numerical Integration

Common name	N	b_{jl}	w_j
Open or explicit Euler	1	All zero	$w_1 = 1$
Improved polygon	2	$b_{21} = \tfrac{1}{2}$	$w_1 = 0$
			$w_2 = 1$
Modified Euler or Heun's method	2	$b_{21} = 1$	$w_1 = \tfrac{1}{2}$
			$w_2 = \tfrac{1}{2}$
Third-order Runge–Kutta	3	$b_{21} = \tfrac{1}{2}$	$w_1 = \tfrac{1}{6}$
		$b_{31} = -1$	$w_2 = \tfrac{2}{3}$
		$b_{32} = 2$	$w_3 = \tfrac{1}{6}$
Fourth-order Runge–Kutta	4	$b_{21} = \tfrac{1}{2}$	$w_1 = \tfrac{1}{6}$
		$b_{31} = 0$	$w_2 = \tfrac{1}{3}$
		$b_{32} = \tfrac{1}{2}$	$w_3 = \tfrac{1}{3}$
		$b_{43} = 1$	$w_4 = \tfrac{1}{6}$

Source: From Ref. 6.

require only one derivative call for each state variable at each time step for comparable accuracy. The disadvantage is that multistep methods are not self-starting since calculations cannot proceed from the initial state alone. Multistep methods must be started, or restarted in the case of discontinuous derivatives, using a single-step method to calculate the first several steps.

The most popular of the multistep methods are the *Adams–Bashforth predictor methods* and the *Adams–Moulton corrector methods*. These methods use the derivative approximation

$$\tilde{f}_i(k) = \sum_{j=0}^{N} b_j f_i[\tilde{x}(k-j), u(k-j)]$$

where the b_j are weighting coefficients. These coefficients are selected such that the error in the approximation is zero when $x_i(t)$ is a specified polynomial. Table 10 gives the values of the weighting coefficients for several Adams–Bashforth–Moulton rules. Note that the predictor methods employ an *open* or *explicit rule* since for these methods $b_0 = 0$ and a prior estimate of $x_i(k)$ is not required. The corrector methods use a *closed* or *implicit rule* since for these methods $b_i \neq 0$ and a prior estimate of $x_i(k)$ is required. Note also that for all of these methods $\sum_{j=0}^{N} b_j = 1$, ensuring unity gain for the integration of a constant.

Predictor–Corrector Methods

Predictor–corrector methods use one of the multistep predictor equations to provide an initial estimate (or "prediction") of $x(k)$. This initial estimate is then used with one of the multistep corrector equations to provide a second and improved (or "corrected") estimate of $x(k)$ before proceeding to the next step. A popular choice is the four-point Adams–Bashforth predictor together with the four-point Adams–Moulton corrector, resulting in a prediction of

$$\tilde{x}_i(k) = \tilde{x}_i(k-1) + \frac{\Delta t}{24}[55\tilde{f}_i(k-1) - 59\tilde{f}_i(k-2)$$
$$+ 37\tilde{f}_i(k-3) - 9\tilde{f}_i(k-4)]$$

for $i = 1, 2, \ldots, n$ and a correction of

$$\tilde{x}_i(k) = \tilde{x}_i(k-1) + \frac{\Delta t}{24}\{9f_i[\tilde{x}(k), u(k)]$$
$$+ 19\tilde{f}_i(k-1) - 5\tilde{f}_i(k-2) + \tilde{f}_i(k-3)\}$$

Predictor–corrector methods generally incorporate a strategy for increasing or decreasing the size of the time step depending on the difference between the predicted and corrected $x(k)$ values. Such *variable time-step methods* are particularly useful if the simulated system possesses local time constants that differ by several orders of magnitude or if there is little a priori knowledge about the system response.

Numerical Integration Errors An inherent characteristic of digital simulation is that the discrete data points generated by the simulation $x(k)$ are only approximations to the exact solution $x(t_k)$ at the corresponding point in time. This results from two types of errors that are unavoidable in the numerical solutions. *Round-off errors* occur because numbers stored in a digital computer have finite word length (i.e., a finite number of bits per word) and therefore limited precision. Because the results of calculations cannot be stored exactly, round-off error tends to increase with the number of calculations performed. For a given total solution interval $t_0 \leq t \leq t_K$, therefore, round-off error tends to increase (a) with increasing integration rule order (since more calculations must be performed at each time step) and (b) with decreasing step size Δt (since more time steps are required).

Truncation errors or *numerical approximation errors* occur because of the inherent limitations in the numerical integration methods themselves. Such

Table 10 Coefficients Commonly Used for Adams–Bashforth–Moulton Numerical Integration

Common name	Predictor or corrector	Points	b_{-1}	b_0	b_1	b_2	b_3
Open or explicit Euler	Predictor	1	0	1	0	0	0
Open trapezoidal	Predictor	2	0	$\frac{3}{2}$	$-\frac{1}{2}$	0	0
Adams three-point predictor	Predictor	3	0	$\frac{23}{12}$	$-\frac{16}{12}$	$\frac{5}{12}$	0
Adams four-point predictor	Predictor	4	0	$\frac{55}{24}$	$-\frac{59}{24}$	$\frac{37}{24}$	$-\frac{9}{24}$
Closed or implicit Euler	Corrector	1	1	0	0	0	0
Closed trapezoidal	Corrector	2	$\frac{1}{2}$	$\frac{1}{2}$	0	0	0
Adams three-point corrector	Corrector	3	$\frac{5}{12}$	$\frac{8}{12}$	$-\frac{1}{12}$	0	0
Adams four-point corrector	Corrector	4	$\frac{9}{24}$	$\frac{19}{24}$	$-\frac{5}{24}$	$\frac{1}{24}$	0

Source: From Ref. 6.

errors would arise even if the digital computer had infinite precision. *Local* or *per-step truncation error* is defined as

$$e(k) = x(k) - x(t_k)$$

given that $x(k-1) = x(t_{k-1})$ and that the calculation at the kth time step is infinitely precise. For many integration methods, local truncation errors can be approximated at each step. *Global* or *total truncation error* is defined as

$$e(K) = x(K) - x(t_K)$$

given that $x(0) = x(t_0)$ and the calculations for all K time steps are infinitely precise. Global truncation error usually cannot be estimated, neither can efforts to reduce local truncation errors be guaranteed to yield acceptable global errors. In general, however, truncation errors can be decreased by using more sophisticated integration methods and by decreasing the step size Δt.

Time Constants and Time Steps As a general rule, the step size Δt for simulation must be less than the smallest local time constant of the model simulated. This can be illustrated by considering the simple first-order system

$$\dot{x}(t) = \lambda x(t)$$

and the difference equation defining the corresponding Euler integration

$$x(k) = x(k-1) + \Delta t \ \lambda x(k-1)$$

The continuous system is stable for $\lambda < 0$, while the discrete approximation is stable for $|1 + \lambda \ \Delta t| < 1$. If the original system is stable, therefore, the simulated response will be stable for

$$\Delta t \le 2 \left| \frac{1}{\lambda} \right|$$

where the equality defines the *critical step size*. For larger step sizes, the simulation will exhibit *numerical instability*. In general, while higher order integration methods will provide greater per-step accuracy, the critical step size itself will not be greatly reduced.

A major problem arises when the simulated model has one or more time constants $|1/\lambda_i|$ that are small when compared to the total solution time interval $t_0 \le t \le t_K$. Numerical stability will then require very small Δt, even though the transient response associated with the higher frequency (larger λ_i) subsystems may contribute little to the particular solution. Such problems can be addressed either by neglecting the higher frequency components where appropriate or by adopting special numerical integration methods for *stiff systems*.

Selecting an Integration Method The best numerical integration method for a specific simulation is the method that yields an acceptable global approximation error with the minimum amount of round-off error and computing effort. No single method is best for all applications. The selection of an integration method depends on the model simulated, the purpose of the simulation study, and the availability of computing hardware and software.

In general, for well-behaved problems with continuous derivatives and no stiffness, a lower order Adams predictor is often a good choice. Multistep methods also facilitate estimating local truncation error. Multistep methods should be avoided for systems with discontinuities, however, because of the need for frequent restarts. Runge–Kutta methods have the advantage that these are self-starting and provide fair stability. For stiff systems where high-frequency modes have little influence on the global response, special stiff-system methods enable the use of economically large step sizes. Variable-step rules are useful when little is known a priori about solutions. Variable-step rules often make a good choice as general-purpose integration methods.

Round-off error usually is not a major concern in the selection of an integration method, since the goal of minimizing computing effort typically obviates such problems. Double-precision simulation can be used where round-off is a potential concern. An upper bound on step size often exists because of discontinuities in derivative functions or because of the need for response output at closely spaced time intervals.

Continuous-System Simulation Languages
Digital simulation can be implemented for a specific model in any high-level language such as FORTRAN or C. The general process for implementing a simulation is shown in Fig. 26. In addition, many special-purpose continuous-system simulation languages are commonly available across a wide range of platforms. Such languages greatly simplify programming tasks and typically provide for good graphical output.

8 MODEL CLASSIFICATIONS

Mathematical models of dynamic systems are distinguished by several criteria that describe fundamental properties of model variables and equations. These criteria in turn prescribe the theory and mathematical techniques that can be used to study different models. Table 11 summarizes these distinguishing criteria. In the following sections, the approaches adopted for the analysis of important classes of systems are briefly outlined.

8.1 Stochastic Systems

Systems in which some of the dependent variables (input, state, output) contain random components are called *stochastic systems*. Randomness may result from

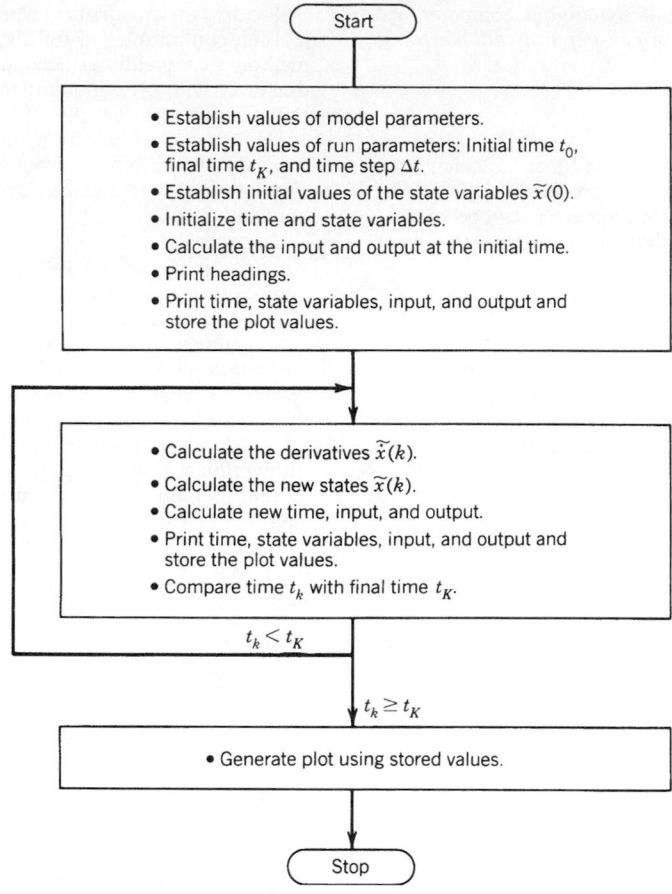

Fig. 26   General process for implementing digital simulation. (Adapted from Ref. 3.)

environmental factors, such as wind gusts or electrical noise, or simply from a lack of precise knowledge of the system model, such as when a human operator is included within a control system. If the randomness in the system can be described by some rule, then it is often possible to derive a model in terms of probability distributions involving, for example, the means and variances of model variables or parameters.

State-Variable Formulation   A common formulation is the fixed, linear model with additive noise

$$\dot{x}(t) = Ax(t) + Bu(t) + w(t)$$

$$y(t) = Cx(t) + v(t)$$

where $w(t)$ is a zero-mean Gaussian disturbance and $v(t)$ is a zero-mean Gaussian measurement noise. This formulation is the basis for many *estimation problems*, including the problem of *optimal filtering*. Estimation

essentially involves the development of a rule or algorithm for determining the best estimate of the past, current, or future values of measured variables in the presence of disturbances or noise.

Random Variables   In the following, important concepts for characterizing random signals are developed. A *random variable x* is a variable that assumes values that cannot be precisely predicted a priori. The likelihood that a random variable will assume a particular value is measured as the *probability* of that value. The probability *distribution function $F(x)$* of a continuous random variable x is defined as the probability that x assumes a value no greater than x, that is,

$$F(x) = \Pr(X \le x) = \int_{-\infty}^{x} f(x)\,dx$$

The probability *density function $f(x)$* is defined as the derivative of $F(x)$.

Table 11 Classification of Mathematical Models of Dynamic Systems

Criterion	Classification	Description
Certainty	Deterministic	Model parameters and variables can be known with certainty. Common approximation when uncertainties are small.
	Stochastic	Uncertainty exists in the values of some parameters and/or variables. Model parameters and variables are expressed as random numbers or processes and are characterized by the parameters of probability distributions.
Spatial characteristics	Lumped	State of the system can be described by a finite set of state variables. Model is expressed as a discrete set of point functions described by ordinary differential or difference equations.
	Distributed	State depends on both time and spatial location. Model is usually described by variables that are continuous in time and space, resulting in partial differential equations. Frequently approximated by lumped elements. Typical in the study of structures and mass and heat transport.
Parameter variation	Fixed or time invariant	Model parameters are constant. Model described by differential or difference equations with constant coefficients. Model with same initial conditions and input delayed by t_d has the same response delayed by t_d.
	Time varying	Model parameters are time dependent.
Superposition property	Linear	Superposition applies. Model can be expressed as a system of linear difference or differential equations.
	Nonlinear	Superposition does not apply. Model is expressed as a system of nonlinear difference or differential equations. Frequently approximated by linear systems for analytical ease.
Continuity of independent variable (time)	Continuous	Dependent variables (input, output, state) are defined over a continuous range of the independent variable (time), even though the dependence is not necessarily described by a mathematically continuous function. Model is expressed as differential equations. Typical of physical systems.
	Discrete	Dependent variables are defined only at distinct instants of time. Model is expressed as difference equations. Typical of digital and nonphysical systems.
	Hybrid	System with continuous and discrete subsystems, most common in computer control and communication systems. Sampling and quantization typical in A/D (analog-to-digital) conversion; signal reconstruction for D/A conversion. Model frequently approximated as entirely continuous or entirely discrete.
Quantization of dependent variables	Nonquantized	Dependent variables are continuously variable over a range of values. Typical of physical systems at macroscopic resolution.
	Quantized	Dependent variables assume only a countable number of different values. Typical of computer control and communication systems (sample data systems).

The *mean* or *expected value* of a probability distribution is defined as

$$E(X) = \int_{-\infty}^{\infty} x f(x)\, dx = \overline{X}$$

The mean is the first moment of the distribution. The *nth moment* of the distribution is defined as

$$E(X^n) = \int_{-\infty}^{\infty} x^n f(x)\, dx$$

The mean square of the difference between the random variable and its mean is the *variance* or *second central*

moment of the distribution,

$$\sigma^2(X) = E(X - \overline{X})^2 = \int_{-\infty}^{\infty} (x - \overline{X})^2 f(x)\, dx$$

$$= E(X^2) - [E(X)]^2$$

The square root of the variance is the *standard deviation* of the distribution:

$$\sigma(X) = \sqrt{E(X^2) - [E(X)]^2}$$

The mean of the distribution therefore is a measure of the average magnitude of the random variable, while the variance and standard deviation are measures of the variability or dispersion of this magnitude.

The concepts of probability can be extended to more than one random variable. The *joint distribution function* of two random variables x and y is defined as

$$F(x, y) = \Pr(X < x \text{ and } Y < y)$$

$$= \int_{-\infty}^{x} \int_{-\infty}^{y} f(x, y) \, dy \, dx$$

where $f(x, y)$ is the joint distribution. The ijth moment of the joint distribution is

$$E(X^i Y^j) = \int_{-\infty}^{\infty} x^i \int_{-\infty}^{\infty} y^j f(x, y) \, dy \, dx$$

The *covariance* of x and y is defined to be

$$E[(X - \overline{X})(Y - \overline{Y})]$$

and the normalized covariance or *correlation coefficient* as

$$\rho = \frac{E[(X - \overline{X})(Y - \overline{Y})]}{\sqrt{\sigma^2(X)\sigma^2(Y)}}$$

Although many distribution functions have proven useful in control engineering, far and away the most useful is the *Gaussian* or *normal distribution*

$$F(x) = \frac{1}{\sigma\sqrt{2\pi}} \exp\left[\frac{(-x - \mu)^2}{2\sigma^2}\right]$$

where μ is the mean of the distribution and σ is the standard deviation. The Gaussian distribution has a number of important properties. First, if the input to a linear system is Gaussian, the output also will be Gaussian. Second, if the input to a linear system is only approximately Gaussian, the output will tend to approximate a Gaussian distribution even more closely. Finally, a Gaussian distribution can be completely specified by two parameters, μ and σ, and therefore a zero-mean Gaussian variable is completely specified by its variance.

Random Processes　A *random process* is a set of random variables with time-dependent elements. If the statistical parameters of the process (such as σ for the zero-mean Gaussian process) do not vary with time, the process is *stationary*. The *autocorrelation function* of a stationary random variable $x(t)$ is defined by

$$\phi_{xx}(\tau) = \lim_{T \to \infty} \frac{1}{2T} \int_{-T}^{T} x(t)x(t + \tau) \, dt$$

a function of the fixed time interval τ. The autocorrelation function is a quantitative measure of the sequential dependence or time correlation of the random variable, that is, the relative effect of prior values of the variable on the present or future values of the variable. The autocorrelation function also gives information regarding how rapidly the variable is changing and about whether the signal is in part deterministic (specifically, periodic). The autocorrelation function of a zero-mean variable has the properties

$$\sigma^2 = \phi_{xx}(0) \geq \phi_{xx}(\tau) \qquad \phi_{xx}(\tau) = \phi_{xx}(-\tau)$$

In other words, the autocorrelation function for $\tau = 0$ is identically the variance and the variance is the maximum value of the autocorrelation function. From the definition of the function, it is clear that (a) for a purely random variable with zero mean, $\phi_{xx}(\tau) = 0$ for $\tau \neq 0$, and (b) for a deterministic variable, which is periodic with period T, $\phi_{xx}(k2\pi T) = \sigma^2$ for k integer. The concept of time correlation is readily extended to more than one random variable. The *cross-correlation function* between the random variables $x(t)$ and $y(t)$ is

$$\phi_{xy}(\tau) = \lim_{T \to \infty} \int_{-\infty}^{\infty} x(t)y(t + \tau) \, dt$$

For $\tau = 0$, the cross correlation between two zero-mean variables is identically the covariance. A final characterization of a random variable is its *power spectrum*, defined as

$$G(\omega, x) = \lim_{T \to \infty} \frac{1}{2\pi T} \left| \int_{-T}^{T} x(t)e^{-j\omega t} \, dt \right|^2$$

For a stationary random process, the power spectrum function is identically the Fourier transform of the autocorrelation function

$$G(\omega, x) = \frac{1}{\pi} \int_{-\infty}^{\infty} \phi_{xx}(\tau)e^{-j\omega t} \, dt$$

with

$$\phi_{xx}(0) = \int_{-\infty}^{\infty} G(\omega, x) \, d\omega$$

8.2　Distributed-Parameter Models

There are many important applications in which the state of a system cannot be defined at a finite number of points in space. Instead, the system state is a continuously varying function of both time and location. When continuous spatial dependence is explicitly accounted for in a model, the independent variables must include spatial coordinates as well as time. The resulting *distributed-parameter model* is described in terms of *partial differential equations*, containing partial derivatives with respect to each of the independent variables.

Distributed-parameter models commonly arise in the study of mass and heat transport, the mechanics of structures and structural components, and electrical transmission. Consider as a simple example the unidirectional flow of heat through a wall, as depicted in Fig. 27. The temperature of the wall is not in general uniform but depends on both the time t and position within the wall x, that is, $\theta = \theta(x, t)$. A distributed-parameter model for this case might be the first-order partial differential equation

$$\frac{d}{dt}\theta(x, t) = \frac{1}{C_t}\frac{\partial}{\partial x}\left[\frac{1}{R_t}\frac{\partial}{\partial x}\theta(x, t)\right]$$

where C_t is the thermal capacitance and R_t is the thermal resistance of the wall (assumed uniform).

The complexity of distributed-parameter models is typically such that these models are avoided in the analysis and design of control systems. Instead, distributed-parameter systems are approximated by a finite number of spatial "lumps," each lump being characterized by some average value of the state. By eliminating the independent spatial variables, the result is a *lumped-parameter (or lumped-element) model* described by coupled ordinary differential equations. If a sufficiently fine-grained representation of the lumped microstructure can be achieved, a lumped model can be derived that will approximate the distributed model to any desired degree of accuracy. Consider, for example, the three temperature lumps shown in Fig. 28, used to approximate the wall of Fig. 27. The corresponding

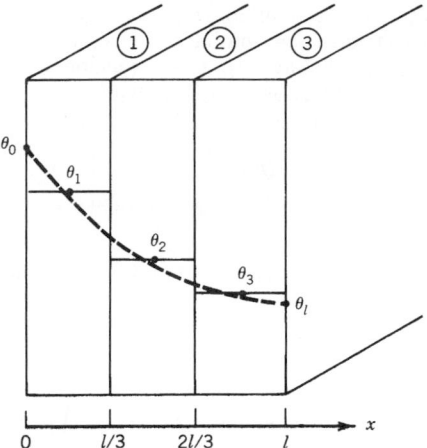

Fig. 28 Lumped-parameter model for uniform heat transfer through a wall.

third-order lumped approximation is

$$\frac{d}{dt}\begin{bmatrix}\theta_1(t)\\\theta_2(t)\\\theta_3(t)\end{bmatrix} = \begin{bmatrix}-\dfrac{9}{C_tR_t} & \dfrac{3}{C_tR_t} & 0\\[2mm] \dfrac{3}{C_tR_t} & \dfrac{6}{C_tR_t} & \dfrac{3}{C_tR_t}\\[2mm] 0 & \dfrac{3}{C_tR_t} & \dfrac{6}{C_tR_t}\end{bmatrix}\begin{bmatrix}\theta_1(t)\\\theta_2(t)\\\theta_3(t)\end{bmatrix}$$
$$+ \begin{bmatrix}\dfrac{6}{C_tR_t}\\[2mm] 0\\[2mm] 0\end{bmatrix}\theta_0(t)$$

If a more detailed approximation is required, this can always be achieved at the expense of adding additional, smaller lumps.

8.3 Time-Varying Systems

Time-varying systems are those with characteristics that change as a function of time. Such variation may result from environmental factors, such as temperature or radiation, or from factors related to the operation of the system, such as fuel consumption. While in general a model with variable parameters can be either linear or nonlinear, the name time varying is most frequently associated with linear systems described by the following state equation:

$$\dot{x}(t) = A(t)x(t) + B(t)u(t)$$

For this linear time-varying model, the superposition principle still applies. Superposition is a great aid in

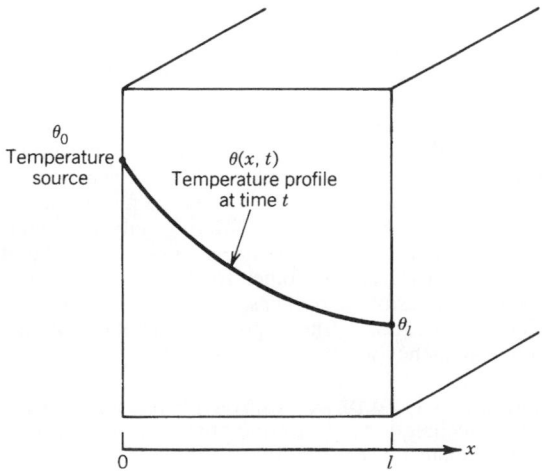

Fig. 27 Uniform heat transfer through a wall.

model formulation but unfortunately does not prove to be much help in determining the model solution.

Paradoxically, the form of the solution to the linear time-varying equation is well known[7]:

$$x(t) = \Phi(t, t_0)x(t_0) + \int_{t_0}^{t} \Phi(t, \tau)B(\tau)u(\tau)\,dt$$

where $\Phi(t, t_0)$ is the time-varying state transition matrix. This knowledge is typically of little value, however, since it is not usually possible to determine the state transition matrix by any straightforward method. By analogy with the first-order case, the relationship

$$\Phi(t, t_0) = \exp\left(\int_{t_0}^{t} A(\tau)\,d\tau\right)$$

can be proven valid *if and only if*

$$A(t)\int_{t_0}^{t} A(\tau)\,d\tau = \int_{t_0}^{t} A(\tau)\,d\tau\ A(t)$$

that is, if and only if $A(t)$ and its integral commute. This is a very stringent condition for all but a first-order system and, as a rule, it is usually easiest to obtain the solution using simulation.

Most of the properties of the fixed transition matrix extend to the time-varying case:

$$\Phi(t, t_0) = I$$
$$\Phi^{-1}(t, t_0) = \Phi(t_0, t)$$
$$\Phi(t_2, t_1)\Phi(t_1, t_0) = \Phi(t_2, t_0)$$
$$\Phi(t, t_0) = A(t)\Phi(t, t_0)$$

8.4 Nonlinear Systems

The theory of fixed, linear, lumped-parameter systems is highly developed and provides a powerful set of techniques for control system analysis and design. In practice, however, all physical systems are nonlinear to some greater or lesser degree. The linearity of a physical system is usually only a convenient approximation, restricted to a certain range of operation. In addition, nonlinearities such as dead zones, saturation, or on–off action are sometimes introduced into control systems intentionally, either to obtain some advantageous performance characteristic or to compensate for the effects of other (undesirable) nonlinearities.

Unfortunately, while nonlinear systems are important, ubiquitous, and potentially useful, the theory of nonlinear differential equations is comparatively meager. Except for specific cases, closed-form solutions to nonlinear systems are generally unavailable. The only universally applicable method for the study of nonlinear systems is *simulation*. As described in

Section 7, however, simulation is an experimental approach, embodying all of the attending limitations of experimentation.

A number of special techniques are available for the analysis of nonlinear systems. All of these techniques are in some sense approximate, assuming, for example, either a restricted range of operation over which nonlinearities are mild or the relative isolation of lower order subsystems. When used in conjunction with more complex simulation models, however, these techniques often provide insights and design concepts that would be difficult to discover through the use of simulation alone.[8]

Linear versus Nonlinear Behaviors There are several fundamental differences between the behavior of linear and nonlinear systems that are especially important. These differences not only account for the increased difficulty encountered in the analysis and design of nonlinear systems, but also imply entirely new types of behavior for nonlinear systems that are not possible for linear systems.

The fundamental property of linear systems is *superposition*. This property states that if $y_1(t)$ is the response of the system to $u_1(t)$ and $y_2(t)$ is the response of the system to $u_2(t)$, then the response of the system to the linear combination $a_1u_1(t) + a_2u_2(t)$ is the linear combination $a_1y_1(t) + a_2y_2(t)$. An immediate consequence of superposition is that the responses of a linear system to inputs differing only in amplitude is qualitatively the same. Since superposition does not apply to nonlinear systems, the responses of a nonlinear system to large and small changes may be fundamentally different.

This fundamental difference in linear and nonlinear behaviors has a second consequence. For a linear system, interchanging two elements connected in series does not affect the overall system behavior. Clearly, this cannot be true in general for nonlinear systems.

A third property peculiar to nonlinear systems is the potential existence of *limit cycles*. A linear oscillator oscillates at an amplitude that depends on its initial state. A limit cycle is an oscillation of fixed amplitude and period, independent of the initial state, that is unique to the nonlinear system.

A fourth property concerns the response of nonlinear systems to sinusoidal inputs. For a linear system, the response to sinusoidal input is a sinusoid of the same frequency, potentially differing only in magnitude and phase. For a nonlinear system, the output will in general contain other frequency components, including possibly harmonics, subharmonics, and aperiodic terms. Indeed, the response need not contain the input frequency at all.

Linearizing Approximations Perhaps the most useful technique for analyzing nonlinear systems is to approximate these with linear systems. While many linearizing approximations are possible, linearization can frequently be achieved by considering small excursions of the system state about a reference trajectory. Consider the nonlinear state equation

$$\dot{x}(t) = f[x(t), \ u(t)]$$

together with a reference trajectory $x^0(t)$ and reference input $u^0(t)$ that together satisfy the state equation

$$\dot{x}^0(t) = f[x^0(t), \ u^0(t)]$$

Note that the simplest case is to choose a static equilibrium or *operating point* $\bar{x}$ as the reference "trajectory" such that $0 = t(\bar{x}, 0)$. The actual trajectory is then related to the reference trajectory by the relationships

$$x(t) = x^0(t) + \delta x(t)$$

$$u(t) = u^0(t) + \delta u(t)$$

where $\delta x(t)$ is some small perturbation about the reference state and $\delta u(t)$ is some small perturbation about the reference input. If these perturbations are indeed small, then applying Taylor's series expansion about the reference trajectory yields the linearized approximation

$$\delta \dot{x}(t) = A(t) \ \delta x(t) + B(t) \ \delta u(t)$$

where the state and distribution matrices are the *Jacobian matrices*

$$A(t) = \begin{bmatrix} \dfrac{\partial f_i}{\partial x_1} & \dfrac{\partial f_1}{\partial x_2} & \cdots & \dfrac{\partial f_1}{\partial x_n} \\[2mm] \dfrac{\partial f_2}{\partial x_1} & \dfrac{\partial f_2}{\partial x_2} & \cdots & \dfrac{\partial f_2}{\partial x_n} \\[2mm] \vdots & \vdots & \ddots & \vdots \\[2mm] \dfrac{\partial f_n}{\partial x_1} & \dfrac{\partial f_n}{\partial x_2} & \cdots & \dfrac{\partial f_n}{\partial x_n} \end{bmatrix}_{x(t)=x^G(t); \ u(t)=u^0(t)}$$

$$B(t) = \begin{bmatrix} \dfrac{\partial f_1}{\partial u_1} & \dfrac{\partial f_1}{\partial u_2} & \cdots & \dfrac{\partial f_1}{\partial u_m} \\[2mm] \dfrac{\partial f_2}{\partial u_1} & \dfrac{\partial f_2}{\partial u_2} & \cdots & \dfrac{\partial f_2}{\partial u_m} \\[2mm] \vdots & \vdots & \ddots & \vdots \\[2mm] \dfrac{\partial f_n}{\partial u_1} & \dfrac{\partial f_n}{\partial u_2} & \cdots & \dfrac{\partial f_n}{\partial u_m} \end{bmatrix}_{x(t)=x^0(t); \ u(t)=u^0(t)}$$

If the reference trajectory is a fixed operating point $\bar{x}$, then the resulting linearized system is time invariant and can be solved analytically. If the reference trajectory is a function of time, however, then the resulting system is linear but time varying.

Describing Functions The describing function method is an extension of the frequency transfer function approach of linear systems, most often used to determine the stability of limit cycles of systems containing nonlinearities. The approach is approximate and its usefulness depends on two major assumptions:

1. All the nonlinearities within the system can be aggregated mathematically into a single block, denoted as $N(M)$ in Fig. 29, such that the equivalent gain and phase associated with this block depend only on the amplitude M_d of the sinusoidal input $m(\omega t) = M \sin(\omega t)$ and are independent of the input frequency ω.
2. All the harmonics, subharmonics, and any direct current (dc) component of the output of the nonlinear block are filtered out by the linear portion of the system such that the effective output of the nonlinear block is well approximated by a periodic response having the same fundamental period as the input.

Although these assumptions appear to be rather limiting, the technique gives reasonable results for a large class of control systems. In particular, the second assumption is generally satisfied by higher order control systems with symmetric nonlinearities, since (a) symmetric nonlinearities do not generate dc terms, (b) the amplitudes of harmonics are generally small when compared with the fundamental term and subharmonics are uncommon, and (c) feedback within a control system typically provides low-pass filtering to further attenuate harmonics, especially for higher order systems. Because the method is relatively simple and can be used for systems of any order, describing functions have enjoyed wide practical application.

The describing function of a nonlinear block is defined as the ratio of the fundamental component of the output to the amplitude of a sinusoidal input. In general, the response of the nonlinearity to the input

$$m(\omega t) = M \ \sin \ \omega t$$

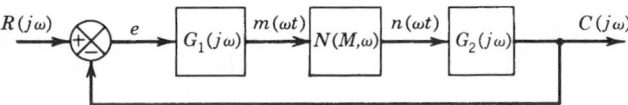

Fig. 29 General nonlinear system for describing function analysis.

is the output

$$n(\omega t) = N_1 \sin(\omega t + \phi_1) + N_2 \sin(2\ \omega t + \phi_2)$$
$$+ N_3 \sin(3\ \omega t + \phi_3) + \cdots$$

and, hence, the describing function for the nonlinearity is defined as the complex quantity

$$N(M) = \frac{N_1}{M} e^{j\phi_1}$$

Derivation of the approximating function typically proceeds by representing the fundamental frequency by the Fourier series coefficients

$$A_1(M) = \frac{2}{T} \int_{-T/2}^{T/2} n(\omega t) \cos \omega t \, d(\omega t)$$

$$B_1(M) = \frac{2}{T} \int_{-T/2}^{T/2} n(\omega t) \sin \omega t \, d(\omega t)$$

The describing function is then written in terms of these coefficients as

$$N(M) = \frac{B_1(M)}{M} + j \frac{A_1(M)}{M} = \left[\left(\frac{B_1(M)}{M} \right)^2 \right.$$
$$\left. + \left(\frac{A_1(M)}{M} \right)^2 \right]^{1/2} \exp \left[j \tan^{-1} \left(\frac{A_1(M)}{B_1(M)} \right) \right]$$

Note that if $n(\omega t) = -n(-\omega t)$, then the describing function is odd, $A_1(M) = 0$, and there is no phase shift between the input and output. If $n(\omega t) = n(-\omega t)$, then the function is even, $B_1(M) = 0$, and the phase shift is $\pi/2$.

The describing functions for a number of typical nonlinearities are given in Fig. 12.30. Reference 9 contains an extensive catalog. The following derivation for a dead-zone nonlinearity demonstrates the general procedure for deriving a describing function. For the saturation element depicted in Fig. 12.30a, the relationship between the input $m(\omega t)$ and output $n(\omega t)$ can be written as

$$n(\omega t) = \begin{cases} 0 & \text{for} \quad -D < m < D \\ K_1 M(\sin \omega t - \sin \omega_1 t) & \text{for} \quad m > D \\ K_1 M(\sin \omega t + \sin \omega_1 t) & \text{for} \quad m < -D \end{cases}$$

Since the function is odd, $A_1 = 0$. By the symmetry over the four quarters of the response period,

$$B_1 = 4 \left[\frac{2}{\pi/2} \int_0^{\pi/2} n(\omega t) \sin \omega t \, d(\omega t) \right]$$
$$= \frac{4}{\pi} \left[\int_0^{\omega t_1} (0) \sin \omega t \, d(\omega t) \right.$$
$$\left. + \int_{\omega t_1}^{\pi/2} K_1 M(\sin \omega t - \sin \omega_1 t) \sin \omega t \, d(\omega t) \right]$$

where $\omega t_1 = \sin^{-1}(D/M)$. Evaluating the integrals and dividing by M yields the describing function listed in Fig. 12.30.

Phase-Plane Method The *phase-plane method* is a graphical application of the state-space approach used to characterize the free response of second-order nonlinear systems. While any convenient pair of state variables can be used, the *phase variables* originally were taken to be the displacement and velocity of the mass of a second-order mechanical system. Using the two state variables as the coordinate axis, the transient response of a system is captured on the *phase plane* as the plot of one variable against the other, with time implicit on the resulting curve. The curve for a specific initial condition is called a *trajectory* in the phase plane; a representative sample of trajectories is called the *phase portrait* of the system. The phase portrait is a compact and readily interpreted summary of the system response. Phase portraits for a sample of typical nonlinearities are shown in Fig. 12.31.

Four methods can be used to construct a phase portrait: (a) direct solution of the differential equation, (b) the graphical *method of isoclines*, (c) transformation of the second-order system (with time as the independent variable) into an equivalent first-order system (with one of the phase variables as the independent variable), and (d) numerical solution using simulation. The first and second methods are usually impractical; the third and fourth methods are frequently used in combination. For example, consider the second-order model

$$\frac{dx_1}{dt} = f_1(x_1, x_2) \qquad \frac{dx_2}{dt} = f_2(x_1, x_2)$$

Dividing the second equation by the first and eliminating the dt terms yield

$$\frac{dx_2}{dx_1} = \frac{f_2(x_1, x_2)}{f_1(x_1, x_2)}$$

This first-order equation describes the phase-plane trajectories. In many cases it can be solved analytically. If not, it always can be simulated.

The phase-plane method complements the describing-function approach. A describing function is an approximate representation of the sinusoidal response for systems of any order, while the phase

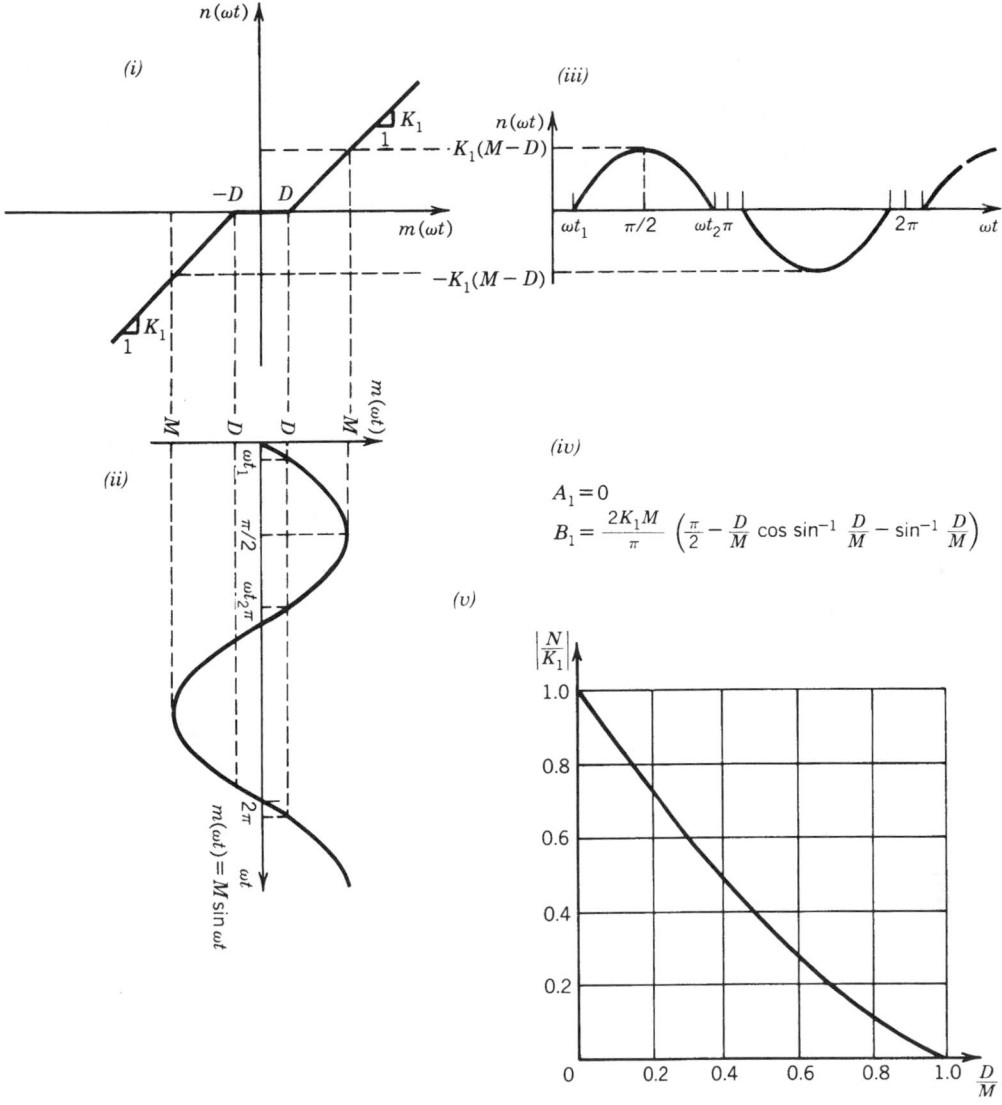

Fig. 30a Describing functions for typical nonlinearities (after Refs. 9 and 10). Dead-zone nonlinearity: (*i*) nonlinear characteristic; (*ii*) sinusoidal input wave shape; (*iii*) output wave shape; (*iv*) describing-function coefficients; (*v*) normalized describing function.

plane is an exact representation of the (free) transient response for first- and second-order systems. Of course, the phase-plane method theoretically can be extended for higher order systems, but the difficulty of visualizing the nth-order state space typically makes such a direct extension impractical. An approximate extension of the method has been used with some considerable success,[8] however, in order to explore and validate the relationships among pairs of variables in complex simulation models. The approximation is based on the assumptions that the paired variables

define a second-order subsystem that, for the purposes of analysis, is weakly coupled to the remainder of the system.

8.5 Discrete and Hybrid Systems

A *discrete-time system* is one for which the dependent variables are defined only at distinct instants of time. Discrete-time models occur in the representation of systems that are inherently discrete, in the analysis and design of digital measurement and control systems, and in the numerical solution of differential equations

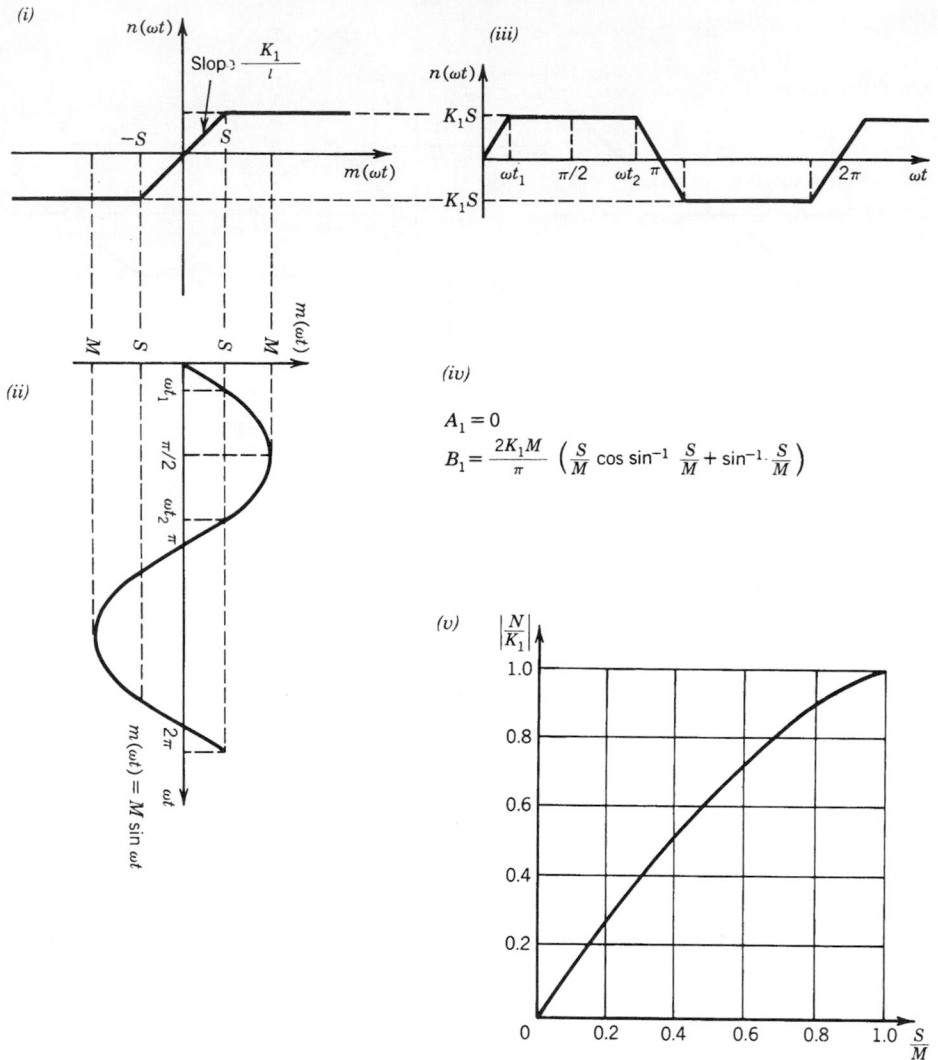

Fig. 30b Saturation nonlinearity: (*i*) nonlinear characteristic; (*ii*) sinusoidal input wave shape; (*iii*) output wave shape; (*iv*) describing-function coefficients; (*v*) normalized describing function.

(see Section 7). Because most control systems are now implemented using digital computers (especially microprocessors), discrete-time models are extremely important in dynamic systems analysis. The discrete-time nature of a computer's sampling of continuous physical signals also leads to the occurrence of *hybrid systems*, that is, systems that are in part discrete and in part continuous. Discrete-time models of hybrid systems are called *sampled-data systems*.

Difference Equations   Dynamic models of discrete-time systems most naturally take the form of *difference equations*. The I/O form of an nth-order difference

equation model is

$$f[y(k+n), y(k+n-1), \ldots, y(k),$$
$$u(k+n-1), \ldots, u(k)] = 0$$

which expresses the dependence of the $(k+n)$th value of the output, $y(k+n)$, on the n preceding values of the output y and input u. For a linear system, the I/O form can be written as

$$y(k+n) + a_{n-1}(k)y(k+n-1) + \cdots + a_1(k)y(k+1)$$
$$+ a_0(k)y(k) = b_{n-1}(k)u(k+n-1) + \cdots + b_0(k)u(k)$$

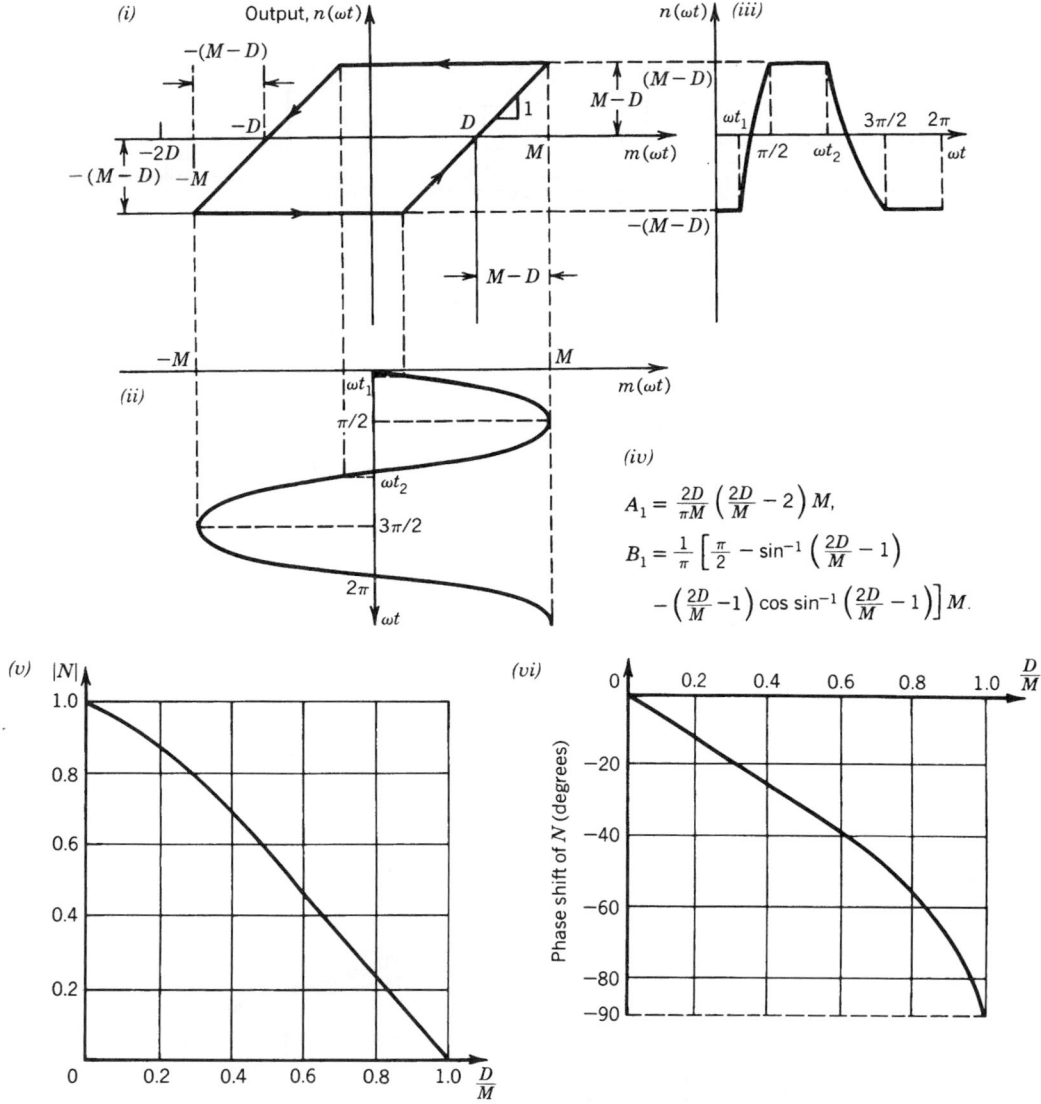

Fig. 30c Backlash nonlinearity: (*i*) nonlinear characteristic; (*ii*) sinusoidal input wave shape; (*iii*) output wave shape; (*iv*) describing-function coefficients; (*v*) normalized amplitude characteristics for the describing function; (*vi*) normalized phase characteristics for the describing function.

In state-variable form, the discrete-time model is the vector difference equation

$$x(k + 1) = f[x(k), u(k)]$$

$$y(k) = g[x(k), u(k)]$$

where x is the state vector, u is the vector of inputs, and y is the vector of outputs. For a linear system, the discrete state-variable form can be written as

$$x(k + 1) = A(k)x(k) + B(k)u(k)$$

$$y(k) = C(k)x(k) + D(k)u(k)$$

The mathematics of difference equations parallels that of differential equations in many important respects. In general, the concepts applied to differential equations have direct analogies for difference equations, although the mechanics of their implementation may vary (see Ref. 11 for a development of dynamic modeling based on difference equations).

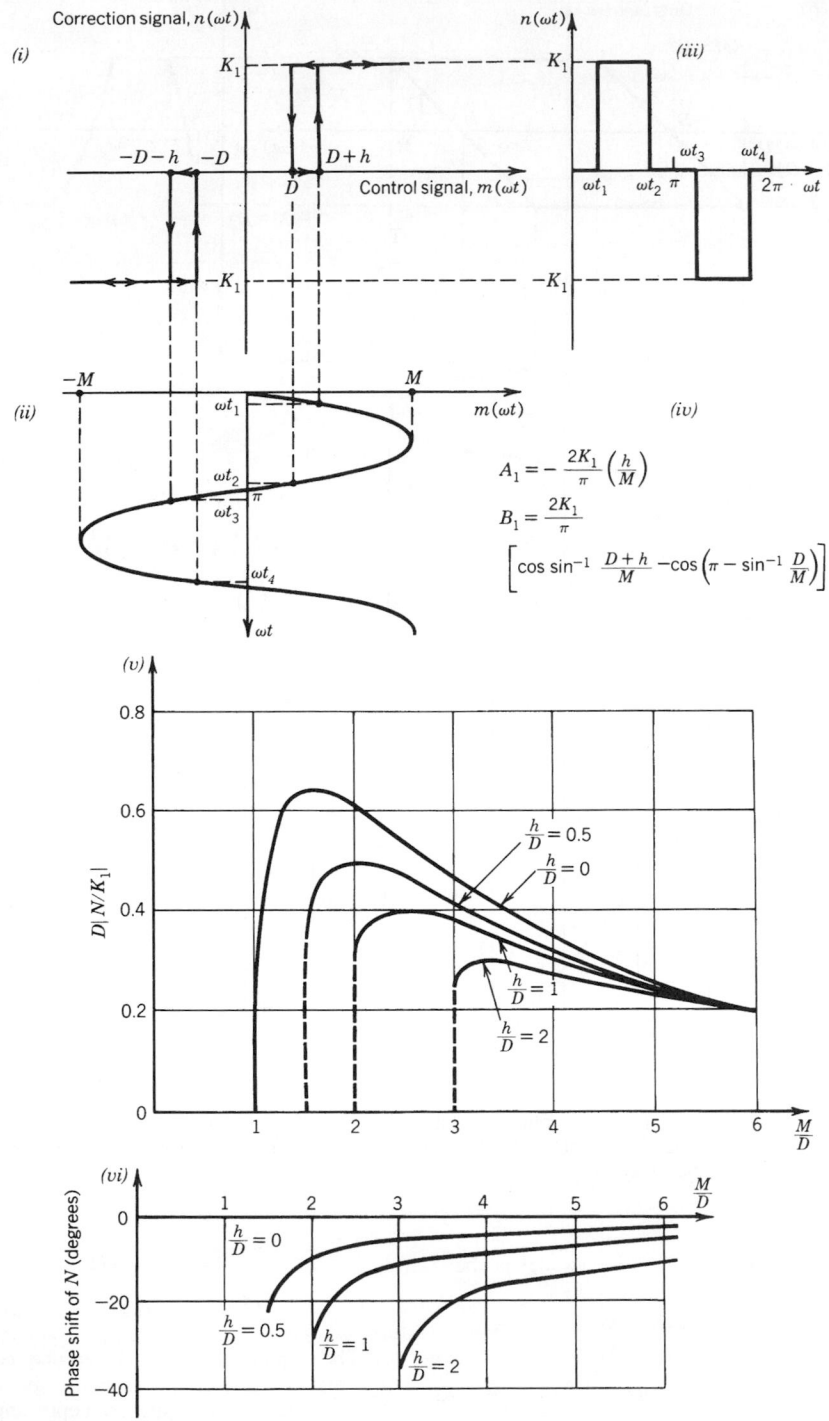

Fig. 30d Three-position on–off device with hysteresis: (*i*) nonlinear characteristic; (*ii*) sinusoidal input wave shape; (*ii*) output wave shape; (*iv*) describing-function coefficients; (*v*) normalized amplitude characteristics for the describing function; (*vi*) normalized phase characteristics for the describing function.

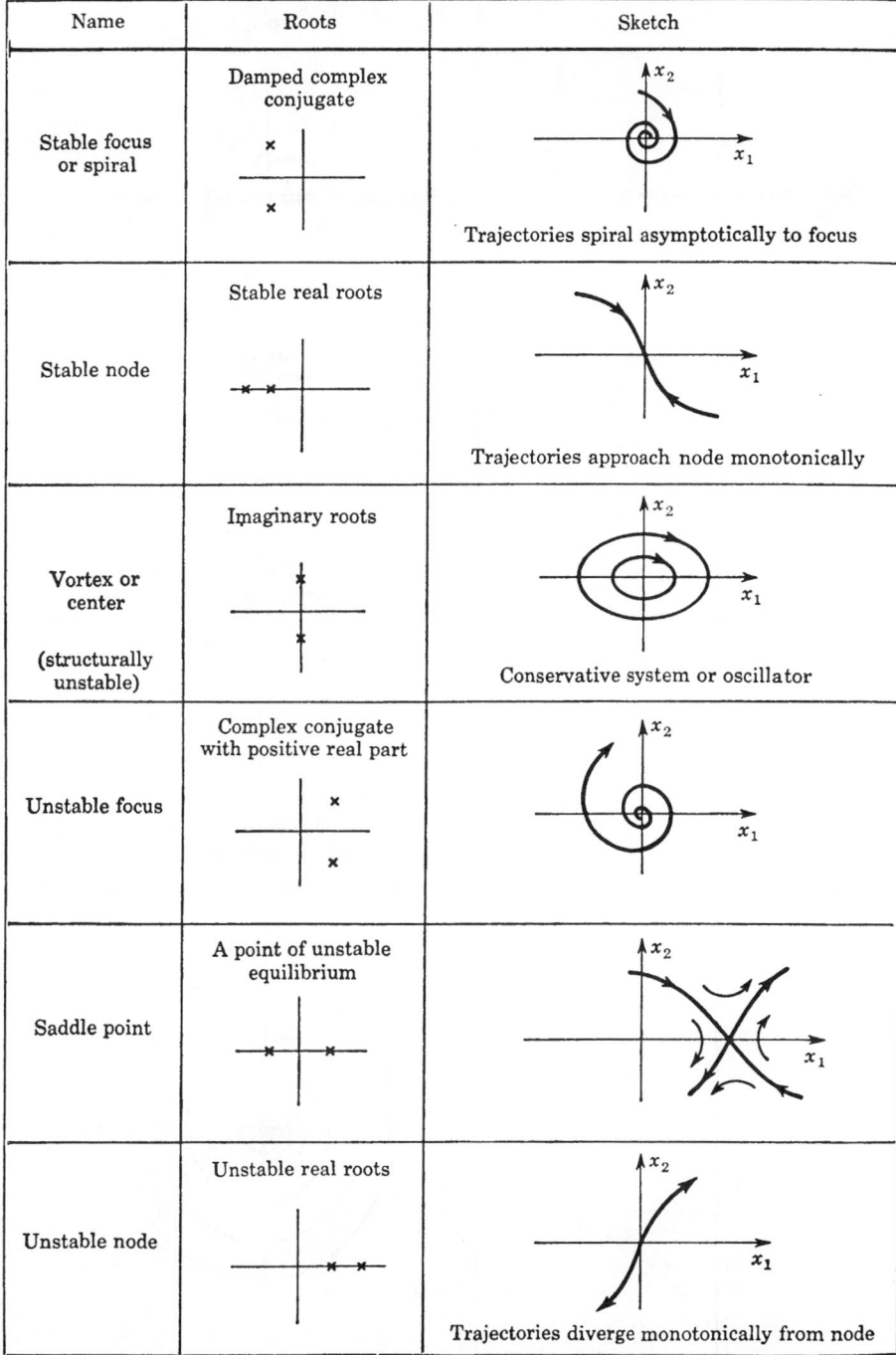

Name	Roots	Sketch
Stable focus or spiral	Damped complex conjugate	Trajectories spiral asymptotically to focus
Stable node	Stable real roots	Trajectories approach node monotonically
Vortex or center (structurally unstable)	Imaginary roots	Conservative system or oscillator
Unstable focus	Complex conjugate with positive real part	
Saddle point	A point of unstable equilibrium	
Unstable node	Unstable real roots	Trajectories diverge monotonically from node

Fig. 31a Typical phase-plane plots for second-order systems.[9]

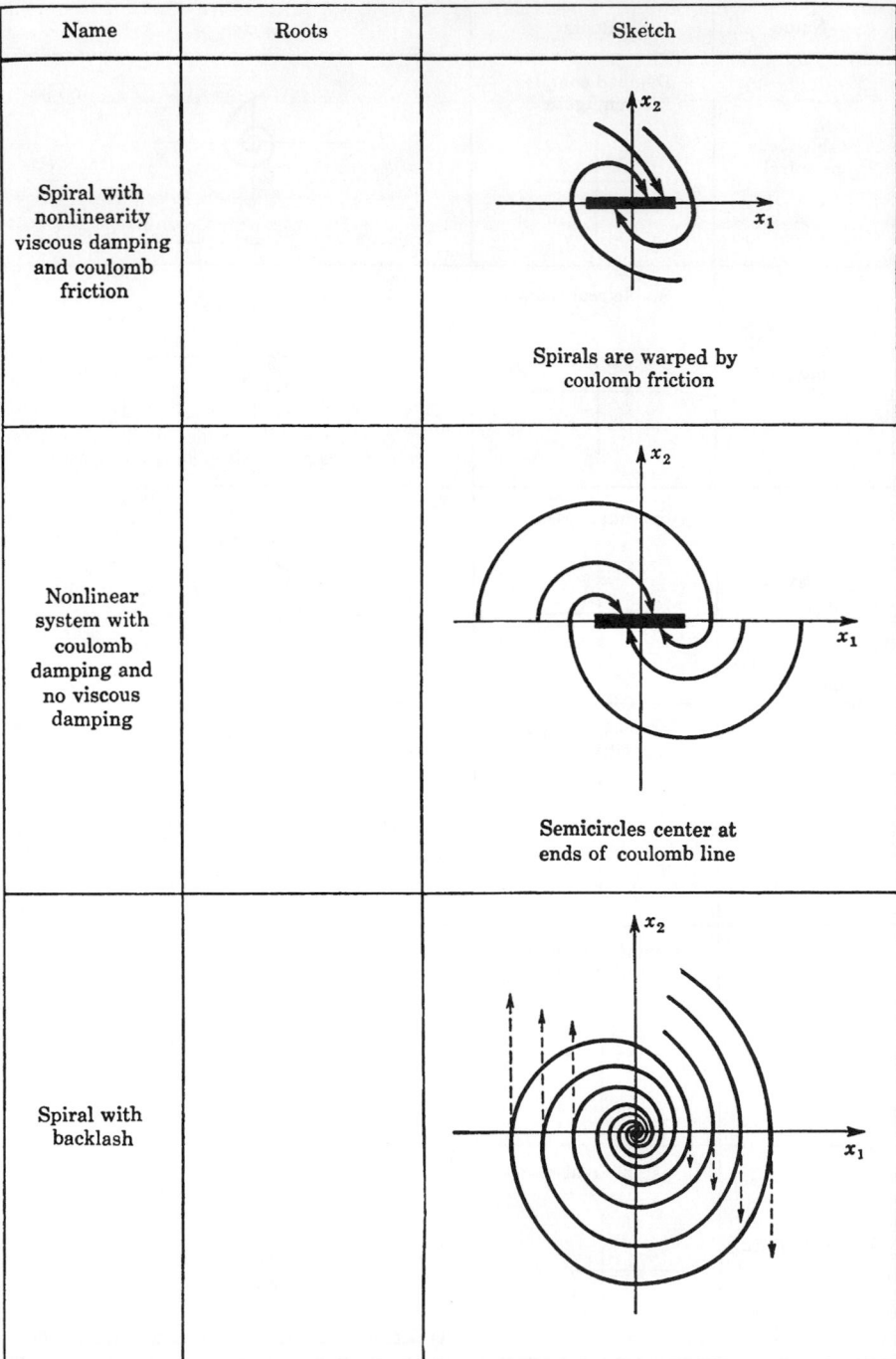

Name	Roots	Sketch
Spiral with nonlinearity viscous damping and coulomb friction		Spirals are warped by coulomb friction
Nonlinear system with coulomb damping and no viscous damping		Semicircles center at ends of coulomb line
Spiral with backlash		

Fig. 31b   (*Continued*)

One important difference is that the general solution of nonlinear and time-varying difference equations can usually be obtained through *recursion*. For example, consider the discrete nonlinear model

$$y(k+1) = \frac{y(k)}{1 + y(k)}$$

Recursive evaluation of the equation beginning with the initial condition $y(0)$ yields

$$y(1) = \frac{y(0)}{1 + y(0)}$$

$$y(2) = \frac{y(1)}{1 + y(1)} = \left[\frac{y(0)}{1 + y(0)}\right] \bigg/ \left[1 + \frac{y(0)}{1 + y(0)}\right]$$

$$= \frac{y(0)}{1 + 2y(0)}$$

$$y(3) = \frac{y(2)}{1 + y(2)} = \frac{y(0)}{1 + 3y(0)}$$

$$\vdots$$

the pattern of which reveals, by induction,

$$y(k) = \frac{y(0)}{1 + ky(0)}$$

as the general solution.

Uniform Sampling *Uniform sampling* is the most common mathematical approach to *A/D conversion*, that is, to extracting the discrete-time approximation $y^*(k)$ of the form

$$y^*(k) = y(t = kT)$$

from the continuous-time signal $y(t)$, where T is a constant interval of time called the *sampling period*. If the sampling period is too large, however, it may not be possible to represent the continuous signal accurately. The *sampling theorem* guarantees that $y(t)$ can be reconstructed from the uniformly sampled values $y^*(k)$ if the sampling period satisfies the inequality

$$T \le \frac{\pi}{\omega_u}$$

where ω_u is the highest frequency contained in the Fourier transform $Y(\omega)$ of $y(t)$, that is, if

$$Y(\omega) = 0 \quad \text{for all} \quad \omega > \omega_u$$

The Fourier transform of a signal is defined to be

$$\mathscr{F}[y(t)] = Y(\omega) = \int_{-\infty}^{\infty} y(t)e^{-j\omega t}\, dt$$

Note that if $y(t) = 0$ for $t \ge 0$, and if the region of convergence for the Laplace transform includes the imaginary axis, then the Fourier transform can be obtained from the Laplace transform as

$$Y(\omega) = [Y(s)]_{s=j\omega}$$

For cases where it is impossible to determine the Fourier transform analytically, such as when the signal is described graphically or by a table, numerical solution based on the *fast Fourier transform (FFT) algorithm* is usually satisfactory.

In general, the condition $T \le \pi/\omega_u$ cannot be satisfied exactly since most physical signals have no finite upper frequency ω_u. A useful approximation is to define the upper frequency as the frequency for which 99% of the signal "energy" lies in the frequency spectrum $0 \le \omega \le \omega_u$. This approximation is found from the relation

$$\int_0^{\omega_u} |Y(\omega)|^2\, d\omega = 0.99 \int_0^{\infty} |Y(\omega)|^2\, d\omega$$

where the square of the amplitude of the Fourier transform $|Y(\omega)|^2$ is said to be the *power spectrum*, and its integral over the entire frequency spectrum is referred to as the "energy" of the signal. Using a sampling frequency 2–10 times this approximate upper frequency (depending on the required factor of safety) and inserting a low-pass filter (called a *guard filter*) before the sampler to eliminate frequencies above the *Nyquist frequency* π/T usually lead to satisfactory results.[4]

The z Transform The z transform permits the development and application of transfer functions for discrete-time systems, in a manner analogous to continuous-time transfer functions based on the Laplace transform. A discrete signal may be represented as a series of impulses

$$y^*(t) = y(0)\delta(t) + y(1)\delta(t - T) + y(2)\delta(t - 2T) + \cdots$$

$$= \sum_{k=0}^{N} y(k)\delta(t - kT)$$

where $y(k) = y^*(t = kT)$ are the values of the discrete signal, $\delta(t)$ is the unit impulse function, and N is the

Table 12   z-Transform Pairs

	$X(s)$	$x(t)$ or $x(k)$	$X(z)$
1	1	$\delta(t)$	1
2	e^{-kTs}	$\delta(t - kT)$	z^{-k}
3	$\dfrac{1}{s}$	$1(t)$	$\dfrac{z}{z-1}$
4	$\dfrac{1}{s^2}$	t	$\dfrac{Tz}{(z-1)^2}$
5	$\dfrac{1}{s+a}$	e^{-at}	$\dfrac{z}{z-e^{-aT}}$
6	$\dfrac{a}{s(s+a)}$	$1 - e^{-at}$	$\dfrac{(1-e^{-aT})z}{(z-1)(z-e^{-aT})}$
7	$\dfrac{\omega}{s^2+\omega^2}$	$\sin \omega t$	$\dfrac{z \sin \omega T}{z^2 - 2z \cos \omega T + 1}$
8	$\dfrac{s}{s^2+\omega^2}$	$\cos \omega t$	$\dfrac{z(z - \cos \omega T)}{z^2 - 2z \cos \omega T + 1}$
9	$\dfrac{1}{(s+a)^2}$	te^{-at}	$\dfrac{Tze^{-aT}}{(z-e^{-aT})^2}$
10	$\dfrac{\omega}{(s+a)^2+\omega^2}$	$e^{-at} \sin \omega t$	$\dfrac{ze^{-aT} \sin \omega T}{z^2 - 2ze^{-aT} \cos \omega T + e^{-2aT}}$
11	$\dfrac{s+a}{(s+a)^2+\omega^2}$	$e^{-at} \cos \omega t$	$\dfrac{z^2 - ze^{-aT} \cos \omega T}{z^2 - 2ze^{-aT} \cos \omega T + e^{-2aT}}$
12	$\dfrac{2}{s^3}$	t^2	$\dfrac{T^2 z(z+1)}{(z-1)^3}$
13		a	$\dfrac{z}{z-a}$
14		$a^k \cos k\pi$	$\dfrac{z}{z+a}$

number of samples of the discrete signal. The Laplace transform of the series is

$$Y^*(s) = \sum_{k=0}^{N} y(k)e^{-ksT}$$

where the shifting property of the Laplace transform has been applied to the pulses. Defining the *shift* or *advance operator* as $z = e^{sT}$, $Y^*(s)$ may now be written as a function of z:

$$Y^*(z) = \sum_{k=0}^{N} \frac{y(k)}{z^k} = \mathscr{Z}[y(t)]$$

where the transformed variable $Y^*(z)$ is called the z transform of the function $y^*(t)$. The inverse of the shift operator $1/z$ is called the *delay operator* and corresponds to a time delay of T.

The z transforms for many sampled functions can be expressed in closed form. A listing of the transforms of several commonly encountered functions is given in Table 12. Properties of the z transform are listed in Table 13.

Pulse Transfer Functions   The transfer function concept developed for continuous systems has a direct analog for sampled-data systems. For a continuous system with sampled output $u(t)$ and sampled input $y(t)$, the *pulse* or *discrete transfer function $G(z)$* is defined as the ratio of the z-transformed output $Y(z)$ to the z-transformed input $U(z)$, assuming zero initial conditions. In general, the pulse transfer function has the form

$$G(z) = \frac{Y(z)}{U(z)} = \frac{b_0 + b_1 z^{-1} + b_2 z^{-2} + \cdots + b_m z^{-m}}{1 + a_1 z^{-1} + a_2 z^{-1} + \cdots + a_n z^{-n}}$$

Table 13 z-Transform Properties

	$x(t)$ or $x(k)$	$\mathscr{Z}[x(t)]$ or $\mathscr{Z}[x(k)]$
1	$ax(t)$	$aX(z)$
2	$x_1(t) + x_2(t)$	$X_1(z) + X_2(z)$
3	$x(t + T)$ or $x(k + 1)$	$zX(z) - zx(0)$
4	$x(t + 2T)$	$z^2X(z) - z^2x(0) - zx(T)$
5	$x(k + 2)$	$z^2X(z) - z^2x(0) - zx(1)$
6	$x(t + kT)$	$z^kX(z) - z^kx(0) - z^{k-1}x(T) - \cdots - zx(kT - T)$
7	$x(k + m)$	$z^mX(z) - z^mx(0) - z^{m-1}x(1) - \cdots - zx(m - 1)$
8	$tx(t)$	$-Tz\dfrac{d}{dz}[X(z)]$
9	$kx(k)$	$-z\dfrac{d}{dz}[X(z)]$
10	$e^{-at}x(t)$	$X(ze^{aT})$
11	$e^{-ak}x(k)$	$X(ze^{a})$
12	$a^kx(k)$	$X\left(\dfrac{z}{a}\right)$
13	$ka^kx(k)$	$-z\dfrac{d}{dz}\left[X\left(\dfrac{z}{a}\right)\right]$
14	$x(0)$	$\lim_{z\to\infty} X(z)$ if the limit exists
15	$x(\infty)$	$\lim_{z\to1}[(z - 1)X(z)]$ if $\dfrac{z - 1}{z}X(z)$ is analytic on and outside the unit circle
16	$\displaystyle\sum_{k=0}^{\infty} x(k)$	$X(1)$
17	$\displaystyle\sum_{k=0}^{n} x(kT)y(nT - kT)$	$X(z)Y(z)$

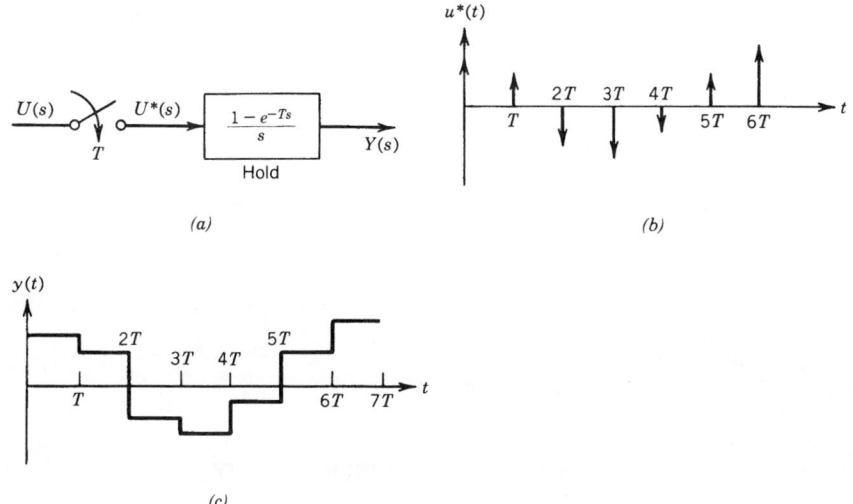

Fig. 32 Zero-order hold: (a) block diagram of hold with a sampler, (b) sampled input sequence, (c) analog output for the corresponding input sequence.[4]

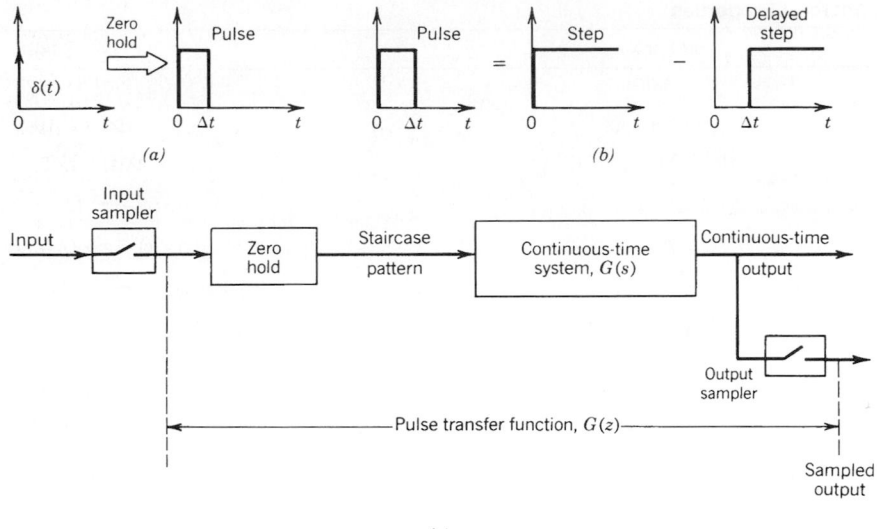

Fig. 33 Pulse transfer function of a continuous system with sampler and zero hold.[12]

Zero-Order Hold The *zero-order data hold* is the most common mathematical approach to *D/A conversion*, that is, to creating a piecewise continuous approximation $u(t)$ of the form

$$u(t) = u^*(k) \quad \text{for} \quad kT \leq t < (k+1)T$$

from the discrete-time signal $u^*(k)$, where T is the period of the hold. The effect of the zero-order hold is to convert a sequence of discrete impulses into a staircase pattern, as shown in Fig. 32. The transfer function of the zero-order hold is

$$G(s) = \frac{1}{s}(1 - e^{-Ts}) = \frac{1 - z^{-1}}{s}$$

Using this relationship, the pulse transfer function of the sampled-data system shown in Fig. 33 can be derived as

$$G(z) = (1 - z^{-1})\mathscr{Z}\left[\mathscr{L}^{-1}\frac{G(s)}{s}\right]$$

The continuous system with transfer function $G(s)$ has a sampler and a zero-order hold at its input and a sampler at its output. This is a common configuration in many computer control applications.

REFERENCES

1. Shearer, J. L., Murphy, A. T., and Richardson, H. H., *Introduction to System Dynamics*, Addison-Wesley, Reading, MA, 1971.

2. Doebelin, E. O., *System Dynamics: Modeling, Analysis, Simulation, and Design*, Merrill, Columbus, OH, 1998.
3. Close, C. M., Frederick, D. K., and Newell, J. C., *Modeling and Analysis of Dynamic Systems*, 3rd ed., Houghton Mifflin, Boston, 2001.
4. Palm, W. J., III, *Modeling, Analysis, and Control of Dynamic Systems*, 2nd ed., Wiley, New York, 2000.
5. Thaler, G. J., and Brown, R. G., *Analysis and Design of Feedback Control Systems*, 2nd ed., McGraw-Hill, New York, 1960.
6. Korn, G. A., and Wait, J. V., *Digital Continuous System Simulation*, Prentice-Hall, Englewood Cliffs, NJ, 1975.
7. Kuo, B. C., and Golnaraghi, F., *Automatic Control Systems*, 8th ed., Prentice-Hall, Englewood Cliffs, NJ, 2002.
8. Thissen, W., "Investigation into the World3 Model: Lessons for Understanding Complicated Models," *IEEE Trans. Syst. Man. Cybernet.*, **SMC-8**(3), 183–193 (1978).
9. Gibson, J. E., *Nonlinear Automatic Control*, McGraw-Hill, New York, 1963.
10. Shinners, S. M., *Modern Control System Theory and Design*, 2nd ed., Wiley, New York, 1998.
11. Luenberger, D. G., *Introduction to Dynamic Systems: Theory, Models, and Applications*, Wiley, New York, 1979.
12. Auslander, D. M., Takahashi, Y., and Rabins, M. J., *Introducing Systems and Control*, McGraw-Hill, New York, 1974.

BIBLIOGRAPHY

Bateson, R. N., *Introduction to Control System Technology*, Prentice-Hall, Englewood Cliffs, NJ, 2001.

Bishop, R. H., and Dorf, R. C., *Modern Control Systems*, 10th ed., Prentice-Hall, Englewood Cliffs, NJ, 2004.

Brogan, W. L., *Modern Control Theory*, 3rd ed., Prentice-Hall, Englewood Cliffs, NJ, 1991.

Cannon, Jr., R. H., *Dynamics of Physical Systems*, McGraw-Hill, New York, 1967.

DeSilva, C. W., *Control Sensors and Actuators*, Prentice-Hall, Englewood Cliffs, NJ, 1989.

Doebelin, E. O., *Measurement Systems*, 5th ed., McGraw-Hill, New York, 2004.

Franklin, G. F., Powell, J. D., and Emami-Naeini, A., *Feedback Control of Dynamic Systems*, 4th ed., Addison-Wesley, Reading, MA, 2001.

Grace, A., Laub, A. J., Little, J. N., and Thompson, C., *Control System Toolbox Users Guide*, The Mathworks, Natick, MA, 1990.

Hartley, T. T., Beale, G. O., and Chicatelli, S. P., *Digital Simulation of Dynamic Systems: A Control Theory Approach*, Prentice-Hall, Englewood Cliffs, NJ, 1994.

Kelton, W. D., Sadowski, R. P., and Sturrock, D. T., *Simulation with Arena*, 3rd ed., McGraw-Hill, New York, 2004.

Kheir, N. A., *Systems Modeling and Computer Simulation*, 2nd ed., Marcel Dekker, New York, 1996.

Lay, D. C., *Linear Algebra and Its Applications*, 3rd ed., Addison-Wesley, Reading, MA, 2006.

Ljung, L., and Glad, T., *Modeling Simulation of Dynamic Systems*, Prentice-Hall, Englewood Cliffs, NJ, 1994.

MathWorks on-line, www.mathworks.com, 1995.

Phillips, C. L., and Harbor, R. D., *Feedback Control Systems*, 4th ed., Prentice-Hall, Englewood Cliffs, NJ, 1999.

Phillips, C. L., and Nagle, H. T., *Digital Control System Analysis and Design*, 3rd ed., Prentice-Hall, Englewood Cliffs, NJ, 1995.

Van Loan, C., *Computational Frameworks for the Fast Fourier Transform*, SIAM, Philadelphia, PA, 1992.

Wolfram, S., *Mathematica Book*, 5th ed., Wolfram Media, Champaign, IL, 2003.

CHAPTER 13

BASIC CONTROL SYSTEMS DESIGN*

William J. Palm III
Department of Mechanical Engineering
University of Rhode Island
Kingston, Rhode Island

*Revised from William J. Palm III, *Modeling, Analysis and Control of Dynamic Systems*, 2nd ed., Wiley, 2000, by permission of the publisher.

1 INTRODUCTION

The purpose of a *control system* is to produce a desired *output*. This output is usually specified by the command *input* and is often a function of time. For simple applications in well-structured situations, *sequencing* devices like timers can be used as the control system. But most systems are not that easy to control, and the controller must have the capability of reacting to disturbances, changes in its environment, and new input commands. The key element that allows a control system to do this is *feedback*, which is the process by which a system's output is used to influence its behavior. Feedback in the form of the room temperature measurement is used to control the furnace in a thermostatically controlled heating system. Figure 1 shows the *feedback loop* in the system's *block diagram*, which is a graphical representation of the system's control structure and logic. Another commonly found control system is the pressure regulator shown in Fig. 2.

Feedback has several useful properties. A system whose individual elements are nonlinear can often be modeled as a linear one over a wider range of its variables with the proper use of feedback. This is because feedback tends to keep the system near its reference operation condition. Systems that can maintain the output near its desired value despite changes in the environment are said to have good *disturbance rejection*. Often we do not have accurate values for some system parameter or these values might change with age. Feedback can be used to minimize the effects of parameter changes and uncertainties. A system that has both good disturbance rejection and low sensitivity to parameter variation is *robust*. The application that resulted in the general understanding of the properties

of feedback is shown in Fig. 3. The electronic amplifier gain A is large, but we are uncertain of its exact value. We use the resistors R_1 and R_2 to create a feedback loop around the amplifier and pick R_1 and R_2 to create a feedback loop around the amplifier and R_1 and R_2 so that $A R_2/R_1 \gg 1$. Then the input–output relation becomes $e_o \approx R_1 e_i/R_2$, which is independent of A as long as A remains large. If R_1 and R_2 are known accurately, then the system gain is now reliable.

Figure 4 shows the block diagram of a *closed-loop* system, which is a system with feedback. An *open-loop* system, such as a timer, has no feedback. Figure 4 serves as a focus for outlining the prerequisites for this chapter. The reader should be familiar with the *transfer function* concept based on the Laplace transform, the *pulse transfer* function based on the z transform, for digital control, and the differential equation modeling techniques needed to obtain them. It is also necessary to understand block diagram algebra, characteristic roots, the final-value theorem, and their use in evaluating system response for common inputs like the step function. Also required are stability analysis techniques such as the Routh criterion and transient performance specifications such as the damping ratio ζ, natural frequency ω_n, dominant time constant τ, maximum overshoot, settling time, and bandwidth. Treatment in depth is given in Refs. 1–4.

2 CONTROL SYSTEM STRUCTURE

The electromechanical position control system shown in Fig. 5 illustrates the structure of a typical control system. A load with an inertia I is to be positioned at some desired angle θ_r. A dc motor is provided for this purpose. The system contains viscous damping, and a disturbance torque T_d acts on the load, in addition to the motor torque T. Because of the disturbance, the

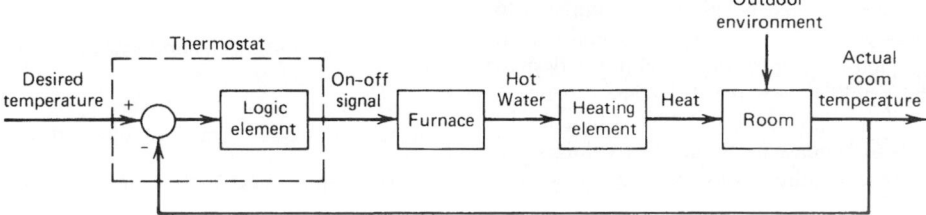

Fig. 1 Block diagram of thermostat system for temperature control.[1]

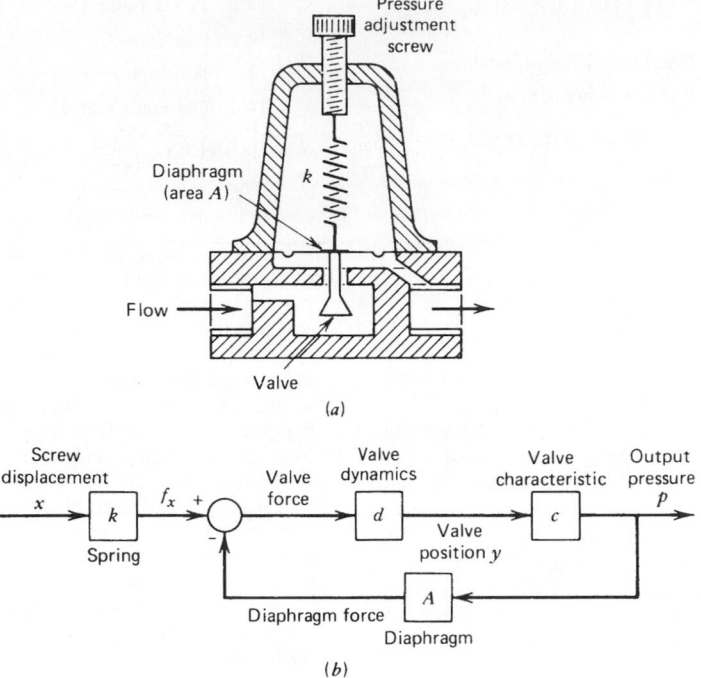

Fig. 2 Pressure regulator: (a) cutaway view; (b) block diagram.[1]

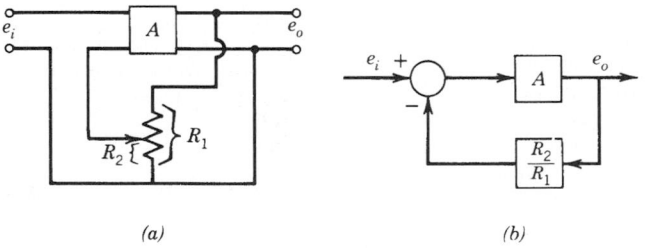

Fig. 3 Closed-loop system.

angular position θ of the load will not necessarily equal the desired value θ_r. For this reason, a potentiometer, or some other sensor such as an encoder, is used to measure the displacement θ. The potentiometer voltage representing the controlled position θ is compared to the voltage generated by the command potentiometer. This device enables the operator to dial in the desired angle θ_r. The amplifier sees the difference e between the two potentiometer voltages. The basic function of the amplifier is to increase the small error voltage e up to the voltage level required by the motor and to supply enough current required by the motor to drive the load. In addition, the amplifier may shape the voltage signal

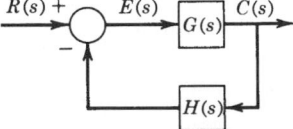

Fig. 4 Feedback compensation of amplifier.

in certain ways to improve the performance of the system.

The control system is seen to provide two basic functions: (a) to respond to a command input that specifies a new desired value for the controlled variable and

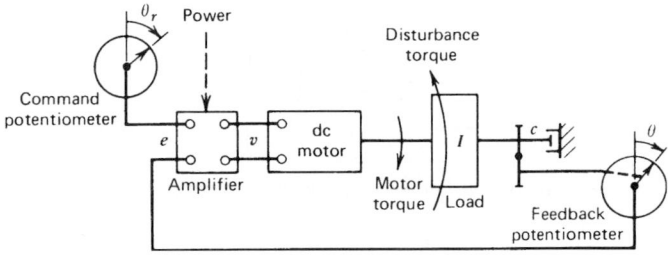

Fig. 5 Position control system using dc motor.[1]

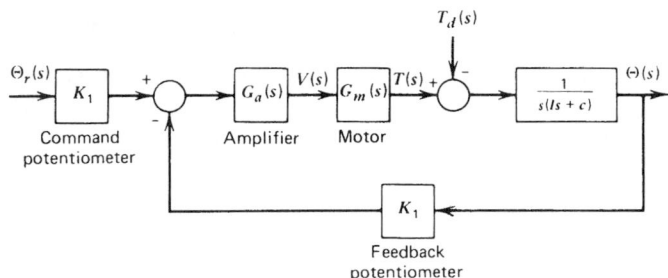

Fig. 6 Block diagram of position control system shown in Fig. 5.[1]

(b) to keep the controlled variable near the desired value in spite of disturbances. The presence of the feedback loop is vital to both functions. A block diagram of this system is shown in Fig. 6. The power supplies required for the potentiometers and the amplifier are not shown in block diagrams of control system logic because they do not contribute to the control logic.

2.1 Standard Diagram

The electromechanical positioning system fits the general structure of a control system (Fig. 7). This figure also gives some standard terminology. Not all systems can be forced into this format, but it serves as a reference for discussion.

The controller is generally thought of as a logic element that compares the command with the measurement of the output and decides what should be done. The input and feedback elements are transducers for converting one type of signal into another type. This allows the error detector directly to compare two signals of the same type (e.g., two voltages). Not all functions show up as separate physical elements. The error detector in Fig. 5 is simply the input terminals of the amplifier.

The control logic elements produce the control signal, which is sent to the *final control elements*. These are the devices that develop enough torque, pressure, heat, and so on to influence the elements under control. Thus, the final control elements are the "muscle" of the system, while the control logic elements are

the "brain." Here we are primarily concerned with the design of the logic to be used by this brain.

The object to be controlled is the *plant*. The *manipulated* variable is generated by the final control elements for this purpose. The disturbance input also acts on the plant. This is an input over which the designer has no influence and perhaps for which little information is available as to the magnitude, functional form, or time of occurrence. The disturbance can be a random input, such as wind gust on a radar antenna, or deterministic, such as Coulomb friction effects. In the latter case, we can include the friction force in the system model by using a nominal value for the coefficient of friction. The disturbance input would then be the deviation of the friction force from this estimated value and would represent the uncertainty in our estimate.

Several control system classifications can be made with reference to Fig. 7. A *regulator* is a control system in which the controlled variable is to be kept constant in spite of disturbances. The command input for a regulator is its *set point*. A *follow-up system* is supposed to keep the control variable near a command value that is changing with time. An example of a follow-up system is a machine tool in which a cutting head must trace a specific path in order to shape the product properly. This is also an example of a *servomechanism*, which is a control system whose controlled variable is a mechanical position, velocity, or acceleration. A thermostat system is not a servomechanism, but a *process control system*, where the controlled variable describes a thermodynamic process.

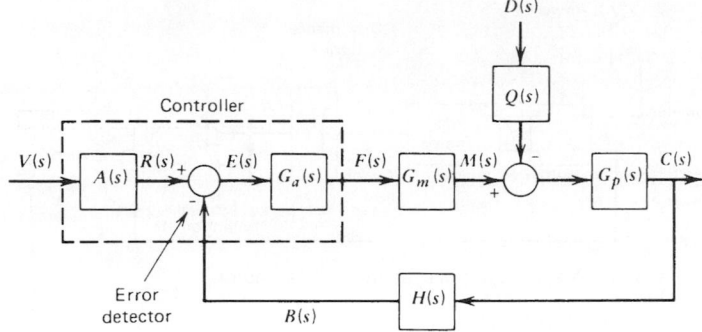

Fig. 7   Terminology and basic structure of feedback control system.[1]

Elements		Signals	
$A(s)$	Input elements	$B(s)$	Feedback signal
$G_a(s)$	Control logic elements	$C(s)$	Controlled variable or output
$G_m(s)$	Final control elements	$D(s)$	Disturbance input
$G_p(s)$	Plant elements	$E(s)$	Error or actuating signal
$H(s)$	Feedback elements	$F(s)$	Control signal
$Q(s)$	Disturbance elements	$M(s)$	Manipulated variable
		$R(s)$	Reference input
		$V(s)$	Command input

Typically, such variables are temperature, pressure, flow rate, liquid level, chemical concentration, and so on.

2.2   Transfer Functions

A transfer function is defined for each input–output pair of the system. A specific transfer function is found by setting all other inputs to zero and reducing the block diagram. The *primary* or *command* transfer function for Fig. 7 is

$$\frac{C(s)}{V(s)} = \frac{A(s)G_a(s)G_m(s)G_p(s)}{1 + G_a(s)G_m(s)G_p(s)H(s)} \tag{1}$$

The *disturbance* transfer function is

$$\frac{C(s)}{D(s)} = \frac{-Q(s)G_p(s)}{1 + G_a(s)G_m(s)G_p(s)H(s)} \tag{2}$$

The transfer functions of a given system all have the same denominator.

2.3   System-Type Number and Error Coefficients

The error signal in Fig. 4 is related to the input as

$$E(s) = \frac{1}{1 + G(s)H(s)}R(s) \tag{3}$$

If the final-value theorem can be applied, the steady-state error is

$$e_{ss} = \lim_{s \to 0} \frac{sR(s)}{1 + G(s)H(s)} \tag{4}$$

The *static error coefficient* c_i is defined as

$$c_i = \lim_{s \to 0} s^i G(s)H(s) \tag{5}$$

A system is of *type n* if $G(s)H(s)$ can be written as $s^n F(s)$. Table 1 relates the steady-state error to the system type for three common inputs and can be used to design systems for minimum error. The higher the

Table 1   **Steady-State Error e_{ss} for Different System-Type Numbers**

$R(s)$	System-Type Number n			
	0	1	2	3
Step $1/s$	$\dfrac{1}{1 + C_0}$	0	0	0
Ramp $1/s^2$	∞	$\dfrac{1}{C_1}$	0	0
Parabola $1/s^3$	∞	∞	$\dfrac{1}{C_2}$	0

system type, the better the system is able to follow a rapidly changing input. But higher type systems are more difficult to stabilize, so a compromise must be made in the design. The coefficients c_0, c_1, and c_2 are called the *position, velocity*, and *acceleration error coefficients*.

3 TRANSDUCERS AND ERROR DETECTORS

The control system structure shown in Fig. 7 indicates a need for physical devices to perform several types of functions. Here we present a brief overview of some available transducers and error detectors. Actuators and devices used to implement the control logic are discussed in Sections 4 and 5.

3.1 Displacement and Velocity Transducers

A *transducer* is a device that converts one type of signal into another type. An example is the potentiometer, which converts displacement into voltage, as in Fig. 8. In addition to this conversion, the transducer can be used to make measurements. In such applications, the term *sensor* is more appropriate. Displacement can also be measured electrically with a *linear variable differential transformer* (LVDT) or a *synchro*. An LVDT measures the linear displacement of a movable magnetic core through a primary winding and two secondary windings (Fig. 9). An ac voltage is applied to the primary. The secondaries are connected together and also to a detector that measures the voltage and phase difference. A phase difference of $0°$ corresponds to a positive core displacement, while $180°$ indicates a negative displacement. The amount of displacement is indicated by the amplitude of the ac voltage in the secondary. The detector converts this

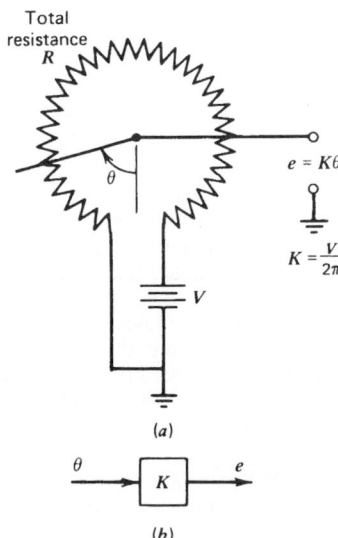

Fig. 8   Rotary potentiometer.[1]

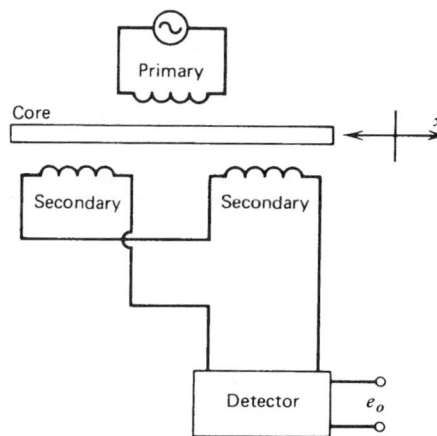

Fig. 9   Linear variable differential transformer.[1]

information into a dc voltage e_o, such that $e_o = Kx$. The LVDT is sensitive to small displacements. Two of them can be wired together to form an error detector.

A synchro is a rotary differential transformer, with angular displacement as either the input or output. They are often used in pairs (a *transmitter* and a *receiver*) where a remote indication of angular displacement is needed. When a transmitter is used with a synchro *control transformer*, two angular displacements can be measured and compared (Fig. 10). The output voltage e_o is approximately linear with angular difference within $\pm 70°$, so that $e_o = K(\theta_1 - \theta_2)$.

Displacement measurements can be used to obtain forces and accelerations. For example, the displacement of a calibrated spring indicates the applied force. The accelerometer is another example. Still another is the strain gage used for force measurement. It is based on the fact that the resistance of a fine wire changes as it is stretched. The change in resistance is detected by a circuit that can be calibrated to indicate the applied force. Sensors utilizing piezoelectric elements are also available.

Velocity measurements in control systems are most commonly obtained with a *tachometer*. This is essentially a dc generator (the reverse of a dc motor). The input is mechanical (a velocity). The output is a generated voltage proportional to the velocity. Translational velocity can be measured by converting it to angular velocity with gears, for example. Tachometers using ac signals are also available.

Other velocity transducers include a magnetic pickup that generates a pulse every time a gear tooth passes. If the number of gear teeth is known, a pulse counter and timer can be used to compute the angular velocity. This principle is also employed in turbine flowmeters.

A similar principle is employed by *optical encoders*, which are especially suitable for digital control purposes. These devices use a rotating disk with

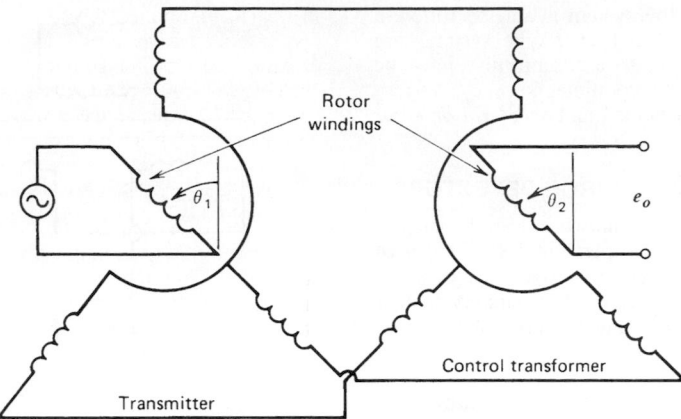

Fig. 10 Synchro transmitter control transformer.[1]

alternating transparent and opaque elements whose passage is sensed by light beams and a photosensor array, which generates a binary (on–off) train of pulses. There are two basic types: the absolute encoder and the incremental encoder. By counting the number of pulses in a given time interval, the incremental encoder can measure the rotational speed of the disk. By using multiple tracks of elements, the absolute encoder can produce a binary digit that indicates the amount of rotation. Hence, it can be used as a position sensor.

Most encoders generate a train of transistor–transistor logic (TTL) voltage level pulses for each channel. The incremental encoder output contains two channels that each produce N pulses every revolution. The encoder is mechanically constructed so that pulses from one channel are shifted relative to the other channel by a quarter of a pulse width. Thus, each pulse pair can be divided into four segments called *quadratures*. The encoder output consists of $4N$ *quadrature counts per revolution*. The pulse shift also allows the direction of rotation to be determined by detecting which channel leads the other. The encoder might contain a third channel, known as the zero, index, or marker channel, that produces a pulse once per revolution. This is used for initialization.

The gain of such an incremental encoder is $4N/2\pi$. Thus, an encoder with 1000 pulses per channel per revolution has a gain of 636 counts per radian. If an absolute encoder produces a binary signal with n bits, the maximum number of positions it can represent is $2n$, and its gain is $2^n/2\pi$. Thus, a 16-bit absolute encoder has a gain of $2^{16}/2\pi = 10{,}435$ counts per radian.

3.2 Temperature Transducers

When two wires of dissimilar metals are joined together, a voltage is generated if the junctions are at different temperatures. If the reference junction is kept at a fixed, known temperature, the thermocouple can be calibrated to indicate the temperature at the other junction in terms of the voltage v. Electrical resistance changes with temperature. Platinum gives a linear relation between resistance and temperature, while nickel is less expensive and gives a large resistance change for a given temperature change. Semiconductors designed with this property are called *thermistors*. Different metals expand at different rates when the temperature is increased. This fact is used in the bimetallic strip transducer found in most home thermostats. Two dissimilar metals are bonded together to form the strip. As the temperature rises, the strip curls, breaking contact and shutting off the furnace. The temperature gap can be adjusted by changing the distance between the contacts. The motion also moves a pointer on the temperature scale of the thermostat. Finally, the pressure of a fluid inside a bulb will change as its temperature changes. If the bulb fluid is air, the device is suitable for use in pneumatic temperature controllers.

3.3 Flow Transducers

A flow rate q can be measured by introducing a flow restriction, such as an orifice plate, and measuring the pressure drop Δp across the restriction. The relation is $\Delta p = Rq^2$, where R can be found from calibration of the device. The pressure drop can be sensed by converting it into the motion of a diaphragm. Figure 11 illustrates a related technique. The Venturi-type flowmeter measures the static pressures in the constricted and unconstricted flow regions. Bernoulli's principle relates the pressure difference to the flow rate. This pressure difference produces the diaphragm displacement. Other types of flowmeters are available, such as turbine meters.

3.4 Error Detectors

The error detector is simply a device for finding the difference between two signals. This function is sometimes an integral feature of sensors, such as with the synchro transmitter–transformer combination. This

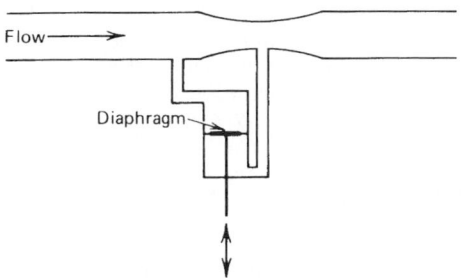

Fig. 11 Venturi-type flowmeter. The diaphragm displacement indicates the flow rate.[1]

concept is used with the diaphragm element shown in Fig. 11. A detector for voltage difference can be obtained, as with the position control system shown in Fig. 5. An amplifier intended for this purpose is a *differential amplifier*. Its output is proportional to the difference between the two inputs. In order to detect differences in other types of signals, such as temperature, they are usually converted to a displacement or pressure. One of the detectors mentioned previously can then be used.

3.5 Dynamic Response of Sensors

The usual transducer and detector models are static models and as such imply that the components respond instantaneously to the variable being sensed. Of course,

any real component has a dynamic response of some sort, and this response time must be considered in relation to the controlled process when a sensor is selected. If the controlled process has a time constant at least 10 times greater than that of the sensor, we often would be justified in using a static sensor model.

4 ACTUATORS

An *actuator* is the final control element that operates on the low-level control signal to produce a signal containing enough power to drive the plant for the intended purpose. The armature-controlled dc motor, the hydraulic servomotor, and the pneumatic diaphragm and piston are common examples of actuators.

4.1 Electromechanical Actuators

Figure 12 shows an electromechanical system consisting of an armature-controlled dc motor driving a load inertia. The rotating armature consists of a wire conductor wrapped around an iron core. This winding has an inductance L. The resistance R represents the lumped value of the armature resistance and any external resistance deliberately introduced to change the motor's behavior. The armature is surrounded by a magnetic field. The reaction of this field with the armature current produces a torque that causes the armature to rotate. If the armature voltage v is used to control the motor, the motor is said to be *armature controlled*. In this case, the field is produced by an electromagnet supplied with a constant voltage or by a permanent

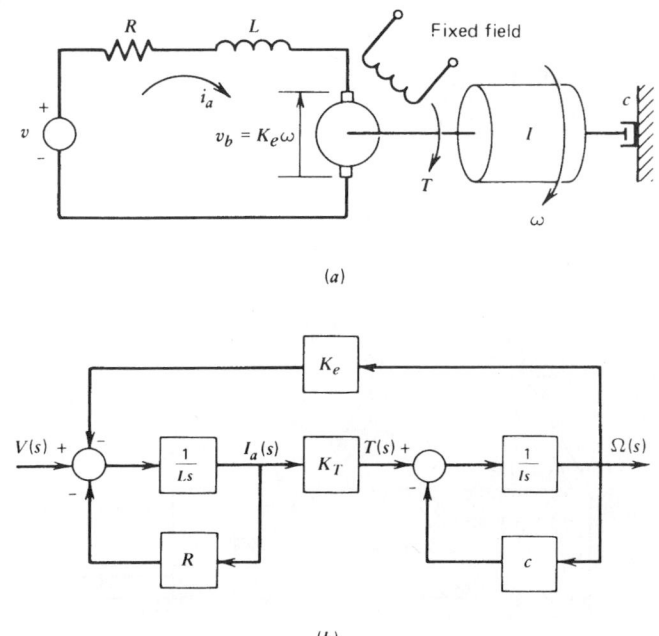

(a)

(b)

Fig. 12 Armature-controlled dc motor with load and the system's block diagram.[1]

magnet. This motor type produces a torque T that is proportional to the armature current i_a:

$$T = K_T i_a \qquad (6)$$

The torque constant K_T depends on the strength of the field and other details of the motor's construction. The motion of a current-carrying conductor in a field produces a voltage in the conductor that opposes the current. This voltage is called the *back emf* (electromotive force). Its magnitude is proportional to the speed and is given by

$$e_b = K_e \omega \qquad (7)$$

The transfer function for the armature-controlled dc motor is

$$\frac{\Omega(s)}{V(s)} = \frac{K_T}{LIs^2 + (RI + cL)s + cR + K_e K_T} \qquad (8)$$

Another motor configuration is the *field-controlled* dc motor. In this case, the armature current is kept constant and the field voltage v is used to control the motor. The transfer function is

$$\frac{\Omega(s)}{V(s)} = \frac{K_T}{(Ls + R)(Is + c)} \qquad (9)$$

where R and L are the resistance and inductance of the field circuit and K_T is the torque constant. No back emf exists in this motor to act as a self-braking mechanism.

Two-phase ac motors can be used to provide a low-power, variable-speed actuator. This motor type can accept the ac signals directly from LVDTs and synchros without demodulation. However, it is difficult to design ac amplifier circuitry to do other than proportional action. For this reason, the ac motor is not found in control systems as often as dc motors. The transfer function for this type is of the form of Eq. (9).

An actuator especially suitable for digital systems is the *stepper motor*, a special dc motor that takes a train of electrical input pulses and converts each pulse into an angular displacement of a fixed amount. Motors are available with resolutions ranging from about 4 steps per revolution to more than 800 steps per revolution. For 36 steps per revolution, the motor will rotate by $10°$ for each pulse received. When not being pulsed, the motors lock in place. Thus, they are excellent for precise positioning applications, such as required with printers and computer tape drives. A disadvantage is that they are low-torque devices. If the input pulse frequency is not near the resonant frequency of the motor, we can take the output rotation to be directly related to the number of input pulses and use that description as the motor model.

4.2 Hydraulic Actuators

Machine tools are one application of the hydraulic system shown in Fig. 13. The applied force f is supplied by the servomotor. The mass m represents that of a cutting tool and the power piston, while k represents the combined effects of the elasticity naturally present in the structure and that introduced by the designer to achieve proper performance. A similar statement applies to the damping c. The valve displacement z is generated by another control system in order to move the tool through its prescribed motion. The spool valve shown in Fig. 13 had two *lands*. If the width of the land is greater than the port width, the valve is said to be *overlapped*. In this case, a dead zone exists in which a slight change in the displacement z produces no power piston motion. Such dead zones

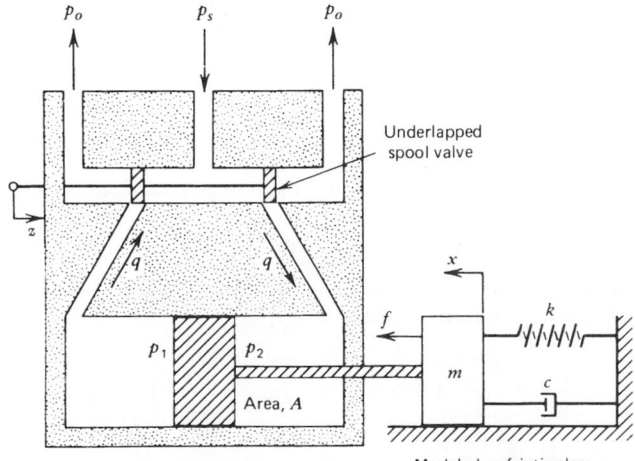

Fig. 13 Hydraulic servomotor with load.[1]

create control difficulties and are avoided by designing the valve to be *underlapped* (the land width is less the port width). For such valves there will be a small flow opening even when the valve is in the neutral position at $z = 0$. This gives it a higher sensitivity than an overlapped valve.

The variables z and $\Delta p = p_2 - p_1$ determine the volume flow rate, as

$$q = f(z, \Delta p)$$

For the reference equilibrium condition ($z = 0$, $\Delta p = 0$, $q = 0$), a linearization gives

$$q = C_1 z - C_2 \, \Delta p \tag{10}$$

The linearization constants are available from theoretical and experimental results.[5] The transfer function for the system is[1,2]

$$T(s) = \frac{X(s)}{Z(s)}$$

$$= \frac{C_1}{(C_2 m/A)s^2 + (cC_2/A + A)s + C_2 k/A} \tag{11}$$

The development of the steam engine led to the requirement for a speed control device to maintain constant speed in the presence of changes in load torque or steam pressure. In 1788, James Watt of Glasgow developed his now-famous flyball governor for this purpose (Fig. 14). Watt took the principle of sensing speed with the centrifugal pendulum of Thomas Mead and used it in a feedback loop on a steam engine. As the motor speed increases, the flyballs move outward and pull the slider upward. The upward motion of the slider closes the steam valve, thus causing the engine to slow down. If the engine speed is too slow, the spring force overcomes that due to the flyballs, and the slider moves down to open the steam valve. The desired speed can be set by moving the plate to change the compression in the spring. The principle of the flyball governor is still used for speed control applications. Typically, the pilot valve of a hydraulic servomotor is connected to the slider to provide the high forces required to move large supply valves.

Many hydraulic servomotors use multistage valves to obtain finer control and higher forces. A *two-stage valve* has a *slave valve*, similar to the pilot valve but situated between the pilot valve and the power piston.

Rotational motion can be obtained with a *hydraulic motor*, which is, in principle, a pump acting in reverse (fluid input and mechanical rotation output). Such motors can achieve higher torque levels than electric motors. A hydraulic pump driving a hydraulic motor constitutes a *hydraulic transmission*.

A popular actuator choice is the *electrohydraulic system*, which uses an electric actuator to control a hydraulic

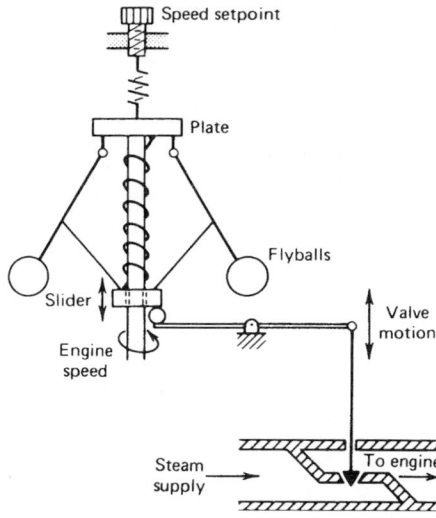

Fig. 14 James Watt's flyball governor for speed control of steam engine.[1]

servomotor or transmission by moving the pilot valve or the swash-plate angle of the pump. Such systems combine the power of hydraulics with the advantages of electrical systems. Figure 15 shows a hydraulic motor whose pilot valve motion is caused by an armature-controlled dc motor. The transfer function between the motor voltage and the piston displacement is

$$\frac{X(s)}{V(s)} = \frac{K_1 K_2 C_1}{As^2(\tau s + 1)} \tag{12}$$

If the rotational inertia of the electric motor is small, then $\tau \approx 0$.

4.3 Pneumatic Actuators

Pneumatic actuators are commonly used because they are simple to maintain and use a readily available working medium. Compressed air supplies with the pressures required are commonly available in factories and laboratories. No flammable fluids or electrical sparks are present, so these devices are considered the safest to use with chemical processes. Their power output is less than that of hydraulic systems but greater than that of electric motors.

A device for converting pneumatic pressure into displacement is the bellows shown in Fig. 16. The transfer function for a linearized model of the bellows is of the form

$$\frac{X(s)}{P(s)} = \frac{K}{\tau s + 1} \tag{13}$$

where x and p are deviations of the bellows displacement and input pressure from nominal values.

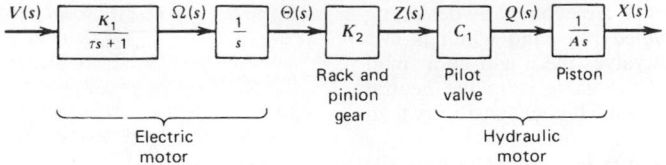

Fig. 15 Electrohydraulic system for translation.[1]

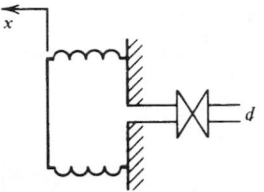

Fig. 16 Pneumatic bellows.[1]

In many control applications, a device is needed to convert small displacements into relatively large pressure changes. The nozzle–flapper serves this purpose (Fig. 17a). The input displacement y moves the flapper, with little effort required. This changes the opening at the nozzle orifice. For a large enough opening, the nozzle back pressure is approximately the same as atmospheric pressure p_a. At the other extreme position with the flapper completely blocking the orifice, the back pressure equals the supply pressure p_s. This variation is shown in Fig. 17b. Typical supply pressures are between 30 and 100 psia. The orifice diameter is approximately 0.01 in. Flapper displacement is usually less than one orifice diameter.

The nozzle–flapper is operated in the linear portion of the back-pressure curve. The linearized back-pressure relation is

$$p = -K_f x \qquad (14)$$

where $-K_f$ is the slope of the curve and is a very large number. From the geometry of similar triangles, we have

$$p = \frac{aK_f}{a+b} y \qquad (15)$$

In its operating region, the nozzle–flapper's back pressure is well below the supply pressure.

The output pressure from a pneumatic device can be used to drive a final control element like the pneumatic actuating valve shown in Fig. 18. The pneumatic pressure acts on the upper side of the diaphragm and is opposed by the return spring.

Formerly, many control systems utilized pneumatic devices to implement the control law in analog form. Although the overall, or higher level, control algorithm is now usually implemented in digital form, pneumatic devices are still frequently used for final control corrections at the actuator level, where the control action must eventually be supplied by a mechanical device. An example of this is the electropneumatic valve positioner used in Valtek valves and illustrated in Fig. 19. The heart of the unit is a pilot valve capsule that moves up and down according to the pressure difference across its two supporting diaphragms. The capsule has a plunger at its top and at its bottom. Each plunger has an exhaust seat at one end and a supply seat at the other. When the capsule is in its equilibrium position, no air is supplied to or

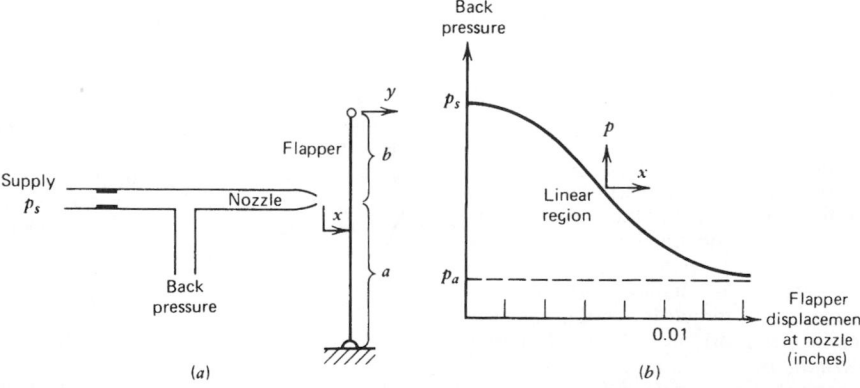

Fig. 17 Pneumatic nozzle–flapper amplifier and its characteristic curve.[1]

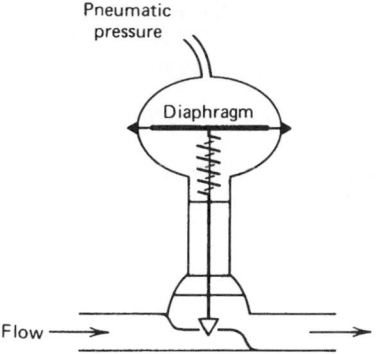

Pneumatic pressure

Diaphragm

Flow

Fig. 18 Pneumatic flow control valve.[1]

exhausted from the valve cylinder, so the valve does not move.

The process controller commands a change in the valve stem position by sending the 4–20-mA dc input signal to the positioner. Increasing this signal causes the electromagnetic actuator to rotate the lever counterclockwise about the pivot. This increases the air gap between the nozzle and flapper. This decreases the back pressure on top of the upper diaphragm and causes the capsule to move up. This motion lifts the upper plunger from its supply seat and allows the supply air to flow to the bottom of the valve cylinder. The lower plunger's exhaust seat is uncovered, thus

decreasing the air pressure on top of the valve piston, and the valve stem moves upward. This motion causes the lever arm to rotate, increasing the tension in the feedback spring and decreasing the nozzle–flapper gap. The valve continues to move upward until the tension in the feedback spring counteracts the force produced by the electromagnetic actuator, thus returning the capsule to its equilibrium position.

A decrease in the dc input signal causes the opposite actions to occur, and the valve moves downward.

5 CONTROL LAWS

The control logic elements are designed to act on the error signal to produce the control signal. The algorithm that is used for this purpose is called the *control law*, the *control action*, or the *control algorithm*. A nonzero error signal results from either a change in command or a disturbance. The general function of the controller is to keep the controlled variable near its desired value when these occur. More specifically, the control objectives might be stated as follows:

1. Minimize the steady-state error.
2. Minimize the settling time.
3. Achieve other transient specifications, such as minimizing the overshoot.

In practice, the design specifications for a controller are more detailed. For example, the bandwidth might also be specified along with a safety margin for stability.

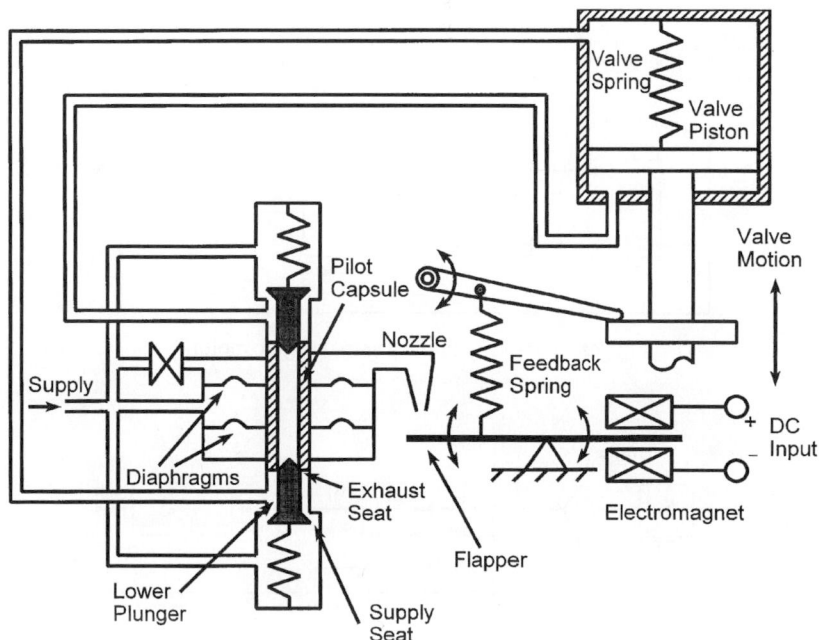

Fig. 19 Electropneumatic valve positioner.

We never know the numerical values of the system's parameters with true certainty, and some controller designs can be more sensitive to such parameter uncertainties than other designs. So a parameter sensitivity specification might also be included.

The following control laws form the basis of most control systems.

5.1 Proportional Control

Two-position control is the most familiar type, perhaps because of its use in home thermostats. The control output takes on one of two values. With the *on–off controller*, the controller output is either on or off (e.g., fully open or fully closed). Two-position control is acceptable for many applications in which the requirements are not too severe. However, many situations require finer control.

Consider a liquid level system in which the input flow rate is controlled by a valve. We might try setting the control valve manually to achieve a flow rate that balances the system at the desired level. We might then add a controller that adjusts this setting in proportion to the deviation of the level from the desired value. This is *proportional control*, the algorithm in which the change in the control signal is proportional to the error. Block diagrams for controllers are often drawn in terms of the deviations from a zero-error equilibrium condition. Applying this convention to the general terminology of Fig. 6, we see that proportional control is described by

$$F(s) = K_P E(s)$$

where $F(s)$ is the deviation in the control signal and K_P is the *proportional gain*. If the total valve displacement is $y(t)$ and the manually created displacement is x, then

$$y(t) = K_P e(t) + x$$

The percent change in error needed to move the valve full scale is the *proportional band*. It is related to the gain as

$$K_P = \frac{100}{\text{band \%}}$$

The zero-error valve displacement x is the *manual reset*.

Proportional Control of a First-Order System

To investigate the behavior of proportional control, consider the speed control system shown in Fig. 20; it is identical to the position controller shown in Fig. 6, except that a tachometer replaces the feedback potentiometer. We can combine the amplifier gains into one, denoted K_P. The system is thus seen to have proportional control. We assume the motor is field controlled and has a negligible electrical time constant. The disturbance is a torque T_d, for example, resulting from friction. Choose the reference equilibrium condition to be $T_d = T = 0$ and $\omega_r = w = 0$. The block diagram is shown in Fig. 21. For a meaningful error signal to be generated, K_1 and K_2 should be chosen to be equal. With this simplification the diagram becomes that

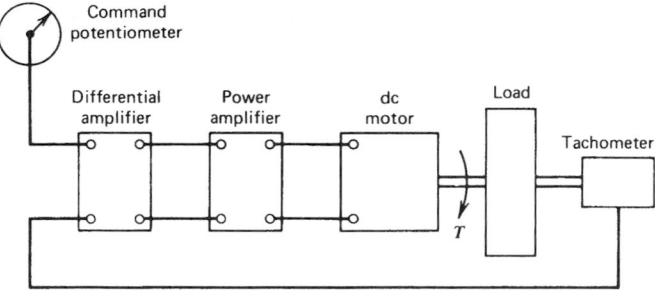

Fig. 20   Velocity control system using dc motor.[1]

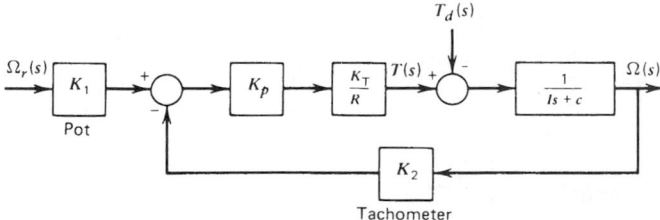

Fig. 21   Block diagram of velocity control system of Fig. 20.[1]

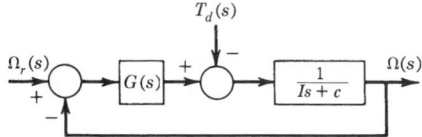

Fig. 22 Simplified form of Fig. 21 for case $K_1 = K_2$.

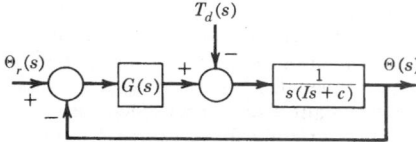

Fig. 23 Position servo.

shown in Fig. 22, where $G(s) = K = K_1 K_P K_T / R$. A change in desired speed can be simulated by a unit step input for ω_r. For $\Omega_r(s) = 1/s$, the velocity approaches the steady-state value $\omega_{ss} = K/(c + K) < 1$. Thus, the final value is less than the desired value of 1, but it might be close enough if the damping c is small. The time required to reach this value is approximately four time constants, or $4\tau = 4I/(c + K)$. A sudden change in load torque can also be modeled by a unit step function $T_d(s) = 1/s$. The steady-state response due solely to the disturbance is $-1/(c + K)$. If $c + K$ is large, this error will be small.

The performance of the proportional control law thus far can be summarized as follows. For a first-order plant with step function inputs:

1. The output never reaches its desired value if damping is present ($c \neq 0$), although it can be made arbitrarily close by choosing the gain K large enough. This is called *offset error*.
2. The output approaches its final value without oscillation. The time to reach this value is inversely proportional to K.
3. The output deviation due to the disturbance at steady state is inversely proportional to the gain K. This error is present even in the absence of damping ($c = 0$).

As the gain K is increased, the time constant becomes smaller and the response faster. Thus, the chief disadvantage of proportional control is that it results in steady-state errors and can only be used when the gain can be selected large enough to reduce the effect of the largest expected disturbance. Since proportional control gives zero error only for one load condition (the reference equilibrium), the operator must change the manual reset by hand (hence the name). An advantage to proportional control is that the control signal responds to the error instantaneously (in theory at least). It is used in applications requiring rapid action. Processes with time constants too small for the use of two-position control are likely candidates for proportional control. The results of this analysis can be applied to any type of first-order system (e.g., liquid level, thermal, etc.) having the form in Fig. 22.

Proportional Control of a Second-Order System
Proportional control of a neutrally stable second-order plant is represented by the position controller of Fig. 6 if the amplifier transfer function is a constant $G_a(s) = K_a$.

Let the motor transfer function be $G_m(s) = K_T/R$, as before. The modified block diagram is given in Fig. 23 with $G(s) = K = K_1 K_a K_T / R$. The closed-loop system is stable if I, c, and K are positive. For no damping ($c = 0$), the closed-loop system is neutrally stable. With no disturbance and a unit step command, $\Theta_r(s) = 1/s$, the steady-state output is $\omega_{ss} = 1$. The offset error is thus zero if the system is stable ($c > 0$, $K > 0$). The steady-state output deviation due to a unit step disturbance is $-1/K$. This deviation can be reduced by choosing K large. The transient behavior is indicated by the damping ratio, $\zeta = c/2\sqrt{IK}$.

For slight damping, the response to a step input will be very oscillatory and the overshoot large. The situation is aggravated if the gain K is made large to reduce the deviation due to the disturbance. We conclude, therefore, that proportional control of this type of second-order plant is not a good choice unless the damping constant c is large. We will see shortly how to improve the design.

5.2 Integral Control

The offset error that occurs with proportional control is a result of the system reaching an equilibrium in which the control signal no longer changes. This allows a constant error to exist. If the controller is modified to produce an increasing signal as long as the error is nonzero, the offset might be eliminated. This is the principle of *integral control*. In this mode the change in the control signal is proportional to the *integral* of the error. In the terminology of Fig. 7, this gives

$$F(s) = \frac{K_I}{s} E(s) \qquad (16)$$

where $F(s)$ is the deviation in the control signal and K_I is the *integral gain*. In the time domain, the relation is

$$f(t) = K_I \int_0^t e(t)\,dt \qquad (17)$$

if $f(0) = 0$. In this form, it can be seen that the integration cannot continue indefinitely because it would theoretically produce an infinite value of $f(t)$ if $e(t)$ does not change sign. This implies that special care must be taken to reinitialize a controller that uses integral action.

Integral Control of a First-Order System Integral control of the velocity in the system of Fig. 20 has the block diagram shown in Fig. 22, where $G(s) =$

K/s, $K = K_1 K_I K_T / R$. The integrating action of the amplifier is physically obtained by the techniques to be presented in Section 6 or by the digital methods presented in Section 10. The control system is stable if I, c, and K are positive. For a unit step command input, $\omega_{ss} = 1$; so the offset error is zero. For a unit step disturbance, the steady-state deviation is zero if the system is stable. Thus, the steady-state performance using integral control is excellent for this plant with step inputs. The damping ratio is $\zeta = c/2\sqrt{IK}$. For slight damping, the response will be oscillatory rather than exponential as with proportional control. Improved steady-state performance has thus been obtained at the expense of degraded transient performance. The conflict between steady-state and transient specifications is a common theme in control system design. As long as the system is underdamped, the time constant is $\tau = 2I/c$ and is not affected by the gain K, which only influences the oscillation frequency in this case. It might be physically possible to make K small enough so that $\zeta \gg 1$, and the nonoscillatory feature of proportional control recovered, but the response would tend to be sluggish. Transient specifications for fast response generally require that $\zeta < 1$. The difficulty with using $\zeta < 1$ is that τ is fixed by c and I. If c and I are such that $\zeta < 1$, then τ is large if $I \gg c$.

Integral Control of a Second-Order System
Proportional control of the position servomechanism in Fig. 23 gives a nonzero steady-state deviation due to the disturbance. Integral control [$G(s) = K/s$] applied to this system results in the command transfer function

$$\frac{\Theta(s)}{\Theta_r(s)} = \frac{K}{Is^3 + cs^2 + K} \tag{18}$$

With the Routh criterion, we immediately see that the system is not stable because of the missing s term. Integral control is useful in improving steady-state performance, but in general it does not improve and may even degrade transient performance. Improperly applied, it can produce an unstable control system. It is best used in conjunction with other control modes.

5.3 Proportional-Plus-Integral Control

Integral control raised the order of the system by 1 in the preceding examples but did not give a characteristic equation with enough flexibility to achieve acceptable transient behavior. The instantaneous response of proportional control action might introduce enough variability into the coefficients of the characteristic equation to allow both steady-state and transient specifications to be satisfied. This is the basis for using *proportional-plus-integral control* (PI control). The algorithm for this two-mode control is

$$F(s) = K_P E(s) + \frac{K_I}{s} E(s) \tag{19}$$

The integral action provides an automatic, not manual, reset of the controller in the presence of a disturbance. For this reason, it is often called *reset action*.

The algorithm is sometimes expressed as

$$F(s) = K_P \left(1 + \frac{1}{T_I s}\right) E(s) \tag{20}$$

where T_I is the *reset time*. The reset time is the time required for the integral action signal to equal that of the proportional term if a constant error exists (a hypothetical situation). The reciprocal of reset time is expressed as repeats per minute and is the frequency with which the integral action repeats the proportional correction signal.

The proportional control gain must be reduced when used with integral action. The integral term does not react instantaneously to a zero-error signal but continues to correct, which tends to cause oscillations if the designer does not take this effect into account.

PI Control of a First-Order System PI action applied to the speed controller of Fig. 20 gives the diagram shown in Fig. 21 with $G(s) = K_P + K_I/s$. The gains K_P and K_I are related to the component gains, as before. The system is stable for positive values of K_P and K_I. For $\Omega_r(s) = 1/s$, $\omega_{ss} = 1$, and the offset error is zero, as with integral action only. Similarly, the deviation due to a unit step disturbance is zero at steady state. The damping ratio is $\zeta = (c + K_P)/2\sqrt{IK_I}$. The presence of K_P allows the damping ratio to be selected without fixing the value of the dominant time constant. For example, if the system is underdamped ($\zeta < 1$), the time constant is $\tau = 2I/(c + K_P)$. The gain K_P can be picked to obtain the desired time constant, while K_I is used to set the damping ratio. A similar flexibility exists if $\zeta = 1$. Complete description of the transient response requires that the numerator dynamics present in the transfer functions be accounted for.[1,2]

PI Control of a Second-Order System Integral control for the position servomechanism of Fig. 23 resulted in a third-order system that is unstable. With proportional action, the diagram becomes that of Fig. 22, with $G(s) = K_P + K_I/s$. The steady-state performance is acceptable, as before, if the system is assumed to be stable. This is true if the Routh criterion is satisfied, that is, if I, c, K_P, and K_I are positive and $cK_P - IK_I > 0$. The difficulty here occurs when the damping is slight. For small c, the gain K_P must be large in order to satisfy the last condition, and this can be difficult to implement physically. Such a condition can also result in an unsatisfactory time constant. The root-locus method of Section 9 provides the tools for analyzing this design further.

5.4 Derivative Control

Integral action tends to produce a control signal even after the error has vanished, which suggests that the controller be made aware that the error is approaching zero. One way to accomplish this is to design the controller to react to the derivative of the error with *derivative control* action, which is

$$F(s) = K_D s E(s) \qquad (21)$$

where K_D is the *derivative gain*. This algorithm is also called *rate action*. It is used to damp out oscillations. Since it depends only on the error rate, derivative control should never be used alone. When used with proportional action, the following proportional-plus-derivative (PD) control algorithm results:

$$F(s) = (K_P + K_D s)E(s) = K_P(1 + T_D s)E(s) \quad (22)$$

where T_D is the *rate time* or *derivative time*. With integral action included, the *proportional-plus-integral-plus-derivative (PID) control law* is obtained:

$$F(s) = \left(K_P + \frac{K_I}{s} + K_D s \right) E(s) \qquad (23)$$

This is called a *three-mode controller*.

PD Control of a Second-Order System   The presence of integral action reduces steady-state error but tends to make the system less stable. There are applications of the position servomechanism in which a nonzero derivation resulting from the disturbance can be tolerated but an improvement in transient response over the proportional control result is desired. Integral action would not be required, but rate action can be added to improve the transient response. Application of PD control to this system gives the block diagram of Fig. 23 with $G(s) = K_P + K_D s$.

The system is stable for positive values of K_D and K_P. The presence of rate action does not affect the steady-state response, and the steady-state results are identical to those with proportional control; namely, zero offset error and a deviation of $-1/K_P$, due to the disturbance. The damping ratio is $\zeta = (c + K_D)/2\sqrt{IK_P}$. For proportional control, $\zeta = c/2\sqrt{IK_P}$. Introduction of rate action allows the proportional gain K_P to be selected large to reduce the steady-state deviation, while K_D can be used to achieve an acceptable damping ratio. The rate action also helps to stabilize the system by adding damping (if $c = 0$, the system with proportional control is not stable).

The equivalent of derivative action can be obtained by using a tachometer to measure the angular velocity of the load. The block diagram is shown in Fig. 24. The gain of the amplifier–motor–potentiometer combination is K_1, and K_2 is the tachometer gain. The advantage of this system is that it does not require signal differentiation, which is difficult to implement if signal noise is present. The gains K_1 and K_2 can be chosen to yield the desired damping ratio and steady-state deviation, as was done with K_P and K_I.

5.5 PID Control

The position servomechanism design with PI control is not completely satisfactory because of the difficulties encountered when the damping c is small. This problem can be solved by the use of the full PID control law, as shown in Fig. 23 with $G(s) = K_P + K_D s + K_I/s$.

A stable system results if all gains are positive and if $(c + K_D)K_P - IK_I > 0$. The presence of K_D relaxes somewhat the requirement that K_P be large to achieve stability. The steady-state errors are zero, and the transient response can be improved because three of the coefficients of the characteristic equation can be selected. To make further statements requires the root-locus technique presented in Section 9.

Proportional, integral, and derivative actions and their various combinations are not the only control laws possible, but they are the most common. PID controllers will remain for some time the standard against which any new designs must compete.

The conclusions reached concerning the performance of the various control laws are strictly true only for the plant model forms considered. These are the first-order model without numerator dynamics and the second-order model with a root at $s = 0$ and no numerator zeros. The analysis of a control law for any other

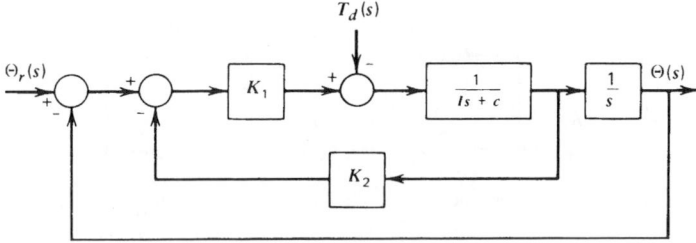

Fig. 24   Tachometer feedback arrangement to replace PD control for position servo.[1]

linear system follows the preceding pattern. The overall system transfer functions are obtained, and all of the linear system analysis techniques can be applied to predict the system's performance. If the performance is unsatisfactory, a new control law is tried and the process repeated. When this process fails to achieve an acceptable design, more systematic methods of altering the system's structure are needed; they are discussed in later sections. We have used step functions as the test signals because they are the most common and perhaps represent the severest test of system performance. Impulse, ramp, and sinusoidal test signals are also employed. The type to use should be made clear in the design specifications.

6 CONTROLLER HARDWARE

The control law must be implemented by a physical device before the control engineer's task is complete. The earliest devices were purely kinematic and were mechanical elements such as gears, levers, and diaphragms that usually obtained their power from the controlled variable. Most controllers now are analog electronic, hydraulic, pneumatic, or digital electronic devices. We now consider the analog type. Digital controllers are covered starting in Section 10.

6.1 Feedback Compensation and Controller Design

Most controllers that implement versions of the PID algorithm are based on the following feedback principle. Consider the single-loop system shown in Fig. 1. If the open-loop transfer function is large enough that $|G(s)H(s)| \gg 1$, the closed-loop transfer function is approximately given by

$$T(s) = \frac{G(s)}{1 + G(s)H(s)} \approx \frac{G(s)}{G(s)H(s)} = \frac{1}{H(s)} \quad (24)$$

The principle states that a power unit $G(s)$ can be used with a feedback element $H(s)$ to create a desired transfer function $T(s)$. The power unit must have a gain high enough that $|G(s)H(s)| \gg 1$, and the feedback elements must be selected so that $H(s) = 1/T(s)$. This principle was used in Section 1 to explain the design of a feedback amplifier.

6.2 Electronic Controllers

The *operational amplifier* (*op amp*) is a high-gain amplifier with a high input impedance. A diagram of an op amp with feedback and input elements with impedances $T_f(s)$ and $T_i(s)$ is shown in Fig. 25. An approximate relation is

$$\frac{E_o(s)}{E_i(s)} = -\frac{T_f(s)}{T_i(s)}$$

The various control modes can be obtained by proper selection of the impedances. A proportional controller can be constructed with a *multiplier*, which uses two resistors, as shown in Fig. 26. An *inverter* is a multiplier circuit with $R_f = R_i$. It is sometimes needed because of the sign reversal property of the op amp. The multiplier circuit can be modified to act as an adder (Fig. 27).

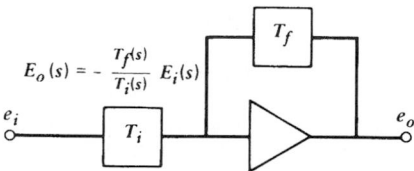

$$E_o(s) = -\frac{T_f(s)}{T_i(s)} E_i(s)$$

Fig. 25 Operational amplifier (op amp).[1]

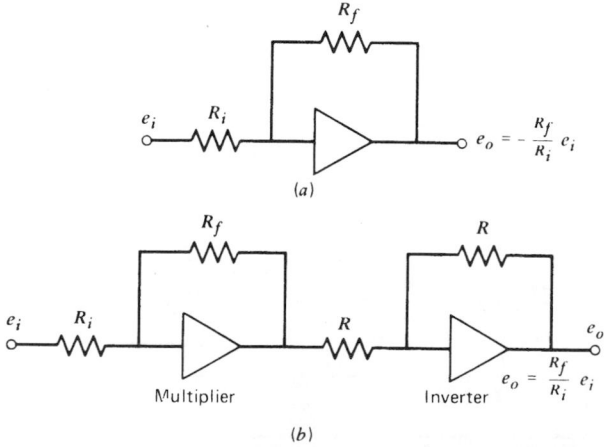

$$e_o = -\frac{R_f}{R_i} e_i$$

(a)

Multiplier

$$e_o = \frac{R_f}{R_i} e_i$$

Inverter

(b)

Fig. 26 Op-amp implementation of proportional control.[1]

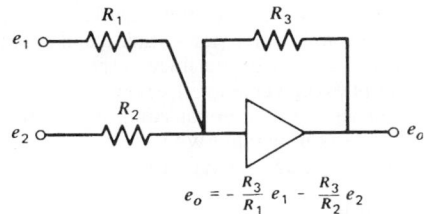

$$e_o = -\frac{R_3}{R_1}e_1 - \frac{R_3}{R_2}e_2$$

Fig. 27 Op-amp adder circuit.[1]

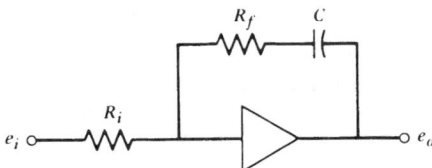

Fig. 28 Op-amp implementation of PI control.[1]

PI control can be implemented with the circuit of Fig. 28. Figure 29 shows a complete system using op amps for PI control. The inverter is needed to

create an error detector. Many industrial controllers provide the operator with a choice of control modes, and the operator can switch from one mode to another when the process characteristics or control objectives change. When a switch occurs, it is necessary to provide any integrators with the proper initial voltages or else undesirable transients will occur when the integrator is switched into the system. Commercially available controllers usually have built-in circuits for this purpose.

In theory, a differentiator can be created by interchanging the resistance and capacitance in the integrating op amp. The difficulty with this design is that no electrical signal is "pure." Contamination always exists as a result of voltage spikes, ripple, and other transients generally categorized as "noise." These high-frequency signals have large slopes compared with the more slowly varying primary signal, and thus they will dominate the output of the differentiator. In practice, this problem is solved by filtering out high-frequency signals, either with a low-pass filter inserted in cascade with the differentiator or by using a redesigned differentiator such as the one shown in Fig. 30. For the ideal PD controller, $R_1 = 0$. The attenuation curve for the ideal controller breaks upward at $\omega = 1/R_2C$ with a slope of 20 dB/decade. The curve for the practical controller does the same but then becomes flat

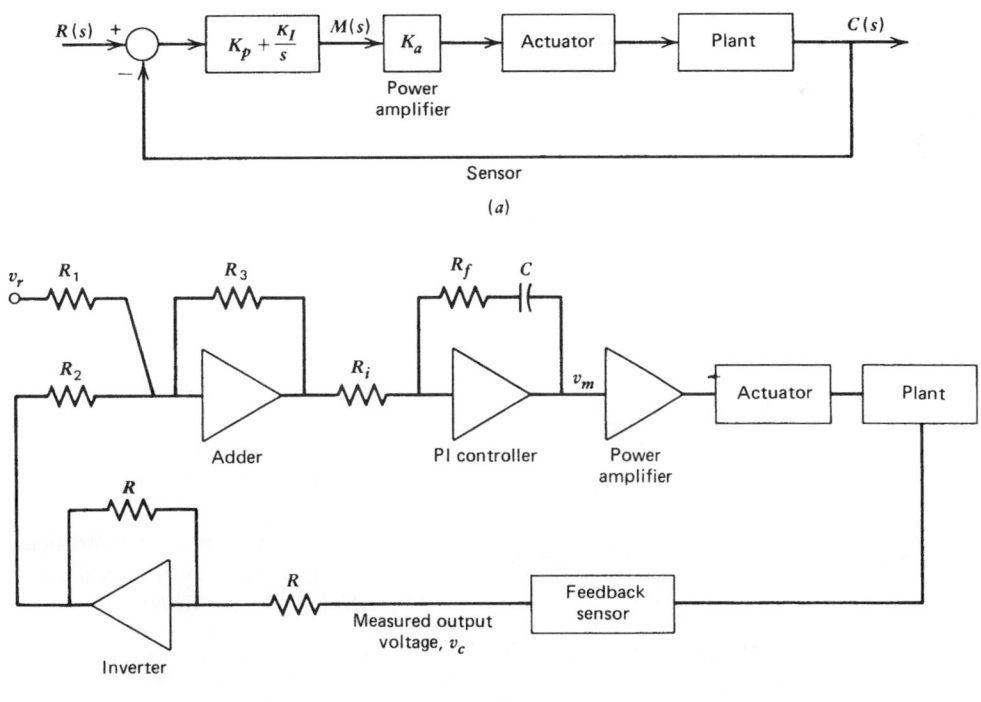

Fig. 29 Implementation of PI controller using op amps. (a) Diagram of system. (b) Diagram showing how op amps are connected.[2]

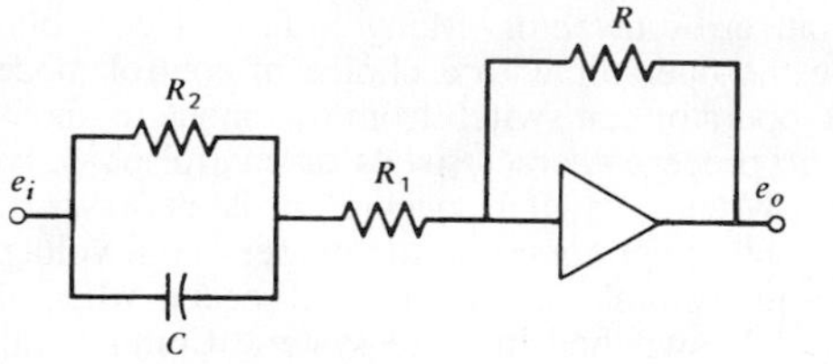

$$e_o = -K_p \left(e_i + T_D \frac{de_i}{dt} \right) - \alpha T_D \frac{de_o}{dt}$$

$$K_p = \frac{R}{R_1 + R_2} \qquad T_D = R_2 C \qquad \alpha = \frac{R_1}{R_1 + R_2}$$

Fig. 30 Practical op-amp implementation of PD control.[1]

for $\omega > (R_1 + R_2)/R_1 R_2 C$. This provides the required limiting effect at high frequencies.

PID control can be implemented by joining the PI and PD controllers in parallel, but this is expensive because of the number of op amps and power supplies required. Instead, the usual implementation is that shown in Fig. 31. The circuit limits the effect of frequencies above $\omega = 1/\beta R_1 C_1$. When $R_1 = 0$, ideal PID control results. This is sometimes called the *noninteractive* algorithm because the effect of each of the three modes is additive, and they do not interfere with one another. The form given for $R_1 \neq 0$ is the *real* or *interactive* algorithm. This name results from the fact that historically it was difficult to implement noninteractive PID control with mechanical or pneumatic devices.

6.3 Pneumatic Controllers

The nozzle–flapper introduced in Section 4 is a high-gain device that is difficult to use without modification.

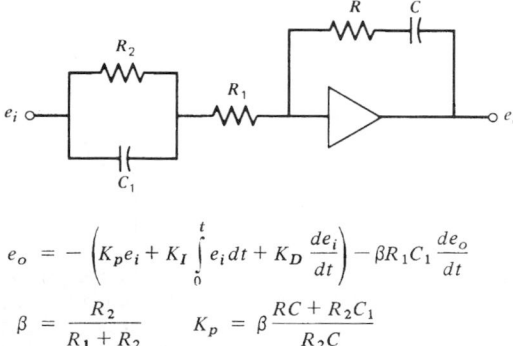

$$e_o = -\left(K_p e_i + K_I \int_0^t e_i \, dt + K_D \frac{de_i}{dt} \right) - \beta R_1 C_1 \frac{de_o}{dt}$$

$$\beta = \frac{R_2}{R_1 + R_2} \qquad K_p = \beta \frac{RC + R_2 C_1}{R_2 C}$$

$$K_I = \frac{\beta}{R_2 C} \qquad K_D = \beta R C_1$$

Fig. 31 Practical op-amp implementation of PID control.[1]

The gain K_f is known only imprecisely and is sensitive to changes induced by temperature and other environmental factors. Also, the linear region over which Eq. (14) applies is very small. However, the device can be made useful by compensating it with feedback elements, as was illustrated with the electropneumatic valve positioner shown in Fig. 19.

6.4 Hydraulic Controllers

The basic unit for synthesis of hydraulic controllers is the hydraulic servomotor. The nozzle–flapper concept is also used in hydraulic controllers.[5] A PI controller is shown in Fig. 32. It can be modified for proportional action. Derivative action has not seen much use in hydraulic controllers. This action supplies damping to the system, but hydraulic systems are usually highly damped intrinsically because of the viscous working fluid. PI control is the algorithm most commonly implemented with hydraulics.

7 FURTHER CRITERIA FOR GAIN SELECTION

Once the form of the control law has been selected, the gains must be computed in light of the performance specifications. In the examples of the PID family of control laws in Section 5, the damping ratio, dominant time constant, and steady-state error were taken to be the primary indicators of system performance in the interest of simplicity. In practice, the criteria are usually more detailed. For example, the rise time and maximum overshoot, as well as the other transient response specifications of the previous chapter, may be encountered. Requirements can also be stated in terms of frequency response characteristics, such as bandwidth, resonant frequency, and peak amplitude. Whatever specific from they take, a complete set of specifications for control system performance generally should include the following considerations for given forms of the command and disturbance inputs:

1. Equilibrium specifications
   (a) Stability
   (b) Steady-state error
2. Transient specifications
   (a) Speed of response
   (b) Form of response
3. Sensitivity specifications
   (a) Sensitivity to parameter variations
   (b) Sensitivity to model inaccuracies
   (c) Noise rejection (bandwidth, etc.)

In addition to these performance stipulations, the usual engineering considerations of initial cost, weight, maintainability, and so on must be taken into account. The considerations are highly specific to the chosen hardware, and it is difficult to deal with such issues in a general way.

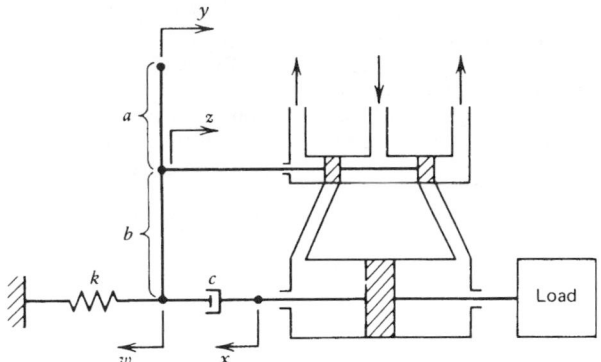

Fig. 32 Hydraulic implementation of PI control.[1]

Two approaches exist for designing the controller. The proper one depends on the quality of the analytical description of the plant to be controlled. If an accurate model of the plant is easily developed, we can design a specialized controller for the particular application. The range of adjustment of controller gains in this case can usually be made small because the accurate plant model allows the gains to be precomputed with confidence. This technique reduces the cost of the controller and can often be applied to electromechanical systems.

The second approach is used when the plant is relatively difficult to model, which is often the case in process control. A standard controller with several control modes and wide ranges of gains is used, and the proper mode and gain settings are obtained by testing the controller on the process in the field. This approach should be considered when the cost of developing an accurate plant model might exceed the cost of controller tuning in the field. Of course, the plant must be available for testing for this approach to be feasible.

7.1 Performance Indices

The performance criteria encountered thus far require a set of conditions to be specified—for example, one for steady-state error, one for damping ratio, and one for the dominant time constant. If there are many such conditions, and if the system is of high order with several gains to be selected, the design process can get quite complicated because transient and steady-state criteria tend to drive the design in different directions. An alternative approach is to specify the system's desired performance by means of one analytical expression called a *performance index*. Powerful analytical and numerical methods are available that allow the gains to be systematically computed by minimizing (or maximizing) this index.

To be useful, a performance index must be selective. The index must have a sharply defined extremum in the vicinity of the gain values that give the desired performance. If the numerical value of the index does not change very much for large changes in the

gains from their optimal values, the index will not be selective.

Any practical choice of a performance index must be easily computed, either analytically, numerically, or experimentally. Four common choices for an index are the following:

$$J = \begin{cases} \displaystyle\int_0^\infty |e(t)|\, dt & \text{(IAE index)} & (25) \\[2ex] \displaystyle\int_0^\infty t|e(t)|\, dt & \text{(ITAE index)} & (26) \\[2ex] \displaystyle\int_0^\infty [e(t)]^2\, dt & \text{(ISE index)} & (27) \\[2ex] \displaystyle\int_0^\infty t[e(t)]^2\, dt & \text{(ITSE index)} & (28) \end{cases}$$

where $e(t)$ is the system error. This error usually is the difference between the desired and the actual values of the output. However, if $e(t)$ does not approach zero as $t \to \infty$, the preceding indices will not have finite values. In this case, $e(t)$ can be defined as $e(t) = c(\infty) - c(t)$, where $c(t)$ is the output variable. If the index is to be computed numerically or experimentally, the infinite upper limit can be replaced by a time t_f large enough that $e(t)$ is negligible for $t > t_f$.

The *integral absolute-error* (IAE) criterion (25) expresses mathematically that the designer is not concerned with the sign of the error, only its magnitude. In some applications, the IAE criterion describes the fuel consumption of the system. The index says nothing about the relative importance of an error occurring late in the response versus an error occurring early. Because of this, the index is not as selective as the *integral-of-time-multiplied absolute-error* (ITAE) criterion (26). Since the multiplier t is small in the early stages of the response, this index weights early errors less heavily than later errors. This makes sense physically. No system can respond instantaneously, and the index is lenient accordingly, while penalizing any design that allows a nonzero error to remain for a long time. Neither criterion allows highly underdamped or highly overdamped systems to be optimum. The ITAE criterion usually results in a system whose step response has a slight overshoot and well-damped oscillations.

The *integral squared-error* (ISE) and *integral-of-time-multiplied squared-error* (ITSE) criteria are analogous to the IAE and ITAE criteria, except that the square of the error is employed for three reasons: (a) in some applications, the squared error represents the system's power consumption; (b) squaring the error weights large errors much more heavily than small errors; (c) the squared error is much easier to handle analytically. The derivative of a squared term is easier to compute than that of an absolute value and does not have a discontinuity at $e = 0$. These differences are important when the system is of high order with multiple error terms.

The closed-form solution for the response is not required to evaluate a performance index. For a given set of parameter values, the response and the resulting index value can be computed numerically. The optimum solution can be obtained using systematic computer search procedures; this makes this approach suitable for use with nonlinear systems.

7.2 Optimal-Control Methods

Optimal-control theory includes a number of algorithms for systematic design of a control law to minimize a performance index, such as the following generalization of the ISE index, called the *quadratic* index:

$$J = \int_0^\infty (\mathbf{x}^T \mathbf{Q} \mathbf{x} + \mathbf{u}^T \mathbf{R} \mathbf{u}) \, dt \qquad (29)$$

where $\mathbf{x}$ and $\mathbf{u}$ are the deviations of the state and control vectors from the desired reference values. For example, in a servomechanism, the state vector might consist of the position and velocity, and the control vector might be a scalar—the force or torque produced by the actuator. The matrices $\mathbf{Q}$ and $\mathbf{R}$ are chosen by the designer to provide relative weighting for the elements of $\mathbf{x}$ and $\mathbf{u}$. If the plant can be described by the linear state-variable model

$$\dot{\mathbf{x}} = \mathbf{A}\mathbf{x} + \mathbf{B}\mathbf{u} \qquad (30)$$

$$\mathbf{y} = \mathbf{C}\mathbf{x} + \mathbf{D}\mathbf{u} \qquad (31)$$

where $\mathbf{y}$ is the vector of outputs—for example, position and velocity—then the solution of this *linear-quadratic* control problem is the linear control law:

$$\mathbf{u} = \mathbf{K}\mathbf{y} \qquad (32)$$

where $\mathbf{K}$ is a matrix of gains that can be found by several algorithms.[1,6,7] A valid solution is guaranteed to yield a stable closed-loop system, a major benefit of this method.

Even if it is possible to formulate the control problem in this way, several practical difficulties arise. Some of the terms in (29) might be beyond the influence of the control vector $\mathbf{u}$; the system is then *uncontrollable*. Also, there might not be enough information in the output equation (31) to achieve control, and the system is then *unobservable*. Several tests are available to check controllability and observability. Not all of the necessary state variables might be available for feedback or the feedback measurements might be noisy or biased. Algorithms known as *observers, state reconstructors, estimators*, and *digital filters* are available to compensate for the missing information. Another source of error is the uncertainty in the values of the coefficient matrices $\mathbf{A}$, $\mathbf{B}$, $\mathbf{C}$, and $\mathbf{D}$. Identification schemes can be used to compare the predicted and the actual system performance and to adjust the coefficient values "online."

7.3 Ziegler–Nichols Rules

The difficulty of obtaining accurate transfer function models for some processes has led to the development of empirically based rules of thumb for computing the optimum gain values for a controller. Commonly used guidelines are the *Ziegler–Nichols rules*, which have proved so helpful that they are still in use 50 years after their development. The rules actually consist of two separate methods. The first method requires the open-loop step response of the plant, while the second uses the results of experiments performed with the controller already installed. While primarily intended for use with systems for which no analytical model is available, the rules are also helpful even when a model can be developed.

Ziegler and Nichols developed their rules from experiments and analysis of various industrial processes. Using the IAE criterion with a unit step response, they found that controllers adjusted according to the following rules usually had a step response that was oscillatory but with enough damping so that the second overshoot was less than 25% of the first (peak) overshoot. This is the *quarter-decay* criterion and is sometimes used as a specification.

The first method is the *process reaction* method and relies on the fact that many processes have an open-loop step response like that shown in Fig. 33. This is the *process signature* and is characterized by two parameters, R and L, where R is the slope of a line tangent to the steepest part of the response curve and L is the time at which this line intersects the time axis. First- and second-order linear systems do not yield positive values for L, and so the method cannot be applied to such systems. However, third- and higher order linear systems with sufficient damping do yield such a response. If so, the Ziegler–Nichols rules recommend the controller settings given in Table 2.

The *ultimate-cycle* method uses experiments with the controller in place. All control modes except

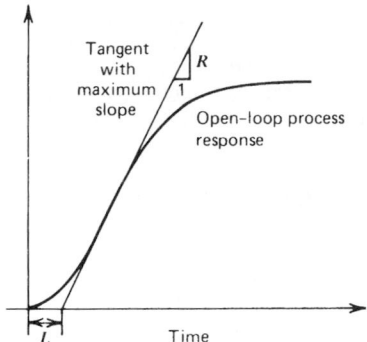

Fig. 33 Process signature for unit step input.[1]

proportional are turned off, and the process is started with the proportional gain K_P set at a low value. The gain is slowly increased until the process begins to exhibit sustained oscillations. Denote the period of this oscillation by P_u and the corresponding *ultimate gain* by K_{Pu}. The Ziegler–Nichols recommendations are given in Table 2 in terms of these parameters. The proportional gain is lower for PI control than for proportional control and is higher for PID control because integral action increases the order of the system and thus tends to destabilize it; thus, a lower gain is needed. On the other hand, derivative action tends to stabilize the system; hence, the proportional gain can be increased without degrading the stability characteristics. Because the rules were developed for a typical case out of many types of processes, final tuning of the gains in the field is usually necessary.

7.4 Nonlinearities and Controller Performance

All physical systems have nonlinear characteristics of some sort, although they can often be modeled as linear systems provided the deviations from the linearization reference condition are not too great. Under certain conditions, however, the nonlinearities have significant effects on the system's performance. One such situation can occur during the startup of a controller if the initial conditions are much different from the reference condition for linearization. The linearized model is then not accurate, and nonlinearities govern the behavior. If the nonlinearities are mild, there might not be much of a problem. Where the nonlinearities are severe, such as in process control, special consideration must be given to startup. Usually, in such cases, the control signal sent to the final control elements is manually adjusted until the system variables are within the linear range of the controller. Then the system is switched into automatic mode. Digital computers are often used to replace the manual adjustment process because they can be readily coded to produce complicated functions for the startup signals. Care must also be taken when switching from manual to automatic. For example, the integrators in electronic controllers must be provided with the proper initial conditions.

7.5 Reset Windup

In practice, all actuators and final control elements have a limited operating range. For example, a motor–amplifier combination can produce a torque proportional to the input voltage over only a limited range. No amplifier can supply an infinite current; there is a maximum current and thus a maximum torque that the system can produce. The final control elements are said to be *overdriven* when they are commanded by the controller to do something they cannot do. Since the limitations of the final control elements are ultimately due to the limited rate at which they can supply energy, it is important that all system performance specifications and controller designs be consistent with the energy delivery capabilities of the elements to be used.

Controllers using integral action can exhibit the phenomenon called *reset windup* or *integrator buildup* when overdriven, if they are not properly designed. For a step change in set point, the proportional term

Table 2 Ziegler–Nichols Rules

Controller transfer function $G(s) = K_p \left(1 + \dfrac{1}{T_I s} + T_D s \right)$		
Control Mode	Process Reaction Method	Ultimate-Cycle Method
P control	$K_p = \dfrac{1}{RL}$	$K_p = 0.5 K_{pu}$
PI control	$K_p = \dfrac{0.9}{RL}$	$K_p = 0.45 K_{pu}$
	$T_I = 3.3L$	$T_I = 0.83 P_u$
PID control	$K_p = \dfrac{1.2}{RL}$	$K_p = 0.6 K_{pu}$
	$T_I = 2L$	$T_I = 0.5 P_u$
	$T_D = 0.5L$	$T_D = 0.125 P_u$

responds instantly and saturates immediately if the set-point change is large enough. On the other hand, the integral term does not respond as fast. It integrates the error signal and saturates some time later if the error remains large for a long enough time. As the error decreases, the proportional term no longer causes saturation. However, the integral term continues to increase as long as the error has not changed sign, and thus the manipulated variable remains saturated. Even though the output is very near its desired value, the manipulated variable remains saturated until after the error has reversed sign. The result can be an undesirable overshoot in the response of the controlled variable.

Limits on the controller prevent the voltages from exceeding the value required to saturate the actuator and thus protect the actuator, but they do not prevent the integral buildup that causes the overshoot. One way to prevent integrator buildup is to select the gains so that saturation will never occur. This requires knowledge of the maximum input magnitude that the system will encounter. General algorithms for doing this are not available; some methods for low-order systems are presented in Ref. 1, Chapter 7; Ref. 2, Chapter 7, and Ref. 4, Chapter 11. Integrator buildup is easier to prevent when using digital control; this is discussed in Section 10.

8 COMPENSATION AND ALTERNATIVE CONTROL STRUCTURES

A common design technique is to insert a *compensator* into the system when the PID control algorithm can be made to satisfy most but not all of the design specifications. A compensator is a device that alters the response of the controller so that the overall system will have satisfactory performance. The three categories of compensation techniques generally recognized are *series compensation, parallel (or feedback) compensation*, and *feedforward compensation*. The three structures are loosely illustrated in Fig. 34,

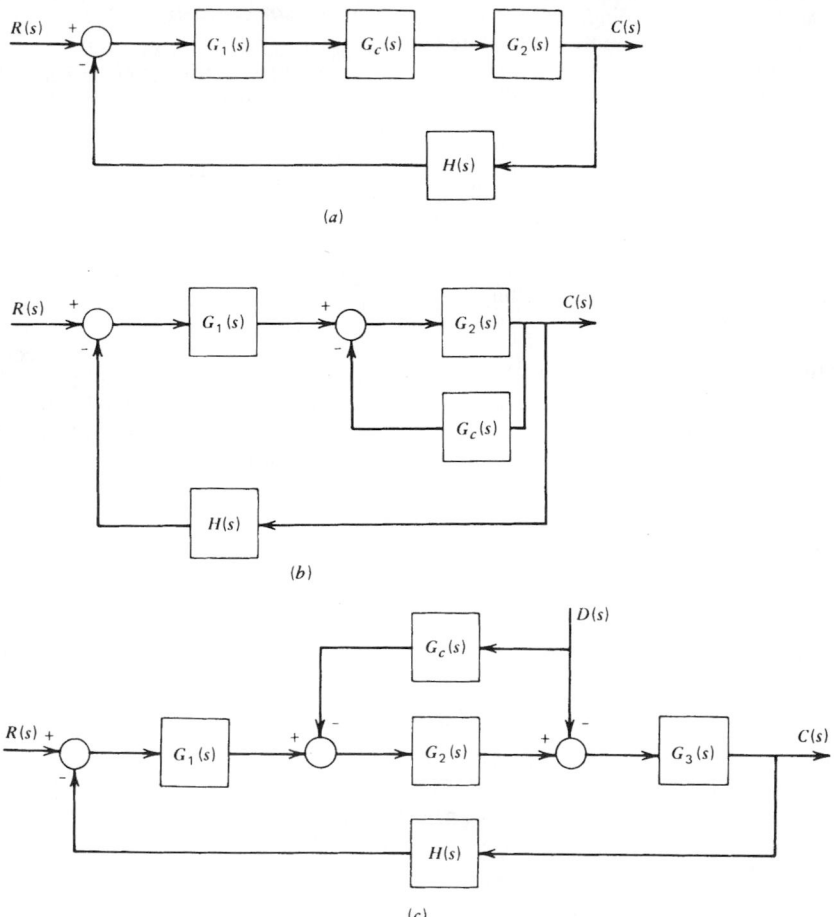

Fig. 34 General structures of three compensation types: (*a*) series; (*b*) parallel (or feedback); (*c*) feedforward. Compensator transfer function is $G_c(s)$.[1]

where we assume the final control elements have a unity transfer function. The transfer function of the controller is $G_1(s)$. The feedback elements are represented by $H(s)$, and the compensator by $G_c(s)$. We assume that the plant is unalterable, as is usually the case in control system design. The choice of compensation structure depends on what type of specifications must be satisfied. The physical devices used as compensators are similar to the pneumatic, hydraulic, and electrical devices treated previously. Compensators can be implemented in software for digital control applications.

8.1 Series Compensation

The most commonly used series compensators are the *lead*, the *lag*, and the *lead–lag* compensators. Electrical implementations of these are shown in Fig. 35. Other physical implementations are available. Generally, the lead compensator improves the speed of response; the lag compensator decreases the steady-state error; and the lead–lag affects both. Graphical aids, such as the root-locus and frequency response plots, are usually needed to design these compensators

(Ref. 1, Chapter 8; Ref. 2, Chapter 9; and Ref. 4, Chapter 11).

8.2 Feedback Compensation and Cascade Control

The use of a tachometer to obtain velocity feedback, as in Fig. 24, is a case of feedback compensation. The feedback compensation principle of Fig. 3 is another. Another form is *cascade control*, in which another controller is inserted within the loop of the original control system (Fig. 36). The new controller can be used to achieve better control of variables within the forward path of the system. Its set point is manipulated by the first controller.

Cascade control is frequently used when the plant cannot be satisfactorily approximated with a model of second order or lower. This is because the difficulty of analysis and control increases rapidly with system order. The characteristic roots of a second-order system can easily be expressed in analytical form. This is not so for third order or higher, and few general design rules are available. When faced with the problem of controlling a high-order system, the designer should first see if the performance requirements can be

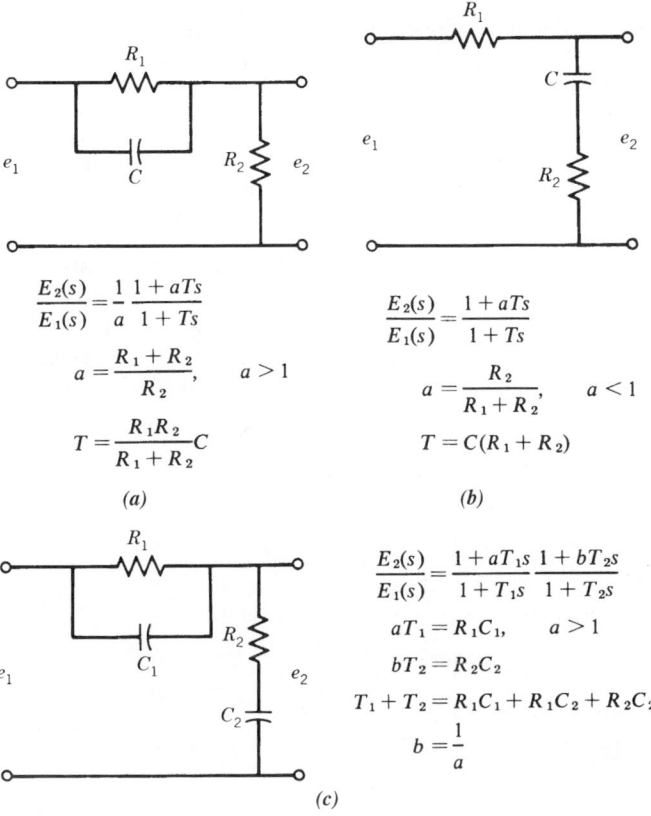

$$\frac{E_2(s)}{E_1(s)} = \frac{1}{a}\frac{1 + aTs}{1 + Ts}$$

$$a = \frac{R_1 + R_2}{R_2}, \qquad a > 1$$

$$T = \frac{R_1 R_2}{R_1 + R_2}C$$

(a)

$$\frac{E_2(s)}{E_1(s)} = \frac{1 + aTs}{1 + Ts}$$

$$a = \frac{R_2}{R_1 + R_2}, \qquad a < 1$$

$$T = C(R_1 + R_2)$$

(b)

$$\frac{E_2(s)}{E_1(s)} = \frac{1 + aT_1s}{1 + T_1s}\frac{1 + bT_2s}{1 + T_2s}$$

$$aT_1 = R_1C_1, \qquad a > 1$$

$$bT_2 = R_2C_2$$

$$T_1 + T_2 = R_1C_1 + R_1C_2 + R_2C_2$$

$$b = \frac{1}{a}$$

(c)

Fig. 35 Passive electrical compensators: (*a*) lead; (*b*) lag; (*c*) lead–lag.

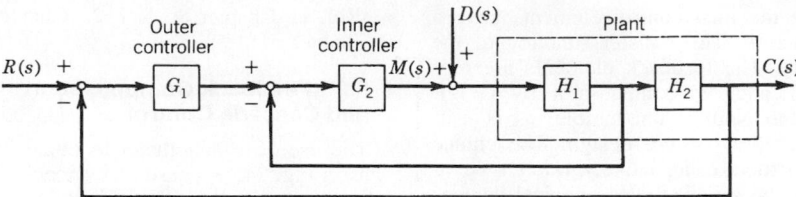

Fig. 36 Cascade control structure.

relaxed so that the system can be approximated with a low-order model. If this is not possible, the designer should attempt to divide the plant into subsystems, each of which is second order or lower. A controller is then designed for each subsystem. An application using cascade control is given in Section 11.

8.3 Feedforward Compensation

The control algorithms considered thus far have counteracted disturbances by using measurements of the output. One difficulty with this approach is that the effects of the disturbance must show up in the output of the plant before the controller can begin to take action. On the other hand, if we can measure the disturbance, the response of the controller can be improved by using the measurement to augment the control signal sent from the controller to the final control elements. This is the essence of feedforward compensation of the disturbance, as shown in Fig. 34c.

Feedforward compensation modified the output of the main controller. Instead of doing this by measuring the disturbance, another form of feedforward compensation utilizes the command input. Figure 37 is an example of this approach. The closed-loop transfer function is

$$\frac{\Omega(s)}{\Omega_r(s)} = \frac{K_f + K}{Is + c + K} \qquad (33)$$

For a unit step input, the steady-state output is $\omega_{ss} = (K_f + K)/(c + K)$. Thus, if we choose the feedforward gain K_f to be $K_f = c$, then $\omega_{ss} = 1$ as desired,

and the error is zero. Note that this form of feedforward compensation does not affect the disturbance response. Its effectiveness depends on how accurately we know the value of c. A digital application of feedforward compensation is presented in Section 11.

8.4 State-Variable Feedback

There are techniques for improving system performance that do not fall entirely into one of the three compensation categories considered previously. In some forms these techniques can be viewed as a type of feedback compensation, while in other forms they constitute a modification of the control law. *State-variable feedback* (SVFB) is a technique that uses information about all the system's state variables to modify either the control signal or the actuating signal. These two forms are illustrated in Fig. 38. Both forms require that the state vector x be measurable or at least derivable from other information. Devices or algorithms used to obtain state-variable information other than directly from measurements are variously termed *state reconstructors, estimators, observers*, or *filters* in the literature.

8.5 Pseudoderivative Feedback

Pseudoderivative feedback (PDF) is an extension of the velocity feedback compensation concept of Fig. 24.[1,2] It uses integral action in the forward path plus an internal feedback loop whose operator $H(s)$ depends on the plant (Fig. 39). For $G(s) = 1/(Is + c)$, $H(s) = K_1$. For $G(s) = 1/Is^2$, $H(s) = K_1 + K_2s$.

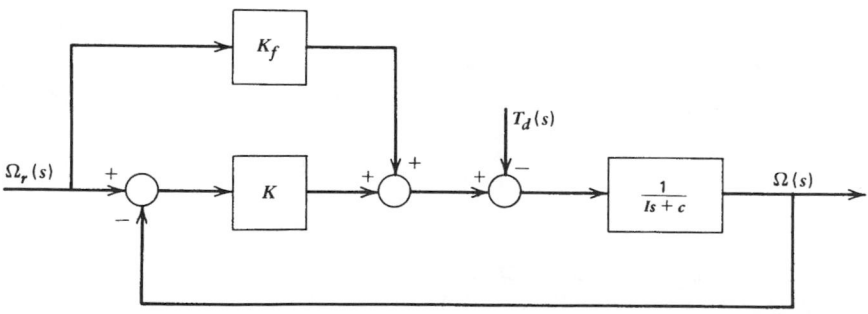

Fig. 37 Feedforward compensation of command input to augment proportional control.[2]

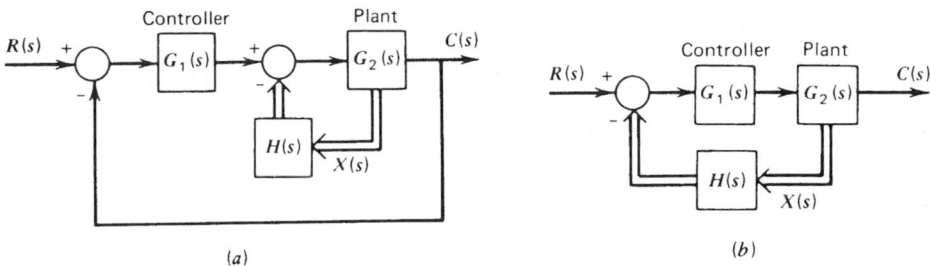

Fig. 38 Two forms of state-variable feedback: (*a*) internal compensation of the control signal; (*b*) modification of the actuating signal.[1]

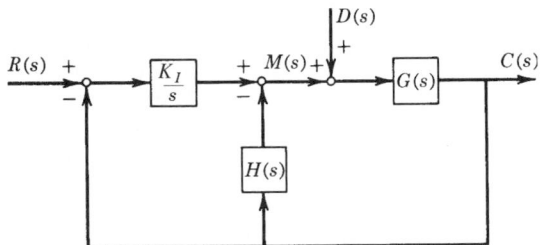

Fig. 39 Structure of pseudoderivative feedback.

The primary advantage of PDF is that it does not need derivative action in the forward path to achieve the desired stability and damping characteristics.

9 GRAPHICAL DESIGN METHODS

Higher order models commonly arise in control systems design. For example, integral action is often used with a second-order plant, and this produces a third-order system to be designed. Although algebraic solutions are available for third- and fourth-order polynomials, these solutions are cumbersome for design purposes. Fortunately, there exist graphical techniques to aid the designer. Frequency response plots of both the open- and closed-loop transfer functions are useful. The *Bode plot* and the *Nyquist plot* present the frequency response information in different forms. Each form has its own advantages. The root-locus plot shows the location of the characteristic roots for a range of values of some parameters, such as a controller gain. The design of two-position and other nonlinear control systems is facilitated by the *describing function*, which is a linearized approximation based on the frequency response of the controller. Graphical design methods are discussed in more detail in Refs. 1–4.

9.1 Nyquist Stability Theorem

The Nyquist stability theorem is a powerful tool for linear system analysis. If the open-loop system has no poles with positive real parts, we can concentrate our attention on the region around the point $-1 + i0$ on the polar plot of the open-loop transfer function. Figure 40 shows the polar plot of the open-loop transfer function of an arbitrary system that is assumed to be open-loop stable. The Nyquist stability theorem is stated as follows: A system is closed-loop stable if and only if the point $-1 + i0$ lies to the left of the open-loop Nyquist plot relative to an observer traveling along the plot in the direction of increasing frequency ω. Therefore, the system described by Fig. 39 is closed-loop stable.

The Nyquist theorem provides a convenient measure of the relative stability of a system. A measure of the proximity of the plot to the $-1 + i0$ point is given by the angle between the negative real axis and a line from the origin to the point where the plot crosses the unit circle (see Fig. 39). The frequency corresponding to this intersection is denoted ω_g. This angle is the *phase margin* (PM) and is positive when measured down from the negative real axis. The phase margin is the phase at the frequency ω_g where the magnitude ratio or "gain" of $G(i\omega)H(i\omega)$ is unity, or 0 decibels (dB). The frequency ω_p, the *phase crossover frequency*, is the frequency at which the phase angle is $-180°$. The *gain margin* (GM) is

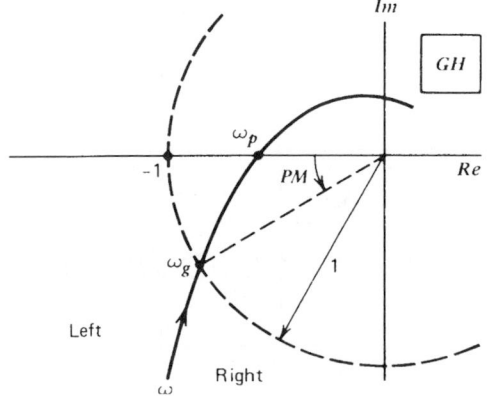

Fig. 40 Nyquist plot for a stable system.[1]

the difference in decibels between the unity gain condition (0 dB) and the value of $|G(\omega_p)H(\omega_p)|$ decibels at the phase crossover frequency ω_p. Thus,

$$\text{Gain margin} = -|G(\omega_p)H(\omega_p)| \quad (\text{dB}) \quad (34)$$

A system is stable only if the phase and gain margins are both positive.

The phase and gain margins can be illustrated on the Bode plots shown in Fig. 41. The phase and gain margins can be stated as safety margins in the design specifications. A typical set of such specifications is as follows:

$$\text{Gain margin} \geq 8 \text{ dB} \quad \text{Phase margin} \geq 30° \quad (35)$$

In common design situations, only one of these equalities can be met, and the other margin is allowed to be greater than its minimum value. It is not desirable to make the margins too large because this results in a low gain, which might produce sluggish response and a large steady-state error. Another commonly used set of specifications is

$$\text{Gain margin} \geq 6 \text{ dB} \quad \text{Phase margin} \geq 40° \quad (36)$$

The 6-dB limit corresponds to the quarter amplitude decay response obtained with the gain settings given by the Ziegler–Nichols ultimate-cycle method (Table 2).

9.2 Systems with Dead-Time Elements

The Nyquist theorem is particularly useful for systems with dead-time elements, especially when the plant is of an order high enough to make the root-locus method cumbersome. A delay D in either the manipulated variable or the measurement will result in an open-loop transfer function of the form

$$G(s)H(s) = e^{-Ds} P(s) \quad (37)$$

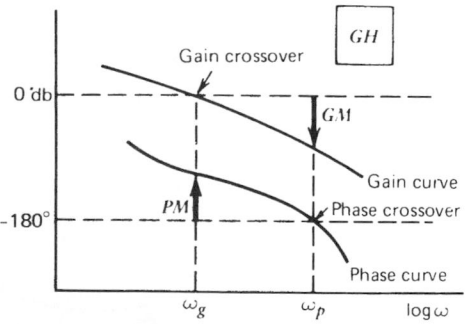

Fig. 41 Bode plot showing definitions of phase and gain margin.[1]

Its magnitude and phase angle are

$$|G(i\omega)H(i\omega)| = |P(i\omega)||e^{-i\omega D}| = |P(i\omega)| \quad (38)$$

$$\angle G(i\omega)H(i\omega) = \angle P(i\omega) + \angle e^{-i\omega D}$$

$$= \angle P(i\omega) - \omega D \quad (39)$$

Thus, the dead time decreases the phase angle proportionally to the frequency ω, but it does not change the gain curve. This makes the analysis of its effects easier to accomplish with the open-loop frequency response plot.

9.3 Open-Loop Design for PID Control

Some general comments can be made about the effects of proportional, integral, and derivative control actions on the phase and gain margins. Proportional action does not affect the phase curve at all and thus can be used to raise or lower the open-loop gain curve until the specifications for the gain and phase margins are satisfied. If integral action or derivative action is included, the proportional gain is selected last. Therefore, when using this approach to the design, it is best to write the PID algorithm with the proportional gain factored out, as

$$F(s) = K_P \left(1 + \frac{1}{T_I s} + T_D s\right) E(s) \quad (40)$$

Derivative action affects both the phase and gain curves. Therefore, the selection of the derivative gain is more difficult than the proportional gain. The increase in phase margin due to the positive phase angle introduced by Derivative action is partly negated by the derivative gain, which reduces the gain margin. Increasing the derivative gain increases the speed of response, makes the system more stable, and allows a larger proportional gain to be used to improve the system's accuracy. However, if the phase curve is too steep near $-180°$, it is difficult to use Derivative action to improve the performance. Integral action also affects both the gain and phase curves. It can be used to increase the open-loop gain at low frequencies. However, it lowers the phase crossover frequency ω_p and thus reduces some of the benefits provided by derivative action. If required, the derivative action term is usually designed first, followed by integral action and proportional action, respectively.

The classical design methods based on the Bode plots obviously have a large component of trial and error because usually both the phase and gain curves must be manipulated to achieve an acceptable design. Given the same set of specifications, two designers can use these methods and arrive at substantially different designs. Many rules of thumb and ad hoc procedures have been developed, but a general foolproof procedure does not exist. However, an experienced designer can often obtain a good design quickly with these techniques. The use of a computer plotting routine greatly speeds up the design process.

9.4 Design with Root Locus

The effect of derivative action as a series compensator can be seen with the root locus. The term $1 + T_D s$ in Fig. 32 can be considered as a series compensator to the proportional controller. The derivative action adds an open-loop zero at $s = -1/T_D$. For example, a plant with the transfer function $1/s(s+1)(s+2)$, when subjected to proportional control, has the root locus shown in Fig. 42a. If the proportional gain is too high, the system will be unstable. The smallest achievable time constant corresponds to the root $s = -0.42$ and is $\tau = 1/0.42 = 2.4$. If derivative action is used to put an open-loop zero at $s = -1.5$, the resulting root locus is given by Fig. 42b. The derivative action prevents the system from becoming unstable and allows a smaller time constant to be achieved (τ can be made close to $1/0.75 = 1.3$ by using a high proportional gain).

The integral action in PI control can be considered to add an open-loop pole at $s = 0$ and a zero at $s = -1/T_I$. Proportional control of the plant $1/(s+1)(s+2)$ gives a root locus like that shown in Fig. 43, with $a = 1$ and $b = 2$. A steady-state error will exist for a step input. With the PI compensator applied to this plant, the root locus is given by Fig. 42b, with $T_I = \frac{2}{3}$. The steady-state error is eliminated, but the response of the system has been slowed because the dominant paths of the root locus of the compensated system lie closer to the imaginary axis than those of the uncompensated system.

As another example, let the plant transfer function be

$$G_P(s) = \frac{1}{s^2 + a_2 s + a_1} \tag{41}$$

where $a_1 > 0$ and $a_2 > 0$. PI control applied to this plant gives the closed-loop command transfer function

$$T_1(s) = \frac{K_P s + K_I}{s^3 + a_2 s^2 + (a_1 + K_P)s + K_I} \tag{42}$$

Note that the Ziegler–Nichols rules cannot be used to set the gains K_P and K_I. The second-order plant, Eq. (41), does not have the S-shaped signature of Fig. 33, so the process reaction method does not apply. The ultimate-cycle method requires K_I to be set to zero and the ultimate gain K_{Pu} determined. With $K_I = 0$ in Eq. (42) the resulting system is stable for all $K_P > 0$, and thus a positive ultimate gain does not exist.

Take the form of the PI control law given by Eq. (42) with $T_D = 0$, and assume that the characteristic roots of the plant (Fig. 44) are real values $-r_1$ and $-r_2$ such that $-r_2 < -r_1$. In this case the open-loop transfer function of the control system is

$$G(s)H(s) = \frac{K_P(s + 1/T_I)}{s(s + r_1)(s + r_2)} \tag{43}$$

One design approach is to select T_I and plot the locus with K_P as the parameter. If the zero at $s = -1/T_I$

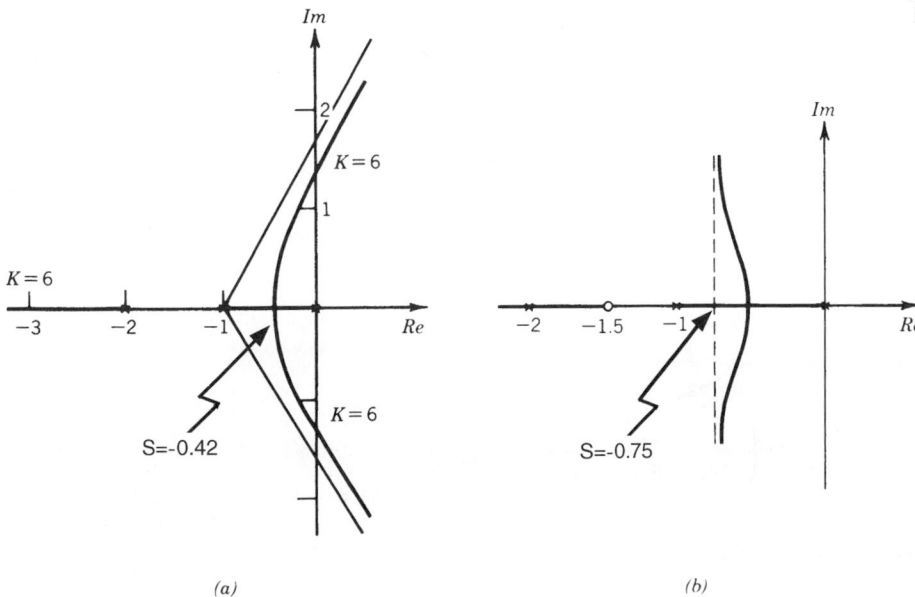

(a)   *(b)*

Fig. 42 (a) Root-locus plot for $s(s+1)(s+2) + K = 0$, for $K \geq 0$. (b) The effect of PD control with $T_D = \frac{2}{3}$.

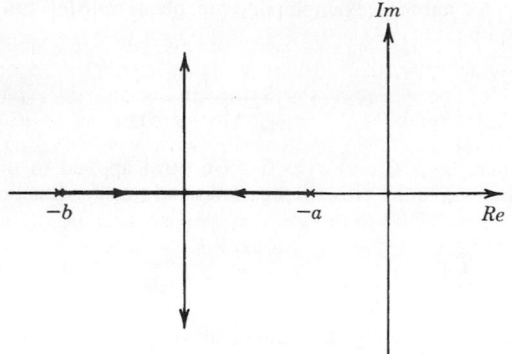

Fig. 43 Root-locus plot for $(s + a)(s + b) + K = 0$.

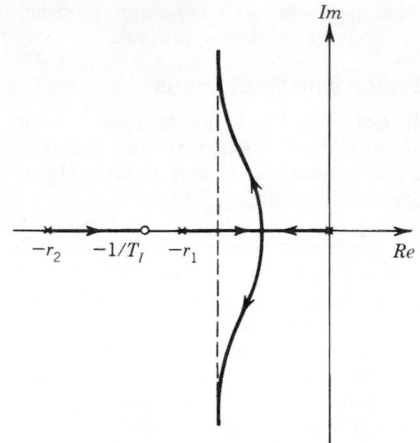

Fig. 44 Root-locus plot for PI control of a second-order plant.

is located to the right of $s = -r_1$, the dominant time constant cannot be made as small as is possible with the zero located between the poles at $s = -r_1$ and $s = -r_2$ (Fig. 44). A large integral gain (small T_I and/or large K_P) is desirable for reducing the overshoot due to a disturbance, but the zero should not be placed to the left of $s = -r_2$ because the dominant time constant will be larger than that obtainable with the placement shown in Fig. 44 for large values of K_P. Sketch the root-locus plots to see this. A similar situation exists if the poles of the plant are complex.

The effects of the lead compensator in terms of time domain specifications (characteristic roots) can be shown with the root-locus plot. Consider the second-order plant with the real distinct roots $s = -\alpha$,

$s = -\beta$. The root locus for this system with proportional control is shown in Fig. 45a. The smallest dominant time constant obtainable is τ_1, marked in the figure. A lead compensator introduces a pole at $s = -1/T$ and a zero at $s = -1/aT$, and the root locus becomes that shown in Fig. 45b. The pole and zero introduced by the compensator reshape the locus so that a smaller dominant time constant can be obtained. This is done by choosing the proportional gain high enough to place the roots close to the asymptotes.

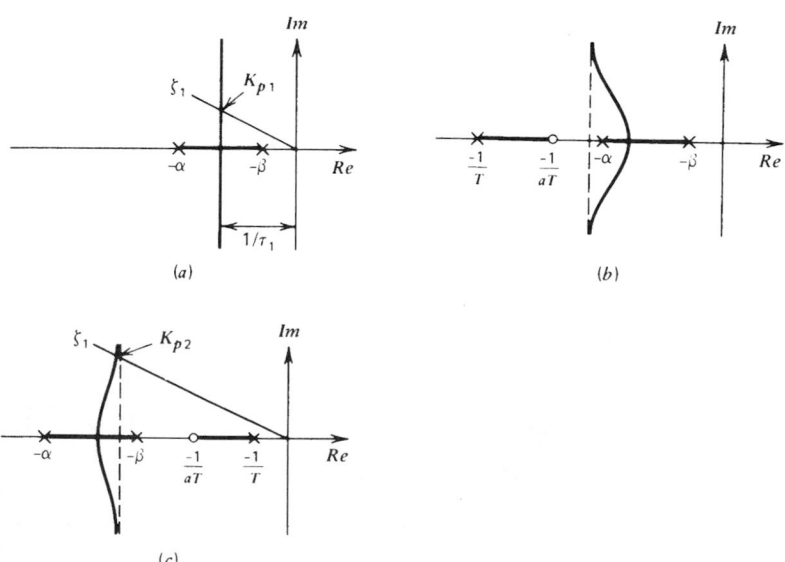

Fig. 45 Effects of series lead and lag compensators: (a) uncompensated system's root locus; (b) root locus with lead compensation; (c) root locus with lag compensation.[1]

With reference to the proportional control system whose root locus is shown in Fig. 45a, suppose that the desired damping ratio ζ_1 and desired time constant τ_1 are obtainable with a proportional gain of K_{P1}, but the resulting steady-state error $\alpha\beta/(\alpha\beta + K_{P1})$ due to a step input is too large. We need to increase the gain while preserving the desired damping ratio and time constant. With the lag compensator, the root locus is as shown in Fig. 45c. By considering specific numerical values, one can show that for the compensated system, roots with a damping ratio ζ_1 correspond to a high value of the proportional gain. Call this value K_{P2}. Thus $K_{P2} > K_{P1}$, and the steady-state error will be reduced. If the value of T is chosen large enough, the pole at $s = -1/T$ is approximately canceled by the zero at $s = -1/aT$, and the open-loop transfer function is given approximately by

$$G(s)H(s) = \frac{aK_P}{(s + \alpha)(s + \beta)} \qquad (44)$$

Thus, the system's response is governed approximately by the complex roots corresponding to the gain value K_{P2}. By comparing Fig. 45a with 45c, we see that the compensation leaves the time constant relatively unchanged. From Eq. (44) it can be seen that since $a < 1$, K_P can be selected as the larger value K_{P2}. The ratio of K_{P1} to K_{P2} is approximately given by the parameter a.

Design by pole–zero cancellation can be difficult to accomplish because a response pattern of the system is essentially ignored. The pattern corresponds to the behavior generated by the canceled pole and zero, and this response can be shown to be beyond the influence of the controller. In this example, the canceled pole gives a stable response because it lies in the left-hand plane. However, another input not modeled here, such as a disturbance, might excite the response and cause unexpected behavior. The designer should therefore proceed with caution. None of the physical parameters of the system are known exactly, so exact pole–zero cancellation is not possible. A root-locus study of the effects of parameter uncertainty and a simulation study of the response are often advised before the design is accepted as final.

10 PRINCIPLES OF DIGITAL CONTROL

Digital control has several advantages over analog devices. A greater variety of control algorithms is possible, including nonlinear algorithms and ones with time-varying coefficients. Also, greater accuracy is possible with digital systems. However, their additional hardware complexity can result in lower reliability, and their application is limited to signals whose time variation is slow enough to be handled by the samplers and the logic circuitry. This is now less of a problem because of the large increase in the speed of digital systems.

10.1 Digital Controller Structure

The basic structure of a single-loop controller is shown in Fig. 46. The computer with its internal clock drives the *digital-to-analog* (D/A) and *analog-to-digital* (A/D) converters. It compares the command signals with the feedback signals and generates the control signals to be sent to the final control elements. These control signals are computed from the control algorithm stored in the memory. Slightly different structures exist, but Fig. 46 shows the important aspects. For example, the comparison between the command and feedback signals can be done with analog elements, and the A/D conversion made on the resulting error signal. The software must also provide for *interrupts*, which are conditions that call for the computer's attention to do something other than computing the control algorithm.

The time required for the control system to complete one loop of the algorithm is the time T, the *sampling time* of the control system. It depends on the time required for the computer to calculate the control algorithm and on the time required for the interfaces to convert data. Modern systems are capable of very high rates, with sample times under $1\,\mu$s.

In most digital control applications, the plant is an analog system, but the controller is a discrete-time system. Thus, to design a digital control system, we must either model the controller as an analog system or model the plant as a discrete-time system. Each approach has its own merits, and we will examine both.

If we model the controller as an analog system, we use methods based on *differential* equations to

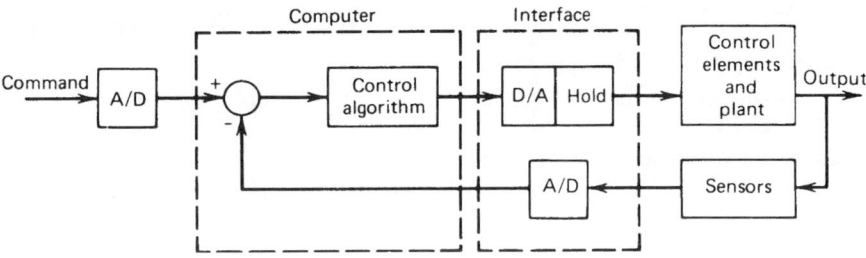

Fig. 46 Structure of digital control system.[1]

compute the gains. However, a digital control system requires *difference* equations to describe its behavior. Thus, from a strictly mathematical point of view, the gain values we will compute will not give the predicted response exactly. However, if the sampling time is small compared to the smallest time constant in the system, then the digital system will act like an analog system, and our designs will work properly. Because most physical systems of interest have time constants greater than 1 ms and controllers can now achieve sampling times less than 1μs, controllers designed with analog methods will often be adequate.

10.2 Digital Forms of PID Control

There are a number of ways that PID control can be implemented in software in a digital control system because the integral and derivative terms must be approximated with formulas chosen from a variety of available algorithms. The simplest integral approximation is to replace the integral with a sum of rectangular areas. With this rectangular approximation, the error integral is calculated as

$$\int_0^{(k+1)T} e(t)\, dt \approx Te(0) + Te(t_1) + Te(t_2)$$

$$+ \cdots + Te(t_k) = T\sum_{i=0}^{k} e(t_i) \qquad (45)$$

where $t_k = kT$ and the width of each rectangle is the sampling time $T = t_{i+1} - t_i$. The times t_i are the times at which the computer updates its calculation of the control algorithm after receiving an updated command signal and an updated measurement from the sensor through the A/D interfaces. If the time T is small, then the value of the sum in (45) is close to the value of the integral. After the control algorithm calculation is made, the calculated value of the control signal $f(t_k)$ is sent to the actuator via the output interface. This interface includes a D/A converter and a *hold* circuit that "holds" or keeps the analog voltage corresponding to the control signal applied to the actuator until the next updated value is passed along from the computer. The simplest digital form of PI control uses (45) for the integral term. It is

$$f(t_k) = K_P e(t_k) + K_I T \sum_{i=0}^{k} e(t_i) \qquad (46)$$

This can be written in a more efficient form by noting that

$$f(t_{k-1}) = K_P e(t_{k-1}) + K_I T \sum_{i=0}^{k-1} e(t_i)$$

and subtracting this from (46) to obtain

$$f(t_k) = f(t_{k-1}) + K_P[e(t_k) - e(t_{k-1})] + K_I T e(t_k)$$
$$(47)$$

This form—called the *incremental* or *velocity* algorithm—is well suited for incremental output devices such as stepper motors. Its use also avoids the problem of integrator buildup, the condition in which the actuator saturates but the control algorithm continues to integrate the error.

The simplest approximation to the derivative is the first-order difference approximation

$$\frac{de}{dt} \approx \frac{e(t_k) - e(t_{k-1})}{T} \qquad (48)$$

The corresponding PID approximation using the rectangular integral approximation is

$$f(t_k) = K_P e(t_k) + K_I T \sum_{i=0}^{k} e(t_i)$$

$$+ \frac{K_D}{T}[e(t_k) - e(t_{k-1})] \qquad (49)$$

The accuracy of the integral approximation can be improved by substituting a more sophisticated algorithm, such as the following trapezoidal rule:

$$\int_0^{(k+1)T} e(t)\, dt \approx T\sum_{i=0}^{k} \frac{1}{2}[e(t_{i+1}) + e(t_i)] \qquad (50)$$

The accuracy of the derivative approximation can be improved by using values of the sampled error signal at more instants. Using the four-point central-difference method (Refs. 1 and 2), the derivative term is approximated by

$$\frac{de}{dt} \approx \frac{1}{6T}[e(t_k) + 3e(t_{k-1}) - 3e(t_{k-2}) - e(t_{k-3})]$$

The derivative action is sensitive to the resulting rapid change in the error samples that follows a step input. This effect can be eliminated by reformulating the control algorithm as follows (Refs. 1 and 2):

$$f(t_k) = f(t_{k-1}) + K_P[c(t_{k-1}) - c(t_k)]$$
$$+ K_I T[r(t_k) - c(t_k)]$$
$$+ \frac{K_D}{T}[-c(t_k) + 2c(t_{k-1}) - c(t_{k-2})] \qquad (51)$$

where $r(t_k)$ is the command input and $c(t_k)$ is the variable being controlled. Because the command input $r(t_k)$ appears in this algorithm only in the integral term, we cannot apply this algorithm to PD control; that is, the integral gain K_I must be nonzero.

11 UNIQUELY DIGITAL ALGORITHMS

Development of analog control algorithms was constrained by the need to design physical devices that could implement the algorithm. However, digital control algorithms simply need to be programmable and are thus less constrained than analog algorithms.

11.1 Digital Feedforward Compensation

Classical control system design methods depend on linear models of the plant. With linearization we can obtain an approximately linear model, which is valid only over a limited operating range. Digital control now allows us to deal with nonlinear models more directly using the concepts of feedforward compensation discussed in Section 8.

Computed Torque Method Figure 47 illustrates a variation of feedforward compensation of the disturbance called the *computed torque method*. It is used to control the motion of robots. A simple model of a robot arm is the following nonlinear equation:

$$I\ddot{\theta} = T - mgL \, \sin\theta \qquad (52)$$

where θ is the arm angle, I is its inertia, mg is its weight, and L is the distance from its mass center to the arm joint where the motor acts. The motor supplies the torque T. To position the arm at some desired angle θ_r, we can use PID control on the angle error $\theta_r - \theta$. This works well if the arm angle θ is never far from the desired angle θ_r so that we can linearize the plant model about θ_r. However, the controller will work for large-angle excursions if we compute the nonlinear gravity torque term $mgL \, \sin\theta$ and add it to the PID output. That is, part of the motor torque will be computed specifically to cancel the gravity torque, in effect producing a linear system for the PID

algorithm to handle. The nonlinear torque calculations required to control multi-degree-of-freedom robots are very complicated and can be done only with a digital controller.

Feedforward Command Compensation Computers can store lookup tables, which can be used to control systems that are difficult to model entirely with differential equations and analytical functions. Figure 48 shows a speed control system for an internal combustion engine. The fuel flow rate required to achieve a desired speed depends in a complicated way on many variables not shown in the figure, such as temperature, humidity, and so on. This dependence can be summarized in tables stored in the control computer and can be used to estimate the required fuel flow rate. A PID algorithm can be used to adjust the estimate based on the speed error. This application is an example of feedforward compensation of the command input, and it requires a digital computer.

11.2 Control Design in *z* Plane

There are two common approaches to designing a digital controller:

1. The performance is specified in terms of the desired continuous-time response, and the controller design is done entirely in the *s* plane, as with an analog controller. The resulting control law is then converted to discrete-time form, using approximations for the integral and derivative terms. This method can be successfully applied if the sampling time is small. The technique is widely used for two reasons. When existing analog controllers are converted to digital control, the form of the control law and the values of its associated gains are known to have been satisfactory. Therefore, the digital version

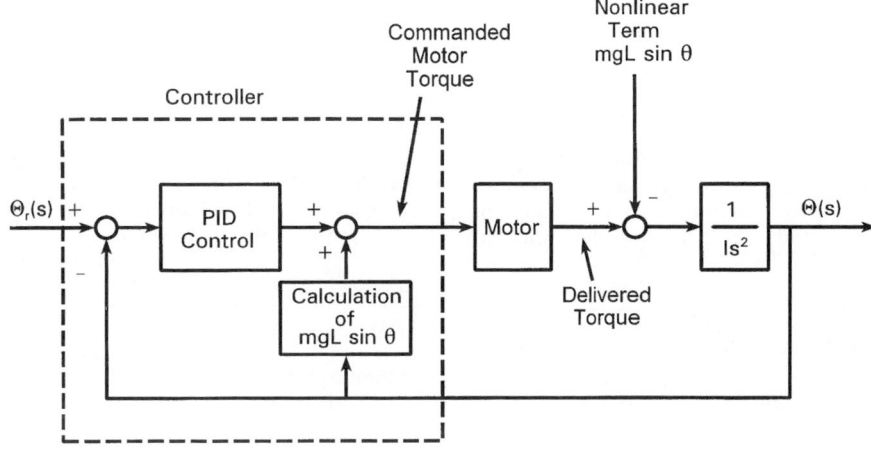

Fig. 47 Computed torque method applied to robot arm control.

Controller

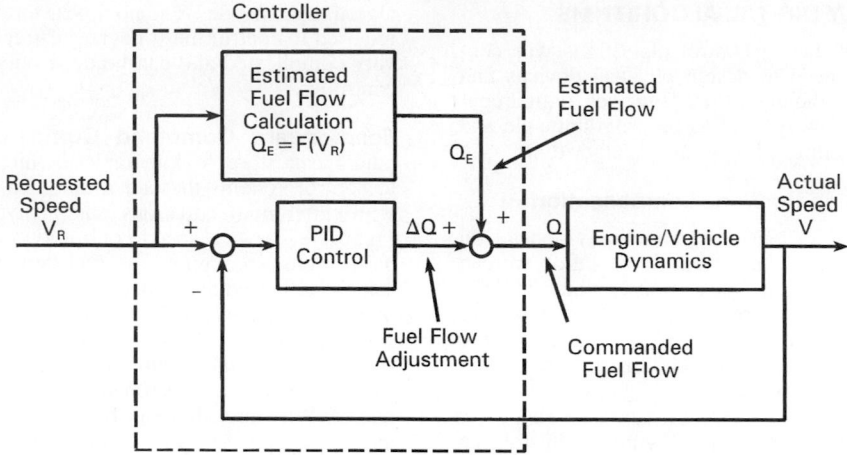

Fig. 48 Feedforward compensation applied to engine control.

can use the same control law and gain values. Second, because analog design methods are well established, many engineers prefer to take this route and then convert the design into a discrete-time equivalent.

2. The performance specifications are given in terms of the desired continuous-time response and/or desired root locations in the s plane. From these the corresponding root locations in the z plane are found and a discrete control law is designed. This method avoids the derivative and integral approximation errors that are inherent in the first method and is the preferred method when the sampling time T is large. However, the algebraic manipulations are more cumbersome.

The second approach uses the z transform and pulse transfer functions. If we have an analog model of the plant, with its transfer function $G(s)$, we can obtain its pulse transfer function $G(z)$ by finding the z transform of the impulse response $g(t) = \mathscr{L}^{-1}[G(s)]$; that is, $G(z) = \mathscr{Z}[g(t)]$. A table of transforms facilitates this process; see Refs. 1 and 2. Figure 49a shows the basic elements of a digital control system. Figure 49b is an equivalent diagram with the analog transfer functions inserted. Figure 49c represents the same system in terms of pulse transfer functions. From the diagram we can find the closed-loop pulse transfer function. It is

$$\frac{C(z)}{R(z)} = \frac{G(z)P(z)}{1 + G(z)P(z)} \tag{53}$$

The variable z is related to the Laplace variable s by

$$z = e^{sT} \tag{54}$$

If we know the desired root locations and the sampling time T, we can compute the z roots from this equation.

Digital PI Control Design   For example, the first-order plant $1/(2s + 1)$ with a zero-order hold has the following pulse transfer function (Refs. 1 and 2):

$$P(z) = \frac{1 - e^{-0.5T}}{z - e^{-0.5T}} \tag{55}$$

Suppose we use a control algorithm described by the following pulse transfer function:

$$G(z) = \frac{F(z)}{E(z)} = \frac{K_1 z + K_2}{z - 1} = \frac{K_1 + K_2 z^{-1}}{1 - z^{-1}} \tag{56}$$

The corresponding difference equation that the control computer must implement is

$$f(t_k) = f(t_{k-1}) + K_1 e(t_k) + K_2 e(t_{k-1}) \tag{57}$$

where $e(t_k) = r(t_k) - c(t_k)$. By comparing (57) with (47), it can be seen that this is the digital equivalent of PI control, where $K_P = -K_2$ and $K_I = (K_1 + K_2)/T$. Using the form of $G(z)$ given by (56), the closed-loop transfer function is

$$\frac{C(z)}{R(z)} = \frac{(1 - b)(K_1 z + K_2)}{z^2 + (K_1 - 1 - b - bK_1)z + b + K_2 - bK_2} \tag{58}$$

where $b = e^{-0.5T}$.

If the design specifications call for $\tau = 1$ and $\zeta = 1$, then the desired s roots are $s = -1, -1$, and the analog PI gains required to achieve these roots are $K_P = 3$ and $K_I = 2$. Using a sampling time of

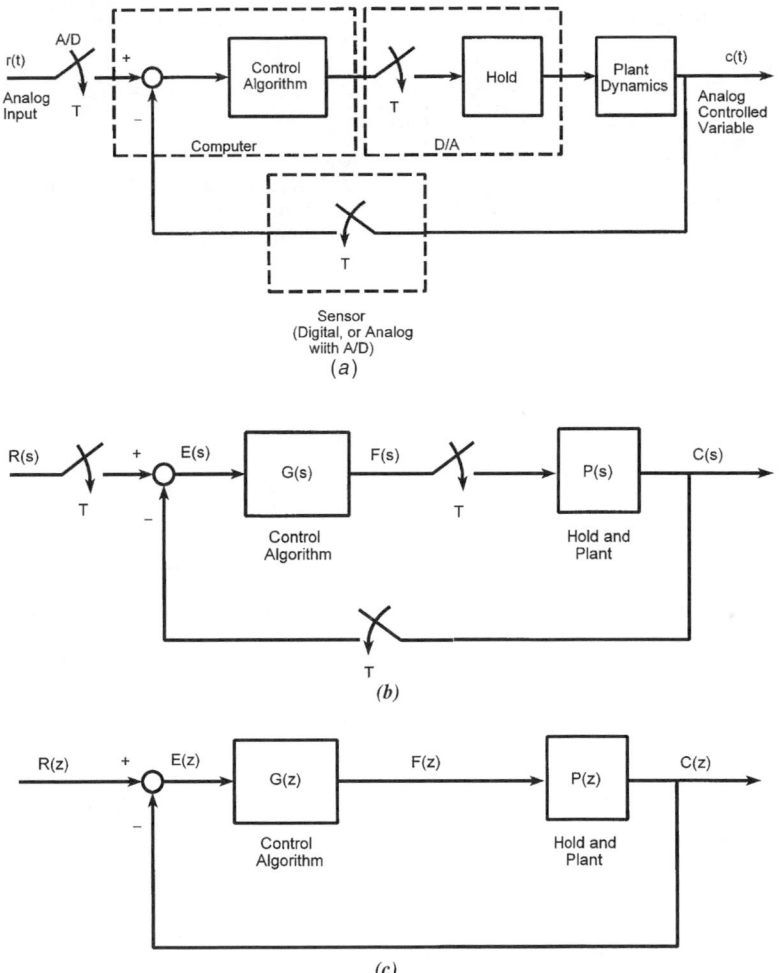

Fig. 49 Block diagrams of typical digital controller. (a) Diagram showing the components. (b) Diagram of the s-plane relations. (c) Diagram of the z-plane relations.

$T = 0.1$, the z roots must be $z = e^{-0.1}, e^{-0.1}$. To achieve these roots, the denominator of the transfer function (58) must be $z^2 - 2e^{-0.1}z + e^{-0.2}$. Thus the control gains must be $K_1 = 2.903$ and $K_2 = -2.717$. These values of K_1 and K_2 correspond to $K_P = 2.72$ and $K_I = 1.86$, which are close to the PI gains computed for an analog controller. If we had used a sampling time smaller than 0.1, say $T = 0.01$, the values of K_P and K_I computed from K_1 and K_2 would be $K_P = 2.97$ and $K_I = 1.98$, which are even closer to the analog gain values. This illustrates the earlier claim that analog design methods can be used when the sampling time is small enough.

Digital Series Compensation Series compensation can be implemented digitally by applying suitable discrete-time approximations for the derivative and integral to the model represented by the compensator's transfer function $G_c(s)$. For example, the form of a lead or a lag compensator's transfer function is

$$G_c(s) = \frac{M(s)}{F(s)} = K\frac{s+c}{s+d} \qquad (59)$$

where $m(t)$ is the actuator command and $f(t)$ is the control signal produced by the main (PID) controller. The differential equation corresponding to (59) is

$$\dot{m} + dm = K(\dot{f} + cf) \qquad (60)$$

Using the simplest approximation for the derivative, Eq. (48), we obtain the following difference equation that the digital compensator must implement:

$$\frac{m(t_k) - m(t_{k-1})}{T} + dm(t_k)$$

$$= K\left[\frac{f(t_k) - f(t_{k-1})}{T} + cf(t_k)\right]$$

In the z plane, the equation becomes

$$\frac{1 - z^{-1}}{T}M(z) + dM(z) = K\left[\frac{1 - z^{-1}}{T}F(z) + cF(z)\right] \tag{61}$$

The compensator's pulse transfer function is thus seen to be

$$G_c(z) = \frac{M(z)}{F(z)} = \frac{K(1 - z^{-1}) + cT}{1 - z^{-1} + dT}$$

which has the form

$$G_c(z) = K_c\frac{z + a}{z + b} \tag{62}$$

where K_c, a, and b can be expressed in terms of K, c, d, and T if we wish to use analog design methods to design the compensator. When using commercial controllers, the user might be required to enter the values of the gain, the pole, and the zero of the compensator. The user must ascertain whether these values should be entered as s-plane values (i.e., K, c, and d) or as z-plane values (K_c, a, and b).

Note that the digital compensator has the same number of poles and zeros as the analog compensator. This is a result of the simple approximation used for the derivative. Note that Eq. (61) shows that when we use this approximation, we can simply replace s in the analog transfer function with $1 - z^{-1}$. Because the integration operation is the inverse of differentiation, we can replace $1/s$ with $1/(1 - z^{-1})$ when integration is used. [This is equivalent to using the rectangular approximation for the integral and can be verified by finding the pulse transfer function of the incremental algorithm (47) with $K_P = 0$.]

Some commercial controllers treat the PID algorithm as a series compensator, and the user is expected to enter the controller's values, not as PID gains, but as pole and zero locations in the z plane. The PID transfer function is

$$\frac{F(s)}{E(s)} = K_P + \frac{K_I}{s} + K_D s \tag{63}$$

Making the indicated replacements for the s terms, we obtain

$$\frac{F(z)}{E(z)} = K_P + \frac{K_I}{1 - z^{-1}} + K_D(1 - z^{-1})$$

which has the form

$$\frac{F(z)}{E(z)} = K_c\frac{z^2 - az + b}{z - 1} \tag{64}$$

where K_c, a, and b can be expressed in terms of K_P, K_I, K_D, and T. Note that the algorithm has two zeros and one pole, which is fixed at $z = 1$. Sometimes the algorithm is expressed in the more general form

$$\frac{F(z)}{F(z)} = K_c\frac{z^2 - az + b}{z - c} \tag{65}$$

to allow the user to select the pole as well.

Digital compensator design can be done with frequency response methods or with the root-locus plot applied to the z plane rather than the s plane. However, when better approximations are used for the derivative and integral, the digital series compensator will have more poles and zeros than its analog counterpart. This means that the root-locus plot will have more root paths, and the analysis will be more difficult. This topic is discussed in more detail in Refs. 1–3 and 8.

11.3 Direct Design of Digital Algorithms

Because almost any algorithm can be implemented digitally, we can specify the desired response and work backward to find the required control algorithm. This is the *direct-design* method. If we let $D(z)$ be the desired form of the closed-loop transfer function $C(z)/R(z)$ and solve for the controller transfer function $G(z)$, we obtain

$$G(z) = \frac{D(z)}{P(z)[1 - D(z)]} \tag{66}$$

We can pick $D(z)$ directly or obtain it from the specified input transform $R(z)$ and the desired output transform $C(z)$, because $D(z) = C(z)/R(z)$.

Finite-Settling-Time Algorithm   This method can be used to design a controller to compensate for the effects of process dead time. A plant having such a response can often be approximately described by a first-order model with a dead-time element; that is,

$$G_P(s) = K\frac{e^{-Ds}}{\tau s + 1} \tag{67}$$

where D is the dead time. This model also approximately describes the S-shaped response curve used with the Ziegler–Nichols method (Fig. 33). When combined with a zero-order hold, this plant has the following pulse transfer function:

$$P(z) = Kz^{-n}\frac{1 - a}{z - a} \tag{68}$$

where $a = \exp(-T/\tau)$ and $n = D/T$. If we choose $D(z) = z^{-(n+1)}$, then with a step command input, the output $c(k)$ will reach its desired value in $n + 1$ sample times, one more than is in the dead time D. This is the fastest response possible. From (66) the required controller transfer function is

$$G(z) = \frac{1}{K(1-a)} \frac{1 - az^{-1}}{1 - z^{-(n+1)}} \tag{69}$$

The corresponding difference equation that the control computer must implement is

$$f(t_k) = f(t_{k-n-1}) + \frac{1}{K(1-a)}[e(t_k) - ae(t_{k-1})] \tag{70}$$

This algorithm is called a *finite-settling-time* algorithm because the response reaches its desired value in a finite, prescribed time. The maximum value of the manipulated variable required by this algorithm occurs at $t = 0$ and is $1/K(1-a)$. If this value saturates the actuator, this method will not work as predicted. Its success depends also on the accuracy of the plant model.

Dahlin's Algorithm   This sensitivity to plant modeling errors can be reduced by relaxing the minimum response time requirement. For example, choosing $D(z)$ to have the same form as $P(z)$, namely,

$$D(z) = K_d z^{-n} \frac{1 - a_d}{z - a_d} \tag{71}$$

we obtain from (66) the following controller transfer function:

$$G(z) = \frac{K_d(1-a_d)}{K(1-a)} \frac{1 - az^{-1}}{1 - a_d z^{-1} - K_d(1-a_d)z^{-(n+1)}} \tag{72}$$

This is *Dahlin's algorithm*.[3] The corresponding difference equation that the control computer must implement is

$$f(t_k) = a_d f(t_{k-1}) + K_d(1-a_d)f(t_{k-n-1})$$
$$+ \frac{K_d(1-a_d)}{K(1-a)}[e(t_k) - ae(t_{k-1})] \tag{73}$$

Normally we would first try setting $K_d = K$ and $a_d = a$, but since we might not have good estimates of K and a, we can use K_d and a_d as tuning parameters to adjust the controller's performance. The constant a_d is related to the time constant τ_d of the desired response: $a_d = \exp(-T/\tau_d)$. Choosing τ_d smaller gives faster response.

Algorithms such as these are often used for system startup, after which the control mode is switched to PID, which is more capable of handling disturbances.

12   HARDWARE AND SOFTWARE FOR DIGITAL CONTROL

This section provides an overview of the general categories of digital controllers that are commercially available. This is followed by a summary of the software currently available for digital control and for control system design.

12.1   Digital Control Hardware

Commercially available controllers have different capabilities, such as different speeds and operator interfaces, depending on their targeted application.

Programmable Logic Controllers (PLCs)   These are controllers that are programmed with relay ladder logic, which is based on Boolean algebra. Now designed around microprocessors, they are the successors to the large relay panels, mechanical counters, and drum programmers used up to the 1960s for sequencing control and control applications requiring only a finite set of output values (e.g., opening and closing of valves). Some models now have the ability to perform advanced mathematical calculations required for PID control, thus allowing them to be used for modulated control as well as finite-state control. There are numerous manufacturers of PLCs.

Digital Signal Processors (DSPs)   A modern development is the *digital signal processor* (DSP), which has proved useful for feedback control as well as signal processing.[9] This special type of processor chip has separate buses for moving data and instructions and is constructed to perform rapidly the kind of mathematical operations required for digital filtering and signal processing. The separate buses allow the data and the instructions to move in parallel rather than sequentially. Because the PID control algorithm can be written in the form of a digital filter, DSPs can also be used as controllers.

The DSP architecture was developed to handle the types of calculations required for digital filters and discrete Fourier transforms, which form the basis of most signal-processing operations. DSPs usually lack the extensive memory management capabilities of general-purpose computers because they need not store large programs or large amounts of data. Some DSPs contain A/D and D/A converters, serial ports, timers, and other features. They are programmed with specialized software that runs on popular personal computers. Low-cost DSPs are now widely used in consumer electronics and automotive applications, with Texas Instruments being a major supplier.

Motion Controllers   *Motion controllers* are specialized control systems that provide feedback control for one or more motors. They also provide a convenient operator interface for generating the commanded trajectories. Motion controllers are particularly well suited for applications requiring coordinated motion of two or more axes and for applications where the commanded trajectory is complicated. A higher level host computer might transmit required distance, speed, and acceleration rates to the motion controller, which then constructs and implements the continuous position profile required for each motor. For example, the host computer would supply the required total displacement, the acceleration and deceleration times, and the desired slew speed (the speed during the zero acceleration phase). The motion controller would generate the commanded position versus time for each motor. The motion controller also has the task of providing feedback control for each motor to ensure that the system follows the required position profile.

Figure 50 shows the functional elements of a typical motion controller, such as those built by Galil Motion Control, Inc. Provision for both analog and digital input signals allows these controllers to perform other control tasks besides motion control. Compared to DSPs, such controllers generally have greater capabilities for motion control and have operator interfaces that are better suited for such applications. Motion controllers are available as plug-in cards for most computer bus types. Some are available as stand-alone units.

Motion controllers use a PID control algorithm to provide feedback control for each motor (some manufacturers call this algorithm a "filter"). The user enters the values of the PID gains (some manufacturers provide preset gain values, which can be changed; others provide tuning software that assists in selecting the proper gain values). Such controllers also have

their own language for programming a variety of motion profiles and other applications. For example, they provide for linear and circular interpolation for two-dimensional coordinated motion, motion smoothing (to eliminate jerk), contouring, helical motion, and electronic gearing. The latter is a control mode that emulates mechanical gearing in software, in which one motor (the slave) is driven in proportion to the position of another motor (the master) or an encoder.

Process Controllers *Process controllers* are designed to handle inputs from sensors, such as thermocouples, and outputs to actuators, such as valve positioners, that are commonly found in process control applications. Figure 51 illustrates the input–output capabilities of a typical process controller such as those manufactured by Honeywell, which is a major supplier of such devices. This device is a stand-alone unit designed to be mounted in an instrumentation panel. The voltage and current ranges of the analog inputs are those normally found with thermocouple-based temperature sensors. The current outputs are designed for devices like valve positioners, which usually require 4–20-mA signals.

The controller contains a microcomputer with built-in math functions normally required for process control, such as thermocouple linearization, weighted averaging, square roots, ratio/bias calculations, and the PID control algorithm. These controllers do not have the same software and memory capabilities as desktop computers, but they are less expensive. Their operator interface consists of a small keypad with typically

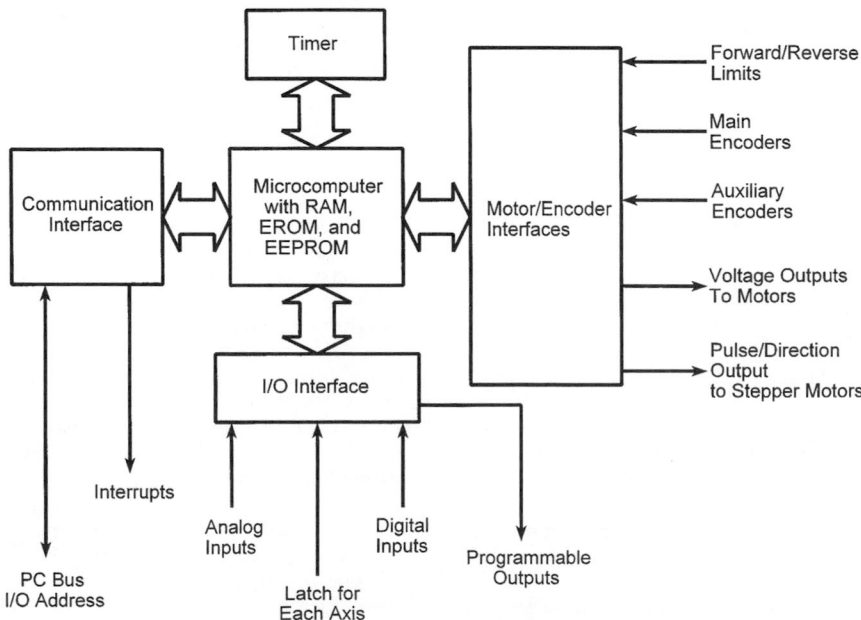

Fig. 50 Functional diagram of motion controller.

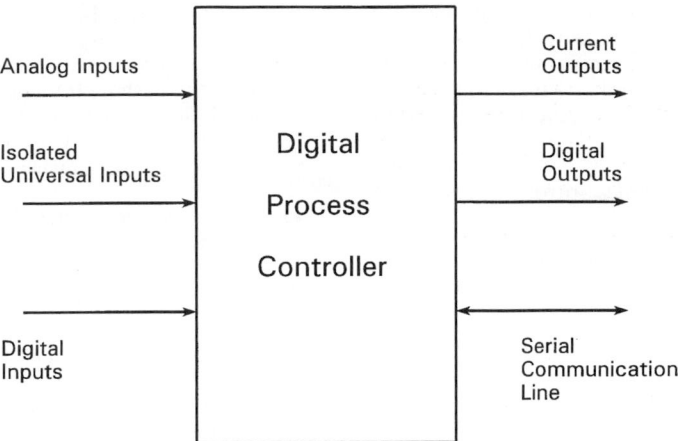

Fig. 51   Functional diagram of digital process controller.

fewer than 10 keys, a small graphical display for displaying bar graphs of the set points and the process variables, indicator lights, and an alphanumeric display for programming the controller.

The PID gains are entered by the user. Some units allow multiple sets of gains to be stored; the unit can be programmed to switch between gain settings when certain conditions occur. Some controllers have an adaptive tuning feature that is supposed to adjust the gains to prevent overshoot in startup mode, to adapt to changing process dynamics, and to adapt to disturbances. However, at this time, adaptive tuning cannot claim a 100% success rate, and further research and development in adaptive control is needed.

Some process controllers have more than one PID control loop for controlling several variables. Figure 52 illustrates a boiler feedwater control application for a controller with two PID loops arranged

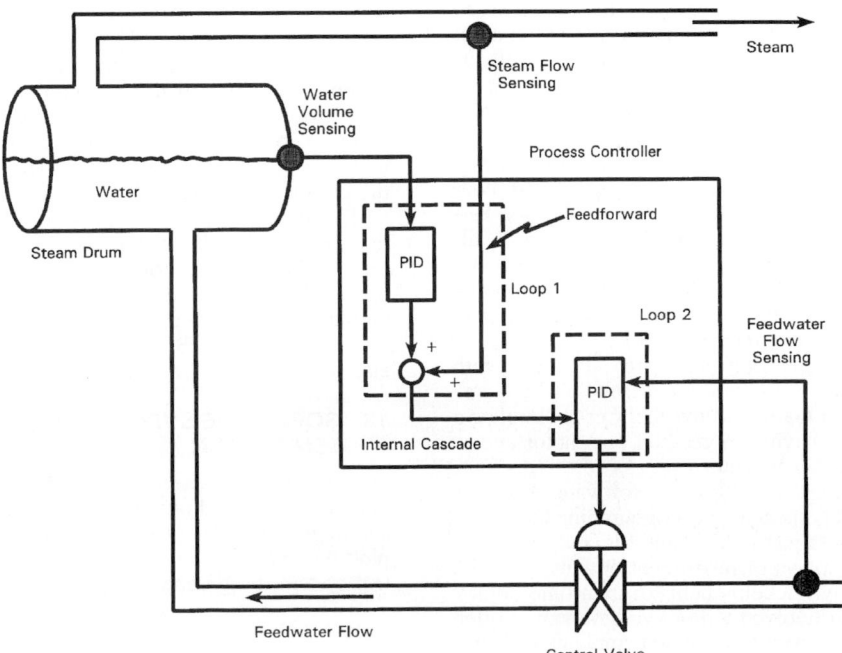

Fig. 52   Application of two-loop process controller for feedwater control.

in a cascade control structure. Loop 1 is the main or outer loop controller for maintaining the desired water volume in the boiler. It uses sensing of the steam flow rate to implement feedforward compensation. Loop 2 is the inner loop controller that directly controls the feedwater control valve.

12.2 Software for Digital Control

The software available to the modern control engineer is quite varied and powerful and can be categorized according to the following tasks:

1. Control algorithm design, gain selection, and simulation
2. Tuning
3. Motion programming
4. Instrumentation configuration
5. Read-time control functions

Many analysis and simulation packages now contain algorithms of specific interest to control system designers. *MATLAB* is one such package that is widely used. It contains built-in functions for generating root-locus and frequency response plots, system simulation, digital filtering, calculation of control gains, and data analysis. It can accept model descriptions in the form of transfer functions or as state-variable equations.[1,4,10]

Some manufacturers provide software to assist the engineer in sizing and selecting components. An example is the *Motion Component Selector* (MCS) sold by Galil Motion Control, Inc. It assists the engineer in computing the load inertia, including the effects of the mechanical drive, and then selects the proper motor and amplifier based on the user's description of the desired motion profile.

Some hardware manufacturers supply software to assist the engineer in selecting control gains and modifying (*tuning*) them to achieve good response. This might require that the system to be controlled be available for experiments prior to installation. Some controllers, such as some Honeywell process controllers, have an autotuning feature that adjusts the gains in real time to improve performance.

Motion programming software supplied with motion controllers was mentioned previously. Some packages, such as Galil's, allow the user to simulate a multiaxis system having more than one motor and to display the resulting trajectory.

Instrumentation configuration software, such as *LabView*, provides specialized programming languages for interacting with instruments and for creating graphical real-time displays of instrument outputs.

Until recently, development of real-time digital control software involved tedious programming, often in assembly language. Even when implemented in a higher level language, such as Fortran or C, programming real-time control algorithms can be very challenging, partly because of the need to provide adequately for interrupts. Software packages are now available that provide real-time control capability, usually a form of the PID algorithm, that can be programmed through user-friendly graphical interfaces. Examples include the Galil motion controllers and the add-on modules for Labview and MATLAB.

12.3 Embedded Control Systems and Hardware-in-the Loop Testing

An *embedded control system* is a microprocessor and sensor suite designed to be an integral part of a product. The aerospace and automotive industries have used embedded controllers for some time, but the decreased cost of components now makes embedded controllers feasible for more consumer and biomedical applications.

For example, embedded controllers can greatly increase the performance of orthopedic devices. One model of an artificial leg now uses sensors to measure in real time the walking speed, the knee joint angle, and the loading due to the foot and ankle. These measurements are used by the controller to adjust the hydraulic resistance of a piston to produce a stable, natural, and efficient gait. The controller algorithms are adaptive in that they can be tuned to an individual's characteristics and their settings changed to accommodate different physical activities.

Engines incorporate embedded controllers to improve efficiency. Embedded controllers in new active suspensions use actuators to improve on the performance of traditional passive systems consisting only of springs and dampers. One design phase of such systems is *hardware-in-the-loop testing*, in which the controlled object (the engine or vehicle suspension) is replaced with a real-time simulation of its behavior. This enables the embedded system hardware and software to be tested faster and less expensively than with the physical prototype and perhaps even before the prototype is available.

Simulink, which is built on top of MATLAB and requires MATLAB to run, is often used to create the simulation model for hardware-in-the-loop testing. Some of the *toolboxes* available for MATLAB, such as the control systems toolbox, the signal-processing toolbox, and the DSP and fixed-point blocksets, are also useful for such applications.

13 SOFTWARE SUPPORT FOR CONTROL SYSTEM DESIGN

Software packages are available for graphical control system design methods and control system simulation. These greatly reduce the tedious manual computation, plotting, and programming formerly required for control system design and simulation.

13.1 Software for Graphical Design Methods

Several software packages are available to support graphical control system design methods. The most popular of these is MATLAB, which has extensive capabilities for generation and interactive analysis of root-locus plots and frequency response plots. Some of these capabilities are discussed in Refs. 1 and 4.

13.2 Software for Control Systems Simulation

It is difficult to obtain closed-form expressions for system response when the model contains *dead time* or nonlinear elements that represent realistic control system behavior. Dead time (also called *transport delay*), rate limiters, and actuator saturation are effects that often occur in real control systems, and simulation is often the only way to analyze their response. Several software packages are available to support system simulation. One of the most popular is Simulink.

Systems having dead-time elements are easily simulated in Simulink. Figure 53 shows a Simulink model for PID control of the plant $53/(3.44s^2 + 2.61s + 1)$, with a dead time between the output of the controller and the plant. The block implementing the dead-time transfer function e^{-Ds} is called the *transport delay* block. When you run this model, you will see the response in the scope block.

In addition to being limited by saturation, some actuators have limits on how fast they can react. This limitation is independent of the time constant of the actuator and might be due to deliberate restrictions placed on the unit by its manufacturer. An example is a flow control valve whose rate of opening and closing is controlled by a *rate limiter*. Simulink has such a block, and it can be used in series with the saturation block to model the valve behavior. Consider the model of the height h of liquid in a tank whose input is a flow rate q_i. For specific parameter values, such a model has the form $H(s)/Q_i(s) = 2/(5s + 1)$. A Simulink model is shown in Figure 54 for a specific PI controller whose gains are $K_P = 4$ and $K_I = \frac{5}{4}$. The saturation block models the fact that the valve opening must be between 0 and 100%. The model enables us to experiment with the lower and upper limits of the rate limiter block to see its effect on the system performance.

An introduction to Simulink is given in Refs. 4 and 10. Applications of Simulink to control system simulation are given in Ref. 4.

14 FUTURE TRENDS IN CONTROL SYSTEMS

Microprocessors have rejuvenated the development of controllers for mechanical systems. Currently, there are several applications areas in which new control systems are indispensable to the product's success:

1. Active vibration control
2. Noise cancellation
3. Adaptive optics
4. Robotics
5. Micromachines
6. Precision engineering

Most of the design techniques presented here comprise "classical" control methods. These methods are

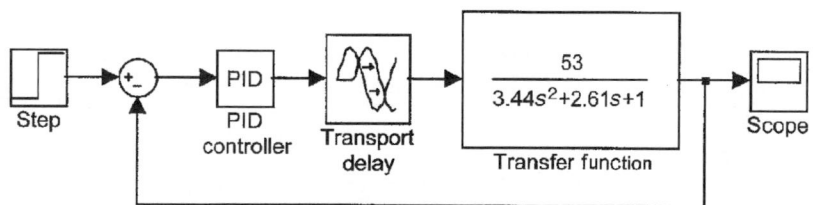

Fig. 53 Simulink model of system with transport delay.

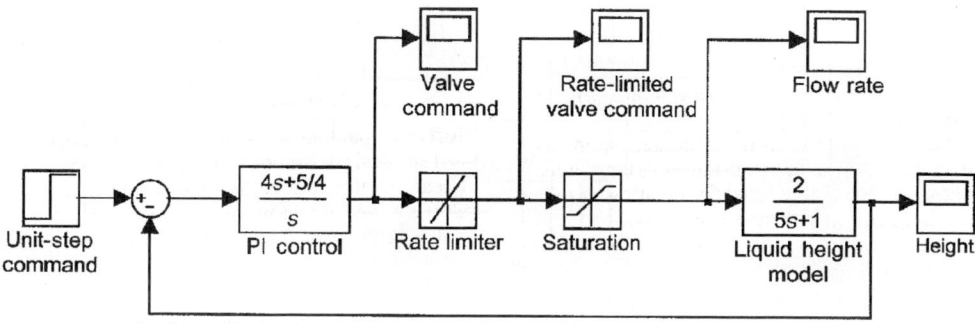

Fig. 54 Simulink model of system with actuator saturation and rate limiter.

widely used because when they are combined with some testing and computer simulation, an experienced engineer can rapidly achieve an acceptable design. Modern control algorithms, such as state-variable feedback and the linear–quadratic optimal controller, have had some significant mechanical engineering applications—for example, in the control of aerospace vehicles. The current approach to multivariable systems like the one shown in Fig. 55 is to use classical methods to design a controller for each subsystem because they can often be modeled with low-order linearized models. The coordination of the various low-level controllers is a nonlinear problem. High-order, nonlinear, multivariable systems that cannot be controlled with classical methods cannot yet be handled by modern control theory in a general way, and further research is needed.

In addition to the improvements, such as lower cost, brought on by digital hardware, microprocessors have allowed designers to incorporate algorithms of much greater complexity into control systems. The following is a summary of the areas currently receiving much attention in the control systems community.

14.1 Fuzzy Logic Control

In classical set theory, an object's membership in a set is clearly defined and unambiguous. *Fuzzy logic control* is based on a generalization of classical set theory to allow objects to belong to several sets with various degrees of membership. Fuzzy logic can be used to describe processes that defy precise definition or precise measurement, and thus it can be used to model the inexact and subjective aspects of human reasoning. For example, room temperature can be described as cold, cool, just right, warm, or hot. Development of a fuzzy logic temperature controller would require the designer to specify the membership functions that describe "warm" as a function of temperature, and so on. The control logic would then be developed as a linguistic algorithm that models a human operator's decision process (e.g., if the room temperature is "cold," then "greatly" increase the heater output; if the temperature is "cool," then increase the heater output "slightly").

Fuzzy logic controllers have been implemented in a number of applications. Proponents of fuzzy logic control point to its ability to convert a human operator's reasoning process into computer code. Its critics argue that because all the controller's fuzzy calculations must eventually reduce to a specific output that must be given to the actuator (e.g., a specific voltage value or a specific valve position), why not be unambiguous from the start, and define a "cool" temperature to be the range between 65° and 68°, for example? Perhaps the proper role of fuzzy logic is at the human operator interface. Research is active in this area, and the issue is not yet settled.[11,12]

14.2 Nonlinear Control

Most real systems are nonlinear, which means that they must be described by nonlinear differential equations. Control systems designed with the linear control theory described in this chapter depend on a linearized approximation to the original nonlinear model. This

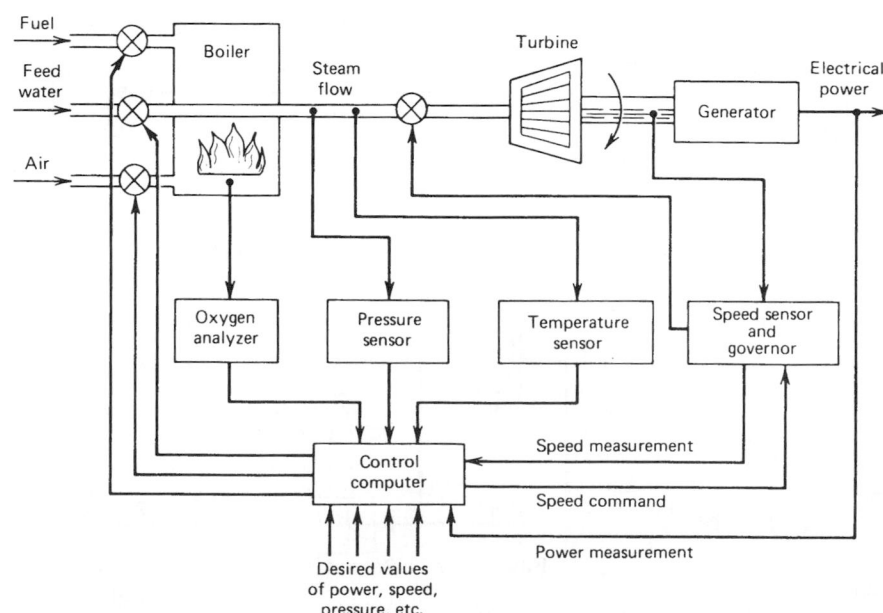

Fig. 55 Computer control system for a boiler-generator. Each important variable requires its own controller. Interaction between variables calls for coordinated control of all loops.[1]

linearization can be explicitly performed, or implicitly made, as when we use the small-angle approximation: $\sin\theta \approx \theta$. This approach has been enormously successful because a well-designed controller will keep the system in the operating range where the linearization was done, thus preserving the accuracy of the linear model. However, it is difficult to control some systems accurately in this way because their operating range is too large. Robot arms are a good example.[13,14] Their equations of motion are very nonlinear, due primarily to the fact that their inertia varies greatly as their configuration changes.

Nonlinear systems encompass everything that is "not linear," and thus there is no general theory for nonlinear systems. There have been many nonlinear control methods proposed—too many to summarize here.[15] *Lyapunov's stability theory* and Popov's method play a central role in many such schemes. Adaptive control is a subcase of nonlinear control (see below).

The high speeds of modern digital computers now allow us to implement nonlinear control algorithms not possible with earlier hardware. An example is the computed-torque method for controlling robot arms, which was discussed in Section 11 (see Fig. 47).

14.3 Adaptive Control

The term *adaptive control*, which unfortunately has been loosely used, describes control systems that can change the form of the control algorithm or the values of the control gains in real time, as the controller improves its internal model of the process dynamics or in response to unmodeled disturbances.[16] Constant control gains do not provide adequate response for some systems that exhibit large changes in their dynamics over their entire operating range, and some adaptive controllers use several models of the process, each of which is accurate within a certain operating range. The adaptive controller switches between gain settings that are appropriate for each operating range. Adaptive controllers are difficult to design and are prone to instability. Most existing adaptive controllers change only the gain values, not the form of the control algorithm. Many problems remain to be solved before adaptive control theory becomes widely implemented.

14.4 Optimal Control

A rocket might be required to reach orbit using minimum fuel or it might need to reach a given intercept point in minimum time. These are examples of potential applications of *optimal control theory*. Optimal control problems often consist of two subproblems. For the rocket example, these subproblems are (a) the determination of the minimum-fuel (or minimum-time) trajectory and the open-loop control outputs (e.g., rocket thrust as a function of time) required to achieve the trajectory and (b) the design of a feedback controller to keep the system near the optimal trajectory.

Many optimal control problems are nonlinear, and thus no general theory is available. Two classes of problems that have achieved some practical successes are the *bang-bang control* problem in which the control variable switches between two fixed values (e.g., on and off or open and closed),[6] and the *linear-quadratic regulator* (LQG), discussed in Section 7, which has proven useful for high-order systems.[1,6]

Closely related to optimal control theory are methods based on stochastic process theory, including *stochastic control theory*,[17] *estimators, Kalman filters*, and *observers*.[1,6,17]

REFERENCES

1. Palm III, W. J., *Modeling, Analysis, and Control of Dynamic Systems*, 2nd ed., Wiley, New York, 2000.
2. Palm III, W. J., *Control Systems Engineering*, Wiley, New York, 1986.
3. Seborg, D. E., Edgar, T. F., and Mellichamp, D. A., *Process Dynamics and Control*, Wiley, New York, 1989.
4. Palm III, W. J., *System Dynamics*, McGraw-Hill, New York, 2005.
5. McCloy, D., and Martin, H., *The Control of Fluid Power*, 2nd ed., Halsted, London, 1980.
6. Bryson, A. E., and Ho, Y. C., *Applied Optimal Control*, Blaisdell, Waltham, MA, 1969.
7. Lewis, F., *Optimal Control*, Wiley, New York, 1986.
8. Astrom, K. J., and Wittenmark, B., *Computer Controlled Systems*, Prentice-Hall, Englewood Cliffs, NJ, 1984.
9. Dote, Y., *Servo Motor and Motion Control Using Digital Signal Processors*, Prentice-Hall, Englewood Cliffs, NJ, 1990.
10. Palm III, W. J., *Introduction to MATLAB 7 for Engineers*, McGraw-Hill, New York, 2005.
11. Klir, G., and Yuan, B., *Fuzzy Sets and Fuzzy Logic*, Prentice-Hall, Englewood Cliffs, NJ, 1995.
12. Kosko, B., *Neural Networks and Fuzzy Systems*, Prentice-Hall, Englewood Cliffs, NJ, 1992.
13. Craig, J., *Introduction to Robotics*, 3rd ed., Addison-Wesley, Reading, MA, 2005.
14. Spong, M. W., and Vidyasagar, M., *Robot Dynamics and Control*, Wiley, New York, 1989.
15. Slotine, J., and Li, W., *Applied Nonlinear Control*, Prentice-Hall, Englewood Cliffs, NJ, 1991.
16. Astrom, K. J., *Adaptive Control*, Addison-Wesley, Reading, MA, 1989.
17. Stengel, R., *Stochastic Optimal Control*, Wiley, New York, 1986.

CHAPTER 14
THERMODYNAMICS FUNDAMENTALS

Adrian Bejan
Department of Mechanical Engineering and Materials Science
Duke University
Durham, North Carolina

1 INTRODUCTION

Thermodynamics describes the relationship between mechanical work and other forms of energy. There are two facets of contemporary thermodynamics that must be stressed in a review such as this. The first is the equivalence of *work* and *heat* as two possible forms of energy exchange. This facet is expressed by the first law of thermodynamics. The second aspect is the one-way character, or irreversibility, of all flows that occur in nature. As expressed by the second law of thermodynamics, irreversibility or entropy generation is what prevents us from extracting the most possible work from various sources; it is also what prevents us from doing the most with the work that is already at our disposal. The objective of this chapter is to review the first and second laws of thermodynamics and their implications in mechanical engineering, particularly with respect to such issues as energy conversion and conservation. The analytical aspects (the formulas) of engineering thermodynamics are reviewed primarily in terms of the behavior of a pure substance, as would be the case of the working fluid in a heat engine or in a refrigeration machine.

Symbols and Units

c specific heat of incompressible substance, $J/(kg \cdot K)$

c_P	specific heat at constant pressure, $J/(kg \cdot K)$
c_T	constant temperature coefficient, m^3/kg
c_v	specific heat at constant volume, $J/(kg \cdot K)$
COP	coefficient of performance
E	energy, J
f	specific Helmholtz free energy $(u - Ts)$, J/kg
$\mathbf{F}$	force vector, N
g	gravitational acceleration, m/s^2
g	specific Gibbs free energy $(h - Ts)$, J/kg
h	specific enthalpy $(u + Pv)$, J/kg
K	isothermal compressibility, m^2/N
m	mass of closed system, kg
$\dot{m}$	mass flow rate, kg/s
m_i	mass of component in a mixture, kg
M	mass inventory of control volume, kg
M	molar mass, g/mol or kg/kmol
n	number of moles, mol
N_0	Avogadro's constant
P	pressure
δQ	infinitesimal heat interaction, J
$\dot{Q}$	heat transfer rate, W
$\mathbf{r}$	position vector, m
R	ideal gas constant, $J/(kg \cdot K)$
s	specific entropy, $J/(kg \cdot K)$
S	entropy, J/K
S_{gen}	entropy generation, J/K
$\dot{S}_{gen}$	entropy generation rate, W/K
T	absolute temperature, K

Reprinted from *Mechanical Engineers' Handbook*, Vol. 4, Wiley, New York, 2006, with permission of the publisher.

u	specific internal energy, J/kg
U	internal energy, J
v	specific volume, m^3/kg
$\bar{v}$	specific volume of incompressible substance, m^3/kg
V	volume, m^3
V	velocity, m/s
δW	infinitesimal work interaction, J
$\dot{W}_{lost}$	rate of lost available work, W
$\dot{W}_{sh}$	rate of shaft (shear) work transfer, W
x	linear coordinate, m
x	quality of liquid and vapor mixture
Z	vertical coordinate, m
β	coefficient of thermal expansion, 1/K
γ	ratio of specific heats, c_P/c_v
η	"efficiency" ratio
η_I	first-law efficiency
η_{II}	second-law efficiency
θ	relative temperature, °C

Subscripts

$()_f$	saturated liquid state (f = "fluid")
$()_g$	saturated vapor state (g = "gas")
$()_s$	saturated solid state (s = "solid")
$()_{in}$	inlet port
$()_{out}$	outlet port
$()_{rev}$	reversible path
$()_H$	high-temperature reservoir
$()_L$	low-temperature reservoir
$()_{max}$	maximum
$()_T$	turbine
$()_C$	compressor
$()_N$	nozzle
$()_D$	diffuser
$()_0$	reference state
$()_1$	initial state
$()_2$	final state
$()_-$	moderately compressed liquid state
$()_+$	slightly superheated vapor state

Definitions

Boundary: The real or imaginary surface delineating the thermodynamic system. The boundary separates the system from its environment. The boundary is an unambiguously defined surface. The boundary has zero thickness and zero volume.

Closed System: A thermodynamic system whose boundary is not crossed by mass flow.

Cycle: The special process in which the final state coincides with the initial state.

Environment: The thermodynamic system external to the thermodynamic system.

Extensive Properties: Properties whose values depend on the size of the system (e.g., mass, volume, energy, enthalpy, entropy).

Intensive Properties: Properties whose values do not depend on the system size (e.g., pressure, temperature). The collection of all intensive properties constitutes the *intensive state*.

Open System: A thermodynamic system whose boundary is permeable to mass flow. Open systems (flow systems) have their own nomenclature: The thermodynamic system is usually referred to as the *control volume*, the boundary of the open system is the *control surface*, and the particular regions of the boundary that are crossed by mass flows are the *inlet* and *outlet ports*.

Phase: The collection of all system elements that have the same intensive state (e.g., the liquid droplets dispersed in a liquid–vapor mixture have the same intensive state, that is, the same pressure, temperature, specific volume, specific entropy, etc.).

Process: The change of state from one initial state to a final state. In addition to the end states, knowledge of the process implies knowledge of the *interactions* experienced by the system while in communication with its environment (e.g., work transfer, heat transfer, mass transfer, and entropy transfer). To know the process also means to know the *path* (the history, or the succession of states) followed by the system from the initial to the final state.

State: The condition (the being) of a thermodynamic system at a particular point in time, as described by an ensemble of quantities called *thermodynamic properties* (e.g., pressure, volume, temperature, energy, enthalpy, entropy). Thermodynamic properties are only those quantities that do not depend on the "history" of the system between two different states. Quantities that depend on the system evolution (path) between states are not thermodynamic properties (examples of nonproperties are the work, heat, and mass transfer; the entropy transfer; the entropy generation; and the destroyed exergy—see also the definition of *process*).

Thermodynamic System: The region or the collection of matter in space selected for analysis.

2 FIRST LAW OF THERMODYNAMICS FOR CLOSED SYSTEMS

The first law of thermodynamics is a statement that brings together three concepts in thermodynamics: work transfer, heat transfer, and energy change. Of these concepts, only energy change, or simply energy, is a thermodynamic property. We begin with a review[1] of the concepts of work transfer, heat transfer, and energy change.

Consider the force F_x experienced by a system at a point on its boundary. The infinitesimal *work transfer* between system and environment is

$$\delta W = -F_x\, dx$$

where the boundary displacement dx is defined as positive in the direction of the force F_x. When the force $\mathbf{F}$ and the displacement of its point of application $d\mathbf{r}$ are not collinear, the general definition of infinitesimal work transfer is

$$\delta W = -\mathbf{F} \cdot d\mathbf{r}$$

The work transfer interaction is considered positive when the system does work on its environment—in other words, when $\mathbf{F}$ and $d\mathbf{r}$ are oriented in opposite directions. This sign convention has its origin in heat engine engineering, because the purpose of heat engines as thermodynamic systems is to deliver work while receiving heat.

For a system to experience work transfer, two things must occur: (1) a force must be present on the boundary and (2) the point of application of this force (hence, the boundary) must move. The mere presence of forces on the boundary, without the displacement or the deformation of the boundary, does not mean work transfer. Likewise, the mere presence of boundary displacement without a force opposing or driving this motion does not mean work transfer. For example, in the free expansion of a gas into an evacuated space, the gas system does not experience work transfer because throughout the expansion the pressure at the imaginary system–environment interface is zero.

If a closed system can interact with its environment only via work transfer (i.e., in the absence of heat transfer δQ discussed later), then measurements show that the work transfer during a change of state from state 1 to state 2 is the same for all processes linking states 1 and 2,

$$-\left(\int_1^2 \delta W \right)_{\delta Q = 0} = E_2 - E_1$$

In this special case the work transfer interaction $(W_{1-2})_{\delta Q=0}$ is a property of the system because its value depends solely on the end states. This thermodynamic property is the *energy change* of the system, $E_2 - E_1$. The statement that preceded the last equation is the first law of thermodynamics for closed systems that do not experience heat transfer.

Heat transfer is, like work transfer, an energy interaction that can take place between a system and its environment. The distinction between δQ and δW is made by the second law of thermodynamics discussed in the next section: Heat transfer is the energy interaction accompanied by entropy transfer, whereas work transfer is the energy interaction taking place in the absence of entropy transfer. The transfer of heat is driven by the *temperature difference* established between the system and its environment.[2] The system temperature is measured by placing the system in thermal communication with a test system called *thermometer*. The result of this measurement is the *relative temperature* θ expressed in degrees Celsius, θ (°C), or

Fahrenheit, θ (°F); these alternative temperature readings are related through the conversion formulas

$$\theta \text{ (°C)} = \tfrac{5}{9}[\theta \text{ (°F)} - 32]$$

$$\theta \text{ (°F)} = \tfrac{5}{9}\theta \text{ (°C)} + 32$$

$$1°\text{F} = \tfrac{5}{9}°\text{C}$$

The boundary that prevents the transfer of heat, regardless of the magnitude of the system–environment temperature difference, is termed *adiabatic*. Conversely, the boundary that is crossed by heat even in the limit of a vanishingly small system–environment temperature difference is termed *diathermal*.

Measurements also show that a closed system undergoing a change of state $1 \to 2$ in the absence of work transfer experiences a heat interaction whose magnitude depends solely on the end states:

$$\left(\int_1^2 \delta Q \right)_{\delta W = 0} = E_2 - E_1$$

In the special case of zero work transfer, the heat transfer interaction is a thermodynamic property of the system, which is by definition equal to the energy change experienced by the system in going from state 1 to state 2. The last equation is the first law of thermodynamics for closed systems incapable of experiencing work transfer. Note that, unlike work transfer, the heat transfer is considered positive when it increases the energy of the system.

Most thermodynamic systems do not manifest purely mechanical ($\delta Q = 0$) or purely thermal ($\delta W = 0$) behavior. Most systems manifest a *coupled* mechanical and thermal behavior. The preceding first-law statements can be used to show that the first law of thermodynamics for a process executed by a closed system experiencing both work transfer and heat transfer is

$$\underbrace{\int_1^2 \delta Q}_{\substack{\text{heat} \\ \text{transfer}}} - \underbrace{\int_1^2 \delta W}_{\substack{\text{work} \\ \text{transfer}}} = \underbrace{E_2 - E_1}_{\substack{\text{energy} \\ \text{change}}}$$

$$\underbrace{\qquad\qquad\qquad\qquad}_{\substack{\text{energy interaction} \\ \text{(nonproperties)}}} \quad \underbrace{\qquad}_{\text{(property)}}$$

The first law means that the net heat transfer into the system equals the work done by the system on the environment plus the increase in the energy of the system. The first law of thermodynamics for a cycle or for an integral number of cycles executed by a closed system is

$$\oint \delta Q = \oint \delta W = 0$$

Note that the net change in the thermodynamic property energy is zero during a cycle or an integral number of cycles.

The energy change term $E_2 - E_1$ appearing on the right-hand side of the first law can be replaced by a more general notation that distinguishes between macroscopically identifiable forms of energy storage (kinetic, gravitational) and energy stored internally,

$$\underbrace{E_2 - E_1}_{\substack{\text{energy}\\\text{change}}} = \underbrace{U_2 - U_1}_{\substack{\text{internal}\\\text{energy}\\\text{change}}} + \underbrace{\frac{mV_2^2}{2} - \frac{mV_1^2}{2}}_{\substack{\text{kinetic}\\\text{energy}\\\text{change}}} + \underbrace{mgZ_2 - mgZ_1}_{\substack{\text{gravitational}\\\text{energy}\\\text{change}}}$$

If the closed system expands or contracts *quasistatically* (i.e., slowly enough, in mechanical equilibrium internally and with the environment) so that at every point in time the pressure P is uniform throughout the system, then the work transfer term can be calculated as being equal to the work done by all the boundary pressure forces as they move with their respective points of application,

$$\int_1^2 \delta W = \int_1^2 P \, dV$$

The work transfer integral can be evaluated provided the path of the quasistatic process, $P(V)$, is known; this is another reminder that the work transfer is path dependent (i.e., not a thermodynamic property).

3 SECOND LAW OF THERMODYNAMICS FOR CLOSED SYSTEMS

A *temperature reservoir* is a thermodynamic system that experiences only heat transfer and whose temperature remains constant during such interactions. Consider first a closed system executing a cycle or an integral number of cycles *while in thermal communication with no more than one temperature reservoir*. To state the second law for this case is to observe that the net work transfer during each cycle cannot be positive,

$$\oint \delta W = 0$$

In other words, a closed system cannot deliver work during one cycle while in communication with one temperature reservoir or with no temperature reservoir at all. Examples of such cyclic operation are the vibration of a spring–mass system or a ball bouncing on the pavement: For these systems to return to their respective initial heights, that is, for them to execute cycles, the environment (e.g., humans) must perform work on them. The limiting case of frictionless cyclic operation is termed *reversible* because in this limit the system returns to its initial state without intervention (work transfer) from the environment. Therefore, the

distinction between reversible and irreversible cycles executed by closed systems in communication with no more than one temperature reservoir is

$$\oint \delta W = 0 \quad \text{(reversible)}$$

$$\oint \delta W < 0 \quad \text{(irreversible)}$$

To summarize, the first and second laws for closed systems operating cyclically in contact with no more than one temperature reservoir are (Fig. 1)

$$\oint \delta W = \oint \delta Q \le 0$$

This statement of the second law can be used to show[1] that in the case of a closed system executing one or an integral number of cycles *while in communication with two temperature reservoirs* the following inequality holds (Fig. 1):

$$\frac{Q_H}{T_H} + \frac{Q_L}{T_L} \le 0$$

where H and L denote the high-temperature and the low-temperature reservoirs, respectively. Symbols Q_H and Q_L stand for the value of the cyclic integral $\oint \delta Q$, where δQ is in one case exchanged only with the H reservoir and in the other with the L reservoir. In the reversible limit, the second law reduces to $T_H/T_L = -Q_H/Q_L$, which serves as definition for the absolute *thermodynamic temperature* scale denoted by symbol T. Absolute temperatures are expressed either in kelvins, T (K), or in degrees Rankine, $T(°R)$; the relationships between absolute and relative temperatures are

$$T(K) = \theta \,(°C) + 273.15K \quad T\,(°R) = \theta \,(°F) + 459.67°R$$
$$1K = 1°C \qquad\qquad 1°R = 1°F$$

A *heat engine* is a special case of a closed system operating cyclically while in thermal communication with two temperature reservoirs, a system that during each cycle receives heat and delivers work:

$$\oint \delta W = \oint \delta Q = Q_H + Q_L > 0$$

The goodness of the heat engine can be described in terms of the heat engine efficiency or the first-law efficiency

$$\eta = \frac{\oint \delta W}{Q_H} \le 1 - \frac{T_L}{T_H}$$

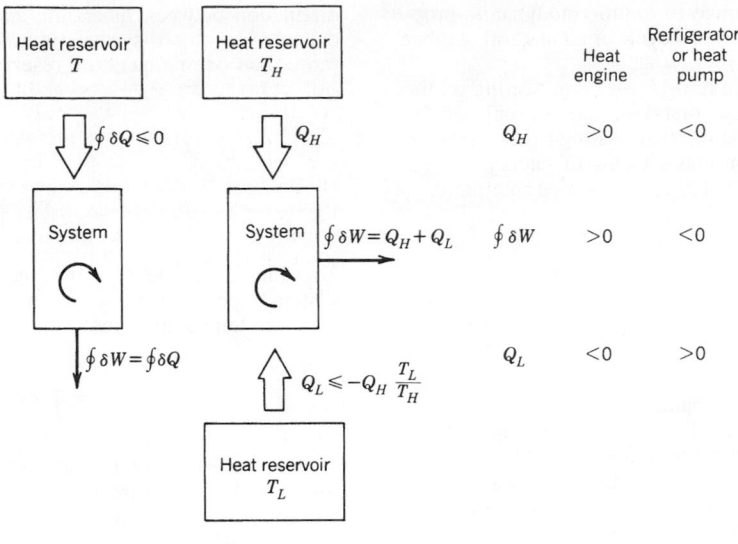

One heat reservoir   Two heat reservoirs

Fig. 1 First and second laws of thermodynamics for a closed system operating cyclically while in communication with one or two heat reservoirs.

Alternatively, the second-law efficiency of the heat engine is defined as[1,3,4]

$$\eta_{II} = \frac{\oint \delta W}{(\oint \delta W)_{\text{maximum (reversible case)}}} = \frac{\eta_I}{1 - T_L/T_H}$$

A *refrigerating machine* or a *heat pump* operates cyclically between two temperature reservoirs in such a way that during each cycle it receives work and delivers net heat to the environment,

$$\oint \delta W = \oint \delta Q = Q_H + Q_L < 0$$

The goodness of such machines can be expressed in terms of a coefficient of performance (COP)

$$\text{COP}_{\text{refrigerator}} = \frac{Q_L}{-\oint \delta W} \le \frac{1}{T_H/T_L - 1}$$

$$\text{COP}_{\text{heat pump}} = \frac{-Q_H}{-\oint \delta W} \le \frac{1}{1 - T_L/T_H}$$

Generalizing the second law for closed systems operating cyclically, one can show that, if during each cycle the system experiences any number of heat interactions Q_i with any number of temperature reservoirs whose respective absolute temperatures are T_i, then

$$\sum_i \frac{Q_i}{T_i} \le 0$$

Note that T_i is the absolute temperature of the boundary region crossed by Q_i. Another way to write the second law in this case is

$$\oint \frac{\delta Q}{T} \le 0$$

where, again, T is the temperature of the boundary pierced by δQ. Of special interest is the reversible cycle limit, in which the second law states $(\oint \delta Q/T)_{\text{rev}} = 0$. According to the definition of thermodynamic property, the second law implies that during a reversible process the quantity $\delta Q/T$ is the infinitesimal change in a property of the system: By definition, that property is the *entropy change*

$$dS = \left(\frac{\delta Q}{T}\right)_{\text{rev}} \quad \text{or} \quad S_2 - S_1 = \left(\int_1^2 \frac{\delta Q}{T}\right)_{\text{rev}}$$

Combining this definition with the second law for a cycle, $\oint \delta Q/T \le 0$, yields the second law of thermodynamics for *any process* executed by a closed system,

$$\underbrace{S_2 - S_1}_{\substack{\text{entropy} \\ \text{change} \\ \text{(property)}}} - \underbrace{\int_1^2 \frac{\delta Q}{T}}_{\substack{\text{entropy} \\ \text{transfer} \\ \text{(nonproperty)}}} \ge 0$$

The entire left-hand side in this inequality is by definition the *entropy generated* by the process,

$$S_{\text{gen}} = S_2 - S_1 - \int_1^2 \frac{\delta Q}{T}$$

The entropy generation is a measure of the inequality sign in the second law and hence a measure of the irreversibility of the process. The entropy generation is proportional to the useful work destroyed during the process.[1,3,4] Note again that any heat interaction (δQ) is accompanied by entropy transfer ($\delta Q/T$), whereas the work transfer δW is not.

4 ENERGY-MINIMUM PRINCIPLE

Consider now a closed system that executes an infinitesimally small change of state, which means that its state changes from $(U, S, \ldots)$ to $(U + dU, S + dS, \ldots)$. The first- and second-law statements are

$$\delta Q - \delta W = dU \qquad dS - \frac{\delta Q}{T} \geq 0$$

If the system is *isolated* from its environment, then $\delta W = 0$ and $\delta Q = 0$, and the two laws dictate that during any such process the energy inventory stays constant ($dU = 0$) and the entropy inventory cannot decrease,

$$dS \geq 0$$

Isolated systems undergo processes when they experience internal changes that do not require intervention from the outside, for example, the removal of one or more of the *internal constraints* plotted qualitatively in the vertical direction in Fig. 2. When all the constraints are removed, changes cease, and, according to $dS \geq 0$, the entropy inventory reaches its highest possible level. This *entropy-maximum principle* is a consequence of the first and second laws. When all the internal constraints have disappeared, the system has reached the *unconstrained equilibrium state*.

Alternatively, if changes occur in the absence of work transfer and at constant S, the first law and the second law require, respectively, $dU = \delta Q$ and $\delta Q \leq 0$, and hence

$$dU \leq 0$$

The energy inventory cannot increase, and when the unconstrained equilibrium state is reached, the system energy inventory is minimum. This *energy-minimum principle* is also a consequence of the first and second laws for closed systems.

The interest in this classical formulation of the laws (e.g., Fig. 2) has been renewed by the emergence of an analogous principle of performance increase (the constructal law) in the search for optimal configurations in the design of open (flow) systems.[5] This analogy is based on the *constructal law* of maximization of flow access.[1,6]

5 LAWS OF THERMODYNAMICS FOR OPEN SYSTEMS

If $\dot{m}$ represents the mass flow rate through a port in the control surface, the principle of *mass conservation*

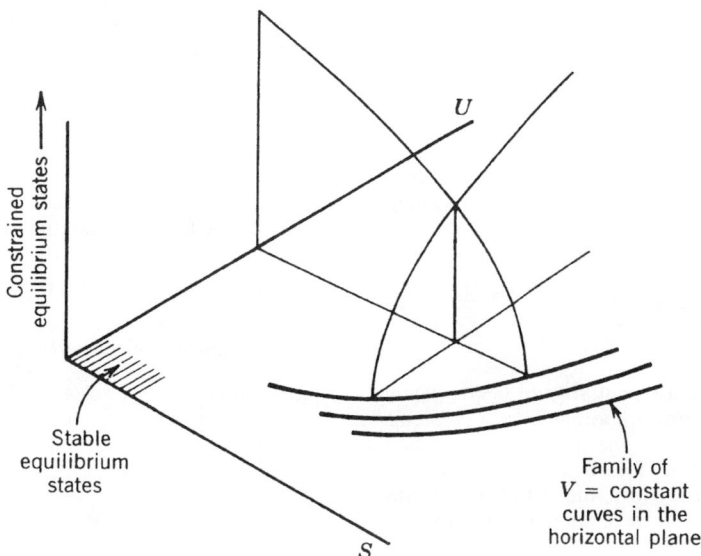

Fig. 2 Energy-minimum principle or entropy-maximum principle.

in the control volume is

$$\underbrace{\sum_{\text{in}} \dot{m} - \sum_{\text{out}} \dot{m}}_{\text{mass transfer}} = \underbrace{\frac{\partial M}{\partial t}}_{\text{mass change}}$$

Subscripts in and out refer to summation over all the inlet and outlet ports, respectively, while M stands for the instantaneous mass inventory of the control volume.

The *first law of thermodynamics* is more general than the statement encountered earlier for closed systems because this time we must account for the flow of energy associated with the $\dot{m}$ streams:

$$\underbrace{\sum_{\text{in}} \dot{m}\left(h + \frac{V^2}{2} + gZ\right) - \sum_{\text{out}} \dot{m}\left(h + \frac{V^2}{2} + gZ\right)}_{} \\ \underbrace{+ \sum_i \dot{Q}_i - \dot{W}}_{\text{energy transfer}}$$

$$= \underbrace{\frac{\partial E}{\partial t}}_{\text{energy change}}$$

On the left-hand side we have the energy interactions: heat, work, and the energy transfer associated with mass flow across the control surface. The specific enthalpy h, fluid velocity V, and height Z are evaluated right at the boundary. On the right-hand side, E is the instantaneous system energy integrated over the control volume.

The *second law of thermodynamics* for an open system assumes the form

$$\underbrace{\sum_{\text{in}} \dot{m}s - \sum_{\text{out}} \dot{m}s + \sum_i \frac{\dot{Q}_i}{T_i}}_{\text{entropy transfer}} \leq \underbrace{\frac{\partial S}{\partial t}}_{\text{entropy change}}$$

The specific entropy s is representative of the thermodynamic state of each stream right at the system boundary. The *entropy generation rate* is defined by

$$\dot{S}_{\text{gen}} = \frac{\partial S}{\partial t} + \sum_{\text{out}} \dot{m}s - \sum_{\text{in}} \dot{m}s - \sum_i \frac{\dot{Q}_i}{T_i}$$

and is a measure of the irreversibility of open-system operation. The engineering importance of $\dot{S}_{\text{gen}}$ stems from its proportionality to the rate of destruction of available work. If the following parameters are fixed—all the mass flows ($\dot{m}$), the peripheral conditions (h, s, V, Z), and the heat interactions (Q_i, T_i) except (Q_0, T_0)—then one can use the first law and the second law to show that the work transfer rate cannot exceed a theoretical maximum[1,3,4]:

$$\dot{W} \leq \sum_{\text{in}} \dot{m}\left(h + \frac{V^2}{2} + gZ - T_0s\right) \\ - \sum_{\text{out}} \dot{m}\left(h + \frac{V^2}{2} + gZ - T_0s\right) - \frac{\partial}{\partial t}(E - T_0s)$$

The right-hand side in this inequality is the maximum work transfer rate $\dot{W}_{\text{sh,max}}$, which would exist only in the ideal limit of reversible operation. The rate of *lost work*, or the rate of exergy (availability) destruction, is defined as

$$\dot{W}_{\text{lost}} = \dot{W}_{\text{max}} - \dot{W}$$

Again, using both laws, one can show that lost work is directly proportional to entropy generation,

$$\dot{W}_{\text{lost}} = T_0 \dot{S}_{\text{gen}}$$

This result is known as the Gouy–Stodola theorem.[1,3,4] Conservation of useful work (exergy) in thermodynamic systems can only be achieved based on the systematic minimization of entropy generation in all the components of the system. Engineering applications of entropy generation minimization as a design optimization philosophy may be found in Refs. 1, 3, and 4.

6 RELATIONS AMONG THERMODYNAMIC PROPERTIES

The analytical forms of the first and second laws of thermodynamics contain properties such as internal energy, enthalpy, and entropy, which cannot be measured directly. The values of these properties are derived from measurements that can be carried out in the laboratory (e.g., pressure, volume, temperature, specific heat); the formulas connecting the derived properties to the measurable properties are reviewed in this section. Consider an infinitesimal change of state experienced by a closed system. If kinetic and gravitational energy changes can be neglected, the first law reads

$$\delta Q_{\text{any path}} - \delta W_{\text{any path}} = dU$$

which emphasizes that dU is path independent. In particular, for a reversible path (rev), the same dU is given by

$$\delta Q_{\text{rev}} - \delta W_{\text{rev}} = dU$$

Note that from the second law for closed systems we have $\delta Q_{\text{rev}} = T\,dS$. Reversibility (or zero-entropy generation) also requires internal mechanical equilibrium at every stage during the process; hence,

$\delta W_{\text{rev}} = P\,dV$, as for a quasistatic change in volume. The infinitesimal change experienced by U is therefore

$$T\,dS - P\,dV = dU$$

Note that this formula holds for an infinitesimal change of state along any path (because dU is path independent); however, $T\,dS$ matches δQ and $P\,dV$ matches δW only if the path is reversible. In general, $\delta Q < T\,dS$ and $\delta W < P\,dV$. The formula derived above for dU can be written for a unit mass: $T\,ds - P\,dv = du$. Additional identities implied by this relation are

$$T = \left(\frac{\partial u}{\partial s}\right)_v \qquad -P = \left(\frac{\partial u}{\partial v}\right)_s$$

$$\frac{\partial^2 u}{\partial s\,\partial v} = \left(\frac{\partial T}{\partial v}\right)_s = -\left(\frac{\partial P}{\partial s}\right)_v$$

where the subscript indicates which variable is held constant during partial differentiation. Similar relations and partial derivative identities exist in conjunction with other derived functions such as enthalpy, Gibbs free energy, and Helmholtz free energy:

- Enthalpy (defined as $h = u + Pv$):

$$dh = T\,ds + v\,dP$$

$$T = \left(\frac{\partial h}{\partial s}\right)_P \qquad v = \left(\frac{\partial h}{\partial P}\right)_s$$

$$\frac{\partial^2 h}{\partial s\,\partial P} = \left(\frac{\partial T}{\partial P}\right)_s = \left(\frac{\partial v}{\partial s}\right)_P$$

- Gibbs free energy (defined as $g = h - Ts$):

$$dg = -s\,dT + v\,dP$$

$$-s = \left(\frac{\partial g}{\partial T}\right)_P \qquad v = \left(\frac{\partial g}{\partial P}\right)_T$$

$$\frac{\partial^2 g}{\partial T\,\partial P} = -\left(\frac{\partial s}{\partial P}\right)_T = \left(\frac{\partial v}{\partial T}\right)_P$$

- Helmholtz free energy (defined as $f = u - Ts$):

$$df = -s\,dT - P\,dv$$

$$-s = \left(\frac{\partial f}{\partial T}\right)_v \qquad -P = \left(\frac{\partial f}{\partial v}\right)_T$$

$$\frac{\partial^2 f}{\partial T\,\partial v} = -\left(\frac{\partial s}{\partial v}\right)_T = -\left(\frac{\partial P}{\partial T}\right)_v$$

In addition to the (P, v, T) surface, which can be determined based on measurements (Fig. 3), the following partial derivatives are furnished by special experiments[1]:

- The specific heat at constant volume, $c_v = (\partial u/\partial T)_v$, follows directly from the constant-volume $(\partial W = 0)$ heating of a unit mass of pure substance.
- The specific heat at constant pressure, $c_P = (\partial h/\partial T)_P$, is determined during the constant-pressure heating of a unit mass of pure substance.
- The Joule–Thompson coefficient, $\mu = (\partial T/\partial P)_h$, is measured during a throttling process, that is, during the flow of a stream through an adiabatic duct with friction (see the first law for an open system in the steady state).
- The coefficient of thermal expansion $\beta = (1/v)(\partial v/\partial T)_P$.
- The isothermal compressibility $K = (-1/v)(\partial v/\partial P)_T$.
- The constant-temperature coefficient $c_T = (\partial h/\partial P)_T$.

Two noteworthy relationships between some of the partial-derivative measurements are

$$c_P - c_v = \frac{Tv\beta^2}{K} \qquad \mu = \frac{1}{c_P}\left[T\left(\frac{\partial v}{\partial T}\right)_P - v\right]$$

The general equations relating the derived properties (u, h, s) to measurable quantities are

$$du = c_v\,dT + \left[T\left(\frac{\partial P}{\partial T}\right)_v - P\right]dv$$

$$dh = c_P\,dT + \left[-T\left(\frac{\partial v}{\partial T}\right)_P + v\right]dP$$

$$ds = \frac{c_v}{T}\,dT + \left(\frac{\partial v}{\partial T}\right)_v\,dv \quad \text{or}$$

$$ds = \frac{c_P}{T}\,dT - \left(\frac{\partial v}{\partial T}\right)_P\,dP$$

These relations also suggest the following identities:

$$\left(\frac{\partial u}{\partial T}\right)_v = T\left(\frac{\partial s}{\partial T}\right) = c_v$$

$$\left(\frac{\partial h}{\partial T}\right)_P = T\left(\frac{\partial s}{\partial T}\right)_P = c_P$$

The relationships between thermodynamic properties and the analyses associated with applying the laws of thermodynamics are simplified considerably in cases

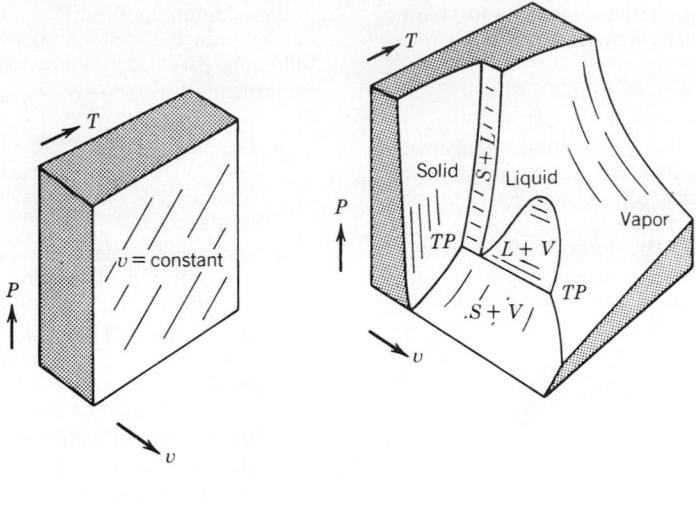

Incompressible substance   Pure substance

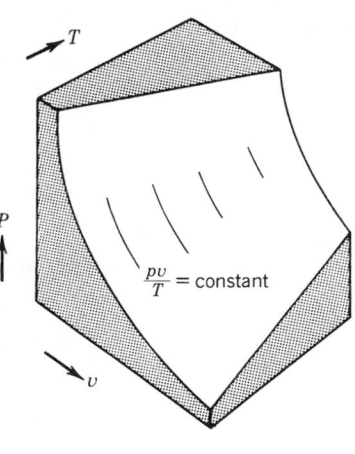

Ideal gas

Fig. 3 The (P, v, T) surface for a pure substance that contracts upon freezing, showing regions of ideal-gas and incompressible fluid behavior: S = solid, V = vapor, L = liquid, TP = triple point.

where the pure substance exhibits *ideal-gas* behavior. As shown in Fig. 3, this behavior sets in at sufficiently high temperatures and low pressures; in this limit, the (P, v, T) surface is fitted closely by the simple expression

$$\frac{Pv}{T} = R \quad \text{(constant)}$$

where R is the ideal-gas constant of the substance of interest (Table 1). The formulas for internal energy,

enthalpy, and entropy, which concluded the preceding section, assume the following form in the ideal-gas limit:

$$du = c_v \, dT \qquad c_v = c_v(T)$$

$$dh = c_P \, dT \qquad c_P = c_P(T) = c_v + R$$

$$ds = \frac{c_v}{T} \, dT + \frac{R}{v} \, dv \quad \text{or} \quad ds = \frac{c_P}{T} \, dT - \frac{R}{P} \, dP \quad \text{or}$$

$$ds = \frac{c_v}{P} \, dP + \frac{c_P}{v} \, dv$$

Table 1   Ideal-Gas Constants and Specific Heats at Constant Volume for Gases Encountered in Mechanical Engineering

Ideal Gas	$R[J/(kg \cdot K)]$	$c_P[J/(kg \cdot K)]$
Air	286.8	715.9
Argon, Ar	208.1	316.5
Butane, C_4H_{10}	143.2	1595.2
Carbon dioxide, CO_2	188.8	661.5
Carbon monoxide, CO	296.8	745.3
Ethane, C_2H_6	276.3	1511.4
Ethylene, C_2H_4	296.4	1423.5
Helium, He_2	2076.7	3152.7
Hydrogen, H	4123.6	10216.0
Methane, CH_4	518.3	1687.3
Neon, Ne	412.0	618.4
Nitrogen, N_2	296.8	741.1
Octane, C_8H_{18}	72.85	1641.2
Oxygen, O_2	259.6	657.3
Propane, C_3H_8	188.4	1515.6
Steam, H_2O	461.4	1402.6

Source: From Ref. 1.

If the coefficients c_v and c_P are constant in the temperature domain of interest, then the *changes* in specific internal energy, enthalpy, and entropy relative to a reference state $(\ )_0$ are given by the formulas

$$u - u_0 = c_v(T - T_0)$$

$$h - h_0 = c_P(T - T_0) \quad \text{(where } h_0 = u_0 + RT_0)$$

$$s - s_0 = \begin{cases} c_v \ln \dfrac{T}{T_0} + R \ln \dfrac{v}{v_0} \\[2mm] c_P \ln \dfrac{T}{T_0} - R \ln \dfrac{P}{P_0} \\[2mm] c_v \ln \dfrac{P}{P_0} + c_P \ln \dfrac{v}{v_0} \end{cases}$$

The ideal-gas model rests on two empirical constants, c_v and c_P, or c_v and R, or c_P and R. The ideal-gas limit is also characterized by

$$\mu = 0 \qquad \beta = \frac{1}{P} \qquad K = \frac{1}{P} \qquad c_T = 0$$

The extent to which a thermodynamic system destroys available work is intimately tied to the system's entropy generation, that is, to the system's departure from the theoretical limit of reversible operation. Idealized processes that can be modeled as reversible occupy a central role in engineering thermodynamics because they can serve as standard in assessing the goodness of real processes. Two benchmark reversible processes executed by closed ideal-gas systems are particularly simple and useful. A *quasistatic adiabatic process* $1 \rightarrow 2$ executed by a closed ideal-gas system has the following characteristics:

$$\int_1^2 \delta Q = 0$$

$$\int_1^2 \delta W = \frac{P_2 V_2}{\gamma - 1}\left[\left(\frac{V_2}{V_1}\right)^{\gamma - 1} - 1\right]$$

where $\gamma = c_P/c_v$.

- Path:

$$PV^\gamma = P_1 V_1^\gamma = P_2 V_2^\gamma \quad \text{(constant)}$$

- Entropy change:

$$S_2 - S_1 = 0$$

Hence the name *isoentropic* or *isentropic* for this process.

- Entropy generation:

$$S_{\text{gen}_{1 \rightarrow 2}} = S_2 - S_1 - \int_1^2 \frac{\delta Q}{T} = 0 \quad \text{(reversible)}$$

A *quasistatic isothermal process* $1 \rightarrow 2$ executed by a closed ideal-gas system in communication with a single temperature reservoir T is characterized by:

- Energy interactions:

$$\int_1^2 \delta Q = \int_1^2 \delta W = mRT \ \ln \frac{V_2}{V_1}$$

- Path:

$$T = T_1 = T_2 \quad \text{(constant)} \qquad \text{or}$$

$$PV = P_1 V_1 = P_2 V_2 \quad \text{(constant)}$$

- Entropy change:

$$S_2 - S_1 = mR \ \ln \frac{V_2}{V_1}$$

- Entropy generation:

$$S_{\text{gen}_{1 \rightarrow 2}} = S_2 - S_1 - \int_1^2 \frac{\delta Q}{T} = 0 \quad \text{(reversible)}$$

Mixtures of ideal gases also behave as ideal gases in the high-temperature, low-pressure limit. If a certain mixture of mass m contains ideal gases mixed in mass proportions m_i, and if the ideal-gas constants of each component are (c_{vi}, c_{Pi}, R_i), then the equivalent ideal-gas constants of the mixture are

$$c_v = \frac{1}{m} \sum_i m_i c_{vi} \qquad c_p = \frac{1}{m} \sum_i m_i c_{Pi}$$

$$R = \frac{1}{m} \sum_i m_i R_i$$

where $m = \sum_i m_i$.

One mole is the amount of substance of a system that contains as many elementary entities (e.g., molecules) as there are in 12 g of carbon 12; the number of such entities is Avogadro's constant, $N_0 \cong 6.022 \times 10^{23}$. The mole is not a mass unit because the mass of 1 mol is not the same for all substances. The *molar mass M* of a given molecular species is the mass of 1 mol of that species, so that the total mass m is equal to M times the number of moles n,

$$m = nM$$

Thus, the ideal-gas equation of state can be written as

$$PV = nMRT$$

where the product MR is the *universal gas constant*

$$\overline{R} = MR = 8.314 \text{ J/(mol} \cdot \text{K)}$$

The equivalent molar mass of a mixture of ideal gases with individual molar masses M_i is

$$M = \frac{1}{n} \sum n_i M_i$$

where $n = \sum n_i$. The molar mass of air, as a mixture of nitrogen, oxygen, and traces of other gases, is 28.966 g/mol (or 28.966 kg/kmol). A more useful model of the air gas mixture relies on only nitrogen and oxygen as constituents, in the proportion 3.76 mol of nitrogen to every mole of oxygen; this simple model is used frequently in the field of combustion.[1]

At the opposite end of the spectrum is the *incompressible substance* model. At sufficiently high pressures and low temperatures in Fig. 3, solids and liquids behave so that their density or specific volume is practically constant. In this limit the (P, v, T) surface is adequately represented by the equation

$$v = \overline{v} \quad \text{(constant)}$$

The formulas for calculating changes in internal energy, enthalpy, and entropy become (see the end of the section on relations among thermodynamic properties)

$$du = c\,dT \qquad dh = c\,dT + \overline{v}\,dP \qquad ds = \frac{c}{T}\,dT$$

where c is the sole specific heat of the incompressible substance,

$$c = c_v = c_P$$

The specific heat c is a function of temperature only. In a sufficiently narrow temperature range where c can be regarded as constant, the finite changes in internal energy, enthalpy, and entropy relative to a reference state denoted by $(\ )_0$ are

$$u - u_0 = c\,(T - T_0)$$

$$h - h_0 = c\,(T - T_0) + \overline{v}\,(P - P_0)$$

$$\text{(where } h_0 = u_0 + P_0 \overline{v})$$

$$s - s_0 = c\,\ln \frac{T}{T_0}$$

The incompressible substance model rests on two empirical constants, c and $\overline{v}$.

As shown in Fig. 3, the domains in which the pure substance behaves either as an ideal gas or as an incompressible substance intersect over regions where the substance exists as a mixture of two phases, liquid and vapor, solid and liquid, or solid and vapor. The two-phase regions themselves intersect along the *triple-point* line labeled TP-TP on the middle sketch of Fig. 3. In engineering cycle calculations, the projections of the (P, v, T) surface on the $P-v$ plane or, through the relations reviewed earlier, on the $T-s$ plane are useful. The terminology associated with two-phase equilibrium states is defined on the $P-v$ diagram of Fig. 4a, where we imagine the isothermal compression of a unit mass of substance (a closed system). As the specific volume v decreases, the substance ceases to be a pure vapor at state g, where the first droplets of liquid are formed. State g is a *saturated vapor state*. It is observed that isothermal compression beyond g proceeds at constant pressure up to state f, where the last bubble (immersed in liquid) is suppressed. State f is a *saturated liquid state*. Isothermal compression beyond f is accompanied by a steep rise in pressure, depending on the compressibility of the liquid phase. The *critical state* is the intersection of the locus of saturated vapor states with the locus of saturated liquid states (Fig. 4a). The temperature and pressure corresponding to the critical state are the *critical temperature* and *critical pressure*. Table 2 contains a compilation of critical-state properties of some of the more common substances.

Figure 4b shows the projection of the liquid and vapor domain on the $T-s$ plane. On the same drawing is shown the relative positioning (the relative slopes) of the traces of various constant-property cuts through the three-dimensional surface on which all the equilibrium states are positioned. In the two-phase region, the temperature is a unique function of pressure. This

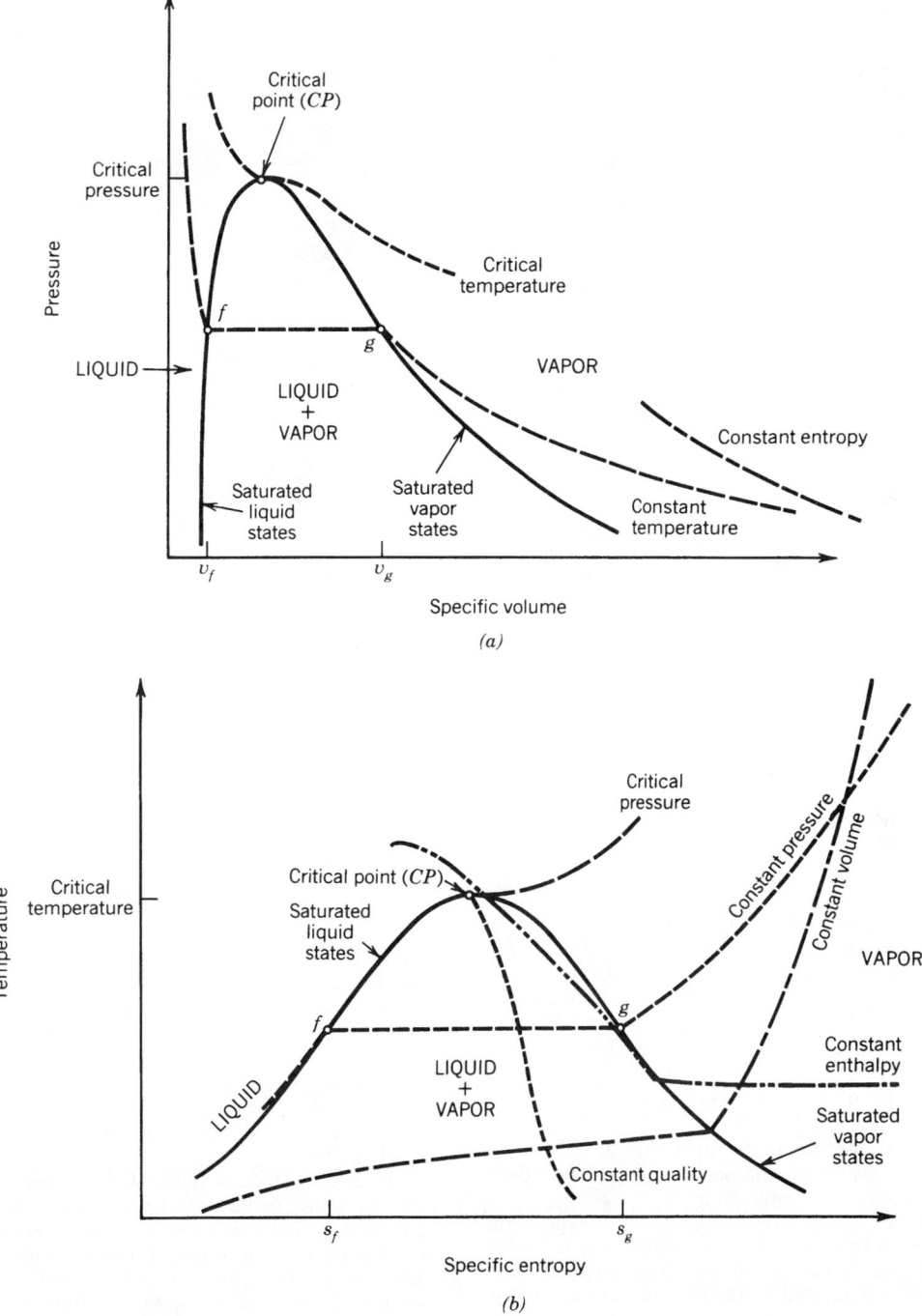

Fig. 4 Locus of two-phase (liquid and vapor) states, projected on (a) P–v plane and (b) T–s plane.

Table 2 Critical-State Properties

Fluid	Critical Temperature [K (°C)]	Critical Pressure [MPa (atm)]	Critical Specific Volume (cm³/g)
Air	133.2(-140)	3.77(37.2)	2.9
Alcohol (methyl)	513.2(240)	7.98(78.7)	3.7
Alcohol (ethyl)	516.5(243.3)	6.39(63.1)	3.6
Ammonia	405.4(132.2)	11.3(111.6)	4.25
Argon	150.9(-122.2)	4.86(48)	1.88
Butane	425.9(152.8)	3.65(36)	4.4
Carbon dioxide	304.3(31.1)	7.4(73)	2.2
Carbon monoxide	134.3(-138.9)	3.54(35)	3.2
Carbon tetrachloride	555.9(282.8)	4.56(45)	1.81
Chlorine	417 (143.9)	7.72(76.14)	1.75
Ethane	305.4(32.2)	4.94(48.8)	4.75
Ethylene	282.6(9.4)	5.85(57.7)	4.6
Helium	5.2(-268)	0.228(2.25)	14.4
Hexane	508.2(235)	2.99(29.5)	4.25
Hydrogen	33.2(-240)	1.30(12.79)	32.3
Methane	190.9(-82.2)	4.64(45.8)	6.2
Methyl chloride	416.5(143.3)	6.67(65.8)	2.7
Neon	44.2(-288.9)	2.7(26.6)	2.1
Nitric oxide	179.2(-93.9)	6.58(65)	1.94
Nitrogen	125.9(-147.2)	3.39(33.5)	3.25
Octane	569.3(296.1)	2.5(24.63)	4.25
Oxygen	154.3(-118.9)	5.03(49.7)	2.3
Propane	368.7(95.6)	4.36(43)	4.4
Sulfur dioxide	430.4(157.2)	7.87(77.7)	1.94
Water	647 (373.9)	22.1(218.2)	3.1

Source: From Ref. 1.

one-to-one relationship is indicated also by the *Clapeyron* relation

$$\left(\frac{dP}{dT}\right)_{\text{sat}} = \frac{h_g - h_f}{T(v_g - v_f)} = \frac{s_g - s_f}{v_g - v_f}$$

where the subscript sat is a reminder that the relation holds for saturated states (such as g and f) and for mixtures of two saturated phases. Subscripts g and f indicate properties corresponding to the saturated vapor and liquid states found at temperature T_{sat} (and pressure P_{sat}). Built into the last equation is the identity

$$h_g - h_f = T(s_g - s_f)$$

which is equivalent to the statement that the Gibbs free energy is the same for the saturated states and their mixtures found at the same temperature, $g_g = g_f$.

The properties of a two-phase mixture depend on the proportion in which saturated vapor, m_g, and saturated liquid, m_f, enter the mixture. The composition of the mixture is described by the property called *quality*,

$$x = \frac{m_g}{m_f + m_g}$$

The quality varies between 0 at state f and 1 at state g. Other properties of the mixture can be calculated in terms of the properties of the saturated states found at the same temperature,

$$u = u_f + x u_{fg} \qquad s = s_f + x s_{fg}$$
$$h = h_f + x h_{fg} \qquad v = v_f + x v_{fg}$$

with the notation $(\ )_{fg} = (\ )_g - (\ )_f$. Similar relations can be used to calculate the properties of two-phase states other than liquid and vapor, namely, solid and vapor or solid and liquid. For example, the enthalpy of a solid and liquid mixture is given by $h = h_s + x h_{sf}$, where subscript s stands for the *saturated solid state* found at the same temperature as for the two-phase state and h_{sf} is the latent heat of melting or solidification.

In general, the states situated immediately outside the two-phase dome sketched in Figs. 3 and 4 do not follow very well the limiting models discussed earlier in this section (ideal gas, incompressible substance). Because the properties of closely neighboring states are usually not available in tabular form, the following approximate calculation proves useful. For a *moderately compressed liquid state*, which is indicated by the subscript $(\ )_*$, that is, for a state situated close to the left of the dome in Fig. 4, the properties may be calculated as slight deviations from those of the saturated liquid state found at the same temperature as the compressed liquid state of interest,

$$h_* \cong (h_f)_{T^*} + (v_f)_{T^*}[P_* - (P_f)_{T^*}] \qquad s \cong (s_f)_{T^*}$$

For a *slightly superheated vapor state*, that is, a state situated close to the right of the dome in Fig. 4, the properties may be estimated in terms of those of the saturated vapor state found at the same temperature:

$$h_+ \cong (h_g)_{T+} \quad s_+ \cong (s_g)_{T+} + \left(\frac{P_g v_g}{T_g}\right)_{T+} \ln \frac{(P_g)_{T+}}{P_+}$$

In these expressions, subscript $()_+$ indicates the properties of the slightly superheated vapor state.

7 ANALYSIS OF ENGINEERING SYSTEM COMPONENTS

This section contains a summary[1] of the equations obtained by applying the first and second laws of thermodynamics to the components encountered in most engineering systems, such as power plants and refrigeration plants. It is assumed that each component operates in *steady flow*:

- *Valve* (throttle) or adiabatic duct with friction (Fig. 5a):

  First law: $\quad h_1 = h_2$

  Second law: $\quad \dot{S}_{\text{gen}} = \dot{m}(s_2 - s_1) > 0$

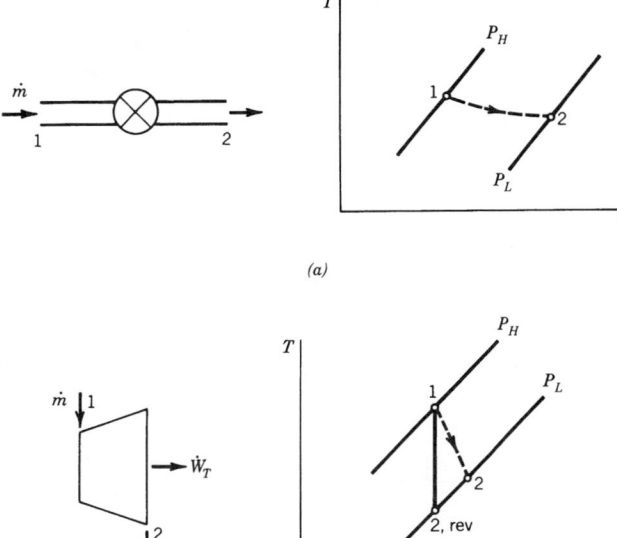

(a)

(b)

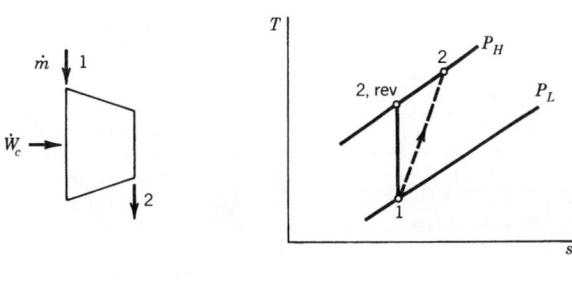

(c)

Fig. 5 Engineering system components and their inlet and outlet states on the $T–s$ plane: P_H = high pressure; P_L = low pressure.

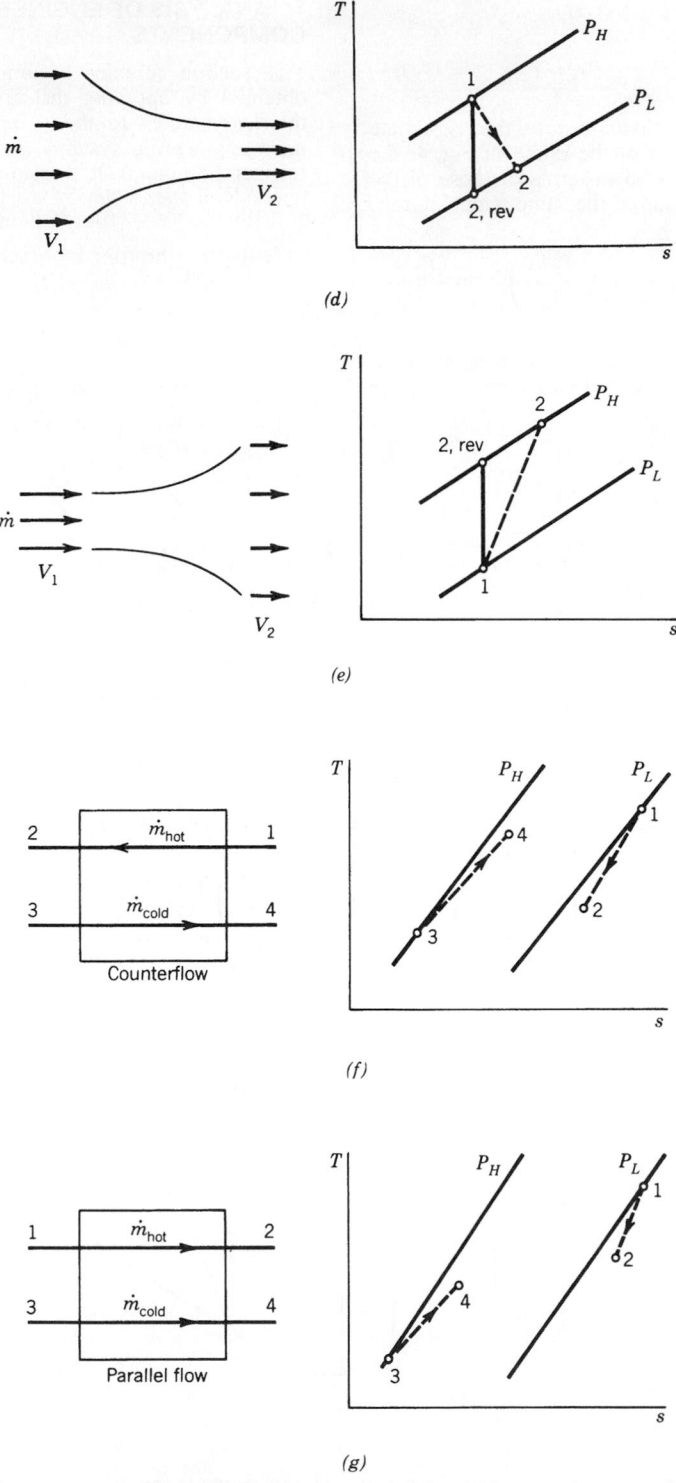

Fig. 5 (*Continued*)

- *Expander* or *turbine* with negligible heat transfer to the ambient (Fig. 5*b*):

  First law: $\dot{W}_T = \dot{m}(h_1 - h_2)$

  Second law: $\dot{S}_{gen} = \dot{m}(s_2 - s_1) \geq 0$

  Efficiency: $\eta_T = \dfrac{h_1 - h_2}{h_1 - h_{2,rev}} \leq 1$

- *Compressor* or *pump* with negligible heat transfer to the ambient (Fig. 5*c*):

  First law: $\dot{W}_c = \dot{m}(h_2 - h_1)$

  Second law: $\dot{S}_{gen} = \dot{m}(s_2 - s_1) \geq 0$

  Efficiency: $\eta_c = \dfrac{h_{2,rev} - h_1}{h_2 - h_1} \leq 1$

- *Nozzle* with negligible heat transfer to the ambient (Fig. 5*d*):

  First law: $\frac{1}{2}(V_2^2 - V_1^2) = h_1 - h_2$

  Second law: $\dot{S}_{gen} = \dot{m}(s_2 - s_1) \geq 0$

  Efficiency: $\eta_N = \dfrac{V_2^2 - V_1^2}{V_{2,rev}^2 - V_1^2} \leq 1$

- *Diffuser* with negligible heat transfer to the ambient (Fig. 5*e*):

  First law: $h_2 - h_1 = \frac{1}{2}(V_1^2 - V_2^2)$

  Second law: $\dot{S}_{gen} = \dot{m}(s_2 - s_1) \geq 0$

  Efficiency: $\eta_D = \dfrac{h_{2,rev} - h_1}{h_2 - h_1} \leq 1$

- *Heat exchangers* with negligible heat transfer to the ambient (Figs. 5*f* and *g*):

  First law: $\dot{m}_{hot}(h_1 - h_2) = \dot{m}_{cold}(h_4 - h_3)$

  Second law: $\dot{S}_{gen} - \dot{m}_{hot}(s_2 - s_1)$
  $+ \dot{m}_{cold}(s_4 - s_3) \geq 0$

Figures 5*f* and *g* show that a pressure drop always occurs in the direction of flow in any heat exchanger flow passage.

REFERENCES

1. Bejan, A., *Advanced Engineering Thermodynamics*, 2nd ed., Wiley, New York, 1997.
2. Bejan, A., *Heat Transfer*, Wiley, New York, 1993.
3. Bejan, A., *Entropy Generation through Heat and Fluid Flow*, Wiley, New York, 1982.
4. Bejan, A., *Entropy Generation Minimization*, CRC Press, Boca Raton, FL, 1996.
5. Bejan, A., and Lorente, S., "The Constructal Law and the Thermodynamics of Flow Systems with Configuration," *Int. J. Heat Mass Transfer*, **47**, 3203–3214 (2004).
6. Bejan, A., *Shape and Structure, from Engineering to Nature*, Cambridge University Press, Cambridge, UK, 2000.

CHAPTER 15

HEAT TRANSFER FUNDAMENTALS

G. P. Peterson
Rensselaer Polytechnic Institute
Troy, New York

SYMBOLS AND UNITS

A Area of heat transfer

Bi   Biot number, hL/k, dimensionless

C Circumference, m, constant defined in text

C_p Specific heat under constant pressure, J/kg·K

D Diameter, m

e Emissive power, W/m^2

f Drag coefficient, dimensionless

F Cross-flow correction factor, dimensionless

F_{i-j} Configuration factor from surface i to surface j, dimensionless

Fo   Fourier number, $\alpha t A^2/V^2$, dimensionless

$F_{o-\lambda T}$ Radiation function, dimensionless

G Irradiation, W/m^2; mass velocity, kg/m^2·s

g Local gravitational acceleration, 9.8 m/s^2

g_c Proportionality constant, 1 kg·m/N·s^2

Gr   Grashof number, $gL^3\beta\,\Delta T/v^2$ dimensionless

h Convection heat transfer coefficient, equals $q/A\,\Delta T$, W/m^2·K

h_{fg} Heat of vaporization, J/kg

J Radiosity, W/m^2

k Thermal conductivity, W/m·K

K   Wick permeability, m^2

L Length, m

Ma   Mach number, dimensionless

N   Screen mesh number, m^{-1}

Nu   Nusselt number, $\mathrm{Nu}_L = hL/k$, $\mathrm{Nu}_D = hD/k$, dimensionless

$\overline{\mathrm{Nu}}$ Nusselt number averaged over length, dimensionless

P Pressure, N/m^2, perimeter, m

Pe   Peclet number, RePr, dimensionless

Pr   Prandtl number, $C_p\mu/k$, dimensionless

q Rate of heat transfer, W

q'' Rate of heat transfer per unit area, W/m^2

R Distance, m; thermal resistance, K/W

r Radial coordinate, m; recovery factor, dimensionless

Ra   Rayleigh number, GrPr; $\mathrm{Ra}_L = \mathrm{Gr}_L\mathrm{Pr}$, dimensionless

Reprinted from *Mechanical Engineers' Handbook*, Vol. 4, Wiley, New York, 2006, with permission of the publisher.

HEAT TRANSFER FUNDAMENTALS **819**

Re	Reynolds Number, $\mathrm{Re}_L = \rho V L/\mu$, $\mathrm{Re}_D = \rho V D/\mu$, dimensionless
S	Conduction shape factor, m
T	Temperature, K or °C
t	Time, s
$T_{\alpha s}$	Adiabatic surface temperature, K
T_{sat}	Saturation temperature, K
T_b	Fluid bulk temperature or base temperature of fins, K
T_e	Excessive temperature, $T_s - T_{\mathrm{sat}}$, K or °C
T_f	Film temperature, $(T_\infty + T_s)/2$, K
T_i	Initial temperature; at $t = 0$, K
T_0	Stagnation temperature, K
T_s	Surface temperature, K
T_∞	Free-stream fluid temperature, K
U	Overall heat transfer coefficient, W/m²·K
V	Fluid velocity, m/s; volume, m³
w	Groove width, m; or wire spacing, m
We	Weber number, dimensionless
x	One of the axes of Cartesian reference frame, m

GREEK SYMBOLS

α	Thermal diffusivity, $k/\rho C_p$, m²/s; absorptivity, dimensionless
β	Coefficient of volume expansion, 1/K
Γ	Mass flow rate of condensate per unit width, kg/m·s
γ	Specific heat ratio, dimensionless
ΔT	Temperature difference, K
δ	Thickness of cavity space, groove depth, m
$\in$	Emissivity, dimensionless
ε	Wick porosity, dimensionless
λ	Wavelength, μm
η_f	Fin efficiency, dimensionless
μ	Viscosity, kg/m·s
ν	Kinematic viscosity, m²/s
ρ	Reflectivity, dimensionless; density, kg/m³
σ	Surface tension, N/m; Stefan–Boltzmann constant, 5.729×10^{-8} W/m²·K⁴
τ	Transmissivity, dimensionless, shear stress, N/m²
Ψ	Angle of inclination, degrees or radians

SUBSCRIPTS

a	Adiabatic section, air
b	Boiling, blackbody
c	Convection, capillary, capillary limitation, condenser
e	Entrainment, evaporator section
eff	Effective
f	Fin
i	Inner
l	Liquid

m	Mean, maximum
n	Nucleation
o	Outer
O	Stagnation condition
p	Pipe
r	Radiation
s	Surface, sonic or sphere
w	Wire spacing, wick
v	Vapor
λ	Spectral
∞	Free stream
$-$	Axial hydrostatic pressure
$+$	Normal hydrostatic pressure

Transport phenomena represents the overall field of study and encompasses a number of subfields. One of these is heat transfer, which focuses primarily on the energy transfer occurring as a result of an energy gradient that manifests itself as a temperature difference. This form of energy transfer can occur as a result of a number of different mechanisms, including *conduction*, which focuses on the transfer of energy through the direct impact of molecules; *convection*, which results from the energy transferred through the motion of a fluid; and *radiation*, which focuses on the transmission of energy through electromagnetic waves. In the following review, as is the case with most texts on heat transfer, *phase change heat transfer*, that is, *boiling* and *condensation*, will be treated as a subset of convection heat transfer.

1 CONDUCTION HEAT TRANSFER

The exchange of energy or heat resulting from the kinetic energy transferred through the direct impact of molecules is referred to as *conduction*, and takes place from a region of high energy (or temperature) to a region of lower energy (or temperature). The fundamental relationship that governs this form of heat transfer is *Fourier's law of heat conduction*, which states that in a one-dimensional system with no fluid motion, the rate of heat flow in a given direction is proportional to the product of the temperature gradient in that direction and the area normal to the direction of heat flow. For conduction heat transfer in the x direction this expression takes the form

$$q_x = -kA\frac{\partial T}{\partial x}$$

where q_x is the heat transfer in the x direction, A is the area normal to the heat flow, $\partial T/\partial x$ is the temperature gradient, and k is the thermal conductivity of the substance.

Writing an energy balance for a three-dimensional body and utilizing Fourier's law of heat conduction yields an expression for the transient diffusion occurring within a body or substance:

$$\frac{\partial}{\partial x}\left(k\frac{\partial T}{\partial x}\right) + \frac{\partial}{\partial y}\left(k\frac{\partial T}{\partial y}\right) + \frac{\partial}{\partial z}\left(k\frac{\partial T}{\partial z}\right) + \dot{q} = \rho c_p \frac{\partial}{\partial x}\frac{\partial T}{\partial t}$$

This expression, usually referred to as the *heat diffusion equation* or heat equation, provides a basis for most types of heat conduction analyses. Specialized cases of this equation can be used to solve many steady-state or transient problems. Some of these specialized cases are as follows:

Thermal conductivity is a constant:

$$\frac{\partial^2 T}{\partial x^2} + \frac{\partial^2 T}{\partial y^2}\frac{\partial^2 T}{\partial z^2} + \frac{\dot{q}}{k} = \frac{\rho c_p}{k}\frac{\partial T}{\partial t}$$

Steady state with heat generation:

$$\frac{\partial}{\partial x}\left(k\frac{\partial T}{\partial x}\right) + \frac{\partial}{\partial y}\left(k\frac{\partial T}{\partial y}\right) + \frac{\partial}{\partial z}\left(k\frac{\partial T}{\partial z}\right) + \dot{q} = 0$$

Steady-state, one-dimensional heat transfer with no heat sink (i.e., a fin):

$$\frac{\partial}{\partial x}\left(\frac{\partial T}{\partial x}\right) + \frac{\dot{q}}{k} = 0$$

One-dimensional heat transfer with no internal heat generation:

$$\frac{\partial}{\partial x}\left(\frac{\partial T}{\partial x}\right) = \frac{\rho c_p}{k}\frac{\partial T}{\partial t}$$

In the following sections, the heat diffusion equation will be utilized for several specific cases. However, in general, for a three-dimensional body of constant thermal properties without heat generation under steady-state heat conduction the temperature field satisfies the expression

$$\nabla^2 T = 0$$

1.1   Thermal Conductivity

The ability of a substance to transfer heat through conduction can be represented by the constant of proportionality, k, referred to as the thermal conductivity. Figure 1 illustrates the characteristics of the thermal conductivity as a function of temperature for several solids, liquids, and gases. As shown, the thermal conductivity of solids is higher than liquids, and liquids higher than gases. Metals typically have higher thermal conductivities than nonmetals, with pure metals having thermal conductivities that decrease with increasing temperature, while the thermal conductivity of nonmetallic solids generally increases with increasing temperature and density. The addition of other metals to create alloys, or the presence of impurities, usually decreases the thermal conductivity of a pure metal.

In general, the thermal conductivity of liquids decreases with increasing temperature. Alternatively, the thermal conductivity of gases and vapors, while lower, increases with increasing temperature and decreases with increasing molecular weight. The thermal conductivities of a number of commonly used metals and nonmetals are tabulated in Tables 1 and 2, respectively. Insulating materials, which are used to prevent or reduce the transfer of heat between two substance or a substance and the surroundings, are listed in Tables 3 and 4, along with the thermal properties. The thermal conductivities for liquids, molten metals, and gasses are given in Tables 5, 6 and 7, respectively.

1.2   One-Dimensional Steady-State Heat Conduction

The steady-state rate of heat transfer resulting from heat conduction through a homogeneous material can be expressed in terms of the rate of heat transfer, q, or $q = \Delta T/R$, where ΔT is the temperature difference and R is the *thermal resistance*. This thermal resistance is the reciprocal of the *thermal conductance* ($C = 1/R$) and is related to the thermal conductivity by the cross-sectional area. Expressions for the thermal resistance, the temperature distribution, and the rate of heat transfer are given in Table 8 for a plane wall, a cylinder, and a sphere. For a plane wall, the heat transfer is typically assumed to be one dimensional (i.e., heat is conducted in only the x direction) and for a cylinder and sphere, only in the radial direction.

Aside from the heat transfer in these simple geometric configurations, other common problems encountered in practical applications is that of heat transfer through layers or composite walls consisting of N layers, where the thickness of each layer is represented by Δx_n and the thermal conductivity by k_n for $n = 1, 2, \ldots, N$. Assuming that the interfacial resistance is negligible (i.e., there is no thermal resistance at the contacting surfaces), the overall thermal resistance can be expressed as

$$R = \sum_{n=1}^{N} \frac{\Delta x_n}{k_n A}$$

Similarly, for conduction heat transfer in the radial direction through a number of N concentric cylinders with negligible interfacial resistance, the overall thermal resistance can be expressed as

$$R = \sum_{n=1}^{N} \frac{\ln(r_{n+1}/r_n)}{2\pi k_n L}$$

where $r_1 =$ inner radius, $r_{N+1} =$ outer radius.

For N *concentric spheres* with negligible interfacial resistance, the thermal resistance can be expressed as

$$R = \sum_{n=1}^{N} \frac{1/r_n - 1/r_{n+1}}{4\pi k}$$

where $r_1 =$ inner radius, $r_{N+1} =$ outer radius.

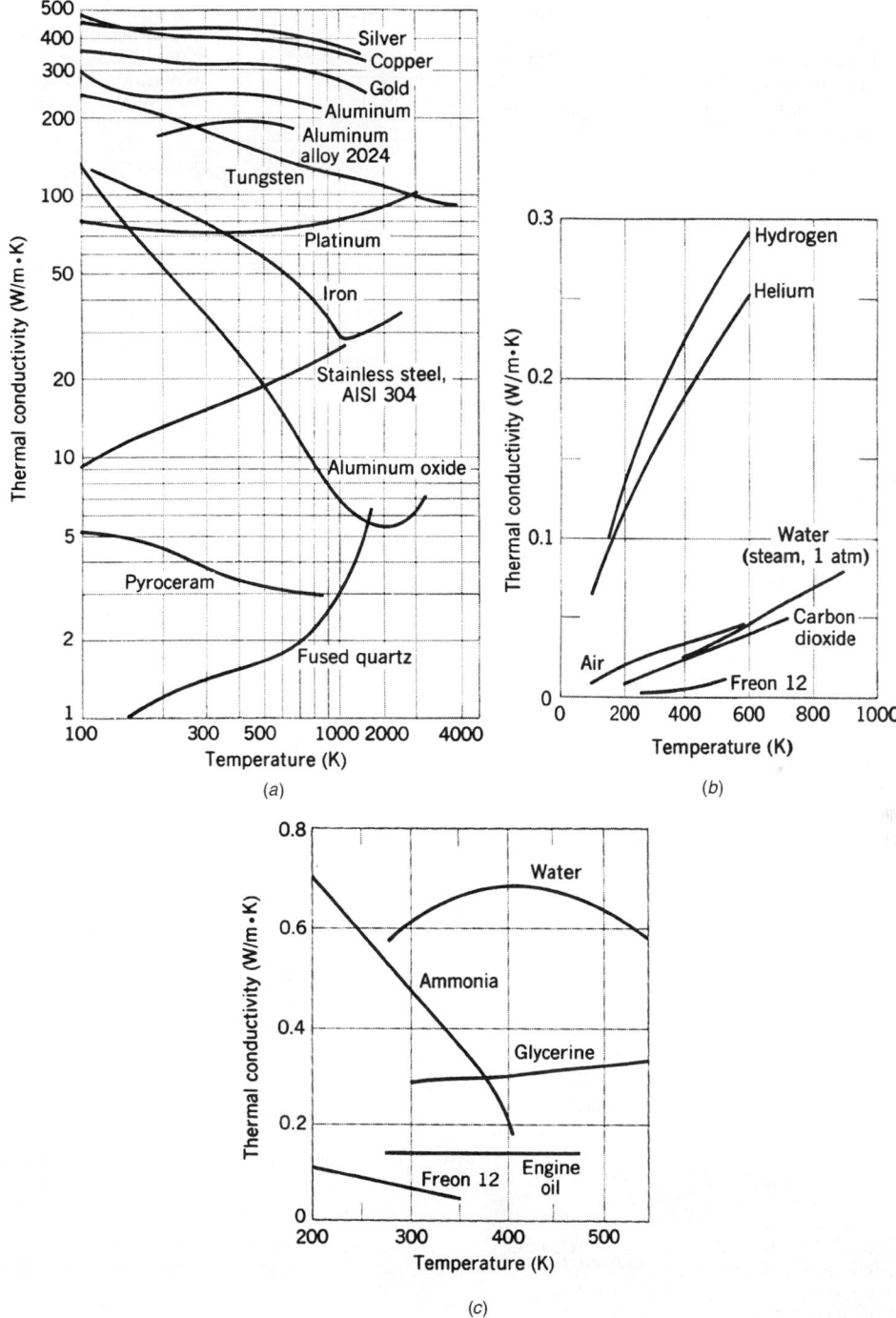

Fig. 1 Temperature dependence of thermal conductivity of (a) selected solids, (b) selected nonmetallic liquids under saturated conditions, and (c) selected gases at normal pressures.[1]

Table 1   Thermal Properties of Metallic Solids[a]

Composition	Melting Point (K)	Properties at 300 K				Properties at Various Temperatures (K) k(W/m·K); C_p(J/kg·K)		
		ρ (kg/m³)	C_p (J/kg·K)	k (W/m·K)	$\alpha \times 10^6$ (m²/s)	100	600	1200
Aluminum	933	2702	903	237	97.1	302; 482	231; 1033	
Copper	1358	8933	385	401	117	482; 252	379; 417	339; 480
Gold	1336	19300	129	317	127	327; 109	298; 135	255; 155
Iron	1810	7870	447	80.2	23.1	134; 216	54.7; 574	28.3; 609
Lead	601	11340	129	35.3	24.1	39.7; 118	31.4; 142	
Magnesium	923	1740	1024	156	87.6	169; 649	149; 1170	
Molybdenum	2894	10240	251	138	53.7	179; 141	126; 275	105; 308
Nickel	1728	8900	444	90.7	23.0	164; 232	65.6; 592	76.2; 594
Platinum	2045	21450	133	71.6	25.1	77.5; 100	73.2; 141	82.6; 157
Silicon	1685	2330	712	148	89.2	884; 259	61.9; 867	25.7; 967
Silver	1235	10500	235	429	174	444; 187	412; 250	361; 292
Tin	505	7310	227	66.6	40.1	85.2; 188		
Titanium	1953	4500	522	21.9	9.32	30.5; 300	19.4; 591	22.0; 620
Tungsten	3660	19300	132	174	68.3	208; 87	137; 142	113; 152
Zinc	693	7140	389	116	41.8	117; 297	103; 436	

[a]Adapted from Ref. 1.

Table 2   Thermal Properties of Nonmetals

Description/Composition	Temperature (K)	Density, ρ(kg/m³)	Thermal Conductivity, k (W/m·K)	Specific Heat, C_p(J/kg·K)	$\alpha \times 10^6$ (m²/s)
Bakelite	300	1300	0.232	1465	0.122
Brick, refractory					
Carborundum	872	—	18.5	—	—
Chrome-brick	473	3010	2.32	835	0.915
Fire clay brick	478	2645	1.0	960	0.394
Clay	300	1460	1.3	880	1.01
Coal, anthracite	300	1350	0.26	1260	0.153
Concrete (stone mix)	300	2300	1.4	880	0.692
Cotton	300	80	0.059	1300	0.567
Glass, window	300	2700	0.78	840	0.344
Rock, limestone	300	2320	2.15	810	1.14
Rubber, hard	300	1190	0.160	—	—
Soil, dry	300	2050	0.52	1840	0.138
Teflon	300	2200	0.35	—	—
	400	—	0.45	—	—

1.3   Two-Dimensional Steady-State Heat Conduction

Two-dimensional heat transfer in an isotropic, homogeneous material with no internal heat generation requires solution of the heat diffusion equation of the form $\partial^2 T/\partial X^2 + \partial T/\partial y^2 = 0$, referred to as the *Laplace equation*. For certain geometries and a limited number of fairly simple combinations of boundary conditions, exact solutions can be obtained analytically. However, for anything but simple geometries or for simple geometries with complicated boundary conditions, development of an appropriate analytical solution can be difficult and other methods are usually employed. Among these are solution procedures involving the use of *graphical* or *numerical* approaches. In the first of these, the rate of heat transfer between two isotherms, T_1 and T_2, is expressed in terms of the conduction shape factor, defined by

$$q = kS(T_1 - T_2)$$

Table 9 illustrates the shape factor for a number of common geometric configurations. By combining these shape factors, the heat transfer characteristics for a wide variety of geometric configurations can be obtained.

Table 3 Thermal Properties of Building and Insulating Materials (at 300 K)[a]

Description/Composition	Density ρ(kg/m³)	Thermal Conductivity, k (W/m·K)	Specific Heat, C_p(J/kg·K)	$\alpha \times 10^6$ (m²/s)
Building boards				
Plywood	545	0.12	1215	0.181
Acoustic tile	290	0.058	1340	0.149
Hardboard, siding	640	0.094	1170	0.126
Woods				
Hardwoods (oak, maple)	720	0.16	1255	0.177
Softwoods (fir, pine)	510	0.12	1380	0.171
Masonry materials				
Cement mortar	1860	0.72	780	0.496
Brick, common	1920	0.72	835	0.449
Plastering materials				
Cement plaster, sand aggregate	1860	0.72	—	—
Gypsum plaster, sand aggregate	1680	0.22	1085	0.121
Blanket and batt				
Glass fiber, paper faced	16	0.046	—	—
Glass fiber, coated; duct liner	32	0.038	835	1.422
Board and slab				
Cellular glass	145	0.058	1000	0.400
Wood, shredded/cemented	350	0.087	1590	0.156
Cork	120	0.039	1800	0.181
Loose fill				
Glass fiber, poured or blown	16	0.043	835	3.219
Vermiculite, flakes	80	0.068	835	1.018

[a]Adapted from Ref. 1.

Table 4 Thermal Conductivities for Some Industrial Insulating Materials[a]

Description/Composition	Maximum Service Temperature (K)	Typical Density (kg/m³)	Typical Thermal Conductivity, k (W/m·K), at Various Temperature (K)			
			200	300	420	645
Blankets						
Blanket, mineral fiber, glass; fine fiber organic bonded	450	10		0.048		
		48		0.033		
Blanket, alumina-silica fiber	1530	48				0.105
Felt, semirigid; organic bonded	480	50–125		0.038	0.063	
Felt, laminated; no binder	920	120			0.051	0.087
Blocks, boards, and pipe insulations						
Asbestos paper, laminated and corruagated, 4-ply	420	190		0.078		
Calcium silicate	920	190			0.063	0.089
Polystyrene, rigid						
Extruded (R-12)	350	56	0.023	0.027		
Molded beads	350	16	0.026	0.040		
Rubber, rigid foamed	340	70		0.032		
Insulating cement						
Mineral fiber (rock, slag, or glass)						
With clay binder	1255	430			0.088	0.123
With hydraulic setting binder	922	560			0.123	
Loose fill						
Cellulose, wood, or paper pulp	—	45		0.039		
Perlite, expanded	—	105	0.036	0.053		
Vermiculite, expanded	—	122		0.068		

[a]Adapted from Ref. 1.

Table 5 Thermal Properties of Saturated Liquids[a]

Liquid	T (K)	ρ (kg/m³)	C_p (kJ/kg·K)	$v \times 10^6$ (m²/s)	$k \times 10^3$ (W/m·K)	$\alpha \times 10^7$ (m²/s)	Pr	$\beta \times 10^3$ (K⁻¹)
Ammonia, NH_3	223	703.7	4.463	0.435	547	1.742	2.60	2.45
	323	564.3	5.116	0.330	476	1.654	1.99	2.45
Carbon dioxide, CO_2	223	1156.3	1.84	0.119	85.5	0.402	2.96	14.0
	303	597.8	36.4	0.080	70.3	0.028	28.7	14.0
Engine oil (unused)	273	899.1	1.796	4280	147	0.910	47,000	0.70
	430	806.5	2.471	5.83	132	0.662	88	0.70
Ethylene glycol, $C_2H_4(OH)_2$	273	1130.8	2.294	57.6	242	0.933	617.0	0.65
	373	1058.5	2.742	2.03	263	0.906	22.4	0.65
Clycerin, $C_3H_5(OH)_3$	273	1276.0	2.261	8310	282	0.977	85,000	0.47
	320	1247.2	2.564	168	287	0.897	1,870	0.50
Freon (Refrigerant-12), CCl_2F_2	230	1528.4	0.8816	0.299	68	0.505	5.9	1.85
	320	1228.6	1.0155	0.190	68	0.545	3.5	3.50

[a]Adapted from Ref. 2. See Table 22 for H_2O.

Table 6 Thermal Properties of Liquid Metals[a]

Composition	Melting Point (K)	T (K)	ρ (kg/m³)	C_p (kJ/kg·K)	$v \times 10^7$ (m²/s)	k (W/m·K)	$\alpha \times 10^5$ (m²/s)	Pr
Bismuth	544	589	10,011	0.1444	1.617	16.4	0.138	0.0142
		1033	9,467	0.1645	0.8343	15.6	1.001	0.0083
Lead	600	644	10,540	0.159	2.276	16.1	1.084	0.024
		755	10,412	0.155	1.849	15.6	1.223	0.017
Mercury	234	273	13,595	0.140	1.240	8.180	0.429	0.0290
		600	12,809	0.136	0.711	11.95	0.688	0.0103
Potassium	337	422	807.3	0.80	4.608	45.0	6.99	0.0066
		977	674.4	0.75	1.905	33.1	6.55	0.0029
Sodium	371	366	929.1	1.38	7.516	86.2	6.71	0.011
		977	778.5	1.26	2.285	59.7	6.12	0.0037
NaK (56%/44%)	292	366	887.4	1.130	6.522	25.6	2.55	0.026
		977	740.1	1.043	2.174	28.9	3.74	0.0058
PbBi (44.5%/55.5%)	398	422	10,524	0.147	—	9.05	0.586	—
		644	10,236	0.147	1.496	11.86	0.790	0.189

[a]Adapted from *Liquid Metals Handbook*, The Atomic Energy Commission, Department of the Navy, Washington, DC, 1952.

Prior to the development of high-speed digital computers, shape factor and analytical methods were the most prevalent methods utilized for evaluating steady-state and transient conduction problems. However, more recently, solution procedures for problems involving complicated geometries or boundary conditions utilize the finite-difference method (FDM). Using this approach, the solid object is divided into a number of distinct or discrete regions, referred to as *nodes*, each with a specified boundary condition. An energy balance is then written for each nodal region and these equations are solved simultaneously. For interior nodes in a two-dimensional system with no internal heat generation, the energy equation takes the form of the Laplace equation discussed earlier. However, because the system is characterized in terms of a nodal network, a finite-difference approximation must be used. This approximation is derived by substituting the following

equation for the x-direction rate of change expression

$$\left.\frac{\partial^2 T}{\partial x^2}\right|_{m,n} \approx \frac{T_{m+1,n} + T_{m-1,n} - 2T_{m,n}}{(\Delta x)^2}$$

and for the y-direction rate of change expression:

$$\left.\frac{\partial^2 T}{\partial y^2}\right|_{m,n} \quad \frac{T_{m,n+1} + T_{m,n-1} + T_{m,n}}{(\Delta y)^2}$$

Assuming $\Delta x = \Delta y$ and substituting into the Laplace equation and results in the following expression:

$$T_{m,n+1} + T_{m,n-1} + T_{m+1,n} + T_{m-1,n} - 4T_{m,n} = 0$$

which reduces the exact difference to an approximate algebraic expression.

Table 7 Thermal Properties of Gases at Atmospheric Pressure[a]

Gas	T (K)	ρ (kg/m^3)	c_p (kJ/kg·K)	$v \times 10^6$ (m^2/s)	k (W/m·K)	$\alpha \times 10^4$ (m^2/s)	Pr
Air	100	3.6010	1.0266	1.923	0.009246	0.0250	0.768
	300	1.1774	1.0057	16.84	0.02624	0.2216	0.708
	2500	0.1394	1.688	543.0	0.175	7.437	0.730
Ammonia, NH$_3$	220	0.3828	2.198	19.0	0.0171	0.2054	0.93
	473	0.4405	2.395	37.4	0.0467	0.4421	0.84
Carbon dioxide	220	2.4733	0.783	4.490	0.01081	0.0592	0.818
	600	0.8938	1.076	30.02	0.04311	0.4483	0.668
Carbon monoxide	220	1.5536	1.0429	8.903	0.01906	0.1176	0.758
	600	0.5685	1.0877	52.06	0.04446	0.7190	0.724
Helium	33	1.4657	5.200	3.42	0.0353	0.04625	0.74
	900	0.05286	5.200	781.3	0.298	10.834	0.72
Hydrogen	30	0.8472	10.840	1.895	0.0228	0.02493	0.759
	300	0.0819	14.314	109.5	0.182	1.554	0.706
	1000	0.0819	14.314	109.5	0.182	1.554	0.706
Nitrogen	100	3.4808	1.0722	1.971	0.009450	0.02531	0.786
	300	1.1421	1.0408	15.63	0.0262	0.204	0.713
	1200	0.2851	1.2037	156.1	0.07184	2.0932	0.748
Oxygen	100	3.9918	0.9479	1.946	0.00903	0.02388	0.815
	300	1.3007	0.9203	15.86	0.02676	0.2235	0.709
	600	0.6504	1.0044	52.15	0.04832	0.7399	0.704
Steam (H$_2$O vapor)	380	0.5863	2.060	21.6	0.0246	0.2036	1.060
	850	0.2579	2.186	115.2	0.0637	1.130	1.019

[a]Adapted from Ref. 2.

Combining this temperature difference with Fourier's law yields an expression for each internal node

$$T_{m,n+1} + T_{m,n+1} + T_{m-1,n} + T_{m-1,n} + \frac{\dot{q}\ \Delta x\ \Delta y}{k}\frac{1}{} - 4T_{m,n} = 0$$

Similar equations for other geometries (i.e., corners) and boundary conditions (i.e., convection) and combinations of the two are listed in Table 10. These equations must then be solved using some form of matrix inversion technique, Gauss–Seidel iteration method or other method for solving large numbers of simultaneous equations.

1.4 Heat Conduction with Convection Heat Transfer on Boundaries

In physical situations where a solid is immersed in a fluid, or a portion of the surface is exposed to a liquid or gas, heat transfer will occur by convection (or when there is a large temperature difference, through some combination of convection and/or radiation). In these situations, the heat transfer is governed by *Newton's law of cooling*, which is expressed as

$$q = hA\ \Delta T$$

where h is the *convection heat transfer coefficient* (Section 2), ΔT is the temperature difference between

the solid surface and the fluid, and A is the surface area in contact with the fluid. The resistance occurring at the surface abounding the solid and fluid is referred to as the *thermal resistance* and is given by $1/hA$, that is, the *convection resistance*. Combining this resistance term with the appropriate conduction resistance yields an *overall heat transfer coefficient U*. Usage of this term allows the overall heat transfer to be defined as $q = UA\ \Delta T$.

Table 8 shows the overall heat transfer coefficients for some simple geometries. Note that U may be based either on the inner surface (U_1) or on the outer surface (U_2) for the cylinders and spheres.

Critical Radius of Insulation for Cylinders A large number of practical applications involve the use of insulation materials to reduce the transfer of heat into or out of cylindrical surfaces. This is particularly true of steam or hot water pipes where concentric cylinders of insulation are typically added to the outside of the pipes to reduce the heat loss. Beyond a certain thickness, however, the continued addition of insulation may not result in continued reductions in the heat loss. To optimize the thickness of insulation required for these types of applications, a value typically referred to as the critical radius, defined as $r_{cr} = k/h$, is used. If the outer radius of the object to be insulated is less than r_{cr} then the addition of insulation will increase the heat loss, while for cases where the outer radii is greater than r_{cr} any additional increases in insulation thickness will result in a decrease in heat loss.

Table 8 One-Dimensional Heat Conduction

Geometry	Heat Transfer Rate and Temperature Distribution	Heat Transfer Rate and Overall Heat Transfer Coefficient with Convection at the Boundaries
Plane wall	$$q = \frac{T_1 - T_2}{(x_2 - x_1)/kA}$$ $$T = T_1 + \frac{T_2 - T_1}{x_x - x_1}(x - x_1)$$ $$R = (x_x - x_1)/kA$$	$$q = UA(T_{\infty,1} - T_{\infty,2})$$ $$U = \left(\frac{1}{h_1} + \frac{x_2 - x_2}{k} + \frac{11}{h_2}\right)^{-1}$$
Hollow cylinder	$$q = \frac{T_1 - T_2}{[\ln(r_2/r_1)]/2\pi kL}$$ $$T = \frac{T_2 - T_1}{\ln(r_2/r_1)}\ln\frac{r}{r_1}$$ $$R = \frac{\ln(r_2/r_1)}{2\pi kL}$$	$$q = 2\pi r_1 L U_1(T_{\infty,1} - T_{\infty,2})$$ $$= 2\pi r_1 L U_2(T_{\infty,1} - T_{\infty,2})$$ $$U_1 = \left(\frac{1}{h_1} + \frac{r_1 \ln(r_2/r_1)}{k} + \frac{r_1}{r_2}\frac{1}{h_2}\right)^{-1}$$ $$U_2 = \left[\left(\frac{r_2}{r_1}\right)\frac{1}{h_1} + \frac{r_2 \ln(r_2/r_1)}{k} + \frac{1}{h_2}\right]^{-1}$$
Hollow sphere	$$q = \frac{T_1 - T_2}{(1/r_1 - 1/r_2)/4\pi k}$$ $$T = \frac{1}{(1 - r_1/r_2)}\left[\frac{r_1}{r}(T_1 - T_2) + \left(T_2 - T_1\frac{r_1}{r_2}\right)\right]$$ $$R = \frac{1/r_1 - 1/r_2}{4\pi k}$$	$$q = 4\pi r_1^2 U_1(T_{\infty,1} - T_{\infty,2})$$ $$= 4\pi r_2^2 U_2(T_{\infty,1} - T_{\infty,2})$$ $$U_1 = \left[\frac{1}{h_1} + \frac{r_1^2(1/r_1 - 1/r_2)}{k} + \left(\frac{r_1}{r_2}\right)^2\frac{1}{h_2}\right]^{-1}$$ $$U_2 = \left[\left(\frac{r_1}{r_2}\right)^2\frac{1}{h_1} + \frac{r_2^2(1/r_1 - 1/r_2)}{k} + \frac{1}{h_2}\right]^{-1}$$

Table 9 Conduction Shape Factors

System	Schematic	Restrictions	Shape Factor
Isothermal sphere buried in a semi-infinite medium having isothermal surface		$z > D/2$	$\dfrac{2\pi D}{1 - D/4z}$
Horizontal isothermal cylinder of length L buried in a semi-infinite medium having isothermal surface		$\left.\begin{array}{l} L \gg D \\ L \gg D \\ z > 3D/2 \end{array}\right\}$	$\dfrac{2\pi L}{\cosh^{-1}(2z/D)}$ $\dfrac{2\pi L}{\ln(4z/D)}$
The cylinder of length L with eccentric bore		$L \gg D_1, D_2$	$\dfrac{2\pi L}{\cosh^{-1}\left(\dfrac{D_1^2 + D_2^2 - 4\varepsilon^2}{2D_1 D_2}\right)}$
Conduction between two cylinders of length L in infinite medium		$L \gg D_1, D_2$	$\dfrac{2\pi L}{\cosh^{-1}\left(\dfrac{4W^2 - D_1^2 - D_2^2}{2D_1 D_2}\right)}$
Circular cylinder of length L in a square solid		$\begin{array}{l} L \gg W \\ w > D \end{array}$	$\dfrac{2\pi L}{\ln(1.08\, w/D)}$
Conduction through the edge of adjoining walls		$D > L/5$	$0.54\, D$
Conduction through corner of three walls with inside and outside temperature, respectively, at T_1 and T_2		$L \ll$ length and width of wall	$0.15\, L$

Extended Surfaces In examining Newton's law of cooling, it is clear that the rate of heat transfer between a solid and the surrounding ambient fluid may be increased by increasing the surface area of the solid that is exposed to the fluid. This is typically done through the addition of extended surfaces or fins to the primary surface. Numerous examples often exist, including the cooling fins on air-cooled engines, that is, motorcycles or lawn mowers or the fins attached to automobile radiators.

Figure 2 illustrates a common uniform cross-section extended surface, fin, with a constant base temperature, T_b, a constant cross-sectional area, A, a circumference of $C = 2W + 2t$, and a length, L, which is much larger

Table 10   **Summary of Nodal Finite-Difference Equations**

Configuration	Finite-Difference Equation for $\Delta x = \Delta y$

Case 1. Interior node

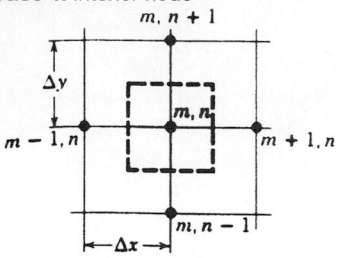

$$T_{m,n+1} + T_{m,n-1} + T_{m-1,n} - 4T_{m,n} = 0$$

Case 2. Node at an internal corner with convection

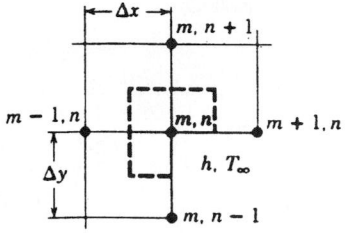

$$2(T_{m-1,n} + T_{m,n+1}) + (T_{m+1,n} + T_{m,n-1})$$

$$+2\frac{h\,\Delta x}{k}T_\infty - 2\left(3 + \frac{h\,\Delta x}{k}\right)T_{m,n} = 0$$

Case 3. Node at a plane surface with convection

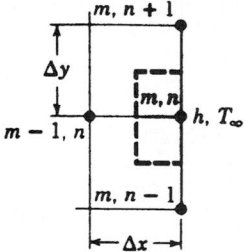

$$2(T_{m-1,n} + T_{m,n+1} + T_{m,n-1}) + \frac{2h\,\Delta x}{k}T_\infty$$

$$-2\left(\frac{h\,\Delta x}{k} + 2\right)T_{m,n} = 0$$

Case 4. Node at an external corner with convection

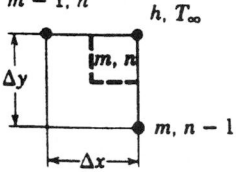

$$(T_{m,n-1} + T_{m-1,n}) + 2\frac{h\,\Delta x}{k}T_\infty$$

$$-2\left(\frac{h\,\Delta x}{k} + 1\right)T_{m,n} = 0$$

Case 5. Node near a curved surface maintained at a nonuniform temperature

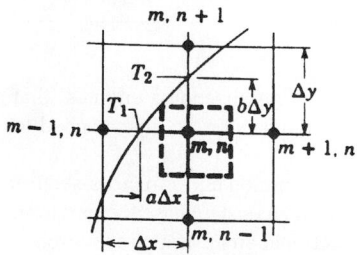

$$\frac{2}{a+1}T_{m+1,n} + \frac{2}{b+1}T_{m,n-1}$$

$$+\frac{2}{a(a+1)}T_1 + \frac{2}{b(b+1)}T_2$$

$$-\left(\frac{2}{a} + \frac{2}{b}\right)T_{m,n} = 0$$

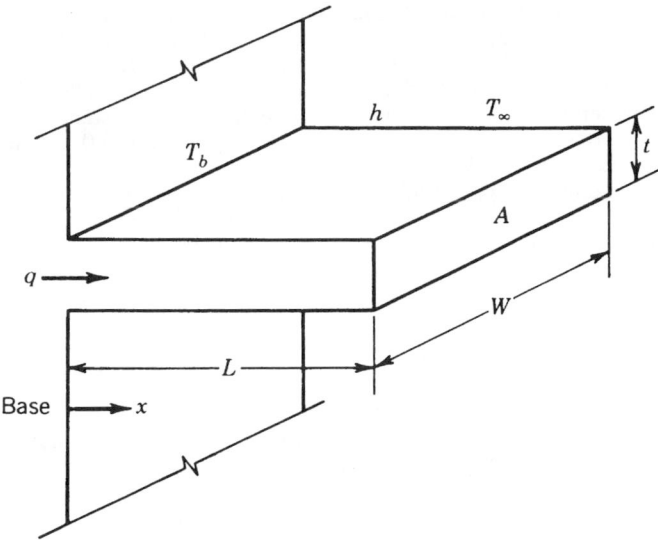

Fig. 2 Heat transfer by extended surfaces.

than the thickness, t. For these conditions, the temperature distribution in the fin must satisfy the following expression:

$$\frac{d^2T}{dx^2} - \frac{hC}{kA}(T - T_\infty) = 0$$

The solution of this equation depends on the boundary conditions existing at the tip, that is, at $x = L$. Table 11 shows the temperature distribution and heat transfer rate for fins of uniform cross section subjected to a number of different tip conditions, assuming a constant value for the heat transfer coefficient, h.

Two terms are used to evaluate fins and their usefulness. The first of these is the *fin effectiveness*, defined as the ratio of the heat transfer rate with the fin to the heat transfer rate that would exist if the fin were not used. For most practical applications, the use of a fin is justified only when the fin effectiveness is significantly greater than 2. A second term used to evaluate the usefulness of a fin is the *fin efficiency*, η_f, This term represents the ratio of actual heat transfer rate from a fin to the heat transfer rate that would occur if the entire fin surface could be maintained at a uniform temperature equal to the temperature of the base of the fin. For this case, Newton's law of cooling can be written as

$$q = \eta_f h A_f (T_b - T_\infty)$$

Table 11 Temperature Distribution and Heat Transfer Rate at the Fin Base ($m = \sqrt{hc/kA}$)

Condition at $x = L$	$\dfrac{T - T_\infty}{T_b - T_\infty}$	Heat Transfer Rate $q/mkA\,(T_b - T_\infty)$
$h(T_{x=L} - T_\infty) = -k\left(\dfrac{dT}{dx}\right)_{x=L}$ (convection)	$\dfrac{\cosh m(L - x) + [h/(mk)]\sinh m(L - x)}{\cosh mL + [h/(mk)]\sinh mL}$	$\dfrac{\sinh mL + [h/(mk)]\cosh mL}{\cosh mL + [h/(mk)]\sinh mL}$
$\left(\dfrac{dT}{dx}\right)_{x=L} = 0$ (insulated)	$\dfrac{\cosh m(L - x)}{\cosh mL}$	$\tanh mL$
$T_{x=L} = T_L$ (prescribed temperature)	$\dfrac{(T_L - T_\infty)/(T_b - T_\infty)\sinh mx + \sinh m(L - x)}{\sinh ml}$	$\dfrac{\cosh mL - (T_L - T_\infty)/(T_b - T_\infty)}{\sinh ml}$
$T_{x=L} = T_\infty$ (infinitely long fin, $L \to \infty$)	e^{-mx}	1

where A_f is the total surface area of the fin and T_b is the temperature of the fin at the base. The application of fins for heat removal can be applied to either forced or natural convection of gases, and while some advantages can be gained in terms of increasing the liquid–solid or solid–vapor surface area, fins as such are not normally utilized for situations involving phase change heat transfer, such as boiling or condensation.

1.5 Transient Heat Conduction

Given a solid body, at a uniform temperature, $T_{\infty i}$, immersed in a fluid of different temperature T_∞, the surface of the solid body will be subject to heat losses (or gains) through convection from the surface to the fluid. In this situation, the heat lost (or gained) at the surface results from the conduction of heat from inside the body. To determine the significance of these two heat transfer modes, a dimensionless parameter referred to as the *Biot number* is used. This dimensionless number is defined as $\mathrm{Bi} = hL/k$, where $L = V/A$ or the ratio of the volume of the solid to the surface area of the solid, and really represents a comparative relationship of the importance of convections from the outer surface to the conduction occurring inside. When this value is less than 0.1, the temperature of the solid may be assumed uniform and dependent on time alone. When this value is greater than 0.1, there is some spatial temperature variation that will affect the solution procedure.

For the first case, $\mathrm{Bi} < 0.1$, an approximation referred to as the *lumped heat capacity* method may be used. In this method, the temperature of the solid is given by

$$\frac{T - T_\infty}{T_i - T_\infty} = \exp\left(\frac{-t}{\tau_t}\right) = \exp(-\mathrm{Bi}\ \mathrm{Fo})$$

where τ_t is the *time constant* and is equal to $\rho C_p V / hA$. Increasing the value of the time constant, τ_t, will result in a decrease in the thermal response of the solid to the environment and hence, will increase the time required for it to reach thermal equilibrium (i.e., $T = T_\infty$). In this expression, Fo represents the dimensionless time and is called the *Fourier number*, the value of which is equal to $\alpha t A^2 / V^2$. The Fourier number, along with the Biot number, can be used to characterize transient heat conduction problems. The total heat flow through the surface of the solid over the time interval from $t = 0$ to time t can be expressed as

$$Q = \rho V C_p (T_i - T_\infty)\left[1 - \exp\left(\frac{-t}{\tau_t}\right)\right]$$

Transient Heat Transfer for Infinite Plate, Infinite Cylinder, and Sphere Subjected to Surface Convection   Generalized analytical solutions to transient heat transfer problems involving infinite plates, cylinders, and finite diameter spheres subjected to surface convection have been developed. These solutions can be presented in graphical form through the use of the *Heisler charts*,[3] illustrated in Figs. 3–11 for plane walls, cylinders, and spheres. In this procedure, the solid is assumed to be at a uniform temperature, T_i, at time $t = 0$ and then is suddenly subjected to or immersed in a fluid at a uniform temperature T_∞. The convection heat transfer coefficient, h, is assumed to be constant, as is the temperature of the fluid. Combining Figs. 3 and 4 for plane walls, Figs. 6 and 7 for cylinders, and Figs. 9 and 10 for spheres allows the resulting time-dependent temperature of any point within the solid to be found. The total amount of energy, Q, transferred to or from the solid surface from time $t = 0$ to time t can be found from Figs. 5, 8, and 11.

1.6 Conduction at Microscale

The mean free path of electrons and the size of the volume involved has long been recognized as having a pronounced effect on electron transport phenomena. This is particularly true in applications involving thin metallic films or wires where the characteristic length may be close to the same order of magnitude as the scattering mean free path of the electrons.[4a] The first notable work in this area was performed by Tien et al.,[4b] where the thermal conductivity of thin metallic films and wires were calculated at cryogenic temperatures. Because the length of the mean free path in these types of applications is shortened near the surface, due to termination at the boundary, a reduction in transport coefficients, such as electrical and thermal conductivities, was observed. Tests at cryogenic temperatures were first performed because the electron mean free path increases as temperature decreases, and the size effects were expected to become especially significant in this range. The primary purpose of this investigation was to outline in a systematic manner a method by which the thermal conductivity of such films and wires at cryogenic temperatures could be determined. The results indicated that, particularly in the case of thin metallic films, size effects may become an increasingly important part of the design and analysis required for application. Due to the increased use of thin films in optical components and solid-state devices and systems, there has been an increasing interest in the effect of decreasing size on the transport properties of thin solid films and wires.

The most common method for calculating the thermal conductivities in thin films and wires consists of three essential steps:

1. Identifying the appropriate expression for the electrical conductivity size effect
2. Determining the mean free path for electrical conductivity, which is essential in calculations of all electron transport properties
3. Applying the electrical–thermal transport analogy for calculating the thermal conductivity size effect[4a]

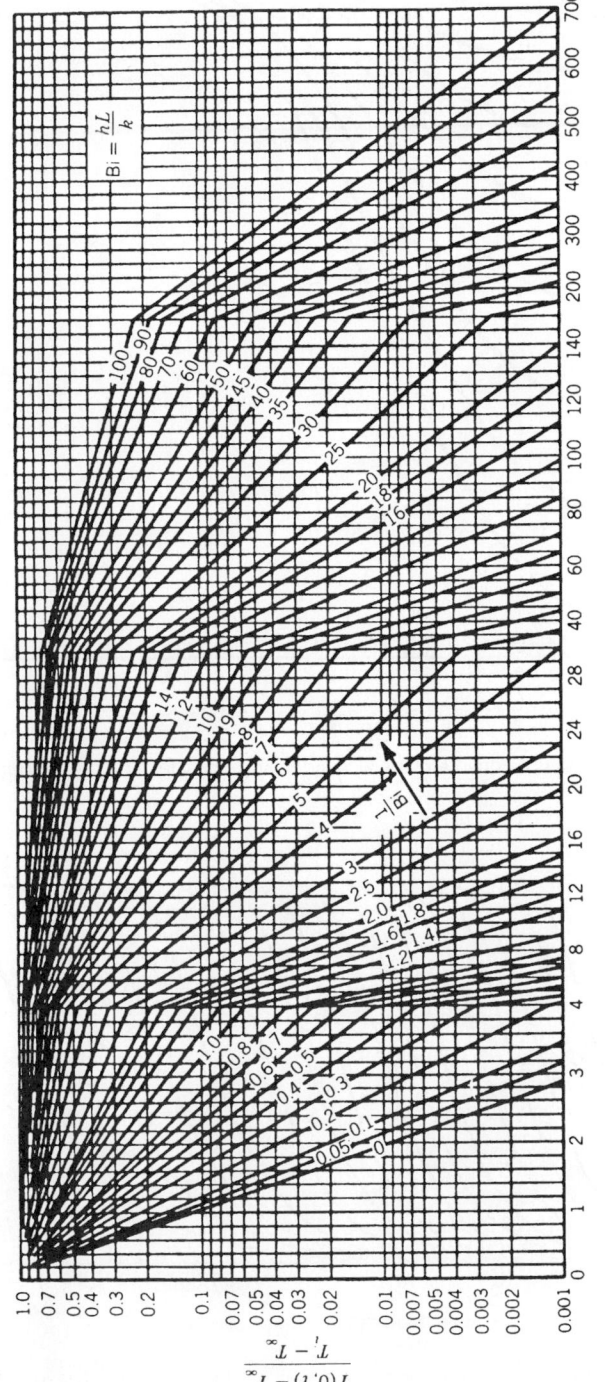

Fig. 3 Midplane temperature as a function of time for a plane wall of thickness 2L. (Adapted from Ref. 3.)

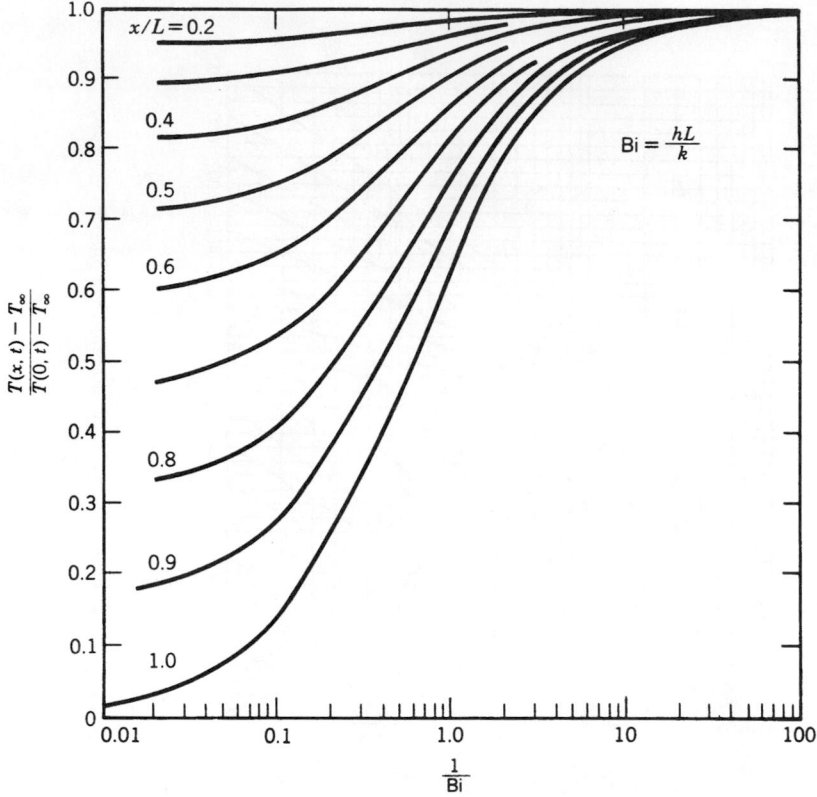

Fig. 4 Temperature distribution in a plane wall of thickness $2L$. (Adapted from Ref. 3.)

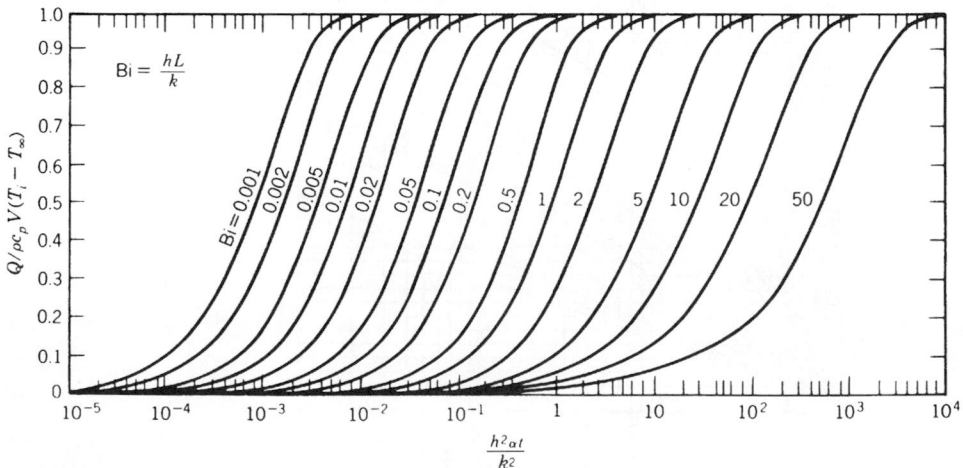

Fig. 5 Internal energy change as a function of time for a plane wall of thickness $2L$.[4] (Used with the permission of McGraw-Hill Book Company.)

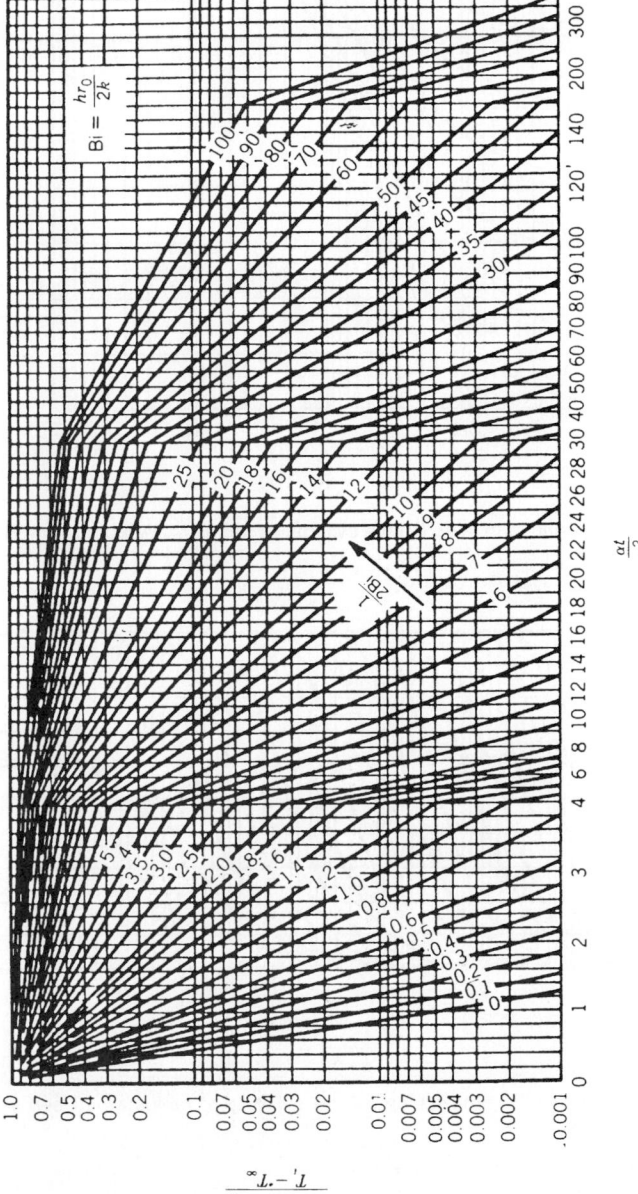

Fig. 6 Centerline temperature as function of time for an infinite cylinder of radius r_o. (Adapted from Ref. 3.)

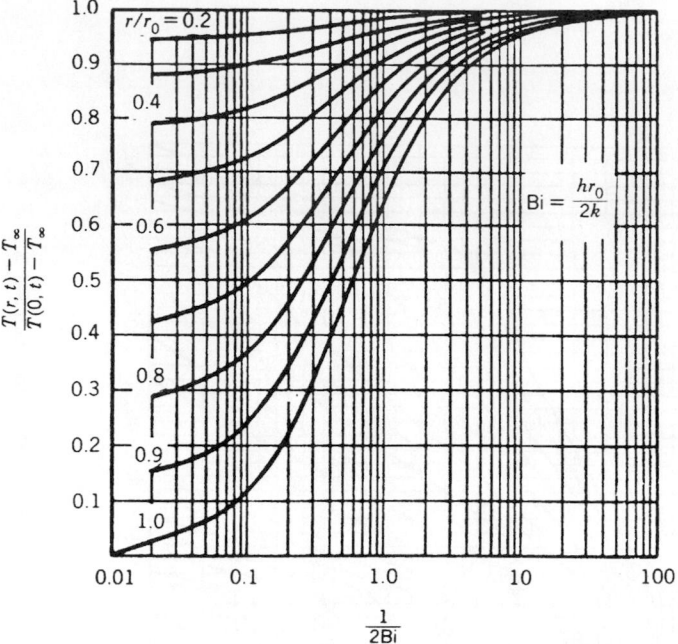

Fig. 7 Temperature distribution in an infinite cylinder of radius r_o. (Adapted from Ref. 3.)

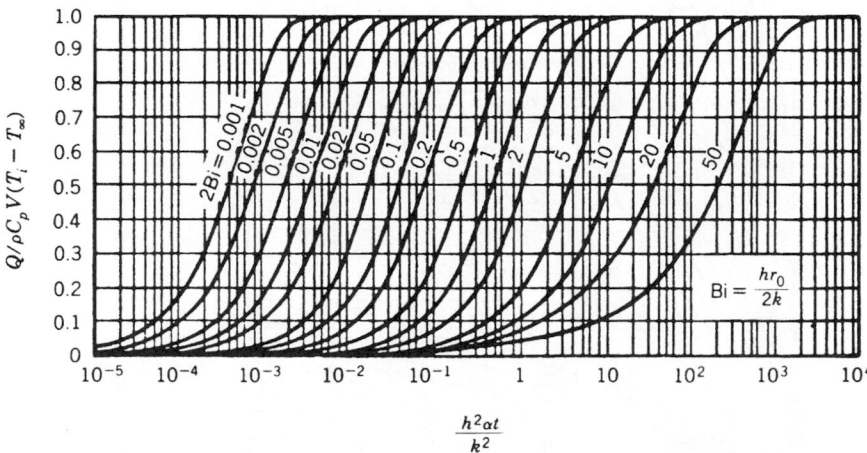

Fig. 8 Internal energy change as function of time for an infinite cylinder of radius r_o.[4] (Used with the permission of McGraw-Hill Book Company.)

For domain thicknesses on the order of the carrier mean free path, jump boundary conditions significantly affect the solution of the conduction problem. This problem can be resolved through the solution of the hyperbolic heat equation-based analysis, which is generally justifiable engineering applications.[4c]

2 CONVECTION HEAT TRANSFER

As discussed earlier, convection heat transfer is the mode of energy transport in which the energy is transferred by means of fluid motion. This transfer can be the result of the random molecular motion or bulk motion of the fluid. If the fluid motion is

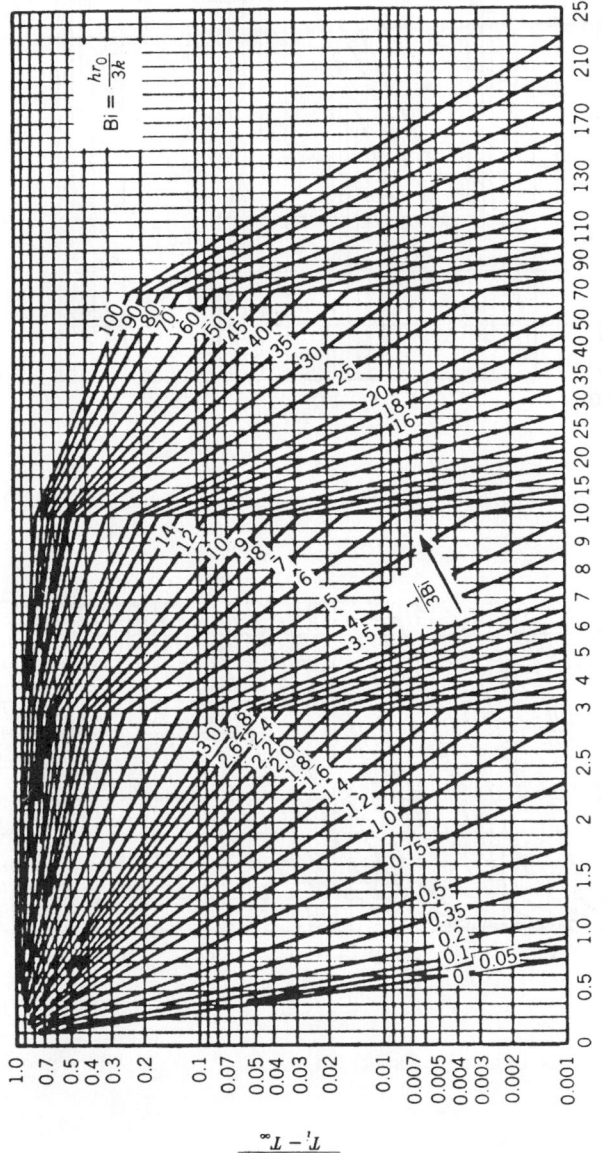

Fig. 9 Center temperature as function of time in a sphere of radius r_o. (Adapted from Ref. 3.)

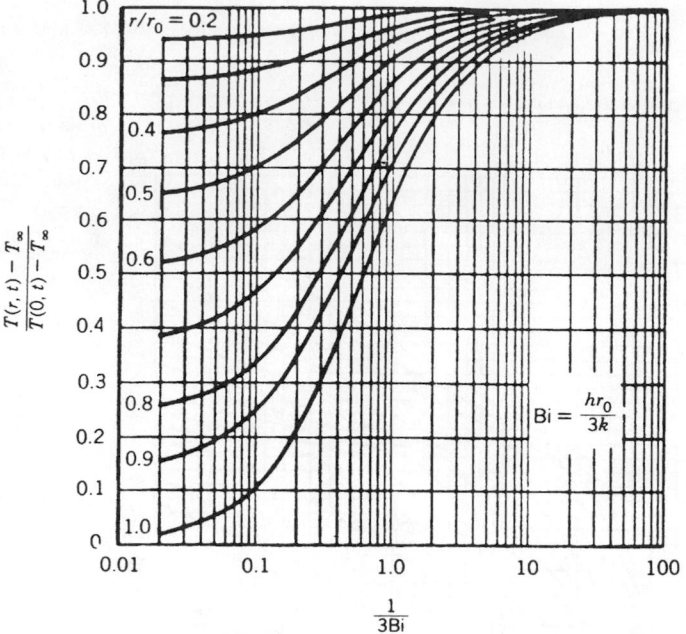

Fig. 10   Temperature distribution in sphere of radius r_o. (Adapted from Ref. 3.)

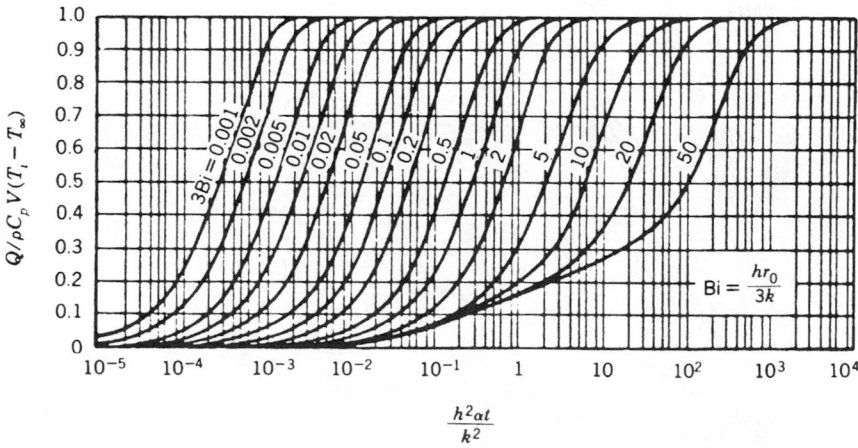

Fig. 11   Internal energy change as function of time for a sphere of radius r_o.[4] (Used with the permission of McGraw-Hill Book Company.)

caused by external forces, the energy transfer is called *forced convection*. If the fluid motion arises from a buoyancy effect caused by density differences, the energy transfer is called *free convection* or *natural convection*. For either case, the heat transfer rate, q, can be expressed in terms of the surface area, A, and the temperature difference, ΔT, by Newton's law of cooling:

$$q = hA\,\Delta T$$

In this expression, h is referred to as the convection heat transfer coefficient or film coefficient and a function of the velocity and physical properties of the fluid, and the shape and nature of the surface. The nondimensional heat transfer coefficient $\text{Nu} = hL/k$ is called the *Nusselt number*, where L is a characteristic length and k is the thermal conductivity of the fluid.

2.1 Forced Convection — Internal Flow

For internal flow in a tube or pipe, the convection heat transfer coefficient is typically defined as a function of the temperature difference existing between the temperature at the surface of the tube and the *bulk* or *mixing-cup temperature*, T_b, that is, $\Delta T = T_s - T_b$ can be defined as

$$T_b = \frac{\int C_p T \ d\dot{m}}{\int C_p \ d\dot{m}}$$

where $\dot{m}$ is the axial flow rate. Using this value, heat transfer between the tube and the fluid can be written as $q = hA(T_s - T_b)$.

In the entrance region of a tube or pipe, the flow is quite different from that occurring downstream from the entrance. The rate of heat transfer differs significantly, depending on whether the flow is *laminar* or *turbulent*. From fluid mechanics, the flow is considered to be turbulent when $\text{Re}_D = V_m D/v > 2300$ for a smooth tube. This transition from laminar to turbulent, however, also depends on the roughness of tube wall and other factors. The generally accepted range for transition is $200 < \text{Re}_D < 4000$.

Laminar Fully Developed Flow For situations where both the thermal and velocity profiles are fully developed, the Nusselt number is constant and depends only on the thermal boundary conditions. For *circular tubes* with $\text{Pr} \geq 0.6$, and $x/D \ \text{Re}_D \ \text{Pr} > 0.05$, the Nusselt numbers have been shown to be $\text{Nu}_D = 3.66$ and 4.36, for constant temperature and constant heat flux conditions, respectively. Here, the fluid properties are based on the mean bulk temperature.

For *noncircular tubes*, the hydraulic diameter, $D_h = 4 \times$ the flow cross-sectional area/wetted perimeter, is used to define the Nusselt number Nu_D and the Reynolds number Re_D. Table 12 shows the Nusselt numbers based on hydraulic diameter for various cross-sectional shapes.

Laminar Flow for Short Tubes At the entrance of a tube, the Nusselt number is infinite, and decreases asymptotically to the value for fully developed flow as the flow progresses down the tube. The Sieder–Tate equation[5] gives good correlation for the combined entry length, that is, that region where the thermal and velocity profiles are both developing or for short tubes:

$$\frac{\overline{\text{Nu}}_D = \bar{h}D}{k} = 1.86(\text{Re} \ D \ \text{Pr})^{1/13} \left(\frac{D}{L}\right)^{1/3} \left(\frac{\mu}{\mu_s}\right)^{0.14}$$

for $T_s = \text{constant}$, $0.48 < \text{Pr} < 16{,}700$, $0.0044 < \mu/\mu_s < 9.75$, and $(\text{Re}_D \ \text{Pr} \ D/L)^{1/3} \ (\mu/\mu_s)^{0.14} > 2$.

Table 12 Nusselt Numbers for Fully Developed Laminar Flow for Tubes of Various Cross Sections[a]

Geometry ($L/DH > 100$)	Nu_{H1}	Nu_{H2}	Nu_r
$\frac{2b}{2a} = 1$	3.608	3.091	2.976
$\frac{2b}{2a} = \frac{1}{2}$	4.123	3.017	3.391
$\frac{2b}{2a} = \frac{1}{4}$	5.099	4.35	3.66
$\frac{2b}{2a} = \frac{1}{8}$	6.490	2.904	5.597
$\frac{2b}{2a} = 0$	8.235	8.235	7.541
$\frac{b}{a} = 0$	5.385	—	4.861
(circle)	4.364	4.364	3.657

[a]Nu_{H1} = average Nusselt number for uniform heat flux in flow direction and uniform wall temperature at particular flow cross section.
Nu_{H2} = average Nusselt number for uniform heat flux both in flow direction and around periphery.
Nu_{Hrr} = average Nusselt number for uniform wall temperature.

In this expression, all of the fluid properties are evaluated at the mean bulk temperature except for μ_s, which is evaluated at the wall surface temperature. The average convection heat transfer coefficient $\overline{h}$ is based on the arithmetic average of the inlet and outlet temperature differences.

Turbulent Flow in Circular Tubes   In turbulent flow, the velocity and thermal entry lengths are much shorter than for a laminar flow. As a result, with the exception of short tubes, the fully developed flow values of the Nusselt number are frequently used directly in the calculation of the heat transfer. In general, the Nusselt number obtained for the constant heat flux case is greater than the Nusselt number obtained for the constant temperature case. The one exception to this is the case of liquid metals, where the difference is smaller than for laminar flow and becomes negligible for $Pr > 1.0$. The Dittus–Boelter equation[6] is typically used if the difference between the pipe surface temperature and the bulk fluid temperature is less than $6°C$ ($10°F$) for liquids or $56°C$ ($100°F$) for gases:

$$Nu_D = 0.023 \ Re_D^{0.8} \ Pr^n$$

for $0.7 \leq Pr \leq 160$, $Re_D \geq 10{,}000$, and $L/D \geq 60$, where

$$n = 0.4 \text{ for heating, } T_s > T_b$$
$$= 0.3 \text{ for cooling, } T_s < T_b$$

For temperature differences greater than specified above, use[5]

$$Nu_D = 0.027 \ Re_D^{0.8} \ Pr^{1/3} \left(\frac{\mu}{\mu_s}\right)^{0.14}$$

for $0.7 \leq Pr \leq 16{,}700$, $Re_D \geq 10{,}000$, and $L/D \geq 60$. In this expression, the properties are all evaluated at the mean bulk fluid temperature with the exception of μ_s, which is again evaluated at the tube surface temperature.

For *concentric tube annuli*, the hydraulic diameter $D_h = D_o - D_i$ (outer diameter − inner diameter) must be used for Nu_D and Re_D, and the coefficient h at either surface of the annulus must be evaluated from the Dittus–Boelter equation. Here, it should be noted that the foregoing equations apply for smooth surfaces and that the heat transfer rate will be larger for rough surfaces and are not applicable to liquid metals.

Fully Developed Turbulent Flow of Liquid Metals in Circular Tubes   Because the Prandtl number for liquid metals is on the order of 0.01, the Nusselt number is primarily dependent on a dimensionless parameter number referred to as the *Peclet number*, which in general is defined as $Pe = RePr$:

$$Nu_D = 5.0 + 0.025 \ Pe_D^{0.8}$$

which is valid for situations where $T_s =$ a constant and $Pe_D > 100$ and $L/D > 60$.

For $q'' =$ constant and $3.6 \times 10^3 < Re_D < 9.05 \times 10^5$, $10^2 < Pe_D < 10^4$, and $L/D > 60$, the Nusselt number can be expressed as

$$Nu_D = 4.8 + 0.0185 \ Pe_D^{0.827}$$

2.2   Forced Convection — External Flow

In forced convection heat transfer, the heat transfer coefficient, h, is based on the temperature difference between the wall surface temperature and the fluid temperature in the free stream outside the thermal boundary layer. The total heat transfer rate from the wall to the fluid is given by $q = hA(T_s - T_\infty)$. The Reynolds numbers are based on the free-stream velocity. The fluid properties are evaluated either at the free-stream temperature T_∞ or at the film temperature $T_f = (T_s + T_\infty)/2$.

Laminar Flow on a Flat Plate   When the flow velocity along a constant temperature semi-infinite plate is uniform, the boundary layer originates from the leading edge and is laminar and the flow remains laminar until the local Reynolds number $Re_x = U_\infty x/v$ reaches the *critical Reynolds number*, Re_c. When the surface is smooth, the Reynolds number is generally assumed to be $Re_c = 5 \times 10^5$, but the value will depend on several parameters, including the surface roughness.

For a given distance x from the leading edge, the *local Nusselt number* and the *average Nusselt number* between $x = 0$ and $x = L$ are given below (Re_x and $Re_L \leq 5 \times 10^5$):

For $Pr \geq 0.6$:

$$Nu_x = \frac{hx}{k} = 0.332 \ Re_x^{0.5} \ Pr^{1/3}$$
$$\overline{Nu_L} = \frac{hL}{k} = 0.664 \ Re_L^{0.5} \ Pr^{1/3}$$

For $Pr \leq 0.6$:

$$Nu_x = 0.565(Re_x \ Pr)^{0.5} \qquad \overline{Nu_L} = 1.13(Re_L \ Pr)^{0.5}$$

Here, all of the fluid properties are evaluated at the mean or average film temperature.

Turbulent Flow on Flat Plate When the flow over a flat plate is turbulent from the leading edge, expressions for the *local Nusselt number* can be written as

$$\mathrm{Nu}_x = 0.0292\ \mathrm{Re}_x^{0.8}\ \mathrm{Pr}^{1/3} \quad \overline{\mathrm{Nu}}_L = 0.036\ \mathrm{Re}_L^{0.8}\ \mathrm{Pr}^{1/3}$$

where the fluid properties are all based on the mean film temperature and $5 \times 10^5 \leq \mathrm{Re}_x$ and $\mathrm{Re}_L \leq 10^8$ and $0.6 \leq \mathrm{Pr} \leq 60$.

Average Nusselt Number between $x = 0$ and $x = L$ with Transition For situations where transition occurs immediately once the critical Reynolds number Re_c has been reached[7]

$$\overline{\mathrm{Nu}}_L = 0.036\ \mathrm{Pr}^{1/3}[\mathrm{Re}_L^{0.8} - \mathrm{Re}_c^{0.8} + 18.44\ \mathrm{Re}_c^{0.5}]$$

provided that $5 \times 10^5 \leq \mathrm{Re}_L \leq 10^8$ and $0.6 \leq \mathrm{Pr} \leq 60$. Specialized cases exist for this situation, that is,

$$\overline{\mathrm{Nu}}_L = 0.036\ \mathrm{Pr}^{1/3}(\mathrm{Re}_L^{0.8} - 18,700)$$

for $\mathrm{Re}_c = 4 \times 10^5$, or

$$\overline{\mathrm{Nu}}_L = 0.036\ \mathrm{Pr}^{1/3}(\mathrm{Re}_L^{0.8} - 23,000)$$

for $\mathrm{Re}_c = 5 \times 10^5$. Again, all fluid properties are evaluated at the mean film temperature.

Circular Cylinders in Cross Flow For circular cylinders in cross flow, the Nusselt number is based upon the diameter and can be expressed as

$$\overline{\mathrm{Nu}}_D = (0.4\ \mathrm{Re}_D^{0.5} + 0.06\ \mathrm{Re}^{2/3})\mathrm{Pr}^{0.4}\left(\frac{\mu_\infty}{\mu_s}\right)^{0.25}$$

for $0.67 < \mathrm{Pr} < 300$, $10 < \mathrm{Re}_D < 10^5$, and $0.25 < 5.2$. Here, the fluid properties are evaluated at the free stream temperature except μ_s, which is evaluated at the surface temperature.[8]

Cylinders of Noncircular Cross Section in Cross Flow of Gases For noncircular cylinders in cross flow, the Nusselt number is again based on the diameter, but is expressed as

$$\overline{\mathrm{Nu}}_D = C(\mathrm{Re}_D)^m\ \mathrm{Pr}^{1/3}$$

where C and m are listed in Table 13, and the fluid properties are evaluated at the mean film temperature.[9]

Flow Past a Sphere For flow over a sphere, the Nusselt number is based on the sphere diameter and can be expressed as

$$\overline{\mathrm{Nu}}_D = 2 + (0.4\ \mathrm{Re}_D^{0.5} + 0.06\ \mathrm{Re}_D^{2/3})\mathrm{Pr}^{0.4}\left(\frac{\mu_\infty}{\mu_s}\right)^{0.25}$$

for the case of $3.5 < \mathrm{Re}_D < 8 \times 10^4$, $0.7 < \mathrm{Pr} < 380$, and $1.0 < \mu_\infty/\mu_s < 3.2$. The fluid properties are calculated at the free-stream temperature except μ_s, which is evaluated at the surface temperature.[8]

Table 13 Constants and *m* for Noncircular Cylinders in Cross Flow

Geometry	Re_D	C	m
Square $V \rightarrow \diamond \updownarrow D$	5×10^3–10^5	0.246	0.588
	5×10^3–10^5	0.102	0.675
$V \rightarrow \square \dfrac{D}{\uparrow}$			
Hexagon $V \rightarrow \updownarrow D$	5×10^3–1.95×10^4	0.160	0.538
	1.95×10^4–10^5	0.0385	0.782
$V \rightarrow \dfrac{D}{\uparrow}$			
Vertical plate $V \rightarrow \square \updownarrow D$	5×10^3–10^5	0.153	0.638
	4×10^3–1.5×10^4	0.228	0.731

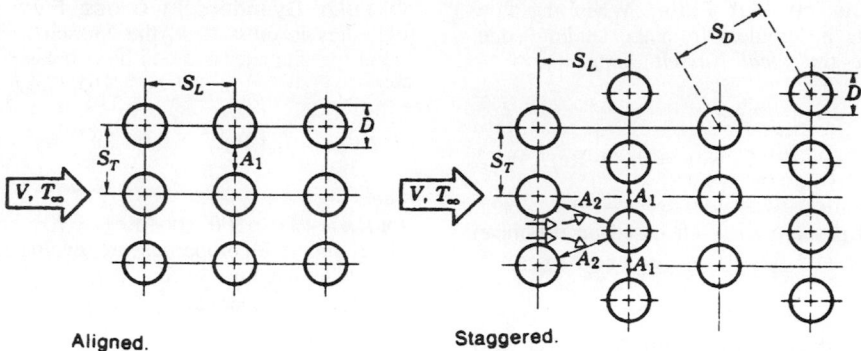

Fig. 12 Tube arrangement.

Flow across Banks of Tubes For banks of tubes, the tube arrangement may be either *staggered* or *aligned* (Fig. 12), and the heat transfer coefficient for the first row is approximately equal to that for a single tube.

In turbulent flow, the heat transfer coefficient for tubes in the first row is smaller than that of the subsequent rows. However, beyond the fourth or fifth row, the heat transfer coefficient becomes approximately constant. For tube banks with more than 20 rows, $0.7 < \text{Pr} < 500$, and $1000 < \text{Re}_{D,\max} < 2 \times 10^6$, the average Nusselt number for the entire tube bundle can be expressed as[10]

$$\overline{\text{Nu}}_D = C(\text{Re}_{D,\max})^m \, \text{Pr}^{0.36} \left(\frac{\text{Pr}_\infty}{\text{Pr}_s} \right)^{0.25}$$

where all fluid properties are evaluated at T_∞ except Pr_s, which is evaluated at the surface temperature. The constants C and m used in this expression are listed in Table 14, and the Reynolds number is based on the maximum fluid velocity occurring at the minimum free flow area available for the fluid. Using the nomenclature shown in Fig. 12, the maximum fluid velocity can be determined by

$$V_{\max} = \frac{S_T}{S_T - D} V$$

for the aligned or staggered configuration provided

$$\sqrt{S_L^2 + (S_T/2)^2} > \frac{S_T + D}{2}$$

or as

$$V_{\max} = \frac{S_T}{\sqrt[2]{S_L^2 + (S_T/2)^2}} V$$

for staggered if

$$\sqrt{S_L^2 + (S_T/2)^2} < \frac{S_T + D}{2}$$

Liquid Metals in Cross Flow over Banks of Tubes
The average Nusselt number for tubes in the inner rows can be expressed as

$$\overline{\text{Nu}}_D = 4.03 + 0.228(\text{Re}_{D,\max} \, \text{Pr})^{0.67}$$

which is valid for $2 \times 10^4 < \text{Re}_{D,\max} < 8 \times 10^4$ and $\text{Pr} < 0.03$ and the fluid properties are evaluated at the mean film temperature.[11]

Table 14   Constants C and m of Heat Transfer Coefficient for the Banks in Cross Flow

Configuration	$\text{Re}_{D,\max}$	C	m
Aligned	$10^3 – 2 \times 10^5$	0.27	0.63
Staggered ($S_T/S_L < 2$)	$10^3 – 2 \times 10^5$	$0.35(S_T/S_L)^{1/5}$	0.60
Staggered ($S_G/S_L > 2$)	$10^3 – 2 \times 10^5$	0.40	0.60
Aligned	$2 \times 10^5 – 2 \times 10^6$	0.21	0.84
Staggered	$2 \times 10^5 – 2 \times 10^6$	0.022	0.84

High-Speed Flow over a Flat Plate When the free stream velocity is very high, the effects of viscous dissipation and fluid compressibility must be considered in the determination of the convection heat transfer. For these types of situations, the convection heat transfer can be described as $q = hA(T_s - T_{as})$, where T_{as} is the *adiabatic surface temperature* or *recovery temperature*, and is related to the *recovery factor* by $r = (T_{as} - T_\infty)/(T_0 - T_\infty)$. The value of the *stagnation temperature*, T_0, is related to the free-stream static temperature, T_∞, by the expression

$$\frac{T_0}{T_\infty} = 1 + \frac{\gamma - 1}{2} M_\infty^2$$

where γ is the specific heat ratio of the fluid and M_∞ is the ratio of the free-stream velocity and the acoustic velocity. For the case where $0.6 < \mathrm{Pr} < 15$,

$$r = \begin{cases} \mathrm{Pr}^{1/2} & \text{for laminar flow } (\mathrm{Re}_x < 5 \times 10^5) \\ \mathrm{Pr}^{1/3} & \text{for turbulent flow } (\mathrm{Re}_x > 5 \times 10^5) \end{cases}$$

Here, all of the fluid properties are evaluated at the reference temperature $T_{\mathrm{ref}} = T_\infty + 0.5(T_s - T_\infty) + 0.22(T_{as} - T_\infty)$. Expressions for the local heat transfer coefficients at a given distance x from the leading edge are given as[2]

$$\mathrm{Nu}_x = \begin{cases} 0.332\ \mathrm{Re}_x^{0.5}\ \mathrm{Pr}^{1/3} & \text{for } \mathrm{Re}_x < 5 \times 10^5 \\ 0.0292\ \mathrm{Re}_x^{0.8}\ \mathrm{Pr}^{1/3} & \text{for } 5 \times 10^5 < \mathrm{Re}_x < 10^7 \\ 0.185\ \mathrm{Re}_x(\log \mathrm{Re}_x)^{-2.584} & \text{for } 10^7 < \mathrm{Re}_x < 10^9 \end{cases}$$

In the case of gaseous fluids flowing at very high free-stream velocities, dissociation of the gas may occur, and will cause large variations in the properties within the boundary layer. For these cases, the heat transfer coefficient must be defined in terms of the enthalpy difference, that is, $q = hA(i_s - i_{as})$, and the recovery factor will be given by $r = (i_s - i_{as})/(i_0 - i_\infty)$, where i_{as} represents the enthalpy at the adiabatic wall conditions. Similar expressions to those shown above for Nu_x can be used by substituting the properties evaluated at a reference enthalpy defined as $i_{\mathrm{ref}} = i_\infty + 0.5(i_s - i_\infty) + 0.22(i_{as} - i_\infty)$.

High-Speed Gas Flow Past Cones For the case of high-speed gaseous flows over conical-shaped objects the following expressions can be used:

$$\mathrm{Nu}_x = \begin{cases} 0.575\ \mathrm{Re}_x^{0.5}\ \mathrm{Pr}^{1/3} & \text{for } \mathrm{Re}_x < 10^5 \\ 0.0292\ \mathrm{Re}_x^{0.8}\ \mathrm{Pr}^{1/3} & \text{for } \mathrm{Re}_x > 10^5 \end{cases}$$

where the fluid properties are evaluated at T_{ref} as in the plate.[12]

Stagnation Point Heating for Gases When the conditions are such that the flow can be assumed to behave as *incompressible*, the Reynolds number is based on the free-stream velocity and $\bar{h}$ is defined as $q = \bar{h}A(T_s - T_\infty)$.[13] Estimations of the Nusselt can be made using the following relationship:

$$\mathrm{Nu}_D = C\ \mathrm{Re}_D^{0.5}\ \mathrm{Pr}^{0.4}$$

where $C = 1.14$ for cylinders and 1.32 for spheres, and the fluid properties are evaluated at the mean film temperature. When the flow becomes *supersonic*, a bow shock wave will occur just off the front of the body. In this situation, the fluid properties must be evaluated at the stagnation state occurring behind the bow shock and the Nusselt number can be written as

$$\overline{\mathrm{Nu}}_D = C\ \mathrm{Re}_D^{0.5}\ \mathrm{Pr}^{0.4} \left(\frac{\rho_\infty}{\rho_0}\right)^{0.25}$$

where $C = 0.95$ for cylinders and 1.28 for spheres; ρ_∞ is the free-stream gas density and ρ_0 is the stagnation density of stream behind the bow shock. The heat transfer rate for this case, is given by $q = \bar{h}A(T_s - T_0)$.

2.3 Free Convection

In free convection the fluid motion is caused by the buoyant force resulting from the density difference near the body surface, which is at a temperature different from that of the free fluid far removed from the surface where velocity is zero. In all free convection correlations, except for the enclosed cavities, the fluid properties are usually evaluated at the mean film temperature $T_f = (T_1 + T_\infty)/2$. The thermal expansion coefficient β, however, is evaluated at the free fluid temperature T_∞. The convection heat transfer coefficient h is based on the temperature difference between the surface and the free fluid.

Free Convection from Flat Plates and Cylinders For free convection from flat plates and cylinders, the average Nusselt number $\overline{\mathrm{Nu}}_L$ can be expressed as[4]

$$\overline{\mathrm{Nu}}_L = C(\mathrm{Gr}_L\ \mathrm{Pr})^m$$

where the constants C and m are given as shown in Table 15. The *Grashof Prandtl number* product, $(\mathrm{Gr}_L\ \mathrm{Pr})$ is called the *Rayleigh number* (Ra_L) and for certain ranges of this value, Figs. 13 and 14 are used instead of the above equation. Reasonable approximations for other types of *three-dimensional shapes*, such as short cylinders and blocks, can be made for $10^4 < \mathrm{Ra}_L < 10^9$, by using this expression and $C = 0.6$, $m = 1/4$, provided that the characteristic length, L, is determined from $1/L = 1/L_{\mathrm{hor}} + 1/L_{\mathrm{ver}}$, where

Table 15   Constants for Free Convection from Flat Plates and Cylinders

Geometry	$Gr_K Pr$	C	m	L
Vertical flat plates and cylinders	10^{-1}–10^4	Use Fig. 12	Use Fig. 12	Height of plates and cylinders; restricted to $D/L \geq 35/Gr_L^{1/4}$ for cylinders
	10^4–10^9	0.59	$1/4$	
	10^9–10^{13}	0.10	$1/3$	
Horizontal cylinders	0–10^{-5}	0.4	0	Diameter D
	10^{-5}–10^4	Use Fig. 13	Use Fig. 13	
	10^4–10^9	0.53	$1/4$	
	10^9–10^{13}	0.13	$1/3$	
Upper surface of heated plates or lower surface of cooled plates	2×10^4–8×10^6	0.54	$1/4$	Length of a side for square plates, the average length of the two sides for rectangular plates
	8×10^6–10^{11}	0.15	$1/3$	
Lower surface of heated plates or upper surface of cooled plates	10^5–10^{11}	0.58	$1/5$	$0.9D$ for circular disks

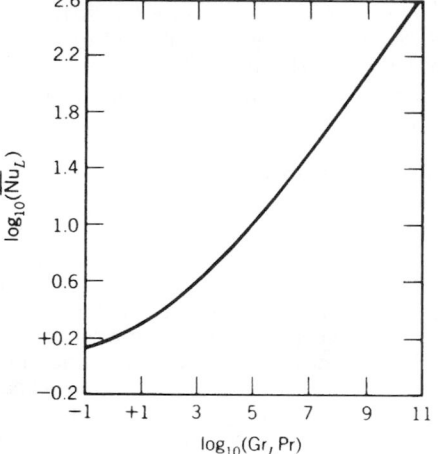

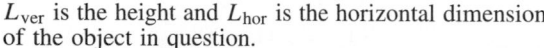

Fig. 13 Free-convection heat transfer correlation for heated vertical plates and cylinders. (Adapted from Ref. 14. Used with permission of McGraw-Hill Book Company.)

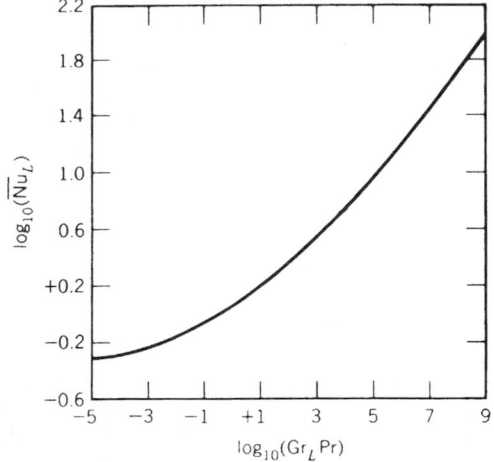

Fig. 14 Free-convection heat transfer correlation from heated horizontal cylinders. (Adapted from Ref. 14. Used with permission of McGraw-Hill Book Company.)

L_{ver} is the height and L_{hor} is the horizontal dimension of the object in question.

For *unsymmetrical horizontal* square, rectangular, or circular surfaces, the characteristic length L can be calculated from the expression $L = A/P$, where A is the area and P is the wetted perimeter of the surface.

Free Convection from Spheres   For free convection from spheres, the following correlation has been developed:

$$\overline{Nu}_D = 2 + 0.43(Gr_D\ Pr)^{0.25} \quad \text{for } 1 < Gr_D < 10^5$$

Although this expression was designed primarily for gases, $Pr \approx 1$, it may be used to approximate the values for liquids as well.[15]

Free Convection in Enclosed Spaces   Heat transfer in an enclosure occurs in a number of different situations and with a variety of configurations. Then a temperature difference is imposed on two opposing walls that enclose a space filled with a fluid, convective heat transfer will occur. For small values of the Rayleigh number, the heat transfer may be dominated by conduction, but as the Rayleigh number increases, the contribution made by free convection will increase. Following are a number of correlations, each designed for a specific geometry. For all of these, the fluid properties are evaluated at the average temperature of the two walls.

Cavities between Two Horizontal Walls at Temperatures T_1 and T_2 Separated by Distance δ (T_1 for Lower Wall, $T_1 > T_2$)

$$q'' = \overline{h}(T_1 - T_2)$$

$$\overline{Nu}_\delta = \begin{cases} 0.069 \, Ra_\delta^{1/3} \, Pr^{0.074} & \text{for } 3 \times 10^5 < Ra_\delta < 7 \times 10^9 \\ 1.0 & \text{for } Ra_\delta < 1700 \end{cases}$$

where $Ra_\delta = g\beta \, (T_1 - T_2) \, \delta^3/\alpha v$; δ is the thickness of the space.[16]

Cavities between Two Vertical Walls of Height H at Temperature by Distance T_Δ and T_Θ Separated by Distance δ[17,18]

$$q'' = \overline{h}(T_1 - T_2)$$

$$\overline{Nu}_\delta = 0.22 \left(\frac{Pr}{0.2 + Pr} Ra_\delta \right)^{0.28} \left(\frac{\delta}{H} \right)^{0.25}$$

for $2 < H/\delta < 10$, $Pr < 10^5$ $Ra_\delta < 10^{10}$;

$$\overline{Nu}_\delta = 0.18 \left(\frac{Pr}{0.2 + Pr} Ra_\delta \right)^{0.29}$$

for $1 < H/\delta < 2$, $10^3 < Pr < 10^5$, and $10^3 < Ra_\delta Pr/(0.2 + Pr)$; and

$$\overline{Nu}_\delta = 0.42 \, Ra_\delta^{0.25} \, Pr^{0.012}(\delta/H)^{0.3}$$

for $10 < H/\delta < 40$, $1 < Pr < 2 \times 10^4$, and $10^4 < Ra_\delta < 10^7$.

2.4 Log-Mean Temperature Difference

The simplest and most common type of heat exchanger is the *double-pipe heat exchanger* illustrated in Fig. 15. For this type of heat exchanger, the heat transfer between the two fluids can be found by assuming a constant overall heat transfer coefficient found from Table 8 and a constant fluid specific heat. For this type, the heat transfer is given by

$$q = UA \, \Delta T_m$$

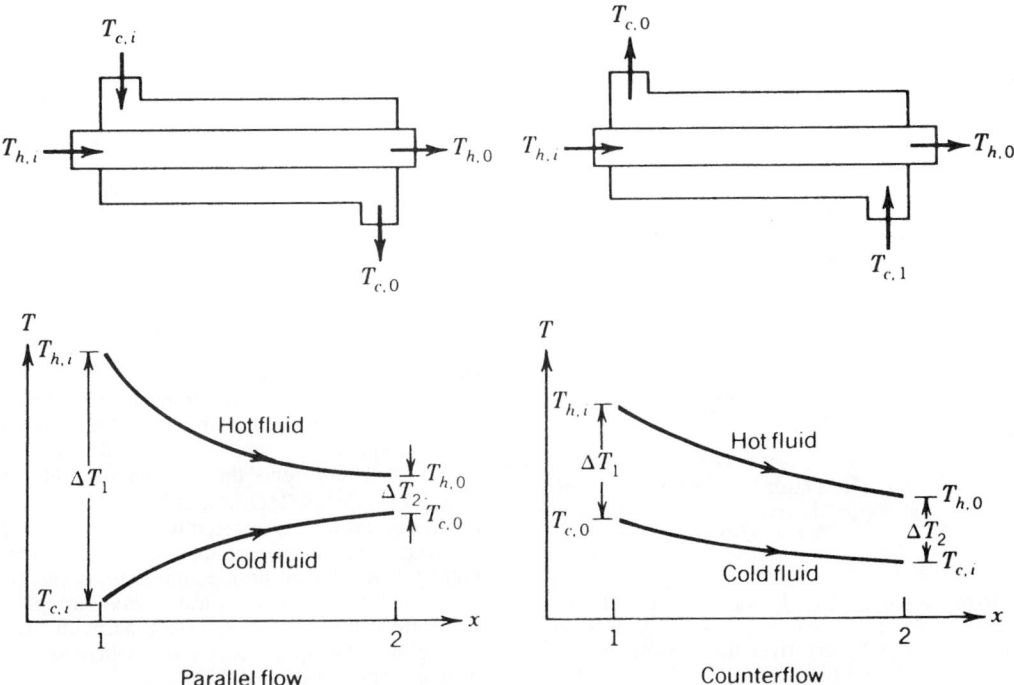

Fig. 15 Temperature profiles for parallel flow and counterflow in double-pipe heat exchanger.

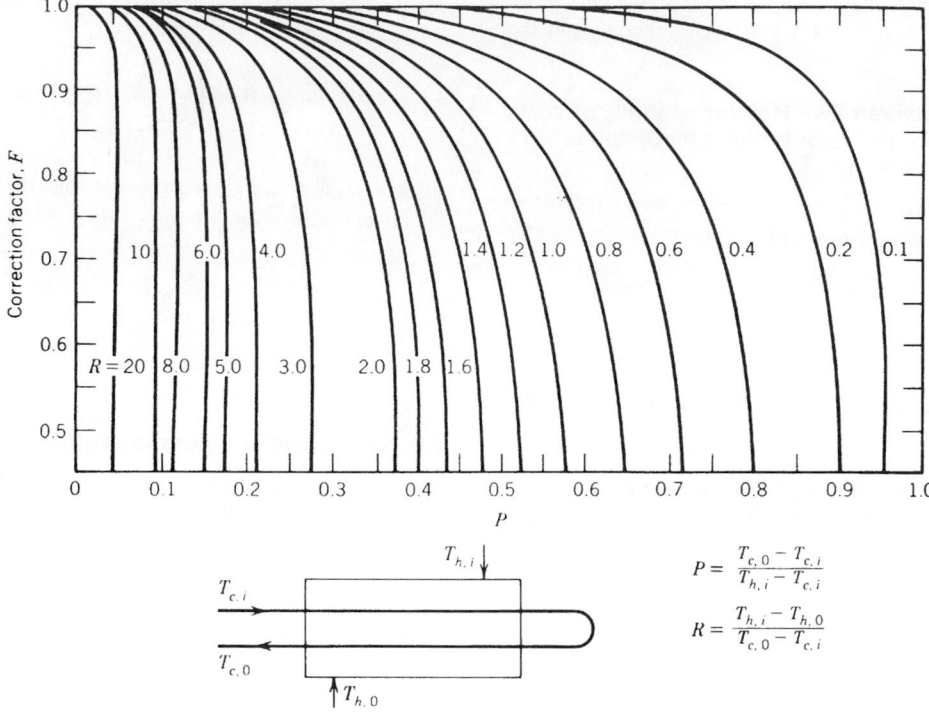

Fig. 16 Correction factor for shell-and-tube heat exchanger with one shell and any multiple of two tube passes (two, four, etc., tube passes.)

where

$$\Delta T_m = \frac{\Delta T_2 - \Delta T_1}{\ln(\Delta T_2 / \Delta T_1)}$$

In this expression, the temperature difference, ΔT_m, is referred to as the *log-mean temperature difference* (LMTD); ΔT_1 represents the temperature difference between the two fluids at one end and ΔT_2 at the other end. For the case where the ratio $\Delta T_2 / \Delta T_1$ is less than two, the *arithmetic mean temperature difference*, $(\Delta T_2 + \Delta T_1)/2$, may be used to calculate heat transfer rate without introducing any significant error. As shown in Fig. 15,

$\Delta T_1 = T_{h,i} - T_{c,i}$ $\Delta T_2 = T_{h,o} - T_{c,o}$ for parallel flow

$\Delta T_1 = T_{h,i} - T_{c,o}$ $\Delta T_2 = T_{h,o} - T_{c,i}$ for counterflow

Cross-Flow Coefficient In other types of heat exchangers, where the values of the overall heat transfer coefficient, U, may vary over the area of the surface, the LMTD may not be representative of the actual average temperature difference. In these cases, it is necessary to utilize a correction factor such that the

heat transfer, q, can be determined by

$$q = U A F \; \Delta T_m$$

Here the value of ΔT_m is computed assuming counterflow conditions, that is, $\Delta T_1 = T_{h,i} - T_{c,i}$ and $\Delta T_2 = T_{h,o} - T_{c,o}$. Figures 16 and 17 illustrate some examples of the *correction factor F* for various multiple-pass heat exchangers.

3 RADIATION HEAT TRANSFER

Heat transfer can occur in the absence of a participating medium through the transmission of energy by electromagnetic waves, characterized by a wavelength, λ, and frequency, v, which are related by $c = \lambda v$. The parameter c represents the velocity of light, which in a vacuum is $c_o = 2.9979 \times 10^8$ m/s. Energy transmitted in this fashion is referred to as *radiant energy* and the heat transfer process that occurs is called radiation heat transfer or simply *radiation*. In this mode of heat transfer, the energy is transferred through electromagnetic waves or through photons, with the energy of a photon being given by hv, where h represents Planck's constant.

In nature, every substance has a characteristic wave velocity that is smaller than that occurring in a vacuum.

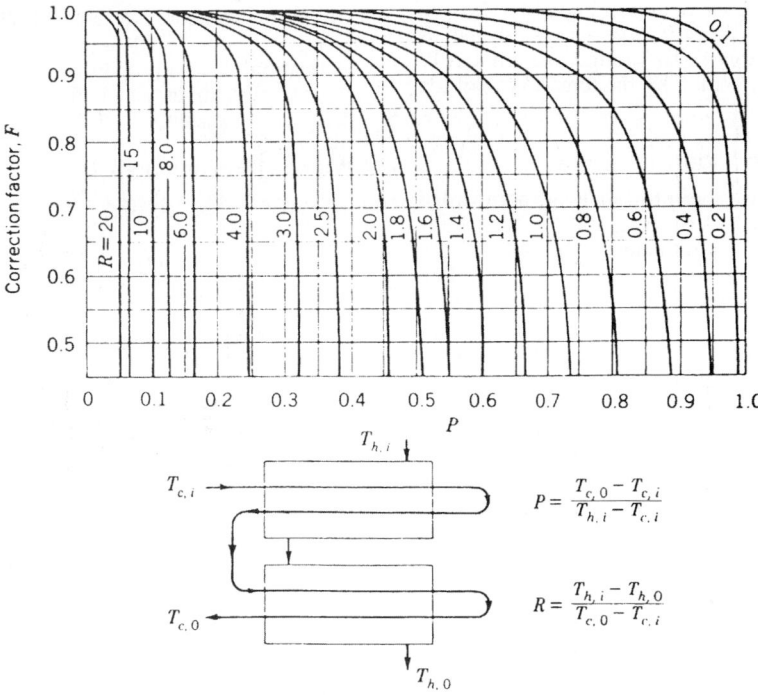

Fig. 17 Correction factor for shell-and-tube heat exchanger with two shell passes and any multiple of four tubes passes (four, eight, etc., tube passes.)

$$P = \frac{T_{c,0} - T_{c,i}}{T_{h,i} - T_{c,i}}$$

$$R = \frac{T_{h,i} - T_{h,0}}{T_{c,0} - T_{c,i}}$$

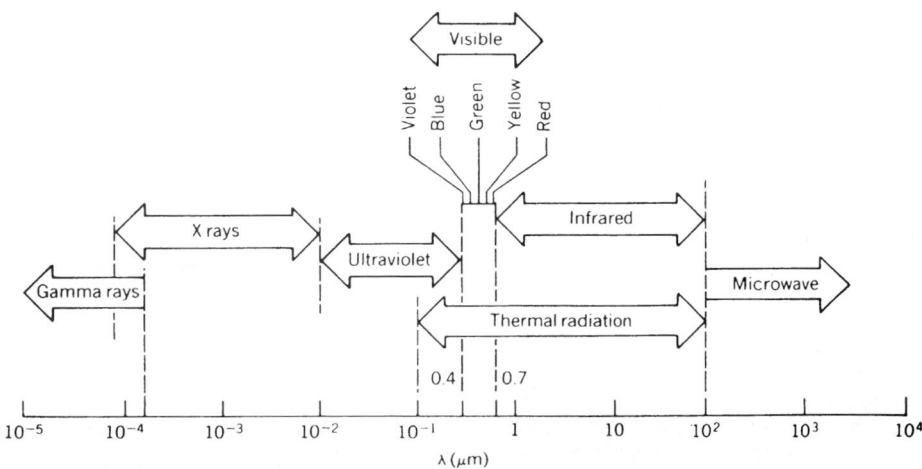

Fig. 18 Electromagnetic radiation spectrum.

These velocities can be related to c_o by $c = c_o/n$, where n indicates the refractive index. The value of the refractive index, n, for air is approximately equal to 1. The wavelength of the energy given or for the radiation that comes from a surface depends on the nature of the source and various wavelengths sensed in different ways. For example, as shown in Fig. 18, the electromagnetic spectrum consists of a number of different types of radiation. Radiation in the visible spectrum occurs in the range $\lambda = 0.4 - 0.74 \ \mu m$, while radiation in the wavelength range 0.1–100 μm is classified as *thermal radiation* and is sensed as heat. For radiant energy in this range, the amount of energy given off is governed by the temperature of the emitting body.

3.1 Blackbody Radiation

All objects in space are continuously being bombarded by radiant energy of one form or another and all of this energy is either absorbed, reflected, or transmitted. An ideal body that absorbs all the radiant energy falling upon it, regardless of the wavelength and direction, is referred to as a *blackbody*. Such a body emits maximum energy for a prescribed temperature and wavelength. Radiation from a blackbody is independent of direction and is referred to as a *diffuse emitter*.

Stefan–Boltzmann Law The Stefan–Boltzmann law describes the rate at which energy is radiated from a blackbody and states that this radiation is proportional to the fourth power of the absolute temperature of the body,

$$e_b = \sigma T^4$$

where e_b is the *total emissive power* and σ is the Stefan–Boltzmann constant, which has the value 5.729×10^{-8} W/m^2·K^4 (0.173×10^{-8} Btu/h·ft^2·$^\circ$R^4).

Planck's Distribution Law The temperature amount of energy leaving a blackbody is described as the *spectral emissive power*, $e_{\lambda b}$, and is a function of wavelength. This function, which was derived from quantum theory by Planck, is

$$e_{\lambda b} = \frac{2\pi C_1}{\lambda^5 [\exp(C_2/\lambda T) - 1]}$$

where $e_{\lambda b}$ has a unit W/m^2·μm (Btu/h·ft^2·μm).

Values of the constants C_1 and C_2 are 0.59544×10^{-16} W·m^2 (0.18892×10^8 Btu·μm^4/h·ft^2) and $14{,}388$ μm·K ($25{,}898$ μm·$^\circ$R), respectively. The distribution of the spectral emissive power from a blackbody at various temperatures is shown in Fig. 19, which shows that the energy emitted at all wavelengths increases as the temperature increases. The maximum or peak values of the constant temperature curves illustrated in Fig. 20 shift to the left for shorter wavelengths as the temperatures increase.

The fraction of the emissive power of a blackbody at a given temperature and in the wavelength interval between λ_1 and λ_2 can be described by

$$F_{\lambda_1 T - \lambda_2 T} = \frac{1}{\sigma T^4} \left(\int_0^{\lambda_1} e_{\lambda b} \, d\lambda - \int_0^{\lambda_2} e_{\lambda b} \, d\lambda \right)$$

$$= F_{o-\lambda_1 T} - F_{o-\lambda_2 T}$$

where the function $F_{o-\lambda T} = (1/\sigma T^4) \int_o^{\lambda} e_{\lambda b} \, d\lambda$ is given in Table 16. This function is useful for the evaluation of total properties involving integration on the wavelength in which the spectral properties are piecewise constant.

Wien's Displacement Law The relationship between these peak or maximum temperatures can be described by *Wien's displacement law*,

$$\lambda_{\max} T = 2897.8 \ \mu\text{m·K}$$

or

$$\lambda_{\max} T = 5216.0 \ \mu\text{m·}^\circ\text{R}$$

3.2 Radiation Properties

While to some degree all surfaces follow the general trends described by the Stefan–Boltzmann and Planck laws, the behavior of real surfaces deviates somewhat from these. In fact, because blackbodies are ideal, all real surfaces emit and absorb less radiant energy than a blackbody. The amount of energy a body emits can be described in terms of the emissivity and is, in general, a function of the type of material, the temperature, and the surface conditions, such as roughness, oxide layer thickness, and chemical contamination. The emissivity is, in fact, a measure of how well a real body radiates energy as compared with a blackbody of the same temperature. The radiant energy emitted into the entire hemispherical space above a real surface element, including all wavelengths is given by $e = \varepsilon \sigma T^4$, where ε is less than 1.0 and is called the *hemispherical emissivity* (or *total hemispherical emissivity* to indicate averaging over the total wavelength spectrum). For a given wavelength the *spectral hemispherical emissivity* ε_λ of a real surface is defined as

$$\varepsilon_\lambda = \frac{e_\lambda}{e_{\lambda b}}$$

where e_λ is the hemispherical emissive power of the real surface and $e_{\lambda b}$ is that of a blackbody at the same temperature.

Spectral irradiation, G_λ, (W/m^2·μm), is defined as the rate at which radiation is incident upon a surface per unit area of the surface, per unit wavelength about the wavelength λ, and encompasses the incident radiation from all directions.

Spectral hemispherical reflectivity, ρ_λ, is defined as the radiant energy reflected per unit time, per unit area of the surface, per unit wavelength per G_λ.

Spectral hemispherical absorptivity, α_λ, is defined as the radiant energy absorbed per unit area of the surface per unit wavelength about the wavelength per G_λ.

Spectral hemispherical transmissivity is defined as the radiant energy transmitted per unit area of the surface, per unit wavelength about the wavelength per G_λ.

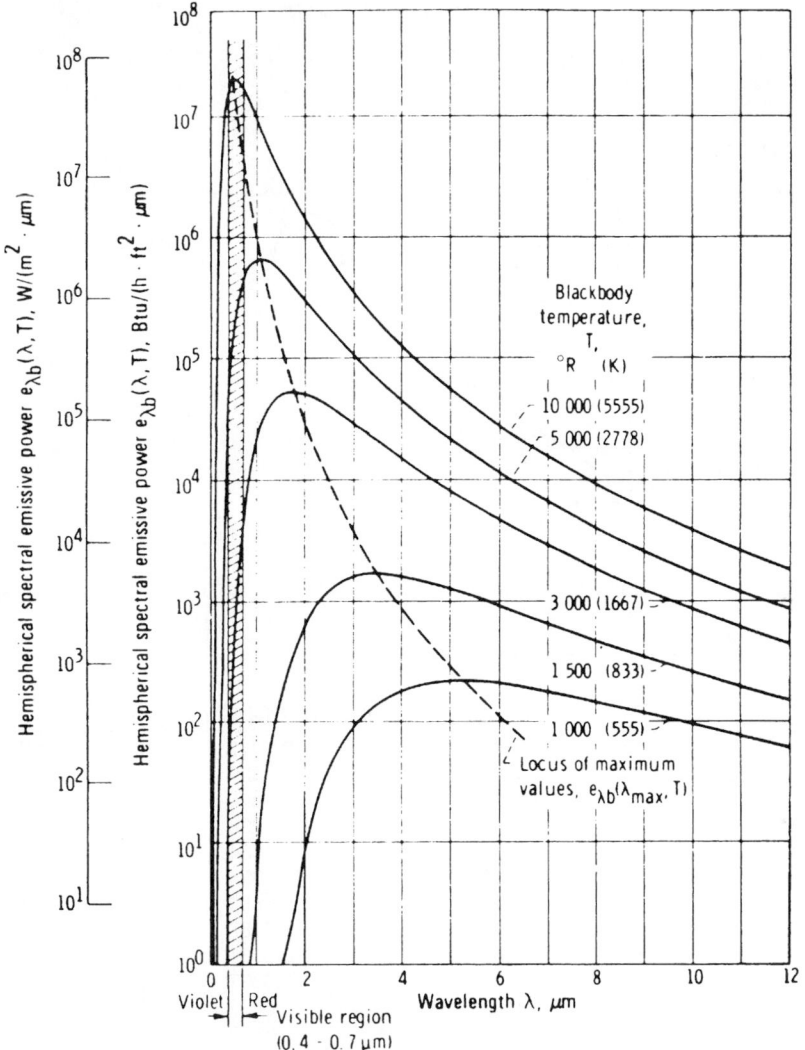

Fig. 19 Hemispherical spectral emissive power of a blackbody for various temperatures.

For any surface, the sum of the reflectivity, absorptivity, and transmissivity must equal unit, that is,

$$\alpha_\lambda - \rho_\lambda \tau_\lambda = 1$$

When these values are averaged over the entire wavelength from $\lambda = 0$ to ∞, they are referred to as *total* values. Hence, the *total hemispherical reflectivity, total hemispherical absorptivity,* and *total hemispherical transmissivity* can be written as

$$\rho = \int_0^\infty \rho_\lambda G_\lambda \frac{d\lambda}{G} \qquad \alpha = \int_0^\infty \alpha_\lambda G_\lambda \frac{d\lambda}{G}$$

and

$$\tau = \int_0^\infty \tau_\lambda G_\lambda \frac{d\lambda}{G}$$

respectively, where

$$G = \int_0^\infty G_\lambda \, d\lambda$$

As was the case for the wavelength-dependent parameters, the sum of the total reflectivity, total absorptivity,

$$dA_i \, dF_{di-dj} = dA_j \, dF_{dj-di}$$

$$dA_i F_{di-j} = A_j \, dF_{j-di}$$

$$A_i F_{i-j} = A_j F_{j-i}$$

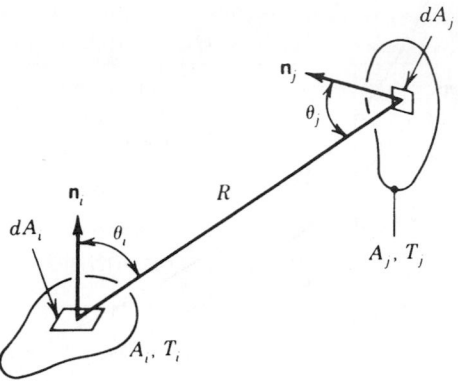

Fig. 20   Configuration factor for radiation exchange between surfaces of area dA_i and dA_j.

Table 16   Radiation Function $F_{o-\lambda T}$

λT			λT			λT		
μm·K	μm·°R	$F_{o-\lambda T}$	μm·K	μm·°R	$F_{o-\lambda T}$	μm·K	μm·°R	$F_{o-\lambda T}$
400	720	0.1864×10^{-11}	3400	6120	0.3617	6400	11,520	0.7692
500	900	0.1298×10^{-8}	3500	6300	0.3829	6500	11,700	0.7763
600	1080	0.9290×10^{-7}	3600	6480	0.4036	6600	11,880	0.7832
700	1260	0.1838×10^{-5}	3700	6660	0.4238	6800	12,240	0.7961
800	1440	0.1643×10^{-4}	3800	6840	0.4434	7000	12,600	0.8081
900	1620	0.8701×10^{-4}	3900	7020	0.4624	7200	12,960	0.8192
1000	1800	0.3207×10^{-3}	4000	7200	0.4809	7400	13,320	0.8295
1100	1980	0.9111×10^{-3}	4100	7380	0.4987	7600	13,680	0.8391
1200	2160	0.2134×10^{-2}	4200	7560	0.5160	7800	14,040	0.8480
1300	2340	0.4316×10^{-2}	4300	7740	0.5327	8000	14,400	0.8562
1400	2520	0.7789×10^{-2}	4400	7920	0.5488	8200	14,760	0.8640
1500	2700	0.1285×10^{-1}	4500	8100	0.5643	8400	15,120	0.8712
1600	2880	0.1972×10^{-1}	4600	8280	0.5793	8600	15,480	0.8779
1700	3060	0.2853×10^{-1}	4700	8460	0.5937	8800	15,840	0.8841
1800	3240	0.3934×10^{-1}	4800	8640	0.6075	9000	16,200	0.8900
1900	3420	0.5210×10^{-1}	4900	8820	0.6209	10,000	18,000	0.9142
2000	3600	0.6673×10^{-1}	5000	9000	0.6337	11,000	19,800	0.9318
2100	3780	0.8305×10^{-1}	5100	9180	0.6461	12,000	21,600	0.9451
2200	3960	0.1009	5200	9360	0.6579	13,000	23,400	0.9551
2300	4140	0.1200	5300	9540	0.6694	14,000	25,200	0.9628
2400	4320	0.1402	5400	9720	0.6803	15,000	27,000	0.9689
2500	4500	0.1613	5500	9900	0.6909	20,000	36,000	0.9856
2600	4680	0.1831	5600	10,080	0.7010	25,000	45,000	0.9922
2700	4860	0.2053	5700	10,260	0.7108	30,000	54,000	0.9953
2800	5040	0.2279	5800	10,440	0.7201	35,000	63,000	0.9970
2900	5220	0.2505	5900	10,620	0.7291	40,000	72,000	0.9979
3000	5400	0.2732	6000	10,800	0.7378	45,000	81,000	0.9985
3100	5580	0.2958	6100	10,980	0.7461	50,000	90,000	0.9989
3200	5760	0.3181	6200	11,160	0.7541	55,000	99,000	0.9992
3300	5940	0.3401	6300	11,340	0.7618	60,000	108,000	0.9994

and total transmissivity must be equal to unity, that is,

$$\alpha + \rho + \tau = 1$$

It is important to note that while the emissivity is a function of the material, temperature, and surface conditions, the absorptivity and reflectivity depend on both the surface characteristics and the nature of the incident radiation.

The terms *reflectance, absorptance*, and *transmittance* are used by some authors for the real surfaces and the terms reflectivity, absorptivity, and transmissivity are reserved for the properties of the ideal surfaces (i.e., those optically smooth and pure substances perfectly uncontaminated). Surfaces that allow no radiation to pass through are referred to as *opaque*, that is, $\tau_\lambda = 0$, and all of the incident energy will be either reflected or absorbed. For such a surface,

$$\alpha_\lambda + \rho_\lambda = 1 \qquad \alpha + \rho = 1$$

Light rays reflected from a surface can be reflected in such a manner that the incident and reflected rays are symmetric with respect to the surface normal at the point of incidence. This type of radiation is referred to as *specular*. The radiation is referred to as *diffuse* if the intensity of the reflected radiation is uniform over all angles of reflection and is independent of the incident direction, and the surface is called a *diffuse surface* if the radiation properties are independent of the direction. If they are independent of the wavelength, the surface is called a *gray surface*, and a *diffuse-gray surface* absorbs a fixed fraction of incident radiation from any direction and at any wavelength, and $\alpha_\lambda = \varepsilon_\lambda = \alpha = \varepsilon$.

Kirchhoff's Law of Radiation The directional characteristics can be specified by the addition of a prime to the value; for example, the spectral emissivity for radiation in a particular direction would be denoted by α'_λ. For radiation in a particular direction, the spectral emissivity is equal to the directional spectral absorptivity for the surface irradiated by a blackbody at the same temperature. The most general form of this expression states that $\alpha'_\lambda = \varepsilon'_\lambda$. If the incident radiation is independent of angle or if the surface is diffuse, then $\alpha_\lambda = \varepsilon_\lambda$ for the hemispherical properties. This relationship can have various conditions imposed on it, depending on whether spectral, total, directional, or hemispherical quantities are being considered.[19]

Emissivity of Metallic Surfaces The properties of pure smooth metallic surfaces are often characterized by low emissivity and absorptivity values and high values of reflectivity. The spectral emissivity of metals tends to increase with decreasing wavelength, and exhibits a peak near the visible region. At wavelengths $\lambda > \sim 5$ μm the spectral emissivity increases with increasing temperature, but this trend reverses at shorter wavelengths ($\lambda < \sim 1.27$ μm). Surface roughness has a pronounced effect on both the hemispherical emissivity and absorptivity, and large *optical roughnesses*, defined as the mean square roughness of the surface divided by the wavelength, will increase the hemispherical emissivity. For cases where the optical roughness is small, the directional properties will approach the values obtained for smooth surfaces. The presence of impurities, such as oxides or other nonmetallic contaminants, will change the properties significantly and increase the emissivity of an otherwise pure metallic body. A summary of the normal total emissivities for metals are given in Table 17. It should be noted that the hemispherical emissivity for metals is typically 10–30% higher than the values normally encountered for normal emissivity.

Emissivity of Nonmetallic Materials Large values of total hemispherical emissivity and absorptivity are typical for nonmetallic surfaces at moderate temperatures and, as shown in Table 18, which lists the normal total emissivity of some nonmetals, the temperature dependence is small.

Absorptivity for Solar Incident Radiation The spectral distribution of solar radiation can be approximated by blackbody radiation at a temperature of approximately 5800 K (10, 000°R) and yields an average solar irradiation at the outer limit of the atmosphere of approximately 1353 W/m^2 (429 Btu/ft^2·h). This solar irradiation is called the *solar constant* and is greater than the solar irradiation received at the surface of the earth, due to the radiation scattering by air molecules, water vapor, and dust, and the absorption by O_3, H_2O, and CO_2 in the atmosphere. The absorptivity of a substance depends not only on the surface properties but also on the sources of incident radiation. Since solar radiation is concentrated at a shorter wavelength, due to the high source temperature, the absorptivity for certain materials when exposed to solar radiation may be quite different from that which occurs for low-temperature radiation, where the radiation is concentrated in the longer wavelength range. A comparison of absorptivities for a number of different materials is given in Table 19 for both solar and low-temperature radiation.

3.3 Configuration Factor

The magnitude of the radiant energy exchanged between any two given surfaces is a function of the emissivity, absorptivity, and transmissivity. In addition, the energy exchange is a strong function of how one surface is viewed from the other. This aspect can be defined in terms of the *configuration factor* (sometimes called the *radiation shape factor, view factor, angle factor, or interception factor*). As shown in Fig. 20, the configuration factor, F_{i-j}, is defined as that fraction of the radiation leaving a black surface, i, that is intercepted by a black or gray surface, j, and is based on the relative geometry, position, and shape

Table 17 Normal Total Emissivity of Metals[a]

Materials	Surface Temperature (K)	Normal Total Emissivity
Aluminum		
Highly polished plate	480–870	0.038–0.06
Polished plate	373	0.095
Heavily oxidized	370–810	0.20–0.33
Bismuth, bright	350	0.34
Chromium, polished	310–1370	0.08–0.40
Copper		
Highly polished	310	0.02
Slightly polished	310	0.15
Black oxidized	310	0.78
Gold, highly polished	370–870	0.018–0.035
Iron		
Highly polished, electrolytic	310–530	0.05–0.07
Polished	700–760	0.14–0.38
Wrought iron, polished	310–530	0.28
Cast iron, rough, strongly oxidized	310–530	0.95
Lead		
Polished	310–530	0.06–0.08
Rough unoxidized	310	0.43
Mercury, unoxidized	280–370	0.09–0.12
Molybdenum, polished	310–3030	0.05–0.29
Nickel		
Electrolytic	310–530	0.04–0.06
Electroplated on iron, not polished	293	0.11
Nickel oxide	920–1530	0.59–0.86
Platinum, electrolytic	530–810	0.06–0.10
Silver, polished	310–810	0.01–0.03
Steel		
Polished sheet	90–420	0.07–0.14
Mild steel, polished	530–920	0.27–0.31
Sheet with rough oxide layer	295	0.81
Tin, polished sheet	310	0.05
Tungsten, clean	310–810	0.03–0.08
Zinc		
Polished	310–810	0.02–0.05
Gray oxidized	295	0.23–0.28

[a]Adapted from Ref. 19.

of the two surfaces. The configuration factor can also be expressed in terms of the differential fraction of the energy or dF_{i-dj}, which indicates the differential fraction of energy from a finite area A_i that is intercepted by an infinitesimal area dA_j. Expressions for a number of different cases are given below for several common geometries:

Infinitesimal area dA_j to infinitesimal area dA_j:

$$dF_{di-dj} = \frac{\cos\theta_i \cos\theta_j}{\pi R^2} dA_j$$

Infinitesimal area dA_j to finite area A_j:

$$F_{di-j} = \int_{Aj} \frac{\cos\theta_i \cos\theta_j}{\pi R^2} dA_j$$

Finite area A_i to finite area A_j:

$$F_{i-j} = \frac{1}{A_i} \int_{Aj} \int_{Aj} \frac{\cos\theta_i \cos\theta_j}{\pi R^2} dA_i \, dA_j$$

Analytical expressions of other configuration factors have been found for a wide variety of simple geometries, and a number of these are presented in Figs. 21–24 for surfaces that emit and reflect diffusely.

Reciprocity Relations The configuration factors can be combined and manipulated using algebraic rules referred to as configuration factor geometry. These expressions take several forms, one of which is the reciprocal properties between different configuration

Table 18 Normal Total Emissivity of Nonmetals[a]

Materials	Surface Temperature (K)	Normal Total Emissivity
Asbestos, board	310	0.96
Brick		
White refractory	1370	0.29
Rough red	310	0.93
Carbon, lampsoot	310	0.95
Concrete, rough	310	0.94
Ice, smooth	273	0.966
Magnesium oxide, refractory	420–760	0.69–0.55
Paint		
Oil, all colors	373	0.92–0.96
Lacquer, flat black	310–370	0.96–0.98
Paper, white	310	0.95
Plaster	310	0.91
Porcelain, glazed	295	0.92
Rubber, hard	293	0.92
Sandstone	310–530	0.83–0.90
Silicon carbide	420–920	0.83–0.96
Snow	270	0.82
Water, deep	273–373	0.96
Wood, sawdust	310	0.75

[a]Adapted from Ref. 19.

factors, which allow one configuration factor to be determined from knowledge of the others:

$$dA_i \, dF_{di-dj} = dA_j \, dF_{dj-di}$$

$$dA_i \, dF_{di-j} = A_j \, dF_{j-di}$$

$$A_i F_{i-j} = A_j F_{j-i}$$

These relationships can be combined with other basic rules to allow the determination of the configuration of an infinite number of complex shapes and geometries form a few select, known geometries. These are summarized in the following sections.

Additive Property For a surface A_i subdivided into N parts $(A_{i_1}, A_{i_2}, \ldots, A_{i_N})$ and a surface A_j subdivided into M parts $(A_{j_1}, A_{j_2}, \ldots, A_{j_M})$,

$$A_i F_{i-j} = \sum_{n=1}^{N} \sum_{m=1}^{M} A_{i_n} F_{i_n - j_m}$$

Table 19 Comparison of Absorptivities of Various Surfaces to Solar and Low-Temperature Thermal Radiation[a]

Surface	Absorptivity	
	For Solar Radiation	For Low-Temperature Radiation (~300 K)
Aluminum, highly polished	0.15	0.04
Copper, highly polished	0.18	0.03
Tarnished	0.65	0.75
Cast iron	0.94	0.21
Stainless steel, No. 301, polished	0.37	0.60
White marble	0.46	0.95
Asphalt	0.90	0.90
Brick, red	0.75	0.93
Gravel	0.29	0.85
Flat black lacquer	0.96	0.95
White paints, various types of pigments	0.12–0.16	0.90–0.95

[a]Adapted from Ref. 20 after J. P. Holman.[27]

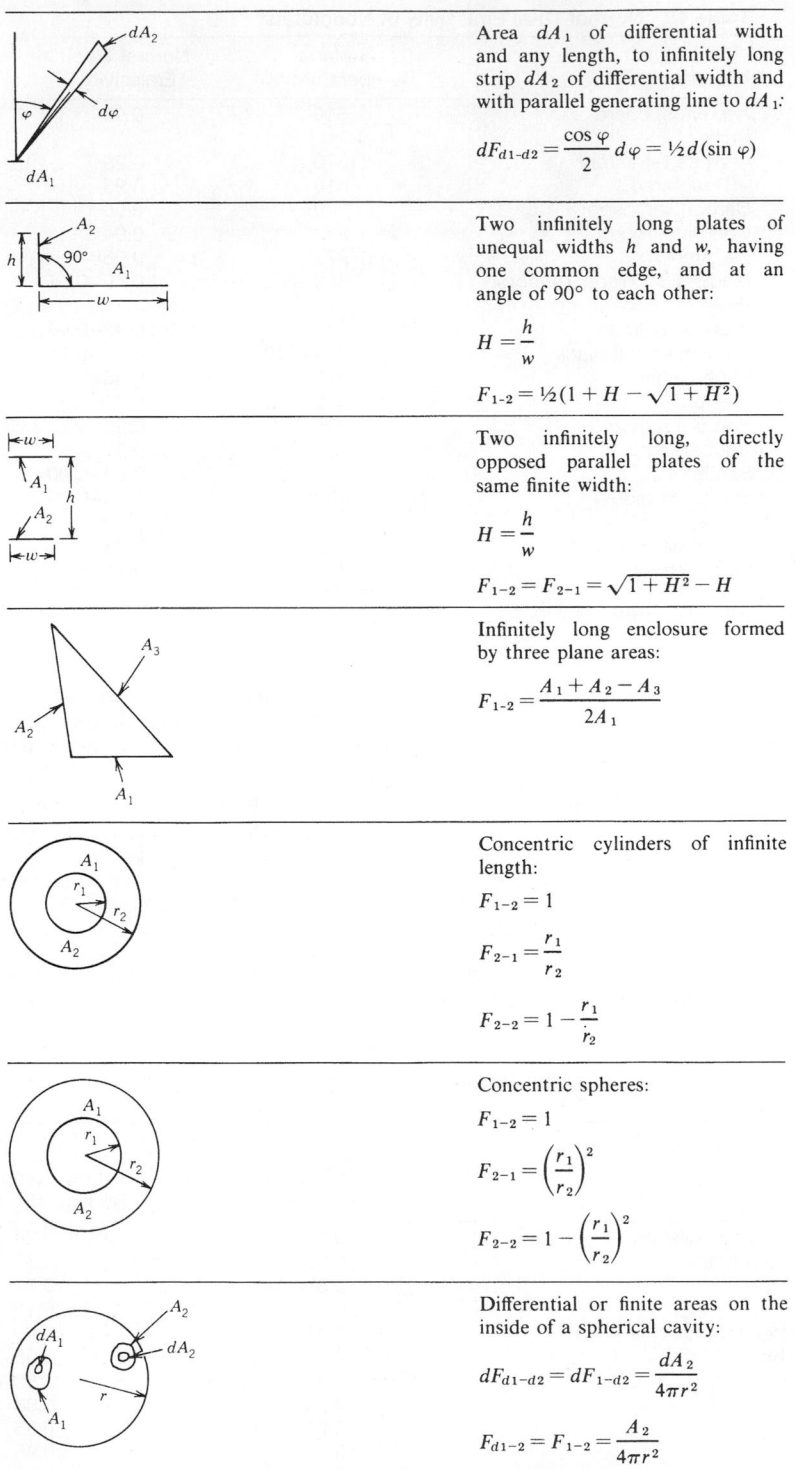

Area dA_1 of differential width and any length, to infinitely long strip dA_2 of differential width and with parallel generating line to dA_1:

$$dF_{d1\text{-}d2} = \frac{\cos \varphi}{2} \, d\varphi = \tfrac{1}{2} d (\sin \varphi)$$

Two infinitely long plates of unequal widths h and w, having one common edge, and at an angle of 90° to each other:

$$H = \frac{h}{w}$$

$$F_{1\text{-}2} = \tfrac{1}{2}(1 + H - \sqrt{1 + H^2})$$

Two infinitely long, directly opposed parallel plates of the same finite width:

$$H = \frac{h}{w}$$

$$F_{1\text{-}2} = F_{2\text{-}1} = \sqrt{1 + H^2} - H$$

Infinitely long enclosure formed by three plane areas:

$$F_{1\text{-}2} = \frac{A_1 + A_2 - A_3}{2A_1}$$

Concentric cylinders of infinite length:

$$F_{1\text{-}2} = 1$$

$$F_{2\text{-}1} = \frac{r_1}{r_2}$$

$$F_{2\text{-}2} = 1 - \frac{r_1}{r_2}$$

Concentric spheres:

$$F_{1\text{-}2} = 1$$

$$F_{2\text{-}1} = \left(\frac{r_1}{r_2}\right)^2$$

$$F_{2\text{-}2} = 1 - \left(\frac{r_1}{r_2}\right)^2$$

Differential or finite areas on the inside of a spherical cavity:

$$dF_{d1\text{-}d2} = dF_{1\text{-}d2} = \frac{dA_2}{4\pi r^2}$$

$$F_{d1\text{-}2} = F_{1\text{-}2} = \frac{A_2}{4\pi r^2}$$

Fig. 21   Configuration factors for some simple geometries.[19]

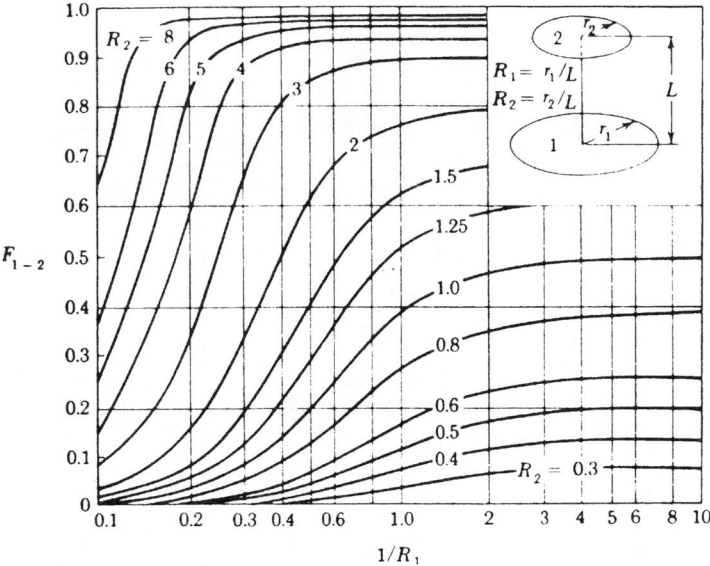

Fig. 22 Configuration factor for coaxial parallel circular disks.

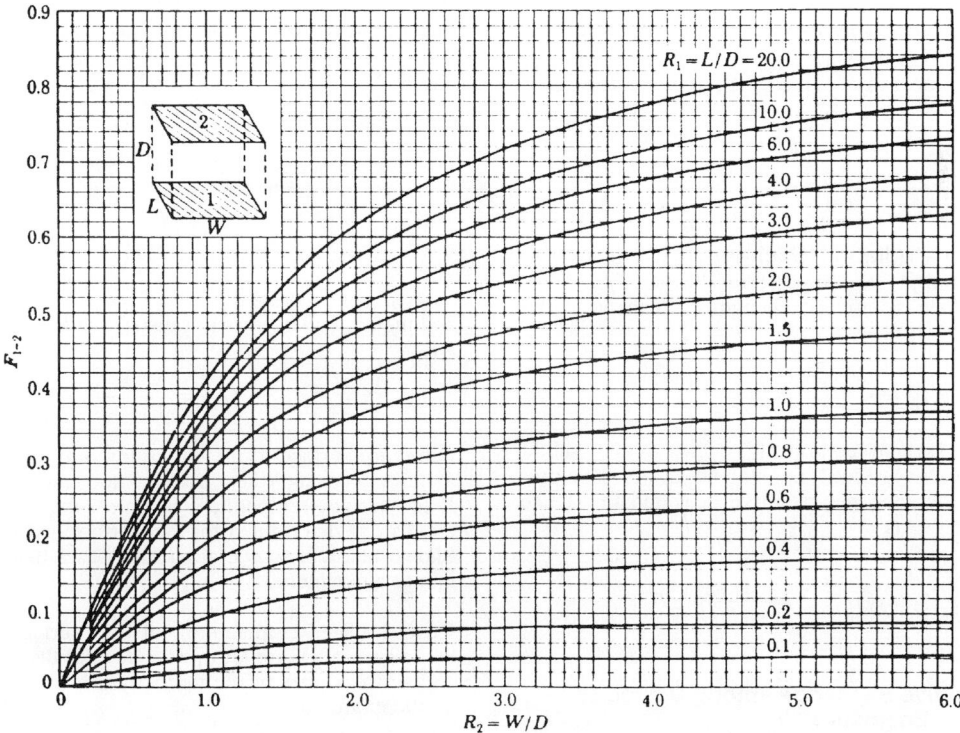

Fig. 23 Configuration factor for aligned parallel rectangles.

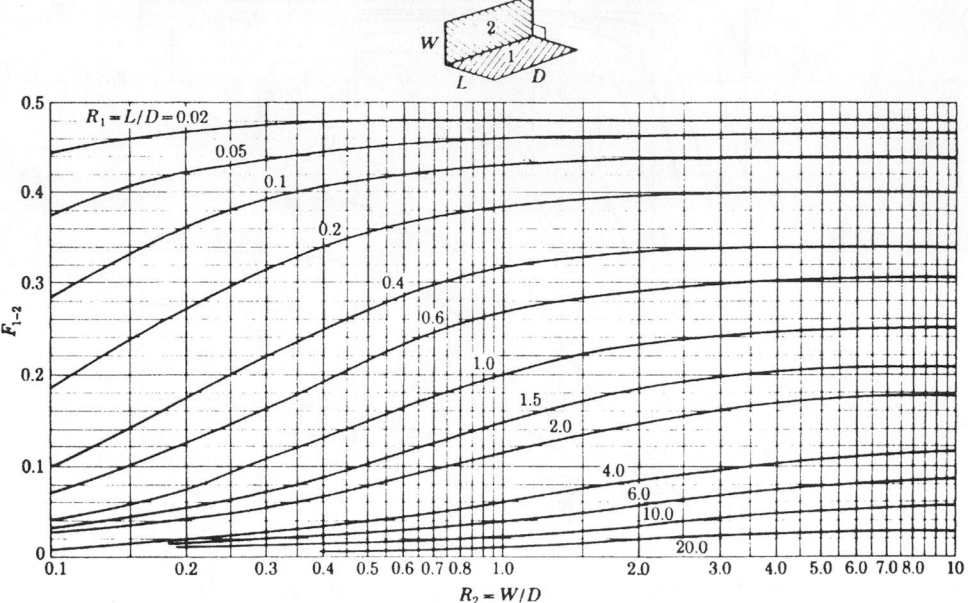

Fig. 24 Configuration factor for rectangles with common edge.

Relation in an Enclosure When a surface is completely enclosed, the surface can be subdivided into N parts having areas $A_1, A_2, \ldots, A_N$, respectively, and

$$\sum_{j=1}^{N} F_{i-j} = 1$$

Blackbody Radiation Exchange For black surfaces A_i, and A_j at temperatures T_i and T_j, respectively, the net radiative exchange, q_{ij}, can be expressed as

$$q_{ij} = A_i F_{i-j} \sigma (T_i^4 - T_j^4)$$

and for a surface completely enclosed and subdivided into N surfaces maintained at temperatures $T_1, T_2, \ldots, T_N$, the net radiative heat transfer, q_i, to surface area A_i is

$$q_i = \sum_{j=1}^{N} A_i F_{i-j} \sigma (T_i^4 - T_j^4) = \sum_{j=1}^{N} q_{ij}$$

3.4 Radiative Exchange among Diffuse Gray Surfaces in Enclosure

One method for solving for the radiation exchange between a number of surfaces or bodies is through the use of the *radiocity, J*, defined as the total radiation

that leaves a surface per unit time and per unit area. For an opaque surface, this term is defined as

$$J = \varepsilon \sigma T^4 + (1 - \varepsilon) G$$

For an enclosure consisting of N surfaces, the irradiation on a given surface i can be expressed as

$$G_i = \sum_{j=1}^{N} J_j F_{i-j}$$

and the net radiative heat transfer rate at given surface i is

$$q_i = A_i (J_i - G_i) = \frac{\varepsilon_i A_i}{1 - \varepsilon_i} (\sigma T_i^4 - J_i)$$

For every surface in the enclosure, a uniform temperature or a constant heat transfer rate can be specified. If the surface temperature is given, the heat transfer rate can be determined for that surface and vice versa. Shown below are several specific cases that are commonly encountered.

Case I. The temperatures of the surfaces, $T_i (i = 1, 2, \ldots, N)$, are known for each of the N surfaces and the values of the radiocity, J_i, are solved from the expression

$$\sum_{j=1}^{N} \{\delta_{ij} - (1 - \varepsilon_i) F_{i-j}\} J_i = \varepsilon_i \sigma T_i^4 \qquad 1 \le i \le N$$

The net heat transfer rate to surface i can then be determined from the fundamental relationship

$$q_i = A_i \frac{\varepsilon_i}{1 - \varepsilon_i}(\sigma T_i^4 - J_i) \qquad 1 \le i \le N$$

where $\delta_{ij} = 0$ for $i \ne j$ and $\delta_{ij} = 1$ for $i = j$.

Case II. The heat transfer rates, $q_i (i = 1, 2, \ldots, N)$, to each of the N surfaces are known and the values of the radiocity, J_i, are determined from

$$\sum_{j=1}^{N} \{\delta_{ij} - F_{i-j}\} J_j = \frac{q_i}{A_i} \qquad 1 \le i \le N$$

The surface temperature can then be determined from

$$T_i = \left[\frac{1}{\sigma} \left(\frac{1 - \varepsilon_i}{\varepsilon_i} \frac{q_i}{A_i} + J_i \right) \right]^{1/4} \qquad 1 \le i \le N$$

Case III. The temperatures, $T_i (i = 1, \ldots, N_1)$, for N_i surfaces and heat transfer rates $q_i (i = N_1 + 1, \ldots, N)$ for $(N - N_i)$ surfaces are known and

the radiocities are determined by

$$\sum_{j=1}^{N} \{\delta_{ij} - (1 - \varepsilon_i) F_{i-j}\} J_j = \varepsilon_i \alpha T_i^4 \quad 1 \le i \le N_1$$

$$\sum_{j=1}^{N} \{\delta_{ij} - F_{i-j}\} J_j = \frac{q_i}{A_i} \quad N_1 + 1 \le i \le N$$

The net heat transfer rates and temperatures van be found as

$$q_i = A_i \frac{\varepsilon_i}{1 - \varepsilon_i}(\sigma T_i^4 - J_i) \qquad 1 \le i \le N_1$$

$$T_i = \left[\frac{1}{\sigma} \left(\frac{1 - \varepsilon_i}{\varepsilon_i} \frac{q_i}{A_i} + J_i \right) \right]^{1/4} \qquad N_1 + 1 \le i \le N$$

Two Diffuse Gray Surfaces Forming an Enclosure The net radiative exchange, q_{12}, for two diffuse gray surfaces forming an enclosure are shown in Table 20 for several simple geometries.

Radiation Shields Often in practice, it is desirable to reduce the radiation heat transfer between two surfaces. This can be accomplished by placing a highly reflective surface between the two surfaces. For this configuration, the ratio of the net radiative exchange

Table 20   Net Radiative Exchange between Two Surfaces Forming an Enclosure

Large (infinite) parallel planes A_1, T_1, ϵ_1 ─────── A_2, T_2, ϵ_2	$A_1 = A_2 = A$	$q_{12} = \dfrac{A\sigma(T_1^4 - T_2^4)}{\dfrac{1}{\varepsilon_1} + \dfrac{1}{\varepsilon_2} - 1}$
Long (infinite) concentric cylinders 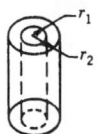	$\dfrac{A_1}{A_2} = \dfrac{r_1}{r_2}$	$q_{12} = \dfrac{\sigma A_1(T_1^4 - T_2^4)}{\dfrac{1}{\varepsilon_1} + \dfrac{1 - \varepsilon_2}{\varepsilon_2} \left(\dfrac{r_1}{r_2} \right)}$
Concentric sphere 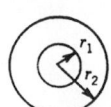	$\dfrac{A_1}{A_2} = \dfrac{r_1^2}{r_2^2}$	$q_{12} = \dfrac{\sigma A_1(T_1^4 - T_2^4)}{\dfrac{1}{\varepsilon_1} + \dfrac{1 - \varepsilon_2}{\varepsilon_2} \left(\dfrac{r_1}{r_2} \right)^2}$
small convex object in a large cavity 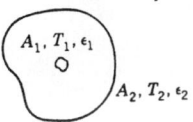	$\dfrac{A_1}{A_2} \approx 0$	$q_{12} = \sigma A_1 \varepsilon_1 (T_1^4 - T_2^4)$

with the shield to that without the shield can be expressed by the relationship

$$\frac{q_{12 \text{ with shield}}}{q_{12 \text{ without shield}}} = \frac{1}{1 + \chi}$$

Values for this ratio, χ, for shields between parallel plates, concentric cylinders, and concentric spheres are summarized in Table 21. For the special case of parallel plates involving more than one or N shields, where all of the emissivities are equal, the value of χ equals N.

Radiation Heat Transfer Coefficient The rate at which radiation heat transfer occurs can be expressed in a form similar to Fourier's law or Newton's law of cooling, by expressing it in terms of the temperature difference $T_1 - T_2$, or as

$$q = h_r A (T_1 - T_2)$$

where h_r is the radiation heat transfer coefficient or *radiation film coefficient*. For the case of radiation between two large parallel plates with emissivities, respectively, of ε_1 and ε_2,

$$h_r \quad \frac{\sigma(T_1^4 - T_2^4)}{T_1 - T_2(1/\varepsilon_1 + 1/\varepsilon_2 - 1)}$$

3.5 Thermal Radiation Properties of Gases

All of the previous expressions assumed that the medium present between the surfaces did not affect the radiation exchange. In reality, gases such as air, oxygen (O_2), hydrogen (H_2), and nitrogen (N_2) have a symmetrical molecular structure and neither emit nor absorb radiation at low to moderate temperatures. Hence, for most engineering applications, such *non-participating gases* can be ignored. However, polyatomic gases such as water vapor (H_2O), carbon dioxide (CO_2), carbon monoxide (CO), sulfur dioxide (SO_2), and various hydrocarbons emit and absorb significant amounts of radiation. These *participating gases* absorb and emit radiation in limited spectral ranges, referred to as spectral *bands*. In calculating the emitted or absorbed radiation for a gas layer, its thickness, shape, surface area, pressure, and temperature distribution must be considered. Although a precise method for calculating the effect of these participating media is quite complex, an approximate method developed by Hottel[21] will yield results that are reasonably accurate.

The effective total emissivities of carbon dioxide and water vapor are a function of the temperature and the product of the partial pressure and the mean beam length of the substance as indicated in Figs. 25 and 26, respectively. The *mean beam length*, L_e, is the characteristic length that corresponds to the radius of a hemisphere of gas, such that the energy flux radiated to the center of the base is equal to the average flux radiated to the area of interest by the actual gas volume. Table 22 lists the mean beam lengths of several simple shapes. For a geometry for which L_e has not been determined, it is generally approximated

Table 21 Values of X for Radiative Shields

Geometry		X
	$\dfrac{\dfrac{1}{\varepsilon_{s1}} + \dfrac{1}{\varepsilon_{s2}} - 1}{\dfrac{1}{\varepsilon_1} + \dfrac{1}{\varepsilon_2} - 1}$	Infinitely long parallel plates
	$\dfrac{\left(\dfrac{r_1}{r_2}\right)^2 \left(\dfrac{1}{\varepsilon_{s1}} + \dfrac{1}{\varepsilon_{s2}} - 1\right)}{\dfrac{1}{\varepsilon_1} + \left(\dfrac{1}{\varepsilon_2} - 1\right)\left(\dfrac{r_1}{r_2}\right)^2}$	$n = 1$ for infinitely long concentric cylinders $n = 2$ for concentric spheres

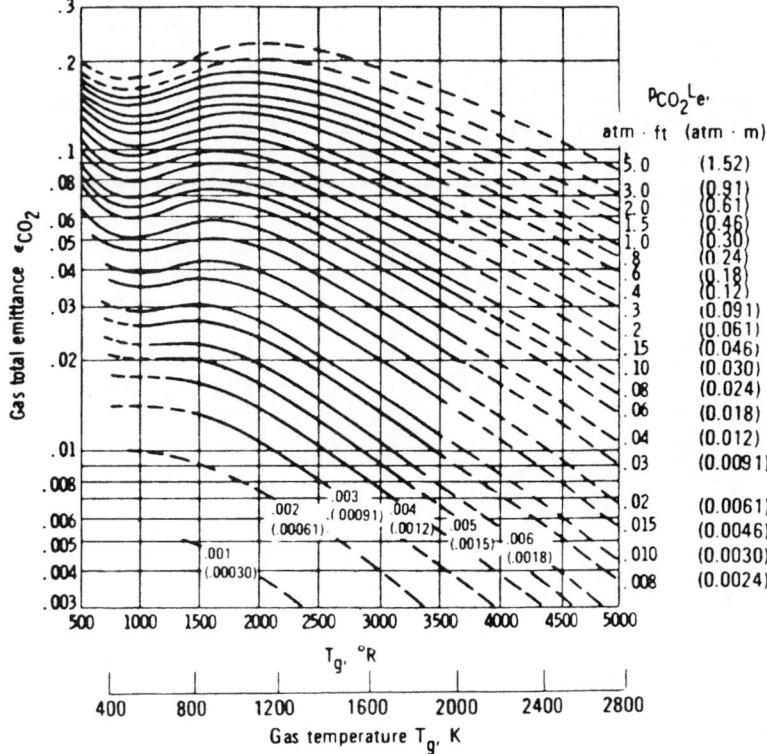

Fig. 25 Total emissivity of CO_2 in a mixture having a total pressure of 1 atm. (From Ref. 21. Used with the permission of McGraw-Hill Book Company.)

by $L_e = 3.6 V/A$ for an entire gas volume V radiating to its entire boundary surface A. The data in Figs. 25 and 26 were obtained for a total pressure of 1 atm and zero partial pressure of the water vapor. For other total and partial pressures the emissivities are corrected by multiplying C_{CO_2} (Fig. 27) and C_{H_2O} (Fig. 28), respectively, to ε_{CO_2} and ε_{H_2O} which are found from Figs. 25 and 26.

These results can be applied when water vapor or carbon dioxide appear separately or in a mixture with other nonparticipating gases. For mixtures of CO_2 and water vapor in a nonparticipating gas, the total emissivity of the mixture, ε_g, can be estimated from the expression

$$\varepsilon_g = C_{CO_2}\varepsilon_{CO_2} + C_{H_2O}\varepsilon_{H_2O} - \Delta\varepsilon$$

where $\Delta\varepsilon$ is a correction factor given in Fig. 29.

Radiative Exchange between Gas Volume and Black Enclosure of Uniform Temperature When radiative energy is exchanged between a gas volume and a black enclosure, the exchange per unit area, q'', for a gas volume at uniform temperature, T_g, and a uniform wall temperature, T_w, is given by

$$q'' = \varepsilon_g(T_g)\sigma T_g^4 - \alpha_g(T_w)\sigma T_w^4$$

where $\varepsilon_g(T_g)$ is the gas emissivity at a temperature T_g and $\alpha_g(T_w)$ is the absorptivity of gas for the radiation from the black enclosure at T_w. As a result of the nature of the band structure of the gas, the absorptivity, α_g, for black radiation at a temperature T_w is different from the emissivity, ε_g, at a gas temperature of T_g. When a mixture of carbon dioxide and water vapor is present, the empirical expression for α_g is

$$\alpha_g = \alpha_{CO_2} + \alpha_{H_2O} - \Delta\alpha$$

where

$$\alpha_{CO_2} = C_{CO_2}\varepsilon'_{CO_2}\left(\frac{T_g}{T_w}\right)^{0.65}$$

$$\alpha_{H_2O} = C_{H_2O}\varepsilon'_{H_2O}\left(\frac{T_g}{T_w}\right)^{0.45}$$

where $\Delta\alpha = \Delta\varepsilon$ and all properties are evaluated at T_w.

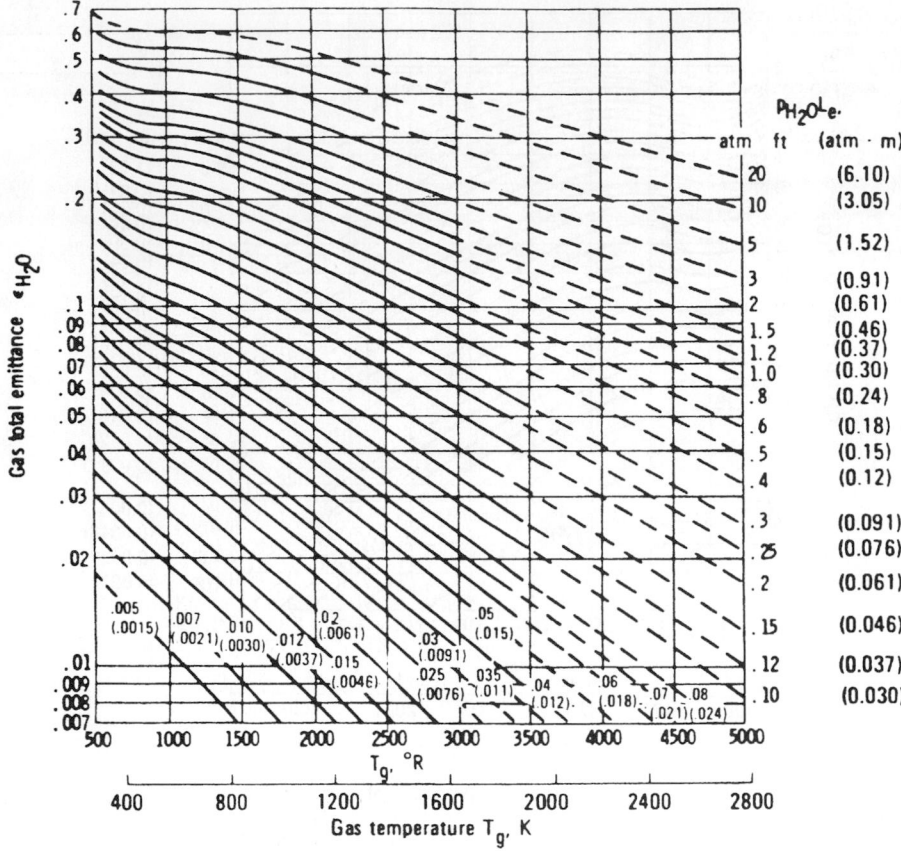

Fig. 26 Total emissivity of H_2O at 1 atm total pressure and zero partial pressure (From Ref. 21. Used with the permission of McGraw-Hill Book Company.)

In this expression, the values of ε'_{CO_2} and ε'_{H_2O} can be found from Figs. 25 and 26 using an abscissa of T_w, but substituting the parameters $p_{CO_2} L_e T_w / T_g$ and $p_{H_2O} L_e T_w / T_g$ for $p_{CO_2} L_e$ and $p_{H_2O} L_e$, respectively.

Radiative Exchange between a Gray Enclosure and a Gas Volume   When the emissivity of the enclosure, ε_w, is larger than 0.8, the rate of heat transfer may be approximated by

$$q_{gray} = \left(\frac{\varepsilon_w + 1}{2} \right) q_{black}$$

where q_{gray} is the heat transfer rate for gray enclosure and q_{black} is that for black enclosure. For values of $\varepsilon_w < 0.8$, the band structures of the participating gas must be taken into account for heat transfer calculations.

4   BOILING AND CONDENSATION HEAT TRANSFER

Boiling and condensation are both forms of convection in which the fluid medium is undergoing a change of phase. When a liquid comes into contact with a solid surface maintained at a temperature above the saturation temperature of the liquid, the liquid may vaporize, resulting in boiling. This process is always accompanied by a change of phase, from the liquid to the vapor state, and results in large rates of heat transfer from the solid surface, due to the latent heat of vaporization of the liquid. The process of condensation is usually accomplished by allowing the vapor to come into contact with a surface at a temperature below the saturation temperature of the vapor, in which case the liquid undergoes a change in state from the vapor state to the liquid state, giving up the latent heat of vaporization.

The heat transfer coefficients for condensation and boiling are generally larger than that for convection

Table 22   Mean Beam Length[a]

Geometry of Gas Volume	Characteristic Length	L_e
Hemisphere radiating to element at center of base	Radius R	R
Sphere radiating to its surface	Diameter D	$0.65D$
Circular cylinder of infinite height radiating to concave bounding surface	Diameter D	$0.95D$
Circular cylinder of semi-infinite height radiating to:		
Element at center of base	Diameter D	$0.90D$
Entire base	Diameter D	$0.65D$
Circular cylinder of height equal to diameter radiating to:		
Element at center of base	Diameter D	$0.71D$
Entire surface	Diameter D	$0.60D$
Circular cylinder of height equal to two diameters radiating to:		
Plane end	Diameter D	$0.60D$
Concave surface	Diameter D	$0.76D$
Entire surface	Diameter D	$0.73D$
Infinite slab of gas radiating to:		
Element on one face	Slab thickness D	$1.8D$
Both bounding planes	Slab thickness D	$1.8D$
Cube radiating to a face	Edge X	$0.6X$
Gas volume surrounding an infinite tube bundle and radiating to a single tube:		
Equilateral triangular array:		
$S = 2D$	Tube diameter D and spacing between tube centers, S	$3.0(S - D)$
$S = 3D$		$3.8(S - D)$
Square array:		$3.5(S - D)$
$S = 2D$		

[a] Adapted from Ref. 19.

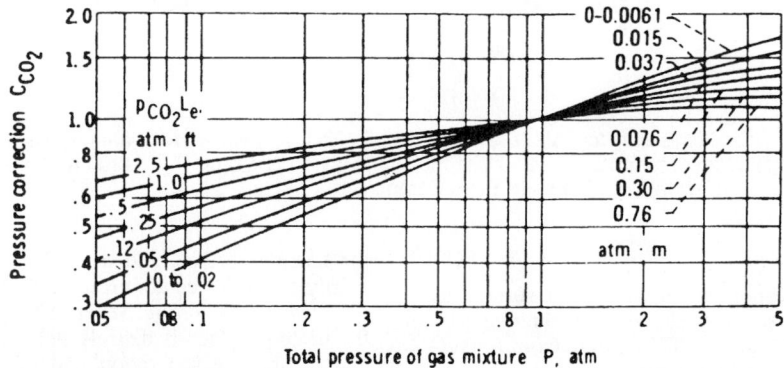

Fig. 27   Pressure correction for CO_2 total emissivity for values of P other than 1 atm. (Adapted from Ref. 21. Used with the permission of McGraw-Hill Book Company.)

without phase change, sometimes by as much as several orders of magnitude. Application of boiling and condensation heat transfer may be seen in a closed-loop power cycle or in a device referred to as a *heat pipe*, which will be discussed in the following section.

In power cycles, the liquid is vaporized in a boiler at high pressure and temperature. After producing work by means of expansion through a turbine, the vapor is condensed to the liquid state in a condenser and then returned to the boiler where the cycle is repeated.

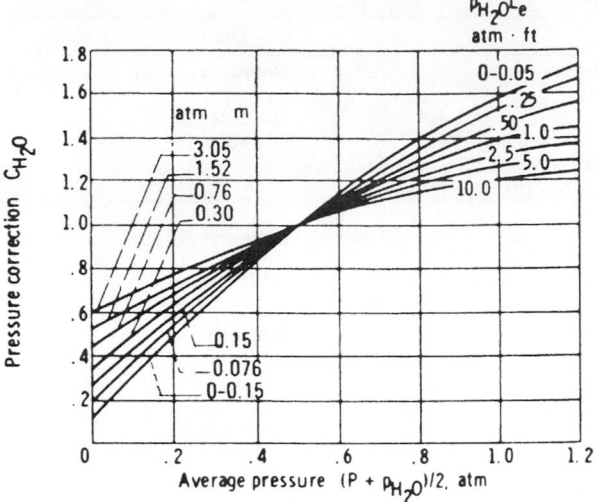

Fig. 28 Pressure correction for water vapor total emissivity for values of p_{H_2O} and P other than 0 and 1 atm. (Adapted from Ref. 21. Used with the permission of McGraw-Hill Book Company.)

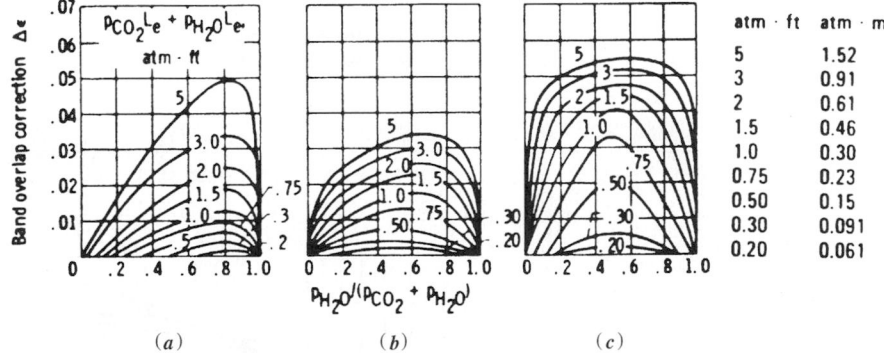

Fig. 29 Correction on total emissivity for band overlap when both CO_2 and water vapor are present: (a) gas temperature $T_g = 400$ K (720°R); (b) gas temperature $T_g = 810$ K (1460°R); (c) gas temperature $T_g = 1200$ K (2160°R). (Adapted from Ref. 21. Used with the permission of McGraw-Hill Book Company.)

4.1 Boiling

The formation of vapor bubbles on a hot surface in contact with a quiescent liquid without external agitation is called *pool boiling*. This differs from *forced-convection boiling* in which forced convection occurs simultaneously with boiling. When the temperature of the liquid is below the saturation temperature, the process is referred to as *subcooled boiling*. When the liquid temperature is maintained or exceeds the saturation temperature, the process is referred to as *saturated or saturation boiling*. Figure 30 depicts the surface heat flux, q'', as a function of the excess temperature, $\Delta T_e = T_s - T_{sat}$, for typical pool boiling of water using an electrically heated wire. In the region $0 < \Delta T_e < \Delta T_{e,A}$ bubbles occur only on selected spots of

the heating surface, and the heat transfer occurs primarily through free convection. This process is called *free-convection boiling*. When $\Delta T_{e,A} < \Delta T_e < \Delta T_{e,C}$, the heated surface is densely populated with bubbles, and the bubble separation and eventual rise due to buoyancy induce a considerable stirring action in the fluid near the surface. This stirring action substantially increases the heat transfer from the solid surface. This process or region of the curve is referred to as *nucleate boiling*. When the excess temperature is raised to $\Delta T_{e,C}$, the heat flux reaches a maximum value, and further increases in the temperature will result in a decrease in the heat flux. The point at which the heat flux is at a maximum value, is called the *critical heat flux*.

Boiling regimes

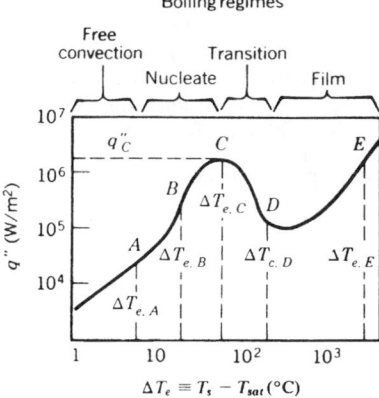

Fig. 30 Typical boiling curve for a wire in a pool of water at atmospheric pressure.

Film boiling occurs in the region where $\Delta T_e > \Delta T_{e,D}$, and the entire heating surface is covered by a vapor film. In this region the heat transfer to the liquid is caused by conduction and radiation through the vapor. Between points C and D, the heat flux decreases with increasing ΔT_e. In this region, part of the surface is covered by bubbles and part by a film. The vaporization in this region is called *transition boiling* or *partial film boiling*. The point of maximum heat flux, point C, is called the *burnout point* or the *Linden frost point*. Although it is desirable to operate the vapor generators at heat fluxes close to q_c'', to permit the maximum use of the surface area, in most engineering applications it is necessary to control the heat flux and great care is taken to avoid reaching this point. The primary reason for this is that, as illustrated, when the heat flux is increased gradually, the temperature rises steadily until point C is reached. Any increase of heat flux beyond the value of q_c'', however, will dramatically change the surface temperature to $T_s = T_{sat} + T_{e,E}$, typically exceeding the solid melting point and leading to failure of the material in which the liquid is held or from which the heater is fabricated.

Nucleate Pool Boiling   The heat flux data are best correlated by[26]

$$q'' = \mu_l h_{fg} \left[\frac{g(\rho_l - \rho_v)}{g_c \sigma} \right]^{1/2} \left(\frac{c_{p,l}\ \Delta T_e}{C h_{fg}\ \mathrm{Pr}_l^{1.7}} \right)^3$$

where the subscripts l and v denote saturated liquid and vapor, respectively. The surface tension of the liquid is σ (N/m). The quantity g_c is the proportionality constant equal to 1 kg·m/N·s². The quantity g is the local gravitational acceleration in m/s². The values of C are given in Table 23. The above equation may be applied to different geometries, such as plates, wire, or cylinders.

The *critical heat flux* (point C of Fig. 30) is given by[28]

$$q_c'' = \frac{\pi}{24} h_{fg}\rho_v \left[\frac{\sigma g g_c (\rho_l - \rho_v)}{\rho_v^2} \right]^{0.25} \left(1 + \frac{\rho_v}{\rho_l} \right)^{0.5}$$

For a water–steel combination, $q_c'' \approx 1290$ kW/m² and $\Delta T_{e,c} \approx 30°$C. For water–chrome-plated copper, $q_c'' \approx 940-1260$ KW/m² and $\Delta T_{e,c} \approx 23-28°$C.

Film Pool Boiling   The heat transfer from the surface to the liquid is due to both convection and radiation. A total heat transfer coefficient is defined by the combination of convection and radiation heat transfer coefficients of the following form[29] for the outside surfaces of horizontal tubes:

$$h^{4/3} = h_c^{4/3} + h_r h^{1/3}$$

where

$$h_c = 0.62 \left[\frac{k_v^3 \rho_v (\rho_l - \rho_v) g (h_{fg} + 0.4 c_{p,v}\ \Delta T_e)}{\mu_v D\ \Delta T_e} \right]^{1/4}$$

and

$$h_r = \frac{5.73 \times 10^{-8} \varepsilon (T_s^4 - T_{sat}^r)}{T_s - T_{sat}}$$

The vapor properties are evaluated at the film temperature $T_f = (T_s + T_{sat})/2$. The temperatures T_s and T_{sat} are in kelvins for the evaluation of h_r. The emissivity of the metallic solids can be found from Table 17. Note that $q = hA\ (T_s - T_{sat})$.

Nucleate Boiling in Forced Convection   The total heat transfer rate can be obtained by simply superimposing the heat transfer due to nucleate boiling and forced convection:

$$q'' = q''_{\text{boiling}} + q''_{\text{forced convection}}$$

Table 23   Values of the Constant C for Various Liquid–Surface Combinations

Fluid-Heating Surface Combinations	C
Water with polished copper, platinum, or mechanically polished stainless steel	0.0130
Water with brass or nickel	0.006
Water with ground and polished stainless steel	0.008
Water with Teflon-plated stainless steel	0.008

For forced convection, it is recommended that the coefficient 0.023 be replaced by 0.014 in the Dittus–Boelter equation (Section 2.1). The above equation is generally applicable to forced convection where the bulk liquid temperature is subcooled (*local forced convection boiling*).

Simplified Relations for Boiling in Water For *nucleate boiling*,[30]

$$h = C(\Delta T_e)^n \left(\frac{p}{p_a}\right)^{0.4}$$

where p and p_a are, respectively, the system pressure and standard atmospheric pressure. The constants C and n are listed in Table 24.

For *local forced convection boiling inside vertical tubes*, valid over a pressure range of 5–170 atm,[31]

$$h = 2.54(\Delta T_e)^3 e^{p/1.551}$$

where h has the unit $W/m^2 \cdot {}^\circ C$, ΔT_e is in $^\circ C$, and p is the pressure in 10^6 N/m^3.

4.2 Condensation

Depending on the surface conditions, the condensation may be a *film condensation* or a *dropwise condensation*. Film condensation usually occurs when a vapor, relatively free of impurities, is allowed to condense on a clean, uncontaminated surface. Dropwise condensation occurs on highly polished surfaces or on surfaces coated with substances that inhibit wetting. The condensate provides a resistance to heat transfer between the vapor and the surface. Therefore, it is desirable to use short vertical surfaces or horizontal cylinders to prevent the condensate from growing too thick. The heat transfer rate for dropwise condensation is usually an order of magnitude larger than that for film condensation under similar conditions. Silicones, Teflon, and certain fatty acids can be used to coat the surfaces to promote dropwise condensation. However, such coatings may lose their effectiveness owing to oxidation or outright removal. Thus, except under carefully controlled conditions, film condensation may be expected to occur in most instances, and the condenser design calculations are often based on the assumption of film condensation.

For condensation on the surface at temperature T_s the total heat transfer rate to the surface is given by $q = \bar{h}_L A (T_{sat} - T_s)$, where T_{sat} is the saturation temperature of the vapor. The mass flow rate is determined by $\dot{m} = q/h_{fg}$; h_{fg} is the latent heat of vaporization of the fluid (see Table 25 for saturated water). Correlations are based on the evaluation of liquid properties at $T_f = (T_s + T_{sat})/2$, except h_{fg}, which is to be taken at T_{sat}.

Film Condensation on a Vertical Plate The Reynolds number for *condensate flow* is defined by $\mathrm{Re}_\Gamma = \rho_l V_m D_h/\mu_l$, where ρ_l and μ_l are the density and viscosity of the liquid, V_m is the average velocity of condensate, and D_h is the hydraulic diameter defined by $D_h = 4 \times$ condensate film cross-sectional area/wetted perimeter. For the condensation on a vertical plate $\mathrm{Re}_\Gamma = 4\Gamma/\mu_l$, where Γ is the mass flow rate of condensate per unit width evaluated at the lowest point on the condensing surface. The condensate flow is generally considered to be laminar for $\mathrm{Re}_\Gamma < 1800$, and turbulent for $\mathrm{Re}_\Gamma > 1800$. The average Nusselt number is given by[22]

$$\overline{\mathrm{Nu}}_L = \begin{cases} 1.13 \left[\dfrac{g\rho_l(\rho_l - \rho_v)h_{fg}L^3}{\mu_l k_l(T_{sat} - T_s)}\right]^{0.25} & \text{for } \mathrm{Re}_\Gamma < 1800 \\[3mm] 0.0077 \left[\dfrac{g\rho_l(\rho_l - \rho_v)L^3}{\mu_l^2}\right]^{1/3} \mathrm{Re}_\Gamma^{0.4} & \text{for } \mathrm{Re}_\Gamma > 1800 \end{cases}$$

Film Condensation on the Outside of Horizontal Tubes and Tube Banks

$$\overline{\mathrm{Nu}}_D = 0.725 \left[\frac{g\rho_l(\rho_l - \rho_v)h_{fg}D^3}{N\mu_l k_l(T_{sat} - T_s)}\right]^{0.25}$$

where N is the number of horizontal tubes placed one above the other; $N = 1$ for a single tube.[23]

Film Condensation Inside Horizontal Tubes For low vapor velocities such that Re_D based on the vapor velocities at the pipe inlet is less than 3500[24]

$$\overline{\mathrm{Nu}}_D = 0.555 \left[\frac{g\rho_l(\rho_l - \rho_l)h'_{fg}D^3}{\mu_l k_l(T_{sat} - T_s)}\right]^{0.25}$$

Table 24   Values of *C* and *n* for Simplified Relations for Boiling in Water

Surface	q''(kW/m^2)	C	n
Horizontal	$q'' < 16$	1042	1/3
	$16 < q'' < 240$	5.56	3
Vertical	$q'' < 3$	5.7	1/7
	$3 < q'' < 63$	7.96	3

Table 25 Thermophysical Properties of Saturated Water

Temperature, T (K)	Pressure, P (bar)	Specific Volume (m³/kg)		Heat of Vaporization, h_{fg} (kJ/kg)	Specific Heat (kJ/kg·K)		Viscosity (N·s/m²)		Thermal Conductivity (W/m·K)		Prandtl Number		Surface Tension $\sigma_l \times 10^3$ (N/m)	Expansion Coefficient, $\beta_l \times 10^6$ (K⁻¹)
		$v_f \times 10^3$	v_u		$C_{p,l}$	$C_{p,u}$	$\mu_l \times 10^6$	$\mu_v \times 10^3$	$k_l \times 10^3$	$k_v \times 10^3$	Pr_l	Pr_v		
273.15	0.00611	1.000	206.3	2502	4.217	1.854	1750	8.02	659	18.2	12.99	0.815	75.5	−68.05
300	0.03531	1.003	39.13	2438	4.179	1.872	855	9.09	613	19.6	5.83	0.857	71.7	276.1
320	0.1053	1.011	13.98	2390	4.180	1.895	577	9.89	640	21.0	3.77	0.894	68.3	436.7
340	0.2713	1.021	5.74	2342	4.188	1.930	420	10.69	660	22.3	2.66	0.925	64.9	566.0
360	0.6209	1.034	2.645	2291	4.203	1.983	324	11.49	674	23.7	2.02	0.960	61.4	697.9
380	1.2869	1.049	1.337	2239	4.226	2.057	260	12.29	683	25.4	1.61	0.999	57.6	788
400	2.455	1.067	0.731	2183	4.256	2.158	217	13.05	688	27.2	1.34	1.033	63.6	896
450	9.319	1.123	0.208	2024	4.40	2.56	152	14.85	678	33.1	0.99	1.14	42.9	
500	26.40	1.203	0.0766	1825	4.66	3.27	118	16.59	642	42.3	0.86	1.28	31.6	
550	61.19	1.323	0.0317	1564	5.24	4.64	97	18.6	580	58.3	0.87	1.47	19.7	
600	123.5	1.541	0.0137	1176	7.00	8.75	81	22.7	497	92.9	1.14	2.15	8.4	
647.3	221.2	3.170	0.0032	0	∞	∞	45	45	238	238	∞	∞	0.0	

where $h'_{fg} + \frac{3}{8}C_{p,l}(T_{\text{sat}} - T_s)$. For higher flow rate,[25] $\text{Re}_G > 5 \times 10^4$,

$$\overline{\text{Nu}}_D = 0.0265\,\text{Re}_G^{0.8}\,\text{Pr}^{1/3}$$

where the Reynolds number $\text{Re}_G = GD/\mu_l$ is based on the equivalent mass velocity $G = G_l + G_v(\rho_l/\rho_v)^{0.5}$. The mass velocity for the liquid G_l and that for vapor G_v are calculated as if each occupied the entire flow area.

Effect of Noncondensable Gases If noncondensable gas such as air is present in a vapor, even in a small amount, the heat transfer coefficient for condensation may be greatly reduced. It has been found that the presence of a few percent of air by volume in steam reduces the coefficient by 50% or more. Therefore, it is desirable in the condenser design to vent the noncondensable gases as much as possible.

4.3   Heat Pipes

Heat pipes are two-phase heat transfer devices that operate on a closed two-phase cycle[32] and come in a wide variety of sizes and shapes.[33,34] As shown in Fig. 31, they typically consist of three distinct regions: the evaporator or heat addition region, the condenser or heat rejection region, and the adiabatic or isothermal region. Heat added to the evaporator region of the container causes the working fluid in the evaporator wicking structure to be vaporized. The

high temperature and corresponding high pressure in this region result in flow of the vapor to the other, cooler end of the container where the vapor condenses, giving up its latent heat of vaporization. The capillary forces existing in the wicking structure then pump the liquid back to the evaporator section. Other similar devices, referred to as two-phase thermosyphons have no wick, and utilize gravitational forces to provide the liquid return. Thus, the heat pipe functions as a nearly isothermal device, adjusting the evaporation rate to accommodate a wide range of power inputs, while maintaining a relatively constant source temperature.

Transport Limitations The transport capacity of a heat pipe is limited by several important mechanisms. Among these are the capillary wicking limit, viscous limit, sonic limit, entrainment, and boiling limits. The capillary wicking limit and viscous limits deal with the pressure drops occurring in the liquid and vapor phases, respectively. The sonic limit results from the occurrence of choked flow in the vapor passage, while the entrainment limit is due to the high liquid vapor shear forces developed when the vapor passes in counterflow over the liquid saturated wick. The boiling limit is reached when the heat flux applied in the evaporator portion is high enough that nucleate boiling occurs in the evaporator wick, creating vapor bubbles that partially block the return of fluid.

To function properly, the net capillary pressure difference between the condenser and the evaporator in a heat pipe must be greater than the pressure losses throughout the liquid and vapor flow paths. This

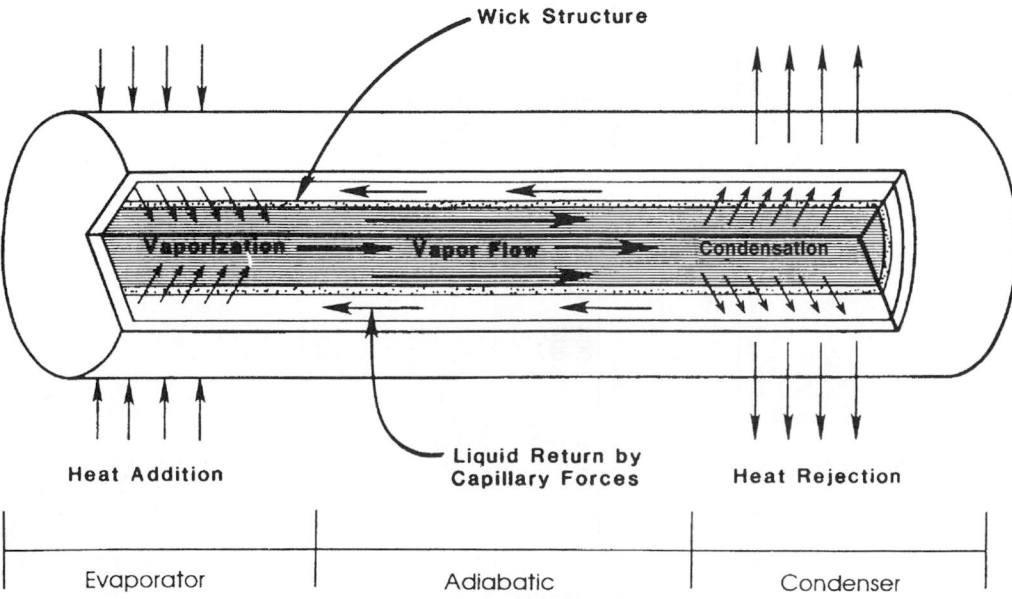

Fig. 31   Typical heat pipe construction and operation.[35]

relationship can be expressed as

$$\Delta P_c \geq \Delta P_+ + \Delta P_- + \Delta P_l + \Delta P_v$$

where ΔP_c = net capillary pressure difference
ΔP_+ = normal hydrostatic pressure drop
ΔP_- = axial hydrostatic pressure drop
ΔP_l = viscous pressure drop occurring in the liquid phase
ΔP_v = viscous pressure drop occurring in the vapor phase

If these conditions are not met, the heat pipe is said to have reached the *capillary limitation*.

Expressions for each of these terms have been developed for steady-state operation, and are summarized below.

Capillary pressure: $\quad \Delta P_{c,m} = \left(\dfrac{2\sigma}{r_{c,e}}\right)$

Values for the effective capillary radius, r_c, can be found theoretically for simple geometries or experimentally for pores or structures of more complex geometry. Table 26 gives values for some common wicking structures.

Normal and axial hydrostatic pressure drop

$$\Delta P_+ + \rho_l g d_v \cos \psi$$

$$\Delta P_- = \rho_l g L \sin \psi$$

In a gravitational environment, the axial hydrostatic pressure term may either assist or hinder the capillary pumping process, depending on whether the tilt of the heat pipe promotes or hinders the flow of liquid back to the evaporator (i.e., the evaporator lies either below or above the condenser). In a zero-g environment, both this term and the normal hydrostatic pressure drop term can be neglected because of the absence of body forces.

Liquid pressure drop $\quad \Delta P_l = \left(\dfrac{\mu_l}{K A_w h_{fg} \rho_l}\right) L_{\text{eff}} q$

where L_{eff} is the effective heat pipe length defined as

$$L_{\text{eff}} = 0.5 L_e + L_a + 0.5 L_c$$

and K is the liquid permeability as shown in Table 27.

Vapor pressure drop $\quad \Delta P_v = \left[\dfrac{C(f_v \text{Re}_v)\mu_v}{2(r_{h,v})^2 A_v \rho_v h_{fg}}\right] L_{\text{eff}} q$

Although during steady-state operation the liquid flow regime is always laminar, the vapor flow may be either laminar or turbulent. It is therefore necessary to determine the vapor flow regime as a function of the heat flux. This can be accomplished by evaluating the local axial Reynolds and Mach numbers and substituting the values as shown below:

$\text{Re}_v < 2300 \qquad \text{Ma}_v < 0.2$

$(f_v \text{Re}_v) = 16$

$C = 1.00$

$\text{Re}_v < 2300 \quad \text{Ma}_v > 0.2$

$(f_v \, \text{Re}_v) = 16$

$C = \left[1 + \left(\dfrac{\gamma_v - 1}{2}\right)\text{Ma}_v^2\right]^{1/2}$

$\text{Re}_v > 2300 \quad \text{Ma}_v < 0.2$

$(f_v \text{Re}_v) = 0.038\left[\dfrac{2(r_{h,v})q}{A_v \mu_v h_{fg}}\right]^{3/4}$

$C = 1.00$

$\text{Re}_v > 2300 \qquad \text{Ma}_v > 0.2$

$(f_v \text{Re}_v) = 0.038\left[\dfrac{2(r_{h,v})q}{A_v \mu_v h_{fg}}\right]^{3/4}$

$C = \left[1 + \left(\dfrac{\gamma_v - 1}{2}\right)\text{Ma}_v^2\right]^{-1/2}$

Table 26 Expressions for the Effective Capillary Radius for Several Wick Structures

Structure	r_c	Data
Circular cylinder (artery or tunnel wick)	r	r = radius of liquid flow passage
Rectangular groove	ω	ω = groove width
Triangular groove	$\omega/\cos \beta$	ω = groove width; β = half-included angle
Parallel wires	ω	ω = wire spacing
Wire screens	$(\omega + d_\omega)/2 = \frac{1}{2}N$	d = wire diameter; N = screen mesh number; ω = wire spacing
Packed spheres	$0.41 r_s$	r_s = sphere radius

Table 27   Wick Permeability for Several Wick Structures

Structure	K	Data
Circular cylinder (artery or tunnel wick)	$r^2/8$	r = radius of liquid flow passage
Open rectangular grooves	$2\varepsilon(r_{h,l})^2/(f_l \mathrm{Re}_l) = \omega/s$	ε = wick porosity ω = groove width s = groove pitch δ = groove depth $(r_{h,l}) = 2\omega\delta/(\omega + 2\delta)$
Circular annular wick	$2(r_{h,l})^2/(f_l \mathrm{Re}_l)$	$(r_{h,l}) = r_1 - r_2$
Wrapped screen wick	$1/22\, d_\omega^2 \varepsilon^3/(1-\varepsilon)^2$	d_ω = wire diameter $\varepsilon = 1 - (1.05\pi N d\omega/4)$ N = mesh number
Packed sphere	$1/37.5 r_s^2 \varepsilon^3/(1-\varepsilon)^2$	r_s = sphere radius ε = porosity (dependent on packing mode)

Since the equations used to evaluate both the Reynolds number and the Mach number are functions of the heat transport capacity, it is necessary to first assume the conditions of the vapor flow. Using these assumptions, the maximum heat capacity, $q_{c,m}$, can be determined by substituting the values of the individual pressure drops into Eq. (1) and solving for $q_{c,m}$. Once the value of $q_{c,m}$ is known, it can then be substituted into the expressions for the vapor Reynolds number and Mach number to determine the accuracy of the original assumption. Using this iterative approach, accurate values for the capillary limitation as a function of the operating temperature can be determined in units of W-m or watts for $(qL)_{c,m}$ and $q_{c,m}$, respectively.

The *viscous limitation* in heat pipes occurs when the viscous forces within the vapor region are dominant and limit the heat pipe operation:

$$\frac{\Delta P_v}{P_v} < 0.1$$

for determining when this limit might be of a concern. Due to the operating temperature range, this limitation will normally be of little consequence in the design of heat pipes for use in the thermal control of electronic components and devices.

The *sonic limitation* in heat pipes is analogous to the sonic limitation in a converging–diverging nozzle and can be determined from

$$q_{s,m} = A_v \rho_v h_{fg} \left[\frac{\gamma_v R_v T_v}{2(\gamma_v + 1)} \right]^{1/2}$$

where T_v is the mean vapor temperature within the heat pipe.

Since the liquid and vapor flow in opposite directions in a heat pipe, at high enough vapor velocities, liquid droplets may be picked up or entrained in the vapor flow. This entrainment results in excess liquid accumulation in the condenser and, hence, dryout of the evaporator wick. Using the Weber number, We, defined as the ratio of the viscous shear force to the force resulting from the liquid surface tension, an expression for the *entrainment limit* can be found as

$$q_{e,m} = A_v h_{fg} \left[\frac{\sigma \rho_v}{2(r_{h,w})} \right]^{1/2}$$

where $(r_{h,w})$ is the hydraulic radius of the wick structure, defined as twice the area of the wick pore at the wick–vapor interface divided by the wetted perimeter at the wick–vapor interface.

The *boiling limit* occurs when the input heat flux is so high that nucleate boiling occurs in the wicking structure and bubbles may become trapped in the wick, blocking the liquid return and resulting in evaporator dryout. This phenomenon, referred to as the boiling limit, differs from the other limitations previously discussed in that it depends on the evaporator heat flux as opposed to the axial heat flux. This expression, which is a function of the fluid properties, can be written as

$$q_{b,m} = \left[\frac{2\pi L_{\mathrm{eff}} k_{\mathrm{eff}} T_v}{h_{fg} \rho_v \ln(r_i/r_v)} \right] \left(\frac{2\sigma}{r_n} - \Delta P_{c,m} \right)$$

where k_{eff} is the effective thermal conductivity of the liquid–wick combination, given in Table 28, r_i is the inner radius of the heat pipe wall, and r_n is the nucleation site radius. After the power level associated with each of the four limitations is established, determination of the maximum heat transport capacity is only a matter of selecting the lowest limitation for any given operating temperature.

Heat Pipe Thermal Resistance   The *heat pipe thermal resistance* can be found using an analogous electrothermal network. Figure 32 illustrates the

Table 28 Effective Thermal Conductivity for Liquid-Saturated Wick Structures

Wick Structures	k_{eff}
Wick and liquid in series	$\dfrac{k_l k_w}{\varepsilon k_w + k_l(1-\varepsilon)}$
Wick and liquid in parallel	$\varepsilon k_l + k_w(1-\varepsilon)$
Wrapped screen	$\dfrac{k_l[(k_l + k_w) - (1-\varepsilon)(k_l - k_w)]}{(k_l + k_w) + (1-\varepsilon)(k_l - k_w)]}$
Packed spheres	$\dfrac{k_l[(2k_l + k_w) - 2(1-\varepsilon)(k_l - k_w)]}{(2k_l + k_w) + (1-\varepsilon)(k_l - k_w)}$
Rectangular grooves	$\dfrac{(w_f k_l k_w \delta) + w k_l(0.185\, w_f k_w + \delta k_l)}{(w + w_f)(0.185\, w_f k_f + \delta k_l)}$

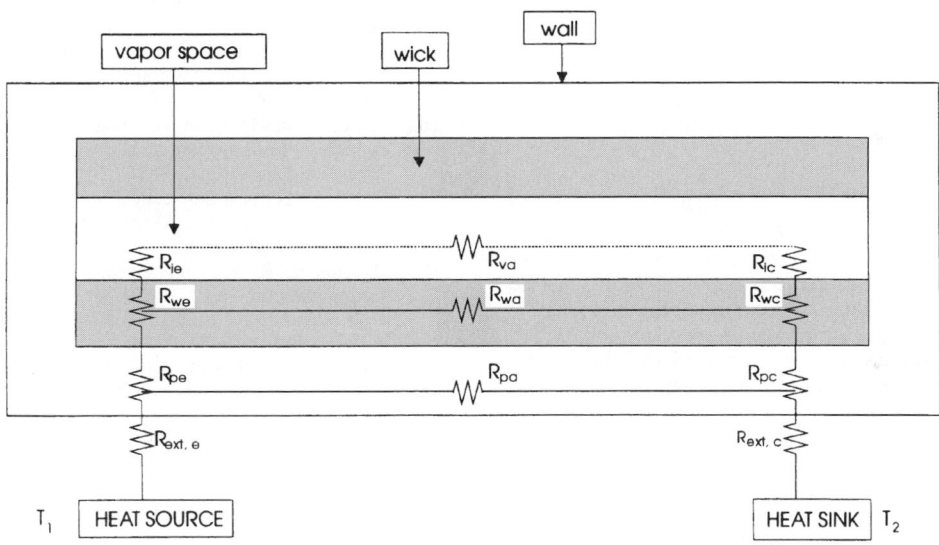

Fig. 32 Equivalent thermal resistance of heat pipe.

electrothermal analog for the heat pipe illustrated in Fig. 31. As shown, the overall thermal resistance is composed of nine different resistances arranged in a series/parallel combination, which can be summarized as follows:

R_{pe} Radial resistance of pipe wall at evaporator
R_{we} Resistance of liquid–wick combination at evaporator
R_{ie} Resistance of liquid–vapor interface at evaporator
R_{ya} Resistance of adiabatic vapor section
R_{pa} Axial resistance of pipe wall
R_{wa} Axial resistance of liquid–wick combination
R_{ic} Resistance of liquid–vapor interface at condenser

R_{wc} Resistance of liquid–wick combination at condenser
R_{pc} Radial resistance of pipe wall at condenser

Because of the comparative magnitudes of the resistance of the vapor space and the axial resistances of the pipe wall and liquid–wick combinations, the axial resistance of both the pipe wall and the liquid–wick combination may be treated as open circuits and neglected. Also, because of the comparative resistances, the liquid–vapor interface resistances and the axial vapor resistance can, in most situations, be assumed to be negligible. This leaves only the pipe wall radial resistances and the liquid–wick resistances at both the evaporator and condenser. The radial resistances at the pipe wall can be computed from Fourier's law as

$$R_{pe} = \frac{\delta}{k_p A_e}$$

for flat plates, where δ is the plate thickness and A_e is the evaporator area, or

$$R_{pe} = \frac{\ln(D_o/D_i)}{2\pi L_e k_p}$$

for cylindrical pipes, where L_e is the evaporator length. An expression for the equivalent thermal resistance of the liquid–wick combination in circular pipes is

$$R_{we} = \frac{\ln(D_o/D_i)}{2\pi L_e k_{\text{eff}}}$$

where values for the effective conductivity, k_{eff}, can be found in Table 28. The adiabatic vapor resistance, although usually negligible, can be found as

$$R_{va} = \frac{T_v(P_{v,e} - P_{v,c})}{\rho_v h_{fg} q}$$

where $P_{v,e}$ and $P_{v,c}$ are the vapor pressures at the evaporator and condenser. Combining these individual resistances provides a mechanism by which the overall thermal resistance can be computed and hence the temperature drop associated with various axial heat fluxes can be computed.

REFERENCES

1. Incropera, F. P., and Dewitt, D. P., *Fundamentals of Heat Transfer*, Wiley, New York, 1981.
2. Eckert, E. R. G., and Drake, R. M., Jr., *Analysis of Heat and Mass Transfer*, McGraw-Hill, New York, 1972.
3. Heisler, M. P., "Temperature Charts for Induction and Constant Temperature Heating," *Trans. ASME*, **69**, 227 (1947).
4. Grober, H., and Erk, S., *Fundamentals of Heat Transfer*, McGraw-Hill, New York, 1961.
4a. Duncan, A. B., and Peterson, G. P., "A Review of Microscale Heat Transfer," invited review article, *Appl. Mechan. Rev.*, **47**(9), 397–428 (1994).
4b. Tien, C. L., Armaly, B. F., and Jagannathan, P. S., "Thermal Conductivity of Thin Metallic Films," in *Proc. 8th Conference on Thermal Conductivity*, October 7–10, 1968.
4c. Bai, C., and Lavine, A. S., "Thermal Boundary Conditions for Hyperbolic Heat Conduction," *ASME HTD*, **253**, 37–44 (1993).
5. Sieder, E. N., and Tate, C. E., "Heat Transfer and Pressure Drop of Liquids in Tubes," *Ind. Eng. Chem.*, **28**, 1429 (1936).
6. Dittus, F. W., and Boelter, L. M. K., *Univ. Calif., Berkeley, Engineering Publication* **2**, 443 (1930).
7. Chapman, A. J., *Heat Transfer*, Macmillan, New York, 1974.
8. Whitaker, S., "Forced Convection Heat Transfer Correlations," *AICHE J.*, **18**, 361 (1972).
9. Jakob, M., *Heat Transfer*, Vol. 1, Wiley, New York, 1949.
10. Zhukauska, A., "Heat Transfer from Tubes in Cross Flow," in *Advances in Heat Transfer*, Vol. 8, J. P. Hartnett and T. F. Irvine, Jr. (Eds.), Academic, New York, 1972.
11. Kreith, F., *Principles of Heat Transfer*, Harper & Row, New York, 1973.
12. Johnson, H. A., and Rubesin, M. W., "Aerodynamic Heating and Convective Heat Transfer," *Trans. ASME*, **71**, 447 (1949).
13. Lin, C. C. (Ed.), *Turbulent Flows and Heat Transfer, High Speed Aerodynamics and Jet Propulsion*, Vol. V, Princeton University Press, Princeton, NJ, 1959.
14. McAdams, W. H., *Heat Transmission*, McGraw-Hill, New York, 1954.
15. Yuge, T., "Experiments on Heat Transfer from Spheres Including Combined Natural and Forced Convection," *J. Heat Transfer*, **82,** 214 (1960).
16. Globe, S., and Dropkin, D., "Natural Convection Heat Transfer in Liquids Confined between Two Horizontal Plates," *J. Heat Transfer*, **81C**, 24 (1959).
17. Catton, I., "Natural Convection in Enclosures," in *Proc. 6th International Heat Transfer Conference*, Vol. 6, Toronto, Canada, 1978.
18. MacGregor, R. K., and Emery, A. P., "Free Convection through Vertical Plane Layers: Moderate and High Prandtl Number Fluids," *J. Heat Transfer*, **91**, 391(1969).
19. Siegel, R., and Howell, J. R., *Thermal Radiation Heat Transfer*, McGraw-Hill, New York, 1981.
20. Gubareff, G. G., Janssen, J. E., and Torborg, R. H., *Thermal Radiation Properties Survey*, 2nd ed., Minneapolis Honeywell Regulator Co., Minneapolis, MN, 1960.
21. Hottel, H. C., in *Heat Transmission*, W. C. McAdams (Ed.), McGraw-Hill, New York, 1954, Chapter 2.
22. McAdams, W. H., *Heat Transmission*, 3rd ed., McGraw-Hill, New York, 1954.
23. Rohsenow, W. M., "Film Condensation" in *Handbook of Heat Transfer*, W. M. Rohsenow and J. P. Hartnett (Eds.), McGraw-Hill, New York, 1973.
24. Chato, J. C., "Laminar Condensation inside Horizontal and Inclined Tubes," *ASHRAE J.*, **4,** 52 (1962).
25. Akers, W. W., Deans, H. A., and Crosser, O. K., "Condensing Heat Transfer within Horizontal Tubes," *Chem. Eng. Prog., Sym. Ser.*, **55**(29), 171 (1958).
26. Rohsenow, W. M., "A Method of Correlating Heat Transfer Data for Surface Boiling Liquids," *Trans. ASME*, **74,** 969 (1952).
27. Holman, J. P., *Heat Transfer*, McGraw-Hill, New York, 1981.
28. Zuber, N., "On the Stability of Boiling Heat Transfer," *Trans. ASME*, **80,** 711 (1958).
29. Bromley, L. A., "Heat Transfer in Stable Film Boiling," *Chem. Eng. Prog.*, **46**, 221 (1950).
30. Jacob, M., and Hawkins, G. A., *Elements of Heat Transfer*, Wiley, New York, 1957.

31. Jacob, M., *Heat Transfer*, Vol. 2, Wiley, New York, 1957, p. 584.
32. Peterson, G. P., *An Introduction to Heat Pipes: Modeling, Testing and Applications*, Wiley, New York, 1994.
33. Peterson, G. P., Duncan, A. B., and Weichold, M. H., "Experimental Investigation of Micro Heat Pipes Fabricated in Silicon Wafers," *ASME J. Heat Transfer*, **115**(3), 751 (1993).
34. Peterson, G. P., "Capillary Priming Characteristics of a High Capacity Dual Passage Heat Pipe," *Chem. Eng. Commun.*, **27**, 1, 119 (1984).
35. Peterson, G. P., and Fletcher, L. S., "Effective Thermal Conductivity of Sintered Heat Pipe Wicks," *AIAA J. Thermophys. Heat Transfer*, **1**(3), 36 (1987).

BIBLIOGRAPHY

American Society of Heating, Refrigerating and Air Conditioning Engineering, *ASHRAE Handbook of Fundamentals*, 1972.

Arpaci, V. S., *Conduction Heat Transfer*, Addison-Wesley, Reading, MA, 1966.

Carslaw, H. S., and Jager, J. C., *Conduction of Heat in Solid*, Oxford University Press, London, 1959.

Chi, S. W., *Heat Pipe Theory and Practice*, McGraw-Hill, New York, 1976.

Duffie, J. A., and Beckman, W. A., *Solar Engineering of Thermal Process*, Wiley, New York, 1980.

Dunn. P. D., and Reay, D. A., *Heat Pipes*, 3rd ed., Pergamon, New York, 1983.

Gebhart, B., *Heat Transfer*, McGraw-Hill, New York, 1971.

Hottel, H. C., and Saroffin, A. F., *Radiative Transfer*, McGraw-Hill, New York, 1967.

Kays, W. M., *Convective Heat and Mass Transfer*, McGraw-Hill, New York, 1966.

Knudsen, J. G., and Katz, D. L., *Fluid Dynamics and Heat Transfer*, McGraw-Hill, New York, 1958.

Ozisik, M. N., *Radiative Transfer and Interaction with Conduction and Convection*, Wiley, New York, 1973.

Ozisik, M. N., *Heat Conduction*, Wiley, New York, 1980.

Peterson, G. P., *An Introduction to Heat Pipes: Modeling, Testing and Applications*, Wiley, New York, 1994.

Planck, M., *The Theory of Heat Radiation*, Dover, New York, 1959.

Rohsenow, W. M., and Choi, H. Y., *Heat, Mass, and Momentum Transfer*, Prentice-Hall, Englewood Cliffs, NJ, 1961.

Rohsenow, W. M., and Hartnett, J. P., *Handbook of Heat Transfer*, McGraw-Hill, New York, 1973.

Schlichting, H., *Boundary-Layer Theory*, McGraw-Hill, New York, 1979.

Schneider, P. J., *Conduction Heat Transfer*, Addison-Wesley, Reading, MA, 1955.

Sparrow, E. M., and Cess, R. D., *Radiation Heat Transfer*, Wadsworth, Belmont, CA, 1966.

Tien, C. L., "Fluid Mechanics of Heat Pipes," *Annu. Rev. Fluid Mechan.*, **7**, 167 (1975).

Turner, W. C., and Malloy, J. F., *Thermal Insulation Handbook*, McGraw-Hill, New York, 1981.

Vargafik, N. B., *Table of Thermophysical Properties of Liquids and Gases*, Hemisphere, Washington, DC, 1975.

Wiebelt, J. A., *Engineering Radiation Heat Transfer*, Holt, Rinehart & Winston, New York, 1966.

CHAPTER 16

ELECTRIC CIRCUITS

Albert J. Rosa
Professor Emeritus
University of Denver
Denver, Colorado

1 INTRODUCTION

1.1 Overview

The purpose of this chapter is to introduce the analysis and design of linear circuits. Circuits are important in electrical engineering because they process electrical signals that carry energy and information. For the present a *circuit* is defined as an interconnection of electrical devices and a *signal* as a time-varying electrical quantity. A modern technological society is intimately dependent on the generation, transfer, and conversion of electrical energy. Recording media like CDs, DVDs, thumb drives, hard drives, and tape-based products; communication systems like radar, cell phones, radio, television, and the Internet; information systems like computers and the world wide web; instrumentation and control systems; and the national electrical power grid X all involve circuits that process

and transfer signals carrying either energy or information or both.

This chapter will focus on *linear circuits*. An important feature of a linear circuit is that the amplitude of the output signal is proportional to the input signal amplitude. The proportionality property of linear circuits greatly simplifies the process of circuit analysis and design. Most circuits are only linear within a restricted range of signal levels. When driven outside this range, they become nonlinear and proportionality no longer applies. Hence only circuits operating within their linear range will be studied.

An important aspect of this study involves interface circuits. An *interface* is defined as a pair of accessible terminals at which signals may be observed or specified. The interface concept is especially important with integrated circuit (IC) technology. Integrated circuits involve many thousands of interconnections, but only

Table 1   Some Important Quantities, Symbols, and Unit Abbreviations

Quantity	Symbol	Unit	Unit Abbreviation
Time	t	Second	s
Frequency	f	Hertz	Hz
Radian frequency	ω	Radians per second	rad/s
Phase angle	θ, φ	Degree or radian	° or rad
Energy	w	Joule	J
Power	p	Watt	W
Charge	q	Coulomb	C
Current	i	Ampere	A
Electric field	$\mathscr{E}$	Volt per meter	V/m
Voltage	v	Volt	V
Impedance	Z	Ohm	Ω
Admittance	Y	Siemen	S
Resistance	R	Ohm	Ω
Conductance	G	Siemens	S
Reactance	X	Ohm	Ω
Susceptance	B	Siemen	S
Inductance, self	L	Henry	H
Inductance, mutual	M	Henry	H
Capacitance	C	Farad	F
Magnetic flux	n	Weber	Wb
Flux linkages	λ	Weber-turns	Wb-t
Power ratio	$\log_{10}(p_2/p_1)$	Bel	B

a small number are accessible to the user. Creating systems using ICs involves interconnecting large circuits at a few accessible terminals in such a way that the circuits are compatible. Ensuring compatibility often involves relatively small circuits whose purpose is to change signal levels or formats. Such interface circuits are intentionally introduced to ensure that the appropriate signal conditions exist at the connections between two larger circuits.

In terms of signal processing, *analysis* involves determining the output signals of a given circuit with known input signals. Analysis has the compelling feature that a unique solution exists in linear circuits. Circuit analysis will occupy the bulk of the study of linear circuits, since it provides the foundation for understanding the interaction of signals and circuits. *Design* involves devising circuits that perform a prescribed signal-processing function. In contrast to analysis, a design problem may have no solution or several solutions. The latter possibility leads to *evaluation*. Given several circuits that perform the same basic function, the alternative designs are rank ordered using factors such as cost, power consumption, and part counts. In reality the engineer's role involves analysis, design, and evaluation, and the boundaries between these functions are often blurred.

There are some worked examples to help the reader understand how to apply the concepts needed to master the concepts covered. These examples describe in detail the steps needed to obtain the final answer. They usually treat analysis problems, although design examples and application notes are included where appropriate.

Symbols and Units   This chapter uses the International System (SI) of units. The SI units include six fundamental units: meter (m), kilogram (kg), second (s), ampere (A), kelvin (K), and candela (cd). All the other units can be derived from these six. Table 1 contains the quantities important to this chapter.

Numerical values encountered in electrical engineering range over many orders of magnitude. Consequently, the system of standard decimal prefixes in Table 2 is used. These prefixes on the unit abbreviation of a quantity indicate the power of 10 that is applied to the numerical value of the quantity.

Circuit Variables   The underlying physical quantities in the study of electronic systems are two basic

Table 2   Standard Decimal Prefixes

Multiplier	Prefix	Abbreviation
10^{18}	Exa	E
10^{15}	Peta	P
10^{12}	Tera	T
10^{9}	Giga	G
10^{6}	Mega	M
10^{3}	Kilo	k
10^{-1}	Deci	d
10^{-2}	Centi	c
10^{-3}	Milli	m
10^{-6}	Micro	μ
10^{-9}	Nano	n
10^{-12}	Pico	p
10^{-15}	Femto	f
10^{-18}	Atto	a

variables, *charge* and *energy*. The concept of electrical charge explains the very strong electrical forces that occur in nature. To explain both attraction and repulsion, we say there are two kinds of charge—positive and negative. Like charges repel while unlike charges attract. The symbol q is used to represent charge. If the amount of charge is varying with time, we emphasize the fact by writing $q(t)$. In SI charge is measured in *coulombs* (abbreviated C). The smallest quantity of charge in nature is an electron's charge ($q_E = 1.6 \times 10^{-19}$ C). There are 6.24×10^{18} electrons in 1 C.

Electrical charge is a rather cumbersome variable to work with in practice. Moreover, in many situations the charges are moving, and so it is more convenient to measure the amount of charge passing a given point per unit time. To do this in differential form, a signal variable i called *current* is defined as follows:

$$i = \frac{dq}{dt} \qquad (1)$$

Current is a measure of the flow of electrical charge. It is the time rate of change of charge passing a given point. The physical dimensions of current are coulombs per second. The unit of current is the *ampere* (abbreviated A). That is,

1 coulomb per second = 1 ampere

In electrical engineering it is customary to define the direction of current as the direction of the net flow of *positive* charges, that is, the opposite of electron flow.

A second signal variable called *voltage* is related to the change in energy that would be experienced by a charge as it passes through a circuit. The symbol w is commonly used to represent energy. Energy carries the units of *joules* (abbreviated J). If a small charge dq were to experience a change in energy dw in passing from point A to point B, then the voltage v between A and B is defined as the change in energy per unit charge. One can express this definition in differential form as

$$v = \frac{dw}{dq} \qquad (2)$$

Voltage does not depend on the path followed by the charge dq in moving from point A to point B. Furthermore, there can be a voltage between two points even if there is no charge motion (i.e., no current), since voltage is a measure of how much energy dw would be involved if a charge dq were moved. The dimensions of voltage are joules per coulomb. The unit of voltage is the *volt* (abbreviated V). That is,

1 joule per coulomb = 1 volt

A third signal variable, *power*, is defined as the time rate of change of energy:

$$p = \frac{dw}{dt} \qquad (3)$$

The dimensions of power are joules per second, which is called a *watt* (abbreviated W). In electrical situations, it is useful to have power expressed in terms of current and voltage. Using the chain rule, Eq. (3) and Eqs. (1) and (2) can be combined as

$$p = \left(\frac{dw}{dq}\right)\left(\frac{dq}{dt}\right) = v \cdot i \qquad (4)$$

This shows that the electrical power associated with a situation is determined by the product of voltage and current.

Signal References   The three signal variables (current, voltage, and power) are defined in terms of two basic variables (charge and energy). Charge and energy, like mass, length, and time, are basic concepts of physics that provide the scientific foundation for electrical engineering. However, engineering problems rarely involve charge and energy directly but are usually stated in terms of the signal variables because current and voltage are much easier to measure.

A signal can be either a current or a voltage, but it is essential that the reader recognize that current and voltage, while interrelated, are quite different variables. Current is a measure of the time rate of charge passing a point. Since current indicates the direction of the flow of electrical charge, one thinks of current as a *through* variable. Voltage is best thought as an *across* variable because it inherently involves two points. Voltage is a measure of the net change in energy involved in moving a charge from one point to another. Voltage is measured not at a single point but rather between two points or across an element.

Figure 1 shows the notation used for assigning reference directions to current and voltage. The reference mark for current [the arrow below $i(t)$] does not indicate the actual direction of the current. The actual direction may be reversing a million times per second. However, when the actual direction coincides with the reference direction, the current is positive. When the opposite occurs, the current is negative. If the net flow of positive charge in Fig. 1 is to the right, the current $i(t)$ is positive. Conversely, if the current $i(t)$ is

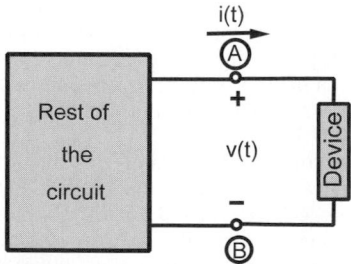

Fig. 1 Voltage and current reference marks for two-terminal device.[1]

positive, then the net flow of positive charge is to the right.

Similarly, the voltage reference marks (+ and B symbols) in Fig. 1 do not imply that the potential at the positive terminal is always higher than the potential at the **B** terminal. However, when this is true, the voltage across the device is positive. When the opposite is true, the voltage is negative.

The importance of relating the reference directions (the plus and minus voltage signs and the current arrows) to the actual direction of the current (in and out) and voltage (high and low) can be used to determine the power associated with a device. That is, if the actual direction of the current is the same as the reference arrow drawn on the device, the current goes "in" and comes "out" of the device in the same direction as the reference arrow. Also, the voltage is "high" at the positive reference and "low" at the negative reference. If the actual and reference directions agree and i and v have the same sign, the power associated with this device is positive since the product of the current and voltage is positive. A positive sign for the associated power indicates that the device absorbs or consumes power. If the actual and reference direction disagrees for either voltage or current so i and v have opposite signs, $p = i \cdot v$ is negative and the device provides power. This definition of reference marks is called the *passive-sign convention*. Certain devices such as heaters (e.g., a toaster) can only absorb power. On the other hand, the power associated with a battery is positive when it is charging (absorbing power) and negative when it is discharging (delivering power). The passive-sign convention is used throughout electrical engineering. It is also the convention used by computer circuit simulation programs.

Ground   Voltage as an across variable is defined and measured between two points. It is convenient to identify one of the points as a reference point commonly called *ground*. This is similar to measuring elevation with respect to mean sea level. The heights of mountains, cities, and so on, are given relative to sea level. Similarly, the voltages at all other points in a circuit are defined with respect to ground. Circuit references are denoted using one of the "ground" symbols shown in Fig. 2. The voltage at the ground point is always taken to be $0\ V$.

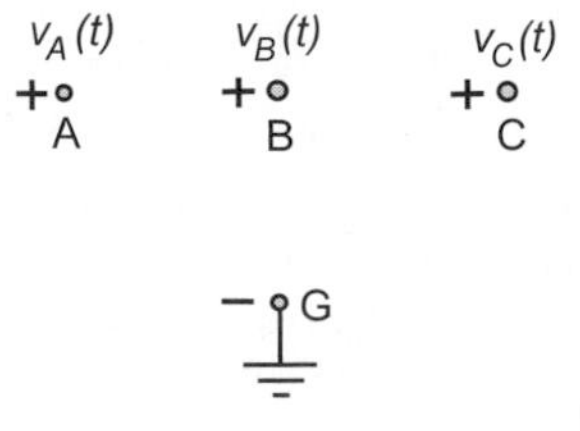

Fig. 2   Ground symbol indicates a common voltage reference point.[1]

1.2   Fundamentals

A *circuit* is a collection of interconnected electrical devices that performs a useful function. An electrical *device* is a component that is treated as a distinct entity.

Element Constraints   A two-terminal device is described by its i–v *characteristic*, that is, the relationship between the voltage across and current through the device. In most cases the relationship is complicated and nonlinear so we use simpler linear models which adequately approximate the dominant features of a device.

Resistor   A resistor is a linear device described by a simple i–v characteristic as follows:

$$v = Ri \qquad \text{or} \qquad i = Gv \qquad (5)$$

where R and G are positive constants related as $G = 1/R$. The power rating of the resistor determines the range over which the i–v characteristic can be represented by this linear relation. Equations (5) are collectively known as *Ohm's law*. The parameter R is called *resistance* and has the unit *ohms* (Ω). The parameter G is called *conductance* with the unit *siemens* (S).

The power associated with the resistor can be found from $p = v \cdot i$. Using Eqs. (5) to eliminate v or i from this relationship yields

$$p = i^2 R = v^2 G = \frac{v^2}{R} \qquad (6)$$

Since the parameter R is positive, these equations state that the power is always nonnegative. Under the passive-sign convention this means the resistor always absorbs power.

Example 1.   A resistor functions as a linear element as long as the voltage and current are within the limits defined by its power rating. Determine the maximum current and voltage that can be applied to a 47-kΩ resistor with a power rating of 0.25 W and remain within its linear operating range.

Solution.   Using Eq. (6) to relate power and current, we obtain

$$I_{\text{MAX}} = \sqrt{\frac{P_{\text{MAX}}}{R}} = \sqrt{\frac{0.25}{47 \times 10^3}} = 2.31\ \text{mA}$$

Similarly, using Eq. (6) to relate power and voltage,

$$V_{\text{MAX}} = \sqrt{RP_{\text{MAX}}} = \sqrt{47 \times 10^3 \times 0.25} = 108\ \text{V}$$

A resistor with infinite resistance, that is, $R = 4\ \Omega$, is called an *open circuit*. By Ohm's law no current can flow through such a device. Similarly, a resistor with

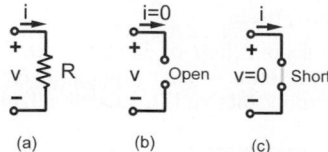

Fig. 3 (*a*) Resistor symbol; (*b*) open circuit; (*c*) short circuit.

no resistance, that is, $R = 0\ \Omega$, is called a *short circuit*. The voltage across a short circuit is always zero. In circuit analysis the devices in a circuit are assumed to be interconnected by zero-resistance wire, that is, by short circuits. Figure 3 shows the circuit symbols for a resistor and open and short circuits.

Ideal Sources The signal and power sources required to operate electronic circuits are modeled using two elements: voltage sources and current sources. These sources can produce either constant or time-varying signals. The circuit symbols of an ideal voltage source and an ideal current source are shown in Fig. 4.

The *i–v* characteristic of an *ideal voltage source* in Fig. 4 is described by the element equations

$$v = v_S \quad \text{and} \quad i = \text{any value} \quad (7)$$

The element equations mean the ideal voltage source produces v_S volts across its terminals and will supply whatever current may be required by the circuit to which it is connected.

The *i–v* characteristic of an *ideal current source* in Fig. 4 is described by the element equations

$$i = i_S \quad \text{and} \quad v = \text{any value} \quad (8)$$

The ideal current source supplies i_S amperes in the direction of its arrow symbol and will furnish whatever voltage is required by the circuit to which it is connected.

In practice, circuit analysis involves selecting an appropriate model for the actual device. Figure 5 shows the practical models for the voltage and current sources. These models are called practical because they more accurately represent the properties of real-world

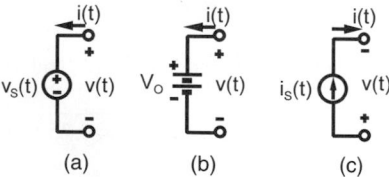

Fig. 4 (*a*) Voltage source; (*b*) battery (traditional symbol); (*c*) current source.

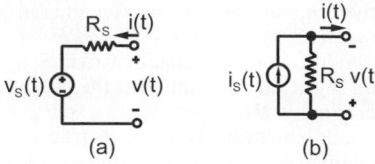

Fig. 5 Circuit symbols for practical independent sources: (*a*) practical voltage source; (*b*) practical current source.[1]

sources than do the ideal models. The resistances R_S in the practical source models in Fig. 5 do not represent physical resistors but represent circuit elements used to account for resistive effects within the devices being modeled.

Connection Constraints The previous section dealt with individual devices and models while this section considers the constraints introduced by interconnections of devices to form circuits. *Kirchhoff's laws* are derived from conservation laws as applied to circuits and are called *connection constraints* because they are based only on the circuit connections and not on the specific devices in the circuit.

The treatment of Kirchhoff's laws uses the following definitions.

A *circuit* is any collection of devices connected at their terminals.
A *node* is an electrical juncture of two or more devices.
A *loop* is a closed path formed by tracing through a sequence of devices without passing through any node more than once.

While it is customary to designate a juncture of two or more elements as a node, it is important to realize that a node is not confined to a point but includes all the wire from the point to each element.

Kirchhoff's Current Law Kirchhoff's first law is based on the principle of conservation of charge. *Kirchhoff's current law* (KCL) states that the algebraic sum of the currents entering a node is zero at every instant. In forming the algebraic sum of currents, one must take into account the current reference directions associated with the devices. If the current reference direction is into the node, a positive sign is assigned to the algebraic sum of the corresponding current. If the reference direction is away from the node, a negative sign is assigned.

There are two signs associated with each current in the application of KCL. The first is the sign given to a current in writing a KCL connection equation. This sign is determined by the orientation of the current reference direction relative to a node. The second sign is determined by the actual direction of the current relative to the reference direction.

The following general principle applies to writing KCL equations: *In a circuit containing N nodes there are only N − 1 independent KCL connection equations.* In general, to write these equations, we select one node as the reference or ground node and then write KCL equations at the remaining $N − 1$ nonreference nodes.

Kirchhoff's Voltage Law The second of Kirchhoff's circuit laws is based on the principle of conservation of energy. *Kirchhoff's voltage law* (KVL) states that the algebraic sum of all of the voltages around a loop is zero at every instant. There are two signs associated with each voltage. The first is the sign given the voltage when writing the KVL connection equation. The second is the sign determined by the actual polarity of a voltage relative to its assigned reference polarity.

The following general principle applies to writing KVL equations: *In a circuit containing E two-terminal elements and N nodes there are only E − N + 1 independent KVL connection equations.* Voltage equations written around $E − N + 1$ different loops contain all of the independent connection constraints that can be derived from KVL. A sufficient condition for loops to be different is that each contains at least one element that is not contained in any other loop.

Parallel and Series Connections Two types of connections occur so frequently in circuit analysis that they deserve special attention. Elements are said to be connected in *parallel* when they share two common nodes. In a parallel connection KVL forces equal voltages across the elements. The parallel connection is not restricted to two elements.

Two elements are said to be connected in *series* when they have one common node to which no other current-drawing element is connected. A series connection results in equal current through each element. Any number of elements can be connected in series.

Combined Constraints The usual goal of circuit analysis is to determine the currents or voltages at various places in a circuit. This analysis is based on constraints of two distinctly different types. The element constraints are based on the models of the specific devices connected in the circuit. The connection constraints are based on Kirchhoff's laws and the circuit connections. The element equations are independent of the circuit in which the device is connected. Likewise, the connection equations are independent of the specific devices in the circuit. But taken together, the *combined constraints* from the element and connection equations provide the data needed to analyze a circuit.

The study of combined constraints begins by considering the simple but important example in Fig. 6. This circuit is driven by the current source i_S and the resulting responses are current/voltage pairs (i_X, v_X) and (i_O, v_O). The reference marks for the response pairs have been assigned using the passive-sign convention.

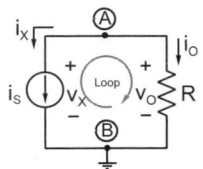

Fig. 6 Circuit used to demonstrate combined constraints.[1]

To solve for all four responses, four equations are required. The first two are the element equations:

$$\begin{aligned}\text{Current source:} \quad & i_X = i_S \\ \text{Resistor:} \quad & v_O = R \cdot i_O\end{aligned} \tag{9}$$

The first element equation states that the response current i_X and the input driving force i_S are equal in magnitude and direction. The second element equation is Ohm's law relating v_O and i_O under the passive-sign convention.

The connection equations are obtained by applying Kirchhoff's laws. The circuit in Fig. 6 has two elements $(E = 2)$ and two nodes $(N = 2)$; hence for a total solution $E − N + 1 = 1$ KVL equation and $N − 1 = 1$ KCL equation are required. Selecting node B as the reference or ground node, a KCL at node A and a KVL around the loop yield

$$\begin{aligned}\text{KCL:} \quad & -i_X - i_O = 0 \\ \text{KVL:} \quad & -v_X + v_O = 0\end{aligned} \tag{10}$$

With four equations and four unknowns all four responses can be found. Combining the KCL connection equation and the first element equations yields $i_O = -i_X = -i_S$. Substituting this result into the second element equations (Ohm's law) produces $v_O = -Ri_S$. The minus sign in this equation does not mean v_O is always negative. Nor does it mean the resistance is negative since resistance is always positive. It means that when the input driving force i_S is positive, the response v_O is negative, and vice versa.

Example 2. Find all of the element currents and voltages in Fig. 7 for $V_O = 10$ V, $R_1 = 2$ kΩ, and $R_2 = 3$ kΩ.

Solution. Substituting the element constraints into the KVL connection constraint produces

$$-V_O + R_1 i_1 + R_2 i_2 = 0$$

This equation can be used to solve for i_1 since the second KCL connection equation requires that $i_2 = i_1$. Hence

$$i_1 = \frac{V_O}{R_1 + R_2} = \frac{10}{2000 + 3000} = 2 \text{ mA}$$

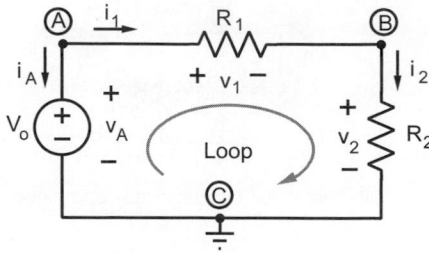

Fig. 7 (From Ref. 1.)

By finding the current i_1, all currents can be found from

$$-i_A = i_1 = i_2$$

since all three elements are connected in series. Substituting all of the known values into the element equations gives

$$v_A = 10 \text{ V} \quad v_1 = R_1 i_1 = 4 \text{ V} \quad v_2 = R_2 i_2 = 6 \text{ V}$$

Assigning Reference Marks In all previous examples and exercises the reference marks for the element currents (arrows) and voltages (+ and −) were given. When reference marks are not shown on a circuit diagram, they must be assigned by the person solving the problem. Beginners sometimes wonder how to assign reference marks when the actual voltage polarities and current directions are as yet unknown. It is important to remember that reference marks do not indicate the actual polarities and directions. They are benchmarks assigned in an arbitrary way at the beginning of the analysis. If it turns out the actual direction and reference direction agree, then the numerical value of the response will be positive. If they disagree, the numerical value will be negative. In other words, the sign of the answer together with arbitrarily assigned reference marks tells us the actual voltage polarity or current direction. When assigning reference marks in this chapter the passive-sign convention will always be used. By always following the passive-sign convention any confusion about the direction of power flow in a device will be avoided. In addition, Ohm's law and other device $i–v$ characteristics assume the voltage and current reference marks follow the passive-sign convention. Always using this convention follows the practice used in all SPICE-based computer circuit analysis programs.

Equivalent Circuits The analysis of a circuit can often be simplified by replacing part of the circuit with one which is equivalent but simpler. The underlying basis for two circuits to be equivalent is contained in their $i–v$ relationships: *Two circuits are said to be equivalent if they have identical i–v characteristics at a specified pair of terminals.*

Equivalent Resistance Resistances connected in series simply add, while conductances connected in parallel also simply add. Since conductance is not normally used to describe a resistor, two resistors R_1 and R_2 connected in parallel result in the expression

$$R_1 \| R_2 = R_{EQ} = \frac{1}{G_{EQ}} = \frac{1}{G_1 + G_2} = \frac{1}{1/R_1 + 1/R_2}$$

$$= \frac{R_1 R_2}{R_1 + R_2} \tag{11}$$

where the symbol $\|$ is shorthand for "in parallel." The expression on the far right in Eq. (11) is called the product over the sum rule for *two* resistors in parallel. The product-over-the-sum rule only applies to two resistors connected in parallel. When more than two resistors are in parallel, the following must be used to obtain the equivalent resistance:

$$R_{EQ} = \frac{1}{G_{EQ}} = \frac{1}{1/R_1 + 1/R_2 + 1/R_3 + \cdots} \tag{12}$$

Example 3. Given the circuit in Fig. 8:

(a) Find the equivalent resistance R_{EQ1} connected between terminals A and B.
(b) Find the equivalent resistance R_{EQ2} connected between terminals C and D.

Solution. First resistors R_2 and R_3 are connected in parallel. Applying Eq. (11) results in

$$R_2 \| R_3 = \frac{R_2 R_3}{R_2 + R_3}$$

(a) The equivalent resistance between terminals A and B equals R_1 and the equivalent resistance

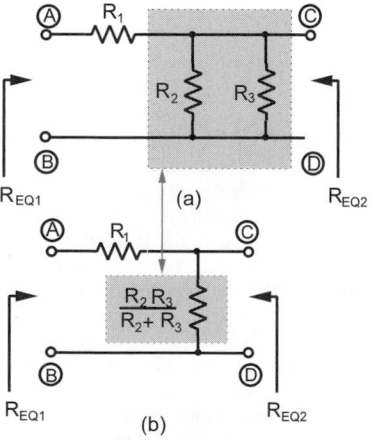

Fig. 8 (From Ref. 1.)

$R_2 \| R_3$ connected in series. The total equivalent resistance R_{EQ1} between terminals A and B thus is

$$R_{EQ1} = R_1 + (R_2 \| R_3)$$

$$R_{EQ1} = R_1 + \frac{R_2 R_3}{R_2 + R_3}$$

$$R_{EQ1} = \frac{R_1 R_2 + R_1 R_3 + R_2 R_3}{R_2 + R_3}$$

(b) Looking into terminals C and D yields a different result. In this case R_1 is not involved, since there is an open circuit (an infinite resistance) between terminals A and B. Therefore only $R_2 \| R_3$ affects the resistance between terminals C and D. As a result R_{EQ2} is simply

$$R_{EQ2} = R_2 \| R_3 = \frac{R_2 R_3}{R_2 + R_3}$$

This example shows that equivalent resistance depends upon the pair of terminals involved.

Equivalent Sources The practical source models shown in Fig. 9 consist of an ideal voltage source in series with a resistance and an ideal current source in parallel with a resistance.

If $R_1 = R_2 = R$ and $v_S = i_S R$, the two practical sources have the same i–v relationship, making the two sources equivalent. When equivalency conditions are met, the rest of the circuit is unaffected regardless if driven by a practical voltage source or a practical current source.

The source transformation equivalency means that either model will deliver the same voltage and current to the rest of the circuit. It does not mean the two models are identical in every way. For example, when the rest of the circuit is an open circuit, there is no current in the resistance of the practical voltage source and hence no $i^2 R$ power loss. But the practical current source model has a power loss because the open-circuit voltage is produced by the source current in the parallel resistance.

Y–Δ Transformations The Y–Δ connections shown in Fig. 10 occasionally occur in circuits and are especially prevalent in three-phase power circuits. One can transform from one configuration to the other by the following set of transformations:

$$R_A = \frac{R_1 R_2 + R_2 R_3 + R_1 R_3}{R_1} \qquad R_1 = \frac{R_B R_C}{R_A + R_B + R_C}$$

$$R_B = \frac{R_1 R_2 + R_2 R_3 + R_1 R_3}{R_2} \qquad R_2 = \frac{R_A R_C}{R_A + R_B + R_C}$$

$$R_C = \frac{R_1 R_2 + R_2 R_3 + R_1 R_3}{R_3} \qquad R_3 = \frac{R_B R_A}{R_A + R_B + R_C}$$

$$\tag{13}$$

Solving Eqs. (13) for R_1, R_2, and R_3 yields the equations for a Δ-to-Y transformation while solving Eqs. (13) for R_A, R_B, and R_C yields the equations for a Y-to-Δ transformation. The Y and Δ subcircuits are said to be *balanced* when $R_1 = R_2 = R_3 = R_Y$ and $R_A = R_B = R_C = R_\Delta$. Under balanced conditions the transformation equations reduce to $R_Y = R_\Delta / 3$ and $R_\Delta = 3 R_Y$.

Voltage and Current Division These two analysis tools find wide application in circuit analysis and design.

Voltage Division Voltage division allows us to solve for the voltage across each element in a series circuit. Figure 11 shows a circuit that lends itself to solution by voltage division. Applying KVL around the loop in Fig. 11 yields $v_S = v_1 + v_2 + v_3$. Since all resistors are connected in series, the same current i exists in all three. Using Ohm's law yields $v_S = R_1 i + R_2 i + R_3 i$. Solving for i yields $i = v_S / (R_1 + R_2 + R_3)$. Once the current in the series circuit is

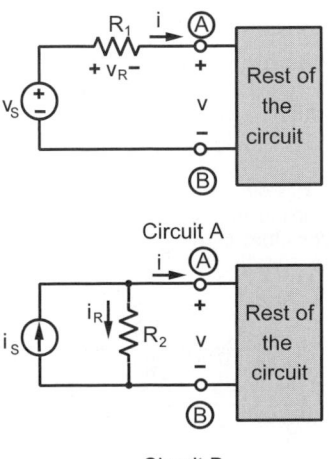

Circuit A

Circuit B

Fig. 9 Equivalent practical source models.[1]

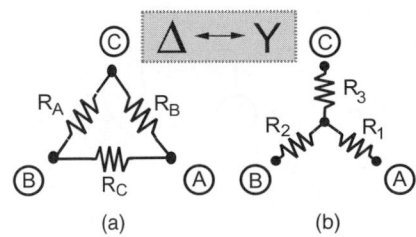

(a)   (b)

Fig. 10 Y–Δ transformation.[2]

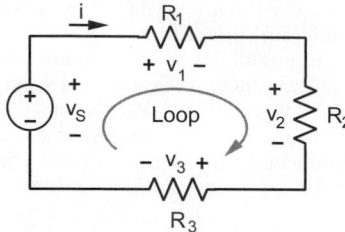

Fig. 11   Voltage divider circuit.[1]

found, the voltage across each resistor is found using Ohm's law:

$$v_1 = R_1 i = \left(\frac{R_1}{R_1 + R_2 + R_3} \right) v_S$$

$$v_2 = R_2 i = \left(\frac{R_2}{R_1 + R_2 + R_3} \right) v_S \qquad (14)$$

$$v_3 = R_3 i = \left(\frac{R_3}{R_1 + R_2 + R_3} \right) v_S$$

In each case the element voltage is equal to its resistance divided by the equivalent series resistance in the circuit times the total voltage across the series circuit. Thus, the general expression of the *voltage division rule* is

$$v_k = \left(\frac{R_k}{R_{EQ}} \right) v_{\text{total}} \qquad (15)$$

The operation of a potentiometer is based on the voltage division rule. The device is a three-terminal element which uses voltage (potential) division to meter out a fraction of the applied voltage. Figure 12 shows the circuit symbol of a potentiometer. Simply stated, a potentiometer is an adjustable voltage divider.

The voltage v_O in Fig. 12 can be adjusted by turning the shaft on the potentiometer to move the wiper arm contact. Using the voltage division rule, v_O is found as

$$v_O = \left(\frac{R_{\text{total}} - R_1}{R_{\text{total}}} \right) v_S \qquad (16)$$

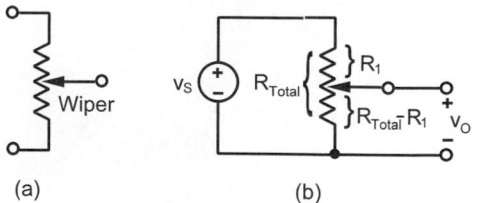

Fig. 12   Potentiometer: (*a*) circuit symbol; (*b*) an application.

Moving the movable wiper arm all the way to the top makes R_1 zero, and voltage division yields v_S. In other words, 100% of the applied voltage is delivered to the rest of the circuit. Moving the wiper to the other extreme delivers zero voltage. By adjusting the wiper arm position we can obtain an output voltage anywhere between zero and the applied voltage v_S. Applications of a potentiometer include volume control, voltage balancing, and fine-tuning adjustment.

Current Division   Current division is the dual of voltage division. By duality current division allows for the solution of the current through each element in a parallel circuit. Figure 13 shows a parallel circuit that lends itself to solution by current division. Applying KCL at node A yields $i_S = i_1 + i_2 + i_3$. The voltage v appears across all three conductances since they are connected in parallel. So using Ohm's law we can write $i_S = vG_1 + vG_2 + vG_3$ and solve for v as $v = i_S/(G_1 + G_2 + G_3)$. Given the voltage v, the current through any element is found using Ohm's law as

$$i_1 = vG_1 = \left(\frac{G_1}{G_1 + G_2 + G_3} \right) i_S$$

$$i_2 = vG_2 = \left(\frac{G_2}{G_1 + G_2 + G_3} \right) i_S$$

$$i_3 = vG_3 = \left(\frac{G_3}{G_1 + G_2 + G_3} \right) i_S$$

These results show that the source current divides among the parallel resistors in proportion to their conductances divided by the equivalent conductances in the parallel connection. Thus, the general expression for the *current division rule* is

$$i_k = \left(\frac{G_k}{G_{EQ}} \right) i_{\text{total}} = \frac{1/R_k}{1/R_{EQ}} i_{\text{total}} \qquad (17)$$

Circuit Reduction   The concepts of series/parallel equivalence, voltage/current division, and source transformations can be used to analyze *ladder circuits* of the type shown in Fig. 14. The basic analysis strategy is to reduce the circuit to a simpler equivalent in which the desired voltage or current is easily found using voltage and/or current division and/or source transformation and/or Ohm's law. There is no fixed pattern to

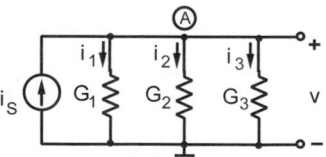

Fig. 13   Current divider circuit.[1]

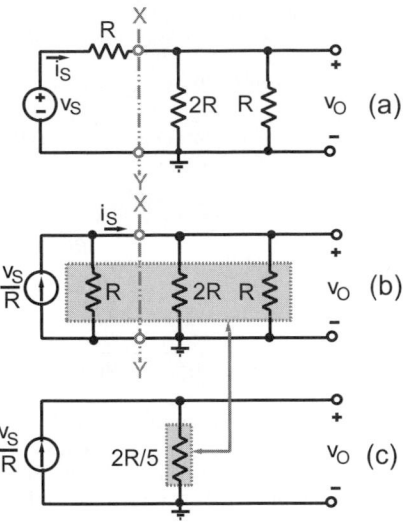

Fig. 14 (From Ref. 1.)

the reduction process, and much depends on the insight of the analyst.

When using circuit reduction it is important to remember that the unknown voltage exists between two nodes and the unknown current exists in a branch. The reduction process must not eliminate the required node pair or branch; otherwise the unknown voltage or current cannot be found. The next example will illustrate circuit reduction.

Example 4. Find the output voltage v_O and the input current i_S in the ladder circuit shown in Fig. 14a.

Solution. Breaking the circuit at points X and Y produces voltage source v_S in series with a resistor R: Using source transformation this combination can be replace by an equivalent current source in parallel with the same resistor, as shown in Fig. 14b. Using current division the input current i_S is

$$ i_S = \frac{R}{(2/3)R + R} \times \frac{v_S}{R} = \frac{v_S}{(5/3)R} = \frac{3}{5}\frac{v_S}{R}, $$

The three parallel resistances in Fig. 14b can be combined into a single equivalent resistance without eliminating the node pair used to define the output voltage v_F:

$$ R_{\text{EQ}} = \frac{1}{1/R + 1/(2R) + 1/R} = \frac{2R}{5} $$

which yields the equivalent circuit in Fig. 14c. The current source v_S/R determines the current through the

equivalent resistance in Fig. 14c. The output voltage is found using Ohm's law:

$$ v_O = \left(\frac{v_S}{R}\right) \times \left(\frac{2R}{5}\right) = \frac{2}{5}v_S $$

Several other analysis approaches are possible.

2 DIRECT-CURRENT (DC) CIRCUITS

This section reviews basic DC analysis using traditional circuits theorems with application to circuit analysis and design.

2.1 Node Voltage Analysis

Using node voltage instead of element voltages as circuit variables can greatly reduce the number of equations that must be treated simultaneously. To define a set of node voltages, a reference node or ground is first selected. The *node voltages* are then defined as the voltages between the remaining nodes and the reference node. Figure 15 shows a reference node indicated by the ground symbol as well as the notation defining the three nonreference node voltages. The node voltages are identified by a voltage symbol adjacent to the nonreference nodes. This notation means that the positive reference mark for the node voltage is located at the node in question while the negative mark is at the reference node. Any circuit with N nodes involves $N - 1$ node voltages.

The following is a fundamental property of node voltages: *If the Kth two-terminal element is connected*

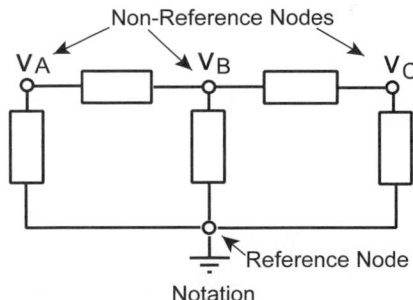

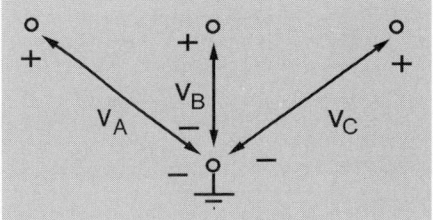

Fig. 15 Node voltage definition and notation.[1]

between nodes X and Y, then the element voltage can be expressed in terms of the two node voltages as

$$v_K = v_X - v_Y \qquad (18)$$

where X is the node connected to the positive reference for element voltage v_K.

Equation (18) is a KVL constraint at the element level. If node Y is the reference node, then by definition $v_Y = 0$ and Eq. (18) reduces to $v_K = v_X$. On the other hand, if node X is the reference node, then $v_X = 0$ and therefore $v_K = -v_Y$. The minus sign here comes from the fact that the positive reference for the element is connected to the reference node. In any case, the important fact is that the voltage across any two-terminal element can be expressed as the difference of two node voltages, one of which may be zero.

To formulate a circuit description using node voltages, device and connection analysis is used, except that the KVL connection equations are not explicitly written down. Instead the fundamental property of node analysis is used to express the element voltages in terms of the node voltages.

The circuit in Fig. 16 will demonstrate the formulation of node voltage equations. The ground symbol identifies the reference node, six element currents (i_0, i_1, i_2, i_3, i_4, and i_5), and three node voltages (v_A, v_B, and v_C).

The KCL constraints at the three nonreference nodes are

$$\text{Node A:} \quad -i_0 + i_1 + i_2 = 0$$
$$\text{Node B:} \quad -i_1 + i_3 - i_5 = 0$$
$$\text{Node C:} \quad -i_2 + i_5 + i_4 = 0$$

Using the fundamental property of node analysis, device equations are used to relate the element currents to the node voltages:

$$R_1: \quad i_1 = \frac{v_A - v_B}{R_1} \qquad R_2: \quad i_2 = \frac{v_A - v_C}{R_2}$$

$$\text{Voltage source:} \quad v_S = v_A$$

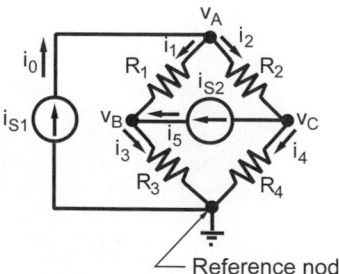

Fig. 16 Bridge circuit for node voltage example.[1]

$$R_3: \quad i_3 = \frac{v_B - 0}{R_3} \qquad R_4: \quad i_4 = \frac{v_C - 0}{R_4}$$

$$\text{Current source:} \quad i_5 = i_{S2}$$

Substituting the element currents into the KCL equations yields

$$\text{Node A:} \quad -i_0 + \frac{v_A - v_B}{R_1} + \frac{v_A - v_C}{R_2} = 0$$
$$\text{Node B:} \quad \frac{v_B - v_A}{R_1} + \frac{v_B}{R_3} - i_{S2} = 0$$
$$\text{Node C:} \quad \frac{v_C - v_A}{R_2} + \frac{v_C}{R_4} + i_{S2} = 0$$

But since the reference is connected to the negative side of the voltage source, $v_A = v_S$. Thus at node A the voltage is already known and this reduces the number of equations that need to be solved. The equation written above can be used to solve for the current through the voltage source if that is desired. Writing these equations in standard form with all of the unknown node voltages grouped on one side and the independent sources on the other yields

$$\text{Node A:} \qquad v_A = v_S$$
$$\text{Node B:} \quad v_B \left(\frac{1}{R_1} + \frac{1}{R_3} \right) = i_S + \frac{v_S}{R_1}$$
$$\text{Node C:} \quad v_C \left(\frac{1}{R_2} + \frac{1}{R_4} \right) = -i_S + \frac{v_S}{R_2}$$

Using node voltage analysis there are only two equations and two unknowns (v_B and v_C) to be solved. The coefficients in the equations on the left side depend only on circuit parameters, while the right side contains the known input driving forces.

Supernodes When neither node of a voltage source can be selected as the reference, a *supernode* must be used. The fact that KCL applies to the currents penetrating a boundary can be used to write a node equation at the supernode. Then node equations at the remaining nonreference nodes are written in the usual way. This process reduces the number of available node equations to $N - 3$ plus one supernode equation, leaving us one equation short of the $N - 1$ required. The voltage source inside the supernode constrains the difference between the node voltages to be the value of the voltage source. The voltage source constraint provides the additional relationship needed to write $N - 1$ independent equations in the $N - 1$ node voltages.

Example 5. For the circuit in Fig. 17:

(a) Formulate node voltage equations.
(b) Solve for the output voltage v_O using $R_1 = R_4 = 2 \text{ k}\Omega$ and $R_2 = R_3 = 4 \text{ k}\Omega$.

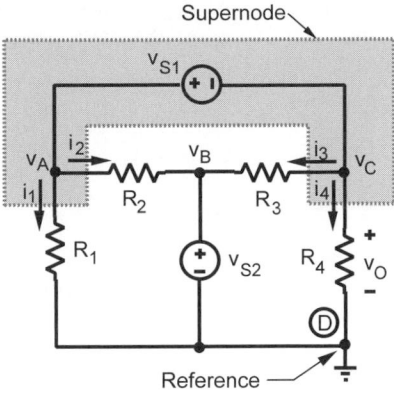

Fig. 17 (From Ref. 1.)

Solution. (a) The voltage sources in Fig. 17 do not have a common node so a reference node that includes both sources cannot be selected. Choosing node D as the reference forces the condition $v_B = v_{S_2}$ but leaves the other source v_{S1} ungrounded. The ungrounded source and all wires leading to it are encompassed by a supernode boundary as shown in the figure. Kirchhoff's current law applies to the four-element currents that penetrate the supernode boundary and we can write

$$i_1 + i_2 + i_3 + i_4 = 0$$

These currents can easily be expressed in terms of the node voltages:

$$\frac{v_A}{R_1} + \frac{v_A - v_B}{R_2} + \frac{v_C - v_B}{R_3} + \frac{v_C}{R_4} = 0$$

But since $v_B = v_{S2}$, the standard form of this equation is

$$v_A \left(\frac{1}{R_1} + \frac{1}{R_2} \right) + v_C \left(\frac{1}{R_3} + \frac{1}{R_4} \right) = v_{S2} \left(\frac{1}{R_2} + \frac{1}{R_3} \right)$$

We have one equation in the two unknown node voltages v_A and v_C. Applying the fundamental property of node voltages inside the supernode, we can write

$$v_A - v_C = v_{S1}$$

That is, the ungrounded voltage source constrains the difference between the two unknown node voltages inside the supernode and thereby supplies the relationship needed to obtain two equations in two unknowns.

(b) Substituting in the given numerical values yields

$$\left(7.5 \times 10^{-4} \right) v_A + \left(7.5 \times 10^{-4} \right) v_C = \left(5 \times 10^{-4} \right) v_{S2}$$

$$v_A - v_C = v_{S1}$$

To find the output v_O, we need to solve these equations for v_C. The second equation yields $v_A = v_C + v_{S1}$, which when substituted into the first equation yields the required output:

$$v_O = v_C = \tfrac{1}{3} v_{S2} - \tfrac{1}{2} v_{S1}$$

Node voltage equations are very useful in the analysis of a variety of electronic circuits. These equations can always be formulated using KCL, the element constraints, and the fundamental property of node voltages. The following guidelines summarize this approach:

1. Simplify the circuit by combining elements in series and parallel wherever possible.
2. If not specified, select a reference node so that as many dependent and independent voltage sources as possible are directly connected to the reference.
3. Label a node voltage adjacent to each nonreference node.
4. Create supernodes for dependent and independent voltage sources that are not directly connected to the reference node.
5. Node equations are required at supernodes and all other nonreference nodes except op amp outputs and nodes directly connected to the reference by a voltage source.
6. Write symmetrical node equations by treating dependent sources as independent sources and using the inspection method.
7. Write expressions relating the node and source voltages for voltage sources included in supernodes.
8. Substitute the expressions from step 7 into the node equations from step 6 and place the result in standard form.

2.2 Mesh Current Analysis

Mesh currents are an alternative set of analysis variables that are useful in circuits containing many elements connected in series. To review terminology, a loop is a sequence of circuit elements that forms a closed path that passes through each element just once. A mesh is a special type of loop that does not enclose any elements.

The following development of mesh analysis is restricted to planar circuits. A *planar circuit* can be drawn on a flat surface without crossovers in a "window pane" fashion. To define a set of variables, a *mesh current* $(i_A, i_B, i_C, \ldots)$ is associated with each window pane and a reference direction assigned customarily in a clockwise sense. There is no momentous reason for this except perhaps tradition.

Mesh currents are thought of as circulating through the elements in their respective meshes; however, this viewpoint is not based on the physics of circuit behavior. There are not different types of electrons flowing that somehow get assigned to mesh currents i_A or i_B. Mesh currents are variables used in circuit analysis. They are only somewhat abstractly related to the physical operation of a circuit and may be impossible to measure directly. Mesh currents have a unique feature that is the dual of the fundamental property of node voltages. In a planar circuit any given element is contained in at most two meshes. When an element is in two meshes, the two mesh currents circulate through the element in opposite directions. In such cases KCL declares that the net element current through the element is the difference of the two mesh currents.

These observations lead to the fundamental property of mesh currents: *If the Kth two-terminal element is contained in meshes X and Y, then the element current can be expressed in terms of the two mesh currents as*

$$i_K = i_X - i_Y \qquad (19)$$

where X is the mesh whose reference direction agrees with the reference direction of i_K.

Equation (19) is a KCL constraint at the element level. If the element is contained in only one mesh, then $i_K = i_X$ or $i_K = -i_Y$ depending on whether the reference direction for the element current agrees or disagrees with the mesh current. The key fact is that the current through every two-terminal element in a planar circuit can be expressed as the difference of at most two mesh currents.

Mesh currents allow circuit equations to be formulated using device and connection constraints, except that the KCL constraints are not explicitly written down. Instead, the fundamental property of mesh currents is used to express the device constraints in terms of the mesh currents, thereby avoiding using the element currents and working only with the element voltages and mesh currents.

For example, the planar circuit in Fig. 18 can be analyzed using the mesh current method. In the figure two mesh currents are shown as well the voltages across each of the five elements. The KVL constraints around each mesh using the element voltages yield

$$\text{Mesh A:} \quad -v_0 + v_1 + v_3 = 0$$

$$\text{Mesh B:} \quad -v_3 + v_2 + v_4 = 0$$

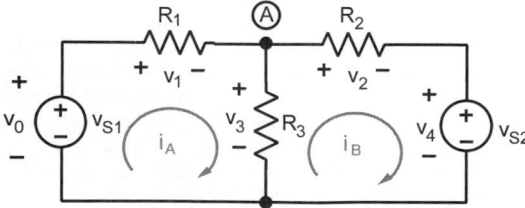

Fig. 18   Circuit demonstrating mesh current analysis.[1]

Using the fundamental property of mesh currents, the element voltages in terms of the mesh currents and input voltages are written as

$$v_1 = R_1 i_A \qquad v_0 = v_{S1}$$

$$v_2 = R_2 i_B \qquad v_4 = v_{S2}$$

$$v_3 = R_3 (i_A - i_B)$$

Substituting these element equations into the KVL connection equations and arranging the result in standard form yield

$$(R_1 + R_3) i_A - R_3 i_B = v_{S1}$$

$$-R_3 i_A + (R_2 + R_3) i_B = -v_{S2}$$

This results in two equations in two unknown mesh currents. The KCL equations $i_1 = i_A$, $i_2 = i_B$ and $i_3 = i_A - i_B$ are implicitly used to write mesh equations. In effect, the fundamental property of mesh currents ensures that the KCL constraints are satisfied. Any general method of circuit analysis must satisfy KCL, KVL, and the device i–v relationships. Mesh current analysis appears to focus on the latter two but implicitly satisfies KCL when the device constraints are expressed in terms of the mesh currents.

Solving for the mesh currents yields

$$i_A = \frac{(R_2 + R_3) v_{S1} - R_3 v_{S2}}{R_1 R_2 + R_1 R_3 + R_2 R_3} \qquad \text{and}$$

$$i_B = \frac{R_3 v_{S1} - (R_1 + R_3) v_{S2}}{R_1 R_2 + R_1 R_3 + R_2 R_3}$$

The results for i_A and i_B can now be substituted into the device constraints to solve for every voltage in the circuit. For instance, the voltage across R_3 is

$$v_A = v_3 = R_3(i_A - i_B) = \frac{R_2 R_3 v_{S1} + R_1 R_3 v_{S2}}{R_1 R_2 + R_1 R_3 + R_2 R_3}$$

Example 6.   Use mesh current equations to find i_O in the circuit in Fig. 19a.

Solution.   The current source in this circuit can be handled by a source transformation. The 2-mA source in parallel with the 4-kΩ resistor in Fig. 19a can be replaced by an equivalent 8-V source in series with the same resistor as shown in Fig. 19b. In this circuit the total resistance in mesh A is 6 kΩ, the total resistance in mesh B is 11 kΩ, and the resistance contained in both meshes is 2 kΩ. By inspection the mesh equations for this circuit are

$$(6000) i_A - (2000) i_B = 5$$

$$-(2000) i_A + (11000) i_B = -8$$

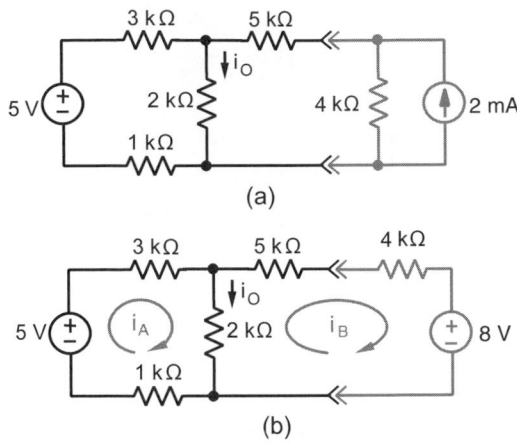

(a)

(b)

Fig. 19 (From Ref. 1.)

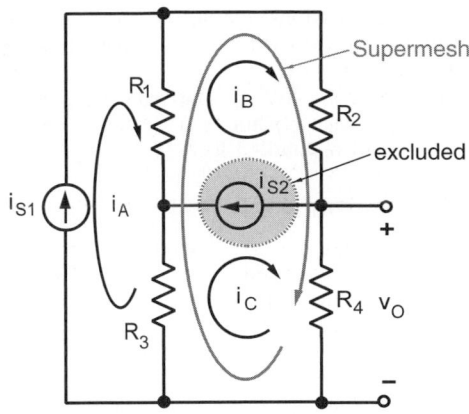

Fig. 20 Example of supermesh.[1]

Solving for the two mesh currents yields

$$i_A = 0.6290 \text{ mA} \qquad \text{and} \qquad i_B = 0.6129 \text{ mA}$$

By KCL the required current is

$$i_0 = i_A - i_B = 1.2419 \text{ mA}$$

Supermesh If a current source is contained in two meshes and is not connected in parallel with a resistance, then a *supermesh* is created by excluding the current source and any elements connected in series with it. One mesh equation is written around the supermesh using the currents i_A and i_B. Then mesh equations of the remaining meshes are written in the usual way. This leaves the solution one equation short because parts of meshes A and B are included in the supermesh. However, the fundamental property of mesh currents relates the currents i_S, i_A, and i_B as

$$i_S = i_A - i_B$$

This equation supplies the additional relationship needed to get the requisite number of equations in the unknown mesh currents. This approach is obviously the dual of the supernode method for modified node analysis. The following example demonstrates the use of a supermesh.

Example 7. Use mesh current equations to find the v_O in Fig. 20.

Solution. The current source i_{S_2} is in both mesh B and mesh C, so we exclude this element and create the supermesh shown in the figure. The sum of voltages around the supermesh is

$$R_1 (i_B - i_A) + R_2 (i_B) + R_4 (i_C) + R_3 (i_C - i_A) = 0$$

The supermesh voltage constraint yields one equation in the three unknown mesh currents. Applying KCL to each of the current sources yields

$$i_A = i_{S1} \qquad i_B - i_C = i_{S2}$$

Because of KCL, the two current sources force constraints that supply two more equations. Using these two KCL constraints to eliminate i_A and i_B from the supermesh KVL constraint yields

$$(R_1 + R_2 + R_3 + R_4) i_C = (R_1 + R_3) i_{S1} \\ - (R_1 + R_2) i_{S2}$$

Hence, the required output voltage is

$$v_O = R_4 i_C = R_4 \times \left[\frac{(R_1 + R_3) i_{S1} - (R_1 + R_2) i_{S2}}{R_1 + R_2 + R_3 + R_4} \right]$$

Mesh current equations can always be formulated from KVL, the element constraints, and the fundamental property of mesh currents. The following guidelines summarize an approach to formulating mesh equations for resistance circuits:

1. Simplify the circuit by combining elements in series or parallel wherever possible.
2. Assign a clockwise mesh current to each mesh.
3. Create a supermesh for dependent and independent current sources that are contained in two meshes.
4. Write symmetrical mesh equations for all meshes by treating dependent sources as independent sources and using the inspection method.
5. Write expressions relating the mesh and source currents for current sources contained in only one mesh.

6. Write expressions relating the mesh and source currents for current sources included in super-meshes.
7. Substitute the expressions from steps 5 and 6 into the mesh equations from step 4 and place the result in standard form.

2.3 Linearity Properties

This chapter treats the analysis and design of *linear circuits*. A circuit is said to be linear if it can be adequately modeled using only linear elements and independent sources. The hallmark feature of a linear circuit is that outputs are linear functions of the inputs. Circuit *inputs* are the signals produced by independent sources and *outputs* are any other designated signals. Mathematically a function is said to be linear if it possesses two properties—homogeneity and additivity. In terms of circuit responses, *homogeneity* means the output of a linear circuit is proportional to the input. *Additivity* means the output due to two or more inputs can be found by adding the outputs obtained when each input is applied separately. Mathematically these properties are written as follows:

$$f(Kx) = Kf(x) \qquad \text{and}$$
$$f(x_1 + x_2) = f(x_1) + f(x_2) \qquad (20)$$

where K is a scalar constant. In circuit analysis the homogeneity property is called *proportionality* while the additivity property is called *superposition*.

Proportionality Property   The *proportionality property* applies to linear circuits with one input. For linear resistive circuits proportionality states that every input–output relationship can be written as

$$y = K \cdot x$$

where x is the input current or voltage, y is an output current or voltage, and K is a constant. The block diagram in Fig. 21 describes a relationship in which the input x is multiplied by the scalar constant K to produce the output y. Examples of proportionality abound. For instance, using voltage division in Fig. 22 produces

$$v_O = \left(\frac{R_2}{R_1 + R_2} \right) v_S$$

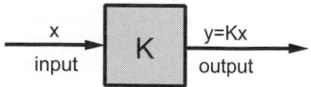

Fig. 21   Block diagram representation of proportionality property.[1]

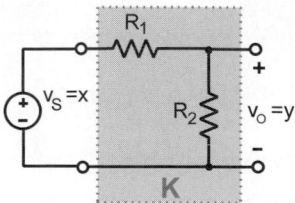

Fig. 22   Example of circuit exhibiting proportionality.[1]

which means

$$x = v_S \qquad y = v_O$$
$$K = \frac{R_2}{R_1 + R_2}$$

In this example the proportionality constant K is dimensionless because the input and output have the same units. In other situations K could carry the units of ohms or siemens when the input or output does not have the same units.

Example 8.   Given the bridge circuit of Fig. 23:

(a) Find the proportionality constant K in the input–output relationship $v_O = Kv_S$.
(b) Find the sign of K when $R_2R_3 > R_1R_4$, $R_2R_3 = R_1R_4$, and $R_2R_3 < R_1R_4$.

Solution.   (a) Note that the circuit consists of two voltage dividers. Applying the voltage division rule to each side of the bridge circuit yields

$$v_A = \frac{R_3}{R_1 + R_3} v_S \qquad v_B = \frac{R_4}{R_2 + R_4} v_S$$

The fundamental property of node voltages allows us to write $v_O = v_A - v_B$. Substituting the equations for v_A and v_B into this KVL equation yields

$$v_o = \begin{cases} \left(\dfrac{R_3}{R_1 + R_3} - \dfrac{R_4}{R_2 + R_4} \right) v_S \\ \left(\dfrac{R_2R_3 - R_1R_4}{(R_1 + R_3)(R_2 + R_4)} \right) v_S \\ (K)v_S \end{cases}$$

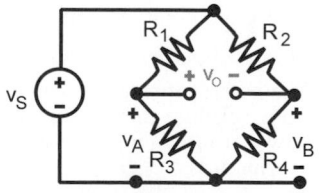

Fig. 23   (From Ref. 1.)

(b) The proportionality constant K can be positive, negative, or zero. Specifically:

If $R_2 R_3 > R_1 R_2$, then $K > 0$.
If $R_2 R_3 = R_1 R_2$, then $K = 0$.
If $R_2 R_3 < R_1 R_2$, then $K < 0$.

When the product of the resistances in opposite legs of the bridge are equal, $K = 0$ and the bridge is said to be balanced.

Superposition Property   The *additivity property*, or superposition, states that any output current or voltage of a linear resistive circuit with multiple inputs can be expressed as a linear combination of several inputs:

$$y = K_1 x_1 + K_2 x_2 + K_3 x_3 + \cdots$$

where $x_1, x_2, x_3, \ldots$ are current or voltage inputs and $K_1, K_2, K_3 \ldots$ are constants that depend on the circuit parameters.

Since the output y above is a linear combination, the contribution of each input source is independent of all other inputs. This means that the output can be found by finding the contribution from each source acting alone and then adding the individual response to obtain the total response. This suggests that the output of a multiple-input linear circuit can be found by the following steps:

Step 1: "Turn off" all independent input signal sources except one and find the output of the circuit due to that source acting alone.

Step 2: Repeat the process in step 1 until each independent input source has been turned on and the output due to that source found.

Step 3: The total output with all sources turned on is then a linear combination (algebraic sum) of the contributions of the individual independent sources.

A voltage source is turned off by setting its voltage to zero ($v_S = 0$) and replacing it with a short circuit. Similarly, turning off a current source ($i_S = 0$) entails replacing it with an open circuit. Figure 24a shows that the circuit has two input sources. Figure 24b shows the circuit with the current source set to zero. The output of the circuit v_{O1} represents that part of the total output caused by the voltage source. Using voltage division yields

$$v_{O1} = \frac{R_2}{R_1 + R_2} v_S$$

Next the voltage source is turned off and the current source is turned on, as shown in Fig. 24c. Using Ohm's law, $v_{O2} = i_{O2} R_2$. Then using current division to express i_{O2} in terms of i_S yields

$$v_{O2} = i_{O2} \times R_2 = \left[\frac{R_1}{R_1 + R_2} i_S \right] R_2 = \frac{R_1 R_2}{R_1 + R_2} i_S$$

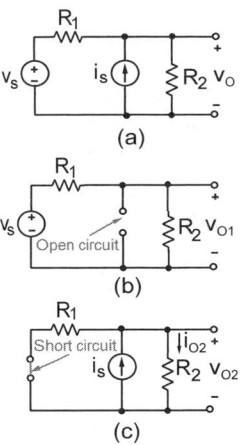

Fig. 24   Circuit analysis using superposition: (a) current source off; (b) voltage source off.[1]

Applying the superposition theorem, the response with both sources "turned on" is found by adding the two responses v_{O1} and v_{O2}:

$$v_O = v_{O1} + v_{O2}$$

$$v_O = \left[\frac{R_2}{R_1 + R_2} \right] v_S + \left[\frac{R_1 R_2}{R_1 + R_2} \right] i_S$$

Superposition is an important property of linear circuits and is used primarily as a conceptual tool to develop other circuit analysis and design techniques. It is useful, for example, to determine the contribution to a circuit by a certain source.

2.4   Thevenin and Norton Equivalent Circuits

An *interface* is a connection between circuits that perform different functions. Circuit interfaces occur frequently in electrical and electronic systems so special analysis methods are used to handle them. For the two-terminal interface shown in Fig. 25, one circuit can be considered as the source S and the other as the load L. Signals are produced by the source circuit and delivered to the load. The source–load interaction at an interface is one of the central problems of circuit analysis and design.

The Thevenin and Norton equivalent circuits shown in Fig. 25 are valuable tools for dealing with circuit interfaces. The conditions under which these equivalent circuits exist can be stated as a theorem: *If the source circuit in a two-terminal interface is linear, then the interface signals v and i do not change when the source circuit is replaced by its Thevenin or Norton equivalent circuit.*

The equivalence requires the source circuit to be linear but places no restriction on the linearity of

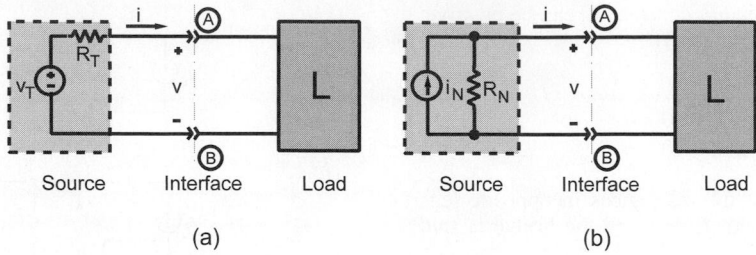

Fig. 25 Equivalent circuits for source circuit: (*a*) Thevenin equivalent; (*b*) Norton equivalent.[1]

the load circuit. The Thevenin equivalent circuit consists of a voltage source (v_T) in series with a resistance (R_T). The Norton equivalent circuit is a current source (i_N) in parallel with a resistance (R_N). The Thevenin and Norton equivalent circuits are equivalent to each other since replacing one by the other leaves the interface signals unchanged. In essence the Thevenin and Norton equivalent circuits are related by the source transformation covered earlier under equivalent circuits.

The two parameters can often be obtained using open-circuit and short-circuit loads. If the actual load is disconnected from the source, an open-circuit voltage v_{OC} appears between terminals A and B. Connecting an open-circuit load to the Thevenin equivalent produces $v_{OC} = v_T$ since the open circuit causes the current to be zero, resulting in no voltage drop across R_T. Similarly, disconnecting the load and connecting a short circuit as shown produce a current i_{SC}. Connecting a short-circuit load to the Norton equivalent produces $i_{SC} = i_N$ since all of the source current i_N is diverted through the short-circuit load.

In summary, the parameters of the Thevenin and Norton equivalent circuits at a given interface can be found by determining the open-circuit voltage and the short-circuit current:

$$v_T = v_{OC} \qquad i_N = i_{SC} \qquad R_N = R_T = \frac{v_{OC}}{i_{SC}} \quad (21)$$

General Applications Since even complex linear circuits can be replaced by their Thevenin or Norton equivalent, the chore of designing circuits that interface with these complex circuits is greatly simplified. Suppose a load resistance in Fig. 26*a* needs to be chosen so the source circuit to the left of the interface A–B delivers 4 V to the load.

The Thevenin and Norton equivalents v_{OC} and i_{SC} are first found. The open-circuit voltage v_{OC} is found by disconnecting the load at terminals A–B as shown in Fig. 26*b*. The voltage across the 15-Ω resistor is zero because the current through it is zero due to the open circuit. The open-circuit voltage at the interface is the same as the voltage across the 10-Ω resistor.

Using voltage division, this voltage is

$$v_T = v_{OC} = \frac{10}{10 + 5} \times 5 = 10 \text{ V}$$

Then the short-circuit current i_{SC} is calculated using the circuit in Fig. 26*c*. The total current i_X delivered by the 15-V source is $i_x = 15/R_{EQ}$, where R_{EQ} is the equivalent resistance seen by the voltage source with a short circuit at the interface:

$$R_{EQ} = 5 + \frac{10 \times 15}{10 + 15} = 11 \text{ }\Omega$$

The source current i_X can now be found: $i_X = 15/11 = 1.36$ A. Given i_X, current division is used to obtain the short-circuit current,

$$i_N = i_{SC} = \frac{10}{10 + 15} \times i_X = 0.545 \text{ A}$$

Finally, we compute the Thevenin and Norton resistances:

$$R_T = R_N = \frac{v_{OC}}{i_{SC}} = 18.3 \text{ }\Omega$$

The resulting Thevenin and Norton equivalent circuits are shown in Figs. 26*d,e*.

It now is an easy matter to select a load R_L so 4 V is supplied to the load. Using the Thevenin equivalent circuit, the problem reduces to a voltage divider,

$$4 \text{ V} = \frac{R_L}{R_L + R_T} \times V_T = \frac{R_L}{R_L + 18.3} \times 10$$

Solving for R_L yields $R_L = 12.2$ Ω.

The Thevenin or Norton equivalent can always found from the open-circuit voltage and short-circuit current at the interface. Often they can be measured using a multimeter to measure the open-circuit voltage and the short-circuit current.

Application to Nonlinear Loads An important use of Thevenin and Norton equivalent circuits is

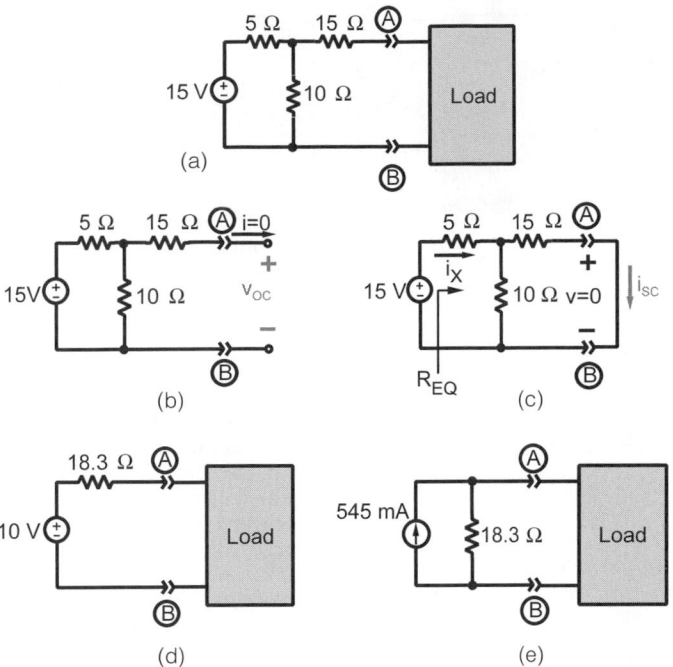

Fig. 26 Example of finding Thevenin and Norton equivalent circuits: (a) given circuit; (b) open circuit yields Thevenin voltage; (c) short circuit yields Norton current; (d) Thevenin equivalent circuit; (e) Norton equivalent circuit.[1]

finding the voltage across, current through, and power dissipated in a two-terminal nonlinear element (NLE). The method of analysis is a straightforward application of device i–v characteristics. An interface is defined at the terminals of the nonlinear element and the linear part of the circuit is reduced to the Thevenin equivalent in Fig. 27a. Treating the interface current i as the dependent variable, the i–v relationship of the Thevenin equivalent is written in the form

$$i = \left(-\frac{1}{R_T}\right) v + \left(\frac{v_T}{R_T}\right)$$

This is the equation of a straight line in the i–v plane shown in Fig. 27b. The line intersects the i axis ($v = 0$) at $i = i_{SC} = v_T/R_T$ and intersects the v axis ($i = 0$) at $v = v_{OC} = v_T$. This line is called the *load line*.

The nonlinear element has the i–v characteristic shown in Fig. 27c. Mathematically this nonlinear characteristic has the form $i = f(v)$. Both the nonlinear equation and the load line equation must be solved simultaneously. This can be done by numerical methods when $f(v)$ is known explicitly, but often a graphical solution is adequate. By superimposing the load line on the i–v characteristic curve of the nonlinear element in Fig. 27d, the point or points of intersection represent the values of i and v that satisfy the source constraints given in the form of the Thevenin equivalent above,

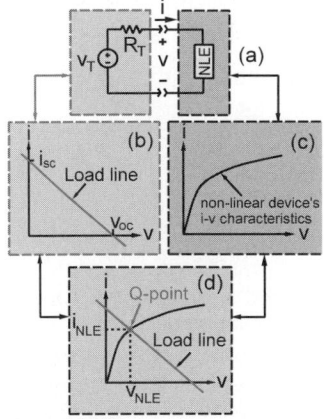

Fig. 27 Graphical analysis of nonlinear circuit: (a) given circuit; (b) load line; (c) nonlinear device i–v characteristics; (d) Q point.[1]

and nonlinear element constraints. In the terminology of electronics the point of intersection is called the operating point, or *Q point*, or the quiescent point.

2.5 Maximum Signal Transfer

Circuit interfacing involves interconnecting circuits in such a way that they are compatible. In this regard an

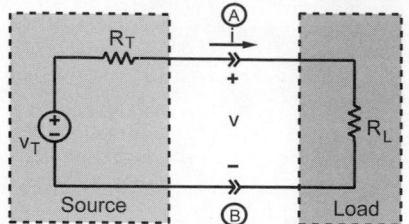

Fig. 28 Two-terminal interface for deriving maximum signal transfer conditions.[1]

important consideration is the maximum signal levels that can be transferred across a given interface. This section defines the maximum voltage, current, and power available at an interface between a *fixed source* and an *adjustable load*.

The source can be represented by its Thevenin equivalent and the load by an equivalent resistance R_L, as shown in Fig. 28. For a fixed source the parameters v_T and R_T are given and the interface signal levels are functions of the load resistance R_L. By voltage division, the interface voltage is

$$v = \frac{R_L}{R_L + R_T} v_T$$

For a fixed source and a variable load, the voltage will be a maximum if R_L is made very large compared to R_T. Ideally R_L should be made infinite (an open circuit), in which case

$$v_{MAX} = v_T = v_{OC}$$

Therefore, the maximum voltage available at the interface is the source open-circuit voltage v_{OC}. The current delivered at the interface is

$$i = \frac{v_T}{R_L + R_T}$$

Again, for a fixed source and a variable load, the current will be a maximum if R_L is made very small

compared to R_T. Ideally R_L should be zero (a short circuit), in which case

$$i_{MAX} = \frac{v_T}{R_T} = i_N = i_{SC}$$

Therefore, the maximum current available at the interface is the source short-circuit current i_{SC}.

The power delivered at the interface is equal to the product vi. Using interface voltage, and interface current results found above, the power is

$$p = v \times i = \frac{R_L v_T^2}{(R_T + R_L)^2}$$

For a given source, the parameters v_T and R_T are fixed and the delivered power is a function of a single variable R_L. The conditions for obtaining maximum voltage ($R_L \to \infty$) or maximum current ($R_L = 0$) both produce zero power. The value of R_L that maximizes the power lies somewhere between these two extremes. The value can be found by differentiating the power expression with respect to R_L and solving for the value of R_L that makes $dp/dR_L = 0$. This occurs when $R_L = R_T$. Therefore, *maximum power transfer* occurs when the load resistance equals the Thevenin resistance of the source. When $R_L = R_T$ the source and load are said to be *matched*. Substituting the condition $R_L = R_T$ back into the power equation above shows the maximum power to be

$$p_{MAX} = \frac{v_T^2}{4R_T} = \frac{i_N^2 R_T}{4}$$

These results are consequences of what is known as the maximum power transfer theorem: *A fixed source with a Thevenin resistance R_T delivers maximum power to an adjustable load R_L when $R_L = R_T$*.

Figure 29 shows plots of the interface voltage, current, and power as functions of R_L/R_T. The plots of v/v_{OC}, i/i_{SC}, and p/p_{MAX} are normalized to the maximum available signal levels so the ordinates in Fig. 29 range from 0 to 1.

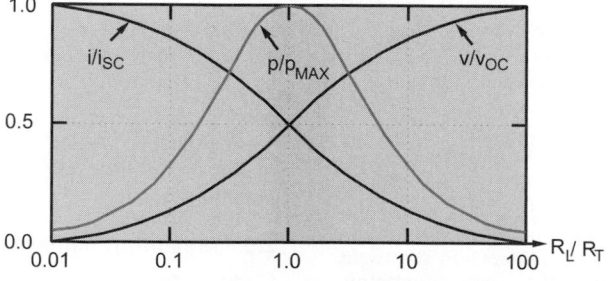

Fig. 29 Normalized plots of current, voltage, and power versus R_L/R_T.[1]

The plot of the normalized power p/p_{MAX} in the neighborhood of the maximum is not a particularly strong function of R_L/R_T. Changing the ratio R_L/R_T by a factor of 2 in either direction from the maximum reduces p/p_{MAX} by less than 20%. The normalized voltage v/v_{OC} is within 20% of its maximum when $R_L/R_T = 4$. Similarly, the normalized current is within 20% of its maximum when $R_L/R_T = 3$. In other words, for engineering purposes maximum signal levels can be approached with load resistances that only approximate the theoretical requirements.

2.6 Interface Circuit Design

The maximum signal levels discussed in the previous section place bounds on what is achievable at an interface. However, those bounds are based on a fixed source and an adjustable load. In practice there are circumstances in which the source or the load or both can be adjusted to produce prescribed interface signal levels. Sometimes it is necessary to insert an interface circuit between the source and load. Figure 30 shows the general situations and some examples of resistive interface circuits. By its very nature the inserted circuit has two terminal pairs, or interfaces, at which voltage and current can be observed or specified. These terminal pairs are also called *ports,* and the interface circuit is referred to as a *two-port network.* The port connected to the source is called the input and the port connected to the load the output. The purpose of this two-port

network is to ensure that the source and load interact in a prescribed way.

Basic Circuit Design Concepts   This section introduces a limited form of circuit design, as contrasted with circuit analysis. Although circuit analysis tools are essential in design, there are important differences. A linear circuit analysis problem generally has a unique solution. A circuit design problem may have many solutions or even no solution. The maximum available signal levels found above provide bounds that help test for the existence of a solution. Generally there will be several ways to meet the interface constraints, and it then becomes necessary to evaluate the alternatives using other factors such as cost, power consumption, or reliability. Currently only the resistor will be used to demonstrate interface design. In subsequent sections other useful devices such as op amps and capacitors and inductors will be used to design suitable interfaces. In a design situation the engineer must choose the resistance values in a proposed circuit. This decision is influenced by a host of practical considerations such as standard values, standard tolerances, manufacturing methods, power limitations, and parasitic elements.

Example 9.   Select the load resistance in Fig. 31 so the interface signals are in the range defined by $v \geq 4$ V and $i \geq 30$ mA.

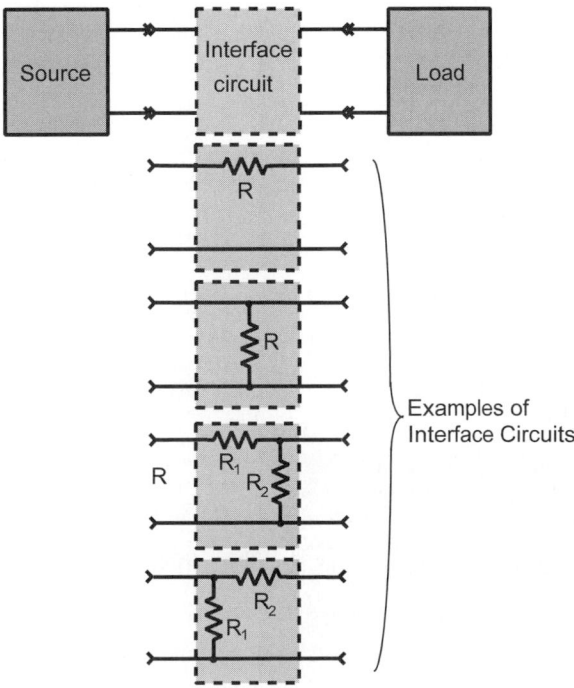

Fig. 30   General interface circuit and some examples.[1]

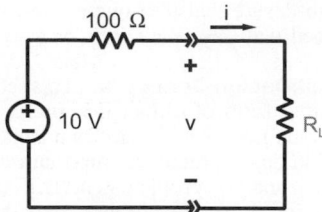

Fig. 31   (From Ref. 1.)

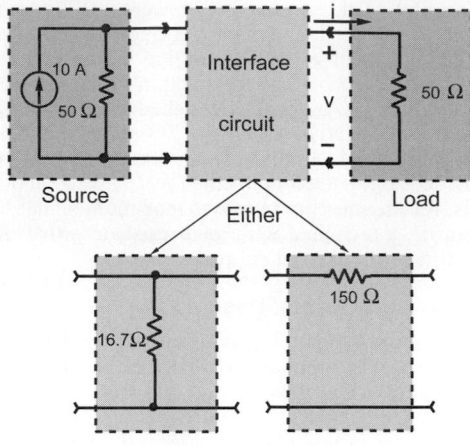

Fig. 32   (From Ref. 1.)

Solution. In this design problem the source circuit is given and a suitable load needs to be selected. For a fixed source the maximum signal levels available at the interface are

$$v_{\text{MAX}} = v_T = 10 \text{ V} \qquad i_{\text{MAX}} = \frac{v_T}{R_T} = 100 \text{ mA}$$

The bounds given as design requirements are below the maximum available signal levels a suitable resistor can be found. Using voltage division, the interface voltage constraint requires

$$v = \frac{R_L}{100 + R_L} \times 10 \geq 4 \quad \text{or} \quad v = 10R_L \geq 4R_L + 400$$

This condition yields $R_L \geq 400/6 = 66.7 \ \Omega$. The interface current constraint can be written as

$$i = \frac{10}{100 + R_L} \geq 0.03 \quad \text{or} \quad i = 10 \geq 3 + 0.03R_L$$

which requires $R_L \leq 7/0.03 = 233 \ \Omega$. In principle any value of R_L between 66.7 and 233 Ω will work. However, to allow for circuit parameter variations, choose $R_L = 150 \ \Omega$ because it lies at the arithmetic midpoint of allowable range and is a standard value.

Example 10. Design the two-port interface circuit in Fig. 32 so that the 10-A source delivers 100 V to the 50-Ω load.

Solution. The problem requires that the current delivered to the load is $i = 100/50 = 2$ A, which is well below the maximum available from the source. In fact, if the 10-A source is connected directly to the load, the source current divides equally between two 50-Ω resistors producing 5 A through the load. Therefore, an interface circuit is needed to reduce the load current to the specified 2-A level. Two possible design solutions are shown in Fig. 32. Applying current division to the parallel-resistor case yields the following constraint:

$$i = \frac{1/50}{1/50 + 1/50 + 1/R_{\text{PAR}}} \times 10$$

For the $i = 2$-A design requirement this equation becomes

$$2 = \frac{10}{2 + 50/R_{\text{PAR}}}$$

Solving for R_{PAR} yields

$$R_{\text{PAR}} = \frac{50}{3} = 16.7 \ \Omega$$

Applying the two-path current division rule to the series-resistor case yields the following constraint:

$$i = \frac{50}{50 + (50 + R_{\text{SER}})} \times 10 = 2 \text{ A}$$

Solving for R_{SER} yields

$$R_{\text{SER}} = 150 \ \Omega$$

Both these two designs meet the basic $i = 2$-A requirement. In practice, engineers evaluate alternative designs using additional criteria. One such consideration is the required power ratings of the resistors in each design. The voltage across the parallel resistor is $v = 100$ V, so the power loss is

$$p_{\text{PAR}} = \frac{100^2}{50/3} = 600 \text{ W}$$

The current through the series-resistor interface is $i = 2$ A so the power loss is

$$p_{\text{SER}} = 2^2 \times 150 = 600 \text{ W}$$

In either design the resistors must have a power rating of at least 600 W. The series resistor is a standard value whereas the parallel resistor is not. Other factors besides power rating and standard size could determine which design should be selected.

Example 11. Design the two-port interface circuit in Fig. 33 so the load is a match to 50 Ω between terminals C and D, while simultaneously the source matches to a load resistance of 300 Ω between A and B.

Solution. No single-resistor interface circuit could work. Hence try an interface circuit containing two resistors. Since the load must see a smaller resistance than the source, it should "look" into a parallel resistor. Since the source must see a larger resistance than the load, it should look into a series resistor. A configuration that meets these conditions is the L circuit shown in Figs. 33b,c.

The above discussion can be summarized mathematically. Using the L circuit, the design requirement at terminals C and D is

$$\frac{(R_1 + 300)\, R_2}{R_1 + 300 + R_2} = 50\ \Omega$$

At terminals A and B the requirement is

$$R_{11} + \frac{50 R_2}{R_2 + 50} = 300\ \Omega$$

The design requirements yield two equations in two unknowns—what could be simpler? It turns out that

solving these nonlinear equations by hand analysis is a bit of a chore. They can easily be solved using a math solver such as MATLAB or MathCad. But a more heuristic approach might serve best.

Given the L circuits in Fig. 33b, such an approach goes as follows. Let $R_2 = 50\ \Omega$. Then the requirement at terminals C and D will be met, at least approximately. Similarly, if $R_1 + R_2 = 300\ \Omega$, the requirements at terminals A and B will be approximately satisfied. In other words, try $R_1 = 250\ \Omega$ and $R_2 = 50\ \Omega$ as a first cut. These values yield equivalent resistances of $R_{CD} = 50||550 = 45.8\ \Omega$ and $R_{AB} = 250 + 50||50 = 275\ \Omega$. These equivalent resistances are not the exact values specified but are within ±10%. Since the tolerance on electrical components may be at least this high, a design using these values could be adequate. The exact values found by a math solver yields $R_1 = 273.861\ \Omega$ and $R_2 = 54.772\ \Omega$.

3 LINEAR ACTIVE CIRCUITS

This section treats the analysis and design of circuits containing active devices such as transistors or operational amplifiers (op amps). An *active device* is a component that requires an external power supply to operate correctly. An *active circuit* is one that contains one or more active devices. An important property of active circuits is that they are capable of providing signal amplification, one of the most important signal-processing functions in electrical engineering. Linear active circuits are governed by the proportionality property so their input–output relationships are of the form $y = Kx$. The term *signal amplification* means the proportionality factor $K > 1$ when the input x and output y have the same dimensions. Thus, active circuits can deliver more signal voltage, current, and power at their output than they receive from the input signal. The passive resistance circuits studied thus far cannot produce voltage, current, or power gains greater than unity.

3.1 Dependent Sources

When active devices operate in a linear mode, they can be modeled using resistors and one or more of the four dependent source elements shown in Fig. 34.

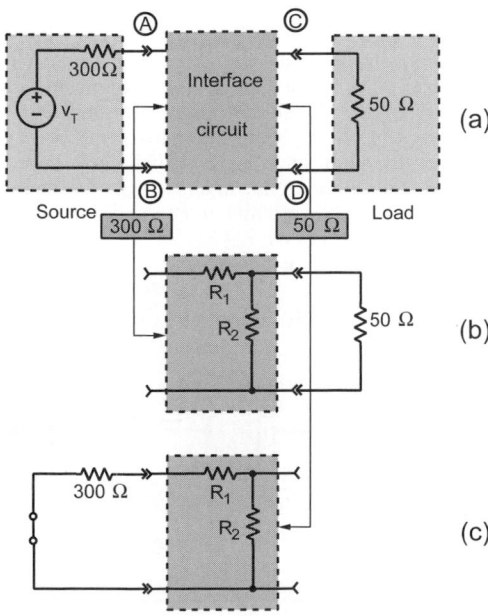

Fig. 33 (From Ref. 1.)

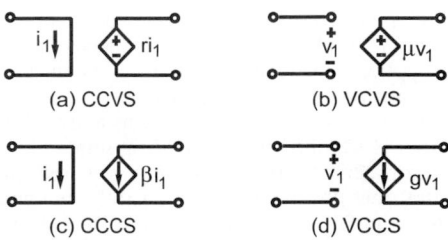

Fig. 34 Dependent source circuit symbols: (a) current-controlled voltage source; (b) voltage-controlled voltage source; (c) current-controlled current source; (d) voltage-controlled current source.[1]

The dominant feature of a dependent source is that the strength or magnitude of the voltage source (VS) or current source (CS) is proportional to—that is, controlled by—a voltage (VC) or current (CC) appearing elsewhere in the circuit. For example, the dependent source model for a current-controlled current source (CCCS) is shown in Fig. 34c. The output current βi_1 is dependent on the input current i_1 and the dimensionless factor β. This dependency should be contrasted with the characteristics of the independent sources studied earlier. The voltage (current) delivered by an independent voltage (current) source does not depend on the circuit to which it is connected. Dependent sources are often but not always represented by the diamond symbol, in contrast to the circle symbol used for independent sources.

Each dependent source is characterized by a single parameter, μ, β, r, or g. These parameters are often called simply the *gain* of the controlled source. Strictly speaking, the parameters μ and β are dimensionless quantities called the *open-loop voltage gain* and *open-loop current gain*, respectively. The parameter r has the dimensions of ohms and is called the *transresistance*, a contraction of transfer resistance. The parameter g is then called *transconductance* and has the dimensions of siemens.

In every case the defining relationship for a dependent source has the form $y = Kx$, where x is the controlling variable, y is the controlled variable, and K is the element gain. It is this linear relationship between the controlling and controlled variables that make dependent sources linear elements.

Although dependent sources are elements used in circuit analysis, they are conceptually different from the other circuit elements. The linear resistor and ideal switch are models of actual devices called resistors and switches. But dependent sources are not listed in catalogs. For this reason dependent sources are more abstract, since they are not models of identifiable physical devices. Dependent sources are used in combination with other resistive elements to create models of active devices.

A voltage source acts like a short circuit when it is turned off. Likewise, a current source behaves like an open circuit when it is turned off. The same results apply to dependent sources, with one important difference. Dependent sources cannot be turned on and off individually because they depend on excitation supplied by independent sources. When applying the superposition principle or Thevenin's theorem to active circuits, the state of a dependent source depends on excitation supplied by independent sources. In particular, for active circuits the superposition principle states that the response due to all independent sources acting simultaneously is equal to the sum of the responses due to each independent source acting one at a time.

Analysis with Dependent Sources With certain modifications the circuit analysis tools developed for passive circuits apply to active circuits as well. Circuit reduction applies to active circuits, but the control variable for a dependent source must not be eliminated. Applying a source transformation to a dependent source is sometimes helpful. Methods like node and mesh analysis can be adapted to handle dependent sources as well. But the main difference is that the properties of active circuits can be significantly different from those of the passive circuits.

In the following example the objective is to determine the current, voltage, and power delivered to the 500-Ω output load in Fig. 35. The control current i_X is found using current division in the input circuit:

$$i_X = \left(\frac{50}{50 + 25}\right) i_S = \frac{2}{3} i_S$$

Similarly the output current i_O is found using current division in the output circuit:

$$i_O = \left(\frac{300}{300 + 500}\right) i_Y = \frac{3}{8} i_Y$$

But at node A KCL requires that $i_Y = -48 i_X$. Combining this result with the equations for i_X and i_O yields the output current:

$$i_O = \left(\tfrac{3}{8}\right)(-48) i_X = -18 \left(\tfrac{2}{3} i_S\right) = -12 i_S \qquad (22)$$

The output voltage v_O is found using Ohm's law:

$$v_O = i_O \times 500 = -6000 i_S \qquad (23)$$

The input–output relationships in Eqs. (22) and (23) are of the form $y = Kx$ with $K < 0$. The proportionality constants are negative because the reference direction for i_O in Fig. 35 is the opposite of the orientation of the dependent source reference mark. Active circuits often produce negative values of K. As a result the input and output signals have opposite signs, a result called *signal inversion*. In the analysis and design of active circuits it is important to keep track of signal inversions.

The delivered output power is

$$p_O = v_O i_O = (-6000 i_S)(-12 i_S) = 72{,}000 i_S^2$$

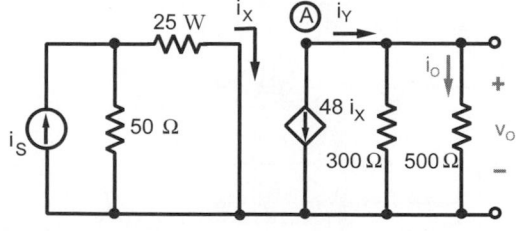

Fig. 35 Circuit with dependent source.[1]

The input independent source delivers its power to the parallel combination of 50 and 25 Ω. Hence, the power supplied by the independent source is

$$p_S = (50\|25)\, i_S^2 = \left(\tfrac{50}{3}\right) i_S^2$$

Given the input power and output power, we find the power gain in the circuit:

$$\text{Power gain} = \frac{p_O}{p_S} = \frac{72{,}000 i_S^2}{(50/3)\, i_S^2} = 432$$

A power gain greater than unity means that the circuit delivers more power at its output than it receives from the input source. At first glance, this appears to be a violation of energy conservation, but dependent sources are models of active devices that require an external power supply to operate. In general, circuit designers do not show the external power supply in circuit diagrams. Control source models assume that the external supply and the active device can handle whatever power is required by the circuit. With real devices this is not the case, and in circuit design engineers must ensure that the power limits of the device and external supply are not exceeded.

Node Voltage Analysis with Dependent Sources

Node voltage analysis of active circuits follows the same process as for passive circuits except that the additional constraints implied by the dependent sources must be accounted for. For example, the circuit in Fig. 36 has five nodes. With node E as the reference both independent voltage sources are connected to ground and force the condition $v_A = v_{S_1}$ and $v_B = v_{S2}$.

Node analysis involves expressing element currents in terms of the node voltages and applying KCL at each unknown node. The sum of the currents *leaving* node C is

$$\frac{v_C - v_{S1}}{R_1} + \frac{v_C - v_{S2}}{R_2} + \frac{v_C}{R_B} + \frac{v_C - v_D}{R_p} = 0$$

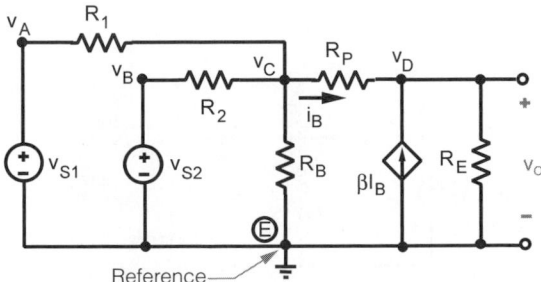

Fig. 36 Circuit used for node voltage analysis with dependent sources.[1]

Similarly, the sum of currents leaving node D is

$$\frac{v_D - v_C}{R_P} + \frac{v_D}{R_E} - \beta i_B = 0$$

These two node equations can be rearranged into the forms

Node C: $\quad v_C \left(\dfrac{1}{R_1} + \dfrac{1}{R_2} + \dfrac{1}{R_B} + \dfrac{1}{R_P} \right) - \dfrac{1}{R_P} v_D$

$$= \frac{1}{R_1} v_{S1} + \frac{1}{R_2} v_{S2}$$

Node D: $\quad -\dfrac{1}{R_P} v_C + \left(\dfrac{1}{R_P} + \dfrac{1}{R_E} \right) v_D$

$$= \beta i_B$$

Applying the fundamental property of node voltages and Ohm's law, the current i_B can be written in terms of the node voltages as

$$i_B = \frac{v_C - v_D}{R_P}$$

Substituting this expression for i_B into the above node equation and putting the results in standard form yield

Node C: $\quad \left(\dfrac{1}{R_1} + \dfrac{1}{R_2} + \dfrac{1}{R_B} + \dfrac{1}{R_P} \right) v_C - \dfrac{1}{R_P} v_D$

$$= \frac{1}{R_1} v_{S1} + \frac{1}{R_2} v_{S2}$$

Node D: $\quad -(\beta + 1) \dfrac{1}{R_P} v_C$

$$+ \left[(\beta + 1) \frac{1}{R_P} + \frac{1}{R_E} \right] v_D = 0$$

The final result involves two equations in the two unknown node voltages and includes the effect of the dependent source.

This example illustrates a general approach to writing node voltage equations for circuits with dependent sources. Dependent sources are initially treated as if they are independent sources and node equations written for the resulting passive circuit. This step produces a set of symmetrical node equations with the independent and dependent source terms on the right side. Next the dependent source terms are expressed in terms of the unknown node voltages and moved to the left side with the other terms involving the unknowns. The last step destroys the coefficient symmetry but leads to a set of equations that can be solved for the active circuit response.

Mesh Current Analysis with Dependent Sources

Mesh current analysis of active circuits follows the same pattern noted for node voltage analysis. Treat the dependent sources initially as independent sources and write the mesh equations of the resulting passive circuit. Then account for the dependent sources by expressing their constraints in terms of unknown mesh currents. The following example illustrates the method.

Example 12

(a) Formulate mesh current equations for the circuit in Fig. 37.
(b) Use the mesh equations to find v_O and R_{IN} when $R_1 = 50\ \Omega$, $R_2 = 1\ k\Omega$, $R_3 = 100\ \Omega$, $R_4 = 5\ k\Omega$, and $g = 100\ mS$.

Solution. (a) Applying source transformation to the parallel combination of R_3 and gv_X in Fig. 37a produces the dependent voltage source $R_{3gv_X} = \mu v_X$ in Fig. 37b. In the modified circuit we have identified two mesh currents. Initially treating the dependent source $(gR_3)v_x$ as an independent source leads to two symmetrical mesh equations:

Mesh A: $(R_1+R_2+R_3)i_A - R_3i_B = v_S - (gR_3)v_X$
Mesh B: $-R_3i_A + (R_3+R_4)i_B = (gR_3)v_X$

The control voltage v_x can be written in terms of mesh currents as

$$v_X = R_2 i_A$$

Substituting this equation for v_x into the mesh equations and putting the equations in standard form yield

$$(R_1 + R_2 + R_3 + gR_2R_3)\,i_A - R_3 i_B = v_S$$

$$-(R_3 + gR_2R_3)\,i_A + (R_3 + R_4)\,i_B = 0$$

The resulting mesh equations are not symmetrical because of the controlled source.

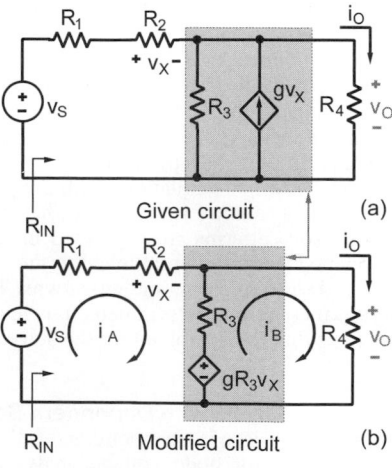

Given circuit   (a)

Modified circuit   (b)

Fig. 37 (From Ref. 1.)

(b) Substituting the numerical values into the mesh equations gives

$$(1.115 \times 10^4)i_A - 10^2 i_B = v_S$$

$$-(1.01 \times 10^4)i_A + (5.1 \times 10^3)i_B = 0$$

Using Cramer's rule the mesh currents are found to be

$$i_A = (0.9131 \times 10^{-4})v_S \quad \text{and}$$

$$i_B = (1.808 \times 10^{-4})v_S$$

The output voltage and input resistance are found using Ohm's law:

$$v_O = R_4 i_B = 0.904 v_S \qquad R_{IN} = \frac{v_S}{i_A} = 10.95\ k\Omega$$

Thevenin Equivalent Circuits with Dependent Sources

To find the Thevenin equivalent of an active circuit, the independent sources are left on or else one must supply excitation from an external test source. This means that the Thevenin resistance can not be found by the "look-back" method, which requires that all independent sources be turned off. Turning off the independent sources deactivates the dependent sources as well and can result in a profound change in input and output characteristics of an active circuit. Thus, Thevenin equivalents of active circuits can be found using the open-circuit voltage and short-circuit current at the interface.

Example 13. Find the Thevenin equivalent at the output interface of the circuit in Fig. 38.

Solution. In this circuit the controlled voltage v_X appears across an open circuit between nodes A and B. By the fundamental property of node voltages, $v_X = v_S - v_O$. With the load disconnected and the input source turned off, $v_x = 0$, the dependent voltage source μv_X acts like a short circuit, and the Thevenin resistance looking back into the output port is R_O. With the load connected and the input source turned on, the sum of currents leaving node B is

$$\frac{v_O - \mu v_X}{R_O} + i_O = 0$$

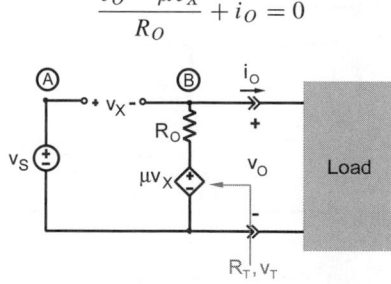

Fig. 38 (From Ref. 1.)

Using the relationship $v_X = v_S - v_O$ to eliminate v_X and then solving for v_O produce the output $i–v$ relationship of the circuit as

$$v_O = \frac{\mu v_S}{\mu + 1} - i_O \left[\frac{R_O}{\mu + 1} \right]$$

The $i–v$ relationship of a Thevenin circuit is $v = v_T - iR_T$. By direct comparison, the Thevenin parameters of the active circuit are found to be

$$v_T = \frac{\mu v_S}{\mu + 1} \quad \text{and} \quad R_T = \frac{R_O}{\mu + 1}$$

The circuit in Fig. 38 is a model of an op amp circuit called a voltage follower. The resistance R_O for a general-purpose op amp is on the order of 100 Ω, while the gain μ is about 10^5. Thus, the active Thevenin resistance of the voltage follower is not 100 Ω, as the look-back method suggests, but around a milliohm!

3.2 Operational Amplifier

The operational amplifier is the premier linear active device made available by IC technology. John R. Ragazzini apparently first used the term operational amplifier in a 1947 paper and his colleagues who reported on work carried out for the National Defenses Research Council during World War II. The paper described high-gain dc amplifier circuits that perform mathematical operations (addition, subtraction, multiplication, division, integration, etc.)—hence the name "operational" amplifier. For more than a decade the most important applications were general- and special-purpose analog computers using vacuum tube amplifiers. In the early 1960s general-purpose, discrete-transistor, op amp became readily available and by the mid-1960s the first commercial IC op amps entered the market. The transition from vacuum tubes to ICs resulted in a decrease in size, power consumption, and cost of op amps by over three orders of magnitude. By the early 1970s the IC version became the dominant

active device in analog circuits. The device itself is a complex array of transistors, resistors, diodes, and capacitors all fabricated and interconnected on a single silicon chip. In spite of its complexity, the op amp can be modeled by rather simple $i–v$ characteristics.

Op Amp Notation Certain matters of notation and nomenclature must be discussed before developing a circuit model for the op amp. The op amp is a five-terminal device, as shown in Fig. 39a. The "+" and "−" symbols identify the input terminals and are a shorthand notation for the noninverting and inverting input terminals, respectively. These "+" and "−" symbols identify the two input terminals and have nothing to do with the polarity of the voltages applied. The other terminals are the output and the positive and negative supply voltage, usually labeled $+V_{CC}$ and $-V_{CC}$. While some op amps have more than five terminals, these five are always present. Figure 39b shows how these terminals are arranged in a common eight-pin IC package.

While the two power supply terminals in Fig. 39 are not usually shown in circuit diagrams, they are always there because the external power supplies connected to these terminals make the op amp an active device. The power required for signal amplification comes through these terminals from an external power source. The $+V_{CC}$ and $-V_{CC}$ voltages applied to these terminals also determine the upper and lower limits on the op amp output voltage.

Figure 40a shows a complete set of voltage and current variables for the op amp, while Fig. 40b shows the typical abbreviated set of signal variables. All voltages are defined with respect to a common reference node, usually ground. Voltage variables v_P, v_N, and v_O are defined by writing a voltage symbol beside the corresponding terminals. This notation means the "+" reference mark is at the terminal in question and the "−" reference mark is at the reference or ground terminal. The reference directions for the currents are directed in at input terminals and out at the output. A global KCL equation for the complete set of variable in Fig. 40a is $i_O = I_{C+} + I_{C-} + i_P + i_N$, NOT $i_O = i_N + i_P$, as might be inferred from Fig. 40b, since it does not

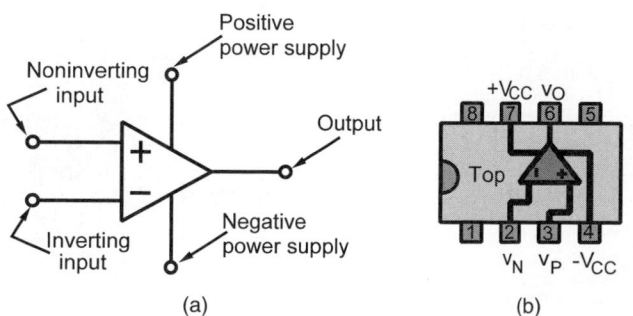

Fig. 39 Op amp: (a) circuit symbol; (b) pin out diagram for eight-pin package.[1]

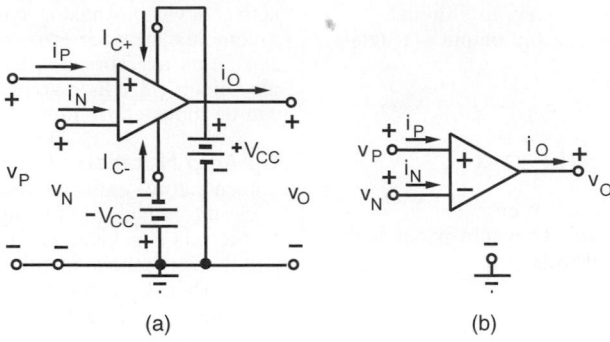

Fig. 40 Op amp voltage and current definitions: (a) complete set; (b) shorthand set.[1]

include all of the currents. More importantly, it implies that the output current comes from the inputs. In fact, this is wrong. The input currents are very small, ideally zero. The output current comes from the supply voltages even though these terminals are not shown on the abbreviated circuit diagram.

Transfer Characteristics The dominant feature of the op amp is the transfer characteristic shown in Fig. 41. This characteristic provide the relationships between the *noninverting input* v_P, the *inverting input* v_N, and the *output voltage* v_O. The transfer characteristic is divided into three regions or modes called *+saturation*, *−saturation*, and *linear*. In the linear region the op amp is a *differential amplifier* because the output is proportional to the difference between the two inputs. The slope of the line in the linear range is called the *open-loop gain*, denoted as μ. In the linear region the input–output relation is $v_O = \mu(v_P - v_N)$. The open-loop gain of an op amp is very large, usually greater than 10^5. As long as the net input $v_P - v_N$ is very small, the output will be proportional to the input. However, when $\mu|v_P - v_N| > V_{CC}$, the op amp is saturated and the output voltage is limited by the supply voltages (less some small internal losses).

The op amp has three operating modes:

1. + Saturation mode when $\mu(v_P - v_N) > +V_{CC}$ and $v_O = +V_{CC}$.
2. − Saturation mode when $\mu(v_P - v_N) < -V_{CC}$ and $v_O = -V_{CC}$.
3. Linear mode when $\mu|v_P - v_N| < V_{CC}$ and $v_O = \mu(v_P - v_N)$.

Usually op amp circuits are analyzed and designed using the linear mode model.

Ideal Op Amp Model A controlled source model of an op amp operating in its linear range is shown in Fig. 42. This model includes an input resistance (R_I), an output resistance (R_O), and a voltage-controlled voltage source whose gain is the open-loop gain μ. Some typical ranges for these op amp parameters are given in Table 3, along with the values for the ideal op amp. The high input and low output resistances and high open-loop gain are the key attributes of an op amp. The ideal model carries these traits to the extreme limiting values.

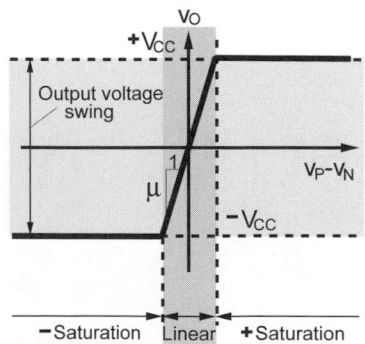

Fig. 41 Op amp transfer characteristics.[1]

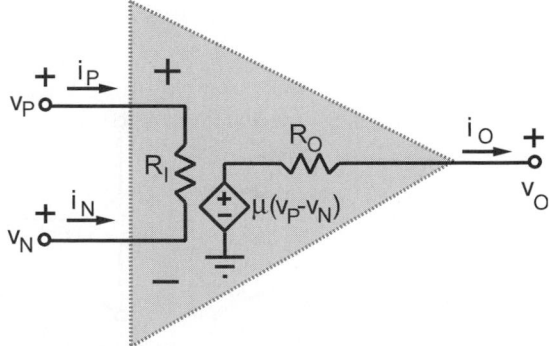

Fig. 42 Dependent source model of op amp operating in linear mode.[1]

Table 3   Typical Op Amp Parameters

Name	Parameter	Range	Ideal Values
Open-loop gain	μ	10^5–10^8	∞
Input resistance	R_I	10^6–10^{13} Ω	∞ Ω
Output resistance	R_O	10–100 Ω	0 Ω
Supply voltages	V_{CC}	± 5 to ± 40 V	

The controlled source model can be used to develop the i–v relationships of the ideal model. This discussion is restricted to the linear region of operation. This means the output voltage is bounded as

$$-V_{CC} \le v_o \le +V_{CC} \qquad -\frac{V_{CC}}{\mu} \le (v_P - v_N) \le +\frac{V_{CC}}{\mu}$$

The supply voltages V_{CC} are most commonly ± 15 V although other supply voltages are available, while μ is a very large number, usually 10^5 or greater. Consequently, linear operation requires that $v_P \cdot v_N$. For the ideal op amp the open-loop gain is infinite ($\mu \to \infty$), in which case linear operation requires $v_P = v_N$. The input resistance R_I of the ideal op amp is assumed to be infinite, so the currents at both input terminals are zero. In summary, the i–v relationships of the *ideal model* of the op amp are

$$v_P = v_N \qquad i_P = i_N = 0 \qquad (24)$$

At first glance the element constraints of the ideal op amp appear to be fairly useless. They actually look more like connection constraints and are totally silent about the output quantities (v_O and i_O), which are usually the signals of interest. In fact, they seem to say that the op amp input terminals are simultaneously a short circuit ($v_P = v_N$) and an open circuit ($i_P = i_N = 0$). The ideal model of the op amp is useful because in linear applications feedback is always present. That is, in order for the op amp to operate in a linear mode, it is necessary that there be feedback paths from the output to one or both of the inputs. These feedback paths ensure that $v_P = v_N$ and allow for analysis of op amp circuits using the ideal op amp element constraints.

Op Amp Circuit Analysis   This section introduces op amp circuit analysis using circuits that are building blocks for analog signal-processing systems. The key to using the *building block* approach is to recognize the feedback pattern and to isolate the basic circuit as a building block.

Noninverting Op Amp   To illustrate the effects of feedback, consider the circuit in Fig. 43. This circuit has a feedback path from the output to the inverting input via a voltage divider. Since the ideal op amp draws no current at either input ($i_P = i_N = 0$),

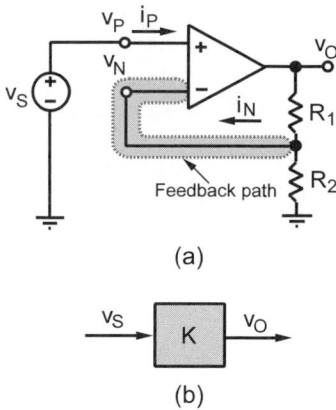

(a)

(b)

Fig. 43   Noninverting amplifier circuit.[1]

voltage division determines the voltage at the inverting input as

$$v_N = \frac{R_2}{R_1 + R_2} v_O$$

The input source connection at the noninverting input requires the condition $v_P = v_S$. But the ideal op amp element constraints demand that $v_P = v_N$; therefore, the input–output relationship of the overall circuit is

$$v_O = \frac{R_1 + R_2}{R_2} v_S \qquad (25)$$

The circuit in Fig. 43a is called a *noninverting amplifier*. The input–output relationship is of the form $v_O = K v_S$, a linear relationship. Figure 43b shows the functional building block for this circuit, where the proportionality constant K is

$$K = \frac{R_1 + R_2}{R_2} \qquad (26)$$

The constant K is called the *closed-loop* gain, since it includes the effect of the feedback path. When discussing op amp circuits, it is necessary to distinguish between two types of gains. The first is the open-loop gain μ provided by the op amp device. The gain μ is a large number with a large uncertainty tolerance. The second type is the closed-loop gain K of the op amp circuit with a feedback path. The gain K must be smaller than μ, typically no more than $1/100$ of μ, and its value is determined by the resistance elements in the feedback path.

For example, the closed-loop gain in Eq. (26) is really the voltage division rule upside down. The uncertainty tolerance assigned to K is determined by the quality of the resistors in the feedback path, and not the uncertainty in the actual value of the closed-loop gain. In effect, feedback converts a very large but imprecisely known open-loop gain into a much smaller but precisely controllable closed-loop gain.

Example 14. Design an amplifier with a closed-loop gain $K = 10$.

Solution. Using a noninverting op amp circuit, the design problem is to select the values of the resistors in the feedback path. From Eq. (26) the design constraint is

$$10 = \frac{R_1 + R_2}{R_2}$$

This yields one constraint with two unknowns. Arbitrarily selecting $R_2 = 10\ \text{k}\Omega$ makes $R_1 = 90\ \text{k}\Omega$. These resistors would normally have high precision ($\pm 1\%$ or less) to produce a precisely controlled closed-loop gain.

Comment: The problem of choosing resistance values in op amp circuit design problems deserves some discussion. Although resistances from a few ohms to several hundred megohms are commercially available, generally designers limit themselves to the range from about 1 kΩ to perhaps 2.2 M Ω. The lower limit of 1 k Ω exists in part because of power dissipation in the resistors and to minimize the effects of loading (discussed later). Typically resistors with 3 W power ratings or less are used. The maximum voltages in op amp circuits are often around ± 15 V although other values exist, including single-sided op amps, with a 0–5 V V_{CC} for use in digital applications. The smallest 3-W resistance we can use is $R_{\text{MIN}} > (15)^2/0.25 = 900\ \Omega$, or about 1 k$\Omega$. The upper bound of 2.2 MΩ exists because it is difficult to maintain precision in a high-value resistor because of surface leakage caused by humidity. High-value resistors are also noisy, which leads to problems when they are in the feedback path. The range 1 kΩ to 2.2 MΩ should be used as a guideline and not an inviolate design rule. Actual design choices are influenced by system-specific factors and changes in technology.

Voltage Follower The op amp in Fig. 44a is connected as *voltage follower* or *buffer*. In this case the feedback path is a direct connection from the output to the inverting input. The feedback connection forces the condition $v_N = v_O$. The input current $i_P = 0$ so there is no voltage across the source resistance R_S. Applying KVL results in $v_P = v_S$. The ideal op amp model requires $v_P = v_N$, so that $v_O = v_S$. By inspection the closed-loop gain is $K = 1$. The output exactly equals the input, that is, the output follows the input, and hence the name voltage follower.

The voltage follower is used in interface circuits because it isolates the source and load—hence its other name, *buffer*. Note that the input–output relationship $v_O = v_S$ does not depend on the source or load resistance. When the source is connected directly to the load as in Fig. 44b, the voltage delivered to the load depends on R_S and R_L. The source and load interaction limits the signals that can transfer across the interface.

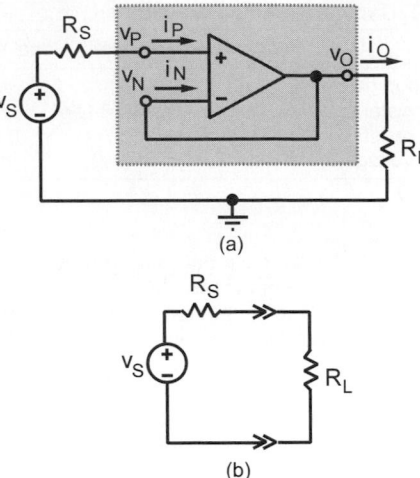

Fig. 44 (a) Source–load interface with voltage follower; (b) interface without voltage follower.[1]

When the voltage follower is inserted between the source and load, the signal levels are limited by the capability of the op amp.

Inverting Amplifier The circuit in Fig. 45 is called an *inverting amplifier*. The key feature of this circuit is that the input signal and the feedback are both applied at the inverting input. Note that the noninverting input is grounded, making $v_P = 0$. Using the fundamental property of node voltages and KCL, the sum of currents entering node A can be written as

$$\frac{v_s - v_N}{R_1} + \frac{v_O - v_N}{R_2} - i_N = 0$$

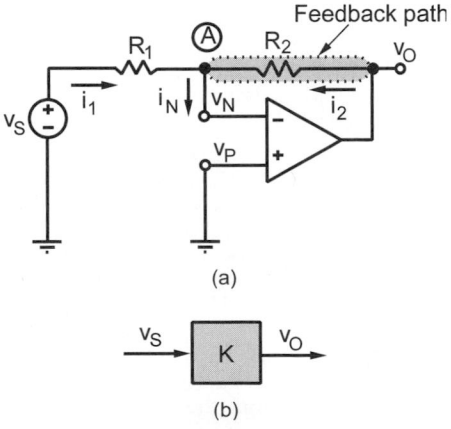

Fig. 45 Inverting amplifier circuit.[1]

The element constraints for the op amp are $v_P = v_N$ and $i_P = i_N = 0$. Since $v_P = 0$, it follows that $v_N = 0$. Substituting the op amp constraints and solving for the input–output relationship yield

$$v_O = -\left(\frac{R_2}{R_1}\right) v_S \qquad (27)$$

This result is of the form $v_O = K v_S$, where K is the closed-loop gain. However, in this case the closed-loop gain $K = -R_2/R_1$ is negative, indicating a signal inversion—hence the name inverting amplifier. The block diagram symbol shown in Fig. 45b is used to indicate either the inverting or noninverting op amp configuration, since both circuits provide a gain of K.

The op amp constraints mean that the input current i_1 in Fig. 43a is

$$i_1 = \frac{v_S - v_N}{R_1} = \frac{v_S}{R_1}$$

This in turn means that the input resistance seen by the source v_S is

$$R_{IN} = \frac{v_S}{i_1} = R_1 \qquad (28)$$

In other words, the inverting amplifier has as finite input resistance determined by the external resistor R_1. This finite input resistance must be taken into account when analyzing circuits with op amps in the inverting amplifier configuration.

Summing Amplifier The *summing amplifier* or *adder* circuit is shown in Fig. 46a. This circuit has two inputs connected at node A, which is called the *summing point*. Since the noninverting input is

grounded, $v_P = 0$. This configuration is similar to the inverting amplifier; hence a similar analysis yields the circuit input–output relationship

$$v_O = \begin{cases} \left(-\dfrac{R_F}{R_1'}\right) v_1 + \left(-\dfrac{R_F}{R_2}\right) v_2 \\[2mm] (K_1)v_1 + (K_2)v_2 \end{cases} \qquad (29)$$

The output is a weighted sum of the two inputs. The scale factors, or gains as they are called, are determined by the ratio of the feedback resistor R_F to the input resistor for each input: that is, $K_1 = -R_F/R_1$ and $K_2 = -R_F/R_2$. In the special case $R_1 = R_2 = R$, Eq. (29) reduces to

$$v_O = K (v_1 + v_2)$$

where $K = -R_F/R$. In this special case the output is proportional to the negative sum of the two inputs—hence the name inverting summing amplifier or simply adder. A block diagram representation of this circuit is shown in Fig. 46b.

Example 15. Design an inverting summer that implements the input–output relationship $v_O = -(5v_1 + 13v_2)$.

Solution. The design problem involves selecting the input and feedback resistors so that

$$\frac{R_F}{R_1} = 5 \quad \text{and} \quad \frac{R_F}{R_2} = 13$$

One solution is to arbitrarily select $R_F = 65$ kΩ, which yields $R_1 = 13$ kΩ and $R_2 = 5$ kΩ. The resulting circuit is shown in Fig. 47a. The design can be modified to use standard resistance values for resistors with $\pm5\%$ tolerance. Selecting the standard

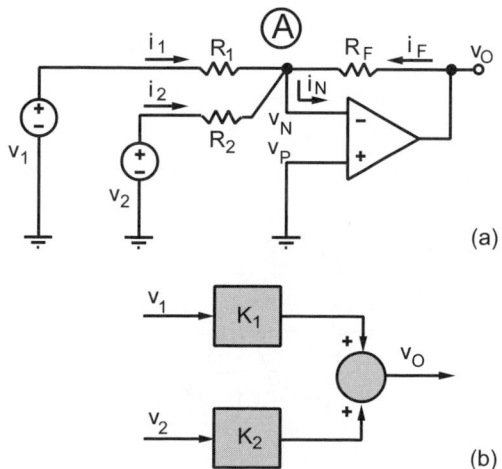

(a)

(b)

Fig. 46 Inverting summer.[1]

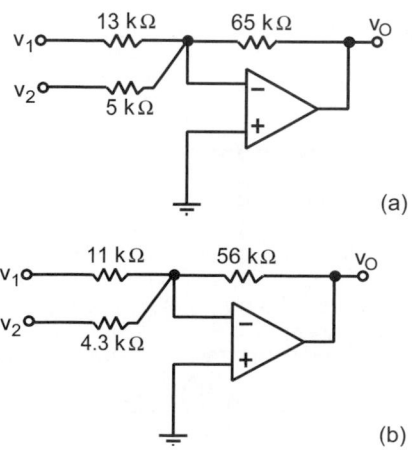

(a)

(b)

Fig. 47 (From Ref. 1.)

value $R_F = 56 \text{ k}\Omega$ requires $R_1 = 11.2 \text{ k}\Omega$ and $R_2 = 4.31 \text{ k}\Omega$. The nearest standard values are 11 and 4.3 kΩ. The resulting circuit shown in Fig. 47b uses only standard value resistors and produces gains of $K_1 = 56/11 = 5.09$ and $K_2 = 56/4.3 = 13.02$. These nominal gains are within 2% of the values in the specified input–output relationship.

Differential Amplifier   The circuit in Fig. 48a is called a *differential amplifier* or *subtractor*. Like the summer, this circuit has two inputs, but unlike the summer, one is applied at the inverting input and one at the noninverting input of the op amp. The input–output relationship can be obtained using the superposition principle.

With source v_2 off there is no excitation at the non-inverting input and $v_P = 0$. In effect, the noninverting input is grounded and the circuit acts like an inverting amplifier with the result that

$$v_{O1} = -\frac{R_2}{R_1}v_1$$

Now turning v_2 back on and turning v_1 off, the circuit looks like a noninverting amplifier with a voltage divider connected at its input. Thus

$$v_{O2} = \left[\frac{R_4}{R_3 + R_4}\right]\left[\frac{R_1 + R_2}{R_1}\right]v_2$$

Using superposition the two outputs are added to obtain the output with both sources on:

$$v_O = \begin{cases} v_{O1} + v_{O2} \\ -\left[\dfrac{R_2}{R_1}\right]v_1 + \left[\dfrac{R_4}{R_3 + R_4}\right]\left[\dfrac{R_1 + R_2}{R_1}\right]v_2 \\ -[K_1]v_1 + [K_2]v_2 \end{cases}$$

$$(30)$$

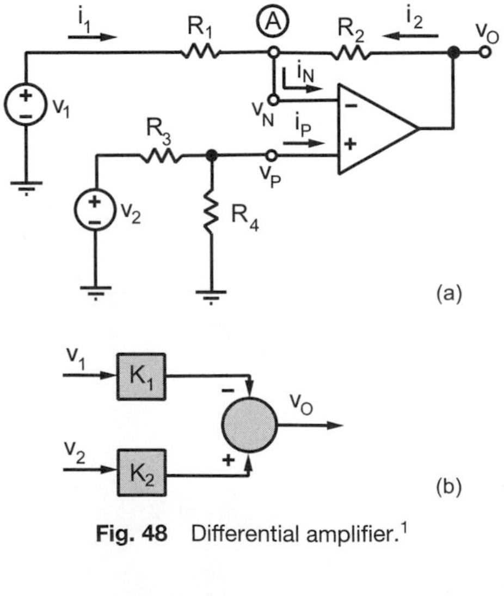

where K_1 and K_2 are the inverting and noninverting gains. Figure 48b shows how the differential amplifier is represented in a block diagram. In the special case of $R_1 = R_2 = R_3 = R_4$, Eq. (30) reduces to $v_O = v_2 - v_1$. In this case the output is equal to the difference between the two inputs—hence the name differential amplifier or subtractor.

Noninverting Summer   The circuit in Fig. 49 is an example of a *noninverting summer*. The input–output relationship for a general noninverting summer is

$$v_O = K\left[\left(\frac{R_{EQ}}{R_1}\right)v_1 + \left(\frac{R_{EQ}}{R_2}\right)v_2\right.$$

$$\left. + \cdots + \left(\frac{R_{EQ}}{R_n}\right)v_n\right] \qquad (31)$$

where R_{EQ} is the Thevenin resistance looking to the left at point P with all sources turned off (i.e., $R_{EQ} = R_1 \| R_2 \| R_3 \cdots \| R_n$) and K is the gain of the noninverting amplifier circuit to the right of point P. Comparing this equation with the general inverting summer result in Eq. (29), we see several similarities. In both cases the weight assigned to an input voltage is proportional to a resistance ratio in which the denominator is its input resistance. In the inverting summer the numerator of the ratio is the feedback resistor R_F and in the noninverting case the numerator is the equivalent of all input resistors R_{EQ}.

Design with Op Amp Building Blocks   The block diagram representation of the basic op amp circuit configurations were developed in the preceding section. The noninverting and inverting amplifiers are represented as gain blocks. The summing amplifier and differential amplifier require both gain blocks and the summing symbol. One should exercise care when translating from a block diagram to a circuit, or vice versa, since some gain blocks may involve negative gains. For example, the gain of the inverting amplifier is negative, as are the gains of the common inverting summing amplifier and the K_1 gain of the differential amplifier. The minus sign is sometimes moved to the summing symbol and the gain within the block changed to a positive number. Since there is

(a)

(b)

Fig. 48   Differential amplifier.[1]

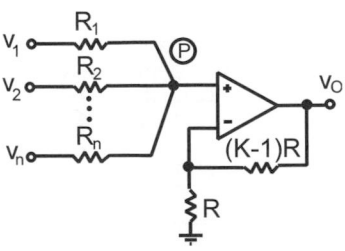

Fig. 49   Noninverting summer.[2]

no standard convention for doing this, it is important to keep track of the signs associated with gain blocks and summing point symbol.

Operational amplifier circuit design generally requires that a given equation or block diagram representation of a signal-processing function be created to implement that the function. Circuit design can often be accomplished by interconnecting the op amp, summer, and subtractor building blocks. The design process is greatly simplified by the near one-to-one correspondence between the op amp circuits and the elements in a block diagram. However, the design process is not unique since often there are several ways to use basic op amp circuits to meet the design objective. Some solutions are better than others are. The following example illustrates the design process.

Example 16. Design an op amp circuit that implements the block diagram in Fig. 50.

Solution. The input–output relationship represented by the block diagram is $v_O = 5v_1 + 10v_2 + 20v_3$. An op amp adder can implement the summation required in this relationship. A three-input adder implements

the relationship

$$v_O = - \left[\frac{R_F}{R_1} v_1 + \frac{R_F}{R_2} v_2 + \frac{R_F}{R_3} v_3 \right]$$

The required scale factors are realized by first selecting $R_F = 100$ kΩ and then choosing $R_1 = 20$ kΩ, $R_2 = 10$ kΩ, and $R_3 = 5$ kΩ. However, the adder involves a signal inversion. To correctly implement the block diagram, we must add an inverting amplifier ($K = -R_2/R_1$) with $R_1 = R_2 = 100$ kΩ. The final implementation is shown in Fig. 51a. An alternate solution avoiding the second inverting op amp by using a non-inverting summer is shown in Fig. 51b.

Digital-to-Analog Converters Operational amplifiers play an important role in the interface between digital systems and analog systems. The parallel 4-bit output in Fig. 52 is a digital representation of a signal. Each bit can only have two values: (a) a high or 1 (typically +5 V) and (b) a low or 0 (typically 0 V).

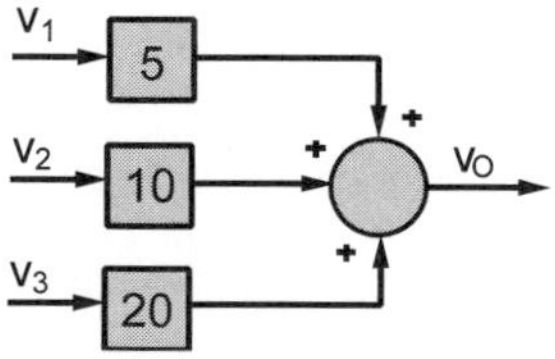

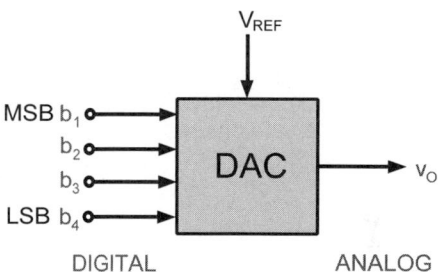

Fig. 50 (From Ref. 2.)

Fig. 52 A DAC.[1]

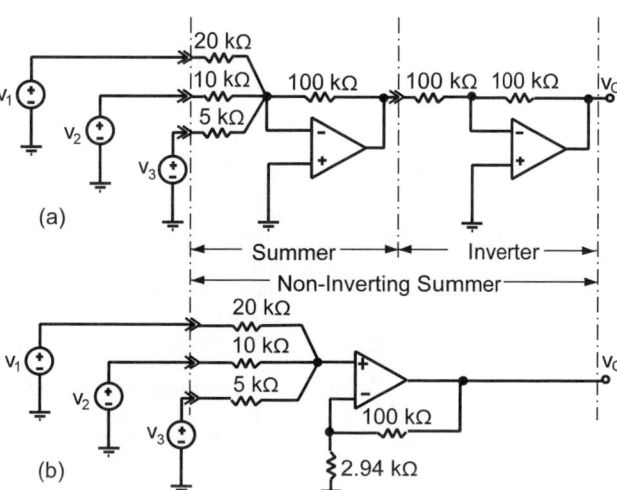

Fig. 51 (From Ref. 2.)

The bits have binary weights so that v_1 is worth $2^3 = 8$ times as much v_4, v_2 is worth $2^2 = 4$ times as much as v_4, v_3 is worth $2^1 = 2$ times as much v_4, and v_4 is equal to $2^0 = 1$ times itself. In a 4-bit DAC v_4 is the least significant bit (LSB) and v_1 the most significant bit (MSB). To convert the digital representation of the signal to analog form, each bit must be weighted so that the analog output v_O is

$$v_O = \pm K(8v_1 + 4v_2 + 2v_3 + 1v_4) \qquad (32)$$

where K is a scale factor or gain applied to the analog signal. Equation (32) is the input–output relationship of a 4-bit digital-to-analog converter (DAC).

One way to implement Eq. (32) is to use an inverting summer with binary-weighted input resistors. Figure 53 shows the op amp circuit and a block diagram of the circuit input–output relationship. In either form, the output is seen to be a binary-weighted sum of the digital input scaled by $-R_F/R$. That is, the output voltage is

$$v_O = \frac{-R_F}{R}(8v_1 + 4v_2 + 2v_3 + v_4)$$

The R–$2R$ ladder in Fig. 54a also implements a 4-bit DAC. The resistance seen looking back into the R–$2R$ ladder at point A with all sources turned off is seen to be $R_T = R$. A Thevenin equivalent circuit of the R–$2R$ network is shown in Fig. 54b, where

$$v_T = \tfrac{1}{2}v_1 + \tfrac{1}{4}v_2 + \tfrac{1}{8}v_3 + \tfrac{1}{16}v_4$$

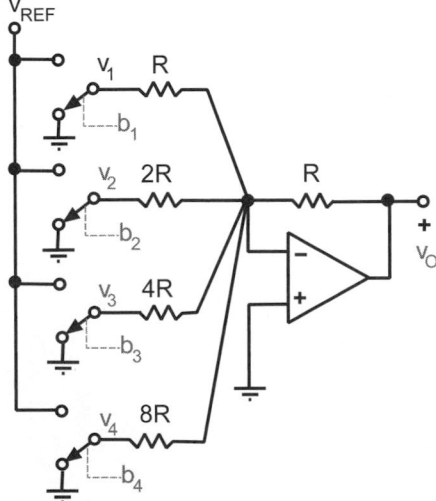

Fig. 53 Binary-weighted summer DAC.[1]

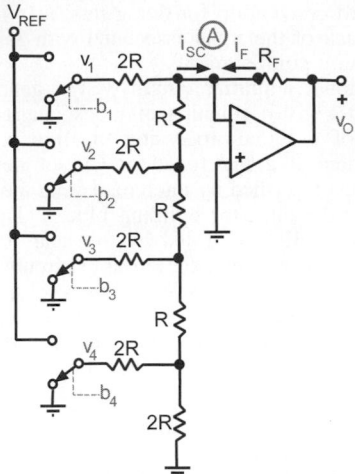

Fig. 54 An R–$2R$ ladder DAC.[1]

The output voltage is found using the inverting amplifier gain relationship:

$$v_O = \frac{-R_F}{R}v_T = \frac{-R_F}{R}\left(\frac{v_1}{2} + \frac{v_2}{4} + \frac{v_3}{8} + \frac{v_4}{16}\right)$$

Using $R_F = 16R$ yields

$$v_O = -(8v_1 + 4v_2 + 2v_3 + v_4)$$

which shows the binary weights assigned to the digital inputs.

In theory the circuits in Figs. 53 and 54 perform the same signal-processing function—4-bit digital-to-analog conversion. However, there are important practical differences between the two circuits. The inverting summer in Fig. 53 requires precision resistors with four different values spanning an 8 : 1 range. A more common 8-bit converter would require eight precision resistors spanning a 256 : 1 range. Moreover, the digital voltage sources in Fig. 53 see input resistances that span an 8 : 1 range; therefore, the source–load interface is not the same for each bit. On the other hand, the resistances in the R–$2R$ ladder converter in Fig. 54 span only a 2 : 1 range regardless of the number of digital bits. The R–$2R$ ladder also presents the same input resistance to each binary input. The R–$2R$ ladder converters are readily made on integrated or thin-film circuits and are the preferred DAC type.

Instrumentation Systems One of the most interesting and useful applications of op amp circuits is in instrumentation systems that collect and process data about physical phenomena. In such a system an *input transducer* (a device that converts some physical quantity, such as temperature, strain, light intensity, acceleration, wavelength, rotation, velocity, pressure, or whatever, into an electrical signal) generates an electrical signal that describes some ongoing physical process. In a simple system the transducer signal is processed by op amp circuits and displayed on an *output transducer* such as a meter or an oscilloscope or more commonly sent into a DAC for further processing or analysis by a microprocessor or digital computer. The output signal can also be used in a feedback control system to monitor and regulate the physical process itself or to control a robotic device.

The block diagram in Fig. 55 shows an instrumentation system in its simplest form. The objective of the system is to deliver an output signal that is directly proportional to the physical quantity measured by the input transducer. The input transducer converts a physical variable x into an electrical voltage v_{TR}. For many transducers this voltage is of the form $v_{TR} = mx + b$, where m is a calibration constant and b is a constant

offset or bias. The transducer voltage is often quite small and must be amplified by the gain K, as indicated in Fig. 55. The amplified signal includes both a signal component $K(mx)$ and a bias component $K(b)$. The amplified bias $K(b)$ is then removed by subtracting a constant electrical signal. The resulting output voltage $K(mx)$ is directly proportional to the quantity measured and goes to an output transducer for display. The required gain K can be found from the relation

$$K = \frac{\text{desired output range}}{\text{available input range}} \tag{33}$$

Example 17. Design a light intensity detector to detect $5-20$ lm of incident light using a photocell serving as the input transducer. The system output is to be displayed on a $0-10$-V voltmeter. The photocell characteristics are shown in Fig. 56a. The design

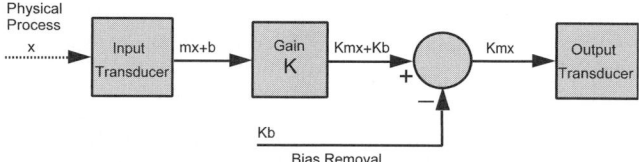

Fig. 55 Block diagram of instrumentation system.[2]

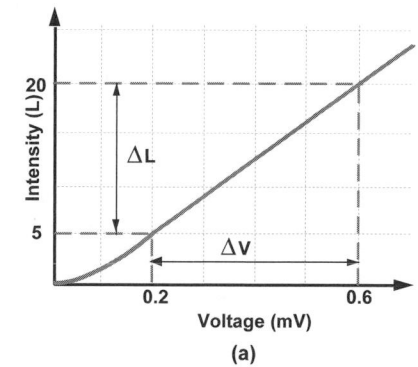

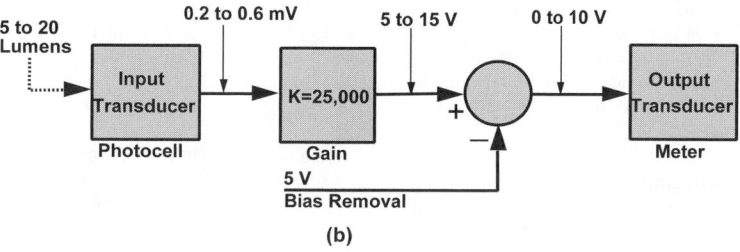

Fig. 56 (From Ref. 2.)

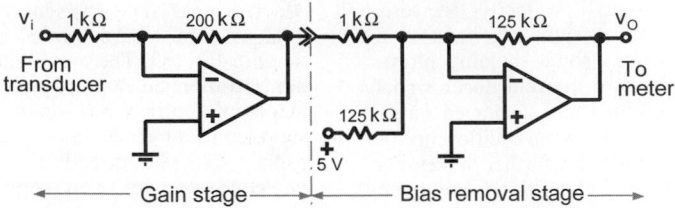

Fig. 57 (From Ref. 2.)

requirements are that 5 lm indicates 0 V and 20 lm indicates 10 V on the voltmeter.

Solution. From the transducer's characteristics the light intensity range $\Delta L = 20 - 5 = 15$ lm will produce an available range of $\Delta v = (0.6 \text{ m} - 0.2 \text{ m}) = 0.4$ mV at the system input. This 0.4-mV change must be translated into a 0–10-V range at the system output. To accomplish this, the transducer voltage must be amplified by a gain of

$$K = \frac{\text{desired output range}}{\text{available input range}}$$

$$= \frac{10 - 0}{0.6 \times 10^{-3} - 0.2 \times 10^{-3}} = 2.5 \times 10^4$$

When the transducer's output voltage range (0.2–0.6 mV) is multiplied by the gain K found above, we obtain a voltage range of 5–15 V. This range is shifted to the required 0–10-V range by subtracting the 5-V bias from the amplified signal. A block diagram of the required signal-processing functions is shown in Fig. 56b.

A cascade connection of op amp circuits is used to realize the signal-processing functions in the block diagram. Figure 57 shows one possible design using an inverting amplifier and an inverting adder. This design includes two inverting circuits in cascade so the signal inversions cancel in the output signal. Part of the overall gain of $K = 2.5 \times 10^4$ is realized in the inverting amplifier ($K_1 = -200$) and the remainder by the inverting summer ($K_2 = -125$). Dividing the overall gain between the two stages avoids trying to produce too large of a gain in a single stage. A single-stage gain of $K = 25,000$ is not practical since the closed-loop gain is not small compared to the open-loop gain μ of most op amps. The high gain would also require a very low input resistance that could load the input and an uncommonly large feedback resistance, for example, 100 Ω and 2.5 MΩ.

Example 18. A strain gauge is a resistive device that measures the elongation (strain) of a solid material caused by applied forces (stress). A typical strain gauge consists of a thin film of conducting material deposited on an insulating substrate. When bonded to a member under stress, the resistance of the gauge changes by an amount

$$\Delta R = 2R_G \frac{\Delta L}{L}$$

where R_G is the resistance of the gage with no applied stress and $\Delta L/L$ is the elongation of the material expressed as a fraction of the unstressed length L. The change in resistance ΔR is only a few tenths of a milliohm, far too little to be measured with an ohmmeter. To detect such a small change, the strain gage is placed in a Wheatstone bridge circuit like the one shown in Fig. 58. The bridge contains fixed resistors R_A and R_B, two matched strain gages R_{G1} and R_{G2}, and a precisely controlled reference voltage v_{REF}. The values of R_A and R_B are chosen so that the bridge is balanced ($v_1 = v_2$) when no stress is applied. When stress is applied, the resistance of the stressed gage changes to $R_{G2} + \Delta R$ and the bridge is unbalanced ($v_1 \neq v_2$). The differential signal ($v_2 - v_1$) indicates the strain resulting from the applied stress.

Design an op amp circuit to translate strains on the range $0 < \Delta L/L < 0.02\%$ into an output voltage on the range $0 < v_O < 4$ for $R_G = 120 \ \Omega$ and $v_{\text{REF}} = 25$ V.

Solution. With external stress applied, the resistance R_{G2} changes to $R_{G2} + \Delta R$. Applying voltage division to each leg of the bridge yields

$$v_2 = \frac{R_{G2} + \Delta R}{R_{G1} + R_{G2}} V_{\text{REF}} \qquad v_1 = \frac{R_B}{R_A + R_B} V_{\text{REF}}$$

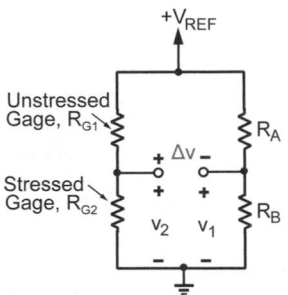

Fig. 58 (From Ref. 2.)

The differential voltage ($\Delta v = v_2 - v_1$) can be written as

$$\Delta v = v_2 - v_1 = V_{\text{REF}} \left[\frac{R_{G1} + \Delta R}{R_{G1} + R_{G2}} - \frac{R_A}{R_A + R_B} \right]$$

By selecting $R_{G1} = R_{G2} = R_A = R_B = R_G$, a balanced bridge is achieved in the unstressed state, in which case the differential voltage reduces to

$$\Delta v = v_2 - v_1 = V_{\text{REF}} \left[\frac{\Delta R}{2R_G} \right] = V_{\text{REF}} \left[\frac{\Delta L}{L} \right]$$

Thus, the differential voltage Δv is directly proportional to the strain $\Delta L/L$. However, for $V_{\text{REF}} = 25$ V and $\Delta L/L = 0.02\%$ the differential voltage is only $(V_{\text{REF}})(\Delta L/L) = 25 \times 0.0002 = 5$ mV. To obtain the required 4-V output, a voltage gain of $K = 4/0.005 = 800$ is required.

The op amp subtractor is specifically designed to amplify differential signals. Selecting $R_1 = R_3 = 10$ kΩ and $R_2 = R_4 = 8$ MΩ produces an input–output relationship for the subtractor circuit of

$$v_O = 800 \, (v_2 - v_1)$$

Figure 59 shows the selected design.

The input resistance of the subtractor circuit must be large to avoid loading the bridge circuit. The Thevenin resistance look-back into the bridge circuit is

$$R_T = R_{G1} \| R_{G2} + R_A \| R_B$$
$$= R_G \| R_G + R_G \| R_G$$
$$= R_G = 120 \ \Omega$$

which is small compared to 10-kΩ input resistance of the subtractor's inverting input.

Comment. The transducer in this example is the resistor R_{G2}. In the unstressed state the voltage across

this resistor is $v_2 = 12.5$ V. In the stressed state the voltage is $v_2 = 12.5$ V plus a 5-mV signal. In other words, the transducer's 5-mV signal component is accompanied by a very large bias. It is important to amplify the 12.5-V bias component by $K = 800$ before subtracting it out. The bias is eliminated at the input by using a bridge circuit in which $v_1 = 12.5$ V and then processing the differential signal $v_2 - v_1$. The situation illustrated in this example is actually quite common. Consequently, the first amplifier stage in most instrumentation systems is a differential amplifier.

4 AC CIRCUITS

4.1 Signals

Electrical engineers normally think of a signal as an electrical current $i(t)$, voltage $v(t)$, or power $p(t)$. In any case, the time variation of the signal is called a waveform. More formally, a *waveform* is an equation or graph that defines the signal as a function of time.

Waveforms that are constant for all time are called *dc signals*. The abbreviation dc stands for direct current, but it applies to either voltage or current. Mathematical expressions for a dc voltage $v(t)$ or current $i(t)$ take the form

$$v(t) = V_0 \qquad i(t) = I_0 \quad \text{for} -\infty < t < \infty$$

Although no physical signal can remain constant forever, it is still a useful model, however, because it approximates the signals produced by physical devices such as batteries. In a circuit diagram signal variables are normally accompanied by reference marks ($+$, $-$, $\rightarrow$ or $\leftarrow$). It is important to remember that these reference marks *do not* indicate the polarity of a voltage or the direction of current. The marks provide a baseline for determining the sign of the numerical value of the actual waveform. When the actual voltage polarity or current direction coincides with the reference directions, the signal has a positive value. When the opposite occurs, the value is negative.

Since there are infinitely many different signals, it may seem that the study of signals involves the uninviting task of compiling a lengthy catalog of waveforms. Most of the waveforms of interest can be addressed using just three basic signal models: the step, exponential, and sinusoidal functions.

Step Waveform   The first basic signal in our catalog is the step waveform. The general step function is based on the *unit step function* defined as

$$u(t) \equiv \begin{cases} 0 & \text{for } t < 0 \\ 1 & \text{for } t \geq 0 \end{cases}$$

Mathematically, the function $u(t)$ has a jump discontinuity at $t = 0$. While it is impossible to generate a true step function since signal variables like current and voltage cannot transition from one value to

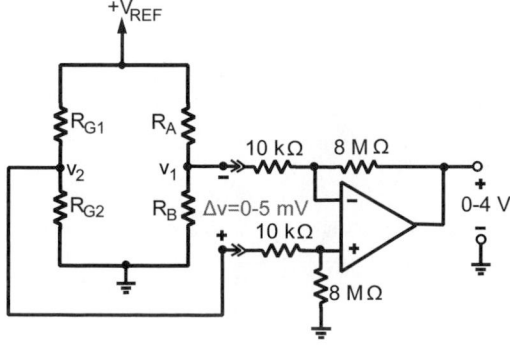

Fig. 59   (From Ref. 2.)

another in zero time, it is possible to generate very good approximations to the step function. What is required is that the transition time be short compared with other response times in the circuit. The step waveform is a versatile signal used to construct a wide range of useful waveforms. It often is necessary to turn things on at a time other then zero and with an amplitude different from unity. Replacing t by $t - T_S$ produces a waveform $V_A u(t - T_S)$ which takes on the values

$$V_A u(t - T_S) = \begin{cases} 0 & \text{for } t < T_S \\ V_A & \text{for } t \geq T_S \end{cases} \qquad (34)$$

The *amplitude* V_A scales the size of the step discontinuity and the *time shift* parameter T_S advances or delays the time at which the step occurs.

Amplitude and time shift parameters are required to define the general step function. The amplitude V_A carries the units of volts. The amplitude of the step function in an electric current is I_A and carries the units of amperes. The constant T_S carries the units of time, usually seconds. The parameters V_A (or I_A) and T_S can be positive, negative, or zero, as shown in Fig. 60.

Example 19. Express the waveform in Fig. 61a in terms of step functions.

Solution. The amplitude of the pulse jumps to a value of 3 V at $t = 1$ s; therefore, $3u(t - 1)$ is part of the equation for the waveform. The pulse returns to zero at $t = 3$ sec, so an equal and opposite step must

occur at $t = 3$ sec. Putting these observations together, we express the rectangular pulse as

$$v(t) = 3u(t - 1) - 3u(t - 3)$$

Figure 61b shows how the two step functions combine to produce the given rectangular pulse.

Impulse Function   The generalization of Example 19 is the waveform

$$v(t) = V_A [u(t - T_1) - u(t - T_2)] \qquad (35)$$

This waveform is a rectangular pulse of amplitude V_A that turns on at $t = T_1$ and off at $t = T_2$. Pulses that turn on at some time T_1 and off at some later time T_2 are sometimes called *gating functions* because they are used in conjunction with electronic switches to enable or inhibit the passage of another signal.

A rectangular pulse centered on $t = 0$ is written in terms of step functions as

$$v_1(t) = \frac{1}{T} \left[u\left(t + \frac{T}{2}\right) - u\left(t - \frac{T}{2}\right) \right] \qquad (36)$$

The pulse in Eq. (36) is zero everywhere except in the range $-T/2 \leq t \leq T/2$, where its amplitude is $1/T$. The area under the pulse is 1 because its amplitude is inversely proportional to its duration. As shown in Fig. 62a, the pulse becomes narrower and higher as T decreases but maintains its unit area. In the limit as $T \to 0$ the amplitude approaches infinity but the area remains unity. The function obtained in the limit is called a *unit impulse*, symbolized as $\delta(t)$. The graphical representation of $\delta(t)$ is shown in Fig. 62b. The impulse is an idealized model of a large-amplitude, short-duration pulse.

A formal definition of the unit impulse is

$$\delta(t) = 0 \quad \text{for } t \neq 0 \qquad \text{and} \qquad \int_{-\infty}^{t} \delta(x)\, dx = u(t)$$

$$(37)$$

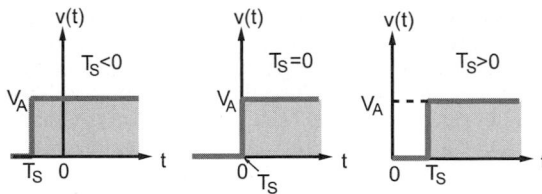

Fig. 60   Effect time shifting on step function waveform.[1]

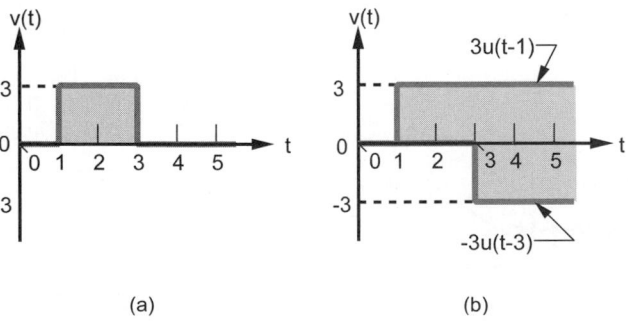

(a)   (b)

Fig. 61   (From Ref. 1.)

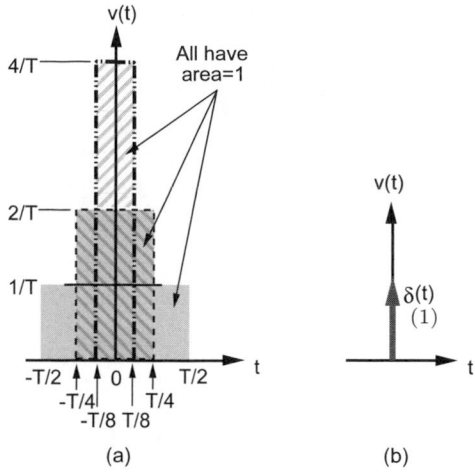

Fig. 62   Rectangular pulse waveforms and impulse.[1]

$$u(t) = \int_{-\infty}^{t} \delta(x)\, dx \qquad \delta(t) = \frac{du(t)}{dt} \qquad (39)$$

$$r(t) = \int_{-\infty}^{t} u(x)\, dx \qquad u(t) = \frac{dr(t)}{dt}$$

These signals are used to generate other waveforms and as test inputs to linear systems to characterize their responses. When applying the singularity functions in circuit analysis, it is important to remember that $u(t)$ is a dimensionless function. But $\delta(t)$ carries the units of reciprocal seconds and $r(t)$ carries units of seconds. Here, $\delta'(t)$ is called a doublet and is included for completeness. It is the derivative of an impulse function and caries the units of reciprocal seconds squared.

The first condition says the impulse is zero everywhere except at $t = 0$. The second condition implies that the impulse is the derivative of a step function although it cannot be justified using elementary mathematics since the function $u(t)$ has a discontinuity at $t = 0$ and its derivative at that point does not exist in the usual sense. However, the concept can be justified using limiting conditions on continuous functions as discussed in texts on signals and systems.

The strength of an impulse is defined by its area since amplitude is infinite. An impulse of strength K is denoted $K\delta(t)$, where K is the *area* under the impulse. In the graphical representation of the impulse the value of K is written in parentheses beside the arrow, as shown in Fig. 62*b*.

Ramp Function   The *unit ramp* is defined as the integral of a step function:

$$r(t) = \int_{-\infty}^{t} u(x)\, dx = tu(t) \qquad (38)$$

The unit-ramp waveform $r(t)$ is zero for $T \le 0$ and is equal to t for $t > 0$. The slope of $r(t)$ is unity. The general ramp waveform is written $Kr(t - T_S)$. The general ramp is zero for $t \le T_S$ and equal to $K(t - T_S)$ for $t > 0$. The scale factor K defines the slope of the ramp for $t > 0$. By adding a series of ramps the triangular and sawtooth waveforms can be created.

Singularity Functions   The impulse, step, and ramp form a triad of related signals that are referred to as *singularity functions*. They are related by integration or by differentiation as

$$\delta(t) = \int_{-\infty}^{t} \delta'(x)\, dx \qquad \delta'(t) = \frac{d\delta(t)}{dt}$$

Exponential Waveform   The *exponential signal* is a step function whose amplitude gradually decays to zero. The equation for this waveform is

$$v(t) = \left[V_A e^{-t/T_C} \right] u(t) \qquad (40)$$

A graph of $v(t)$ versus t/T_C is shown in Fig. 63. The exponential starts out like a step function. It is zero for $t < 0$ and jumps to a maximum amplitude of V_A at $t = 0$. Thereafter it monotonically decays toward zero versus time. The two parameters that define the waveform are the *amplitude* V_A (in volts) and the *time constant* T_C (in seconds). The amplitude of a current exponential would be written I_A and carry the units of amperes.

The time constant is of special interest, since it determines the rate at which the waveform decays to zero. An exponential decays to about 37% of its initial amplitude $v(0) = V_A$ in one time constant because, at $t = T_C$, $v(T_C) = V_A e^{-1}$ or approximately $0.368 V_A$. At $t = 5T_C$, the value of the waveform is $V_A e^{-5}$

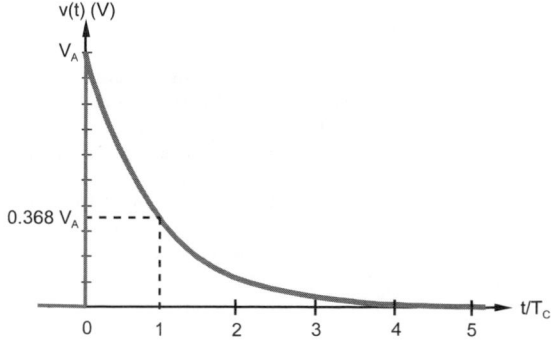

Fig. 63   Exponential waveform.

or approximately $0.00674V_A$. An exponential signal decays to less than 1% of its initial amplitude in a time span of five time constants. In theory an exponential endures forever, but practically speaking after about $5T_C$ the waveform amplitude becomes negligibly small. For this reason we define the *duration* of an exponential waveform to be $5T_C$.

Decrement Property of Exponential Waveforms

The *decrement property* describes the decay rate of an exponential signal. For $t > 0$ the exponential waveform is given by $v(t) = V_A e^{-t/T_C}$. At time $t + \Delta t$ the amplitude is

$$v(t + \Delta t) = V_A e^{-(t+\Delta t)/T_C} = V_A e^{-t/T_C} e^{-\Delta t/T_C}$$

The ratio of these two amplitudes is

$$\frac{v(t + \Delta t)}{v(t)} = \frac{V_A e^{-t/T_C} e^{-\Delta t/T_C}}{V_A e^{-t/T_C}} = e^{-\Delta t/T_C} \quad (41)$$

The decrement ratio is independent of amplitude and time. In any fixed time period Δt, the fractional decrease depends only on the time constant. The decrement property states that the same percentage decay occurs in equal time intervals.

Slope Property of Exponential Waveforms

The slope of the exponential waveform (for $t > 0$) is found by differentiating Eq. (40) with respect to time:

$$\frac{dv(t)}{dt} = -\frac{V_A}{T_C} e^{-t/T_C} = -\frac{v(t)}{T_C} \quad (42)$$

The *slope property* states that the time rate of change of the exponential waveform is inversely proportional to the time constant. Small time constants lead to large slopes or rapid decays, while large time constants produce shallow slopes and long decay times.

Sinusoidal Waveform

The cosine and sine functions are important in all branches of science and engineering. The corresponding time-varying waveform in Fig. 64 plays an especially prominent role in electrical engineering.

In contrast with the step and exponential waveforms studied earlier, the sinusoid extends indefinitely in time in both the positive and negative directions. The sinusoid in Fig. 64 is an endless repetition of identical oscillations between positive and negative peaks. The *amplitude* V_A defines the maximum and minimum values of the oscillations. The *period* T_0 is the time required to complete one cycle of the oscillation. Using these two parameters, a voltage sinusoid can be expressed as

$$v(t) = V_A \cos\left(\frac{2\pi t}{T_0}\right) \quad \text{V} \quad (43)$$

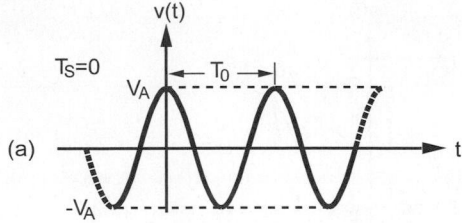

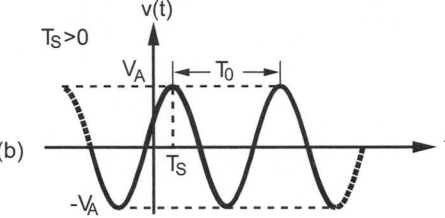

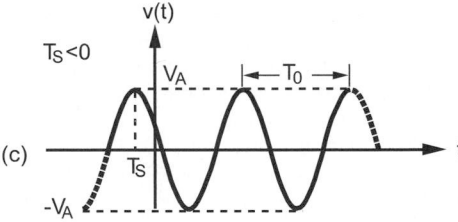

Fig. 64   Effect of time shifting on sinusoidal waveform.[1]

The waveform $v(t)$ carries the units of V_A (volts in this case) and the period T_0 carries the units of time t (usually seconds). Equation (43) produces the waveform in Fig. 64 which has a positive peak at $t = 0$ since $v(0) = V_A$.

As in the case of the step and exponential functions, the general sinusoid is obtained by replacing t by $t - T_S$. Inserting this change in Eq. (43) yields a general expression for the sinusoid as

$$v(t) = V_A \cos\left[\frac{2\pi (t - T_S)}{T_0}\right] \quad (44)$$

where the constant T_S is the time shift parameter. The sinusoid shifts to the right when $T_S > 0$ and to the left when $T_S < 0$. In effect, time shifting causes the positive peak nearest the origin to occur at $t = T_S$.

The time-shifting parameter can also be represented by an angle:

$$v(t) = V_A \cos\left[\frac{2\pi t}{T_0} + \phi\right] \quad (45)$$

The parameter ϕ is called the *phase angle*. The term phase angle is based on the circular interpretation of the cosine function where the period is divided into

2π radians or $350°$. In this sense the phase angle is the angle between $t = 0$ and the nearest positive peak. The relation between T_S and ϕ is

$$\phi = -2\pi \frac{T_S}{T_0} \tag{46}$$

An alternative form of the general sinusoid is obtained by expanding Eq. (45) using the identity $\cos(x + y) = \cos(x)\cos(y) - \sin(x)\sin(y)$. This results in the general sinusoid being written as

$$v(t) = a \cos\left(\frac{2\pi t}{T_0}\right) + b \sin\left(\frac{2\pi t}{T_0}\right) \tag{47}$$

The two amplitude-like parameters a and b have the same units as the waveform (volts in this case) and are called Fourier coefficients. By definition the Fourier coefficients are related to the amplitude and phase parameters by the equations

$$\begin{array}{ll} a = V_A \cos\phi & V_A = \sqrt{a^2 + b^2} \\ b = -V_A \sin\phi & \phi = \tan^{-1}\dfrac{-b}{a} \end{array} \tag{48}$$

It is customary to describe the time variation of the sinusoid in terms of a frequency parameter. *Cyclic frequency* f_0 is defined as the number of periods per unit time. By definition the period T_0 is the number of seconds per cycle; consequently the number of cycles per second is

$$f_0 = \frac{1}{T_0} \tag{49}$$

where f_0 is the cyclic frequency or simply the frequency. The unit of frequency (cycles per second) is the *hertz* (Hz). Because there are 2π radians per cycle, the *angular frequency* ω_0 in radians per second is related to cyclic frequency by the relationship

$$\omega_0 = 2\pi f_O = \frac{2\pi}{T_0} \tag{50}$$

In summary, there are several equivalent ways to describe the general sinusoid:

$$v(t) = \begin{cases} V_A \cos\left[\dfrac{2\pi(t - T_S)}{T_0}\right] = V_A \cos\left(\dfrac{2\pi t}{T_0} + \phi\right) \\[2mm] = a \cos\left(\dfrac{2\pi t}{T_0}\right) + b \sin\left(\dfrac{2\pi t}{T_0}\right) \\[2mm] V_A \cos[2\pi f_0(t - T_S)] = V_A \cos(2\pi f_0 t + \phi) \\[2mm] = a \cos(2\pi f_0 t + \phi) + b \sin(2\pi f_0 t + \phi) \\[2mm] V_A \cos[\omega_0(t - T_S)] = V_A \cos(\omega_0 t + \phi) \\[2mm] = a \cos(\omega_0 t) + b \sin(\omega_0 t) \end{cases} \tag{51}$$

Additive Property of Sinusoids The *additive property* of sinusoids states that summing two or more sinusoids with the same frequency yields a sinusoid with different amplitude and phase parameters but the same frequency.

Derivative and Integral Property of Sinusoids The *derivative* and *integral* properties of the sinusoid state that a sinusoid maintains its wave shape when differentiated or integrated. These operations change the amplitude and phase angle but do not change the basic sinusoidal wave shape or frequency. The fact that the wave shape is unchanged by differentiation and integration is a key property of the sinusoid. No other periodic waveform has this shape-preserving property.

Waveform Partial Descriptors An equation or graph defines a waveform for all time. The value of a waveform $v(t)$, $i(t)$, or $p(t)$ at time t is called the *instantaneous value* of the waveform. Engineers often use numerical values or terminology that characterizes a waveform but do not give a complete description. These waveform *partial descriptors* fall into two categories: (a) those that describe temporal features and (b) those that describe amplitude features.

Temporal Descriptors Temporal descriptors identify waveform attributes relative to the time axis. A signal $v(t)$ is *periodic* if $v(t + T_0) = v(t)$ for all t, where the period T_0 is the smallest value that meets this condition. Signals that are not periodic are called *aperiodic*. The fact that a waveform is periodic provides important information about the signal but does not specify all of its characteristics. The period and periodicity of a waveform are partial descriptors. A sine wave, square wave, and triangular wave are all periodic. Examples of aperiodic waveforms are the step function, exponential, and damped sine.

Waveforms that are identically zero prior to some specified time are said to be *causal*. A signal $v(t)$ is *casual* if $v(t)/0$ for $t < T$; otherwise it is *noncausal*. It is usually assumed that a causal signal is zero for $t < 0$, since time shifting can always place the starting point of a waveform at $t = 0$. Examples of causal waveforms are the step function, exponential, and damped sine. An infinitely repeating periodic waveform is noncausal.

Causal waveforms play a central role in circuit analysis. When the input driving force $x(t)$ is causal, the circuit response $y(t)$ must also be causal. That is, a physically realizable circuit cannot anticipate and respond to an input before it is applied. Causality is an important temporal feature but only a partial description of the waveform.

Amplitude Descriptors Amplitude descriptors are generally positive scalars that identify size features of the waveform. Generally a waveform's amplitude varies between two extreme values denoted as V_{MAX}

and V_{MIN}. The *peak-to-peak* value (V_{pp}) describes the total excursion of $v(t)$ and is defined as

$$V_{pp} = V_{MAX} - V_{MIN} \qquad (52)$$

Under this definition V_{pp} is always positive even if V_{MAX} and V_{MIN} are both negative. The *peak value* (v_p) is the maximum of the absolute value of the waveform. That is,

$$V_P = \max \{|V_{MAX}|, |V_{MIN}|\} \qquad (53)$$

The peak value is a positive number that indicates the maximum absolute excursion of the waveform from zero. Figure 65 shows examples of these two amplitude descriptors.

The *average value* (v_{avg}) smoothes things out to reveal the underlying waveform baseline. Average value is the area under the waveform over some period of time T divided by that time period:

$$V_{avg} = \frac{1}{T} \int_{t_0}^{t_0+T} v(x) \, dx \qquad (54)$$

For periodic signals the averaging interval T equals the period T_0. The average value measures the waveform's baseline with respect to the $v = 0$ axis. In other words, it indicates whether the waveform contains a constant, non-time-varying component. The average value is also called the *dc component* of the waveform because dc signals are constant for all t.

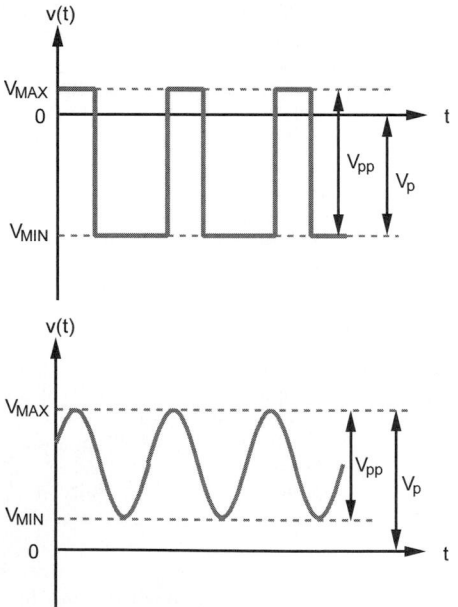

Fig. 65 Peak value (V_P) and peak-to-peak value (V_{pp}).[1]

Root-Mean-Square Value   The *root-mean-square value* (v_{rms}) of a waveform is a measure of the average power carried by the signal. The instantaneous power delivered to a resistor R by a voltage $v(t)$ is

$$p(t) = \frac{1}{R} [v(t)]^2$$

The average power delivered to the resistor in time span T is defined as

$$P_{avg} = \frac{1}{T} \int_{t_0}^{t_0+T} p(t) \, dt$$

Combining the above equations yields

$$P_{avg} = \frac{1}{R} \left[\frac{1}{T} \int_{t_0}^{t_0+T} [v(t)]^2 \, dt \right]$$

The quantity inside the large brackets is the average value of the square of the waveform. The units of the bracketed term are volts squared. The square root of this term defines the amplitude descriptor v_{rms}:

$$V_{rms} = \sqrt{\frac{1}{T} \int_{t_0}^{t_0+T} [v(t)]^2 \, dt} \qquad (55)$$

For periodic signals the averaging interval is one cycle since such a waveform repeats itself every T_0 seconds. The average power delivered to a resistor in terms of v_{rms} is

$$P_{avg} = \frac{1}{R} V_{rms}^2 \qquad (56)$$

The equation for average power in terms of v_{rms} has the same form as the instantaneous power. For this reason the rms value is also called the *effective value*, since it determines the average power delivered to a resistor in the same way that a dc waveform $v(t) = v_{dc}$ determines the instantaneous power. If the waveform amplitude is doubled, its rms value is doubled, and the average power is quadrupled. Commercial electrical power systems use transmission voltages in the range of several hundred kilovolts (rms) to transfer large blocks of electrical power.

4.2   Energy Storage Devices

Capacitor   A capacitor is a dynamic element involving the time variation of an electric field produced by a voltage. Figure 66a shows the parallel-plate capacitor, which is the simplest physical form of a capacitive device, and two common circuit symbols for the capacitor are shown in Fig. 66b.

Electrostatics shows that a uniform electric field $\mathscr{E}(t)$ exists between the metal plates when a voltage exists across the capacitor. The electric field produces

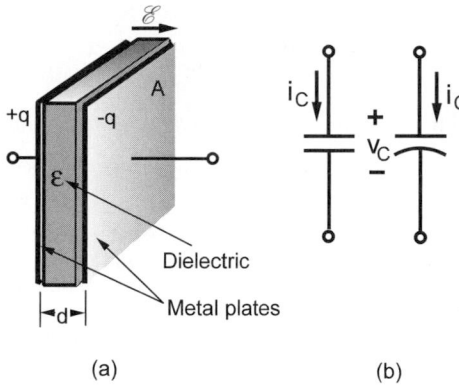

Fig. 66 Capacitor: (a) parallel-plate device; (b) circuit symbol.[1]

charge separation with equal and opposite charges appearing on the capacitor plates. When the separation d is small compared with the dimension of the plates, the electric field between the plates is

$$\mathscr{E}(t) = \frac{q(t)}{\varepsilon A}$$

where ε is the permittivity of the dielectric, A is the area of the plates, and $q(t)$ is the magnitude of the electric charge on each plate. The relationship between the electric field and the voltage across the capacitor $v_C(t)$ is given by

$$\mathscr{E}(t) = \frac{v_C(t)}{d}$$

Setting both equations equal and solving for the charge $q(t)$ yields

$$q(t) = \left[\frac{\varepsilon A}{d}\right] v_C(t) = C v_C(t) \qquad (57)$$

The proportionality constant inside the bracket in this equation is the *capacitance C*. The unit of capacitance is the farad (F), a term that honors the British physicist Michael Faraday. Values of capacitance range from picofarads (10^{-12} F) in semiconductor devices to tens of millifarads (10^{-3} F) in industrial capacitor banks. Differentiating Eq. (57) with respect to time t and realizing that $i_C(t)$ is the time derivative of $q(t)$ result in the capacitor i–v relationship

$$i_C(t) = \frac{dq(t)}{dt} = \frac{d\,[C v_C(t)]}{dt} = C \frac{d v_C(t)}{dt} \qquad (58)$$

The time derivative in Eq. (58) means the current is zero when the voltage across the capacitor is constant, and vice versa. In other words, the capacitor acts like an open circuit ($i_C = 0$) when dc excitations are applied. The capacitor is a dynamic element because the current is zero unless the voltage is changing. However, a discontinuous change in voltage requires an infinite current, which is physically impossible. Therefore, the capacitor voltage must be a continuous function of time.

Equation (59a) is the integral form of the capacitor i–v relationship where x is a dummy integration variable:

$$v_C(t) = v_C(0) + \frac{1}{C} \int_0^t i_C(x)\,dx \qquad (59a)$$

With the passive-sign convention the power associated with the capacitor is

$$p_C(t) = i_C(t) \times v_C(t) \qquad (59b)$$

This equation shows that the power can be either positive or negative because the capacitor voltage and its time rate of change can have opposite signs. The ability to deliver power implies that the capacitor can store energy. Assuming that zero energy is stored at $t = 0$, the capacitor energy is expressed as

$$w_C(t) = \tfrac{1}{2} C v_C^2(t) \qquad (60)$$

The stored energy is never negative, since it is proportional to the square of the voltage. The capacitor absorbs power from the circuit when storing energy and returns previously stored energy when delivering power to the circuit.

Inductor The inductor is a dynamic circuit element involving the time variation of the magnetic field produced by a current. Magnetostatics shows that a magnetic flux φ surrounds a wire carrying an electric current. When the wire is wound into a coil the lines of flux concentrate along the axis of the coil as shown in Fig. 67a. In a linear magnetic medium the flux is proportional to both the current and the number of turns in the coil. Therefore, the total flux is

$$\phi(t) = k_1 N i_L(t)$$

where k_1 is a constant of proportionality involving the permeability of the physical surroundings and dimensions of the wire.

The magnetic flux intercepts or links the turns of the coil. The flux linkages in a coil is represent by the symbol λ, with units of webers (Wb), named after the German scientist Wilhelm Weber (1804–1891). The number of flux linkages is proportional to the number of turns in the coil and to the total magnetic flux, so λ is given as

$$\lambda(t) = N\phi(t)$$

Substituting for $\phi(t)$ gives

$$\lambda(t) = \left[k_1 N^2\right] i_L(t) = L i_L(t) \qquad (61)$$

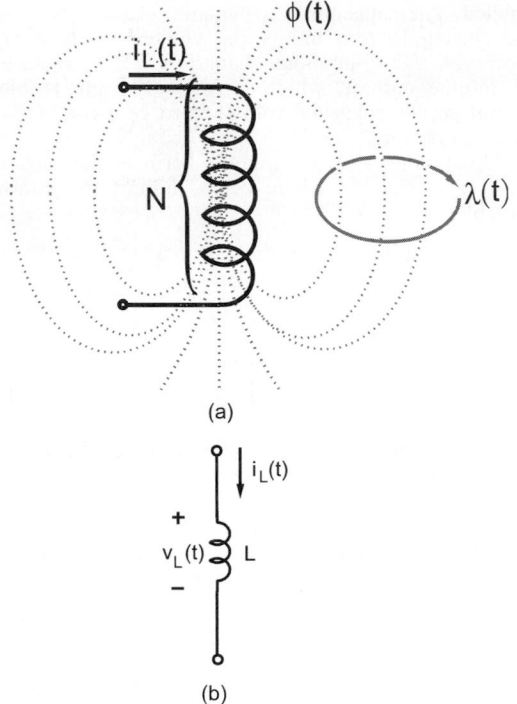

(a)

(b)

Fig. 67 (a) Magnetic flux surrounding current-carrying coil. (b) Circuit symbol for inductor.[1]

The $k_1 N^2$ inside the brackets in this equation is called the *inductance* L of the coil. The unit of inductance is the henry (H) (plural henrys), a name that honors American scientist Joseph Henry. Figure 67b shows the circuit symbol for an inductor.

Equation (61) is the inductor element constraint in terms of current and flux linkages. Differentiating Eq. (61) with respect to time t and realizing that according to Faraday's law $v_L(t)$ is the time derivative of $\lambda(t)$ result in the inductor i–v relationship

$$v_L(t) = \frac{d\,[\lambda(t)]}{dt} = \frac{d\,[Li_L(t)]}{dt} = L\frac{di_l(t)}{dt} \quad (62)$$

The time derivative in Eq. (62) means that the voltage across the inductor is zero unless the current is time varying. Under dc excitation the current is constant and $v_L = 0$ so the inductor acts like a short circuit. The inductor is a dynamic element because only a changing current produces a nonzero voltage. However, a discontinuous change in current produces an infinite voltage, which is physically impossible. Therefore, the current $i_L(t)$ must be a continuous function of time t.

Equation (63) is the integral form of the inductor i–v relationship where x is a dummy integration

variable:

$$i_L(t) = i_L(0) + \frac{1}{L}\int_0^t v_L(x)\,dx \quad (63)$$

With the passive-sign convention the inductor power is

$$p_L(t) = i_L(t) \times v_L(t) \quad (64)$$

This expression shows that power can be positive or negative because the inductor current and its time derivative can have opposite signs. The ability to deliver power indicates that the inductor can store energy. Assuming that zero energy is stored at $t = 0$, the inductor energy is expressed as

$$w_L(t) = \tfrac{1}{2}Li_L^2(t) \quad (65)$$

The energy stored in an inductor is never negative because it is proportional to the square of the current. The inductor stores energy when absorbing power and returns previously stored energy when delivering power.

Equivalent Capacitance and Inductance Resistors connected in series or parallel can be replaced by equivalent resistances. The same principle applies to connections of capacitors and inductors. N capacitors connected in parallel can be replaced by a single capacitor equal to the sum of the capacitance of the parallel capacitors, that is,

$$C_{EQ} = C_1 + C_2 + \cdots + C_N \quad \text{(parallel connection)} \quad (66)$$

The initial voltage, if any, on the equivalent capacitance is $v(0)$, the common voltage across all of the original N capacitors at $t = 0$. Likewise, N capacitors connected in series can be replaced by a single capacitor equal to

$$C_{EQ} = \frac{1}{1/C_1 + 1/C_2 + \cdots + 1/C_N}$$

$$\text{(series connection)} \quad (67)$$

The equivalent capacitance of a parallel connection is the sum of the individual capacitances. The reciprocal of the equivalent capacitance of a series connection is the sum of the reciprocals of the individual capacitances. Since the capacitor and inductor are dual elements, the corresponding results for inductors are found by interchanging the series and parallel equivalence rules for the capacitor. That is, in a series connection the equivalent inductance is the sum of the individual inductances:

$$L_{EQ} = L_1 + L_2 + \cdots + L_N \quad \text{(series connection)} \quad (68)$$

For the parallel connection the reciprocals add to produce the reciprocal of the equivalent inductance:

$$L_{EQ} = \frac{1}{1/L_1 + 1/L_2 + \cdots + 1/L_N}$$

(parallel connection) (69)

Example 20. Find the equivalent capacitance and inductance of the circuit in Fig. 68a.

Solution. The circuit contains both inductors and capacitors. The inductors and the capacitors are combined separately. The 5-pF capacitor in parallel with the 0.1-μF capacitor yields an equivalent capacitance of 0.100005 μF. For all practical purposes the 5-pF capacitor can be ignored, leaving two 0.1-μF capacitors in series with equivalent capacitance of 0.05 μF. Combining this equivalent capacitance in parallel with the remaining 0.05-μF capacitor yields an overall equivalent capacitance of 0.1 μF. The parallel 700- and 300-μH inductors yield an equivalent inductance of $1/(1/700 + 1/300) = 210$ μH. This equivalent inductance is effectively in series with the 1-mH inductor at the bottom, yielding $1000 + 210 = 1210$ μH as the overall equivalent inductance. Figure 68b shows the simplified equivalent circuit.

Mutual Inductance The $i{-}v$ characteristics of the inductor result from the magnetic field produced by current in a coil of wire. The magnetic flux spreads out around the coil forming closed loops that cut or link with the turns in the coil. If the current is changing, then Faraday's law states that voltage across the coil is equal to the time rate of change of the total flux linkages.

Now suppose that a second coil is brought close to the first coil. The flux from the first coil will link with the turns of the second coil. If the current in the first

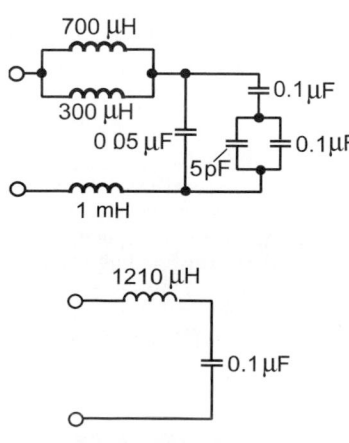

Fig. 68 (From Ref. 1.)

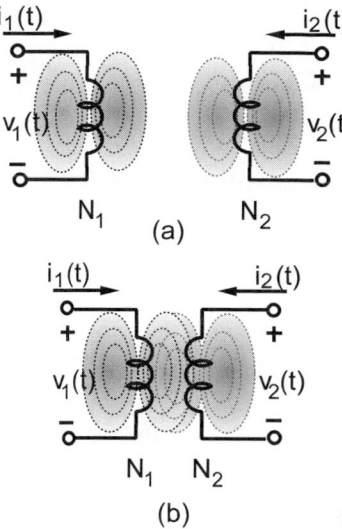

Fig. 69 (a) Inductors separated, only self-inductance present. (b) Inductors coupled, both self- and mutual inductance present.[1]

coil is changing, then these flux linkages will generate a voltage in the second coil. The coupling between a changing current in one coil and a voltage across a second coil results in *mutual inductance*.

If there is coupling between the two coils in Fig. 69, there are two distinct effects occurring in the coils. First there is the self-inductance due to the current flowing in each individual coil and the voltage induced by that current in that coil. Second, there are the voltages occurring in the second coil caused by current flowing through the first coil and vice versa. A double-subscript notation is used because it clearly identifies the various cause-and-effect relationships. The first subscript indicates the coil in which the effect takes place and the second identifies the coil in which the cause occurs. For example, $v_{11}(t)$ is the voltage across coil 1 due to causes occurring in coil 1 itself, while $v_{12}(t)$ is the voltage across coil 1 due to causes occurring in coil 2. The self-inductance is

$$\text{Coil 1:} \quad v_{11}(t) = \frac{d\lambda_{11}(t)}{dt} = N_1 \frac{d\phi_1(t)}{dt}$$

$$= \left[k_1 N_1^2 \right] \frac{di_1(t)}{dt}$$

$$\text{Coil 2:} \quad v_{22}(t) = \frac{d\lambda_{22}(t)}{dt} = N_2 \frac{d\phi_2(t)}{dt}$$

$$= \left[k_2 N_2^2 \right] \frac{di_2(t)}{dt} \quad (70)$$

Equations (70) provide the $i{-}v$ relationships for the coils when there is no mutual coupling. The mutual inductance is

Coil 1: $v_{12}(t) = \dfrac{d\lambda_{12}(t)}{dt} = N_1 \dfrac{d\phi_{12}(t)}{dt}$

$= [k_{12}N_1N_2]\dfrac{di_2(t)}{dt}$

Coil 2: $v_{21}(t) = \dfrac{d\lambda_{21}(t)}{dt} = N_2 \dfrac{d\phi_{21}(t))}{dt}$

$= [k_{21}N_1N_2]\dfrac{di_1(t)}{dt}$ (71)

The quantity $\varphi_{12}(t)$ is the flux intercepting coil 1 due to the current in coil 2 and $\varphi_{21}(t)$ is the flux intercepting coil 2 due to the current in coil 1. The expressions in Eq. (71) are the i–v relationships describing the cross coupling between coils when there is mutual coupling.

When the magnetic medium supporting the fluxes is linear, the superposition principle applies, and the total voltage across the coils is the sum of the results in Eqs. (70) and (71):

Coil 1: $v_1(t) = v_{11}(t) + v_{12}(t)$

Coil 2: $v_2(t) = v_{21}(t) + v_{22}(t)$

There are four inductance parameters in these equations. Two *self-inductance* parameters $L_1 = k_1 N_1^2$ and $L_2 = k_2 N_2^2$ and two *mutual inductances* $M_{12} = k_{12}N_1N_2$ and $M_{21} = k_{21}N_2N_1$. In a linear magnetic medium $k_{12} = k_{21} = k_M$, there is a single mutual inductance parameter M defined as $M = M_{12} = M_{21} = k_M N_1 N_2$. Putting these all together yields

Coil 1: $v_1(t) = L_1 \dfrac{di_1(t)}{dt} \pm M \dfrac{di_2(t)}{dt}$

$\qquad\qquad\qquad\qquad\qquad\qquad$ (72)

Coil 2: $v_2(t) = \pm M \dfrac{di_1(t)}{dt} + L_2 \dfrac{di_2(t)}{dt}$

The coupling across coils can be additive or subtractive. This gives rise to the $\pm$ sign in front of the mutual inductance M. Additive (+) coupling means that a positive rate of change of current in coil 2 induces a positive voltage in coil 1, and vice versa for subtractive coupling (−).

When applying these element equations, it is necessary to know when to use a plus sign and when to use a minus sign. Since the additive or subtractive nature of a coupled-coil set is predetermined by the manufacturer of the windings, a dot convention is used. The dots shown near one terminal of each coil are special reference marks indicating the relative orientation of the coils. Figure 70 shows the dot convention.

The correct sign for the mutual inductance term hinges on how the reference marks for currents and voltages are assigned relative to the coil dots: *Mutual*

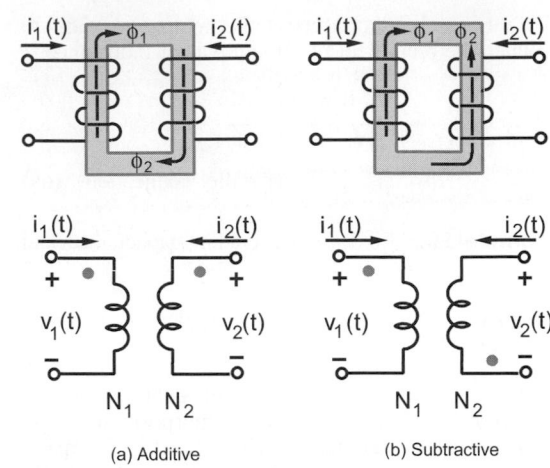

Fig. 70 Winding orientations and corresponding reference dots: (a) additive; (b) subtractive.

inductance is additive when both current reference directions point toward or both point away from dotted terminals; otherwise, it is subtractive.

Ideal Transformer A *transformer* is an electrical device that utilizes mutual inductance coupling between two coils. Transformers find application in virtually every type of electrical system, especially in power supplies and commercial power grids.

In Fig. 71 the transformer is shown as an interface device between a source and a load. The coil connected to the source is called the *primary winding* and the coil connected to the load the *secondary winding*. In most applications the transformer is a coupling device that transfers signals (especially power) from the source to the load. The basic purpose of the device is to change voltage and current levels so the signal conditions at the source and load are compatible.

Transformer design involves two primary goals: (a) to maximize the magnetic coupling between the two windings and (b) to minimize the power loss in the windings. The first goal produces near-perfect coupling ($k \cong 1$) so that almost all of the flux in one winding

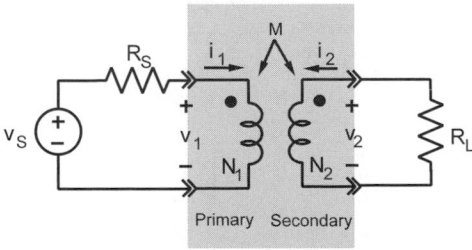

Fig. 71 Transformer connected at source–load interface.[1]

links the other. The second goal produces nearly zero power loss so that almost all of the power delivered to the primary winding transfers to the load. The *ideal transformer* is a circuit element in which coupled coils are assumed to have perfect coupling and zero power loss.

Perfect coupling assumes that all the coupling coefficients are equal to each other, that is, $k_{11} = k_{22} = k_{12} = k_{21} = k_M \cong 1$. Dividing the two equations in Eq. (72) and using the concept of perfect coupling result in the equation

$$\frac{v_2(t)}{v_1(t)} = \pm \frac{N_2}{N_1} = \pm n \tag{73}$$

where n is the *turns ratio*. With perfect coupling the secondary voltage is proportional to the primary voltage so they have the same wave shape. For example, when the primary voltage is $v_1(t) = V_A \sin \omega t$, the secondary voltage is $v_2(t) = \pm n V_A \sin \omega t$. When the turns ratio $n > 1$, the secondary-voltage amplitude is larger than the primary and the device is called a *step-up transformer*. Conversely, when $n < 1$, the secondary voltage is smaller than the primary and the device is called a *step-down transformer*. The ability to increase or decrease ac voltage levels is a basic feature of transformers. Commercial power systems use transmission voltages of several hundred kilovolts. For residential applications the transmission voltage is reduced to safer levels (typically 220/110 V_{rms}) using step-down transformers.

The $\pm$ sign in Eq. (73) depends on the reference marks given the primary and secondary currents relative to the dots indicating the relative coil orientations. The rule for the ideal transformer is a corollary of the rule for selecting the sign of the mutual inductance term in coupled-coil element equations.

The ideal transformer model also assumes that there is no power loss in the transformer. With the passive-sign convention, the power in the primary winding and secondary windings is $v_1(t)i_1(t)$ and $v_2(t)i_2(t)$, respectively. Zero power loss requires

$$v_1(t)i_1(t) + v_2(t)i_2(t) = 0$$

which can be rearranged in the form

$$\frac{i_2(t)}{i_1(t)} = -\frac{v_1(t)}{v_2(t)}$$

But under the perfect-coupling assumption $v_2(t)/v_1(t) = \pm n$. With zero power loss and perfect coupling the primary and secondary currents are related as

$$\frac{i_2(t)}{i_1(t)} = \mp \frac{1}{n} \tag{74}$$

The correct sign in this equation depends on the orientation of the current reference directions relative to the dots describing the transformer structure.

With both perfect coupling and zero power loss, the secondary current is inversely proportional to the turns ratio. A step-up transformer ($n > 1$) increases the voltage and decreases the current, which improves transmission line efficiency because the $i^2 R$ losses in the conductors are smaller.

Using the ideal transformer model requires some caution. The relationships in Eqs. (73) and (74) state that the secondary signals are proportional to the primary signals. These element equations appear to apply to dc signals. This is of course wrong. The element equations are an idealization of mutual inductance, and mutual inductance requires time-varying signals to provide the coupling between two coils.

Equivalent Input Resistance Because a transformer changes the voltage and current levels, it effectively changes the load resistance seen by a source in the primary circuit. Consider the circuit shown in Fig. 72. The device equations are

$$
\begin{aligned}
\text{Resistor:} \quad & v_2(t) = R_L i_L(t) \\
\text{Transformer:} \quad & v_2(t) = n v_1(t) \\
& i_2(t) = -\frac{1}{n} i_1(t)
\end{aligned}
$$

Dividing the first transformer equation by the second and inserting the load resistance constraint yield

$$\frac{v_2(t)}{i_2(t)} = \frac{i_L(t)R_L}{i_2(t)} = -n^2 \frac{v_1(t)}{i_1(t)}$$

Applying KCL at the output interface tells us $i_L(t) = -i_2(t)$. Therefore, the equivalent resistance seen on the primary side is

$$R_{\text{EQ}} = \frac{v_1(t)}{i_1(t)} = \frac{1}{n^2} R_L \tag{75}$$

The equivalent load resistance seen on the primary side depends on the turns ratio and the load resistance.

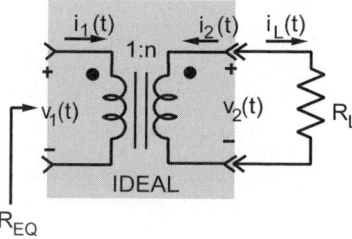

Fig. 72 Equivalent resistance seen in primary winding.[1]

Adjusting the turns ratio can make R_{EQ} equal to the source resistance. Transformer coupling can produce the resistance match condition for maximum power transfer when the source and load resistances are not equal.

4.3 Phasor Analysis of Alternating Current Circuits

Those ac circuits that are excited by a single frequency, for example, power systems, can be easily and effectively analyzed using sinusoidal steady-state techniques. Such a technique was first proposed by Charles Steinmetz (1865–1923) using a vector representation of sinusoids called *phasors*.

Sinusoids and Phasors The phasor concept is the foundation for the analysis of linear circuits in the sinusoidal steady state. Simply put, a *phasor* is a complex number representing the amplitude and phase angle of a sinusoidal voltage or current. The connection between sine waves and complex numbers is provided by Euler's relationship:

$$e^{j\theta} = \cos\theta + j\sin\theta$$

To develop the phasor concept, it is necessary to adopt the point of view that the cosine and sine functions can be written in the form

$$\cos\theta = \text{Re}\{e^{j\theta}\} \qquad \text{and} \qquad \sin\theta = \text{Im}\{e^{j\theta}\}$$

where Re stands for the "real part of" and Im for the "imaginary part of." Development of the phasor concept begins with reference of phasors to the cosine function as

$$v(t) = V_A \cos(\omega t + \phi) = V_A \text{Re}\left\{e^{j(\omega t + \phi)}\right\}$$
$$= V_A \text{Re}\left\{e^{j\omega t} e^{j\phi}\right\} = \text{Re}\left\{\left(V_A e^{j\phi}\right) e^{j\omega t}\right\} \quad (76)$$

Moving the amplitude V_A inside the real-part operation does not change the final result because it is real constant. By definition, the quantity $V_A e^{j\varphi}$ in Eq. (76) is the *phasor representation* of the sinusoid $v(t)$. The phasor **V**—a boldface V—or sometimes written with a tilde above the variable, $\tilde{V}$, can be represented in polar or rectangular form as

$$\mathbf{V} = \underbrace{V_A e^{j\varphi}}_{\text{polar form}} = \underbrace{V_A\left(\cos\phi + j\sin\phi\right)}_{\text{rectangular form}} \quad (77)$$

Note that **V** is a complex number determined by the amplitude and phase angle of the sinusoid. Figure 73 shows a graphical representation commonly called a phasor diagram. An alternative way to write the polar form of a phasor is to replace the exponential $e^{j\varphi}$ by the shorthand notation $\angle\varphi$, that is, $\mathbf{V} = V_A\angle\varphi$,

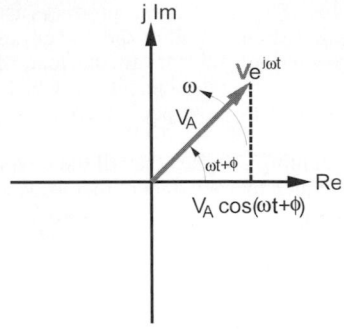

Fig. 73 Complex exponential $\mathbf{V}e^{j\omega t}$.[1]

which is equivalent to the polar form in Eq. (77). It is important to realize that a phasor is determined by its amplitude and phase angle and does not contain any information about the frequency of the sinusoid.

The first feature points out that signals can be described in different ways. Although the phasor **V** and waveform $v(t)$ are related concepts, they have quite different physical interpretations and one must clearly distinguish between them. The absence of frequency information in the phasors results from the fact that in the sinusoidal steady state all currents and voltages are sinusoids with the same frequency. Carrying frequency information in the phasor would be redundant, since it is the same for all phasors in any given steady-state circuit problem.

In summary, given a sinusoidal waveform $v(t) = V_A \cos(\omega t + \varphi)$, the corresponding phasor representation is $\mathbf{V} = V_A e^{j\varphi}$. Conversely, given the phasor $\mathbf{V} = V_A e^{j\varphi}$, the corresponding sinusoid waveform is found by multiplying the phasor by $e^{j\omega t}$ and reversing the steps in Eq. (76) as follows:

$$v(t) = \text{Re}\left\{\mathbf{V}e^{j\omega t}\right\} = \text{Re}\left\{\left(V_A e^{j\phi}\right) e^{j\omega t}\right\}$$
$$= V_A \text{Re}\left\{e^{j(\omega t + \phi)}\right\} = V_A \cos\left(\omega t + \phi\right)$$

The frequency ω in the complex exponential $\mathbf{V}e^{j\omega t}$ in Eq. (76) must be expressed or implied in a problem statement, since by definition it is not contained in the phasor. Figure 73 shows a geometric interpretation of the complex exponential $\mathbf{V}e^{j\omega t}$ as a vector in the complex plane of length v_A, which rotates counterclockwise with a constant angular velocity ω. The real-part operation projects the rotating vector onto the horizontal (real) axis and thereby generates $v(t) = V_A \cos(\omega t + \varphi)$. The complex exponential is sometimes called a *rotating phasor*, and the phasor **V** is viewed as a snapshot of the situation at $t = 0$.

Properties of Phasors Phasors have two properties. The *additive property* states that the phasor representing a sum of sinusoids of the same frequency is obtained by adding the phasor representations of the component sinusoids. To establish this property, we write the expression

$$v(t) = v_1(t) + v_2(t) + \cdots + v_N(t)$$

$$v(t) = \text{Re}\{\mathbf{V}_1 e^{j\omega t}\} + \text{Re}\{\mathbf{V}_2 e^{j\omega t}\} + \cdots + \text{Re}\{\mathbf{V}_N e^{j\omega t}\} \tag{78}$$

where $v_1(t)$, $v_2(t), \ldots,$ $v_N(t)$ are sinusoids of the same frequency whose phasor representations are $\mathbf{V}_1$, $\mathbf{V}_2, \ldots,$ $\mathbf{V}_N$. The real-part operation is additive, so the sum of real parts equals the real part of the sum. Consequently, Eq. (78) can be written in the form

$$v(t) = \text{Re}\{\mathbf{V}_1 e^{j\omega t} + \mathbf{V}_2 e^{j\omega t} + \cdots + \mathbf{V}_N e^{j\omega t}\}$$

$$= \text{Re}\{(\mathbf{V}_1 + \mathbf{V}_2 + \cdots + \mathbf{V}_N)e^{j\omega t}\} \tag{79}$$

Hence the phasor $\mathbf{V}$ representing $v(t)$ is

$$\mathbf{V} = \mathbf{V}_1 + \mathbf{V}_2 + \cdots + \mathbf{V}_N \tag{80}$$

The result in Eq. (80) applies only if the component sinusoids all have the same frequency so that $e^{j\omega t}$ can be factored out as shown in the last line in Eq. (79).

The *derivative property* of phasors allows us to easily relate the phasor representing a sinusoid to the phasor representing its derivative. Differentiating Eq. (76) with respect to time t yields

$$\frac{dv(t)}{dt} = \frac{d}{dt}\text{Re}\{\mathbf{V}e^{j\omega t}\} = \text{Re}\left\{\mathbf{V}\frac{d}{dt}e^{j\omega t}\right\}$$

$$= \text{Re}\{(j\omega\mathbf{V})e^{j\omega t}\} \tag{81}$$

From the definition of a phasor we see that the quantity $j\omega\mathbf{V}$ on the right side of this equation is the phasor representation of the time derivative of the sinusoidal waveform.

In summary, the *additive property* states that adding phasors is equivalent to adding sinusoidal waveforms of the same frequency. The *derivative property* states that multiplying a phasor by $j\omega$ is equivalent to differentiating the corresponding sinusoidal waveform.

Phasor Circuit Analysis   Phasor circuit analysis is a method of finding sinusoidal steady-state responses directly from the circuit without using differential equations.

Connection Constraints in Phasor Form   Kirchhoff's laws in phasor form are as follows:

KVL: The algebraic sum of phasor voltages around a loop is zero.

KCL: The algebraic sum of phasor currents at a node is zero.

Device Constraints in Phasor Form   The device constraints of the three passive elements are

$$\begin{aligned}
\text{Resistor:} \quad & v_R(t) = Ri_R(t) \\
\text{Inductor:} \quad & v_L(t) = L\frac{di_L(t)}{dt} \\
\text{Capacitor:} \quad & i_C(t) = C\frac{dv_C(t)}{dt}
\end{aligned} \tag{82}$$

Now in the sinusoidal steady state all of these currents and voltages are sinusoids. In the sinusoidal steady state the voltage and current of the resistor can be written in terms of phasors as $v_R(t) = \text{Re}\{\mathbf{V}_R e^{j\omega t}\}$ and $i_R(\text{t}) = \text{Re}\{\mathbf{I}_R e^{j\omega t}\}$. Consequently, the resistor i–v relationship in Eq. (82) can be expressed in terms of phasors as follows:

$$\text{Re}\left\{\mathbf{V}_R e^{j\omega t}\right\} = R \times \text{Re}\left\{\mathbf{I}_R e^{j\omega t}\right\}$$

Moving R inside the real-part operation on the right side of this equation does not change things because R is a real constant:

$$\text{Re}\left\{\mathbf{V}_R e^{j\omega t}\right\} = \text{Re}\left\{R\mathbf{I}_R e^{j\omega t}\right\}$$

This relationship holds only if the phasor voltage and current for a resistor are related as

$$\mathbf{V}_R = R\mathbf{I}_R \tag{83a}$$

If the current through a resistor is $i_R(t) = I_A\cos(\omega t + \varphi)$. Then the phasor current is $\mathbf{I}_R = I_A e^{j\varphi}$ and, according to Eq. (83a), the phasor voltage across the resistor is

$$\mathbf{V}_R = RI_A e^{j\phi} \tag{83b}$$

This result shows that the voltage has the same phase angle (φ) as the current. Phasors with the same phase angle are said to be *in phase*; otherwise they are said to be *out of phase*.

In the sinusoidal steady state the voltage and phasor current for the inductor can be written in terms of phasors as $v_L(t) = \text{Re}\{\mathbf{V}_L e^{j\omega t}\}$ and $i_L(t) = \text{Re}\{\mathbf{I}_L e^{j\omega t}\}$. Using the derivative property of phasors, the inductor i–v relationship can be expressed as

$$\text{Re}\{\mathbf{V}_L e^{j\omega t}\} = L \times \text{Re}\{j\omega\mathbf{I}_L e^{j\omega t}\}$$

$$= \text{Re}\{j\omega L\mathbf{I}_L e^{j\omega t}\} \tag{84}$$

Moving the real constant L inside the real-part operation does not change things, leading to the conclusion that phasor voltage and current for an inductor are related as

$$\mathbf{V}_L = j\omega L\mathbf{I}_L \tag{85}$$

When the current is $i_L(t) = i_A \cos(\omega t + \varphi)$, the corresponding phasor is $\mathbf{I}_L = I_A e^{j\varphi}$ and the $i-v$ constraint in Eq. (85) yields

$$\mathbf{V}_L = j\omega L\mathbf{I}_L = (\omega L e^{j90^\circ})(I_A e^{j\phi})$$

$$= \omega L I_A e^{j(\phi+90^\circ)}$$

The resulting phasor diagram in Fig. 74 shows that the inductor voltage and current are 90° out of phase. The voltage phasor is advanced by 90° counterclockwise, which is in the direction of rotation of the complex exponential $e^{\omega t}$. When the voltage phasor is advanced counter clockwise, that is, ahead of the rotating current phasor, the voltage phasor *leads* the current phasor by 90° or equivalently the current *lags* the voltage by 90°.

Finally, the capacitor voltage and current in the sinusoidal steady state can be written in terms of phasors as $v_C(t) = \text{Re}\{\mathbf{V}_C e^{j\omega t}\}$ and $i_C(t) = \text{Re}\{\mathbf{I}_C e^{j\omega t}\}$. Using the derivative property of phasors, the $i-v$ relationship of the capacitor becomes

$$\text{Re}\left\{\mathbf{I}_C e^{j\omega t}\right\} = C \times \text{Re}\left\{j\omega \mathbf{V}_C e^{j\omega t}\right\}$$
$$= \text{Re}\left\{j\omega C\mathbf{V}_C e^{j\omega t}\right\} \tag{86}$$

Moving the real constant C inside the real-part operation does not change the final results, so we conclude that the phasor voltage and current for a capacitor are related as

$$\mathbf{I}_C = j\omega C\mathbf{V}_C \quad \text{or} \quad \mathbf{V}_C = \frac{1}{j\omega C}\mathbf{I}_C \tag{87}$$

When $i_C(t) = I_A \cos(\omega t + \varphi)$, then Eq. (87) the phasor voltage across the capacitor is

$$\mathbf{V}_C = \frac{1}{j\omega C}\mathbf{I}_C = \left(\frac{1}{\omega C}e^{-j90^\circ}\right)(I_A e^{j\phi})$$

$$= \frac{I_A}{\omega C}e^{j(\phi-90^\circ)}$$

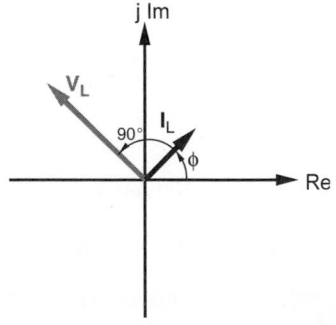

Fig. 74 Phasor $i-v$ characteristics of inductor.[1]

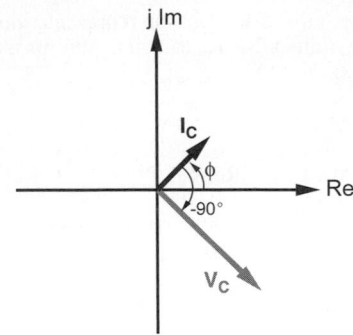

Fig. 75 Phasor $i-v$ characteristics of capacitor.[1]

The resulting phasor diagram in Fig. 75 shows that voltage and current are 90° out of phase. In this case the voltage phasor is retarded by 90° clockwise, which is in a direction opposite to the rotation of the complex exponential $e^{j\omega t}$. When the voltage is retarded clockwise, that is, behind the rotating current phasor, we say the voltage phasor *lags* the current phasor by 90° or equivalently the current *leads* the voltage by 90°.

Impedance Concept The **I–V** constraints in Eqs. (83a), (85), and (87) are all of the form

$$\mathbf{V} = Z\mathbf{I} \tag{88}$$

where Z is called the impedance of the element. Equation (88) is analogous to Ohm's law in resistive circuits. *Impedance* is the proportionality constant relating phasor voltage and phasor current in linear, two-terminal elements. The impedances of the three passive elements are

$$\begin{aligned} \text{Resistor:} \quad & Z_R = R \\ \text{Inductor:} \quad & Z_L = j\omega L \\ \text{Capacitor:} \quad & Z_C = \frac{1}{j\omega C} = -\frac{j}{\omega C} \end{aligned} \tag{89}$$

Since impedance relates phasor voltage to phasor current, it is a complex quantity whose units are ohms. Although impedance can be a complex number, it is not a phasor. Phasors represent sinusoidal signals while impedances characterize circuit elements in the sinusoidal steady state.

Basic Circuit Analysis in Phasor Domain The phasor constraints have the same format as the constraints for resistance circuits; therefore, familiar tools such as series and parallel equivalence, voltage and current division, proportionality and superposition, and Thevenin and Norton equivalent circuits are applicable to phasor circuit analysis. The major difference is that the circuit responses are complex numbers (phasors) and not waveforms.

Series Equivalence and Voltage Division Consider a simple series circuit with several impedances connected to a phasor voltage. The same phasor responses **V** and **I** exist when the series-connected elements are replaced by equivalent impedance Z_{EQ}:

$$Z_{EQ} = \frac{V}{I} = Z_1 + Z_2 + \cdots + Z_N$$

In general, the equivalent impedance Z_{EQ} is a complex quantity of the form

$$Z_{EQ} = R + jX$$

where R is the real part and X is the imaginary part. The real part of Z is called *resistance* and the imaginary part (X, not jX) is called *reactance*. Both resistance and reactance are expressed in ohms. For passive circuits resistance is always positive while reactance X can be either positive or negative. A positive X is called an *inductive* reactance because the reactance of an inductor is ωL, which is always positive. A negative X is called a *capacitive* reactance because the reactance of a capacitor is $-1/\omega C$, which is always negative.

The phasor voltage across the kth element in the series connection is

$$\mathbf{V}_k = Z_k \mathbf{I}_k = \frac{Z_k}{Z_{EQ}} \mathbf{V} \qquad (90)$$

Equation (90) is the phasor version of the voltage division principle. The phasor voltage across any element in a series connection equals the ratio of its impedance to the equivalent impedance of the connection times the total phasor voltage across the connection.

Example 21. The circuit in Fig. 76*a* is operating in the sinusoidal steady state with $v_S(t) = 35 \cos 1000t$ volts.

(a) Transform the circuit into the phasor domain.
(b) Solve for the phasor current **I**.

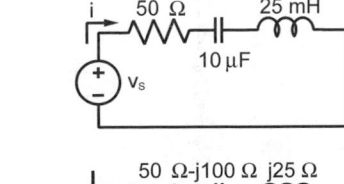

Fig. 76 (From Ref. 1.)

(c) Solve for the phasor voltage across each element.
(d) Find the waveforms corresponding to the phasors found in (b) and (c).

Solution

(a) The phasor representing the input source voltage is $\mathbf{V}_S = 35\angle 0°$. The impedances of the three passive elements are

$$Z_R = R = 50 \ \Omega$$

$$Z_L = j\omega L = j1000 \times 25 \times 10^{-3} = j25 \ \Omega$$

$$Z_C = \frac{1}{j\omega C} = \frac{1}{j1000 \times 10^{-5}} = -j100 \ \Omega$$

Using these, results we obtain the phasor domain circuit in Fig. 76*b*.

(b) The equivalent impedance of the series connection is

$$Z_{EQ} = 50 + j25 - j100 = 50 - j75$$

$$= 90.1\angle - 56.3° \ \Omega$$

The current in the series circuit is

$$\mathbf{I} = \frac{\mathbf{V}_S}{Z_{EQ}} = \frac{35\angle 0°}{90.1\angle - 56.3°} = 0.388\angle 56.3° \ \text{A}$$

(c) The current **I** exists in all three series elements so the voltage across each passive element is

$$\mathbf{V}_R = Z_R \mathbf{I} = 50 \times 0.388\angle 56.3°$$

$$= 19.4\angle 56.3° \ \text{V}$$

$$\mathbf{V}_L = Z_L \mathbf{I} = j25 \times 0.388\angle 56.3°$$

$$= 9.70\angle 146.3° \ \text{V}$$

$$\mathbf{V}_C = Z_C \mathbf{I} = -j100 \times 0.388\angle 56.3°$$

$$= 38.8\angle - 33.7° \ \text{V}$$

(d) The sinusoidal steady-state waveforms corresponding to the phasors in (b) and (c) are

$$i(t) = \text{Re}\{0.388e^{j56.3°} e^{j1000t}\}$$

$$= 0.388 \cos(1000t + 56.3°) \ \text{A}$$

$$v_R(t) = \text{Re}\{19.4e^{j56.3°} e^{j1000t}\}$$

$$= 19.4 \cos(1000t + 56.3°) \ \text{V}$$

$$v_L(t) = \text{Re}\{9.70e^{j146.3°} e^{j1000t}\}$$

$$= 9.70 \cos(1000t + 146.3°) \ \text{V}$$

$$v_C(t) = \text{Re}\{38.8e^{j-33.7°}e^{j1000t}\}$$

$$= 38.8\cos(1000t - 33.7°) \text{ V}$$

Parallel Equivalence and Current Division   Consider a number of impedances connected in parallel so the same phasor voltage **V** appears across them. The same phasor responses **V** and **I** exist when the parallel-connected elements are replaced by equivalent impedance Z_{EQ}:

$$\frac{1}{Z_{EQ}} = \frac{\mathbf{I}}{\mathbf{V}} = \frac{1}{Z_1} + \frac{1}{Z_2} + \cdots + \frac{1}{Z_N}$$

These results can also be written in terms of admittance Y, which is defined as the reciprocal of impedance:

$$Y = \frac{1}{Z} = G + jB$$

The real part of Y is called *conductance* and the imaginary part B is called *susceptance*, both of which are expressed in units of siemens.

The phasor current through the kth element of the parallel connection is

$$\mathbf{I}_k = Y_k \mathbf{V}_k = \frac{Y_k}{Y_{EQ}}\mathbf{I} \qquad (91)$$

Equation (91) is the phasor version of the current division principle. The phasor current through any element in a parallel connection equals the ratio of its admittance to the equivalent admittance of the connection times the total phasor current entering the connection.

Example 22.   For the circuit in Fig. 77 solve for the phasor voltage **V** and for the phasor current through each branch.

Solution

(a) The admittances of the two parallel branches are

$$Y_1 = \frac{1}{-j500} = j2 \times 10^{-3} \text{ S}$$

$$Y_2 = \frac{1}{500 + j1000} = 4 \times 10^{-4} - j8 \times 10^{-4} \text{ S}$$

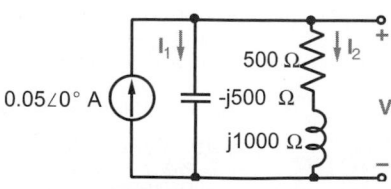

Fig. 77   (From Ref. 1.)

The equivalent admittance of the parallel connection is

$$Y_{EQ} = Y_1 + Y_2 = 4 \times 10^{-4} + j12 \times 10^{-4}$$

$$= 12.6 \times 10^{-4} \angle 71.6° \text{ S}$$

and the voltage across the parallel circuit is

$$\mathbf{V} = \frac{\mathbf{I}_S}{Y_{EQ}} = \frac{0.05\angle 0°}{12.6 \times 10^{-4}\angle 71.6°}$$

$$= 39.7\angle -71.6° \text{ V}$$

(b) The current through each parallel branch is

$$\mathbf{I}_1 = Y_1\mathbf{V} = j2 \times 10^{-3} \times 39.7\angle -71.6°$$

$$= 79.4\angle 18.4° \text{ mA}$$

$$\mathbf{I}_2 = Y_2\mathbf{V} = (4 \times 10^{-4} - j8 \times 10^{-4})$$

$$\times 39.7\angle -71.6° = 35.5\angle -135° \text{ mA}$$

Y–Δ Transformations   In section 1.2 in the discussion of equivalent circuits the equivalence of Δ- and Y-connected resistors to simplify resistance circuits with no series- or parallel-connected branches was covered. The same basic concept applies to the Δ- and Y-connected impedances (see Fig. 78). The equations for the Δ–Y transformation are

$$Z_1 = \frac{Z_B Z_C}{Z_A + Z_B + Z_C} \qquad Z_2 = \frac{Z_C Z_A}{Z_A + Z_B + Z_C}$$

$$Z_3 = \frac{Z_A Z_B}{Z_A + Z_B + Z_C} \qquad (92)$$

The equations for a Y–Δ transformation are

$$Z_A = \frac{Z_1 Z_2 + Z_2 Z_3 + Z_1 Z_3}{Z_1}$$

$$Z_B = \frac{Z_1 Z_2 + Z_2 Z_3 + Z_1 Z_3}{Z_2}$$

$$Z_C = \frac{Z_1 Z_2 + Z_2 Z_3 + Z_1 Z_3}{Z_3} \qquad (93)$$

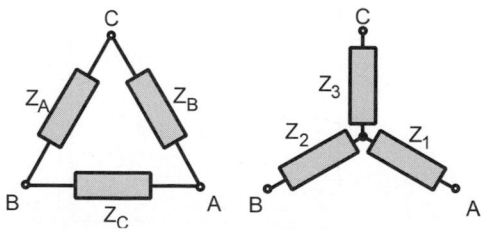

Fig. 78   Y–Δ impedance transformation.[2]

The equations have the same form except that here they involve impedances rather than resistances.

Example 23. Find the phasor current $\mathbf{I}_{IN}$ in Fig. 79a.

Solution. One cannot use basic reduction tools on the circuit because no elements are connected in series or parallel. However, by replacing either the upper Δ (A, B, C) or lower Δ (A, B, D) by an equivalent Y subcircuit, series and parallel reduction methods can be applied. Choosing the upper Δ because it has two equal resistors simplifies the transformation equations. The sum of the impedance in the upper Δ is $100 + j200\,\Omega$. This sum is the denominator in the expression in $\Delta - Y$ transformation equations. The three Y impedances are found to be

$$Z_1 = \frac{(50)(j200)}{100 + j200} = 40 + j20 \ \Omega$$

$$Z_2 = \frac{(50)(j200)}{100 + j200} = 40 + j20 \ \Omega$$

$$Z_3 = \frac{(50)(50)}{100 + j200} = 5 - j10 \ \Omega$$

Figure 79b shows the revised circuit with the equivalent Y inserted in place of the upper Δ. Note that the transformation introduces a new node labeled N. The revised circuit can be reduced by series and parallel equivalence. The total impedance of the path NAD is $40 - j100\ \Omega$. The total impedance of the path NBD is $100 + j20\ \Omega$. These paths are connected in parallel so

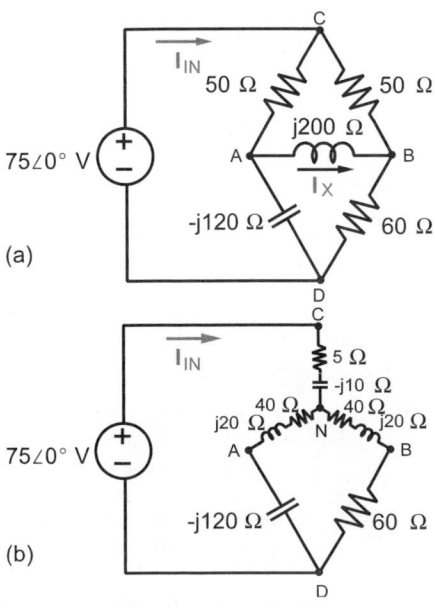

Fig. 79 (From Ref. 2.)

the equivalent impedance between nodes N and D is

$$Z_{ND} = \frac{1}{1/(40 - j100) + 1/(100 + j20)}$$

$$= 60.6 - j31.1 \ \Omega$$

The impedance Z_{ND} is connected in series with the remaining leg of the equivalent Y, so the equivalent impedance seen by the voltage source is

$$Z_{EQ} = 5 - j10 + Z_{ND} = 65.6 - j41.1 \ \Omega$$

The input current then is

$$\mathbf{I}_{IN} = \frac{\mathbf{V}_S}{Z_{EQ}} = \frac{75\angle 0^\circ}{65.6 - j41.1} = 0.891 + j0.514$$

$$= 968\angle 32.0^\circ \ \text{mA}$$

Circuit Theorems in Phasor Domain Phasor analysis does not alter the linearity properties of circuits. Hence all of the theorems that are applied to resistive circuits can be applied to phasor analysis. These include proportionality, superposition, and Thevenin and Norton equivalence.

Proportionality The *proportionality* property states that phasor output responses are proportional to the input phasor. Mathematically proportionality means that $\mathbf{Y} = K\mathbf{X}$, where $\mathbf{X}$ is the input phasor, $\mathbf{Y}$ the output phasor, and K the proportionality constant. In phasor circuit analysis the proportionality constant is generally a complex number.

Superposition Care needs to be taken when applying superposition to phasor circuits. If the sources all have the same frequency, then one can transform the circuit into the phasor domain (impedances and phasors) and proceed as in dc circuits with the superposition theorem. If the sources have different frequencies, then superposition can still be used but its application is different. With different frequency sources each source must be treated in a separate steady-state analysis because the element impedances change with frequency. The phasor response for each source must be changed into waveforms and then superposition applied in the time domain. In other words, the superposition principle always applies in the time domain. It also applies in the phasor domain when all independent sources have the same frequency. The following example illustrates the latter case.

Example 24. Use superposition to find the steady-state current $i(t)$ in Fig. 80 for $R = 10\ \text{k}\Omega$, $L = 200$ mH, $v_{S1} = 24 \cos 20{,}000t$ V, and $v_{S2} = 8 \cos(60{,}000t + 30^\circ)$ V.

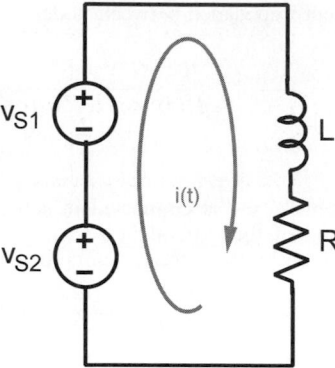

Fig. 80 (From Ref. 1.)

Solution. In this example the two sources operate at different frequencies. With source 2 off, the input phasor is $\mathbf{V}_{S1} = 24\angle 0°$ V at a frequency $\omega = 20$ krad/sec. At this frequency the equivalent impedance of the inductor and resistor is

$$Z_{EQ1} = R + j\omega L = (10 + j4) \text{ k}\Omega$$

The phasor current due to source 1 is

$$\mathbf{I}_1 = \frac{\mathbf{V}_{S1}}{Z_{EQ1}} = \frac{24\angle 0°}{10,000 + j4000} = 2.23\angle - 21.8° \text{ mA}$$

With source 1 off and source 2 on, the input phasor $\mathbf{V}_{S2} = 8\angle 30°$V at a frequency $\omega = 60$ krad/sec. At this frequency the equivalent impedance of the inductor and resistor is

$$Z_{EQ2} = R + j\omega L = (10 + j12) \text{ k}\Omega$$

The phasor current due to source 2 is

$$\mathbf{I}_2 = \frac{\mathbf{V}_{S2}}{Z_{EQ2}} = \frac{8\angle 30°}{10,000 + j12,000}$$

$$= 0.512\angle - 20.2° \text{ mA}$$

The two input sources operate at different frequencies so the phasors responses $\mathbf{I}_1$ and $\mathbf{I}_2$ cannot be added to obtain the overall response. In this case the overall response is obtained by adding the corresponding time domain waveforms:

$$i(t) = \text{Re}\{\mathbf{I}_1 e^{j20,000t}\} + \text{Re}\{\mathbf{I}_2 e^{j60,000t}\}$$

$$i(t) = 2.23 \cos(20,000t - 21.8°)$$

$$+ 0.512 \cos(60,000t - 20.2°) \text{ mA}$$

Thevenin and Norton Equivalent Circuits In the phasor domain a two-terminal circuit containing linear

elements and sources can be replaced by Thevenin or Norton equivalent circuits. The general concept of Thevenin's and Norton's theorems and their restrictions are the same as in the resistive circuit studied earlier. The important difference here is that the signals $\mathbf{V}_T$, $\mathbf{I}_N$, $\mathbf{V}$, and $\mathbf{I}$ are phasors and $Z_T = 1/Y_N$ and Z_L are complex numbers representing the source and load impedances.

Thevenin equivalent circuits are useful to address the maximum power transfer problem. Consider the source–load interface as shown in Fig. 81. The source circuit is represented by a Thevenin equivalent circuit with source voltage $\mathbf{V}_T$ and source impedance $Z_T = R_T + jX_T$. The load circuit is represented by an equivalent impedance $Z_L = R_L + jX_L$. In the maximum-power-transfer problem the source parameters $\mathbf{V}_T$, R_T, and X_T are given, and the objective is to adjust the load impedance R_L and X_L so that average power to the load is a maximum.

The average power to the load is expressed in terms of the phasor current and load resistance:

$$P = \tfrac{1}{2} R_L |\mathbf{I}|^2$$

Then, using series equivalence, the magnitude of the interface current is

$$|\mathbf{I}| = \left| \frac{\mathbf{V}_T}{Z_T + Z_L} \right| = \frac{|\mathbf{V}_T|}{|(R_T + R_L) + j(X_T + X_L)|}$$

$$= \frac{|\mathbf{V}_T|}{\sqrt{(R_T + R_L)^2 + (X_T + X_L)^2}}$$

Combining the last two equations yields the average power delivered across the interface:

$$P = \frac{1}{2} \frac{R_L |\mathbf{V}_T|^2}{(R_T + R_L)^2 + (X_T + X_L)^2}$$

Since the quantities $|\mathbf{V}_T|$, R_T, and X_T are fixed, P will be maximized when $X_L = -X_T$. This choice of X_L always is possible because a reactance can be positive or negative. When the source Thevenin equivalent has an inductive reactance ($X_T > 0$), the load is selected

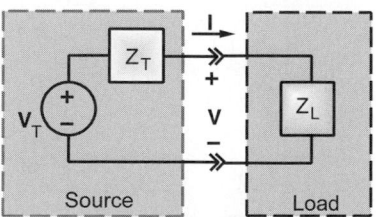

Fig. 81 Source–load interface in the sinusoidal steady state.[1]

to have a capacitive reactance of the same magnitude and vice versa. This step reduces the net reactance of the series connection to zero, creating a condition in which the net impedance seen by the Thevenin voltage source is purely resistive. In summary, to obtain maximum power transfer in the sinusoidal steady state, we select the load resistance and reactance so that $R_L = R_T$ and $X_L = -X_T$. The condition for maximum power transfer is called a *conjugate match*, since the load impedance is the conjugate of the source impedance $Z_L = Z_T^*$. Under conjugate-match conditions the maximum average power available from the source circuit is

$$P_{\text{MAX}} = \frac{|\mathbf{V}_T|^2}{8R_T}$$

where $|\mathbf{V}_T|$ is the peak amplitude of the Thevenin equivalent voltage.

It is important to remember that conjugate matching applies when the source is fixed and the load is adjustable. These conditions arise frequently in power-limited communication systems. However, conjugate matching does not apply to electrical power systems because the power transfer constraints are different.

Node Voltage and Mesh Current Analysis in Phasor Domain
The previous sections discuss basic analysis methods based on equivalence, reduction, and circuit theorems. These methods are valuable because they work directly with element impedances and thereby allow insight into steady-state circuit behavior. However, node and mesh analysis allows for solution of more complicated circuits than the basic methods can easily handle. There general methods use node voltage or mesh current variables to reduce the number of equations that must be solved simultaneously. These solution approaches are identical to those in resistive circuits except that phasors are used for signals and impedances in lieu of only resistors. The following are examples of node voltage and mesh current problems.

Example 25. Use node analysis to find the node voltages $\mathbf{V}_A$ and $\mathbf{V}_B$ in Fig. 82a.

Solution. The voltage source is connected in series with an impedance consisting of a resistor and inductor connected in parallel. The equivalent impedance of this parallel combination is

$$Z_{\text{EQ}} = \frac{1}{1/50 + 1/(j100)} = 40 + j20 \ \Omega$$

Applying a source transformation produces an equivalent current source of

$$\mathbf{I}_{\text{EQ}} = \frac{10\angle -90°}{40 + j20} = -0.1 - j0.2 \ \text{A}$$

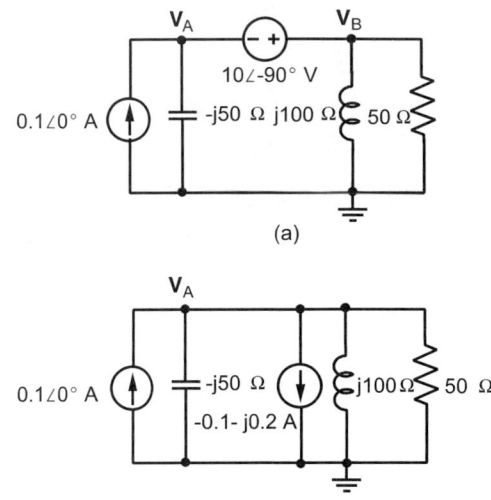

(a)

(b)

Fig. 82   (From Ref. 1.)

Figure 82b shows the circuit produced by the source transformation. The node voltage equation at the remaining nonreference node in Fig. 82b is

$$\left(\frac{1}{-j50} + \frac{1}{j100} + \frac{1}{50} \right) \mathbf{V}_A = 0.1\angle 0° - (-0.1 - j0.2)$$

Solving for $\mathbf{V}_A$ yields

$$\mathbf{V}_A = \frac{0.2 + j0.2}{0.02 + j0.01} = 12 + j4 = 12.6\angle 18.4° \ \text{V}$$

Referring to Fig. 82a, KVL requires $\mathbf{V}_B = \mathbf{V}_A + 10\angle -90°$. Therefore, $\mathbf{V}_B$ is found to be

$$\mathbf{V}_B = (12 + j4) + 10\angle -90° = 12 - j6$$

$$= 13.4\angle -26.6° \text{V}$$

Example 26. The circuit in Fig. 83 is an equivalent circuit of an ac induction motor. The current $\mathbf{I}_S$ is called the stator current, $\mathbf{I}_R$ the rotor current, and $\mathbf{I}_M$ the magnetizing current. Use the mesh current method to solve for the branch currents $\mathbf{I}_S$, $\mathbf{I}_R$, and $\mathbf{I}_M$.

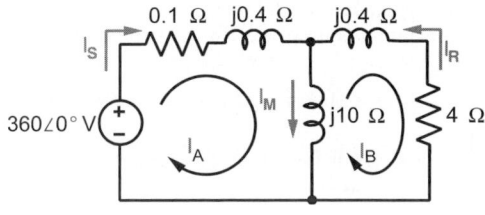

Fig. 83   (From Ref. 1.)

Solution. Applying KVL to the sum of voltages around each mesh yields

$$\text{Mesh A:} \quad -360\angle 0° + [0.1 + j0.4]\mathbf{I}_A$$
$$+ j10[\mathbf{I}_A - \mathbf{I}_B] = 0$$
$$\text{Mesh B:} \quad j10[\mathbf{I}_B - \mathbf{I}_A]$$
$$+ [4 + j0.4]\mathbf{I}_B = 0$$

Solving these equations for $\mathbf{I}_A$ and $\mathbf{I}_B$ produces

$$\mathbf{I}_A = 79.0 - j48.2 \text{ A} \qquad \mathbf{I}_B = 81.7 - j14.9 \text{ A}$$

The required stator, rotor, and magnetizing currents are related to these mesh currents as follows:

$$\mathbf{I}_S = \mathbf{I}_A = 92.5\angle 31.4° \text{ A}$$
$$\mathbf{I}_R = -\mathbf{I}_B = -81.8 + j14.9 = 83.0\angle 170° \text{A}$$
$$\mathbf{I}_M = \mathbf{I}_A - \mathbf{I}_B = -2.68 - j33.3 = 33.4\angle -94.6° \text{A}$$

4.4 Power in Sinusoidal Steady State

Average and Reactive Power In power applications it is normal to think of one circuit as the source and the other as the load. It is important to describe the flow of power across the interface between source and load when the circuit is operating in the sinusoidal steady state. The interface voltage and current in the time domain are sinusoids of the form

$$v(t) = V_A \cos(\omega t + \theta) \qquad i(t) = I_A \cos \omega t$$

where v_A and i_A are real, positive numbers representing the peak amplitudes of the voltage and current, respectively. The forms of $v(t)$ and $i(t)$ above are completely general. The positive maximum of the current $i(t)$ occurs at $t = 0$ whereas $v(t)$ contains a phase angle θ to account for the fact that the voltage maximum may not occur at the same time as the current's. In the phasor domain the angle $\theta = \varphi_V - \varphi_I$ is the angle between the phasors $\mathbf{V} = V_A\angle\varphi_V$ and $\mathbf{I} = i_A\angle\varphi_I$. In effect, choosing $t = 0$ at the current maximum shifts the phase reference by an amount $-\varphi_I$ so that the voltage and current phasors become $\mathbf{V} = v_A\angle\theta$ and $\mathbf{I} = I_A\angle 0°$.

The instantaneous power in the time domain is

$$p(t) = v(t) \times i(t) = V_A I_A \cos(\omega t + \theta) \cos \omega t \text{W}$$

This expression for instantaneous power contains both dc and ac components. Using the identities $\cos^2 x = 2(1 + \cos 2x)$ and $\cos x \sin x = 2 \sin 2x$, $p(t)$ can be written as

$$p(t)$$
$$= \underbrace{\left[\tfrac{1}{2}V_A I_A \cos \theta\right]}_{\text{dc component}}$$
$$+ \underbrace{\left[\tfrac{1}{2}V_A I_A \cos \theta\right] \cos 2\omega t - \left[\tfrac{1}{2}V_A I_A \sin \theta\right] \sin 2\omega t}_{\text{ac component}}$$
$$(94)$$

The instantaneous power is the sum of a dc component and a double-frequency ac component. That is, the instantaneous power is the sum of a constant plus a sinusoid whose frequency is 2ω, which is twice the angular frequency of the voltage and current. The instantaneous power in Eq. (94) is periodic and its average value is

$$P = \frac{1}{T} \int_0^T p(t)\, dt$$

where $T = 2\pi/2\omega$ is the period of $p(t)$. Since the average value of a sinusoid is zero, the *average value* of $p(t)$, denoted P, is equal to the constant or dc term in Eq. (94):

$$P = \tfrac{1}{2}V_A I_A \cos \theta \qquad (95)$$

The amplitude of the $\sin 2\omega t$ term in Eq. (94) has a form much like the average power in Eq. (95), except it involves $\sin \theta$ rather than $\cos \theta$. This amplitude factor is called the *reactive power* of $p(t)$, where reactive power Q is defined as

$$Q = \tfrac{1}{2}V_A I_A \sin \theta \qquad (96)$$

The instantaneous power in terms of the average power and reactive power is

$$p(t) = \underbrace{P(1 + \cos 2\omega t)}_{\text{unipolar}} - \underbrace{Q \sin 2\omega t}_{\text{bipolar}} \qquad (97)$$

The first term in Eq. (97) is said to be unipolar because the factor $1 + \cos 2\omega t$ never changes sign. As a result, the first term is either always positive or always negative depending on the sign of P. The second term is said to be bipolar because the factor $\sin 2\omega t$ alternates signs every half cycle.

The energy transferred across the interface during one cycle $T = 2\pi/2\omega$ of $p(t)$ is

$$W = \int_0^T p(t)\, dt$$

$$W = P\underbrace{\int_0^T (1 + \cos 2\omega t)\, dt}_{\text{net energy}} - Q\underbrace{\int_0^T \sin 2\omega t\, dt}_{\text{no net energy}} \quad (98)$$

$$W = P \times T \qquad - \qquad 0$$

Only the unipolar term in Eq. (97) provides any net energy transfer and that energy is proportional to the average power P. With the passive-sign convention the energy flows from source to load when $W > 0$. Equation (98) shows that the net energy will be positive if the average power $P > 0$. Equation (95) points out that the average power P is positive when $\cos\theta > 0$, which in turn means $|\theta| < 90°$.

The bipolar term in Eq. (97) is a power oscillation which transfers no net energy across the interface. In the sinusoidal steady state the load borrows energy from the source circuit during part of a cycle and temporarily stores it in the load's reactance, namely its inductance or capacitance. In another part of the cycle the borrowed energy is returned to the source unscathed. The amplitude of the power oscillation is called reactive power because it involves periodic energy storage and retrieval from the reactive elements of the load. The reactive power can be either positive or negative depending on the sign of $\sin\theta$. However, the sign of Q says nothing about the net energy transfer, which is controlled by the sign of P.

Consumers are interested in average power since this component carries net energy from source to load. For most power system customers the basic cost of electrical service is proportional to the net energy delivered to the load. Large industrial users may also pay a service charge for their reactive power as well. This may seem unfair, since reactive power transfers no net energy. However, the electric energy borrowed and returned by the load is generated within a power system that has losses. From a power company's viewpoint the reactive power is not free because there are losses in the system connecting the generators in the power plant to the source–load interface at which the lossless interchange of energy occurs.

In ac power circuit analysis, it is necessary to keep track of both the average power and reactive power. These two components of power have the same dimensions, but because they represent quite different effects, they traditionally are given different units. The average power is expressed in watts while reactive power is expressed in volt-amperes reactive (VARs).

Complex Power   It is important to relate average and reactive power to phasor quantities because ac circuit analysis is conveniently carried out using phasors. The magnitude of a phasor represents the peak amplitude of a sinusoid. However, in power circuit analysis it is convenient to express phasor magnitudes in rms values. In this chapter phasor voltages and currents are expressed as

$$\mathbf{V} = V_{\text{rms}}e^{j\phi_V} \qquad \text{and} \qquad \mathbf{I} = I_{\text{rms}}e^{j\phi_I}$$

Equations (95) and (96) express average and reactive power in terms of peak amplitudes v_A and i_A. The peak and rms values of a sinusoid are related by $V_{\text{rms}} = V_A/\sqrt{2}$. The expression for average power can be easily converted to rms amplitudes, Eq. (95), as

$$P = \frac{V_A I_A}{2}\cos\theta = \frac{V_A}{\sqrt{2}}\frac{I_A}{\sqrt{2}}\cos\theta$$
$$P = V_{\text{rms}}I_{\text{rms}}\cos\theta \tag{99}$$

where $\theta = \varphi_V - \varphi_I$ is the angle between the voltage and current phasors. By similar reasoning, Eq. (96) becomes

$$Q = V_{\text{rms}}I_{\text{rms}}\sin\theta \tag{100}$$

Using rms phasors, we define the *complex power* (S) at a two-terminal interface as

$$S = \mathbf{VI}^* = V_{\text{rms}}e^{j\phi_V}I_{\text{rms}}e^{-j\phi_I} = [V_{\text{rms}}I_{\text{rms}}]e^{j(\phi_V-\phi_I)} \tag{101}$$

That is, the complex power at an interface is the product of the voltage phasor times the conjugate of the current phasor. Using Euler's relationship and the fact that the angle $\theta = \phi_V - \phi_I$, complex power can be written as

$$S = [V_{\text{rms}}I_{\text{rms}}]e^{j\theta} = [V_{\text{rms}}I_{\text{rms}}]\cos\theta$$
$$+ j[V_{\text{rms}}I_{\text{rms}}]\sin\theta = P + jQ \tag{102}$$

The real part of the complex power S is the average power while the imaginary part is the reactive power. Although S is a complex number, it is not a phasor. However, it is a convenient variable for keeping track of the two components of power when voltage and current are expressed as phasors.

The power triangles in Fig. 84 provide a convenient way to remember complex power relationships and terminology. Considering those cases in which net energy is transferred from source to load, $P > 0$ and the power triangles fall in the first or fourth quadrant.

The magnitude $|S| = v_{\text{rms}}I_{\text{rms}}$ is called *apparent power* and is expressed using the unit volt-ampere (VA). The ratio of the average power to the apparent power is called the *power factor* (pf):

$$\text{pf} = \frac{P}{|S|} = \frac{V_{\text{rms}}I_{\text{rms}}\cos\theta}{V_{\text{rms}}I_{\text{rms}}} = \cos\theta$$

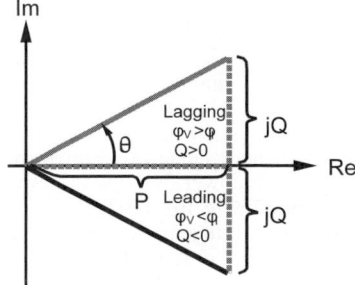

Fig. 84   Power triangles.[1]

Since pf $= \cos\theta$, the angle θ is called the *power factor angle*.

When the power factor is unity, the phasors **V** and **I** are in phase ($\theta = 0°$) and the reactive power is zero since $\sin\theta = 0$. When the power factor is less than unity, the reactive power is not zero and its sign is indicated by the modifiers lagging or leading. The term *lagging power factor* means the current phasor lags the voltage phasor so that $\theta = \phi_V - \phi_I > 0$. For a lagging power factor S falls in the first quadrant in Fig. 84 and the reactive power is positive since $\sin\theta > 0$. The term *leading power factor* means the current phasor leads the voltage phasor so that $\theta = \varphi_V - \varphi_I < 0$. In this case S falls in the fourth quadrant in Fig. 84 and the reactive power is negative since $\sin\theta < 0$. Most industrial and residential loads have lagging power factors.

The apparent power rating of electrical power equipment is an important design parameter. The ratings of generators, transfomers, and transmission lines are normally stated in kilovolt-amperes. The rating of most loads is stated in kilowatts and power factor. The wiring must be large enough to carry the required current and insulated well enough to withstand the rated voltage. However, only the average power is potentially available as useful output, since the reactive power represents a lossless interchange between the source and device. Because reactive power increases the apparent power rating without increasing the available output, it is desirable for electrical devices to operate as close as possible to unity power factor (zero reactive power).

In many cases power circuit loads are described in terms of their power ratings at a specified voltage or current level. In order to find voltages and current elsewhere in the circuit, it is necessary to know the load impedance. In general, the load produces the element constraint $\mathbf{V} = Z\mathbf{I}$. Using this constraint in Eq. (101), we write the complex power of the load as

$$S = \mathbf{V} \times \mathbf{I} = Z\mathbf{I} \times \mathbf{I}^* = Z|\mathbf{I}|^2$$

$$= (R + jX)I_{rms}^2$$

where R and X are the resistance and reactance of the load, respectively. Since $S = P + jQ$, we conclude that

$$R = \frac{P}{I_{rms}^2} \qquad \text{and} \qquad X = \frac{Q}{I_{rms}^2} \qquad (103)$$

The load resistance and reactance are proportional to the average and reactive power of the load, respectively.

The first condition in Eq. (103) demonstrates that resistance cannot be negative, since P cannot be negative for a passive circuit. The second condition points out that when the reactive power is positive the load is inductive, since $X_L = \omega L$ is positive. Conversely, when the reactive power is negative the load

is capacitive, since $X_C = -1/\omega C$ is negative. The terms inductive load, lagging power factor, and positive reactive power are synonymous, as are the terms capacitive load, leading power factor, and negative reactive power.

Example 27. At 440 V (rms) a two-terminal load draws 3 kVA of apparent power at a lagging power factor of 0.9. Find i_{rms}, P, Q, and the load impedance.

Solution

$$I_{rms} = \frac{|S|}{V_{rms}} = \frac{3000}{440} = 6.82 \text{ A (rms)}$$

$$P = V_{rms}I_{rms}\cos\theta = 3000 \times 0.9 = 2.7 \text{ kW}.$$

For $\cos\theta = 0.9$ lagging, $\sin\theta = 0.436$ and $Q = v_{rms} I_{rms}\sin\theta = 1.31$ kVAR.

$$Z = \frac{P + jQ}{(I_{rms})^2} = \frac{2700 + j1310}{46.5} = 58.0 + j28.2 \ \Omega$$

Three-Phase Circuits The three-phase system shown in Fig. 85 is the predominant method of generating and distributing ac electrical power. The system uses four lines (A, B, C, N) to transmit power from the source to the loads. The symbols stand for the three phases A, B, and C and a neutral line labeled N. The three-phase generator in Fig. 85 is modeled as three independent sources, although the physical hardware is a single unit with three separate windings. Similarly, the loads are modeled as three separate impedances, although the actual equipment may be housed within a single container.

The terminology Y connected and Δ connected refers to the two ways the source and loads can be electrically connected. In a Y connection the three elements are connected from line to neutral, while in the Δ connection they are connected from line to line.

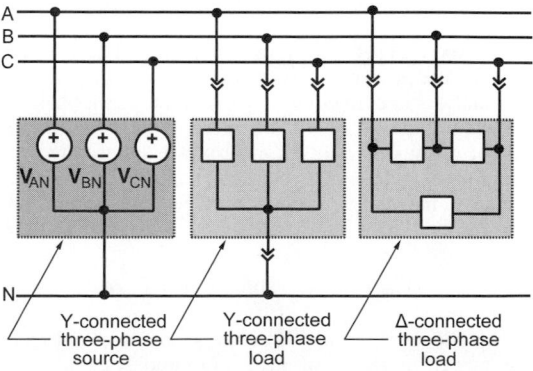

Fig. 85 Three-phase source connected to three-phase Y connection and to three-phase Δ connection.[2]

In most systems the source is Y connected while the loads can be either Y or Δ, although the latter is more common.

Three-phase sources usually are Y connected because the Δ connection involves a loop of voltage sources. Large currents may circulate in this loop if the three voltages do not exactly sum to zero. In analysis situations, a Δ connection of ideal voltage sources is awkward because it is impossible to uniquely determine the current in each source.

A double-subscript notation is used to identify voltages in the system. The reason is that there are at least six voltages to deal with: three line-to-line voltages and three line-to-neutral voltages. The two subscripts are used to define the points across which a voltage is defined. For example, $\mathbf{V}_{AB}$ means the voltage between points A and B with an implied plus reference mark at the first subscript (A) and an implied minus at the second subscript (B).

The three line-to-neutral voltages are called the *phase voltages* and are written in double-subscript notation as $\mathbf{V}_{AN}$, $\mathbf{V}_{BN}$, and $\mathbf{V}_{CN}$. Similarly, the three line-to-line voltages, called simply the *line voltages*, are identified as $\mathbf{V}_{AB}$, $\mathbf{V}_{BC}$, and $\mathbf{V}_{CA}$. From the definition of the double-subscript notation it follows that $\mathbf{V}_{XY} = -\mathbf{V}_{YX}$. Using this result and KVL we derive the relationships between the line voltages and phase voltages:

$$\mathbf{V}_{AB} = \mathbf{V}_{AN} + \mathbf{V}_{NB} = \mathbf{V}_{AN} - \mathbf{V}_{BN}$$

$$\mathbf{V}_{BC} = \mathbf{V}_{BN} + \mathbf{V}_{NC} = \mathbf{V}_{BN} - \mathbf{V}_{CN} \qquad (104)$$

$$\mathbf{V}_{CA} = \mathbf{V}_{CN} + \mathbf{V}_{NA} = \mathbf{V}_{CN} - \mathbf{V}_{AN}$$

A balanced three-phase source produces phase voltages that obey the following two constraints:

$$|\mathbf{V}_{AN}| = |\mathbf{V}_{BN}| = |\mathbf{V}_{CN}| = V_P$$

$$\mathbf{V}_{AN} + \mathbf{V}_{BN} + \mathbf{V}_{CN} = 0 + j0$$

That is, the phase voltages have equal amplitudes (v_P) and sum to zero. There are two ways to satisfy these constraints:

Positive Phase Sequence

$$\mathbf{V}_{AN} = V_P \angle 0°$$
$$\mathbf{V}_{BN} = V_P \angle -120°$$
$$\mathbf{V}_{CN} = V_P \angle -240°$$

Negative Phase Sequence

$$\mathbf{V}_{AN} = V_P \angle 0°$$
$$\mathbf{V}_{BN} = V_P \angle -240°$$
$$\mathbf{V}_{CN} = V_P \angle -120°$$

$$(105)$$

Figure 86 shows the phasor diagrams for the positive and negative phase sequences. It is apparent that both sequences involve three equal-length phasors that are separated by an angle of 120°. As a result, the sum of any two phasors cancels the third. In the positive sequence the phase B voltage lags the phase A voltage by 120°. In the negative sequence phase B lags by 240°. It also is apparent that one phase sequence can be converted into the other by simply interchanging the labels on lines B and C. From a circuit analysis viewpoint there is no conceptual difference between the two sequences.

However, the reader is cautioned that "no conceptual difference" does not mean phase sequence is unimportant. It turns out that three-phase motors run in one direction when the positive sequence is applied and in the opposite direction for the negative sequence. In practice, it is essential that there be no confusion about which line A, B, and C is and whether the source phase sequence is positive or negative.

A simple relationship between the line and phase voltages is obtained by substituting the positive-phase-sequence voltages from Eq. (105) into the phasor sums in Eq. (104):

$$\mathbf{V}_{AB} = \mathbf{V}_{AN} - \mathbf{V}_{BN} = \sqrt{3}V_P \angle 30°$$

$$\mathbf{V}_{BC} = \sqrt{3}V_P \angle -90°$$

$$\mathbf{V}_{CA} = \sqrt{3}V_P \angle -210°$$

Figure 87 shows the phasor diagram of these results. The line voltage phasors have the same amplitude and are displaced from each other by 120°. Hence, they obey equal-amplitude and zero-sum constraints like the phase voltages.

If the amplitude of the line voltages is v_L, then $V_L = \sqrt{3}V_P$. In a balanced three-phase system the

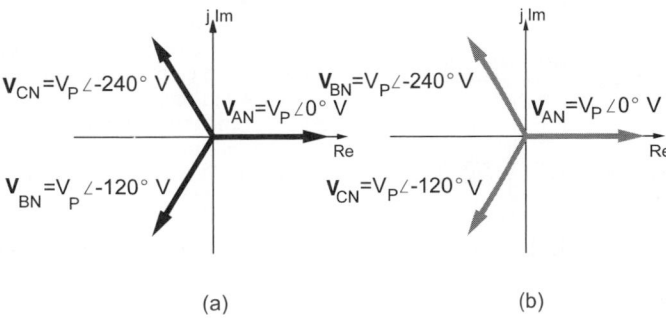

Fig. 86 Two possible phase sequences: (a) positive; (b) negative.[1]

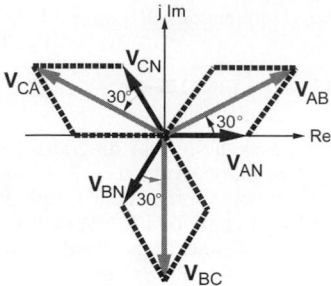

Fig. 87 Phasor diagram showing phase and line voltages for positive phase sequence.[1]

line voltage amplitude is $\sqrt{3}$ times the phase voltage amplitude. This ratio appears in equipment descriptions such as 277/480 V three-phase, where 277 is the phase voltage and 480 the line voltage. It is necessary to choose one of the phasors as the zero-phase reference when defining three-phase voltages and currents. Usually the reference is the line A phase voltage (i.e., $\mathbf{V}_{AN} = V_P\angle 0°$), as illustrated in Figs. 86 and 87.

5 TRANSIENT RESPONSE OF CIRCUITS

5.1 First-Order Circuits

First-order RC and RL circuits contain linear resistors and a single capacitor or a single inductor. Figure 88 shows RC and RL circuits divided into two parts: (a) the dynamic element and (b) the rest of the circuit containing only linear resistors and sources.

Dealing first with the RC circuit in Fig. 88a, a KVL equation is

$$R_T i(t) + v(t) = v_T(t)$$

The capacitor i–v constraint is

$$i(t) = C\frac{dv(t)}{dt}$$

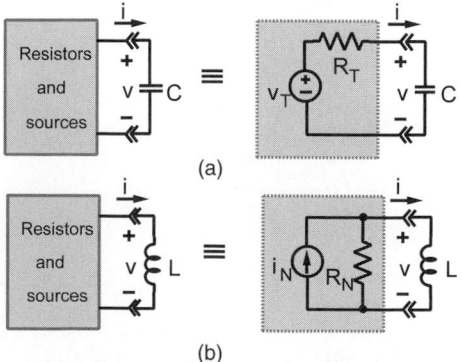

(a)

(b)

Fig. 88 First-order circuits: (a) RC circuit; (b) RL circuit.[1]

Substituting the i–v constraint into the source constraint produces the equation governing the RC series circuit:

$$R_T C\frac{dv(t)}{dt} + v(t) = v_T(t) \qquad (106)$$

The unknown in Eq. (106) is the capacitor voltage $v(t)$, which is called the *state variable* because it determines the amount or state of energy stored in the capacitive element.

Writing a KCL equation for the RL circuit in Fig. 88b yields

$$\frac{1}{R_N}v(t) + i(t) = i_N(t)$$

The element constraint for the inductor is

$$v(t) = L\frac{di(t)}{dt}$$

Combining the element and source constraints produces the differential equation for the RL circuit:

$$\frac{L}{R_N}\frac{di(t)}{dt} + i(t) = i_N(t) \qquad (107)$$

The unknown in Eq. (107) is the inductor current, also called the state variable because it determines the amount or state of energy stored in the inductive element.

Note that Eqs. (106) and (107) have the same form. In fact, interchanging the following quantities converts one equation into the other:

$$G \leftrightarrow R \qquad L \leftrightarrow C \qquad i \leftrightarrow v \qquad i_N \leftrightarrow v_T$$

This interchange is an example of the principle of duality. Because of duality there is no need to study the RC and RL circuits as independent problems. Everything learned solving the RC circuit can be applied to the RL circuit as well.

Step Response of RL and RC Circuits For the RC circuit the response $v(t)$ must satisfy the differential equation (106) and the initial condition $v(0)$. The initial energy can cause the circuit to have a nonzero response even when the input $v_T(t) = 0$ for $t \geq 0$.

When the input to the RC circuit in Fig. 88 is a step function, the source can be written as $v_T(t) = v_A u(t)$. The circuit differential equation (106) then becomes

$$R_T C\frac{dv(t)}{dt} + v(t) = V_A u(t)$$

The step response of this circuit is a function $v(t)$ that satisfies this differential equation for $t \geq 0$ and meets the initial condition $v(0)$. Since $u(t) = 1$ for $t \geq 0$

$$R_T C \frac{dv(t)}{dt} + v(t) = V_A \quad \text{for } t \geq 0 \qquad (108)$$

The solution $v(t)$ can be divided into two components:

$$v(t) = v_N(t) + v_F(t)$$

The first component $v_N(t)$ is the *natural response* and is the general solution equation (108) when the input is set to zero. The natural response has its origin in the physical characteristic of the circuit and does not depend on the form of the input. The component $v_F(t)$ is the *forced response* and is a particular solution of Eq. (108) when the input is the step function.

Finding the natural response requires the general solution of Eq. (108) with the input set to zero:

$$R_T C \frac{dv_N(t)}{dt} + v_N(t) = 0 \quad \text{for } t \geq 0$$

But this is the homogeneous equation that produces the zero-input response. Therefore, the form of the natural response is

$$v_N(t) = K e^{-t/(R_T C)} \qquad t \geq 0$$

This is a general solution of the homogeneous equation because it contains an arbitrary constant K. To evaluate K from the initial condition the total response is needed since the initial condition applies to the total response (natural plus forced).

Turning now to the forced response, a particular solution of the equation needs to be found:

$$R_T C \frac{dv_F(t)}{dt} + v_F(t) = V_A \quad \text{for } t \geq 0 \qquad (109)$$

The equation requires that a linear combination of $v_F(t)$ and its derivative equal a constant v_A for $t \geq 0$. Setting $v_F(t) = K_F$ meets this condition since $dv_F/dt = dV_A/dt = 0$. Substituting $v_F = K_F$ into Eq. (109) results in $K_F = v_A$.

Now combining the forced and natural responses yields

$$v(t) = v_N(t) + v_F(t)$$

$$v(t) = K e^{-t/(R_T C)} + V_A \qquad t \geq 0$$

This equation is the general solution for the step response because it satisfies Eq. (106) and contains an arbitrary constant K. This constant can now be evaluated using the initial condition, $v(0) = V_0 = K e^0 + V_A = K + V_A$. The initial condition requires that $K = V_0 - V_A$. Substituting this conclusion into the

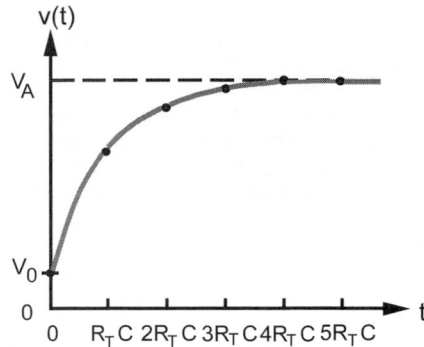

Fig. 89　Step response of first-order RC circuit.[1]

general solution yields the step response of the RC circuit:

$$v(t) = (V_0 - V_A)e^{-t/(R_T C)} + V_A \qquad t \geq 0 \qquad (110)$$

A typical plot of the waveform of $v(t)$ is shown in Fig. 89. The RL circuit in Fig. 88 is the dual of the RC circuit, so the development of its step response is similar. The result is

$$i(t) = (I_0 - I_A)e^{-R_N t/L} + I_A \qquad t \geq 0 \qquad (111)$$

The RL circuit step response has the same form as the RC circuit step response in Eq. (110). At $t = 0$ the starting value of the response is $i(0) = I_0$ as required by the initial condition. The final value is the forced response $i(\infty) = i_F = i_A$, since the natural response decays to zero as time increases.

Initial and Final Conditions　The state variable responses can be written in the form

$$v_c(t), i_L(t) = [\text{IC} - \text{FC}]e^{-t/T_C} + \text{FC} \qquad t \geq 0 \qquad (112)$$

where IC stands for the initial condition $(t = 0)$ and FC for the final condition $(t = 4)$. To determine the step response of any first-order circuit, only three quantities, IC, FC, and T_C, are needed.

The final condition can be calculated directly from the circuit by observing that for $t > 5T_C$ the step responses approach a constant, or dc, value. Under the dc condition a capacitor acts like an open circuit and an inductor acts like a short circuit, so the final value of the state variable can be calculated using resistance circuit analysis methods.

Similarly the dc analysis method can be used to determine the initial condition in many practical situations. One common situation is a circuit containing a switch that remains in one state for a period of time that is long compared with the circuit time constant. If the switch is closed for a long period of time, then the state variable approaches a final value determined by the dc input. If the switch is now opened at $t = 0$, a transient occurs in which the state variable is driven to a new final condition.

The initial condition at $t = 0$ is the dc value of the state variable for the circuit configuration that existed before the switch was opened at $t = 0$. The switching action cannot cause an instantaneous change in the initial condition because capacitor voltage and inductor current are continuous functions of time. In other words, opening a switch at $t = 0$ marks the boundary between two eras. The dc condition of the state variable for the $t < 0$ era is the initial condition for the $t > 0$ era that follows. The parameters IC, FC, and T_C in switched dynamic circuits are found using the following steps:

Step 1: Find the initial condition IC by applying dc analysis to the circuit configuration for $t < 0$.

Step 2: Find the final condition FC by applying dc analysis to the circuit configuration for $t \geq 0$.

Step 3: Find the time constant T_C of the circuit with the switch in the position for $t \geq 0$.

Step 4: Write the step response directly using Eqs. (112) without formulating and solving the circuit differential equation.

Example 28. For the circuit shown in Fig. 90a the switch has been closed for a long time. At $t = 0$ it

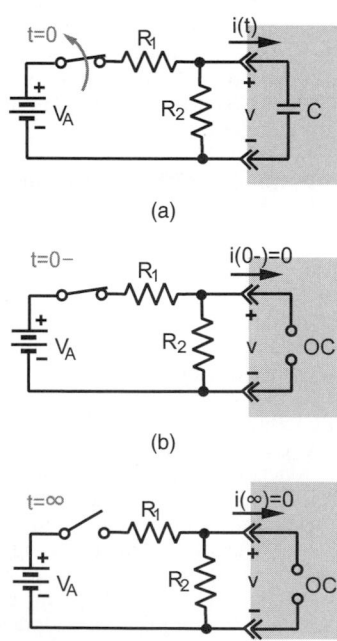

(a)

(b)

(c)

Fig. 90 Solving switched dynamic circuit using initial and final conditions.[1]

opens. Find the capacitor voltage $v(t)$ and current $i(t)$ for $t \geq 0$.

Solution

Step 1: The initial condition is found by dc analysis of the circuit configuration in Fig. 90b where the switch is closed. Using voltage division the initial capacitor voltage in found to be

$$v_C(0-) = \text{IC} = \frac{R_2 V_0}{R_1 + R_2}$$

Step 2: The final condition is found by dc analysis of the circuit configuration in Fig. 90c where the switch is open. Five time constants after the switch is opened the circuit has no practical dc excitation, so the final value of the capacitor voltage is zero.

Step 3: The circuit in Fig. 90c also is used to calculate the time constant. Since R_1 is connected in series with an open switch, the capacitor sees an equivalent resistance of only R_2. For $t \geq 0$ the time constant is $R_2 C$. Using Eq. (112) the capacitor voltage for $t \geq 0$ is

$$v_C(t) = (\text{IC} - \text{FC})e^{-t/T_C} + FC \qquad t \geq 0$$

$$v_C(t) = \frac{R_2 V_A}{R_1 + R_2} e^{-t/(R_2 C)} t \geq 0$$

This result is a zero-input response, since there is no excitation for $t \geq 0$. To complete the analysis, the capacitor current is found by using its element constraint:

$$i_C(t) = C \frac{dv_C}{dt} = -\frac{V_0}{R_1 + R_2} e^{1/(R_2 C)} \qquad t \geq 0$$

For $t < 0$ the initial-condition circuit in Fig. 90b points out that $i_C(0-) = 0$ since the capacitor acts like an open circuit.

Example 29. The switch in Fig. 91a has been open for a "long time" and is closed at $t = 0$. Find the inductor current for $t > 0$.

Solution. The initial condition is found using the circuit in Fig. 91b. By series equivalence the initial current is

$$i(0-) = \text{IC} = \frac{V_0}{R_1 + R_2}$$

The final condition and the time constant are determined from the circuit in Fig. 91c. Closing the switch shorts out R_2 and the final condition and time constant for $t > 0$ are

$$i(\infty) = \text{FC} = \frac{V_0}{R_1} \qquad T_C = \frac{L}{R_N} = \frac{L}{R_1}$$

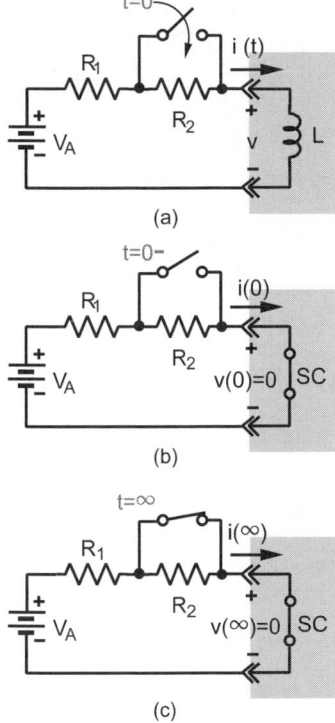

Fig. 91 (From Ref. 1.)

Using Eq. (112) the inductor current for $t \geq 0$ is

$$i(t) = (IC - FC)e^{-t/T_C} + FC \qquad t \geq 0$$

$$i(t) = \left[\frac{V_0}{R_1 + R_2} - \frac{V_0}{R_1} \right] e^{-R_1 t/L} + \frac{V_0}{R_1} \qquad A \quad t \geq 0$$

First-Order Circuit Response to Other Than dc Signals The response of linear circuits to a variety of signal inputs is an important concept in electrical engineering. Of particular importance is the response to a step reviewed in the previous section to the exponential and sinusoid. If the input to the RC circuit in Fig. 88 is an exponential or a sinusoid, then the circuit differential equation is written as

Exponential input: $\quad R_T C \dfrac{dv(t)}{dt} + v(t)$

$$= V_A e^{-\alpha t} u(t)$$

Sinusoidal input: $\quad R_T C \dfrac{dv(t)}{dt} + v(t)$ $\qquad$ (113)

$$= V_A \cos \omega t \, u(t)$$

The inputs on the right side of Eq. (113) are signals that start at $t = 0$ through some action such as closing a

switch. A solution function $v(t)$ is needed that satisfies Eq. (113) for $t \geq 0$, and that meets the prescribed initial condition $v(0) = V_0$.

As with the step response, the solution is divided into two parts: natural response and forced response. The natural response is of the form

$$v_N(t) = K e^{-t/(R_T C)}$$

The natural response of a first-order circuit always has this form because it is a general solution of the homogeneous equation with input set to zero. The form of the natural response depends on physical characteristics of the circuit and is independent of the input.

The forced response depends on both the circuit and the nature of the forcing function. The forced response is a particular solution of the equation

Exponential input: $\quad R_T C \dfrac{dv_F(t)}{dt} + v_F(t)$

$$= V_A e^{-\alpha t} \qquad t \geq 0$$

$\qquad\qquad\qquad\qquad\qquad\qquad\qquad$ (114)

Sinusoidal input: $\quad R_T C \dfrac{dv_F(t)}{dt} + v_F(t)$

$$= V_A \cos \omega t \qquad t \geq 0$$

This equation requires that $v_F(t)$ plus $R_T C$ times its first derivative add to produce either an exponential or a sinusoidal waveform for $t \geq 0$. The only way this can happen is for $v_F(t)$ and its derivative to be either an exponential of the same decay or sinusoids of the same frequency. This requirement brings to mind the derivative property of the exponential or the sinusoid. Hence one chooses a solution in the form of

Exponential: $\quad v_F(t) = K_F e^{-\alpha t}$

Sinusoidal: $\quad v_F(t) = K_A \cos \omega t + K_B \sin \omega t$

In this expression the constant K_F or the Fourier coefficients K_A and K_B are unknown. The approach we are using is called the method of undetermined coefficients. The unknown coefficients are found by inserting the forced solution $v_F(t)$ into the differential equation and equating the coefficients of the exponential in that case or of the sine and cosine terms. This yields the following:

Exponential: $\quad K_F = \dfrac{V_A}{1 - \alpha R_T C}$

Sinusoidal: $\quad K_A = \dfrac{V_A}{1 + (\omega R_T C)^2}$

$$K_B = \dfrac{\omega R_T C V_A}{1 + (\omega R_T C)^2}$$

The undetermined coefficients are now known, since these equations express constants in terms of known circuit parameters $(R_T C)$ and known input signal parameters (ω and v_A).

The forced and natural responses are combined and the initial condition used to find the remaining unknown constant K:

$$\text{Exponential:} \quad K = V_0 + \frac{V_A}{R_T C\alpha - 1}$$

$$\text{Sinusoidal:} \quad K = V_0 - \frac{V_A}{1 + (\omega R_T C)^2}$$

Combining these together yields the function $v(t)$ that satisfies the differential equation and the initial conditions:

$$
v(t) = \begin{cases}
\underbrace{\left[V_0 + \dfrac{V_A}{R_T C\alpha - 1}\right] e^{-t/(R_T C)}}_{\text{Natural response}} \\[2em]
\left. \underbrace{- \dfrac{V_A}{R_T C\alpha - 1} e^{-\alpha t}}_{\text{Forced response}} \right\} u(t) \quad V \\[2.5em]
\overbrace{\left[V_0 - \dfrac{V_A}{1 + (\omega R_T C)^2}\right] e^{-t/(R_T C)}}^{\text{Natural response}} \\[2em]
+ \overbrace{\dfrac{V_A}{1 + (\omega R_T C)^2}(\cos \omega t + \omega R_T C \sin \omega t)}^{\text{Forced response}}
\end{cases}
$$

$$\times u(t) \quad V$$

(115)

Equations (115) are the complete responses of the *RC* circuit for an initial condition V_0 and either an exponential or a sinusoidal input.

5.2 Second-Order Circuits

Second-order circuits contain two energy storage elements that cannot be replaced by a single equivalent element. They are called *second-order circuits* because the circuit differential equation involves the second derivative of the dependent variable. The series *RLC* circuit will illustrate almost all of the basic concepts of second-order circuits.

The circuit in Fig. 92*a* has an inductor and a capacitor connected in series. The source–resistor circuit can be reduced to the Thevenin equivalent shown in Fig. 92*b*. Applying KVL around the loop on the right side of the interface and the two *i–v* characteristics of

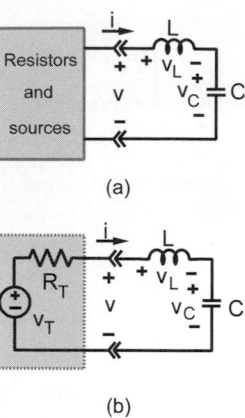

(a)

(b)

Fig. 92 Series *RLC* circuit.[1]

the inductor and capacitor yields

$$LC\frac{d^2 v_C(t)}{dt^2} + R_T C\frac{dv_C(t)}{dt} + v_C(t) = v_T(t)$$

$$v_L(t) + v_R(t) + v_C(t) = v_T(t) \quad (116)$$

In effect, this is a KVL equation around the loop in Fig. 92*b*, where the inductor and resistor voltages have been expressed in terms of the capacitor voltage.

The Thevenin voltage $v_T(t)$ is a known driving force. The initial conditions are determined by the values of the capacitor voltage and inductor current at $t = 0$, that is, V_0 and I_0:

$$v_C(0) = V_0 \quad \text{and} \quad \frac{dv_C}{dt}(0) = \frac{1}{C}i(0) = \frac{I_0}{C}$$

The circuit dynamic response for $t \geq 0$ can be divided into two components: (1) the zero-input response caused by the initial conditions and (2) the zero-state response caused by driving forces applied after $t = 0$. With $v_T = 0$ (zero input) Eq. (116) becomes

$$LC\frac{d^2 v_C(t)}{dt^2} + R_T C\frac{dv_C(t)}{dt} + v_C(t) = 0$$

This result is a second-order homogeneous differential equation in the capacitor voltage. Inserting a trial solution of $v_{CN}(t) = Ke^{st}$ into the above equation results in the following *characteristic equation* of the series *RLC* circuit:

$$LCs^2 + R_T Cs + 1 = 0$$

In general, the above quadratic characteristic equation has two roots:

$$s_1, s_2 = \frac{-R_T C \pm \sqrt{(R_T C)^2 - 4LC}}{2LC}$$

The roots can have three distinct possibilities:

Case A: If $(R_T C)^2 - 4LC > 0$, the discriminant is positive and there are two real, unequal roots ($s_1 = -\alpha_1 \neq s_2 = -\alpha_2$).

Case B: If $(R_T C)^2 - 4LC = 0$, the discriminant vanishes and there are two real, equal roots ($s_1 = s_2 = -\alpha$).

Case C: If $(R_T C)^2 - 4LC < 0$, the discriminant is negative and there are two complex conjugate roots ($s_1 = -\alpha - j\beta$ and $s_2 = -\alpha + j\beta$).

Second-Order Circuit Zero-Input Response

Since the characteristic equation has two roots, there are two solutions to the homogeneous differential equation:

$$v_{C1}(t) = K_1 e^{s_1 t} \qquad \text{and} \qquad v_{C2}(t) = K_2 e^{s_2 t}$$

Therefore, the general solution for the zero-input response is of the form

$$v_C(t) = K_1 e^{s_1 t} + K_2 e^{s_2 t} \qquad (117)$$

The constants K_1 and K_2 can be found using the initial conditions:

$$v_c(t) = \frac{s_2 V_0 - I_0/C}{s_2 - s_1} e^{s_1 t}$$

$$+ \frac{-s_1 V_0 + I_0/C}{s_2 - s_1} e^{s_2 t} \qquad t \geq 0 \qquad (118)$$

Equation (118) is the general zero-input response of the series *RLC* circuit. The response depends on two initial conditions, V_0 and I_0, and the circuit parameters R_T, L, and C since s_1 and s_2 are the roots of the characteristic equation $LCs^2 + R_T Cs + 1 = 0$. The response has different waveforms depending on whether the roots s_1 and s_2 fall under case A, B, or C.

For case A the two roots are real and distinct. Using the notation $s_1 = -\alpha_1$ and $s_2 = -\alpha_2$, the form of the zero-input response for $t \geq 0$ is

$$v_c(t) = \left[\frac{\alpha_2 V_0 + I_0/C}{\alpha_2 - \alpha_1} \right] e^{-\alpha_1 t}$$

$$- \left[\frac{\alpha_1 V_0 + I_0/C}{\alpha_2 - \alpha_1} \right] e^{-\alpha_2 t} \qquad t \geq 0$$

This form is called the *overdamped response*. The waveform has two time constants $1/\alpha_1$ and $1/\alpha_2$.

With case B the roots are real and equal. Using notation $s_1 = s_2 = -\alpha$, the general form becomes

$$v_C(t) = V_0 e^{-\alpha t} + \left(\alpha V_0 + \frac{I_0}{C} \right) t e^{-\alpha t} \qquad t \geq 0$$

This special form is called the *critically damped response*. The critically damped response includes an exponential and a damped ramp waveform.

Case C produces complex-conjugate roots of the form $s_1 = -\alpha - j\beta$ and $s_2 = -\alpha + j\beta$. The form of case C is

$$v_C(t) = V_0 e^{-\alpha t} \cos \beta t$$

$$+ \left(\frac{\alpha V_0 + I_0/C}{\beta} \right) e^{-\alpha t} \sin \beta t \qquad t \geq 0$$

This form is called the *underdamped response*. The underdamped response contains a damped sinusoid waveform where the real part of the roots (α) provides the damping term in the exponential, while the imaginary part (β) defines the frequency of the sinusoidal oscillation.

Second-Order Circuit Step Response

The general second-order linear differential equation with a step function input has the form

$$a_2 \frac{d^2 y(t)}{dt^2} + a_1 \frac{dy(t)}{dt} + a_0 y(t) = Au(t)$$

where $y(t)$ is a voltage or current response, $Au(t)$ is the step function input, and a_2, a_1, and a_0 are constant coefficients. The step response is the general solution of this differential equation for $t \geq 0$. The step response can be found by partitioning $y(t)$ into forced and natural components:

$$y(t) = y_N(t) + y_F(t)$$

The natural response $y_N(t)$ is the general solution of the homogeneous equation (input set to zero), while the forced response $y_F(t)$ is a particular solution of the equation

$$a_2 \frac{d^2 y_F(t)}{dt^2} + a_1 \frac{dy_F(t)}{dt} + a_0 y_F(t) = A \qquad t \geq 0$$

The particular solution is simply $y_F = A/a_0$.

In a second-order circuit the zero-state and natural responses take one of the three possible forms: overdamped, critically damped, or underdamped. To describe the three possible forms, two parameters are used: ω_0, the *undamped natural frequency*, and ζ, the *damping ratio*. Using these two parameters, the general homogeneous equation is written in the form

$$\frac{d^2 y_N(t)}{dt^2} + 2\zeta\omega_0 \frac{dy_N(t)}{dt} + \omega_0^2 y_N(t) = 0$$

The above equation is written in *standard form* of the second-order linear differential equation. When a second-order equation is arranged in this format, its damping ratio and undamped natural frequency can be readily found by equating its coefficients with those in the standard form. For example, in the standard form the homogeneous equation for the series RLC circuit is

$$\frac{d^2 v_C(t)}{dt^2} + \frac{R_T}{L}\frac{dv_c(t)}{dt} + \frac{1}{LC}v_C(t) = 0$$

Equating like terms yields

$$\omega_0^2 = \frac{1}{LC} \quad \text{and} \quad 2\zeta\omega_0 = \frac{R_T}{L}$$

for the series RLC circuit. Note that the circuit elements determine the values of the parameters ω_0 and ζ. The characteristic equation is $s^2 + 2\zeta\omega_0 s + \omega_0^2 = 0$ and its roots are

$$s_1, s_2 = \omega_0\left(-\zeta \pm \sqrt{\zeta^2 - 1}\right)$$

The expression under the radical defines the form of the roots and depends only on the damping ratio ζ:

Case A: For $\zeta > 1$ the discriminant is positive and there are two unequal, real roots

$$s_1, s_2 = -\alpha_1, -\alpha_2 = \omega_0\left(-\zeta \pm \sqrt{\zeta^2 - 1}\right)$$

and the natural response is of the form

$$y_N(t) = K_1 e^{-\alpha_1 t} + K_2 e^{-\alpha_2 t} \quad t \geq 0 \quad (119)$$

Case B: For $\zeta = 1$ the discriminant vanishes and there are two real, equal roots,

$$s_1 = s_2 = -\alpha = -\zeta\omega_0$$

and the natural response is of the form

$$y_N(t) = K_1 e^{-\alpha t} + K_2 t e^{-\alpha t} \quad t \geq 0 \quad (120)$$

Case C: For $\zeta < 1$, the discriminant is negative leading to two complex, conjugate roots $s_1, s_2 = -\alpha \pm j\beta$, where $\alpha = \zeta\omega_0$ and $\beta = \omega_0\sqrt{1 - \zeta^2}$ and the natural response is of the form

$$y_N(t) = e^{-\alpha t}(K_1 \cos \beta t + K_2 \sin \beta t) \quad t \geq 0 \quad (121)$$

In other words, for $\zeta > 1$ the natural response is overdamped, for $\zeta = 1$ the natural response is critically damped, and for $\zeta < 1$ the response is underdamped.

Combining the forced and natural responses yields the step response of the general second-order differential equation in the form

$$y(t) = y_N(t) + \frac{A}{a_0} \quad t \geq 0$$

The factor A/a_0 is the forced response. The natural response $y_N(t)$ takes one of the forms in Eqs. (119)–(121) depending on the value of the damping ratio. The constants K_1 and K_2 in the natural response can be evaluated from the initial conditions.

Example 30. The series RLC circuit in Fig. 93 is driven by a step function and is in the zero state at $t = 0$. Find the capacitor voltage for $t \geq 0$.

Solution. This is a series RLC circuit so the differential equation for the capacitor voltage is

$$10^{-6}\frac{d^2 v_C(t)}{dt^2} + 0.5 \times 10^{-3}\frac{dv_C(t)}{dt}$$
$$+ v_C(t) = 10 \quad t \geq 0$$

By inspection the forced response is $v_{CF}(t) = 10$ V. In standard format the homogeneous equation is

$$\frac{d^2 v_{CN}(t)}{dt^2} + 500\frac{dv_{CN}(t)}{dt} + 10^6 v_{CN}(t) = 0 \quad t \geq 0$$

Comparing this format, the standard form yields

$$\omega_0^2 = 10^6 \quad \text{and} \quad 2\zeta\omega_0 = 500$$

so that $\omega_0 = 1000$ and $\zeta = 0.25$. Since $\zeta < 1$, the natural response is underdamped (case C) and has the form

$$\alpha = \zeta\omega_0 = 250 \text{ Np}$$

$$\beta = \omega_0\sqrt{1 - \zeta^2} = 968 \text{ rad/sec}$$

$$v_{CN}(t) = K_1 e^{-250t}\cos 968t + K_2 e^{-250t}\sin 968t$$

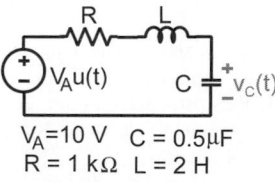

$V_A = 10$ V $C = 0.5\mu$F
$R = 1$ kΩ $L = 2$ H

Fig. 93 (From Ref. 1.)

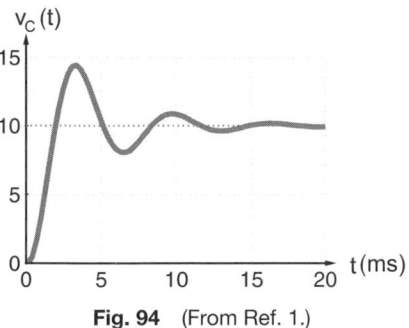

Fig. 94 (From Ref. 1.)

The general solution of the circuit differential equation is the sum of the forced and natural responses:

$$v_C(t) = 10 + K_1 e^{-250t} \cos 968t$$

$$+ K_2 e^{-250t} \sin 968t \qquad t \geq 0$$

The constants K_1 and K_2 are determined by the initial conditions. The circuit is in the zero state at $t = 0$, so the initial conditions are $v_C(0) = 0$ and $i_L(0) = 0$. Applying the initial-condition constraints to the general solution yields two equations in the constants K_1 and K_2:

$$v_C(0) = 10 + K_1 = 0$$

$$\frac{dv_C}{dt}(0) = -250\,K_1 + 968K_2 = 0$$

These equations yield $K_1 = -10$ and $K_2 = -2.58$. The step response of the capacitor voltage step response is

$$v_C(t) = 10 - 10e^{-250t} \cos 968t$$

$$- 2.58e^{-250t} \sin 968t \qquad \text{V} \qquad t \geq 0$$

A plot of $v_C(t)$ versus time is shown in Fig. 94. The waveform and its first derivative at $t = 0$ satisfy the initial conditions. The natural response decays to zero so the forced response determines the final value of $v_C(\infty) = 10$ V. Beginning at $t = 0$ the response climbs rapidly but overshoots the final value several times before eventually settling down. The damped sinusoidal behavior results from the fact that $\zeta < 1$, producing an underdamped natural response.

6 FREQUENCY RESPONSE

Linear circuits are often characterized by their behavior to sinusoids, in particular, how they process signals versus frequency. Audio, communication, instrumentation, and control systems all require signal processing that depends at least in part on their frequency response.

6.1 Transfer Functions and Input Impedance

The proportionality property of linear circuits states that the output is proportional to the input. In the *phasor* domain the proportionality factor is a rational function of $j\omega$ called a *transfer function*. More formally, in the phasor domain a transfer function is defined as the ratio of the output phasor to the input phasor with *all initial conditions set to zero*:

$$\text{Transfer function} = \frac{\text{Output phasor}}{\text{Input phasor}} = H(j\omega)$$

To study the role of transfer functions in determining circuit responses is to write the phasor domain input–output relationship as

$$Y(j\omega) = H(j\omega) \cdot X(j\omega) \qquad (122)$$

where $H(j\omega)$ is the transfer function, $X(j\omega)$ is the input signal transform (a voltage or a current phasor), and $Y(j\omega)$ is the output signal transform (also a voltage or current phasor). Figure 95 shows a block diagram representation of the phasor domain input–output relationship.

In an *analysis* problem the circuit defined by $H(j\omega)$ and the input $X(j\omega)$ are known and the response $Y(j\omega)$ is sought. In a *design* problem the circuit is unknown. The input and the desired output or their ratio $H(j\omega) = Y(j\omega)/X(j\omega)$ are given, and the objective is to devise a circuit that realizes the specified input–output relationship. A linear circuit analysis problem has a unique solution, but a design problem may have one, many, or even no solution. Choosing the best of several solutions is referred to as an *evaluation* problem.

There are two major types of functions that help define a circuit: input impedance and transfer functions. *Input impedance* relates the voltage and current at a pair of terminals called a port. The input impedance $Z(j\omega)$ of the one-port circuit in Fig. 96 is defined as

$$Z(j\omega) = \frac{V(j\omega)}{I(j\omega)} \qquad (123)$$

When the one port is driven by a current source, the response is $V(j\omega) = Z(j\omega)I(j\omega)$. On the other hand, when the one port is driven by a voltage source, the response is $I(j\omega) = [Z(j\omega)]^{-1} V(j\omega)$.

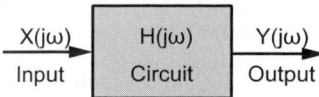

Fig. 95 Block diagram for phasor domain input–output relationship.[1]

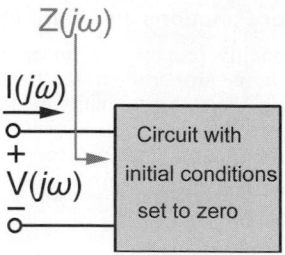

Fig. 96 One-port circuit.[1]

The term input impedance means that the circuit is driven at one port and the response is observed at the same port. The impedances of the three basic circuit elements $Z_R(j\omega)$, $Z_L(j\omega)$, and $Z_C(j\omega)$ are elementary examples of input impedances. The equivalent impedances found by combining elements in series and parallel are also effectively input impedances. The terms input impedance, *driving-point impedance*, and *equivalent impedance* are synonymous. Input impedance is useful in impedance-matching circuits at their interface and to help determine if *loading* will be an issue.

Transfer functions are usually of greater interest in signal-processing applications than input impedances because they describe how a signal is modified by passing through a circuit. A *transfer function* relates an input and response (or output) at different ports in the circuit. Since the input and output signals can be either a current or a voltage, four kinds of transfer functions can be defined:

$$H_V(j\omega) = \text{voltage transfer function} = \frac{V_2(j\omega)}{V_1(j\omega)}$$

$$H_Y(j\omega) = \text{transfer admittance} = \frac{I_2(j\omega)}{V_1(j\omega)}$$

$$H_I(j\omega) = \text{current transfer function} = \frac{I_2(j\omega)}{I_1(j\omega)}$$

$$H_Z(j\omega) = \text{transfer impedance} = \frac{V_2(j\omega)}{I_1(j\omega)}$$

$$(124)$$

The functions $H_V(j\omega)$ and $H_I(j\omega)$ are dimensionless since the input and output signals have the same units. The function $H_Z(j\omega)$ has units of ohms and $H_Y(j\omega)$ has unit of siemens.

Transfer functions always involve an input applied at one port and a response observed at a different port in the circuit. It is important to realize that a transfer function is only valid for a given input port and the specified output port. They cannot be turned upside down like the input impedance. For example, the voltage transfer function $H_V(j\omega)$ relates the voltage $V_1(j\omega)$ applied at the input port to the voltage response $V_2(j\omega)$ observed at the output port in Fig. 95. The voltage transfer function for signal transmission in the opposite direction is usually *not* $1/H_V(j\omega)$.

Determining Transfer Functions The divider circuits in Fig. 97 occur so frequently that it is worth taking time to develop their transfer functions in general terms. Using phasor domain analysis the voltage transfer function of a voltage divider circuit is

$$H_V(j\omega) = \frac{V_2(j\omega)}{V_1(j\omega)} = \frac{Z_2(j\omega)}{Z_1(j\omega) + Z_2(j\omega)}$$

Similarly, using phasor domain current division in Fig. 97b results in the current transfer function of a current divider circuit:

$$H_I(j\omega) = \frac{I_2(j\omega)}{I_1(j\omega)} = \frac{1/[Z_2(j\omega)]}{1/[Z_1(j\omega)] + 1/[Z_2(j\omega)]}$$

$$= \frac{Z_1(j\omega)}{Z_1(j\omega) + Z_2(j\omega)}$$

By series equivalence the driving-point impedance at the input of the voltage divider is $Z_{EQ}(j\omega) = Z_1(j\omega) + Z_2(j\omega)$. By parallel equivalence the driving-point impedance at the input of the current divider is $Z_{EQ}(j\omega) = 1/[1/Z_1(j\omega) + 1/Z_2(j\omega)]$.

Two other useful circuits are the inverting and noninverting op amp configurations shown in Fig. 98.

The voltage transfer function of the inverting circuit in Fig. 98a is

$$H_V(j\omega) = \frac{V_2(j\omega)}{V_1(j\omega)} = -\frac{Z_2(j\omega)}{Z_1(j\omega)}$$

The input impedance of this circuit is simply $Z_1(j\omega)$ since $v_B(j\omega) = 0$. The effect of $Z_1(j\omega)$ should be studied when connecting it to another circuit or a nonideal source since it can cause undesired loading.

For the noninverting circuit in Fig. 98b the voltage transfer function is

$$H_V(j\omega) = \frac{V_2(j\omega)}{V_1(j\omega)} = \frac{Z_1(j\omega) + Z_2(j\omega)}{Z_1(j\omega)}$$

The ideal op amp draws no current at its input terminals, so theoretically the input impedance of the noninverting circuit is infinite; in practice it is quite high, upward of $10^{10}\ \Omega$.

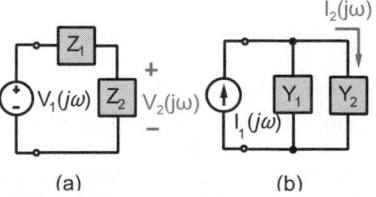

Fig. 97 Basic divider circuits: (a) voltage divider; (b) current divider.[1]

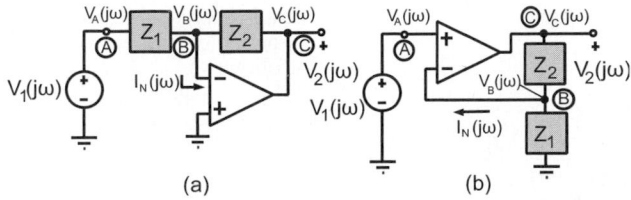

Fig. 98 Basic op amp circuits: (a) inverting amplifier; (b) noninverting amplifier.[1]

Example 31. For the circuit in Fig. 99, find (a) the input impedance seen by the voltage source and (b) the voltage transfer function $H_V(j\omega) = V_2(j\omega)/V_1(j\omega)$.

Solution

(a) The circuit is a voltage divider. First find the equivalent impedances of the two legs of the divider. The two elements in parallel combine to produce the series leg impedance $Z_1(j\omega)$:

$$Z_1(j\omega) = \frac{1}{C_1 j\omega + 1/R_1} = \frac{R_1}{R_1 C_1 j\omega + 1}$$

The two elements in series combine to produce shunt (parallel) leg impedance $Z_2(j\omega)$:

$$Z_2(j\omega) = R_2 + \frac{1}{C_2 j\omega} = \frac{R_2 C_2 j\omega + 1}{C_2 j\omega}$$

Using series equivalence, the input impedance seen at the input is

$$Z_{EQ}(j\omega) = Z_1(j\omega) + Z_2(j\omega)$$

$$= \frac{R_1 C_1 R_2 C_2 (j\omega)^2 + (R_1 C_1 + R_2 C_2 + R_1 C_2) j\omega + 1}{C_2 j\omega (R_1 C_1 j\omega + 1)}$$

(b) Using voltage division, the voltage transfer function is

$$H_V(j\omega) = \frac{Z_2(j\omega)}{Z_{EQ}(j\omega)}$$

$$= \frac{(R_1 C_1 j\omega + 1)(R_2 C_2 j\omega + 1)}{R_1 C_1 R_2 C_2 (j\omega)^2 + (R_1 C_1 + R_2 C_2 + R_1 C_2) j\omega + 1}$$

Example 32

(a) Find the driving-point impedance seen by the voltage source in Fig. 100.
(b) Find the voltage transfer function $H_V(j\omega) = V_2(j\omega)/V_1(j\omega)$ of the circuit.
(c) If $R_1 = 1\ k\Omega$, $R_2 = 10\ k\Omega$, $C_1 = 10\ nF$, and $C_2 = 1\ \mu F$, evaluate the driving-point impedance and the transfer function.

Solution. The circuit is an inverting op amp configuration. The input impedance and voltage transfer function of this configuration are

$$Z_{IN}(j\omega) = Z_1(j\omega) \quad \text{and} \quad H_V(j\omega) = -\frac{Z_2(j\omega)}{Z_1(j\omega)}$$

(a) The input impedance is

$$Z_1(j\omega) = R_1 + \frac{1}{j\omega C_1} = \frac{R_1 C_1 j\omega + 1}{j\omega C_1}$$

(b) The feedback impedance is

$$Z_2(j\omega) = \frac{1}{j\omega C_2 + 1/R_2} = \frac{R_2}{R_2 C_2 j\omega + 1}$$

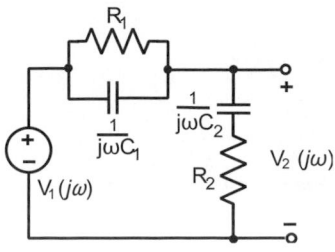

Fig. 99 (From Ref. 1.)

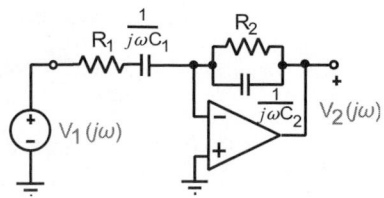

Fig. 100 (From Ref. 1.)

and the voltage transfer function is

$$H_V(j\omega) = -\frac{Z_2(j\omega)}{Z_1(j\omega)}$$

$$= -\frac{R_2 C_1 j\omega}{(R_1 C_1 j\omega + 1)(R_2 C_2 j\omega + 1)}$$

(c) For the values of R's and C's given

$$Z_1(j\omega) = \frac{1000\left(j\omega + 10^5\right)}{j\omega}$$

and

$$H_V(j\omega) = -\frac{1000 j\omega}{(j\omega + 100)\left(j\omega + 10^5\right)}$$

6.2 Cascade Connection and Chain Rule

Signal-processing circuits often involve a *cascade connection* in which the output voltage of one circuit serves as the input to the next stage. In some cases, the overall voltage transfer function of the cascade can be related to the transfer functions of the individual stages by a *chain rule*:

$$H_V(j\omega) = H_{V1}(j\omega) \times H_{V2}(j\omega) \times \cdots \times H_{Vk}(j\omega)$$
$$(125)$$

where $H_{V1}, H_{V2}, \ldots, H_{Vk}$ are the voltage transfer functions of the individual stages when operated separately. It is important to understand when the chain rule applies since it greatly simplifies the analysis and design of cascade circuits.

Figure 101 shows two *RC* circuits or stages connected in cascade at an interface. When disconnected

and operated separately, the transfer functions of each stage are easily found using voltage division as follows:

$$H_{V1}(j\omega) = \frac{R}{R + 1/j\omega C} = \frac{Rj\omega C}{Rj\omega C + 1}$$

$$H_{V2}(j\omega) = \frac{1/j\omega C}{R + 1/j\omega C} = \frac{1}{Rj\omega C + 1}$$

When connected in cascade the output of the first stage serves as the input to the second stage. If the chain rule applies, the overall transfer function would be expected to be

$$H_V(j\omega) = \frac{V_3(j\omega)}{V_1(j\omega)} = \left(\frac{V_2(j\omega)}{V_1(j\omega)}\right)\left(\frac{V_3(j\omega)}{V_2(j\omega)}\right)$$

$$= H_{V1}(j\omega) \times H_{V2}(j\omega)$$

$$= \underbrace{\left(\frac{Rj\omega C}{Rj\omega C + 1}\right)}_{\text{1st stage}}\underbrace{\left(\frac{1}{Rj\omega C + 1}\right)}_{\text{2nd stage}}$$

$$= \underbrace{\frac{Rj\omega C}{(Rj\omega C)^2 + 2Rj\omega C + 1}}_{\text{combined}}$$

However, the overall transfer function of this circuit is actually found to be

$$H_V(j\omega) = \frac{Rj\omega C}{(Rj\omega C)^2 + 3Rj\omega C + 1}$$

which disagrees with the chain rule result.

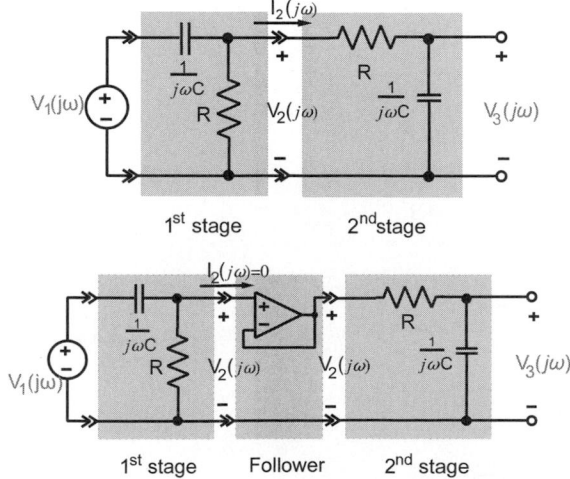

Fig. 101 (a) Two-port circuits connected in cascade. (b) Cascade connection with voltage follower isolation.[1]

The reason for the discrepancy is that when they are connected in cascade the second circuit "loads" the first circuit. That is, the voltage divider rule requires the current $I_2(j\omega)$ in Fig. 101a to be zero. The no-load condition $I_2(j\omega) = 0$ is valid when the stages operate separately, but when connected together the current is no longer zero. The chain rule does not apply here because loading caused by the second stage alters the transfer function of both stages.

The loading problem goes away when an op amp voltage follower is inserted between the RC circuit stages (Fig. 101b). With this modification the chain rule in Eq. (125) applies because the voltage follower isolates the two RC circuits. Recall that ideally a voltage follower has infinite input resistance and zero output resistance. Therefore, the follower does not draw any current from the first RC circuit $[I_2(j\omega) = 0]$ and its transfer function of "1" allows $V_2(j\omega)$ to be applied directly across the input of the second RC circuit.

The chain rule in Eq. (125) applies if connecting a stage does not change or load the output of the preceding stage. Loading can be avoided by connecting an op amp voltage follower between stages. More importantly, loading does not occur if the output of the preceding stage is the output of an op amp or controlled source unless the load resistance is very low. These elements act very close to ideal voltage sources whose outputs are unchanged by connecting the subsequent stage.

For example, in the top representation the two circuits in Fig. 102 are connected in a cascade with circuit C1 appearing first in the cascade followed by circuit C2. The chain rule applies to this configuration because the output of circuit C1 is an op amp that can handle the load presented by circuit C2. On the other hand, if the stages are interchanged so that the op amp circuit C1 follows the RC circuit C2 in the cascade, then the chain rule would not apply because the input impedance of circuit C1 would then load the output of circuit C2.

6.3 Frequency Response Descriptors

The relationships between the input and output sinusoids are important to frequency-sensitive circuits and can be summarized in the following statements.

Realizing the circuit transfer function is usually a complex function of ω, its effect on the sinusoidal steady-state response can be found through its *gain* function $|H(j\omega)|$ and *phase* function $\angle H(j\omega)$ as follows:

$$\text{Magnitude of } H(j\omega) = |H(j\omega)| = \frac{\text{output amplitude}}{\text{input amplitude}}$$

$$\text{Angle of } H(j\omega) = \angle H(j\omega)$$

$$= \text{output phase} - \text{input phase}$$

Taken together the gain and phase functions show how the circuit modifies the input amplitude and phase angle to produce the output sinusoid. These two functions define the *frequency response* of the circuit since they are frequency-dependent functions that relate the sinusoidal steady-state input and output. The gain and phase functions can be expressed mathematically or

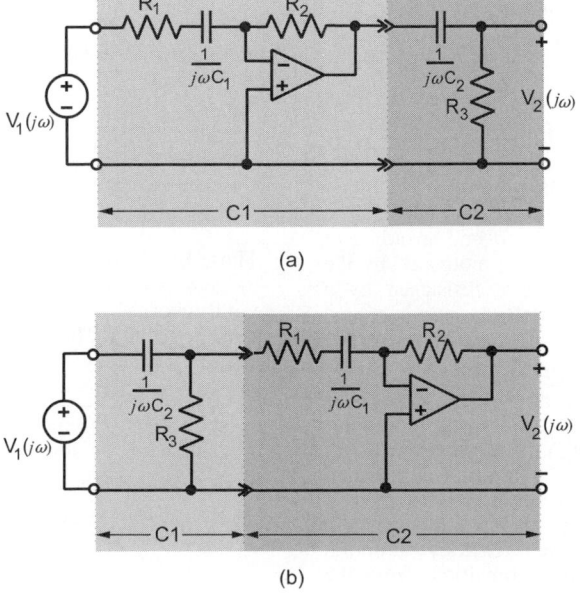

(a)

(b)

Fig. 102 Effects of stage location on loading.[1]

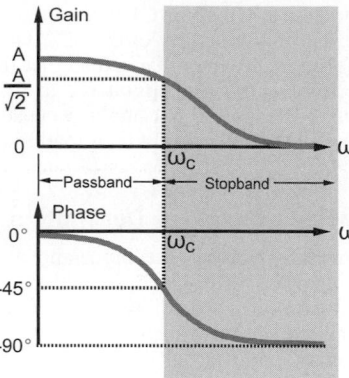

Fig. 103   Frequency response plots.[1]

presented graphically as in Fig. 103, which shows an example frequency response plot called a Bode diagram. These diagrams can be constructed by hand but are readily and more accurately produced by simulation and mathematical software products.

The terminology used to describe the frequency response of circuits and systems is based on the form of the gain plot. For example, at high frequencies the gain in Fig. 103 falls off so that output signals in this frequency range are reduced in amplitude. The range of frequencies over which the output is significantly attenuated is called the *stopband*. At low frequencies the gain is essentially constant and there is relatively little attenuation. The frequency range over which there is little attenuation is called a *passband*. The frequency associated with the boundary between a passband and an adjacent stopband is called the *cutoff frequency* ($\omega_C = 2\pi f_C$). In general, the transition from the passband to the stopband is gradual so the precise location of the cutoff frequency is a matter of definition. The most widely used definition specifies the cutoff frequency to be the frequency at which the gain has decreased by a factor of $1/\sqrt{2} = 0.707$ from its maximum value in the passband.

Again this definition is arbitrary, since there is no sharp boundary between a passband and an adjacent stopband. However, the definition is motivated by the fact that the power delivered to a resistance by a sinusoidal current or voltage waveform is proportional to the square of its amplitude. At a cutoff frequency the gain is reduced by a factor of $1/\sqrt{2}$ and the square of the output amplitude is reduced by a factor of $\frac{1}{2}$. For this reason the cutoff frequency is also called the *half-power frequency*. In filter design the region from where the output amplitude is reduced by 0.707 and a second frequency wherein the output must have decayed to some specified value is called the transition region. This region is where much of the filter design attention is focused. How rapidly a filter transitions from the cutoff frequency to some necessary attenuation is what occupies much of the efforts of filter designers.

Additional frequency response descriptors are based on the four prototype gain characteristics shown in Fig. 104. A *low-pass* gain characteristic has a single passband extending from zero frequency (dc) to the cutoff frequency. A *high-pass* gain characteristic has a single passband extending from the cutoff frequency to infinite frequency. A *bandpass* gain has a single passband with two cutoff frequencies neither of which is zero or infinite. Finally, the *bandstop* gain has a single stopband with two cutoff frequencies neither of which is zero or infinite.

The *bandwidth* of a gain characteristic is defined as the frequency range spanned by its passband. The bandwidth (BW) of a low-pass circuit is equal to its cutoff frequency (BW $= \omega_C$). The bandwidth of a high-pass characteristic is infinite since passband extends to infinity. For the bandpass and bandstop cases in Fig. 104 the bandwidth is the difference in the two cutoff frequencies:

$$\text{BW} = \omega_{C2} - \omega_{C1} \tag{126}$$

For the bandstop case Eq. (126) defines the width of the stopband rather than the passband.

The gain responses in Fig. 104 have different characteristics at zero and infinite frequency:

Prototype	Gain at $\omega = 0$	Gain at $\omega = 4$
Low pass	Finite	0
High pass	0	Finite
Bandpass	0	0
Bandstop	Finite	Finite

Since these extreme values form a unique pattern, the type of gain response can be inferred from the values of $|H(0)|$ and $|H(\infty)|$. These endpoint values in turn are usually determined by the impedance of capacitors and inductors in the circuit. In the sinusoidal steady state the impedances of these elements are

$$Z_C\,(j\omega) = \frac{1}{j\omega C} \quad \text{and} \quad Z_L\,(j\omega) = j\omega L$$

These impedances vary with respect to frequency. An inductor's impedance increases linearly with increasing frequency, while that of a capacitor varies inversely with frequency. They form a unique pattern at zero and infinite frequency:

Element	Impedance (Ω) at $\omega = 0$ (dc)	Impedance (Ω) at $\omega = 4$
Capacitor $(1/j\omega C)$	Infinite (open circuit)	0 (short circuit)
Inductor $(j\omega L)$	0 (short circuit)	Infinite (open circuit)
Resistor (R)	R	R

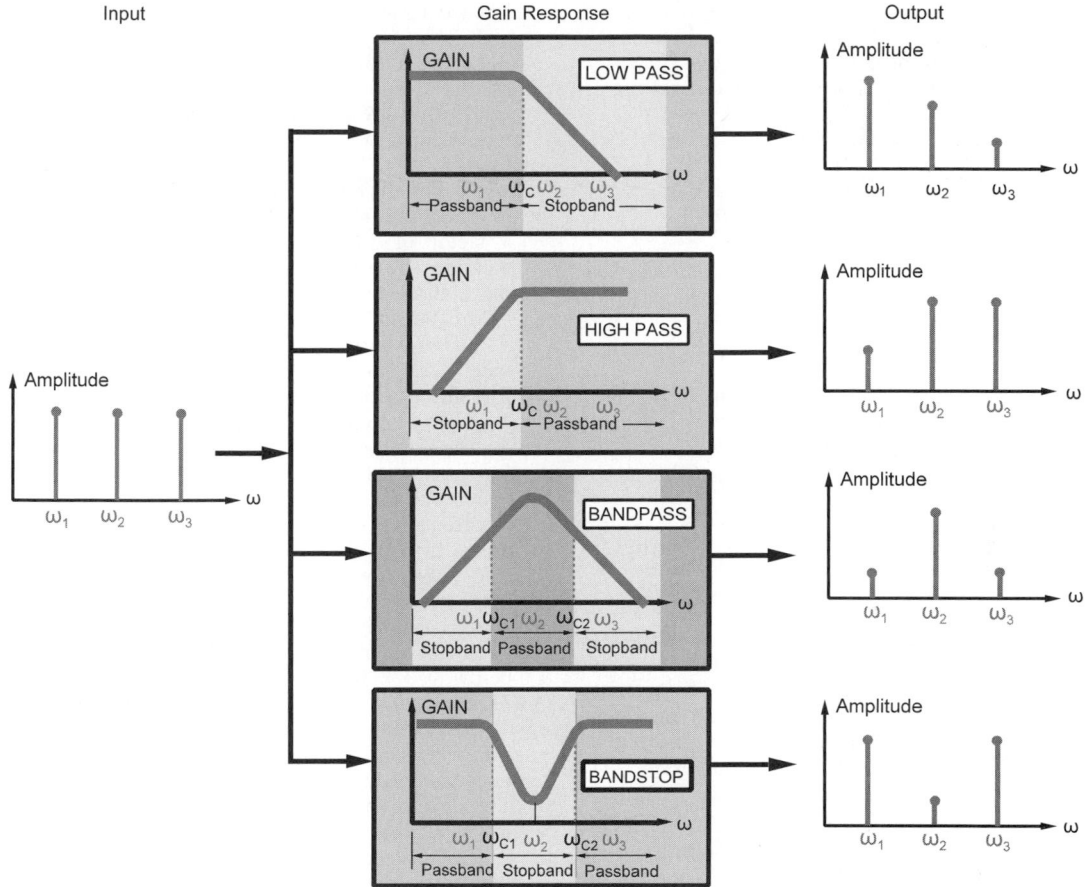

Fig. 104 Four basic gain responses.[1]

These observations often allow one to infer the type of gain response and hence the type of filter directly from the circuit itself without finding the transfer functions.

Frequency response plots are almost always made using logarithmic scales for the frequency variable. The reason is that the frequency ranges of interest often span several orders of magnitude. A logarithmic frequency scale compresses the data range and highlights important features in the gain and phase responses. The use of a logarithmic frequency scale involves some special terminology. Any frequency range whose endpoints have a 2 : 1 ratio is called an *octave*. Any range whose endpoints have a 10 : 1 ratio is called a *decade*. For example, the frequency range from 10 to 20 Hz is one octave, as is the range from 20 to 40 MHz. The standard UHF (ultrahigh frequency) band spans one decade from 0.3 to 3 GHz.

In frequency response plots the gain $|H(j\omega)|$ is often expressed in *decibels* (dB), defined as

$$|H(j\omega)|_{\mathrm{dB}} = 20\log_{10}|H(j\omega)|$$

The gain in decibels can be positive, negative, or zero. A gain of 0 dB means that $|H(j\omega)| = 1$; that is, the input and output amplitudes are equal. Positive decibel gains mean the output amplitude exceeds the input since $|H(j\omega)| > 1$ and the circuit is said to *amplify* the signal. A negative decibel gain means the output amplitude is smaller than the input since $|H(j\omega)| < 1$ and the circuit is said to *attenuate* the signal. A cutoff frequency occurs when the gain is reduced from its maximum passband value by a factor $1/\sqrt{2}$ or 3 dB. For this reason the cutoff is also called the *3-dB down frequency*.

6.4 First-Order Frequency Response and Filter Design

Frequency-selective circuits are fundamental to all types of systems. First-order filters are simple to design and can be effective for many common applications.

First-Order Low-Pass Response A first-order low-pass transfer function can be written as

$$H(j\omega) = \frac{K}{j\omega + \alpha}$$

The constants K and α are real. The constant K can be positive or negative, but α must be positive so that the natural response of the circuit is stable.

The gain and phase functions are given as

$$|H(j\omega)| = \frac{|K|}{\sqrt{\omega^2 + \alpha^2}}$$

$$\angle H(j\omega) = \angle K - \tan^{-1}(\omega/\alpha)$$

(127)

The gain function is a positive number. Since K is real, the angle of $K(\angle K)$ is either $0°$ when $K > 0$ or $\pm 180°$ when $K < 0$. An example of a negative K occurs in an inverting op amp configuration where $H(j\omega) = -Z_2(j\omega)/Z_1(j\omega)$.

Figure 105 shows the gain and phase functions versus normalized frequency ω_c/α. The maximum passband gain occurs at $\omega = 0$ where $|H(0)| = |K|/\alpha$. As frequency increases, the gain gradually decreases until at $\omega = \alpha$:

$$|H(j\alpha)| = \frac{|K|}{\sqrt{\alpha^2 + \alpha^2}} = \frac{|K| \to \alpha}{\sqrt{2}} = \frac{|H(0)|}{\sqrt{2}}$$

That is, the cutoff frequency of the first-order low-pass transfer function is $\omega_C = \alpha$. The graph of the gain function in Fig. 105a displays a low-pass characteristic with a finite dc gain and zero infinite frequency gain.

The low- and high-frequency gain asymptotes shown in Fig. 105a are especially important. The low-frequency asymptote is the horizontal line and the high-frequency asymptote is the sloped line. At low frequencies ($\omega \ll \alpha$) the gain approaches $|H(j\omega)| \to |K|/\alpha$. At high frequencies ($\omega \gg \alpha$) the gain approaches $|H(j\omega)| \to |K|/\omega$. The intersection of the two asymptotes occurs when $|K|/\alpha = |K|/\omega$. The intersection forms a "corner" at $\omega = \alpha$, so the cutoff frequency is also called the *corner frequency*.

The high-frequency gain asymptote decreases by a factor of 10 (-20 dB) whenever the frequency increases by a factor of 10 (one decade). As a result the high-frequency asymptote has a slope of -1 or -20 dB/decade and the low-frequency asymptote has a slope of 0 or 0 dB/decade. These two asymptotes provide a straight-line approximation to the gain response that differs from the true response by a maximum of 3 dB at the corner frequency.

The semilog plot of the phase shift of the first-order low-pass transfer function is shown in Fig. 105b. At $\omega = \alpha$ the phase angle in Eq. (127) is $\angle K - 45°$. At low frequency ($\omega < \alpha$) the phase angle approaches $\angle K$ and at high frequencies ($\omega > \alpha$) the phase approaches $\angle K - 90°$. Almost all of the $-90°$ phase change occurs in the two-decade range from $\omega/\alpha = 0.1$ to $\omega/\alpha = 10$. The straight-line segments in Fig. 105b provide an approximation of the phase response. The phase approximation below $\omega/\alpha = 0.1$ is $\theta = \angle K$ and above $\omega/\alpha = 10$ is $H\angle\theta = \angle K - 90°$. Between these values the phase approximation is a straight line that begins at $H\angle\theta = \angle K$, passes through $H\angle\theta = \angle K - 45°$ at the cutoff frequency, and reaches $H\angle\theta = \angle K - 90°$ at $\omega/\alpha = 10$. The slope of this line segment is $-45°$/decade since the total phase change is $-90°$ over a two-decade range.

To construct the *straight-line approximations* for a first-order low-pass transfer function, two parameters are needed, the value of $H(0)$ and α. The parameter α defines the cutoff frequency and the value of $H(0)$ defines the passband gain $|H(0)|$ and the low-frequency phase $\angle H(0)$. The required quantities $H(0)$ and α can be determined directly from the transfer function $H(j\omega)$ and can often be estimated by inspecting the circuit itself.

Example 33.   Design a low-pass filter with a passband gain of 4 and a cutoff frequency of 100 rad/sec.

Solution.   See Fig. 106. Start with an inverting amplifier configuration since a gain is required:

$$H(j\omega) = -\frac{Z_2(j\omega)}{Z_1(j\omega)}$$

$$Z_1(j\omega) = R_1 \qquad \text{and}$$

$$Z_2(j\omega) = \frac{1}{j\omega C_2 + 1/R_2} = \frac{R_2}{R_2 C_2 j\omega + 1}$$

$$H(j\omega) = -\frac{R_2}{R_1} \times \frac{1}{R_2 C_2 j\omega + 1}$$

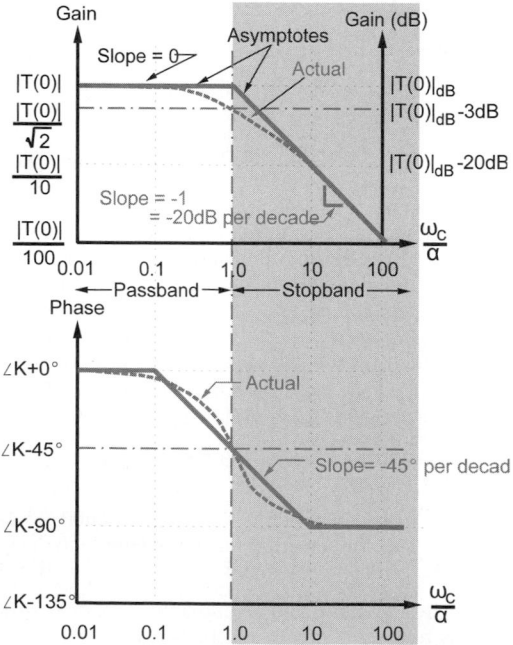

Fig. 105   First-order low-pass Bode plots.[1]

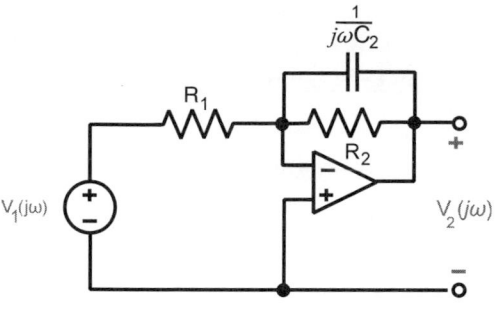

Fig. 106 (From Ref. 1.)

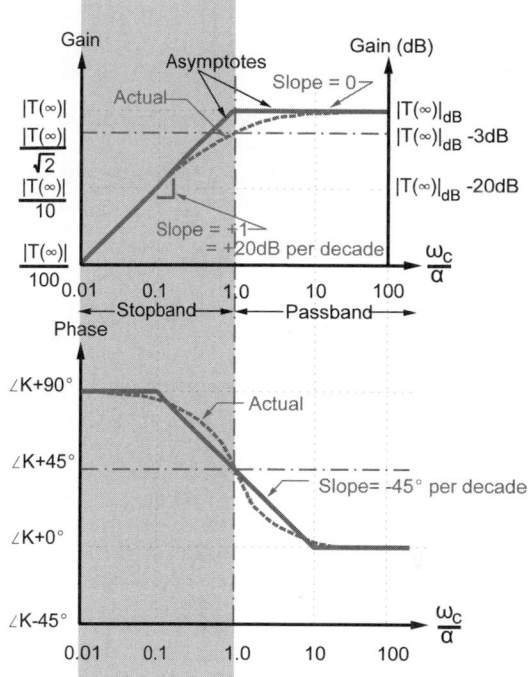

Fig. 107 First-order high-pass Bode plots.[1]

Rearrange the standard low-pass form as

$$H(j\omega) = \frac{K/\alpha}{j\omega/\alpha + 1}$$

$$\omega_C = \alpha = \frac{1}{R_2 C_2} \quad \text{and} \quad H(0) = -\frac{R_2}{R_1}$$

The design constraints require that $\omega_C = 1/R_2 C_2 = 100$ and $|H(0)| = R_2/R_1 = 4$. Selecting $R_1 = 10\ \text{k}\Omega$ requires $R_2 = 40\ \text{k}\Omega$ and $C = 250\ \text{nF}$.

First-Order High-Pass Response A first-order high-pass transfer function is written as

$$H(j\omega) = \frac{Kj\omega}{j\omega + \alpha}$$

The high-pass function differs from the low pass case by the introduction of a $j\omega$ in the numerator, resulting in the function becoming zero at $\omega = 0$. Solving for the gain and phase functions yields

$$|H(j\omega)| = \frac{|K|\,\omega}{\sqrt{\omega^2 + \alpha^2}}$$

$$\angle H(j\omega) = \angle K + 90° - \tan^{-1}\left(\frac{\omega}{\alpha}\right) \qquad (128)$$

Figure 107 shows the gain and phase functions versus normalized frequency ω/α. The maximum gain occurs at high frequency ($\omega > \alpha$) where $|H(j\omega)|6|K|$. At low frequency ($\omega < \alpha$) the gain approaches $|K|\omega/\alpha$. At $\omega = \alpha$ the gain is

$$|H(j\alpha)| = \frac{|K|\alpha}{\sqrt{\alpha^2 + \alpha^2}} = \frac{|K|}{\sqrt{2}}$$

which means the cutoff frequency is $\omega_C = \alpha$. The gain response plot in Fig. 107a displays a high-pass characteristic with a passband extending from $\omega = \alpha$ to infinity and a stopband between zero frequency and $\omega = \alpha$.

The low- and high-frequency gain asymptotes approximate the gain response in Fig. 107a. The high-frequency asymptote ($\omega > \alpha$) is the horizontal line whose ordinate is $|K|$ (slope $= 0$ or 0 dB/decade). The low-frequency asymptote ($\omega < \alpha$) is a line of the form $|K|\omega/\alpha$ (slope $= +1$ or $+20$ dB/decade). The intersection of these two asymptotes occurs when $|K| = |K|\omega/\alpha$, which defines a corner frequency at $\omega = \alpha$.

The semilog plot of the phase shift of the first-order high-pass function is shown in Fig. 107b. The phase shift approaches $\angle K$ at high frequency, passes through $\angle K + 45°$ at the cutoff frequency, and approaches $\angle K + 90°$ at low frequency. Most of the 90° phase change occurs over the two-decade range centered on the cutoff frequency. The phase shift can be approximated by the straight-line segments shown in the Fig. 107b. As in the low-pass case, $\angle K$ is 0° when K is positive and $\pm 180°$ when K is negative.

Like the low-pass function, the first-order high-pass frequency response can be approximated by straight-line segments. To construct these lines, we need two parameters, $H(\infty)$, and α. The parameter α defines the cutoff frequency and the quantity $H(\infty)$ gives the passband gain $|H(\infty)|$ and the high-frequency phase angle $\angle H(\infty)$. The quantities $H(\infty)$ and α can be determined directly from the transfer function or estimated directly from the circuit in some cases. The straight line shows the first-order high-pass response

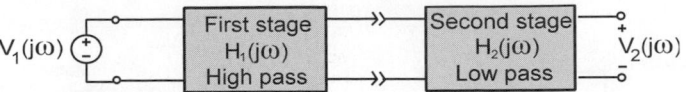

Fig. 108 Cascade connection of high- and low-pass circuits.[1]

can be characterized by calculating the gain and phase over a two-decade band from one decade below to one decade above the cutoff frequency.

Bandpass and Bandstop Responses Using First-Order Circuits

The first-order high- and low-pass circuits can be used in a building block fashion to produce a circuit with bandpass and bandstop responses. Figure 108 shows a cascade connection of first-order high- and low-pass circuits. When the second stage does not load the first, the overall transfer function can be found by the chain rule:

$$H(j\omega) = H_1(j\omega) \times H_2(j\omega)$$

$$= \underbrace{\left(\frac{K_1 j\omega}{j\omega + \alpha_1}\right)}_{\text{high pass}} \underbrace{\left(\frac{K_2}{j\omega + \alpha_2}\right)}_{\text{low pass}}$$

Solving for the gain response yields

$$|H(j\omega)| = \underbrace{\left(\frac{|K_1|\,\omega}{\sqrt{\omega^2 + \alpha_1^2}}\right)}_{\text{high pass}} \underbrace{\left(\frac{|K_2|}{\sqrt{\omega^2 + \alpha_2^2}}\right)}_{\text{low pass}}$$

Note the gain of the cascade is zero at $\omega = 0$ and at infinite frequency.

When $\alpha_1 < \alpha_2$ the high-pass cutoff frequency is much lower than the low-pass cutoff frequency, and the overall transfer function has a bandpass characteristic. At low frequencies ($\omega < \alpha_1 < \alpha_2$) the gain approaches $|H(j\omega)| \to |K_1 K_2|\omega/\alpha_1\alpha_2$. At midfrequencies ($\alpha_1 < \omega < \alpha_2$) the gain approaches $|H(j\omega)| \to |K_1 K_2|/\alpha_2$. The low- and midfrequency asymptotes intersect when $|K_1 K_2|\omega/\alpha_1\alpha_2 = |K_1 K_2|/\alpha_2$ at $\omega = \alpha_1$, that is, at the cutoff frequency of the high-pass stage. At high frequencies ($\alpha_1 < \alpha_2 < \omega$) the gain approaches $|H(j\omega)| \to |K_1 K_2|/\omega$. The high- and midfrequency asymptotes intersect when $|K_1 K_2|/\omega = |K_1 K_2|/\alpha_2$ at $\omega = \alpha_2$, that is, at the cutoff frequency of the low-pass stage. The plot of these asymptotes in Fig. 109 shows that the asymptotic gain exhibits a passband between α_1 and α_2. Input sinusoids whose frequencies are outside of this range fall in one of the two stopbands.

In the bandpass cascade connection the input signal must pass both a low- and a high-pass stage to reach the output. In the parallel connection in Fig. 110 the

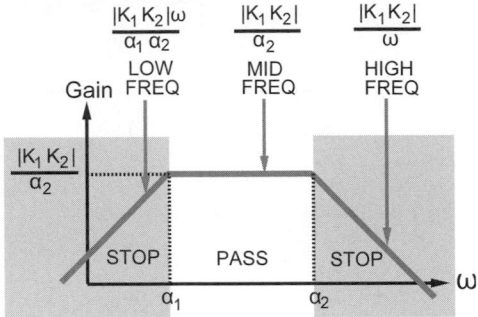

Fig. 109 Bandpass gain characteristic.[1]

input can reach the output via either a low- or a high-pass path. The overall transfer function is the sum of the low- and high-pass transfer functions:

$$|H(j\omega)| = \underbrace{\left(\frac{|K_1|\,\omega}{\sqrt{\omega^2 + \alpha_1^2}}\right)}_{\text{high pass}} + \underbrace{\left(\frac{|K_2|}{\sqrt{\omega^2 + \alpha_2^2}}\right)}_{\text{low pass}}$$

Any sinusoid whose frequency falls in either passband will find its way to the output unscathed. An input sinusoid whose frequency falls in both stopbands will be attenuated.

When $\alpha_1 > \alpha_2$, the high-pass cutoff frequency is much higher than the low-pass cutoff frequency, and the overall transfer function has a bandstop gain response as shown in Fig. 111. At low frequencies ($\omega < \alpha_2 < \alpha_1$) the gain of the high-pass function is negligible and the overall gain approaches $|H(j\omega)| \to |K_2|/\alpha_2$, which is the passband gain of the low-pass function. At high frequencies ($\alpha_2 < \alpha_1 < \omega$) the low-pass function is negligible and the overall gain approaches $|H(j\omega)| \to |K_1|$, which is the passband gain of the high-pass function. With a bandstop function the two passbands normally have the same gain, hence $|K_1| = |K_2|/\alpha_2$. Between these two passbands there is a stopband. For $\omega > \alpha_2$ the low-pass asymptote is $|K_2|/\omega$, and for $\omega < \alpha_1$ the high-pass asymptote is $|K_1|\omega/\alpha_1$. The asymptotes intersect at $\omega^2 = \alpha_1|K_2|/|K_1|$. But equal gains in the two passband frequencies requires $|K_1| = |K_2|/\alpha_2$, so the intersection frequency is $\omega = \sqrt{\alpha_1\alpha_2}$. Below this frequency the stopband attenuation is determined by the low-pass

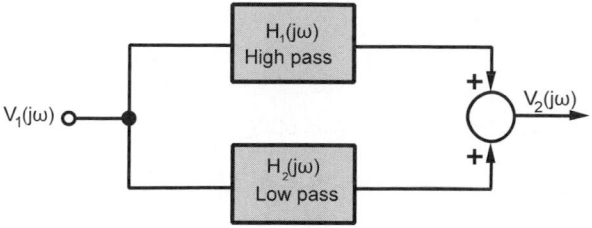

Fig. 110   Parallel connection of high- and low-pass circuits.[1]

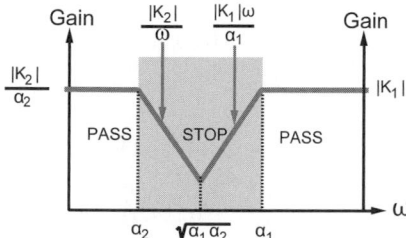

Fig. 111   Bandstop gain characteristic.[1]

function and above this frequency the attenuation is governed by the high-pass function.

Analysis of the transfer functions illustrates that the asymptotic gain plots of the first-order functions can help one understand and describe other types of gain response. The asymptotic response in Figs. 109 and 111 are a reasonably good approximation as long as the two first-order cutoff frequencies are widely separated. The asymptotic analysis gives insight to see that to study the passband and stopband characteristics in greater detail one needs to calculate gain and phase responses on a frequency range from a decade below the lowest cutoff frequency to a decade above the highest. This frequency range could be very wide, since the two cutoff frequencies may be separated by several decades. Mathematical and simulation software packages can produce very accurate frequency response plots.

Example 34.   Design a first-order bandpass circuit with a passband gain of 10 and cutoff frequencies at 20 Hz and 20 kHz.

Solution.   A cascade connection of first-order low- and high-pass building blocks will satisfy the design. The required transfer function has the form

$$H(j\omega) = H_1(j\omega) \times H_2(j\omega)$$

$$= \underbrace{\left(\frac{K_1 j\omega}{j\omega + \alpha_1}\right)}_{\text{high pass}} \underbrace{\left(\frac{K_2}{j\omega + \alpha_2}\right)}_{\text{low pass}}$$

with the following constraints:

Lower cutoff frequency: $\alpha_1 = 2\pi(20)$
$= 40\pi$ rad/s

Upper cutoff frequency: $\alpha_2 = 2\pi(20 \times 10^3)$
$= 4\pi \times 10^4$ rad/s

Midband gain: $\dfrac{|K_1 K_2|}{\alpha_2} = 10$

With numerical values inserted, the required transfer function is

$$H(j\omega) = \underbrace{\left[\frac{j\omega}{j\omega + 40\pi}\right]}_{\text{high pass}} \underbrace{[10]}_{\text{gain}} \underbrace{\left[\frac{40\pi \times 10^4}{j\omega + 40\pi \times 10^4}\right]}_{\text{low pass}}$$

This transfer function can be realized using the high-pass/low-pass cascade circuit in Fig. 112. The first stage is a passive RC high-pass circuit and the third stage is a passive RL low-pass circuit. The noninverting op amp second stage serves two purposes: (a) It isolates the first and third stages, so the chain rule applies, and (b) it supplies the midband gain. Using the chain rule, the transfer function of this circuit is

$$H(j\omega) = \underbrace{\left[\frac{j\omega}{j\omega + 1/RC}\right]}_{\text{high pass}} \underbrace{\left[\frac{R_1 + R_2}{R_1}\right]}_{\text{gain}} \underbrace{\left[\frac{R/L}{j\omega + R/L}\right]}_{\text{low pass}}$$

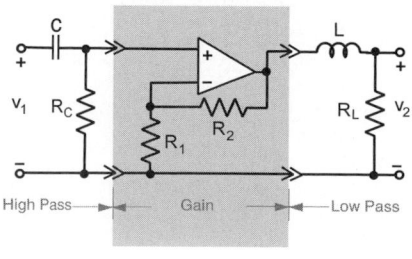

Fig. 112   (From Ref. 1.)

Comparing this to the required transfer function leads to the following design constraints:

High-Pass Stage: $R_C C = 1/40\pi$. Let $R_C = 100$ kΩ. Then $C = 79.6$ nF.

Gain Stage: $(R_1 + R_2)/R_1 = 10$. Let $R_1 = 10$ kΩ. Then $R_2 = 90$ kΩ.

Low-Pass Stage: $R_L/L = 40000\pi$. Let $R_L = 200$ kΩ. Then $L = 0.628$ H.

6.5 Second-Order *RLC* Filters

Simple second-order low-pass, high-pass, or band pass filters can be made using series or parallel *RLC* circuits. Series or parallel *RLC* circuits can be connected to produce the following transfer functions:

$$H(j\omega)_{\text{LP}} = \frac{K}{-\omega^2 + 2\zeta\omega_0 j\omega + \omega_0^2}$$

$$H(j\omega)_{\text{HP}} = \frac{-K\omega^2}{-\omega^2 + 2\zeta\omega_0 j\omega + \omega_0^2} \qquad (129)$$

$$H(j\omega)_{\text{BP}} = \frac{Kj\omega}{-\omega^2 + 2\zeta\omega_0 j\omega + \omega_0^2}$$

where for a series *RLC* circuit

$$\omega_0 = \sqrt{LC} \qquad \text{and} \qquad \zeta = \frac{R}{2}\sqrt{\frac{C}{L}}.$$

The undamped natural frequency ω_0 is related to the cutoff frequency in the high- and low-pass cases and is the center frequency in the band pass case. Zeta (ζ) is the damping ratio and determines the nature of the roots of the equation that translates to how quickly a transition is made from the passband to the stopband. In the band pass case ζ helps define the bandwidth of the circuit, that is, $B = 2\zeta\omega_0$. Figure 113 shows how a series *RLC* circuit can be connected to achieve the transfer functions given in Eq. (129). The gain $|H(j\omega)|$ plots of these circuits are shown in Figs. 114–116.

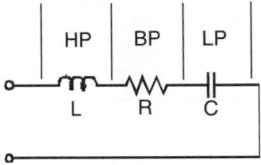

Fig. 113 Series *RLC* connected as low-pass (LP), high-pass (HP), or bandpass (BP) filter.

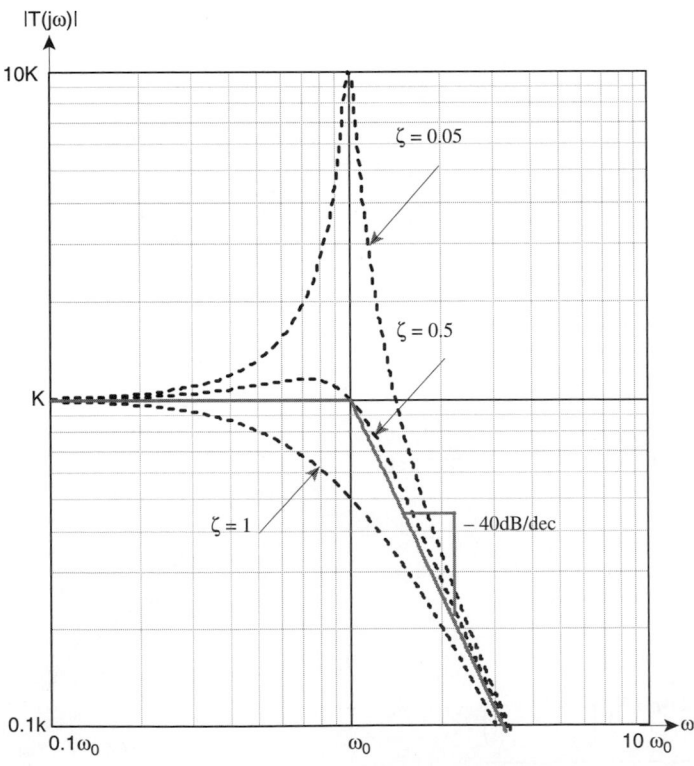

Fig. 114 Second-order low-pass gain responses.[1]

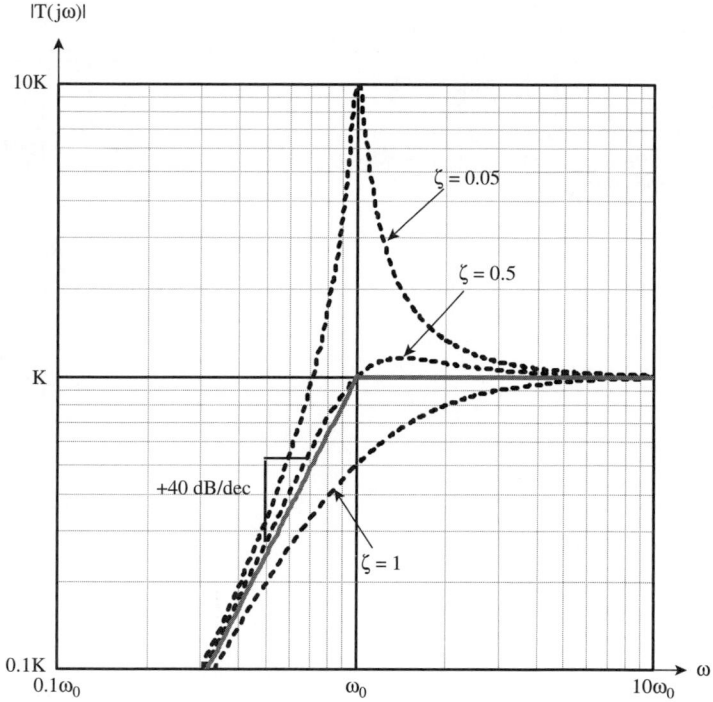

Fig. 115   Second-order high-pass gain responses.[1]

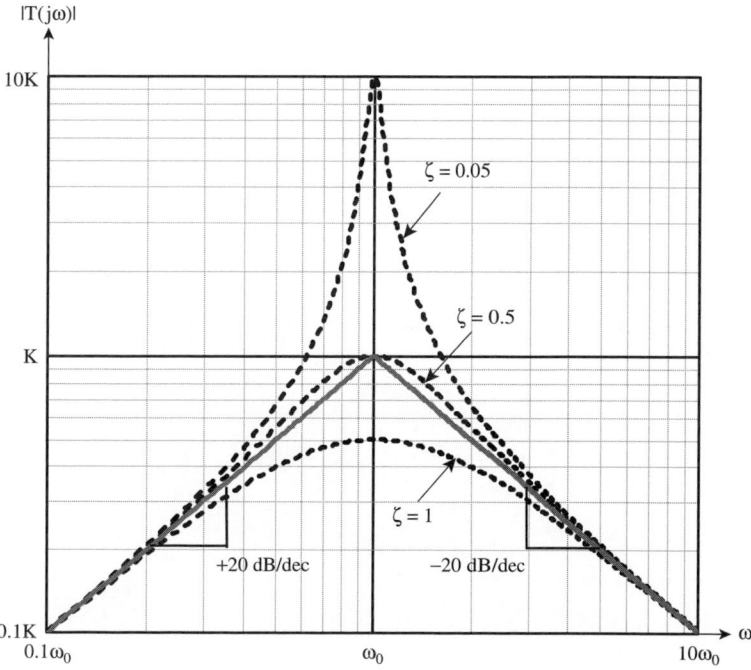

Fig. 116   Second-order bandpass gain responses.[1]

6.6 Compound Filters

Compound filters are higher order filters obtained by cascading lower order designs. *Ladder circuits* are an important class of compound filters. Two of the more common passive ladder circuits are the *constant-k* and the *m-derived* filters, either of which can be configured using a *T section, π section, L section* (Fig. 117), or combinations thereof, the bridge-T network and parallel-T network.

Active filters are generally designed using first-, second-, or third-order modules such as the *Sallen–Key* configurations shown in Fig. 118*a* and *b*, and the Delyannies–Friend configurations shown in Fig. 118*c* and *d*. The filters are then developed using an algorithmic approach following the *Butterworth, elliptical,* or *Tchebycheff* realizations.

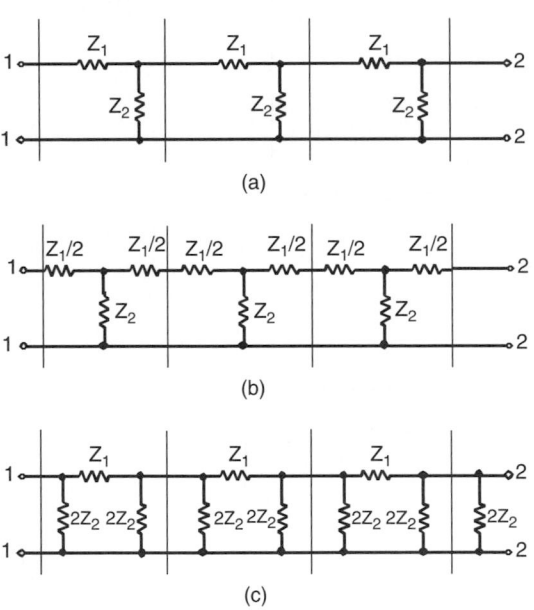

Fig. 117 Passive cascaded filter sections: (*a*) L section; (*b*) T section; (*c*) π section.

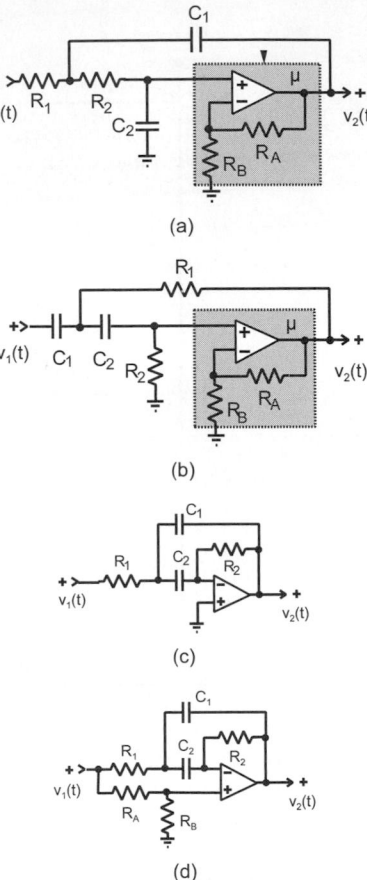

Fig. 118 Second-order configurations: (*a*) LP; (*b*) HP; (*c*) tuned; (*d*) Notch.[1]

REFERENCES

1. Thomas, R., and Rosa, A. J., *The Analysis and Design of Linear Circuits*, 5th ed., Wiley, 2005.
2. Thomas, R., and Rosa, A. J., *The Analysis and Design of Linear Circuits*, 2nd ed., Prentice-Hall, Englewood Cliffs, NJ, 1998.

CHAPTER 17
ELECTRONICS*

John D. Cressler
Georgia Institute of Technology
Atlantita, Georgia

**Kavita Nair, Chris Zillmer,
Dennis Polla, and Ramesh Harjani**
University of Minnesota
Minneapolis, Minnesota

Arbee L. P. Chen and Yi-Hung Wu
National Tsing Hua University
Hsinchu, Taiwan
Republic of China

Konstantinos Misiakos
NCSR "Demokritos"
Athens, Greece

Clarence W. de Silva
University of British Columbia
Vancouver, British Columbia
Canada

**Georges Grinstein
and Marjan Trutschl**
University of Massachusetts Lowell
Lowell, Massachusetts

Halit Eren
Curtin University of Technology
Bentley, Western Australia
Australia

**N. Ranganathan
and Raju D. Venkataramana**
University of South Florida
Tampa, Florida

Robert P. Colwell
Intel Corporation
Hillsboro, Oregon

Andrew Rusek
Oakland University
Rochester, Michigan

Alex Q. Huang and Bo Zhang
Virginia Polytechnic Institute
and State University
Blacksburg, Virginia

*Reprinted from *Wiley Encyclopedia of Electrical and Electronics Engineering*, Wiley, New York, 1999, with permission of the publisher.

1   BIPOLAR TRANSISTORS

John D. Cressler

The basic concept of the bipolar junction transistor (BJT) was patented by Shockley in 1947[1], but the BJT was not experimentally realized until 1951.[2] Unlike the point contact transistor demonstrated earlier in 1947, the BJT can be completely formed *inside* the semiconductor crystal, and thus it proved to be more manufacturable and reliable and better suited for use in integrated circuits. In a real sense, the BJT was the device that launched the microelectronics revolution and, hence, spawned the *Information Age*. Until the widespread emergence of complementary metal−oxide−semiconductor (CMOS) technology in the 1980s, the BJT was the dominant semiconductor technology in microelectronics, and even today represents a significant fraction of the global semiconductor market.

At its most basic level the BJT consists of two back-to-back *pn* junctions (*p–n–p* or *n–p–n* depending on the doping polarity), in which the intermediate *n* or *p* region is made as thin as possible. In this configuration the resultant three-terminal (emitter−base−collector) device exhibits current amplification (current gain) and thus acts as a "transistor" that can be used to build a wide variety of electronic circuits. Modern applications of the BJT are varied and range from high-speed digital integrated circuits in mainframe computers, to precision analog circuits, to radio frequency (*RF*) circuits found in radio communications systems.

Compared to CMOS, the BJT exhibits higher output current per unit length, larger transconductance (g_m) per unit length, faster switching speeds (particularly under capacitive loading), and excellent properties for many analog and RF applications (e.g., lower $1/f$ and broadband noise). Today, frequency response above 50 GHz and circuit switching speeds below 20 ps are readily attainable using conventional fabrication techniques. The primary drawback of BJT circuits compared to CMOS circuits lies in their larger dc power dissipation and increased fabrication complexity, although in applications requiring the fastest possible switching speeds, the BJT remains the device of choice. Figure 1 shows unloaded emitter-coupled logic (ECL) gate delay for today's technology

and indicates that state-of-the-art BJT technology is rapidly approaching 10 ps switching times.

In this section we review the essentials of modern bipolar technology, the operational principles of the BJT, second-order high-injection effects, issues associated with further technology advancements, and some future directions. Interested readers are referred to Refs. 3−5 for review articles on modern BJT technology, and to Ref. 6 for an interesting historical perspective on the development of the BJT.

1.1 Double-Polysilicon Bipolar Technology

In contrast to the depictions commonly found in many standard electronics textbooks, BJT technology has evolved radically in the past 15 years, from double-diffused, large geometry, non-self-aligned structures to very compact, self-aligned, "double-polysilicon" structures. Figure 2 shows a schematic cross section of a modern double-polysilicon BJT. This device has deep-trench and shallow-trench isolation to separate one transistor from the next, a p^+ polysilicon extrinsic base contact, an n^+ polysilicon emitter contact, and an ion-implanted intrinsic base region. The two polysilicon layers (hence the name *double-polysilicon*) act as both diffusion sources for the emitter and extrinsic base dopants as well as low-resistance contact layers. In addition, to form the active region of the transistor, a "hole" is etched into the p^+ polysilicon layer, and afterwards a thin dielectric "spacer" oxide is formed. In this manner, the emitter and extrinsic base regions are fabricated without the need of an additional lithography step ("self-aligned"), thereby dramatically reducing the size of the transistor and hence the associated parasitic resistances and capacitances of the structure. The first double-polysilicon BJT structures appeared in the early 1980s[7,8] and today completely dominate the high-performance BJT technology market. The reader is referred to Refs. 9−15 for specific BJT technology examples in the recent literature.

The doping profile from the intrinsic region of a state-of-the-art double-polysilicon BJT is shown in Fig. 3. The transistor from which this doping profile was measured has a peak cutoff frequency of about 40 GHz[14], and is typical of the state of the art. The emitter polysilicon layer is doped as heavily as possible with arsenic or phosphorus, and given a sort rapid-thermal-annealing (RTA) step to out-diffuse the dopants from the polysilicon layer. Typical metallurgical emitter−base junction depths range from 25 to 45 nm in modern BJT technologies. The collector region directly under the active region of the transistor is formed by local ion implantation of phosphorus. A collector doping of about 1×10^{17} cm^{-3} at the base−collector junction is adequate to obtain a peak cutoff frequency of 40 GHz at a collector-to-emitter breakdown voltage (BV$_{CEO}$) of about 3.5 V, consistent with the needs of digital ECL circuits. The intrinsic

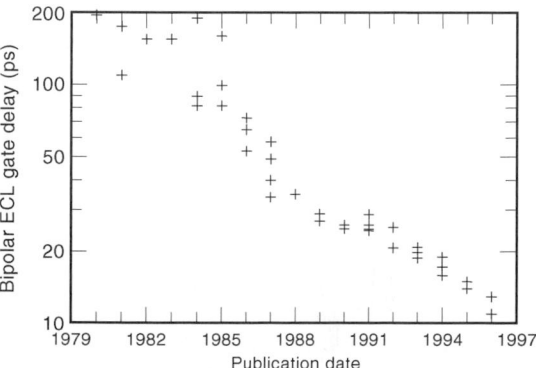

Fig. 1 Unloaded ECL gate delay (as function of publication date) showing rapid decrease in delay with technology evolution.

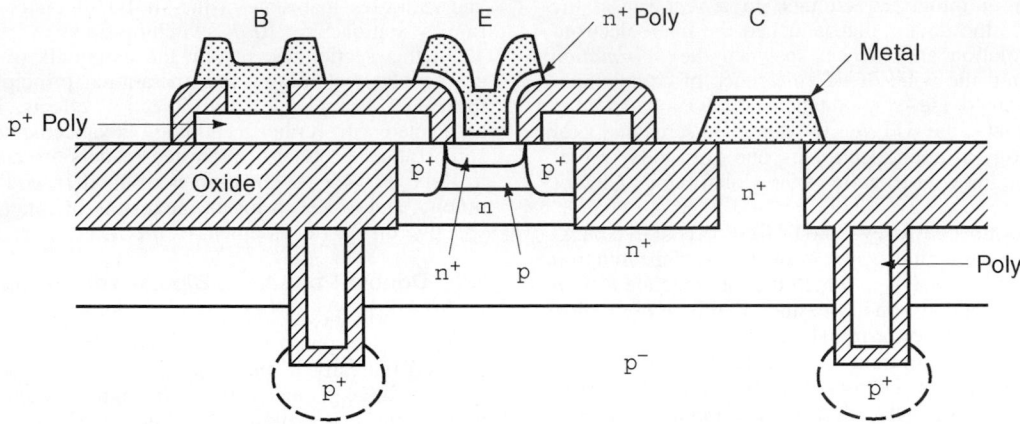

Fig. 2 Schematic device cross section of modern double-polysilicon self-aligned bipolar transistor.

base region is also formed by low-energy ion implantation of boron. Resultant base widths range from about 60 to 150 nm at the state of the art, with peak base doping levels in the range of $3-5 \times 10^{18}$ cm^{-3}. A traditional (measurable) metric describing the base profile in a BJT is the intrinsic base sheet resistance (R_{bi}), which can be written in terms of the integrated base

doping (N_{ab}) according to

$$R_{bi} = \left[q \int_0^{W_b} \mu_{pb}(x) N_{ab}(x)\, dx \right]^{-1} \qquad (1)$$

In Eq. (1), μ_{pb} is the position-dependent hole mobility in the base and W_b is the neutral base width. Typical

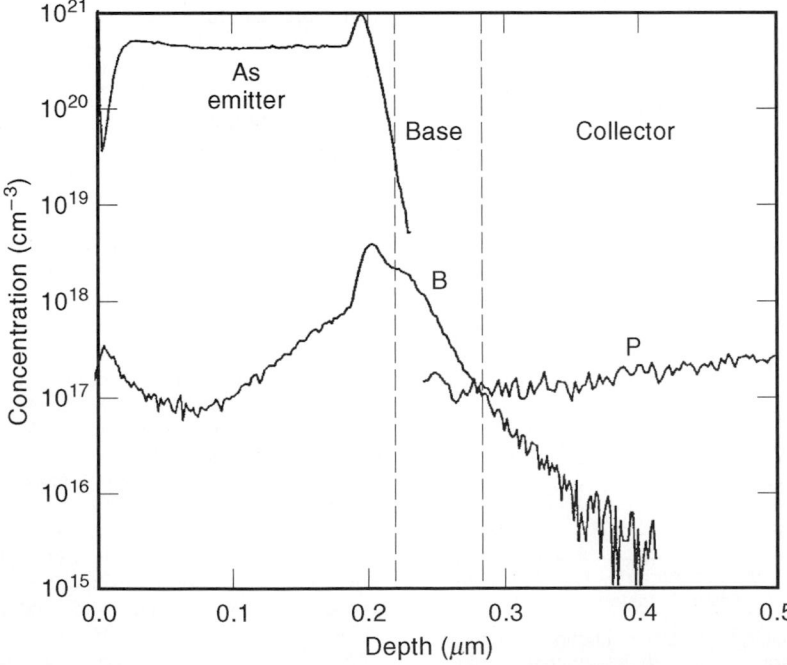

Fig. 3 Measured secondary ion mass spectroscopy (SIMS) doping profile from ion-implanted base bipolar technology with 40-GHz peak cutoff frequency.[14]

R_{bi} values in modern BJT technologies range from 10 to 15 kW/f.

1.2 Theory of Operation

Basic Physics The BJT is in essence a barrier-controlled device. A voltage bias is applied to the emitter–base junction such that we modulate the size of the potential barrier seen by the electrons moving from emitter to base, and thus can (exponentially) modulate the current flowing through the transistor. To best illustrate this process, we have used a one-dimensional device simulator called SCORPIO.[16] SCORPIO is known as a "drift diffusion" simulator because it solves the electron and hole drift diffusion transport equations self-consistently with Poisson's equation and the electron and hole current-continuity equations (see, e.g., Ref. 6 for a formulation of these equations and the inherent assumptions on their use). These five equations, together with the appropriate boundary conditions completely describe the BJT.

Figure 4 depicts a "toy" doping profile of the ideal BJT being simulated. Both the layer thicknesses and doping levels are consistent with those found in modern BJTs, although the constancy of the doping profile in each region is idealized and hence unrealistic. Figure 5 shows the resultant electron energy band diagram of this device at zero bias (equilibrium). The base potential barrier seen by the electrons in the emitter is clearly evident. The equilibrium carrier concentrations for each region are shown in Fig. 6. The majority carrier densities are simply given by the doping level in each region, while the minority carrier densities are obtained by use of the "law of mass action" according to the following:

$$p_{e0} = \frac{n_{ie}^2}{N_{de}} \qquad n_{e0} = N_{de} \quad \text{(emitter)} \qquad (2)$$

$$p_{b0} = N_{ab} \qquad n_{b0} = \frac{n_{ib}^2}{N_{ab}} \quad \text{(base)} \qquad (3)$$

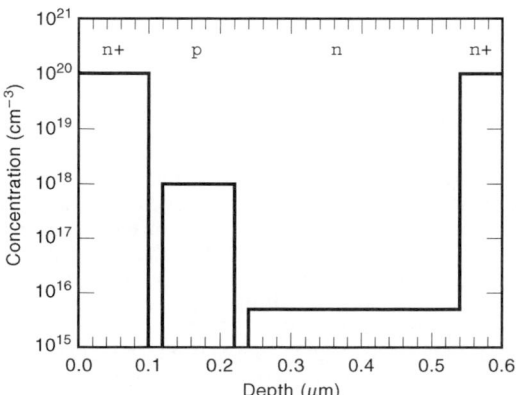

Fig. 4 Doping profile of hypothetical bipolar transistor used in one-dimensional SCORPIO simulations.

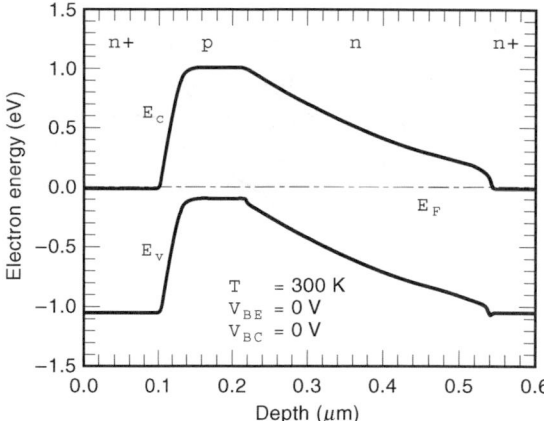

Fig. 5 Simulated zero-bias energy band diagram of hypothetical bipolar transistor depicted in Fig. 4.

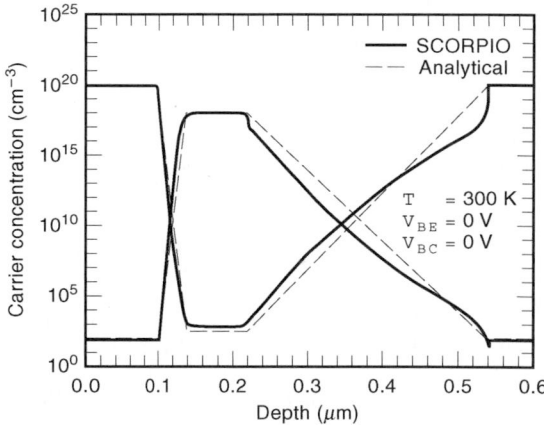

Fig. 6 Simulated electron and hole concentrations of hypothetical bipolar transistor depicted in Fig. 4. Also shown are analytical calculations.

$$p_{c0} = \frac{n_{i0}^2}{N_{dc}} \qquad n_{c0} = N_{dc} \quad \text{(collector)} \qquad (4)$$

where n_{i0} is the intrinsic carrier density, the subscripts e, b, and c represent the emitter, base, and collector regions, respectively, N is the doping density, and,

$$n_{ie}^2 = n_{i0}^2 e^{\Delta E_{ge}^{\mathrm{app}}/kT} = N_C N_V e^{-E_g/kT} e^{\Delta E_{ge}^{\mathrm{app}}/kT} \qquad (5)$$

$$n_{ib}^2 = n_{i0}^2 e^{\Delta E_{gb}^{\mathrm{app}}/kT} = N_C N_V e^{-E_g/kT} e^{\Delta E_{gb}^{\mathrm{app}}/kT} \qquad (6)$$

where $\Delta E_{ge}^{\mathrm{app}}$ and $\Delta E_{gb}^{\mathrm{app}}$ represent the heavy-doping-induced apparent bandgap narrowing.[17] The resultant

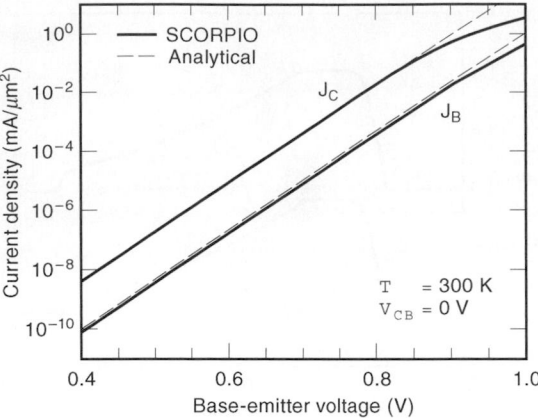

Fig. 7 Simulated collector and base current densities as function of emitter–base bias. Also shown are analytical calculations.

collector current density (J_C) and base current density (J_B) from this structure are shown in Fig. 7. Observe that the BJT exhibits useful current gain ($\beta = J_C/J_B$) over a wide operating range.

The basic operational principles of the BJT can be described as follows. If we imagine forward biasing the emitter–base junction, and reverse biasing the base–collector junction (i.e., forward-active mode), electrons from the heavily doped emitter are injected into and diffuse across the base region and are collected at the collector contact, thereby giving rise to a useful collector current. At the same time, if the base region is thin enough, the base current consists primarily of the back-injected hole current from base to emitter. Because the emitter is doped heavily with respect to the base, the ratio of forward-injected (emitter-to-base) electron current to back-injected (base to emitter) hole current is large (roughly equal to the ratio of emitter-to-base doping), and the BJT exhibits useful current gain. It is critical that the intermediate base region be kept as thin as possible because (a) we do not want electrons traversing the base to have sufficient time to recombine with holes before they reach the collector contact, and (b) the transit time of the electrons through the base typically limits the frequency response and, hence, the speed of the transistor. In the forward-active mode, a schematic representation of the magnitude of the various currents flowing in an ideal BJT is illustrated in Fig. 8.[6]

Current–Voltage Characteristics For simplicity, we will limit this discussion to the currents flowing in the BJT under forward-active bias. Other bias regimes (e.g., saturation) are not typically encountered in high-speed circuits such as ECL. The reader is referred to Refs. 17–19 for a discussion of other operating regimes. In this case, for a BJT with a position-dependent base doping profile, the collector current

density can be expressed as[20]

$$J_c = \frac{q[e^{qV_{BE}/kT} - 1]}{\int_0^{W_b} \frac{N_{ab}(x)\,dx}{D_{nb}(x)n_{ib}^2(x)}} \tag{7}$$

We see then that the collector current density in a BJT depends on the details of the base doping profile [more specifically the integrated base charge, and, hence, R_{bi} given in Eq. (1)]. The base current density can be obtained in a similar manner, except that the physics of the polysilicon emitter contact must be properly accounted for.[21,22] For the "transparent emitter domain" in which the holes injected from the base to emitter do not recombine before reaching the emitter contact, the base current density can be written as

$$J_B = \frac{q[e^{qV_{BE}/kT} - 1]}{\int_0^{W_e} \frac{N_{de}(x)\,dx}{D_{pe}(x)n_{ie}^2} + \frac{N_{de}(W_e)}{S_{pe}n_{ie}^2(W_e)}} \tag{8}$$

where S_{pe} is the "surface recombination velocity" characterizing the polysilicon emitter contact.[21] More detailed base current density expressions can be found in Refs. 21 and 22. Observe that in this transparent domain, the base current density depends on the specifics of the emitter doping profile as well as the influence of the polysilicon emitter contact.

For position-independent base and emitter doping profiles, with no polysilicon emitter contact, Eqs. (8) simplify to their familiar forms:

$$J_c \cong \left[\frac{q D_{nb} n_{ib}^2}{W_b N_{ab}} \right] e^{qV_{BE}/kT} \tag{9}$$

$$J_B \cong \left[\frac{q D_{pe} n_{ie}^2}{L_{pe} N_{de}} \right] e^{qV_{BE}/kT} \tag{10}$$

from which the ideal BJT current gain can be obtained

$$\beta \cong \frac{D_{nb} L_{pe} N_{de}}{D_{pe} W_b N_{ab}} e^{(\Delta E_{gb}^{app} - \Delta E_{gb}^{app})/kT} \propto \frac{N_{de}}{N_{ab}} \tag{11}$$

Thus, the current gain of the BJT depends on the ratio of emitter-to-base doping level. Given this fact, it is not surprising that the actual ratio of emitter-to-base doping level is typically found to be 100 (refer to Fig. 4), a common value for β in modern technologies. Note as well, however, from Eq. (11) that the ideal current gain in a BJT is reduced by the exponential dependence of the heavy-doping-induced bandgap narrowing parameters (the exponent is negative because the emitter is more heavily doped than the base). This latter dependence is also responsible for determining the temperature dependence of β in a BJT.

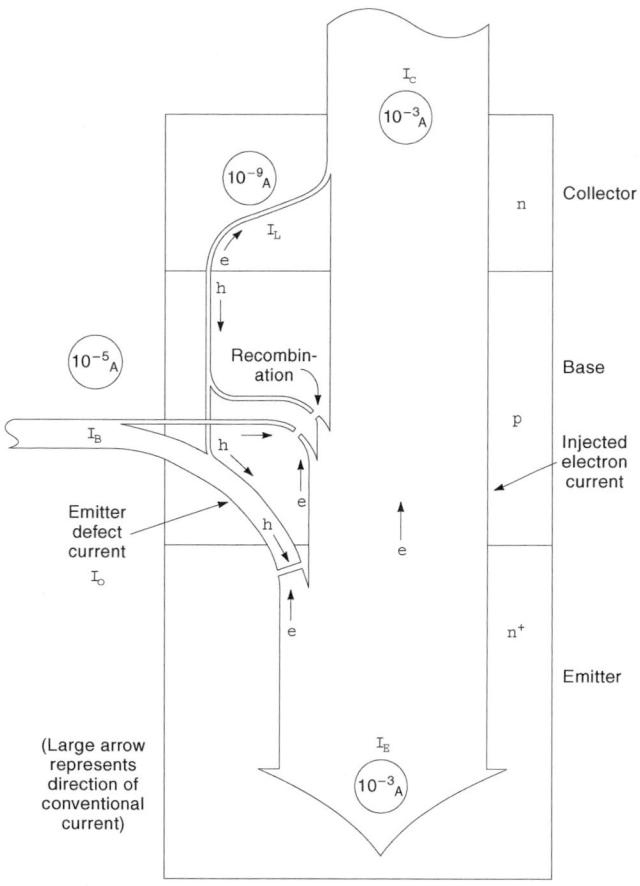

Fig. 8 Schematic current flow distributions in realistic bipolar transistor.

If one compares the measured $I–V$ characteristics of a BJT with those expected from Eqs. (9)–(11), substantial deviations are typically observed, as depicted schematically in Figs. 9 and 10 (the dashed lines represent the ideal results). Referring to Fig. 9, at low current levels, base current nonideality is the result of emitter–base space-charge region recombination effects; at high current levels, the deviations are the result of various "high-injection" effects (discussed in what follows). Only over an intermediate bias range are ideal characteristics usually observed. Figure 11 shows typical measured $I–V$ characteristics (a so-called Gummel plot) from the same 40-GHz profile depicted in Fig. 3.[14] The inset of Fig. 3 shows the linear "output characteristics" of the BJT. The shape and doping level of the collector profile controls the breakdown characteristics of the device. In this case, the collector-to-emitter breakdown voltage (BV_{CEO}) is approximately 3.3 V, typical for a high-performance digital BJT technology.

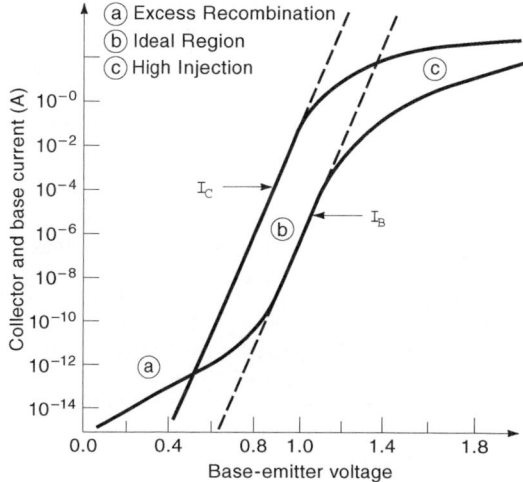

Fig. 9 Schematic Gummel characteristics for realistic bipolar transistor.

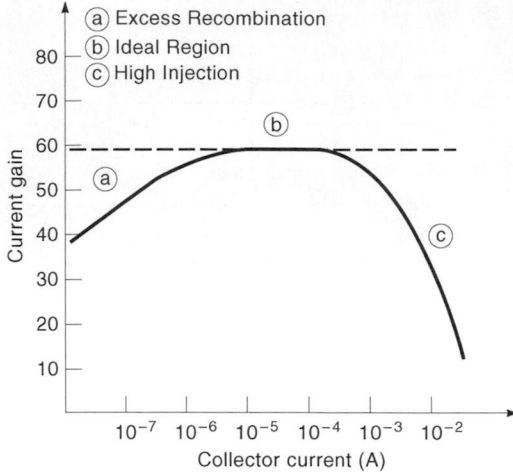

Fig. 10 Schematic current gain versus bias for realistic bipolar transistor.

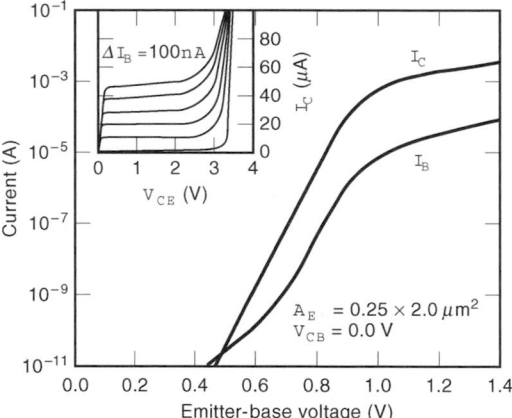

Fig. 11 Measured Gummel characteristics for scaled 0.25-μm double-polysilicon bipolar technology.[14] Inset shows common-emitter breakdown characteristics of transistor.

Frequency Response The frequency response of a BJT is determined by both the intrinsic speed of the carriers through the device (transit time), as well as the parasitic resistances and capacitances of the transistor. Two primary figures of merit are used to characterize the frequency response of a BJT, the unity gain cutoff frequency (f_T) and the maximum oscillation frequency (f_{max}). Using a small-signal hybrid-π model both f_T and f_{max} can be derived (17), yielding

$$f_T = \frac{1}{2\pi\tau_{ec}} = \left[\frac{1}{g_m}(C_{be} + C_{bc}) + \tau_b + \tau_e + \tau_c \right]^{-1} \tag{12}$$

$$f_T = \left[\frac{kT}{qI_c}(C_{be} + C_{bc}) + \frac{W_b^2}{\eta \tilde{D}_{nb}} \right.$$
$$\left. + \frac{1}{\beta_{ac}}\left(\frac{W_e}{S_{pe}} + \frac{W_e^2}{2D_{pe}} \right) + \frac{W_{bc}}{2v_s} + r_c C_{bc} \right]^{-1} \tag{13}$$

and

$$f_{max} = \sqrt{\frac{f_T}{8\pi C_{bc} R_b}} \tag{14}$$

In Eqs. (12)–(14), g_m is the transconductance ($\partial I_C/\partial V_{BE}$), C_{be} and C_{bc} are the base–emitter and base–collector capacitances, τ_b, τ_e, and τ_c are the base, emitter, and collector transit times, respectively, v_s is the saturation velocity (1×10^7 cm/s), η accounts for any doping-gradient-induced electric fields in the base, and R_b is the base resistance; f_T and, hence, f_{max} is typically limited by τ_b in conventional Si–BJT technologies. A major advantage of ion-implanted base, double-polysilicon BJT technology is that the base width can be made very small (typically < 150 nm), and thus the intrinsic frequency response quite large. Figure 12 shows measured f_T data as a function of bias current for a variety of device sizes for the doping profile shown in Fig. 3.[14]

ECL Gate Delay Due to its nonsaturating properties and high logical functionality, the ECL is the highest speed bipolar logic family and is in widespread use in the high-speed digital bipolar world. Figure 13 shows a simplified two-phase ECL logic gate. A common large-signal performance figure of merit is the unloaded

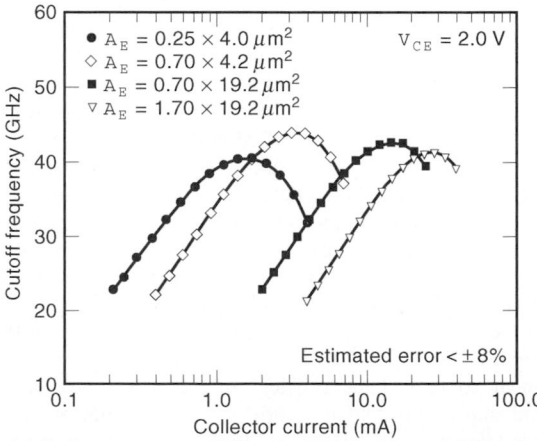

Fig. 12 Measured cutoff frequency as function of collector[14] current for scaled 0.25-μm double-polysilicon bipolar technology. Shown are a variety of device geometries.

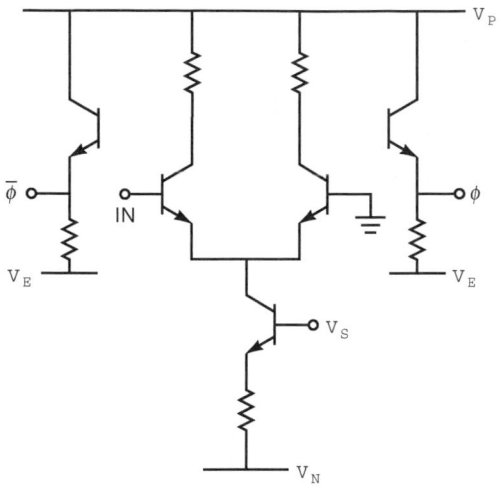

Fig. 13 Circuit schematic of two-phase. ECL gate.

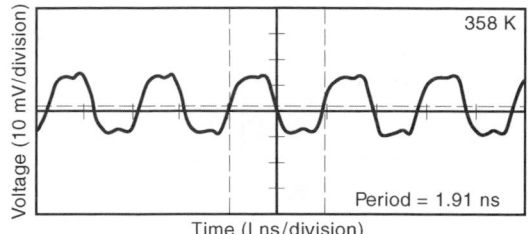

Fig. 15 Measured output waveform from ECL ring oscillator.

$$= \frac{V_L}{I_{CS}}[a_1 C_{bc} + a_2 C_{be}$$
$$+ a_3 C_{cs} + a_4 C_w + \cdots] \quad (16)$$

$$\alpha \frac{1}{I_{CS}} \propto \frac{1}{\text{power}} \quad (17)$$

while under high current (or power) conditions the ECL gate delay can be written as

$$\tau_{\text{ECL}}(\text{high power}) \cong C_{\text{diff}} \sum_{k=1}^{n} b_k R_k \quad (18)$$

$$= \frac{q \tau_{ec} I_{CS}}{kT}[b_1 R_{bi} + b_2 B_{bx}$$
$$+ b_3 R_e + b_4 R_c + \cdots] \quad (19)$$

$$\alpha \ I_{CS} \alpha \ \text{power} \quad (20)$$

In Eqs. (15)–(20), R_{CC} is the circuit pull-up resistor, V_L is the logic swing, a_k and b_k are delay "weighting factors," I_{CS} is the switch current, and C_{diff} is the transistor diffusion capacitance. We see then that at low currents, the parasitic capacitances dominate the ECL delay with a delay that is reciprocally proportional to the power dissipation, whereas at high currents, the parasitic resistances dominate the ECL delay, yielding a delay that is proportional to power dissipation. It is thus physically significant to plot the log of the ECL delay as a function of the log of the power (or current), as shown in Fig. 16. Also shown in Fig. 16 are large-signal circuit simulation results using the compact model depicted in Fig. 17, which confirm the stated dependence of delay on power.

ECL gate delay, which can be measured using a "ring oscillator." A ring oscillator is essentially a delay chain of ECL inverters with output tied back to its input, thus rendering the resultant circuit unstable (Fig. 14). From the period of the oscillation (Fig. 15), the average gate delay can be determined for a given bias current. Multiple ring oscillators can then be configured to operate at various bias currents, and hence the "power delay" characteristics of the BJT technology determined (average gate delay is plotted as a function of average power dissipation—or current in this case, because the supply voltage is constant). Figure 16 shows a typical measured ECL power delay curve.[14] A minimum ECL gate delay of 20.8 ps is achieved with this technology. Observe that the speed of the ECL gate becomes faster as the average switch current increases, until some minimum value of delay is reached. To better understand the functional shape of the power delay curve, asymptotic expressions can be developed using a weighted time constant approach.[23] Under low-current (or low-power) conditions, the ECL gate delay is given by

$$\tau_{\text{ECL}}(\text{low power}) \cong R_{CC} \sum_{k=1}^{n} a_k C_k \quad (15)$$

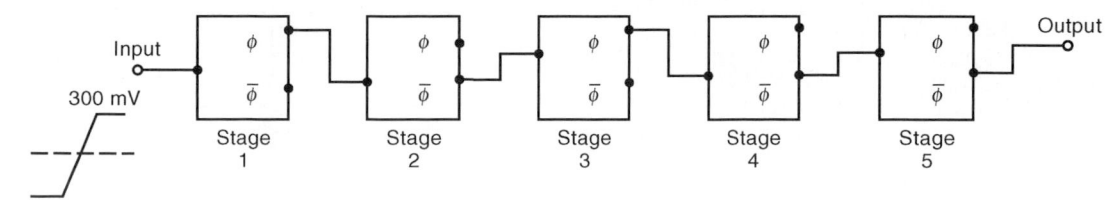

Fig. 14 Schematic representation of ECL ring oscillator circuit configuration.

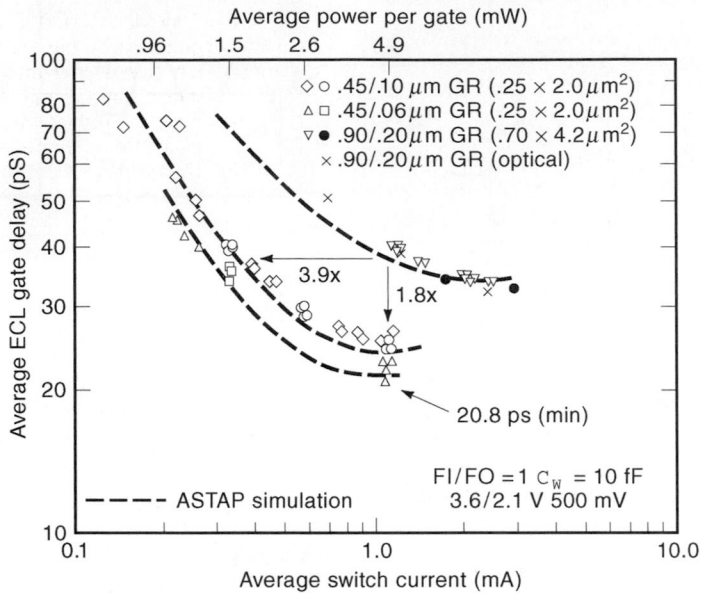

Fig. 16 The ECL power delay characteristics for scaled 0.25-μm double-polysilicon bipolar technology.[14] Minimum delay of 20.8 ps is achieved. The ECL circuits were operated on 3.6/2.1-V power supplies at 500-mV logic swing. Fan-in (FI) and fan-out (FO) of one was used. Impact of transistor scaling from 0.90/0.20-μm lithography to 0.45/0.06-μm lithography is indicated. Also shown are circuit simulations calibrated to data using compact circuit model implemented in ASTAP.

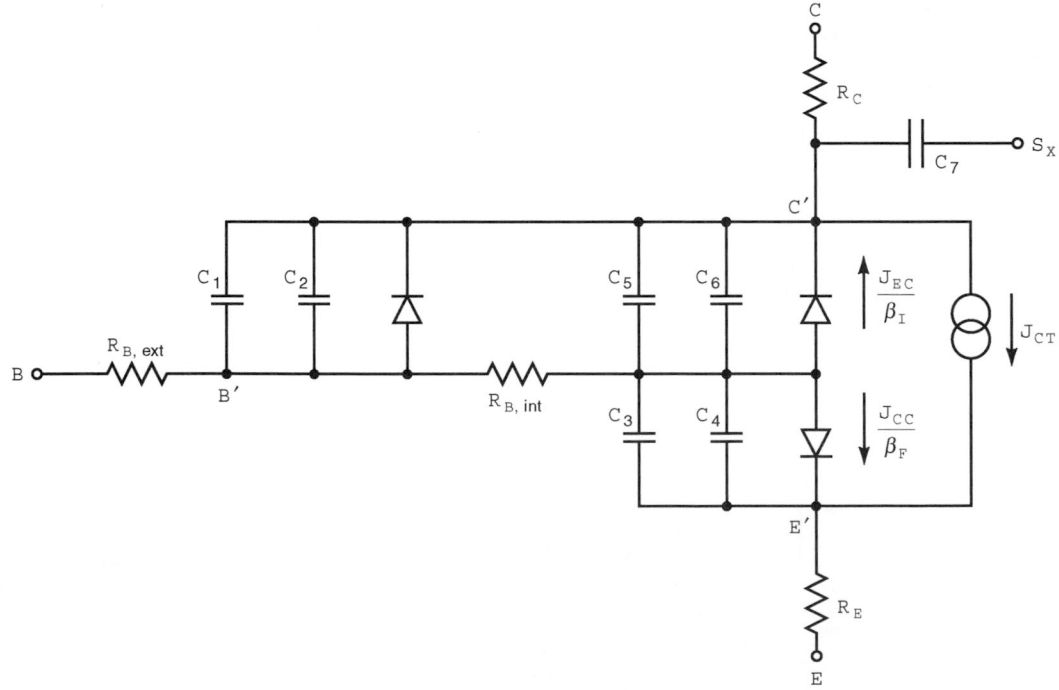

Fig. 17 Compact circuit model used in ASTAP circuit simulations.

1.3 High-Injection Effects

Substantial deviations from ideal behavior occur for BJTs operating at high current densities (as a rule of thumb, for $J_C \sim 1.0$ mA/μm^2 in a modern high-performance technology). This deviation from simple theory can be observed in the premature roll-off of both the current gain and the cutoff frequency at high current densities, as shown in Figs. 10–12. These so-called high-injection effects are particularly important because most high-performance BJT circuits will be biased at high current densities in order to achieve maximum transistor performance. High injection in a BJT can generally be defined as that current density at which the injected minority carrier density (e.g., electrons in the base) becomes comparable to the local doping density. High-injection effects are generally the result of a number of competing physical mechanisms in the collector, base, and emitter regions and are thus difficult to analyze together theoretically. In this work we will simply emphasize the physical origin of each high-injection phenomenon region by region, discuss their impact on device performance, and give some rule-of-thumb design guidelines. The interested reader is referred to Ref. 6 for a more in-depth theoretical discussion.

Collector Region Collector region high-injection effects in BJTs can be divided into two separate phenomena: (a) Kirk effect, sometimes referred to as "base push-out"[24]; and (b) quasi-saturation. The physical origin of the Kirk effect is as follows. As the collector current density continues to rise, the electron density in the base–collector space-charge region is no longer negligible and modifies the electric field distribution in the junction. At sufficiently high current density, the (positive) background space charge

due to the donor doping in the collector ($N+_{dc}$) is compensated by the injected electrons, and the electric field in the junction collapses, thereby "pushing" the original base region deeper into the collector (Figs. 18 and 19). Because both β and f_T depend reciprocally on W_b, this injection-induced increase in effective base width causes a strong degradation in both parameters. Approximate theoretical analysis can be used to determine the critical current density at which the Kirk effect is triggered, resulting in a BJT design equation

$$J_{\text{Kirk}} \cong q v_s N_{\text{dc}} \left(1 + \frac{2\varepsilon V_{BC}}{q W_{epi}^2 N_{\text{dc}}} \right) \qquad (21)$$

From Eq. (21) it is apparent that increasing the collector doping level is the most efficient method of delaying the onset of the Kirk effect, although this will have a detrimental impact on the BV$_{CEO}$ and collector–base capacitance of the transistor. As the Kirk effect is typically the limiting high-injection phenomenon in modern high-performance BJTs, a fundamental trade-off thus exists between peak f_T and BV$_{CEO}$.

The second major collector region high-injection phenomenon is called "quasi-saturation." At a basic level, quasi-saturation is the result of the finite collector resistance of the n-type epi layer separating the base from the heavily doped subcollector in a BJT. At sufficiently high current levels, the infrared (IR) drop associated with the collector epi becomes large enough to internally forward bias the base–collector junction, even though an external reverse bias on the collector is applied. For instance, for a collector resistance of 1 kΩ and a collector current of 2 mA, an internal voltage drop of 2 V is obtained. If the BJT

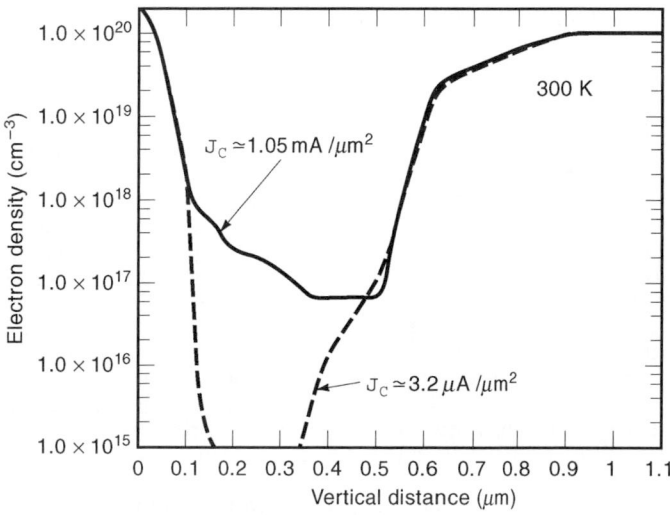

Fig. 18 Simulated electron profile in bipolar transistor at both low injection (3.2 μA/μm^2) and high injection (1.05 mA/μm^2).

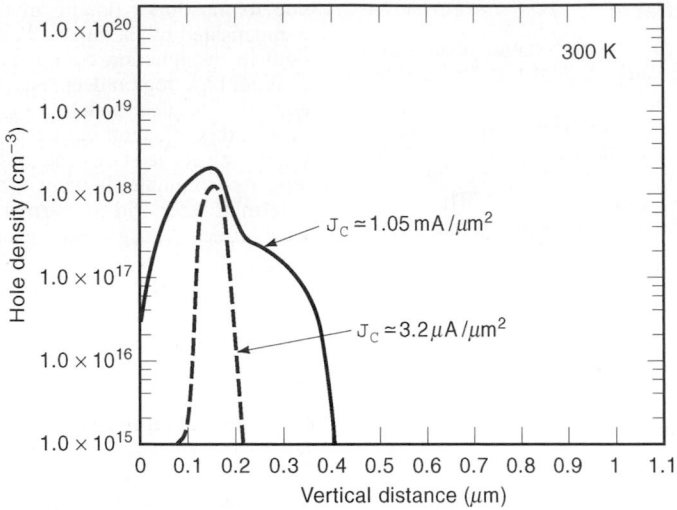

Fig. 19 Simulated hole profile in bipolar transistor at both low injection (3.2 $\mu A/\mu m^2$) and high injection (1.05 mA/μm^2). Observe that at high-injection levels hole profile in base exceeds local doping level (as indicated by low-injection result), and holes are present in (*n*-type) collector region.

were biased at a base–collector reverse voltage of 1 V, then the internal base–collector junction would be forward-biased by 1 V, artificially saturating the transistor. With both base–emitter and base–collector junctions forward biased, the dc signature of quasi-satuation is a strong increase in base current together with a "clipping" of the collector current. Dynamically, quasi-saturation has a strong negative impact on the f_T and, hence, circuit speed because excess minority charge is injected into the base region under saturation. Theoretically, quasi-saturation is difficult to model because the resistance of the epi layer is strongly bias-dependent and the collector doping profile in real devices is highly position dependent. In a well-designed high-performance BJT, the Kirk effect is much more important than quasi-saturation.

Base Region High injection in the base region of a BJT leads to two major degradation mechanisms: (a) the Webster–Rittner effect,[25,26] sometimes known as "base conductivity modulation," and (b) emitter current crowding. In the Webster–Rittner effect, the large electron density in the base region under high injection is no longer small compared to the doping in the base. To maintain charge neutrality in the neutral base, the hole density must therefore rise (refer to Figs. 18 and 19), changing the (low-injection) Shockley boundary condition at the emitter–base junction, and effectively doubling the electron diffusivity in the base. The result is a different voltage dependence of the collector current, which changes to one-half the slope of the exponential low-injection collector current

according to

$$J_C(\text{Webster–Rittner}) \cong \left[\frac{q2D_{nb}n_{ib}(0)}{W_b} \right] e^{qV_{BE}/2kT}$$

(22)

This slope change of J_C has a detrimental impact on the current gain, although in practice for high-performance BJTs, the Kirk effect typically onsets before the Webster–Rittner effect because the base is much more heavily doped than the collector.

Emitter current crowding is the result of the finite lateral resistance associated with the intrinsic base profile (i.e., R_{bi}). Because the collector current depends on the actual base–emitter voltage applied at the junction itself, rather than that applied at the base and emitter terminals, large base currents flowing at high-injection levels can produce a lateral voltage drop across the base. This yields a lateral distribution in the actual base–emitter voltage at the junction, resulting in higher bias at the emitter periphery than in the center of the device. In essence, then, the collector current "crowds" to the emitter edge where the static and dynamic properties of the device are generally worse, and can even produce "thermal runaway" and catastrophic device burnout. This is typically only a problem in large geometry power transistors, not high-speed digital technologies. In addition, as the base current is a factor of β smaller than the collector current, emitter current crowding is not generally a problem unless there is very large base resistance in the device.

Emitter Region Because it is very heavily doped, the emitter region in modern BJTs always operate in

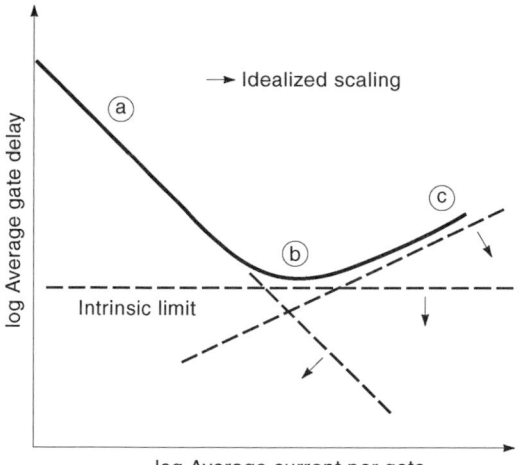

Fig. 20 The ECL power delay characteristics showing impact of idealized scaling.

low injection. Thus, the only significant emitter region high-injection effect is the result of the finite emitter resistance of the transistor. Because polysilicon emitter contacts in fact exhibit reasonably high specific contact resistance (e.g., 20–60 $\Omega \ \mu m^2$), however, emitter resistance (R_E) can be a serious design constraint. Emitter resistance degrades the collector and base currents exponentially as it decreases the applied base–emitter voltage according to

$$I_C = I_{C0}e^{q(V_{BE}-I_E R_E)/kT} \qquad (23)$$

$$I_B = I_{B0}e^{q(V_{BE}-I_E R_E)/kT} \qquad (24)$$

For instance, for a 1.0-μm^2 emitter area transistor operating at a collector current of 1.0 mA, a specific emitter contact resistance of 60 $\Omega \ \mu m^2$ results in an emitter–base voltage loss of 60 mV, yielding a 10× decrease in collector current. Proper process optimization associated with the polysilicon emitter contact is key to obtaining a robust high-speed BJT technology, particularly as the emitter geometry shrinks.

1.4 Scaling Issues

Device miniaturization ("scaling") has been a dominant theme in bipolar technology over the past 15 years, and has produced a monotonic decrease in circuit delay over that period (refer to Fig. 1). In general, optimized BJT scaling requires a coordinated reduction in both lateral and vertical transistor dimensions, as well as a change in circuit operating point.[23] Unlike in CMOS technology, BJT circuit operating voltages (for conventional circuits such as ECL) cannot be scaled because the junction built-in voltage is only weakly dependent on doping. The

evolution of BJT technology from non-self-aligned, double-diffused transistor structures to self-aligned, ion-implanted, double-polysilicon transistor structures was the focus for BJT scaling in the 1980s. During the 1990s more emphasis has been placed on vertical profile scaling and a progression toward both forms of advanced lithography [e.g., deep ultraviolet (UV) or electron beam lithography], low-thermal budget processing, and structural innovation to continue the advances in circuit speed over time.

Figure 20 represents an idealized ECL power delay curve and indicates the three principle regions that require attention during optimized scaling. In region (a), which is dominated by parasitic transistor capacitances [see Eqs. (15)–(17)], a reduction in lithography, and hence decrease in transistor size, is effective in reducing circuit delay at low current levels. Region (b) is dominated by the intrinsic speed of the transistor (i.e., τ_{ec}). Thinning the vertical profile, particularly the base width, is key to reducing the ECL delay at intermediate current levels. The evolution of ion implantation has proven key to realizing viable sub-150-nm metallurgical base widths in modern BJT technologies. In region (c), the ECL delay is dominated by base resistance and high-injection roll-off of the frequency response of the device [Eqs. (18)–(20)]. Doping the base and collector regions more heavily is successful in improving the delay at very high current levels, although tradeoffs exist. For instance, doping the base more heavily decreases the peak f_T of the transistor (due to a lower electron mobility), and, hence, degrades the speed in region (b) at intemediate current levels. In addition, increasing the collector doping level to improve the high-injection performance in region (c) effectively increases the collector–base capacitance, degrading the ECL delay in region (a) at low-current levels. Optimized scaling is thus a complex tradeoff between many different profile design issues.

Clever solutions to certain scaling tradeoffs have emerged over the years and include, for instance, the now pervasive use of the so-called self-aligned, implanted collector (SIC) process. In an SIC process (see Ref. 10), phosphorus is implanted through the emitter window in the base polysilicon layer (either before or after sidewall spacer formation) to increase the collector doping level locally under the intrinsic device without increasing the collector–base capacitance in the extrinsic transistor.

Figure 21 and 22 show the results of a recent BJT lithographic scaling experiment.[14] In this study a comparison was made between BJTs fabricated using three different lithographies (0.09-μm/0.20-μm lithographic linewidth/lithographic overlay, 0.45 μm/0.10 μm, and 0.45 μm/0.06 μm). The latter two processes used advanced electron-beam lithography. As can be seen, the impact of scaling on device parameters is dramatic, resulting in an expected improvement in ECL delay across the entire power delay characteristic, and a minimum ECL gate delay of 20.8 ps (Fig. 16).

Lithographic GR Min image/OL (μm)	0.90/0.20	0.45/0.10	0.45/0.06
A_E (μm^2)	0.7×4.2	0.25×2.0	0.25×2.0
β_{MAX}	80	87	90
R_{BI} (kΩ/□)	15.2	15.2	15.2
ρ_{p+poly} (Ω/□)	176	176	176
BV_{CEO} (V)	3.4	3.4	3.4
BV_{CBO} (V)	11.2	11.3	11.4
R_E (Ω)	14.3	101	93
C_{CBi} (fF/μm^2)	1.3	1.3	1.3
C_{CBx} (fF/μm^2)	0.70	0.70	0.70
C_{EB} (fF/μm^2)	4.64	4.64	4.64
Lumped ASTAP parameters, $I_{CS} = 0.5$ mA			
C_{CB} (fF)	10.2	3.5	2.7 (.2×)
C_{EB} (fF)	13.8	2.7	2.7 (.2×)
C_{CS} (fF)	11.7	7.0	5.6 (.4×)
R_{BX} (Ω)	203	73	66 (.3×)

Fig. 21 Comparison of measured device parameters as function of scaling for: (1) 0.90/0.20 μm (lithographic image/overlay); (2) 0.45/0.10 μm; and (3) 0.45/0.06 μm transistors.[14] Lumped ASTAP parameters are extracted from calibrated simulations of ECL ring oscillator data.

Nonetheless, practical limits do exist for conventional ion-implanted, double-polysilicon BJT technology. Obtaining metallurgical basewidths below 80–100 nm with reasonable base resistance using low-energy ion implantation is very difficult and places a practical limit of about 40–50 GHz on the resultant f_T of such transistors (see Fig. 12, which corresponds to the doping profile shown in Fig. 2). In addition, circuit operating voltages limit the useful BV_{CEO} of the transistor to about 3.0 V, and thus place a practical limit on collector doping levels of about 1×10^{17} cm^{-3} and a consequent maximum operating current density of about 1–2 mA/μm^2. The emitter junction depth (and, hence, the thermal process associated with the polysilicon emitter) is limited to about 25–30 nm because the emitter–base space charge region must lie inside the single-crystal emitter region to avoid the generation/recombination centers associated with the heavily defective polysilicon region. More advanced profiles can be obtained using epitaxial growth techniques, as will be discussed in the next section.

1.5 Future Directions

Despite the continual improvements in speed that BJT technology has enjoyed over the past 15 years, and the inherent superiority of the analog and digital properties of BJTs compared to field-effect transistors (FETs), the world market for BJT integrated circuits (ICs) has steadily eroded. This is due to both the improved performance of FET technology as gate lengths are scaled into the submicron domain, the widespread emergence of CMOS with its low power delay product, and the decreased cost associated with CMOS ICs compared to competing bipolar technologies. To confront this situation, many bipolar + CMOS (BiCMOS) technologies have been developed that seek to combine low-power CMOS with high-performance BJTs. The reader is referred to Ref. 4 for an examination of the process integration issues associated with modern BiCMOS technologies.

In addition, there are several areas of current research with the potential to extend BJT technology well into the twenty-first century; they include: (a) complementary bipolar technology, (b) Silicon-on-insulator (SOI) bipolar technology, and (c) silicon–germanium (SiGe) bipolar technology. Each of these three research areas seeks to improve either the power dissipation associated with conventional BJT circuit families such as ECL, or improve the transistor performance to levels not possible in Si BJTs and thus capture new and emerging IC markets.

Complementary Bipolar Technology Complementary bipolar (C-bipolar) technology, which combines n–p–n and p–n–p transistors on the same chip, has been used for decades. In conventional usage, the n–p–n BJT is a standard, vertical high-performance

(a) 0.9/0.2

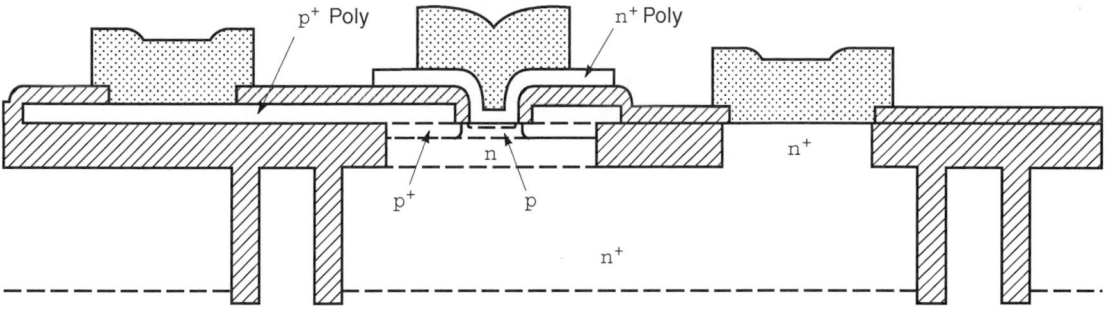

(b) 0.45/0.06

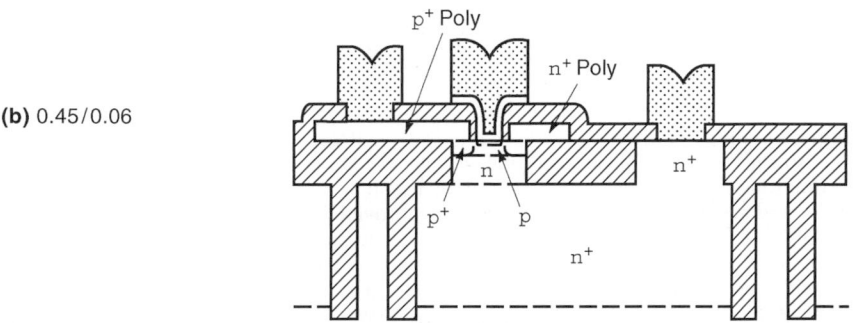

Fig. 22 Scaled comparison of (a) a 0.90/0.20-μm (lithographic image/overlay) transistor with (b) a 0.90/0.06-μm transistor.

transistor, while the $p-n-p$ BJT is typically a slow-speed lateral device used only in analog circuits such as current sources where high speed is unnecessary. Modern implementations of C-bipolar technology, on the other hand, combine a high-performance vertical $n-p-n$ BJT and a high-performance vertical $p-n-p$ BJT (see, e.g., Refs. 27 and 28). The resulting IC technology, though inherently more complex than a traditional $n-p-n$ only BJT technology, opens many new possibilities for novel high-speed, low-power circuit families. New C-bipolar circuit families such as accoupled push–pull emitter-coupled logic (ACPP-ECL) and nonthreshold logic with complementary emitter–follower (NTL-CEF) offer dramatic improvements in power delay product compared to conventional ECL (Fig. 23).

Silicon-on-Insulator Bipolar Technology Silicon-on-insulator IC technologies have existed since the 1960s but have emerged recently as a potential scaling path for advanced CMOS technologies. In SOI technology, a buried oxide dielectric layer is placed below the active Si region, either by ion implantation (SIMOX) or by wafer bonding (BESOI). For the

CMOS implementation, the active Si region is made thin, so that it is fully depleted during normal device operation, resulting in improved subthreshold slope, better leakage properties at elevated temperatures, and improved dynamic performance due primarily to the reduction in parasitic source/drain capacitance. Given this development, it is natural to implement a lateral BJT together with the SOI-CMOS to form an SOI-BiCMOS technology. While lateral BJTs are not generally considered high-speed transistors, the reduction in parasitic capacitance in the lateral BJT, together with clever structural schemes which allow very aggressive base widths to be realized, have resulted in impressive performance.[29]

SiGe Bipolar Technology Attempts to reduce the base widths of modern BJT technologies below 100 nm typically rely on epitaxial growth techniques. A recent high-visibility avenue of research has been the incorporation of small amounts of germanium (Ge) into these epitaxial films to tailor the properties of the BJT selectively while maintaining compatibility with conventional Si fabrication techniques. The resultant

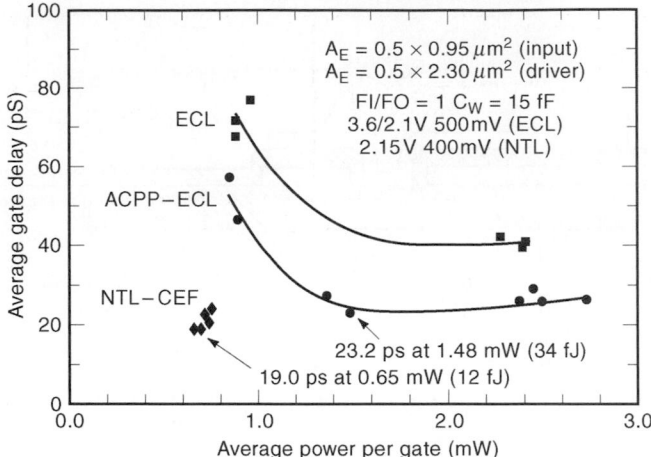

Fig. 23 Measured power delay characteristics from an advanced complementary bipolar technology.[27] Three circuit families are compared: (1) conventional (*npn*-only) ECL, (2) ACPP-ECL, and (3) nonthreshold logic with complementary emitter–(NTL-CEF). The NTL-CEF circuit achieved a minimum power delay product of 12 fJ.

device, called an SiGe heterojunction bipolar transistor (HBT), involves introducing strained epitaxial SiGe alloys into the base region of the transistor, and represents the first practical bandgap-engineered device in Si technology (refer to Refs. 30–32, and references contained within, for reviews of SiGe HBTs).

Compared to an Si BJT with an identical doping profile, the SiGe HBT has significantly enhanced current gain, cutoff frequency, Early voltage (output conductance), and current gain Early voltage product, according to Refs. 31 and 32,

$$\frac{J_{C\text{SiGe}}}{J_{C\text{Si}}} = \frac{\beta_{\text{SiGe}}}{\beta_{\text{Si}}}$$

$$= \gamma\eta\frac{\Delta E_{g,\text{Ge}}(\text{grade})/kT\,e^{\Delta E_{g,\text{Ge}}(0)/kT}}{1 - e^{-\Delta E_{g,\text{Ge}}(\text{grade})/kT}} \tag{25}$$

$$\frac{\tau_{b,\text{SiGe}}}{\tau_{b,\text{Si}}}\alpha\frac{f_{T,\text{Si}}}{f_{T,\text{SiGe}}} = \frac{2}{\eta}\left(\frac{kT}{\Delta E_{g,\text{Ge}}(\text{grade})}\right)$$

$$\left[1 - \frac{1 - e^{-\Delta E_{g,\text{Ge}}(\text{grade})/kT}}{\Delta E_{g,\text{Ge}}(\text{grade})/kT}\right] \tag{26}$$

$$\frac{V_{A,\text{SiGe}}}{V_{A,\text{Si}}} = e^{\Delta E_{g,\text{Ge}}(\text{grade})/kT}\left[\frac{1 - e^{-\Delta E_{g,\text{Ge}}(\text{grade})/kT}}{\Delta E_{g,\text{Ge}}(\text{grade})/kT}\right] \tag{27}$$

$$\frac{\beta V_A|_{\text{SiGe}}}{\beta V_A|_{\text{Si}}} = \gamma\eta e^{\Delta E_{g,\text{Ge}}(0)/kT}e^{\Delta E_{g,\text{Ge}}\text{grade}/kT} \tag{28}$$

where $\Delta E_{g,\text{Ge}}(0)$ is the Ge-induced band offset at the emitter–base junction, $\Delta E_{g,\text{Ge}}(\text{grade}) = \Delta E_{g,\text{Ge}}$

$(W_b) - \Delta E_{g,\text{Ge}}(0)$ is the base bandgap grading factor, and γ, η are the strain-induced density-of-states reduction and mobility enhancement factors, respectively. With its improved transistor performance compared to Si BJTs and compatibility with standard Si fabrication processes, SiGe HBT technology is expected to pose a threat to more costly compound semiconductor technologies such as GaAs for emerging high-speed communications applications. Figure 24 shows a representative SiGe doping profile. Observe that the Ge is introduced only in the base region of the transistor. Experimental results comparing a SiGe HBT and a Si BJT having identical layout and doping profile are shown in Figs. 25 and 26 and indicate that significant enhancements compared to comparably designed Si devices are possible. It is now clear that cutoff frequencies well above 300 GHz are possible using SiGe HBT technology, and thus SiGe represents the next evolutionary step in Si BJT technology.

2 DATA ACQUISITION AND CONVERSION

Kavita Nair, Chris Zillmer, Dennis Polla, and Ramesh Hargani

Data acquisition and conversion pertain to the generation of signals from sensors, their conditioning, and their conversion into a digital format. In this section we describe typical sensors that generate signals and examples of data converter topologies suitable for sensor interfaces. We restrict ourselves to integrated implementations of sensors and sensor interface circuits. In particular, we target sensors and sensor interfaces that are compatible with CMOS fabrication technologies.

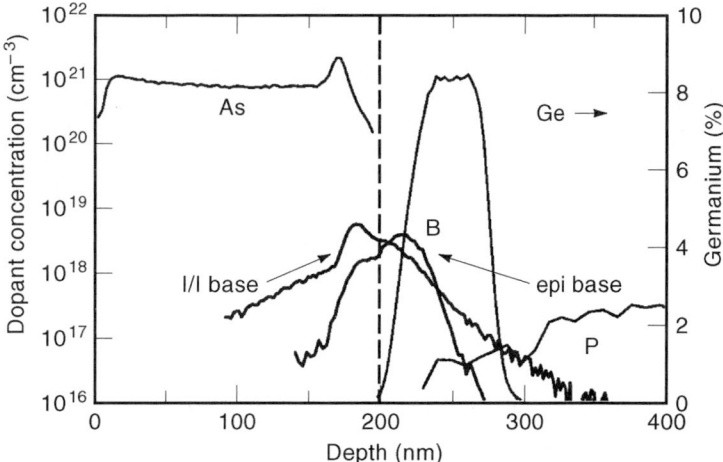

Fig. 24 Measured secondary ion mass spectroscopy (SIMS) doping profile comparing a 60-GHz cutoff frequency epitaxial SiGe base bipolar technology with an aggressive (40-GHz cutoff frequency) ion-implanted (*I/I*) base bipolar technology.

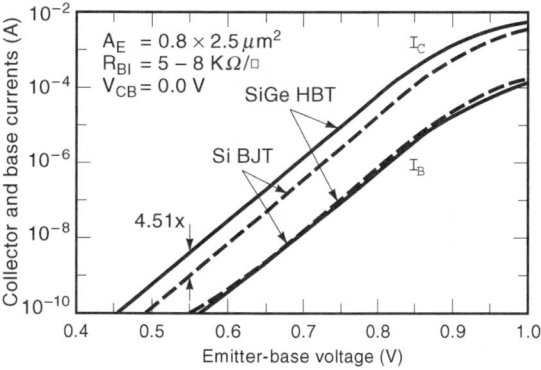

Fig. 25 Measured Gummel characteristics for SiGe and Si transistors with comparable doping profiles. Expected enhancement in collector current (4.51×) can be observed.

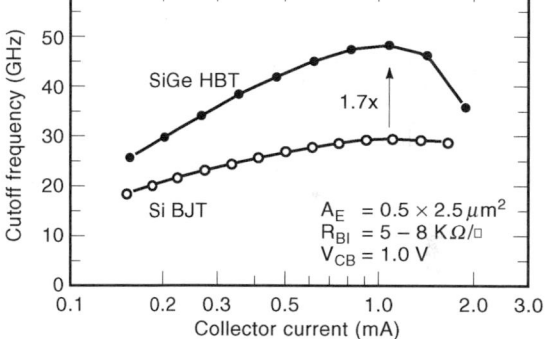

Fig. 26 Measured cutoff frequency as a function of collector current for SiGe and Si transistors with comparable doping profiles. The expected enhancement in collector current (1.71×) can be observed.

This section is organized as follows. First, we describe some examples of sensors and sensor interfaces; then we describe some sample data converter topologies. After that, we provide two complete design examples.

2.1 Sensors

Sensors are devices that respond to a physical or chemical stimulus and generate an output that can be used as a measure of the stimulus. The sensed inputs can be of many types: chemical, mechanical, electrical, magnetic, thermal, and so on. The input signal sensed by the sensor is then processed (amplified, converted from analog to digital, etc.) by some signal conditioning electronics, and the output transducer converts this processed signal into the appropriate output form. The primary purpose of interface electronics is to convert the sensor's signal into a format that is more compatible with the electronic system that controls the sensing system. The electric signals generated by sensors are usually small in amplitude. In addition to this, sensors often exhibit errors, such as offsets, drift, and nonlinearities that can be compensated for with the correct interface circuitry. Analog elements have been improved substantially to achieve high speed and high accuracy; however, for many applications digital is still the preferred format. The sensors yield a wide variety of electric output signals: voltages, currents, resistances, and capacitances. The signal conditioning

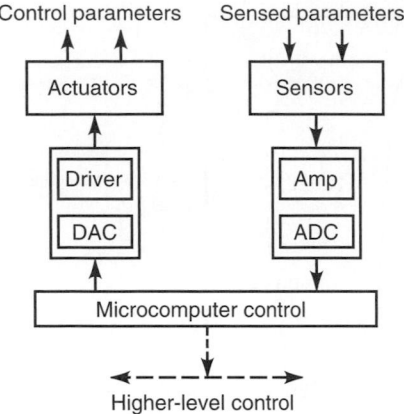

Fig. 27 Overall system architecture of a sensor–actuator control system.

circuitry modifies the input signal into a format suitable for the follow-on data converter.

Figure 27 shows the system architecture for a sensor–actuator-based control system. The sensor(s) senses the external physical and chemical parameters and converts them into an electrical format. The sensed data are processed and digitized using integrated circuitry and transmitted to the host controller. The host uses this information to make the appropriate decisions, and information is fed back to the external environment through a set of actuators.[33] These microprocessor-based controllers have revolutionized the design and use of instrumentation systems by allowing system operation to be defined in software, thus permitting a substantial increase in signal-processing and user–interface features. In general, a power supply is also connected to these blocks but is not explicitly shown in Fig. 27. If a sensor can provide a signal without a power supply, it is referred to as a self-generating sensor.

Integrated sensors are used in many applications, including automotive, manufacturing, environmental monitoring, avionics, and defense. In the past few years, integrated sensors that monolithically combine the sensor structure and some signal-processing interface electronics on the same substrate have begun to emerge. By combining microsensors and circuits, integrated smart sensors increase accuracy, dynamic range, and reliability and at the same time reduce size and cost. Some examples of semiconductor sensors are pressure sensors used in pneumatic systems, magnetic sensors used in position control, temperature sensors used in automotive systems, chemical sensors used in biological diagnostic systems, and acoustic emission sensors used in structural diagnostics.

We now illustrate the use of sensors and sensor interfaces with the two most common types of sensors: resistive and capacitive sensors. We then describe

two complete sensor systems that include an acoustic emission sensor and a temperature sensor.

Resistive Sensors Sensors based on the variation of electric resistance are called *resistive* sensors. They can be further classified according to the physical quantity that they measure: thermal, magnetic, optical, and so on.

A *potentiometer* is a simple resistance measurement device in which the resistance is proportional to its length. However, the linearity of a potentiometer is limited because its resistance is not perfectly uniform. The resistance value also drifts with temperature. Applications of potentiometers are in the measurement of linear or rotary displacements.

Another simple and commonly used resistive sensor is the *strain gauge*, which is based on the variation of the resistance of a conductor or semiconductor when subjected to a mechanical stress. The variation in the resistance of a metal is given by[34]

$$R = R_0(1 + G\varepsilon) \tag{29}$$

where R_0 is the resistance when there is no applied stress, G is the *gauge factor*, and ε is the strain. There are a number of limitations on strain gauges, such as temperature dependence, light dependence, and inaccuracies in the measurement of a nonuniform surface; but in spite of these limitations, they are among the most popular sensors because of their small size and linearity.

Some of the applications of the strain gauge are in measuring force, torque, flow, acceleration, and pressure. Figure 28 shows a micromachined piezoresistive cantilever beam used as a strain gauge sensor. Strain gauges are capable of detecting deformations as small as 10 μm or lower.

A resistance temperature detector (RTD) is a temperature detector based on the variation in electric resistance. An increase in temperature increases the vibrations of atoms around their equilibrium positions,

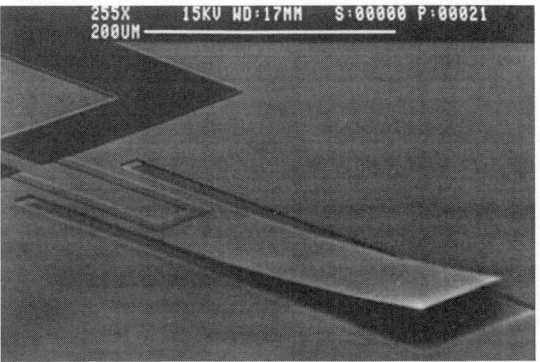

Fig. 28 Micromachined piezoresistive cantilever beam used as strain gauge sensor.

and this increases the resistance in a metal: Thus there is a positive temperature coefficient of resistance. The complete temperature dependence can be expressed[34] as

$$R = R_0(1 + \alpha_1 T + \alpha_2 T^2 + \cdots + \alpha_n T^n) \qquad (30)$$

where T is the temperature difference from the reference and R_0 is the resistance at the reference temperature.

The main advantages of these sensors are their high sensitivity, repeatability, and low cost. There are some limitations too. First, to avoid destruction through self-heating, the RTD cannot measure temperatures near the melting point of the metal. Second, the change in temperature may cause physical deformations in the sensor. Additionally, for each metal there is only a small range over which the RTD is linear. The most common metals used for RTDs are platinum, nickel, and copper.

Thermistors are also temperature-dependent resistors but are made of semiconductors rather than metals. The temperature dependence of the resistance of a semiconductor is due to the variation in the available charge carriers. Semiconductors have a negative temperature coefficient, as the resistance is inversely proportional to the number of charge carriers. The temperature dependence of thermistors is given by[34]

$$R_T = R_0 \ \exp\left[B\left(\frac{1}{T} - \frac{1}{T_0} \right) \right] \qquad (31)$$

where T_0 is the reference temperature, R_0 is the resistance at T_0, and B is the characteristic temperature of the material, which itself is temperature dependent. The limitations and advantages of a thermistor are similar to those of an RTD, except that the thermistor is less stable. There are many types of thermistors available, and each type has its own applications. The foil and bead types are suitable for temperature measurement, whereas the disk and rod types are suitable for temperature control. Some of the applications of thermistors are in the measurement of temperature, flow, level, and time delay. Two simple applications of thermistors are discussed below.

Light-dependent resistors, or LDRs, are devices whose resistance varies as a function of the illumination. LDRs are also known as photoconductors. The conductivity is primarily dependent on the number of carriers in the conduction band of the semiconductor material used. The basic working of the photoconductor is as follows. The valence and conduction bands in a semiconductor are quite close to each other. With increased illumination, electrons are excited from the valence to the conduction band, which increases the conductivity (reduces the resistance). The relation between resistance and optical radiation or illumination is given by[34]

$$R = AE^{-\alpha} \qquad (32)$$

where A and α are process constants, R is the resistance, and E is the illumination.

An important limitation of LDRs is their nonlinearity. Also, their sensitivity is limited by fluctuations caused by changes in temperature. Finally, the spectral response of LDRs is very narrow and primarily depends on the type of material used.

Some of the most common LDRs are made of PbS, CdS, and PbSe. Some applications of LDRs are shutter control in cameras and contrast and brightness control in television receivers.

Measurement Techniques for Resistive Sensors. Various measurement techniques can be used with resistive sensors. The basic requirement for any measurement circuitry is a power supply to convert the change in resistance into a measurable output signal. In addition, it is often necessary to custom-build interface circuits for some sensors. For example, we may be required to add a linearization circuit for thermistors.

Resistance measurements can be made by either the deflection method or the nulling method. In the deflection method the actual current through the resistance or the voltage across the resistance is measured. In the nulling method a bridge is used.

The two-readings method is a fundamental approach to resistance measurement. A known resistance is placed in series with the unknown resistance as shown in Fig. 29. The voltage is then measured across each of them. The two voltages can be written as

$$V_K = \frac{V}{R_K + R_U} R_K \qquad (33)$$

$$V_U = \frac{V}{R_K + R_U} R_U \qquad (34)$$

where V is the supply voltage, V_K and R_K are the known voltage and resistance, and V_U and R_U are the unknown voltage and resistance. Thus from the above equations R_U can be written as follows:

$$R_U = R_K \frac{V_U}{V_K} \qquad (35)$$

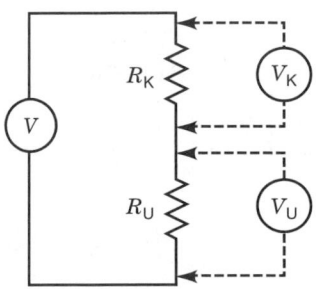

Fig. 29 Two-readings method for resistance measurement.

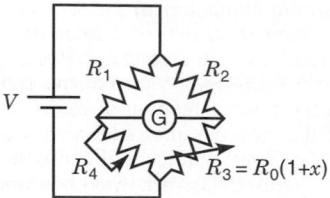

Fig. 30 Simple Wheatstone bridge measurement method.

A similar method is the *voltage divider* in which the unknown resistance is once again calculated from known voltages and resistances. It is easier to resolve small voltage changes for low voltages than it is for high voltages. Thus to measure small changes in resistance, another voltage divider is placed in parallel to the one with the sensor. The parallel voltage dividers are designed to give the same voltage for no input. Thus the signal obtained by taking the difference between their output signals is totally dependent on the measured signal. This method of measuring small changes using parallel voltage dividers is called the *Wheatstone bridge method*.[34,34a,35,36] A simple Wheatstone bridge measurement method is shown in Fig. 30.

The Wheatstone bridge is balanced with the help of a feedback system, which adjusts the value of the standard resistor until the current through the galvanometer is zero. Once this is done, the value for R_3 is given by

$$R_3 = R_4 \frac{R_2}{R_1} \qquad (36)$$

Thus the resistance R_3 is directly proportional to the change required in R_4 in order to balance the circuit.

The Wheatstone bridge can also be used for deflection measurement. In this case, instead of measuring the change needed to balance the bridge, the voltage difference between the bridge outputs is measured or the current through the center arm is measured. This method is shown in Fig. 30. When the bridge is completely balanced (i.e., $x = 0$), k is defined as follows:

$$k = \frac{R_1}{R_4} = \frac{R_2}{R_0} \qquad (37)$$

Thus the voltage difference between the outputs can be written as follows:

$$V_0 = V\left(\frac{R_3}{R_2+R_3} - \frac{R_4}{R_1+R_4}\right) = V\frac{kx}{(k+1)(k+1+x)} \qquad (38)$$

The maximum sensitivity for very small changes in x is obtained when $k = 1$.

Capacitive Sensors Recently capacitive sensors have gained popularity. They generally exhibit lower temperature sensitivity, consume less power, and provide an overall higher sensor sensitivity with higher resolution than resistive sensors. For these reasons they have begun to show up in areas where resistive sensors were the norm. They are used in many applications such as pressure sensors and accelerometers. Capacitive sensors typically have one fixed plate and one moving plate that responds to the applied measurand. The capacitance between two plates separated by a distance d is given by $C = \varepsilon A/d$, where ε is the dielectric constant and A is the area of the plate. It is easily seen that the capacitance is inversely proportional to the distance d.

For capacitive sensors there are several possible interface schemes. Figure 31 shows one of the most common capacitive sensor interfaces. The circuit is simply a charge amplifier, which transfers the difference of the charges on the sensor capacitor C_s and the reference capacitor C_{ref} to the integration capacitor C_I. If this interface is used in a pressure sensor, the sensing capacitor C_s can be written as the sum of the sensor capacitor value C_{s0} at zero pressure and the sensor capacitor variation $\Delta C_s(p)$ with applied pressure: $C_s = C_{s0} + \Delta C_s(p)$. In many applications C_{s0} can be 5 to 10 times larger than the full-scale sensor capacitance variation $\Delta C_s(p)_{max}$; the reference capacitor C_{ref} is used to subtract the nominal value of the sensor capacitor at half the pressure range, which is $C_{ref} = C_{s0} + \Delta C_s(p)_{max}/2$. This ensures that the transferred charge is the charge that results from the change in the capacitance. This results in a smaller integration capacitor and increased sensitivity.

This type of capacitive interface is insensitive to the parasitic capacitance between the positive and negative terminals of the opamp, since the opamp maintains a virtual ground across the two terminals of the parasitic capacitor. This type of interface is also much faster than most other capacitive interfaces; its speed of operation is determined by the opamp's settling time. This technique also allows for the amplifier's offset and flicker noise to be removed very easily by using correlated double sampling or chopper stabilization. The resolution of this interface is in most cases limited by kT/C noise and charge injection due to the switches.

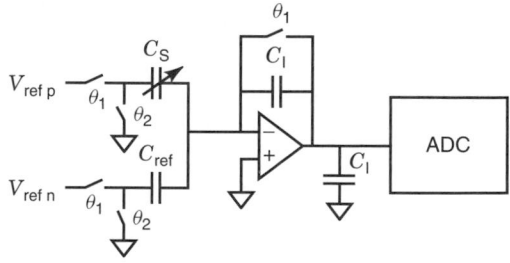

Fig. 31 Capacitive sensor interface.

There are a number of other sensor types, and two more will be discussed later in this section. However, we first describe the most common data converters that are used as part of sensor interfaces.

2.2 Data Converters

The analog signals generated and then conditioned by the signal conditioning circuit are usually converted into digital form via an analog-to-digital converter (ADC). In general, most of the signals generated by these sensors are in the low-frequency region. For this reason, certain data converter topologies are particularly well suited as sensor interface subblocks. These include the charge redistribution implementation of the successive approximation converter, along with incremental and sigma–delta converters.[37] In the following we shall briefly describe successive approximation (incremental) and sigma–delta converters. Incremental and sigma–delta converters are very similar, and the details of the former are later described extensively as part of a sample system design.

Successive Approximation Converter A block diagram for the successive approximation converter is shown in Fig. 32. The successive approximation topology requires N clock cycles to perform an N-bit conversion. For this reason, a sample-and-held (S/H) version of the input signal is provided to the negative input of the comparator. The comparator controls the

digital logic circuit that performs the binary search. This logic circuit is called the successive approximation register (SAR). The output of the SAR is used to drive the digital-to-analog converter (DAC) that is connected to the positive input of the comparator.

During the first clock period, the input is compared with the most significant bit (MSB). For this, the MSB is temporarily raised high. If the output of the comparator remains high, then the input lies somewhere between zero and $V_{ref}/2$ and the MSB is reset to zero. However, if the comparator output is low, then the input signal is somewhere between $V_{ref}/2$ and V_{ref} and the MSB is set high. During the next clock period the MSB-1 bit is evaluated in the same manner. This procedure is repeated so that at the end of N clock periods all N bits have been resolved.

The charge redistribution implementation of the successive approximation methodology is the most common topology in metal–oxide–semiconductor (MOS) technologies.[38] The circuit diagram for a 4-bit charge redistribution converter is shown in Fig. 33. In this circuit the binary weighted capacitors $\{C, C/2, \ldots, C/8\}$ and the switches $\{S_1, S_2, \ldots, S_5\}$ form the 4-bit scaling DAC. For each conversion the circuit operates as a sequence of three phases. During the first phase (sample), switch S_0 is closed and all the other switches $S_1, S_2, \ldots, S_6$ are connected so that the input voltage V_{in} is sampled onto all the capacitors. During the next phase (hold), S_0 is open and the bottom plates of all the capacitors are connected to ground, that is, switches $S_1, S_2, \ldots, S_5$ are switched to ground. The voltage V_x at the top plate of the capacitors at this time is equal to $-V_{in}$, and the total charge in all the capacitors is equal to $-2CV_{in}$. The final phase (redistribution) begins by testing the input voltage against the MSB. This is accomplished by keeping the switches $S_2, S_3, \ldots, S_5$ connected to ground and switching S_1 and S_6 so that the bottom plate of the largest capacitor is connected to V_{ref}. The voltage at the top plate of the capacitor is equal to

$$V_x = \frac{V_{ref}}{2} - V_{in} \qquad (39)$$

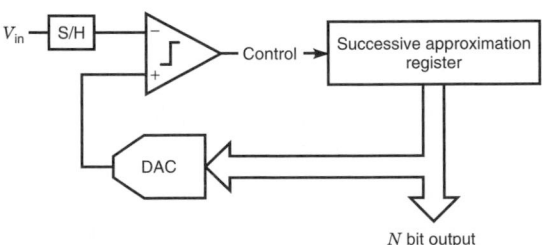

Fig. 32 Successive approximation converter: block diagram.

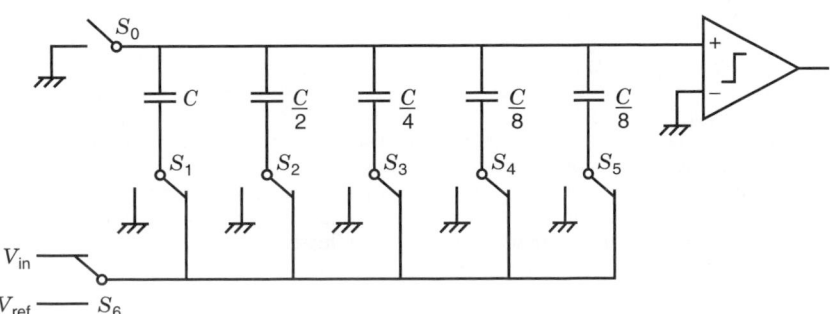

Fig. 33 Charge redistribution implementation of successive approximation architecture.

If $V_x > 0$, then the comparator output goes high, signifying that $V_{in} < V_{ref}/2$, and switch S_1 is switched back to ground. If the comparator output is low, then $V_{in} > V_{ref}/2$, and S_1 is left connected to V_{ref} and the MSB is set high. In a similar fashion the next bit, MSB-1, is evaluated. This procedure is continued until all N bits have been resolved. After the conversion process the voltage at the top plate is such that

$$V_x = -V_{in} + \left(b_3 \frac{V_{ref}}{2^1} + b_2 \frac{V_{ref}}{2^2} + b_1 \frac{V_{ref}}{2^3} + b_0 \frac{V_{ref}}{2^4} \right)$$
(40a)

$$-\frac{V_{ref}}{2^4} < V_x < 0$$
(40b)

where b_i is 0 or 1 depending on whether bit i was set to zero or one, and LSB is the least significant bit.

One of the advantages of the charge redistribution topology is that the parasitic capacitance from the switches has little effect on its accuracy. Additionally, the clock feed-through from switch S_0 only causes an offset, and those from switches $S_1, S_2, \ldots, S_5$ are independent of the input signal because the switches are always connected to either ground or V_{ref}. However, any mismatch in the binary ratios of the capacitors in the array causes nonlinearity, which limits the accuracy to 10 or 12 bits. Self-calibrating[39] techniques have been introduced that correct for errors in the binary ratios of the capacitors in charge redistribution topologies. However, these techniques are fairly complex, and for higher resolutions sigma–delta converters are the preferred topology. We now briefly describe sigma–delta converters.

Sigma–Delta Data Converters Oversampling converters sample the input at a rate larger than the Nyquist frequency. If f_s is the sampling rate, then $f_s/2f_0 = $ OSR is called the *oversampling ratio*. Oversampling converters have the advantage over Nyquist rate converters that they do not require very tight tolerances from the analog components and that they simplify the design of the antialias filter. Sigma–delta converters[40] are oversampling single-bit converters that use frequency shaping of the quantization noise to increase resolution without increasing the matching requirements for the analog components.

Figure 34 shows a block diagram for a general noise-shaping oversampled converter. In a sigma–delta converter both the ADC and DAC shown in Fig. 34 are single-bit versions and as such provide perfect linearity. The ADC, a comparator in the case of a sigma–delta converter, quantizes the output of the loop filter, H_1. The quantization process approximates an analog value by a finite-resolution digital value. This step introduces a quantization error Q_n. Further, if we assume that the quantization noise is not correlated to the input, then the system can be modeled as a linear

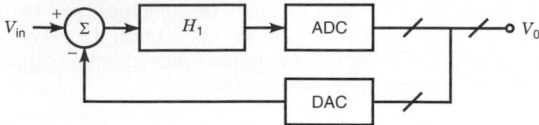

Fig. 34 Figure for general noise-shaping oversampled converter.

system. The output voltage for this system can now be written as

$$V_0 = \frac{Q_n}{1 + H_1} + \frac{V_{in} H_1}{1 + H_1}$$
(41)

For most sigma–delta converters H_1 has the characteristics of a low-pass filter and is usually implemented as a switched-capacitor integrator. For a first-order sigma–delta converter H_1 is realized as a simple switched-capacitor integrator, $H_1 = z^{-1}/(1 - z^{-1})$. Making this substitution in Eq. (41), we can write the transfer function for the first-order sigma–delta converter as

$$V_0 = V_{in} z^{-1} + Q_n (1 - z^{-1})$$
(42)

As can be seen from Eq. (44) below, the output is a delayed version of the input plus the quantization noise multiplied by the factor $1 - z^{-1}$. This function has a high-pass characteristic with the result that the quantization noise is reduced substantially at lower frequencies and increases slightly at higher frequencies. The analog modulator shown in Fig. 34 is followed by a low-pass filter in the digital domain that removes the out-of-band quantization noise. Thus, we are left with only the in-band $(0 < f < f_0)$ quantization noise. For simplicity the quantization noise is usually assumed to be white with a spectral density equal to $e_{rms}\sqrt{2/f_s}$. Further, if the OSR is sufficiently large, then we can approximate the root-mean-square (rms) noise in the signal band by

$$N_{f_0} \approx e_{rms} \frac{\pi}{3} \left(\frac{2f_0}{f_s} \right)^{3/2}$$
(43)

As the oversampling ratio increases, the quantization noise in the signal band decreases; for a doubling of the oversampling ratio the quantization noise drops by $20(\log 2)^{3/2} \approx 9$ dB. Therefore, for each doubling of the oversampling ratio we effectively increase the resolution of the converter by an additional 1.5 bits.

Clearly, H_1 can be replaced by other, higher order functions that have low-pass characteristics. For example, in Fig. 35 we show a second-order modulator. This modulator uses one forward delay integrator and one feedback delay integrator to avoid stability

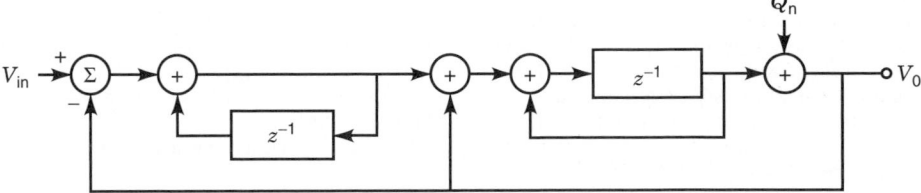

Fig. 35 Modulator for second-order oversampled converter.

problems. The output voltage for this figure can be written as

$$V_0 = V_{in}z^{-1} + Q_n(1 - z^{-1})^2 \qquad (44)$$

The quantization noise is shaped by a second-order difference equation. This serves to further reduce the quantization noise at low frequencies, with the result that the noise power in the signal bandwidth falls by 15 dB for every doubling of the oversampling ratio. Alternatively, the resolution increases by 2.5 bits for every doubling of the oversampling ratio. In general, increasing the order of the filter will reduce the necessary oversampling ratio for a given resolution. However, for stability reasons, topologies other than the simple Candy-style[41] modulator discussed above are required for filter orders greater than two. Topologies that avoid this stability problem include the multistage delta–sigma (MASH) and interpolative topologies.[37]

For low-frequency inputs, the white-noise assumption for the quantization noise breaks down. This results in *tones* that reduce the effective resolution of lower order sigma–delta converters. Incremental converters utilize this observation to simplify the low-pass filter that follows the sigma–delta converter. Details for the incremental converter are discussed below.

We now consider two system design examples. The first is an acoustic emission sensor system and the second is a temperature measurement system.

2.3 System Design Examples

We illustrate the sensor and sensor interface scenario with two examples. The first uses a piezoelectric acoustic emission sensor interfaced with a charge amplifier and a data converter. The second describes an integrated temperature sensor.

Acoustic Emission Sensing System Acoustic emission sensors are microsensors that are used for the detection of acoustic signals. These devices use elastic acoustic waves at high frequencies to measure physical, chemical, and biological quantities. Typically, integrated acoustic sensors can be made to be extremely sensitive and also to have a large dynamic range. The output of these sensors is usually a frequency, a charge, or a voltage.

The piezoelectric effect is one of the most convenient ways to couple elastic waves to electrical circuits. Piezoelectricity is caused by the electric polarization produced by mechanical strain in certain crystals. Conversely, an electric polarization will induce a mechanical strain in piezoelectric crystals. As a consequence, when a voltage is applied to the electrodes of a piezoelectric film, it elongates or contracts depending on the polarity of the field. Conversely, when a mechanical force is applied to the film, a voltage develops across the film. Some properties of a good piezoelectric film are wide frequency range, high elastic compliance, high output voltage, high stability in wet and chemical environments, high dielectric strength, low acoustic impedance, and low fabrication costs. Piezoelectric materials are anisotropic, and hence their electrical and mechanical properties depend on the axis of the applied electric force. The choice of the piezoelectric material depends on the application.

Crystalline quartz (SiO_2) is a natural piezoelectric substance. Some other commonly used piezoelectric materials are ferroelectric single-crystal lithium niobate ($LiNbO_3$) and thin films of ZnO and lead zirconium titanate (PZT). Recently, advances have been made in sensor technology with ultrasonic sensor configurations such as the surface acoustic wave (SAW) and acoustic plate mode (APM). In SAW devices the acoustic waves travel on the solid surface, and in an APM arrangement they bounce off at an acute angle between the bounding planes of a plate. The main types of acoustic wave sensors are shown in Fig. 36.[42]

Piezoelectric thin films are particularly well suited for microsensor applications that require high reliability and superior performance. When prepared under optimal conditions piezoelectric thin films have a dense

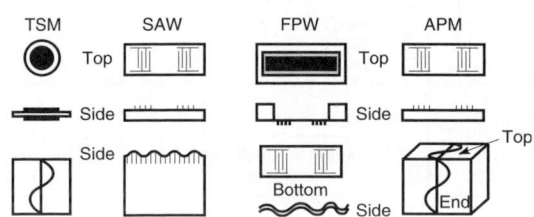

Fig. 36 Types of acoustic wave sensors.

microstructure without cracks and holes, good adherence, and good electrical properties. The three most popular materials used for thin films include ZnO (zinc oxide), AIN (aluminum nitride), and PZT (lead zirconium titalate). Deposition, sputtering, and sol–gel are some of the methods used for preparing piezo films; the choice depends on the material and substrate used. ZnO thin films are prepared using laser-assisted evaporation and are often doped with lithium. Such films have excellent orientation. AIN thin films maintain a high acoustic velocity and are able to withstand extremely high temperatures. PZT thin films have a much higher piezoelectric coefficient than ZnO and AIN.

Recently, it has become possible to generate piezoelectric thin films with extremely good properties through the sol–gel process. This process consists of the following steps: synthesis of a metal–organic solution, deposition of this solution by spin coating, and a final heating that helps to crystallize the ceramic film. A cross-sectional view of a thin-film PZT sensor is shown in Fig. 37. The advantages of thin-film PZT sensors include their small size, which allows them to be positioned virtually anywhere, and their ability to operate at high frequencies.

Measurement Techniques. The different modes of use for an acoustic sensor are summarized in Fig. 38. Using either a resonator-transducer or a delay line, measurements can be made on the device itself or incorporated into an oscillator circuit. There are basically two ways to implement this measurement technique: active or passive. In the case of passive bulk-wave resonators, we measure the resonant frequency to infer the wavelength and hence the velocity. Likewise, for passive delay lines the phase shift between the input and the output of the transducer, which are separated by a known distance, yields the velocity. On the other hand, for active resonators or delay-line oscillators, the frequency can be directly measured with the help of a digital counter.

As an example, let us consider the complete design and implementation of an integrated acoustic emission sensor with low-power signal-conditioning circuitry for the detection of cracks and unusual wear in aircraft and submarines. Within a health and usage monitoring system, it is necessary by some means, either directly or indirectly, to monitor the condition of critical components, for example, airframe, gearboxes, and turbine blades. The overall aim is to replace the current practice of planned maintenance with a regime of required maintenance. Typical parameters used include stress (or strain), pressure, torque, temperature, vibration, and crack detection. In this example, acoustic emission sensors are used for crack detection. The thin-film piezoelectric sensor, coupled to an aircraft component, senses the outgoing ultrasonic waves from any acoustic emission event as shown in Fig. 39. The magnitude of the output signal is proportional to the magnitude of the acoustic emission event. For our example design, the acoustic emission signal bandwidth varies from 50 kHz to approximately 1 MHz. Mixed in with the desired acoustic emission signal is vibration noise due to fretting of the mechanical parts. However, this noise is limited to about 100 kHz and is easily filtered out.

Due to the acoustic emission event, the piezoelectric sensor generates a charge on the top and bottom plates of the sensor. There are two basic methods of interfacing to this sensor. We can use either a voltage amplifier (Fig. 40) or a charge amplifier (Fig. 41).

In general, the charge amplifier interface provides a number of advantages. First, it is not affected by parasitic capacitances at the input of the amplifier. Second, the output voltage at the piezoelectric sensor is very small. This is because the piezoelectric material, PZT, that is used for its high piezoelectric coefficient

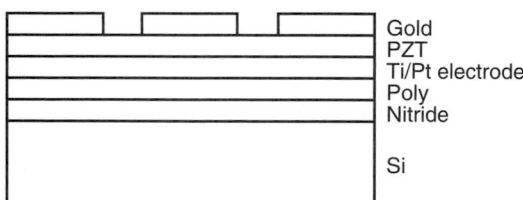

Fig. 37 Cross-sectional view of thin-film PZT sensor.

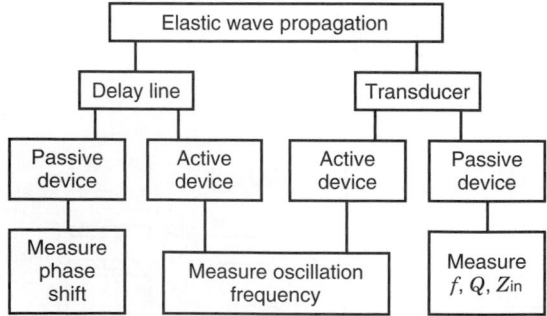

Fig. 38 Different measurement techniques for acoustic sensors.

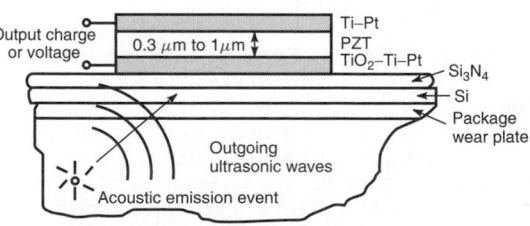

Fig. 39 Acoustic emission sensor.

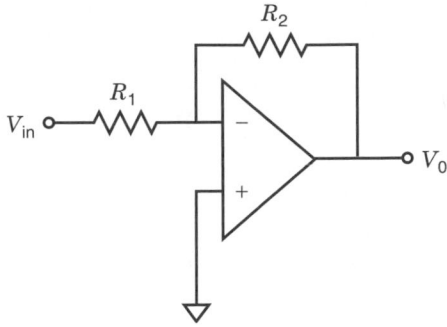

Fig. 40 Voltage amplifier.

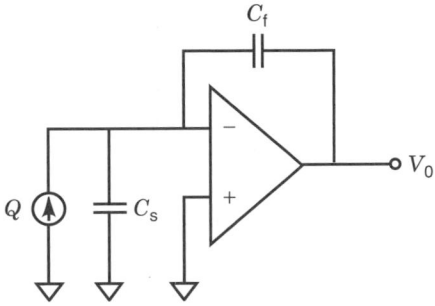

Fig. 41 Charge amplifier.

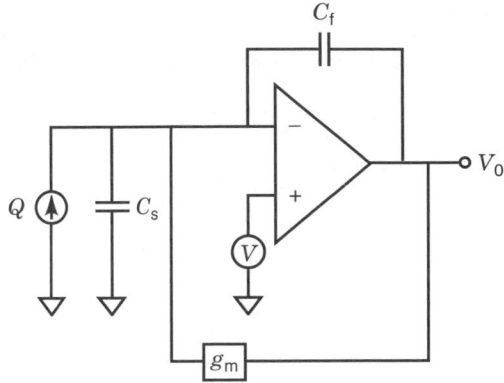

Fig. 42 Modified charge amplifier circuit.

also has a very high dielectric constant. As shown below, the output voltage is proportional to the charge and inversely proportional to the dielectric constant:

$$V = \frac{Q}{C} = \frac{Q}{\varepsilon A/d} = \frac{eSA}{\varepsilon A/d} = \frac{eSd}{\varepsilon} \qquad (45)$$

[The output voltage can also be written in terms of the strain S, the distance d, the electron charge e, and the dielectric constant ε as shown in Eq. (47).] For these and other reasons the charge amplifier interface was selected for our design example.

The charge amplifier circuit shown in Fig. 41 is in its simplest form. The charge Q and capacitance C_s are used to model the sensor charge and sensor capacitance. The inverting terminal of the operational amplifier is a virtual ground, and no charge flows into the operational amplifier inputs. Therefore, any charge that is generated across the sensor has to flow into the feedback capacitance C_f. The output voltage developed across the feedback capacitor is inversely proportional to the value of this capacitance. The voltage gain of the circuit is given by the ratio of C_s to C_f, and hence, to obtain high gain, C_f can be made much smaller than C_s. This basic topology has a number of limitations, including low-frequency flicker

noise of the amplifier, operational amplifier offset, and long-term drift. Traditionally, correlated double sampling and chopper stabilization are used to remove low-frequency noise and offset. However, as noted earlier, our signal band does not include the frequencies from dc to 50 kHz, and our maximum signal frequencies are fairly high. Therefore, an alternative design topology shown in Fig. 42 was selected to circumvent the problem.

Here, low-frequency feedback is provided to reduce the effects of offset, long-term drift, and low-frequency noise. In the modified circuit, a transconductor is connected in negative feedback. The transfer function of the modified circuit is given by

$$\frac{V_0(s)}{Q_{in}(s)} = -\frac{s(g_{ma} - g_m - C_f s)}{C_s C_f s^2 + s(C_s g_m + g_{m_a} C_f) + g_{m_a} g_m} \qquad (46)$$

In this equation, C_s is the sensor capacitance, C_f is the feedback capacitance of the operational amplifier, g_{m_a} and g_m are the transconductances of the operational amplifier and the transconductor. If the higher order terms are neglected, then Eq. (46) can be simplified to

$$\frac{V_0(s)}{Q_{in}(s)} = -\left(\frac{s}{g_m}\right)\left(\frac{1}{1 + \frac{C_s s}{g_{m_a}}}\right) \qquad (47)$$

From Eq. (47) it is clear that the circuit has the characteristics of a high-pass filter, that is, none of the low-frequency noise or offsets affect the circuit performance.

Next, we perform a power analysis to analyze the effects of different design tradeoffs. Both MOS and bipolar transistor technologies are considered, and power and noise analysis and design tradeoffs for both technologies are presented.

Power Analysis. If MOS transistors in strong inversion (SI) are used to implement the operational amplifier, then the minimum power requirement is given by

$$P = VI = \frac{V(2\pi BWC)^2}{2K(W/L)} \quad (48)$$

where BW is the signal bandwidth, C is the sensor capacitance, K is the transconductance factor, V is the output voltage, I is the supply current, and W/L is the aspect ratio of the transistor. From this equation it is clear that the power is proportional to the square of the signal bandwidth and sensor capacitance.

If, however, bipolar transistors are used to implement the operational amplifier, the minimum power requirement is given by

$$P = VI = V2\pi \text{ BW } U_T C \quad (49)$$

Here, U_T is the thermal voltage, which is equal to 26 mV at room temperature. From this equation it is clear that in the case of bipolar transistors, the power is linearly proportional to the signal bandwidth and sensor capacitance. This difference in the power consumption between bipolar and MOS implementations for a signal frequency of 1 MHz is shown in Fig. 43. Here we note that the power consumption for both MOS and bipolar implementations increases with increased sensor capacitance. However, for very low frequencies, the MOS devices can be operated in weak inversion (WI). In WI, MOS devices behave very similarly to bipolar devices, and hence the slopes for weak inversion and bipolar devices are initially very similar. However, at higher frequencies MOS devices are forced to operate in strong inversion and hence consume more power for the same performance.

Next, we consider the design tradeoffs in connection with device noise.

Noise Analysis. The power spectral density for the wideband gate-referred noise voltage for MOS transistors is given by

$$V_{nT}^2 = \frac{8}{3}\frac{kT}{g_m} \quad (50)$$

Here, k is Boltzmann's constant, T is the temperature, g_m is the transconductance. Likewise, for bipolar transistors the power spectral density for the wide-band input-referred noise voltage is given by

$$V_{nT}^2 = 2q I_C \quad (51)$$

For both MOS and bipolar implementations the total rms input-referred noise is independent of frequency and inversely proportional to the sensor capacitance as shown in Fig. 44. Here, we note that the ratio of the noise spectral density for the MOS and the bipolar implementations is a constant equal to 4.

In summary we note that: For an MOS implementation the power consumption is proportional to the square of the sensor capacitance, whereas for a bipolar implementation it is linearly proportional to the sensor capacitance. On the other hand, the input-referred noise for both the MOS and bipolar implementations is inversely proportional to the sensor capacitance. Thus, there is a clear tradeoff between the minimum power consumption and the maximum input-referred noise. If the sensor capacitance is increased, then the input-referred noise decreases, but the power increases, and vice versa. Using the equation above, we can calculate the minimum bound on the power requirements for our application. For 10 bits of accuracy and a signal bandwidth of 1 MHz, the minimum sensor capacitance size is 5 pF and the minimum power consumption is around 500 μW.

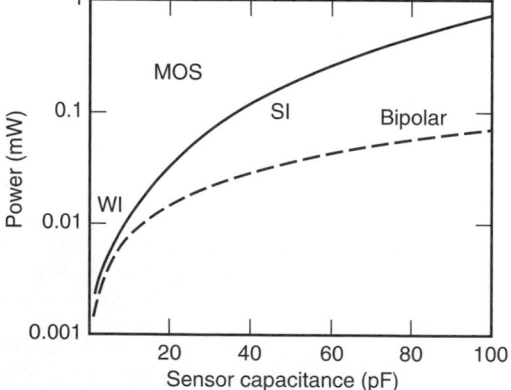

Fig. 43 Minimum power requirements versus sensor capacitance for a MOS or bipolar design.

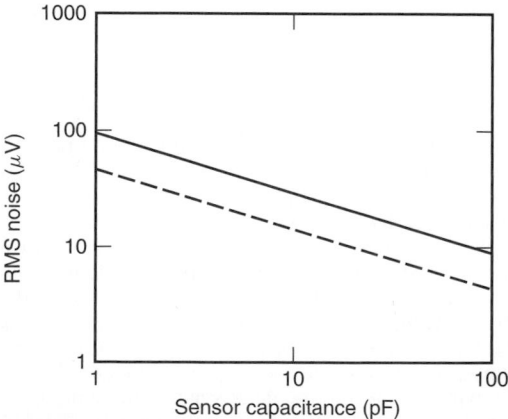

Fig. 44 Noise power spectral density versus capacitance.

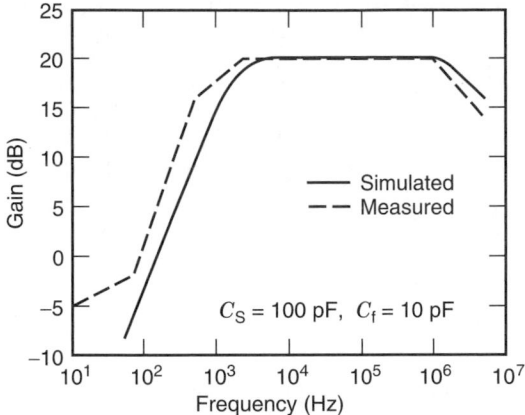

Fig. 45 Small-signal frequency response of charge amplifier.

Next, we provide some simulation and measurement results for our acoustic emission sensor system.

Results. Simulation and measurement results for the charge amplifier with a sensor capacitance of 100 pF and a feedback capacitance of 10 pF are shown in Fig. 45. For this measurement, discrete versions of the sensor and feedback capacitors were used. As expected, the signal band gain is given by the ratio of the sensor to the feedback capacitance, which is equal to 20 dB. Both measurement and simulation results agree fairly well with this value. The primary difference between the measurement and simulation results is in the low-frequency and high-frequency poles. It is expected that this is largely a result of parasitic capacitances and possibly a lower realized transconductance in comparison with the simulated value.

The charge amplifier circuit design just described converts the sensor charge into a voltage. This amplified signal voltage is then converted to digital form using an ADC. For our implementation a 10-bit fourth-order sigma–delta implemented as a MASH topology was used. The fourth-order topology was used to keep the oversampling ratio low, as the signal frequency is fairly high. Details of this implementation are not included here; interested readers are referred to Ref. 37 for more information.

Next, we describe a complete temperature sensor system.

Temperature-Sensing System In many control systems, temperature sensors are used as the primary sensor. Additionally, as most electronic components and circuits are affected by temperature fluctuations, temperature sensors are often needed in microsensor systems to compensate for the temperature variations of the primary sensor or sensors.

Because integrated sensors can be manufactured on the same substrate as the signal-processing circuitry, most recent temperature measurement schemes concentrate on integrated silicon temperature sensors. The resulting smart sensor is extremely small and is also able to provide extremely high performance, as all the signal processing is done on chip before the data is transmitted. This avoids the usual signal corruption that results from data transmission. The disadvantage of the smart sensor is that since all the processing is done on chip, it is no longer possible to maintain the signal preprocessing circuits in an isothermal environment. The on-chip sensor interface electronics must therefore be temperature insensitive or be compensated to provide a temperature-insensitive output.

A *smart* temperature sensor is a system that combines on the same chip all the functions needed for measurement and conversion into a digital output signal. A smart temperature sensor includes a temperature sensor, a voltage reference, an ADC, control circuitry, and calibration capabilities. A block diagram for a smart temperature sensor is shown in Fig. 46. The use of $p-n$ junctions as temperature sensors and for the generation of the reference voltage signals has been reported extensively.[43,44] A bandgap voltage reference can be generated with the help of a few $p-n$ junctions. The basic principle for the operation of a bandgap voltage reference is illustrated in Fig. 47.

The base–emitter voltage V_{be} of a bipolar transistor decreases almost linearly with increasing temperature. The temperature coefficient varies with the applied current, but is approximately -2 mV/°C. It is also well known that the difference between the base–emitter voltages of two transistors, ΔV_{be}, operated at a constant ratio of their emitter current densities, possesses a positive temperature coefficient. At an emitter

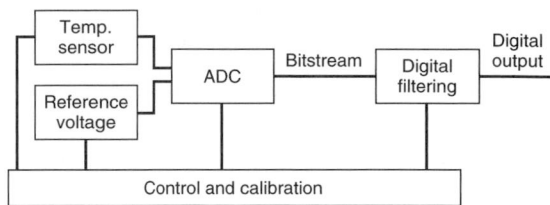

Fig. 46 Smart temperature sensor.

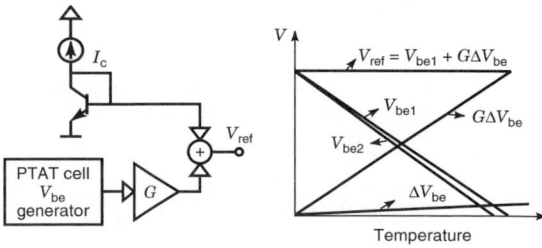

Fig. 47 Principle of bandgap reference.

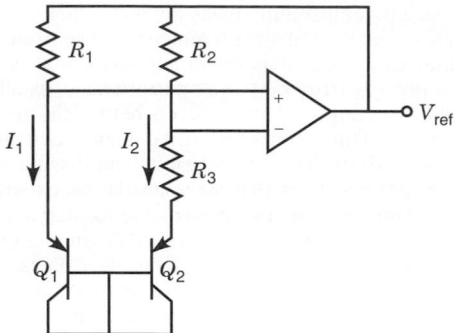

Fig. 48 Example bandgap voltage reference circuit.

current density ratio of 8, the temperature coefficient of this PTAT (proportional to absolute temperature) source is approximately 0.2 mV/°C. Amplifying this voltage ($G\Delta V_{be}$) and adding it to a base–emitter voltage V_{be} produces a voltage reference that is independent of temperature. Many circuits have been developed to realize bandgap voltage references using this principle.[45,46] A circuit diagram for one of the early bandgap reference implementations is shown in Fig. 48.[47]

For an ideal operational amplifier, the differential input voltage is equal to zero, so that resistors R_1 and R_2 have equal voltages across them. Since the voltage across the resistors is the same, the two currents I_1 and I_2 must have a ratio that is determined solely by the ratio of the resistances R_1 and R_2. The base–emitter voltage of a diode-connected bipolar transistor is given by Eq. (52), where T is the absolute temperature of the junction, I_s is the reverse saturation current, I_d is the current through the junction, k is Boltzmann's constant, q is the electronic charge, and n is a constant that depends on the junction material and fabrication technique. To see this, we write

$$V_{be} = \frac{nkT}{q} \ln \frac{I_d + I_s}{I_s} \approx \frac{nkT}{q} \ln \frac{I_d}{I_s} \qquad (52)$$

Therefore, the difference between the two base–emitter voltages (ΔV_{be}) is given by

$$\Delta V_{be} = V_{be1} - V_{be2} = \frac{nkT}{q} \ln \frac{I_1 I_{s2}}{I_2 I_{s1}} = \frac{nkT}{q} \ln \frac{R_2 I_{s2}}{R_1 I_{s1}} \qquad (53)$$

This voltage appears across R_3. Since the same current that flows through R_3 also flows through R_2, the voltage across R_2 is given by

$$V_{R_2} = \frac{R_2}{R_3} \Delta V_{be} = \frac{R_2}{R_3} \frac{nkT}{q} \ln \frac{R_2 I_{s2}}{R_1 I_{s1}} \qquad (54)$$

as desired.

The output voltage is the sum of the voltage across R_1 and the voltage across Q_1. Since the voltage across R_1 is equal to the voltage across R_2, the output voltage is equal to

$$V_{out} = V_{be1} + \frac{R_2}{R_3} \frac{nkT}{q} \ln \frac{R_2 I_{s2}}{R_1 I_{s1}} = V_{be1} + G\Delta V_{be} \qquad (55)$$

Therefore, this circuit behaves as a bandgap reference, where the gain factor G is set by the ratios R_2/R_3, R_2/R_1, and I_{s2}/I_{s1}. In many designs $R_2 = R_1$ and $I_{s2} = 8I_{s1}$. Since the reverse saturation current I_s is proportional to the emitter area, to make $I_{s2} = 8I_{s1}$ we let the emitter area of Q_2 be 8 times as large as the emitter area of Q_1.

The operational amplifier's input-referred voltage offset is the largest error source in this type of voltage reference. This voltage offset is highly temperature dependent and nonlinear, making an accurate calibration of such a reference virtually impossible. It is therefore necessary to use some type of offset cancellation technique such as autozero or chopper stabilization.[48]

Another source of error is the nonzero temperature coefficient of the resistors. Usually, on-chip resistors are used in the form of polysilicon resistors or well resistors. Both of these resistor implementations tend to occupy very large amounts of chip area if low power is desired. Low-power implementations demand the use of large-value resistors, which unfortunately require large areas. Though well resistors have a much larger resistivity than polysilicon resistors, they also have a very nonlinear temperature coefficient, which makes for difficult calibration.

A solution to these problems is to use switched-capacitor circuits to implement the resistors in the voltage reference circuit. A switched-capacitor implementation makes offset removal simple and also reduces the power consumption, as the area occupied by large-value switched-capacitor resistors is significantly smaller than the area occupied by continuous-time resistors. In fact, the area occupied by switched-capacitor resistors is inversely proportional to the value of the resistance desired. Another advantage is that the temperature coefficient of on-chip poly–poly capacitors is much smaller than that of on-chip resistors, making design and calibration easier. A switched-capacitor implementation of the bandgap voltage reference is shown in Fig. 49.

The structure of this voltage reference is similar to the one shown in Fig. 48, except that the continuous time resistors have been replaced by switched-capacitor resistors, and capacitors C_T and C_F have been added. The switched capacitors emulate resistors with an effective resistance value given by

$$R_{eff} = \frac{1}{f_C C} \qquad (56)$$

where f_C is the clock frequency of the switch. The feedback capacitor C_F is designed to be very small

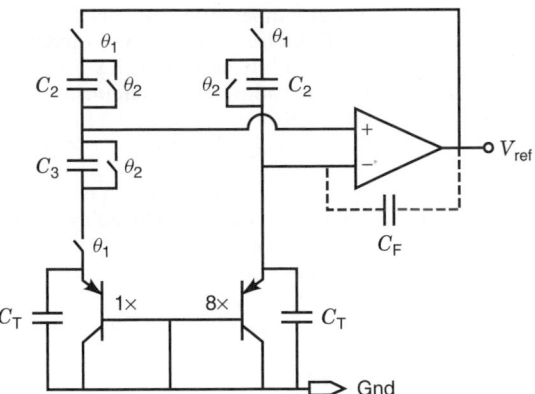

Fig. 49 Switched-capacitor implementation of bandgap reference.

and is added to ensure the operational amplifier is never in an open-loop mode of operation. The capacitors located in parallel with the diodes act as tank capacitors to ensure that current is constantly supplied to the diodes. The output of this voltage reference can similarly be calculated and is given by

$$V_{\text{ref}} = V_{be1} + \frac{C_3}{C_2} \frac{nkT}{q} \ln \frac{C_1 I_{s2}}{C_2 I_{s1}} = V_{be1} + G\Delta V_{be} \quad (57)$$

which is the desired bandgap voltage reference.

Most temperature-sensing devices also use the difference between two diodes (ΔV_{be}) as the sensing element of the system. Since the temperature coefficient of ΔV_{be} is small (≈ 0.2 mV/$^{\circ}$C), it is almost always amplified to a much larger value (≈ 10 mV/$^{\circ}$C) for increased sensitivity. Since we already have an amplified value of ΔV_{be} in the voltage reference ($G\Delta V_{be}$), all that needs to be done is to subtract V_{be1} from the voltage reference to obtain an amplified value of ΔV_{be}.

If more sensitivity is needed, the additional amplification can be incorporated in the ADC by simply adjusting the capacitor ratio of C_A and C_B as shown in Fig. 50. Additionally, the subtraction of V_{be1} from the voltage reference can be easily accomplished with the circuit shown in Fig. 51, where V_{in1} is the output of the voltage reference, V_{in2} is equal to V_{be1}, and V_G is the negative input of the operational amplifier in the follow-on data converter. During clock cycle θ_1 the capacitor C is charged to the input voltage V_{in2}. During clock cycle θ_2, the charge $(V_{in1} - V_{in2})/C$ is transferred. This circuit effectively does the voltage subtraction that is needed to obtain the amplified temperature-dependent output voltage ($G\Delta V_{be}$).

Incorporating the voltage reference and temperature-sensing circuitry shown in Figs. 49 and 51 into a smart temperature sensor system involves some additional circuitry. Since switched capacitors are already being used for the voltage reference and the sensing circuitry, it makes sense to use switched-capacitor technology for the ADC. A simple ADC that utilizes oversampling techniques is the incremental converter.[49] The advantage of this data converter topology, shown in Fig. 50, is its low power consumption, small area, and insensitivity to component mismatch. Additionally, in comparison with sigma–delta converters the postquantization digital low-pass filter is much simpler. It consists of just an up–down counter instead of a more complicated decimation filter. Unfortunately, the first-order incremental ADC has a relatively long conversion time, making this converter suitable only for very slow signals such as temperature.

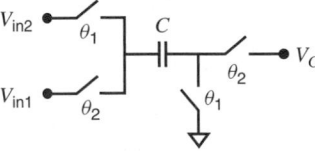

Fig. 51 Switched-capacitor subtraction circuit.

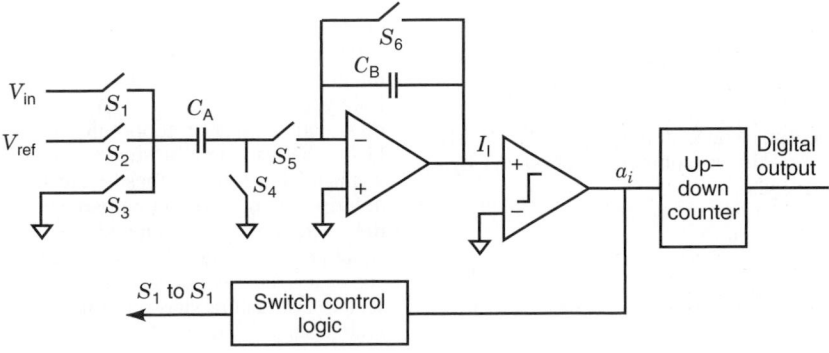

Fig. 50 Incremental ADC.

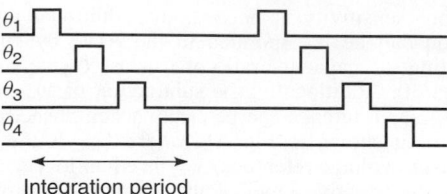

Fig. 52 Four-phase nonoverlapping clock.

The first-order incremental ADC shown in Fig. 50 is composed of a stray-insensitive switched-capacitor integrator, a comparator, switch control logic, and an up−down counter. A four-phase nonoverlapping clock as shown in Fig. 52 constitutes an integration period. The integrator output voltage is designated by $V_I[i, j]$, where i corresponds to the current integration period and j to the clock cycle (1, 2, 3, or 4).

During clock cycle θ_1, S_1 and S_4 are closed, charging C_A to the input voltage V_{in}. During θ_2, S_3 and S_5 are closed, transferring the charge that was stored on C_A to C_B. At the end of the charge transfer from C_A to C_B the comparator output is denoted by

$$a_i = \begin{cases} 1 & \text{if} \quad V_I[i, 2] > 0 \\ -1 & \text{if} \quad V_I[i, 2] < 0 \end{cases}$$

During θ_3, S_4 is closed, and if:

$$a_i = \begin{cases} 1 & S_3 \text{ is closed} \\ -1 & S_2 \text{ is closed} \end{cases}$$

During θ_4, S_5 is closed, and if:

$$a_i = \begin{cases} 1 & S_2 \text{ is closed} \\ -1 & S_3 \text{ is closed} \end{cases}$$

Also during θ_4, the integrator output voltage $V_I[i, 4]$ is given by

$$V_I[i, 4] = V_I[i, 1] + \frac{C_A}{C_B}(V_{in} - a_i V_{ref}) \qquad (58)$$

The final N-bit output code, denoted by D_{out}, that results from the up−down counter is obtained by evaluating the quantity

$$D_{out} = \frac{1}{n} \sum_{i=1}^{n} a_i \qquad (59)$$

Here n is the number of integration periods, and is a function of the resolution that is required of the ADC.

The complete smart temperature sensor is shown in Fig. 53. The subtraction circuit of Fig. 51 is incorporated into the ADC by simply adding switch S_{sub}. The only difference in the operation of the incremental converter shown in Fig. 53 from the one shown in Fig. 50 is that now during θ_2, S_3 is not closed but instead S_{sub} is closed.

The calibration of this system is done in two steps. First the voltage reference is calibrated by adjusting the ratio of C_3 and C_2; next the amplified sensor voltage is calibrated by adjusting the ratio of C_A and C_B. Adjusting the ratios of the capacitors is done with the use of a capacitor array that is controlled digitally. The output is an N-bit digital word.

In Fig. 54 we show measurement results for the voltage reference and final temperature output. For these results a first-pass design of the circuit in Fig. 53 was used. This design was not completely integrated and included external resistors to obtain gain. We expect final integrated results to behave similarly. Figure 54a shows the reference voltage obtained as a sum of a V_{be} and an amplified ΔV_{be} as described in Eq. (57). The x axis shows the temperature in kelvin and the y axis shows the measured output reference voltage in volts. The measured value is fairly close to the expected value except for some small experimental variations. We suspect these variations are a result of the length of time used to stabilize the temperature between temperature output measurements. The graph in Fig. 54b shows the output voltage, which is $V_{ref} - V_{be}$. As expected, this voltage varies linearly with temperature. Figure 55 shows the expected 1-bit output stream (α_i shown in Fig. 54) of the sigma−delta converter before the digital low-pass filter. This output corresponds to an input voltage equal to one-eighth of the reference voltage.

We have provided detailed designs for two complete data acquisition systems, namely an acoustic emission sensor system and a smart temperature sensor system. We provide both measurement and simulation results to show their performance.

2.4 Conclusion

In this section we have provided brief descriptions of data acquisition and data conversion systems. In particular, we provided some general descriptions of integrated capacitive and resistive sensors. This was followed by descriptions of two of the most common data converter topologies used in sensor interface systems, namely successive approximation and sigma−delta. Finally, these were followed by detailed descriptions of two complete acquisition systems. The first system was based on a piezoelectric acoustic emission sensor interfaced to a charge amplifier and data converter. The second system was a smart temperature sensor. As feature sizes continue to decrease and integrated sensor technologies progress, it is likely that extremely smart and high-performance systems will be integrated on single chips. Additionally, significant

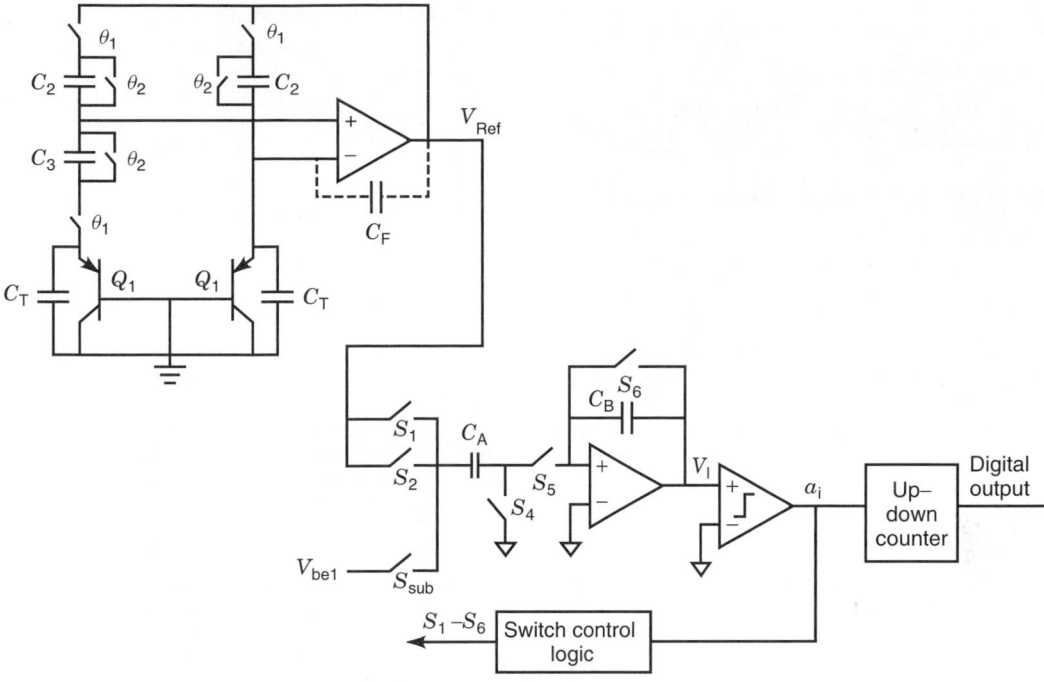

Fig. 53 Smart temperature sensor circuit.

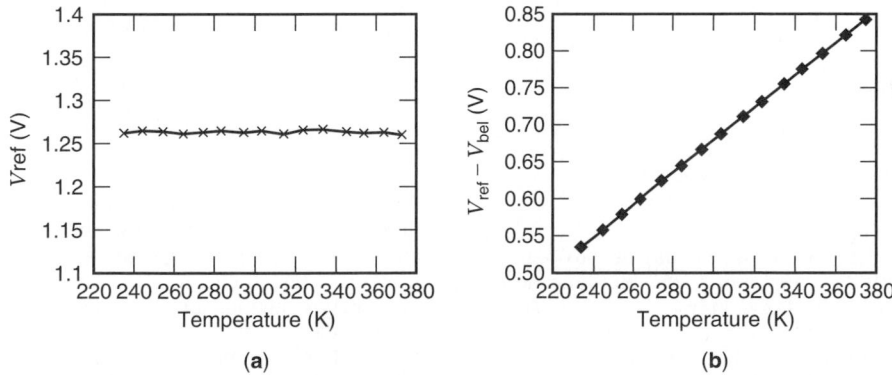

Fig. 54 Measurement results for the (a) voltage reference and (b) temperature sensor.

reduction in power and area as a result of smaller feature sizes will make such systems ubiquitous.

3 DATA ANALYSIS

Arbee L. P. Chen and Yi-Hung Wu

What is data analysis? Nolan[50] gives a definition that is a way of making sense of the patterns that are in, or can be imposed on, sets of figures. In concrete terms, data analysis consists of an observation and an investigation of the given data, and the derivation of characteristics from the data. Such characteristics, or features as they are sometimes called, contribute to the insight of the nature of data. Mathematically, the features can be regarded as some variables, and the data are modeled as a realization of these variables with some appropriate sets of values. In traditional data analysis,[51] the values of the variables are usually numerical and may be transformed into symbolic representation. There are two general types of variables: discrete and continuous.

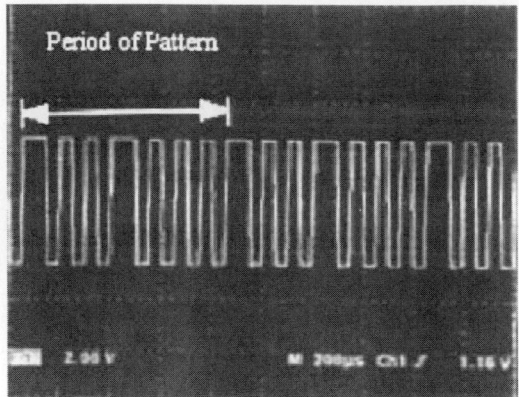

Fig. 55 Measurement results for analog-to-digital converter.

Discrete variables vary in units, such as the number of words in a document or the population in a region. In contrast, continuous variables can vary in less than a unit to a certain degree of accuracy. The stock price and the height of people are examples of this type. The suitable method for collecting values of discrete variables is counting, and for continuous ones it is measurement.

The task of data analysis is required among various application fields, such as agriculture, biology, economics, government, industry, medicine, military, psychology, and science. The source data provided for different purposes may be in various forms, such as text, image, or wave form. There are several basic types of purposes for data analysis:

1. Obtain the implicit structure of data
2. Derive the classification of data
3. Search particular objects in data

For example, the stockbroker would like to get the future trend of the stock price, the biologist needs to divide animals into taxonomies, and the physician tries to find the related symptoms of a given disease. The techniques to accomplish these purposes are generally drawn from statistics that provide well-defined mathematical models and probability laws. In addition, some theories, such as fuzzy-set theory, are also useful for data analysis in particular. This section is an attempt to give a brief description of these techniques and concepts of data analysis. In Section 3.1, a variety of data analysis methods are introduced and illustrated by examples. We first give two categories of data analysis according to its initial conditions and resultant uses. Next, we show two well-known methods based on different mathematical models. In Section 3.2, an approach to data analysis for Internet applications is proposed. Some improvements of the data analysis methods are discussed in Section 3.3. Finally, we give a brief summary.

3.1 Data Analysis Methods

In data analysis, the goals are to find significant patterns in the data and apply this knowledge to some applications. Analysis is generally performed in the following stages:

1. Feature selection
2. Data classification
3. Conclusion evaluation

The first stage consists of the selection of the features in the data according to some criteria. For instance, features of people may include their height, skin color, and fingerprints. Considering the effectiveness of human recognition, the fingerprint, which is the least ambiguous, may get the highest priority for selection. In the second stage, the data are classified according to the selected features. If the data consist of at least two features, for example, the height and the weight of people, which can be plotted in a suitable coordinate system, we can inspect so-called scatter plots and detect clusters or contours for data grouping. Furthermore, we can investigate ways to express data similarity. In the final stage, the conclusions drawn from the data would be compared with the actual demands. A set of mathematical models has been developed for this evaluation. In the following sections, we first divide the study of data analysis into two categories according to different initial conditions and resultant uses. Then, we introduce two famous models for data analysis. Each method will be discussed first, followed by examples. Because the feature selection depends on the actual representations of data, we postpone the discussion about this stage until the next section. In this section, we focus on the classification procedure based on the given features.

Categorization of Data Analysis There are a variety of ways to categorize the methods of data analysis. According to the initial conditions and the resultant uses, there are two categories, supervised data analysis and unsupervised data analysis. The term *supervised* means that human knowledge has to be provided for the process. In supervised data analysis, we specify a set of classes called a *classification template* and select some samples from the data for each class. These samples are then labeled by the names of the associated classes. Based on this initial condition, we can automatically classify the other data termed *to-be-classified* data. In *unsupervised* data analysis, there is no classification template, and the resultant classes depend on the samples. Following are descriptions of supervised and unsupervised data analysis with an emphasis on their differences.

Supervised Data Analysis. The classification template and the well-chosen samples are given as an initial state and contribute to the high accuracy of data classification. Consider the K nearest-neighbor classifier, which is a typical example of supervised data analysis. The input to the classifier includes a set of labeled samples S, a constant value K, and a to-be-classified datum X. The output after the classfication is a label denoting a class to which X belongs. The classification procedure is as follows.

1. Find the K nearest neighbors (K NNs) of X from S.
2. Choose the dominant classes by K NNs.
3. If there exists only one dominant class, label X by this class; otherwise, label X by any dominant class.
4. Add X to S, and the process terminates.

The first step selects K samples from S such that the values of the selected features (also called patterns) of these K samples are closest to those of X. Such a similarity may be expressed in a variety of ways. The measurement of distances among the patterns is one of the suitable instruments, for example, the Euclidean distance as shown in Eq. (60). Suppose the K samples belong to a set of classes; the second step is to find the set of dominant classes C'. A dominant class is a class that contains the majority of the K samples. If there is only one element in C', say class C_i, we assign X to C_i. On the other hand, if C' contains more than one element, X is assigned to an arbitrary class in C'. After deciding on the class of X, we label it and add it into the set S.

$$\delta(X, Y) = \sqrt{\sum_{k=1}^{m}(X_k - Y_k)^2} \qquad (60)$$

where each datum is represented by m features.

Example. Suppose there is a data set about the salaries and ages of people. Table 1 gives such a set of samples S and the corresponding labels. There are three labels that denote three classes: rich, fair, and poor. These classes are determined based on the assumption that richness depends on the values of the salary and age. In Table 1, we also append the rules for assigning labels for each age. From the above, we can get the set membership of each class.

$C_{\text{rich}} = \{Y_1, Y_4, Y_8\}$ $C_{\text{fair}} = \{Y_2, Y_5, Y_6, Y_{10}\}$

$C_{\text{poor}} = \{Y_2, Y_7, Y_9\}$

If there is a to-be-classified datum X with age 26 and salary \$35,000 (35k), we apply the classification procedure to classify it. Here we let the value of K be 4 and use the Euclidean distance as the similarity measure.

1. The set of 4 NNs is $\{Y_4, Y_5, Y_6, Y_9\}$.

Table 1   Set of Samples with Salary and Age Data

Sample	Age	Salary	Label	Assumed Rules to Assign Labels
Y_1	20	25k	Rich	rich, >20k; poor, <10k
Y_2	22	15k	Fair	rich, >26k; poor, <13k
Y_3	24	15k	Poor	rich, >35k; poor, <16k
Y_4	24	40k	Rich	
Y_5	28	25k	Fair	rich, >44k; poor, <22k
Y_6	30	40k	Fair	rich, >50k; poor, <25k
Y_7	30	20k	Poor	
Y_8	32	60k	Rich	rich, >56k; poor, <28k
Y_9	36	30k	Poor	rich, >68k; poor, <34k
Y_{10}	40	70k	Fair	rich, >80k; poor, <40k
X	26	35k	Fair	rich, >38k; poor, <19k

2. The dominant class is the class C_{fair} because $Y_6, Y_5 \in C_{\text{fair}}$, $Y_4 \in C_{\text{rich}}$, and $Y_9 \in C_{\text{poor}}$.
3. Label X by C_{fair}.
4. New sample S contains an updated class $C_{\text{fair}} = \{Y_2, Y_5, Y_6, Y_{10}, X\}$.

We can also give an assumed rule to decide the corresponding label for the age of X as shown in Table 1. Obviously, the conclusion drawn from the above classification coincides with such an assumption from human knowledge.

Unsupervised Data Analysis.   Under some circumstances, data analysis consists of a partition of the whole data set into a number of subsets. Moreover, the data within each subset have to be similar to a high degree, whereas the data between different subsets have to be similar to a very low degree. Such subsets are called clusters, and the way to find a good partition is sometimes also called cluster analysis. There are a variety of methods developed to handle this problem. A common characteristic among them is the iterative nature of the algorithms.

The C-mean clustering algorithm is representative in this field. The input contains the sample set S and a given value C, which denotes the number of clusters in the final partition. Notice that no labels are assigned to the samples in S in advance. Before classification, we must specify an initial partition W_0 with C clusters. The algorithm terminates when it converges to a stable situation in which the current partition remains the same as the previous one. Different initial partitions can lead to different final results. One way to get the best partition is to apply this algorithm with all different W_0's. To simplify the illustration, we only consider a given W_0 and a fixed C. The classification procedure is as follows.

1. Let W be W_0 on S.
2. Compute the mean of each cluster in W.
3. Evaluate the nearest mean of each sample and move a sample if its current cluster is not the one corresponding to its nearest mean.

4. If any movement occurs, go to step 2; otherwise, the process terminates.

The first step sets the current partition W to be W_0. Then we compute a set of means M in W. In general, a mean is a virtual sample representing the whole cluster. It is straightforward to use averaging as the way to find M. Next, we measure the similarities between each sample in S and every mean M. Suppose a sample Y_j belongs to a cluster C_i in the previous partition W, while another cluster C_k has a mean nearest to Y_j. Then we move Y_j from C_i to C_k. Finally, if there exists such a sample movement, the partition W would become a new one and requires more iterations. On the other hand, if no such movement occurs during an iteration, the partition would become stable and produce the final clustering.

Example. Consider the data in Table 1 again. Suppose there is no label on each sample and only the salary and the age data are used as the features for analysis. For clarity, we use a pair of values on the two features to represent a sample, for instance, the pair (20, 25k) refers to the sample Y_1. Suppose there is an initial partition containing two clusters C_1 and C_2. Let the means of these clusters be M_1 and M_2, respectively. The following shows the iterations for the clustering.

1. For the initial partition W: $C_1 = \{Y_1, Y_2, Y_3, Y_4, Y_5\}$, $C_2 = \{Y_6, Y_7, Y_8, Y_9, Y_{10}\}$.

First iteration

1. <label>2.</label>$M_1 = (23.6, 24k)$, $M_2 = (33.6, 44k)$.
2. <label>3.</label>Move Y_4 from C_1 to C_2; move Y_7 and Y_9 from C_2 to C_1.
3. <label>4.</label>For the new partition W: $C_1 = \{Y_1, Y_2, Y_3, Y_5, Y_7, Y_9\}$, $C_2 = \{Y_4, Y_6, Y_8, Y_{10}\}$.

Second iteration <list1 type = "custom">

1. <label>2.</label>$M_1 = (26.6, 21.6k)$, $M_2 = (31.5, 52.5k)$.
2. <label>3.4.</label>There is no sample movement; the process terminates.

We can easily find a simple discriminant rule behind this final partition. All the samples with salaries lower than 40k belong to C_1, and the others belong to C_2. Hence we may conclude with a discriminant rule that divides S into two clusters by checking the salary data. If we use another initial partition, say W', where $C_1 = \{Y_1, Y_3, Y_5, Y_7, Y_9\}$ and $C_2 = \{Y_2, Y_4, Y_6, Y_8, Y_{10}\}$, the conclusion is the same. The following process yields another partition with three clusters.

1. For the initial partition W: $C_1 = \{Y_1, Y_4, Y_7\}$, $C_2 = \{Y_2, Y_5, Y_8\}$, $C_3 = \{Y_3, Y_6, Y_9, Y_{10}\}$.

First iteration

1. <label>2.</label>$M_1 = (24.6, 28.3k)$, $M_2 = (27.3, 33.3k)$, $M_3 = (32.5, 38.7k)$.
2. <label>3.</label>Move Y_4 from C_1 to C_2, move Y_2 and Y_5 from C_2 to C_1, move Y_8 from C_2 to C_3, move Y_3 from C_3 to C_1, move Y_9 from C_3 to C_2.
3. <label>4.</label>For the new partition W: $C_1 = \{Y_1, Y_2, Y_3, Y_5, Y_7\}$, $C_2 = \{Y_4, Y_9\}$, $C_3 = \{Y_6, Y_8, Y_{10}\}$.

Second iteration

1. <label>2.</label>$M_1 = (24.8, 20k)$, $M_2 = (30, 35k)$, $M_3 = (34, 56.6k)$.
2. <label>3.</label>Move Y_6 from C_3 to C_2.
3. <label>4.</label>For the new partition W: $C_1 = \{Y_1, Y_2, Y_3, Y_5, Y_7\}$, $C_2 = \{Y_4, Y_6, Y_9\}$, $C_3 = \{Y_8, Y_{10}\}$.

Third iteration

1. <label>2.</label>$M_1 = (24.8, 20k)$, $M_2 = (30, 36.6k)$, $M_3 = (36, 65k)$.
2. <label>3.4.</label>There is no sample movement; the process terminates.

After three iterations, we have a stable partition and also conclude with the discriminant rule that all the samples with salaries lower than 30k belong to C_1, the other samples with salaries lower than 60k belong to C_2, and the remainder belongs to C_3. The total number of iterations depends on the initial partition, the number of clusters, the given features, and the similarity measure.

Methods for Data Analysis In the following, we introduce two famous techniques for data analysis. One is Bayesian data analysis based on probability theory, and the other is fuzzy data analysis based on fuzzy-set theory.

Bayesian Data Analysis. Bayesian inference, as defined In Ref. 52, is the process of fitting a probability model to a set of samples, which results in a probability distribution to make predictions for to-be-classified data. In this environment, a set of samples is given in advance and labeled by their associated classes. Observing the patterns contained in these samples, we can obtain not only the distributions of samples for the classes but also the distributions of samples for the patterns. Therefore, we can compute a distribution of classes for these patterns and use this distribution to predict the classes for the to-be-classified data based on their patterns. A typical process of Bayesian data analysis contains the following stages:

1. Compute the distributions from the set of labeled samples.

2. Derive the distribution of classes for the patterns.

3. Evaluate the effectiveness of these distributions.

Suppose a sample containing the pattern a on some features is labeled class C_i. First, we compute a set of probabilities $P(C_i)$ that denote a distribution of samples for different classes and let each $P(a|C_i)$ denote the conditional probability of a sample containing the pattern a, given that the sample belongs to the class C_i. In the second stage, the conditional probability of a sample belonging to class C_i, given that the sample contains the pattern a, can be formulated as follows:

$$P(C_i|a) = \frac{P(a|C_i)P(C_i)}{P(a)} \qquad (61)$$

where

$$P(a) = \sum_i P(a|C_i)P(C_i)$$

From Eq. (62), we can derive the probabilities of a sample belonging to classes according to the patterns contained in the sample. Finally, we can find a way to determine the class by using these probabilities. The following is a simple illustration of data analysis based on this probabilistic technique.

Example. Consider the data in Table 1. We first gather the statistics and transform the continuous values into discrete ones as in Table 2. Here we have two discrete levels, young and old, representing the age data, and three levels, low, median, and high, referring to the salary data. We collect all the probabilities and derive the ones for prediction based on Eq. (62):

$P(\text{young}, \text{low}|C_{\text{rich}}) = \frac{1}{3}$ $\qquad P(\text{young}, \text{low}|C_{\text{fair}}) = \frac{1}{2}$

$P(\text{young}, \text{low}|C_{\text{poor}}) = \frac{1}{3},$

$P(\text{young}, \text{median}|C_{\text{rich}}) = \frac{1}{3}$ $\quad P(\text{young}, \text{median}|C_{\text{fair}}) = 0$

$P(\text{young}, \text{median}|C_{\text{poor}}) = 0, \ldots$

$P(\text{young}, \text{low}) = \frac{4}{10}$ $\qquad P(\text{young}, \text{median}) = \frac{1}{10}$

$P(\text{young}, \text{high}) = 0, \ldots$

$P(C_{\text{rich}}) = \frac{3}{10}$ $\qquad P(C_{\text{fair}}) = \frac{2}{5}$ $\qquad P(C_{\text{poor}}) = \frac{3}{10}$

$P(C_{\text{rich}}|\text{young}, \text{low}) = \frac{1}{4}$ $\qquad P(C_{\text{fair}}|\text{young}, \text{low}) = \frac{1}{2}$

$P(C_{\text{poor}}|\text{young}, \text{low}) = \frac{1}{4}$

$P(C_{\text{rich}}|\text{young}, \text{median}) = 1$ $\quad P(C_{\text{fair}}|\text{young}, \text{median}) = 0$

$P(C_{\text{poor}}|\text{young}, \text{median}) = 0, \ldots$

Because there are two features representing the data, we compute the joint probabilities instead of the individual probabilities. Here we assume that the two features have the same degree of significance. At this

Table 2 Summary of Probability Distribution for Data in Table 1

Sample	Rich	Fair	Poor	Expressions of New Condensed Features
Young	2	2	1	Age is lower than 30
Old	1	2	2	Other ages
Low	1	2	3	Salary is lower than 36k
Median	1	1	0	Other salaries
High	1	1	0	Salary is higher than 50k

point, we have constructed a model to express the data with their two features. The derived probabilities can be regarded as a set of rules to decide the class of any to-be-classified datum.

If there is a to-be-classified datum X whose age is 26 and salary is 35k, we apply the derived rules to label X. We transform the pattern of X to indicate that the age is young and the salary is low. To find the suitable rules, we can define a penalty function $\lambda(C_i|C_j)$, which denotes the payment when a datum belonging to C_j is classified into C_i. Let the value of this function be 1 if C_j is not equal to C_i and 0 if two classes are the same. Furthermore, we can define a distance measure $\iota(X, C_i)$ as in Eq. (64), which represents the total amount of payments when we classify X into C_i. We conclude that the lower the value of $\iota(X, C_i)$, the higher the probability that X belongs to C_i. In this example, we label X by C_{fair} because $\iota(X, C_{\text{fair}})$ is the lowest.

$$\iota(X, C_i) = \sum_j \lambda(C_i|C_j)P(C_j|X) \qquad (62)$$

$$\iota(X, C_{\text{rich}}) = 0 \times \tfrac{1}{4} + 1 \times \tfrac{1}{2} + 1 \times \tfrac{1}{4} = \tfrac{2}{4}$$

$$\iota(X, C_{\text{fair}}) = \tfrac{1}{2} \qquad \iota(X, C_{\text{poor}}) = \tfrac{3}{4}$$

Fuzzy Data Analysis. Fuzzy-set theory, established by Zadeh,[53] allows a gradual membership $MF_A(X)$ for any datum X on a specified set A. Such an approach more adequately models the data uncertainty than using the common notion of set membership. Take cluster analysis as an example. Each datum belongs to exactly one cluster after the classification procedure. Often, however, the data cannot be assigned exactly to one cluster in the real world, such as the jobs of a busy person, the interests of a researcher, or the conditions of the weather. In the following, we replace the previous example for supervised data analysis with the fuzzy-set notion to show its characteristic.

Consider a universe of data U and a subset A of U. Set theory allows to express the membership of A on U by the characteristic function $F_A(X) : U \to \{0, 1\}$.

$$F_A(X) = \begin{cases} 1 & X \in A \\ 0 & X \notin A \end{cases} \qquad (63)$$

From the above, it can be clearly determined whether X is an element of A or not. However, many real-world phenomena make such a unique decision impossible. In this case, expressing membership is more suitable. A fuzzy set A on U can be represented by the set of pairs that describe the membership function $MF_A(X) : U \rightarrow [0, 1]$ as defined in Ref. 54:

$$A = \{(X, MF_A(X)) | X \in U, MF_A(X) \in [0, 1]\} \quad (64)$$

Example. Table 3 contains a fuzzy-set representation of the data set in Table 1. The membership function of each sample is expressed in a form of possibility that stands for the degree of the acceptance that a sample belongs to a class. Under the case of supervised data analysis, the to-be-classified datum X needs to be labeled using an appropriate classification procedure. All the distances between each sample and X are calculated using the two features and Euclidean distance.

1. Find the K NNs of X from S.
2. Compute the membership function of X for each class.
3. Label X by the class with a maximal membership.
4. Add X to S and stop the process.

The first stage in finding K samples with minimal distances is the same, so we have the same set of four nearest neighbors $\{Y_4, Y_5, Y_6, Y_9\}$ when the value of $K = 4$. Let $\delta(X, Y_j)$ denote the distance between X and the sample Y_j. In the next stage, we calculate the membership function $MF_{C_i}(X)$ of X for each class C_i as follows:

$$MF_{C_i}(X)$$
$$= \frac{\sum_j MF_{C_i}(Y_j)\delta(X, Y_j)}{\sum_j \delta(X, Y_j)} \qquad \forall Y_j \in k \text{ NNs of } X \qquad (65)$$

Table 3 Fuzzy-Set Membership Functions for Data in Table 1

Sample	Rich	Fair	Poor	Estimated Distances between the Sample and X
Y_1	0.5	0.2	0.3	11.66
Y_2	0.1	0.5	0.4	20.39
Y_3	0	0.2	0.8	20.09
Y_4	0.6	0.3	0.1	5.38
Y_5	0.2	0.5	0.3	10.19
Y_6	0.2	0.5	0.2	6.4
Y_7	0	0	1	15.52
Y_8	0.9	0.1	0	25.7
Y_9	0	0.3	0.7	11.18
Y_{10}	0.4	0.6	0	37.69
X	0.2	0.42	0.38	

$$MF_{C_{\text{rich}}}(X)$$
$$= \frac{0.6 \times 5.38 + 0.2 \times 10.19 + 0.2 \times 6.4 + 0 \times 11.18}{5.38 + 10.19 + 6.4 + 11.18} \approx 0.2$$

$$MF_{C_{\text{fair}}}(X)$$
$$= \frac{0.3 \times 5.38 + 0.5 \times 10.19 + 0.6 \times 6.4 + 0.3 \times 11.18}{5.38 + 10.19 + 6.4 + 11.18} \approx 0.42$$

$$MF_{C_{\text{poor}}}(X)$$
$$= \frac{0.1 \times 5.38 + 0.3 \times 10.19 + 0.2 \times 6.4 + 0.7 \times 11.18}{5.38 + 10.19 + 6.4 + 11.18} \approx 0.38$$

Because the membership of X for class C_{fair} is higher than all others, we label X by C_{fair}. The resultant membership directly gives a confidence measure of the classification.

3.2 Data Analysis on Internet Data

The dramatic growth of information systems over the past years has brought about the rapid accumulation of data and an increasing need for information sharing. The World Wide Web (WWW) combines the technologies of the uniform resource locator (URL) and hypertext to organize the resources on the Internet into a distributed hypertext system.[54] As more and more users and servers register on the WWW, data analysis on its rich content is expected to produce useful results for various applications. Many research communities such as network management, information retrieval, and database management have been working in this field.[54]

Many tools for Internet resource discovery[55] use the results of data analysis on the WWW to help users find the correct positions of the desired resources. However, many of these tools essentially keep a keyword-based index of the available resources (Web pages). Owing to the imprecise relationship between the semantics of keywords and the Web pages,[56] this approach clearly does not fit the user requests well.

The goal of Internet data analysis is to derive a classification of a large amount of data, which can provide a valuable guide for the WWW users. Here the data are the Web pages produced by the information providers of the WWW. In some cases, data about the browsing behaviors of the WWW users are also interesting to the data analyzers, such as the most popular sites browsed or the relations among the sites in a sequence of browsing. Johnson and Fotouhi[57] propose a technique to aid users to roam through the hypertext environment. They gather and analyze all the browsing paths of some users to generate a summary as a guide for other users. Many efforts have been made to apply the results of such data analysis.[57] In this section, we focus on the Web pages that are the core data of the WWW. First, we present a study on the nature of Internet data. Then we show the feature selection stage and enforce a classification procedure to group the data at the end.

Each site within the Web environment contains one or more Web pages. Under this environment, any WWW user can make a request to any site for any Web page in it. Moreover, the user can also roam through the Web by means of the anchor information provided in each Web page. Such an approach has resulted in several essential difficulties for data analysis.

1. Huge amounts of data
2. Frequent changes
3. Heterogeneous presentations

Basically, Internet data originate from all over the world, and the amount of data is huge. As any WWW user can create, delete, and update the data, and change the locations of the data at any time, it is difficult to get a precise view of the data. Furthermore, the various forms of expressing the same data also reveal the status of the chaos on the WWW. As a whole, Internet data analysis should be able to handle the large amount of data and control the uncertainty factors in a practical way. The data analysis procedure consists of the following stages:

1. Observe the data.
2. Collect the samples.
3. Select the features.
4. Classify the data.
5. Evaluate the results.

In the first stage, we observe the data and conclude with a set of features that may be effective for classifying the data. Next, we collect a set of samples based on a given scope. In the third stage, we estimate the fitness of each feature for the collected samples to determine a set of effective features. Then, we classify the to-be-classified data according to the similarity measure on the selected features. At last, we evaluate the classified results and find a way for further improvement.

Data Observation   In the following, we provide two directions for observing the data.

Semantic Analysis.   We may consider the semantics of a Web page as potential features. Keywords contained in a Web page can be analyzed to determine the semantics such as which fields it belongs to or what concepts it provides. There have been many efforts at developing techniques to derive the semantics of a Web page. The research results of information retrieval[58,59] can also be applied for this purpose.

Observing the data formats of Web pages, we can find several parts expressing the semantics of the Web pages to some extent. For example, the title of a Web page usually refers to a general concept of the Web page. An anchor, which is constructed by the home-page designer, provides a URL of another Web page and makes a connection between the two Web pages. As far as the home-page designer is concerned, the anchor texts must sufficiently express the semantics of the whole Web page to which the anchor points. As to the viewpoint of a WWW user, the motivation to follow an anchor is based on the fact that this anchor expresses desired semantics for the user. Therefore, we can make a proper connection between the user's interests and those truly relevant Web pages. We can group the anchor texts to generate a corresponding classification of the Web pages pointed to by these anchor texts. Through this classification we can relieve the WWW users of the difficulties on Internet resource discovery through a query facility.

Syntactic Analysis.   Syntactic analysis is based on the syntax of the Web pages to derive a rough classification. Because the data formats of Web pages follow the standards provided on the WWW, for example, hypertext markup language (HTML) we can find potential features among the Web pages. Consider the features shown in Table 4. The white pages, which mean the Web pages with a list of URLs, can be distinguished from the ordinary Web pages by a large number of anchors and the short distances between two adjacent anchors within a Web page. Note that here the distance between two anchors means the number of characters between them. For publication, the set of the headings has to contain some specified keywords, such as "bibliography" or "Publications." The average distance between two adjacent anchors has to be lower than a given threshold and the placement of anchors has to center to the bottom of the Web page.

According to these features, some conclusions may be drawn in the form of classification rules. For instance, the Web page is designed for publication if it satisfies the requirements of the corresponding features. Obviously, this approach is effective only when the degree of support for such rules is high enough. Selection of effective features is a way to improve the precision of syntactic analysis.

Sample Collection   It is impossible to collect all the Web pages, and thus choosing a set of representative samples becomes a very important task. On the Internet, we have two approaches to gather these samples.

Table 4   Potential Features for Some Kinds of Web Pages

Kind of Home Page	Potential Feature
White page	Number of anchors, average distance between two adjacent anchors
Publication	Headings, average distance between two adjacent anchors, anchor position
Person	Title, URL directory
Resource	Title, URL filename

1. Supervised sampling
2. Unsupervised sampling

Supervised sampling means the sampling process is based on human knowledge that specifies the scope of the samples. In supervised data analysis, there exists a classification template that consists of a set of classes. The sampling scope can be set based on the template. The sampling is more effective when all classes of the template contain at least one sample. On the other hand, we consider unsupervised sampling if there is not enough knowledge about the scope, as in the case of unsupervised data analysis. The most trivial way to get samples is to choose any subset of Web pages. However, this arbitrary sampling may not fit the requirement of random sampling well. We recommend the use of search engines that provide different kinds of Web pages in a form of directory.

Feature Selection In addition to collecting enough samples, we have to select suitable features for the subsequent classification. No matter how good the classification scheme is, the accuracy of the results would not be satisfactory without effective features. A measure for the effectiveness of a feature is to estimate the degree of class separability. A better feature implies a higher class separability. This measure can be formulated as a criterion to select effective features.

Example. Consider the samples shown in Table 5. From Table 4, there are two potential features for white pages, the number of anchors (F_0) and the average distance between two adjacent anchors (F_1). We assume that $F_0 \geq 30$ and $F_1 \leq 3$ when the sample is a white page. However, a sample may actually belong to the class of white pages although it does not satisfy the assumed conditions. For example, Y_6 is a white page although its $F_0 < 30$. Therefore, we need to find a way to select effective features.

From the labels, the set membership of the two classes is as follows, where the class C_1 refers to the

class of white pages.

$$C_0 = \{Y_1, Y_2, Y_3, Y_4, Y_5\} \quad C_1 = \{Y_6, Y_7, Y_8, Y_9, Y_{10}\}$$

We can begin to formulate the class separability. In the following formula, we assume that the number of classes is c, the number of samples within class C_j is n_j, and Y_k^i denotes the kth sample in class C_i. First, we define the interclass separability D_b, which represents the ability of a feature to distinguish the data between two classes. Next, we define the intraclass separability D_w, which expresses the power of a feature to separate the data within the same class. The two measures are formulated in Eqs. (69) and (67) based on the Euclidean distance defined in Eq. (60). Since a feature with larger D_b and smaller D_w implies a better class separability, we define a simple criterion function D_{F_j} [Eq. (71)] as a composition of D_b and D_w to evaluate the effectiveness of a feature F_j. Based on this criterion function, we get $D_{F_0} = 1.98$ and $D_{F_1} = 8.78$. Therefore, F_1 is more effective than F_0 due to its higher class separability.

$$D_b = \frac{1}{2} \sum_{i=1}^{c} P_i \sum_{j \neq i} P_j \frac{1}{n_i n_j} \sum_{k=1}^{n_i} \sum_{m=1}^{n_j} \delta(Y_k^i, Y_m^j) \quad (66)$$

where

$$P_i = \frac{n_i}{\sum_{j=1}^{c} n_j}$$

$$D_w = \frac{1}{2} \sum_{i=1}^{c} P_i \sum_{j=i} P_j \frac{1}{n_i n_j} \sum_{k=1}^{n_i} \sum_{m=1}^{n_j} \delta(Y_k^i, Y_m^j) \quad (67)$$

where

$$P_i = \frac{n_i}{\sum_{j=1}^{c} n_j}$$

and

$$D_{F_j} = D_b - D_w \quad (68)$$

We have several ways to choose the most effective set of features:

1. Ranking approach
2. Top-down approach
3. Bottom-up approach
4. Mixture approach

Ranking approach selects the features one by one according to the rank of their effectiveness. Each time we include a new feature from the rank, we compute the joint effectiveness of the features selected so far by Eqs. (69)–(71). When the effectiveness degenerates,

Table 5   Set of Samples with Two Features. The Labels Come from Human Knowledge

Sample	F_0^a	F_1^b	White Page
Y_1	8	5	No
Y_2	15	3.5	No
Y_3	25	2.5	No
Y_4	35	4	No
Y_5	50	10	No
Y_6	20	2	Yes
Y_7	25	1	Yes
Y_8	40	2	Yes
Y_9	50	2	Yes
Y_{10}	80	8	Yes

[a] F_0 denotes the number of anchors.
[b] F_1 denotes the average distance for two adjacent anchors.

the process terminates. Using a top-down approach, we consider all the features as the initial selection and drop the features one by one until the effectiveness degenerates. On the contrary, the bottom-up approach adds a feature at each iteration. The worse case of the above two approaches occurs if we choose the bad features earlier in the bottom-up approach or the good features earlier in the top-down approach. The last approach allows us to add and drop the features at each iteration by combining the above two approaches. After determining the set of effective features, we can start the classification process.

Data Classification   In the following, we only consider the anchor semantics as the feature, which is based on the dependency between an anchor and the Web page to which the anchor points. As mentioned previously, the semantics expressed by the anchor implies the semantics of the Web page to which the anchor points, and also describes the desired Web pages for the users. Therefore, grouping the semantics of the anchors is equivalent to classifying the Web pages into different classes. The classification procedure consists of the following stages:

1. Label all sample pages.
2. For each labeled pages, group the texts of the anchors pointing to it.
3. Record the texts of the anchors pointing to the to-be-classified page.
4. Classify the to-be-classified page based on the anchor information.
5. Refine the classification process.

In the beginning, we label all the samples and record all the anchors pointing to them. Then we group together the anchor texts contained in the anchors pointing to the same sample. In the third stage, we group the anchor texts contained in the anchors pointing to the to-be-classified page. After the grouping, we determine the class of the to-be-classified page according to the corresponding anchor texts. At last, we can further improve the effectiveness of the classification process. There are two important measures during the classification process. One is the similarity measure of two data, and the other is the criterion for relevance feedback.

Similarity Measure.   After the grouping of samples, we have to measure the degree of membership between the to-be-classified page and each class. Considering the Euclidean distance again, there are three kinds of approaches for such measurement:

1. Nearest-neighbor approach
2. Farthest-neighbor approach
3. Mean approach

The first approach finds the sample in each class nearest to the to-be-classified page. Among these representative samples, we can choose the class containing the one with a minimal distance and assign the page to it. On the other hand, we can also find the farthest sample in each class from the page. Then we assign the page to the class that contains the representative sample with a minimal distance. The last approach is to take the mean of each class into consideration. As in the previous approaches, the mean of each class represents a whole class, and the one with a minimal distance from the page would be chosen. An example follows by using the mean approach.

Example.   Inspect the data shown in Table 6. There are several Web pages and anchor texts contained in some anchors pointing to the Web pages. Here we consider six types of anchor texts, $T_1, T_2, \ldots, T_6$. The value of an anchor text for a Web page stands for the number of the anchors pointing to the Web page, which contain the anchor text. The labeling is the same as in the previous example. We can calculate the means of the two classes:

$$M_0 = (0, 4, 1, 1, 1, 0.2, 1)$$

$$M_1 = (4.2, 3.4, 2.6, 1.4, 2, 1.4)$$

Suppose there is a Web page X to be classified as shown in Table 6. We can compute the distances between X and the two means. They are $\delta(X, M_0) = 6.94$ and $\delta(X, M_1) = 4.72$. Thus we assign X to class C_1.

Relevance Feedback.   The set of samples may be enlarged after a successful classification by including the classified Web pages. However, the distance

Table 6   Set of Web Pages with Corresponding Anchor Texts and Labels. The Labels Come from Human Knowledge

Sample	T_1^a	T_2^b	T_3^c	T_4^d	T_5^e	T_6^f	White Page
Y_1	0	0	0	1	1	2	No
Y_2	0	1	2	0	0	2	No
Y_3	0	2	0	4	0	0	No
Y_4	0	0	3	0	0	1	No
Y_5	2	2	0	0	0	0	No
Y_6	1	3	0	0	2	3	Yes
Y_7	3	3	1	6	3	0	Yes
Y_8	4	2	5	0	1	0	Yes
Y_9	5	5	3	0	0	2	Yes
Y_{10}	8	4	4	1	4	2	Yes
X	5	2	0	0	5	0	Yes

$^a T_1 =$ "list."
$^b T_2 =$ "directory."
$^c T_3 =$ "classification."
$^d T_4 =$ "bookmark."
$^e T_5 =$ "hot."
$^f T_6 =$ "resource."

between a to-be-classified page and the nearest mean may be very large, which means that the current classification process does not work well on this Web page. In this case, we reject the classification of such a Web page and wait until more anchor texts for this Web page are accumulated. This kind of rejection not only expresses the extent of the current ability to classify Web pages, but also promotes the precision of the classified results. Furthermore, by the concept of class separability formulated in Eqs. (69)–(71), we can define a similar criterion function D_S to evaluate the performance of the current set of samples.

$$D_S = D_F(S) \qquad (69)$$

where F is the set of all effective features and S is the current set of samples.

Example. Reconsider the data shown in Table 6. Before we assign X to C_1, the initial $D_S = 0.75$. When C_1 contains X, $D_{S \cup \{X\}}$ yields a smaller value 0.16. On the other hand, $D_{S \cup \{X\}}$ becomes 1.26 if we assign X to C_0. Hence, although X is labeled C_1, it is not suitable to become a new sample for the subsequent classification. The set of samples can be enlarged only when such an addition of new samples gains a larger D_s value, which means the class separability is improved.

3.3 Improvement of Data Analysis Methods

Although the previous procedures are able to fit the requirements of data analysis well, there are still problems, such as speed or memory requirements and the complex nature of real-world data. We have to use some heuristic techniques to improve the classification performance. For example, the number of clusters given in unsupervised data analysis has significant impact on the time spent at each iteration and the quality of the final partition. Notice that the initial partition may contribute to a specific sequence of adjustments and then a particular solution. Therefore, we have to find an ideal number of clusters during the analysis according to the given initial partition. The bottom-up approach for decreasing the number of clusters at each iteration is a way to determine the final partition. Given a threshold of similarity among the clusters, we can merge two clusters that are similar enough to become a new single cluster at each iteration. We can find a suitable number of clusters when there are no more similar clusters to be merged. In the following sections, we introduce two more techniques to improve the work of data analysis.

Rough-Set-Based Data Analysis The approach to classifying Internet data by anchor semantics requires a large amount of anchor texts. These anchor texts may be contained in the anchors pointing to the Web pages in different classes. An anchor text is said to be indiscernible when it cannot be used to distinguish the Web pages in different classes. We employ the rough-set theory[60,61] to find the indiscernible anchor texts, which will then be removed. The remaining anchor texts will contribute to a higher degree of accuracy for the subsequent classification. In addition, the cost of distance computation can also be reduced. In the following, we introduce the basic idea of the rough-set theory and an example for the reduction of anchor texts.

Rough-Set Theory. By the rough-set theory, an information system is modeled in the form of a 4-tuple (U, A, V, F), where U represents a finite set of objects, A refers to a finite set of attributes, V is the union of all the domains of the attributes in A, and F is a binary function $(U \times A: \rightarrow V)$. The attribute set A often consists of two subsets, one refers to condition attributes $\overline{C}$ and the other stands for decision attributes $\overline{D}$. In the approach of classification on Internet data, U stands for all the Web pages, A is the union of the anchor texts $(\overline{C})$ and the class of Web pages $(\overline{D})V$ is the union of all the domains of the attributes in A, and F handles the mappings. Let B be a subset of A. A binary relation called indiscernibility relation is defined as

$$\text{IND}_B = \{(X_i, X_j) \in U \times U | \forall p \in B, p(X_i) = p(X_j)\} \qquad (70)$$

That is, X_i and X_j are indiscernible by the set of attributes B if $p(X_i)$ is equal to $p(X_j)$ for every attribute p in B. IND_B is an equivalence relation that produces an equivalence class denoted $[X_i]_B$ for each sample X_i. With regard to the Internet data, two Web pages X_i and X_j, which have the same statistics for each anchor text in $\overline{C}$ belong to the same equivalence class $[X_i]\overline{C}$ (or $[X_j]\overline{C}$). Let U' be a subset of U. A lower approximation $\text{LOW}_{B,U'}$, which contains all the samples in each equivalence class $[X_i]_B$ contained in U', is defined as

$$\text{LOW}_{B,U'} = \{X_i \in U | [X_i]_B \subset U'\} \qquad (71)$$

Based on Eq. (75), $\text{LOW } \overline{C}, [X_i]\overline{D}$ contains the Web pages in the equivalence classes produced by $\text{IND } \overline{C}$, and these equivalence classes are contained in $[X_i]\overline{D}$ for a given X_i. A positive region $\text{POS } \overline{C}, \overline{D}$ is defined as the union of $\text{LOW } \overline{C}, [X_i]\overline{D}$ for each equivalence class produced by $\text{IND } \overline{D}$. $\text{POS } \overline{D}, \overline{D}$ refers to the samples that belong to the same class when they have the same anchor texts. As defined in Ref. 62, $\overline{C}$ is independent of $\overline{D}$ if each subset $\overline{C_i}$ in $\overline{C}$ satisfies the criterion that $\text{POS } \overline{C}, \neq \text{POS } \overline{C_i}, \overline{D}$; otherwise, $\overline{C}$ is said to be dependent on $\overline{D}$. The degree of dependency $\gamma \overline{C}, \overline{D}$ is defined as

$$\gamma_{\overline{C},\overline{D}} = \frac{\text{card}(\text{POS}_{\overline{C},\overline{D}})}{\text{card}(U)} \qquad (72)$$

where card denotes set cardinality;

$$\mathrm{CON}_{p,\gamma_{\overline{C},\overline{D}}} = \gamma_{\overline{C},\overline{D}} - \gamma_{\overline{C}-|p|,\overline{D}} \tag{73}$$

From these equations, we define the contribution $\mathrm{CON}_{p,\gamma}\overline{C}, \overline{D}$ of an anchor text p in $\overline{C}$ to the degree of dependency γ $\overline{C}$, CIDbar; by using Eq. (73). According to Eq. (73), we say an anchor text p is dispensable if γ $\overline{C} - \{p\}, \overline{D} = \gamma$ $\overline{C}, \overline{D}$. That is, the anchor text p makes no contribution to γ $\overline{C}, \overline{D}$ and the value of $\mathrm{CON}_{p,\gamma}\overline{C}, \overline{D}$ equals 0. The set of indispensable anchor texts is the core of the reduced set of anchor texts. The remaining task is to find a minimal subset of $\overline{C}$ called a reduct of $\overline{C}$ that satisfies Eq. (74) and the condition that the minimal subset is independent of $\overline{D}$.

$$\mathrm{POS}_{\overline{C},\overline{D}} = \mathrm{POS}_{\text{minimal subset of } \overline{C},\overline{D}} \tag{74}$$

Reduction of Anchor Texts. To employ the concepts of the rough-set theory for the reduction of anchor texts, we transform the data shown in Table 6 into those in Table 7. The numerical value of each anchor text is transformed into a symbol according to the range in which the value falls. For instance, a value in the range between 0 and 2 is transformed into the symbol L. This process is a generalization technique usually used for a large database.

By Eq. (73), we can compute $\mathrm{CON}_{p,\gamma}\overline{C}, \overline{D}$ for each anchor text p and sort them in ascending order. In

this case, all $\mathrm{CON}_{p,\gamma}\overline{C}, \overline{D}$ are 0 except $\mathrm{CON}_{T_1}, \gamma \overline{C}, \overline{D}$. That is, only the anchor text T_1 is indispensable, which becomes the unique core of $\overline{C}$ Next, we use a heuristic method to find a reduct of $\overline{C}$ because such a task has been proved to be NP-complete in Ref. 63. Based on an arbitrary ordering of the dispensable anchor texts, we check the first anchor text to see whether it is dispensable. If it is, then remove it and continue to check the second anchor text. This process continues until no more anchor texts can be removed.

Example. Suppose we sort the dispensable anchor texts as the sequence $\{T_2, T_3, T_4, T_5, T_6\}$, we then check one at a time to see whether it is dispensable. At last, we obtain the reduct $\{T_1, T_6\}$. During the classification process, we only consider these two anchor texts for similarity measure. Let the symbols used in each anchor text be transformed into three discrete values, 0, 1, and 2. The means of the two classes are $M_0 = (0, 0)$ and $M_1 = (1, 0.8)$. Therefore, we classify X into the class C_1 due to its minimum distance. When we use the reduct $\{T_1, T_6\}$ to classify data, the class separability $D_{\{T_1, T_6\}}$ is 0.22. Different reducts may result in different values of class separability. For instance, the class separability becomes 0.27 if we choose the reduct $\{T_1, T_2\}$.

Hierarchical Data Analysis Consider the 1-nearest-neighbor classifier for supervised data analysis. We may not want to compute all the distances each time a to-be-classified datum X arrives. We can organize the set of samples into a hierarchy of subsets and record a mean M_i for each subset S_i and the farthest distance d_i from M_i to any sample in S_i. If there exists a nearest neighbor of X in a subset other than S_i, we do not need to compute the distances between X and all the samples in S_i as the triangular inequality [Eq. (75)] holds. Such techniques can reduce the computation time to find the nearest neighbor.

$$\delta(X, M_i) - d_i \geq \delta(X, Y) \tag{75}$$

where Y is the nearest neighbor of X.

3.4 Summary

In this section, we describe the techniques and concepts of data analysis. A variety of data analysis methods are introduced and illustrated by examples. Two categories, supervised data analysis and unsupervised data analysis, are presented according to their different initial conditions and resultant uses. Two methods for data analysis are also described, which are based on probability theory and fuzzy-set theory, respectively. An approach of data analysis Internet data is presented. Improvements for the data analysis methods are also discussed.

Table 7   Set of Data in Symbolic Values Transformed from Table 6

Sample	T_1	T_2	T_3	T_4	T_5	T_6	White Page
Y_1	L[a]	L	L	L	L	L	No
Y_2	L	L	L	L	L	L	No
Y_3	L	L	L	M[b]	L	L	No
Y_4	L	L	M	L	L	L	No
Y_5	L	L	L	L	L	L	No
Y_6	L	M	L	L	L	M	Yes
Y_7	M	M	L	H[c]	M	L	Yes
Y_8	M	L	M	L	L	L	Yes
Y_9	M	M	M	L	L	L	Yes
Y_{10}	H	M	M	L	M	L	Yes
X	M	L	L	L	M	L	Yes

[a] L = [0, 2].
[b] M = [3, 5].
[c] H = [6, 8].

4 DIODES

Konstantinos Misiakos

The word *diode* originates from the Greek word Διοδος meaning passage or way through. In electronics terminology, in fact, diode refers to a two-terminal device that allows current to flow in one direction while it blocks the flow of current in the opposite direction. Such devices usually employ semiconductor junctions or metal–semiconductor junctions. There are also diodes made of vacuum tubes or metal—purely ionic crystal contacts. This section deals with semiconductor *p–n* junction diodes because they are the most widely used in practice due to their versatility, reproducibility, stability, and compatibility with integrated circuit technology. Additionally, an insight into the operation of the *p–n* junction diode is the basis for understanding the device physics of other semiconductor devices, the majority of which use the *p–n* junction as the building block. The semiconductor of our choice will be silicon because almost all diodes, discrete or integrated, are made of this element. Extensions to other semiconductors will be made to generalize theoretical results or to set limits to the validity of certain equations.

Figure 56 shows the electrical symbol of a diode. The arrow-type symbol indicates the conduction direction. For a diode to conduct an appreciable electric current, the voltage on the left side of the symbol must be a little higher than the voltage on the right side. If this polarity is reversed, the current drops to negligible values even for a large bias. The two previous polarity modes are known as forward and reverse bias, respectively. In Fig. 56*b* the very basic material structure of a diode is shown. The starting material is a high-purity silicon crystal, the properties of which are properly modified by selectively introducing dopants (elements) from either the third or the fifth column of the periodic table. The third-column elements, when introduced into the silicon lattice, behave as acceptors: They trap electrons from the valence band, thereby creating positively charged holes in the valence band and negatively charged immobile acceptor ions. The acceptor-doped part of the diode is called the *p* side. On the other hand, fifth-column elements behave as donors: They give up their fifth electron, creating a population of conduction band electrons and positively charged immobile donor ions. The donor-doped part of the device is the *n* side. The introduction of acceptor and donor dopants into silicon creates the two polarity sides of the diode, as shown in Fig. 56*a*. Schematically speaking, when applying a forward bias, the higher voltage on the *p* side makes the electron and hole gases move into each other. Thus, an electric current is created through electron–hole pair recombination. On the contrary, a lower voltage on the *p* side moves the charge carriers away, thus preventing recombination and eliminating the current. In terms of dopants, the previous account of how the *p–n* diode is formed also holds for germanium diodes, which also is a fourth-column elemental

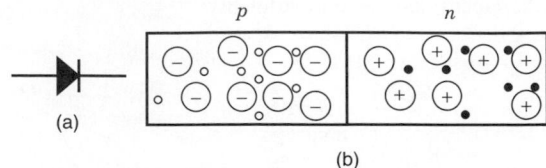

Fig. 56 Electrical symbol of a diode (*a*) and illustration of a semiconductor *p–n* junction (*b*). In (*b*) the large circles with the minus and the plus signs are the acceptor and the donor ions, whereas the small circles are the holds (empty) and the electrons (dark).

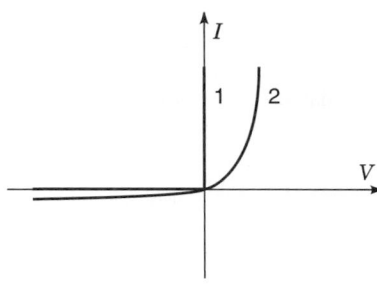

Fig. 57 Current–voltage characteristics of an ideal diode, curve 1, and a realistic one, curve 2.

semiconductor. For compound semiconductors (e.g., GaAs, InP, CdTe), the chemical origin of donor and acceptor dopants is more complex in relation to the semiconductor elements themselves.

In Fig. 57 the current–voltage (*I–V*) characteristic of an ideal diode as well as of a realistic one is shown. The ideal diode would behave as a perfect switch when forward biased: Unlimited current flows without any voltage drop across the device. The same ideal diode would allow no current in the reverse direction, no matter what the magnitude of the reverse bias is. Now, a realistic semiconductor diode would exhibit a resistance to current flow in the forward direction, whereas in reverse bias a small current would always be present due to leakage mechanisms. The disagreement between the ideal and the actual electrical behavior is not restricted only to the static *I–V* characteristics shown in Fig. 57. It extends to the transient response obtained when applying a time-dependent terminal excitation. The response of a realistic diode cannot follow at exactly the same speed as the terminal excitation of an ideal diode would. When designing a diode to be used as a switching device, care is taken to bring the device electrical characteristics as close to the ideal ones as possible. This is done by choosing both the geometrical features and the fabrication process steps in a way to suppress the parasitic components of the diode. As a result of the semiconductor electronic band structure as well as technological constraints, material limitations impose certain basic restrictions on the device performance and create the subsequent deviation from the ideal performance. In the following sections, these restrictions will be investigated, and the deviation from the ideal performance will be analyzed in terms of the basic device physics, material constants, and geometry considerations.

Before considering the device physics of the diode, we will briefly discuss the steps in the basic fabrication process employed when making a silicon diode. These steps determine its basic geometrical and technological characteristics, which in turn determine the device electrical behavior. Today, almost all silicon diodes are made through the standard planar process of the silicon integrated circuit technology. A silicon wafer is first oxidized at temperatures in the vicinity of $1000°C$. Such oxidation creates a silicon dioxide (SiO_2) cover layer with a thickness on the order of a micron. This layer is used as a mask for the subsequent technological steps. The SiO_2 film is then patterned by lithographic techniques and through etching, which allows windows of exposed silicon to be opened. Then, either by diffusion or by ion implantation, dopants are introduced into the exposed areas. The dopants are of a type opposite to the one already existing in the original wafer. In this way, $p–n$ junctions are created in the exposed areas. In the rest of the wafer, the SiO_2 layer stops the ions and prevents diffusion into the silicon bulk. On the back surface, another diffusion or implantation of the same dopants as in the bulk is usually applied for reasons that will become clear in the next sections. At the end, metal contacts are evaporated on the front and the back. Lithography, again, on the front side defines the contacts of the individual diodes. The metal contacts are required for the diodes to interact with the external world in terms of terminal excitation (voltage or current) and terminal response (current or voltage, respectively). Similar methods are used for germanium diodes, whereas the compound semiconductor devices are usually made by epitaxial growth on proper substrates and by in situ doping.

4.1 Fundamentals of $p–n$ Junctions

The basic $p–n$ junction device physics was proposed by Shockley.[64] He derived the current–voltage characteristics, considering the electron and hole current continuity equations and the relationship between the carrier quasi-Fermi levels and the externally applied potentials. Here, we rederive the general current–voltage relation of a $p–n$ junction based on Shockley's classic work[64] and its later extension.[65]

Basic Equations and Assumptions To formulate the electron and hole transport in a semiconductor device mathematically, we can always start by expressing the carrier densities and currents in terms of the carrier quasi-Fermi potentials under uniform temperature conditions:

$$J_n = \begin{cases} -e\mu_n n\,\nabla F_n & \text{(76a)} \\[2mm] e\mu_n n E + e D_n\,\nabla n & \text{(76b)} \end{cases}$$

$$n = n_i \exp\left(\frac{-eF_n - E_i}{kT}\right) \tag{76c}$$

$$J_p = \begin{cases} -e\mu_p p\,\nabla F_p & \text{(77a)} \\[2mm] e\mu_p p E - e D_p\,\nabla p & \text{(77b)} \end{cases}$$

$$p = n_i \exp\left[\frac{-(-eF_p) + E_i}{LT}\right] \tag{77c}$$

In Eqs. (76a) and (77a), J_n and J_p are the electric current densities of electrons and holes, respectively. Equations (76b) and (77b) express the currents in terms of drift and diffusion, where μ_n, D_n, n and μ_p, D_p, p are the mobilities, diffusivities, and volume densities of electrons and holes, respectively. Finally, F_n and F_p are the electron and hole quasi-Fermi potentials, E_i is the intrinsic energy level, and E the electric field density. Figure 58 shows the energy band diagram of a $p–n$ junction under forward bias and illustrates the space dependence of the quasi-Fermi potentials, of the bottom of the conduction band E_c, and of the top of the valence band E_v. Equations (76c) and (77c) hold provided that the differences $E_c - (eF_n)$ and $(eF_p) - E_v$ are positive and at least several times the thermal energy kT. Equations (76a) and (77a) are borrowed from thermodynamics and hold provided that the bias is such that perturbations from equilibrium are small. Small, here, implies that the energy distribution of electrons and holes in the conduction and the valence band, respectively, continue (within a good approximation) to follow the Boltzmann statistics. Additionally, we assume that the mean free paths of the carriers are negligible compared to the physical dimensions of the device. Finally, Eqs. (76) and (77) hold provided quantum mechanical tunneling of carriers across potential barriers is not important. Such a constraint is relaxed in the last section of nonconventional transport diodes.

The second set of equations to be considered is the electron and hole continuity equations:

$$\frac{\partial n}{\partial t} = \frac{1}{e}\nabla J_n - U(n, p) + G \tag{78}$$

$$\frac{\partial p}{\partial t} = -\frac{1}{e}\nabla J_p - U(n, p) + G \tag{79}$$

where U is the electron–hole net recombination rate either by band-to-band transitions or through traps, whereas G is the band-to-band generation rate resulting from ionizing radiation or impact ionization processes. For the sake of simplicity, we assume that U

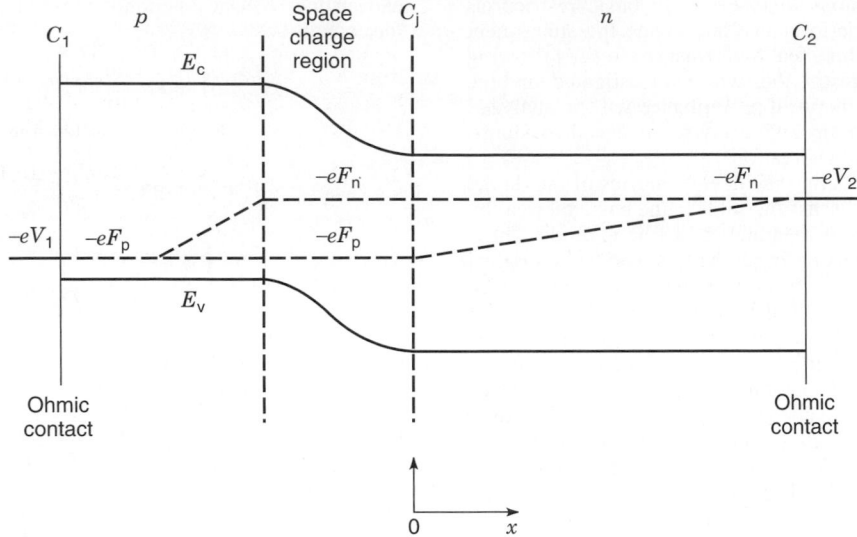

Fig. 58 Band diagram of forward-biased p–n junction. Boundaries C_1 and C_2 are the ohmic contacts, whereas C_j is the base-injecting boundary.

and G are the same for both carriers. Equations (78) and (79) are more general than the previous ones because they do the "bookkeeping" by equating the increase in the rate of carrier density to minus the carrier losses resulting from carrier out-fluxing ($\nabla J_p/e$ and $-\nabla J_n/e$) and recombination.

Next is the Poisson equation, which relates the electric field to the charge density caused by both mobile and immobile charges:

$$\nabla E = \frac{e}{\varepsilon}[-n + p + N_D - N_A] \qquad (80)$$

where ε is the semiconductor dielectric constant and N_D and N_A are the donor and acceptor densities, respectively. The charge density resulting from donors and acceptors is not carrier density dependent, unless the temperature drops to the cryogenic region.

The final equation to be considered is the one that equates the electric field to the gradient of the electrostatic potential:

$$E = \frac{1}{e}\nabla E_c = \frac{1}{e}\nabla E_v = \frac{1}{e}\nabla E_i \qquad (81)$$

Equation (81) implies that the electrostatic potential is determined by conduction and the valence band edges because the carriers there have only potential energy. The last equation assumes that the separation in the energy scale of the three levels (E_c, E_v, and E_i) is space independent. So, in Eq. (81), as well as in eqs. (76b), (76c), (77b), and (77c), we neglect band distortion resulting from heavy doping or other effects

(e.g., mechanical stain). This is discussed in a later section on heavy doping effects.

Boundary Conditions. Equations (76)–(81) form a system of six relations with six unknown variables: F_n, F_p, E_i, J_n, J_p, and E. They apply, within the range of their validity, to any semiconductor device. In this sense, any semiconductor device understanding, design, operation, and performance is based on this set of six equations. What distinguishes a device of a particular kind is its boundary conditions, as well as the doping and trap density and type.

Ohmic Contacts. For a diode, a two-terminal device, the boundary conditions necessarily include two ohmic contacts that will supply the charge to be transported through the device. The voltage across and the current through the two ohmic contacts are, interchangeably, the excitation or the response of the device. The ohmic contacts are realized by depositing metals (e.g., Ti or Al) on heavily doped regions of the semiconductor. An ideal ohmic contact should, by definition, establish thermodynamic equilibrium between the metal and the semiconductor at all the contact points. In analytical terms, this is expressed by equating the carrier quasi-Fermi potentials to the metal Fermi potential:

$$F_n(C_1) = F_p(C_1) = V_1 \qquad (82a)$$

$$F_n(C_2) = F_p(C_2) = V_2 \qquad (82b)$$

where V_1 and V_2 are the voltages of the metal contacts ($V_1 - V_2$ is the terminal voltage), whereas C_1

and C_2 are the contact areas of the first and the second ohmic contact, respectively. The pinning of the Fermi potentials at the externally applied voltages is illustrated in Fig. 58. If $V_1 = V_2$ and G in Eqs. (78) and (79) were zero, the device would be in equilibrium. Then the solution of the previous system of six equations would be zero currents and equal and flat Fermi potentials throughout the diode. When an external voltage is applied, the splitting of the Fermi potential values between the two ohmic contacts drives the device out of equilibrium. Such a boundary value split enforces a separation of the electron and hole quasi-Fermi potentials through the device, as shown in Fig. 58. The separation of the two potentials implies that the nonequilibrium conditions mainly refer to the interaction between electrons and holes. At any point, excluding ohmic contacts, electrons are out of equilibrium with respect to holes because the relaxation time of interband transitions (recombination and generation mechanisms) required to bring them into equilibrium are too slow (milliseconds or microseconds for germanium and silicon and nanoseconds for most compound semiconductors). On the contrary, the intraband transitions resulting from scattering have short relaxation times (picoseconds) so that electrons or holes are nearly in equilibrium within their band. This is required for the carrier Fermi potentials to have a meaning, as mentioned in the discussion following Eqs. (76) and (77).

Semiconductor–Insulator Interfaces. The surface that bounds the device includes, in addition to the ohmic contact, the semiconductor–vacuum or semiconductor–insulator interface. The exposed semiconductor surface is usually covered by an insulating film (SiO_2 in silicon) to reduce recombination. If we assume that there is no injection in the insulator, then at the interface the boundary conditions for Eqs. (78) and (79) are

$$\frac{\partial n_s}{\partial t} = -\frac{1}{e} J_n^n - U_s \qquad (83)$$

$$\frac{\partial p_s}{\partial t} = \frac{1}{e} J_p^n - U_s \qquad (84)$$

where the subscript s refers to surface densities and the superscript n refers to the normal component looking into the insulator. The boundary conditions for Eq. (80) are dictated by the lows of electrostatics. The discontinuity of the normal component of the dielectric displacement vector must be equal to the surface-charge density, whereas the tangential component of the electric field must be continuous.

Because the boundary conditions have been set, the system of six equations [Eqs. (76)–(81)] can be solved, in principle. As it turns out, the solution of such a nonlinear system of coupled equations can be found only numerically even for one-dimensional p–n junctions with uniform acceptor and donor densities. To derive analytical approximations, we need to make certain assumptions regarding the physical makeup of the device and the degree of bias. These analytical expressions help predict the device response under reasonable bias, whereas the appreciation of their validity range provides an insight into the diode device physics.

4.2 Doping Carrier Profiles in Equilibrium and Quasi-Neutral Approximation

As mentioned earlier, a p–n junction diode consists of an acceptor-doped p region in contact with a donor-doped n region. The two-dimensional area where the donor and acceptor densities are equal is called the metallurgical junction. Let us assume, for the moment, equilibrium conditions. In such a case, the currents are zero and the quasi-Fermi potentials are equal and spatially independent, $F_n = F_p = F$. Therefore, from Eqs. (76) and (77), $pn = n_i^2$, where n_i is the intrinsic-carrier density. Now, the six equations reduce to the Poisson equation, which, with the help of Eqs. (76c), (77c), and (81), takes the form

$$\Delta^2[E_i - (-eE)] = \frac{e^2}{\in}\left[-n_i \exp\left(\frac{(-eF) - E_i}{kT}\right)\right.$$

$$\left. +n_i \exp\left(\frac{E_i - (-eF)}{kT}\right) + N_D - N_A\right] \qquad (85)$$

The last equation is known as the Boltzmann–Poisson equation. Approximate analytical solutions are possible when the donor and acceptor densities are uniform in the n and p regions, respectively. In this case, the field is zero, and the bands are flat everywhere except at and near the metallurgical junction. The finite-field region around the metallurgical junction is called the space-charge region, whereas the zero-field regions are called quasi-neutral regions, for reasons to be explained shortly. In the n and p quasi-neutral regions, electrons and holes are the majority carriers, respectively. The majority-carrier densities are equal to the respective doping densities. With reference to the metallurgical junction, the space-charge region extends W_A and W_D within the p and the n regions. At zero bias, and in one dimension, an approximate solution of Eq. (85) gives

$$W_A = \sqrt{\frac{2\varepsilon}{e} V_{bi} \frac{N_D}{N_A(N_A + N_D)}} \qquad (86a)$$

$$W_D = \sqrt{\frac{2\varepsilon}{e} V_{bi} \frac{N_A}{N_D(N_A + N_D)}} \qquad (86b)$$

$$V_{bi} = \frac{kT}{e} \ln\left(\frac{N_A N_D}{n_i^2}\right) \qquad (86c)$$

In Eq. (86c) V_{bi} is the zero bias electrical potential difference, or barrier, between the p and the n side reflected in the level differences of the flat bands of each side. Such a barrier prevents majority carriers from diffusing into the other side. These approximations result by assuming that the electron and hole densities are zero in the space-charge region. This is the depletion approximation, which reduces Eq. (85) to a linear second-order differential equation with constant terms and coefficients.

The zero-field condition for the rest of the n and p sides, outside the space-charge region, apparently justifies the term quasi-neutral regions. This term also applies when the n and p regions have gradually changing dopant profiles in the sense that the net space charge is much less than the majority-carrier charge. Here, by gradually changing we mean that the doping profile $N(x)$ in the quasi-neutral region is such that[66]

$$\frac{\varepsilon kT}{e^2} \left| \nabla^2 \left[\ln \frac{N(x)}{n_i} \right] \right| \ll N(x) \qquad (87)$$

In such regions, the zero-bias majority-carrier density continues to be nearly the same as the net dopant density, but the electric field is not zero as in the uniform doping case.

4.3 Forward- and Reverse-Bias Conditions

The quasi-neutrality condition of the n and p regions is preserved even under bias, but now the boundaries with the space-charge region move appropriately to accommodate the new boundary conditions. This neutrality condition can be expressed as

$$n \approx p + N_D - N_A \qquad (88)$$

Under a small forward bias, the applied voltage changes the electric field preferentially at the space-charge region, because it is the region with the fewest carriers, has the highest resistance, and is in series with more conductive n and p regions. The equilibrium barrier height V_{bi} lowers under forward bias, and the majority electrons overcoming the repulsive field diffuse from the n side to the p side, whereas the holes are doing the opposite. The carrier quasi-Fermi potentials are no longer equal, as shown in Fig. 58. The diffusion process, through the space-charge region and inside the quasi-neutral regions, increases dramatically the minority-carrier population on either side and gives rise to an appreciable electric current.

For forward voltages, the degree of bias defines three injection-level regimes distinguished by how the minority-carrier density compares to the majority one in the quasi-neutral regions. These regimes are the low-level, the moderate-level, and the high-level injection condition. In the low-level injection regime, the minority-carrier density is well below the majority-carrier density, and the electric field in the quasi-neutral regions is practically unaffected by the bias. As a result, the applied voltage drops across the space-charge region and reduces the barrier height from V_{bi}

to $V_b = V_{bi} - V$. Provided the depletion approximation still holds, Eq. (86) still applies with V_{bi} being replaced by V_b. In low-level injection, the majority-carrier density is the same as at zero bias, as Eq. (88) points out, and is nearly equal to the net doping density. In the high-level injection regime, the minority-carrier injection is so intense that the injected carriers have densities far exceeding the dopant densities. Now, both carrier densities are about the same, $n = p$, to preserve neutrality in the quasi-neutral region. In other words, there is no real distinction between minority and majority carriers in terms of concentrations, but we obtain an electron–hole plasma having densities well above that of the dopant densities instead. In the moderate injection, the minority-carrier density approaches the order of magnitude of the majority-carrier density causing the majority-carrier density to start to increase, as Eq. (88) implies.

Under reverse bias, the built-in barrier increases in the space-charge region, the repulsive forces on the majority carriers coming from the quasi-neutral regions increase, and injection of minority carriers is not possible. The space-charge region is now totally depleted from both carriers, and a small leakage current exists as a result of thermal generation of electron and hole pairs in the depletion region.

Recombination Currents in the Steady State

Here, we will introduce the base and emitter terms as well as a general expression for the terminal current as the sum of recombination components. Between the two quasi-neutral regions, emitter is the one that is heavily doped, usually by diffusion or implantation, whereas the base is more lightly doped and occupies most of the substrate on which the device is made, at least in silicon. The heavy doping of the emitter excludes the possibility of moderate- or high-level injection conditions in this region. At forward bias, majority carriers from the emitter diffuse as minority carriers to the base where they recombine. Simultaneously, recombination occurs in the emitter because minority carriers are back-injected from the base, as well as in the space-charge region. At steady state, $\partial n/\partial t = \partial p/\partial t = 0$, and in the dark $G = 0$. Now, the continuity equations [Eqs. (78) and (79)] become after volume integration:

$$I_D = I_e + I_b + I_{SCR} \qquad (89)$$

where I_D is the terminal current and I_e, I_b, and I_{SCR} are the net recombination currents in the emitter, the base, and the space-charge region, respectively. Equation (89) expresses the total current as the sum of the recombination currents in the three regions of the device. Therefore, excess carrier recombination along with diffusion are the two basic transport mechanisms that determine the diode current at a given bias. The carrier recombination occurs either at the ohmic contacts, at the surface, or in the bulk. The minority carriers that arrive at the ohmic contact are supposed to recombine simultaneously there, to preserve the boundary condition, Eq. (82b). The bulk recombination occurs either through traps or through band-to-band transitions. In terms of trap-mediated recombination, the Shockley–Read–Hall mechanism[67,68] is the most common:

$$U_{SRH} = \frac{(pn - n_i^2)N_t}{\frac{1}{\sigma_p v_{th}}\left[n + n_i \exp\left(\frac{E_t - E_i}{kt}\right)\right]} + \frac{1}{\sigma_n v_{th}}\left[p + n_i \exp\left(-\frac{E_t - E_i}{kt}\right)\right] \quad (90)$$

where N_t is the trap density, σ_n and σ_p are the electron and hole capture cross sections, respectively, E_t, is the trap energy level in the gap, and v_{th} is the carrier thermal velocity. The band-to-band recombination is discussed in the heavy-doping effects section.

4.4 Approximate Analytical Expressions in Steady State

As previously mentioned, the set of Eqs. (76)–(81) has no analytical solution in the general case. Approximate closed-form expressions, though, are possible when low-level injection conditions dominate in the quasi-neutral base region. Without loss of generality, we assume that we are dealing with a p–n diode with a heavily doped p emitter and an n base. The steady-state situation results when a terminal bias, say a terminal voltage V, is steadily applied on the terminals, and we wait long enough for the initial transient to disappear. The steady-state version of the continuity equations [Eqs. (78) and (79)] is simplified because the time derivatives are set equal to zero. First, we will derive the expressions for the base current, and then extensions will be made for the recombination current in the emitter and the space-charge regions. If low-level injection conditions prevail in the base, then, to a good approximation, the original system of equations [Eqs. (76)–(81)] reduces to the minority-carrier equations [Eqs. (77) and (79)], which are now decoupled from Eq. (80) (the Poisson equation). This decoupling results because, as previously mentioned, at low-level injection the electric field is practically bias independent. Any small field variations would affect only the drift current of the majority carriers because of their high density; the minority carriers would not be influenced. That is why we focus on the minority-carrier transport to exploit the Poisson equation decoupling. Another reason for focusing on the minority carriers is the fact that the recombination in low-level injection, where $p \ll n$, can always be written as a linear function of their density:

$$U = \frac{p - p_0}{\tau} \quad (91)$$

where p_0 is the equilibrium carrier density, whereas the variable τ, called minority-carrier lifetime, is the

inverse of the derivative of the recombination rate with respect to the minority-carrier density. In the case of Shockley–Read–Hall recombination, $\tau = 1/\sigma_p v_{th} N_t$. Therefore, from Eqs. (77b), (79), and (91), we end up at

$$-\nabla(E\mu_p p' - D_p \nabla p') - \frac{p'}{\tau} + G = 0 \quad (92)$$

where $p' = p - p_0$ is the excess minority-carrier density. Because of the field independence on p', Eq. (92) is linear and becomes homogenous if $G = 0$. In the later case, the solution is proportional to $p'(C_j)$, the excess minority-carrier density at the injecting boundary (Fig. 58).

Forward Bias and Low-Level Injection   Under forward bias, a basic assumption will be made. This assumption allows the coupling of the minority-carrier density to the externally applied terminal voltage: the Fermi levels are flat in the regions where the carriers are a majority and also in the space-charge region. Under this condition and from Eqs. (76c), (77c), and (82),

$$p(C_j)n(C_j) = n_i^2 \exp\left(\frac{eV}{kT}\right) \quad (93)$$

$$p'(C_j) = \frac{n_i^2[eV/kT) - 1]}{N_D(C_j)} \quad (94)$$

Equation (93) holds under any injection level, provided that the flat Fermi potential assumption holds, whereas Eq. (94) for the excess minority-carrier density holds only in low-level injection. The proportionality of the solution with respect to $p'(C_j)$ forces all carrier densities and currents to become proportional to the term $\exp(eV/kT) - 1$. Here, we note that the surface recombination is also a linear function of the excess minority-carrier density when $p \ll n$. More analytically, Eq. (84) becomes

$$J_p^n = eS_p p' \quad (95)$$

where S_p is called surface recombination velocity. Therefore, the total base recombination current in Eq. (89) is proportional to the term $\exp(eV/kT) - 1$. The same is true for the quasi-neutral emitter recombination. Thus, Eq. (89) becomes

$$I_D = (I_{0e} + I_{0b})\left[\exp\left(\frac{eV}{kT}\right) - 1\right] + I_{SCR} \quad (96)$$

where the preexponential factors I_{0b} and I_{0e} are called base and emitter saturation currents, respectively. Equation (94) points out that the saturation currents are proportional to n_i^2.

The space-charge region recombination is a current component that is hard to express in analytical terms. This difficulty arises because in this region the field depends on the bias and there is no such entity as a minority carrier. Consequently, the linearity conditions that allowed us to derive Eq. (96) no longer hold. To derive an approximate expression for the bias dependence of I_{SCR}, certain simplifications must be made throughout the space-charge region regarding the integral of Eq. (90). These simplifications result in a bias dependence of the form $\exp(eV/nkT) - 1$, where n, the ideality or slope factor, takes values from 1 to 2.[65] The specific value depends on the trap position in the energy gap, the doping profiles, and the cross section for hole capture relative to the cross section for electron capture. This range for n holds provided that the capture coefficients do not depend on the electric field. Now the preexponential factor I'_{CSCR} is proportional to n_i. Finally, the expression for the forward current of a diode in the base of which low-level injection conditions prevail becomes

$$I_D = (I_{0e} + I_{0b}) \left[\exp\left(\frac{eV}{kT}\right) - 1 \right]$$
$$+ I'_{SCR} \left[\exp\left(\frac{eV}{nkT}\right) - 1 \right] \qquad (97)$$

For voltages higher than $3nkT/e$, the unity can be dropped from Eqs. (96) and (97). Because of a better slope factor, the emitter and base recombination will dominate the diode current for voltages above a certain level. Below this level, the space-charge region recombination must be considered too. Such trends are seen in Fig. 59. Curve 1 is the $I-V$ characteristic of a diode with a base doping of 5.5×10^{14} cm^{-3} and has an ideality factor of 1 in the bias range from 0.2 up to 0.4 V. For lower voltages, the space-charge region recombination slightly increases the ideality factor and makes the measured current deviate from the $\exp(eV/kT) - 1$ dependence. The ideality factor also increases for voltages above 0.4 V because of high-injection effects, which are discussed in the next subsection. The device of curve 2 has a very light doping density in the base, 4.5×10^{11} cm^{-3} and is driven in high injection at even smaller bias. As discussed in the next subsection, curve 2 exhibits unity slope factor even at very low voltages.

Curve 3 shows what happens if the temperature is reduced to 78 K. The sharp reduction of the intrinsic-carrier density due to its $\cong \exp(-E_g/2kT)$ dependence requires much higher voltages to reach the same current as at 300 K. In fact, to reach a current density of 10 mA/cm^2, a voltage in excess of 1 V is required. The reduction of n_i reduces the recombination in the base and the emitter is much faster than in the space-charge region because the proportionality constants are n_i^2 and n_i, respectively. Therefore, at low temperatures, the bias regions with higher than 1 ideality factor are

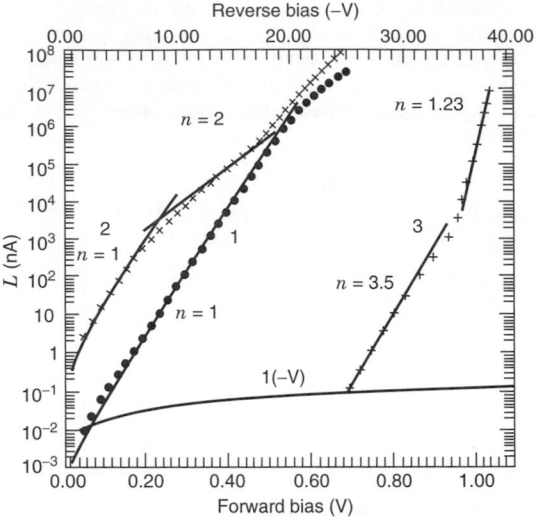

Fig. 59 Experimental $I-V$ characteristics of two different diodes. Diode 1 has a base thickness of 250 μm, a base-doping density of 5.5×10^{14} cm^{-3}, and an area of 2.9×2.9 mm^2. Diode 2 has a base thickness of 300 μm, a base-doping density of 4.5×10^{11} cm^{-3}, and an area of 5×5 mm^2. The base in both devices is of n type. Plot 3 is the $I-V$ characteristic of diode 2 at 78 K. The other plots are at 300 K. Curve 1(−V) is the reverse bias characteristic, with reversed sign, of diode 1 (top axis). The straight lines in curves 1, 2, and 3 are the exponential $\exp(eV/nkT) - 1$ fits to the experimental points. The slope factor n is also shown.

expected to be wider. This is evidenced in curve 3 of Fig. 59, where the ideality factor is 3.5 for voltages below 950 mV. The increase of the ideality factor above 2 is a result of the Poole–Frenkel effect, which reduces the effective energy separation of the traps from the bands.[69] The influence of small values of n_i on the ideality factor is evident not only when the temperature drops but also when the bandgap increases. In several compound semiconductor devices, their large bandgap, compared to 1.1 eV of silicon, results in an intrinsic-carrier density, which is several orders of magnitude smaller than the 10^{10} cm^{-3} value for silicon at 300 K.[70,71] Consequently, their $I-V$ characteristics show slope factors substantially larger than 1 for the entire range of bias. On the contrary, germanium diodes have slope factors of 1 even at reduced temperatures because of the smaller gap, 0.66 eV, of the semiconductor.

One-Dimensional Case. Equation (97) holds for any three-dimensional geometry and doping profiles because no assumption, except for low-level injection, was made so far regarding doping profiles and device topology. If, however, we want to express in close form the saturation values of the emitter and base recombination currents, then one-dimensional devices with uniform doping profiles must be considered. In such a case, the one-dimensional, homogenous, and constant-coefficient version of Eq. (92) becomes

$$\frac{d^2 p'}{dx^2} = \frac{p'}{L_p^2} \tag{98}$$

where $L_p = \sqrt{D_p \tau}$ is the minority-carrier diffusion length. The first boundary condition for Eq. (98) is Eq. (94) applied at the injecting boundary. The other one refers to the ohmic contact. If it is an ideal ohmic contact deposited directly on the uniformly doped base, then the second boundary condition becomes, from Eq. (82), $p'(l) = 0$. Here, l is the base length and the coordinate origin is at the injecting boundary, as shown in Fig. 58. In many cases, between the ohmic contact and the uniformly doped base, a thin and heavily doped region intervenes.

This region has thickness on the order of a micrometer and a doping of the same type as the rest of the base. The purpose of such a layer, called back-surface field, is to provide a better ohmic contact and to isolate the contact from the lightly doped base so that carrier recombination generation is reduced.[72] Such a back-surface field terminates the lightly doped base of diode 2 in Fig. 59 making it a $p-i-n$ diode, where i stands for intrinsic. Therefore, in the presence of this contact layer, the base ends at a "low/high" $n-n^+$ junction. In terms of minority-carrier recombination, this interface is characterized by an effective recombination velocity S_{pe}, experienced by the minority carriers at the low side of the junction. The expression for S_{pe} is

$$S_{pe} = \frac{I_{0c} N_D}{e n_i^2 S} \tag{99}$$

where I_{0c} is the saturation value of the recombination current in the back-surface field and S is the device cross section. Equation (99) can be derived from Eq. (95), by applying Eq. (94) at the $n-n^+$ junction and by equating the minority current at the low/high junction to the recombination in the heavily doped region.

Under the previous boundary conditions, the solution of Eq. (98) yields for the base saturation current:

$$I_{0b} = S \frac{e n_i^2}{N_D} \frac{D_p}{L_p} \frac{1 + [D_p/(S_{pe} L_p)] \tanh(\ell/L_p)}{\tanh(\ell/L_p) + D_p/(S_{pe} L_p)} \tag{100}$$

Equation (100) shows that, in terms of the one-dimensional geometry, the quantity that matters is the ratio l/L_p. Values of this ratio much less than one define the short base, whereas values above 3 define the long base. In the long base case, Eq. (100) becomes $I_{0b} = S e n_i^2 D_p / (N_D L_p)$. Similar equations hold for a uniform

emitter, too, but now the heavy doping effects could modify the value of n_i^2, as will be discussed in the section on heavy doping effects. If the base doping is very light, as in a $p-i-n$ diode, then the increased value of I_{0b} will make the base recombination dominate the current components in Eq. (97). Accordingly, the influence of space-charge region recombination current on the slope factor will be suppressed even for voltages as low as a few kT/e, as shown in Fig. 59, curve 2. Also, by extrapolating the $\exp(eV/kT)$ fit of curve 2 at zero voltage, a base recombination current of 0.7 nA is obtained. This corresponds to a 300-K saturation current density of 2.8 nA/cm^2 compared to emitter saturation current densities on the order of pA/cm^2. On the other hand, diode 1, with a base-doping density three orders of magnitude higher than that in diode 2, exhibits a saturation current of 24 pA/cm^2. This saturation current comes mainly from the base recombination as a result of its relatively light doping density and the absence of a back-surface field which gives S_{pe} very high values.

Equation (100) applies to uniformly doped regions. If the doping is nonuniform, close form expressions are not possible, in the general case. This is the case because the one-dimensional version of Eq. (92) is still an ordinary differential equation with nonconstant coefficients. However, analytical approximations can be derived based on iterative techniques.[73]

Diffusion in Three Dimensions. Equation (98) holds provided the cross-sectional dimensions of the diode are much larger than the diffusion length. Otherwise, lateral diffusion of minority carriers in the base becomes important. In such a case, the three-dimensional version of Eq. (98) takes the form

$$\nabla^2_{p'} = \frac{p'}{L_p^2} \tag{101}$$

The last equation can be solved very accurately by semianalytical techniques based on the two-dimensional Fourier transform.[74] Simulation results are as shown in Fig. 60. As illustrated, in the case of a point contact diode having emitter dimensions of $0.1 L_p$, the base recombination is expected to increase by a factor of 25 as a result of the lateral carrier diffusion.

High-Level Injection So far, our analytical approaches were based on the low-level injection assumption. In high-level injection, where $n = p$, an equation similar to Eq. (98) can be derived where now the hole diffusion length is replaced by the ambipolar diffusion length.[75] The boundary conditions, however, are not linear and depend on the electric field, which, now, is a function of bias. If the quasi-Fermi potentials are flat in the quasi-neutral base, then the electron–hole plasma density p is space independent and equals n_i $\exp(eV/2kT)$, as can be derived from Eq. (93). In such

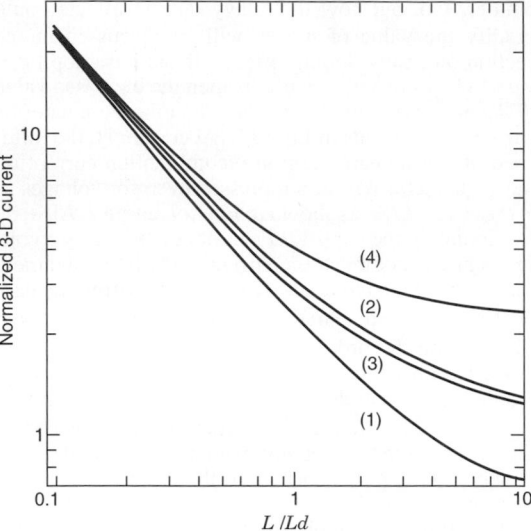

Fig. 60 Three-dimensional diffusion base saturation current of a planar $p-n$ junction with a square emitter having a side length L. The top surface of the base is supposed to have zero recombination velocity. The current is normalized with respect to the one-dimensional diffusion current $I_{0b} = en_i^2 L^2 D_p / N_D L_p$. In curve 2, $S_{pe} = 0$ and $l = L_p/2$. In curve 2, l is assumed to be infinite. In curve 3, $S_{pe} = D_p/L_p$ and $l = L_p/2$, whereas in curve 4, S_{pe} is assumed to be infinite and $l = L_p/2$.

a case, by integrating the recombination current in the emitter, the base, and the back-surface field, we obtain

$$I_D = (I_{0e} + I_{0c}) \exp\left(\frac{eV}{kT}\right) + \left(\frac{eSl}{\tau'}\right) \exp\left(\frac{eV}{2kT}\right) n_i \tag{102}$$

where τ' is the high-injection lifetime defined as the ratio of the recombination rate divided by the plasma density. The flat Fermi level condition can easily be satisfied in $p-i-n$ diodes where the light base doping density makes the high-level injection possible even at a bias of 0.4 V. In Fig. 59, curve 2 shows the $\exp(eV/2kT)$ dependence, or slope factor of 2, for voltages of about 0.4 V, which drive the $p-i-n$ device to high-level injection. For even higher voltages, the emitter and back-surface field recombination in Eq. (102) starts dominating the current, and the slope factor drops again. For higher base doping densities, as in curve 1 of diode 1, the required voltage for high-level injection conditions could exceed 0.5 V at 300 K. Now, the heavily doped region recombination in Eq. (102) competes with the bulk recombination, and the slope factor of 2 does not appear. The bent of both curves 1 and 2 at voltages near 0.6 V is a result of series resistance effects, which invalidate the assumption of flat Fermi levels across the base. In

such a case, the simulation is possible only by device simulators that solve the complete system of the transport equations.

Reverse Bias   Under reverse bias where $V < 0$, the assumption of flat Fermi levels across the space-charge region that led to Eq. (97) no longer hold. On the other hand, however, the space-charge region can be considered to be fully depleted from free carriers. In such a case, Eq. (86) holds with V_{bi} replaced by $V_{bi} + |V|$. Therefore, the depleted space-charge region will expand toward the base according to the square root of the bias for $|V| > 5$ V. In this region, the Shockley–Read–Hall Eq. (90) predicts a negative recombination or generation of electron–hole pairs. This generation current is the basic component of the leakage current in reverse bias. The contribution of the diffusion components from the base and the emitter, $-I_{0b} - I_{0e}$, is usually negligible unless the base is very lightly doped. The bottom line in Fig. 59 shows the reverse-bias current for diode 1. The square-root dependence on voltage is not exactly obeyed because of the Poole–Frenkel effect, which increases the generation rate at higher fields.

4.5 Transient Response of Diodes

If a diode is subjected to a transient terminal bias, then in addition to currents due to carrier diffusion and recombination, we also have the dielectric displacement current resulting from the time dependence of the electric field. If low-level injection is observed in the quasi-neutral regions, the displacement current is restricted in the space-charge region. At the same time, low-level injection ensures that linearity holds in the base and the emitter, and Eq. (92) still applies with $\partial p'/\partial t$ replacing zero in the right-hand side of the relation. The solution of the time-dependent edition of Eq. (92) provides the minority-carrier currents at the injecting boundaries of the base and the emitter, $I_b(t)$ and $I_e(t)$, respectively. These currents have now two components: the minority-carrier recombination and the minority-carrier storage current $\partial Q'/\partial t$, where Q' is the total excess minority-carrier charge. To calculate the total transient current, reconsider Eq. (89) in its transient version. Therefore, in addition to $I_b(t)$ and $I_e(t)$, the transient space-charge region current is required. Unlike the base and the emitter, this current in addition to the recombination and storage component also includes the displacement current.[76] Insofar as the displacement current is concerned, the space-charge region behaves as a parallel plate capacitor with a plate distance $W = W_A + W_D$, Eq. (86), a dielectric constant ε and a capacitance $C_{SCR} = \varepsilon S/W$. During transit, the dielectric displacement current is supplied by the majority carriers from either side of the junction.

To calculate the transient currents in the base and the emitter, the boundary conditions must be defined. Boundary condition Eq. (95) holds because of linearity. The other condition at the injecting boundary depends on the kind of transient to be considered (89). Here we will assume that the device is in equilibrium for $t < 0$, whereas at $t = 0$ a constant voltage V is applied. We can now assume that Eq. (94) applies with $p'(C_j)$ replaced by $p'(C_j, t)$ for $t > 0$. This assumption has a validity range depending on how fast the flat quasi-Fermi potential condition can be established across the space-charge region. As a matter of fact, even in the absence of series resistance effects, it takes a short time for this condition to be established. This short time relates to the dielectric response time of the majority carriers and the minority-carrier diffusion time across the space-charge region.[77] For almost all practical cases, the delay in establishing a fixed minority-carrier density at the edge of the quasi-neutral region will not exceed the limit of a few tens of a picosecond[77] in the absence of series resistance effects. Therefore, if the time granularity used in solving the time-dependent version of Eq. (92) is restricted to about a nanosecond, then the solutions will be accurate. In practical cases, however, the very first part of the transient current, following the sudden application of a voltage, will be determined by charging C_{CSR} through the series resistance of the majority carriers in the base and the emitter. The respective time constant could be on the order of a nanosecond. In such a case, the minority-carrier transport in the base will determine the transient only after several nanoseconds have elapsed since the application of the voltage. The transient base transport can be expressed in semianalytical forms using Laplace transform techniques,[77] especially in the case of uniform and one-dimensional quasi-neutral regions. In a long-base diode, the transition will last for about a minority-carrier lifetime. In a short-base device with an ohmic contact at the base end, the transient will last approximately $l^2/2D_p$, which is the minority-carrier diffusion time through the base.

Small-Signal Response In many cases, the device operates under sinusoidal small-signal excitation superimposed on a steady-state excitation. In such cases, Eq. (92) still holds, but now $1/\tau$ will have to be replaced by $1/\tau + j\omega$, where j is the imaginary unit and ω is the angular frequency of the excitation. This is the case because the time derivative of the small-signal carrier density is the carrier density amplitude times $j\omega$. Having done the complex lifetime replacement, the analysis that followed Eq. (92) still holds. Now, however, the small-signal value of the excess minority-carrier density at the injecting boundary will be the steady-state value in Eq. (94) times ev/kT. Here, v is the small-signal terminal voltage, which is supposed to be much less than kT/e. Under low-level injection and in view of the previous transient response discussion, the small-signal version of Eq. (97) will refer to a terminal current I_D^* having a real and an imaginary component:

$$I_D^* = (I_{0e}^* + I_{0b}^*)\frac{ev}{kT}\left[\exp\left(\frac{eV}{kT}\right) - 1\right]$$

$$+ I_{SCR}^*\frac{ev}{nkT}\left[\exp\frac{eV}{nkT} - 1\right]$$

$$+ j\omega C_{SCR}v = v(G + j\omega C) \tag{103}$$

The star exponents denote the complex values of the saturation currents as a result of the complex lifetime. In Eq. (103), G and C are the diode small-signal parallel conductance and capacitance, respectively. These two components are of great importance because their frequency dependence can reveal minority-carrier properties, such as diffusivity and lifetime[78] and allow the device circuit representation when the diode is part of a greater small-signal circuit. For uniformly doped quasi-neutral regions, I_{0e}^* and I_{0b}^* can be obtained from Eq. (100) by replacing the diffusion length $L = (D\tau)^{1/2}$ with the complex diffusion length $L^* = L/(1 + j\omega\tau)^{1/2}$. For frequencies sufficiently high, the magnitude of the complex diffusion length will become much shorter than the base thickness. Then, the complex version of Eq. (100) predicts that the base current would change as $1/L^*$. If the base component in Eq (103) were to dominate, then beyond a certain frequency, C would change as $\omega^{-1/2}$ while G would change as $\omega^{1/2}$.

This frequency dependence is confirmed in Fig. 61, which shows the frequency response of diode 1, from

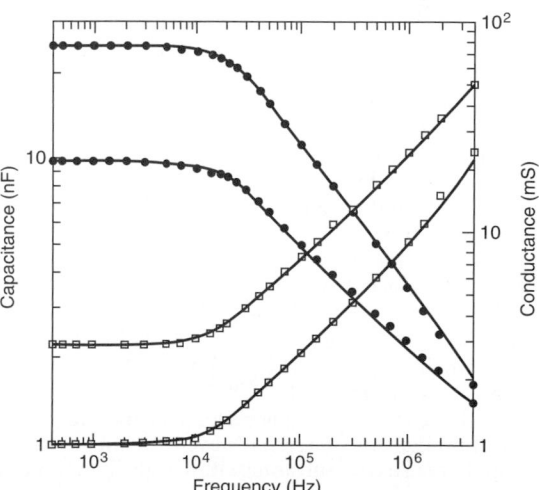

Fig. 61 Experimentally measured capacitance (dots) and conductance (squares) at 300 K for diode 1. The bottom and the top curves correspond to two different bias points: 420 and 450 mV, respectively. The solid curves are the theoretical fits from the equivalent circuit of Fig. 62.

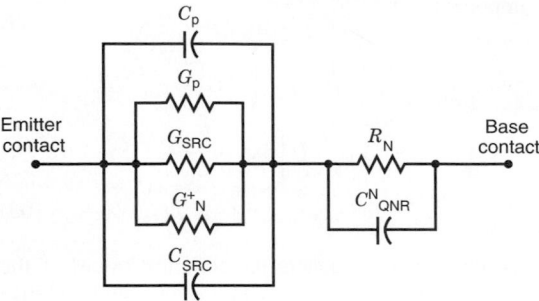

Fig. 62 The equivalent circuit model of a diode. The base injection currents in Eq. (103) corresponds to C_p, imaginary part, and G_p, real part. The injection in the emitter is represented by G_N^+. The space-charge region recombination is represented by G_{SCR}, whereas C_{SRC} is the space-charge region capacitance. The rest of the components are accounted for in the text.

Fig. 59, at two bias points. The theoretical fit to the experimental results was obtained on the basis of the diode equivalent circuit shown in Fig. 62. This circuit includes all the components relating to carrier injection and storage in the device's three regions in accordance with Eq. (103). It also includes the base resistance R_N, which has been ignored in Eq. (103). In Fig. 61, the square root law is better obeyed at the higher bias point and for frequencies less than 1 MHz, especially for the capacitance. This is a combined result of the space-charge region capacitance, the relative contribution of which increases at lower bias, and the series resistance, the influence of which is stronger at high frequencies. The corner frequencies of the conductance and the capacitance depend on the base thickness and the lifetime. The fit shown in Fig. 61 gave a minority-hole lifetime in the n-type base of about 30 μs. Such a lifetime and Eq. (100) imply that the saturation current density of 24 pA/cm^2 at 300 K, as shown in Fig. 59, is 90% due to base recombination. The emitter contribution of 10% is discussed in the section on heavy doping effects.

The series resistance R_N becomes the bulk majority-carrier resistance under reverse bias or even under forward bias, provided that the frequency is high ($|L^*| \ll l$). The capacitance C_{QNR}^N in parallel with R_N, as shown in Fig. 62, is the geometrical capacitance of the quasi-neutral base.[79,80] For ordinary resistivity devices, it can be ignored unless the frequency is in the gigahertz range. However, for diodes made on high resistivity substrates, this capacitance must be considered especially at reverse bias and high frequencies.[80] From Fig. 62 and in the limit of very high frequencies under forward bias, the parallel conductance saturates at $1/R_N$ whereas the parallel capacitance does so at C_{QNR}^N. This is because of the combination of the increasing injection conductancies and

the space-charge region capacitance. Then, the product $R_N C_{QNR}^N$ becomes the dielectric response time of the majority carriers in a uniform base. At high injection, the parallel conductance will saturate at the sum of the two carrier conductances.[81] Under reverse bias, the circuit of Fig. 62 reduces to the space-charge region capacitance in series with the parallel combination of C_{QNR}^N and R_N. Unlike the forward-bias case, where the circuit parameters depend roughly exponentially on the terminal voltage V, in reverse bias the voltage dependence would be restricted to $V^{-1/2}$. In the sense of the voltage dependence, the circuit of Fig. 62 is the circuit of a varactor.

4.6   Heavy Doping Effects in Emitter

In the previous subsection, the emitter saturation current density was estimated to be about 2 pA/cm^2. From Eq. (100) and by assuming microsecond lifetimes, we would expect saturation currents on the order of a fA/cm^2 from an emitter doped in the range 10^{19}–10^{20} cm^{-3}. Such a discrepancy by three orders of magnitude is due to the heavy doping effects, namely the short lifetime resulting from Auger recombination and the effective increase of n_i due to bandgap narrowing. In the Auger recombination process, a minority carrier recombines directly with a majority one, and the energy is transferred to another majority carrier. Because of such kinetics, the Auger minority-carrier lifetime is inversely proportional to the square of the majority-carrier density. The proportionality constant is $\sim 10^{-31}$ cm^6/s for minority electrons in p^+ emitters and 3×10^{-31} cm^6/s for minority holes in n^+ emitters.[82] In heavily doped regions, the Auger recombination rate is by far higher than the Shockley–Read–Hall rate and determines the lifetime. Therefore, nanosecond lifetimes are expected, especially for holes, in emitters doped in the vicinity of 10^{20} cm^{-3}.

In a heavily doped region, every minority carrier interacts strongly with the majority carriers because of their high density. The minority-majority carrier attraction along with the carrier-dopant interaction and the semiconductor lattice random disruption by the dopant atoms reduces the banggap and changes the density of states in both bands.[83,84] The net result is an effective shrinkage of the gap depending on the doping type and density.[85-87] This shrinkage changes the intrinsic-carrier density n_i to a much higher effective n_{ie}. The result of the band distortion is that the original system of transport equations [Eqs. (76)–(81)] no longer holds. More specifically, Eqs. (76b), (76c), (77b), and (77c) are not valid for the majority carriers even if n_{ie} substitutes n_i because Boltzmann statistics must be replaced by Fermi–Dirac statistics. Also, Eq. (81) no longer holds in a nonuniform region because the band edges are not parallel any more and each carrier experiences a different field. However, the minority carriers still follow the Boltzmann statistics, and Eq. (92) holds for the minority carriers. Now E is the minority-carrier field $(1/e\nabla E_c$ for electrons), and the boundary condition Eq. (94) is valid with n_i replaced by n_{ie}. Therefore, Eq. (100) still applies for the minority-carrier recombination in a uniformly doped emitter. For an emitter doped at about 10^{20} cm^{-3}, a gap narrowing of about 100 meV is expected,[15-87] which makes n_{ie} several tens higher than n_i. If such an n_{ie} as well as nanosecond lifetimes replace n_i and microsecond lifetimes in Eq. (100), an emitter saturation current on the order of pA/cm^2 is predicted, in accordance with the experimental results of the previous section.

4.7 Diodes of Nonconventional Transport

So far in this section, devices based on the drift and diffusion model of Eqs. (76) and (77) were studied. Charge carriers can be transported from one region to another by tunneling. Also, they can be temporarily trapped in energy-gap states, atom clusters, or crystallites imbedded in insulating films, thereby affecting the tunneling or the conventional transport of the free carriers.

In this respect, the first device to be examined is the $p-i-n$ diode 2 of Fig. 59, operating at cryogenic temperatures. Around 4.2 K, the equilibrium Fermi level in the lightly doped n^- region is pinned at the donor level. These levels, now, are not ionized except for a fraction to compensate the charge of the unintentionally introduced acceptor ions. At such low temperatures, there are no free carriers in the base, and no measurable conduction is possible unless the voltage is raised enough to achieve the flat-band condition.[88,89] For silicon, this voltage V_0 would be about 1.1 V. For even higher voltages, conduction is possible only if electrons and holes can be injected in the frozen substrate from the n and p regions, respectively. In this sense, Eq. (99) based on the assumption of flat majority-carrier Fermi levels no longer holds. For $T < 10$ K, injection is possible by carrier tunneling through the small potential barrier existing at each of the $p-i$ and $i-n$ interfaces.[59] These barriers exist because of the band distortion in the heavily doped regions and the smaller gap there, as outlined in the previous section. For $V > V_0$, electrons tunnel in the i layer, and the higher the forward bias, the higher the current due to a field-induced effective lowering of the barriers.

As shown in Fig. 63, for temperatures below 10 K it takes at least several volts to establish a current of few nanoamps. The injected electrons in the i layer are trapped by the ionized donors and built a space charge and a subsequent potential barrier. For even higher voltages approaching 10 V, the barrier at the $i-p$ interface lowers, holes now enter the i layer in large numbers. Their charge neutralizes the trapped electron charge and causes the voltage breakdown and the negative differential resistance that appears in Fig. 63 for $T < 10$ K. The negative resistance persists

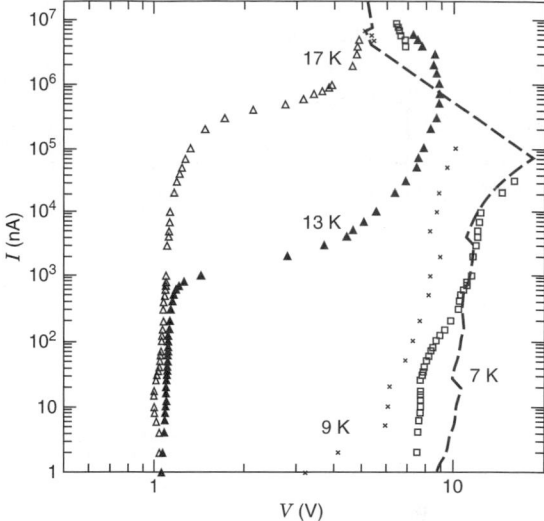

Fig. 63 Measured $I-V$ characteristics of diode 2 at cryogenic temperatures. The square points correspond to 4.2 K. The $T < 10$ K plots exhibit a distinct voltage breakdown. Reprinted from K. Misiakos, D. Tsamakis, and E. Tsoi, Measurement and modeling of the anomalous dynamic response of high resistivity diodes at cryogenic temperatures, *Solid State Electronics*, **41**: 1099–1103, 1997, with kind permission from Elsevier Science Ltd., The Boulevard, Langford Lane, Kidlington 0X5 1 GB, UK.

and beyond breakdown as a result of new carrier generation by the impact ionization of occupied shallow donors by the injected carriers. The interaction of free and trapped carriers through impact ionization gives rise to a negative dynamic conductance and capacitance which for frequencies high enough change as ω^{-2}.[90] For $T > 10$ K the injection mechanism changes to thermion emission over the interface potential barriers, whereas the space-charge effects are now less pronounced.

Another example of tunneling injection mechanism is the breakdown effect in zener diodes. Here, the base is quite heavily doped ($\approx 10^{18}$ cm^{-3}), and the strong electric field in the space-charge region increases even further by applying a reverse bias. For fields approaching 10^6 V/cm, a valence band electron can tunnel to a conduction band state of the same energy. This way, electron–hole pairs are created, and the reverse current sharply increases. Another diode structure based on tunneling is a new metal–insulator–semiconductor device having silicon nanocrystals imbedded in the thin insulating film.[91] One way to realize such diodes is by depositing an aluminum electrode on a thin (on the order of 10 nm) SiO$_2$ layer containing silicon nanocrystals. The substrate is n-type crystalline silicon. The silicon nanocrystals can be created either by oxidizing deposited amorphous silicon layers[91] or by low-energy silicon ion implantation in the SiO_2 film.[92] In the absence of the nanocrystals, by applying a negative voltage of a few volts on the aluminum electrode relative to the n-type silicon substrate, only a small tunneling current would be present.

When the nanocrystals are introduced, much higher currents are observed while the conductance curve exhibits characteristic peaks. Such peaks are shown in Fig. 64 showing the reverse current and conductance of a quantum dot diode formed by low-energy implantation of silicon in a 10-nm SiO_2 layer.[92] The conductance peaks appear when the metal Fermi level is swept across the discrete energy states of the nanocrystals, thus enabling resonant tunneling from the metal to the semiconductor.[91] The three-dimensional confinement of electrons in the quantum-box crystallites creates a large separation between energy states, which along with the Coulomb blockade effect of the occupied states explains the large voltage separation of the three first conductance peaks in Fig. 64.[91,92] Such quantum dot devices hold the

promise of single-electron transistors[93] and silicon-based light-emitting diodes.[94]

As we end this section, we would like to mention the basic uses of the diode as a device. The most frequent use of the diode is the protection of CMOS integrated circuits from electrostatic discharges by clamping the output pads to the power-supply voltages through reverse-biased $p-n$ junctions. In analog integrated circuits, forward-biased diodes are used for voltage shifting. Such diodes usually come from properly wired bipolar transistors (e.g., emitter–base diodes with base–collector short circuited). Diodes, as discrete devices, find applications mainly as rectifying elements in power circuits. The breakdown effect of zener diodes makes these devices useful as voltage reference sources in power supplies. Photodiodes are widely used for detecting photons or charge particles. Finally, large area diodes with exposed front surface and proper design and engineering can efficiently convert solar light into electricity and are used as solar cells.[95]

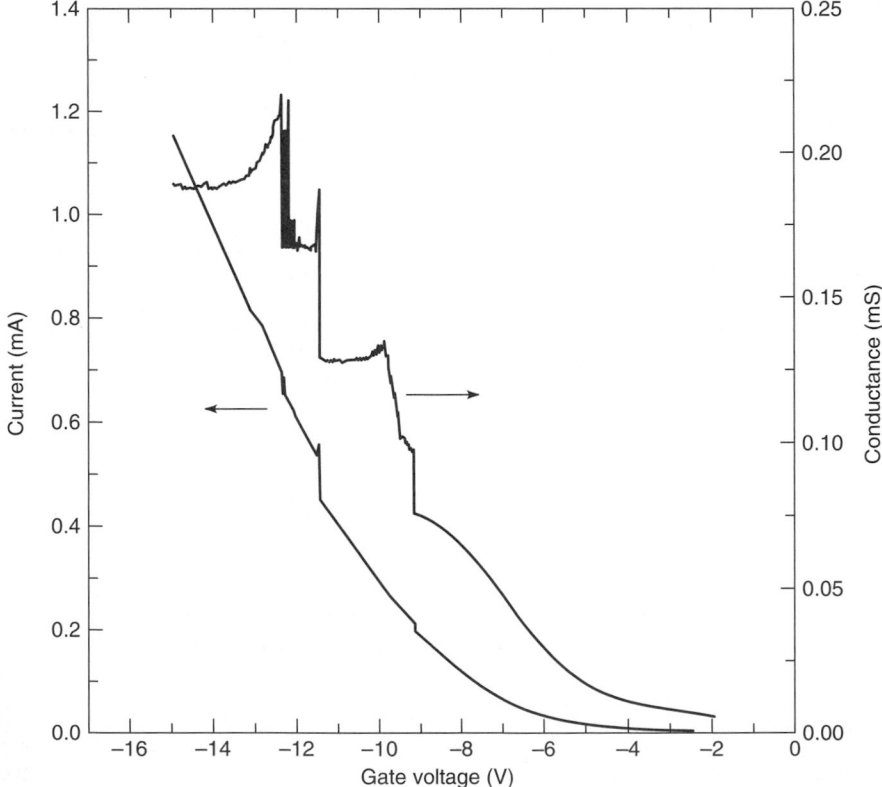

Fig. 64   Current and conductance plots of a reverse-biased quantum-dot diode. The conductance peaks correspond to steps in current curve. Reprinted from P. Normand et al., Silicon nanocrystal formation in thin thermal-oxide films by very low energy Si[+] ion implantation, *Microelectronic Engineering*, **36**(1–4): 79–82, 1997, with kind permission of Elsevier Science-NL, Sara Burgerharstraat 25, 1055 KV Amsterdam, The Netherlands.

5 ELECTRONIC COMPONENTS
Clarence W. de Silva

5.1 Materials and Passive Components
Conductive Material and Components

Conductance and Resistance. When a voltage is applied across a conductor, a current will flow through the conductor. For a given voltage v (volts), the current i (amperes) will increase with the conductance G of the conductor. In the linear range of operation, this characteristic is expressed by Ohm's law:

$$i = Gv$$

Resistance $R(\Omega)$ is the inverse of conductance:

$$R = \frac{1}{G}$$

Silver, copper, gold, and aluminum are good conductors of electricity.

Resistivity. For a conductor, resistance increases with the length (L) and decreases with the area of cross section (A). The corresponding relationship is

$$R = \frac{\rho L}{A}$$

The constant of proportionality ρ is the resistivity of the conducting material. Hence, resistivity may be defined as the resistance of a conductor of unity length and unity cross-sectional area. It may be expressed in the units $\Omega \cdot cm^2 / cm$ or $\Omega \cdot cm$. A larger unit would be $\Omega \, m^2/m$ or $\Omega \, m$.

Alternatively, resistivity may be defined as the resistance of a conductor of unity length and unity diameter. According to this definition,

$$R = \frac{\rho L}{d^2}$$

where d represents the wire diameter. If the wire diameter is 1 mil (or 1/1000 in), the wire area would be 1 circular mil (or cmil). Furthermore, if the wire length is 1 foot, the units of ρ would be $\Omega \cdot cmil/ft$. Resistivities of several common materials are given in Table 8.

Effect of Temperature. Electrical resistance of a material can change with many factors. For example, the resistance of a typical metal increases with temperature, and the resistance decreases with temperature for many nonmetals and semiconductors. Typically, temperature effects on hardware have to be minimized in precision equipment, and temperature compensation or calibration would be necessary. On the other hand, high-temperature sensitivity of resistance in

Table 8 Resistivities of Some Materials

Material	Resistivity ρ ($\Omega \cdot m$) at 20°C(68°F)
Aluminum	2.8×10^{-8}
Copper	1.7×10^{-8}
Ferrite (manganese-zinc)	20.0
Gold	2.4×10^{-8}
Graphite carbon	775.0×10^{-8}
Lead	9.6×10^{-8}
Magnesium	45.8×10^{-8}
Mercury	20.4×10^{-8}
Nichrome	112.0×10^{-8}
Polyester	1×10^{10}
Polystyrene	1×10^{16}
Porcelain	1×10^{16}
Silver	1.6×10^{-8}
Steel	15.9×10^{-8}
Tin	11.5×10^{-8}
Tungsten	5.5×10^{-8}

Note: Multiply by 6.0×10^8 to obtain the resistivity in $\Omega \cdot emil/ft$.

some materials is exploited in temperature sensors such as RTDs and thermistors. The sensing element of an RTD is made of a metal such as nickel, copper, platinum, or silver. For not too large variations in temperature, the following linear relationship could be used:

$$R = R_0(1 + \alpha \, \Delta t)$$

where R is the final resistance, R_0 is the initial resistance, ΔT is the change in temperature, and α is the temperature coefficient of resistance.

Values of α for several common materials are given in Table 9. These values can be expressed in ppm/°C (parts per million per degree centigrade) by multiplying each value by 10^6. Note that graphite has a negative temperature coefficient, and nichrome has

Table 9 Temperature Coefficients of Resistance for Several Materials

Material	Temp. Coeff. Resistance α(per °C) at 20°C(68°F)
Aluminum	0.0040
Brass	0.0015
Copper	0.0039
Gold	0.0034
Graphite carbon	−0.0005
Iron	0.0055
Lead	0.0039
Nichrome	0.0002
Silver	0.0038
Steel	0.0016
Tin	0.0042
Tungsten	0.0050

a very low temperature coefficient of resistance. A platinum RTD can operate accurately over a wide temperature range and possesses a high sensitivity (typically 0.4 $\Omega/^\circ$C).

Thermistors are made of semiconductor material such as oxides of cobalt, copper, manganese, and nickel. Their resistance decreases with temperature. The relationship is nonlinear and is given approximately by

$$R = R_0 e^{-\beta(1/T_0 - 1/T)}$$

where the temperatures T and T_0 are in absolute degrees (kelvins or rankines), and R and R_0 are the corresponding resistances. The parameter β is a material constant.

Effect of Strain. The property of resistance change with strain in materials, or piezoresistivity, is used in strain gauges. The foil strain gauges use metallic foils (e.g., a copper–nickel alloy called constantan) as their sensing elements. The semiconductor strain gauges use semiconductor elements (e.g., silicon with the trace impurity boron) in place of metal foils. An approximate relationship for a strain gauge is

$$\frac{\Delta R}{R} = S_s \varepsilon$$

where ΔR is the change in resistance due to strain ε, R is initial resistance, and S_s is the sensitivity (gauge factor) of the strain gauge.

The gauge factor is of the order of 4.0 for a metal-foil strain gauge and can range from 40.0 to 200.0 for a semiconductor strain gauge.[96]

Temperature effects have to be compensated for in high-precision measurements of strains. Compensation circuitry may be employed for this purpose. In semiconductor strain gauges, self-compensation for temperature effects can be achieved due to the fact that the temperature coefficient of resistance varies nonlinearly with the concentration of the dope material.[96] The temperature coefficient curve of a p-type semiconductor strain gauge is shown in Fig. 65.

Superconductivity. The resistivity of some materials drops virtually to zero when the temperature is decreased close to absolute zero, provided that the magnetic field strength of the environment is less than some critical value. Such materials are called superconducting materials. The superconducting temperature T (absolute) and the corresponding critical magnetic field strength H are related through

$$H = H_0 \left(1 - \frac{T}{T_c}\right)^2$$

where H_0 is the critical magnetic field strength for a superconducting temperature of absolute zero, and T_c

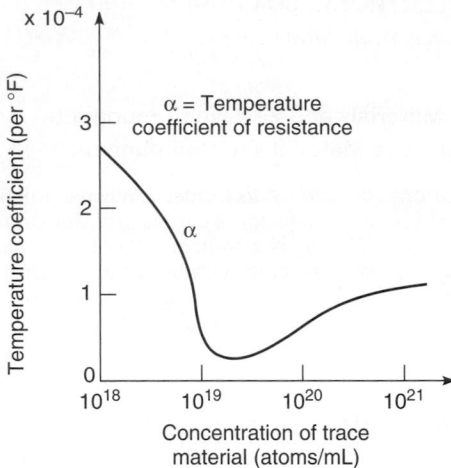

Fig. 65 Temperature coefficient of resistance of p-type semiconductor strain gauge.

is the superconducting temperature at zero magnetic field. The constants H_0 and T_c for several materials are listed in Table 10.

Superconducting elements can be used to produce high-frequency (e.g., 1×10^{11} Hz) switching elements (e.g., Josephson junctions) that can generate two stable states (e.g., zero voltage and a finite voltage, or zero magnetic field and a finite magnetic field). Hence, they are useful as computer memory elements. Other applications of superconductivity include powerful magnets with low dissipation (for medical imaging, magnetohydrodynamics, fusion reactors, particle accelerators, etc.), actuators (for motors, magnetically levitated vehicles, magnetic bearings, etc.), sensors, and in power systems.

Color Code for Fixed Resistors. Carbon, wound metallic wire, and conductive plastics are commonly used as commercial resistors. A wire-wound resistor element is usually encapsulated in a casing made of an insulating material such as porcelain or bakelite. Axial

Table 10 Superconductivity Constants for Some Materials

Material	T_c(K)	H_0(A/m)
Aluminum	1.2	0.8×10^4
Gallium	1.1	0.4×10^4
Indium	3.4	2.3×10^4
Lead	7.2	6.5×10^4
Mercury	4.0	3.0×10^4
Tin	3.7	2.5×10^4
Vanadium	5.3	10.5×10^4
Zinc	0.9	0.4×10^4

Table 11 Color Code for Fixed Resistors

Color	First Stripe, First Digit	Second Stripe, Second Digit	Third Stripe, Multiplier	Fourth Stripe, Tolerance (%)
Silver	—	—	10^{-2}	±10
Gold	—	—	10^{-1}	±5
Black	0	0	1	—
Brown	1	1	10	±1
Red	2	2	10^2	±2
Orange	3	3	10^3	
Yellow	4	4	10^4	
Green	5	5	10^5	
Blue	6	6	10^6	
Violet	7	7	10^7	
Gray	8	8	10^8	
White	9	9	10^9	

or radial leads are provided for external connection. The outer surface of a fixed resistor is color coded for the purpose of its specification. Four stripes are used for coding. The first stripe gives the first digit of a two-digit number, and the second stripe gives the second digit. The third stripe specifies a multiplier, which should be included with the two-digit number to give the resistance value in ohms. The fourth stripe gives the percentage tolerance of the resistance value. This color code is given in Table. 11

Dielectric Material and Components

Dielectrics and Capacitors. Dielectric materials are insulators, having resistivities larger than 1×10^{12} $\Omega \cdot m$ and containing less than 1×10^6 mobile electrons per cubic meter. When a voltage is applied across a medium of dielectric material sandwiched between two electrode plates, a charge polarization takes place at the two electrodes. The resulting charge depends on the capacitance of the capacitor formed in this manner. In the linear range, the following relationship holds:

$$q = Cv$$

where v is applied voltage (in volts), q is stored charge (in coulombs), and C is capacitance (farads). Since current (i) is the rate of change of charge (dq/dt), we can write

$$i = C\frac{dv}{dt}$$

Hence, in the frequency domain (substitute $j\omega$ for the rate of change operator), we have

$$i = Cj\omega v$$

and the electrical impedance (v/i in the frequency domain) of a capacitor is given by

$$\frac{1}{j\omega C}$$

where ω is the frequency variable, and $j = \sqrt{-1}$.

Permittivity. Consider a capacitor made of a dielectric plate of thickness d sandwiched between two conducting plates (electrodes) of common (facing) area A. Neglecting the fringe effect, its capacitance is given by

$$C = \frac{\varepsilon A}{d}$$

where ε is the permittivity of the dielectric material. The relative permittivity (or dielectric constant) ε_r is defined as

$$\varepsilon_r = \frac{\varepsilon}{\varepsilon_0}$$

where ε_0 = permittivity of vacuum (approx. 8.85×10^{-12} F/m). Relative permittivities of some materials are given in Table 12.

Capacitor Types. The capacitance of a capacitor is increased by increasing the common surface area of the electrode plates. This increase can be achieved, without excessively increasing the size of the capacitor, by employing a rolled-tube construction. Here, a dielectric sheet (e.g., paper or a polyester film) is placed between two metal foils, and the composite is rolled into a tube. Axial or radial leads are provided

Table 12 Dielectric Constants of Some Materials

Material	Relative Permittivity ε_r
Air	1.0006
Carbon dioxide gas	1.001
Ceramic (high permittivity)	8000.0
Cloth	5.0
Common salt	5.9
Diamond	5.7
Glass	6.0
Hydrogen (liquid)	1.2
Mica	6.0
Oil (mineral)	3.0
Paper (dry)	3.0
Paraffin wax	2.2
Polythene	2.3
PVC	6.0
Porcelain	6.0
Quartz (SiO_2)	4.0
Vacuum	1.0
Water	80.0
Wood	4.0

for external connection. If the dielectric material is not flexible (e.g., mica), a stacked-plate construction may be employed in place of the rolled construction to obtain compact capacitors having high capacitance. High-permittivity ceramic disks are used as the dielectric plates in miniature, single-plate, high-capacitance capacitors. Electrolytic capacitors can be constructed using the rolled-tube method, using a paper soaked in an electrolyte in place of the dielectric sheet. When a voltage is applied across the capacitor, the paper becomes coated with a deposit of dielectric oxide that is formed through electrolysis. This becomes the dielectric medium of the capacitor. Capacitors having low capacitances of the order of 1×10^{-12} F (1 pF), and high capacitances of the order of 4×10^{-3} F are commercially available.

An important specification for a capacitor is the breakdown voltage, which is the voltage at which discharge will occur through the dielectric medium (i.e., the dielectric medium ceases to function as an insulator). This is measured in terms of the dielectric strength, which is defined as the breakdown voltage for a dielectric element of thickness 1 mil (1×10^{-3} in). Approximate dielectric strengths of several useful materials are given in Table 13.

Color Code for Fixed Capacitors.
Color codes are used to indicate the specifications of a paper or ceramic capacitor. The code consists of a colored end followed by a series of four dots printed on the outer surface of the capacitor. The end color gives the temperature coefficient of the capacitance in parts per million per degree centigrade (ppm/°C). The first two dots specify a two-digit number. The third dot specifies a multiplier which, together with the two-digit number, gives the capacitance value of the capacitor in picofarads. The fourth dot gives the tolerance of the capacitance. This code is shown in Table 14.

Piezoelectricity.
Some materials, when subjected to a stress (strain), produce an electric charge. These are termed piezoelectric materials, and the effect is called piezoelectricity. Most materials that posses a nonsymmetric crystal structure are known to exhibit the piezoelectric effect. Examples are barium titanate, cadmium sulfide, lead zirconate titanate, quartz, and rochelle salt. The reverse piezoelectric effect (the

Table 14 Color Code for Ceramic and Paper Capacitors

Color	End Color, Temp. Coeff. (ppm/°C)	First Dot, First Digit	Second Dot, Second Digit	Third Dot, Multiplier	Fourth Dot Tolerance For ≤10 pF	Fourth Dot Tolerance For >10 pF
Black	0	0	0	1	±2 pF	±20%
Brown	−30	1	1	10	±0.1 pF	±1%
Red	−80	2	2	1×10^2	−	±2%
Orange	−150	3	3	1×10^3	−	±2.5%
Yellow	−220	4	4	1×10^4	−	−
Green	−330	5	5	−	±0.5 pF	±5%
Blue	−470	6	6	−	−	−
Violet	−750	7	7	−	−	−
Gray	30	8	8	0.01	±0.25 pF	−
White	100	9	9	0.1	±1 pF	±10%

material deforms in an electric field) is also useful in practice.

The piezoelectric characteristic of a material may be represented by its piezoelectric coefficient, k_p, which is defined as

$$k_p = \frac{\text{change in strain(m/m)}}{\text{change in electric field strength(V/m)}}$$

with no applied stress. Piezoelectric coefficients of some common materials are given in Table 15.

Applications of piezoelectric materials include actuators for ink-jet printers, miniature step motors, force sensors, precision shakers, high-frequency oscillators, and acoustic amplifiers. Note that large k_p values are desirable in piezoelectric actuators. For instance, PZT is used in microminiature step motors.[96] On the other hand, small k_p values are desirable in piezoelectric sensors (e.g., quartz accelerometers).

Magnetic Material and Components

Magnetism and Permeability.
When electrons move (or spin), a magnetic field is generated. The combined effect of such electron movements is the cause of magnetic properties of material.

Table 13 Approximate Dielectric Strengths of Several Materials

Material	Dielectric Strength (V/mil)
Air	25
Ceramics	1000
Glass	2000
Mica	3000
Oil	400
Paper	1500

Table 15 Piezoelectric Coefficients of Some Materials

Material	Piezoelectric Coefficient k_p (m/V)
Barium titanate	2.5×10^{-10}
PZT	6.0×10^{-10}
Quartz	0.02×10^{-10}
Rochelle salt	3.5×10^{-10}

In the linear range of operation of a magnetic element, we can write

$$B = \mu H$$

where B is the magnetic flux density (webers per meter squared or teslas), H is magnetic field strength (amperes per meter), and μ is the permeability of the magnetic material. The relative permeability μ_r of a magnetic material is defined as

$$\mu = \frac{\mu}{\mu_0}$$

where μ_0 is the permeability of a vacuum (approx. $4\pi \times 10^{-7}$ H/m). (Note: 1 T = 1 Wb/m^2; 1 H = 1 Wb/A.)

Hysteresis Loop. The B versus H curve of a magnetic material is not linear and exhibits a hysteresis loop as shown in Fig. 66. It follows that μ is not a constant. Initial values (when magnetization is started at the demagnetized state of $H = 0$ and $B = 0$) are usually specified. Some representative values are given in Table 16.

Properties of magnetic materials can be specified in terms of parameters of the hysteresis curve. Some important parameters are shown in Fig. 66:

H_c = coercive field or coercive force (A/m)

B_r = remnant flux density (Wb/m^2 or T)

B_{sat} = saturation flux density (T)

Magnetic parameters of a few permanent-magnetic materials are given in Table 17. Note that high values

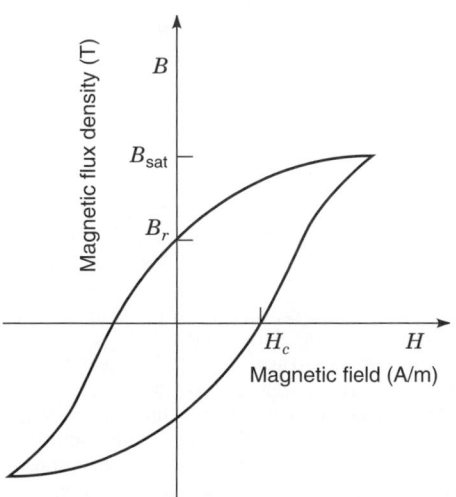

Fig. 66 Hysteresis curve (magnetization curve) of magnetic material.

Table 16 Initial Relative Permeability (Approximate) of Some Materials

Material	Relative Permeability μ_r
Alnico (Fe$_2$ Ni Al)	6.5
Carbon steel	20
Cobalt steel (35% Co)	12
Ferrite (manganese-zinc)	800–10,000
Iron	200
Permalloy (78% Ni, 22% Fe)	3000
Silicon iron (grain oriented)	500–1500

Table 17 Parameters of Some Magnetic Materials

Material	H_c (A/m)	B_r(Wb/m^2)
Alnico	4.6×10^4	1.25
Ferrites	14.0×10^4	0.65
Steel (carbon)	0.4×10^4	0.9
Steel (35% Co)	2.0×10^4	1.4

of H_c and B_r are desirable for high-strength permanent magnets. Furthermore, high values of μ are desirable for core materials that are used to concentrate magnetic flux.

Magnetic Materials. Magnetic characteristics of a material can be imagined as if contributed by a matrix of microminiature magnetic dipoles. Paramagnetic materials (e.g., platinum and tungsten) have their magnetic dipoles arranged in a somewhat random manner. These materials have a μ_r value approximately equal to 1 (i.e., no magnetization). Ferromagnetic materials (e.g., iron, cobalt, nickel, and some manganese alloys) have their magnetic dipoles aligned in one direction (parallel) with virtually no cancellation of polarity. These materials have a high μ_r (of the order of 1000) in general. At low H values, μ_r will be correspondingly low. Antiferromagnetic materials (e.g., chromium and manganese) have their magnetic dipoles arranged in parallel, but in an alternately opposing manner thereby virtually canceling the magnetization ($\mu_r = 1$). Ferrites have parallel magnetic dipoles arranged alternately opposing, as in antiferromagnetic materials, but the adjacent dipoles have unequal strengths. Hence, there is a resultant magnetization (μ_r is of the order of 1000).

Applications of magnets and magnetic materials include actuators (e.g., motors, magnetically leviated vehicles, tools, magnetic bearings), sensors and transducers, relays, resonators, and cores of inductors and transformers. Also, see the applications of superconductivity.

Piezomagnetism. When a stress (strain) is applied to a piezomagnetic material, the degree of magnetization of the material changes. Conversely, a piezomagnetic material undergoes deformation when the magnetic field in which the material is situated is changed.

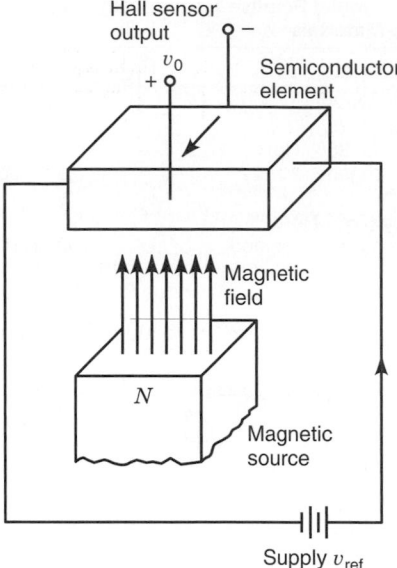

Fig. 67   Hall effect sensor.

Hall Effect Sensors.   Suppose that a dc voltage v_{ref} is applied to a semiconductor element that is placed in a magnetic field in an orthogonal direction, as shown in Fig. 67. A voltage v_0 is generated in the third orthogonal direction, as indicated in the figure.[96] This is known as the Hall effect. Hall effect sensors use this phenomenon. For example, the motion of a ferromagnetic element can be detected in this manner since the magnetic field in which the sensor is mounted would vary as a result of the motion of the ferromagnetic element. Hall effect sensors are useful as position sensors, speed sensors, commutation devices for motors, and instrument transformers for power transmission systems.

Magnetic Bubble Memories.   Consider a film of magnetic material such as gadolinium gallium oxide ($Gd_3\,Ga_5O_{12}$) deposited on a nonmagnetic garnet layer (substrate). The direction of magnetization will be perpendicular to the surface of the film. Initially, some regions of the film will be N poles, and the remaining regions will be S poles. An external magnetic field can shrink either the N regions or the S regions, depending on the direction of the field. The size of the individual magnetic regions can be reduced to the order of 1 μm in this manner. These tiny magnetic bubbles are the means with which information is stored in a magnetic bubble memory.

Inductance.   Suppose that a conducting coil having n turns is placed in a magnetic field of flux ϕ (in webers). The resulting flux linkage is $n\phi$. If the flux linkage is changed, a voltage is induced in the coil.

This induced voltage (v) is given by

$$v = \frac{d(n\phi)}{dt} = n\frac{d\phi}{dt}$$

If the change in magnetic flux is brought about by a change in current (i), we can write

$$v = L\frac{di}{dt}$$

where L is the inductance of the coil (in henries).

In the frequency domain, we have

$$v = Lj\omega i$$

where $\omega = $ frequency and $j = \sqrt{-1}$.

It follows that the electrical impedance of an inductor is given bv $j\omega L$.

5.2   Active Components

Active components made of semiconductor junctions and field effect components are considered in this section. Junction diodes, bipolar junction transistors, and field-effect transistors are of particular interest here. Active components are widely used in the monolithic (integrated-circuit) form as well as in the form of discrete elements.

pn Junctions   A pure semiconductor can be doped to form either a p-type semiconductor or an n-type semiconductor. A pn junction is formed by joining a p-type semiconductor element and an n-type semiconductor element.

Semiconductors.   Semiconductor materials have resistivities that are several million times larger than those of conductors and several billion times smaller than those of insulators. Crystalline materials such as silicon and germanium are semiconductors. For example, the resistivity of pure silicon is about 5×10^{10} times that of silver. Similarly, the resistivity of pure germanium is about 5×10^7 times that of silver. Typically, semiconductors have resistivities ranging from 10^{-4} to $10^7 \, \Omega \cdot$ m. Other examples of semiconductor materials are gallium arsenide, cadmium sulfide, and selenium.

A pure (intrinsic) semiconductor material has some free electrons (negative charge carriers) and holes (positive charge carriers). Note that a hole is formed in an atom when an electron is removed. Strictly, the holes cannot move. But suppose that an electron shared by two atoms (a covalent electron) enters an existing hole in an atom, leaving behind a hole at the point of origin. The resulting movement of the electron is interpreted as a movement of a hole in the direction opposite to the actual movement of the covalent electron.

The number of free electrons in a pure semiconductor is roughly equal to the number of holes.

The number of free electrons or holes in a pure semiconductor can be drastically increased by adding traces of impurities in a controlled manner (doping) into the semiconductor during crystal growth (e.g., by alloying in a molten form, and by solid or gaseous diffusion of the trace). An atom of a pure semiconductor that has four electrons in its outer shell will need four more atoms to share in order to form a stable covalent bond. These covalent bonds are necessary to form a crystalline lattice structure of atoms that is typical of semiconductor materials. If the trace impurity is a material such as arsenic, phosphorus, or antimony whose atoms have five electrons in the outer shell (a donor impurity), a free electron will be left over after the formation of a bond with an impurity atom. The result will be an *n*-type semiconductor having a very large number of free electrons. If, on the other hand, the trace impurity is a material such as boron, gallium, aluminum, or indium whose atoms have only three electrons in the outer shell (an acceptor impurity), a hole will result on formation of a bond. In this case, a *p*-type semiconductor, consisting of a very large number of holes, will result. Doped semiconductors are termed extrinsic.

Depletion Region.

When a *p*-type semiconductor is joined with an *n*-type semiconductor, a *pn* junction is formed. A *pn* junction exhibits the diode effect, much larger resistance to current flow in one direction than in the opposite direction across the junction. As a *pn* junction is formed, electrons in the *n*-type material in the neighborhood of the common layer will diffuse across into the *p*-type material. Similarly, the holes in the *p*-type material near the junction will diffuse into the opposite side (strictly, the covalent electrons will diffuse in the opposite direction). The diffusion will proceed until an equilibrium state is reached. But, as a result of the loss of electrons and the gain of holes on the *n* side and the opposite process on the *p* side, a potential difference is generated across the *pn* junction, with a negative potential on the *p* side and a positive potential on the *n* side. Due to the diffusion of carriers across the junction, the small region surrounding the common area will be virtually free of carriers (free electrons and holes). Hence, this region is called the depletion region. The potential difference that exists in the depletion region is mainly responsible for the diode effect of a *pn* junction.

Biasing.

The forward biasing and the reverse biasing of a *pn* junction are shown in Fig. 68. In the case of forward biasing, a positive potential is connected to the *p* side of the junction, and a negative potential is connected to the *n* side. The polarities are reversed for reverse biasing. Note that in forward biasing, the external voltage (bias voltage *v*) complements the potential difference of the depletion region (Fig. 68*a*). The free electrons that crossed over to the *p* side from the *n* side will continue to flow toward the positive terminal of

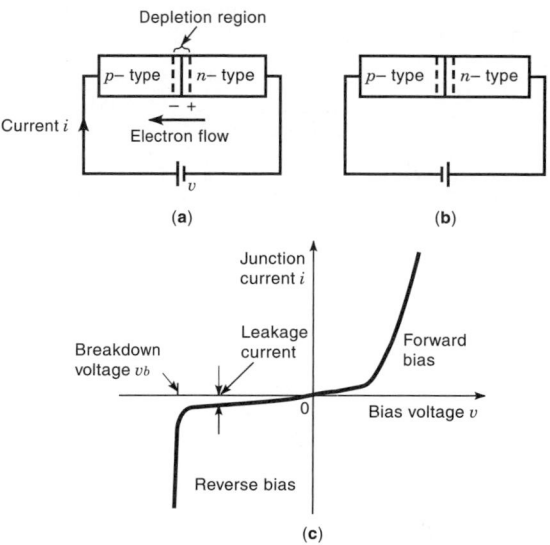

Fig. 68 A *pn* junction diode: (*a*) forward biasing; (*b*) reverse biasing; (*c*) characteristic curve.

the external supply, thereby generating a current (junction current *i*). The junction current increases with the bias voltage, as shown in Fig. 68*c*.

In reverse biasing, the potential in the depletion region is opposed by the bias voltage (Fig. 68*b*). Hence, the diffusion of free electrons from the *n* side into the *p* side is resisted. Since there are some (very few) free electrons in the *p* side and some holes in the *n* side, the reverse bias will reinforce the flow of these minority electrons and holes. This will create a very small current (about 10^{-9} A for silicon and 10^{-6} A for germanium at room temperature), known as the *leakage current*, in the opposite direction to the forward-bias current. If the reverse bias is increased, at some voltage (breakdown voltage v_b in Fig. 68*c*) the junction will break down, generating a sudden increase in the reverse current. There are two main causes of this breakdown. First, the intense electric field of the external voltage can cause electrons to break away from neutral atoms in large numbers. This is known as Zener breakdown. Second, the external voltage will accelerate the minority free electrons on the *p* side (and minority holes on the *n* side), creating collisions that will cause electrons on the outer shells of neutral atoms to break away in large numbers. This is known as the avalanche breakdown. In some applications (e.g., rectifier circuits), junction breakdown is detrimental. In some other types of applications (e.g., as constant voltage sources and in some digital circuits), the breakdown state of specially designed diodes is practically utilized. Typical breakdown voltages of *pn* junctions made of three common semiconductor materials are given in Table 18. Note that the breakdown voltage decreases with the concentration of the trace material.

Table 18   Typical Breakdown Voltage of *pn* Junction at Room Temperature

| Semiconductor | Breakdown Voltage (V) | |
	Dope Concentration $= 10^{15}$ atoms/cm^3	Dope Concentration $= 10^{17}$ atoms/cm^3
Germanium	400	5.0
Silicon	300	11.0
Gallium arsenide	150	16.0

The current through a reverse-biased *pn* junction will increase exponentially with temperature. For a forward-biased *pn* junction, current will increase with temperature at low to moderate voltages and will decrease with temperature at high levels of voltage.

Diodes   A semiconductor diode is formed by joining a *p*-type semiconductor with an *n*-type semiconductor. A diode offers much less resistance to current flow in one direction (forward) than in the opposite direction (reverse). There are many varieties of diodes. Zener diodes, voltage variable capacitor (VVC) diodes, tunnel diodes, microwave power diodes, pin diodes, photodiodes, and light-emitting diodes (LED) are examples. The last two varieties will be discussed in separate sections.

Zener Diodes.   Zener diodes are a particular type of diodes that are designed to operate in the neighborhood of the reverse breakdown (both Zener and avalanche breakdowns). In this manner, a somewhat constant voltage output (the breakdown voltage) can be generated. This voltage depends on the concentration of the trace impurity. By varying the impurity concentration, output voltages in the range of 2–200 V may be realized from a Zener diode. Special circuits would be needed to divert large currents that are generated at the breakdown point of the diode. The rated power dissipation of a Zener diode should take into consideration the current levels that are possible in the breakdown region. Applications of Zener diodes include constant voltage sources, voltage clipper circuits, filter circuits for voltage transients, digital circuits, and two-state devices.

VVC Diodes.   VVC diodes use the property of a diode that, in reverse bias, the capacitance decreases (nonlinearly) with the bias voltage. The depletion region of a *pn* junction is practically free of carriers (free electrons and holes) and, hence, behaves like the dielectric medium of a capacitor. The adjoining *p* region and *n* region serve as the two plates of the capacitor. The width of the depletion region increases with the bias voltage. Consequently, the capacitance of a reverse-biased *pn* junction decreases as the bias voltage is increased. The obtainable range of capacitance can be varied by changing the dope concentration and also by distributing the dope concentration nonuniformly along the diode. For example, a capacitance variation of 5–500 pF may be obtained in this manner (note: 1 pF $= 1 \times 10^{-12}$ F). VVC diodes are also known as varactor diodes and varicaps and are useful in voltage-controlled tuners and oscillators.

Tunnel Diodes.   The depletion of a *pn* junction can be made very thin by using very high dope concentrations (in both the *p* and *n* sides). The result is a tunnel diode. Since the depletion region is very narrow, charge carriers (free electrons and holes) in the *n* and *p* sides of the diode can tunnel through the region into the opposite side on application of a relatively small voltage. The voltage–current characteristic of a tunnel diode is quite linear at low (forward and reverse) voltages. When the forward bias is further increased, however, the behavior will become very nonlinear; the junction current will peak, then drop (a negative conductance) to a minimum (valley), and finally rise again, as the voltage is increased. Due to the linear behavior of the tunnel diode at low voltages, almost instantaneous current reversal (i.e., very low reverse recovery time) can be achieved by switching the bias voltage. Tunnel diodes are useful in high-frequency switching devices, sensors, and signal conditioning circuits.

pin Diodes.   The width of the depletion region of a conventional *pn* junction varies with many factors, primarily the applied (bias) voltage. The capacitance of a junction depends on this width and will vary due to such factors. A diode with practically a constant capacitance is obtained by adding a layer of silicon in between the *p* and *n* elements. The sandwiched silicon layer is called the intrinsic layer, and the diode is called a *pin* diode. The resistance of a *pin* diode varies inversely with junction current. *Pin* diodes are useful as current-controlled resistors at constant capacitance.

Schottky Barrier Diodes.   Most diodes consist of semiconductor–semiconductor junctions. An exception is a Schottky barrier diode, which consists of a metal–semiconductor (*n*-type) junction. A metal such as gold, silver, platinum, or palladium and a semiconductor such as silicon or gallium arsenide may be used in the construction. Since no holes exist in the metal, a depletion region cannot be formed at the metal–semiconductor junction. Instead, an electron barrier is formed by the free electrons from the *n*-type semiconductor. Consequently, the junction capacitance will be negligible, and the reverse recovery time will be very small. For this reason. Schottky diodes can handle very high switching frequencies (10^9 Hz range). Since the electron barrier is easier to penetrate than a depletion region, by using a reverse bias, Schotky diodes exhibit much lower breakdown voltages. Operating noise is also lower than for semiconductor–semiconductor diodes.

Thyristors.   A thyristor, also known as a silicon-controlled rectifier, a solid-state controlled rectifier, a

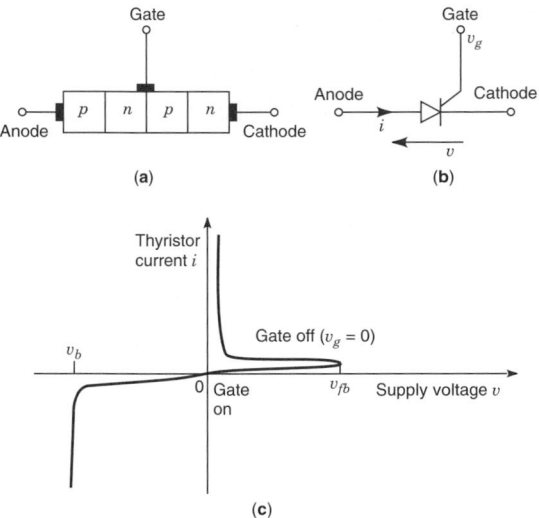

Fig. 69 Thyristor: (a) schematic representation; (b) circuit symbol; (c) characteristic curve.

Table 19   Characteristic Variables and Parameters for Diodes

Diode Variable/Parameter	Description
Forward bias (v_f)	Positive external voltage at p with respect to n
Reverse bias (v_r)	Positive external voltage at n with respect to p
Breakdown voltage (v_b)	Minimum reverse bias that will break down the junction resistance
Junction current (i_f)	Forward current through a forward-biased diode
Leakage current (i_r)	Reverse current through a reverse-biased diode
Transition capacitance (C_t)	Capacitance (in the depletion region) of a reverse-biased diode
Diffusion capacitance (C_d)	Capacitance exhibited while a forward-biased diode is switched off
Forward resistance (R_f)	Resistance of a forward-biased diode
Reverse recovery time (t_{rr})	Time needed for the reverse current to reach a specified level when the diode is switched from forward to reverse
Operating temperature range (T_A)	Allowable temperature range for a diode during operation
Storage temperature range (T_{srg})	Temperature that should be maintained during storage of a diode
Power dissipation (P)	Maximum power dissipation allowed for a diode at a specified temperature

semiconductor-controlled rectifier, or simply an SCR, possesses some of the characteristics of a semiconductor diode. It consists of four layers (*pnpn*) of semiconductor and has three terminals—the anode, the cathode, and the gate—as shown in Fig. 69a. The circuit symbol for a thyristor is shown in Fig. 69b. The thyristor current is denoted by i, the external voltage is v, and the gate potential is v_g. The characteristic curve of a thyristor is shown in Fig. 69c. Note that a thyristor cannot conduct in either direction (i almost zero) until either the reverse voltage reaches the reverse breakdown voltage (v_b) or the forward voltage reaches the forward breakover voltage (v_{fb}). The forward breakover is a bistable state, and once this voltage is reached, the voltage drops significantly, and the thyristor begins to conduct like a forward-biased diode. When v_g is less than or equal to zero with respect to the cathode, v_{fb} becomes quite high. When v_g is made positive, v_{fb} becomes small, and v_{fb} will decrease as the gate current (i_g) is increased. A small positive v_g can make v_{fb} very small, and then the thyristor will conduct from anode to cathode but not in the opposite direction (i.e., it behaves like a diode). It follows that a thyristor behaves like a voltage-triggered switch; a positive firing signal (a positive v_g) will close the switch. The switch will be opened when both i and v_g are made zero. When the supply voltage v is dc and nonzero, the thyristor will not be able to turn itself off. In this case a commutating circuit that can make the trigger voltage v_g slightly negative has to be employed. Thyristors are commonly used in control circuits for dc and ac motors.

Parameter values for diodes are given in data sheets provided by the manufacturer. Commonly used variables and characteristic parameters in association with

diodes are described in Table 19. For thyristors, as mentioned before, several other quantities such as v_{fb}, v_g, and i_g should be included. The time required for a thyristor to be turned on by the trigger signal (turn-on time) and the time for it to be turned off through commutation (turn-off time) determine the maximum switching frequency (bandwidth) for a thyristor. Another variable that is important is the holding current or latching current, which denotes the small forward current that exists at the breakover voltage.

Bipolar Junction Transistors   A bipolar junction transistor (BJT) has two junctions that are formed by joining p regions and n regions. Two types of transistors, *npn* and *pnp*, are possible with this structure. A BJT has three terminals, as indicated in Fig. 70a. The middle (sandwiched) region of a BJT is thinner than the end regions, and this region is known as the base. The end regions are termed the emitter and the collector. Under normal conditions, the emitter–base junction is forward biased, and the collector–base junction is reverse biased, as shown in Fig. 70b.

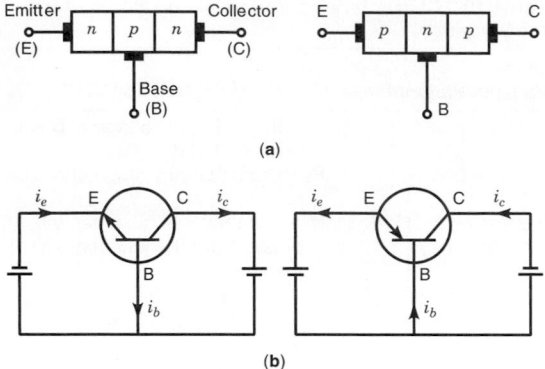

Fig. 70 Bipolar junction transistors: (a) *npn* and *pnp* transistors; (b) circuit symbols and biasing.

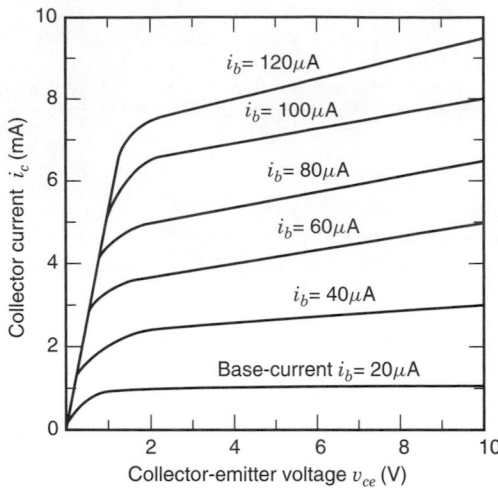

Fig. 71 Characteristic curves of common emitter BJT.

To explain the behavior of a BJT, consider an *npn* transistor under normal biasing. The forward bias at the emitter–base junction will cause free electrons in the emitter to flow into the base region, thereby creating the emitter current (i_e). The reverse bias at the collector–base junction will increase the depletion region there. The associated potential difference at the collector–base junction will accelerate the free electrons in the base into the collector and will form the collector current (i_c). Holes that are created in the base, for recombination with some free electrons that entered the base, will form the base current (i_b). Usually, i_c is slightly smaller than i_e. Furthermore, i_b is much smaller than i_c.

Transistor Characteristics. The common-emitter connection is widely used for transistors in amplifier applications. In this configuration, the emitter terminal will be common to the input side and the output side of the circuit. Transistor characteristics are usually specified for this configuration. Figure 71 shows typical characteristic curves for a junction transistor in the common-emitter connection. In this configuration, both voltage gain (output voltage/input voltage) and current gain (collector current/base current) will be greater than unity, thereby providing a voltage amplification as well as a current amplification. Note from Fig. 71 that the control signal is the base current (i_b), and the characteristic of the transistor depends on i_b. This is generally true for any bipolar junction transistor; a BJT is a current-controlled transistor. In the common-base configuration, the base terminal is common to both input and output.

Maximum frequency of operation and allowable switching rate for a transistor are determined by parameters such as rise time, storage time, and fall time. These and some other useful ratings and characteristic parameters for bipolar junction transistors are defined in Table 20. Values for these parameters are normally given in the manufacturer's data sheet for a particular transistor.

Table 20   Rating Parameters for Transistors

Transistor Parameter	Description
Collector-to-base voltage (v_{cb})	Voltage limit across collector and base with emitter open
Collector-to-emitter voltage (v_{ce})	Voltage limit across collector and emitter with base connected to emitter
Emitter-to-base voltage (v_{eb})	Voltage limit across emitter and base with collector open
Collector cutoff current (i_{co})	Reverse saturation current at collector with either emitter open (i_{cbo}) or base open (i_{co})
Transistor dissipation (P_T)	Power dissipated by the transistor at rated conditions
Input impedance (h_i)	Input voltage/input current with output voltage = 0 (Defined for both common emitter and common base configurations, h_{ie}, h_{ib})
Output admittance (h_o)	Output current/output voltage with input current = 0 (h_{oe}, h_{ob} are defined)
Forward current transfer ratio (h_f)	Output current/input current with output voltage = 0 (h_{fe}, h_{fb} are defined)
Reverse voltage transfer ratio (h_r)	Input voltage/output voltage with input current = 0 (h_{re}, h_{rb} are defined)
Rise time (t_r)	Time taken to reach the full current level for the first time when turned on
Storage time (t_s)	Time taken to reach the steady current level when turned on
Fall time (t_f)	Time taken for the current to reach zero when turned off

Fabrication Process. The actual manufacturing process for a transistor is complex and delicate. For example, an *npn* transistor can be fabricated by starting with a crystal of *n*-type silicon. This starting element is called the wafer or substrate. The *npn* transistor is formed, by using the planar diffusion method, in the top half of the substrate as follows: The substrate is heated to about 1000°C. A gas stream containing a donor-type impurity (which forms *n*-type regions) is impinged on the crystal surface. This produces an *n*-type layer on the crystal. Next the crystal is oxidized by heating to a high temperature. The resulting layer of silicon dioxide acts as an insulating surface. A small area of this layer is then dissolved off using hydrofluoric acid. The crystal is again heated to 1000°C, and a gas stream containing acceptor-type impurity (which forms *p*-type regions) is impinged on the window thus formed. This produces a *p* region under the window on top of the *n* region, which was formed earlier.

Oxidation is repeated to cover the newly formed *p* region. Using hydrofluoric acid, a smaller window is cut on the latest silicon dioxide layer, and a new *n* region is formed, as before, on top of the *p* region. The entire manufacturing process has to be properly controlled so as to control the properties of the resulting transistor. Aluminum contacts have to be deposited on the uppermost *n* region, the second *p* region (in a suitable annular window cut on the silicon dioxide layer), and on the *n* region below it or on the crystal substrate. A pictorial representation of an *npn* transistor fabricated in this manner is shown in Fig. 72.

Field-Effect Transistors An FET, unlike a BJT, is a voltage-controlled transistor. The electrostatic field generated by a voltage applied to the gate terminal of an FET controls the behavior of the FET. Since the device is voltage controlled at very low input current levels, the input impedance is very high, and the input power is very low. Other advantages of an FET over a BJT are that the former is cheaper and requires significantly less space on a chip in the monolithic form. FETs are somewhat slower (in terms of switching rates) and more nonlinear than BJTs, however.

There are two primary types of FETs: metal–oxide–semiconductor field-effect transistor (MOSFET) and junction field-effect transistor (JFET). Even though the

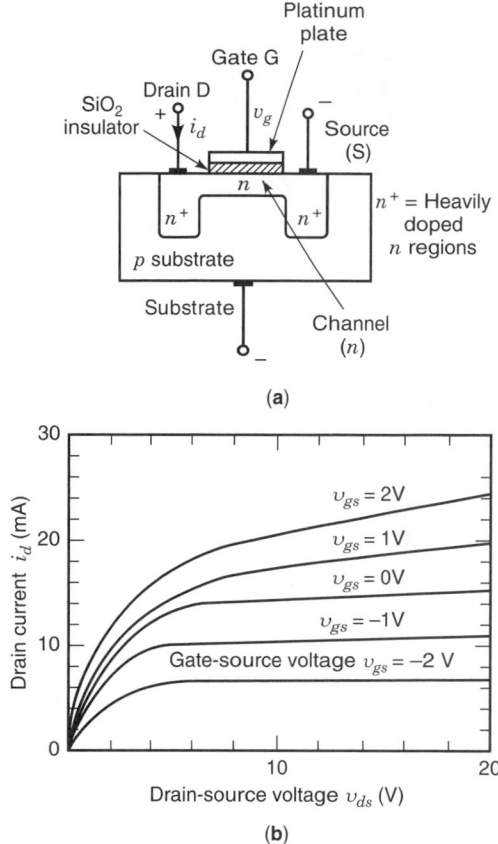

(a)

(b)

Fig. 73 MOSFET: (a) an *n*-channel depletion-type MOS-FET; (b) D-MOSFET characteristics.

physical structure of the two types is somewhat different, their characteristics are quite similar. Insulated-gate FET (or IGFET) is a general name given to MOSFETs.

MOSFET. An *n*-channel MOSFET is produced using a *p*-type silicon substrate, and a *p*-channel MOSFET by an *n*-type substrate. An *n*-channel MOSFET is shown in Fig. 73*a*. During manufacture, two heavily doped *n*-type regions are formed on the substrate. One region is termed source (S) and the other region drain (D). The two regions are connected by a moderately doped and narrow *n* region called a channel. A metal coating deposited over an insulating layer of silicon dioxide, which is formed on the channel, is the gate (G). The source lead is usually joined with the substrate lead. This is a depletion-type MOSFET (or D-MOSFET). Another type is the enhancement-type MOSFET (or E-MOSFET). In this type, a channel linking the drain and the source is not physically present in the substrate but is induced during operation of the transistor.

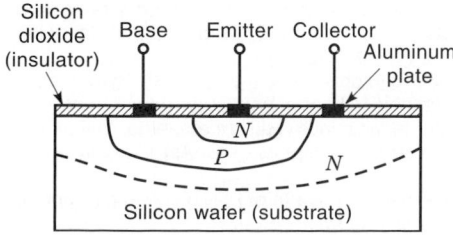

Fig. 72 An *npn* transistor manufactured by the planar diffusion method.

Consider the operation of the *n*-channel D-MOSFET shown in Fig. 73*a*. Under normal operation, the drain is positively biased with respect to the source. Drain current i_d is considered the output of a MOSFET (analogous to the collector current of a BJT). The control signal of a MOSFET is the gate voltage v_{gs} with respect to the source (analogous to the base current of a BJT). It follows that a MOSFET is a voltage-controlled device. Since the source terminal is used as the reference for both input (gate voltage) and output (drain), this connection is called the common-source configuration. Suppose that the gate voltage is negative with respect to the source. This will induce holes in the channel, thereby decreasing the free electrons there through recombination. This, in turn, will reduce the concentration of free electrons in the drain region and, hence, will reduce the drain current i_d. Clearly, if the magnitude of the negative voltage at the gate is decreased, the drain current will increase, as indicated by the characteristic curves in Fig. 73*b*. A positive bias at the gate will further increase the drain current of an *n*-channel MOSFET as shown. The opposite will be true for a *p*-channel MOSFET.

The JFET. A junction field-effect transistor (JFET) is different in physical structure to a MOSFET but similar in characteristics. The structure of an *n*-channel JFET is shown in Fig. 74. It consists of two *p*-type regions formed inside an *n*-type region. The two *p* regions are separated by a narrow *n* region called a

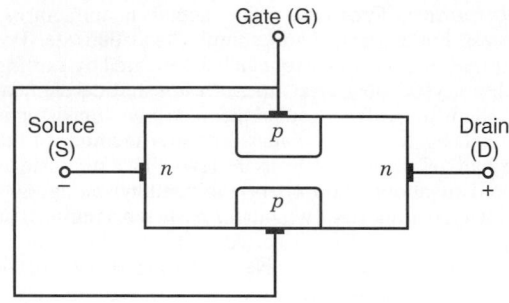

Fig. 74   An *n*-channel JFET.

channel. The channel links two *n*-type regions called source (S) and drain (D). The two *p* regions are linked by a common terminal and form the gate (G). As for a MOSFET, drain current i_d is considered the output of the JFET, and gate voltage v_{gs}, with respect to the source, is considered the control signal. For normal operation, the drain is positively biased with respect to the source, as for an *n*-channel MOSFET, and the common-source configuration is used.

To explain the operation of a JFET, consider the *n*-channel JFET shown in Fig. 74. Depletion regions are present at the two *pn* junctions of the JFET (as

Table 21   Common Transistor Types

Abbreviation	Transistor Type — Name	Description
BJT	Bipolar junction transistor	Three-layer device (*npn* or *pnp*) Current controlled Control = base current Output = collector current
FET	Field-effect transistor	Physical or induced channel (*n*-channel or *p*-channel) voltage controlled Control = gate voltage Output = drain current
MOSFET	Metal–oxide–semiconductor FET	*n* channel or *p* channel
D-MOSFET	Depletion-type MOSFET	Channel is physically present
E-MOSFET	Enhancement-type MOSFET	Channel is induced
VMOS	V-shaped Gate MOSFET or VFET	An E-MOSFET with increased power-handling capacity
DG-MOS	Dual-gate MOSFET	Secondary gate is present between main gate and drain (lower capacitance)
D-MOS	Double-diffused MOSFET	Channel layer is formed on a high-resistivity substrate and then source and drain are formed (by diffusion). High breakdown voltage
CMOS	Complementary symmetry MOSFET	Uses two E-MOSFETs (*n* channel and *p* channel). Symmetry is used to save space on chip. Cheaper and lower power consumption.
GaAs	Gallium arsenide MOSFET	Uses gallium arsenide, aluminum gallium arsenide, (AlGaAs), indium gallium arsenide phosphide (InGaAsP), etc. in place of silicon substrate. Faster operation
JFET	Junction FET	*p* channel or *n* channel. Has two (*n* or *p*) regions in a (*p* or *n*) region linked by a channel (*p* or *n*) Control = gate voltage Output = drain current

for a semiconductor diode). If the gate voltage is made negative, the resulting field will weaken the p regions. As a result, the depletion regions will shrink. Some of the free electrons from the drain will diffuse toward the channel to occupy the growing n regions due to the shrinking depletion regions. This will reduce the drain current. It follows that drain current decreases as the magnitude of the negative voltage at the gate is increased. This behavior is similar to that of a MOSFET.

A p-channel JFET has two n regions representing the gate and two p regions forming the source and the drain, which are linked by a p-channel. Its characteristic is the reverse of an n-channel JFET.

Common types of transistor are summarized in Table 21. Semiconductor devices have numerous uses. A common use is as switching devices or as two-state elements. Typical two-state elements are schematically illustrated in Fig. 75.

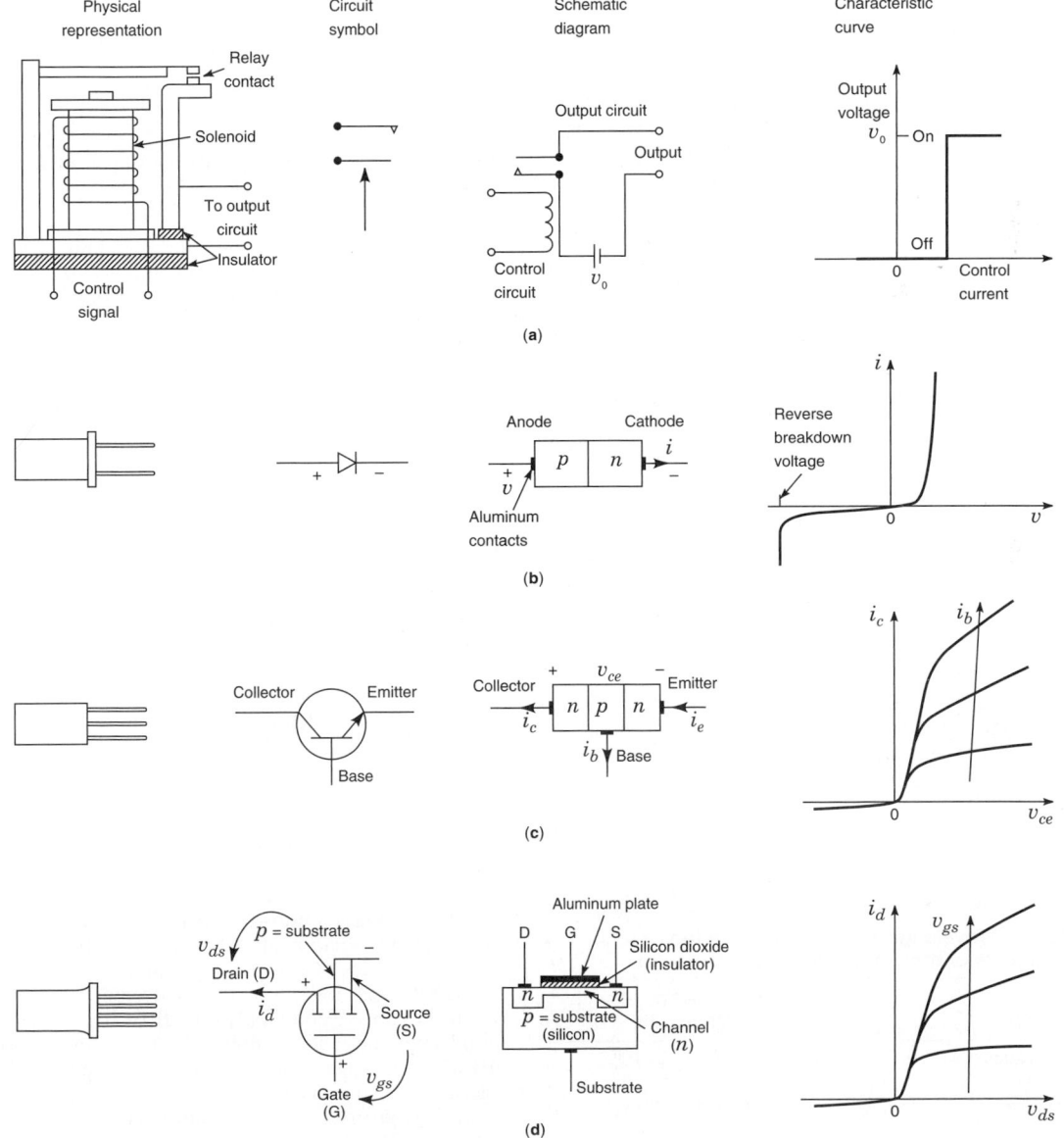

Fig. 75 Discrete switching (two-state) elements: (*a*) electromagnetic relay; (*b*) Zener diode; (*c*) BJT (*npn*); (*d*) *n*-channel MOSFET.

5.3 Light Emitters and Displays

Visible light is part of the electromagnetic spectrum Electromagnetic waves in the wave length range of 390–770 nm (Note: $1\,nm = 1 \times 10^{-9}$m) form the visible light. Ultraviolet rays and X rays are also electromagnetic waves but have lower wavelengths (higher frequencies). Infrared rays, microwaves, and radio waves are electromagnetic waves having higher wavelengths. Table 22 lists wavelengths of several types of electromagnetic waves. Visible light occupies a broad range of wavelengths. For example, in optical coupling applications, the narrower the wave spectrum, the clearer (noise free) the coupling process. Consequently, it is advantageous to use special light sources in applications of that type. Furthermore, since visible light can be contaminated by environmental light, thereby introducing an error signal into the system, it is also useful to consider electromagnetic waves that are different from what is commonly present in operating environments in applications such as sensing, optical coupling, and processing.

Incandescent Lamps   Tungsten-filament incandescent lamps that are commonly used in household illumination emit visible light in a broad spectrum. Furthermore, they are not efficient because they emit more infrared radiation than useful visible light. Ionizing lamps filled with gases such as halogens, sodium vapor, neon, or mercury vapor have much narrower spectra, and they emit very pure visible light (with negligible infrared radiation). Hence, these types of incandescent lamps are more efficient for illumination purposes. Regular fluorescent lamps are known to create a line-frequency (60 or 50 Hz) flicker but are quite efficient and durable. All these types of light sources are usually not suitable in many applications primarily because of the following disadvantages:

1. They are bulky.
2. They cannot be operated at high switching rates (from both time constant and component life points of view).
3. Their spectral bandwidth can be very wide.

Note that a finite time is needed for an incandescent lamp to emit light once it is energized. That is, it has a

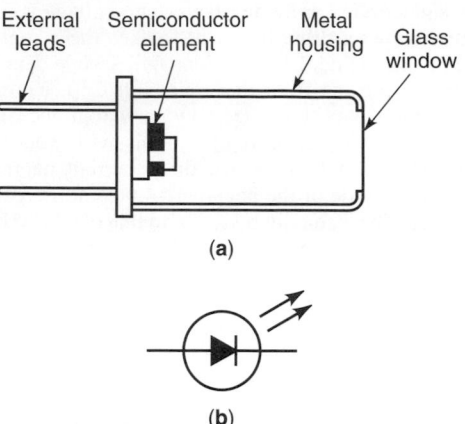

Fig. 76   LED: (*a*) physical construction; (*b*) circuit symbol.

large time constant. This limits the switching speed to less than 100 Hz. Furthermore, lamp life will decrease rapidly with increasing switching frequency.

Light-Emitting Diodes   The basic components of an LED are shown in Fig. 76*a*. The element symbol that is commonly used in electrical circuits is shown in Fig. 76*b*. The main component of an LED is a semiconductor diode element, typically made of gallium compounds (e.g., gallium arsenide or GaAs and gallium arsenide phosphide or GaAsP). When a voltage is applied in the forward-bias direction to this semiconductor element, it emits visible light (and also other electromagnetic wave components, primarily infrared). In the forward-bias configuration, electrons are injected into the *p* region of the diode and recombined with holes. Radiation energy (including visible light) is released spontaneously in this process. This is the principle of operation of an LED. Suitable doping with trace elements such as nitrogen will produce the desired effect. The radiation energy generated at the junction of a diode has to be directly transmitted to a window of the diode in order to reduce absorption losses. Two types of construction are commonly used; edge emitters emit radiation along the edges of the *pn* junction, and surface emitters emit radiation normal to the junction surface.

     Infrared light-emitting diodes (IRED) are LEDs that emit infrared radiation at a reasonable level of power. Gallium arsenide (GaAs), gallium aluminum arsenide (GaAlAs), and indium gallium arsenide phosphide (InGaAsP) are the commonly used IRED material. Gallium compounds and not silicon or germanium are used in LEDs for reasons of efficiency and intensity characteristics. (Gallium compounds exhibit sharp peaks of spectral output in the desired frequency bands.) Table 23 gives wavelength characteristics of common LED and IRED types ($1\,\mathring{A} = 1 \times 10^{-10}$ m $= 0.1$ nm). Note that $\mathring{A}$ denotes the unit angstrom.

Table 22   Wavelengths of Several Selected Components of the Electromagnetic Spectrum

Wave Type	Approximate Wavelength Range (μm)
Radio waves	$1 \times 10^6 - 5 \times 10^6$
Microwaves	$1 \times 10^3 - 1 \times 10^6$
Infrared rays	$0.8 - 1 \times 10^3$
Visible light	$0.4 - 0.8$
Ultraviolet rays	$1 \times 10^{-2} - 0.4$
X rays	$1 \times 10^{-6} - 5 \times 10^{-2}$

Table 23 Wavelength Characteristics of Common LEDs (1 Å = 1×10⁻¹⁰ m)

LED Type	Wavelength at Peak Intensity (Å)	Color
Gallium arsenide	5500	Green
	9300	Infrared
Gallium arsenide phosphide	5500	Green
	7000	Red
Gallium phosphide	5500	Green
Gallium aluminum arsenide	8000	Red
	8500	Infrared
Indium gallium arsenide phosphide	13000	Infrared

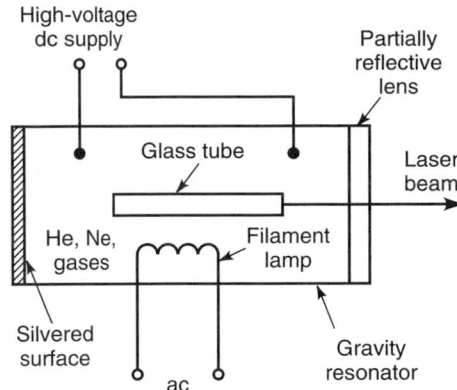

Fig. 77   Helium–neon (He–Ne) laser.

Light-emitting diodes are widely used in optical electronics because they can be constructed in miniature sizes, they have small time constants and low impedances, they can provide high switching rates (typically over 1000 Hz), and they have much longer component life than incandescent lamps. They are useful as both light sources and displays.

Lasers   Laser (light amplification by stimulated emission of radiation) is a light source that emits a concentrated beam of light that will propagate typically at one or two frequencies (wavelengths) and in phase. Usually, the frequency band is extremely narrow (i.e., monochromatic), and the waves in each frequency are in phase (i.e., coherent). Furthermore, the energy of a laser is highly concentrated (power densities of the order of one billion watts/cm²). Consequently, a laser beam can travel in a straight line over a long distance with very little dispersion. Hence, it is useful in gauging and aligning applications. Lasers can be used in a wide variety of sensors (e.g., motion sensors, tactile sensors, laser-doppler velocity sensors) that employ photosensing and fiber optics. Also, lasers are used in medical applications, microsurgery in particular. Lasers have been used in manufacturing and material removal applications such as precision welding, cutting, and drilling of different types of materials, including metals, glass, plastics, ceramics, leather, and cloth. Lasers are used in inspection (detection of faults and irregularities) and gauging (measurement of dimensions) of parts. Other applications of lasers include heat treatment of alloys, holographic methods of nondestructive testing, communication, information processing, and high-quality printing.

Lasers may be classified as solid, liquid, gas, and semiconductor. In a solid laser (e.g., ruby laser, glass laser), a solid rod with reflecting ends is used as the laser medium. The laser medium of a liquid laser (e.g., dye laser, salt-solution laser) is a liquid such as an organic solvent with a dye or an inorganic solvent with dissolved salt compound. Very high peak power levels are possible with liquid lasers. Gas lasers (e.g., helium–neon or He–Ne laser, helium–cadmium or He–Cd laser, carbon dioxide or CO₂ laser) use a gas as

the laser medium. Semiconductor lasers (e.g., gallium arsenide laser) use a semiconductor diode similar to an edge-emitting LED. Some lasers have their main radiation components outside the visible spectrum of light. For example, a CO_2 laser (wavelength of about 110,000 Å) primarily emits infrared radiation.

In a conventional laser unit, the laser beam is generated by first originating an excitation to create a light flash. This will initiate a process of emitting photons from molecules within the laser medium. This light is then reflected back and forth between two reflecting surfaces before the light beam is finally emitted as a laser. These waves will be limited to a very narrow frequency band (monochromatic) and will be in phase (coherent). For example, consider the He–Ne laser unit schematically shown in Fig. 77. The helium and neon gas mixture in the cavity resonator is heated by a filament lamp and ionized using a high dc voltage (2000 V). Electrons released in the process will be accelerated by the high voltage and will collide with the atoms, thereby releasing photons (light). These photons will collide with other molecules, releasing more photons. This process is known as lasing. The light generated in this manner is reflected back and forth by the silvered surface and the partially reflective lens (beam splitter) in the cavity resonator, thereby stimulating it. This is somewhat similar to a resonant action. The stimulated light is concentrated into a narrow beam by a glass tube and emitted as a laser beam through the partially silvered lens.

A semiconductor laser is somewhat similar to an LED. The laser element is typically made of a *pn* junction (diode) of semiconductor material such as gallium arsenide (GaAs) or indium gallium arsenide phosphide (InGaAsP). The edges of the junction are reflective (naturally or by depositing a film of silver). As a voltage is applied to the semiconductor laser, the ionic injection and spontaneous recombination that take place near the *pn* junction will emit light as in an LED. This light will be reflected back and forth between the reflective surfaces, passing along

Table 24   Properties of Several Types of Lasers
($1\ \text{Å} = 1 \times 10^{-10}$ m)

Laser Type	Wavelength (Å)	Output Power (W/cm^2)
Solid		
Ruby	7,000	0.1–100
Glass	1,000	0.1–500
Liquid		
Dye	4,000–10,000	0.001–1
Gas		
Helium–neon	6,330	0.001–2
Helium–cadmium	4,000	0.001–1
Carbondioxide	110,000	$1–1 \times 10^4$
Semiconductor:		
GaAs	9,000	0.002–0.01
InGaAsP	13,000	0.001–0.005

the depletion region many times and creating more photons. The stimulated light (laser) beam is emitted through an edge of the *pn* junction. Semiconductor lasers are often maintained at very low temperatures in order to obtain a reasonable component life. Semiconductor lasers can be manufactured in very small sizes. They are lower in cost and require less power in comparison to the conventional lasers. Wave length and power output characteristics of several types of lasers are given in Table 24.

Liquid Crystal Displays   A liquid crystal display (LCD) consists of a medium of liquid crystal material (e.g., organic compounds such as cholesteryl nonanote and *p*-azoxyanisole) trapped between a glass sheet and a mirrored surface, as shown in Fig. 78. Pairs of transparent electrodes (e.g., indium tin oxide), arranged in a planar matrix, are deposited on the inner surfaces of the sandwiching plates. In the absence of an electric field across an electrode pair, the atoms of liquid crystal medium in that region will have a parallel orientation. As a result, any light that falls on the glass sheet will first travel through the liquid crystal, then will be reflected back by the mirrored surface, and finally will return unscattered. Once an electrode pair is energized, the molecular alignment of the entrapped

medium will change, causing some scattering. As a result, a dark region in the shape of the electrode will be visible. Alphanumeric characters and other graphic images can be displayed in this manner by energizing a particular pattern of electrodes.

Other types of LCD construction are available. In one type, polarized glass sheets are used to entrap the liquid crystal. In addition, a special coating is applied on the inner surfaces of the two sheets that will polarize the liquid crystal medium in different directions. This polarization structure is altered by an electric field (supplied by an electrode pair), thereby displaying an image element. LCDs require external light to function. But they need significantly low currents and power levels to operate. For example, an LED display might need a watt of power, whereas a comparable LCD might require just a small fraction of a milliwatt. Similarly, the current requirement for an LCD will be in the microampere range. LCDs usually need an ac biasing, however. An image resolution on the order of 5 lines/mm is possible with an LCD.

Plasma Displays   A plasma display is somewhat similar to an LCD in construction. The medium used in a plasma display is an ionizing gas (e.g., neon with traces of argon or xenon). A planar matrix of electrode pairs is used on the inner surfaces of entrapping glass. When a voltage above the ionizing voltage of the medium is applied to the electrode pair, the gas will break down, and a discharge will result. The electron impacts that are generated at the cathode as a result will cause further release of electrons to sustain the discharge. A characteristic orange glow will result. The pattern of energized electrodes will determine the graphic image.

The electrodes could be either dc coupled or ac coupled. In the case of the latter, the electrodes are coated with a layer of dielectric material to introduce a capacitor at the gas interface. The power efficiency of a plasma display is higher than that of an LED display. A typical image resolution of 2 lines/mm is obtainable.

Cathode Ray Tubes   A schematic representation of a cathode ray tube (CRT) is given in Fig. 79. In a CRT,

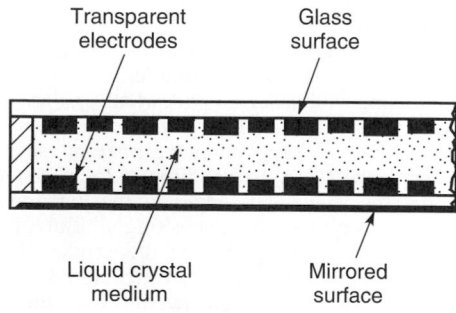

Fig. 78   LCD element.

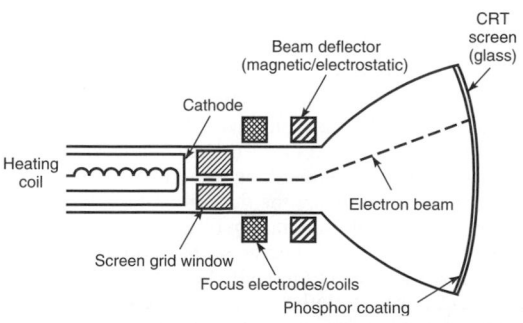

Fig. 79   Schematic of CRT.

an electron beam is used to trace lines, characters, and other graphic images on the CRT screen. The electron beam is generated by an electron gun. A cathode made of a metal such as nickel coated with an oxide such as barium strontium calcium oxide forms the electron gun and is heated (say, using a tungsten coil heater) to generate electrons. Electrons are accelerated toward the inner surface of the CRT screen using a series of anodes, biased in increasing steps. The CRT screen is made of glass. Its inner surface is coated with a crystalline phosphor material. The electrons that impinge on the screen will excite the phosphor layer, which will result in the release of additional electrons and radiation. As a result, the point of impingement will be illuminated. The electron beam is focused using either electrostatic (a pair of electrode plates) or magnetic (a coil) means. The position of the luminous spot on the screen is controlled using a similar method. Two pairs of electrodes (or two coils) will be needed to deflect the electron to an arbitrary position on the screen.

Different types of phosphor material will provide different colors (red, green, blue, white, etc.). The color of a monochrome display is determined by this. Color displays employ one of two common techniques. In one method (masking), three guns are used for the three basic colors (red, green, and blue). The three beams pass through a small masking window and fall on the faceplate. The faceplate has a matrix of miniature phosphor spots (e.g., at 0.1-mm spacing). The matrix consists of a regular pattern of R–G–B phosphor elements. The three electron beams fall on three adjacent spots of R–G–B phosphor. A particular color is obtained as a mixture of the three basic colors by properly adjusting the intensity of the three beams. In the second method (penetration), the faceplate has several layers of phosphor. The color emitted will depend on the depth of penetration of the electron beam into the phosphor.

Flicker in a CRT display, at low frequencies, will strain the eye and also can deteriorate dynamic images. Usually, a minimum flicker frequency of 40 Hz will be satisfactory, and even higher frequencies can be achieved with most types of phosphor coatings. Flicker effect worsens with the brightness of an image. The efficiency of a phosphor screen is determined by the light flux density per unit power input (measured in lumens/watt). A typical value is 40 lm/W. Time constant determines the time of decay of an image when power is turned off. Common types of phosphor and their time constants are given in Table 25.

CRTs have numerous uses. Computer display screens, television picture tubes, radar displays, and oscilloscope tubes are common applications. The raster-scan method is a common way of generating an image on a computer or television screen. In this method, the electron beam continuously sweeps the screen (say, starting from the top left corner of the screen and tracing horizontal lines up to the bottom right corner, continuously repeating the process). The spot is turned on or

Table 25   Time Constants of CRT Phosphor

Phosphor	Color	Time Constant (ms)
P1	Green	30.0
P4	White	0.1
P22	Red	2.0
	Green	8.0
	Blue	6.0
RP20	Yellow–green	5.0

off using a controller according to some logic that will determine the image that is generated on the screen. In another method used in computer screens, the beam is directly moved to trace the curves that form the image. In oscilloscopes, the horizontal deflection of the beam can be time sequenced and cycled in order to enable the display of time signals.

5.4   Light Sensors

A light sensor (also known as a photodetector or photosensor) is a device that is sensitive to light. Usually, it is a part of an electrical circuit with associated signal conditioning (amplification, filtering, etc.) so that an electrical signal representative of the intensity of light falling on the photosensor is obtained. Some photosensors can serve as energy sources (cells) as well. A photosensor may be an integral component of an optoisolator or other optically coupled system. In particular, a commercial optical coupler typically has an LED source and a photosensor in the same package, with a pair of leads for connecting it to other circuits, and perhaps power leads.

By definition, the purpose of a photodetector or photosensor is to sense visible light. But there are many applications where sensing of adjoining bands of the electromagnetic spectrum, namely infrared radiation and ultraviolet radiation, would be useful. For instance, since objects emit reasonable levels of infrared radiation even at low temperatures, infrared sensing can be used in applications where imaging of an object in the dark is needed. Applications include infrared photography, security systems, and missile guidance. Also, since infrared radiation is essentially thermal energy, infrared sensing can be effectively used in thermal control systems. Ultraviolet sensing is not as widely applied as infrared sensing.

Typically, a photosensor is a resistor, diode, or transistor element that brings about a change (e.g., generation of a potential or a change in resistance) into an electrical circuit in response to light that is falling on the sensor element. The power of the output signal may be derived primarily from the power source that energizes the electrical circuit. Alternatively, a photocell can be used as a photosensor. In this latter case, the energy of the light falling on the cell is converted into electrical energy of the output signal. Typically, a photosensor is available as a tiny cylindrical element with a sensor head consisting of a circular window (lens). Several types of photosensors are described below.

Photoresistors A photoresistor (or photoconductor) has the property of decreasing resistance (increasing conductivity) as the intensity of light falling on it increases. Typically, the resistance of a photoresistor could change from very high values (megohms) in the dark to reasonably low values (less than 100 Ω) in bright light. As a result, very high sensitivity to light is possible. Some photocells can function as photoresistors because their impedance decreases (output increases) as the light intensity increases. Photocells used in this manner are termed photoconductive cells. The circuit symbol of a photoresistor is given in Fig. 80a.

A photoresistor may be formed by sandwiching a photoconductive crystalline material such as cadmium sulfide (CdS) or cadmium selenide (CdSe) between two electrodes. Lead sulfide (PbS) or lead selenide (PbSe) may be used in infrared photoresistors.

Photodiodes A photodiode is a *pn* junction of semiconductor material that produces electron–hole

pairs in response to light. The symbol for a photodiode is shown in Fig. 80b. Two types of photodiodes are available. A photovoltaic diode generates a sufficient potential at its junction in response to light (photons) falling on it. Hence, an external bias source is not necessary for a photovoltaic diode. A photoconductive diode undergoes a resistance change at its junction in response to photons. This type of photodiode is usually operated in reverse-biased form; the *p* lead of the diode is connected to the negative lead of the circuit, and *n* lead is connected to the positive lead of the circuit. The breakdown condition may occur at about 10 V, and the corresponding current will be nearly proportional to the intensity of light falling on the photodiode. Hence, this current can be used as a measure of the light intensity. Since the current level is usually low (a fraction of a milliampere), amplification might be necessary before using it in the subsequent application (e.g., actuation, control, display). Semiconductor materials such as silicon, germanium, cadmium sulfide, and cadmium selenide are commonly used in photodiodes. A diode with an intrinsic layer (a pin diode) can provide faster response than with a regular *pn* diode.

Phototransistor Any semiconductor photosensor with amplification circuitry built into the same package (chip) is popularly called a phototransistor. Hence, a photodiode with an amplifier circuit in a single unit might be called a phototransistor. Strictly, a phototransistor is manufactured in the form of a conventional bipolar junction transistor with base (B), collector (C) and emitter (E) leads.

Symbolic representation of a phototransistor is shown in Fig. 80c. This is an *npn* transistor. The base is the central (*p*) region of the transistor element. The collector and the emitter are the two end regions (*n*) of the element. Under operating conditions of the phototransistor, the collector–base junction is reverse biased (i.e., a positive lead of the circuit is connected to the collector, and a negative lead of the circuit is connected to the base of an *npn* transistor). Alternatively, a phototransistor may be connected as a two-terminal device with its base terminal floated and the collector terminal properly biased (positive for an *npn* transistor). For a given level of source voltage (usually applied between the emitter lead of the transistor and load, the negative potential being at the emitter load), the collector current (current through the collector lead) i_c is nearly proportional to the intensity of the light falling on the collector–base junction of the transistor. Hence, i_c can be used as a measure of the light intensity. Germanium or silicon is the semiconductor material that is commonly used in phototransistors.

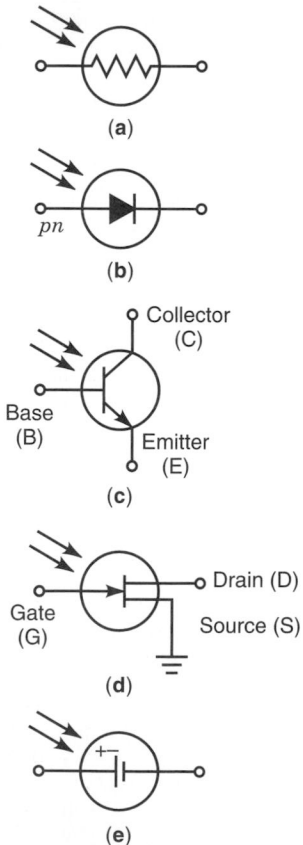

Fig. 80 Circuit symbols of some photosensors: (*a*) photoresistor; (*b*) photodiode; (*c*) phototransistor (*npn*); (*d*) photo-FET (*n*-channel); (*e*) photocell.

Photo-FET A photo–field-effect transistor is similar to a conventional FET. The symbol shown in Fig. 80d is for an *n*-channel photo-FET. This consists of an *n*-type semiconductor element (e.g., silicon doped with boron), called channel. A much smaller element of *p*-type material is attached to the *n*-type element. The lead on the p-type element forms the gate (G). The drain (D) and the source (S) are the two leads on the channel. The operation of an FET depends on the electrostatic fields created by the potentials applied to the leads of the FET.

Under operating conditions of a photo-FET, the gate is reverse biased (i.e., a negative potential is applied to the gate of an n-channel photo-FET). When light is projected at the gate, the drain current i_d will increase. Hence, drain current (current at the D lead) can be used as a measure of light intensity.

Photocells Photocells are similar to photosensors except that a photocell is used as an electricity source rather than a sensor of radiation. Solar cells, which are more effective in sunlight are commonly available. A typical photocell is a semiconductor junction element made of a material such as single-crystal silicon, polycrystalline silicon, and cadmium sulfide. Cell arrays are used in moderate-power applications. Typical power output is 10 mW/cm^2 of surface area, with a potential of about 1.0 V. The circuit symbol of a photocell is given in Fig. 80e.

Charge-Coupled Device A charge-coupled device (CCD) is an integrated circuit (a monolith device) element of semiconductor material. A CCD made from silicon is schematically represented in Fig. 81. A silicon wafer (p type or n type) is oxidized to generate a layer of SiO$_2$ on its surface. A matrix of metal electrodes is deposited on the oxide layer and is linked to the CCD output leads. When light falls onto the CCD element, charge packets are generated within the substrate silicon wafer. Now if an external potential is applied to a particular electrode of the CCD, a potential well is formed under the electrode, and a charge packet is deposited here. This charge packet can be moved across the CCD to an output circuit by sequentially energizing the electrodes using pulses of external voltage. Such a charge packet corresponds to a pixel (a picture element). The circuit output is the video signal. The pulsing rate could be higher than 10 MHz. CCDs are commonly used in imaging application, particularly in video cameras. A typical CCD element with a facial area of a few square centimeters may detect 576×485 pixels, but larger elements

(e.g., 4096×4096 pixels) are available for specialized applications. A charge injection device (CID) is similar to a CCD. In a CID, however, there is a matrix of semiconductor capacitor pairs. Each capacitor pair can be directly addressed through voltage pulses. When a particular element is addressed, the potential well there will shrink, thereby injecting minority carriers into the substrate. The corresponding signal, tapped from the substrate, forms the video signal. The signal level of a CID is substantially smaller than that of a CCD, as a result of higher capacitance.

Applications of Optically Coupled Devices One direct application is in the isolation of electric circuitry. When two circuits are directly connected through electrical connections (cables, wires, etc.), a two-way path is created at the interface for the electrical signals. In other words, signals in circuit A will affect circuit B and signals in circuit B, will affect circuit A. This interaction means that noise in one circuit will directly affect the other. Furthermore, there will be loading problems; the source will be affected by the load. Both these situations are undesirable. If the two circuits are optically coupled, however, there is only a one-way interaction between the two circuits (see Fig. 82). Variations in the output circuit (load circuit) will not affect the input circuit. Hence, the input circuit is isolated from the output circuit. The connecting cables in an electrical circuit can introduce noise components such as electromagnetic interference, line noise, and ground-loop noise. The likelihood of these noise components affecting the overall system is also reduced by using optical coupling. In summary, isolation between two circuits and isolation of a circuit from noise can be achieved by optical coupling. Optical coupling is widely used in communication networks (telephones, computers, etc.) and in circuitry for high-precision signal conditioning (e.g., for sophisticated sensors and control systems) for these reasons.

The medium through which light passes from the light source to the photosensor can create noise problems, however. If the medium is open (see Fig. 82), then ambient lighting conditions will affect the output circuit, resulting in an error. Also, environmental impurities (dust, smoke, moisture, etc.) will affect the light received by the photosensor. Hence, a more controlled medium of transmission would be desirable. Linking the light source and the photosensor using

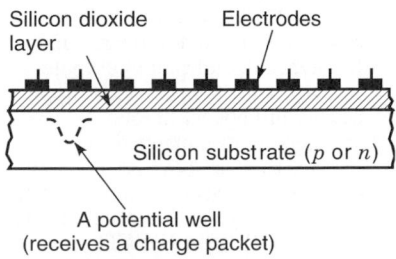

Fig. 81 A CCD.

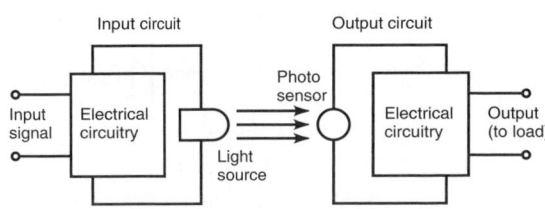

Fig. 82 An optically coupled device.

optical fibers is a good way to reduce problems due to ambient conditions in optically coupled systems.

Optical coupling may be used in relay circuits where a low-power circuit is used to operate a high-power circuit. If the relay that operates the high-power circuit is activated using an optical coupler, reaction effects (noise and loading) on the low-power circuit can be eliminated. Optical coupling is used in power electronics and control systems in this manner.

Many types of sensors and transducers that are based on optical methods do, indeed, employ optical coupling (e.g., optical encoders, fiberoptic tactile sensors). Optical sensors are widely used in industry for parts counting, parts detection, and level detection. In these sensors, a light beam is projected from a source to a photodetector, both units being stationary. An interruption of the beam through the passage of a part will generate a pulse at the detector, and this pulse is read by a counter or a parts detector. Furthermore, if the light beam is located horizontally at a required height, its interruption when the material filled into a container reaches that level could be used for filling control in the packaging industry. Note that the light source and the sensor could be located within a single package if a mirror is used to reflect light from the source back onto the detector. Further applications are within computer disk drive systems, for example, to detect the write protect notch as well as the position of the recording head.

6   INPUT DEVICES

George Grinstein and Marjan Trutschl

Human–computer interaction (HCI) is now a multidisciplinary area focusing on the interface and interactions between people and computer systems. Figure 83 presents a conceptual view of HCI: A user interacts with a system (typically a processor or device) using one or multiple input devices.

Input devices convert some form of energy, most often kinetic or potential energy, to electric energy. In this section we consider analog and digital input devices. Analog input devices generate voltages that vary over a continuous range ($R = V_{max} - V_{min}$) of values and are converted to binary values by an ADC. Digital input devices are based on binary digits. An input device that generates logical 0's and 1's, on and off, respectively, is called a *binary switch*. A binary switch generates the binary digit 1 when the input voltage is equal to or greater than a specified threshold value and the binary digit 0 otherwise. A second type of digital input device approximates an analog signal and provides a binary stream. Thus, any device that produces an electrical signal or responds to an electrical signal can be used as an input device. Preprocessed analog (digitized) and digital signals generated by an input device are passed on to the processor/device for processing. Once processed, the processor/device may, and often does, generate a new signal or a series of signals. These signals can be used to trigger events on some attached output device.

Figure 84 shows examples of a signal produced by an analog input device. To be used with a digital computer, the analog signal can be processed to mimic an on/off switch or it can be digitized using an ADC. The performance of an ADC depends on its architecture. The more bits the ADC operates with, the better the resolution of the signal approximation.

Input devices can be further classified as acoustic, inertial, mechanical, magnetic, and optical input devices.

6.1   Devices

Based on their basic operation, input devices can be classified as 2-D, 3-D, 6-D, or *n*-D (degrees of freedom) input devices. Table 26 lists some of the most popular input devices and degrees of freedom associated with each. Many devices can fit in several categories. Also, as any device can emulate another, this table is to be used simply as a guide. Finally, there are other forms of input technologies that are described elsewhere in this encyclopedia.

Many of the aforementioned devices can be used in combinations with other input devices, thus providing the notion of either two-handed input or multimodal input. For example, the use of two data gloves is considered two-handed input, as is the use of a mouse along with a Spaceball, whereas the use of a mouse along with speech recognition is considered to be multimodal input.

6.2   Commonly Used Input Devices

Keyboard   The keyboard is now considered the most essential input device and is used with the majority of computers. Keyboards provide a number of keys (typically more than 100) labeled with a letter or a function that the key performs. Keyboards manufactured for use with notebooks and palm computers or those designed for users with special needs typically provide a reduced set of keys. Different alphabets require different characters to be mapped to each key on the keyboard (i.e., English QWERTY versus German QWERTZ keyboard). Such mappings are achieved by reprogramming the keyboard's instruction set. Certain keys (e.g., ALT, CTRL, and SHIFT) can be used in conjunction with other keys, thus permitting one key to map to several different functions.

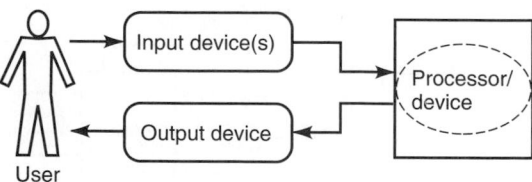

Fig. 83 Fundamental human–computer interaction model.

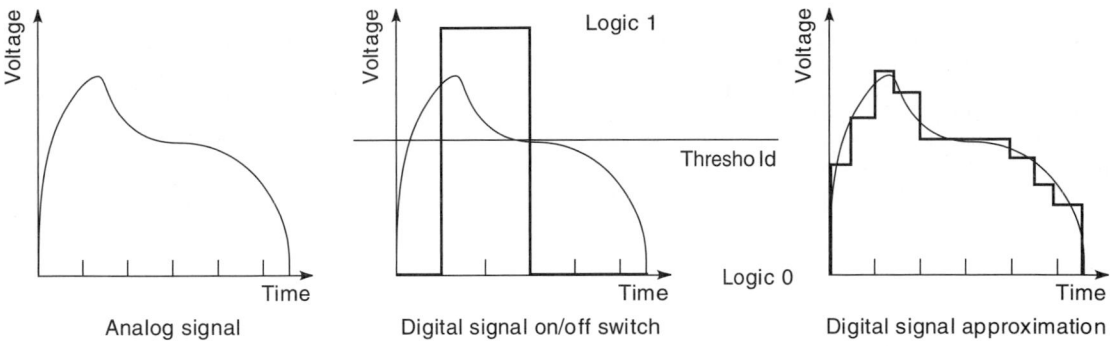

Fig. 84   Input signals.

Table 26   Input Device Classes

Input Device	1-D	2-D	3-D	4-D	6-D	n-D
Data glove						×
Digitizer		×	×			
Eye tracker		×				
Graphic tablet		×				
Trackpoint device		×				
Joystick		×	×	×	×	
Lightpen		×				
Monkey						×
Mouse		×	×	×		
Position tracker	×	×	×	×	×	×
Scanner		×	×			
Slider	×	×				
Spaceball					×	
Touch screen		×				
Touchpad		×				
Trackball		×				

Mouse   Since its creation at Xerox Palo Alto Research Center (PARC), the mouse has become the most popular 2-D input device and has a wide number of variants. Regardless of the variation, each mouse has one, two, or three buttons. For most mice, the motion of a ball, located underneath the mouse, is converted to planar motion—a set of x and y values—using a photoelectric switch as an input transducer. The photoelectric switch contains an LED as a source, a phototransistor as a sensor, and a circular perforated disk as a switch. When the light emitted from the diode reaches the sensor, a pulse (logic 1) is generated and passed on to the interface electronics. The frequency of pulses is interpreted as the velocity of the mouse. There are two such input transducers built in a mouse—one for the x and one for the y axis. Figure 85 shows the principle of motion-to-electric energy conversion.

The majority of mice use this principle of motion conversion. Optical mice take advantage of the reflective properties of mouse pads that have a grid of thin lines printed on their smooth and reflective surface. As the mouse passes across the line of a grid, a portion of the light emitted from the LED is diffracted, resulting in a slight drop of a voltage on the sensor's side (Fig. 86). These drops of voltage are used to determine the direction and speed of movement of the mouse.

Trackball   A trackball can be described as an inverted mouse. To move a cursor on the screen, the user moves

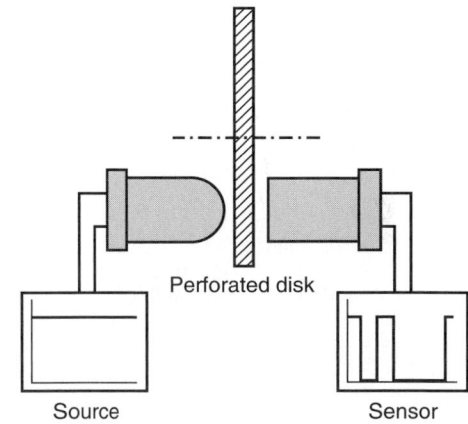

Fig. 85   Motion-to-energy conversion.

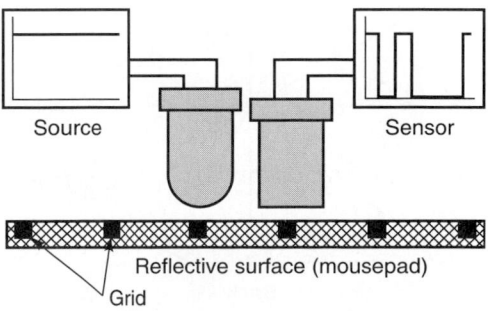

Fig. 86   Optical mouse structure.

the ball in the desired direction. The motion of the ball is translated to electric signals using a set of perforated disks (one for the x and the other for the y direction). Trackballs, like mice, are equipped with one or more buttons that are pressed to perform a desired operation. Many notebooks and portable computers provide built-in trackballs, as these require much less space than a mouse.

Joystick The joystick made its first major appearance in arcade machines in the early 1980s. The basic joystick is a 2-D input device that allows users to move a cursor or an object in any direction on a plane. Typically, a joystick consists of two major parts—a vertical handle (the stick) and a base—each providing one or more buttons that can be used to trigger events. To move the cursor or an object, the stick is moved in the desired direction. Figure 87 shows a major limitation imposed by the resolution of a joystick. The resolution in this example joystick makes it impossible to move in the indicated direction (desired direction), making navigation a bit difficult.

There are two major types of joysticks—isotonic and isometric. *Isotonic* joysticks are precision position-sensitive devices, used in animation, special-effects development, and games. These joysticks are equipped with a set of springs, which return the joystick to the center position when released. A stream of x and y values is generated based on and proportional to the angle between the initial and the current position of the control stick. Some implementations of isotonic joysticks are insensitive to the angle α. These use switches to provide information on direction. *Isometric* joysticks provide no spring action—the control stick does not move. The x and y values generated by the joystick are proportional to the force applied to the control stick. Some newer joysticks also have been provided with tactile and force feedback.

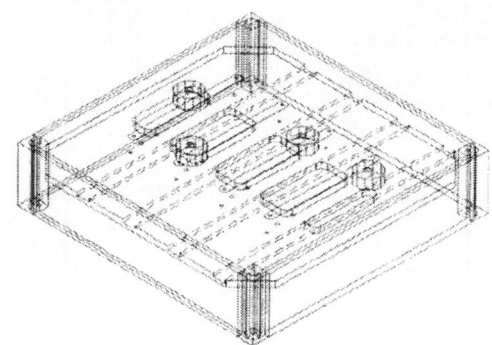

Fig. 88 Slider box. (Image courtesy of Simulation Special Effect, LLC).

Slider A slider is a 1-D input device (Fig. 88). Although sliders are usually implemented in software as part of a graphical user interface (GUI), slider boxes are available as input devices in applications requiring a large number of independent parameters to be controlled [as in musical instrument digital interface (MIDI) applications requiring multiple channels to be manipulated independently]. Most windowing systems incorporate sliders to support panning of the window's content or for color scale value selections.

Spaceball The Spaceball is a 6-D input device used primarily in computer-aided design and engineering, animation, virtual reality, and computer games. It enables users to manipulate a 3-D model with 6-degrees-of-freedom control (simultaneous x, y, z, translations and rotations) and as easily as if they were holding it in their hands. A Spaceball is often used in conjuction with the mouse. Spaceballs made their appearances initially with high-end graphic workstations, but this is not the case anymore. As desktop computers have become more powerful, many applications make use of the Spaceball and its derivatives.

Touchpad A touchpad is a 2-D input device developed for use in areas with limited space. Touchpads provide precise cursor control by using a fingertip moving on a rectangular area. Buttons located on the side of the rectangular input area can be programmed to perform specific operations as modifier keys on keyboards. Touchpads are usually located under the SPACE bar or the cursor keys, or they can be attached to a computer through a serial port.

Input Tablet An input tablet is a variation of a touchpad. It is larger than a touchpad, and instead of a finger, a penlike device with a button to perform specific operations is used. A coil in the pen generates a magnetic field, and a wire grid in the tablet transmits the signal to the tablet's microprocessor. The output data include the pen's location, the pressure of the pen

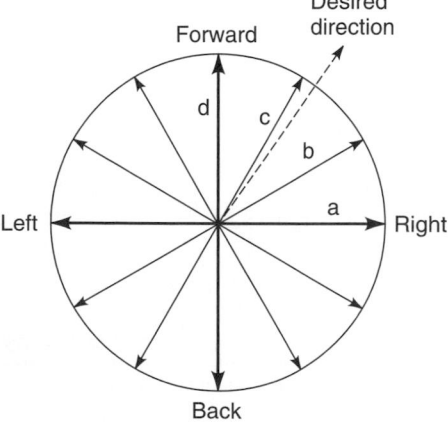

Fig. 87 Directional limitation of Joysticks.

on the tablet, and the tilt of the pen in relation to tablet. Input tablets are mostly used in the design arts and in mechanical and engineering computer-aided design.

Integrated Pointing Device — Stick A 2-D integrated pointing device, called a stick, is a miniature isometric joystick embedded between the keys on the keyboard. To move the cursor, the user pushes or pulls the stick in desired direction. The buttons associated with the stick are located under the SPACE bar on the keyboard.

Lightpen A lightpen is a penlike 2-D device attached to a computer through one of the communications ports or through a dedicated controller board. It is used to draw or select objects directly on the screen. Lightpens may be optically or pressure driven. An optically driven lightpen receives light from the refresh update on the screen; the x, y position of the refreshed pixel is then available for processing. A pressure-driven lightpen is triggered by pressing the lightpen on the screen or by pushing a button.

Touch Screen A touch screen is a special type of a 2-D hybrid device because it can both display and acquire information at the same time. On the input side, a touch screen contains a set of sensors in the x and y directions. These sensors may be magnetic, optical, or pressure. Users simply touch the screen, and the sensors in both x and y directions detect an event at some x and y coordinate. Since users tend to use a finger to interact with the touch screen, the resolution of the input device is not fully utilized. In fact, it is often limited to the size of a fingertip. Touch screens are very popular in menu-driven environments such as information booths, fast-food restaurants, and control rooms.

Scanner A scanner is a 2-D input device used to capture pictures, drawings, or text. Images, color or black and white, can be captured and stored in digital form for analysis, manipulation, or future retrieval. Associated application software is typically bundled with scanners. This includes imaging software, photo manipulation software, vector graphics conversion software, or text creation (using optical character recognition) software. Three major scanners are available: handheld, flatbed, and sheet scanners. Handheld scanners are suitable for small-scale scanning, flatbed scanners usually handle up to legal-size documents, and sheet scanners usually handle documents of fixed width but arbitrary length. Some engineering firms and geographers use special large-scale scanners for digitizing blueprints and maps.

Digitizer A digitizer can be considered either a 2-D or a 3-D input device. There are numerous kinds of digitizers available. Many older and less expensive systems require a great deal of manual work to acquire the data points. For example, the user may need to draw a grid on the object to be digitized to enable the acquisition of coordinates for every point on that grid. This is both time consuming and error prone.

3-D Laser Digitizers Nonmanual digitizers can automate several parts of the digitization process. These are primarily laser-based scanners. An object is positioned on a podium and the scanner rotates the podium while the digitization takes place. Some digitizers revolve around the object when the object is too big or to heavy to be rotated easily around its axes. Such scanners project a beam of laser light onto the model. The intersection of the laser beam and the surface of the object creates a contour of the model captured by a camera and displayed on the screen. This can be done in real time, and a color camera can be used to generate a color model. Most laser scanners use laser triangulation to reconstruct the object.

Position Trackers Position trackers are used to detect motion and are often attached to objects or body parts. Trackers perform reasonably well. Newer trackers have removed the tethering limitation of older trackers. Newer technologies are also solving the line-of-sight problem (the receiver's requiring an unobstructed view of the sensors). Some trackers need to be recalibrated often to maintain a high degree of accuracy.

Mechanical. Mechanical position trackers use a rigid jointed structure with a known geometry. Such a structure has one fixed and one active end, with the position of the active and available in real time. Mechanical tracking devices are very fast (less than 5 ms response time) and very accurate. The accuracy depends on the accuracy of joint angle encoders. A tracker with a full-color head-coupled stereoscopic display can provide high-quality, full-color stereoscopic images and full 6 degrees of freedom (translation along x, y, and z as well as roll, pitch, yaw).

Magnetic. Magnetic trackers use a source that generates three fields of known strength. Detectors are attached to the object to be tracked and measure the magnetic field strengths at a given point. These values are used to determine 6 degrees of freedom in space. Magnetic trackers do not experience any line-of-sight problems and are scalable to many detectors. However, the amount of wiring increases as the number of detectors increases. Magnetic trackers do not operate well around ferrous materials.

Ultrasonic. Ultrasonic trackers are often attached to a virtual reality (VR) headset. The tracker consists of three receivers and three transmitters. The position and orientation of the object is calculated based on the time required for each transmitted signal to reach a receiver. Ferrous materials do not affect such trackers. However, ultrasonic trackers are affected by the line-of-sight problem and may be affected by other sources of ultrasonic harmonics.

High-Speed Video. High-speed video along with fiducial markings on a tracked object is used to determine the location of an object in space. A single picture or a series of pictures are acquired and later processed using image-processing techniques. Fiducial markings can also be located in the space (i.e., scene or walls) and the camera can be attached to the object itself. Such devices can then be used to control the navigation of a robot between two given locations. High-speed video is used for work in a large space because no extra wiring is necessary. Video is unaffected by ferrous and other metals, ultrasonic sound, and light. However, the line-of-sight problem does affect video-tracking systems.

Inertial. Inertial position trackers are used to measure orientation and velocity. They are untethered and are not limited by the range or the size of the volume they operate in. Inertial position trackers provide almost complete environmental immunity. Such trackers are sensitive to vibrations and can thus result in inaccurate readings.

Biological. Eye tracking is a relatively old technology although not in common use. Eye tracking can be used for control or monitoring. For example, a pilot can control various instruments by simply looking at them. A low-powered infrared (IR) beam is used to illuminate the eye, which in turn is captured using a small camera. The image is processed to track pupil and corneal reflection. Today's eye tracking devices operate at one degree of resolution. It takes approximately one-third of a second to select, acquire, and fix on an image. Modern applications of eye tracking include its use as an input device for the disabled.

Digital Whiteboard A digital whiteboard is a 2-D input device designed to replace traditional blackboards and whiteboards. Everything written on the digital whiteboard with a standard dry-erase marker can be transmitted to a computer. That information can then be used by any application, such as e-mail, fax, or teleconferencing.

Data Glove A data glove is an input device that uses properties of leaky fiber-optic cables or resistive strain gauges to determine the amount of movement of fingers and wrists. Leaky fiber-optic cables provide good data, but it is the resistive strain-based input gloves that provide more accurate data. Each data glove is often combined with a 3-D tracker and with 10 strain gauges—at least one for each finger joint—which provides a very high degree of freedom. The latest data gloves also have been extended to provide tactile/force feedback using pneumatic pistons and air bladders. Data gloves can be used along with gestures to manipulate virtual objects or to perform other tasks.

Microphone/Speech Recognition and Understanding The microphone has proved to be one of the most useful input devices for digitizing voice and sound input or for issuing short commands that need to be recognized by a computer. Longer commands cannot be handled by simple recognition. Most sophisticated systems available today still cannot guarantee 100% understanding of human speech.

Monkeys or Mannequins The first monkeys were humanlike input devices with a skeleton and precision rheostats at the joints to provide joint angles. Monkeys can be used to set up and capture humanlike motions and offer much better degree-of-freedom match than other devices. Since the first monkeys, a series of animal-like input devices and building blocks have been created that allow users to create their own creatures.

Game Input Devices There are a number of other specialized input devices designed to make playing games a more exciting and more realistic experience. Most of these input devices offer additional degrees of freedom and can be used along with other input devices.

6.3 Conclusions

There are a large number of input devices, and the technology is rapidly changing. It is expected that speech recognition and command interpretation, gesture recognition for highly interactive environments (game and virtual), and real-time imaging will become more prominent in the next decade. These will increase the level of human participation in applications and the bandwidth of the data transferred.

7 INSTRUMENTS

Halit Eren

Measurement is essential for observing and testing scientific and technological investigations. It is so fundamental and important to science and engineering that the whole science can be said to be dependent on it. Instruments are developed for monitoring the conditions of physical variables and converting them into symbolic output forms. They are designed to maintain prescribed relationships between the parameters being measured and the physical variables under investigation. The physical parameter being measured is known as the *measurand*. The sensors and transducers are the primary sensing elements in the measuring systems that sense the physical parameters to produce an output. The energy output from the sensor is supplied to a transducer, which converts energy from one form to another. Therefore, a transducer is a device capable of transferring energy between two physical systems.

Measurement is a process of gathering information from a physical world and comparing this information with agreed standards. Measurement is carried out with instruments that are designed and manufactured to fulfill given specifications. After the sensor generates the signals, the type of signal processing depends on the information required from it. A diverse range of sensors and transducers may be available to meet the measurement requirements of a physical system. The sensors and transducers can be categorized in a number of ways depending on the energy input and output, input variables, sensing elements, and electrical or physical principles. For example, from an energy input and output point of view, there are three fundamental types of transducers: modifiers, self-generators, and modulators.

In modifiers, a particular form of energy is modified rather than converted; therefore, the same form of energy exists at the input and the output. In self-generators, electrical signals are produced from nonelectric inputs without the application of external energy. These transducers produce very small signals, which may need additional conditioning. Typical examples are piezoelectric transducers and photovoltaic cells. Modulators, on the other hand, produce electric outputs from nonelectric inputs, but they require an external source of energy. Strain gauges are typical examples of such devices.

The functionality of an instrument can be broken into smaller elements, as illustrated in Fig. 89. Most measurement systems have a sensor or transducer stage, a signal-conditioning stage, and an output or termination stage. All instruments have some or all of these functional blocks. Generally, if the behavior of the physical system under investigation is known, its performance can be assessed by means of a suitable method of sensing, signal conditioning, and termination.

In the applications of instruments, the information about a physical variable is collected, organized, interpreted, and generalized. Experiments are conceived, performed, and repeated; as we acquire confidence in the results, they are expressed as scientific laws. The application of instruments ranges from laboratory conditions to arduous environments such as inside nuclear reactors or on satellite systems and spaceships. In order to meet diverse application requirements of high complexity and capability, many manufacturers have developed a large arsenal of instruments. Some of these manufacturers are listed in Table 27.

In recent years, rapid growth of integrated circuit (IC) electronics and the availability of cheap analog-to-digital and microprocessors have led to progress in the instrumentation field, with the development of instruments, measuring techniques, distributed architectures, and standards aimed to improve performance.

Instruments are applied for static or dynamic measurements. The static measurements are relatively easy since the physical quantity (e.g., fixed dimensions and weights) does not change in time. If the physical quantity is changing in time, which is often the case, the measurement is said to be dynamic. In this case, steady-state and transient behavior of the physical variable must be analyzed so that it can be matched with the dynamic behavior of the instrument.

7.1 Design, Testing, and Use of Instruments

Instruments are designed on the basis of existing knowledge, which is gained either from the experiences of people about the physical process or from our structured understanding of the process. In any case, ideas conceived about an instrument must be translated into hardware and/or software that can perform well within the expected standards and easily be accepted by the end users.

Usually, the design of instruments requires many multidisciplinary activities. In the wake of rapidly changing technology, instruments are upgraded often to meet the demands of the marketplace. Depending on the complexity of the proposed instrument, it may take many years to produce an instrument for a relatively short commercial lifetime. In the design and production of instruments, we must consider such factors as simplicity, appearance, ease and flexibility of use, maintenance requirements, lower production costs, lead time to product, and positioning strategy in the marketplace.

In order to design and produce instruments, a firm must consider many factors. These include sound business plans, suitable infrastructure, plant, equipment, understanding of technological changes, skilled and trained personnel, adequate finance, marketing and distribution channels, and a clear understanding about worldwide instrument and instrumentation system trends. It is important to choose the right product that is very likely to be in demand in the years to come. Here entrepreneurial management skills may be an important factor.

The design process itself may follow well-ordered procedures from idea to marketing stages. The process may be broken down into smaller tasks such as identifying specifications, developing possible solutions for these specifications, modeling, prototyping,

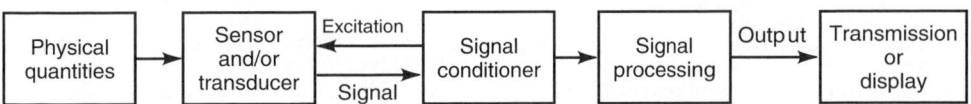

Fig. 89 An instrument has a number of relatively independent components that can be described as functional elements. These functional elements are the sensors and transducers, signal conditioners, and output or terminations. In general, if the behavior of the physical system is known, its performance is measured by a suitable arrangement and design of these components.

Table 27   List of Manufacturers

ABB, Inc.
501 Merritt 7, P.O. Box 5308
Norwalk, CT 06856-5308
Tel: 800-626-4999
Fax: 203-750-2263

Allied Signal, Inc.
101 Columbia Road
Morristown, NY 07962
Tel: 800-707-4555
Fax: 608-497-1001

Bailey-Fisher and Porter Company
125 E County Line Road
Wanminster, PA 18974
Tel: 800-268-8520
Fax: 215-674-7183

Consolidated Instrument, Inc.
510 Industrial Avenue
Teterboro, NC 07608
Tel: 800-240-3633
Fax: 201-288-8006

Davies Instrument Manufacturing
    Company, Inc.
4701 Mt. Hope Drive
Baltimore, MD 21215
Tel: 800-548-0409
Fax: 410-358-0252

Dwyer Instrument, Inc.
P.O. Box 373-T
Michigan City, IN 46361-0373
Tel: 219-879-8000
Fax: 219-872-9057

Fuji Corporation of America
Park 80 West, Plaza Two
Saddlebrook, NJ 07663
Tel: 201-712-0555
Fax: 201-368-8258

Hanna Instrument, Inc.
Highland Industrial Park
584 Park East Drive
Woonscocket, RI 02895-0849
Tel: 800-999-4144
Fax: 401-765-7575

Hewlett-Packard Company
5301 Stevens Creek Boulevard
Santa Clara, CA 95052-8059
Fax: 303-756-6800

Industrial Instruments
    and Supply, Inc.
P.O. Box 416
12 County Line Industrial Park
Southampton, PA 18966
Tel: 800-523-6079
Fax: 215-396-0833

Instrument and Control Services
    Company
1351-T Cedar Lake Road
Lake Villa, IL 60046
Tel: 800-747-8367
Fax: 847-356-9007

Keithley Instrument, Inc.
28775-T Aurora Road
Cleveland, OH 44139-1891
Tel: 800-552-1115
Fax: 440-248-6168

MCS Calibration, Inc.
Engineering Division
1533 Lincoln Avenue
Halbrook, NY 11741
Tel: 800-790-0512
Fax: 512-471-6902

MSC Industrial Supply Company
151-T Sunnyside Boulevard
Plainview, NY 11803
Tel: 800-753-7937
Fax: 516-349-0265

National Instruments
6504 Bridge Point Parkway
Austin, TX 78730-7186
Tel: 512-794-0100; 888-217-7186
Fax: 512-794-8411

Omega Engineering, Inc.
P.O. Box 4047
Stamford, CT 06907
Tel: 800-826-6342
Fax: 203-359-7700

Rosemount Analytical
600 S. Harbor Boulevard, Dept TR
La Habra, CA 90631-6166
Tel: 800-338-8099
Fax: 562-690-7127

Scientific Instruments, Inc.
518 W Cherry Street
Milwaukee, WI 53212
Tel: 414-263-1600
Fax: 415-263-5506

Space Age Control, Inc.
38850 20th Street East
Palmdale, CA 93550
Tel: 800-366-3408
Fax: 805-273-4240

Tektronix, Inc.
P.O. Box 500
Beaverton, OR 97077
Tel: 503-627-7111

Texas Instrument, Inc.
34 Forest Street, MS 23-01
P.O. Box 2964
Attleboro, MA 02703
Tel: 508-236-3287
Fax: 508-236-1598

Warren-Knight Instrument Company
2045 Bennett Drive
Philadelphia, PA 19116
Tel: 215-464-9300
Fax: 215-464-9303

Yokogawa Corporation of America
2 Dart Road
Newnon, GA 30265-1040
Tel: 800-258-2552
Fax: 770-251-2088

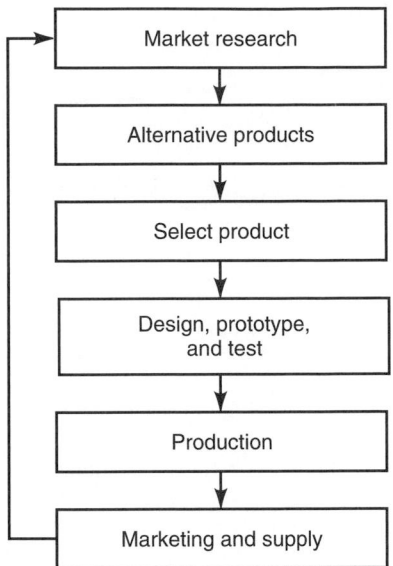

Fig. 90 Design process from the conception of ideas to marketing follows carefully considered stages. The proper identification and effective implementation of these stages is important in the success of a specific instrument in the marketplace.

installing and testing, making modifications, manufacturing, planning marketing and distribution, evaluating customer feedback, and making design and technological improvements. Figure 90 illustrates the stages for the design and marketing of an instrument. Each one of these stages can be viewed in detail in the form of subtasks. For example, many different specifications may be considered for a particular product. These specifications include but are not limited to operational requirements, functional and technological requirements, quality, installation and maintenance, documentation and servicing, and acceptance level determination by the customers.

In recent years, computers have been used extensively in the instrument manufacturing industry in the form of computer-aided design (CAD), automated testing, and in other applications. The computer enables rapid access to knowledge-based information and makes design time considerably shorter thus enabling manufacturers to meet rapid demand.

In CAD systems, mechanical drafting software, electronic circuit design tools, control analysis tools, and mathematical and word processing tools are integrated to assist the design procedure. Design software is available from various manufacturers listed in Table 27.

Testing and Use of Instruments After the instrument is designed and prototyped, various evaluation tests may be conducted. These tests may be made under reference conditions or under simulated environmental conditions. Some examples of reference condition tests are accuracy, response time, drift, and warmup time. Simulated environmental tests may be compulsory, being regulated by governments and other authorities. Some simulated environment tests include climatic test, drop test, dust test, insulation resistance test, vibration test, electromagnetic compatibility (EMC) tests, and safety and health hazard tests. Many of these tests are strictly regulated by national and international standards.

Adequate testing and proper use of instruments is important to achieve the best results out of them. When the instruments are installed, a regular calibration is necessary to ensure the consistency of the performance over the time period of operation. Incorrect measurements can cost a considerable amount of money or even result in the loss of lives.

For maximum efficiency, an appropriate instrument for the measurement must be selected. Users should be fully aware of their application requirements, since instruments that do not fit their purpose will deliver false data resulting in wasted time and effort. When selecting the instrument, users must evaluate many factors such as accuracy, frequency response, electrical and physical loading effects, sensitivity, response time, calibration intervals, power supply needs, spare parts, technology, and maintenance requirements. They must ensure compatibility with their existing equipment.

Also, when selecting and implementing instruments, quality becomes an important issue from both quantitative and qualitative perspectives. The quality of an instrument may be viewed differently depending on the people involved. For example, quality as viewed by the designer may be an instrument designed on sound physical principles, whereas from the user's point of view quality may be reliability, maintainability, cost, and availability.

For the accuracy and validity of information collected from the instruments, correct installation and proper use become very important. The instruments must be fully integrated with the overall system. Sufficient background work must be conducted prior to installation to avoid a possible shutdown of the process that is longer than necessary.

Once the system is installed, the reliability of the instrument must be assessed, and its performance must be checked regularly. The *reliability* of the system may be defined as the probability that it will operate at an agreed level of performance for a specified period of time. The reliability of instruments follows a bath tub shape against time. Instruments tend to be unreliable in the early and later stages of their lives. During normal operations, if the process conditions change (e.g., installation of large machinery nearby), calibrations must be conducted to avoid possible performance deterioration of the instrument. Therefore, the correct operations of the instruments must be assured at all times throughout the lifetime of the device.

Once the instruments are installed, they may be left alone and expected to operate reliably. They may be communicating with other devices, and their performance may affect the performance of the rest of the system, as in the case of the process industry. In some applications, the instruments may be part of a large instrumentation system, taking a critical role in monitoring and/or controlling the process and operations. However, in many applications, instruments are used on a stand-alone basis for laboratory and experimental work, and the success of the experiments may entirely depend on their correct performance. In these cases, the experiments must be designed and conducted carefully by identifying the primary variables, controlling, selecting the correct instruments, assessing the relative performances, validating the results, and using the data effectively by employing comprehensive data analysis techniques. Set procedures for experimental designs can be found in various sources (e.g., see Ref. 97 as well as the Bibliography).

After having performed the experiments, the data must be analyzed appropriately. This can be done at various stages by examining the consistency of the data, performing appropriate statistical analyses, estimating the uncertainties of the results, relating the results to the theory, and correlating the data. Details of statistical data analysis can be found in many books; also many computer software programs are available for the purpose of analysis including common packages such as Microsoft Excel.

7.2   Instrument Response and Drift

Instruments respond to physical phenomena by sensing and generating signals. Depending on the type of instrument used and the physical phenomenon observed, the signals may be either slow or fast to change, and may also contain transients. The response of the instruments to the signals can be analyzed in a number of ways by establishing static and dynamic performance characteristics. Although, the static performances are relatively simple, the dynamic performances may be complex.

Static Response   Instruments are often described by their dynamic ranges and full-scale deflections (span). The *dynamic range* indicates the largest and smallest quantities that can be measured. The *full-scale deflection* of an instrument refers to the maximum permissible value of the input quoted in the units of the particular quantity to be measured.

In instruments, the change in output amplitude resulting from a change in input amplitude is called the *sensitivity*. System sensitivity often is a function of external physical variables such as temperature and humidity. The relative ratio of the output signal to the input signal is the *gain*. Both, the gain and sensitivity are dependent on the amplitude of the signals and the frequency, which will be discussed in the section on dynamic response.

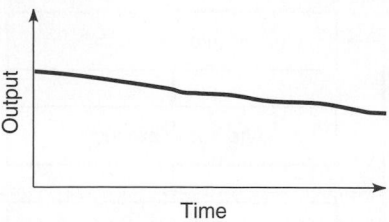

Fig. 91   Drift in the output of an instrument. The main causes of the drift are aging, temperature, ambient conditions, and component deterioration. The drift in an instrument may be predicted by performance analysis of components, past experience, environmental tests, and so on.

In the design stages or during manufacturing, there might be small differences between the input and output, which is called the *offset*. In other words, when the input is zero the output is not zero or vice versa. The signal output also may change in time, which is known as *drift*. The drift can occur for many reasons including temperature and aging. Fortunately, drift usually occurs in a predictable manner. A typical drift curve of an instrument against time is illustrated in Fig. 91.

During practical applications, readings taken from an instrument under the same conditions may not be repeatable. In this case, a repeatability test may be conducted, and statistical techniques must be employed to evaluate the repeatability of the instrument.

Dynamic Response   The dynamic response of an instrument is characterized by its natural frequency, amplitude, frequency response, phase shift, linearity and distortions, rise and settling times, slew rate, and the like. These characteristics are a common theme in many instrumentation, control, and electronics books. Although sufficient analysis will be given here, the detailed treatment of the topic can be very lengthy and complex; Hence the full treatment of this tonic is not within the scope of this section. Interested readers should refer to the literature (e.g., Ref. 98).

The dynamic response of an instrument can be linear or nonlinear. Fortunately, most instruments exhibit linear characteristics, leading to simple mathematical modeling by using differential equations such as

$$a_n \frac{d^n y}{dt^n} + a_{n-1} \frac{d^{n-1} y}{dt^{n-1}} + \cdots + a_0 y = x(t) \qquad (104)$$

where x is the input variable or the forcing function, y is the output variable, and $a_n, a_{n-1}, \ldots, a_0$ are the coefficients or the constants of the system.

The dynamic response of instruments can be categorized as zero-order, first-order, or second-order responses. Although higher order instruments may exist, their behaviors can be understood adequately in the form of a second-order system. From Eq. (104)

$$a_0 y = x(t) \quad \text{zero order} \quad (105)$$

$$a_1 \frac{dy}{dt} + a_0 y = x(t) \quad \text{first order} \quad (106)$$

$$a_2 \frac{d^2 y}{dt^2} + a_1 \frac{dy}{dt} + a_0 y = x(t) \quad \text{second order} \quad (107)$$

Equations (105)–(107) can be written as Laplace transforms, thus enabling analysis in the frequency domain,

$$\frac{Y(s)}{X(s)} = \begin{cases} 1 & (108) \\ \dfrac{1}{\tau_1 s + 1} & (109) \\ 0 & \\ \dfrac{1}{(\tau_1 s + 1)(\tau_2 s + 1)} & (110) \end{cases}$$

where s is the Laplace operator and τ is the coefficient also called time constant.

In zero-order instruments, there is no frequency dependence between the input and output. The amplitude change is uniform across the spectrum of all possible frequencies. In practice, such instruments are difficult to obtain, except in a limited range of operations.

In first-order instruments, the relation between the input and the output is frequency dependent. Figure 92 illustrates the response of a first-order instrument for a unit step input in the time domain. Mathematically, the output may be written as

$$y(t) = K e^{-t/\tau} \quad (111)$$

where K and τ are constants determined by the system parameters. In many cases, the input signals may be a complex rather than a simple step input. In the analysis, we need to multiply the transfer function, the second member of Eq. (109), by the Laplace transform of the input signal and then transform it back to the time domain if we are to understand the nature of transient and steady-state responses. Also, if the first-order

systems are cascaded, the relative magnitudes of the time constants become important; some may be dominant, and others may be neglected. Second-order systems exhibit the laws of simple harmonic motion, which can be described by linear wave equations. Equation (110) may be rearranged as

$$\frac{X(s)}{Y(s)} = \frac{1/a_0}{s^2/\omega_n^2 + 2\zeta s/\omega_n + 1} \quad (112)$$

where ω_n is the natural or undamped frequency (radians per second) and ζ is the damping ratio.

As can be seen, the performance of instruments become a function of natural frequency and the damping ratio of the system. The natural frequency and damping ratios are related to the physical parameters of the devices, such as mass and dimensions. In the design stages, these physical parameters may be selected, tested, and modified to obtain a desired response from the system.

Typical time response of a second-order system to unit step inputs is illustrated in Fig. 93. The response here indicates that a second-order system can either resonate or be unstable. Furthermore, we can deduce that, since the second-order system is dependent on time, wrong readings can be made depending on the time that the results are taken. Clearly, recording the output when the instrument is still under transient conditions will give an inadequate representation of the physical variable. The frequency compensation, selection of appropriate damping, acceptable time responses, and rise time settling time of instruments may need careful attention in both the design and

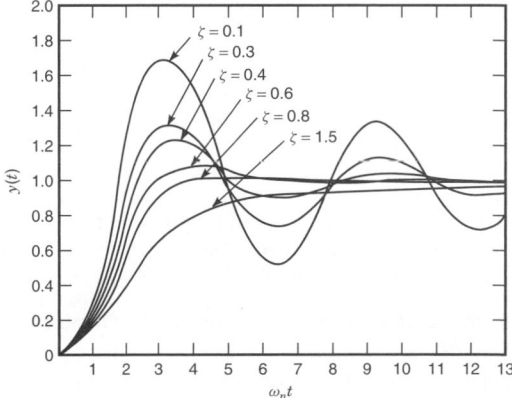

Fig. 93 Unit step time responses of a second-order system with various damping ratios. The maximum overshoot, delay, rise, settling times, and frequency of oscillation depend on the damping ratio. A smaller damping ratio gives a faster response but larger over shot. In many applications, a damping ratio of 0.707 is prefered.

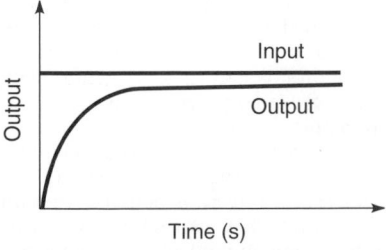

Fig. 92 First-order-hold instrument responds to a step input in an exponential form. For a good response the time delay must be small. Drift is usually expressed in percentage of output.

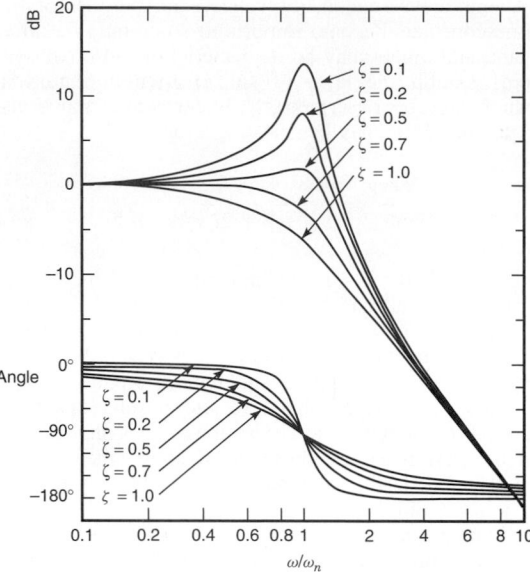

Fig. 94 Bode plots of gains and phase angles against frequency of a second-order system. Curves are functions of frequencies as well as damping ratios. These plots can be obtained theoretically or by practical tests conducted in the frequency range.

application stages of an instrument. In these systems, system analysis is essential to ensure that they can measure the input measurand adequately.

A typical frequency dependence of gain and phase angle between input and output is illustrated in Fig. 94 in the form of Bode diagrams. Here, the *bandwidth*, which is the frequencies over which the gain is reasonably constant, is also shown. Usually, half power point (3 dB), which symbolizes 70.7% of the maximum value, is taken as the bandwidth.

An important concept in instruments is *response time*, which can be described as the time required for the instrument to respond to an input signal change. For automatic measurements, the response time is an indication of how many readings can be done per second. Response time is affected by many factors such as analog-to-digital (A/D) conversion time, settling time, delays in electronic components, and delays in sensors.

7.3 Measurement Errors and Error Control Systems

The performance of an instrument depends on its static and dynamic characteristics. The performance may be indicated by its *accuracy*, which may be described as the closeness of measured values to the real values of the variable. The total response is a combination of dynamic and static responses. If the signals generated by the physical variable are changing rapidly, then the dynamic properties of the instrument become important. For slowly varying systems the dynamic errors

may be neglected. In order to describe the full relationships between the inputs and outputs, differential equations can be used, as discussed previously.

The performance of an instrument may also be decided by other factors, such as the magnitudes of errors; the *repeatability*, which indicates the closeness of sets of measurements made in the short term; and the *reproducibility* of the instrument. The reproducibility is the closeness of sets of measurements when repeated in similar conditions over a long period of time.

The ideal or perfect instrument would have perfect sensitivity, reliability, and repeatability without any spread of values and would be within the applicable standards. However, in many measurements, there will be imprecise and inaccurate results as a result of internal and external factors. The departure from the expected perfection is called the *error*. Often, sensitivity analyses are conducted to evaluate the effect of individual components that are causing these errors. Sensitivity to the affecting parameter can be obtained by varying that one parameter and keeping the others constant. This can be done practically by using the developed instruments or mathematically by means of appropriate models.

When determining the performance of an instrument, it is essential to appreciate how errors arise. There may be many sources of errors; therefore, it is important to identify these sources and draw up an error budget. In the error budget, there may be many factors, such as (a) imperfections in electrical and mechanical components (e.g., high tolerances and noise or offset voltages), (b) changes in component performances (e.g., shift in gains, changes in chemistry, aging, and drifts in offsets), (c) external and ambient influences (e.g., temperature, pressure, and humidity), and (d) inherent physical fundamental laws (e.g., thermal and other electrical noises, Brownian motion in materials, and radiation).

In instrumentation systems, errors can be broadly classified as systematic, random, or gross.

Systematic Errors   Systematic errors remain constant with repeated measurements. They can be divided into two basic groups as instrumental errors and environmental errors. Instrumental errors are inherent within the instrument, arising because of the mechanical structure, electronic design, improper adjustments, wrong applications, and so on. They can also be subclassified as loading error, scale error, zero error, and response time error. The environmental errors are caused by environmental factors such as temperature and humidity. Systematic errors can also be viewed as static or dynamic errors.

Systematic errors can be quantified by mathematical and graphical means. They can be caused by the nonlinear response of the instrument to different inputs as a result of hysteresis. They also emerge from wrong biasing, wear and aging, and other factors such as modifying the effects environment (e.g., interference). Typical systematic error curves are illustrated in Fig. 95.

Because of the predictability of systematic errors, deterministic mathematics can be employed. In the

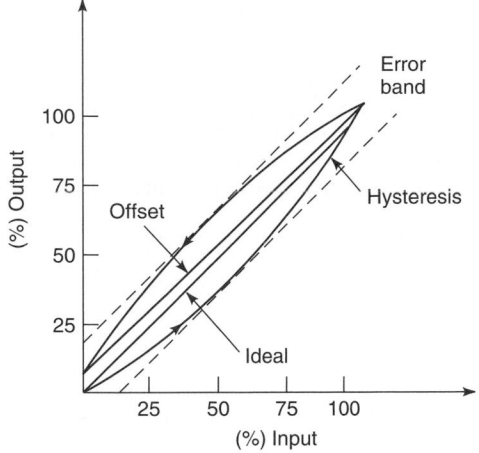

Fig. 95 Systematic errors are static errors and they can be quantified theoretically or experimentally. There are many different types, including hysteresis, linearity, and offset. They are contained within an error band typical to particular instrument.

simplest form, the error of a measurement may be expressed as

$$\Delta x(t) = x_m(t) - x_r(t) \qquad (113)$$

where $\Delta x(t)$ is the absolute error, $x_\tau(t)$ is the correct reference value, and $x_m(t)$ is the measured value.

From Eq. (113), the relative error $r_e(t)$ may be calculated as

$$r_e(t) = \frac{\Delta x(t)}{x_r(t)} \qquad (114)$$

However, in complex situations, correction curves obtained either empirically or theoretically may be used. Manufacturers usually supply correction curves, especially if their products embrace wide ranging and different applications (e.g., slurries with changing characteristics in time).

In many applications, the measurement system is made up of many components that have errors in their own rights. The deterministic approach may be adapted to calculate the overall propagated error of the system, as

$$y = f(x_1, x_2, x_3, \ldots, x_n) \qquad (115)$$

where y is the overall output and $x_1, x_2, \ldots$ are the components affecting the output.

Each variable affecting the output will have its own absolute error of Δx_i. The term Δx_i indicates the mathematically or experimentally determined error of each component under specified operating conditions. The overall performance of the overall system with the errors may be expressed as

$$y \pm \Delta y = f(x_1 \pm \Delta x_1, x_2 \pm \Delta x_2, \ldots, x_n \pm \Delta x_n) \qquad (116)$$

For an approximate solution, the Taylor series may be applied to Eq. (116). By neglecting the higher order terms of the series, the total absolute error Δy of the system may be written as

$$\Delta y = \left| \frac{\Delta x_1 \delta y}{\delta x_1} \right| + \left| \frac{\Delta x_2 \delta y}{\delta x_2} \right| + \cdots + \left| \frac{\Delta x_n \delta y}{\delta x_n} \right| \qquad (117)$$

The absolute error is predicted by measuring or calculating the values of the errors of each contributing component.

Slight modification of Eq. (116) leads to uncertainty analysis, where

$$w_y = [(w_1 \delta y / \delta x_1)^2 + (w_2 \delta y / \delta x_2)^2 + \ldots \\ + (w_n \delta y / \delta x_n)^2]^{1/2} \qquad (118)$$

where w_y is the uncertainty of the overall system and $w_1, w_2, \ldots, w_n$ are the uncertainties of affecting the component.

Uncertainty differs from error in that it involves such human judgmental factors as estimating the possible values of errors. In measurement systems, apart from the uncertainties imposed by the instruments, experimental uncertainties also exist. In evaluating the total uncertainty, several alternative measuring techniques should be considered and assessed, and estimated accuracies must be worked out with care.

Random and Gross Errors Random errors appear as a result of rounding, noise and interference, backlash and ambient influences, and so on. In experiments, the random errors vary by small amounts around a mean value. Therefore, the future value of any individual measurement cannot be predicted in a deterministic manner. Random errors may not easily be offset electronically; therefore, in the analysis and compensation, stochastic approaches are adapted by using the laws of probability.

Depending on the system, the random error analysis may be made by applying different probability distribution models. But, most instrumentation systems obey normal distribution laws; therefore, the Gaussian model can broadly be applied enabling the determination of the mean values, standard deviations, confidence intervals, and the like, depending on the number of samples being taken. A typical example of a Gaussian curve is given in Fig. 96. The mean value $\bar{x}$ and the standard deviation σ may be found by

$$\bar{x} = \frac{\sum x_i}{n} \qquad (119)$$

and

$$\sigma = \frac{\sum (x_i - \bar{x})^2}{n - 1} \qquad (120)$$

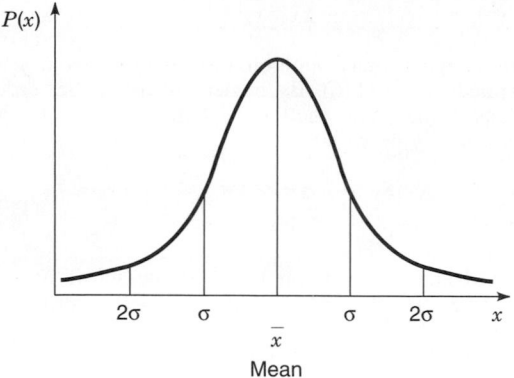

Fig. 96 Random errors of instruments can be analyzed by using probability methods. In many instruments the errors can be described by a Gaussian distribution curve.

Discussions relating to the application of stochastic theory in error analysis are very lengthy and will not be repeated here. Interested readers should refer to the literature (e.g., Ref. 98a).

Gross errors are the result of human mistakes, equipment fault, and the like. Human errors may occur in the process of observations or during the recording and interpretation of experimental results. A large number of errors can be attributed to carelessness, the improper adjustment of instruments, the lack of knowledge about the instrument and the process, and so on. These errors cannot be treated mathematically and eliminated completely, but they can be minimized by having different observers repeat the experiments.

Error Reduction Techniques   Controlling errors is an essential part of instruments and instrumentation systems. Various techniques are available to achieve this objective. The error control begins in the design stages by choosing the appropriate components, filtering, and bandwidth selection, by reducing the noise, and by eliminating the errors generated by the individual subunits of the complete system. In a good design, the errors of the previous group may be compensated adequately by the following groups.

The accuracy of instruments can be increased by postmeasurement corrections. Various calibration methods may be employed to alter parameters slightly to give correct results. In many cases, calibration graphs, mathematical equations, tables, the experiences of the operators, and the like are used to reduce measurement errors. In recent years, with the application of digital techniques and intelligent instruments, error corrections are made automatically by computers or the devices themselves.

In many instrumentation systems, the application of compensation strategy is used to increase static and dynamic performances. In the case of static characteristics, compensations can be made by many methods

including introducing opposing nonlinear elements in the system, using isolation and zero environmental sensitivity, opposing compensating environmental inputs, using differential systems, and employing feedback systems. On the other hand, the dynamic compensation can be achieved by applying these techniques as well as by reducing harmonics, using filters, adjusting bandwidth, using feedback compensation techniques, and the like.

Open-loop and close-loop dynamic compensations are popular methods employed in both static and dynamic error corrections. For example, using high-gain negative feedback can reduce the nonlinearity generated by the system. A recent and fast developing trend is the use of computers for estimating measured values and providing compensation during the operations if any deviations occur from the estimated values.

7.4   Standards and Reference Materials

Standards of fundamental units of length, time, weight, temperature, and electrical quantities have been developed for measurements to be consistent all over the world. The length and weight standards—the meter and the kilogram—are kept in the International Bureau of Weights and Measures in Sèvres, France. Nevertheless, in 1983 the meter was defined as the length of the path traveled by light in a vacuum in the fraction 1/299,792,458 of a second, which was adopted as the standard meter. The standard unit of time—second—is established in terms of known oscillation frequencies of certain devices, such as the radiation of the cesium-133 atom. The standards of electrical quantities are derived from mechanical units of force, mass, length, and time. Temperature standards are established as international scale by taking 11 primary fixed points. If different units are involved, the relationship between different units are defined in fixed terms. For example, 1 lbm = 453.59237 g.

Based on these standards, primary international units, SI (Système International d'Unités), are established for mass, length, time, electric current, luminous intensity, and temperature, as illustrated in Table 28. From these units, SI units of all physical quantities can be derived as exemplified in Table 29. The standard multiplier prefixes are illustrated in Table 30.

Table 28   Basic SI Units

Quantity	Unit	Symbol
Length	meter	m
Mass	kilogram	kg
Time	second	s
Electric current	ampere	A
Temperature	kelvin	K
Amount of substance	mole	mol
Luminous intensity	candela	cd
Plane angle	radian	rad
Solid angle	steradian	sr

Table 29   Fundamental, Supplementary, and Derived Units

Quantity	Symbol	Unit Name	Unit Symbol
Mechanical Units			
Acceleration	a	Meter/second^2	m/s^2
Angular acceleration	α	Radian/second^2	rad/s^2
Angular frequency	ω	Radian/second	rad/s
Angular velocity	ω	Radian/second	rad/s
Area	A	Square meter	m^2
Energy	E	Joule	J(kg $\cdot$ m^2/s^2)
Force	F	Newton	N(kg $\cdot$ m/s^2)
Frequency	f	Hertz	Hz
Gravitational field strength	g	Newton/kilogram	N/kg
Moment of force	M	Newton $\cdot$ meter	N $\cdot$ m
Plane angle	$\alpha, \beta, \theta, \phi$	Radian	Rad
Power	P	Watt	W(J/s)
Pressure	p	Newton/meter^3	N/m^3
Solid angle	ω	Steradian	Sr
Torque	T	Newton meter	N $\cdot$ m
Velocity	v	Meter/second	m/s
Volume	V	Cubic meter	m^3
Volume density	ρ	Kilogram/meter^3	kg/m^3
Wavelength	λ	Meter	M
Weight	W	Newton	N
Weight density	γ	Newton/cubic meter	N/m^3
Work	w	Joule	J
Electrical Units			
Admittance	Y	Mho (siemen)	mho (S)
Capacitance	C	Farad	F(A $\cdot$ s/V)
Conductance	G	Mho (siemen)	mho(S)
Conductivity	γ	Mho/meter	mho/m(S/m)
Current density	J	Ampere/meter^2	A/m^2
Electric potential	V	Volt	V
Electric field intensity	E	Volt/meter	V/m
Electrical energy	W	Joule	J
Electrical power	P	Watt	W
Impedance	Z	Ohm	Ω
Permittivity of free space	ε	Farad/meter	F/m
Quantity of electricity	Q	Coulomb	C(A $\cdot$ s)
Reactance	X	Ohm	Ω
Resistance	R	Ohm	Ω
Resistivity	ρ	Ohm $\cdot$ meter	$\Omega \cdot$ m
Magnetic Units			
Magnetic field intensity	H	Ampere/meter	A/m
Magnetic flux	Φ	Weber	Wb
Magnetic flux density	B	Tesla (weber/meter^2)	T (Wb/m^2)
Magnetic permeability	μ	Henry/meter	H/m
Mutual inductance	M	Henry	H
Permeability of free space	μ_0	Henry/meter	H/m
Permeance	P	Henry	H
Relative permeability	μ_τ	—	—
Reluctance	R	Henry^{-1}	H^{-1}
Self-inductance	L	Henry	H
Optical Units			
Illumination	lx	Lux	cd $\cdot$ sr/m^2
Luminous flux	lm	Lumen	cd $\cdot$ sr
Luminance	cd	Candela/meter^2	cd/m^2
Radiance	L_e	Watt/steradian $\cdot$ meter^3	W/sr $\cdot$ m^3
Radiant energy	W	Joule	J
Radiant flux	P	Watt	W
Radiant intensity	I_c	Watt/steradian	W/sr

Table 30 Decimal Multiples

Name	Symbol	Equivalent
Exa	E	10^{18}
Peta	P	10^{15}
Tera	T	10^{12}
Giga	g	10^{9}
Mega	M	10^{6}
Kilo	k	10^{3}
Hecto	h	10^{2}
Deca	da	10
Deci	d	10^{-1}
Centi	c	10^{-2}
Milli	m	10^{-3}
Micro	μ	10^{-6}
Nano	n	10^{-9}
Pico	p	10^{-12}
Femto	f	10^{-15}
Atto	a	10^{-18}

In addition to primary international standards, standard instruments are available having stable and precisely defined characteristics that are used as references for other instruments that are performing the same function. Hence, the performance of an instrument can be cross-checked against a known device. At a global level, checking is done by using an international network of national and international laboratories, such as the National Bureau of Standards (NBS), the National Physical Laboratory (NPL), and the Physikalisch-Technische Bundesanstalt of Germany. A treaty between the world's national laboratories regulates the international activity and coordinates development, acceptance, and intercomparisons. Basically, standards are kept in four stages:

1. *International standards* represent certain units of measurement with maximum accuracy possible within today's available technology. These standards are under the responsibility of an international advisory committee and are not available to ordinary users for comparison or calibration purposes.

2. *Primary standards* are the national standards maintained by national laboratories in different parts of the world for verification of secondary standards. These standards are independently calibrated by absolute measurements that are periodically made against the international standards. The primary standards are compared against each other.

3. *Secondary standards* are maintained in the laboratories of industry and other organizations. They are periodically checked against primary standards and certified.

4. *Working standards* are used to calibrate general laboratory and field instruments.

Another type of standard is published and maintained by the Institute of Electrical and Electronics Engineer (IEEE) in New York. These standards are for test procedures, safety rules, definitions, nomenclature, and so on. The IEEE standards are adopted by many organizations around the world. Many nations also have their own standards for test procedures, instrument usage procedures, safety, and the like.

7.5 Calibration, Calibration Conditions, and Linear Calibration Model

The calibration of all instruments is essential for checking their performances against known standards. This provides consistency in readings and reduces errors, thus validating the measurements to be valid universally. After an instrument is calibrated, future operation is deemed to be error bounded for a given period of time for similar operational conditions. The calibration procedure involves comparison of the instrument against primary or secondary standards. In some cases, it may be sufficient to calibrate a device against another one with a known accuracy.

Many nations and organizations maintain laboratories with the primary functions of calibrating instruments and field measuring systems that are used in everyday operations. Examples of these laboratories are National Association of Testing Authorities (NATA) of Australia and the British Calibration Services (BCS).

Calibrations may be made under static or dynamic conditions. A typical calibration procedure of a complex process involving many instruments is illustrated in Fig. 97. In an ideal situation, for an instrument that responds to a multitude of physical variables, a commonly employed method is to keep all inputs constant except one. The input is varied in increments in

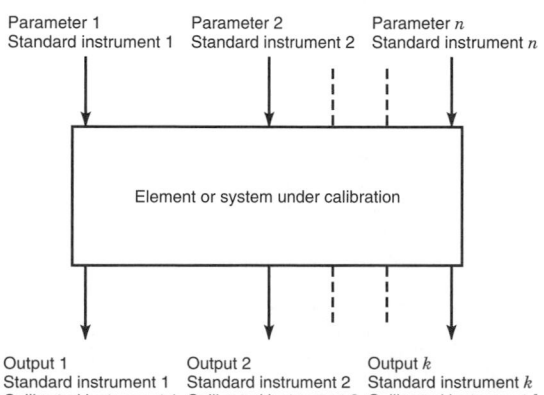

Fig. 97 Instruments are frequently calibrated sequentially for all affected inputs. Calibrations are made under static or dynamic conditions, usually keeping all inputs constant and varying only one and observing the output. Calibration continues until all other inputs are covered.

increasing and decreasing directions over a specified range. The observed output then becomes a function of that single input. The calibration is continued in a similar manner until all other inputs are covered. For better results, this procedure may be repeated by varying the sequences of inputs, thus developing a family of relationships between the inputs and outputs. As a result of these calibration readings, the input and output relation usually demonstrates statistical characteristics. From these characteristics, appropriate calibration curves can be obtained, and other statistical techniques can be applied.

In many instruments, the effect of a single input may not represent the true output values when one input is varied and all others are kept constant. In these cases, calibration is conducted by varying several inputs simultaneously. Throughout the calibration procedure, the n number of variables of the system are monitored by appropriate standard instruments. The rule of thumb is that each calibrated variable must have a traceable ladder starting from laboratory standards and secondary standards leading to primary standards. This is known as the linear calibration model or traceability.

Most instrument manufacturers supply calibrated instruments and reliable information about their products. But their claims of accuracy and reliability must be taken at face value. Therefore, in many cases, application-specific calibrations must be made periodically within the recommended calibration intervals. Usually, manufacturers supply calibration programs. In the absence of such programs, it is advisable to conduct frequent calibrations in the early stages of installation and lengthen the period between calibrations as the confidence builds based on satisfactory performance. Recently, with the wide applications of digital systems, computers can make automatic and self-calibrations as in the case of many intelligent instruments. In these cases, postmeasurement corrections are made, and the magnitudes of various errors are stored in the memory to be recalled and used in laboratory and field applications.

7.6   Analog and Digital Instruments

Instruments can be analog or digital or a combination of the two. Nowadays, most instruments are produced to be digital because of the advantages that they offer. However, the front end of majority of instruments are still analog; that is, the majority of sensors and transducers generate analog signals. Initially, the signals are conditioned by analog circuits before they are put into digital form for signal processing. It is important

to mention that digital instruments operating purely on digital principles are developing fast. For instance, today's smart sensors contain the complete signal condition circuits in a single chip integrated with the sensor itself. The output of smart sensors can be interfaced directly with other digital devices.

In analog instruments, the useful information is conveyed by changes in amplitudes, phases, or frequencies or a combination of the three. These signals can be deterministic or nondeterministic. In all analog or digital instruments, as in the case with all signal-bearing systems, there are useful signals that respond to the physical phenomena and unwanted signal resulting from various forms of noise. In the case of digital instruments, additional noise is generated in the process of A/D conversion.

Analog signals can also be nondeterministic; that is, the future state of the signal cannot be determined. If the signal varies in a probabilistic manner, its future can be foreseen only by statistical methods. The mathematical and practical treatment of analog and digital signals having deterministic, stochastic, and nondeterministic properties is a very lengthy subject and a vast body of information can be found in the literature; therefore, they will not be treated here.

As is true of all instruments, when connecting electronic building blocks, it is necessary to minimize the loading effects of each block by ensuring that the signal is passed without attenuation, loss, or magnitude and phase alterations. It is also important to ensure maximum power transfer between blocks by appropriate impedance-matching techniques. Impedance matching is very important in all instruments but particularly at a frequency of 1 MHz and above. As a rule of thumb, output impedances of the blocks are usually kept low, and input impedances are kept high so that the loading effects can be minimized.

Analog Instruments   Analog instruments are characterized by continuous signals. A purely analog system measures, transmits, displays, and stores data in analog form. The signal conditioning is usually made by integrating many functional blocks such as bridges, amplifiers, filters, oscillators, modulators, offsets and level converters, buffers, and the like, as illustrated Fig. 98. Generally, in the initial stages, the signals produced by the sensors and transducers are conditioned mainly by analog electronics, even if they are configured as digital instruments later. Therefore, we pay more attention to analog instruments, keeping in mind

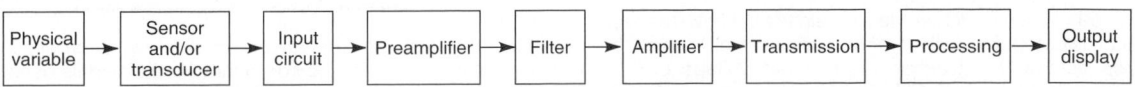

Fig. 98   Analog instruments measure, transmit, display, and store data in analog form. Signal conditioning usually involves such components as bridges, amplifiers, filters, oscillators, modulators, offsets and level converters, buffers, and so on. These components are designed and tested carefully to suit the characteristics of particular instruments.

that much of the information given here also may be used in various stages of the digital instruments.

Instrument bridges are commonly used to measure such basic electrical quantities as resistance, capacitance, inductance, impedance, and admittance. Basically, they are two-port networks in which the component to be measured is connected to one of the branches of the network. There are two basic groups, ac and dc bridges. Also, there are many different types in each group, such as Wheatstone and Kelvin dc bridges and Schering, Maxwell, Hay, and Owen ac bridges. In a particular instrument, the selection of the bridge to be employed and the determination of values and tolerances of its components is very important. It is not our intent to cover all bridges here; however, as typical example of an ac bridge, a series RC bridge is given in Fig. 99. We also offer some analysis to illustrate briefly their typical operational principles. At balance,

$$Z_1 Z_3 = Z_x Z_z \qquad (121)$$

Substitution of impedance values gives

$$R_3 \left(R_1 - \frac{j}{\omega} C_1 \right) = \left(R_x - \frac{j}{\omega} C_x \right) R_2 \qquad (122)$$

Equating the real and imaginary terms gives the values of unknown components as

$$R_x = \frac{R_1 R_3}{R_2} \qquad (123)$$

and

$$C_x = \frac{C_1 R_2}{R_3} \qquad (124)$$

In instruments, the selection and use of amplifiers and filters are also very important since many transducers generate extremely weak signals in comparison to the noise existing in the device. Today, operational amplifiers and high-precision instrumentation amplifiers are the building blocks of modern instruments.

The operation amplifiers may be used as inverting and noninverting amplifiers, and by connecting suitable external components, they can be configured to perform many other functions, such as multipliers, adders, limiters, and filters. Instrumentation amplifiers are used in situations where operational amplifiers do not meet the requirements. They are essentially high-performance differential amplifiers consisting of several closed-loop operational amplifiers. The instrumentation amplifiers have improved common-mode rejection ratios (CMRR) (up to 160 dB), high input impedances (up to 500 MΩ), low output impedance, low offset currents and voltages, and better temperature characteristics. To illustrate amplifiers in instrumentation systems, a typical current amplifier used in charge amplification is illustrated in Fig. 100. In this circuit, if the input impedance of the operational amplifier is high, output is not saturated, and the differential input voltage is small, it is possible to write

$$\frac{1}{C_f} \int i_f \ dt = e_{ex} - e_{ai} - e_{ex} \qquad (125)$$

$$\frac{1}{C_x} \int i_x \ dt = e_0 - e_{ai} = e_0 \qquad (126)$$

$$i_f + i_x - i_{ai} = 0 = i_f + i_x \qquad (127)$$

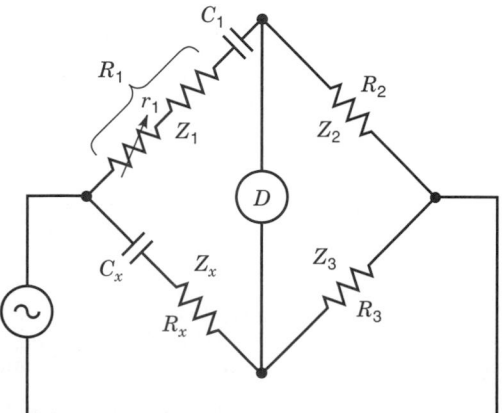

Fig. 99 A series RC bridge wherein the unknown capacitance is compared with a known capacitance. The voltage drop across R_1 balances the resistive voltage drop in branch Z_2. The bridge balance is achieved relatively easily when capacitive branches have substantial resistive components. The resistors R_1 and either R_2 or R_3 are adjusted alternately to obtain the balance.

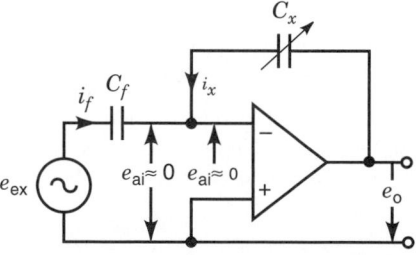

Fig. 100 Using an operational amplifier signal processor is useful to eliminate the nonlinearity in the signals generated by capacitive sensors. With this type of arrangement, the output voltage can be made to be directly proportional to variations in the signal representing the nonlinear operation of the device.

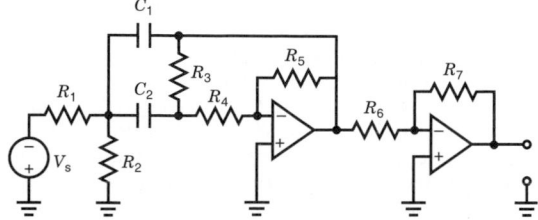

Fig. 101 Filtering is used in various stages of signal processing to eliminate unwanted components of signals. They can be designed and constructed to eliminate or pass signals at certain frequency ranges. Suitable arrangements of components yield to bandpass, high-pass, bandpass, bandstop, and notch filters. Filters can be classified as active and passive.

Manipulation of these equations gives

$$e_0 = \frac{-C_f e_{ex}}{C_x} \qquad (128)$$

However, a practical circuit requires a resistance across C_f to limit output drift. The value of this resistance must be greater than the impedance of C_f at the lowest frequency of interest.

Filtering is used to reject unwanted components of signals. For example, by using a filter that narrows the bandwidth, the broadband noise energy is reduced, and unwanted signals outside the passband are rejected. Analog filters can be designed by using various techniques, such as Butterworth, Chebyshev, and Bessel–Thomson filters. They can be low-pass, high-pass, bandpass, bandstop, and notch filters. Filters can be classified as active and passive. Active filters involve active components such as operational or instrumentation amplifiers, whereas passive filters are configured completely by inductive, capacitive, and resistive components. The choice of active or passive filters depends on the available components, the precision required, and the frequency of operations. A typical filter used in instrument is given in Fig. 101.

Digital Instruments In modern instruments, the original data acquired from the physical variables are usually in analog form. This analog signal is converted to digital before being passed on to the other parts of the system. For conversion purposes, analog-to-digital converters are used together with appropriate

sample-and-hold devices. In addition, analog multiplexers enable the connection of a number of transducers to the same signal-processing media. The typical components of a digital instrument are illustrated in Fig. 102. The digital systems are particularly useful in performing mathematical operations and storing and transmitting data.

Analog-to-digital conversion involves three stages: sampling, quantization, and encoding. The Nyquist sampling theorem must be observed during sampling; that is, "the number of samples per second must be at least twice the highest frequency present in the continuous signal." As a rule of thumb, depending on the significance of the high frequencies, the sampling must be about 5–10 times the highest frequency of the signal. The next stage is the quantization, which determines the resolution of the sampled signals. The quantization error decreases as the number of bits increases. In the encoding stage, the quantized values are converted to binary numbers to be processed digitally. Figure 103 illustrates a typical A/D sampling process of an analog signal.

After the signals are converted to digital form, the data can be further processed by employing such various techniques as fast Fourier transform (FFT) analysis, digital filtering, sequential or logical decision making, correlation methods, spectrum analysis, and so on.

Virtual Instruments (VIs) Traditional instruments have three basic components—acquisition and control,

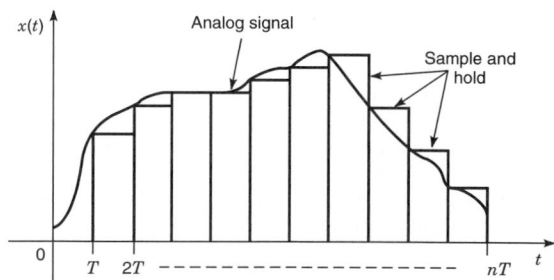

Fig. 103 Analog-to-digital converters involve three stages: sampling, quantization, and encoding. However, the digitization introduces a number of predictable errors. After the conversion, the data can be processed by techniques such as FFT analysis, discrete Fourier transform (DFT) analysis, digital filtering, sequential or logical decision making, correlation methods, spectrum analysis, and so on.

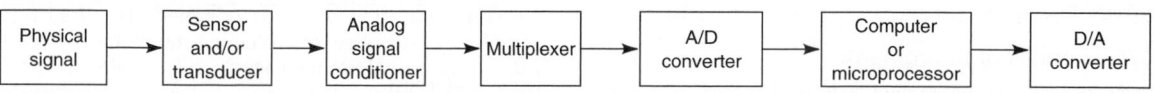

Fig. 102 Digital instruments have more signal-processing components than analog instruments. Usually, analog signals are converted to digital form by analog-to-digital (A/D) converters. The digital instruments have the advantage of processing, storing, and transmitting signals more easily than their analog counterparts.

data analysis, and data presentation. In VIs, the use of digital techniques, software, and computers replace the display and processing capabilities of most traditional instruments. In this technology, plug-in data acquistion (DAQ) boards, PC cards (PCMCIA), and parallel-port input–output (I/O) devices are used to interface sensors and transducers of the system under investigation to computers. There are standard interface buses such as VXIbus, which stands for VMEbus Extensions for Instrumentation (also known as the IEEE Standard 1155–1992).

Once the system is interfaced, the computer can be programmed to act just like a stand-alone instrument, but offering additional benefits of flexibility of the processing, display, and storage. In VIs, the data can be saved or loaded in memory to be processed in popular spreadsheet programs and word processors, and a report generation capability complements the raw data storage by adding timestamps, measurements, user names, and comments.

VI technology allows the user to build test systems that fit specific applications. The VI software can be programmed to resemble familiar instrument panels, including buttons and dials. The user interface tools include knobs, meters, gauges, dials, tank displays, thermometers, graphs, strip charts, and the like to simulate the appearance of traditional instruments. Computer displays can show more colors and allow users to quickly change the way they display test data and controls as required. The software also contains analysis libraries with high-powered statistics, curve fitting, signal processing, and filtering to standard dynamic link libraries (DLLs).

Designing a VI system is similar to designing a test system with stand-alone instruments. The first step is to determine what types of signals are needed to measure, including their frequencies, amplitudes, and other signal characteristics together with the level of accuracy expected from these signals. To develop the software for the test application, a programming language or test development software package needs to be selected such as C or Microsoft Visual Basic. Since the display is not fixed, as on a stand-alone instrument, it can be as complex or as simple as the application requires.

Nowadays, users can configure their VIs to update front panels and display real-time, animated VIs over the Internet. The toolkits let applications be published over the web and viewed with a standard web browser with little additional programming. With these tools, developers can monitor VIs running in remote locations, publish experiment results on the Web, and automatically notify operators of alarm conditions or status information.

7.7   Control of Instruments

Instruments can be manual, semiautomatic, or fully automatic. Manual instruments need human intervention for adjustment, parameter setting, and interpreting readings. Semiautomatic instruments need limited

intervention such as the selection of operating conditions and so on. In the fully automatic instruments, however, the variables are measured either periodically or continuously without human intervention. The information is either stored or transmitted to other devices automatically. Some of these instruments can also measure the values of process variables and regulate their deviations from preset points.

It is often necessary to measure many parameters of a process by using two or more instruments. The resulting arrangement for performing the overall measurement function is called the measurement system. In measurement systems, instruments operate in an autonomously but coordinated manner. The information generated by each device is communicated between instruments themselves, or between the instrument and other devices such as recorders, display units, and computers. The coordination of instruments can be done in three ways: analog to analog, analog to digital, and digital to digital.

Analog systems consist of instruments that generate continuous current and voltage waveforms in response to the physical variations. The signals are processed by using analog electronics; therefore, signal transmission between the instruments and other devices is also done in the analog form. In assembling these devices, the following characteristics must be considered:

Signal transmission and conditioning
Loading effects and buffering
Proper grounding and shielding
Inherent and imposed noises
Ambient conditions
Signal level compatibility
Impedance matching
Proper display units
Proper data storage media

Offset and level conversion is used to convert the output signal of an instrument from one level to another, compatible with the transmission medium in use. In analog systems, signals are usually transmitted at suitable current levels (4–20 mA). In this way, change in impedance does not affect the signal levels, and standard current signal levels can easily be exchanged.

In digital instrumentation systems, analog data are converted and transmitted in digital form. The transmission of data between digital devices can be done relatively easily, by using serial or parallel transmission techniques. However, as the measurement system becomes large by the inclusion of many instruments, the communication becomes complex. To avoid this complexity, message interchange standards are used for digital signal transmission such as RS-232 and IEEE-488 VXIbus.

Many instruments are manufactured with output ports to pass measurement data and various control signals. The IEEE-488 (also known as the GPIB) bus is one of the established industry standard instrumentation interfacings. It enables simultaneous measurements by interfacing up to 15 instruments together at the same time. It has 16 signal lines distributed as 8 data lines, 3 control lines, and 5 general interface management lines. The line configuration of an IEEE-488 bus is given in Fig. 104. Once connected, any one device can transfer data to one or more other devices on the bus. All devices must be able to perform at least one of the following roles: talker, listener, controller. The minimum device consists of one talker and one listener without a controller. The length of cables connected to the bus cannot exceed 20 m, and the maximum data rate is restricted to 250 kbytes per second.

RS-232 is issued by the Electronic Industries Association (EIA). It uses serial binary data interchange and applies specifically to the interconnection of data communication equipment (DCE) and data terminal equipment (DTM). Data communications equipment may include modems, which are the devices that convert digital signals suitable for transmission through telephone lines. The RS-232 uses standard DB-25 connectors, the pin connection is given in Table 31. Although 25 pins are assigned, a complete data transmission is possible by using only three pins—2, 3, and 7. The transmission speed can be set to certain baud rates such as 19,200 bits per second and can be used for synchronous or non-synchronous communication purposes. The signal voltage levels are very flexible, with any voltage between -3 and -25 V representing logic 1 and any voltage between $+3$ and $+25$ V representing logic 0.

In many industrial applications, the current loop digital communication is used. This communication is similar to analog current loop systems, but the signal is transmitted in digital form, with 20 mA signifying logic 1 and 0 mA representing logic 0. Depending on the external noise sources in the installation environment, the current loop can be extended up to 2 km.

When data are transmitted distances greater than those permitted by the RS-232 or current loop, the modem, microwave, or RF transmissions are used. In this case, various signal modulation techniques are necessary to convert digital signals to suitable formats. For example, most modems, with medium-speed asynchronous data transmission, use frequency-shift keyed (FSK) modulation. The digital interface with modems uses various protocols such as MIL-STD-188C to transmit signals in simplex, half-duplex, or full-duplex forms depending on the directions of the data flow. The simplex interface transmits data in one direction, whereas full duplex transmits it in two directions simultaneously.

Table 31   RS-232 Pin Connections

Pin Number	Direction	Function
1	—	Frame ground
2	Out	Transmitted data (−TxD)
3	In	Received data (−RxD)
4	Out	Request to send (RTS)
5	In	Clear to send (CTS)
6	In	Data set ready (DSR)
7	—	Signal ground (SG)
8	In	Received line signal detector (DCD)
9	Out	+ Transmit current loop data
11	Out	− Transmit current loop data
18	In	+ Receive current loop data
20	Out	Data terminal ready (DTR)
22	In	Ring indicator (RI)
25	In	− Receive current loop return

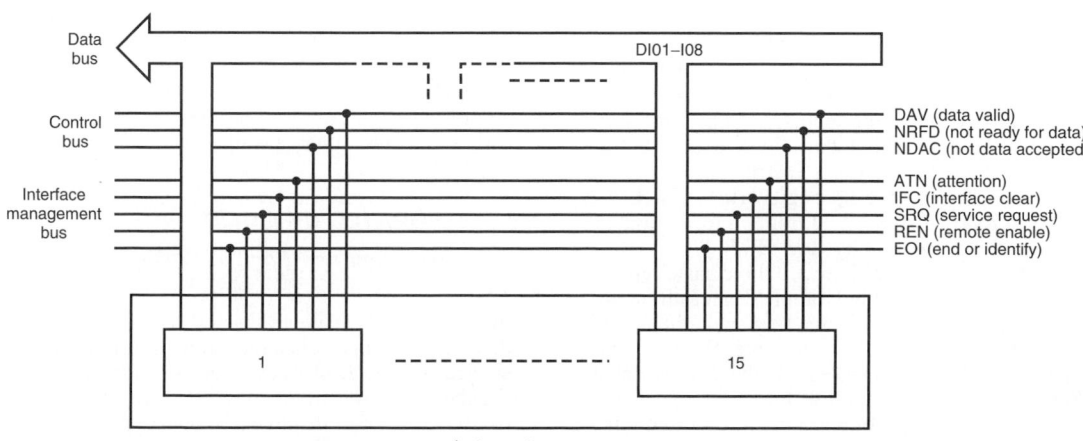

Fig. 104   The IEEE-488 or the GPIB bus is an industry standard for interface medium. It has 8 data lines, 3 control lines, and 5 general interface management lines. In noisy environments the maximum length of cable is recommended to be not more than 20 m.

As far as industrial applications are concerned, several standards for digital data transmission are available, commonly known as *field buses* in the engineering literature. For example, WorldFIP and Profibus have been developed and Foundation Fieldbus is under development to increase the performance of the 20-mA current loop. New devices allow for an increase in the data rates (e.g., National Instruments chips and boards operating with high-speed protocol HS488 for 8 Mbytes/s transfer rate). A new standard is under discussion at the IEEE by the working group for higher performance IEEE Std. 488.1, with a very high increase in the data rate.

Concerning the design software, there are important tools that help implement control (application) software for automatic measuring equipment, such as Lab-Windows and LabVIEW from National Instruments and VEE from Hewlett-Packard.

In many applications, many instruments (say over 1000) may be used to monitor and control the process as in the case of computer-integrated manufacturing (CIM). In these cases, instruments are networked either in groups or as whole via a center computer or group of computers. Appropriate network topologies (e.g., star, ring, field bus) may be employed to enable the signal flow between the instruments and computers, among the instruments themselves, or between instruments and control panels.

7.8 Industrial Measuring Instruments

In industry, instruments are used to sense and maintain the functions of the process. Because the requirements of diverse industries are different, the instruments are made quite differently to suit applicational differences from one industry to another. Here, instruments specific to some industries will be discussed briefly.

The process industry uses instruments extensively for on-line monitoring and off-line analysis. Specific instruments are used commonly for sensing variables such as temperature, pressure, volumetric and mass flow rate, density, weight, displacement, pH levels, color, absorbency, viscosity, material flow, dew point, organic and inorganic components, turbidity, solid and liquid level, humidity, and particle size distribution. The selection and use of these instruments constitute an important part of process engineering, which is a discipline in its own right. Additional information can be found in the Bibliography (e.g., Ref. 97).

In medical technology, there are three basic types of instruments—imaging, physiological measurements, and laboratory analysis. In imaging and physiological measurements, the instruments are closely linked with patients. Some examples of these instruments are X-ray tomography, nuclear magnetic resonance (NMR) and nuclear spin tomography, ultrasound imaging, thermography, brain and nervous system sensors, and respiratory sensors. Many instruments are based on the radiation and sound, force and tactile sensing, electromagnetic sensing, and chemical and bioanalytical sensors.

Power plants are instrumented for maximum availability, operational safety, and environmental planning. Therefore, their measurements must be as accurate as possible and reliable. Instruments are used for temperature, pressure, flow, level, vibration measurements, and water, steam, and gas analysis. For example, gas analysis requires instruments to measure carbon compounds, sulfur and nitrogen compounds, and dust and ash contents.

Environmental monitoring requires a diverse range of instruments for air, water, and biological monitoring. Instruments are used for measuring various forms of radiation, chemicals hazards, air pollutants, and organic solvents. Many sophisticated instruments are also developed for remote monitoring via satellites, and they operate on optical, microwave, and RF electromagnetic radiation principles.

In automobiles, instruments are used to assist drivers by sensing variables such as cooling, braking, fuel consumption, humidity control, speed, travel route monitoring, and position sensing. Instruments also find applications for safety and security purposes, such as passenger protection and locking and antitheft systems. Recently, with the advent of micromachined sensors, many diverse instruments such as engine control, fuel injection, air regulation, and torque sensing are developed.

The manufacturing industry, especially automated manufacturing, requires a diverse range of instruments. Machine diagnosis and process parameters are made by instruments based on force, torque, pressure, speed, temperature, and electrical parameter-sensing instruments. Optics, tactile arrays, and acoustic scanning instruments are used for pattern recognition. Distance and displacement measurements are made by many methods (e.g., inductive, capacitive, optical, and acoustic techniques).

Aerospace instrumentation requires an accurate indication of physical variables and the changes in these variables. Instruments are designed to suit specific conditions of operations. Some of the measurements are gas temperature and pressure, fluid flow, aircraft velocity, aircraft icing, thrust and acceleration, load, strain and force, position, altitude sensing, and direction finding.

8 INTEGRATED CIRCUITS

N. Ranganathan and Raju D. Venkataramana

The invention of the transistor in 1947 by William Shockley and his colleagues John Bardeen and Walter Brattain at Bell Laboratories, Murray Hill, NJ, launched a new era of ICs. The transistor concept was based on the discovery that the flow of electric current through a solid semiconductor material like silicon can be controlled by adding impurities appropriately through the implantation processes. The transistor replaced the vacuum tube due to its better reliability, lesser power requirements, and, above all, its much smaller size. In the late 1950s, Jack Kilby of Texas Instruments developed the first integrated circuit. The ability to develop flat or planar ICs, which allowed the interconnection of circuits on a single substrate (due to Robert Noyce and Gordon Moore), began the microelectronics revolution. The substrate is the supporting semiconductor material on which the various devices that form the integrated circuit are attached. Researchers developed sophisticated photolithography techniques that helped in the reduction of the minimum feature size, leading to larger circuits being implemented on a chip. The miniaturization of the transistor led to the development of integrated circuit technology in which several hundreds and thousands of transistors could be integrated on a single silicon die. IC technology led to further developments, such as microprocessors, mainframe computers, and supercomputers.

Since the first integrated circuit was designed following the invention of the transistor, several generations of integrated circuits have come into existence: SSI (small-scale integration) in the early 1960s, MSI (medium-scale integration) in the latter half of the 1960s, and LSI (large-scale integration) in the 1970s. The VLSI (very large scale integration) era began in the 1980s. While the SSI components consisted on the order of 10–100 transistors or devices per integrated circuit package, the MSI chips consisted of anywhere from 100 to 1000 devices per chip. The LSI components ranged from roughly 1000 to 20,000 transistors per chip, while the VLSI chips contain on the order of up to 3 million devices. When the chip density increases beyond a few million, the Japanese refer to the technology as ULSI (ultra large scale integration), but many in the rest of the world continue to call it VLSI. The driving factor behind integrated circuit technology was the scaling factor, which in turn affected the circuit density within a single packaged chip. In 1965, Gordon Moore predicted that the density of components per integrated circuit would continue to double at regular intervals. Amazingly, this has proved true, with a fair amount of accuracy.[99]

Another important factor used in measuring the advances in IC technology is the minimum feature size or the minimum line width within an integrated circuit (measured in microns). From about 8 μm in the early 1970s, the minimum feature size has decreased steadily, increasing the chip density or the number of devices that can be packed within a given die size. In the early 1990s, the minimum feature size decreased to about 0.5 μm, and currently 0.3, 0.25, and 0.1 μm technologies (also called deep submicron technologies) are becoming increasingly common. IC complexity refers, in general, to the increase in chip area (die size), the decrease in minimum feature size, and the increase in chip density. With the increase in IC complexity, the design time and the design automation complexity increase significantly. The advances in IC technology are the result of many factors, such as high-resolution lithography techniques, better processing capabilities, reliability and yield characteristics, sophisticated design automation tools, and accumulated architecture, circuit, and layout design experience.

8.1 Basic Technologies

The field of integrated circuits is broad. The various basic technologies commonly known are shown in Fig. 105. The inert substrate processes, further divided as thin- and thick-film processes, yield devices with good resistive and temperature characteristics. However, they are mostly used in low-volume circuits and in hybrid ICs. The two most popular active substrate materials are silicon and gallium arsenide (GaAs). The silicon processes can be separated into two classes: MOS (the basic device is a metal–oxide–semiconductor field-effect transistor) and bipolar (the basic device is bipolar junction transistors). The bipolar process was commonly used in the 1960s and 1970s and yields high-speed circuits with the overhead of high-power dissipation and the disadvantage of low density. The transistor–transistor logic (TTL) family of circuits constitutes the most popular type of bipolar and is still used in many high-volume applications. The ECL devices are used for high-speed parts that form the critical path delay of the circuit. The MOS family of processes consists of PMOS, NMOS, CMOS, and BiCMOS. The term *PMOS* refers to a MOS process that uses only *p*-channel transistors, and *NMOS* refers to a MOS process that uses only *n*-channel transistors. PMOS is not used much due to its electrical characteristics, which are not as good as the *n*-channel FETs, primarily since the mobility of the *n*-channel material is almost twice compared to the mobility of the *p*-channel material. Also, the NMOS devices are smaller than the PMOS devices, and thus PMOS do not give good packing density.

CMOS was introduced in the early 1960s; however, it was only used in limited applications, such as watches and calculators. This was primarily due to the fact that CMOS had slower speed, less packing density, and latchup problems although it had a high noise margin and lower power requirements. Thus, NMOS was

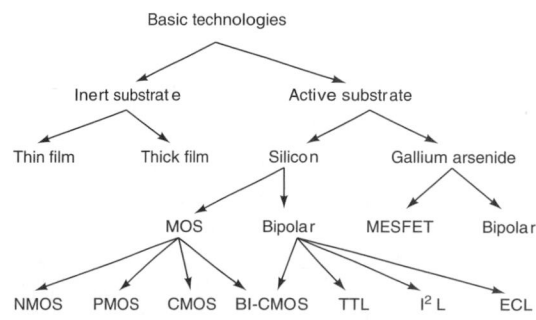

Fig. 105 Overview of basic technologies.

preferred over CMOS, in general, until the p-channel devices developed began to have similar characteristics as the nMOS and both the p-channel and n-channel transistors started delivering close to equal amounts of currents with similar transistor sizes. In the 1980s and the 1990s, the need for lower power consumption was the driving factor, and thus CMOS emerged as the leading IC technology.[100] The BiCMOS technology combines both bipolar and CMOS devices in a single process. While CMOS is preferred for logic circuits, BiCMOS is preferred for input/output (I/O) and driver circuits due to its low input impedance and high current driving capability.

Since the 1980s, efforts have been directed toward designing digital ICs using GaAs devices. In many high-resolution radar systems, space systems, high-speed communication circuits, and microwave circuits, the integrated circuits need to operate at speeds beyond several gigahertz. In silicon technology, it is possible to obtain speeds on the order of up to 10 GHz using ECL circuits, which is almost pushing the limits of the silicon technology. In GaAs technology, the basic device is the metal–semiconductor (Schottky gate) field-effect transistor, called the GaAs MESFET. Given similar conditions, the electrons in n-type GaAs material travel twice faster than in silicon. Thus, the GaAs circuits could function at twice the speed than the silicon ECL circuits for the same minimum feature size. The GaAs material has a larger bandgap and does not need gate oxide material, as in silicon, which makes it immune to radiation effects. Also, the GaAs

material has very high resistivity at room temperatures and lower parasitic capacitances, yielding high-quality transistor devices. However, the cost of fabricating large GaAs circuits is significantly high due to its low reliability and yield characteristics (primarily due to the presence of more defects in the material compared to silicon). The fabrication process is complex, expensive, and does not aid scaling. Also, the hole mobility is the same as in silicon, which means GaAs is not preferable for complementary circuits. Thus, the GaAs technology has not been as successful as initially promised. Since CMOS has been the most dominant technology for integrated circuits, we examine the MOS transistor and its characteristics as a switch in the next section.

8.2 MOS Switch

The MOSFET is the basic building block of contemporary CMOS circuits, such as microprocessors and memories. A MOSFET is a unipolar device; that is, current is transported by means of only one type of polarity (electrons in an n type and holes in a p type). In this section, we describe the basic structure of MOSFETS and their operation and provide examples of gates built using MOS devices.

Structure The basic structure of a MOSFET (n and p type) is shown in Fig. 106. We describe the structure of an n-type MOSFET.[101,102] It consists of four terminals with a p-type substrate into which two n^+ regions are implanted. The substrate is a silicon

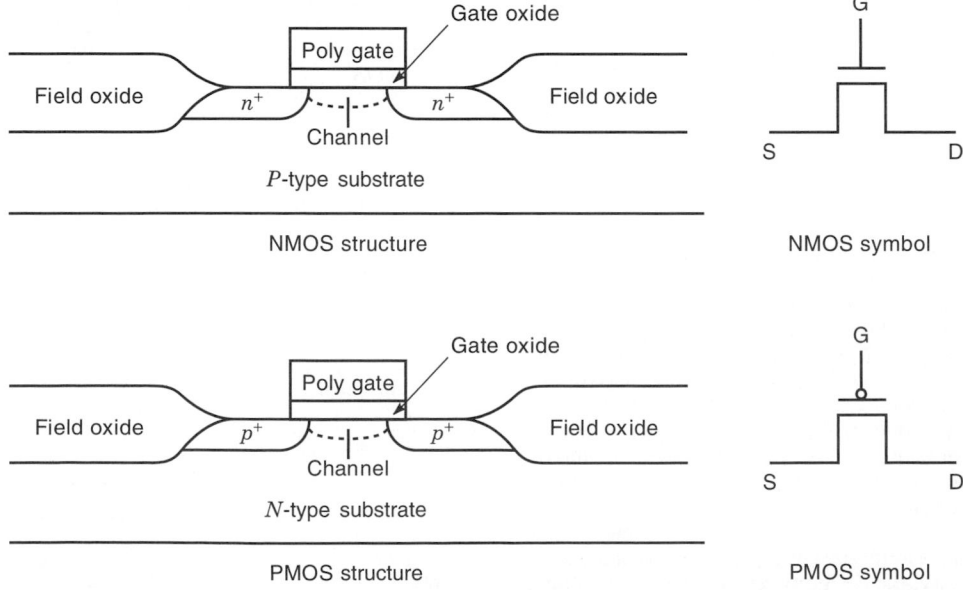

Fig. 106 Structure of n- and p-type MOSFET.

wafer that provides stability and support. The region between the two n^+ regions is covered by an insulator, typically polysilicon and a metal contact. This contact forms the gate of the transistor. The insulating layer is required to prevent the flow of current between the semiconductor and the gate. The two n^+ regions form the source and the drain. Due to the symmetry of the structure, the source and the drain are equivalent. The gate input controls the operation of the MOSFET. A bias voltage on the gate causes the formation of a channel between the n^+ regions. This channel causes a connection between the source and drain and is responsible for the flow of the current. The MOSFET is surrounded by a thick oxide, called the field oxide, which isolates it from neighboring devices. Reversal of n and p types in the discussion will result in a p-type MOSFET. Typical circuit symbols for n-type and p-type MOSFETS are also shown in Fig. 106.

Operation When no gate bias is applied, the drain and the source behave as two pn junctions connected in series in the opposite direction. The only current that flows is the reverse leakage current from the source to the drain. When a positive voltage is applied to the gate, the electrons are attracted and the holes are repelled. This causes the formation of an inversion layer or a channel region. The source and the drain are connected by a conducting n channel through which the current can flow. This voltage-induced channel is formed only when the applied voltage is greater than the threshold voltage, V_t. MOS devices that do not conduct when no gate bias is applied are called *enhancement mode* or normally OFF transistors. In nMOS enhancement mode devices, a gate voltage greater than V_t should be applied for channel formation. In pMOS enhancement mode devices, a negative gate voltage whose magnitude is greater than V_t must be applied. MOS devices that conduct at zero gate bias are called normally ON or *depletion mode* devices. A gate voltage of appropriate polarity depletes the channel of majority carriers and hence turns it OFF.

Considering an enhancement mode n-channel transistor, when the bias voltage is above the predefined threshold voltage, the gate acts as a closed switch between the source and drain, the terminals of which become electrically connected. When the gate voltage is cut off, the channel becomes absent, the transistor stops conducting, and the source and the drain channels get electrically disconnected. Similarly, the p-channel transistor conducts when the gate voltage is beneath the threshold voltage and stops conducting when the bias voltage is increased above the threshold. The behavior of the MOS transistor as a switch forms the fundamental basis for implementing digital Boolean circuits using MOS devices.

Output Characteristics We describe the basis output characteristics[103,104] of a MOS device in this subsection. There are three regions of operation for a MOS device:

1. Cutoff region
2. Linear region
3. Saturation region

In the cutoff region, no current flows and the device is said to be off. When a bias, V_{gs}, is applied to the gate such that $V_g > V_t$, the channel is formed. If a small drain voltage, V_{ds}, is applied, drain current, I_{ds}, flows from source to drain through the conducting channel. The channel acts like a resistance, and the drain current is proportional to the drain voltage. This is the linear region of operation. As the value of V_{ds} is increased, the channel charge near the drain decreases. The channel is pinched off when $V_{ds} = V_{gs} - V_t$. An increase in V_{ds} beyond the pinchoff value causes little change in the drain current. This is the saturation region of operation of the MOS device. The output characteristics of n- and p-type devices is shown in Fig. 107. The equations that describe the regions of operation can be summarized as follows:

$$I_{ds} = \begin{cases} 0 \quad \text{if } V_{gs} \leq V_t & \text{(cutoff)} \\ k/2[2(V_{gs} - V_t)V_{ds} - V_{ds}^2] \\ \quad \text{if } V_g > V_t, V_{ds} \leq (V_{gs} - V_t) & \text{(linear)} \\ k/2(V_{gs} - V_t)^2 \\ \quad \text{if } V_g > V_t, V_{ds} > (V_{gs} - V_t) & \text{(saturation)} \end{cases}$$

where k is the transconductance parameter of the transistor. A detailed analysis of the structure and operation of MOS devices is described in Refs. 101, 103, 105, and 106.

CMOS Inverter The basic structure of an inverter is shown in Fig. 108, and the process cross section is shown in Fig. 109. The gates of both the NMOS and the PMOS transistors are connected. The PMOs transistor is connected to the supply voltage V_{dd}, and the NMOS transistor is connected to G_{nd}. When a logical 0 is applied at the input V_{in}, then the PMOS

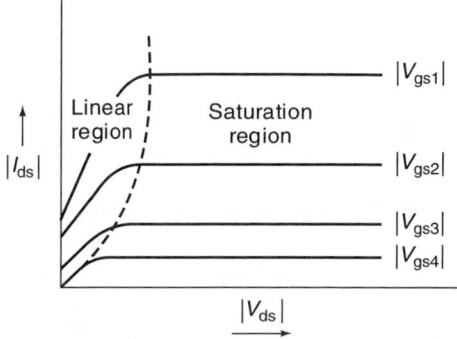

Fig. 107 Output characteristics of MOS transistor.

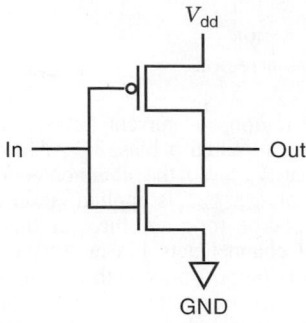

Fig. 108   Circuit schematic of inverter.

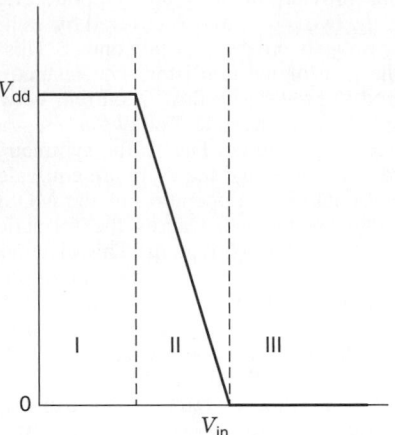

Fig. 110   Operating regions of transistor.

device is on and the output is pulled to V_{dd}. Hence the output is a logical 1. On the other hand, when a logical 1 is applied at the input, then the NMOS transistor is on and the output is pulled to the ground. Hence we have a logical 0. The operating regions of the transistor are shown in Fig. 110. In region I, the n device is off and the p device operates in the linear region. Hence the output is pulled to V_{dd}. In region II, the n and p devices operate in the linear and saturation region depending on the input voltage. In region III, the p device is cut off and the n device is operating in the linear region. The output is pulled to the ground. In region II, when both the transistors are on simultaneously, a short is produced between V_{dd} and G_{nd}. This accounts for the short circuit power dissipation in CMOS logic.

Transmission Gate   Consider the device shown in Fig. 111, which represents an NMOS or a PMOS device. By suitably controlling the gate bias, the device can be made to turn on or off. It behaves as an electrical switch that either connects or disconnects the points s and d. An NMOS device is a good switch when it passes a logical 0, and a PMOS is a good switch when it passes a logical 1. In CMOS logic, both the NMOS and PMOS devices operate together. In general, the NMOS transistor pulls down the output

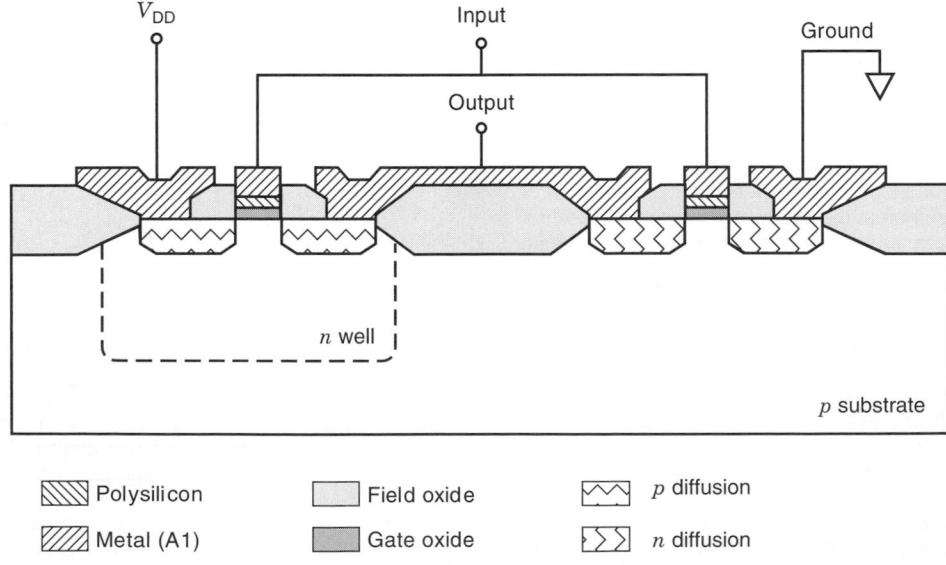

Fig. 109   Process cross section of n-well inverter.

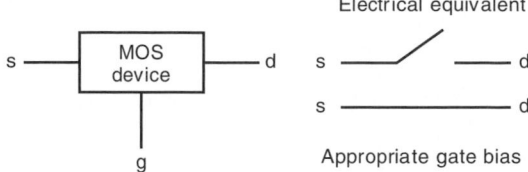

Fig. 111   A MOS device as switch.

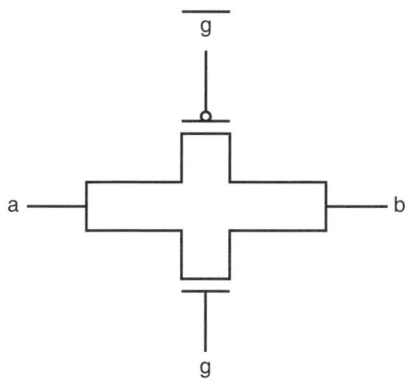

Fig. 112   Transmission gate.

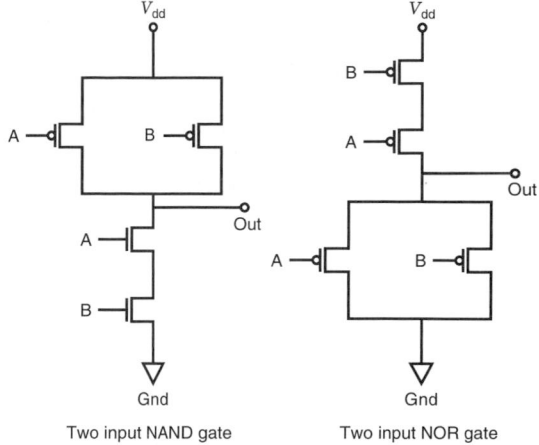

Fig. 113   Two-input NAND and NOR gate.

node to logical 0, and the PMOS device pulls up a node to logical 1. A transmission gate is obtained by connecting the two in parallel, as shown in Fig. 112. The control signal (say, g) applied to the n-type device is complemented and applied to the p-type device. When g is high, both the transistors are on and hence a good 1 or a 0 is passed. When g is low, both the devices are off. This is also called a complementary switch, or a C SWITCH.[103]

NAND and NOR Gates   CMOS combinational gates are constructed by connecting the PMOS and NMOS devices in either series or parallel to generate different logical functions. The structures for a two-input NAND and NOR gate are shown in Fig. 113.

NAND Gate.   The p devices are connected in parallel, and the n devices are connected in series. When either of the inputs A or B is a logical 0, the output is pulled high to V_{dd}. When both A and B are high, then the output is pulled to the ground. Hence this structure implements the operation $f = (AB)'$.

NOR Gate.   Similarly, in the NOR gate, the p devices are connected in series and the n devices are connected in parallel. When either of the inputs A or B is a logical 1, then the output is pulled to the ground. When both A and B are low, then the output is pulled

to V_{dd}. Hence this structure implements the operation $f = (A + B)'$. The p structure is the logical dual of the n structure. An n input NAND and NOR gate can be constructed in a similar fashion.

8.3   IC Design Methodology

To design and realize VLSI circuits, several factors play key roles. The goal of an IC designer is to design a circuit that meets the given specifications and requirements while spending minimum design and development time avoiding design errors. The designed circuit should function correctly and meet the performance requirements, such as delay, timing, power, and size. A robust design methodology has been established over the years, and the design of complex integrated circuits has been made possible essentially due to advances in VLSI design automation. The various stages in the design flow are shown in Fig. 114. The design cycle ranges from the system-level specification and requirements to the end product of a fabricated, packaged, and tested integrated circuit. The basic design methodology is briefly described here, and the various stages are discussed in detail in the following sections using simple examples.

The first step is to determine the system-level specifications, such as the overall functionality, size, power, performance, cost, application environment, IC fabrication process, technology, and chip-level and board-level interfaces required. There are several tradeoffs to be considered. The next step is the functional design and description, in which the system is partitioned into functional modules and the functionality of the different modules and their interfaces to each other are considered. The issues to be considered are regularity and modularity of structures, subsystem design, data flow organization, hierarchical design approach, cell types, geometric placements, and communication between the different blocks.

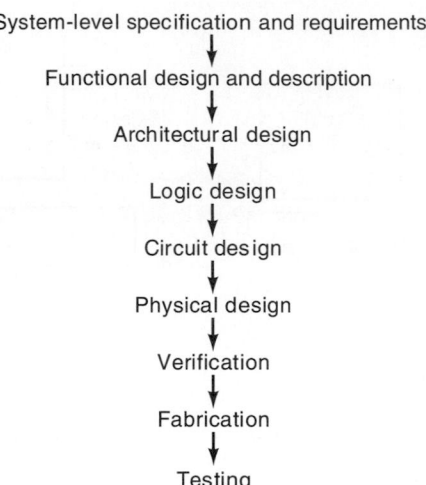

System-level specification and requirements

↓

Functional design and description

↓

Architectural design

↓

Logic design

↓

Circuit design

↓

Physical design

↓

Verification

↓

Fabrication

↓

Testing

Fig. 114 IC design methodology.

Once the functionality of the various modules is determined, the architectural design of the modules is pursued. Many design alternatives are considered toward optimization. This stage also includes the design of any hardware algorithms to be mapped onto architectures. A behavioral-level description of the architecture is obtained and verified using extensive simulations, often with an iterative process. This stage is critical in obtaining an efficient circuit in the end and for simplifying the steps in some of the following stages. In the logic design stage, the architectural blocks are converted into corresponding gate-level logic designs, Boolean minimization is performed, and logic simulation is used to verify the design at this level. In some design flows, the circuit could be synthesized from the logic level by using gate-level libraries (this is referred to as logic synthesis). The logic design usually includes a conventioinal logic design approach and a nontraditional design, such as precharge logic. At this stage, gate delays are considered and timing diagrams are derived to verify the synchronization of the various logic modules. The next step is the circuit design stage, which essentially involves converting the logic design modules into a circuit representation. At this stage, the essential factors considered are clocking, switching speeds or delays, switching activity and power requirements, and other electrical characteristics (e.g., resistance, capacitance).

The most complex step in VLSI design automation is the physical design, which includes floor planning, partitioning, placement, routing, layout, and compaction. This process converts the given circuit design or description into a physical layout that is a geometric representation of the entire circuit. Each step of the physical design by itself is complex and takes significant amounts of iterations and time. The various types

of transistors, the interconnecting wires, and contacts between different wires and transistors are represented as different geometric patterns consisting of many layers placed according to several design rules that govern a given fabrication technology and process. The floor-planning step involves higher level planning of the various components on the layout. The partitioning step converts the overall circuit into smaller blocks to help the other steps. It is usually impractical to synthesize the entire circuit in one step. Thus, logic partitioning is used to divide the given circuit into a smaller number of blocks, which can be individually synthesized and compacted. This step considers the size of the blocks, the number of blocks, and the interconnections between the blocks and yields a netlist for each block that can be used in the further design steps.

During the next step, which is the placement of the blocks on the chip layout, the various blocks are placed such that the routing can be completed effectively and the blocks use minimum overall area, avoiding any white spaces. The placement task is iterative in that an initial placement is obtained first and evaluated for area minimization and effective routing possibility, and alternate arrangements are investigated until a good placement is obtained. The routing task completes the routing of the various interconnections, as specified by the netlists of the different blocks. The goal is to minimize the routing wire lengths and minimize the overall area needed for routing. The routing areas between the various blocks are referred to as channels or switchboxes. Initially, a global routing is performed in which a channel assignment is determined based on the routing requirements, and then a detailed routing step completes the actual point-to-point routing.

The last step in the physical design is the compaction step, which tries to compact the layout in all directions to minimize the layout area. A compact layout leads to less wire lengths, lower capacitances, and more chip density since the chip area is used effectively. The compaction step is usually an interactive and iterative process in which the user can specify certain parameters and check if the compaction can be achieved. The goal of compaction, in general, is to achieve minimum layout area. The entire physical design process is iterative and is performed several times until an efficient layout for the given circuit is obtained.

Once the layout is obtained, design verification needs to be done to ensure that the layout produced functions correctly and meets the specifications and requirements. In this stage, design rule checking is performed on the layout to make sure that the geometric placement and routing rules and the rules regarding the separation of the various layers, the dimensions of the transistors, and the width of the wires are followed correctly. Any design rule violations that occurred during the physical design steps are detected and removed. Then circuit extraction is performed to complete the functional verification of the layout. This step verifies the correctness of the layout produced by the physical design process. After layout verification, the circuit layout is ready to be submitted for fabrication, packaging, and testing. Usually, several dies are produced on a single wafer and the wafer is tested for faulty dies. The correct ones are diced out and packaged in the form of a pin grid array (PGA), dual in-line package (DIP), or any other packaging technology. The packaged chip is tested extensively for functionality, electrical and thermal characteristics, and performance. The process of designing and building an integrated circuit[107] that meets the performance requirements and functions perfectly depends on the efficiency of the design automation tools.

8.4   Circuit Design

To create performance-optimized designs, two areas have to be addressed to achieve a prescribed behavior: (a) circuit or structural design, and (b) layout or physical design. While the layout design is discussed in a later section, this section focuses on the former.

A logic circuit must function correctly and meet the timing requirements. There are several factors that can result in the incorrect functioning of a CMOS logic gate: (a) incorrect or insufficient power supplies, (b) noise on gate inputs, (c) faulty transistors, (d) faulty connections to transistors, (e) incorrect ratios in ratioed logic, and (f) charge sharing or incorrect clocking in dynamic gates. In any design, there are certain paths, called *critical paths*, that require attention to timing details since they determine the overall functional frequency. The critical paths are recognized and analyzed using timing analyzer tools and can be dealt with at four levels:

1. Architecture
2. RTL/logic level
3. Circuit level
4. Layout level

Designing an efficient overall functional architecture helps to achieve good performance. To design an efficient architecture, it is important to understand the characteristics of the algorithm being implemented as the architecture. At the register transfer logic (RTL)/logic level, pipelining, the type of gates, and the fan-in and the fan-out of the gates are to be considered. Fan-in is the number of inputs to a logic gate, and fan-out is the number of gate inputs that the output of a logic gate drives. Logic synthesis tools can be used to achieve the transformation of the RTL level. From the logic level, the circuit level can be designed to optimize a critical speed path. This is achieved by using different styles of CMOS logic, as explained later in this section. Finally, the speed of a set of logic can be affected by rearranging the physical layout. The following techniques can be used for specific design constraints.

The various CMOS logic structures that can be used to implement circuit designs are as follows:

1. *CMOS Complementary Logic.* The CMOS complementary logic gates are designed as ratioless circuits. In these circuits, the output voltage is not a fraction of the V_{dd} (supply), and the gates are sized to meet the required electrical characteristics of the circuits. The gate consists of two blocks, and n block and a p block, that determine the function of the gate. The p block is a dual of the n block. Thus, an n-input gate will consist of $2n$ transistors.

2. *Pseudo-NMOS Logic.* In this logic, the load device is a single p transistor with the gate connected to $V_{dd}^{(103,108)}$. This is equivalent to replacing the depletion NMOS load in a conventional NMOS gate by a p device. The design of this style of gate[109,110] involves ratioed transistor sizes to ensure proper operation and is shown in Fig. 115. The static power dissipation that occurs whenever the pull-down chain is turned on is a major drawback of this logic style.

3. *Dynamic CMOS Logic.* In the dynamic CMOS logic style, an n-transistor logic structure's output node is precharged to V_{dd} by a p transistor and conditionally discharged by an n transistor connected to V_{ss}.[103] The input capacitance of the gate is the same as the pseudo-NMOS gate. Here, the pull-up time is improved by virtue of the active switch, but the pull-down time is increased due to the ground. The disadvantage of this logic structure is that the inputs can only change during the precharge phase and must be stable during the evaluate portion of the cycle. Figure 116 depicts this logic style.

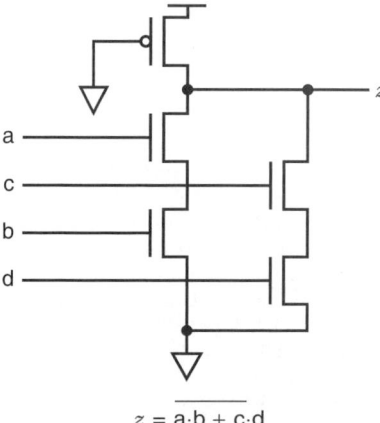

$$z = \overline{a \cdot b + c \cdot d}$$

Fig. 115   Pseudo-NMOS logic.

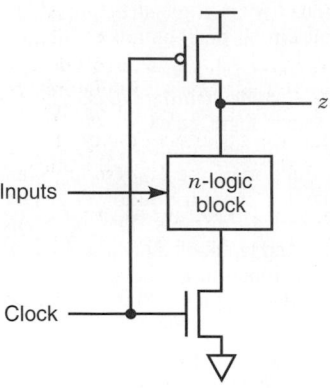

Fig. 116   Dynamic CMOS logic.

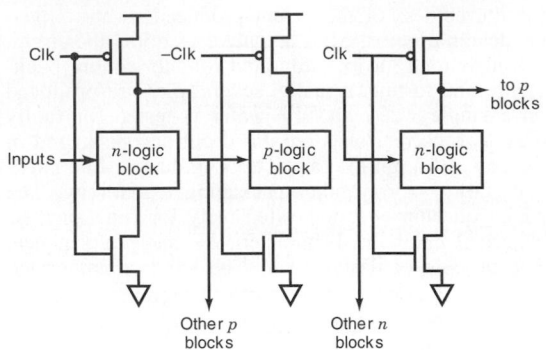

Fig. 118   NP domino logic.

4. *Clocked CMOS Logic.* To build CMOS logic gates with low-power dissipation,[111] this logic structure was proposed. The reduced dynamic power dissipation is realized due to the metal gate CMOS layout considerations. The gates have larger rise and fall times because of the series clocking transistors, but the capacitance is similar to the CMOS complementary gates. This is a recommended strategy for "hot electron" effects because it places an additional *n* transistor in series with the logic transistors.[112]

5. *CMOS Domino Logic.* This is a modification of the clocked CMOS logic, in which a single clock is used to precharge and evaluate a cascaded set of dynamic logic blocks. This involves incorporating a static CMOS inverter into each logic gate,[113] as shown in Fig. 117. During precharge, the output node is charged high and hence the output of the buffer is low. The transistors in the subsequent logic blocks

will be turned off since they are fed by the buffer. When the gate is evaluated, the output will conditionally go low (1–0), causing the buffer to conditionally go high (0–1). Hence, in a cascaded set of logic blocks, each state evaluates and causes the next stage to evaluate, provided the entire sequence can be evaluated in one clock cycle. Therefore, a single clock is used to precharge and evaluate all logic gates in a block. The disadvantages of this logic are that (a) every gate needs to be buffered, and (b) only noninverting structures can be used.

6. *NP Domino Logic (Zipper CMOS).* This is a further refinement of the domino CMOS.[114–116] Here, the domino buffer is removed, and the cascading of logic blocks is achieved by alternately composed *p* and *n* blocks, as is shown in Fig. 118. When the clock is low, all the *n*-block stages are precharged high while all the *p*-block stages are precharged low. Some of the advantages of the dynamic logic styles include (a) smaller area than fully static gates, (b) smaller parasitic capacitances, and (c) glitch-free operation if designed carefully.

7. *Cascade Voltage Switch Logic (CVSL).* The CVSL is a differential style of logic requiring both true and complement signals to be routed to gates.[117] Two complementary NMOS structures are constructed and then connected to a pair of cross-coupled *p* pull-up transistors. The gates here function similarly to the domino logic, but the advantage of this style is the ability to generate any logic expression involving both inverting and noninverting structures. Figure 119 gives a sketch of the CVSL logic style.

8. *Pass Transistor Logic.* Pass transistor logic is popular in NMOS-rich circuits. Formal methods for deriving pass transistor logic for NMOS are presented in Ref. 118. Here, a set of control signals and a set of pass signals are applied to the gates and sources of the *n* transistor,

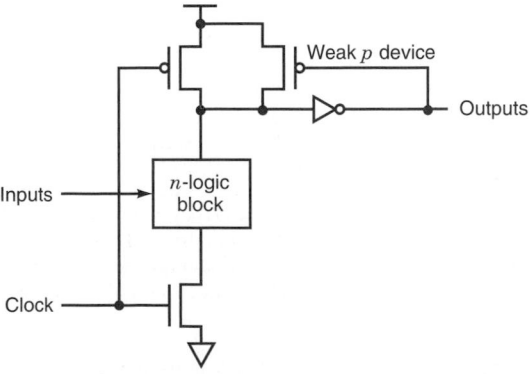

Fig. 117   CMOS domino logic.

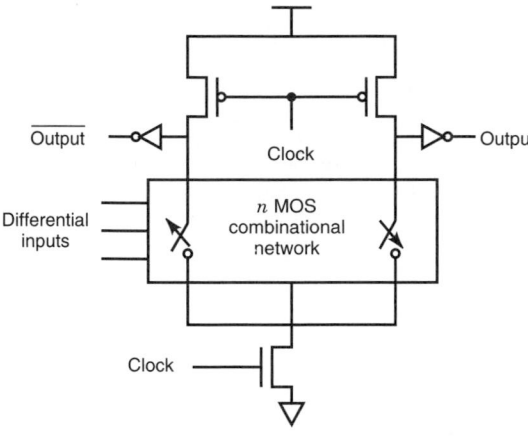

Fig. 119 Cascade voltage switch logic.

correspondingly. From these signals, the truth table for any logic equation can be realized.

9. *Source Follower Pull-Up Logic (SFPL).* This is similar to the pseudo-NMOS gate except that the pull-up is controlled by the inputs.[119] In turn, this leads to the use of smaller pull-down circuits. The SFPL gate style reduces the self-loading of the output and improves the speed of the gate. Therefore, it shows a marked advantage in high fan-in gates.

Using the various design styles, any circuit design can be built in a hierarchical fashion. The basic gates are first built, from which functional blocks like a multiplexer or an adder circuit can be realized. From these basic blocks, more complex circuits can be constructed. Once a design for a specific application has been designed, the functionality of the circuit needs to be verified. Also, other constraints, like the timing and electrical characteristics, have to be studied before the design can be manufactured. The techniques and tools to achieve this are the subject of the next section.

8.5 Simulation

Simulation is required to verify if a design works the way it should. Simulation can be performed at various levels of abstraction. A circuit can be simulated at the logic level, the switch level, or with reference to the timing. Simulation is a critical procedure before committing the design to silicon. The simulators themselves are available in a wide variety of types.[120]

Logic-Level Simulation Logic-level simulation occurs at the highest level of abstraction. It uses primitive models of NOT, OR, AND, NOR, and NAMD gates. Virtually all digital logic simulators are event driven (i.e., a component is evaluated based on when an event occurs on its inputs). Logic simulators are

categorized according to the way the delays are modeled in the circuit: (a) unit-delay simulators, in which each component is assumed to have a delay of one time unit, and (b) variable-delay simulators, which allow components to have arbitrary delays. While the former helps in simulating the functionality of the circuit, the latter allows for more accurate modeling of the fast-changing nodes.

The timing is normally specified in terms of an inertial delay and a load-dependent delay, as follows:

$$T_{\text{gate}} = T_{\text{intrinsic}} + C_{\text{load}} \times T_{\text{load}}$$

where T_{gate} = delay of the gate
$T_{\text{intrinsic}}$ = intrinsic gate delay
C_{load} = actual load in some units (pF)
T_{load} = delay per load in some units (ns/pF)

Earlier, logic simulators used preprogrammed models for the gates, which forced the user to describe the system in terms of these models. In modern simulators, programming primitives are provided that allow the user to write models for the components. The two most popular digital simulation systems in use today are VHDL and Verilog.

Circuit-Level Simulation The most detailed and accurate simulation technique is referred to as *circuit analysis*. Circuit analysis simulators operate at the circuit level. Simulation programs typically solve a complex set of matrix equations that relate the circuit voltages, currents, and resistances. They provide accurate results but require long simulation times. If N is the number of nonlinear devices in the circuit, then the simulation time is typically proportional to N^m, where m is between 1 and 2. Simulation programs are useful in verifying small circuits in detail but are unrealistic for verifying complex VLSI designs. They are based on transistor models and hence should not be assumed to predict accurately the performance of designs. The basic sources of error include (a) inaccuracies in the MOS model parameters, (b) an inappropriate MOS model, and (c) inaccuracies in parasitic capacitances and resistances. The circuit analysis programs widely used are the SPICE program, developed at the University of California at Berekely,[121] and the ASTAP program from IBM.[122] HSPICE [123] is the commercial variant of these programs. The SPICE program provides various levels of modeling. The simple models are optimized for speed, while the complex ones are used to get accurate solutions. As the feature size of the processes is reduced, the models used for the transistors are no longer valid and hence the simulators cannot predict the performance accurately unless new models are used.

Switch-Level Simulation Switch-level simulation is simulation performed at the lowest level of abstraction. These simulators model transistors as switches to merge the logic-level and circuit-level simulation techniques. Although logic-level simulators also model transistors as switches, the unidirectional logic gate model cannot model charge sharing, which is a result of the bidirectionality of the MOS transistor. Hence, we assume that all wires have capacitance since we need to locate charge-sharing bugs. RSIM[124] is an example of a switch-level simulator with timing. In RSIM, CMOS gates are modeled as either pull-up or pull-down structures, for which the program calculates a resistance to power or ground. The output capacitance of the gate is used with the resistance to predict the rise and the fall times of a gate.

Timing Simulators Timing simulators allow simple nonmatrix calculations to be employed to solve for circuit behavior. This involves making approximations about the circuit, and hence accuracy is less than that of simulators like SPICE. The advantage is the execution time, which is over two orders of magnitude less than for SPICE. Timing simulator implementations typically use MOS-model equations or table look-up methods. Examples of these simulators are in Ref. 125.

Mixed-Mode Simulators Mixed-mode simulators are available commercially today and merge the aforementioned different simulation techniques. Each circuit block can be simulated in the appropriate mode.

The results of the simulation analysis are fed back to the design stage, where the design is tuned to incorporate the deviations. Once the circuit is perfected and the simulation results are satisfactory, the design can be fabricated. To do this, we need to generate a geometric layout of the transistors and the electrical connections between them. This has been a subject of intense research over the last decade and continues to be so. The following section introduces this problem and presents some of the well-known techniques for solving it.

8.6 Layout

The layout design is considered a prescription for preparing the photomasks used in the fabrication of ICs.[103] There is a set of rules, called the design rules, used for the layout; these serve as the link between the circuit and the process engineer. The physical design engineer, in addition to knowledge of the components and the rules of the layout, needs strategies to fit the layouts together with other circuits and provide good electrical properties. The main objective is to obtain circuits with optimum yield in as small an area as possible without sacrificing reliability.

The starting point for the layout is a circuit schematic. Figure 106 depicts the schematic symbols for an n-type and p-type transistor. The circuit schematic is treated as a specification for which we must implement the transistors and connections between them in the layout. The circuit schematic of an inverter is shown in Fig. 108. We need to generate the exact layout of the transistors of this schematic, which can then be used to

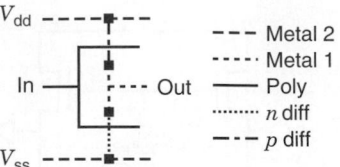

Fig. 120 Stick diagram of inverter.

build the photomask for the fabrication of the inverter. Generating a complete layout in terms of rectangles for a complex system can be overwhelming, although at some point we need to generate it. Hence designers use an abstraction between the traditional transistor schematic and the full layout to help organize the layout for complex systems. This abstraction is called a *stick diagram*. Figure 120 shows the stick diagram for the inverter schematic. As can be seen, the stick diagram represents the rectangles in the layout as lines, which represent wires and component symbols. Stick diagrams are not exact models of the layouts but let us evaluate a candidate design with relatively little effort. Area and aspect ratio are difficult to estimate from stick diagrams.

Design Rules Design rules for a layout[126] specify to the designer certain geometric constraints on the layout artwork so that the patterns on the processed wafer will preserve the topology and geometry of the designs. These help to prevent separate, isolated features from accidentally short circuiting, or thin features from opening, or contacts from slipping outside the area to be contacted. They represent a tolerance that ensures very high probability of correct fabrication and subsequent operation of the IC. The design rules address two issues primarily:

1. The geometrical reproduction of features that can be reproduced by the mask-making and lithographical process
2. The interaction among the different layers

Several approaches can be used to descibe the design rules. These include the micron rules, stated at some micron resolution, and the lambda (λ)-based rules. The former are given as a list of minimum feature sizes and spacings for all masks in a process, which is the usual style for the industry. Mead–Conway[127] popularized the λ-based approach, where λ is process dependent and is defined as the maximum distance by which a geometrical feature on any one layer can stray from another feature. The advantage of the λ-based approach is that by defining λ properly the design itself can be made independent of both the process and fabrication house, and the design can be rescaled. The goal is to devise rules that are simple, constant in time, applicable to many processes,

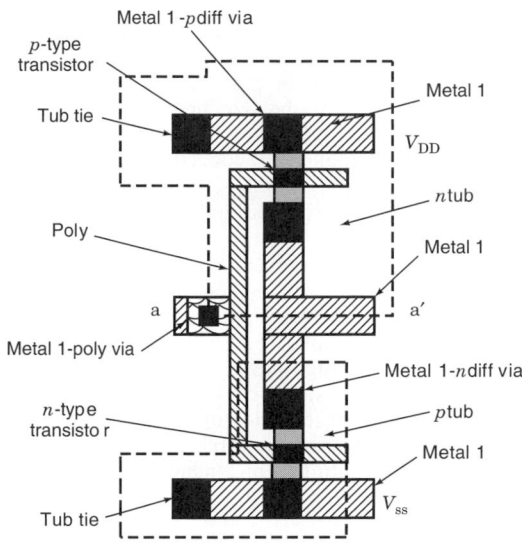

Fig. 121 Transistor layout of inverter.

standardized among many institutions, and have a small number of exceptions for specific processes. Figure 121 gives the layout of the inverter, with the design rules specified.

To design and verify layouts, different CAD tools can be used. The most important of these are the layout editors,[128] design rule checkers, and circuit extractors. The editor is an interactive graphic program that allows us to create and delete layout elements. Most editors work on hierarchical layouts, but some editors, like Berekely's

Magic tool,[129] work on a symbolic layout. The latter include somewhat more detail than the stick diagram but are still more abstract than the pure layout. The design rule checker, or DRC, programs look for design rule violations in the layouts. Magic has an online design rule checking. The circuit extractor is an extension of the DRC programs. While the DRC must identify transistors and vias to ensure proper checks, the circuit extractor performs a complete job of component and wire extraction. It produces a netlist, which lists the transistors in the layouts and the electrical nets that connect their terminals.

Physical Design From the circuit design of a certain application and the design rules of a specific process, the physical design problem is to generate a geometric layout of the transistors of the circuit design conforming to the specified design rules. From this layout, photomasks can be generated that will be used in the fabrication process. To achieve this, the different modules of the design need to be placed first and then electrical connections between them realized through the metal layers. For instance, a two-layer

metallization would allow the designer to lay out metal both vertically and horizontally on the floorplan. Whenever the wire changes direction, a via can be used to connect the two metal layers. Due to the complexity of this problem, most authors treat module placement and the routing between modules as two separate problems, although they are related critically. Also, in former days, when designs were less complex, design was done by hand. Now we require sophisticated tools for this process.

Placement. Placement is the task of placing modules adjacent to each other to minimize area or cycle time. The literature consists of a number of different placement, algorithms that have been proposed.[130–133] Most algorithms partition the problem into smaller parts and then combine them, or start with a random placement solution and then refine it to reach the optimal. The modules are usually considered as rectangular boxes with specified dimensions. The algorithms then use different approaches to fit these boxes in a minimal area or to optimize them to certain other constraints. For instance, consider a certain number of modules with specific dimensions and a given area in which to fit them. This is similar to the bin-packing algorithm. After the placement step, the different modules are placed in an optimal fashion and the electrical connections between them need to be realized.

Routing. Once the modules have been placed, we need to create space for the electrical connections between them. To keep the area of the floorplan minimal, the first consideration is to determine the shortest path between nodes, although a cost-based approach may also be used. The cost is defined to include an estimate of the congestion, number of available wire tracks in a local area, individual or overall wire length, and so on. Since the problem is a complex one, the strategy is to split the problem into global or loose routing and local or detailed routing. Global routing is a preliminary step, in which each net is assigned to a particular routing area, and the goal is to make 100% assignment of nets to routing regions while minimizing the total wire length. Detailed routing then determines the exact route for each net within the global route. There are a number of approaches to both of these problems.

Global Routing. Global routing[134] is performed using a wire-length criterion because all timing critical nets must be routed with minimum wire length. The routing area itself can be divided into disjoint rectangular areas, which can be classified by their topology. A two-sided channel is a rectangular routing area with no obstruction inside and with pins on two parallel sides. A switch box is a rectangular routing area with no obstructions and signals entering and leaving through all four sides.[135] The focus in this problem is only to create space between the modules for all the nets and not to determine the exact route. The algorithms proceed by routing one net at a time, choosing the shortest possible path. Since there is a lot of dependency among the nets, different heuristics are used to generate the least possible routing space in which to route the nets. Once space is created for all the nets, the exact route of each net can be determined.

Detailed Routing. Detailed routing is usually done by either a maze-search or a line-search algorithm. The maze-running algorithm[136,137] proposed by Lee–Moore finds the shortest path between any two points. For this, the layout is divided into a grid of nodes, in which each node is weighted by its distance from the source of the wire to be routed. The route that requires the smallest number of squares is then chosen. If a solution exists, this algorithm will find it, but an excessive amount of memory is required to achieve this. In the linesearch algorithm, vertical and horizontal lines are drawn from the source and the target, followed by horizontal or vertical lines that intersect the original lines. This is repeated until the source and target meet. There are also a number of other heuristic algorithms that exploit different characteristics of the design to generate optimal routing solutions. Genetic algorithms and simulated annealing approaches to this problem have gained importance in recent years.

An introduction to the various algorithms that have been proposed for layouts can be found in Ref. 138. Once the layout has been determined and the photomasks made, the circuit can go to the fabrication plant for processing.

8.7 Fabrication

The section describes the approach used in building integrated circuits on monolithic pieces of silicon. The process involves the fabrication of successive layers of insulating, conducting, and semiconducting materials, which have to be patterned to perform specific functions. The fabrication therefore must be executed in a specific sequence, which constitutes an IC process flow. The manufacturing process itself is a complex interrelationship of chemistry, physics, material science, thermodynamics, and statistics.

Semiconducting materials, as the name suggests, are neither good conductors nor good insulators. While there are many semiconducting elements, silicon is primarily chosen for manufacturing ICs because it exhibits few useful properties. Silicon devices can be built with a maximum operating temperature of about $150°C$ due to the smaller leakage currents as a result of the large bandgap of silicon (1.1 eV). IC planar processing requires the capability to fabricate a passivation layer on the semiconductor surface. The oxide of silicon, SiO_2, which could act as such a layer, is easy to form and is chemically stable. The controlled addition of specific impurities makes it possible to alter the characteristics of silicon. For these reasons, silicon is almost exclusively used for fabricating microelectronic components.

Silicon Material Technology Beach sand is first refined to obtain semiconductor-grade silicon. This is then reduced to obtain electronic-grade polysilicon in the form of long, slender rods. Single-crystal silicon is grown from this by the Czochralski (CZ) or float-zone (FZ) methods. In CZ growth, single-crystal ingots are pulled from molten silicon contained in a crucible. For VLSI applications, CZ silicon is preferred because it can better withstand thermal stresses[139] and offers an internal gettering mechanism than can remove unwanted impurities from the device structures on wafer surfaces.[140] FZ crystals are grown without any contact to a container or crucible and hence can attain higher purity and resistivity than CZ silicon. Most high-voltage, high-power devices are fabricated on FZ silicon. The single-crystal ingot is then subjected to a complex sequence of shaping and polishing, known as wafering, to produce starting material suitable for fabricating semiconductor devices. This involves the following steps:

1. The single-crystal ingot undergoes routine evaluation of resistivity, impurity content, crystal perfection size, and weight.
2. Since ingots are not perfectly round, they are shaped to the desired form and dimension.
3. The ingots are then sawed to produce silicon slices. The operation defines the surface orientation, thickness, taper, and bow of the slice.
4. To bring all the slices to within the specified thickness tolerance, lapping and grinding steps are employed.
5. The edges of the slices are then rounded to reduce substantially the incidence of chipping during normal wafer handling.
6. A chemical-mechanical polishing[141] step is then used to produce the highly reflective and scratch- and damage-free surface on one side of the wafer.
7. Most VLSI process technologies also require an epitaxial layer, which is grown by a chemical vapor deposition process.

The most obvious trend in silicon material technology is the increasing size of the silicon wafers. The use of these larger-diameter wafers presents major challengers to semiconductor manufacturers. Several procedures have been investigated to increase axial impurity uniformity. These include the use of double crucibles, continuous liquid feed (CLF) systems,[142] magnetic Czochralski growth (MCZ),[142,143] and controlled evaporation from the melt.

Epitaxial Layer The epitaxial growth process is a means of depositing a single-crystal film with the same crystal orientation as the underlying substrate. This can be achieved from the vapor phase, liquid phase, or solid phase. Vapor-phase epitaxy has the widest acceptance in silicon processing, since excellent control of the impurity concentration and crystalline perfection can be achieved. Epitaxial processes are used for the fabrication of advanced CMOS VLSI circuits because epitaxial processes minimize latch-up effects. Also in the epitaxial layer, doping concentration can be accurately controlled, and the layer can be made oxygen and carbon free. Epitaxial deposition is a chemical vapor deposition process.[144] The four major chemical sources of silicon used commercially for this deposition are (a) silicon tetrachloride ($SiCl_4$), (b) trichlorosilane ($SiHCl_3$), (c) dichlorosilane (SiH_2Cl_2), and (d) silane (SiH_4). Depending on particular deposition conditions and film requirements, one of these sources can be used.

Doping Silicon The active circuit elements of the IC are formed within the silicon substrate. To construct these elements, we need to create localized *n*-type and *p*-type regions by adding the appropriate dopant atoms. The process of introducing controlled amounts of impurities into the lattice of the monocrystalline silicon is known as *doping*. Dopants can be introduced selectively into the silicon using two techniques: diffusion and ion implantation.

Diffusion. The process by which a species moves as a result of the presence of a chemical gradient is referred to as diffusion. Diffusion is a time- and temperature-dependent process. To achieve maximum control, most diffusions are performed in two steps. The first step is predeposition,[145] which takes place at a furnace temperature and controls the amount of impurity that is introduced. The second step, the drive-in step,[145] controls the desired depth of diffusion.

Predeposition. In predisposition, the impurity atoms are made available at the surface of the wafer. The atoms of the desired element in the form of a solution of controlled viscosity can be spun on the wafer, in the same manner as the photoresist. For these spin-on dopants, the viscosity and the spin rate are used to control the desired dopant film thickness. The wafer is then subjected to a selected high temperature to complete the predeposition diffusion. The impurity atoms can also be made available by employing a low-temperature chemical vapor deposition process in which the dopant is introduced as a gaseous compound—usually in the presence of nitrogen as a diluent. The oxygen concentration must be carefully controlled in this operation to prevent oxidation of the silicon surface of the wafer.

Drive-In. After predeposition the dopant wafer is subjected to an elevated temperature. During this step, the atoms further diffuse into the silicon crystal lattice. The rate of diffusion is controlled by the temperature employed. The concentration of the dopant atoms is maximum at the wafer surface and reduces as the silicon substrate is penetrated further. As the atoms migrate during the diffusion, this concentration changes. Hence a specific dopant profile can be achieved by controlling the diffusion process. The dopant drive-in is usually performed in an oxidizing temperature to grow a protective layer of SiO_2 over the newly diffused area.

Ion Implantation. Ion implantation is a process in which energetic, charged atoms or molecules are directly introduced into the substrate. Ion implantation[146,147] is superior to the chemical doping methods discussed previously. The most important advantage of this process is its ability to control more precisely the number of implanted dopant atoms into substrates. Using this method, the lateral diffusion is reduced considerably compared to the chemical doping methods. Ion implantation is a low-temperature process, and the parameters that control the ion implantation are amenable to automatic control. After this process the wafer is subjected to annealing to activate the dopant electrically. There are some limitations to this process. Since the wafer is bombarded with dopant atoms, the material structure of the target is damaged. The throughput is typically lower than diffusion doping process. Additionally, the equipment used causes safety hazards to operating personnel.

Photolithography Photolithography is the most critical step in the fabrication sequence. It determines the minimum feature size that can be realized on silicon and is a photoengraving process that accurately transfers the circuit patterns to the wafer. Lithography[148,149] involves the patterning of metals, dielectrics, and semiconductors. The photoresist material is first spin coated on the wafer substrate. It performs two important functions: (a) precise pattern formation and (b) protection of the substrate during etch. The most important property of the photoresist is that its solubility in certain solvents is greatly affected by exposure to ultraviolet radiation. The resist layer is then exposed to ultraviolet light. Patterns can be transferred to the wafer using either positive or negative masking techniques. The required pattern is formed when the wafer undergoes the development step. After development, the undesired material is removed by wet or dry etching.

Resolution of the lithography process is important to this process step. It specifies the ability to print minimum size images consistently under conditions of reasonable manufacturing variation. Therefore, lithographic processes with submicron resolution must be available to build devices with submicron features. The resolution is limited by a number of factors, including (a) hardware, (b) optical properties of the resist material, and (c) process characteristics.[150]

Most IC processes require 5–10 patterns. Each one of them needs to be aligned precisely with those already on the wafer. Typically, the alignment distance between two patterns is less than 0.2 μm across the entire area of the wafer. The initial alignment is made with respect to the crystal lattice structure of the wafer, and subsequent patterns are aligned with the existing ones. Earlier, mask alignment was done using *contact printing*,[151,152] in which the mask is held just off the wafer and visually aligned. The mask is then pressed into contact with the wafer and impinged with ultraviolet light. There is a variation of this technique called *proximity printing*, in which the mask is held slightly above the wafer during exposure. Hard contact printing was preferred because it reduced the diffraction of light, but it led to a number of yield and production problems. So the projection alignment and exposure system was developed, in which the mask and wafer never touch and an optical system projects and aligns the mask onto the wafer. Since there is no damage to the mask or photoresist, the mask life is virtually unlimited. VLSI devices use projection alignment as the standard production method.

Junction Isolation   When fabricating silicon ICs, it must be possible to isolate the devices from one another. These devices can then be connected through specific electrical paths to obtain the desired circuit configuration. From this perspective, the isolation technology is one of the critical aspects of IC fabrication. For different IC types, like NMOS, CMOS, and bipolar, a variety of techniques have been developed for device isolation. The most important technique developed was termed LOCOS, for LOCal Oxidation of Silicon. This involves the formation of semirecessed oxide in the nonactive or field areas of the substrate. With the advent of submicron-size device geometries, alternative approaches for isolation were needed. Modified LOCOS processes, trench isolation, and selective epitaxial isolation were among the newer approaches adopted.

LOCOS.   To isolate MOS transistors, it is necessary to prevent the formation of channels in the field regions. This implies that a large value of V_T is required in the field region, in practice about 3–4 V above the supply voltage. Two ways to increase the field voltage are to increase the field oxide thickness and raise the doping beneath the field oxide. Thicker field oxide regions cause high enough threshold voltages but unfortunately lead to step coverage problems, and hence thinner field oxide regions are preferred. Therefore, the doping under the field oxide region is increased to realize higher threshold voltages. Nevertheless, the field oxide is made 7–10 times thicker than the gate oxide. Following this, in the channel-stop implant step, ion implantation is used to increase the doping under the field oxide. Until about 1970, the thick field oxide was grown using the *grow-oxide-and-etch* approach in which the oxide is grown over the entire wafer and then etched over the active regions. Two disadvantages of this approach prevented it from being used for VLSI applications: (a) Field-oxide steps have sharp upper corners, which poses a problem to the subsequent metallization steps; and (b) channel-stop implant must be performed before

the oxide is grown. In another approach, the oxide is selectively grown over the desired field regions. This process was introduced by Appels and Kooi in 1970[153] and is widely used in the industry. This process is performed by preventing the oxidation of the active regions by covering them with a thin layer of silicon nitride. After etching the silicon nitride layer, the channel-stop dopant can be implanted selectively. The process has a number of limitations for submicron devices. The most important of these is the formation of the "bird's beak," which is a lateral extension of the field oxide into the active areas of the device. The LOCOS bird's beak creates other problems as junctions become shallower in CMOS ICs. The LOCOS process was therefore modified in several ways to overcome these limitations: (a) etched-back LOCOS, (b) polybuffered LOCOS, and (c) sealed-interface local oxidation (SILO).[154]

Non-LOCOS Isolation Technologies.   There have been non-LOCOS isolation technologies for VLSI and ultra-large-scale integration (ULSI) applications. The most prominent of these is *trench* isolation technology. Trench technologies are classified into three categories: (a) shallow trenches (<1 μm), (b) moderate depth trenches (1–3 μm), and (c) deep, narrow trenches (>3 μm deep, <2 μm wide). Shallow trenches are used primarily for isolated devices of the same type and hence are considered a replacement to LOCOS. The buried oxide (BOX)[155] isolation technology uses shallow trenches refilled with a silicon dioxide layer, which is etched back to yield a planar surface. This technique eliminates the bird's beak of LOCOS. The basic BOX technique has certain drawbacks for which the technique is modified.

Metallization   This subsection describes the contact technology to realize the connections between devices and how the different metal layers are connected to realize the circuit structure.

Contact Technology.   Isolated active-device regions in the single-crystal substrate are connected through high-conductivity, thin-film structures that are fabricated over the silicon dioxide that covers the surface. An opening in the SiO_2 must be provided to allow *contacts* between the conductor film and the Si substrate. The technology involved in etching these contacts is referred to as contact technology. These contacts affect the circuit behavior because of the parasitic resistances that exist in the path between the metal-to-Si substrate and the region where the actual transistor action begins. Conventional contact fabrication involves the fabrication of a contact to silicon at locations where the silicon dioxide has been etched to form a window. It involves the following steps:

1. In regions where contacts are to be made, the silicon substrate is heavily doped.

2. A window or *contact hole* is etched in the oxide that passivates the silicon surface.

3. The silicon surface is cleaned to remove the thin native-oxide layer that is formed rapidly when the surface is exposed to an oxygen-containing ambient.

4. The metal film is deposited on the wafer and makes contact with silicon wherever contact holes were created. Aluminum is the most commonly used metal film.

5. After depositing the metal, the contact structure is subjected to a thermal cycle known as *sintering* or *annealing*. This helps in bringing the Si and metal into intimate contact.

Aluminum is desired as an interconnect material because its resistivity, 2.7 $\mu\Omega$-cm, is very low, and it offers excellent compatibility with SiO_2. Al reacts with SiO_2 to form Al_2O_3, through which the Al can diffuse to reach the Si, forming an intimate Al–Si contact. But using pure Al has certain disadvantages. Since Al is polycrystalline in nature, its grain boundaries offer very fast diffusion paths for the Si at temperatures above 400°C. Hence, if a large volume of Al is available, a significant quantity of the Si can diffuse into Al. As a result, the Al from the film moves rapidly to fill in the voids created by the migrating Si, which leads to large leakage currents or electrically shorting the circuit. This effect is referred to as *junction spiking*.[156] To prevent junction spiking, different techniques are used:

1. Add approximately 1% of Si to Al.

2. Add a diffusion barrier to prevent Si from diffusing into Al.

3. Decrease sintering temperature, but this increases contact resistance.

4. Add a "barrier" metal to the contact hole.[157]

Of these techniques, the most commonly used is the barrier metal. The idea is to block or hinder Al from intermixing with Si. There are three main types of contact barrier metallurgies: (a) sacrificial barrier, (b) passive barrier, and (c) stuffed barrier.

The use of Al has its own problems, the most important being its high resistivity and electromigration. There is also the problem with "hillock" formation. Hillocks are formed due to the thermal expansion mismatch among Al, SiO_2, and Si. As the wafer is cooled, thermal expansion mismatch forms stresses (usually compressive), which forms these hillocks. Therefore, copper metallization has been gaining importance. Copper is preferred over Al because it has a low resistivity (1.2 $\mu\Omega$-cm) and is resistant to electromigration. In fact, copper is added in small quantities to Al to reduce the electromigration problem of Al. However, there are some real problems with copper that need to be addressed before it can replace Al:

1. Cu is a terrible contaminant in Si. It has a very high diffusivity in Si and causes junction leakage, which degrades the gate oxide integrity (GOI).

2. Cu diffuses and drifts easily through SiO_2. Hence, Cu needs to be encapsulated for use in metallization.

3. Cu oxidizes to CuO easily.

4. The etch chemistry for Cu is highly contaminated, and the wafers need to be held at higher temperatures.

Typical process steps involved in the fabrication of a 0.8-μm LOCOS n-well inverter are as follows:

1. Wafer: 1×10^{15} to 1×10^{16} CZ(p) with $\langle 100 \rangle$ crystal orientation. Epitaxial layer required because of latch-up. The thickness is 2–16 μm with 5×10^{15}

2. Grown screen oxide layer, with the thickness in the range 400–1000.

3. Expose the n-well photo on the wafer.

4. An n-well ion implant. Use $1 \times 10^{13}/cm^2$ phosphorus. The voltage used is 60 keV to 2 MeV.

5. An n-well drive-in. This step is carried out at 1050 to 1100°C for 2–6 h. This activates the dopant atoms. The drive-in depth is around 1–10 μm.

6. Perform LOCOS process.

   6.1. Strip wafer.

   6.2. Pad oxide. Thickness is 100–400.

   6.3. Pad nitride. Thickness is 1000–2000. Low-pressure chemical vapor deposition (LPCVD) silicon nitride is used.

   6.4. Expose the diffusion photo on the wafer.

   6.5. Etch the nitride layer.

   6.6. Expose the block field (BF) photo. This is the inverse of the n-well photo and prevents the formation of the parasitic transistors between adjacent transistors.

   6.7. Field ion implantation. $1 \times 10^{13}/cm^2$ boron at 60 keV.

   6.8. Strip the BF and the resist layer.

   6.9. Grow the field oxide. The thickness is about 4000–6000 of SiO_2. The process used is a pyro process at 900–1050°C for 3–6 h.

   6.10. Strip the pad nitride layer by dipping the wafer in H_3PO_4.

   6.11. Strip the pad oxide layer by dipping the wafer in 50:1 HF.

7. Grow a sacrificial oxide layer and strip it. The thickness is about 600–1000. The sacrificial oxide layer eats into the bare silicon, thus exposing fresh silicon area on which the device can be grown.

8. Grow a sacrificial gate oxide layer. Here the thickness is about 80–130. This layer protects the gate when the V_T implant is done.

9. V_T implant. Two masks, one for the n region and one for the p region, are used. The concentration is 1×10^{11} to $1 \times 10^{12}/cm^2$ at 5–30 keV.

10. Strip the sacrificial gate oxide layer using a 50:1 HF solution.

11. Grow the gate oxide layer. Typical thickness is 80–130. The gate oxide layer is grown at 800–900°C for 20 min.

12. Polysilicon is deposited all over the wafer. LPCVD silane is used at 620°C for a blanket deposition. The typical thickness is 2500–4000.

13. Polysilicon doping is done by ion implantation using 5×10^{15} phosphorus.

14. The polysilicon etch is a very critical photo/etch process.

    14.1. Polysilicon photo is exposed on the wafer.

    14.2. Reactive ion etch (RIE) is used to etch the polysilicon.

    14.3. The resist is stripped.

15. Diffusion processing

    15.1. Mask the p^+ regions.

    15.2. Perform n^+ source/drain ion implantation using $5 \times 10^{15}/ cm^2 As^{75}$ at 40 keV. Arsenic is used because it is slow and does not diffuse deep into the silicon substrate.

    15.3. Perform n^+ anneal at 900°C for 15 min to activate the dopant.

    15.4. Strip the resist.

    15.5. Mask the n^+ regions.

    15.6. Perform p^+ source/drain ion implantation using $1 \times 10^{15}/ cm^2 BF_2/B^{11}$ at 5–20 keV.

    15.7. Source/drain anneal at 900°C for 30 min in an oxidizing atmosphere. This is a rapid thermal process.

    15.8. Strip the resist off.

16. Interlevel dielectric. Boro-phospho silicon glass (BPSG) is used because it flows well. Typical thickness is 5000–8000. A 900°C reflow anneal is also performed.

17. The contact photo is exposed on the wafer. This is critical to the layout density.

18. The contacts are etched using RIE.

19. After etching, the contact resist is stripped off.

20. Metallization. Ti barrier metallurgy is used. The actual contact is made with an alloy of Al/Cu/Si with percentages 98, 1, and 1%, respectively. The Al alloy is sputtered onto the wafer. The typical thickness is about $1.2\mu m$.

21. The metal-1 layer photo is exposed.

22. Metal-1 etch.

23. Strip resist.

24. Foaming gas anneal is performed to improve the mobility of the electrons and relieve stress on the wafer.

The inverter is finally fabricated. Figure 109 describes the process cross section of this inverter after the various steps have been performed.

8.8 Testing

Testing is a critical factor in the design of circuits. The purpose of testing is to verify conformance to the product definition. To understand the complexity of this problem, consider a combinational circuit with n inputs. A sequence of 2^n inputs must be applied and observed to test the circuit exhaustively. Since the number of inputs are high for VLSI circuits, testing the chip exhaustively is impossible. Hence, this becomes an area of importance to circuit design. There are three main areas that need to be addressed to solve this problem:

1. Test generation
2. Test verification
3. Design for testability

Test generation corresponds to the problem of generating a minimum number of tests to verify the behavior of a circuit. The problem of test verification, which is commonly gauged by performing fault simulations, involves evaluating measures of the effectiveness of a given set of test vectors. Circuits can also be designed for testability.

Test Generation Test generation[158] involves the search for a sequence of input vectors that allow faults to be detected at the primary device outputs. VLSI circuits are typically characterized by buried flip-flops, asynchronous circuits, indeterminate states, complex clock conditions, multiple switching of inputs simultaneously, and nonfunctional inputs. Due to these factors, an intimate knowledge of the internal circuit details is essential to develop efficient test strategies. The goal of a test generation strategy[159,160] is multifold: (a) chip design verification in conjunction with the designer, (b) incorporation of the customer's specification and patterns into the manufacturing test program, and (c) fault detection by fault simulation methods.

Test Verification Test verification[161] involves calculating measures for how efficient the test vectors for a given circuit are. This is often accomplished by using fault models.[162] Fault simulation requires a good circuit simulator to be efficient and is hence closely related to logic simulation and timing analysis. While the logic simulator verifies the functionality of a design and ensures that the timing constraints are met, the fault simulator tells the designer if enough analysis has been performed to justify committing the design to silicon. In fault simulation, the true value of a circuit and its behavior under possible faults is simulated. The fault model is a hypothesis based on an abstract model of the circuit, conformed to some precise real physical defects. To begin with, the simulator generates a fault list that identifies all the faults to be modeled. Then a set of test vectors is simulated against the fault-free and faulty models. Those faults that cause an erroneous signal at an output pin are considered as detected faults. Now the fault coverage of the test vector set can be computed as the number of faults detected over the total number of faults modeled.

The most widely used fault model is the single stuck-at fault model. This model assumes that all faults occur due to the shorting of a signal node with the power rail. A number of faults can be detected by this model, but a major disadvantage of this model is its assumption that all faults appear as stuck-at faults. The limitations of this model have led to the increased use of other models, like the stuck-open[163] and bridging fault models.[164] The former can occur in a CMOS transistor or at the connection to a transistor. The bridging faults are short circuits that occur between signal lines. These represent a frequent source of failure in CMOS ICs. A majority of the random defects are manifested as timing delay faults in static CMOS ICs. These are faults in which the increased propagation delay causes gates to exceed their rated specifications. The statically designed circuits have a transient power supply that peaks when the gates are switching and then settles to a low current value in the quiescent state. The quiescent power supply current,[165] known as I_{DDQ}, can be used as an effective test to detect leakage paths due to defects in the processing. The measured I_{DDQ} of a defect-free CMOS IC is approximately 20 nA. Most physical defects will elevate I_{DDQ} by one to five orders of magnitude. Thus the I_{DDQ} testing approach can be used to detect shorts not detectable by the single stuck-at fault model.

There are several other ways of applying logic and fault simulation to testing:

1. *Toggle Testing.* This is the cheapest, simplest, and least time-consuming method of applying simulation to testing. Toggle testing provides a testability measure by tracking the activity of circuit nodes. From a set of vectors, the method identifies those parts of the network that exhibit no activity. Since the test vectors do not affect these nodes, faults occurring here cannot be detected by the fault simulator.
2. *Fault Simulation of Functional Tests.* The outputs of the functional simulator can be used in the design process as an effective design analysis tool. The lists of detectable and undetectable faults generated by the simulator can be used to

locate problems in the design and correct them. This results in substantial savings in development and manufacturing.
3. *Fault Simulation after New Test Vector Generation.* High-quality testing in a reasonable time frame would require an efficient test pattern generation system and a fault simulator. Test vectors are first generated to detect specific faults, and the fault simulator determines the quality of the vector set. In this scenario, it becomes important to fault simulate after every new test vector is generated in order to catch multiple faults. Accelerated fault simulation is faster than test pattern generation.

Design for Testability *Design for testability* commonly refers to those design techniques that produce designs for which tests can be generated by known methods. The advantage of these techniques are (a) reduced test generation cost, (b) reduced testing cost, (c) high-quality product, and (d) effective use of computer-aided design tools. The key to designing circuits that are testable are two concepts, *controllability* and *observability*. Controllability is defined as the ability to set and reset every node that is internal to the circuit. Observability is defined as the ability to observe either directly or indirectly the state of any node in the circuit. There are programs like SCOAP[166] that, given a circuit structure, can calculate the ability to control or observe internal circuit nodes. The concepts involved in design for testability can be categorized as follows: (a) ad hoc testing, (b) scan-based test techniques, and (c) built-in self-test (BIST).

Ad Hoc Testing. Ad hoc testing comprises techniques that reduce the combinational explosion of testing. Common techniques partition the circuit structure and add test points. A long counter is an example of a circuit that can be partitioned and tested with fewer test vectors. Another technique is the use of a bus in a bus-oriented system for testing. The common approaches can be classified as (a) partitioning techniques, (b) adding test points, (c) using multiplexers, and (d) providing for easy state reset.

Scan-Based Test Techniques. Scan-based approaches stem from the basic tenets of controllability and observability. The most popular approach is the level-sensitive scan design, or LSSD, approach, introduced by IBM.[167] This technique is illustrated in Fig. 122. Circuits designed based on this approach operate in two modes—namely normal mode and test mode. In the normal mode of operation, the shift register latches act as regular storage latches. In the test mode, these latches are connected sequentially and data can be shifted in or out of the circuit. Thus, a known sequence of data (controllability) can be input to the circuit and the results can be shifted out of the circuit using the registers (observability). The primary disadvantage of this approach is the increased

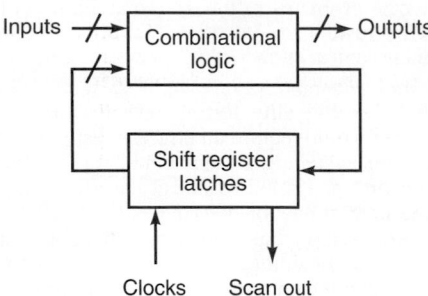

Fig. 122 Level-sensitive scan design.

complexity of the circuit design and the increased external pin count. The serial scan approach is similar to the LSSD, but the design of the shift register latch is simplified to obtain faster circuits. For most circuit designs, only the input and output register need be made scannable. This technique makes the designer responsible for deciding which registers need to be scanned and is called the partial serial scan technique.[168] The parallel scan[169] approach is an extension of the serial scan in which the registers are arranged in a sort of a grid, where on a particular column all the registers have a common read/write signal. The registers that fall on a particular row have common data lines. Therefore, the output of a register can be observed by enabling the corresponding column and providing the appropriate address. Data can also be written into these registers in a similar fashion.

Built-In Self-Test. Signature analysis[170] or cyclic redundancy checking can be used to incorporate a built-in self-test module in a circuit. This involves the use of a linear feedback shift register, as depicted in Fig. 123. The value in the register will be a function of the value and number of latch inputs and the counting function of the signature analyzer. The good part of the circuit will have a particular number or signature in the register, while the faulty portion will have a different number. The signature analysis approach can be merged with the level-sensitive scan design approach to create a structure called a built-in logic block observation, or BILBO.[171] Yet another approach

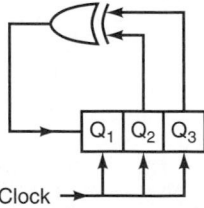

Fig. 123 Linear feedback shift register.

to built-in test is called a design for autonomous test, in which the circuit is partitioned into smaller structures that are tested exhaustively. The partitioning method involves the use of multiplexers. The syndrome testing method, in which all possible inputs are applied and the number of 1's at the outputs is counted, is also a test method that requires exhaustive testing. The resultant value is compared to that of a known good machine.

Other Tests So far we have discussed techniques for testing logic structures and gates. But we need testing approaches at the chip level and the system level also. Most approaches for testing chips rely on the aforementioned techniques. Memories, for instance, can use the built-in self-test techniques effectively. Random logic is usually tested by full serial scan or parallel scan methods. At the system level, traditionally the "bed-of-nails" testers have been used to probe points of interest. But with the increasing complexity of designs, system designers require a standard to test chips at the board level. This standard is the IEEE *1149 boundary scan*[172] architecture. ICs that are designed based on this standard enable complete testing of the board. The following types of tests can be performed in a unified framework: (a) connectivity test between components, (b) sampling and setting chip I/Os, and (c) distribution and collection of built-in self-test results.

9 MICROPROCESSORS
Robert P. Colwell

In 1972, Intel Corporation sparked an industrial revolution with the world's first microprocessor, the 4004. The 4004 replaced the logic of a numeric calculator with a general-purpose computer, implemented in a single silicon chip. The 4004 is shown in Fig. 124. The 4004 integrated 2300 transistors and ran at a clock rate of 108 kHz (108,000 clock cycles per second). In 1997, Intel introduced the Pentium II processor, running at 300 MHz (300 million clock cycles per second) and incorporating nearly 8 million transistors. The Pentium II processor is shown in Fig. 125. In 2000, Intel introduced the Pentium 4.

From the 4004's humble beginning, the microprocessor has assumed an importance in the world's economy similar to that of the electric motor or the internal combustion engine. Microprocessors now supply more than 90% of the world's computing needs, from small portable and personal desktop computers to large-scale supercomputers such as Intel's Teraflop machine, which contains over 9000 microprocessors. A variant of the microprocessor, the microcontroller, has become the universal controller in machines from automobile engines to audio systems to wristwatches.

9.1 Microprocessors and Computers
Microprocessors are the processing units or the "brains" of the computer system. Every action that microprocessors perform is specified by a computer program that

Fig. 124 World's first microprocessor, the Intel 4004, ca. 1971. Originally designed to be a less expensive way to implement the digital logic of a calculator, the chip instead spawned a computing revolution that still shows no signs of abating.

has been encoded into "object code" by a software program known as a compiler. Directed by another software program known as the operating system (e.g., Microsoft's Windows), the microprocessor locates the desired application code on the hard drive or compact disk and orders the drive to begin transferring the program to the memory subsystem so that the program can be run.

Digital electronic computers have at least three major subsystems:

- A memory to hold the programs and data structures
- An I/O subsystem
- A central processor unit (CPU)

A microprocessor is the central processor subsystem, implemented on a single chip of silicon.

In microprocessor-based computer systems, the I/O subsystem moves information into and out of the computer system. I/O subsystems usually include some form of nonvolatile storage, which is a means of remembering data and programs even when electrical

Fig. 125 The 1998 successors to the line of microprocessors started by the 4004, Intel's Pentium II processor, mounted within its Single-Edge Cartridge Connector (SECC). This picture shows the cartridge with its black case removed. On the substrate within the cartridge, the large octagonal package in the center is the Pentium II CPU itself. The rectangular packages to the right and left of the CPU are the cache chips. The small components mounted on the substrate are resistors and capacitors needed for power filtering and bus termination.

power is not present. Disk drives, floppy drives, and certain types of memory chips fulfill this requirement in microprocessor-based systems. Keyboards, trackballs, and mice are common input devices. Networks, modems, and compact discs are also examples of I/O devices. The memory subsystem, a place to keep and quickly access programs or data, is usually random-access memory (RAM) chips.

Microprocessors and microcontrollers are closely related devices. The differences are in how they are used. Essentially, microcontrollers are microprocessors for embedded control applications. They run programs that are permanently encoded into read-only memories and optimized for low cost so that they can be used in inexpensive appliances (printers, televisions, power tools, and so on). The versatility of a micro-controller is responsible for user-programmable VCRs and microwave ovens, the fuel-savings of an efficiently managed automobile engine, and the convenience of sequenced traffic lights on a highway and of automated bank teller machines.

Microprocessor software is typically created by humans who write their codes in a high-level language such as C or Fortran. A compiler converts that source code into a machine language that is unique to each particular family of microprocessors. For instance, if the program needs to write a character to the screen, it will include an instruction to the microprocessor that specifies the character, when to write it, and where to put it. Exactly how these instructions are encoded into the 1's and 0's (bits) that a computer system can use determines which computers will be able to run the program successfully. In effect, there is a contract between the design of a microprocessor and the compiler that is generating object code for it. The compiler and microprocessor must agree on what every computer instruction does, under all circumstances of execution, if a program is to perform its intended function. This contract is known as the computer's instruction set. The instruction set plus some additional details of implementation such as the number of registers (fast temporary storage) are known as the computer's instruction set architecture (ISA). Programs written or compiled to one ISA will not run on a different ISA. During the 1960s and 1970s, IBM's System/360 and System/370 were the most important ISAs. With the ascendancy of the microprocessor, Intel's x86 ISA vied with Motorola's MC68000 for control of the personal computer market. By 1997, the Intel architecture was found in approximately 85% of all computer systems sold.

Early microprocessor instruction set architectures, such as the 4004, were designed to operate on 8-bit data values (operands). Later microprocessors migrated to 16-bit operands, including the microprocessor in the original IBM PC (the Intel 8088). Microprocessors settled on 32-bit operands in the 1980s, with the Motorola 68000 family and Intel's 80386. In the late 1980s, the microprocessors being used in the fastest servers and high-end workstations began to run into the intrinsic

addressability limit of 4 GB (four gigabytes, or four billion bytes, which is 2 raised to the power 32). These microprocessors introduced 64-bit addressing and data widths. It is likely that 64-bit computing will eventually supplant 32-bit microprocessors. It also seems likely that this will be the last increase in addressability that the computing industry will ever need because 2 raised to the power 64 is an enormous number of addresses.

Prior to the availability of microprocessors, computer systems were implemented in discrete logic, which required the assembly of large numbers of fairly simple digital electronic integrated circuits to realize the basic functions of the I/O, memory, and central processor subsystems. Because many (typically thousands) of such circuits were needed, the resulting systems were large, power-hungry, and costly. Manufacturing such systems was also expensive, requiring unique tooling, hand assembly, and a large amount of human debug effort to repair the inevitable flaws that accumulate during the construction of such complex machinery. In contrast, the fabrication process that underlies the microprocessor is much more economical. As with any silicon-integrated circuit, microprocessor fabrication is mainly a series of chemical processes performed by robots. So the risk of introducing human errors that would later require human debugging is eliminated. The overall process can produce many more microprocessors than discrete methods could.

9.2 Moore's Law

In 1964, Gordon Moore made an important observation regarding the rate of improvement of the silicon-integrated circuit industry. He noted that the chip fabrication process permitted the number of transistors on a chip to double every 18 months. This resulted from the constantly improving silicon process that determines the sizes of the transistors and wiring on the integrated circuits. Although he made the initial observation on the basis of experience with memory chips, it has turned out to be remarkably accurate for microprocessors as well. Moore's law has held for well over 30 years. Figure 126 plots the number of transistors on each Intel microprocessor since the 4004.

These improvements of the underlying process technology have fueled the personal computer industry in many different ways. Each new process generation makes the transistors smaller. Smaller transistors are electrically much faster, allowing higher clock rates. Smaller wires represent less electrical capacitance, which also increases overall clock rates and reduces power dissipation. The combination of both permits far more active circuitry to be included in new design. Constant learning in the silicon fabrication plants have also helped drive up the production efficiency, or yield, of each new process to be higher than its predecessor, which also helps support larger die sizes per silicon chip.

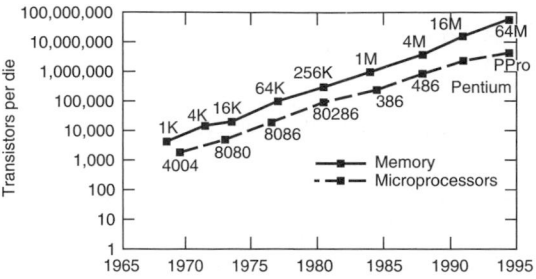

Fig. 126 Moore's law has accurately predicted the number of transistors that can be incorporated in microprocessors for over 25 years. Since this transistor count strongly influences system performance, this remarkable "law" has become one of the central tenets in the field of computers and integrated electronics. It guides the design of software, hardware, manufacturing production capacity, communications, and corporate planning in nearly every major area.

The impact of this progression has been profound for the entire industry. The primary benefit of a new microprocessor is its additional speed over its predecessors, at ever better price points. The effect of Moore's law has been for each new microprocessor to become obsolete within only a few years after its introduction. The software industry that supplies the applications to run on these new microprocessors expects this performance improvement. The industry tries to design so that its new products will run acceptably on the bulk of the installed base but can also take advantage of the new performance for the initially small number of platforms that have the new processor. The new processor's advantages in price/performance will cause it to begin to supplant the previous generation's volume champion. The fabrication experience gained on the new product allows its price to be driven ever downward until the new design completely takes over. Then an even more advanced processor on an even better process technology is released, and the hardware/software spiral continues.

9.3 Microprocessor Architectures

Another factor in the performance improvement of microprocessors is its microarchitecture. Microarchitecture refers to *how* a microprocessor's internal systems are organized. The microarchitecture is not to be confused with its instruction set architecture. The ISA determines what kind of software a given chip can execute. The earliest microprocessors (e.g., Intel 4004, 4040, 8008, 8080, 8086) were simple, direct implementations of the desired ISA. But as the process improvements implied by Moore's law unfolded, microprocessor designers were able to borrow many microarchitectural techniques from the mainframes that preceded them, such as caching (Intel's 486, MC68010), pipelining (i486 and all subsequent chips),

parallel superscalar execution (Pentium processor), superpipelining (Pentium Pro processor), and out-of-order and speculative execution (Pentium Pro processor, MIPS R10000, DEC Alpha 21264).

Microprocessor designers choose their basic microarchitectures very carefully because a chip's microarchitecture has a profound effect on virtually every other aspect of the design. If a microarchitecture is too complicated to fit a certain process technology (e.g., requires many more transistors than the process can economically provide), then the chip designers may encounter irreconcilable problems during the chip's development. The chip development may need to wait for the next process technology to become available. Conversely, if a microarchitecture is not aggressive enough, then it could be very difficult for the final design to have a high enough performance to be competitive.

Microarchitectures are chosen and developed to balance efficiency and clock rate. All popular microprocessors use a synchronous design style in which the microarchitecture's functions are subdivided in a manner similar to the way a factory production line is subdivided into discrete tasks. And like the production line, the functions comprising a microprocessor's microarchitecture are pipelined, such that one function's output becomes the input to the next. The rate at which the functions comprising this pipeline can complete their work is known as the pipeline's clock rate. If the functions do not all take the same amount of time to execute, then the overall clock rate is determined by the slowest function in the pipeline.

One measure of efficiency for a microarchitecture is the average number of clock cycles required per instruction executed (CPI). For a given clock rate, fewer clocks per instruction implies a faster computer. The more efficient a microarchitecture is, the fewer the number of clock cycles it will need to execute the average instruction. Therefore, it will need fewer clock cycles to run an entire program. However, the desire for high microarchitectural efficiency is often in direct conflict with designing for highest clock rate. Generally, the clock rate is determined by the time it takes a signal to traverse the slowest path in the chip, and adding transistors to a microarchitecture to boost its efficiency usually makes those paths slower.

Figure 127 illustrates the functional block diagram of the Intel 486, a very popular microprocessor of the early 1990s. (The microarchitectures of microprocessors that followed the 486, such as the Pentium processor or the Pentium Pro processor, are too complex to be described here.) The prefetch unit of the 486 fetches the next instruction from the instruction cache at a location that is either the next instruction after the last instruction executed or some new fetch address that was calculated by a previous branch instruction. If the instruction requested is not present in the cache, then the bus interface unit generates an access to main memory across the processor bus, and the memory sends the missing instruction back to the cache. The requested instruction is sent to the instruction decode

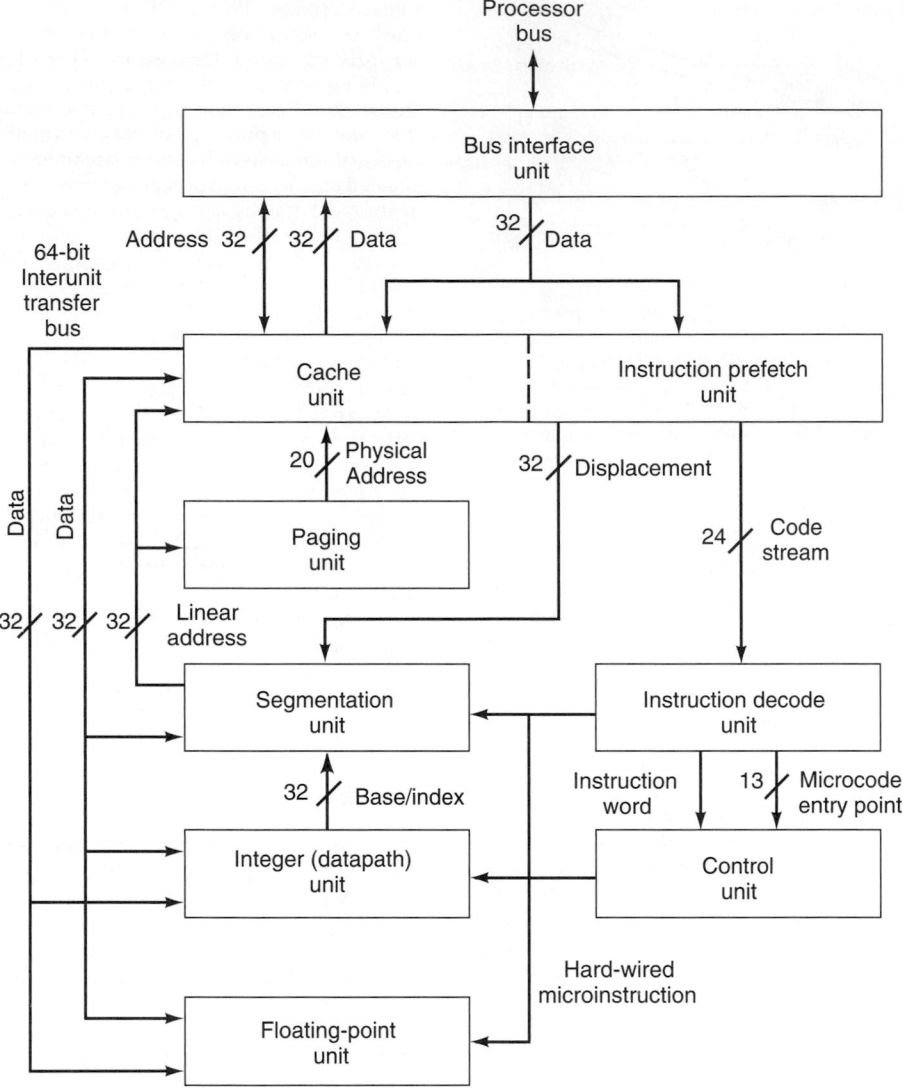

Fig. 127 Block diagram of the most popular microprocessor of the early 1990s, the Intel i486. The various blocks shown work together to execute the Intel Architecture instruction set with approximately 1.1M transistors. Newer designs, such as the Pentium processor, or the P6 microarchitecture at the core of the latest Pentium II processor, are much more complicated.

unit, which extracts the various fields of the instruction, such as the opcode (the operation to be performed), the register or registers to be used in the instruction, and any memory addresses needed by the operation. The control unit forwards the various pieces of the instruction to the places in the microarchitecture that need them (register designators to the register file, memory addresses to the memory interface unit, opcode to the appropriate execution unit).

Certain very complex instructions are implemented in an on-chip read-only memory called the microcode. When the instruction decoder encounters one of these, it signals a microcode entry point for the microcode unit to use in supplying the sequence of machine operations that correspond to that complex macroinstruction.

Although it is not obvious from the block diagram, the Intel 486 microarchitecture is pipelined, which allows the machine to work on multiple instructions at any given instant. While one instruction is being decoded, another instruction is accessing its registers, a third can be executing, and a fourth is writing the results of an earlier execution to the memory subsystem.

See Ref. 173–176 for sources of more details on designing microarchitectures.

9.4 Evolution of ISAs

Although microprocessor ISAs are crucial in determining which software will run on a given computer system, they are not static and unchangeable. There is a constant urge to develop the ISA further, adding new instructions to the instruction set or (much more rarely) removing old obsolete ones. Almost all old ISAs have many instructions, typically hundreds, some of which are quite complicated and difficult for compilers to use. Such architectures are known as complex instruction set computers (CISC). In the early 1980s, substantial academic research was aimed at simplifying ISAs [reduced instruction set computers (RISC)], and designing them with the compiler in mind, in the hopes of yielding much higher system performance. Some important differences remain, such as the number of registers, but with time the differences in implementations between these two design philosophies have diminished. RISC ISAs have adopted some of the complexity of the CISC ISAs, and the CISC designers borrowed liberally from the RISC research. Examples of the CISC design style are the Intel x86, the Motorola MC68000, the IBM System/360 and/370, and the DEC VAX. RISC ISAs include MIPS, PowerPC, Sun's Sparc, Digital Equipment Corp. Alpha, and Hewlett-Packard PA-RISC.

9.5 Coprocessors and Multiple Processors

Some microprocessor systems have included a separate chip known as a coprocessor. This coprocessor was intended to improve the system's performance at some particular task that the main microprocessor was unsuited for. For example, in the Intel 386 systems, the microprocessor did not implement the floating-point instruction set; that was relegated to a separate numerics coprocessor. (In systems that lacked the coprocessor, the microprocessor would emulate the floating-point functions, albeit slowly, in software.) This saved die size and power on the microprocessor in those systems that did not need high floating-point performance, yet it made the high performance available in systems that did need it, via the coprocessor. However, in the next processor generation, the Intel 486, enough transistors were available on the microprocessor, and the perceived need for floating-point performance was large enough, that the floating-point functions were directly implemented on the microprocessor.

Floating-point coprocessors have not reappeared, but less integrated hardware for providing audio (sound generation cards) and fast graphics are quite common in personal computers of the 1990s, which are similar to the coprocessors of the past. As the CPUs get faster, they can begin to implement some of this functionality in their software, thus potentially saving the cost of the previous hardware. But the audio and graphics hardware also improves, offering substantially faster functionality in these areas, so that buyers are tempted to pay a small amount extra for a new system.

9.6 High-End Microprocessor Systems

Enough on-chip cache memory and external bus bandwidth is now available that having multiple microprocessors in a single system has become a viable proposition. These microprocessors share a common platform, memory, and I/O subsystem. The operating system attempts to balance the overall computing workload equitably among them. Dedicated circuits on the microprocessor's internal caches monitor the traffic on the system buses, in a procedure known as "snooping" the bus, to keep each microprocessor's internal cache consistent with every other microprocessor's cache. The system buses are designed with enough additional performance so that the extra microprocessors are not starved.

In the late 1990s, systems of 1, 2, and 4 microprocessors became more common. Future high-end systems will probably continue that trend, introducing 8, 16, 32, or more microprocessors organized into clusters. As of the mid-1990s, the fastest computers in the world no longer relied on exotic specialized logic circuits but were composed of thousands of standard microprocessors.

9.7 Future Prospects for Microprocessors

From their inception in 1971, microprocessors have been riding an exponential growth curve in the number of transistors per chip, delivered performance, and growth in the installed base. But no physical process can continue exponential growth forever. It is of far more than academic interest to determine when microprocessor development will begin to slow and what form such a slowdown will take.

For example, it is reasonable to surmise that the process technology will eventually hit fundamental limitations in the physics of silicon electronic devices. The insulators most commonly used in an integrated circuit are layers of oxide, and these layers are only a few atoms thick. To keep these insulators from breaking down in the presence of the electric fields on an integrated circuit, designers try to lower the voltage of the chip's power supply. At some point, the voltage may get so low that the transistors no longer work.

Power dissipation is becoming an increasingly important problem. The heat produced by fast microprocessors must be removed so that the silicon continues to work properly. As the devices get faster, they also generate more heat. Providing the well-regulated electrical current for the power supply, and then removing the heat, means higher expense in the system. With the 486 generation, aluminum blocks with large machined surface areas, known as heat sinks, became commonplace. These heat sinks help transfer the heat from the microprocessor to the ambient air inside the computer; a fan mounted on the chassis transfers this ambient air outside the chassis. With the Pentium processor generation, a passive aluminum block was no longer efficient enough, and a fan was mounted directly on the heat sink itself. Future microprocessors must find ways to use less power, transfer the heat more efficiently and inexpensively to the outside, and modulate their operations to their circumstances more adroitly. This may involve slowing down when high performance is temporarily unnecessary, changing their power supply voltages in real time, and managing the program workload based on each program's thermal characteristics.

Microprocessor manufacturers face another serious challenge: complexity, combined with the larger and less technically sophisticated user base. Microprocessors are extremely complicated, and this complexity will continue to rise commensurate with, among other things:

- Higher performance
- Higher transistor counts
- The increasing size of the installed base (which makes achieving compatibility harder)
- New features to handle new workloads
- Larger design teams
- More difficult manufacturing processes

This product complexity also implies a higher risk that intrinsic design or manufacturing flaws may reach the end user undetected. In 1994, such a flaw was found in Intel's Pentium processor, causing some floating-point divides to return slightly wrong answers. A public relations debacle ensued, and Intel took a $475 million charge against earnings, to cover the cost of replacing approximately 5 million microprocessors. In the future, if existing trends continue, microprocessor manufacturers may have tens or even hundreds of millions of units in the field. The cost of replacing that silicon would be prohibitive. Design teams are combating this problem in a number of ways, most notably by employing validation techniques such as random instruction testing, directed tests, protocol checkers, and formal verification.

What really sets microprocessors apart from the other tools that humankind has invented is the chameleonlike ability of a computer to change its behavior completely under the control of software. A computer can be a flight simulator, a business tool for calculating spreadsheets, an Internet connection engine, a household tool to balance the checkbook, and a mechanic to diagnose problems in the car. The faster the microprocessor and its supporting chips within the computer, the wider the range of applicability across the problems and opportunities that people face. As microprocessors

continue to improve in performance, there is ample reason to believe that the computing workloads of the future will evolve to take advantage of the new features and higher performance, and applications that are inconceivable today will become commonplace.

Conversely, one challenge to the industry could arise from a saturated market that either no longer needs faster computers or can no longer afford to buy them. Or perhaps the ability of new software to take advantage of newer, faster machines will cease to keep pace with the development of the hardware itself. Either of these prospects could conceivably slow the demand for new computer products enough to threaten the hardware/software spiral. Then the vast amounts of money needed to fund new chip developments and chip manufacturing plants would be unavailable.

However, negative prognostications about computers or microprocessors have been notoriously wrong in the past. Predictions such as "I think there is a world market for maybe five computers" (Thomas Watson, chairman of IBM, 1943) or "photolithography is no longer useful beyond one micron line widths" have become legendary for their wrongheadedness. It is usually far easier to see impending problems than to conceive ways of dealing with them, but computer history is replete with examples of supposedly immovable walls that turned out to be tractable.

In its short life, the microprocessor has already proven itself to be a potent agent of change. It seems a safe bet that the world will continue to demand faster computers and that this incentive will provide the motivation for new generations of designers to continue driving the capabilities and applications of microprocessors into areas as yet unimagined.

ACKNOWLEDGMENTS

Various trademarks are the property of their respective owners.

10 OSCILLOSCOPES

Andrew Rusek

The oscilloscope, or the *scope*, is an important tool in engineering. The oscilloscope displays a graph of an electrical signal on a screen. In its basic operational mode, the oscilloscope graphs instantaneous signal values versus time. Displayed waveforms have three basic components: Y, the vertical or amplitude component; X, the horizontal or time component; Z, the signal brightness or intensity. Single channel and multichannel oscilloscopes are available. The most popular is a two-channel instrument for which a single trace display with channel multiplexing is available. In this case not only instantaneous signal values versus time, but also amplitude waveform component versus the other, can be observed.

Oscilloscopes are used in physics, electronics, instrumentation and measurements, communications, control, mechanical engineering, and many other areas where nonelectrical signal can be converted into electrical signals. The graphical display allows a user to determine many signal parameters, such as:

- Voltage and time values
- Frequency, period, and rise and fall times
- Phase shift
- Noise level, jitter
- Frequency ratio of two signals
- Calculated Fourier spectrum
- Calculated statistical parameters
- Calculated mathematical functions of signals, such as the integral, derivative, sum, difference, and product

Two major types of oscilloscopes have emerged over the last decade: analog and digital. Within these two categories, a number of operational modes can be distinguished. There are real-time data acquisition and display scopes, real-time acquisition and equivalent or transformed-time display scopes, and random data acquisition and random display scopes. A typical analog scope directly acquires signal and displays them on the screen in real time. An analog sampling scope can operate in real time and low signal frequencies, and in equivalent time at high frequencies. A digital scope can operate in real time or in equivalent time to acquire signals, but the display shows the reconstructed signal in the equivalent-time mode, which allows for great reduction of the display unit speed, which in turn achieves great reduction of the scope cost. Currently, the major oscilloscope manufacturers, such as Agilent Technologies, Tektronix, and LeCroy, do not list analog scopes in their catalogs but many analog scopes have been used recently and will be used in the near future. The digital scopes, also called digitizing or digital storage oscilloscopes, are considerably more powerful and flexible for signal inspection and analysis. Signal digitization has opened unlimited possibilities for signal storage and processing.

10.1 Analog Scopes

Analog scopes without sampling acquire and display waves in real time. The most critical part of the analog oscilloscope is a CRT, whose maximum speed of wave tracing limits the speed of the entire scope. Figure 128 shows the simplified structure of a CRT. The beam of electrons generated by the cathode is directed toward the fluorescent screen by a set of electrodes. The beam makes a bright spot on the screen at a position controlled by the voltages applied to the vertical and horizontal deflection plates. The tube can have a single or a double beam system. The latter has two separate beam-forming and deflection structures. The tube shown in Fig. 128 employs a two-stage electron acceleration process. The electrons emitted from the cathode are formed into the beam, whose density is controlled by the cylindrical electrode. Later, the beam is focused and deflected, passing through the vertical and horizontal deflection plates. Final acceleration, called postdeflection acceleration (which is necessary to produce a visible spot on the screen), is achieved by applying very high voltage, at least several kilovolts, to a distributed electrode located inside the tube between the deflection plates and the screen. The inner surface of the screen is covered with a phosphor, which emits visual radiation on being bombarded by electrons. The length of time during which the phosphor emits radiation until its level decays to 10% of the initial value is called the persistence. Several types of phosphor inner coatings are applied in CRTs. The most popular phosphor, P 31, exhibiting high luminance and reasonable persistence of about 32 ms,[177] is found in most CRTs.

In analog scopes, an image of a nonrepetitive signal fades quickly or slowly, depending upon the tube persistence. Fast single-shot events may not be registered by a classical CRT whose screen directly reacts to the

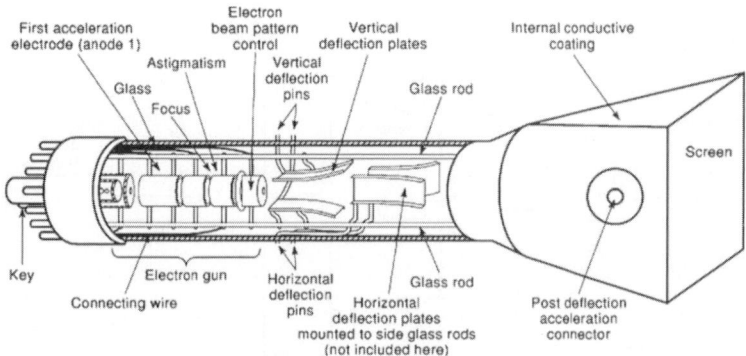

Fig. 128 Internal structure of a CRT.

electron current delivered to it. In order to enhance the quality of writing, microchannel plate CRTs are used in some analog scopes.[178] Inside the tube, an array of hollow glass fibers is located. The fibers internally coated with a semiconductor, multiply the number of electrons through secondary emission when activated by an adequately accelerated electron beam. The beam deflection can be affected by the proximity of power supplies, especially transformers, and chokes. To avoid interference, the tubes and potentially interfering components are well shielded.

The block diagram of a typical analog scope in shown in Fig. 129. The vertical channels, A and B,

include input coupling circuits, attenuators, preamplifiers, delay line, and differential output amplifiers driving the vertical plates of the CRT, as shown in Fig. 130. Quite often, voltage or current probes are connected to the inputs to provide better interfacing with the tested circuits. The horizontal section of the oscilloscope is composed of triggering unit with its coupling circuits, trigger shaping unit, trigger level control, and trigger holdoff units; a sweep generator, which defines the main time base of the scope; and a differential output amplifier driving the horizontal deflection plates of the tube (Fig. 131). The main sweep generator can be supported by a fast time-base

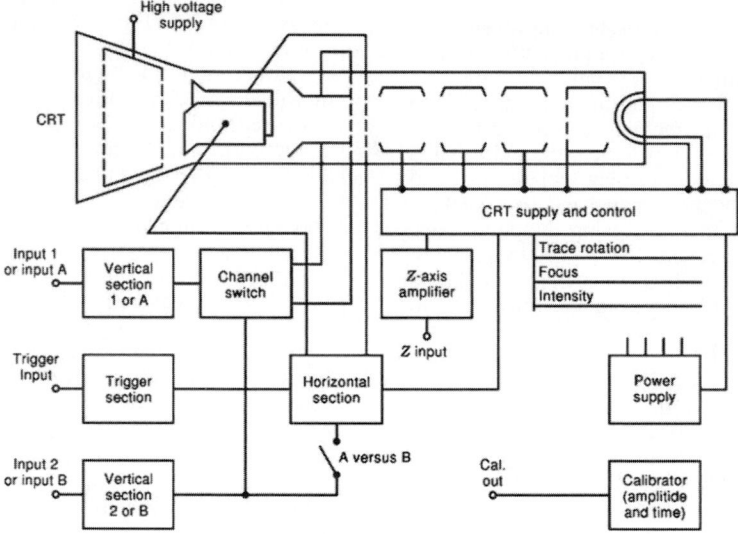

Fig. 129　General diagram of an analog oscilloscope.

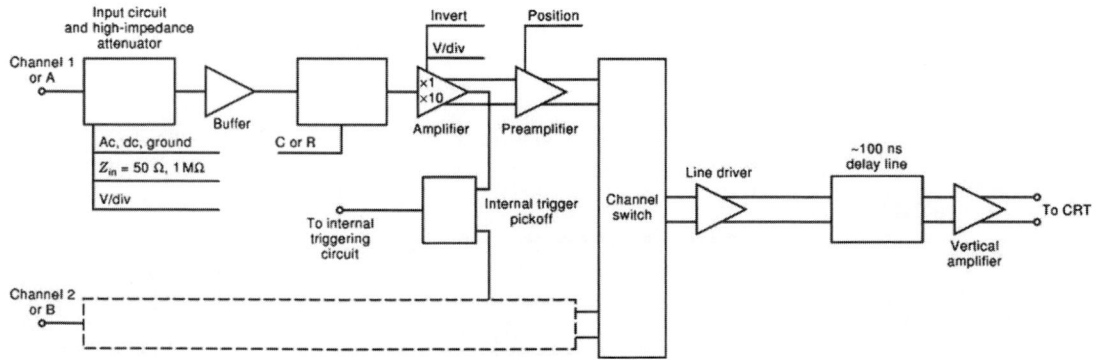

Fig. 130　Vertical channel of an analog oscilloscope.

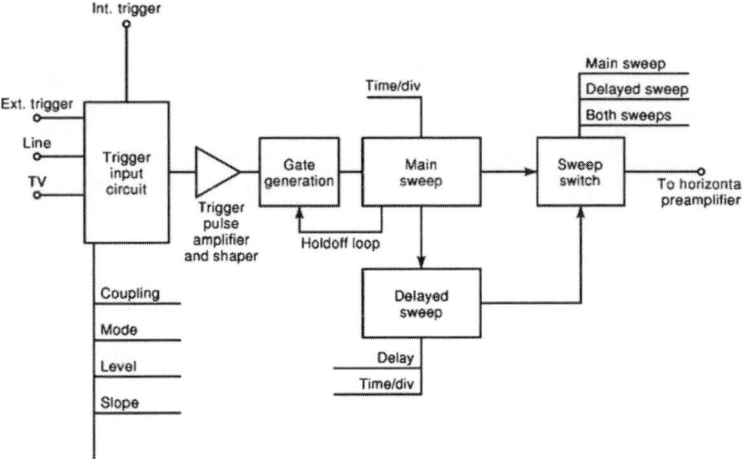

Fig. 131   Horizontal channel of analog oscilloscope.

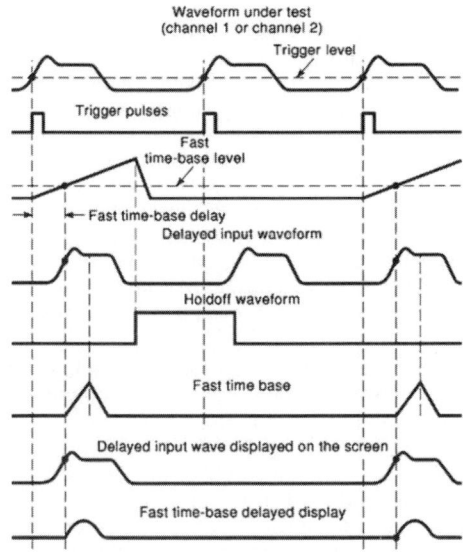

Fig. 132   Waveforms in analog oscilloscope.

unit triggered at a preselected level to expand desired parts of the tested waves. Figure 132 illustrates typical waveforms of analog scopes. Two time bases are activated to show distinguished sections of the signals under test with the help of the fast ramp, which is called the delayed time base. The waves observed on the tube screen are delayed in relation to the input signals due to the action of the delay lines forming parts of the vertical amplifier. In this way the trigger and the time-base circuit delays and nonlinearities

are compensated to create good conditions for tests of the wave leading edges without introducing distortions. Figure 132 also shows internally generated triggering pulses initiated by the input signal of channel 1 or 2.

To stabilize the horizontal wave position on the screen, the trigger level is selected within the frame of the vertical span of the incoming waves. At the same time an auxiliary system, called *holdoff*, has to be adjusted to make the time-base generator insensitive to excessively frequent triggering. If more than one channel is used and a single-trace tube is applied, then a channel switch can direct only one channel to the screen at a time for the duration of the time base, especially at high rates of the time base (alternating mode). At lower time-base rates small segments of the individual channels can be displayed one by one (chopped mode). The waves for both modes of the display are shown in Fig. 133. Multichannel data acquisition allows one to perform basic hardware arithmetic operations on signals, such as addition and subtraction. The process of subtraction, which can disclose a differential signal component, is corrupted by interchannel interference due to the common ground pathways and nonideal channel isolation of the channel switch. Many analog oscilloscopes have internal amplitude calibrators, which help check the scope amplitude calibration and verify or adjust the compensation of the voltage probes.

10.2   Sampling Scopes

Typical real-time analog oscilloscopes have operating bandwidths up to several hundred megahertz. The limits are mainly imposed by the CRT writing speed. One of the means to reduce the speed of observed waves has been the time transformation by means of coherent sampling. The method can be only applied to repetitive waveforms. A basic set of waveforms for a

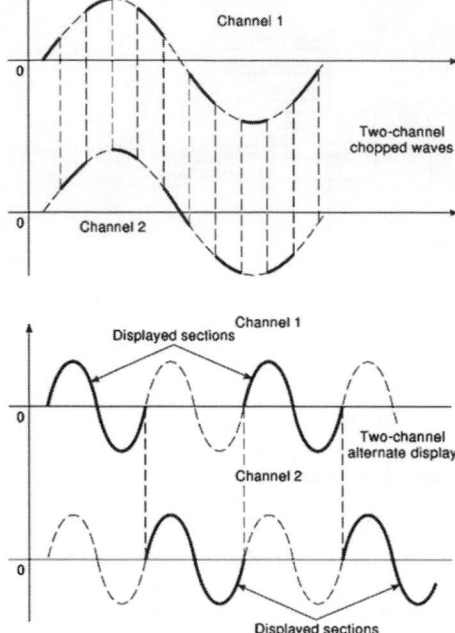

Fig. 133 Modes of display of analog oscilloscope.

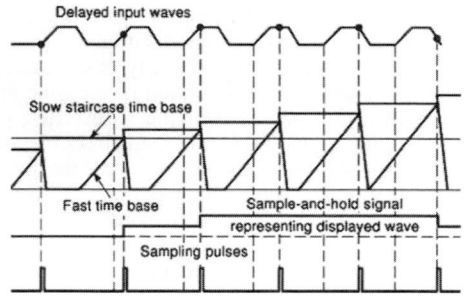

Fig. 134 Time transformation in analog sampling scope.

sampling scope is shown in Fig. 134; the scope block diagram is depicted in Fig. 135. The signal under test applied to the input is delayed by a broadband transmission line, and then sampled by very narrow pulses in a sampling gate. If the repetition frequency is less than several hundred kilohertz, which is the maximum repetition frequency of the trigger circuitry, a single sample is taken from each pulse of a frame defined by the time base. For higher repetition rates the trigger circuits divide the frequency so that every nth one sample is taken from a single pulse of a repetitive wave. When the samples are taken, they are stretched,

amplified, and memorized in a sample-and-hold circuit. Later, the samples are applied to low-frequency vertical amplifiers driving the tube. The sampling scope system uses a track-and-hold feature that enables the scope to process the small samples formed by the differences between the two consecutive values of the sampled signal. Generation of subsequent sampling pulses is achieved by comparison of two fully synchronized time bases. One of them is a fast time base, and the other is a staircase signal. The horizontal channel provides the timing of all samples. It also delivers trigger signal to the sampling pulse generator and to the time bases. Sampling oscilloscopes can operate up to frequencies above 70 GHz of an equivalent bandwidth, which corresponds to the rise time of 5 ps. Both, vertical and horizontal channels of such, scopes have been digitized and new generations of sampling scopes have been available for the last few years.[179] They are applied not only in the typical scope measurements but also in time domain reflectometry (TDR).

10.3 Digital Scopes

Digital oscilloscopes have overcome many disadvantages of analog scopes, especially the necessity for using long persistence CRTs and delay lines. A digital scope display can be adjusted after wave acquisition and storage. The input circuitry of a typical digital scope does not differ from that of analog scope. After preliminary attenuation or amplification, analog signals are sampled and digitized. The digitized samples are stored in a memory. The display receives the reconstructed analog samples of vertical and horizontal information through DACs. The process of data display is delayed and stretched in time, so the samples of the signal are displayed at a slower rate than they are acquired. Three major types of digitizers are applied in digital scopes:

- Real-time digitizers operating at high sampling and conversion rates, which can capture nonperiodic short transient processes.
- Coherent sampling digitizers working in the same way as the sampling oscilloscopes with time transformation described before. The A/D conversion is applied to much slower waves after the sampling and time transformation. In this approach, a pretrigger pulse is required; it is obtained either from the tested signal or from a separate source whose output wave follows the trigger.
- Random interleaved sampling digitizers, which sample repetitive signals in a pseudorandom manner. Amplitude and timing data are digitized and memorized in corresponding memories and later displayed to follow the stored timing sequence.

All three digitizers can equally well acquire periodic signals. In the last two approaches, only a single sample can be taken during the input signal cycle, especially at low frequencies of very short input

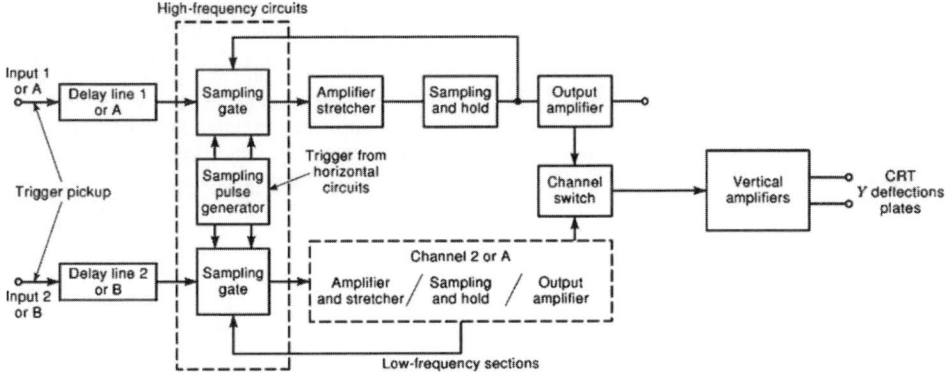

Fig. 135 Vertical system of coherent sampling oscilloscope.

pulses. The real-time signal capture capability of most digital scopes is limited by the maximum sampling rate, which is directly associated with the maximum speed of the analog-to-digital converters. The inherent storage capability of digital scopes, especially the possibility of storing long records of a number of channels, allows for processing the signals applying complex mathematical operation, including Fourier analysis and statistical functions.[180] When the oscilloscope calculates the FFT of the input signal, the frequency spectrum can be displayed separately from the time domain signal image. A typical block diagram of a two-channel digital storage scope DSO is shown in Fig. 136. The channels acquire and digitize the signals independently, and then the digital signals are accumulated and processed according to the chosen function. In the simplest case, the digitized signals are retrieved

from the memory, converted back into their analog forms by the DACs, and displayed. The reduced speed of the signal display processes eliminates the need for high-frequency CRTs. Many scope displays utilize LCD tubes, which are directly controlled by addressing the picture elements (pixels) of a display matrix. Analog scopes with CRTs can update their pictures about 10^5 times per second. Time changes in the signals are then recorded immediately. The display process of a typical digital storage scope is usually a memory reconstruction of the acquired samples taken from many periods, so the entire image can be acquired only 5–10 times per second. High quality of the picture display of analog scopes requires rather high frequency of image updates. Such update frequencies are not available when the signal is not repeated frequently. Some digital storage scopes with high-speed digitizers and

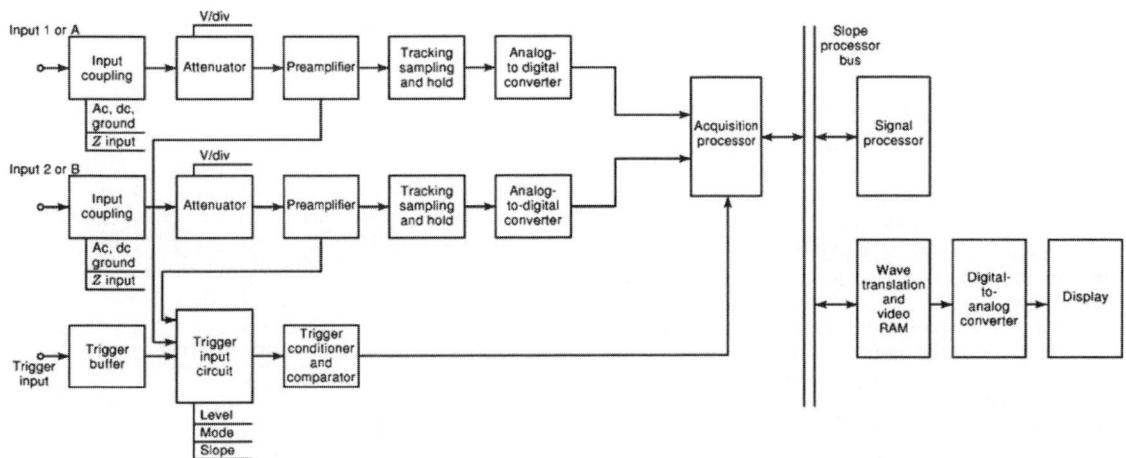

Fig. 136 General block diagram of digital oscilloscope.

processors, and with smart display algorithms, update their displays by creating high-quality analog persistence images over a broad range of signal frequencies. The number of samples displayed on the screen is fixed independently of the time-base settings, which leads to changes of the sampling rates for different time bases. In order to reproduce periodic waves, taking 5–10 samples per period of the highest harmonic is regarded as the minimum to preserve sufficient signal details. The Nyquist sampling rate is too low to reconstruct periodic signals in all cases, especially when the signals are sampled near zero crossings. Major features of the digital storage scopes can be summarized as follows:

- Large amounts of pretrigger information can be acquired without delay lines.
- Single or repetitive waves can be stored permanently.
- Computer or printer interfacing to store or print more data can be easily achieved.
- One may reliably capture unpredictable events.
- One may reliably capture glitches shorter than the resolution time by means of peak detectors.
- Within the memory capacity, there is no dead time between acquired events.
- Measurements can be automated.
- Currently acquired waves can be compared with reference waves.
- Wave tolerance tests are possible, including go-no-go tests.
- Acquired waves can be processed mathematically, including statistical operations.
- Finite resolution and a small number of display sweeps per second can mask the time distribution of waveforms.
- The limited number of samples and the constant sample intensity commonly used in displaying simple repetitive waves can distort composite repetitive signals, including waves of various speeds.

10.4 Technical Parameters and Limitations

This section describes basic parameters of the oscilloscopes to help select the right one for an application. In the selection process, the scope parameters must be related to the waves under test. The major specifications of analog scopes are number of channels, bandwidth, voltage sensitivity, maximum time-base speed, amplitude noise, time jitter, and overall accuracy. From these parameters other scope features can be derived. For digital scopes, the same set of parameters is also important, but maximum sampling rate, vertical and horizontal resolutions, and memory length should be considered.

Number of Channels Common-purpose single-channel scopes are not very popular in measurement

laboratories. If the number of channels is greater than one, the interchannel interference becomes important and a high CMRR is desired. For example, if CMRR = 40 dB, then with one channel receiving a signal of 1 V and the other channel receiving a signal of 10 mV, the lower amplitude channel will measure the amplitude with an error of about 10%, which is well above scope error specifications.

Bandwidth The BW is usually defined as the frequency range between the dc and the frequency at which a sinusoidal signal amplitude observed on the screen is reduced 3 dB in relation to the value at dc or low frequencies. The input sinusoidal signal is supplied from a source having very low internal resistance, normally 50 Ω. The bandwidth of an analog scope is primarily determined by the frequency response parameters of its vertical amplification channel, including attenuators and the tube or other display. The magnitude response error introduced by the frequency limitations is shown in Table 32. The frequency response bandwidth relates to an equivalent time domain parameter, namely, the rise time, usually considered as the time the step response changes between 10 and 90% of the steady-state level. For the simplest frequency response approximation, the relationship between the rise time and the bandwidth is as follows:

$$T_B = \frac{0.35}{\text{BW}}$$

For instance, for a 100-MHz scope the rise time is 3.5 ns. This leads to time domain measurement error, whose effects can be corrected for using the following expression:

$$T_T = \sqrt{T_B^2 + T_R^2}$$

where T_T = actual rise time
T_0 = measured rise time
T_R = scope's own rise time

Table 33 shows the time domain error calculated without corrections.

Both tables suggest that in order to maintain reasonable measurement accuracy, the scope should be at least several times faster than the fastest tested wave. Frequency domain or time domain speed limitations of digital scopes are additionally imposed by the sampler

Table 32 Magnitude Response Error

Signal Freq. for 3 dB BW (MHz)	Error (%)
0.1	1
0.2	2
0.5	11
1.0	29

Table 33 Time Domain Error

$T_R{}^a$ (ns)	T_T (ns)	T_0 (ns)	Error (%)
3.5	1	3.6	260
3.5	2	4	100
3.5	5	6.1	22
3.5	10	10.6	6
3.5	20	20.3	1.5

^{a} BW 100 MHz.

sampling speed and the ADC speed. The sampler includes the sample-and-hold unit, whose sampling pulse width determines the fastest change that can be captured and later converted in the ADC. Majority of digital scopes use flash or parallel ADCs. With ideal sampling pulses, the sampling rate should be several times greater than the expected maximum frequency of the tested signals. The sampling rate also determines the scope horizontal resolution.

Vertical Sensitivity This parameter expresses the scope ability to accommodate small signals as well as large signals within the same space of the screen. Typical ranges of scope sensitivities are between several millivolts per division and 20 V per division. The sensitivity range is a compromise between the user's desired values and the technical limitations imposed by the noise level of the scope amplifiers and the maximum available attenuation of high-frequency compensated attenuators.

Amplitude Noise Level The scope noise mainly includes the broadband noise generated by the input stages of the scope and by the source resistance. The observed noise levels are usually on the order of 1 mV. Bandwidth reduction decreases the noise level and allows increasing the scope sensitivity. In digital scopes, the input analog amplifiers introduce their noise, thus degrading the resolution of the ADC conversion process. For example, 1 mV noise on a 1-V signal limits the resolution of the conversion to less than 12 bits.

Time Jitter Time jitter, or jitter, is caused by threshold instabilities in triggering and time-base circuits. The jitter is observed on the screen as a spread of vertical parts of the tested signals. In digital scopes, the clock jitter adds its component to this time domain instability.[181] The levels of jitter are on the order of picoseconds for fast sampling scopes and on the order of nanoseconds for low-frequency scopes.

Accuracy The inaccuracy of a scope is mainly caused by vertical and horizontal calibration and reading errors. The reading errors of digital scopes, with built-in digital amplitude and time meters, are negligible in comparison with calibration errors of amplifiers, voltage attenuators, and input resistors. The total

amplitude error can reach several percent, but the dc voltage error can be as low as 1%. The time-base errors in digital scopes are comparable with the errors of digital counters, which are below 0.1%,[182] while the time-base errors in analog scopes are usually 2–5%. The process of selecting a scope is primarily determined by the information about the waves under test: their shapes, repetitivity, frequency and amplitude, drift, and speed. Additional conditions that should be considered are the length of the record, possibility of test automation, signal statistics, and wave storage. The scope capabilities and parameters should be also chosen with regard to future applications and affordability.

10.5 Oscilloscope Probes

Almost all standard oscilloscopes use different voltage probes to connect to a circuit or a device under test, chosen to minimize the effects of the scope input impedance and electromagnetic interference. Some probes convert a nonelectrical quantity or another electrical quantity, such as current, into a voltage, which is measured by the scope. Two major types of voltage probes are used. The first type is a passive probe, which yields high input resistance and low input capacitance but introduces significant signal attenuation, usually 10 times. The second type is an active probe, which does not attenuate the signal but provides high input resistance and low input capacitance. Unfortunately, the supply voltage required for active probe devices also introduces undesired noise and offset. Active probes usually employ FETs and broadband operational amplifiers.[183] A typical voltage probe has two input terminals: a ground terminal and a signal terminal. In many cases, tests of differential signals cannot be done by means of two-channel differential measurement, due to the limitations of the CMRR. For these tests, differential active probes, which have very high CMRR, are recommended.

Passive Probes Passive probes include low-impedance resistive voltage divider probes and compensated high-impedance voltage divider probes. There are also high-impedance probes with a very large voltage division factor, which are used in high-voltage measurements. Figure 137 shows a low-impedance passive probe. This type of probe has very low capacitive loading and wide bandwidth. It can be used in testing low-impedance circuits. High-impedance passive probes are shown in Figs. 138 and 139. In both cases,

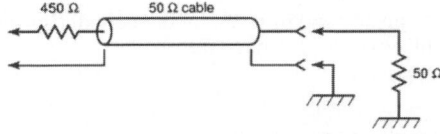

Fig. 137 Low-impedance passive probe.

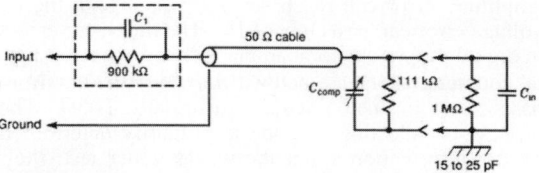

Fig. 138 Passive probe, 1 MΩ input resistance.

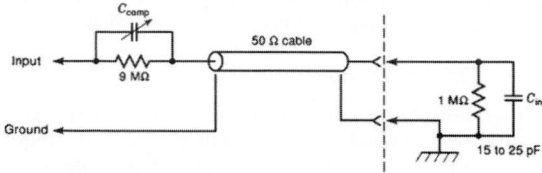

Fig. 139 Passive probe, 10 MΩ input resistance.

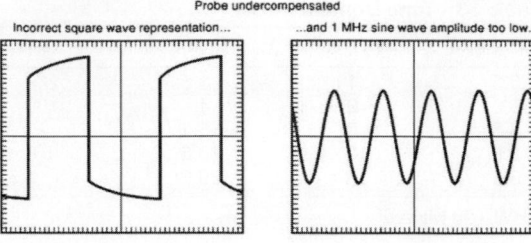

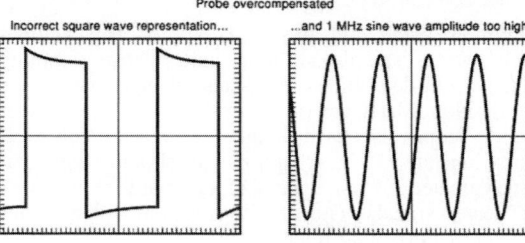

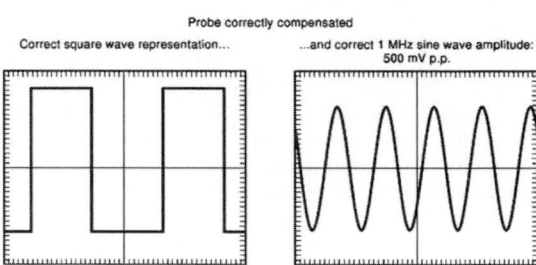

Fig. 140 Compensated probe waveforms.

the probes require compensation before the measurement process starts. The compensation is performed by observing a standard square wave signal generated by the scope calibrator. The effects of the pulse compensation on the square wave and the sine wave are shown in Fig. 140. High-voltage probes have similar structure to that in Fig. 140. The series resistor is of much higher value and larger physical size, to withstand voltages higher than 500 V. A high-voltage probe circuit is shown in Fig. 141.

The parameters of passive probes are:

- Voltage attenuation (e.g., 10× or 1×)
- Bandwidth for 50-Ω source resistance
- Maximum input voltage (e.g., 500 V)
- Input resistance (e.g., 1 MΩ)
- Input capacitance (e.g., 2 pF)

Active Probes Figure 142 is a diagram of an active probe with a single input. A differential probe is shown in Fig. 143. Both probes have very large input resistance, broadband, and no attenuation. The differential probe also has high CMRR, greater than 60 dB.

10.6 Oscilloscope Measurements

Scopes can measure the following basic quantities:

- Voltage
- Current (with a current probe)
- Frequency and time, including period, time delay, pulse width, and duty factor
- Phase shift
- Rise and fall times
- Amplitude modulation index

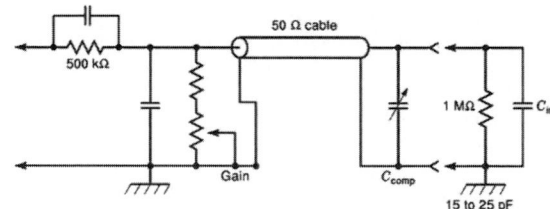

Fig. 141 High-input-resistance high-voltage probe.

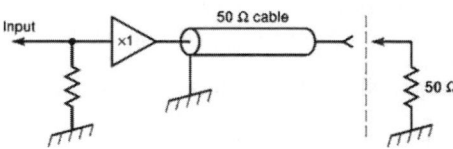

Fig. 142 Diagram of active probe.

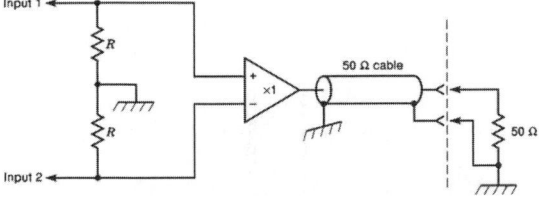

Fig. 143 Differential probe.

- Jitter
- Eye diagram to evaluate quality of communication signals

Voltage Voltage measurements are elementary tests done with oscilloscopes. Direct current voltages are measured by observing the level shift of the horizontal line on the screen when the input of the scope is dc coupled to the source. The screen graticule and the vertical sensitivity setting indicate the amount of shift. Modern scopes have internal digital voltmeters indicating numerical values of the beam position and the positions of the different markers. In older scopes, ac voltages are measured in terms of the voltage differences, which are calculated with the aid of the graticule. In modern scopes, the marked levels are measured digitally and displayed numerically on the screen.

Current Current measurement requires the use of current-to-voltage converting devices, such as Hall effect sensors for dc currents, current transformers clamping sensors for ac currents, or combinations of both. The Hall effect current sensors are supplied from the scope or from a separate power supply, which also incorporates additional amplifiers.

Frequency and Time With older scopes, the time domain parameters can be measured using the screen reading in relation to the current time-base setting. Modern scopes, especially digital, have digital readout and they display the time domain parameters with reference markers. Rise and fall times are measured with the help of horizontal levels marked at 10 and 90%. Figure 144 illustrates the way the rise time is evaluated.

Phase Shift The phase shift between two waves is the amount of time, expressed in degrees, that passes between the beginnings or other selected points of the two waves. The phase shift is thus obtained from two time measurements in which the relative position and the period of the waves are found and phase angles are calculated. Another method, which can be used only for coherent sinusoidal signals, involves the XY display mode. One signal is applied to the horizontal

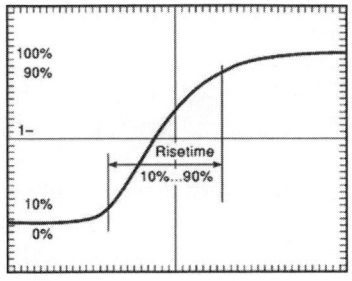

Fig. 144 Measurement of rise time.

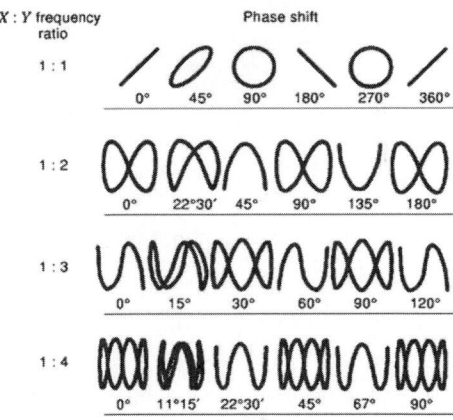

Fig. 145 Phase measurement applying XY display.

channel, the other to the vertical channel. The XY patterns displayed on the screen are called Lissajous graphs, and from the shape of the pattern the phase shift and frequency can be determined. Figure 145 shows different Lissajous patterns.

Amplitude Modulation (AM) Index Figure 146 shows a time domain AM wave displayed on the screen while the external triggering of the scope is synchronized by means of a modulating lower frequency wave. The modulation index is calculated as a ration of M to C. It is also possible to display the XY pattern in which the modulated wave is applied to the scope vertical system, and the modulating signal is applied to the horizontal system. Figure 147 shows a typical AM wave pattern for this method.

Eye Diagram The eye diagram is a very useful visual method, which is often applied in digital communication systems, to evaluate quality of the information data. The vertical channel of the scope displays the demodulated and filtered bipolar symbol stream,

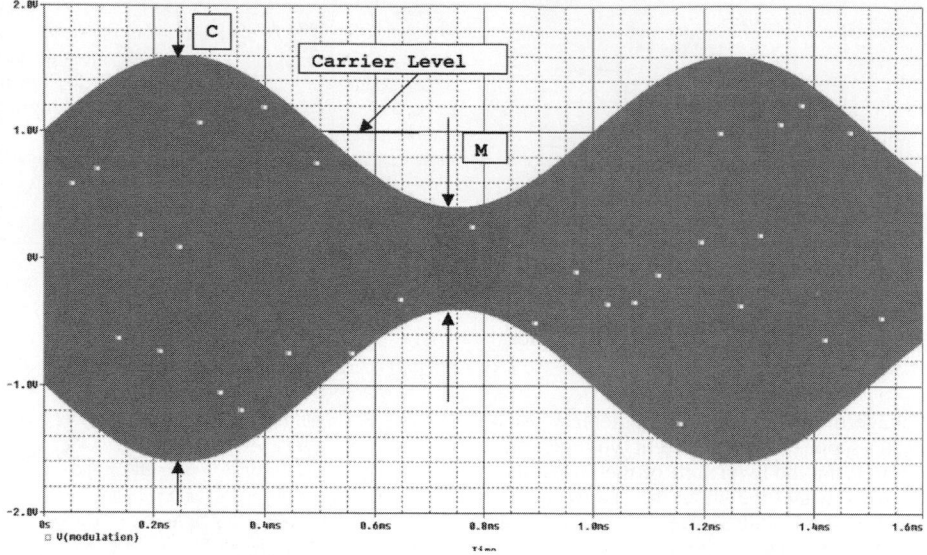

Fig. 146   Time domain measurement of AM depth (index).

but the horizontal channel is triggered at every symbol period or every multiple of the symbol period (Fig. 148). Thanks to the screen persistence the consecutive symbols of various durations can be observed and amplitude and time domain distortions evaluated.

10.7 Programmability of Oscilloscopes

Most modern digital scopes can be controlled by computers. The computer provides more flexibility in storing data and in programming and controlling experiments. Advance interfacing techniques involve applications of GPIB or IEE-488 cards and connections controlled by the dedicated software, like VEE of Agilent Technologies or LAB VIEW of National Instruments.[184,185]

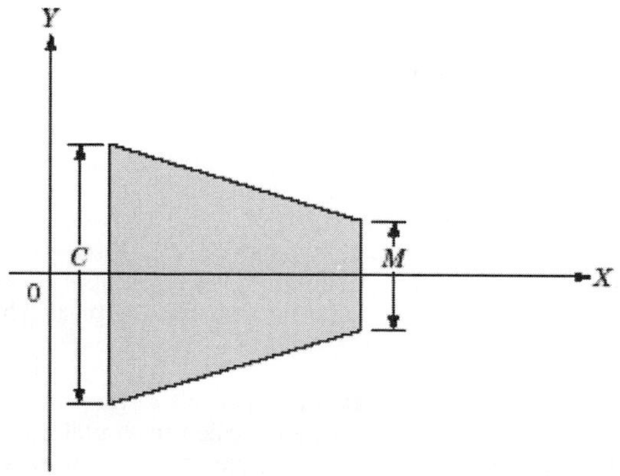

Fig. 147   Measurement of modulation depth (index) applying *XY* display.

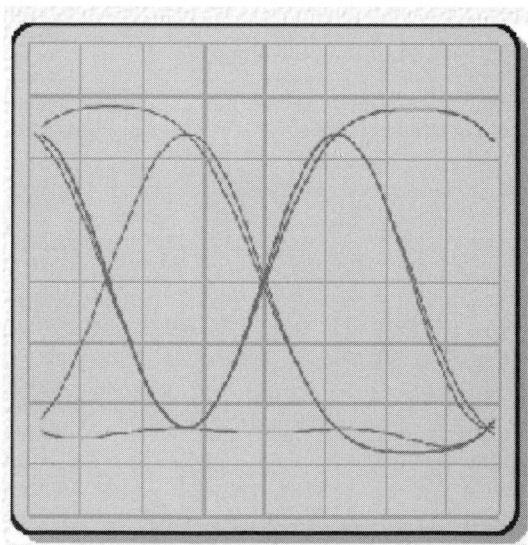

Fig. 148   Eye diagram.

11 POWER DEVICES

Alex Q. Huang and Bo Zhang

11.1 Rectifiers

The semiconductor rectifier[186] was the first semiconductor device developed for power circuit applications. It is a semiconductor device specifically designed to rectify alternating current; that is, to exhibit a very low resistance to current flow in one direction and a very high resistance in the other direction.

P–i–N Rectifier   In the past, only $P–i–N$ rectifiers were available for use in power circuits. The first devices were made by using germanium. However, the high leakage current in germanium devices associated with its small energy bandgap (0.66 eV) led to their replacement by silicon $P–i–N$ rectifiers. Since the 1950s, the performance of silicon $P–i–N$ rectifiers has been continually improving due to the optimization of the device structure and lifetime control[187] that is used to adjust the switching speed. The basic $P–i–N$ rectifier structure is shown in Fig. 149. The doping concentration and the thickness of the i region (N^- region) are designed to support the required reverse blocking voltage. In order to support large reverse blocking voltage, it is necessary to use a low doping concentration as well as a large thickness for the i region. In the on state, minority carrier holes are injected from the P^+ anode and electrons injected from the N^+ cathode; they are equal in number to maintain charge neutrality. This phenomenon of injecting a high concentration of holes and electrons into the i region, called conductivity modulation, is an extremely important effect

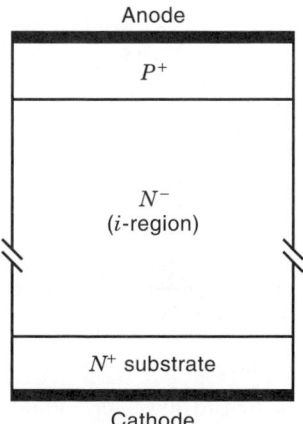

Fig. 149 Basic structure of the $P–i–N$ rectifier. The lightly doped thick i region is designed to support the required reverse voltage. The heavily doped p^+ anode and n^+ cathode inject carriers into the i region to modulate its conductivity during forward conduction.

that allows the transporting of high currents through the $P–i–N$ rectifiers with breakdown voltages of up to 5000 V. In the steady-state reverse blocking state, the reverse current is only the leakage current due to minority carrier generation in the depletion region, and the $P–i–N$ rectifier exhibits a very high resistance to current flow.

As described, the injection of a high concentration of minority carriers into the i region can increase the conductivity of the i region. However, the injected carriers also create problems during switching of the $P–i–N$ rectifier. When the voltage across the device reverses polarity, the injected electron–hole population (called stored charge) must be removed before the formation of a depletion region can occur to support the reverse blocking voltage. This leads to a reverse recovery current. The presence of such a reverse recovery transient leads to power dissipation that limits the maximum switching frequency of $P–i–N$ rectifiers and degrades the reliability of the applied circuits, an additional concern is the large voltage overshoot caused by the di/dt of the reverse recovery current flowing through the stray inductance in the circuit. When the switching frequency of a power circuit increases, the turn-off di/dt must be increased. It has been found that this causes an increase in both the peak reverse recovery current and the ensuing reverse recovery di/dt. If this reverse recovery di/dt is large, an increase in the breakdown voltage of all the circuit components becomes essential. Raising the breakdown voltage capability causes an increase in the forward voltage drop of power devices, which in turn degrades the system efficiency because of a higher conduction loss. Many methods of lifetime control have been developed to reduce the minority carrier lifetime in the i region to decrease the switching loss of the $P–i–N$ rectifier, but they also lead to an increase in the on-state voltage drop. It is therefore customary to perform a tradeoff between on-state and turn-off losses when designing $P–i–N$ rectifiers.

Another drawback of the $P–i–N$ rectifier is the forward voltage overshoot during its turn on. The forward voltage overshoot in a $P–i–N$ rectifier arises from the existence of the high resistance i region. Under steady-state current conduction, the i-region resistance is drastically reduced by hole–electron population injected by the N^+ and the P^+ regions. However, during turn on under high di/dt condition, the current rises at a faster rate than the diffusion of the minority carriers injected from the junction. A high voltage drop develops across the i region for a short period of time until the minority carriers can swamp out the i-region resistance.

Schottky Barrier Diode   In order to eliminate the reverse recovery problem associated with $P–i–N$ rectifiers, the Schottky barrier diode (SBD, Schottky rectifier) was developed in the 1970s. The basic structure of the SBD is shown in Fig. 150. It consists of a metal–semiconductor rectifying contact with

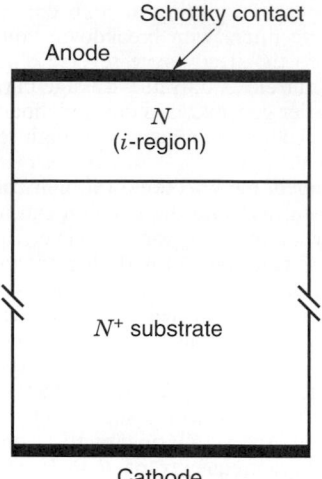

Fig. 150 Basic structure of the Schottky barrier diode. A specially selected anode metalization (such as tungsten, aluminum) forms a Schottky barrier with the lightly doped N region.

an N-region designed to support the required reverse blocking voltage. When a positive bias is applied to the metal with respect to the N-type semiconductor, forward conduction in the SBD occurs by thermal emission of majority carrier electrons across a lowered metal–semiconductor barrier. The on-state voltage drop of the SBD therefore consists of the sum of the voltage drop across the barrier and the ohmic voltage drop across the N region. There is no conductivity modulation of the N region in the SBD structure because the minority carrier injection is negligible. The N region resistance depends on the reverse blocking voltage. For low breakdown voltages (<100 V), the doping concentration of the N region lies between 5×10^{15} and 1×10^{16} cm^{-3} and its thickness can be made to less than 10 μm. This leads to a relatively small voltage drop in the N region. For the case of a typical SBD with a reverse breakdown voltage of 50 V and a barrier height of 0.8 V, the forward voltage drop is about 0.554 V at a forward conduction current density of 100 A/cm^2. This lower forward voltage drop compared to a $P–i–N$ rectifier (whose forward voltage drop at 100 A/cm^2 is about 0.9 V) and its faster switching speed (because of the absence of large reverse recovery current observed in a $P–i–N$ rectifier) make the SBD attractive in low-voltage switching applications.

With the increase of the reverse blocking capability, the forward voltage drop of the SBD will increase rapidly and will approach that of a $P–i–N$ rectifier when the reverse blocking capability is increased to 200 V. In addition, the Schottky rectifier has a larger reverse leakage current, which also increases with

the increase of the temperature, and a soft blocking characteristic. These make the silicon Schottky rectifier generally unacceptable for use in high-voltage applications.

As discussed previously, the silicon Schottky rectifier eliminates the reverse recovery problem that limits its high-frequency application, but the forward voltage drop of the silicon Schottky rectifier increases rapidly with the increase of the reverse blocking capability. A much superior power rectifier can be created by using SBD contact if the resistance of the N region can be reduced while achieving the same blocking capability. An approach for achieving this is to replace silicon with a wide-bandgap semiconductor. Based on this fundamental analysis, it was demonstrated that a gallium arsenide (GaAs) Schottky rectifier has a better forward voltage drop than a silicon $P–i–N$ diode for blocking voltages of up to 500 V.

The mobility of electrons in gallium arsenide at low field is larger than in silicon by a factor of 5.6. In addition, because of the larger energy bandgap, the critical electric field for breakdown in gallium arsenide is higher than in silicon. These two facts result in a reduction in the specific on resistance (the on resistance per unit area) of the N region by a factor of 13. For a typical Schottky barrier height of 0.8 V, the gallium arsenide Schottky rectifiers are expected to have a lower forward drop than silicon $P–i–N$ rectifiers for breakdown voltages of up to about 500 V at a typical operating current density of 100–200 A/cm^2. In this voltage range, the gallium arsenide rectifiers offer a clear advantage over $P–i–N$ rectifiers due to faster switching speed because of the absence of the reverse recovery current. The fabrication of gallium arsenide Schottky barrier power rectifiers can be accomplished by using aluminum or titanium Schottky barrier contacts. These devices are now commercially available.

An even more promising material for power rectifiers is silicon carbide (SiC) because of its much higher critical electric field at breakdown. The on-state voltage drop of the silicon carbide Schottky barrier power rectifier is superior to that of the silicon $P–i–N$ power rectifier for a blocking voltage up to 3000 V. The silicon carbide Schottky barrier power rectifiers have also been found to have excellent reverse recovery and reverse bias leakage characteristics even at high operating temperatures. They are likely to replace silicon $P–i–N$ rectifiers in high-voltage power electronic circuits in the next decade.

11.2 JBS Rectifiers

With the trend toward lower operating voltages for VLSI chips, there is an increasing demand to reduce the forward voltage drop in rectifiers. The forward voltage drop of a Schottky rectifier can be reduced by decreasing the Schottky barrier height. Unfortunately, a low barrier height results in a severe increase in leakage current and a reduction in maximum operating

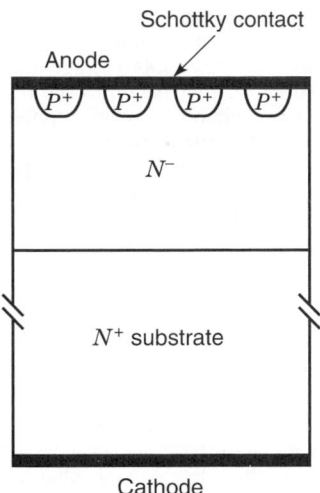

Fig. 151 Cross-sectional view of JBS rectifier structure. The p^+ junction grid is designed so that its depletion layers do not pinch-off under zero and forward bias conditions of the rectifier, but intersect with each other underneath the Schottky contact when the reverse bias exceeds a few volts.

temperature. Further, Schottky power rectifiers fabricated with barrier heights of less than 0.7 eV have been found to exhibit an extremely soft breakdown characteristic, which makes them prone to failure.

The junction barrier-controlled Schottky (JBS)[188] rectifier is a Schottky rectifier structure with a $P–N$ junction grid integrated into its N region. A cross section of the JBS structure is provided in Fig. 151. The junction grid is designed so that its depletion layers do not pinch-off under zero and forward bias conditions of the rectifier, but intersect with each other under the Schottky contact when the reverse bias exceeds a few volts.

Under reverse blocking states, after the depletion layer pinches off, a potential barrier is formed in the channel between the two grids, and further increase of the applied reverse voltage is supported by it with the depletion layer extending toward the N^+ substrate. Therefore, the potential barrier shields the Schottky contact from the applied voltage. This shielding prevents the Schottky barrier lowering phenomenon and eliminates the large increase in leakage current observed in conventional Schottky rectifiers. During on-state operation, there are multiple conductive channels under the Schottky contact through which current can flow. Because of the suppressed leakage current, the Schottky barrier height used in JBS rectifiers can be decreased compared to that of conventional Schottky rectifiers. This has allowed a reduction in the forward voltage drop while maintaining acceptable reverse blocking characteristics. For the same leakage current, the JBS rectifier has been found to provide a

forward voltage of 0.25 V, compared with about 0.5 V for the Schottky rectifier.

In the design of the JBS rectifier, the lowest on-state voltage drop can be obtained by making the width of the junction diffusion window as small as possible. This minimizes the dead space below the junction where the current does not flow. The best JBS rectifier characteristics can therefore be expected when submicron lithography is used to pattern the diffusion windows for the P^+ regions.

MPS Rectifiers The merged $P–i–N$/Schottky (MPS)[189] rectifier has a structure similar to that of the JBS rectifier, as shown in Fig. 152. However, the operating physics and applications of the two rectifiers are quite different. In the JBS rectifier, there is no injection of minority carrier holes from the $P–N$ junction, and the on-state voltage drop is less than 0.5 V. In the MPS rectifier, the N region is designed for supporting a high reverse blocking voltage, and the forward biasing $P–N$ junction becomes necessary in the on state. The forward bias of the $P–N$ junction produces the injection of holes into the N region and results in conductivity modulation in the N region in a manner similar to the $P–i–N$ rectifier, which drastically reduces the resistance of the N region to current flow. This also allows larger current flow via the Schottky region due to a lower series resistance. Because of the existence of the Schottky region, the injection level required to reduce the resistance in the MPS is not as large as that observed in the $P–i–N$ rectifier. As a consequence, the stored charge in the MPS rectifier is much smaller than that in the $P–i–N$ rectifier when they have the

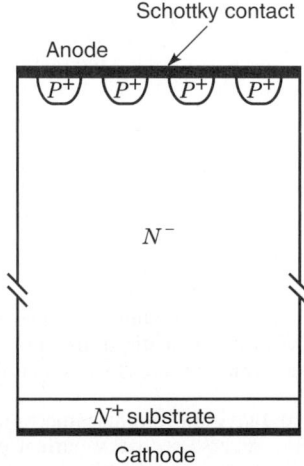

Fig. 152 Basic structure of MPS rectifier. The N^- region is designed to support a high reverse blocking voltage. The purpose of the p^+ region is to enhance the minority carrier injection when the forward voltage is higher than 0.7 V. This p^+ region also reduces the leakage current.

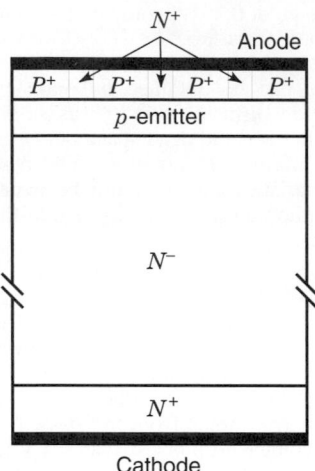

Fig. 153 Basic structure of ESD. Its N^+ short regions reduce the emitter injection efficiency of holes by clamping the peak values of the electron concentration underneath these N^+ regions.

same on-state voltage drop. Consequently, MPS rectifiers have a better reverse recovery characteristic and a superior tradeoff curve between the on state and the turn-off loss than those for $P-i-N$ rectifiers.

Emitter Short Diode The emitter short diode (ESD)[190,191] is an improved $P-i-N$ diode structure proposed for high-voltage and high-speed applications. Figure 153 shows the basic structure of the ESD.

In the ESD structure, additional N^+ short regions are formed on a lightly doped p-emitter layer to reduce the hole injection efficiency and attain asymmetric carrier profiles for the N^- region in the on state by control of the junction depth of the N^+ regions. To block a high reverse voltage, a high impurity concentration for the p emitter is needed. However, to reduce emitter injection efficiency, a shallow junction and low impurity concentration for the p emitter are needed. Consequently, the minimum impurity dose of the shallow p emitter should be chosen carefully to block the desired high reverse voltage and avoid the depletion region reaching the N^+ short layers. To reduce the lateral resistance in the p emitter layer under N^+ short regions and avoid the parasitic transistor effect during reverse recovery, the N^+ areas should be formed in a fine pattern.

The ESD reduces the emitter injection efficiency and controls the reverse recovery current behavior by using a shallow p emitter combined with the emitter short structure. It attains half of the reverse recovery current, half of the reverse recovery time, and one-fourth of the reverse charge, when compared to a conventional $P-i-N$ diode. Its leakage current is also as low as a conventional $P-i-N$ diode, even at $125°C$.

Synchronous Rectifier With the increasing requirement in applications such as computers for power supplies for even lower supply voltage, the conduction loss in the output rectifier becomes the biggest source of power loss in switching power supplies. Even the commonly used Schottky diodes have a relatively large voltage drop and, hence, a large power loss in such low-output-voltage applications. Consequently, low-voltage MOSFETs, which operate in the third quadrant, with a very low on-state resistance and fast switching speed can be used to replace the diodes in the output stage. Because the gate signal to the low-voltage power MOSFET is provided in synchronism with the drain–source voltage to maintain low on resistance in one direction and blocking state in another direction, the low-voltage power MOSFET in these applications is called a synchronous rectifier (SR).[186,192,193] Low-voltage power MOSFETs are successfully used as SR, because of their linear $V-I$ characteristic. The conduction loss can therefore be reduced to a very low value by paralleling more MOSFETs. The SR is also fast because it is a majority carrier device.

11.3 Switches

A switch, in semiconductor terms, is a device that has two states for current flow in the same direction—a low-impedance state (ON state) and a high-impedance state (OFF state). Switching between these two states can be controlled by voltage, current, temperature, or light.

Bipolar Power Transistor Bipolar power transistors have been commercially available for more than 30 years. They were favored for low and medium power applications because of their faster switching capability. The rating for bipolar power transistors grew steadily until the end of the 1970s, when power MOSFET started to appear in the market.

Figure 154 shows the cross-sectional view of a bipolar power transistor cell and its circuit symbol. Although the operating physics for bipolar power transistors is essentially the same as that for signal transistors, their characteristics differ because they need to support a high collector voltage in the forward blocking mode. The high voltage capability of the bipolar power transistor is obtained by incorporating a high resistive, thick N^- region into the collector structure. In addition, the base region must be carefully designed to prevent punch-through breakdown. These differences strongly influence the current gain of the device.

Another distinguishing feature of bipolar power transistors is that they operate at relatively high current densities in the saturation region when both the emitter and collector junction are forward biased. This produces high-level injection not only in the base but also in the collector region. The high-level injection results in severe degradation of the current gain. In general, bipolar power transistors have a low current gain at typical operating current levels. Since the

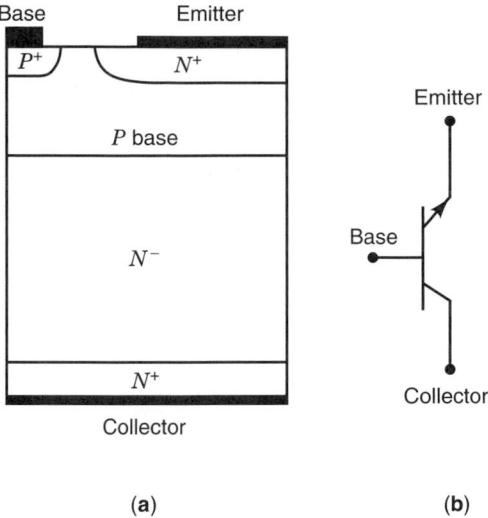

Fig. 154 (a) Cross-sectional view of a bipolar power transistor cell and (b) its circuit symbol.

bipolar power transistor is fundamentally a current-controlled device, with the magnitude of the collector current determined by the base drive current, a low current gain means a bulky and expensive control circuit requiring many discrete components in applications. Consequently, one of the most critical design goals has been to improve the current gain in order to reduce the complexity, size, and weight of the base control circuit. Unfortunately, achieving a high current gain conflicts with the achievement of high breakdown voltage. Further, the fall-off in current gain at typical operating current densities due to high injection level leads to a gain of less than 10. The current gain can be improved by using the Darlington power transistor, but this has the disadvantage of considerably increasing the on-state voltage drop.

In addition, the devices are prone to failure due to the second breakdown phenomenon. This occurs because of the affinity for the formation of a local region in the emitter through which the current tends to constrict itself. It appears on the output characteristic of the power bipolar transistor as a precipitous drop in the collector–emitter voltage at large collector currents. As the collector voltage drops, there is a significant increase in the collector current and a substantial increase in the power dissipation. What makes this situation particularly dangerous for the power bipolar transistor is that the dissipation is concentrated in highly localized regions where the local temperature may grow very quickly to unacceptably high values because of the positive feedback relationship between the current and the temperature within the power bipolar transistor. If this situation is not terminated quickly, device destruction results. This positive feedback relationship between the current and the temperature also

means that power bipolar transistors are difficult to parallel. For these reasons, the bipolar power transistor has been displaced by the power MOSFET for high-speed, low-power applications in the 1980s, and for medium-power applications by the IGBT in the 1990s.

Darlington Power Transistor Figure 155 shows the cross-sectional view of a monolithic Darlington power transistor[193] and its equivalent circuit. In this structure, two transistors have a common collector connection, and the emitter of the drive transistor (T_1) is connected to the base terminal of the output transistor (T_2). The base drive current (I_B) is supplied to the drive transistor. This current turns "on" the transistor T_1, which then provides the base drive current for transistor T_2. Consequently, the current gain of the Darlington power transistor, β, is approximately equal to the product of the current gain of the drive transistor (β_1) and the current gain of the output transistor (β_2). That is

$$\beta = \beta_1 \beta_2$$

However, in order to turn on the output transistor T_2, it is necessary to raise the potential of the collector. Thus, the on-state voltage drop of the Darlington power transistor is higher than that of a single bipolar power transistor.

The Darlington power transistor was the only switching power device with a fast switching speed to deliver several hundred amperes of current and up to 1000 V before the appearance of insulated-gate bipolar transistors (IGBTs). These power Darlington transistors were also called giant transistors (GTRs). Darlington power transistor can be in the form of a monolithic device, such as that shown in Fig. 155; it can also be formed by multichip packaging technique, commonly known as power module.

Thyristor A thyristor is loosely defined as a device having a four-layer $P-N-P-N$ structure, leading to bistable behavior, that can be switched between a high-impedance, low-current OFF state and a low-impedance, high-current ON state. In the past, the thyristor was also commonly called the semiconductor-controlled rectifier or the silicon-controlled rectifier (SCR). The basic structure of the thyristor and its equivalent circuit are illustrated in Fig. 156.

As shown in Fig. 156a, a thyristor consists of four semiconductor layers ($P-N-P-N$). The N^- layer is lightly doped and supports a high voltage when the device is in its blocking state, with junction J_1 or J_2 reverse biased. Thyristors hence offer both forward and reverse blocking capability of comparable magnitude. This makes them well suited for ac circuit applications. Thyristors are now available with ratings of 10 kV and 6 kA. These devices are manufactured from single 10–12.5-cm diameter wafers by using a matured deep diffusion process with gallium and aluminum as dopants. High breakdown voltages are

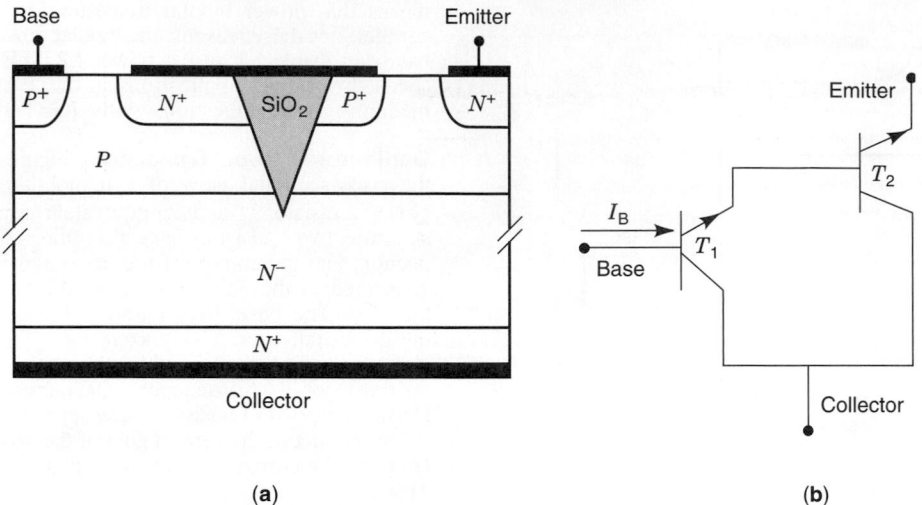

Fig. 155   (*a*) Cross-sectional view of the monolithic Darlington power transistor and (*b*) its equivalent circuit.

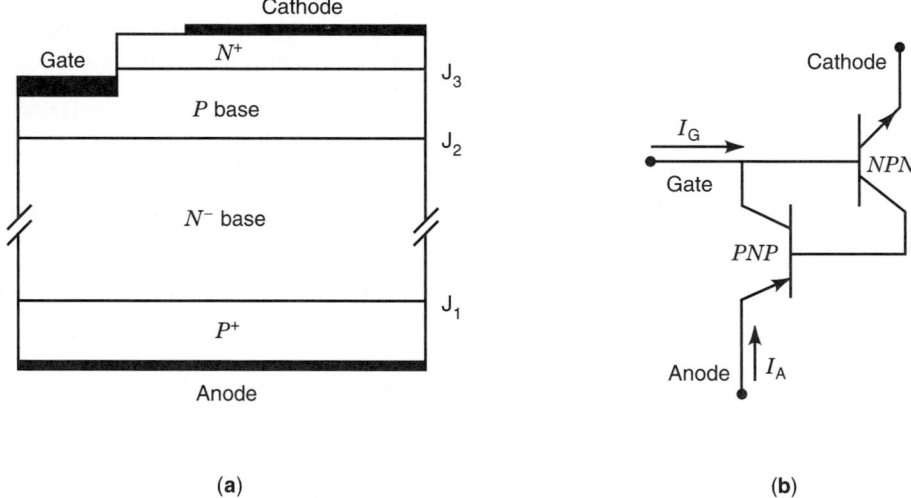

Fig. 156   (*a*) Basic structure of the thyristor and (*b*) its equivalent circuit.

realized by using positive and negative bevel etching techniques at the edge of the wafer. An SCR typically only has a few gate fingers, so device turn off relies on forced commutation in ac circuits.

From Fig. 156*b* it is clear that a thyristor can be bisected into a P–N–P and an N–P–N bipolar transistor, with each base connected to the other's collector. Consequently, the anode current I_A can be described by

$$I_A = \frac{\alpha_{NPN} I_G + (I_{\mathrm{co1}} + I_{\mathrm{co2}})}{1 - (\alpha_{NPN} + \alpha_{PNP})}$$

where α_{NPN} and α_{PNP} are current gains of the N–P–N and the P–N–P bipolar transistors, respectively. I_{co1} and I_{co2} are leakage currents of the N–P–N and the P–N–P bipolar transistors, respectively. In the blocking state, the sum of $\alpha_{NPN} + \alpha_{PNP}$ is much smaller than unity and I_G is zero so that the anode current can be kept quite small. If the sum of $\alpha_{NPN} + \alpha_{PNP}$ approaches unity, the anode current will be arbitrarily large.

For turning on the SCR, a small triggering current is required at the gate. This gate drive current serves to turn on the N–P–N bipolar transistor and

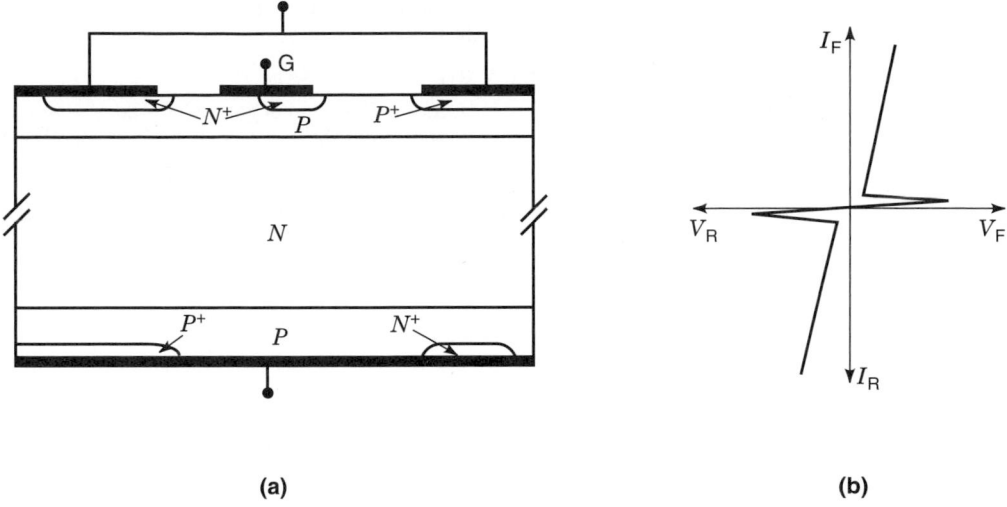

(a)   (b)

Fig. 157   (a) Basic structure of a TRIAC and (b) its output characteristic.

increases the current gain. Once the current gains of the two transistors within the thyristor structure become sufficiently large, the two transistors can provide the base drive currents for each other and the thyristor enters self-sustaining mode. This mode is referred to as regenerative mode, or as latch-up state.

Because thyristors operate in the latch-up mode, they have very low conduction loss. However, it is difficult to turn off a thyristor. In the case of a conventional thyristor structure, the device is primarily used in ac circuits where the anode voltage periodically reverses to force the current to decrease to zero.

A light-triggered thyristor is a thyristor that can be directly triggered via an optical signal. It is useful for high-power systems, such as the high-voltage direct-current (HVDC) transmission system, because the control system is isolated from the power stage.

TRIAC   A TRIAC (triode ac switch)[186] is a semiconductor device with bi-directional voltage blocking capability and bi-directional current conduction capability. Figure 157 shows a basic TRIAC structure with a single control gate electrode and its output characteristic. It is clear that this structure has two back-to-back thyristors integrated monolithically in an antiparallel configuration.

GTO Thyristor   The gate turn-off (GTO) thyristor is similar to the SCR in device structure but has gate current turn-off capability. Figure 158 shows the vertical cross-sectional view of the GTO. It has the basic $P-N-P-N$ four-layer structure of the SCR. In order to have more efficient gate-controlled turn off, during the on state the base current of the $N-P-N$ bipolar transistor and the collector current of the $P-N-P$ bipolar transistor have to be minimized. This demands that $\alpha_{NPN} \gg \alpha_{PNP}$. Consequently, in the GTO

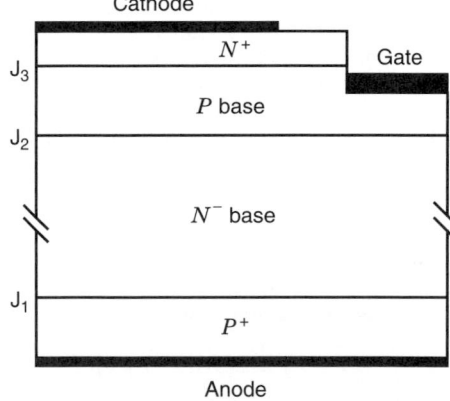

Fig. 158   Vertical cross-sectional view of the GTO structure. Structure of the GTO is similar to that of the thyristor, but it has a much narrower emitter cell width than that of the thyristor so as to improve its gate-controlled turn-off capability.

structure, the thickness of the P base layer is generally somewhat smaller than that in an SCR structure. Another significant difference between a GTO and an SCR is that a lot of gate fingers are placed next to the cathode emitter in an interdigital fashion. The basic goal is to maximize the periphery of the cathode and minimize the distance from the gate to the center of the cathode region.

In the GTO, a negative gate current, I_G, alone is able to turn off the thyristor without forced commutation. A large reverse gate drive current extracts charges from the base region of the upper transistor and disrupts the self-sustaining current conduction mechanism in the GTO. I_A/I_G is defined as the turn-off gain, β_{off}, where I_A is the anode current. For successful turn off, β_{off} has to satisfy

$$\beta_{\text{off}} \leq \frac{\alpha_{NPN}}{\alpha_{PNP} + \alpha_{NPN} - 1}$$

Typical values for turn-off gain range between 5 and 10.

The main advantage of the GTO thyristor is the elimination of an external circuit for forced commutation, which provides increasing flexibility in circuit applications. Another advantage is a smaller turn-off time and the capability for the higher speed operation compared to that of the SCR. The disadvantage is the large gate currents required for turn on and turn off of the GTO. The GTO is the only commercially available device today with the ability to block 8 kV and control over 6 kA for applications such as traction control. The fabrication process of the GTO is similar to that of the SCR.

The main problem that limits the application of GTOs is that they require a complicated and expensive gate drive to turn off GTOs. Not only is a large negative gate current required but also a high di_G/dt for that negative gate current. The latter is crucial to ensure that each segment in the GTO turns off uniformly. Otherwise, the GTO could be destroyed permanently due to a process called current filamentation, in which a slow gate turn-off current only turns off some of the GTO cells, forcing all of the anode current to crowd around a few cells or even a single cell, hence destroying the device through a very high localized power

dissipation. A large snubber is therefore routinely used in GTO applications to solve this problem.

SIT The static induction transistor (SIT)[194] was introduced in 1972 and began to be produced in the market in the mid-1980s as a switching power device. Several structures of the SIT are shown in Fig. 159. The buried gate structure is the original proposed scheme, while the planar gate structure and the recessed gate structure are more popular. In an SIT structure, the gate and source regions are highly interdigital. Thousands of these basic gate–source cells are connected in parallel to make up a single SIT. The most critical parameters in an SIT are the spacing between gates and the channel doping level. Since most SITs are designed as "normally on" devices, the doping is chosen such that the depletion regions from the gates do not merge, and there exists a narrow neutral channel opening at zero gate bias. The gates in SITs are formed normally by $P-N$ junctions, but the SIT operations can also be generalized to include metal (Sckottky) gates, or even metal–insulator–semiconductor (MIS) gates. Normally off SITs are also being fabricated by using very narrow channel design.

The SIT is basically a JFET or MESFET with supershort channel length and with multiple channels connected in parallel. As a result of short channel length, punch-through occurs with high drain bias even if the transistor is originally turned off (static induction is equivalent to punch-through). The output characteristics of a normally on SIT are shown in Fig. 160. These characteristics are quite different from those of a bipolar transistor and are often referred to as

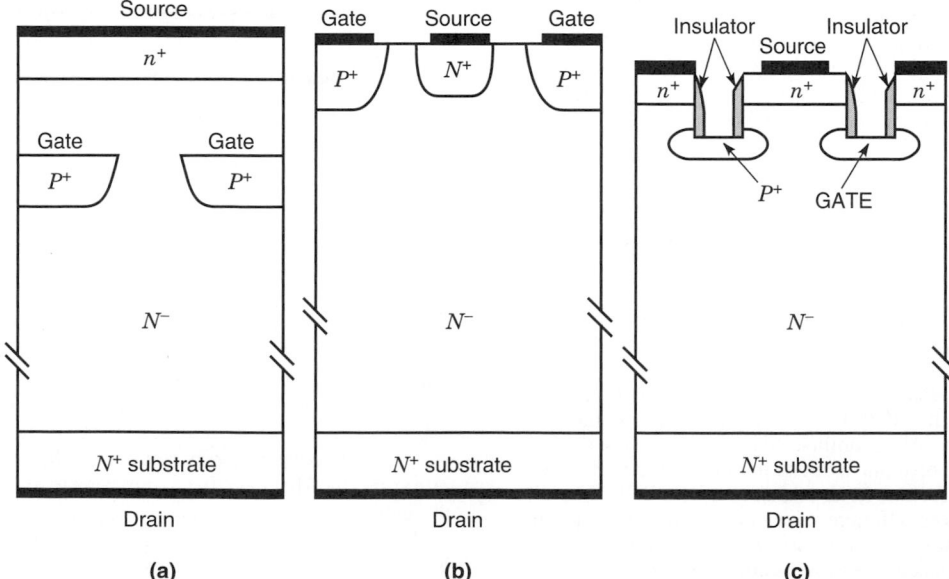

Fig. 159 Structure of SIT with (a) buried gate, (b) planar gate, and (c) recessed gate.

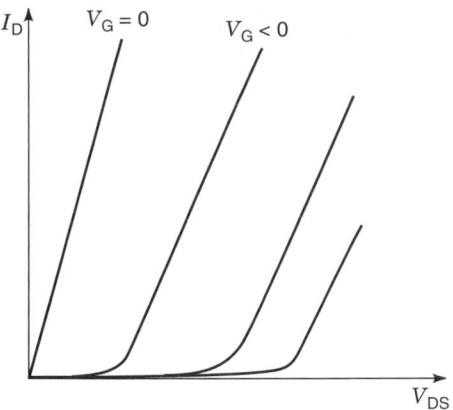

Fig. 160 Output characteristics of the normally on SIT. These *I–V* curves are very different from those of a bipolar transistor and similar to a triode. The SIT typically conducts current when the gate voltage is zero. This type of device is called a "normally on" device.

triodelike characteristics because of their resemblance to the *I–V* characteristics of a vacuum triode.

As shown in Fig. 160, when a positive bias is applied to the drain, the normally on SIT is in the on state when the gate–source voltage is zero. The current conduction is adrift in nature and is similar to a JFET. When a reverse bias is applied to the gate, the depletion layers widen, and pinch off the channel. The depletion layers set up a potential barrier to the flow of the drain electron current. As a consequence, there will be no flow of current between drain and source as long as the drain–source voltage is kept small. This is the off state of the SIT. As the drain-to-source bias voltage, V_{DS}, increases, the potential barrier to the drain current flow gets smaller and smaller. When V_{DS} is large enough to suppress the potential barrier set up by the gate–source bias voltage, current begins to flow again and increases with the increase of the V_{DS}.

The main attractiveness of an SIT is the combination of high-voltage and high-speed capability. Its cut-off frequency can be up to 2 GHz.[194] As an audio power amplifier, the SIT has low noise, low distortion, and low output impedance. It can be used in high-power oscillators of microwave equipment, such as broadcasting transmitters and microwave ovens. As switching power devices, SITs are limited by their normally on characteristic and because power MOSFETs, developing concurrently with SITs, are superior to SITs in switching power applications due to their fast speed, high input impedance, and normally off characteristic. SITs are also difficult in scaling up to high voltages, not only because the conduction loss will increase but also because a large negative gate bias is needed to block higher voltage. A factor commonly defined as forward-blocking voltage gain is defined as the change of the blocking voltage V_{DS} induced by the change

of V_G for the same drain current. High-voltage SITs typically have a blocking gain of less than 20.

SITH   The static induction thyristor (SITH) is also called the field-controlled thyristor (FCT). The SITH was introduced in the mid-1970s with the aim to reduce the conduction loss of high-voltage SITs. Although the SITH is commercially available as a power device, its performance is superseded by the IGBTs.

The structure of the SITH is similar to that of the SIT with a P^+ anode replacing the N^+ drain. The basic structures of the SITH with planar gates, buried gates, and double gates are shown in Figs. 161*a*, *b*, and *c*, respectively. It is clear that the SITH consists of a $P–i–N$ diode with part of the channel surrounded by closely spaced junction grids or gates. There are two types of SITH—normally on SITH and normally off SITH. In the normally on SITH, pinch-off of the channel does not occur with zero gate voltage, and a high current can flow. In the normally off SITH, the depletion regions of the nearby gate merge, and the pinch-off occurs at zero gate voltage. The output characteristics for a normally on SITH are shown in Fig. 162.

In the normally on SITH, at zero gate bias or small positive gate bias, the depletion regions around the gates do not pinch off the gap completely. The current conduction from anode to cathode is similar to that of a $P–i–N$ diode. At a forward-biased voltage V_{AK}, electrons are injected from the cathode and holes are injected from the anode, and they are equal in number to maintain charge neutrality. These excess electrons and holes increase the conductivity of the N^- layer. Note that although the output characteristics are similar in shape to those of the SIT, the P^+ anode can inject holes and enable conductivity modulation, resulting in a lower forward voltage drop or lower on resistance.

With a larger reverse gate bias, the depletion layers extend, pinch-off of the channel is introduced, and a barrier for electrons is formed. This barrier limits the electron supply and becomes the controlling factor for the overall current. Without an ample electron supply, the hole current reduces to leakage generation current and becomes insignificant. The SITH enters the forward blocking state.

In the SITH structure, the channel barrier height can be influenced by the gate voltage as well as by the anode voltage. A large forward anode bias V_{AK} can lower this barrier height. This dependence of the barrier height on the forward anode bias is called static induction. Static induction current is basically a punch-through current due to the thin and small barrier in the direction of current flow.

One useful parameter for the SITH is the forward-blocking voltage gain, which is defined as the change of the blocking voltage V_{AK} induced by the change of V_G for the same anode current. The forward-blocking voltage gain depends on the structure of the gates and the channel doping.

One of the advantages of the SITH when compared with GTOs is its higher speed of operation due to a

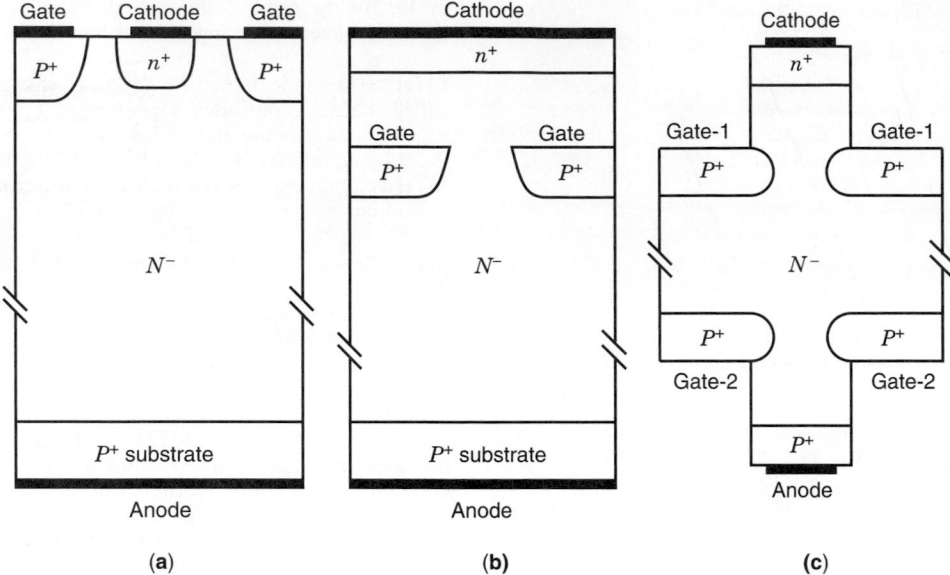

Fig. 161 Structure of SITH with (*a*) planar gates, (*b*) buried gates, and (*c*) double gates.

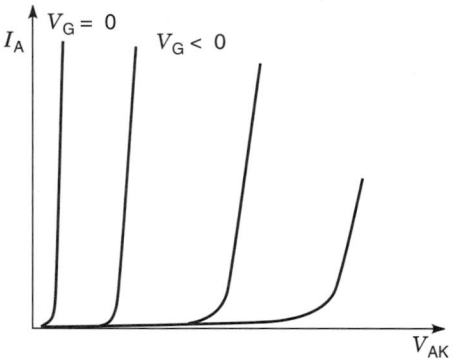

Fig. 162 Output characteristics of normally on SITH.

faster turn-off process. During turn off, the reverse gate bias can extract the excess minority carriers (holes) quickly. The excess electrons, being majority carriers in the N^- region, can be swept away quickly by the drift process. The hole current contributes to an instantaneously large gate current, and a small gate resistance is critical to avoid gate debiasing. An alternate technique to reduce the turn-off time is to reduce the minority carrier lifetime by lifetime control technique. The penalty for using this technique is a larger forward voltage drop.

In the SITH structure, the planar gate structure has a lower gate resistance since a metal contact can be deposited directly over it. This results in a smaller debiasing effect during the turn-off process

when there is a substantial current through the gate. The advantage of the buried gates structure is a higher forward blocking voltage gain resulting from a more efficient use of the cathode area and a more effective gate control of the current. The double-gate SITH is capable of higher speed than the single-gate structure, but it has a more complicated fabrication process.

Because of the fast turn-off capability, SITHs with an operating frequency up to 100 kHz are possible. With a high forward-blocking voltage gain of up to 700, stable operation at high temperature, and large dI/dt and dV/dt capabilities the SITHs have been applied mainly in power source conversion such as ac-to-dc converters, dc-to-ac converters, and chopper circuits. Other applications of the SITH include pulse generation, induction heating, lighting of fluorescent lamps, and driving pulsed lasers.

The main problem for the SITHs is similar to that of the GTOs, in that a large and expensive gate drive circuit has to be provided. Because of the nature of their planar and shallow junction process, the SITHs have not reached the power ratings of GTOs. They are, therefore, seriously challenged by the IGBTs in the medium power range, because IGBTs have many of the SITHs advantages plus a simple control interface owing to their high input impedance.

Power MOSFET   Prior to the development of the power metal–oxide–semiconductor field-effect transistors (power MOSFETs), the most favorable device available for high-speed, medium-power switching applications was the power bipolar transistor. But the power bipolar transistors exhibit several fundamental drawbacks in their operating characteristics, such as that they are current-controlled devices and are difficult to parallel. In order to suppress these performance limitations, the power MOSFET was developed in the 1970s due to the advancement of VLSI technology.

In the power MOSFET, the control signal is applied to a metal (or polysilicon) gate electrode that is separated from the semiconductor surface by an intervening insulator (typically silicon dioxide). Thus, the power MOSFET has a very high input impedance in steady state, and it is classified as a voltage-controlled device that can be controlled using integrated circuits because of the small gate currents that are required to charge and discharge the input gate capacitance. Even during the switching of the devices between the ON and OFF states, the gate current is small at typical operating frequencies of less than 100 kHz because it serves only to charge and discharge the input gate capacitance. When the operating frequency becomes high (>100 kHz), this capacitance current can become significant, but it is still possible to integrate the control circuit due to the low gate bias voltages (typically 5–15 V) required to drive the device into its on state with a low forward voltage drop.

In comparison with the bipolar transistor, the power MOSFET is a unipolar device, it therefore has a very fast switching speed due to the absence of minority carrier injection. The switching time for the MOSFET is dictated by the ability to charge and discharge the input capacitance rapidly. This feature is particularly attractive in circuits operating at high frequency, where switching power loss is dominant. Further, the power MOSFET has superior ruggedness and has been found to display an excellent safe-operating area (i.e., they can withstand the simultaneous application of high current and voltage without undergoing destructive failure).

These characteristics of power MOSFETs make them important candidates for many applications such as high-frequency power conversion and lamp ballasts. Three discrete vertical channel power MOSFET structures are described.

VVMOS (VMOS).

The vertical V-shaped groove MOSFET (VVMOS or VMOS) was the first commercial structure of power MOSFET developed in the 1970s. The VVMOS structure is shown in Fig. 163. This structure is based on the V-shaped groove by anisotropically etching in a ⟨100⟩ silicon substrate within which the gate is located along the ⟨111⟩ planes extending through the P layer. It can be fabricated by first performing an unpatterned P-region diffusion followed by the N^+ source region diffusion. A V-shaped groove extending through these diffusions is then formed by using preferential etching with potassium-hydroxide-based solutions. For an aluminum gate VMOS, the gate oxidation layer is grown and the gate electrode is then deposited and patterned. The channel region for this structure is formed along the walls of the V-groove. This structure can therefore

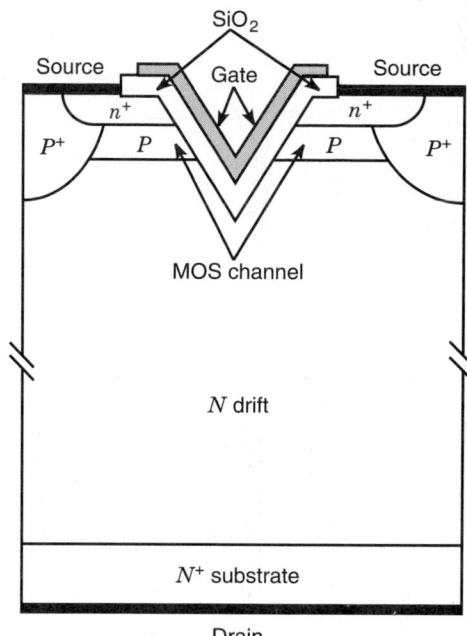

Fig. 163 Cross section of the VVMOS structure. The V-shaped groove is formed by anisotropically etching a ⟨100⟩ silicon substrate. The MOSFET channel region is formed along the walls of the V-groove and can therefore provide short channel length without fine lithography capability. The vertical current flow also maximizes the total current for a given surface area. The drawback of this structure is that a strong electric field peak exists at the V groove corner when a high drain voltage is applied.

provide short channel length without fine lithography capability, and the vertical current flow also maximizes the total current for a given surface area.

For the N-channel structure shown, when a positive bias larger than the threshold voltage of the MOSFET is applied to the gate electrode, an inversion layer forms along the V-groove and the channel is turned on. When a positive bias voltage is applied to the drain, electrons flow from the N^+ source via the MOSFET channel into the drift region and are then collected by the N^+ drain. When the gate voltage is lower than the threshold voltage of the MOSFET, the device is in the forward blocking state and can support a high drain voltage across the P-body/N-drift junction.

Because the gate of the VVMOS is located in the V-groove formed by the preferential etching, there are instabilities in the threshold voltage during manufacturing. In addition, the sharp tip at the bottom of the V-groove creates a high electric field during the forward blocking state, which degrades its breakdown voltage. Furthermore, the mobility of the carrier on a ⟨111⟩ etched slope is somewhat lower. For these reasons, the VVMOS has been displaced by the VDMOS

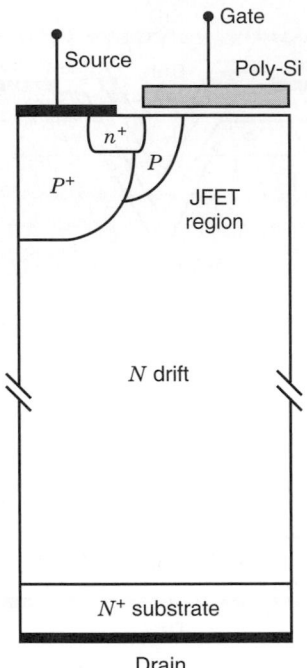

Fig. 164 Cross section of the VDMOS structure. This VDMOS is fabricated by using planar diffusion technology with a refractory gate as a mask. The double-diffusion process forms the channel laterally and enables fabrication of submicrometer channel length without resorting to high-resolution lithography.

structure based on the double-diffusion MOS (DMOS) process.

VDMOS. The vertical double-diffusion MOSFET (VDMOS) is the most popular power MOSFET structure. Figure 164 shows the cross section of one-half of a VDMOS cell structure. This VDMOS is fabricated by using planar diffusion technology with a refractory gate, such as polysilicon, as a mask. The P region and the N^+ source region are diffused through a common window defined by the edge of the polysilicon gate. The name for this device is derived from this *double-diffusion process*. The P region is driven deeper than the N^+ source region, hence defining the surface channel region. This process enables fabrication of a submicrometer channel length without resorting to high-resolution lithography. A commercial VDMOS chip contains millions of such VDMOS cells in parallel, with a common gate control.

For an N-channel VDMOS, the device operates with a positive voltage applied to the drain. When the gate voltage is lower than the threshold voltage of the DMOS, the device can support a high drain voltage across the P-region/N-drift region junction. The forward blocking capability is determined by the breakdown characteristic

of the P-region/N-drift junction. The voltage blocking capability of the VDMOS is better than that of the VVMOS because the field distribution in the cell region is essentially one dimensional. When the gate voltage is higher than the threshold voltage of the DMOS, the surface channel is turned on, and electrons will flow from the source via the channel into the drift region and then bend 90° before being collected by the drain.

The conduction loss of the VDMOS is specified by the on resistance. The on resistance is an important device parameter because it determines the maximum current rating. The specific on resistance, defined as the on resistance per unit area, is a preferable parameter in the design of the VDMOS. The on resistance of the VDMOS consists of the N^+ source resistance, the channel resistance, the accumulation layer resistance under the polysilicon gate, the JFET region resistance, the drift region resistance, the substrate resistance, and the contact resistance.

A tradeoff between the forward blocking voltage and the specific on resistance exists for the VDMOS and other unipolar devices such as the SIT. A high forward blocking voltage needs a lightly doped and thick drift region, hence creating a large specific on resistance. This tradeoff relationship can be described by:

$$R_{\text{drift, specific}} = \frac{4\text{BV}^2}{\varepsilon_s \mu_n E_C^3}$$

where ε_s is the dielectric constant of the silicon, μ_n is the electron mobility, BV is the breakdown voltage, and E_c is the critical electric field at avalanche breakdown. Because of this tradeoff, the VDMOS usually are designed to operate at high-voltage, low-current levels or low-voltage, large-current levels.

The forward voltage drop or the on resistance in the VDMOS increases with the increase of temperature; therefore, VDMOSs can be easily paralleled. This characteristic of VDMOSs makes them important candidates for many applications. The VDMOS is a unipolar device; current conduction occurs via transport of majority carriers in the drift region without minority carrier injection required in bipolar transistor operation. Thus, the VDMOS has a faster switching speed than bipolar transistors. This feature is particularly attractive in circuits operating at high frequencies, where switching power losses are dominant.

UMOS (Trench-Gate MOSFET, UMOSFET). The UMOS structure is shown in Fig. 165. The name for this structure is derived from the U-shaped groove formed in the gate region by using RIE. The fabrication of this structure can be performed by following the same sequence as the VVMOS structure with the V-groove replaced by the U-groove. The U-groove structure has a higher channel density than either the VMOS or DMOS structures, which allows significant reduction in the on resistance of the device. The technology for the fabrication of this structure was derived

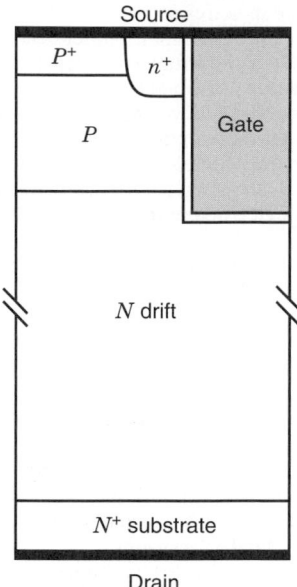

Fig. 165 Cross section of the UMOS structure. A U-shaped gate region is formed by using reactive ion etching (RIE) followed by gate oxidation and polysilicon refill. The UMOSFET has a higher channel density than either the VMOS or DMOS structure, resulting in the lowest on-resistance per unit silicon area.

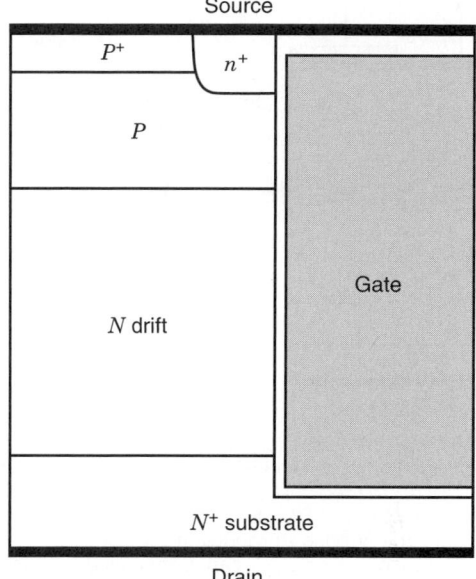

Fig. 166 Cross section of deep-trench UMOS. This structure extends the trench down to the N^+ substrate, and the drift region resistance component of the on resistance is reduced by the parallel current flow path of an accumulation layer on the sidewall of the trench.

from the trench etching technique developed for memory cells in DRAMs.

In the UMOS structure, the UMOS cell size can be made relatively small (6 μm) when compared with the DMOS cell (20 μm) for the same design rules. This results in an increase in the channel density (channel width per square centimeter of device area). In the UMOS structure, no JFET region exists. The on resistance of the UMOS consists of the N^+ source resistance, the channel resistance, the drift region resistance, the N^+ substrate resistance, and the contact resistance. Unlike the DMOS structure, there is no optimum design for the UMOS cell. In this case, it is beneficial to reduce the mesa and trench width as much as possible. As these dimensions becomes smaller, the channel resistance contribution becomes smaller because the channel density increases. Therefore, the UMOS has the lowest specific on resistance in power MOSFETs. Using UMOS structure, however, will not reduce the on resistance of the drift region, hence, UMOS structure is only beneficial for low-voltage power MOSFETs.

Figure 166 shows a modified UMOS structure with a deep trench, which has a very low specific on resistance approaching the limits for silicon FET performance.[195] In this structure, the trench extends down to the N^+ substrate, and the drift region resistance component of the on resistance is reduced by the

parallel current flow path created by the formation of an accumulation layer on the sidewall of the trench. But it must be noted that the blocking voltage of this structure is limited to less than 30 V by the high electric field created in the gate oxide by the extension of the trench into the N^+ substrate.

Low breakdown voltage power MOSFETs (<30 V) are also successfully used as synchronous rectifiers because their conduction losses can be reduced to even lower than those of the SBDs because of their linear V–I relationships.[190]

IGBT The name insulated-gate bipolar transistor (IGBT) comes from its operation based on an internal interaction between an insulated-gate FET (IGFET) and a bipolar transistor. It has also been called previously an IGT (insulated-gate transistor), an IGR (insulated-gate rectifier), a COMFET (conductivity-modulated field-effect transistor), a GEMFET (gain-enhanced MOSFET), a BiFET (bipolar FET), and an injector FET. IGBTs have been successfully used since they were first demonstrated in 1982.

The IGBT is an important power switch used in converters with ratings up to several hundred kilowatts. A cross section of the planar DMOS-technology-based IGBT structure is shown in Fig. 167a. It is clear from Fig. 167a that the IGBT structure is similar to that of the VDMOS from the fabrication point of view. This has made its manufacturing relatively easy

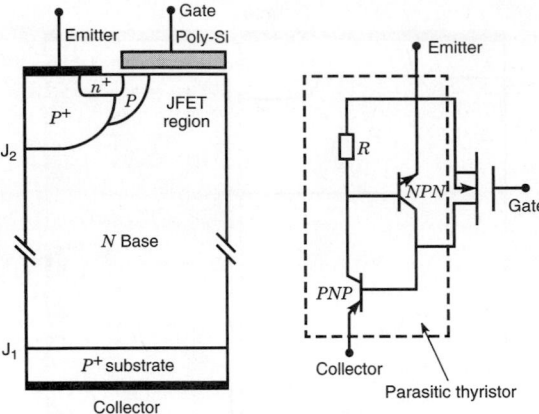

Fig. 167 (a) Cross section of the IGBT structure and (b) its equivalent circuit. The IGBT structure is similar to that of the VDMOS except that a p^+ collector is used to replace the n^+ drain in the VDMOS. This p^+ collector can inject holes into the N base to modulate the conductivity and improve the forward current capability during the forward conduction. The introduction of such p^+ layer results in a vertical $P–N–P$ transistor, which in turn forms a parasitic thyristor with the parasitic $N–P–N$ transistor.

immediately after conception and its power rating has grown at a rapid pace due to the ability to scale up both the current and the blocking voltage.

The equivalent circuit for the IGBT, shown in Fig. 167b consists of a wide-base $P–N–P$ bipolar transistor driven by a short-channel MOSFET. This $P–N–P$ transistor has a long base region, and therefore a very low current gain. A parasitic $N–P–N$ transistor also exists in the IGBT, which forms a parasitic thyristor with the $P–N–P$ transistor. In the IGBT structure, when a positive bias voltage larger than the threshold voltage of the DMOS is applied to the gate electrode, an inversion layer is formed along the P-base surface of the DMOS, and the DMOS channel is turned on. When a positive bias is applied to the collector, electrons flow from the N^+ emitter via the DMOS channel into the N^- region. This provides the base drive current for the wide-base vertical $P–N–P$ transistor in the IGBT structure. Since the emitter junction (J_1) for this bipolar transistor is forward biased, the P^+ substrate injects holes into the N^- base region. When the positive bias on the collector terminal of the IGBT is increased, the injected hole concentration increases and reduces the resistance of the N^- region. Consequently, the IGBT can operate at much higher current densities than the VDMOS even when it is designed to support high blocking voltages.

As long as the gate bias is sufficiently large to produce a strong inversion layer charge of electrons at the N^- base region surface, the IGBT's forward conduction characteristic looks like that of a $P–i–N$ diode. Therefore, the IGBT can also be considered as

a $P–i–N$ diode in series with a MOSFET. However, if the DMOS channel or JFET channel becomes pinched-off and the electron current saturates, the hole current also saturates due to the saturation of the base drive current for the $P–N–P$ transistor. Consequently, the device operates with current saturation in its active region with a gate-controlled output current. This current saturation characteristic is useful for applications in which the device is required to sustain a short-circuit condition.

When the gate voltage is lower than the threshold voltage of the DMOS, the inversion layer cannot sustain and the electron current via the DMOS channel is terminated. The IGBT then operates in the forward blocking mode. A large voltage can then be supported by the reverse-biased P-base/N-base region junction (J_2). Figure 168 shows the typical output characteristics of the IGBT. Because of the existence of the P^+ collector junction, IGBT also has a reverse blocking capability. Commercial IGBTs, however, are mostly using asymmetric structures, and their reverse blocking voltages are typically very low.

The IGBT was the first commercially successful device based on combining the physics of MOS-gate control with bipolar current conduction. Because of the injection of a high concentration of holes from the P^+ substrate into the N^- drift, the conductivity of the long N^- region is modulated and the IGBT exhibits $P–i–N$ diodelike on-state characteristic with a low forward voltage drop. Thus, the IGBT exhibits excellent current-carrying capability with forward conduction current densities 20 times higher than that of a power MOSFET and 5 times greater than that of a bipolar transistor operating at a current gain of 10. Since the input signal for the IGBT is a voltage applied to the MOS gate, the IGBT has the high input impedance of the power MOSFET and can be classified as a voltage-controlled device. However, unlike the power MOSFET, the switching speed of the IGBT is limited by the time taken to remove the stored charges in the N region due to the injection of holes during on-state current conduction. The turn-off time for the

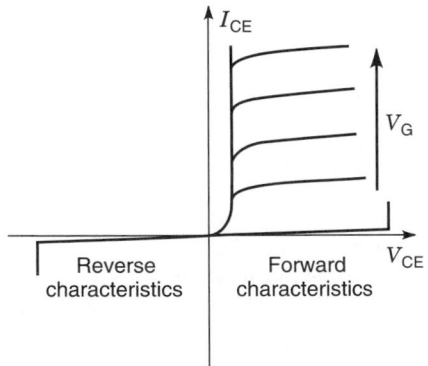

Fig. 168 Output characteristics of the IGBT.

IGBT is dictated by the conduction modulation of the N region and the minority-carrier lifetime. The latter can be controlled by a lifetime control process, such as electron irradiation. Although the lifetime control process can be successful in reducing the turn-off time, it was found that there is a tradeoff between the on-state voltage drop (conduction loss) and the turn-off time (switching loss). A shorter minority-carrier lifetime makes the switching loss of the IGBT lower, but the shorter minority-carrier lifetime also results in a higher conduction loss.

One of the problems encountered when operating the IGBT at high current levels has been the latch-up of the parasitic $P-N-P-N$ thyristor structure inherent in the device structure. Latch-up of this thyristor can occur, causing losses of gate-controlled current conduction. Since the current gains of the $N-P-N$ and $P-N-P$ transistors increase with increasing temperature, the latching current decreases with increasing temperature. This effect is also aggravated by an increase in the resistance of the P base with temperature due to a decrease in the mobility of holes. Many methods have been explored to suppress the latch-up of the parasitic thyristor, such as the use of a deep P^+ diffusion (see p. 454 of Ref. 186), a shallow P^+ diffusion (see p. 456 of Ref. 186), or a self-aligned sidewall diffusion of N^+ emitter.[196,197] The objective of these methods is to reduce the gain of the parasitic $N-P-N$ transistor to minimal.

Traditionally, IGBTs are fabricated on a lightly doped epitaxial substrate, such as the one shown in Fig. 167a. Because of the difficulty of growing the lightly doped epitaxial layer, the breakdown voltage of this type of IGBT is limited to below 1200 V.

To benefit from such a design, an N buffer layer is normally introduced between the P^+ substrate and the N^- epitaxial layer, so that the whole N^- region can be depleted when the device is blocking the off-state voltage, and the electric field shape inside the N^- region is close to rectangular. This type of design is referred to as punch-through IGBT (PT IGBT), as shown in Fig. 169a. The PT structure allows it to support the same forward blocking voltage with about half the thickness of the N^- base region of the $P-N-P$ transistor, resulting in a greatly improved tradeoff relationship between the forward voltage drop and the turn-off time. Thus, the PT structure together with lifetime control is preferred for IGBTs with forward blocking capabilities of up to 1200 V.

For higher blocking voltages, the thickness of the N-base region becomes too large for cost-effective epitaxial growth. Another type of design, the non-punch-through IGBT (NPT IGBT, as shown in Fig. 169b), is gaining popularity.[198] In the NPT IGBTs, devices are built on an N^- wafer substrate that serves as the N^- base region. The collector is implanted from the backside of the wafer and no field stopping N buffer layer is applied to the NPT IGBT. In this concept, the shape of the electric field is triangular in the forward blocking state, which makes a longer N^- base region necessary to achieve the same breakdown voltage as compared with the PT IGBT. However, the NPT IGBT offers some advantages over the PT IGBT. For instance, the injection efficiency from the collector side can be more easily controlled and devices with voltage ratings as high as 4 kV can be realized. Further, by optimizing the injection efficiency of carriers from the P^+ collector layer and the transport factor of carriers in the N^-

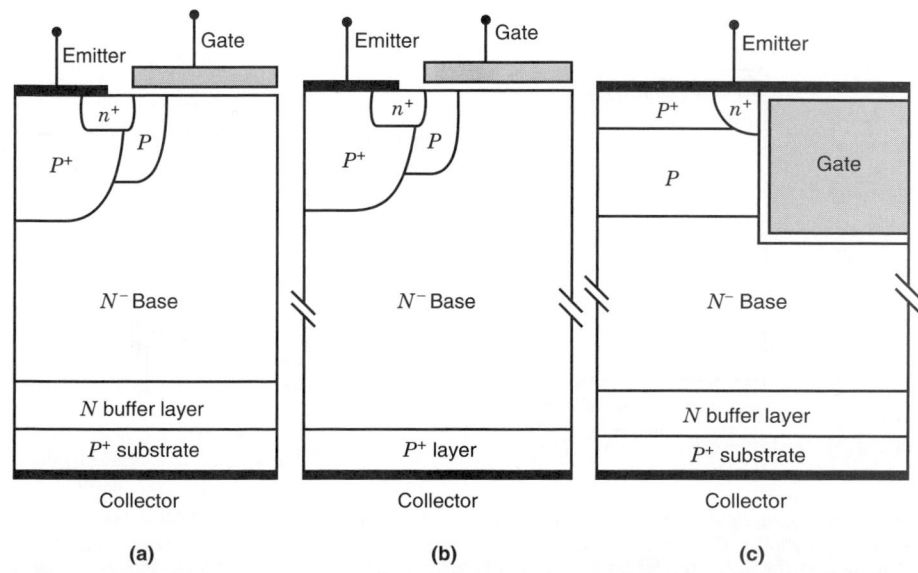

Fig. 169 (a) PT IGBT structure, (b) NPT IGBT structure, and (c) UMOS gate PT IGBT structure.

base, the tradeoff between the forward voltage drop and the turn-off time for the NPT IGBT can be improved to become similar to that of the PT-type IGBT.[199] NPT IGBT is now widely used in high-voltage IGBT design where no lifetime control is used.

Generally speaking, the current tail in the NPT IGBT is longer than the PT IGBT, but the NPT IGBT is more robust than the PT IGBT, particularly under a short-circuit condition.

The trench gate IGBT (trench IGBT, TIGBT, and UMOS gate IGBT) structure[200] is shown in Fig. 169c. With the UMOS structure in place of the DMOS gate structure in the IGBT, the channel density is greatly increased and the JFET region between the two adjacent P-base region is eliminated. In addition, the electron–hole concentration is enhanced at the bottom of the trench because an N-type accumulation layer forms. This creates a catenary-type carrier distribution profile in the IGBT that resembles that obtained in a thyristor or P–i–N diode.[201] These improvements lead to a large reduction in the on-state voltage drop until it approaches that of a P–i–N diode, hence approaching the theoretical limitation of a silicon device. The latching current density of the UMOS IGBT structure is superior to that of the DMOS structure. This is attributed to the improved hole current flow path in the UMOS structure. As shown in Fig. 169c, the hole current flow can take place along a vertical trajectory in the UMOS structure, while in the DMOS structure hole current flow occurs below the N^+ emitter in the lateral direction. The resistance for the hole current that causes the latch-up is determined only by the depth of the N^+ emitter region. A shallow P^+ layer can be used, as shown in the figure, to reduce this resistance. As a consequence, the

safe operating area (SOA) of the UMOS IGBT structure is superior to that of the DMOS IGBT structure. Further, because of a very strong percentage of electron current flow in the trench gate IGBT, the turn-off speed of the trench-based IGBT is generally faster than the DMOS-based IGBT. To obtain better tradeoff between conduction and switching losses, several improved structures and technologies, such as the carrier injection enhancement,[202] local lifetime control by proton irradiation,[203] and p^+/p^- collector region,[204] have been proposed for the high-voltage trench IGBT. It can be anticipated that trench gate IGBTs will replace the DMOS IGBT structures in the future.

MCT   The MOS-controlled thyristor (MCT)[205] is a newer commercially available semiconductor power switch that also combines the physics of MOSFET and bipolar conduction. It is basically a thyristor with two MOSFETs built into the gate structure. One of the two MOSFETs, the ON-FET, is responsible for turning the MCT on, and the other MOSFET, the OFF-FET, is responsible for turning the MCT off. There are two types of MCTs, the N-MCT and the P-MCT, and both combine the low on-state losses and large current-handling capability of a thyristor structure with the advantages of MOSFET-controlled turn on and turn off and relatively fast switching speed. MCTs provide both an easy gate drive, due to the high input impedance of the MOS gate, and a low forward voltage drop, due to the strong conductivity modulation effect of the thyristor structure. They are expected to compete with IGBTs and GTOs in high-power applications.

A cross-sectional view of a single cell of an N-MCT is shown in Fig. 170a. A complete N-MCT is

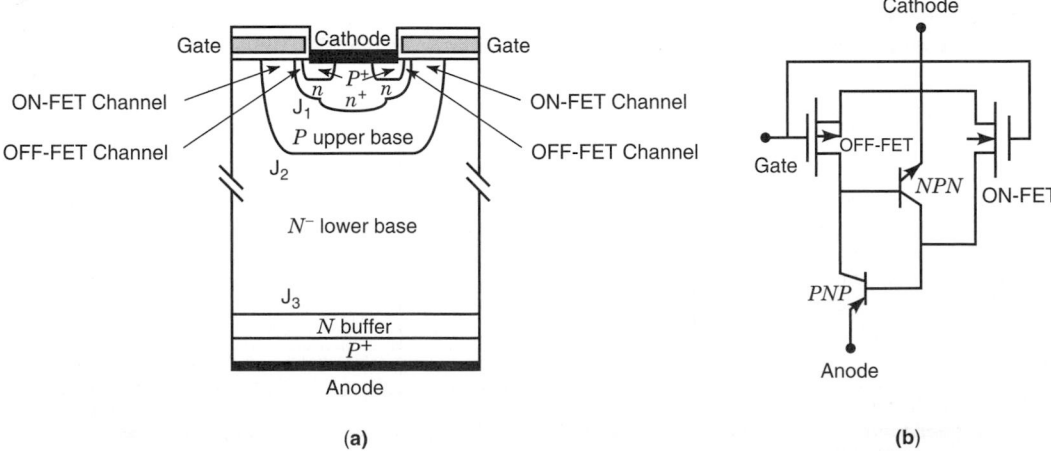

(a)   (b)

Fig. 170   (a) Cross-sectional view of the MCT cell and (b) its equivalent circuit. The MCT is a five-layer semiconductor device. It is basically a thyristor with two MOSFET channels underneath the gate. One of the two MOSFETs, the ON-FET, is responsible for turning the MCT on, when the gate voltage is positive (such as +15 V), and the other MOSFET, the OFF-FET, is responsible for turning the MCT off, when the gate voltage is negative (such as −15 V).

composed of thousands of these cells fabricated integrally on the same silicon wafer connected in parallel to achieve the desired current rating. The ON-FET density in an MCT can also be adjusted to a suitable percentage of the overall cell density. From Fig. 170, it is clear that the MCT has a four-layer $P–N–P–N$ thyristor structure and a MOS gate controlling both the turn-on and turn-off FETs. When the gate bias is zero or negative, the OFF-FET is turned on to short the upper transistors in the emitter junction. The MCT exhibits a high forward blocking voltage by supporting the voltage across the reverse biased junction J_2. When a positive bias is applied to the gate electrode, the ON-FET channel turns on and electrons are supplied to the N^- base of the $P–N–P$ transistor. This results in the injection of holes from the P^+ anode into the N^- base region, and they are collected at the reverse-biased junction J_2. The current in the P upper base created by the collection of holes across junction J_2 acts as the base drive current of the $N–P–N$ transistor, which turns on the $N–P–N$ transistor then triggers the regenerative feedback mechanism between the two coupled transistors within the thyristor structure. The thyristor can therefore be turned on by the application of a positive gate voltage to the MOS electrode.

The MCT can be truned off when a negative bias is applied to the gate electrode because a P channel is formed by the inversion of the N-emitter surface. This provides a path for the flow of holes from the P-base region into the cathode contact that bypasses the N^+-emitter/P-base junction. The holes that are flowing into the P-base region when the thyristor was operating in its on state can then be diverted via the P-channel MOSFET into the cathode electrode. This will reduce the current gain of the $N–P–N$ transistor. If the resistance of the P-channel MOSFET is significantly low, a sufficient number of holes are diverted to the cathode by the P-channel MOSFET. The latch-up condition would then be broken and the thyristor would be turned off successfully.

As mentioned previously, there are two types of MCTs, the N-MCT and the P-MCT. The MCT with an N-channel ON-FET is called an N-MCT and the MCT with a P-channel ON-FET is called a P-MCT. A P-MCT can turn off higher currents because of a higher electron mobility in the turn-off N-channel MOSFET compared to that of the N-MCT. Thus, the first two generations of MCTs developed were P-MCTs. On the other hand, applications normally require an N-MCT, because its SOA is larger than the P-MCT's and its bias configuration is compatible with an N-channel IGBT.

An alternative N-MCT structure that uses an N-channel MOSFET for turn off is shown in Fig. 171.[206] It has an N-channel OFF-FET and its bias configuration is compatible with an N-channel IGBT. In this N-MCT, a floating ohmic contact (FOC) is used to form the bridge of transferring hole current to electron channel current in the event of device turn-off.

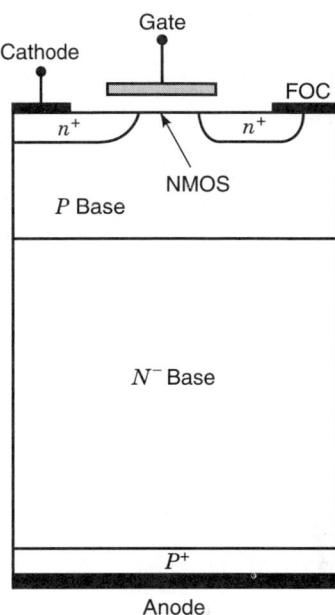

Fig. 171 Cross-sectional view of the N-MCT cell structure. The N-channel MOSFET shorts the upper base via the floating ohmic contact (FOC) to the cathode, resulting in MOS-controlled turn-off.

Figure 172 shows the basic structure of a trench MCT (TMCT) cell based on trench technology.[207] The TMCT can have a much smaller cell pitch than its planar counterpart, so it has a much more uniform turn off across the chip. The TMCT has a turn-on-cell–turn-off-cell ratio of 1 without losing silicon area. Thus, the turn-on capability of the TMCT is also expected to be much better than the planar MCT.

Although MCTs combine the advantages of the high input impedance of the MOS gate and the low forward voltage drop of the thyristor, there are some drawbacks limiting the development of the MCT. One is that the MCT has the current filamentation problem during device turn-off that can destroy the device because of internal regenerative action and the negative temperature coefficient for the on-state voltage within the thyristor. Even if a uniform turn-off is possible, the OFF-FET channel resistance will limit the maximum turn-off current and hence the SOA of the MCT. Further, the MCT lacks the ability to saturate the anode current level, making it a fundamentally different device than the IGBT from an application viewpoint. Consequently, a number of new MOS-gated thyristors are currently being developed and studied, and commercialization of some of these newer devices is expected in the future.[208,209]

Table 34 lists the major electric characteristics of the switches.[210] These switches are all based on silicon material.

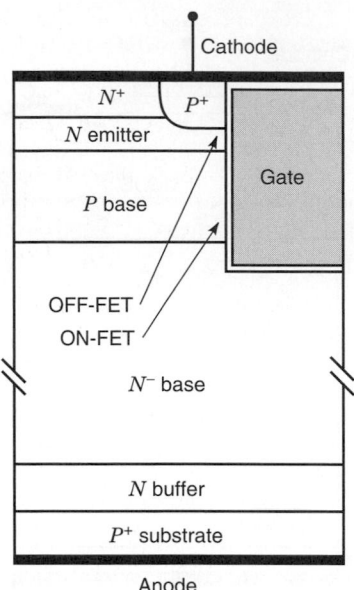

Fig. 172 Basic structure of trench MCT cell.

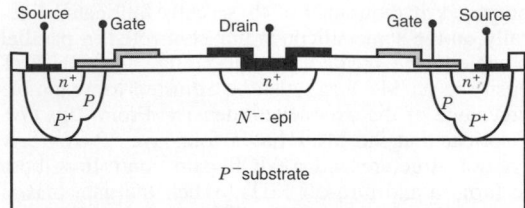

Fig. 173 Cross-sectional view of the RESURF LDMOS. The LDMOS is a lateral version of the VDMOS, with all three electrodes on the substrate surface. There is no need of excessive epitaxial growth and the high voltage is supported laterally. The LDMOS is used mostly in PICs.

11.4 Integrable Lateral Power Devices

LDMOS The lateral double-diffusion MOSFET (LDMOS) was one of the first integrable lateral power devices developed and finds wide use in power ICs (PICs).[211,212] A schematic diagram of the cross section of the LDMOS transistor is shown in Fig. 173. The use of the self-aligned double-diffusion process results in a relatively short channel. A lightly doped, thin drift region is used to support a high drain-to-source breakdown voltage. Although the operation mechanism of the LDMOS is the same as that of the VDMOS, the LDMOS has all three electrodes on the substrate surface and there is no need of excessive epitaxial growth. In addition, it is also easily integrated with CMOS circuitry. These reasons make the LDMOS attractive in monolithic PICs, where interconnection between the LDMOS and low-voltage analog and digital ICs can be easily achieved.

Significant efforts have been directed toward increasing the breakdown voltage and reducing the on resistance of the LDMOS. The important parameters related to a required breakdown voltage are the charge

Table 34 Comparison among Commercially Available Power Semiconductor Switches

Switch	Number of Junctions	Control Type	Carrier Conduction Type	Highest Voltage Rating (V)	Largest Current Rating (A)	Typical Switching Frequency (Hz)	Major Limiation Factor
BJT	2	Current	Bipolar	1,500	800	15,000	• Low current gain • Second breakdown • Difficult to parallel
Darlington	3	Current	Bipolar	1,500	150	10,000	• High forward voltage drop • Second breakdown
SCR	3	Current	Bipolar	10,000	6,000	1,000	• No turn-off capability • Low operating frequency
TRIAC	3	Current	Bipolar	1,000	300	400	• No turn-off capability • Low operating frequency
GTO	3	Current	Bipolar	8,000	6,000	5,000	• Complex gate drive • Current filamentation
SIT	1	Current	Unipolar	1,200	300	200,000	• High on resistance • Normal on device • Low blocking gain
SITH	2	Current	Bipolar	1,500	500	20,000	• Normal on device • Complex gate drive
Power MOS	2	Voltage	Unipolar	1,600	400	200,000	• High on reistance
IGBT	3	Voltage	Bipolar	3,500	1.2	30,000	• Latch-up
MCT	4	Voltage	Bipolar	1,200	1,000 100	20,000	• Maximum turn-off current • Current filamentation • No current saturation

in the drift layer per unit area, the length of the drift region, and the substrate doping density. The use of charge control technology, namely, RESURF (reduced surface field)[213] enabled an increase in breakdown voltage or a corresponding reduction in its on resistance per unit area (specific on resistance) by almost a factor of 2.

Figure 173 also illustrates the RESURF principle. As shown in Fig. 173, a thin, lightly doped N^--type epitaxial layer is located on a P^- substrate. When the total charge in the N^- region between the P^+ source region and the N^+ drain region is large, the surface electric field near the channel reaches the critical electric field before the N^- epitaxial layer is fully depleted. Therefore, the surface-limited breakdown occurs. In the opposite case, when the charge of the N^- epitaxial layer is carefully controlled so that the surface electric field is always lower than that in the bulk, and the superficial N^- layer is fully depleted prior to reaching the critical electric field, the full bulk breakdown value is achieved. The RESURF technology permits realization of lateral power transistors with breakdown voltages of up to 1200 V.[214]

In high-frequency applications, the LDMOS offers the desired high switching speed, in the order of a few tens of nanoseconds, with no significant storage time. This is because current transport occurs solely by majority carriers. As an RF power device (for power amplification, not for power conversion), submicrometer channel LDMOS with very small gate-to-source is being used due to its low cost and high efficiency.[215]

LIGBT   The lateral insulated-gate bipolar transistor (LIGBT),[211,212] the lateral version of the IGBT, is another promising integrable power device. Its cross-sectional view is shown in Fig. 174a.

In the LIGBT structure, a lightly doped drift region is also needed to support a high forward blocking voltage. The use of the RESURF technology enabled an increase in breakdown voltage similar to that of an LDMOS. An N-buffer layer is needed in the LIGBT to prevent vertical punch-through breakdown of the vertical $P-N-P$ transistor formed by the P^+ collector and the P^- substrate. The operating mechanism of the LIGBT is similar to that of the IGBT. Because a high concentration of holes is injected from the P^+ collector into the N^- region, the conductivity of the long N^- region is modulated and the LIGBT has a specific on resistance that is lower by about a factor of 5–10 than that of the LDMOS transistor. As a consequence, the LIGBT results in substantial reduction of die size for the same power handling capability in comparison with the LDMOS, which is an important factor in power ICs.

However, during turn off, the minority carriers injected by the P^+ collector at the on state, called the storage charge, reduce the switching speed compared to the LDMOS. Storage charge effect improves the dc performance of the LIGBT, but degrades its switching performance and limits its usefulness at the high operating frequencies. Another disadvantage of the LIGBT is the existence of substrate current due to its bipolar current conduction mechanism, which may cause interference with the neighboring analog and digital ICs.

Similar to the IGBT, the LIGBT can improve its switching speed by lifetime control technique[187] at the expense of higher on resistance. But in the PIC, reducing the switching loss of the LIGBT by lowering the minority-carrier lifetime is difficult because it can also degrade the characteristics of other devices on the chip. An alternative is to use the shorted collector structure, as shown in Fig. 174b, which provides an efficient way to remove excess carriers at turn off, hence reducing the switching loss.[216]

In a shorted collector LIGBT, the device operates like a conventional LDMOS at low drain currents. As

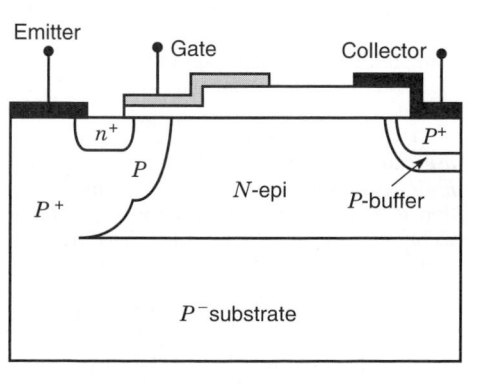

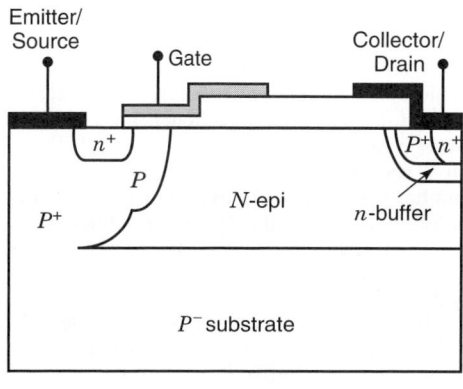

(a)   (b)

Fig. 174   Cross-sectional view of (a) an LIGBT and (b) a shorted collector LIGBT.

the current increases, the voltage drop across the N-buffer layer resistance underneath the P^+ collector increases, and eventually the P^+ collector becomes forward biased. Holes will then be injected into the N^--base region from the P^+ collector. As soon as the injected hole density becomes comparable to the N^--base region doping, conductivity modulation takes place and the on resistance is reduced. Because the P^+ region is shorted to the N^+ region, the emitter efficiency of the P^+ collector is lower than that of the conventional LIGBT, and the minority carrier concentration in the N^--base region is also lower. Further, because the N^- base is connected to the collector contact by the N^+ region, excess electrons can be removed rapidly by the N^+ drain, resulting in a faster turn-off process.

The LIGBT is susceptible to latch-up in the same way as the discrete IGBT because of an inherent parasitic $P-N-P-N$ thyristor in the device. At high current levels, sufficient voltage drop occurs across the emitter/body junction and the parasitic $N-P-N$ transistor turns on. The collector current of the $N-P-N$ constitutes the base current of the lateral $P-N-P$ transistor. When the sum of the current gain of the two transistors reaches unity, latch-up occurs and gate control is lost. In the RESURF LIGBT, a parasitic vertical $P-N-P$ transistor also exists that diverts some of the hole current into the substrate, and the latch-up threshold is increased. Another method to suppress latch-up is to lower the gain of the $P-N-P$ transistor by using an N-buffer layer. The buffer layer can also be used to limit carrier injection by controlling the collector emitter efficiency. This increases both latch-up current and switching speed and provides a convenient way to trade off speed and forward voltage drop.

Latch-up can be either static, as discussed previously, or dynamic, that is, during switching. Dynamic latch-up occurs because of a rapid increase in the displacement current as well as the change in electron and hole current components near the emitter side. Consequently, the rate at which the device is turned on is critical to dynamic latch-up.

LMCT   The lateral MOS-controlled thyristor (LMCT) is another integrable lateral power device, the lateral type of the MCT. The cross-sectional view of an LMCT is shown in Fig. 175.[217]

The LMCT shown in Fig. 175 is built on a P^-/P^+ substrate and has a P-channel DMOS transistor to turn it off and an N-channel lateral MOS transistor to turn it on. An efficient utilization of area is obtained by using the RESURF technique to achieve high breakdown voltage. By controlling the charge in the ion-implanted N-RESURF base layer, it becomes fully depleted at maximum blocking voltage and the breakdown occurs in the bulk at the N-RESURF/P^- substrate junction (J_2). The N-buffer layer has a higher doping concentration than the N-RESURF layer and is used to prevent punch-through breakdown between the anode and the P^- substrate in the off state.

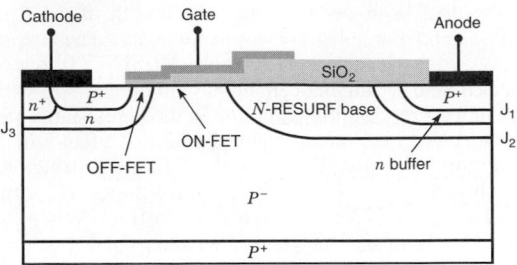

Fig. 175   Cross-sectional view of an LMCT structure.

The LMCT structure turns on by biasing the gate positively with respect to the cathode and turning on the lateral NMOS. Electrons flow into the N-RESURF base layer via the NMOS channel. These electrons form the base current of the $P-N-P$ transistor and turn the anode junction (J_1) on. Because of the $P-N-P-N$ thyristor in the LMCT, at a certain current level, regenerative action takes place and the LMCT latches up. Turn-off is achieved by biasing the gate negatively with respect to the cathode and turning the P-channel DMOS transistor on, which effectively shorts the base–emitter junction (J_3) of the $N-P-N$ transistor and diverts its base current to the cathode contact.

Because the LMCT operates at the latch-up state in its on state, it has much better current-carrying capability than the LDMOS and the LIGBT. However, the LMCT has a poorer current turn-off capability. Its maximum controllable turn-off current is limited by the P-channel DMOS transistor's channel resistance. Other variations of the LMCT also exist, which are being studied to improve performance and process compatibility.[218]

11.5   Isolation Technologies for PICs

In order to obtain high performance, low cost, small size, and high reliability of electronic equipment, the development of monolithic PICs has been promoted for several years. For integrating both power devices and low-voltage control ICs on the same silicon substrate, isolating the high- and low-voltage components is essential.

Junction Isolation   The junction isolation (JI, or $P-N$ isolation) is the most commonly used isolation technology for PICs. With the junction isolation technique, the silicon islands where the various components are integrated are separated through reverse-biased junctions realized by the diffusion of P regions through the entire depth of the N-type epitaxial layer grown on a P-type substrate (Fig. 176a). This technique is the most widely used because it offers the best compromise between cost and versatility. For thick epitaxial layers, up and down diffusion is used to form the isolation islands (Fig. 176b). Because the isolation diffusion must extend through the entire epitaxial layer

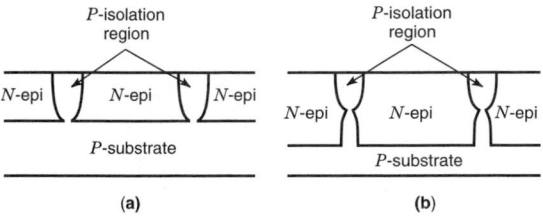

Fig. 176 Junction isolation techniques: (a) down diffusion in thin epitaxial layer and (b) up and down diffusion in thick epitaxial layer.

in the junction isolation, the thickness of the epitaxial layer is limited. However, blocking voltages of up to 1200 V are possible by applying RESURF technique.

The major drawback of junction isolation is that it uses a significant amount of silicon space. Moreover, it introduces an extra $P-N$ junction, which has capacitance and causes leakage current to the substrate. Under unfavorable conditions, JI can also introduce parasitic components such as a $P-N-P$ transistor.

Dielectric Isolation   Dielectric isolation (DI) allows the realization of silicon islands completely surrounded by oxide. DI has advantages such as low parasitic capacitance to the substrate, the absence of leakage current, and a reduction in the size of high-voltage components. However, this method requires a more complex and costly manufacturing process. Moreover, since oxide is a poor conductor of heat, it limits the integration of high-power devices.

The conventional dielectric isolation process (epitaxial passivated integrated circuit, EPIC) is shown in Fig. 177. Although EPIC is the dielectric isolation technology for power ICs in mass production, it has a problem with wafer warpage caused by its thick polysilicon layer, which must be solved before the wafer size can be increased and the minimum device patterning size can be lowered. Therefore, many advanced DI technologies are also being developed.

SOI Isolation Technology.   The SOI isolation technology is one of the dielectric isolation technologies

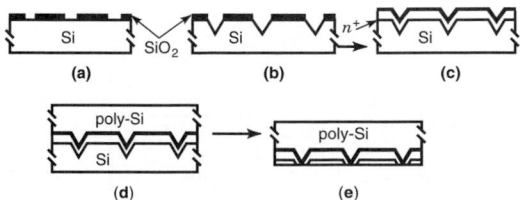

Fig. 177 Fabrication process flow of an EPIC-type wafer: (a) oxidation and photoetching, (b) V-groove etching, (c) N^+ deposition and isolation oxidation, (d) polysilicon epitaxial growth, and (e) grinding and polishing.

developed for ICs aimed at high-speed, high-level integration, and low power consumption. Although the origin of the SOI can be traced back to the 1934 patent of Oscar Heil on MIS structure, it is only in the 1980s that the SOI material became an evolution of the silicon and not an exotic revolutionary material. SOI material has been successfully introduced in production at the begin of the 1990s in some applications where limited volumes of wafer are required. This first IC market helped the SOI development. Among all the techniques proposed in the 1980s to perform the SOI structure, only three are still competing and are serious challengers of standard silicon: SDB (silicon direct bonding),[212] SIMOX (synthesis by implanted oxygen),[219] and smart cut.[220]

Power devices fabricated on an SOI substrate have attracted a lot of attention in the area of smart power integrated circuits. The reason for this is the many advantages offered by SOI over a conventional bulk substrate. V-groove etching, trench, or LOCOS (local oxidation of the silicon) isolation processes between adjacent devices on SOI offer true dielectric isolation and allow simple integration of power and logic devices on the same substrate. It is also possible to achieve significant improvements in breakdown voltage and switching speed with an SOI substrate. These advantages can be attributed to the excellent insulating properties of silicon dioxide in these devices. The buried oxide helps sustain a high electric field, which results in high breakdown voltage, and confines the carriers, which reduces the switching time of minority carrier devices.

On the other hand, the buried oxide underneath the device is also a good thermal insulator. The thermal conductivity of silicon dioxide is only 1.4 W/K · m compared to 140 W/K · m for bulk silicon. This significant difference impedes the dissipation of the heat generated inside the device. Therefore, the temperature rise inside an SOI device can be much higher than that in a bulk device.

In the SOI technologies, the SDB dielectric isolation technology and the SIMOX are promising for high-voltage power IC applications.

SDB Dielectric Isolation Technique.   The SDB (silicon direct bonding) dielectric isolation technique, which is also called BESOI (bond and etch back SOI) or DISDB (dielectric isolation by silicon wafer direct bonding) is widely used to prepare starting SOI material. The use of SDB technology has made SOI technology more viable and cost effective. The key process steps are shown in Fig. 178. Starting from two silicon wafers, at least one with an oxide layer on top, these two wafers are bonded together using van der Walls forces. Subsequent annealing increases the mechanical strength of the bonded interface by the chemical reaction which can occur at this interface. One of the substrates is then thinned down to the required thickness from several hundred microns by mechanical grinding and polishing.

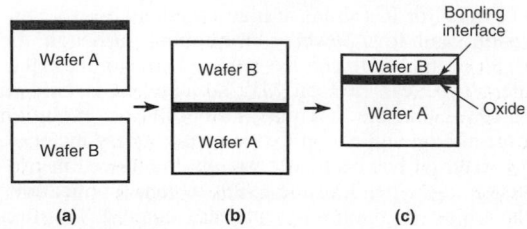

Fig. 178 Fabrication process flow of a SDB wafer: (a) oxidation of slice A, (b) cleaning and bonding, and (c) annealing, then grinding and polishing of wafer B.

The SDB dielectric isolation technique is a promising candidate for power ICs because thick silicon islands, which are required for handling high current and high voltage, can be easily fabricated. The adjacent devices isolation on the SDB wafer can be provided by etching V-grooves, by RIE (reactive ion etching) trench isolation, or even by LOCOS if the SOI layer is thin enough.

The SDB dielectric isolation technique is used not only in the isolation between lateral power devices such as LDMOSs and low-voltage integrated circuits, it is also used in the isolation between vertical power devices such as IGBTs and low-voltage integrated circuits. There are two methods for this application, as shown in Figs. 179 and 180.

Figure 179 shows the process flow of the first method for the SDB wafer to be used in isolating PICs. First, in the SDB wafer, the silicon and oxide films where the power device is to be formed are removed by wet etching. Then, the etched place is buried with Si epitaxial growth and the lapping and polishing are performed. Next, lateral isolation regions are formed by the conventional steps of groove etching (trench etching or V-groove etching), thermal oxidation for isolation film formation, and refilling the groove with polysilicon.

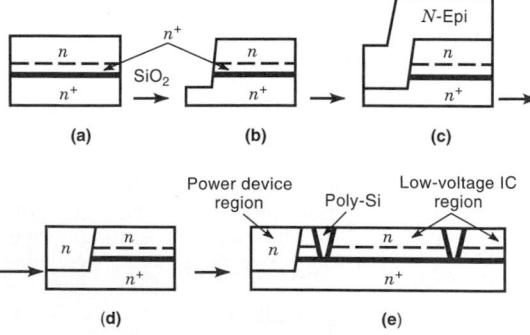

Fig. 179 Method 1 for using SDB in PICs: (a) SDB wafer, (b) silicon and oxide film etching, (c) epitaxial growth, (d) lapping and surface polishing, and (e) lateral isolation.

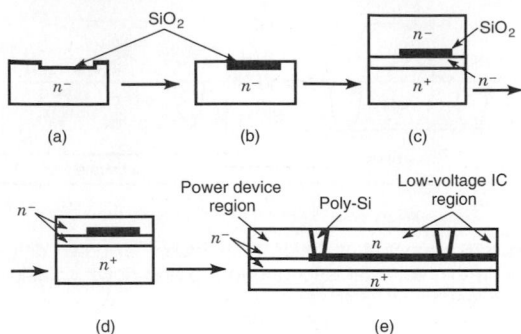

Fig. 180 Method 2 for using SDB in PICs: (a) etching and oxidation, (b) grinding and polishing, (c) bonding, (d) grinding and polishing, and (e) lateral isolation.

Another method is shown in Fig. 180. First, in the wafer A, the region where the low-voltage controlling circuit is formed is masked and etched. After thermal oxidation is carried out to form an isolation film, the wafer is polished until the optically flat Si–SiO$_2$ coexistent surface is exposed. Then, wafer bonding is performed. After that, the side of the bonded wafer on which devices are fabricated is ground and polished to a thickness of several microns. Finally, lateral isolation regions are formed by conventional steps of groove etching, thermal oxidation for isolation film formation, and refilling of the groove with polysilicon.

SIMOX. The SIMOX (separation by implanted oxygen, or selective implantation with oxygen) technique is considered to be one of the most advanced and promising SOI technologies for high-density CMOS circuits. The key processes of the SIMOX technology are shown in Fig. 181. First, oxygen ions are implanted into the silicon underneath the initial silicon surface. Then, a postimplantation annealing regenerates the crystalline quality of the silicon layer remaining over the oxide. This annealing also drives the chemical reaction that forms the stoichiometric oxide buried in the silicon wafer. Although the ideal annealing conditions are not fully identified yet, it is known that good SIMOX must be annealed at about 1320°C, for 6 h, in argon ambient containing 1% of oxygen.

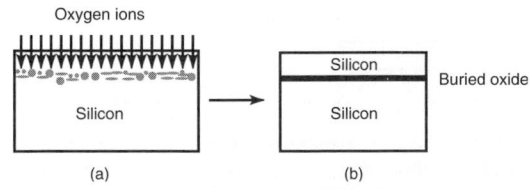

Fig. 181 Fabrication process of SIMOX wafer: (a) oxygen ion implantation and (b) high-temperature annealing.

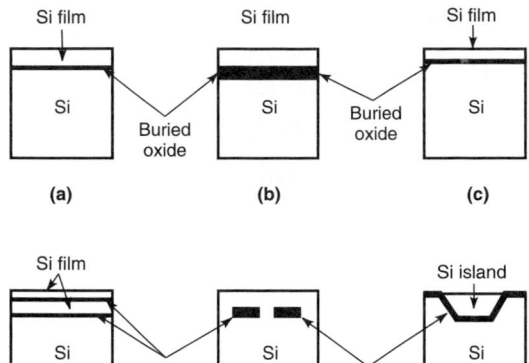

Fig. 182 Various types of SIMOX structure: (a) thin BOX, (b) thick BOX, (c) thin Si film, (d) double SIMOX, (e) interrupted BOX, and (f) totally isolated island SIMOX.

A number of SIMOX variants have been explored, as shown in Fig. 182. The thin or thick buried oxide (BOX) can be fabricated by using a lower implanted dose or a higher implanted dose. The thin and thick silicon film over the BOX can be obtained by using lower implanted energy or higher implanted energy. The double SIMOX structure is fabricated by two sequential oxygen implants. The thicker silicon layer can be achieved by using epitaxy technology. The interrupted BOX and the totally isolated island SIMOX can be processed by masked implantation.

The main disadvantage of the SIMOX technology is the need of >1300°C annealing, which could be a limitation for the 300-mm wafers.

The SIMOX technology is also considered to be one of the most promising dielectric isolation technologies for PICs. Figure 183 shows a PIC using SIMOX technology. As shown in Fig. 183, the wafer can be

selectively implanted with oxygen ions and forms local dielectric isolation. This local SIMOX technology with the epitaxy and the trench technology offers the unique opportunity to integrate monolithic devices with vertical current pathlike VDMOS or IGBT with lateral low-voltage control circuits by means of rather standard VLSI process steps. In contrast to other dielectric isolation technologies, no "exotic" process steps like selective epitaxy or mechanical back-lapping and surface polishing are necessary. Further, since the SIMOX technology involves forming a local buried oxide layer, improved smart power discrete devices are possible using SIMOX technology and they are being studied and developed.

SPSDB Technique. The single-silicon polysilicon direct bonding (SPSDB) technique,[221–223] which is also called laminated dielectric isolation (LDI or laminated DI) is a new SDB isolation method used for power ICs. Figure 184 shows the cross-sectional view of the SPSDB wafer. The SPSDB wafer has inverse V-groove isolation regions with a narrow isolation width of about 5 μm, which is independent of breakdown voltage. Consequently, the SPSDB technique has a very high packing density. Further, the SPSDB wafer has a simple fabrication process and is suited to mass production. Furthermore, the same design rules as those for the EPIC DI wafer can be utilized.

The process of the SPSDB is based on those of the EPIC and the SDB. Similar to the process of the conventional EPIC, first the ⟨100⟩ silicon wafer is oxide-masked and the V-grooves are etched with a preferential etching solution like KOH. When the grooves are completed, the etching stops automatically. The wafer is then oxidized and subsequently covered with a polysilicon layer about $80 - \mu$m thick. Next, the polysilicon layer is lapped and polished to about 10 μm thickness. The polished wafer is treated with $NH_4OH-H_2O_2$ solution at 70°C, rinsed with deionized water, and dried using a spin dryer. Then, a single-crystal silicon wafer is placed onto the polished polysilicon surface. These wafers are bonded at 1100°C for 2 h in an oxidizing atmosphere. Finally, the slice is inverted and the original single-crystal substrate is lapped and polished until the silicon islands are isolated from each other.

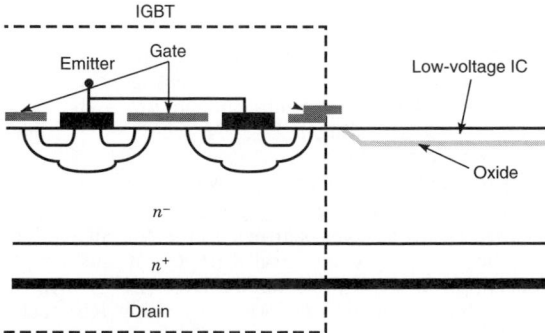

Fig. 183 One possible way of using SIMOX technology in PIC.

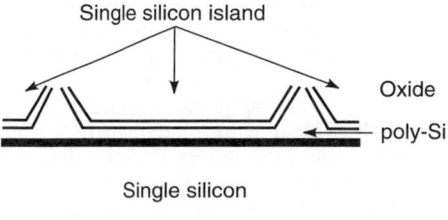

Fig. 184 Cross-sectional view of SPSDB wafer.

The SPSDB wafer has an unchanging warpage height during high-temperature heat treatments and has a high bonding strength comparable to that of the thermal oxidizing layer interface.

11.6 SiC-Based Power Devices

In recent years, silicon carbide (SiC) has received increased attention as a potential material for power devices operating at high temperatures, high power levels, and high frequencies due to its unique material properties.[224,225] Silicon carbide has a bandgap about three times wider than that of silicon (3.0 eV for 6H–SiC and 3.25 eV for 4H–SiC), high avalanche breakdown electric field of $2-4 \times 10^6$ V/cm, high saturated electron drift velocity of 2×10^7 cm/s, and high thermal conductivity of 4.9 W/cm · K. The high breakdown electric field allows the use of much higher doping and thinner layers for a given blocking voltage than silicon devices, resulting in much lower specific on resistance for unipolar devices. Further, high thermal conductivity and high saturated electron drift velocity also make SiC especially attractive in the power device arena.

With the arrival of commercial single-crystal substrates of 6H– and 4H–SiC and the ability to grow high-quality SiC epitaxial layers, the silicon carbide process has developed rapidly and the fabrication of power devices has become viable.

Among the SiC substrates, the two SiC polytypes 6H and 4H are having the biggest impact on power devices. Although the 6H–SiC has the best single-crystal quality of the established polytypes, the 4H–SiC is more attractive for power devices than the 6H–SiC. The reason for this is that the electron mobility in the 4H–SiC is two times that of the 6H–SiC in the direction perpendicular to the c axis and almost 10 times that of the 6H–SiC in the direction parallel to the c axis.

Compared with silicon and gallium arsenide materials, SiC has a lower mobility in the inversion layer and very small diffusion rates for dopants. These factors are limiting the pace of SiC power device development.

Schottky and *P–i–N* Junction Diodes High-voltage SiC Schottky rectifiers are already commercially available. Figure 185 shows the cross-sectional view of a SiC Schottky barrier diode (SBD, or Schottky diode) with high-resistance edge termination. This structure consists of an N^+-doped substrate with backside ohmic contact, a lightly doped epitaxial layer, and a topside Schottky contact with a high-resistance termination. The Schottky diode is fabricated by evaporating a high work function metal, such as titanium, nickel, or gold, onto the lightly doped epitaxial layer to form the Schottky contact and by depositing a metal onto the back of the N^+ substrate to form the back ohmic contact. The high resistivity edge termination is achieved by implanting argon, which damages the exposed semiconductor to create a high-resistance

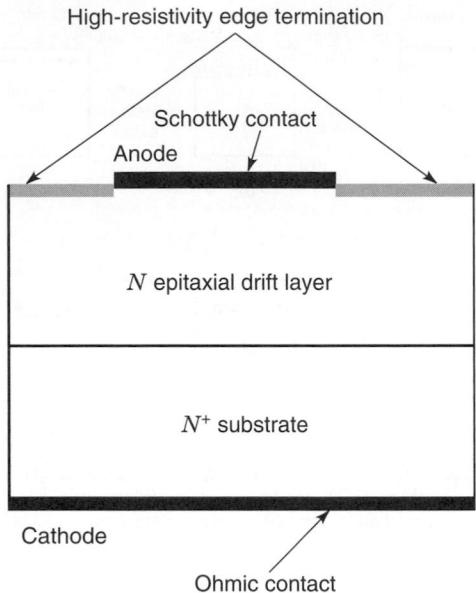

Fig. 185 Cross-sectional view of SiC Schottky diode with high-resistivity edge termination. The SiC Schottky diode consists of an N^+-doped substrate with backside ohmic contact, a lightly doped N epitaxial layer, and a topside Schottky contact surrounded by a high-resistance termination.

region. This process is self-aligned to the Schottky contact because the Schottky metal acts as a mask preventing damage under the contact.[226]

Because of the higher breakdown electric field, the epitaxial layer of the SiC SBD can have a higher doping and thinner drift layer at the same blocking capability when compared with that of a gallium arsenide (GaAs) SBD and Si SBD. The specific on resistance of a 1000-V 4H–SiC SBD is 15 times lower than that of a 1000-V GaAs SBD and over 200 times lower than that of a 1000-V silicon SBD. However, due to the high electron mobility of the GaAs material, a GaAs SBD has a lower specific on resistance than an SiC SBD at block voltages lower than 200 V.

SBDs have also been found to have excellent reverse recovery and reverse bias leakage characteristics even at high operating temperatures. They are likely to replace silicon $P–i–N$ rectifiers in high-voltage power electronic circuits.

A somewhat more complex device is a SiC $P–i–N$ diode, as shown in Fig. 186.[227] A high concentration N^+ SiC epitaxial layer is grown on the N^+ SiC wafer, and then an N-type epitaxial drift region and a high concentration P^+ thin epitaxial layer are grown. A mesa edge termination is formed by using RIE technology to block reverse voltage. This etching process was self-aligned in that the aluminum etch mask also acts as the top contact to the P^+ layer in the device.

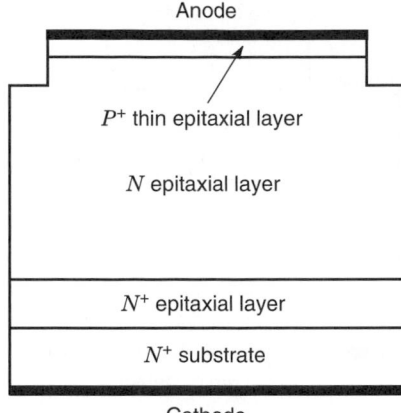

Fig. 186 Cross-sectional view of SiC $P-i-N$ diode. In the SiC $P-i-N$ diode structure, a high concentration N^+ SiC epitaxial layer is grown on the N^+ SiC wafer, and then an N-type epitaxial drift region and a high concentration P^+ thin epitaxial layer are grown. A mesa edge termination is formed by using RIE technology to block reverse voltage.

Thyristors For very high voltage (5–10 kV) applications, such as traction control and high-voltage dc transmission, silicon bipolar devices have much lower on-state losses than silicon unipolar devices. The same is expected to be true for SiC bipolar devices. At these very high voltages, a single SiC thyristor could replace a stack of silicon thyristors and thereby achieve a lower forward voltage drop.[228] In addition it is expected that properly designed SiC bipolar devices, which take advantage of the high breakdown field of SiC, will have lower voltage drops than silicon bipolar devices. The most promising SiC thyristor structure to date has been an $N-P-N-P$ device in 4H–SiC, as shown in Fig. 187. This structure utilized a mesa structure, with all of the doping being done in situ during epitaxy. The device periphery was terminated using an RIE mesa.

MOSFET SiC vertical power MOSFETs have a strong advantage over those made in silicon because the drift layer may use a 10 times higher doping level and one-tenth the thickness for a given breakdown voltage because of the much higher breakdown electric field of the SiC material. Ultimately, this could translate into specific on resistances as low as 1/300th that of an equivalent Si device. Because dopant diffusion rates in SiC material are very small, the UMOS process was considered to be the most suitable for making SiC power MOSFETs because the UMOS process can rely on epitaxy to form the channel region.[229]

The cross-sectional view of an SiC UMOSFET structure is shown in Fig. 188. An N^- epitaxial drift layer is grown on the N^+ substrate, and then a P-type channel layer is grown. N^+ source regions are formed by using implantation into the P-type channel layer.

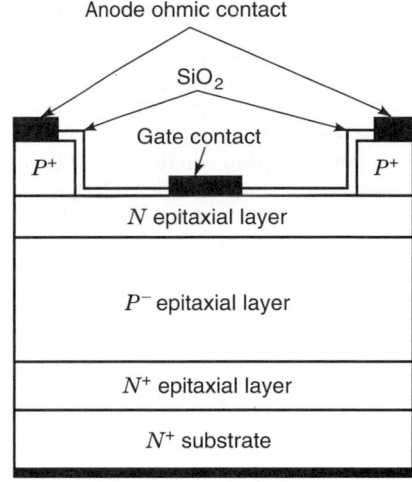

Fig. 187 Cross-sectional view of the SiC $N-P-N-P$ thyristor. In the SiC thyristor structure, epitaxy is used to grow all semiconductor layers. The RIE is used to define the gate contact. Because an N^+ substrate is used, the resulting thyristor is a p-type thyristor.

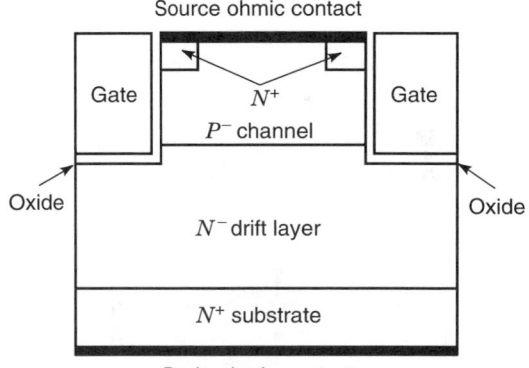

Fig. 188 Cross-sectional view of SiC UMOSFET. An N^- epitaxial drift layer is grown on the N^+ substrate, and then a P-type channel layer is epitaxially grown. N^+ source regions are formed by using implantation into the p-type channel layer. RIE is used to form the trenches on either side of p-type layer.

RIE is used to form the trenches on either side of the P-type layer. After the gate oxide (SiO_2) is grown and annealed, ohmic contacts are formed on the source and drain areas. Finally, the gate metal and interconnect metal are defined. During operation, current flows from the N^+ source contacts through an inversion channel layer to the N^- drift layer and they are collected by the N^+ drain. The current flow from source to drain is controlled by the voltage on the gate electrode.

It was found, however, that there is a high electric field at the corners of the trenches, which restricts the breakdown voltage of the UMOSFET far lower than its theoretical breakdown voltage. Further, the side-wall inversion channel mobilities in the SiC UMOSFET are lower than those in planar SiC MOSFETs, which leads to a severe increase in the on resistance of the device. A planar high-voltage SiC MOSFET using double implants has also been reported,[230] which avoids both of these problems by forming the inversion channel on the silicon. Shown in Fig. 189, the DMOS structure is formed by using multiple energy boron and nitrogen implants. Both implants are activated simultaneously at 1600°C for 30 min in an argon ambient. Then, the wafer is thermally oxidized to obtain the gate oxide, and the polysilicon is deposited or aluminum is thermally evaporated to form the gate electrode.

6H–SiC UMOS IGBT has also been experimentally demonstrated.[229] Its structure is similar to that of the SiC UMOS MOSFET, shown in Fig. 188, except that a p^+ collector substrate is used to replace the n^+ substrate of the MOSFET. The SiC IGBT has better forward conduction capability than that of the SiC MOSFET at high blocking voltage (>1000 V) and high operating temperature (>200°C) due to the conductivity modulation.

SIT   Although the SiC SIT (static induction transistor) has a structure resembling that of the UMOSFET, as shown in Fig. 190,[231] the operation mechanism is significantly different. The SiC SIT is a vertical device with an ohmic source contact on the top and an ohmic drain contact on the back of the wafer. Between the

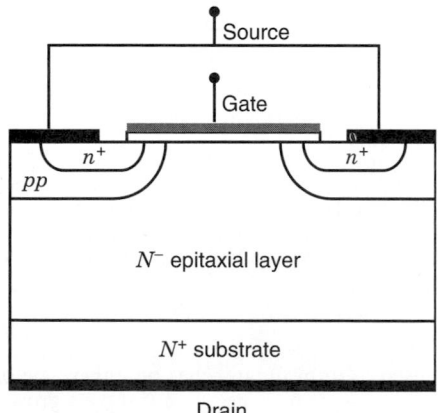

Fig. 189   Structure of planar SiC MOSFET. An N^- epitaxial drift layer is first grown on the N^+ substrate. The DMOS structure is formed by using multiple high-energy boron (*p* region) and nitrogen implants (*n* region).

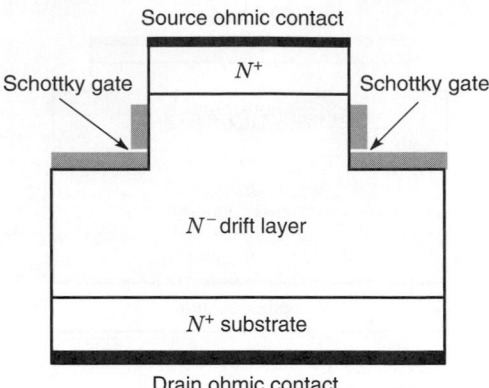

Fig. 190   Cross-sectional view of SiC SIT. An N^- epitaxial drift layer is grown on the N^+ substrate, and then an N^+ layer is grown. Trenches are etched to define the channel region, and Schottky gate contacts are formed in the bottom and along the sidewalls of the trench.

N^+ source and N^+ drain regions is an N^- epitaxial drift layer whose doping is one of the factors that determines the device breakdown voltage and pinch-off voltage. Trenches are etched to define the channel region, and Schottky gate contacts are formed in the bottom and along the side walls of the trench. Majority carriers flow from the source contact to the drain contact through the N-type channel region. By applying a negative voltage to the gate contact, the current flow can be modulated and even decreased to zero when depletion regions under each gate contact meet in the middle of the channel.

The SiC SIT is ideally suited to high-power microwave devices owing to the remarkable transport properties, very high breakdown field strength, and thermal conductivity of SiC. The SiC SIT is being developed as a discrete power microwave transistor for operation at frequencies up to S-band.

RF MESFET   The cross-sectional view of an RF SiC MESFET is shown in Fig. 191.[232] This device is a lateral device with both source and drain contacts on the top surface of the wafer. The MESFET epitaxial structure consists of an undoped P-buffer layer, N-type channel layer, and N^+ contact layer. The majority of carriers flow laterally from source to drain, confined to the N-type channel by the P^- buffer layer and controlled by the Schottky gate electrode.

For RF Si LDMOS, GaAs MESFET, and SiC MESFET, the device parameters that are important in different power densities are low field electron mobility, breakdown electric field, and electron saturation velocity. At a doping density of 1×10^{17} cm^{-3} the electron mobility of 4H–SiC is 560 cm^2/V · s, which is slightly lower than that of Si (800 cm^2/V · s) and significantly lower than that of GaAs (4900 cm^2/V · s). On the

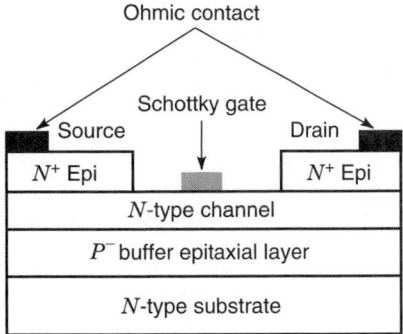

Fig. 191 Cross-sectional view of RF SiC MESFET. In the SiC RF MESFET structure, all semiconductor layers are epitaxially grown. The RIE is used to define the Schottky gate.

other hand, the breakdown electric field of 4H–SiC is about 10 times that of Si and GaAs, and the saturated drift velocity is 2 times that of Si and GaAs. Consequently, at low voltages, GaAs MESFETs, which have the highest electron mobility, have the highest power density. The higher power density of SiC MESFETs is only achieved at drain voltages higher than those normally used with either Si or GaAs devices.

RF JFET High-frequency SiC JFETs are of interest for high-temperature RF applications because a much lower gate leakage current can be obtained with a $P-N$ junction at high temperature than with a Schottky gate.[233] The cross-section of a SiC RF JFET (shown in Fig. 192) is similar to that of the RF MESFET, except a P^+. SiC epitaxial region with an ohmic contact on top is used in place of a Schottky contact,

and ion-implanted N^+ source and drain contact regions are used in place of the N^+ epitaxial region.

REFERENCES

1. Shockley, W., U.S. Patent 2,569,347, filed June 26, 1947; issued September 25, 1951.
2. Shockley, W., Sparks, M., and Teal, G. K., "*pn* Junction Transistors," *Phys. Rev.*, **83**, 151 (1951).
3. Ning, T. H., and Tang, D. D., "Bipolar Trends," *Proc. IEEE*, **74**, 1669 (1986).
4. Warnock, J. D., "Silicon Bipolar Device Structures for Digital Applications: Technology Trends and Future Directions," *IEEE Trans. Electron Devices*, **42**, 377 (1995).
5. Nakamura, T., and Nishizawa, H., "Recent Progress in Bipolar Transistor Technology," *IEEE Trans. Electron Devices*, **42**, 390 (1995).
6. Warner, R. M., Jr., and Grung, B. L., *Transistors: Fundamentals for the Integrated Circuit Engineer*, Wiley, New York, 1983.
7. Nakashiba, H., et al., "An Advanced PSA Technology for Highspeed Bipolar LSI," *IEEE Trans. Electron Devices*, **27**, 1390 (1980).
8. Tang, D. D., et al., "1.25 μm Deep-Groove-Isolated Self-Aligned Bipolar Circuits," *IEEE J. Solid-State Circuits*, **17**, 925 (1982).
9. Chen, T. C., et al., "A Submicron High-Performance Bipolar Technology," *Symp. VLSI Technol. Tech. Dig.*, 87 (1989).
10. Konaka, S., et al., "A 20-ps Si Bipolar IC Using Advanced Super-Self-Aligned Process Technology with Collector Ion Implantation," *IEEE Trans. Electron Devices*, **36**, 1370 (1989).
11. Shiba, T., et al., "A 0.5 μm Very-High-Speed Silicon Bipolar Technology U-Groove Isolated SICOS," *IEEE Trans. Electron Devices*, **38**, 2505 (1991).
12. de la Torre, V., et al., "MOSAIC V—A very high performance bipolar technology," paper presented at the Bipolar Circuits and Technology Meeting Tech. Dig., 21, 1991.
13. Warnock, J. D., et al., "High-Performance Bipolar Technology for Improved ECL Power-Delay," *IEEE Electron Device Lett.*, **12**, 315 (1991).
14. Cressler, J. D., et al., "A Scaled 0.25 μm Bipolar Technology Using Full E-Beam Lithography," *IEEE Electron Device Letters*, **13**, 262 (1992).
15. Uchino, T., et al., "15-ps ECL/74 GHz f_T Si Bipolar Technology," paper presented at the Int. Electron Device Meeting Tech. Dig., 67, 1993.
16. Richey, D. M., Cressler, J. D., and Joseph, A. J., "Scaling Issues and Ge Profile Optimization in Advanced UHV/CVD SiGe HBTs," *IEEE Trans. Electron Devices*, **44**, 431 (1997).
17. Roulston, D. J., *Bipolar Semiconductor Devices*, McGraw-Hill, New York, 1990.
18. Yang, E. S., *Microelectronic Devices*, McGraw-Hill, New York, 1988.
19. Pierret, R. F., *Semiconductor Device Fundamentals*, Addison-Wesley, New York, 1996.
20. Moll, J. L., and Ross, I. M., "The Dependence of Transistor Parameters on the Distribution of Base Layer Resistivity," *Proc. IRE*, **44**, 72 (1956).

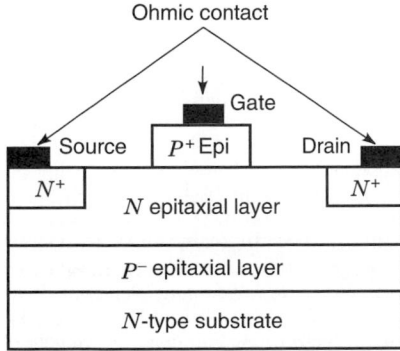

Fig. 192 Cross-sectional view of SiC JFET. In the SiC RF JFET structure, a p^- epitaxial layer is grown on the N-type substrate, and then an N-type epitaxial layer and high concentration P^+ epitaxial layer are grown. The P^+ mesa is formed by using RIE technology and N^+ source and drain regions are formed by using ion implantation.

21. Kapoor, A., and Roulston, D. (Eds.), *Polysilicon Emitter Bipolar Transistors*, IEEE Press, New York, 1989.

22. Post, I. R. C., Ashburn, P., and Wolstenholme, G., "Polysilicon Emitters for Bipolar Transistors: A Review and Re-evaluation of Theory and Experiment," *IEEE Trans. Electron Devices*, **39**, 1717 (1992).

23. Solomon, P. M., and Tang, D. D., "Bipolar circuit scaling," paper presented at the Int. Solid-State Circuits Conf. Tech. Dig., 86, 1979.

24. Kirk, C. T., Jr., "A Theory of Transistor Cutoff Frequency (ft) Falloff at High Current Densities," *IRE Trans. Electron Devices*, **9**, 164 (1962).

25. Rittner, E. S., "Extension of the Theory of the Junction Transistor," *Phys. Rev.*, **94**, 1161 (1954).

26. Webster, W. M., "On the Variation of Junction-Transistor Current-Amplification Factor with Emitter Current," *Proc. IRE*, **42**, 914 (1954).

27. Cressler, J. D., et al., "A High-Speed Complementary Silicon Bipolar Technology with 12-fJ Power-Delay Product," *IEEE Electron Device Lett.*, **14**, 523 (1993).

28. Onai, T., et al., "An npn 30 GHz, pnp 32 GHz f_T Complementary Bipolar Technology," paper presented at the Int. Electron Device Meeting Tech. Dig., 63, 1993.

29. Dekker, R., van den Einden, W. T. A., and Maas, H. G. R., "An Ultra Low Power Lateral Bipolar Polysilicon Emitter Technology on SOI," paper presented at the Int. Electron Device Meeting Tech. Dig., 75, 1993.

30. Cressler, J. D., "Re-Engineering Silicon: Si-Ge Heterojunction Bipolar Technology," *IEEE Spectrum*, pp. 49–55 (1995).

31. Cressler, J. D., and Niu, G., *Silicon-Germanium Heterojunction Bipolar Transistors*, Artech House, Boston, MA, 2003.

32. Cressler, J. D. (Ed.), *Silicon Heterostructure Handbook—Materials, Fabrication, Devices, Circuits, and Applications of SiGe and Si Strained-Layer Epitaxy*, CRC Press, Taylor & Francis Group, Boca Raton, FL, 2006.

33. Gopel, W., "Sensors in Europe and Eurosensors: State-of-the-Art and the Science in 1992," *Sensors Actuators A*, **37–38**, 1–5 (1993).

34. Pallás-Areny, R., and Webster, J. G., *Sensors and Signal Conditioning*, Wiley-Interscience, New York, 1991.

34a. Najafi, K., Wise, K. D., and Najafi, N., "Integrated Sensors," in S. M. Sze (Ed.), *Semiconductor Sensors*, Wiley, New York, 1994.

35. Sheingold, D. H. (Ed.), *Transducer Interfacing Handbook*, Analog Devices, Norwood, MA, 1980.

36. van de Plassche, R. J., Huijsing, J. H., and Sansen, W. M. C., *Analog Circuit Design—Sensor and Actuator Interfaces*, Kluwer, Norwell, MA, 1994.

37. Harjani, R., "Analog to Digital Converters," in W.-K. Chen (Ed.), *The Circuits and Filters Handbook*, IEEE/CRC Press, New York, 1995.

38. McCreary, J. L., and Gray, P. R., "All-MOS Charge Redistribution Analog-to-Digital Conversion Techniques—Part I," *IEEE J. Solid-State Circuits*, **10**, 371–379 (1975).

39. Lee, H. S., Hodges, D. A., and Gray, P. R., "A Self-Calibrating 15 Bit CMOS A/D Converter," *IEEE J. Solid-State Circuits*, **19**(6), 813–819 (1984).

40. Wang, F., and Harjani, R., *Design of Modulators for Oversampled Converters*, Kluwer, Norwell, MA, 1998.

41. Candy, J. C., and Temes, G. C. (Eds.), *Oversampling Delta-Sigma Data Converters—Theory, Design and Simulation*, IEEE Press, New York, 1992.

42. Sze, S. M. (Ed.), *Semiconductor Sensors*, Wiley, New York, 1994.

43. Bakker, A., and Huijsing, J., "Micropower CMOS Temperature Sensor with Digital Output," *IEEE J. Solid-State Circuits*, **SC-31**(7), 933–937 (1996).

44. Meijer, G., "An IC Temperature Transducer with an Intrinsic Reference," *IEEE J. Solid-State Circuits*, **SC-15**(3), 370–373 (1980).

45. Lin, S., and Salama, C., "A $V_{be}(T)$ Model with Applications to Bandgap Reference Design," *IEEE J. Solid-State Circuits*, **SC-20**(6), 1283–1285 (1985).

46. Song, B., and Gray, P., "A Precision Curvature-Compensated CMOS Bandgap References," *IEEE J. Solid-State Circuits*, **SC-18**(6), 634–643 (1983).

47. Kuijk, K., "A Precision Reference Voltage Source," *IEEE J. Solid-State Circuits*, **SC-8**(3), 222–226 (1973).

48. Enz, C., and Temes, G., "Circuit Techniques for Reducing the Effects of Op-Amp Imperfections: Autozeroing, Correlated Double Sampling, and Chopper Stabilization," *Proc. IEEE*, **84**(111), 1584–1614 (1996).

49. Robert, J., and Deval, P., "A Second-Order High-Resolution Incremental A/D Converter with Offset and Charge Injection Compensation," *IEEE J. Solid-State Circuits*, **23**(3), 736–741 (1988).

50. Nolan, I. B., *Data Analysis: An Introduction*, Polity Press, Cambridge, 1994.

51. Tukey, J. W., *Exploratory Data Analysis*, Addison-Wesley, Reading, MA, 1977.

52. Gelman, A., et al., *Bayesian Data Analysis*, Chapman & Hall, London, 1995.

53. Zadeh, L. A., "Fuzzy Sets," *Information Control*, **8**, 338–353 (1965).

54. Bandemer, H., Nather, W., *Fuzzy Data Analysis*, Kluwer, Dordrech, 1992; Berners-Lee, T., Cailliau, R., Luotonen, A., Nielsen, H. F., and Secret, A., "The World Wide Web," *Commun. the ACM*, **37**(8), 76–82 (1994); Baentsch, M., Baum, L., Molter, G., Rothkugel, S., and Sturm, P., "Enhancing the Web's Infrastructure: From Caching to Replication," *IEEE Internet Computing*, **1**(2), 18–27 (March/April 1997); Gudivada, V. N., Raghavan, V. V., Grosky, W. I., and Kasanagottu, R., "Information Retrieval on the World Wide Web," *IEEE Internet Computing*, **1**(5), 58–68 (September/October 1997); Florescu, D., Levy, A., and Mendelzon, A., "Database Techniques for the World Wide Web: A Survey," *ACM SIGMOD Record*, **27**(3), 59–74 (September 1998).

55. Obraczka, K., Danzig, P. B., and Li, S. H., "Internet Resource Discovery Services," *IEEE Comput. Mag.*, **26**(9), 8–22 (1993).

56. Chang, C. S., and Chen, A. L. P., "Supporting Conceptual and Neighborhood Queries on WWW," *IEEE Trans. Syst. Man Cybernet.*, **28**(2), 300–308 (1998); Chakrabarti, S., Dom, B., and Indyk, P., "Enhanced Hypertext Categorization Using Hyperlinks," in *Proceedings of ACM SIGMOD Conference on Management of Data*, 1998, pp. 307–318.

57. Johnson, A., and Fotouhi, F., "Automatic Touring in Hypertext Systems," in *Proc. IEEE Phoenix Conf. Comput. Commun.*, Phoenix, 1993, pp. 524–530; Buchner, A., and Mulvenna, M. D., "Discovering Internet Marketing Intelligence through Online Analytical Web Usage Mining," *ACM SIGMOD Record*, **27**(4), 54–61 (December 1998); Yan, T. W., Jacobsen, M., Garcia-Molina, H., and Dayal, U., "From User Access Patterns to Dynamic Hypertext Linking," *Computer Networks ISDN Syst.*, **28**, 1007–1014 (1996).

58. Salton, G., and McGill, M. J., *Introduction to Modern Information Retrieval*, McGraw-Hill, New York, 1983.

59. Salton, G., *Automatic Text Processing*, Addison Wesley, Reading, MA, 1989.

60. Pawlak, Z., "Rough Set," *Commun. ACM*, **38**(11), 88–95 (1995).

61. Pawlak, Z., *Rough Sets: Theoretical Aspects of Reasoning about Knowledge*, Kluwer, Norwell, MA, 1991.

62. Hu, X., and Cercone, N., "Mining Knowledge Rules from Databases: A Rough Set Approach," in *Proc. 12th Int. Conf. Data Eng.*, Ed. Y. W. Su Stanley (Ed.), New Orleans, LA, 1996, pp. 96–105.

63. Slowinski, R. (Ed.) *Handbook of Applications and Advances of the Rough Sets Theory*, Norwell, MA. Kluwer 1992.

64. Shockley, W., *Electrons and Holes in Semiconductors*, Van Nostrand, Princeton, NJ, 1950.

65. Sah, C. T., Noyce, R. N., and Shockley, W., "Carrier Generation and Recombination in p-n Junction and p-n Junction Characteristics," *Proc. IRE*, **45**, 1228–1243 (1957).

66. del Alamo, J. A., "Charge Neutrality in Heavily Doped Emitters," *Appl. Phys. Lett.*, **39**, 435–436 (1981).

67. Shockley, W., and Read, W. T., "Statistics of the Recombination of Holes and Electrons," *Phys. Rev.*, **87**, 835–842 (1952).

68. Hall, R. N., "Electron-Hole Recombination in Germanium," *Phys. Rev.*, **87**, 387 (1952).

69. Woo, J. C. S., Plummer, J. D., and Stork, J. M. C., "Non-Ideal Base Current in Bipolar Transistors at Low Temperatures," *IEEE Trans. Electron Devices*, **34**, 131–137 (1987).

70. Sproul, A. B., and Green, M. A., "Intrinsic Carrier Concentration and Minority Carrier Mobility from 77 K to 300 K," *J. Appl. Phys.*, **74**, 1214–1225 (1993).

71. Misiakos, K., and Tsamakis, D., "Accurate Measurements of the Intrinsic Carrier Density from 78 to 340 K," *J. Appl. Phys.*, **74**, 3293–3297 (1993).

72. Fossum, J. G., "Physical Operation of Back Surface Field Solar Cells," *IEEE Trans. Electron Devices*, **24**, 322–325 (1977).

73. Park, J. S., Neugroschel, A., and Lindholm, F. A., "Systematic Analytical Solution for Minority-Carrier Transport in Semiconductors with Position Dependent Composition with Application to Heavily Doped Silicon," *IEEE Trans. Electron Devices*, **33**, 240–249 (1986).

74. Kavadias, S., and Misiakos, K., "Three-Dimensional Simulation of Planar Semiconductor Diodes," *IEEE Trans. Electron Devices*, **40**, 1875–1878 (1993).

75. Sze, S. M., *Physics of Semiconductor Devices*, 2nd ed., Wiley, New York, 1981, p. 87.

76. Lindholm, F. A., "Simple Phenomenological Model of Transition Region Capacitance of Forward Biased p-n Junction Diodes or Transistor Diodes," *J. Appl. Phys.*, **53**, 7606–7608 (1983).

77. Jung, T., Lindholm, F. A., and Neugroschel, A., "Unifying View of Transient Responses for Determining Lifetime and Surface Recombination Velocity in Silicon Diodes and Back-Surface Field Solar Cells with Application to Experimental Short Circuit Current Decay," *IEEE Trans. Electron Devices*, **31**, 588–595 (1984).

78. Neugroschel, A., et al., "Diffusion Length and Lifetime Determination in p-n Junction Solar Cells and Diodes by Forward Biased Capacitance Measurements," *IEEE Trans. Electron Devices*, **25**, 485–490 (1978).

79. Vul, B. M., and Zavatitskaya, E. I., "The Capacitance of p/n Junctions at Low Temperatures," *Sov. Phys.—JETP (Engl. Transl.)*, **11**, 6–11 (1960).

80. Kavadias, S., et al., "On the Equivalent Circuit Model of Reverse Biased Diodes Made on High Resistivity Substrates," *Nucl. Instrum. Methods Phys. Res.*, **A322**, 562–565 (1992).

81. Misiakos, K., and Tsamakis, D., "Electron and Hole Mobilities in Lightly Doped Silicon," *Appl. Phys. Lett.*, **64**, 2007–2009 (1994).

82. Dziewior, J., and Schmid, W., "Auger Coefficients for Lightly Doped and Highly Excited Silicon," *Appl. Phys. Lett.*, **31**, 346–348 (1977).

83. Mahan, G. D., "Energy Gap in Si and Ge: Impurity Dependence," *J. Appl. Phys.*, **51**, 2634–2646 (1980).

84. Landsberg, P. T., et al., "A Model for Band-p Shrinkage in Semiconductors with Application to Silicon," *Phys. Status Solidi B*, **130**, 255–266 (1985).

85. Slotboom, J. W., and de Graaff, H. C., "Measurements of Band Gap Narrowing in Si Bipolar Transistors," *Solid-State Electron*, **19**, 857–862 (1976).

86. Wieder, A. W., "Emitter Effects in Shallow Bipolar Devices: Measurements and Consequences," *IEEE Trans. Electron Devices*, **27**, 1402–1408 (1980).

87. del Alamo, J. A., and Swanson, R. M., "Measurement of Steady-State Minority-Carrier Recombination in Heavily Doped n-Type Silicon," *IEEE Trans. Electron Devices*, **34**, 1580–1589 (1987).

88. Jonscher, A. K., "p-n Junctions at Very Low Temperatures," *Br. J. Appl. Phys.*, **12**, 363–371 (1961).

89. Yang, Y. N., Coon, D. D., and Shepard, P. F., "Thermionic Emission in Silicon at Temperatures Below 30 K," *Appl. Phys. Lett.*, **45**, 752–754 (1984).

90. Misiakos, K., Tsamakis, D., and Tsoi, E., "Measurement and Modeling of the Anomalous Dynamic

Response of High Resistivity Diodes at Cryogenic Temperatures," *Solid-State Electronics*, **41**, 1099–1103 (1997).

91. Nicollian, E. H., and Tsu, R., "Electrical Properties of a Silicon Quantum Dot Diode," *J. Appl. Phys.*, **74**, 4020–4025 (1993).

92. Normand, P., et al., "Silicon Nanocrystal Formation in Thin Thermaloxide Films by Very-Low Energy Si Ion Implantation," *Microelectron. Eng.*, **36**(1–4), 79–82 (1997).

93. Yano, K., et al., "Room-Temperature Single-Electron Memory," *IEEE Trans. Electron Devices*, **41**, 1628–1638 (1994).

94. Dimaria, D. J., et al., "Electroluminescence Studies in Silicon Dioxide Films Containing Tiny Silicon Islands," *J. Appl. Phys.*, **56**, 410 (1984).

95. Zhang, F., Wenham, S., and Green, M. A., "Large Area, Concentrator Buried Contact Solar Cells," *IEEE Trans. Electron Devices*, **42**, 145–149 (1995).

96. de Silva, C. W., *Control Sensors and Actuators*, Prentice-Hall, Englewood Cliffs, NJ, 1989.

97. Sydenham, P. H., Hancock, N. H., and Thorn, R., *Introduction to Measurement Science and Engineering*, Wiley, New York, 1989.

98. Doebelin, E. O., *Measurement Systems: Application and Design*, 4th ed., McGraw-Hill, New York, 1990.

98a. Holman, J. P., *Experimental Methods for Engineers*, 5th ed., McGraw-Hill, New York, 1989.

99. Schaller, R. S., "Moore's Law: Past, Present and Future," *IEEE Spectrum*, **34**(6), 52–59 (June 1997).

100. Chen, J. Y., "CMOS—The Emerging Technology," *IEEE Circuits Devices Mag.*, **2**(2), 16–31 (1986).

101. Wolf, S., and Tauber, R. N., *Silicon Processing for the VLSI Era: Process Integration*, Vol. 2, Lattice Press, Sunset Beach, CA, 1986.

102. Kahng, D., and Atalla, M. M., "Silicon-Silicon Dioxide Field Induced Surface Devices," paper presented at the IRE Solid State Devices Res. Conf., Carnegie Inst. Technol., Pittsburgh, PA, 1960.

103. Weste, N. H. E., and Esharaghian, K., *Principles of CMOS VLSI Design*, 2nd ed., Addison-Wesley, Reading, MA, 1993.

104. Pao, H. C., and Shah, C. T., "Effects of Diffusion Current on Characteristics of Metal-Oxide (Insulator)-Semiconductor Transistors (MOST)," *Solid State Electron.*, **9**, 927–937 (1966).

105. Sze, S. M., *Physics of Semiconductor Devices*, Wiley, New York, 1981.

106. Hodges, D. A., and Jackson, H. G., *Analysis and Design of Digital Integrated Circuits*, McGraw-Hill, New York, 1983.

107. Chaterjee, P. K., "Gigabit Age Microelectronics and Their Manufacture," *IEEE Trans. VLSI Syst.*, **1**, 7–21 (1993).

108. Wu, C. Y., Wang, J. S., and Tsai, M. K., "The Analysis and Design of CMOS Multidrain Logic and Stacked Multidrain Logic," *IEEE JSSC*, **SC-22**, 47–56 (1987).

109. Johnson, M. G., "A Symmetric CMOS NOR Gate for High Speed Applications," *IEEE JSSC*, **SC-23**, 1233–1236 (1988).

110. Schultz, K. J., Francis, R. J., and Smith, K. C., "Ganged CMOS: Trading Standby Power for Speed," *IEEE JSSC*, **SC-25**, 870–873 (1990).

111. Susuki, Y., Odagawa, K., and Abe, T., "Clocked CMOS Calculator Circuitry," *IEEE JSSC*, **SC-8**, 462–469 (1973).

112. Sakurai, T., et al., "Hot-Carrier Generation in Submicrometer VLSI Environment," *IEEE JSSC*, **SC-21**, 187–191 (1986).

113. Krambeck, R. H., Lee, C. M., and Law, H. S., "High-Speed Compact Circuits with CMOS," *IEEE JSSC*, **SC-17**, 614–619 (1982).

114. Friedman, V., and Liu, S., "Dynamic Logic CMOS Circuits," *IEEE JSSC*, **SC-19**, 263–266 (1984).

115. Gonclaves, N. F., and DeMan, H. J., "NORA: A Race-free Dynamic CMOS Technique for Pipelined Logic Structures," *IEEE JSSC*, **SC-18**, 261–266 (1983).

116. Lee, C. M., and Szeto, E. W., "Zipper CMOS," *IEEE Circuits Devices*, **2**(3), 101–107 (1986).

117. Heller, L. G., et al., "Cascade Voltage Switch Logic: A Differential CMOS Logic Family," in *Proc. IEEE Int. Solid State Circuits Conf.*, San Francisco, CA, February 16–17, 1984.

118. Simon, T. D., "A Fast Static CMOS NOR Gate, in *Proc. 1992 Brown/MIT Conf. Advanced Res. VLSI Parallel Syst.*, T. Knight and J. Savage (Eds.), MIT Press, Cambridge, MA, 1992, pp. 180–192.

119. Radhakrishnan, D., Whitaker, S. R., and Maki, G. K., "Formal Design Procedures for Pass Transistor Switching Circuits," *IEEE JSSC*, **SC-20**, 531–536 (1985).

120. Terman, C. J., in *Simulation Tools for VLSI, VLSI CAD Tools and Applications*, W. Fichtner and M. Morf (Eds.), Kluwer, Norwell, MA, 1987, Chapter 3.

121. Nagel, L. W., "SPICE2: A Computer Program to Simulate Semiconductor Circuits, Memo ERL-M520, Dept. Electr. Eng. Comput. Sci., Univ. California, Berkeley, May 9, 1975.

122. Weeks, W. T., et al., "Algorithms for ATSAP—A Network Analysis Program," *IEEE Trans. Circuit Theory*, **CT-20**, 628–634 (1973).

123. *HSPICE User's Manual H9001*, Meta-Software, Campbell, CA, 1990.

124. Terman, C., "Timing Simulation for Large Digital MOS Circuits," in *Advances in Computer-Aided Engineering Design*, Vol. 1, A. Sangiovanni-Vincentelli (Ed.), JAI Press, Greenwich, CT, 1984, pp. 1–91.

125. White, J., and Sangiovanni-Vincentelli, A., *Relaxation Techniques for the Simulation of VLSI Circuits*, Kluwer, Hingham, MA, 1987.

126. Lyon, R. F., "Simplified Design Rules for VLSI Layouts," *LAMBDA*, **II**(1), 54–59 (1981).

127. Mead, C. A., and Conway, L. A., *Introduction to VLSI Systems*, Addison-Wesley, Reading, MA, 1980.

128. Rubin, S. M., *Computer Aids for VLSI Design*, Addison-Wesley, Reading, MA, 1987, Chapter 11.

129. Ousterhout, J. K., et al., "Magic: A VLSI Layout System," in *Proc. 21st Design Autom. Conf.*, 1984, pp. 152–159.

130. Lauther, U., "A Min-Cut Placement Algorithm for General Cell Assemblies Based on a Graph," in *Proc. 16th Design Autom. Conf.*, 1979, pp. 1–10.

131. Kuh, E. S., "Recent Advances in VLSI Layouts," *Proc. IEEE*, **78**, 237–263 (1990).

132. Kirkpatrick, S., Gelatt, C., and Vecchi, M., "Optimization by Simulated Annealing," *Science*, **220**(4598), 671–680 (1983).

133. Sechen, C., and Sangiovanni-Vincentelli, A., "TimberWolf 3.2: A new Standard Cell Placement and Global Routing Package," in *Proc. 23rd Design Autom. Conf.*, Las Vegas, NV, 1986, pp. 432–439.

134. Clow, G. W., "A Global Routing Algorithm for General Cells," in *Proc. 21st Design Autom. Conf.*, Albuquerque, NM, 1984, pp. 45–50.

135. Dupenloup, G., "A Wire Routing Scheme for Double Layer Cell-Layers," in *Proc. 21st Design Autom. Conf.*, Albuquerque, NM, 1984, pp. 32–35.

136. Moore, E. F., "The Shortest Path through a Maze," in *Proc. Int. Symp. Switching Theory*, Vol. 1, Harvard University Press, 1959, pp. 285–292.

137. Lee, C. Y., "An Algorithm for Path Connection and Its Applications," *IRE Trans. Electron. Comput.*, 346–365 (September 1961).

138. Lengauer, T., *Combinatorial Algorithms for Integrated Circuit Layouts*, Wiley, New York, 1981.

139. Doerschel, J., and Kirscht, F. G., "Differences in Plastic Deformation Behavior of CZ and FZ Grown Si Crystals," *Phys. Status Solid*, **A64**, K85–K88 (1981).

140. Zuhlehner, W., and Huber, D., *Czochralski Grown Silicon, Crystals*, Vol. 8, Springer-Verlag, Berlin, 1982.

141. Biddle, D., "Characterizing Semiconductor Wafer Topography," *Microelectron. Manuf. Testing*, **15**, 15–25 (1985).

142. Fiegl, G., "Recent Advances and Future Directions in CZ-Silicon Crystal Growth Technology," *Solid State Technol.*, **26**(8), 121–131 (1983).

143. Suzuki, T., et al., "CZ Silicon Growth in a Transverse Magnetic Field, in *Semiconductor Silicon 1981*, Electrochemical Society, Pennington, NJ, 1981, pp. 90–93.

144. Bloem, J., and Gilling, L. J., "Epitaxial Growth by Chemical Vapor Deposition," in *VLSI Electronics*, Vol. 12, N. G. Einspruch and H. Huff (Eds.), Academic, Orlando, FL, 1985, Chapter 3, pp. 89–139.

145. Wolf, S., and Tauber, R. N., *Silicon Processing for the VLSI Era: Process Technology*, Lattice Press, Sunset Beach, CA, 1986.

146. Burggraff, P., "Ion Implantation in Wafer Fabrication," *Semiconductor Int.*, **39**, 39–48 (1981).

147. Glawishnig, H., and Noack, N., "Ion Implantation System Concepts," in *Ion Implantation, Science and Technology*, J. F. Ziegler (Ed.), Academic, Orlando, FL, 1984, pp. 313–373.

148. Thompson, L. F., and Bowden, M. J., "The Lithographic Process: The Physics," in *Introduction to Microlithography*, L. F. Thompson, C. G. Willson, and M. S. Bowden (Eds.), Advances in Chemistry Series, Vol. 219, American Chemical Society, Washington, DC, 1983, pp. 15–85.

149. King, M. C., "Principles of Optical Lithography," in *VLSI Electronics Micro Structure Science*, Vol. 1, N. G. Einspruch (Ed.), Academic, New York, 1981.

150. Gwozdz, P. S., "Positive vs. Negative: A Photoresist Analysis," *SPIE Proc., Semicond. Lithography VI*, **275**, 156–182 (1981).

151. Elliot, D. J., *Integrated Circuit Fabrication Technology*, McGraw-Hill, New York, 1982, Chapter 8.

152. Braken, R. C., and Rizvi, S. A., "Microlithography in Semiconductor Device Processing," in *VLSI Electronics—Microstructure Science*, Vol. 6, N. G. Einspruch and G. B. Larabee (Eds.), Academic, Orlando, FL, 1983, pp. 256–291.

153. Kooi, E., and Appels, J. A., "Semiconductor Silicon 1973," in The Electrochem. Symp. Ser., H. R. Huff and R. Burgess (Eds.), Electrochemical Society, Princeton, NJ, 1973, pp. 860–876.

154. Deroux-Dauphin, P., and Gonchond, J. P., "Physical and Electrical Characterization of a SILO Isolation Structure," *IEEE Trans. Electron Devices*, **ED-32**(11), 2392–2398 (1985).

155. Mikoshiba, M., "A New Trench Isolation Technology as a Replacement of LOCOS," *IEDM Tech. Dig.*, 1984, pp. 578–581.

156. Pauleau, Y., "Interconnect Materials for VLSI Circuits: Part II: Metal to Silicon Contacts," *Solid-State Technol.*, **30**(4), 155–162 (1987).

157. Nicolet, M. A., and Bartur, M., "Diffusion Barriers in Layered Contact Structures," *J. Vacuum Sci. Technol.*, **19**(3), 786–793 (1981).

158. Agrawal, V. D., and Seth, S. C., *Tutorial: Test Generation for VLSI Chips*, IEEE Computer Society Press, Los Alamitos, CA, 1988.

159. Chakradhar, S. T., Bushnell, M. L., and Agrawal, V. D., "Toward Massively Parallel Automatic Test Generation," *IEEE Trans. CAD*, **9**, 981–994 (1990).

160. Calhoun, J. D., and Brglez, F., "A Framework and Method for Hierarchical Test Generation," *IEEE Trans. CAD*, **11**, 598–608 (1988).

161. Reghbati, H. K., *Tutorial: VLSI Testing and Validation Techniques*, IEEE Computer Society Press, Los Alamitos, CA, 1985.

162. Malay, W., "Realistic Fault Modeling for VLSI Testing," in *IEEE/ACM Proc. 24th IEEE Design Autom. Conf.*, Miami Beach, FL, 1987, pp. 173–180.

163. Jayasumana, A. P., Malaiya, Y. K., and Rajsuman, R., "Design of CMOS Circuits for Stuck-Open Fault Testability," *IEEE JSSC*, **26**(1), 58–61 (1991).

164. Acken, J. M., "Testing for Bridging Faults (Shorts) in CMOS Circuits," in *Proc. 20th IEEE/ACM Design Autom. Conf.*, Miami Beach, FL, 1983, pp. 717–718.

165. Lee, K., and Breuer, M. A., "Design and Test Rules for CMOS Circuits to Facilitate IDDQ Testing of Bridging Faults," *IEEE Trans CAD*, **11**, 659–670 (1992).

166. Goldstein, L. H., and Thigpen, E. L., "SCOAP: Sandia Controllability/Observability Analysis Program," in *Proc. 17th Design Autom. Conf.*, 1980, pp. 190–196.

167. Eichelberger, E. B., and Williams, T. W., "A Logic Design Structure for LSI Testing," *J. Design Autom. Fault Tolerant Comput.*, **2**(2), 165–178 (1978).

168. Gupta, R., Gupta, R., and Breuer, M. A., "An Efficient Implementation of the BALLAST Partial Scan Architecture," in *IFIP Proc. Int. VLSI'89 Conf.*, Munich, 1990, pp. 133–142.

169. Ando, H., "Testing VLSI with Random Access Scan," *IEEE/ACM Dig. Papers COMPCON 80*, February 1980, pp. 50–52.

170. Frohwerk, R. A., "Signature Analysis—A New Digital Field Service Method," *Hewlett Packard J.*, **28**(9), 2–8 (1977).

171. Koenemann, B., Mucha, J., and Zwiehoff, G., "Built-in Logic Block Observation Techniques," *Dig. 1979 IEEE Test Conf.*, October 1979, pp. 37–41.

172. *IEEE Standard 1149.1-1990: IEEE Standard Test Access Port and Boundary-Scan Architecture*, IEEE Standards Board, New York, p. 19.

173. Patterson, D. A., and Hennessy, J. L., *Computer Architecture: A Quantitative Approach*, 2nd ed., Morgan Kaufmann, San Francisco, 1996.

174. Blaauw, G. A., and Brooks, F. P., Jr., *Computer Architecture Concepts and Evolution*, Addison-Wesley, Reading, MA, 1997.

175. Siewiorek, D. P., Bell, C. G., and Newell, A., *Computer Structures: Principles and Examples*, McGraw-Hill, New York, 1981.

176. Malone, M. S., *The Microprocessor: A Biography*, Springer-Verlag, Santa Clara, CA, 1995.

177. Helfrick, D. A., and Cooper, W. D., *Modern Electronic Instrumentation and Measurement Techniques*, Prentice-Hall, Englewood Cliffs, NJ, 1990.

178. *TDS 210 and TDS 220 Digital Real-Time Oscilloscopes*, 070-8483-02, Tektronix, Beaverton, OR, 1997.

179. "Digital Serial Analyzer Sampling Oscilloscope," http://www.tektronix.com.

180. White, A., "Low-Cost, 100-MHz Digitizing Oscilloscopes," *Hewlett-Packard J.*, 43(1), 6–11 (February 1992).

181. "Measuring Random Jitter on a Digital Sampling Oscilloscope," Application Note, HFAN-04.5.1, Rev 0,08/02, Maxim, http://www.maxim-ic.com.

182. "XYZ of Oscilloscopes," http://www.tektronix.com.

183. "ABC's of Probes," http://tektronix.com.

184. "VEE," http://adn.tm.agilent.com/index.cgi?CONTENT_ID=830.

185. Khalid, S. F., *Lab Windows/CVI Programming for Beginners*, Prentice Hall, http://www.phptr.com, 2000.

186. Baliga, B. J., *Power Semiconductor Devices*, PWS Publishing, Boston, 1996.

187. Baliga, B. J., and Sun, E., "Comparison of Gold, Platinum, and Electron Irradiation for Controlling Lifetime in Power Rectifier," *IEEE Trans. Electron Devices*, **ED-24**, 685–688 (1977).

188. Mehrotra, M., and Baliga, B. J., "Very Low Forward Drop JBS Rectifiers Fabricated Using Submicron Technology," *IEEE Trans. Electron Devices*, **ED-40**, 2131–2132 (1993).

189. Tu, L., and Baliga, B. J., "Controlling the Characteristics of the MPS Rectifier by Variation of Area of Schottky Region," *IEEE Trans. Electron Devices*, **ED-40**, 1307–1315 (1993).

190. Kitagawa, M., Matsushita, K., and Nakagawa, A., "High-Voltage Emitter Short Diode (ESD)," *Japan, J. Appl. Phys.*, **35**, 5998–6002 (1997).

191. Schlangenotto, H., et al., "Improved Recovery of Fast Power Diodes with Self-Adjusting, *p* Emitter Efficiency," *IEEE Electron Device Lett.*, **10**, 322–324 (1989).

192. Blanc, J., "Practical Application of MOSFET Synchronous Rectifiers," paper presented at the 13th Int. Telecommun. Energy Conf., INTELEC-91, 1991, pp. 495–501.

193. Mohan, N., Undeland, T. M., and Robbins, W. P., *Power Electronics*, 2nd ed., Wiley, New York, 1995.

194. Ng, K. K., *Complete Guide to Semiconductor Devices*, McGraw-Hill, New York, 1995.

195. Syau, T., Venkatraman, P., and Baliga, B. J., "Comparison of Ultralow Specific On-Resistance UMOSFET Structure: The ACCUFET, EXTFET, INVFET, and Conventional UMOSFET's," *IEEE Trans. Electron Devices*, **ED-41**, 800–808 (1994).

196. Mori, M., Nakano, Y., and Tanaka, T., "An Insulated Gate Bipolar Transistor with a Self-Aligned DMOS Structure," *IEEE Int. Electron Devices Meet. Dig.*, **IEDM-88**, 813–816 (1988).

197. Chow, T. P., et al., "A Self-Aligned Short Process for Insulated-Gate Bipolar Transistor," *IEEE Trans. Electron Devices*, **ED-39**, 1317–1321 (1992).

198. Miller, G., and Sack, J., "A New Concept for Non-Punch Through IGBT with MOSFET Like Switching Characteristics," paper presented at the Conf. Rec. IEEE Power Electron. Specialists, 1989, pp. 21–25.

199. Laska, T., Miller, G., and Niedermeyer, J., "2000-V Non-Punch Through IGBT with High Ruggedness," *Solid-State Electron*, **35**, 681–685 (1992).

200. Chang, H. R., et al., "Insulated Gate Bipolar Transistor (IGBT) with Trench Gate Structure," *IEEE Int. Electron Devices Meet. Dig.*, **IEDM-87**, 674–677 (1987).

201. Harada, M., et al., "600-V Trench IGBT in Comparison with Planar IGBT," *Int. Symp. 1994 IEEE Int. Symp. Power Semicond. Devices and IC's*, **ISPSD-94**, 411–416 (1994).

202. Omura, I., et al., "Carrier Injection Enhancement Effect of High Voltage MOS Devices," *1997 IEEE Int. Symp. Power Semicond. Devices and ICs*, **ISPSD-97**, 217–220 (1997).

203. Eicher, S., et al., "Advanced Lifetime Control for Reducing Turn-Off Switching Loss of 4.5 kV IEGT Devices," *1998 IEEE Int. Symp. Power Semicond. Devices and ICs*, **ISPSD-98**, 39–42 (1998).

204. Suekawa, E., et al., "High Voltage IGBT (HV-IGBT) Having p^+/p^- Collection Region, "*1998 IEEE Int. Symp. Power Semicond. Devices and ICs*, **ISPSD-98**, 249–252 (1998).

205. Temple, V. A. K., "MOS Controlled Thyristors (MCT's)," *IEEE Int. Electron Devices Meet. Dig.*, **IEDM-84**, 282–285 (1984).

206. Huang, Q., et al., "Analysis of *n*-Channel MOS Controlled Thyristors," *IEEE Trans. Electron Devices*, **ED-38**, 1612–1618 (1991).

207. Huang, A. Q., "Analysis of the Inductive Turn-Off of Double Gate MOS Controlled Thyristor," *IEEE Trans. Electron Devices*, **ED-43**, 1029–1032 (1996).

208. Baliga, B. J., "Trends in Power Semiconductor Devices," *IEEE Trans. Electron Devices*, **ED-43**, 1727–1731 (1996).

209. Huang, A. Q., "A Unified View of the MOS Gated Thyristors," *Solid-State Electronics*, **42**(10), 1855–1865 (1998).

210. *Power Semiconductors*, ed. 36, DATA Digest, an IHS group company, Englewood, 1996.

211. Hart, P. A. H. (Ed.), *Bipolar and Bipolar-MOS Integration*, Elsevier Science, Amsterdam, The Netherlands, 1994.

212. Murari, B., Bertotti, F., and Vignola, G. A., (Eds.), *Smart Power ICs*, Springer, New York, 1995.

213. Appels, J. A., and Vaes, H. M. J., "High Voltage Thin Layer Devices (RESURF Devices)," *IEEE Int. Electron Devices Meet. Dig.*, **IEDM-79**, 238–241 (1979).

214. Ludikuize, A. W., "A Versatile 700–1200 V IC Process for Analog and Switching Applications," *IEEE Trans. Electron Devices*, **ED-38**, 1582–1589 (1991).

215. Wood, A., Dragon, C., and Burger, W., "High Performance Silicon LDMOS Technology for 2-GHz RF Power Amplifier Applications," *IEEE Int. Electron Devices Meet. Dig.*, **IEDM-96**, 87–90 (1996).

216. Simpson, M. R., et al., "Analysis of the Lateral Insulated Gate Transistor," *IEEE Int. Electron Devices Meet. Dig.*, **IEDM-85**, 740–743 (1985).

217. Darwish, M. N., "A New Lateral MOS Controlled Thyristor," *IEEE Electron Device Lett.*, **11**, 256–257 (1990).

218. Huang, A. Q., "Lateral Insulated Gate $P–i–N$ Transistor (LIGPT)—A New MOS Gate Lateral Power Device," *IEEE Electron Device Lett.*, **17**, 297–299 (1996).

219. Haddara, H. (Ed.), *Characterization Methods for Submicron MOS-FETs*, Kluwer, Boston, 1995.

220. Bruel, M., Aspar, B., and Auberton-Herve, A. J., "Smart Cut: A New Silicon on Insulator Material Technology Based on Hydrogen Implantation and Wafer Bonding," *Jpn. J. Appl. Phys., Part 1*, **36**, 1636–1641 (1997).

221. Inoue, Y., Sugawara, Y., and Kurita, S., "Characteristics of New Dielectric Isolation Wafers for High Voltage Power ICs by Single-Si Poly-Si Direct Bonding (SPSDB) Technique," *IEEE Trans. Electron Devices*, **ED-42**, 356–358 (1995).

222. Easier, W. G., et al., "Polysilicon to Silicon Bonding in Laminated Dielectrically Isolated (LDI) Wafers," in *Proc. 1st Int. Symp. Semicond, Wafer Bonding*, 1991, pp. 223–229.

223. Sugawara, Y., Inoue, Y., and Kurita, S., "New Dielectric Isolation for High Voltage Power ICs by Single Silicon Poly Silicon Direct Bonding (SPSDB) Technique," paper presented at the 1992 IEEE Int. Symp. Power Semicond. Devices and ICs, ISPSD-92, 1992, pp. 316–319.

224. Weitzel, C. E., et al., "Silicon Carbide High-Power Devices," *IEEE Trans. Electron Devices*, **ED-43**, 1732–1739 (1996).

225. Palmour, J. W., et al., "Silicon Carbide for Power Devices," paper presented at the 1997 IEEE Int. Symp. Power Semicond. Devices and ICs, ISPSD-97, 1997, pp. 25–32.

226. Bhatnagar, M., Mclarty, P., and Baliga, B. J., "Silicon-Carbide High-Voltage (400 V) Schottky Barrier Diodes," *IEEE Electron Device Lett.*, **13**, 501–503 (1992).

227. Neudeck, P. G., and Fazi, C., "Positive Temperature Coefficient of Breakdown Voltage in 4H–SiC $P–N$ Junction Rectifiers," *IEEE Electron Device Lett.*, **18**, 96–98 (1997).

228. Palmour, J. W., et al., "Silicon Carbide Substrates and Power Devices," in *Compound Semiconductors 1994*, H. Goronkin and U. Mishra (Eds.), IOP Publishing, Bristol, UK; *Inst. Phys. Pub.*, **141**, 377–382 (1994).

229. Ramungul, N., et al., "A Fully Planarized 6H-SiC UMOS Insulated-Gate Bipolar-Transistor," paper presented at the 54th Annu. Device Res. Conf., 1996, pp. 24–26.

230. Shenoy, J. N., Cooper, J. A., and Melloch, M. R., "High-Voltage Double-Implanted Power MOSFET's in 6H-SiC," *IEEE Electron Device Lett.*, **18**, 93–95 (1997).

231. Siergiej, R. R., et al., "High Power 4H–SiC Static Induction Transistors," *IEEE Int. Electron Devices Meet. Dig.*, **IEDM-95**, 353–356 (1995).

232. Weitzel, C. E., "Comparison of Si, GaAs, and SiC RF MESFET Power Densities," *IEEE Electron Device Lett.*, **16**, 451–453 (1995).

233. Weitzel, C. E., et al., "SiC Microwave Power MESFET's and JFET's," in *Compound Semiconductors 1994*, H. Goronkin and U. Mishra (Eds.), IOP Publishing, Bristol, UK, **141**, 389–394 (1994).

BIBLIOGRAPHY

Abele, M. G., *Structures of Permanent Magnets*, Wiley, New York, 1993.

"About Oscilloscope," http://www.hobbyprojects.com/oscilloscope_tutorial.html.

"Advances in Oscilloscope Technology," LeCroy, White Paper, http://www.lecroy.com.

Annaratone, M., *Digital CMOS Circuit Design*, Kluwer, Norwell, MA, 1986.

Baecker, R. M., and Buxton, W. S. (Eds.), *Readings in Human-Computer Interaction: A Multidisciplinary Approach*, Morgan Kaufmann, San Mateo, CA, 1987.

Balakrishnan, et al., R., "The Rockin' Mouse: Integral 3D Manipulation on a Plane," in *CHI97 Conf. Proc.*, Atlanta, GA, ACM, 1997.

Barfield, W., and Furness III, T. A. (Eds.), *Virtual Environments and Advanced Interface Design*, Oxford University Press, Oxford, 1995.

Belove, C. (Ed.), *Handbook of Modern Electronics and Electrical Engineering*, Wiley, New York, 1986.

Bentley, J. P., *Principles of Measurement Systems*, 2nd ed., Longman Scientific and Technical, Burnt Mill, UK, 1988.

Bogart, T. F., *Electronic Devices and Circuits*, 3rd ed., Macmillan, New York, 1993.

Brey, B. B., *Microprocessors and Peripherals: Hardware, Software, Interfacing, and Applications*, 2nd ed., Macmillan, New York, 1988.

Chang, C. Y., and Sze, S. M., *ULSI Technology*, McGraw-Hill, New York, 1996.

Dahl, P. F., *Superconductivity*, American Institute of Physics, New York, 1992.

Dix, A., et al., *Human-Computer Interaction*, Prentice-Hall, Englewood Cliffs, NJ, 1993.

Esposito, C., *User Interfaces for Virtual Reality Systems*, Tutorial Notes, CHI'96, Vancouver, British Columbia, Canada, 1996.

Fink, D. G., and Christiansen, D. (Eds.), *Electronics Engineers' Handbook*, McGraw-Hill, New York, 1982.

Fishbane, P. M., Gasiorowicz, S., and Thornton, S. T., *Physics for Scientists and Engineers*, Prentice-Hall, Upper Saddle River, NJ, 1996.

Gallagher, R. S., *Computer Visualization*, CRC Press, Boca Raton, FL, 1995.

Giancoli, D. C., *Physics Principles and Applications*, Prentice-Hall, Englewood Cliffs, NJ, 1991.

Gieger, R. L., Allen, P. E., and Strader, N. R., *VLSI Design Techniques for Analog and Digital Circuits*, McGraw-Hill, New York, 1990.

Glasser, L. A., and Dobberpuhl, D. W., *The Design and Analysis of VLSI Circuits*, Addison-Wesley, Reading, MA, 1985.

Gopel, W., Hesse, J., and Zemel, J. N., *Sensors—A Comprehensive Survey*, WCH, Weinheim, Germany, 1989.

Hartson, H. R., and Hix, D., *Advances in Human-Computer Interaction*, Vol. 4, Ablex, Norwood, NJ, 1993.

Herbst, L. J., *Integrated Circuit Engineering*, Oxford University Press, London, 1996.

Holmes-Siedle, A., and Adams, L., *Handbook of Radiation Effects*, Oxford University Press, New York, 1993.

Interrante, L. V., Casper, L. A., and Ellis, A. B. (Eds.), *Materials Chemistry*, American Chemical Society, Washington, DC, 1995.

Kang, S. M., and Leblebici, Y., *CMOS Digital Integrated Circuits*, McGraw-Hill, New York, 1996.

Kaufaman, M., and Seidman, A. H. (Eds.), *Handbook for Electronic Engineering Technicians*, McGraw-Hill, New York, 1984.

Metzger, D., *Electronic Components, Instruments, and Troubleshooting*, Prentice-Hall, Englewood Cliffs, NJ, 1981.

Mukherjee, A., *Introduction to nMOS and CMOS VLSI Systems Design*, Prentice-Hall, Englewood Cliffs, NJ, 1986.

Pucknell, D. A., and Eshraghian, K., *Basic VLSI Design: Systems and Circuits*, Prentice-Hall of Australia, Sydney, 1988.

Rosenstein, M., and Morris, P., *Modern Electronic Devices: Circuit Design and Application*, Reston Publishing Company, Reston, VA, 1985.

Sadiku, M. N. O., *Elements of Electromagnetics*, Saunders College Publishing, Orlando, FL, 1994.

Schroeter, J., *Surviving the ASIC Experience*, Prentice-Hall, Englewood Cliffs, NJ, 1992.

Seymour, J., *Electronic Devices and Components*, Pitman Publishing, London, 1981.

Sherwani, N., *Algorithms for VLSI Physical Design Automation*, Kluwer, Boston, 1993.

Shoji, M., *CMOS Digital Circuit Technology*, Prentice-Hall, Englewood Cliffs, NJ, 1988.

Smith, M. S., *Application Specific Integrated Circuits*, Addison-Wesley, Reading, MA, 1997.

Solymar, L., and Walsh, D., *Lectures on the Electrical Properties of Materials*, 3rd ed., Oxford University Press, Oxford, 1984.

Thomson, C. M., *Fundamentals of Electronics*, Prentice-Hall, Englewood Cliffs, NJ, 1979.

Tocci, R. J., *Digital Systems Principles and Applications*, 5th ed., Prentice-Hall, Englewood Cliffs, NJ, 1991.

Tompkins, W. J., and Webster, J. G., *Interfacing Sensors to the IBM PC*, Prentice-Hall, Englewood Cliffs, NJ, 1988.

"User's and Service Guide, 3000 Series Oscilloscopes," Agilent Technologies, http://www.agilent.com.

Wolf, S., and Smith, R. F. M., *Student Reference Manual for Electronic Instrumentation Laboratories*, 2nd ed, Prentice-Hall, Upper Saddle River, NJ, 2004.

Wolf, W., *Modern VLSI Design: A System Approach*, Prentice-Hall, Englewood Cliffs, NJ, 1994.

CHAPTER 18

LIGHT AND RADIATION

M. Parker Givens
Institute of Optics
University of Rochester
Rochester, New York

1 INTRODUCTION

Radiation is the transfer of energy through space without requiring any intervening medium; for example, the energy reaching Earth from the sun is classified as radiation.

The majority of this is in the form of electromagnetic waves, which have a wide range of frequencies. Of this, a relatively narrow frequency band between 4×10^{14} and 8×10^{14} Hz is capable of stimulating the visual system; this is light. Our attention will be directed primarily toward the visible part of the spectrum, but most of the principles are valid in other parts of the spectrum.

Electromagnetic radiation propagates through empty space with velocity c, which is one of the fundamental constants of nature; its approximate value is $2.998\ldots \times 10^8$ m/s. This velocity is independent of frequency. In any material medium, the velocity of propagation v is less than c; the ratio $c/v \equiv n$ is called the *index of refraction* of the medium. The velocity v (and therefore n) depends upon the frequency of the radiation; this variation of velocity (or index) with frequency is known as dispersion.

The simplest wave to discuss is one in which some physical quantity varies sinusoidally with time at any point in space and this variation propagates with velocity v. Such a wave is represented by an equation of the form

$$b = A \sin \frac{2\pi}{\lambda}(x \pm vt) \qquad (1)$$

where b represents the value of the quantity at position x and at time t and A is the maximum value of b and is called the *amplitude* of the wave. This equation represents a *plane wave*, that is, the quantity b is constant over a plane surface perpendicular to the x axis. The minus sign gives a wave propagating in the positive x direction; the plus sign gives a wave propagating in the negative x direction. The wavelength λ represents the smallest, nonzero distance for which $b(x + \lambda) = b(x)$ for all x; alternately, λ may be defined as the distance between adjacent crests of the wave. See Fig. 1, which is a plot of b as a function of x for some fixed t. Equation (1) is meaningful for all values of x and t; in this sense, it represents a wave of infinite extent in time and space. We can also define the period T of the wave as the time required to execute one cycle, or the smallest, nonzero, time for which $b(t + T) = b(t)$. Frequency v is the reciprocal of the period; for period in seconds, the frequency unit is the hertz. Equation (1) is called a monochromatic wave since it contains only one frequency. These quantities are interrelated as follows:

$$v = \lambda v = \frac{\lambda}{T} \quad \text{or} \quad \lambda = vT$$

As the radiation passes from one medium to another, such as from glass to air, the frequency remains constant, but the velocity and wavelength change.

No real wave extends indefinitely in time but must begin and end; real waves cannot be monochromatic in the strictest interpretation of the term.

The methods of Fourier analysis enable us to construct a finite wave train as a sum of appropriately selected infinite wave trains of the proper phase. The amplitude as a function of time in the finite wave train and the amplitude as a function of frequency for the infinite components form a Fourier transform pair. Although the details in each case will depend upon the manner in which the wave builds up initially and dies away at the end, some "rule-of-thumb" statements are often helpful. These are:

$$\frac{\Delta v}{\overline{v}} \approx \frac{\overline{T}}{\Delta t} \approx \frac{\overline{\lambda}}{v \Delta t} \approx \frac{\Delta \lambda}{\lambda}$$

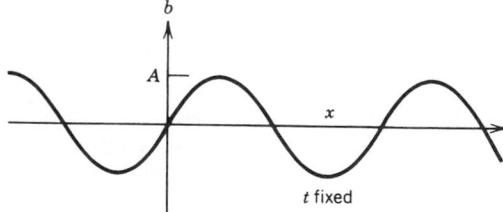

Fig. 1　Simple harmonic wave of the form $y = A \sin(2\pi / \lambda)$ $(x - vt)$. Plot is for fixed $t = 0$. Over time, disturbance moves to right.

where Δt is the duration of the wave train; Δv is the frequency spread of the infinite, or monochromatic, components making up the wave train; $\overline{v}$, $\overline{T}$, and $\overline{\lambda}$ are the average frequency, period, and wavelength of the finite wave. These are rule-of-thumb or *order-of-magnitude* statements. A wave for which $\Delta v \ll \overline{v}$ is properly called *quasi-monochromatic* but is frequently called *monochromatic*. For many classical (i.e., prelaser) light sources, $\Delta v / \overline{v} \approx 10^{-5}$ and the wave is, for most practical purposes, monochromatic. Sunlight, on the other hand, has a very broad spectral range (Δv comparable to $\overline{v}$) and may be described equally well as a series of randomly spaced short pulses or a broad spectrum of randomly phased monochromatic waves. The two descriptions are equally valid and interchangeable.

As already mentioned, light is electromagnetic radiation. The quantities described by Eq. (1) are electric and magnetic fields $\mathbf{E}$ and $\mathbf{B}$. These two fields each obey Eq. (1) and are in phase with each other. They are perpendicular to each other in space and each is perpendicular to the direction of propagation. This is illustrated in Fig. 2, which shows $\mathbf{E}$ in the y direction and $\mathbf{B}$ in the z direction for a wave propagated in the positive x direction. Here, $\mathbf{E}$, $\mathbf{B}$, and v form a right-handed orthogonal system so that in order to have a wave propagated in the negative x direction, either $\mathbf{E}$ or $\mathbf{B}$ must be reversed.

Within the constraint that it remain in a plane perpendicular to the direction of propagation, $\mathbf{E}$ may have any direction. Usually the direction of $\mathbf{E}$ changes in a random way, and the light is called unpolarized. If the direction of $\mathbf{E}$ remains constant, the light is called linearly polarized. Also, $\mathbf{B}$ is always perpendicular to $\mathbf{E}$.

Theoretical considerations indicate that in vacuum $c = (\mu_0 \varepsilon_0)^{-1/2}$, which experiment confirms. Theory also predicts that in a medium $v = (\mu \varepsilon)^{-1/2}$ or $n = (\mu \varepsilon / \mu_0 \varepsilon_0)^{1/2}$. Here μ and ε are the permeability and permittivity of the medium; μ_0 and ε_0 are the corresponding quantities for vacuum. The prediction for the velocity in a real medium cannot be experimentally confirmed since μ and ε are frequency dependent and at optical frequencies the only available measurements are the measurements of n or v; there are no direct measurements of μ and ε. The velocity v in Eq. (1) is the phase velocity. For a finite wave train or pulse, the envelope of the pulse moves forward with the *group velocity* U. The value of U may be expressed in a variety of forms, including

$$U = \frac{c}{(d / dv)(nv)} = \frac{c}{n} \left(1 + \frac{\lambda}{n} \frac{dn}{d\lambda} \right) \qquad (2)$$

For common transparent materials, $dn / d\lambda < 0$ and $U < v$. In nondispersive media (e.g., vacuum), $U = v$; λ is the wavelength in the medium.

One common aspect of wave propagation, as observed with water waves, is the tendency of the

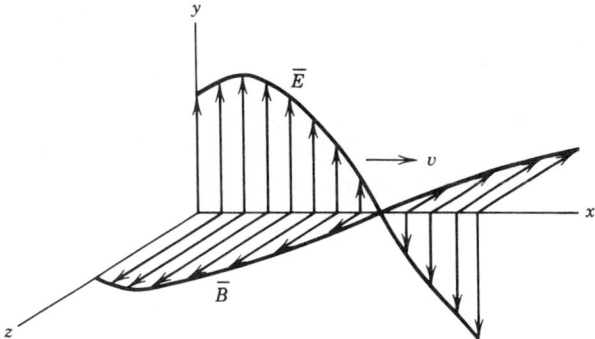

Fig. 2 Electromagnetic wave, where **E**, **B**, and v form right-handed orthogonal system as shown. Plot is for some fixed time; wave moves to right; E_y and B_z have maximum values for same value of x.

wave to spread into the shadow region behind barriers. This phenomenon is known as diffraction. Light also exhibits diffraction, but the effects are much smaller than for water waves because the wavelengths of light waves are so small ($\sim 5 \times 10^{-7}$ m). Diffraction effects are important if we attempt to pass light through openings only a few wavelengths wide or to focus the light into a very small spot. Otherwise, we may describe the light in terms of "rays" that represent the direction of energy flow and coincide with the direction of propagation. In a homogeneous isotropic medium, the rays are straight. That part of optics that may be treated by tracing rays is called geometric optics.

2 GEOMETRIC OPTICS

If a ray of light strikes a boundary separating two homogeneous isotropic media such as air and glass (see Fig. 3), a simple wave calculation will show and experiment will confirm the following statements:

1. The incident ray will be partially reflected at the boundary and partially transmitted (refracted) into the second medium.
2. The incident ray, the reflected ray, the refracted ray, and the normal to the surface (erected at the point of incidence) are coplanar.
3. The angle of reflection θ' is equal to the angle of incidence θ; these angles are measured between the surface normal and the rays, as shown in Fig. 3.
4. The angle of refraction ϕ and the angle of incidence θ are related by the following equation, which is known as Snell's law:

$$\frac{\sin \theta}{\sin \phi} = \frac{v_1}{v_2} = \frac{n_2}{n_1} \qquad (3)$$

where v_1 and v_2 are the velocities of propagation in medium 1 and medium 2 and n_1 and n_2 are the indices of refraction.

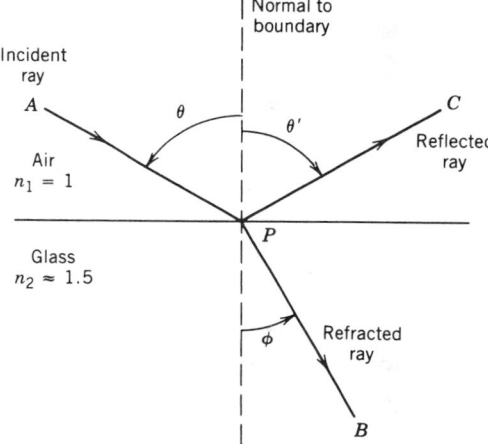

Fig. 3 Refraction at plane boundary separating two media with different indices of refraction. According to Snell's law, $n_1 \sin \theta = n_2 \sin \phi$. In the special case for which θ_B satisfies condition $\tan \theta_B = n_2/n_1$, reflected light is linearly polarized.

These statements can be proven without postulating that the light wave is electromagnetic. However, one must recognize the electromagnetic nature of the wave in order to calculate the fraction of the light that is reflected or transmitted. The path of light is reversible; that is, light will travel from B to A along the same path BPA.

It is a straightforward exercise in calculus to show that if A and B (of Fig. 3) are fixed points, one in each medium, and if P is an arbitrary point on the boundary, then the location of P that minimizes the propagation time from A to P to B is the same location of P for which Snell's law is satisfied. Also, for A and C as fixed points in the same medium, the choice of P that produces a minimum of $AP + PC$ is the same P for which $\theta = \theta'$. These are examples of Fermat's principle, which states that the path of an actual ray from one point to another is a path for which the transit time is stationary. By *stationary* we mean that the derivative of the transit time with respect to small changes in the path (such as small changes in the location of P) must be zero.

For a ray such as APB in Fig. 3, which passes through more than one medium, it is convenient to define the *optical pathlength* from A to B as $n_1 AP + n_2 PB$, or in case the ray passes through many media, the optical pathlength is $\sum_i l_i n_i$, where l_i is the pathlength in medium i and n_i is the index of medium i. The optical pathlength between two points A and B is the distance in vacuum that light could travel during the time required to propagate from A to B through the intervening media.

Snell's law and the law of reflection are sufficient to explain the image-forming properties of lenses and mirrors. We first turn our attention to lenses and to the case in which the same medium (air) is on both sides of the lens. The lens will have spherical surfaces and rotational symmetry about some line called the axis; distances are measured along and perpendicular to this axis. The following results are usually derived using small-angle approximations, $\sin \alpha = \alpha = \tan \alpha$, and are called *paraxial calculations*. Consider first a ray that is parallel to the axis at a distance h above the axis (see Fig. 4). Upon passing through the lens, the ray will be refracted according to Snell's law and cross the axis at the point F in the figure. For a good lens, the point F is independent of h; it is called the focal point of the lens. As the ray passes through F, it has a slope of u; in the figure, u is negative. The ratio $h/(-u)$ is the focal length f. For a thin lens, f is the distance from the lens to F, where F is the second (or back) focal point. There is another point, F', in front

of the lens called the first (or front) focal point. Any ray that passes through F' (with slope u) and strikes the lens will be refracted to be parallel to the axis (at height h). The front focal length, $h/u \equiv f'$, will be equal to f provided there is the same medium on both sides of the lens. For a thin lens in air

$$\frac{1}{f'} = \frac{1}{f} = (n - 1)\left(\frac{1}{r_1} - \frac{1}{r_2}\right) \qquad (4)$$

where n is the index of refraction of the lens material and r_1 and r_2 are the radii of curvature of the first and second surfaces of the lens; r_1 and r_2 are considered positive (negative) if the center of curvature of the surface is to the right or downstream (left or upstream) relative to the surface; and $1/f$ is the power of the lens. In Fig. 4, r_1 is positive and r_2 is negative; for this lens, the focal length f is positive. This lens is *convergent*.

Figure 5 shows the application of these definitions to a *negative*, or divergent, lens. In this case, r_1 is negative and r_2 is positive, making f negative. Notice that F, the *second* focal point, is to the left of the lens; the refracted ray does not pass through F but must be extended backward to intersect the axis (at F). In Fig. 5b, the incident ray is headed toward F', the *first* focal point, but is refracted by the lens to be parallel to the axis.

For both positive and negative lenses, a ray that crosses the axis at the center of the lens continues undeviated into the region beyond. This is called a chief ray.

It follows from Eq. (4) that positive, or convergent, lenses are thicker on axis than at the edge, whereas negative, or divergent, lenses are thinner on axis than at the edge. They may have a variety of shapes, as illustrated in Fig. 6.

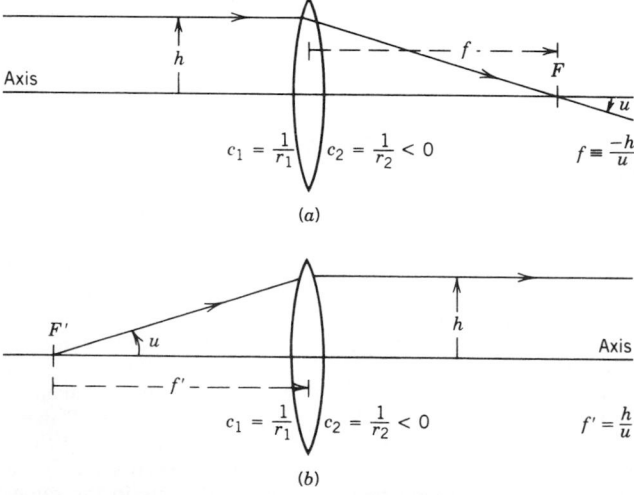

Fig. 4 Two focal points of *positive* lens. In (a) F is second (or back) focal point. In (b) F' is first (or front) focal point. For lens shown, r_1 (radius of curvature of front surface) is positive; r_2 (radius of curvature of second surface) is negative.

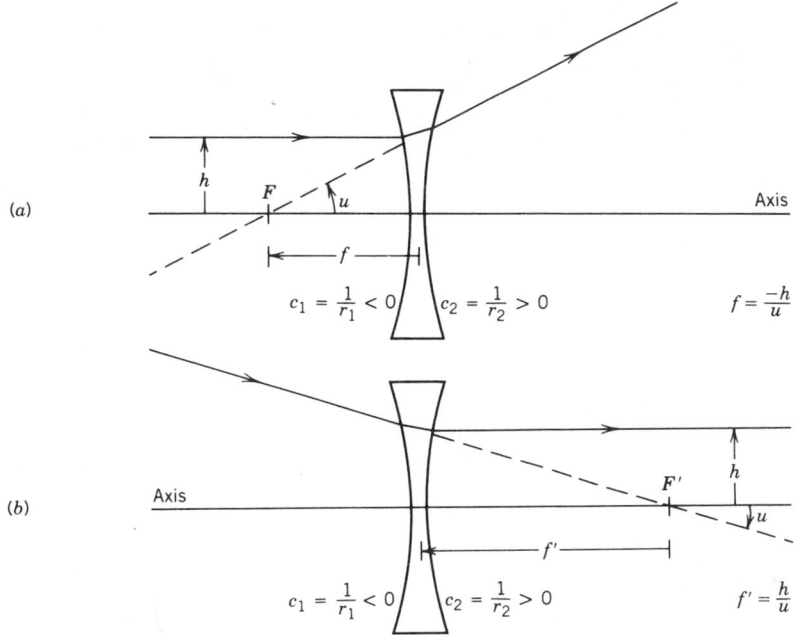

Fig. 5 Two focal points of negative lens. In (a) F, the second focal point, is in front of lens. In (b) F', first focal point, is behind lens. For lens shown, r_1 is negative and r_2 is positive.

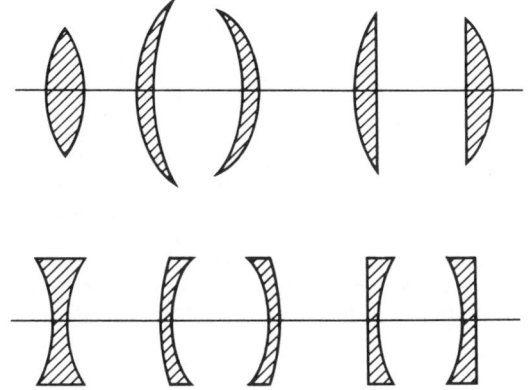

Fig. 6 Variety of positive lenses (upper group) and negative lenses (lower group).

Figure 7 shows a positive lens forming an image of point A at point B. All rays from A that pass through the lens converge to B, but only three are shown. It is assumed that the locations of F and F' are known; rays 1 and 3 are drawn to satisfy the definitions of these points. Ray 2 passes undeviated through the center of the lens. The image is inverted and the lateral magnification m is defined as y'/y or $B'B/A'A$; it is negative in the cases shown, indicating an inverted

image. The image is real since the rays actually arrive at point B.

It is simple to calculate the image position by the equation

$$xx' = f^2 \tag{5}$$

where x is the distance from the object to the first focal point; it is taken as positive if (as shown in Fig. 7) the object-to-focal-point direction is the same as the direction of light propagation. On the image side x' is the distance from the second focal point to the image; it is positive in Fig. 7. From Eq. (5), we see that x and x' always have the same sign. Since the product xx' is constant, moving the object to the right (toward F') moves the image to the right (away from F). The lateral magnification m is given as

$$m \equiv \frac{y'}{y} = -\frac{f}{x} = -\frac{x'}{f} \tag{6}$$

Another pair of equations may be used

$$\frac{1}{p} + \frac{1}{q} = \frac{1}{f} \quad \text{and} \quad m = -\frac{q}{p} \tag{7}$$

where p is the distance from the object to the lens and q is the distance from the lens to the image; they are considered positive if they are in the same

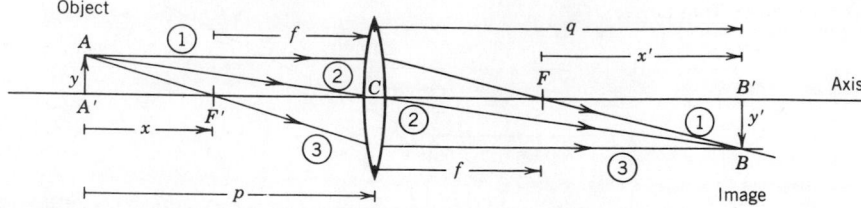

Fig. 7 Image formation by positive lens, illustrating quantities that appear in Eqs. (5)– (7). Also shown are three rays easily used in graphical ray tracing.

direction as the light propagation. In Fig. 7, both p and q are positive. Equation (7) is very convenient for use with thin lenses. For thick lenses or lenses consisting of several elements, it is not obvious what point (or points) in the lens should be used for measuring p and q. By reversing the rays, A becomes the image of B, and A and B are said to be conjugate points.

Figure 8 shows the corresponding situation for a negative lens. Here the first focal point F' is to the right of the lens and the second focal point F is to the left of the lens. Ray 1 is parallel to the axis until it strikes the lens and is refracted along a line that appears to have come from F. Ray 3 is headed for F' but is refracted to be parallel to the axis. Ray 2 passes straight through the center of the lens. These rays do not intersect anywhere to the right of the lens but if extended backward appear to have intersected at B. Only ray 2 actually passes through B. Point B is a *virtual* image of A (in contrast to the real image formed in Fig. 7). Equations (5) and (6) or Eq. (7) work for this case, but notice the following: f is negative; x is measured from A' to F' as before, but F' is to the right of the lens; x' is measured from F to B', but F is to the left of the lens; p is positive; Eq. (7) gives a negative value for q, indicating that B' is to the left of the lens and therefore virtual; m is positive but less than 1, so the image is upright or erect and smaller than the object.

Equations (5)–(7) may be used to establish the information in Table 1.

The *focal length* of a thin positive lens may be calculated from Eq. (4) if the curvatures and the index of the glass are known. It may also be measured in the laboratory by setting up on an optical bench an experiment similar to Fig. 7, measuring the appropriate distances and calculating f. A small luminous source, such as the filament of an unfrosted lightbulb, might serve as a suitable object. A ground glass screen is used to locate the image. Negative lenses cannot be measured in this way because the image in Fig. 8 is virtual and virtual images cannot be caught upon a screen. There are two ways around this problem.

Two thin lenses of focal lengths f_1 and f_2 when placed in contact are equivalent to a single lens of focal length f_c, given by the equation

$$\frac{1}{f_c} = \frac{1}{f_1} + \frac{1}{f_2} \qquad (8)$$

If f_1 is a negative lens under test, it may be combined with a positive lens of known focal length f_2. If $f_2 < |f_1|$, then the combined focal length f_c will be positive and can be measured by the experiment of Fig. 7; f_1 is then calculated from Eq. (8).

An alternate method is shown in Fig. 9. A positive lens is used to form a real image at $A'A$; its position is determined and recorded by observing the image on

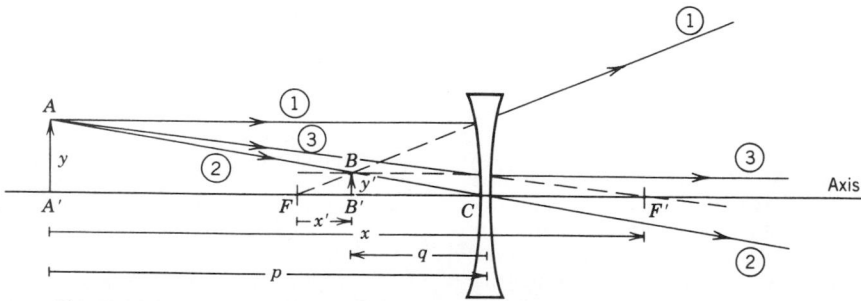

Fig. 8 Image formation by negative lens illustrating quantities from (5)–(7). Also shown are the rays easily used in graphical ray tracing.

Table 1 Images Formed by Thin Lenses

Lens	Object Position		Image Position		Nature of Image
Positive, or convex, $f > 0$	$x = \infty$	$p = \infty$	$x' = 0$	$q = f$	Real
	$\infty > x > f$	$\infty > p > 2f$	$0 < x' < f$	$f < q < 2f$	Real, inverted, $|m| < 1$
	$x = f$	$p = 2f$	$x' = f$	$q = 2f$	Real, inverted, $|m| = 1$
	$f > x > 0$	$2f > p > f$	$f < x' < \infty$	$2f < q < \infty$	Real, inverted, $|m| > 1$
	$0 > x > -f$	$f > p > 0$	$-\infty < x' < -f$	$-\infty < q < 0$	Virtual, erect, $|m| > 1$
Negative, or concave, $f < 0$	$x = \infty$	$p = \infty$	$x' = 0$	$q = f = -|f|$	Virtual
	$\infty > x > -f$	$\infty > p > 0$	$0 < x' < -f$	$f < q < 0$	Virtual, erect, $|m| < 1$

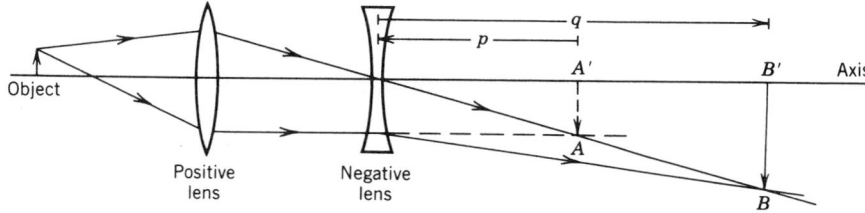

Fig. 9 Measuring focal length of negative lens using image AA' formed by positive lens as virtual object for negative lens (p is negative).

a screen. Being careful not to move the object or the positive lens, the negative lens to be measured is placed between the positive lens and the image $A'A$. The image now has a new location, $B'B$, which may be adjusted to some convenient position by moving the negative lens to the left or right. The initial image $A'A$ is the object for the negative lens; it is called a *virtual object* because the light is intercepted by the negative lens before it reaches point A. The distance from object to lens, p, is negative since it points upstream. The image distance q is the distance from the negative lens to the image $B'B$; in the figure it is positive (i.e., downstream). The focal length f of the negative lens may be calculated from Eq. (7) and the values of p and q just obtained.

A *spherical mirror*, concave toward the light source, forms images and behaves in many ways as a positive

lens. There is, however, only one focal point F (see Fig. 10). A ray that is initially parallel to the axis is reflected to pass through F. A ray that passes through F before striking the mirror will be parallel to the axis after reflection. A ray that passes through the center of curvature C will strike the mirror normally and be reflected back on itself. These three rays are shown in the figure starting from A and intersecting at B, which is the image of A. The focal point F is midway between the center of curvature and the mirror surface.

Equations (5)–(7) may be used to calculate image locations and magnification. The focal length of a concave mirror is positive. The downstream direction is always positive, but this reverses when the rays reflect from the mirror. In Fig. 10, rays from the object that have not reached the mirror (called rays

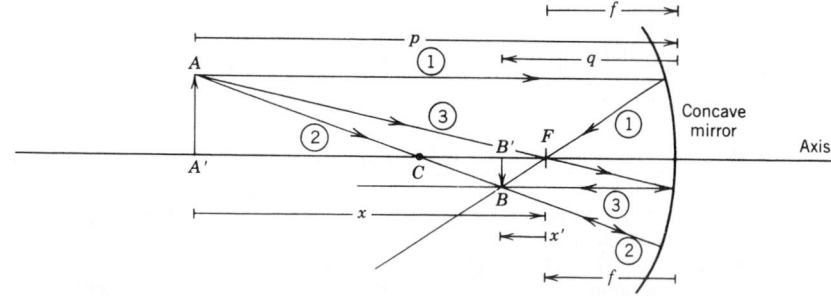

Fig. 10 Image formation by concave mirror. Focal point F is halfway between center of curvature C and mirror surface.

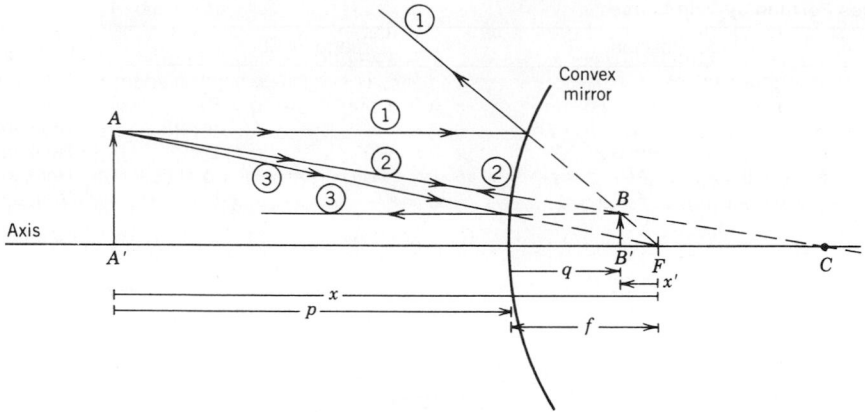

Fig. 11   Image formation by convex mirror.

in object space) are downstream to the right, so this is the positive direction for p and x. Rays that have been reflected from the mirror (called rays in image space) are downstream to the left, so right to left is the positive direction for q and x', which locate the image. In the figure, x, x', p, and q are all positive.

A mirror that is *convex* toward the light source is similar to a negative lens. The focal point is behind the mirror surface and $|f| = \frac{1}{2}|r|$, where r is the radius of curvature. Image formation for this case is illustrated in Fig. 11. In the equations, f is negative; x and p are in object space and positive in the figure. Distances x' and q are part of image space where downstream is to the left, so x' is positive and q is negative in the figure.

2.1 Aberrations

In first-order optical calculations, it is found that all the rays from a point object that pass through the optical system converge upon the image point. If the object lies in a plane that is perpendicular to the axis of the system, the image will lie in a plane perpendicular to the axis of the system and be similar in shape to the object.

If we abandon first-order approximations ($\sin \theta \approx \theta$) and use more accurate calculations, we discover that the preceding statements are not true. The differences between the actual behavior of the optical system and the ideal (or first-order) behavior are called *aberrations*, which are generally classified as follows.

2.2 Chromatic Aberration

According to Eq. (4), the focal length of a lens depends upon the index of refraction of the glass from which it was made. Since the index is wavelength (or frequency) dependent, the focal length is also wavelength dependent. This is *chromatic aberration*. Usually the focal length is longer for red light than for blue light. By combining two lens elements (such as a positive

lens of crown glass and a negative lens of flint glass), it is possible to create a "doublet" that has the same focal length for two specified wavelengths, such as a wavelength in the red and a wavelength in the blue. It will then have nearly the same focal length for the intervening wavelengths. Such a doublet is said to be *achromatic*. Objectives for small telescopes are usually of this design.

2.3 Spherical Aberration

See Fig. 4a. If the point F at which the ray crosses the axis depends upon h, the lens has spherical aberration. For the simple lens shown, F moves slightly toward the lens as h increases. Likewise, in Figs. 7 and 10, all the rays from A' do not cross the axis at B'; rather, the crossing point depends to some extent upon height (or off-axis distance) at which the ray strikes the lens or mirror.

Spherical aberration and chromatic aberration exist for source and image points that are on axis or off axis. The following aberrations do not exist for object and image points on axis; they exist only for off-axis points and increase in magnitude as the object and image points are more off axis.

A lens is said to exhibit *field curvature* if the image of an off-axis point does not lie upon the paraxial image plane but departs from that plane by a distance proportional to the square of the angle off axis.

For *astigmatism*, the image of an off-axis point consists of two short line segments. These two line segments are at different distances from the lens. One line segment (called the sagittal image) is directed radially outward from the axis; the other (called the tangential image) is perpendicular to the radial line through its center. If a lens suffering from astigmatism is used to image an object consisting of radial lines and concentric circles, they will not be sharply imaged on the same plane. The radial lines will be imaged at the sagittal position; the concentric circles will be imaged at the tangential position. Midway between these two images is the position of "least confusion." The separation between the sagittal and tangential lines is a measure of astigmatism; it increases with the square of the angle off axis.

For a lens with *coma*, an image of a point source is formed at the paraxially predicted position by the rays that pass through the central portion of the lens (Fig. 12). Rays that pass through the lens at some fixed distance from the axis (i.e., through a circular zone) do not come to the paraxial image point. Instead, they intersect the image plane in a circle, the center of which is displaced from the paraxial image point by twice its radius. The image of a point produced by all the rays through the lens is a bright spot at the expected location with a $60°$ flare extending from it. It resembles a comet (hence the name *coma*). The size of the coma pattern increases linearly with angle off axis.

A lens with *distortion* produces a plane image of a plane object, and the rays that pass through the lens all reach the image point, but the image of an extended object (such as a square) is not the same shape as the object. Instead of having constant magnification m so that

$$y' = my$$

there is an additional term to give

$$y' = my + Cy^3$$

The second term on the right is small compared with the first term, but y'/y is no longer independent of y.

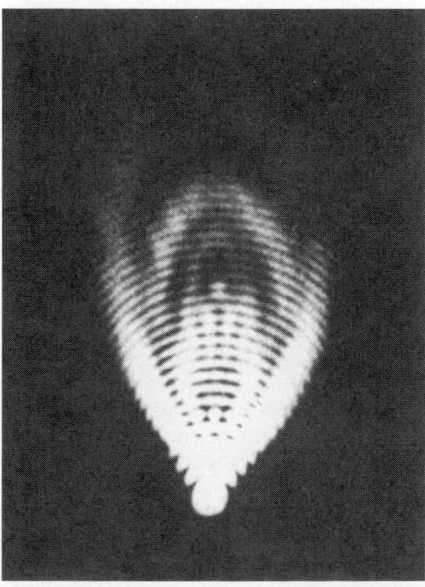

Fig. 12 Image of point source as formed by lens with coma. (Photograph by David Dutton.)

In case $m < 0$ (real images) and positive C, we find that $|y'/y|$ decreases as $|y|$ increases; the image of a square has a barrel shape. Conversely, a lens for which C is negative images a square into a figure resembling a pin cushion.

The monochromatic aberrations (spherical, astigmatism, coma, field curvature, and distortion) can be calculated using third-order approximations, that is, approximations consistent with $\sin \theta \approx \theta - \frac{1}{6}\theta^3$. These are called Seidel, or third-order, aberrations. In third-order calculations there exist only the aberrations listed in the preceding (and linear combinations of them). We have also assumed the lens has rotational symmetry about its axis.

Most practical lenses consist of several elements or components, each of which is a simple lens. The lens designer reduces the aberrations by adjusting the following: (a) the distribution of the power among the elements; (b) the shapes, thicknesses, and spacings of the elements; (c) the position of the stop (or limiting aperture); and (d) the choice of glass in each element. Usually it is not possible to minimize all of the aberrations, and the designer must know the application for which the lens is intended and make some judgment about the relative importance of the various aberrations. A lens that is optimally corrected for one pair of conjugate points will not be optimally corrected for another pair of conjugate points; for this reason the tube length of a compound microscope is important (and has been standardized).

A few lenses have been produced in which one or more elements are made of inhomogeneous glass that contains a deliberately introduced gradient of the index of refraction. This is a recent method of controlling the aberrations.

The "ray" description of image formation by lenses and mirrors, which has occupied our attention in the previous sections, takes no account of the wave nature of light. We now recognize several situations in which the wave properties of light are important; these are part of a field known as *physical optics*.

3 PHYSICAL OPTICS

We set up the experiment diagrammed in Fig. 13. A source of light L is placed behind a slit S_0 on axis. Light from S_0 passes through two slits S_1 and S_2 in an opaque screen and continues on to a viewing screen. The slits S_1 and S_2 are symmetrically located about the axis at $\pm a/2$. The point P is an arbitrary point on the viewing screen at a distance y from the axis, R is the distance from S_0 to the slits S_1 and S_2, and D is the distance from these slits to the viewing screen. In a typical laboratory experiment, a would be in the range $0.5-1.0$ mm and R and D in the range $50-100$ cm. The three slits should be less than 0.1 mm wide.

The path difference Δ for the two possible paths is

$$\Delta \equiv S_0 S_2 P - S_0 S_1 P = S_2 P - S_1 P = \frac{ay}{D} \qquad (9)$$

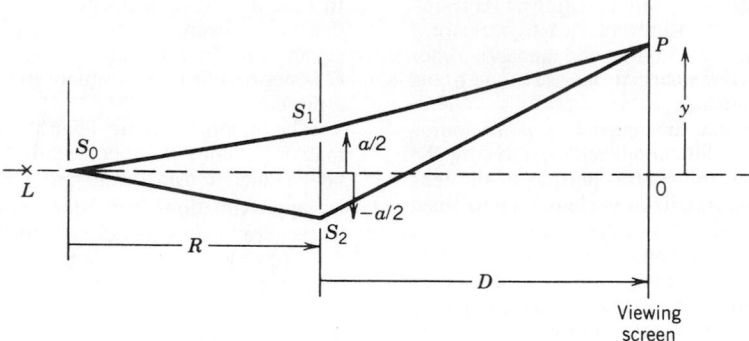

Fig. 13 Young's double-slit interference experiment. Light from the source slit S_0 passes through slits S_1 and S_2 and continues on to viewing screen. Interference pattern is formed on viewing screen. Intensity at point P depends upon path difference for two possible paths.

The light reaching P by way of S_2 is out of phase with the light reaching P by way of S_1. The phase difference δ is

$$\delta = \frac{2\pi}{\lambda}\Delta = \frac{2\pi}{\lambda}\frac{ay}{D} \qquad (10)$$

Depending upon the value of δ, the two waves will interfere at P either constructively or destructively.

If the slits S_1 and S_2 are of equal width so that they contribute equally to the irradiance at P, then the irradiance at P is I_p:

$$I_p = 4I_0 \cos^2 \frac{\delta}{2} = 2I_0(1 + \cos\delta) \qquad (11)$$

where I_0 is the irradiance that would be produced at P by light from only one slit. From Eqs. (10) and (11), we see that I_p will have maxima at values of y given by

$$y = \frac{m\lambda D}{a} \qquad (12)$$

where $m = 0, \pm 1, \pm 2, \pm 3, \ldots$; m is the order number or order of interference. Halfway between each pair of consecutive maxima is a minimum at which $I_p = 0$.

This pattern of fluctuating irradiance is called a pattern of *interference fringes*, or a *fringe pattern*.

The minima in the fringe pattern are not always zero. In this case, we define the fringe contrast or visibility V as

$$V = \frac{I_{max} - I_{min}}{I_{max} + I_{min}} \qquad (13)$$

where I_{max} and I_{min} are the maximum and minimum values of I_p. For the fringe pattern described by Eq. (11), $V = 1$; this is its maximum possible value. The case $V = 0$ corresponds to uniform irradiance, that is, no fringe pattern, or $I_{min} = I_{max}$.

If the two slits do not contribute equally to the irradiance at P, the fringe visibility will be reduced. Let I_1 represent the irradiance at P due to slit S_1 and I_2 the irradiance due to S_2; then Eq. (11) becomes

$$I_p = I_1 + I_2 + 2\sqrt{I_1 I_2} \cos\delta \qquad (14)$$

giving a fringe pattern with visibility

$$V = \frac{2\sqrt{I_1 I_2}}{I_1 + I_2} \qquad (15)$$

If we keep S_1 and S_2 of equal width so that they contribute equally to the irradiance at P, the fringe visibility will be reduced if we increase the width of the source slit S_0. If S_0 is centered on axis and has width w, the fringe pattern is described by

$$I_p = 2I_0(1 + V\cos\delta) \qquad (16)$$

where the fringe visibility V is

$$V = \frac{\sin\beta}{\beta} \qquad \text{where} \quad \beta = \frac{\pi wa}{\lambda R} \qquad (17)$$

The visibility V decreases from 1 (for very small values of w) to zero for $\beta = \pi$ or $w = \lambda R/a$. The visibility as a function of w is often identified with γ, the degree of coherence of the light emerging from slits S_1 and S_2. Here, γ is the normalized cross-correlation function of the waves emerging from the two slits. Loosely speaking, $|\gamma|$ represents the reliability with which the phase at S_2 could be predicted from a hypothetical determination of the phase at S_1. If β lies in the range $\pi < \beta < 2\pi$, the value of V (or γ) is negative. For the specific case $w = 1.4\lambda R/a$ or $\beta = 4.4$, we obtain from (17) that $V = -0.2$ or $\gamma = -0.2$. Fringes of contrast 0.2 are not very good but are easily recognizable. The negative value of γ indicates that there is a minimum rather than a maximum at the axial point ($y = 0$) of the viewing screen. The waves emerging from S_1 and S_2 are poorly phase correlated (only 0.2), and the most probable phase difference is 180° (minus sign). Since the separation of S_1 and S_2 is perpendicular to the direction of propagation, this is called spatial coherence.

Figure 14 illustrates the irradiance profile for fringe patterns with three different values of V.

Another interference experiment is illustrated in Fig. 15. A broad source of light is used to illuminate two partially reflecting mirrors M_1 and M_2. The reflected light is viewed by the eye; in order to have both illumination and viewing directions nearly normal to the mirrors M_1 and M_2, a third mirror, M_0, is inserted as shown. Here, M_0 is partially reflecting and partially transmitting.

In order to have interference effects, the two rays must be coherent. Since the source is broad, we have coherence only between rays reflected from M_1 at A_1 and from M_2 at A_2, where points S, A_1, and A_2 lie on a straight line, where S is an arbitrary point of source. The light reflected at A_1 will be reflected at B_1 into the eye by mirror M_0. Light reflected at A_2 reaches the eye by a similar path. In order for an interference pattern to be seen, both of these rays must enter the eye. Interference will take place where these rays intersect on the retina of the eye, but the eye (or the brain) will interpret this as an interference at P, the intersection

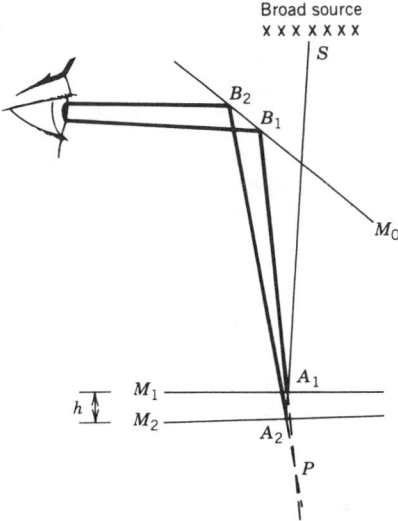

Fig. 15 Interference produced by light reflected from thin wedge. Effect may be observed in oil slick or in soap bubble.

point of lines $A_1 B_1$ and $A_2 B_2$. Interference such as this is observed in an oil film on wet pavement or sometimes in a soap bubble.

Both rays must enter the pupil of the eye; in order to obtain this, the mirrors M_1 and M_2 must be nearly parallel and *either* h (the separation of the mirrors) must be small *or* the ray SA_1 must be at near normal incidence. The mirror M_0 permits near normal illumination and viewing.

The path difference and therefore the phase difference between the two rays will be determined by the mirror separation h at the point A_1. If the space between the mirrors M_1 and M_2 is slightly wedged or if the mirrors are not exactly plane, then the locus of a maximum (or a minimum) in the fringe pattern represents the locus of points of constant h. A change from one bright (or dark) fringe to the next bright (or dark) fringe corresponds to a change of h by one-half wavelength.

Fringes of good contrast [corresponding to Eq. (11)] are obtained if the illumination is nearly monochromatic, if h is small, and if the two reflecting surfaces M_1 and M_2 have small but equal reflectances. The last condition ensures that the two rays $A_1 B_1$ and $A_2 B_2$ are of nearly equal irradiance. A clean glass surface in air reflects 4% and would serve very well as M_1 or M_2.

If the mirror M_2 is moved away from M_1 so as to increase h, this serves to introduce a time delay in the ray $A_2 B_2$ relative to $A_1 B_1$. If air separates the mirrors M_1 and M_2, the time delay τ is given by

$$\tau = \frac{2h}{c} \tag{18}$$

As h (and therefore τ) is increased, the fringe visibility gradually decreases, indicating a decrease in

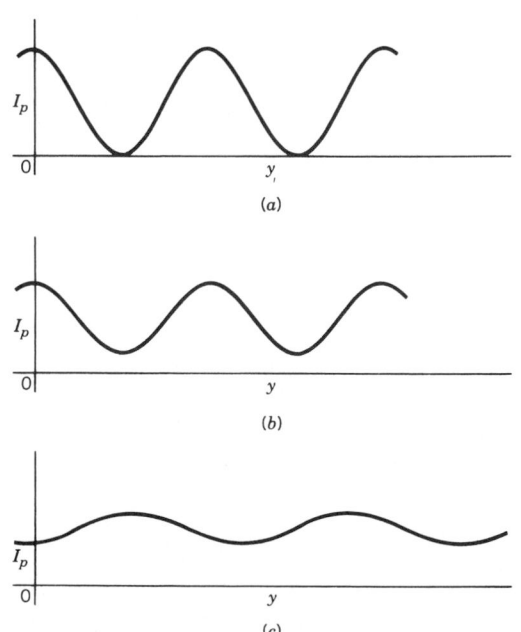

Fig. 14 Plots of I_p vs. y for fringes with different visibility V: (a) $V = 1.0$; (b) $V = 0.5$; (c) $V = -0.2$.

the coherence between the two waves or rays. In this case, the coherence decreases due to a time difference between the rays and is called temporal coherence. If a dispersive medium separates the two mirrors, the time delay should be calculated using the group velocity [U of Eq. (2)] instead of c. The largest value of τ for which fringes of reasonable visibility can be observed is called τ_c, the coherence time of the source; $c\tau_c$ is called the coherence length; τ_c is the same as Δt in the discussion following Eq. (1).

Aside from the interference fringes produced in Young's experiment (Fig. 13), there is another aspect of the experiment that is contrary to the ideas of ray optics. If light reaches P by two paths $S_0 S_1 P$ and $S_0 S_2 P$, then at least one (and usually both) of these paths must be bent. This ability of light to bend into regions that geometrical optics would call shadow regions is known as *diffraction*. Diffraction is a property of all wave propagation.

Practical diffraction problems can be solved by the Huygens–Fresnel method. The method is outlined in Fig. 16. Light from a small source S_0 spreads out to pass through an aperture A in an opaque screen. A wave front (surface of constant phase) reaches the aperture and is transmitted through the open part but is blocked by the opaque portion of the screen. Each elemental area on the transmitted wave front acts as a secondary source radiating spherical waves into the region to the right of A. To calculate the disturbance (i.e., electric field) at P, an arbitrary point on the viewing screen, we must sum (i.e., integrate) the contributions due to all the elemental areas of the aperture. The summation process must take into account the fact that the disturbances reaching P from the various elemental areas of the aperture are not in phase at P because contributions from the various elemental areas have traveled unequal distances from S_0 to P. Mathematically this is written

$$U(P) = -\frac{iA}{2\lambda} \iint_A \frac{e^{ik(r_1+r_2)}}{r_1 r_2} \, dS \qquad (19)$$

where $U(P)$ is the disturbance at P; dS is an element of area in the aperture; $k = 2\pi/\lambda$; and r_1 and r_2 are distances from S_0 to dS and from dS to P. The integration is over the open aperture. In many practical cases, the product $r_1 r_2$ in the denominator may be considered as a constant during the integration. The values of r_1 and r_2 in the exponent may not be considered as constant because the exponential is periodic and k is large; A is the open area of the aperture. (We have also assumed that r_1 and r_2 are nearly perpendicular to the plane of the opaque screen.) A complex quantity, $U(P)$ represents the amplitude and phase of the disturbance at P. The irradiance at P, $I(P) = |U(P)|^2 = U(P)U^*(P)$, where $U^*(P)$ is the complex conjugate of $U(P)$.

The integral in Eq. (19) is generally difficult to evaluate, but results are available for some simple cases.

If the aperture is a slit with length much larger than its width, we can consider this as a two-dimensional problem (as illustrated in Fig. 17), ignoring the dimension along the length of the slit. The point source may be replaced by S_0, a narrow slit (or line) source provided this is parallel to the diffracting slit S_1; use of the slit source rather than a point source will noticeably increase the irradiance on the viewing screen.

In discussing this problem, we take as axis the line from the source S_0 through the center of the diffracting slit S_1; it intersects the viewing screen at O. The slit S_1 is of width w and is symmetrically located about the axis so that the two edges D and D' of the slit are at $\pm \frac{1}{2} w$ from the axis. The distance $S_0 D O$ is larger than the axial distance $S_0 O$, and on the assumption that $w \ll a$ or $w \ll b$, the difference is

$$K \equiv S_0 D O - S_0 O = \frac{a+b}{2ab}\left(\frac{w}{2}\right)^2 \qquad (20)$$

where a and b are the distances from source to diffracting slit and from diffracting slit to viewing screen, as shown in Fig. 17.

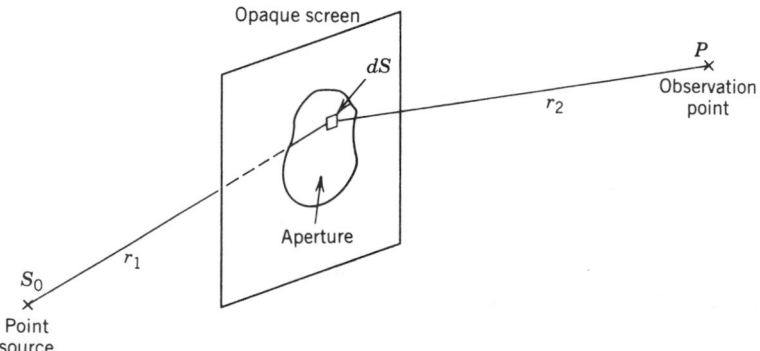

Opaque screen

dS

P
Observation
point

r_2

Aperture

r_1

S_0

Point
source

Fig. 16 Geometry of general diffraction problem. Here, S_0 is point source of light; disturbance at P calculated using Eq. (9) where r_1 and r_2 are distances illustrated.

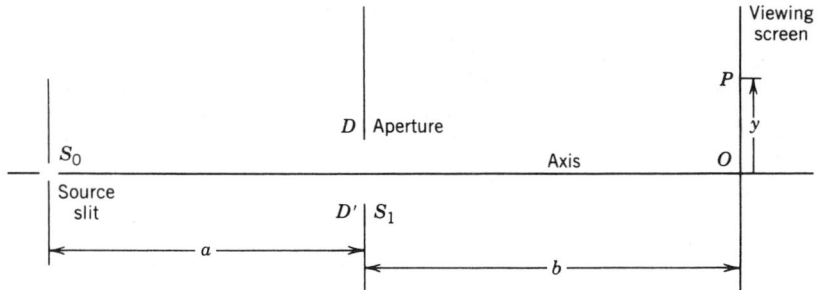

Fig. 17 Geometric quantities used in calculating diffraction pattern produced by slit.

Some generality may be achieved by replacing w by the dimensionless parameter u:

$$u = w\sqrt{\frac{2(a+b)}{ab\lambda}} \qquad (21)$$

and replacing y (the distance along the viewing screen from the axis to the point of observation P) by the dimensionless variable v:

$$v = y\sqrt{\frac{2a}{\lambda b(a+b)}} \qquad (22)$$

With these changes $K = \frac{1}{16}u^2\lambda$.

Figure 18 shows the irradiance at P as a function of v for several values of the parameter u. In each case, the edge of the geometrical shadow is indicated by vertical lines.

Notice that for the larger values of u, the curves have several maxima and minima and that there is very little irradiance outside the geometrical shadow region. For the smaller values of u (e.g., $u = 1.5, 0.5$), the curves show less structure and for values of $u < 1$ spread out well beyond the geometrical shadow region.

The experiment of Fig. 17 may be repeated using a point source and a small circular aperture in place of the slit. The observed diffraction pattern on the viewing screen will have circular symmetry and the minima will be deeper than those observed with the slit. For $u > 1$, most of the light will be within the region of

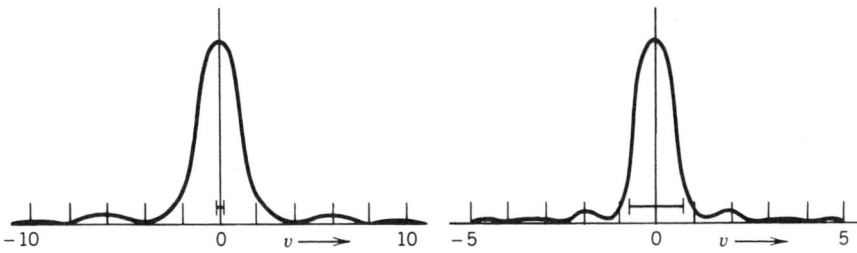

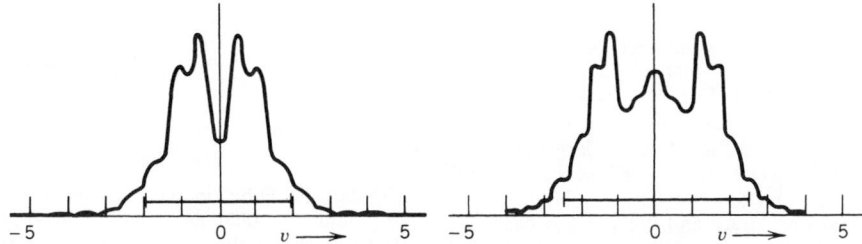

Fig. 18 Diffraction patterns of four slits of different widths. Each figure is plot of irradiance at P as function of v. (a) For narrow slit, $u = 0.5$. Other figures are for successively wider slits: (b) $u = 1.5$; (c) $u = 3.8$; (d) $u = 5.0$. In each case geometrically predicted width is indicated by horizontal line segment just above v axis.

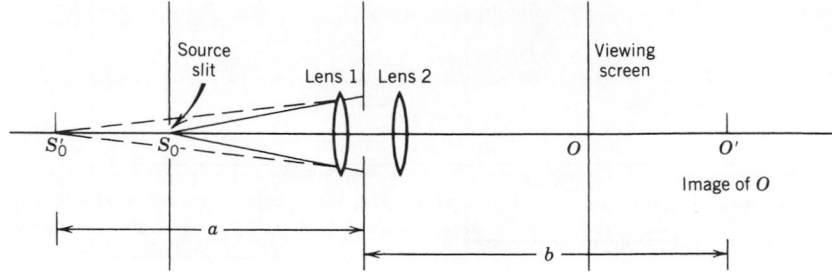

Fig. 19 Effect of using lenses in diffraction experiment. Source distance measured from S_0', image formed by lens 1 of real source slit. Screen distance is measured to O', image formed by lens 2 of actual viewing screen.

the geometrical shadow; and for $u < 1$, the light will spread out into the shadow region.

Diffraction problems are generally divided into two classes, known as *Fresnel diffraction* and *Fraunhofer diffraction*. Fresnel diffraction, or near-field diffraction, covers the cases for which u [of Eq. (21)] is larger than 1. Fraunhofer (or far-field) diffraction covers the cases for which $u < 1$. For a given aperture and wavelength, you can move from the Fresnel to the Fraunhofer region by increasing the distances a and b.

A lens may be placed on each side of the aperture as indicated in Fig. 19. Lens 1 forms an image of the source S_0 at S_0' and lens 2 images point O of the viewing screen to O'. In this situation, the distance a is measured from S_0' to the aperture and b is measured from the aperture to O'. If S_0 is at the front focal point of lens 1 and the viewing screen is at the second focal plane of lens 2, then a and b are infinite and $u = 0$ independent of the aperture size. Fraunhofer diffraction calculations assume that the preceding conditions are satisfied and some mathematical simplifications result therefrom. Even though the calculations are for the $u = 0$ case, the results are good approximations to any case for which $u \leq 1$. The irradiance distribution for the Fraunhofer diffraction pattern of a slit is

$$I(y) = I_0 \left(\frac{\sin \beta}{\beta} \right)^2 \qquad (23)$$

In this equation, I_0 is the irradiance on axis; $I(y)$ is the irradiance at P that is at a distance y from the axis; $\beta = (\pi w y)/(\lambda f)$. Here w is the slit width and f is the focal length of lens 2. Figure 20a is a plot of Eq. (23); notice that it has a large central maximum with small secondary maxima on each side. The irradiance is zero at $\beta = \pi, 2\pi, \ldots$. Notice the similarity between this figure and the $u = 0.5$ case of Fig. 18.

For a circular aperture and a point source on axis, the Fraunhofer diffraction pattern will have circular symmetry. Calculations show that the irradiance $I(r)$

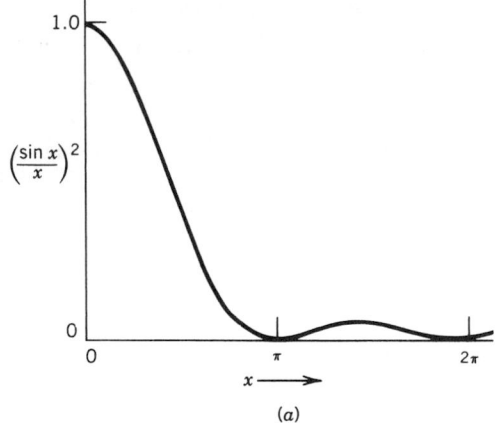

(a)

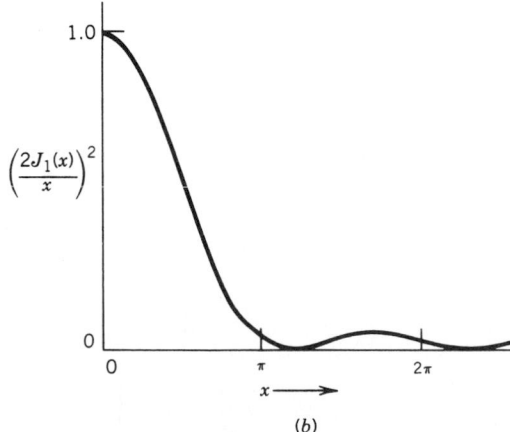

(b)

Fig. 20 (a) Function $(\sin x/x)^2$ as function of x. This is Fraunhofer diffraction pattern of single slit. (Pattern is symmetrical about $x = 0$.) (b) Function $[2J_1(x)/x]^2$ as function of x. This is Fraunhofer diffraction pattern of circular aperture. (Pattern has rotational symmetry about axis.)

at a distance r from the axis in the rear focal plane of lens 2 is given by

$$I(r) = I_0 \left(\frac{2J_1(x)}{x}\right)^2 \qquad (24)$$

where J_1 is the first-order Bessel function and $x = (\pi w r)/(\lambda f)$. In this case w represents the *diameter* of the circular aperture. Figure 20*b* is a plot of Eq. (24). The pattern has rotational symmetry about the $x = 0$ axis. For the circular aperture the zeros of irradiance are not quite equally spaced, occurring at $x = 1.220\pi, 2.23\pi, 3.328\pi, \ldots$. The first dark ring is at $r = 1.22\lambda f/w$. Of all the optical power that passes through the aperture, 84% is focused inside this first dark ring, forming what is known as the Airy disk.

For many optical instruments (e.g., telescopes and cameras), the light enters through a circular aperture and the image of a distant point source such as a star is not a bright point but an Airy disk that has the size just discussed. It follows that two separate sources will not be recognized as separate (we say they will not be "resolved") unless their images are separated by a distance at least equal to the radius of the Airy disk. The angular separation θ of two just resolved stars is

$$\theta = 1.22\frac{\lambda}{w} \qquad (25)$$

As a rule of thumb the 1.22 is often ignored.

Photographers often express the size of the camera aperture in terms of the f number, which we represent as $F^{\#}$. It is defined as $F^{\#} = f/w$. In terms of the f number, the diameter d of the Airy disk is

$$d = 2\frac{1.22\lambda f}{w} = 2.44\lambda F^{\#} \qquad (26)$$

Using a typical value of $\lambda = 0.55$ μm, we obtain

$$d = 1.33F^{\#} \quad (\mu\text{m})$$

The diameter of the Airy disk (in micrometers) is about equal to (or 33% larger than) the f number used. This assumes the lens has no aberrations; the performance of such a lens is said to be *diffraction limited*.

Microscopists generally speak of the *numerical aperture* (NA) of the microscope:

$$\text{NA} \equiv n \, \sin i \qquad (27)$$

where i is the half angle of the cone of light that enters the microscope from the object (see Fig. 21) and n is the index of refraction of the region between the object and the microscope objective. Usually $n = 1.0$, but in oil immersion microscopes, n is the index of the oil that fills the space between the object and the lens.

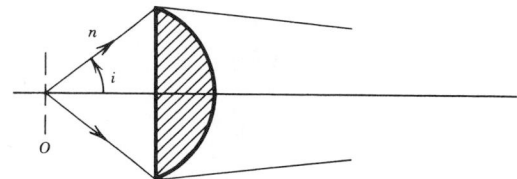

Fig. 21 Numerical aperture (NA) of microscope objective defined as NA = n sin i. Here i is half-angle of cone of light entering objective from object O; n is index of refraction of medium between object and objective.

Microscopes generally cover a small field of view and are diffraction limited. The separation s of two just resolved points is

$$s = \frac{\lambda}{2n \, \sin i} = \frac{\lambda}{2(\text{NA})} \qquad (28)$$

where λ is the wavelength in a vacuum.

Equation (24) describes the irradiance as a function of r, the distance off axis, in a plane through the rear focal point and perpendicular to the axis. The *irradiance on axis* as a function of x, the distance to the right of the focal point, is given by

$$I(x) = I_0 \left(\frac{f}{f+x}\right)^2 \left(\frac{\sin(U_N/4)}{U_N/4}\right)^2 \qquad (29)$$

where I_0 is the irradiance at $x = 0$ and

$$U_N \equiv 2\pi N \frac{x}{f+x} \qquad (30)$$

and $N \equiv w^2/(4\lambda f)$. See Fig. 22 for a sketch of this experiment. An examination of Eq. (29) shows that the maximum value of $I(x)$ is not at $x = 0$ but is shifted toward the lens (i.e., to negative x) by the amount Δf given by

$$\frac{\Delta f}{f} = -\frac{1}{1 + N^2\pi^2/12} \qquad (31)$$

This shift is insignificant unless N is small ($N < 10$), which corresponds to a very large $F^{\#}$ or a very small cone angle in the converging light beam.

In spite of the shift just calculated, if one is given a fixed aperture size w with a distant *fixed* object on which it is required to produce maximum irradiance and the available variable is the focal length of the lens (or the curvature of the wave front emerging from the lens), the optimum choice is to select the lens for which the distant object will be at the center of curvature of the emerging wave front.

A beam with *Gaussian* irradiance profile, that is, a beam for which

$$I_p(r) = I_0 e^{-r^2/w^2} \qquad (32)$$

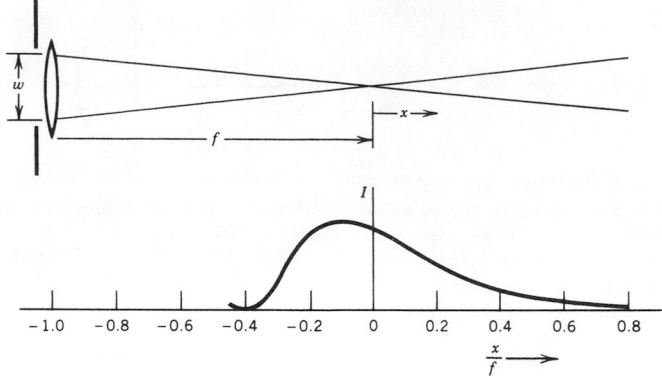

Fig. 22 Position of maximum irradiance shifted toward lens from geometrical focal point. Effect becomes significant only if w^2 is comparable to $4\lambda f$ or less. Lower curve of irradiance vs. x/f: $N = w^2/4\lambda f = 3$.

is unusual in the sense that as it propagates, the irradiance profile remains Gaussian. The curvature of the wave front and the width w will vary as the beam propagates, but it will remain of the form given by Eq. (32), that is, it will remain Gaussian.

A *diffraction grating* may be used in spectroscopy to measure the wavelength of light or to study the distribution of optical power throughout the spectrum.

In its simplest form, the diffraction grating is an opaque screen in which a large number N of transparent slits have been made. The slits should be parallel, evenly spaced, and of the same width. We represent the width of each slit by w and the center-to-center spacing of adjacent slits by d. (Since gratings are routinely produced with d in the range of 3–5 μm and sometimes even smaller, their production requires great care and skill.)

The grating is used in Fraunhofer diffraction, as shown in Fig. 23. This device is called a spectroscope. Light enters the collimator through slit S_0, which is at the front focal point of lens L_c. Parallel rays strike the grating G; the diffracted light is viewed through the telescope, which is focused for infinity. The telescope consists of an objective L_t and an eyepiece E_p. The light is incident upon the grating at angle θ and is viewed from the direction ϕ; both of these angles are measured from the normal to the plane of the grating and with a sign convention that makes both angles positive in Fig. 23. Since w and d are only a few times the wavelength of light, the diffraction angles will be so large that we cannot use small-angle approximations for $\sin\theta$ and $\sin\phi$.

A calculation applying Eq. (19) to this problem gives

$$I = I_0 \frac{\sin^2\beta}{\beta^2} \frac{\sin^2 N_\gamma}{\sin^2\gamma} \qquad (33)$$

where

$$\beta = \frac{\pi w}{\lambda}(\sin\theta + \sin\phi) \qquad (34a)$$

and

$$\gamma = \frac{\pi d}{\lambda}(\sin\theta + \sin\phi) \qquad (34b)$$

Equation (33) contains two factors, one of which depends upon β and therefore w; the other depends upon γ and therefore d. Here, I_0 is a normalizing factor and is equal to the value of I when $\theta = -\phi$ (i.e., the straight-through irradiance).

The factor involving γ has maxima (equal to N^2) when $\gamma = m\pi$, where $m = 0, \pm 1, +2, \ldots$. By way of Eq. (34b), this corresponds to maxima of I when

$$m\lambda = d(\sin\theta + \sin\phi) \qquad (35)$$

The integer m is called the *order number*.

For large values of N, these maxima of I are very sharp (or narrow) functions of θ and ϕ. If γ changes from the value $m\pi$, which produces a maximum, by the amount $\pm\Delta\gamma = \pi/N$, the factor goes to zero. For good commercially available gratings, N may be in the range from 10^4 to 10^5 and $\Delta\gamma$ is correspondingly quite small.

If the source slit of the spectrometer (Fig. 23) is illuminated with monochromatic light and if d is known and θ and ϕ can be measured, then Eq. (35) enables us to calculate the wavelength λ. Conversely, if λ is known, we can calculate the grating space d. If the source slit of the spectrometer is sufficiently narrow, the half width $\Delta\phi$ of the image line may be calculated from Eq. (34b) using $\Delta\gamma = \pi/N$. Two spectral lines of wavelength λ and $\lambda + \Delta\lambda$ are said to be just resolved (Rayleigh criterion) when their angular separation has this value of $\Delta\phi$. It can be shown that

$$\frac{\lambda}{\Delta\lambda} = \frac{\gamma}{\Delta\gamma} = mN \qquad (36)$$

The ratio $\lambda/\Delta\lambda$ is known as the resolving power of the grating.

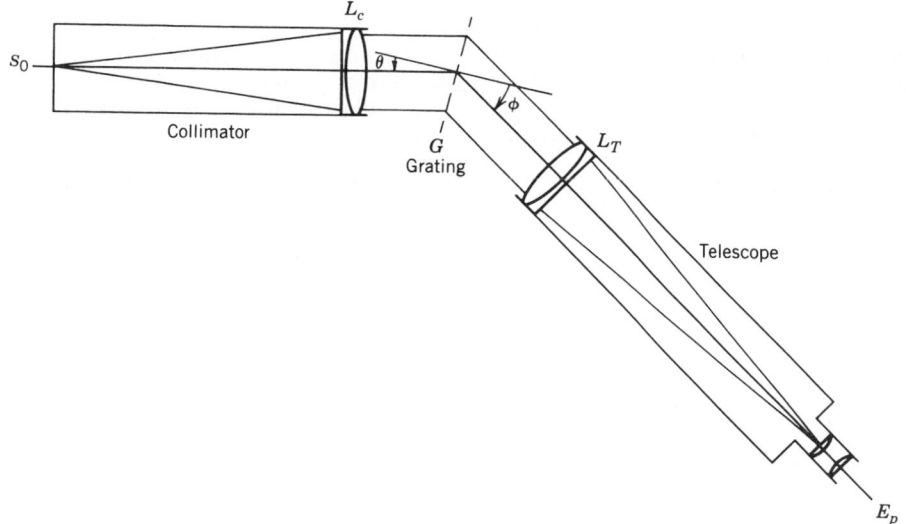

Fig. 23 Use of grating in spectroscopy. If grating space is known, measurement of angles θ and ϕ provides sufficient information to calculate λ, wavelength of light.

If the source of light contains several colors, the deviation $|\theta + \phi|$ of the light will, in any given order, be larger in the red than in the blue.

In Eq. (33), the factor involving β is the same factor found in Eq. (23); it describes the diffraction pattern of a single slit of width w. It does not alter the value of ϕ for any of the maxima, but it determines the relative irradiance of the various orders. For example, if we make $w = \frac{1}{2}d$, then the orders $m = \pm 2, \pm 4, \ldots$ will be missing because the zeros of this factor fall at the same values of ϕ as the even-numbered maxima given by Eq. (35).

The grating need not be composed of alternate opaque and transparent lines. It may be everywhere transparent but have a thickness variation that is periodic with period d in one dimension. One surface of such a grating would represent a small-scale copy of a sheet of corrugated steel. Alternately, the grating might be a transparent, uniformly thick sheet in which the index of refraction is a periodic function of one dimension with period d. In all these cases, Eqs. (34) and (35) are still valid, but the factor involving β has to be changed to represent the diffraction pattern of one period of the grating structure.

If the tool that cuts the grooves or lines on the grating is deliberately shaped to increase the irradiance in some order (say, order $m = 1$) at the expense of the irradiance in other orders (such as $m = 0, -1$), the resulting grating is said to be "blazed." In measuring weak spectral lines, this is desirable.

In Fraunhofer diffraction, the *amplitude distribution* in the observation plane is the Fourier transform of the *amplitude transmittance function* of the aperture (with a suitable scaling factor, which is wavelength dependent). Irradiance is proportional to the square of the amplitude. With this information, many Fraunhofer diffraction patterns are predictable.

3.1 Holography

Interference and diffraction form the basis of holography. Consider Fig. 24. A photographic plate P that is to become a hologram is exposed to two coherent beams of monochromatic light O and R that strike the

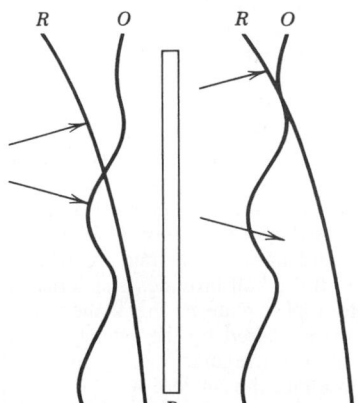

Fig. 24 Process of holography. Hologram is formed when photographic plate P records interference pattern produced by two overlapping wave fronts O and R. If plate is processed, returned to its original position, and illuminated by *one* wave front (e.g., R), diffraction will reproduce copy of other wave front (i.e., O).

plate from different directions. A wave front for each beam is sketched in the figure. An interference pattern is formed on the plate. The fringe pattern formed will be irregular in shape and spacing unless *both* wave fronts have some simple form, such as plane, spherical, or cylindrical. The photographic plate is exposed to this interference pattern for a suitable time and is then processed in the darkroom in the usual way. The photographic density as a function of position on the plate is a record of the irradiance in the interference pattern. The photographic plate is now a hologram. In some respects it resembles a grating, but the lines are unequally spaced and distorted as a means of carrying information about the shape of the two wave fronts.

If the hologram is now replaced in its original position (as in Fig. 24) and illuminated from the left by *either* of the beams (O or R), then to the right of P there will be observed both beams (O and R) as if they had been propagated through a window at P. This process is called wave front reconstruction. The reconstructed wave front will correspond in direction to the first-order beam from a diffraction grating, but the irregular spacing and distortion of the lines will cause the wave front to duplicate, in amplitude distribution and phase, the wave front used to produce the hologram. There will usually be a third beam, called the conjugate beam, corresponding to order -1 of the grating. In some cases the form of the conjugate beam is easily predicted.

It is customary for one wave, R (called the reference wave), to be of some simple form, such as plane or spherical, that can be easily reproduced for use in the reconstruction process. The other wave, O, is from some object to be recorded; it carries information about the object and usually has rapid spatial variations of amplitude and phase. If such a complicated wave were used as the reference wave, then in the reconstruction process, the hologram would have to be repositioned very accurately. (Error from the exposure position must be small compared to the spatial scale of the amplitude and phase variations in the wave front.)

The preceding statements may be derived from the interference and diffraction equations. Let us look at a simple example as illustrated in Fig. 25a. A coherent, monochromatic beam of light such as an expanded laser beam falls upon an object (represented by a mug) and upon a polished steel ball. The light reflected from the ball provides a spherical wave front at the photographic plate P; this is the reference wave. Light is also scattered by the object and reaches the plate P as a very irregular object wave front carrying information about the object.

The plate is processed in the darkroom and becomes a hologram. This hologram is returned to its original position and is illuminated by light reflected by the ball, which is also in its original position. An eye looking through the hologram will see the object at its former position even though it has been removed. The hologram and the reference wave have reconstructed the object wave front. This reconstructed object wave front has entered the eye and produced therein the same image that would have been produced by the original object wave front. The eye is using only that part of the hologram that is between the eye pupil and the reconstructed image. By moving the eye so as to look through a different part of the hologram, the observer sees the image from a different perspective and thereby observes that the image is three dimensional.

In this case, the conjugate wave gives rise to a conjugate image, as shown in Fig. 25b. It may be seen by looking through the hologram. In this case, both the image and the conjugate image are virtual. Real images may be obtained at these same positions if the reconstructing reference wave is replaced by its "conjugate," a spherical wave that strikes the hologram from the left and is converging toward B.

If the reference wave is plane, then the image in the reconstruction is a duplicate of the object and is virtual. In this case, the conjugate image is real and located on the opposite side of the hologram from the object position. The conjugate reference wave is a wave of the same curvature as the reference wave but propagating in the reverse direction. In this case, it is plane and produces a real image with virtual conjugate image.

The hologram is a photographic record of an interference pattern. The average fringe space d is given by

$$d = \frac{\lambda}{\sin \theta} \qquad (37)$$

where θ is the angle between the reference and object beams; see Fig. 25. It is necessary that the photographic emulsion (or whatever recording medium is used) be able to resolve lines of this spacing.

The recording system is very sensitive to vibration. The relative motion of one wavelength of light of the various parts during exposure will reduce the contrast of the interference pattern to nearly zero and make the "hologram" useless. It is usually necessary to isolate the hologram-recording system from building vibrations.

By using reasonably large values of $\theta (\approx 30°)$, it is possible to make the fringe spacing much less than the thickness of the photographic emulsion. In this case, the hologram has some of the properties of a blazed grating; that is, the reconstructed image is much brighter than the reconstructed conjugate image. It also makes the reconstructing angle of incidence more critical.

The main advantage of holography is its ability to record and reproduce the phase distribution in a wave front. Other recording methods record only the irradiance distribution.

4   LIGHT SOURCES

If we consider the light sources that we encounter in our daily lives, these would probably include

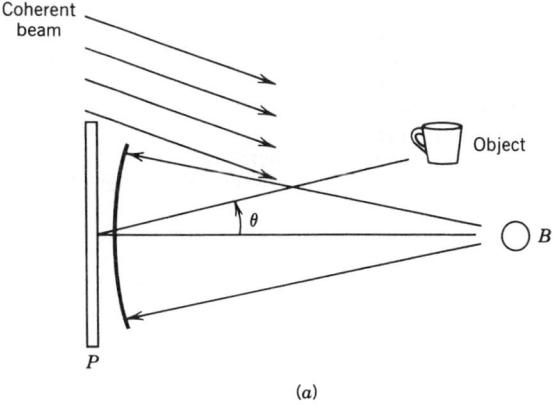

(a)

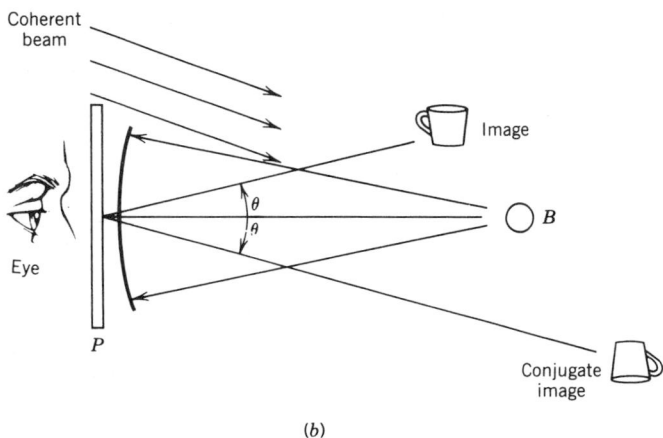

(b)

Fig. 25 Simple hologram experiment. Hologram is produced with arrangement of upper figure. Light reflected from polished steel ball *B* forms reference wave. Object wave is light scattered diffusely by object. In lower figure, object has been removed; reference wave falls on hologram. Eye, looking through hologram, sees image of object in its original position; conjugate image is also seen.

(a) tungsten filament lamps and fluorescent lamps used to illuminate our homes and workplaces (b) mercury or sodium arcs frequently used for highway lighting, and (c) in recent years, lasers, which have become fairly common but must be used with caution.

The tungsten filament lamp is the most common example of light sources that are solids and produce light because they are hot (i.e., incandescent). The oil lamp of years past also belongs in this class, producing light because small specks of carbon (i.e., soot) are heated to incandescence by the flame.

Discussion of these sources usually begins with a consideration of *blackbody radiation*. This is because the radiation of light from a blackbody is understood from a theoretical view and provides a standard of comparison for other sources of radiation. The subject is treated in many textbooks of thermodynamics or

modern physics. We shall be content to summarize the major results of the theory.

A blackbody is an object that absorbs all the radiation that falls upon it. In order to remain in thermal equilibrium with its surroundings, it must also emit radiation. A laboratory blackbody would consist of a cavity, the walls of which are maintained at some temperature *T*. There is thermal equilibrium between the walls and the radiation in the cavity except for a small hole in one wall that allows radiation to enter and leave the cavity. This hole is black in the sense that any radiation that enters the cavity through the hole has a negligible chance of finding its way back out. If the area of the hole is only a small part of the total wall area of the cavity, then the radiation escaping from the cavity through the hole is blackbody radiation at temperature *T*. The properties of this radiation are independent of the shape of the cavity and the material of which the walls are made. It is also assumed that the cavity is filled with a transparent medium of index of refraction $n = 1.0$.

Experimental investigations support the following theoretically derived equations that describe the radiation emitted from a blackbody:

$$\frac{d\Phi}{dA} \equiv M = \sigma T^4 \tag{38}$$

This is known as the Stefan–Boltzmann law, where M is the total radiant flux Φ per unit area (watts per square meters) emitted by the blackbody, T is the temperature (degrees Kelvin), σ is the Stefan–Boltzmann constant, which has the value 5.67×10^{-8} W/m^2 K^4. Here, M is called the *radiant exitance*.

It is also customary to define the radiance L, which is the radiant flux (or power) per unit solid angle per unit *projected* area. In Fig. 26, the solid angle $d\Omega$, in steradians, is defined as dA_2/r^2, where the area dA_2 is perpendicular to r and dA_1 is an element of area of the source (blackbody in our present discussion) but projected area means apparent area as seen from the direction θ; so projected area is $dA_1 \cos\theta$:

$$L \equiv \frac{d^2\Phi}{d\Omega\, dA_1 \cos\theta} \tag{39}$$

A source for which L is independent of θ is called a Lambertian source; it looks equally "bright" from all viewing directions. A blackbody is a Lambertian source; some other sources are approximately Lambertian. For a Lambertian source that radiates into a hemisphere,

$$M = \pi L \tag{40}$$

The spectral distribution of blackbody radiation is described by the Planck radiation law, which may be written in either of two forms. The first form is

$$L(\nu) = \frac{2h\nu^3}{c^2} \frac{1}{e^{h\nu/kT} - 1} \tag{41}$$

where $L(\nu)\,\Delta\nu$ is the radiance in the spectral region between frequencies ν and $\nu + \Delta\nu$; h is Planck's constant and k is the Boltzmann constant.

If the spectral distribution is to be given in terms of wavelength, then Eq. (41) may be converted to

$$L(\lambda) = \frac{2hc^2}{\lambda^5} \frac{1}{e^{hc/\lambda kT} - 1} \tag{42}$$

Here $L(\lambda)\,\Delta\lambda$ is the radiance in the spectral region between wavelengths λ and $\lambda + \Delta\lambda$. The velocity of light in free space is represented by c.

If we divide both sides of the preceding equation by T^5, we see that $L(\lambda)/T^5$ depends upon λ and T only through the product λT. We can therefore plot a single graph of $L(\lambda)T^{-5}$ versus λT, which will serve for all wavelengths and temperatures. Figure 27 is such a plot.

The maximum value of $L(\lambda)T^{-5}$ is 4.095×10^{-6} if $L(\lambda)\,\Delta\lambda$ is in watts per square meter per steradian and $\Delta\lambda$ is measured in meters. The maximum occurs at $\lambda T = 2.898 \times 10^{-3}$ m K. The radiance $L = \int_0^\infty L(\lambda)\,d\lambda = \int_0^\infty L(\nu)\,d\nu$.

The surfaces of real solid objects are not black since they absorb only a fraction α of the radiation that falls upon them. The unabsorbed radiation will be reflected or scattered or, if the object is not too thick, transmitted. The fraction absorbed, α, will be a function of wavelength, temperature, and angle of incidence θ.

Heated solids radiate, but the spectral radiance $L(\lambda)$ is less than that of a blackbody at the same

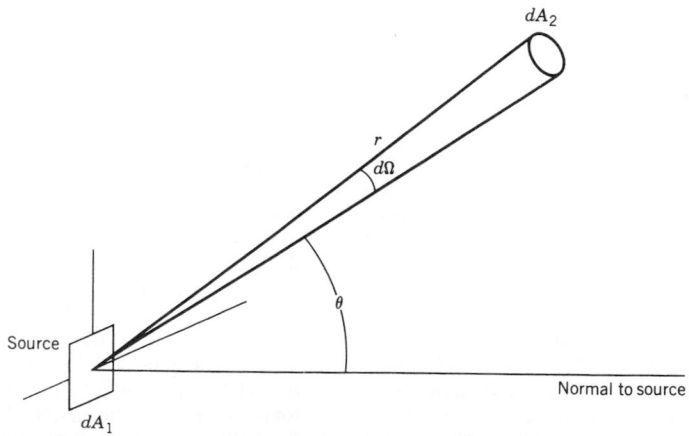

Fig. 26 Geometric quantities used in defining radiance L.

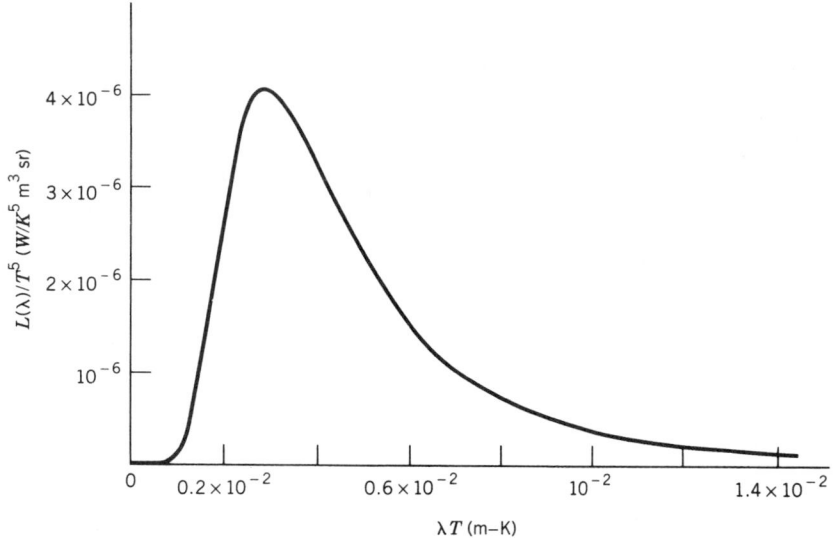

Fig. 27 Blackbody spectral distribution curve $L(\lambda)/T^5$ vs. λT. Power in watts; all lengths (including λ) in meters, and temperatures are degrees Kelvin. Curve has maximum value of 4.095×10^{-6} at $\lambda T = 2.898 \times 10^{-3}$.

temperature. The emissivity of the material, $\varepsilon(\lambda)$, is the ratio of the radiance $L(\lambda)$ for the material at hand to the radiance $[L(\lambda)]_{BB}$ of a blackbody at the same temperature; $\varepsilon(\lambda)$ is not only a function of wavelength but depends also upon the temperature and the angle of viewing, θ. For given temperature, wavelength, and angle

$$\varepsilon(\lambda, \theta, T) = \alpha(\lambda, \theta, T) \qquad (43)$$

Solids for which $\varepsilon(\lambda)$ varies slowly over the spectral range are known as *gray bodies*. For example, the emissivity of tungsten at 2500 K varies in the visible spectrum from about 0.49 in the blue to about 0.40 in the red. It is considered a gray body. As a function of temperature, the value of $\varepsilon(\lambda)$ of tungsten at $\lambda = 800$ nm is 0.414 for $T = 2000$ K and 0.396 for $T = 3000$ K.

Radiation from the sun closely resembles, in spectral distribution, radiation from a blackbody at 6500 K. Selective absorption by the atmosphere removes the ultraviolet ($\lambda < 350$ nm) and various regions of the infrared so that the sunlight reaching Earth is different from blackbody radiation.

If light passes through an optical system, we can measure the radiance L_1 before it enters the system and also measure L_2 after it leaves the optical system. [Recall the definition of L as given by Eq. (39).] It can be shown that if there are no losses (e.g., by absorption or by reflection at surfaces), then

$$\frac{L_1}{n_1^2} = \frac{L_2}{n_2^2} \qquad (44)$$

where n_1 and n_2 are the indices of refraction of the medium before and the medium after the optical system. In many cases, the same medium (e.g., air) exists on both sides of the optical system; in these cases $L_2 = L_1$. If there are losses in the system, then L_2 will be less than the value given by this calculation.

Consider the situation presented in Fig. 28. A source S of area dA is imaged by an optical system to S', which may serve as an apparent source for the region to the right of S'. The radiance of the beam to the left of the lens is L_1; neglecting losses and assuming air on both sides of the lens, the radiance to the right of the lens (both left and right of the image S') is $L_2 = L_1$. This is true only within the solid angle subtended by the lens at the image S'. Outside this solid angle L_2 is zero since no light reaches S' from the region beyond the circumference of the lens. As a secondary source, S' has radiance equal to L_1 out to this limiting angle and zero radiance for larger angles. No passive optical system can produce $L_2 > L_1$.

The irradiance (incident power per unit area) onto S' is E, given by

$$E = \frac{d\phi}{dA} = \int L_2 \, \cos\theta \, d\Omega$$

where the integration is over the solid angle subtended at S' by the lens. This is the same irradiance that would be produced at S' if the source S filled the exit pupil of the lens. The value of E depends upon the solid angle subtended. In photography, the solid angle is usually represented through the f number [see Eq. (26)] so that irradiance is proportional to $(F^{\#})^{-2}$.

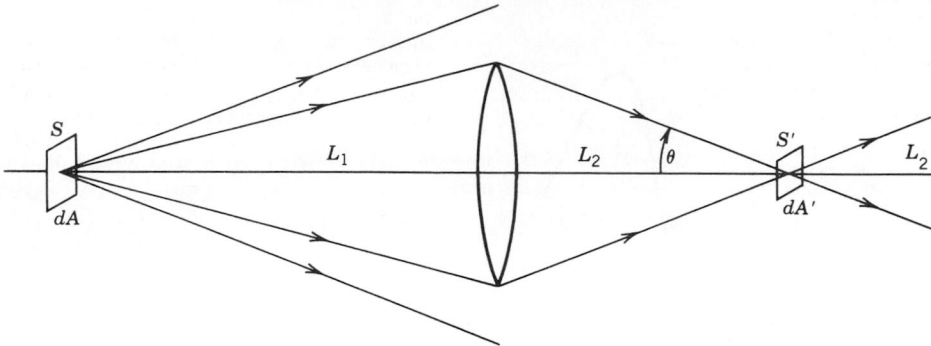

Fig. 28 Neglecting losses in lens, $L_2 = n_2^2 L_1/n_1^2$. Assuming there is no diffusing screen at image position S', L_2 remains the same to right of image; L_2 cannot be greater than L_1 if $n_2 = n_1$.

If one were to place a small luminous source, such as the filament of a small lightbulb, at S and then place the pupil of his eye at S' while looking back at the lens, he would see the exit pupil of the lens entirely filled with light. Only if the pupil of the eye is located at S' will the exit pupil be entirely filled with light.

5 LASERS

In gases, the intermolecular spacing is generally so large that the molecules radiate independently of each other, and the spectrum produced is a line structure for monatomic molecules or a band structure of fine lines for polyatomic molecules. We comment briefly about the atomic (or monatomic) case. As an example, sodium vapor may be observed to emit light by introducing a little salt (NaCl) into a bunsen flame; alternatively, radiation from sodium vapor may be observed from the sodium arcs, which are widely used for highway lighting.

In the heat of the bunsen flame, the salt dissociates into sodium and chlorine atoms. Most of these atoms are in their lowest energy (or "ground") state. The atoms can, however, be put into more energetic (or "excited") states; the energy of the excited states is specific, depending upon the atom. The number of atoms per unit volume, N_1, in the excited state and the number of atoms per unit volume, N_0, in the ground state will be in the ratio

$$\frac{N_1}{N_0} = e^{-\Delta E/kT} \qquad (45)$$

where ΔE is the energy difference between an atom in the excited state and an atom in the ground state, T is the temperature (in Kelvin), and k is the Boltzmann constant. Clearly $N_1 < N_0$. The atom will have several excited states, each with its own ΔE, and there will be an equation similar to (45) for each state. The atoms in the more energetic states are said to be thermally excited. An excited atom may give up its excess energy

in the form of light and return to the ground state. The frequency and wavelength of the emitted light are given by the equation

$$h\nu = \frac{hc}{\lambda} = \Delta E \qquad (46)$$

This bundle of light energy is called a *photon* or a *quantum*. Since each atom (e.g., sodium, potassium, mercury, and hydrogen) is characterized by its own set of excited states, it is also characterized by its own set of spectral lines.

In the case of the sodium arc lamp and other gaseous discharge lamps, the atomic excitations take place by inelastic collisions of the atoms with the moving electrons or ions that make up the electric current. Equation (45) is no longer an adequate description of the ratio N_1/N_0, but in most cases it remains true that $N_1 \ll N_0$; that is, only a small fraction of the atoms are excited.

If there are N_1 excited atoms per unit volume, there will be a rate of spontaneous return to the ground state. That rate will be

$$\frac{dN_1}{dt} = -A_{01}N_1 \qquad (47)$$

where A_{01} is a constant determined by the nature of the two states. Aside from the value of this constant, the rate of spontaneous return depends only upon N_1. Each atom that returns from the excited state to the ground state emits one quantum of light.

If light of the resonant frequency [i.e., the frequency given by Eq. (46)] passes through the gas, some of the atoms in the ground state will absorb a quantum and be excited into the more energetic state. This process is called absorption, and the rate at which it takes place is given by

$$\frac{dN_0}{dt} = -B_{10}N_0 L \qquad (48)$$

where L is the radiance at the resonant frequency and B_{10} is a constant determined by the nature of the two states. The rate at which atoms are excited (or quanta absorbed) is proportional to L.

The atoms in the excited state are also affected by the light, causing some of them to emit a quantum and return to the ground state. This process is called stimulated emission. The rate at which it takes place is

$$\frac{dN_1}{dt} = -B_{01}N_1 L \qquad (49)$$

This is in addition to the spontaneous emission. It can be shown that $B_{10} = B_{01}$ and $A_{01} = (h\nu^3/\pi c^2)B_{10}$. The photon emitted by stimulated emission is indistinguishable from the photon that stimulated it. There are now two photons instead of one; they have the same direction of propagation, the same frequency, the same phase, and the same polarization.

In most cases, $N_0 \gg N_1$, so that absorption predominates and stimulated emission is of little consequence. However, if it can be arranged so that $N_1 > N_0$, then the stimulated emission will exceed the absorption and the light can increase in L as it propagates. This might be called negative absorption. This condition must exist in order to produce a laser. The condition in which $N_1 > N_0$ is called a *population inversion*. It may be created in a variety of ways.

In the helium–neon laser, for example, the population inversion is produced as follows: The medium is a mixture of helium and neon atoms in the ratio of about $4 : 1$; the gas pressure of the mixture is about 1.0 torr to obtain a stable discharge. Electrical discharge in this gas mixture has little direct effect upon the neon but serves to excite some of the helium atoms into metastable states known as the $2^1 S$ and the $2^3 S$ states. These states are metastable in the sense that there are no allowed radiative transitions by which the atoms can return to the ground state; these helium atoms remain excited long enough to experience inelastic collisions with neon atoms. Fortunately, the excitation energy of the $3s_2$ state of a neon atom is the same as the $2^1 S$ state of a helium atom; collision between an excited helium atom and an unexcited neon atom can result in energy transfer, producing a neon atom in the $3s_2$ state and a helium atom in the ground state. In the same way, a helium atom in the $2^3 S$ state can excite a neon atom to

the $2s_2$ state. In this way, a small but useful fraction of the neon atoms are excited into these two states even though the neon was not directly involved in the electrical discharge. At a lower energy than the two states we have been discussing is a state of the neon atom known as $2p_4$. There are essentially no neon atoms in this state, so there is a population inversion between states $3s_2$ and $2p_4$ and between states $2s_2$ and $2p_4$. These two population inversions can produce lasing at $\lambda = 632.8$ nm and $\lambda = 1152.3$ nm, respectively. For the process to run continuously, the neon atoms in the $2p_4$ state must return to the ground state. This involves a radiative transition to the $1s_5$ state and finally an inelastic collision of the neon atom with the walls of the tube.

Since the rate of stimulated emission is proportional to the spectral irradiance (or to the spectral energy density) and the rate of spontaneous emission is independent of the spectral irradiance, it follows that stimulated emission will become the dominant process when the spectral irradiance is large. To bring this about, the lasing medium [in our case the helium–neon (He–Ne) gas mixture] is placed between two mirrors as indicated in Fig. 29. These mirrors should have high reflectivity; it is common for the reflectivities to exceed 99.5%. One of the mirrors should have a slight transmissivity (a few tenths of a percent) so that some of the light may escape from the space between the mirrors (called the cavity) into outside space.

If the mirror separation l and the wavelength λ_m are such that a round-trip path equals an integral number m of wavelengths, that is,

$$2l = m\lambda_m \qquad (50)$$

there will be constructive interference for light reflected from the mirrors upon successive round trips. The *cavity* is said to be resonant at wavelength λ_m and at the corresponding frequency $\nu_m = c/\lambda_m$. The frequency spacing $\Delta\nu$ between ν_m and ν_{m+1} is $\Delta\nu = c/2l$. Cavity resonances are called *modes*. For the typical He–Ne laser, l may be about 30 cm so that m is a very large number and $\Delta\nu$ is about 500 MHz.

If a resonance of the cavity exists at the same wavelength as the resonance of the neon states $3s_2$ and $2p_4$ and there is also a population inversion for these two states sufficiently large that the gain per pass exceeds

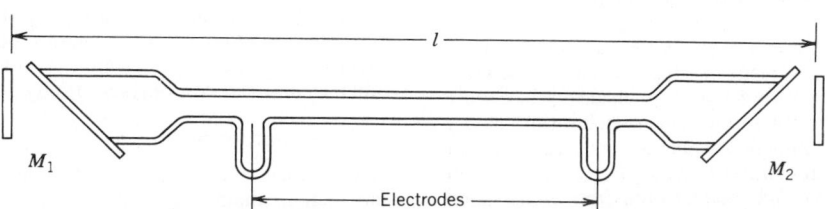

Fig. 29 Helium–neon laser with concave end mirrors and Brewster angle windows.

the losses, then there will be lasing (oscillation at optical frequencies) within the cavity. The losses from the cavity include (a) the light that escapes through the partially transparent mirrors, (b) absorption by the mirrors, (c) diffraction and scattering from the beam inside the cavity, and (d) any reflection losses at the end windows of the discharge tube.

The spectral lines of neon have some width, primarily because of the Doppler broadening due to thermal motion. For an atom in motion, its Doppler-shifted resonant frequency must match the incident photon frequency in order to produce stimulated emission; that is, the Doppler-shifted frequency of the atom must match the cavity-resonant frequency for the system to lase. The spectral width of the neon resonance due to Doppler broadening is temperature dependent, but in a typical laser it is roughly 1.0 GHz. This width is sufficient to cover two, or sometimes three, cavity modes of a 30-cm laser cavity. These two or three modes will lase simultaneously.

In the first He–Ne lasers, the mirrors M_1 and M_2 were plane mirrors. It was necessary that they be accurately parallel to each other; if they were not, a ray reflected back and forth between them would not remain in the discharge tube to be amplified by the active medium. It was later discovered that curved mirrors could be used, in which case the alignment is less critical. It can be shown by diffraction calculations or by ray tracing that a ray from M_1 to M_2 that is slightly off axis will remain trapped near the axis if the center of curvature of M_1 is to the right of M_2 and the center of curvature of M_2 is to the left of M_1. The axis is a line through the two centers of curvature; the axis should pass through the discharge tube.

A limiting case is to use one plane mirror, M_1, and let the center of curvature of the other (M_2) lie slightly to the left of M_1; this is called a hemispherical cavity. There are other mirror arrangements that form stable cavities (stable in the sense that rays are trapped near the axis), but the ones just mentioned are commonly used.

In contrast, there are unstable cavities in which a ray starting slightly off axis diverges from the axis and escapes from the cavity. As an example, we give a cavity for which the center of curvature of M_1 lies slightly to the left of M_2 and the center of curvature of M_2 lies slightly to the left of M_1. (One center is inside the cavity and the other outside the cavity.) The rays are not trapped in this cavity but escape after a few round trips; the losses in this cavity are very large. It does not lase.

In the earliest He–Ne lasers, the reflection losses at the windows on the ends of the discharge tube were so great that the system could be made to lase only by eliminating these windows and attaching the cavity mirrors directly onto the discharge tube. Later it was realized that by attaching the windows at the Brewster angle (see Fig. 3), one polarization (the polarization with E parallel to the plane of incidence) would experience no reflection loss and the cavity mirrors could be mounted independently of the discharge tube. The light from such a laser is linearly polarized since the gain exceeds the losses only for the polarization, which experiences no loss at the Brewster angle windows.

The mirrors M_1 and M_2, which form the resonant cavity, are usually multilayer dielectric coatings since most metals do not have sufficiently high reflectivity. Also, with multilayer dielectric coatings the mirrors may be spectrally selective in their reflectivity. For example, the mirrors may be highly reflecting at $\lambda = 632.8$ nm with much lower reflectivity at $\lambda = 1152.3$ nm. In this case, the He–Ne laser will lase at $\lambda = 632.8$ nm but not at 1152.3 nm. By exchanging the mirrors for a pair with high reflectivity at 1152.3 nm, we can cause the system to lase at that wavelength.

The modes considered a few paragraphs back are properly called longitudinal modes. A laser may also have several "transverse" modes, but the manufacturers of lasers usually suppress all the transverse modes except the TEM_{00} mode. In this mode the amplitude of the electric field at the output mirror as a function of distance off axis r is given by

$$E = E_0 e^{-r^2/w^2} \qquad (51)$$

where E_0 is the amplitude on axis, E is the amplitude at a distance r from the axis, and w is a constant depending upon the geometry of the cavity. The surface of the output mirror is a surface of constant phase (i.e., a wave front) for the emerging wave; the beam width w may be only a few millimeters.

We have given our attention to the He–Ne laser because it is readily available and illustrates the principles involved. There are many other media in which population inversion can be produced and which can provide lasing if used in a suitable cavity.

6 THE EYE AND VISION

The eye is important because most of the information obtained in a lifetime is brought to the brain through the eye. For the student of optics, the eye is important because many optical instruments, for example, microscopes and telescopes, are used in conjunction with the eye so that the eye becomes a part of the optical system. The pupil of the eye may become the aperture stop of the system or in some cases the eye may limit the spatial frequency response or resolution of the optical system. A geometrical or physical description is inadequate because the eye is a living, functioning organ that should be considered in terms of physiology and neurology, but these fields are beyond the scope of this chapter and can be considered only superficially.

6.1 Structure of the Eye

The human eye is an almost spherical organ about an inch in diameter. It is shown in cross section in Fig. 30. Six muscles, two of which are shown in the figure as Z_1 and Z_2, hold the eye in place and rotate

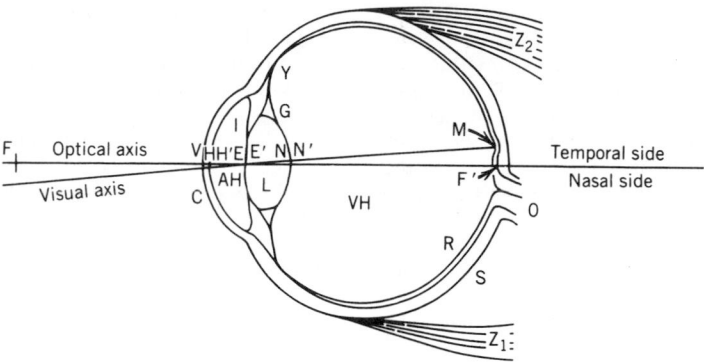

Fig. 30 Horizontal section of right human eye according to Helmholtz.

it relative to the head. These muscles are attached to the sclera S, which is a tough white skin covering most of the eye. At the front of the eye the sclera is replaced by the cornea C, which is a transparent membrane through which light enters the eye. After entering through the cornea, light passes through the aqueous humor AH, the crystalline lens L, the vitreous humor VH, and finally reaches the retina R. The aqueous humor is a weak salt solution; the vitreous humor is a soft jelly consisting primarily of water. The fluids of the eye are slightly ($\sim$25 torr) above atmospheric pressure. This pressure helps to maintain the shape of the eyeball. The crystalline lens is a fibrous jelly contained in a thin membrane or sac; it is hard at the center and progressively softer toward the outside. The lens is held in place and attached to the ciliary muscle Y by the suspensory ligament G. When the ciliary muscle is relaxed, the second focal point is at the retina and distant objects are in focus. To view nearby objects, the ciliary muscle contracts, allowing the lens to become more nearly spherical. This is known as accommodation; with age, the lens becomes less elastic, and the ability to accommodate gradually decreases. The lens of the eye is not transparent to ultraviolet light.

The retina is the interior lining for a large part of the eyeball. It consists of rods and cones that are light-sensitive nerve endings, along with a delicate network of nerve fibers connecting the rods and cones to the optic nerve O and a network of capillary blood vessels that supply the necessary oxygen and nutrients. The yellow spot, or macula lutea M, which contains many cones and relatively few rods, is a slight depression in the retina; the central region, called the fovea centralis, contains cones exclusively, no rods. The macula lutea is about 2 mm in diameter, and the fovea centralis is about 0.25 mm in diameter. Cones in the fovea centralis are about 1.5 μm in diameter, increasing in size to about 5.5 μm in the outer portion of the macula lutea and several times this size in other portions of the retina. See Fig. 31. In the outer portion of the retina, the rods outnumber the cones by 10:1.

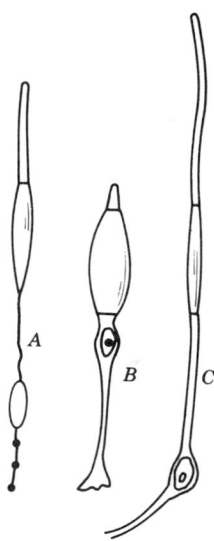

Fig. 31 Rods and cones of retina: *A*, rod; *B*, cone from extrafoveal region; *C*, cone from central fovea.

Each human eye contains roughly 7 million cones and 120 million rods.

Vision in the fovea centralis is so much more acute than in the extra foveal region that the muscles surrounding the eye involuntarily rotate the eyeball until the object of interest is imaged upon the fovea centralis. The angle in *object space* covered by the fovea centralis is less than 1°; it is only a little more than sufficient to cover one letter of this printed page when the book is held at the usual reading distance of 25 cm. In reading or examining an extended object, the eye must move frequently. Extra foveal vision is not useful in observing details but enables one to be aware of objects around him. For a healthy eye, the total field is about 128°; an early sympton of glaucoma is the shrinking of the field of view.

The mosaic structure of the cones in the fovea centralis limits the resolution of the eye. Considering the size of the cones, this varies from 0.3 to 1.0 min for the angular resolution in object space. These numbers should be slightly larger because the cones are separated by a small amount of inactive tissue. It is interesting to observe that this is comparable to the resolution limit set by diffraction at the pupil and also comparable to the limits produced by aberrations, primarily spherical aberration of the optical system. One minute is a good round number representing the overall resolution of the eye.

Part of the blood supply to the retina is provided by a network of blood vessels on the front of the retina. If one stares at a blue sky (or a white wall illuminated by blue or violet light), the red blood cells coursing through these blood vessels can be seen since they cast a shadow on the retina. Unlike the specks of dust that float upon the front of the eye or in the vitreous humor when one is tired, these shadows follow definite paths; that is, they are confined to the blood vessels. These shadows are called *muscae volitantes*, which means flying flies. Red blood cells are about 8 μm in diameter, so each one can cast a shadow over several cones of the fovea.

6.2 Adaptation of Eye to Light

The iris diaphragm, I, is a ring-shaped involuntary muscle that controls the amount of light entering the eye. It is located just in front of the lens, and the diaphragm opening or pupil is the aperture stop of the eye. It varies in diameter from 2 to 8 mm. This is a factor of 4^2, or 16, in the area of the entrance pupil. The eye functions under illumination conditions that vary by a factor of $\sim10^9$. Variation in pupil size is certainly not sufficient to account for this wide range; most of the adaptation to light and dark is accomplished by changing the sensitivity of the retina. The photosensitive chemicals (or pigments) in the rods and cones are bleached or altered by light and must be constantly reconstituted. Due to the lower rate at which the pigment is consumed in low illumination, the steady-state concentration of the pigments is higher and the retina more sensitive in low illumination than in high illumination.

6.3 Scotopic Vision

When the eye has been dark adapted (i.e., kept for half an hour or more in darkness comparable to outdoor illumination by a moonless night sky), the eye becomes sufficiently sensitive to see a small source of 2×10^{-8} cd at a distance of 3 m. Neglecting atmospheric absorption, this is equivalent to seeing a standard candle at a distance of 13 miles. Astronomers observe that except under unusually good conditions, stars of sixth magnitude represent the limit of vision of the unaided eye. This corresponds to seeing a standard candle at a distance of about 6.6 miles through the atmosphere. However one describes it, the dark-adapted eye is incredibly sensitive.

Vision by the dark-adapted eye is called *scotopic* vision and takes place in the rods of the eye, not in the cones. Since there are no rods in the fovea, the dark-adapted eye has no central vision, and in order to see an object in subdued light, one must look not at the object of interest but to the side so that the object of interest will be imaged on the outer part of the retina, which contains rods. There is no color in scotopic vision.

In the rods, the pigment that absorbs the light and somehow triggers the signal along the nerves to the brain is called rhodopsin. The chemical composition and structure is known to be a protein molecule combined with a molecule of retinal. Retinal is closely related to the compounds known as retinol (vitamin A) and carotene (the yellow pigment of carrots and many other yellow vegetables).

The spectral sensitivity of rod vision is shown in curve B of Fig. 32. The ordinate at each wavelength is inversely proportional to the minimum amount of energy that is just perceptable (i.e., to the threshold of vision). The curve is normalized to 1 at its peak. This closely matches the absorption curve of rhodopsin.

6.4 Photopic Vision

For conditions of ordinary illumination, the rhodopsin in the rods is almost completely bleached and vision is by the cones. This is called *photopic* vision, or cone vision. The spectral sensitivity for cone vision is shown by curve A of Fig. 32. Notice the shift of this curve toward the red relative to the scotopic curve B. Because of this shift, two nonluminous objects of different colors (e.g., yellow and blue-green) that appear "equally bright" in ordinary illumination will not appear equally bright in subdued illumination (e.g., twilight), the blue-green becoming much more conspicuous than the yellow. This shift in the spectral sensitivity and the resulting change in relative brightness of various colors is known as the Purkinje effect. It is a source of trouble in making visual comparisons of light sources of different colors.

The level of illumination at which the eye changes from photopic to scotopic vision (or vice versa) with the attendant change in spectral sensitivity, loss of color discrimination, and foveal vision is about the illumination level produced by the full moon on a clear night, or 0.16 lux.

6.5 Color Vision

Color vision takes place in the cones. There are three different types of cones in the eye; the three differ in that they contain different photosensitive pigments and have distinct spectral response curves. There is no observable physical structure that enables one to distinguish between the three types; the photosensitive pigments are present in such low concentrations that it is difficult to distinguish even on this basis.

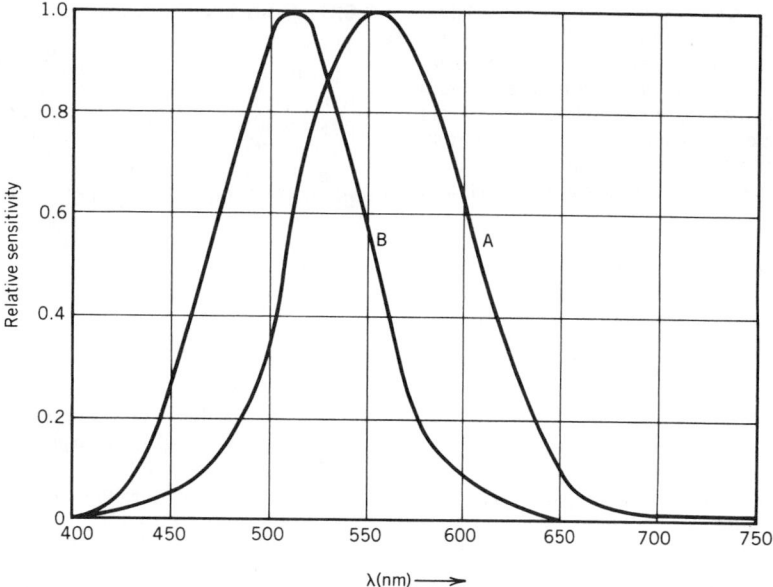

Fig. 32 Spectral sensitivity curves for normal human eye: *A*, light-adapted (photopic) eye; *B*, dark-adapted (scotopic) eye. Each curve is normalized to 1 at its maximum.

The three pigments are probably three different protein molecules, each in combination with a molecule of retinal. Because of the chemical similarity of the three dyes to each other and to rhodopsin (which is much more abundant), they are difficult to isolate and identify. The spectral sensitivity of the three cone types is given in Fig. 33. The output of each cone is determined by the intensity reaching it, the wavelength of the light, and the spectral sensitivity of the cone for that wavelength; the same output signal could be obtained by use of a lower intensity at a wavelength closer to the peak of the sensitivity curve. Each cone is color blind (just as the rods are color blind); the sense of color is derived from the relative response of the three types of cones.

As shown in Fig. 33, the three cone types have peak sensitivities in the blue (~440 nm), green (~535 nm), and orange (~565 nm); they are labeled *C*, *B*, and *A*,

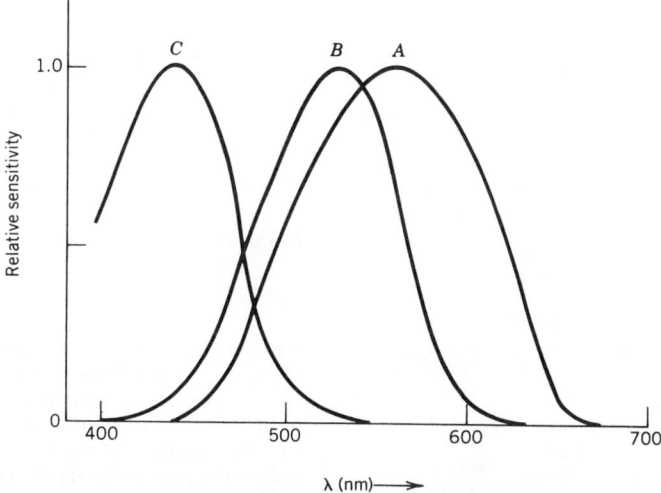

Fig. 33 Spectral sensitivity curves for three cone types of human eye. Each curve is normalized to 1 at its maximum.

respectively. For each spectral wavelength, the relative response of these three cone types is unique and determines the color sensation. If several wavelengths are present, each wavelength evokes a response in each cone type, and the relative size of the total response in each of the three cones determines the color sensation. Curve A peaks at 565 nm, which is in the orange; at this wavelength, the other curves, in particular B, still have nonzero values and the sensation of orange is produced by a signal from A and a weaker signal from B. At longer wavelengths, cones of type A respond less than they did at 565 nm, but the response from cones of type B decreases even more rapidly so that the signal from A makes up a larger fraction of the total output and the color sensation changes from orange to red.

Common forms of color blindness result from the absence of type A or type B cones. Protonopes are persons color blind due to lack of type A cones; duteranopes lack type B cones.

In ordinary vision, the output of type A cones is added to the output of type B cones (with perhaps a weak contribution from type C cones), and this sum is transmitted to the brain along the optic nerves. This sum encoded as nerve pulses per second is interpreted by the brain as luminosity (white) without color information. Color information is transmitted in two channels as the difference $(A - B)$ between the output of type A cones and type B cones and the difference $(A - C)$ between the output of type A cones and type C cones. The data is processed into these sum and differences in or near the eye and then transmitted along nerve fibers to the brain in the form of an increase or a decrease of the pulse frequency from the spontaneous value of the pulse frequency that exists when the eye is in the dark. Type C cones have little effect upon the sensation of brightness but are effective in producing color discrimination.

Although the system just described is believed by many to be the usual one, others are possible and sometimes effective because one can cover the right eye with a red filter and the left eye with a green filter (or vice versa) and obtain color vision. In this case, some of the data processing that usually takes place at the eye appears to be deferred to some later stage of the visual process, perhaps in the brain or perhaps at the optic chiasma, the point at which the two optic nerves (one from each eye) come together on their way to the brain.

For a person with normal vision, the colors associated with various portions of the spectrum are as shown in Table 2.

Table 2 Colors and Associated Wavelengths

Color	Wavelength (nm)
Violet	<450
Blue	450–500
Green	500–570
Yellow	570–590
Orange	590–610
Red	>610

6.6 Colorimetry

The word *color* has several definitions. In one, it is associated with the properties of a dye; in another, it is a property of light; and in yet another, it is a physiological sensation produced in the brain by light entering the eye. In an earlier section, we have given a brief description of color vision. In this section, we use the word color as descriptive of the light entering the eye and present the methods used to give a quantitative description of the color. The branch of optics that deals with the quantitative specification of color is called colorimetry.

6.7 Color Mixing

In order to understand colorimetry, we must first establish the basic facts of color mixing, which are illustrated by the following experiment. We attempt to match all possible colors by mixing three "primaries." The selected primaries are monochromatic (or spectral) colors of wavelength 450 nm (blue), 550 nm (green), and 620 nm (red). We identify them as α, β, and γ, respectively. There is nothing unique about these particular wavelengths that entitle them to be primaries; we select them because the experimental data using these three primaries was carefully determined in early color-mixing experiments. We now allow the eye to look at a white diffusing card. Two adjacent areas of the card are illuminated (a) by light of arbitrary or unknown color and (b) by a mixture of the three primaries. Area 2 is illuminated by all three primaries, and the amount of each primary is adjusted to obtain a match with the unknown. Most colors can be matched by this mixing process; a few cannot. In cases for which the unknown cannot be matched by the preceding process, a match can be obtained by moving one (or very rarely two) of the primaries from area 2 to area 1 and then adjusting the amount of each primary; this is equivalent to subtracting or using a negative amount of the moved primary in area 2. The use of three primaries widely spaced in the spectrum, as are the ones suggested here, reduces the number of cases in which a negative amount of any of the primaries is required. Neither the unknown nor the primaries need be monochromatic (spectrally pure) colors; a match can always be made. If the unknown is represented by U and the amount of each primary by A, B, and C, respectively, the experimental results may be represented by the equation

$$U = A + B + C \qquad (52)$$

which is interpreted to mean that the sensation of light and color produced by the unknown may be duplicated by the mixture of the three primaries. The values of A, B, and C are unique if U is given. The eye sees the overall effect of the mixture; it is not aware of the individual primaries that make up the mixture.

If we now restrict our unknown to spectrally pure (i.e., monochromatic) light and keep the power of the

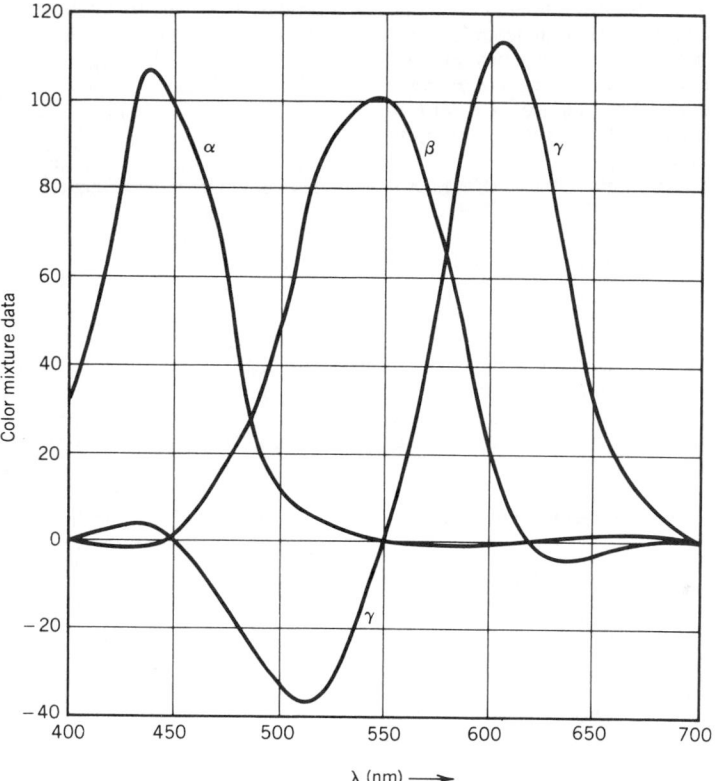

Fig. 34 Color mixture curves for matching spectrally pure colors by mixing primaries having wavelengths 450, 550, and 620 nm.

unknown constant but vary its wavelength, we can at each wavelength determine experimentally the power of each primary required to produce a match. The results of this experiment are given in Fig. 34, which gives the amount of each primary α, β, and γ required to match each spectral color. The curves have been normalized to $\beta = 100$ at 550 nm; α and γ are zero at this wavelength, which corresponds to the β primary. Each curve is normalized to 100 when the unknown wavelength is the same as that primary. For example, the curves indicate that a match is obtained for an unknown at 500 nm by combining 47.5 units of α (light at 450 nm) with 125 units of β (light at 550 nm) and subtracting (i.e., adding to the unknown) 30.0 units of γ (light at 620 nm). Notice that the primaries do not add to 100; this is because the spectral sensitivity of the eye for each of the primaries differs from its sensitivity at the unknown wavelength. In this case, the most significant difference is a factor of about 3 between the sensitivity of the eye to the β primary and its sensitivity to the 500 nm unknown.

Similar curves for the mixing of other sets of monochromatic primaries could be determined experimentally, but it is unnecessary to do so because it is possible to deduce them from the curves already given. The process is straightforward but tedious, and we shall not describe it. It is also possible to specify a new set of primaries by giving the curves α', β', and γ', which give the mixing data required to match spectral colors using the new primaries. As long as the new curves α', β', and γ' (as functions of wavelength) are a linear combination of the experimental curves α, β, and γ (which were given in Fig. 34), the new system will give a satisfactory system of color specification. The requirement of algebraic linearity means that

$$\alpha' = K_{11}\alpha + K_{12}\beta + K_{13}\gamma$$

$$\beta' = K_{21}\alpha + K_{22}\beta + K_{23}\gamma$$

$$\gamma' = K_{31}\alpha + K_{32}\beta + K_{33}\gamma \tag{53}$$

where the K_{ij} are real and independent of wavelength but are otherwise subject to no restriction except that the determinant

$$\begin{vmatrix} K_{11} & K_{12} & K_{13} \\ K_{21} & K_{22} & K_{23} \\ K_{31} & K_{32} & K_{33} \end{vmatrix} \neq 0 \tag{54}$$

With such a wide choice of primaries and with the possibility of algebraic transformation from one set to another, color-mixing data could not provide any insight into the spectral sensitivity curves of Fig. 33, but this did not prevent the development of colorimetry in advance of a detailed understanding of color vision.

With so much freedom in the choice of the K_{ij} (or the curves α', β', and γ'), it is probably not surprising that some of the possible sets of primaries so described contain primaries that are not spectral colors (e.g., purples) or not spectrally pure (e.g., pinks). It is also true that many acceptable sets of primaries contain primary colors that are not real; by this we mean that they exist mathematically in terms of the mixing curves α', β', and γ', which produce real colors; in this sense, they are entirely satisfactory primaries, and yet they do not exist in the sense that the single primary alone cannot be seen as light and color.

6.8   Tristimulus Values and Trichromatic Coefficients

Since color specification is commercially important and since there is so much freedom in the choice of primaries, it was inevitable that there should develop some agreement on what set of primaries would be used. In 1931, the International Commission on Illumination (ICI) [also known by its French name, Commission Internationale de l'Eclairage (CIE)] agreed to express all color specifications in terms of three primaries defined by the color-mixing curves of Fig. 35. The letters $\bar{x}$, $\bar{y}$, and $\bar{z}$ have become standard, replacing the α, β, and γ used by earlier workers. The ordinates, called *tristimulus values*, are in arbitrary units and have been adjusted so that the areas under the three curves are equal. The shape of curve $\bar{y}$ was arbitrarily chosen to be the same as curve A of Fig. 32. Curves $\bar{x}$ and $\bar{z}$ have shapes selected so that the three primaries satisfy Eq. (53). For computational convenience it was also required that none of the curves is ever negative. None of the primaries defined by this set of mixing curves is real; they form a satisfactory base for the quantitative specification of color, but only real colors can be produced and mixed in the laboratory.

For monochromatic light with wavelength 500 nm (green), the tristimulus values are

$$\bar{x} = 0.0049 \qquad \bar{y} = 0.3230 \qquad \bar{z} = 0.2720$$

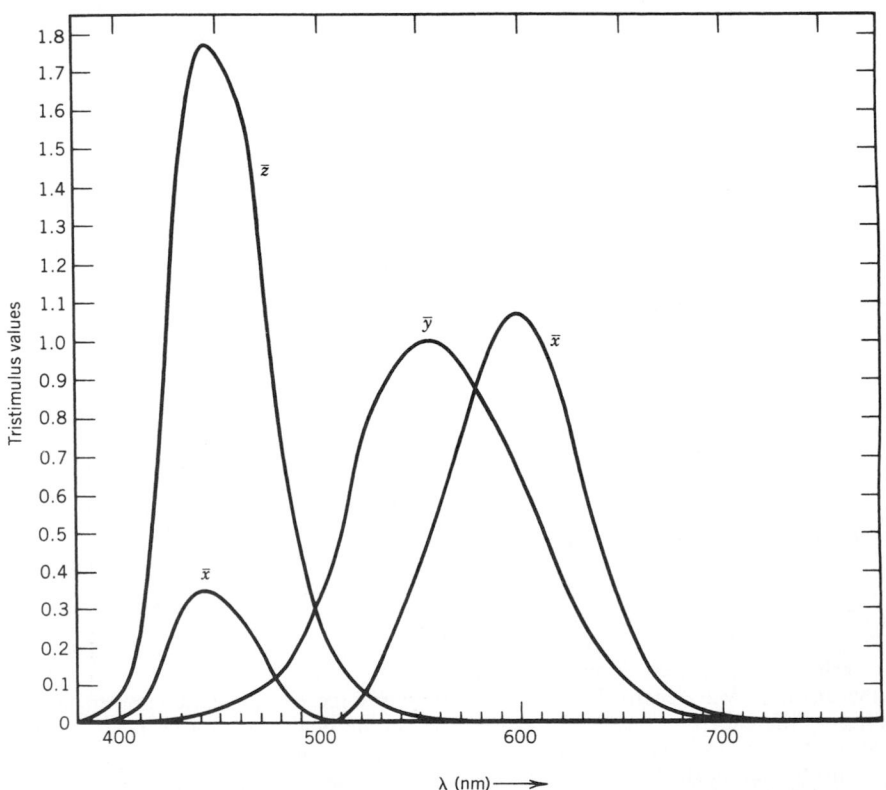

Fig. 35   Standard ICI (or CIE) tristimulus curves $\bar{x}, \bar{y}$, and $\bar{z}$ for unit power at indicated wavelength. Numerical values for these curves may be found in Refs. 1 and 2.

Define three new quantities x, y, and z such that

$$x \equiv \frac{\bar{x}}{\bar{x} + \bar{y} + \bar{z}} \qquad y \equiv \frac{\bar{y}}{\bar{x} + \bar{y} + \bar{z}} \qquad z \equiv \frac{\bar{z}}{\bar{x} + \bar{y} + \bar{z}}$$

These new quantities are called *trichromatic coefficients* and by definition have the property that $x + y + z = 1$; any two of the three quantities are sufficient to specify the color. For 500-nm light, the values are

$$x = 0.0082 \qquad y = 0.5384 \qquad z = 0.4534$$

This system cannot contain any intensity information, only color information.

In this system, any spectral color may be specified by giving any two of the trichromatic coefficients; the values of x and y are usually given. If we plot on ordinary graph paper the values of x and y for the spectral colors, we obtain the curve of Fig. 36, where the wavelength (in nanometers) is shown at various places along the curve. A diagram such as this in which color information is plotted using the trichromatic coefficients is called a *chromaticity diagram*; the curve is known as the spectrum locus.

6.9 Trichromatic Coefficients for Nonmonochromatic Light

In the previous section we defined the trichromatic coefficients of any monochromatic light using the ICI (or CIE) primaries. In very few cases is the light reaching the eye monochromatic; it is usually a mixture or distribution of spectral colors. If we represent the spectral distribution by the function $f(\lambda)$ defined so that $f(\lambda)\, d\lambda$ is the power (e.g., in watts) in the spectral interval between λ and $\lambda + d\lambda$, then we calculate the *tristimulus values* of the light by the equations

$$X \equiv \int_0^\infty \bar{x} f(\lambda)\, d\lambda \qquad (55a)$$

$$Y \equiv \int_0^\infty \bar{y} f(\lambda)\, d\lambda \qquad (55b)$$

$$Z \equiv \int_0^\infty \bar{z} f(\lambda)\, d\lambda \qquad (55c)$$

where $\bar{x}$, $\bar{y}$, and $\bar{z}$ are the functions represented in Fig. 35, the ICI color mixture curves. This process

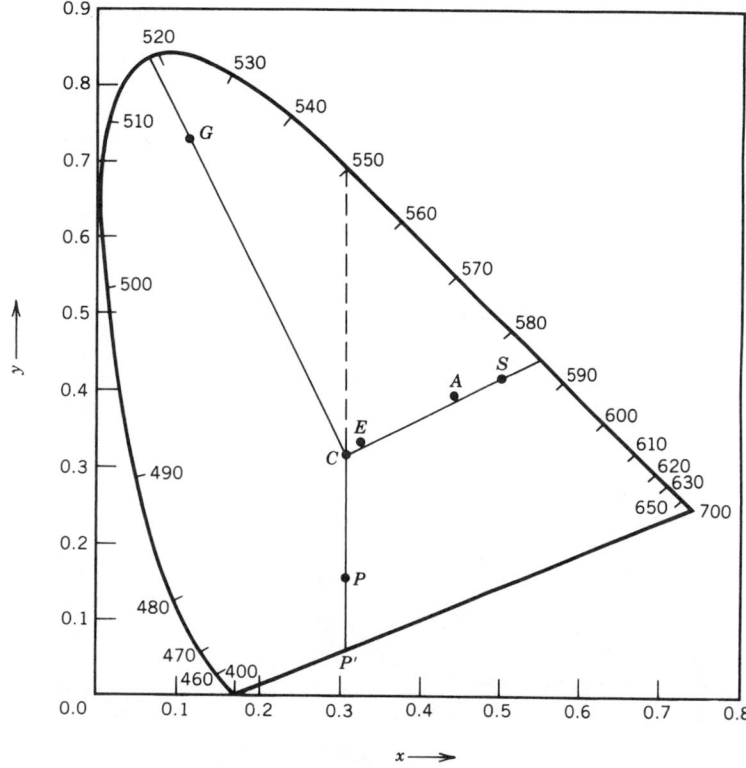

Fig. 36 Chromaticity diagram: Horseshoe curve, spectrum locus; *E*, source for which $f(\lambda)$ is constant; *C*, illuminant C, approximately daylight; *A*, illuminant A, illumination from tungsten filament lamp; *S*, light reflected from orange skin illuminated by illuminant C.

amounts to treating each spectral color in the light by the methods of the previous section and then adding (integrating) all of these effects together. By definition, the integrals are from zero to infinity; but since the functions $\bar{x}$, $\bar{y}$, and $\bar{z}$ are zero outside of the visible range, the integration is effectively limited to the visible-wavelength interval. The functions involved cannot be integrated by elementary methods; the integration is carried out numerically. The numerical data represented by the curves $\bar{x}$, $\bar{y}$, and $\bar{z}$ may be found in the original ICI report[1] or in any textbook on colorimetry.[2]

The three tristimulus values calculated by Eqs. (55) are converted to *trichromatic coefficients* by the equations

$$x = \frac{X}{X + Y + Z} \qquad y = \frac{Y}{X + Y + Z}$$

$$z = \frac{Z}{X + Y + Z} \qquad (56)$$

We again have the property that

$$x + y + z = 1 \qquad (57)$$

and any two of these may be used to specify the color of the light; x and y are usually used. Information about the total intensity of the light has been lost, but all color information is retained.

As an example, c onsider light for which the spectral distribution $f(\lambda)$ is a constant. This is light for which, at any wavelength, a small wavelength interval $d\lambda$ contains the same power as an equal interval $d\lambda$ located at any other wavelength. Since the three curves x, y, and z of Fig. 35 have equal areas under them, the integrals of Eqs. (55) will, for this example, be equal, that is, $X = Y = Z$. When these are converted to the trichromatic coefficients x, y, and z, we obtain

$$x = y = z = 0.3333$$

On the chromaticity diagram (Fig. 36), this is represented by the point E at $\left(\frac{1}{3}, \frac{1}{3}\right)$.

Another important example is light from a source known as illuminant C. Illuminant C is intended to have the same spectral distribution as average daylight, at least in the visible. It consists of a gas-filled tungsten lamp operated at the color temperature 2848 K combined with filters designed to alter the spectral distribution of the lamp to that of daylight. The spectral distribution of illuminant C is in Fig. 37. From this distribution, one can evaluate numerically the integrals of Eqs. (55) and then the trichromatic coefficients x, y, and z of Eqs. (56). The results of these calculations give point C at (0.3101, 0.3163) on the chromaticity diagram in Fig. 37. Light from illuminant C is generally considered to be "white," although the term *white light* has no universally accepted definition.

6.10 Color of an Orange Skin

In Fig. 37, there is shown the reflectance of an orange skin as a function of wavelength, $R(\lambda)$. This curve may be obtained by illuminating the orange skin successively at several different wavelengths; at each

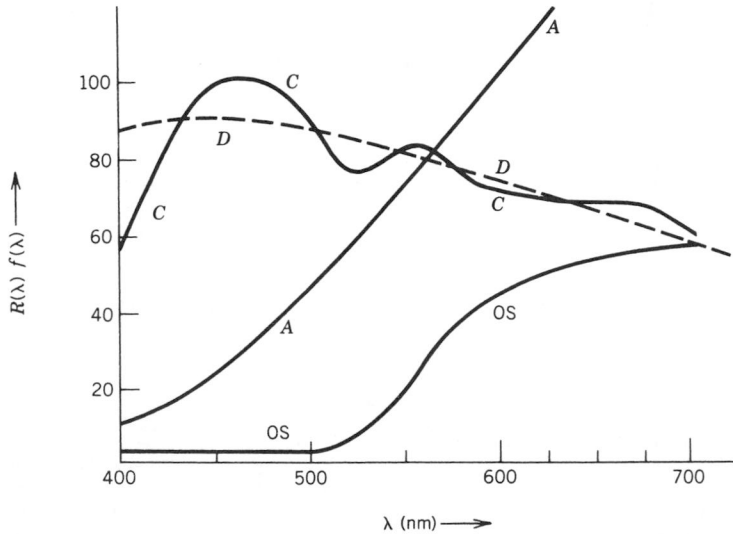

Fig. 37 Dashed curve D, curve A, and curve C are spectral distributions $f(\lambda)$ of average daylight, light from illuminant A, and light from illuminant C. (Vertical scale is arbitrary and not the same for the three curves.) Lower curve OS, spectral reflectance $R(\lambda)$ of orange skin; magnesium carbonate powder is taken as 100%.

wavelength, the reflected radiance is measured for the orange skin and for some white object. The reflectance $R(\lambda)$ at each wavelength is the ratio of these two measurements. Since the orange skin is a diffuse reflector, the white comparison object should be a diffuse reflector also. Freshly fallen snow is a good white diffuse reflector, but in the laboratory a powder of magnesium carbonate is more practical.

The observed color of the orange skin depends not only on its spectral reflectance but also on the spectral distribution of the illuminating light. Let us assume that illuminant C is used and we represent its spectral distribution by $C(\lambda)$. The spectral distribution of the light reflected from the orange skin is the product $C(\lambda)R(\lambda)$. We calculate the tristimulus values of this light from Eqs. (55):

$$X \equiv \int_0^\infty \overline{x} C(\lambda) R(\lambda)\, d\lambda = 341$$

$$Y \equiv \int_0^\infty \overline{y} C(\lambda) R(\lambda)\, d\lambda = 277$$

$$Z \equiv \int_0^\infty \overline{z} C(\lambda) R(\lambda)\, d\lambda = 50$$

and when these are normalized to the trichromatic coefficients, we have

$$x = 0.511 \qquad y = 0.414 \qquad z = 0.075$$

These locate a point (marked S) on the chromaticity diagram (see Fig. 36). This point is fairly close to the spectral locus for 586 nm. The light reflected from this orange skin is therefore close to the orange-yellow color of the sodium D lines.

If we used another illuminant instead of illuminant C, the location of point S representing the chromaticity of the light reflected by the orange skin would have to be recalculated and would probably have changed.

Two pieces of cloth that have the same spectral reflectance will always look alike, that is, have the same coordinates on the chromaticity diagram, as long as the same illuminant is used on each piece no matter what illuminant is used. It is possible, and sometimes happens, that two pieces of cloth that have different spectral reflectance curves may look alike when illuminant C is used but will be noticeably different when another illuminant, such as illuminant A, is used. Illuminant A is the gas-filled tungsten lamp operated at the color temperature 2848 K and used without filters; it is typical of the illumination produced by tungsten filament lamps. The chromaticity of illuminant A is represented by point A in Fig. 36. It more frequently happens that two pieces of cloth look alike under illuminant A but are noticeably different under illuminant C. Illuminant A is relatively weak in the short-wavelength region so that a match using this illuminant is relatively insensitive to the reflectance

of the cloth for blue light. Illuminant C is slightly stronger at the shorter wavelengths than it is at the longer wavelength (see Fig. 37).

6.11 Chromaticity Diagram as Aid to Color Mixing

From the definitions of the trichromatic coefficients [Eqs. (55)], it follows that if we have two colors represented by points such as G and R (Fig. 38) of the chromaticity diagram, any additive mixture of these two colors will be represented by a point lying on the line GR. If each component (G and R) is assigned a weight proportional to the sum of its tristimulus values ($X + Y + Z$), the point representing the chromaticity of the mixture will lie at the center of gravity of these weights. For example, if the mixture contains more of light G than of light R such that the sum $X + Y + Z$ for light G is twice the corresponding sum for light R, the mixture will have color represented by the point D on the line GR located so that the distance DR is twice the distance GD. Any color on the line from G to R may be obtained by additive mixing properly selected amounts of lights G and R. After obtaining light D in this way, light D may be mixed with some other light, such as that represented by B, to obtain any color along the line BD. It follows that by additive mixing properly selected amounts of the three lights represented by points G, R, and B, one can obtain any color within the triangle GRB. Colors outside this triangle cannot be produced by *additive* mixing of colors GRB.

Since all the real colors are mixtures of the spectral colors, they must lie in the area enclosed by the horseshoe-shaped spectrum locus curve and the straight line connecting the violet and red ends of the horseshoe.

If the triangle GRB is to enclose most of the real colors, the point G should lie close to the spectrum locus point for 520-nm (green) light, and the points R and B should be near the red and violet ends of the spectrum locus curve. In this sense, red, green, and blue are desirable primaries.

Equations (55) and (56) give us a means of calculating the trichromatic coefficients (and therefore the location on the chromaticity diagram) for light with any given spectral distribution; the answer is unique. The reverse process is not unique. Given a light represented by a point such as G that has a specific set of trichromatic coefficients, this light may be matched by a mixture of two monochromatic colors with wavelengths 500 and 530 nm, by a pair with wavelengths 510 and 550 nm, or by several other pairs of monochromatic colors or a variety of continuously variable spectral distributions. These various matches are easily distinguished with the aid of a spectrometer, but to the unaided eye all look the same. There is no unique spectral distribution associated with a given point on the chromaticity diagram.

Color television is an example of additive color mixing. The screen consists of a mosaic (dots) of three different phosphors that can be excited independently;

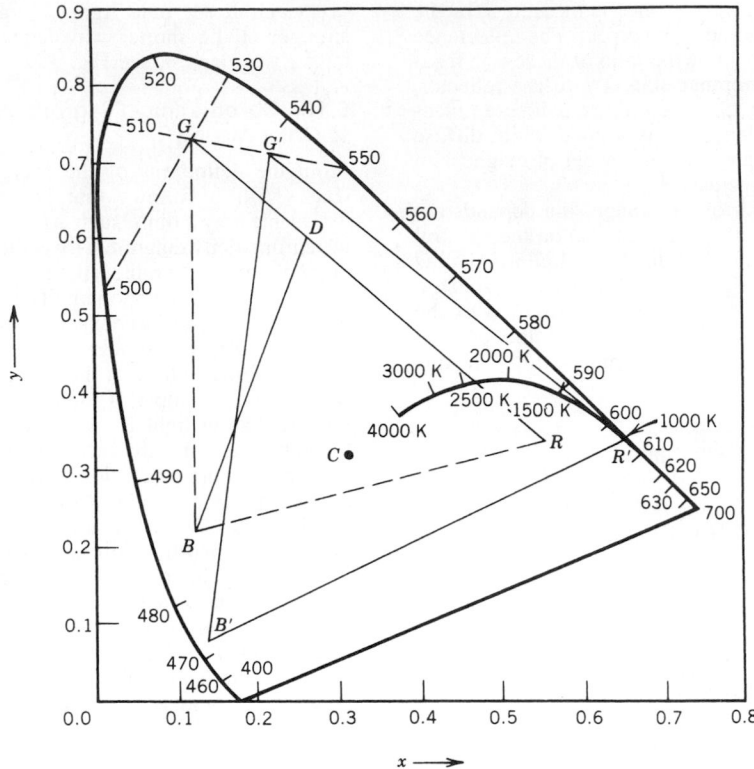

Fig. 38 Chromaticity diagram as aid to color mixing. Any color along line *GR* may be produced by adding colors *G* and *R*. Any color within triangle *GBR* may be produced by adding properly selected amounts of light of colors *G*, *B*, and *R*. Color television uses primaries *G′*, *B′*, and *R′*. Curve entering horseshoe near 610-nm locus gives color of blackbody radiation for several temperatures.

the three phosphors emit three different colors: red, green, and blue. At the customary viewing distance, the spacing of the dots is too small to be resolved by the eye and so the light from several dots is added together to give the sensation of color. The phosphors used give the colors represented by R', G', and B' in Fig. 38. The coordinates of these points are $R' = (0.670, 0.330)$, $G' = (0.210, 0.710)$, and $B' = (0.140, 0.080)$. By exciting these three phosphors in the proper ratio, any color within the triangle $R'G'B'$ can be produced.

In Fig. 38, we have added a curve representing the color of a blackbody at temperatures from 1000 to 4000 K. It is "cherry red" at 1000 K and progresses through orange toward white as the temperature rises. At very high temperatures (e.g., 15,000 K), the blackbody color is on the blue side of illuminant C.

6.12 Dominant Wavelength and Purity

Dominant wavelength and purity are physical properties of light that evoke the physiological sensations called hue and saturation.

As an illustration, consider again the orange skin. The chromaticity of the light scattered from it is represented by the point S in Fig. 36. Continuing to use C as the white point, we see that a line C to S may be extended to intersect the spectrum locus at $\lambda = 587$ nm, which is labeled D and has coordinates $(0.56, 0.44)$. It follows that the light from the orange skin may be color matched by a mixture of white light (illuminant C) and monochromatic light of wavelength 587 nm. The light from the orange skin is said to have a *dominant wavelength* of 587 nm. The distance from C to S divided by the distance from C through S to D is 0.83; light from the orange skin is said to have spectral purity p of 83%.

Similar procedures show that light represented by the point G has dominant wavelength of 519 nm and spectral purity of 80%. Specifying dominant wavelength and spectral purity is an alternate method of locating a point on the chromaticity diagram. For most people, these quantities are easier to interpret than the trichromatic coefficients x and y.

One runs into trouble for colors in the lower part of the diagram, for example, the color represented by the point P. A line from C to P, if extended, does not intersect the spectrum locus but intersects at P', the straight line closing the bottom of the horseshoe. If the line is extended backward, it intersects the spectrum locus at P'', or 550 nm. The color represented by P is said to have dominant wavelength of -550 nm, or complementary, 550 nm. The spectral purity is the ratio of the distances $\overline{CP}/\overline{CP'}$, which is about 65% for this case.

Colors commonly called "pastel colors" are of low spectral purity. The color pink is a red of low spectral purity; but in every-day language, low purity is often indicated by some adjectives (e.g., "baby" blue and "apple" green).

Two colors are said to be complementary if they may be added to make white. In terms of the chromaticity diagram, two colors are complementary if the line joining the two points representing them passes through C. The negative, or complementary, greens are called magenta or purple; frequently they are incorrectly called red.

6.13 Average Reflectance

We have seen that the dominant wavelength and purity of light reflected by an object depends upon the spectral distribution of the illuminant. The average reflectance also depends upon the illuminant; the average reflectance will be large if the spectral distribution of the illuminant is large at the wavelengths for which the reflectance of the object is also large. The average reflectance depends in this same way upon the spectral sensitivity of the detector. For the light-adapted (photopic) eye, the spectral sensitivity is represented by curve A of Fig. 32, which is the same as curve $\overline{y}$ of Fig. 35. Since the eye has its maximum sensitivity in the wavelength interval near 550 nm, the averaging process must be weighted in favor of these wavelengths. The average reflectance r_a is calculated as

$$r_a = \frac{\int_0^\infty r(\lambda)\overline{y}(\lambda)C(\lambda)\,d\lambda}{\int_0^\infty \overline{y}(\lambda)C(\lambda)\,d\lambda} \qquad (58)$$

where $r(\lambda)$ is the spectral reflectance of the object; $r(\lambda)$ for an orange skin was given in Fig. 37. The spectral distribution of the illuminant is $C(\lambda)$, and $\overline{y}(\lambda)$ is the spectral sensitivity of the photopic eye. If some other detector were used, its spectral sensitivity would replace $\overline{y}(\lambda)$ in the equation.

For the orange skin, illuminant C, and the photopic eye we obtain an average reflectance

$$r_a = 0.26 \quad \text{or } 26\%$$

6.14 Subtractive Color Mixing

If a white paper is used as a background for water colors, light must pass through the water color to get to the paper and after diffuse reflection from the paper again pass through the water color. Selective absorption by the dye in the water color gives the scattered light its color. Consider a dye that is absorbent and transmits only a little (say, 10%) for wavelengths shorter than 500 nm but is only slightly absorbent, transmitting 80 or 90%, for wavelengths longer than 500 nm. If this dye is painted on white paper, it will produce a yellow color. Another dye may transmit well for wavelengths shorter than 550 nm and absorb most of the light of longer wavelength; this dye will produce a blue color. If these two dyes do not react chemically and are used one on top of the other (or mixed together) so that light must pass through both of them, then most of the light between 500 and 550 nm will emerge but only a little of the light outside this wavelength interval will emerge. The resultant is a green color. This is a subtractive process in which yellow and blue give green; it must not be confused with the additive processes discussed earlier.

The principles in the water color experiment just described may be illustrated using a slide projector and two pieces of cellophane, one yellow and the other blue. Light projected onto a white wall through one piece of cellophane appears either yellow or blue, but when it passes through both pieces in series, it appears green.

In Fig. 39 are the transmission curves of a yellow filter (A), a blue filter (B), and the two in series (G). At each wavelength, the transmission represented by curve G is the product of the transmission for A and the transmission for B.

Color pictures and color slides (i.e., transparencies) use the subtractive method of producing colors. Three dyes are sufficient, and those that are most effective (i.e., produce the widest range of colors) are dyes that control the red, green, and blue. The dye that subtracts the red is blue in color and often is described as cyan. The dye that subtracts the green, leaving the red and blue, is magenta. The third dye subtracts the blue, leaving the green and longer wavelengths unaffected; it is yellow. By varying the concentration of each of these dyes, one can produce all real colors except the highly saturated ones. The available colors are sufficient for all ordinary use since spectrally pure colors are rare outside of the laboratory.

In color printing, it is often necessary to include a black and white image in addition to the three subtractive colors described. The use of black controls the average reflectance of a given area of the picture, not its color.

Good color reproduction in prints or in slides for projection necessitates prior knowledge of the illuminant used in viewing them. Slides are usually projected using a tungsten filament lamp so the dyes are adjusted on the assumption of illuminant A. In many cases, the color purity is increased deliberately and the blues emphasized because it is pleasing to have "bright colors" and "nice blue skys." Color pictures are more likely to be viewed in daylight and therefore are processed for use with illuminant C. If the pictures were produced photographically, the illuminant used for the initial exposure affects the final color.

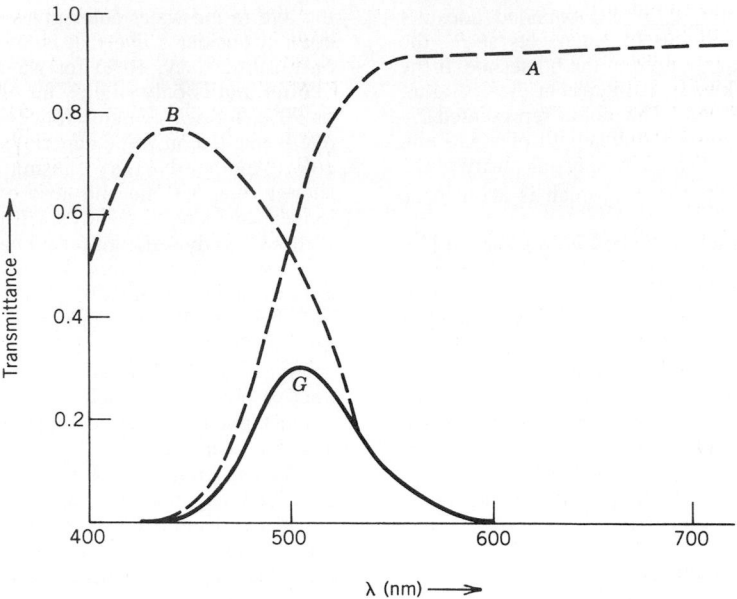

Fig. 39 Subtractive combination of yellow filter, *A*, and blue filter, *B*, to produce green, *G*. At each wavelength, *G* is product of *A* and *B* (*A* and *B* are Wratten filters 9 and 47A).

6.15 Munsell System

In matching or specifying paints, the average reflectance is as important as the dominant wavelength and spectral purity. In the Munsell system, paint samples are assigned a "value" from 0 to 10. Zero is black (i.e., nonreflecting) and 10 is white (100% diffuse reflecting); the intermediate shades of gray are equally spaced *subjectively*. At each value level is a plane polar arrangement in which the distance from the center indicates saturation or the sensation called *chroma* in subjectively equal steps, from zero or neutral at the center to 12 for a saturated color. Dominant waelength, or its subjective equivalent *hue*, is represented by the horizontal direction on the polar plot; five principal hues—red, yellow, green, blue, and purple—are recognized with the intermediate hues—yellow-red, green-yellow, and so on—making a total of 10 equally spaced hue segments. Each of these is divided into 10 numbered subdivisions (see Fig. 40). The Munsell quantities value, hue, and chroma are subjective, corresponding roughly to the physical quantities average reflectance, dominant wavelength, and purity. About a thousand distinguishable paints have been prepared and classified in this way. These samples preserved as an atlas are used to specify paints.

The average reflectance and ICI color specification (under illuminant C) for these samples have been measured, but equating the subjective and the physical quantities is difficult. The Munsell system predates the ICI system.

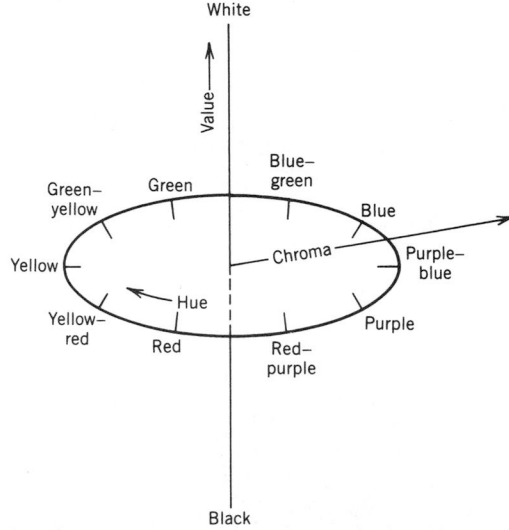

Fig. 40 Munsell representation of color data.

6.16 Photometric Units

In our earlier discussions of sources, such as the blackbody, we measured the radiated energy in physical units (i.e., in watts). The units used are called radiometric units. Long before it became possible to make such measurements, light sources and levels of illumination were compared and measured using the eye as the detector. The eye is quite good at judging the equality of illumination on two adjacent areas. A whole set of units evolved around this process and are still in use; these are known as photometric units, often distinguished from the corresponding radiometric units by including the word *luminous* in the name and the subscript v on the symbol.

The unit of luminous flux is the *lumen*. At the wavelength 555 nm, 1 W produces 683 lm; but since the sensitivity of the photopic eye follows curve A of Fig. 32 (or curve $\overline{y}$ of Fig. 35), the number of lumens per watt at other wavelengths is smaller, as indicated by this curve. For a nonmonochromatic source, the luminance (or luminous radiance) is

$$L_v = 683 \int_0^\infty L(\lambda)\overline{y}(\lambda)\,d\lambda \qquad (59)$$

and has units of nits. A nit is one lumen per square meter per steradian.

A small source that radiates 1 lm into each steradian is said to have a luminous intensity I_v of 1 candela. (It was formerly called a "standard candle.") Common units of illuminance are the foot-candle (1 lm/ft^2), the lux (1 lm/m^2), and the phot (1 lm/cm^2).

7 DETECTORS OR OPTICAL TRANSDUCERS

Aside from the eye, there are numerous devices that are used to detect and measure radiant flux. These are usually separated into two groups: (a) thermal detectors and (b) quantum detectors. In thermal detectors, the radiation is absorbed and converted into heat, which raises the temperature of the detector. The temperature change causes a measurable change in some other physical property of the detector (e.g., its resistance). Thermal detectors are sensitive throughout the spectrum. For quantum detectors, the incident light (photons) affect the detector directly (i.e., without heating it); the best known of these is the photoelectric detector in which light causes electrons to be emitted from a surface.

Two thermal detectors are in common use. (a) In the *thermocouple*, two different materials (usually metals) are connected to form a closed circuit. One junction is exposed to the radiation and thereby heated slightly while the other junction is shielded from the radiation and remains at ambient temperature. The temperature difference between the two junctions produces an electromotive force (emf) in the circuit that may be measured. (This is known as the Peltier effect.) (b) The **bolometer** depends upon the change of electrical resistance with temperature. Two identical small detectors are arranged in a Wheatstone bridge circuit; again one detector is exposed to the radiation and the other shielded from the radiation. Small changes in the resistance of the exposed detector are taken as a measure of the incident flux.

If the bolometer element is small and thermally well insulated (except for the thin connecting wires) from its surroundings, then a very small optical or radiant power will produce a relatively large temperature rise; in this sense, the detector is very sensitive. However, it will cool slowly and be unable to respond to rapid fluctuations of the incident flux. For rapid response, the thermal isolation of the bolometer should be reduced. For any given application, one must find the optimum compromise between good sensitivity and rapid response.

Bolometer elements may be small flakes or ribbons of metal; nickel or platinum are commonly used. For metals, the resistance **increases** with increasing temperature, as represented by the equation

$$R = R_0[1 + \alpha(T - T_0)] \qquad (60)$$

where R is the resistance at temperature T and R_0 is the resistance at ambient temperature T_0. The constant α depends upon the metal used, but values of 0.003–0.004 per degree Kelvin are typical.

Semiconductor bolometer elements (known as thermistors) are also available. For semiconductors, the resistance *decreases* with increasing temperature according to the equation

$$\frac{R(T)}{R(T_0)} = \frac{e^{\beta/T}}{e^{\beta/T_0}} \qquad (61)$$

A typical value of β is 3600 K, which gives dR/dT equivalent to $\alpha = -0.04$ per degree Kelvin in Eq. (60). In this sense, the thermistor is about 10 times as sensitive as a metal bolometer. Since the resistance decreases with increasing temperature, it must be used with a suitably large series resistance to prevent self-burnout.

In detecting or measuring weak signals, the signal-to-noise ratio becomes important. It is important to realize that a small object (such as a bolometer element) that is in thermal equilibrium with its surroundings is not at a constant temperature but is constantly exchanging energy with its surroundings and fluctuating in temperature. It will experience a root-mean-square random fluctuation of temperature ΔT given by

$$\overline{\Delta T^2} = \frac{kT^2}{C} \qquad (62)$$

where k is Boltzmann's constant, T is the absolute temperature of the surroundings, and C is the thermal capacity of the small object. Even if all the amplifying and/or measuring circuits could be noise free, the random temperature fluctuations given by this equation represent unavoidable noise. If the incoming radiation in this ideal case produces a temperature rise equal to ΔT, it is said to have *noise equivalent power* (NEP) and a signal-to-noise ratio of 1.0. In this respect, the thermistor has no advantage over the metal bolometer; it gives larger response to both signal and noise but does not improve the signal-to-noise ratio.

The simplest quantum detector is an evacuated glass tube containing two electrodes. The anode (or positive electrode) collects electrons emitted from the cathode (or negative electrode). Light striking the cathode, called the photocathode, causes the emission of electrons from the cathode; these are collected by the anode and are measured as current in an external circuit. Not every photon causes the emission of an electron, and the term *quantum efficiency* is used to represent the ratio of the number of electrons emitted to the number of photons incident on the cathode. Quantum efficiencies of 10–15% are typical of good photocathode surfaces.

Even in a light beam from a well-stabilized source, the photons do not arrive on the cathode at equally spaced times but at random time intervals. Also, the photons that produce electrons are randomly selected from those that arrive. The subject is usually treated by Poisson statistics, giving the result that if many measurements are made of n (the number of observed electrons in some constant time interval) and the average value of n is $\bar{n}$, then the departures of the individual measurements from the average, $\bar{n}$, will be given by

$$\overline{(n - \bar{n})^2} = \bar{n} \qquad (63)$$

This statistical fluctuation of n about its average value $\bar{n}$ is known as photon noise; it arises from the same statistical considerations as "shot noise" in electric circuits. The signal-to-noise ratio is $\sqrt{\bar{n}}$, which may be increased by making $\bar{n}$ larger by (a) increasing the rate at which photons arrive or (b) increasing the observation time for each measurement. The time allotted to each measurement is often built into the associated amplifier, that is, the reciprocal of its bandwidth.

Quantum detectors are wavelength selective. The long-wavelength (low-frequency) limit is determined by the equation

$$\frac{hc}{\lambda} = h\nu \geq \phi e$$

where e is the electronic charge, ϕ is the "work function" of the photo cathode. Here, ϕe is the minimum energy required to remove an electron from the cathode into the vacuum, where ϕ is this energy expressed in electron volts and e is the electronic charge. The short-wavelength (high-frequency) limit is usually determined by the absorption of light by the glass walls of the vacuum tube. The sensitivity of photocathodes is wavelength dependent, and a variety of photocathodes are available having peak sensitivities at different regions of the spectrum.

At room temperature, there will be some emission from the cathode even in the dark. This is known as "dark current" and is due mainly to thermionic emission; it may be reduced by refrigerating the detector.

Some photodetectors realize an amplification of about 10 in the current by having a few torrs of gas in the tube. The electrons are accelerated toward the anode and gain enough energy to ionize some of the gas; the ions then contribute to the current. The recommended cathode-to-anode potential difference must be maintained. Too little will not provide the specified gain; too much will result in a glow discharge independent of light input (and damaging to the cathode). Recommended potential differences are usually 50–100 V.

The photomultiplier is a vacuum tube in which the photocathode is followed by several other electrodes called dynodes. Electrons emitted by the cathode are accelerated to the first dynode, which has a positive potential relative to the cathode on the order of 100 V. Each electron striking the dynode gives up its kinetic energy, thereby causing emission from the dynode of several (e.g., four) slow-moving electrons; this process is called secondary emission. These secondary electrons are accelerated to the next dynode where the process is repeated. It is repeated at each dynode until the electrons are finally collected on the anode and measured in some external circuit. If there is a gain of 4 electrons at each dynode and there are 10 dynodes, there will be $\sim 10^6$ electrons at the anode for each electron that left the cathode; under these conditions, one can observe individual photoemissive events and count their number. The photon noise is determined by the number of electrons leaving the cathode; the large gain makes the individual events easier to count; it does not improve the signal-to-noise ratio. The anode is usually at or near ground potential; to get 100 V for each of 10 dynodes requires the cathode to be −1000 V. The gain is sensitive to this voltage. The photomultiplier is useful primarily at low levels of illumination.

Advances in semiconductor science and technology have provided a number of solid quantum detectors that are more rugged and easier to use than the vacuum tube detectors of earlier years.

The electrical behavior of solids is usually described in terms of allowed energy bands for the electrons. For intrinsic semiconductors, the valence band contains all of the valence electrons of the solid and is filled by these electrons. There is no room for any net motion of these electrons. Above the valence band there is an energy region in which no electrons can exist. The width of this forbidden region is called the bandgap ϕ (usually expressed in electron volts). Above the bandgap is an energy band in which electrons are permitted and in which they are free to move; this is called the conduction band. Normally there are no electrons in the conduction band except for a negligible few that may be thermally excited there from the valence band. Light of wavelength shorter than that given by Eq. (63) can be absorbed by the semiconductor. A photon so absorbed can excite an electron from the valence band to the conduction band. An empty space, or hole, is left in the valence band; this hole acts as a small positive charge and can move through the solid in the valence band.

If the semiconductor just described is connected to a current meter and a source of small emf, the observed current is due to the motion of electrons and holes; it will depend upon the irradiance. This process is called photoconductivity, and a semiconductor used in this way is called a photoconductor. There are a number of photoconductors available, each having its characteristic bandgap. A semiconductor made by mixing mercury telluride and cadmium telluride will have a bandgap dependent upon the relative concentration of the components. In practice, one seeks a detector with bandgap a little less than the quantum energy $h\nu$ of the radiation to be detected; in this way the detector becomes blind to undesired radiation at lower frequencies.

The electron has a mean lifetime before it recombines with a hole, after which it no longer contributes to the current. The lifetime is a random variable. This randomness contributes to the noise and is known as generation recombination noise; it is larger than the photon noise of photoemissive detectors.

An extrinsic semiconductor is an intrinsic semiconductor (such as silicon) into which a small concentration of "impurity" has been introduced. Silicon, as carbon, has four valence electrons, and these are just sufficient to fill the valence band. The bandgap for silicon is 1.14 eV. If a small concentration of an element with five valence electrons (e.g., phosphorus or arsenic) is included, each impurity atom will contribute four electrons to the valence band and the fifth electron will be loosely bound to its parent atom. A little energy, <0.1 eV, will be sufficient to ionize this electron into the conduction band where it can move about and contribute to a measurable current. The ionization energy can be supplied by a photon. Impurities of this type are called donor impurities, and the semiconductor is known as n type since negative charges (electrons) produce the current.

If the impurity is trivalent, such as boron or aluminum, then there is an electron missing (a hole). This hole is loosely bound to its parent atom. A little energy can move an electron from the valence band into this hole, leaving in the valence band a hole that is free to move. Impurities of this type are known as acceptor impurities, and the semiconductor is p type since the hole behaves as a positive charge. Photoconductive detectors can be made of either p-type, n-type, or intrinsic semiconductors.

A short length of semiconductor that is p type at one end but n type at the other is called a ***p−n* junction diode**. This device is commonly used as a rectifier because it conducts electricity much better if a small potential difference (a few volts) is applied positive to the p type and negative to the n type than if the polarity is reversed. The first case is called forward biased (good conductivity); the second case is reverse biased (poor conductivity).

This device may be used as an optical detector. If the device is reverse biased, it will show a small dark current and, in addition, a current that is nearly linear with the irradiance at the junction between the p region and the n region. It is characterized by shot or photon noise, not generation recombination noise. This is known as the photoconductive mode. These $p−n$ junction photodiodes usually show faster response to changes of irradiance than photoconductors.

Junction photodiodes are also used in the photovoltaic mode. In this case, there is no external bias provided. A voltmeter (or sometimes an ammeter) is connected to the two ends of the diode. A potential difference (or current) is observed that is dependent on the irradiance at the junction. Usually the output is not a linear function of the irradiance.

The $p−i−n$ photodiode is a $p−n$ junction with an intrinsic region separating the p region from the n region. The radiation is absorbed in the intrinsic region. This device has a low internal capacitance, allowing it to respond quickly to changes of irradiance. It can respond in times shorter than 10^{-9} s. It also functions over a very large dynamic range of irradiance.

This is certainly not an exhaustive list of the available photodetectors, but perhaps it gives a flavor of the subject. For details, one should consult the manufacturers' specification sheets.

The most commonly used figure of merit for detectors is the specific detectivity, or D^*. By definition,

$$D^* = \frac{\sqrt{A \, \Delta f}}{P_N}$$

where P_N is the incident radiant power that produces a signal just equal to the noise signal, A is the area of the detector, and Δf is the bandwidth of the detector and its amplifying circuits. (These are low-frequency circuits used to monitor fluctuations in radiant power.) Theoretical considerations show that if A and Δf are changed, P_N will change in such a way as to keep D^* constant. Therefore, D^* is characteristic of the process and material being used for detection and is not dependent upon detector geometry or bandwidth.

REFERENCES

1. Judd, D. B., "The 1931 I.C.I. Standard Observer and Coordinate System for Colorimetry," *J. Opt. Soc. Am.* **23**, 359 (1933).
2. Billmeyer, F. W., and Saltzman, M., *Principles of Color Technology*, 2nd ed., New York, Wiley, 1981.

BIBLIOGRAPHY

General Optics Textbooks

Born, M., and Wolf, E., *Principles of Optics*, 6th ed., Pergamon, Oxford, England, 1980.
Hecht, E., and Zajac, A., *Optics*, 2nd ed., Addison-Wesley, Reading, MA, 1979.
Jenkins, F. A., and White, H. E., *Fundamentals of Optics*, 4th ed., McGraw-Hill, New York, 1976.
Klein, M. V., *Optics*, Wiley, New York, 1970.

Lens Design and Testing

Malacara, D., *Optical Shop Testing*, Wiley, New York, 1978.

Smith, W. J., *Modern Optical Engineering: The Design of Optical Systems*, McGraw-Hill, New York, 1966.

Interference, Diffraction, and Holography

Collier, R. J., Burckhardt, C. B., and Lin, L. H., *Optical Holography*, Academic, New York, 1971.

Françon, M., *Optical Interferometry*, Academic, New York, 1966.

Givens, M. P., "Focal Shifts in Diffracted Converging Spherical Waves," *Opt. Commun.*, **41**, 145 (1982).

Givens, M. P., "Introduction to Holography," *Am. J. Phys.*, **35**, 1056 (1967).

Givens, M. P., "Image Location and Magnification in Holography," *Am. J. Phys.*, **40**, 1311 (1972).

Goodman, J. W., *Introduction to Fourier Optics*, McGraw-Hill, New York, 1968.

Li, Y., and Wolf, E., "Focal Shifts in Diffracted Converging Spherical Waves," *Opt. Commun.*, **39**, 211 (1981).

Smith, H. M., *Principles of Holography*, Wiley, New York, 1969.

Solymar, L., and Cooke, D. J., *Volume Holography and Volume Gratings*, Academic, New York, 1981.

Steel, W. H., *Interferometry*, Cambridge University Press, London, England, 1967.

Lasers

Lengyel, B. A., *Lasers*, 2nd ed., Wiley, New York, 1971.

Siegman, A. E., *An Introduction to Lasers and Masers*, McGraw-Hill, New York, 1968.

Svelto, O., *Principles of Lasers*, 2nd ed., Plenum, New York, 1982.

Yariv, A., *Introduction to Optical Electronics*, 3rd ed., Holt, Rinehart, and Winston, New York, 1985.

Color and Vision

Evans, R. M., *An Introduction to Color*, Wiley, New York, 1948.

Wright, W. D., *The Measurement of Color*, Van Nostrand, New York, 1969.

Light Sources and Detectors

Boyd, R. W., *Radiometry and the Detection of Optical Radiation*, Wiley, New York, 1983.

Kingston, R. H., *Detection of Optical and Infrared Radiation*, Springer, New York, 1978.

Kruse, P. W., McGlauchlin, L. D., and McQuistan, R. B., *Elements of Infrared Technology*, Wiley, New York, 1962.

Smith, R. A., Jones, F. E., and Chasmar, R. P., *The Detection and Measurement of Infrared Radiation*, Clarendon Press, Oxford, England, 1957.

CHAPTER **19**

ACOUSTICS

Jonathan Blotter, Scott Sommerfeldt, and Kent L. Gee
Department of Mechanical Engineering
Brigham Young University
Provo, Utah

1   INTRODUCTION

Acoustics is the term used to describe the science of sound. It is derived from the Greek word *akoustos*, which means "hearing." Acoustics is a very broad field of science that covers the generation, reduction, control, transmission, and reception aspects of sound. The earliest known studies of sound were made by Pythagoras during the sixth century BC. Pythagoras was the first to identify the musical consonances described as the octave, the fifth, and the fourth. He also initiated the theory of the inverse proportionality between pitch and the length of a vibrating string. In the Renaissance era of the late fifteenth century, the famous artist Leonardo do Vinci (1452–1519) also made many significant contributions to the theory of mechanics and wave motion. The works of Galileo Galilei (1564–1642) have had a larger impact on theoretical mechanics than any other before Newton. Galileo is recognized as having built on Leonardo's speculative work and developed it into quantitative logic. Some likely consider Galileo as formalizing the science of sound. However, most generally recognized as the first significant contributor to the quantitative study of acoustics is Marin Mersenne (1588–1648). Mersenne performed and developed significant experimental techniques that laid foundations in many areas of acoustics. From the time of Mersenne to the twentieth century, many significant contributions to the area of acoustics were made by those such as Newton, Euler, Green, Stokes, Helmholtz, Kirchhoff, Rayleigh, and Sabine. However, recent contributions during the twentieth century have also had a tremendous impact in the area of acoustics today. One of these developments was active noise control by Paul Leug in Germany in 1932. Other developments include computational sound field modeling, measurement methods, and measurement hardware, including computers, digital signal processors, and various transducers.[1–3]

2   SOUND POWER, SOUND INTENSITY, AND SOUND PRESSURE

In this section we discuss three fundamental measures of sound. These are sound power, sound intensity, and sound pressure. These three terms are discussed in most books on acoustics.[1,3]

2.1   Sound Power

Sound power is the term that describes the rate at which sound energy is radiated per unit time. The units of sound power are joules per second or watts. As an example of sound power, assume that there is a small spherical sound source. The sound power is a measure of the total amount of sound that the source can produce. Assuming that the source has constant sound power output and that unobstructed radiation occurs, the sound power output of the source is the same at any radius away from the source. A balloon analogy is often used to describe sound power. A balloon has the same amount of material surrounding the internal gas whether the balloon is partially or fully inflated. This is the same concept in that the sound power radiating from the source is a constant independent of distance from the source.

The range of sound powers encountered can be extremely large. For example, the sound power of a large rocket engine is on the order of 1,000,000 W while the sound power of a soft whisper is approximately 0.000,000,001 W. Because of this large range in power values, a logarithmic scale known as the decibel scale is widely used in acoustic applications.

When sound power is expressed in terms of decibels, it is called a sound power level and is computed as

$$L_w = 10 \log \left(\frac{W}{W_{\text{ref}}} \right) \qquad (1)$$

In Eq. (1), L_w represents the sound power level, W is the power of the sound energy source, and W_{ref} is the standard reference power, usually taken to be 1×10^{-12} W, or 1 pW. Table 1 provides an example of the range of sound power and sound power levels.

Sound power may be measured in various ways depending on the environment in which the measurements are to be made (free field, diffuse field, hemianechoic, etc). The two basic techniques of measuring sound power consist of using an array of pressure microphones placed around the source or by using a sound intensity probe. Other comparative techniques also exist where measurements are compared to calibrated sources. The international standards that outline the processes for measuring sound power are International Organization for Standardization (ISO) 3741, 3742, 3744, 3745.[4]

2.2   Sound Intensity

Sound intensity is defined as the amount of sound power for a given area of the wave front. Relating this back to the balloon analogy, sound intensity would be a measure of the balloon wall thickness. For a given balloon there will be more balloon material due to the thickness of the balloon wall if the balloon is partially inflated as compared to fully inflated where a thinner balloon wall will occur. This is the same as sound

Table 1   Sound Power and Sound Power Level

Sound Source	Sound Power (W)	Sound Power Level, L_w (dB re 1×10^{-12} W)
Saturn rocket	25–40 million	195
Rocket engine	1,000,000	180
Turbojet engine	10,000	160
Siren	1,000	150
Heavy-truck engine/rock concert	100	140
75-Piece orchestra pipe organ	10	130
Piano/small aircraft, jackhammer	1	120
Excavator/trumpet	0.3	115
Chain saw/blaring radio	0.1	110
Helicopter	0.01	100
Auto on highway, loud speech/shouting	0.001	90
Normal speech	1×10^{-5}	70
Refrigerator	1×10^{-7}	50
Auditory threshold	1×10^{-12}	0

intensity, in that the closer to the source, the higher the intensity for a given area. As previously mentioned, sound power and sound intensity are related by an area. For a spherical wave front, the relationship is expressed as

$$I = \frac{W}{4\pi r^2} \qquad (2)$$

where I represents the sound intensity, W represents the sound power, and $4\pi r^2$ is the surface area of the sphere where r is the radius of the sphere. From Eq. (2), it should be noted that the decay of the sound intensity is directly proportional to the square of the radius of the sphere. This relationship is often referred to as the inverse-square law.

Sound intensity values are directly related to sound power values by an area term and hence also have a very wide range of values. It is therefore again common to express the sound intensity using a decibel scale and then refer to it as a sound intensity level. Sound intensity level is expressed as shown by Eq. (3), where L_i is the sound intensity level, I is the measured intensity at some distance from the source, and I_{ref} is the reference sound intensity usually taken as 1×10^{-12} W/m^2:

$$L_I = 10 \log \left(\frac{I}{I_{\text{ref}}} \right) \qquad (3)$$

Sound intensity is the time-averaged product of the pressure and particle velocity. The pressure can simply be measured with a pressure microphone. However, the particle velocity is somewhat more difficult to measure. One common method of measurement is to place two similar microphones a known distance apart. By doing this, the pressure gradient can be measured. Then using Euler's equation the pressure gradient can be used to compute the particle velocity.

2.3 Sound Pressure

Sound pressure refers to the small deviations of pressure from the ambient value that are propagated by sound waves. To better understand this wave phenomenon, take, for example, a U-shaped tuning fork as shown in Fig. 1a. The tuning fork is a device used

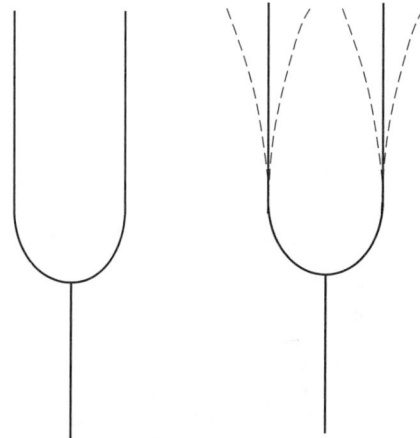

Fig. 1   Tuning fork: (a) static position; (b) vibrating motion showing tine motion.

to create a tone at a specific frequency that can then be used to tune various musical instruments. The tone is created by striking one of the tines with a hammer causing the forks to freely vibrate, as shown in Fig. 1b. As the tine moves outward from equilibrium, the air adjacent to the outward moving tine is compressed and forced outward. As the air adjacent to the tine is pushed outward, the pressure at that point increases as does the density of the air molecules. As the tine returns back to equilibrium and displacement in the opposite direction, the air molecules spread back out, leaving a lower pressure and air molecule density as they return to their original position. The expansion and subsequent return to equilibrium are commonly called rarefaction.

As with sound power and sound intensity, sound pressures encountered also have a very wide range of values. For example, a typical jet engine is on the order of 630 Pa while the threshold of hearing is approximately 2×10^{-5} Pa at 2 kHz. Therefore, as with sound power and sound intensity, sound pressure level L_p has been developed as

$$L_p = 20 \log \left(\frac{p}{p_{ref}} \right) \qquad (4)$$

where p is the measured rms sound pressure and p_{ref} is the reference sound pressure usually taken as 20 μPa (2×10^{-5} Pa), which is the threshold of hearing for an undamaged ear and also corresponds closely with the reference power of 1×10^{-12} W. Table 2 shows the relationship between sound pressure and sound pressure level.

3   DECIBEL AND OTHER SCALES

The decibel scale is a unitless logarithmic scale that is widely used in many fields of science, including acoustics, electronics, signals, and communication. The decibel scale is used to describe a ratio of values which among others may be sound power, sound intensity, sound pressure, and voltage. One main advantage of a logarithmic scale is that a very wide range of values can be reduced to a much smaller range which provides the user with a better feel for the particular values of interest.

In the field of acoustics, the decibel (dB), which is one-tenth of a bel, is almost exclusively used. Decibels are used over bels to simply increase the sensitivity of the values, similar to using inches instead of feet or centimeters instead of meters. The term bel (in honor of Alexander Graham Bell) refers to the basic logarithm of the ratio of two values. The definition of the base 10 logarithm is shown as

$$10^{\log p} = p \qquad (5)$$

The log of p is the value to which 10 must be raised in order to obtain the value p. The decibel equation can then be expressed as

$$N(\text{dB}) = 10 \log \frac{A}{A_{ref}} \qquad (6)$$

Two examples using the decibel scale are now provided to better illustrate the significance of the relationship.

Example 1.   An increase in sound power of a ratio of 2 : 1 results in what gain in decibels?

Solution

$$N(\text{dB}) = 10 \log \ 2/1 = 10 \times (0.3010) = 3.01 \, \text{dB}$$

Therefore a doubling of the sound power results in an increase of 3 dB.

Example 2.   The noise level at a factory property line caused by 12 identical air compressors running simultaneously is 60 dB. If the maximum sound pressure level permitted at this location is 57 dB, how many compressors can run simultaneously?

Table 2   Sound Pressure and Sound Pressure Level

Sound Source	Sound Pressure (Pa)	Sound Pressure Level, L_p (dB re 20 μPa)
Krakatoa (at 160 km)	20,000	180
M1 Garand being fired	12,000	176
Jet engine (at 30 m)	630	150
Threshold of pain	100	130
Hearing damage (short-term exposure)	20	120
Jet (at 100 m)	6–200	110–140
Jack hammer (at 1 m)	2	100
Major road (at 10 m)	0.2–0.6	80–90
Passenger car (at 10 m)	0.02–0.2	60–80
Normal talking (at 1 m)	0.002–0.02	40–60
Very calm room	0.0002–0.0006	20–30

Solution For 1 compressor the L_p is $L_{p1} = 20 \log(P_{rms1}/P_{ref})$. For 12 compressors the L_p is $P^2_{rms12} = 12P^2_{rms1}$; therefore $P_{rms12} = \sqrt{12}P_{rms1}$:

$$L_{p12} = 20\log\left(\frac{\sqrt{12}P_{rms1}}{P_{ref}}\right) = 20\log\left(\frac{P_{rms1}}{P_{ref}}\right)$$
$$+ 20\log\sqrt{12}$$

Therefore, $L_{p12} - L_{p1} = 20\log\sqrt{12} = 10.79$ and

$$L_{p1} = L_{p12} - 10.79 = 60 - 10.79 = 49.21\,\text{dB}$$

Now for x compressors

$$L_{p,max} = 20\log\sqrt{x} + L_{p1}$$

Therefore,

$$\log\sqrt{x} = \frac{L_{p\,max} - L_{p1}}{20} = \frac{57 - 49.21}{20} = 0.3895$$
$$x = (10^{(0.3895)})^2 = 6.01$$

Therefore 6 compressors, at most, can be operated in order to meet the 57-dB limit.

Besides the conventional decibel scale, there are other scales that have been developed which filter or weight the sound. This chapter continues by presenting and discussing several of these filtered scales.

In 1933, Fletcher and Munson[5] presented what is typically called the equal-loudness curves (see Fig. 2). This set of curves represents what is considered to be tones of equal loudness over the frequency range from 20 Hz to 15 kHz. The unit of measure for loudness is the phon. A phon is a unit of apparent loudness. The value for the unit of phon comes from the intensity level in decibels of a 1-kHz tone with the same perceived loudness. Therefore, a decibel value and a phon value are the same for a 1-kHz tone. These curves were determined experimentally. The experiment consisted of playing pure tones at a specific intensity level along with a 1-kHz reference signal. The intensity of the reference signal was increased until the two tones were perceived to be of the same loudness by the listener. Other work on the development of equal-loudness curves has been generated since Fletcher and Munson. The most recent standards can be found in ISO 226:2003.[4]

4 WEIGHTING FILTERS

The first weighting filter to be discussed is called A weighting and usually carries the associated symbol dBA or dB(A). This weighting was developed to apply the same filter to a tone that the human ear applies. As a result, this weighting is based on the 40-phon pure-tone equal-loudness curve. The 40-phon pure tone is a relatively quiet tone. It should be noted that the A-weighted curve roughly corresponds to the inverse of the 40-phon equal-loudness curve. The A-weighted curve tends to amplify signals in the region of 1–6 kHz. Sound outside this frequency range is reduced

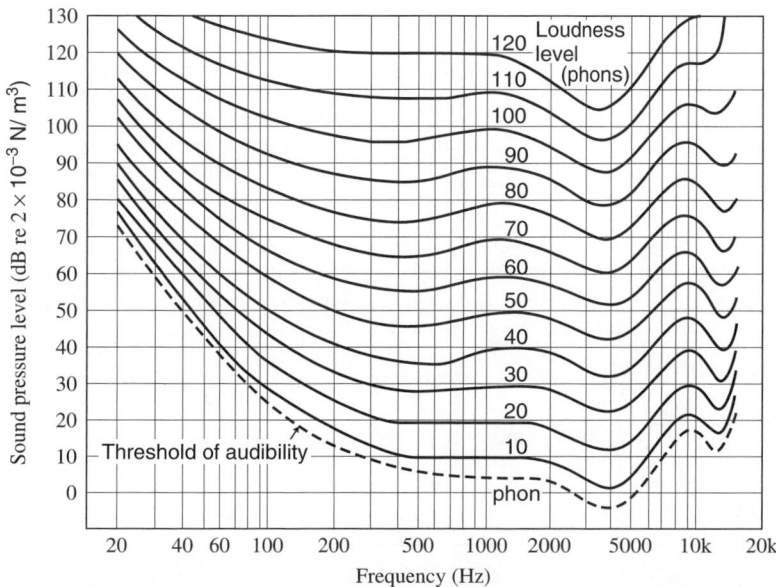

Fig. 2 Equal-loudness curves.

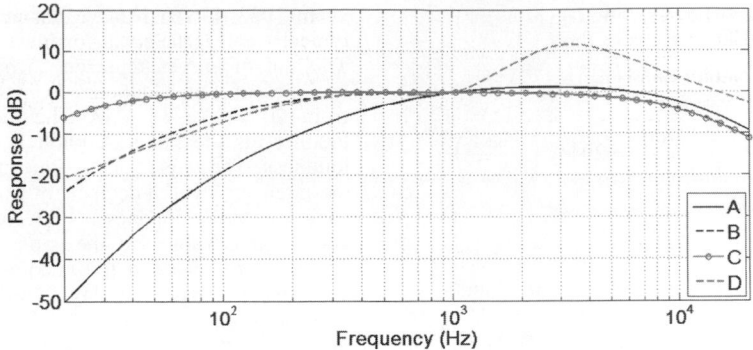

Fig. 3 A, B, C, and D weighting filters.

by the A weighting. The A-weighted filter is shown in Fig. 3.

The A-weighted scale is very popular and is used in many applications. These include environmental noise, noise control, construction noise, and community noise standards. The main violation of usage of the A-weighted scale is that it was developed for relatively quiet pure tones.

The dB(B) and dB(C) scales were developed along similar methods as the dB(A) scale except that they were developed for tones with higher sound pressure levels. The dB(B) scale is seldom used but was developed for tones between the low sound pressure level of the A-weighted scale and the high sound pressure level of the C-weighted scale. The dB(C) scale is used for the highest sound pressure levels and, as shown in Fig. 3, is nearly linear over a large frequency range. The dB(D) scale is often used for aircraft noise. The equations for the weighting scales are shown in Eqs. (7), where $s = j\omega$, $j = \sqrt{-1}$ and ω is the frequency:

$$G_A(s) = \frac{k_A s^1}{(s+129.4)^2(s+676.7)(s+4636)(s+76,655)^2}$$

(7)

$$G_B(s) = \frac{k_B s^3}{(s+129.4)^2(s+995.9)(s+76,655)^2}$$

$$G_C(s) = \frac{k_C s^2}{(s+129.4)^2(s+76,655)^2}$$

$$G_D(s) = \frac{k_D s(s^2+6532s+4.0975\times10^7)}{(s+1776.3)(s+7288.5)(s^2+21,514s+3.8836\times10^8)}$$

where $k_A \approx 7.39705 \times 109$, $k_B \approx 5.99185 \times 109$, $k_C \approx 5.91797 \times 109$, and $k_D \approx 91104.32$.

Another unit of measure for perceived loudness is the sone. The sone was developed experimentally to relate perceived loudness to phons. In the experiment, volunteers were asked to adjust the loudness of a tone until it was perceived to be twice as loud as the original tone. As a result, 1 sone is arbitrarily equal to 40 phons, and a doubling of the loudness (in sones) corresponds to a doubling of the perceived loudness. A graph relating sones and phons is shown in Fig. 4.

Octave Bands in Audio Range   In reference to the audio range, the term octave corresponds to a doubling or a halving of a frequency. For example, 200 Hz is one octave above 100 Hz and 4000 Hz is one octave above 2000 Hz. The 400 Hz would be considered to be two octaves above 100 Hz and so on.

In audio acoustics, the term "octave band" is commonly used to represent a range of frequencies where the highest frequency is twice the lowest frequency. These octave bands are commonly used to represent what is going on in the acoustic signal over a band or range of frequencies as compared to discrete frequencies. Most sound level meters will output the data in some form of a standardized octave band.

Octave bands are typically defined by a center frequency f_C. The upper frequency f_U and the lower frequency f_L of the band are computed as

$$f_L = \frac{f_C}{\sqrt{2}} \qquad f_U = \sqrt{2}f_C \qquad \text{where } f_C = \sqrt{f_L f_U}$$

(8)

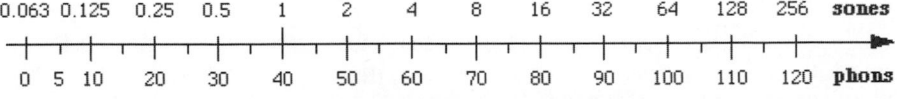

Fig. 4 Relationship of sones and phons.

Table 3   Octave-Band Lower, Center, and Upper Band Frequencies (Hz)

Lower Band Frequency, f_L	Center Band Frequency, f_C	Upper Band Frequency, f_U
22.4	31.5	45
45	63	90
90	125	180
180	250	355
355	500	710
710	1,000	1,400
1,400	2,000	2,800
2,800	4,000	5,600
5,600	8,000	11,200
11,200	16,000	22,400

Table 4   One-Third-Octave-Band Lower, Center, and Upper Band Frequencies (Hz)

Lower Band Frequency, f_L	Center Band Frequency, f_C	Upper Band Frequency, f_U
18.0	20	24.4
22.4	25	28.0
28.0	31.5[a]	35.5
35.5	40	45
45	50	56
56	63[a]	71
71	80	90
90	100	112
112	125[a]	140
140	160	180
180	200	224
224	250[a]	280
280	315	355
355	400	450
450	500[a]	560
560	630	710
710	800	900
900	1,000[a]	1,120
1,120	1,250	1,400
1,400	1,600	1,800
1,800	2,000[a]	2,240
2,240	2,500	2,800
2,800	3,150	3,550
3,550	4,000[a]	4,500
4,500	5,000	5,600
5,600	6,300	7,100
7,100	8,000[a]	9,000
9,000	10,000	11,200
11,200	12,500	14,000
14,000	16,000[a]	18,000
18,000	20,000	22,400

[a]Octave-band center frequencies.

Furthermore, the bandwidth for a given octave can be computed by as

$$\text{Bandwidth} = f_U - f_L = \frac{f_C}{\sqrt{2}} \qquad (9)$$

In the field of acoustics standard octave bands have been defined. The lower frequency, center frequency, and upper frequency, for the standard acoustic octave bands are given in Table 3

In many applications, a higher resolution on the frequency axis is desired. In these cases, the octave band is split into smaller groups. Another very common scale is the one-third octave band, in which each of the octave bands is split into three smaller bands. The one-third-octave-band scale is commonly used for environmental, noise control, and building acoustics, among others. The lower, center, and upper frequencies for the one-third octave bands are given in Table 4 and calculated using the equations

$$f_L = \frac{f_C}{\sqrt[6]{2}} \qquad f_U = \sqrt[6]{2}\, f_C \qquad \text{where } f_C = \sqrt{f_L f_U}$$

$$\text{Bandwidth} = f_C \left(\sqrt[6]{2} - \sqrt[-6]{2} \right) \qquad (10)$$

5   IMPEDANCE

Mathematical models for mechanical, electrical, fluid, and thermal systems can be developed essentially independent of the discipline based on concepts of system similarity. In each discipline, the variables that are used to express the differential equations can be classified as either effort, flow, or combinations of effort and flow variables.[6]

Effort variables describe the effort which is put on a component. Typical effort variables are force, voltage, pressure, and temperature. Flow variables are those that describe the rate of change, such as velocity, current, volume flow rate, and heat flow. Impedance of a component is defined as the ratio of the effort variable to the flow variable. For a mechanical system this is force divided by velocity.

In acoustic systems, the effort variable is pressure p and the flow variable is velocity. However, there are two velocities often considered in acoustics: volume velocity U and particle velocity u. With these two velocities there are two acoustic impedance terms. Acoustic impedance is defined as the acoustic pressure divided by the volume velocity as

$$Z = \frac{p}{U} \qquad (11)$$

Specific acoustic impedance is defined as the acoustic pressure divided by the particle velocity as

$$z = \frac{p}{u} \qquad (12)$$

Radiation impedance is a term often used to represent the coupling between acoustic waves and a driving source or driven load. The radiation impedance for a vibrating surface with area A is given by the equation

$$Z_r = zA \tag{13}$$

This can also be computed as the ratio of the force f acting on the fluid and the particle velocity:

$$Z_r = \frac{f}{u} \tag{14}$$

The radiation impedance is part of the overall mechanical impedance of the system. The magnitude of the contribution of the radiation impedance to the overall mechanical impedance is generally small but is, of course, related to the magnitudes of the values in the impedance equations.

6 THEORY OF SOUND

In this section, some of the fundamental theoretical developments of acoustics in a fluid are summarized. For theoretical developments of acoustic propagation in solids, see Ref. 7. Of foremost importance in this section is a discussion of the acoustic wave equation and forms of solutions.

6.1 Constitutive Equations

Acoustic disturbances are ultimately an unsteady portion of the overall dynamics of the fluid. Consequently, the equations that describe the motion of a fluid—the equations of continuity, force, state, and so on—can be applied to describe acoustic pressure fluctuations. However, with the exception of Section 14 on non-linear acoustics, we will consider the fluctuating pressure term in the overall description of the fluid to be "small." In terms of the acoustic Mach number, which is a dimensionless quantity that describes the velocity of acoustic fluctuations, $\mathbf{u}$, relative to the speed of sound in the fluid, c_0, this assumption of "smallness" may be expressed as $|\mathbf{u}|/c_0 \gg 1$. This assumption applies to the vast majority of applications of acoustics and allows for considerable simplification of the constitutive equations of fluid mechanics to first-order accuracy.

The constitutive equations of the wave equation in fluid acoustics are used to relate the acoustic pressure p, the acoustic density ρ, and the acoustic "particle" velocity $\mathbf{u}$, which despite its name, describes the motion of a small volume or continuum of fluid. These acoustic variables describe the unsteady fluctuations in the fluid that are acoustic waves. On the other hand, variables that describe the ambient pressure or density will be denoted with the subscript 0. Note that both the mean velocity of the fluid and losses (e.g., thermoviscous) are considered to be zero in the equations that follow.

The first equation is the equation of continuity, which relates ρ to $\mathbf{u}$ and connects the compression

of the fluid with its motion. In its linearized form, the equation of continuity may be written as

$$\frac{\partial \rho}{\partial t} + \rho_0 \nabla \cdot \mathbf{u} = 0 \tag{15}$$

where the first term, $\partial \rho / \partial t$, describes the time rate of change of the acoustic density for a small volume fixed in space. For continuity of the fluid to hold, $\partial \rho / \partial t$ must be equal to the net flux of fluid into the volume, approximated by $-\rho_0 \nabla \cdot \mathbf{u}$. (The minus sign is needed because the divergence operator describes flow of fluid away from the volume.)

The second equation relates $\mathbf{u}$ to p and is an expression of Newton's second law, $\sum F = ma$. For a fluid element that moves with the surrounding fluid, the linearized Euler equation (a lossless form of the Navier–Stokes equation) is written as

$$\rho_0 \frac{\partial \mathbf{u}}{\partial t} = -\nabla p \tag{16}$$

Note that because $\mathbf{u}$ is a vector that describes the velocity of the fluid in three-dimensional space, Eq. (16) is a composite vector equation.

The third constitutive equation necessary in the development of a small-amplitude wave equation relates the two scalars p and ρ. If we neglect losses and treat the acoustic processes in air as perfectly adiabatic, the equation of state between p and ρ may be approximated to first order as

$$p = c_0^2 \rho \tag{17}$$

where

$$c_0^2 = \frac{B}{\rho_0} \tag{18}$$

where B is the adiabatic bulk modulus of the fluid. In air, $B = \gamma p_0$, where γ is the ratio of specific heats ($\gamma = 1.402$). By use of the ideal gas law, the sound speed in air may be expressed as a function of temperature as

$$c_0^2 = \gamma r T \tag{19}$$

where r is the specific gas constant (287 J/kg·K) and T is the temperature in kelvin. An alternative expression is to write the sound speed relative to some reference value, which is 331 m/sec at 0°C (273.15 K). Relative to 331 m/sec, c_0 is expressed as

$$c_0 = 331 \sqrt{\frac{T}{273.15}} \qquad \text{m/s} \tag{20}$$

Equation (19) may be used to calculate sound speed in other gases given the appropriate values of γ and r, whereas in fluids that cannot be considered ideal gases, Eq. (18) can be used. For the sake of completeness, the speed of compressional waves in solids may be calculated as

$$c_0 = \sqrt{\frac{Y}{\rho}} \qquad (21)$$

where Y is Young's modulus and, in this case, ρ is the density of the solid in kilograms per meters cubed. A list of sound speed values in different media is given in Table 5. A more complete listing may be found in the literature.[7]

6.2 Wave Equation

Although not repeated here, the constitutive equations may be manipulated to form a wave equation for any of the acoustic variables, p, ρ, and $\mathbf{u}$. The acoustic wave equation for the pressure p may be expressed as

$$\nabla^2 p - \frac{1}{c_0^2} \frac{\partial^2 p}{\partial t^2} = 0 \qquad (22)$$

If we assume a one-dimensional wave equation in the x direction, Eq. (22) may be divided up as

$$\left(\frac{\partial}{\partial x} - \frac{1}{c_0} \frac{\partial}{\partial t} \right) \left(\frac{\partial}{\partial x} + \frac{1}{c_0} \frac{\partial}{\partial t} \right) p = 0 \qquad (23)$$

Equation (23) demonstrates that the wave equation may be viewed as right-going and left-going wave operators that act on p. In fact, the general solution to the one-dimensional wave equation for p is

$$p(x, t) = p_1 \left(t - \frac{x}{c_0} \right) + p_2 \left(t + \frac{x}{c_0} \right) \qquad (24)$$

where the function p_1 is a wave that travels in the $+x$ direction and p_2 is a wave that travels in the $-x$ direction. The forms of functions p_1 and p_2 depend on source, boundary, and initial conditions. Specific types of solutions to the acoustic wave equation are now discussed.

Plane Waves One possible solution to the wave equation is a monofrequency plane wave with angular frequency ω. A plane wave's amplitude and phase are constant on any plane perpendicular to the direction of travel. The most common application of plane wave propagation is in ducts at low frequencies, where the acoustic wavelength is much greater than the cross-sectional dimensions of the duct. In addition, at large distances from an acoustic source, spherical waves look locally planar. Expressed in complex exponential form, the plane wave solution for the acoustic pressure in one dimension (x) is

$$\boldsymbol{p}(x, t) = A e^{j(\omega t - kx)} + \boldsymbol{B} e^{j(\omega t + kx)} \qquad (25)$$

where the boldface italic text denotes a complex quantity and k is the wavenumber, where $k = \omega/c_0$. Complex notation is frequently used for mathematical convenience in this section and throughout the remainder of this section. The actual scalar pressure is

$$p(x, t) = \mathbf{Re}\{\boldsymbol{p}(x, t)\} \qquad (26)$$

The above result may be extended for a plane wave of frequency ω that is propagating in an arbitrary direction. If we consider just a single forward-propagating plane wave, the solution to the wave equation may be expressed as

$$\boldsymbol{p}(\mathbf{r}, t) = A e^{j(\omega t - \mathbf{k} \cdot \mathbf{r})} \qquad (27)$$

where $\mathbf{r}$ is the position vector from the origin and $\mathbf{k}$, the wavenumber propagation vector with magnitude $k = \omega/c_0$, is written as

$$\mathbf{k} = k_x \hat{x} + k_y \hat{y} + k_z \hat{z} \qquad (28)$$

An example of a two-dimensional plane wave is shown in Fig. 5.

One final point regarding plane waves is the relationship between p and $\mathbf{u}$. Use of the linearized Euler

Table 5 Speed of Sound in Various Media

Medium	Sound Speed (m/sec)	
Gases (20°C and 1 atm)		
Air	343	
Helium	1007	
Oxygen (O_2)	317	
Carbon dioxide	258	
Liquids		
Fresh water	1481	
Sea water	1500	
Glycerin	1980	
Solids	Bar	Bulk
Aluminum	5150	6300
Steel	5050	6100
Hard rubber	1450	2400
Glass (Pyrex)	5200	5600
Lead	1200	2050

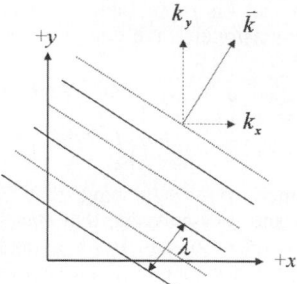

Fig. 5 Representation of plane wave propagation perpendicular to z axis. The lines represent surfaces of constant phase.

equation for a wave traveling in the positive direction results in the expression

$$\frac{p}{|\mathbf{u}|} = \rho_0 c_0 \tag{29}$$

which has units of specific acoustic impedance. This quantity is referred to as the characteristic impedance of the medium and is approximately equal to 415 Pa· sec/m in air.

Spherical Waves For sources of sound whose dimensions are much less than a wavelength or for distances much greater than the source dimensions, the acoustic radiation in free space is spherically diverging. For radial symmetry, the wave equation in spherical coordinates may be written as

$$\frac{\partial^2 p}{\partial r^2} + \frac{2}{r}\frac{\partial p}{\partial r} = \frac{1}{c_0^2}\frac{\partial^2 p}{\partial t^2} \tag{30}$$

A different form of Eq. (30) lends insight into the physical behavior of spherical waves. Rather than p as the dependent variable, rewriting Eq. (30) in terms of rp results in

$$\frac{\partial^2 (rp)}{\partial r^2} = \frac{1}{c_0^2}\frac{\partial^2 (rp)}{\partial t^2} \tag{31}$$

Comparison of Eq. (31) with the general wave equation (24) shows that the general solution to this wave equation for spherical waves can be written as

$$p(r, t) = \frac{1}{r}p_1\left(t - \frac{r}{c_0}\right) + \frac{1}{r}p_2\left(t + \frac{r}{c_0}\right) \tag{32}$$

The first term in Eq. (32) represents outgoing waves and the second term, incoming waves. For diverging

monofrequency waves of frequency ω, the solution to Eq. (30) in complex form is

$$p(r, t) = \frac{A}{r}e^{j(\omega t - kr)} \tag{33}$$

From this expression for p, particle velocity, intensity, and sound power can all be calculated. However, of particular interest are the ratios of acoustic pressure between r_1 and r_2. For spherical spreading between r_1 and r_2,

$$\frac{p(r_1)}{p(r_2)} = \frac{r_2}{r_1}e^{-jk(r_2-r_1)} \tag{34}$$

For $r_2 = 2r_1$,

$$\left|\frac{p(r_2)}{p(r_1)}\right| = \frac{1}{2} \tag{35}$$

6.3 Absorptive Processes

Although losses were neglected in the theoretical development of the wave equation, they are essential to everyday acoustic analyses. Losses may be readily incorporated into the solution of a propagating wave by making the wavenumber k complex:

$$k = k - j\alpha \tag{36}$$

The variable α is known as the absorption coefficient, with units of nepers per meter, and its value depends on both frequency and the propagation environment (the Neper is a dimensionless unit). The solution for a traveling plane wave takes the form

$$p(\mathbf{r}, t) = Ae^{-\alpha r}e^{j(\omega t - \mathbf{k}\cdot\mathbf{r})} \tag{37}$$

The term $e^{-\alpha r}$ represents an exponential decay in amplitude as a function of distance. The absorption coefficient can also be directly considered with regards to its impact on sound pressure level. For plane wave propagation from the source ($x = 0$) to some distance x,

$$L_p(x) = L_p(0) - 8.686\alpha x \tag{38}$$

Two types of absorption are briefly considered, atmospheric absorption and boundary layer absorption in pipes. Note that the absorptive processes in the atmosphere and near pipe walls give rise to not only absorption but dispersion as well. However, because dispersion does not affect amplitude at a given frequency, the phenomenon is not discussed further in this section. Instead the reader is referred to Refs. 8 and 9, which contain more in-depth discussions on the absorptive processes.

The absorption of sound by the atmosphere is dominated by thermoviscous (tv) losses and vibrational relaxation losses due to the diatomic molecules of oxygen (O_2) and nitrogen (N_2). The combined effects of these processes represent the total atmospheric absorption coefficient

$$\alpha = \alpha_{tv} + \alpha_{O_2} + \alpha_{N_2} \tag{39}$$

To calculate atmospheric absorption as given below, values of p_0 in atmospheres, ambient temperature (T_0) in kelvin, and relative humidity (RH) in percent are needed as inputs. Although other units are possible, they result in different expressions for α. With ambient pressure, temperature, and relative humidity expressed in the appropriate units, α is calculated as

$$\alpha = p_0 F^2 \left\{ 1.84 \times 10^{-11} \left(\frac{T_0}{293.15} \right)^{1/2} \right.$$
$$+ \left(\frac{T_0}{293.15} \right)^{-5/2} \left[0.01275 \frac{e^{-2239.1/T_0}}{F_{r,O} + F^2/F_{r,O}} \right.$$
$$\left. \left. + 0.1068 \frac{e^{-3352/T_0}}{F_{r,N} + F^2/F_{r,N}} \right] \right\} \tag{40}$$

where

$$F = \frac{f}{p_0} \qquad F_{r,O} = \frac{f_{r,O}}{p_0} \qquad F_{r,N} = \frac{f_{r,N}}{p_0} \tag{41}$$

The normalized relaxation frequencies F_r for oxygen and nitrogen are respectively

$$F_{r,O} = 24 + 4.04 \times 10^4 h \frac{0.02 + h}{0.391 + h} \tag{42}$$

and

$$F_{r,N} = \left(\frac{293.15}{T_0} \right)^{1/2}$$
$$\times \left(9 + 280 h \exp \left[-4.17 \left\{ \left(\frac{293.15}{T_0} \right)^{1/3} - 1 \right\} \right] \right) \tag{43}$$

where h is the absolute humidity, or molar concentration of water vapor, in percent. Absolute humidity can be calculated from relative humidity as

$$h = \text{RH} \frac{p_{sat}}{p_0} \tag{44}$$

Finally, the saturation vapor pressure p_{sat} can be calculated from

$$\log_{10}(p_{sat}) = -6.8346 \left(\frac{273.16}{T_0} \right)^{1.261} + 4.6151 \tag{45}$$

In Fig. 6, curves of the absorption coefficient for 1 atm and 20°C (293.15 K) are displayed as a function of frequency for three values of relative humidity. The figure shows that variation in water vapor content in the atmosphere can significantly affect the absorption over the audio range (20 Hz–20 kHz). At low frequencies, the dominant absorption mechanism is due to nitrogen relaxation. Above a few hundred hertz, oxygen relaxation becomes the dominant process. It is not until high frequencies (50–100 kHz) that the diatomic molecules are considered frozen and thermoviscous processes dominate.

The second type of absorption is boundary layer absorption, which is relevant to propagation in pipes or ducts. The additional absorption near the walls of the duct is due to the viscous and thermal boundary layers. For a duct with a given hydraulic diameter (HD = $4S/C$, where S is the area and C is the perimeter

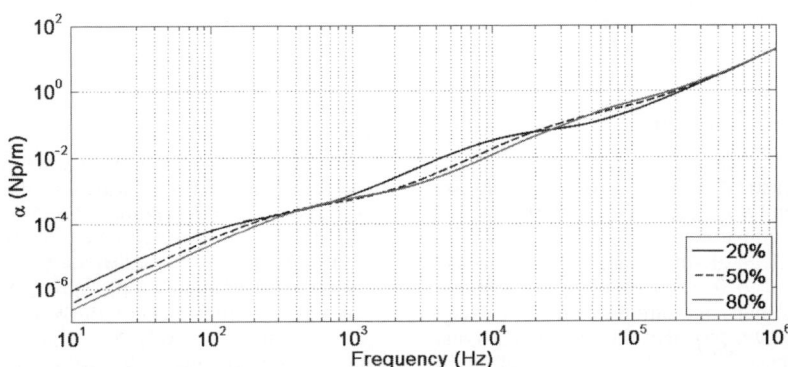

Fig. 6 Atmospheric absorption coefficient as function of frequency for 1 atm, 20°C, and various values of relative humidity.

of the cross section), α for an air-filled pipe may be calculated as

$$\alpha = \frac{2}{\text{HD}} \sqrt{\frac{\pi f \mu}{\rho_0 c_0^2}} \left(1 + \frac{\gamma - 1}{\sqrt{\text{Pr}}} \right) \qquad (46)$$

where μ is the shear viscosity coefficient, γ is the ratio of specific heats, and Pr is the Prandtl number. Unlike atmospheric absorption, which approaches an f^2 dependence well above the relaxation frequencies, boundary layer absorption only increases as $\sqrt{f}$.

7 REFLECTION, TRANSMISSION, AND ABSORPTION

When a wave propagating in a medium encounters a change in the impedance presented to the wave, reflection and transmission of the wave occur. There are two general ways in which this could occur: (i) the medium could change, such as when a wave propagates from air into water, or (ii) the geometry could change even though the medium does not, such as when a wave propagates from a pipe into an expansion chamber.

The physical concepts that govern the reflection and transmission are not difficult to understand, although the algebra that results in analyzing the situation can quickly become very cumbersome. Thus, for all but the simplest cases, one would typically solve the resulting expressions with the aid of a computer.

There are several reflection and transmission coefficients of interest. The *pressure* reflection and transmission coefficients ($\mathbf{R}$, $\mathbf{T}$) provide the ratio of the reflected and transmitted pressure to the incident pressure. The *intensity* reflection and transmission coefficients (R_I, T_I) provide the ratio of the reflected and transmitted acoustic intensity to the incident acoustic intensity. Note that these two sets of coefficients are not generally the same, as they involve two different physical quantities. Finally, the *power* reflection and transmission coefficients (R_Π, T_Π) provide the ratio of the reflected and transmitted acoustic power to the incident acoustic power. These coefficients will, in general, be different than the intensity reflection and transmission coefficients, except in the case of normal incidence.

7.1 Reflection/Transmission from a Single Interface

Consider first the case of a plane wave impinging on an interface between two fluid media. One of the most common examples considered for this case is that of sound propagating from air into water (or vice versa), although it can be used for any two fluids in contact with each other. The general configuration can be seen in Fig. 7. There are two physical conditions that must be met. First, the pressure in fluid 1 at the interface will be the same as the pressure in fluid 2 at the interface.

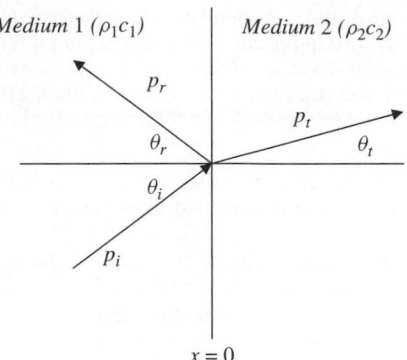

Fig. 7 Reflection and transmission at interface of two fluid media. The vectors represent the propagation directions of the incident, reflected, and transmitted waves.

Second, the displacement normal to the interface (or equivalently, the particle velocity) in fluid 1 will be the same as the normal displacement (particle velocity) in fluid 2 at the boundary. If the interface lies in the $y-z$ plane, the waves in the two media can be expressed as

$$p_i = \boldsymbol{p}_i e^{j(\omega t - k_1 x \cos \theta_i - k_1 y \sin \theta_i)}$$
$$p_r = \boldsymbol{p}_r e^{j(\omega t + k_1 x \cos \theta_r - k_1 y \sin \theta_r)} \qquad (47)$$
$$p_t = \boldsymbol{p}_t e^{j(\omega t - k_2 x \cos \theta_t - k_2 \sin \theta_t)}$$

where, θ_i is the angle of incidence of the wave, θ_r is the angle of reflection, and θ_t is the angle of transmission. The wavenumber k is specific to the medium of interest and will differ for the two media, since the phase speed c differs in the two media. Analysis of the conditions at the boundary leads to the following important result:

$$\theta_i = \theta_r \qquad (48)$$

This result is known as the law of reflection and simply indicates that the angle of reflection must equal the angle of incidence. This is the same result as for light reflecting from a plane mirror.

Equation (49) is known as Snell's law and governs the refraction of sound waves as they propagate from one medium into another. Note that the refraction depends on the speed of sound in the two media. This is also manifest in the atmosphere when there is a temperature gradient in the air. In this case, there is a continuous rather than a discrete change in the sound speed, and as a result the sound waves bend either upward or downward as the wave propagates. This phenomenon is responsible for being able to clearly hear a cricket that is a long distance away on a cool evening, for example:

$$\frac{\sin \theta_i}{c_1} = \frac{\sin \theta_t}{c_2} \qquad (49)$$

Equations (50) are the pressure reflection and transmission coefficients. In these expressions, $\mathbf{Z}_n$ refers to the normal specific acoustic impedance of either medium 1 or medium 2. For a fluid, this is given by $\mathbf{Z}_n = \rho c / \cos \theta$, where the angle θ is the angle of incidence in medium 1 and the transmitted angle in medium 2 and the ρc of the appropriate medium is used. For normal incidence, $\cos \theta = 1$. It is also of note that $\mathbf{T} = 1 + \mathbf{R}$ for the pressure coefficients, but for the power coefficients (and intensity for normal incidence), $R_\Pi + T_\Pi = 1$ (conservation of energy). Equations (51) are the intensity and power reflection and transmission coefficients:

$$\mathbf{R} = \frac{\mathbf{Z}_{n2} - \mathbf{Z}_{n1}}{\mathbf{Z}_{n2} + \mathbf{Z}_{n1}} \qquad \mathbf{T} = 1 + \mathbf{R} = \frac{2\mathbf{Z}_{n2}}{\mathbf{Z}_{n1} + \mathbf{Z}_{n2}} \qquad (50)$$

$$R_I = R_\Pi = \left(\frac{\mathbf{Z}_{n2} - \mathbf{Z}_{n1}}{\mathbf{Z}_{n2} + \mathbf{Z}_{n1}}\right)^2$$

$$T_I = \frac{\rho_1 c_1}{\rho_2 c_2} \left(\frac{2\mathbf{Z}_{n2}}{\mathbf{Z}_{n2} + \mathbf{Z}_{n1}}\right)^2 \qquad (51)$$

$$T_\Pi = \frac{4\mathbf{Z}_{n1}\mathbf{Z}_{n2}}{(\mathbf{Z}_{n2} + \mathbf{Z}_{n1})^2}$$

An analysis of these results yields several important special cases:

1. Angle of Intromission. This occurs when $\mathbf{R}$ in Eq. (50) equals zero and corresponds to complete transmission through the interface. Although the material properties of density and speed of sound change, the normal specific impedance does not so that the wave does not "see" any discontinuity.

2. Grazing Incidence. In this case, θ_i goes to $90°$ and $\mathbf{Z}_{n1}$ goes toward infinity, resulting in a pressure reflection coefficient of $\mathbf{R} = -1$. Thus, the boundary between the two fluids acts as a pressure release boundary.

3. Critical Angle for $c_1 < c_2$. If $c_1 < c_2$, there is an angle of incidence θ_i where $\sin \theta_t = 1$ [using Eq. (49)]. This gives $\theta_t = 90°$. For angles of incidence greater than this, θ_t must become complex to satisfy the mathematical condition. Physically, this means that the transmitted waves become "evanescent," meaning that the waves decay exponentially in the x direction, propagate in the y direction, and cannot effectively penetrate into the second medium.

7.2 Reflection/Transmission from a Solid Surface

In the case of a wave in a fluid impinging upon a solid surface, the nature of the reflection/transmission phenomena depends upon the properties of the solid. If the solid has properties consistent with a bulk isotropic solid (no significant elastic response), the transmission of the wave into the solid will obey the equations given above, and the analysis can proceed as if the solid behaved as a fluid. If the solid behaves as a *locally reacting* material, the analysis must be modified slightly. One example of such a material would be an anisotropic solid, where waves propagate much faster in a direction perpendicular to the solid surface than they do in a direction parallel to the surface. Examples of such materials include many sound absorption materials such as acoustic tile, perforated panels, and other materials with a honeycomb structure. In these cases, the expressions given above can be used if one lets the transmission angle θ_t go to zero, that is, $\mathbf{Z}_{n2} = \rho_2 c_2$.

Reflection/Transmission through a Fluid Layer
One can often encounter situations where sound waves propagate through one or more fluid layers. These fluid layers may be small or large relative to an acoustic wavelength. The expressions will be developed here for a single layer with normal incidence. The methods can be extended in a straightforward manner to multiple layers and/or oblique incidence, but the algebra is such that use of a computer would be desirable. Use of these techniques can allow one to design a single or multilayer to have desired reflection/transmission characteristics. It also can be used for the analysis needed for noise control problems such as transmission through walls and other barriers.

The general configuration for understanding transmission through a layer can be seen in Fig. 8. With normal incidence, the acoustic waves involved can be

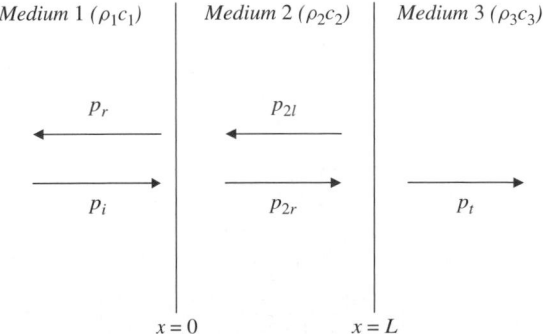

Fig. 8 Reflection and transmission through a fluid layer. The vectors represent the propagation directions of the incident, reflected, and transmitted waves in the three fluids.

represented as

$$p_i = \boldsymbol{P}_i e^{j(\omega t - k_1 x)}$$
$$p_r = \boldsymbol{P}_r e^{j(\omega t + k_1 x)}$$
$$p_{2r} = \boldsymbol{P}_{2r} e^{j(\omega t - k_2 x)} \quad (52)$$
$$p_{2l} = \boldsymbol{P}_{2l} e^{j(\omega t + k_2 x)}$$
$$p_t = \boldsymbol{P}_t e^{j(\omega t - k_3 x)}$$

The new terms in these equations represent the right- and left-going waves in the fluid layer. Again using continuity of acoustic pressure and acoustic particle velocity at the two interfaces ($x = 0$ and $x = L$) leads to the following result for the pressure reflection coefficient:

$$\mathbf{R} = \frac{\begin{aligned}[1 - \rho_1 c_1/(\rho_3 c_3)]\cos(k_2 L) + j[\rho_2 c_2/(\rho_3 c_3) \\ - \rho_1 c_1/(\rho_2 c_2)]\sin(k_2 L)\end{aligned}}{\begin{aligned}[1 + \rho_1 c_1/(\rho_3 c_3)]\cos(k_2 L) + j[\rho_2 c_2/(\rho_3 c_3) \\ + \rho_1 c_1/(\rho_2 c_2)]\sin(k_2 L)\end{aligned}}$$

$$(53)$$

For transmission through the layer, the intensity transmission coefficient is given by

$$T_I = \frac{4}{\begin{aligned}2 + [\rho_3 c_3/(\rho_1 c_1) + \rho_1 c_1/(\rho_3 c_3)]\cos^2(k_2 L) \\ + [(\rho_2 c_2)^2/(\rho_1 c_1 \rho_3 c_3) + \rho_1 c_1 \rho_3 c_3/(\rho_2 c_2)^2]\sin^2(k_2 L)\end{aligned}}$$
$$(54)$$

It is of note that this expression is symmetric, implying that the same energy propagates through the layer regardless of the direction of incidence. Many specific cases could be analyzed. However, one case that is often of practical interest is when fluid 1 is the same as fluid 3, such as transmission through a wall with air on both sides. In this case,

$$T_I = \frac{1}{1 + (1/4)[\rho_2 c_2/(\rho_1 c_1) - \rho_1 c_1/(\rho_2 c_2)]^2 \sin^2(k_2 L)}$$
$$(55)$$

From this expression, it can be seen that for low frequencies and/or thin walls ($k_2 L \gg 1$) there will be nearly perfect transmission through the layer. There will also be perfect transmission at discrete frequencies where $\sin(k_2 L) = 0$. Also, if the specific acoustic impedance of the layer ($\rho_2 c_2$) is large (as is the case with most walls, for example) and $k_2 L \gg 1$, the expression in Eq. (55) reduces to the well-known mass law, expressed as

$$T_I = \left(\frac{2\rho_1 c_1}{\omega \rho_2 L}\right)^2 \quad (56)$$

Thus, at low frequencies, one must increase the mass per unit area of the intermediate layer in order to reduce transmission for this case.

7.3 Reflection/Transmission at Discontinuities in Pipes

Another important class of problems involves the reflection and transmission that occurs when waves propagating in pipes or vents encounter a geometric discontinuity, such as a branch in the pipe or an expansion chamber or constriction, as shown in Fig. 9. For this analysis, continuity of pressure still holds at the discontinuity. However, instead of continuity of particle velocity, there is now continuity of volume velocity, given as the product of the particle velocity and the cross-sectional area ($\mathbf{U} = \mathbf{u}S$). For this case, the pressure reflection coefficient is given by

$$\mathbf{R} = \frac{\mathbf{Z}_0 - \rho c / S_1}{\mathbf{Z}_0 + \rho c / S_1} \quad (57)$$

where $\mathbf{Z}_0$ is the acoustic impedance at $x = 0$ and S_1 is the cross-sectional area of the pipe for $x > 0$. The power reflection and transmission coefficients are given by $R_\Pi = |\mathbf{R}|^2$ and $T_\Pi = 1 - R_\Pi$. For the case of a single pipe with a different cross-sectional area, for $x > 0$, $\mathbf{Z}_0 = \rho c / S_2$. For branches in the pipe or for

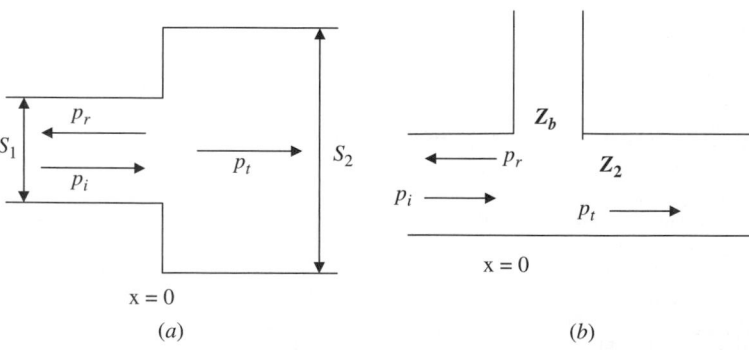

Fig. 9 Reflection and transmission from geometric discontinuities: (a) pip expansion; (b) side branch with impedance $\mathbf{Z}_b$.

more complex impedances such as expansion chambers and constrictions, the measured or calculated acoustic impedance at $x = 0$ should be used.

Side Branch A special case of practical interest is where a side branch exists at $x = 0$. This could be a branching of the pipe or it could be a noise reduction device such as a Helmholtz resonator (see Section 9). Let the acoustic impedance of the side branch be given by $\mathbf{Z}_b = R_b + j X_b$ and the impedance of the continuing pipe be $\rho c / S_1$. Using these values leads to the power reflection coefficient, the power transmission coefficient for waves propagating further down the pipe, and the power transmission coefficient for waves propagating into the side branch:

$$R_\Pi = \frac{(\rho c / 2 S_1)^2}{(\rho c / 2 S_1 + R_b)^2 + X_b^2}$$

$$T_\Pi = \frac{R_b^2 + X_b^2}{(\rho c / 2 S_1 + R_b)^2 + X_b^2} \quad (58)$$

$$T_{\Pi b} = \frac{(\rho c / 2 S_1) \, R_b}{(\rho c / 2 S_1 + R_b)^2 + X_b^2}$$

8 HEARING LOSS

8.1 Hearing Protection

Because excess noise can be damaging to the human ear, precautions must be taken to protect one's hearing. Sensorineural hearing loss, or permanent damage to the inner ear, can occur through sudden trauma, for example, caused by an explosion at close range. It can also occur if chemicals that are damaging to the cilia (called ototoxic chemicals) are not handled properly. However, the most common cause of sensorineural hearing loss is prolonged exposure to significant noise levels. To help workers safeguard their hearing, the Occupational Safety and Health Administration (OSHA) act (see http://www.osha.gov/dts/osta/otm/noise/standards.html) set forth permissible exposure limits in the workplace. Table 6 summarizes daily exposure limits for various A-weighted sound levels. Note that if noise exposure occurs at various levels throughout the day, in order to be under the cumulative noise exposure limit, the condition of $\sum t_i / T_i \leq 1$, where t_i is the exposure time at the ith level and T_i is the daily limit for that level that must be met. However, because a typical workday may consist of many different sound levels for as many different exposure periods, a noise dosimeter, which consists of a microphone and a sound level logger, can be worn by workers to track total daily noise exposure. Other noise and vibration exposure criteria are summarized by von Gierke and Ward.[10]

There are underlying assumptions regarding the OSHA act that should be made clear. First, the act is only intended to protect hearing for the purposes of understanding speech. Hearing loss above 4 kHz

Table 6 Daily Noise Exposure Limits (L_{pA}) in dBA

Hours	(dBA)
8	90
6	92
4	95
3	97
2	100
1.5	102
1	105
0.5	110
<0.25	115

is likely to occur and is deemed acceptable. Second, an 8-hr workday and a 5-day workweek are assumed. It is further assumed that leisure time is not spent in high-noise-level activities that could be damaging to the ear. Finally, and perhaps most importantly, the OSHA criteria are designed to protect only 85% of the population exposed at the limits. The remaining, more damage-prone 15% are to be financially compensated for the hearing loss suffered as part of their jobs.

The risk of hearing impairment can be summarized another way, as a function of both frequency and level. Shown in Fig. 10 are different zones which represent different amounts of potential for damage to the ear. In zone I, sounds are inaudible. The levels in zone II are everyday audible sounds that present no risk. In zone III, there is a qualified risk, especially for prolonged exposure periods, and in zone IV, there is a high risk of hearing damage. The division between zones II and III roughly follows the equal-loudness contours, meaning the ear is most susceptible to damage at those frequencies where it is most sensitive.

8.2 Hearing Protection Devices

To preserve hearing, in addition to limiting exposure time, hearing protectors can be worn to reduce the noise levels to which the ear is exposed. Figures 11 and 12 summarize many classes of hearing protectors and ranges for attenuation provided by each type as a function of frequency. In general, the effectiveness of hearing protectors increases as a function of frequency, but to achieve sufficient attenuation, the user should take care to wear them properly. For simple formable earplugs, the attenuation can be quite high, but care must be taken to insert them sufficiently deep into the ear canal. For ear muffs to provide the proper attenuation, the wearer needs to ensure that the cushioned cups that enclose the pinnae (outer ears) fit snugly enough that an airtight seal is formed.

9 PASSIVE NOISE CONTROL

There are numerous situations where it is desirable to reduce unwanted sound, or noise. There are many approaches for accomplishing this, with some

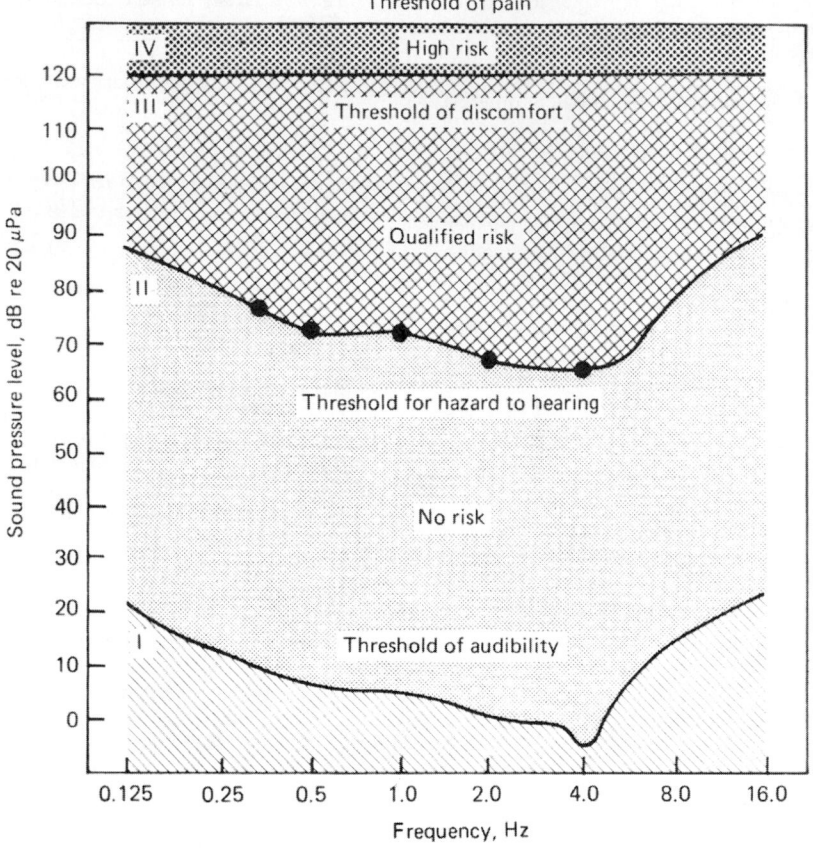

Fig. 10 Graphical representation of hearing damage risk as function of frequency and level.[11]

approaches being more appropriate for a given situation than others. This section will outline some general concepts associated with the use of passive noise control for reducing unwanted noise and the next section will outline concepts associated with active noise control. Much more information regarding these control methods can be found using the references at the end of the chapter (see particularly Refs. 13 and 14). The distinguishing feature between these two approaches is that passive noise control relies on the inherent acoustic properties of the noise control material or device being used to reduce the unwanted noise, while active noise control involves the active generation of additional sound or vibration to interact with the existing noise or vibration in a manner that attenuates the field.

9.1 Source/Path/Receiver Considerations

For any given noise control application, one should consider that three components exist: (a) there is an acoustic source that is responsible for generating the noise, (b) there is an acoustic path that the sound

energy propagates through, and (c) there is a receiver (often the human ear) that receives the acoustic energy. Where possible, this also gives the hierarchy of how to most effectively attenuate unwanted noise. The most effective solution is to modify the source characteristics so that less acoustic energy is generated. If that is not possible, the next most effective approach would be to modify the acoustic path. This can be done, for example, by using walls, barriers, absorptive materials, acoustic mufflers, and so forth. The final possibility is to modify the receiver. There is little possibility for doing much in this regard when the receiver is the human ear (except that the ear has a protective mechanism that reduces the sensitivity of the ear when exposed to loud sounds; however, this mechanism has a finite response time, and one should never rely on this ear protection as a means of making noise exposure acceptable). (Depending on one's point of view, hearing protectors such as ear muffs could be viewed as part of the receiver or part of the acoustic path.) If the receiver is not the human ear, there may be the possibility of modifying its response characteristics.

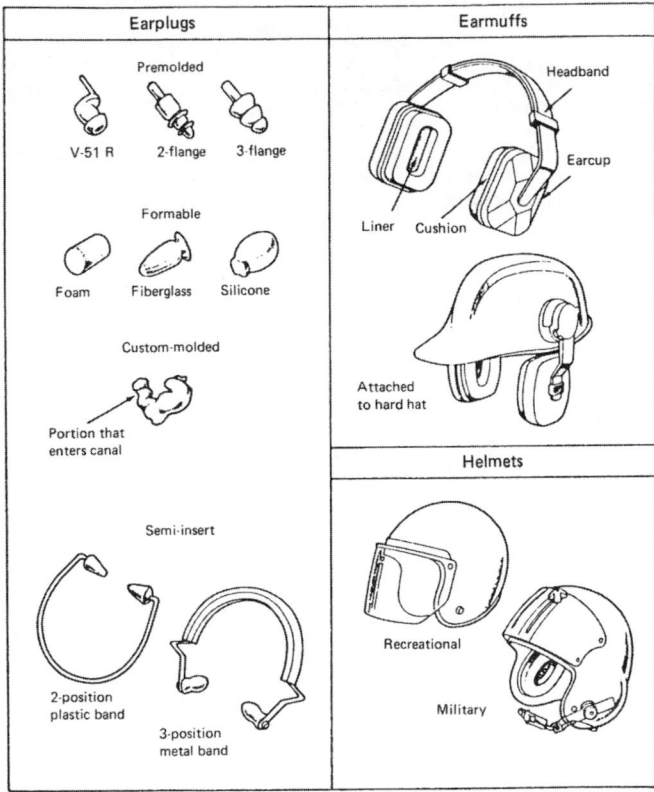

Fig. 11 Various types of hearing protectors.[12]

Most typical noise control approaches are based on the assumption that the source has been optimized and the receiver characteristics are fixed. Thus, the focus then needs to be on modifying the acoustic path.

9.2 Definitions

A number of terms are commonly used in the context of noise control, and an understanding of their meaning is beneficial:

Transmission Coefficient (τ): The fraction of incident energy that is transmitted through a noise control element.

Transmission Loss (TL): Expresses the transmission coefficient on a logarithmic scale and is obtained as $TL = -10 \log_{10} \tau$ (dB).

Insertion Loss (IL): The decrease in sound power level measured at a receiver location after a noise control element is inserted into the acoustic transmission path.

Absorption Coefficient (α): The fraction of energy that is randomly incident on a given material that is absorbed.

Noise Reduction Coefficient (NRC): Given by the arithmetic mean of the sound absorption coefficients for a material at 250, 500, 1000, and 2000 Hz and used as a single-number rating of the absorption characteristics of the material.

9.3 Impedance Considerations: Vibration Isolation Mounts

Many noise control situations involve a source that converts mechanical vibration energy into acoustic energy. Such examples could be a vibrating motor or generator that transmits energy to a support structure which then radiates acoustic energy or a vibrating wall or casing which in turn radiates acoustic energy. In general, there are three properties of the source that can be modified to attenuate the noise: mass, stiffness, and damping. Modifying each of these properties will be effective under certain conditions but completely ineffective under other conditions. This can best be seen by considering a vibration isolation mount, such as might be used for mounting a motor or generator. The general configuration is shown in Fig. 13. The

Type of protection	One-third-octave-band center frequencies, Hz						
	125	250	500	1000	2000	4000	8000
 Earplugs (premolded, user formable)	10–30	10–30	15–35	20–35	20–40	30–45	25–45
 Foam earplugs (attenuation varies with depth of insertion)	20–35	20–35	25–40	25–40	30–40	40–45	35–45
 Earplugs (custom-molded)	5–20	5–20	10–25	10–25	20–30	25–40	25–40
 Semi-insert earplugs (also called semiaural devices or canal caps)	10–25	10–25	10–30	10–30	20–35	25–40	25–40
 Earmuffs (with or without communications components)	5–20	10–25	15–30	25–40	30–40	30–40	25–40
 Earplugs and earmuffs (in combination)	20–40	25–45	25–50	30–50	35–45	40–50	40–50
Active noise reduction headsets				Identical to earmuffs above 1000 Hz			
	15–25	15–30	20–45	25–40	30–40	30–40	25–40
 Military helmets	0–15	5–15	15–25	15–30	25–40	30–50	20–50
 Motorcycle helmets	0–5	0–5	0–10	0–15	5–20	10–30	15–35

Fig. 12 Expected attenuation (dB) for various types of hearing protectors.[12]

mechanical impedance of this mount is given by

$$\mathbf{Z}_m = R_m + j\left(\omega m - \frac{s}{\omega}\right) \qquad (59)$$

Looking at this expression in various frequency regions indicates how to most effectively modify the

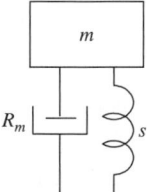

Fig. 13 Single-stage vibration isolation mount showing mass m to be isolated, spring element with stiffness s, and mechanical damping R_m.

vibration response. At low frequencies, the stiffness term dominates, and the impedance is given as $\mathbf{Z}_m \approx -js/\omega$. This is referred to as the stiffness-controlled region, and one should alter the stiffness of the mount to most effectively modify the response of the structure. Around the resonance frequency of the mount, the impedance is given as $\mathbf{Z}_m \approx R_m$. This is referred to as the damping-controlled region, and increased damping will be most effective in this frequency region. At higher frequencies above the resonance frequency, the impedance is given by $\mathbf{Z}_m \approx j\omega m$. This is referred to as the mass-controlled region, and one should alter the mass to most effectively modify the response of the structure.

Although this was developed for a vibration isolation mount, the same concepts hold for all vibrating sources—at any given frequency the system response will be dominated by the stiffness, damping, or mass properties of the structure. An investigation of the phase of the impedance can reveal which property is dominant. The stiffness-controlled region exhibits an impedance phase near $-90°$, the damping-controlled region has a phase near $0°$, and the mass-controlled region has a phase near $90°$.

9.4 Transmission Loss for Isolation Mounts

Isolation mounts are normally designed for operation above the resonance frequency of the mount. For the single-stage isolation mount shown in Fig. 13, the resonance frequency is given by

$$f_r = \frac{1}{2\pi}\sqrt{\frac{s}{m}} \tag{60}$$

The transmission loss for the mount is given by

$$\text{TL} = -10 \log_{10}\left[\frac{1 + \left[2\left(\dfrac{R}{4\pi m f_r}\right)\left(\dfrac{f}{f_r}\right)\right]^2}{\left[1 - \left(\dfrac{f}{f_r}\right)^2\right]^2 + \left[2\left(\dfrac{R}{4\pi m f_r}\right)\left(\dfrac{f}{f_r}\right)\right]^2}\right] \tag{61}$$

In this expression, the quantity $4\pi m f_r$ is commonly called "critical damping" and represents the damping that results in the fastest decay time for a disturbance with no oscillation. It can be seen that at frequencies above the resonance frequency the transmission loss increases by 12 dB/octave (doubling of frequency). If additional isolation is required, a two-stage isolation mount can be implemented, as shown in Fig. 14. For this type of mount, the transmission loss increases by 24 dB/octave above the resonance frequencies of the mount. The trade-off is that an additional resonance frequency is introduced at lower frequencies, so the isolation performance is degraded at lower frequencies. However, if operation will always be at frequencies above the resonance frequencies, this type of mount design can be desirable when greater isolation is required.

9.5 Acoustic Filters

Acoustic filters can be designed effectively when the wavelength at the frequencies of interest is significantly larger than the dimensions of the acoustic elements used to construct the acoustic filters. It can be shown that a tube whose length is small relative to a wavelength behaves as a mass, and a cavity (enclosed volume of fluid) behaves as a spring. These, along with damping materials, allow one to develop low-pass, high-pass, and bandstop filters in much the same way as one designs LRC electrical circuits. These three types of filters are reviewed briefly.

Bandstop Filter The design of a bandstop filter is typically dependent on the use of a Helmholtz resonator. A Helmholtz resonator consists of a tube connected to a cavity, typically with some damping included. A common example we are familiar with is

a soda bottle, where the neck of the bottle behaves as a tube and the portion below the neck behaves as the cavity. A Helmholtz resonator has a resonance frequency associated with it, and when used as a filter, it is capable of attenuating that frequency. The use of such a filter is shown in Fig. 15. The reactance of the side branch (Helmholtz resonator) is given by

$$X_b = \rho_0\left(\frac{\omega L_{\text{eff}}}{S_b} - \frac{c^2}{\omega V}\right) \tag{62}$$

where L_{eff} is the effective length of the tube (including end corrections), S_b is the cross-sectional area of the tube, and V is the volume of the cavity. The effective length is given by

$$L_{\text{eff}} = \begin{cases} L + 1.7a & \text{(outer end flanged)} \\ L + 1.4a & \text{(outer end unflanged)} \end{cases} \tag{63}$$

where a is the radius of the tube. This configuration will give a stopband centered at the frequency

$$\omega_{\text{sb}} = c\sqrt{\frac{S_b}{L_{\text{eff}}V}} \tag{64}$$

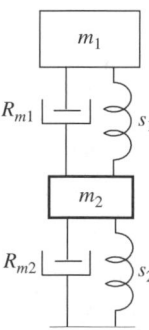

Fig. 14 Two-stage vibration isolation mount showing mass m_1 to be isolated, spring elements with stiffness s_1 and s_2, mechanical damping R_{m1} and R_{m2}, and intermediate mass (raft) m_2.

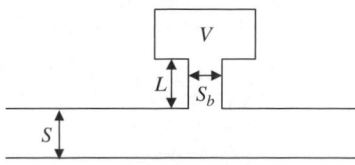

Fig. 15 Bandstop filter using Helmholtz resonator as side branch. The stopband is centered at the resonance frequency of the Helmholtz resonator.

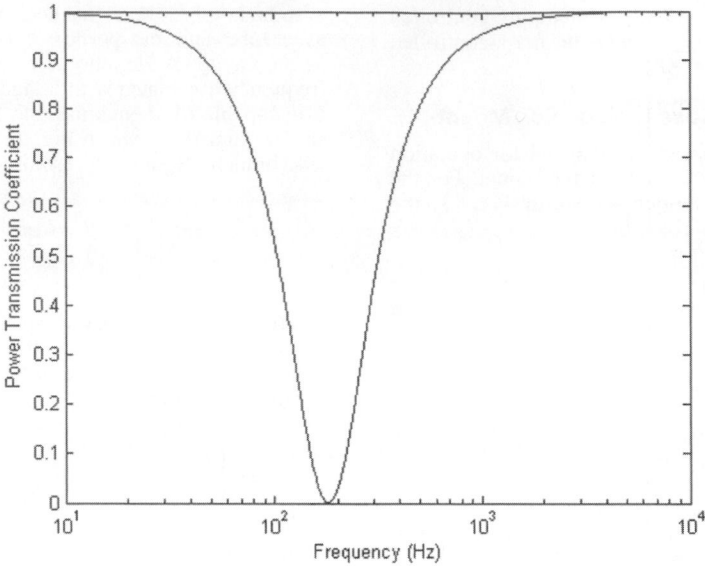

Fig. 16 Power transmission coefficient for bandstop filter using Helmholtz resonator. The resonator has a volume of 2000 cm^3, a neck length of 0.8 cm, and a cross-sectional area of 7.5 cm^2. The main pipe has a cross-sectional area of 28 cm^2.

and the power transmission coefficient for this filter is given by

$$T_\Pi = \frac{1}{1 + [(c/2S)/(\omega L_{\text{eff}}/S_b - c^2/\omega V)]^2} \quad (65)$$

where S is the cross-sectional area of the main duct. Figure 16 shows the characteristics of the power transmission coefficient for a typical bandstop filter.

Low-Pass Filter An acoustic low-pass filter can be constructed using an enlarged section of pipe or a constriction in the pipe. An enlarged section of pipe is used, for example, in the basic design of mufflers and will be considered here, as shown in Fig. 17. The enlarged section behaves acoustically as a cavity, and

the acoustic impedance of the cavity is given by

$$Z_c \approx \frac{-j\rho_o c^2}{\omega (S_1 - S) L} \quad (66)$$

The resulting power transmission coefficient is given by

$$T_\Pi \approx \frac{1}{1 + \{[(S_1 - S)/2S]/kL\}^2} \quad (67)$$

A typical response for this low-pass filter is shown in Fig. 18. It should be remembered that this acoustic response is only valid for low frequencies (wavelength significantly larger than the acoustic elements), and the response shown in Fig. 18 will not be valid for $kL > 1$.

High-Pass Filter An acoustic high-pass filter can be constructed using a side branch consisting of a short length of unflanged pipe (with radius a), such as is used for toneholes in musical instruments like a flute or clarinet, as shown in Fig. 19. The acoustic impedance of the side branch is given by

$$\mathbf{Z}_{\text{sb}} = \frac{\rho_o c k^2}{4\pi} + j\omega \left(\frac{\rho_o L_{\text{eff}}}{\pi a^2} \right) \quad (68)$$

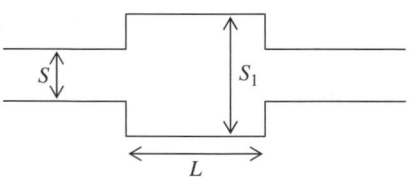

Fig. 17 Low-pass filter using expansion chamber.

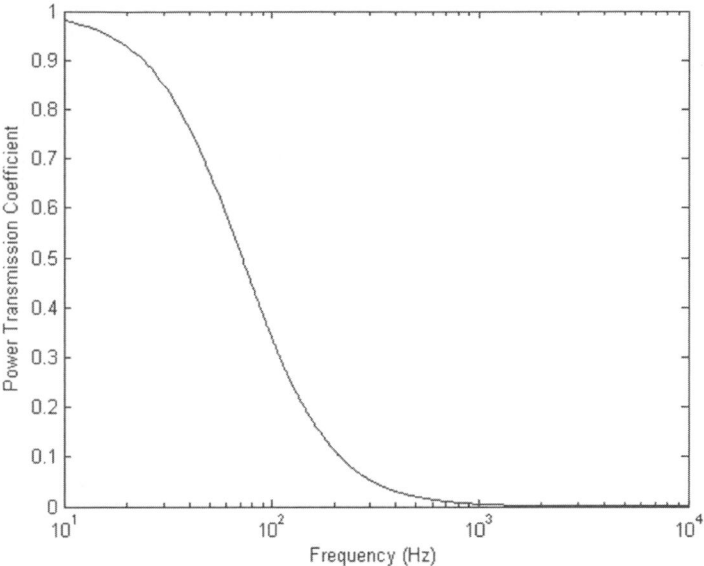

Fig. 18 Power transmission coefficient for low-pass filter using enlarged section of pipe. The main pipe has a radius of 2.54 cm, and the enlarged section has a length of 30.5 cm and an area six times that of the main pipe.

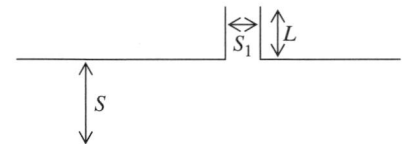

Fig. 19 High-pass filter using open side branch.

The resulting power transmission coefficient is given by

$$T_\Pi = \frac{1}{1 + [S_1/(2SL_{\text{eff}}k)]^2} \qquad (69)$$

A typical response for this high-pass filter is shown in Fig. 20. It should again be remembered that this acoustic response is only valid for frequencies where the wavelength is significantly larger than the side branch.

9.6 Lined Ducts

Ducts lined with absorbing material are often used as dissipative muffling devices to muffle fans in heating and air conditioning systems. The liner material generally consists of a porous material such as fiberglass or rockwool, usually covered with a protective facing. The protective facing may be a thin layer of acoustically

transparent material, such as a lightweight plastic sheet, or it may be a perforated heavy-gage metal facing. If a perforated facing is used, it should have a minimum open area of 25% to ensure proper performance.

The performance of a lined duct with liner thickness l and airway width $2h$ is shown in Fig. 21 for the case of zero mean flow. This figure shows some of the dependencies on the ratio of liner thickness to airway width as well as the flow resistivity.

9.7 Single- and Double-Leaf Partitions

Partitions (such as walls) are often used to separate a noise source from a receiving space. When partitions are used, flanking paths and leakage are important to check for. Small openings or flanking paths with low impedance can easily reduce the effectiveness of partitions significantly. Single-leaf partitions exist when there is a single surface or when both surfaces of the wall vibrate as a unit. Double-leaf partitions consist of two unconnected walls separated by a cavity.

At low frequencies for single-leaf partitions, the transmission through the partition is governed by the mass of the partition, and the mass law governs this behavior. The intensity transmission coefficient is given in Eq. (56), and the resulting transmission loss can be expressed as

$$\text{TL} = 20 \ \log(f\rho_s) - 47(\text{dB}) \qquad (70)$$

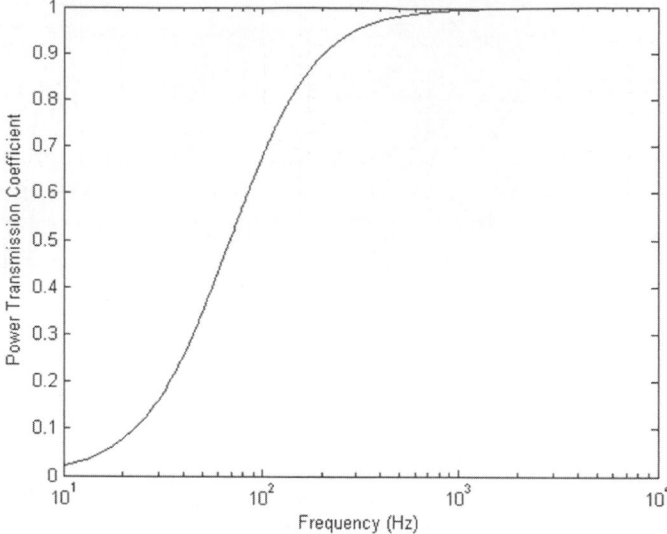

Fig. 20 Power transmission coefficient for high-pass filter using open side branch. The main pipe has a cross-sectional area of 28 cm^2, and the side branch has a length of 8 cm and a cross-sectional area of 7.5 cm^2.

where ρ_s is the surface density of the partition, given as the product of the density and thickness of the partition. Thus, at low frequencies, one must increase the density of the partition in order to increase the transmission loss; doubling the density of the partition increases the transmission loss by 6 dB.

For a single-leaf partition, wave effects in the partition lead to a coincidence frequency, which corresponds to the condition where the flexural wavelength in the partition matches the acoustic wavelength along the direction of the partition. When this condition is met, the transmission loss drops significantly, with the decrease being governed by the damping in the partition. Above the coincidence frequency, the partition becomes stiffness controlled, and the transmission loss increases at a rate higher than the mass law (theoretically 18 dB/octave for a single angle of incidence or about 9 dB/octave for diffuse-field incidence). The coincidence frequency separates these two behaviors and is given by

$$f_{co} = \frac{1}{2\pi}\sqrt{\frac{\rho_p h}{D}}\left(\frac{c}{\sin\phi}\right)^2 \qquad (71)$$

In this equation, c is the speed of sound in the fluid, ρ_p is the density of the partition, h is the thickness of the partition, and D is the bending rigidity of the partition, given by $D = Eh^3/[12(1-\nu^2)]$, where E is Young's modulus of the partition and ν is Poisson's ratio for the partition.

The transmission loss through a double-leaf partition is noticeably higher than through a single-leaf partition with the same mass density. While the behavior is too complex to be covered extensively here, the following characteristics are generally associated with double-leaf partitions. (a) At low frequencies, the transmission loss follows the mass law, with the combined mass of the two leaves being used to determine the surface density. (b) A mass–air–mass resonance exists where the air cavity between the two leaves behaves as a spring between two masses. (c) Above the mass–air–mass resonance, the transmission loss increases sharply (18 dB/octave) until resonance effects in the cavity become important. (d) The transmission loss oscillates as the cavity resonance effects become important. At resonances of the cavity (cavity depth $\approx n\lambda/2$), the transmission loss drops to values consistent with the mass law, while the peaks in the transmission loss occur at antiresonances of the cavity and continue to rise at about 12 dB/octave. Experimental transmission loss measurements typically show some variation from these predicted trends, but the results are generally consistent with predicted behavior.

9.8 Enclosures

For noisy equipment, one can install an enclosure around the piece of equipment. The insertion loss of the enclosure can be estimated using

$$\text{IL} = \text{TL} - C \qquad (\text{dB}) \qquad (72)$$

where TL is the transmission loss associated with the walls of the enclosure and

$$C = 10\,\log\left(0.3 + \frac{S_E\,(1-\overline{\alpha}_i)}{S_i\overline{\alpha}_i}\right) \qquad \text{dB} \qquad (73)$$

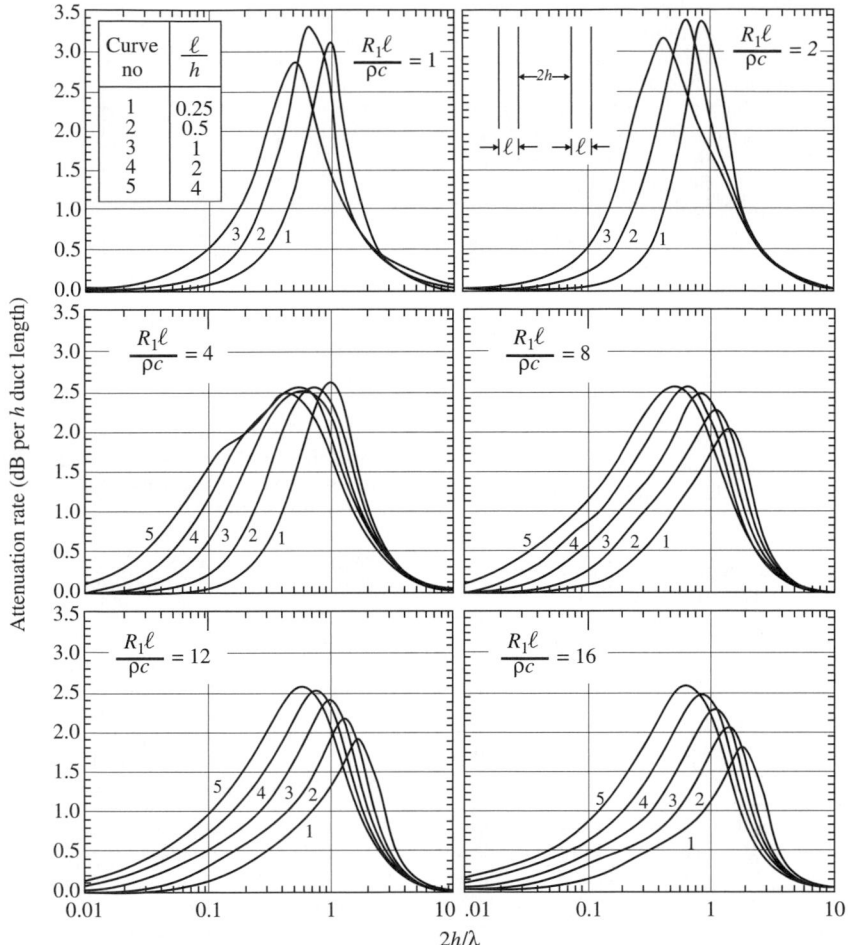

Fig. 21 Predicted octave-band attenuations for rectangular duct lined on two opposite sides. Lined circular ducts or square ducts lined on all four sides give twice the attenuation shown here. The quantity ρ is the density of fluid flowing in the duct, c is the speed of sound in the duct, ℓ is the liner thickness, h is the half width of the airway, and R_1 is the liner flow resistivity. For these results, a bulk reacting liner with no limp membrane covering and zero mean flow is assumed.[14]

$\overline{\alpha_i}$ is the mean Sabine absorption coefficient of the interior of the enclosure, S_i is the interior surface area of the enclosure, and S_E is the external surface area of the enclosure.

10 ACTIVE NOISE CONTROL

Over the last couple of decades, there has been considerable interest in the use of active noise control to address noise control applications. In many cases, it has been a technology that has not been well understood. There are many applications for which active noise control is not a good solution. In such cases, active noise control will be quite ineffective, and as a result, people can be easily disappointed. Thus, active noise control as a noise control solution should be

chosen with care. However, for applications where active noise control is appropriate, it works *very* well and can produce impressive results. While active noise control can be very effective for proper applications, it is generally not a straightforward "off-the-shelf" solution. Thus, the focus of this section is to give an overview of how active noise control works and the direction in identifying proper applications. If active noise control is a viable solution, expertise should be sought in implementing the solution.

In deciding whether active noise control is a viable solution, there are several characteristics of active noise control that should be understood. First, active noise control is inherently a low-frequency solution. Implementation at high frequencies has several difficulties associated with it. First, effective control requires precise phase and amplitude matching. For example, if one wishes to achieve 20 dB of attenuation, the control signal must have a phase error of less than 4.7° (assuming perfect amplitude matching) or a magnitude error of less than 0.9 dB (assuming perfect phase matching). This tight tolerance in phase and magnitude matching is significantly easier to achieve at lower frequencies than at higher frequencies.

It is also easier to achieve significant spatial control of the acoustic field at lower frequencies than it is at higher frequencies. In active noise control, one can achieve localized control or global control of the field, depending on the physical configuration of the problem. In those cases where local control is achieved, the spatial volume where significant attenuation occurs scales according to the wavelength. The diameter of the sphere where at least 10 dB of attenuation is achieved is about one-tenth of a wavelength. Thus, lower frequencies result in larger volumes of control. If global control is to be achieved, good spatial matching must be achieved, which is also easier to achieve at lower frequencies than at higher frequencies.

In a number of applications, it is desirable to achieve global control of the acoustic field. In order to accomplish this, there must be a good acoustic coupling between the primary noise source and the secondary control source used to control the field. This can be achieved using one of two mechanisms. The first is to have the spacing between the primary noise source and the secondary control source be significantly less than an acoustic wavelength. For an extended noise source, this would require multiple control sources that will acoustically couple to the primary noise source. For an enclosed noise field (such as in rooms or cabs), it is also possible to achieve acoustic coupling without the control source being less than a wavelength away from the noise source. Instead, the coupling occurs through the acoustic modes of the enclosed field. The acoustic modes have a distinct spatial response, and by exciting the secondary control source properly, it is possible to achieve the same spatial (modal) response but with opposite phase, thus resulting in global attenuation of the field. However, it should be noted that this approach is effective only for low modal density fields. If too many modes are excited, global control can rarely be achieved and local control is the result.

It is also important to understand that discrete tonal noise is significantly easier to control than broadband noise. When multiple frequencies must be controlled, it is necessary to control the precise phase and amplitude matching at all frequencies. This is easier to accomplish with a small number of discrete tones than it is with many frequencies or broadband noise. In addition, when controlling broadband noise, *causality* also becomes an important issue. When controlling tonal noise, if it is not possible to generate the control signal to arrive with the noise signal at the error sensor at exactly the same time, it is still possible to

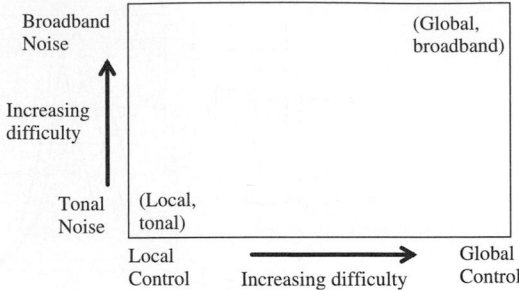

Fig. 22 Level of difficulty for various classes of noise control problems.

achieve effective control by delaying the control signal to match the noise signal one period later. However, with broadband noise, this approach is not possible. If broadband control is to be achieved, the control signal and the noise signal must be temporally aligned, which cannot occur if it takes longer for the control signal to get to the error sensor than the primary noise signal.

To summarize, when considering a particular application, the lower the frequencies involved, the more effective the control can be, in general. In addition, the difficulty of the solution depends on the frequency content of the noise and the spatial extent of the control needed, as shown in Fig. 22.

10.1 Control Architectures

Active control can be implemented in either an adaptive mode or a nonadaptive mode. In a nonadaptive mode, the control filter is fixed such that if the filter is designed properly, good attenuation results. However, if the acoustic system changes, reduced effectiveness can result. In adaptive mode, the control filter has the ability to adjust itself to a changing acoustic environment, based on the response of one or more error sensors. Most active noise control solutions are based on an adaptive mode solution that is based on a digital signal processing (DSP) platform.

There are two general architectures that can be implemented with an active noise control system. These are referred to as feedforward and feedback. While both adaptive control and nonadaptive control have been used with both architectures, adaptive control has generally been used for feedforward control systems. To understand the basic working of an adaptive feedforward control system, consider the control of a plane wave propagating in a duct, as shown in Fig. 23. This is a prototypical application and allows one to understand the basic configuration in a straightforward manner. The noise to be controlled is detected by a "reference sensor," which could be a microphone in the duct or some other sensor whose output is correlated to the noise to be attenuated. It should be understood that the control system will only be capable of attenuating noise that is correlated with the signal

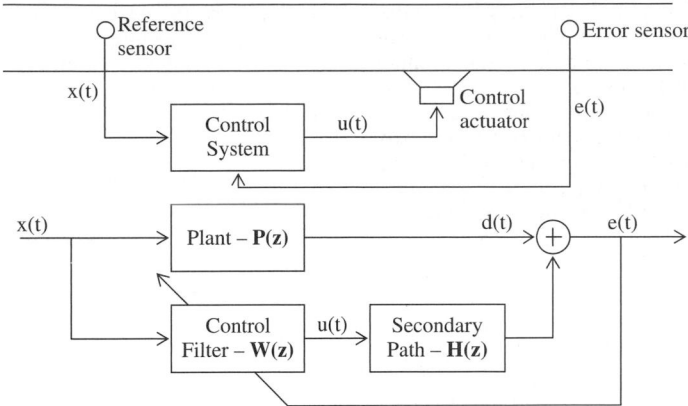

Fig. 23 Adaptive feedforward control for a duct. The upper figure shows the physical layout of the system, while the lower figure shows a block diagram of the control system implementation.

from this reference sensor. This signal is then used as the input to an adaptive control filter that determines the control signal output. The output signal is passed to a control actuator (such as one or more loudspeakers), where it generates an acoustic response that combines with the uncontrolled field and is measured by one or more error sensors. The error sensor response is then used to update the adaptive control filter. For implementation of a feedforward control system, care must also be taken to account for any possible response from the control output at the reference sensor. In our example, not only does the control signal propagate "downstream" in the duct to provide the desired attenuation, but it also propagates "upstream" in the duct where it can alter the reference signal if a microphone is being used. This can be accounted for either by using a nonacoustic reference signal (such as a tachometer signal if the noise is being created by some piece of rotating equipment such as a fan) or by modeling the feedback contribution from the control output to the reference sensor and compensating for that feedback component in the control system.

Causality is a potentially important issue in implementing feedforward control. There is an acoustic delay that exists as the noise to be controlled propagates from the reference sensor to the error sensor. If the reference signal can be processed and the response from the control actuator can arrive at the error sensor at the same time as the uncontrolled noise, the system will be causal. In this case, both random and periodic noise could be effectively controlled, since it will be possible to achieve the precise time alignment needed. If the noise to be controlled is periodic, the causality constraint can be relaxed. If the control signal does not arrive in time to be perfectly aligned in time, the DSP will adjust itself to effectively line up the signal properly one cycle later, thus still achieving the desired attenuation.

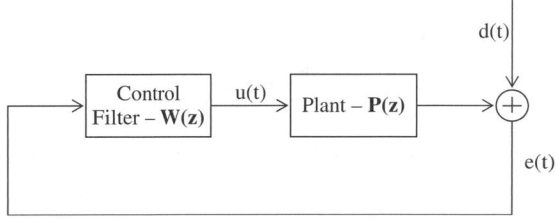

Fig. 24 Block diagram of feedback control implementation. The uncontrolled signal is given by $d(t)$.

Feedback control systems have used both adaptive and nonadaptive control configurations. Perhaps the most common example involving feedback control is with active headsets that are commercially available to reduce noise (although some headsets are now also implementing feedforward control). For a feedback control system (Fig. 24), the noise to be controlled is detected by a reference sensor that is used as the input to the control filter. The output of the control filter is again sent to a control actuator, and the sound generated combines with the uncontrolled field. The result is measured at the reference sensor and thus creates the "feedback loop." Feedback control systems are generally more tolerant of model errors in the control system implementation. Thus, implementation of feedforward control in a nonadaptive mode is rarely effective in a practical application. However, feedback control systems can also easily become unstable if designed improperly. Staying within stability constraints often leads to a solution that does not achieve as much attenuation as with feedforward control. The stability constraint also often determines the frequency bandwidth that can be effectively controlled.

It should also be understood that there is inherently an acoustic delay present in the feedback loop when implementing feedback control. This delay results from the finite response time of the control actuator, the acoustic propagation time from the control actuator to the reference sensor, and any delays associated with the electronics of the system. As a result, it is impossible to achieve perfect time alignment of the uncontrolled noise and the noise generated by the active control system. In general, the shorter the delay time in the feedback loop, the greater the attenuation that can be achieved and the wider the frequency bandwidth that can be controlled.

10.2 Attenuation Limits

Good estimates of the maximum attenuation achievable can be determined for both feedforward and feedback control implementations. Because of the structure of feedforward systems, the control system will only attenuate noise that is correlated with the reference sensor signal. Thus, it is imperative that the error signal sensor and the reference signal sensor be correlated. This is embodied in the expression that gives the maximum obtainable attenuation,

$$\Delta L_{\max} = 10 \, \log[1 - \gamma_{xd}^2(\omega)] \tag{74}$$

where ΔL_{max} is the maximum obtainable attenuation and $\gamma_{xd}^2(\omega)$ is the coherence (frequency dependent) between the reference signal, $x(t)$, and the uncontrolled error signal, $d(t)$. A quick method of determining if active noise control would be effective would be to measure the coherence between a proposed reference sensor and a proposed error sensor. It should be remembered that this gives a prediction of the attenuation at the error sensor. If local control is being achieved, this will give no indication as to how effective the control will be at locations removed from the error sensor location.

For feedback control, the autocorrelation of the reference signal can be used to predict the attenuation that can be achieved if the time delay associated with the feedback loop is known. Conversely, if the desired attenuation is known, this autocorrelation can also be used to determine how short the time delay must be. The predicted attenuation is given by

$$\Delta L_{\max} = -10 \, \log\left(1 - \frac{E_p}{E_0}\right) \tag{75}$$

where E_0 is the autocorrelation level at zero delay and E_p is the largest magnitude of the autocorrelation that exists in the autocorrelation at any time greater than the group delay of the feedback loop.

10.3 Filtered-x Algorithm

The most common adaptive algorithm in current use for active noise control is the filtered-x algorithm or some variation of that algorithm. A brief review of the algorithm is helpful in understanding the general

architecture of most active noise control systems. In the DSP architecture, the control filter is implemented as a finite impulse response filter, whose response can be represented by a vector of the filter coefficients. Thus,

$$\mathbf{W} = \begin{bmatrix} w_0, w_1, w_2, \dots, w_{I-1} \end{bmatrix}^{\mathrm{T}} \tag{76}$$

Similarly, the secondary path transfer function (represented by $\mathbf{H}$ in Fig. 23) can be represented by a vector of filter coefficients:

$$\mathbf{H} = \begin{bmatrix} h_0, h_1, h_2, \dots, h_{J-1} \end{bmatrix}^{\mathrm{T}} \tag{77}$$

The output of a digital filter is the convolution sum of the input signal with the filter response vector. Thus, if the reference input and control output signals are represented as vectors,

$$\begin{aligned} \mathbf{X}(t) &= [x(t)x(t-1)x(t-2)\cdots x(t-I+1)]^{\mathrm{T}} \\ \mathbf{U}(t) &= [u(t)u(t-1)u(t-2)\cdots u(t-J+1)]^{\mathrm{T}} \end{aligned} \tag{78}$$

the control output signal is given by $u(t) = \mathbf{W}^{\mathrm{T}}\mathbf{X}(t)$ and the error signal is given by $e(t) = d(t) + \mathbf{H}^{\mathrm{T}}\mathbf{U}(t)$. Most active noise control systems are based on quadratic minimization techniques. For the filtered-x algorithm, the algorithm updates its coefficients according to the negative gradient (with respect to the control filter coefficients) of the squared instantaneous error signal. Calculating the gradient of the squared error signal leads to

$$\mathbf{W}(t+1) = \mathbf{W}(t) - \mu\mathbf{R}(t)e(t) \tag{79}$$

which gives the control filter coefficients for the next iteration of the algorithm. In this expression, μ is a convergence parameter chosen to maintain stability and $\mathbf{R}(t)$ is the "filtered-x" signal vector, whose components are given by

$$r(t) = \hat{\mathbf{H}}^{\mathrm{T}}\mathbf{X}(t) \tag{80}$$

where $\hat{\mathbf{H}}$ is a vector of filter coefficients that models the physical secondary path transfer function $\mathbf{H}$.

10.4 System Identification

In order to achieve stable, effective control, it is necessary to have a reasonable model of the secondary path transfer function $\mathbf{H}$. It has been shown that the phase of the model is the primary concern in achieving good control. While phase errors of up to $\pm 90°$ can be tolerated in order to maintain stability, the performance of the control system degrades seriously as the phase errors approach this limit. Thus, an accurate model with minimal phase errors will result in substantially improved control results.

There are several methods that have been used to obtain a good model of **H**. The most straightforward method is to obtain a model of **H** a priori. This is done by injecting broadband noise into the secondary path (typically from the DSP used for the control) and measuring the response at the error signal with the primary noise source turned off. In this manner, a straightforward adaptive system identification routine can be used to obtain the coefficients of **H**.

A second method implements an adaptive online secondary-path estimation technique by injecting low-level broadband noise, $n(t)$, along with the control signal, $u(t)$. This broadband signal is uncorrelated with the primary noise, $d(t)$, and with the control signal, $u(t)$. Thus, the error signal can be used as an output, with the noise signal $n(t)$ as an input in a typical adaptive system identification routine. Since the primary noise and control signal are not correlated with $n(t)$, they do not affect the system identification and the process proceeds similar to the offline approach. The difficulty with this approach is that the injected noise must be high enough in level to achieve good system identification and yet kept low enough in level to not affect the overall noise level at the error sensor. In many cases, this can involve adaptive gain control to maintain the correct balance between the control signal and the injected noise.

A final method that has been used is also adaptive in nature. This method performs system identification not only for the secondary path but also for the plant, P. In other words, it implements a model of the entire system which is unlike the previous two methods. One of the results of this approach is that the model of the secondary path is not unique, unless the excitation of the system can be characterized as "persistent excitation," which essentially means broadband excitation. For excitation signals that are narrowband in nature, although there is not a unique solution for the secondary path, it has been shown that the solution obtained leads to stable, effective control. For more information on both adaptive system identification methods, the reader is referred to Refs. 15 and 16.

10.5 Control Applications

This section briefly outlines application areas where active noise control may be applicable:

(i) Active Control in Ducts. Active control of ducts has been implemented in a number of commercial applications, such as in exhaust stacks at industrial plants and in heating, ventilation, and air-conditioning (HVAC) ducts. The most successful applications have been at low frequencies, where only plane waves propagate in the duct. At higher frequencies, higher order modes propagate in the duct. While control of such fields can be effective, it requires sensing and actuating configurations that are able to sense and control those higher order modes. Another consideration is that even in

the frequency range where only plane waves propagate the control actuator will generate evanescent higher order modes. Thus, the control system must be configured so that all evanescent modes have effectively decayed by the time they reach the error sensor and/or the reference sensors.

(ii) Active Control of Free-Field Radiation. Active control has been investigated for applications such as radiation from transformers and even as part of noise barriers for highways. For these applications, the control configuration significantly impacts the amount of control that can be achieved and whether the control is local or global. The control source configuration must be carefully selected for these applications if control in a desired direction or even global control is to be achieved.

(iii) Active Control in Enclosures. The most successful application currently is for active headsets. These implement control in a small confined volume surrounding the ears and a number of active headsets are commercially available. Other applications include active control in automobile cabins, aircraft fuselages, other vehicles, and rooms. The active control will be more effective at lower frequencies and is dependent to a large extent on the modal density in the enclosure. In a number of applications, local control is achieved, although if the modal density is low, it can be possible to achieve global control, or at least control extended over a much broader portion of the volume. In general, if global control is desired, the number of control actuators used must be at least as great as the number of modes to be controlled.

(iv) Active Vibration Isolation Mounts. This approach uses active vibration control to minimize the transmission of vibration energy through isolation mounts associated with engines, generators, and so forth. Active mounts have been investigated for automobile engine mounts and aircraft engine mounts, among others. Depending on the mount configuration, it may be necessary to control multiple degrees of freedom in the mount in order to achieve the desired isolation, and it may be necessary to use active mounts on most, if not all, of the engine mounts.

(v) Active Control of Transmission Loss. This approach is focused on increasing the transmission loss through a partition, such as an aircraft fuselage or a partition in a building. There are multiple possible approaches, including controlling the structural response of the partition using structural actuators or directly controlling the acoustic field (on either the source or receiver side) through the use of acoustic actuators. For these applications, a thorough understanding of the physics associated with the structural response of the partition and its coupling with the incident and transmitted acoustic fields is essential in developing an effective solution.

With all of these applications, it is important to do a careful analysis of the noise reduction requirements in order to assess whether active noise control is an appropriate solution. Several applications using active noise control are currently commercially available and others are nearing commercialization. Nonetheless, if one does not carefully consider the application, it is easy to be disappointed in active noise control when it is not as effective as hoped. In review, active noise control is better suited for low frequencies. It is easier to achieve success for tonal noise than it is for broadband noise. It will generally be more effective for compact noise sources than for complex extended sources. If these criteria are met, active noise control could be a very effective and viable solution, although currently it would still generally require the involvement of someone knowledgeable in the field.

11   ARCHITECTURAL ACOUSTICS

This section defines the four principal, physical measures used to determine performance hall listening quality. Several perceptual attributes correlated with a physical measure are also listed.

The four main measures used to qualify concert halls are the binaural quality index (BQI), the early decay time (EDT), the strength (G), and the initial time delay gap (ITDG). The BQI is defined as BQI = 1 − IACC, where IACC is the average of the interaural cross correlation in octave bands of 500, 1000, and 2000 Hz. The BQI and IACC are measures of the relative sound reached at each ear. If a listener's two ears receive identical reflections such as from a ceiling, floor or back wall, the BQI would equal zero and the IACC would equal unity. If reflection at both ears are received from side walls or such that the reflections at both ears are not identical the BQI would be greater than zero and the IACC would be less than unity. BQI values of 0.65–0.71 are representative of the best concert halls.

The second measure is EDT. This is a measure of the time required for a 10-dB decay to occur in the signal. This time is then multiplied by a factor of 6 that provides an extrapolated comparison to a 60-dB decay time that is a similar measure of the reverberation time ($\frac{T}{60}$). Because of the 10-dB decay, the abbreviation of EDT10 is often used. EDT typically has a linear relationship in frequency between occupied and unoccupied halls. This can simplify the process of gathering data. The better concert halls have EDT values in the range from 1.7 to 2.1 sec.

The third measure is G, defined as $G = L_p - L_w + 31$ dB, where L_p is the sound pressure level measured at the point of interest and L_w is the power level of the source. Typically, L_p will decrease as the room volume and room absorption increase. However, if the reverberation time (RT) also increases with room volume, the L_p can be held constant. The better concert halls have G values that range from 3–6 dB while relatively large concert halls with less sound quality have G values in the range of 0–3 dB.

The fourth measure is the ITDG, which is defined as the time interval in milliseconds from the direct sound to the first reflected sound. ITDG values are functions of walls, balconies, and other obstacles which provide a reflective surface for the sound. ITDG values should not exceed 35 msec for best results.

There are also several perceptual attributes which are used to describe concert halls. These attributes are listed and defined by providing a measure in Table 7.

Table 7   Perceptual Attributes, Physical Measures, and Optimal Values for Concert Halls

Perceptual Attribute	Physical Measure	Optimal Values
Spaciousness	Binaural quality index, lateral fraction	BQI > 0.64
Reverberance	Early decay time	EDT and RT depend on type of music
Dynamic loudness	Strength	G > 3 dB, low background level
Intimacy	Initial time delay gap	ITDG (15–35 msec)
Clarity	Early/late ratio	Large early/late ratio for speech, depends on music
Envelopment	Highly diffuse, reverberant sound	Similarity of EDT and RT
Warmth	Frequency dependence of EDT	$EDT_{low} > EDT_{high}$
Ensemble	Early (15–35-msec) stage reflections, at frequencies above 500 Hz	Stage with ample reflecting surfaces

This table was produced from class notes provided by William Strong at Brigham Young University.[17] It is important to note that halls that provide good speech intelligibility are not necessarily the best halls for music.

12 COMMUNITY AND ENVIRONMENTAL NOISE

Measurement and analysis of the impact of noise on both individuals and communities represent a major subfield within acoustics. They are also topics inherently fraught with debate, because human perception of noise is ultimately a subjective phenomenon. To introduce this section, some of the basic principles of outdoor sound propagation are summarized. This is done to demonstrate the impact the propagation environment can have on the noise at a receiver.

12.1 Outdoor Sound Propagation

Many phenomena can affect the propagation of noise from source to receiver and therefore have a direct impact on community noise issues. Some of these phenomena are:

- Geometric spreading
- Atmospheric absorption
- Ground effect
- Refraction
- Atmospheric turbulence
- Barriers

The effects that each of these can have on sound pressure level are now reviewed.

Geometric spreading for distances much larger than the characteristic dimensions of the source will be spherical. For every doubling of distance, ΔL_p will be -6 dB. There are situations, such as supersonic aircraft or steady traffic near a roadway, where the spreading will be cylindrical, which reduces ΔL_p to -3 dB for every doubling of distance.

Because of its frequency-dependent nature, absorption causes high-frequency energy to decay much more rapidly than low-frequency energy. In addition to overall level, absorption can play an important role in changing spectral shape and therefore community response to the noise.

Another effect is that of reflections of nonplanar waves off a finite-impedance ground. If we consider the basic setup in Fig. 25, where the distance from the source to the receiver is r_1 and the distance from the image source to the receiver is r_2, the complex pressure amplitude at the receiver may be expressed as

$$p = \frac{Ae^{-jkr_1}}{r_1} + Q\frac{Ae^{-jkr_2}}{r_2} \qquad (81)$$

where the quantity Q is the spherical wave reflection coefficient and may be calculated from Appendix D.4

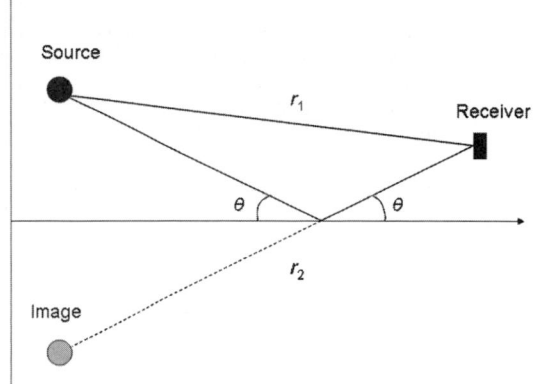

Fig. 25 Direct and ground-reflected paths from source to receiver.

in (Ref. 18). This coefficient is generally complex and accounts for the amplitude and phase changes encountered when the nonplanar sound wave reflects off the finite-impedance ground.

An example of the significant effect that the impedance of the ground can have is shown in Fig. 26, where the impedance values chosen are representative of grass, gravel, and asphalt. In this example, for which one-third octave bands are displayed, the source and receiver are both at a height of 6 ft (1.8 m) and are separated by a distance of 500 ft (152 m). At low frequencies, the wavelengths of sound are such that the direct and reflected sound waves arrive in phase, resulting in constructive interference and a doubling of pressure ($+6$ dB ΔL_p). The first interference null varies significantly in frequency for the three surfaces. Note that asphalt begins to approximate a rigid ground surface.

Atmospheric refraction can also have a significant impact on the propagation of sound from source to receiver. Refraction is caused by variations in sound speed. The first cause of a variable sound speed is wind. For sound propagation upwind, upward refraction occurs. For sound propagation downwind, downward refraction occurs, as illustrated in Fig. 27. The second cause of sound speed variation is a temperature gradient. During the day, solar radiation causes the ground to warm up and a temperature lapse to occur, meaning that the temperature decreases as a function of height. This condition causes upward refraction to occur as the sound waves bend toward where the sound speed is slower and can create a "shadow region" near the ground where the sound (theoretically) does not reach. At night, however, temperature inversions can occur as the ground cools more quickly than the surrounding air. In this case, downward refraction occurs. This also occurs at the surface of a body of water, where the air temperature just above the water is cooler than the surrounding air. This condition makes

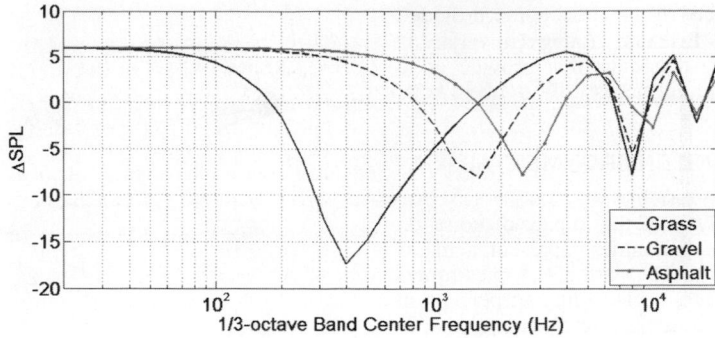

Fig. 26 Change in sound pressure level relative to free-field propagation due to ground reflections. The source and receiver are both at a height of 6 ft and the distance between them is 500 ft.

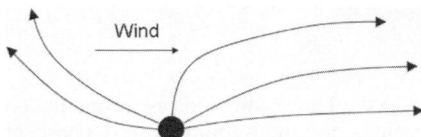

Fig. 27 Effect of wind on direction of sound rays radiating from a source.

sound propagated over large distances near the ground more readily audible and can greatly impact community noise issues.

Atmospheric turbulence may be viewed as small-scale refraction. Small-scale inhomogeneities in temperature or air velocity can cause sound to be scattered (diffracted). The main effect of turbulence is to generally lessen the impact of other propagation effects. For example, in Fig. 28, ΔL_p has been calculated for propagation over grass with and without atmospheric turbulence. Turbulence minimizes the interference nulls at high frequencies because the scattered sound takes slightly different paths to the receiver. The effect for a refracting atmosphere is similar. Although the shadow zone can readily occur near the ground for upward refraction, turbulence causes some sound to be scattered into the region.

The final phenomenon in outdoor sound propagation that is discussed is the behavior of acoustic waves when a barrier is encountered. For the case of natural barriers, such as hills, acoustic propagation over a hill can often be treated as propagation through an upward-refracting atmosphere over a flat plane. For the case of man-made barriers, such as sound walls, analytical methods may be used to account for the sound that reaches a receiver. If the length of the barrier is much greater than the height, there are four basic paths that need to be accounted for, which are depicted in Fig. 29. The first path is a direct path from the source to barrier and then, due to diffraction, from the top of the barrier to the receiver. The other paths involve one or two ground reflections before reaching the receiver.

Although more sophisticated analytical methods exist (e.g., Ref. 18), the basic effects of the multipath problem can be included in the following equation,[19] which describes the insertion loss, in decibels, of a thin barrier for a point source and for ranges less than 100 m, where atmospheric effects are ignored:

$$IL = -\Delta SPL = 10 \log_{10}[3 + 10N] - A_{ground} \quad (82)$$

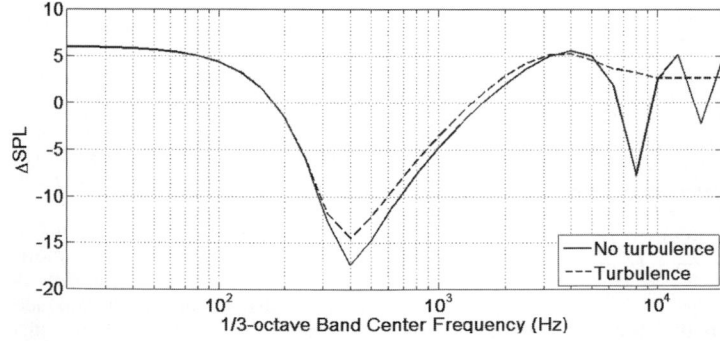

Fig. 28 Change in sound pressure level for propagation over grass with and without atmospheric turbulence.

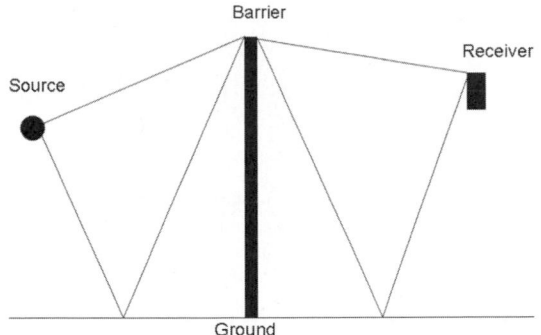

Fig. 29 Four different paths that the barrier-diffracted sound can take between the source and receiver.

where A_{ground} is the absorption due to the ground (in decibels) before the barrier is installed, JPL stands for sound pressure level, and N is the Fresnel number, which for sound of wavelength λ can be calculated as

$$N = \frac{2}{\lambda}(d_{\text{SB}} + d_{\text{BR}} - d_{\text{SR}}) \qquad (83)$$

The distances are the distances from the source to the top of the barrier (SB), from the barrier to the receiver (BR), and the direct line-of-sight distance from the source to the receiver (SR). One point to make is that although the barrier effectiveness does generally increase as a function of frequency, the diffraction from the top of the barrier can play a significant role and result in diminished performance. This is especially true for pathlengths for which the interference is constructive. More general analytical techniques, applicable to thick barriers or diffraction due to gradual structures like hills, do exist and can be found in Refs. 14 and 18. However, explicit inclusion of atmospheric effects (e.g., refraction) is usually accomplished with numerical models.

12.2 Representations of Community Noise Data

There are numerous ways of representing the noise to which communities are subjected. One way is to simply display the A-weighted sound pressure level (L_A) as a function of time. One example, displayed in Fig. 30, is the emptying of several trash dumpsters during the early morning at an apartment complex in Provo, Utah. Before the garbage truck arrived, major noise sources were due to intermittent traffic from the nearby street. The most significant noise events were due to the dumpsters being shaken by the hydraulic arms on the truck before being noisily set back down.

Other representations of community noise are statistical in nature. Using the same garbage truck example, the estimated probability density function of the A-weighted level is displayed in Fig. 31. The broad tail of large values is caused primarily by the garbage truck noise events. Another statistical representation is a cumulative distribution, which displays the percentage of time that the noise levels exceed a given value. This is shown for the same garbage truck data in Fig. 32. Finally, statistical moments can also be calculated from the time series. For example, the mean level during the 15-min sampling period was 62.4 dBA and the skewness of the data was 1.1. This latter moment emphasizes the non-Gaussian characteristics of the noise distribution because skewness is zero for Gaussian distributions.

In addition to A-weighted sound pressure level, there are many other single-number metrics that are used to describe community noise. These metrics have been the result of attempts to correlate subjective response with objective, albeit empirical measures. Some of the commonly used metrics are as follows:

- Equivalent Continuous Sound Level (L_{eq}). The A-weighted level of the steady sound that has the same time-averaged energy as the noise event. Common averaging times include hourly levels, day levels (7 AM–10 PM), evening levels (7–10 PM), and night levels (10 PM–7 AM).

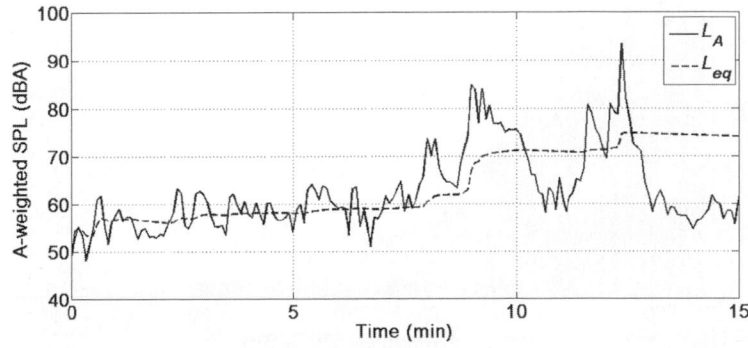

Fig. 30 A-weighted L_p and L_{eq} as function of time before, during, and after garbage truck arrival.

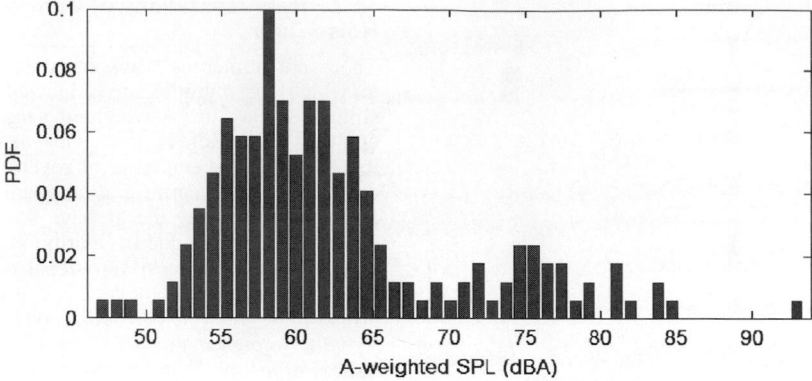

Fig. 31 Probability density function (PDF) of L_p for time series shown in Fig. 30.

For the time interval T which runs between T_1 and T_2, L_{eq} is calculated as

$$L_{eq} = 10 \, \log_{10} \left\{ \frac{1}{4 \times 10^{-10}} \frac{1}{T} \int_{T_1}^{T_2} p_A^2(t) \, dt \right\} \tag{84}$$

where p_A is the instantaneous A-weighted sound pressure. The L_{eq} as a function of time was shown for the previous garbage truck example in Fig. 30.

- Day–Night Level (DNL or L_{dn}). The L_{eq} obtained for a 24-hr period after a 10-dBA penalty is added to the night levels (10 PM–7 AM). For individual L_{eq} calculations carried out over 1-hr intervals (L_{1h}), L_{dn} may be expressed as

$$L_{dn} = 10 \, \log_{10} \left\{ \frac{1}{24} \left[\sum_{i=0100}^{0700} 10^{0.1[L_{1h}(i)+10]} \right. \right.$$

$$\left. \left. + \sum_{i=0800}^{2200} 10^{0.1 L_{1h}(i)} + \sum_{i=2300}^{2400} 10^{0.1[L_{1h}(i)+10]} \right] \right\} \tag{85}$$

- Community Noise Equivalent Level (CNEL). The L_{eq} obtained for a 24-hr period after 5 dBA is added to the evening levels (7–10 PM) and 10 dBA is added to the night levels. It can be calculated similar to L_{dn}, with the appropriate penalty given during the evening (between 2000 and 2200 hours).

- X-Percentile-Exceeded Sound Level (L_X). Readily calculated from the cumulative distribution (e.g., see Fig. 32), L_X is the level exceed X percent of the time. Common values are L_{10}, L_{50}, and L_{90}. In Fig. 33, L_x values are shown as bars for L_{99}, L_{10}, L_{50}, L_{90}, and L_1 for representative noise environments.

- C-Weighted Sound Pressure Level (L_C). Similar to A weighting, but designed to mimic the

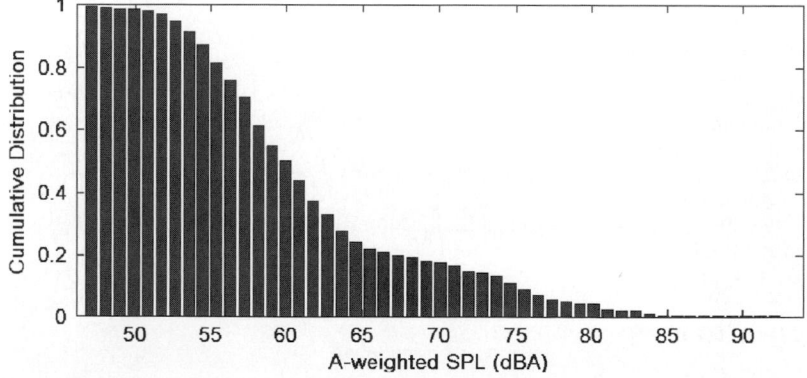

Fig. 32 Cumulative distribution of time series in Fig. 30 showing fraction of time sound level exceeds given L_p.

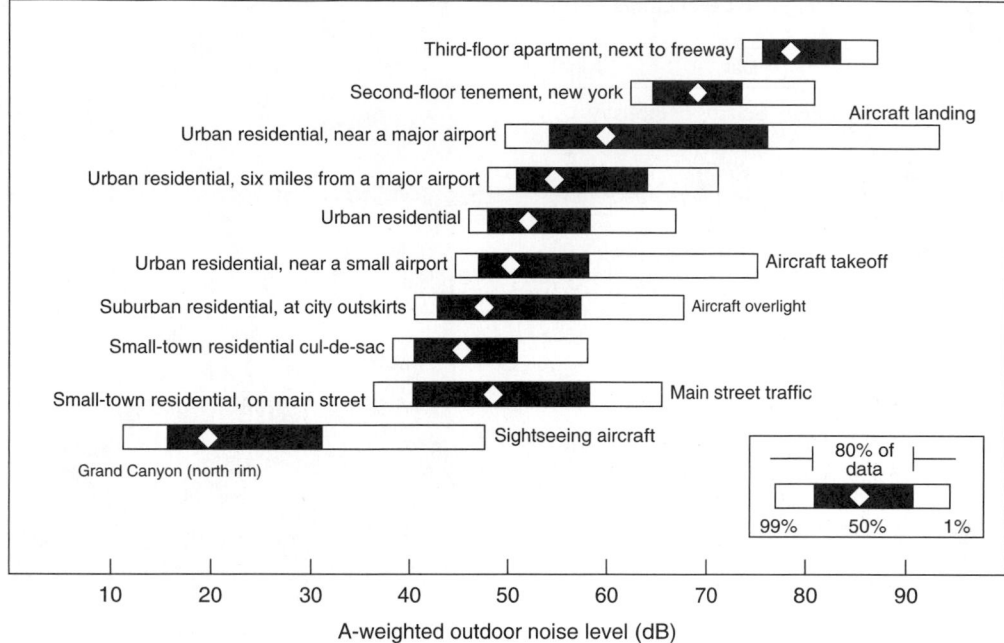

Fig. 33 Community noise data showing L_{99}, L_{90}, L_{50}, L_{10}, and L_1 data points for various noise events.

90-phon equal-loudness contour. Consequently, C weighting is more appropriate than A weighting for louder sounds. The equation for the C-weighting filter was given previously in Eq. (7).

- D-Weighted Sound Pressure Level (L_D). Developed for assessing the auditory impact of aircraft noise. The weighting curve heavily penalizes high frequencies to which the ear is most sensitive (see Fig. 3). The equation for the D-weighting filter was given previously in Eq. (7).

- Effective Perceived Noise Level (EPNL). This metric was designed for characterizing aircraft noise impact and is used by the Federal Aviation Administration (FAA; see FAR Part 36, Sec. A.36) in the certification of commercial aircraft. The metric accounts for (a) the nonuniform response of the human ear as a function of frequency (i.e., the perceived noise level), (b) the additional annoyance due to significant tonal components of the spectrum (the tone-corrected perceived noise level), and (c) the change in perceived noisiness due to the duration of the flyover event. Too involved to be repeated here, calculation procedures for EPNL may be found in FAR Part 36, Sec. A.36.4, or Ref. 20.

12.3 Community Noise Criteria

Because of increased awareness regarding community noise issues, city noise ordinances are becoming more commonplace. Many of these ordinances are based on maximum allowable A-weighted sound pressure level, broken down into land usage and day or night. In addition, consideration can be given to the nature of the noise source (e.g., is it essential to commerce/industry) and its duration (e.g., is it intermittent or continuous). As an example, portions of the Provo, Utah, noise ordinance, which is representative of many cities, are summarized in Table 8. Continuous sounds are those that have a duration greater than 6 min, intermittent sounds last between 2 sec and 6 min, and impulse sounds last less than 2 sec. The level listed is not to be exceeded at the property line of interest.

Noise levels are an important consideration when considering land use. Guidelines for outdoor DNL (for structures in 24-hr/day use) or L_{eq} (for structures being used only part of the day) have been put forth in a land use compatibility report published by the FAA. If the appropriately measured outdoor levels for a yearly average are <65 dB, then the land is compatible for all the uses listed below. In Table 9, if a number is listed, then the guidelines indicate that the transmission loss between outdoors and indoors needs to meet the value listed.

In the case of airports, DNL contour maps showing community noise impact are used to make land usage decisions. Shown in Fig. 34 is an example of a DNL map for the Colorado Springs Airport superimposed on a photograph of the area. The figure shows appropriate land usage in that urban development has occurred

Table 8   Summary of A-Weighted Level Limits (dBA) in Provo, Utah

District	Day	Night
Continuous and Intermittent Sounds of Industry and Commerce		
Residential/agricultural	85	55
Commercial	85	65
Industrial	85	85
Continuous Public Disturbances		
Residential/agricultural	65	55
Commercial	70	65
Industrial	75	75
Intermittent Public Disturbances		
Residential/agricultural	70	60
Commercial	75	65
Industrial	80	80
Impulse Noises		
Residential/agricultural	75	60
Commercial	80	65
Industrial	85	85

Table 9   Land Use Compatibility

Structure	Outdoor DNL or L_{eq} (dB)				
	65–70	70–75	75–80	80–85	<85
Home	25	30	N	N	N
Church, school, hospital, library	25	30	N	N	N
Business offices, retail, restaurant	Y	25	30	N	N
Manufacturing	Y	25[a]	30[a]	35[a]	N
Agriculture	Y	Y	Y[NB]	Y[NB]	Y[NB]
Livestock breeding	Y	Y	N	N	N
Mining	Y	Y	Y	Y	Y
Nature exhibits/zoos	Y	N	N	N	N
Amusement parks	Y	Y	N	N	N
Outdoor music, amphitheaters	N	N	N	N	N
Outdoor sports arenas	Y[SR]	Y[SR]	N	N	N

Source: Adapted from FAA AC 150/2050-1, "Noise Control and Compatibility Planning for Airports."
Note: NB denotes residential buildings not permitted and SR means sound reinforcement may be required.
[a]Applies to areas sensitive to noise.

outside the 65-dB DNL contour. The land usage that occurs closest to the airport is not a residential district, but rather an Air Force base.

12.4   Community Response to Noise

The response of an individual to a given noise event is a highly subjective, variable measure. Noise surveys

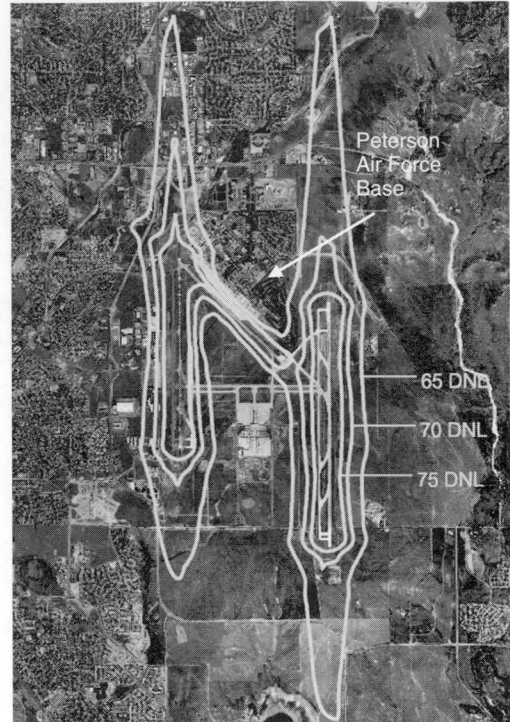

Fig. 34 DNL contour map for Colorado Springs Airport. Urban development has occurred outside the 65-dB DNL contour.

are conducted to obtain the average response of citizens to different noise sources. For transportation noise (e.g., traffic, aircraft), these surveys have resulted in an empirical relationship between transportation noise events and the percentage of people that were highly annoyed.[21] If this percentage is defined as %HA, it may be expressed in terms of the DNL (L_{dn}) as

$$\%HA = 0.0360L_{dn}^2 - 3.265L_{dn} + 78.9 \qquad (86)$$

The uncertainty for this measure is approximately ±5 dBA between 45 and 85 dBA. The percent highly annoyed is plotted as a function of DNL in Fig. 35. Note that for 65-dB DNL 20% of the population will still be highly annoyed. This curve suggests the difficulty that noise control engineers and land use planners encounter when trying to balance the needs and desires of the community and industry.

13   SOUND QUALITY ANALYSIS

Good sound quality has traditionally been synonymous with quiet sounds. However, over the last three decades, the definition of sound quality has changed.

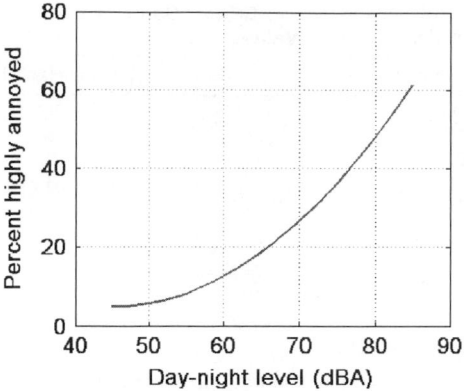

Fig. 35 Percentage of people highly annoyed for given DNL. Calculated from Eq. (86).

Blauert and JeKosch[22] define sound quality as "the adequacy of a sound in the context of a specific technical goal and/or task." The term "compatibility" has also been used in this context, especially with regard to sounds accompanying actions of product users.[23] With this definition in mind, sound quality now represents the "sensory pleasantness" of the sound, which is a combination of the perceived loudness, roughness, and pitch.[24] The difference in definition contrasts the one-dimensional approach initially used to the current multifaceted technique that includes aspects of psychology and anatomy as well as the physical parameters engineers are accustomed to using.

While much of what is now considered sound quality analysis originated in the automotive industry, there are a myriad of applications in nearly all consumer products and industrial processes. Sound quality analysis has generally focused on many automotive aspects, including exterior vehicle passing noise,[25] holistic vehicle sounds and research,[26,27] engine noise, and exhaust noise.[28,29] Sound quality analysis techniques have also been used to improve the acoustic signatures of everything from helicopter main rotor blades and aircraft interior noise[30,31] to hairdryers and vacuum cleaners.[32,33] Computer hard drives, HVAC systems in buildings,[34] construction machinery, and boxes at Italian opera houses[35] have also been improved through psychoacoustic studies.

As sound quality is dependent on the opinions of consumers, a common and effective approach is to form listening juries and survey their responses to different sounds. The primary reason for this method is the disparity between what a sound level meter measures and what an individual reports regarding the loudness of a given sound as well as the instrument's inability to describe other parameters of sound that a consumer hears and recognizes. This inconsistency results from the nonlinear nature of the human auditory system and the inability of traditional sound measurement techniques to approximate this behavior. A significant drawback to this technique is the requirement of large groups of research subjects to perform the listening tests. However, approximations using empirical equations have allowed the calculation of metrics that closely approximate the human subject responses.

These metrics are grouped into four categories: loudness, sharpness, grating, and tonality. Loudness characterizes the response of the cochlea given the sound's frequency, bandwidth, and amplitude in decibels. Sharpness is a metric that describes the amount of high-frequency energy present in a sound. It is a weighted average, or area moment, of the loudness with more weight given to frequencies above ~3000 Hz. Grating describes two metrics: fluctuation strength and roughness. Fluctuation strength is a metric which characterizes amplitude modulation of the sound that is partially to entirely audibly perceptible. The modulation frequencies that cause the sensation of fluctuation strength are frequencies up to ~20 Hz. Roughness represents amplitude modulation of the sound that is too rapid to be even partially audibly perceptible; that is, it is perceived that the sound is modulated but the degree of modulation is unclear to the human ear. Frequencies of modulation that cause a sensation of roughness fall between ~20 and 300 Hz. Tonality characterizes the tonal presence in a sound. The question it answers is, "Are there distinct tones present in a seemingly broadband sound? If so, how strong are they?" These five metrics can then be combined into a relative-pleasantness measure when comparing two or more sounds. This pleasantness metric, called sensory pleasantness, is convenient as it consolidates these five independent metrics into a single response in the same way a juror does subconsciously.

13.1 Mathematical Metrics Procedures

This section continues by presenting an overview of the procedures developed for using mathematical metrics to assess sound quality. The complexity of using humans as instruments for sound quality evaluation has inspired several attempts to quantify human responses, hence creating parameters representing the acts of auditory perception, cognition, and judgment. These parameters can be measured using traditional sound analysis instruments and are quantified based on empirical equations to represent a juror's response regarding the quality of a sound. The process consists of using standard microphone recordings in either a free field (with frontal incidence from the noise source) or a binaural field (microphones set up in a dummy head to approximate how the ears receive sound). The measured sound is then processed mathematically to calculate the sound quality metrics. These results can then be compared to jury listener tests and correlated for the given sound source. The result is a process that can accurately predict the sensory pleasantness of a sound with a minimum of human subjects, allowing sound quality analysis to be performed earlier in the product development and often with fewer jurors during the product life cycle. To better understand the mathematical metric method, a few common terms and each of the sound quality metrics will be briefly reviewed (see Refs. 36–38 for more details).

13.2 Critical Band Rate

The human auditory systems response is dependent on both the frequency and amplitude of a given sound. Therefore, it is useful to alter the frequency scaling used in sound quality metric calculations. Characteristic frequency bands are defined as bandwidths that produce the same perceived loudness in a tone and in narrowband noise spectrum within that band when the tone is just masked. This scale of critical band rates is used to approximate the frequency dependence of each of the metrics, as it represents the bandpass nature of the auditory system. These critical band rates are given the name barks, named after Heinrich Barkhausen, an acoustician whose pioneering work in loudness approximations has produced two subjective-loudness amplitude scales. Figure 36 and Table 10 illustrate the relationship between barks (critical band rate) and center frequencies in hertz. All of the metrics will use critical band rate as the abscissa in all of their plots and calculations.

The following sections give derivations as to the method for calculating each metric as well as examples to help clarify the necessity for each of the individual metrics. The overall combination metric is called sensory pleasantness. Sensory pleasantness describes the holistic result similar to the response a juror would give when evaluating a sound.

Table 10   Frequency to Critical Band Rate for Selected Frequency Values

Frequency (Hz)	Critical Band Rate
100	1
510	5
1,000	8.5
2,000	13
4,800	18.5
10,500	22.5
15,500	24

13.3 Loudness

The loudness of a sound is a perceptual measure of the effect of the energy content of sound on the ear. It is related to but not the same as the sound pressure level (L_p), a logarithmic scale used to quantify the pressure amplitude of a sound. Loudness, however, is also dependent on the frequency content of a sound. For example, a low-frequency sound such as a 100-Hz tone at 40 dB would be perceived to be significantly quieter to a normal-hearing person than a 1-kHz tone at 40 dB; in fact, the 1-kHz tone would sound nearly four times as loud.

The fundamental assumption of this model of loudness is that it is not a product of spectral lines or obtained from the spectral distribution of the sound directly but that the total loudness is the sum of the specific loudness from each critical band. Empirical studies have shown that this process yields the best equivalent psychoacoustic values.

Hence, using the threshold of quiet as a base or minimum level of excitation, the specific loudness (N')

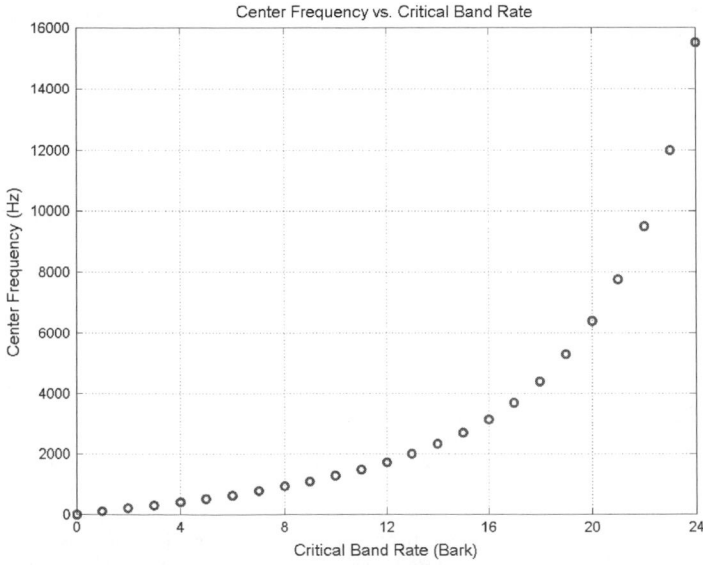

Fig. 36   Frequency versus critical band rate.

of a sound can be determined as

$$N' = N_0' \left(\frac{E_{TQ}}{s \cdot E_0} \right)^k \left[\left(1 + \frac{s \cdot E}{E_{TQ}} \right)^k - 1 \right] \quad (87)$$

where E_{TQ} is the excitation level at the threshold of quiet, E_0 is the excitation level that corresponds to the reference intensity $I_0 = 10^{-12}$ W/m², and N_0 is a reference-specific loudness. The variable s, which is the ratio between the intensity of a just audible test tone and the intensity of broadband noise appearing within the same critical band as the test tone, is determined experimentally. The exponent k is also determined experimentally.

The result of numerous jury tests using specific tones and broadband noise yields an approximation for the specific loudness in each critical band formulated by Zwicker and Fastl[36] and presented here as

$$N' = 0.08 \left(\frac{E_{TQ}}{E_0} \right)^{0.23} \left[\left(0.5 + \frac{E}{2E_{TQ}} \right)^{0.23} - 1 \right]$$
$$\times \text{ sone/bark} \quad (88)$$

The calculation of the total loudness is then a summation of all of the specific loudnesses across all of the critical bands, as shown in the equation

$$N = \sum_{z=1}^{24 \text{ bark}} N_z' \quad (89)$$

This results in a mathematical representation of the sensation of loudness in the human auditory system. This model accounts for the effects of masking as well as the nonlinear relationship between loudness and frequency. Extensive empirical studies have verified the accuracy of this model and provided a foundation for its use in many applications.

The process for calculating the loudness of a given sound is as follows. Assuming this is being done digitally in a computer, the first step is to convert the amplitude from volts to a calibrated decibel value. The array of sound pressure levels is subsequently multiplied by a series of 24 one-third-octave filters to decompose the spectral response into the critical bands. The only unknown in the above equations is the excitation level (E) of the sound in a particular critical band, where E is approximated using the sound pressure level. The result is an array that can be interpreted to represent the excitation level of the cochlear duct over the audible spectrum. This array is then summed to give a value representing the total loudness that a listener would perceive when exposed to this sound.

13.4 Sharpness

Sharpness is a measure of the high-frequency energy content of a sound. Sharpness is therefore very similar to a weighted loudness ratio, with emphasis on the high-frequency sounds. A plot of the weighting function versus critical band rate is shown in Fig. 37. The unit of measure for sharpness is the *acum*, which is the Latin word for sharp.

This metric is normalized to the sound pressure level and frequency scales by the following relationship; 1 acum corresponds to a narrowband noise with a bandwidth of one critical band bandwidth wide at a center frequency of 1 kHz with a level of 60 dB. The model for sharpness is simply a weighted first moment of the critical band rate distribution of specific loudness, as seen in Eq. (90), where $g(z)$ is the weighting

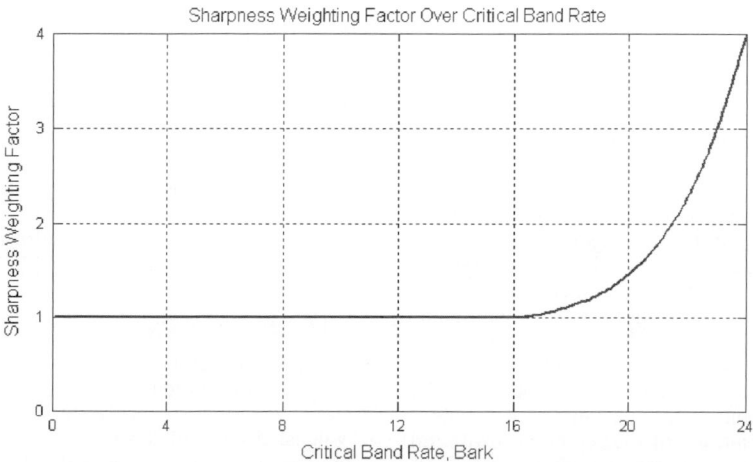

Fig. 37 Sharpness weighting factor versus critical band rate.

function given in Fig. 37:

$$S = 0.11 \frac{\int_0^{24 \text{ bark}} N_0' g(z) z \, dz}{\int_0^{24 \text{bark}} N' dz} \quad (90)$$

This model of sharpness takes into account that the sharpness of a narrowband noise increases significantly at high center frequencies. Although the model is relatively simple, empirical data show that the agreement between the model and test subject responses is very good.

13.5 Fluctuation Strength

Modulated sounds cause two different kinds of hearing sensations. At low modulation frequencies, fluctuation strength is produced. At higher modulation frequencies, a sensation of roughness is produced. The sensation of fluctuation strength occurs for modulation frequencies between 0 and about 20 Hz. This metric represents the modulation of the sound that is strongly audible.

This leads to a relationship for fluctuation strength. Empirical studies have shown that the total fluctuation for a fluctuating sound is approximated by Zwicker's model, given here as

$$F = \frac{0.008 \cdot \int_0^{24 \text{ bark}} \left[4 \log \left(N_{\text{max}}' / N_{\text{min}}' \right) \text{ dB (dB/bark)} \right] dz}{f_{\text{mod}}/4 \text{Hz} + 4Hz/f_{\text{mod}}} \quad (91)$$

where f_{mod} is the modulation frequency, which is determined for a given sound by examining its spectral content and evaluating what frequency is modulating the sound. The unit for fluctuation strength is 'vacil', which is Latin for oscillate. The relationship describing vacils is that 1 vacil corresponds to a 60-dB, 1-kHz tone that is 100% amplitude modulated at a frequency of 4 Hz.

13.6 Roughness

Roughness is similar to fluctuation strength. However, it quantifies the subjective perception of rapid (15–300-Hz) amplitude modulation of a sound. This metric is often considered to be the property of a sound that would be described as grating, as the temporal masking effects of the human auditory system do not allow the total recognition of changes in amplitude at these frequencies. Roughness, like fluctuation strength, is therefore influenced by two main factors: frequency resolution and temporal resolution of the human auditory system.

For very low frequencies of modulation, roughness is also small. For midfrequencies, around 70 Hz, roughness is at its maximum. At very high frequencies, due to the restrictions of the temporal resolution of the human auditory system, roughness falls off. As the value of ΔL is also dependent on the critical

band rate, a more accurate approximation of roughness would include this dependence, which leads to a fairly accurate proportionality, given as

$$R \sim f_{\text{mod}} \int_0^{24 \text{ bark}} \Delta L(z) \, dz \quad (92)$$

Using the boundary conditions of one *asper* (a Latin word meaning rough, which will be the unit for this metric), corresponding to a 60-dB, 1-kHz tone that is 100% amplitude modulated at a frequency of 70 Hz, Zwicker and Fastl[36] propose the following model of roughness, again based on significant empirical data:

$$R = 0.3 \frac{f_{\text{mod}}}{1\text{kHz}} \int_0^{24 \text{ bark}} \frac{20 \log \left(N_{\text{max}}' / N_{\text{min}}' \right) \text{ dB}}{\text{dB/bark}} dz \quad (93)$$

where z is the critical band rate variable and the parameter dB/bark is simply a unit conversion factor. Similar to the rest of the metrics, roughness is a summation of effects across all of the critical bands.

13.7 Tonality

Tonality is concerned with the tonal prominence of a sound (i.e., are there tones present/absent in the sound). Several models of tonality have been developed, each with unique strengths and weaknesses. However, there is one common weakness with each of the currently accepted models of tonality. As the noise floor around the tone is increased in bandwidth to values greater than 100 Hz, the model for describing the tonality falls off at such a rate that it becomes inaccurate. Tonality can also be estimated using fast Fourier transform (FFT) analysis, where a time-aliased (frequency-averaged) FFT versus a non-time-aliased FFT are compared. Plotting the two FFTs in the same window, the position and amplitude of peaks can be compared versus averaging. If the peaks move, they are not tonal; however, if they remain the same, there is a tone at that frequency. The magnitude of tonality can be estimated from a comparison in amplitude of the "noise floor" versus the peak amplitude of the tone. This estimate is expressed in Eq. (94) as the ratio of amplitudes of the tone and the broadband noise. This is, however, a subjective approach to determining tonality:

$$T \approx \frac{\text{amplitude of tone}}{\text{amplitude of noise floor}} \quad (94)$$

13.8 Sensory Pleasantness

The relationship between sensory pleasantness and the sensations of loudness, sharpness, roughness, fluctuation, and tonality can be modeled based on relative values of each of the metrics. The resulting relationship proposed by Zwicker and Fastl[36] is presented in Eq. (95). It should be noted that because the influence of tonality T is small, the tonality term [the term in the parenthesis in Eq. (95)] reduces to an approximate value of 0.24:

$$\frac{P}{P_0} = e^{-[0.023(N/N_0)]^2} e^{-1.08(S/S_0)} e^{-0.7(R/R_0)}$$

$$\times \left(1.24 - e^{-2.43(T/T_0)}\right) \qquad (95)$$

where P_0, N_0, S_0, R_0, and T_0 represent the sensory pleasantness, loudness, sharpness, roughness, and tonality of the reference sound to which the current sound is being compared. In other words, the model is a model of relative sensory pleasantness measured against a benchmark sound.

One of the key factors of Eq. (95) is that sensory pleasantness is not dominated by loudness. In fact, the effect of loudness on pleasantness is only strongly evident for sounds with significantly strong loudness values. Zwicker and Fastl[36] suggest that the role of loudness in sensory pleasantness is only important when a sound has a calculated loudness greater than 15 sones.

13.9 Limitations

Each sound quality metric is based on using "curve-fit" functions designed around sets of tests using jury listening evaluation of specific tones and developing empirical equations to match the juror response in each of the five categories. They are based on assumptions that approximations for each of the immeasurable factors can be made using one-third-octave filters and weighting functions to represent the cochlear response to auditory inputs. The question then arises as to whether this method can be applied to other sounds in general as they are not specific tones well defined by their frequency and amplitude. This brings to light the importance of at least some jury testing to compare with the results from the sound quality metrics.

14 NONLINEAR ACOUSTICS

This section provides an overview of nonlinear acoustics. All the theory and applications treated in this chapter thus far have relied on the fundamental assumption put forth in the wave equation section, which was that the acoustic perturbations are very small ($|\mathbf{u}|/c_0 \ll 1$). This supposition allowed us to keep only the first-order terms in the constitutive equations. However, there are certain situations in which this small-amplitude assumption breaks down and second-order (nonlinear) phenomena result. This section is a brief introduction into nonlinear acoustics theory and applications. It concludes with the description of data analysis tools that can be used to help in characterizing nonlinear phenomena.

14.1 Theory

It is first relevant to consider the basic effects of nonlinearity. In a linear system, an amplitude and phase change may occur, as is shown in Fig. 38,

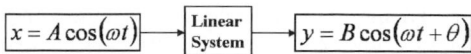

Fig. 38 Example of linear system that induces amplitude and phase change.

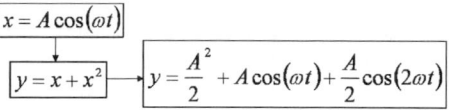

Fig. 39 Example of quadratic nonlinear system that produces dc and 2ω.

but the acoustic excitation remains at frequency ω. In a quadratically nonlinear system, however, such as $y = x + x^2$ shown in Fig. 39, the output contains energy not only at ω but also at zero frequency (dc) and at 2ω. If an input waveform that contains energy at ω_1 and at a greater frequency ω_2 is passed through $y = x + x^2$, trigonometric relationships demonstrate that the output of the nonlinear system will contain energy at dc, ω_1, ω_2, $2\omega_1$, $2\omega_2$, $\omega_1 + \omega_2$, and $\omega_2 - \omega_1$. The nonlinear generation of harmonics is called harmonic distortion and generation of sum and difference frequencies is called intermodulation distortion.

There are other classes of nonlinearities (e.g., cubic), but quadratic nonlinearities are emphasized here because they appear in the constitutive equations for sound propagation before they are neglected in first-order (i.e., linear) approximations. These nonlinearities culminate in an amplitude-dependent sound speed, expressed for one-dimensional propagation as

$$c = c_0 + \beta u \qquad (96)$$

where u is the particle velocity, c_0 is the small-signal sound speed, and β is the coefficient of nonlinearity in the fluid ($\beta = 1.201$ in air). For general fluids or biological tissues, $\beta = 1 + B/2A$, where B/A is the "parameter of nonlinearity" of the medium. Measured values of B/A in a few other media are listed in Table 11. The tabulated values reveal that most media are significantly more nonlinear than air.

The significance of Eq. (96) is that the portions of the waveforms with positive particle velocities (or pressures) travel faster than those with negative particle velocities or pressures. For $|\mathbf{u}/c_0| \ll 1$, $c \approx c_0$ and

Table 11 Parameter of Nonlinearity for Selected Media

Medium	B/A
Distilled water, 20°C	5.0
Sea water, 20°C	5.25
Liquid nitrogen, 90 K	9.00
Human liver, 30°C	7.6($\pm$0.8)

the linear approximation is recovered. If we consider an initially sinusoidal waveform that propagates as a plane wave, the waveform will eventually steepen to the point that an acoustic shock forms. This distance is known as the shock formation or discontinuity distance and may be written as

$$\overline{x} = \frac{\rho_0 c_0^2}{\beta p_{\text{init}} k} \tag{97}$$

where p_{init} is the pressure amplitude of the initial sinusoid. We can also express propagation distance in terms of $\overline{x}$ and define a dimensionless distance, $\sigma = x/\overline{x}$. The nonlinear evolution of an initially sinusoidal waveform is shown in Fig. 40 between $\sigma = 0$ and $\sigma = 3$. The time waveforms are expressed both as traditional pressure waveform traces in Fig. 40a and as longitudinal wave representations in Figs. 40b–d. At $\sigma = 1$ a discontinuity has just formed, and by $\sigma = 3$ the waves is essentially a fully developed sawtooth wave. The waveform steepening represents a transfer of energy from ω to higher harmonics that can eventually be expressed as the Fourier series of a sawtooth waveform that, in principle, contains an infinite number of harmonics to account for the discontinuous slope at the shock front.

The shock formation distance is an important parameter in determining the degree of nonlinearity of the propagation. However, another essential phenomenon to consider is the absorptive losses that occur in the propagation. The losses, which increase as a function of frequency, would "compete" with the nonlinear steepening that transfers energy to high harmonics. The importance of nonlinearity relative to absorption is expressed in the Gol'dberg number, a dimensionless quantity that can be written for an initially

sinusoidal plane wave in a thermoviscous medium as

$$\Gamma = \frac{1}{\alpha \overline{x}} \tag{98}$$

If $\Gamma \gg 1$, absorption and/or the shock formation distance is very small, which means that the propagation is highly nonlinear and that shocks will form. On the other hand, if $\Gamma \ll 1$, absorption and/or $\overline{x}$ is very large, which means that nonlinear waveform steepening is negligible. Finally, $\Gamma \sim 1$ represents an intermediate case, where steepening is likely but an acoustic shock may not develop as the higher harmonic energy is absorbed. As a final note, the above description of the Gol'dberg number is only qualitative for any acoustic waveform besides a pure sinusoid in a thermoviscous medium; however, Γ demonstrates qualitatively the processes (e.g., amplitude, frequency, absorption) that determine when nonlinearity plays a significant role in sound propagation. In all cases, as propagation occurs, the effective Gol'dberg number gradually lessens as a function of distance as acoustic amplitudes are reduced through losses and possibly geometric spreading. This will eventually cause steepened waveforms to gradually unsteepen as the rate of absorption at high frequencies begins to exceed the nonlinear energy transfer rate.

14.2   Radiation Pressure and Streaming

Although their derivations are beyond the scope of this discussion, there are physical phenomena directly related with the nonlinear energy transfer to dc. The first is called acoustic streaming, which is a mean velocity that is set up in the fluid by the acoustic fluctuations. The second phenomenon is radiation pressure, which is a time-averaged pressure exerted on an object in the fluid or on a boundary. (Acoustic radiation pressure is analogous to the radiation pressure exerted by

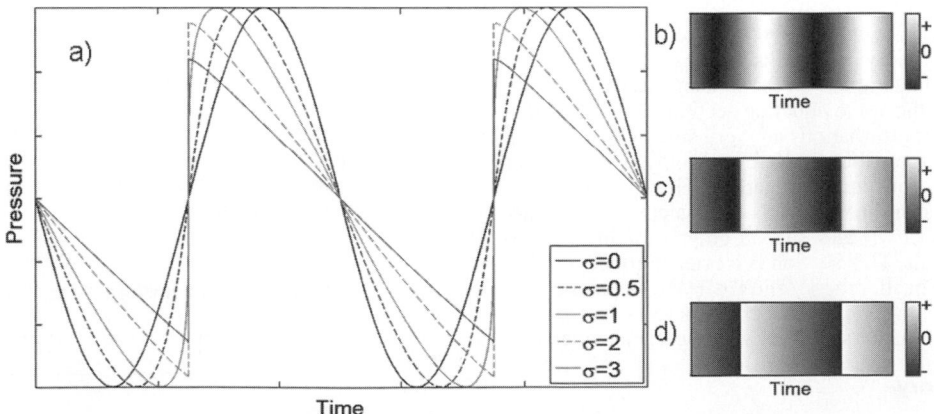

Fig. 40 (a) Nonlinear evolution of sinusoidal waveform into sawtooth wave. Longitudinal wave representation at (b) $\sigma = 0$, (c) $\sigma = 1$, and (d) $\sigma = 3$.

light waves.) One or both of these phenomena can be very relevant to applications of nonlinear acoustics, particularly in enclosed systems. For more detailed discussions, see Refs. 39 and 40.

14.3 Applications of Nonlinear Acoustics

Nonlinearity is often a factor in the atmospheric propagation of high-amplitude sources. For example, the nonlinear evolution of sonic boom waveforms is a problem that has been investigated in a variety of ways. The pressure fluctuations created by the supersonic aircraft travel at different speeds and the shocks that form often coalesce into two single shocks that form a wave called an N wave (see Fig. 41). The propagation of noise from explosions is also nonlinear, but unlike the sonic boom, there is typically a single shock front associated with the blast waveform. The last example of nonlinear atmospheric propagation is the noise radiated by military fighter jets and by rockets. The nonlinear interactions between spectral components as the shaped broadband noise waveform propagates result in a transfer of energy to high frequencies. The waveform steepening can lead to the formation of significant acoustic shocks. The excess high-frequency energy (relative to that linearly predicted) and its arrival in the form of shocks at discrete instances in time appear to have a significant impact on human perception of these sounds.[41]

Ultrasound is used in many applications of nonlinear acoustics that include the fields of audio, biomedical, and industrial acoustics. Parametric arrays have been investigated for audio applications for several decades. In its simplest form, the parametric array uses high-intensity ultrasonic transducers configured so as to produce a narrow primary beam of sound (in general, the larger the dimensions of the array, the narrower the beam). The array is driven at two closely spaced ultrasonic frequencies, ω_1 and ω_2, at high enough amplitudes to produce the difference frequency, $\omega_2 - \omega_1$. For proper selection of source frequencies, $\omega_2 - \omega_1$ will be in the audio range. Harmonics and sum frequencies are also generated, but atmospheric absorption at any of these ultrasonic frequencies is much greater than at the difference frequency. Consequently, the parametric array is unique in its ability to generate a narrow beam of low-frequency sound. Potential applications include audio complements to displays in museums, individualized stereo systems in vehicles, and other situations where spatially narrow audio is desired.

Applications of nonlinear acoustics in biomedical ultrasound are many, but a few are briefly summarized. The harmonic imaging technique in diagnostic ultrasound relies on harmonic distortion as waves propagate through human tissue (which is generally a nonlinearity-prone medium.) The ultrasonic transducer transmits a wave at ω, but then the ultrasonic sensors "listen" for the scattered signal at 2ω. The result is better spatial resolution of the tissue to be imaged. Other applications of nonlinear ultrasound include for high-intensity focused ultrasound (HIFU). Shock wave lithotripsy uses focused ultrasonic waves that nonlinearly steepen into a shock wave to break up kidney stones. Although the physical mechanisms by which this occurs are still being researched, it is believed that cavitation, or the formation and subsequent violent collapse of bubbles due to large pressure changes in a fluid, plays a significant role in breaking up of the stones. Because HIFU can lead to significant heating of internal tissue (which is ordinarily avoided), other applications of HIFU that are being investigated include ultrasonic treatment for cancer and stoppage of internal bleeding by wound cauterization.

Industrial applications of high-intensity ultrasound include ultrasonic cleaning, ultrasonic welding, and nondestructive evaluation (NDE).[43] There is a wealth

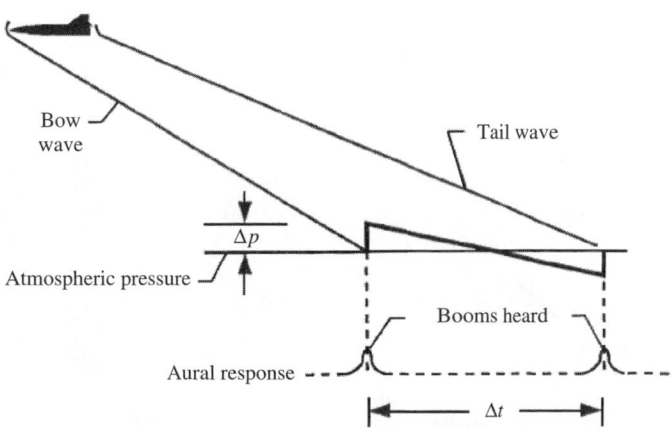

Fig. 41 N wave, characteristic of sonic booms.[42]

of information on each of these topics that can be consulted. However, due to the brevity of the discussion, general principles of operation will be given. An ultrasonic cleaner is a fluid-filled chamber that is activated using one or more ultrasonic transducers. Although all the physical mechanisms behind the cleaning are not totally understood, microjets created by cavitating bubbles are believed to be largely responsible for the cleaning. One principal advantage of ultrasonic cleaning is that milder chemicals may be used in the cleaning process than other cleaning techniques. The use of ultrasonic cleaning is widespread and can be readily found in, for example, automotive, pharmaceutical, and medical applications.

Ultrasonic welding, which has been around since the 1960s, utilizes directional, high-intensity sound energy with typical frequencies ranging from 20 to 40 kHz. Although ultrasonic welding is most extensively used in the welding of plastics, it is also used in small-scale welding of metals (e.g., wires). The principle of ultrasonic welding is different for thermoplastics and metals. In thermoplastics, as the materials are pressed together and ultrasound is applied, the acoustic energy absorbed by the materials is transformed into heat, resulting in high temperatures in a localized area. This causes localized melting of the plastic surfaces and the formation of a bond. In ultrasonic welding of metals surfaces, studies have shown that the heat generated is insufficient to melt the metals at the interface. Rather, the mechanism is due to abrasive shear forces excited at the interface between the metals that disperse oxides and other contaminants. If the exposed metal surfaces are held together with sufficient pressure, a solid-state bonding takes place. Because the temperatures involved in an ultrasonic weld are

relatively low, embrittlement of the metals, which can occur with other welding techniques, is avoided.

Finally, ultrasound is a viable method within the broad field of NDE for testing material properties. Although some ultrasonic NDE, which typically involves frequencies in the megahertz range, employ linear acoustic techniques and processing, nonlinear phenomena can be used to readily identify defects in solids. Cracks and other defects exhibit high nonlinearity such that the amount of harmonic and intermodulation distortion increases significantly. Therefore, harmonic imaging or measurements of the amount of harmonic and intermodulation distortion in the scattered wave can be used to assess the properties of the solid. Examples of ultrasonic NDE include weld inspection, pipe wall inspection, assessment of concrete strength, and even recent examinations of the Space Shuttle wings and external fuel tanks.

15 THE HUMAN EAR AND HEARING

The human auditory system is complex from the standpoint of both how the nervous system responds to auditory inputs and the physical makeup of the human ear. This section first will discuss the structure of the ear, outlining the physical aspects of how humans hear, followed by a brief discussion of the nervous system response to auditory inputs and the role of the brain in hearing.[44]

The human ear can be subdivided into three main parts, as illustrated in Fig. 42: the outer, middle, and inner ear. The outer ear consists of the pinna, which serves as a horn that collects sound and directs it into the auditory canal. The pinna is comparatively ineffective in humans. The auditory canal is essentially

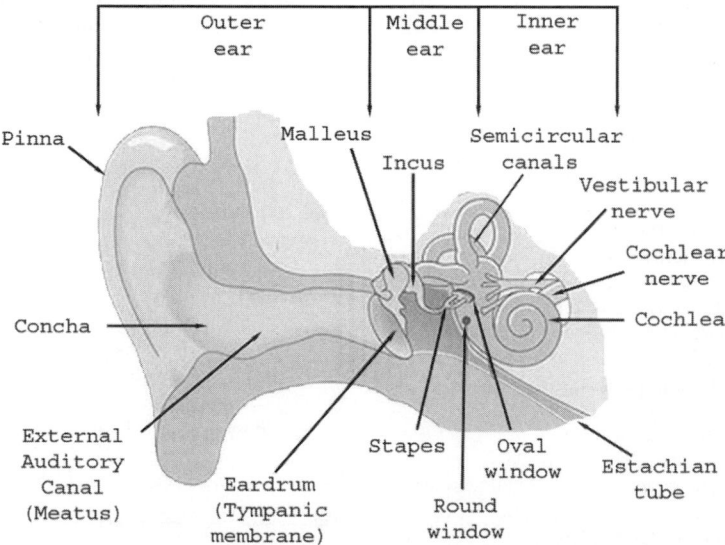

Fig. 42 Diagram of human ear.

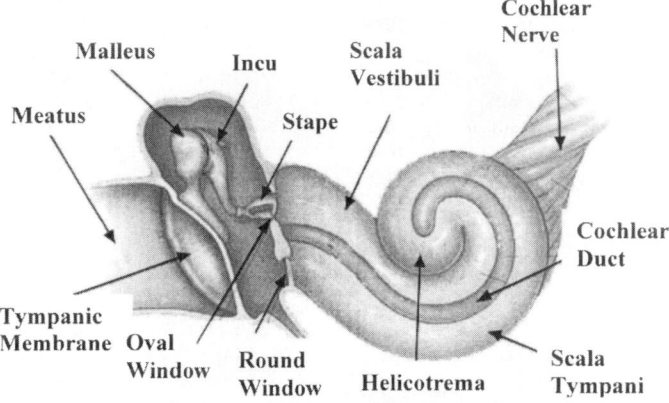

Fig. 43 Middle-to-inner-ear transition.

a straight tube about 0.7 cm in diameter and 2.5 cm long, open at one end and closed at the other, as seen in Fig. 43 in the section labeled outer ear.

There are resonances associated with this tube, the lowest resonance occuring at about 3 kHz. At this resonance frequency, the sound pressure level is about 10 dB higher at the closed end than at the open end. This is the most sensitive frequency region of the ear. Above this frequency, resonance phenomena can be observed, but resonance peaks tend to be relatively broad and flat.

The outer ear is connected to the middle ear by the tympanic membrane, commonly known as the eardrum. This membrane forms a flattened cone with its apex facing inward. It is flexible in the center and attached to the end of the auditory canal at its edges. The middle ear is an air-filled cavity, with a volume of approximately 2 cm³, which contains three ossicles (bones): the malleus (hammer), the incus (anvil), and the stapes (stirrup). These bones form a mechanical linkage to amplify the force transmitted to the inner ear from the outer ear. The stapes is connected to the inner ear via the "oval window." The area ratio of the eardrum to the oval window is about 30:1. This system combines to create an approximate impedance match between the auditory canal and the inner ear. This is controlled to some extent to protect the ear from high-intensity sounds, known as the acoustic reflex. Acoustic reflex takes about one-half of a millisecond to respond. Therefore, it is ineffective at protecting the ear from impulsive noise, for example, explosions, sonic booms, and gunshots. The middle ear is connected to the throat via the eustachian tube, also shown in Fig. 42. The purpose of the eustachian tube is to equalize the pressure on both sides of the eardrum. Normally it is closed, but it opens during yawning and swallowing.

The inner ear contains three significant parts, the vestibule (entrance chamber), semicircular canals, and the cochlea, as illustrated in Fig. 43. The semicircular canals give humans their sense of balance and do not affect hearing. The vestibule is connected to the middle ear through the oval window and the round window. Both windows are sealed to prevent the inner ear fluids from escaping. Bone surrounds the remainder of the inner ear. The cochlea is a tube with an approximately circular cross section, which is curled in a shape similar to a snail shell. The tube makes approximately two and one-half turns with a length of approximately 3.5 cm. The total volume of the cochlear tube is approximately 0.05 cm³, with the cross section decreasing from entrance to termination. However, the average diameter is approximately 1.3 mm. The cochlear tube is divided by the cochlea partition into two channels, the upper gallery (scala vestibule) and the lower gallery (scala tympani). The two galleries arc joined at the apex by the helicotrema. The other ends of the galleries are connected to the oval (upper gallery) and round (lower gallery) windows.

The cross section of the cochlear duct is presented in Fig. 44. There is a "bony ledge" which projects (from the right in the figure) into the fluid-filled tube, which carries the auditory nerve. At the termination of the bony ledge, the nerve fibers enter the basilar membrane, which continues across the tube and is attached to the spiral ligament on the opposite side of the tube. Above the basilar membrane is the tectorial membrane, which projects into the fluid in the scala media. Reissner's membrane (vestibular membrane) runs diagonally across the tube from the bony ledge to the opposite wall, forming the pie-shaped cochlea sack, which is completely sealed. The cochlea sack is filled with endolymphatic fluid and the two galleries are filled with perilymphatic fluid. The organ of Corti is attached to the top of the basilar membrane. It contains four rows of hair cells, which span the entire length of the cochlear duct, giving roughly 30,000 cells. Several small hairs extend from each hair cell to the undersurface of the tectorial membrane.

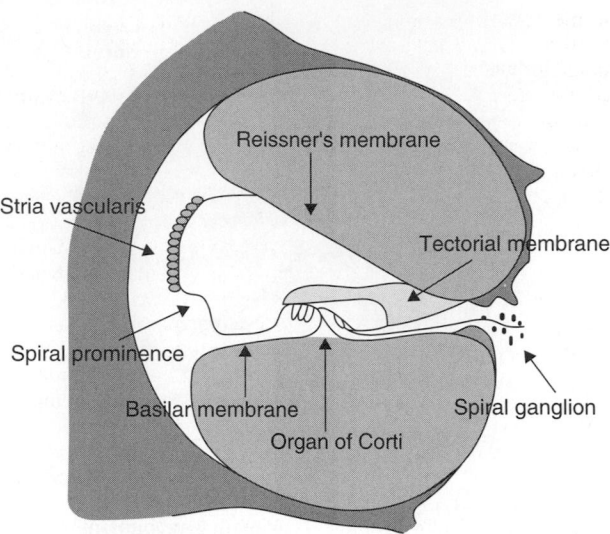

Fig. 44   Cross section of cochlear ducts.

When the ear is exposed to a sound, the eardrum's vibration is transmitted to the inner ear via the bones in the middle ear. The fluid of the inner ear in the upper gallery is disturbed by the motion of the stapes against the oval window. This fluid disturbance travels down the length of the upper gallery, through the helicotrema, then back down the lower gallery to the round window. The round window acts like a pressure release termination to the tube. The basilar membrane is driven by this fluid motion into highly damped motion with the peak amplitude increasing slowly with distance from the oval window. Once it reaches a maximum, it decreases rapidly. The location of the maximum amplitude is a function of frequency, with low frequencies causing the peak to occur close to the helicotrema and high frequencies reaching a peak much closer to the oval window. It is because of this behavior that low frequencies mask higher frequencies better than vice versa. As the organ of Corti is attached to the basilar membrane and the tectorial membrane is attached to the bony ledge, relative motion between the two flex the hair follicles, which excite the nerve endings attached to these hair cells. This in turn creates an electrical impulse that is sent up the nerve fibers. As may be implicated by the complex physical nature of the inner ear, the individual nerves do not fire at the same frequency as that of the excitation sound. In fact, the nerves tend to fire in a quasi-random frequency that is dependent upon the stress on the individual hair cells, which is more closely related to the sound intensity. These pulses then travel to the auditory center in the brain. Here a complex decoding and processing process formulates the mental "picture" of the sound.

15.1 Psychoacoustic Effects

From the previous discussion, it is apparent that when a sound excites the basilar membrane a small group of cells at the point of maximum deviation has the maximum excitation, causing it to send the largest electrical impulse up the nerve, whereas the neighboring cells to either side of this point are also disturbed but to a lesser degree. As a result, they also fire impulses, but not as strongly. Each point of the basilar membrane is the point of maximum excitation for some frequency but will also join in the excitation when a different frequency excites one of its neighbors. Therefore, a sound at a given frequency excites nerves belonging to a range of frequencies. Any sound at a high level centered at a given frequency raises the hearing level threshold for frequencies in its vicinity. This phenomenon of some sounds rendering other sounds inaudible (or less audible) is termed masking. Studies have shown that tones or narrowband noise at a given frequency raise the audibility threshold of tones at neighboring frequencies. A broadband noisy sound, like a vacuum, a loud ventilation system, or a jet airplane engine, can therefore raise the hearing threshold of just about everything.[44]

The segment of the basilar membrane that joins the excitement is called the critical band. The frequency range spanned by this section is called the critical bandwidth. It is important to note the difference between the two because they do not maintain a constant relationship. A frequency of 350 Hz stimulates a band of cells having a bandwidth of 100 Hz (300–400 Hz). However, a frequency of 4 kHz excites a critical band of cells having a bandwidth of 700 Hz (3.7–4.4 Hz). Thus, the critical bandwidth is much wider for higher frequencies than it is for lower frequencies.

Precisely where along the membrane this point of excitement occurs is another important aspect of the auditory system. As the frequencies of single-tone or narrowband noise are doubled, the point of excitement moves in equal increments. An equal length of basilar membrane is traversed to reach the points excited by 500 Hz, 1 kHz, 2 kHz, 4 kHz, 8 kHz, and so on. This corresponds to the human perception of pitch; individuals hear doubling of frequencies as a change of an octave. Other musical intervals are also based on the ratio of any two given frequencies, not the absolute distance between them. For example, a perfect fifth above some note is always a frequency ratio of 3:2: the E above A (440 Hz) is 660 Hz, an increase of 220 Hz and a ratio of 3 (660 Hz) to 2 (440 Hz). An additional fifth above 660 Hz is not, however, at 880 Hz (660 + 220), but rather at 990 Hz (660 × $\frac{3}{2}$). Such a relationship, based on multiplication rather than addition, is logarithmic rather than linear.

From the example above, it is evident that the logarithmic relationship of the basilar membrane to the spectrum applies to the perception of pitch for human beings. A logarithmic relationship is also behind the changes in the sensitivity to frequency differences over the frequency spectrum. When two tones are played consecutively, the minimum frequency difference they must have in order for listeners to notice that difference is called the "just noticeable difference." The just noticeable difference depends on a variety of factors, including frequency range and suddenness of the change. However, the just noticeable difference below 1 kHz is about 3 Hz, with the just noticeable difference for tones from 1 to about 4 kHz being 0.5% of the frequency. The just noticeable difference becomes larger above 5 kHz. Sine wave melodies transposed into that range tend to melt into a bunch of screaming, high beeps.

The auditory system is not completely egalitarian, however. Thus, the minimum sound pressure level required for a sound to be audible, called the threshold of audibility, is different for different frequencies. Lower frequencies must be produced at much greater sound pressure levels than higher frequencies in order for them to be perceived as having equal loudness, if perceived at all. The threshold is lowest for frequencies in the region of greatest importance for speech—hence giving the finest resolution of amplitude differences in this range. High frequencies excite the basilar membrane at points near the oval window, leaving points that are more distant relatively undisturbed. Low frequencies, however, excite the membrane at points more distant from the oval window, creating waves in the membrane that have to travel past those closer points excited by higher frequencies. Therefore, when high and low frequencies are heard together, the low frequencies can, in some circumstances, interfere with the high frequencies.

In review, the main psychoacoustic effects are masking, critical bands, critical bandwidths, pitch, threshold of audibility, and just noticeable differences. Masking is caused by a high level of excitation at a particular point along the basilar membrane. This high level of excitation results in a change in the threshold of audibility for frequencies in the neighborhood of the masking tone. Critical bands are the bands of frequencies that humans recognize as an interference zone, which is nominally one-third-octave frequency band spacing. Pitch is related to the critical bands and the location along the basilar membrane where the maximum excitation occurs. Threshold of audibility is the sound pressure level where a tone is just audible; it is important to reiterate that it is not equal for all frequencies. This, however, is good because it prevents humans from hearing their own heartbeat and other low-frequency, low-amplitude sounds that would be disturbing but allows them to hear the spoken word very well. Just noticeable differences are the distances between two frequencies that are perceivable. It is this psychoacoustic effect, as well as that of masking, that is the basis of the principles of music compression for the MP3 format. The tones that would be masked or are not noticeably different from another simultaneously played tone are removed, allowing the file size to be greatly reduced without sacrificing acoustic quality.

16 MICROPHONES AND LOUDSPEAKERS

A microphone is a transducing device which converts acoustic energy into electrical energy in the form of a time-varying voltage. The voltage should fluctuate so as to accurately represent the vibration of the air. The basic operation of a microphone can be represented as shown in Fig. 45. Sound impinging on a diaphragm causes the diaphragm to vibrate. The mechanical motion of the diaphragm is coupled to the transducer, producing the varying voltage. The original acoustic vibrations are thus preserved in the form of a time-varying voltage.

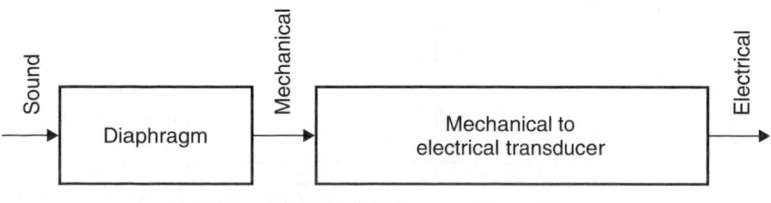

Fig. 45 Basic processes in a microphone.

A microphone can be classified as pressure-sensing, velocity-sensing, or some combination of these two types. In the pressure-sensing type sound is allowed to strike only one side of the diaphragm. Thus, the diaphragm is sensitive to pressure changes. Pressure microphones are said to be omnidirectional or nondirectional because they respond to sound coming from any direction. In velocity microphones both sides of the diaphragm are exposed to the sound. Velocity microphones respond mostly to sounds striking one side or the other of the diaphragm in a head-on fashion. Sound approaching the edge of the diaphragm exerts equal but oppositely directed pressures on each side of the diaphragm so that no motion is produced. A velocity microphone is said to be bidirectional.

The piezoelectric effect is the transducing principle employed in crystal microphones and ceramic microphones. Ceramic materials are more commonly used because they are less sensitive than crystal materials to mechanical shock, humidity, and temperature. Typical elements of a ceramic microphone are shown in Fig. 46. One end of the ceramic is attached to the microphone casing and the other end is coupled to the diaphragm by a drive pin. Conducting material is deposited on the two surfaces of the ceramic. When the diaphragm moves, the ceramic element is bent and a positive or negative (depending on the direction of bending) voltage appears on the electrical lead attached to one side of the ceramic. The principal disadvantage of a ceramic microphone is that relatively small output voltages are obtained for normal acoustic pressures. However, in modern designs this disadvantage is usually overcome by mounting a small preamplifier in the microphone casing. Microphones of this type have a fairly wide (20 Hz–10 kHz) but somewhat uneven frequency response. They are small and relatively inexpensive and are widely used in hearing aids, public address systems, portable sound equipment, and tape recorders.

Typical elements of a *condenser microphone* are shown in Fig. 47. One conducting plate is fixed to the microphone casing, although it is electrically isolated from the casing. A metal or metallized plastic diaphragm is used and acts as the second (and movable) conducting plate of the capacitor. When the diaphragm vibrates, an oscillating voltage appears on the electrical leads attached to the plates. The principal disadvantage of these microphones is the need to have a bias voltage (200–400 V) across the plates and a preamplifier in close proximity to the plates. Condenser microphone designs have typically incorporated a preamplifier in the microphone casing. Condenser microphones are used in research, to produce high-fidelity recordings, and in hearing aids and portable sound equipment.

Electret condenser microphones (see Fig. 48) are now in wide use because they offer the advantages of a condenser microphone without the disadvantage of needing a polarizing voltage. The diaphragm consists of an electret plastic foil that has been permanently electrically charged and overlayed with a thin metallic layer. The electret condenser microphone typically employs a preamplifier. It is slightly less sensitive than the condenser microphone.

A *ribbon microphone* (see Fig. 49) consists of a very thin metallic ribbon set in a magnetic field. A current is caused to flow in the ribbon as it is pushed and pulled back and forth in the magnetic field by acoustic pressures. Maximum response occurs when the pressure is perpendicular to the face of the ribbon. Pressure coming from the edge pushes oppositely on the two

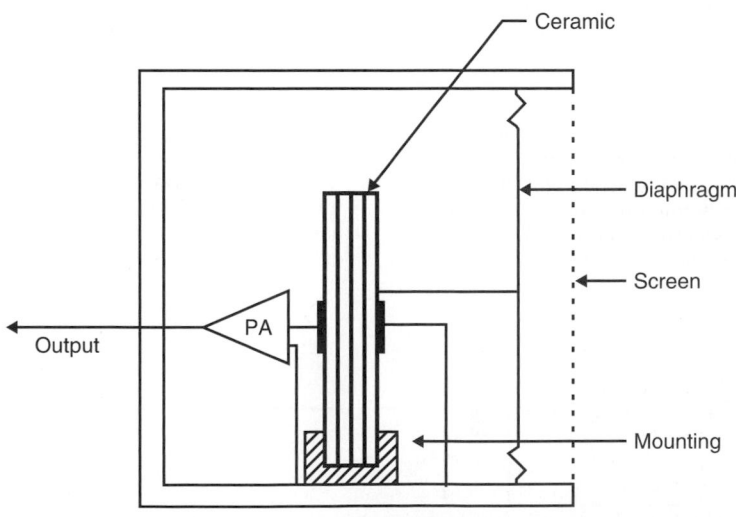

Fig. 46 Elements of a ceramic microphone.

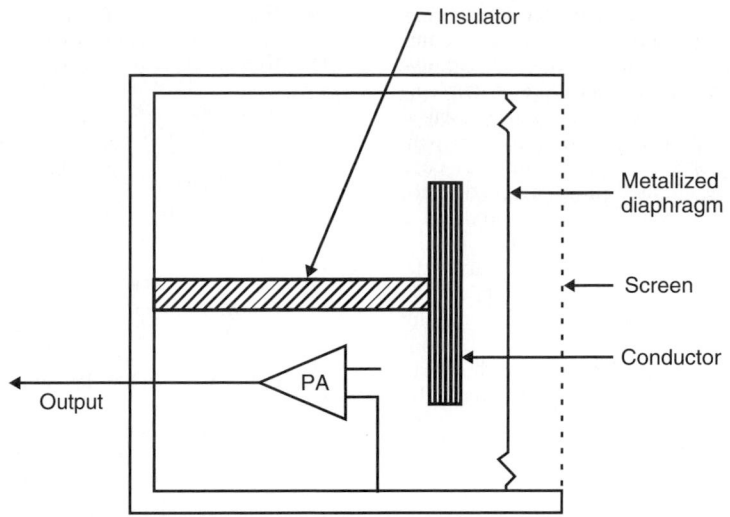

Fig. 47   Elements of a condenser microphone.

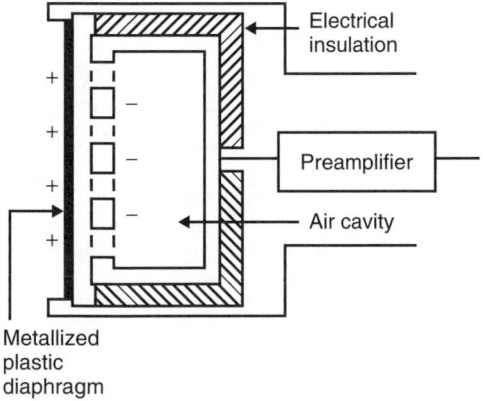

Fig. 48   Elements of an electret condenser microphone.

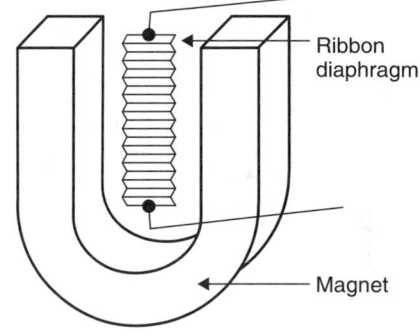

Fig. 49   Elements of a ribbon microphone.

sides of the ribbon and so little response results. A ribbon microphone is bidirectional in response and is an example of a velocity microphone.

A *dynamic microphone* has a coil of wire attached to its diaphragm as shown in Fig. 50. The transducing element is a coil that is free to move between the poles of a magnet. The electrons in the coil move with the coil and experience a force (as described by the magnetic force law) which produces a varying electric voltage across the ends of the coil. Dynamic microphones are capable of a relatively high power output. They are rugged and capable of a broad frequency response over a wide dynamic range. Because they are able to withstand the high-intensity sound levels often associated with popular music, they are widely used for live performances and for recording sessions.

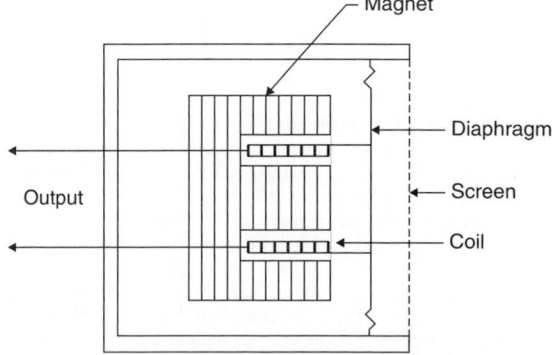

Fig. 50   Elements of a dynamic microphone.

A *loudspeaker* is a transducing device which converts electrical energy into acoustic energy. One common type of loudspeaker is the direct radiator dynamic loudspeaker. It operates on the same principle as the dynamic microphone, only in the opposite sense. Figure 50 illustrates a dynamic loudspeaker as well as a dynamic microphone. A voltage source is connected to a coil of wire situated in the magnetic field of a permanent magnet. A varying voltage produces a varying current in the coil. The coil experiences a force (because a magnetic field is present) which causes the coil to vibrate. The coil is attached to a diaphragm so the motion of the coil "drives" the diaphragm, causing the air to vibrate. Because of the simplicity of construction, the small space requirements, and the fairly uniform frequency response, this type of speaker is the most widely used.

REFERENCES

1. Hunt, F. V., *Origins in Acoustics, The Science of Sound from Antiquity to the Age of Newton*, Acoustical Society of America, Woodbury, NY, 1992.
2. Speaks, C. E., *Introduction to Sound: Acoustics for the Hearing and Speech Sciences*, 3rd ed., Singular Publications Group, San Diego, CA, 1999.
3. Raichel, D. R., *The Science and Applications of Acoustics*, Springer-Verlag, New York, NY, 2000.
4. International Organization for Standardization (ISO), 1, ch. de la Voie-Creuse, Case postale 56, CH-1211 Geneva 20, Switzerland.
5. Fletcher, H. and Munson, N. A. "Loudness; Its Definition, Measurement, and Calculation," *J. Acoust. Soc. Am.*, **5**, 82–108 (1933).
6. Woods R. L., Lawrence, K. L., *Modeling and Simulation of Dynamic Systems*, Prentice-Hall, Englewood Cliffs, NJ, 1997.
7. Kinsler, L. E., Frey, A. R., Coppens, A. R., and Sanders, J. V., *Fundamentals of Acoustics*, 4th ed., Wiley, New York, 2000.
8. Blackstock, O., *Fundamentals of Physical Acoustics*, Wiley, New York, 2000.
9. Pierce, A. D., *Acoustics: An Introduction to Its Physical Principles and Applications*, Acoustical Society of America, Woodbury, NY, 1989.
10. von Gierke, H. E., and Dixon Ward, W., "Criteria for Noise and Vibration Exposure," in *Handbook of Acoustical Measurements and Noise Control*, 3rd ed., C. M. Harris (Ed.), Acoustical Society of America, Woodbury, NY, 1998.
11. Melnick, W., "Hearing Loss from Noise Exposure," in *Handbook of Acoustical Measurements and Noise Control*, 3rd ed., C. M. Harris (Ed.), Acoustical Society of America, Woodbury, NY, 1998.
12. Nixon, C. W., and Berger, E. W., "Hearing Protection Devices," in *Handbook of Acoustical Measurements and Noise Control*, 3rd ed., C. M. Harris (Ed.), Acoustical Society of America, Woodbury, NY, 1998.
13. Harris, C. M. (Ed.), *Handbook of Acoustical Measurements and Noise Control*, Acoustical Society of America, Woodbury, NY, 1998.
14. Hansen, C., *Noise Control: From Concept to Application*, Taylor & Francis, London, 2005.
15. Kuo, S. M., and Morgan, D. R., *Active Noise Control Systems: Algorithms and DSP Implementations*, Wiley, New York, 1996.
16. Sommerfeldt, S. D., "Multi-Channel Adaptive Control of Structural Vibration," *Noise Control Eng. J.*, **37**(2), 77–89 (1991).
17. Strong, W. J., and Plitnik, G. R., *Music Speech High-Fidelity*, 2nd ed., Soundprint, Provo, UT, 1983.
18. Salomons, E. M., *Computational Atmospheric Acoustics*, Kluwer Academic, Dorbrecht, 2001.
19. Piercy, J. E., and Daigle, G. A., "Sound Propagation in the Open Air," in *Handbook of Acoustical Measurements and Noise Control*, 3rd ed., C. M. Harris (Ed.), Acoustical Society of America, Woodbury, NY, 1998.
20. Raney, J. P., and Cawthorn, J. M., "Aircraft Noise," in *Handbook of Acoustical Measurements and Noise Control*, 3rd ed., C. M. Harris (Ed.), Acoustical Society of America, Woodbury, NY, 1998.
21. Fidell, S., Barber, D. S., and Schultz, T. J., "Updating a Dosage Effect Relationship for the Prevalence of Annoyance Due to General Transportation Noise," *J. Acoust. Soc. Am.*, **89**, 221–233 (1991).
22. Blauert, J., and Jekosch, U., "Sound–Quality Evaluation—A Multi-Layered Problem," *Acustica*, **83**(5), 747–753 (1997).
23. Blauert, J., "Product-Sound Assessment: An Enigmatic Issue from the Point of View of Engineering," *Proc. Internoise 94*, **2**, 857–862 (1994).
24. Guski, R., "Psychological Methods for Evaluating Sound-Quality and Assessing Acoustic Information," *Acustica*, **83**(5), 765–774 (1997).
25. Blauert, J., and Jekosch, U., "A Semiotic Approach towards Product Sound Quality," *Proc. Internoise 96*, 2283–2286 (1996).
26. Lyon, R. H., "Designing for Sound-Quality," *Proc. Internoise 94 J-Yokohama*, **2**, 863–868 (1994).
27. Vastfjall, D., Gulbol, M. A., and Kleiner, M., "Wow, What Car is That?': Perception of Exterior Vehicle Sound Quality," *Noise Control Eng. J.*, **51**(4), 253–261 (2003).
28. Otto, N. C., Feng, B. J., and Wakefield, G. H., "Sound Quality Research at Ford—Past, Present and Future," *Sound Vib.*, **32**(5), 20–24 (1998).
29. Ishihama, M., "Sound Quality R and D in the Japanese Automotive Industry," *Noise Control Eng. J.*, **51**(4), 191–194 (2003).
30. Otto, N. C., and Wakefield, G. H., "Design of Automotive Acoustic Environments. Using Subjective Methods to Improve Engine Sound Quality," *Proc. Human Factors Society*, **1**, 475–479 (1992).
31. Gonzalez, A., Ferrer, M., DeDiego, M., Pinero, G., and Barcia-Bonito, J. J., "Sound Quality of Low-Frequency and Car Engine Noises after Active Noise Control," *J. Sound Vib.*, **265**(3), 663-679 (2003).
32. Brentner, K. S., Edwards, B. D., Riley, R., and Schillings, J., "Predicted Noise for a Main Rotor With Modulated Blade Spacing," *J. Am. Helicopter Soc.*, **50**(1), 18–25 (2005).

33. Vecchio, A., Polito, T., Janssens, K., and Van Der Auwearaer, H., "Real-Time Sound Quality Evaluation of Aircraft Interior Noise," *Acustica*, **89** (Suppl.), S53 (2003).

34. Gerges, S. N. Y., and Zmijevski, T. R. L., "Hairdryer Sound Quality," *Acustica*, **89**, (Suppl.), S90 (2003).

35. Jiang, L., and Macioce, P., "Sound Quality for Hard Drive Applications," *Noise Control Eng. J.*, **49**(2), 65–67 (1996).

36. Zwicker, E., and Fastl, H., *Psychoacoustics Facts and Models*, Springer-Verlag, Berlin, 1990.

37. Hastings, A., and Davies, P., "An Examination of Aures's Model of Tonality," *Proc. Internoise 02*, Dearborn, MI, 2002.

38. Hastings, A., Lee, K. H., Davies, P., and Surprenant, A. M., "Measurement of the Attributes of Complex Tonal Components Commonly Found in Product Sound," *Noise Control Eng. J.*, **51**(4), 195–209 (2003).

39. Beyer, R. T., *Nonlinear Acoustics*, Acoustical Society of America, Woodbury, NY, 1997.

40. Hamilton M. F., and Blackstock, D. T., *Nonlinear Acoustics*, Academic, San Diego, CA, 1998.

41. Gee, K. L., Swift, S. H., Sparrow, V. W., Plotkin, K. P., and Downing, J. M., "On the Potential Limitations of Conventional Sound Metrics in Quantifying Perception of Nonlinearly Propagated Noise," *J. Acoust. Soc. Am.*, **121**, EL1–EL7 (2007).

42. Maglieri, D. J., and Plotkin, K. J., *Aeroacoustics of Flight Vehicles: Theory and Practice*, Vol. 1: *Noise Sources*, H. H. Hubbard (Ed.), NASA Reference Publication 1258, 1991.

43. Shoh, A., "Industrial Applications of Ultrasound—A Review: I. High Power Ultrasound," *IEEE Trans. Sonics Ultratsonics*, **SU-22**, 60–71 (1975).

44. Kawaguchi, M., and Nishimura, M., "Sound Quality Evaluation, Based on Hearing Characteristics, for Construction Machinery," *Nippon Kikai Gakkai Ronbunshu, C Hen/Trans. Jpn. Soc. Mech. Eng., Pt. C*, **61**(584), 1509–1515 (1995).

SUGGESTED FURTHER READING

Blazier, Jr., W. E., "Sound Quality Considerations in Rating Noise from Heating, Ventilating, and Air-Conditioning (HVAC) Systems in Buildings," *Noise Control Eng. J.*, **43**(3), 53–63 (1995).

Cocchi, A., Garai, M., and Tavernelli, C., "Boxes and Sound Quality in an Italian Opera House," *J. Sound Vib.*, **232**(1), 171–191 (2000).

Fahy, F. J., *Sound and Structural Vibration: Radiation, Transmission, and Response*, Academic, San Diego, CA, 1985.

Hamilton, M. F., and Blackstock, D. T., *Nonlinear Acoustics*, Academic, San Diego, 1998.

Ih, J. G., Lim, D. H., Shin, S. H., and Park, Y., "Experimental Design and Assessment of Product Sound Quality: Application to a Vacuum Cleaner," *Noise Control Eng. J.*, **51**(4), 244–252 (2003).

Junger, M. C., and Feit, D., *Sound Structures and Their Interaction*, Massachusetts Institute of Technology, Cambridge, MA, 1986.

Letowski, T. R., "Guidelines for Conducting Listening Tests on Sound Quality," in *Proc. National Conference on Noise Control Engineering*, Fort Lauderdale, FL, May 1–4, 1994, 987–992.

Nelson, P. A., and Elliot, S. J., *Active Control of Sounds*, Academic, London, 1992.

Snyder, S. D., *Active Noise Control Primer*, Springer-Verlag, New York, 2000.

Sommerfeldt, S. D., *Fundamentals of Acoustics Course Notes*, Brigham Young University, Provo, UT, 2003.

Susini, P., McAdams, S., Winsberg, S., Perry, I., Vieillard, S., and Rodet, X, "Characterizing the Sound Quality of Air-Conditioning Noise," *Appl. Acoust.*, **65**(8), 763–790 (2004).

Vastfjall, D., "Contextual Influences on Sound Quality Evaluation," *Acustica*, **90**(6), 1029–1036 (2004).

Whitfield, I. C., *The Auditory Pathway*, Edward Arnold, London, 1967.

Yost, W. A., *Fundamentals of Hearing: An Introduction*, Academic, San Diego, CA, 2000.

CHAPTER 20
CHEMISTRY

D. A. Kohl
The University of Texas at Austin
Austin, Texas

1 ATOMIC STRUCTURE AND THE PERIODIC TABLE

Atoms.   Atoms contain a dense, positively charged nucleus surrounded by a cloud of negatively charged orbital electrons that occupy discrete energy levels and orbital configurations. The nucleus consists of Z positively charged protons and $A - Z$ uncharged neutrons. Atomic number Z determines the name of the element and the number of orbital electrons in a neutral atom. Mass number A equals the number of protons plus neutrons in the nucleus. Gain or loss of electrons results in ionized atoms with a net charge. Chemical reactions are concerned only with electronic structure changes, while nuclear reactions involve changes in the constitution of the nucleus. Isotopes are atoms with the same atomic number, the same number of electrons, and the same chemical reactions but differing in mass number because the nucleus contains a different $(A - Z)$ number of neutrons.

Atomic Weights of Elements.   The mass of an element's naturally occurring distribution of isotopes is its atomic weight. This may be expressed in atomic mass units (amu) or in grams for an Avogadro number (6.023×10^{23}) of atoms. Atomic weights are relative to 12.000 for the $_{6}^{12}C$ isotope of carbon, an atom whose $Z = 6$ and $A = 12$. Examples: Naturally occurring chlorine contains 75.53 atom % of $_{17}^{35}Cl$ and 24.47% of $_{17}^{37}Cl$; it is assigned an atomic weight of 35.453. Naturally occurring tin consists of a presumably geographically invariant distribution of 10 stable isotopes ranging from mass number 112 to 124. These yield a weighted average atomic weight of 118.69.

Atomic Radii.   The radii of atoms, though nebulous and difficult to estimate, range from 0.4 to 2.5 Å $(1 \text{ Å} = 10^{-8} \text{ cm})$. The radii may be deduced from quantum mechanical calculations or from experimentally determined crystal radii. Ionic radius differs from atomic radius due to a greater or lesser occupancy of the outer electron orbitals.

Periodicity of Elements.   The periodic table is an arrangement of elements in rows and columns in order of increasing atomic number according to similarity of chemical behavior. Differences in electron energy levels, deduced from transitions that absorb or emit

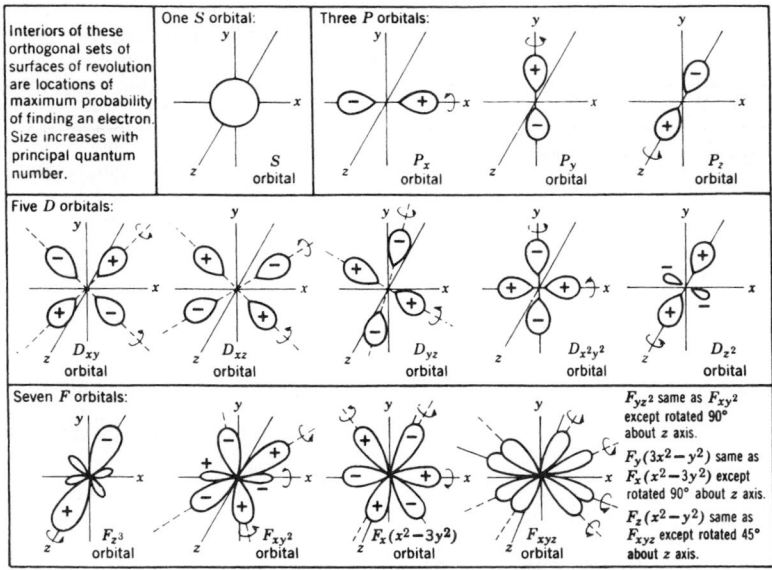

Fig. 1 Azimuthal characteristics of electronic orbitals.

spectral energy, yield a similar and consistent pattern of electron buildup in numbers and energy levels.

Electronic Orbitals.   The rationalization of spectroscopic data via quantum mechanics leads to four types of quantum numbers for describing the building up of shells of electron orbitals with discrete energy levels for each orbital. This buildup, with electrons, individually occupying the lowest vacant energy level, results in calculated spectra that closely match the experimental. These four types of quantum numbers are:

(a) *Principal quantum number n* represents a major grouping of energy levels, where n can be $1, 2, 3, \ldots$, with 1 representing the lowest energy level.

(b) *Azimuthal quantum number l* denotes different shapes of orbitals within the major grouping. These shapes are labeled *s, p, d,* and *f* for $l = 0, 1, 2,$ and 3, respectively, where l may range from 0 to $n - 1$.

(c) *Magnetic quantum number m_l* has $2l + 1$ possible values: $-l, -l + 1, -l + 2, \ldots, 0, \ldots, l - 1, l$.

(d) *Spin magnetic quantum number m_S* may have a value of $+\frac{1}{2}$ or $-\frac{1}{2}$ depending on the direction of electronic spin.

Azimuthal Characteristics of s, p, d, and f Electronic Orbitals.   Directionality of the most probable electronic position about the nucleus is provided by sets of mathematically symmetrical and orthogonal orbitals shown in Fig. 1. Each orbital can contain two electrons of opposite spin. An electron has the highest probability of being located inside the surfaces of the revolution shown. The size of these surfaces (distance from the nucleus) increases with the principal quantum number.

Electronic Structure of Elements in the Ground State.   As atomic number increases through the periodic table, the buildup of electronic structure follows a regular pattern. One *s*, three *p*, five *d*, and seven *f* orbitals can contain 2, 6, 10, and 14 electrons (with different magnetic and spin quantum numbers), respectively. Electron energy levels associated with quantum number designations are sequentially filled, beginning with the lowest energy level available. Figure 2 shows the pattern of electron occupancy of orbitals for elements from hydrogen ($Z = 1$) to lawrencium ($Z = 103$) in their lowest, or ground, energy state.

Oxidation States of the Elements.   The oxidation state of the elements is the atomic charge attainable by chemical methods. This is correlated with the systematic buildup of electrons in orbitals. The addition or loss of electrons is subject to energy barriers that become essentially insurmountable with filled shells exemplified by the noble-gas structures. In the discussion of groups of elements in each vertical column of the periodic chart, the known oxidation states are cited and related to electronic structure. This does not mean that all can exist as stable states in contact with oxygen, environmental moisture, or aqueous solution, for many of the oxidation states are particularly reactive. Electronic structure notation uses the structure of the

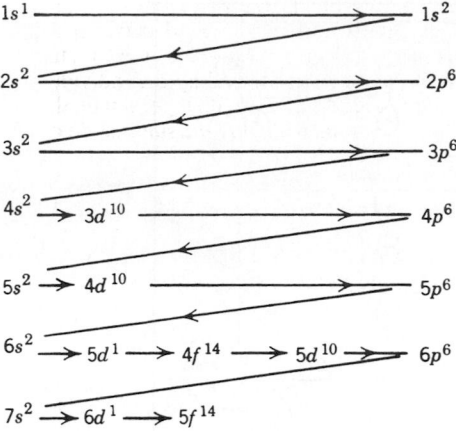

Fig. 2 Progressive filling of electron orbitals: $3d^{10}$ indicates 10-electron (complete) occupancy of the five d orbitals existent in principal quantum number 3 energy levels. These are filled after the $4s$ orbital is occupied by two electrons. Interruption of progressive buildup of the $5d$ sequence by 14-electron filling of $4f$ orbitals is partially explained by coulombic shielding of the positively charged nucleus by inner portions of the electron cloud.

prior noble gas plus the additional electronic orbitals that are filled for each element.

(a) *Inert gases He, Ne, Ar, Kr, Xe,* and *Rn* have filled p orbitals, and this maximum-stability, minimum-energy configuration exists as a neutral atom with nuclear charge equaling orbital electron charge. The energy barrier for gain or loss of electrons is very high and the oxidation state is zero although a small number of compounds with fluorine and oxygen have been prepared such as $XeOF_4$. With the exception of xenon (XeF_6) and radon compounds, where inner electrons shield the nuclear charge and reduce this potential barrier, stable compound formation has not been accomplished.

(b) *F, Cl, Br, I,* and *At,* lacking one electron of filled p orbitals, readily gain one electron; can lose one p orbital electron, reaching stability of a half-filled orbital; can lose five p orbital electrons, leaving a filled s orbital below; or may further lose both of those s electrons. Oxidation states of $-1, 0, +1, +3, +5,$ or $+7$ are possible.

(c) *O, S, Se, Te,* and *Po* lack two p electrons of the inert-gas configuration and easily reach the oxidation state of -2. Peroxides are a covalent combination of two oxygen atoms with -2 for the pair. Higher atomic number members of the group also can lose all four p electrons and may also lose the two s electrons lying beneath. Oxidation states of $-2, +4,$ and $+6$ result.

(d) *N, P, As, Sb,* and *Bi* with three p electrons can either gain or lose three and may also lose two s electrons beneath (exception: NO or NO_2 gases). Oxidation states of ±3 and $+5$ exist for the group.

(e) *C, Si, Ge, Sn,* and *Pb* with two p electrons can lose these plus two s electrons beneath, resulting in $+2$ and $+4$ oxidation states for the group. Carbon forms numerous covalent bonds with an oxidation state of -4.

(f) *B, Al, Ga, In,* and *Tl* lose their single p electron plus the two s electrons below for an oxidation state of $+3$. Thallium has an additional stable $+1$ oxidation state.

(g) *Zn, Cd,* and *Hg,* with 10 electrons filling the d orbitals in stable configurations, lose two s electrons, forming $+2$. Exception: mercurous dimer Hg_2^{2+}.

(h) *Cu, Ag,* and *Au,* with filled d orbitals, can lose their one s electron, forming $+1$ oxidation state. Exceptions for the column are Cu^{2+} and Au^{3+} due to removal of additional d electrons over low energy barriers.

(i) *Ni, Pd,* and *Pt,* with some overlap of d and s orbital energies, exhibit oxidation states of $+2, +4,$ and $+6$.

(j) *Co, Rh, Ir,* and the *Fe, Ru,* and *Os* transition groups tend to have two electrons in the s orbital at the expense of vacant d orbitals. Oxidation states of $+2, +3, +4,$ and $+6$ occur, with the $+8$ oxidation state reached in Ru and Os.

(k) *Mn, Tc,* and *Re,* with five-electron, half-filled d orbitals and filled s orbital, tend to lose the s orbital electrons for $+2$ and any in the d orbitals for a maximum oxidation state of $+7$. Exception: Re also forms a -1 ion; it is the only transition element to do so.

(l) *Cr, Mo,* and *W,* with four electrons in the d orbitals and two in the s orbital, can have oxidation states ranging from $+2$ to $+6$.

(m) The *V, Nb,* and *Ta* group, with three or four electrons in the d orbitals and one or two in the s orbital, can have oxidation states ranging from $+2$ to $+5$.

(n) The *Ti, Zr,* and *Hf* groups, with two d and two s electrons, tends to have an oxidation state of $+4$. Exception: Ti^{3+} also occurs.

(o) The *Sc, Y, La,* and *Ac* group, with one d and two s electrons, can lose these for an oxidation state of $+3$.

(p) The 14 *rare earths (lanthanides),* following lanthanum and beginning with cerium, have the lanthanum $+3$ oxidation state and structure with approximately progressive filling of the seven $5f$ orbitals. Ce, Nd, Pr, and Nd form $+4$ ions while Sm, Eu, Tm, and Yb also form $+2$ ions.

(q) The 14 *actinides*, beginning with thorium, paralleling cerium above, approximately repeat the structure of the rare earths with 14 electrons filling of $5f$ orbitals. Some exceptions occur, forming the half-filled orbital situation, and the $+3$ oxidation state can be increased to $+6$ due to much electronic shielding of the nuclear positive charge.

(r) *Beryllium* and the *alkaline earth* group have two electrons filling the s orbital. Loss of these electrons results in a $+2$ oxidation state.

(s) *Hydrogen* and the *alkali metals* have a single electron in the s orbital. Its loss results in a $+1$ oxidation state. Exception: Gain of an electron by hydrogen to fill the s orbital results in the hydride H^{1-}.

Excited Electronic States and Ionization Energy.

Excited electronic states are produced through energy absorption, which raises an electron's energy level to that of an unfilled orbital. Loss of energy by photoemission may be very fast (10^{-8} s) or very slow depending on the particular state. Energy absorption of magnitude greater than the highest available energy level is synonomous with having achieved an electronic escape velocity and ionization occurs. This ionization energy ranges from 3.9 eV (90 kcal/g atom) for cesium to 24.5 eV for helium.

Electron Affinity.

The counterpart to ionization potential, electron affinity is energy released when an isolated monatomic gaseous atom gains an electron and becomes an anion. For halogens this energy is on the order of 3–4 eV; experimental problems have precluded measurement for many of the other elements that form negative ions.

Electronegativity.

The electron-attracting power of the elements has been estimated from chemical bond energies and from average electron densities. The result is a consistent series of relative numbers based on zero for inertness of the noble gases and ranging from a maximum of 4.0 for fluorine (which readily forms the anion F^-) to 2.5 for carbon (which forms equally shared covalent bonds) to a minimum of 0.7 for cesium (which forms the cation Cs^+). Large differences in electronegativities of two chemically bonded atoms indicate an ionic bond with the more electronegative atom monopolizing the electron pair. Small differences in electronegativity indicate a covalent bond with approximately equal sharing of the electron pair.

2 MOLECULAR STRUCTURE AND CHEMICAL BONDING

Molecules.

Molecules consist of groups of atoms held together in geometric arrangement by chemical bonds. Chemical bonds are based on electronic configuration of the individual atoms and electron occupancy of vacancies existent in the atomic orbitals. Mutual occupancy of orbitals of two atoms by an electron pair constitutes a chemical bond between atoms and results in joint configuration of greater stability, or lower energy, than possessed by the individual atoms. The total charge on a molecule is zero, with the number of orbital electrons matching the summation of positive nuclear charges.

Chemical Bonds.

Mutual occupancy of orbitals, the shared electron pair, leads to two general classifications of chemical bonds. These are ionic and covalent. An ionic bond exists where the shared electron pair is monopolized by one atom, giving that portion of the molecule a net negative charge and leaving an electron deficiency, or net positive charge, on the other. A covalent bond results when both atoms have approximately the same electronegativity or electron affinity; the electron pair is shared without monopoly by either atom. One shared electron pair constitutes one chemical bond, commonly designated by a dash between symbols of the elements. The spatial geometry of molecules is related to orientation of the atomic orbitals involved in the bond.

Bond Energy.

Energy liberated on bond formation, or energy required to rupture a bond, is the bond energy. This varies with the parent atoms and the particular molecular configuration involved (see Table 10 in Section 9). The summation of all bond energies is related to the heat of formation of a compound from its elements.

Chemical Compounds and Ions.

Compounds can be classified into categories according to their ions in aqueous solution. Ionic compounds are those with ionic bonds between major atomic groupings, and these ionize in aqueous solution to form cations that are positive and anions that are negative. The naturally occurring combinations are used for inorganic chemical nomenclature. Metals, having low ionization potential, readily form positive cations, in agreement with their electronic configuration. Examples: Na^+, Ca^{2+}, Al^{3+}, Fe^{2+} and Fe^{3+}, Zn^{2+}, Ag^+, and so on. Halogens, having high electron affinity, form halide ions: F^-, Cl^-, Br^-, and I^-. Oxygen forms oxides and readily combines with other atoms to form MnO_4^-, $Cr_2O_7^{2-}$, SO_4^{2-}, SO_3^{2-}, NO_2^-, NO_3^-, PO_4^{-3}, CO_3^{-2}, SiO_3^{-2}, and similar ions.

Electrical Neutrality of Compounds.

This results from summation of component oxidation states and permits assignment of an oxidation state to component atoms in the molecule.

Example 1. Determine oxidation state of Cr in $NaCrO_4$.

Sodium's electronic structure, revealed by its position on the periodic table, definitely assigns its ion as Na^+. Oxygen, except in peroxides, assumes -2 oxidation state. Chromium's usual oxidation states are

+2, +3, and +6, as shown on the periodic table. Therefore, the oxidation state of Cr is +6, as required for compound electrical neutrality.

Inorganic Chemical Nomenclature.
Systematic and unambiguous nomenclature for inorganic and organic compounds and ions is necessitated by proliferation of known species. Traditional (often trivial) names of common compounds are paralleled by more cumbersome names following rules of the International Union of Pure and Applied Chemistry (IUPAC). Traditional nomenclature for inorganic compounds (Table 1) uses simple prefix and suffix terminology.

Acids and Bases.
Three acid–base definitions are recognized:

(a) The traditional Arrhenius definition that acids yield H^+ and bases yield OH^- suffices for inorganic reactions in aqueous solution.

(b) Brönsted introduced the broader definition that acids are proton (H^+) donors and bases are proton acceptors. Since H^+ in aqueous solution exists hydrated as the hydronium ion H_3O^+, water is by definition a base because it can accept a proton. In nonaqueous media this definition recognizes hydrides, alcoholates, alkyllithium compounds, and ammonia as the powerful bases they are. Brönsted broadens the traditional definition because OH^- does accept a proton.

(c) Lewis defined acids as molecules or ions capable of accepting electron pairs and bases as molecules or ions capable of donating electron pairs. Thus $AlCl_3$, BF_3, $FeCl_3$, $ZnCl_2$, and $SnCl_4$ are Lewis acids because they lack pairs of electrons from filled orbitals. These strong acids are important in nonaqueous organic reactions. Lewis bases have an uncoordinated electron pair available; for example, on the nitrogen atom of ammonia, $:NH_3$.

The Brönsted definition of proton donors and acceptors, broadened by the Lewis definition of electron pair donors and acceptors, is necessary to explain reactions in nonaqueous media. The remaining sections on inorganic chemistry utilize the traditional simplicity of acid H^+ and base OH^- because these ions do result when any of the more broadly defined acids and bases react with water.

Strong and Weak Acids and Bases.
The strong or weak designation refers to extent of ionization; this does not refer to concentration in solution. Strong acids and bases are completely dissociated into their ions, in water, while weak ones are only slightly dissociated. A 1-M solution of a strong acid such as HCl yields a hydrogen ion concentration, H^+, of 1 mol/l, while a 1-M solution of a weak acid such as acetic, CH_3COOH, yields a H^+ of only 4×10^{-3} mol/l. A similar classification of bases is made based on the extent of ionization in the solvent employed.

Complex Ions and Coordination Compounds.
Unfilled electron orbitals of transition metal elements, those occurring in the center of the periodic table,

Table 1 Common Terminology

Cations (+ ions)	Anions (− ions)
-ic. Higher usual oxidation state of cation. Examples: Fe^{3+} is ferric and Cr^{3+} is chromic.	*-ide.* Usual oxidation state of anion in binary compounds. Examples: oxide O^{2-}, chloride Cl^-, and hydride H^-.
-ous. Lower usual oxidation state of cation. Examples: Au^+ is aurous and Cr^{2+} is chromous.	*-ate.* Higher usual positive oxidation state of major element in oxygenated acid anions. Examples: ZnO_2^{2-} is zincate, SO_4^{2-} is sulfate, and ClO_3^- is chlorate.
-yl. Oxygenated cation containing the -ic or higher usual oxidation state of the major element. Examples UO_2^{2+} is uranyl, TiO^{2+} is titanyl.	*-ite.* Lower usual positive oxidation state of major element in oxygenated acid anions. Examples: NO_2^- is nitrite, SO_3^{-2} is sulfite, and ClO_2^- is chlorite.
No suffix is used for cations that form only one ion. Examples: Zn^{2+} is zinc and Na^+ is sodium.	*hypo -ite* or *-ous.* Lower oxidation state than -ite or -ous implies. Example: ClO^- is hypochlorite.
Instead of the prefix or suffix notation, oxidation states are also denoted by Roman numerals following the name, such as chromium(III), instead of Cr^{3+} or chromic.	*per- -ate* or *-ic.* Higher oxidation state than -ate or -ic implies. Example: ClO_4^- is perchlorate and MnO_4^- is permanganate.
	pyro-. Infers a dimeric structure (a di- -ate) of the oxygenated acid anion. Example: $P_2O_7^{4-}$ is pyrophosphate.
	thio-. Means single sulfur substitution for oxygen in oxygenated acid anions. Example: $S_2O_2^{-2}$ is thiosulfate (SO_4^{-2} with S replacing one O).

Table 2   Coordination Compound Ligands

F^-	NH_3, PH_3	$S_2O_3{}^{2-}$ thiosulfate
Cl^-	$NH_2-CO-NH_2$ (urea)	$C_2O_4{}^{2-}$ oxalate
Br^-		
CN^-	H_2O	$S_x{}^{2-}$ polysulfide
OH^-	CO, NO	
SCN$^-$	Various unsaturated hydrocarbons, including acetylide ion $(HC{\equiv}C)^-$ and cyclopentadienyl ion $(C_5H_5)^-$	

permit forming complex ions and coordination compounds that appear inconsistent with the usual oxidation state of the metallic element. But complex ions are consistent with the metal's electronic structure; its unfilled orbitals are filled by the sharing of electron pairs to reach a stable configuration. Ligands (see Table 2) are groups attached to the central metal atom. Each ligand contributes an electron pair to be shared with and occupy vacant orbitals. Coordination number is the number of ligands surrounding the central metal atom. Solvation in water (hydration) or in liquid ammonia (ammoniation) and formation of hydrated and double salts and the numerous complex ions of inorganic and metal organic chemistry are examples of complex formation. The geometry of the complexes is spatially related to the hybridized orbitals involved. In the following examples, carbon monoxide, ammonia, and the cyanide ion are ligands:

$Ni(CO)_4$	Nickel carbonyl
$Cr(NH_3)_6{}^{3+}$	Hexamminechromium III ion
$Fe(CN)_6{}^{3-}$	Ferricyanide, or hexacyanoiron III ion

In nickel carbonyl, nickel has an elemental, or zero, oxidation state with one $4s$ and three $4p$ orbitals unfilled before reaching the next noble-gas structure, Kr. An oxygen from each of the four CO molecules provides an electron pair that effectively occupies one of the four equal sp^3 hybrid orbitals to generate a stable compound.

The hexamminechromium III ion consists of Cr^{3+} symmetrically surrounded by six ammonia molecules, each providing an electron pair from its nitrogen atom. The resultant complex ion has gained stability by having all the $4d$ orbitals filled and the $4p$ orbitals half-filled. Further electron sharing with three negative ions will yield a hexamminechromium III salt; this effectively gives the chromium atom the stable electronic structure of Kr.

The ferricyanide ion consists of Fe^{3+} surrounded by six CN^- ions, each N of which shares an electron pair to occupy the six d^2sp^3 hybrid orbitals. Three of these same cyanide ions further share an electron to fill the remaining three vacancies and give the iron atom an effective Kr electronic structure. Charge on the remaining three CN^- ions gives the very stable ferricyanide ion its 3− charge.

A similar rationale can be applied to other coordination compounds composed of a central metal atom or ion and mixtures of ligands.

Chelate Compounds.   Chelating (clawlike) ligands have a stereospecific configuration that permits multiple attachement to the metallic ion to form complexes. Examples are EDTA (ethylenediamine tetraacetate ion), the acetylacetonate ion, and ethylenediamine. Multiple attachment can consist of a chemical bond plus coordination with oxygen or nitrogen atoms of the ligands.

Dissociation of Complex Ions.   The variable stability of complexes is represented by equilibrium constants for their dissociation. Many complexes are very stable, while others can exist only in specific environments.

3   CHEMICAL REACTIONS AND STOICHIOMETRY

Chemical Reactions.   Chemical reactions fit into two broad categories: equilibrium-controlled conversions of reactants to products without changes in oxidation state and equilibrium control of oxidation–reduction reactions where changes in oxidation state occur.

Conversion of reactants to products can proceed only to an equilibrium condition because no driving force exists to go beyond equilibrium. Products can be maximized by continuously removing product from the reaction system so conversion can proceed without reaching equilibrium and stopping.

Le Chatelier's Principle.   This principle is basic to the understanding of all chemical equilibria. It is universally applicable and states that systems adjust to reduce applied stress. For example, equilibrium of the reaction $aA + bB \rightleftarrows cC + dD$ can be shifted to the right when excess concentration of reactants A or B are present, or when products C or D are removed from the reaction mixture. Conversely, equilibria may be shifted to the left by adjustment of reactant or product concentrations.

Equilibrium-controlled   Conversions.   Many chemical operations use processes that consist of a mixture of chemical species that proceed toward equilibrium and whose conversion to a desired product is governed by product removal from the reaction system. Product removal from the reaction system,

Table 3 Equilibrium-controlled Conversions

Product Removal Method	Example	Net Reaction of Example	Equilibrium Governing Conversion Completeness
Slightly ionized product	Acid–base neutralization, formation of weak acid or base	$H^+ + OH^- \rightleftharpoons H_2O$	Dissociation of water
		$H^+ + CH_3COO^- \rightleftharpoons$ CH_3COOH	Dissociation of weak acid or base
Slightly soluble products (least soluble product of several alternatives will prevail)	Precipitation reactions leaving uninvolved ions remaining in solution	$Ca^{2+} + SO_4^{2-} \rightleftharpoons CaSO_4$	Solubility product of precipitate
Complete ion formation (most stable complex of several alternatives will prevail)	Removal of ions from solution (convert them to soluble species that do not participate in reaction)	$Ag^+ + 2CN^- \rightleftharpoons Ag(CN)_2^-$	Dissociation of complex ion
Volatile product formation	Loss of CO_2 from aqueous solution	$H_2CO_3 \rightleftharpoons H_2O + CO_2$	Solubility of gas at reaction temperature

as shown in Table 3, occurs through removal of ions by (a) forming slightly ionized products, (b) forming slightly soluble products, (c) complex ion formation, and (d) gas evolution.

Oxidation–Reduction (Redox) Reactions. Some chemical reactions change element oxidation states as molecular configurations and chemical bondings are rearranged. Redox terminology is given in Table 4. Electron transfer takes place from an oxidation source to a reduction sink. Electrons are not lost, only transferred. The number of electrons transferred determines the relative stoichiometry between their oxidation half-reaction source and their reduction half-reaction sink. Redox reactions are recognized by finding changes in oxidation state. Except in disproportionation, oxidation occurs for one atom and reduction occurs for another.

Changes in oxidation state are determined by considering the following:

(a) Elemental state of the elements; that is, S, O_2, H_2, Fe, and so on; oxidation state is 0.

(b) Oxygen in other molecules: Oxidation state is -2 (exception, when covalently bonded in peroxide, -1).

(c) Hydrogen in other molecules: Oxidation state is $+1$ (exception in hydride, -1).

(d) Other elements according to position on periodic table and list of usual oxidation states.

(e) Summation of oxidation states in any molecule is 0, and in any ion it equals the net charge on the ion.

Table 4 Redox Terminology

Oxidation	Process producing electrons; electron source; occurs at anode in electrochemistry; electron loss produces increase in oxidation state.
Oxidizing agent	Species that is reduced is electron sink and consumes electrons produced by oxidation of other species.
Oxidation half-reaction	Charge-balanced equation for electron source (reversing equation makes it reduction half-reaction).
Oxidation potential	Energetics of oxidation half-reaction, expressed thermally as free energy change or electrically as voltage, under standard conditions.
Disproportionation	Self-redox; reaction occurring when metastable oxidation state simultaneously oxidizes and reduces atom to more stable higher and lower oxidation states.
Reduction	Process consuming electrons; electron sink; occurs at cathode in electrochemistry; electron gain causes reduction in oxidation state.
Reducing agent	Species that is oxidized is electron source and produces electrons for reduction of other species.
Reduction half-reaction	Charge-balanced equation for electron sink.
Reduction potential	Energetics of reduction half-reaction. Positive sign means reaction can proceed spontaneously as written. Reversal of reduction half-reaction changes it to oxidation half-reaction and changes algebraic sign of potential.

(f) Consult a table of reduction potentials (Table 10 later in the chapter) for typical half-reactions of the elements in acidic or neutral and in basic aqueous media.

Stoichiometric Equations. Stoichiometric equations can be written for reactions relating initial reactants and final products without regard to details such as equilibria, steps in the reaction mechanism, kinetics, percentage of conversion in a closed system, intermediate products, excess reactants required in the process, process variables, solvents, catalysts, and so. The balanced stoichiometric equation relates mass and moles of each product species to mass and moles of each of the reactants, assuming that the reaction goes to completion as written. The stoichiometric equation is the balanced chemical equation.

Balanced chemical equations involve three basic principles: (a) a *mass balance*, based on conservation of mass, which requires that total mass in equal total mass out; (b) an *atom balance*, based on constant mass of each atomic species involved in ordinary nonnuclear reactions, which requires that there is no change in the numbers of each atomic species; and (c) an *electron balance*, based on conservation of the number of electrons involved in oxidation–reduction reactions.

Example 2

1. *Reaction not involving change in oxidation states* in a mixture of ions that may proceed to complete conversion because a product is removed from the reaction system. The stoichiometric equation $CaCl_2 + Na_2SO_4 \rightarrow CaSO_4 + 2NaCl$ can be written as a net ionic reaction: $Ca^{2+} + SO_4^{2-} \rightarrow CaSO_4$ for the formation of slightly soluble $CaSO_4$, whose solubility is quantified by its solubility product (see Section 6). Atom balance and mass balance apply; but electron balance, although valid, is not required since no oxidation–reduction occurs.

2. *Reaction involving redox* requires atom, mass, and electron balances. A gross reaction for dissolving copper in concentrated nitric acid (with liberation of brown N_2O_4 gas resulting from air oxidation of NO, the reaction product) can be written as two half-reactions:

(a) $NO_3^- + 4H^+ + 3e^- \rightarrow NO + 2H_2O$

(b) $Cu \rightarrow Cu^{2+} + 2e^-$

Charge balance requires electron balance so twice (a) provides $6e^-$ that match the requirement of three times (b) for the net result $2NO_3^- + 8H^+ + 3Cu \rightarrow 2NO + 4H_2O + 3Cu^{2+}$. It is apparent that $8H^+$ must have come from $8HNO_3$, and the product $3Cu^{2+}$ must be accompanied by $6NO_3^-$ for charge balance. The final stoichiometric equation is $3Cu + 8HNO_3 \rightarrow$ $2NO + 4H_2O + 3Cu(NO_3)_2$. Equations (a) and (b) can be found among the equations in the table of reduction potentials (Table 10).

3. *Balance an unbalanced redox equation given* $Ag_2S + CN^- + O_2 \rightarrow S + Ag(CN)_2^- + OH^-$. Atom, mass, and electron balances are required. Water can participate in the reaction, but H^+ cannot exist in alkaline solution. Determine which one element is oxidized and between which oxidation states. Do the same for the element reduced. (a) Reduction of O_2 from zero to -2 oxidation state in basic solution. (b) Oxidation of S from -2 to zero oxidation state in basic solution. Write net reactions for (a) and (b) or find those reactions in the table of reduction potentials.

(a) $O_2 + 2H_2O + 4e^- \rightarrow 4OH^-$

Note the need for H_2O inclusion to balance this half-reaction.

(b) $S^{2-} \rightarrow S + 2e^-$ or $Ag_2S \rightarrow S + 2Ag^+ + 2e^-$

Equation (a) supplies $4e^-$, meeting the electron needs of twice Eq. (b) for the net result:

$$2Ag_2S + O_2 + 2H_2O \rightarrow 4OH^- + 2S + 4Ag^+$$

This matches that given, except it needs the CN^- added to give the $Ag(CN)_2^-$ complex. The final stoichiometric equation is

$$2Ag_2S + O_2 + 2H_2O + 8CN^- \rightarrow 4OH^- + 2S$$
$$+ 4Ag(CN)_2^-$$

A redox equation cannot be correctly balanced unless the electron balance requirement is first met. Trial and error on the basis of an atom balance usually leads to erroneous results.

4. *Balance the disproportionation*: $ClO^- \rightarrow ClO_3^- + Cl^-$. The oxidation state of chlorine changes from $+1$ to $+5$ and -1. The simplified oxidation half-reaction is $Cl^+ \rightarrow Cl^{5+} + 4e^-$. The simplified reduction half-reaction is $Cl^+ + 2e^- \rightarrow Cl^-$. Electron balance equates the oxidation half-reaction with twice the reduction half-reaction. Addition produces a net skeleton reaction of $3Cl^+ \rightarrow Cl^{5+} + 2Cl^-$. The final redox equation is $3ClO^- \rightarrow ClO_3^- + 2Cl^-$ after oxygens are appropriately inserted to obtain the required mass and atom balance for the equation.

Molar Volume of Gases. At standard conditions of $0°C$ and 1 atm ($32°F$, 14.7 psia), the molar volume of ideal gases is approximately 22.4 l/g mol or 359 ft^3/lb mol. Application of the ideal-gas law ($Pv = RT$) using ratios of absolute pressure and absolute temperature permits estimation of molar volumes at other temperatures and pressures. These molar volumes of gases can usefully elaborate the mass and molar quantities of gases directly available from stoichiometric equations.

Material Balances. Material balances for processes can be written on many bases, for example, per ton mole of product, per hour of operation, and per batch. All are based on a stoichiometric equation for the process and on a definition of the system such that mass or molar balances may be made at the steady state using "out minus in equals no accumulation" as the guiding principle. Selection of the element(s) used in the material balance depends upon the system; for example, the nitrogen of air in combustion processes passes through unchanged and quantities of other gases may be related to it via gas analyses.

Example 3

1. Determine the amount of CaO required to neutralize a short ton of 37 wt % H_2SO_4 waste.

   (a) Establish a balanced stoichiometric equation. Add the molecular weights of the species involved:

$$CaO + H_2SO_4 \rightarrow CaSO_4 + H_2O$$
$$56 \qquad 98 \qquad\quad 136 \qquad 18$$

   Note that a material balance exists.

   (b) Determine the weight of 100% H_2SO_4 that must be neutralized:

$$2000(0.37) = 740 \text{ lb}$$

   (c) Ratio reactants and products according to the stoichiometric equation:

$$\frac{\text{wt CaO}}{56} = \frac{740}{98} = \frac{\text{wt CaSO}_4}{136} = \frac{\text{wt H}_2\text{O}}{18}$$

   $\text{wt CaO} = 423 \text{ cb}$

2. Per 1000 standard cubic feet (scf) of methane, calculate maximum carbon black available from partial combustion, scf dry air required for that combustion, and composition of wet flue gas.

   (a) Establish balanced stoichiometric equation. Add atomic and molecular weights of all reactants and products:

$$CH_4 + O_2 \rightarrow C + 2H_2O$$
$$16 \qquad 32 \quad\; 12 \quad 2(18)$$

   Note that 1 lb mol of CH_4 can produce 1 lb atom of carbon black.

   (b) Convert 1000 scf CH_4 to pound moles and pounds (359 scf gas = 1 lb mol); therefore,

$$\frac{1000}{359} = 2.79 \text{ lb mol} \quad \text{or} \quad 44.7 \text{ lb, } CH_4$$

   (c) Ratio reactants and products according to the stoichiometric equation:

$$\frac{44.7}{16} = \frac{\text{lb O}_2}{32} = \frac{\text{lb mol O}_2}{1} = \frac{\text{lb C}}{12}$$
$$= \frac{\text{lb mol H}_2\text{O}}{2}$$

   Obtain 2.79 lb mol O_2, 2.79 lb mol (or 33.5 lb) C, and 5.58 lb mol H_2O.

   (d) Dry air composition is 21 vol% = 21 mol% oxygen and 79 mol % nitrogen and inerts:

$$\text{Dry air required} = 2.79\frac{100}{21}$$
$$= 13.3 \text{ lb mol, or}$$
$$13.3(359), = 4770 \text{ scf}$$

   (e) Assume complete electrostatic precipitation of carbon black and no water condensation. Then the flue gases consist of nitrogen, inerts, and H_2O vapor (nitrogen + inerts in = nitrogen + inerts out):

$$13.3(0.79) = 10.5 \text{ lb mol N} + \text{inerts}$$
$$\underline{5.58 \text{ lb mol H}_2\text{O}}$$
$$\text{Total: } 16.08 \text{ lb mol}$$

$$N_2 = \frac{10.5}{16.08} = 65.3 \text{ vol\%}$$

$$H_2O = \frac{5.58}{16.08} = 34.7 \text{ vol\%}$$

   (Ideal-gas behavior has been assumed in the moles-to-volume conversion.)

3. Products of hydrocarbon combustion analyzed on a dry basis are 10.34 vol % CO_2, 0.80 vol % CO, 5.16 vol % O_2, and the remainder N_2. Water is undetermined. Calculate the carbon/hydrogen weight ratio of the hydrocarbon fuel, assuming all hydrogen burns to water before any carbon is converted to other products, and the percentage of excess air entering the combustion zone.

   (a) Establish 100 lb mol dry product as the basis of calculation. On a molar basis, the products are

$$10.34 \text{ lb mol CO}_2$$
$$0.80 \text{ lb mol CO}$$
$$5.16 \text{ lb mol O}_2$$
$$\underline{83.70 \text{ lb mol N}_2}$$
$$\text{Total: } {\scriptstyle 100.00} \text{ lb mol dry products}$$

   (b) Use a nitrogen balance to determine the O_2 entering the combustion zone:

$$83.70 \text{ lb mol N}_2 \text{ in} = 83.70 \text{ lb mol N}_2 \text{ out}$$

$$O_2 \text{ of air entering} = 83.70\frac{21}{79}$$

$$= 22.25 \text{ lb mol}$$

(c) Determine the pound moles of oxygen accounted for in known products:

$$\text{in } CO_2\text{: } 10.34 \text{ lb mol}$$

$$\text{in CO: } \frac{0.80}{2} = 0.40 \text{ lb mol}$$

$$\text{as } O_2\text{: } \underline{5.16 \text{ lb mol}}$$

$$\text{Total: } \overline{15.90 \text{ lb mol}}$$

(d) Remainder of oxygen must exist as undetermined water:

$$22.25 - 15.90 = 6.35 \text{ lb mol } O_2 \text{ in water}$$

Therefore

$$2(6.35) = 12.70 \text{ lb mol } H_2O \text{ are in products}$$

(e) Hydrogen in products $= 12.70(1) = 12.70$ lb mol $= 25.4$ lb.
(f) Carbon in products $= 10.34(1) + 0.80(1) = 11.14$ lb atom $= 133.5$ lb; therefore

$$\text{C/H wt ratio} = \frac{133.5}{25.4} = 5.26$$

(g) Percentage of excess air:

$$100 \times \frac{\text{(total } O_2 \text{ entering)} - (O_2 \text{ required for complete combustion)}}{O_2 \text{ required for complete combustion}}$$

(h) Alternate method for calculating percentage of excess air:

Free oxygen in combustion products

$$= 5.16 \text{ lb mol, oxygen required to burn CO to } CO_2$$

$$= \frac{0.80}{2} = 0.40 \text{ lb mol}$$

Therefore excess O_2 beyond that required for complete combustion is 4.76 lb mol:

Percentage of excess O_2
$$= \text{percentage of excess air} = 100 \times \frac{4.76}{17.49}$$

$$= 27.2\%$$

Mixture and Dilution Calculations. Mixtures with properties linearly dependent on mass, volume, or mole fractions of components in the mixture can be handled by a simple summation:

$$P = \sum_i X_i P_i \qquad (1)$$

where

$$1 = \sum_i X_i \qquad (2)$$

X_i is the mass, volume, or mole fraction, as appropriate, of component i, and P_i is the property P of pure i. For example,

$$P \text{ of a binary mixture} = X_i P_i + (1 - X_i)P_i \qquad (3)$$

Example 4

1. A mixture of density 3.6 is desired from components having densities of 2.6 and 7.5:

$$1(3.6) = (x)2.6 + (1 - x)7.5$$

$$x = \frac{3.9}{4.9} = 0.796$$

$$1 - x = 0.204$$

Since density is weight per volume, this is a weight summation when x is the volume fraction.

2. Dilute 37 wt % solution with a 3 wt % solution to make a 20 wt % solution:

$$1(20) = x(37) + (1 - x)3 \qquad x = 0.50$$

$$(1 - x) = 0.50$$

Since (wt) $\times$ (wt%) = wt, this is also a weight summation when x is mass or weight fraction. Densities and assumption of volume additivity of ideal solutions are not required unless volume fractions are sought.

4 CHEMICAL THERMODYNAMICS

Chemical thermodynamics deals with work–energy relationships and the driving forces of chemical reactions. It is based on a wider variety of energy terms than the expansion–compression work, thermal, flow, kinetic, and potential energy terms of engineering thermodynamics. Basic relationships presented in the engineering thermodynamics section are augmented with other forms of work and energy that are applicable to the systems considered.

Energy Terms of Thermodynamics. Energy terms are the product of an intensive property (independent of quantity) and an extensive property (varies with amount) as shown in Table 5. When any of these

Table 5   Energy Terms of Thermodynamics

Kind of Energy	Intensive Property	Extensive Property	Energy Term
Expansion-contraction work	P, pressure	dV, volume	$P\,dV$
Thermal	T, temperature	dS, entropy	$T\,dS$
Mechanical	F, force	dX, displacement	$F\,dX$
Electrical	E, voltage	dQ, charge	$E\,dQ$ or $EF\,dn$
Magnetic	H, magnetic field intensity	dM, magnetization	$H\,dM$, work of magnetization
Surface	γ, surface tension	dA, area	$\gamma\,dA$, work of increasing area
Chemical	μ, chemical potential (partial molal free energy)	dN, number of moles	$\mu\,dN$, energy due to undergoing reaction or crossing phase boundaries

terms are involved in a chemical thermodynamic system, they are included in the basic equations of engineering thermodynamics.

Chemical Potential μ. For a multicomponent system with N_i moles of component i,

$$\mu_i = \left(\frac{\partial G}{\partial N_i}\right)_{T, P, N_{j \neq i}} = \left(\frac{\partial A}{\partial N_i}\right)_{T, V, N_{j \neq i}} \tag{4}$$

where G and A are the Gibbs free energy and the Helmholtz function, respectively.

Chemical Energy. As other forms of energy, all products of an intensive and an extensive property, chemical energy added to a system is $\mu_i\,dN_i$ for one component and $\Sigma \mu_i\,dN_i$ for all components of the system.

Chemical Thermodynamic Equations. Basic differential equations of engineering thermodynamics are expanded by adding the chemical energy and other appropriate energy terms. The thermal energy sign convention is used for chemical energy; positive terms are energy added and negative terms are energy removed from the system:

Relation of chemical potential to internal energy:

$$dU = T\,dS - P\,dV + \sum_i \mu_i\,dN_i \tag{5}$$

Relation to enthalpy:

$$dH = T\,dS + V\,dP + \sum_i \mu_i\,dN_i \tag{6}$$

Relation to Gibbs free energy:

$$dG = -S\,dT + V\,dP + \sum_i \mu_i\,dN_i \tag{7}$$

Relation to Helmholtz work function:

$$dA = -S\,dT - P\,dV + \sum_i \mu_i\,dN_i \tag{8}$$

For constant-temperature $(dT = 0)$ and constant-pressure $(dP = 0)$ reaction systems, Eq. (7) reduces to the most widely used relationship:

$$dG = \sum_i \mu_i\,dN_i \tag{9}$$

Standard Thermodynamic Data. Augmenting usual engineering sources of thermodynamic data (steam tables, gas tables, and special compilations for working fluids), useful tabulations of thermochemical data are available at 25°C (298 K).

$\Delta H_f{}^\circ =$ heat of formation from elements, kcal/g mol ($\Delta H_f{}^\circ$ of all elements in standard state arbitrarily taken as zero)

$\Delta G_f{}^\circ =$ Gibbs free energy of formation from elements, kcal/g mol (some references use ΔF°)

$S^\circ =$ entropy, cal/g mol K

$C_p{}^\circ =$ heat capacity, cal/g mol K

New compilations may be expressed in joules rather than calories, where 1 cal = 4.184 J. The degree superscript designates standard state at 25°C and unit activity or 1 atm fugacity. For a pure compound with more than one allotropic form, this refers to the stable modification. Aqueous solution data are given at 1 molal activity. The standard state is independent of pressure or concentration variables because activity (as a thermodynamically corrected concentration) and fugacity (as a corrected pressure to give ideal-gas behavior) have been introduced as substitute variables.

Activity a. Activity is a thermodynamically effective concentration used in lieu of actual concentrations to compensate for deviations of gases from ideality, incomplete ionization of strong electrolytes in solution, and other discrepancies between calculated and experimental behavior.

Fugacity f. Fugacity deviations from the model behavior of a gas equals its activity. Fugacity is an effective partial pressure expressed in atmospheres.

For ideal gases fugacity f exactly equals partial pressure. Since real gases in the standard state (in their usual phase at 1 atm pressure and 25°C) deviate slightly from ideality, the activity of real gases is defined as the ratio of fugacity f to fugacity f° in a standard state where $f^\circ = 1$ atm. For all practical purposes, $f^\circ = 1$ atm pressure. Nonidealities of real gases at other temperatures and pressures are handled by an activity coefficient γ.

Activity Coefficient γ. Activity coefficient is the ratio of fugacity to partial pressure or the ratio of activity to molal concentration.

For ideal gases $\gamma = 1$.

For real gases γ varies with T and P and is the ratio f/p, where f is the fugacity of the gas and p is its partial pressure.

For ions in solution $\gamma = a/m$, where a is the activity of the ion and m is its molality in solution.

Activity of Gases in Mixtures. The activity of any gas in a mixture (Lewis and Randall rule) is calculated as

$$a_i = p_i \gamma_i \tag{10}$$

where p_i is the partial pressure in atmospheres and γ_i is the fugacity coefficient of pure i at the temperature and total pressure of the mixture.

Calculation of ΔG° at Elevated Temperatures. The effect of temperature on ΔG° at constant pressure and at constant number of moles can be calculated from thermodynamic data at another temperature, which can be taken to be 298 K. At any temperature T,

$$\Delta G_T^\circ = \Delta H_T^\circ - T\,\Delta S_T^\circ \tag{11}$$

$$\Delta H_T^\circ = \Delta H_{298}^\circ + \int_{298}^{T} \Delta C_P \, dT' \tag{12}$$

$$\Delta S_T^\circ = \Delta S_{298}^\circ + \int_{298}^{T} \frac{\Delta C_P \, dT'}{T'} \tag{13}$$

where ΔC_P is the difference in heat capacity between products and reactants. If any phase changes occur between T and 298 K, there are additional contributions to Eqs. (12) and (13). In the absence of any phase changes, Eqs. (11)–(13) can be combined to obtain

$$\Delta G_T^\circ = \Delta H_{298}^\circ - T\,\Delta S_{298}^\circ + \int_{298}^{T}\left(1 - \frac{T}{T'}\right)\Delta C_P\,dT' \tag{14}$$

If ΔC_P is represented by a power series in T', the required integration in the last term is straightforward. (If ΔC_P is small and/or T is close to 298 K, the value of the integral is approximately zero.) An example is: Calculate ΔG_{298}° for the reaction: $CO + \frac{1}{2}O_2 \rightleftharpoons CO_2$ at unit fugacity and activity of all products and reactants at (a) 298 K and (b) 800 K:

(a) At standard conditions 25°C (298 K) and 1 atm

	CO	O_2	CO_2
ΔH_f° (kcal/mol)	-26.42	0	-94.05
ΔS° (cal/K mol)	47.30	49.00	51.06

Therefore,

$$\Delta H_{298}^\circ = \Delta H_f^\circ \text{ (products)} - \Delta H_f^\circ \text{ (reactants)}$$

$$= -94.05 - (-26.42) - \tfrac{1}{2}(0)$$

$$= -67.63 \text{ kcal/mol}$$

$$\Delta S_{298}^\circ = 51.06 - 47.30 - \tfrac{1}{2}(49.00)$$

$$= -20.74 \text{ cal/K mol}$$

and

$$\Delta G_{298}^\circ = \Delta H_{298}^\circ - 298\Delta S_{298}^\circ$$

$$= -67.63 - 298(-20.74/100)$$

$$= -61.45 \text{ kcal/mol}$$

(b) The heat capacity data are

For CO: $C_p = 6.3424 + 1.8363 \times 10^{-3} T$
$$- 0.2801 + 10^{-6} T^2,$$
$$\text{cal/g mol K}$$

For O_2: $G_p = 6.0954 + 3.2533 \times 10^{-3} T$
$$- 1.0171 + 10^{-6} T^2$$

For CO_2: $C_p = 6.393 + 10.100 \times 10^{-3} T$
$$- 3.405 \times 10^{-6} T^2$$

Here, ΔC_p for the reaction equals the C_p for products $-C_p$ for the reactants:

$$\Delta C_p = 1(C_{p,CO_2}) - \tfrac{1}{2}(C_{p,O_2})$$

$$= -2.997 + 6.637 + 10^{-3} T$$

$$- 2.716 \times 10^{-6} T^2$$

$$= \alpha + \beta T + \gamma T^2$$

At $T = 800$ K, one finds ΔG_{800}° from Eq. (14) and

$$\Delta G_{800}^\circ = -67.63 - 800\left(-\frac{20.74}{1000}\right) + 0.19$$

$$= -50.85 \text{ kcal/mol}$$

Note that in this case the value of the integral, 0.19 kcal/mol, made a minor contribution to the net result.

Calculation of ΔG at Elevated Pressure*

At constant temperature and constant number of moles, $dT = 0$ and $dN_i = 0$, Eq. (7) reduces to $dG = +V\,dP$. The calculation of ΔG for ideal gases, real gases, solids, and liquids at any pressure is dependent only on an expression for V as a function of P.

For *ideal gases* $PV = RT$, and $dG = (RT/P)\,dP = RT\,d(\ln P)$. Integrating between limits gives

$$\Delta G = RT \ \ln \frac{P_2}{P_1} \tag{15}$$

For *real gases* more complicated equations of state relating V and P can be substituted for the ideal-gas law and the integration performed. Deviations from ideality can also be incorporated by the use of the fugacity f_i and

$$\Delta G = RT \ \ln \frac{f_2}{f_1} \tag{16}$$

For *incompressible liquids and solids* V is constant, so

$$\Delta G = V(P_2 - P_1) \tag{17}$$

using appropriate units. The magnitude of this change for all condensed phases is very small.

For *compressible liquids and solids* V is dependent on isothermal compressibility β, $V = V_0(1 - \beta P)$:

$$dG = V_0(1 - \beta P)\,dP = V_0\,dP - \beta V_0 P\,dP \tag{18}$$

or

$$\Delta G = V_0(P_2 - P_1) - \frac{\beta V_0}{2}(P_2^2 - P_1^2)$$

Standard Gibbs Free-Energy Change and Reactivity. The ΔG_T° for a reaction is

$$\Delta G_T^\circ = \sum_i \nu_i G_{T_i}^\circ \tag{19}$$

where ν_i is the stoichiometric coefficient of species i, taken as positive for products and negative for reactants. Here, ΔG_T° is the standard Gibbs free-energy change for a reaction at temperature T when all reactants and products are at unit activity (unit fugacity for gases). The ΔG_T° is independent of pressure; its magnitude as a driving force for predicting reactivity is very important:

---

*Note: ΔG is not the same as the pressure-independent ΔG° introduced in the preceding.

If ΔG_T° is negative	Reaction can occur as written; a forward driving force or chemical potential exists
If ΔG_T° is zero	No driving force exists; system is at equilibrium
If ΔG_T° is positive	Reverse reaction can occur; a reverse driving force or chemical potential exists

$$\tag{20}$$

Since the rate of reaction is dependent on temperature and activation energy, the value of ΔG° provides no information about the rate.

Example 5

1. Reaction solid 1 CaO + gaseous 1 CO_2 = solid 1 $CaCO_3$

$\Delta G_f^\circ =$	$\Delta G_f^\circ =$	$\Delta G_f^\circ =$
-144.4	-94.26	-269.78
kcal/gmol	kcal/gmol	kcal/gmol

   where ΔG° for the reaction equals $(1)(-269.78) -(1)(-144.4) - (1)(-94.26) = -31.1$ kcal. For the reaction as written at 298 K, the process is spontaneous.

2. Phase change:

$$C_{graphite} = C_{diamond}$$
$$\Delta G_f^\circ = 0 \qquad \Delta_f^\circ = +0.68$$

   $\Delta G_{298}^\circ = 1(+0.68) - 1(0) = +0.68$ kcal for the reaction as written. The positive value of ΔG° indicates that the reaction as written cannot proceed at the standard condition, but the reverse reaction of diamond transition to graphite would be spontaneous if kinetic factors were favorable.

Thermodynamics of Equilibrium. At equilibrium the capability of any system for doing work is zero because there is no net energy available as the driving force to modify the status quo. In chemical systems at constant temperature and pressure, the change in the Gibbs free energy, ΔG, is zero at equilibrium between reactants and products; it is negative when a thermodynamic driving force for forward reaction is existent. (For equilibrium at constant temperature and constant volume, the change in the Helmholtz work function ΔA is zero.) Since most significant chemical equilibria are at constant temperature and pressure, this section contains only the relationships involving Gibbs free energy. The conditions for chemical equilibrium are

$$dG = \begin{cases} 0 & \text{at constant } T, \ P \\ \sum_i \mu_i \, dN_i = 0 \end{cases}$$

$$\tag{21}$$

$$dA = \begin{cases} 0 & \text{at constant } T, \ V \\ \sum_i \mu_i \, dN_i = 0 \end{cases}$$

Standard Gibbs Free-Energy Change and Equilibrium Constant.

The magnitude of the driving force ΔG_T° at unit activity or unit fugacity of reactants and products, available to cause the reaction to proceed, was calculated for temperature T. The standard free energy ΔG_T° was considered pressure independent because unit activities or fugacities were involved.

Here, ΔG_T° is related to the equilibrium constant K, where K is expressed using activities (or fugacities of gases). This K is pressure independent because the product of any set of equilibrium fugacities, each to the power of its stoichiometric coefficient, is a constant:

$$\Delta G_T^\circ = -RT \ln K \qquad (22)$$

where K is expressed using activities (or fugacities) of gases.

Temperature Dependence of Equilibrium Constant.

The temperature dependence of ΔG_T° directly affects K. The temperature dependence of K can be obtained by combining Eqs. (11) and (22) to yield

$$\ln K = -\frac{\Delta H_T^\circ}{RT} + \frac{\Delta S_T^\circ}{R} \qquad (23)$$

If T is close to 298 K and/or ΔC_P for the reaction is approximately zero,

$$\ln K = -\frac{\Delta H_{298}^\circ}{RT} + \frac{\Delta S_{298}^\circ}{R} \qquad (24)$$

Pressure Dependence of Equilibrium Constant.

For an equilibrium constant expressed in activities and fugacity, the effect of pressure is introduced through change of activity and fugacity terms.

For *gases*: Fugacity equals the partial pressure of ideal gases and corrects for nonlinearities in the partial pressure of real gases.

For *liquids* or *solids*: Activities of solids and liquids are essentially invariant. The change in ΔG with pressure for these condensed phases, calculated from Eq. (17) $[\Delta G = V_0(P_2 - P_1)]$ is of small magnitude.

For *solutes*: The changes of activity coefficient and equilibrium constant with pressure are given by

$$\frac{d(\ln \gamma_i)}{dP} = \frac{\overline{v}_i}{RT} \qquad (25)$$

where $\overline{v}_i$ is the partial molal volume of component i. The magnitude of pressure dependence of the equilibrium constant is extremely small:

$$\frac{d(\ln K)}{dP} = \frac{\sum_i \nu_i \overline{v}_i}{RT} \qquad (26)$$

where $\sum \nu_i \overline{v}_i$ is the difference in the partial molal volumes of products and reactants.

Thermodynamics and Phase Equilibria.

Phase equilibrium exists, as with other equilibria, when change in Gibbs free energy $\Delta G = 0$. Since the chemical potential, μ_i of component i is defined as $(\partial G / \partial N_i)_{T,P,N_{j \neq i}}$, μ is the partial molal free energy G_i. At constant temperature and pressure, $\Delta G = \Sigma_i \mu_i \, dN_i$. Phase equilibrium exists (with dN_i moles of component i transferred between phases) when the partial molal free energy $\overline{G}_i$ or chemical potential μ_i of each component is the same in all phases:

$$\mu_i \text{ (phase 1)} = \mu_i \text{ (phase 2)} \qquad (27)$$

Since $\overline{G}_i = \overline{G}_i^\circ + RT \ln a_i$, or $\mu_i = \mu_i^\circ + RT \ln a_i$, it follows that activities in each phase can be used with standard free energies to determine chemical potentials. For liquid phases, activities can be related to concentrations, and for vapor phases, activities are related to partial pressures. These relationships allow calculation for the effects of changes in temperature, pressure, and concentrations.

Partial Molal Quantities.

Mixture nonidealities are often empirically treated by summing partial molal quantities contributed by each component. Partial molal quantities, such as molal volume or molal free energy, are of interest in phase equilibria:

$$dV = \sum_i \overline{v}_i \, dN_i \qquad dG = \sum_i \overline{G}_i \, dN_i \qquad (28)$$

where the overbar indicates partial molal quantities.

5 THERMOCHEMISTRY

ΔH° Standard Heats of Reaction and Formation.

Energy is transferred in all reactions and phase changes. The enthalpy change during an endothermic reaction is positive, in agreement with the basic sign convention of thermodynamics. Conversely, the enthalpy change in an exothermic process is negative.

Data at a standard condition of 25°C (298 K) and 1 atm (or at unit fugacity of gases and unit activity of solutes) are designated by the ° superscript. Standard heats of formation of compounds from their elements are available, all based on a reference state of $\Delta H^\circ = 0$ for elements in their stable phase at the standard condition. These data are given in kilocalories per gram mole.

Example 6

$$H_2(g) + \tfrac{1}{2}O_2(g) \rightarrow H_2O(g)$$

$$\Delta H_f^\circ = -57.80 \text{ kcal/g mol water(g)}$$

$$H_2(g) + \tfrac{1}{2}O_2(g) \rightarrow H_2O(l)$$

$$\Delta H_f^\circ = -68.32 \text{ kcal/g mol water(l)}$$

Parentheses enclose the phase of each species involved.

Additivity of Extensive State Functions. Extensive properties that are state functions, such as H, G, and S, are additive over successive reactions carried out at constant temperature and pressure according to the principle known as Hess's law:

$$\Delta X = \sum_i (v_i X_i)_1 + \sum_i (v_i X_i)_2 + \cdots \quad (29)$$

where v_i is the stoichiometric coefficient of species i. Thus ΔX for a reaction may be obtained from the summation of a series of simpler or better known reactions that give the same overall chemical balance. Thus the heat of a reaction may be obtained from the heats of combustion or formation of the compounds involved in the reaction. If X is any extensive property of the system and X_i is the corresponding molal property of pure i for the reaction,

$$\Delta X = \sum_i v_i X_i \quad (30)$$

In simultaneous or successive reactions, where the numerical subscripts refer to the different reactions,

$$dN_i = (dN_i)_1 + (dN_i)_2 + \cdots \quad (31)$$

and

$$N_i = N_i^\circ + (v_i \Delta N)_1 + (v_i \Delta N)_2 + \cdots \quad (32)$$

Estimating Heat of Formation and Reaction from Bond Energies. In the absence of heat-of-formation data for organic compounds, the standard heat of formation can be estimated by summing average nonpolar energies for each of the bonds of the compound. These energies are for gaseous molecules and are to be used with caution since bond strengths are influenced by molecular configuration, being reduced in resonance structures and increased in polar compounds. Heats of formation can be expected within $\pm 10\%$ of the experimentally determined values using the values of Table 6.

Example 7. Heat of formation of acetaldehyde

$$
\begin{array}{ccc}
\text{H} & \text{H} & \\
| & / & \\
\text{H}-\text{C}-\text{C} & = & \text{O}: \\
| & & \\
\text{H} & &
\end{array}
$$

$$2\text{C(s)} + 2\text{H}_2 + \tfrac{1}{2}\text{O}_2 \rightarrow \text{CH}_3\text{CHO}$$

Energy evolved to make products bonds:

4 C–H bonds at 98.7	$= -394.8$
1 C–C bond	$= -82.6$
1 C=O aldehyde type bond	$= -176$
	$\overline{-653.4}$ kcal/mol

Energy added to break reactant bonds:

2 H–H bonds at 104.2	$= +208.4$
$\tfrac{1}{2}$O=O bond at 119.1	$= +59.6$

Table 6　Bond Energies, kcal/g mol at 25°C[a]

H–H	104.2	C–N	72.8
O=O	119.1	C=N	147
N≡N	225.8	C≡N	212.6
C=O (carbon monoxide type)	255.8	C–O	85.5
C–H	98.7	C=O (carbon dioxide type)	192
N–H	93.4	C=O (aldehyde type	176
O–H	110.6	C=O (ketone type)	179
S–H	83	H–F	134.6
P–H	76	H–Cl	103.2
N–N	39	H–B	87.5
N=N	100	H–I	71.4
O–O	35	C–F	116
S–S	54	C–Cl	81
N–O	53	C–Br	68
N=O	145	C–I	51
F–F	36.6	C–S	65
Cl–Cl	58.0	N–F	65
Br–Br	46.1	N–Cl	46
I–I	36.1	O–F	45
C–C	82.6	O–Cl	52
C=C	145.8	O–Br	48
C≡N	199.6		

[a]These bond energies apply to energies of formation of gaseous molecules from gaseous atoms. For graphitic carbon to gaseous carbon atoms the heat of sublimation and atomization is 172 kcal/g mol. Heat of vaporization of water is 10.5 kcal/g mol.

2(heat of sublimation and
  atomization of graphite at
  172) $= \dfrac{+344}{+612.0}$ kcal/mol

Net enthalpy–energy
  change for reaction $= -653.4 + 612.0$
                                $= -41.4$ kcal/mol

Compare with -39.8 actually evolved.

Estimating Heats of Reaction.

Standard heats of reaction can be estimated by considering only the energy added to break specific bonds of reactants and energy released in forming the new product bonds. Where reactants and products are liquid or solid, enthalpies of vaporization or sublimation must be used to get reactants and products into and out of the vapor phase where the bond energies are valid.

ΔH for Changes of Phase.

Thermal energy changes during phase transitions are significant factors in the overall energetics of chemical reaction systems. The numerical value of enthalpy of vaporization, sublimation, fusion, and other phase transitions is $f(T, P)$.

Trouton's Rule for Estimating ΔHᵥ.

Based on an estimated average entropy change of 21 cal/K g mol occurring at the normal boiling point at constant pressure with no change in chemical potential between phases,

$$dH = T\,ds + V\,dP + \sum_i \mu_i\,dN_i \qquad (33)$$

reduces to $\Delta H_v = 21\ (T\ \text{K})$. This generality applies for a number of nonpolar materials but is grossly inaccurate for others where $\Delta S \neq 21$.

Clausius–Clapyron Equation for Phase Transitions

$$\frac{dP}{dT} = \frac{\Delta H}{T\,\Delta V} = \frac{\Delta S}{\Delta V} \qquad (34)$$

where ΔH and ΔS are enthalpy and entropy changes of phase transitions at temperature T, and ΔV is the associated molal volume change. An integrated form, assuming ideal-gas behavior of the vapor phase and negligible volume for condensed phases, is applicable to sublimation and vaporization:

$$\frac{d\ln P}{dT} = \frac{\Delta H}{RT^2} \quad \text{or} \quad \frac{d(\ln P)}{d(1/T)} = -\frac{\Delta H}{R} \qquad (35)$$

Enthalpies of Solution, Dilution, Solvation, Adsorption, Crystallization, and Mixing.

Very limited published data usually require direct experimental measurement in all but the most widely used industrial systems. Energies involved in solvation, association, and hydration are small relative to those normally associated with chemical bond rupture and formation.

Heat Capacity.

The basic relationship that $\Delta H = \int C_p dT$ for temperature changes of a system plus enthalpy of phase transitions provides the basis for thermal calculations at other than the standard temperature. Further thermodynamic use of heat capacity data is in calculation of entropy changes via $\Delta S = \int (C_p/T)dT$ and calculation of Gibbs free-energy changes using the definitive relationship $G = H - TS$.

Heat Capacity Equations.

Power functions of temperature best fit molar C_p data over a wide temperature range. These are of the form $C_p = \alpha + \beta T + \gamma T^2$. Since C_p is expressed as either cal/g mole K or Btu/lb mole $^\circ R$, care must be used in either inserting T in Kelvins in the heat capacity equation or else using degrees Rankine divided by 1.8 to obtain the proper numerical value of C_p. Typical heat capacity equations are given in Table 7.

Mean Molar Heat Capacity.

Integration of heat capacity equations between 25°C (298 K, 77°F, 537°R) and temperature T yields a mean molar heat capacity over the range 25°C to T that grossly simplifies thermal calculations. Graphs of these results for the usual gases are given by Smith and Van Ness.[1] Mean molar heat capacities between any pair of temperatures can be obtained by difference.

Heat Capacities of Ideal Gases.

The C_p molar heat capacity at a constant pressure is independent of pressure. The C_v molar heat capacity at constant volume is independent of volume. $C_p - C_v = R$, where $R = 1.985$ cal/g mol K. The approximate values C_v of $\frac{3}{2}R$ for monatomic, $\frac{5}{2}R$ for diatomic, and $\frac{7}{2}R$ for triatomic ideal gases are inadequate for most engineering purposes.

Zero-Pressure Heat Capacity of Real Gases, C_p°.

Zero-pressure heat capacities for an ideal-gas state can be calculated from spectroscopic data. Since C_p for ideal gases is independent of pressure and is a function of temperature only, it can be used for real gases under T, P conditions where the real gas does not grossly deviate from ideal-gas behavior. For conditions near critical and near the vapor–liquid phase change, recourse is made to more detailed compilations of specific thermal data for that system (steam tables, gas tables, etc.).

C_p of Liquids and Solids.

The relative incompressibility of liquids and solids usually reduces the difference between C_p and C_v to less than the experimental error in their measurement. The specific heat of liquids and solids increases with temperature but to a lesser extent than for gases. Experimental data are reported on both molar and unit mass bases at one temperature, average over a temperature range, or as power functions of T.

Kopp's Rule.

The molar C_p° of solid molecules at 25°C is approximated by summation of contributory

Table 7 Molar Heat Capacities of Gases in Ideal Gaseous State[a]

Compound	Formula	a	$b \times 10^3$	$c \times 10^6$	$d \times 10^9$	Range (K)	Accuracy (%)
Bromine	Br_2	8.4228	0.9379	−0.3555	—	300–1500	0.5
Chlorine	Cl_2	7.5755	2.4244	−0.9650	—	300–1500	0.5
Fluorine	F_2	6.115	5.864	−4.186	0.9797	273–2000	0.5
Hydrogen	H_2	6.9469	−0.1999	0.4808	—	300–1500	0.3
Iodine	I_2	8.504	1.3135	−1.0684	0.3125	273–1773	0.1
Nitrogen	N_2	6.4492	1.4125	−0.0807	—	300–1500	1
Oxygen	O_2	6.0954	3.2533	−1.0171	—	300–1500	0.5
Sulfur	S_2	6.499	5.298	−3.888	0.9520	273–1773	0.5
Air		6.557	1.477	−0.2148	—	273–3773	1
Ammonia	NH_3	6.189	7.887	−0.728	—	273–1000	0.5
Carbon dioxide	CO_2	6.393	10.100	−3.405	—	300–1500	1
Carbon monoxide	CO	6.3424	1.8363	−0.2801	—	300–1500	0.5
Cyanogen	$(CN)_2$	9.892	14.484	−6.207	—	291–1000	0.5
Hydrogen bromide	HBr	5.5776	0.9549	0.1581	—	300–1500	0.5
Hydrogen chloride	HCl	6.732	0.4325	0.3697	—	300–1500	0.5
Hydrogen cyanide	HCN	5.974	10.208	−4.317	—	300–1000	0.5
Hydrogen iodide	HI	6.702	0.4546	1.216	−0.4813	273–1873	0.5
Hydrogen sulfide	H_2S	6.385	5.704	−1.210	—	298–1500	1
Nitric oxide	NO	7.020	−0.370	2.546	−1.087	298–1500	0.5
Nitrous oxide	N_2O	6.529	10.515	−3.571	—	298–1500	1
Phosgene, carbonyl chloride	$COCl_2$	10.35	1.653	−8.408	—	273–973	0.5
Phosphine	PH_3	4.496	14.372	−4.072	—	298–1500	0.5
Phosphorus pentachloride	PCl_5	4.739	107.329	−119.2	—	298–500	2
Sulfur dioxide	SO_2	6.147	13.844	−9.103	2.057	273–1773	0.5
Sulfur trioxide	SO_3	6.077	23.537	−0.687	—	298–1200	1.5
Stannic chloride	$SnCl_4$	21.72	6.33	—	—	273–573	1
Water	H_2O	7.219	2.374	0.267	—	298–1500	0.5
Methane	CH_4	3.381	18.044	−4.300	—	298–1500	1
Ethane	C_2H_6	2.247	38.201	−11.049	—	298–1500	0.5
Propane	C_3H_8	2.410	57.195	−17.533	—	298–1500	1.5
n-Butane	C_4H_{10}	4.453	72.270	−22.214	—	298–1500	1.5
2-Methyl propane	C_{10}	3.332	75.214	−23.743	—	298–1500	1.5
n-Pentane	C_5H_{12}	5.910	88.449	−27.388	—	298–1500	1.5
n-Hexane	$C_6H_{13}4$	7.477	104.422	−32.471	—	298–1500	1.5
2,2-Dimethyl butane	C_6H_{14}	0.593	133.001	−52.878	—	298–1000	0.5
n-Heptane	C_7H_{16}	9.055	120.352	−37.528	—	298–1500	1.5
n-Octane	C_8H_{18}	10.626	136.398	−42.592	—	298–1500	1.5
2,2,4-Trimethyl pentane	C_8H_{18}	−3.2	152.5	—	—	400–500	2
Ethene	C_2H_4	2.830	28.601	−8.726	—	298–1500	1.3
Propene	C_3H_6	2.253	45.116	−13.740	—	298–1500	1
1-Butene	C_4H_8	5.132	61.760	−19.322	—	298–1500	1.5
cis-2-Butene	C_4H_8	1.625	64.836	−20.047	—	298–1500	1.3
trans-2-Butene	C_4H_8	4.967	59.961	−18.147	—	298–1500	0.8
Ethyne	C_2H_2	7.331	12.622	−3.889	—	298–1500	1.5
Propyne	C_3H_4	6.334	30.990	−9.457	—	298–1500	1
2-Butyne	C_4H_6	5.700	48.207	−14.479	—	298–1500	0.5
Benzene	C_6H_6	−0.409	77.621	−26.429	—	298–1500	3
Toluene	$C_6H_5CH_3$	0.576	93.493	−31.227	—	298–1500	2.5
Cyclopropane	C_3H_6	−6.481	82.06	−55.77	15.61	273–973	0.5
Cyclopentane	C_5H_{10}	−5.763	97.377	−31.328	—	298–1500	2.4
Cyclohexane	C_6H_{12}	−7.701	125.675	−41.584	—	298–1500	2
Methyl cyclohexane	$C_6H_{11}CH_3$	−4.624	140.87	−46.698	—	298–1500	2
Formaldehyde	HCHO	4.498	13.953	−3.730	—	291–1500	0.5
Acetaldehyde	CH_3CHO	7.422	29.029	−8.742	—	298–1500	1.5
Acetone	$(CH_3)_2CO$	5.371	49.227	−15.182	—	298–1500	1.5
Methanol	CH_3OH	4.398	24.274	−6.855	—	273–1000	2
Ethanol	C_2H_5OH	6.990	39.741	−11.926	—	298–1500	0.8
2-Propanol	$(CH_3)_2CHOH$	0.7936	85.02	−50.16	11.56	273–1473	0.2
n-Propanol	$C_2H_5CH_2OH$	−1.307	92.35	−58.00	14.14	273–1473	0.5

Table 7 *(Continued)*

Compound	Formula	a	$b \times 10^3$	$c \times 10^6$	$d \times 10^9$	Range (K)	Accuracy (%)
Ethyl ether	$(C_2H_5)_2O$	−24.83	338.7	−593	—	300–400	2
Ethylene oxide	$(CH_2)_2O$	−1.12	4.925	−23.89	3.149	273–973	0.3
Bromomethane	CH_3Br	4.184	22.445	−7.496	—	300–1200	1
Chloromethane	CH_3Cl	3.563	22.998	7.571	—	273–773	0.5
Fluoromethane	CH_3F	3.616	18.239	−2.035	—	298–600	0.5
Iodomethane	CH_3I	4.105	24.487	−9.733	—	300–600	0.5
Dichloromethane	CH_2Cl_2	4.309	31.67	−16.35	—	250–600	0.5
Fluorochloromethane	CH_2FCl	4.292	27.025	−10.605	—	250–600	0.5
Tribromomethane, bromoform	$CHBr_3$	9.356	32.319	−21.272	—	300–600	0.1
Trichloromethane, chloroform	$CHCl_3$	7.052	35.598	−21.686	—	273–773	0.5
Methyl cyanide	CH_3CN	5.018	27.935	−9.302	—	291–1200	0.5

$^a C_p^\circ = a + bT + cT^2 + dT^3$, where T is in Kelvins.

Source: Reprinted with permission from Wilson and Ries, *Principles of Chemical Engineering Thermodynamics*, New York, McGraw-Hill, 1956.

Table 8   Contributory Atomic C_p° Values for Kopp's Rule^a

C	1.8
H	2.3
B	2.7
Si	3.5
O	4.0
F	5.0
N	5.0
P	5.4
S	5.4
Mg	5.7
Al	5.8

^{a}For all elements of higher molecular weight use an average value of 6.4 cal/g atom K
Example: $Ag_2CO_3 = 2(6.4) + 1.8 + 3(4.0) =$ 26.6 cal/g mol K, (actual 26.8).

atomic C_p° values for constituents. Average values are given in Table 8.

C_p of Solutions and Mixtures.   In lieu of experimental data, the heat capacity of a solution or mixture can be estimated by

$$C_{p,\text{mixture}} = \sum_i y_i C_{pi} \tag{36}$$

where y_i is the mole fraction and C_{pi} is the molar heat capacity of component i.

Effect of Temperature on Heat of Reaction. Consider the reactants as one state and the products as another and that thermodynamically one can proceed from one state function to another regardless of the path. From reactants at temperature T, one can change temperature (using heat capacities and the enthalpies of any phase changes) to 25°C, then utilize the standard heat of reaction at 25°C and 1 atm, and then change products back to temperature T.

Example 8.   Calculate the heat of the ammonia synthesis reaction, $N_2 + 3H_2 \rightarrow 2\,NH_3$, as a function of temperature T. Given is the standard heat of reaction $\Delta H^\circ = -22.08$ kcal for the equation as written at 25°C (298 K) and 1 atm.

The heat capacity equations for reactants and products are

H_2: $6.62 + 0.0081\,T\,\text{(K)cal/g mol K}$ or

$$6.62 + \frac{0.00081}{1.8}T(^\circ R)\frac{\text{Btu}}{\text{lb mol}^\circ R}$$

N_2: $6.50 + 0.00100\,T$ K

NH_3: $6.70 + 0.00630\,T$ K

1. Cool reactants from T down to 298°K:

$$\Delta H_1 = \sum \int_T^{298} C_P\,dt = 1\left[(6.50)(298 - T)\right.$$

$$\left. + 0.00100\left(\frac{298^2 - T^2}{2}\right)\right]$$

$$+ 3\left[(6.62)(298 - T) + 0.00081\right.$$

$$\left. \times \left(\frac{298^2 - T^2}{2}\right)\right]$$

$$= 26.36(298 - T) + 0.00343$$

$$\times \left(\frac{298^2 - T^2}{2}\right)$$

$$= -26.36(T - 298) - 0.00343$$

$$\times \left(\frac{T^2 - 298^2}{2}\right)$$

2. Standard heat of reaction $\Delta H° = -22.08$ kcal/g mol quantities for $N_2 + 3H_2 = 2\,NH_3$ at the 298 K. Given is $\Delta H° = -11.04$ kcal for $\frac{1}{2}N_2 + \frac{3}{2}H_2 = NH_3$ in tables. Doubling the stoichiometric quantities doubles heat of reaction. Multiply by 1000 to convert the Btu/lb mol quantities.

3. Heat products from 298 K up to T K:

$$\Delta H_2 = \int_{298}^{T} C_p \, dt$$

$$= 2\left[6.70(T - 298) + 0.00630 \times \left(\frac{T^2 - 298^2}{2}\right) \right]$$

$$= 13.40(T - 298) + 0.01260 \times \left(\frac{T^2 - 298^2}{2}\right)$$

4. Algebraically sum the three ΔH values to get ΔH_T for the reaction at temperature T K:

$$\Delta H_1 = -26.36(T - 298) \\ -0.00343\left(\frac{T^2 - 298^2}{2}\right)$$
$$\Delta H° = -22.08$$
$$\Delta H_2 = +13.40(T - 298) \\ +0.01260\left(\frac{T^2 - 298^2}{2}\right)$$
$$\overline{\Delta H_T = -22.08 - 12.96(T - 298) \\ +0.00917\left(\frac{T^2 - 298^2}{2}\right)}$$

kcal/g mol quantities for the equation as written.

5. Convert to Btu and lb mol units for T °R since

$$298 \text{ K} = 537°R$$

$$\Delta H_T = -22.08(1000) - \frac{12.96}{1.8}(T - 537) \\ + \frac{0.00917}{1.8}\left(\frac{T^2 - 537^2}{2}\right)$$

Btu/lb mol quantities for the equation as written.

Combustion Calculations

1. For purposes of combustion calculations *dry air* is assumed to have a molecular weight of 29 and the following composition:

Oxygen, 21 vol% = 21 mol%

Nitrogen, 79 vol% = 79 mol%

(The nitrogen includes 1% argon and traces of carbon dioxide.) This composition allows combustion calculation on the basis of 100 lb mol (2900 lb) dry air entering a reaction zone with 21 lb mol oxygen available for reaction and 79 lb mol nitrogen and inerts that pass through unreacted.

2. The *moisture* in ambient air entering a combustion zone can often be ignored in approximate combustion calculations; however, it cannot be ignored in the exhaust gases. Gases saturated with water vapor contain a partial pressure of water vapor equal to the vapor pressure of water at that temperature. Saturation is 100% relative humidity; and the dew point, being the temperature of condensation initiation, is then the ambient temperature. The vapor pressure of water is available from steam tables.

Example 9. At a dew point of 77°F, the vapor pressure of water from steam tables (saturated steam, temperature table) is 0.459 psia. Applying Dalton's law of partial pressures, the mole fraction of water vapor in gases at 1 atm is 0.459 psia/14.7 psia = 0.031. At a dew point of 160°F, the vapor pressure is 4.739 psia and the mol fraction of water vapor in gases at 28 in. Hg pressure is

$$\frac{4.739 \text{ psia}}{\frac{28}{30}(14.7 \text{ psia})} = 0.347$$

3. *Psychrometric charts*, useful for air conditioning and humidification calculations, express moisture content in grains of water per pound of dry air (7000 grains = 1 lb) and utilize wet-bulb (essentially dew point) and dry-bulb (ambient) temperatures as variables for determining percentage relative humidity.

4. *Percentage excess air.* Stoichiometric air is the amount of air theoretically necessary to burn all hydrogen in fuels to water and to burn all carbon to carbon dioxide. With insufficient air, hydrogen is preferentially oxidized to water and carbon may be released as elemental carbon (soot), carbon monoxide, or mixtures of carbon monoxide and dioxide. Significant amounts of excess air and residence time in the combustion zone are required for complete combustion of fuel-contained carbon to carbon dioxide.

5. *Basis for calculation.* All stoichiometric calculations should be preceded by a basis of calculation in order to readily permit scale-up or scale-down of the result. Examples of usual bases are per 100 lb of fuel, per 100 lb mol of a reactant, and per ton of a product.

6. *Heat balances.* Thermal calculations for reaction systems are most easily made for steady-state conditions involving an accounting calculation that equates all forms of energy in plus heat of reaction to all forms of energy out plus losses. Forms of energy involved are thermal, chemical, electrical, and mechanical in the form of shaft work, kinetic, potential, and flow energies plus enthalpies of all phase transitions

involved. Heat balances are necessarily predicated on valid material balances, and vice versa. Calculation can be based on mass flow of some component, a unit of operating time, or any other useful basis chosen to simplify calculation and utilize available thermal data.

6 CHEMICAL EQUILIBRIUM

Equilibrium Constant. The equilibrium condition for any reaction $aA + bB + \cdots = cC + dD + \cdots$ is defined by the expression

$$K = \frac{[C]^c[D]^d \cdots}{[A]^a[B]^b \cdots} \tag{37}$$

The exponents are stoichiometric coefficients of the equation as written, and bracketed terms are the activities of each species existent at equilibrium. The wide numerical range of K encountered has promoted usage of $pK = -\log_{10} K$ for tabulated data. The equilibrium constant is an expression relating activities of reactants and products such that the net chemical potential is zero at the temperature and pressure of reaction. The standard free energy of a reaction is based on the unit activity of all reactants and products. The K varies with temperature and has a pressure dependence introduced via fugacity of gaseous species, being derived from the relationship $\Delta \bar{G}_T^\circ = -RT \ln K$ [see Eq. (22)].

For reactions where equilibrium constant data are unavailable, combining K's from other known reactions will yield the one desired. This is a consequence of the additivity of reactions and their thermodynamic state functions.

Example 10 See the tabulation for Example 10 below.

The activities of solids and solvents are substantially invariant, each in its own phase; they are assigned numerical values of 1. The activity of gases is their fugacity in atmospheres. Fugacity is related to partial pressure via a fugacity coefficient $\gamma = f/p$ that varies with pressure and temperature. The fugacity coefficient of ideal gases is 1, but the deviations of real gases necessitate its consideration. For real gases

it is expedient to divide K, based on activities, into two parts:

$$K = K_\gamma K_p \tag{38}$$

where K_γ is an activity coefficient product and K_p is the pressure product.

Equilibrium Composition of Reaction Mixtures. The effects of pressure and temperature on equilibrium composition, qualitatively predicted by Le Chatelier's principle (Section 3), are increased conversion with pressure rise if moles of reactants exceeds moles of products and decreased conversion with temperature rise in an exothermic reaction. Quantitatively, equilibrium composition may be calculated from the equilibrium constant.

Sample Calculation of Equilibrium Composition for Gaseous Reactions

$$\tfrac{3}{2}H_2(g) + \tfrac{1}{2}N_2(g) = NH_3(g)$$

$$K = \frac{[NH_3]^1}{[H_2]^{3/2}[N_2]^{1/2}}$$

$$= 0.0431 \text{ atm}^{-1} \quad \text{at } 600 \text{ K}$$

Note that the equilibrium constant is dimensional, and its units are useful in reconstructing the equilibrium constant expression and the reaction equation on which it is based.

Example 11

$$K = \frac{(f_{NH_3})^1}{(f_{H_2})^{3/2}(f_{N_2})^{1/2}} = 0.0431 \text{ atm}^{-1}$$

applies for NH_3 synthesis. For a stoichiometric ratio of hydrogen and nitrogen without inerts, calculate the mole fraction of ammonia in the equilibrium mixture at (a) 1 atm and (b) 100 atm.

At 1 atm total pressure, where the assumption of ideal-gas behavior for this system may be valid, the

---

Tabulation for Example 10

$$CO_2 = CO + \tfrac{1}{2}O_2$$

$$H_2O = H_2 + \tfrac{1}{2}O_2 \quad K = \frac{[H_2][O_2]^{1/2}}{[H_2O]} = 7.25 \times 10^{-1} \text{ at } 1000 \text{ K}$$

$$CO_2 + H_2 = CO + H_2O \quad K = \frac{[CO][H_2O]}{[CO_2][H_2]} = 8.70 \times 10^{-11} \text{ at } 1000 \text{ K}$$

$$\therefore CO_2 = CO + \tfrac{1}{2}O_2 \quad K = \frac{[CO][O_2]^{1/2}}{[CO_2]} = (7.25 \times 10^{-1})(8.70 \times 10^{-11})$$

$$= 6.3 \times 10^{-11} \text{ at } 1000 \text{ K}$$

Tabulation for Example 11

	P_C (atm)	T_C (K)	T_R	P_R at $P = 100$ atm	$\gamma = f/p$ at $P = 100$ atm	P_R at $P = 1$ atm	$\gamma = f/p$ at $P = 1$ atm
NH_3	111.3	405.5	0.90	1.48	ca~0.92	0.009	~1
H_2	12.8	33.3	7.8	18.	ca~1.1	0.078	~1
N_2	33.5	126.2	2.98	4.76	ca~1.0	0.030	~1

$$K_\gamma = \frac{(0.92)^1}{(1.1)^{3/2}(1.0)^{1/2}} = \frac{0.92}{(1.16)(1.0)}$$

$$= 0.79 \text{ at } 100 \text{ atm}$$

$$= 1.02 \text{ at } 1 \text{ atm}$$

activity coefficient γ is assumed and confirmed to be 1 for all species. At 100 atm γ should be estimated for each species by entering the $\gamma = f/p$ chart (see Section 7) using reduced temperature and pressure based on the reaction temperature and total pressure. Significant deviation of K_γ from 1 signals the necessity of its inclusion in the calculation. See the tabulation for Example 11 above.

Calculation of mole fraction NH_3: At total $P = 1$ atm, $K_\gamma = 1$. Let

$$X = \text{partial pressure of } N_2 \text{ in the reaction mixture}$$

$$3X = \text{partial pressure of } H_2$$

$$1 - 4X = \text{partial pressure of } NH_3$$

$$K = K_\gamma K_p = 0.0431 \text{ atm}^{-1} = \frac{1 - 4X}{(3X)^{3/2}(X)^{1/2}}$$

$$1 - 4X = 0.0431(5.2X^2)$$

This is solved by the quadratic formula or successive approximation to give $X \cong 0.246$:

$$\text{Mole fraction } y_{NH_3} = \frac{p_{NH_3}}{p_{total}} = \frac{1 - 4X}{1} = 0.014$$

At total $P = 100$ atm, $K = 0.79$. Let

$$X = p_{N_2}$$

$$3X = p_{H_2}$$

$$K = 0.0431 = \frac{0.79(100 - 4X)}{(3X)^{3/2}(X)^{1/2}}$$

$$100 - 4X = p_{NH_3}$$

$$100 - 4X = \frac{0.0431}{0.79}(5.2X^2)$$

$$X = 13.0$$

$$\text{Mole fraction } y_{NH_3} = \frac{p_{NH_3}}{p_{total}} = \frac{100 - 4X}{100} = 0.48$$

Applications to Solution Chemistry. Equilibrium constants have important quantitative applications in solution chemistry, and specific K's are given appropriate subscripts:

$K_A =$ ionization constant for weak acids

$K_B =$ ionization constant for weak bases

$K_W =$ dissociation constant for water

$K_D =$ dissociation constant for complex ions

$K_{sp} =$ solubility product for slightly soluble species

Solutions and Solution Concentrations. Solutes are smaller amounts of any phase dissolved in larger amounts of liquid or solid solvents, resulting in single-phase solutions.

Solution concentrations are expressed in three ways:

1. *Molar* concentration M is gram moles of solute per liter of solution. Small temperature dependence exists due to density changes.
2. *Molal* concentration m is gram moles of solute per kilogram of solvent. Molal concentrations are not temperature dependent. Since volume of solution = (grams solute + grams solvent)/(solution density), molality m approximately equals molarity M only in dilute aqueous solutions.
3. *Mole fraction* x is the ratio of moles of component i to total moles of all species present in that phase:

$$x_i \text{ or } y_i = \frac{N_i}{N_{total}} \tag{39}$$

where N is number of moles.

Activity–Concentration Relationship for Solutes.
Activity coefficient γ is the ratio of activity a to molality m. At infinite dilution $\gamma = 1$ for all solutes, but γ varies with temperature, concentration, and the species involved (see Section 9). Activity is correctly used for equilibrium calculations, although molar concentrations are widely used as approximations.

Hydrogen Ion Concentration pH.
The pH is an exponential notation using the operator p meaning $-\log_{10}$ to express a wide range of $[H^+]$. At $10^{-7}\ M$ $[H^+]$, resulting from dissociation of water at ambient temperature, pH = 7.

7 PHASE EQUILIBRIA

Temperature–Pressure Phase Diagrams.
Pure chemical species, in the absence of thermal decomposition, have temperature–pressure phase diagrams of the same general shape exemplified by Fig. 3. In this figure, changes of state occur along transition lines, that is, sublimation along AB, fusion along BC, and vaporization along BD. Each phase transition is accompanied by volume, enthalpy, and entropy changes. Phase equilibrium exists at any T, P condition on a transition line. If multiple solid phases exist, transition occurs at T, P along lines such as EF.

Excepting material (H_2O, Bi) which expand on solidification, BC slopes only slightly to the right. For different species the coordinates of triple and critical points differ, but the diagram retains the same general form. Logarithmic compression of scales is often helpful. The normal boiling point occurs along transition curve BD at $P = 1$ atm. The normal freezing point occurs approximately at the temperature of the triple point because of the almost vertical slope of line BC.

Equations of State.
The PVT relations for gases are most simply expressed by the ideal-gas law $Pv = RT$, where v is molar volume, applicable with small error at pressures well below the critical pressure and at temperatures well above the condensation temperature. More precise equations of state are necessary for gases near their condensation conditions where atomic size becomes important and interaction must be considered. Several modifications are used, and all contain additional terms to improve precision near the liquid state. One of the several historical and more precise equations of state is the van der Waals equation

$$\left(P - \frac{a}{v^2}\right)(v - b) = RT \tag{40}$$

where a and b are dimensional and specific van der Waals constants applicable to individual gases. Constant a is a measure of the attractive force between molecules, and constant b is due to incompressibility and finite molecular volume.

The Redlich–Kwong equation is another example:

$$P = \frac{RT}{v - b} - \frac{a}{T^{1/2}v(v - b)} \tag{41}$$

where a and b are constants calculated from P_c and T_c available in Table 9.

Virial coefficients are used in more easily computed power series expansion used as equations of state:

$$Pv = RT + \frac{\beta}{v} + \frac{\gamma}{v^2} + \frac{\delta}{v^3} \quad \text{and}$$

$$Pv = RT + BP + CP^2 + DP^3 + \cdots \tag{42}$$

where β, γ, δ, and B, C, D are second, third, and fourth virial coefficients.

Reduced Equations of State.
Coefficients of van der Waals equation of state, a and b, vary for individual gases and also vary to some extent with temperature. Coefficients a and b and the gas constant R can be generalized and expressed in terms of the critical constants for all gases using consistent units:

$$a = 3V_c^2 P_c = \left(\frac{3}{4}\right)^3 \frac{R^2 T_c^2}{P_c} \tag{43}$$

$$b = \frac{V_c}{3} = \frac{RT_c}{8P_c} \tag{44}$$

$$R = \frac{8}{3} \frac{P_c V_c}{T_c} \tag{45}$$

Using reduced variables P_R, V_R, and T_R, the van der Waals equation of state yields a dimensionless reduced equation of state:

$$\left(P_R + \frac{3}{V_R^2}\right)(3V_R - 1) = 8T_R \tag{46}$$

This infers that all gases have the same V_R at equal values of P_R and of T_R. This reduced equation of state is generally applicable at elevated pressures and near the critical point.

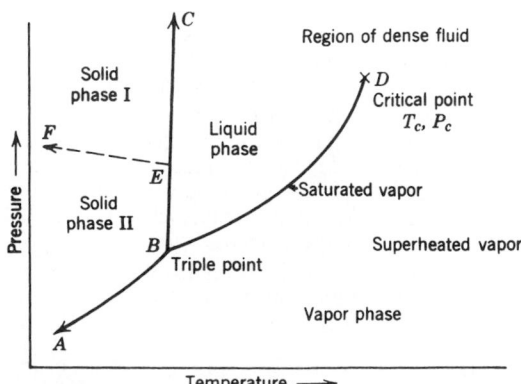

Fig. 3 Generalized temperature–pressure phase diagram.

Table 9 Critical Constants; Latent Heat of Vaporization at Normal Boiling Point

Compound	Formula	T_B (K)	P_c (atm)	ρ_c (g/cm^3)	T_c (K)	ΔH_v at T_B (cal/g mol)
Bromine	Br_2	—	102	1.18	584	—
Chlorine	Cl_2	239.1	76.1	0.573	417	4878
Hydrogen	H_2	20.39	12.797	0.0310	33.24	216
Nitrogen	N_2	77.36	33.5	0.311	126.0	1333
Oxygen	O_2	90.19	50.14	0.430	154.78	1630
Sulfur	S_2	717.8	120	0.4	1313	2500
Air	—	—	37.2	0.35	132.5	—
Ammonia	NH_3	239.8	111.5	0.235	405.6	5581
Carbon dioxide	CO_2	194.7	72.9	0.459	304.1	6100
Carbon monoxide	CO	81.7	34.53	0.301	133.0	1444
Carbon disulfide	CS_2	319.4	76	0.4	546.2	6400
Hydrogen bromide	HBr	206.4	84		363	4210
Hydrogen chloride	HCl	188.1	81.6	0.421	324.6	3860
Hydrogen cyanide	HCN	298.86	50.0	0.20	456.7	6027
Hydrogen sulfide	H_2S	212.8	88.9	0.349	373.6	4463
Nitric oxide	NO	121.4	65	0.52	179	3292
Nitrous oxide	N_2O	184.7	71.7	0.45	309.7	3956
Phosgene	$COCl_2$	280.7	56	0.52	455.0	5832
Phosphine	PH_3	185.4	64.5	0.30	324.5	3490
Sulfur dioxide	SO_2	263.1	77.7	0.518	430.4	5950
Sulfur trioxide	SO_3	316.5	83.8	0.633	491.5	9990
Stannic chloride	$SnCl_4$	386	37	0.74	522	8300
Water	H_2O	373.2	218.4	0.323	647.3	9717
Methane	CH_4	111.67	45.8	0.162	191.0	1955
Ethane	C_2H_4	184.5	48.2	0.203	305.5	3517
Propane	C_3H_8	231.1	42.0	0.022	370.0	4487
n-Butane	C_4H_{10}	272.7	37.5	0.228	425.2	5350
2-Methyl propane	C_4H_{10}	261.4	36.0	0.221	408.15	5089
n-Pentane	C_5H_{12}	309.2	33.3	0.232	470.1	6160
n-Hexane	C_6H_{14}	341.9	29.9	0.234	507.9	6900
2,2-Dimethyl butane	C_6H_{14}	322.9	30.7	0.242	489.4	6290
n-Heptane	C_7H_{16}	371.6	27.0	0.235	540.2	7580
n-Octane	C_8H_{18}	398.8	24.6	0.233	569.4	8215
2,2,4-Trimethyl pentane	C_8H_{18}	372.4	25.4	0.243	544	7410
Ethene	C_2H_4	169.5	50.0	0.227	282.5	3237
Propene	C_3H_6	225.5	45.6	0.233	364.9	4405
1-Butene	C_4H_8	266.9	39.6	0.232	419.7	5240
cis-2-Butene	C_4H_8	276.9	40.8	0.239	428.2	5580
trans-2-Butene	C_4H_8	274.05	40.8	0.239	428.2	5440
Ethane	C_2H_2	184.7	61.6	0.231	308.7	4270
Propyne	C_2H_4	—	52.8	—	401	—
2-Butyne	C_4H_6	—	60	—	489	—
Benzene	C_6H_6	353.3	48.3	0.304	562.1	7350
Toluene	C_7H_8	383.8	41.6	0.291	593.8	8000
Cyclopropane	C_3H_6	240.2	54	—	398	—
Cyclopentane	C_5H_{10}	322.4	44.6	0.270	511.8	6525
Cyclohexane	C_6H_{12}	353.9	40	0.272	553.7	7190
Methyl cyclohexane	C_7H_{14}	374.1	34.32	0.285	572.3	7580
Acetaldehyde	CH_3CHO	293.3	44	0.26	461	6500
Acetone	$(CH_3)_2CO$	329.35	46.6	0.273	508.7	7100
Methanol	CH_3OH	337.9	78.5	0.272	513.2	8430
Ethanol	C_2H_5OH	351.7	63.0	0.2755	516	9220
i-Propanol	$(CH_3)_2CHOH$	355.36	53	0.27	509	9650
n-Propanol	C_3H_7OH	370.5	50.2	0.273	537	9890
Methyl ether	$(CH_3)_2O$	248.3	53	0.271	400.1	5141
Ethyl ether	$(C_2H_5)_2O$	307.8	35.6	0.263	467.0	6220
Ethylene oxide	$(CH_2)_2O$	283.7	71	0.31	469	6101
Chloromethane	CH_3Cl	248.9	65.9	0.35	416.3	5150
Fluoromethane	CH_3F	195.1	58.0	0.300	317.8	4230
Trichloromethane, chloroform	$CHCl_3$	334.4	55	0.516	536	7020
Methyl cyanide, acetonitrile	CH_3CN	354.7	47.7	0.240	547.9	7830

Source: Kobe and Lynn, *Chem. Rev.*, Vol. 52, p. 117, 1953.

Corresponding States. The reduced equation of state, similar shapes of the basic T, P phase diagrams, maximum ΔH_v and ΔS_v at the triple point, and the disappearance of ΔH_v and ΔS_v at the critical point have led to the use of reduced pressure, volume, and temperature for correlation of other properties. These reduced variables P_R, V_R, and T_R are defined as ratios of the existent to the critical variable:

$$P_R = \frac{P}{P_c} \qquad V_R = \frac{V}{V_c} \qquad T_R = \frac{T}{T_c} \qquad (47)$$

Two gases are in corresponding states when they have the same values for two of the three reduced variables. Corresponding states permit property estimation by relating gases or liquids to comparable species where more complete PVT data are available.

Compressibility. The PVT relations of any gas may be expressed as

$$Pv = ZRT \qquad (48)$$

where Z is the compressibility factor or ratio of the volume of a real gas to that of an ideal gas. If the parameters P and T are replaced by their ratios to the corresponding critical values, called reduced properties, all gases are in corresponding states and can be approximately represented by a general chart (Fig. 4) of Z versus P_R on lines of constant T_R.

Most individual gases deviate from such a general graph by 2–7%. Hydrogen and helium, however, deviate so much that the reduced properties for them are computed by adding 8 atm to the critical pressure and 8 K to the critical temperature.

Generalized Fugacity Chart. A generalized fugacity chart can be constructed from the compressibility chart by integration, along each isotherm, of

$$\ln \frac{f}{p} = \int_0^{P_R} (Z - 1)\, d(\ln P_R) \qquad (49)$$

The result of these integrations is shown in Fig. 5.

Gas Mixtures. Mixtures of ideal gas have a compressibility $Z = 1$ at low pressures; however, mixtures of real gases deviate from ideality as pressure is increased. A compressibility factor for the mixture can be determined to simplify calculation. At pressures below 50 atm Dalton's law of additive partial pressures can be used for calculating mixture pressure, while above 300 atm Amagat's law of additive molal volumes allows better calculation of mixture volume. The pseudocritical point method may be used to estimate mixture compressibility between 50 and 300 atm.

Dalton's Law of Additive Partial Pressures. For real gases below 50 atm, where the actual volume of the molecules is small relative to the volume occupied and where molecular interaction does not have a large effect, Dalton's law is valid:

$$P = \sum_i p_i \qquad (50)$$

where $p_i = P y_i$ and P is total pressure, p_i is partial pressure of species i, and y_i is the mole fraction of species i.

Amagat's Law of Additive Molal Volumes. Amagat's law is valid at pressures above 300 atm:

$$V = \sum_i v_i \qquad (51)$$

where $v_i = V y_i$ and v_i is the molal volume of species i.

Molal Average Compressibility

$$Z = \sum_i y_i Z_i \qquad (52)$$

where Z_i is determined using T_R and P_R for species i at the temperature of the mixture and at p_i for low total pressures (50 atm maximum) or at P for high total pressures (300 atm minimum).

Pseudocritical Point. In the absence of experimentally determined critical conditions for gas mixtures, pseudocritical temperature and pressures can be estimated as molal average values derived from components:

$$T_{pc} = \sum_i y_i T_{ci} \quad \text{and} \quad P_{pc} = \sum_i y_i P_{ci} \qquad (53)$$

where T_{ci} and P_{ci} are critical temperature and pressure of pure component i and the subscript pc refers to pseudocritical. These values may be used for determining a reduced temperature and pressure for the mixture, from which a compressibility may be estimated.

Examples of Gas Mixture Calculations. Calculate molal volume at $100°C$ (373 K) of a mixture containing 30 mol % CO_2 and 70% N_2 at (a) 20 atm, (b) 70 atm, and (c) 400 atm.

(a) At 20 atm use Dalton's law and molal average compressibility:

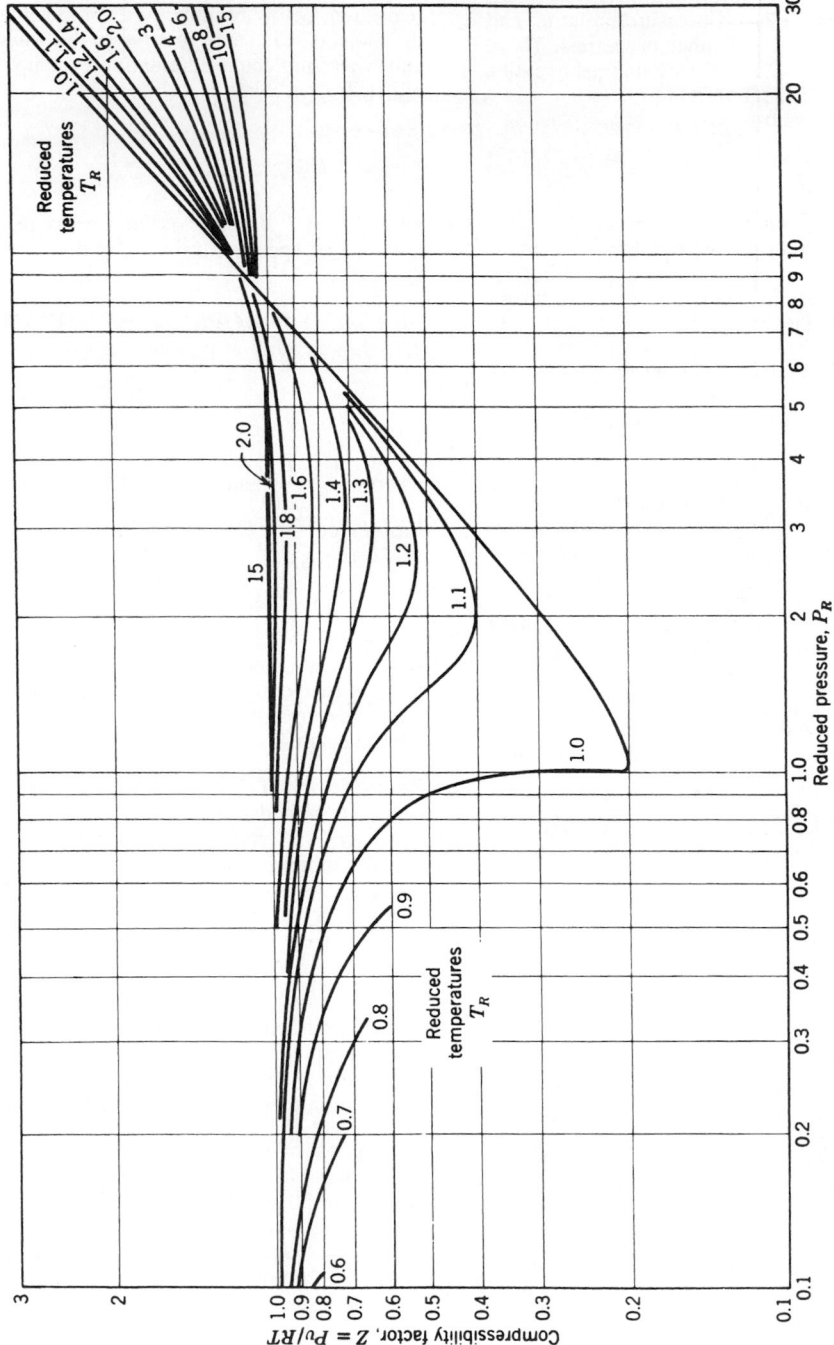

Fig. 4 Compressibility factor. (Reprinted by permission from M. Souders, *Engineers Companion*, Wiley, New York, 1966).

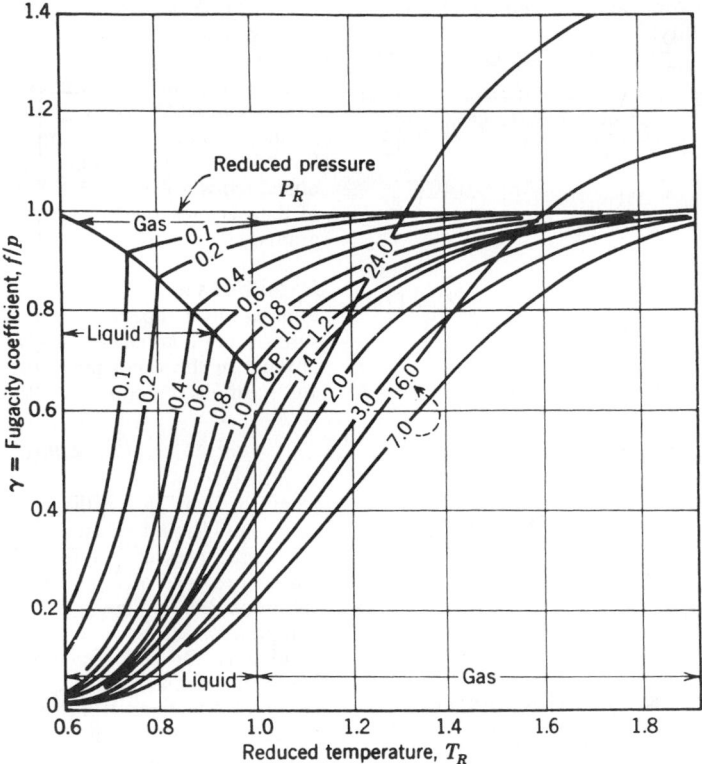

Fig. 5 Fugacity of gases and liquids. (Reprinted from O. A. Hougen, K. M. Watson, and R. A. Ragatz, *Chemical Process Principles*, Wiley, New York, 1959.)

	For CO_2	For N_2
	$y_i = 0.3$	$y_i = 0.7$
	$p_i = 20(0.3)$	$p_j = 20(0.7)$
	$= 6$ atm^a	$= 14$ atm

$P_c = 72.9$ atm $P_c = 33.5$ atm

$T_R = \dfrac{373}{304}$ $T_R = \dfrac{373}{126}$

$= 1.23$ $= 2.96$

$P_R = \dfrac{6 \text{ atm}}{72.9}$ $P_R = \dfrac{14 \text{ atm}}{33.5}$

$= 0.082$ $= 0.42$

$Z_i = 0.98$ $Z_j = 0.99$

$Z_{\text{mixture}} = (0.3)(98) + (0.7)(99)$

$= 0.987$

$v = \dfrac{ZRT}{P}$

$= \dfrac{(0.987)(0.0821)}{(20) \text{ atm}} 1 \text{ atm } (373) \text{ K/g mol K}$

$= 1.5$ l/mol mixture

^{a}This assumes that Dalton's law holds.

(b) At 70 atm use pseudocritical point to determine mixture compressibility:

Mixture $T_{\text{pc}} = 0.3(304) + 0.7(126) = 179$ K

Mixture $P_{\text{pc}} = 0.3(72.9) + 0.7(33.5) = 45.3$ atm

$T_R = \dfrac{373}{179} = 2.08$

$P_R = \dfrac{70}{45.3} = 1.54$

$Z_{\text{mixture}} = 0.980$

$v = \dfrac{ZRT}{P} = \dfrac{(0.980)(0.0821)(373)}{70}$

$= 0.43$ l/mol mixture

(c) At 400 atm use Amagat's law and molal average compressibility:

	For CO_2	For N_2
$T_R =$	$\dfrac{373}{304}$	$T_R = \dfrac{373}{126}$
	$= 1.23$	$= 2.96$
$P_R =$	$\dfrac{400}{72.9}$	$P_R = \dfrac{400}{33.5}$
	$= 5.49$	$= 8.38$
	$Z_i = 0.76$	$Z_i = 1.08$

$$Z_{mixture} = (0.3)(0.76) + (0.7)(1.08)$$
$$= 0.984$$

$$v = \frac{ZRT}{P} = \frac{(0.984)(0.0821)(373)}{400}$$
$$= 0.075 \ 1/\,mol\ mixture$$

Ideal Solutions. The activity of each constituent of ideal-liquid solutions is equal to its mole fraction under all conditions of temperature, pressure, and concentration. The solution volume exactly equals the summation of component volumes. The enthalpy of mixing of components is zero. The total vapor pressure is the summation of the contribution of individual components following Raoult's law. This also applies to solutions containing nonvolatile components. The freezing point of solvent in ideal solutions occurs at that temperature where the vapor pressure of the solution equals the vapor pressure of the solid solvent.

Real Solutions. Actual liquid solutions deviate from the preceding conditions of ideality. Most significant are positive or negative vapor pressure deviations from direct summation of component contributions; these affect behavior on distillation for separating the components. Deviations from ideality increase with solute concentration; that is, dilute solutions behave reasonably ideally.

Henry's Law. The solubility, and hence activity, of gaseous solute in liquid solvent is directly proportional to the partial pressure (fugacity) of the solute vapor phase in equilibrium with the solution. If the solution and vapor phases behave ideally, the model fraction of the gas solute in solution at low concentration equals solute activity ($X_i = a_i$). Henry's law is

$$P y_i = h_T x_i \tag{54}$$

where y_i and x_i are mole fractions of i in the vapor and liquid phases, P is total pressure, and h_T is Henry's law of coefficient for i at temperature T. The solubility of gases in liquids is inversely proportional to temperature, and frequently chemical similarity between solute and solvent leads to a higher solubility. Henry's law applies to the same molecular species of solute in the solution and in the gas phases. For example, for

NH_3, CO_2, SO_2, H_2S, Cl_2, HCl, and so on, in water it applies to the unhydrated, undissociated species in equilibrium.

For nonreactive, unhydrated, and undissociated gases such as O_2, N_2, H_2, CO, and CH_4 in water, h_T is on the order of 10^{-5} atm^{-1} between 0 and 80°C. Similar data of different magnitudes are existent for nonaqueous solvents, but a wider range of this type of data is presented in vapor–liquid equilibrium diagrams.

Raoult's Law. For ideal-vapor and ideal-liquid solutions at any temperature, the equilibrium partial pressure of any component of a liquid solution equals the product of the vapor pressure of that pure component and its mole fraction in solution:

$$P y_i = p_i^\circ x_i = p_i \tag{55}$$

where y_i and x_i are the mole fractions of i in the vapor and liquid phase, P is total pressure, p_i° is the vapor pressure of pure i, and p_i is the actual partial pressure of i, all at temperature T. For ideal solutions Raoult's law holds for all mole fractions of a component, while for real solutions it can be assumed valid only for mole fractions near unity. Interaction between dissimilar species in the liquid phase leads to large deviations from ideal solution behavior and hence divergence from conformity with Raoult's law. Raoult's law can be considered a special case of Henry's law where Henry's law coefficient h_T is equal to the vapor pressure of pure component i, p_i° at that temperature.

Nernst Distribution Law. The equilibrium distribution of solute between two immiscible liquid phases at temperature T is constant regardless of concentration and is equal to the ratio of solute activities in the two phases:

$$K = \frac{a_1}{a_2} \tag{56}$$

where a_1 is the activity in phase 1 and a_2 is the activity in phase 2. In dilute solutions molar concentrations can be used to replace activities.

Gibbs Phase Rule. The phase rule gives the relationship to define heterogeneous phase equilibria:

$$P + F = C + 2 \tag{57}$$

where P is the number of separate phases involved, C is the number of components (minimum number of chemical species necessary to define composition of all phases), and F is the number of degrees of freedom or number of variables, such as temperature, pressure, or concentration, of each component that can be varied independently without changing the number of phases.

Example 12

1. *Water at the triple point.* Solid, liquid, and vapor phases are in equilibrium. A pure compound is a single component, and its concentration in each phase is determined to be 100 mol %. Therefore $P = 3, C = 1$, and F is to be determined by substituting these values in Eq. (57):

$$3 + F = 1 + 2$$

Hence $F = 0$, and the system is invariant; temperature and pressure are fixed, and a shift in either will reduce the number of phases in equilibrium.

2. *For a three-component system, defined by T, P but not composition*, determine the number of phases present; $F = 2, C = 3$, and P is to be determined:

$$P + 2 = 3 + 2$$

Therefore $P = 3$ phases. Further defining the concentration of one of the three components raises F to 3 and reduces the number of phases to 2. Defining the concentration of two components also establishes the concentration of the third, raises F to 4, and reduces the number of phases to 1.

3. *Two-component system at liquid–vapor equilibrium (two phases).* The concentrations in the two phases are not independent; $P = 2, C = 2$, and F is to be determined:

$$2 + F = 2 + 2$$

Two of the three independent variables (T and P, or concentration of only one component) are required to define the system; the third is fixed.

Colligative Properties of Solutions. Colligative properties depend on the number, rather than the nature, of particles in solution: nonelectrolytes, being undissociated, yield one particular per molecules unless they are dimerized. Strong electrolytes ionized to an extent indicated by their activity yield several ions. Colligative properties are (a) boiling point elevation as a direct consequence of vapor pressure reduction; (b) freezing point depression, also as a direct consequence of vapor pressure reduction; and (c) osmotic pressure. Since ideal solutions are expected at low concentrations, changes in these colligative properties are predictable.

Binary Solution Vapor–Liquid Equilibria. In the vapor phase

$$P_{total} = p_i + p_j$$

$$N_{total} = N_i + N_j$$

$$y_i = \frac{p_i}{P_{total}}$$

$$y_i + y_j = 1 \tag{58}$$

where p_i is the partial pressure, y_i is the mole fraction, and N_i is the number of moles of i.

In the liquid phase

$$x_i = \frac{N_i}{N_{total}}$$

$$x_i + x_j = 1$$

$$N_{total} = N_i + N_j \tag{59}$$

where x_i is the mole fraction of i.

(a) *Ideal Solutions.* Each component of an ideal solution obeys Raoult's law [Eq. (55)], relating concentrations in vapor and liquid phases.

(b) *Real Solutions.* Numerical distillation calculations often use a vapor–liquid equilibrium ratio K for each component:

$$K_i = \frac{y_i}{x_i} = \frac{f \text{ of pure } i \text{ at its vapor pressure at } T \text{ of system}}{f \text{ of pure } i \text{ at } T, P \text{ of system}} \tag{60}$$

The volatility ratio between components of binary solutions is

$$\alpha = \frac{k_i}{k_j} = \frac{y_i x_i}{x_i y_y} \tag{61}$$

where α is the volatility ratio. For ideal binary solutions, where α is constant, manipulation of Eqs. (60) and (61) relates the mole fraction in the vapor phase y_i to the volatility ratio α and the mole fraction in the liquid phase x_i:

$$y_i = \frac{\alpha x_i}{1 + x_i(\alpha - 1)} \tag{62}$$

Binary Solution Vapor Pressure–Composition Diagrams

1. Ideal solutions follow Raoult's law [Eq. (55)]. As shown in Fig. 6a, the vapor pressure of each component is linear and proportional to the mole fraction, and the vapor pressure of the mixture is the simple sum of the component vapor pressures. The diagrams shown represent one fixed temperature.

2. Real solutions show significant deviations from linearity of individual vapor pressures with the mole fraction. At low mole fractions in the liquid phase. Henry's law [Eq. (54)] is followed, while at high mole fractions, Raoult's law tends to be followed. These deviations from linearity are shown in Figs. 6b, c.

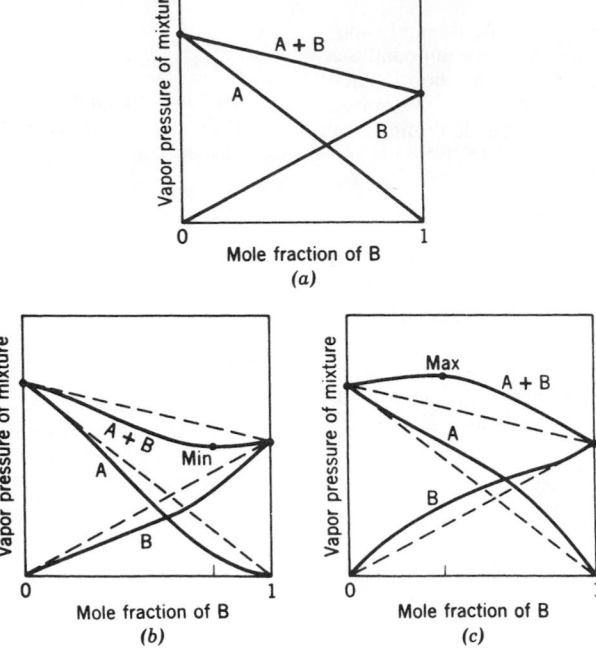

Fig. 6 Binary solution vapor pressure–composition diagrams.

Boiling Point–Composition Diagrams. Figure 7 shows three types of boiling point–composition diagrams. These are drawn for a single pressure. Figure 7a is the usual case for ideal solutions; it corresponds with the vapor pressure diagram 6a. At any given temperature, vapor composition y is in equilibrium with liquid composition x. This ideal type of boiling point diagram exists for many combinations of chemically similar materials.

Figures 7b,c correspond to the vapor pressure diagrams in Figs. 6b,c.

Azeotropes (constant-boiling mixtures) exist at 1 and 2. Their composition can be altered by changing pressure. These arise from nonideality of the solutions. For collected data on azeotropes, see *Handbook of Chemistry and Physics*.[2]

Freezing Point–Composition Diagrams. The freezing point is essentially independent of pressure, and the binary freezing point–composition diagrams of Fig. 8 apply over an extended pressure range. The usual case of two similar and mutually soluble components is to form a continuous series of solid and liquid solutions as shown in Fig. 8a. At any one temperature the composition of the liquid phase differs from that of the solid phase in equilibrium with it. Nonidealities in the liquid and solid solutions can give rise to a composition of minimum freezing point typified by Fig. 8b. Where components of the binary mixture have limited mutual solubility, a eutectic composition exists, and this is a mechanical mixture of the two components. A phase diagram idealized by Fig. 8c then occurs. Maximum-melting-point mixtures are rare, and the existence of a maximum is due to formation of a definite compound. Figure 8d shows the existence of an equimolar compound A_1B_1. The resultant phase diagram then shows two eutectic diagrams placed side by side. One of these eutectics exists for component A and compound A_1B_1; the other exists for compound A_1B_1 and component B.

Numerous proliferations of these basic types of phase diagrams exist. All may be interpreted on the basis of the temperature and composition range for mutual solubility, the existence of nonideal solutions, and the formation of compounds stable over an often limited temperature range.

8 CHEMICAL REACTION RATES

Process Kinetics and Conversion. The kinetics of chemical reactions vary widely, ranging from a rapid approach to equilibrium for ionic reactions in aqueous media (primarily limited by diffusion and mixing) to a much slower approach to equilibrium for organic reactions involving covalent bonds. The net reaction rate at equilibrium is zero, and further conversion of reactants to products depends upon the rate of product removal from the reaction system so the

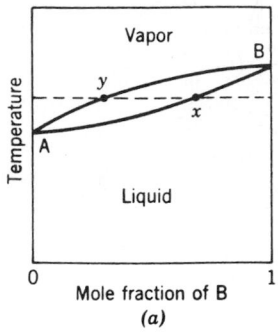

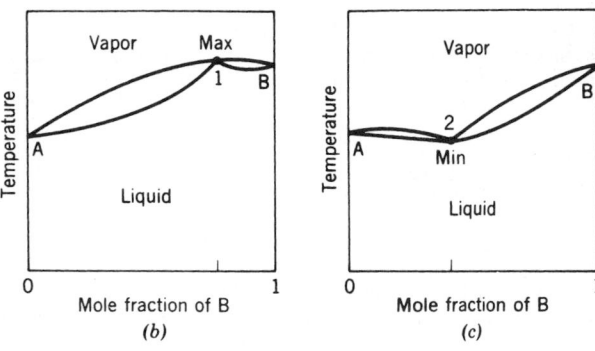

Fig. 7   Binary solution boiling point–composition diagrams.

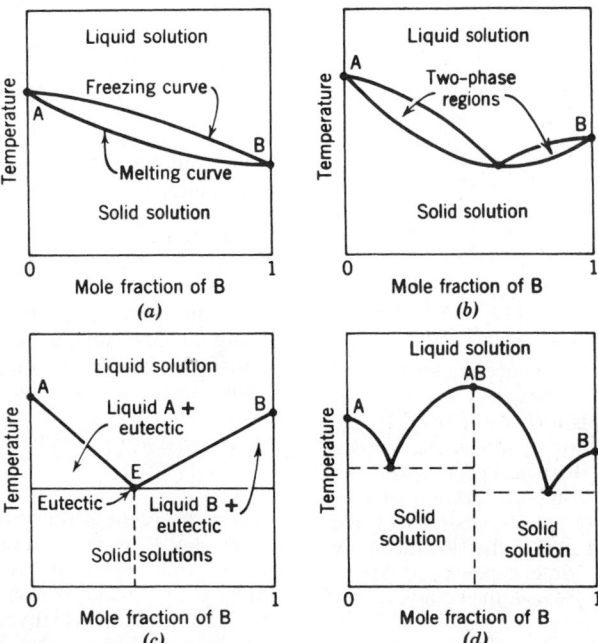

Fig. 8   Binary solution freezing point–composition diagrams.

reaction can again proceed at a finite rate toward equilibrium. Only temperature can shift the equilibrium constant of the reaction system; however, the physical control of reactant and product concentrations can allow a reaction to proceed to complete conversion of reactants to product despite an equilibrium situation that may permit a maximum of a few percent of product in the system at any instant.

Kinetics of Reaction. This is concerned with the rate of approach to a temperature-dependent equilibrium condition. The rate varies with the following:

(a) Displacement from Equilibrium. This is dependent on the reactant and product concentrations.

(b) Temperature. In addition to changing the numerical value of the equilibrium constant and altering the diffusional mobility, increased temperature raises the fraction of molecules with enough energy to react.

(c) Catalysis. This modifies the reaction mechanism and may provide a lower energy path from reactants to products and vice versa. Catalysis has no effect on equilibrium.

(d) Inhibition. Inhibitors block an existent low-energy path between reactants and products, for example, by an essentially irreversible reaction with a catalytic surface.

Reaction Mechanism and Reaction Rate. Most reactions proceed via a sequence of several reversible steps. For example, the net reaction $aA + bB \rightleftharpoons cC$ might actually involve the following elementary steps:

$$\begin{array}{ccc} k_1 & k_2 & k_3 \\ A \rightleftharpoons 2D & D + B \rightleftharpoons E & E + A \rightleftharpoons C \\ K_{-1} & k_{-2} & k_{-3} \end{array}$$

where D and E are intermediates. The rate law for each of the species follow directly from the mechanism, for example,

$$\frac{d[E]}{dt} = k_2[B][D] - k_{-2}[E] - k_3[A][E] + k_{-3}[C]$$

where the brackets denote the concentration of the species. (There are analogous expressions for, for example, $d[A]/dt$.) Determination of the overall rate of reaction requires the solution of the set of coupled differential equations for all of the species, and the resulting rate law will be a complex function of [A], [B], [C], and the k's. While the expression for the equilibrium constant K is related to the stoichiometry of just the net reaction, the time dependence of the approach to equilibrium depends on the details of the mechanism.

Activation Energy. An enthalpy-versus-reaction diagram (Fig. 9) can be ideally drawn for any reac-

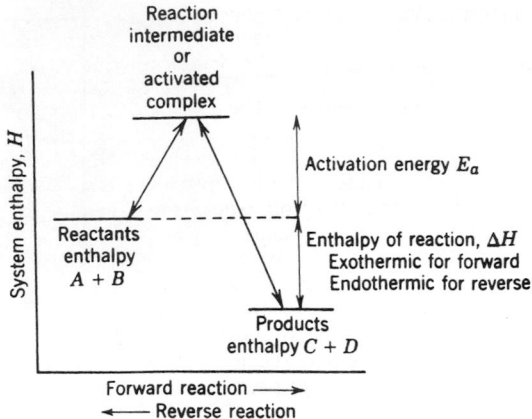

Fig. 9 Enthalpy–reaction diagram.

tion. An activation energy barrier exists between reactants and products, and enthalpy of reaction ΔH separates the enthalpies of reactants and of products. Reactant molecules must receive a minimum activation energy E_a before they are energetic enough to react. For the reverse reaction, product molecules must receive $E_a + \Delta H$ before they can react in the reverse direction. Catalysts modify the intermediate condition and result in a reduced activation energy by providing a lower energy path between reactants and products; by so doing, they effectively speed *both* forward and reverse reactions without affecting the equilibrium condition.

Collision Theory. Energy of activation results from the statistical probability that some molecules, after collision with others, conserving momentum, have energy enough for reaction to occur. An elevated temperature increases the Maxwell–Boltzmann energy distribution of molecules and raises the number of molecules having the energy required for the reaction.

Rate Dependence on Temperature. The population of high-energy molecules is directly correlated with the temperature dependence on the reaction rate using the Arrhenius equation (which has a form similar to the dependence of the equilibrium constant on temperature)

$$\frac{d \ln k}{dT} = \frac{E_a}{RT^2} \tag{63}$$

where E_a is the activation energy, R is the gas constant, and T is the absolute temperature. Integration of Eq. (63) yields $\ln k = -E_a/RT + \text{const}$. A plot of $\ln k$ versus $1/T$ is a straight line with slope equal to $-E_a/R$, from which the activation energy can be obtained. It shows that a rate at $25°C$ will double for a $10°C$ temperature rise when the activation energy is 12.5 kcal/g mol. A lower activation energy gives a smaller rate increase, and a higher energy gives a larger rate increase. *Caution*: Changes of mechanism, via catalysis or otherwise, will alter the E_a of a reaction.

Reaction Mechanisms. The low statistical probability of trimolecular collisions essentially precludes the possibility that any reaction can proceed in one step according to the stoichiometric equation. More probable is a sequence of bimolecular reactions forming intermediate species, some of which may be too short-lived to isolate. Each of these stepwise reactions has its equilibrium constant and rates dependent on species concentrations. Usually one step is limiting and becomes the rate-determining factor for the entire reaction. The intermediate species may be an activated complex that rearranges to produce the reaction products.

Different reaction mechanisms have been postulated and proved for many reactions:

1. Simple bimolecular collision.
2. Chain reaction for polymerization, with initiation, propagation, and termination steps.
3. Free-radical, photoinitiated reactions.
4. Acid- or base-catalyzed intermediate steps.
5. Activated complex on a heterogeneous catalyst surface, with adsorption–desorption contributing to the overall rate.
6. Solvolysis participation in the reaction sequences.

Many processes have been studied in detail, with reaction sequences, proven or postulated, consistent with experimentally determined kinetics.

Initial Rate Equation. A reaction rate equation can be experimentally deduced and, from this, a reaction mechanism postulated. The initial rate equation has the generalized form

$$-\frac{dC_A}{dt} = kC_A^a C_B^b \cdots \tag{64}$$

where $-dC_A/dt$ is the rate of decrease of reactant A and is dependent upon a product of concentration terms. The order of a reaction is the sum of the exponents of concentration terms in the rate equation. It need not be a whole number; it may be 0, $\frac{1}{2}$, 1, $\frac{3}{2}$, or 2. There is no necessary relationship between the reaction order and the form of the overall stoichiometric relation. Where one of the reactants is a solvent, its concentration does not effectively change, and it will not appear in the rate equation.

Heterogeneous (Contact) Catalysis. Contact catalysis for gas-phase reactions involves surface adsorption and reaction at active sites. Adsorption can be either physical, with a mono- or multimolecular gas layer, or chemical (chemisorption), with a monomolecular layer in some activated state. A large surface area is required. Steps involved in a catalyzed reaction are

1. Adsorption of reactants (favored at low temperature)
2. Activation
3. Reaction
4. Desorption of products (favored at high temperature)

For monolayer coverage, the amount of adsorption from the gas phase can be represented by the Langmuir adsorption isotherm

$$y = \frac{ap}{1 + bp} \tag{65}$$

where y is the fraction of area covered, p is pressure, and a and b are empirical constants applicable at one temperature only.

Adsorption from the liquid phase is represented by the Freundlich equation

$$y = aC^{1/b} \tag{66}$$

where C is the concentration of the solute and a and b are empirical and applicable at one temperature only.

Catalyst poisoning involves blocking active sites on the catalyst surface by adsorbed material that is not readily desorbed.

Photochemical Reactions. Many reactions proceed via ultraviolet (UV) energy absorption. Photons from mercury vapor at 2536 Å $\cong$ 113 kcal/g mol are absorbed, one maximum per molecule, and have energy sufficient to disrupt chemical bonds (usual range of 80–100 kcal) and form reactive free-radical species. Photons of infrared (IR) wavelength 9000 Å $\cong$ 31.8 kcal/g mol do not have sufficient energy for bond disruption, although they do increase the overall thermal energy of a system.

An Einstein is an Avogadro number (6.02×10^{23}) of photons.

9 ELECTROCHEMISTRY

Chemical Energy of Redox Reactions. The chemical energy of redox reactions can be equated at constant T, P with Gibbs free energy using Eq. (7):

$$dG = -S\,dT + V\,dP + \sum_i \mu_i\,dN_i$$

Since dT and dP are both zero at constant T, P, then $dG = \Sigma_i \mu_i dN_i$. Gibbs free energy then can be manifest as electrical energy by an electron flow through an external conductor provided there is electrical isolation of the oxidation reaction (electron source) and reduction reaction (electron sink). The external conductor transfers electrons from source to sink, and a liquid junction permits internal migration of ions for preservation of charge neutrality of the system as a whole.

Chemical energy at constant T, P can be considered convertible to electrical and/or thermal energy, for it can be shown by conservation of energy that $\Sigma_i \mu_i dN_i = -E\,dq - T\,dS$. In the absence of $T\,dS$ thermal energy,

$$dG = -E\,dq \qquad (67)$$

where E is in volts, dq is in coulombs per gram mole, and dG is energy in joules per gram mole. Joules are convertible to calories by $J = 4.185\ \text{J/cal}$.

Without source–sink electrical isolation, an internal short circuit results in energy conversion to thermal. Thus at constant T, P

$$dG = -T\,dS \qquad (68)$$

where T is in degrees Kelvin and dS is in calories per gram mole per degrees Kelvin.

Faraday F. This is the 96,500 C charge carried by 1 g equivalent or an Avogadro number of electrons. Hence

$$dq = F\,dn \qquad (69)$$

where dn is the number of gram equivalents.

Standard Potential, E° Volts. The superscript refers to the standard conditions of $25°C$ (298 K), 1 atm fugacity (pressure), and 1 M activity of reactants and products, the same conditions as for $\Delta G°$, thus establishing dimensionally correct conversion of energy terms at the standard thermodynamic reaction condition:

$$dG° = -\frac{dn\,F\,E°}{J} \quad \text{or} \quad \Delta G° = -\frac{n\,F\,E°}{J} \quad (70)$$

in integrated form. If the equilibrium constant for the reaction is known at 298 K, $\Delta G_T° = -RT \ln K$ can be used to calculate $\Delta G°$, which can then be related to $E°$ via Eq. (70). $\Delta G°$ indicates the magnitude of displacement from equilibrium that exists at 298 K when all reactants and products are at unit activity:

$$\Delta G° = RT \ln(1) - RT \ln K = -RT \ln K = -\frac{n\,F\,E°}{J} \qquad (71)$$

Potential $E_T°$ at Temperatures Other Than 298 K. The $\Delta G_T°$ calculated for reactions at other temperatures [via. Eq. (11)] can be directly related to $E_T°$. Similarly, experimental values of the equilibrium constant K can be related to $\Delta G_T°$ [via Eq. (23)] and then to $E_T°$ using

$$\Delta G_T° = -RT \ln K = -\frac{n\,F\,E_T°}{J} \qquad (72)$$

Potential E at Other Concentrations — Nernst Equation. No driving force exists at equilibrium where $\Delta G = 0$, and the driving force is directly related to displacement from equilibrium. Therefore, an E value can be corrected for reactant and product activities other than 1 at 298 K by using the general equation

$$G = RT \ln K' - RT \ln K = \frac{-nFE}{J} \qquad (73)$$

where K' has the form and exponents of the reaction equilibrium constant but uses actual activities of products and reactants in the system rather than those existent at equilibrium.

Combining Eqs. (71) and (73) yields the Nernst equation

$$E = E_0 \frac{-JRT}{nF} \ln K' \qquad (74)$$

At $T = 298$ K, this equation reduces to the usual equation

$$E = E° - \frac{0.0592}{n} \log_{10} K' \qquad (75)$$

Both equations reduce E to zero at equilibrium concentrations. At temperatures other than 298 K, $E_T°$ instead of $E°$ may be used in the general equation to obtain E actual.

Standard Reduction Potential E°. Being related to both $\Delta G°$ and to equilibrium constant K at standard conditions, $E°$ is a measure of the chemical energy available from redox reactions. Tabulated potentials for a number of half-reactions in aqueous solution can be combined to give the standard potential of all possible combinations of these half-reactions. Table 10 gives the standard potentials for reduction half-reactions in 1 M acid and neutral media and in 1 M base media because the reactions and their potentials change in many cases. The table is reversible because a reduction half-reaction (with its $E°$), on reversal, becomes an oxidation half-reaction (with the same numerical $E°$ but of opposite sign).

Example 13 See the tabulation for Example 13 on page 1236.

(a) *Sign Convention.* Since $E°$ for $2H^+ + 2e^- = H_2$ is taken as a zero reference potential, all positive $E°$ values indicate the reaction can proceed as written, and all negative $E°$ values indicate that the reverse reaction (with a positive $E°$ value) can occur.

(b) *Oxidation Potential.* Reversal of any reduction half-reaction and change of sign of the reduction potential generates an oxidation half-reaction with its oxidation potential.

Table 10 Reduction Potentials

Acid Solution	E^0	Basic Solution	E_B°
Aluminum			
$Al^{3+} + 3e^- \rightarrow Al$	-1.66	$H_2AlO_3^- + H_2O + 3e^- \rightarrow Al + 4OH^-$	-2.35
Antimony			
$Sb + 3H^+ + 3e^- \rightarrow SbH_3$	-0.51	$Sb + 3H_2O + 3e^- \rightarrow 3bH_3 + 3OH^-$	~ -1.3
$SbO^+ + 2H^+ + 3e^- \rightarrow Sb + 3H_2O$	$+0.21$	$SbO_2^- - 2H_2O + 3e^- \rightarrow Sb + 4OH^-$	-0.66
$Sb_2O_5 + 6H^+ + 4e^- \rightarrow 2SbO^+ + 3H_2O$	$+0.58$	$H_3SbO_6^{4-} + H_2O + 2e^- \rightarrow SbO_2^- + 5OH^-$	~ -0.4
Arsenic			
$As + 3H^+ + 3e^- \rightarrow AsH_3$	-0.60	$As + 3H_2O + 3e^- \rightarrow AsH_3 + 3OH^-$	-1.43
$HAsO_2 + 3H^+ + 3e^- \rightarrow As + 2H_2O$	$+0.25$	$AsO_2^- + 2H_2O + 3e^- \rightarrow As + 4OH^-$	-0.68
$H_3AsO_4 + 2H^+ + 2e^- \rightarrow HAsO_2 + 2H_2O$	$+0.56$	$AsO_4{}^{3-} + 2H_2O + 2e^- \rightarrow AsO_2^- + 4OH^-$	-0.67
Barium			
$Ba^{2+} + 2e^- \rightarrow Ba$	-2.90	$Ba(OH)_2 \cdot 8H_2O + 2e^- \rightarrow Ba + 2OH^- + 8H_2O$	-2.97
Beryllium			
$Be^{2+} + 2e^- \rightarrow Be$	-1.85	$Be_2O_3{}^{2-} + 3H_2O + 4e^- \rightarrow 2Be + 6OH^-$	-2.62
Bismuth			
$BiO^+ + 2H^+ + 3e^- \rightarrow Bi + H_2O$	$+0.32$	$Bi_2O_3 + 3H_2O + 6e^- \rightarrow 2Bi + 6OH^-$	-0.46
Boron			
$H_3BO_3 + 3H^+ + 3e^- \rightarrow B + 3H_2O$	-0.87	$H_2BO_3^- + 3e^- \rightarrow B + 4OH^-$	-1.79
Bromine			
$Br_2 + 2e^- \rightarrow 2Br^-$	$+1.07$	$Br_2 + 2e^- \rightarrow 2Br^-$	$+1.07$
$HBrO + H^+ + e^- \rightarrow Br_2 + H_2O$	$+1.59$	$BrO^- + H_2O + 2e^- \rightarrow Br^- + 2OH^-$	$+0.76$
$BrO_3^- + 6H^+ + 5e^- \rightarrow \frac{1}{2}Br_2 + 3H_2O$	$+1.52$	$BrO_3^- + 3H_2O + 6e^- \rightarrow Br^- + 6OH^-$	$+0.61$
Cadmium			
$Cd^{2+} + 2e^- \rightarrow Cd$	-0.40	$Cd(OH)_2 + 2e^- \rightarrow Cd + 2OH^-$	-0.81
Calcium			
$Ca^{2+} + 2e^- \rightarrow Ca$	-2.87	$Ca(OH)_2 + 2e^- \rightarrow Ca + 2OH^-$	-3.03
Cerium			
$Ce^{4+} + e^- \rightarrow Ce^{3+}$	$+1.61$		
$Ce^{3+} + 3e^- \rightarrow Ce$	-2.48 or -2.34		
Cesium			
$Cs^+ + e^- \rightarrow Cs$	$+2.92$	$Cs^+ + e^- \rightarrow Cs$	-2.92
Chlorine			
$Cl_2 + 2e^- \rightarrow 2Cl^-$	$+1.36$		
$HClO + H^+ + e^- \rightarrow \frac{1}{2}Cl_2 + H_2O$	$+1.63$	$ClO + H_2O + e^- \rightarrow \frac{1}{2}Cl_2 + 2OH^-$	$+0.40$
$ClO_3^- + 6H^+ + 5e^- \rightarrow \frac{1}{2}Cl_2 + 3H_2O$	$+1.47$	$ClO_3^- + 2H_2O + 4e^- \rightarrow ClO^- + 4OH^-$	$+0.50$
$ClO_4^- + 2H^+ + 2e^- \rightarrow ClO_3^- + H_2O$	$+1.20$	$ClO_4^- + H_2O + 2e^- \rightarrow ClO_3^- + 2OH^-$	$+0.36$
Chromium			
$Cr^{3+} + 3e^- \rightarrow Cr$	-0.74	$CrO_4^{2-} + 4H_2O + 3e^- \rightarrow Cr(OH)_3 + 5OH^-$	-0.13
$Cr^{3+} + e^- \rightarrow Cr^{2+}$	-0.41	$CrO_2^- + H_2O + 3e^- \rightarrow Cr + 4OH^-$	-1.2
$Cr_2O_7^{2-} + 14H^+ + 6e^- \rightarrow 2Cr^{3+} + 7H_2O$	$+1.33$	$Cr(OH)_3 + 3e^- \rightarrow Cr + 3OH^-$	-1.3
Cobalt			
$Co^{2+} + 2e^- \rightarrow Co$	-0.28	$Co(OH)_2 + 2e^- \rightarrow Co + 2OH^-$	-0.73
$Co^{3+} + e^- \rightarrow Co^{2+}$	$+1.82$	$Co(OH)_3 + e^- \rightarrow Co(OH)_2 + OH^-$	$+0.14$
Copper			
$Cu^{2+} + 2e^- \rightarrow Cu$	$+0.34$	$Cu_2O + H_2O + 2e^- \rightarrow 2Cu + 2OH^-$	-0.36
$Cu^{2+} + e^- \rightarrow Cu^+$	$+0.15$	$2Cu(OH)_2 + e^- \rightarrow Cu_2O + 4OH^-$	-0.08
Fluorine			
$F_2 + 2e^- \rightarrow 2F^-$	$+2.85$		
Gallium			
$Ga^{3+} + 3e^- \rightarrow Ga$	-0.53	$H_2GaO_3^- + H_2O + 3e^- \rightarrow Ga + 4OH^-$	-1.22
Germanium			
$GeO_2 + 4H^+ + 4e^- \rightarrow Ge + 2H_2O$	-0.15	$HGeO_3^- + 2H_2O + 4e^- \rightarrow Ge + 5OH^-$	-1.0

(Continues)

Table 10 (*Continued*)

Acid Solution	E^0	Basic Solution	E_B°
Gold			
$Au^{3+} + 3e^- \rightarrow Au$	+1.50	$H_2AuO_3^- + H_2O + 3e^- \rightarrow Au + 4OH^-$	+0.7
$Au^{3+} + 2e^- \rightarrow Au^+$	+1.41		
Hafnium			
$Hf^{4+} + 4e^- \rightarrow Hf$	−1.70	$HfO(OH)_2 + H_2O + 4e^- \rightarrow Hf + 4OH^-$	−2.50
Hydrogen			
$H_2 + 2e^- \rightarrow 2H^-$	−2.23	$2H_2O + 2e^- \rightarrow H_2 + 2OH^-$	−0.83
$2H^+ + 2e^- \rightarrow H_2$	0 reference	$H_2 + 2e^- \rightarrow 2H^-$	−2.23
		$2H^+(\text{at pH 7}) + 2e^- \rightarrow H_2$	−0.41
Indium			
$In^{3+} + 3e^- \rightarrow In$	−0.34	$In(OH)_3 + 3e^- \rightarrow In + 3OH^-$	−1.0
Iodine			
$I_2 + 2e^- \rightarrow 2I^-$	+0.54	$I_2 + 2e^- \rightarrow 2I^-$	+0.54
$HIO + H^+ + e^- \rightarrow \frac{1}{2}I_2 + H_2O$	+1.45	$2IO^- + 2H_2O + 2e^- \rightarrow \frac{1}{2}I_2 + 4OH^-$	+0.45
$IO_3^- + 5H^+ + 4e^- \rightarrow HIO + 2H_2O$	+1.14	$IO_3^- + 2H_2O + 4e^- \rightarrow IO^- + 4OH^-$	+0.14
Iridium			
$Ir^{3+} + 3e^- \rightarrow Ir$	+1.15	$Ir_2O_3 + 3H_2O + 6e^- \rightarrow Ir + 6OH^-$	+0.1
$IrO_2 + 4H^+ + E^- \rightarrow Ir^{3+} + H_2O$	+0.7	$2IrO_2 + H_2O + 2e^- \rightarrow Ir_2O_3 + 2OH^-$	+0.1
Iron			
$Fe^{2+} + 2e^- \rightarrow Fe$	−0.44	$Fe(OH)_2 + 2e^- \rightarrow Fe + 2OH^-$	−0.88
$Fe^{3+} + e^- \rightarrow Fe^{2+}$	+0.78	$Fe(OH)_3 + e^- \rightarrow Fe(OH)_2 + OH^-$	−0.56
Lanthanum			
$La^{3+} + 3e^- \rightarrow La$	−2.52	$La(OH)_3 + 3e^- \rightarrow La + 3OH^-$	−2.90
Lead			
$Pb^{2+} + 2e^- \rightarrow Pb$	−0.13	$PbO_2 + H_2O + 2e^- \rightarrow PbO + 2OH^-$	+0.25
$PbO_2 + 4H^+ + 2e^- \rightarrow Pb^{2+} + 2H_2O$	+1.46	$HPbO_2^- + H_2O + 2e^- \rightarrow Pb + 3OH^-$	−0.54
Lithium			
$Li^+ + e^- \rightarrow Li$	−3.05	$Li^+ + e^- \rightarrow Li$	−3.05
Magnesium			
$Mg^{2+} + 2e^- \rightarrow Mg$	−2.37	$Mg(OH)_2 + 2e^- \rightarrow Mg + 2OH^-$	−2.69
Manganese			
$Mn^{2+} + 2e^- \rightarrow Mn$	−1.18	$Mn(OH)_2 + 2e^- \rightarrow Mn + 2OH^-$	−1.55
$MnO_2 + 4H^+ + 2e^- \rightarrow Mn^{2+} + 2H_2O$	+1.23	$Mn(OH)_3 + e^- \rightarrow Mn(OH)_2 + OH^-$	−0.4
$MnO_4^- + 4H^+ + 3e^- \rightarrow MnO_2 + 2H_2O$	+1.70	$MnO_2 + H_2O + 2e^- \rightarrow Mn(OH)_2 + 2OH^-$	−0.05
		$MnO_4^- + 2H_2O + 3e^- \rightarrow MnO_2 + 4OH^-$	+0.59
Mercury			
$Hg^{2+} + 2e^- \rightarrow Hg$	+0.85	$HgO + H_2O + 2e^- \rightarrow Hg + 2OH^-$	+0.10
$2Hg^{2+} + 2e^- \rightarrow Hg_2^{2+}$	+0.92		
Molybdenum			
$Mo^{3+} + 3e^- \rightarrow Mo$	−0.20	$MoO_2 + 2H_2O + 4e^- \rightarrow Mo + 4OH^-$	−0.87
$MoO_2^+ + 4H^+ + 2e^- \rightarrow Mo^{3+} + 2H_2O$	0.0	$MoO_4^{2-} + 2H_2O + 2e^- \rightarrow MoO_2 + 4OH^-$	−1.40
$H_2MoO_4 + 2H^+ + e^- \rightarrow MoO_2^+ + 2H_2O$	+0.4	$MoO_4^{2-} + 4H_2O + 6e^- \rightarrow Mo + 8OH^-$	−1.05
Nickel			
$Ni^{2+} + 2e^- \rightarrow Ni$	−0.25	$Ni(OH_2) + 2e^- \rightarrow Ni + 2OH^-$	−0.72
$NiO_2 + 4H^+ + 2e^- \rightarrow Ni^{2+} + 2H_2O$	+1.78	$NiO_2 + 2H_2O + 2e^- \rightarrow Ni(OH)_2 + 2OH^-$	+0.49
Niobium			
$Nb^{3+} + 3e^- \rightarrow Nb$	−1.10		
$Nb_2O_5 + 10H^+ + 10e^- \rightarrow 2Nb + 5H_2O$	−0.65		
Nitrogen			
$2NO_3^- + 4H^+ + 2e^- \rightarrow N_2O_4 + 2H_2O$	+0.80	$2NO_3^- + 2H_2O + 2e^- \rightarrow N_2O_4 + 4OH^-$	−0.86
$NO_3^- + 4H^+ + 3e^- \rightarrow NO + 3H_2O$	+0.96	$NO_3^- + H_2O + 2e^- \rightarrow NO_2^- + 2OH^-$	+0.01
$NO_3^- + 2H^+ + e^- \rightarrow HNO_2 + H_2O$	+0.94		
Osmium			
$OsO_4 + 8H^+ + 8e^- \rightarrow Os + 4H_2O$	+0.85	$HOsO_5^- + 4H_2O + 8e^- \rightarrow Os + 90H^-$	+0.02

Table 10 (*Continued*)

Acid Solution	E^0	Basic Solution	E_B°
Oxygen			
$O_2 + 4H^+ + 4e^- \rightarrow 2H_2O$	+1.23	$O_3 + H_2O + 2e^- \rightarrow O_2 + 2OH^-$	+1.24
$O_3 + 2H^+ + 2e^- \rightarrow O_2 + H_2O$	+2.07	$O_2 + 2H_2O + 4e^- \rightarrow 4OH^-$	+0.40
		$O_2 + 4H^+ (\text{at pH 7}) + 4e^- = 2H_2O$	+0.81
Palladium			
$Pd^{2+} + 2e^- \rightarrow Pd$	+0.99	$Pd(OH)_2 + 2e^- \rightarrow Pb + 2OH^-$	+0.07
$Pd^{4+} + 2e^- \rightarrow Pd^{2+}$	+1.60	$Pd(OH)_4 + 2e^- \rightarrow Pd(OH)_2 + 2OH^-$	+0.73
Phosphorus			
$P + 3H^+ + 3e^- \rightarrow PH_3$	−0.07	$P + 3H_2O + 3e^- \rightarrow PH_3 + 3OH^-$	−0.89
$H_3PO_2 + H^+ + e^- \rightarrow P + 2H_2O$	−0.51	$H_2PO_2^- + e^- \rightarrow P + 2OH^-$	−2.05
$H_3PO_3 + 2H^+ + 2e^- \rightarrow H_3PO_2 + H_2O$	−0.51	$HPO_3^{2-} + 2H_2O + 2e^- \rightarrow H_2PO_2^- + 3OH^-$	−1.57
$H_3PO_4 + 2H^+ + 2e^- \rightarrow H_3PO_3 + H_2O$	−0.28	$PO_4^{3-} + 2H_2O + 2e^- \rightarrow HPO_3^{2-} + 3OH^-$	−1.12
Platinum			
$Pt^{2+} + 2e \rightarrow Pt$	+1.20	$Pt(OH)_2 + 2e^- \rightarrow Pt + 2OH^-$	+0.15
$Pt(OH)_2 + 2H^+ + 2e^- \rightarrow Pt + 2H_2O$	+0.98	$Pt(OH)_6^{2-} + 2e^- \rightarrow Pt(OH)_2 + 4OH^-$	+0.20
Potassium			
$K^+ + e^- \rightarrow K$	−2.93	$K^+ + e^- \rightarrow K^+$	−2.93
Rhenium			
$ReO_2 + 4H^+ + 4e^- \rightarrow Re + 2H_2O$	+0.25	$ReO_2 + H_2O + 4e^- \rightarrow Re + 4OH^-$	−0.58
$ReO_4^- + 4H^+ + 3e^- \rightarrow ReO_2 + 2H_2O$	+0.51	$ReO_4^- + 2H_2O + 3e^- \rightarrow ReO_2 + 4OH^-$	−0.59
Rhodium			
$Rh^{3+} + 3e^- \rightarrow Rh$	+0.8	$Rh_2O_3 + 3H_2O + 6e^- \rightarrow 2Rh + 6OH^-$	+0.04
Rubidium			
$Rb^+ + 3e \rightarrow Rb$	−2.93	$Rb^+ + e^- \rightarrow Rb$	−2.93
Ruthenium			
$RuO_2 + 4H^+ + 4e^- \rightarrow Ru + 2H_2O$	+0.79	$RuO_2 + 2H_2O + 4e^- \rightarrow Ru + 4OH^-$	−0.04
		$RuO_4 + H_2O + 4e^- \rightarrow RuO_2 + 4OH^-$	+0.58
Scandium			
$Sc^{3+} + 3e^- \rightarrow Sc$	−2.08	$Sc(OH)_3 + 3e^- \rightarrow Sc + 3OH^-$	$\sim$ −0.26
Selenium			
$Se + 2H^+ + 2e^- \rightarrow H_2 Se$	−0.40	$Se + 2e^- \rightarrow Se^{2-}$	−0.92
$H_2SeO_3 + 4H^+ + 4e^- \rightarrow Se + 3H_2O$	+0.74	$SeO_3^{2-} + 3H_2O + 4e^- \rightarrow Se + 6OH^-$	−0.37
$SeO_4^{2-} + 4H^+ + 2e^- \rightarrow H_2SeO_3 + H_2O$	+1.15	$SeO_4^{2-} + H_2O + 2e^- \rightarrow SeO_3^{2-} + 2OH^-$	+0.05
Silicon			
$H_2SiO_3 + 4H^+ + 4e^- \rightarrow Si + 3H_2O$	−0.87	$SiO_3^{2-} + 3H_2O + 4e^- \rightarrow Si + 6OH^-$	−1.73
Silver			
$Ag^+ + e^- \rightarrow Ag$	+0.80	$Ag_2O + H_2O + 2e^- \rightarrow 2Ag + 2OH^-$	+0.34
Sodium			
$Na^+ + e^- \rightarrow Na$	−2.71	$Na^+ + e^- \rightarrow Na$	−2.71
Strontium			
$Sr^{2+} + 2e^- \rightarrow Sr$	−2.89	$Sr(OH)_2 \cdot 8H_2O + 2e^- \rightarrow Sr + 2OH^- + 8H_2O$	−2.99
Sulfur			
$S + 2H^+ + 2e^- \rightarrow H_2S$	+0.14	$S + 2e^- \rightarrow S^{2-}$	−0.51
$S_2O_3^{2-} + 6H^+ + 4e^- \rightarrow 2S + 3H_2O$	+0.50	$S_2O_3^{2-} + 3H_2O + 4e^- \rightarrow 2S + 6OH^-$	−0.74
$S_4O_6^{2-} + 2e^- \rightarrow 2S_2O_3^{2-}$	+0.08	$S_4O_6{2-} + 2e^- \rightarrow 2S_2O_3^{2-}$	+0.08
$4H_2SO_3 + 4H^+ + 6e^- \rightarrow S_4O_6^{2-} + 6H_2O$	+0.51	$2SO_3^{2-} + 3H_2O + 4e^- \rightarrow S_2O_3^{2-} + 6OH^-$	−0.58
$SO_4^{2-} + 4H^+ + 2e^- \rightarrow H_2SO_3 + H_2O$	+0.17	$SO_4^{2-} + H_2O + 2e^- \rightarrow SO_3^{2-} + 2OH^-$	−0.93
Tantalum			
$Ta_2O_5 + 10H^+ + 10e^- \rightarrow 2Ta + 5H_2O$	−0.81		
Tellurium			
$Te + 2H^+ + 2e^- \rightarrow H_2Te$	−0.72	$Te + 2e^- \rightarrow Te^{2-}$	−1.14
$TeO_2 + 4H^+ + 4e^- \rightarrow Te + 2H_2O$	+0.53	$TeO_3^{2-} + 3H_2O + 4e^- \rightarrow Te + 6OH^-$	−0.57
$H_6TeO_6 + 2H^+ + 2e^- \rightarrow TeO_2 + 4H_2O$	+1.02	$TeO_4^{2-} + H_2O + 2e^- \rightarrow TeO_3^{2-} + 2OH^-$	+0.4
Tallium			
$Tl^+ + e^- \rightarrow Tl$	−0.34	$Tl(OH) + e^- \rightarrow Tl + OH^-$	−0.34
$Tl^{3+} + 2e^- \rightarrow Tl^+$	+1.25	$Tl(OH)_3 + 2e^- \rightarrow Tl(OH) + 2OH^-$	−0.05

(*Continues*)

Table 10 *(Continued)*

Acid Solution	E^0	Basic Solution	E_B°
Thorium			
$Th^{4+} + 4e^- \rightarrow Th$	−1.90	$Th(OH)_4 + 4e^- \rightarrow Th + 4OH^-$	−2.48
Tin			
$Sn^{2+} + 2e^- \rightarrow Sn$	−0.14	$HSnO_2^- + H_2O + 2e^- \rightarrow Sn + 3OH^-$	−0.93
$Sn^{4+} + 2e^- \rightarrow Sn^{2+}$	+0.15	$Sn(OH)_6^{2-} + 2e^- \rightarrow HSnO_2^- + H_2O + 3OH^-$	−0.90
Titanium			
$Ti^{2+} + 2e^- \rightarrow Ti$	−1.63	$TiO_2(xH_2O) + 4e^- \rightarrow Ti + 4OH^- + (x-2)H_2O$	
$Ti^{3+} + e^- \rightarrow Ti^{2+}$	−0.37		−1.69
$TiO^{2+} + 2H^+ + e^- \rightarrow Ti^{3+} + H_2O$	+0.10		
Tungsten			
$WO_2 + 4H^+ + 4e^- \rightarrow W + 2H_2O$	−0.12	$WO_42- + 4H_2O + 6e^- \rightarrow W + 8OH^-$	−1.05
$W_2O_6 + 2H^+ + 2e^- \rightarrow 2WO_2 + H_2O$	−0.04		
$2WO_3 + 2H^+ + 2e^- \rightarrow W_2O_5 + H_2O$	−0.03		
Uranium			
$U^{3+} + 3e^- \rightarrow U$	−1.80	$U(OH)_3 + 3e^- \rightarrow U + 3OH^-$	−2.17
$U^{4+} + e^- \rightarrow U^{3+}$	−0.61	$U(OH)_4 + e^- \rightarrow U(OH)_3 + OH^-$	−2.14
$UO_2^+ + 4H^+ + e^- \rightarrow U^{4+} + 2H_2O$	+0.62	$Na_2UO_4 + 4H_2O + 2e^- \rightarrow U(OH)_4$	
$UO_2^{2+} + e^- \rightarrow UO_2^+$	+0.05	$+2Na^+ + 4OH^-$	−1.61
Vanadium			
$V^{2+} + 2e^- \rightarrow V$	−1.18	$VO_3^- + 3H_2O + 5e^- \rightarrow V + 6OH^-$	−1.15
$V^{3+} + e^- \rightarrow V^{2+}$	−0.26		
$VO_{2+} + 2H^+ + e^- \rightarrow V^{3+} + H_2O$	+0.36		
$VO_2^+ + 2H^+ + e^- \rightarrow VO^{2+} + H_2O$	+1.00		
Yttrium			
$Y^{3+} + 3e^- \rightarrow Y$	−2.37	$Y(OH)_3 + 3e^- \rightarrow Y + 3OH^-$	−2.8
Zinc			
$Zn^+ + 2e^- \rightarrow Zn$	−0.76	$Zn(OH)_2 + 2e^- \rightarrow Zn + 2OH^-$	−1.24
		$ZnO_2^{2-} + 2H_2O + 2e^- \rightarrow Zn + 4OH^-$	−1.22
Zirconium			
$Zr^{4+} + 4e^- \rightarrow Zr$	−1.53	$H_2ZrO_3 + H_2O + 4e^- \rightarrow Zr + 4OH^-$	−2.36

Tabulation for Example 13

Reduction	$Zn^2 + 2e^- \rightarrow Zn$	$E^\circ = -0.76$ V in neutral or acid
Oxidation	$Zn \rightarrow Zn^{2+} + 2e^-$	$E^\circ = +0.76$ V in neutral or acid
Reduction	$ZnO_2^{2-} + 2H_2O + 2e^- \rightarrow Zn + 4OH^-$	$E_B^\circ = -1.22$ V in base
Oxidation	$Zn + 4OH^- \rightarrow ZnO_2^{2-} + 2H_2O + 2e^-$ $E_B^\circ = +1.22$ V in base

---

Reduction Potentials for Other Than Tabulated Half-Reactions.

These may be obtained by adding or subtracting known half-reactions provided correction is made for the different numbers of electrons involved. This is done by summing voltage equivalents nE° and dividing by n, the number of electrons involved in the final derived equation.

Example 14

$$\frac{Fe^{2+} + 2e^- \rightarrow Fe \quad E^\circ = -0.44 \text{ V} \quad nE^\circ = -0.88}{Fe^{3+} + e^- \rightarrow Fe^{2+} \quad E^\circ = +0.77 \text{ V} \quad nE = +0.77}$$
$$Fe^{3+} + 3e^- \rightarrow Fe \qquad\qquad nE^\circ = -0.11$$

Therefore, $E^\circ = -0.11/3 = -0.04$ V.

Doubling stoichiometric coefficients of an equation does not change E° since E° is a molal quantity.

Relationship of Tabulated Reduction Potentials to "Electromotive Series." A compilation of standard reduction potentials, in acid, of metal-to-ion reduction half-reactions arranged in order of decreasing E° values comprises the usual abbreviated electromotive series tabulations.

Table of Standard Reduction Potentials in Aqueous Solution.

The potentials listed are derived from Latimer with additions from the recent literature. The E° values in volts for the reduction equations as written are based on a zero-reference voltage for the hydrogen couple at 25°C. Not all reactions are physically reversible or achievable as written under usual laboratory conditions. They are for 1 M activities for water-soluble species, for 1 atm fugacity for gases, and do not take into account overvoltages necessary for gas evolution at electrodes. Within the limitations imposed, these remain one of the most reliable sources of relative potential for reaction, although they infer nothing about reaction rate.

Significance of Reduction Half-Reactions and Potentials for Predicting Chemical Reactivity

1. Reaction with acids, water, and bases. Stability of solutions to air oxidation.
2. Redox reactions. A source of balanced half-reactions that may be added to give balanced redox reactions.
3. A systematic guide to descriptive inorganic chemistry and usual oxidation states.
4. Compound disproportionation (self-oxidation and reduction).

Reactions with Acids, Bases, Water, or Air.
Such reactions add complexity to all chemical reactions and offer competing reactions. If a potential exists to oxidize or reduce water (consider the pH of the system), that oxidation or reduction can proceed. Atmospheric oxygen is easily reduced (is a moderately strong oxidizing agent), and its reactions in systems exposed to air may be significant. These important potentials, listed under hydrogen and oxygen in the reduction potential table, are pH dependent and are summarized in Table 11.

Example 15

1. Air can oxidize Sn^{2+} to Sn^{4+} in neutral solution:

$$\begin{array}{ll} O_2 + 4H^+ + 4e^- & \to 2H_2O \\ 2[Sn^{2+} & \to Sn^{4+} + 2e^-] \\ \hline 2Sn^{2+} + O_2 + 4H^+ & \to 2Sn^{4+} + 2H_2O \end{array}$$

$$\begin{array}{l} E^\circ = +0.81 \text{ V} \\ E^\circ = -0.15 \text{ V} \\ \hline E^\circ = +0.66 \text{ V} \end{array}$$

2. Gold is dissolved by dilute cyanide solution with air oxidation in alkaline solution:

$$\begin{array}{ll} O_2 + 2H_2O + 4e^- & \to 4OH^- \\ 4[2CN^- + Au & \to Au(CN)_2^- + e^-] \\ \hline 4Au + 8CN^- + O_2 + 4H^+ & \to 2H_2O + 4Au(CN)_2^- \end{array}$$

$$\begin{array}{l} E_B^\circ = +0.40 \text{ V} \\ E_B^\circ = +0.60 \text{ V} \\ \hline E_B^\circ = +1.00 \text{ V} \end{array}$$

3. Iron rusting in neutral solution:

$$\begin{array}{ll} O_2 + 4H^+ (\text{at pH 7}) + 4e^- & \to 2H_2O \\ Fe + 2OH^- & \to Fe(OH)_2 + 2e^- \\ \hline Fe(OH)_2 + OH^- & \to Fe(OH)_3 + e^- \end{array}$$

$$\begin{array}{ll} E^\circ = 0.81 & \text{(a)} \\ E^\circ = (+0.88) \text{ (this } E^\circ \text{ is reduced at} \\ \quad \text{pH 7 to} \sim +0.6 \text{ V)} & \text{(b)} \\ E_B^\circ = +0.56 \text{ V} & \text{(c)} \end{array}$$

The reduction of oxygen reaction (a) and the oxidation of iron reaction (b) are followed by further reactions (a) and (c). An alkaline solution reduces potential (a) to +0.40 and an acid solution increases it to +1.23. Many reaction schemes are postulated for rusting; this one is plausible. Overall corrosion rate is increased at low pH and is dependent on oxygen diffusion rate.

4. Aluminum dissolves in a base to liberate hydrogen:

$$\begin{array}{ll} 2[Al + 4OH^- & \to H_2AlO_3^- + H_2O + 3e^-] \\ 3[2H_2O + 2e^- & \to H_2 + 2OH^-] \\ \hline 2Al + 2OH^- + 4H_2O & \to 2H_2AlO_3^- + 3H_2 \end{array}$$

$$\begin{array}{l} E_B^\circ = +2.35 \text{ V} \\ E_B^\circ = -0.83 \text{ V} \\ \hline E_B^\circ = +1.52 \text{ V} \end{array}$$

5. Copper will not dissolve in a nonoxidizing acid:

$$\begin{array}{ll} Cu & \to Cu^{2+} + 2e^- \\ 2H^+ + 2e^- & \to H_2 \\ \hline Cu + 2H^+ & \to Cu^{2+} + H_2 \end{array}$$

$$\begin{array}{l} E^\circ = -0.34 \text{ V} \\ E^\circ = 0 \text{ V} \\ \hline E^\circ = -0.34 \text{V} \end{array}$$

and reaction cannot proceed spontaneously

Table 11 Summary of Air and Water Reduction Potentials

Acidic Solution, pH 0	Neutral, pH 7	Basic Solution, pH 14
$O_2 + 4H^+ + 4e^- \to 2H_2O$	$O_2 + 4H^+ + 4e^- \to 2H_2O$	$O_2 + 2H_2O + 4e^- \to 4OH^-$
$E^\circ = +1.23$ V	$E^\circ = +0.81$ V	$E_B^\circ = +0.40$ V
$2H^+ + 2e^- \to H_2$	$2H^+ + 2e^- \to H_2$	$2H_2O + 2e^- \to H_2 + 2OH^-$
$E^\circ = 0$ V	$E^\circ = -0.41$ V	$E_B^\circ = -0.83$ V

Tabulation for Example 17

Reactions at Anode		Reactions Possible at Cathode
		At low concentrations:
$Fe \rightarrow Fe^{2+} + 2e^-$	$E° = +0.44$ V	$Fe^2 + 2e^- \rightarrow Fe, E° = -0.44$ V
$Cu \rightarrow Cu^{2+} + 2e^-$	$E° = -0.34$ V	At high concentrations:
$Ag \rightarrow Ag^+e^-$	$E° = -0.80$ V	$Cu^{2+} + 2e^- \rightarrow Cu, \ E° = 0.34$ V

6. Copper will dissolve in oxidizing acids (example: nitric):

$$3[Cu \quad\quad\quad \rightarrow \quad Cu^{2+} + 2e^-]$$
$$\underline{2[NO_3^- + 4H^+ + 3e^- \rightarrow 2H_2O + NO]}$$
$$3Cu + 2NO_3^- + 8H^+ \rightarrow 3Cu^{2+} + 4H_2O + 2NO$$

$$E° = -0.34 \text{ V}$$
$$\underline{E° = +0.96 \text{ V}}$$
$$E° = +0.62 \text{ V}$$

Electrolytic Cells. Electrolysis cells and the charging of batteries are comparable because an external applied voltage is used, depending on polarity, to augment the rate or to reverse the direction of spontaneous reaction. For example, charging a lead storage battery requires an opposing voltage greater than that of the spontaneous discharge reaction to reverse the direction of the oxidation and reduction reactions. Electrolysis cells can be operated using either molten salts or aqueous solutions, both of which are conductive.

(a) *Molten Salt Cells.* These cells are simple and straightforward in their operation.

Example 16. Electrolysis of fused $MgCl_2$ (with NaCl added to reduce melting point) liberates chlorine at the anode and magnesium metal at the cathode:

Oxidation at anode:
$$2Cl^- \quad\quad \rightarrow \quad Cl_2 + 3e^- \quad E° = -1.36 \text{ V}$$
Reduction at cathode:
$$Mg^{2+} + 2e^- \quad \rightarrow \quad Mg \quad\quad \underline{E° = -2.34 \text{ V}}$$
$$E° = -3.70 \text{ V}$$

Possible competitive reactions do not occur because the minimum applied potential of 3.70 V (plus *IR* losses) reduces the magnesium ion more easily than the sodium ion:

$$Na^+ + e^- \rightarrow Na \quad\quad E° = -2.71 \text{ V}$$

Although $E°$ is not directly applicable because of the nonaqueous media, elevated temperature, and high ionic concentrations, it can be helpful as a very crude approximation.

(b) *Aqueous Solution Electrolysis Cells.* These cells are used for electroplating, electrolytic copper purification, and many chemical reactions. Competing reactions such as electrolysis of water and overvoltages

required for oxygen and hydrogen bubble formation and evolution are complications encountered.

Example 17. Electrolysis in Aqueous Solution of Electrolytic Copper. Purification of impure copper requires its separation from Fe, Ag, Ni, Sb, and As metals. An electrolytic cell with an impure copper anode, $CuSO_4$ solution as electrolyte, and a pure copper cathode is used. See the tabulation for Example 17 above.

By application of a low voltage, Cu and the more easily oxidized metals Fe, Ni, Sb, and As are selectively dissolved at the anode, leaving Ag and metals that are more difficult to oxidize. Copper metal is deposited at the cathode and the more easily oxidized (less easily reduced) ions Fe^{2+}, Ni^{2+}, Sb^{3+}, and As^{3+} remain and accumulate in solution, replacing the Cu^{2+} of the electrolyte.

Overvoltage for Gas Evolution. Gaseous products can be formed and react with an anode at voltages consistent with their $E°$ values; however, bubble formation and evolution from inert anodes require an overvoltage. Overvoltage is experimentally dependent upon anode surface and current density. The approximate values in Table 12 show the large potentials required and the facts that hydrogen and halogen have very low overvoltages on platinum black and that oxygen, though high, is minimum on that particular surface. Reaction with noninert electrodes occurs whenever possible. The effect of overvoltage on the selection of alternate reactions that may occur is shown in the following example: Electrolysis of NaCl (basic aqueous solution) forms H_2 and Cl_2 on graphite electrodes:

$$2[Na^+ + e^- \rightarrow Na]E° \quad\quad = -2.71 \text{ V} \quad (a)$$
$$2Cl^- \rightarrow Cl_2 + 2e^- E° \quad\quad = -1.36 \text{ V} \quad (b)$$
$$2[2H_2O + 2e^- \rightarrow H_2 + 2OH^-]E°_B = -0.83 \text{ V} \quad (c)$$
$$4OH^- \rightarrow O_2 + 2H_2O + 4e^- E°_B = -0.40 \text{ V} \quad (d)$$

Reduction reactions (a) or (c) can occur; (c), the easiest reaction, occurs and liberates hydrogen (despite the hydrogen evolution overvoltage). Oxidation reactions (b) or (d) can occur; they are dependent on the low concentration of OH^-, the high concentration of Cl^-, and the much lower overvoltage for Cl_2 than for O_2 evolution, and Cl_2 is preferentially evolved. The net result is that reactions (b) and (c) occur.

Table 12   Approximate Overvoltages for Gas Evolution

	At 0.01 A/cm^2			At 0.1 A/cm^2			At 1 A/cm^2		
	O_2	H_2	Cl_2	O_2	H_2	Cl_2	O_2	H_2	Cl_2
Graphite	0.80	0.70	—	1.09	0.89	0.25	1.24	1.17	0.50
Pt black	0.40	0.03	0.02	0.64	0.04	0.03	0.79	0.05	0.08
Pt smooth	0.72	0.07	0.03	1.28	0.29	0.05	1.38	0.68	0.24
Ni	0.35	0.74	—	0.73	1.05	—	0.87	1.24	—
Cu	0.42	0.58	—	0.66	0.80	—	0.84	1.25	—
Ag	0.58	0.76	—	0.98	0.98	—	1.14	1.10	—
Au	0.67	0.39	—	1.24	0.59	—	1.68	0.80	—
Fe	—	0.56	—	—	0.82	—	—	1.29	—

Definitions

Anode.

Oxidation occurs at this electrode; site where reducing agent is oxidized; source of electrons to external circuit whether battery or electrolytic cell; electrode labeled negative terminal on batteries (physically labeled for condition of discharge); anode of a discharging battery becomes its cathode during charging (or acting as an electrolytic cell); name of electrode connected to positive terminal of external voltage source for charging a battery or for electrolytic cell; anions (negative ions) migrate toward it in an electrolytic cell.

Cathode.

Reduction occurs at this electrode. Its characteristics are reversed from those of the anode.

Refer to redox terminology of Table 4.

Batteries.   Batteries are redox cells that are physically arranged for external flow and internal ion mobility. When different electrolytes are used, gel structures or a porous membrane prevent mixing. The electrolyte(s) must permit ionization of the reactant species and be conductive. Mobility of ions in solution provides the mechanism for maintaining a charge balance and electrical neutrality within the electrolyte. By convention, batteries are labeled negative at the anode of the spontaneous discharge reaction.

(a) *Voltage.* Open-circuit terminal voltage is concentration and temperature dependent due to changes in chemical potential and is reduced by internal IR drop under current flow conditions. The current available depends on the reaction rate, which is a function of temperature and is not directly predictable from $E°$ values. For practical battery configurations $E°$ provides a reasonable voltage estimate.

(b) *Polarization.* Polarization is a localized accumulation of reaction products at an electrode until their concentration is reduced by diffusion, precipitation, formation of complexes, or further reaction to form new species not involved in the electrode reaction. Depolarization is the process of reducing the localized high concentration of reaction products; if depolar-ization is by a diffusional process, the cell may be reversible because the products retain their chemical identity and physical availability.

(c) *Irreversibility.* Although all redox reactions are ideally reversible near equilibrium, many cells operate far from equilibrium and with secondary reactions or physical configurations that render the cell partially, if not completely, irreversible for practical purposes. This situation results in having many cells that can convert their chemical energy to electrical and a limited few that are reversible for use as practical and economic storage batteries.

(d) *Battery Reactions.* Battery reactions on discharge are listed in what follows. Only a few are designed for reversibility.

1. *Lead Storage Battery* (H_2SO_4 Electrolyte) (Reversible). Oxidation reaction at lead plates, labeled electrically negative:

$$Pb + HSO_4^- \rightarrow PbSO_4 + 2e^- + H^+ \quad E° = 0.36 \text{ V}$$

Reduction reaction at lead dioxide plates, labeled electrically positive:

$$PbO_2 + 3H^+ + HSO_4^- + 2e^- \rightarrow PbSO_4 + 2H_2O$$

$$E° = +1.68 \text{ V}$$

2. *Mercury Cell* (KOH Electrolyte Saturated with ZnO). Oxidation (negative terminal):

$$Zn + 4OH^- \rightarrow ZnO_2^{2-} + 2H_2O + 2e^-$$

$$E°_B = +1.22 \text{ V}$$

Reduction (positive terminal):

$$HgO + H_2O + 2e^- \rightarrow Hg + 2OH^-$$

$$E°_B = +0.10 \text{ V}$$

3. *Zinc–Silver Peroxide Cell* (KOH Electrolyte Saturated with ZnO). Oxidation (negative terminal):

$$Zn + 4OH^- \rightarrow ZnO_2^{2-} + 2H_2O + 2e^-$$

$$E°_B = +1.22 \text{ V}$$

Reduction (positive terminal):

$$Ag_2O + H_2O + 3e^- \rightarrow 2Ag + 2OH^-$$

$$E_B^\circ = +0.34 \text{ V}$$

4. *LeClanche Cell* (Flashlight Battery, NH_4Cl Electrolyte). Oxidation (negative terminal):

$$Zn \rightarrow Zn^{2+} + 2e^-$$

$$E^\circ = +0.76 \text{ V}$$

Reduction (positive terminal):

$$2MnO_2 + 2\,NH_4^+ + 2e^- \rightarrow Mn_2O_3 + H_2O + 2\,NH_3$$

$$\text{(for complexing } Zn^{2+}) \quad E^\circ = +0.74 \text{ V}$$

5. *Alkaline Zinc Manganese Dioxide* "Alkaline Flashlight Battery" (KOH Electrolyte). Oxidation (negative terminal):

$$Zn + 4OH^- \rightarrow ZnO_2^{2-} + 2H_2O + 2e^-$$

$$E_B^\circ = +1.22 \text{ V}$$

Reduction (positive terminal):

$$MnO_2 + 2H_2O + e^- \rightarrow Mn(OH)_3 + OH^-$$

$$E_B^\circ = +0.35 \text{ V}$$

6. *Nickel–Cadmium Storage Battery* (KOH Electrolyte) (Reversible). Oxidation (negative terminal):

$$Cd + 2OH^- \rightarrow Cd(OH)_2 + 2e^- \quad E_B^\circ = +0.81 \text{ V}$$

Reduction (positive terminal):

$$NiO_2 + 2H_2O + 2e^- \rightarrow Ni(OH)_2 \quad E_B^\circ = +0.49 \text{ V}$$

10　ORGANIC CHEMISTRY

Tetrahedral Carbon. The electronic orbitals of carbon are consistent with the formation of four covalent bonds. Symmetry leads to a tetrahedral spatial orientation. The ability of carbon to bond to carbon leads to a multiplicity of organic compounds based on linear, branched chain, and ring sequences of carbon atoms.

Carbon–Carbon Bonds. Saturation refers to carbon–carbon single bonds, with the remaining three bonds to another carbon, to hydrogen, or to substituent groups. Double and triple bonds between carbon atoms are strained and quite reactive; compounds containing these are termed *unsaturated* because each double bond replaces two substituents.

10.1　Classes of Compounds Based on Structure of Carbon Chain

Aliphatic Compounds. Chains and branched chains of tetrahedral carbon atoms covalently bonded to each other give rise to a proliferation of aliphatic compounds. Bonds between carbon atoms forming the backbone chain may be single, double, or triple, and the remaining bonds available at each carbon atom are made to hydrogen or other substituent groups. Compounds have backbone and branch combinations ranging from C_1 to C_{40} or more. Since substituent groups other than hydrogen are possible at the carbon atoms and can be placed in different sequences along the backbone, the numbers of organic compounds become very large—before considering the effect of spatial configuration that often permits one part of the molecule to combine with another reactive part of the same or a kindred molecule.

Cyclic Compounds. Ring structures of aliphatic compounds exist subject to bond strain and the possible three-dimensional configurations of the carbon atom chain. Cyclohexane, C_6H_{12} (Fig. 10a), and cyclopentane, C_5H_{10} (Fig. 10b), most commonly exist as the backbone structures. These consist of tetrahedral carbon atoms connected by single bonds and with two hydrogen atoms existent at each apex. Double bonds between carbon atoms can occur: cyclohexene, C_6H_{10} (Fig. 10c) and cyclopentadiene, C_5H_5 (Fig. 10d), are examples.

Note that these can be variously oriented on paper just as the molecules can be in space, so there is no correct or preferred orientation. The presence of a double bond limits the number of hydrogen atoms or substituent groups to one at the double-bonded carbon atoms.

Aromatic Compounds. A particularly stable minimum-energy configuration of six carbon atoms (Fig. 10e) consists of equivalent carbon atoms in planar configuration with single bonds to a substituent at each apex of the hexagon. The absence of other substituent groups infers the presence of hydrogen, and the figure drawn signifies benzene, C_6H_6. Other groups may replace any or all hydrogen atoms subject to steric (space) considerations; two or three large groups or six small groups may each replace a hydrogen. Higher analogs of benzene (e.g., Figs. 10f–(h) are condensed planar ring structures containing H atoms at the protruding corners that are subject to similar space-limited substitutions.

Heterocyclic Compounds. Heterocyclic compounds are ring structures with O, S, or N substituted for one or more of the carbon atoms. Bonds at these substituent atoms are those expected from their electronic structure. Examples are furan, C_4H_4O (Fig. 10i), thiophene, C_4H_4S (Fig. 10j), pyridine, C_5H_5N (Fig. 10k), a nitrogen analog of the aromatic hydrocarbon benzene, and piperidine, C_5H_{10} NH (Fig. 10l), an alicyclic amine.

Cyclic Compounds

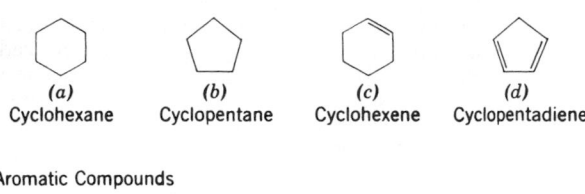

(a) Cyclohexane (b) Cyclopentane (c) Cyclohexene (d) Cyclopentadiene

Aromatic Compounds

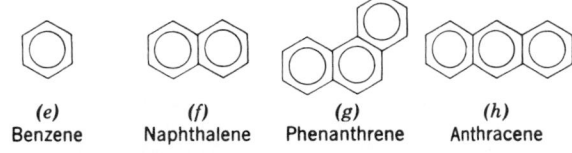

(e) Benzene (f) Naphthalene (g) Phenanthrene (h) Anthracene

Heterocyclic Compounds

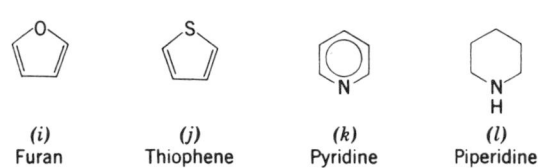

(i) Furan (j) Thiophene (k) Pyridine (l) Piperidine

Fig. 10 Cyclic, aromatic, and heterocyclic compounds.

10.2 Abbreviated Organic Chemical Nomenclature

Organic nomenclature is systematic and dependent upon an extensive set of rules of the IUPAC that discourages use of older trivial names. Use of the structural formula is encouraged because it is unambiguous. A few of the simpler rules are helpful: Ring compounds are considered the basic unit and all appendages as secondary. If required, positions around ring compounds are numbered sequentially. Carbon atoms of linear and branched-chain compounds are numbered along the longest string, starting at the end nearest a substituent group or branch. The positions of substituents and side chains are located by the number of the carbon atom where they are attached. Multiple bonds are located by the lowest numbered of the two numbered carbon atoms involved. Some of the prefix and suffix terms used in organic nomenclature are listed in Table 13.

Common usage of trivial names that are unrelated to structure will continue. The only recourse in a practical situation is to locate the compound and its physical constants in the *Merck Index*[3] or the *Handbook of Chemistry and Physics*.[2]

10.3 Classification of Compounds into Unreactive and Functional Groups

Although the number of known organic compounds is astronomic, the number of inexpensive and commer-

cially available ones is drastically limited. Their chemistry ranges from complete oxidation (during combustion) to conversion into specific compounds with academic or commercial application.

Reactivity of organic compounds occurs only at selected sites on the molecule. Synthetic procedures attack substituent groups, more reactive-than-normal hydrogen atoms, and multiple bonds in the carbon chains; the remainder of the molecule remains unchanged in the process. This permits classification of compounds into a reactive portion of the molecule and one that is unaffected by the synthetic process. The reactive portion is a substituent or functional group. The unaffected portion is the remainder of the compound and is called an aliphatic (alkyl), cyclic aromatic (aryl), or a heterocyclic R group according to its structure; this R group is the backbone that carries the reactive substituent group (Tables 14 and 15).

10.4 Organic Reactions

Organic reactions make changes in the bonding to carbon atoms; their course is greatly influenced by reaction conditions that are experimentally established on the basis of reaction mechanism studies. Complications arise because several different reactions may occur simultaneously accompanied by cyclization, molecular rearrangements, and oxidation–reduction. All occur in the direction of minimum energy and increased stability. An encyclopedic literature is replete with tens of thousands of reactions, several hundred of which are generally useful in synthesis and are "name reactions"

Table 13 Organic Chemical Nomenclature

-ane	Saturated hydrocarbon compound
-ene	Unsaturated hydrocarbon compound with double bond
-yne	Unsaturated hydrocarbon compound with triple bond
-yl	Hydrocarbon group as substituent (see next section)
cis-	Refers to stereochemistry
trans-	Refers to stereochemistry
ortho-	1,2 positions on benzene ring (abbreviated *o-*)
meta-	1,3 positions on benzene ring (abbreviated *m-*)
para-	1,4 positions on benzene ring (abbreviated *p-*)
d-	Dextro-rotatory optical isomer
l-	Levo-rotatory optical isomer
r-	Racemic, mixture of *d* and *l* isomers
n-	Normal or linear chain
iso-	Branched chain; usually with isopropyl group
t-	Tertiary, three carbons attached to one
sec-	Secondary, two carbons attached to one
neo-	Four carbons attached to one
di-	Two
tri-	Three
cyclo-	Ring structure

Three parts of a simple compound name are:
1. A position number
2. Name of substituent group (see next section)
3. Name of parent structure

Examples:

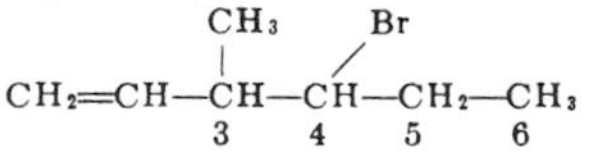

$$CH_2{=}CH\underset{3}{-}\overset{\overset{\textstyle CH_3}{|}}{CH}\underset{4}{-}\overset{\overset{\textstyle Br}{\diagup}}{CH}\underset{5}{-}CH_2\underset{6}{-}CH_3 \qquad \text{3-Methyl-4-bromo-hex-l-ene}$$

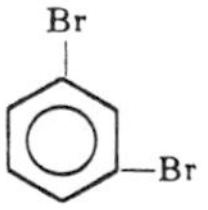

1,3-Di-bromo-benzene

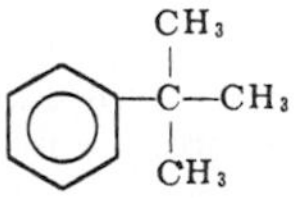

t-Butyl-benzene

1,3-Cyclo-hexa-di-ene

honoring their early investigators. Studies of rate, catalysis, stepwise mechanism, and stereochemistry have coalesced much of the accumulated information into a few generalized categories of related mechanisms.

Substitution. In substitution reactions a group attached to a carbon atom is removed and another enters in its place. Reactions are designated S_{N_1}, S_{N_2}, S_{E_1}, or S_{E_2} dependent on the nucleophilic or electrophilic nature of the reagent and the unimolecular or bimolecular dependence of the reaction rate. Nonpolar substitutions are dependent on a free-radical mechanism. Solvolysis reactions, such as hydrolysis of esters or amides, and organic acid–base neutralizations are substitution reactions. Other reactions occur at both saturated and unsaturated carbon atoms. Intra- and intermolecular substitutions are the mechanism for rearrangements and cyclization (ring formation).

Addition. Addition of functional groups occurs at unsaturated carbon atoms, increasing the number of groups attached at those positions. The reagent species may be nucleophilic, electrophilic, or free radical.

Elimination. Reducing the number of functional groups bound to one carbon atom necessitates the formation of an unsaturated carbon bond to the adjacent carbon. Elimination thus involves removal of two functional groups from contiguous carbon atoms and is the reverse of addition reactions. Elimination reactions often occur simultaneously and competitively with substitution reactions at saturated carbon atoms.

Rearrangement. Many reactions are encountered where functional groups migrate within the molecule and result in products other than anticipated. These migrations tend to occur at adjacent carbon atoms; they

Table 14   Alkyl and Aryl R Groups

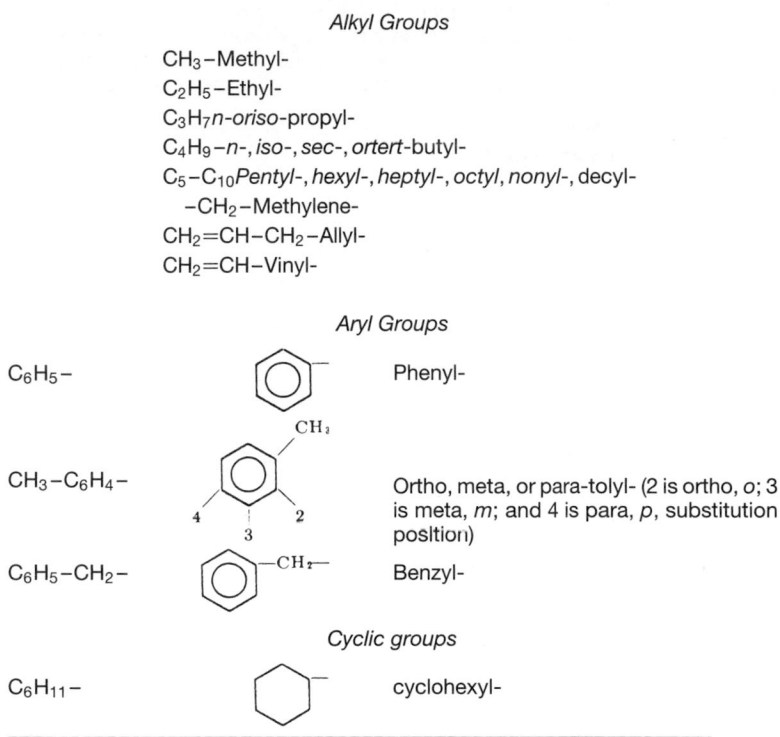

Alkyl Groups

CH_3–Methyl-
C_2H_5–Ethyl-
C_3H_7 *n*-or*iso*-propyl-
C_4H_9 –*n*-, *iso*-, *sec*-, or*tert*-butyl-
C_5–C_{10} *Pentyl*-, *hexyl*-, *heptyl*-, *octyl*, *nonyl*-, decyl-
   –CH_2–Methylene-
CH_2=CH–CH_2–Allyl-
CH_2=CH–Vinyl-

Aryl Groups

C_6H_5 –   Phenyl-

CH_3–C_6H_4 –   Ortho, meta, or para-tolyl- (2 is ortho, *o*; 3
                                                  is meta, *m*; and 4 is para, *p*, substitution
                                                  position)

C_6H_5–CH_2 –   Benzyl-

Cyclic groups

C_6H_{11} –   cyclohexyl-

---

may involve five- and six-membered heterocyclic ring intermediates or neighboring group participation in the reaction. The driving force for any rearrangement, as for any reaction, is increased stability of the resultant product species.

Oxidation–Reduction.   Redox reactions are selectively applied to carbon–hydrogen bonds and carbon–carbon multiple bonds, but few oxidative procedures allow selective breaking of a specific carbon–carbon single bond without disrupting the entire molecule. The removal of hydrogen to form multiple carbon–carbon bonds or to make new bonds between carbon and oxygen, sulfur, nitrogen, or halogens is called oxidation. Primary alcohols are oxidized to carboxylic acids, and acids are reduced to alcohols.

10.5   Reagents of Organic Reactions

Further classification is made according to classes of reagent, the species which attacks a substrate to produce products. Reagents are collected into three classes based on their electronic structure:

(a) Nucleophilic. Lewis bases can donate unshared electron pairs, reducing agents.
(b) Electrophilic. Lewis acids can accept unshared electron pairs, oxidizing agents.

(c) Free Radical. A reactive transient fragment that has one or more unpaired electrons. Examples: $Cl·$, $≡C·$, and $=C$: are chlorine, carbon, and carbene radicals, respectively.

The more common reagents are summarized in Table 16.

10.6   Catalysis

Catalysis is absolutely dependent on the reaction mechanism; other than transition metal participation in intermediate species, acid and base catalyses are most important:

1. Acid catalysis facilitates production of electrophilic reagent species. Lewis acids provide acid catalysis.
2. Base catalysis facilitates production of nucleophilic reagent species. Lewis bases provide base catalysis.
3. Free-radical initiation: Heat, light, or peroxide can initiate free-radical formation, independent of both solvent and acid or base catalysis.

10.7   Solvents

The solvent for reactions may be one of the reactants or be introduced to facilitate mutual contact via solution.

Table 15   Functional Groups

—OH	Alcohol when attached to alkyl R, phenol when attached to aryl R		Sulfoxide
—O—	Ether		Sulfone
—O—O—	Peroxide		
—O—OH	Hydroperoxide		
—O—N=O	Nitrite		Sulfinic acid
—O—N (=O, O)	Nitrate		Sulfonic acid
—N (O, =O)	Nitro compound		Carboxylic acid
—NH₂	Primary amine		Acid anhydride
—N (H)	sec-Amine		Ester
—N	tert-Amine		
=NH	Imide		Amide
—C≡N	Nitrile		
—X	Halide, where X = F, Cl, Br, or I		Acid halide
—H	Hydrogen		
C=C	Double bond		Ketone
—C≡C—	Triple bond		Aldehyde
—SH	Thioalcohol, mercaptan		
—S—	Sulfide		
—S—S—	Disulfide		

Table 16   Nucleophilic and Electrophilic Reagents

Nucleophilic Reagent, Bases and Reducing Agents, Donors of Unshared e^- Pair		Electrophilic Reagent, Acids and Oxidizing Agents, Acceptors of e^- Pair	
I$^-$	Iodide ion	H_2O^+	Hydronium ion
OH$^-$	Hydroxyl ion	$\begin{cases} R_2C=\overset{+}{O}H \\ R_2\overset{+}{C}-OH \end{cases}$	These occur after protonation of ketones
RO$^-$	Alcoholate ion		
RS$^-$	Mercaptide ion	BF_3	Boron trifluoride
CN$^-$	Cyanide ion	$AlCl_3$	Aluminum chloride
H$_2$Ö:	Water	Cl_2	Their cleavage forms X^+ and X^-; X^+ is actual electrophilic reagent
		Br_2	
RÖH	Alcohol	I_2	
:NH$_2$	Ammonia	$NO_2{}^+$	Nitronium ion (formed after protonation of nitric acid)
Br$^-$	Bromide ion		
Cl$^-$	Chloride ion		
—C$^-$:	Carbanion	—C$^+$	Carbonium ion

Table 17 Solvents for Organic Reactions

Polar	Nonpolar
Water	Hydrocarbons
Formic acid	Acetone — will dissolve some ionic species
Dimethylsulfoxide	Ethers
Dimethylformamide	Ethylene glycol dimethyl ether (diglyme)
Methanol	Tetrahydrofuran (THF)
Ethanol	Dioxane
Acetic acid	
Nitromethane	
Acetonitrile	
Liquid ammonia	

The choice is necessarily dependent on reaction conditions to avoid undesired solvent participation:

(a) Polar solvents are chosen to solubilize ionic species.
(b) Nonpolar solvents dissolve un-ionized molecules.

Table 17 summarizes the more widely used polar and nonpolar solvents.

REFERENCES

1. Smith, J. M. and Van Ness, H. C., *Introduction to Chemical Engineering Thermodynamics*, 3rd ed., McGraw-Hill, New York, 1975.
2. R. C. Weast, Ed. *Handbook of Chemistry and Physics*, 66th ed., CRG Press, Boca Raton, FL, 1985.
3. Windholz, M., *The Merck Index*, 10th ed., Merck, Rahway, NJ, 1983.

BIBLIOGRAPHY

Atkins, P. W., *Physical Chemistry*, 2nd ed., Freeman, San Francisco, 1982.
Bard, A. J., and Faulkner, L. R., *Electrochemical Methods*, Wiley, New York, 1980.
Davis, R. E., Gailey, K. D., and Whitten, K. W., *Principles of Chemistry*, Saunders, New York, 1984.
Dean, J. A., ed., *Lange's Handbook of Chemistry*, 13th ed., McGraw-Hill, New York, 1985.
Morrison, R. T., and Boyd, R. N., *Organic Chemistry*, 5th ed., Allyn and Bacon, Boston, 1987.
Perry, R. H., and Green, D. W., *Perry's Chemical Engineers Handbook*, 6th ed., McGraw-Hill, New York, 1984.
Selected Values of Chemical Thermodynamic Properties, NBS Circular 500, U.S. Government Printing Office, Washington, DC, 1952.
Stryer, L., *Biochemistry*, 2nd ed., Freeman, San Francisco, 1981.
Willard, H. H., Merritt, L. L., and Dean, J. A., *Instrumental Methods of Analysis*, 5th ed., Van Nostrand, New York, 1974.

CHAPTER 21

ENGINEERING ECONOMY

Kate D. Abel
School of Systems and Enterprises
Stevens Institute of Technology
Hoboken, New Jersey

1 INTRODUCTION

Engineering economics is a necessary field for practicing engineers and those studying to become engineers because it helps students prepare for real-world situations and provides professionals with valuable conceptual and mathematical tools for implementing cost analysis. Cost analysis and project selection in a cost-efficient manner are important factors impacting economies and business competitiveness. Companies must respond to increased competition with lower cost operations. Thus, the purpose of engineering economics is to formulate economic studies for the proper allocation of financial resources. Just as operations research or other management sciences deal with the allocation of human or material resources in a methodical and calculated manner, engineering economics focuses on the distribution of financial resources in the best manner to maximize financial payback. This type of analysis is intricately intertwined with engineering since many problems can be easily solved with engineering solutions, but it is only when we add

the economic constraint that we realize the solution is not necessarily feasible. As Eugene Grant[1] said, engineering "involves a realization that quite as definite a body of principles governs the economic aspects of an engineering decision as governs its physical aspects." For example, the problem of excessive carbon dioxide emissions could easily be solved by giving everyone an alternate-fuel car. However, it is only when we add the economic constraint that alternate-fuel cars are too expensive for most people that the real engineering begins. Thus, this chapter on engineering economics discusses the kinds of entrepreneurial decisions that engineers need to make in choosing among competitive design alternatives or in deciding which business case or project they or their company should pursue.

2 CASH FLOWS AND TIME VALUE OF MONEY

Money makes (or loses) value as it changes hands over time. Transactions, where money moves from one entity to another, are called cash flows and can happen at any point in time. Whether these transactions, or cash flows, are from lender to borrower or personal investor to bank, the details of cash flow cost one person some part of the wealth and increase the other's wealth. In contrast to taking dollars and putting them under a mattress, called hoarding, in a transaction, or cash flow, scenario, one is modifying wealth via the transaction, whereas in the hoarding example one is not modifying wealth.

In the hoarding scenario, money, or the future value of money, can only be lost. This is so since inflation causes the purchasing power of today's dollars to decrease over time. Add the interest lost by putting the money under the mattress instead of a simple savings account, and we can see that under the mattress money is not standing still but instead is losing value at a rather quick rate. Contrast that to funds and put in a bank account, certificate of deposit, or money market account. This investment would earn some money through the interest rate on the account—this might at least keep pace with inflation.

But what is the value of money used to purchase something such as a home or a car? For example, a person takes out a mortgage for a specified time period (e.g., 30 years) at some rate of interest. To pay back the mortgage, the person generally pays the same mortgage payment in year 1 as in year 30. But what is this payment really worth in year 1? And what is it worth in year 30? Are they the same value? The answer is that the values of the two amounts, though the same in dollar amount, are not the same in value. The first payment is made in today's dollars and the last payment, 30 years later, is made when the value of those future dollars is less. Using numbers to illustrate: When your parents first took out their mortgage in the last quarter of the twentieth century, a monthly payment of $1000 was a hefty monthly mortgage payment for a young couple to meet. By the time you were out of college and your parents were making their last mortgage payment, that same $1000 monthly payment was just a drop in the bucket. This is because inflation went up while purchasing power went down, making the future value of that same $1000 less after 30 years.

Thus, not only is a transaction, or cash flow, necessary to modify the value of money, but the time line on which the transaction occurs also needs to be known. For, as can be seen above, a dollar in a future transaction is not worth as much as a dollar today. To equate the two without correcting for the "time value of money" would be as incorrect as adding pounds to kilograms without first converting the items into the same units. This is the basis for engineering economics—equating sums of money from different periods of time (i.e., the *time value of money*) based on the risk associated with the investment, which is normally expressed in terms of the interest rate.

3 EQUIVALENCE

Economics is basically the moving of money from some point in time to some other point in time. For example, if one loans money to someone else, money could be made later. However, there are a many things to consider. For example, there is a risk the borrower will not pay back the loan, the borrower is getting to use the money when you are not, there is a risk of inflation over the life of the loan, and the dollars you get back, even if more than you loaned, may not buy as much as they would have if you had just kept the money. In addition to these concepts surrounding the loan, one must consider what else could be done with the money if it was not lent. By considering these factors, an agreement between lender and borrower will say that if one is going to lend money, the borrower will have to pay back more than what was borrowed. One might think of this extra amount as rental income charged for the use of the money. To account for these types of issues, we use an interest rate to indicate the value of money on the time continuum. Thus, the concept of *equivalence* within economics is that today's and tomorrow's dollars are related to each other by some interest rate and time period.

In order to make sensible comparisons between present and future sums on this time continuum, we must bring all the sums being compared into the same moment of time. This moment can be either present or future, but not both. The value of present money increases as we move it into the future, while the value of future income falls as we move it back to the present.

The central notion here is that of equivalence. Present and future sums of money are equivalent if a rational person would be indifferent as to which he or she received. Thus, for example, if one won the lottery, the payout is typically either one cash payment today or multiple payments over a specified time period in the future. At some interest rate, both of these payouts are equivalent. Assuming the winner could only reinvest the winnings at the same interest rate at which the lottery is paying out, the winner would be indifferent to the options. Or put more simply, if the best available interest rate is 5%, one should be indifferent between receiving $100 now or $105 this time next year. Similarly, we should now see the principle that $100 this time next year is worth less than $100 today. (These concepts do not include specific exceptions such as when income taxes may affect a decision to accept money today.)

3.1 Cash Flow Diagrams

Sometimes one has to use money now and pay it back later or vice versa. In order to move money from one in point in time to another, the units of the time line need to be specified along the time line, or the planning horizon of the project, on a cash flow diagram. Similarly, the equations need to be specified. Within these equations are standard notations for the quantities involved. The five most needed quantities are P, present worth; F, future worth (or "compound amount"); A, annual worth; i, interest rate per period; and N, number of time periods we are considering.

Cash flow diagrams are a way to define and visualize the problem. Similar to other engineering and science concepts which use concepts such as the free-body diagram, cash flow diagrams are an important step in taking a problem from words to one that can be easily analyzed using mathematical techniques. And within the time line, or cash flow diagram, standard conventions are similarly employed. On a cash flow diagram the x axis represents the time period up to the duration of the project, N. The y axis represents the monetary amount where up is a positive inflow of money (i.e., revenues) and down is a negative outflow of money (i.e., expenditures). Also note that an arrow means that money is changing hands. However, a key assumption within the cash flow diagram is that all interest occurs at the end of the time period. See Figs. 1a, b for examples of typical cash flow diagrams as well as the information usually presented on the diagram. It should be noted that the *end-of-year* convention is the generally accepted format for cash flow diagrams. What this means is that any cash flows within the year will be shown as occurring at the end of the year.

3.2 Solving the Problem

The steps to solving a typical cash flow problem are simple. First one would want to draw a picture, or cash flow diagram, in order to visualize the entire cash flow within the problem. Second, one should identify all the known components and unknowns within the problem. Third, one needs to convert all known components to the same units of time. Lastly, one needs to identify the correct equation and solve the problem using engineering economic techniques. The general equations used in engineering economics are shown in Table 1.

Factor notation can be used to represent the necessary equations (see the notation column of Table 1) and is available in tabular form, such as Table 2, in the literature,[2] or on the Internet.[3] The tables used will depend upon which factor is used (designated in columns within the tables) and which interest rate is needed (designated by separate tables). Normally each table is based on one interest rate or minimum attractive rate of return (MARR). Thus, one would need to find the table with the appropriate interest rate for the problem at hand. Common mathematical operations would tell one to multiply by a factor which would "cancel out" the variable on the right side of the equation and leave the variable you are looking for on the left. This is not what actually happens. However, it allows one to easily determine which factor should be used as located across the top row of the table. In addition to determining the factor and interest rate, one would also need to determine the period, N, which would be located down the left side of the table. The tables all work this way, and when used in place of the formulas, they make calculations very simple. Interest table use is not always possible if the interest rate is not listed in the provided tables. (Tables normally run only in whole numbers from 1 to 50%.) In situations where the interest table is not available, one needs to use the formulas to calculate the answer the long way. See Table 1 for factor functional formats as well as formulas.

In addition, it should be noted that the interest rate is assumed to be *compound interest* where the money within the account is compounded monthly, quarterly, or yearly. Also the interest rate used must be the effective rate per period, N. Thus the basis of the rate, whether monthly, quarterly, or yearly, must agree with the period used for N. So, for example, one cannot use an annual interest rate when the time period N is in months.

Lastly, there are different kinds of interest rates. The effective interest rate is the rate that would yield the accrued interest at the end of the year. The nominal rate (r) equals the rate per period times the number of periods in the year (m). (For example, $m = 12$ for monthly compounding.) Thus, if the interest rate were 1% per month, the nominal rate would be 12% per year. Thus, the effective interest rate is simply r/m. The nominal and effective rates differ as the effective rate is the yield from the *continuous* compounding process. Thus, for the effective rate, if the interest rate was 1% per month, the effective rate would actually be greater than 12% due to the power of compounding. Generally speaking, one can assume the rate given in a problem is the effective rate if the compounding is annual. If, however, the compounding is not annual, the rate given is the nominal rate.

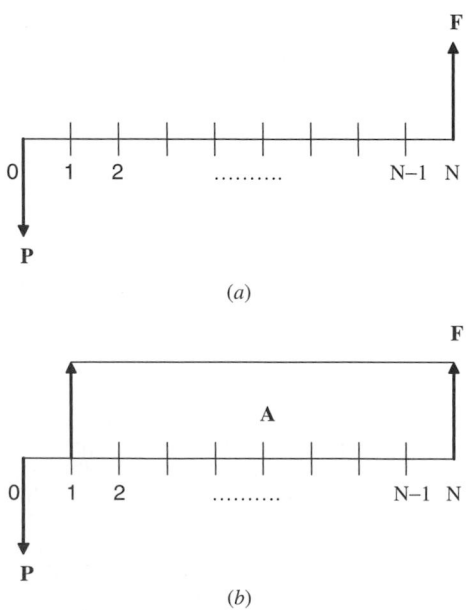

Fig. 1 Typical cash flow diagrams: (a) single sum; (b) uniform series.

Table 1 Engineering Economy Symbols, Notation, and Formulas

Type of Factor	Factor Symbol	Factor Notation	Formula
Present Worth			
Single payment	P/F	P/F, i, N	$\dfrac{1}{(1+i)^N}$
Annuity	P/A	P/A, i, N	$\dfrac{(1+i)^N - 1}{i(1+i)^N}$
Gradient	P/G	P/G, i, N	$\dfrac{(1+i)^N - 1}{i^2(1+i)^N} - \dfrac{N}{i(1+i)^N}$
Geometric gradient		P/A, i, g, N	$\dfrac{1 - [(1+g)/(1+i)]^N}{i-g}$
Future Worth			
Single payment	F/P	F/P, i, N	$(1+i)^N$
Annuity	F/A	F/A, i, N	$\dfrac{(1+i)^N - 1}{i}$
Gradient	F/G	F/G, i, N	$\dfrac{(1+i)^N - 1}{i^2} - \dfrac{N}{i}$
Annuity			
Capital recovery	A/P	A/P, i, N	$\dfrac{i(1+i)^N}{(1+i)^N - 1}$
Sinking fund	A/F	A/F, i, N	$\dfrac{i}{(1+i)^N - 1}$
Gradient	A/G	A/G, i, N	$\dfrac{1}{i} - \dfrac{N}{(1+i)^N - 1}$

Table 2 6% Interest Rate

N	Compound Amount Factor (F/P, i, N)	Present-Worth Factor (P/F, i, N)	Compound Amount Factor (F/A, i, N)	Sinking-Fund Factor (A/F, i, N)	Present-Worth Factor (P/A, i, N)	Capital Recovery Factor (A/P, i, N)	Uniform Gradient Series Factor (A/G, i, N)
1	1.0600	0.9434	1.0000	1.0000	0.9434	1.0600	0.0000
2	1.1236	0.8900	2.0600	0.4854	1.8334	0.5454	0.4854
3	1.1910	0.8396	3.1836	0.3141	2.6730	0.3741	0.9612
4	1.2625	0.7921	4.3746	0.2286	3.4651	0.2886	1.4272
5	1.3382	0.7473	5.6371	0.1774	4.2124	0.2374	1.8836
6	1.4185	0.7050	6.9753	0.1434	4.9173	0.2034	2.3304
7	1.5036	0.6651	8.3938	0.1191	5.5824	0.1791	2.7676
⋮	⋮	⋮	⋮	⋮	⋮	⋮	⋮
100	339.3021	0.0029	5638.3681	0.0002	16.6175	0.0602	16.3711

4 SINGLE SUM AND UNIFORM, GRADIENT, AND GEOMETRIC SERIES

When one is considering two investment opportunities, if you are indifferent to one or the other in making a choice between them, they are said to be equivalent. For example, two values, P occurring in the present and F occurring at some future time N, are equivalent at some interest rate. The most common conversion is that which reduces the cash flow pattern to a single sum at year zero. This is referred to as the present worth (P).

But another common conversion reduces the cash flow to a future single sum. This single sum can occur at year N, the end of the cash flow, or at any time in the middle. This is referred to as the future worth (F). (*Note:* There is a special function of F called salvage value (S) which occurs only at year N at the end of a project.) The parameters P, F, and S are all equivalent single sums within a given time period. See Fig. 1a.

Another common conversion is to reduce a cash flow pattern to a uniform series that runs without change in amount from year 1 to year N. This is the annual worth (A). This is an equivalent uniform series within a given time period. For an example, see Fig. 1b.

Another common conversion is a series which increases or decreases in magnitude by a uniform amount, G, from one period to the next. This is called a gradient, where G increases starting with the end of the second period until the end of year N, at which time its cash flow is calculated as $(n-1)G$. Note that the value of the gradient G at year 1 is zero. This kind of gradient calculation would be common in replacement studies where expenses such as maintenance are assumed to increase by a fixed amount each year. See Fig. 2a. A final common conversion is a series which increases or decreases in magnitude at a constant rate or percentage from one period to the next. This is called a geometric gradient (g), where g increases starting with the end of the second period. A common example of a geometric gradient is inflation. See Fig. 2b.

The conversion patterns, singly or in combination, represent most of the situations where money is exchanged whether in one's personal finance or within a company (e.g., selection, insurance, pensions, loans, funding, and replacements). These conversions are critical since they allow the comparison of investment opportunities whose cash flow patterns differ. With the conversions, all the cash flows are reduced to a common level in which direct comparisons can be made. A common calculation is switching between present dollars (PW), annual dollars (AW), and future dollars (FW). Table 3 shows which factors to multiply by to yield a needed worth conversion. Thus, any worth can be converted to any other worth by multiplying it by the proper factor, as shown in the table. For example, a present worth can be converted to a future worth simply by multiplying by F/P or to an annuity by multiplying by A/P. In essence, any of the "worths" (A, F, P) can be readily converted into the others.

Up to this point we have discussed all the notation in the equations in Table 1 except for i, the interest rate. Present and future values, annual and present values, and so on, are all related through this interest rate or the MARR one will accept. To determine whether or not a project should be undertaken, its rate of return should be compared to the MARR or the required rate of return for the project. The more risk one is willing to accept, the greater the MARR will be. Stated another way, MARR is the rate of return that one will accept for giving up the money now in favor of future purchasing power. If after engineering economic calculations the rate of return is shown to be higher than the MARR, the investment should be undertaken. Conversely, if the rate of return is shown to be lower than the MARR, the investment should not be undertaken since it is assumed that the money can be invested elsewhere at a rate at least equal to the MARR.

Finally, while working to make cash flows equivalent, a few items must be kept in mind. Cash flows that occur at the same point in time can be added or subtracted from one another. Time horizons must be aligned so that every problem starts at year 0 with income or expenses starting at year 1. If expenses are incurred prior to the start of the project, or time zero, they are called sunk costs and are not relevant to the problem unless they have tax consequences. Lastly, gradients may need to be manipulated.

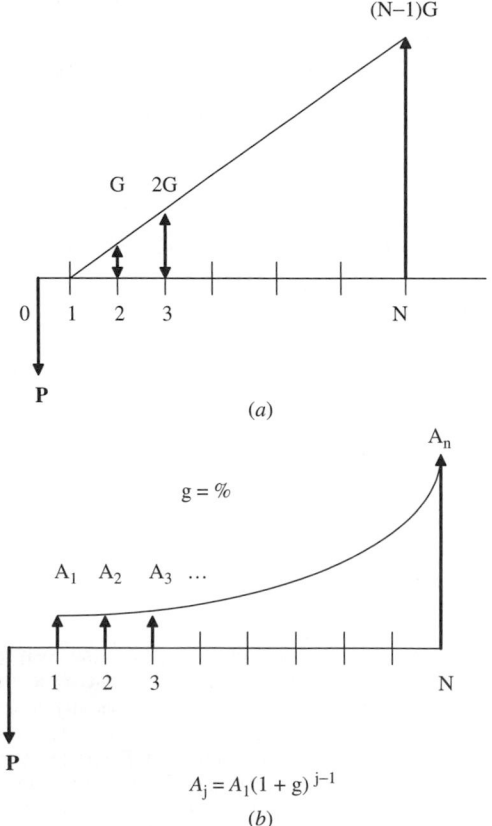

(a)

$$A_j = A_1(1+g)^{j-1}$$

(b)

Fig. 2 (a) Arithmetic gradient cash flow Diagram. (b) Geometric gradient cash flow diagram.

Table 3 Conversion of Worth Factors

If you are given →		P	F	A	G
	P	P	P/F	P/A	P/G
To solve for	F	F/P	F	F/A	F/G
Use factors in table	A	A/P	A/F	A	A/G

P = PW = present worth (NPV = net present value)
F = FW = future worth
A = AW = annual worth

Note: There is commonly no P/G factor. Therefore to get P/G, simply multiply $(A/G)(P/A) = P/G$.

Example 1 provides a problem encompassing the basics of engineering economics described thus far.

Example 1 Basic Engineering Economics Problem.

A company wants to purchase equipment today which costs $5 million. During each year of ownership of the equipment, the company estimates that an additional $800,000 in revenues will be brought in. The company plans to hold the equipment for seven years and estimates it will be able to sell the equipment in year 7 for $1 million. The interest rate is estimated at 6%.

(a) Draw a cash flow diagram.
(b) Is this a profitable enterprise and should it be undertaken?

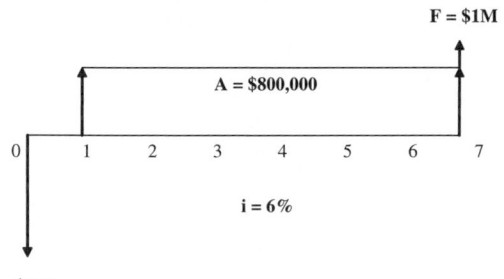

Identify the items from the word problem and list them with appropriate notation:

$P = -\$5,000,000$
$A = \$800,000$
S (also called F) $= \$1,000,000$
$i = 0.06$
$N = 7$

Solution

1. Solving the problem using the equations in Table 1 gives

$$\text{PW} = -P + A\left(\frac{P}{A}, i, N\right) + F\left(\frac{P}{F}, i, N\right)$$

$$= -5,000,000 + 800,000\left(\frac{P}{A}, 6, 7\right)$$

$$+ 1,000,000\left(\frac{P}{F}, 6, 7\right)$$

$$= -5,000,000 + 800,000 \times \frac{(1+i)^N - 1}{i(1+i)^N}$$

$$+ 1,000,000 \times \frac{1}{(1+i)^N}$$

$$= -5,000,000 + 800,000 \times \frac{(1+0.06)^7 - 1}{0.06(1+0.06)^7}$$

$$+ 1,000,000 \times \frac{1}{(1+0.06)^7}$$

$$= -5,000,000 + 800,000 \times 5.5824$$

$$+ 1,000,000 \times 0.6651$$

$$= -5,000,000 + 4,465,920 + 665,100$$

$$= \$131,020$$

This is a profitable project which should be undertaken since the present worth of the project is positive.

2. Solving the problem using factors from interest tables such as Table 2 yields

$$\text{PW} = -P + A\left(\frac{P}{A}, i, N\right) + F\left(\frac{P}{F}, i, N\right)$$

$$= -5,000,000 + 800,000\left(\frac{P}{A}, 6, 7\right)$$

$$+ 1,000,000\left(\frac{P}{F}, 6, 7\right)$$

$$= -5,000,000 + 800,000 \times 5.5824$$

$$+ 1,000,000 \times 0.6651$$

$$= \$131,020$$

This is a profitable project which should be undertaken since the present worth of the project is positive.

5 COMPARING ALTERNATIVES: DEFINING OPTIONS

The object of an economic evaluation is to select the most cost-effective solution among several alternatives. Comparing engineering design alternatives can take many forms: assessment of technology, risk, quality, safety, serviceability, and so on. But in engineering economics the comparison of engineering design alternatives revolves around the monetary aspects and the economic impact of the various projects to determine which project best serves the financial interest of the firm or individual. Thus, it is often worthwhile to be able to rank engineering projects or investment opportunities not only from a technological perspective but also from a monetary perspective. Engineering economics provides us with this ability through its financial criteria, called figures of merit, numerical quantities based on the monetary attributes of the projects. Thus, through engineering economics, one can rank and compare investment opportunities, solve design problems, and make business decisions based on their financial impact as expressed through the figures of merit.

There are two steps to ranking projects: technological and financial. For both steps one needs to determine the set of alternatives from which a project will be chosen. There are many stories of how "pet" projects are chosen. But to be a truly adequate selection process, tests of feasibility, both technical and financial, need to occur. Under technical feasibility one might consider technical limitations (e.g., of material or production capability) or political, administrative, or other physical hindrances that disqualify alternatives from further review. For financial feasibility one must consider the amount of available funding and how that funding will be allocated to the projects under review.

As mentioned, the steps for calculating an engineering economic problem are simple. First draw a cash flow diagram for each project to outline what is occurring in the project. The cash flow diagram, by means of its layout, would indicate the time line on which the entire problem was occurring from time zero to N, the cash flows and the points of time of their occurrence, and the specified interest rate, or MARR. In many instances a "do nothing" alternative will also be examined that represents the concept of passing on the project and reinvesting the project funds as would usually be done.

Cash flows often need to be estimated because, just as stock analysts cannot accurately predict the future stock market level, economists cannot always predict the future of the economy. Variations in cash flow can be understood from concepts such as whether projected income from a project is only what is anticipated and, if the salvage value of an asset depends heavily on what is occurring in the economy N years later at the end of the project life, whether inflation can impact the variability of the direct and overhead costs.

In addition to the time line (N) and the cash flows (P, F, and/or A), interest rate i, also called MARR, must be estimated for each project. As already discussed, the MARR indicates the level of risk one is willing to accept to undertake the project. Or put more simply, MARR is the rate of return that one will accept for giving up the money now in favor of future payout from the project. This interest rate is used to convert the cash flows into figures of merit that allow one to compare the engineering alternatives. The selection or estimation of MARR can be based on various factors, including but not limited to the current interest rate, the rate of inflation, the risk of defaulting on the loan, and the risk of failing to achieve the project outcome. In essence, the greater these factors, the higher the MARR should be for the project. Another way of calculating MARR is to simply estimate the current return in the market for other investments of similar profile and risk.

6 COMPARING ALTERNATIVES THROUGH FIGURES OF MERIT

The most common figures of merit for comparing and ranking alternative investment opportunities are as follows:

Present value or present worth or net present worth (PV, PW, NPW)

Annual worth (AW), which if negative is also known as equivalent uniform annual cost (EUAC) or annual cost (AC)

Future value or future worth (FV or FW)

Internal Rate of Return (IRR)

Benefit–cost ratio (BCR)

The first three items are commonly called "the three worths." As mentioned previously, any worth can be converted to any other worth simply by multiplying it by the correct factor (see Table 3). These worths are often useful in problems that involve comparing projects directly at present costs (present worth), mortgages or auto loans (annual worth), or life insurance or pensions (future worth), for example.

The last two items, IRR and BCR, cannot be compared directly or converted into each other. IRR is used by investors to ease the comparison of outcomes to market rates or other investment opportunities such as standard bank options (i.e., CDs, bonds, and money market funds). BCR is used mostly in the public sector in major investment projects involving capital expenditures with little or no projected profits from the undertaking (i.e., roads, bridges, and municipal buildings).

6.1 Present Worth

The present worth of a cash flow is the discounted value of future cash flows at time zero at some interest rate. To maximize benefits of the projects under consideration, one would choose the projects with the highest present worth. However, when the projects all entail costs (with no benefits), then one would want to minimize costs and choose the alternative with the least cost. (i.e., the smallest negative number). See Example 2 for a demonstration of calculating the present worth.

6.2 Annual Worth

The annual worth converts the cash flows in a project into an equivalent uniform annual series based on the interest rate. The rules for choosing the best alternatives are the same as those for the present worth. In a maximum-benefits case, one wants the highest annual worth. In a least-cost case, one wants to choose the alternative with the smallest negative number. Example 2 converts worths from present to annual.

6.3 Future Worth

The future worth is the compounded value of cash flows in a project at time N at some interest rate. As previously stated for the other worths, the project having the greatest future worth is the one to be selected in a maximum-benefits problem, while the project having the lowest future worth is the one to be selected in a least-cost problem. Example 2 converts worths from present to future.

Example 2 American Mirror Company is looking into expanding its operations and incorporating another site. The two business alternatives, sites A and B, are as follows:

Description	Site A	Site B
First cost, P	$1,500,000	$1,250,000
Annual revenue, $+A$	$600,000	$475,000
Annual expense, $-A$	$125,000	$100,000
Salvage value, S	$150,000	$80,000
Life, N	10	10
MARR, i	12%	12%

(a) Draw a cash flow diagram for each opportunity:

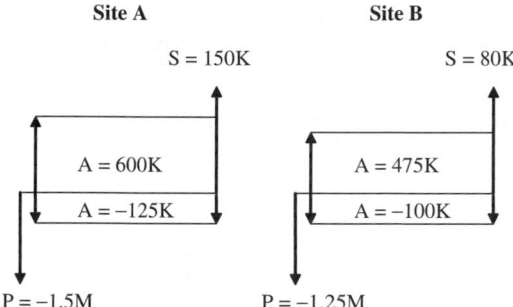

Site A S = 150K

A = 600K

A = −125K

P = −1.5M

Site B S = 80K

A = 475K

A = −100K

P = −1.25M

(b) Calculate the present worth for both sites:

$$PW = A\left(\frac{P}{A}, i, N\right) + S\left(\frac{P}{F}, i, N\right) - P$$

PW site A $= (600,000 - 125,000)$

$$\times \left(\frac{P}{A}, 12\%, 10\right) + 150,000$$

$$\times \left(\frac{P}{F}, 12\%, 10\right) - 1,500,000$$

$$= (600,000 - 125,000) \times 5.6502$$

$$+ 150,000 \times 0.322 - 1,500,000$$

$$= \$1,232,145$$

PW site B $= (475,000 - 100,000)$

$$\times \left(\frac{P}{A}, 12\%, 10\right) + 80,000$$

$$\times \left(\frac{P}{F}, 12\%, 10\right) - 1,250,000$$

$$= (475,000 - 100,000) \times 5.6502$$

$$+ 80,000 \times 0.322 - 1,250,000$$

$$= \$894,585$$

(c) Calculate the annual worth for both sites:

$$AW = A + S\left(\frac{A}{F}, i, N\right)$$

$$- P\left(\frac{A}{P}, i, N\right)$$

AW site A $= (600,000 - 125,000)$

$$+ 150,000\left(\frac{A}{F}, 12\%, 10\right)$$

$$- 1,500,000\left(\frac{A}{P}, 12\%, 10\right)$$

$$= (600,000 - 125,000) + 150,000$$

$$\times 0.0569 - 1,500,000 \times 0.1769$$

$$= \$218,185$$

or

$$AW \text{ site A} = P\left(\frac{A}{P}, i, N\right) = 1,232,145$$

$$\times 0.1769 = \$217,966$$

Note the small error from $218,185 due to rounding.

AW site B $= (475,000 - 100,000)$

$$+ 80,000\left(\frac{A}{F}, 12\%, 10\right)$$

$$- 1,250,000\left(\frac{A}{P}, 12\%, 10\right)$$

$$AW = (475,000 - 100,000) + 80,000$$

$$\times 0.0569 - 1,250,000 \times 0.1769$$

$$AW = \$158,427$$

or

$$AW \text{ site B} = P\left(\frac{A}{P}, i, N\right)$$

$$= 894,585 \times 0.1769 = \$158,252$$

Note the small error from $158,427 due to rounding.

(d) Calculate the future worth for both sites:

$$FW = S + A\left(\frac{F}{A}, i, N\right) - P\left(\frac{F}{P}, i, N\right)$$

$$\text{FW site A} = 150{,}000 + 475{,}000 \left(\frac{F}{A}, 12\%, 10\right)$$

$$- 1{,}500{,}000 \left(\frac{F}{P}, 12\%, 10\right)$$

$$= 150{,}000 + 475{,}000 \times 17.5487$$

$$- 1{,}500{,}000 \times 3.1059$$

$$= 3{,}826{,}783$$

or

$$\text{FW site A} = P \left(\frac{F}{P}, i, N\right) = 1{,}232{,}145$$

$$\times 3.1059 = \$3{,}826{,}919$$

Note the small error from \$3,826,783 due to rounding.

$$\text{FW site B} = 80{,}000 + 375{,}000 \left(\frac{F}{A}, 12\%, 10\right)$$

$$- 1{,}250{,}000 \left(\frac{F}{P}, 12\%, 10\right)$$

$$= 80{,}000 + 375{,}000 \times 17.5487$$

$$- 1{,}250{,}000 \times 3.1059$$

$$= \$2{,}778{,}388$$

or

$$\text{FW site B} = P \left(\frac{F}{P}, i, N\right) = 894{,}585$$

$$\times 3.1059 = \$2{,}778{,}491$$

Note the small error from \$2,778,388 due to rounding.

(e) Which alternative should be chosen? Why? Choose site A because it has a much greater group of values or worths.

6.4 Internal Rate of Return

The (IRR) is the rate for which the sum of the worths of all cash flows equals zero:

$$\sum \text{PW} = 0 \qquad (1)$$

The IRR has many names such as target rate, hurdle rate, and profitability index. These descriptive names are accurate because the IRR is the return an investor will get if the cash flows used in the engineering economics problem prove true. Thus, the IRR is often compared to

the MARR since the MARR was chosen by an investor as the minimally acceptable level of return. The rules for how to compare IRR to MARR follow:

If IRR is greater than MARR, the project is a good investment.

If IRR is less than MARR, the project is a bad investment.

These rules can be easily explained. If the IRR (the return expected from the project) is more than the MARR, then one is making more than expected or hoped for, and thus the project is a good investment. Conversely, if the IRR (the return expected) is less than the MARR, then one is making less than hoped or expected, and one should not proceed with the investment.

The drawback is that IRRs of different projects cannot be compared directly. In order to compare the IRR of more than one project, one must perform incremental analysis on each set of projects and then choose the best project based on the following rules:

If $\text{IRR}_{\text{beta-alpha}} > \text{MARR}$, then beta is the best choice.
If $\text{IRR}_{\text{beta-alpha}} < \text{MARR}$, then alpha is the best choice.

Example 3 provides details on solving an IRR problem.

Example 3. A company is deciding between three different projects with properties as listed below. All projects have a MARR of 15% and will occur over 15 years.

(a) Draw the generic cash flow diagram for the projects involved.
(b) What is the best project using MARR?
(c) What is the best project using IRR?

The data for this example are as follows:

Projects	P	Annual	IRR
B	500	110	20.7
C	725	149	19
D	885	170	12

Solution

(a) The generic cash flow diagram is as follows:

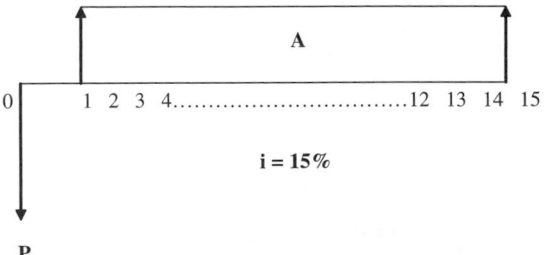

(b) To determine the best project using MARR, calculate the net present value (NPV), also called present worth, for each alternative:

$$NPV = -P + A\left(\frac{P}{A}, i, N\right) + S\left(\frac{P}{F}, i, N\right)$$

$$NPV_B = -500 + 110\left(\frac{P}{A}, 15, 15\right)$$

$$= -500 + 110 \times 5.847 = 143$$

$$NPV_C = -725 + 149\left(\frac{P}{A}, 15, 15\right)$$

$$= -725 + 149 \times 5.847 = 146$$

$$NPV_D = -885 + 170\left(\frac{P}{A}, 15, 15\right)$$

$$= -885 + 170 \times 5.847 = 109$$

Project C is the best since it has the greatest NPV.

(c) IRRs cannot be compared directly. Thus, the information provided in the problem statement about IRR is immaterial. Use the NPV equation of part (a) to solve for IRR, only this time set NPV equal to zero as as stated in Eq. 1:

$$NPV = 0 = -P + A\left(\frac{P}{A}, i^*, 15\right)$$

$$NPV_{dc} = 0 = -885 - (-725)$$

$$+ (170 - 149)\left(\frac{P}{A}, i^*, 15\right)$$

$$160 = 21\left(\frac{P}{A}, i^*, 15\right)$$

$$7.62 = \left(\frac{P}{A}, i^*, 15\right)$$

Use interest rate tables (see Refs. 2 and 3) to search for P/A at $N = 15$ nearest to 7.62. *Note:* Interpolate between 8 and 10% in the table to find i^*:

$$i^* = 9.97\% = IRR$$

The rule states that, for NPV, alpha − beta = $0 = -P + A(P/A, i^*, 15)$:
  Choose alpha if $i^* > MARR$.
  Choose beta if $i^* < MARR$.
Project C is the best since IRR < MARR. Now compare project C to B:

$$NPV = 0 = -P + A\left(\frac{P}{A}, i^*, 15\right)$$

$$NPV_{C-B} = 0 = -725 - (-500)$$

$$+ (149 - 110)\left(\frac{P}{A}, i^*, 15\right)$$

$$225 = 39\left(\frac{P}{A}, i^*, 15\right)$$

$$5.77 = \left(\frac{P}{A}, i^*, 15\right)$$

Use Table 2 to search for P/A at $N = 15$ nearest to 5.77. *Note:* Interpolate between 15 and 18% in the table to find i^*:

$$i^* = 15.28\% = IRR$$

The rule states that, for NPV, alpha − beta = $0 = -P + A\left(\frac{P}{A}, i^*, 15\right)$:
  Choose alpha if $i^* > MARR$.
  Choose beta if $i^* < MARR$.
Project C is the best since IRR>MARR. Therefore, C is the overall best project.

6.5   Benefit–Cost Analysis

Benefit–cost analysis provides a ratio of benefits to costs, which indicates a measure of worthiness for competing projects. Generally governments and other public entities use this figure of merit for determining between projects involving capital expenditures with minimal projected profits. If the BCR is greater than 1, the benefits exceed the costs, which would indicate a good investment. If, however, the BCR ratio is less than 1, than the costs would exceed the benefits, which would indicate a poor investment choice. Just as with IRR, the drawback is that BCRs of different alternatives cannot be compared with one another.

The equation for BCR is listed in Eq. (2) below. Benefits B are favorable monetary consequences. Disbenefits D are unfavorable monetary consequences. The initial cost is I. Cash disbursements C are costs such as operating and maintenance. Cash receipts R are fees such as tolls. Lastly, the key to BCR analysis is to have all units be consistent. Thus, all costs must either be annualized or brought to the present worth:

$$\text{Conventional BCR} = \frac{B - D}{I + (C - R)} \qquad (2)$$

The first step in doing a comparison among projects using BCR is to calculate BCR for each project. If any project has a BCR ratio less than 1 (indicating a poor investment), that project should be eliminated from further analysis. Then, just as was done with IRR, each set of projects is compared and evaluated according to a set of rules until only one project remains:

When looking at individual projects: If $B/C < 1$, eliminate that project.

When comparing projects:

If $B/C_{\text{X-Y}} > 1$, choose Y.

If $B/C_{\text{X-Y}} < 1$, choose X.

Example 4. A company is examining projects W, X, Y, and Z. The details of each project are outlined below. Determine which project should be picked based on the BCR as the figure of merit.

Projects	PW of Benefits	PW of Costs	B/C ratio
W	60	80	0.75
X	150	110	1.36
Y	70	25	2.80
Z	120	73	1.64

The first step is to examine the BCR ratios to see if any fall below the critical limit of 1. Since the BCR of project W is 0.75 and thus below 1, project W can be eliminated from further study. The next step is to compare the remaining alternatives using Eq. (2) and the rules governing benefit–cost analysis to determine the best project:

$$B/C_{\text{Z-Y}} = \frac{120 - 70}{73 - 25} = 1.04 > 1$$

Therefore select project Z. Now compare project Z to a remaining alternative, project X:

$$B/C_{\text{X-Z}} = \frac{150 - 120}{110 - 73} = 0.81 < 1$$

Therefore select project Z. Thus, of projects W, X, Y and Z, the BCR analysis showed that project Z was the best alternative to pursue.

7 ADDITIONAL ANALYSES IN SELECTION PROCESS

After calculating the figures of merit for alternative projects, one might also want to determine the breakeven point, sensitivity analysis, or risk analysis of each project.

7.1 Breakeven Analysis

Breakeven analysis shows the point at which if we go to one side of the point we choose one alternative, but if we go to the other side we choose the competing alternative. Thus at the breakeven point, total revenues equal total costs, and an increase in sales will lead to profit. Conversely, if sales fail to meet expectations and fall below the breakeven point, the company will experience a loss. An example of breakeven analysis might be to determine how many "widgets" need to be produced and sold before the company breaks even and shifts from the debt-making to the profit-making

side of the breakeven point. Breakeven analysis could also be used to determine a lease-versus-buy solution as well. In this case, an analysis of the expenses associated with ownership is compared with those fees associated with leasing in order to determine the optimal solution. This type of analysis would be commonly used when the choice among alternatives is dependent mainly upon a single factor such as utilization of the equipment.

7.2 Sensitivity Analysis

Engineering economy studies are about the future, and as such uncertainty about prospective project results cannot be avoided. Which factors are of specific concern for analysis will vary from project to project, but generally one or more of them will need to be further analyzed in order to come to the best decision. Thus, *sensitivity analysis* can be used to determine which parameters are the most sensitive to a project's economic feasibility. By knowing which items are the most sensitive, one can examine more closely the details of the project in order to have greater confidence in the calculated figures of merit. Therefore, this technique makes clear the impact of the uncertainty when two or more project factors are of concern on the figure of merit (e.g., present worth).

The first step in a sensitivity analysis is to calculate the base-case figure of merit (i.e., present worth), which is developed using the most likely value for each input. The next step is to change one variable of interest by some percentage above and below that used in the base case while holding the remaining inputs constant. The last step is to calculate the new figure of merit for the new input values. Results of the sensitivity analyses can be displayed in both tabular and graphic forms. In tabular form, the item with the greatest numerical value as a sensitivity would have the greatest impact on the pertinent figure of merit. In graphic form, the item with the largest slope would have the greatest impact. Thus, the items highest in sensitivity would be the inputs that would affect the final outcome the most.

7.3 Risk Analysis

Risk analysis is founded on the application of probability theory where one compares the risk propensity of alternative investment opportunities. These kinds of risk analysis activities occur when there is a lack of precise knowledge regarding the future business conditions surrounding, or involved in, the project under consideration. In these cases, the probabilities of the occurrences are estimated. There are risk problems where probabilities of various possible outcomes can be estimated using decision rules such as dominance, aspiration level, and most probable future value. Decision trees and laws of expected value and variance share similar concepts of describing risk and the estimates of various possible outcomes. Lastly, simulation techniques, such as the Monte Carlo method, can be used as ways of imitating real-life situations and occurrences with models to produce risk profiles for projects under review.

The discussions on risk analysis as well as the specific details on breakeven and sensitivity analysis are beyond the scope of this handbook. For more information the reader is referred to Ref. 4.

8 CAPITAL RECOVERY, CAPITAL COST, AND REPLACEMENT STUDIES

Many decisions in business are based around equipment and their *retirement and replacement studies*. For example, when should a company retire a facility, or specific equipment, or when is the technical limit of an asset reached? Perhaps a more important question surrounds when the economics of keeping and maintaining an item becomes more expensive than purchasing new and replacing the item. How does one determine this critical time period? These types of analyses are called retirement and replacement studies and are probably the largest class of cost studies performed by companies and organizations.

Since there is usually no profit from replacing an item, only savings over time, replacement problems are considered "least-cost" problems, where one wants to minimize costs and choose the alternative with the least cost (i.e., the smallest negative number). The two major cost elements in replacement studies are capital recovery costs and operating and maintenance costs.

The capital recovery (CR) of an investment is the cost of recovering the initial cost of investment. It is a uniform series representing the difference between the equivalent annual cost of the first cost and the equivalent annual worth of the salvage value. This annualized return usually declines over time:

$$\text{CR} = P\left(\frac{A}{P}, i, N\right) - S_N\left(\frac{A}{F}, i, N\right) \quad (3)$$

Operating and maintenance costs behave in the opposite direction to capital recovery costs. See Fig. 3. While capital recovery costs normally decrease over time, operation and maintenance costs generally increase over the lifespan of a facility or piece of equipment. This happens as parts get harder to find and labor skilled on the workings of the equipment disappears. Breakdowns also become more frequent as an object ages and complete overhauls become necessary. Thus, one must calculate the EUAC, or AC, (which is simply a negative annual worth) of the operations and maintenance costs in addition to the capital recovery costs for each year of possible ownership until time period N. Factors used for the EUAC include the present worth $(P/F, i, j)$; the cumulative sum of the present worth, $\sum A_j(P/F, i, j)$, and the annualized cost to convert the present worth to an annuity $(A/P, i, n)$. Note that n is the total years for the asset in question and j runs from 1 to n.

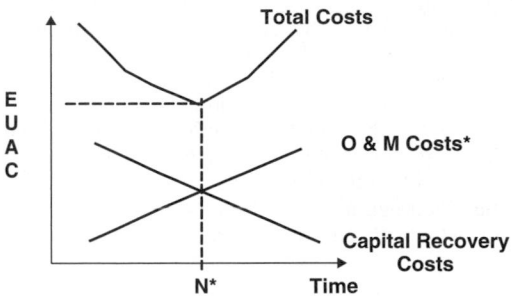

Fig. 3 EUAC versus time to determine total costs.

As mentioned, retirement and replacement problems are least-cost calculations with the goal being to minimize the EUAC. Since capital recovery decreases over time and operation and maintenance costs increase over time, there is usually an intersection of the two costs. This is the point (N^*) at which it is economically ideal to retire the item or replace it. See Fig. 3. The detailed calculations of these studies are beyond the topics in this handbook. For more details the reader should refer to Ref. 5.

8.1 Capital Cost

Related to capital recovery [Eq. (3)] is capitalized cost (CC). Capitalized cost analysis can be used for investments whose life is very long, generally greater than 30 years. Items with long lives include bridges, tunnels, and highways and therefore have lives that can be assumed to be infinite, as opposed to finite. While capital recovery was an annual worth, capitalized cost is a present worth. Capitalized cost is the equivalent present worth of the initial cost plus a series of periodic cash flows (maintenance, tolls, operating costs, major overhauls, etc.) extended to infinity. Thus, capitalized cost can be thought of as the amount of money that one needs to set aside now (at some interest rate) to provide the funds needed to operate the project indefinitely. And when one can assume that the project runs indefinitely, then N runs to infinity. This assumption of infinity introduces negligible errors to the problem and allows the cash flows to simplify and reduce to the equation

$$P = A\left(\frac{P}{A}, i, \infty\right) = \frac{A}{i} \quad (4)$$

which yields the equation for CC:

$$\text{CC} = P + \frac{A}{i} \quad (5)$$

For further discussion of capital cost see Ref. 5.

9 CONCLUSION

In addition to being part of the fundamentals in engineering (FE) exam and the professional engineering

(PE) exam, engineering economics is a topic commonly encountered by engineers faced with design decisions dealing with alternative designs, methods, or materials; and the choice of these influences the majority of the costs of manufacturing or construction. So beyond designs, method materials, function, and performance, alternative solutions must also be viable economically. It is for these reasons that Arthur Wellington[6] defined engineering as "the art of doing well with one dollar what any bungler can do with two."

In many instances what an engineer designs has the requirements of being designed and built at some expense for the benefits or revenues that may occur from the design over time. The design alternatives and their comparison, therefore, can maximize benefits or minimize costs and show which projects are worthwhile and which are not or which projects should be given higher priority than others simply based on their economic impact. This chapter provides the tools to choose a suitable economic criterion for the problem and analyze and solve these decisions from an economic perspective. If one understands the decision-making process and can make accurate economic comparisons between alternatives, one can make better, more fiscally responsible decisions.

Engineering economics is therefore the application of economic analysis techniques in the comparison of engineering design alternatives. As such, one can only practice engineering if one also knows engineering economics. The Forty-Fifth Annual Report of the Engineers Council for Professional Development defined engineering as a profession in which knowledge of mathematics and science is applied to economically solve problems, create products, and so on. And the U.S. Department of Labor[7] states that engineers should apply the principles of science and mathematics to develop economical solutions to technical problems. Thus, engineering by definition is a trade-off among science, engineering, and economic constraints to efficiently use capital to minimize costs and maximize benefits.

REFERENCES

1. Grant, E. L., *Principles of Engineering Economy*, Ronald, New York, 1930.
2. Newman, D. G., Lavelle, J. P., and Eschenbach, T. G., *Engineering Economic Analysis*, 9th ed., Oxford University Press, New York, 2004.
3. Oxford University Press, Compound interest tables, www.oup.com/pdf/ca/compoundstudent.pdf, accessed Feb. 1, 2007.
4. Park, C. S., *Fundamentals of Engineering Economics*, 1st ed., Prentice-Hall, Englewood Cliffs, NJ, 2004.
5. Lang, H. J., and Merino, D. N., *The Selection Process for Capital Projects*, 6th ed., Wiley, New York, 1993.
6. Wellington, A. M., *The Economic Theory of the Location of Railways: An Analysis*, Wiley, New York, 1887.
7. U. S. Department of Labor, Bureau of Labor Statistics, http://www.bls.gov/oco/ocos027.htm, accessed Jan. 25, 2007.

CHAPTER **22**

SOURCES OF MATERIALS DATA

J. G. Kaufman
Kaufman Associates, Inc.
Columbus, Ohio

1　INTRODUCTION AND SCOPE

It is the purpose of this chapter to aid engineers and materials scientists in locating reliable sources of high-quality materials property data. While sources in hard-copy form are referenced, the main focus is on electronic sources that provide well-documented searchable property data.

To identify useful sources of materials data, it is important to have clearly in mind at the outset (a) the intended use of data, (b) the type of data required, and (c) the quality of data required. These three factors are key in narrowing a search for property data and improving its efficiency. Therefore, as an introduction to the identification of some specific potentially useful sources of materials data, we will discuss those three factors in some detail and then describe the options available in types of data sources.

It is beyond the scope of this chapter to attempt to provide a comprehensive list of all of the several thousand sources of materials data in various forms and formats. Readers interested in a more comprehensive list of sources and of more discussion of the technology of

material property data technology and terminology are referred to Westbrook's extensive treatment of these subjects in Refs. 1 and 2 and the *ASM International Directory of Material Property Databases*.[3]

2　INTENDED USES FOR DATA

Numeric material property data are typically needed for one of the following purposes by individuals performing the respective functions as part of their jobs:

- Mathematical modeling of material or product performance
- Materials selection (finding candidate materials for specific applications)
- Analytical comparisons (narrowing the choices)
- Preliminary design (initially sizing the components)
- Final design (assuring performance; setting performance specifications)

- Material specification (defining specifications for purchase)
- Manufacturing process control (assuring processes to achieve desired product)
- Quality assurance (monitoring manufacturing quality)
- Maintenance (repairing deterioration/damage)
- Failure analysis (figuring out what went wrong)

It is useful to note some of the differing characteristics of data needed for these different functions.

2.1 Mathematical modeling of Material and/or Product Performance

To an increasing extent, mathematical modeling is used to establish the first estimates of the required product performance and material behavior and even in some cases the optimum manufacturing process that should be used to achieve the desired performance. The processes and/or performance analyzed and represented may include any of the issues addressed in the following paragraphs, and so the types of data described under the various needs are the same as those needed for the modeling process itself.

2.2 Materials Selection

The needs of materials specialists and engineers looking for numeric materials data to aid in the selection of a material for some specific application are likely to be influenced by whether they (a) are in the early stages of their process or (b) have already narrowed the options down to two or three candidates and are trying to make the final choice. The second situation is covered in Section 2.3 on analytical comparisons.

If the materials engineers are in the early stages of finding candidate materials for the application, they are likely to be looking for a wide variety of properties for a number of candidate materials. More often, however, they may decide to focus on two or three key properties that most closely define the critical performance requirements for that application and search for all possible materials providing relatively favorable combinations of those key properties. In either case, they may not be as much concerned about the quality and statistical reliability of the data at this stage as much as the ability to find a wide variety of candidates and to make direct comparisons of the performance of those candidates.

In cases where there is interest in including relatively newly developed materials in the survey, it may be necessary to be satisfied with only a few representative test results or even educated "guesstimates" of how the new materials may be expected to perform. The engineers will need to be able to translate these few data into comparisons with the more established materials, but at this stage they are probably most concerned with not missing out on important new materials.

Thus, at this early stage of materials selection, the decision makers may be willing to accept data rather widely ranging in type and quality, with few restrictions on statistical reliability. They may even be satisfied with quite limited data to identify a candidate that may merit further evaluation.

2.3 Analytical Comparisons

If, on the other hand, the task is to make a final decision on which of two or three candidate materials should be selected for design implementation (the process defined here as "analytical comparison"), the quality and reliability of the data become substantially more important, particularly with regard to the key performance requirements for the application. It will be important that all of those key properties, for example, density, tensile yield strength, and plane strain fracture toughness, are available for all of the candidate materials that may be the next cut in the list.

The search will also be for data sources where the background of the data is well defined in terms of the number of tests made, the number of different lots tested, and whether the numbers included in the data source are averages or the result of some statistical calculation, for example, that to define three σ limits. It would not be appropriate at this stage to be uncertain whether the available data represent typical, average values or statistically minimum properties: it may not be important which they are, but the same quality and reliability must be available for all of the final candidates for a useful decision to be made.

In addition, the ability to make direct comparisons of properties generated by essentially the same, ideally standard, methods is very important. The decision maker will want to be able to determine if the properties reported were determined from the same or similar procedures and whether or not those procedures conformed to American Society for Testing and Materials (ASTM) International, International Organization for Standardization (ISO),[4] or other applicable standard test methods.

One final requirement is added at this stage: The materials themselves for which the data are presented must all represent to the degree possible comparable stages of material production history. It would be unwise to base serious decisions on comparisons of data for a laboratory sample on one hand and a commercial-size production lot on the other. Laboratory samples have a regrettable history of promising performance seldom replicated in production-size lots.

Thus, for analytical comparisons for final candidate material selection, specialists need databases for which a relatively complete background of metadata (i.e., data about the data) are included and readily accessible.

Incidentally, it is not unusual at this point in the total process to decide that more data are needed for a particular candidate than are available in any existing database, and so a new series of tests are needed to increase confidence in the comparisons being made.

2.4 Preliminary Design

Once a decision is made on a candidate material (or sometimes two) for an application, the task of designing a real component out of that material begins. The requirement for statistical reliability steps up, and the importance of the availability of a data source that provides applicable metadata covering quality, reliability, and material history becomes even more important.

At this stage, the statistical reliability required includes not only a minimum value but one based upon a statistically significant sample size, ideally something comparable to the standards required in the establishment of MMPDS (previously know as MIL-HDBK-5) A or B values.[5] In MMPDS terminology, an A value is one that would be expected to be equaled or exceeded by 99% of the lots tested with 95% confidence; the B value provides for 90% of lots tested equaling or exceeding the value with 95% confidence. Furthermore, the MMPDS guidelines require that A and B values be based upon predefined sample sizes, representing a minimum number of lots (normally 100 or more) and compositions (normally at least 3) of a given alloy. The provision of such statistical levels needs to be a part of data sources used for design purposes, and the description of the statistical quality needs to be readily available in the data source.

For preliminary design, then, the data sources sought will include both statistically reliable data and well-defined metadata concerning the quality and reliability of the data.

2.5 Final Design

Setting the final design parameters for any component or structure typically requires not only data of the highest level of statistically reliability but also, in many cases, data that have been sanctioned by some group of experts for use for the given purpose. It is also not unusual that at this stage the need is identified for additional test data generated under conditions as close as possible to the intended service conditions, conditions perhaps not available from any commercial database.

Databases providing the level of information required at this stage often contain what are characterized as "evaluated" or "certified" values. Evaluated data are those that, in addition to whatever analytical or statistical treatment they have been given, have been overviewed by an expert or group of experts who make a judgment as to whether or not the data adequately and completely represent the intended service conditions and, if necessary, incorporate their own analysis into the final figures. This technique has been widely used in digesting and promulgating representative physical property data for many years; examples are the thermophysical property data provided by Thermophysical Properties Research Center (TPRC).[6]

Other databases may be said to provide "certified" data. In this case, the database or set of data going into a database have been evaluated by a group of experts and certified as the appropriate ones to be used for the design of a particular type of structure. Two examples of this are the aforementioned MMPDS values,[5] which are approved for aircraft design by the MMPDS Coordination Committee, consisting of aerospace materials experts, and the American Society of Mechanical Engineers (ASME) *Boiler and Pressure Vessel Code*,[7] with properties certified by materials experts in that field for the design of pressure vessels and companion equipment for high-temperature chemical processes.

As noted, it is often the case when designers reach this stage (if not the earlier preliminary design stage) that they find it necessary to conduct additional tests of some very specialized type to ensure adequate performance under the specific conditions the component or structure will see in service but for which reliable databases have not previously been identified or developed. The net result is the creation of new materials databases to meet highly specialized needs in a manner that provides the appropriate level of statistical confidence.

2.6 Material Specification

Material specifications typically include specific property values that must be equaled or exceeded in tests of those materials that are being bought and sold. The properties that one requires in this case may differ from those needed for other purposes in two respects. First and foremost, they must be properties that will ensure that the material has been given the desired mechanical and thermal processing to consistently achieve the desired performance. The second requirement is that, while in most cases there may be only one or a very few properties required (most often tensile properties), they are required at a very high level of precision and accuracy, similar to or better than that required for the MMPD5 A properties defined earlier.

Examples are the material specifications required for the purchase of commercial aluminum alloy products.[8] These are usually only the chemical composition and the tensile properties. So while many of the other properties needed for design are not required as part of the purchase specification, those properties that are required are needed with very high reliability. In the case of aluminum alloys, the requirements for tensile strength, yield strength, and elongation are normally that 99% of lots produced must have properties that equal or exceed the published purchase specification values with 95% confidence, and they must have been defined from tests of more that 100 different production lots from two or more producers.

In many cases, the databases needed to generate material specification properties are proprietary and are contained within individual companies or within the organizations that set industry specifications. However, the resultant statistically reliable specification properties are resident and more readily available in industry or ASTM material standards.[4,8]

2.7 Manufacturing Process Control

The properties required for manufacturing purposes may be the most difficult to find in commercially available databases because they typically involve the specialized treatments or processes utilized by specific producers or suppliers of the specific products in question. Sometimes these processes are proprietary and closely held for competitive advantage. An added complication is that once some semifabricated component (e.g., aircraft sheet) has been purchased, it will require forming to very tight tolerances or finishing at some relatively high temperature. The fabricator may require data to enable the process to be carried through without otherwise damaging or changing the properties of the component but may have difficulty representing the fabricating conditions in meaningful tests. That fabricator may well have to carry out its own tests and build the needed database to provide the desired assurance of quality and to provide a source of information to which its own employees can refer to answer specific questions. Typically such databases never become commercially available, and new situations will require compiling new data sources.

Some processing data sources are available, of course. The Aluminum Association, for example, provides to all interested parties a data source defining standard solution heat treating, artificial aging, and annealing treatments for aluminum alloys that will assure the proper levels of properties will be obtained.[8] In some cases, ASTM and American National Standards Institute (ANSI) material specifications will also contain such information.

2.8 Quality Assurance

Quality assurance may be considered to be the flip side of material specifications, and so the types of data and the data sources themselves required for the two functions are essentially the same.

Purchasers of materials, for example, may choose to do their own testing of the materials once delivered to their facilities by materials producers. If so, they will use exactly the same tests and refer to exactly the same data sources to determine compliance. The one difference may be that such purchasers may choose to gradually accumulate the results of such tests and build their own databases for internal use, not only by their quality control experts, but also by their designers and materials experts who must establish safe levels of performance of the structures. These types of databases also tend to be proprietary, of course, and are seldom made available to the outside world, especially competitors.

2.9 Maintenance

The principal value of material data in connection with maintenance concerns is for reference purposes when problems show up with either deterioration of surface conditions by local corrosion or the suspicion of the development of fatigue cracks at local stress raisers.

In both cases, the important features of the types of data desired to address such issues are more likely to be those based upon exposures representing service experience, and so the user may be more concerned about the degree of applicability than upon data quality and statistical reliability (though both features would be desired if available). Typically such data are hard to find in any event and once again are more likely to be buried in proprietary files than in published databases.

In the case of engineers and technicians needing databases comprised of service experience, they may well be faced with building their own data sources based upon their organization's production and service experience than expecting to be able to locate applicable external sources.

2.10 Failure Analysis

The occurrence of unexpected failure of components in a structure usually calls for follow-up study to determine the cause and possible ways to avoid further loss. In such cases it is inevitable that such failure analysis will involve both (a) a review of the old databases used to design the part and (b) a search for or the development of new data sources that may shed more light on the material's response to conditions that developed during the life of the structure that had not been anticipated beforehand.

The types of databases sought in this case will likely be those containing statistically reliable data, but recognizing the unexpected nature of some problems, an interest in a wider range of data sources and a willingness to consider a lower level of data quality may result. Databases for failure analysis studies may need to be wider in scope and to cover subjects such as corrosion that are not always easily treated by statistical means. In fact, sources covering failure experience may be the most valuable, though hardest to find because historically engineers and scientists do not publish much detail about their mistakes.

The net result is that when dealing with failure analysis, the search may be quite broad in terms of data quality, and the focus most likely will be more on applicability to the problem than on the quality and structure of the compilation. As in the case of maintenance engineers and technicians needing databases comprised of service experience, failure analysts may well be faced with building new data sources based upon their organization's production and service experience.

3 TYPES OF DATA

It is useful at this stage to note that there are several basic types of materials databases, that is databases containing significantly different types of information and, hence, different data formats. Note that this is different from the type of platform or presentation format (e.g., hard copy, CD, online, etc.); these will be discussed in Section 6.

The two fundamental types of databases discussed here in are *textual data* and *numeric data*. In fact, many databases represent a combination of both types, but there are some basic differences worth noting as in the next two paragraphs. The concept of metadata will also be described in more detail in this section.

3.1 Textual Data

The terminology *textual data* is generally applied to data entries that are purely alphabetic in nature with numbers used only as necessary to complete the thoughts. Textual databases are typically searched with alphabetical strings, for example, by searching for all occurrences of a term such as *aluminum* or *metallurgy* or whatever subject is of interest.

The subjects of textual databases are predominantly bibliographies and abstracts of publications, but they may reflect other specific subjects such as descriptions of failures of components. Bibliographic databases seldom reflect the final answers to whatever inquiry is in mind but rather references to sources where the answers may be found.

The majority of all databases in existence in any form (see Section 6) are textual in nature, even many of those purporting to be property databases. Searchers of such textual property databases are searching based upon strings of alphanumeric characters reflecting their interest, not on numeric values of the properties except as they are expressed as strings. This is quite different from the case for numeric property databases, as we shall see in Section 3.2.

3.2 Numeric Databases

Databases classed as numeric (a) have data stored in them in numeric format and (b) are searched numerically. For example, numeric databases may be searched for all materials having a specific property equal to or greater than a certain value or within a certain specific range; this would not be possible in a textual database. To provide such searchability, almost all numeric databases are electronic in nature (see Section 4) though many hard-copy publications also contain extensive amounts of numeric data.

To accommodate numeric searching, the properties must be entered into a database digitally as numbers not simply as alphanumeric strings. And to be useful to material specialists and designers, they must have meaningful precision (i.e., numbers of significant figures) and units associated with each number. The numeric number of a property is of no value if it does not have the applicable unit(s) associated with it.

3.3 Metadata

The concept of metadata as "data about data" was introduced earlier, and it is appropriate at this stage to describe the concept and its importance in greater detail.[9] It is vital, especially for numeric databases and independent of their platform, to have ample background information on the numeric properties included in the database and closely associated with the individual properties.

Examples of metadata include:

- Original source of data (e.g., from tests at ABC laboratories)
- Test methods by which data were obtained (e.g., ASTM methods, size and type of specimen)
- Production history of the samples for which the data are applicable (e.g., annealed, heat treated, cold worked, special handling?)
- Exposure experienced by the samples tested, if any, prior to the test (e.g., held 1000 h at 500°F)
- Conditions under which the properties were determined (e.g., tested at 300°F, 50% humidity)
- Number of individual tests represented by and statistical precision of the values presented (e.g., individual test results; averages of x number of tests, statistical basis)
- Any subsequent evaluation or certification of the data by experts (e.g., ASME Boiler & Pressure Vessel Committee)

It should be clear that the value of information in any database, but especially in a numeric database, is greatly diminished by the absence of some and potentially all of these types of metadata. For example, a listing of the properties of an alloy is of no value at all if it is not clear at what temperature they were determined. It should be clear that metadata are an integral part of every material property database. Similarly, the properties of a material are of virtually no value if it is not clear how the materials were mechanically and thermally processed before they were tested.

4 SUBJECTS OF DATA SOURCES

It is next useful to note some of the major categories of data available and illustrate the manner in which they are likely to be classified or structured. From this point on, the discussion will focus primarily upon numeric materials property data of interest to the engineer and scientist.

The total breadth of properties and characteristics may reasonably be illustrated by the following four categories:

- Fundamental (atomic level) properties
- Physical properties (atomic and macro/alloy levels)
- Mechanical properties (macro/alloy level)
- Application performance (macro/alloy or component levels)

To provide greater detail on the first three categories, it is helpful to utilize the taxonomy of materials information developed by Westbrook[1] and illustrated

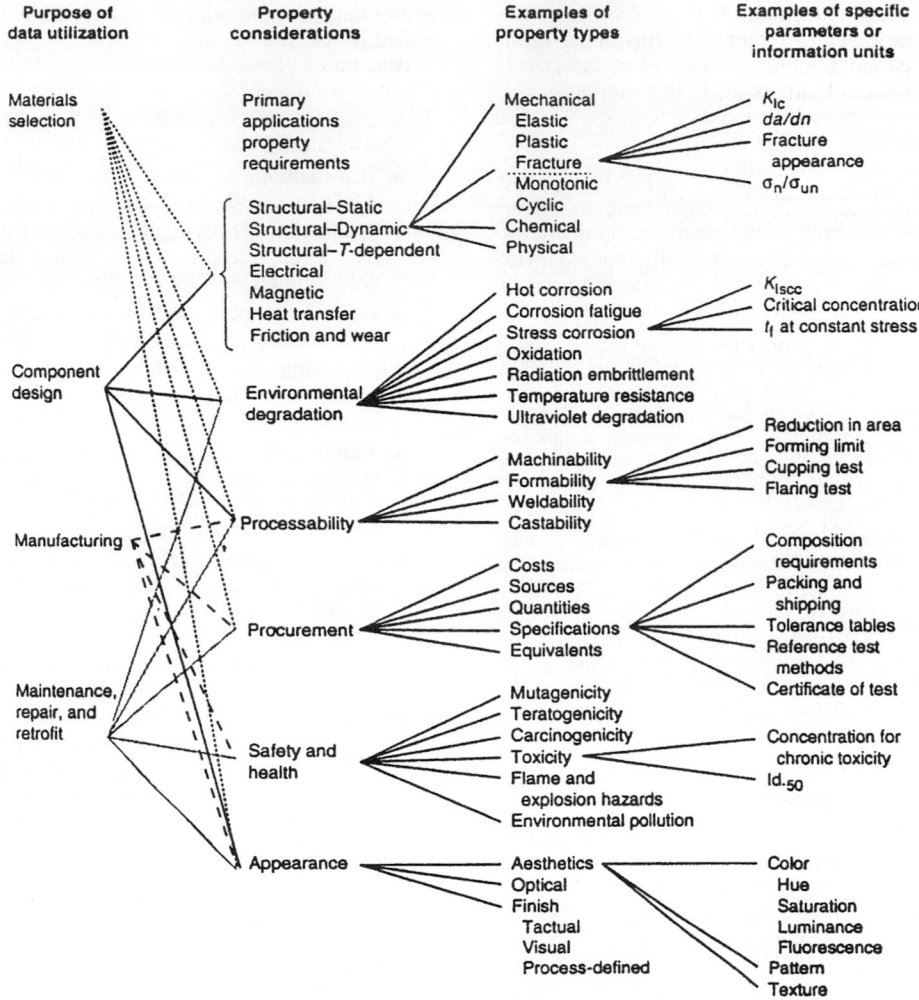

Fig. 1 Taxonomy of materials information. From *ASM Handbook*, Vol. 20: *Materials Selection and Design* (1997), ASM International, Materials Park, OH, Fig. 1, p. 491.

in Fig. 1. While the entire list of potential uses of data described above is not included, the taxonomy in Fig. 1 illustrates the variety of subject matter quite well.

The fourth major category identified above, referred to as application performance, incorporates several individual types of information:

- Fabrication characteristics (sometimes called the "ilities")
  - (a) Fabricability ("workability")
  - (b) Forming characteristics ("formability")
  - (c) Joining characteristics ("joinability")
  - (d) Finishing characteristics ("finishability")

- Service experience
  - (a) Exposure conditions
  - (b) Service history
  - (c) Failures observed and their causes

What is necessary is not to discuss these categories individually but simply to recognize that specific databases may well focus on only one or a few of these subjects and that rarely if ever would all subjects be included in any one database.

5 DATA QUALITY AND RELIABILITY

It was noted in Section 2 that individuals doing preliminary material selection or screening may have different needs with regard to quality and reliability than those doing design functions. It is appropriate at this stage to review the major factors that go into judgments of data quality and reliability.[10] There are two parts to such a discussion: (1) the two major factors affecting the data themselves and (2) the degree to which those factors are reflected in database content.

The two major factors affecting quality and reliability include (a) the manner in which the data were obtained and (b) the statistical reliability of the data presented.

First and foremost, the users of a database for whatever purpose will want to know that consistent standards were applied in assembling the data for that database and that the properties of one material may reliably be compared with those for another. They would also prefer that those properties have been generated by well-known standard methods such as those prescribed by organizations such as ASTM and ISO.[4] They may also be quite interested in knowing the specific source of the original data; realistically some laboratories [e.g., the National Institute of Standards and Technology (NIST)] have a more widely recognized reputation than others.

The second major factor affecting how the user applies information from a specific database is the statistical reliability of the values included therein. The user will be interested in which of the following categories best describes the values presented:

- Raw data (the results of individual tests)
- Average of multiple tests (how many tests represented?)
- Statistically analyzed (at what statistical definition and with what confidence?)
- Evaluated/certified (by whom? for what purpose?)

Of equal importance to the user of any database for whatever reason is the degree to which those factors affecting quality and reliability as noted above are presented in the database itself and, therefore, can be fully understood by the user. In some instances there may be one or several screens of background information laying out the general guidelines upon which the database was generated. This is particularly effective if the entire database represents properties of a common lineage and character. On the other hand, if the user fails to consult this upfront information, some important delimiters may be missed and the data misinterpreted.

Another means of presenting the metadata concerning quality and reliability is by incorporating them within the database itself. This is especially true for such factors as time–temperature parameters that delimit the applicability of the data and units and other elements of information that may restrict its application (see Section 3.3). The incorporation of such information is also especially important in instances where the individual values may vary with respect to their statistical reliability. While the later case may seem unlikely, it is actually quite common, as when "design" data are presented; in such cases, the strength values are likely to be statistical minimum values while moduli of elasticity and physical properties are likely to be average values, and the difference should be made clear in the database.

6 PLATFORMS: TYPES OF DATA SOURCES

The last feature we will consider before identifying specific data sources is the variety of platforms available for databases today. While it is not necessary to discuss them in detail, it will be obvious that the following choices exist:

- Hard copy (published books, monographs, etc.)
- Self-contained electronic (CDs, DVDs, etc.)
- Internet sites (online availability; perhaps downloadable)

The only amplification needed on these obvious options is that the last one, the Internet, has become an interesting and sometimes challenging resource from which to identify and locate specific sources of materials data, and that trend will likely continue to increase. Two large caveats go with the use of Internet sources however: (a) a great deal of "junk" (i.e., unreliable, undocumented data) may be found on Internet websites and (b) even those containing more reliable data seldom meet all users' needs with respect to covering the metadata. It is vital that the users themselves apply the guidelines listed earlier to judge the quality and reliability of sources located on the Internet.

7 SPECIFIC DATA SOURCES

It is beyond the scope of this chapter to provide an exhaustive list of data sources because there are thousands of them of varying presentation platforms, styles, and content. What we will do here is highlight a few of the potentially most useful and highest quality sources, in the sense that their coverage is relatively broad and/or they represent good places to go to look for new and emerging sources of materials data.

The descriptions below will include the following four groups:

- ASM International, a technical society for materials engineers and specialists that produce and provide a wide range of textual resources and materials databases in hard copy and several electronic formats (see Section 7.1)
- STN International, a service of the American Chemical Society, providing (for-fee) access to several numeric databases as well as a great many textual/bibliographic databases on materials-related subjects (see Section 7.2).

- Knovel, a division of William Andrew, providing online access via www.knovel.com to more than 800 published scientific texts and databases.
- Other online Internet websites. (see Section 7.3).

For readers interested in more extensive listings, reference is made to the article "Sources of Materials Property Data and Information" by Jack Westbrook in Volume 20 of the *ASM Handbook*[1] and to the *ASM Directory of Databases*.[3]

7.1 ASM International

1 ASM International has emerged as one of strongest providers of both textual and numeric data, and those sources are usually available in at least three formats: hard copy, CD, and online. For example, one of the most extensive sources of high- and low-temperature data for aluminum alloys is available in book form, *Properties of Aluminum Alloys—Tensile, Creep, and Fatigue Data at High and Low Temperatures*,[11] on a CD, and online as part of the Alloy Center.

Most impressive is the emergence of the ASM International website, www.asm-international.org, as a portal to a wealth of materials data, including but not limited to the following:

- The *ASM Handbook*[12]—covering the 20-volume handbook; versions are also available for single or networked workstations.
- A research library of thousands technical publications, including technical books, CDs, and DVDs, articles from magazines and journals, and life-long learning support.
- The Alloy Center, including:
  - (a) Data sheets and diagrams—include thousands of data documents from publications such as *Alloy Digest* and *Heat Treater's Guide*, along with time–temperature curves, creep curves, and fatigue curves.
  - (b) Alloy finder—includes alloy designations and trade names from around the world, enabling users to locate compositions, producers, and tensile properties.
  - (c) Materials property data—include numeric databases of mechanical and physical properties and processing information for most industrially important alloys; include graphs of data as functions of time and temperature.
  - (d) Coatings data—feature detailed information on commercial coating systems that can be searched by trade name, manufacturer, or process of coating type.
  - (e) Corrosion data—feature corrosion information for alloys in specific environments, searchable by either category.

- Materials for medical devices—a comprehensive and authoritative set of mechanical, physical, biological response, and drug compatibility properties for the materials and coatings used in medical devices, including carbonaceous materials, ceramics, metals, and polymers.
- Materials for micro electro mechanical system (MEMS) packaging—a comprehensive and authoritative set of mechanical and physical properties, processing, and component data to facilitate materials selection and design for MEMS packages. Includes capability to search, select, and report features for composites, metals, and polymers as well as getters and solders.
- Alloy phase diagrams—About 28,000 binary and ternary phase diagrams, also available in hard-copy or CD versions
- Micrograph center—A comprehensive collection of micrograph images and associated data for industrially important alloys, including material designation and composition, processing history, service history, metallographic preparation/technique, magnification, significance of the structures shown, selected materials properties data, and other relevant data.
- Failure analysis center—Features over 1000 case histories along with authoritative handbook information on failure mechanisms and analysis methods, enabling the user to find specific information to help solve failure analysis or materials performance issues.

The access and availability of the ASM sources is more cost-effective with membership in the society. For more information of any aspect of the ASM International resources, readers are referred to www.asm-international.org.

7.2 STN International

STN International is the online worldwide scientific and technical information network, providing one of the most extensive sources of numeric materials property data.[13] Built and operated by the American Chemical Society at its Chemical Abstract Service site in Columbus, Ohio, STN International includes about 25 numeric databases, including several developed during the collaboration with the Materials Property Data Network.[14–16] Of great interest to professional searchers, STN International has the most sophisticated numeric data search software available anywhere online. Thus the data sources may be searched numerically, that is, using numeric values as ranges or with "greater or less than" types of operators, making it possible to search for alloys that meet required performance needs.

The disadvantages of using STN International to search for numeric data are twofold: (a) the primary search software is keyed to a command system best known to professional searchers and engineers and scientists will need patience and some training to master the technique and (b) use of the STN International system is billed via a time- and content-based cost accounting system, and so a private account is needed.

For those able to deal with those conditions, a number of valuable databases exist on STN International, including the following representative sources:

- Aluminium—Aluminum Industry Abstracts (textual: 1868–present)
- ASMDATA—ASM Materials Databases online version
- BEILSTEIN—Beilstein Organic Compound Files (1779–1999)
- COPPERLIT—Copper and Copper Alloy Standards & Data
- CORROSION—Cambridge Scientific Abstracts databases of corrosion science and engineering (current)
- DETHERM—Thermophysical Properties Database (1819–present)
- GMELIN—GMELIN Handbook of Inorganic Chemistry (1817–present)
- ICSD—Inorganic Crystal Structure Database (1912–present)
- MDF—Metals Datafile (1982–1993)
- PIRA—PIRA and PAPERBASE Database (1975–present)
- RAPRA—Rubber, Plastics, Adhesives, and Polymer Composites (1972–present)
- WELDASEARCH–Literature about joining of metals, polymers, and ceramics

For more information on STN International, readers are referred to the website www.cas.org.

7.3 Knovel

Another online source of materials information that has grown significantly over the last decade is www.knovel.com, a product of William Andrew Publishing in cooperation with a number of publishers of scientific information, including John Wiley & Sons, McGraw-Hill, and Elsevier. On the knovel.com website, the sources are enhanced with time-saving analytical tools to help analyze and manipulate the data, so users can analyze and reorganize the data from over 800 full-text engineering and scientific reference works, handbooks, and databases. They can sort, filter, and export data from "live tables," resolve equations, plot graphs, capture values from existing graphs, and perform "what if" experiments on the data. Knovel's unique and powerful search engine has a user-friendly browse capability to allow the user to discover and analyze data in relatively intuitive ways.

The specific content areas recognized on knovel.com include:

- Adhesives, coatings, sealants, and inks
- Aerospace and radar technology
- Biochemistry, biology, and biotechnology
- Ceramics and ceramic engineering
- Chemistry and chemical engineering
- Civil engineering and construction materials
- Earth sciences
- Electrical and power engineering
- Electronics and semiconductors
- Environment and environmental engineering
- Food science
- General engineering and engineering management
- Mechanics and mechanical engineering
- Metals and metallurgy
- Oil and gas engineering
- Pharmaceuticals, cosmetics, and toiletries
- Plastics and rubber
- Textiles

7.4 Other Internet Websites

As indicated earlier, there are many websites on the Internet with materials information content. The challenge is to determine which have useful, reliable, and relatively easily accessible data. In the interests of readers, we will focus on guidance on these latter points developed by Fran Cverna, Director of Electronic and Reference Data Sources at ASM International, who produced and presented a survey of the scope and quality of materials data content found on Internet websites.[17]

Among the most useful websites of the several hundred screened by Cverna and her ASM resources are the following:

www.about.com	Internet search engine—search materials and properties
www.nist.gov/public affairs/database.htm	NIST—standard reference database
www.tprl.com	TPRL—thermophysical properties data
www.campusplastics.com	Campus consortium—plastics database
www.matweb.com	Matweb databases by Automation Creations
www.copper.org	Copper Development Association—copper alloys database
www.brushwellman.com/homepage.htm	Brush Wellman—supplier materials database
www.timet.com/overviewframe.html	Timet—supplier materials database

www.special-metals.com	Special Metals—supplier materials database
www.cartech.com	Cartech—supplier materials database (compositions)
www.macsteel.com/mdb	McSteel supplier materials database (limited)
www.aluminum.org	Aluminum Association applications—publications' limited data

Of the sources above, three merit special mention: www.about.com, the National Institute for Science and Technology databases, and www.matweb.com

The about.com site provides an excellent means of locating materials data sites and provides a subcategory called "Materials Properties and Data" if you search for such information. Many sites are identified, some of which overlap the ASM survey, but others are unique to that site. Many included in the ASM survey are not included here, so the two are supplementary in scope. A wide range of properties may be searched, ranging froms physical and mechanical properties, and support calculations for densities, unit changes, and currencies are provided. In addition there are links from about.com to many other materials sites, including one called aluminum.com that also provides materials data for a variety of metals, but often without adequate citation and metadata.

The NIST database site www.wst.gov/public affairs/database.htm provides direct online access to the highest quality, carefully evaluated numeric data from the following databases, among others:

- Standard Reference Data—reliable scientific and technical data extracted from the world's literature assessed for reliability and critically evaluated
- Ceramics WebBook—evaluated data and access to data centers as well as tools and resources
- Chemistry WebBook—chemical and physical property data for specific compounds
- Fundamental Physical Constants—internationally recommended values of a wide range of often used physical constants
- Thermophysical Properties of Gases for the Semiconductor Industry.
- MatWeb—designed, developed, and maintained by Automation Creations, the website www.matweb.com provides access to the properties of over 65,000 materials, including metals, polymers, ceramics, and composite materials. The search engine is relatively flexible, permitting searches by property, composition, and/or material, and in the latter case, material designations ranging from trade names to Unified Numbering System (UNS) designations may be used.

To summarize the Internet discussion, there are many sources of numeric materials data available from Internet sites. It remains for the potential users of the data, however, to approach each site with caution, look for the pedigree of the data in terms of quality and reliability, and make certain that the source used meets the requirement of the intended use.

ACKNOWLEDGMENTS

The contributions of Jack Westbrook (Brookline Technologies) and Fran Cverna (ASM International) are acknowledged.

REFERENCES

1. Westbrook, J. H., "Sources of Materials Property Data and Information," in *ASM Handbook*, Vol. 20, ASM International, Materials Park, OH, 1997, pp. 491–506.
2. Westbrook, J. H., and Reynard, K. W., "Data Sources for Materials Economics, Policy, and Management," in *Concise Encyclopedia of Materials Economics, Policy, and Management*, M. B. Bever (Ed.), Pergamon, New York, 1993, pp. 35–42.
3. *ASM International Directory of Materials Property Databases*, ASM International, Materials Park, OH, published periodically.
4. ASTM International and ANSI/ISO Standards: *Annual Book of ASTM Standards*, published annually, ASTM, Philadelphia; and American National Standards Institute (ANSI) and International Standards Organization (ISO) Standards, published periodically, ISO, Brussels.
5. Metallic Materials Properties Development and Standardization MMPDS (previously known as MIL-HDBK-5H, *Metallic Materials and Elements for Aerospace Vehicle Structures*), published periodically by Battelle for the Federal Aeronautic Administration (FAA), Washington, DC.
6. Publications of the Thermophysical Properties Research Center (TPRC, previously known as CINDAS), Lafayette, IN.
7. *ASME Boiler & Pressure Vessel Code*, Section 2, Material–Properties, American Society of Mechanical Engineers, New York, published periodically.
8. *Aluminum Standards & Data, 2000*, and *Aluminum Standards & Data 1998 Metric SI*. The Aluminum Association, Washington, DC, published periodically.
9. Westbrook, J. H., and Grattidge, W., "The Role of Metadata in the Design and Operation of a Materials Database," in *Computerization and Networking of Materials Databases*. ASTM STP 1106, J. G. Kaufman and J. S. Glazman (Eds.), ASTM, West Conshohocken, PA, 1991.
10. Kaufman, J. G., "Quality and Reliability Issues in Materials Databases," in ASTM Committee E49.05, *Computerization and Networking of Materials Databases*, Vol. 3, ASTM STP 1140, T. I. Barry and K. W. Reynard (Eds.), ASTM, West Conshohocken, PA, 1992, pp. 64–83.
11. Kaufman, J. G., *Properties of Aluminum Alloys—Tensile, Creep, and Fatigue Data at High and Low*

Temperatures, ASM International, Materials Park, OH, 1999.

12. *ASM Handbook*, Vols. 1 and 2. *Properties and Selection*, ASM International, Materials Park, OH, published periodically.

13. STNews, a newsletter of STN International, the Worldwide Scientific and Technical Information Network. Chemical Abstract Services (CAS), a Division of the American Chemical Society, Columbus, OH. published bimonthly.

14. Kaufman, J. G., "The National Materials Property Data Network Inc., The Technical Challenges and the Plan." in *Materials Property Data: Applications and Access*, J. G. Kaufman (Ed.). MPD-Vol. PVP-Vol. 111 ASME, New York, 1986, pp. 159–166.

15. Kaufman, J. G., "The National Materials Property Data Network, Inc.—A Cooperative National Approach to Reliable Performance Data," in *Computerization and Networking of Materials Data Bases*, ASTM STP 1017, J. S. Glazman and J. R. Rumble, Jr. (Eds.), ASTM, West Conshohocken, PA, 1989, pp. 55–62.

16. Westbrook, H., and Kaufman, J. G., "Impediments to an Elusive Dream," in *Modeling Complex Data for Creating Information*, J. E. DuBois and N. Bershon (Eds.), Springer-Verlag, Berlin, 1996.

17. Cverna, F., "Overview of Commercially Available Material Property Data Collections" (on the Internet), presented at the 2000 ASM Materials Solutions Conference on Materials Property Databases, ASM International, St Louis, MO. October 10–12, 2000.

INDEX